2011 IEEE International Reliability Physics Symposium

(IRPS 2011)

Monterey, California, USA
10 - 14 April 2011

Pages 1 - 508

IEEE Catalog Number: CFP11RPS-PRT
ISBN: 978-1-4244-9113-1

Copyright © 2011 by the Institute of Electrical and Electronic Engineers, Inc
All Rights Reserved

Copyright and Reprint Permissions: Abstracting is permitted with credit to the source. Libraries are permitted to photocopy beyond the limit of U.S. copyright law for private use of patrons those articles in this volume that carry a code at the bottom of the first page, provided the per-copy fee indicated in the code is paid through Copyright Clearance Center, 222 Rosewood Drive, Danvers, MA 01923.

For other copying, reprint or republication permission, write to IEEE Copyrights Manager, IEEE Service Center, 445 Hoes Lane, Piscataway, NJ 08854. All rights reserved.

***This publication is a representation of what appears in the IEEE Digital Libraries. Some format issues inherent in the e-media version may also appear in this print version.

IEEE Catalog Number: CFP11RPS-PRT
ISBN 13:   978-1-4244-9113-1
ISSN: 1541-7026

Additional Copies of This Publication Are Available From:

Curran Associates, Inc
57 Morehouse Lane
Red Hook, NY 12571 USA
Phone: (845) 758-0400
Fax: (845) 758-2633
E-mail: curran@proceedings.com
Web: www.proceedings.com

2011 IEEE International Reliability Physics Symposium (IRPS 2011)

Monterey, California, USA
10-14 April 2011

IEEE Catalog Number: CFP11RPS-POD
ISBN: 978-1-42449-113-1

TABLE OF CONTENTS

SESSION 2A: GATE DIELECTRICS:

Post-Breakdown Statistics and Acceleration Characteristics in High-K Dielectric Stacks 1
E. Wu, J. Suñé, B. Linder, R. Achanta, B. Li, S. Mittl

Methodologies for Sub-1nm EOT TDDB Evaluation 7
Thomas Kauerauf, Robin Degraeve, Lars-Åke Ragnarsson, Philippe Roussel, Sahar Sahhaf, Guido Groeseneken, Robert O'Connor

Frequency Dependent TDDB Behaviors and Its Reliability Qualification in 32nm High-k/Metal Gate CMOSFETs 17
Kyong Taek Lee, Jongik Nam, Minjung Jin, Kidan Bae, Junekyun Park, Lira Hwang, Jungin Kim, Hyunjin Kim, Jongwoo Park

Re-Investigation of Gate Oxide Breakdown on Logic Circuit Reliability 22
Y. C. Huang, T. Y. Yew, W. Wang, Y.-H. Lee, R. Ranjan, N. K. Jha, P. J. Liao, J. R. Shih, K. Wu

SESSION 2B: CIRCUIT RELIABILITY:

In Situ Screening Techniques for Defective Oxides in Devices for Automotive Applications 28
V. Malandruccolo, M. Ciappa, W. Fichtner, H. Rothleitner

A TDC-Based Test Platform for Dynamic Circuit Aging Characterization 36
Min Chen, Vijay Reddy, John Carulli, Srikanth Krishnan, Vijay Rentala, Venkatesh Srinivasan, Yu Cao

Fast Characterization of the Static Noise Margin Degradation of Cross-Coupled Inverters and Correlation to BTI Instabilities in MG/HK Devices 41
A. Kerber, N. Pimparkar, S. Balasubramanian, T. Nigam, W. McMahon, E. Cartier

Reliability Monitoring Ring Oscillator Structures for Isolated/Combined NBTI and PBTI Measurement in High-K Metal Gate Technologies 47
Jae-Joon Kim, Barry P. Linder, Rahul M. Rao, Tae-Hyoung Kim, Pong-Fei Lu, Keith A. Jenkins, Chris H. Kim, Aditya Bansal, Saibal Mukhopadhyay, Ching-Te Chuang

Negative Bias Temperature Instability "Multi-Mode" Compact Model Based on Threshold Voltage and Mobility Degradation 51
D. Varghese, R. Higgins, S. Dunn, A. T. Krishnan, V. Reddy, S. Krishnan

A New Smart Device Array Structure for Statistical Investigations of BTI Degradation and Recovery 56
C. Schlünder, J. M. Berthold, M. Hoffman, J.-M. Weigmann, W. Gustin, H. Reisinger

SESSION 2C: FABLESS AND PRODUCTION RELIABILITY:

Characterization and Challenge of TDDB Reliability in Cu/Low K Dielectric Interconnect 61
Feng Xia, Jun He, Prad Prabhumirashi, Anthony Schmitz, Anthony Lowrie, Jeff Hicks, Yuriy Shusterman, Ruth Brain

Backend Low-k TDDB Chip Reliability Simulator 65
Muhammad Bashir, Dae Hyun Kim, Krit Athikulwongse, Sung Kyu Lim, Linda Milor

Determination of CPU Use Conditions 75
Robert Kwasnick, Athanasios E. Papathanasiou, Matthew Reilly, Al Rashid, Bashir Zaknoon, John Falk

Si3N4 Extrinsic Defects and Capacitor Reliability 81
John Scarpulla, Everett E. King, Jon V. Osborn

SESSION 2D: MEMORY:

AC-DC Factor Sensitivity for DRAM Components Lifetime under Hot-Carrier Injection 91
Sanghyeon Baeg, Hyeowoo Nam, Pierre Chia, ShiJie Wen, Richard Wong

STI Stress-Induced Degradation of Data Retention Time in DRAM and a New Characterizing Method for Mechanical Stress 95
Tae-Su Jang, Kyung-Do Kim, Min-Soo Yoo, Yong-Taik Kim, Seon-Yong Cha, Jae-Goan Jeong, Sung-Joo Hong

Hot Hole Induced Damage in 1T-FBRAM on Bulk FinFET...99
M. Aoulaiche, N. Collaert, A. Mercha, M. Rakowski, B. De Wachter, G. Groeseneken, L. Altimime, M. Jurczak, Z. Lu

Effects of BTI during AHTOL on SRAM VMIN...105
Sun-Me Lim, Heebum Hong, Sunil Yu, Ming Zhang, Jongwoo Park, Yongshik Kim

SESSION 2E: THIN FILM TRANSISTOR:

Flexible Biomedical Devices for Mapping Cardiac and Neural Electrophysiology.......................111
Dae-Hyong Kim, John A. Rogers, Jonathan Viventi, Brian Litt

Low-Energy UV Effects on Organic Thin-Film-Transistors...113
N. Wrachien, A. Cester, D. Bari, G. Meneghesso, J. Kovac, J. Jakabovic, M. Sokolsky, D. Donoval, J. Cirak

A New Method for Predicting the Lifetime of Highly Stable Amorphous-Silicon Thin-Film Transistors from Accelerated Tests..121
T. Liu, S. Wagner, J. C. Sturm

Investigation of Ultra Thin Polycrystalline Silicon Channel for Vertical NAND Flash...............126
Bio Kim, Seung-Hyun Lim, Dong Woo Kim, Toshiro Nakanishi, Sangryol Yang, Jae-Young Ahn, HanMei Choi, Kihyun Hwang, Yongsun Ko, Chang-Jin Kang

SESSION 2F: BEOL DIELECTRICS:

Electrical Reliabilities of Porous Silica Low-k Films...130
Takamaro Kikkawa, Yasuhisa Kayaba, Kazuo Kohmura, Shinichi Chikaki

Invasion Percolation Model for Abnormal TDDB Characteristic of ULK Dielectrics with Cu Interconnect at Advanced Technology Nodes...134
F. Chen, M. Shinosky, B. Li, J. Aitken, S. Cohen, G. Bonilla, A. Simon, P. McLaughlin, R. Achanta, F. Baumann, C. Parks, M. Angyal

Comparison between Intrinsic and Integrated Reliability Properties of Low-k Materials..............142
K. Croes, M. Pantuovaki, L. Carbonell, L. Zhao, G. P. Beyer, Zs. Tökei

Statistics of Breakdown Field and Time-Dependent Dielectric Breakdown in Contact-to-Poly Modules................149
Shinji Yokogawa, Satoshi Uno, Ichiro Kato, Hideaki Tsuchiya, Tatsuo Shimizu, Mitsuhiro Sakamoto

Reliability Limitations to the Scaling of Porous Low-K Dielectrics..155
Shou-Chung Lee, A. S. Oates

A Comprehensive Process Engineering on TDDB for Direct Polishing Ultra-Low K Dielectric Cu Interconnects at 40nm Technology Node and Beyond..160
W. C. Lin, T. C. Tsai, H. K. Hsu, Jack Lin, W. C. Tsao, Willis Chen, C. M. Cheng, C. L. Hsu, C. C. Liu, C. M. Hsu, J. F. Lin, C. C. Huang, J. Y. Wu

SESSION 2G: EXTREME ENVIRONMENTS:

Space Radiation Effects and Reliability Considerations for the Proposed Jupiter Europa Orbiter.......165
Allan Johnston

Reliability and Performance Characterization of a MEMS-Based Non-Volatile Switch.................171
Roberto Gaddi, Cor Schepens, Charles Smith, Cristian Zambelli, Andrea Chimenton, Piero Olivo

Microanalysis for Tin Whisker Risk Assessment...177
Maribeth Mason, Genghmun Eng, Martin Leung, Gary Stupian, Terence Yeoh

SESSION 3A: GATE DIELECTRICS:

Random Telegraph Noise Reduction in Metal Gate High-kappa Stacks by Bipolar Switching and the Performance Boosting Technique...182
W. H. Liu, K. L. Pey, N. Raghavan, X. Wu, M. Bosman, T. Kauerauf

Correlation of I_d- and I_g-Random Telegraph Noise to Positive Bias Temperature Instability in Scaled High-kappa/Metal Gate n-Type MOSFETs...190
Chia-Yu Chen, Qiushi Ran, Hyun-Jin Cho, Andreas Kerber, Yang Liu, Ming-Ren Lin, Robert W. Dutton

SILC-Based Reassignment of Trapping and Trap Generation Regimes of Positive Bias Temperature Instability...196
J. Q. Yang, M. Masuduzzaman, J. F. Kang, M. A. Alam

A New Interface Defect Spectroscopy Method 202
J. T. Ryan, L. C. Yu, J. H. Han, J. J. Kopanski, K. P. Cheung, F. Zhang, C. Wang, J. P. Campbell, J. S. Suehle, V. Tilak, J. Fronheiser

Experimental Identification of Unique Oxide Defect Regions by Characteristic Response of Charge Pumping 207
Muhammad Masuduzzaman, Ahmad Islam, Robin Degraeve, Moonju Cho, Mohammed Zahid, Muhammad Alam

SESSION 3B: MEDICAL ELECTRONICS:

The Implications of RoHS on Active Implantable Medical Devices 213
Timothy Scott Savage

Implantable Microtechnologies for the Brain: Challenges and Strategies for Reliable Operation 220
Jit Muthuswamy, Sindhu Anand, Jemmy Sutanto, Michael Baker, Murat Okandan

A Swallowable Diagnostic Capsule with a Direct Access Sensor Using Anisotropic Conductive Adhesive 224
Pio Jesudoss, Alan Mathewson, Karen Twomey, Frank Stam, William M. D. Wright

Application Based Reliability Assessment and Qualification Methodology for Medical ICs 231
Xiaowei Zhu, Karthik Vasanth, Xiaochen Xu, Charles Smyth, Brent Rhoton

SESSION 3C: SOFT ERRORS:

Quantification and Mitigation Strategies of Neutron Induced Soft-Errors in CMOS Devices and Components - The Past and Future 239
Eishi Ibe, Ken-ichi Shimbo, Hitoshi Taniguchi, Tadanobu Toba, Koji Nishii, Yoshio Taniguchi

Effects of Scaling on Muon-Induced Soft Errors 247
Brian D. Sierawski, Robert A. Reed, Marcus H. Mendenhall, Robert A. Weller, Ronald D. Schrimpf, Shi-Jie Wen, Richard Wong, Nelson Tam, Robert C. Baumann

Neutron Induced Single Event Multiple Transients with Voltage Scaling and Body Biasing 253
Ryo Harada, Yukio Mitsuyama, Masanori Hashimoto, Takao Onoye

Double-Pulse-Single-Event Transients in Combinational Logic 258
J. R. Ahlbin, T. D. Loveless, D. R. Ball, B. L. Bhuva, A. F. Witulski, L. W. Massengill, M. J. Gadlage

SESSION 3D: THERMO-MECHANICAL AND MEMS:

Thermomechanical Reliability of Through-Silicon Vias in 3D Interconnects 264
Kuan-Hsun Lu, Suk-Kyu Ryu, Jay Im, Rui Huang, Paul S. Ho

Characterization of Steady and Transient Heating of Interconnects - A Review 271
Banafsheh Barabadi, Yogendra Joshi, Satish Kumar

Impact of Scaling on the Performance and Reliability Degradation of Metal-Contacts in NEMS Devices 280
Hamed F. Dadgour, Muhammad M. Hussain, Alan Cassell, Navab Singh, Kaustav Banerjee

The Effect of Temperature on Dielectric Charging of Capacitive MEMS 290
M. Koutsoureli, L. Michalas, G. Papaioannou

SESSION 3E: ELECTROMIGRATION/VOIDING:

Electromigration Induced Void Kinetics in Cu Interconnects for Advanced CMOS Nodes 297
L. Arnaud, P. Lamontagne, R. Galand, E. Petitprez, D. Ney, P. Waltz

Formation of Highly Reliable Cu/Low-k Interconnects by Using CVD Co Barrier in Dual Damascene Structures 307
Hye Kyung Jung, Hyun-Bae Lee, Matsuda Tsukasa, Eunji Jung, Jong-Ho Yun, Jong Myeong Lee, Gil-Heyun Choi, Siyoung Choi, Chilhee Chung

Electromigration-Resistance Enhancement with CoWP or CuMn for Advanced Cu Interconnects 312
Cathryn Christiansen, Baozhen Li, Matthew Angyal, Terence Kane, Vincent McGahay, Yun Yu Wang, Shaoning Yao

A Study of Via Depletion Electromigration with Very Long Failure Times 317
Baozhen Li, Cathryn Christiansen, Kaushik Chanda, Matt Angyal, Jennifer Oakley

Study of Void Formation Kinetics in Cu Interconnects Using Local Sense Structures 321
K. Croes, M. Lofrano, C. J. Wilson, L. Carbonell, Y. K. Siew, G. P. Beyer, Zs. Tökei

SESSION 3F: PROCESS INTEGRATION AND 3D/TSV:

3D Integration Technology and Reliability... 328
Mitsumasa Koyanagi

Impact of Air-Induced Poly-SI/Oxynitride Interface Layer Degradation on Gate-Edge Leakage............. 335
Ziyuan Liu, Shuu Ito, Tomoya Saito, Soon W. Chang, Arito Ogawa, Sadayoshi Horii, Tsuyoshi Horikawa, Markus Wilde, Katsuyuki Fukutani, Toyohiro Chikyow

On the Thermal Failure in Nanoscale Devices: Insight Towards Heat Transport Including Critical BEOL and Design Guidelines for Robust Thermal Management & EOS/ESD Reliability.................................... 342
Mayank Shrivastava, Manish Agrawal, Jasmin Aghassi, Harald Gossner, Wolfgang Molzer, Thomas Schulz, V. Ramgopal Rao

Resistance Increase Due to Electromigration Induced Depletion under TSV 347
T. Frank, C. Chappaz, P. Leduc, L. Arnaud, F. Lorut, S. Moreau, A. Thuaire, R. El Farhane, L. Anghel

SESSION 4A: TRANSISTORS/ORGANICS TFTS:

Reliability- and Process-Variation Aware Design of Integrated Circuits - A Broader Perspective............................ 353
Muhammad A. Alam, Kaushik Roy, Charles Augustine

Response of a Single Trap to AC Negative Bias Temperature Stress... 364
M. Toledano-Luque, B. Kaczer, Ph. J. Roussel, T. Grasser, G. I. Wirth, J. Franco, C. Vrancken, N. Horiguchi, G. Groeseneken

PBTI under Dynamic Stress: From a Single Defect Point of View.. 372
K. Zhao, J. H. Stathis, B. P. Linder, E. Cartier, A. Kerber

Understanding of Traps Causing Random Telegraph Noise Based on Experimentally Extracted Time Constants and Amplitude.. 381
Kenichi Abe, Akinobu Teramoto, Shigetoshi Sugawa, Tadahiro Ohmi

Mechanistic Understanding of Breakdown and Bias Temperature Instability in High-K Metal Devices Using Inline Fast Ramped Bias Test 387
Siddarth A. Krishnan, Eduard Cartier, James Stathis, Michael Chudzik, Andreas Kerber

SESSION 4C: ESD AND LATCH-UP:

Design of Modified ESD Protection Structure with Low-Trigger and High-Holding Voltage in Embedded High Voltage CMOS Process.. 392
Tai-Hsiang Lai, Lu-An Chen, Tien-Hao Tang, Kuan-Cheng Su

An EOS-Free PNP-Enhanced Cascoded NMOSFET Structure for High Voltage Application 396
Shih-Yu Wang, Yao-Wen Chang, Yan-Yu Chen, Chieh-Wei He, Guan-Wei Wu, Tao-Cheng Lu, Kuang-Chao Chen, Chih-Yuan Lu

Latch-Up Free ESD Protection Design with SCR Structure in Advanced CMOS Technology 401
Chang-Tzu Wang, Tien-Hao Tang, Kuan-Cheng Su

Transient Latchup in Power Analog Circuits.. 405
V. A. Vashchenko, D. LaFonteese, A. Concannon

WCDM2 - Wafer-Level Charged Device Model Testing with High Repeatability.. 409
Nathan Jack, Timothy J. Maloney, Bruce Chou, Elyse Rosenbaum

SESSION 4E: COMPOUND OPTO-ELECTRONICS:

Reliability of GaN-HEMTs for High-Voltage Switching Applications.. 417
Wataru Saito

Time Evolution of Electrical Degradation under High-Voltage Stress in GaN High Electron Mobility Transistors.. 422
Jungwoo Joh, Jesús A. del Alamo

Reliability-Limiting Defects in AlGaN/GaN HEMTs 426
Tania Roy, En Xia Zhang, Daniel M. Fleetwood, Ronald D. Schrimpf, Yevgeniy S. Puzyrev, Sokrates T. Pantelides

250 GHz Heterojunction Bipolar Transistor: From DC to AC Reliability 430

Malick Diop, Salim Ighilahriz, Florian Cacho, Vincent Huard

SESSION 5A: HIGH VOLTAGE/RF:

Intrinsic Reliability of RF Power LMDOS FETs 435

David C. Burdeaux, Wayne R. Burger

**Investigation of Multistage Linear Region Drain Current Degradation and Gate-Oxide Breakdown
under Hot-Carrier Stress in BCD HV PMOS** 444

Yu-Hui Huang, J. R. Shih, C. C. Liu, Y.-H. Lee, R. Ranjan, Puo-Yu Chiang, Dah-Chuen Ho, Kenneth Wu

**New Investigation of Hot Carrier Degradation of RF Small-Signal Parameters in High-k/Metal Gate
nMOSFETs** 449

*Hyun Chul Sagong, Chang Yong Kang, Chang-Woo Sohn, Min Sang Park, Do-Young Choi, Eui-Young Jeong, Jack
C. Lee, Yoon-Ha Jeong*

**Design-in Reliability Approach for Hot Carrier Injection Modeling in the Context of AMS/RF
Applications** 454

Vincent Huard, Thomas Quemerais, Florian Cacho, Laurence Moquillon, Sebastien Haendler, Xavier Federspiel

**Impact of Source/Drain Contact and Gate Finger Spacing on the RF Reliability of 45-nm RF
nMOSFETs** 461

*Rajan Arora, Sachin Seth, John Chung Hang Poh, John D. Cressler, Akil K. Sutton, Hasan M. Nayfeh, Giuseppe
L. Rosa, Greg Freeman*

SESSION 5B: SOFT ERRORS:

Soft-Error Testing at Advanced Technology Nodes 467

B. Bhuva, B. Narasimham, A. Oates, K. Patterson, N. Tam, M. Vilchis, S.-J. Wen, R. Wong, Y. Xu

Measurement of Neutron-Induced SET Pulse Width Using Propagation-Induced Pulse Shrinking 471

Jun Furuta, Chikara Hamanaka, Kazutoshi Kobayashi, Hidetoshi Onodera

**Multicenter Comparison of Alpha Particle Measurements and Methods Typical of Semiconductor
Processing** 476

*Jeffrey D. Wilkinson, Brett M. Clark, Richard Wong, Charles Slayman, Barry Carroll, Michael Gordon, Yi He,
Olivier Lauzeral, Keith Lepla, Jennifer Marckmann, Brendan McNally, Philippe Roche, Mike Tucker, Tommy Wu*

The Impact of New Technology on Soft Error Rates 486

Anand Dixit, Alan Wood

SESSION 5C: FAILURE ANALYSIS:

**Quantitative, Nanoscale Free-Carrier Concentration Mapping Using Terahertz Near-Field
Nanoscopy** 493

J. Wittborn, R. Weiland, A. J. Huber, F. Keilmann, R. Hillenbrand

**Rapid and Automated Grain Orientation and Grain Boundary Analysis in Nanoscale Copper
Interconnects** 500

K. J. Ganesh, S. Rajasekhara, D. Bultreys, P. J. Ferreira

High Reliable Strain Measurement for Power Devices Using STEM-CBED Method 503

N. Nakanishi, H. Arie, H. Maeda, Y. Hirose, N. Hattori, T. Koyama, E. Murakami

**Electron Beam Induced Current Characterization of Dark Line Defects in Failed and Degraded High
Power Quantum Well Laser Diodes** 509

Maribeth Mason, Nathan Presser, Yongkun Sin, Brendan Foran, Steven C. Moss

Spectral Resolution of Photon Emission from SiGe:C Heterojunction Bipolar Transistors (HBTs) 514

Ulrike Kindereit, Oana-Mihaela Mutihac, Christian Boit, Bernd Tillack

Isolating Light-Sensitive Defects Using C-AFM 520

Hung Sung Lin, Mong Sheng Wu

SESSION 5D: PROCESS INTEGRATION AND 3D/TSV:

A Holistic Approach to Process Co-Optimization for Through-Silicon Via 524

Sesh Ramaswami

Thermal and Spatial Profiling of TSV-Induced Stress in 3DICs 527
  Colin McDonough, Benjamin Backes, Wei Wang, Robert E. Geer

Reliability Studies of a 32nm System-on-Chip (SoC) Platform Technology with 2nd Generation High-K/Metal Gate Transistors 533
  A. Rahman, M. Agostinelli, P. Bai, G. Curello, H. Deshpande, W. Hafez, C.-H. Jan, K. Komeyli, J. Park, K. Phoa, C. Tsai, J.-Y. Yeh, J. Xu

VTH Shift Mechanism in Dysprosium (Dy) Incorporated HfO₂ Gate nMOS Devices 539
  Tackhwi Lee, Sanjay K. Banerjee

SESSION 5E: PHOTOVOLTAIC DEVICES:

Physics of Instability of Thin Film Si and (Si,Ge) Alloy Solar Cells 546
  Vikram L. Dalal, Zhao Li

Reliability Testing beyond Qualification as a Key Component in Photovoltaic's Progress toward Grid Parity 551
  John H. Wohlgemuth, Sarah Kurtz

Identification, Characterization and Implications of Shadow Degradation in Thin Film Solar Cells 557
  Sourabh Dongaonkar, Muhammad A. Alam, Karthik Y., Souvik Mahapatra, Dapeng Wang, Michel Frei

Metastability of Hydrogenated Amorphous Silicon Passivation on Crystalline Silicon and Implication to Photovoltaic Devices 562
  Bahman Hekmatshoar, Davood Shahrjerdi, Marinus Hopstaken, Devendra Sadana

Optical Stress and Reliability Study of Ruthenium-Based Dye-Sensitized Solar Cells (DSSC) 566
  Daniele Bari, Nicola Wrachien, Andrea Cester, Gaudenzio Meneghesso, Roberto Tagliaferro, Stefano Penna, Thomas M. Brown, Andrea Reale, Aldo Di Carlo

SESSION 5F: CPI/ELECTROMIGRATION/VOIDING:

The Role of Elastic and Plastic Anisotropy of Sn on Microstructure and Damage Evolution in Lead-Free Solder Joints 573
  Thomas R. Bieler, Bite Zhou, Lauren Blair, Amir Zamiri, Payam Darbandi, Farhang Pourboghrat, Tae-Kyu Lee, Kuo-Chuan Liu

Robust Pad Layout to Improve Wire Bonding Reliability 582
  Kyoung-Hwan Kim, Hong Kook Min, Se Yeoul Park, So Ra Park, Seung Jin Yang, Byung Sup Shim, Yong Tae Kim, Jeong-Uk Han

Electromigration Characterization of Lead-Free Flip-Chip Bumps for 45nm Technology Node 588
  Christine Hau-Riege, You-Wen Yau, Nick Yu

Electromigration Failure Mechanisms for Different Flip Chip Bump Configurations 592
  Riet Labie, Tomas Webers, Christophe Winters, Vladimir Cherman, Kristof Croes, Bart Vandevelde, Franck Dosseul

SESSION 6A: TRANSISTORS/ORGANICS TFTS:

Understanding and Modeling AC BTI 597
  Hans Reisinger, Tibor Grasser, Karsten Ermisch, Heiko Nielen, Wolfgang Gustin, Christian Schluender

The 'Permanent' Component of NBTI: Composition and Annealing 605
  T. Grasser, Th. Aichinger, G. Pobegen, H. Reisinger, P.-J. Wagner, J. Franco, M. Nelhiebel, B. Kaczer

A Critical Re-Evaluation of the Usefulness of R-D Framework in Predicting NBTI Stress and Recovery 614
  S. Mahapatra, A. E. Islam, S. Deora, V. D. Maheta, K. Joshi, A. Jain, M. A. Alam

On the Recoverable and Permanent Components of Hot Carrier and NBTI in Si pMOSFETs and Their Implications in Si₀.₄₅Ge₀.₅₅ pMOSFETs 624
  J. Franco, B. Kaczer, G. Eneman, Ph. J. Roussel, M. Cho, J. Mitard, L. Witters, T. Y. Hoffmann, G. Groeseneken, F. Crupi, T. Grasser

Impact of HK / MG Stacks and Future Device Scaling on RTN 630
  Naoki Tega, Hiroshi Miki, Zhibin Ren, Christopher P. D'Emic, Yu Zhu, David J. Frank, Michael A. Guillorn, Dae-Gyu Park, Wilfried Haensch, Kazuyoshi Torii

SESSION 6B: MEMORY:

Split-Gate Flash Memory for Automotive Embedded Applications.. 636
Y. S. Chu, Y. H. Wang, C. Y. Wang, Y. H. Lee, A. C. Kang, R. Ranjan, W. T. Chu, T. C. Ong, H. W. Chin, K. Wu

Novel Negative Vt Shift Program Disturb Phenomena in 2X~3X nm NAND Flash Memory Cells........................... 641
Soonok Seo, Hyungseok Kim, Sungkye Park, Seokkiu Lee, Seiichi Aritome, Sungjoo Hong

Junction Optimization for Reliability Issues in Floating Gate NAND Flash Cells.................................. 645
C. H. Lee, I. C. Yang, Chienying Lee, C. H. Cheng, L. H. Chong, K. F. Chen, J. S. Huang, S. H. Ku, N. K. Zous, I. J. Huang, T. T. Han, M. S. Chen, W. P. Lu, K. C. Chen, Tahui Wang, Chih-Yuan Lu

Charge Diffusion in Silicon Nitrides: Scalability Assessment of Nitride Based Flash Memory 650
Seung Jae Baik, Koeng Su Lim, Wonsup Choi, Hyunjun Yoo, Hyunjung Shin

The Effect of Crystallinity of HfO2 on the Resistive Memory Switching Reliability 656
Min Gyu Sung, Wan Gee Kim, Jong Hee Yoo, Sook Joo Kim, Jung Nam Kim, Byung Gu Gyun, Jun Young Byun, Taeh Wan Kim, Won Kim, Moon Sig Joo, Jae Sung Roh, Sung Ki Park

BEOL DIELECTRICS:

A Novel Pre-Clean Process of BEOL Barrier-Seed Process to Enhance Reliability Performance of Advanced 40nm Node.. 661
Chun-Min Cheng, Chi-Mao Hsu, W. C. Lin, Hsin-Fu Huang, Yan-Chun Liu, Kun-Hsien Lin, Jin-Fu Lin, C. C. Huang, J. Y. Wu

A Charge Transport Based Acceleration Model for Interlevel Dielectric Breakdown 665
Ravi Achanta, Paul McLaughlin

A Model for Post-CMP Cleaning Effect on TDDB 670
Chia-Lin Hsu, Wen-Chin Lin, Teng-Chun Tsai, Climbing Huang, J.-Y. Wu

COMPOUND OPTO-ELECTRONICS:

Performance and Structure Degradations of SiGe HBT after Electromagnetic Field Stress............................ 674
A. Alaeddine, M. Kadi, K. Daoud

Reliability Testing of AlGaN/GaN HEMTs under Multiple Stressors 680
Bradley D. Christiansen, Ronald A. Coutu Jr., Eric R. Heller, Brian S. Poling, G. David Via, Rama Vetury, Jeffrey B. Shealy

CPI / ELECTROMIGRATION / VOIDING:

Long Term Isothermal Reliability of Copper Wire Bonded to Thin 6.5 μm Aluminum 685
F. C. Classe, Sesha Gaddamraja

A New ESD Model Induced Yield Loss during Chip-On-Film Package Process and It's Failure Mechanism 690
Jian-Hsing Lee, J. R. Shih, Yu-Hui Huang, C. P. Lin, David Su, Kenneth Wu

CIRCUIT RELIABILITY:

A Robust Reliability Methodology for Accurately Predicting Bias Temperature Instability Induced Circuit Performance Degradation in HKMG CMOS.. 696
D. P. Ioannou, K. Zhao, A. Bansal, B. Linder, R. Bolam, E. Cartier, J.-J. Kim, R. Rao, G. La Rosa, G. Massey, M. Hauser, K. Das, J. H. Stathis, J. Aitken, D. Badami, S. Mittl

Bias Temperature Instability Model for Digital Circuits – Predicting Instantaneous FET Response...................... 700
Aditya Bansal, Kai Zhao, Jae-Joon Kim, Rahul Rao

Soft Oxide Breakdown Impact on the Functionality of a 40 nm SRAM Memory...................................... 704
Saad Cheffah, Vincent Huard, Remy Chevallier, Alain Bravaix

The Relationship between Transistor-Based and Circuit-Based Reliability Assessment for Digital Circuits.. 706
Balaji Vaidyanathan, Shawn Bai, Anthony S. Oates

The Impact of RTN on Performance Fluctuation in CMOS Logic Circuits.. 710
Kyosuke Ito, Takashi Matsumoto, Shinichi Nishizawa, Hiroki Sunagawa, Kazutoshi Kobayashi, Hidetoshi Onodera

ESD AND LATCH-UP:

The Modified P+ Electrode Layout Schemes to Enhance ESD Robustness of SCR Structure for PMIC Applications 714
Lu-An Chen, Chang-Tzu Wang, Tai-Hsiang Lai, Tien-Hao Tang, Kuan-Cheng Su

Impact of Shielding Line on CDM ESD Robustness of Core Circuits in a 65-nm CMOS Process 717
Ming-Dou Ker, Chun-Yu Lin, Tang-Long Chang

Test Chip Design for Study of CDM Related Failures in SoC Designs 719
Nicholas Olson, Vrashank Shukla, Elyse Rosenbaum

Nanosecond Transient Thermoreflectance Imaging of Snapback in Semiconductor Controlled Rectifiers 725
Kerry Maize, Dustin Kendig, Ali Shakouri, Vladislav Vashchenko

ELECTROMIGRATION/VOIDING:

Degradation and Failure Analysis of Polysilicon Resistor Connecting with Tungsten Contact and Copper Line 731
Clement Huang, Mingte Lin, James W. Liang, Alex Juan, K. C. Su

A Practical Modeling for Transient Thermal Characteristics of Multilevel Interconnects 734
Seung-Man Choi, Dong-Cheon Baek, Tae-Young Jeong, Myung-Soo Yeo, Miji Lee, Andrew T. Kim, Jongwoo Park

Electromigration of Cu Interconnects under AC, Pulsed-DC and DC Test Conditions: Ramifications on Accelerated Testing 740
Roey Shaviv, Gregory J. Harm, Sangita Kumari, Robert R. Keller, David T. Read

Improving Lifetime of Cu Interconnects with Adding Compressive Stress at Cathode End 746
L. Arnaud, P. Lamontagne, E. Petitprez, R. Galand

EXTREME ENVIRONMENTS:

A Study on the Short- and Long-Term Effects of X-Ray Exposure on NAND Flash Memories 751
S. Gerardin, M. Bagatin, A. Paccagnella, A. Visconti, S. Beltrami, M. Bertuccio, L. T. Czeppel

Application of Reliability Test Standards to SiC Power MOSFETs 756
Ronald Green, Aivars Lelis, Daniel Habersat

FAILURE ANALYSIS:

A Novel and Low-Cost Method to Detect Delay Variation by Dynamic Thermal Laser Stimulation 765
Chunlei Wu, Masuda Motohiko, Winter Wang, Grace Song, Jinglong Li, Joe Yu, Li Tian, Miao Wu

A Study of the Influence of High Voltage Device Characteristics by Electron Beam Irradiation during Nanoprobing 770
Hung Sung Lin

Detecting Laser Beam Reflectance Modulated by Electronic Device Operation with a Simple Setup 774
Carlo Pagano, Christian Boit, Yoshiyuki Yokoyama

Backside Reflectance Modulation of Microscale Metal Interconnects 780
J. K. J. Teo, C. M. Chua, L. S. Koh, J. C. H. Phang

GATE DIELECTRICS:

Nanoscale Electrical and Physical Study of Polycrystalline High-kappa Dielectrics and Proposed Enhancement Techniques 786
K. Shubhakar, K. L. Pey, S. S. Kushvaha, M. Bosman, S. J. O'Shea, N. Raghavan, M. Kouda, K. Kakushima, Z. R. Wang, H. Y. Yu, H. Iwai

Investigation of Progressive Breakdown and Non-Weibull Failure Distribution of High-k and SiO_2 Dielectric by Ramp Voltage Stress 792
Nilufa Rahim, Ernest Y. Wu, Durgamadhab Misra

Oxide Defects Generation Modeling and Impact on BD Understanding 798
Y. Mamy Randriamihaja, V. Huard, A. Zaka, S. Haendler, X. Federspiel, M. Rafik, D. Rideau, D. Roy, A. Bravaix

Comprehensive Analysis of Charge Pumping Data for Trap Identification 802
D. Veksler, G. Bersuker, A. Koudymov, C. D. Young, M. Liehr, B. Taylor

A Physics-Based Model of the Dielectric Breakdown in HfO₂ for Statistical Reliability Prediction 807
Luca Vandelli, Andrea Padovani, Gennadi Bersuker, Jung Yum, Paolo Pavan, Luca Larcher

HIGH VOLTAGE / RF:

Advanced 45nm MOSFET Small-Signal Equivalent Circuit Aging under DC and RF Hot Carrier Stress 811
L. Negre, D. Roy, S. Boret, P. Scheer, D. Gloria, G. Ghibaudo

MEMORY:

Characterization of Hexagonal Rare-Earth Aluminates for Application in Flash Memories 815
M. B. Zahid, R. Degraeve, M. Toledano-Luque, J. Van Houdt

Charge Gain, NBTI Recovery and Random Telegraph Noise in Localized-Trapping NVM Devices 819
Meir Janai, Ilan Bloom, Yael Shur

A Highly Reliable Embedded P-Channel SONOS Memory Using Dynamic Programming Method 824
Ying-Je Chen, Cheng-Jye Liu, Chun-Yuan Lo, Yun-Jen Ting, T. H. Hsu, Wein-Town Sun

Analysis of Edge Wordline Disturb in Multimegabit Charge Trapping Flash NAND Arrays 828
Cristian Zambelli, Andrea Chimenton, Piero Olivo

Investigation of the Programming Accuracy of a Double-Verify ISPP Algorithm for Nanoscale NAND Flash Memories 833
Carmine Miccoli, Christian Monzio Compagnoni, Alessandro S. Spinelli, Andrea L. Lacaita

Precise Understanding of Data Retention Mechanisms for MONOS Memories: Toward Simultaneous Improvement of Retention and Endurance Performances by SiN Engineering 839
Shosuke Fujii, Ryota Fujitsuka, Katsuyuki Sekine, Naoki Yasuda

Variability of Resistive Switching Memories and Its Impact on Crossbar Array Performance 843
An Chen, Ming-Ren Lin

Statistical Analysis of Retention Behavior and Lifetime Prediction of HfOₓ-Based RRAM 847
Lijie Zhang, Ru Huang, Yen-Ya Hsu, Frederick T. Chen, Heng-Yuan Lee, Yu-Sheng Chen, Wei-Su Chen, Pei-Yi Gu, Wen-Hsing Liu, Shun-Min Wang, Chen-Han Tsai, Ming-Jinn Tsai, Pang-Shiu Chen

PROCESS INTEGRATION AND 3D / TSV:

Behaviors and Physical Degradation of HfSiON MOSFET Linked to Strained CESL Performance Booster 852
Kidan Bae, Minjung Jin, Hajin Lim, Lira Hwang, Dongseok Shin, Junekyun Park, Jinchul Heo, Jongho Lee, Jinho Do, Ilchan Bae, Chulhee Jeon, Jongwoo Park

Experimental Study on Origin of V_TH Variability under NBT Stress 857
Yuichiro Mitani, Akira Toriumi

Multiple Cell Upsets Tolerant Content-Addressable Memory 863
Syed Mohsin Abbas, Sanghyeon Baeg, Sungju Park

An Automated Approach to Isolate Dominant SER Susceptibilities in Microcircuits 868
James Castillo, David Mavis, Paul Eaton, Mike Sibley, Don Elkins, Rich Floyd

Bit Error and Soft Error Hardenable 7T/14T SRAM with 150-nm FD-SOI Process 876
Shusuke Yoshimoto, Takuro Amashita, Shunsuke Okumura, Kosuke Yamaguchi, Masahiko Yoshimoto, Hiroshi Kawaguchi

Pulsed Laser-Induced Transient Currents in Bulk and Silicon-On-Insulator FINFETs 882
F. El-Mamouni, E. X. Zhang, R. D. Schrimpf, R. A. Reed, K. F. Galloway, D. McMorrow, E. Simoen, C. Claeys, S. Cristoloveanu, W. Xiong

Neutron- and Alpha-Particle Induced Soft-Error Rates for Flip Flops at a 40 nm Technology Node 886
Srikanth Jagannathan, T. D. Loveless, Z. Diggins, B. L. Bhuva, S.-J. Wen, R. Wong, L. W. Massengill

Analysis of Multiple Cell Upsets Due to Neutrons in SRAMs for a Deep-N-Well Process 891
Nihaar Mahatme, Bharat Bhuva, Y.-P. Fang, Anthony Oates

Impact of Ion-Induced Transients on High-Speed Dual-Complementary Flip-Flop Designs 897
Dolores A. Black, Robert A. Reed, William H. Robinson, Jeffrey D. Black, Daniel B. Limbrick, Kevin D. Dick

THIN FILMS:

Stability Improvement of a-ZIO TFT Circuits Using Low Temperature Anneal 904
Aritra Dey, David R. Allee

TRANSISTORS:

Low-Frequency Noise Behavior of La-Doped HfSiON/Metal Gate nMOSFETs 908
Do-Young Choi, Min Sang Park, Chang Woo Sohn, Hyun Chul Sagong, Eui-Young Jung, Jeong-Soo Lee, Yoon-Ha Jeong, Chang Yong Kang

Neutral Interface Traps for Negative Bias Temperature Instability 913
Z. Chen, X. Zhou, Y. Z. Hu, K. S. Machavolu

Atomistic Approach to Variability of Bias-Temperature Instability in Circuit Simulations 915
B. Kaczer, S. Mahato, V. Valduga de Almeida Camargo, M. Toledano-Luque, Ph. J. Roussel, T. Grasser, F. Catthoor, P. Dobrovolny, P. Zuber, G. Wirth, G. Groeseneken

Probabilistic Defect Occupancy Model for NBTI 920
J. Martin-Martinez, B. Kaczer, M. Toledano-Luque, R. Rodriguez, M. Nafria, X. Aymerich, G. Groeseneken

MOSFET's Hot Carrier Degradation Characterization and Modeling at a Microscopic Scale 926
Y. Mamy Randriamihaja, A. Zaka, V. Huard, M. Rafik, D. Rideau, D. Roy, A. Bravaix

Simultaneous Extraction of Threshold Voltage and Mobility Degradation from On-The-Fly NBTI Measurements 929
Rodolf W. Herfst, Jurriaan Schmitz, Andries J. Scholten

Analysis of Recoverable and Non-Recoverable NBTI and PBTI Using AC and DC Stresses 933
Frederic Monsieur, Eduard Cartier, James Stathis

On the Evolution of the Recoverable Component of the SiON, HfSiON and HfO$_2$ P-MOSFETs under Dynamic NBTI 935
Y. Gao, A. A. Boo, Z. Q. Teo, D. S. Ang

New Observations on the Physical Mechanism of Vth-Variation in Nanoscale CMOS Devices after Long Term Stress 941
E. R. Hsieh, Steve S. Chung, C. H. Tsai, R. M. Huang, C. T. Tsai, C. W. Liang

On the Cyclic Threshold Voltage Shift of Dynamic Negative-Bias Temperature Instability 943
Z. Q. Teo, A. A. Boo, D. S. Ang, K. C. Leong

Author Index

April 10-14, 2011 • HYATT REGENCY Monterey Resort & Spa • Monterey, CA http://www.irps.org/

General Chair
J.H. Stathis—IBM Research
914-945-2559 Fax:...2141
stathis@us.ibm.com

Vice General Chair
E. Ogawa—Broadcom
949-926-5507 Fax: 9240
etogawa1@yahoo.com

Secretary
C. Slayman—Ops a la Carte
charlies@opsalacarte.com

Technical Program
S. Krishnan—Texas Instruments
972-995-9606 Fax:...1724
s-krishnan1@ti.com

Finance
P. Chaparala—Alta Devices
408-721-8985
prasadc@altadevices.com

Tutorial Program
G. Meneghesso, U. of Padova
39-498-277-653 Fax: ...699
gauss@dei.unipd.it

Workshop Program
M. Porter—Medtronic
480-929-5661
Mark.Porter@Medtronic.com

Registration
G. La Rosa—IBM
845-892-3179..Fax:...6850
larosa@us.ibm.com

Arrangements
A. Haggag—Freescale
512-665-2571 Fax:933-6962
Amr.Haggag@freescale.com

Publicity
C.L. Henderson—Semitracks Inc
505-858-0454 Fax:...9813
henderson@semitracks.com

Audio Visual
E. Rosenbaum—Univ. of Illinois
217-333-6754..Fax:244-1946
elyse@illinois.edu

Presentations
R. Kaplar—Sandia National Labs
rjkapla@sandia.gov

Equipment Exhibits
E. Ogawa—Broadcom
949-926-5507 Fax: 9240
etogawa1@yahoo.com

Publications
Y. Chen—JPL
818-393-0940
yuan.chen@jpl.nasa.gov

Year-In-Review
E. Ogawa—Broadcom
949-926-5507 Fax: 9240
etogawa1@yahoo.com

Consultants
Scien-Tech Associates, Inc.
828-898-6375 Fax:...6379
dbarbsta@aol.com

Widerkehr and Associates
301-527-0900 ext. 2
phyllism@widerkehr.com

Historian
Bernie Pietrucha
pietrucha@rowan.edu

PREFACE

On behalf of the IRPS 2011 Management Committee and the IRPS Board of Directors, it is my pleasure to present the 49th edition of the International Reliability Physics technical proceedings. Within these pages are the manuscripts detailing the oral and poster technical presentations that are the heart of this annual symposium.

These proceedings and the symposium are the result of the dedication and tireless efforts of many individuals. First, I wish to express my deepest thanks to my management committee, who volunteered many hours to organize this year's symposium; next, to our consultants whose professionalism and guidance in managing the logistics of symposium, including registration, local arrangements, exhibits, publications, and finances are essential to the success of the meeting.   Finally, this volume would not be possible without the authors whose work was selected for its outstanding technical quality. The 2011 technical program consists of more than 85 contributed talks and 60 poster presentations, selected from over 210 submissions, plus an additional 25 invited papers.

This year the International Reliability Physics Symposium introduces a technical focus on Electronics for Medical and Healthcare Applications, and is pleased to have a keynote talk on this topic delivered by Paul Gerrish, Technical Fellow and Director of Technology Development, Medtronic.

Our signature "Virtual IRPS" will be available shortly after the symposium to provide the video, audio, and presentation material for all of the platform presentations given at all of the technical sessions, including the keynote talk.

In addition to the papers included in these proceedings, the symposium also includes 24 stimulating and educational tutorials, four Reliability Year-in-Review talks, informal workshops and panel discussions, an equipment exhibit, and a lively poster reception. There are many opportunities to network and to increase your knowledge of this technically important and ever changing field.

We hope that you have an enriching and valuable experience at IRPS and look forward to your participation and attendance at many future IRPS.

James H. Stathis
2011 IRPS General Chair

Reliability Year In Review

COMPOUND DEVICES – Reliability of GaN Based HEMTs Devices – Gaudenzio Meneghesso, University of Padova

GaN High Electron Mobility Transistors are excellent devices for high power and high frequency applications. Thanks to the advantageous material properties of GaN-based semiconductors, such devices will operate at very high drain voltages, where the extremely high electric fields drive high current densities in the two-dimensional channel electron gas at fairly high channel temperatures. The resulting operating conditions are by far more severe than those encountered by any other semiconductor developed so far. Even though significant improvements in the quality of GaN-related substrates and epitaxial structures have been achieved in the past few years, GaN based devices still have material-related issues which need to be resolved. This review will present the main parasitic and reliability issues reported in the last year on GaN-based HEMTs devices.

MEMORIES - Review of the State-of-the-Art Flash Memory Devices and Post-Flash Emerging Memories - Hang-Ting Lue, Macronix

Although conventional floating gate (FG) Flash memory has recently gone into the 2Xnm node, the technology challenges are formidable below 20nm. Charge-trapping (CT) devices are promising to scale beyond 20nm but below 10nm both CT and FG devices hold too few electrons for robust MLC (Multi-level Cell, or more than one bit storage per cell) storage. However, due to the simpler structure and its more robust storage (not sensitive to tunnel oxide defects since charges are stored in deep trap levels), CT is much more desirable than FG in 3D stackable Flash memory. Optimistically, 3D CT Flash memory may allow the density increase to continue for at least another decade beyond the 1Xnm node. In this paper, we review the current status of FG devices, their scaling and reliability challenges, and the operation principles of CT devices and several variations such as MANOS and BE-SONOS. We will then discuss various 3D memory architectures, technology challenges and address the poly-silicon thin film transistor (TFT) issues.

Devices that do not rely on charge storage are naturally not limited by the number of electrons, thus promise further scaling below 10nm. Several of the most promising post-flash era devices, their operation principle and critical issues are reviewed. (One of them, phase change memory, will be covered by other speaker and thus not included here.) Their potential applications and challenges for 3D stacking are critically examined.

CIRCUIT RELIABILITY - Circuit Reliability: Cross-layer Resilience Challenges and Solutions - Yu Cao, ASU

Reliability issues that start from the technology level need to be propagated into circuit and system levels for design protection. This session presents recent advancements in design for reliability (DFR), addressing modeling, silicon characterization, on-chip sensing, and simulation techniques. It focuses on reliability analysis in CMOS logic and memory units, with the outreach to emerging technologies, such as flexible electronics.

PACKAGING - Package Reliability - Darvin R. Edwards, Texas Instruments

With the increasing pace of new package technology introduction and the reduction of development time for learning cycles, rapid evaluation of device reliability and failure modes has become ever more important to ensure optimum field lifetimes. New packaging technologies which are currently being prepared or which are in early production include Through-Silicon-Via (TSV) package solutions, Cu wire bonding, embedded die substrates, and fan-out packages. This year's publications surfaced a range of potential concerns for TSVs, including delamination of the Cu from the liner oxides, pump-out of the Cu material, and ILD cracking above the Cu. Wire bonded packages are undergoing a radical technology shift with the move from Au wires to Cu to reduce costs. Much attention focused on Cu wire bond corrosion issues and the mechanisms behind the corrosion. Acceleration of artificial failure modes in life tests which are inconsistent with field failures is a concern for the new technologies, and a number of groups have been studying the applicability of existing and new test methods to weed out artificial fails or to identify gaps in the testing regimes. These and more topics gleaned from this year's literature will be reviewed with the goal of providing a practical summary of the problems and potential solutions.

2011 IRPS Officers and Committees

General Chair
Jim Stathis
IBM

Vice General Chair
Ennis Ogawa
Broadcom

FINANCE
Prasad Chaparala, *Chair*
Alta Devices

ARRANGEMENTS
Amr Haggag, Chair
Freescale

Audio Visual
Elyse Rosenbaum, Chair
University of Illinois

Presentations
Robert Kaplar, *Chair*
Sandia National Labs

Publications
Yuan Chen, *Chair*
NASA Langley

Publicity & Web
Chris Henderson, *Chair*
Semitracks

Registration
Giuseppe La Rosa, *Chair*
IBM

Secretary
Charles Slayman, *Chair*
Ops a la Carte

Tutorials
Gaudenzio Meneghesso, *Chair*
University of Padova

Workshop
Mark Porter, *Chair*
Medtronic

TECHNICAL PROGRAM COMMITTEE

Srikanth Krishnan,
Chair
Texas Instrument

Koji Eriguchi
Co-Chair
University of Kyoto

BEOL Dielectrics

Fen Chen, Chair
IBM

Baozhen Li, Co-Chair
 IBM
Walter Yao
 Global Foundries
Jim Lloyd
 SUNY Albany
Guo Qiang
 SMIC
Jim Hsu
 UMC
Feng Xia
 Intel
A.S. Oates
 TSMC
Tae-Young Jeong
 Samsung
Zsolt Tokei
 IMEC
Naohito Suzumura
 Renesas
Changsoo Hong
 Freescale
Guoyong Yang
 Global Foundries
 Singapore

Compound/Opto-electronics

Jose Jimenez, Chair
TriQuint Semiconductor

Martin Kuball. Co-Chair
University of Bristol

Jesus del Alamo
 MIT
Mitsuo Fukuda
 Tokohashi University
Gaudenzio Meneghesso
 University of Padova
Rama Vetury
 RFMD
Giovanni Verzellesi
 Universita di Modena
Michael Damman
 IAF Fraunhofer
Brian Skromme
 Arizona State University
Aris Christou
 University of Maryland

Chip-Package Interaction

Choong-un Kim, Chair
University of Texas

Li Li, Co-Chair
CISCO

Chang, Je-Young
 Intel
JungWoo Pyun
 Samsung Electronics
Kejun Zeng
 Texas Instruments
Heidi Raynolds
 Sun Microsystems
Ingrid De Wolf
 IMEC
Charlie Zhai,
 nVidia

Circuit Reliability

Vijay Reddy, Chair
Texas Instruments

Chris Kim, Co-Chair
University of Minnesota

Jim Tschanz
 Intel
Kevin Cao
 Arizona State University
Vincent Huard
 ST Microelectonics
Subhashish Mitra
 Stanford University
Manjul Bhushan
 IBM
Karl Hofmann
 Infineon
LeRoy Winemberg
 Freescale
Tony Oates
 TSMC
Georgios Konstadinidis
 Oracle Corporation

ESD and Latchup

Gianluca Boselli, Chair
Texas Instruments

Junjun Li, Co-Chair
 IBM
Guido Notermans
 ST-Ericsson
Dimitri Linten
 IMEC
Michael Khazhinsky
 Freescale
Vladislav Vashchenko
 National Semiconductors

Electromigration/ Voiding

Oliver Aubel, Chair
Globalfoundries

Armin Fischer, Co-Chair
Infineon

Kristof Croes
 IMEC
Yeow Kheng LIM
 Globalfoundries
Guillaume Ribes
 STM
Baozhen Li
 IBM
Christine Hau-Riege
 QualComm
Young-Joon Park
 Texas Instruments
Shinji Yokogawa
 Renesas
Tony Oates
 TSMC
Jim Lloyd
 SUNY
Wen Wu
 Novellus

Extreme Environments

Mark White, Chair
NASA

Failure Analysis

David Su, Chair
TSMC

Fabless & Product Reliability

Tom Anderson, Chair

Gate Dielectrics

Paul Nicollian, Chair
Texas Instruments

Yuichiro Mitani, Co-chair
 Toshiba
Thomas Kauerauf
 IMEC
Andreas Kerber
 Global Foundries
Wing Lai
 IBM
Ziyuan Liu
 Renesas
Enrigue Miranda
 University Autonoma
 Barcelona
Sangwoo Pae
 Intel
Mustapha Rafik
 STMicroelectronics
JR Shih

TSMC
Kin-Leong Pey
 National Taiwan Univ.
Thomas Pompl
 Infineon

High Voltage/RF

Merlyne De Souza, Chair
University of Sheffield

Jong Mun Park, Co-Chair
Austria Microsystems

Young Chung
 Freescale
Sameer Pendarkar
 Texas Instruments
Paul Van Der Wel
 NXP

Medical Electronics

Scott Hareland, Chair
Medtronic

Jit Muthuswamy
 Arizona State University
Yan Liu
 Medtronic
Doug Barlage
 University of Alberta
Suman Datta
 Penn State University
Borna Obradovic
 Texas Instruments

Thermo-mechanical/MEMS

Ganesh Subarrayan, Chair
Purdue University

Memory

Sanjay Rangan, Chair
Intel

Susumu Shuto, Co-Chair
Toshiba

Souvik Mohapatra
 IIT Mumbai
Luca Larcher
 Univ of Modena
Chandra Mouli
 Micron
Robert Gleixner
 Numonynx/Micron
Hiroshi Watanabe
 National Chiao Tung
University
Luca Perniola
 LETI
Yasuhiro Taniguchi
 Renesas

PI Integration/3D/TSV

Jongwoo Park, Chair
Samsung

Chris Connor, Co-Chair
Intel

Matt Nowak
 Qualcomm
Larry Smith
 SEMATECH
Motoyuki Sato
 Toshiba
Vivian Ryan
 GlobalFoundries
Timothy D. Sullivan
 IBM
Kristof Croes
 IMEC
 Henley Liu
 Xilinx

Photovoltaic Devices

S. Ashok, Chair
Pennsylania State University

Bhushan Sopori, Co-Chair
NREL

Norma Sosa
 IBM
Ken Wu
 TSMC
M. Ashraf Alam
 Purdue University
Siva Sivanathan
 University of Illinois -
 Chicago

Soft Errors

Jeff Wilkinson, Chair
Medtronic

Thin Films

Andrea Cester, Chair
University of Padova

Ryoichi Ishihara
Delft Technical University

Genoe Jan
 IMEC
Yue Kuo
Thomas Riedl
 University of Wuppertal
Hsiao-Wen Zan
 National Chiao Tung
 University

Transistor/TFTs

Tanya Nigam, Chair
Global Foundries

BOARD OF DIRECTORS

Thomas M. Moore,
Chair
Omniprobe, Inc.

Marsha Abramo,
IBM, Retired

Edward I. Cole, Jr.,
Sandia National Labs

Lucien A. Kasprzak,
Siemens

Ronald C. Lacoe,
The Aerospace
Corporation

Neal R. Mielke,
Intel

Joe W. McPherson,
Texas Instruments,
Emeritus

Ennis Ogawa,
Broadcom

James Stathis,
IBM Research

CONSULTANTS

SITE SELECTION/ARRANGEMENTS/
EQUIPMENT DEMONSTRATIONS/AUDIO-
VISUAL/SIGNS

David Barber
Scien-Tech Assoc., Banner Elk, NC

PUBLICATIONS/REGISTRATION /
TECHNICAL PROGRAM SUPPORT/WEB
PAGE

Phyllis Mahoney and Wendy Walker
Widerkehr and Associates
Montgomery Village, MD

HISTORIAN

Bernie Petruchka

2009 Paper Awards

Best Paper Award

THE EFFECT OF A THRESHOLD FAILURE TIME AND BIMODAL BEHAVIOR ON THE ELECTROMIGRATION LIFETIME OF COPPER INTERCONNECTS

R. G. Filippi, P.-C. Wang, A. Brendler,
P. S. McLaughlin, J. Poulin, and B. Redder,
IBM Systems and Technology Group, Hopewell Junction, NY, USA
J. R. Lloyd, IBM Research Division, Yorktown Heights, NY USA
J. J. Demarest, IBM at Albany Nano Tech, Albany, NY USA

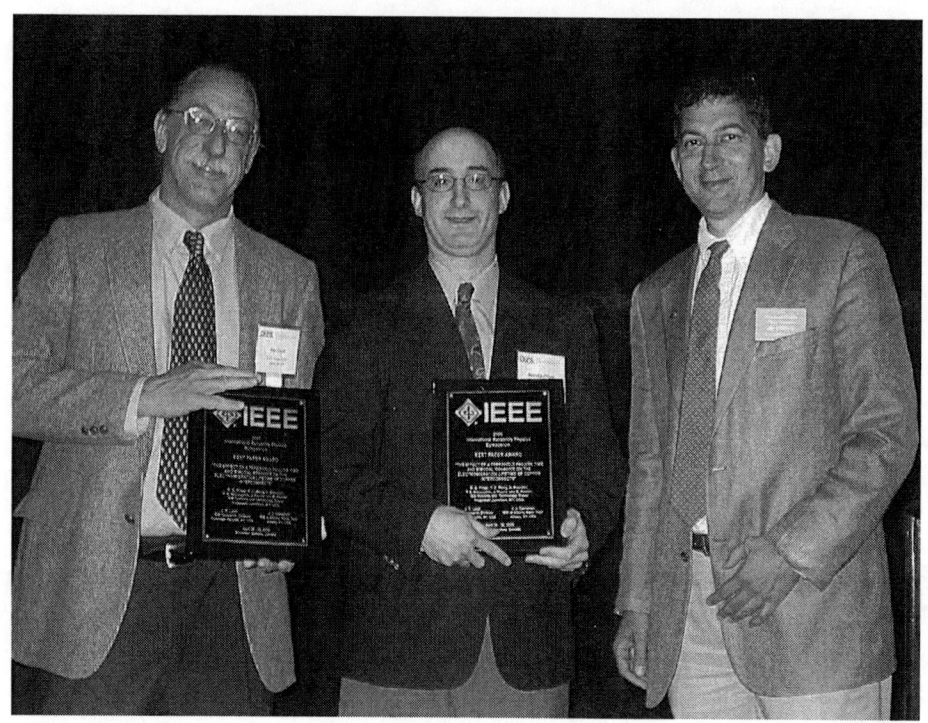

(left to right) Jim Lloyd and Ron Filippi accepting the award from Jim Stathis, 2009 Technical Program Chair

Outstanding Paper Award

Random Telegraph Noise in Highly Scaled nMOSFETs

J.P. Campbell[1], J. Qin[1,2], K.P. Cheung[1], L.C. Yu[1,3],
J.S. Suehle[1], A. Oates[4], and K. Sheng[3]

[1]Semiconductor Electronics Division, NIST, Gaithersburg, MD, USA

[2]Department of Mechanical Engineering,
University of Maryland, College Park, MD USA

[3]Department of Electrical and Computer Engineering,
Rutgers University, Piscataway, NJ USA

[4]TSMC Ltd., Hsin-Chu, Taiwan, R.O.C

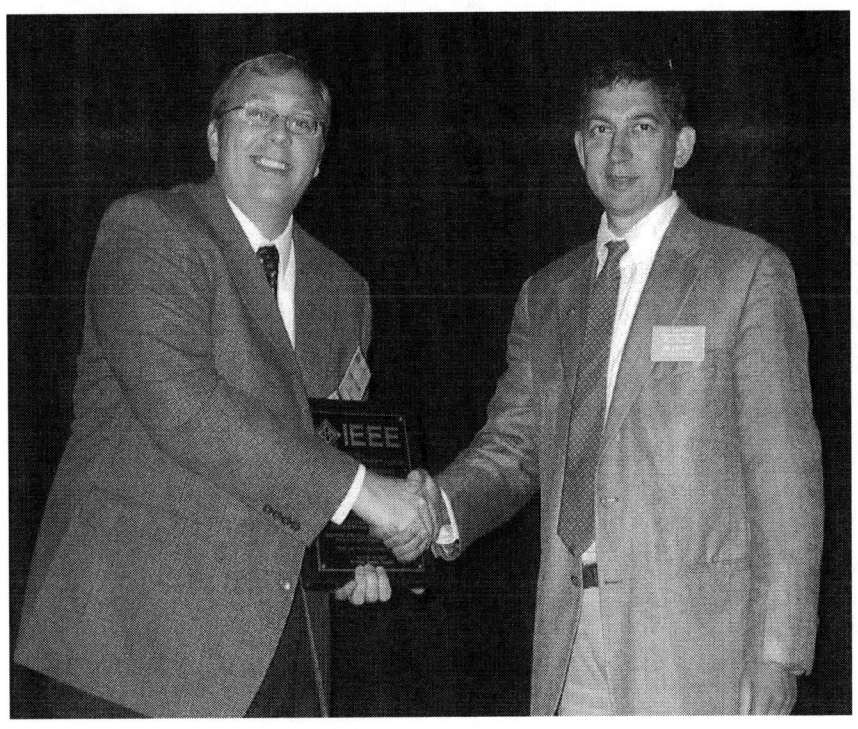

Jason Campbell (left) accepting the award from Jim Stathis, 2009 Technical Program Chair

Best Poster Award

VERY FAST TRANSIENT SIMULATION AND MEASUREMENT METHODOLOGY FOR ESD TECHNOLOGY DEVELOPMENT"

Slavica Malobabic, David F. Ellis, and Juin J. Liou
School of Electrical Engineering and Computer Science
University of Central Florida, Orlando, FL, USA

Javier A. Salcedo, Jean-Jacques Hajjar, and
Yuanzhong Zhou, Analog Devices, Wilmington, MA, USA

Dr. Andrew Olney, Director of Reliability,
Product Analysis, Calibration & ESD,
at Analog Devices, accepting the award
on behalf of the authors from Jim Stathis,
2009 Technical Program Chair.

(left to right) Jim
Stathis, 2009
Technical Program
Chair, with poster
authors Slavica
Malobabic and Juin
J. Liou.

2010 IRPS Paper Awards to be Recognized at the 2011 IRPS

Best Paper Award

To: T. Grasser, H. Reisinger, P. J. Wagner, F. Schanovsky, W. Goes, and B. Kaczer

For the paper entitled: "The Time Dependent Defect Spectroscopy (TDDS) for the Characterization of the Bias Temperature Instability"

Best Poster Award

To: Nathan Jack and Elyse Rosenbaum

For the paper entitled: "ESD Protection for High-Speed Receiver Circuits"

Best Student Paper Award

To: J. T. Ryan, P. M. Lenahan, T. Grasser, and H. Enichlmair

For the paper entitled: "Recovery-Free Electron Spin Resonance Observations of NBTI Degradation"

Keynote Address

Implantable Medical Electronics: Race to the bottom or race to the future?

PAUL GERRISH, DIRECTOR TECHNOLOGY DEVELOPMENT, TECHNICAL FELLOW

IMPLANTABLE MICROSYSTEMS TECHNOLOGY GROUP
MEDTRONIC TEMPE CAMPUS, TEMPE ARIZONA USA

Abstract

Implantable Medical Electronics have made tremendous strides improving the lives of millions of people over the last half century, yet realizing the full promise of this application to treat chronic disease across the globe in both developed and emerging markets has been limited by a number of constraints such as health care access, therapy delivery, total cost of care, system reliability, time and cost to prove clinical evidence, cultural attitudes towards chronic disease, and lack of information and innovation to address prevention of disease rather than just palliative care. New approaches for overcoming these constraints with innovation in ideas and application of technology can make meaningful impacts helping people worldwide live fuller lives with improved health care system outcomes in both cost and quality of care. This innovation is needed to create a better future for patients by engaging medical device makers, health care providers, reimbursement and regulatory bodies in a race to the future – a competitive landscape of innovative technologies that improve patient care, rather than a race to the bottom – a highly constrained, commodity device landscape with lowered incentives that slow innovation in the practice of medicine.

Paul has served in his present role as Medtronic Tempe Campus Technology Director, since January 2008. In this role, he is responsible for technology development, design automation and product development for the New Therapies and Diagnostics organization within the Cardiac Rhythm Disease Management business unit of Medtronic. The Tempe campus technology organization includes over 90 personnel with competencies in developing and supporting next generation technologies and design environments for integrated circuits, electronic packaging and implantable sensors, brought together to enable system solutions across multiple business units of Medtronic. They are energized by the belief that there is still tremendous opportunity for hardware solutions to contribute toward making a difference in improving the lives of people worldwide.

Biography

Paul has been with Medtronic in Tempe for the last 24 years. During that time he has served in a variety of positions in product, IC design, semiconductor device and technology development, where he has made contributions to all IC fabrication processes developed for Medtronic over the past 20 years. This included serving as the primary liaison between Medtronic's IC design community to both internal and external foundries for more than a decade and work as a scout for new sensors, components, and IC technology joint development opportunities prior to taking on the role of technology director.

In 2004, Paul was recognized for his technical contributions to the corporation in developing reliable, ultra low power IC processes and design environments, as a Medtronic technical fellow, and in 2005 with his election to the Bakken society, Medtronic's highest technical award. He holds 3 issued US patents and several pending.

Paul holds a BSEE in Electrical Engineering from the University of New Mexico, and an MSE degree in Electrical Engineering from Arizona State University, with an emphasis in solid state physics. He lives in Phoenix, AZ with his wife and three children.

2011 IRPS Tutorial Program

Gaudenzio Meneghesso, University of Padova

[111] ELECTRICAL CHARACTERIZATION METHODS AND THEIR APPLICATION TO METAL GATE / HIGH-K CMOS RELIABILITY EVALUATION - Andreas Kerber, GLOBALFOUNDRIES - Since the 45nm CMOS process technology node metal gate (MG) / high-k (HK) is used in semiconductor manufacturing. The introduction of high-k as a new gate dielectric in combination with a metal electrode brings additional reliability challenges for qualification of advanced technologies nodes previously not encountered with conventional poly-Si / SiON gate stacks. In addition to negative bias temperature instability (NBTI) in pFET devices, positive bias temperature instability (PBTI) and stress-induced leakage currents (SILC) in nFET devices as well as dielectric breakdown are in the focus of MG/HK reliability. This tutorial summarizes recent achievements in the electrical characterization of MG/HK stacks towards understanding the reliability physics for CMOS applications. In particular time resolved characterization techniques are being discussed and various stress modes are being explored to address the reliability impact. Unipolar and bipolar AC characterization techniques are considered beside the conventional DC device reliability characterization methods to assess BTI, SILC and TDDB. Finally, applicability of Voltage Ramp Stress is explored for MG/HK process screening and monitoring.

[112] FORECASTING BTI IMPACT IN CIRCUITS - IT'S SUNNY AND HUMID WITH CHANCES OF RAIN - Aditya Bansal and Jae-Joon Kim, IBM T.J. Watson Research - In current high-k metal gate technologies, NFETs and PFETs weaken with time causing circuit performance to deviate from post-fabrication test specifications. Accurate prediction of BTI induced degradation before shipping a product is important to eliminate chip failures at customer site. Prediction of BTI impact on circuits can be classified as model-based or direct. Model based prediction involves generating BTI models for FETs, circuit representation and simulation under usage conditions (voltage, temperature etc.). FET level models are typically extensively verified with hardware under various usage conditions. Alternatively, in direct prediction, representative circuits are implemented to monitor BTI induced degradation. These circuits can be used for BTI characterization during technology ramp-up and/or for monitoring BTI shift in the field.

We will discuss in depth the nature of functional failures which differ for each design block; for example, a combination block may suffer from increased delay whereas a memory cell may suffer from stability failures.Further, we'll give a detailed review of the process flow for model-based prediction and existing state-of-the-art BTI stress-recovery characterization circuits. Finally, we will discuss the possible circuit and architecture based solutions to reduce the impact of BTI.

[113] RTN ANALYSIS FOR DEFECT IDENTIFICATION IN ADVANCED GATE STACKS - Gennadi Bersuker, SEMATECH - In highly scaled devices and, especially, devices fabricated on high mobility substrates characterized by a relatively low carrier density, fluctuations of the drive current caused by carrier trapping/detrapping at the defects in the dielectrics presents a serious challenge to meeting performance requirements. Indeed, advanced gate stacks, such as metal/high-k dielectrics with a variety of cap layers, barrier-engineered tunnel oxides in charge trapping memories, etc., exhibit complex compositional profiles, which are usually prone to defect formation. Due to the compositional complexity of these stacks, identifying the nature of traps that contribute to RTN is critically important to process improvement efforts. In the most widely used elastic trapping description, the time of an injected electron capture by a bulk defect is controlled by the electron tunneling from/to the substrate. However, this model was recently shown to be inadequate since it results in RTN times many orders of magnitude shorter than measured ones. A more comprehensive analysis needs to consider that the bulk oxide traps change the atomic configuration of the surrounding lattice when capturing/releasing electrons. These structural changes may control, to a great degree, the characteristic trapping/detrapping times. Therefore, analysis based on explicit consideration of the trap atomic properties provides an opportunity to identify the nature of the defect from the RTN data. The model considering structural relaxation processes is applicable to analyzing charge pumping, trap-assisted tunneling transport, and other measurements probing defects in the dielectrics.

[114] HOT-CARRIER DEGRADATION IN ADVANCED CMOS NODES: FROM THE NBTI SHADOW BACK TO THE FRONT SCENE - A. Bravaix, ISEN-IM2NP - For the last ten years, Negative Bias Temperature Instability (NBTI) has eclipsed all other CMOS device degradation modes, including Hot Carrier (HC) degradation. The difficulties in optimizing last CMOS technologies in convergence with aggressive CMOS scaling have recently seen the Hot-Carrier (HC) phenomena back to the front of topics for device and circuit reliability challenges. Besides, Among all these CMOS nodes, the impact of Channel Hot Carriers (CHC), Cold Carriers (CCC) and Non Conductive Hot-Carriers (NCHC) are indicative of the quality of gate and drain processes and the ability to resist to wearout mechanisms with operating time. This makes the HC examination mandatory to qualify actual CMOS nodes.

This tutorial presents the underlying HC mechanisms through basics to advanced analysis focusing the distinction between the channel energetic 'hot' carriers (CHC), the channel thermalized 'cold' carriers (CCC) and finally the non conductive hot carriers (NCHC) mechanisms. A thorough review of the existing HC modeling yields to a composite lifetime modeling (in three distinctive modes) for the generation of interface traps in advanced CMOS nodes. Oxide charge trapping in thicker gate-oxides is added on top of the generation of neutral traps. All these elements are required to allow a complete modeling function which could be used for both accurate DC and AC device lifetimes. Additionally, temperature effects have a strong effect on the PMOS side through NBTI - and to a lesser magnitude in NMOS under positive bias - we will show that both phenomena are closely interlinked during digital application, while finally, the HC degradation in last CMOS nodes is compared to Self Heating (SH) mechanism which is found much larger in FD SOI technologies than in last silicon bulk 40nm node. This opens new perspectives for an accurate lifetime evaluation of the future scaled CMOS nodes at high frequency operation.

[121] FLASH MEMORY RELIABILITY - A. S. Spinelli and C. Monzio Compagnoni - This tutorial will cover the operation principles and the main reliability constraints which affect the performance of scaled floating-gate non-volatile memories. After

introducing the operating principle of the memory cell and the array architecture, the key reliability issues and their underlying physical concepts are described. The tutorial will then address P/E endurance, charge detrapping and SILC phenomena, random telegraph noise and charge injection effects, and will eventually describe some recently-found phenomena that have become important for Flash reliability. The emphasis of the tutorial will be on physical comprehension rather than quantitative modeling, and will mostly rely on published experimental data rather than theoretical extrapolations.

[122] RELIABILITY ISSUES ON PCM MEMORIES - Matthew J. Breitwisch, IBM T.J. Watson Research Center - This tutorial will review the reliability issues associated with the phase change memory (PCM) technology and will cover the following topics: an introduction to PCM technology including physical structure and electrical characterization, a review of key metrics for performance and functionality, and a review of reliability issues including endurance, resistance drift, retention, and write disturb.

[123] FROM DEVICE TO LIBRARY RELIABILITY IN ADVANCED CMOS NODES - Vincent Huard, STMicroelectronics - The continuous scaling of CMOS technologies down to sub-micron range inevitably yields to increased reliability challenges, such as Negative Bias Temperature Instability (NBTI), Time-Dependent Dielectric Breakdown (TDDB) and Hot Carrier Injection (HCI). All these effects, which contribute to degrade transistor temporal performances, cannot be handled anymore by process means only. Reliability mitigation should then be addressed by design strategies of information processing and knowledge ordering.

A top-down approach is essentially the breaking down of a system to gain insight into in many compositional sub-systems down to libraries elements (standard cells, IPs) often specified with the assistance of "black boxes". However, black boxes may fail to elucidate elementary mechanisms or be detailed enough to realistically validate the model.

A bottom-up approach is the piecing together of systems to give rise to grander systems. In this approach the individual libraries elements of the system are first specified in great detail. These libraries are then linked together to form larger subsystems, which then in turn are linked, sometimes in many levels, until a complete top-level system is formed.

Independently of the design strategy chosen for reliability mitigation, libraries elements are always at the heart of the approach. This tutorial will address the need to move from device to library level reliability to handle coming challenges in advanced CMOS nodes. Both basics and advanced topics on reliability modeling and simulation tools will be presented including: transistor-level reliability compact modeling, library-level modeling approach, ageing effects on various kinds of libraries and simulation approaches for hierarchical reliability analysis.

[124] TIME-DEPENDENT DIELECTRIC BREAKDOWN (TDDB) IN HIGH-K DIELECTRICS - Robin Degraeve, IMEC - The introduction of Hf-based high-k dielectrics as a low-leakage alternative for the SiO_2 gate insulator has raised many questions on the reliability issues of these materials. Concerning Time-Dependent Dielectric Breakdown (TDDB), the physical and statistical models developed for SiO_2 served historically as a

starting point for the research on high-k TDDB. An important change when moving from SiO_2 to high-k is the increased defect density, which poses significant extra reliability challenges even in process-optimized layers. Further complications arise from the change from a single-layer oxide to a multi-layer stack consisting of an interface SiO_2 layer and a high-k. On top of this, the gate electrode has been changed from poly-Si to metal (usually TiN). This tutorial discusses the theory and applicability of some electrical defect characterization methods in high-k (like charge pumping, charge and sense methods, Stress-Induced Leakage Current,...) and demonstrates how they helped in understanding the details of high-k stack degradation and breakdown. The multi-layer aspect of breakdown is discussed with emphasis on the latest developments in the theoretical and statistical models on this subject. The phenomenon of soft, hard and progressive breakdown in high-k is explained and critically evaluated. It is shown how reliability criteria can be constructed in presence of soft and progressive breakdown. Finally, guidelines and boundary conditions for fast TDDB evaluation techniques are presented. For this tutorial, it is assumed that the attendant has basic knowledge of TDDB statistics (Weibull distribution, area scaling,...) and of SiO_2 degradation and breakdown concepts (accelerated testing, percolation model,...).

[131] SEMICONDUCTOR CHIP QUALIFICATION AND VARIABILITY - Pascal Nsame, IBM STG - The robust functional operation of semiconductor chips over their specified lifetime is a requirement for products in the current and future nanotechnology era. The robustness of key IP with a wide range of functions must be realized in the presence of variability. This tutorial compares and contrasts variability sources such as: design, test, analytics, and manufacturing environments. We analyze the impact of the variability of these components on the reliability models; relative to the target product functional robustness across semiconductor technology nodes from 90nm to 32nm with examples from recent high performance compute systems including advanced processors and custom logic.

[132] LATEST DEVELOPMENTS IN FAILURE MODELING OF ELECTROMIGRATION AND ILD TDDB - Jim Lloyd, SUNY Albany College - Recently there has been some progress in the modeling of both electromigration and TDDB for low-k interlevel dielectrics that may impact how these problems are dealt with in engineering. For electromigration, the concept of damage nucleation and growth provides both a more satisfying conceptual framework as well as a more logical method to project lifetimes from test to use conditions than the classical Black Model (n=2) or the modified Black model (which has serious conceptual problems). In addition, the time dependent dielectric breakdown (TDDB) in low-k interlevel dielectrics (ILD) has been shown to behave quite differently than in SiO_2 based dielectrics. Here we have seen that several "root-E" models provide a better fit to the empirical data than the classic but controversial E or 1/E models. These new models will be described in detail and compared with the traditional methods and the physical interpretations discussed in detail.

[133] ELECTROMIGRATION, THERMOMIGRATION, AND STRESS-MIGRATION IN FLIP CHIP SOLDER JOINTS - K. N. Tu, Dept. of Materials Science and Engineering, UCLA, Los Angeles, CA - Owing to the line-to-bump configuration in flip chip solder joints, electromigration is affected by current crowding

and has a unique mode of failure by growing a pancake-type of void across the contact area at the cathode. Joule heating in interconnect line can induce a temperature gradient across flip chip solder joints, so thermomigration accompanies electromigration. A temperature gradient of 1000 °C/cm or a temperature difference of 10 °C across a solder joint of 100 μm in diameter is sufficient for thermomigration. Phase separation occurs in composite flip chip solder joints driven by thermomigration. It has been found that Sn moves to the hot side. On stress-migration, current crowding drives Sn atoms to the anode and leads to Sn whisker growth.

[134] BEOL RELIABILITY CHALLENGES AND ITS INTERACTION WITH PROCESS INTEGRATION - Oliver Aubel, GLOBALFOUNDRIES - In former technologies nodes the reliability investigation was a central part of the process qualification but mainly served as verification for the technology development success only. In recent technologies, latest since 65nm technology node, the reliability characterization is becoming one major part of the technology development itself. It has strong impact on the choice of process options or necessary process changes. Future technology nodes are very challenging with respect to meeting reliability targets. Only with carefully chosen unit process options and an optimized balance between design and reliability demands, technology nodes of 32nm dimensions and beyond can be successfully introduced.

In this tutorial the interaction between process options and reliability requirements will be covered and several critical aspects to ensure BEOL reliability in 32nm technology nodes and beyond will be discussed.

A common key work for intrinsic reliability performance loss is for example the electromigration crisis. Here the reliability robustness is reduced due to scaling leading to reduced critical void volume yielding a resistance increase failure. Aspects such as barrier via side wall coverage or potential etch back issues are not cover is this physics based model. The tutorial will focus on the process related items which overlay the intrinsic reliability performance.

[141] BASICS OF RELIABILITY PHYSICS - J.W. McPherson, Texas Instruments - All devices are expected to degrade with time so device reliability is of great practical importance. Reliability investigations generally start with measuring the degradation rate for a material/device and then modeling the time-to-failure versus the applied stress. The term "stress" is very general ---- any external agent (electrical, mechanical, chemical, thermal, electrochemical, etc.) which is capable of producing material/device degradation. Time-to-failure occurs when the amount of degradation reaches some critical threshold level. Time-to-failure (TF) models generally assume either a power law or exponential stress-dependence and with either an Arrhenius or Erying-like activation energy. From these TF models, the all important acceleration factors can be established and serve as the foundation for accelerated testing.

During this tutorial, the basics of reliability physics and accelerated testing methods are discussed: degradation rate modeling, TF model generation, TF statistical distribution determination, and acceleration factor development. Several TF models, which are commonly used for common IC failure mechanisms, will be discussed. These failure mechanisms include: Electromigration (EM), Stress Migration (SM), Time-Dependent Dielectric Breakdown (TDDB), Negative-Biased Temperature Instability (NBTI), Hot-Carrier Injection (HCI), Surface Inversion/Mobile-Ions, Plasma-Induced Damage (PID), Thermal Cycling (TC) , Energy-Density Issues (EDI), and Single-Event Upsets (SEU). This tutorial will provide the attendee the basics of reliability physics ---- which should serve as a solid foundation for a better understanding of the papers presented at the IRPS.

[142] PHOTOVOLTAIC MODULE RELIABILITY: ENDURING A STORM - Glenn Alers, University of California at Santa Cruz - In contrast to integrated circuits, solar panels need to withstand large variations in environment and weather conditions. Photovoltaic modules must withstand 0-100% humidity at -20C to 100C with voltages up to 1000V and thousands of thermal cycles. Therefore, predicting lifetime is no better than predicting the weather. Yet lifetime estimates directly impact the levelized cost of a power generation system and therefore must be quantified. This tutorial will review the qualification procedures for assuring reliability of photovoltaic panels. The wide range of failure modes for solar panels will be summarized along with the failure analysis techniques that are commonly used. Thermal and emission images are the dominate method for identifying degradation in a solar module.

[143] FAILURE MECHANISMS AND THE USE OF ACCELERATED TESTS IN THE DEVELOPMENT OF RELIABLE PV MODULES - John Wohlgemuth, National Renewable Energy Laboratory - Development of reliable PV modules requires an understanding of potential failure mechanisms. The most straightforward way to determine these failure mechanisms is to observe them in the field, but we can't wait 20 or 25 years (the warranty lifetime of today's commercial PV module) to see:

- What failure mechanisms a module type might suffer from;
- To get an estimate of lifetime or degradation rate (durability) to see if the product can meet the warranty;
- To determine if a change in materials or design will have an effect on the safety, lifetime and durability of the subsequent modules.

Therefore we try to develop stress tests that accelerate the same failure mechanisms that have been seen in the field.
This tutorial will start with a discussion of observed field failure modes for PV modules. From this list and knowledge of the environment in which PV modules operate, a number of accelerated stress tests have been developed. As the tests became more sophisticated and their use more widespread the lifetime and reliability of PV modules increased.

Accelerated stress tests can be utilized in a number of different ways. PV module Qualification tests (like IEC 61215 for crystalline Si or 61646 for Thin Films) have become a commercial tool used by purchasers of PV modules to assure a minimum level of performance and by module manufacturers as both marketing and quality assurance tools. Details of these qualification tests and their strengths and weaknesses will be discussed.

Reliability testing goes beyond Qualification testing to actually cause failures. It is used to:

- Make long term predictions about module lifetime;
- Establish the correct set of accelerated tests to be utilized on new PV technologies; and
- Support efforts in cost reduction by verifying that inherently lower cost designs or use of lower cost materials do not negatively impact the long term safety, reliability, lifetime or durability of PV modules.

[144] RELIABILITY ISSUES IN OPTOELECTRONICS DEVICES - Matteo Meneghini, Gaudenzio Meneghesso, and Enrico Zanoni, University of Padova - With this tutorial we describe in detail the most important physical mechanisms that limit the reliability of optoelectronic devices, focusing on the case of LEDs and laser diodes. Starting with a brief introduction on the operating principles of advanced LED and laser structures, we will describe in detail a number of case studies focused on the analysis of the following degradation mechanisms:

(i) degradation of the active layer due to the generation of non-radiative defects;
(ii) degradation of the ohmic contacts of optoelectronic devices;
(iii) rapid degradation of laser diodes due to the generation of dark-line defects;
(iv) catastrophic degradation of the laser facets;
(v) sudden degradation of LEDs and lasers due to Electrostatic Discharge events.

This tutorial will give a general overview on the critical factors that limit the lifetime of optoelectronic devices, an on the analytical techniques that can be adopted for identifying the degradation mechanisms. Part of the presentation will be devoted to advanced device technologies such as visible LEDs (for lighting and automotive applications) and Blu-Ray laser diodes.

[211] NBTI: RECENT FINDINGS AND CONTRO-VERSIAL TOPICS - Hans Reisinger, Infineon Technologies - Recent investigations of the properties of single defects in the insulators of MOSFETs have created new impact on the controversy about the physical origin of NBTI. Basics and results of these new experiments and corresponding physical models will be explained in other tutorials. This tutorial will be divided in three main parts:

Part 1 will take the main findings from the defect spectroscopy experiments and use them to establish an empirical, but physics-based NBTI model. It will be shown how such a model explains DC-degradation and recovery and also the response to dynamic NBTI stress with arbitrary stress/recovery sequences. Of special interest for combinational logic is high frequency AC-stress. Thus a focus topic will be the understanding of the special features of AC-BTI, for example the duty cycle and frequency dependence and the recovery after AC-stress.

Part 2 will present a critical review of the contradicting findings and claims made regarding NBTI. An example for such a claim - often said to be generally accepted - is that there is a fast, T-independent precursor to NBTI. It will be carefully checked if the most popular claims are in agreement with hard experimental facts or may be believed only as a consequence of experimental artifacts.

Part 3 will give an overview over fast measuring techniques which have been presented in the literature. Only the techniques being able to provide insight into the NBTI-physics - in contrast to fast measurements just done for qualification - will be treated. Some of these techniques are based on commercial instruments; other ones require home-made components. Compared will be their performances and limitations with respect to time and their resolution in extracting $\Box VT$ values.

[212] CHARGE TRAPPING IN OXIDES: FROM RTN TO NBTI - Tibor Grasser, TU Wien - Even under stationary bias conditions, fluctuations in the terminal currents of MOSFETs can be observed, a phenomenon which has become known as random telegraph noise (RTN). The magnitude of these fluctuations increases with decreasing device area. The commonly accepted interpretation explains the noise as a result of stochastic trapping of charge carriers into oxide or interface defects. Experimental data show that the average capture and emission times of this trapping process exhibit pronounced temperature and bias dependences.

Of particular importance is the exponential bias dependence of the capture times, which naturally links RTN to the bias temperature instability (BTI): application of large electric fields results in a dramatic decrease of the capture times, thereby upsetting the dynamic equilibrium typical for RTN. Since the defects present in a MOSFET have a wide distribution of time constants, not all defects capture their charge at the same time. Rather, one defect after the other captures its charge, resulting in slow drifts in the terminal characteristics of the MOSFET. The opposite is observed once the stress field is removed, producing long recovery transients. In nanoscale MOSFETs these degradation and recovery transients proceed in discrete steps, with each step being due to charge exchange of a particular defect with the substrate or the gate.

This tutorial starts with a review of the stochastic properties of charge capture and emission, which can be described by a Markov process. Fundamental properties like the probability distribution of the capture and emission times, and their expectation values and variance will be discussed. A special focus will be also put on multi-state defects (switching traps), as have been observed recently. These multi-state defects are of fundamental importance, as they explain not only charge trapping but also defect creation.

Next, physical models for the capture and emission times required in the Markov model will be discussed. Conventional models based on elastic tunneling or on an extended Shockley-Read-Hall mechanism can neither explain the bias nor the temperature dependence of the data. In contrast, a non-radiative multiphonon processes appears to be consistent with the data. The basic difference between these models is that the latter also considers the additional thermal energy required to accommodate the distortion of the lattice after a charge capture or emission event.

Finally, the implications of these stochastic charge trapping events on the lifetime of nanoscale MOSFETs will be discussed. It will become important to realize that each transistor is unique in various aspects: each transistor will have a random number of defect precursors (scaling with area), each defect will have different properties, and each defect contributes in a stochastic manner to the measured degradation. As a consequence of the stochastic nature of these processes, the lifetime of nanoscale MOSFETs becomes a stochastic quantity, requiring different qualification procedures.

[221] RELIABILITY AND MICROSTRUCTURE OF POLY-SI TFTS - Ryoichi Ishihara, Delft University of Technology - Low-temperature processed polycrystalline-silicon (LT poly-Si or LTPS) has been used as both pixel TFTs and driver circuits for active-matrix liquid-crystal displays (AM-LCDs), because of the much higher mobility than that of the amorphous silicon TFTs. LTPS TFTs have been widely investigated as a potentially suitable material for organic light-emitting diode (OLED) displays as well. If the mobility of poly-Si TFTs is further increased, this technology will enable realization of system on panel (SOP) that will integrate memory, CPU, and display with wireless data communication. For realization of those future applications, improvements of the mobility, uniformity and reliability of LTPS TFTs are very important.

This tutorial will provide an overview of reliability of LTPS TFTs under various static and dynamic operations and its relation with microstructure of LT poly-Si. After introducing fabrication technology for LTPS TFTs, degradations on mobility, flat-band voltage and subthreshold characteristic will be explained in terms of physics of trap generation in LT poly-Si and injection of charges into the gate insulator. In addition, effects of grain boundaries on mobility, uniformity and the bias stabilities will be addressed. Finally, reliability of TFTs fabricated inside a single, orientation-controlled Si grain will be discussed.

[222] RELIABILITY IN THE IMPLANTABLE MEDICAL ARENA - Scott Hareland and Dennis Scranton, Medtronic - It is impossible to separate the reliability of a circuit from the application of the circuit. What is very reliable in one application may not work well at all in another. The implantable medical environment is an application which puts unique demands on electronic circuits and their manufacturing processes. For example, the ultra-low current application raises the significance of Iddq testing. The implantable environment requires a hermetic enclosure for the electronics which in turn establishes a unique atmosphere for the electronics. This internal atmosphere can have impacts on the performance of components like ceramic capacitors. Even supplier relationships and failure reporting take on unique aspects for implantable electronics.

Implantable medical devices must also be designed for a wide variety of operating modes, including sensing signals in the microvolt range, all the way to delivering electrical therapies that operate with energy levels in excess of 30J. All of this is done with the same integrated electrical system which poses unique challenges for the design as well as the reliability of electrical components. These constraints drive, and the relatively low volume of medical electronics permit, design techniques that provide more latitude to increase reliability margin compared to general commercial electronic products. It is important to recognize the operating space and constraints in which implantable medical electronics operate in order to optimize the design and manufacture of medical grade electronic products.

[231] DIELECTRIC CHARGING IN MEMS. A REVIEW OF PRESENT KNOWLEDGE ON ASSESSMENT METHODS AND MATERIALS - George Papaioannou, University of Athens - Dielectric charging constitutes a major problem that still inhibits the commercialization of RF MEMS capacitive switches. During last decade several methods have been applied to assess the charging process involving MEMS switches, MIM capacitors and simple dielectric films on metal or/and insulating substrates.
The device temperature has been used extensively because accelerates the charging and discharging processes by providing energy to trapped charges and to dipoles to overcome potential barriers and randomize orientation and reduce the time of charge collection.

Materials such as SiO_2, Si_3N_4, AlN, Al_2O_3, Ta_2O_5 have been used due to deposition method maturity and high dielectric constant. These materials consist of covalent or ionic bonds, including one piezoelectric, that significantly affects the charging processes. The presence or absence of dielectric film as well as its expansion on the insulating substrate constitute a key issue parameter that influences the charging process.

Finally several theoretical models have been proposed to describe the macroscopic effects of dielectric charging. Aim of the tutorial is to discuss these approaches showing the present available knowledge.

[232] RELIABILITY CHALLENGES RELATED TO TSV-INTEGRATION AND 3D-STACKING - Kristof Croes, IMEC - 3D integration is a technology that allows for the vertical connection of basic electronic components. This technology allows better performance and smaller and cheaper systems, linking various designs and applications (logic, memory, analog, passives, sensors, etc.) together in 3D.

This will however induce unknown reliability issues related to TSV-integration, impact of the TSV on the FEOL and BEOL performance and reliability, the wafer thinning process and the stacking of these thinned chips.

In this course various reliability related challenges encountered in the 3D-stacked IC process which is under development in imec will be discussed. This includes problems related to:

- TSV-reliability itself: stress and stability of the Cu in the TSV, stress induced in the silicon and TSV barrier/liner integrity, etc.
- Possible impact on FEOL and BEOL performance and reliability.
- Backside processing: thinning induced damage in the Si, effect of released copper nail, backside passivation, etc.
- Bonding: Sn Cu micro-bumps reliability, IMC growth, Cu-Cu bonding issues, bonding induced stresses and damage, particles, co-planarity issues, etc.

[241] ESD DESIGN CHALLENGES IN STATE-OF-THE-ART ANALOG TECHNOLOGIES - Gianluca Boselli, Texas Instruments Inc. - The relevance of Analog technologies has rapidly increased over recent years by virtue of the phenomenal success of portable consumer electronics (MP3 readers, smart-phones, navigation devices...), which require a variety of analog functions integrated into the same device.

From an ESD standpoint, the large components portfolio typical of state-of-the-art Analog technologies poses significant development challenges.

The focus of this seminar is the discussion of the technical challenges (high current behavior of analog components, applications requirements, design methodology,) encountered in the ESD design in state-of-the-art Analog technologies.

Peculiar challenges of ESD design for Analog technologies as opposed to Digital technologies will be discussed as well.

[242] ESD PROTECTION CONCEPTS AND METHODOLOGY FOR MODERN SOC DESIGNS - Harald Gossner, Infineon Technologies - Beyond the underlying base ESD elements the overall ESD protection network of a modern system-on chip (SoC) design poses a significant challenge. This class of IC designs require the integration of digital logic, large SRAM or NVM memory blocks, high voltage circuitry, RF high frequency IOs and sensitive analog circuits on one piece of silicon - many of them working on different VDD levels, most of them belonging to separate power domains. The overall ESD network has to establish an efficient ESD protection path between any pin combination across all supply domains. At the same time the ESD circuits at the RF and high voltage IOs need to comply with high performance requirements of IOs exploiting the limits of technology. Advanced circuit solutions and their verification by a SoC compatible EDA verification flow will be presented. The tutorial will highlight the critical issue of CDM. In addition, the specific limitations and novel approaches of failure analysis to support troubleshooting of these designs will be discussed.

BIOGRAPHIES

Abbas, Syed Mohsin
Syed Mohsin Abbas received the B.Sc degree in computer engineering from University of Engineering and Technology Taxila, Pakistan in 2007. He is currently a master's candidate in computer engineering at Hanyang University, Ansan, Korea. His research interests include reliability issue such as Soft Error for memory devices and Design-for-Testability(DFT).

Abe, Kenichi
Kenichi Abe (S'07) was born in Miyagi, Japan in 1983. He received the B.S., M.S., and Ph.D. degrees in electronic engineering from Tohoku University, Sendai, Japan, in 2006, 2008, and 2011, respectively. His current research topics are low-frequency noise and random telegraph noise in deep submicron devices and variability of FET characteristics in ULSI. Dr. Abe received the IEEE Electron Devices Society Japan Chapter Student Award in 2008.

Achanta, Ravi
Ravi Achanta received his MS and PhD(2008) in chemical engineering from Rensselaer Polytechnic Institute(RPI). His Phd research dealt with the reliability of the copper/low-k interconnect system and was funded by Semiconductor Research Corporation (SRC). He is currently a reliability engineer with IBM Microelectronics in Hopewell Junction, NY where he is involved in the technology reliability evaluation of both back-end (TDDB and SM) and front-end (high-k/MG TDDB) material processes. His research interests include microelectronics material reliability and transport processes involved in understanding materials behavior.

Ahlbin, Jonathan
Jonathan R. Ahlbin received B.E. and M.S. degrees in electrical engineering from Vanderbilt University in 2005 and 2009 respectively, and is currently pursuing his Ph.D. there. His research interests include microelectronic circuit analysis and design, and the effects of radiation on integrated circuits, specifically the modeling of circuit-level soft errors.

Aitken, John
John Aitken joined IBM in 1974 at the T.J. Watson Research Center. He is a currently a Senior Technical Staff Member at IBM Burlington managing the Semiconductor Technology Reliability Engineering Department evaluating reliability of new technologies and materials for advanced semiconductor device technologies . He is a Senior Member of the IEEE and past committee member and General Chairman of the International Electron Devices Meeting . He received a MS and Ph.D. in Physics /Materials Science from Rensselaer Polytechnic Institute 1972 and a BS in Physics from Fordham University. John is an Adjunct Professor at University of Vermont.

Alam, Muhammad A
Muhammad Ashraful Alam is a Professor of Electrical and Computer Engineering where his research and teaching focus on physics, simulation, characterization and technology of classical and emerging electronic devices. From 1995 to 2003, he was with Bell Laboratories, Murray Hill, NJ, where he made important contributions to reliability physics of electronic devices, MOCVD crystal growth, and performance limits of semiconductor lasers. At Purdue, Alam's research has broadened to include nanocomposite flexible electronics, organic solar cells, and nanobiosensors. He is a fellow of the IEEE and APS and received the 2006 IEEE Kiyo Tomiyasu Award for contributions to device technology.

Ali, Alaeddine
Ali Alaeddine was born in Zabboud (Lebanon), in April 10, 1984. He received master's degrees in electronics, electrotechnic, automatic from the Lebanese University, Beirut, in 2006, and research master's degrees in microtechnology, architecture, and systems from INSA of Rennes, France, in 2007. Since then, he has been working toward the Ph.D degree in collaboration with the IRSEEM/ESIGELEC, Rouen, France, and GPM/University of Rouen, France, with the partnership of THALES AIR SYSTEMS company, France, dealing with reliability study of the silicon/germanium components under electromagnetic disturbance. His research interests include electromagnetic compatibility and microelectronics reliability.

Allee, David
David R. Allee received a B.S. in Electrical Engineering from the University of Cincinnati in 1984 and the M.S. and Ph.D. in Electrical Engineering from Stanford University in 1986 and 1990, respectively. He was a post-doctoral fellow at Cambridge University in 1990 and 1991. He is currently the director of research, backplane electronics at FDC. He has co authored over 80 archival publications.

Amashita, Takuro
received the B.E. degree in Computer and Systems Engineering from Kobe University, Hyogo, Japan, in 2010. He is currently working in the master course at the same university. His current research is soft-error tolerant SRAM designs.

Andrea, Padovani
Andrea Padovani graduated in Electronics Engineering at the University of Modena and Reggio Emilia, Italy, in 2005. He received his Ph.D. in 2009 from the University of Ferrara, Italy. He is currently an Assistant Professor at the University of Modena and Reggio Emilia, Italy. His research activity focuses on the reliability and modeling of logic transistors based on high-k/metal gate technology, and of innovative non-volatile memories, such as resistive rams (RRAM) and charge-trapping devices (NROM, TANOS). He authored and co-authored more that 40 technical papers in international journals and conference proceedings. He serves as reviewer for several international journals.

Ang, Diing Shenp
D. S. Ang received the B.Eng. (Hons) and Ph.D. degrees in Electrical Engineering from National University of Singapore. He joined Nanyang Technological University, Singapore in July 2002 as an assistant professor. From April 2008 onwards, he has been an associate professor. His research interests include bias-temperature instability, reliability physics and characterization of advanced CMOS and novel devices. He has authored/co-authored more than 100 papers in international refereed journals and conferences. Dr. Ang has served on the technical sub-committees of the International Reliability Physics Symposium and the International Symposium on the Physical and Failure Analysis of Integrated Circuits (IPFA).

Anghel, Lorena
Lorena Anghel got her MS (97) from in Electrical Engineering and Telecommunication Department of Polytechnic Institute of Bucharest, and her PhD in 2000 from INPG, Grenoble. She is currently Professor at National Polytechnic Institute at Grenoble (INPG), France and member of the research staff of TIMA Laboratory. Her research interests include VLSI testing, fault tolerance, soft errors, reliable design, timing optimization, power

analysis and optimization. She has been an Organizing Committee member of Design Automation and test in Europe, IEEE VLSI Test Symposium and IEEE On-Line Test Symposium. She was General Chair of IEEE On-Line Test Symposium in 2005, and Program Chair of DCIS Conference in 2008 and 2009, SERESSA from 2006 to 2008. She is the University Booth chair of DATE 2011 and General Chair of European Test Symposium 2012. She has been involved in European FP and Eureka Projects, as well as French national projects.

Angyal, Matthew
Matthew Angyal received his B. S. in Electrical Engineering from the University of Maryland in 1991 and his Ph. D. in Electrical Engineering from Cornell University in 1996. From 1996 through 2001 he worked at Motorola's Advanced Products Research and Development Lab (Austin, TX) where he contributed to the development of first and second generations of Motorola products utilizing copper interconnect technology. He joined the IBM Semiconductor Research and Development Center in 2001, where he has contributed to copper / low-k interconnect technology development for multiple advanced bulk and SOI product technologies.

Arie, Hiroyuki
Hiroyuki Arie received the B.S. (1994) in faculty of engineering from Okayama University, Okayama, Japan, the M.S. (1996) degrees in engineering from Tsukuba University, Ibaraki, Japan, and the Ph.D. (2000) degree of physics from Tokyo Institute of Technology, Tokyo, Japan. He joined Process Development Dept., Hitachi Ltd. in 2000. He joined Renesas Electronics Corporation in 2010. He has been engaged in designing Si wafer, evaluation of Si crystal characteristics and device failure analysis concerning crystallography.

Aritome, Seiichi
He received the M.E. degrees from Hiroshima University, Japan, in 1985. He joined the Toshiba R&D Center, Kawasaki, Japan, in 1985. He had worked for several companies, Micron, ID USA (2003~), Powerchip, Taiwan (2007~), and Hynix, Korea (2009~).

Arnaud, Lucile
Lucile Arnaud joined CEA-LETI in 1984. She first covered design and characterization of magnetic and electromagnetic passives devices. She has started studies in electromigration physics since 1995 . She was first involved in Al interconnect reliability where her field of interest was failure analysis under pulsed current. She moved to Cu interconnect reliability studies in 2000 where she was active in stress effects related to electromigration degradation. During this time she also chaired many european projects / subprojects for BEOL development. Since 2007 , she is assigned at ST Microelectronics for interconnect reliability expertise of most advanced CMOS technology.

Arora, Rajan
Rajan Arora (S'04) received his B. E. degree in Electronics and Communication Engineering from Punjab Engineering College, India in 2007. He received his M.S degree in Electrical Engineering from Vanderbilt Universty, Nashville, TN in 2009 working on reliability of Germanium and SiC based FETs. Currently he is working towards his Ph. D. degree in Electrical Engineering in the SiGe devices and circuits team at Georgia Institute of Technology, Atlanta, GA. His research interests include RFCMOS, SiGe HBT device and circuit design and their reliability. Rajan has authored or co-authored over 10 refereed publications in addition to conference presentations and

proceedings. He is the co-recipient of the outstanding conference paper award at IEEE Nuclear and Space Radiations Effects Conference (NSREC), 2010.

Athikulwongse, Krit
Krit Athikulwongse received the B.Eng and M.Eng degrees from the Department of Electrical Engineering, Chulalongkorn University, Bangkok, Thailand, in 1995 and 1997, respectively. He also received the M.S. degree in 2005, and is currently working toward the Ph.D. degree from the School of Electrical and Computer Engineering, Georgia Institute of Technology, Atlanta. He joined the Georgia Tech Computer Aided Design Laboratory in 2008. His research focus is on physical design for 3D integrated circuits

Aymerich, Xavier
Xavier Aymerich, Ph.D. from Universitat Autònoma de Barcelona (1980), with a thesis on tunnel devices performed in collaboration with the LAAS in Toulouse. Full Professor from 1991, moved in 1996 to Department of Electronic Engineering. His research interest are the reliability in micro and nanoelectronics, from device failure modeling and device and circuit reliability simulation, including ultra thin gate oxides, high-k dielectrics, characterization of devices by combining SPM tools and conventional microelectronic techniques.. He is member of several national and international committees, and authored more than 350 research papers in scientific journals and conferences.

Bae, Kidan
Technical Quality & Reliability, System LSI division, Samsung Electronics, Korea

Baeg, Sanghyeon
Sanghyeon Baeg received the Ph.D. degrees in electrical and computer engineering from the University of Texas at Austin, Texas, U.S.A in 1994. From 1994 to 1997 he was a Staff Researcher at Samsung Electronics Co., Korea. In 1997, he joined Cisco Systems, Inc. San Jose, CA and worked as a Hardware Manager. Since 2004, he has been working as Professor at Hanyang University, Korea in the School of Electrical Engineering and Computer Science. His work has focused on reliability issue such as HCI, and Soft Error for memory devices, low power SRAM/CAM, and VLSI DFT implementation and methodologies.

Baek, Dong-Cheon
Dong-Cheon Baek received the B.S., M.S., and Ph.D. degrees in mechanical engineering from Korea Advanced Institute Science and Technology (KAIST), Daejeon, Korea, in 2001, 2003 and 2009, respectively. Since 2009, he is with Technology Quality Reliability in SYSTEM LSI, Samsung Electronics as a senior engineer responsible for interconnect reliability qualification of advanced CMOS process and product.

Bagatin, Marta
Marta Bagatin received the Laurea degree (cum laude) in Electronic Engineering from the University of Padova, Italy, in 2006. Since January, 2007, she has been following the Ph.D. School in Information Science and Technology at the Department of Information Engineering, University of Padova. Her research interests concern radiation and reliability effects on electronic devices, especially on volatile and non-volatile semiconductor memories. The results of her work were recognized with the Outstanding Student Paper Award at NSREC 2008 and NSREC

2009, the Best Student Presentation Award at RADECS 2008, and the NPSS Phelps Award 2009.

Bai, Shawn
Shawn Bai received his Ph.D. in Electrical Engineering Department, University of Cincinnati, Ohio in 1998. He is currently working in TSMC, Taiwan focused on device reliability, device modeling and characterization. Before he joined TSMC, he worked for Texas Instruments, Agere Systems/Lucent Technologies and Motorola engaged on CMOS and BJT device modeling, device RF modeling.

Baik, Seung Jae
Seung Jae Baik received the B.S., M.S., and Ph.D. degree in electrical engineering from the Korea Advanced Institute of Science and Technology (KAIST), Daejeon, Korea, in 1994, 1996, and 2001, respectively. From 2001 to 2009, he was with Samsung Electronics Co. as a senior engineer and principle engineer, where he has contributed to novel Si devices and high density flash memories. From 2009 to present he is with KAIST as a research professor. His current research interests include high density memory devices, solar cells, nano-structured devices, and devices based on colloidal quantum dots.

Balasubramanian, Sriram
Sriram Balasubramanian received his B.Tech. degree from the Indian Institute of Technology, Madras, in 1998 and the M.S. and the Ph.D. degree from the University of California, Berkeley in 2002 and 2006 respectively. From 2006 to 2008, he worked as a Device Engineer with Advanced Micro Devices focused on SRAM bitcell optimization and scaling for AMD's high-performance technology. He is currently a Member of Technical Staff at GLOBALFOUNDRIES, Sunnyvale CA, and continues to work on SRAM scaling across multiple technologies. His interests include CMOS devices, device variability, device-design interactions and SRAM bitcell scaling into the nanoscale regime.

Ball, Dennis
Dennis Ball is an engineer at the Institute for Space and Defense Electronics.

Banerjee, Sanjay
His research area is the development of the process integration and electrical characterization of metal gate high-k stack and its reliability, including the charge trapping characteristics and dielectric breakdown mechanism of the thin high-dielectric MOS device. He also explored GaAs MOSFET, charge trap layer Flash memory and Graphene nano-device intensively during his graduate study.

Bansal, Aditya
Aditya Bansal received the B.Tech. degree in Electrical Engineering from the Institute of Technology, BHU, India, in 2001; M.S. and the PhD degrees in Electrical and Computer Engineering from Purdue University, West Lafayette, IN, in 2003 and 2007, respectively. He is currently a Research Staff Member at IBM T. J. Watson Research Center, Yorktown Heights, NY. His current research interests are in design of technology characterization circuits for analysis and optimization of technological challenges in extremely scaled silicon technologies with primary focus on process immunity and temporal reliability. In addition, he is involved in the design of IBM's high performance servers. He has (co-)authored over 35 research papers in refereed journals and conferences.

Bari, Daniele
Daniele Bari was born in Rovigo, Italy, in 1984 and received the degree cum laude in Electronic Engineering from the University of Padova (Italy) on December 2009 with the thesis "Thermal characterization and Reliability study of Organic semiconductor LED". At present he is working as Ph.D student at MOST at Departement of Information Engineering - University of Padova, working on OLED (organic LED) and on DSSC (dye sensitized solar cells). He is also a IEEE Student Member.

Barry, Linder
Barry P. Linder received his B.S. from Pennsylvania State University in 1993, and an M.S and Ph.D. in Electrical Engineering from the University of California at Berkeley in 1999. His doctoral thesis dealt with plasma processing, plasma implantation, and plasma charging damage. Since graduation Dr. Linder has been employed as a Research Staff Member at the IBM T. J. Watson Research Center, Yorktown Heights, NY. Initially his work centered on the breakdown of ultra-thin gate oxides, including the statistics of breakdown phenomenon, post-breakdown conduction mechanisms, and the interaction between oxide breakdown and circuit operation. This work formed the basis for the paper that received an "Outstanding Paper Award" at the 2003 International Reliability Physics Symposium. After 2003, his focus switched to electrical characterization and integration of metal gates and high-k materials. He has studied the full array of advanced gate stack materials including their integration and their effect on effective work function, channel mobility, gate leakage, and inversion layer thickness scaling. He specialized on the interaction between cap layers, interface layers, and metal gate composition on the final electrical properties of the gate stack. As manufacturing of high-k/metal gate stacks approached, he concentrated on all reliability aspects with emphasis on dielectric breakdown and bias temperature instability. More recently, he has added focus on circuit reliability, integrating device level reliability with circuit level functionality.

Bart, Vandevelde
Dr. Bart Vandevelde received his Masters degree in mechanical engineering from the Catholic University of Leuven (Belgium) in June 1994. In March 2002, he received a PhD degree at IMEC in the field of thermo-mechanical modelling for electronic packages. Currently, he is responsible for the packaging level reliability research team at IMEC. He has many publications in the field of thermal and thermo-mechanical modelling and characterisation for advanced IC packaging technologies. He is also co-founder and member of the organisation committee for the IEEE conference EUROSIME.

Bashir, Muhammad
Muhammad Bashir received the B.S degree from the Electrical Engineering Department, University of Engineering and Technology, Lahore, Pakistan in 2006 and the M.S degree in ECE from the Georgia Institute of Technology, Atlanta, GA in 2008. He is working towards his Ph.D. degree at the Georgia Institute of Technology. His research interests include reliability and yield modeling.

Baumann, Robert
Dr. Baumann heads the Texas Instruments radiation effects program and is a TI and IEEE Fellow and EDS Distinguished Lecturer with 21 years of experience in semiconductor reliability. He co-led the SIA panel tasked with mitigating the impact of ITAR export restrictions responsible for changes to ITAR that reduced the risk of U.S. commercial electronics becoming inadvertently controlled. He led and was one of the authors of the JEDEC JESD89 test standard for radiation testing of commercial

microelectronics and was awarded the JEDEC Chairman's Award. He has published > 55 papers, two book chapters, and holds eight U.S. patents.

Beltrami, Silvia
Silvia Beltrami was born in Bergamo, Italy, in 1978. She received the Laurea degree in Electronic Engineering from Politecnico di Milano, Italy, in 2003. She joined STMicroelectronics in 2003 and now she is working in the R&D Technology Center of Numonyx. Her work is about the reliability improvement of flash memory and, in particular, NAND and multilevel devices. She is a co-author of various scientific papers about flash memories.

Bersuker, Gennadi
Gennadi Bersuker completed his M.S. and Ph.D. in Physics at the Leningrad State University and Kishinev State University, respectively. After graduation, he joined Moldavian Academy of Sciences, and then worked at Leiden University and the University of Texas at Austin. Since 1994, he has been working at SEMATECH on electrical characterization of Cu/low-k interconnect, high-k gate stacks, advanced memory, and CMOS process development. He is an editor of IEEE Transactions on Device Materials and Reliability and has been involved in organizing, chairing, or serving as a committee member in a number of technical conferences, including IRW, IRPS, IEDM, ULSI-TFT, ISAGST, LEC, NGCM, APS. He is a SEMATECH Fellow and has published over 200 papers on the electronic properties of dielectrics and semiconductor processing and reliability.

Berthold, Jörg M.
Joerg Berthold received the Diplom and Dr.rer. nat. degrees in physics from the University of Heidelberg, Germany. From 1985 to 2007, he was involved in the development of processes for semiconductor fabrication , in the design of low-power digital circuits, and in investigations of the interaction between digital circuits and <100nm-technologies, at Siemens Semiconductor, Infineon and at Qimonda. Since 2007, at Infineon and Intel Mobil Communications, he is working on the power-management of complex SoCs, and on the power-efficient implementation of communication macros.

Bertuccio, Massimo
Massimo Bertuccio was born in Vizzolo Predabissi (Milano, Italy), in 1983. He received his Master Degree in Electronic Engineering from Politecnico di Milano, Italy, in 2008. He joined Numonyx in the same year and he is currently working in the Micron R&D Division of Agrate Brianza, Italy, dealing with reliability improvement of NAND Flash and Phase Change Memory devices.

Beyer, Gerald P.
Gerald Beyer is the program manager of Cu/low-k program. His background is sputter deposition. He earned a PhD in Materials Science from Imperial College, London.

Bhuva, Bharat
Bharat L. Bhuva reeived his Ph.D. degrees in electrical engineering from North Carolina State University, Raleigh. He is currently a Professor of electrical engineering with the Department of Electrical Engineering and Computer Science, Vanderbilt University, Nashville, TN. His research interests include criuit design, radiation effects on microelectronics, CAD tools development, and biosensors, .

Black, Dolores
Dolores A. Black receied her B.S. from the University of New Mexico in 1991, M.S from Vanderbilt University in 2006 and is currently a Research Assistant and PhD candidate at Vanderbilt in electrical engineering. After completing her B.S degree she worked in the aerospace industry in the areas of circuit design, FPGA and ASIC design for 15 years before returning to graduate school. Her research interest is in soft error rate prediction as it applies to complex digital IC's, FPGA's and ASICS.

Black, Jeffrey
Jeffrey D. Black is a Research Associate Professor for Electrical Engineering at Vanderbilt University. He received a BSEE at the United States Air Force Academy in 1988, a MSEE at the University of New Mexico in 1991, and a Ph.D. at Vanderbilt University in 2008. Jeff's areas of specialty and interest are single event effects and mitigation approaches. Prior to joining Vanderbilt University in 2004, Jeff worked for Mission Research Corporation, now ATK Mission Research, in Albuquerque, NM for 11 years. He spent most of his time in the Microelectronics Division and was its manager for the last 20 months. Efforts performed included program management; microelectronic design, layout, and testing; radiation testing; computer architecture identification; and consultations on navigation and communications systems. From 1988 to 1993, Jeff was assigned to the Phillips Laboratory, now Air Force Research Laboratory. At the laboratory, Jeff worked on advanced space communications projects, specifically on technology for laser and RF crosslinks, and on survivable space communications.

Bloom, Ilan
Dr. Bloom is Fellow of Spansion Israel Ltd., specializing in device and process technology, product algorithms, bit statistics and reliability. he received his B.A. in Physics (1987) and. B.Sc., M.Sc. and Ph.D. in Microelectronics (EE) from the Technion, Israel Institute of Technology (1992). From 1998-2003 he was Technology Development Director at Saifun Semiconductors, Ltd., developing the MirrorBit NVM technology. From 1992-1994 he was a Postdoctoral Fellow at the University of Illinois at Urbana-Champaign, and from 1994-1998 he was EE Faculty member at the Technion, IIT. Ilan published 33 papers, 36 conference papers, 7 patents and 2 book chapters.

Boit, Christian
Prof. Christian Boit received the diploma of physics and the Ph.D. of electrical engineering at Berlin University of Technology, Germany. He joined Siemens Research Laboratories, Munich, Germany, in 1986, as expert of photon emission and laser stimulation. He participated IBM-Siemens 64M DRAM project at East Fishkill, NY. Later, he became Director of Failure Analysis at Infineon Technologies. He currently holds a chair of Semiconductor Devices at Berlin University of Technology. Dr. Boit co-founded EDFAS and the European EUFANET. He was General Chair of ISTFA 2002 and is member of IEEE, VDE and acatech, the German Academy of Science and Engineering.

Bonilla, Griselda
Dr. Griselda Bonilla is the Manager of the Materials and Reliability Sciences group at IBM Research in Yorktown Heights, NY, where she is responsible for extending IBM's microelectronics technology and manufacturing leadership through innovations in materials, processes, and reliability methodology. Specifically, her work deals with a detailed understanding of the impact of scaling and material properties on reliability performance of semiconductor devices. Prior to joining IBM, Griselda earned a Ph.D. in Chemical Engineering from the

University of Massachusetts Amherst, and was awarded the "Best PhD in Particle Technology" by the American Institute of Chemical Engineers. She has authored or coauthored over 40 papers and presentations, including a paper in Science.

Boo, Ann Ann
A. A. Boo received the B. Eng. (Hons.) degree from the School of Electrical and Electronic Engineering, Nanyang Technological University, Singapore in August 2009. She has been working towards the Ph.D. degree in the same school under the GLOBALFOUNDRIES Singapore-NTU Graduate Research Programme since then. Her research interests are mainly in device reliability physics and characterization, focusing on the bias-temperature instability problem.

Boret, Samuel
Samuel Boret was born in Malo-les-Bains, France, on December 23, 1972. He received the Ph.D. degree from the University of Lille, Lille, France, in 1999. In 1996, he joined the Centre Hyperfréquences et Semiconducteurs, University of Lille. As part of his graduate studies, he was involved with monolithic integrated circuits in coplanar technology for applications of reception up to 110 GHz. He is currently with Central Research and Development, RF Electrical Characterization Group, STMicroelectronics, Crolles, France. His main interests include design, characterization and modeling of RF devices in advanced silicon technologies.

Bosman, Michel
Michel Bosman received his M.Sc. degree in Materials Science from the Delft University of Technology, Delft, Netherlands, and Ph.D. degree in Electron Microscopy from the University of Sydney, Australia. He is currently with the Institute of Microelectronics, A*STAR, Singapore. His research interests include nanoscale optics and atomic-resolution spectroscopy using transmission electron microscopes.

Brain, Ruth
Ruth A. Brain is a Principal Engineer with the Technology and Manufacturing Group based in Hillsboro, Oregon. She is currently the interconnect integration group leader responsible for the SoC interconnect technology development. Ruth joined Intel in 1995 working on Intel's 0.25um process technology in the metal deposition area. In 1997, she worked to develop the processes for Intel's first Cu interconnect process at the 130nm technology node, and has subsequently worked on the 90nm, 65nm, and 32nm process technologies. Ruth earned her bachelor's degree in physics and mathematics from Iowa State University in 1990. She received her M.S. and Ph.D. in physics from the California Institute of Technology in 1993 and 1995, respectively.

Bravaix, Alain
Alain BRAVAIX graduated from the University of Sciences of Paris and received the Ph.D. degree in microelectronics in 1991, a post doctorate fellowship at the Institut d'Electronique et de Microélectronique du Nord (IEMN). Since 1994 he is teaching for Engineering and Master Degree at the Institut Supérieur d'Electronique et du Numérique (ISEN-Toulon) developing research activities in collaboration with STMicroelectronics (Crolles) on the reliability and optimization of CMOS and BICMOS nodes for Design in Reliability. Since 2000 he is a member of the Institut Matériaux Microélectronique Nanosciences de Provence (IM2NP) UMR 6242.

Brown, Thomas M.
Thomas Brown graduated in Physics from Rome University with a thesis on a-Si:H. He investigated poly-Si TFTs as research assistant at Cambridge University Engineering Department (1996-97) and polymer LEDs for his PhD at the Cavendish Laboratory. From 2001-05 he developed E-paper at Plastic Logic Ltd as Senior Engineer. In 2005 he won a "Re-entry" Fellowship awarded by the Italian Ministry of University and Research to carry out research on organic photovoltaics within the Electronic Engineering Department of the University of Rome – Tor Vergata. In 2006 he co-founded the Centre for Hybrid and Organic Solar Energy and became Associate Professor in 2007. He has authored over 50 publications and 11 patents

Burdeaux, David
David Burdeaux graduated from Michigan State University with a B.S. in Chemical Engineering in 1984. He received a M.S. in Manufacturing Management from GMI Engineering & Management Institute in 1989. He worked for The Dow Chemical Company between 1984 and 1994 in various research functions. In 1994 he joined Motorola's Semiconductor Product Sector (which later became Freescale Semiconductor). At Motorola / Freescale he has held various positions in Process and Device Engineering for silicon based Bipolar and MOSFET device manufacturing and development. He has been in his current Device Design position since 2001 working on high voltage LDMOS.

Burger, Wayen
Wayne Burger graduated from MIT with a Ph.D. in Electrical Engineering in 1987, with his thesis work focusing on the deposition and characterization of low temperature silicon epitaxial films. After working on BiCMOS SRAM's at National Semiconductor for two years, he joined Motorola's Semiconductor Product Sector (which later became Freescale Semiconductor) in 1990. Dr. Burger has been manager of the RF-LDMOS Device Development team at Freescale Semiconductor since 1994. During this time LDMOS has grown to rapidly become the dominant PA technology in cellular infrastructure, and is expanding into adjacent RF power markets.

Cacho, Florian
Florian Cacho was born in Belfort, France. He received the PhD degree in Material Science from l'Ecole des Mines de Paris in 2005. Since 2005, he has been employed by STMicroelectronics, Crolles. He is in charge of reliability models in Design-in-Reliability team.

Campbell, Jason
Jason P. Campbell received his B.S. and Ph.D. in Engineering Science from the Pennsylvania State University, University Park, PA in 2001 and 2007, respectively. In 2007, he was awarded a National Research Council (NRC) post-doctoral fellowship which he spent in the Semiconductor Electronics Division at the National Institute of Standards and Technology (NIST) where he is currently still employed as a member of the technical staff. He has contributed to more than 40 refereed papers and conference presentations at national and international conferences and has been involved in the technical and managerial committees of both the IIRW and IRPS conferences. His research interests involve the fundamentals of NBTI, random telegraph noise in highly scaled devices, galvanomagnetic effects, and alternative magnetic resonance measurements.

Cao, Yu

Yu Cao received the B.S. degree in physics from Peking University in 1996. He received the M.A. degree in biophysics and the Ph.D. degree in electrical engineering from University of California, Berkeley, in 1999 and 2002, respectively. He is now an Associate Professor of Electrical Engineering at Arizona State University, Tempe, Arizona. He has published numerous articles and coauthored one book on nano-CMOS physical and circuit design. His research interests include physical modeling of nanoscale technologies, design solutions for variability and reliability, and reliable integration of post-silicon technologies.

Carbonell, Laureen

Laureen Carbonell received the diploma in chemistry from the Ecole Nationale Supérieure de Chimie de Toulouse (ENSCT), France, in 1996. From 1997 until 2000 she carried out her Ph.D. in materials science at the Laboratoire de Minéralogie-Cristallographie de Paris (LMCP), France, focusing on the epitaxial growth and characterization of ferromagnetic layers deposited on ZnSe-based semiconductors. Since 2000 she has been with the Interuniversity Microelectronics Center (IMEC), Leuven, Belgium, where her R&D activity has been focusing on the metallization of interconnects. She has authored and co-authored more than 60 publications. In 2009 she received the Michel Lerme Best Paper Award of the International Interconnect Technology Conference for her work on metallization of sub- 30 nm interconnects.

Carbonell, Laureen

Laureen Carbonell received the diploma in chemistry from the Ecole Nationale Supérieure de Chimie de Toulouse (ENSCT), France, in 1996. From 1997 until 2000 she carried out her Ph.D. in materials science at the Laboratoire de Minéralogie-Cristallographie de Paris (LMCP), France, focusing on the epitaxial growth and characterization of ferromagnetic layers deposited on ZnSe-based semiconductors. Since 2000 she has been with the Interuniversity Microelectronics Center (IMEC), Leuven, Belgium, where her R&D activity has been focusing on the metallization of interconnects. She has authored and co-authored more than 60 publications. In 2009 she received the Michel Lerme Best Paper Award of the International Interconnect Technology Conference for her work on metallization of sub- 30 nm interconnects.

Cartier, Eduard

Eduard Albert Cartier was born in Switzerland in 1951. He earned a bachelors degree (1971), a masters degree (ETH, Zurich, Switzerland, 1977) and Ph.D degree (Dr.sc.nat., granted with honors) from ETH in 1982. From 1984 till 1987 he worked as a research staff member at the ASEA Brown Boweri (ABB) Research Center in Baden-Dattwil, Switzerland. Since 1988, he works as a research staff member of the IBM Research Division at the T.J. Watson Research Center in Yorktown Heigths, NY, USA. Over the last 10 years he has been working on the development of alternative, high-k gate dielectrics for CMOS applications.

Carulli, John

John M. Carulli Jr. is a distinguished member of the technical staff in the Analog Engineering Operations division of Texas Instruments. He was previously the manager of design reliability activities for new technology development in the External Development and Manufacturing division. His research interests include outlier analysis, product reliability modeling, and performance modeling. He has an MSEE from the University of Vermont. He is a member of the IEEE and the Material Research Society.

Castillo, James

James A. Castillo received his B.S. Degree in Electrical Engineering from the New Mexico Institute of Mining and Technology in 2010. While at Micro-RDC, he has taken the technical lead of the Milli-BeamTM heavy-ion test program and is also leading a project to emulate single event effects through laser irradiation. Prior to joining Micro-RDC, his major interests were working with laser control systems for the detection, monitoring, and mapping of biochemical contamination and working on data generation and processing techniques for the Las Alamos Neutron Science Center's proton storage ring.

Catthoor, Francky

Francky Catthoor received the engineering degree and a Ph.D. in electrical engineering from the Katholieke Universiteit Leuven, Belgium in 1982 and 1987 respectively. Between 1987 and 2000, he has headed several research domains in the area of high-level and system synthesis techniques and architectural methodologies, including related application and deep submicron technology aspects, all at the Inter-university Micro-Electronics Center (IMEC), Heverlee, Belgium. Currently he is an IMEC fellow. He is part-time full professor at the EE department of the K.U.Leuven. In 1986 he received the Young Scientist Award from the Marconi International Fellowship Council. He has been associate editor for several IEEE and ACM journals, like Trans. on VLSI Signal Procsesing, Trans. on Multi-media, and ACM TODAES. He was the program chair of several conferences including ISSS'97 and SIPS'01. He has been elected an IEEE fellow in 2005.

Cester, Andrea

Andrea Cester received the degree (magna cum laude) in electronic engineering and the Ph.D. degree in electronic and telecommunication engineering from the Padova University, Italy, in 1998 and 2002, respectively. He is currently an Assistant Professor with the Department of Information Engineering, Padova University. He is the author of more than 120 papers published in international journals and conference proceedings. His previous research interest included the reliability issues of deep-submicrometer CMOS devices and advanced nonvolatile memories. His current research interests are characterization, reliability, and modeling of organic electronic devices such as OTFT, OLED, and organic and hybrid solar cells.

Cha, Seon-Yong

Seon Yong Cha received the B.S. in 1990, M.S. in 1995, and Ph.D. in 2000, in the electrical engineering from Korea Advanced Institute of Science and Technology (KAIST), Taejeon, Korea. Since 2005, He has been with Hynix Semiconductor Inc. in developing DRAM. Currently, he is a senior member of technical staff in Hynix R&D Division and leading the Device Technology Team. His current interests are to develop cell and peripheral transistors for next generation DRAM.

Chanda, Kaushik

Kaushik Chanda received the B.Tech in Metallurgical Engineering and Materials Science (1999) from Indian Institute of Technology, Bombay, India, and the M.S. in Materials Science and Engineering (2001) from Rensselaer Polytechnic Institute, Troy, New York. Since then he has been working as an Advisory Engineer at IBM's Systems and Technology Group (STG) focusing on BEOL technology reliability issues including dielectric TDDB, electromigration, stress voiding and thermal cycling.

Chang, Soon W.
Soon W. Chang received his BS in electrical engineering from University of Washington and a graduate certificate from Stanford University. He is currently the manager of product engineering at Renesas Electronics America. He has 9 years experience in diffusion and CVD process engineering and 9 years experience in product engineering specializing mirco-controller with embedded flash.

Chappaz, Cédrick
Cedrick Chappaz received the M.S. in electrical and optronics engineering from the Ecole Nationale Superieure d'Ingenieur (ENSI CAEN) in 1998. Afterwards, he pursued a Ph.D. in the field of Microsystems. He obtained its Ph.D. in 2003. its expertise in MEMS, integrated optics and characterization lead him to do a post-doctorate year in the Commissariat a l'Energie Atomique (CEA) in LETI laboratories from 2003 to 2004. Then, he joined STMicroelectronics as an expert in Reliability and Characterization fields in 2004. Since this date, he's in charge of the qualification of BEOL of advanced technologies and specific 3D integration.

Cheffah, Saad
Saad Cheffah received the Engineer degree in electronics from the Ecole Nationale Superieure d'Ingenieurs de Caen (ENSICAEN), Caen, France. and the M.S degree in solid state physics from the University of Caen-Basse Normandie, Cean,France, in 2007. In 2008 he worked as an application engineer at the Commissariat a l'Energie Atomique (CEA),Saclay,France. He is currently working toward the Ph.D. degree in micro and nano-electronics at the university of Aix-Marseille,France, with close collaboration between the IM2NP loboratory and STMicroelectronics, Crolles,France.

Chen, Yu-Sheng
Yu-Sheng Chen is a device and testing engineer in ITRI, Taiwan. His research interests include CMOS technology, non-volatile memory technology, and circuit design for memory. Chen is a PhD student in Institute of Electronics Engineering, National Tsing Hua University, Taiwan.

Chen, Chia-Yu
Chia-Yu Chen (S'07) received the B.S. degree from National Central University, Taiwan, R.O.C., in 2003; he received M.S. and Ph.D. degrees from Stanford University, CA, USA in 2008 and 2011, all in electrical engineering. He is currently a postdoc associate in Massachusetts Institute of Technology, MA, USA. He has held positions at KLA-Tencor, Berkeley Design Automation, AMD, TSMC, Global Foundries, and Intel Research, where he worked on device scaling, noise, distortion, power efficiency, and reliability.

Chen, Fen
Fen Chen received his Ph.D. degree in Electrical Engineering in 1998 from University of Delaware. From 1997 to 1998, he was with IBM System Group at Rochester, MN and Intel Component Research at Santa Clara, CA as a graduate intern working on system stress and IC interconnect reliability. He joined IBM microelectronics at Essex Junction, VT in 1998 and has worked on semiconductor technology reliability issues since that time. During the past several years he has focused on low-k ILD TDDB issue for various IBM and IBM Alliance development programs.

Chen, Frederick T
Dr. Chen received his B.S., M.S. and PhD degrees in applied and engineering physics from Cornell University in 1990, 1992 and

1996, respectively. Following graduation, he worked for nine years at Intel Corporation, in the areas of advanced photomask development and lithography. Since 2005, he has been at the Industrial Technology Research Institute in Hsinchu, Taiwan. He holds 10 US patents and has 12 others pending. He has authored more than 10 papers in optics, lithography and advanced memories. His research interests include advanced three-dimensional memories, nanoelectronics reliability, and sub-22 nm patterning.

Chen, Kuan-Fu
Kuan-Fu Chen was born in Taipei, Taiwan, R.O.C. on September 21, 1978. He received the B.S. degree in physics from National Taiwan University, Taipei, Taiwan, R.O.C. in 2000 and the M.S. degree in electrical engineering from National Taiwan University, Taipei, Taiwan, R.O.C. in 2002. He joined Macronix International Company, Ltd., Hsinchu, in 2002 as a Process Integration Engineer. He worked on process integration and devices characteristics analysis for nonvolatile floating-gate-type and nitride storage flash memory.

Chen, Kuang-Chao
Kuang-Chao Chen received the M.S. degrees in chemistry from the National Chong-Shan University, Taiwan, in 1987. From 1989 to 1995, he joined Electronic Research and Service Organization (ERSO), Hsinchu, Taiwan, where he has been involved in the development of BEOL planarization process technology. From 1995 to 1998, he was with Mosel-Vitelic International Co., Ltd, Hsinchu, Taiwan. He performed yield improvement in manufacturing line. In 1998, he joined Vanguard International Semiconductor Co., Ltd, Hsinchu, Taiwan, as a department manager. He was responsible for thin film module development. In 2000, he joined Macronix International Semiconductor Co., Ltd, Hsinchu, Taiwan, where he worked on advanced module development. He is currently Executive Director of Technology Development Center.

Chen, Lu-An
Lu-An Chen received the B.S. degree in Electrical Engineering from the National Sun Yat-Sen University, Kaohsiung, Taiwan, in 2007 and the M.S. degree in the Institute of Nanotechnology, National Chiao-Tung University, Hsinchu, Taiwan, in 2009. He joined United Microelectronic Corp. (UMC) in 2009 and his main research interests in ESD protection design for high-voltage (HV) technology.

Chen, Min
Min Chen reveived the B.S. Degree in electical engineering from Huazhong University of Technology and Science In 1997. He joined Huawei Tech. as a product engineer. He received his Ph.D. degree in electrical engineering from Arizona State University and joined Texas Instruments in 2010. He currently work on sensor developement and design method intergration for circuit reliability.

Chen, Ming-Shiang
Ming-Shiang Chen was born in Tainan, Taiwan, R.O.C. on September 22, 1970. He received the B.S. and M.S. degrees in electronics engineering from National Chiao-Tung University, Hsinchu, Taiwan, in 1992 and 1994, respectively. He joined Macronix International Company, Ltd., Hsinchu, Taiwan, in 1996 as a Device Engineer. From 1996 to 1998, he has worked on nonvolatile memory devices characteristics analysis. Since 1998, he has been engaged in the development floating-gate Flash memory technology and Nitride-based NBit technology.

Chen, Pang-Shiu
Pang-Shiu Chen received the Ph.D. degree in material science and engineering from National Chiao-Tung University, Hsinchu, Taiwan, R.O.C., in 2000. He is with MingHsin University. His current research interests include the fabrication and characterization of heterostructure of SiGe:C, strained Si MOSFET devices, Ge nano-structure, and resistive memory based on transition metal oxide.

Chen, Wei-Su
Wei-Su Chen received the B.S. and M.S. degrees in electronic engineering from National Tsin-Hua University, Taiwan, ROC, in 1988 and 1990, respectively, and studied the Ph.D. grade in National Tsin-Hua University, Taiwan, ROC, from 1991 to 1997 but not graduated. Since 1997, he has been working in EOL/ITRI, where he is currently a project manager of the Department II of Non-Volatile Memory. He got ITRI awards in 2005 and 2007. He is the holder of 25 patents and the First Author of 10 papers in SPIE Advanced Lithography and 1 paper in IEDM with Author/Coauthor of 22 papers in many fields. His research interests include new non-volatile memories, e-beam/DUV lithography, Organic LED/memory, RF-MEMS, semiconductor/MEMS manufacturing processes, GaAs MESFET, scientific equipments and bacteria photoresponse.

Chen, Ying-Je
Ying-Je Chen was born in Taiwan on October 18, 1983. He received the M.S. degree in electronics engineering from the National Tsing-Hua University, Hsinchu, Taiwan, in 2008. He is currently with the Technology Development Division, eMemory Technology Inc., Hsinchu. His research focuses on VLSI design, process integration, and embedded nonvolatile memory design.

Chen, Zuhui
Zuhui Chen received the B.S. degree in physics from Fujian Normal University and M.S. degree in solid-state physics from Xiamen University, Fujian, China, respectively in 1998 and 2001, and Ph.D. in engineering science from University of Florida, FL, USA, in 2005. In 2006, he was a faculty of the Pen-Tung Sah MEMS Research Center, Xiamen University, China. In 2007, he joined Nanyang Technological University as a research fellow with Lee Kuan Yew Postdoctoral Fellowship. His current interests include NBTI and interface-trap modeling in silicon MOSFETs.

Chen, An
An Chen works as a Member of Technical Staff in the Strategic Technology Group at GLOBALFOUNDRIES. He received Ph.D. in Electrical Engineering from Yale University.

Cheng, Cheng-Hsien
Cheng-Hsien Cheng was born in Yunlin, Taiwan, ROC., on December 11, 1982. He recieved the BS and MS degree in Engineering and System Science from National Tsing Hua University, Hsinchu, Taiwan, in 2005 and 2007. In 2007, he joined the Advanced Device Department of Macronix International Company Ltd., Hsinchu, Taiwan. His research interests include the device characterization of Flash memory devices.

Cheung, Kin
Kin P. Cheung obtained the Ph.D. degree in physical chemistry from the New York University in 1983. From 1983 to 1985 he was a post doc at Bell Laboratories during which he pioneered Terahertz Spectroscopy. From 1985 to 2001 he was a member of the technical staff in Bell Laboratories at Murray Hill. From 2001 to 2006, he was an associate professor at Rutgers University. He is currently a project leader at the National Institute of Standards & Technology, Semiconductor Electronics Division. Dr. Cheung has published over 150 refereed journal and conference papers. He authored a book on plasma charging damage, a book chapter and edited three conference proceedings. He served on the committee of a number of international conferences and has given tutorials at 10 international conferences. His area of interest covers VLSI technology/devices, and MEMS/NEMS. He is an editor of the IEEE Transaction on Device & Material Reliability. He chairs the Reliability sub-working group of the Process-Integration, Devices and Structures Technical Working Group of the International Technology Roadmap for Semiconductor (ITRS).

Chevallier, Remy
Rémy Chevallier was born in 1977 in France. He obtained degrees in microelectronics and computer science in 2000. Employed after his studies by STMicroelectronics as verification engineer, Remy worked at the beginning on equivalence checking on microprocessor projects. In a second step, Remy was involved in the improvement of functional verification methods by developing internal tool. Since 2004, he has worked on the reliability issues faced at library level. In close collaboration with reliability experts in one side and with CAD providers in the other side, he is specifying and providing verification flows and methodologies to forecast and prevent reliability effects (ESD, Electromigration, NBTI, PBTI, HCI) at design phase.

Chia, Pierre Chor-Fung
Pierre Chor-Fung Chia received both of his M.S. and B.S. degree in Chemical Engineering from University of Texas at Austin in 1995 and 1993 respectively. In 2005 he joined Cisco Systems Inc., San Jose, CA where he focuses on technology quality and reliability aspects of semiconductor components. He also engaged in various subjects of improving system reliability. Prior to Cisco he had worked in Cypress Semiconductor in Process R&D, and high-speed synchronous SRAM products developments and manufacturing.

Chiang, Puo-Yu
Puo-Yu Chiang received the M.S. degree in Physics from National Taiwan Normal University, Taiwan, R.O.C., in 2002. He has been working in Taiwan Semiconductor Manufacturing Company (TSMC), Hsinchu, Taiwan, R.O.C., as an High Voltage Process Integration Engineer from 2004 to 2006, and a High Voltage Device R&D Engineer now.

Chieh-Wei, He
Chieh-Wei He was born in Kaohsiung, Taiwan in 1984. He received the B.S in physics from National Cheng Kung University in 2006 and M.S. degrees in Electro-Optical Engineering from National Chiao Tung University in 2008, respectively. In 2008, he joined Macronix International Co., Ltd. (MXIC), and has been with the Device Engineering Department. Where he has engaged in the ESD/EOS protective scheme of NOR flash memory process technologies.

Chih-Yuan, Lu
Chih-Yuan Lu received B.S. degree from National Taiwan University in 1972, and Ph.D. degree in physics from Columbia University, NYC, in 1977. Dr. Lu has been a professor in National Chiao-Tung Univ. and with AT&T Bell Labs from 1984-1989; later joined ERSO/ITRI in 1989 as a Deputy General Director responsible for the MOEA grand Submicron Project. This project successfully developed Taiwan first 8-inch manufacturing technology with high density DRAM/SRAM. He was therefore granted the highest honor prize--National Science & Technology

Achievement Award by the Prime Minister of ROC, due to his leadership and achievement in this Submicron Project. In 1994, Dr. Lu becomes the co-founder of Vanguard International Semiconductor Corporation, which is a spin-off memory IC Company from ITRI's Submicron Project. He was the VP of Operation, VP of R&D, and later President from 1994-99. Dr. Lu now is the founding chairman and CEO of Ardentec Corp. a VLSI testing service company; and also serves Macronix International as a Senior VP/CTO, and now the President. Dr. Lu led MXIC's technology development team to successfully achieve the state of the art nonvolatile memory technology and now responsible for MXIC's overall operation.

Chikyow, Toyohiro
Prof. Toyohiro Chikyow was born in Fukuoka, Japan in 1959. He graduated from the School of Science and Engineering, Electronics and Communication, Waseda University as a Bachelor in1983, Master in 1985 and Ph.D in 1989. He currently investigates gate stack materials including metal gate and higher-k dielectric for future nano device. In this study, he has been developing combinatorial synthesis systems and high throughput characterization tools to screen the candidate materials in a short time. He also studies, memory device materials such as ReRAM or atomic switch, transparent conductive oxide and fusion with organic materials on Si device

Chimenton, Andrea
Andrea Chimenton received the PhD degree in information engineering from the University of Ferrara in 2004. In 2002 he joined the Electronic Engineering Department of the University of Ferrara as assistant professor working in the field of electronic system reliability and device physics. In 2003–2004 he was a research scientist at Intel, Santa Clara (CA) working on Flash scaling issues. In 2010 he joined Intel, Boise (ID). His scientific interests are in the area of non-volatile memory characterization, reliability, compact and statistical modeling.

Cho, Moonju
Moonju Cho received the M.Sc. and the Ph.D. degrees in materials science and engineering from Seoul National University, Korea, in 2003 and 2007, respectively. Her study was focused on the physical properties of ALD-HfO2 films and reliability problems based on chlorine/carbon residues. In 2005, she stayed at IMEC, Leuven, Belgium, working on TDDB and charge pumping. From 2007 till February 2009, she did postdoctoral study at IMEC in theoretical modeling for oxide trap characterization. She is currently working at the Reliability Group, IMEC. Her current interest is on hot carrier injection and trap characterization study on ultrathin high-k gate oxide.

Choi, Do-Young
received the B.S. degree (2007) in electronic and electrical engineering and computer science from the Kyungpook National Uiversity (KNU), Daegu, Korea, and the M.S. degree (2009) in electronic and electrical engineering from the Pohang University of Science and Technology (POSTECH), Pohang, Korea, where he is currently working toward the Ph.D. degree. His main research interests include bias temperature instability, low-frequency noise, and time-dependent dielectric breakdown of high-k dielectric CMOS devices.

Choi, Seung-Man
Seung-Man Choi received a B.S. degree in Chemical Engineering from Korea University, Seoul, Korea, in 1997. He joined Unit Process Development team in Samsung Electronics in 1997 and moved to 45nm CMOS BEoL integration team as a senior member

of technical staff in 2006. Since 2008, he is with Technology Quality and Reliability in System LSI, Samsung Electronics as a manager of Interconnect Reliability group responsible for reliability qualification of advanced CMOS BEoL process and product.

Choi, Wonsup
Wonsup Choi received a B.S and M.S degree in School of Advance Materials Engineering from Kookmin University, Seoul, Korea, in 2008 and 2010. His research interests include nonvolatile memory devices, SONOS devices, ReRAM devices, and atomic force microscopy.

Chong, Lit-Ho
was born in Johor Bahru, Malaysia, in 1976. He received the PhD degree in Microelectronics from the University of Southampton in 2006. He joined Macronix International Company, Ltd., Hsinchu, Taiwan, in 2006 as a device engineer. He has been working on the research and development of the nitride storage Flash memory.

Chou, Bruce
Bruce Chia-Te Chou received the B.S. degrees in Electrical Engineering and Mathematics from the University of Florida in 2004, and an M.S. degree in Electrical Engineering from the University of Florida 2006. He joined Intel Corp., Santa Clara, CA in 2007 as a component design engineer after interned in 2005 and 2006. He has been with Portland Technology Development group since 2007 and is the owner of ESD & I/O protection circuit design kits in Intel's 32, 22, and 15 nm technologies. His main research interests are SPICE simulation of ESD systems and distributed ESD protection network.

Christiansen, Cathryn
Cathryn J. Christiansen received her B.S. in Broad Field Science Eduation at Marquette University in Milwaukee, WI, and her Ph. D. in Physics at the University of Minnesota, Minneapolis in 2001. She immediately joined IBM Systems and Technology Group in Essex Junction, VT. She is currently working on BEOL related technology reliability issues, including electromigration, stress voiding, thermal cycling and dielectric breakdown for advanced CMOS logic technologies.

Christophe, Winters
Ing. Christophe Winters obtained a degree in electrical engineering, with a specialisation in micro-electronics, in 1997 at GROUP T Leuven Institute of Technology. In 1997 he joined imec in the High-Density Interconnection and advanced Packaging group within the Materials and Packaging Division (MCP/HDIP). He's involved in research and design in the field of high-density packaging and processing related topics and the development of several System-in-Package (SiP) applications and coordination of assembly activities. Also design and activities related to Cu Back-end / 3D TSV / MEMS processing are part of his work. For the FP6 European ENCAST / Good-Die Co-ordinated Action Project Mr. Winters ran for 8 years the Network Office which took care of dissemination of Known Good Die and Microelectronic Packaging related topics by publishing the Newsletters, maintenance of webpages and organizing international workshops.

Chu, Wen-Ting
Dr. Wen-Ting Chu is a manager of Non-volatile Memory Development Department at Taiwan Semiconductor Manufacturing Company (TSMC). He joined TSMC in 1997, where he has been primarily engaged in the development of nonvolatile memory, especially dedicated to embedded applications from the 0.25-mm generation to the 40-nm

generation. His research interests mainly focus on deep submicron nonvolatile memory technology development, flash memory reliability studies, and MOS device physics. He has published more than 8 papers on the field of semiconductor technology. He received the B.S. degree in physics and the M.S. and Ph.D. degrees in electrical engineering, all from the National Taiwan University, Taipei, Taiwan, in 1991, 1993 and 2005, respectively.

Chu, Yi-Shin
Yi-Shin Chu received her M.S. in EE from National Taiwan University in 2005. She has served as the Flash memory process integration engineer at Taiwan Semiconductor Manufacturing Company for 3 years, and currently served as the engineer of Technology Quality & Reliability Department (TSMC) focusing on NVM reliability.

Chua, Choon Meng
Choon Meng Chua received both his Bachelor and Masters Degree in Engineering from the National University of Singapore in 1988 and 1990 respectively. He is the Chief Executive Officer of SEMICAPS Corporation and current responsibility includes overseeing the general operations of the companies within the SEMICAPS Corporation Group.

Chuang, Ching-Te
Ching-Te Chuang received B.S.E.E. from National Taiwan University, Taipei, Taiwan in 1975 and Ph.D. in Electrical Engineering from University of California, Berkeley, CA in 1982. He worked at IBM T. J. Watson Research Center, Yorktown Heights, NY from 1982 to Jan. 2008, holding technical and management positions in areas ranging from bipolar devices, circuits, and technologies, BiCMOS logic and memory, CMOS microprocessor and SRAM design. He joined National Chiao-Tung University, Hsinchu, Taiwan as a Chair Professor in the Department of Electronics Engineering in Feb. 2008. He is a Fellow of IEEE, and has authored or coauthored over 290 papers.

Chudzik, Michael
Dr. Chudzik manages process development of FEOL, MOL and BEOL high-k and metallization processes for advanced technology nodes at IBM SRDC in East Fishkill, NY. Dr. Chudzik joined IBM in 2000 as a development engineer working on SiON films and high-k dielectrics for DRAM, eDRAM technologies and transistor gate dielectric development. Prior to his current position, Dr. Chudzik spent two years at IBM's T.J. Watson Research Center working on atomic layer deposition (ALD) of dielectric and metal films. At IBM Research, he served as a manager of the advanced gate dielectric research group and as the technical assistant to the Vice President of Science and Technology.

Ciappa, Mauro
Mauro Ciappa received the MSc degree in physics from the University Zurich, Zurich, Switzerland, and the Ph.D. degree in engineering sciences from the Swiss Federal Institute of Technology (ETH), Zurich, Switzerland. He joined the Reliability Laboratory, ETH, in 1986, where he was Head of the Failure Analysis and Reliability Physics Laboratory and Lecturer for reliability physics and failure analysis techniques until 1997. He is currently a Member of the Integrated Systems Laboratory, ETH, where he leads the group for physical characterization of semiconductor devices and he is Lecturer for Smart Power technologies.

Claeys, Cor
Cor Claeys received his masters Ph.D. degree from the KULeuven in Belgium, where he is Professor since 1990. He is also at imec Director of Advanced Semiconductor Technologies responsible for strategic relations. His main interests are, in general, silicon technology for ULSI; device physics, low frequency noise phenomena, and radiation effects; and defect engineering and material characterization. He authored and coauthored 14 book chapters and more than 900 technical papers. Prof. Claeys is a Fellow of the Electrochemical Society and of IEEE. In 2004, he received the Electronics Division Award of the Electrochemical Society.

Clark, Brett
Dr. Brett Clark manages the Analytical and Metrology Laboratory at Honeywell's Electronic Materials Division in Spokane, Washington. His research areas include liquid particle counting, low level radiation measurement, and materials science of high purity metals and alloys. He received a Ph.D. degree in physical/analytical chemistry from Brigham Young University, and has authored/coauthored seven journal publications and holds ten patents.

Classe, Francis
Francis is a Member of the Technical Staff in the Quality and Reliability Engineering division at Spansion. He has been working in the semiconductor industry for almost eight years, and has focused on a variety of quality and reliability issues in the past, most recently studying and analyzing the reliability of Spansion's copper wire bonding process. His Bachelor's degree came from the Univeristy of Virginia, and he obtained his Master's degree at the Georgia Institute of Technology.

Cressler, John
John D. Cressler received his Ph.D. in applied physics from Columbia University in 1990. He was on the research staff at IBM Research (1984-1992), the faculty of Auburn University (1992 to 2002), and currently is Ken Byers Professor of Electrical and Computer Engineering at Georgia Tech. His research interests center on silicon-based heterostructure devices and circuits, and he and his team have published over 500 papers in this area. He is the co-author of Silicon-Germanium Heterojunction Bipolar Transistors (2003), the editor of Silicon Heterostructure Handbook: Materials, Fabrication, Devices, Circuits, and Applications of SiGe and Si Strained-Layer Epitaxy (2006), and author of Silicon Earth: Introduction to the Microelectronics and Nanotechnology Revolution (2009). He has received a number of awards for both his teaching and research, and he was elected Fellow of the IEEE in 2001.

Cristoloveanu, Sorin
Sorin Cristoloveanu received the PhD in Electronics and the French Doctorat ès-Sciences in Physics from Grenoble Polytechnic Institute, France. He is currently Director of Research CNRS. He also worked at JPL, Motorola, and the Universities of Maryland, Florida, Vanderbilt, Western Australia, and Kyungpook. He served as the director of the LPCS Laboratory and the Center for Advanced Projects in Microelectronics, initial seed of Minatec center. He authored more than 700 technical journal papers and communications at international conferences. He is the author or the editor of 24 books, and he has organized 20 international conferences.

Croes, Kristof

Kristof Croes received his BSc in physics at the Catholic University of Louvain (Belgium) in 1993 and his MSc in biostatistics at the Limburgs Universitair Centrum (LUC) in 1994. In 1999, he obtained his PhD, concerning the development of statistical techniques for planning reliability experiments. After that, he joined the reliability business unit of XPEQT, first as the software responsible and than as the manager of the R&D. From 2003 till end 2006, he was product and application manager of the package level reliability products of Chiron holdings. Beginning 2007, he went back to research, working on BEOL and 3D-reliability problems at imec, where he is currently heading the wafer level interconnect reliability team.

Crupi, Felice

Felice Crupi received the M.Sc. degree in electronic engineering from the University of Messina, Messina, Italy, in 1997, and the Ph.D. degree from the University of Firenze, Firenze, Italy, in 2001. Since 2002, he has been with the University of Calabria, Rende, Italy, where he is currently an Associate Professor of electronics. Since 1998, he has been a repeat Visiting Scientist with the Interuniversity Micro-Electronics Center, Leuven, Belgium. In 2000, he was a Visiting Scientist with the IBM Thomas J. Watson Research Center, Yorktown Heights, NY. In 2006, he was a Visiting Scientist with the Universitat Autonoma de Barcelona, Barcelona, Spain. His main research interests include reliability of very large scale integration CMOS devices, electrical characterization techniques for solid-state electronic devices, and the design of ultralownoise electronic instrumentation. He has authored or coauthored more than 100 publications in international scientific journals and in international conference proceedings.

Czeppel, Laura

Laura Tatiana Czeppel was born in Milan, Italy, in 1973. She received the Master Degree in Physics at "Universita' Statale di Milano", in 1998. She joined STMicroelectronics in 2000 and now she is working in R&D Technology Division in Micron. After joining the Flash Reliability Group, working on NAND memory devices, now she is a Sr Device Engineer for PCM memory development.

Degraeve, Robin

Robin Degraeve received the M.Sc. degree in electrical engineering from the University of Gent, Belgium, in 1992, and the Ph.D. degree from the Catholic University of Leuven, Belgium, in 1998. He joined imec, Leuven, in 1992 in the CMOS Reliability and Characterization group, where he is working as a Senior Researcher. His work currently focuses on advanced and novel characterization techniques for studying electrical defects in dielectrics and related reliability aspects. This includes characterization and reliability of high-k material in transistor and memory applications, hot-carrier related issues and breakdown physics.

del Alamo, Jesús

Jesús A. del Alamo obtained a Telecommunications Engineer degree from the Polytechnic University of Madrid in 1980 and MS and PhD degrees in Electrical Engineering from Stanford University in 1983 and 1985, respectively. From 1985 to 1988 he was with NTT LSI Laboratories in Atsugi (Japan) and since 1988 he has been with the Department of Electrical Engineering and Computer Science of Massachusetts Institute of Technology where he is currently Donner Professor and MacVicar Faculty Fellow. He is a member of the Royal Spanish Academy of Engineering and Fellow of the IEEE. He currently serves as Editor of IEEE Electron Device Letters.

Deora, Shweta

Shweta Deora received the B.E. degree in Electronics and communication engineering from Sardar Patel University, Gujarat, India, in 2004. From 2005-2006, she was at IIT Bombay working on development of radiation sensor. Since 2006, she is doing her PhD in electrical engineering department, IIT Bombay. Her research interests are in the field of semiconductor device reliability, characterization and modeling. She is currently working on NBTI in SiON, high-k/MG and strained devices.

Dey, Aritra

Aritra Dey received his B.Tech from the West Bengal University of Technology in 2006 and his MS (by research) from Indian Institute of Technology (IIT), Madras in 2008. He is currently pursuing his PhD at the Flexible Display Center at ASU. His research interests include developing novel analog and mixed analog circuits on flexible substrates. He has already co-authored 9 research papers in international journals and conferences. He currently has 2 pending patents.

Di Carlo, Aldo

Aldo Di Carlo is associate professor of optoelectronics and leader of the Nano&Optoelectronic research group (http://www.optolab.uniroma.it <http://www.optolab.uniroma.it/>) of the University of Rome "Tor Vergata" (Italy). Since 2006 he is co-director of the Centre for Hybrid and Organic Solar Energy (CHOSE) of the Lazio Region- Italy. CHOSE involves more that 30 researchers for the development and industrialization of the DSC technology. Di Carlo is author/coauthor of more than 190 scientific publications on international journals, 6 patents, several book chapters and co-author of two books of optoelectronics.

Dick, Kevin

Kevin Dick is a graduate student at Vanderbilt University. He received his Bachelor's Degree from Western Kentucky University in Electrical Engineering and Mathematics. His graduate research focuses on understanding and mitigating radiation effects on a system level.

Diggins, Zachary

Zachary Diggins is an undergraduate at Vanderbilt University pursuing a B.E. in electrical engineering. He is involved in radiation effects research, including testing ICs with different radiation sources and performing data analysis. He plans to pursue a Ph.D in electrical engineering.

Diop, Malick

Malick DIOP received his phD degree in 2009 from Institut National Polytechnique Grenoble (INPG) on reliability physics and low frequency noise in high speed SiGe: C HBTs. He his with STMicroelectronics working on CMOS and BiCMOS reliability modeling

Dixit, Anand

Anand Dixit received the B.Tech degree from Indian Institute of Technology, Kanpur, India, and M.S. degree from Carnegie Mellon University, Pittsburgh, PA, both in electrical engineering, in 1996 and 1998, respectively. From 1998 to 2000, he worked at National Semiconductor, Santa Clara, CA, where he worked on touch screen controller, PLL and other analog designs. Since 2000, he has been with Oracle (formerly Sun Microsystems), where he has been responsible for the I/O designs on UltraSparc III and UltraSparc IV processors. His recent work includes high-speed interface design, advance circuit development, soft error modeling and processor test/debug.

Dobrovolny, Petr

Petr Dobrovolný received his MSc and PhD degree from the Technical University of Brno, Dept. Microelectronics, in 1987 and 1998, respectively. During his Ph.D. studies he investigated the problem of the symbolic analysis of large analog circuits. This research was done in the cooperation with the KUL, Dept. ESAT-MICAS. Since 1999, he has been with IMEC, where is currently a senior research in the VAM (Variability Aware Modeling) team. He focuses on the research and development of CAD tools for circuit characterization under process variability and in the presence of time dependent degradation mechanisms.

Dongaonkar, Sourabh

Sourabh Dongaonkar completed his B.Tech. in Electrical Engineering from the Indian Institute of Technology Kanpur, India in 2007. In 2007-08 he worked at the Global Market Center, Deutsche Bank, Mumbai, India as a quantitative analyst. Since 2008, he is a graduate student in the School of Electrical and Computer Engineering at Purdue University. His research interests lie in the areas of thin film solar cells, and transport in amorphous and random materials. He was the recipient of 2008 Ross Fellowship from Purdue University Graduate School.

Dosseul, Franck

Franck received his Engineering Degree in Materials Physics from INSA, Toulouse, France in 1995. He then joined ST Microelectronics in the Packaging Engineering & Development group in Tours, France. Franck has developed and industrialized the first WLCSP line in ST Microelectronics. He has been working since more than 12 years in the field of System In Package development and Advanced Chip Scale Package programs, focusing on R&D programs management and process bricks development and Integration. Franck is now Advanced Programs Manager in charge of Advanced Collaborative Programs in the field of 3D Wafer Level Packaging and Flexible Printed Electronics.

Dutton, Robert

Robert W. Dutton is Professor of Electrical Engineering at Stanford University and director of Integrated Circuits Laboratory. He received the B.S., M.S., and Ph.D. degrees from the University of California, Berkeley, in 1966, 1967, and 1970, respectively. He has held summer staff positions at Fairchild, Bell Telephone Laboratories, Hewlett-Packard, IBM Research, and Matsushita during 1967, 1973, 1975, 1977, and 1988 respectively. His research interests focus on Integrated Circuit process, device, and circuit technologies--especially the use of Computer-Aided Design (CAD) in device scaling and for RF applications. Dr. Dutton has published more than 200 journal articles and graduated more than four dozen doctorate students. He was Editor of the IEEE CAD Journal (1984-1986), winner of the 1987 IEEE J. J. Ebers and 1996 Jack Morton Awards, 1988 Guggenheim Fellowship to study in Japan, was elected to the National Academy of Engineering in 1991, honored with the C&C Prize (Japan) in 2000, received Career Achievement Award (2005) from the Semiconductor Industry Association (SIA) for sustained contributions in support of research that is critical to SIA needs, and the Phil Kaufman Award in 2006.

Eaton, Paul

Paul Eaton received his bachelor's degree and master's degree from the Texas Tech University in 1994 and 1996 respectively. He is one of the co-founders and the Chief Engineer of Micro-RDC where he has been since June 2005. While at Micro-RDC, he has lead all test and analysis efforts as well as the development of FPGA based high-speed test boards.

Elkins, Donald

Donald Elkins received advanced avionics training while enlisted in the US Navy and has over 20 years experience as an engineering technician. Currently his major interests are involved with the implementation and testing of several radiation hardened high speed microprocessors within a structured ASIC program.

El-Mamouni, Farah

Farah El Mamouni is a Ph.D student in Electrical Engineering and Computer Science departement at Vanderbilt University. She received her B.E.E and M.S.E.E degrees from the University of Montpellier II in France in 2005 and 2007. Farah received her second M.S.E.E degree in Electrical engineering and computer science in 2009 from Vanderbilt University. Farah's research activity focuses on studying reliability of emerging SOI devices in particular, FinFETs.

Eneman, Geert

Geert Eneman received the B.S. and M.S. degrees in electrical engineering and the Ph.D. degree on the topic of "Design, fabrication, and characterization of strained silicon transistors" from the Catholic University of Leuven, Leuven, Belgium, in 1999, 2002, and 2006, respectively. His Ph.D. work was done in the Interuniversity MicroElectronics Center (IMEC), Leuven. He is currently with the CMOS Technology Department, IMEC. He also holds a postdoctoral position at the Catholic University of Leuven. He is a Postdoctoral Fellow with the Fund for Scientific Research-Flanders, Belgium. His current research interests include the characterization and modeling of Ge MOS devices, Ge junction analysis, and modeling of alternative device structures.

Eng, Genghmun

Dr. Genghmun Eng is a Senior Scientist in the Microelectronics Technology Department at The Aerospace Corporation, studying device reliability and the physics and chemistry associated with device failures. His work includes tin whiskers, tin plasmas, electron beams, electron-beam systems, advanced cathodes, RF power transistors, thin-layer interfaces, and multi-layer diffusion. He holds a PhD in Physics (1978) from the University of Illinois at Urbana-Champaign, where he studied nonlinear diffusion in ionic crystals.

Ermisch, Karsten

Karsten Ermisch received is diploma in physics in 1998 from the University of Hanover and finished his PhD in 2003 in nuclear physics in Groningen, Netherlands. After working as a Post-Doc in the field of detector characterization for space applications he started at Infineon Technologies as a test program engineer for DRAM modules. Later on he switched to soft-error rate measurements of DRAMs. His current activities include aging simulations and integration of agemos models into Cadence RelXpert, measurements of soft-error rates of logic chips and simulations of single-event upsets using Synopsys TCAD.

Falk, John

John Falk is Power Validation Technical Lead for high volume client platforms at Intel Corporation. He has a B.S. in Physics from the University of Washington and graduate work at the Oregon Graduate Institute of Science and Technology. His career spans semiconductor process and CPU product development including fault isolation, failure analysis, micro analytics, yield improvement, test quality, product reliability, validation and qualification of CPUs. His current interests are optimizing power of PC platforms.

Federspiel, Xavier
Xavier Federspiel received his PhD in Microelectronics from the Institut National Polytechnique de Grenoble in 2001. Afterwards, he went to work for Delphi Automotive working in the failure analysis group. In 2002 he joined Philips Semiconductors Crolles R&D labs where he worked in the reliability of interconnects for advanced CMOS technology. In 2007, he worked on developpement of DRAM interconnect for Qimonda as Process Integration Engineer. Since 2009 he is with Dolphin Integration where he works on CMOS device reliability.

Fichtner, Wolfgang
Wolfgang Fichtner a physicist, graduated at the Technical University of Vienna, Austria, in 1978. From 1979 through 1985, he worked at AT&T Bell Laboratories, Murray Hill, NJ. Since 1985 he is full Professor at ETH Zurich, Switzerland. In 1993, he founded ISE AG, a company in the field of Technology CAD (now the TCAD Business Unit of Synopsys, Inc.). He is a Fellow of the IEEE and a member of the Swiss National Academy of Engineering. In 2000, he received the IEEE Andrew S. Grove Award for his contributions to Technology CAD.

Fleetwood, Daniel
Dan Fleetwood received his Ph.D. from Purdue University in 1984, and then joined Sandia National Laboratories. In 1990, he was named a Distinguished Member of Technical Staff in the Radiation Technology and Assurance Department. He moved to Vanderbilt University as professor of electrical engineering in 1999, and now serves as Chairman of Vanderbilt's Electrical Engineering and Computer Science Department. Dan is author or co-author of more than 375 publications on radiation effects and low frequency noise. He is a Fellow of the IEEE and the American Physical Society. In 2009 he received the IEEE Nuclear and Plasma Sciences Merit Award.

Floyd, Richard
Richard Floyd was employed by Micro-RDC as a senior technician. His electronics career has spanned over 20 years with experience in flight avionics electronic hardware assembly and test, RFID transit system test and assembly and radiation hardened microelectronics testing and hardware assembly. He received an associate's degree in Applied Science from Albuquerque Technical-Vocational Institute. Rich currently holds a Technician Class amateur radio operators license.

Foran, Brendan
Brendan Foran is a section manager within the Physical Sciences Laboratories at The Aerospace Corporation. His current technical focus is analytical and high resolution transmission electron microcopy of materials and devices for microelectronics and optoelectronics to aid reliability and physics of failure studies. Prior to his current position, Dr. Foran led the transmission electron microscopy group for ten years at SEMATECH supporting research and development for semiconductor processing. Dr. Foran earned a Ph.D. from The University of Michigan with a thesis focused on relationships between crystal structure and electronic band structure for low-dimensional semi-metal phases controlled by charge density waves.

Franco, Jacopo
Jacopo Franco received the B.Sc. and M.Sc. degrees in Electronic Engineering from the University of Calabria - Italy, in 2005 and 2008 respectively. His M.Sc. thesis was developed at imec, Leuven - Belgium, and it is related to reliability issues in advanced Silicon and Germanium MOSFETs. He is currently working toward a Ph.D. degree in the reliability group of imec and at the Katholieke Universiteit Leuven, on the topic "Interface stability and reliability of Ge and III-V transistors for future CMOS applications". He has authored or co-authored more than 30 publications and he received the IEEE SISC Ed Nicollian Award for the best student paper in 2009.

Frank, Thomas
Thomas Frank received the engineering degree in Physic, Electronic and Materials from the Institut National Polytechnique de Grenoble (INP Grenoble Phelma), France and M.S. degree of Micro and Nanoelectronic from the Université Joseph Fourier de Grenoble, France in 2009. He is currently working toward the Ph.D. degree with the Electrical Characterization and Reliability Group at STMicroelectronics, Crolles, France. His research field is the reliability issues of 3D Integration Circuits.

Freeman, Greg
Greg Freeman received the Ph.D. degree from Stanford University in 1991, and has been with IBM's East Fishkill, NY, facility since 1991 where he has been involved in characterization, HBT design, and SOI CMOS technology development. Dr. Freeman has authored or coauthored over 60 publications and has to his name approximately 26 patents. Currently he is manager of the 22nm SOI device design department at IBM.

Frei, Michel
Michel Frei leads the Cell Analysis group in the Solar division of Applied Materials in Santa Clara, CA. Previously, he was with Bell Laboratories in Murray Hill, NJ, where he developed technologies for III-V HBTs, SiGe BiCMOS, and Si RF LDMOS. He holds a Diploma from the Swiss Federal Institute of Technology in Lausanne, Switzerland, and a PhD in Electrical Engineering from Princeton University.

Fronheiser, Jody
Jody Fronheiser received his B.S. degree in chemical engineering from Clarkson University, Potsdam, in 2000 and the M.S. degree in electrical engineering from Union Graduate College, Schenectady, NY in 2010. He has since been a field service and process engineer for wide-bandgap materials growth and characterization at Emcore Corporation. Since 2003, he has been at General Electric Global Research, Niskayuna, NY where he develops SiC wide-bandgap power transistors.

Fujii, Shosuke
Shosuke Fujii received the B.S. (2005) and M.S. (2007) in materials science and engineering from Kyoto University, Japan. He joined Advanced LSI Technology Lab, Toshiba Corp., Yokohama, Japan, in 2007, where he has been engaged in the research on reliability physics of non-volatile memories.

Fujitsuka, Ryota
Ryota Fujitsuka received the B.S. (2003) and M.S. (2005) degrees, from Nagoya University, Nagoya, Japan. In 2005, He joined Process and Manufacturing Engineering Center, Toshiba Corporation, Semiconductor Company, Yokohama, Japan, where he worked on development of dielectric film for advanced nonvolatile memory device. He is currently engaged in development of nonvolatile memory device in Advanced Memory Development Center, Semiconductor Company, Toshiba Corporation, Yokkaich, Japan

Fukutani, Atsuyuki

Atsuyuki Fukutani received D.Sc. from the University of Tokyo in 1990. From 1990 to 1995, he worked as a research associate in the Surface Science group at Institute for Solid State Physics of the University of Tokyo. Since 1995, he has been at Institute of Industrial Science of the University of Tokyo where he is a Professor of the Department of Fundamental Engineering. His major is surface and interface physics, and he is currently interested in chemistry and physics of hydrogen at solid surfaces and interfaces.

Furuta, Jun

Jun Furuta received the B.E. and M.E. degrees in Electrical and Electronic Engineering from Kyoto University, Kyoto, Japan, in 2009, 2011, respectively. He is presently a doctor's course student at Kyoto University.

Gaddamraja, Sesha

Sesha is a Senior Member of Technical Staff at the Final Manufacturing (packaging) division of Spansion, responsible for new product/ process introductions. He has been working in the high tech semiconductor manufacturing industry for over ten years, with a special focus on wirebonding. Recent accomplishment involve the development, qualification and implementation of Cu wire to mass production at Spansion's packaging facilities. Currently, his focus is on qualifying and implementing Cu wire in laminate substrate-based packages, including multi-chip packages. He obtained his Master's degree at Louisiana Tech University, and his MBA from San Jose State University.

Gaddi, Roberto

Roberto Gaddi received the Laurea degree in electronic engineering from University of Parma, Italy, in 1997 and the Ph.D. degree from University of Wales, Cardiff, UK, in 2002. In 2002 he joined as Post Doc the RF-IC design group at ARCES, University of Bologna, Italy. He lead the modeling, design and characterization of RF MEMS devices and reconfigurable RF circuits. From April 2007 he is with Cavendish Kinetics ,'s-Hertogenbosch, Netherlands. He is now leading the RF product modeling and design activity and is involved with MEMS device reliability.

Gadlage, Matthew

Matthew J. Gadlage received his B.S. in electrical engineering from the University of Evansville in 2002 and his M.S. in electrical engineering from Vanderbilt University in 2004. Since 2002, he has been a member of the Radiation Sciences Branch at Crane Naval Surface Warfare Center. His primary research interests include the effects of radiation on electronic circuits and soft errors. He has authored or co-authored over 30 papers in these research areas. He finished his Ph.D. in electrical engineering at Vanderbilt University in the spring of 2010.

Galloway, Kenneth

Is a Professor of Electrical Engineering and Dean of Engineering at Vanderbilt University. He has worked in radiation effects on semiconductor devices for almost forty years.

Gao, Yuan

Y. Gao received the B. Eng. (Hons.) degree in Electrical Engineering from National University of Singapore, Singapore. He is currently working toward to the Ph.D. degree in the School of Electrical and Electronic Engineering, Nanyang Technological University (NTU), Singapore. He had worked as a process engineer in United Microelectronics Corporation (Singapore Branch) for about two years before joining NTU. His research project is on the characterization of bias-temperature instability in the metal/high-k gate dielectric MOSFETs.

Geer, Robert

Bio: Robert Geer is a Professor of Nanoscience and Vice President for Academic Affairs at the College of Nanoscale Science and Engineering at the University at Albany, SUNY. His research has focused on materials, structures, and processing for advanced interconnects in current and emerging nanoelectronics architectures – especially with respect to nanoscale mechanics. Recent research efforts have focused on stress-induced effects in 3DICs, strained SOI, and 3D Si-based logic devices. In addition, he leads the Electrical Interconnect Theme for the SRC/DARPA Interconnect Focus Center. His current research is actively supported by NSF, Sematech, SRC-FRCP, SRC-NRI, NIST, DOE, and the AFOSR.

Gennadi, Bersuker

Gennadi Bersuker completed his M.S. and Ph.D. in Physics at the Leningrad State University and Kishinev State University, respectively. After graduation, he joined Moldavian Academy of Sciences, and then worked at Leiden University and the University of Texas at Austin. Since 1994, he has been working at SEMATECH on electrical characterization of Cu/low-k interconnect, high-k gate stacks, advanced memory, and CMOS process development. He is an editor of IEEE Transactions on Device Materials and Reliability and has been involved in many technical conferences, including IRW, IRPS, IEDM, ULSI-TFT, ISAGST. He is a SEMATECH Fellow and has published over 200 papers.

George, Papaioannou

George Papaioannou received the BSc degree in physics from University of Athens, the MSc degree from University College London, England, and the PhD degree in solid state physics from the University of Athens. He is associate professor at the Solid State Physics Section of Athens University, leading a team on the transport properties and the radiation effects in semiconductors devices. Presently his research work is focused on the investigation of polarization/charging effects in thin insulating films and insulating materials for MEMS capacitive and ohmic switches. He has been contributed 110 publications and more than 180 conference presentations.

Gerardin, Simone

Simone Gerardin received the Laurea degree (cum laude) in Electronics Engineering in 2003, and a Ph.D. in Electronics and Telecommunications Engineering in 2007, both from the University of Padova - Italy. He is currently a research assistant at the same university. His research is focused on soft and hard errors induced by ionizing radiation in advanced CMOS technologies, and on their interplay with device aging and ESD. Simone has authored or co-authored about 50 papers published in international journals, more than 60 conference presentations, two book chapters, and two tutorials at international conferences.

Ghibaudo, Gérard

Gérard Ghibaudo was born in France in 1954. He graduated from Grenoble Institute of Technology in 1979, obtained the PhD degree in Electronics in 1981 and the State Thesis degree in Physics from the same University in 1984. He became associate researcher at CNRS in 1981 where he is now Director of Research at CNRS and Director of IMEP-LAHC Laboratory located at MINATEC-INPG center. During the academic year 1987-1988 he spent a sabbatical year at Naval Research Laboratory in

Washington, DC (USA) where he worked on the characterization of MOSFETs. His main research activities were or are in the field of electronics transport, oxidation of silicon, MOS device physics, fluctuations and low frequency noise and dielectric reliability. Dr. G. Ghibaudo has surpervised over 66 PhD students in his career.Dr. G. Ghibaudo is also member of the Editorial board of Solid State Electronics and associate editor of Microelectronics Reliability Journals. During his career he has been author or co-author of over 312 articles in International Refereed Journals, 513 communications and 60 invited presentations in International Conferences and of 21 book chapters.

Gloria, Daniel
Daniel Gloria received in 1995 the engineering degree in electronics from the Ecole Nationale Supérieure d'Electronique et de Radioélectricité and the M.S.E.E. in optics, optoelectronics and microwaves design systems from the Institut National de Grenoble (INPG). He spent two years, from 1995 to 1997, in ALCATEL Bell Network System Labs, in Charleroi, Belgium, as an RF designer engineer and was involved in the development of the Cablephone RF front end and its integration in Hybrid-Fiber-Coax telecommunication networks. Since 1997, he has been working for ST Microelectronics, Technology R&D Crolles, TPS Laboratory where he is in charge of HF characterization and RF passives modeling group. His interests are in optimization of active and passives devices for HF applications in BiCMOS and CMOS advanced technologies.

Gordon, Michael
Michael Gordon is a Research Staff Member at the IBM TJ Watson Research Center. At IBM, he initially worked in the field of Electron Beam Lithography before joining the Research Laboratory. Dr. Gordon received his B.S. degree in 1982 in Engineering Physics from University of Colorado, Boulder, and his Ph.D. in 1989 in Experimental Nuclear Physics from SUNY Stony Brook. Dr. Gordon's research interests are in applications of accelerator-based ions including materials analysis and single-event upsets. Dr. Gordon has 32 Patents issued and has coauthored about 50 technical articles. In 2007 Dr. Gordon received an IBM Outstanding Technical Achievement Award.

Grasser, Tibor
Tibor Grasser received his Ph.D. degree in technical sciences from the TU Wien where he is currently employed as an Associate Professor. In 2003 he was appointed head of the Christian Doppler Laboratory for TCAD in Microelectronics. Dr. Grasser is the co-author or author of nearly 300 scientific articles, editor of a book on advanced device simulation, a senior member of IEEE, has been involved in the program committees of SISPAD, IWCE, ESSDERC, IRPS, IIRW, and ISDRS, and is a recipient of the Best Paper Awards at IRPS and ESREF. He was also a chairman of SISPAD 2007.

Green, Ronald
Ronald Green received the B.S. degree in electrical engineering in 1994, and the Doctor of Engineering degree in electrical engineering in 2010 from Morgan State University located in Baltimore, Maryland. He has been with the Army Research Laboratory (ARL) since 2005, working in the Power Components Branch of the Sensors and Electron Devices Directorate (SEDD) as an electronics engineer. His dissertation research focused on the performance and reliability of SiC power MOSFET technology for continuous power conversion applications. Mr. Green's research interests include device characterization, modeling, and reliability testing and application of wide band gap semiconductors.

Groeseneken, Guido
Dr. Guido Groeseneken received the M.Sc. degree in electrical engineering (1980) and the Ph.D degree in applied sciences (1986), both from the Katholieke Universiteit Leuven, Belgium. In 1987 he joined the R&D Laboratory of IMEC (Interuniversity Microelectronics Center) in Leuven, Belgium, where he is responsible for research in reliability physics for deep submicron CMOS technologies. From October 2005 until April 2007 he was also responsible for the IMEC Post CMOS Nanotechnology program within IMEC's core partner research program. Since 2001 he is Professor at the KU Leuven,where he is Program Director of the Master in Nanoscience and Nanotechnology, and where he is also coordinating a European Erasmus Mundus Master program in Nanoscience and nanotechnology. He became an IEEE Fellow in 2005 and an IMEC Fellow in 2007. He has made contributions to the fields of non-volatile semiconductor memory devices and technology, reliability physics of VLSI-technology, hot carrier effects in MOSFET's, time-dependent dielectric breakdown of oxides, Negative-Bias-Temperature Instability effects, ESD-protection and –testing, plasma processing induced damage, electrical characterization of semiconductors and characterization and reliability of high k dielectrics. Recently he has also interest in nanotechnology for post-CMOS applications, such as carbon nanotubes for interconnect applications, tunnel FET's for alternative nanowire devices etc. He has served as a technical program committee member of several international scientific conferences, among which the IEEE International Electron Device Meeting (IEDM), the European Solid State Device Research Conference (ESSDERC), the International Reliability Physics Symposium (IRPS), the IEEE Semiconductor Interface Specialists Conference (SISC) and the EOS/ESD Symposium. From 2000 until 2002 he also acted as European Arrangements Chair of IEDM. In 2005 he was the General Chair of the Insulating Films on Semiconductor (INFOS) conference, organized in Leuven, Belgium. Finally from 1999 until 2006 he acted as an editor of IEEE Transactions on Electron Devices. He has authored or co-authored more than 500 publications in international scientific journals and in international conference proceedings, 6 book chapters and 10 patents in his fields of expertise.

Gu, Pei-Yi
Pei-Yi Gu received her master degree in applied chemistry from NCTU, Hsin Chu, Taiwan, in 2006.After graduation, she joined the Non-volatile Memories Development Department, Electronics and Opto-Electronics Research Laboratories, ITRI (Industrial Technology Research Institute), Taiwan, where she has been developing etching process technologies including GeSbTe, high-k dielectric film,

Guan-Wei, Wu
Guan-Wei Wu was born in Kaohsiung, Taiwan in 1983. He received the B.S and M.S. degrees in Department of Electrical Engineering from National Chung Hsing University in 2005 and 2006, respectively. In 2006, he joined Macronix International Co., Ltd. (MXIC), and has been with the Device Engineering Department. Where he has engaged in the TCAD simulation of CMOS devices and memory cells.

Gustin, Wolfgang
Wolfgang Gustin received the diploma in physics (1990) from the University of Stuttgart, and the Ph.D. (1994) from the Max-Planck-Institut Stuttgart. From 1994-1998, he had been with Philips & IBM, working on Integration and Unit Process Issues for Logic and DRAM Technologies. In 1998 he joined the DRAM Development Group at Infineon Technologies. Currently he is the

manager of the Device Reliability Group at Infineon Technologies and responsible for Design for Reliability and Radiation induced Reliability.

Habersat, Daniel
Daniel Habersat earned his B.S. degree in physics from the University of Maryland, College Park in 2001, and a M.S. degree in Applied Physics from Johns Hopkins University in 2007. Since 2002, he has been working for the Sensors and Electron Devices Directorate at the Army Research Laboratory in Adelphi, MD. Mr. Habersat is currently the lead technical engineer for the Wide Bandgap Device Evaluation Team of the Power Components Branch. His research interests are focused on the evaluation of MOS interface defects and their influences on device performance and reliability.

Hamanaka, Chikara
Chikara Hamanaka received the B.E. and M.E. degrees from Kyoto Institute of of Technology, Kyoto, Japan, in 2009, 2011, respectively.

Han, Jae
Jae Ho Han received B.S. and M.S. degrees in electronic engineering and radio engineering from Korea University, Seoul, Korea, in 1998 and 2000, respectively and the Ph.D. degree in electrical and computer engineering from Johns Hopkins University, Baltimore, MD in 2010. From 2000 to 2005, he was with the LG R&D Center, Anyang, where he was engaged in the development of high-speed optoelectronic devices. Since 2010, he has been with the National Institute of Standards and Technology, Gaithersburg, MD, working on electrical and optical characterization of novel semiconductor devices.

Han, Tzung-Ting
Tzung-Ting Han was born in I-Lan, Taiwan, ROC., on November 18, 1973. He received the B.S. degree in engineering science from National Cheng Kung University, Tainan, Taiwan, R.O.C. in 1997 and the M.S. degree in electrical engineering from National Cheng Kung University, Tainan, Taiwan, R.O.C. in 1999. In 1999, he joined Macronix International Co., Ltd., Hsinchu, Taiwan, to work on advanced diffusion module process development. Since 2002, he has worked on process integration of advanced non-volatile memory.

Hattori, Nobuyoshi
Nobuyoshi Hattori received the B.S. (1983) and the M.S. (1986) degrees in faculty of engineering from Kyoto University, Kyoto, Japan. Since 1986 he had been engaged in development of production engineering for the LSI Research & Development Laboratory, Mitsubishi Electric Corp. In 2010 he joined Renesas Electronics Corporation and now is interested in mechanical characteristics of silicon crystal for LSI manufacturing.

Hau-Riege, Christine
Dr. Hau-Riege received S.B. and Ph.D. degrees in material science and engineering from MIT in 1996 and 2000, respectively. She has since held positions at Intel (Portland, 2000 - 2001) and AMD (Sunnyvale, 2001 – 2008), and is currently with Qualcomm (Santa Clara, since 2009) in the area of interconnect reliability. She is the first-author of more than 15 papers and holds 19 US patents. She has served on the IRPS committee for most years since 2004 and was a committee chair in 2008.

He, Jun
Jun He received his B. S. degree in Physics and Ph.D. in Material Science from the University of California, Santa Barbara in 1991

and 1996, respectively. Since joined Intel, Jun has held numerous technical and management positions in Logic Technology Development Q&R. In 2007, Jun took on additional responsibility as reliability program manager overseeing all Si quality & reliability of Intel's 32nm logic technology. He is current managing the Assembly TD Q&R department. He has 24 technical publications and holds 12 U.S. /E.U. patents with 16 Patents pending.

He, Yi
Yi He received a B.Sc. degree in physics from Sichuan University, Chengdu, China; a M.Sc. degree in physics from University of Waterloo, Waterloo, Ontario, Canada; and a PhD in physics from University of Virginia, Charlottesville, VA. He is a Sr. Staff Engineer at Intel, Chandler, AZ, working in the Assembly Test & Technology Development. His main focus is on physical characterization of electronic packaging materials.

Heebum,
Hee Bum Hong received the B.S., M.S. and Ph.D. degrees in materials science and engineering from Seoul National University in 1998, 2000 and 2005, respectively. Since 2008 he has been working as a senior engineer in samsung electronics. His research interests include CMOS modeling and characterization.

Hekmatshoar, Bahman
Bahman Hekmatshoar received his Ph.D. from Princeton University in 2010, where he worked on the reliability of amorphous Si thin-film transistor backplanes for realizing active-matrix organic light-emitting diode displays on flexible clear plastic substrates. He is currently a research staff member at IBM T.J. Watson Research Center, Yorktown Heights, NY, where his research focus is on high-efficiency Si-based heterojunction and/or tandem-junction photovoltaic devices. He has authored or co-authored over 50 journals and conference papers.

Herfst, Rodolf W.
Rodolf Herfst received the M.Sc. degree in applied physics from the Eindhoven University of Technology, The Netherlands, in 2004. He received his Ph.D. degree in electrical engineering from the University of Twente in 2008. Since then he has worked at Epcos Netherlands B.V. on reliability of RF MEMS capacitive switches, and studied NBTI in CMOS at the University of Twente. In 2010 he joined NXP Research and his current research interests include RF characterization, reliability simulation and reliability characterization.

Hicks, Jeff
Jeffrey Hicks is a Senior Principle Engineer in Intel's Technology Development Quality and Reliability Group. He received his BS in Applied Physics from Caltech in 1980 and joined Intel's Technology Development Group in Santa Clara CA in that year as a reliability engineer working on Bipolar and EPROM non-volatile memory technologies. In his career at Intel to date Mr. Hicks has served in numerous Quality and Reliability functions spanning Technology Development, Manufacturing, Product, and Customer Quality and Reliability. He has several patents received or pending and has published a number of technical papers including Best paper at the 1999 International Reliability Physics symposium and is the recipient of four Intel achievement awards.

Hidetoshi, Onodera
His inteHidetoshi Onodera received the B.E., M.E., and Dr. Eng. degrees in Electronic Engineering from Kyoto University, Kyoto, J apan, in 1978, 1980, 1984, respectively. Since 1983 he has been an Instructor(1983-1991), an Associate Professor (1992-1998), a Professor (1999-) in the Department of Communications and

Computer Engineering, Graduate School of Informatics, Kyoto University. His research interests include computer-aided-design for integrated circuits, and analog and mixed analog-digital circuits design. He is a member of the IPSJ, ACM and IEEE.rests are in reconfigurable architectures utilizing device variations, architectures and implementations of parallel computers.

Higgins, Robert
Robert Higgins received his B.S. in Engineering Physics (1991), and M.S. in Chemical Engineering (1995) from the Colorado School of Mines, and joined Texas Instruments in 1996. His focus has been on analog technology reliability development, including high-voltage and bipolar transistor hot carrier degradation, gate oxides, NBTI, and electromigration. He is a senior member of the technical staff at Texas Instruments, and a Certified Reliability Engineer (CRE). He is also an active member of the JEDEC 14.2 Wafer Level Reliability Standards Committee.

Hillenbrand, Rainer
Rainer Hillenbrand is Ikerbasque Research Professor and Nanooptics Group Leader at the nanoscience research center CIC nanoGUNE in San Sebastian, Spain. He is also co-founder of the company Neaspec GmbH (Martinsried, Germany), which develops and manufactures near-field optical microscopes. From 1998 to 2007 he worked at the Max-Planck-Institut fuer Biochemie (Martinsried, Germany), where he led the Nano-Photonics Research Group from 2003 to 2007. He obtained his PhD degree in physics from the Technical University of Munich in 2001. Hillenbrand's research activities include the development of infrared and terahertz near-field nanoscopy and its applications in material sciences, semiconductor technology and nano-photonics.

Hiroki, Sunagawa
Hiroki Sunagawa received the B.E. and M.E. degrees in Electrical and Electronic Engineering from Kyoto University, Kyoto, Japan, in 2008, 2010, respectively.

Hirose, Yukinori
Yukinori Hirose received the B.S. (1985) and the M.S. (1987) degrees in faculty of engineering from Tohoku University, Sendai, Japan, and received the Ph.D. (2007) degree of engineering from Osaka City University, Osaka, Japan. Since 1987 he had been engaged in development of advanced process evaluation and failure analysis technology for the LSI Research & Development Laboratory, Mitsubishi Electric Corp. In 2010 he joined Renesas Electronics Corporation. Also, he is a member of the Japanese Society of Microscopy, the Japan Society of Applied Physics, and the Japan Institute of Metals.

Ho, Dah-Chuen
Dah-Chuen Ho received the M.S. degree in Physics from National Tsing-Hua University, Hsinchu, Taiwan, R.O.C. in 1996. He joined the Taiwan Semiconductor Manufacturing Company (TSMC), Hsinchu, Taiwan, R.O.C. in 2000. He works in Division of Anolog Power IC and Special Technology. He is currently the Section Manager for the development on power IC and BCD

Ho, Paul S.
Dr. Paul S. Ho is the Director of the Laboratory for Interconnect and Packaging at The University of Texas at Austin. He received his Ph.D. degree in physics from Rensselaer Polytechnic Institute. In 1972, he joined the IBM T.J. Watson Research Center and became Senior Manager of the Interface Science Department in 1985. In 1991, he joined the faculty at the University of Texas and was appointed the Cockrell Family Regents Chair in Materials Science and Engineering. His current research is in the areas of materials, processing and reliability for interconnect and packaging.

Hoffmann, Marcel
Marcel Hoffmann joined Infineon Technologies in 2005. He received his diploma in electrical engineering (2008) from the Baden-Württemberg Cooperative State University. From 2006 he has been with Device Department of the Central Reliability Methodology Group where he completed his diploma thesis on NBTI-Characterization methodology .
Since 2008 he is working in the Application Engineering Team of Infineon Technologies on UMTS physical layer firmware. 2011 this business was transferred to Intel Mobile Communications GmbH within the Intel Cooperation.

Hoffmann, Thomas Y.
Thomas Y. Hoffmann received a Ph.D. degree from Lille University, France, in 2000. He then joined Intel Corporation's R&D group in Hillsboro, Oregon, as a TCAD engineer for sub-90nm technologies. In 2004, he moved to Intel's technology development group as a device engineer for 45nm process development. In 2005, he joined IMEC in Leuven, Belgium, to lead the electrical characterization group for advanced silicon technologies. He has authored or co-authored approximately 50 technical papers for publication in journals and presentations at conferences.

Hong, Sung-Joo
Sung-Joo Hong received the B.S. in 1985 in the department of Physics from Seoul National University (SNU) and the M.S. in 1987, and Ph.D. in 1992 in the department of physics from Korea Advanced Institute of Science and Technology (KAIST), Taejeon, Korea. Since he joined Hynix Semiconductor inc. in 1992, He has developed many DRAM and FLASH technologies. Currently, he is the head of R&D division in Hynix Semiconductor Inc.

Hopstaken, Marinus
Marinus Hopstaken obtained his PhD from the Eindhoven University of Technology, Eindhoven, the Netherlands, in 2000. He has worked at Philips Research Labs, Eindhoven, The Netherlands, Philips/NXP Semiconductors, Crolles, France, and CEA-LETI, Grenoble, France during 2001-2007 in various capacities involving physical characterization of semi-conductor materials, processes and devices. He joined IBM Research, Yorktown Heights, NY in 2008 where he is currently a Senior Staff Member in the Advanced Materials Characterization group. His publications include nearly 80 papers in various scientific journals and conference proceedings, 2 issued patents, and 1 book chapter.

Horii, Sadayoshi
Sadayoshi Horii received the B.S. degree in precision engineering from Kanazawa University, Ishikawa, Japan, in 1987, and the M.S. and Ph.D. degrees in material science from Japan Advanced Institute of Science and Technology, Ishikawa, Japan in 1998 and 2001, respectively. In 1987, he joined in Toyama Works, Kokusai Electric Ltd., where he had been developing the software system for the information equipment and the semiconductor equipment. From October 1996 to September 2001, he studied the ferroelectric materials and devices in Japan Advanced Institute of Science and Technology on leave from Kokusai Electric Ltd. In 2001, he was with Semiconductor Equipment System Laboratory, Hitachi Kokusai Electric Inc., engaging in development of thin-film deposition for advanced CMOS and DRAM devices. Dr. Horii is a member of the Japan Society of Applied Physics and the IEEE Electron Devices Society.

Horikawa, Tsuyoshi
Tsuyoshi Horikawa received the B.S. degree and the M.S. degree in chemistry from Shizuoka University (1983) and from Kyoto University (1985), respectively. He joined Mitsubishi Electric Corp. in 1985. In 2003, he moved to AIST, Japan, where he has been investigating material science and process technology for CMOS and silicon photonics. He has also been with Millennium Research for Advanced Information Technology (MIRAI) project, and with Photonics-Electronics Convergence System Technology (PECST) project. He is an associate editor of Japanese Journal of Applied Physics, and a member of the IEEE Electron Device Society and the Japan Society of Applied Physics.

Hsu, T. H.
Te-Hsun Hsu was born in Kaohsiung, Taiwan. He received the M.S. degree in Department of Electric Engineering from the National Cheng-Kung University, Tainan, Taiwan. In 2000, he joined Taiwan Semiconductor Manufacturing Company, Ltd., Hsinchu, where he was engaged in the research and development of nonvolatile memory device. In 2009, he was with the eMemory Technology Inc., Hsinchu, where he was engaged in the research and development of embedded nonvolatile memory device.

Hsu, Yen-Ya
Yen-Ya Hsu received the B.S. degrees in Physics from National Sun Yat-sen University, Taiwan, in 2003 and M.S. degrees in electronic engineering from National Chia-Tung University, Taiwan, in 2007, respectively. Since 2007, she has been with Industrial Technology Research Institute, Taiwan, and she is a Associate Engineer. Her research interests include RRAM device design, fabrication and measurement.

Hu, Youzhou
Youzhou Hu, received the B.S. and M.S. degrees in Applied Physics from TianJin University, China, in 2001 and 2004. And he worked as a Process Integration Engineer in SMIC of China and Chartered Silicon for two and half years. He is currently pursuing the Ph.D. degree in the School of Electrical and Electronic Engineering, Nanyang Technological University. His research interest is the BTI of MOS transistors with high-K oxides.

Huang, Ru
Ru Huang received the B.S. (with highest honors) and M.S. degrees in electronic engineering from Southeast University, Nanjing, China, in 1991 and 1994, respectively, and the Ph.D. degree from Peking University, Beijing, China, in 1997. Since 1997, she has been with Peking University, where she is currently a Professor and the Chairman of the Department of Microelectronics. She is the holder of 11 patents and the Author/Coauthor of 4 books and more than 150 papers. Her research interests include nanoscale CMOS device design and fabrication, flash memory devices, modeling and simulation, and RF IC technology.

Huang, Clement
Clement Huang received his B.S. (1995) in Chemical Engineering from Tatung institute of Technology and M.S. (1997) in Environment Engineering from National Chiao Tung University Taiwan. He joined United Microelectronics Corp. in 2000 and worked with thin-film module division for over 6 years. Currently he is a staff engineer of reliability department for three years.

Huang, I-Jen
I-Jen Huang received the B.S. degree in electrical engineering from National Taiwan University, Taiwan, in 1992, and the M.S. degree in electrical engineering from National Taiwan University

in 1994. He joined Macronix International Co., Ltd., Hsinchu, Taiwan, in 1996 as a process integration engineer. He has worked on process integration of non-volatile memory and SRAM for several generations. He is presently engaged in array characterization of NVM products.

Huang, Jyun-Siang
Jyun-Siang Huang was born in Chiayi, Taiwan,R.O.C., in 1981. He received the M.S. degree from Electrophysics Department, National Chiao Tung University, Hsinchu, Taiwan, in 2006. In 2006, he joined the Advanced Device Department of Macronix International Company Ltd., Hsinchu, Taiwan. His research interests include the device characterization of Flash memory devices

Huang, Yen-Chieh
Yen-Chieh Huang received his M.S. in Physics from National Taiwan University in 2006. He is the senior engineer of Mainstream Technology Quality and Reliability Department (TSMC). Prior to joining TSMC in 2008, he served as DRAM process integration engineer in Nanya Technology Corporation for 2 years.

Huang, Yu-Hui
Yu-Hui Huang received the B.S. degree in Physics from National Central University, Taoyuan, Taiwan, R.O.C., in 2002, and the M.S. degree in electronics engineering from National Chiao Tung University, Hsinchu, Taiwan, R.O.C., in 2004. She has been working in Taiwan Semiconductor Manufacturing Company (TSMC), Hsinchu, Taiwan, R.O.C., as an High Voltage Process Integration Engineer from 2004 to 2008 and a HV Reliability Engineer now.

Huard, Vincent
Vincent Huard received the B.S degree in physics and the M.S. degree in electrical engineering from the Institut National Polytechnique de Grenoble (INPG), Grenoble, France, in 1996 and 1997, respectively, and the Ph.D. degree in Physics from the University of Grenoble, Grenoble, in 2000. He is currently with STMicroelectronics, Crolles, France. His current research interests include oxide reliability, hot carrier injection, and NBTI degradation.

Hwang, Lira
Technical Quality & Reliability, System LSI division, Samsung Electronics, Korea

Ighilahriz, Salim
Salim IGHILAHRIZ received his Master's degree in Physics for microelectronics and nanotechnologies from Montpellier's sciences university in 2010. He is currently working with STmicroelectronics and IMEP Grenoble toward the PhD degree. His research focuses on modelization of the HBTs RF integrated circuits reliability.

Islam, Ahmed Ehteshamul
Ahmad Ehteshamul Islam received the B.S. Degree from Bangladesh University of Engineering and Technology (2004) and the Ph.D. Degree from Purdue University (2010), all in Electrical Engineering. Currently, he is working as Postdoctoral Research Associate in Prof. John Rogers group at University of Illinois at Urbana-Champaign. His research interests include the reliability, variability of high-κ/strained-CMOS and CNT-based flexible-electronic applications, and characterization, performance improvement of flexible-electronic devices. He has (co)-authored more than 30 journals and conference papers. He is a member of

the IEEE EDS and also serves as a reviewer for several IEEE, Elsvier, APS and ECS journals. His academic record has been recognized with a number of awards, namely the Kintar-Ul-Haque Gold Medal (2005) tfor his undergraduate result, and IEEE EDS PhD Fellowship (2008), Intel PhD Fellowship (2009-2010) for his work on CMOS reliability.

Ito, Shuu
Shuu Ito received the B.S. (1990) and M.S. (1992) in mathematical science from Osaka Prefecture University, Osaka, Japan. In 1992, he joined NEC Corp. He is currently engaged in the research and development of embedded flash memory LSIs in Renesas Electronics. He is a member of the Physical Society of Japan and the Japan Society of Applied Physics.

Iwai, Hiroshi
Hiroshi Iwai is a professor in Frontier Research Center and also in Interdisciplinary Graduate School of Science and Engineering, Tokyo Institute of Technology. He received the B.E. and Ph.D. degrees in electrical engineering from the University of Tokyo. He has been worked in the R & D for LSI and nano-device technologies for more than 35 years; at Toshiba for 26 years, and at Tokyo Institute of Technology for 10 years. His current main research fields are Si-nanowire FET and high-k gate insulator below 0.5 nm.

Jack, Nathan
Nathan Jack received a B.S. degree from Utah State University in 2007 and a M.S. degree in 2009 from the University of Illinois at Urbana-Champaign, both in Electrical and Computer Engineering. He is currently pursuing a Ph.D. at UIUC, also in ECE. His research interests include on-chip ESD protection design and ESD test methods. He has completed five summer internships at Micron Technology, Inc, in Boise, Idaho, of which two were in the ESD/LUP R&D Reliability Group. During the summer of 2010, Nathan was an intern at Intel Corp. in Hillsboro, OR, also in the ESD Reliability Group.

Jagannathan, Srikanth
Srikanth Jagannathan (S'07) received the B.E. degree in electrical engineering from the University of Madras, Chennai, India, in 2003, and the M.S. degree in electrical engineering from Vanderbilt University, Nashville, TN, in 2009. He is currently working toward the Ph.D. degree at Vanderbilt University. He has worked with Intel Corporation, Hillsboro, OR, as an Intern. His research interests include digital/analog circuit design, soft errors, and reliability of semiconductors. Mr. Jagannathan is a member of the Nuclear and Plasma Sciences Society.

Jain, Ankit
Ankit Jain received the B.Tech degree in Electrical Engineering (EE) from Indian Institute of Technology (IIT) Kanpur, India in May 2008. Currently he is working towards the PhD degree in the Department of Electrical and Computer Engineering (ECE), Purdue University, West Lafayette, IN, USA under the guidance of Prof. M. A. Alam. His research interest mainly focuses on simulation, modeling and reliability of nanoscale devices. Currently he is working on reliability of RF-MEMS capacitive switch and NEMFET.

James, Sturm
James C. Sturm is a professor of Electrical Engineering and Director of the Princeton Institute for the Science and Technology of Materials (PRISM) at Princeton University. He was educated at Princeton and Stanford Universities, and previous experience in industry includes microprocessor design at Intel Corp (Santa Clara, CA) and Siemens (Munich, Germany). He is co-founder of Aegis-Lightwave, which makes advanced fiber optics devices and systems, and a fellow of IEEE . He has received over a dozen awards for teaching excellence. Current research interests include silicon-based nanodevices, large area and flexible electronics, photovoltaics, and the nanotechnology-biology interface.

Janai, Meir
Dr. Janai received his B.Sc in Physics and Mathematics from the Hebrew University of Jerusalem (1968) and his M.Sc and D.Sc in Physics from the Technion, Israel Institute of Technology (1977). He is a Senior Fellow of Spansion Inc., Sunnyvale, CA. Prior to joining Spansion he held positions of VP of Q&R at Saifun Semiconductors and Chief Scientist at Chip Express Corporation. From 1972 to 1985 he was member of the Faculty of Physics at the Technion. Dr. Janai served on Israel National Microelectronics Committee and is a member and former Chaiman of JEDEC NVM reliability task-group.

Jang, Tae-Su
Tae Su Jang received the M.S. and Ph. D. degrees in electrical engineering from Pohang University of Science and Technology (POSTECH), Pohang, Kyungbook, Korea, in 2002 and 2006, respectively. In 2006, Jang joined the R&D division of Hynix Semiconductor Inc. where Jang has worked on the development of DRAM devices including 3-dimensional cell transistor and capacitor-less 1T DRAM as well as on the reliability characterization of plasma process-induced damages.

Javier, Martin-Martinez
Javier Martin-Martinez received the M.S. degree in physics from the Universidad de Zaragoza, in 2004 and his Ph.D. from Universitat Autònoma de Barcelona (UAB) in 2009. He is currently assistant professor with the Departament d'Enginyeria Electrònica in UAB. During his Ph.D. studies, he has been with the Università degli Studi di Padova, Padova, Italy, and IMEC, Leuven, Belgium, where he has work on electrical modelling for NBTI and dielectric breakdown. His main research interests include characterization and modelling of failure mechanisms in MOSFETs and their impact on circuits.

Jenkins, Keith
Keith A. Jenkins received a PhD in physics from Columbia University, for experimental work in high energy physics. At the IBM Research Division, he has worked in a variety of device and circuit subjects, including high frequency measurement techniques, electron beam circuit testing, radiation-device interactions, low temperature electronics, SOI self-heating and history effects, and substrate signal coupling. His current activities include evaluating the frequency response of non-silicon nanoscale devices, and replacing bench instruments with on-chip circuits to measure timing jitter, power supply transients, device variability, and circuit reliability. He is a senior member of the IEEE.

Jeong, Eui-Young
received the B.S. degree (2010) in electronic and electrical engineering from the Pohang University of Science and Technology (POSTECH), Pohang, Korea, where he is currently working toward the M.S. degree. His main research interests include characterization and modeling of CMOS devices.

Jeong, Jae-Goan
Since 1998, He has been with Hynix Semiconductor Inc. in developing DRAM. He has developed many DRAM technologies

including sub-50nm technology. Currently, he is leading the Device Technology Group of DRAM in R&D division.

Jeong, Tae-Young
Tae-Young Jeong studied the electrical and electronic engineering in college is located in Seoul, Korea. He joined Technology Quality Reliability in SYSTEM LSI, Samsung Electronics in 2004. Since then, he has been engaged in interconnect reliability of logic process technology.

Jeong, Yoon-Ha
received the Ph.D. degree (1987) in electronic engineering from the University of Tokyo, Tokyo, Japan. From 1976 and 1981, he was an Assistant Professor of electrical engineering with the Kyungnam College of Technology, Pusan, Korea. From 1982 to 1987, he was a Research Assistant with the Department of Electronic Engineering, University of Tokyo, where he pioneered in situ vapor phase deposition and the development of photo-chemical vapor deposition technology for InP metal–insulator–semiconductor field-effect transistors. In 1987, he joined the Pohang University of Science and Technology (POSTECH), Pohang, Korea, where he is currently a Professor with the Department of Electronic and Electronical Engineering and a Senior Research Vice President. He is also an Adjunct Professor with the Department of Electrical Engineering, University of Texas at Dallas, Richardson. From 2001 to 2009, he was the Director of Nanotechnology Research Center, POSTECH. Since 2004, he has been responsible for the National Center for Nanomaterials Technology (NCNT), Pohang. Dr. Jeong is a Fellow of the Institution of Engineering and Technology and the Institution of Electrical Engineers. He is the Chair of the IEEE Electron Devices Society Yeongnam Chapter in Korea. He was the Conference Chair of the IEEE Nanotechnology Materials and Devices Conference (NMDC) in 2006. He was nominated on Who's Who in Science and Technology.

Jeong, Yoon-Ha
received the Ph.D. degree (1987) in electronic engineering from the University of Tokyo, Tokyo, Japan. In 1987, he joined the Pohang University of Science and Technology (POSTECH), Pohang, Korea, where he is currently a Professor with the Department of Electronic and Electronical Engineering and a Senior Research Vice President. He is also an Adjunct Professor with the Department of Electrical Engineering, University of Texas at Dallas, Richardson. From 2001 to 2009, he was the Director of Nanotechnology Research Center, POSTECH. He is the Chair of the IEEE Electron Devices Society Yeongnam Chapter in Korea.

Jesudoss, Pio
Pio is a PhD student at Tyndall National Institute. He did a Bachelor of Science (BSc) degree in Physics in 2004 from University Louis Pasteur, Strasbourg, France. He received an MSc degree in Elaboration and Characterisation technique of surface materials and thin films from University Louis Pasteur, Strasbourg, France. During this course, he worked on a project titled 'Characterisation of low field GMR Sensors'. He is currently working on his PhD in Electrical Engineering, under Dr. Alan Mathewson, Dr. William M. D. Wright and Frank Stam. For his PhD, he is working on Adhesive Flip Chip on flex for in vivo diagnostics using swallowable capsule

Jha, Neeraj
Dr. Neeraj K. Jha received the M.S. degree in physics from IIT Delhi and the Ph.D. degree in microelectronics from the Indian Institute of Technology (IIT) Bombay, India, in 1998 and 2005,

respectively. His interests include MOS physics and technology, and characterization and modeling of MOSFET degradation under mixed-signal circuit conditions. He has also worked on the characterization and modeling of hot-carrier and bias temperature reliability issues in MOSFET. In August 2005, he joined Taiwan Semiconductor manufacturing Company (TSMC), Ltd., Hsinchu Taiwan, R.O.C., where he has been engaged in CHC, NBTI and TDDB modelling for circuit and product reliability

Jin, Minjung
Technical Quality & Reliability, System LSI division, Samsung Electronics, Korea

Joh, Jungwoo
Jungwoo Joh received the B.S. degree in Electrical Engineering from Seoul National University, Korea, in 2002, and the M.S. and Ph.D. degrees in Electrical Engineering from the Massachusetts Institute of Technology, Cambridge, MA, in 2007 and 2009, respectively. From 2002 to 2005, he worked with Alticast in Seoul, Korea as a software engineer. Since 2005, he has been conducting research on reliability, modeling, and characterization of GaN HEMTs.

Johnston, Allan
B. S. and M. S. degrees from the University of Washington. He joined Boeing Aerospace in 1965, specializing in radiation effects in microelectronics. He joined JPL in 1992, and is currently a Principal Staff Engineer. Interests include opto-electronics, single-event upset in microelectronics, low dose-rate effects, latchup, and device scaling. He was Technical Chairman of the Nuclear and Space Radiation Effects conference in 1997, General Chairman in 2003, and received the NSREC Outstanding Paper award in 1999. He has published more than 90 papers, and authored a recent book, Reliability and Radiation Effects in Compound Semiconductors, published by World Scientific.

Park, Jongwoo
Jongwoo Park received the Ph. D. from Lehigh University in 1998. After post doctoral research at Lehigh University in 1999, he joined Lucent Technologies as a Member of Technical Staff. His research projects were packaging, reliability, polymer characterization and interfacial failure mechanism. In 2002, he joined Princeton Optronics as a Manager of Quality/Reliability. He was involved in hermetic/nonhermetic package development and reliability focused on tunable laser module including pump laser, MEMS and VCSEL. Since 2003, he is with Technology Quality Reliability in SYSTEM LSI, Samsung Electronics as a Vice President responsible for reliability qualification of advanced CMOS process and product. engineering from Tokyo Institute of Technology in 2005 with Samsung scholarship program.

Joo, Moonsig
Moon Sig Joo, Ph.D. received a B.S. degree in physics education from Seoul Nation University, Korea, the M.S. degree in physics from Korea Advanced Institute of Science and Technology (KAIST), and the Ph.D. degree in electrical engineering from National University of Singapore (NUS), Singapore, in 1989, 1991 and 2006, respectively. Joo has been Principal Research Engineer at Hynix Semiconductor since 1991 with advanced gate stack development for both DRAM and NAND Flash Device. Currently, Joo is responsible for process development and process integration for ReRAM device.

Joshi, Kaustubh
Kaustubh Joshi received the B.E. degree from Chhattisgarh Swami Vivekananda Technical University, Bhilai in 2009. Since 2009 he

is pursuing master's degree in Electrical Engineering at Indian Institute of Technology (IIT) Bombay. His research interests include device reliability characterization and modeling. Currently he is working on NBTI in high-k/MG devices.

Juan, Alex

Alex Juan received the B.S. degree (1992) and M.S. degree (1994) in Electronics Engineering from National Chiao Tung university, Taiwan, ROC. He joined United Microelectronic Corp. (UMC) in 1998 and worked with platform process development for over 10 years. He is currently working for reliability department and is focusing on the technology & methodology development.

Jung, Eui-Young

received the B.S. degree (2010) in electronic and electrical engineering from the Pohang University of Science and Technology (POSTECH), Pohang, Korea, where he is currently working toward the M.S. degree. His main research interests include characterization and modeling of CMOS devices.

Kaczer, Ben

Ben Kaczer is a Senior Reliability Scientist at IMEC, Belgium. He received the M. S. degree in Physical Electronics from Charles University, Prague, Czech Republic, in 1992 and the M. S. and Ph. D. degrees in Physics from The Ohio State University, Columbus, in 1996 and 1998, respectively. For his Ph. D. research on the ballistic-electron emission microscopy of SiO2 and SiC films he received the OSU Presidential Fellowship and support from Texas Instruments, Inc. In 1998 he joined the reliability group of IMEC, Leuven, Belgium where his activities have included the research of the degradation phenomena and reliability assessment of SiO2, SiON, high-k, and ferroelectric films, planar and multiple-gate FETs, circuits, and characterization of Ge/III-V and MIM devices. He has authored or co-authored more than 250 journal and conference papers, presented invited papers and tutorials at several international conferences, and received three Best and one Outstanding Paper Awards at IRPS and the Best Paper Award at IPFA. He has served or is serving at various functions at the IEDM, IRPS, SISC, INFOS, and WoDiM conferences.

Kakushima, Kuniyuki

Kuniyuki Kakushima received the Ph.D. degree in engineering from the University of Tokyo in 2003. Currently, he is an assistant professor at Tokyo Institute of Technology. His research interests include high-k gate dielectric. He is a memeber of IEEE.

Kalya, Shubhakar

SHUBHAKAR K received his B.E (Electronics and Communication, 2000) and M.E (Microelectronics, 2007) from NMAMIT Nitte, Mangalore and Indian Institute of Science (I.I.Sc), Bangalore respectively. He is currently pursuing his Ph.D. at the Division of Microelectronics, School of EEE, Nanyang Technological University working on reliability physics and degradation mechanisms in novel high-κ dielectric materials using scanning tunneling microscopy (STM) and conductive atomic force microscopy (CAFM) technique.

Kane, Terence

A graduate of University of California, Berkeley, Terence Kane is a Senior Engineer in the IBM Semiconductor Research Design Center in East Fishkill, New York, responsible for electrical and physical characterization of 45nm, 32nm and 22nm node technologies. Terence has introduced atomic force probing and nanoprobe capacitance spectroscopy to IBM in characterizing discrete SRAM bulk silicon and SOI MOSFET devices as well as SOI embedded DRAM devices. Terence is a holder of more than 26 patents and has published over 35 technical papers. He is member of IEEE and EDFAS. He has worked for IBM for over 26 years.

Kang, Chang Yong

received the B.S. degree (1993) and the M.S. degree (1995) from Hanyang University, Seoul, Korea and the Ph.D (2005) in electrical and computer engineering from the University of Texas at Austin. He worked at Hynix, former LG Semicon (1995-2001), where he was involved in DRAM, high performance and low standby power logic devices design and he covered device characterization, TCAD/Spice modeling and process integration. Since 2005, he is with SEMATECH and now leads Simulation & Modeling group. He has authored and co-authored over 130 peer-reviewed journal and conference papers in the various semiconductor research areas including gate oxide reliability, SOI FinFET device and process, strained silicon devices, alternative channel device, and high-k/ metal gate process and devices. Dr. Kang is the recipient of the 2008 SSDM Best Paper Award. He is a senior member of IEEE and his name was listed on Marquis Who's Who in Science and Engineering and Marquis Who's Who in the World. Also, he serves as a reviewer for various technical journals such as TED, EDL, APL, JAP, Thin Solid Film, etc. Currently, he is a technical committee member of IEEE ICMTS.

Kang, Jinfeng

Jinfeng Kang received his B.S. degree in physics from Dalian University of Technology in 1984, and M.S. and Ph.D degrees in solid-state electronics from Peking University in 1992 and 1995 respectively. He is Full Professor of Electronics Engineering Computer Science School, Peking University. He has published more than 100 journals and conferences papers. Now His research interest focuses on nano devices and materials for computing and data storage, high-k/metal gate technology, and solar cell.

Kaouther, Daoud

Kaouther Daoud was born in Tunis (Tunisia), in March 18, 1960. She was a graduate from IEF, Paris XI University. Her Ph.D. research activities in France Telecom Laboratories (CNET,1984–1987) include optimizing a selfaligned technology of GaAs/GaAlAs heterojunction bipolar transistors for microwave applications. Since 1989, she is a lecturer and a professor at INSA of Rouen. She is in charge of a microelectronic team in the GPM laboratory, Rouen, France. She is mainly involved in fundamental and applied research with microelectronic critical processes related to device miniaturization and failure analysis: defects and diffusion of p- and n-type dopants in ultra-shallow junctions, segregation in semiconductor–oxide interfaces, activation anomalies, and failure analysis. She is also currently collaborating in academic and industrial partnerships in French and European programs.

Kato, Ichiro

Ichiro Kato received the B.S.(2000) degree in engeneering from Kogakuin University, Tokyo, Japan. He joined NEC Corp., Japan, in 2000. He has been with Renesas Electronics Corp., Japan, since 2010. He has been working on 90/40nm-node technology reliability.

Kauerauf, Thomas

Thomas Kauerauf received his degree in electrical engineering from the Technical University of Ilmenau, Germany, in 2001 and the Ph.D. degree from the Katholieke Universiteit Leuven, Belgium, in 2007. In 2006 he joined imec, Leuven, Belgium, where he is currently working in the Device Reliability and

Electrical characterization group focusing on high-k gate stacks and the impact of Cu contacts on the FEOL reliability. From 1999 to 2000 he stayed several months at Bell Laboratories, Murray Hill, USA. Thomas Kauerauf has authored or co-authored more than 50 publications and received the IEEE SISC Ed Nicollian Award for the best student paper in 2001.

Kawaguchi, Hiroshi
received B.E. and M.E. degrees in Electronic Engineering from Chiba University, Chiba, Japan, respectively, in 1991 and 1993. He received a Ph.D. degree in Engineering from the University of Tokyo, Tokyo, Japan, in 2006. Since 2007, he has been an Associate Professor at the Department of Computer Science and Systems Engineering, Kobe University. Dr. Kawaguchi received the IEEE ISSCC 2004 Takuo Sugano Outstanding Paper Award. He has served as a Program Committee Member for IEEE Symposium on Low-Power and High-Speed Chips, and as a Guest Associate Editor of IEICE Transactions on Fundamentals of Electronics.

Kazutoshi, Kobayashi
Kazutoshi Kobayashi received the B.E., M.E. and Dr. Eng. degrees in Electronic Engineering from Kyoto University, Kyoto, Japan, in 1991, 1993, 1999, respectively. His interests are in reconfigurable architectures utilizing device variations, architectures and implementations of parallel computers. He was an Instructor (1993-2001), an Associate Professor(2001-2002, 2004-2009) in Kyoto University. From 2002 to 2004, he was an Associate Professor of VLSI Design and Education Center (VDEC) at the University of Tokyo. Since 2009, he is a Professor in Kyoto Institute of Technology He is a member of IEEE, IEICE and IPSJ.

Keller, Robert
Bob holds a B.S. and Ph.D. in Materials Science and Engineering from the University of Minnesota. He has worked at the University of Erlangen-Nürnberg and the Max-Planck-Institute for Metals Research in Germany. He has been with NIST since 1993, in various roles, including post-doctoral fellow, project leader, and group leader. His interests include deformation and reliability of metals and semiconductors, and development of electron microscopy methods. He has twice served as an MRS Symposium organizer, and organized two NIST-NSF workshops on reliability and characterization of nanoscale materials. He received a U.S. Department of Commerce Bronze Medal in 2008.

Kerber, Andreas
Andreas Kerber was born in Schnann, Austria, in 1973. He received his Diploma in physics from the University of Innsbruck, Austria, in 2001 and a PhD in electrical engineering from the TU-Darmstadt, Germany in 2004 (granted with honors). He worked as intern at Bell Laboratories, Lucent Technologies, Murray Hill, NJ, USA (1999-2000), at IMEC in Leuven, Belgium (2001-03) as Infineon Technologies assignee to International SEMATECH, for the Reliability Methodology Department at Infineon Technologies in Munich, Germany (2004-06), for AMD in Yorktown Heights, NY (2006-09), and for GLOBALFOUNDRIES in Yorktown Heights, NY (since 2009). Much of his work centered around Front-End-Of-Line (FEOL) reliability research with focus on metal gate / high-k CMOS technologies. He has co-authored 65 papers in Journals and Conferences.

Kerber, Andreas
Andreas Kerber was born in Schnann, Austria, in 1973. He received his Diploma in physics from the University of Innsbruck, Austria, in 2001 and a PhD in electrical engineering from the TU-

Darmstadt, Germany in 2004 (granted with honors). He worked as intern at Bell Laboratories, Lucent Technologies, Murray Hill, NJ, USA (1999-2000), at IMEC in Leuven, Belgium (2001-03) as Infineon Technologies assignee to International SEMATECH, for the Reliability Methodology Department at Infineon Technologies in Munich, Germany (2004-06), for AMD in Yorktown Heights, NY (2006-09), and for GLOBALFOUNDRIES in Yorktown Heights, NY (since 2009). Much of his work centered around Front-End-Of-Line (FEOL) reliability research with focus on metal gate / high-k CMOS technologies. He has co-authored 65 papers in Journals and Conferences.

Kim, Andrew
Andrew Kim received a B.S. degree from California State University, Fullerton, in May 1995, and M.S. and Ph.D. from Rensselaer Polytechnic Insitute, NY, all in Mechanical Enginnering, in 1997 and 2001, respectiely. He was a TCAD engineer at Texas Instruments from 2002 to 2004. In 2004, he moved to Samsung Electronics, South Korea, and stayed until 2010 as the manager of Interconnect Reliability group. He is now a senior member of technical staff (SMTS) at Globalfoundries, NY, focusing on sub-20nm advanced BEoL integration.

Kim, Chris
Chris H. Kim received his B.S. and M.S. degrees from Seoul National University and a Ph.D. degree from Purdue University. He spent a year at Intel Corporation and is currently an Associate Professor at the University of Minnesota. Prof. Kim is the recipient of an NSF CAREER Award, a Mcknight Foundation Land-Grant Professorship, a 3M Non-Tenured Faculty Award, DAC/ISSCC Student Design Contest Awards, IBM Faculty Partnership Awards, an IEEE Circuits and Systems Society Outstanding Young Author Award, ISLPED Low Power Design Contest Awards, and an Intel Ph.D. Fellowship. His research interests are in digital, mixed-signal, and memory circuit design.

Kim, Dae Hyun
Dae Hyun Kim received the B.S. degree in electrical engineering from Seoul National University, Seoul, Korea, in 2002, and received the M.S. degree in electrical and computer engineering from Georgia Institute of Technology, Atlanta, in 2007. Currently, he is working toward the Ph.D. degree in the School of Electrical and Computer Engineering, Georgia Institute of Technology, Atlanta. His research interests are physical design algorithms and methodologies for 3D integrated circuits.

Kim, Hyung-Seok
He received the B.S. from Dongguk University, Korea, in 2003. He joined Hynix. in 2003 and currently he is participating in the next generation of device reliability.

Kim, Jae-Joon
Jae-Joon Kim received the B.S. and M.S. degrees in Electronics Engineering from Seoul National University, Seoul, Korea and Ph.D. degree from the School of Electrical and Computer Engineering of Purdue University at West Lafayette, IN, USA in 1994, 1998, and 2004, respectively. Since 2004, he has been with IBM T. J. Watson Research Center as a Research Staff Member and has contributed to POWER6 and POWER7 microprocessor design. His current research interest includes circuit design for exploratory devices, 3D VLSI integration, robust memory design and on-chip monitoring circuit design for reliability/process variations.

Kim, Kyung-Do

Kyung Do Kim received the B.S. and M.S. degrees in electrical engineering from Seoul National University, Seoul, Korea, in 1998 and 2000,respectively. In 2000, he joined Hynix Semiconductor Inc. ,Icheon, Korea, where he was engaged in deveolping 512M and 1G DDR2 DRAM. His research interest is the reliability of DRAM.

Kim, Tae-Hyoung
Prof. Kim received the B.S. and M.S. degrees in electrical engineering from Korea University, Korea, in 1999 and 2001, respectively. He received the Ph.D. degree in electrical and computer engineering from the University of Minnesota, Minneapolis, USA in 2009. From 2001 to 2005, he worked for Samsung Electronics where he performed research on the design of high-speed SRAMs, and IO circuits. In 2007 ~ 2009 summer, he was with IBM T. J. Watson Research Center and Broadcom Corporation where he conducted research on circuit reliability and ultra-low power memory design. In November 2009, he joined Nanyang Technological University as an assistant professor.

Kim, Yong-Taik
Yong-Taik Kim joined Hynix Semiconductor Inc. in 1996. He has developed many of DRAM technologies. Especially, Kim played a leading role in developing the 60nm DRAM technology. Kim also involved in the development of W-Gate and Fin cell transistor. He is now developing sub-30nm DRAM technology.

Kim, Hyunjin
Technical Quality & Reliability, System LSI division, Samsung Electronics, Korea

Kim, Jungin
Technical Quality & Reliability, System LSI division, Samsung Electronics, Korea

Kindereit, Ulrike
Dr. Ulrike Kindereit focused on diagnostics of semiconductor devices early, with a master thesis on Laser Stimulation, supported by Robert Bosch GmbH. She received the diploma of electrical engineering from Berlin University of Technology in 2005. After being engaged in a research project with DCG Systems, CA, with a thesis on Laser Voltage Probing, she received the Ph.D. in 2008. From 2008-2010 Dr. Kindereit was working on diagnostics and reliability of IHP's Si/SiGe technology in Frankfurt (Oder), Germany. Currently she is working in the Circuit Test and Diagnostics Technology group at the IBM T.J. Watson Research Center, Yorktown Heights, NY.

Kobayashi, Kazutoshi
Kazutoshi Kobayashi received the B.E., M.E. and Dr. Eng. degrees in Electronic Engineering from Kyoto University, Kyoto, Japan, in 1991, 1993, 1999, respectively. His interests are in reconfigurable architectures utilizing device variations, architectures and implementations of parallel computers. He was an Instructor (1993-2001), an Associate Professor(2001-2002, 2004-2009) in Kyoto University. From 2002 to 2004, he was an Associate Professor of VLSI Design and Education Center (VDEC) at the University of Tokyo. Since 2009, he is a Professor in Kyoto Institute of Technology He is a member of IEEE, IEICE and IPSJ.

Koh, Lian Ser
Koh Lian Ser received his B.Eng(Hons) in 1991 and M.Eng in 1993 from the National University of Singapore. He joined SEMICAPS Pte Ltd in 1993 and is now the R&D Section Manager responsible for the development of technologies for photon emission and laser scanning microscopy.

Kopanski, Joseph
Joseph Kopanski received a B. S. degree in Applied Physics in 1982 and an M. S. degree in Electrical Engineering in 1985, both from Case Western Reserve University in Cleveland, OH. He joined the National Institute of Standards and Technology in Gaithersburg, MD in 1985 where he is currently a member of the technical staff. His research interest are focused on using scanning probe techniques, such as scanning capacitance microscopy and scanning Kelvin force microscopy for electrical transport characterization of nanostructures. He is the first author of over 40 publications, a member of ECS, APS, and a senior member of IEEE.

Koudymov, Alex
Alexei Koudymov received the M.S. degree in Physics from the St-Petersburg State technical University, Russia, in 1997 and Ph.D. in Electrical Engineering from the University of South Carolina in 2003. He was in Research positions at the University of South Carolina in 2003-2006, and at the Rensselaer Polytechnic Institute in 2006-2010. Currently with Sensitron Semiconductor, a division of RSM Electron Power, Inc., of Deer Park, NY. Area of expertize includes Nitride-based electronics and optoelectronics, Si electronics and emerging semiconductor applications.

Koyama, Toru
Toru Koyama received the B.S. degrees in Chemical Engineering from the Kyoto Institute of Technology, Kyoto, Japan, in 1984. He later obtained the Ph. D. in material and life science at the Osaka University in 2002. He joined Mitsubishi Electric Corporation in 1984, where he has been engaged in development of failure analysis and fault isolation technique for ULSI devices. In 2010, he joined Renesas Electronics Corporation. He is a member of the Reliability Engineering Association of Japan.

Krishnan, Anand
Anand T. Krishnan received his Bachelor of Technology degree in Metallurgical Engineering from Institute of Technology, BHU, Varanasi, India in 1994, and M.S. and Ph.D degrees in Materials from The Pennsylvania State University in 1997 and 2000, respectively. In 2000, he joined the silicon technology development group at Texas Instruments. His interests and activities are in the areas of negative bias temperature instability and plasma charging damage. He has served in the technical program committee for the International Electron Devices Meeting (IEDM), International Reliability Physics Symposium (IRPS), Integrated Reliability Workshop (IRW) and for the Plasma Process-Induced Damage Symposium (P2ID). He has authored or co-authored more than 50 papers and 15 patents. He is the co-receeipient of IRPS outstanding paper awards in 2002, 2004 and 2006, and the Mehboob Khan award of the SRC in 2010.

Krishnan, Siddarth
Siddarth Krishnan is the High-K Dielectric Team Lead in the Unit Process team in IBM SRDC, in East Fishkill. He joined IBM in October 2006, as a characterization and reliability engineer for the 32nm and 22nm development programs. He is currently focused on unit process, device analysis and fundamental learning in setting work function and EOT scaling. Prior to joining IBM, he was a postdoctoral researcher with SEMATECH, where he worked on High-k, Hybrid Orientation Technology and Wafer Level Reliability. He has a Masters and a Ph.D in ECE from the University of Texas, at Austin and a bachelor's degree in MSE from IIT in Madras.

Kristof, Croes
Kristof Croes received his MSc in physics at the Catholic University of Louvain (Belgium) in 1993 and his MSc in biostatistics at the Limburgs Universitair Centrum (LUC) in 1994. In 1999, he obtained his PhD, concerning the development of statistical techniques for planning reliability experiments. After that, he joined the reliability business unit of XPEQT, first as the software responsible and than as the manager of the R&D. From 2003 till end 2006, he was product and application manager of the package level reliability products of Chiron holdings. Beginning 2007, he went back to research, working on BEOL and 3D-reliability problems at imec, where he is currently heading the wafer level interconnect reliability team.

Ku, Kuan-Cheng
K C Su received the B.S. degree (1986) and M.S. degree (1988) in Electrical Engineering from National Cheng Kung university, Taiwan, ROC. Then, he joined United Microelectronic Corp. (UMC) and worked with process technology development division for over 10 years. He is currently focusing on reliability technology & methodology development. His main interests are reliability engineering, device engineering and logic technology development.

Ku, Shaw-Hung
Shaw-Hung Ku was born in Taipei, Taiwan, R.O.C. 1977. He received the M.S. and Ph.D. degree in electronics engineering from the National Chiao-Tung University, Hsinchu, Taiwan, R.O.C., in 2001 and in 2006. Then, he joined Macronix International Company, Ltd. and is responsible for the developement of nitride-based and floating-gate storage memories.

Kuang-Chao, Chen
Kuang-Chao Chen received the M.S. degrees in chemistry from the National Chong-Shan University, Taiwan, in 1987. From 1989 to 1995, he joined Electronic Research and Service Organization (ERSO), Hsinchu, Taiwan, where he has been involved in the development of BEOL planarization process technology. From 1995 to 1998, he was with Mosel-Vitelic International Co., Ltd, Hsinchu, Taiwan. He performed yield improvement in manufacturing line. In 1998, he joined Vanguard International Semiconductor Co., Ltd, Hsinchu, Taiwan, as a department manager. He was responsible for thin film module development. In 2000, he joined Macronix International Semiconductor Co., Ltd, Hsinchu, Taiwan, where he worked on advanced module development. He is currently Executive Director of Technology Development Center.

Kumari, Sangita
Sangita kumari is a PhD student at University of Arizona, US, in the Department of Material Science and Engineering and holds the degrees of B.Sc and M.Sc in Physics from Goa University, India. Sangita's dissertation work involves investigating the role of dissolved gases in the generation of Sonoluminescence (SL) and its relation to wafer damage and cleaning efficiencies. SL is a phenomenon where light is generated in sound-irradiated fluids by the process of cavitation. Sangita's research aims at reducing the extent of damage to wafer features caused by unstable cavities formed in process fluids during megasonic cleaning of wafers.

Kushvaha, Sunil Singh
Dr. Sunil Singh Kushvaha obtained his Masters degree in Physics from Banaras Hindu University (BHU), Varanasi India in 2000, M.Tech. degree in Solid State Materials from Indian Institute of Technology (IIT) Delhi, India in 2002 and PhD in Physics from National University of Singapore (NUS) in 2008. He is currently working as a Research Engineer at Institute of Materials Research and Engineering (IMRE), Singapore. Dr. Kushvaha has worked on different scanning probe microscopy and surface analysis techniques such as scanning tunneling microscopy (STM), atomic force microscopy (AFM), four nanoprobes, low energy electron diffraction (LEED), Auger electron spectroscopy (AES), scanning electron microscopy with polarization analysis (SEMPA) in ultra-high vacuum systems. He has 16 publications in international journal papers and 2 invited book chapters.

Kwasnick, Robert
Robert Kwasnick is Q&R Manager of the Hard IP Group at Intel Corporation. He has a B.A. in Physics (High Honors) from Swarthmore College, and a M.S. and Ph.D. in Physics from MIT. Dr. Kwasnick worked at GE on digital x-ray imaging technology and works at Intel on design for reliability and knowledge-based qualification. He has 74 issued patents and 20 publications. Dr. Kwasnick has served on IRPS subcommittees, and presented IRPS tutorials in 2007 and 2008 and a Year in Review topic in 2009. Dr. Kwasnick was elected a Fellow of the IEEE in 2001.

Kyosuke, Ito
Kyosuke Ito received the B.E. and M.E. degrees in Electrical and Electronic Engineering from Kyoto University, Kyoto, Japan, in 2009, 2011, respectively.

Labie, Riet
Riet Labie obtained the M.Sc. and Ph.D. degrees in Materials Science from the Katholieke Universiteit Leuven, Belgium, in 1999 and 2007, respectively. She joined Imec in 1999 where she first worked as process and research engineer in advanced packaging and flip chip interconnections. After her PhD, she started working as a reliability engineer for packaging and interconnects, with particular focus on 3D integration. In 2010, she moved to the Si PV department of Imec where she is currently working on the development and reliability of PV modules.

Lacaita, Andrea L.
Andrea L. Lacaita is a Full Professor of Electronics at the Politecnico di Milano. Scientist since 1987, he has been Visiting Professor at the AT&T Bell Laboratories, Murray Hill, NJ (1989-90), IBM T.J. Watson Research Center, Yorktown Heights, NY (1999). 2009 IEEE Fellow. He has contributed to study quantum effects as well as experimental characterization techniques and numerical models of non-volatile memories, both Flash and emerging (PCM,RRAM). He is co-author of more than 200 papers, patents and several educational books in Electronics.

Lai, Tai-Hsiang
Tai-Hsiang Lai received the B.S. degree from the Department of Electrical Engineering, Yuan Ze University, Taoyuan, Taiwan, in 2004 and the M.S. degree in the Institute of Electronics, National Chiao-Tung University, Hsinchu, Taiwan, in 2006. He joined United Microelectronic Corp. (UMC) in 2006 and he is currently an Assistant Manager of ESD Engineering Dept., Reliability Technology & Assurance Division. His main research interests include ESD and Latch-up protection designs for high-voltage (HV) and power management ICs (PMIC) technologies.

Lai, Tai-Hsiang
Tai-Hsiang Lai received the B.S. degree from the Department of Electrical Engineering, Yuan Ze University, Taoyuan, Taiwan, in 2004 and the M.S. degree in the Institute of Electronics, National Chiao-Tung University, Hsinchu, Taiwan, in 2006. He joined United Microelectronic Corp. (UMC) in 2006 and he is currently

an Assistant Manager of ESD Engineering Dept., Reliability Technology & Assurance Division. His main research interests include ESD and Latch-up protection designs for high-voltage (HV) and power management ICs (PMIC) technologies.

Lamontagne, Patrick
Patrick Lamontagne received the engineering degree in Materials and Nanotechnologies from the Institut National des Sciences Appliquées de Rennes, Rennes, France and M.S. degree of Physic from the Université de Rennes 1, Rennes, France in 2007. He is currently working toward the Ph.D. degree with the Electrical Characterization and Reliability Group at STMicroelectronics, Crolles, France. His research field is the electromigration issues in advanced copper interconnects.

Leduc, Patrick
Patrick Leduc received his M. Sc. degree in physics from the Polytechnics Institute of Grenoble (INPG-Phelma), France, in 1998 and joined the CEA-LETI in 2000. His current position is project leader in 3-D integration for CMOS applications. His fields of expertise are 3-D integration and CMOS interconnects technologies. During 5 years, he has been working on process development (Chemical Mechanical Planarization) for CMOS interconnects. Since 2006, he is in charge of 3D integration projects. He participated in European projects dedicated to interconnects (NanoCMOS, Pullnano, ELITE). During his carrier, he has been author or co-author of about 30 technical papers, communications or invited presentations in international conferences related to the above fields.

Lee, Kyong Taek
Kyong Taek Lee received a Ph.D. degree in electronics and electrical engineering from Pohang University of Science and Technology (POSTECH), Pohang, Korea, in 2010 and is currently working at Technical Quality & Reliability, System LSI division, Samsung Electronics, Korea.His current research interests include device characterization and reliability physics of below 28nm high-k/metal gate CMOSFETs.

Lee, Chien-Ying
Chien-Ying Lee was born in Taipei, Taiwan, R.O.C. on June 10, 1984. He received the MS degree in Electronics Engineering from National Chiao-Tung University, Hsinchu, Taiwan, R.O.C. in 2008. He joined Macronix International Company, Ltd., Hsinchu, in 2008, as a Process Integration Engineer in technology development center. Since 2009, he has been engaged in the development and integration for the floating-gate-type flash memory at Macronix.

Lee, Chih-Hsiung
Chih-Hsiung Lee was born in Chiayi, Taiwan, R.O.C., on November 15, 1982. He received the B.S. degree and the M.S. degree in electronics engineering from the National Chiao-Tung University, Hsinchu, Taiwan, in 2005 and 2007. He jointed Macronix International Company, Ltd., Hsinchu, in 2009, as a Process Integration Engineer in technology development center. He has been working on the development and integration for the floating-gate-type NAND flash memory at Macronix.

Lee, Heng-Yuan
Hengyuan Lee received the B.S. degrees in material science and engineering from National Tsing Hua University, Hsinchu, Taiwan, in 2002, and the M.S. degrees in electronics engineering from National Chiao Tung University, Hsinchu, Taiwan, in 2004,

and the Ph.D. degree in electronics engineering from National Tsing Hua University, Hsinchu, Taiwan, in 2010. Since 2005, he has been with Industrial Technology Research Institute, where he is currently a deputy project manager of Department II of Non-Volatile Memory. His research interests include emerging memory technology, nanoscale CMOS device, DRAM and flash device.

Lee, Jack
received the B.S. degree (1980) and M.S. degree (1981) in electrical engineering from the University of California, Los Angeles, and the Ph.D. degree(1988) in electrical engineering from the University of California, Berkeley. Currently, he is Professor with the Electrical and Computer Engineering Department and holds the Cullen Trust For Higher Education Endowed Professorship in Engineering at the University of Texas at Austin. From 1981 to 1984, he was a Member of the technical staff at the TRW Microelectronics Center in the High-Speed Bipolar Device Program. He worked on bipolar circuit design, fabrication, and testing. In 1988, he joined the faculty of The University of Texas at Austin. His current research interests include thin dielectric breakdown and reliability, high-K gate dielectrics and gate electrode, high-K thin films for semiconductor memory applications, electronic materials and semiconductor device fabrication processes, characterization, and modeling. He has published over 250 journal publications and conference proceedings. Dr. Lee has been awarded two Best Paper Awards, numerous teaching awards, and several patents.

Lee, Jeong-Soo
received the Ph.D. degree (1996) in electronic and electrical engineering from the Pohang University of Science and Technology (POSTECH), Pohang, Korea. From 1996 to 2001, he was a senior engineer of semiconductor R&D division in Samsung electronics Co. and he was research assistant professor of University of Texas at Dallas from 2002 to 2005. In 2008, he joined the Pohang University of Science and Technology (POSTECH), Pohang, Korea, where he is currently a assistant professor of the department of electronic and electrical engineering.

Lee, Miji
After graduated from Korea Aerospace University in Korea with a bachelor's degree in materials engineering in 2006, she joined Quality/Reliability team in SYSTEM LSI, Samsung Electronics . She is involved in 28nm, 45nm BEOL qualification of semiconductor as a reliability engineer.

Lee, Shou-Chung
Shou-Chung Lee received the B.S. (1994) in physics from National Sun Yet-Sen University, Kaohsiung, Taiwan and M. S. (1996) from Institute of Electro-Optical Engineering of National Chiao Tung University, Hsinchu, Taiwan. In 1998, he joined TSMC where he work on Cu/Low-K interconnect reliability. His current research interests are in the reliability physics of Copper electromigration and low-k dielectric breakdown.

Lee, Tackhwi
Tackhwi LEE received the B. S. and M. S. degrees in Physics from Yonsei University, Seoul, Korea, in 1996 and 1998, respectively. He obtained the Ph. D degree in Electrical and Computer Engineering at the University of Texas, Austin, in 2010.

Lee, Yung-Huei
Yung-Huei Lee (S'82-M'86-SM'06) received Ph.D. in EE from Ohio State University. He is a 23 years veteran of Intel and has

worked in the development of various CPU, RF/analog, and Flash technology generations. He is currently a Technical Director at TSMC. Dr. Lee holds 2 US patents and has published over 60 technical papers.

Lee, Yung-Huei
Yung-Huei Lee (S'82-M'86-SM'06) received Ph.D. in EE from Ohio State University. He is a 23 years veteran of Intel and has worked in the development of various CPU, RF/analog, and Flash technology generations. He is currently a Technical Director at TSMC. Dr. Lee holds 2 US patents and has published over 60 technical papers.

Lee, Seok-Kiu
He received the Ph.D. from SNU, Korea, in 2000. He joined Hynix Semiconductor inc. in 1986 and now he is in charge of managing the development of NAND flash device development in R&D.

Lelis, Aivars
Aivars Lelis received his MS degree in Electrical Engineering from the Johns Hopkins University in 2000 and is presently completing his doctoral work in reliability engineering at the University of Maryland, focusing on the causes of the threshold-voltage instability in SiC MOSFETs and their effect on device reliability. He is currently the Device Physics and Evaluation Team Leader in the Power Components Branch at the U.S. Army Research Laboratory. His research has focused on the device physics effects of charge trapping in gate insulators, first in regard to the effects of ionizing radiation in Si MOSFETs, and more recently in regard to the effects of processing and operational stresses in SiC MOSFETs.

Leong, Kam Chew
K.C. Leong received his B. Eng (Hons) and MEng (Microelectronics) from Nanyang Technological University in 1996 and 1998 respectively. He joined the then-CHARTERED SEMICONDUCTOR MANUFACTURING (now GLOBALFOUNDRIES) in 1998 in the R&D Department working on 0.25mm, 0.18mm and 0.13mm CMOS technologies as well as the 0.18mm SiGe BiCMOS technology. He is currently holding the post of Program Manager in charge of close to 50 postgraduate students in GLOBALFOUNDRIES, on top of being a technical trainer for advanced technologies.

Lepla, Keith
Keith Lepla: B.Sc. Chemistry University of Manitoba, 83, Ph.D. Analytical Chemistry University of Alberta 89. He is the Senior Analytical Chemist at Teck Resources Zinc-Lead Operation in Trail BC. His interests are atomic spectroscopy and laboratory automation and database applications.

Leung, Martin
Dr. Martin Leung received his Ph.D. in physical chemistry from the University of California at Los Angeles in 1974. He worked for Union Carbide Research Center after graduation and joined The Aerospace Corporation in 1979. He is currently an Associate Director of the corporation's Microelectronics Technology Department. His research interests encompass the use of various analytical techniques to investigate and resolve reliability issues related to space hardware.

Li, Baozhen
Baozhen Li received the B. S and M. S. degrees in Metallurgy from Northeastern University in China, and the Ph. D degree in Materials Science and Engineering from University of Notre

Dame, IN., in 1990. He was with the Science and Technology Center of Westinghouse Electric Corp., where his research was centered on the fuel cell technology development. Since 1996, he has been with IBM Microelectronics in the Systems and Technology Group, Essex Junction, Vermont. Currently he serves as a leading engineer for advanced semiconductor technology qualifications for IBM alliances.

Li, Baozhen
Baozhen Li received the B. S and M. S. degree in Metallurgy from Northeastern University in China, and the Ph. D degree in Materials Science and Engineering from University of Notre Dame, IN., in 1990. He was with the Science and Technology Center of Westinghouse Electric Corp., where his research was centered on the fuel cell technology development. Since1996, he has been with IBM Microelectronics in the Systems and Technology Group, Essex Junction, Vermont. Currently he serves as a leading engineer for advanced semiconductor technology qualifications for IBM alliances.

Li, Jinglong
I am a failure analysis engineer in Quality Department in Freescale Semiconductor (China) Limited since 2008. In 2008, I graduated form Tianjin University of Technology in Tianjin China and got master degree in automation Engineering.

Liang, James
James Laing received his B.S. (1992) in Ceramic Engineering from Rutgers University and M.S. (1994) in Material Science from New Jersey Institute of Technology. He joined United Microelectronics Corp. in 2002 and has been involved in the quality and reliability of CMOS process. Currently he is a staff engineer of reliability department.

Liao, Pei-Jean
Pei-Jean Liao received her M.S. degree in electrophysics from the National Chiao Tung University, Hsinchu, Taiwan, in 1998. She was then with the Technology Quality and Reliability Division at Taiwan Semiconductor Manufacturing Company (TSMC), where her research centered on theory and mechanism in CMOS gate oxide integrity. During this time, she engaged several TSMC technology reliability qualifications and published 2 technical papers regarding TDDB category. She is currently working with SRAM product level reliability.

Lim, Koeng Su
Koeng Su Lim received the B.S. degree in electrical engineering and the M.S. degree in energy materials engineering from Yokohama National University, Yokohama, Japan, in 1977 and 1979, respectively, and the Ph.D. degree in physical electronics from the Tokyo Institute of Technology, Tokyo, Japan, in 1984. From 1984 to present, he has been an assistant, associate, and full professor in the department of electrical engineering, Korea Advanced Institute of Science and Technology. He has authored more than 280 journal/conference papers and more than 30 domestic/international patents. His research interests include amorphous silicon solar cells, memory devices, and optical/display devices.

Lim, Sung Kyu
Sung Kyu Lim received the B.S., M.S., and Ph.D. degrees from the Computer Science Department, University of California, Los Angeles, in 1994, 1997, and 2000, respectively. He joined the School of Electrical and Computer Engineering, Georgia Institute of Technology in 2001, where he is currently an Associate Professor. His research focus is on 3-D IC architecture, design,

and computer-aided design. Dr. Lim received the National Science Foundation Faculty Early Career Development Award in 2006. He has served the Technical Program Committee of several conferences on electronic design automation including ACM Design Automation Conference and IEEE International Conference on Computer-Aided Design.

Limbrick, Daniel
Daniel Limbrick is currently a PhD student in the Department of Electrical Engineering and Computer Science at Vanderbilt University under the advisement of Dr. William H. Robinson. He received a Bachelor of Science degree in Electrical Engineering at Texas A&M University in May 2007 and his Masters of Science degree in the same field at Vanderbilt University in December 2009. His research interests include computer architecture, logic synthesis, and reliability of microelectronics.

Lin, Hung Sung
Hung-Sung Lin was born in Hsinchu, Taiwan in 1972. He received the M.S. in MEMS from National Tsing Hua University, Hsinchu, Taiwan in 1997. He joined Micro System Laboratory, ITRI, Hsinchu, Taiwan in 1999 where he worked on the research of advanced MEMS devices and process development. In March 2000, he joined Product Engineering Division, UMC, Hsinchu, Taiwan. He is a section manager of Logic&MM group and presently in charge of failure analysis of Logic and MM products. He has published 18 papers at international conferences. Mr. Lin is an inventor with 2 granted U.S. patents, and he received UMC ten best composing invention disclosure Award in 2006 and 2008 for the contribution on composing patents in the fields of CMOS.

Lin, Ming-Ren
Ming-Ren Lin is a Senior Fellow and the director of the Technology Research Group at GLOBALFOUNDRIES.

Lin, Mingte
Mingte Lin received his B.Eng. (1987) in Power Mechanical engineering and M.S. (1992) in Physics from Tsing Hua University Taiwan. He joined United Microelectronics Corp. in 1997 and has been involved in the quality and reliability of CMOS process. Currently he is a staff engineer of reliability department.

Linder, Barry
Barry P. Linder received his B.S. from Pennsylvania State University in 1993, and an M.S and Ph.D. in Electrical Engineering from the University of California at Berkeley in 1999. His doctoral thesis dealt with plasma processing, plasma implantation, and plasma charging damage. Since graduation Dr. Linder has been employed as a Research Staff Member at the IBM T. J. Watson Research Center, Yorktown Heights, NY. Initially his work centered on the breakdown of ultra-thin gate oxides, including the statistics of breakdown phenomenon, post-breakdown conduction mechanisms, and the interaction between oxide breakdown and circuit operation. This work formed the basis for the paper that received an "Outstanding Paper Award" at the 2003 International Reliability Physics Symposium. After 2003, his focus switched to electrical characterization and integration of metal gates and high-k materials. He has studied the full array of advanced gate stack materials including their integration and their effect on effective work function, channel mobility, gate leakage, and inversion layer thickness scaling. He specialized on the interaction between cap layers, interface layers, and metal gate composition on the final electrical properties of the gate stack. As manufacturing of high-k/metal gate stacks approached, he concentrated on all reliability aspects with emphasis on dielectric breakdown and bias temperature instability. More recently, he has added focus on circuit reliability, integrating device level reliability with circuit level functionality.

Liu, Cheng-Jye
Cheng-Jye Liu was born in Tao-Yuan, Taiwan, R.O.C. in 1975. He received the B.S. and M.S. degrees from the Electrical Engineering Department of National Tsing-Hua University, Hsinchu, Taiwan in 1997 and 1999 respectively. From 1999 to 2006, he worked in Macronix International Co., Ltd. and was engaged in the technology development of 0.35um/0.25um/75nm SONOS type flash memory. From 2006 to 2009, he was with PowerFlash where he was engaged in the product development of 70nm/50nm/40nm FG flash memory. Now he works in eMemory Technology Inc. and leads the device team to develop NeoFlash technology on various process platforms.

Liu, Chien-Chih
Chien-Chih Liu received the M.S. degree in Materials Science and Engineering from National Chiao-Tung University, Hsinchu, Taiwan, R.O.C.in 1995. He joined the Taiwan Semiconductor Manufacturing Company (TSMC) , Hsinchu, Taiwan, R.O.C. in 2000. He has worked on the development for several logic technologies in R&D, and then moved to FEOL reliability department for HV device reliability qualification. Currently, he is the HV reliability section manager in Technology Quality and Reliability Division.

Liu, Wen-Hsing
Wen Hsing Liu received the B.S. degree from the department of electronic engineering, Feng Chia University, Taichung, Taiwan, and M.S. degrees in electronic engineering from National Tsing Hua University, Hsinchu, Taiwan, in 2006 and 2008. Since 2008, she has been with Industrial Technology Research Institute (ITRI), where she is currently a RRAM device engineer for the department of Electronics and Optoelectronics Research Laboratory. Her researches include semiconductor device testing development, device reliability testing.

Liu, Wenhu
Liu Wenhu was born in Shandong, China in 1986. He received his B.Eng, 1st Class Honors (Electrical and Electronics Engineering) from Nanyang Technological University (NTU) in 2008. He is currently pursuing his Ph.D. at the Division of Microelectronics, School of Electrical and Electronic Engineering in NTU. His project is focusing on electrical characterization and reliability of novel high-κ gate dielectrics in nano-scale MOSFETs. He is currently a Graduate Student Member of IEEE (2008-present).

Liu, Ziyuan
Ziyuan Liu received the B. S. (1982) and the M. S. (1985) in Materials science and engineering from Beijing Institute of Aeronautics, Beijing, China, and the Dr. Sc. (1994) in Material Science from Tokyo Institute of Technology, Tokyo, Japan. She has worked as a frontier researcher in RIKEN (The Institute of Physical and Chemical Research). In 1997 she joined NEC Corporation, and engaged in the development of device analysis technology. She is currently interested in understand of hydrogen behavior at surfaces and interfaces of the MOS stacks and their relation to the reliability issue in Renesas Electronics.

Liu, Yang
Yang Liu (M'03) received the B.S. degree from the University of Science and Technology, Hefei, China, in 1998 and the Ph.D. degree in physics from the University of Illinois, Urbana, in 2002.

In 2003, he joined the Stanford TCAD Group, Stanford University, Stanford, CA, as a Research Associate. His current research interests include the modeling and design of novel biological sensors, digital and analog electronic devices, and optoelectronic LEDs and lasers.

Lo, Chun-Yuan

Chun Yuan Lo was born in Kaohsiung, Taiwan on February 2, 1979. He received the M.S. degree in Department of Engineering and System Science from the National Tsing-Hua University, Hsinchu, Taiwan, in 2003. In 2003, he was with the Macronix International Company, Ltd., Hsinchu, where he was engaged in the research and development of nonvolatile memory device. In 2009, he was with the eMemory Technology Inc., Hsinchu, where he was engaged in the research and development of embedded nonvolatile memory device.

Lofrano, Melina

Melina Lofrano received the BSc degree in Physics in 2001 and the MSc in Mechanical Engineering in 2003 at the University of São Paulo, Brazil. From 2003 until 2006 she worked in research and product development, where she was responsible for the mechanical modeling and analysis. From 2006 she was involved with several CAE research projects at Katholieke Universiteit Leuven, Belgium. In 2008 she joined IMEC modeling and reliability team.

Loukas, Michalas

Loukas Michalas received the B.Sc in Physics, the M.Sc. in Solid State Physics and the Ph.D. in Solid State Electronics from Physics Department of University of Athens in 2002, 2004 and 2009 respectively. His research interests include the electrical characterization and reliability of semiconductor devices. Recently, his research interests have been extended on the study of dielectric charging in Metal-Insulator-Metal (MIM) structures and RF Micro-Electro-Mechanical-Systems (MEMS) capacitive switches. He has published 9 articles in international journals and contributed more than 20 conferences presentations

Loveless, Daniel

Dr. Daniel Loveless received the B.S. degree from Georgia Institute of Technology in 2004 and the M.S. and Ph.D. degrees from Vanderbilt University in 2007 and 2009, respectively.
He is currently a staff research engineer at the Institute for Space and Defense Electronics at Vanderbilt University and an adjunct assistant professor in the Department of Electrical Engineering and Computer Science at Vanderbilt University. His work has included the design of single-event-hardened phase-locked loops, and the analysis of high-speed analog and digital circuits in radiation environments. His research interests include CMOS, mixed-signal circuit design, and radiation effects in microelectronics.

Lowrie, Anthony

Tony Lowrie is a Staff Engineer with the LTD Q&R based in Hillsboro, Oregon. He is currently working on the reliability concerns for Intel 32 and 22nm Logic technologies focusing on Interconnect reliability. Tony joined Intel in 2002 working in Intel's F24 site as the 65nm certification lead and focusing on Cu interconnect reliability. In 2008 Tony moved to Oregon focusing again on Interconnect reliability on 32nm in particular BTS concerns. Tony developed a robust BTS monitor and screening methodology for high volume manufacturing ensuring effective screening prior to shipping from Intel fabs. Tony earned his Bachelor of Engineering with honors degree from Glasgow University before starting a PHD in Optoelectronics in 1994

before the draw of industry pulled him in to work for National Semiconductor from 1994 to 1999. Tony then moved to Chartered Semiconductor in Singapore where he held various technical and management roles in Chartered Technology Development department.

Lu, Chih-Yuan

Chih-Yuan Lu received B.S. from National Taiwan University (1972), and Ph.D. (1977) in physics from Columbia University, NYC. Dr. Lu has been a professor in National Chiao-Tung Univ. and with AT&T Bell Labs from 1984- 1989; later joined ERSO/ITRI in 1989 as a Deputy General Director responsible for the MOEA grand Submicron Project. This project successfully developed Taiwan first 8-inch manufacturing technology with high density DRAM/ SRAM. He was therefore granted the highest honor prize—National Science & Technology Achievement Award by the Prime Minister of ROC, due to his leadership and achievement in this Submicron Project. In 1994, Dr. Lu becomes the co-founder of Vanguard International Semiconductor Corp., which is a spin-off memory IC Company from ITRI's Submicron Project. He was the VP of Operation, VP of R&D, and later President from 1994-99. Dr. Lu now is the chairman and CEO of Ardentec Corp. a VLSI testing service company; and also serves Macronix International as a Senior VP/CTO, and now the President. Dr. Lu led Macronix's technology development team to successfully achieve the state of the art nonvolatile memory technology and now responsible for Macronix's overall operation. Dr. Lu has published more than 200 papers and has been granted 130 worldwide patents, and was elected a Fellow of IEEE, and a Fellow of APS. He also received IEEE Millennium Medal, and the most prestige semiconductor R&D Award in Taiwan from Pan Wen Yuan Foundation.

Lu, Pong-Fei

Pong-Fei Lu received his Ph.D. degree in electrical engineering from Princeton University, Princeton, N.J., in 1986. He joined IBM T. J. Watson Research Center, Yorktown Heights, N.Y., in 1985 as a Research Staff Member. His early research work involved high-speed bipolar device design and exploratory bipolar SRAMs. Since 1989 he has been with the High-Performance Circuit Group working on the circuit design for IBM microprocessors. He received two Outstanding Technical Awards for his contribution to IBM S390 and Power5 microprocessors. He is currently the manager of Advanced RISC Design group working on next generation of IBM Power server design

Lu, Wen-Pin

Wen-Pin Lu was born in I-Lan, Taiwan, R.O.C. on December 20, 1967. He received the B.S. degree in electronics engineering from National Chiao-Tung University, Hsinchu, Taiwan, R.O.C. in 1990 and the M.S. degree in electrical engineering from National Taiwan University, Taipei, Taiwan, R.O.C. in 1992. He joined Macronix International Co., Ltd., Hsinchu, Taiwan, in 1994 as a device engineer. From 1994 to 1999, he has worked on device analysis of non-volatile memory, especially in floating-gate flash memory. Since 2000, he has been engaged in the development of PACAND Flash memory technology, and has accomplished 0.18mm and 0.15mm of delivering. He is presently responsible for the Nitrite-based NBit technology and also floating gate NOR flash development at Macronix.

Luca, Larcher

Luca Larcher graduated in Electronics Engineering from University of Padova in 1998. He received the PhD degree in 2001 from the University of Modena and Reggio Emilia, where he is currently Associate Professor of Electronics. His research interests are twofold. He focused on the experimental characterization, reliability and modeling non-volatile memories and logic devices. He worked on the characterization, reliability and design of both RF Integrated Circuits for telecommunications and circuits for energy harvesting from renewable sources in CMOS technology. He authored and co-authored a book, more than 85 technical papers published on international journals and conferences.

Luca, Vandelli
Vandelli Luca was born in Reggio Emilia, Italy, in 1985. He received the Master degree in electronic engineering from the University of Modena and Reggio Emilia Italy, in 2009. Since 2010, he has been pursuing the Ph.D. degree at the Department of Engineering Sciences and Methods of the University of Modena and Reggio Emilia. His work focuses on modeling and characterization of thin gate oxides in nonvolatile memory devices and MOS transistors.

Machavolu, Srikanth
Srikanth Machavolu is pusuing the Ph.D. degree in the School of Electrical and Electronic Engineering, Nanyang Technology University. His research interests are nano device physics and modeling.

Maeda, Hitoshi
Hitoshi Maeda received the B.S. (1992) degree in physics from Kobe University, Hyogo, Japan. Since 1992 he had been engaged in development of advanced process evaluation and failure analysis technology for the LSI Research & Development Laboratory, Mitsubishi Electric Corp. He is currently working on process evaluation and failure analysis in Renesas Electronics Corp. Also, he is a member of the Japan Society of Applied Physics.

Mahapatra, Souvik
Souvik Mahapatra received his PhD in Electrical Engineering from IIT Bombay, India, in 1999. He was at Bell Laboratories, Murray Hill, NJ, USA during 2000-2001. Since 2002 he is with the Department of Electrical Engineering, IIT Bombay, India, and presently holds the position of Professor. His research interests are in the area of characterization, modeling and simulation of CMOS, Flash memory and thin film PV devices, and device reliability. He has published more than 100 papers in international journals and conferences, delivered invited talks at leading international conferences in the USA, Europe and Asia-pacific including at IEEE IEDM, delivered reliability tutorials at IEEE IRPS, and acted as a reviewer of several international journals and conferences. He is a distinguished lecturer of IEEE EDS and senior member of the IEEE

Mahapatra, Souvik
Souvik Mahapatra received his PhD in Electrical Engineering from IIT Bombay, India, in 1999. He was at Bell Laboratories, Murray Hill, NJ, USA during 2000-2001. Since 2002 he is with the Department of Electrical Engineering, IIT Bombay, India, and presently holds the position of Professor. His research interests are in the area of characterization, modeling and simulation of CMOS, Flash memory and thin film PV devices, and device reliability. He has published more than 100 papers in international journals and conferences, delivered invited talks at leading international conferences in the USA, Europe and Asia-pacific including at IEEE IEDM, delivered reliability tutorials at IEEE IRPS, and acted as a reviewer of several international journals and conferences. He is a distinguished lecturer of IEEE EDS and senior member of the IEEE

Mahato, Swaraj
Swaraj Bandhu Mahato was born in Bandwan, West Bengal, India, in 1984. He received the BTech (Bachelor of Technology) degree of electronics and communication engineering from West Bengal University of Technology (WBUT), India in 2007. He is currently doing MSc in Integrated Circuit Design (ICD) at TU Munich, Germany and NTU, Singapore. From 2007 to 2009, he was a Design Engineer in new medical diagnostic equipment design & development as well as Embedded Software design & development at D&D, Larsen & Toubro Limited, India. He has filed two patents for "Temperature Control and Protection System for Patient Monitor System" and "Auto Debugging System for Patient Monitoring System". Currently he is doing his Master Thesis at Interuniversity MicroElectronics Centre (IMEC), Leuven, Belgium. His research work involves workload dependent circuit and architecture level NBTI modelling based on device level trapping model.

Maheta, Vrajesh D
Vrajesh D. Maheta received the B.E. degree in Electronics Engineering from Sardar Patel University, Gujarat, India, in 1993, the M.E. degree in Microelectronics from Birla Institute of Technology and Science, Pilani, India, in 2002, and the Ph.D. degree in electrical engineering from Indian Institute of Technology (IIT) Bombay, Mumbai, India, in 2009. From 1998 to 2009, he was with G. H. Patel College of Engineering & Technology, as a regular faculty. Since 2010, he has been with Middle East College of Information Technology, Muscat, Oman, where he is currently working as an Associate Professor in the department of electronics and communication engineering. His research interests are in the field of semiconductor device physics and simulation, modeling and characterization of CMOS silicon devices.

Maize, Kerry
Kerry Maize is a doctoral student at the Baskin School of Engineering at the University of California, Santa Cruz. Kerry received his B.S. in Electrical Engineering and Computer Science from UC Berkeley in 2002. His current research is focused on thermal characterization of microelectronic power ICs and thermoelectric devices and materials using high thermal and spatial resolution methods. His research interests are in the areas of applied quantum electronics and nanoelectronics.

Malandruccolo, Vezio
Vezio Malandruccolo received his MSc with honors in electronic engineering, from the University of Rome "La Sapienza", in 2004. In 2005, he joined ANSOFT as a specialist of high frequency simulation tools. From 2006 to 2007 he was in Infineon Technologies, Villach, Austria, where he worked on the design and simulation of mixed signal circuits for telecommunications products. Currently, he is working towards his Ph.D. at the ETH Zurich. His research interests include the realization of theoretical and experimental tools for the implementation of built-in reliability program, mainly based on in-situ reliability monitors, and indicators.

Maloney, Timothy
Timothy J. Maloney has a bachelor's degree in physics from MIT (1971), and M.S. (physics, 1973) and PhD. (EE, 1976) degrees from Cornell, where he was an NSF Fellow. Following postdoctoral work at Cornell he worked in semiconductor research at Varian Associates, Palo Alto, CA, until 1984. Then he joined Intel Corp., Santa Clara, CA, where he has been concerned with IC ESD protection and testing, CMOS latchup testing, signal integrity, system ESD, and other topics. Dr. Maloney is now a Senior Principal Engineer at Intel. He has 31 patents issued, is co-

author of a book, "Basic ESD and I/O Design" (Wiley, 1998), and is a Fellow of the IEEE. He has many publications at the EOS/ESD Symposium, IRPS, and other IEEE-connected entities.

Mamy Randriamihaja, Yoann
Yoann Mamy Randriamihaja received the M.S. degree in micro- and nanoelectronic from University Joseph Fourier and the Engineering degree in microelectronics from the Ecole Nationale Supérieure d'Electronique et de Radioélectricité de Grenoble (ENSERG) of the Institut National Polytechnique de Grenoble (INPG), in 2009. Since October 2009, he is pursuing the Ph.D. degree in micro- and nanoelectronic in the frame of a collaboration between the Institut Matériaux Microélectronique Nanosciences de Provence (IM2NP), Toulon, and the Electrical Characterization and Reliability Departement of STMicroelectronics, Crolles. His Ph.D. work focuses on the study of the oxide degradation.

Mason, Maribeth
Maribeth Mason received a Ph.D. in Applied Physics from the California Institute of Technology in 2004. She is currently a Research Scientist in the Microelectronics Technology Department of The Aerospace Corporation, specializing in defect characterization and failure analysis of microelectronic and optoelectronic devices.

Massengill, Lloyd
Lloyd Massengill received the Ph.D. degree in solid state circuits from North Carolina State University, Raleigh, in 1987. He is currently a Professor with the Department of Electrical Engineering and Computer Science, Vanderbilt University, Nashville, TN, where he teaches microelectronic circuit analysis and design, and studies the effects of radiation on the operation of integrated circuits, particularly the modeling of circuit-level soft errors. He also serves as the Director of Engineering for the Vanderbilt Institute for Space and Defense Electronics, Nashville.

Masuduzzaman, Muhammad
Muhammad Masuduzzaman received the B.S. degree in Electrical and Electronic Engineering (EEE) from Bangladesh University of Engineering and Technology (BUET), Dhaka, Bangladesh, in 2004. During 2005-2006, he worked as a Lecturer in the Department of EEE, BUET. He is currently enrolled in the direct Ph.D. program at the Department of Electrical and Computer Engineering, Purdue University, West Lafayette, IN, USA. His research interest includes physics, simulation and characterization of nanoscale devices. Currently he is working on reliability physics of ferro-electric, high-κ, and other novel dielectric materials.

Mathewson, Alan
Dr. Alan Mathewson is a senior member of technical staff at the Tyndall National Institute. He is responsible for systems integration in the institute with particular emphasis on the integration of autonomous wireless sensor systems. He has been involved in the development of three dimensional integrated circuits and systems for twenty years and worked on one of the first three dimensional integrated circuits fabricated under the European ESPRIT programme. Dr. Mathewson has been responsible for a large number of innovations in silicon integrated circuit technology, especially relating to novel devices and processes and their characterization and modelling.

Matroni, Koutsoureli

Matroni Koutsoureli received the B.S. degree in physics from the National and Kapodistrian University of Athens (NKUA), Athens, Greece, in 2006, and the M.Sc. degree in solid state physics from NKUA in 2008. She is currently working toward the Ph.D. degree in solid state physics at the NKUA. Her current research is focused on the polarization effects in dielectric materials of (RF) microelectromechanical (MEMS) switches.

Mavis, David
David G. Mavis is co-founder and Chief Scientist of Micro-RDC. He has over 20 years of experience in the research, development, and design of nuclear and space radiation hardened microelectronics. He received his bachelor's degree in Physics from the University of Wisconsin, Madison, and his doctorate in Nuclear Physics from Stanford University. His major interests are currently in the area of Single Event Effects testing of ultra-deep submicron microcircuits.

McGahay, Vincent
Vincent McGahay received his Bachelor and PhD degrees in Materials Engineering from Rensselaer Polytechnic Institute. He joined IBM in 1992 as a staff engineer in Logic BEOL Development, initially with responsibilities in insulator development and subsequently moving into BEOL integration. Vince was appointed Senior Technical Staff Member and Master Inventor in 2008. He is currently 1x Module Lead for the 32nm technology node.

McMahon, William
William McMahon received S.B. degrees in Physics and Electrical Engineering/Computer Science from MIT, Cambridge, Massachusetts in 1997 and a Ph.D. degree in physics from the University of Illinois, Urbana in 2002. He was with the Data Storage Institute, Singapore for two years working on spin transport in ferromagnetic materials. He worked for several years at Intel/Numonyx on NOR Flash reliability. He joined GlobalFoundries in 2009 and is currently working as part of the ISDA/ASTA alliances to enable various high-k metal gate bulk and SOI technologies. His research currently focuses on intrinsic reliability of semiconductor devices.

McMorrow, Dale
Dale McMorrow received the Ph.D. degree in Physical Chemistry from The Florida State University, Tallahassee, FL under the direction of Dr. Michael Kasha. His graduate studies focused on the role of intermolecular structure and dynamics in shaping the excitation and relaxation processes of molecular species. After a postdoctoral fellowship at the University of Toronto he joined the technical staff at the Naval Research Laboratory, and presently is head of the Radiation Effects Section. His current research interests include the development, characterization and application of linear- and nonlinear-optical techniques for simulating single-event phenomena in microelectronic devices and complex integrated circuits.

Mendenhall, Marcus
Marcus Mendenhall received his PhD from Caltech in 1983. He has been involved with ion-beam analytical techniques and computational methods for ion scattering and transport. He served as the associate director for operations of the Vanderbilt Free Electron Laser. His current work is in computer modeling of radiation effects in solids, numerical computing methods, and development of new xray sources for radiobiological applications.

Meneghesso, Gaudenzio

Gaudenzio Meneghesso graduated in Electronics Engineering at the University of Padova in 1992 working on the failure mechanism induced by hot-electrons in GaAs MESFETs and HEMTs. His research interests include Electrical characterization, modeling and reliability of microwave and optoelectronic devices like compound semiconductors HEMTs and MESFETs, RF-MEMS switches, and organic semiconductors devices. He is also developing ESD protection structures. Within these activities he published over 400 technical papers (of which more than 50 Invited). He is reviewer of several international journals and he is Associate Editor of the IEEE Electron Device Letter for the compound semiconductor devices area since 2007.

Miccoli, Carmine
Carmine Miccoli was born in Cantù, Italy, in 1984. He received the Bachelor (BS) and the Master (MS) degrees with full marks (cum laude) in Electronics Engineering from the Politecnico di Milano, Milan, Italy, in 2006 and 2009, respectively. Since 2009 he has been with the Dipartimento di Elettronica e Informazione, Politecnico di Milano, where he is currently pursuing the Ph.D. degree in Information Technology. His research activities include characterization and modeling of ultra-scaled Flash memories.

Milor, Linda
Linda Milor is an associate professor of electrical and computer engineering at the Georgia Institute of Technology. Her research interests include yield and reliability modeling, testing, and design-for-testability of analog and digital circuits. She has a PhD in electrical engineering from the University of California, Berkeley.

Min, Hong Kook
1996, B.S. Hanyang Univ./ 1996. Samsung Electronics, Dev. of embedded NVM

Ming, Zhang
Ming Zhang received his B.S. in Physics from Peking University, China, and Ph.D. in EE from University of Illinois at Urbana-Champaign. He is currently a design engineer at Samsung Electronics America Headquarter, responsible for design-for-reliability/design-for-yield development. Prior to joining Samsung, he was a senior staff architect in the Atom and SOC Development group at Intel and led corporate-wide strategic pathfinding for next-generation power-efficient mobile devices. He has more than 15 US patents and 30 papers in the areas of SER, cache VMIN, power management, DFT, and MEMS.

Mitani, Yuichiro
Yuichiro Mitani received the B. E. and M. E. in material science and engineering from Tohoku University, Sendai, Japan, in 1990 and 1992, respectively. He received the Ph.D. from the University of Tokyo in 2009. He joined the R&D Center, Toshiba Corporation in 1992. His primary works were concerned in the Si-CVD and the ultra-shallow junction process technology. Since 1999, he has been with the Advanced LSI Technology Laboratory, Corporate R&D Center, Toshiba Corporation, Yokohama, Japan. His present research interests and activities cover the ultra-thin oxide process technology and the study of the reliability of ultra-thin gate dielectrics (SiO2, SiON and High-k) for ULSI technology. He serves (or served) on the technical committees of International Conference on IC Design & Technology (ICICDT) and IEEE International Reliability Physics Symposium (IRPS). He is a member of the JSAP.

Mitard, Jerome

Jérôme Mitard received, in 2003, the M.S. degree in Microelectronic Engineering from the Polytechnic University School of Marseille, France, and the Ph.D. degree in Microelectronics from the MINATEC Center (CEA-LETI-IMEP), Grenoble, France. For more than three years, he acted as an STMicroelectronics assignee at CEA-LETI, Grenoble, where he was involved in the advanced electrical characterization of hafnium-based dielectrics and metal gate, for sub-70-nm CMOS technologies. Subsequently, he joined imec, Leuven, Belgium, as a Device Engineer, working on the integration of High-Mobility channel materials for the 16-nm CMOS node and beyond.

Moncef, Kadi
Moncef Kadi was born in Constantine (Algeria), in March 17, 1974. He received the Electr. Eng. Dipl. from the University of Constantine, Constantine, in 1996, the master's research degree (D.E.A.) in optoelectronic, optics, and microwaves from the National Polytechnic Institute of Grenoble (INPG), Grenoble, France, in 2001, and the Ph.D. degree in RF and optics from the University Joseph Fourier, Grenoble, in 2004. In October 2004, he joined the IRSEEM/ESIGELEC, Rouen, France, as a postdoctoral fellow, and is currently a lecturer/researcher. His current research interests include electromagnetic compatibility, antenna design, probe characterisation, and susceptibility of integrated circuits.

Montse, Nafira
received the Ph.D. in Physics in 1993 from the Universitat Autònoma de Barcelona, where, currently, she is an Associate Professor at the Department of Electronic Engineering. Her major research interest is in the area of CMOS devices and circuits reliability. Currently, she is working on the reliability of advanced MOS devices, including their nanoscale characterization using AFM related techniques, and the modelling of gate dielectric related failure mechanisms (breakdown, BTI, hot carriers) for circuit reliability simulation.

Monzio Compagnoni, Christian
Christian Monzio Compagnoni received the Laurea degree (cum laude) in Electronics Engineering and the Ph.D. degree in Information Technology from the Politecnico di Milano, Milan, Italy, in 2001 and 2005, respectively. Since 2002, he has been with the Dipartimento di Elettronica e Informazione, Politecnico di Milano, where he became an Assistant Professor in 2006. His research activities include characterization and modeling of advanced non-volatile memories and MOS devices. Dr. Monzio Compagnoni received the Outstanding Paper Award at the IRPS in 2008 and was a member of the memory committee of the IRPS in 2009 and 2010.

Moquillon, Laurence
Laurence Moquillon received the M.S. and Ph.D. degrees from the University of Limoges, France, in 2001. Her doctoral research with the Microwave and Optical Communication Research Institute (now XLIM), Limoges, France, concerned the study of microwave planar ring resonator multipole active filters. Since 2001, she has been with STMicroelectronics, Crolles, France. Her principal research interests are RF and millimeter-wave circuit designs for wireless communication using SiGe BiCMOS and advancedCMOS technologies.

Moreau, Stéphane
Stéphane MOREAU received the Ph.D. degree from François-Rabelais University, Tours, France, in 2005 on the environmental reliability of TRIAC (semiconductor switch). In 2006, he joined the CEA, LETI, Grenoble, France. His researches include BEOL

reliability issues (EM, SIV, TCT, THS) and multiphysics simulations.

Moss, Steven C.
Steven C. Moss is Director of the Microelectronics Technology Department at The Aerospace Corporation in Los Angeles, CA. He received a BS Degree (physics, mathematics) from Arkansas A&M College (1970), an MS Degree (physics) from Purdue University (1972), and a Ph. D. Degree (physics) from North Texas State University (1981). His research interests include the use of ultrashort optical pulses to simulate the transient effects of ionizing radiation on microelectronic devices, transient photoluminescence from materials and advanced devices, permanent damage produced by radiation effects in microelectronic/optoelectronic devices, and reliability and physics of failure of advanced microelectronic/optoelectronic devices.

Motohiko, Masuda
I am a failure analysis engineer in Quality Department in Freescale Semiconductor (China) Limited since 2005. In 1983, I graduated form Nagaoka University of Technology in Japan and got master degree in Electronic devices Engineering.

Mukhopadhyay, Saibal
Dr. Saibal Mukhopadhyay has received PhD in Electrical and Computer Engineering from Purdue University, West Lafayette, IN, in 2006. Currently he is an Assistant Professor at School of ECE at Georgia Institute of Technology. Prior to joining Georgia Tech, he was a Research Staff Member at IBM T. J. Watson Research Center, Yorktown Heights, NY. His research interests include low-power and robust circuit and system design and 3D integrated circuits. He has co-authored over 100 papers in refereed journals and conferences. He has received NSF Career Award in 2010 and IBM Faculty Award in 2009 and 2010

Murakami, Eiichi
Eiichi Murakami received the B.S. (1981), M.S. (1983) and Ph.D. (1995) degrees in applied physics from Waseda University, Tokyo, Japan. He joined the Central Research Laboratory, Hitachi, Ltd. Japan, in 1983. He worked on Si-SPE, SiGe, ultrashallowjunction, and MOSFET's design & characterization studies. He is now the manager of Analysis and Evaluation Technology Department in Renesas Electronics Corp. His current interest is in reliability and failure physics in Si-LSI. Dr. Murakami is a member of the Japan Society of Applied Physics and the IEEE EDS. He served as a sub-committee member of CMOS and Interconnect Reliability in IEDM 2004, 05.

Mutihac, Oana-Mihaela
Dr.-Ing. Oana-Mihaela Mutihac received the diploma in electrical engineering from "Politehnica" University of Bucharest, Romania in 2004 and the Ph.D. in microelectronics from Darmstadt University of Technology, Germany in 2009. Oana has been a scholarship holder and graduate member of the Graduate College "Tunable Integrated Components in Microwave Technology and Optics", funded by the German Research Foundation. Currently, she is working in the Department of Semiconductor Devices at Berlin University of Technology being focused on diagnostic of Si/SiGe HBT semiconductor devices by photon emission and laser stimulation methods.

Nagarajan, Raghavan
NAGARAJAN RAGHAVAN was born in Bangalore, India in 1985. He received his B.Eng, 1st Class Honors, (Electronics Engineering, 2007), S.M. (Advanced Materials for Micro & Nano

Systems, 2008) and M.Eng (Materials Science and Engineering, 2008) from Nanyang Technological University (NTU), National University of Singapore (NUS) and Massachusetts Institute of Technology (MIT) respectively. He was the recipient of the prestigious Nanyang Scholarship, NTU President Research Scholar and Singapore-MIT Alliance (SMA) Graduate Fellowship awards. He is also one of the five recipients to be bestowed with the IEEE Reliability Society Graduate Scholarship award in 2008 for his research accomplishments in reliability and its application to nanoelectronics. He is currently pursuing his Ph.D at the Division of Microelectronics, School of EEE, NTU focusing on reliability modeling and statistical characterization of novel high-κ dielectric materials in nanodevices. He serves on the review committee for IEEE Transactions on Device and Materials Reliability (TDMR). He is currently a Graduate Student Member of IEEE (2005-present).

Nakanishi, Nobuto
Nobuto Nakanishi received the B.S. (2001), M.S. (2003) and Ph.D. (2006) degrees in physics from Tokyo University of Science, Tokyo, Japan. He joined Renesas Technology Corp., Japan, in 2006. He is currently working on process evaluation and failure analysis in Renesas Electronics Corp. His current interests are material characterizations of semiconductor devices using advanced TEM techniques. He is a member of Japanese Society of Microscopy.

Nam, Hyeonwoo
Hyeonwoo Nam received B.S. degree in electronics and commuication engineering from the University of Hanyang, Korea in 2010. He is currently working for M.S. degree at the same university. His current research interest is in HCI issues in circuit design for emerging technilogies. He is currently working on stressing and measuring test chip for HCI analysis.

Nam, Jongik
Technical Quality & Reliability, System LSI division, Samsung Electronics, Korea

Narasimham, Balaji
Balaji Narasimham received the B.E. degree in electrical engineering from the University of Madras, Chennai, India, in 2003 and the M.S. and Ph.D. degrees in electrical engineering from Vanderbilt University, Nashville, TN, in 2005 and 2008, respectively. He is currently a Senior Staff Reliability Scientist with Broadcom Corporation, Irvine, CA, where his work focuses on device- and circuit-level reliability and characterization of soft errors for memory and logic circuits. He was with Intel Corporation, Hillsboro, OR, and IBM T. J. Watson Research Center, Yorktown Heights, NY, where he held a graduate level cooperative position. Dr. Narasimham's research interests include CMOS circuit design, radiation effects and reliability of semiconductor devices and circuits. He has authored or co-authored over 30 papers related to his research and has authored a book chapter on single-event transients. He has served in the technical committees of IRPS and NSREC and is the recipient of the Best Paper Award at the 2007 RADECS conference. He has also served as a Reviewer for the IEEE Transactions on Electron Devices, IEEE Transactions on Nuclear Science and the IET Circuits, Devices, and Systems.

Nayfeh, Hasan
Hasan Nayfeh is a Senior Engineer at the IBM Semiconductor R&D Center in East Fishkill, NY. He received his Ph.D. (2003) in Electrical Engineering in the area of strained silicon devices from MIT. After joining IBM in 2003, he has worked on SOI device

design that has resulted in the successful deployment of 65nm and 45nm technology nodes. Since early 2010 his focus is on technology definition for the 22nm node. He has over 30 technical publications and 4 patents and is a senior member of the IEEE.

Negre, Laurent
Laurent Negre received the engineering degree in Physics with a specialization in Microtechnology and Microsystems from the Institut National des Sciences Appliquées de Toulouse, France, in 2008. He is currently working toward the Ph.D. degree at the Institute of Microelectronics, Electromagnetism and Photonics (IMEP-LAHC) at Minatec, Grenoble. His Ph.D. work, which is devoted to the study of MOSFET RF reliability, is carried out in the reliability group of ST Microelectronics, Crolles, France.

Nielen, Heiko
Heiko Nielen received his diploma thesis in 1994 on mechanical stress in thin films at the research center Jülich. He finished his PhD thesis in 1998 at the technical university in Aachen (RWTH) on X-ray investigations of mechanical stress in passivated metal lines. After that he joined the central reliability department of Siemens semiconductor group, now Infineon Technologies. Since then he's been responsible for numerous device reliability qualification projects and is currently focused on degradation model development and flow implementation topics for circuit reliability simulations.

Nigam, Tanya
Tanya Nigam received her Bachelor's degree in Physics (Hons.) from St. Stephens College, Delhi University. She obtained a M.Sc in Physics from IIT Kanpur and a M.Sc in Electrical Engineering from the Katholieke Universiteit Leuven in 1995. Between 1995 and 1999, she obtained Ph.D in the area of ultra-thin gate oxides at IMEC, Belgium. From 1999 until 2001, Tanya was a Member of Technical Staff at Bell Labs where she worked on novel device geometries to overcome sub-50nm device challenges. From 2001 until 2005, she was with Agere Systems, formerly the Microelectronics Division of Lucent Technology. At Agere, she worked on reliability issues for power LDMOS devices, and HCI/NBTI reliability concerns for CMOS. From October 2005 till 2007 Tanya worked as a Senior Staff at Cypress Semiconductor involved in the optimization of 65nm CMOS. In 2008 she was with AMD and since 2009 she is with GLOBALFOUNDRIES as SMTS working on the correlation between device and product level degradation. She has co-authored 30 papers in Journals and Conferences.

O'Connor, Robert
Robert O'Connor received a Degree in Applied Physics from Dublin City University (DCU) in 2001 and a PhD in Semiconductor Physics from the same institution in 2005, which was focused on the electrical and chemical characterisation of transistor gate dielectrics. In 2006 he was at Intel Ireland working as a process engineer. He was awarded a Marie Curie Fellowship to work on charges and defects in the gate dielectric at imec, Belgium. He is currently at DCU and the Tyndall National Institute working on the characterisation of high-k materials on both silicon and III-V substrates.

Oakley, Jennifer
Jennifer Oakley received a B.S. in Physics and Astronomy from the University of Iowa and a M.S and Ph.D. in Physics from the University of Florida. She has worked as an engineer for IBM since 2005 in the Microelectronics division and currently works in Back-End of Line technology development.

Oates, Anthony
Tony Oates received his B.Sc. and PhD degrees in physics from the University of Reading, U.K, in 1982 and 1985 respectively. He is presently responsible for reliability physics development at TSMC. Prior to joining TSMC in 2002, Dr. Oates spent 17 years as a distinguished member of technical staff and technical manager at Bell Labs and Agere Systems, where he was responsible for reliability characterization and process qualification of CMOS technologies. Dr. Oates has over 100 publications, and holds 5 patents. He is a frequent conference speaker and tutorial presenter. Dr. Oates has a long-standing involvement with the International Reliability Physics Symposium, serving as the General Chair of the Symposium in 2001. He has participated in the technical leadership of the International Electron Device Meeting (IEDM) and Materials Research Society (MRS) Meetings.He is currently Editor-in-Chief of the IEEE Transactions on Device and Materials Reliability (TDMR).

Ogawa, Arito
Arito Ogawa received the B.S. and the M.S. degrees in electronics from the Tohoku Inst. of Technol., Sendai, Japan in 1996 and 1998, respectively. He joined the Hitachi Kokusai Electric Inc., Toyama, Japan, in 1998. In 2004, he joined with the High-k Gate Stack Group,Millennium Research for Advanced Information Technology (MIRAI) project, Association of Super-Advanced Electronics Technology (ASET), Tsukuba, Japan, where he investigated metal/high-k MOSFETs. Since April 2007, he has been with Hitachi Kokusai Electric Inc., Toyama, Japan, where he has been engaged in research and development of high-k dielectrics films and metal films for DRAM capacitors.Mr. Ogawa is a member of the Japan Society of Applied Physics.

Ohmi, Tadahiro
Tadahiro Ohmi (M'81–SM'01–F'03) received the B.S., M.S., and Ph.D. degrees in electrical engineering from Tokyo Institute of Technology, Tokyo, Japan, in 1961, 1963, and 1966, respectively. Prior to 1972, he served as a Research Associate in the Department of Electronics, Tokyo Institute of Technology, where he worked on Gunn diodes such as velocity overshoot phenomena, multivalley diffusion and frequency limitation of negative differential mobility due to an electron transfer in the multi-valleys, high-field transport in semiconductor such as unified theory of space-charge dynamics in negative differential mobility materials, Bloch-oscillation-induced, negative mobility and Bloch oscillators, and dynamics in injection lasers. In 1972, he moved to Tohoku University, Sendai, Japan, where he is currently a Professor at the New Industry Creation Hatchery Center. He is engaged in researches on high-performance ULSI such as ultrahigh-speed ULSI based on gas-isolated-interconnect metal-substrate SOI technology, base store image sensor (BASIS) and high-speed flat-panel display, and advanced semiconductor process technologies such as low kinetic-energy particle bombardment processes including high-quality oxidation, high-quality metallization, very-low-temperature Si epitaxy, and crystallinity-controlled film growth technologies from single-crystal, grain-size-controlled polysilicon and amorphous highly selective CVD, highly selective RIE, and high-quality ion implantation with low-temperature annealing capability based on ultraclean technology concept supported by newly developed ultraclean gas supply system, ultrahigh vacuum-compatible reaction chamber with self-cleaning function, and ultraclean wafer surface cleaning technology. His research activities are summarized by the publication of over 1300 original papers and the application of 1600 patents. Dr. Ohmi serves as the President of the Institute of Basic Semiconductor Technology-Development

(Ultra Clean Society). He is a Fellow of the Institute of Electricity, Information and Communication Engineers of Japan. He is a member of the Institute of Electronics of Japan, the Japan Society of Applied physics, and the Electrochemical Society. He received the Ichimura Award in 1979, the Inoue Harushige Award in 1989, the Ichimura Prizes in Industry-Meritorious Achievement Prize in 1990, the Okouchi Memorial Technology Prize in 1991, the Minister of State for Science and Technology Award for the Promotion of Invention (the Invention Prize) in 1993, the IEICE Achievement Award in 1997, the Okouchi Memorial Technology Prize in 1999, the Werner Kern Award in 2001, the ECS Electronics Division Award, the Medal with Purple Ribbon from Government of Japan and the Best Collaboration Award (the Prime Minister's Award) in 2003.

Okumura, Shunsuke
received the B.E. and M.E degree in Computer and Systems Engineering from Kobe University, Hyogo, Japan, in 2008 and 2010, respectively. He is currently working in the doctral course at the same university. His current research is high-performance and low-power SRAM designs.

Olivo, Piero
Piero Olivo was born in Bologna in 1956. He graduated in electronic engineering in 1980 at the University of Bologna, where he received the PhD degree in 1987. Since 1994 he is full professor of electronics at the University of Ferrara (Italy) where, since 2007, he is dean of the Engineering Faculty. The scientific activity concerns the physics, the reliability and the experimental characterization of thin oxide devices and of non-volatile memories. He is author of the first paper describing and analyzing stress-induced leakage current (SILC) in thin oxides.

Olson, Nicholas
Nicholas Olson received his B.S. degree in electrical engineering from Iowa State University of Science and Technology in 2005. He received his MS degree from the University of Illinois at Urbana-Champaign in 2008 and is working towards his Ph.D at the same university. His research is in the field of ESD and he plans to graduate in 2011.

Ong, Tong-Chern
TC Ong received his Master and Ph.D degrees in Electrical Engineering from the University of California at Berkeley in 1979 and 1987, respectively. He worked for National Semiconductor from 1979 to 1981 as a DRAM circuit designer. From 1987 to 1999, he was with Intel Corporation and worked in the area of Flash Memory reliability. He joined TSMC in 1999 and is currently the manager of Embedded Non-Volatile Memory group.

Onodera, Hidetoshi
Hidetoshi Onodera received the B.E., M.E., and Dr. Eng. degrees in Electronic Engineering from Kyoto University, Kyoto, J apan, in 1978, 1980, 1984, respectively. Since 1983 he has been an Instructor(1983-1991), an Associate Professor (1992-1998), a Professor (1999-) in the Department of Communications and Computer Engineering, Graduate School of Informatics, Kyoto University. His research interests include computer-aided-design for integrated circuits, and analog and mixed analog-digital circuits design. He is a member of the IPSJ, ACM and IEEE.

O'Shea, Sean Joseph
Sean O'Shea received his BSc and PhD in Physics from Sydney University, Australia. Subsequently he undertook research in various areas of nanoscale science at Cambridge University, UK from 1989 to 1998, where he was appointed a Royal Society

University Research Fellow. From 1999 he has been at the Institute of Materials Research and Engineering (IMRE), Singapore. His research interests are in Nanotechnology (chiefly scanning probe microscopy), MEMS, and the creation and application of new bio-sensors

Paccagnella, Alessandro
Alessandro Paccagnella is Full Professor of Electronics and Director of the Department of Information Engineering at the University of Padova. He is the author of more than 300 scientific papers, and about 200 of them have been published on international journals. In the past, his research has been directed to the study of different aspects of physics, technology, and reliability of semiconductor devices. At present, he coordinates the activity of a research group focused on the study of ultra-thin gate dielectrics in MOS devices and on Total Ionizing Dose and Single Event Effects induced by ionizing radiation on integrated circuits.

Pagano, Carlo
Carlo Pagano received the M. S. (cum laude) in electronics engineering from the University of Salerno, Salerno, Italy, in 2008, with a thesis on circuit edit by focused ion beam in cooperation with Berlin University of Technology, Germany. He is currently working on his Ph. D. degree at the Berlin University of Technology and a project with Hamamatsu Photonics Deutschland GmbH, Hersching, Germany. His research interests include laser beam modulation intensity mapping and laser stimulation with different wavelengths.

Pantelides, Sokrates
Sokrates T. Pantelides is University Distinguished Professor of Physics and Engineering, McMinn Professor of Physics and Professor Electrical Engineering at Vanderbilt University. He is also a Distinguished Guest Scientist at Oak Ridge National Laboratory. He joined Vanderbilt in 1994 after 20 years in IBM's Research Division as a researcher and manager. At Vanderbilt, Pantelides served as Chair of the Department of Physics. He earned a PhD in Physics at the University of Illinois at Urbana-Champaign in 1973. His research interests span semiconductor physics and devices, radiation effects, quantum transport, complex oxides, catalysis.

Pantouvaki, Marianna
M. Pantouvaki received the MSc degree in Physics from the University of Athens, Greece, and the PhD in Photonics from University College London (UCL), UK. After working for three years as a post-doctoral fellow at the Ultrafast Photonics group at UCL, she joined IMEC in 2006 as a BEOL integration engineer, where she is currently involved in low-k characterization and integration in narrow pitch interconnects."

Paolo, Pavan
Paolo Pavan graduated in Electrical Engineering at the University of Padova, Italy, and received his PhD from the same University in 1994. From 1992 to 1994 he was at the University of California at Berkeley. His research interests are in the characterization, modeling and development of nonvolatile memory cells. He has been active in the IEDM Technical and Executive Committees, and in VLSI-TSA since 2006. He authored and co-authored many technical papers, one book and two chapters in edited books. He is currently Professor of Electronics at the University of Modena and Reggio Emilia, Italy.

Papathanasiou, Athanasios
Athanasios E. Papathanasiou received the BSc. and MSc. degrees in Computer Science from the University of Crete, Greece in 1997

and 1999 respectively and the PhD. degree in Computer Science from University of Rochester in 2005. Dr. Papathanasiou is currently a Systems Architect at Intel Corporation. His research activities focus on the design, implementation and monitoring of computer systems with an emphasis on power management and reliability.

Park, Jongwoo
Jongwoo Park received the Ph. D. from Lehigh University in 1998. After post doctoral research at Lehigh University in 1999, he joined Lucent Technologies as a Member of Technical Staff. His research projects were packaging, reliability, polymer characterization and interfacial failure mechanism. In 2002, he joined Princeton Optronics as a Manager of Quality/Reliability. He was involved in hermetic/nonhermetic package development and reliability focused on tunable laser module including pump laser, MEMS and VCSEL. Since 2003, he is with Technology Quality Reliability in SYSTEM LSI, Samsung Electronics as a Vice President responsible for reliability qualification of advanced CMOS process and product.

Park, Min Sang
received the B.S. degree (2005) and the Ph.D. degree (2011) in electrical and computer engineering from the Pohang University of Science and Technology (POSTECH), Pohang, Korea. He is currently working in the Hynix Semiconductor Inc., Cheongju, Korea. His main research interests include characterization and modeling of CMOS devices.

Park, So Ra
So Ra Park received the M.S. degree in electrical engineering from University of Kookmin, in 2009. She majored in the semiconductor device and integrated circuit and researched the 3-D devices, FinFET and GAA, mainly. She has worked in TCAD team where develops the submicron transistor and suggests the design guide-lines. Her recent work is to research the stress in BEOL by TCAD.

Park, Sungju
Sungju Park received the B.S. degree in electronics engineering from Hanyang University, Korea in 1983 and the M.S. and Ph.D. degrees in electrical and computer engineering from the University of Massachusetts at Amherst in 1988 and 1992 respectively. He was with the Gold Star Company in Seoul, Korea from 1983 to 1986 in charge of developing microcomputer and network interface systems. From 1992 to 1994 he worked for IBM Microelectronics in Endicott, NY in charge of Test Design Automation. He has been a Professor in the Department of Electrical and Computer Engineering, Hanyang University, Ansan, Korea since 1995.

Park, Sung-Kye
He received the Ph.D. from KAIST, Korea, in 1994. He joined Hynix Semiconductor inc. in 1994, and currently he is a research fellow with R&D Center.

Park, Jongwoo
Technical Quality & Reliability, System LSI division, Samsung Electronics, Korea

Park, Junekyun
Technical Quality & Reliability, System LSI division, Samsung Electronics, Korea

Patterson, Ken
J. Ken Patterson (M'10) received his B.S. and Ph.D. degrees from Iowa State University, and University of Illinois Urbana-

Champaign in 1984 and 1994 respectively. He joined Tektronix Inc. in 1984, Hewlett-Packard Co. in 1996, Agilent Technologies in 2000, and Avago Technologies in 2005. He is currently a Member of Technical Staff at Avago Technologies.

Penna, Stefano
Stefano Penna was born in Rome, Italy, on December 18, 1979. He received the B.En., the M.En. and the Ph.D. degrees in Telecommunications Engineering from the University of Rome Tor Vergata, Italy, in 2002, 2005 and 2009, respectively. Since 2009 he is working as a post-doc fellow at the Center for Hybrid and Organic Solar Energy (CHOSE) of the University of Rome Tor Vergata. His work involves large area processing, outdoor and indoor testing of organic photovoltaic technologies, including dye solar cells and panels and bulk-heterojunction polymer solar cells. Dr. Penna is a member of the IEEE Photonics Society.

Petitprez, Emmanuel
Emmanuel Petitprez received the Engineering degree in solid state physics from the Institut National des Sciences Appliquées, Toulouse, France, in 1994, and the Ph.D. degree in materials science from the University of São Paulo, Brazil, in 2001. He was with Serma Technologies, Grenoble, France, as a Characterization Engineer. In 2005 he moved to Freescale Semiconductor, Crolles, France working on interconnect reliability issues in SOI technologies. Since 2007, he has been with the STMicroelectronics Central R&D Laboratory, Crolles, France where he works on the reliability of interconnects for advanced CMOS technology.

Pey, Kin-Leong
Kin Leong Pey received his Bachelor of Engineering (1989) and Ph.D. (1994) in Electrical Engineering from the National University of Singapore (NUS). He has held various research positions in the Institute of Microelectronics, GlobalFoundries Singapore, Agilent Technologies and National University of Singapore. He is currently a Visiting Professor at Nanyang Technological University (NTU) and an Associate Provost of Singapore University of Technology and Design (SUTD), Singapore and also holds a concurrent Fellowship appointment in the Singapore-MIT Alliance (SMA). He has published more than 150 international refereed publications and 160 technical papers at international meetings/conferences and holds 33 US patents. Dr. Pey is a senior member of IEEE and an IEEE EDS Distinguished Lecturer.

Phang, Jacob CH
Jacob Phang received both his BA and PhD degrees from the University of Cambridge in 1975 and 1979 respectively. He joined the National University of Singapore (NUS) in 1979 where he is now Professor at the Centre for Integrated Circuit Failure Analysis and Reliability (CICFAR). His field of research is in integrated circuit failure analysis and solar cell characterization. He is also Chairman of SEMICAPS Corporation Pte Ltd, an NUS spin-off company he co-founded in 1988 to commercialize the technologies developed at CICFAR for world-wide distribution.

Pimparkar, Ninad
Ninad Pimparkar was born in India in 1980. He received B.Tech and M.Tech degrees in electrical engineering from India Institute of Technology (IIT), Bombay, in 2003 from the five year dual degree program. He received M.S and Ph.D. from Purdue University, USA in 2007. His doctoral advison was Prof. M. A. Alam. He was with AMD in 2008 and with GLOBALFOUNDRIES since 2009 working on compact modeling and reliability modeling.

Poh, John
John Chung Hang Poh (S'08) received the B.Eng in Electrical and Electronic Engineering from Nanyang Technological University (NTU), Singapore, in 2005, and the M.S.E.C.E in Electrical and Computer Engineering from Georgia Institute of Technology (Georgia Tech), Atlanta, in 2009. He is currently working toward the Ph.D. degree at Georgia Tech. His research interests include SiGe HBT based RF front-ends, and modeling of RF packages.

Presser, Nathan
Dr. Presser received his B. A in mathematics and chemistry from New York University and his doctorate in physical chemistry from NYU in 1975. As a postdoctoral fellow at MIT and research assistant professor at UIC, he performed research on simple chemical reactions of atmospheric interest using crossed molecular beams, vacuum ultraviolet flash photolysis and laser multiphoton ionization and dissociation. Dr Presser is presently a senior scientist in the Microelectronics Technology Department of the Electronics and Photonics Laboratory of The Aerospace Corporation, where the focus of his work is failure analysis and materials characterization using FIB and SEM based techniques .

Puzyrev, Yevgeniy
Yevgeniy Puzyrev received his B.S. from South Ural State Univ. (2000), and Ph.D. (2005) in physics from Florida Atlantic Univ. He joined X-ray Diffraction and Microscopy group at Oak Ridge National Laboratory in 2006 and has been involved in physical characterization of chemical and magnetic order of metallic alloys. Since 2008 he is a Research Associate at Physics and Astronomy Department of Vanderbilt University. His research interests include physics of defects in semiconductors and reliability physics and characterization of novel devices.

Raghavan, Nagarajan
Nagarajan Raghavan was born in Bangalore, India in 1985. He received his B.Eng, 1st Class Honors, (Electronics Engineering, 2007), S.M. (Advanced Materials for Micro & Nano Systems, 2008) and M.Eng (Materials Science and Engineering, 2008) from Nanyang Technological University (NTU), National University of Singapore (NUS) and Massachusetts Institute of Technology (MIT) respectively.

Ragnarsson, Lars-Åke
Lars-Åke Ragnarsson received a M.S. degree in 1993 and a Ph.D. degree in 1999 in Electrical Engineering from Chalmers University of Technology, Göteborg, Sweden. He did post-doctoral studies at the IBM T.J. Watson Research Center in Yorktown Heights, NY, USA between 2000 and 2002, focusing mainly on electrical characterization of high-k dielectrics. He is since 2002 employed by imec in Leuven, Belgium as a senior scientist on high-k dielectrics and metal gates.

Ramaswami, Sesh
Sesh Ramaswami is Sr. Director, in the Silicon Systems Group at Applied Materials, responsible for program definition and execution, external collaboration and co-leading process integration for wafer level processing for TSV. Over the past 16 years, he has had technical and business responsibilities in product management and development. Prior to joining Applied Materials in 1994, Sesh had thin film process responsibilities for seven years at Advanced Micro Devices and four years at National Semiconductor. He holds 35 US patents. Sesh has undergraduate and graduate degrees in Chemical Engineering from Indian Institute of Technology, Kanpur, and Syracuse University respectively and a MBA

Ran, Qiushi
Qiushi Ran received the B.S. and M.S. degrees in Electrical Engineering from Tsinghua University, Beijing, China, in 2007 and 2009 respectively. She is currently working towards her Ph.D. degree in Electrical Engineering, Stanford University, CA, USA. Her research intestes include device Physics, nanodevice modeling.

Ranjan, Rakesh
Dr. Ranjan Rakesh received his Master of Science (1999) and Master of Technology (2001) from IIT Delhi and Ph.D (2007) in Microelectronics from Nanyang Technological University, Singapore. He has worked as Research Associates in the NTU Singapore and as Research Engineer in IME Singapore before joining TSMC Taiwan. Presently, Dr. Ranjan is working as a Technical Manager in Technology Quality and Reliability Division of TSMC, Taiwan. He has published more than 30 papers in reputed journals (APL, EDL, EL, TDMR etc.) and conferences (IEDM, IRPS, IPFA, INFOS etc.). He is also the IEEE member and reviewer of TDMR.

Rao, Rahul
Rahul Rao received B. E. degree in Electronics Engineering from the University of Bombay, India, in 2000. He received his M. S and Ph.D. degrees in Electrical and Computer Engineering from the University of Michigan in 2002 and 2004, respectively. He is a Research Staff Member at IBM T. J. Watson Research Center, NY. He has been actively involved in the design of processors for IBM's high performance servers and characterizing process variations and device degradation in high performance digital circuits and SRAMs. Dr. Rao holds seven US patents and has (co)-authored over 25 papers in referred journals and conferences.

Rashid, Al
Al Mamunur Rashid received his B.Sc. in Computer Science from Bangladesh University of Science & Technology, and his M.Sc. and Ph.D. from University of Minnesota, Twin Cities. Dr. Rashid's interests include automated learning of user preferences, recommender systems, and machine learning and data mining in general. He has been working at Intel Corporation since 2007 as a data miner.

Read, David T.
David T. Read holds a Ph. D. in physics from the University of Illinois at Urbana; he has been with the National Institute of Standards and Technology (formerly the National Bureau of Standards) from 1974 until the present, researching measurement techniques for evaluation of materials reliability in a variety of contexts, from macro to micro. Now a contractor at NIST, Read's interests include mechanical and electrical measurements of material behavior and modeling of material response to mechanical, thermal, and electrical stresses using molecular dynamics and peridynamics. His publications include over 75 archival papers and many conference proceedings papers.

Reale, Andrea
Andrea Reale received the Laurea degree in electronic engineering and Ph.D. degree from the University of Rome "Tor Vergata," Rome. Since 2004, he has been an Assistant Professor in the University of Rome "Tor Vergata," Rome, Italy. His current research interests include photovoltaics systems based on organic and hybrid organic–inorganic semiconductors, experimental study of nanostructured materials, such as carbon nanotube-based materials, and theoretical and experimental analysis of optical, electrooptical, electrical, and thermal properties of heterostructure devices for electronics and telecommunications.

Reddy, Vijay
After receiving the Ph.D. (1994) in Electrical Engineering from the University of Texas at Austin, Vijay Reddy joined Texas Instruments and has worked on several topics concerning transistor, circuit reliability and product qualification methodologies. He is currently a Distinguished Member Technical Staff and is the CMOS Design Reliability Manager focusing on digital, analog, and RF circuit reliability for SOC products. He has served on the IRPS/IEDM program committees and has presented papers at IRPS/IEDM and invited tutorials at IRPS/ICTMS/VLSI Test. Symposium. He received the 2002 IRPS Outstanding Paper Award, 2004 IRPS Outstanding Paper Award, and the 2002 ESD/EOS Symposium Best Paper/Best Presentation Awards. He has received twelve patents with several pending along with more than thirty publications.

Reddy, Vijay
After receiving the Ph.D. (1994) in Electrical Engineering from the University of Texas at Austin, Vijay Reddy joined Texas Instruments and has worked on several topics concerning transistor, circuit reliability and product qualification methodologies. He is currently a Distinguished Member Technical Staff and is the CMOS Design Reliability Manager focusing on digital, analog, and RF circuit reliability for SOC products. He has served on the IRPS/IEDM program committees and has presented papers at IRPS/IEDM and invited tutorials at IRPS/ICTMS/VLSI Test. Symposium. He received the 2002 IRPS Outstanding Paper Award, 2004 IRPS Outstanding Paper Award, and the 2002 ESD/EOS Symposium Best Paper/Best Presentation Awards. He has received twelve patents with several pending along with more than thirty publications.

Reed, Robert
Robert A. Reed received his M.S. and Ph.D. degrees in Physics from Clemson University in 1993 and 1994. He worked as a post-doctoral fellow at the Naval Research Laboratory and later worked for Hughes Space and Communication. From 1997 to 2004, he was a research physicist at NASA Goddard Space Flight Center where he supported space flight and research programs. He is currently an Associate Professor at Vanderbilt University and a member of the IEEE/NPSS AdCom. His research activities include topics such as single event effect and displacement damage basic mechanisms and on-orbit performance analysis and prediction techniques.

Reisinger, Hans
Hans Reisinger received his diploma in physics (1979) and his Ph.D. (1982) both from the Technical University of Munich. In 1982/83 he was with the IBM T.J.Watson Research Ctr. in Yorktown Heights, NY. Topics of his research were electronic properties of 2d-space charge layers on Si and III-V semiconductors. In 1986 he joined the Siemens Semiconductor Department (now Infineon). His work was focused on the study of thin dielectrics and interfaces in DRAMs and NVMs, including film fabrication and optical and electrical characterization. Currently he is with the Infineon Central Reliability Department and mainly works on the problems of threshold instabilities of MOSFETs with SiO2 and HiK dielectrics.

Rhoton, Brent
Quality Assurance Manager at Texas Instruments Medical and High Reliability Business Group. Involved in several industry and TI committees regarding quality and reliability of ceramic and plastic integrated circuits in medical, military, automotive, High Temp and other high reliability applications. Twenty seven years experience in quality assurance and process development for products used in Defense and Aerospace, Automotive and Extended Temperature Applications.

Robinson, William
William H. Robinson received his B.S. in electrical engineering from the Florida Agricultural and Mechanical University in 1996, and his M.S. and Ph.D. in electrical engineering from the Georgia Institute of Technology in 1998 and 2003, respectively. Currently, Dr. Robinson is an Associate Professor in the Department of Electrical Engineering and Computer Science at Vanderbilt University. His research explores hardware and software tradeoffs to improve system performance, system reliability, and system security. Dr. Robinson's major honors include a National Science Foundation CAREER Award in 2008. Dr. Robinson is a Senior Member of both the IEEE and the ACM.

Roche, Philippe
Philippe Roche received the M.S. (1995) and Ph.D. (1999) in semiconductor physics from the University of Montpellier, France. From 1995 to 1999, he worked consecutively at the University of Eindhoven, the Netherlands, at the French Atomic Energy Commission, Military applications centre at Bruyères-le-Chatel, at the University of Montpellier and in the Radiation Effects Group at Vanderbilt University, USA. Since 1999, he has been with STMicroelectronics, Central CAD and Design Solutions, Crolles France, as senior expert and manager of a group in charge of both SER safety/reliability aspects and subthreshold (~0.3V) IP designs. His primary research activities are Single Event Effects and Total Ionizing Dose, as well as Ultra Low Voltage IPs, on sub-0.25µm commercial technologies down to CMOS 20nm. He has been serving in conferences since 1997, as session chairman and short course instructor, in 10 international conferences, such as IRPS, NSREC, IOLTS, RADECS and SOI conference. Philippe has coauthored +100 papers and has filed +20 patents and 3 trade marks in radiation hardening.

Rosa, Giuseppe
Giuseppe La Rosa is a Senior Technical Staff Member at the IBM Semiconductor R&D Center, IBM Microelectronics, Hopewell Junction, NY. Since he joined IBM, he has been leading projects in the Technology Reliability of advanced DRAM/Logic submicron technologies. His main interests are in the area of advanced submicron transistor reliability mechanism including BTI and Hot Carrier Effects. He is a worldwide known expert in BTI instabilities and recently Chair of across industry working groups in BTI benchmarking (RTAB Sematech, Jedec 14.2). He has reached the 3rd IBM Invention Achievement Plateau and coauthored 16 articles and given several invited talks and tutorials in the area of FEOL Reliability Physics. He has organized and chaired several Transistor Reliability committees at IRPS from 1997 to present as well as being part of the transistors committee at several conferences (IRPS, IEDM, IRW etc). Recently he has been member of IRPS management committee. He also has been reviewer for several journals (TDMR, EDL, etc) and Editor of two TDMR special issues on NBTI. He holds a Doctor Degree in Physics from the University of Catania (Italy) and an M.S. in Electronic Materials from MIT and an M.S. in Physics from Northeastern University. He is a Senior Member of IEEE and member of the Electron Device Society.

Rosana, Rodriguez
Rosana Rodríguez received the Ph. D. in Electrical Engineering from the Universitat Autònoma de Barcelona in 2000. Funded by the Fulbright program, she worked on devices and circuits reliability at the IBM Thomas J. Watson Research Center (USA). Currently, she is associate professor at the Universitat Autònoma

de Barcelona. Her main research interests are focused on the effect of CMOS failures such as dielectric breakdown, BTI and Channel Hot Carriers on the performance of devices and digital and analogical circuits. She also analyses the process-related variability impact on devices and circuits functionality and the resistive switching effect for non-volatile memories.

Rosenbaum, Elyse

Elyse Rosenbaum completed a Ph.D. degree in electrical engineering at the University of California, Berkeley in 1992. She is currently a Professor in the Department of Electrical and Computer Engineering at the University of Illinois at Urbana-Champaign. Dr. Rosenbaum's present research interests include design, testing, modeling and simulation of ESD protection circuits, latch-up, gate oxide degradation, and substrate noise coupling. She has authored or co-authored over 100 technical papers and is an editor for IEEE Transactions on Device and Materials Reliability. She has presented tutorials at the International Reliability Physics Symposium, the EOS/ESD Symposium, and the RFIC Symposium. She was the keynote lecturer at the 2004 Taiwan ESD Conference, and has given invited lectures at many universities and industrial laboratories. She is a Fellow of the IEEE.

Rothleitner, Hubert

Hubert Rothleitner graduated in electronic engineering at the Technical University of Graz, Austria in 1982. Contributing to the up growing smart power technology development, he worked on High Side and Low Side switches as well as designs for safety applications (e.g. airbag or breaking systems) for the automotive industry. As a Senior Principal for smart power design at Infineon his fields of interest are novel system architectures with new technologies and device concepts combined with reliability related aspects for integrated smart power devices, especially MOS-surface- and bipolar- parasitics, latch-up immunity and ESD-protection.

Roussel, Ph.J

Philippe J. Roussel was born in Bruges, Belgium, on May 18, 1955. He received the diploma in Electrical Engineering from the Industriële Hogeschool of Ghent, Belgium, in 1983. In 1984, he joined the ESAT laboratory of the KULeuven, Belgium as an assistant in a reliability research project on plastic encapsulation. From 1987, he assisted in a government funded project on electromigration at imec, Leuven, which also resulted in a maximum likelihood fitting program for the assessment of statistical distributions and acceleration models from reliability test data. From 1991 till 2000, he worked on analysis techniques like microprobe, TEM and Spectroscopic Ellipsometry (SE) in the PT/MCA group, while continuing his research on Reliability Statistics, SPC and Data Analysis in the PT/DRE group. At present he also works in a multidisciplinary team on the statistics required for Technology Aware Design (TAD). He co-authored papers in fields as diverse as scientific handheld programming, virology, SE, ESD, oxide and interconnect reliability, and TAD.

Roussel, Philippe

Philippe J. Roussel was born in Bruges, Belgium, on May 18, 1955. He received the diploma in Electrical Engineering from the Industriële Hogeschool of Ghent, Belgium, in 1983. In 1984, he joined the ESAT laboratory of the KULeuven, Belgium as an assistant in a reliability research project on plastic encapsulation. From 1987, he assisted in a government funded project on electromigration at imec, Leuven, which also resulted in a maximum likelihood fitting program for the assessment of statistical distributions and acceleration models from reliability

test data. From 1991 till 2000, he worked on analysis techniques like microprobe, TEM and Spectroscopic Ellipsometry (SE) in the PT/MCA group, while continuing his research on Reliability Statistics, SPC and Data Analysis in the PT/DRE group. At present he also works in a multidisciplinary team on the statistics required for Technology Aware Design (TAD). He co-authored papers in fields as diverse as scientific handheld programming, virology, SE, ESD, oxide and interconnect reliability, and TAD.

Roy, David

David Roy received the BS (1997) in physics, the M.S.(1998) in Physics (from the Institut National Polytechnique de Grenoble (INPG)) and the Magistere(1998) of Physics Research from University Joseph Fourier de Grenoble. He worked for the CEA-Grenoble on the 3D-optical micro-system in the "Laboratoire d Electronique et des Technologies de l information" (LETI) in 1999. In 1999, he joined STMicroelectronics as a reliability engineer, working on oxide and device reliability. Since 2007, he is in charge of the Front-end Reliability team. His current research interests include transistor reliability as well as low k interconnect reliability

Roy, Tania

Tania Roy received her B.E. (Hons.) degree in Electrical and Electronics Engineering from BITS, Pilani in India in 2006. She completed her Masters in Electrical Engineering at Vanderbilt University in 2008. She is working towards her PhD in Electrical Engineering on the reliability of GaN HEMTs. Her research interests include semiconductor device physics, modeling, characterization and reliability of devices, particularly heterostructure devices.

Ryan, Jason

Jason T. Ryan received the B.S. degree in Physics from Millersville University, Millersville, PA in 2004. He received the M.S. degree in Engineering Science and the Ph.D in Materials Science and Engineering from The Pennsylvania State University, University Park, PA in 2006 and 2010 respectively. He is currently employed as an NRC post-doctoral fellow in the Semiconductor Electronics Division at the National Institute of Standards and Technology, Gaithersburg, MD. Dr. Ryan has serves on the management committees of both the IRPS and IIRW conferences.

Sadana, Devendra

Devendra Sadana obtained his PhD from the Indian Institute of Technology, New Delhi in 1975. He has worked at the University of Oxford, England, University of California, Berkeley, Microelectronics Center of North Carolina, and Philips Research Labs, Sunnyvale, CA during 1975-87 in various capacities. He joined IBM Research in 1987 where he is currently a Senior Staff/Manager of Advanced Substrate Research group. His publications include nearly 200 papers in various scientific journals and conference proceedings, over 75 issued patents, 5 book chapters, multiple industrial courses on ion implantation, numerous invited talks at international conferences, and co-editing of several conference proceedings.

Sagong, Hyun Chul

received the B.S. degree (2008) in electrical engineering from the Pusan National University (PNU), Pusan, Korea. He is currently in integrated M.S./Ph.D. program cource at the Pohang University of Science and Technology (POSTECH), Pohang, Korea. His main research interests include characterization and modeling of CMOS devices.

Sahhaf, Sahar
Sahar Sahhaf was born on 25 July 1981 in Mashad, Iran. In 2006, she received the M.Sc. degree in ICT-Microelectronic engineering from the Katholieke Universiteit Leuven (K.U.L) in Belgium. Later on, in September 2006, she started her doctoral studies in Advanced Devices Reliability at imec and K.U.Leuven (Belgium) on the topic of "Electrical defects in high-k/Metal gate MOS transistors" and she graduated in September 2010.

Saito, Tomoya
Tomoya Saito received the B.S. (1997) in Electronics Engineering, M.S. (1999) and Ph.D. (2005) in Electronics and Information Systems from Osaka University, Osaka, Japan. In 1999, he joined Halo LSI Inc, where he was involved in Flash memory technology development. In 2005, he joined NEC Electronics Corp., and currently engaged in Embedded Flash memory Process development in Renesas Electronics.

Saito, Wataru
Wataru Saito received Ph. D. degrees in electrical and electronics engineering from Tokyo Institute of Technology, Tokyo, Japan, in 1999. He joined Discrete Semiconductor Division, Toshiba Corporation Semiconductor Company, Kawasaki, Japan, in 1999, where he has been engaged in the development of high-voltage power MOSFETs. His current interest is basic researches on the next generation power semiconductor devices including wide band-gap semiconductor materials applications.

Sakamoto, Mitsuhiro
Mitsuhiro Sakamoto received the B.S.(1997) and M.S.(1999) degree in engeneering from Kyushu Institute of Technology, Fukuoka, Japan. He joined NEC Corp., Japan, in 1999. He has been with Renesas Electronics Corp., Japan, since 2010. He has been working on reliability of eDRAM/SRAM technology.

Savage, Scott
Scott Savage currently works for Medtronic as a Senior Principal Reliability Engineer for active implantable medical device circuit assemblies. Scott received a BS degree in Chemistry from Illinois State University and an MS degree in Materials Science from Northwestern University and is an ASQ Certified Reliability Engineer. He started his career at Motorola Semiconductor with 8 years in failure analysis and 3 years in component reliability engineering. Major projects included FCBGA and QFN package development and the overall Pb-free transition. Scott joined Medtronic in 2004 supporting stacked die package development, ceramic and tantalum capacitor reliability, and circuit assembly development.

Scheer, Patrick
Patrick Scheer was born in Grenoble, France, in 1970. He received the engineering degree in electronics from the Ecole Nationale Supérieure d'Electronique et de Radioélectricité de Grenoble and the M.S. degree in optics, optoelectronics and microwaves from the Institut National Polytechnique de Grenoble in 1993. He received the Ph.D. degree in optics and optoelectronics from the Ecole Nationale Supérieure de l'Aéronautique et de l'Espace, Toulouse, France, in 1998. He joined the central R&D site of STMicroelectronics in Crolles, France, in 1998 to develop high frequency models for MOS transistors in advanced CMOS and BiCMOS technologies. His interests are in the small-signal, noise and large-signal behavior of active devices, device physics, compact modeling and parameter extraction methodologies. He is currently leading a modeling group focused on analog, high frequency and high voltage devices and applications.

Schepens, Cor
Cor Schepens received his M.Sc in Electronics from the Technical University of Eindhoven in 1987 and has more than 20 years of experience in the semiconductor industry in various roles. He started as an ASIC designer, moved into project management, worked as an advisor for Dutch SMEs and had marketing and sales responsibilities. He was founder and CEO of Adveda, a startup in HW/SW co-simulations tools. He joined CK in December 2005 and has a role as project manager as well as marketing manager and has filed patents related to the Nanomech technology.

Schlünder, Christian
Christian Schlünder has received his Dipl.-Ing. (1999) in electrical engineering and his doctoral degree in engineering science (2006) accompanying his regular work (both from the Technical University of Dortmund, Germany). From 1998-1999 he worked in a cooperative program between Siemens Corporate Research Labs in Munich and the Technical University of Dortmund. 1999 he joined Infineon as a member of the Corporate Research Department, where he was active in research on hot carrier stress in analogue and mixed signal applications. In the year 2000 he changed to the Corporate Reliability Methodology Group. There he evaluates device reliability of new in-house and silicon foundry technologies. Today, as a Senior Staff Engineer, he manages technology qualification and quality assurance for various state-of-the-art CMOS-Technologies. Furthermore he leads the device reliability research activities for Infineon. His current work is focussed on BTI recovery phenomena. Christian Schlünder has published some 40 papers in various conference proceedings and microelectronic journals. Additionally, he has presented invited talks and tutorials at many conferences such as 'IRPS', 'ESSDERC' or "ZuE". He is frequently a member of the Technical Program Committee of the IEEE-conferences 'IRPS', 'IRW' and referee of several microelectronic journals.

Schmitz, Anthony
Anthony Schmitz received the BSEE (1995) and MSEE (1997) from the University of Illinois at Urbana-Champaign. His thesis research was in generating high quality Schottky contacts to GaN for device applications at elevated temperatures. After joining Intel in 1997, he worked on various reliability mechanisms in the manufacturing environment in the technology nodes spanning 0.35um to 90nm. In 2004, he joined the Logic Technology Development Q&R department in Hillsboro, OR, USA with a deepened focus on interconnect reliability mechanisms for each of Intel's latest microprocessor technologies.

Schmitz, Jurriaan
Jurriaan Schmitz (M '02/SM '06) Received his M.Sc. (1990, cum laude) and Ph.D. (1994) degrees in Experimental Physics at the University of Amsterdam. He then joined Philips Research, studying CMOS transistor scaling, characterization and reliability. Since 2002, he is a full professor at the University of Twente, leading the Group of Semiconductor Components. He serves as a TPC member of the IEDM, ESSDERC and ICMTS conferences, being General Chairman of the latter in 2011. He (co-)authored over 180 publications and holds 16 US patents. His research interests include CMOS post-processing, novel silicon device concepts, and electrical characterization of devices.

Scholten, Andries J.
Andries J. Scholten received the M.Sc. and Ph.D. degrees in experimental physics from Utrecht University, The Netherlands, in 1991 and 1995, respectively. In 1996, he joined Philips Research Laboratories (now NXP Research), Eindhoven, The Netherlands,

where he has worked on compact MOS modeling for circuit simulation, with a focus on the modeling of thermal noise and non-quasi-static effects. He has contributed to the development and industrialization of well-known compact MOSFET models such as MOS Model 9, MOS Model 11, and the world-standard PSP model. His current research is directed towards RF CMOS reliability and reliability simulation.

Schrimpf, Ronald

Ron Schrimpf is the Orrin Henry Ingram Professor of Electrical Engineering at Vanderbilt University and the Director of Vanderbilt's Institute for Space and Defense Electronics (ISDE). He received his B.E.E., M.S.E.E., and Ph.D. degrees from the University of Minnesota in 1981, 1984, and 1986, respectively, and served as a Professor of Electrical and Computer Engineering at the University of Arizona prior to joining Vanderbilt in 1996. Ron's research activities focus on semiconductor device physics, particularly radiation effects and reliability in microelectronics and semiconductor devices.

Sekine, Katsuyuki

Katsuyuki Sekine was born in Saitama, Japan, in 1970. He received the B.S., M.S., and Ph.D. degrees in electronic engineering from Tohoku University, Sendai, Japan in 1995, 1997, and 2000, respectively. In 2000, He joined Process and Manufacturing Engineering Center, Toshiba Corporation, Semiconductor Company, Yokohama, Japan, where he worked on development of advanced gate dielectric film formation process for CMOS device.
He is currently engaged in development of dielectric film for nonvolatile memory device in Advanced Memory Development Center, Semiconductor Company, Toshiba Corporation, Yokkaich, Japan

Seo, Soonok

She received the M.S. from GIST, Korea, in 2007. She joined Hynix Semiconductor inc. in 2007 and currently she is participating in the Flash memory reliability.

Seth, Sachin

Sachin Seth (S, 08) received the B.Eng in Electronics and Communications Engineering from Delhi University, Delhi, in 2007, and the M.S.E.C.E in Electrical and Computer Engineering from Georgia Institute of Technology (Georgia Tech), Atlanta, in 2009. He is currently working toward the Ph.D. degree at Georgia Tech. His research interests include SiGe HBT based RF frontends, compact modeling and device physics of HBTs, and low-power circuit design.

Shahrjerdi, Davood

Davood Shahrjerdi finished his PhD in electrical engineering at The University of Texas at Austin in 2008, where he worked on novel high-mobility channel materials, such as III-Vs and Ge for future CMOS technologies. He is currently a Research Staff Member at IBM T J Watson Research Center, working on advanced high-efficiency solar cell structures and materials. He has authored or coauthored more than 50 publications in scientific journals and conferences.

Shakouri, Ali

Ali Shakouri is a Professor of Electrical Engineering at University of California Santa Cruz. He received his Ph.D. from the California Institute of Technology in 1995. His current research is on nanoscale heat and current transport in semiconductor devices, high resolution thermal imaging, micro refrigerators on a chip, and waste heat recovery. He has initiated an international summer school on practical renewable energies. He is the director of the multi-university Thermionic Energy Conversion center. He received the Packard Fellowship in Science and Engineering (1999), the NSF Career award (2000), and the UCSC School of Engineering FIRST Professor Award in 2004.

Shaviv, Roey

Roey Shaviv received his Ph.D. in chemistry in 1988 from the University of Michigan, in Ann Arbor. He is a senior integration and reliability technologist at Novellus Systems Inc. He joined Novellus in 2001 and is specializing in reliability and integration issues associated with copper / low k interconnects. Prior it this, he was the Etch Engineering Section Head, and earlier the Thin Films and CMP Engineering Section Head, at Tower Semiconductors. He was a Lecturer at the Chemistry Department of the University of Michigan and a Research Associate in the Chemistry Department of University of Illinois, Chicago

Shih, Jiaw-Ren

Jiaw-Ren Shih received his M.S. and Ph.D. from National Tsing-Hua University, Taiwan, in 1992 and 2000, respectively. All are with Electrical Engineering. Dr. Shih is the Academician in TSMC Academy now, and also serves as the manager of Advanced Reliability Development Program and Mainstream Technology Quality and Reliability Department at Technology Quality and Reliability Division. Prior to joining reliability division in 2000, he served as the transistor designer at device department of R&D since 1992. Dr. Shih has served on the IRPS Technical Committee in 2002, 2004 ~ 2009 for Transistor, Process Integration and Dielectric Sections, respectively. He holds 53 U.S. patents and 40 Taiwan patents, and also published more than 41 technical papers.

Shih-Yu, Wang

Shih-Yu Wang was born in Taipei, Taiwan in 1974. He received the B.S and M.S. degrees in physics from Tamkang University in 1996 and National Taiwan University in 1998, respectively. In 2000, he joined Macronix International Co., Ltd. (MXIC) as a reliability engineer. And in 2004, he transfered to Device Engineering Department as an ESD engineer, where he has engaged in the ESD protection desgin of NOR flash memory process technologies.

Shimizu, Tatsuo

Tatsuo Shimizu received the B.S.(2002) and M.S.(2006) in apllied physics from Waseda Univ., Tokyo, Japan. In 2006, he joined NEC Electronics Corp., Kawasaki, Japan. Since 2010, he has been with Renesas Electronics Corp., Sagamihara, Japan, where he has been working on MOSFET and gate oxide reliability for advanced CMOS technology.

Shin, Hyunjung

Hyunjung Shin received his B.S. degree in ceramic engineering from Yonsei University, Seoul, Korea, in 1991, M.S. and Ph.D. degrees in Materials Science and Engineering from Case Western Reserve University, Cleveland, OH in 1994 and 1996, respectively. He spent one year at Max-Planck Institute fur Metallforschung in Stuttgart, Germany, and was a member of the research staff in Samsung Advanced Institute of Technology from 1997 to 2002. He is currently an associate professor of School of Advanced Materials Engineering at Kookmin University in Seoul, Korea. His research interests include functional nanoscale materials, ferroelectric nanostructures, transition metal oxide, scanning probe microscope, and self-assembly.

Shinich, Nishizawa
Shinichi Nishizawa received the B.E. and M.E. degrees in Electrical and Electronic Engineering from Kyoto University, Kyoto, Japan, in 2009, 2011, respectively. He is presently a doctor's course student at Kyoto University.

Shinosky, Mike
Mike Shinosky received an Associate of Engineering degree in Electronic Engineering from A.T.E.S. Technical Institute of Niles, Ohio in 1982. Mike joined IBM in 1982, where he was responsible for building and running Soft Error Rate testers for DRAM Memory products. In 1985 he began T2 Functionality Qualification testing of DRAM memory cards for large scale server applications. In 1999 Mike transferred to an ASICS department doing book level Physical Design Mask Layout work. In 2003 Mike joined Technology Reliability Engineering, where he is currently and primarily responsible for TDDB testing of BEOL Dielectric technologies.

Shukla, Vrashank
Vrashank Shukla received his B.E. degree from the National Institute of Technology, Karnataka in 2001. He is currently working towards a PhD degree in Electrical Engineering at University of Illinois Urbana Champaign. He worked at Texas Instruments, Bangalore, India from 2001 to 2007 in the ASIC CAD group. He worked on the development of EDA tools for reliability analysis of circuits. His current research focus is on full-chip and package modeling for electro-static discharge (ESD) simulations, in particular, simulation of charged device model of ESD. He works in the ESD research group of Co-ordinated Science Laboratory at University of Illinois Urbana Champaign.

Shur, Yael
Received a B.Sc. degree in physics and astronomy in 2005 and M.Sc. degrees in electrical engineering (physical electronics) in 2008 from the Tel Aviv University, Israel. The MSc. thesis focused on unified retention model for localized charge trapping NROM nonvolatile memory device She joined Saifun Semiconductors Ltd. in 2004 and currently holds the position of Member of technical staff in Technology Development. Published 2 scientific papers and 4 conference papers. Yael's main experience is in MirrorBit cell technology, she has participated and lead NROM development programs, product algorithms, reliability topics, lab experimentations and more.

Shusterman, Yuriy
Yuriy Shusterman has been a Process Engineer with Intel (Portland, OR) since 2001. He spent initial five years on Defect Metrology development and last five years on Interconnect technology, development. Prior to joining Intel, he received PhD in Applied physics from RPI (Troy, NY) for work studying growth and electron-transport properties of epitaxial metal films on insulators. Before RPI, he did research on epitaxial fluoride-on-silicon interface at Ioffe Physico-Technical Institute (S. Petersburg, Russia). During his academic career, he published over fifteen articles.

Sibley, Michael
Michael Sibley received his bachelor's degree and master's degree from the University of New Mexico in 1998 and 2001 respectively. He is one of the founders of Micro-RDC where he has worked since December 2005. His major interests at Micro-RDC have been design of circuits for SEU/SET characterization and performing heavy ion and laser experiments of the circuits. He also has interests in investigating the TID hardness of different RHBD techniques through TID testing.

Sierawski, Brian
Brian D. Sierawski received his M.S.E degree in computer science and Engineering from the University of Michigan in 2004 where studied formal verification methods for EDA. He is pursuing a Ph.D. in electrical engineering at Vanderbilt University. In 2002 he held an internship at Intel. He currently works at the Institute for Space and Defense Electronics where his efforts include modeling and evaluation of single event effects in semiconductor devices for the purposes of error rate predictions. His research interests are in the simulation of soft errors at the device, gate, and register-transfer levels.

Siew, Yong Kong
Yong Kong SIEW received his B.A.Sc in materials engineering in 1998 and his PhD, concerning integration and characterization of low dielectric constant materials for interconnect application in 2003, both from Nanyang Technological University (NTU), Singapore. From 2002 to 2008 he worked in Chartered Semiconductor Manufacturing (now part of GlobalFoundries), where he was involved in technology development and process integration of 90nm to 40nm technologies. From 2008, he joined imec, working on advanced interconnect integration.

Sigurd, Wagner
Sigurd Wagner received a Ph.D. from the University of Vienna in 1968, came to the U.S. as a postdoctoral fellow at Ohio State University, and from 1970 to 1978 worked at Bell Telephone Laboratories on semiconductor memories and heterojunction solar cells. He is co-inventor of two solar cells that are in commercial production. From 1978 to 1980 he was founding Chief of Photovoltaic Research at the Solar Energy Research Institute (now NREL) in Golden, CO. Since 1980 as Professor of Electrical Engineering at Princeton University he is developing materials, processes and components for flexible large-area electronics, electrotextiles, and electronic skin.

Simoen, Eddy
Is a senior researcher at imec working on defects in semiconductor materials and devices, with focus on low-frequency noise and Deep-Level Transient Spectroscopy (DLTS) characterisation. Since the early 90ties he has been working on radiation effects in semiconductor devices and technologies. In these fields, he has published over 1000 papers in technical Journals and Conference Proceedings and 5 Books, two of which have been translated in Chinese.

Sin, Yongkun
Yongkun Sin received a Ph.D. in electrical engineering from North Carolina State Univ in Raleigh, NC. He was involved in research and development of optical components including semiconductor lasers, detectors, and modulators for telecom and chemical sensing applications at Oki, Hyundai, JPL, and TRW. He is currently at The Aerospace Corporation, where he is studying reliability and degradation mechanisms in GaAs-based high power lasers, HBTs, and HEMTs. He has published over forty journal and conference papers.

Slayman, Charles
Charles Slayman received his B.A. in Physics and Ph.D. in Electrical Engineering and Computer Science in from the University of California at Berkeley. He is currently a Senior Reliability Consultant at Ops A La Carte. He has held positions at Hughes Research Labs in optical/microwave device research, Cypress Semiconductor as R&D Pilot Line Manager, Alliance Semiconductor as Foundry Operations Manager and Sun

Microsystems as SPARC Supply Engineering Manager and Senior Member of the Technical Staff in memory reliability. He has also participated in various task groups responsible for the JEDEC soft error standards and served as the chair of the Soft Error Committee for IRPS.

Smith, Charles
Charles G. Smith received his degree in Physics in 1983 from the University of St Andrews Scotland. He spent a year a the University of Oregon in the USA before completing a PhD at the University of Cambridge in 1987, where he worked on nanomechanical devices and on quantized thermal conductivity in sub-micron free-standing metal structures. He has been a Professor of Physics at the University of Cambridge since 20005 where he works on quantum transport in nanodevices. He is the founder of Cavendish Kinetics and is now the Chief Scientific Officer.

Smyth, Charles
Charles Smyth received a Bachelor of Science in Electrical Engineering from the University of Florida in 2005. He continued on to complete a Master's of Science in EE at the University of Florida in 2008 with a specialization in analog integrated circuit design with some focus on device physics. He has been working at Texas Instruments in Medical Business Unit since 2008 as a characterization engineer and specializes in medical imaging systems, working in MRI devices and Ultrasound transmission and front-end devices.

Sohn, Chang Woo
received the B.S. degree (2007) in electrical and computer engineering from the Hanyang University, Seoul, Korea, and the M.S. degree (2009) in electronic and electrical engineering from the Pohang University of Science and Technology (POSTECH), Pohang, Korea, where he is currently working toward the Ph.D. degree. His main research interests include characterization and modeling of CMOS devices.

Song, Grace
I am a failure analysis engineer in Quality Department in Freescale Semiconductor (China) Limited since 2005. In 2005, I graduated form Tianjin University in Tianjin China and got master degree in microelectronic Engineering.

Spinelli, Alessandro S.
Alessandro S. Spinelli is a Full Professor of electronics with the Politecnico di Milano. He has conducted experimental and theoretical research in electronics instrumentation and microelectronics, co-authoring more than 130 papers published in international journals or presented at international conferences, and serving in the technical committees of the IEDM and IRPS conferences. His current research interests include experimental characterization and modeling of non-volatile memory cells performance and reliability, development of innovative non-volatile memory technologies and circuit design for biological signal readout.

Stam, Frank
Frank Stam received his MEngSc (Mechanical Engineering) from University Twente, The Netherlands in 1989. He was with Digital Equipment Corporation (DEC), Galway, Ireland as a project engineer from 1989 until 1992 working on advanced manufacturing processes for Multi Chip Assemblies. He then joined the Tyndall National Institute and provided consultancy and failure analysis services to industry. In 1996 he became a senior research scientist and got involved in various Irish and EU

microelectronics research projects. In 2001 he setup a team to develop Biomedical Microsystems applications, and more recently his interest converged to biomedical implants.

Stathis, James
Jim Stathis received the bachelor's in physics from Washington University in St. Louis (1980), and the Ph.D. in physics from the Massachusetts Institute of Technology (1986), joining the IBM Research Division the same year. At IBM the focus of his work has been the electrical properties of point defects in SiO2, including basic studies of defect structure using magnetic resonance and electrical measurement techniques, and the role of defects in wearout and breakdown. He is the author or coauthor of more than 100 research papers and over 60 invited talks. From November 2005 to February 2007 he served as Technical Assistant to the Vice President for Science and Technology, IBM Research Division. In February 2007 he became manager of High-k/Metal-Gate Characterization and Reliability, IBM Research. Jim has served on technical program committees for SISC, INFOS, IRPS, ESREF, IPFA, MIEL and other conferences, served as Chair of the dielectrics sub-committee for the 2003 IRPS in Dallas, and is on the IRPS management committee. He was the Technical Program Chair for IRPS 2009 and is Vice-General Chair for IRPS 2010. He has presented tutorials on CMOS reliability at IRPS, ESREF, MRS, and IPFA and is an Associate Editor of the journal Microelectronics Reliability. He is a Senior Member of IEEE and a Fellow of the American Physical Society.

Stupian, Gary
Dr. Gary Stupian received his B.S. in physics from the California Institute of Technology in 1961 and his Ph.D. in physics from the University of Illinois Urbana-Champaign in 1967. He joined The Aerospace Corporation in 1969 after post-doctoral research at Cornell University. He is currently a Distinguished Scientist in the corporation's Electronics and Photonics Laboratory. His research interests center about the application of advanced microanalytic techniques to the reliability of space systems.

Su, Kuan-Cheng
K C Su received the B.S. degree (1986) and M.S. degree (1988) in Electrical Engineering from National Cheng Kung university, Taiwan, ROC. Then, he joined United Microelectronic Corp. (UMC) and worked with process technology development division for over 10 years. He is currently focusing on reliability technology & methodology development. His main interests are reliability engineering, device engineering and logic technology development.

Suehle, John
John S. Suehle received his B.S., M.S., and Ph.D. degrees in electrical engineering from the University of Maryland, College Park, in 1980, 1982, and 1988 respectively. Since 1982 he has been at the National Institute of Standards and Technology (NIST) where he is leader of the CMOS and Novel Devices Group. Dr. Suehle has authored or co-authored over 200 technical papers or conference proceedings and holds 5 U.S. patents. He has served as General Chair of the 2008 IRPS and also serves on the executive management committee of the IEEE IEDM. He is currently an Editor of the IEEE Transactions on Electron Devices.

Sugawa, Shigetoshi
Shigetoshi Sugawa (M'86) received his M.S. in 1982 in Physics from Tokyo Institute of Technology and his Ph.D. in 1996 in Electrical Engineering from Tohoku University. In 1982-1999 he worked in Canon Inc., where he researched high S/N ratio solid-state imaging devices, high performance amorphous silicon

devices, high-speed low power SOI devices and high-density liquid crystal display devices. In 1999 he moved to Tohoku University and he is presently a professor at Graduate School of engineering, Tohoku University. He is currently engaged in researches on CMOS image sensors, high-performance ULSI's and advanced displays such as high-performance, high-speed and low-power circuits/devices, and advanced semiconductor process technologies related to high-quality low-temperature oxidation, nitridation, CVD and etching process using microwave-exited high-density plasma. Dr. Sugawa is a member of the IEEE, the Institute of Image Information and Television Engineering of Japan.

Sun, Wein-Town
Wein-Town Sun was born in Kaohsiung, Taiwan, R.O.C., in 1970. He received the B.S. and Ph.D. degrees in electrical engineering from National Tsing-Hua University, Hsinchu, Taiwan, in 1992 and 1998, respectively. He was with Process Integration Department, eMemory Technology Inc., Hsinchu, in 2006, working in the area of SONOS Flash nonvolatile memory. His research interests are device physics of sub-micron MOSFETs and TFTs, and process development of gate and drain engineering for sub-micron MOSFETs. Currently, his research involves in developing high P/E endurance SONOS cells and next-generation embedded SONOS Flash.

Yu, Sunil
Sunil Yu received the Ph. D. degree in EECS from Korea Advanced Institute of Science and Technology in 2006. In 1991, he joined Semiconductor Division, Samsung Electronics. He has worked on process development of standard CMOS technologies for various generation. His interests include SRAM characterization and yield improvement.

Lim, Sun-Me
Sun-Me Lim received the B.S. in physics form Korea University in 2000. She also received the M.S. and Ph.D. degrees in solid state physics from Korea Advanced Institute of Science and Technology in 2002, and 2006, repectively. Since 2006 she has been working for SRAM technology as a senior engineer in System LSI, Samsung Electronics.

Sutton, Akil
Akil K. Sutton (M'01) received the B.S. degree in mathematics from Morehouse College, Atlanta GA in 2003 and the M.S. and Ph.D. degrees in electrical engineering from the Georgia Institute of Technology in 2005 and 2009, respectively. His doctoral research involved the study of the effects of total ionizing dose and single-event effects on the radiation response of SiGe HBT devices and circuits. In 2009 he joined the IBM Semiconductor Research and Development Center, Hopewell Junction, NY. At IBM, he has been involved in the power performance analysis of analog and mixed signal blocks in highly scaled bulk and SOI CMOS technology nodes.

Tagliaferro, Roberto
Roberto Tagliaferro was born in Rome, Italy, on May 31, 1984. He graduated cum laude in Electronic Engineering for Telecommunications at the University of Rome "Roma Tre" on 23th July 2009. He did a thesis on: "Advanced techniques of Ray Tracing to study electromagnetic field in indoor/outdoor environments". Since November 2009, Roberto is a Ph.D. student at C.H.O.S.E. (Centre for Hybrid and Organic Solar Energy), in OLAB (Opto&Nano Electronics) group, at University of Rome "Tor Vergata" and is working on Dye solar cells. Particularly he deals with light management in DSSC to improve performance of cell test.

Takashi, Matsumoto
Takashi Matsumoto received the B.E. and M.E. degrees in Applied Physics from The University of Tokyo, Tokyo, Japan, in 1999, 2001, respectively. From 2001 to 2009, he was with Fujitsu Laboratories. He is presently a doctor's course student at Kyoto University.

Tam, Nelson
Dr. Nelson Tam joined Intel Corporation in 1991, where he worked on the development of phase shifting mask. In 1997, he joined the Enterprise Processor Division as a Quality and Reliability Engineer focusing on processor pre-Si reliability verification. In 2006, he jointed Marvell Technology Group as a Principal Reliability Engineer focusing on soft error issues. His research interests include simulation and experimental techniques for determining radiation effects on integrated electronic devices. Dr. Nelson Tam received the B.S. degree in chemical engineering and the M.S. and Ph.D. degrees in electrical engineering and computer science from the University of California, Berkeley.

Tang, Tien-Hao
Howard Tien-Hao Tang received the B.S. degree in Electrical Engineering from National Cheng Kung University and the M.S. degree in Electrical Engineering from University of Missouri-Columbia. He joined UMC in 1997 and currently a senior department manager of ESD Engineering Department. He has participated the Technical Program Committee of EOS/ESD Symposium, IRPS, IPFA, and IEW. He is author or co-author of 70 technical papers and has issued 100 patents. He won the National Invention and Creation Award in year 2006. In year 2011, he is selected as one of the Top Ten Distinguished Inventors in Taiwan.

Tang, Tien-Hao
Howard Tien-Hao Tang received the B.S. degree in Electrical Engineering from National Cheng Kung University and the M.S. degree in Electrical Engineering from University of Missouri-Columbia. He joined UMC in 1997 and currently a senior department manager of ESD Engineering Department. He has participated the Technical Program Committee of EOS/ESD Symposium, IRPS, IPFA, and IEW. He is author or co-author of 70 technical papers and has issued 100 patents. He won the National Invention and Creation Award in year 2006. In year 2011, he is selected as one of the Top Ten Distinguished Inventors in Taiwan.

Tao-Cheng, Lu
Tao Cheng Lu was born in Taiwan in 1967. He received B.S., M.S. and Ph.D. in electrical engineering in 1989, 1991 and 1993 respectively, all from National Taiwan University. He joined the Macronix International Co., Ltd. in 1995, where he engaged in the simulation and modeling of CMOS and nonvolatile memory devices. From 1997 to 1999, he set up the ESD group in device department and became the manager of the device department in 2000. Currently, he is the Director of Device Technology Division and is in charge of Process/Device Simulation, Device Modeling, ESD/Latchup and Nonvolatile Memory Device Design, including the development of nitride trapping storage flash devices. His research interests include device/flash cell reliability and scaling, device modeling, interconnect capacitance measurement and modeling, and on-chip ESD protection. He has published more than 50 papers in technical journals and conferences and has authored and co-authored more than 100 U.S. patents. He has served as the scientific committee member of International Memory Workshop and Non-Volatile Semiconductor Memory Workshop since 2009 and 2005.

Teo, Jason KJ
Jason Teo is a fourth year PhD student at the National University of Singapore (NUS). He is a receipient of the NUS scholarship. His field of research is in integrated circuit failure analysis, working in depth on the physics and development of laser reflectance techniques.

Teo, Zhiqiang
Z. Q. Teo received the B. Eng. (Hons.) degree from the School of Electrical and Electronic Engineering, Nanyang Technological University, Singapore. He is currently working towards the Ph.D. degree in the same school under a GLOBALFOUNDRIES Singapore-NTU Graduate Research Programme. His research project is on the characterization of bias-temperature instability in state-of-the-art p-MOSFETs.

Teramoto, Akinobu
Akinobu Teramoto (M'02) received the B.S. and M.S. degrees in electronic engineering from Tohoku University in 1990 and 1992, respectively and Ph.D. in 2001 in Electrical Engineering from Tohoku University. In 1992-2002, he worked Mitsubishi Electric Corporation, Hyogo, Japan, where he has been engaged in the research and development of thin silicon dioxide films. In 2002, he moved to Tohoku University and he is presently an associate professor at New Industry Creation Hatchery Center, Tohoku University. He is currently engaged in an advanced semiconductor device technologies and process technologies, such as SOI MOS transistors, accumulation-mode transistors, variation and noise of transistors, high-quality low-temperature oxidation, nitridation, and CVD process using microwave-exited high-density plasma. Dr. Teramoto is a member of the IEEE, the Electrochemical Society, the Japan Institute of Electronics Packaging, Information and Communication Engineers of Japan and the Japan Society of Applied Physics.

Tian, Li
I am a failure analysis engineer in Quality Department in Freescale Semiconductor (China) Limited since 2010. In 2010, I graduated form Tianjin University of Technology in Tianjin China and got master degree in Physical electronic Engineering.

Tilak, Vinayak
Vinayak Tilak was born in Chennai, India. He received M.Sc in Physics and B.E. Electrical Engineering from Birla Institute of Technology and Science, Pilani in 1997. He received his PhD in Applied and Engineering Physics from Cornell University, Ithaca, N.Y in 2002 where his doctoral work focused on development of high frequency AlGaN/GaNF FETs for RF applications. He is currently a member of technical staff at GE Global Research at Niskayuna, NY. He is working on Silicon Carbide and Gallium Nitride based devices and integrated circuits for various applications including power switching and sensors. He is a senior member of IEEE.

Tillack, Bernd
Prof. Bernd Tillack received the PhD degree from the University Halle-Merseburg in 1980. In 1981 he joined the IHP Frankfurt (Oder), Germany, as a staff member of the process technology. He had been the project leader of different IHP Si/SiGe technology projects. His research interests include SiGe BiCMOS technology development following the "More than Moore" strategy for embedded system applications. Since 2004 he is in charge of the Si/SiGe process and device technology in the IHP. In 2008 he got a professorship for Si based high frequency technologies at the Berlin Institute of Technology (TU Berlin).

Ting, Liu
Ting Liu received the B.E. degree from Nanjing University of Science and Technology, China, in 2005 and the M.E. degree from Shanghai Jiao Tong University, China, in 2008, both in electrical engineering. She is currently working towards her Ph.D. degree in electrical engineering at Princeton University. Her research is the fabrication of amorphous-silicon thin-film transistors with very high stability for application as current sources. She is evaluating and testing these transistors, and simulates their long-term performance in terms of physics-based models.

Ting, Yun-Jen
Yun-Jen Ting received the M.S. degree in electrical engineering from National Chi Nan University, Nantou, Taiwan, R.O.C., in 2002. From 2002 to 2007, he was with Winbond Electronics Corp., Hsinchu, Taiwan, where he was engaged in development of 0.18um-0.09um Flash memory technology including floating gate and nitride based flash. Since 2007, he has been with the Device Engineering Department of eMemory Technology Inc., Hsinchu, Taiwan, Where he is currently engaged in the development of nitride based embedded flash IP technology.

Tőkei, Zsolt
Zsolt Tőkei is program director for the advanced interconnect program. He joined IMEC in 1999 and since then he is working in the field of copper low-k interconnects. He obtained his M.S. (1994) in physics from the University Kossuth in Debrecen, Hungary. In the framework of a co-directed thesis between the Hungarian University Kossuth and the French University Aix Marseille-III, he earned his PhD (1997) in materials science. From 1998 he worked at the Max-Planck Institute of Düsseldorf, Germany, as a post-doctorate researcher. In 1999, he joined the Interuniversity Microelectronics Center (IMEC), Belgium, where he has worked on a range of interconnect issues such as metallization, process development, electrical performance, interconnect scaling and dielectric reliability. He has authored or co-authored more than 100 publications in international scientific journals and in international scientific proceedings.

Toledano-Luque, Maria
María Toledano Luque was born in Madrid, Spain, in 1980. She received the M.Sc. degree in electrical engineering in 2003, M.Sc. degree in physics in 2009, and Ph.D. degree in 2008 from the Universidad Complutense de Madrid (UCM), Madrid, Spain. In 2003, she became a member of the Thin Films and Microelectronics research group (UCM). In 2007, she joined the Department of Applied Physics (UCM) as Assistant Professor. Currently, she is a Postdoctoral Researcher at imec, Leuven, Belgium, working in the field of advance electrical characterization.

Toriumi, Akira
Akira Toriumi received the B.S. degree in physics, the M.S. and Ph.D. degrees in applied physics from the University of Tokyo in Japan, 1978, 1980 and 1983, respectively. He joined R&D Center of Toshiba Corporation in 1983. In May 2000, he moved to the University of Tokyo. His research interests have been on silicon device physics and related materials science throughout his professional carrier. He is currently studying the materials science of high-k dielectrics, and physics and technology of Ge as well as organic electron devices. He is a member of the JSAP, JPS, APS, ECS and IEEE.

Tsai, Chen-Han

Chen-Han Tsai received the B.S. and M.S. degrees in electronic engineering from National Chia-Tung University, Taiwan, in 2007 and 2009, respectively. Since 2009, he has been with Industrial Technology Research Institute, Taiwan, and he is currently an Associate Engineer. He is the Author/Coauthor of 4 papers. His research interests include RRAM device design, fabrication and measurement.

Tsai, Ming-Jinn

Ming-Jinn Tsai received his Ph.D. in Materials Science and Engineering from MIT before he joined Industrial Technology Research Institute (ITRI) in Taiwan. Since 1994 he has been working with Electronics Research and Service Organization of ITRI mainly on the semiconductor device and process technology, such as CCD, IRCCD, Power and RF devices and High K capacitors. Currently he is a research director of EOL of ITRI and responsible for the Nanoelectronics program, which includes the development of new non-volatile memory technologies, such as MRAM and PCM, RRAM and other new devices. He has published ~180 papers and holds 15 patents.

Tsuchiya, Hideaki

Hideaki Tsuchiya received the B.S. (2000) and M.S. (2002) in physics from Tohoku University, Japan. He joined NEC Corp. and transfered to NEC Electronics Corp., in 2002. He has been with Renesas Electronics Corp., since 2010, where he has been working on reliability engineering of Cu interconnects.

Twomey, Karen

Karen Twomey obtained her degree in electrical and electronic engineering from University College Cork in 1999. She obtained her PhD in portable sensing systems from University of Limerick in 2002. She is employed as a staff research scientist at the Molecular Microsystems Group, Tyndall National Institute, Cork. Her research interests include swallowable capsule technology, portable and miniaturized sensor systems, electrochemical microsensor fabrication and characterisation, and methods for signal processing and data interpretation.

Uno, Satoshi

Satoshi Uno received the B.S.(1999) and M.S.(2001) degree in engeneering from the University of Shinshu, Nagano, Japan. He joined NEC Corp., Japan, in 2001. He has been with Renesas Electronics Corp., Japan, since 2010. He has been working on 40nm-node technology reliability and solder bump EM.

Vaidyanathan, Balaji

Balaji Vaidyanathan received his Ph.D. in Computer Science and Engineering from Pennsylvania State University in 2010. He has been working in TSMC, Taiwan since 2007 focusing on device reliability characterization, circuit reliability modeling and characterization for soft-errors, NBTI, HCI, and PBTI in advanced technologies.

Van Houdt, Jan

Jan Van Houdt (SM '02) was born in Leuven, Belgium, on June 20, 1963. He received the M.Sc. degree in electrical and mechanical engineering in 1987 and the Ph.D. degree in applied sciences in 1994, both from the Katholieke Universiteit Leuven, Belgium. His M.S. thesis dealt with the degradation of short-channel MOS transistors under hot-carrier injection conditions. His Ph.D. work concentrated on the physics and characteristics of HIMOS™ Flash memory devices. He received the Best Student

Paper Award at the 22nd European Solid-State Device Research Conference in 1992 and the Scientific Award of the Royal Academy for Science, Literature and Fine Arts of Belgium in 1995. After his M.S. thesis, he joined the Interuniversity Micro-Electronics Center (IMEC), Leuven, Belgium. In 1990, he invented the High Injection MOS (HIMOS™) transistor, a novel fast-programmable Flash EEPROM cell that has led to a high performance, cost-effective nonvolatile memory technology, on which he holds numerous international patents. In 1996, he became responsible for the development and dissemination of Flash memory technology based on IMEC's proprietary concepts, including the licensing and the transfer of these technologies toward 4 industrial product lines. Since 1999, he manages the memory group at IMEC. From 2000 on, he also managed IMEC's Industrial Affiliation Program on Advanced Memory Technology and expanded it to become one of IMEC's largest research programs today. His research interests are physics of semiconductor devices, hot-carrier injection and degradation phenomena in MOS structures, thin dielectrics, modeling and optimization of floating gate and nitride nonvolatile memory devices, physics, reliability and design aspects of memories in general. He has published more than 160 papers in international journals, wrote 2 book chapters and accumulated more than 140 conference contributions (including more than 25 invitations and 5 best paper awards). He has filed more than 50 patent applications worldwide in the area of nonvolatile memories, out of which 27 patents have been granted so far. He serves (or served) on the program and/or organizational committees of the IEEE Nonvolatile Semiconductor Memory Workshop (NVSMW), the IEEE Reliability Physics Symposium (IRPS), the European Solid-State Device Research Conference (ESSDERC), the International Conference on Memory Technology and Design (ICMTD), the IEEE International Workshop on Memory Technology, Design and Testing (MTDT, Taiwan) , the Solid-State Devices and Materials (SSDM) conference, the MRS symposium on non-volatile memory technologies, the International Memory Workshop (IMW), the IEEE International Electron Devices Meeting (IEDM) and the IEEE Semiconductor Interface Specialists Conference (SISC). In 2007 he was the General Chairman of the International Conference on Memory Technology and Design.

Varghese, Dhanoop

Dhanoop Varghese received the B.Tech. degree in electronics and communication engineering from REC Calicut, India in 2002, M.Tech degree in electrical engineering from Indian Institute of Technology (IIT), Bombay, India in 2005 and Ph.D. in electrical engineering from Purdue University, IN, in 2009. Since 2009 he is with Texas Instruments, Dallas working on the reliability of high voltage analog components. His research interests are in the field of semiconductor device physics, simulation, modeling and characterization of various transistor degradation mechanisms. He has worked on bias temperature, and hot carrier reliability issues in MOSFETs and high-k gate dielectrics.

Vasanth, Karthik

Karthik Vasanth received the Bachelor of Technology degree in Electronics and Communication Engineering from the Indian Institute of Technology Madras (Chennai) in 1991. He received his Ph.D degree in Electrical Engineering from Princeton University in 1995. He joined the Silicon Technology Development group at Texas Instruments in 1995 and worked on compact process and device simulation models. From 2003-2007 was the RF Design manager for the High Performance Wireless Infrastructure Products group at Texas Instruments. He was also elected as a Distinguished Member of the Technical Staff at Texas

Instruments in 2005. In 2007 he became the product line manager of the Medical business Unit at Texas Instruments. In 2010 he became the general manager for the medical and high reliability business unit at Texas Instruments. He has published over 30 papers and authored/co-authored several patents.

Veksler, Dmitry
Dmitry Veksler received his B.S. in 1996 and M.S. in 1998 from Nizhniy Novgorod State University and Ph.D. in Physics in 2007 from Rensselaer Polytechnic Institute, Troy, NY. From 1998 till 2003 he was with Institute for Physics of Microstructures, Nizhniy Novgorod (Russian Academy of Sciences) investigating optical and transport phenomena in low-dimensional semiconductor hetero structures. During 2003-2008 he was with Rensselaer Polytechnic Institute working on plasma wave electronics for THz spectroscopy and imaging applications. In 2009 he joined SEMATECH Inc., Albany, NY. His current research focus is on electrical characterization/reliability of high-k and alternative substrate MOSFETs.

Vijay, Rentala
Vijay Rentala obtained his bachelor's in electrical and electronics engineering from Birla Institute of Technology, Pilani in 1998, and his masters in Electrical engineering from Iowa State University in 2000. He has been with Texas Instruments. He is currently working as an RF/analog circuit designer in Wireless Business Unit of Texas Instruments. His research interests include RF/analog circuit design in advanced digital processes, data converters and IC design for wireless communication systems.

Vilchis, Miguel
Miguel A. Vilchis-Cruz. Received the M.Sc. degree in physics from the University of Wisconsin, Milwaukee, in 1978, and the M.Sc. degree in materials science from Rensselaer Polytechnic Institute, Troy, NY, in 1980. Between 1980 and 1997 he worked at Amdahl Corporation, Sunnyvale, CA, where he was involved with intra-board level optical interconnects. In 1997 he joined LSI Corporation, Milpitas, CA, where his work concentrates on reliability assurance of storage CMOS devices and soft errors mitigation solutions in system-on-chip applications.

Visconti, Angelo
Angelo Visconti (M'07) received the Laurea degree in physics (cum laude) from the University of Milano, in 1997. The same year he joined the Non-Volatile Memory Process Development Group, R&D department of STMicroelectronics (now Numonyx), Agrate Brianza, Italy. Since that, he has been involved in the developments of ten-generations of Flash Memory Process. His current research interests include reliability, radiation effects, and multilevel applications of Flash cells. He is co-author of more than 100 scientific publications and 14 patents. He is a Lecturer on NVM Reliability and Radiation Effects on Flash Memory with the University of Padova and Politecnico di Milano. Since 2005, he has been the ST-rapresentative, and chair from 2009, within the JEDEC 14.3 task-group on NVM reliability. Finally, he was member of the IEDM-CIR committee, IRPS-PIR committee, IRPS-Memory committee and co-chair in 2010. Mr. Visconti won the IRPS 2008 "Outstanding Paper Award" and the JEDEC 2010 Team Award.

Vladimir, Cherman
Vladimir Cherman received his M.Sc. and Ph.D. degrees in electronic engineering from Saint-Petersburg Electrotechnical University (ETU) Russia, in 1994 and 1999, respectively. From 1997 to 2000, he was with Morion Inc. St.-Petersurg, Russia as an electronics engineer. From 2000 to 2007 he was with Ceramics

Laboratory at Swiss Federal Institute of Technology-Lausanne (EPFL) where his research was focused on microwave properties of ferroelectric materials. He joined the Reliability and Modeling group of IMEC as a researcher in 2007, where he is involved in reliability study of MEMS devices and electronic packages.

Waltz, Patrice
Patrice Waltz received the Engineering degree in solid state physics in 1994 and the Ph.D. degree in materials science in 1998 from the Institut National des Sciences Appliquées, Lyon, France. He joined beginning of 2000 the Electrical Characterization & Reliability Group from STMicroelectronics, Crolles, France where he first worked on High Voltage transistors and Non Volatile CMOS Embedded devices characterization. He is leading the 8' Back-End Reliability Activity since 2004 and 8' & 12' Back-End Reliability Activity since 2007.

Wang, Chang-Tzu
Chang-Tzu Wang (S'06-M'11) received the B.S. and Ph.D. degrees from the National Chiao Tung University, Hsinchu, Taiwan, in 2005 and 2010, respectively. In 2006, he joined the Electrostatic Discharge Engineering Department, Reliability Technology and Assurance Division, United Microelectronics Corp. (UMC), where he is currently an Assistant Manager and in charge of supervising the electrostatic discharge (ESD) development for advanced CMOS processes. His technical area of expertise includes on-chip ESD protection circuit design for advanced CMOS processes, mixed voltage I/O circuits, and high-speed/low-power applications.

Wang, Chen
Chen Wang received her B.S. degree in electrical engineering from the Zhengzhou University, China, in 2004, and her M.S. degree in electrical engineering from Xiamen University, China, in 2007. Since 2007, she has been a Ph.D. candidate based on the Fudan University-Purdue University Joint Training program. From winter 2007 to summer 2009, she was a guest researcher in the Semiconductor Electronics Division, National Institute of Standards and Technology, where she was engaged in research on the reliability of III-V Transistors. Now, she is at Purdue University to finish her Ph.D. topic on the electrical characterization of III-V MOSFETs.

Wang, Chia-Yu
Chia-Yu Wang received his M.S. in Material Science Engineering from National Tsing-Hua University in 2006. He has served as the engineer of Technology Quality & Reliability Department at Taiwan Semiconductor Manufacturing Company (TSMC) focusing on NVM reliability for 3 years.

Wang, Shun-Min
Shun Min Wang received the B.S. degree from the department of chemical engineering, National Cheng Kung University, Tainan, Taiwan, and M.S. degrees in material science and engineering from National Chiao Tung University, Hsinchu, Taiwan, in 2004 and 2006. Since 2007, he has been with Industrial Technology Research Institute (ITRI), where he is currently a RRAM integration engineer for the department of Electronics and Optoelectronics Research Laboratory. His researches include semiconductor module process development, device fabrication, and 3DIC technology.

Wang, Tahui
Tahui Wang (S'84-M'85-SM'94) was born in Taoyuan, Taiwan, R.O.C., on May 3, 1958. He received the B.S.E.E. degree from National Taiwan University, Taipei, Taiwan, in 1980, and the Ph.D. degree in electrical engineering from the University of Illinois at Urbana–Champaign, Urbana, in 1985. From 1985 to

1987, he was with Hewlett-Packard Laboratories, Palo Alto, CA, where he was engaged in the development of GaAs HEMT devices and circuits. Since 1987, he has been with the Department of Electronics Engineering, National Chiao-Tung University, Hsinchu, Taiwan, where he is currently a Professor. His research interests include hot-carrier phenomena characterization and reliability physics in very large scale integration (VLSI) devices, RF CMOS devices, and nonvolatile semiconductor devices. Dr. Wang was given the Best Teacher Award by Taiwan's Ministry of Education. He has served as technical committee member of many international conferences, among them International Electron Devices Meeting, International Reliability Physics Symposium, and VLSI Technology, Systems, and Applications.

Wang, Wayne
Wayne Wang received his M.S. degree in EE from National Chiao Tung University, Taiwan. He is the section manager of Mainstream Technology Quality and Reliability Department at TSMC. Prior to joining reliability division in 2007, he served as embeded flash circuit designer since 1999. He holds 8 U.S. patents and 9 Taiwan patents.

Wang, Winter
I am a failure analysis engineer in Quality Department in Freescale Semiconductor (China) Limited since 2000. In 1998, I graduated form Tianjin University of Technology in Tianjin China and got bachelor degree in Electronic Engineering.

Wang, Yun Yu
Yun-Yu Wang attended Fudan University in Shanghai, China, to study laser physics. He obtained a Ph.D. in physics at Virginia Tech in 1990. He worked at the University of Virginia and Northwestern University as a research associate. He joined IBM in 1997 and is now a senior engineer specializing in TEM characterization of semiconductor devices. He has authored and co-authored 80 papers and obtained 56 US patents. His invention on dual-lens operation for electron holography has been able to obtain spatial resolution down to 1 nm scale, which allows junction profiling and strain imaging for semiconductor device in great detail.

Wang, Zhongrui
Wang Zhongrui received his bachelors degree from School of EEE, Nanyang Technological University, Singapore in 2009. He is currently pursuing his Ph.D at the Division of Microelectronics, School of EEE, NTU focusing on fabrication and characterization of high-κ dielectric based switching materials.

Weigmann, Jan-Michael
Jan-Michael Weigmann received his bachelor in electrical engineering (2010) from the Baden-Württemberg Cooperative State University. 2008 he joined Qimonda as a student. 2009 he has changed to the Device Department of the Central Reliability Methodology Group of Infineon where he completed his bachelor thesis on statistal investigation methodologies for NTBI. Since 2010 he is working in the Design Enabling Service Team for Infineon Technologies .

Weiland, Rainer
Rainer Weiland (56) received his diploma and his Ph.D. in chemistry from the University of Kaiserslautern, Germany, in 1977 and 1983, respectively. His thesis was in the field of optical spectroscopy of oriented molecules. In 1986 he joined Siemens and worked primarily on the development of thermal ink jet printers. 1995 he took over new responsibilities in the Semiconductor Division of Siemens and is currently Senior Manager of the Physical Failure Analysis Department PTA 5 of Infineon in Munich.

Weller, Robert
Robert A. Weller received his Ph.D. in physics in the Kellogg Radiation Laboratory at Caltech in 1978, and is now a Professor of Electrical Engineering, Materials Science, and Physics at Vanderbilt University. Throughout his career, he has been active in the field of radiation interactions with materials, with an emphasis on the application of the techniques of nuclear physics. Most recently, he has shifted the primary focus of his activities from experimental research to the computational analysis of radiation effects in electronics. He is the principle author of Vanderbilt's radiation effects code, MRED.

Wen, Shi-Jie
Shi-Jie Wen received his Ph.D in Material Engineering from University of Bordeaux I in 1993. He joined Cisco Systems Inc., San Jose, CA in 2004, where he has been engaged in IC component technology reliability assurance. His main interest is in silicon technology reliability, such as SEU, WLR, and complex failure analysis, etc. He is a member of DFR, SEU core teams in Cisco. Before Cisco, he worked in Cypress Semiconductor where he was involved in the area of product reliability qualification with technology in 0.35u, 0.25u, 0.18u, 0.13u and 90nm.

Wilde, Markus
Markus Wilde obtained master and doctoral degrees in Physical Chemistry from the University of Bochum, Germany, and is currently Associate Professor at the Institute of Industrial Science, University of Tokyo (Japan). He investigates the reactive behavior of hydrogen (H) on surfaces, in shallow surface-bulk transition regions and at thin film interfaces, combining surface science techniques with quantitative high-resolution H depth profiling by nuclear reaction analysis (NRA). Current research focuses on H-absorption and surface/subsurface H exchange at transition metal single crystals and nanoparticles. Other interests include H impurities at MOS electrode/dielectric interfaces, surface hydroxylation and hydrophilicity of oxides, and semiconductor H-passivation.

Wilkinson, Jeff
Jeff Wilkinson joined Medtronic, Inc. in 1983 and is now a Technical Fellow and Senior Principal R&D Engineer, worked in product development on a variety of internal and external medical devices. Since 1999 he has been a part of the Device Technology group, developing and adapting technologies to meet the unique challenges of the implanted medical device. This work has included microprocessor design, ultralow power DSP, implantable sensors, and high reliability non-volatile memory. Since 2003 he has concentrated on ionizing radiation effects with a particular emphasis on soft errors in implanted devices. Mr. Wilkinson is a member of the AAAS, senior member of the IEEE and holds a B.A. degree in Mathematics, Physics and Computer Science from St. Olaf College. He holds 10 US and international patents related to implantable medical device design.

Wilson, Christopher J.
Christopher J. Wilson (S'07) received the M.Eng. degree (first class honors) in Electronics and Ph.D. degree in interconnect reliability and mechanical stress in thin films from Newcastle University, Newcastle upon Tyne, U.K. in 2006 and 2009 respectively. Between 2007 and 2008 he spent 16 months as Marie Curie APROTHIN Fellow and Visiting Scholar in the Advanced Interconnect Program at the Interuniversitair Micro-Elektronica Centrum (IMEC). He was an Institution of Electrical

Engineers Robinson Research Scholar and held an Institution of Electrical and Electronic Engineers Reliability Society Scholarship. Dr. Wilson is now the International Mobility Post-Doctoral Fellow in IMEC's Interconnect Process Technology unit, where he is working on advanced metallization integration and reliability.

Wirth, Gilson
Gilson I. Wirth received the B.S.E.E and M.Sc. degrees from UFRGS, Brazil, in 1990 and 1994, respectively. In 1999 he received the Dr.-Ing. degree in Electrical Engineering from the University of Dortmund, Dortmund, Germany. He is a IEEE Senior Member and Distinguished Lecturer of the IEEE Circuits and Systems Society. He is currently a full professor at the Electrical Engineering Department at UFRGS. In July, August and December 2001 he was at Motorola, Austin, Texas, leading the team working in CMOS process technology transfer to CEITEC, Porto Alegre, Brazil. In February and March 2002 he was at the Corporate Research Department of Infineon Technologies, Munich, Germany, working as guest researcher on low-frequency noise in deep submicron MOS devices. His research interests include low-frequency noise, ionizing radiation effects, bias temperature instability (BTI), reliability and design for yield of digital, analog and mixed-signal circuits.

Wittborn, Jesper
Jesper Wittborn received a M.Sc. in physics from Stockholm University, Sweden, and a Ph.D. in material physics from the Royal Institute of Technology, Stockholm, Sweden, in 1992 and 2000, respectively. His thesis was in the field of scanning probe microscopy on biological, magnetic and ferroelectric materials. From 2001 to 2004 he worked at Ericsson Microelectronics doing physical failure analysis. In 2004 he joined Infineon Technologies where, besides working as a failure analysis expert, he leads the development of scanning probe microscopy (SPM) methods. His research interests include SPM use for failure analysis of microelectronics and measurement of doping in semiconductors.

Witters, Liesbeth
Liesbeth Witters received the B.Sc. and M.Sc. degrees in chemical engineering from Katholieke Universiteit Leuven, Belgium and Institut Francais du Petrole, Paris, France in 1992 and 1993, respectively. From 1994 till 1995, she did graduate research work at the civil engineering department of University of California, Irvine. In 1995, she joined Rockwell Semiconductor Systems, later Conexant, in Newport Beach, California where she worked as a CMP process development engineer. In 2001, she joined IMEC, Leuven, Belgium as a BiCMOS process integration engineer. Since 2004 she has been working on different applications in the CMOS process development department.

Witulski, Arthur
Arthur F. Witulski received the B.S., M.S., and Ph.D. degrees in electrical engineering from the University of Colorado, Boulder in 1981, 1986, and 1988 respectively. He has worked in analog electronics, power electronics, and microelectronics in industry and academia. He has worked as a design engineer at Storage Technology in Colorado, Associate Professor at the University of Arizona, and is currently Senior Research Engineer at the Institute for Space and Defense Electronics at Vanderbilt University, where he also holds an appointment as a Research Associate Professor. He is author or co-author of more than 60 technical papers in these areas.

Wong, Richard
Richard Wong received his M.S. degree in electrical engineering from Santa Clara University and his B.S. degree in chemical engineering from UC Berkeley. He joined Cisco Systems Inc., San Jose, CA in 2006. He is engaged in IC component technology reliability assurance in issues such as Single Event Upset, Electro Static Discharge, Wafer Level Reliability, failure analysis and reliability modeling. Prior to Cisco, he had worked on ASICs, FPGAs, TCAMs and memories.

Wood, Alan
Alan Wood is a Consulting Engineer in Oracle Labs. He is developing fault-tolerant data base software and simulating data base performance. He is also working on energy-efficient computing and software reliability. Previously, as a Distinguished Engineer at Sun Microsystems, he was the fault tolerance architect for a DARPA-funded next generation supercomputer. Prior to joining Sun, Alan was a Distinguished Technologist at HP, in charge of fault-tolerant modeling and architecture for NonStop systems. He has a PhD in Operations Research from Stanford and has been an officer of the local IEEE Reliability chapter for the past several years.

Wrachien, Nicola
Nicola Wrachien was born in Treviso, Italy, in 1982. He received the degree (magna cum laude) in electronic engineering in 2006 from the University of Padova, Italy, and the PhD in Information Engineering in 2010, working on advanced nonvolatile memories. He currently holds a post-doc position working on organic semiconductor devices and on III-V MOSFETs.

Wright, William M.D.
William M. D. Wright received his B.Eng. and Ph.D. degrees in engineering from the University of Warwick, England in 1991 and 1996, respectively. He continued to work there as a postdoctoral researcher until 1997, when he joined the Department of Electrical and Electronic Engineering in University College Cork, Ireland where he is currently Senior Lecturer in Mechanical Engineering. His research interests include non-destructive evaluation, ultrasonic MEMS sensors, applications of non-contact ultrasound, tomographic imaging, and ultrasonic flow metering. He is a member of the Acoustical Society of America, and a Senior Member of the IEEE

Wu, Chunlei
I am a failure analysis engineer in Quality Department in Freescale Semiconductor (China) Limited since 2006. In 2006, I graduated form Hebei University of Technology in Tianjin China and got master degree in Theory and New Technology of Electrical Engineering.I have published 3 papers in IPFA2009 and IPFA2010.

Wu, Ernest
Ernest Y. Wu is a senior technical staff member in Technology Reliability Department at Semiconductor Research and Development Center (SRDC) in IBM Microelectronics Division in IBM System and Technology Group. He received M.S. and Ph.D. degrees in physics from University of Kansas in 1986 and 1989, respectively. Dr. Wu joined IBM Microelectronics Division in 1994 at Essex Junction, Vermont. He is responsible for technology qualification and development of dielectric reliability methodologies including advanced high-k dielectrics. Dr. Wu has served on the device dielectric committee as chair and co-chair for 2007 and 2005 International Reliability Physics Symposium (IRPS), respectively. He is a member of CMOS and Interconnect Reliability committee of International Electron Device Meeting

(IEDM) for 1999 and 2000. He has authored and co-authored more than 100 papers in technical journals and international conferences with several invited papers and tutorials as well as twelve IEDM papers. He has co-authored two books on gate dielectric reliability entitled "Reliability Wearout Mechanisms in Advanced CMOS Technologies" and "Defects in Semiconductors". In 2004, he received IBM Outstanding Technical Achievement Award for his contribution to ultra-thin gate reliability in advanced CMOS technology. His research interests include dielectric reliability physics, device physics and simulation, and carrier transport phenomena.

Wu, Kenneth
Dr. Kenneth Wu is the Director of Technology Quality & Reliability at Taiwan Semiconductor Manufacturing Company (TSMC), responsible for technology reliability characterization and qualification. Prior to join TSMC in 2002, Dr. Wu worked at Intel Corporation since 1982 in technology development and reliability qualification. Dr. Wu has published more than 50 papers on a variety of technology and reliability subjects; and received IRPS The Best Paper Award in 1990. Dr. Wu received his B.S from National Taiwan University in 1975, an M.S. from Northwestern University in 1978 and his Ph.D. from Princeton University in 1982; all are in Electrical Engineering.

Wu, Miao
I am a failure analysis engineer in Quality Department in Freescale Semiconductor (China) Limited since 2010. In 2006, I graduated form Tianjin University in Tianjin China and got doctor degree in microelectronic Engineering.

Wu, Mong Sheng
M.S. Wu, born in Chia-yi in 1982, is a engineer in UMC Corporation. Wu received a bachelor degree from National Tsing Hua University in Taiwan. Now, he has been a FA engineer for 4 years.

Wu, Tommy
Tommy Wu is an Assistant Supervisor at SGS Taiwan Ltd., Kaohsiung, working in the Material & Engineering Lab. He is also a radiation protection expert serving on the Atomic Energy Council(AEC) of Taiwan. His current research interests include radiation detection and testing of physical and chemical properties for metallic and polymeric materials.

Wu, Xing
Xing Wu was born in Xi'an, China in 1984. She received the B.E. degree in Electronic Science and Technology in 2008 from Xi'an Jiaotong University, People's Republic of China. She was an international exchange student at Osaka University from 2006 to 2007. She is currently pursuing her Ph.D. degree at the Division of Microelectronics, School of Electrical and Electronic Engineering, Nanyang Technological University (NTU), Singapore. Her current research interests include reliability study of novel high-κ dielectric materials in nanodevices using atomistic simulations and transmission electron microscopy (TEM) analysis. She is currently a graduate student member of the IEEE (2008-present).

Xia, Feng
Feng Xia received his B.S. degree in Material Science and Engineering from Tsinghua University in 1994, and Ph.D. in Electronic Materials and Devices from Xi'an Jiaotong University in China in 1999. He received 2nd Ph.D. degree in Electrical Engineering from the Penn State University, University Park in 2004. His research areas include electronic ceramics and electroactive polymers and their applications in sensors, actuators,

MEMs and microfluidic devices. Since joined Intel, he worked on reliability of polymer and flash memory, his current focus is interconnect reliability. He has more than 60 technical publications and contributed to two book chapters.

Xiong, Weize
Weize Xiong received his PhD in EE from UC Davis in 2001. From 2000 –2002, he led Cypress Semi-Conductor's 90nm CMOS development. From 2003 to 2006 he was in charge of multi-gate transistors and nano-wire transistors research. From 2007 to 2010 Wade led TI's 28nm node low power CMOS transistor development. Currently he is manager of SEMATECH Emerging Technology department. His research interests, including TFET, NEMS, RRAM and III-V devices. Dr. Xiong has authored and co-authored over 80 journal/conference papers and wrote a book chapter on multi-gate MOSFET technologies.

Xu, Xiaochen
Xiaochen Xu was born in Hangzhou, China, in 1976. He received the B.A.Sc. in Electrical Engineering from Nanjing University, Nanjing, China, in 1999, and the M.Sc. degree in Acoustics from the same university in 2002. He received his Ph.D. degree in Biomedical Engineering from the University of Southern California, Los Angeles, in 2007. His research interests included the design of beamformers for high-frequency ultrasonic arrays and high-frequency Doppler ultrasound for medical applications. Dr. Xu joined the Medical Business Unit at Texas Instruments in June, 2007. Currently he is a systems and applications engineer at TI and developing highly integrated circuit for ultrasound applications.

Xu, Yanzhong
Yanzhong Xu received B.S., M.S. and Ph.D. from Huazhong Univ. of Science and Technology, China and De Montfort Univ., England, all in electric engineering. He worked in Huazhong University as a Lecturer. Later, he worked in De Montfort University and Cypress Semiconductor. Currently, he is with Altera Corporation, in charge of device modeling and SEU. He has published about 40 conference and journal papers and holds 7 US patents.

Y, Karthik
Karthik Y completed his Bachelor of Engineering in Electronics and Communication in 2007 from Sri Jayachamarajendra College of Engineering, Mysore, India. After completing his bachelors he worked in Microelectronics Lab, ECE Department, Indian Institute of Science, Bangalore as a Project Assistant during 2007-08. Currently he is a graduate research assistant in Electrical Department, IIT Bombay pursuing his masters (Master of Technology) in Microelectronics. His research interests are semiconductor device physics, characterization, simulation and modeling of new and innovative materials and semiconductor devices for various applications.

Yamaguchi, Kosuke
received the B.E. degree in Computer and Systems Engineering from Kobe University, Hyogo, Japan, in 2009. He is currently working in the master course at the same university. His current research is low-power SRAM designs.

Yang, I-Chen
I-Chen Yang was born in Changhwa, Taiwan, R.O.C., in 1976. She received the B.S. degree in electrical engineering and M.S. degree in photonics from National Sun Yat-Sen University, Kaohsiung, Taiwan, in 1999 and 2002, respectively. She joined Macronix International Company Ltd., Hsinchu as a device

engineer in 2002. She worked on the process and device simulations of floating-gate type and nitride trap flash memory devices.

Yang, Jiaqi
Jiaqi Yang received her Bachelor's degree in Electronic Science and Technology from Dalian University of Technology, China, in 2006. She is a direct Ph.D student at the Institute of Microelectronics, Peking University, Beijing, China. Since Nov. 2009, she joined Electrical and Computer Engineering, Purdue University, West Lafayette, as a visiting Ph.D scholar. Her research interests include characterization and modeling of reliability issues in nanoscale transistors. She is currently working on NBTI/PBTI and low frequency noise in MG/HK dielectrics.

Yang, Seung Jin
1998, B.S. Yonsei Univ./ 2000, M.S. KAIST/ 2004. Ph.D of Material Sci.&Eng. KAIST / 2004. Samsung Electronics, Dev. of embedded NVM

Yan-Yu, Chen
Yan-yu Chen was born in PingTung, Taiwan in 1978. He received the B.S degree in electrical engineering from Yuan Ze University in 2002 and M.S. degree in Photonics from National Sun Yat-sen University in 2004, respectively. In 2005, he joined Macronix International Co., Ltd. (MXIC), and has been with the Device Engineering Department. Where he has engaged in the ESD/EOS protection design and characterization of NOR flash memory process technologies.

Yao, Shaoning
Shaoning Yao received her PhD degree in Materials Science from University of New Hampshire in 2006. She joined IBM Semiconductor Research and Development Center after she graduated from UNH. She has been working as an advisory engineer in BEOL integration team across 65nm, 45nm, 32nm and 22nm technology nodes since 2006. She is currently working in the area of 32nm thin wire module and process integration with alloy seed.

Yao-Wen, Chang
Yao-wen Chang was born in Kaohsiung, Taiwan in 1972. He received the B.S and M.S. degrees in electrical engineering from National Taiwan University in 1994 and 1996, respectively, and Ph.D. degree from National Tsing-Hua University in 2010. In 1998, he joined Macronix International Co., Ltd. (MXIC), and has been with the Device Engineering Department. Where he has engaged in the device design, characterization, and modeling of CMOS devices and memory cells.

Yasuda, Naoki
Naoki Yasuda received the B.S., M.S., and Ph.D. degrees in electronic engineering from Osaka University, Japan. He joined Toshiba Corporation in 1992, where he was engaged in the research of SiO2 in Si MOSFETs. From 1998 to 1999, he was visiting scientist at Rutgers University, NJ. When he returned to Toshiba, he was engaged in the research on silicon oxynitride in CMOS devices. From 2001 to 2004, he worked for MIRAI Project, Tsukuba, Japan, as researcher in High-k Gate Stack Group. Since 2005, he has been back in Toshiba to work in the area of non-volatile memory devices.

Yau, You-Wen
Dr. Yau received the B.S. degree in physics from the National Taiwan University in 1976, and the Ph.D. degrees in applied physics from Stanford University in 1983. He held various

engineering and management positions at Honeywell Solid State Electronics Division (Plymouth, Minnesota, 1983 – 1985), IBM Microelectronics Division (East Fishkill, New York, 1985 – 1993), and TSMC (Hsinchu, Taiwan, 1993 – 2008). He is with Qualcomm CDMA Technologies (San Diego, California) since 2008. He served as the IRPS Publication Committee Chair in 2004 and holds 6 US patents.

Yeo, Myung-Soo
After graduated from Hanyang University in Korea with a bachelor's degree in material engineering in 2005, He joined Samsung Electronics as an assistant engineer. He was involved in 90nm BEOL process integration for the first 3 years and then in charge of 32nm BEOL process integration for the next 2 years. He transferred to Technology Quality Reliability team in 2010 after 5 years experience in process development team. Currently, He is responsible for 28nm Interconnect and TSV Reliability.

Yeoh, Terence
Terence Yeoh received a Ph.D in Materials Science and Engineering from the University of Illinois at Urbana-Champaign in 2003. He is currently a Senior Scientist at the Aerospace Corporation, specializing in microelectronic failure analysis using the focused ion beam and transmission electron microscopy.

Yew, Tsung-Yi
Tsung-Yi Yew received M.S in EE from National Tsing-Hua University, Taiwan. He served as the engineer of Technology Quality & Reliability at Taiwan Semiconductor Manufacturing Company (TSMC). His work is focus on device and NVM memory reliability.

Yokogawa, Shinji
Shinji Yokogawa recieved the B.S.(1992), M.S.(1994), and Ph.D(2008) in engineering from The University of Electro-Communications, Tokyo, Japan. In 1994, he joined NEC Corp., Japan. He is currently with Renesas Electronics Corporation, Sagamihara, Japan. Currently, he serves as a leading engineer of advanced technology qualification in Renesas Electronics. Dr. Yokogawa is a member of the Japan Society of Applied Physics and Reliability Engineering Association of Japan.

Yokoyama, Yoshiyuki
Yoshiyuki Yokoyama started working at Hamamatsu Photonics K.K. in Japan in 2002. He has worked as development and manufacturing engineer of emission microscopy system for semiconductor. Since 2008, he has been working at Hamamatsu Photonics Deutschland GmbH in Germany. He currently serves as a technical and application support engineer on emission microscopy in Europe.

Yongshik Kim,
Yongshik Kim received the B.S. and M.S. degrees in materials science and engineering from Yonsei University in 1988 and Korea Advanced Institute of Science and Technology in 1990. He also received Ph. D. degree in applied electrical In 1990, he joined MOS process development group, Samsung electronics, starting with 1.2um OTP process development. While with Samsung, he has worked on EEPROM/SRAM/DRAM-embedded and standard CMOS logic technologies from 1.2um to 20nm. Since 2005, he has led SRAM and Design Rule teams in foundry business center. His interets are CMOS modeling, technology-aware design of SRAM and CMOS DR, statitistical yield modeling, and advanced SRAM development.

Yoo, Hyunjun

Hyunjun Yoo received B.S. and M.S. degree in Physics from Sogang University, Seoul, Korea, in 2003 and 2005, respectively. He is a currently Ph. D. candidate of Advanced Materials Engineering at Kookmin University in Seoul, Korea. His research interests include fabrication of oxide nanotubes, energy applications such as dye-sensitized solar cells, and scanning probe microscopy.

Yoo, Min-Soo

Yoo received the M.S. degree in Materials Science and Engineering in 1999 and the Ph.D. degree in the same field in 2003, respectively, from POSTECH, Pohang, KyungBuk, Korea. Since 2003, Yoo has been with the R&D Division of Hynix Semiconductor Inc. responsible for DRAM device development. Until 2007, Yoo was responsible for the development of the 90nm and 54nm DRAM cell transistor. Yoo also involved in the development of W-Gate and Fin cell transistor. His current research interests are sub 30nm DRAM device and, especially, advanced 3-D cell transistor development for the next generation.

Yoshimoto, Masahiko

received a B.S. degree in Electronic Engineering from the Nagoya Institute of Technology, Nagoya, Japan, in 1975, and an M.S. degree in Electronic Engineering from Nagoya University, Nagoya, Japan, in 1977. He received a Ph.D. degree in Electrical Engineering from Nagoya University, Nagoya, Japan in 1998. He holds 70 registered patents. He served on the Program Committee of the IEEE International Solid State Circuit Conference during 1991-1993. In addition, he has served as a Guest Editor for special issues on Low-Power System LSI, IP, and Related Technologies of IEICE Transactions in 2004.

Yoshimoto, Shusuke

received the B.E. degree in Computer and Systems Engineering from Kobe University, Hyogo, Japan, in 2009. He is currently working in the master course at the same university. His current research is soft-error tolerant and low-power SRAM designs.

Young, Chadwin

Chadwin D. Young received his B.S. degree in Electrical Engineering from the University of Texas at Austin in 1996. He then went on to receive a M.S. and Ph.D. from the North Carolina State University in 1998 and 2004, respectively. He has held several internships during this time until 2001 when he joined SEMATECH. Here, he completed his dissertation research on high-k gate stacks. He now continues this research at SEMATECH as a Member of the Technical Staff working on electrical characterization and reliability methodologies for the evaluation of high-k gate stacks, and he has authored or co-authored 150+ journal and conference papers.

Yu, Hong Yu

Prof. Yu Hong Yu is currently in the School of EEE, NTU Singapore. He received a B.Eng. degree from Tsinghua University, a M. ASc. degree from University of Toronto, and a Ph.D. degree from National University of Singapore. From 06/2004 to 01/2008, he was with IMEC (Leuven Belgium), as a senior researcher in the Si process and device technology division. He has authored or co-authored more than 190 publications in the peer-reviewed international journals and conferences, including 24 IEDM / Tech. VLSI Sym. conference presentations, >40 IEEE electron device letters. He was awarded with an inaugural Nanyang Assistant Professorship in 2008, and Tan Chin Tuan Academic Exchange Fellowship in 2009.

Yu, Joe

I am a failure analysis engineer in Quality Department in Freescale Semiconductor (China) Limited since 2006. In 2004, I graduated form Tianjin University in Tianjin China and got bachelor degree in microelectronic Engineering.

Yu, Liangchun

Liangchun C. Yu received the B.S. degree in physics from Nanjing University, China, in 2003; and the M.S. degree in material science from National University of Singapore (Singapore-MIT Alliance) in 2004; and the Ph.D. degree in electrical engineering from Rutgers University, NJ, in 2010. From 2008 to 2010, she was a guest researcher at the National Institute of Standards and Technology (NIST), where she evaluated oxide reliability of SiC MOS devices and developed a wafer-level Hall mobility measurement technique. She is currently with General Electric Global Research, Niskayuna, NY. Her research interests include reliability of semiconductor devices, development of wide-bandgap power devices and development of novel characterization techniques. Dr. Yu has authored and co-authored over 20 journal and conference papers. She served as the vice arrangements chair of the 2009 IEEE International Integrated Reliability Workshop.

Yu, Nick

Nick Yu is a Vice President of Engineering at Qualcomm's CDMA Technologies Division. He is currently responsible for setting Qualcomm's semiconductor technology roadmaps including wafer fab process node, backend interconnect and packaging technologies. He manages engineering teams that are involved with our supply chain partners for Qualcomm's chipset products. Nick has 18 years of experience with Qualcomm on low power wireless chipset and SoC development, including managing chipset design, advanced semiconductor technology, deep submicron circuit design, advanced semiconductor R&D and packaging development. He is one of the architects of many Qualcomm chipset products. Nick has an MSEE degree from Georgia Institute of Technology.

Yum, Jung Hwan

Jung Hwan Yum received the B.S degree from Korea University and M.S degree from University of Sothern California. He is currently working toward the PhD degree in University of Texas at Austin. His PhD work is the III-V MOS devices fabrication and characterization. Since 2009, He has been working as an intern at Seamtech. He has authored and co-authored over 50 journal and conference papers in the various semiconductor research areas including high-k gate oxides and III-V alternative semiconductors.

Zahid, Mohammed B

Mohammed B Zahid received the University Diploma in Electrical and Computer science in 2001 from the University of Technologies (IUT), Cachan, France. He received the Bachelor in Electrical Engineering, Master degree in Microelectronics and PhD in "Characterization of high-k layers as the gate dielectric for MOSFETs" from Liverpool John Moores University (LJMU), Liverpool, U.K, in 2002 , 2003 and 2007 respectively. He worked as a Post-Doc Researcher on the topic of characterization of Non-Volatile Memory (NVM) with alternative high-k dielectrics at the Interuniversity Microelectronics Center (IMEC) with the group of Device Reliability and Electrical Characterization (DRE), Leuven, Belgium. His currently working in the field of High Power based GaN devices and his research interest includes the technology, Novel electrical characterization techniques and reliability for NVM, flash memory and GaN High Power devices

Zaknoon, Bashir
Bashir Zaknoon is a product reliability engineer at Intel Corporation. He has a B.S. and M.Sc. in Electrical Engineering from the Technion - Israel Institute of Technology. During his M.Sc. studies, Bashir's research focused on characterization of the quantum electronic properties of silicon nanocrystals. Currently, Bashir's work includes lab-based and simulation aspects of product reliability.

Zambelli, Cristian
Cristian Zambelli received the Laurea Triennale degree (B.Sc.) in Electronic and Telecommunications Engineering and the Laurea Magistrale degree (M.Sc.) in Technology for Telecommunications and Electronic Engineering from the Università degli Studi di Ferrara respectively in 2006 and 2008. From 2009 he started pursuing the Ph.D. title in Engineering Science within the same institution. His research interests are focused on the electrical characterization, physics and modeling of non-volatile memories such as PCM (Phase Change Memories), NOR/NAND Charge Trapping Flash and MEMS-based memories. Recently he started to work on memory hardware/software co-design and ad-hoc algorithmic solutions for smart memory management.

Zambelli, Cristian
Cristian Zambelli received the Laurea Triennale degree (B.Sc.) in Electronic and Telecommunications Engineering and the Laurea Magistrale degree (M.Sc.) in Technology for Telecommunications and Electronic Engineering from the Università degli Studi di Ferrara respectively in 2006 and 2008. From 2009 he started pursuing the Ph.D. title in Engineering Science within the same institution. His research interests are focused on the electrical characterization, physics and modeling of non-volatile memories such as PCM (Phase Change Memories), NOR/NAND Charge Trapping Flash and MEMS-based memories. Recently he started to work on memory hardware/software co-design and ad-hoc algorithmic solutions for smart memory management.

Zhang, Enxia
Enxia Zhang received her Ph.'D from Shanghai Institute of Microsystem and Information Technology (SIMIT), China Academic of Sciences in 2006. She joined Shanghai University of Engineering Science in 2007 as an associate professor. In 2008, she joined Radiation Effect and Reliability Research Group at Vanderbilt University as a research associate. She is currently s a research assistant professor in Electrical Engineering and Computer Science Department at Vanderbilt University.

Zhang, Enxia
Enxia Zhang received her Ph. D from Shanghai Institute of Microsystem and Information Technology (SIMIT), China Academic of Sciences in 2006, and joined Shanghai University of Engineering Science in 2007 as an associate professor. In 2008, she joined the Radiation Effect and Reliability Research Group at Vanderbilt University as a research associate, and now is a research assistant professor of the Department of Electrical Engineering and Computer Science at Vanderbilt University.

Zhang, Fei
Fei Zhang received the B.S. and Ph.D. degrees from the Department of Physics, Wuhan University, Wuhan, China, in 2002 and 2007, respectively. Currently, he is with the Department of Electrical and Computer Engineering, Michigan State University, East Lansing, USA. His research interests include biomedical sensors, signal and image processing, semiconductor devices and RF circuits. He is author or coauthor of over 70 scientific publications in peer-reviewed journals and conference proceedings.

Zhang, Lijie
Lijie Zhang received the B.S. (with highest honors) in Hebei University of Technology, Tianjin, China, in 2006. Currently she is pursuing her PhD degree in Peking University, Beijing, China, and will graduate in July 1st, 2011. Her major is Microelectronics. Her research interests include nonvolatile memory, power devices, device reliability.

Zhao, Kai
Kai Zhao received the bachelor's and master 's in physics from University of Science and Technology of China in 2000 and 2002 respectively and the Ph.D. from Physics department at University of California, San Diego in 2008. In the same year, he joined IBM research division as a postdoctroal researcher, working in the reliability and characterization group. At IBM research, he is focused on understanding the fundamental reliability aspects of advanced Field Effect Transistors, including HKMG FETs and Carbon Nanotube Transistors. In 2010, Kai transferred to IBM SRDC, where he is working on SOI technology development.

Zhoa, Larry
Larry Zhao joined Intel Corporation in 2002 and is currently working at IMEC as an assignee from Intel. He has more than 10 years of experience in the research and development of Cu interconnects and holds 6 U.S. patents. He has also co-authored more than 50 scientific publications. Larry Zhao received his B.S. degree in Materials Science from Shanghai Jiaotong University in 1984 and M.S. degree in Engineering from Dartmouth College, Hanover, NH in 1993. He studied for Ph.D. at Columbia University, NYC from 1993 to 1997.

Zhou, Xing
Xing Zhou received the B.E. degree from Tsinghua University in 1983 and the M.S. and Ph.D. degrees in electrical engineering from the University of Rochester in 1987 and 1990, respectively. Since 1992 he has been a faculty of the School of Electrical and Electronic Engineering, Nanyang Technological University, Singapore. He was a visiting professor with Stanford University in 1997 and 2001 and with Hiroshima University in 2003. He is a distinguished lecturer of the IEEE EDS and an editor for the IEEE Electron Device Letters. His research focuses on development of compact models for circuit simulation for nanoscale MOS devices.

Zhu, Xiaowei
Xiaowei (Vivian) Zhu received the B.S. (1994) and M.S. (1997) in Applied Physics from South China University of Technology, Guangzhou, China, and the M.S. (2000), Ph.D (2002) in Electrical Engineering from Vanderbilt University, Nashville, Tennessee. She joined the silicon technology development group of Texas Instruments, Inc. in 2002, and she worked in the area of characterization and modeling of radiation induced soft error rates and statistical modeling of gate oxide reliability in advanced CMOS technologies from 2002 to 2007. She joined high reliability and medical business group within TI in Sep 2007, since then her focus has been the quality and reliability in medical ICs. Dr. Zhu has authored and co-authored several papers in the field of electronic device reliability, presented at her work at major conferences, including IEDM, IRPS and NSREC, and she has contributed one chapter in JEDEC soft error rate testing specification (JESD-89). Dr. Zhu is currently holding 3 US patents.

Zous, Nian-Kai

Nian-Kai Zous received the Ph.D. degree in electronics engineering from National Chiao-Tung University (NCTU), Hsin-chu, Taiwan, R.O.C., in 2002. He joined the Macronix International Co., Ltd. (MXIC) in Hsinchu, Taiwan, in 2003, working in the advanced device group of the Device Technology Division and his research activities include the study of reliability and scaling issues in nitride storage devices. From 2005 to 2007, he is working in the NBIT Flash Tech. Dept. II of MXIC and is responsible for the reliability issue on NBIT product. Since 2007, he has been engaged in the methodologies for technology development and yield enhancement .

Post-Breakdown Statistics and Acceleration Characteristics in High-K Dielectric Stacks

E. Wu, J. Suñé*, B. Linder**, R. Achanta, B. Li, and S. Mittl

IBM System and Technology Group, Essex Junction, VT (USA)
*Departament d'Enginyeria Electrònica, Universitat Autònoma de Barcelona (SPAIN)
** IBM Research Division, Yorktown Heights, NY USA)

Abstract

Contrary to recent claims, experimental results obtained in thin and thick Hafnium-based high-K gate dielectric stacks demonstrate that progressive breakdown is relevant in these insulators. For thin and thick stacks and both in NFETs and PFETs, the residual time distributions are found to be non-Weibull with two regions: a universally shallower slope at long times and a steeper slope at short times. The shallow distributions favour the coexistence of single-spot BD and multiple competing spots in different samples. Contrary to what happens in the case of SiON dielectrics, the final failure distribution is reported to be strongly dependent on the threshold current I_F used to define device failure. Also contrary to what found for SiON single-layer dielectrics, the voltage acceleration and temperature activation energy of the residual time is reported to be much stronger than that of the first breakdown time. All these results emphasize the important role of progressive breakdown for high-K reliability assessment methodology.

I. Introduction

The topic of non-Weibull failure distribution remains controversial for high-K dielectric stacks [1,2] since some authors have ruled out the existence of progressive breakdown (PBD) and have attributed the bending at low percentiles (LP) exclusively to the fact that the defect generation rates in the layers of the stack are very different [1]. In this regard, a recent formulation of the percolation model for stack dielectrics shows that the first breakdown (BD) distribution would be non-Weibull if the defect generation rate in the high-K layer is different from that in the interfacial layer, with larger Weibull slope at LP [1,3]. Nevertheless, while the validity of percolation model is well established, it is worthwhile to point out that failure-current based TDDB methodology including PBD is a more general formulation for the treatment of dielectric failure as illustrated in Appendix A. In this context, the first BD framework is only a special case with PBD time being negligibly small. Consequently, as discussed in the Appendix, it is impossible to conclusively rule out the existence of progressive BD as claimed in [1].

In the case of SiO_2-based single-layer thin gate oxides, it is well established that considering the final failure distribution which includes PBD is crucial for circuit reliability [4] because first BD events do not necessarily destroy circuit operation. For high-K stacks, a recent report has shown that the 1st BD model prediction fails to explain product burn-in data with a significant discrepancy, thus indicating that the first BD

methodology is unrealistic [5]. In this work, we present a comprehensive investigation of post-BD in high-K stacks. While we have found that the LP bending effect due to first BD is not always pronounced at least for the data reported here, PBD and multiple BD are unequivocally observed in thin and thick high-K stacks and for all stress conditions investigated. We have also found out that the progressive BD time distribution is non-Weibull for high-K stacks with a shallow region at high percentiles (HP) and a rapid decrease at low percentiles. The flat region is much shallower than those reported for SiON films, exhibiting very small slopes ($\beta_{PBD}<1$). This observation is found to be universal and it can explain why PBD might be sometimes apparently absent [1]. By means of long-term module stress (~ 4 months), we report an increased voltage acceleration factor and temperature activation energy for the residual time, T_{RES}. In contrast to what is found for SiON films, voltage acceleration is found to be larger for T_{RES} than for the time to first BD, T_{BD} for high-K stacks. Our results are critical to eventually resolve the discrepancy between product data and 1st BD model prediction reported recently for high-K stacks [5] with the implementation of a realistic circuit TDDB methodology.

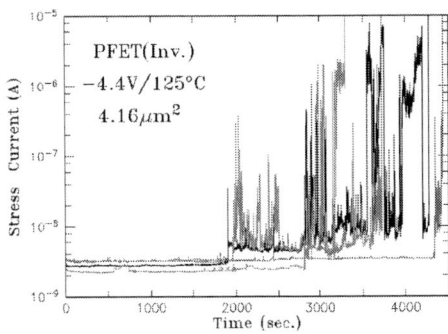

Fig. 1. Typical current transients measured during constant-voltage stress of pFETs with *thick* gate stack dielectric. The blue and red lines represent single PBD events with $t_{PBD}/t_{BD}<1$ while the pink and black lines reveal the multiple competing PBD spots due to $t_{PBD}/t_{BD}>1$.

II. Evidence of progressive breakdown in high-K gate dielectric stack.

In this work, two sets of Hf-based high-K stacks with different equivalent thickness are considered. We reference them as *thick* and *thin*, corresponding to an interfacial layer thickness of ~20Å and ~10Å and a high-K layer thickness >25Å and <25Å, respectively. Fig. 1 shows the typical current transients for *thick* stack PFETs stressed in inversion at 125°C. Due to the

978-1-4244-9113-1/11 $26.00 © 2011 IEEE

reduced leakage currents through these *thick* high-K stacks, a clear identification of the first BD event is possible and after this, a pronounced PBD phase can be observed. Figure 2 displays the individual distributions of first BD times, the times-to-failure specified at $I_{FAIL}=0.1\mu A$, and the residual time, $t_{RES}=t_F(I_{FAIL})-t_{BD}$, for *thick* PFET inversion. The t_{RES} distribution associated to PBD deviates from the Weibull model at LP while at HP a very shallow slope ($\beta\sim0.3$) can be seen over many orders of magnitude in time, in comparison to what is usually found in SiON gate dielectrics as discussed later.

The first BD distributions normalized to the area of $4.16\mu m^2$ by Poisson area scaling are shown in Fig. 3. These data clearly show a single Weibull distribution capable of fitting a large range of percentiles, $Ln(-Ln(1-F))$, varying from -12 to 2. Only a very limited number of data points below -12 show some bending at LP, probably due to dual-layer percolation statistics [1-3].

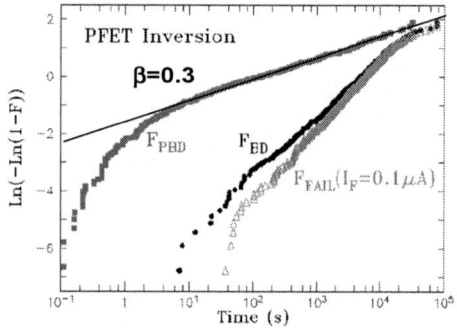

Fig. 2. Measured first BD (F_{BD}), final failure (F_{FAIL}) and residual time ($F_{RES}=F_{PBD}$) distributions for *thick* PFETs with an area of $4.16\mu m^2$ stressed under constant-voltage conditions at 125°C. A local current change of $\Delta I=I(t+dt)-I(t)$ of 0.6nA is used for 1st BD detection. A failure current (I_F) of $0.1\mu A$ is chosen for single-spot PBD detection. Single spot is ensured by the fact that $T_{PBD}\ll T_{BD}$ in this particular case.

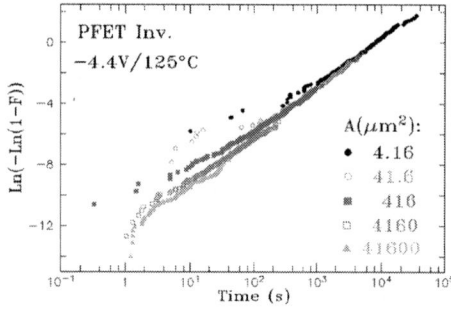

Fig. 3. The normalized first BD distributions of *thick* pFETs following Poisson area scaling with some indication of the bending at low percentiles as suggested by the percolation model of dual layer statistics [1-3]. The reference area is $4.16\mu m^2$. The data is obtained by proper screening to eliminate non-Poisson area scaling at high percentiles caused by sample-to-sample current variations.

At this point it is worth to precisely define the PBD time and the experimentally measurable residual time. The PBD time (t_{PBD}) for a particular sample is defined as

the time required for the current of a particular BD spot to reach the threshold current that causes the device failure (I_F). On the other hand, the residual time (t_{RES}) has been defined as the time elapsed between the first BD detection (t_{BD}) and the final device failure (t_F). If a single BD spot is created and the current of this spot progressively grows and finally causes the device failure, the residual time and the PBD time coincide for all the samples and thus the associated distributions also coincide as in Fig. 2, $F_{RES}=F_{PBD}$. The first BD times are shown to be uncorrelated with PBD time as given in Fig. 4, suggesting that defect generation process in the 1st BD phase is independent of progressive BD phase.

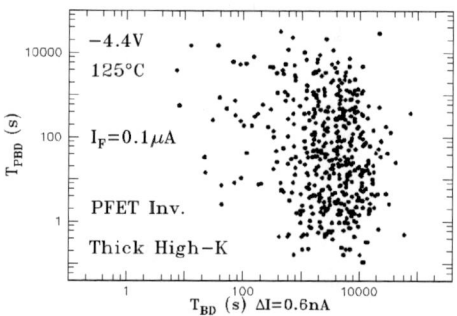

Fig. 4. Progressive BD times vs. 1st BD times suggesting the defect generation phase (1st BD) and post BD phase are independent.

Fig. 5 depicts the I-t traces measured in *thin* PFETs with smaller area. As can be seen, the magnitude of leakage jump due to 1st BD can be very small, down to ~1 μA or lower. The corresponding t_{RES} and t_{FAIL} distributions are shown in Fig. 6 (a) and (b) revealing a failure current dependence of the PBD distributions and time-to-failure distributions consistent with the PBD effect. The larger is the current level that defines the device failure, the longer is the residual time for the first BD spot to reach this current level.

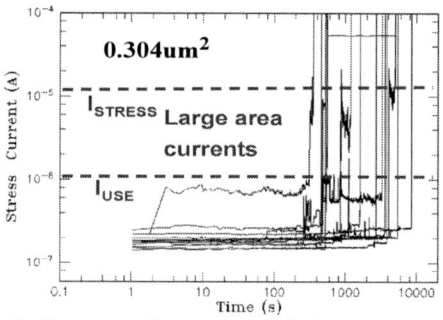

Fig. 5. Current transients measured during stress at -2.5V and 125°C for *thin*-gate-stack pFETs ($0.304\mu m^2$). The dashed lines represent the current levels measured at stress and use conditions in larger areas as discussed in the text in relation to Fig. 7. Notice the small jump due to 1st BD indicating the detection difficulties.

The current-transients of *thin* pFETs measured on larger areas are shown in Fig. 7 (a) and also display a clear PBD phase. In addition, Fig. 7 (b), which shows the same data in a linear-linear plot, clearly revealing that $t_{BD}\ll t_{PBD}$ so that multiple BD spots might compete during the PBD phase according to the successive BD theory [6,7]. It is well known that larger areas favour the occurrence of multiple BD events while single-spot BD tends to dominate for smaller areas. This is because the PBD time does not change with area while the scale factor of the first BD distribution (T_{BD}) increases when the area is reduced. As illustrated in Fig. 5, the current levels from large areas are much higher than the small jumps of 1st BD evens. Even at low use-voltages (V_{USE}), *ie*. operating voltages, the current measurement may not reliably detect the first BD events which remain hidden under the background level.

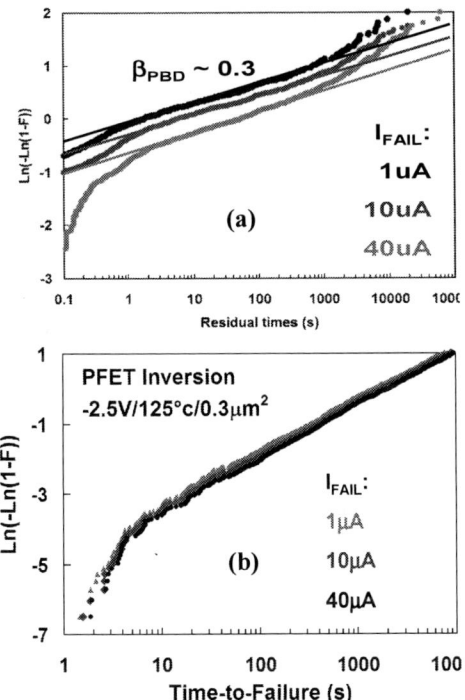

Fig. 6. Experimental residual time distributions **(a)** and time-to-failure distributions **(b)** measured in *thin* pFETs with $0.304\mu m^2$ stressed at -2.5V and 125°C as a function of failure currents.

Nevertheless, the time-to-failure (t_{FAIL}) at specified failure currents is more relevant than t_{BD} for circuit applications in microelectronics products. To this end, we have investigated the area scaling properties from large ($\sim160\mu m^2$) to small areas ($\sim0.1\mu m^2$) for *thin*-stack pFETs and we have found the t_{FAIL} distributions at a specified current (I_F), always follow Poisson area scaling law as shown in Fig. 8, in agreement with basic theory and with previous results reported for SiO_2-based insulators [7]. This means that the statistics of multiple competing BD events or those of single-spot BD (despite the difficulties of detecting the first BD) both obey the weakest-link characteristics as a universal scaling property, thus simplifying the TDDB evaluation with structures of different areas. It is worthwhile mentioning that even for small areas, multiple competing BDs cannot be completely excluded (Fig. 5). It is well established that the current variation effect due to thickness non-uniformity can cause non-Poisson area

scaling at high percentiles for SiO_2 dielectrics in T_{BD} distributions while Q_{BD} distributions are less prone to this effect [8]. Similarly, to extend the Q_{BD} approach to high-K stacks, we define $Q_{FAIL}=T_{FAIL}xJ_0$ where J_0 is the current density at t=0. As shown in Fig. 8, not only do Q_{FAIL} eliminate non-Poisson area scaling at high percentiles but it also yields a steeper Weibull slope at LP. These results reveal the important role of current variation, and the effectiveness of the Q_{BD} (or Q_{FAIL}) methodology.

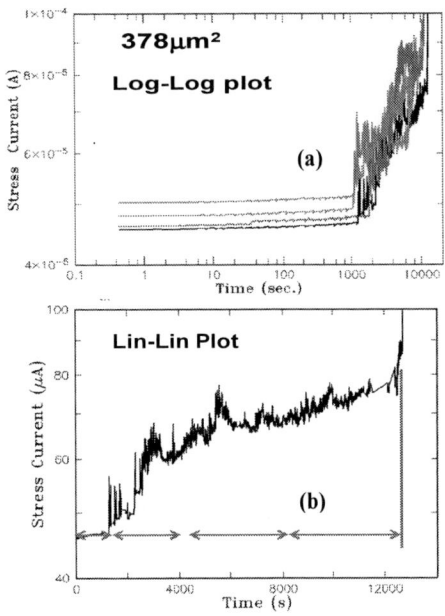

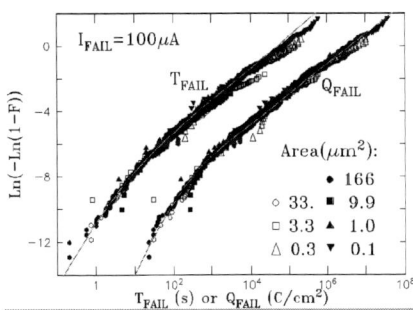

Fig. 7. (a) Current transients measured on *thin*-stack PFETs with large areas at -1.8V and 170°C, clearly revealing the PBD phase. **(b).** Linear-linear plot revealing very long residual times between first BD and time-to-failure (I_F=100μA) suggests the occurrence of multiple BD events as dictated by the successive BD statistics [6,7].

Fig. 8. The scaled T_{FAIL} and Q_{FAIL} distributions (I_F=100μA) using Poisson statistics showing a universal overlay of distributions from larger area to very small areas despite the 1st BD detection difficulties associated with larger areas. $Q_{FAIL}=T_{FAIL}xJ(t=0)$. The red lines represent the fit of using the 5-parameter proposed in Ref. [9]. The obtained values of LP beta are 1.2 and 2.4 for T_{FAIL} and Q_{FAIL}.

The I_{FAIL} (or I_F) dependence of the T_{FAIL} distributions is shown in Fig. 9. In this case, a small change of failure current from 60μA to 70μA, is found to cause the T_{FAIL}

978-1-4244-9113-1/11 $26.00 © 2011 IEEE

distribution (measured for ~1000 samples) to change from single-Weibull to non-Weibull distribution with a large β_{LP} of ~2.4. Such strong sensitivity to the fail current contrasts to that found in SiON single-layer dielectrics as compared in Fig. 10 (a). In SiON, it was found that the low-percentile T_{FAIL} shows a very weak dependence on the failure current whereas for high-K dielectrics T_{FAIL} significantly increases with the increase of I_F. The current-transients in Figs. 1, 5, and 7 and the strong failure-current-dependence of T_{RES} and T_{FAIL} distributions (Figs. 6 and 10) provide direct evidence that the PBD phase can be unequivocally identified.

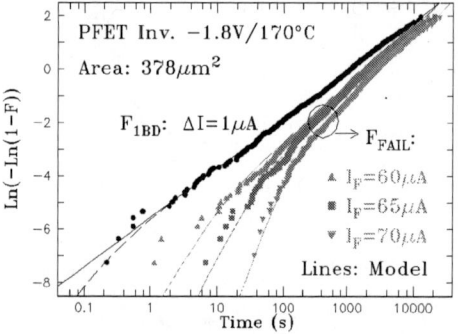

Fig. 9. The final failure distributions as a function of failure current for *thin*-stack PFETs. The first BD data was fitted by both a single Weibull (black solide line) and a percolation model for dual layer statistics (black dashed line) using the analytic model in [3] while the final failure distributions were fitted with a model including PBD [9].

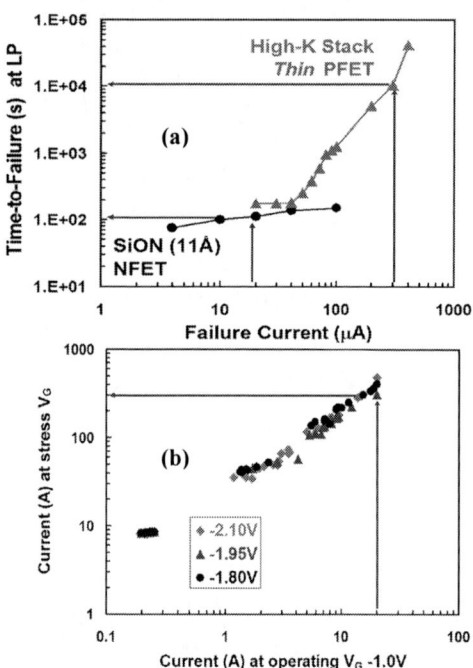

Fig. 10. Low-percentile T_{FAIL} vs. failure current for high-K stacks as compared to that of SiON [4] **(a)**. The strong I_F dependence coupled to the relation of currents measured at stress-voltage and use-voltage **(b)** yields extra TDDB margin in contrast with SiON due to its weak sensitivity to I_F. The data are taken from the samples used in Fig. 14.

Moreover, the failure currents obtained from circuit simulation and failure analysis [4] are generally specified at operating-voltages which are much lower than stress voltages. To account for this difference, Fig. 10 (b) shows the relationship between stress currents and operating-voltage currents, indicating that I_F ~20µA at V_{USE} corresponds to about I_F ~400µA at the stress voltage. In the SiON case, this difference yields only a 2X change in the time to failure while for the high-K stack dielectrics, the coupling effect leads to a 100X improvement in the LP T_{FAIL} estimation due to the observed post-BD effect. This provides a significant increase in reliability margin from the perspective of circuit reliability.

We have conducted similar investigations of PBD distributions for nFETs using *thin* and *thick* high-K stacks and the experimental results are compiled in Fig. 11. At long times or high percentiles, very broad T_{RES} distributions (β~0.3) are observed for all the studied high-K stacks, thus suggesting that this is a universal result. At one limit, for very short t_{PBD}, the first BD events tend to appear as very hard events and one might conclude that PBD is absent. At other limit, with very long t_{PBD} (>>T_{BD}), multiple competing BD events tend to dominate.

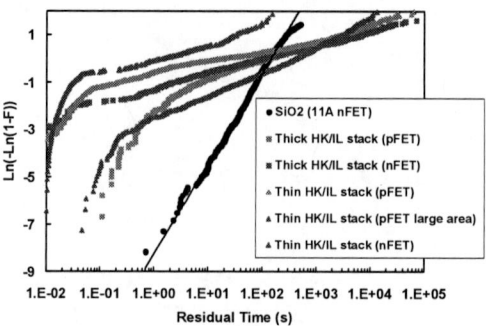

Fig. 11. Comparison of PBD distributions of thin and thick stacks of PFETs and NFETs with NFET data of SiON. An universal observation of a very shallow PBD distribution, i.e. very small Weibull slope (β_{RES} ranging from 0.3 to 0.6) while the Weibull slope of T_{PBD} for SiON dielectrics is ~1.6

Until now, we have been discussing on the shallow region of the distributions (Figs. 2, 6(a), and 11) of both PBD time and residual time to failure, assuming a Weibull distribution. Furthermore, it is know that at low percentile limit, single-spot BD dominates [7] with $\beta_{FAIL}=\beta_{1BD}+\beta_{PBD}$ [12]. It is unclear how to explain a value of ~2.4 for β_{FAIL} with a shallower slope of β_{PBD} ~ 0.3. On the other hand, experimental data in these figures clearly suggest that the Weibull distribution model is not adequate for progressive breakdown times. For SiON films, it has been shown [10] that the PBD times are Weibull-distributed at low failure currents and then gradually approach the lognormal distribution at high failure currents as the BD spots continue to grow. A phenomenological PBD model was developed and provided much insight in the case of SiON dielectrics [10,11]. No work has been reported regarding the validity of PBD time distributions for high-K stacks,

978-1-4244-9113-1/11 $26.00 © 2011 IEEE

either experimentally or theoretically. Nevertheless, the choice of Weibull distributions can have significant impact on TDDB reliability projection as discussed in Fig. 12. The area ratio involved in the TDDB projection is the ratio between the relatively small device areas used in stress experiments and the total gate dielectric area in the entire chip. Depending on the area ratio, the choice of one distribution or the other can have an effect of as many as four orders of magnitude in time or more for large area ratios.

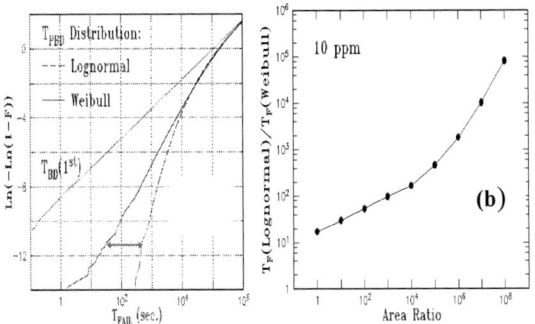

Fig. 12. (a). The final failure distributions for small dimension devices obtained from MC simulation considering a Weibull and a Lognormal distribution of t_{PBD}, the projection to low percentiles strongly depends on the distribution of choice. **(b).** Ratio of the failure times projected to low failure percentiles (10 ppm) using either lognormal and Weibull function of t_{PBD} distribution. Area ratio of 1 corresponds to the unit device of small dimension as shown in (a). In this figure, a single Weibull function for 1st BD is used for simplicity. A non-Weibull distribution of 1st BD according to dual layer statistics [1,3] can be also made.

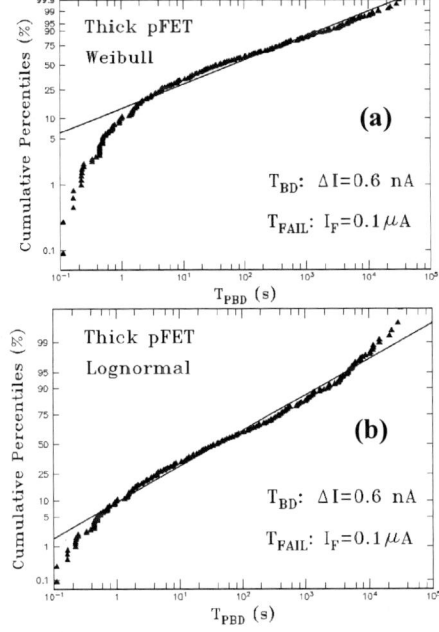

Fig. 13. Progressive BD times are plotted in Weibull **(a)** and lognormal **(b)** distributions. The lines represent the maximum likelihood fit with 95% confidence bound.

Fig. 13 represents the same PBD statistical data of Fig. 2 in the Weibull and Lognormal functional plots for comparison. As discussed earlier, this is the case of single spot BD ($T_{PBD}<T_{BD}$) for $T_{PBD}=T_{RES}$. It is evident that the PBD distribution is non-Weibull as indicated by the deviation at failure percentiles below ~25%. On the other hand, the use of the lognormal distribution gives a better fit over a much wider range of percentiles although at both ends some deviations can be seen. These results clearly reveal that a single Weibull distribution cannot provide a realistic description for progressive BD times, consequently its pessimistic TDDB projection as pointed out in Fig. 12 remain questionable.

III. Voltage and Temperature Acceleration

Voltage-acceleration and temperature-acceleration are the key issues for reliability assessment methodology and technology qualification. The voltage acceleration of "first BD", residual time and final time-to-failure are shown in Fig. 14 while "first BD time", residual time, and final time-to-failure are given in Fig. 15 as a function of temperature. Both T_{RES} and T_{FAIL} show much higher accelerations than that of first BD. It is also interesting to note that at high voltages or temperatures, T_{BD} values approach those of T_{FAIL}. However, at low voltages or temperatures, the residual times become comparable to those of T_{FAIL} as post-BD phase plays much dominant role. In the case of SiON films, comparable voltage acceleration factors were found for 1st BD and for PBD times and this was interpreted as an indication that the same mechanisms (defect generation) are responsible for both degradation phases [7].

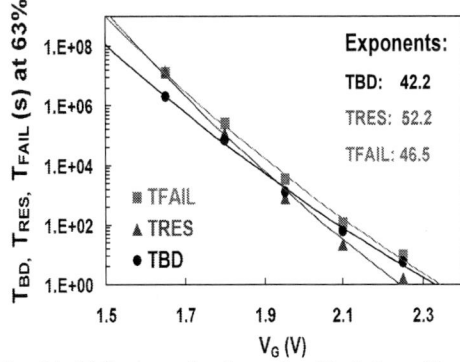

Fig. 14. Voltage acceleration of residual time (T_{RES}), failure time (T_{FAIL}) and first BD time (T_{BD}) at $I_F=40\mu A$ for thin pFET inversion. The results demonstrated a much larger voltage acceleration of PBD (implying a larger acceleration of T_{RES} and T_{FAIL}) than that of "first BD". The area is $48\mu m^2$ and stress temperature is $170°C$.

In contrast, our results for high-K stacks suggest that defect generation processes in post-BD phase can be different from those responsible for the generation of the percolation path that triggers the first BD. This might be an indication that the first BD in controlled by one layer of the stack (probably the interfacial layer, where the defect generation rate is known to be much smaller) and the PBD phase might correspond to propagation of the

978-1-4244-9113-1/11 $26.00 © 2011 IEEE

damage in the high-K layer. In this regard, we have recently shown that, assuming the validity of the percolation picture of stack dielectrics [1,3], the cumulative distribution of the propagation of damage into the high-K layer is quite analogous to the PBD phase and that the slope of the propagation time is very broad due to the damage induced in the high-K layer during the formation of the percolation path in the interfacial layer [13] .

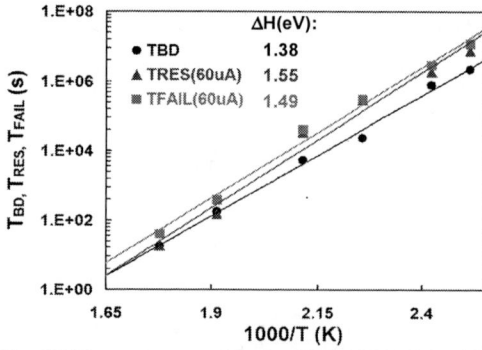

Fig. 15. Temperature acceleration (150°C-290°C) for thin pFET inversion of residual time and time-to-failure as well as "first BD" time, showing a much larger activation energies for T_{RES} and T_{FAIL} than that of "1st BD". The area is 370μm².

IV. Conclusions.

It has been demonstrated that progressive BD with both single-spot and multiple-spot BD modes is a universal phenomenon and plays a dominant role in high-K TDDB reliability, contrary to what recently claimed [1]. Unlike SiON single-layer films, high-K stacks exhibit very wide T_{RES} (T_{PBD}) distributions (with Weibull slopes between 0.3 and 0.6 at high percentiles). These wide residual time distributions are responsible for the observation of these two BD modes: single spot BD at low failure percentiles (short times to failure) and stronger competition among several BD spots at higher percentiles (longer times to failure). More importantly, we have found either PBD or residual time distributions are non-Weibull or better fitted with a lognormal distribution. In contrast to SiON, the voltage acceleration factor and temperature activation energy of T_{RES} are found to be much larger than those of the time to first BD, thus emphasizing the dominant role of PBD in high-K TDDB reliability assessment. All these findings are critical to resolve the discrepancy between product stress data [5] and the first BD TDDB methodology.

Appendix A

In Figure A, we show the first BD distribution (F_{BD}) and the final failure distribution (F_{FAIL}) for different ratios of T_{PBD} to T_{BD}. As can be seen, the transition point for the change in Weibull slopes moves towards low percentiles or short times as this ratio becomes smaller. As the ratio T_{PBD}/T_{BD} decreases, more and more samples are required to discern these two distributions. At $T_{PBD}/T_{BD}=10^{-10}$, it requires at least one trillion samples to

discern the F_{BD} and F_{FAIL} distributions at $W=Ln(-Ln(1-F))= -27$. This illustrates that it is practically impossible to rule out the existence of PBD. On the contrary, the first BD distribution can be always viewed as a special case with T_{PBD} being negligibly small so that the F_{BD} and F_{FAIL} distributions coincide for certain range of percentiles. Although the Weibull distribution is assumed for both F_{BD} and F_{PBD} to obtain F_{FAIL} for simplicity of this illustration, this is a general conclusion for any choice of distribution.

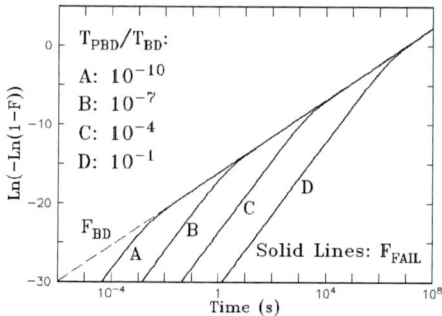

Fig. A. Final failure distributions (F_{FAIL}) for several values of the T_{PBD}/T_{BD} ratio. For simplicity, Weibull functions are used for F_{BD} and F_{PBD} distributions with $\beta_{BD}=\beta_{PBD}=1$ to obtain the F_{FAIL} distributions in solving the convolution integral. The dashed line is for F_{BD} with $T_{BD}(63\%)=10^7 s$.

Acknowledgements

J. Suñé acknowledges funding by the MICINN under contract TEC2009-09350 (partially funded by the EU FEDER program), and by the DURSI of the Generalitat de Catalunya under project no. 2009SGR783. J.S. also acknowledges the ICREA ACADEMIA award and IBM funding. The experimental work and hardware fabrication was performed by the Research Alliance Teams at various IBM Research and Development facilities.

References

[1]. T. Nigam, A. Kerber and P. Peumans, IRPS, p.523-530, 2009.
[2] G. Bersuker, N. Chowdhury, C. Young, D. Heh, D. Misra, and R. Choi, IRPS, pp. 49-54, 2007.
[3] J. Suñé, S. Tous, and E.Y. Wu, IEEE Electron Device Letters, 30 (12), pp. 1359, 2009.
[4] E. Wu, G. Braceras, D. Turner, A. Swift, M. Johnson, J. Suñé, S. Tous, and M. Khare, IEDM pp.397-400, 2009
[5] S. Pae, et al. IRPS, pp. 287-292, 2010.
[6] J. Suñé and E. Y. Wu, IEEE Electron Device Letters, vol. 24, pp. 272-274, 2003.
[7] E. Wu and J. Suñé, IRPS, pp. 54-62, 2006
[8] E. Wu, E. Nowak, R. Vollertsen, and K. Han, IEEE Tran. Electron Device, vol. 47, n 12, pp. 2301-9, 2000.
[9] J. Suñé, E. Wu, and S. Tous, IEDM pp. 86-89, 2010.
[10] E.Wu, S. Tous, and J. Suñé, IEDM pp. 493-496, 2007.
[11] J. Suñé, E. Wu, and S. Tous, INFOS-2007, Micro. Eng. 84, pp.1917-1920, (2007).
[12] E. Wu and J. Suñé, IRPS, pp. 36-45, 2006
[13] S. Tous, E. Y. Wu, E. Miranda, and J. Suñé, (unpublished).

978-1-4244-9113-1/11 $26.00 © 2011 IEEE

Methodologies for sub-1nm EOT TDDB evaluation

Thomas Kauerauf, Robin Degraeve, Lars-Åke Ragnarsson,
Philippe Roussel, Sahar Sahhaf, Guido Groeseneken*

imec
Kapeldreef 75, B-3001 Leuven, Belgium
phone: +32 1628 1148, e-mail: kauerauf@imec.be
* also at KU Leuven, ESAT Department, Leuven, Belgium

Robert O'Connor
Tyndall National Institute, Lee Maltings, Prospect Row
Cork City, Ireland

Abstract – **Measuring and understanding TDDB reliability in sub-1nm EOT dielectrics is both a practical and scientific challenge. We present three different methods for the experimental determination of the SBD and wearout parameters needed to construct an all-in-one TDDB reliability prediction consisting of a SBD-free region, a leakage current-dominated region and a HBD-limited region. We demonstrate these methods on several sub-1nm EOT high-k/metal gate nMOS and pMOS devices and evaluate their advantage and disadvantages. We also discuss the validity and interpretation of the SBD/wearout model, confronting it with experiments that demonstrate how SBD paths can be annealed by reversing the stress polarity.**

Keywords-component: MOS, TDDB, reliability, high-k, gate dielectric, soft breakdow,hard breakdown, Weibull slope

I. INTRODUCTION

For sub-1nm ultra thin (UT) EOT high-k/metal gate MOS devices, the conventional time-dependent dielectric breakdown (TDDB) lifetime extrapolation method based on soft breakdown (SBD) after applying constant voltage stress (CVS) is challenged for mainly two reasons: First, due to the low breakdown voltage and the low time-to-breakdown (t_{BD}) Weibull slope β_{tBD}, the extrapolated SBD maximum applicable voltage for 10 years is significantly below the operating voltage [1, 2]. In Fig. 1 this is illustrated for pMOS, where the SBD lifetimes of various high-k/metal gate combinations over a wide EOT range clearly confirm the trend of insufficient TDDB lifetime for UT-EOT devices. So in order to guarantee the 10 years TDDB device lifetime at V_{DD}, the wearout phase (WO) [3] before final hard breakdown (HBD) [4] has to be included in the extrapolation (Fig. 2). Secondly, already the determination of the SBD parameters from the I-t traces is not trivial. For nMOS devices the gate leakage tunneling current J_G is intrinsically so high and the current increase through the generated percolation path is relatively low, and it is almost impossible to observe SBD and to trigger on it. This can be solved by periodically interrupting the stress and switch to a lower sense voltage as suggested by some researchers [5, 6]. For pMOS, J_G is less of an issue [7], but the pre-HBD features in the gate current are also much smaller in magnitude and

hardly detectable. For both nMOS and pMOS, the selection of the adequate SBD trigger criterion is therefore of crucial importance. In this paper, we present possible methodologies to extract the soft and hard breakdown parameters in UT-EOT oxides. We demonstrate these with examples and elaborate on their physical understanding and statistical framework.

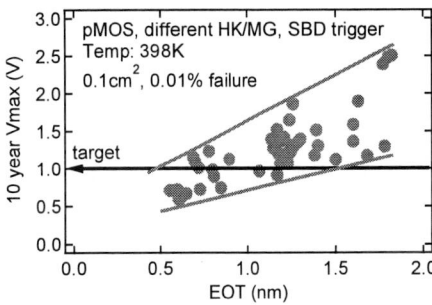

Fig. 1. The extrapolated 10 years maximum applicable voltage for different pMOS high-k gate stacks when triggering on soft breakdown. Below 1nm EOT the V_{DD} target is not met and soft breakdown will occur within the 10 years lifetime.

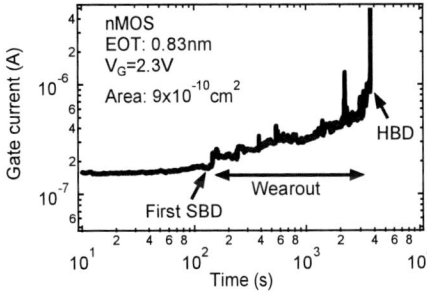

Fig. 2. Gate current vs. time of an nMOS under CVS. The first SBD occurs after 150 s, initiating the wearout phase of multiple SBD's and enhanced localized degradation. The dielectric looses its isolating properties when HBD is formed.

978-1-4244-9113-1/11 $26.00 © 2011 IEEE

This paper is divided into three main parts: In the first part, section II: "The all-in-one TDDB lifetime extrapolation", we present a slightly modified version of the TDDB lifetime extrapolation model previously presented in [8] which combines both SBD and HBD. With the help of some example extrapolations we motivate why such a 6-parameter analysis is absolutely necessary for sub-1nm EOT devices. Finally some computational guidance is given.

In the second part, section III: "Reliability parameter extraction" we present three different methods to experimentally obtain the six intrinsic SBD and HBD reliability parameters necessary for a complete sub-1nm EOT high-k MOS TDDB lifetime extrapolation. We discuss the requirements for each method as well as their practical advantages and disadvantages.

In section IV, we discuss the validity of the soft breakdown/ wearout model in the context of recent publications that challenge this concept.

Note that all measurements in this work were done in inversion and the voltages mentioned for the pMOS are absolute values.

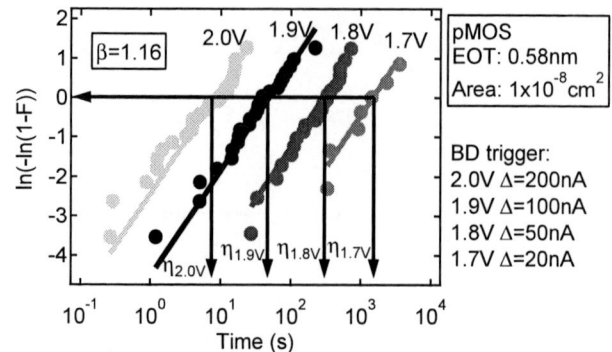

Fig. 3. TDDB Weibull distributions of a 0.58 nm EOT pMOS high-k metal gate stack. The t_{BD} was triggered at current steps between 20 and 200 nA (SBD trigger).

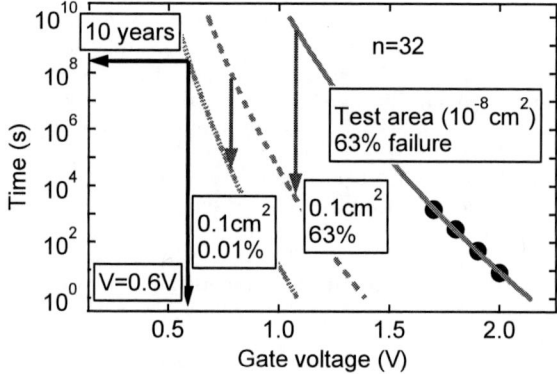

Fig. 4. TDDB lifetime extrapolation of the 0.58 nm EOT pMOS high-k metal gate stack. The power-law fit of the η-values results in $n = 32$. After scaling to 0.1 cm^2 and 0.01% failure, for this dielectric a gate voltage of 0.59 V is extrapolated to guarantee 10 years without SBD.

II. THE ALL-IN-ONE TDDB LIFETIME EXTRAPOLATION

The conventional TDDB lifetime extrapolation is based on sets of I-t traces measured at different V_G (and optional on different areas), where the t_{BD} is determined with an appropriate failure criterion which is typically the first significant current increase. The area-scaled t_{BD} Weibull distributions are then fitted together, yielding the Weibull slope (β_{BD}) and the 63% failure values (η_{BD}) for the different V_G as shown in Fig. 3. Calculating the lifetime at specification conditions is done in three steps. First, by using a power law [9], η_{BD} is extrapolated to specification V_G, second, η_{BD} is scaled to the specification area of 0.1cm^2 [10], and third, η_{BD} is extrapolated to low failure percentiles (usually 0.01%). The scaling and extrapolation formulas for voltage, area and percentiles are given in (1), (2) and (3), respectively. Alternatively, the maximum voltage to guarantee 10 years lifetime can be determined as done in Fig. 4.

$$\eta_1 = \eta_2 \left(\frac{V_{G1}}{V_{G2}}\right)^{-n} \quad (1)$$

$$\eta_1 = \eta_2 \left(\frac{A_1}{A_2}\right)^{-\frac{1}{\beta}} \quad (2)$$

$$\ln[-\ln(1-F)] = \beta \ln(t) - \beta \ln \eta \quad (3)$$

Our new concept for a combined TDDB extrapolation including both SBD and HBD was first presented in [8]. The t_{BD} lifetime was no longer plotted as a function of the applied gate voltage, but a dielectric area/gate voltage plot was split into three regions: 1) the SBD free region, confined by the 10 years and 63% SBD probability, 2) the region of multiple SBD's, wearout and runaway, and 3) the HBD region confined by the 0.01% probability of having HBD after 10 years [11].

Since it is more consistent to define all the regions by the 0.01% probability, in this paper we present a modified version of the all-in-one reliability extrapolation plot as shown in Fig. 5. To directly access the number of SBD's for each area/voltage combination of the dielectric tested, region 2 now indicates the higher percentile contours and from 63% on the number of SBD's. Knowing the current through an average individual SBD path at a certain gate voltage, it is possible to calculate the total SILC after 10 years in the entire chip. The all-in-one TDDB lifetime plot of a 0.63 nm EOT HfO$_2$/TiN pMOS MIPS gate stack [12] shown in Fig. 5, predicts no SBD

978-1-4244-9113-1/11 $26.00 © 2011 IEEE

on 0.1 cm^2 after 10 years if $V_G < 0.52V$. If $V_G = 0.9$ V more than 1000 SBD's will be created after 10 years, and for $V_G = 1$ V, 0.01 % of the chips will fail due to HBD.

To generate the all-in-one TDDB reliability prediction in Fig. 5, the following intrinsic reliability parameters from the SBD and the wearout distributions are required: the Weibull slopes β_{SBD} & β_{WO}, the 63% values η_{SBD} & η_{WO} and the voltage acceleration power-law exponents n_{SBD} & n_{WO}.

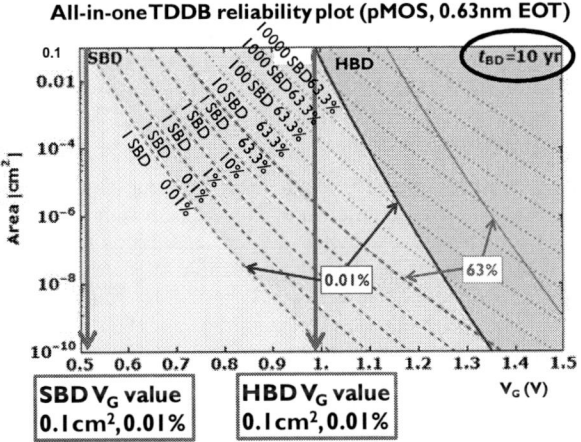

Fig. 5. All-in-one TDDB reliability for a 0.63 nm EOT pMOS. For a fixed lifetime of 10 years three different regions of area/V_G combinations can be distinguished: SBD free, SBD & wearout, and HBD. For this wafer a $V_{G,SBD} = 0.5$ V and $V_{G,HBD} = 1$ V is extracted.

It is important to point out that the HBD distribution is not a simple convolution of the SBD and wearout distributions. As was shown in [13], properly accounting for the competition between the multiple SBD paths leads to the following expression for $F_{HBD}(t)$ (4):

$$F_{HBD}[t] = 1 - Exp\left[-\left(\frac{\eta_{wo}}{\eta_{SBD}}\right)^{\beta_{SBD}} \cdot \int_0^{\left(\frac{t}{\eta_{wo}}\right)^{\beta_{wo}}} \left(\frac{t}{\eta_{wo}} - u^{\frac{1}{\beta_{SBD}}}\right)^{\beta_{SBD}} e^{-u} du\right]$$

(4)

Equation 4 is computationally rather challenging. Fortunately, the lower tail of this distribution can be approximated rather well with an asymptotic Weibull distribution, with lower asymptotic parameters η_{la}, β_{la} and n_{la} given by:

$$\eta_{la} = \left(\frac{\Gamma[\beta_{SBD} + \beta_{WO} + 1] \cdot \eta_{SBD}^{\beta_{SBD}} \cdot \eta_{WO}^{\beta_{wo}}}{\Gamma[\beta_{SBD} + 1] \cdot \Gamma[\beta_{WO} + 1]}\right)^{\frac{1}{\beta_{SBD} + \beta_{WO}}}$$

(5)

In this equation $\Gamma(x)$ represents the gamma function [14].

$$\beta_{la} = \beta_{SBD} + \beta_{WO} \tag{6}$$

$$n_{la} = \frac{n_{SBD} \cdot \beta_{SBD} + n_{WO} \cdot \beta_{WO}}{\beta_{SBD} + \beta_{WO}} \tag{7}$$

In practice, this approximation can often be used for the extrapolation to low percentiles with sufficient accuracy.

The all-in-one reliability plot (Fig. 5) is constructed as follows. Both for the SBD and the lower percentile Weibull approximation of the HBD, (8) holds:

$$Ln\left[\frac{A}{A_r}\right] = Ln[-Ln[1 - F]] - \beta \cdot Ln\left[\frac{t}{\eta_r}\right] - \beta \cdot n \cdot Ln\left[\frac{V_G}{V_{G,r}}\right] \quad (8)$$

In more detail, for the plot in Fig. 5, t is fixed to 10 years, F is chosen 0.01%, 0.1%, 1%, etc. and A is calculated as a function of V_G resulting in the dashed and full lines, respectively. The SBD curves are calculated using $\beta=\beta_{SBD}$, $\eta=\eta_{SBD}$, $n=n_{SBD}$ while the HBD curves are calculated using β_{la}, η_{la} and n_{la} from Eq. (5), (6) and (7). When the curvature of the HBD distribution remains non-negligible upon extrapolation, (9) has to be applied instead of (8):

$$A = \frac{A_r \cdot Ln[1 - F]}{Ln[1 - F_{HBD}[t, V_G, A_r]]} \tag{9}$$

, wherein both $\eta_{SBD,r}$ & $\eta_{WO,r}$ in (4) have been rescaled with (1). The r subscripts in (8) & (9) stand for the parameters at reference conditions. $V_{G,r}$ and A_r are the reference voltage and area that were chosen to fit the measured hard breakdown distribution and obtain the reference η_r. Best choice for $V_{G,r}$ and A_r are in the range of the test voltage and test device area.

To calculate the 63 % time to k^{th} soft BD curves, the inverse of the Gamma Regularized function [14] has to be applied onto the zero Weibit fail fraction. On the Weibit scale, this leads to the following offset w_0:

$$w_0(k) = Ln\left[\Gamma^{[-1]}\left[k, \frac{1}{e}\right]\right] \tag{10}$$

$\Gamma^{[-1]}[k, 1/e]$ represents the solution in x of $\frac{\Gamma[k,x]}{\Gamma[k]} = 1/e$, wherein $\Gamma[k, x]$ is the incomplete Gamma function [14]. The kth soft breakdown curves in Fig. 5 are calculated using (11):

$$Ln\left[\frac{A}{A_r}\right] = w_0[k] - \beta_{SBD} \cdot Ln\left[\frac{t}{\eta_r}\right] - \beta_{SBD} \cdot n_{SBD} \cdot Ln\left[\frac{V_G}{V_{G,r}}\right] \quad (11)$$

III. RELIABILITY PARAMETER EXTRACTION

A. Test structures

We first address the issue of test structures that can be used for sub-1nm EOT dielectric breakdown evaluations. In order to observe extrinsic tails and to stress at gate voltages as close as possible to the target operating voltage, typically devices larger than the transistors in the final CMOS product are used for t_{BD} and V_{BD} tests. On average, the total series resistance R_S seen by the gate current in a test of the dielectric is around 150 Ω in an isolated device. From a gate current around 200 µA on, the voltage drop over the R_S is larger than 30mV and cannot be neglected anymore. This limitation is less stringent for pMOS than for nMOS. For an 0.8 nm EOT nMOS gate stack which meets the ITRS high performance logic device leakage spec of J_G (1 V) = 1 A/cm^2, at 125°C the (pre-BD) gate current of a 1×10^{-8} cm^2 device at V_{BD} is larger than 200 µA (Fig. 6), and also at lower gate voltage during a CVS the gate current prior HBD can exceed 200 µA due to excessive Stress-Induced Leakage Current. For nMOS around 0.6 nm EOT the gate current is even higher, and the only practical solution is to limit the experiment to devices significantly smaller than 1×10^{-9} cm^2.

EOT's below 1 nm are achieved by the scavenging technique which reduces the interfacial layer SiO$_2$ thickness [12]. We, however, revealed electrically and by physical analysis that the scavenging process can be less efficient close to the device edges, and a dependence of the average EOT on channel length for L < 1 µm caused by a slightly thicker interfacial layer can be observed. The area-scaling of the gate leakage current I_G and consequently that of the t_{BD} is not always guaranteed. Combining the results of different areas for UT-EOT devices needs to be done with caution.

With the proper choice of test structures, we now describe three different methods to experimentally obtain the six intrinsic reliability parameters (the Weibull slopes β_{SBD} & β_{WO}, the 63% values η_{SBD} & η_{WO} and the voltage acceleration power-law exponents n_{SBD} & n_{WO}) needed to reach a complete reliability prediction as in Fig. 5.

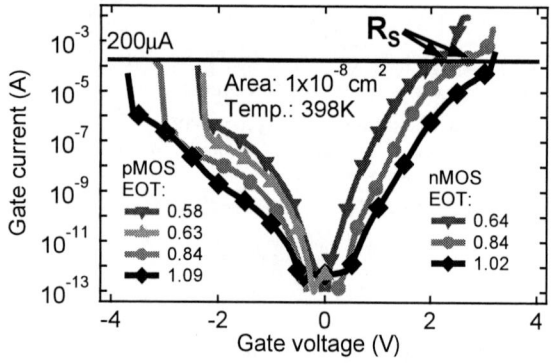

Fig. 6. The gate leakage current for some of the gate stacks discussed in this work (the symbol code corresponds to the same gate stack). The I_G for pMOS is decades lower than for nMOS where at CVS condition I_G easily exceeds 200 µA.

B. Method 1: From HBD-only

Since (4) describes the shape of the entire HBD distribution, the most obvious way to reach our goal is the measurement of the HBD over a range of stress voltages. From an experimental point of view, this is also the easiest approach,

and this method can be applied for both nMOS and pMOS over a wide area range.

The HBD distribution given by (4) does not appear as a straight line on a Weibit scale (Fig. 7). As was discussed in section II, (6), the low asymptote of the HBD distribution is Weibull with slope $\beta_{SBD} + \beta_{WO}$. The asymptote at high percentiles is also Weibull [11] with slope β_{SBD}.

Extracting all parameters from a set of HBD-distributions is, however, a mathematical challenge. Maximum likelihood fits need to be done for all voltage conditions simultaneously in order to extract the six parameters with satisfactory accuracy. It is mandatory to have both low and high percentile tails in the data set in order to stabilize the fit. In practice, this requires a very large number of samples at different V_G resulting in a very long total measurement time.

This issue can partially be alleviated by combining different areas. Indeed, the sample size and measurement time can significantly be reduced when the test conditions are chosen such that mainly the low or high percentile tails are measured. Measurements on large areas (>1×10^{-8} cm^2 if resistance allows) at low V_G result in the low percentile SBD + WO dominated HBD distribution, while measuring on small areas (<1×10^{-10}cm^2) at higher V_G results in negligible wearout and therefore the HBD distribution reveals the SBD parameters. Furthermore, the combination of small area/high voltage with large area/low voltage also keeps the experimental time window limited. Note that combining different areas requires I_G area-scaling.

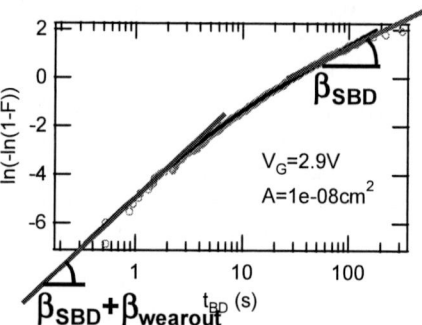

Fig. 7. Large sample size t_{HBD} distributions do not appear as a straight line on a Weibit scale but exhibits a certain curvature due to the convolution of SBD and wearout and the competition effect between multiple SBD's.

C. Method 2: From SBD and HBD distributions

The mathematical and experimental challenges of having to fit a set of distributions with (4) can be greatly simplified if we can determine part of the model parameters in a different way. When it is possible to distinguish both SBD and HBD separately after the analysis of the current-time traces, we can determine the SBD parameters first and use them as *input* for the fit of the HBD distribution, effectively reducing the number of fitting parameters from six to three.

While identifying and triggering on HBD with a typical step current increase > 10μA is straightforward, the correct determination of the time-to-soft breakdown (t_{SBD}) is a major challenge. In UT-EOT nMOS devices, SBD and wearout are often completely hidden in the tunneling current. In pMOS devices, J_G is lower due to the asymmetry of the high-k gate stack [7], but the current increase due to SBD is also lower. To detect SBD, the current increase due to an individual SBD path has to be clearly above the noise level of the tunneling current. For nMOS this is only the case for EOT's > 0.8 nm and/or when measuring on very small area devices ($<1\times10^{-10}cm^2$).

Fig. 8 illustrates the SBD trigger issue. On the $2.5\times10^{-9}cm^2$ device, two abrupt current increases of different magnitude can be observed. The first one has a magnitude in the order of 20 to 100 nA, the second one is in the range of 2 to 10 μA. Based on statistical arguments, the different magnitudes of both events is due to a different number of defects in the created percolation path. On a $1\times10^{-6}cm^2$ device, no individual abrupt current increases are detectable, but instead a more gradual current increase is observed. This SILC is the result of *multiple* soft breakdowns. The example in Fig. 8 also demonstrates that SBD is not properly *defined*. Naively, in the $2.5\times10^{-9}cm^2$ sample, the first current jump could be taken as SBD. However, suppose a sample with initial gate current of 10 μA had been chosen. A 20-100 nA current increase would be hard to detect, but the 2-10 μA increase would remain visible and would therefore be taken as the first SBD. In conclusion, the definition of SBD is never unambiguous and the determination of t_{SBD} depends on the sample choice, measurement conditions and trigger level.

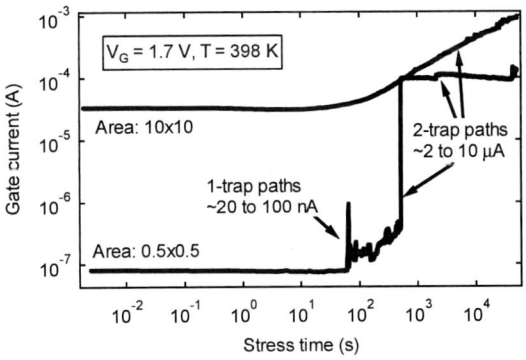

Fig. 8. The gate leakage current under CVS measured on two different areas. On the large 10x10 transistor the current through 1-trap paths disappears in the tunneling current and multiple 2-trap paths cause the observed SILC increase [15]. On small area devices, however, it is possible to distinguish between low conductive 1-trap paths and the several μA current increase through 2-trap paths.

In this work, we use a spike and transient-insensitive detection algorithm [16] for finding the first current step ΔI in each current-time trace. The trigger magnitude is chosen such that the smallest steps detectable above the noise level are considered SBD's. A more systematic approach involves the determination of t_{BD} with a variable trigger magnitude. After constructing for each ΔI the corresponding t_{BD}-distribution and fitting it assuming Weibull, it is observed how at sufficiently

low ΔI, the Weibull slope of the distributions saturates at its SBD value [3, 17]. In UT-EOT, this systematic approach is often not possible since the required small current step is already below the gate current noise level.

If it is possible to detect t_{SBD} at different V_G, the parameters β_{SBD}, η_{SBD} and n_{SBD} can be extracted directly from the SBD distributions. These numbers serve as input for (4). Subsequently, the missing wearout parameters can be obtained through a maximum likelihood fit of the HBD distributions. With only three parameters still remaining, the fit is stable even if a small amount of data is used.

In the (rare) special case that only one SBD is observed before HBD, as can occur in (very) small area devices, the wearout parameters are also directly determinable, since then $t_{WO} = t_{HBD} - t_{SBD}$.

The major advantage of the SBD/HBD method is that only a relatively small sample size is required (e.g. 20 devices per V_G). The drawback, however, is the selection of the 'correct' SBD trigger ΔI. To facilitate the SBD detection, V_G can be lowered, but then extremely long t_{HBD} is found. On the other hand, it is not a necessity to determine t_{SBD} and t_{HBD} in the same voltage range. A scheme optimized for minimum measurement time selects high V_G for HBD only and low V_G for SBD only.

This approach is shown in Fig. 9a for the 0.63 nm pMOS MIPS wafer where on 1x1 μm² transistors CVS in inversion was applied until HBD at $V_G = 2.2$ to 1.9 V and only until SBD at $V_G = 1.8$ to 1.6 V. A 10 μA current trigger was combined with a 1 nA ΔI trigger for the detection of t_{HBD} and t_{SBD}, respectively.

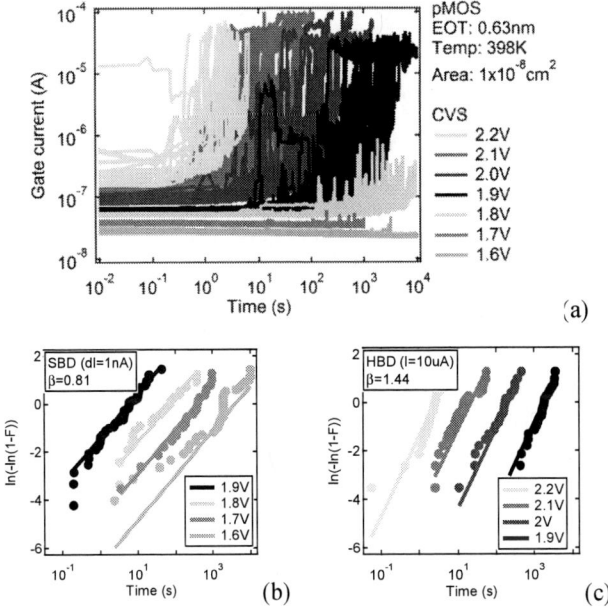

Fig. 9. I-t traces measured on a 0.63 nm EOT pMOS gate stack over a wide voltage range (9a). For low V_G a 1nA ΔI trigger is applied to obtain the SBD distributions (9b), while the high V_G data yield the HBD distributions (9c). Note that all voltages are absolute values measurements were done in inversion.

978-1-4244-9113-1/11 $26.00 © 2011 IEEE

The SBD Weibull distributions are shown in Fig. 9b and after fitting them together with maximum likelihood estimation, β_{SBD}, η_{SBD} and n_{SBD} are extracted. The HBD distributions are sufficiently straight to be approximated with a Weibull, and its parameters correspond to the low asymptote distribution characterized by (5-7). With the SBD parameters as input, the wearout parameters can be calculated. In general, the three wearout parameters are extracted with a maximum likelihood fit using the complete HBD distribution formula (4). Having all SBD and wearout parameters it is then possible to perform the full all-in-one TDDB lifetime extrapolation for the 0.63 nm EOT HfO$_2$/TiN pMOS gate stack as shown in Fig. 5.

Further optimization of the measurement time vs parameter accuracy balance can be obtained by (as in method 1) combining different device areas. Small area devices allow more accurate t_{SBD} determination, while large area devices can be used for finding t_{HBD}. This approach can only be used if I_G scales exactly with area for all device dimensions and is limited by the maximum tolerable voltage drop over the series resistance R_S during stress.

D. Method 3: From SILC and HBD distributions

The third method attempts to avoid the difficulty of finding the first SBD. The SBD parameters (β_{SBD}, η_{SBD}, n_{SBD}) are not obtained from t_{SBD} Weibull distributions, but from the SILC generated during CVS measurements. As we already pointed out when discussing Fig. 8, once the device area is sufficiently large, a gradual increase of the gate current due to multiple SBD's is observed rather than individual SBD's.

As has been shown in [18, 19], that the SILC vs. time slope of multiple parallel SBD paths is equal to the Weibull slope β_{SBD} of the corresponding t_{SBD} breakdown distribution. Moreover, one measurement at each stress voltage is sufficient to determine the voltage acceleration n_{SBD} [19]. In this way, two out of the six required reliability parameters are extracted without the need of a SBD trigger algorithm, making this approach ideal for UT-EOT nMOS devices where the individual SBD events are hidden in the overall current. In this paper we extend this approach to obtain η_{SBD} as well. In order to perform the complete TDDB lifetime prediction, the wearout parameters are extracted in the same way as in method 2 from HBD distributions.

We will now discuss by means of an example, the technical difficulties of extracting the SBD parameters from SILC. Single I-t traces at different V_G on the largest possible area (R_S-limited, typically 1x1 μm^2) were measured at 125°C for nMOS gate stacks with EOT's ranging from 0.6 to 1 nm [13]. In Fig. 10, the traces for the thinnest gate stack are shown. The current increase was monitored up to 250 μA only in order to avoid any R_S effects, or the measurement was interrupted before HBD after 20 ks because of the low V_G. Similar to the approach shown in [15], the SILC component can be extracted by normalizing the gate current to its initial value and compensating for the BTI component caused by charge trapping in initially present and stress-generated traps in the high-k. We perform this compensation by fitting the initial current decay and subtracting the extrapolated fit from the entire current-time trace (Fig. 11). The voltage dependence of the charging needs to be taken into account. At the high stress

voltages (V_G >1.7V in the example shown) the SILC might dominate from t = 0s (Fig. 11) while the overall slope of the current increase is still reduced by the hidden charging component. Therefore, the slope of charging component at high voltage is extrapolated from the low voltage values.

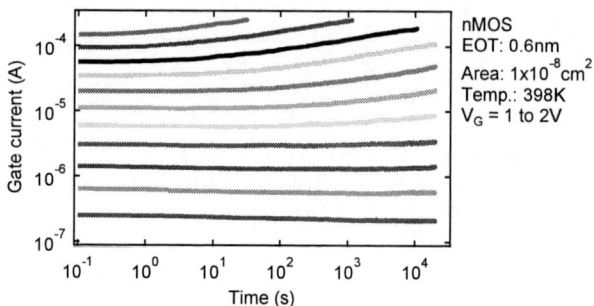

Fig. 10. The I-t traces of a 0.6 nm EOT nMOS gate stack measured on 1x10^{-8}cm^2 transistors at 398 K for gate voltages from 2 to 1 V.

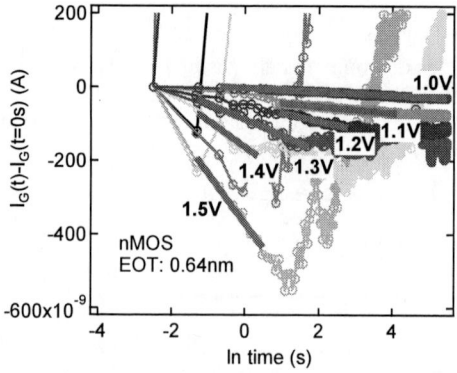

Fig. 11. The initial part of the normalized gate currents reveals a strong voltage dependent of the charging component. The measured current has to be correct for this charging component in order to obtain the true SILC.

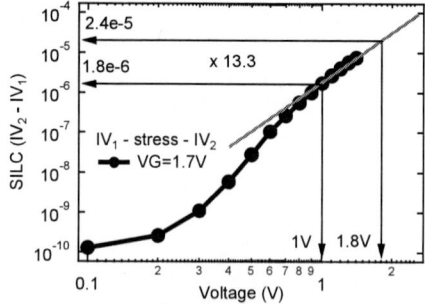

Fig. 12. Above 0.9 V the SILC I_G-V_G curve can be fitted with a power law to determine the ratios at which the true I_G-t SILC curves have to be scaled to a common sense voltage.

As a last step, the SILC still has to be scaled to a sense voltage [15]. To achieve this, the I_G-V_G characteristics were measured before and after each stress. As an example, the V_G dependence of the SILC after a stress at $V_{stress} = 1.7$ V (data in Fig. 10) is shown in Fig. 12. For V_G above 0.9 V the SILC can be fitted with a power law. The slope of the line is found to be independent of the stress condition and is used to scale the SILC at each stress voltage to a chosen sense voltage of 1 V.

After normalization, correcting for the charging component and scaling to a sense voltage, we are able to determine the SBD parameters (Fig. 13). The initial part of the normalized SILC is still affected by discrete SBD events [18] and the high current levels are affected by series resistance. The most accurate power function exponent ($=\beta_{SBD}$) of the fit results from the middle part of the curve. A common exponent is forced to all SILC traces. The intersect of the fits of the different gate voltages with a chosen SILC level (e.g. 1 µA) can be interpreted as η_{SBD} (Fig. 13), and using (1), n_{SBD} can be calculated. The choice of the SILC-level corresponds to the choice of the SBD trigger level in method 2. However, since η_{SBD} is extracted from the power law fit, the statistical spread on the first SBD is eliminated, making this methodology superior to method 2 which requires a complete distribution of SBD's on different samples.

The main advantages of this method are that a) triggering on an individual SBD is avoided making this method well suited for the extreme low EOT nMOS down to 0.5 nm EOT, b) only a single trace per V_G has to be measured and c) the accuracy of β_{SBD} is high. The remaining wearout parameters however still have to be obtained from a set of t_{HBD} distributions as in method 2.

Measuring SILC and HBD on the same area is difficult. While for SILC, large areas and low V_G are required, HBD measurements on large areas are affected by R_S issues or, if low stress voltages are used, are extremely lengthy. The combination of low V_G SILC analysis in large area devices with high V_G HBD measurements in small area devices results in very accurate SBD parameters and a time-optimized determination of the wearout parameters. Therefore, method 3 is in many cases the best approach to obtain for nMOS devices the all-in-one reliability plot of Fig. 5.

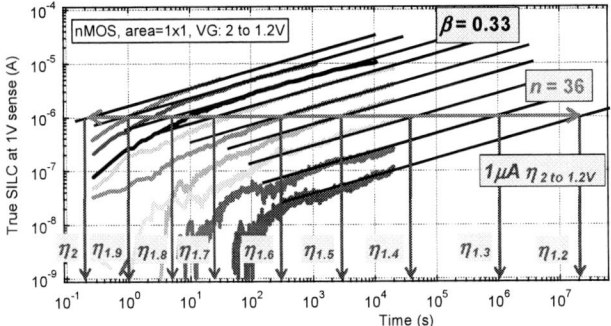

Fig. 13. A set of normalized SILC traces measured on a 0.64 nm EOT HfO$_2$/TiN nMOS gate stack at different V_G. All SBD parameters (β_{SBD}, η_{SBD} and n_{SBD}) are obtained from this graph and β=0.33 indicates the dominance of 1-trap SBD paths.

IV. DISCUSSION

In previous sections, we showed how to optimize the extraction of the SBD and wearout parameters of TDDB for UT-EOT oxide layers. All the presented methods use CVS. Some authors [6] claim that ramped voltage measurements provide all necessary parameters even faster. To a certain extend extent this is true. The ramp rate can indeed take the role of stress time. However, in [6] the conversion formula between t_{BD} and V_{BD} assumes the same voltage acceleration exponent for both SBD and wearout. The authors have experimentally justified this assumption in [20]. Also on our own samples we observed in some cases only a small difference between n_{SBD} and n_{WO}, but this observation cannot be generalized as is clear from the examples shown in this work. In particular for pMOS, a significant difference between n_{SBD} and n_{WO} is found (Fig. 5).

Even more fundamental than assumptions on the parameters, is the validity of the soft breakdown/wearout concept as such. This concept was originally developed for SiO$_2$ and SiON layers. Separate experimental determination of SBD and wearout parameters was proven to yield the same results as the extraction of those parameters from the overall HBD distribution using (4) [11]. One should however be very cautious in applying these results to high-k stacks that are inherently asymmetric since they consist of a thin SiO$_2$ or SiON interface layer with a thicker high-k layer on top.

In particular, it has been demonstrated [6] that nMOS SBD's can be 'annealed' by switching for a short time to a negative polarity. Yet, even when all SBD's are systematically annealed as soon as they appear, t_{HBD} remains unaffected. On large area samples the entire SILC can be annealed [21], consistent with the picture that SILC is merely caused by multiple SBD's. As a consequence, no SILC is observed during AC stress, while the AC t_{HBD} is identical to the DC t_{HBD} after proper correction for the differences in oxide charging [22]. All these experimental observations suggest that SBD and HBD are not directly related as is assumed in the SBD/wearout concept.

In order to verify these claims, we carefully created a single SBD in a WxL=0.15x1µm^2 nMOS transistor with 1/1.8 nm SiO$_2$/HfO$_2$ and TiN gate using a very slow positive ramped current stress. The I_G-V_G characteristic at negative polarity is presented in Fig. 14. The current ratio $I_S/(I_S+I_D) = \sim 0.4$ indicates the position of the breakdown spot [23]. Subsequently, the negative gate bias was slowly increased and at ~-1.5V the SBD path was abruptly annealed (Fig. 15) with post-anneal I_G-V_G also shown in Fig. 14. During the following positive voltage ramp, the SBD current reappeared at ~1.7V, but since both the I_G-V_G characteristic and the BD position significantly changed compared to the 1st BD, it is concluded that a new 2nd SBD path was created. The procedure was repeated, yielding also a 3rd SBD.

The observation in Fig. 15 is identical to the reset behavior in Resistive RAM (ReRAM) cells. Indeed, HfO$_2$ has been shown to be suited as switching material for the fabrication of this type of exploratory memories [24]. The switching in HfO$_2$ is explained as an anneal of oxide vacancies in the conductive filament (='breakdown path' in TDDB terminology).

978-1-4244-9113-1/11 $26.00 © 2011 IEEE

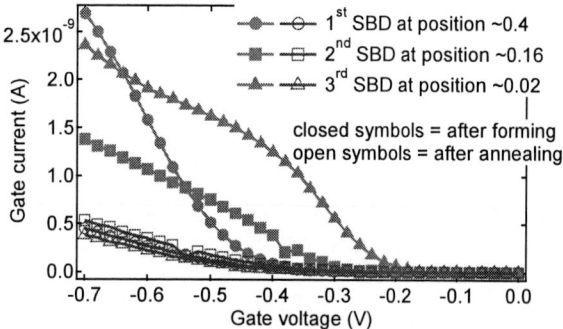

Fig. 14. The I_G-V_G characteristics of three SBD's in the same transistor that were created at positive gate bias and subsequently annealed at negative gate bias. The shape of the curve and the position factor indicate three separate SBD's.

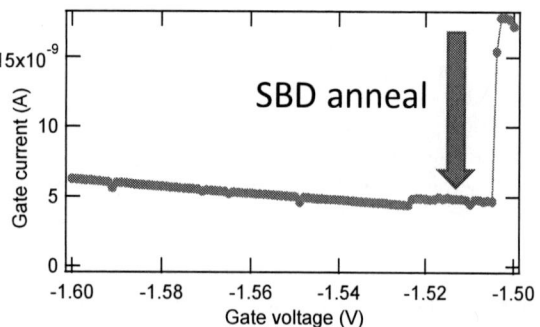

Fig. 15. The abrupt annealing of a SBD path in an nMOS after applying negative gate bias. In the subsequent I_G-V_G no SBD path is observed.

It is concluded that SILC and SBD's in thin high-k stacks are caused by leakage paths through stress-generated defects in the high-k layer (Fig. 16). Negative polarity can anneal the SBD conduction paths in a similar way as in ReRAM. Note, however, that annealing of a SBD path requires the removal of only 1 trap and the total defect density hardly changes. In other words, the SBD's and SILC can be annealed, but the oxide degradation not. This explains why the SILC quickly returns to its pre-annealed value upon continuation of the stress [21].

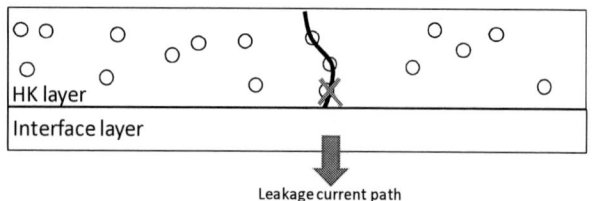

Fig. 16. In thin high-k gate stacks SILC and SBD's are caused by leakage paths through stress-generated defects. An anneal of an oxide vacancy in the high-k can cut off a local conduction path, but the total trap density hardly changes.

The slope of the SBD distribution, or equivalently, the exponent of the SILC increase indicates how many defects N_{def} are in the conduction path, since:

$$\beta_{SBD} = m \cdot N_{def} \qquad (12)$$

with m the defect generation rate (between 0.35 and 0.4) [25]. For UT-EOT devices typically 1 or 2 well-placed traps are sufficient to create a SBD path and β_{SBD} is quantized to either ~0.35-0.4 or ~0.7-0.8 [15] as shown in Fig. 17. This result is confirmed in several other publications [6, 26, 27, 28].

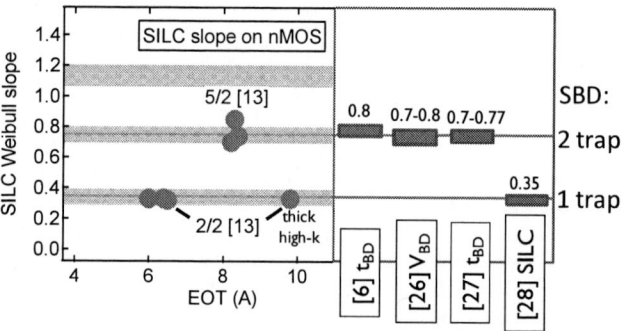

Fig. 17. The quantization of the nMOS SILC slopes indicating either a 1 or 2-trap conduction path. Similar results extracted from either t_{BD}, V_{BD} or SILC are found in other references.

The confrontation of all these results inevitably leads to the conclusion that the concept of SBD followed by a localized wearout phase becomes questionable. Instead, a two-layer breakdown model [29] seems a more likely explanation for the curved HBD-distribution in high-k dielectrics. In this model, it is assumed that the degradation in the interface and the high-k layer happen independently at two different rates, and HBD is triggered when a percolation path through both layers is formed. From the analytic elaboration of this model by Tous et al. [30], it follows that the lower and upper asymptotes (subscript la and ua, respectively) of the HBD distribution are again Weibull distributed with

$$\beta_{la} = \beta_{slow} + \beta_{fast} \ and \ \beta_{ua} = \beta_{slow} \qquad (13)$$

with β_{slow} and β_{fast} being the Weibull slopes of the t_{BD} distribution of the slowest and fastest degrading layer, respectively. The mathematical resemblance with (6) and Fig. 7 is striking. Indeed, the two-layer model of [30] is a competition model in space, while the SBD/wearout model is a competition model in time.

Following [27], we can introduce the concept of correlation between the percolation path in the high-k and the interface layer. Correlation means that the presence of a percolation path in the high-k layer induces an *accelerated* local degradation in the interface layer. This is the concept of wearout. If no correlation at all exists, a soft breakdown path in the high-k layer will be completed in the interface layer by pure chance.

Yet, *mathematically* this can still be interpreted as wearout. There is one important difference between the two-layer model in [29] and the SBD/wearout model: wearout can only start when the SBD path is created while in the two layer model both layers start degrading at t=0. If t_{SBD} is significantly smaller than t_{HBD}, this difference becomes negligible.

In summary, irrespective of the conceptual model, SBD's still introduce a leakage current-dominated transition region as in Fig. 5. In thin SiO_2 and SiON, one can be certain that a SBD/wearout approach is correct [11], but in SiO_2/high-k stacks, future experimental and modeling work is needed to further unravel the exact breakdown mechanism.

V. CONCLUSION

In this paper we present 3 different methods for the experimental determination of the SBD and wearout parameters needed to construct an all-in-one TDDB reliability prediction consisting of a SBD-free region, a leakage current-dominated region and a HBD-limited region. The first method relies on the fitting of the curved HBD distribution. In the second method, we investigated the possibility of extracting the t_{SBD} separately, an in the third method, we show how the SILC can be used to extract the SBD parameters. We illustrated these methods on several sub-1nm EOT high-k/metal gate nMOS and pMOS devices.

Since for UT-EOT pMOS devices the detection of SBD is still possible while there is insufficient SILC before HBD, the best suited method is based on t_{SBD} & t_{HBD} TDDB distributions. For nMOS in terms of measurement time and precision, however, the SILC & t_{HBD} method is the best.

We also discuss the validity and interpretation of the SBD/wearout model, confronting it with experiments that demonstrate how SBD paths can be annealed by reversing the stress polarity. The introduction of the concept of correlation between breakdown paths in the two layers of the dielectric stack might prove mandatory to completely understand the subtle breakdown statistics of these layers.

ACKNOWLEDGMENT

The authors want to thank imec's core partners and the members of the CMOST department for their support, the imec p-line for the processing of the samples and amsimec for the assistance with the electrical measurements. Robert O'Connor acknowledges the Science Foundation Ireland FORME SRC for funding.

REFERENCES

[1] T. Sakura, H. Utsunomiya, Y. Kamakura, K. Taniguchi, "A Detailed Study of Soft- and Pre-Soft-Breakdowns in Small Geometry MOS Structures", IEDM Technical Digest, pp. 183–186, 1998.

[2] J.S. Suehle, B. Zhu, Y. Chen, J.B. Bemstein, "Acceleration factors and mechanistic study of progressive breakdown in small area ultra-thin gate oxides", Proceedings IRPS, pp. 95-101, 2004.

[3] B. Kaczer, R. Degraeve, R. O'Connor, Ph. Roussel, G. Groeseneken, "Implications of progressive wear-out for lifetime extrapolation of ultra-thin (EOT - 1 nm) SiON films",IEDM Technical Digest, pp. 713-716, 2004.

[4] T. Kauerauf, R. Degraeve, M. B. Zahid, M. Cho, B. Kaczer, P. Roussel, and G. Groeseneken, "Abrupt breakdown in dielectric/metal gate stacks:

A potential reliability limitation?" IEEE Electron Device Lett., Volume 26, Issue 10, pp. 773–775, 2005.

[5] A. Kerber, E. Cartier, B.P. Linder, S.A. Krishnan, T. Nigam, "TDDB failure distribution of metal gate/high-k CMOS devices on SOI substrates", Proceedings IRPS, pp. 505-509, 2009.

[6] A. Kerber, E.A. Cartier, "Reliability Challenges for CMOS Technology Qualifications With Hafnium Oxide/Titanium Nitride Gate Stacks", IEEE TDMR, Volume 9, Issue 2, pp. 147-162, 2009.

[7] T. Kauerauf, B. Govoreanu, R. Degraeve, G. Groeseneken, H. Maes, "Scaling CMOS: Finding the gate stack with the lowest leakage current", Solid-State Electronics 49, pp 695-701, 2005.

[8] S. Sahhaf, R. Degraeve, P. J. Roussel, T. Kauerauf, B. Kaczer, and G. Groeseneken, "TDDB reliability prediction based on the statistical analysis of hard breakdown including multiple soft breakdown and wearout", IEDM Technical Digest, pp. 501–504, 2007.

[9] E.Y. Wu, A. Vayshenker, E. Nowak, J. Sune, R.P. Vollertsen, W. Lai, D. Harmon, "Experimental evidence of TBD power-law for voltage dependence of oxide breakdown in ultrathin gate oxides", IEEE Trans. Electron Devices, Volume 49, Issue 12, pp. 2244-2253, 2002.

[10] T. Nigam, R. Degraeve, G. Groeseneken, M.M. Heyns, H.E. Maes, "Constant current charge-to-breakdown: Still a valid tool to study the reliability of MOS structures?", Proceedings IRPS, pp. 62-69, 1998.

[11] S. Sahhaf, R. Degraeve, P.J. Roussel, B. Kaczer, T. Kauerauf, G. Groeseneken, "A New TDDB Reliability Prediction Methodology Accounting for Multiple SBD and Wear Out", IEEE Trans. Electron Devices, Volume 56, Issue 7, pp. 1424-1432, 2009.

[12] P. Roussel, R. Degraeve, S. Sahhaf, and G. Groeseneken, "A consistent model for the hard breakdown distribution including digital soft breakdown: The noble art of area scaling," Microelectron. Eng., Volume. 84, Issue 9/10, pp. 1925–1928, 2007.

[13] L.-Å. Ragnarsson, Z. Li, J. Tseng, T. Schram, E. Rohr, M.J. Cho, T. Kauerauf, T. Conard, Y. Okuno, B. Parvais, P. Absil, S. Biesemans, T.Y. Hoffmann, "Ultra Low-EOT (5Å) Gate-First and Gate-Last High Performance CMOS Achieved by Gate-Electrode Optimization", IEDM Technical Digest, pp. 1–4, 2009.

[14] Mathematica™ manual, Wolfram research inc.

[15] R. Degraeve, T. Kauerauf, M. Cho, M. Zahid L-Å. Ragnarsson, D.P. Brunco, B. Kaczer, Ph. Roussel, S De Gendt, G. Groeseneken, "Degradation and breakdown of 0.9 nm EOT SiO2/ALD HfO2/metal gate stacks under positive Constant Voltage Stress", IEDM Technical Digest, pp. 408–411, 2005.

[16] T. Kauerauf, "Degradation and breakdown of MOS gate stacks with high permittivity dielectrics", PhD thesis, 2007.

[17] T. Kauerauf, R. Degraeve, F. Crupi, B. Kaczer, G. Groeseneken, H. Maes, "Trap generation and progressive wearout in thin HfSiON", Proceedings IRPS, pp. 45-49, 2005.

[18] E. Farrés, M. Nafria, J. Suñé, and X. Aymerich, "The statistical distribution of breakdown from multiple breakdown events in one sample," J. Phys. D: Appl. Phys. 24, pp. 407-414, 1991.

[19] M., Alam, R.K. Smith, "A Phenomenological Theory of Correlated Multiple Soft-Breakdown Events in Ultra-thin Gate Dielectrics", Proceedings IRPS, pp. 406-411, 2003.

[20] A. Kerber, M. Röhner, T. Pompl, R. Duschl, M. Kerber, "Lifetime prediction for CMOS devices with ultrathin gate oxides based on progressive breakdown", Proceedings IRPS, pp. 217-220, 2007.

[21] E. Cartier, A. Kerber, "Stress-induced leakage current and defect generation in nFETs with HfO2/TiN gate stacks during positive-bias temperature stress", Proceedings IRPS, pp. 486-492, 2009.

[22] A. Kerber, A. Vayshenker, D. Lipp, T. Nigam, E. Cartier, "Impact of charge trapping on the voltage acceleration of TDDB in metal/high-k n-channel MOSFETs", Proceedings IRPS, pp. 369-372, 2010.

[23] R. Degraeve, B. Kaczer, A. De Keersgieter, G. Groeseneken, "Relation between breakdown mode and breakdown location in short channel NMOSFETs and its impact on reliability specifications", Proceedings IRPS, pp. 360-366, 2001.

[24] L. Goux, P. Czarnecki, Y.Y. Chen, L. Pantisano, X.P. Wang, R. Degraeve, B. Govoreanu, M. Jurczak, D.J. Wouters, L. Altimime,

978-1-4244-9113-1/11 $26.00 © 2011 IEEE

"Evidences of oxygen-mediated resistive-switching mechanism in TiN\HfO$_2$\Pt cells ", Appl. Phys. Lett. 97(24), 2010.

[25] R. Degraeve, B. Kaczer, Ph. Roussel, G. Groeseneken, "On the trap generation rate in ultrathin SiON under constant voltage stress", Microelectronic Engineering, vol. 80, pp. 440-443, 2005.

[26] B.P. Linder, E. Cartier, S. Krishnan, J.H. Stathis, "The effect of interface thickness of high-k/metal gate stacks on NFET dielectric reliability", Proceedings IRPS, pp. 510-513, 2009.

[27] G. Ribes, P. Mora, F. Monsieur, M. Rafik, F. Guarin, G. Yang, D, Roy, W.L. Chang, J. Stathis, "High-k gate stack breakdown statistics modeled by correlated interfacial layer and high-k breakdown path", Proceedings IRPS, pp. 364-368, 2010.

[28] W.L. Chang, J.H. Stathis, E. Cartier, "Role of interface layer in stress-induced leakage current in high-k/metal-gate dielectric stacks", Proceedings IRPS, pp. 787-791, 2010.

[29] T. Nigam, A. Kerber, P. Peumans, "Accurate model for time-dependent dielectric breakdown of high-k metal gate stacks", Proceedings IRPS, pp. 523-530, 2009.

[30] S. Tous, E.Y. Wu, J. Suñé, "A compact analytic model for the breakdown distribution of gate stack dielectrics", Proceedings IRPS, pp. 792-798, 2010.

Frequency Dependent TDDB Behaviors and Its Reliability Qualification in 32nm High-k/Metal Gate CMOSFETs

Kyong Taek Lee*, Jongik Nam, Minjung Jin, Kidan Bae, Junekyun Park, Lira Hwang, Jungin Kim, Hyunjin Kim and Jongwoo Park

Technology Reliability, System LSI division, Samsung Electronics
San #24 Nongseo-Dong Giheung-Gu, Yongin-City, Gyeonggi-Do, Korea 446-711
82-31-209-5733 (phone), 82-31-209-4312 (fax), kyongtaek.lee@samsung.com

Abstract

The TDDB failure mechanism of high-k dielectric/metal gate (HK/MG) CMOSFETs on DC and AC stress conditions are investigated in comparison to poly-Si/SiON. All devices under unipolar AC stress exhibit longer failure time (t_{bd}) as frequency increases. In case of HK/MG, the SILC behavior has been attributed to the bulk transient charge trapping by pre-existing defects in HK. Since trapped charges in HK can easily be detrapped once a relaxation bias is applied, t_{bd} is increased as frequency becomes higher. Unlike unipolar AC bias condition, HK/MG nMOSFETs with bipolar AC stress exhibit shorter t_{bd} than with DC at a lower frequency. This is attributed to hole trapping into IL as V_g is at the gate injection bias since HK/MG stack has higher probability of electron injection than poly-Si/SiON due to relatively lower barrier height. However, bipolar AC TDDB in high frequency shows longer t_{bd} than DC TDDB because of lack of time to generate enough holes in the IL. In bipolar AC bias condition, the higher power-law time exponent (n) appears because G_m degradation by hole generation is aggravated at the gate injection bias in nMOSFET, while pMOSFET SILC is generated by bulk charge trapping at the substrate injection bias.

I. INTRODUCTION

To date Hf-based high-k dielectric/metal gate stack (HK/MG) CMOSFETs have enabled very aggressive equivalent -oxide-thickness (EOT) scaling due mainly to less gate leakage achieved by relatively thicker physical thickness than conventional silicon oxynitride (SiON) dielectrics for the same EOT [1-2]. However, such aggressive gate stack scaling often suffers from time dependent dielectric breakdown (TDDB) and, in effect, an accurate lifetime estimation becomes a crucial reliability concern in terms of test methodology and modeling. Understanding the breakdown mechanism of HK/MG is complicated by multilayer structures, which include, in addition to HK dielectric, an interfacial SiON layer (IL) [3]. Furthermore, inside a circuit, CMOSFET gate stack experiences time-varying bias. Hence, both DC and AC characterizations for HK/MG TDDB are inevitable for technology process reliability [4].

In this study, we present TDDB behaviors of HK/MG CMOSFETs compared with poly-Si/SiON devices under DC and AC stress conditions. The propensity of dielectric breakdown mechanism of HK/MG CMOSFET on various stress bias conditions will be discussed.

II. EXPERIMENTAL

32nm HK/MG CMOSFETs with a Hf-based gate dielectric were fabricated by utilizing a conventional CMOS process. The T_{inv} values in HK/MG and P-SiON MOSFETs are 1.4nm and 2.2 nm, respectively. TDDB measurements were performed at 125°C on transistors with 2.496μm² (W/L=2/0.04μm, 5×6 array) and W/L=1/0.03μm using three different bias conditions (Fig.1). With the AC stress method, the gate voltage was altered between the stress bias (+V_g) and a fixed recovery bias such as 0V (unipolar) and -V_g (bipolar) in a frequency range of 1hz ~ 100khz with a duty cycle of 50%. The SILC at a lower sense voltage (V_g= 1V) was monitored periodically with a minimal stress interruption as an aid for BD detection (Fig.2). All represented stress times in this paper are mean effective stress times. Rise and fall times of pulse generator were optimized at all bias conditions to minimize over- or undershoots problem during unipolar and bipolar AC bias conditions. In order to investigate over- or undershoot issue, DC was generated by AC pulse generator with the duty cycle of 99.99% under unipolar and bipolar AC bias. In comparison, no difference in TDDB failure time was observed between unipolar and bipolar AC with duty cycle of 99.99% at all frequency which means that no problem with over- or undershoot is observed.

To examine the soft breakdown (SBD) positions, the amount of I_d / (I_d + I_s) at V_g = −1V (V_d = V_s = 0V) was measured [5]. Moderate changes in I_d / (I_d + I_s) were observed after SBD of all samples (0.3 ~ 0.5), which indicates that the SBD location is in the middle of the channel rather than the S/D edge.

III. RESULTS AND DISCUSSION

TDDB failure distributions under DC and unipolar AC stress for poly-Si/SiON and HK/MG nMOSFETs are shown in Fig. 3 and Fig. 4, respectively. All devices under unipolar AC stress exhibit longer failure time (t_{bd}) as frequency increases.

978-1-4244-9113-1/11 $26.00 © 2011 IEEE

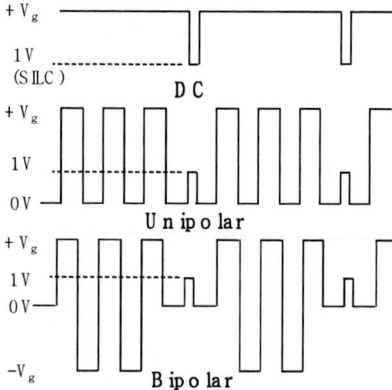

Fig.1. Three different bias conditions were used. With the AC stress method, the gate voltage was altered between the stress bias ($+V_g$) and a fixed recovery bias such as 0V (unipolar), $-V_g$ (bipolar).

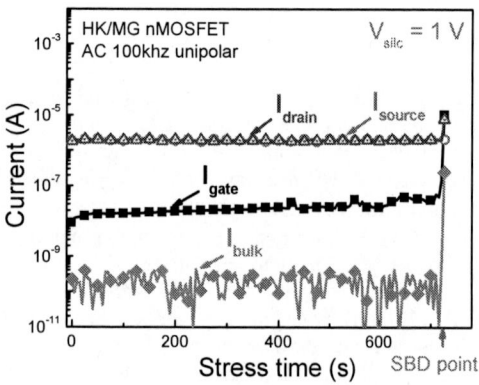

Fig.2. Current time trace during unipolar AC stress. Moderate changes in $I_d / (I_d + I_s)$ were observed after SBD of all samples ($0.3 \sim 0.5$), which indicates that the SBD location is in the middle of the channel rather than the S/D edge.

In poly-Si/SiON, hole generation and trapping in oxide are primary causes to breakdown at substrate injection bias condition ($+V_g$). Trapped holes may recombine with injected electrons to form neutral electron traps. Therefore, the increase of t_{bd} in high frequency is related to less net hole generation when short stress pulses are applied and detrapping of holes when the signal is zero [6]. In case of HK/MG, the SILC behavior has been attributed to the bulk transient charge trapping by pre-existing defects in HK. Normally, SILC on HK is employed for SBD failure time projection. Since trapped charges in HK can easily be detrapped once a relaxation bias is applied, t_{bd} is increased as frequency becomes higher. Neither significant subthreshold slope nor transconductance (G_m) degradation is observed after SBD, which indicates that most of the SILC increase is related to the transient charge trapping within the bulk of HK (Fig.5).

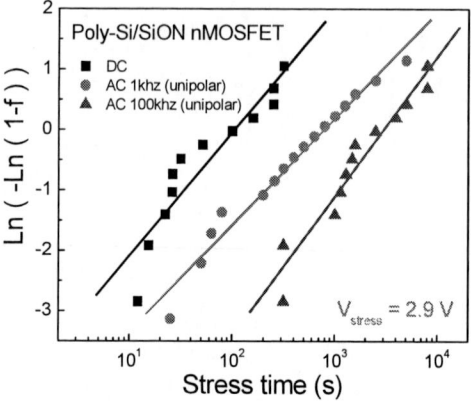

Fig.3. TDDB failure distribution under DC and unipolar AC stress for poly-Si/SiON nMOSFETs. All devices under unipolar AC stress exhibit longer failure time (t_{bd}) as frequency increases

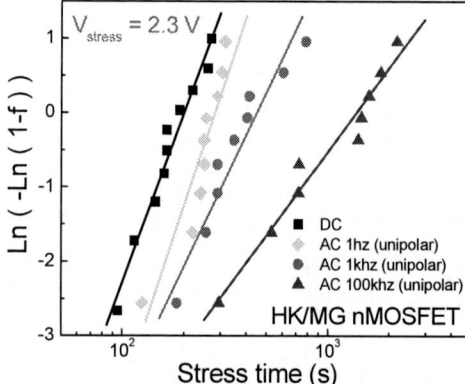

Fig.4. TDDB failure distribution under DC and unipolar AC stress for HK/MG nMOSFETs. The unipolar AC conditions exhibit longer t_{bd} than DC conditions regardless of frequency because of trapped charges in HK can easily be detrapped once a relaxation bias is applied.

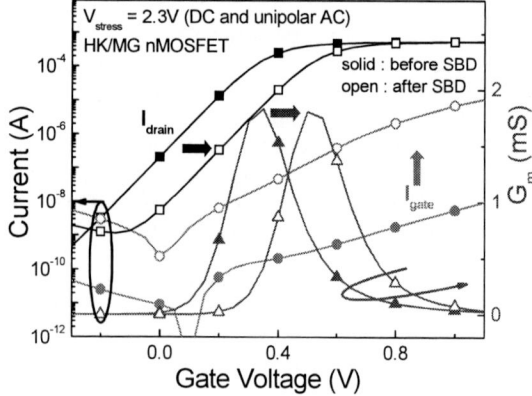

Fig.5. Transistor current-voltage (I-V) characteristics and G_m (measured in the linear region with V_{ds}=50mV) before and after SBD in HK/MG nMOSFET under DC and unipolar AC stress. Neither significant subthreshold slope nor G_m degradation is observed after SBD. Therefore, SILC increase is related to the transient charge trapping within the bulk of HK

978-1-4244-9113-1/11 $26.00 © 2011 IEEE

Figures 6 and 7 show TDDB failure distribution under DC, unipolar AC and bipolar AC stress for poly-Si/SiON and HK/MG nMOSFETs, respectively. The poly-Si/SiON devices under bipolar AC stress show longer t_{bd} than DC and unipolar AC bias condition because deeper traps are also detrapped as V_g is reverse bias. Unlike poly-Si/SiON, HK/MG nMOSFETs under bipolar AC stress in low frequency exhibit shorter t_{bd} than DC condition. During bipolar AC stress, SILC increase is suppressed by fixed recovery V_g bias due to more detrapping in HK (Fig. 8). Thereby, V_{th} shift is lower than unipolar AC stress (Fig. 9). However, G_m is degraded after SBD (Fig. 9). Furthermore, interface states density (D_{it}) increase is observed in charge pumping measurement during bipolar AC stress which means that IL is degraded (not shown). Therefore, contrary to unipolar AC stress, the major cause of shorter t_{bd} in bipolar AC stress is IL degradation by applied reverse V_g bias.

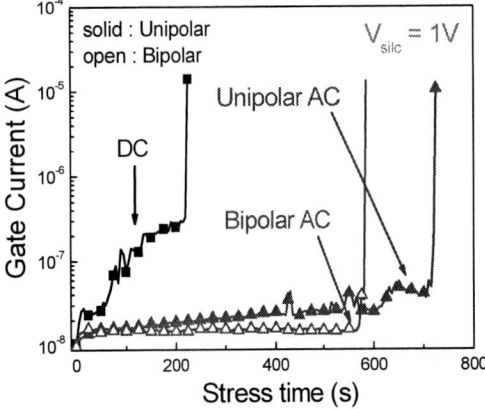

Fig.8. SILC behaviors during DC, unipolar and bipolar AC bias stress. SILC increase is suppressed by fixed recovery V_g bias due to more detrapping in HK

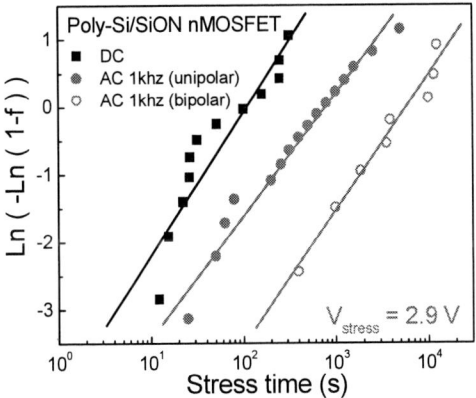

Fig.6. TDDB failure distribution under DC and bipolar AC stress for poly-Si/SiON nMOSFETs. The devices under bipolar AC stress show longer t_{bd} than unipolar AC and DC conditions because deeper traps are also detrapped as V_g is reverse bias.

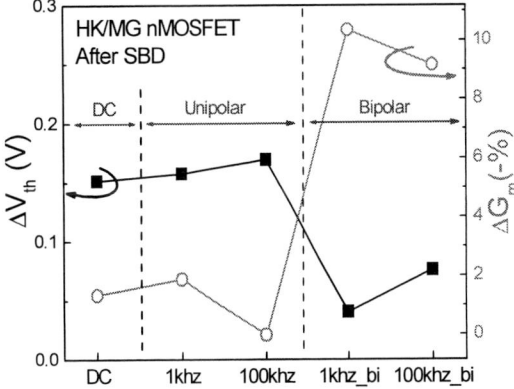

Fig.9. ΔV_{th} and ΔG_m degradation after SBD for HK/DG nMOSFETs on various stress bias conditions. During bipolar AC stress, G_m degradation were observed due to hole traps in IL

In substrate injection ($+V_g$), electrons injected from the substrate can generate new oxygen vacancies undergo trap assistant tunneling (TAT) or direct tunneling (DT) and Fowler-Nordheim (FN) tunneling (Fig. 10(a)). Once an electron is trapped, the defect also appears to become a center for TAT, contributing to the SILC [4]. In a gate injection case ($-V_g$), bulk transient charge trapping in HK is mostly reversible due to its shallow trap level, which is attributed to SILC recovery (Fig. 8). In contrast, high energy electron from the gate can generate hot holes at the interface or at the surface of silicon substrate because HK layer has higher probability of electron injection due to a lower barrier height than that of SiON (Fig. 10(b)) [7]. As a result, the interfacial layer (IL) is degraded by holes which in turn led to G_m degradation. However, bipolar AC TDDB in high frequency shows longer t_{bd} than DC TDDB because of insufficient time to generate enough holes.

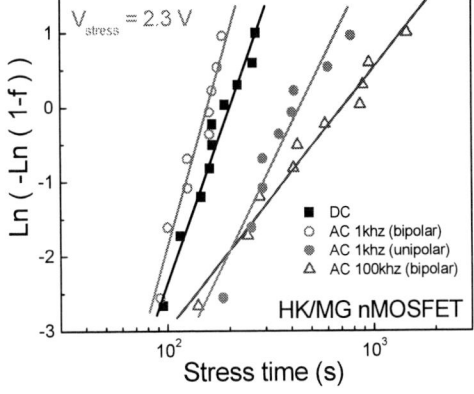

Fig.7. TDDB failure distribution under DC and bipolar AC stress for HK/MG nMOSFETs. Unlike poly-Si/SiON, the bipolar AC conditions in low frequency exhibit shorter t_{bd} than DC conditions. The major cause of shorter t_{bd} in bipolar AC stress is IL degradation by applied reverse V_g bias.

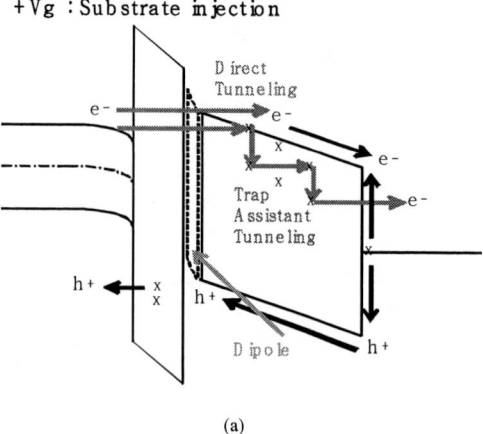

+Vg : Substrate injection

(a)

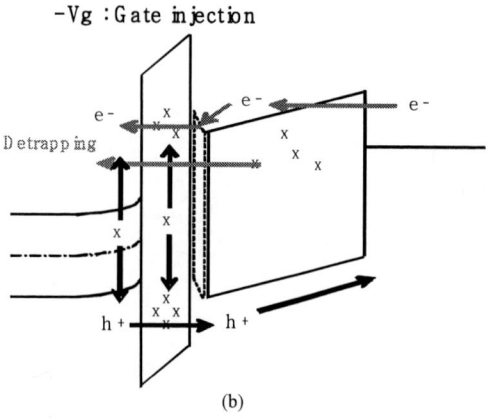

−Vg : Gate injection

(b)

Fig.10. Schematic band diagram describing the models for TDDB of high-k dielectrics. In a gate injection case, high energy electron from the gate can generate hot holes at the interface or at the surface of silicon substrate because HK layer has higher probability of electron injection due to a lower barrier height than that of SiON.

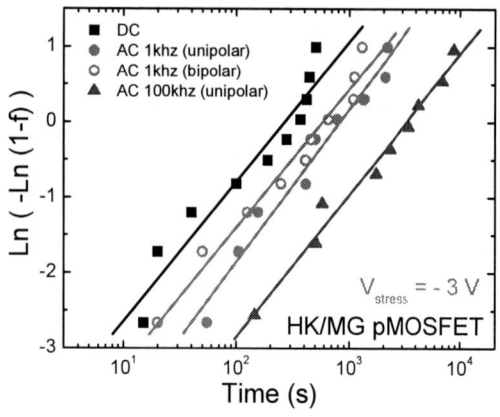

Fig.11. TDDB failure distribution for HK/MG pMOSFETs on various stress bias conditions. Unlike nMOSFETs, bipolar AC conditions also show higher t_{bd} compared to DC conditions.

Figure 11 shows TDDB failure distribution for HK/MG pMOSFETs on various stress bias conditions. Unlike nMOSFET, the IL degradation by hole traps is primary cause to breakdown of pMOSFET at gate injection bias [1]. Such that after BD severe G_m degradation was observed without SILC generation. In case of bipolar AC stress, SILC increase during substrate injection bias ($+V_g$) although hole generation is reduced. Thus, the t_{bd} of bipolar AC stress is located between that of DC and unipolar AC stress.

Figure 12 and 13 show the comparison of ΔI_{dsat} degradation for HK/MG nMOSFETs and pMOSFETs with various stress bias conditions. All bipolar AC devices exhibit lower ΔI_{dsat} degradation at initial time due to electron detrapping in HK or hole detrapping in the IL. However, the higher power-law time exponent (n) appears because G_m degradation by hole generation is aggravated at the gate injection bias in nMOSFET, while pMOSFET SILC is generated by bulk charge trapping at the substrate injection bias.

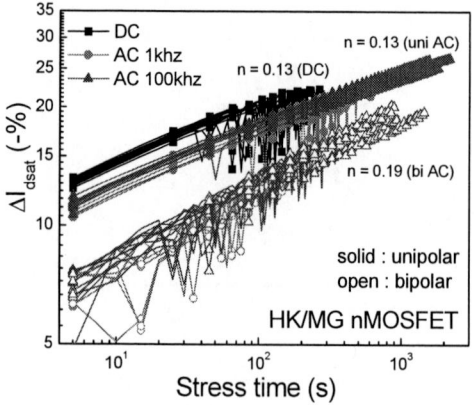

Fig.12. Comparison of ΔI_{dsat} degradation for HK/MG nMOSFETs on various stress bias conditions. Due to electron detrapping in bulk HK, bipolar AC devices exhibit lower ΔI_{dsat} degradation at initial time

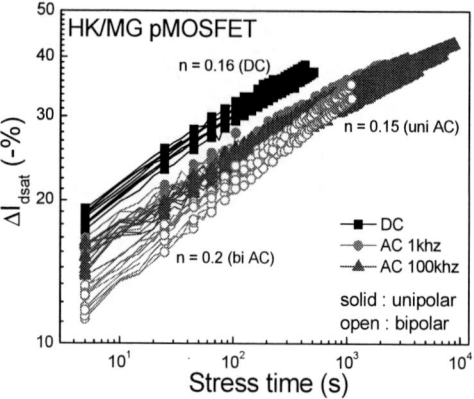

Fig.13. Comparison of ΔI_{dsat} degradation for HK/MG pMOSFETs on various stress bias conditions. The bipolar AC conditions show higher n because SILC is generated by bulk charge trapping at $+V_g$

IV. CONCLUSION

In this study, the propensity of TDDB behaviors of HK/MG CMOSFETs on DC and AC stress conditions is considered in comparison to poly-Si/SiON. Unlike poly-Si/SiON, HK/MG nMOSFETs with bipolar AC stress exhibit shorter t_{bd} than with DC at a lower frequency. This is attributed to hole trapping into IL as V_g is at the gate injection bias since HK/MG stack has higher probability of electron injection than poly-Si/SiON due to relatively lower barrier height. However, bipolar AC TDDB in high frequency shows longer t_{bd} than DC TDDB because of lack of time to generate enough holes in the IL. As such AC based TDDB would not be a major concern for technology process qualification for HK/MG technology. Since most of circuit, in reality, is subjected to operate at high frequency range with unipolar AC bias conditions, DC based TDDB methodology is more conservative for HK/MG technology qualification.

REFERENCES

[1] Y. Yasuda, N. Kimizuka, T. Iwamoto, S. Fujieda, T. Ogura, H. Watanabe, T. Tatsumi, I. Yamamoto, K. Ito, H. Watanabe, Y. Yamagata, and K. Imai, "A 65 nm-node LSTP (low standby power) poly-Si/a–Si/HfSiON transistor with high Ion–Istandby□ ratio and reliability," in *VLSI Symp. Tech. Dig.*, 2004, pp. 40–41.

[2] S. Bae, A. Ashok, J. Choi, T. Ghani, J. He, S. Lee, K. Lemay, M. Liu, R. Lu, P. Packan, C. Parker, R. Purser, A. St. Amour and B. Woolery, "Reliability Characterization of 32nm High-K and Metal-Gate Logic Transistor Technology," in *International Reliability Physics Symp. Proc.*, 2010, p. 287-292.

[3] G. Bersuker, D. Heh, C. D. Young, L. Morassi, A. Padovani, L. Larcher, K. S. Yew, Y. C. Ong, D. S. Ang, K. L. Pey, W. Taylor, in *International Reliability Physics Symp. Proc.*, 2010, p. 373-378.

[4] A. Kerber, E. A. Cartier, "Reliability Challenges for CMOS Technology Qualifications With Hafnium Oxide/Titanium Nitride Gate Stacks," *IEEE Trans. Device Mater. Rel.*, vol. 9, no. 2, 2009, p. 147-162.

[5] H.S. Choi, S. H. Hong, R. H. Baek, K. T. Lee, C. Y. Kang, R. Jammy, B. H. Lee, S. W. Jung and Y. H. Jeong, "Low-Frequency Noise After Channel Soft Oxide Breakdown in HfLaSiO Gate Dielectric," , *IEEE Electron Device Lett*, vol. 30, no. 5, 2009, p. 523-525.

[6] E. Rosenbaum, Z. Liu, C. Hu, "Silicon Dioxide Breakdown Lifetime Enhancement Under Bipolar Bias Conditions," *IEEE Trans. Electron Devices*, vol. 40, no. 12, 1993, p. 2287-2295.

[7] B. H. Lee, "Unified TDDB Model for Stacked High-k Dielectrics," ICICDT, 2009, p. 83-87.

Re-investigation of Gate Oxide Breakdown on Logic Circuit Reliability

Y.C. Huang, T.Y. Yew, W. Wang, Y.-H. Lee, R. Ranjan, N.K. Jha, P.J. Liao, J.R. Shih, and K. Wu

Taiwan Semiconductor Manufacturing Company, Ltd. 121, Park Ave. 3, Hsinchu Science Park, Hsinchu, Taiwan 300-77, R.O.C.

E-mail: huangycw@tsmc.com

Abstract-**Gate oxide breakdown has been studied in the circuit-like patterns, i.e. e-Fuse arrays and two-stage inverter circuit. It is observed that time-dependent dielectric breakdown (TDDB) lifetime of eFuse chip is larger compared to discrete devices. Gate oxide breakdown study using two-stage inverter circuit (1st-stage I/O N/PMOS worked as current limiting transistors and 2nd-stage core N/PMOS is stressed transistors) reveals that, even by applying a significant high voltage stress ($\leq$ 3xVdd) on stressed device, the stress device will suffer only soft breakdown not a hard breakdown and it is independent with the current drive capability of current limiting transistors. Soft breakdown results in very small voltage drop across the current limiting device (i.e. between source and drain terminals), which will have negligible impact on the circuit functionality. It suggests circuit functionality will be immune from gate oxide breakdown in normal circuit operating condition, i.e. V_{dd} of ~1V, and designers will get extra reliability margin. Our HSPICE simulation results on ring oscillator (RO) also suggest the logic circuit functionality immunity with gate oxide breakdown.**

Keywords – (Breakdown (BD), Hard BD (HBD), Soft BD (SBD), time-depedent dielectric BD (TDDB), e-Fuse, 2-stage inverter circuit, Ring oscillator (RO))

I. INTRODUCTION

Linder *et al.* [1] reported that the transistor-limited constant voltage stress (CVS) significantly reduces post-breakdown (BD) conduction as compared to standard CVS, and severity of gate oxide BD is dependent on limiting transistors current drive capability. It is also reported that the logic circuits can function even with hard BD (HBD) suffered gate oxide [1]-[2]. But most of the circuit level oxide BD studies were performed on very high voltage stress (e.g. > 3x V_{dd}) and circuit functionality has been predicted at operational voltage on the basis of high voltage oxide BD behavior. It may not be practical as it is reported that the growth of severity of discrete transistor's oxide BD is voltage dependent, i.e. at higher voltage oxide BD will be more severe [3]. In addition, Lo *et al.* [4] reported that nature of oxide BD of discrete transistors depends on the critical voltage, i.e. below critical voltage stress leads the device to soft BD (SBD) or digital BD only but above critical voltage stress leads the device to HBD. Hence, higher voltage stress may cause HBD of the stressed device in the circuit and results in higher voltage drop across the limiting transistor and leads the transistor to operate in saturation region. It may give higher degradation of the circuit functionality. But in real circuit where device operates at

nominal voltage, the stressed transistor will suffer only SBD and results in very small voltage drop across the limiting transistor and leads it to operate in the linear region, which gives negligible impact on circuit functionality. Therefore, lifetime prediction at operational voltage on the basis of high voltage stress study may mislead us. In this study, we report that the gate oxide suffers only SBD with moderate voltage stress on the basis of e-fuse and two-stage inverters circuit's studies. Hence, at nominal operating voltage circuit's functionality will not be impacted by gate oxide BD, as oxide degradation will be negligible. HSPICE simulation has been performed to understand the impact of oxide BD on circuit's functionality.

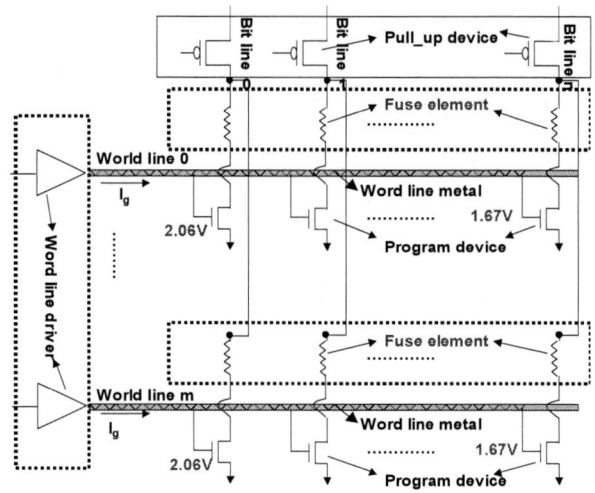

Figure 1. Schematic of an eFuse array shows program voltage (V_g of program device) decreases from 2.06V to 1.67V along word line. Nevertheless, program devices on the same bit line see the same program voltage.

II. EXPERIMENTAL

The device and circuit used in this study are fabricated using 28nm technology having electrical T_{ox} of ~20Å. Figs. 1 and 2 show schematics of e-fuse array and two-stage inverter circuits, respectively used in this study. The program devices in e-fuse array are core NMOSFET (Fig. 1). In Fig.2, the PMOS (P1) in the 1st-stage of inverter circuit will serve as the current limiting transistor for the NMOS (N2) (stressed device)

978-1-4244-9113-1/11 $26.00 © 2011 IEEE

in the 2nd-stage of inverter circuit. The 1st-stage inverter is fabricated using IO N/PMOS having thicker oxide to ensure the immunity of current limiting transistors from oxide BD, while 2nd-stage inverter is made by core N/PMOS to study oxide BD. During N2 stress, the clamp current, i.e. the drain current of P1, can be set at different level by adjusting V_{in} and vice-versa during P2 stress (Fig. 2). In this work, clamp current is chosen to be the typical drive current I_d of a standard logic device.

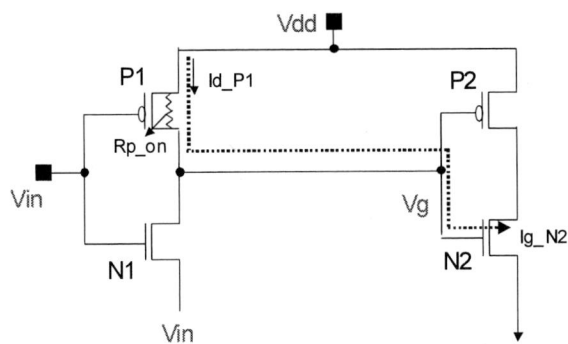

Figure 2. Schematic shows 2-stage inverters used in this work. N2 is the stressed NMOS and P1 is the clamp. Clamp current is defined as the I_{dsat} of P1 for a given $|\Delta V_{gs}|$ (i.e. $V_{dd} - V_{in}$). N1 and P2 are off during stress. I_g of N2 is actually the I_d of P1. R_{p_on} resulted IR drop from V_{dd} to V_g after SBD. $|\Delta V_g|$ alleviates TDDB stress of N2.

TABLE I. Calculated TDDB lifetime of array and periphery circuits based on discrete device TDDB model. Core devices except for power switch constitute periphery circuit. Lifetime is normalized to the discrete model in 0.1% cumulative failure for compassion. The stress during programming was performed under a highly accelerated voltage compared to the read operation. There is a significant TDDB lifetime discrepancy between single device and circuit level.

Program at 2.58xVdd	Array		Periphery
Program time ~ 10us	WL w/o Vg drop	WL w/i Vg drop	
Stressed area (um^2)	44236.8	44236.8	4.3
0.1% TDDB lifetime	1x	1y	1z
0.5% TDDB lifetime	4.3x	4.3y	4.3z
actual stress time in real chip	202.8x	4.2y	43.3z

III. RESULT AND DISCUSSION

Table I lists the predicted lifetime (based on discrete transistor model) and actual stress time of e-fuse array stressed under moderate voltage (i.e. program voltage, $V_{ddq-max} > 2xV_{dd}$) at short period of time. The actual stress time for e-fuse is used for comparison because there has been no detectable failure signal during the time of chip stress. Lifetime is normalized to the discrete model in 0.1% statistical failure for the sake of compassion. It is better to notify that e-fuse has zero tolerance (0.1% lifetime is already aggressive) of any program device HBD after programming as program path and read path share the same program device for each bit. Core devices in both array and periphery circuits are experiencing very high voltage stress only for very short period of time

during programming. From Table I, it can be seen that e-Fuse TDDB lifetime is much better compared to discrete devices. On the other hand, if we compare the predicted lifetime listed in Table I with actual stress time of chip, it can be said that program devices of e-fuse should suffer HBD and results in random single word line failure. However, experimental data (a few hundreds of e-fuse chips have been stressed) does not show such kind of function fail and/or any increase of standby current. It suggests that program devices experienced only SBD and circuit's lifetime is much longer compared to predicted lifetime extrapolated from the oxide TDDB model using high voltage stress on discrete devices.

Gate oxide BD study has been performed on the two-stage inverter circuit to further clarify the stress voltage effect on the nature of gate oxide BD and its impact on circuit reliability. Fig. 3 shows lifetime and V_g drop dependence on stress voltage. It indicates that severity of gate oxide BD (i.e. I_g change and V_g drop after breakdown) depends on stress voltage. It is also observed that although I_g is quite large (>100μA for 3.0V stress), stressed devices still have transistor characteristics with a small I_d degradation, e.g. < 5% degradation.

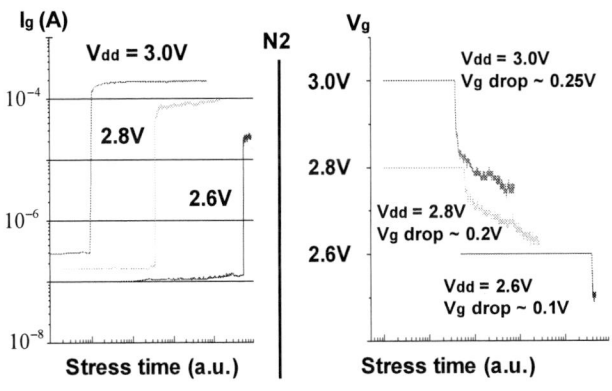

Figure 3. I_g (left) and V_g (right) of stress-N2 device in Fig. 2 measured under different V_{dd} with the same clamp current (250uA). TDDB and I_g after SBD have clear dependence on stress voltage (V_{dd}). Larger V_{dd} corresponds to larger V_g drop (for a given $|V_{gs}|$, smaller I_{ds} accompanies smaller $|V_{ds}|$ from P1's perspective).

On the other hand, I_g after SBD measured at same stress voltage (2.8V) with different clamp current between 50uA to 300uA shows the independency of I_g increase just after SBD (i.e. BD severity) on clamp current, as illustrated in Fig. 4. These devices have similar oxide damage. So, neither I_g just after SBD nor SBD lifetime depends on the clamp current. In addition, Fig. 4 shows I_g is always lower than clamp current, which indicates that at moderate voltage stress, devices suffered only SBD and not reach to HBD. This is different than the literature's report on the basis of high voltage stress [1], where it is claimed that severity of breakdown (i.e. I_g increment) is dependent on clamp current limit. It may be due to different nature of oxide BD with high and moderate voltage stress. It is speculated that the I_g increment will be large enough to be clamped by the saturation current of current limiting transistor (current drive capability) under high voltage stress. But, I_g increment is not big enough under moderate voltage stress and cannot be limited by the current limiting device, which results in independency of the current drive capability with I_g increment just after SBD. So, I_g increment just after SBD under the same moderate stress voltage will be close to each other.

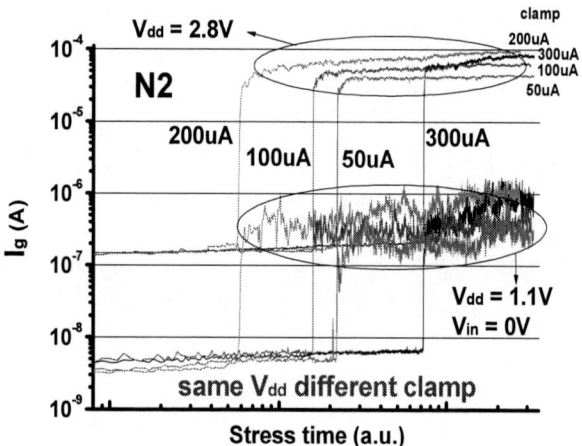

Figure 4. I_g of stressed-N2 device (Fig. 2) measured with different clamp currents. The two V_{dd} groups belong to the same set of stressed devices during TDDB test. I_g under 1.1V V_{dd} was also measured during test to compare with I_g under 2.8V V_{dd}. After SBD, I_g is smaller than 1uA under operation condition and each device still has normal I_d-V_d characteristic. 300uA, the largest clamp current, has the longest lifetime and second largest I_g just after SBD. These devices have similar oxide damage (under the same but not so high V_{dd}, $I_g \leq$ clamp current). So, neither I_g just after SBD nor SBD lifetime depends on the clamp current.

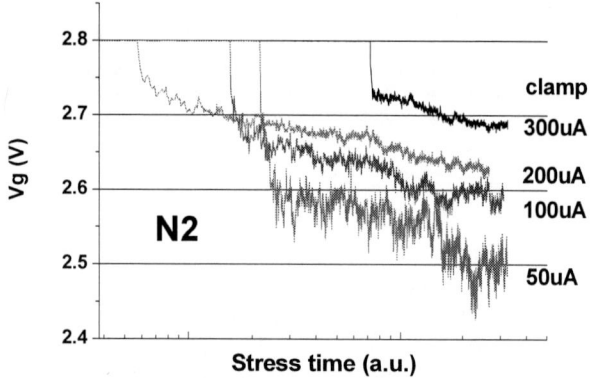

Figure 5. V_g measured for different clamp current at 2.8V V_{dd} Stress. Since, all the stressed devices have the similar damage under same stress condition (i.e. similar I_g of N2 (~ <100e-6A) in Fig. 4), larger $|V_{gs}|$ has to accompany smaller $|V_{ds}|$ (i.e. ΔV_g) to achieve the similar I_d (i.e. I_g of N2) from P1's perspective (Fig. 2). The larger the clamp current, the V_g drop will be smaller after SBD. Note that a smaller clamp current at smaller $|V_{gs}|$ results in larger R_{on}, which magnifies V_g (i.e. $I_g * R_{on}$) fluctuation.

Fig. 5 shows stressed device V_g with stress time under 2.8V CVS stress. It can be seen that the larger the current driving capability (i.e. clamp current) of current limiting device the V_g drop of stressed device is smaller after SBD. It is speculated that for a given width/length of the limiting transistors, the driving capability (i.e. clamp current I_d) of the limiting devices are controlled by $|V_{gs}|$ (i.e. V_{in} in Fig. 2). $|V_{gs}|$ will be fixed to larger value for the higher driving capability of the limiting transistor. It is also worth mentioning that the stressed device will be suffered by similar damage (i.e. similar I_g of N2 (30uA ~ 50uA) in Figs. 2 and 4) under same stress voltage condition. Hence, with higher driving capbility (i.e. larger $|V_{gs}|$) of the limiting transistor, the voltage drop across the limiting transistor (i.e. P1 in Fig. 2) will be lower to achieve the similar I_d (i.e. I_g of N2) after SBD under same stress voltage. This hypothesis correlates well with the experimental results as shown in Fig. 5, where it can be seen that the V_g drop on the stressed device is lower with higher clamp current (i.e. higher driving capability of the limiting transistor). It explains post-SBD I_g with clamp current of 300uA climbs faster (smallest V_g drop; highest V_g) than 200uA (Fig. 4). It can be also seen that post-SBD I_g with clamp current of 50uA and 100uA remains relatively flat (Fig. 4) due to higher voltage drop as shown in Fig. 5.

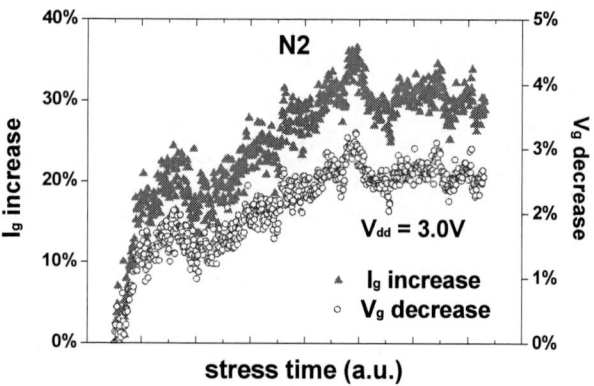

Figure 6. I_g (I_d of P1) increase tracks V_g decrease ($|\Delta V_g|$, i.e. $|\Delta V_{ds}|$ of P1) perfectly and vise versa. Cross stage interaction between P1 and N2 resulted in this seemingly counterintuitive relationship of N2. This negative feedback mechanism keeps I_g in a relative stable region instead of thermal runaway.

From Figs. 4 and 5, it can be also seen that the device progressive BD lifetime will depend on the voltage drop across the limiting transistor, i.e. with higher I_g after BD results in higher V_g drop across the gate oxide and that will again arrest the further BD (i.e. I_g increase or thermal runway) and keeps the device only in SBD region. It implies I_g (I_d of P1) increase tracks V_g decrease ($|\Delta V_g|$, i.e. $|\Delta V_{ds}|$ of P1) perfectly and vise versa. We can see this phenomenon clearly in Fig. 6. This negative feedback mechanism suppresses huge sudden I_g increment and keeps stressed device away from HBD. Hence, it can be concluded that as long as there is less probability of the occurrence of HBD, then the lifetime of the logic circuits will be much longer than the predicted lifetime based on discrete device model.

HSPICE simulations are conducted by letting $I_g = I\_clamp$ to verify the impact of current limiting device on TDDB lifetime improvement of logic circuit compared to lifetime extracted by discrete device model. From Fig. 7, it can be seen that simulation results not matching well with the experimental data. Simulation results show dependence of the BD severity (i.e. I_g increase) with the clamp current limit similar to literature report [1], but experimentally it is not observed. This discrepancy between simulation data and experimental results may be due to hardness of breakdown considered in simulation is much severe (i.e. close to HBD) than actual experimental BD hardness as experimentally it is observed that device suffered only SBD under moderate voltage CVS stress due to current limiting transistor operation region.

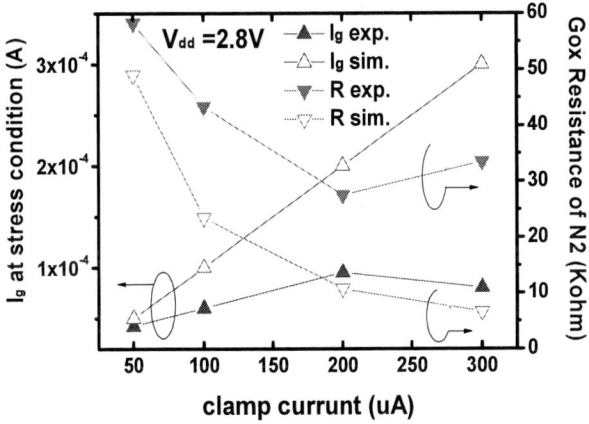

Figure 7. I_g and GOX resistance comparisons of N2 after SBD between experiment and simulation show the clamp current dependence of oxide damage can exist providing the damage is big enough to push I_g reach clamp current limit. Resistance is defined as V_g/I_g. It reflects the oxide damage.

Fig. 8 shows that $|V_{ds}|$ of P1 in Fig. 2 (i.e. $|\Delta V_g|$ of N2 in Fig. 2) behaviors of simulation and experimental results have different trend with the clamp current after breakdown. Experimental $|V_{ds}|$ decreases with clamp current increase, but the simulated result shows that $|V_{ds}|$ increases with clamp current increment. This different trend of $|V_{ds}|$ of the simulation and experimental data with clamp current limit of the current limiting transistor may be attributed to its operation in the different region, i.e. saturation and linear region, respectively. If the oxide damage during experiment is not big enough and I_g is not reaching to I_clamp, then P1 is operating in the linear region due to smaller $|V_{ds}|$ of P1 (i.e. $|\Delta V_g|$ of N2) as V_s and $|V_{gs}|$ of P1 are fixed, so the I_g determines the $|V_{ds}|$, smaller I_g results in smaller $|V_{ds}|$. More over, higher clamp current implies that $|V_{gs}|$ is larger, so $|V_{ds}|$ have to be smaller to achieve the similar I_g as shown in Fig. 4, and results in $|V_{ds}|$ decrement with clamp current (Fig. 8). On

the contrary, simulated $|V_{ds}|$ increases with clamp current because of higher I_g due to consideration of $I_g = I\_clamp$, which results in larger $|V_{ds}|$ and drags P1 to operate in the saturation region by providing the condition of $|V_{gd}| \sim |V_t|$ (Fig. 8).

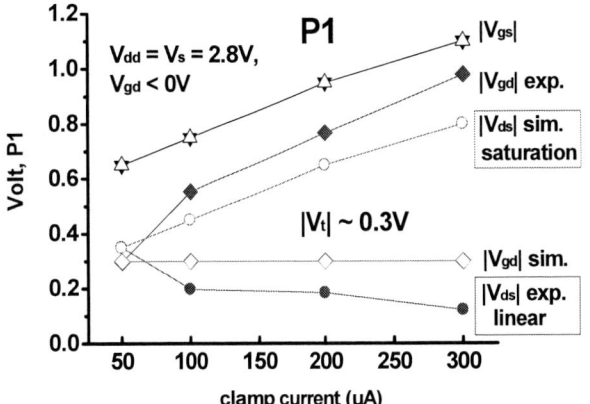

Figure 8. Bias comparison of P1 between experiment and simulation (in simulation I_g is considered to be equal to I_clamp) shows they have opposite $|\Delta V_g|$ (i.e. $|V_{ds}|$ of P1) behaviors as a function of clamp current. If the oxide damage during stress is not big enough and I_g is not reaching to I_clamp, then P1 is operating in linear region due to smaller $|\Delta V_g|$ of N2 (i.e. $|V_{ds}|$ of P1). On the contrary, simulated $|\Delta V_g|$ increases with clamp current because of higher I_g due to consideartion of $I_g = I\_clamp$, which results in larger $|V_{ds}|$.

TABLE II. Observations in Ref [1] (3.4V on 17Å GOX) showed positive correlation (+) between clamp current and I_g. Comparison between experiment in this work and simulation indicates clamp current dependence of I_g do exist providing P1 is in saturation region which is quite likely under even higher stress voltage.

Vdd = 2.8V		experiment	simulation
operation region	Id of P1	linear	saturation
clamp current dependence	Ig of N2	None, depends on Vdd	(+)
	|Vds| of P1 |ΔVg| of N2	(-)	(+)

Table II summaries the impact of the operating region of P1 on the clamp current dependency with severity of BD (i.e. I_g increase in Fig. 7) and V_g drop (Fig. 8) after BD. If the severity of BD of N2 (in Fig. 2) will be higher the voltage drop across P1 will be higher and that will lead P1 to be operated in the saturation region due to larger $|V_{ds}|$ and results in clamp current dependency on I_g. Thus, it can be said that under higher voltage stress devices will suffer severe breakdown and show dependency on the clamp current similar to literature report [1] due to limiting transistor operation in saturation region. But, under moderate voltage stress the transistor N2 will suffer only SBD, which results in lower voltage drop across the P1 and leads the P1 to operate in the linear region only, which will give clamp current independency on I_g just after SBD. Therefore, it can be said that even we get 1st SBD at lower voltage stress, it will take very long time to reach to 2nd BD or HBD due to operation of current limiting transistor in the linear region, which provides limited current and arrests the BD growth of the gate oxide. Hence, it can be concluded that SBD has negligible impact on circuit functionality and results in extra reliability margin to the designer. To further investigate the impact of SBD on digital circuit, HSPICE simulations of two inverter-based oscillators are conducted. Resistances are added to mimic leakage paths in gate oxides (Fig. 9). Fig. 10 shows the output frequency of 51-stage oscillator can increase even SBD occurs in every possible location. Despite the existence of multiple leakage paths (i.e. $I_g \sim 10\mu A$, which is already larger than the post SBD gate leakage of $\sim 1\mu A$ at 1.1V as shown in Fig. 4), there is insignificant impact on oscillator performance.

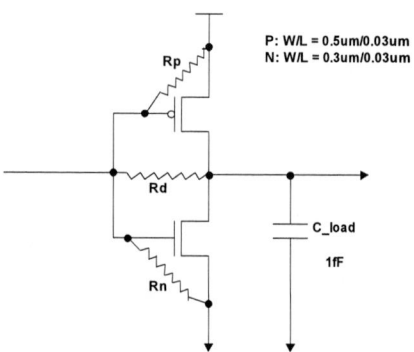

Figure 9. Schematic shows the inverter in our simulation. Resistances mimic the gate laekage paths for N- and P-devices.

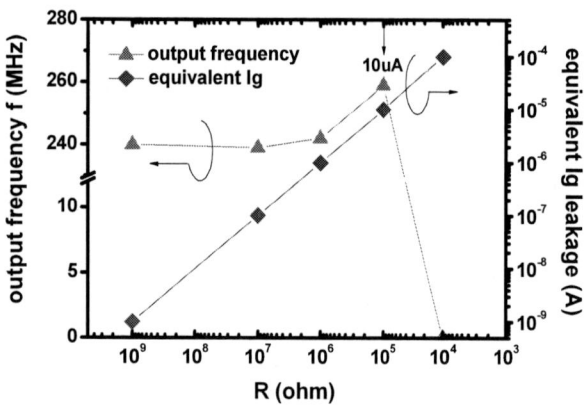

Figure 10. Simulation of output frequency of a 51-stage oscillator. Resistances between gate to drain and gate to source are added in each inverter. Note that gate to well parallel resistance is lumped into R_{gd} and R_{gs}. Each resistance is set to be equal. Simulation result shows the output frequency of 51-stage oscillator can increase even SBD occurs in every possible location.

In addition, from Fig. 11, it can be seen that SBD has negligible impact on circuit functionality, but significant impact on standby leakage current, which results in higher power consumption. It indicates that at lower voltage if the

gate oxide breakdown is not reach to HBD, device functionality will be not impacted but we have to consider the increase of power consumption. Therefore, it can be said that the circuit reliability prediction at nominal operating condition on the basis of high voltage stress data will be much conservative.

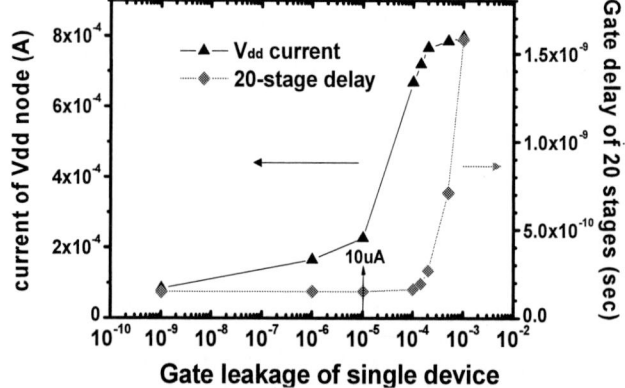

Figure 11. Simulation of gate delay as a function of gate leakage of a 201-stage inverter. Resistance between device gate and drain is added repeatedly in one inverter after every 20 inverters of this 201-stage oscillator. Simulation result shows the gate delay is negligible with a 10μA gate leakage. Nevertheless, current measured at V_{dd} node increases monotonically.

IV. CONCLUSION

Experimental and HSPICE simulation results show that with moderate stress voltage gate oxide only suffer SBD due to the negative feedback relationship between I_g increment and V_g. The possibility of HBD occurrence in logic circuit is very remote. Hence, circuit functionality will be immune from gate oxide BD at nominal operating voltage but we have to take consideration of standby leakage current increment. Our study suggests that circuit will survive much longer at operating condition than the predicted lifetime by the gate oxide BD model on the basis of single device under high voltage stress, which leads the device to HBD. Therefore, designers have to think about the standby current increment only and do not have to worry about the gate oxide BD impact on circuit functionality. The concept of SBD tolerance described in this study can also apply to other circuit types (with different failure criteria). But the feasibility of this concept will be influenced by the I_g sensitivity of other types of circuits. Finally, the circuit-like TDDB test pattern depicted in this study is highly recommended for logic circuit reliability assessment to reflect true oxide BD behaviors.

ACKNOWLEDGMENT

The authors would like to express their gratitude to N. K. Emani for the drawing of test patterns and valuable discussion, and to TSMC manufacturing for the wafer processing.

REFERENCES

[1] B. P. Linder, D. J. Frank, J. H. Stathis and S. A. Cohen, "Transistor-Limited Constant Voltage Stress of Gate Dielectrics." Symposium, on VLSI Technology Digest of Technical Papers, p.93, 2001.

[2] B. Kaczer, R. Degraeve, K. Van de Mieroop, P. J. Roussel and G. Groeseneken, "Impact of MOSFET Gate Oxide Breakdown on Digital Circuit Operation and Reliability." IEEE Transactions on Electron Devices, Vol. 49, No. 3, p. 500, 2002.

[3] B.P. Linder, S. Lombardo, J. H. Stathis, A. Vayshenker and D. J. Frank, "Voltage Dependence of Hard Breakdown Growth and the Reliability Implication in Thin Dielectrics" IEEE Electron Device Letter, Vol. 23, No. 11, p. 661, 2002.

[4] V.L. Lo, K. L. Pey, R. Ranjan, J. R. Shih and K. Wu, "Critical Gate Voltage and Digital Breakdown: Extending Post-Breakdown Reliability Margin in Ultrathin Gate Dielectric with Thickness <1.6nm." IRPS, p. 696, 2009.

In situ Screening Techniques for Defective Oxides in Devices for Automotive Applications

V. Malandruccolo*, M. Ciappa, W. Fichtner
Institute of Technology (ETH) – Integrated Systems Laboratory
Zurich, Switzerland
* vezio@iis.ee.ethz.ch

H. Rothleitner
Infineon Technologies
Villach, Austria

Abstract— Efficient screening procedures for the control of the defectivity are vital to limit early failures especially in critical automotive applications. Traditional strategies based on burn-in and in-line tests are able to provide the required level of reliability but they are expensive and time consuming. This paper presents novel built-in circuitries to screen out oxide defects in integrated circuits for the most important building blocks used in automotive applications. The proposed techniques are based on an embedded circuitry that includes control logic, high voltage generation, and leakage current monitoring. The concept and advantages of the proposed screening procedure are described in very detail and demonstrated experimentally in conjunction with the integration of test-chips.

Keywords: : Built-In Reliability, Gate Oxide Reliability, STI Defects, Capacitor Reliability, Burn-In

I. INTRODUCTION

All modern automotive systems fully depend on electronic systems for their basic operation. The failure of the related equipment has important cost implications for the producers, which could be faced among other with safety issues for the car passengers, as well as expensive call back actions of the unreliable parts [1, 2]

In recent years, due to the implementation of efficient design for reliability strategies, wear-out failures almost disappeared in integrated devices during the typical life cycle of a car ranging from 10 to 15 kilo-hours. Stated otherwise, the majority of the reliability issues that are observed at present in properly designed products, integrated into well matching and well controlled processes, mainly occur due to processing incidents and process-related defects, e.g. in the form of particles, lithography, or defects [3].

The advent of new Smart Power Technologies in the automotive environment enables to design high (HV) and low voltage (LV) devices into the same die. HV devices are often used to match the voltage level of signals from the outside world with the low voltage levels required for the integration of complex analog and digital circuits. Thanks to this variety of devices, a number of additional features can be added to existing circuit solutions in order to extend their functionalities, reduce their cost, and make them more flexible.

Nowadays, most of the area of modern automotive ICs is occupied by HV devices, implemented with LDMOS, for power applications (50%-60% relative area occupation). LV devices are used to implement the control logic (15%-20% relative area occupation) and the analog/mixed-signal part (25%-30% relative area occupation). In critical applications most of the area of the analog/mixed-signal section is occupied by Analog to Digital Converters (ADCs), e.g. a modern four channel airbag driver includes more than twenty ADCs in the regulation loop of the squib current. As a consequence, vital components exhibit a large area of thin and thick gate oxide combined with a large volume of capacitor dielectric material and STI isolation. Since both dielectrics are inherently prone to defects, efficient screening procedures are needed to reach the very low levels of residual defectivity required by the zero-defect strategies.

This paper focuses on the improvement of the procedures to control the defectivity in LDMOS transistors, low voltage transistors, and Vertical Parallel Plate (VPP) Capacitors used in the most advanced processes for automotive devices (Table 1). As a first step, traditional screening procedures will be reviewed, highlighting their intrinsic limitations. In the following sections, the principles and the main advantages of the newly developed built-in screening methodologies will be

TABLE I. PROPOSED BUILT-IN SCREENING METHODOLOGIES DEPENDING ON THE AFFECTED OXIDE

Block	Device	Affected oxide	Traditional screening	Proposed built-in screening
HV	LDMOS	Thick gate Oxide	GST	BI-GST
HV	LDMOS	STI Oxide	DLT	BI-DLT
Mixed signal	ADC	Intermetal Dielectric	None	BI-VPPST (Built-in VPP Stress Test)
Mixed signal	ADC	Thin Gate Oxides	IDDQ ΔIDDQ	Integrated IDDQ sensor + background current compensation

978-1-4244-9113-1/11 $26.00 © 2011 IEEE

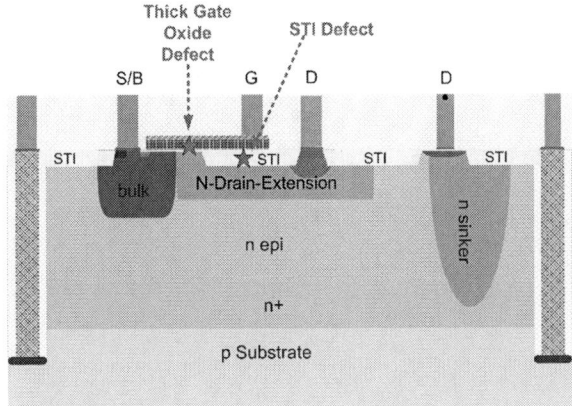

Figure 1. Thick Gate Oxide and STI defects in a LDMOS

presented referring to previous works published by the same authors for the circuit details.

A. Traditional Screening Procedures for Oxides in HV Devices

As shown in Fig. 1 two kinds of oxides are used in HV devices: the Thick Gate Oxide and the Shallow Trench Isolation. Both oxides can be affected by defects introduced during manufacturing phases that could jeopardize the functionality of the device itself.

1) Traditional screening procedures for Thick Gate Oxides

According to the usual defective oxide classification, Class A includes such oxides, which are expected to fail (usually due to pinholes) for an applied field strength (EBD) lower than 1 MV/cm. Class B includes the so-called extrinsic oxides that fail for an EBD lower than the intrinsic value, and Class C includes those oxides, which fail at an EBD above 10 MV/cm (for the thickness range of interest in this paper) due to intrinsic dielectric breakdown. At present the required failure rate during the life cycle of the product is assured by properly designed screening procedures before and after packaging of the chip, which force oxides in the Class A and B to fail, without introducing any substantial pre-damaging of the robust subpopulation [4].

Nowadays, the most common procedure to screen gate oxide defects at chip level is still the so-called gate stress test (GST), where a high voltage pulse is applied to dedicated test pads of the LDMOS gate through an Automatic Test Equipment (ATE). Subsequently, the eventual occurrence of the oxide breakdown is detected by the measurement of the leakage current through the gate oxide [5]. A typical configuration of such a circuitry is shown in Fig. 2.

The popularity of this solution is justified by the fact that is very simple to be implemented, but on the other hand it does not solve a multitude of challenges at different levels.

Design level: The diode D3 in Fig. 2 limits the driving voltage of the gate of the LS-SWITCH. Therefore, in order to get the same Ron performances, the area of the HV-LDMOS has to be increased. Moreover, the resistance R is critical for applications where fast turn-on/off switching time of the transistor is required.

Wafer level test: The wafer-probecard requires extra needles exclusively dedicated to the GST. In addition, one has to rely on external Automatic Test Equipment (ATE) in order to perform a non trivial measurement. This fact impacts the test cost and reduces the capability to screen several devices at the same time (parallelization).

Package Level test: Since the stress introduced by bonding and plastic molding compounds can affect the gate oxide reliability, performing GST is essential after packaging, as well. However, for pin number reduction and in order to avoid electrostatic discharge events, both the GST and the TP1 pad are not bonded so that the gates of the HV-LDMOS switches are no longer accessible.

Burn-in level test: Static temperature stress under constant bias voltage is a way to accelerate the Time Dependent Dielectric Breakdown (TDDB) during burn-in. An issue related to this approach is that the stress is applied at the same time even to non-defective gate oxides. This reduces the residual lifetime of the surviving gate oxides. Moreover, the detection of possible breakdown events can only be performed indirectly through a parametric or functional test, since accurate leakage current measurement is no longer possible.

2) Traditional screening procedures for STI

The Shallow Trench Isolation (STI) processes are getting attention as an essential technology to integrate isolated high-density semiconductor devices. STI is created early during the semiconductor device fabrication process, before transistors are formed. The key steps of the STI process involve etching a pattern of trenches in the silicon, deposition of one or more dielectric materials (such as silicon dioxide) to fill the trenches, and removing the excess dielectric using a technique such as chemical-mechanical planarization [6]. Defects can be either inherently present in the bulk silicon or generated during some critical process steps like epitaxial growth, implantation, and the formation of shallow trench isolations [7-8]. During the lifetime of a device, those defects can coalesce or act as a gettering site for dopant atoms and contaminations. This can result into the formation of local highly conductive paths, which under some circumstances can heavily affect the performance of the integrated circuit. The correlation between the presence of such defects and the increase of the transistor leakage current by various orders of magnitude is reported in literature [9-10]. For this reason, screening for STI defects is

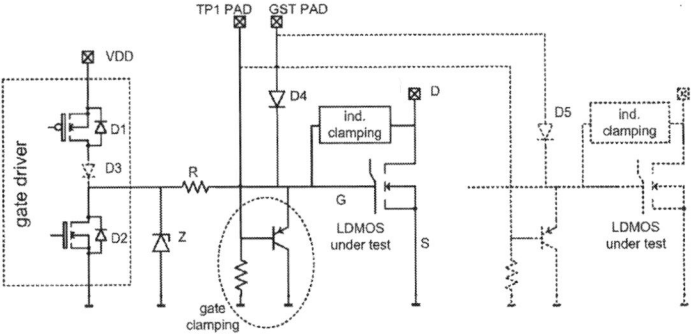

Figure 2. Traditional gate stress test scheme (GST)

978-1-4244-9113-1/11 $26.00 © 2011 IEEE

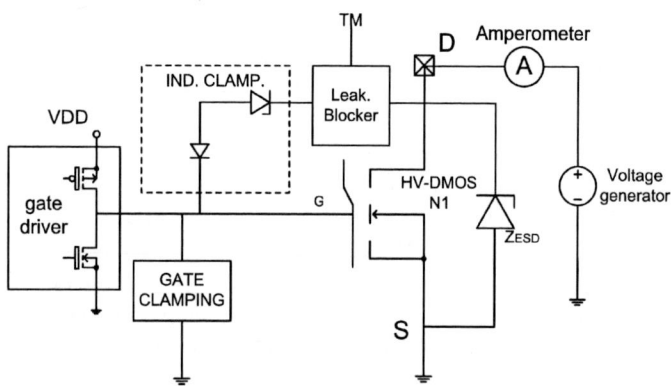

Figure 3.   Traditional Drain Leakage Test scheme (DLT)

usually accomplished by measuring the drain leakage current while biasing the device under sub-threshold conditions, after application of a high voltage pulse to the drain terminal. This procedure is usually defined as Drain Leakage Test (DLT). The configuration used for this test is shown in Fig. 3.

The main issue of the traditional DLT is related to the fact that this procedure can only be carried out if the direct access to the drain contact is directly (and separately) ensured [11]. Actually, the direct contact to the drain contact of LDMOS is just granted through the pins if they are used to drive external loads. Nevertheless, in the case of modern designs using large power transistors for internal purposes, this represent a major limitation, since the design of an addition pin just for testing purposed is rarely viable.

B.   Traditional Screening Procedures for Oxides In ADCs

The vast majority of A/D converters available today in the automotive field are implemented as successive approximation register (SAR) ADCs. Their popularity is deeply rooted in the fact that they offer a good trade off between accuracy, speed, and die size requirements. A major concern in this respect is the integrity of the thick dielectric in Vertical Parallel Plate (VPP) capacitor banks because of their large area occupation. In fact, it has to be noticed that more than 50% of the area consumption related to ADC circuits is due to VPP capacitors. Because of the complexity of the searching algorithm, especially when implementing the Digital Error Correction Algorithm, the control logic is in general very demanding in terms of area. Thus, it is of primary importance to use test techniques, which are able to detect defects that escape to functional tests and may result in early field failures.

1)   Traditional screening procedures for inter-metal dielectric in VPP capacitors

Vertical Parallel Plate Capacitors (VPP) are commonly used to implement the capacitor array (Fig. 4). The capacitor is realized by a stacked comb-comb configuration linked between the thin wire levels by staggered vias. Their popularity is deeply rooted in the fact that the controlling switches and logic can be placed below the capacitors and because this technology is suitable for high integration densities. In fact, the lateral dimensions shrink at every technology step, but not the vertical ones.

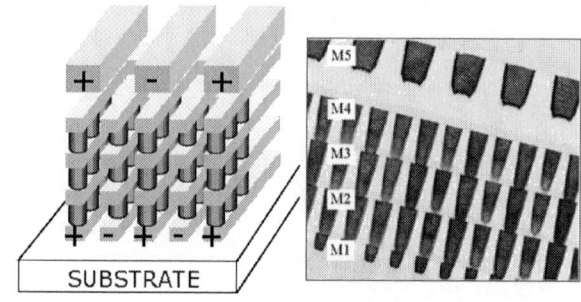

Figure 4.   VPP Capacitor

From manufacturing and reliability point of view these devices can be rather challenging since they may contain millions of vias and meters of metal interconnects. Unfortunately the silicon dielectric is not uniform but contains micro-inhomogeneities, entrapped particulates, metal precipitates and local thinning. These non-uniformities cause breakdown voltage reduction, and are the cause of early failures.

It has been demonstrated in [12] that the direct application of high voltage pulses is very effective for screening, because of the higher acceleration factor related to voltage activated phenomena. However, this approach has never been attempted before, due to the difficult task to apply the required high voltage pulses to the capacitor bank. Three main challenges have to be tackled to implement the new screening procedure. The first problem to be solved refers to the avoidance of parasitics due the additional circuitry for screening. The second problem is to provide a design with a negligible area overhead. Finally, a solution has to be found to supply to the VPP capacitor bank high voltage pulses up to 60V by using low voltage technology.

2)   Traditional screening procedures for thin gate oxide in the SAR logic

IDDQ testing is known as a powerful method for obtaining higher quality and reliability. IDDQ is used profitably in conjunction with High Voltage Stress or Burn-In for decreasing failure rate of a product during early field life [13]. However this methodology presents severe drawbacks which limit its application for new and emerging technologies.

Problems related to the use of IDDQ - Emerging submicron technologies have increased leakage currents, mainly due to sub-threshold conduction of MOSFETs. A large background current makes distinguishing defect current extremely difficult, thus reducing the defect resolution of IDDQ test.

Problems related to the use of ΔIDDQ - Current signatures and ΔIDDQ were developed in the past to overcome this issue. Nevertheless the two approaches are able to detect only active defects (involving switching nodes) but do not cover passive defects (involving non-switching circuit nodes) [13]. Furthermore both techniques require the use of complicated data acquisition and post-processing procedures. This result in high complexity of the test equipment and in increase of the test cost.

Problems due to the instrumentation - IDDQ can be measured only after inputs are stabilized and internal toggling is settled. This makes IDDQ testing comparatively slow. Furthermore, off-chip measurements are degraded by the impedance loading of the tester probe (from 20 to 200 pF), and current leakages into/out of the tester. Also, the high noise of the tester load board is degrading the IDDQ measurement performances.

Problems related to the scan chain introduction - The use of scan design has two types of penalties. The scan hardware increases the chip size (the expected area overhead is in the range between 15% and 20%), and slows the signals down (a scan design can reduce the clock speed by 5 to 10%).

II. PROPOSED BUILT-IN SOLUTIONS FOR THE SCREENING OF OXIDE DEFECTS IN HV-DEVICES

In order to solve the test limitations discussed in the previous section, a new approach to the screening of defective oxides in automotive ICs is presented, which is based on dedicated embedded circuitry to perform on chip the voltage stress and the measurement of the leakage current through the stressed device. Due to the fact that it relies on a built-in circuit, the proposed solution can be applied both at chip level and to packaged devices, targeting directly the point of interest. Furthermore, since the whole process is managed by an internal circuitry, it does not require any additional testing equipment and can be run in parallel on a very large number of devices.

A. Proposed Built-In Screening of Oxides in HV Devices

1) Working Principle

The proposed solution consists of the integration of all the functionality required to perform a GST and DLT into each device. Following this approach, every chip should become responsible for its own stress test. The principles of the implementation of this solution are presented in Fig. 5.

The sections responsible for the Built-In GST (BI-GST) and Built-In (BI-DLT), work completely independently, in a sense that the operation of the BI-GST section is not affected by the BI-DTL, and vice versa. The only restriction is that both tests have to be carried out sequentially.

In order to save additional pads/pins for stressing, the high voltage, which has to be applied to the LDMOS, is forced via the battery pin VS. The stress sequence is enabled by a

command sent via a serial interface. A digital controller asserts the stress controlling signals GS, GS_ISO for the BI-GST and DL and DL_ISO for the BI-DLT according to the basic waveforms show in Fig. 5. GS, GS_ISO signals control respectively the protection switch and the stress switch for the BI-GST. The former is responsible for the disconnection of the gate driver unit from the LDMOS gate. The latter controls the application of the stress voltage on the gate. The signals DL and DL_ISO control the application of the test voltage on the drain and the opening of the inductive clamping.

The protection switch, the gate clamping and the inductive clamping are designed in such a way that no current flows through them during the stress. The currents Ileak_gate and Ileak_drain flowing from the battery pin respectively to the gate during BI-GST and to the drain during BI-DLT, are measured, mirrored by the current mirror and compared with a threshold current Ith by the current comparator. The result of the comparison is a digital signal, COMP_O that is sent to the external world via a serial interface, indicating whether the leakage current is higher than a given threshold current.

A detailed description of the circuit implementing the concept together with circuit simulation results can be found in [11].

Advantages of the BI-GST. The BI-GST approach is beneficial under many aspects. Firstly, it reduces the dependency on external ATE, reducing at the same time the testing time and the resulting testing cost. This is due in particular to the fact that the average testing time required is shorter than in the traditional one, and that it introduces the capability to test more devices in parallel. Furthermore, the BI-GST approach enables the manufacturer to perform a targeted GST after packaging of the device what it has not been possible until nowadays by the use of the traditional GST solution. This fact removes the need to perform burn-in and reduces the test time to some milliseconds as it has been shown in [5].

Advantages of the BI-DLT. The BI-DLT presents major advantages either in the case of it is used as standalone screening technique, or in conjunction with a traditional burn-in procedure. The design solution proposed here for the standalone version refers to a LOW SIDE (LS) SWITCH with direct access to the drain and source through external pins. Obviously, this design could be easily extended to the case of embedded LDMOS transistors without any direct access from outside. This just requires an additional connection to ground of the source terminal, as well as an isolation of the drain node to block any unwanted parasitic leakage current during the test. The BI-DLT has been integrated in a process for temperatures up to 200 °C, it can be used very proficiently in conjunction with traditional burn-in storage under bias. In this case, BI-DLT just requires the high voltage to be supplied through the battery pin.

2) Experimental validation

Before extensive use in commercial products, the proposed scheme has been firstly prototyped and then experimentally validated by implementation on a test-chip.

This solution has been introduced in a typical LS-SWITCH application, designed to obtain RON = 0.4Ω. The additional

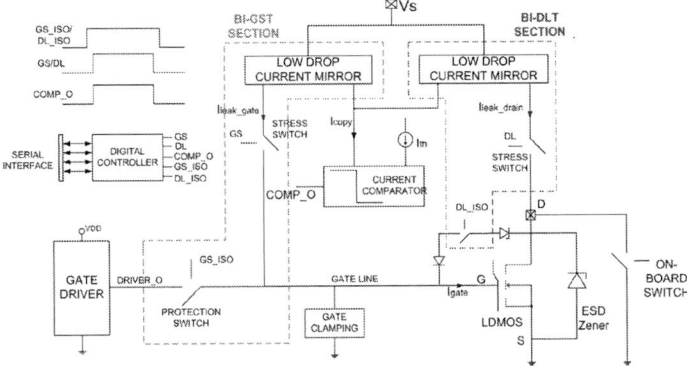

Figure 5. Working principle of the solution implementing BI-GST and BI-DLT

978-1-4244-9113-1/11 $26.00 © 2011 IEEE   31

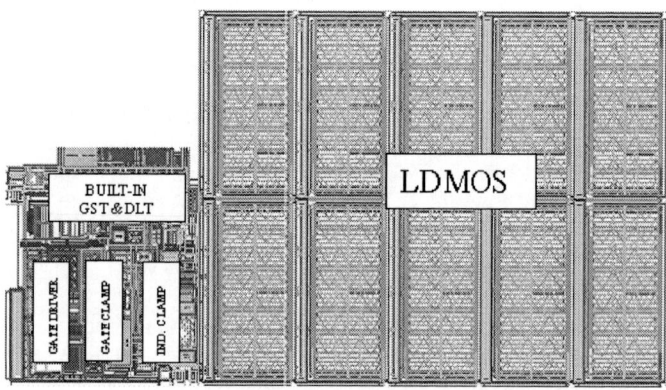

Figure 6. Layout of the solution implementing BI-GST and BI-DLT

area occupied by the built-in test circuitry is 10% of the area occupied by a single LS-SWITCH as shown in Fig. 6. In this case, the reduction in terms of reliability due to the additional circuitry is negligible, since the gate oxide area associated with built-in test unit is a factor of 500 smaller than the gate oxide area in a HV-LDMOS. The relative area overhead is further reduced, if multiple LS-SWITCH can be tested in parallel. The small size of the stress circuitry and the fact that no device is degraded during the stress phase (excepted for the LS-SWITCH) makes the solution attractive with negligible risks for additional reliability problems.

The experimental results reported in Fig. 7, refer to the case of a gate oxide defect screening process by BI-GST in a LDMOS. Measurements are shown that arise from an intact and leaking gate oxides. Fig. 7a, 7b, and 7c show the timing of the most significant control signals introduced in the previous section during the high voltage stress phase for the BI-GST.

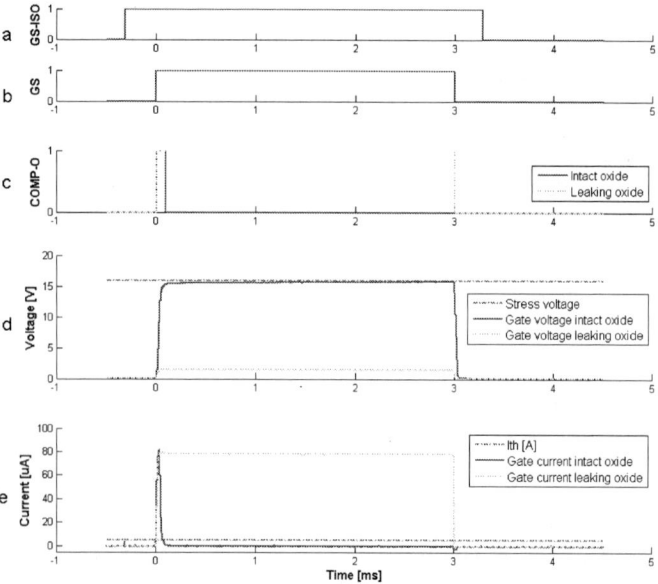

Figure 7. Measurement results in the case of BI-GST application for intact and a leaking oxide. a,b: The timing of BI-GST test; c: The result of test. COMP_O remains high at the end of the test in the case of leaking oxide; d: Voltage applied on the gate; e: Copy of the current flowing into the gate

Fig 7d shows the application of the stress voltage to the gate of the intact and leaking oxide. The voltage on the gate oxide of the leaking devices does not rise up to the full level of the stress voltage because of the current limiting feature of the stress generator. Finally Fig. 7e, shows the case of a leaking oxide, where the leakage current is higher than the chosen threshold. On the contrary, no appreciable current is detected for an intact device.

III. PROPOSED BUILT-IN SOLUTIONS FOR THE SCREENING OF OXIDE DEFECTS IN SAR ADCs

The goal of the new methodology is to screen out defects in the dielectric of Vertical Parallel Plate (VPP) capacitors and MOSFETs used in Successive Approximation Register (SAR) Analog to Digital Converter (ADC). The proposed strategies are implemented through embedded circuits for stressing and characterizing the devices under test.

A. Proposed Technique for the Screening of the VPP Capacitor Array Bank

1) The proposed BI-VPPST concept

A novel approach is proposed to screen out defective capacitors used in SAR ADCs. The BI-VPPST (Built-In VPP stress test) method consists of two steps: in a first phase, the capacitors are stressed with an HV pulse in order to force the breakdown of defective devices, and in a second phase the effect of the stress is measured with an advanced functional test, to check for possible breakdown events. The principles of the implementation of this solution are presented in Fig.8.

In order to save additional pads/pins for stressing, the high voltage to be applied to the capacitors is forced via any HV capable pin (stess_pad), like the battery pin for example. The stress sequence is enabled by a command sent via a serial interface. A digital controller asserts the stress_iso and stress_on signals according to the basic waveforms shown in Fig. 8. The former controls the disconnection of the VDD_HV and VSS_HV pins of the ADC from the default supply rails VDD and VSS respectively. The latter enables the connection to the stress voltage. The same voltage stress generator can be commonly used to stress all the ADCs. The working principles of the ADC under stress and the basics for the voltage stress

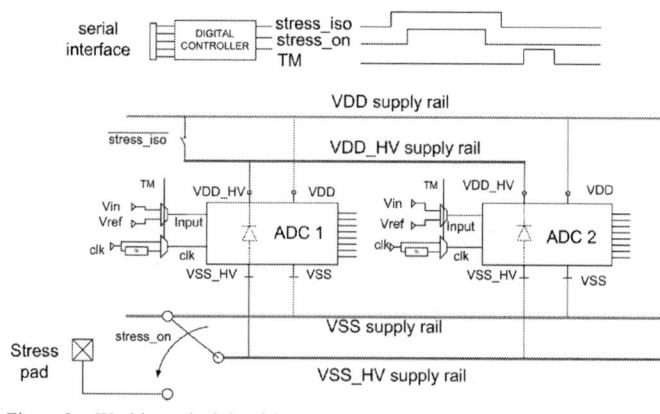

Figure 8. Working principle of the BI-VPPST for the screening of defects in VPP capacitors

978-1-4244-9113-1/11 $26.00 © 2011 IEEE

generator have been described in very detail in [14].

Once the stress phase is completed, a functional test is performed to detect the presence of damages. The same digital controller will force the ADCs to perform a conversion, will connect the inputs of the ADCs to known signals generated internally, and will activate and ADC test condition under reduced clock speed (Fig. 8 TM signal is on). The voltage or charge retention of the DAC capacitors is essential for proper conversion results of a SAR ADC. If a charge loss is introduced by some defect in the dielectric, a wrong data conversion will take place. The idea introduced here makes use of this concept, in a way that the integrity of the capacitor bank of the SAR ADC is assessed by checking whether the conversion is performed correctly, or not. A wrong conversion performed on some reference input signal will be interpreted as a signature of a defect, which changes the charge stored into the capacitor array. To increase the sensitivity of this procedure a lower clock frequency as the nominal value has been used.

The proposed solution is beneficial for different reasons:

- The reliability test of a critical component in modern automotive designs can be covered with a completely Built-In concept.

- The solution requires a very low area overhead.

- No design change.

- No expensive burn-in procedure.

- Low reliability risk.

- The data collected during the production phase could be profitably used to tune the back-end process.

2) Experimental validation

Before extensive use in commercial products, the proposed scheme has been firstly prototyped and then experimentally validated by implementation on a test-chip. The details of the implementations of the test-chip are reported in Fig 9.

The working principle of the block is described based on the measurement results shown in Fig. 10. In this particular case, a 60V pulse with 2ms pulse duration has been used. In general, the voltage can be set to any level, which can be sustained by the HV section of the circuit. The pulse duration can be adjusted in such a way to reach the desired acceleration factor during the stress.

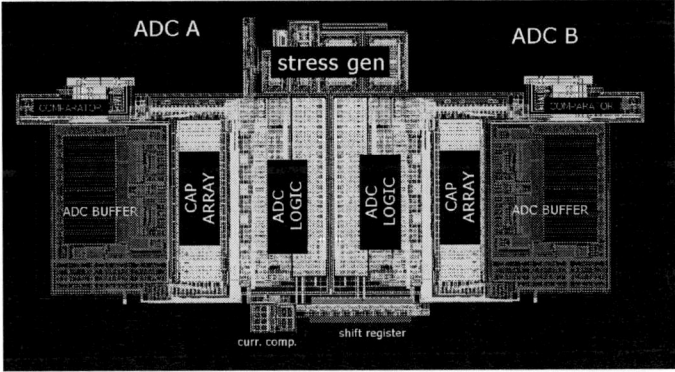

Figure 9. Layout of the proposed BI-VPPST solution as integrated into the test-chip

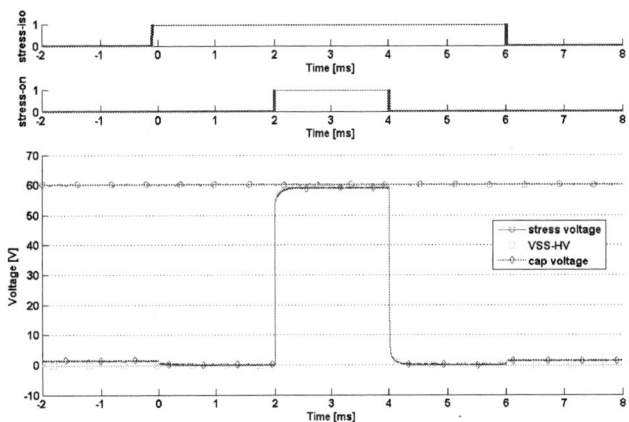

Figure 10. Measurement results showing the controlling signals and the stress voltage applied on the capacitor

In functional mode, stress_iso and stress_on signals are low. In this phase VSS_HV and VDD_HV are connected to the default supply rails VSS and VDD respectively. The stress sequence consists of two different phases. When the stress_iso signal is asserted (time = 0ms), the VSS_HV and VDD_HV node are disconnected from the default supply rails, and subsequently after the rising edge of the stress_on signal (time = 2ms) the stress voltage is injected into the VSS_HV net, which is connected to the pseudo-substrate. Once the pseudo-substrate has been charged, any node of the first stage of the comparator is able to float safely to the stress voltage. In this phase, the capacitors of the capacitor array are stressed. The final phase of the stress is defined by the falling edge of the stress_on signal (time = 4ms), since the connection to the stress_pad (VSTRESS) is removed. Subsequently, at the falling edge of the stress_iso signal (time = 6ms), the connections to the default supply rails are re-established.

The fact that the LV components are not damaged by the HV stress applied on the bank of capacitors is proved by the fact that no ADC performance reduction has been measured.

B. Proposed Technique for the Screening of the Control Logic of SAR ADCS

1) The proposed concept

In this paper, a novel built-in ΔIDDQ testing technique and circuit are presented based on the background current compensation concept proposed in [15]. This technique use spatial correlation between different IDDQ measurements performed on neighboring ADCs in the same die, instead of vector-to-vector correlation as is the case of traditional ΔIDDQ. Neighboring ADCs have highly correlated fault-free parameters because they undergo similar processing. Spatial correlation techniques exploit this feature to take into account background IDDQ.

The idea behind the proposed concepts consists in comparing the difference between two IDDQ measurements, performed at the same time on two identical neighboring ADCs in the same state, with a threshold current. If the absolute value of this difference is higher than the given threshold the die is scraped. The principle works only under the plausible assumption that the two logic blocks are not affected by the same defect at the same position. The request of the availability of at least two ADCs is always satisfied in the targeted applications.

In Fig. 11 the block schematic of the proposed technique is presented, while the main controlling signals are reported in Fig. 12.

The state of the ADC logic can be programmed by a proper choice of signals to use as input to the ADC logic. The peculiarity of the ADC logic is to be particularly active during the conversion phase. For this reason it exists the possibility to make the ADC logic assume many different state during test mode, simply emulating what happens during a conversion in functional mode providing different patterns to the input of the ADC logic receiving the feedback signal from the analog part.

In functional mode (TM=0) the supply nodes of the two ADC logics are connected to the supply rail (VDD) through the use of low resistance PMOS switches. During test mode (TM=1) the supply switches are turned on only during the first half of the clock period. In this phase, the ADC logic changes its state, and a low impedance connection to VDD is required to handle the high current required to charge the parasitic capacitances inside the logic. In the second half of the clock, the ADC logic is settled, and the supply node are connected to the inputs of the low drop current comparator. The comparator

is able to perform the function comp_o=sign(I1-I2-Ith) without generating any significant voltage drop between VDD and the supply node of the logic. In this phase (S1=1) the comp_o signal will rise only if an excessive leakage from logic A is detected (comp_o=1 if IDDQA>IDDQB+Ith).

Once the ADC logic has passed through the number of state changes necessary to obtain the required IDDQ coverage, the inputs of the current comparator are swapped. In this phase (S1=0) the comp_o signal will rise only if an excessive leakage from logic B is detected (comp_o=1 if IDDQB>IDDQA+Ith). In Fig. 12 is reported the case of leaky LOGIC B. The comp_o is high on those periods where the ADC state is appropriate to make IDDQ leakage visible.

2) Experimental validation

The proposed concept has been applied to an existing 8 bit SAR ADC design for validation and performance verification. Subsequently, both ADCs and the additional circuitry for the built-in reliability test have been integrated into a test-chip for experimental validation (Fig. 9).

In order to test the sensitivity of the technique, a dedicated circuit has been designed and implemented, which simulates the presence of defects inside the ADC logic and does not alter the functionality of the original design. This circuit is used for characterization purposes only, and it is not a part of the proposed solution. The circuit shown with a dashed line in Fig. 13 enables to introduce a programmable value of leakage current in given nodes of the ADC logic. The leakage introduced is used to simulate the presence of a defect insisting on the affected net. The leakage current introduced can be accurately measured by an external current meter, which measures the IDDQ current flowing inside the logic through dedicated test-pins. The leakage current introduced can be set to the desired value by mirroring a current forced with an external current generator through N1-N3 and P1-P2. Four nets inside the ADC can be connected to the leakage generator via switches (sw1-sw4) that can be easily turned on-off by a command sent via the serial interface.

The correct functionality of the application is demonstrated by the experimental data collected from the test-chip and reported in Fig. 14, which also shows the controlling digital signals used in the concept discussed before. Reported are the

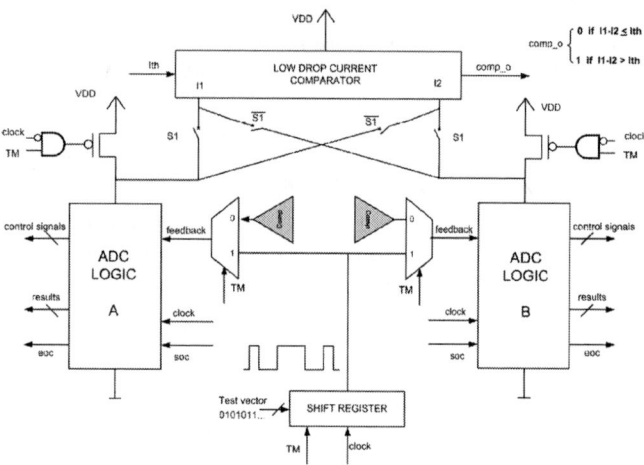

Figure 11. Working principle of the proposed built-in solution for reliability test of SAR logic

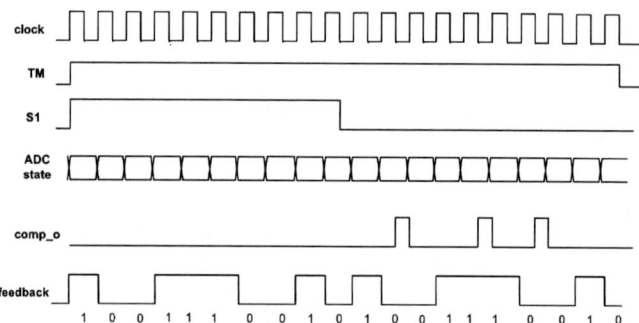

Figure 12. Controlling signals of the built-in Concept. Logic B is assumed to be affected by defects

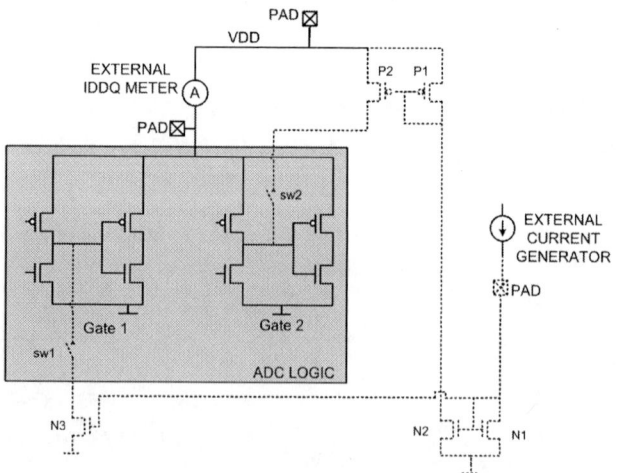

Figure 13. Circuit for the introduction of artificial defects inside the ADC logic

978-1-4244-9113-1/11 $26.00 © 2011 IEEE

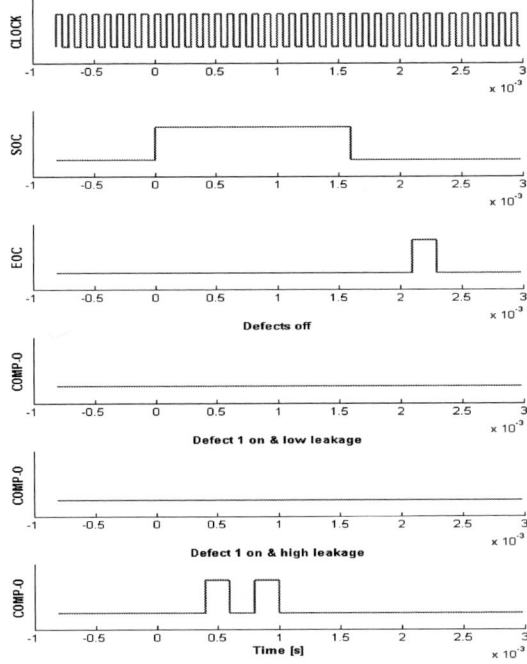

Figure 14. Measurement data showing the controlling signal for the built-in reliability test for the ADC logic

CLOCK, the SOC (start of conversion), the EOC (end of conversion), and the comp_o signals. The conversion required for the test is triggered at the rising edge of the SOC signal and ends at the rising edge of the EOC. The test vector is shifted in the ADC logic during this period. The comp_o signal, which is the result of the test, is reported here for three different cases. In a first case all the defects artificially introduced were turned off, so that the comp_o remains low for the entire duration of the conversion. In a second case as single defect (defect number 1) is activated and the leakage introduced is too small to make the comparator trigger. Also in this case the comp_o remains low for the entire duration of the test. Finally the leakage current introduced is raised above the programmed threshold and in this case the comp_o correctly triggers on those states where the defect is made visible.

IV. CONCLUSIONS

Novel built-in approaches have been proposed to screen out defective oxides in integrated circuits for automotive applications. These techniques are based on a programmable embedded circuitry for built-in screening, which provides high voltage pulse-stressing and accurate quantitative measurement of the leakage current through the oxides. The test sequence is controlled by internal logic and the unit communicates to the outside world through a serial interface. The peculiarities of the circuit design have been discussed with particular focus on the design solutions adopted for high-voltage operation and accurate current measurement by dedicated circuitry.

In opposite to the traditional techniques, the proposed solution can be applied both at chip level, as well as in packaged devices, and do not requires sophisticated ATE to be used. This, in conjunction with the fact that they can be performed in parallel in several devices, results at the same time in a more targeted stress than for the traditional burn-in and in a noticeable decrease of the costs involved with the screening.

REFERENCES

[1] Millington, D.E.; , "Planning for high reliability electronic products [vehicle applications]," Guaranteeing the Reliability of Automotive Electronics, IEE Colloquium on , vol., no., pp.5/1-5/3, 18 May 1988

[2] [F. Kuper, "Automotive IC reliability: elements of the battle towards zero defects", Microelectronics Reliability 48(2008)1459-1463.

[3] J. van der Pol, E. Ooms, T. van't Hof, and F. Kuper, "Impact of screening of latent defects at electrical test on the yield-reliability relation and application to burn-in elimination," in Proc. Int. Reliability Physics Symp., 1998, pp. 370–377.

[4] M. Ciappa, "Some Reliability Aspects of IGBT Modules for High-Power Applications", Chapter 4 pp. 107-140, "Modeling the Gate Oxide Reliability", Hartung-Gorre Editor, Konstanz 2001.

[5] V. Malandruccolo, M.Ciappa, H. Rothleitner, W. Fichtner, "New On-Chip Screening of Gate Oxides in Smart Power Devices for Automotive Applications", Reliability Physics Symposium, 2009. IRPS 2009 pp. 573-578

[6] P Sallagoity, F Gaillard, M Rivoire, M Paoli, M Haond, S McClathie, STI process steps for sub-quarter micron CMOS, Microelectronics Reliability, Volume 38, Issue 2, 27 February 1998, Pages 271-276, ISSN 0026-2714, DOI: 10.1016/S0026-2714(97)00166-2.

[7] Byoung-Ho Kwon; Jong-Hyup Lee; Hee-Jeen Kim; Seoung Soo Kweon; Young-Gyoon Ryu; Jeong-Gun Lee; , "Dishing and erosion in STI CMP," VLSI and CAD, 1999. ICVC '99. 6th International Conference on , vol., no., pp.456-458, 1999

[8] C. -F. Lin, W. -T. Tseng, M. -S. Feng, Y. -L. Wang, A ULSI shallow trench isolation process through the integration of multilayered dielectric process and chemical-mechanical planarization, Thin Solid Films, Volume 347, Issues 1-2, 22 June 1999, Pages 248-252, ISSN 0040-6090, DOI: 10.1016/S0040-6090(99)00029-2.

[9] I. Mica, M. L. Polignano, G. Carnevale, P. Ghezzi, M. Brambilla, F. Cazzaniga, M. Martinelli, G. Pavia and E. Bonera, "Crystal defects and junction properties in the evolution of device fabrication technology" 2002 J. Phys.: Condens. Matter 14 13403-13410

[10] F. Siegelin and A. Stuffer, "Dislocation related Leakage in Advanced CMOS devices", Proceedings of the 31 International Symposium for Testing and Failure Analysis

[11] V. Malandruccolo, M. Ciappa, H. Rothleitner, W. Fichtner, "A new built-in screening methodology to achieve zero defects in the automotive environment", Microelectronics Reliability, Volume 49, Issues 9-11, September-November 2009, Pages 1334-1340

[12] Fischer, A.H.; Lim, Y.K.; Riess, P.; Pompl, T.; Zhang, B.C.; Chua, E.C.; Keller, W.W.; Tan, J.B.; Klee, V.; Tan, Y.C.; Souche, D.; Sohn, D.K.; von Glasow, A.;" TDDB robustness of highly dense 65NM BEOL vertical natural capacitor with competitive area capacitance for RF and mixed-signal applications", Reliability Physics Symposium, 2008. IRPS 2008. IEEE International, Page(s): 126 – 131

[13] Sabade, S.S.; Walker, D.M.H.; , "IDDQ test: will it survive the DSM challenge?," Design & Test of Computers, IEEE , vol.19, no.5, pp. 8- 16, Sep-Oct 2002

[14] Malandruccolo V., Ciappa M., Rothleitner H., Hommel M., Fichtner W., "A new built-in screening methodology for Successive Approximation Register Analog to Digital Converters", Microelectronics Reliability, Volume 50, Issues 9-11, September-November 2010, Pages 1750-1757

[15] Matakias, S.; Tsiatouhas, Y.; Arapoyanni, A.; Haniotakis, T.; , "An embedded IDDQ testing circuit and technique," Electronics, Circuits and Systems, 2005. ICECS 2005. 12th IEEE International Conference on , vol., no., pp.1-4, 11-14 Dec.

978-1-4244-9113-1/11 $26.00 © 2011 IEEE

A TDC-based Test Platform for Dynamic Circuit Aging Characterization

Min Chen, Vijay Reddy, John Carulli, Srikanth Krishnan, Vijay Rentala, Venkatesh Srinivasan
Texas Instruments
Dallas, TX 75243, USA
(214)- 567-9539, min.chen@ti.com

Yu Cao
Department of Electrical Engineering
Arizona State University
Tempe, AZ 85287, USA

Abstract— An on-chip 45nm test platform that directly monitors circuit performance degradation during dynamic operation is demonstrated. In contrast to traditional ring-oscillator (RO) based frequency measurements, it utilizes a Time-to-Digital Converter (TDC) with 2ps resolution to efficiently monitor circuit delay change on-the-fly. This new technique allows the capability of measuring signal edge degradation under various realistic circuit operating scenarios, such as asymmetric aging, dynamic voltage/frequency scaling, dynamic duty cycle factors, and temperature variations.

Keywords-variations; relibility; asymetric aging; NBTI; duty cycle; test structure; RO; TDC

I. INTRODUCTION

Circuit performance degradation due to transistor aging is a significant impediment to high-performance IC design due to increasing concerns of negative-bias-temperature-instability (NBTI) and channel-hot-carrier (CHC) degradation [1]. The degradation rate is strongly affected by dynamic operating conditions, such as the effective duty cycle and the supply voltage (V_{DD}) [1] [2]. Therefore, on-the-fly test techniques to monitor the aging effect in dynamic operating modes are necessary for developing effective design-in-reliability methodologies and reducing timing margins from a worst-case perspective.

There have been intensive studies on test structure design for device and circuit aging. Device level test is important to understand the underlying mechanism of device aging; but it inadequate to provide circuit level information. At the circuit level, traditional test method usually utilizes a ring-oscillator (RO) as the aging sensor, because of its easy implementation and the accessibility to the output frequency [3] [4]. Some RO-based beat circuits are well developed to achieve very high resolution [5] or effectively separate different aging effects [6]. However, due to its closed-loop structure, there are several limitations of RO-based aging test (Fig. 1). A ring oscillator lacks the flexibility to tune the stress frequency, V_{DD}, and duty cycle separately (Fig. 1(a)). In reality, a data path may

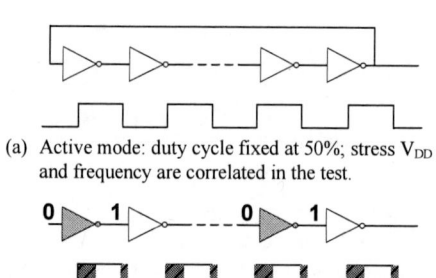

(a) Active mode: duty cycle fixed at 50%; stress V_{DD} and frequency are correlated in the test.

(b) Gated mode: asymmetric stress leads to different degradation rate at different switching edges.

Fig. 1. The RO-based structure is not sufficient for general-purpose circuit aging test because of fixed duty cycle and averaging effects.

experience different operation sequences from the test RO, leading to inaccuracy in aging prediction.

Furthermore, the RO structure cannot capture the difference in degradation rate between the rising and falling edge, a scenario that occurs due to asymmetric aging in low-power clock gating operation (Fig. 1(b)). Asymmetric aging is induced by unbalanced input logic pattern in a data path or a gated clock path. Widely deployed in low-power design, clock gating sets the gated path into standby mode when there is no logic switching activity. Since NBTI mainly affects the PMOS device, certain logic/clock inverting stages experience more stress than their neighbor experiences and thus, leads to unbalanced degradation. When the activity resumes on this path, one edge may suffer more delay degradation than the other edge. Figure 1(b) illustrates how asymmetric aging leads to different amount of edge degradation in a gated path. With the aggressive scaling of clock frequency and the reduction in timing budget, asymmetric aging imposes significant challenges to reliable design with unbalanced critical path and half-cycle path. As reported in [7], clock skew in a clock tree may increase by up to 7X under gated condition due to this effect. To ensure appropriate guard bands, a desirable test structure should be able to differentiate the degradation rate of different signal edges under asymmetric stress. A traditional

978-1-4244-9113-1/11 $26.00 © 2011 IEEE

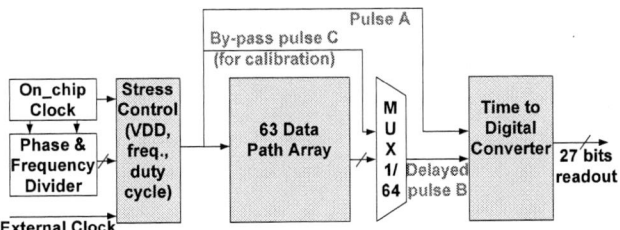

Fig. 2.   The block diagram of proposed test platform, composed of three major blocks (on-chip clock/stress control, data paths and TDC).

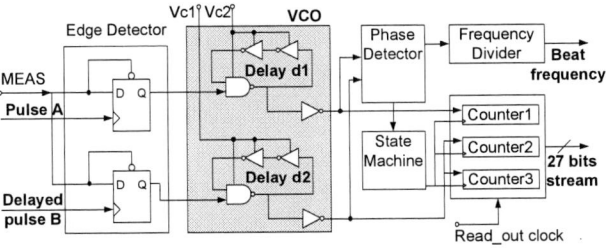

Fig. 3.   The architecture of the cyclic TDC. It measures the edge delay difference between a certain stressed data path and the clock, and converts it into digital readout.

RO structure is not capable in this scenario because its output frequency is an averaged result of both switching edges. Recent data [8] shows similar RO frequency degradation under both DC (static) stress and AC (dynamic) stress conditions. However, under DC stress conditions the signal edge degradation is asymmetric and thus not captured by the RO frequency degradation.

To overcome these drawbacks, we integrate the TDC circuit as a solution to directly sample circuit delay change of a stressed data path. The most significant improvement is that the measurement is performed in an open-loop manner with this platform. Therefore it allows the detection of the signal edge instead of the averaged frequency, consequently avoided the delay averaging over both signal edges. The integrated large array of data paths in this platform further allows the reliability characterization of many types of circuit topologies and operating modes conveniently.

II.   CIRCUIT ARCHITECTURE AND TEST PRINCIPLE

Figure 2 illustrates the block diagram of proposed test structure.   There are three major blocks, the on-chip clock generator and stress control circuit, the array of data paths, and the TDC. A fresh clock signal with desired frequency is generated on-chip and passed to the stress control module. The stress control circuit creates the stress signal (Pulse A) with duty cycles from 10% to 90%. Pulse A is sent to both the data path under the test, as well as the TDC as a reference signal. The delayed signal after the data path (Pulse B) enters the TDC, and is compared with fresh Pulse A. In addition to 63 data paths and the multiplexer (MUX), the test array also contains a by-pass path C, which is used to calibrate the TDC, especially to count the delay shift by the MUX. Finally, the

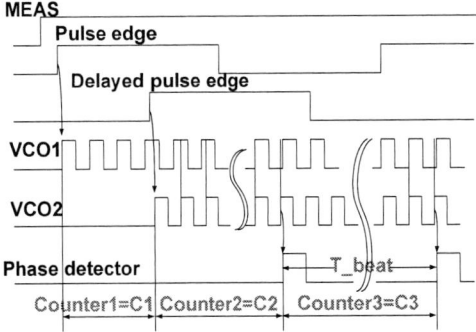

Fig. 4 (a).   Switching waveforms to illustrate the test principle of the TDC detection.

Beat frequency
$$f_{beat} = \frac{1}{T_{beat}} = \frac{1}{C3 \cdot d1} = \frac{1}{(C3+1) \cdot d2}$$
Resolution
$$r = d1 - d2 = \frac{1}{(C3+1) \cdot C3 \cdot f_{beat}}$$
Detected delay
$$D = d1 \cdot C1 + d2 \cdot C2 = \frac{(C1+C2) \cdot C3 + C1}{C3 \cdot (C3+1) \cdot f_{beat}}$$

Fig. 4(b).   The TDC resolution, r, is determined by the delay difference between two low noise voltage controlled oscillators. It is detected through the beat frequency and the read of Counter 3.

TDC measures the delay difference, and converts it into digital readout.

Figure 3 shows the architecture of the cyclic TDC [9] [10]. This design utilizes a single stage delay loop in the low noise voltage controlled oscillator (VCO), minimizing the delay mismatch among multiple stages in the Vernier chain structure [6] while exploiting its area efficiency in the implementation. Once the TDC is activated into the measurement mode, the rising edges of Pulse A and Pulse B are captured by two edge detectors, which consequently activate two VCOs (Fig. 3) to sense the delay difference.

The waveforms in Fig. 4(a) explain the principle of TDC detection. Depending on the bias, one VCO is faster than other other (Fig. 3). Over a period, such slight delay difference of two VCOs generates one cycle difference. This procedure repeats itself and the needed period is therefore detected by a phase detector as the beat frequency. Controlled by a state

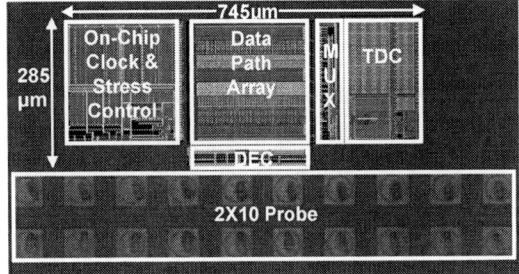

Fig. 5.   The die photo of the 45nm test chip.

978-1-4244-9113-1/11 $26.00 © 2011 IEEE   37

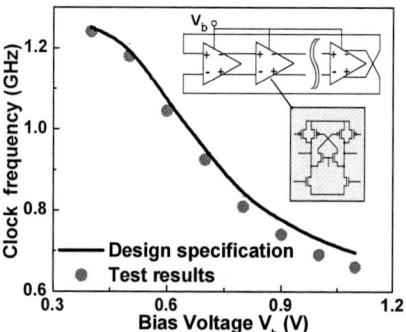

Fig. 6. The output range of 10-stage on-chip clock generator, from 680MHz to 1.23 GHz.

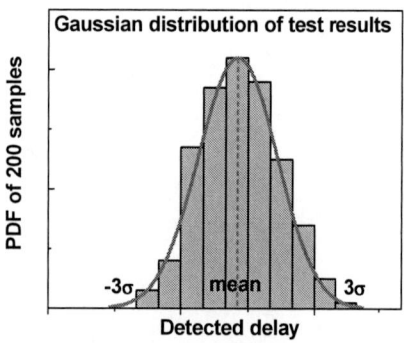

Fig. 7. Measured data path delays follow the Gaussian distribution due to random jitter.

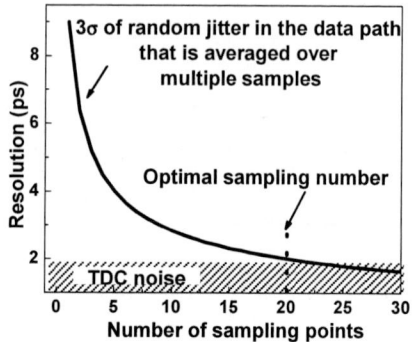

Fig. 8. Oversampling helps improve the resolution to 2ps.

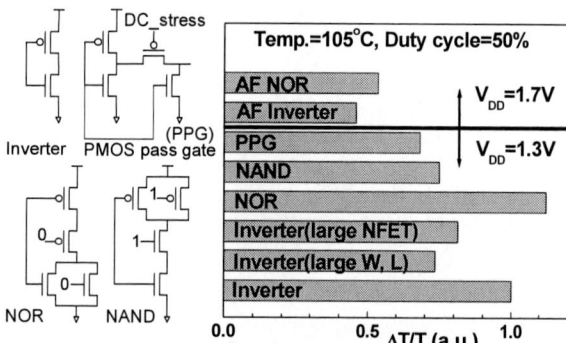

Fig. 9. Examples of various gate structures to form data paths and comparison of speed degradation after 8-hour stress.

machine, the desired delay target is detected by Counter 1 (3 bit) and Counter 2 (12 bit), while Counter 3 (12 bit) picks up the full cycle numbers in a complete beat frequency period. The TDC resolution, r, is measured by the beat frequency output and the count numbers in Counter 3. As shown in Fig. 4(b), the resolution of the TDC is mainly restricted by the noise of the VCO, which is designed and calibrated to be less than 2ps. Such a resolution is corresponding to ~0.5% delay degradation of an 11 stage buffer data path. It is sufficient for monitoring both short-term and long-term aging effect. Overall, the cyclic TDC is efficient for the integration, and accurate for aging test.

To further improve the accuracy, all function blocks along with two VCOs are independently powered. The TDC is turned off during the stress to avoid unnecessary degradation in the measurement circuitry. Figure 5 shows the 45nm die photo. The total die area of this 45nm chip is 0.24mm^2.

III. MEASUREMENT RESULTS AND DISCUSSIONS

Figure 6 presents the frequency range of the on-chip clock generator. Designed with low noise differential cells, it matches the design expectation by reaching gigahertz range. The delay measured by TDC is defined by the rising edge difference between pulse A and pulse B. However, the noise from supply or coupling may lead to different arrival time of pulse B cycle by cycle, which is known as jitter. One test sample at a time may not be enough to contain the correct speed degradtion value. Multiple samples are collected with the TDC for each stress measurement. Figure 7 illustrates the

PDF of detected delay in a data path at time zero with 200 samples. The fact that it follows a Gaussian distribution indicates that the random jitter dominates in the clock propagating path. Therefore, by increasing the sampling number to 20 and measuring their average value, the clock noise is effectively reduced to achieve a better resolution. The measurement resolution is eventually limited by the noise in the TDC (Fig. 7).

The integrated test platform provides a convenient method to characterize circuit aging under realistic dynamic operations. With an array of 63 data paths, the performance degradation rate as a function of both circuit topology and transistor type can be efficiently studied. Figure 8 shows four representative gates in the data path. Other gate types includes analog friendly (AF) NOR, AF inverter and inverter with different transistor sizes. After 8-hour dynamic stress, the relative degradations of eight different data paths are compared in Fig. 9. The results show that the NOR gate degrades the most, due to the sensitivity of the PMOS stack to NBTI degradation. The analog friendly (AF) devices exhibit lower degradation due to their thicker oxide.

Figure 10 shows that increasing V_{DD} and temperature accelerates the delay degradation of an 11 stage buffer chain. Figures 11 and 12 investigate the impact of dynamic operation sequence on circuit aging, for duty cycle and voltage scaling, respectively. During a realistic operation, both duty cycle and V_{DD} vary due to power saving demand, resulting in different amount of delay degradation (Fig. 10 and Fig. 11(a)).

978-1-4244-9113-1/11 $26.00 © 2011 IEEE

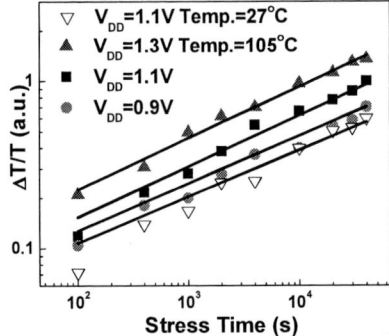

Fig. 10. Delay degradation under various V_{DD} and temperatures.

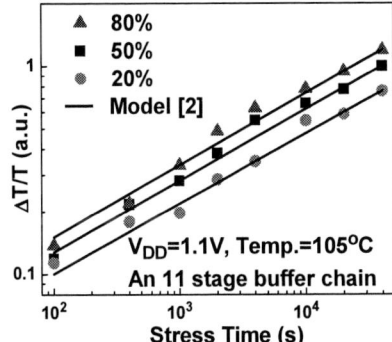

Fig. 11(a). Delay degradation under various duty cycles.

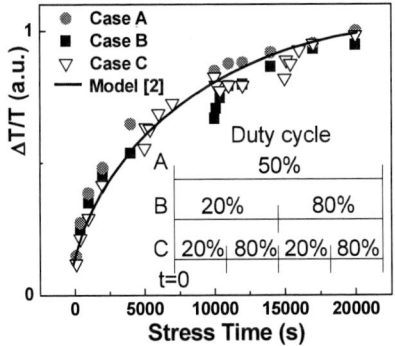

Fig. 11(b). Delay degradation due to NBTI is relatively insensitive to the sequence of duty cycles.

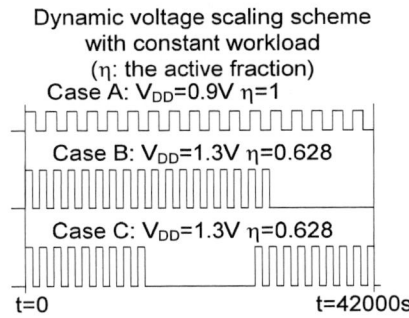

Fig. 12 (a). Three test schemes with same workload.

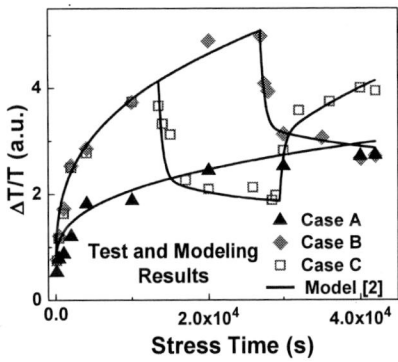

Fig. 12 (b). Test results show strong impact on voltage sequence.

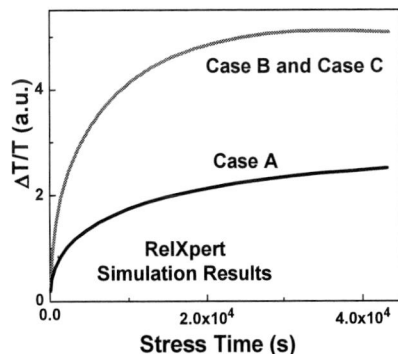

Fig. 12 (c). Conventional aging tool (RelXpert) fails to capture the sleep mode in Case B and C.

Compared to that by V_{DD} and temperature, duty cycle does not appear to have a significant impact on NBTI induced aging under AC stress. Therefore the sequence of duty cycles does not have a pronounced impact (Fig. 11(b)). This is explained by the approximately linear dependence of aging on duty cycle in 10%-90% range [1].

In contrast, speed degradation of different choices of V_{DD} level and the sleep mode lead to dramatic difference in the degradation. Circuit degradation are strongly interacted with dynamic voltage scaling (DVS) while the circuit can be designed either to improve performance under higher V_{DD} or to achieve better power efficiency under lower V_{DD}. As shown in Fig. 12(a), under the same workload, three schemes are evaluated. The first data path switches constantly under 0.9V, while the other two identical data paths runs faster under

1.3V. The latter two paths operate with a sleep phase ($V_{DD}=0$) inserted at different moments; the active fraction in the entire stress period is 0.628 with duty cycle at 50%. The measurement results prove that the exact sequence of voltage scaling, especially the last operation period, does have a significant influence on delay change, as observed in Fig. 12(b) [2].

The test results are employed to calibrate dynamic aging model in [2], which is shown in Fig. 12(b) as the solid line. The new model in [2] well matches the data, especially after the sleep mode. To further evaluate aging tools, simulations using RelXpert are conducted under the same test sequences in three cases. Figure 12(b) presents the result. In Case B and C, the simulation predictions are similar with a flat or even higher degradation after the sleep mode. The conventional tool fails

978-1-4244-9113-1/11 $26.00 © 2011 IEEE

to capture the behavior during the sleep period, leading to a significant over-estimation of circuit aging as shown in Figure 12(c). Such calibration with appropriate silicon data proves the importance of correct understanding and modeling of dynamic circuit operation.

IV.   CONCLUSIONS

A highly efficient aging test platform is demonstrated with 45nm data paths. Close to realistic circuit operation, it directly samples the delay change using a TDC with 2ps resolution. This method is convenient for large-scale aging test, without any interruption to dynamic circuit operations. Measurement data validate the aging model developed in [2]. They illustrate the significant impact of voltage scaling and the sleep duration on delay shift, guiding early design decision for circuit reliability.

REFERENCES

[1]   W. Wang, V. Reddy, A. T. Krishnan, R. Vattikonda, S. Krishnan, Y. Cao, "Compact modeling and simulation of circuit reliability for 65nm CMOS technology," *TDMR*, vol. 7, no. 4, pp. 509-517, 2007.

[2]   R. Zheng, *et al.*, "Circuit aging prediction for low-Power operation," *CICC*, pp. 427-430, 2009.

[3]   T. H. Kim, R. Persaud, C. H. Kim, "Silicon odometer: an on-chip reliability monitor for measuring frequency degradation of digital circuits," *VLSI Circuit Symp.*, pp.122-123, 2007.

[4]   M. B. Ketchen, M. Bhushan and R. Bolam, "Ring oscillator based test structure for NBTI analysis," *ICMTS*, pp. 42-27, 2007.

[5]   J. Keane, W. Zhang and C.H. Kim, "An on_chip monitor for statistically significant circuit aging characterization," *IEDM*, 2010.

[6]   J. Keane, X. F. Wang, D. Persaud and C.H. Kim, "An all-in-one silicon odometer for separately monitoring HCI, BTI, and TDDB," *JSSC*, Vol. 45, pp. 817-829, 2010.

[7]   A. Chakraborty, G. Ganesan, A. Rajaram, and DZ. Pan, " Analysis and optimization of NBTI induced clock skew in gated clock trees," *DATE*, pp. 296-299, 2009

[8]   A. T. Krishnan, et al., " Product drift from NBTI: guardbanding, circuit and statistical effects," *IEDM, 2010.*

[9]   P. Dudek and J. V. Hatfield, "A high-resolution CMOS time-to-digital converter utilizing a Vernier delay line," *JSSC*, vol. 35, pp. 240-247, 2000.

[10] A. H. Chan and G. W. Roberts, "A synthesizable, fast and high-resolution timing measurement device using a component-invariant Vernier delay line," *TEST*, pp. 858-867, 2001.

978-1-4244-9113-1/11 $26.00 © 2011 IEEE

Fast Characterization of the Static Noise Margin Degradation of Cross-Coupled Inverters and Correlation to BTI Instabilities in MG/HK Devices

A. Kerber

Technology Research Group, GLOBALFOUNDRIES Inc.
1101 Kitchawan Road, Yorktown Heights, NY, 10598, USA
phone: +1 (914) 945 1607, **email:** Andreas.Kerber@globalfoundries.com

N. Pimparkar, S. Balasubramanian, T. Nigam, W. McMahon

GLOBALFOUNDRIES Inc.
1050 East Arques Street, Sunnyvale, CA, 94085, USA

E. Cartier

T.J. Watson Research Center, IBM
1101 Kitchawan Road, Yorktown Heights, NY, 10598, USA

Abstract — A fast BTI characterization setup is introduced to study the Static Noise Margin (SNM) of cross-coupled inverters using metal gate / high-k devices. It is shown that static stress leads to significant SNM degradation due to positive bias temperature instability (PBTI) at high stress voltage consistent with device level data. For the dynamic stress mode the bias temperature instability (BTI) induced degradation of the Pull UP and Pull down devices becomes symmetric which can mask the "worst-case" SNM degradation depending on the initial cell symmetry.

Keywords- high-k dielectrics, metal gate, BTI, SRAM

I. INTRODUCTION

Bias temperature instability (NBTI and PBTI) is one of the critical reliability mechanisms in advanced CMOS technology nodes employing metal gate / high-k (MG/HK) stacks. Most of the research has been focusing on extensive device level characterization, discussing in detail the degradation and recovery effects [1, and ref. therein]. Fast characterization methods were developed to minimize the recovery effects and accurately quantify the magnitude of the instability [2, 3]. Product level burn-in testing still lacks such fast characterization methods, potentially leading to an underestimation of the magnitude of the degradation. Therefore, it is of general interest to extend the fast BTI characterization methods to simple CMOS circuits like cross-coupled inverters used for SRAM. In this paper, we discuss the correlation between device level BTI degradation / recovery and the Static Noise Margin (SNM) of cross-coupled inverters employing static and dynamic stress in metal gate / high-k CMOS devices.

II. EXPERIMENTAL

MG/HK CMOS devices with a Hafnium based gate dielectric were fabricated utilizing a conventional process flow on SOI substrates. In this paper, cross-coupled inverter circuits were used for electrical testing at elevated temperature (T = 125 ^{O}C). The schematic layout of the circuit is given in Fig. 1, including the nomenclature used in this paper to describe individual FETs and bias nodes in the circuit. The input terminals are Sense Node Left (SNL) and Sense Node Right (SNR). The pFET devices are called Pull UP Left (PUL) and Pull UP Right (PUR) and the nFET devices are called Pull Down Left (PDL) and Pull Down Right (PDR).

In order to obtain the butterfly transfer characteristics of cross-coupled inverters, the fast PCI card based BTI characterization setup [3] was modified by adding a fast switching capability and a high impedance voltmeter as schematically shown in Fig. 2. The switching was realized by using a standard CMOS switch with Ron < 100 Ω. The switch is controlled by digital outputs of the Peripheral Component Interconnect (PCI) card minimizing overall switching delays to < 10 μs.

The high impedance volt-meter was realized with an operational amplifier configured as unity gain buffer amplifier. The meter readings are recorded by the analog input terminal (ADC) of the PCI card. To avoid charging, the input terminal of the volt-meter was terminated with 20MΩ.

An inverting operational amplifier circuit was used as a combined voltage source and current meter. The value of the source is set by the analog output (DAC) of the PCI

978-1-4244-9113-1/11 $26.00 © 2011 IEEE

card. The current values are calculated from the ADC readings of the PCI card.

The configuration of the switches during a sense measurement is provided in Fig. 2. First SNL is forced in the "normal" configuration and the voltage at SNR is measured. Then the configuration is "altered" to force SNR and measure SNL. The total time for a sense cycle is determined by the number of bias conditions used and the measurement time per condition. A typical number for the bias conditions is 60 (2 x 30) and the measurement time per condition is 1ms, as will be discussed in more detail below.

Typical voltage time traces for the static and dynamic BTI stress-and-sense measurements are shown in Fig. 3. In the static stress mode the SNL terminal is either biased at GND (top panel of Fig. 3) or VDD (center panel of Fig. 3) whereas, in the dynamic mode the bias is altered between GND and VDD using at 100 Hz frequency and 50% duty cycle (bottom panel of Fig. 3). From these voltage traces butterfly characteristics (SNR is plotted versus SNL) were obtained as shown in Fig. 4. The SNM is then derived by placing the maximum squares in the butterfly characteristics SNL_{LO} & SNR_{HI} and SNL_{HI} & SNR_{LO} following the procedure given in [4].

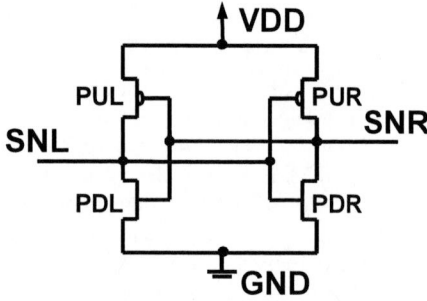

Fig. 1. Schematic showing the cross-coupled inverter circuit. The circuit is biased via the Sense Node Left (SNL) and Sense Node Right (SNR) terminals in addition to the supply (VDD) and ground (GND) terminal. The nomenclature for the transistors are pFET pull up left (PUL), right (PUR) and nFET pull down left (PDL), right (PDR).

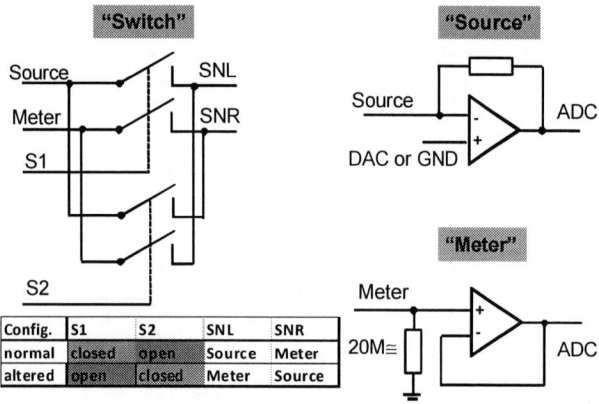

Fig. 2. Schematics of the various components used in the fast cross-coupled inverter characterization setup. The analog CMOS switch is used to terminate SNL (S1 closed) and SNR (S2 closed) with either source and meter or vice versa.

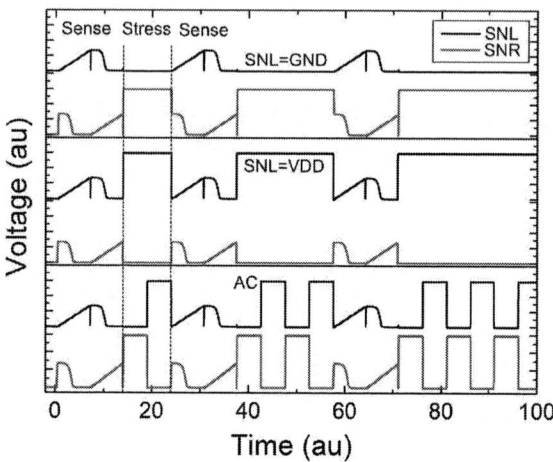

Fig. 3. Measured voltage time traces for static (upper two panels) and dynamic (lower panel) BTI characterization of cross-coupled inverters. In the static stress mode SNL is either forced to GND (upper panel) or VDD (middle panel), whereas in the dynamic mode it is altered between GND and VDD at 100Hz and 50% duty cycle.

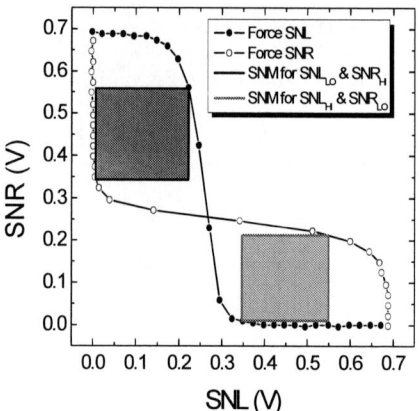

Fig. 4. Butterfly characteristics of unstressed cross-coupled inverters. The static noise margin (SNM) is obtained by placing maximum squares.

To verify the impact of the measurement time on the stability of the butterfly transfer characteristics, the sense measurement sequence shown in Fig. 3 was modified (see in the inset of Fig. 5) and a hysteresis sweep on unstressed cross-coupled inverters was performed. As can be seen from the data in Fig. 5, the forward and reverse traces for short measurement times (~30 μs/pt, ~200 μs/pt) corresponding to fast ramp rates of ~725 V/s and ~120 V/s do not coincide, leading to a hysteresis in the transfer characteristic. Longer measurement time (~1 ms/pt,) or slower ramp rate (24 V/s), however, shows no hysteretic behavior.

The hysteresis for short measurement times or fast ramp rates is caused by charging and discharging of the parasitic cable capacitance through the high output impedance terminal of the cell during the switching event. The hysteresis directly impacts the extraction of the SNM as shown in Fig. 6. For ramp rates of > 50 V/s a significantly different value for the SNM in forward and reverse direction is obtained. To avoid transient charging effects, a measurement time of ~ 1 ms was chosen to investigate the correlation between the SNM and BTI degradation. Therefore, the total sense time is ~ 60 ms.

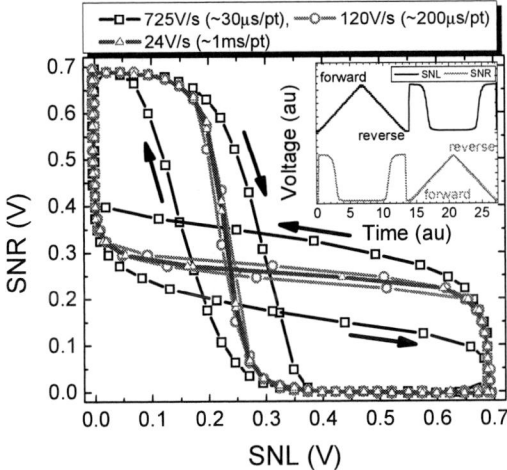

Fig. 5. Butterfly curve for unstressed cross-coupled inverters measured at different ramp rates. Note the hysteretic behavior of the butterfly characteristics for ramp rates > 100 V/s caused by charging of the parasitic capacitance. The inset shows the voltage time traces during the hysteresis measurements.

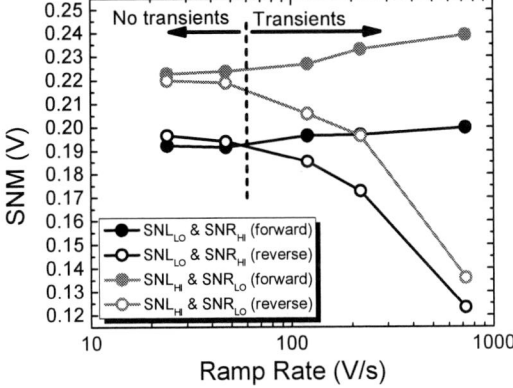

Fig. 6. The static noise margin (SNM) of unstressed cross-coupled inverters plotted versus the ramp rate of the voltage sweep at SNL or SNR terminal. Transients due to charging of the parasitic capacitance diminish for ramp rates of < 50 V/s.

III. RESULTS AND DISCUSSION

Typical butterfly characteristics for the static and dynamic BTI stress modes are compared in Fig. 7. In all cases first a pre-stress characteristic was measured and then the high voltage stress was applied. The stress voltage (VDD_{stress}) was chosen significantly above the sense voltage of $VDD_{sense} = 0.7V$ (2 x VDD_{sense} < VDD_{stress} < 3x VDD_{sense}). During stress the SNL terminal was forced with SNL = GND or SNL = VDD in static mode and altered between GND and VDD in dynamic mode using a frequency of 100 Hz and 50% duty cycle. Intermittent characterization of the transfer characteristic was performed with a measurement time of ~1 ms per data point. The stress cycle was immediately followed by a recovery cycle where VDD and SNL were set to GND and again intermittent characterization of the transfer characteristic was carried out.

As can be seen from the data in Fig. 7, the static mode at high stress bias leads mainly to a shift of one transfer characteristic which is attributed to the PBTI degradation of the nFET pull down devices. The dynamic stress at high stress bias simultaneously degrades both transfer characteristics but remains dominated by PBTI. The time dependence of the SNM during stress and recovery is summarized in Fig. 8. In all cases, the static stress leads to a significant degradation in the SNM for one of the segments and an increase in the SNM for the opposite one. Depending on the pre- stress margin this can either lead to a continuous degradation in the SNM or to an initial improvement followed by degradation. The dynamic stress, however, shows overall less change in the SNM. This suggests that PBTI drift initially improves the cell symmetry [5] and leads to degradation in the SNM only at long stress times.

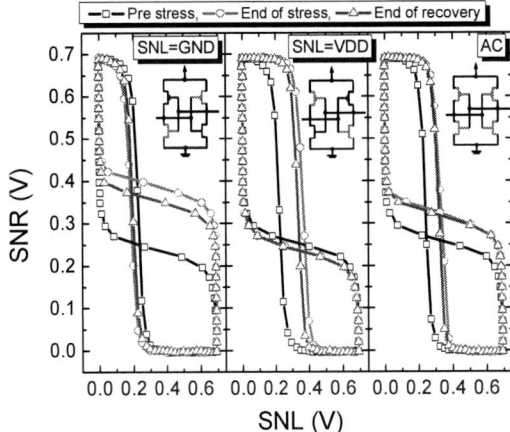

Fig. 7. Pre-stress, end-of-stress and end-of-recovery butterfly characteristics obtained for static stress mode with SNL biased at GND (left) or VDD (middle) and dynamic stress mode (right) altering between GND and VDD. Note that the shift in transfer characteristics at high stress voltages is dominated by PBTI.

978-1-4244-9113-1/11 $26.00 © 2011 IEEE

The post-stress recovery of the SNM is summarized in the lower panel of Fig. 8. For static stress a significant SNM recovery is seen for both segments whereas, for dynamic stress the SNM recovery is minimal. These results are consistent with device level data [6].

In addition to the butterfly characteristics, pre- and post-stress transistor characteristics for all 4 devices in the cross coupled inverter were measured. Typical pre- and post-stress Id-Vg traces are shown in Fig. 9 for Pull Up devices (left) and Pull Down devices right. The stress case SNL = GND is given in the upper, SNL = VDD in the middle and the dynamic stress mode altering between GND and VDD in the lower panel. As mentioned above, the dominant degradation mechanism at high voltage stress condition for MG/HK devices is PBTI in the nFET or Pull Down devices. NBTI on Pull Up devices is found to be less significant than PBTI. These results are consistent with previously published data on discrete CMOS devices [7].

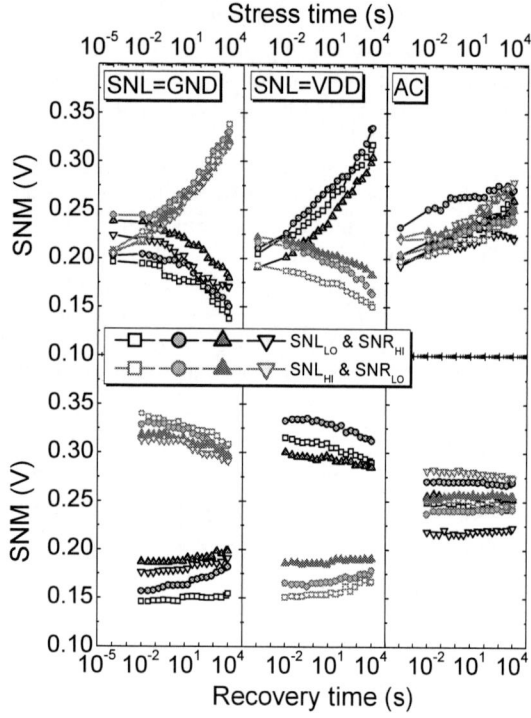

Fig. 8. Static noise margin (SNM) as function of stress (upper panel) and recovery (lower panel) time for static stress mode with SNL biased at GND (left) or VDD (middle) and dynamic stress mode (right) altering between GND and VDD. Much larger degradation and recovery can be seen for the static modes than for the dynamic mode.

From the pre- and post-stress Id-Vg and Id-Vd data (Id-Vd characteristics are not shown), pre- and post-stress SPICE model parameters were derived. To illustrate the accuracy of the SPICE model parameters obtained from the data in Fig. 9, the model currents are plotted versus the measured data

for the stress case SNL = VDD in Fig. 10. As can be seen the model for the Pull Down devices shows good correlation to the measured data. From the inset it can be seen that the PDR device shows significant PBTI degradation whereas the PDL shows no degradation.

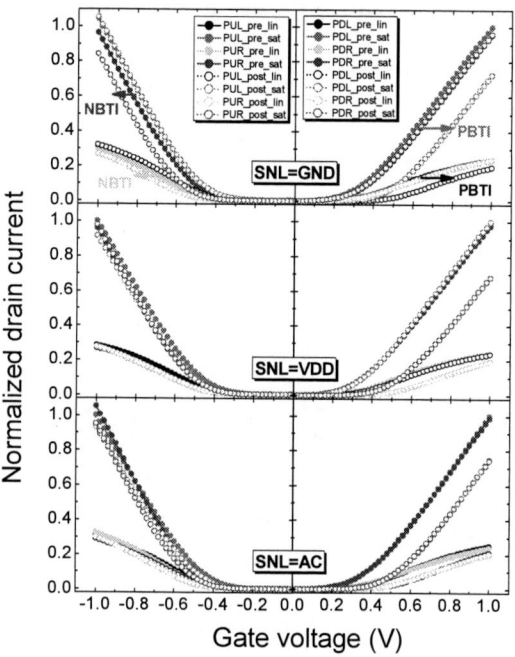

Fig. 9. Measured pre- (solid) and post-stress (open) Id-Vg characteristics with SNL biased at GND (upper panel) or VDD (middle panel) and dynamic stress mode (lower panel) altering between GND and VDD. Pre- and post-stress Id-Vg characteristics were used to determine the SPICE model parameters for circuit simulation.

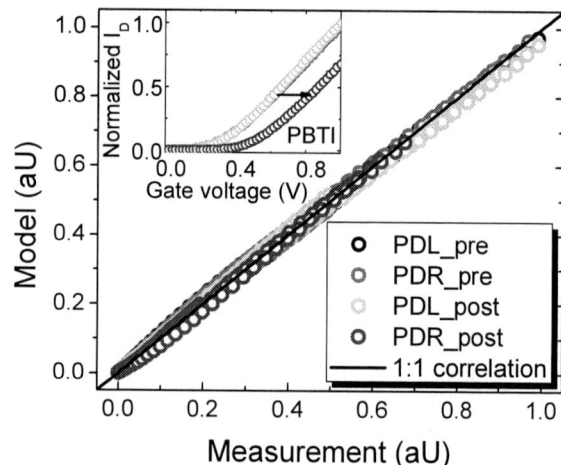

Fig. 10. Comparison of nFET SPICE model currents with the measurements for the stress case SNL = VDD shown in Fig. 9. The model values are typically within 5% of the experimental data. The inset shows the normalized pre- and post-stress drain current in saturation.

The SPICE models were then used to simulate the transfer characteristics of the cross coupled inverters. In Fig. 11, simulated pre- and post-stress characteristics are compared to the experimental traces for SNL = GND, VDD and for the dynamic stress mode, altering between GND and VDD. This demonstrates that the SNM correlates well with the BTI-induced voltage shifts for static and dynamic stress conditions, justifying the application of device level transistor degradation data to simulate SRAM circuits.

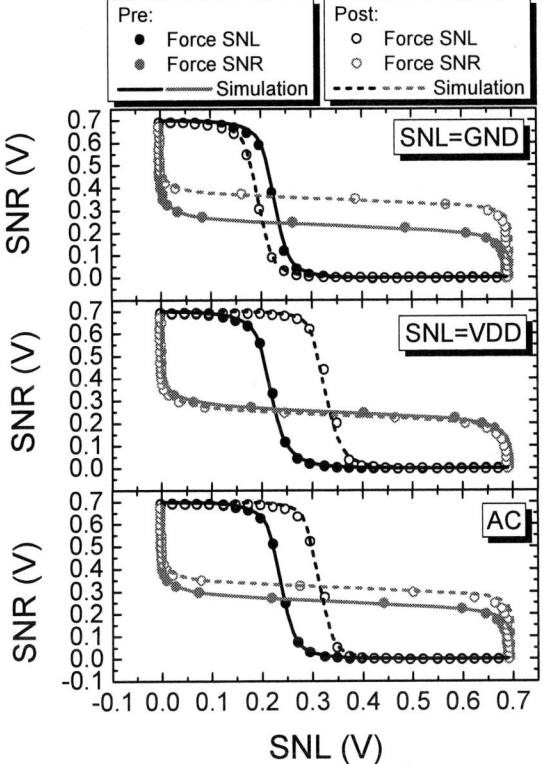

Fig. 11. Pre- (solid symbols) and post-stress (open symbols) transfer characteristics with SNL biased at GND (upper panel) or VDD (middle panel) and dynamic stress mode (lower panel) altering between GND and VDD. Pre- (solid lines) and post-stress (dashed lines) simulations of the transfer characteristics are overlaid on experimental data. Excellent agreement between measured and simulated transfer curves is demonstrated.

The implication of the stress mode and corresponding recovery effects on the SRAM product burn-in test are summarized in Fig. 12. First, the magnitude of SNM degradation and improvement seen at the end of stress for the static stress mode are not symmetric. SNM improvement yields higher values compared to the SNM degradation. Second, for the static stress mode a 20 - 30 mV recovery is observed for SNM degradation or improvement after $t_{Recovery} = 10^4$ s.

For the dynamic stress mode negligible recovery of the SNM is observed after $t_{Recovery} = 10^4$ s. In addition, the dynamic stress mode resulted in an improvement of the SNM for this hardware which is attributed to compensation of V_t mis-match between pFET and nFET devices due to PBTI.

Thus, SRAM burn-in test will overestimate the minimum operation voltage for the static stress mode due to significant recovery. Dynamic burn-in test will show negligible recovery but the SNM degradation will most likely not present a "worst-case" scenario typically considered for reliability testing..

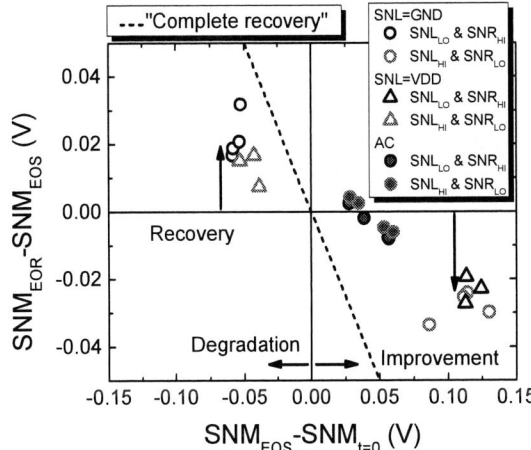

Fig. 12. SNM recovery plotted versus SNM degradation for static and dynamic stress shown in Fig. 8. Note the significant recovery of the SNM for the static stress mode and the negligible recovery for dynamic stress mode. Furthermore note the asymmetry in SNM during stress between SNM degradation and SNM improvement

IV. CONCLUSION

A fast BTI characterization setup was introduced to study the stability of cross-coupled inverters used in SRAM. It is shown that static stress leads to significant SNM degradation and recovery consistent with device level data. Large differences between static and dynamic stress modes are observed. It is also shown that the cell symmetry needs to be considered carefully because it can lead to masking of the "worst-case" degradation.

ACKNOWLEDGMENT

The authors would like to acknowledge the stimulating discussions with Gottfried Kurz and Gernot Krause. This work was performed by the Research Alliance Teams at various IBM Research and Development facilities.

REFERENCES

[1] A. Kerber and E. Cartier, "Reliability challenges for CMOS technologyqualifications with hafnium oxide/titanium nitride gate stacks", IEEE Trans. Device Mater. Rel., vol. 9, no. 2, pp. 147–162, 2009.

[2] H. Reisinger, Reisinger, O. Blank, W. Heinrigs, A. Mühlhoff, W. Gustin, and C. Schlünder, "Analysis of NBTI degradation- and recoverybehavior based on ultra fast V T-measurements," in Proc. IRPS, pp. 448–453, 2006.

[3] A. Kerber, K. Maitra, A. Majumdar, M. Hargrove, R. J. Carter, and E. Cartier, "Characterization of fast relaxation during BTI stress in conventional and advanced CMOS devices with HfO2/TiN gate stacks", IEEE Trans. Electron Devices, vol. 55, no. 11, pp. 3175–3183, 2008.

[4] F. J. List, "The Static Noise Margin of SRAM cells", ESSCIRC, pp. 16-18, 1986.

[5] Jiajing Wang, Satyanand Nalam, Zhenyu(Jerry) Qi, Randy W. Mann, Mircea Stan, and Benton H. Calhoun, " Improving SRAM Vmin and Yield by Using Variation-Aware BTI Stress", IEEE CICC, 2010.

[6] K. Zhao, J. H. Stathis, A. Kerber and E. Cartier, "PBTI Relaxation Dynamics after AC vs. DC Stress in High-K/Metal Gate Stacks", in Proc. IRPS,, pp. 50–54. 2010.

[7] A. Kerber, S.A. Krishnan and E.A. Cartier, "Voltage Ramp Stress for Bias Temperature Instability Testing of Metal Gate/High-k Stacks", IEEE Electron Device Letters, Vol. 30, No. 12, pp. 1347-1349, 2009.

Reliability Monitoring Ring Oscillator Structures for Isolated/Combined NBTI and PBTI Measurement in High-K Metal Gate Technologies

Jae-Joon Kim[1], Barry P. Linder[1], Rahul M. Rao[1], *Tae-Hyoung Kim[1,2], Pong-Fei Lu[1], Keith A Jenkins[1],
Chris H. Kim[2], Aditya Bansal[1], †Saibal Mukhopadhyay[1], ‡Ching-Te Chuang[1]

1. IBM T. J. Watson Research Center, Yorktown Heights, NY,
2. University of Minnesota. Minneapolis, MN.
(E-mail: jjkim2@us.ibm.com)

Abstract— Ring oscillator (RO) structures that separate NBTI and PBTI effects are implemented in a high-k metal gate technology. The measurement results clearly show significant RO frequency degradation from PBTI as well as NBTI. For comparison, RO structures with the same principle are also implemented in a SiO₂/poly-gate technology, where PBTI is negligible. Experimental results show noticeable frequency degradation under NBTI-only stress mode but negligible degradation under PBTI-only mode, which illustrates the validity of the proposed principle and structures.

Keywords- NBTI, PBTI, Ring Oscillator, Circuit

I. INTRODUCTION

Ring oscillator (RO) based monitor circuits have been a mainstay for characterizing Negative Bias Temperature Instability (NBTI) in SiO₂/poly-gate technologies utilizing the fact that Positive Bias Temperature Instability (PBTI) is negligible compared to the dominant NBTI effect in the technologies [1][2]. However, the advent of high-k metal-gates increases PBTI making it significant compared with NBTI [3], which prevents the conventional RO from separating out NBTI and PBTI effects. Fig. 1 shows the RO structure that has been traditionally used for NBTI measurement. During the non-ringing stress period, PMOS and NMOS transistors in alternate stages are stressed. Therefore, the measurement result quantifies the impact of mixed NBTI and PBTI. The PMOS and NMOS devices in an RO structure can be separately

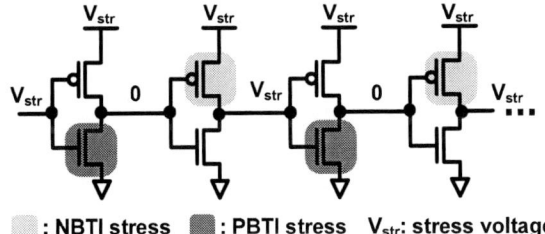

: NBTI stress   : PBTI stress V_{str}: stress voltage

Fig. 1. DC Stress condition for conventional RO used for NBTI measurement in SiO₂/poly-gate technology [1][2].

* T. Kim, †S. Mukhopadhyay, and ‡C. Chuang are currently with Nanyang Technological University, Singapore, †Georgia Institute of Technology, USA and ‡ National Chiao-Tung University, Taiwan, respectively.)

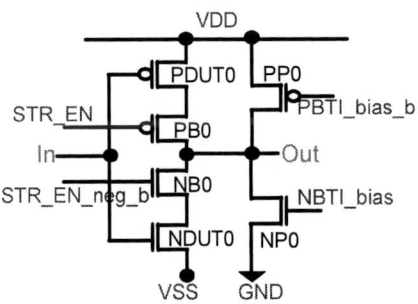

Fig. 2. Proposed ring oscillator (RO4)
(1-stage).

stressed if the input of each stage is selectively biased and isolated from the output of the previous stage [4]. In this paper, several RO structures based on this principle are proposed and implemented in a SiO₂/poly-gate technology and a high-k metal gate technology. Experimental results are presented that highlight the ability of these structures to successfully isolate NBTI and PBTI effects while maintaining the simplicity of RO.

II. ISOLATED/ COMBINED NBTI/PBTI MONITOR IMPLEMENTATION IN A HIGH-K/METAL GATE TECHNOLOGY

Fig. 2 shows a single stage of the proposed multi-staged RO structure for separately measuring the impact of NBTI and PBTI on circuit delay. PMOS (PDUT0 for NBTI-only) or NMOS (NDUT0 for PBTI-only) devices can be selectively stressed based on the control inputs (Table 1). In addition, both PMOS and NMOS devices in each stage can be simultaneously

Table 1. Input signal conditions for circuits in Fig. 2 and 4.

Signal	Stress Mode			Meas. Mode
	NBTI	PBTI	N/PBTI	
VDD	Vstr	Vstr	Vstr	Vnom
STR_EN	Vstr	Vstr	Vstr	GND
STR_EN_neg_b	GND	GND	-Vstr	Vnom
PBTI_bias_b	Vstr	GND	Vstr	Vnom
NBTI_bias	Vstr	GND	Vstr	GND
VSS	GND	GND	-Vstr	GND

978-1-4244-9113-1/11 $26.00 © 2011 IEEE

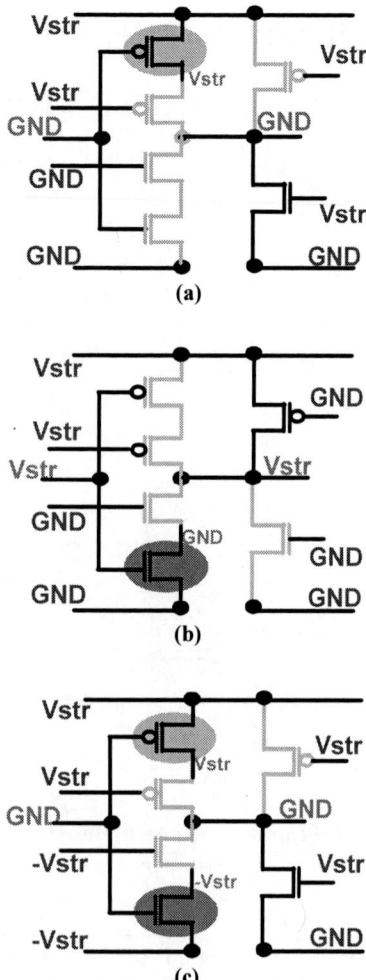

(a)

(b)

(c)

Fig. 3. DC static stress conditions and internal voltages for the proposed RO. (a) NBTI-only stress mode, (b) PBTI-only stress mode. (c) simultaneous N/PBTI stress mode

stressed. During stress, a voltage (Vstr) that is higher than the nominal supply voltage (Vnom) is applied. In static NBTI-only stress mode (Fig. 3a), STR_EN (at Vstr) and STR_EN_neg_b (at GND) turn off PB0 and NB0. The output node (OUT) is forced to GND by NBTI_bias (Vstr) which enables NP0, while PBTI_bias (Vstr) turns off PP0. Thus, the PMOS device-under-test (DUT) in each stage is stressed at DC stress voltage. The stressing of NMOS devices in PBTI-only mode can be explained in complementary fashion (Fig. 3b). In simultaneous N/PBTI stress mode (Fig. 3c), the OUT node is biased to GND (NBTI_bias and PBTI_bias are both at Vstr) while STR_EN_neg_b and VSS receive negative voltage –Vstr. Thus, both NMOS DUTs and PMOS DUTs are simultaneously stressed. In measurement mode, PB0 and NB0 are turned on while NP0 and PP0 are turned off so that the structure becomes equivalent to the conventional RO.

This structure (RO4) was implemented in a high-k/metal gate PD/SOI CMOS technology. In addition, three other

simpler structures shown in Fig. 4 with the same design principle [4] were implemented (RO1: NOR structure for NBTI-only, RO2: NAND structure for PBTI-only, and RO3: Unified structure to measure NBTI or PBTI selectively). During the stress mode, the input of every stage in the NOR-style RO1 is biased to Vstr for NBTI stress and the input of every stage in the NAND-style RO2 is biased to GND for PBTI stress.

The die photo is shown in Fig. 5. Each RO frequency is observable through a dedicated frequency divider at an output pad enabling simultaneous measurement of all 4 ROs. The circuits were stressed at various high stress voltages and an elevated temperature for 10,000 seconds followed by an equal relaxation period at VDD=0V. Frequency degradation is monitored at Vnom by interrupting the stress or relaxation for 20ms at periodic intervals. Note that the measurement time is longer than that reported in [2] (2 μs measurement time) due to the use of simple frequency divider for readout. Our intention was to utilize these ROs early on in the technology development stage, and therefore we chose to use simple and

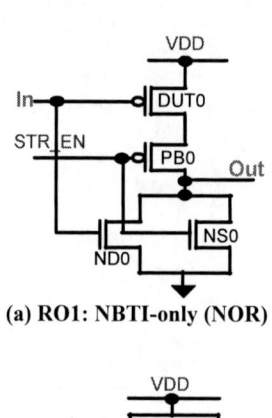

(a) RO1: NBTI-only (NOR)

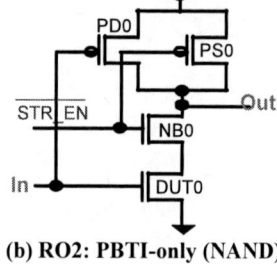

(b) RO2: PBTI-only (NAND)

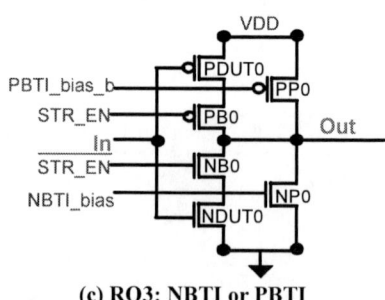

(c) RO3: NBTI or PBTI

Fig. 4. Simpler RO Structures [4] implemented in the test chip in addition to the RO shown in Fig. 2 (RO4). See Table 1 for input conditions.

978-1-4244-9113-1/11 $26.00 © 2011 IEEE

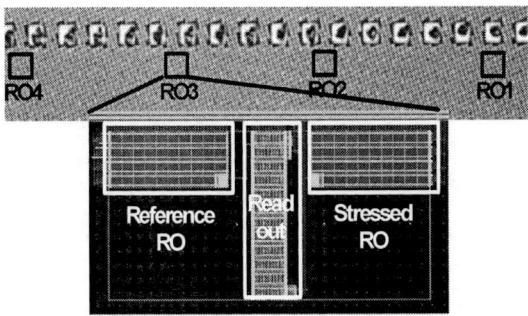

Fig. 5 Die micrograph and layout for BTI monitors for a high-k/metal gate technology. (170x22µm² for 4 RO experiments)

robust read circuitry. The proposed RO structures can be connected to the faster readout circuitry [2] seamlessly if necessary. As shown in Fig. 5, a corresponding reference RO is placed next to each RO. The power supply voltage of the reference RO is biased to GND during stress mode and is raised to the nominal operating VDD during measurement mode only. The RO output frequency change due to the environmental variations such as VDD or temperature fluctuations can be separated from the change due to BTI stress by measuring both the reference RO and stressed RO frequencies [2]. Fig. 6 shows the RO frequency degradation as a function of time for one particular stress voltage. The first data point plotted at 100 second of stress time when $T_{relaxation}/T_{stress}$ = 20ms/100s = 0.02%. This mitigates the effect of BTI recovery on the analysis as the recovery depends on the ratio between stress time and relaxation time [5]. As expected, combined NBTI and PBTI frequency degradation is the highest. RO frequency degradation due to NBTI effect was slightly higher than the one due to PBTI in the technology we used for this study. The frequency degradation due to NBTI in

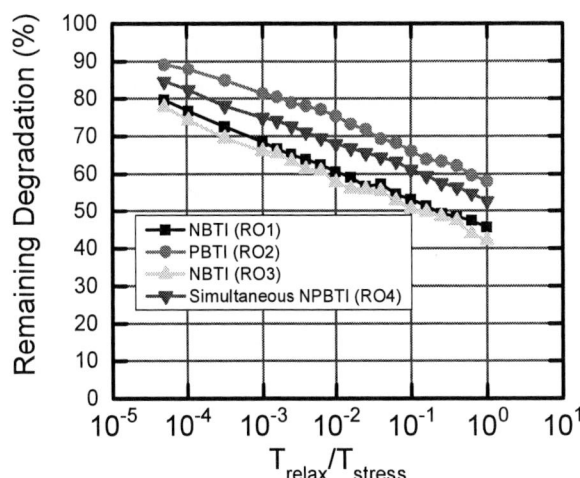

Fig. 7 Remaining Frequency degradation vs. ratio between relaxation time and stress time (relaxation starts after 10,000s stress)

RO3 is larger than in RO1 because device width ratios between PDUT and PB are different (1:1 in RO1 and 1:9 in RO3). The larger PB width makes the overall RO frequency more sensitive to the Vth change in PDUT. The impact of recovery during relaxation is shown in Fig. 7. NBTI relaxes faster than PBTI, which is consistent with previously reported device level data [6].

Fig. 8 shows that frequency degradation from simultaneous N/PBTI stress is close to the sum of degradations from individual NBTI and PBTI as expected. Measurement results for different stress voltages showed that the slope for voltage acceleration for PBTI (~5.6) is steeper than the one for NBTI (~4.3), which is consistent with a device level measurement report [7].

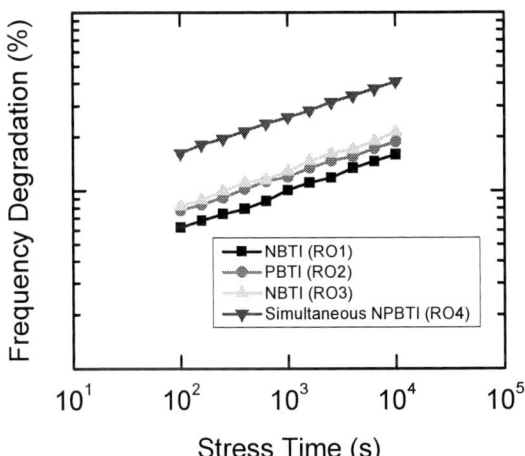

Fig. 6 RO frequency degradation vs. stress time. NBTI results from RO1 and RO3 differ because W_{PDUT}/W_{PB}= 1:1 in RO1 and W_{PDUT}/W_{PB}= 1:9 in RO3.

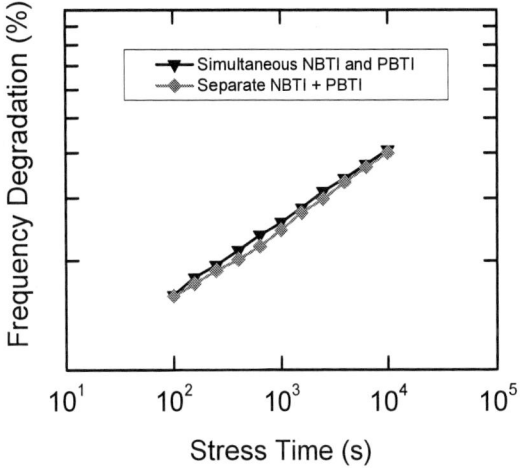

Fig. 8 RO4 frequency degradation comparison between simultaneous N/PBTI stress case and sum of individual NBTI and PBTI stress cases.

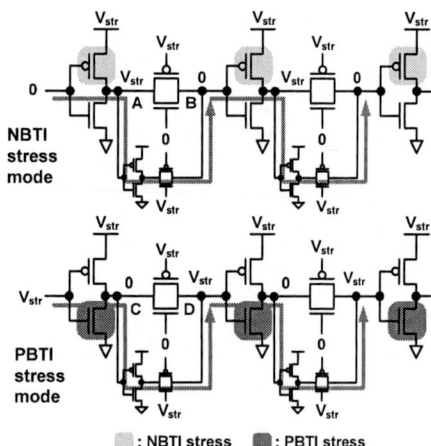

Fig. 9. Alternative circuit topology implemented in a SiO$_2$/poly-gate technology (stress mode)

III. IMPLEMENTATION IN A SiO$_2$/POLY-GATE TECHNOLOGY

Fig. 9 depicts an alternative transmission-gate based structure. The circuit consists of two signal paths; a measurement path for frequency measurements and a control path for applying NBTI or PBTI stress. During stress mode, the transmission gate in the measurement path is cut off while that in the control path is turned on. Signal A is inverted and transferred to B, making the input of each inverter in the measurement path identical. For static NBTI stress, the primary input of the ring oscillator is connected to ground in order to stress all PMOS transistors. Likewise, the input of the ring oscillator is connected to Vstr for PBTI stress. We implemented the circuit in a SiO$_2$/poly-gate technology. As shown in Fig. 10, the RO frequency is degraded noticeably after NBTI-only stress but there is little change in RO frequency after PBTI-only stress. This, combined with measurement results from the high-k/metal gate technology in the previous section, confirms that the proposed RO structures successfully isolate NBTI and PBTI effects. Fig. 11 shows the die photo for the design.

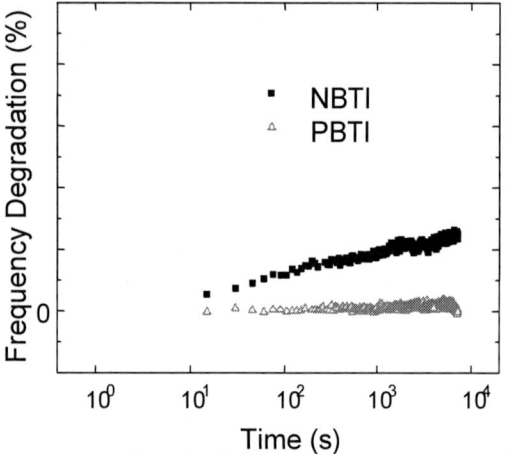

Fig. 10. RO frequency degradation comparison between NBTI stress and PBTI stress for the proposed circuit (Fig. 9) in SiO$_2$/poly-gate technology

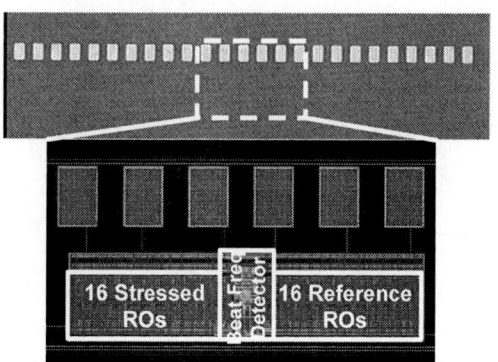

Fig. 11 Die micrograph and layout for BTI monitor in Fig. 9 (372x90μm^2)

IV. SUMMARY

We propose new ring oscillator circuits which monitor NBTI and PBTI separately. In case the conventional ring oscillator circuit is used for characterization of model-to-hardware correlation for NBTI and PBTI, it can only provide overall frequency degradation information due to combined NBTI and PBTI effects. In such a case, the modeled value of each BTI component can be quite different from the real value even when the overall degradation is close to the values from the models. The proposed circuits can provide information for each BTI component and enables more accurate model-to-hardware correlation.

ACKNOWLEDGMENT

This work was partially performed by the Research Alliance Team at various IBM Research and Development facilities.

REFERENCES

[1] V. Reddy, et. al, "Impact of negative bias temperature instability on digital circuit reliability", *International Reliabilty Physics Symposium (IRPS)*, pp. 248-254, 2002.

[2] T. Kim, R. Persaud, C. Kim, "An On-Chip Reliability Monitor for Measuring Frequency Degrdation of Digital Circuits", Proceeding of the Symposium on VLSI Circuits, pp.14-16, June 2007

[3] S. Zafar, et. al, "A Comparative Study of NBTI and PBTI (Charge Trapping) in SiO$_2$/HfO$_2$ Stacks with FUSI, TiN, Re Gates, " Proceeding of the Symposium on VLSI Technology, pp.23-25, June 2006

[4] J.-J. Kim, R. Rao, S. Mukhopadyay, and C-T. Chuang, "Ring oscillator circuit structures for measurement of isolated NBTI/PBTI effects", IEEE International Conference on Integrated Circuit Design and Technology (ICICDT), pp. 163 -166, June 2008.

[5] T. Grasser, W. Gos, V. Sverdlov, B. Kaczer, "The Universality of NBTI Relaxation and Its Implication for Modeling and Characterization", *International Reliabilty Physics Symposium (IRPS)*, pp. 268-280, 2007

[6] S. Ramey et. al., "Frequency and Recovery Effects in High-K BTI Degradation" *International Reliabilty Physics Symposium (IRPS)*, pp. 26-30, 2009

[7] A. Kerber, S. Krishnan, E. Cartier, "Voltage Ramp Stress for Bias Temperature Instability Testing of Metal-Gate/High-k Stacks," IEEE *Electron Device Letters*, vol. 30, no. 12, pp.1347-1349, 2009

978-1-4244-9113-1/11 $26.00 © 2011 IEEE

Negative Bias Temperature Instability "Multi-Mode" Compact Model Based on Threshold Voltage and Mobility Degradation

D. Varghese, R. Higgins, S. Dunn, A. T. Krishnan[#], V. Reddy[#], S. Krishnan

Analog Technology Development, #Technology-Design Integration
Texas Instruments
Dallas, TX, USA
+1–972–995–6001, Email: dhanoop@ti.com

Abstract— **In this paper we have developed a model to obtain drain current (ID) degradation at all transistor operating modes (linear, saturation and sub-threshold) during NBTI stress based on threshold voltage (VT) and mobility (μ) degradation. This model provides a compact way to comprehend NBTI induced drain current degradation for transistors subject to multiple operating modes (e.g., dynamic voltage scaling, active/standby modes).**

Keywords--Negative bias temperature instability; compact model; threshold-voltage degradation; mobility degradation

I. Introduction

Negative Bias Temperature Instability (NBTI) is a reliability issue that limits the lifetime of modern CMOS transistors [1]. Various physical mechanisms have been put forward to explain the transistor degradation during NBTI stress [2]-[4]. One of the unique features of NBTI is its ultra-fast relaxation where close to 50% of the degradation recovers within a few milliseconds after removal of the stress conditions [5]. An accurate characterization of NBTI therefore requires ultra-fast (tmeas ≈ 10 μs) [6] or zero-delay on-the-fly [7] measurements. Estimating the degradation in transistor parameters like threshold voltage (VT) and mobility (μ) based on such measurement data set is challenging and various approaches are proposed in literature [8]-[12]. Moreover, such fast measurements are typically done at constant drain/gate bias conditions, while there exist certain applications (*e.g.*, burn-in, dynamic voltage scaling, active/standby modes) where stress and operating conditions are different and it is necessary to know the ID degradation at various operating VD, VG bias conditions. Further, reliability simulators require operating bias dependent degradation models for accurate circuit design [13], [14]. In this paper we will discuss a measurement scheme to obtain VT and μ degradation during NBTI stress from pre- and post-stress ID-VG characteristics and use them to obtain a compact model for drain current degradation at arbitrary VD, VG operating bias condition.

II. Experimental Details

The experimental methodology used in this work is described in Fig. 1. The measurement starts with a full ID-VG

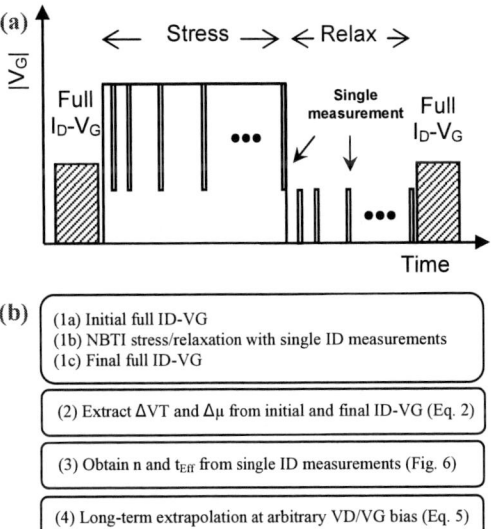

Figure 1. (a) Timing diagram for the NBTI measurements and (b) long-term projection methodology of ID degradation at arbitrary VD/VG bias conditions. Full ID-VG characterizations are made before stress and after sufficient relaxation. Single measurements at constant bias conditions (tmeas<1ms) are made at logarithmic intervals during stress and relaxation.

$$I_D = \mu C_{OX} \frac{W}{L}\left(V_G - V_T - \frac{mV_D}{2}\right)V_D \quad \text{(linear region)} \quad (1a)$$

$$I_D = \mu C_{OX} \frac{W}{L}\frac{(V_G - V_T)^2}{2m} \quad \text{(saturation region)} \quad (1b)$$

$$I_D = \mu C_{OX} \frac{W}{L}(m-1)\left(\frac{kT}{q}\right)^2 e^{\frac{q(V_G-V_T)}{mkT}}\left(1 - e^{\frac{qV_D}{kT}}\right) \quad \substack{\text{(sub-}\\\text{threshold)}} \quad (1c)$$

$$I_D = \mu \times f(V_G - V_T) \quad (1d)$$

$$\Delta I_D = \frac{\partial I_D}{\partial V_T} \times \Delta V_T + \frac{\partial I_D}{\partial \mu} \times \Delta\mu \quad (1e)$$

$$\frac{\partial I_D}{\partial V_T} = -\frac{\partial I_D}{\partial V_G}; \frac{\partial I_D}{\partial \mu} = \frac{I_D}{\mu} \quad (1f)$$

Figure 2. Equations used to model ID degradation during NBTI stress [15]. Degradation in transistor parameters VT and μ are extracted from the measured ID degradation based on these equations.

978-1-4244-9113-1/11 $26.00 © 2011 IEEE

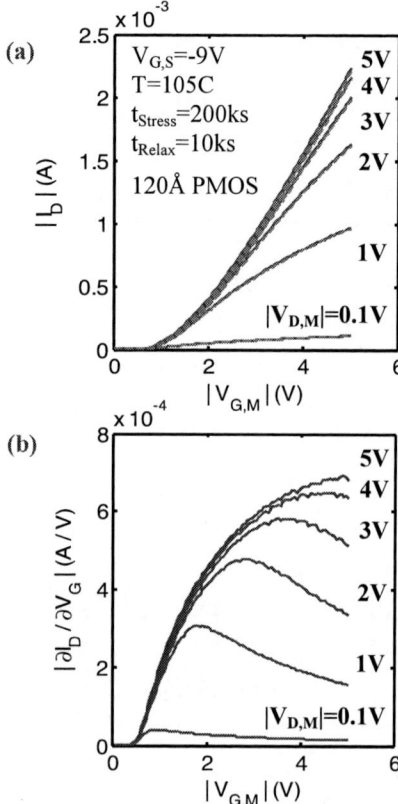

Figure 3. (a) ID-VG curves measured before stress (blue solid) and after relaxation (red dashed) on 120Å I/O PMOS transistor from 180nm node (VG,S=-9V, T=105C, tStress=200ks, tRelax=10ks). (b) $\partial \tilde{I}_D / \partial V_G$ for pre-stress ID-VG curve obtained through numerical differentiation.

characterization of an un-stressed PMOS transistor. The device is then subjected to NBTI stress phase and a subsequent relaxation phase. The stress and relaxation phases are interrupted at logarithmic intervals to perform single-shot fast ID measurements at constant bias condition ($|V_G|>|V_T|$, VD= −0.1V, tmeas < 1ms). Full ID-VG characterizations are also made at the end of relaxation. The relaxation phase is included to allow the fast recovery transient to settle, so that the entire post-stress ID-VG family of curves corresponds to the same level of interface defect density. Single drain current measurements at constant bias condition are used to monitor degradation during stress phase and recovery during the relaxation phase.

The pre- and post-stress ID-VG characteristics provide the VT and μ degradation based on (2): (see Fig. 2):

$$\Delta I_D(V_D, V_G, t) = (-\partial I_D/\partial V_G) \times \Delta V_T(t) + I_D \times (\Delta\mu(t)/\mu). \quad (2)$$

The single ID measurements provide the time exponent of degradation, the effective stress time at end of relaxation phase, voltage and temperature acceleration factors that are required for long-term extrapolation. Measurements were carried out on transistors from 180nm (drain-extended, I/O, core PMOS), 90nm and 45nm (core PMOS) technology nodes.

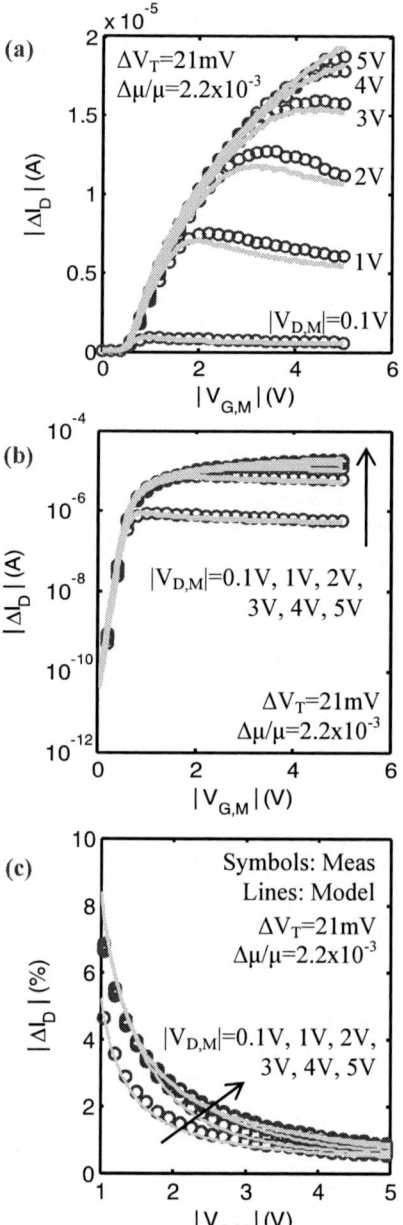

Figure 4. Absolute drain current degradations after NBTI stress and relaxation phases on 120Å I/O PMOS transistor from 180nm node (VG,S=-9V, T=105C, tStress=200ks, tRelax=10ks) are plotted in (a) linear scale and (b) logarithmic scale. The blue symbols represent the measurements and the green lines are the model fit based on (2). V_T and μ degradation are chosen to be 21mV and 2.2x10⁻³ respectively so as to fit the measured degradation. (c) Percentage drain current degradation peaks close to threshold voltage and reduces at higher gate overdrives.

III. RESULTS AND DISCUSSION

A. Extraction of VT and μ degradation

Fig. 3a shows the ID-VG sweeps before and after NBTI stress and relaxation phases on 120Å I/O PMOS from 180nm node. The pre- and post-stress ID-VG curves look almost overlapping since the drain current degradation after the NBTI stress and recovery phases is much smaller compared to the time-zero drain current. Fig. 3b shows the $\partial \tilde{I}_D/\partial V_G$ for pre-stress ID-VG family of curves obtained through numerical differentiation.

Though the pre- and post-stress ID-VG curves in Fig. 3a looks almost overlapping, the differences between these drain currents when plotted in linear (Fig. 4a) and logarithmic (Fig. 4b) show remarkable features. The difference plot in Fig. 4a looks like a combination of $\partial \tilde{I}_D/\partial V_G$ and I_D curves as also predicted by (2). The measured drain current degradation is fitted based on (2) to obtain VT and μ degradation factors. Since the pre-factor $\partial I_D/\partial V_G$ in the first term of (2) peaks close to VT (especially at low measurement VD) while the pre-factor ID in the second term peaks at VG>>VT, a very good sensitivity is obtained in the estimation of VT and μ degradation. The model fits in Fig. 4a and 4b shows that the two parameters explain the drain current degradation at all operating modes of the transistor. Fig. 4c shows the percentage ID degradation which peaks close to threshold voltage and reduces at higher gate overdrive biases.

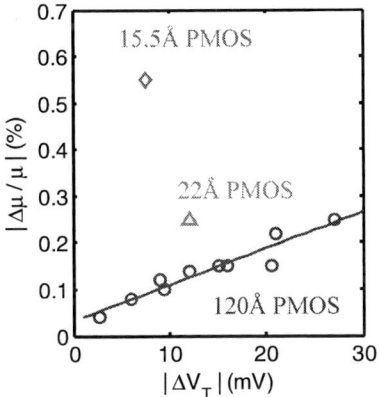

Figure 5. Correlation between V_T and μ degradation for 120Å PMOS transistor. Individual data points were obtained from NBTI measurements carried out at different stress biases and durations. Data from other technology nodes are also included for comparison.

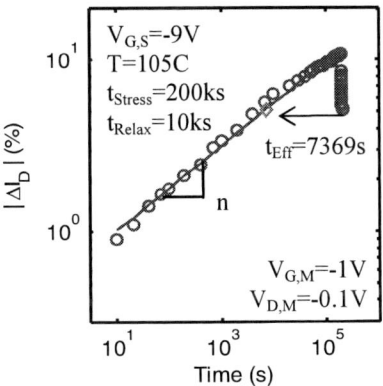

Figure 6. I_D degradation from single fast measurement during stress and relaxation phases for 120Å PMOS transistor. This measurement provides the time exponent of degradation (n) and the effective stress time at the end of relaxation phase (t_{Eff}).

To check for correlation between VT and μ degradation, NBTI stress and relaxation experiments were performed at various stress biases and durations on separate identical transistors to generate different defect densities. The VT and μ degradation extracted from the corresponding pre- and post-stress ID-VG curves are plotted in Fig. 5. Good correlation between ΔVT and Δμ is obtained in agreement with previous findings [8]. Once the correlation between the two parameters is known for a given technology and oxide thickness, a single parameter is sufficient to model NBTI induced ID degradation.

This methodology can also be used to back extract NBTI induced VT degradation from the measured drain current degradation based on the following equation:

$$\Delta V_T(t) = \Delta I_D(V_D, V_G, t)/[-\partial I_D/\partial V_G(V_D, V_G, 0) + k \times I_D(V_D, V_G, 0)], \quad (3)$$

where k is the technology and oxide thickness dependent constant that relates VT and μ degradation (Fig. 5).

B. Long term extrapolation

Long-term extrapolation of ID degradation requires the time exponent (*n*) during stress phase and the effective stress time (t_{Eff}) corresponding to the defect density at the end of relaxation phase. Since drain current degradations for entire ID-VD family of curves are driven by correlated VT and μ degradation, the time exponent of degradation and the recovery transient should be independent of the specific VD, VG measurement bias condition used. We therefore use the single-shot fast drain current measurement to obtain the time exponent and effective stress time for the entire ID-VG family of curves.

Fig. 6 shows the time evolution of drain current degradation measured through single-shot fast drain current measurement during the stress and relaxation phases. The time exponent of degradation is obtained from power-law fit of degradation in the stress phase. The effective stress time is obtained by finding the stress time corresponding to end of recovery degradation level (see Fig. 6). Long term extrapolations at arbitrary VD/VG bias can then be made using the equation:

$$\Delta I_D(VD, VG, t) = \Delta I_D(VD, VG, t_{Eff}) x (t/t_{Eff})^n. \quad (4)$$

The drain current degradation in (4) is for a given stress gate bias and temperature, while it is necessary to obtain the degradation at arbitrary stress gate bias and temperature. The voltage and temperature acceleration factors are once again obtained from single-shot fast drain current measurements and applied to the entire family of curves. Fig. 7a shows the I_D degradation from single-shot measurements at multiple stress gate biases. ID degradation after 50ks of NBTI stress plotted against the stress gate biases is modeled by power-law, though other models proposed in literature [2] can also be used.

Temperature acceleration factor can also be obtained in a similar fashion. The drain current degradation at arbitrary stress and measurement condition is finally given by (5):

$$\Delta I_D(V_{G,S}, T, V_D, V_G, t) \approx \Delta I_D(V_D, V_G, t_{Eff})(t/t_{Eff})^n (V_{G,S}{}^\gamma)(e^{-EA/kT}), \quad (5)$$

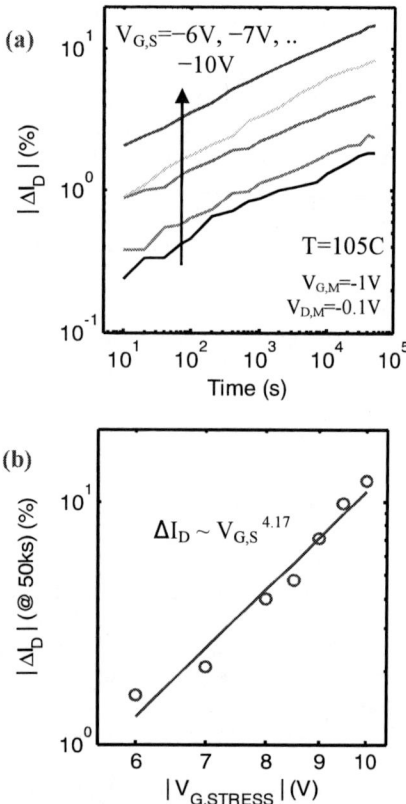

Figure 7. (a) ID degradation from single-shot fast measurements at multiple stress gate biases for 120Å PMOS transistor. (b) ID degradation after 50ks stress plotted agains stress gates biases. NBTI stress gate voltage acceleration is modeled as a power-law fit.

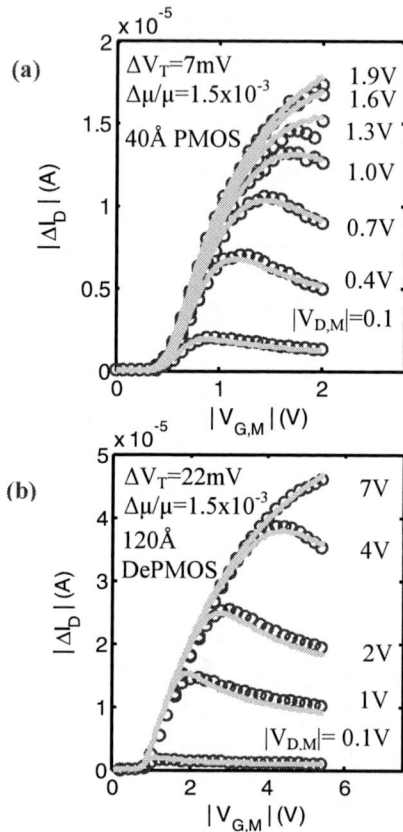

Figure 8. Absolute I_D degradation after NBTI stress/relaxation on (a) 40Å PMOS transistor ($V_{G,S}$=-3V, T=105C, t_{Stress}=30ks, t_{Relax}=1ks) and (b) 120Å drain extended PMOS transistor ($V_{G,S}$=-9V, T=105C, t_{Stress}=200ks, t_{Relax}=10ks) from 180nm node. Blue symbols represent the measured degradation and green lines are the model fits.

where γ and EA are the voltage and temperature acceleration factors respectively.

C. NBTI in different device geometries and technology nodes

In the previous sections, we used the 120Å I/O PMOS from 180nm node to demonstrate extraction of VT and μ degradation from measured ID-VG family of curves and to perform long-term extrapolation at arbitrary stress and measurement bias conditions. Measurements were also carried out on various other device geometries and technology nodes to verify the wide range of applicability of the proposed methodology. The ID-VG difference plots from NBTI measurements in 40Å core PMOS and 120Å drain extended PMOS from 180nm node and the corresponding model fits based on (2) are shown in Fig. 8a-b. The drain current degradation over the entire operating range of the transistor is successfully explained by VT and μ degradation.

NBTI measurements were also carried out on 22Å core PMOS transistor from 90nm node (Fig. 9a) and 15.5Å core PMOS transistor from 45nm node (Fig. 9b). The measured ID degradation in these transistor structures are also successfully explained based on (2). The extracted VT and μ degradation are plotted in Fig. 5 and compared with the 120Å PMOS from 180nm node. The data clearly shows the impact of mobility degradation becoming more significant at lower gate oxide thickness. This is understood from the fact that more interface charge is required to shift threshold voltage by a given amount at lower oxide thickness, resulting in higher mobility degradation.

Fig. 10 shows the percentage ID degradation for the entire VD/VG operating range of the 22Å PMOS transistor after 10 year operation at VG=−1.1V, T=105C, obtained based on (5). The linear and saturation drain current degradations (top left and right corners respectively) are found to be less than 10%. The drain current degradation, however, increases at lower gate biases and can significantly impact circuit performance on operating modes like standby and dynamic voltage scaling. The compact model comprehends the increased degradation at VG ≈ VT and allows accurate modeling of circuit performance.

The ID degradation plot in Fig. 10 also suggests that the OFF-state leakage current of the PMOS transistor (along the bottom edge of Fig. 10) reduces significantly after NBTI stress. This can result in lower stand-by power, a factor to be considered in the design optimizations.

Before concluding, we would like to mention that the discussion in this paper assumes DC NBTI, while in actual designs the transistors mostly undergo AC operation. There

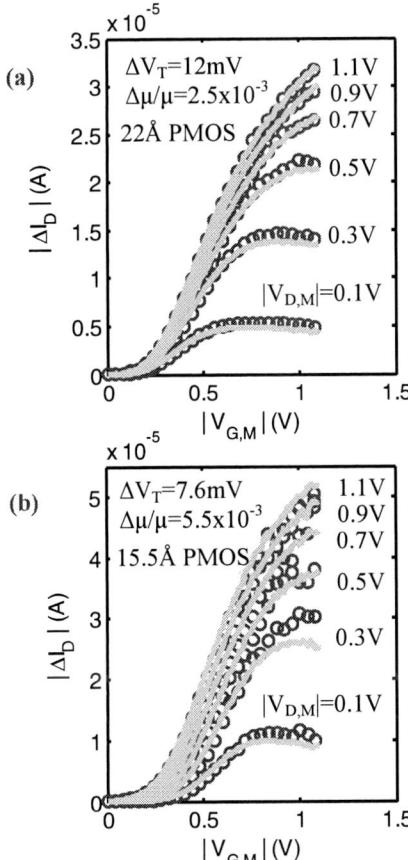

Figure 9. Absolute I_D degradation after NBTI stress/relaxation on (a) 22Å PMOS transistor from 90nm node ($V_{G,S}$=-1.9V, T=105C, t_{Stress}=10ks, t_{Relax}=1ks) and (b) 15.5Å PMOS transistor from 45nm node ($V_{G,S}$=-1.3V, T=105C, t_{Stress}=10ks, t_{Relax}=1ks). Blue symbols represent the measured degradation and green lines are the model fits. The transistors with lower dielectric thickness show higher mobility degradation for similar V_T shifts compared to the 120Å PMOS transistor (see Fig. 5).

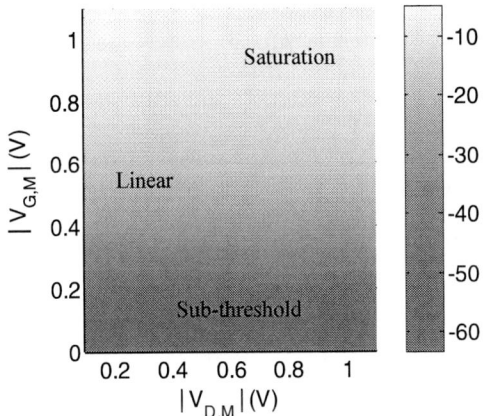

Figure 10. Percentage I_D degradation expected for 22Å PMOS transistor after 10 year operation at VG=-1.1V and T=105C based on (5). Linear and saturation drain current degrade less than 10%, while the degradation increases at lower gate overdrives. The compact model comprehends the ID degradation at various VD/VG operating conditions and allows accurate reliability assessments

are various approaches in the literature [13], [14] to obtain the asymptotic limit of AC NBTI based on the DC degradation. Such frequency and duty cycle dependent models can easily be incorporated to (5) to obtain the asymptotic AC NBTI degradation.

CONCLUSIONS

The drain current reduction for the entire range of VD/VG bias conditions during NBTI is explained based on VT and μ degradation. We have proposed an experimental methodology to extract these parameters and use them in a compact model to perform long-term extrapolation of ID degradation at arbitrary VD/VG bias conditions. This methodology can be applied to develop reliability compact models for accurate degradation assessment during dynamic circuit operation and can be implemented for *e.g.* in Verilog-A.

REFERENCES

[1] V. Reddy *et al.*, "Impact of negative bias temperature instability on digital circuit reliability", Microelectronics Reliability, vol. 45, pp. 31–38, January, 2005.

[2] M. A. Alam, and S. Mahapatra, "A comprehensive model of PMOS NBTI degradation", Microelectronics Reliability, vol. 45, pp. 71–81, January, 2005.

[3] T. Grasser, B. Kaczer, W. Goes, T. Aichinger, P. Hehenberger, and M. Nelhiebel, "A two-stage model for negative bias temperature instability", in Int. Reliability Physics Symposium, pp. 33–44, April, 2009.

[4] V. Huard, M. Denais, and C. Parthasarathy, "NBTI degradation: From physical mechanisms to modelling", Microelectronics and Reliability, vol. 46, pp.1–23, January, 2006.

[5] M. Ershov *et al.*, "Dynamic recovery of negative bias temperature instability in p-type metal–oxide–semiconductor field-effect transistors", Applied Physics Letters, vol. 83, pp. 1647–1649, August, 2003.

[6] H. Reisinger *et al.*, "Analysis of NBTI degradation- and recovery-behavior based on ultra fast VT-measurements", in Int. Reliability Physics Symposium, pp. 448–453, March, 2006.

[7] S. Rangan, N. Mielke, E. C. C. Yeh, "Universal recovery behavior of negative bias temperature instability", in Int. Electron Devices Meeting, pp. 14.3.1–14.3.4, December, 2003.

[8] A. E. Islam, V. D. Maheta, H. Das, S. Mahapatra, and M. A. Alam, "Mobility degradation due to interface traps in plasma oxynitride PMOS devices", in Int. Reliability Physics Symposium, pp. 87–96, April, 2008.

[9] A. E. Islam *et al.*, "Theory and Practice of On-the-fly and ultra-fast VT measurements for NBTI degradation: Challenges and opportunities", in Int. Electron Devices Meeting, pp. 805–808, December, 2007.

[10] M. Denais *et al.*, "On-the-fly characterization of NBTI in ultra-thin gate oxide PMOSFET's", in Int. Electron Devices Meeting, pp. 109–112, December, 2004.

[11] C. Schlunder *et al.*, "A new smart Vth-extraction methodology considering recovery and mobility degradation due to NBTI", in Integrated Reliability Workshop Final Report, pp. 1–5, October, 2007.

[12] J. F. Zhang, Z. Ji, M. H. Chang, B. Kaczer, and G. Groeseneken, "Real Vth instability of pMOSFETs under practical operation conditions", in Int. Electron Devices Meeting, pp. 817–820, December, 2007.

[13] S. Bhardwaj, W. Wang, R. Vattikonda, Y. Cao, and S. Vrudhula, "Predictive modeling of the NBTI effect for reliable design", in IEEE Custom Intergrated Circuits Conference, pp. 189–192, September, 2006.

[14] W. Wang, V. Reddy, A. T. Krishnan, R. Vattikonda, S. Krishnan, and Y. Cao, "Compact modeling and simulation of circuit reliability for 65-nm CMOS technology", IEEE Transactions on Device and Materials Reliability, vol. 7, pp. 509–517, December, 2007.

[15] Y. Taur and T. H. Ning, *"Fundamentals of Modern VLSI Design"*, Cambridge Univerisity Press, 1998.

A new smart device array structure for statistical investigations of BTI degradation and recovery

C. Schlünder, J.M. Berthold, M. Hoffmann, J.-M. Weigmann, W. Gustin and H. Reisinger

Corporate Reliability Department, Infineon Technologies AG, Am Campeon 1-12, 85579 Neubiberg

Introduction

BTI is nowadays the most critical device degradation mechanism and a limiting factor for CMOS technology scaling. Although (N)BTI is under investigation since more than 30 years there is still no consensus regarding origin and mechanism [1-6]. With the method of single defect spectroscopy (SDS) we evaluate new important insights and understanding behind the degradation and recovery behavior of BTI [7,8]. This starts a sequence of several papers investigating BTI based on this method.

Due to the enormous measurement effort and time demand of SDS, all results in the literature are based on a single or very few device measurements. Up to our knowledge a statistical investigation of numerous devices combined with a measurement technique capable to consider recovery is missing. A few publications can be found dealing with device array structures, but all results are impacted by long measurement delays, since standard array structure or SRAM investigations are prone to long and particularly non-uniform recovery times for each device [9, 10].

We develop a new universal smart test-structure solving the typical restrictions of BTI device array investigations. Our integrated structure combined with an adapted measurement methodology ensures very short and most notably uniform recovery times for each device of the entire array. This is the absolutely required precondition for statistical evaluations of the BTI degradation und recovery behavior. Our new smart device array structure approach enables us to investigate the statistic of the parameter degradations itself and the combination of the variability of virgin devices at zero hour together with the additional variations due the non-uniform behavior of BTI degradation.

Restrictions of regular array test structure

Regular test structures with standard pad rows are only capable of offering a restricted number of DUTs (devices under test), since the number of pads is limited by practical aspects like needle-cards, minimal pads size and pitch for reliable contacts. Furthermore the measurements equipment (SMUs) is not arbitrary available in practice. Device array structures offer to integrate a high number of devices within one structure, enabling the characterization of each device by a decoder in combination with selection system on chip.

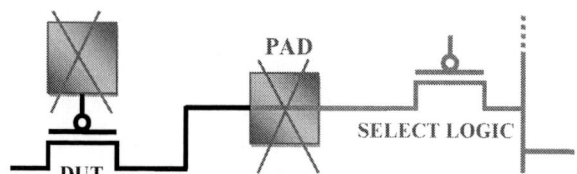

Fig. 1: Schematic to illustrate the direct connection of the transistor nodes to pads in a standard test structure in contrast to array structure with select devies within the signal path prone to IR-drops impacting stress & characterization.

All DUTs can be stressed in parallel and individually characterized in a serial fashion. But reliability investigations with these regular device array test structures are strongly limited due to several severe problems:

Since the DUTs are no longer directly accessible by pads, but have to be selected by decoder logic and select-devices, transistors are within the signal path between DUT and measurement equipment (Fig. 1). These elements impact the applied voltages and consequently the operation points of the DUTs for both stress- and characterization mode. Furthermore the select devices are potentially also prone to degradation. At least the elevated stress temperature is hard to separate for DUT and select devices.

The strongest restriction for BTI origins in the investigation approach itself. After stopping the BTI stress, the DUTs are characterized serially one by one. Long recovery times handicap recovery investigations. In particular non-uniform recovery times avoid statistical investigations at all. The first DUT of the array is characterized directly after end of stress, but all the following DUT have different delay times between stress and characterization.

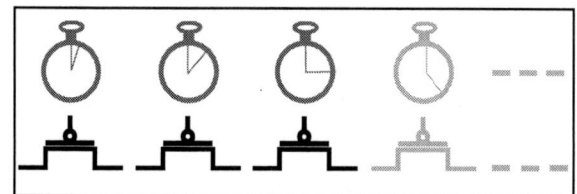

Fig. 2a: Severe restriction in regular array structures: If parallel stress and serial characterization is used, each device got a different recovery time between end of stress and characterization. The stopwatches above the array DUTs symbolize the non-uniform, increasing delay times between stress and characterization As a consequence the results are useless for statistical investigations.

978-1-4244-9113-1/11 $26.00 © 2011 IEEE

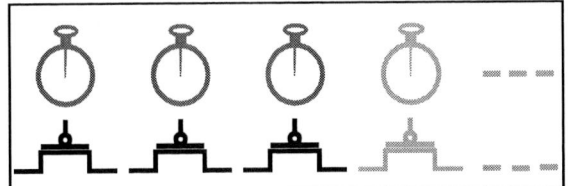

Fig. 2b: Our smart array and measurement approach guarantees the required precondition for statistical investigations of NBTI degradation and recovery: Identical shortest possible recovery times for each device of the entire array. The stopwatches above the array DUTs symbolize the uniform delay times. The achievable delay times depends only on the used measurement technique not on the array test structure.

They recover until the measurements of all previous DUTs are finished. Fig. 2a sketches the situation. The stopwatches above the DUTs symbolizes the different recovery times. The characterization results of each DUT are not comparable and therefore useless.

New smart array test structure

Fig 3 shows a block diagram of our designed test-structure. A decoder controls a selection logic part to connect test transistors of three separate blocks to different pads for stressing or characterization. The decoder contains a robust shifting register supplemented by some buffers to ensure stable signal flanks on the rather long path (>1mm). The decoder actuates the select devices to distinguish between pads for stress potentials and pads for characterization for each DUT.

To cure the selection logic against degradation, due to the elevated DUT stress voltages, transistors with thicker oxides were used. No minimal channel lengths were used, to prevent any possible degradation due to off-state-stress [11, 12]. Typically the I/O devices offered by the CMOS technology were used for this task. Since they have higher operation voltages, level-shifter circuits were used to achieve the necessary signals for the select devices.

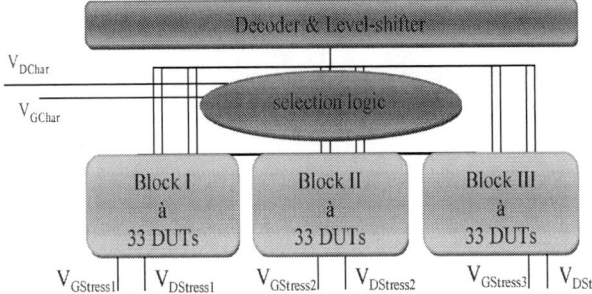

Fig. 3: Block-diagram of the main components of the device array test structure. A decoder controls select devices to switch the DUTs between stress and characterization rails. The structure contains three blocks. In our investigated structure each block houses 33 test devices.

All in all approximately 1300 devices are integrated. Fig. 13 shows the layout of the test-structure in the software tool. The structure contains three array parts of DUTs. In our investigated samples these blocks house 33 devices each, but not restricted to this number. The blocks can be used e.g. for different device geometries or they can be stressed with different stress conditions, e.g. for a lifetime prediction. Therefore each block got separate drain and gate-pads for different stress voltages. Within one block stress conditions are identical. Theoretically even more blocks can be integrated in the test-structure, only limited by available voltage sources and pads. In Fig.4 a part of the structure is drawn to a larger scale to depict the parts in detail (marked with a rectangle in Fig.). The select devices stick out, due to their large width, to reduce IR-drops. Also all paths leading a higher current are drawn wide and stacked.

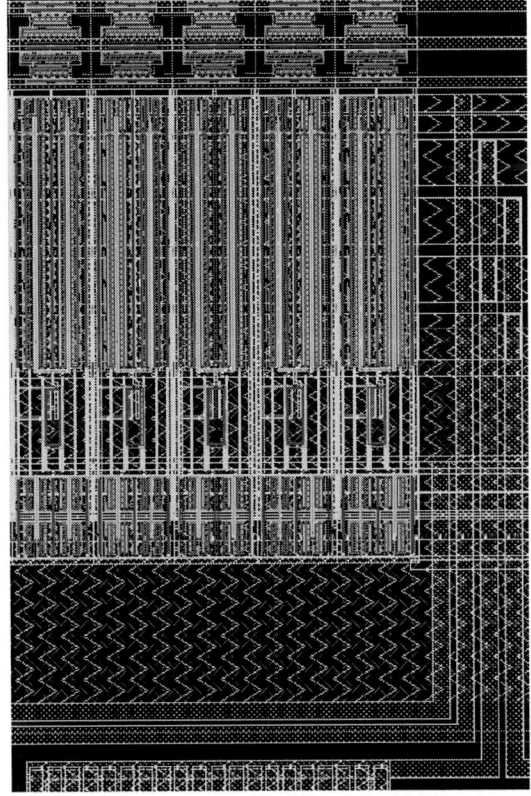

Fig. 4: A part of the test structure to a larger scale. From the top: shifting-register circuits and level-shifter; very wide select devices for drain node, the DUTs itself in a free space for arbitrary geometries, select devices for the gate node and the edge of a pad.

Correct operation points

To ensure correct operation points at all DUTs both for stress conditions and for characterization conditions, the IR-drops impacting these values have to be avoided or compensated. Without special measures even the used wide select transistors in transmission-gate configuration cannot

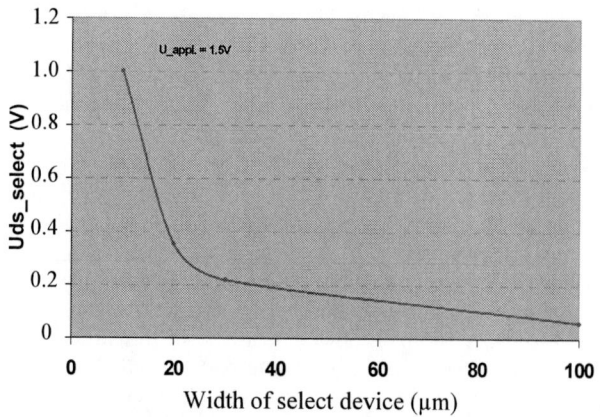

Fig. 5: Simulation of the IR-drop Uds_select at a select device if 1.5V is applied at the DUT as a function of its channel width.

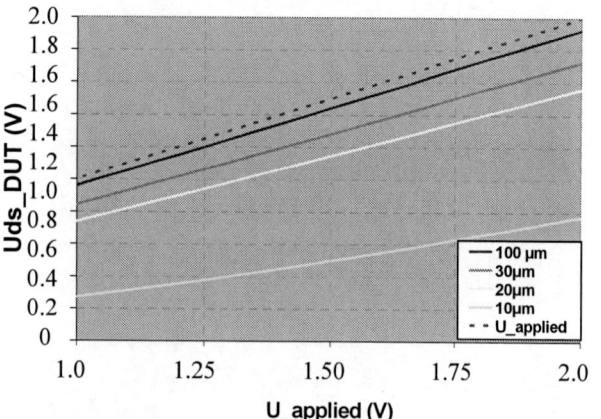

Fig. 6: Voltage directly at the DUT as function of the applied voltage. Even wide select devices clearly impact the DUT voltage.

guarantee that the voltages of the analyzer are accurately applied at the DUTs. Clearly more than 100mV deviation is possible. Fig. 5 and Fig. 6 show simulations of the voltage drop *Uds_select* at select devices and the voltage directly at the DUT *Uds_DUT* for different geometries. But very wide select devices show also disadvantages e.g. higher leakage current in the off-state Therefore we use a force/sense wiring to provide the correct operation voltages for the test-devices.

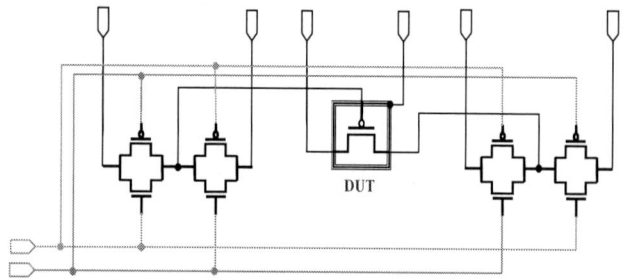

Fig. 7: Schematic of the device under test and doubled transmissions gates as select devices for Force/Sense signal paths.

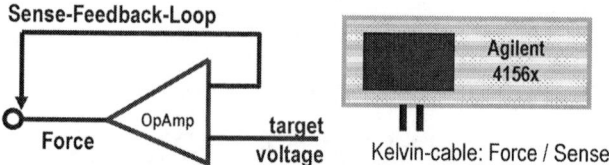

Fig. 8: Operational amplifier with feedback loop to guarantee the correct voltages directly at the DUT. For our test structure we use already optimized opamp of the Kelvin-cable system.

The entire connection paths including all transmission gates for the transistor nodes are doubled. Fig 7 shows the configuration with two transmission-gates each for the test-transistor nodes gate and drain. The force/sense principle adjusts the voltage at the force pad until the voltage directly at the DUT is correct. Every possible IR-drop due to select devices and circuit path will be compensated, even in the unrealistic case that the select logic degrades. The necessary operational amplifier for this task can be attached externally. We use the Kelvin-cable configuration of an Agilent 4556C Analyzer (Fig. 8). The integrated amplifier is developed and optimized to compensate IR-drops on the feed cable.

We expand the feed cable through our integrated structure down to the DUTs. Force and sense path to the DUTs are absolutely identical, but due to highly resistive input of the analyzer (>1MOhm) only negligible current flow through the sense path and according to this no IR-drop impacts the sense-measurement. This measure guarantees correct operation point of the DUTs for stress & measurement. The remaining voltage deviations are clearly below 1mV and therefore negligible.

Uniform recovery times

To prevent non-uniform long recovery times for the DUTs we develop a new smart approach for the characterization phase of our test array. For the V_{th} extraction at a time only one transistors is switched of stress rails to characterization rails, all other DUTs stay on stress. This hot switching is done at stress voltage on both rails. There is no impact on the degradation phases of the devices and no time adder to the measurement delay. We checked the switching event with a digital storage oscilloscope Tektronix TDS5034B and see no glitches or other impacting abnormalities. For this feature we have to double the entire select logic once more (Fig. 9).

After switching the parameter extraction can be done with an arbitrary measurement-routine and –hardware, e.g. with our SIS method with commercial analyzers [13] or with our self-developed fast feedback loop [14]. The achievable delay time depends only on the measurement routine and hardware but not on the test array. In this manner we guarantee shortest possible and particularly uniform recovery times for each DUT (Fig. 2b). Since all other devices stay on stress during the characterization of one device, each device will be measured after different stress times, but this can easily be considered for the evaluations.

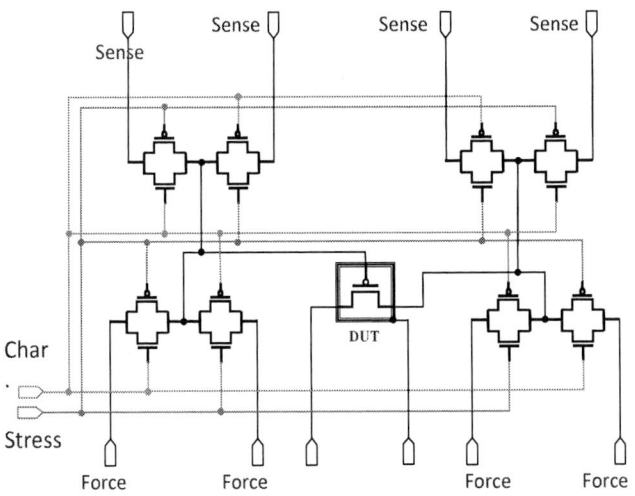

Fig. 9 Schematic of the slice with one DUT and appending select devices. For the switching and force/sense principle 16 select devices per DUT are required.

Results & Discussion

Fig. 10 and Fig. 11 show sample characterizations of the array structure. The diagrams show input characteristics of each DUT in one plot. Fig. 10 show the variability of the identically drawn virgin devices. The curves slightly differ due to parameter mismatch. Fig. 11 shows the input characteristic of all DUT after NBTI stress in one plot. Additional variations are added on top on the variability of the virgin devices before stress. The relatively long measurement leads to recovery.

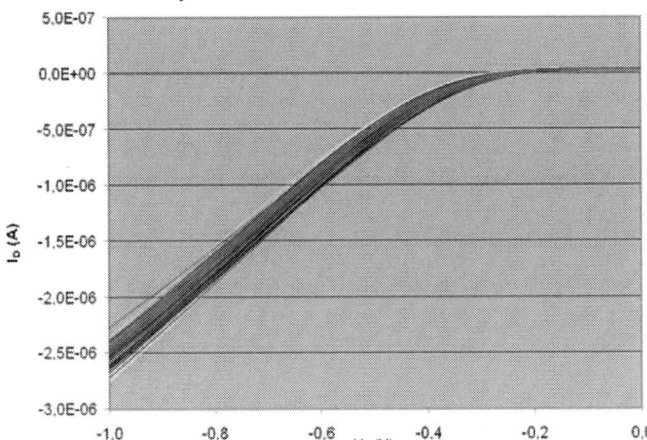

Fig. 10: Input characteristics of all DUTs in one structure before stress in one plot. Due to the variability the identically drawn devices show slight deviations.

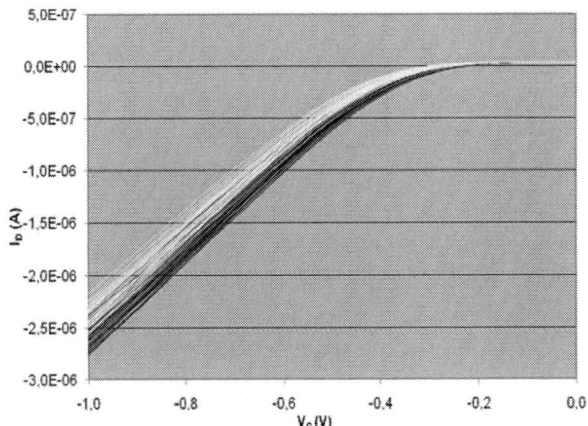

Fig. 11: Input characteristic of all DUT after NBTI stress in one plot. Additional variations are added on top on the variability of the virgin devices before stress.

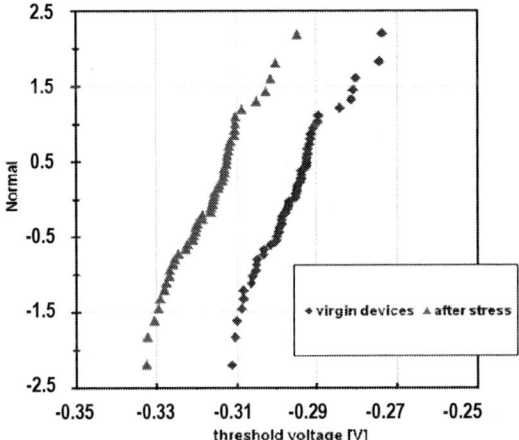

Fig. 12: Example of a distribution of the threshold voltage before and after NBTI stress of a further stress experiment. The threshold voltages are extracted from an array with wide devices. The entire distribution is shifted by the induced threshold voltage drift.

Fig 12 the results of a further experiment with the array test structure. The diagram shows the distribution of the threshold voltage before and after NBTI stress. The array contains productive quality 10μm wide pMOSFETs with 2.2nm PNO. With standard test structure we would need to measure a large number of dies to get the same results but with much larger time demand.

Our new smart array test structure enables us to investigate NBTI degradation and recovery behavior without the restrictions of regular arrays. The special selection circuit guarantees correct operation points and uniform short recovery times.

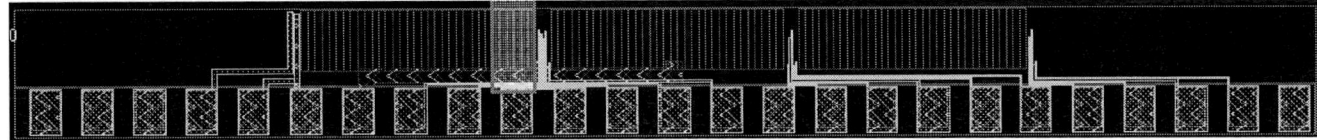

Fig. 13: Layout picture of the new smart device array structure. The test circuit is optimized for 25 pad row. Here it contains three DUT blocks with 33 test devices each

978-1-4244-9113-1/11 $26.00 © 2011 IEEE

References

[1] Jeppson, K.O. and Svensson, C.M., "Negative bias stress of MOS devices at high electric fields and ...", J. Appl. Phys., 48#5, 1977. p. 2004

[2] Harris, H.R.; et al., "Recovery of NBTI degradation in HfSiON/metal gate transistors", IRW 2007, pp. 132-135

[3] Grasser, T., et al., "Switching Oxide Traps as the Missing Link Between Negative Bias Temperature Inst...", IEDM 2009, pp. 729-732

[4] Haggag, A.; et al., "Generalized Models for Optimization of BTI in SiON and High-K Dielectrics", IRPS 2006, pp. 665-666

[5] Nigam, T.; et al., "Accurate product lifetime predictions based on device-level measurements", IRPS 2009. pp. 634-639

[6] Mahapatra; S., et al., "On the Physical Mechanism of NBTI in Silicon Oxynitride p-MOSFETs: Can Differences in Insulator Process..", IRPS 07, p. 1

[7] Grasser, T., et al., "The Time Dependent Defect Spectroscopy (TDDS) for the Characteriza-tion of the Bias Temperature Instability", IRPS 2010

[8] Reisinger, H., et al., "The statistical analysis of individual defects constituting NBTI and its implications for modelling, IRPS 2010

[9] Fischer, T.; et al., „A 65nm test structure for the analysis of NBTI induced statistical variation in SRAM transistors" ESSDERC 2008, pp. 51-54

[10] Drapatz, S.; et al., "Fast stability analysis of large-scale SRAM arrays and the impact of NBTI degradation", ESSDERC 2009, pp. 93-96

[11] Hofmann, K.; et al., "A comprehensive analysis of NFET degradation due to off-state stress", IRW 2004, pp. 94-98

[12] Muehlhoff, A., "An Extrapolation Model for Lifetime Prediction for Off-State Degradation of MOS-FETs, J. Micr. Reliability, 41, 1289 (2001)

[13] Schlünder, C., et al., "A novel multi-point NBTI Characterization Methodology using Smart Intermediate Stress (SIS)", IRPS 2007, pp. 79-86

[14] Reisinger, H., et al., "Analysis of NBTI degradation- and recovery-behavior based on ultra fast VT-measurements", IRPS 2006, pp. 448-453

Characterization and Challenge of TDDB Reliability in Cu/Low K Dielectric Interconnect

Feng Xia, Jun He, Prad Prabhumirashi, Anthony Schmitz, Anthony Lowrie, Jeff Hicks,
Yuriy Shusterman[#], Ruth Brain[#]

Intel Corporation, LTD Quality and Reliability, #Logic Technology Development
5200 NE Elam Young Pkwy, MS: RA3-402, Hillsboro, OR 97124 USA
Phone: 971-214-1887; email: feng.xia@intel.com

Abstract— **Interconnect dielectric reliability challenges increase every generation due to dimension scaling and pursuing of lower K dielectrics for performance. In this paper, TDDB reliability characteration, process innovation, process control and product validation are presented based on Intel 32nm technology node. In definition and technology development phase, extensive characterization and process innovation are needed to enable the proper choices of materials and processes. For any given healthy process and material set, the TDDB reliability is determined by via to line and line to line space distribution, and dominated by the tail distribution of dies with small space. Although different TDDB physical models such as E, root E or other models will project very different failure probability when extrapolating from high Efield to low Efield for the main population, the choice of models makes little difference for product level failure probability from the tail population because the Efield is already in the range of that used for accelerated stressing. Since product failure rate is dominated by tail population, more focus has been given to this area in terms of process development and control. Novel self-aligned via patterning process has been developed for 32nm technology and significantly improved via to line space and thus the low K TDDB performance. In addition, to ensure superior quality and reliability, extensive process window and product level validation through extended life test are necessary to capture and eliminate worse case process and product corners. A scaling trend and potential path to enable continuing scaling are highlighted.**

Keywords- BEOL, low k dielectric, self aligned via (SAV), Cu interconnects, Reliability, TDDB.

I. INTRODUCTION

Low K TDDB is one of the key reliability challenges for Cu/low K Interconnect, and the risk is higher every generation due to dimension scaling and pursuing of lower K dielectrics [1-2]. The typical failure modes are dielectric breakdown which could be hard breakdown or soft breakdown, or increased leakage due to pure dielectric breakdown, Cu migration, or the combination of both (aka. Cu diffusion or migration enhanced dielectric breakdown). Comparing to transitional low K dielectric, ultra low K dielectric materials are more prone to soft breakdown [3]. In recent years, many studies have been done to address the reliability risk for continuing scaling, several models have been proposed such as E, root E, E^n models, etc [4-7]. In many cases, different

processes and materials integration could lead to very different results and complexities when comparing each other's conclusion. There is a tendency to move to root E with recent and more supporting data, however no consistency has been reached yet.

For reliability study and evaluation, many novel test structures have been proposed, such as planar capacitor, metal comb, and metal tip to tip and via comb structures [8]. Voltage ramp and TDDB testing are the main methods to evaluate interconnect dielectric reliability. However most of the studies are focused on much higher voltage and electrical field with the assumption that low use voltage performance over small space shall follow the same trend at the same Efield. TDDB data from low voltage and small space are limited due to long stress times, low temperature acceleration and lack of precise space estimation method at the tightest space ranges that represent the tail in the space distribution. Increased variations in time to failure (TTF) and Efield for small space lead to difficulty and reduced accuracy in modeling and reliability characterization. This challenge may also have played role in the diversity of the TDDB modeling.

Beside fundamental study, process interactions vs TDDB reliability has been reported with foci on CMP, dielectric cap layer and barriers [9-10]. The key component is to reduce or eliminate Cu migration at the interface. New integration schemes were also studied for improving TDDB performance; self-aligned via (SAV) patterning process implemented on Intel's 32nm technology [11] is one of new processes that can significantly improve the low K TDDB performance. It is also no doubt that process variations and process-design interactions play significant roles for reliability. How to evaluate the product level risk and eliminate the process corners are among the key components for low K TDDB reliability.

In this paper, we will present the TDDB characteration, process innovation, process control and product validation that have enabled Intel's 32nm technology node for high volume production. Fundamental characterization method and process innovation such as SAV improvement for TDDB reliability will be discussed in details. In addition, the impact of different physical models on product level failure probability (Pfail), methods for process control and product validation to ensure superior quality and reliability will be covered.

II. EXPERIMENTAL DETAILS

Accelerated stresses were performed on 32nm Cu/low K interconnects on different structures and spaces for reliability performance. Standard voltage ramp method and TDDB stress were used on test structures to get current-voltage (I-V) curve and leakage current vs time (I-t) for time to failure (TTF).

To validate product failure rate and capture worse case layouts and process corners, product level validations were carried out using extended life testing with intentional mis-registration to detect any interconnect dielectric TDDB related failures in product qualification.

III. RESULTS AND DISCUSSION

(1) Test structure and method for intrinsic study and process optimization

Due to line edge roughness, line-to-line is often considered a primary test case for TDDB modeling. However in multi-layer interconnect system; via-to-line is the limiting case for low K dielectric TDDB due to via bulge-out and via-to-line mis-registration. How to accurately know the space between via and adjacent line is one of the key challenges in TDDB modeling, if not the most. To better know and monitor the space, via to line structures are designed to capture the CD variation, etch bias and mis-registration by shifting via closer to the adjacent metal line. Fig. 1 presents the concept of such a structure, via is shifted reference to upper and lower metal layers intentionally in a big range. The shift step can be as small as possible for accuracy. By testing the leakage of this group of structures for shoring or not, the space window and mis-registration can be estimated. Other usages of this kind of test structures include (a) IV curves can be obtained by ramping voltage to get breakdown voltage (Vbd); (2) these structures can be used to get different spaces for TDDB testing to develop TTF vs Efield models.

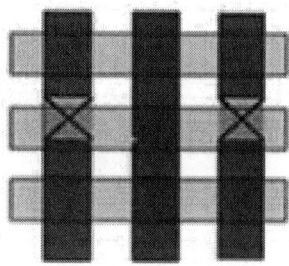

Figure 1 Via to line space structure with shift reference to lower and upper metal layers.

Fig. 2 is the typical I-V curves that tested on the via-line shift structures with different via shift and space. As expected, the breakdown voltage (Vbd) decreases for reducing space and higher leakages were observed for smaller space at same stress voltage. A linear relation between Vbd vs via shift is demonstrated as shown in Fig. 3. The slope is the breakdown electrical field (Ebd), which can be used to measure the intrinsic dielectric properties and for process selection such as liners options and CMP optimization. A similar concept was

independently reported by K. Y. Yiang et al [12]. The advantages of Ebd method are (a) the testing is fast; (b) the comparison is more intrinsic which reduces the noise from space variations. Furthermore, Ebd was collected from small and large space to study the effect of process damage and recovery on low K dielectrics immediately adjacent to metal/via. The results showed that the Ebd is the same for space at a few nm versus large space. A relative high Ebd is a prerequisite for low K dielectric selection. For low K dielectric materials with intrinsic lower Ebd, space will have to be used to compensate lower Ebd to meet the product specs. This will negatively impact pitch scaling.

Another usage of Ebd is to estimate space at sub-nm resolution for accurate TDDB evaluation by combing Vbd and time zero leakage (Io). Fig. 4 shows a universal Vbd vs leakage curve obtained from similar test structures with different spaces. The current was tested at a fixed low voltage, big space structures have lower leakage and higher Vbd. Using Vbd vs Io curve, Vbd can be predicted for TDDB testing. The data have confirmed that the predicted Vbd vs tested Vbd is within 10% error bar. If the via shift step as shown in Fig. 1 is 2 nm for instance, the error is 0.2 nm or less. From Vbd vs Io and Io vs TTF as shown in Fig. 4 and Fig. 5, finer space resolution at nm resolution and more accurate Efield can be extrapolated for intrinsic modeling.

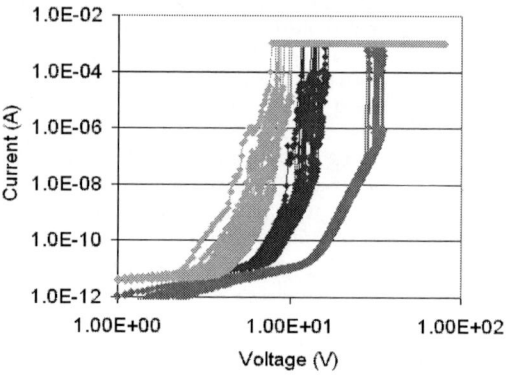

Figure 2 I-V curves for via to line structure with different via shift to modulate space.

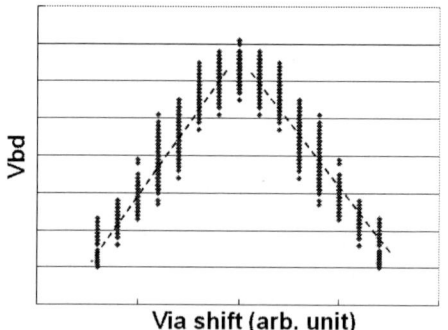

Figure 3 Vbd vs via shift. Via shifts in either direction from nominal space, the slope is Ebd.

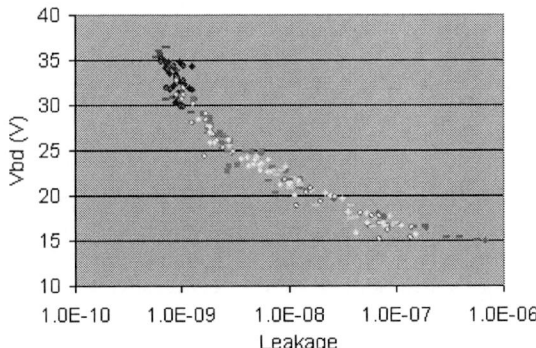

Figure 4 Vbd vs leakage current tested at fixed low voltage from similar structures with different space.

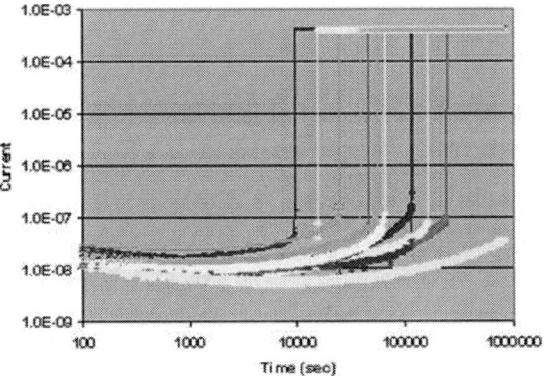

Figure 5 Current vs time at constant stress voltage.

(2) Impact of space distribution and physical models on product Pfail

For a given interconnect system with fixed process and materials, the Pfail is dominated by space distribution at the tight pitch layers. The small space dies at the tail of the distribution will subject to much higher Efield than the ones with large space; shall have significant failure rate for any Efield based models regardless of whether the acceleration model is E-model, root-E, etc.

Fig. 6 showed the fail probability (Pfail) vs space under same voltage using the most conservative E model. The abrupt curve clearly states that for the die population with large space, the probability to fail is zero or negligible, while the Pfail from small space dies is much higher. In other words, the Pfail is dominated by the tail distribution, not the main population. The Pfail from the main population is very low even using the conservative E-model; it will be much lower if root-E or other models applied. Not until the scaling makes the main population near a few nm will different models have a significant contribution to product Pfail. Even with continuous scaling, the main population does become increasingly important each generation; however the tail population will already require more attention in each technology node, and always reach limit earlier than main population. Yet, the value

of the fundamental studies is to understand the physical mechanism and consequently develop better materials and integration schemes.

For the tail populations at the few nm space range, the Efield is relatively high and in the same range of testing Efield. The real TTF vs E field curve can be used and little extrapolation is needed for reliability evaluation. At this Efield range, there is little different in term of fitting using different models such as E, root E and other models as illustrated conceptually in Fig. 7. To this aspect, more focus should be given to the tail distribution no matter which model dominates in the future. Rigorous control and elimination of the tail population is the key for product quality and reliability, although in reality such worse case space occurs very rare only when both metal/via CD and registration hitting the worse case corner simultaneously. The tail population is difficult to be obtained on simple test structures, process skew on actual product or test chip is the right approach and associated reliability evaluation would not be complete without it. With process innovation and rigorous process control, Intel's 32nm node was certified for high volume manufacturing using the conserve E model.

For further scaling, the tail population may cause significant yield loss if no revolutionary process and integration innovations are made. Self-aligned via process is one of those new ideas that will be presented in next session.

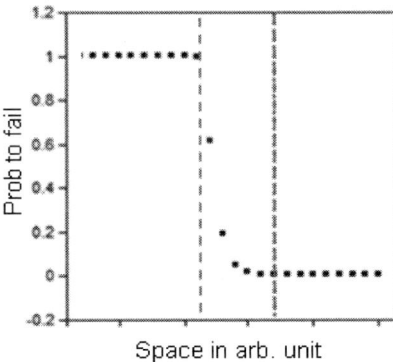

Figure 6 Prob to fail vs space using E model, two dash line highlights the focus area that dominating Pfail.

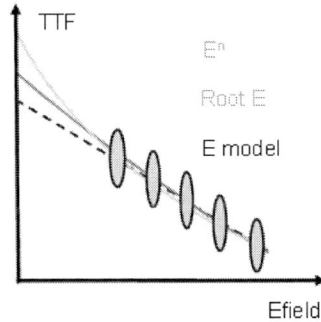

Figure 7 TTF vs Efield curve. The tail population with much higher Efield dominates Pfail and there is little difference for different models at those Efield range.

(3) Self-aligned via (SAV) process and improvement for TDDB

To reduce the tail population in space and have larger space for intrinsic reliability, SAV process was developed. The process details and SEM images have been published previously [11] to show that there is no via bulge and mis-registration comparing to via first process. From reliability point of view, SAV process has led to significant improvement in via-to-line space for Cu/low K TDDB reliability comparing to via fist process, especially for tail distribution with smaller space which dominates Pfail. Fig. 8 showed the distribution of the space ratio for 32 nm over 45 nm node. The ratio varies from 0.85 to 1.50 for large and small space, respectively. What most important is that most of improvement is in the small space regime that dominating product Pfail. The 0.85 ratio for large space is consistent with the 0.70x historical scaling factor from 45nm to 32nm node with 1.2x improvement. The 1.5 x improvement for small space will become 2.1 x if counting the 0.7x scaling factor from 45nm to 32nm node. The improvement in small space regime is due to the elimination of via bulge and mis-registration contributions to space and space variation. As a result, better TDDB performance has been achieved on bigger space and tight distribution through process innovation.

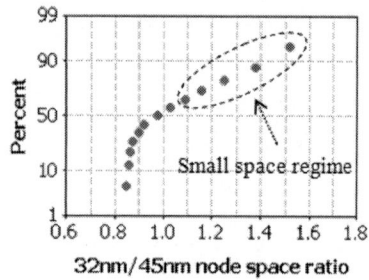

Figure 8 Distribution of space ratio of 32 nm vs 45 nm node; SAV demonstrated clear improvement in space, especially in the small space regime which dominates product fail rate.

(4) Process control and product level validation

To ensure supreme quality and reliability, rigorous process control and product validation are required. Quick turn monitors were designed using time zero leakage vs Vbd relations as presented previously; and good correlation has been demonstrated with yield and inline signals. This allows tools and monitors to be developed and implemented to dynamically monitor process health in high volume production line.

Product is much more complex than the test structures. To ensure full coverage and capture of worse case processes and design corners, extensive validation has been designed to evaluate the process-design interactions using intentionally process skews and extended life testing methods. Both test chip and real product were intentional skewed to worse case corners and much smaller space using mis-registration, even those worst corners are very rare in production. Then extended life test was conducted with statistical sound sampling size. The validated failure rate from extended life testing was compared to the model predictions. Within tolerable conservatism as the nature of any reliability modeling, the Pfail is well matched between modeling and product reality. The extensive validation has ensured high confidence in TDDB reliability characterization and product qualification.

(5) Scaling trend

Continuing scaling and extreme low K dielectrics for better performance with more porous materials shall bring more challenge for TDDB reliability. In our view, fundamental understanding, intrinsic material and interface engineering, process innovation, and rigorous process control are the key components to enable quality and reliability for high volume manufacturing.

IV.   SUMMARY

Although the physical TDDB models are important for intrinsic Cu/low K TDDB reliability, the product level failure rate is determined by space distribution, and dominated by tail distribution. Rigorous process control, validation, and elimination of worse case space corners are the key components to ensure product quality and reliability.

REFERENCES

[1] Jeffrey P. Gambino, et al, IPFA 2009.

[2] E. T. Ogawa et al, pp 166, IRPS 2003.

[3] F. Chen, et al, pp464, IRPS 2009.

[4] F. Chen, Year in Review, IRPS 2010.

[5] K. Croes, et al, pp IRPS 2010.

[6] J. Noguchi, IEEE Trans Electron Devices, pp1743, 2005.

[7] S C Lee, etal, IRPS pp481, 2009.

[8] T. Kamoshima, et al. pp185, IITC 2009.

[9] W. Liu et al, pp613, IRPS 2009.

[10] D. Roest, et al, pp103, AMC 2009.

[11] Ruth Brain et al, IITC, 2009.

[12] K. Y. Yiang, et al. IRPS, 2010.

Backend Low-k TDDB Chip Reliability Simulator

Muhammad Bashir, Dae Hyun Kim, Krit Athikulwongse, Sung Kyu Lim and Linda Milor
School of Electrical and Computer Engineering
Georgia Institute of Technology
Atlanta, GA USA
mbashir@gatech.edu

Abstract—**Backend low-k time-dependent dielectric breakdown degrades reliability of circuits with Copper metallization. We present test data and link it to a methodology to evaluate chip lifetime due to low-k time-dependent dielectric breakdown. Other failure mechanisms can be integrated into our methodology. We analyze several layouts using our methodology and present the results to show that the methodology can enable the designer to consider easy design modifications and their impact on lifetime, separate from the design rules.**

Keywords-Copper interconnects; TDDB; dielectric breakdown; chip lifetime; reliability

I. INTRODUCTION

Copper (Cu) is the choice material for metallization in today's very-large-scale integrated (VLSI) circuits. When Cu metallization is used with materials termed low-k dielectrics, which have a dielectric constant (k) lower than that of Silicon Dioxide (SiO_2), the combination is known as Cu/Low-k interconnect. Cu/Low-k interconnects reduce interconnect delays and coupling capacitances. However, these performance advantages are accompanied by drawbacks critical to reliable chip operation and lifetime.

The primary reason of using low-k dielectrics is the reduction in parasitic capacitance in comparison with SiO_2 dielectrics. However, lower-k dielectrics are formed by increasing porosity, at the possible cost of reducing reliability. Moreover, each technology node reduces the interconnect dimensions without always reducing the supply voltage by the same proportion. This results in the backend dielectric in the newer nodes being subjected to higher electric fields than the previous nodes.

Figure 1(a) shows a cross-section of a conventional Cu/Low-k interconnect stack. The dielectric between the interconnect lines is the one that is most vulnerable to breakdown since it is stressed by the highest electric fields and uses material with the lowest dielectric constant. Hence, test structures used to evaluate backend dielectric breakdown are single-layer test structures that stress the dielectric between the lines in the same layer.

The standard approach to assess backend dielectric reliability is by using process data. The typical test structure is a comb structure, as shown in Figure 1(b) (top view). In testing a comb structure, a voltage difference is applied between the two combs. The current between the combs is monitored to determine the time-to-failure (*TF*).

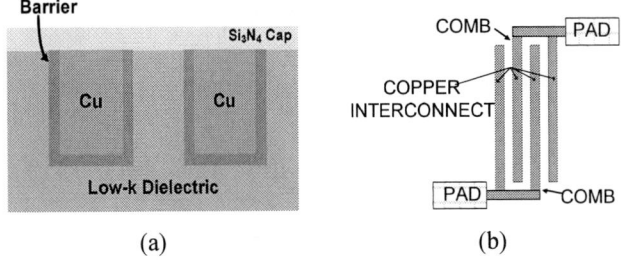

Figure 1. (a) Cross section of a typical Cu/Low-k interconnect (b) top view of a comb test structure.

Test structures are stressed at high voltages and high temperatures to accelerate dielectric breakdown and device aging. Appropriate adjustments, and extrapolations, are made to the test results to scale them to actual operation conditions. In addition, corrections are also needed to account for the difference between the vulnerable area of the chip and the test structure. There is no literature on the method to find the vulnerable area for a chip for backend dielectric breakdown, except for the statement that the vulnerable area is "the total length of such [minimum spaced] lines within a product" [1].

Chip reliability analysis requires techniques to extend the results gathered from small test structures and circuits to large complex chips. Such an endeavor must also be accompanied by solutions to manage and use the deluge of data that comes with analyzing large layouts. The physics describing IC failure mechanisms both in the front-end and in the backend has matured as a result of years of refinement to existing theories. However, the extension of these models to large and complex circuits has not proven to be straightforward and is complex.

The purpose of this paper is to present a methodology to assess chip lifetimes based on low-k TDDB chip lifetimes, by developing the link between data collected from test structures and the chip. We demonstrate the feasibility of our methodology by presenting results from a simulator based on the proposed methodology. Our methodology includes all layers of a chip, which can have different vulnerable areas.

The ultimate purpose of our work is to introduce backend dielectric reliability in design. The onus of meeting this end falls on the designers, if the reliability concerns and the accurate estimation of chip lifetimes can be conveyed to the designer in a designer-friendly manner. Designers ensure reliability, often inadvertently, by strictly adhering to design rules that assume worst case scenarios. Design rules often do not adapt themselves to the complexity of different circuits and

This work has been sponsored by SRC.

978-1-4244-9113-1/11 $26.00 © 2011 IEEE

operating conditions. Instead, design rules are kept general enough to encompass a large number of circuits. A design rule that may be too restrictive for one design can be completely unrealistic for another design.

As chip complexity increases, designers become less aware of the actual physical functioning of their chip. Similarly, extending the models of the physics of failure to a chip requires the consideration of a myriad of factors. The task can be simplified if test results are used incrementally, as proposed in this work.

We start by commenting on the efforts to simulate chip level reliability, followed by laying out the guidelines for our methodology, and consequently the simulator. In section IV, we summarize our test structures and test results. In the following section, we outline our methodology. In section VI, we detail the circuits we have used to run our simulations. In section VII, we study the results from our simulator and present the insights gathered from our results. Section VIII concludes the paper.

II. RELIABILITY SIMULATIONS AND SIMULATORS

The most important reliability concerns for interconnects are electromigration, stress-induced voiding, and TDDB of the backend dielectric [1]. To date, reliability simulators for interconnects have only been developed for electromigration and stress evolution [2], [3]. Our purpose is to consider an additional wear-out mechanism: backend low-k TDDB that is currently not included in reliability simulators.

A. Modeling system lifetime

It should be noted that circuits wear-out for a variety of reasons, both related to devices and interconnect. All of these wear-out mechanisms happen simultaneously. It is common to describe reliability mechanisms with a Weibull distribution

$$P(t) = 1 - \exp\left(-\left(\frac{t}{\eta}\right)^{\beta}\right), \qquad (1)$$

having two parameters: the characteristic lifetime (η) and the shape parameter (β) [4]. The characteristic lifetime is the time-to-failure at the 63% probability point, when 63% of the population has failed, and the shape parameter describes the dispersion of the failure rate population. Typically, the shape parameter is close to one. If we have a collection of n independent wear-out mechanisms modeled with Weibull distributions, having parameters, $\eta_i, i = 1, \dots, n$, and $\beta_i, i = 1, \dots, n$, then the characteristic lifetime of the system, η, i.e. the time when 63% of the population has failed from any mechanism, is the solution of [5]

$$1 = \sum_{i=1}^{n} \left(\frac{\eta}{\eta_i}\right)^{\beta_i}. \qquad (2)$$

If the characteristic lifetime of one mechanism is significantly smaller than others, this mechanism will dominate the failure rate. However, in general, it is prudent to consider

all major sources of wear-out. Hence, as we scale the dimensions of interconnect and lower the dielectric constant, one should no longer neglect the potential reliability failures due to backend low-k dielectric.

B. TDDB Models

We note that models that describe backend TDDB, although they may have been initially developed for device TDDB, are of the general form [6], [7], [8], [9]

$$\ln TF = A - \gamma E^m, \qquad (3)$$

where A is a constant that depends on the material properties of the dielectric, γ is a field acceleration factor, m is one for the E model [6],[7] and $m = 1/2$ for the $\sqrt{E}$ model [8], and TF is the time-to-failure. Note that although these models can be generally represented in this form, they are based upon very different physical mechanisms. This representation is only used for modeling.

Time-to-failure is both a function of the electric field and temperature. Equation (3) provides the correction that takes into account the difference in the electric field between use conditions and during accelerated stress test. The temperature dependence is modeled with an Arrhenius relationship [8]

$$\ln TF = B - \frac{C}{T}, \qquad (4)$$

where B and C are constants. Equation (4) provides a correction between chip operating conditions and accelerated stress conditions. There is a concern that stressing at high temperatures can activate failure modes that are not present during use conditions. Hence, stressing at high electric fields is preferred in comparison with testing at high temperatures. Our tests were conducted at $150^{\circ}C$.

III. GROUND RULES

This work forms an interface among reliability physicists, semiconductor foundry engineers, and designers by suitably partitioning their combined effort, thereby keeping each unburdened with the details of the other's efforts.

Test structures have identified the well known sensitivities of backend dielectric breakdown to area and distance between the lines (electric field in the dielectric) and the less well known sensitivity to metal linewidth. Through our methodology [5], we link test structure data to the entire product die (chips), by extracting the corresponding "vulnerable areas" for a chip, and we characterize the failure rate for the chip by combining the failure rates due to all "vulnerable areas".

The simulator focuses on chip wear-out due to backend dielectric breakdown. The results can be combined with other wear-out mechanisms via (2).

The simulator combines the test results from test structures that stress different vulnerable areas to determine the failure rate for a chip. In other words, our methodology links chip layout geometries to those on the test structures. We assume that these results can be scaled to use conditions with whatever version of (3) and (4) that is deemed to be appropriate.

The focal point of our work is the mapping between the test structures and the chip. We assume that the conductors in the chip are very similar to the conductors in the test structures and that they have similar geometries. Conductor geometry does impact the data that is collected from the test structures and the models developed from these data. The methods that account for conductor geometry will be noted in the following sections.

IV. TEST STRUCTURE DESIGN AND TEST RESULTS

A. Test structures

We have designed test structures to assess the impact of area and linewidth on Cu/low-k TDDB. The details of the test structures, their design and results, are given in [10] and [11]. The test structure set includes comb structures that vary area, linespace, and linewidth.

The test structures were manufactured with an industrial 45nm dual-damascene process and subjected to accelerated stress tests at $150^{o}C$ with electric fields ranging from $0.25MV/cm$ to $1.5MV/cm$. Breakdown was considered as the point of onset of leakage current greater than $100\mu A$. The sample size was 30 and the Weibull failure rate distribution was used to model the failure population.

B. Test results

Test results indicated a strong impact of area. Die-to-die linewidth variation creates curvature in failure rate distributions. This curvature does not impact η. Hence we extract η and determine β by area scaling. Once these parameters have been determined for the unit area, the relationship between characteristic lifetimes for different areas in known.

Note that LER can also be taken into account when extrapolating failure rates. The impact of LER is considered in [11].

Lifetime is also impacted by the linewidth on each side of the dielectric segment. We consider linewidth variation when determining η and β for our test structures [10]. Specifically, linewidth can be taken into account by extracting η as a function of linewidth, since β, which is extracted from the area test structures, is assumed to be constant.

Test results showed a strong impact of linewidth on low-k TDDB, even when the vulnerable area and linespace of the dielectric under stress remained constant. If W_a is the actual linewidth and W_d is the drawn linewidth, then the difference between them is given by $\Delta W = W_a - W_d$. The shift in linespace is $S_a = S_d - \Delta W$, where S_a is the actual linespace and S_d is the drawn linespace. This shift arises because of aspect-ratio-dependent-etching (ARDE) [12]. During etch a protective compound builds up on the sidewalls of wide trenches, preventing lateral etch. This compound does not build up as much in narrow trenches. Hence, narrow trenches suffer from greater lateral etch near the critical CMP interface. Linespaces

with larger positive values of ΔW breakdown faster, since $E = V/S_a$. We can use data directly to determine the relationship between the drawn linespace and η through regression.

Our data indicates a difference in linespace as a function of the widths of lines on the right and left as shown in Figure 2.

If SEM data is available, then Equation (3) along with $E = V/S_a$ can be used to calculate A and γ. The ΔW can also be found for any dielectric with any linewidth on the right and the left. The model in Figure 2 gives an estimate of ΔW and S_a. The constants from Equation (3), in turn, provide an estimate of the corresponding characteristic lifetime.

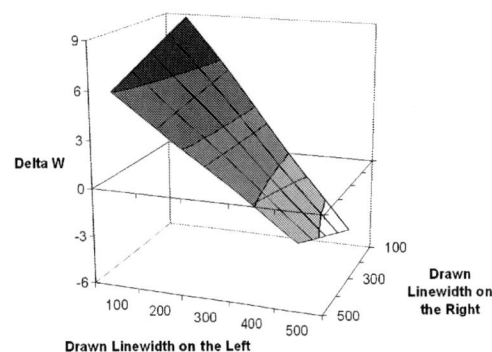

Figure 2. Variation in linespace as a function of the widths of the lines in either side of the dielectric. The data was collected using scanning electron microscopy (SEM).

Figure 3 shows a plot of characteristic lifetime varying with area and linespace, extracted from our test structures.

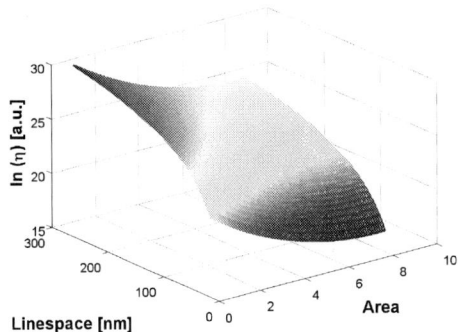

Figure 3. η for the $\sqrt{E}$ model. Area is the ratio of the area of extrapolation to the unit area test strcture.

This plot is obtained by determining the characteristic lifetime for different area ratios, in comparison with the $1X$ test structure, for different linespaces, i.e.,

$$\ln \eta \propto f(S, A), \qquad (5)$$

where S is the linespace and A is the area.

C. Scaling test results to use conditions

Electric field and temperature can affect the relationship between test conditions and use conditions. The relationship between test conditions and use conditions is given in Equation (3). However, the test structure is stressed with DC stress while the chip dielectrics undergo AC stress. Nonetheless, it should be noted that the backend dielectric TDDB under AC stress does not show any recovery [13], as observed in bias temperature instability degradation, and lifetime relaxation or healing, as observed in degradation due to electromigration.

In our analysis we assume a signal activity factor of 0.5. Figures 4 shows our results scaled to use conditions for 45nm technology, with a supply voltage of 0.8V under alternating pulsed stress.

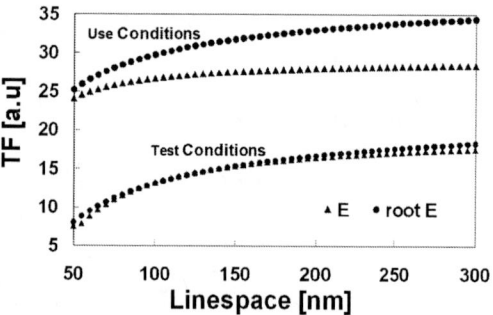

Figure 4. Test results, for structures used to isolate the impact of linewidth, scaled to use conditions.

It should be noted that segments of the circuit may undergo different signal activity factors. In that case, these sectors should be analyzed separately, before combining them with estimates of lifetime of other sectors. Similarly, segments of the circuit may experience different values of average temperature. These sectors should also be analyzed separately, before combination with estimates of lifetime from other sectors.

V. TDDB CHIP LIFETIME

A. Feature extraction

We determine the vulnerable sites, vulnerable area, in a given layout from the layout features. The vulnerable area is a block of dielectric between the two Copper lines separated by linespace S_1 for length L_1 and having an area $S_1 L_1$.

The feature that is extracted from layouts is the vulnerable length between two lines L_i associated with a linespace S_i, which is a function of the widths of the two adjacent lines, $W_{i,L}$ and $W_{i,R}$, determined by the model in Figure 2.

A given layout is analyzed by determining the pairs $\left(S_i(W_L, W_R), L_i \right)$ for each layer for all linespaces surrounded by the linewidths W_L and W_R. When we integrate temperature profiles, we partition the layout with a $5\mu m \times 5\mu m$ grid prior to the extraction of $\left(S_i(W_L, W_R), L_i \right)$.

The details of our methodology can be found in [5]. Here we give its gist in the following subsections.

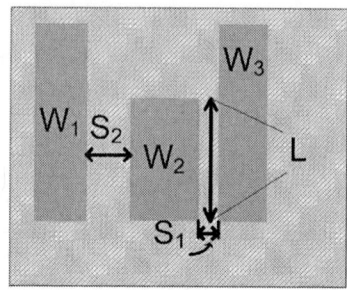

Figure 5. Vulnerable area characterized by the linespace. The rectangles are Cu wires and the shaded area is the low-k dielectric.

B. TF for a layer

Let η_t be the characteristic lifetime associated with vulnerable linespace S_t of length L_t. η_t is determined by stressing a test structure with vulnerable linespace S_t of length L_t. In the layout, if the vulnerable length corresponding to S_t is L_v, then the corresponding characteristic lifetime is

$$\eta_v = \eta_t \left(\frac{L_t}{L_v} \right)^{1/\beta}. \qquad (6)$$

Since a chip has many different linespacings, we combine failure rates for all linespacings by computing a defect count for each linespacing in the layout. The total defect count for a layer is the sum of all these defect counts. The characteristic lifetime for a layer (η_l) at the probability point $P = 0.63$ is

$$\eta_l = \sum_n \left(\frac{1}{\eta_n^\beta} \right)^{-1/\beta}. \qquad (7)$$

C. Chip lifetime

The defect density of the chip is computed from the defect densities of the individual layers that are in turn computed by the defect densities of all the linespace groups present within a layer. The chip lifetime (η_{chip}) is computed as

$$\eta_{chip} = \sum_l \sum_n \left(\frac{1}{\eta_n^\beta} \right)^{-1/\beta}. \qquad (8)$$

Unlike for a single layer, multiple layers of a chip may have different process details. In that case, data would be collected for each layer separately. If β were not common to all layers, then η_{chip} is implicitly defined as

$$1 = \sum_l \sum_n \left(\frac{\eta_{chip}}{\eta_n} \right)^{\beta(l)}. \qquad (9)$$

Since we only have data from one layer, we assume that CMP, etching and photolithography impact all the layers in the same way. This assumption is simplistic, and if data from

978-1-4244-9113-1/11 $26.00 © 2011 IEEE

different layers is available, it can be easily incorporated into Equations (7) - (9).

Reliability is adversely affected by linewidth variation and line edge roughness (LER). It has been shown that large scale linewidth variation impacts β [10], while LER impacts η with little or no impact on β [14]. We consider linewidth variation when determining η and β for our test structures [11]. Any effect of LER is reflected in the characteristic lifetime through η_t in equations (7) and (9) [14], and thus is included in the results.

D. Temperature profile

Including the temperature map in the layout statistics adds another dimension to the problem because now we have to consider the different characteristic lifetimes at different temperatures for every linespace. If we have a collection of m different temperatures for a linespace S_1, then the corresponding characteristic lifetime for the linespace is

$$\eta_{S_1} = \sum_m \left(\frac{1}{\eta_m^\beta} \right)^{-\frac{1}{\beta}}. \quad (10)$$

Characteristic lifetimes for the layer and the chip can be calculated using (7) and (9).

E. Overview

We extract pairs $\left(S_i (W_L, W_R), L_i \right)$ for each layer to determine the vulnerable length associated with each linespace in a layout and then determine the associated η. We also partition the layout by temperature. For the chip, we determine lifetime from (9) after summing the defects on all the layers.

VI. DETAILS OF THE CIRCUITS

A. Process

The NCSU 45nm technology library was used for our experiments [15]. This process has ten metal layers and the details of relevant features are given in Table 1.

TABLE I.   METAL LAYERS IN NCSU 45NM PDK

Metal Layer	Minimum Linewidth [nm]	Minimum Linespace [nm]
1	65	65
2, 3	70	70
4, 5, 6	140	140
7, 8	400	400
9, 10	800	800

B. Circuits

We synthesized the radix-2 pipelined, 256-points and 512-points, Fast Fourier Transform (FFT) HDL source code [16]. The circuit cf_fft_256_8 has 324k gates and 329k nets. The circuit cf_fft_512_8 has 708k gates and 712k nets. Both circuits have precision eight. The names of different instantiations of

cf_fft_256_8 and cf_fft_512_8 start with f1 and f2, respectively. The block diagram of the circuit is shown in Figure 6.

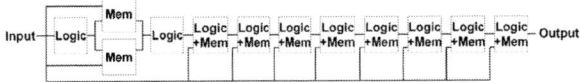

Figure 6.   Block diagram of a FFT circuit.

Synopsys Design Compiler is used for synthesis [17]. Cadence SoC Encounter is used for placement, clock-tree synthesis, routing, optimization, and RC extraction [18]. Synopsys PrimeTime is used for timing analysis [19]. We use seven different instantiations of cf_fft_256_8 and three different instantiations of cf_fft_512_8 .

Our metrics of performance comparison are the number of layers in a circuit and its timing performance. The details of the circuits are given in Table 2. Table 2 shows the timing performance of each circuit, the total wirelength of each circuit, and the percentage of total wirelength in each layer, referred to as wire density.

Circuits f1_M5, f1_M6, f1_M7, and f1_M8 are used to isolate the impact of the number of layers on reliability. Circuits labeled 'M' use Metal1 to Metal'X' during routing. Using more routing layers tends to result in shorter wirelengths and better timing performance, as shown in Table 2.

Circuits f1_RT1, f1_RT2, f1_RT3, f2_RT1, f2_RT2, and f2_RT3 are used to analyze the impact of timing performance on reliability. In RT'Y', we optimize timing using buffer insertion and gate sizing. M'X' does not do this. A higher value of 'Y' means more aggressive timing optimization with a higher clock frequency.

All the circuits have been synthesized using the same technology library. Thus the values in Table 1 are consistent across all the instantiated circuits.

C. Vulnerable Area Extraction

We have developed our layout extraction tool using standard object oriented programming languages. The layout extraction flow is shown in Algorithm 1. We explain the flow for horizontal line segments only. Vertical line segments can be handled in a similar way.

We start by reading in line segments for each metal layer from a given layout (ReadLineSegments in Algorithm 1) and by sorting all the horizontal segments in the ascending order of the y-coordinate of the bottom-left corner of the line segments, and the x-coordinate of the bottom-left corner (tie-breaker for equal y-coordinates). A layout may have a large number of line segments. For instance, there are eight million segments in Metal2 of cf_fft_512_8. Hence, a fast sorting algorithm is required. Therefore we use Bucket sort in the ReadLineSegments process in Algorithm 1.

After the reading-in process, we compare two adjacent line segments. If their vertical spacing is less than or equal to the pre-determined maximum line spacing, there exists a vulnerable area surrounded by these two line segment. We

TABLE II.   Wirelength of individual metal layers, as well as timing performance and reliability, of the used designs. The table shows the percentage of total wirelengths of chips present in each layer. Wirelength (WL) and timing performance, critical path delay (CPD), are also given.

Layer	Design									
	f1_M5	f1_M6	f1_M7	f1_M8	f1_RT1	f1_RT2	f1_RT3	f2_RT1	f2_RT2	f2_RT3
Metal1	1.4	1.4	1.4	1.4	2.59	0.77	2.20	2.23	2.24	1.19
Metal2	18.8	18	18.1	18.1	22.61	14.10	19.56	20.83	23.37	15.80
Metal3	33.9	33.1	33.1	33.1	38.41	25.80	32.01	35.44	37.92	27.90
Metal4	29.5	25.7	25.7	25.6	24.95	18.89	23.11	24.90	25.3	22.75
Metal5	16.2	16.1	15.4	15.3	9.52	19.92	15.49	13.42	9.04	19.83
Metal6		5.4	5.07	5.03	1.86	15.91	6.99	2.77	2.07	12.53
Metal7			0.9	1	0.05	4.60	0.65	0.38	0.04	
Metal8				0.1		0.01			0.004	
Metal9						0.01				
Metal10						0.0001				
Total Wirelength [m]	8.06	8.05	7.99	7.99	6.28	13.56	7.52	15.06	13.91	21.09
CPD [ns]	3.51	3.51	3.33	3.29	3.16	2.9	2.86	2.909	2.96	2.98

Algorithm 1 : Layout extraction flow.

Input: The maximum line spacing S_{max} and a given layout L
Output: A table of vulnerable areas (VulnerableAreaTable)

for each metal layer m do
   LineData $(m) \leftarrow$ **ReadLineSegments** (L); // BucketSort
   $c \leftarrow 1$;
   $n \leftarrow 2$;
   while true **do**
     if $c = N_{line}$ **then** // N_{line} : # lines in LineData
       break;
     end
     $L_1 \leftarrow$ LineData (m,c); // $c-$th line
     $L_2 \leftarrow$ LineData (m,n); // $n-$th line
     if Spacing $(L_1, L_2) \le S_{max}$ **then**
       VulnerableAreaTable $(m) \leftarrow$ **VulnerableArea** (L_1, L_2);
       LineData $(m) \leftarrow$ **Cut** (L_1, L_2);
       Adjust (N_{line}, c, n)
     else
       $c \leftarrow c+1$;
       $n \leftarrow n+1$;
     end
   end
end

insert this into our vulnerable area table. After the extraction of a vulnerable area, we apply two post processes (Cut and Adjust in Algorithm 1). In the Cut process, we remove one of the compared line segments from our data structure (LineData), and insert one or two new segment(s) into the data structure. In the Adjust process, we readjust indexes of line segments for the next comparison.

We illustrate our algorithm with the help of an example shown in Figure 7. In this example, there are three vulnerable segments (between S_1 and S_2, between S_1 and S_3, and between S_2 and S_3) as shown in Figure 7(a). We first compare two line segments S_1 and S_2, and find a vulnerable area. After we store this vulnerable area in our vulnerable area table, we apply the Cut process. In this process, we remove S_1 from LineData, cut the overlapped part from S_1, create two segments (S_{1_1} and S_{1_2}), and insert these segments into LineData (sorting is performed during insertion) as shown in Figure 7(b). Since there are four new line segments, we increase the total number of segments by one and start comparing segments with S_{1_1} (Adjust process). S_{1_1} does not overlap with any other line segment, and therefore we proceed to S_{1_2} and find a vulnerable area between S_{1_2} and S_3. We store this vulnerable area in our vulnerable area table, and cut S_{1_2} as seen in Figure 7(c). In Figure 7(d), we find a vulnerable area between S_2 and S_3. We store it and cut S_2.

D. Runtime

The runtime for the simulator is the sum of the time taken to extract features from the layout and a constant time to

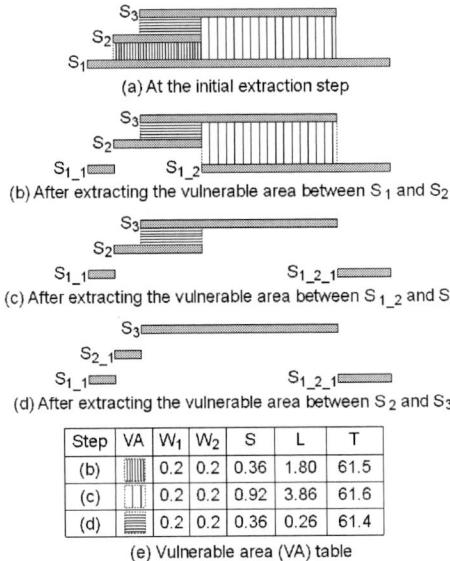

(a) At the initial extraction step

(b) After extracting the vulnerable area between S$_1$ and S$_2$

(c) After extracting the vulnerable area between S$_{1\_2}$ and S$_3$

(d) After extracting the vulnerable area between S$_2$ and S$_3$

Step	VA	W$_1$	W$_2$	S	L	T
(b)		0.2	0.2	0.36	1.80	61.5
(c)		0.2	0.2	0.92	3.86	61.6
(d)		0.2	0.2	0.36	0.26	61.4

(e) Vulnerable area (VA) table

Figure 7. Extraction of vulnerable area (a) at the initial step (b) after extracting the vulnerable area between S1 and S2 (c) After extracting the vulnerable area between S1_2 and S3 (d) After extracting the vulnerable area between S2 and S3 (e) Vulnerable area (VA) table: W1 and W2 denote linewidths, S is the linespace, L is the vulnerable length and T is the temperature

evaluate Equations (7) - (9). Complexity of feature extraction and database extraction is $O(n)$, where n is the number of features, since bucket-sort is used. Complexity of extracting statistics from the features is also $O(n)$, because we scan the bucket from the bottom most element, and the maximum number of features within a fixed distance from an element is constant. Lifetime is estimated in constant time.

VII. EFFECT OF LAYOUT ON TDDB RELIABILITY

A. Results based on geometry

Figure 8 shows η for Metal1-Metal6 for the circuits used in the study and the chip according to the $\sqrt{E}$ Model.

B. Observations

η for chips are more pessimistic than η for individual layers, because in calculating characteristic lifetime for the chip we combine the vulnerable areas for all the layers. Figure 8 does not indicate a trend for reliability with respect to timing performance. Figure 9 shows the lack of correlation between timing performance and reliability. Our results show that increasing the number of layers affects reliability marginally, while decreasing wirelengths increases reliability, as shown in Figure 9.

1) Number of layers

Using more metal layers generally results in a decrease in routing congestion and reduces the need for long detours to avoid routing congestion. This leads to less coupling capacitances between wires, which results in a decrease in the critical path delay. Since a router can spread out wires in several metal layers, we expect this to improve reliability.

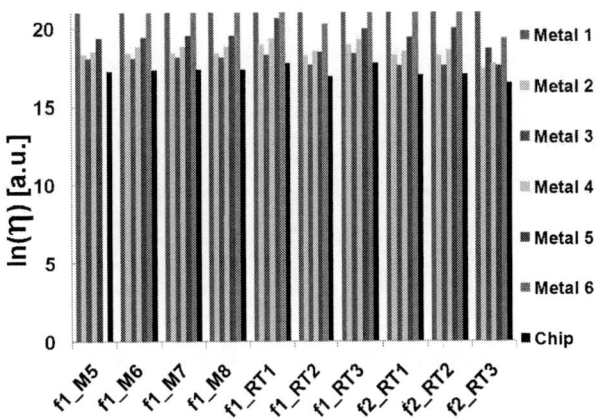

Figure 8. η for each layer and η for the chip for the circuits using the $\sqrt{E}$ Model.

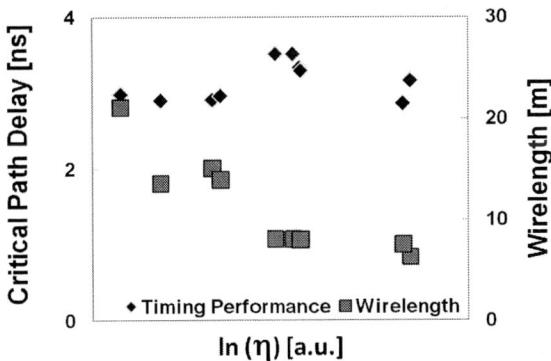

Figure 9. A comparison of reliability, timing performance and wirelength for the circuits under study.

critical path delay. Since a router can spread out wires in several metal layers, we expect this to improve reliability.

As expected, the critical path delay decreases as we increase the number of metal layers. However, as shown in Figure 8, reliability increases only marginally as the number of layers increases, and the layer most critical to lifetime remains the one with the highest wire density. This change of reliability as a function of the number of layers, or lack thereof, can be expected because even though the number of layers increases from five to eight, the percentage of total wirelength in the additional layers is less than 6%. Moreover, a large percentage of the wirelength still remains in a single layer, Metal3. Metal3 has mid-distance interconnects performing vital operations and it is highly unlikely that any optimization would cause major changes in Metal3. An even distribution of wirelengths can lead to an increase in lifetime This will the topic of future research.

2) Critical physical features

Figure 10 shows the characteristic lifetime for each layer of fl_M5, η for the critical linespace, and the linespace group with the smallest η according to (6), for the same layer. The critical linespace group is the most frequent linespace in a layout. It is not necessarily the smallest linespace.

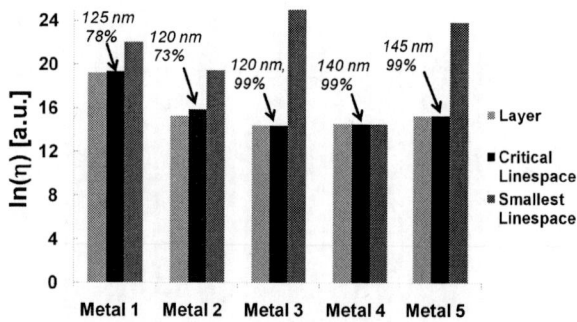

Figure 10. η for each layer of f1_M5, the critical linespace group, and for the smallest linespace group in each layer. The critical linespace, along with its percentage, is also given.

All dielectric area in a layer and in the chip falls in some linespace group, determined by its immediate neighbors. According to (7), η for a layer is dominated by the η of the critical linespace group, i.e. the most frequent linespace group. Figure 10 shows the percentage of dielectric area formed by the critical linespace group. A way to increase η for layers with a critical linespace greater than the smallest linespace is to redistribute linewidths. We could optimize this distribution for reliability

If we were to estimate the characteristic lifetime based on the most frequent (critical) linespacing alone, we only need to determine the area for this single linespace in each layer. Such an approach is simplistic. However, Figure 10 shows that lifetime estimates based on the critical linespace group are reasonably accurate.

Figure 10 also shows the characteristic lifetime, η, for the minimum linespace group for each layer, where the minimum linespace for each layer is given in Table 1. Consider Metal2 in f1_M5, the smallest linespace in Metal2 is 70nm, but the linespace dominating the characteristic lifetime, η, for this particular layer is 120nm for both the E Model and the $\sqrt{E}$ Model. Forty different linespaces are present throughout the layer, with the minimum being 70nm and the maximum being 252.5nm. However, 73% of the dielectric segments in this layer have a linespace of 120nm, with only 0.11% of dielectric segments having a linespace of 70nm between them. Therefore, we cannot just consider the smallest linespace group, as suggested in [1], when the layout is dominated by a linespace group other than the minimum linespace. Such an approach leads to lifetimes that are optimistic by orders of magnitude.

3) Wire density

Figures 11 shows that there is a strong correlation between wire density, the proportion of total wirelength in a layer, and η_l, with layers having the highest wire density dominating η_l in (8). This result is expected because of the nature of the breakdown mechanism. Higher wire coverage is achieved by closely packing the metal lines together, resulting in an increase in E and consequently degrading reliability. Also, as expected, reliability increases with a decrease in wirelength.

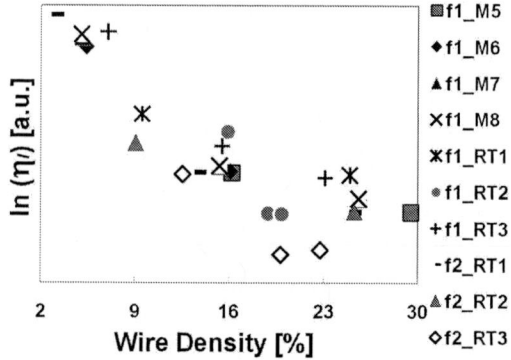

Figure 11. Wire density for Metal4-Metal6 and their characteristic lifetimes according to the $\sqrt{E}$ model.

4) Linewidth and Reliability

Linewidth impacts TF by affecting etching [11] and photolithography. As shown by our previous results, η increases as linewidth increases, for a given linespace, because of the interaction of physical design with etching and photolithography [11]. However, we found that increasing the linespace design rule does not improve reliability because the overall dielectric area increases [20] after re-routing .

5) Timing and Reliability

Timing optimization is achieved through buffer insertion, changing gate locations, and gate sizing. In terms of interconnect, densely routed areas raise coupling capacitance issues, which are addressed by ripping-up and re-routing the nets. Wire sizing is another way to obtain timing performance although we did not use it.

Buffer insertion results in an increase in total wirelength, resulting in a decrease in reliability, as apparent from the results. Gate re-placement needs to be managed carefully from the reliability perspective because re-placing gates in a crowded region can result in higher electric fields. If gate sizing is used for timing optimization, then the goals of timing align with those of reliability. Increasing the gate size increases the degrees of freedom for wire-to-pin connections, and this can be taken advantage of to enhance reliability. Rip-up and re-routing are aimed at reducing wiring congestion and coupling capacitance, factors that are also critical to reliability.

Despite all of these factors we did not observe any relationship between timing and reliability, because timing optimization generally uses heuristic algorithms instead of deterministic algorithms.

The results showed a strong correlation between the coverage in a given layer by the critical linespace group and the TF. For instance, for circuits f1_RT1, f1_RT2, and f1_RT3, 99% of the lines in Metal3 are separated by two linespacing groups, 120nm and 310nm. The lifetime is determined by the 120nm linespace group. Interestingly, for the circuits optimized for timing, in every layer 95% of the lines fall into only three linespace groups, and out of these three, more than 50% of linespaces were from the critical linespace group.

C. Results based on geometry and function

The steady state temperature of a point $p = (x, y, z)$ inside a thermal structure can be obtained by solving the heat equation

$$\nabla \cdot (k(p)\nabla T(p)) + S_h(p) = 0, \qquad (11)$$

where k is the thermal conductivity, T is temperature and S_h is the volumetric heat source. This model can be implemented by meshing the integrated circuit (IC) structure into thermal cells. To perform the thermal analysis, we start with the layout in DEF or GDSII format from Cadence SoC Encounter [18] and then perform static power analysis, for a given circuit frequency (f) and logic cell switching activity (α_i), to determine power dissipation

$$P_i = (\alpha_i C_i V_{DD}^2 f) / 2, \qquad (12)$$

where C_i is the loading capacitance of a logic cell, V_{DD} is the supply voltage, and f is the clock frequency of each logic cell. The layout along with the logic cell power dissipation is then used by our analyzer. The analyzer automatically generates the meshed structure for the IC along with the thermal conductivity and the volumetric heat source of each thermal cell. This information is used to perform thermal analysis using ANSYS FLUENT [21]. Figure 12 shows the thermal map, with an activity factor of 0.5, for Metal3 of *f2_RT3*.

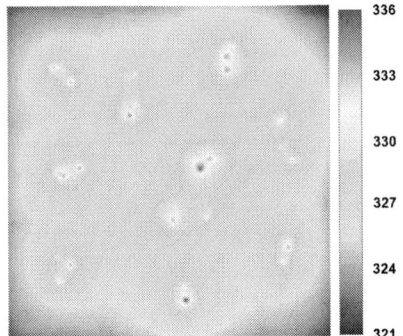

Figure 12. Thermal map of Metal3 of the circuit *f2_RT* for an activity factor of 0.5.

1) Runtime of thermal simulations

The runtime of thermal analysis consists of the runtime for determining the percentage of material in each thermal cell to determine the thermal conductivity, the volumetric heat sources inside each thermal cell of the meshed structure, and the runtime for solving the partial differential equations. The worst case complexity for the former is $O(n^2)$ and the average is $O(n \log n)$, n being the number of layout geometries. FLUENT [21] uses the finite volume method and its runtime varies between $1/40^{th}$ and $1/25^{th}$ of the time it takes to determine the percentage of the material in each thermal cell and the volumetric heat sources. Note that once the thermal analysis has been done, the layout statistics are generated, whose runtime is given in Section VI. They are integrated with the thermal profile of the chip.

2) Results

Figure 13 shows the results for our circuits incorporating their temperature profiles for a given signal activity level.

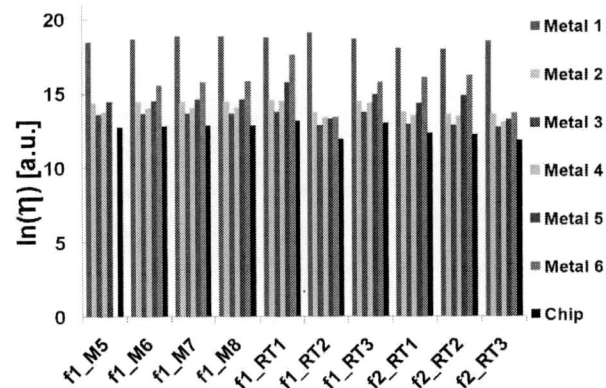

Figure 13. η for each layer and η for the chip for the circuits using the E Model and temperature profiles.

The trend among the models and the circuits remains the same after integrating the temperature profiles. Only the magnitudes of the characteristic lifetimes change. Figure 14 shows lifetimes with and without temperature profiles for the circuits for both the E model and the $\sqrt{E}$ model.

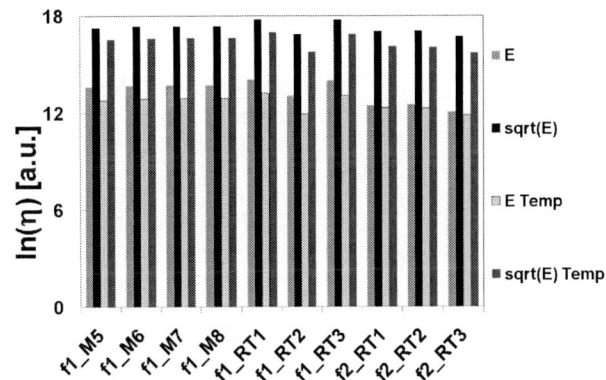

Figure 14. Characteristic lifetimes for the circuits according to the E model and the $\sqrt{E}$ model.

Figure 15 shows the characteristic lifetime for all layers of *f1_M5* for both the E model and the $\sqrt{E}$ model, with and without the thermal map.

D. Temperature Map

Integrating the temperature profile in our methodology takes into account the variation in characteristic lifetime caused by the variation in on-chip temperature. However, for large layouts the best-case complexity for generating the thermal map is greater than the worst-case complexity for generating the layout statistics. Moreover, different input vectors will affect the thermal map differently, thus requiring exhaustive thermal profiling, unless there are formal methods to generate thermal profiles. Hence the efficacy of including thermal maps

978-1-4244-9113-1/11 $26.00 © 2011 IEEE

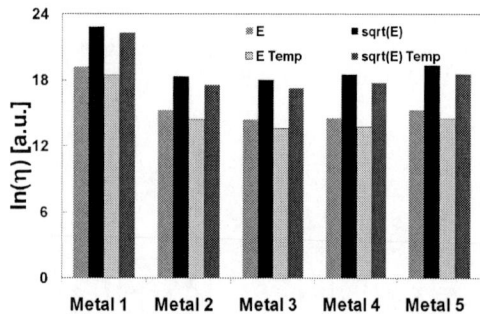

Figure 15. Characteristic lifetime for all layers of *f1_M5* with and without temperature according to the E Model and the $\sqrt{E}$ model.

ultimately depends on the intended use of the simulator. If the simulator is being used for accurate reliability estimates, then thermal maps must be integrated. The same will be the case if the intention is to observe the effect of a particular class of input signals. However, if the intended use for the designer is to get some quick reliability numbers, then results without thermal maps can give somewhat of an accurate guess at best, or describe the range of lifetimes at worst.

VIII. CONCLUSION

A methodology was proposed to assess backend TDDB chip reliability. The methodology has been developed in a way that other failure mechanisms can be integrated into it. Results from the simulator, built upon our methodology, showed the feasibility of our approach. In doing so, we also analyzed the effect of layout on backend TDDB reliability. We showed the absence of any correlation between timing performance and reliability. We also showed that greater wire coverage will result in smaller lifetimes, as expected. We demonstrated that the narrowest linespace group may not impact the chip lifetime critically. Instead, it is the linespace group with the highest coverage that is most instrumental in determining lifetime. We demonstrated that integrating temperature maps result in lower, though accurate, TF estimates. This study is a step towards our eventual goal of developing a tool to help designers build in reliability not just for backend low-k dielectric failures, but also for other failure mechanisms.

Our methodology does not assume the design to be "as drawn" or that the failures are being caused by the layout features. Of course, if such were the case, then we would not have been using an extreme value distribution. We assume that the layout is manufactured from the geometries in our test structures and modeling takes into account the invariance of Weibull statistics to area scaling, thus justifying extrapolations. It is assumed that any failure causing mechanisms that manifest themselves in test structures are reflected in the characteristic lifetimes used in the methodology.

REFERENCES

[1] T. Pompl, *et al.*, "Practical aspects of reliability analysis for IC designs," in *Proc. Design Automation Conf.*, 2006, pp. 193-198.

[2] S. M. Alam, *et al.*, "Reliability computer-aided design tool for full-chip electromigration analysis and comparison with different interconnect metallizations," *Microelectronics Journal*, vol. 38, pp. 463-73, 2007.

[3] Synopsys Inc, "FAMMOS."

[4] E. Y. Wu, *et al.*, "On the Weibull shape factor of intrinsic breakdown of dielectric films and its accurate experimental determination. Part I: theory, methodology, experimental techniques," *IEEE Transactions on Electron Devices*, vol. 49, pp. 2131-2140, 2002.

[5] M. Bashir and L. Milor, "Towards a chip level reliability simulator for copper/low-k backend processes," in *Proc. Design, Automation & Test in Europe (DATE)*, , 2010, pp. 279-282.

[6] G. S. Haase and J. W. McPherson, "Modeling of Interconnect Dielectric Lifetime Under Stress Conditions and New Extrapolation Methodologies for Time-Dependent Dielectric Breakdown," in *Proc.Int. Reliability Physics Symposium (IRPS)*, 2007, pp. 390-398.

[7] J. Kim, *et al.*, "Time Dependent Dielectric Breakdown Characteristics of Low-k Dielectric (SiOC) Over a Wide Range of Test Areas and Electric Fields," in *Proc. Int. Reliability Physics Symposium (IRPS)*, 2007, pp. 399-404.

[8] F. Chen, *et al.*, "A Comprehensive Study of Low-k SiCOH TDDB Phenomena and Its Reliability Lifetime Model Development," in *Proc. Int. Reliability Physics Symposium (IRPS)*, 2006, pp. 46-53.

[9] F. Chen, *et al.*, "Cu/low-k dielectric TDDB reliability issues for advanced CMOS technologies," *Microelectronics Reliability*, vol. 48, pp. 1375-1383, 2008.

[10] M. Bashir and L. Milor, "A methodology to extract failure rates for low-k dielectric breakdown with multiple geometries and in the presence of die-to-die linewidth variation," *Microelectronics Reliability*, vol. 49, pp. 1096-102, 2009.

[11] M. Bashir and L. Milor, "Analysis of the Impact of linewidth variation on Low-k Dielectric Breakdown," in *Proc. Int. Reliability Physiscs Symp.*, 2010, pp. 895 - 902.

[12] R. A. Gottscho, *et al.*, "Microscopic uniformity in plasma etching," *Journal of Vacuum Science and Technology* vol. 10, pp. 2133-47, 1992.

[13] J. Sung-Yup, *et al.*, "The characteristics of Cu-drift induced dielectric breakdown under alternating polarity bias temperature stress," in *Proc. Int. Reliability Physics Symp.*, 2009, pp. 825-827.

[14] M. Vilmay, *et al.*, "Copper line topology impact on the SiOCH low-k reliability in sub 45nm technology node. From the time-dependent dielectric breakdown to the product lifetime," in *Proc Int. Reliability Physics Symp.*, 2009, pp. 606-612.

[15] NCSU EDA. *NCSU Free PDK45*. Available: http://www.eda.ncsu.edu/wiki/FreePDK

[16] Launchbird Design Systems Inc. *CF FFT*. Available: http://www.opencores.org

[17] Synopsys Inc. *Design Compiler*.

[18] Cadence Design Systems Inc, "SoC Encounter RTL-to-GDSII."

[19] Synopsys Inc, "PrimeTime."

[20] M. Bashir, *et al.*, "Methodology to determine the impact of linewidth variation on chip scale copper/low-k backend dielectric breakdown," *Microelectronics Reliability*, vol. 50, pp. 1341-6, 2010.

[21] ANSYS Inc.,"FLUENT."

DETERMINATION OF CPU USE CONDITIONS

Robert Kwasnick, Athanasios E. Papathanasiou, Matthew Reilly, Al Rashid, Bashir Zaknoon, John Falk

Intel Corporation
Santa Clara, CA 95054
(001) –(408)-765-3982, robert.kwasnick@intel.com

Abstract— **Use condition inputs to physics-of-failure models are required to use knowledge-based qualification of ICs. Modern CPUs have multiple voltage-frequency states which vary widely in reliability stress, but it is not obvious what time in the various states to use in product qualification. We present a methodology for developing a time in state model for CPUs which combines large scale user monitoring and lab-based studies. Results for a specific CPU family, including field validation and implications for knowledge-based qualification, are discussed.**

Keywords-product reliablity; knowledge based qualification; use conditions; methodology; CPU

I. INTRODUCTION

Knowledge Based Qualification (KBQ) of ICs using physics-of-failure models is recognized as a complement to standard-based qualification (SBQ) for IC products [1]. KBQ includes the prediction of field reliability based on product-specific failure models and use conditions, which allows reliability predictions more closely aligned to product usage than is possible only by SBQ assessment. For example, to predict time-dependent dielectric breakdown (TDDB) characterized by a voltage acceleration factor and thermal activation energy, the needed use conditions are the time at specific voltages and temperatures [2].

To employ KBQ, those use conditions must be accurately determined. Values for some use conditions have been noted [3], and some have been studied in detail [4], but the literature on this subject seems sparse. Furthermore, given the ongoing evolution of IC technology and product usages, use conditions evolve. In particular, by enabling multiple voltage-frequency states, central processor unit (CPU) power architecture has evolved to better match workload performance demand and be more energy efficient [5, 6]. The implication is that, while some use conditions, such as ambient climate [4], are relatively stable over the years, others may need to be reconsidered in qualifying new products.

In this paper we present a methodology to determine a time in state model for CPUs, focusing on the Intel® Core™ 2 Duo CPU family released in 2007 on a 45 nm process [7], hereafter referred to as "45 nm CPU", and its use in notebook PC. The 45 nm CPU has two processor cores within the IC. We use "CPU" to refer more broadly to all CPU.

We define the time in state (TIS) use model to be the fraction of time each CPU spends in the available power states,

which include S, C and P [8]. The system is active in S0, and the CPU is executing instructions in C0, where P represents discrete voltage-frequency (V-F) operating states within C0. The highest V-F state is designated as P0, while lower states are designated as P1, P2 and so on down to the lowest defined state Pn. For the 45 nm CPU, the highest performance state P0 represents the Intel® Dynamic Acceleration™ mode, commonly known as Turbo mode. Turbo mode (that is, P0) can be entered if only one of the two cores is active. See the Appendix for a full description of the system and processor power states.

For KBQ assessments, for example for TDDB, the time in P0 and P1 contribute the most to device aging, relative to the other P states, because they are associated with higher voltage and temperature. For a given CPU model and platform, the voltages and temperatures for each C and P state need to be modeled to enable a KBQ failure prediction vs. use time.

The methodology has two elements which will be described in detail in the following sections. Field monitoring (Section II.A): The time in S states and C states are assessed by monitoring CPUs already in the field. Since the TIS model needs to be developed before product release to be used in KBQ, the monitored units are older CPU models. These may differ in some aspects from the CPU for which the TIS model is being built, but we observe that user behavior does not change radically over time (Section III.A), so this data is a reasonable basis for setting %C0 in the reference time in state use model. Lab-based studies (Section II.B): Since the power architecture and other influences on TIS may differ for a new CPU, field monitoring is supplemented by monitoring the time in P states for a CPU in a lab system while running various benchmark software applications (Section III.B) which are mapped to models of user behavior (Section IV.A) to build the TIS use model.

The paper includes (Section IV.B) a field monitoring study done in 2009 which reasonably validates the lab-based model of 2007, and concludes with an evaluation of failure prediction sensitivity to variations in the TIS use model and its impact on KBQ (Section V).

II. METHODOLOGY

In this section, we first describe field monitoring of the power states, and then describe the lab-based data collection.

A. Field Monitoring Infrastructure

During the development of the 45 nm CPU use model, we made a significant effort to acquire reliability data from systems in the field in large-scale and in a continuous way (non-stop monitoring). For this purpose, we implemented a low performance impact monitoring agent, called SEMA (System Environment Monitoring Agent) and deployed it in a large corporate pool of systems in 2007.

SEMA is a software infrastructure designed to continuously monitor and log system data in large scale. SEMA keeps track of the time the processor spends in the available power states (S, C and P states) using operating system and processor monitoring hooks [9]. The agent logs data samples periodically, e.g., every 30 minutes, and transmits them to a central database for post-processing and analysis. S state data are monitored by tracking operating system boot, shutdown, suspend and hibernate operations and system crashes. Processor utilization and C state data are monitored through the Microsoft** Windows** Management and Instrumentation classes (WMI).

In the study described in Section IV.B, P state data are collected at high frequency (0.5 seconds sampling period) from the processor using the IA32_MPERF, IA32_APERF, IA32_PERF_STAT model specific registers (MSR) [9, 10]. The IA32_MPERF MSR counter increments in proportion to a fixed (guaranteed or marked) frequency, which is configured when the processor is booted. The I32_APERF MSR counter increments in proportion to actual performance. Thus, the ratio of IA32_APERF to IA32_MPERF provides an estimate of the average performance state during the sampling interval. The IA32_PERF_STAT register captures the P state the processor operates at the time of sampling. Based on these data, we were able to determine the percentage of time each core spends in the C0 state, the distribution of time in each of the supported P states (P0, P1, ..., Pn).

B. Lab-Based Measurements

In the lab, the time in C and P states could have been measured using the same method described above for field monitoring. But, for the 45 nm CPU, we could visually discriminate the high voltage, and hence most wearing, P0 and P1 states and number of cores active by simply plotting the instantaneous voltage and current of the 45 nm CPU over time.

Factors which might influence the time in P states include operating system power algorithms, executing workloads, and power source (AC or battery). It would be challenging to predict the impact of all these factors on the time in P states, especially for a CPU with a new power architecture, thus motivating lab-based measurements of some of the factors.

In this study, the TIS was assessed while running benchmark programs. A 2.4 GHz 45 nm CPU was installed on an Intel® Centrino™ 2 platform, running Microsoft Vista*

operating system with the same power states enabled as used in the field. The system was AC powered.

III. RESULTS

A. Field Monitoring Data

S0 and C0 data were logged from about 29,000 notebook PCs with 45 nm CPUs from April, 2008 to Oct., 2010.

These had a variety of model numbers and were running the Windows XP operating system. Some notebooks had beginning and end dates within the time period, but the data was censored so the minimum time period is 30 days.

Fig. 1 is a normal probability plot of %S0/day for the monitored population. The median 45 nm CPU uptime is 8.2 hr/day (34%), and the 95th percentile is 18.4 hr/day (77%). S0/day values longer than the typical workday suggest some use their notebook computers outside the office. Some of the notebooks which approach 24 hr/day on time may be used in lab settings, which is not the behavior of the typical notebook user.

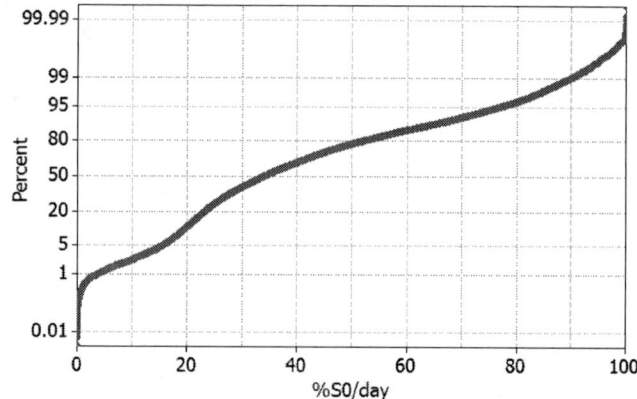

Figure 1. Normal probability plot of %S0/day from 28,621 notebooks with 45 nm CPUs.

Fig. 2 is a normal probability plot of %C0/day for the monitored. For the 45 nm CPU, the time in C0 is reported for each of the two cores independently. However, the 45 nm CPU is in C0 if either core is in C0, and thus the C0 data for the two cores needs to be combined to determine the 45 nm CPU time in C0. For each user, we set the C0 for the 45 nm CPU to the maximum of the two C0 values for its cores plus 50% of the minimum of the two C0 values. That is, we assumed two core operation occurs 50% of the time it is possible. This resulted in median and 95th percentile %C0/day of 14.4% and 27.8%, respectively.

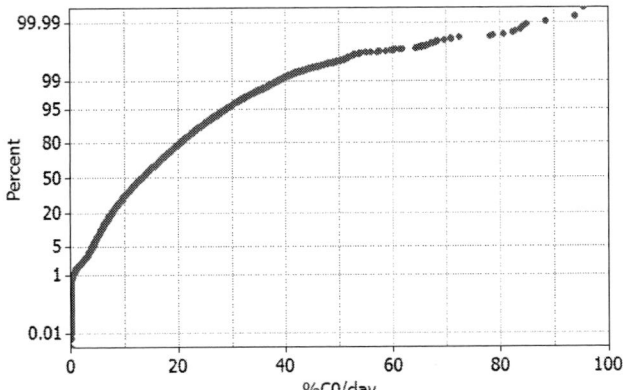

Figure 2. Normal probability plot of %C0/day of 28,621 notebooks with 45 nm CPUs from Apr., 2008 to Oct., 2010.

Fig. 3 is a normal distribution plot of %C0/day for field monitored notebooks from a previous generation with 1 core CPUs. The data was taken on 8,930 systems over about 54 days near the end of 2008. In this case, no post-processing of the %C0 data is needed. The %C0/day median and 95th percentile values are 10.6% and 27.6%, respectively. These values are reasonably close to those of the later 45 nm CPU, which is plausible because while %C0 is reduced in more recent CPU due to technology-driven performance improvements, it may also be increased by new applications, corporate IT requirements, and operating system demands.

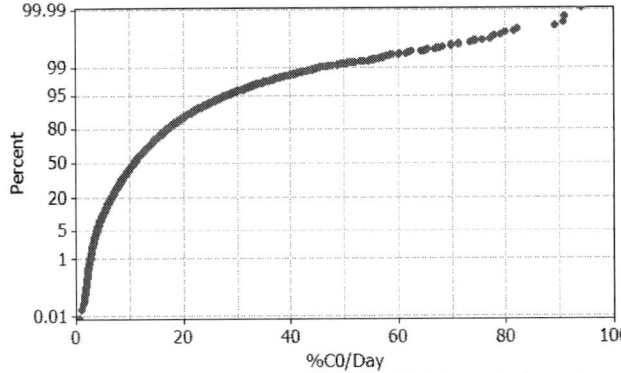

Figure 3. Normal probability plot of %C0/day of 8,930 notebooks with one core CPUs near the end of 2008.

B. Lab-Based Data

Fig. 4 is a contour plot of voltage (Vcc) and current (Icc) of a 45 nm CPU in the lab while exposing the system to the Sysmark* 2007 Preview use case [10]. Sysmark 2007 Preview is a standard benchmark developed by BAPCo**, an IT industry consortium. It is commonly used to assess system performance levels. The work performed by the benchmark was developed to represent the behavior and actions of IT industry professionals. It represents workflows performed on business applications by an IT end-user across four categories—E-Learning, Video Creation, Office Productivity and 3D Modeling.

Each region of high density represents one particular processor state, with the most wearing P0 and P1 states labeled. The density of points is used to determine the relative time in

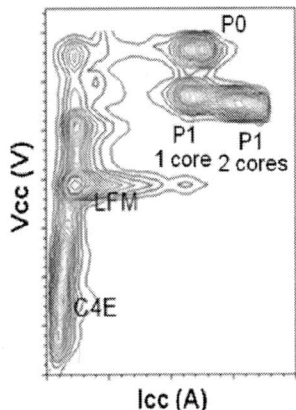

Figure 4. Vcc and Icc contour plot for a 45 nm CPU running on an Intel® Centrino™ 2 platform under AC power, with Windows Vista operating system, and "High Performance" user-selectable power scheme. Power states are noted on the figure. LFM is the CPU's low frequency mode and C4E is an even lower power state.

the power states. As noted in Section I, for this 45 nm CPU only 1 core operation is allowed in P0 but 1 and 2 core operation are allowed in P1, which is apparent in the figure. Other represented use cases were also measured to create more diverse usage scenarios, described in the next section. In particular, we report measurements while exposing the system to 3DMark06*, a graphics intensive industry standard benchmark produced by Futuremark** Corp. This workload stresses the graphics subsystems using the new technologies available at the time of release. We measured with both Windows Vista "High Performance" and "Balanced" power schemes, and observed for most applications a several percent higher time in P0 with the "High Performance" scheme.

IV. ANALYSIS

A. Time In State Use Model Development

The key parameters for a time in state use model for wearout calculations for a 45 nm CPU are %C0/day and the time in P0 and P1. %C0/day was based on monitored S0 and C0 data. As noted, monitored field usage provides a better measure of %C0/day than lab data because it is based on actual behavior over a large set of users, and, for notebooks used in a corporate environment, was not observed to change greatly over time per Figs. 2 and 3. A conservative percentile is used within the %C0/day distribution to account for variation between users and the possible evolution of use behavior.

Time in the P states was determined in three steps:

- First, typical use behaviors, which for notebooks include activities such as office work and gaming, were considered (Table 1, left column), and benchmarks which represent those behaviors assigned (Table 1, second column) [11, 12].

- Second, TIS for those benchmarks were measured in the lab, providing data like that depicted in Fig. 4 (Table 1, remaining columns).

978-1-4244-9113-1/11 $26.00 © 2011 IEEE

- Third, use scenarios were considered by combining the use behaviors. The relative times in P0, P1 and other measured states was derived from the weighted sum of the measured times in those states.

TABLE I.   MEAN TIME IN STATE RESULTS

Use Behavior	Represented By	%P0	%P1	%C0
General Usage	Sysmark* 2007 All	9%	20%	32%
Office Work	Sysmark* 2007 Office Productivity	5%	12%	20%
Gaming	3DMark06* Game Demo	6%	29%	47%
Corporate	Field Monitoring	7%	7%	33%

As an example of defining the time in P states in the third step, assume the use scenario is represented simply by the use behavior "General Usage". Per Table I, the time in %P0 (turbo) is 9/32 and %P1 is 20/32 of the time in C0. If the time in C0 were set to 32%/day, then %P0 is 9%/day, %P1 is 20%/day, and the time in lower P states is 3%/day. Alternative use scenarios might emphasize gaming or office work. A range of plausible scenarios and the relative reliability impact of each were considered in setting the time in the P states for reliability modeling.

B. Time In State Use Model Results

The first three rows of Table I show the measured fraction of time for each use behavior in %P0 (turbo), %P1 and %C0 with "Balanced" power scheme (%C0 is the sum of the time in all P states, including those below P1). Note that the %C0 values of the benchmarks, 20%, 32% and 47%, are similar to those of a typical user's %C0/day in Fig. 2, after the benchmark %C0 values are scaled down to account for typical %S0/day values per Fig. 1.

A reference TIS use model for KBQ can thus be set based on a combination of field data, lab data and consideration of use scenarios. User behavior variation and uncertainty and failure mechanism sensitivity are important considerations in setting the reference TIS use model sufficiently conservatively, discussed further in Section V.

C. Time In State Use Model Validation

In mid-2009, we monitored the time in P states of about 400 randomly selected corporate notebooks with Intel® Core™2 Duo CPUs (Model T7300) for four week periods. The systems had an Intel® Centrino™ 2 platform and were running the Windows XP operating system. About 97% of participants had "Balanced" user-selectable power scheme. The mean results are in the last row of Table I, labeled "Corporate". Note that the %C0 value in the last row is the mean %C0 during each user's S0 time, so the data is directly comparable to the lab data in the first three rows because the lab system is in S0 during the benchmark measurements. The Corporate %P0 90th, 50th, and 10th percentile values are 13.6%, 6.3%, and 0.3%, respectively. We observe the Corporate results are fairly similar to the benchmarks measurements, taken in the lab before the product was released, especially

regarding the time in P0, the most wearing operating state. Note that only a small fraction of 45 nm CPU will have high %P0 over a long use period because P0 is a valid state only when one core is active, which is consistent with the observed %P0 values in Table I.

V. DISCUSSION

Advanced multi-core CPU architectures feature a large number of possible power states designed to maximize performance per watt and to maximize performance when there is power and temperature margin available. These states can vary widely in terms of voltage, frequency and temperature and therefore potentially have very different reliability implications for the processor. For Intel® Core™ 2 Duo processors and beyond, individual processor cores can be placed into idle (C1, C2 and so on) C states while other cores execute code, and in addition each core can operate at multiple frequencies over a range as large as 2 GHz with operating voltages that vary by 100's of mV.

Fig. 5 shows the modeled normalized failure rate for a 45 nm CPU vs. %C0 for 3 yr and 5 yr use times. Normalization here means the fail rate is set to 1.0 at 50% C0 for 5 yr, thus the figure depicts the relative impact of %C0 utilization. For example, if the KBQ assessment were done with a %C0/day of 50%, a unit with %C0/day at 75% has about 2x the modeled failure rate. Note that this high a %C0 would apply to only a small fraction of units, per Figs. 2 and 3.

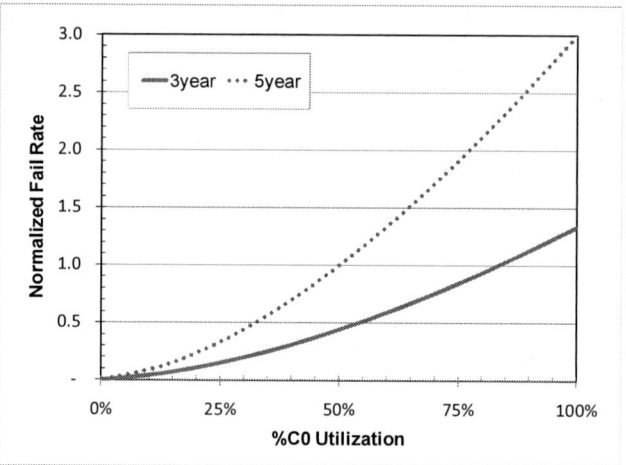

Figure 5.   Modeled normalized failure rate of a 45 nm CPU vs. % C0 for two use times. The curves are normalized to 1.0 at %C0 Utilization of 50% for 5 yr.

For KBQ, it is important to understand the sensitivity of failure mechanisms to variations in the TIS model. Some important reliability mechanisms on modern silicon logic processes, for example TDDB, may be fit to Weibull model, key parameters of which are a shape parameter, β, and voltage acceleration factor (Vaf) [13]. Fig. 6 shows the dependence of a hypothetical failure model on %Turbo Utilization, defined as the P0 fraction of C0 operation time, with the operating state in the other fraction of C0 operation time being P1 with 2 cores active. C0 operation time was set to 2.5 yr. The four curves are for a range of Vaf values relevant to recent CMOS processes [13], and for two different voltage "P0 Boost" values of P0 above that of P1. Each curve is normalized to its value at 50% P0 being 1.0. As %P0 increases, the failure prediction increases only somewhat superlinearly, with increased sensitivity at higher Vaf and higher P0 boost.

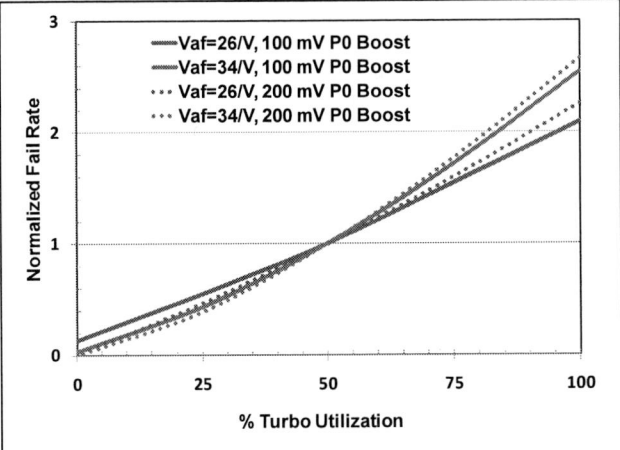

Figure 6. Modeled normalized failure rate vs. %Turbo Utilization for a Weibull failure mechanism with β=1.5, Vaf = 26/V or 34/V, and voltage difference (P0 Boost) of either 100 mV or 200 mV between P1 and P0. Each curve is normalized to 1 at 50% Turbo Utilization.

Fig. 7 shows a related assessment with fixed Vaf and two β values, again chosen to be representative of modern CMOS processes [13]. Similar to Fig. 6, the normalized failure rate is only somewhat superlinear in its dependence on %Turbo Utilization, i.e., %P0, with stronger sensitivity at higher β.

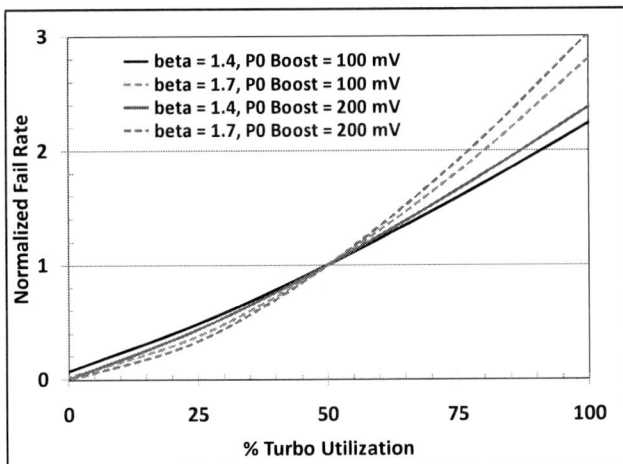

Figure 7. Modeled normalized failure rate vs. %Turbo Utilization for a Weibull failure mechanism with Vaf of 30/V, β of either 1.4 or 1.7, and voltage difference (P0 Boost) of either 100 mV or 200 mV between P0 and P1. Each curve is normalized to 1 at 50% Turbo Utilization.

These figures illustrate the importance of accounting for usage differences in order to ensure acceptable field reliability. This is especially relevant with modern CPU which have a wide range of possible operating states. It is important that the TIS use models employed for reliability prediction account both for the range of CPU utilization and of user behaviors,

and to include sufficient conservatism to protect against uncertainties.

APPENDIX

This appendix provides background information and definitions of system and processor power states.

Processors are equipped today with multiple, both operational and nonoperational, low-power modes that can save significant amounts of power when the system is not working at its peak. Non-operational modes provide the opportunity to power down a processor's subcomponents during periods of inactivity. Operational modes allow the processor to dynamically vary its performance level in an attempt to match that required by the workload. This performance level variation is achieved through processor voltage and frequency scaling.

Non-operational low-power modes of processors are known as C states. The Advanced Configuration and Power Interface Specification (ACPI) [8] defines processor power states as active (operational), in which the processor executes instructions, or non-active (non-operational), in which the processor does not execute instructions. The active state is referred to as C0 and is associated with a higher wear-out impact. Non-active states are designated as C1, C2, ..., Cn. Lower non-active states consume less power and are associated with a higher entry and exit latency. Modern system implementations support three non-operational processor states: C1, C2 and C3.

While in the active (C0) state, modern processors support multiple performance states, known as P states. The ACPI specification designates processor performance states as P0, P1, ..., Pn. P0 represents the highest available performance state, in which the processor executes instructions at its maximum supported frequency and voltage. Lower performance states (P1, ..., Pn) correspond to progressively lower performance levels. In these lower states, the processor executes instructions at reduced frequency and voltage. Higher performance states are generally associated with higher wear-out impact.

Modern Intel® processors support a novel form of processor operation that takes advantage of design headroom to opportunistically increase performance. Specifically, Intel® Core™ 2 CPU, supports Intel® Dynamic Acceleration™ (frequently referred to as Turbo mode) [7], which allows a single core to operate at a higher performance level. In this mode, the processor's voltage and frequency temporarily exceed their marked values to provide increased performance. A core in the Intel® Core™ 2 CPU can enter the Intel® Dynamic Acceleration™ mode only when other cores sharing the same package are in a non-active power state. The ACPI specification maps the Intel® Dynamic Acceleration™ mode (when supported and enabled) to the P0 state.

The ACPI specification defines system power states in a way similar to how it defines processor power states. The working state is designated as G0 or S0. G1 is the sleeping state, in which the system appears to be off from user's perspective. The sleeping state supports multiple sub-states, designated as S1, S2, S3 and S4. Each sleeping sub-state is characterized by reduced power consumption and a higher latency to enter and exit the state. The S3 state is known as the Sleep, Standby or Suspend state, while S4 corresponds to hibernation. Finally the off state is designated as G2 (soft off) or G3 (mechanical off).

ACKNOWLEDGMENT

We thank Arijit Biswas, Greg Kaine, David Pullen and Mohsen Alavi for helpful suggestions.

REFERENCES

[1] JEP148A, "Reliability qualification of semiconductor devices based on physics of failure risk and opportunity assessment", Dec., 2008.

[2] JEP122F, "Failure mechanisms and models for semiconductor devices", Nov., 2010.

[3] JESD94A, "Application specific qualification using knowledge based test methodology, July, 2008.

[4] Chen Gu, Robert F. Kwasnick, Neal R. Mielke, Eric M. Monroe, C. Glenn Shirley "Ambient use-condition models for reliability assessment", IRPS 2006, pp. 299-306.

[5] Mark Weiser, Brent Welch, Alan Demers and Scott Shenker, "Scheduling for reduced CPU energy", Proc. of the 1st USENIX Symposium on Operating Systems Design and Implementation (OSDI'94)", November 1994.

[6] Krisztian Flautner and Trevor Mudge, "Vertigo: Automatic performance-setting for Linux", Proc. of the 5th USENIX Symposium on Operating Systems Design and Implementation (OSDI'02), December 2002.

[7] Jose Allarey, Varghese George, Sanjeev Jahagirdar, "Original 45nm Intel® Core™ 2 processor performance", Intel Technology Journal, http://www.intel.com/technology/itj/2008/.

[8] Advanced Configuration and Power Interface Specification 4.0a, April 5, 2010, at www.acpi.info.

[9] Intel® 64 and IA-32 Architectures Software Developer's Manual Volume 3A: System Programming Guide, Part 1.

[10] Intel® 64 and IA-32 Architectures Software Developer's Manual Volume 3A: System Programming Guide, Part 2.

[11] "SYSmark 2007 An overview of Sysmark 2007 Preview", at http://www.bapco.com/support/technical_documents/SYSmark2007Preview_WhitePaper.pdf.

[12] 3DMark06*, at www.futuremark.com. Note: Sysmark* 2007 Preview and 3DMark06* are platform tests. The processor is only one of many components that make up the platform. While it is clearly influenced by choice of processor, Sysmark* 2007 Preview and 3DMark06* will also be influenced by memory size and configuration, disk choices, and operating system. All of these factors need to be considered and documented when making an evaluation using this benchmark.

[13] Ernest Wu and Jordi Sune, "Power-law voltage acceleration: A key element for ultra-thin gate oxide reliability" Microelectronics Reliability, Vol. 45, pp. 1809-1834, 2005.

** Other names and brands may be claimed as the property of others.

Si_3N_4 Extrinsic Defects and Capacitor Reliability

John Scarpulla, Everett E. King and Jon V. Osborn

Electronics and Photonics Laboratory, Physical Sciences Laboratories

The Aerospace Corporation

El Segundo, CA

310-336-7998, john.scarpulla@aero.org

Abstract - **The capacitors implemented in RF/microwave MMICs seem to dominate the reliability in many cases, rather than the active devices (pHEMTs or HBTs) themselves. We have examined MIMCAP extrinsic defect density by extracting it from published data available in the open literature. The methodology for extracting defect densities from probability plots of times to breakdown or of ramped breakdown voltages is shown. We have noted that the extrinsic densities are quite varied across the dozen sets of data compiled. We also contributed data using Hg-dot-formed capacitors. Using this industry-wide data we provide some design charts for the sizing of capacitors in MMICs based upon their extrinsic reliability.**

Keywords-Silicon Nitride; GaAs MMIC; MIMCAP ; Dielectric Reliability; TDDB; Extrinxic Defects; Effective Thickness; Defect Desity

I. INTRODUCTION

In GaAs and other compound semiconductor RF microwave MMICs (monolithic microwave integrated circuits), the capacitor dielectric of choice is predominantly formed using PECVD (plasma-enhanced chemical vapor deposited) silicon nitride. A MIMCAP (metal-insulator-metal capacitor) is built with the dielectric sandwiched between evaporated layers of Au or Ti/Pt/Au. Silicon nitride is compatible with the GaAs material system, and it is generally deposited at the relatively low temperature of approximately 300°C or less. This minimizes the possibility of thermal damage when pHEMT (pseudomorphic high electron mobility transistor) gates or HBT (heterojunction bipolar transistor) mesa structures are pre-existing during intermediate steps of fabrication.

For MMICs, there are two types of reliability effects that must be covered. The first is the built-in or "intrinsic" reliability of the device. This is the reliability that is associated with the wear-out or end-of-life of the materials or devices. The second is the "extrinsic" or defect-related reliability that results in early to mid-life failures due to process and material defects. In general, the intrinsic reliability of most active and passive devices in a MMIC today is quite satisfactory, being many hundreds or thousands of years at typical usage conditions. Phenomena such as gate sinking, HBT ledge degradation, electromigration and walkout-related phenomena are well understood for the most part. In the GaN and SiC power device field the intrinsic wearout reliability of the basic device is still improving.

The extrinsic or defect-related phenomena are also very important. In mature MMIC GaAs or InP technologies we have found that the typical reliability is dominated no longer by the intrinsic device reliability, but rather by the extrinsic reliability of components such as the Si_3N_4 based MIMCAPs. Ironically it is the lower frequency X- and S- band applications generally requiring larger sized MIMCAPs that tend to suffer the most from extrinsic defects. These lower-frequency applications generally are thought to have much higher reliability since the RF-driven aspects of reliability are believed to be more forgiving at lower frequencies. We posit that the reliability or failure rate in most MMICs is dominated not by the active devices but by the MIMCAPs. This proposition is explored further in the following with a re-evaluation of data from the literature.

II. EXTRINSIC MIMCAP DATA

There is much written on the PECVD Si_3N_4 based MIMCAP fabrication and material quality. Most work has focused on the intrinsic reliability and the maximization of the capacitance per unit area. A small-area, reliable, high capacitance value is the goal, generally achieved by making the Si_3N_4 dielectric thinner or improving the process or the material quality. The literature seems to fall into two categories – (1) papers that deal with the plasma process chemistry, optimization and characterization of the nitride films by various means such as FTIR (Fourier transform infrared spectroscopy, EPR (electron paramagnetic resonance), ellipsometry, etc., and (2) those that assess the electrical and reliability properties of the film. Unfortunately, most deal with the intrinsic film to the exclusion of the extrinsic defects. Very few papers mention the extrinsic defects, and fewer still provide extrinsic data. Nor is data on extrinsic defect density made readily available from GaAs foundries or from the various fabrication companies. It is safe to say that there is a dearth of data on extrinsic MIMCAP reliability.

Recently we have had the need to characterize a Si₃N₄ nitride film used in a foundry process, and have obtained defect densities. We have compared our results to those (few) defect densities that have been published. We surveyed the literature from approximately the last dozen years and discovered a subset of approximately a dozen papers containing extrinsic data. [1-12]. They all utilize similar low temperature PECVD nitride films, and cover the gamut from GaAs, InP, Si, and SiC technologies. All have some extrinsic data displayed whether the focus of the work or not. All provide useful results along with enough information to deduce a defect density characteristic. We used this compilation of data to assess MIMCAP reliability trends in the industry. We provide reliability benchmarks in the form of maximum allowable MIMCAP area or maximum allowable capacitance under typical operating conditions. We found an unexpectedly large variability in defect densities across the published results. The methods used to extract these results from published data are described below.

A. Experimental

As part of this work, we also characterized a nitride film from a commercial foundry having nitride thickness of 1400Å as determined by ellipsometry and cross sectioning. This unpatterned PECVD film was made available on a 6 in. blank GaAs wafer, onto which was first deposited Ti/Pt/Au serving as the lower plate of the MIMCAP. Temporary MIMCAPs were formed using a mercury probe (Materials Development Corp. model MDC 802B-200). Ramped breakdown, charge to breakdown (QBD), and IV measurements were performed using a semiconductor parameter analyzer (Agilent E5270B mainframe with E5280B modules). The mercury dot area was 0.51mm², the ramp rate was 4 V/sec, and the current density

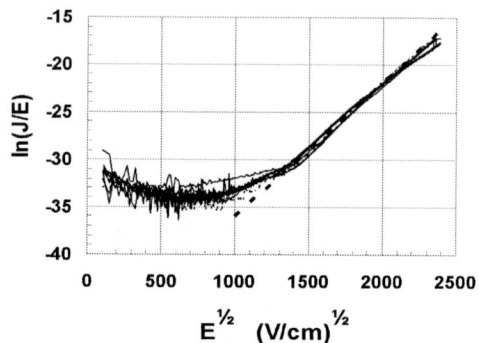

Figure 1. A Frenkel-Poole (F-P) plot of ln(J/E) versus $E^{1/2}$ for one dozen Hg-dot-formed capacitors. The dashed straight line shows that that the F-P conduction mode is valid and is used to extract the current conduction parameters. At low currents, the noise in the measurement setup obscures the conduction in the dielectric

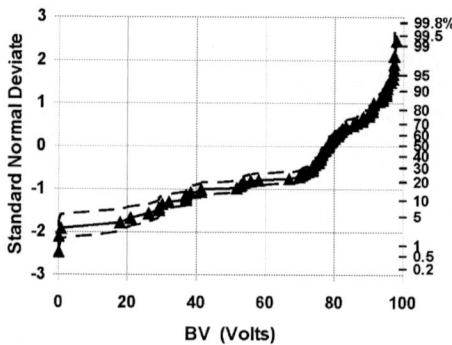

Figure 2. Probability plot of the ramped breakdown voltages of approximately 100 Hg-dot-formed capacitors. The use of a normal probability plot is for convenience only.

for QBD measurements set at 0.1A/cm². The films were annealed at 250°C for 24 hours in order to simulate the stabilization bake and other thermal processing in the actual fab.

The film IV characteristics were good and fit the typical Frenkel-Poole conduction mode [9] for PECVD silicon nitride as shown in Fig. 1 with a current density

$$J(E) = \sigma E \exp \frac{-\left(\phi_t - \beta\sqrt{E}\right)}{kT} \qquad (1)$$

The fitting parameters were: conductivity $\sigma = 7.29\times10^{-9}$ S/cm, trap energy $\phi_t = 0.8$ eV, and field lowering factor $\beta = 3.61\times10^{-4}$ V$^{1/2}$cm$^{1/2}$. The charge to breakdown Q_{BD} was a very typical 20 Coul/cm². The ramped breakdown data was also quite typical, with intrinsic ramped breakdown voltages of 80V – 100V as shown in the probability plot of Fig. 2. The usual extrinsic tail appears in the probability plot and is our focus of attention here.

III. Extrinsic Data Analysis

A probability plot of the extrinsic times to breakdown t_{BD} (in a constant voltage test), or the extrinsic breakdown voltages V_{BD} (in a ramped voltage test) constitutes the data needed to obtain the extrinsic failure probability of a MIMCAP. From the probability axis we extract the cumulative defect density D. From the t_{BD} or V_{BD} values we extract the effective thickness x_{eff}.

A. Effective Thickness

According to the traditional TDDB theory [9], the time to breakdown is proportional to the charge to breakdown Q_{BD}. The charge to breakdown is a constant quantity for a given nitride thickness and process. After a sufficient current for a sufficient duration has passed through, the charge to breakdown is attained and the breakdown occurs. If we assume the current density is constant, the time-to-breakdown is

$$t_{BD} = \frac{Q_{BD}}{J(E)}. \tag{2}$$

If the current density varies because of a time-varying electric field applied to the MIMCAP (for example in ramp voltage tests), then the time averaged current density must be used instead,

$$t_{BD} = \frac{Q_{BD}}{\dfrac{1}{t_{BD}}\displaystyle\int_0^{t_{BD}} J(E(t))dt} \tag{3}$$

or more simply, t_{BD} is found from

$$Q_{BD} = \int_0^{t_{BD}} J(E(t))dt \tag{4}$$

In MIMCAPs with typical nominal dielectric thicknesses $d_{nom} > 500\text{Å}$, the expected or intrinsic times to breakdown are extremely long, exceeding hundreds or thousands of years at normal usage voltages. In actual testing, however, a small fraction of unexpectedly short times to breakdown are found. These are attributed to the random extrinsic defects. The effective thickness concept is a useful way to characterize these defects. It is postulated that at the site of the defect, the thickness of the dielectric is not the nominal d_{nom}, but rather is reduced to a thinner value x_{eff}. Often this effective thinning is perfectly invisible, there being no physically assignable defect such as a particle, a surface roughness, or a pinhole to which to ascribe the origin of the breakdown. (This is not to say that physical defects, such as evaporated metal "spits" or particles can be discounted. However in modern processing, metal spits have been nearly eliminated by the use of Ta in the Au evaporator crucible [13], and cleanroom particle counts today are extremely low.) The unfavorable arrangement of Si - N bonds or the incorporation of other elements such as H in the dielectric may relegate some small fraction(s) of the total MIMCAP area to have a thinner effective thickness. The effective thickness may be deduced from the measured times to breakdown in constant voltage testing, or from the breakdown voltages in ramp voltage testing, two of the most common of the many test techniques. In the extrinsic data literature referenced here, one of these two methods has been used.

Combining equations (1) and (2) with the fact that in the planar MIMCAP, the electric field at a defect is given by $E = V/x_{eff}$ we find for the constant voltage test condition

$$x_{eff}^{1/2} = \frac{\beta V^{1/2}}{\phi_t + kT \ln\left(\dfrac{Q_{BD}}{\sigma t_{BD} V} x_{eff}\right)} \tag{5}$$

This gives the effective thickness x_{eff} that corresponds to a measured time to breakdown t_{BD} for a constant applied voltage V. This equation must be solved iteratively for x_{eff}, first assuming a value, inserting it into the logarithm to find a new x_{eff}, and repeating. Convergence is usually found in just a few iterations.

For a ramped voltage test with ramp rate R, the voltage is $V(t) = Rt$, the electric field is $E = Rt/x_{eff}$, and breakdown occurs at $V = V_{BD}$. The corresponding x_{eff} is found by integrating eq. (4) with (1) to obtain similarly (see [14] for integral):

$$x_{eff}^{1/2} = \frac{\beta V_{BD}^{1/2}}{kT \ln\left[\dfrac{Q_{BD}R}{2\sigma}\exp\dfrac{\phi_t}{kT} - 6g^4 x_{eff}\right] - kT \ln f(x_{eff})} \tag{6}$$

where f is a function of x_{eff} that has dimensions of $V^2 cm^{-1}$ given by

$$f(x_{eff}) = g V_{BD}^{3/2} x_{eff}^{-1/2} - 3g^2 V_{BD} + 6g^3 V_{BD}^{1/2} x_{eff}^{1/2} - 6g^4 x_{eff} \tag{7}$$

and where $g = kT/\beta$ and has dimensions $V^{1/2} cm^{-1/2}$. The solution is then found again by iteration on x_{eff}. In summary, either eq. (5) or (6) can be used to obtain the effective nitride thicknesses from any extrinsic breakdown probability plot.

B. Defect Densities

The next order of business is to extract the actual defect densities. For a certain defect density level D per unit MIMCAP area, the probability of failure caused by n random defects can be obtained by assuming a Poisson distribution,

$$P_f(n) = \frac{DA}{n!}\exp^{-DA} \qquad (8)$$

where A is the area of one MIMCAP. We assume that a failure occurs when one dominant defect is present, $n = 1$. By approximating the exponential as $e^{-x} \approx 1/(1+x)$, we obtain the simple Seeds [15] defect model

$$P_f = \frac{DA}{1+DA} = 1 - \frac{1}{1+DA} \qquad (9)$$

which differs little from the Poisson model for the relatively low probability levels consistent with the extrinsic defects. Inverting this equation provides the cumulative defect density

$$D = \frac{1}{A}\frac{P_f}{1-P_f} \qquad (10)$$

For every cumulative probability point, a cumulative defect value is thus obtained. The area in this expression is the area of a single one of the capacitors tested (not the total area).

IV. EXTRINSIC DATA COMPILATION

We wanted to obtain an industry-wide collection of extrinsic data from as many sources of published data as possible. References [1-12] contain useful extrinsic data that could be put to our purpose. In some of these works, the authors are forthright about characterizing the level of the extrinsic defects. In others, there are efforts described to improve the MIMCAP reliability when it is noted that a significant fraction of breakdown events occur at the edges or corners. By improving the geometry factors, such as the extension of the nitride film beyond the top metal, or by smoothing the rough edge left by liftoff, or by improving the underlying wafer defect density, the extrinsic breakdowns can be minimized. However, there still remains a certain level of extrinsic defects even with the improvements made, apparently with no discernable defect pre-existing. We also can find instances where the intrinsic breakdown improvement is the subject of the work, but

nevertheless an extrinsic tail is still visible in the data plotted. In Table 1 we have summarized all the available references in the open literature where an extrinsic breakdown failure mode is displayed whether acknowledged or not.

Armed with a scanner, digitizing software, equations (5), (6) and (10) and some patience (a lot of patience) we converted the probability plots in references [1-12] into the chart shown in Fig. 3. Sufficient information was needed from the references to proceed, including the ramp rate or the constant test voltage, the area and quantity of test structures, the nominal nitride thickness, the approximate intrinsic breakdown voltage, the charge to breakdown and the Frenkel-Poole (F-P) conduction parameters. Not all of these values were provided in all the data references, so some assumptions were necessary. For example, lacking the charge to breakdown, we assumed $Q_{BD} = 20$ Coul/cm^2 as a standard value. In some cases the film thickness was not specified, but could be inferred by capacitances or by the intrinsic breakdown voltages. When lacking the F-P parameters, we assumed standard values: $\phi_t = 1$ eV, $\beta = 2.771 \times 10^{-4}$ V$^{1/2}$cm$^{1/2}$, and $\sigma = 1 \times 10^{-4}$ S/cm. When necessary, we adjusted the F-P conductivity σ so that the effective thickness at the intrinsic breakdown condition was at least 80% of the nominal thickness. The parameters

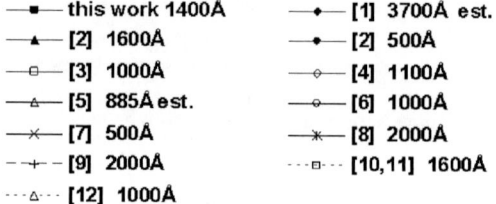

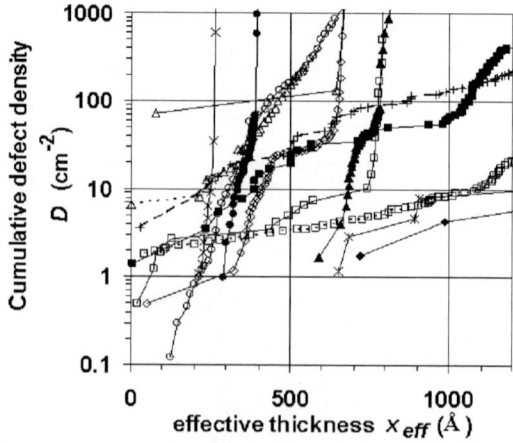

Figure 3. Extrinsic defect density versus effective thickness extracted from 12 different publications, together with our Hg-dots data. The legend provides the nominal nitride thickness for each reference.

Q_{BD} and σ appear in eqs. (5) and (6) as a ratio, so even if we are ignorant of both values, the more important ratio is still likely correct. Where parameters were provided in the references, we used them without prejudice, and adjusted any remaining parameters as needed. Table 1 shows some of the details of the assumptions and other information about our compilation process.

Fig. 3 represents a snapshot of the industry data over about the last dozen years. The data includes GaAs processes, InP processes, one Si processes and one SiC process, all utilizing PECVD nitrides. There seems to be no correlation between defect density and process type or the date of the works. Some of the defect densities are quite low, less than 1 per cm². None are higher than about 100 per cm² disregarding the intrinsic breakdown branches. In the fourth column of Table 1 we provide the area of the individual capacitor test structures, the quantity tested, and their product -- the total area devoted to each investigation. We see that in most cases, the authors have devoted much more than 0.1 cm² (10 mm²) of test structure area to these investigations. This is necessary in order to detect some of the lower defect density numbers

V. PARAMETERIZATION OF THE EXTRINSIC DEFECT DENSITIES

We also were interested in parameterizing the extrinsic defect density tails in order to summarize and condense all the data in Fig. 3. In the past a good model for the cumulative extrinsic defect density that has served well is a simple exponential [9]

$$D = D_0 \exp\frac{x_{eff}}{x_C} \;, \qquad (11)$$

where D_0 is the intercept parameter and x_C is a characteristic thickness. This forms a straight line with slope $1/x_C$ on a typical plot of cumulative defect density versus effective thickness. Unfortunately, in many of the extrinsic probability plots in our compilation [1-12], the defect density of "dead on arrivals" (DOAs) or time-zero failures have been removed. These failures correspond essentially to $x_{eff} = 0$, and are the infant mortalities that can occur when many capacitors are tested. Removing these from the probability plot causes the extrinsic defect tail to become approximately

$$D = D_0\left[\exp\frac{x_{eff}}{x_C} -1 \right] \qquad (12)$$

We used either (11) or (12) to approximate the extrinsic defect densities depending upon whether or not DOAs were included. They usually are *not* included in the references, especially when constant voltage testing is employed. This is because at the elevated constant voltage, the times to fail for the smaller effective thicknesses (weakest spots) are so short as to be indistinguishable from true DOAs. This tends to artificially remove the extrinsics near $x_{eff} = 0$. In our experience, ramp voltage testing does not seem to suffer this ill. Two examples showing how the extrinsic tails can be parameterized in either situation are shown in Fig. 4. In one case we show the extrinsic data collected on 1400Å films as part of this work using a ramp voltage test along with eq. (6). In the other case we show points acquired from [7], a constant voltage test on capacitors with a nitride thickness of 500Å, along with eq. (5). In the former case the DOAs are evident, and in the

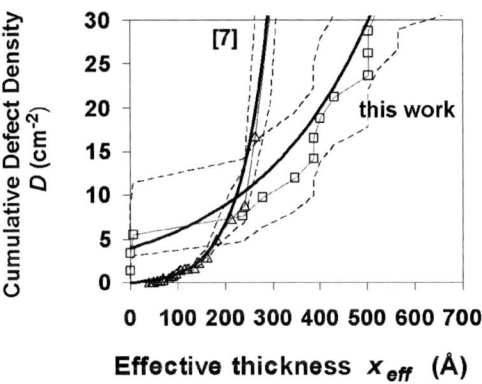

(a)

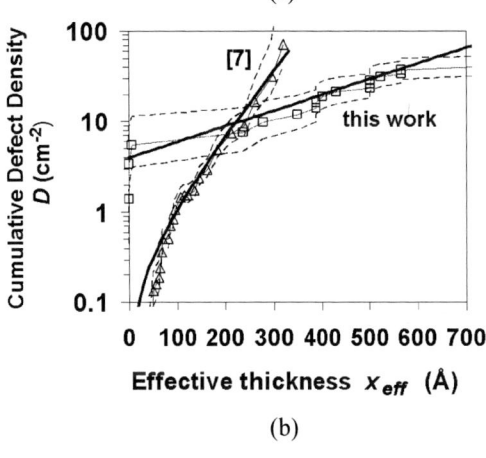

(b)

Figure 4. Defect densities extracted from Hg-dots using a ramped voltage method as part of this work as compared to [7] using a constant voltage method. The same data is plotted in (a) on linear axes and (b) on semilogarithmic axes. The dashed lines are 60% confidence limits, and the heavy lines are the empirical fits.

TABLE 1.   SUMMARY OF COMPILED Si_3N_4 MIMCAP EXTRINSIC DATA.

Ref. and year of pub.	Nominal thickness (Å)	Test Method	Area A(cm²), Qty. tested & Tot. Area tested (cm²)	Φ_c (eV)	β (cm$^{1/2}$V$^{1/2}$)	σ (S/cm)	Q_{BD} (C/cm²)	Notes
this '11	1400	Ramp 4V/s	5.1×10⁻³ 100 0.51	0.8	3.78 ×10⁻⁴	7.29 ×10⁻⁹	20	Hg dots
[1] '10	3700 est.	Ramp 5-20 V/s	1.0×10⁻³ 400 est. 0.4	1.0	2.771 ×10⁻⁴	1.0 ×10⁻³	20	Combined all extrinsic points of Fig. 6 that are indep. of edge process. Avg. ramp rate 12.5V/s assumed.
[2] '08	500	Const 43V	2.04×10⁻⁴ 2000 est. 0.408	1.0	2.771 ×10⁻⁴	1.0 ×10⁻⁶	20	None of the capacitor edge extrinsics are used here, only the remaining areal ones.
	1600	Const 146V	2.04×10⁻⁴ 3500 est. 0.714	1.0	2.0 ×10⁻⁴	1.0 ×10⁻⁵	20	
[3] '06	1000	Ramp 50V/s	1.06×10⁻³ 100 0.106	1.06	4.59 ×10⁻⁴	3.5 ×10⁻⁸	120	Utilize only the conventional MIMCAPs, not MIMCAPs on vias
[4] '06	1100	Ramp 5V/s	3.6×10⁻³ 1000 est. 0.36	1.0	2.0 ×10⁻⁴	5.0 ×10⁻⁴	20	Utilized unprocessed extrinsics in Fig.1 (medium defect density)
[5] '07	885 est.	Ramp 0.5 or 4.3V/s	7.1×10⁻⁶ est. 4000 est. 0.028	1.0	2.771 ×10⁻⁴	5.0 ×10⁻⁶	20	Used Fig.3., converting to a cum. probplot. Assumed the higher ramp rate for conservatism.
[6] '03	1000	Ramp 3V/s	4.0×10⁻⁴ 14,600 5.84	0.96	3.59 ×10⁻⁴	7.5 ×10⁻⁸	127	Used Fig. 9 data from 10 wafers.
[7] '01	500	Const 44V	4.6×10⁻³ 200 est. 0.345	0.7	5.0 ×10⁻⁵	1.0 ×10⁻⁹	5	Used extrinsic branch of Fig. 1, and the nitride consts. from the IV meas.
[8] '00	2000	Ramp 3V/s	4.0×10⁻⁴ 1475 0.59	1.0	3.0 ×10⁻⁴	1.5 ×10⁻⁴	20	Utilized prob. chart on p. 43.
[9] '99	2000	Ramp 25V/s	1.20×10⁻⁴ 820 0.1	0.9	2.771 ×10⁻⁴	7.0 ×10⁻⁴	150	Used Fig. 14. Note that this nitride is composed of two layers
[10] [11] '98	1600	Const 50V	6.4×10⁻² 200 est. 1.28	1.0	2.771 ×10⁻⁴	1.0 ×10⁻⁴	20	Used slide 11 of [10] having largest area, and thickness from [11]
[12] '98	1000	Ramp 2V/s	4.3×10⁻⁴ 1000 est. 0.43	1.0	2.771 ×10⁻⁴	1.0 ×10⁻⁸	20	The 2V/sec data of Fig 4a having an extrinsic tail was utilized

latter case they have been expunged. That they have been expunged is fairly evident in the semilog plot of Fig. 4, where no data points are found within about 40Å of zero. These were probably mistaken for DOAs and removed from the probability plot in the original reference. Also in Fig. 4 we included the light dashed lines showing the 60% two-sided pointwise binomial confidence limits [16] which aid in the fitting of the lines. It should be noted that the important extrinsic defects region lies between approximately 100Å to about 400Å for most capacitor voltages of interest. This is the range of extrinsic defects that affects the mid- to late life of typical MMICs. Defects with x_{eff} smaller than this are infant mortalities or early failures and are

removed during initial test or burn-in. Defects larger than this have long lifetimes beyond our interest. It is the middle region that concerns us most, and this is where the parameterization accuracy is important.

This technique was used to parameterize all the data of Fig. 3. The resulting set of curves in Fig. 5 shows our results. Again we see a wide range of defect densities extending over perhaps two orders of magnitude. A plot of the defect densities at the specific $x_{eff} = 250$Å in the middle of our range of interest versus the nominal dielectric thickness is shown in Fig. 6. Here we see that there is a fair amount of scatter in the data, but that a weak trend exists. The thicker dielectrics seem to have a bit lower defect density. This trend is worth noting, even if it is a weak trend. A reasonable encapsulation of this trend is the modification of (12) to

$$D = D_0 \exp\left(\frac{-d_{nom}}{d_C}\right)\left[\exp\frac{x_{eff}}{x_C} - 1\right] \qquad (13)$$

where d_{nom} is the nominal nitride thickness and d_C is a characteristic thickness. In this way, the defect density is a function of both the effective thickness as well as the nominal thickness of the nitride. We decided to characterize the wide range of defect densities with a "low-mid-high" type of ranking. If defect densities are low, or if one needs an optimistic prediction, the "low" values should be used. If pessimistic, the "high" numbers are in order, and if one needs an industry average trend, the "mid" values are appropriate. The values (ordered as low, middle, high) are: D_0 = (3.5, 4.5, 8) defects/cm^2, x_C = (500, 180, 100) Å, and d_C = 1000Å. The heavy solid, dashed and dotted lines in Fig. 5 show this empirical model for three different nitride thicknesses of 500Å, 1000Å and 2000Å. The model covers the wide range of the defect densities.

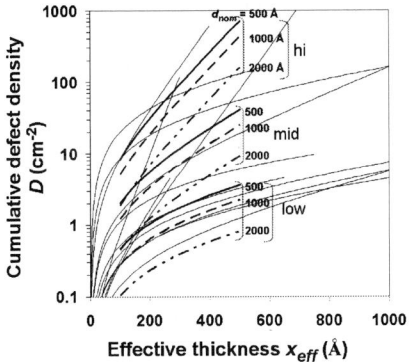

Figure 5. Family of parameterized defect densities extracted from the compilation of data sources.

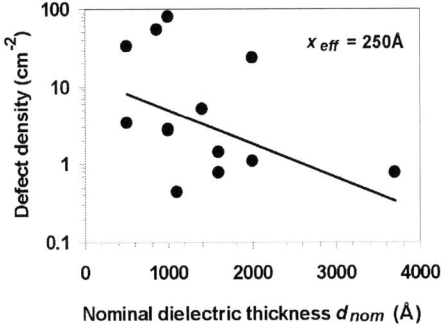

Figure 6. Defect densities determined at the specific x_{eff} value of 250Å versus the nominal dielectric thickness.

VI. RELIABILITY OF MIMCAPs

We can combine eqs. (9) and (13) to provide a probability of failure $P_f(x_{eff}, A)$ as a function of an effective thickness x_{eff} and the product MIMCAP area A, choosing a particular defect density (low-mid-high). In the MIMCAP product, we assume a constant applied voltage and temperature for its useful life, and from eq. (5), we can predict the value of x_{eff} that is associated with that lifetime. In fact we can find the failure probability at any time t. In using eq. (5) the nitride properties must be known. We have used the standard values as described above. Therefore we have a complete model for MMCAP reliability.

One can go further by computing the cumulative hazard value $H(t) = -\ln(P_f - 1)$ as a function of the time up to the desired lifetime, say 10 years. Next, a numerical differentiation produces the hazard rate $h(t)$, which decreases with time. The hazard rate can be averaged over the desired mission or usage time to obtain the AFR (average failure rate) expressed in FITs. This entire procedure has been described previously [9], and a typical result is shown here in Fig. 7. In performing this calculation we have assumed the "mid" defect density, a voltage of 3.3V applied continuously d_{nom} = 2000Å, an area of 0.1 mm^2 and a usage temperature of 125°C. The resulting AFR after 10 years is about 30 FITs in this case. The failures amounting to 0.05% in the first hour were removed as infant mortalities. Using the defect densities model of eq. (13), not many early failures or burn-in failures would be predicted since we have removed the $x_{eff} = 0$ defects from consideration by design. After the first hour, we have $x_{eff} = 103$Å in this case, increasing to 337Å by ten years.

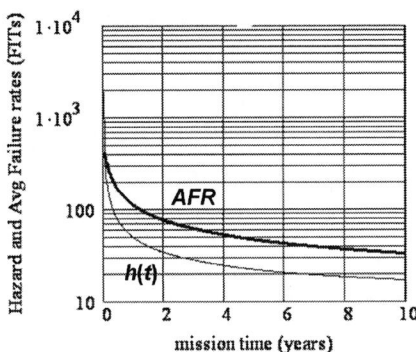

Figure 7. The hazard rate $h(t)$ (light line) and the average failure rate, AFR (heavy line) versus time for a 0.1mm^2 MIMCAP exposed to 3.3V and 125°C.

Note that Fig. 7 shows a decreasing failure rate–initially high and decreasing with time. Contrast this with the failure rate of a typical pHEMT or HBT, as described using the usual lognormal distribution. For a pHEMT/HBT there is essentially a zero failure rate in the first few years (excluding infant mortalities), increasing with wearout. On the other hand the MIMCAP failure rate begins at a moderately high value and decreases in the later years. At moderate temperatures where wearout of the active devices is not of concern, the MIMCAPs dominate the AFR of a MMIC, especially in the shorter-lived applications.

VII. MIMCAP DESIGN SPACE

Now that the defect density has been parameterized, it is possible to go even further. For a given nitride thickness, the defect density (low, mid, or high depending upon one's sense of optimism or pessimism) can be used to determine the maximum allowable area to achieve a certain reliability level, expressed as an AFR over the time of interest. Figures 8-10 show the results. They give the maximum allowable capacitor areas in (a) and corresponding capacitance values in (b) under various usage conditions for the specified reliability level of 10 FITs. These contours are the design space for the sizing of MIMCAPs and show the tradeoff between reliability versus areas. Fig. 8 is for a "stringent commercial" application with 3.3V, 125°C, and 10 FITs over 10 years. The continuous lines are contours of constant average failure rate level of 10 FITs and have been plotted using the approximated defect density models (low, mid and high) along with the standard F-P parameters. To the left of each line is reliable operation with an average failure rate lower than the specified 10 FITs. To the right of each line is unreliable operation where the average failure rate exceeds 10 FITs.

The diamond points in Fig. 8 have been plotted atop the contours for comparison. These points represent the AFRs derived from each of the references computed using the F-P parameters of Table 1 along with the individually measured defect densities for each reference as shown in Fig. 3. The diamond points are therefore based upon more exact TDDB computations from each reference. The contours are approximations for purposes of summarization and design prediction. The contours in Fig. 8 seem to represent the range of reliability results rather well. The most optimistic contours fall on the upper range of the data points, while the middle contour, being more conservative is close to the center of the data

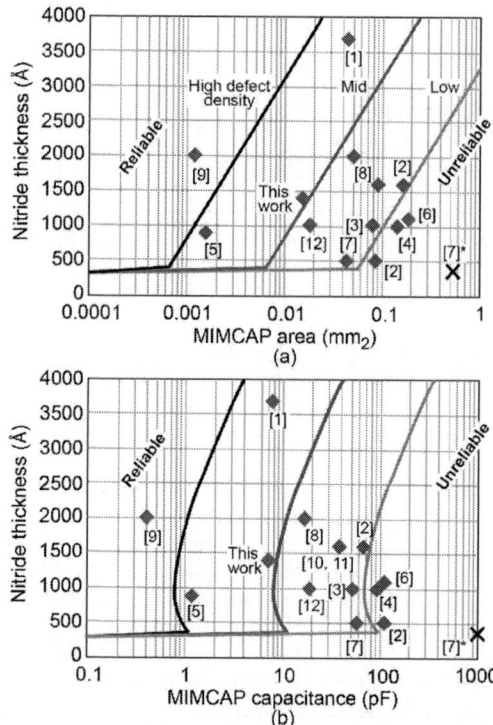

Figure 8. Design space for MIMCAPs under a "stringent commercial" reliability goal of 10 FITs averaged over 10 years with operating conditions 3.3V and 125°C. The contours with respect to the nominal nitride thickness show the industry-wide max. permissible area (a) and max. capacitance (b) for low (optimistic) to high (pessimistic) defect densities. The diamonds show computations from the references numbered using defect densities and F-P parameters as published for comparison. The "×" is a prediction from [7] using a very different method.

points. For extreme conservatism in the face of the highest defect densities, the pessimistic leftmost contour can be used for design purposes.

On Fig. 8 is an "×" showing a prediction found in ref. [7] that a MIMCAP with a 350Å thick nitride and an area of about 0.5mm² operated at 3.3V and 125°C would meet a 10 year life with an AFR of 10 FITs. It is more optimistic than even the "low" defect density industry-wide contour by 10×. It is also even more optimistic than the diamond symbol displaying our more exact TDDB treatment for [7]. The reason for the discrepancy may be because it was obtained by extrapolation of a voltage acceleration factor derived from the F-P field-lowering parameter, rather than going through the computations as described here. Alternatively, the defect densities for that actual capacitor could have been lower than we adduced from [7].

We have also computed the maximum permissible area and capacitance for two more scenarios. Fig. 9 shows a "high rel small signal "

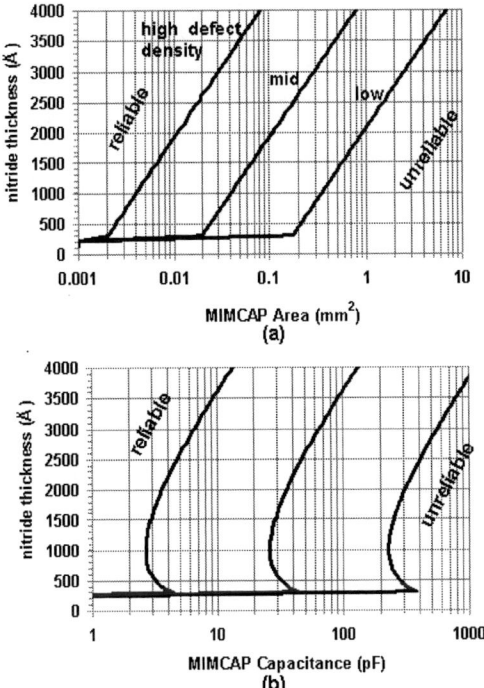

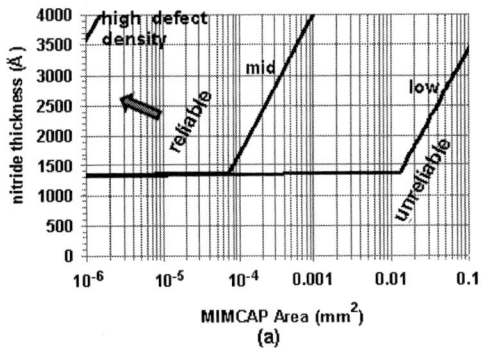

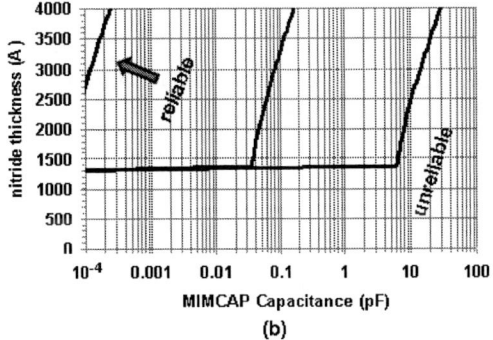

Figure 9. Design space for MIMCAPs under a "hi-rel small signal" goal of 10 FITs averaged over 15 years with operating conditions 5V and 60°C. A burn-in is also assumed consisting of 5V, 125°C for 320 hours.

Figure 10. Design space for MIMCAPs under a "hi-rel power" goal of 10 FITs averaged over 15 years with operating conditions 12V and 125°C. A burn-in is also assumed consisting of 12V, 125°C for 320 hours.

case for a satellite or similar mission with a 10 FITs requirement over 15 years with usage conditions of 5V and 60°C. It was also assumed that a 320 hr. burn-in is performed at 5V and 125°C. These contours are more favorable than those in Fig. 8, mostly owing to the lower temperature. A 60°C usage temperature is typical for LNAs (low noise amplifiers) or other similar small signal MMIC applications. For the "low" defect density range, we can have capacitances up to 200pF or more. In Fig. 10 we show another high-rel case —"hi-rel power" — with the same 15 year 10 FITs requirement. However the operating conditions are more difficult at 12V and 125°C. A burn-in for 320 hrs, 12V and 125°C is also assumed. This resembles the usage conditions for microwave power amplifiers or other higher power MMIC applications. These contours are much less favorable. A nitride thickness less than about 1500Å will not support this reliability goal. The design space above 1500Å is still perfectly usable, with reasonable capacitances up to about 10 pF still feasible. If the defect density is pessimistically graded as "high", it is not feasible to expect a failure rate of 10 FITs even for capacitors of size 10^{-6}mm² (1μm x 1μm) under these operating conditions.

VIII. DISCUSSION

A feature of Figs. 8-10 is that the reliability contours collapse below a certain nominal MIMCAP nitride thickness. This happens at somewhat below 400Å for the "stringent commercial" and "hi-rel small signal" cases, and at about 1400Å for the "hi-rel power" case. The collapse occurs because with sufficient voltage and/or time, the x_{eff} will advance beyond the extrinsic region and begin to enter the intrinsic region. In other words, the end of life or wearout phase of the capacitor looms too quickly to support the 10 FITs reliability goal.

A larger Q_{BD} value or a lower film conductivity would extend the useful lifetime according to eq. (2). If an effort were mounted towards increasing the Q_{BD} values for nitrides, rather than optimizing for maximum dielectric constant, or minimum loss tangent, it might be possible to extend the collapse point to lower nitride thicknesses. This would be of great benefit to the industry. Of course it would have to be done while maintaining low defect densities as well.

The charge to breakdown has been assumed to be a constant in this work. However there is

978-1-4244-9113-1/11 $26.00 © 2011 IEEE

no guarantee that the charge to breakdown measured using intrinsic capacitors (the usual practice, and typically close to 20 C/cm^2) is the correct one for extrinsic capacitors. This is a possible weakness in the TDDB theory as utilized here. To our knowledge, a measurement of the QBD values specifically for extrinsics has never been reported.

The main results of this work show that the industry seems to have a wide variability in the nitride defect densities. We have tried to summarize the wide range of defect densities by the three levels low-mid-high. A weak thickness dependence was included in our empirical model of defect densities. We were expecting to find a much stronger dependence on the nominal nitride thickness buried in the extrinsic data over all of the references. Instead we found that the variability across all the references overcomes any possible thickness dependence. We believe that the effects of plasma chemistry and conditions, growth rate, metal surface roughness, processing conditions, cleanliness and particle counts are all factors affecting these nitride defect densities across the industry. There is a potential for further improvement of nitride reliability by taking stock of the very important extrinsic defects.

IX. CONCLUSIONS

We have provided a data compilation focusing on extrinsic silicon nitride defects that are important for the reliability of MIMCAPs. We have summarized the state of the industry over the last dozen years by characterizing the wide range of defect densities that we found. The characterization takes the form of defect density models for low-mid-high levels of defects depending upon one's degree of optimism. We provided the methodology that we used to extract data on defect densities from probability plots found from the references cited. We have also provided design spaces for MIMCAP reliability that show the tradeoff between the defect density, the nitride thickness, and the maximum permissible area or capacitance that can be implemented for a given specified failure rate. The industry data seem to correlate well with our rank characterization of the defect densities.

ACKNOWLEDGEMENT

This work was supported by The Aerospace Corporation's Independent Research and Development (IR&D) Program.

REFERENCES

[1] J. Gurganus, T. Alcorn, A. Mackenzie and Z. Ring "Investigation and Improvement of Early MIM Capacitor Breakdown with a Focus on Edge-Related Failures" Digest 2010 Compound Semiconductor Mantech Conf., May 17-20, Portland OR, p. 73.

[2] J. Oliver, H. Cramer, and R. Porter "Layout Design Rule Effects on Capacitor Reliability" paper 10.3, Digest 2008 Compound Semiconductor Mantech Conf., April 14-17, Chicago, IL

[3] X.Zeng, M. Barsky, J. Uyeda, D. Farkas, F. Yamada, D. Biendenbender, D. Eng, J. Wang, and R. Lai "The Reliability Study of MIM Capacitor Built on Top of Backside Via in III-V Compound MMIC" Digest 2006 Compound Semiconductor Mantech Conf., April 24-27, Chicago, IL, p. 87

[4] P. van der Wel, J. de Beer, R. van Boxtel, Y. Hsieh, and Y. Wang "Reliability Assessment of Extrinsic Defects in SiNx Metal-Insulator-Metal Capacitors" paper 2.1, Proc. 2006 Reliability of Compound Semiconductors (ROCS) Workshop, Nov. 12, San Antonio, TX

[5] B. De and M. Shokrani "On-Wafer Test Method of Metal-Insulator-Metal Life Using Time Dependent Dielectric Breakdown" ECS Tranactions 6 (3), 2007 p. 415-429

[6] W. Rowe, B. Paine, A. Schmitz, R. Walden, and M. Delaney "Reliability of 100nm Silicon Nitride Capacitors in an InP HEMT MMIC Process" Microelectronics Reliability 43 (2003) p 845-851

[7] K. Allers, M. Schrenk, K. Koller, M. Schwerd, and H. Korner "Reliability of Metal-Insulator-Metal Capacitors" Proc. 2001 Advanced Metallization Conf., MRS Oct. 8-11, Montreal, CAN, p. 447

[8] B. Paine, R. Wong, A. Schmitz, R. Walden, L. Nguyen and M. Delaney "Ka Band InP HEMT MMIC Reliability" Proc. 2000 GaAs Reliability Workshop, Nov. 5, Seattle, WA, p. 21

[9] J. Scarpulla, D. Eng, S. Olson, and CS Wu, "A TDDB Model of Si3N4 – based Capacitors in GaAs MMICs" Proc. 1999 Int. Rel. Physics Symp., San Diego, CA, p. 128

[10] H. Cramer, J. Oliver and G. Dix "MMIC Capacitor Dielectric Reliability" Proc. 1998 GaAs Reliability Workshop, Nov. 1, Atlanta, GA, p. 46.

[11] H. Cramer, J. Oliver, G. Dix and M. Zimmerman "Development of an Improved Capacitor Dielectric" Digest 1998 International Compound Semiconductor Workshop, Seattle WA, April 27-30, 1998, p. 15

[12] B.Yeats "Assessing the Reliability of Silicon Nitride Capacitors in a GaAs IC Process" IEEE Trans. on Electron Dev., 45 (4) April 1998 p. 939

[13] J. Cotronakis, M. Clarke, R. Lawrence, J Campbell, and C. Gaw "Continuous Defectivity Improvements and Impact on High Density Metal-Insulator-Metal (HDMIM) Capacitor Yields" Digest 2004 International Compound Semiconductor Workshop, paper 12.3

[14] H. Dwight Tables of Integrals and Other Mathematical Data (4th ed.) MacMillan 1961 p. 134, eq. 567.3

[15] R. Seeds "Yield and Cost Analysis of Bipolar LSI" Digest Int. Elec. Dev. Meeting, 1967, p. 02

[16] [15] W. Meeker and L. Escobar Statistical Methods for Reliability Data Wiley,1998 p. 50

AC-DC Factor Sensitivity for DRAM Components Lifetime Under Hot-Carrier Injection

Sanghyeon Baeg[1], Hyeonwoo Nam[2], Pierre Chia[3], ShiJie Wen[3], Richard Wong[3]

Hanyang University at ERICA Campus, 426-791, Korea[1,2]; Cisco Systems Inc., San Jose, CA 95134, USA[3]

Email: bau@hanyang.ac.kr[1]

Abstract—**Estimating operating lifetime is critical for dynamic random access memory (DRAM) components with hot-carrier injection (HCI). Using DC device lifetime to substitute a component lifetime can be too pessimistic and can disqualify good DRAM products. This work proposes the DC to AC lifetime ratio factor and its sensitivity over three parameters: device degradation, effective drain voltage in word-line driving circuit, and access frequency. The timing margin in word-line driver signal is directly related to HCI degradation and correlated to DC to AC lifetime ratio. The results are discussed with measured I_{sub} current in three different technologies and DRAM components, and correlated with both simulation and test data.**

Keywords-hot-carrier injection; AC-DC Factor; pumped voltage; DRAM

I. INTRODUCTION

The voltage at dynamic random access memory (DRAM) word-lines is boosted to pumped voltage (V_{CCP}) to store full V_{CC} power supply voltage at a DRAM cell and to compensate for the voltage droop over long word-lines. The V_{CCP} at the driving NMOS circuit causes more hot-carrier injection (HCI) than nominal operating voltage.

Regarding typical HCI device requirements, Cisco asks for a device lifetime of 0.1 year under DC stress; the device lifetime is measured at 10% I_{DSAT} degradation. In many technology nodes used in DRAM products, we have noticed that the 10% degradation lifetime is too strict, and many vendors are unable to conform to the guidelines over wide ranges of technology nodes.

On the other hand, we have observed that the 10% I_{DSAT} degradation due to HCI did not cause component failures in many products. We also observed that the product with a very short device DC lifetime outlived other products with longer device DC lifetimes. Such observations motivated the development of more realistic qualification processes and methodologies to assess a component AC lifetime at a product level, rather than disqualifying a component because of unqualified device DC lifetime under HCI.

The AC lifetime estimation was previously based on the observation that HCI degradation occurs primarily during the switching period [1]. We extended the idea to analyze the sensitivity of AC lifetime based on the circuit design and device degradation parameters. Such sensitivity analysis will eventually provide a vehicle for assessing a product AC lifetime and may result in justifying the insufficient device DC lifetime at a device level. The opposite situation can also happen; higher device DC lifetime may not be able to satisfy the AC lifetime requirement of 10 years operational time.

The rest of this paper is organized as follows. Section II introduces the AC lifetime estimation from DC device lifetime based on the substrate current due to HCI degradations. Section III discusses the sensitivities of AC lifetime over device HCI degradation, effective V_{CCP}, and access frequency to cells. The timing margin in a word-line signal is discussed in Section IV from the perspective of AC lifetime estimation. Finally, Section V concludes this paper.

II. AC-DC Factor

In this work, the NOR type word-line driver circuit is selected for simulation purposes and is shown in Fig. 1 [3] [4]. When the pull-down transistor (MN1) in a word-line driver circuit is degraded by HCI, the fall time of the word-line signal (L-word-line) is erroneously extended and can cause the charge at a cell to be erroneously shared with bit-line charges in the subsequent pre-charging cycle.

The pull-up transistor (MP1) is replaced with NMOS in the case of the bootstrap word-line driver circuit, in which case both rise and fall times can be affected by the HCI degradation. We have seen the different variations of transistor usages depending on V_{CCP} and timing margin values. In the rest of this paper, the pull-down transistor (MN1) will be referred to as pull-down transistor in the word-line driver circuit (PDWD); the pull-up transistor will be referred to as PUWD. PUWD is not necessarily PMOS as in Fig. 1 and can be implemented with NMOS transistor.

The word-line is pre-charged with V_{CCP_sig} and switches to the ground as a memory cell is accessed and de-selected. Let us assume V_{CCP_sig} is the same as V_{CCP} for the time-being. As the drain at PDWD switches between V_{CCP} and ground, it suffers from HCI, resulting in increasing threshold voltage and a reduced I_{DSAT} current. The HCI degradation of the NMOS is proportional to the substrate current as shown in Equation (1) [2].

$$\Delta V_T = C\left(\int_0^t I_{SUB}{}^m dt\right)^n \quad (1)$$

Figure 2 shows the measured I_{sub} with DC bias conditions for the technology used in the simulation; the results and test data in this paper are normalized for confidential reasons. The maximum I_{sub} occurred with the maximum voltage and its half voltages at drain and gate terminals, respectively.

Figure 3 shows the simulated gate and drain signals and I_{sub} of PDWD when a word-line is de-selected. As can be seen from Fig. 3, the bias condition for the maximum I_{sub} does not happen in real circuit operations or, if it does happen, it is only for short periods of time. The I_{sub} exists only when the local word-line switches to ground. It is therefore necessary to consider such timing dynamics for component HCI reliability analysis. The AC-DC factor (ADF) can be represented as in Equation 2 to express the ratio of I_{sub} in both DC and AC operations. The AC lifetime (AC_{life}) is estimated by multiplying the ADF value times the DC lifetime (DC_{LIFE}), as in Equation 3. The objective of this work is to investigate the sensitivity of ADF within various sensitivity parameters.

This work was supported by the GRRC program of Gyeonggi province. [GRRC Hanyang2010-A01, developing sensor network SoC equipped with low power and high reliability characteristics.

$$ADF = \frac{I_{SUB}{}^m}{\sum_{i=1}^{n} I_{sub}(i)^m \Delta t}, n = 1/\Delta t \qquad (2)$$

$$AC_{life} = ADF * DC_{LIFE} \qquad (3)$$

In Equation 2, Δt can be determined to be small enough to properly trace the I_{sub} changes during the switching period.

III. Sensitivity

We used three I_{sub} measurement data (A, B, C) from three different technologies and DRAM products from three different vendors to see the trends of simulated ADF values. The three different products used different V_{CCP} voltages. We believe they are good representatives of DRAM products over the wide ranges of technology nodes.

Figure 4 shows ADF values with I_{DSAT} degradations in PDWD. As expected, the ADF value decreased with increasing degradations; the AC lifetime is reduced with decreased ADF based on Equation (3). Data C had the worst DC lifetime among the three samples. However, it had the highest ADF ratio because of the higher ratio of DC I_{sub} over AC I_{sub} current. Please note that the maximum DC I_{sub} condition is not likely to happen in real circuit operations; the maximum substrate current is the basis for DC lifetime and can cause large discrepancies between device and product lifetime.

Figure 5 shows the failing time measurements with different t_{RP} degradations when a row is continuously accessed for up to 500 hours under a low temperature environment with highly elevated V_{CCP}. The stressed condition can affect various part of DRAM circuits and failures can be caused by multiple different reasons. From the past failure analysis experiences, we believe the t_{RP} increase in the test result is primarily caused by the accelerated device HCI degradation with elevated V_{CCP}.

The I_{DSAT} degradation in Fig. 4 is translated into timing degradation (i.e., fall time degradation of the local word-line) to correlate with the test data in Fig. 5. The corresponding ADF values are plotted over Fig. 4 (see plot *falltime*). The slope in Fig. 4 and the plot in Fig. 5 are close to 0.003. However, the slopes were opposite because the ADF was calculated with fixed DC_{life}, while the other was not.

Figure 2 shows the I_{sub} current increases with increasing drain voltage in PDWD, which prompted the idea of reducing the PDWD drain voltage before deactivating the word-line signal. Figure 1 shows that the supply voltage for the word-line comes from V_{CCP_sig}. The stored charge at the word-line can be initially drained through PUWD, and the rest of the charge is drained through PDWD, which will be referred to as PUWD drainage technique (PDT). If PUWD is implemented by PMOS, such PDT would help reduce the effective V_{CCP} voltage for PDWD with extra timing cost due to the slower PMOS device. If PUWD is implemented with NMOS, both PUWD and PDWD can be exposed to HCI degradation. The PMOS HCI degradation was much smaller the degradation of NMOS. This paper will focus on the HCI degradations at PDWD.

Figure 6 shows I_{sub} plots in PDWD during the transition of a word-line after initially reducing 10% to 30% of the word-line voltage by PDT. It is very clear that I_{sub} is significantly reduced by increasingly draining the word-line charge through PUWD. The corresponding ADF values are plotted in Fig. 7.

The ADF significantly increases with the percentage increases of PUWD drainage. The side effect of PUWD drainage is an increase in the total word-line fall time, which is not the main interest of this work.

Figure 8 shows the measured DC device lifetime for different devices with different V_{CCP} levels over the wide ranges of technology nodes. The slopes in Fig. 7 (data C) and Fig. 8 are 4.4214 and 4.4556, respectively. It was interesting to note that the DC lifetime with various V_{CCP} is more closely related to the effective V_{CCP} voltage than the technology; Fig. 8 is for different technologies up to about 40nm with the corresponding V_{CCP} at each technology node, and Fig. 9 is for one technology with varying effective V_{CCP}.

Figure 9 shows the ADF values with different access rates to a word-line. The X-axis in Fig. 9 is the frequency accessing a word-line. As the access frequency increases, the ADF value approaches 1. The maximum rate in a DRAM component is t_{RC} time, assuming the one word-line is continuously accessed. The continuous access to a word-line was not just the worst case assumption; HCI failure issues were observed in multiple products with back-to-back accesses to one word-line by Cisco system software.

IV. Timing Margin

The device DC lifetime of a vendor DRAM component was less than the DC lifetime requirement of 0.1 years. However, the device did not cause t_{RP} timing failure while satisfying much higher AC lifetime than the estimated AC lifetime based on the device DC lifetime when stress tests were performed with about 35% elevated V_{CCP} stress voltage.

On the other hand, the second component from a different vendor had a similar device DC lifetime as the first one. The V_{CCP} stress test voltage in the second component had a slightly higher V_{CCP} stress percentage compared to the first component. However, the second component had about 10 times less t_{RP} shift compared to the first one with a larger AC stress time; the stress was applied while a word-line was accessed with regular DRAM access operations. The second component did not really meet the device DC lifetime requirement; however, the AC lifetime estimation was within the Cisco product deployment budget, and the component was deployed with the risk factor noted.

The above observations could not be easily explained by device DC lifetime alone. The second component case can be partly explained by the conclusion that the timing margin added to the word-line signal largely compensated for the HCI degradation. It is therefore necessary to add timing margin considerations as a part of AC lifetime requirement evaluation due to HCI reliability. When the timing margin is large enough, the corresponding component can tolerate more than 10% I_{DSAT} degradation, which can lead to relaxing the device DC lifetime requirement.

Equation 3 is now changed to Equation 4, in which the DC lifetime of S% degradation is used instead of 10% degradation. In order to determine the value of S, proper engineering practices should be implemented to make a reasonable decision about the timing margin that can be allocated for the HCI issue.

978-1-4244-9113-1/11 $26.00 © 2011 IEEE

$$AC_{life} = ADF * DC_{LIFE}(S) \quad (4)$$

According to the configuration in Fig. 1, the local word-line signal (L_Wordline) should be deactivated before starting to set up for subsequent memory access operations. If the local word-line signal is not completely turned off due to the HCI degradation in PDWD, the accessed cells could be erroneously affected by the subsequent pre-charging operations. This type of situation will eventually affect memory access timing; one type of timing that belongs to this category is t_{RP}.

On the other hand, it is also possible that the PUWD in the word-line driver can be HCI-degraded, especially when PUWD is implemented with NMOS; as a result, the word-line to cells will not be sufficiently charged within a given access time window, and it will eventually affect the access operation. This can happen in either write or read operation. Typical timing related to this issue can be t_{RCD}, which is the number of clock cycles between row and column addresses.

The timing margin between the deactivation of a word-line drive and the activation of pre-charging kickoff can compensate the HCI degradation in the word-line driver circuits. A larger timing margin will tolerate more HCI degradations without causing failures; it shows that HCI-related word-line failure at a product level does not entirely depend on the device lifetime, but also on the circuit design margin and access rate to a word.

Simulations are performed to see the changes in timing margin with HCI degradations. Figure 10 shows the simulation results for the timing margin shifts for three different technology examples. The X-axis shows the HCI degradations in the PDWD in Fig. 1. The Y-axis shows the normalized timing margin; it is normalized over the timing margin without any degradation. As expected, as the word-line driver is degraded, the timing margin is reduced. The three different technologies showed different margin variations. For fair comparisons in this experiment, we made the three designs have the same design margin.

Typically, the timing margin is reserved for many uncertain random factors such as the skews in clocks and critical signals, process variations, and power and signal noises. However, we rarely found that the timing margin was reserved without the consideration of HCI degradations. It is therefore necessary to cautiously take a reasonable portion of the timing margin to assess the HCI degradation effects at a product level. For example, three different cases in Fig. 10 start to fail at 13.1%, 9.7%, and 10.6% HCI degradations for cases A, B, and C, respectively, when 25% of the timing margin is assumed to be allocated for HCI degradation.

Figure 11 shows the timing margin changes over the change in the ADF values due to the HCI degradations. The difference in timing margin for the three different technologies was not clearly observable in Fig. 10. However, the changes in timing margin are more clearly observable with the change in ADF values as shown in Fig. 11. We observed that the changes in ADF value have a close relationship to the slope of the falling word-line signal; the slope of the falling word-line signal is correspondingly closely related to the timing margin shift.

Case B has the steepest slope in Fig. 11, which can imply an initial conclusion that case B will likely fail first due to HCI if the ADF change rates for all three cases are assumed to be the same. However, Fig. 4 shows that case B had the least ADF change over the HCI degradation. The two factors need to be considered at the same time for proper evaluations.

It is also interesting to see the effects on the timing margin with the PDT. Figure 12 shows the timing margin changes as the drainage through the PUWD increases. The general trend in Fig. 12 shows that the timing margin is enhanced as the drainage through PUWD increases. This is because there is less charge through PDWD, and PDWD is less stressed because of reduced effective V_{CCP} when it turns on. Please note that the time in Fig. 12 is only for the fall time starting when PDWD is enabled; it does not include the time through the drainage through PUWD. In the case experiments, the duration through PUWD was larger than the time benefit in PDWD draining. So, from a timing perspective, there is no benefit in using PMOS PUWD; however, the drainage through PMOS PUWD provides a large benefit in the ADF value and can reduce HCI risks.

V. Conclusions

HCI risk assessment is overly pessimistic when it is evaluated solely by device HCI figures. Product level HCI qualification should consider the real-time application of the devices. The AC-DC factor (ADF) can enable the AC lifetime estimation. It inherently considers device DC lifetime, application and circuit structures. The ADF was most influenced by reducing effective operating V_{CCP} voltage. The timing margin in word-line signal was more related to the change in ADF than the change in degradation. As a result, ADF can serve in evaluating the HCI risk factor at a product level, because it can be related to device HCI degradation, as well as effective V_{CCP} voltage level, circuit operation, and circuit design structures.

REFERENCES

[1] J. Choi et al, "Hot-Carrier-Induced MOSFET Degradation Under AC Stress," IEEE Electron Device Letters vol. EDL-8, no. 8, Aug. 1987

[2] C. Hu et al., "Hot-electron-induced MOSFET degradation – Model, monitor, and improvement," IEEE Trans. Electron Devices, vol. ED-32, p. 375, Feb 1985

[3] Youn et al., "Semiconductor memory device having split word line driver circuit with layout patterns that provide increased integration density," U.S.A patent 7,729,195, June, 2010

[4] B. Keeth et al., "DRAM circuit design – a tutorial," IEEE Press 2001

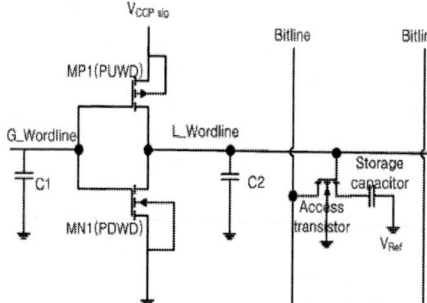

Fig. 1: Word-line driving circuit structure and DRAM cell used for simulation works.

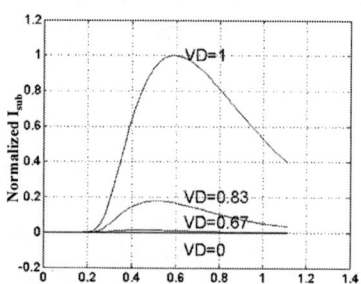

Fig. 2: One of three I_{sub} current measurement data used in the simulations with different gate (X-axis) and drain voltages (VD): I_{sub} current is normalized over the peak current.

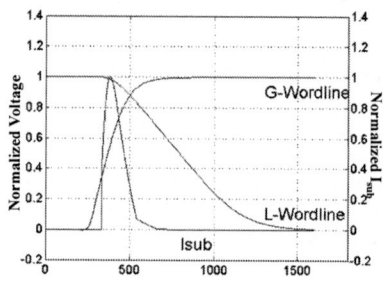

Fig. 3: Waveforms of global and location word-lines and I_{sub} current when a word-line is deactivated; I_{sub} exists only during the switching period.

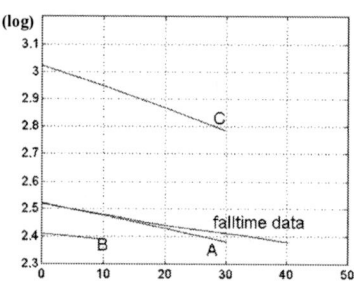

Fig 4: The ADF factor changes with the I_{DSAT} degradations in percentage at the driving PDWD. The plot *falltime_data* shows ADF values with the fall time degradation. (X-axis: PDWD degradation; fall time degradation for the plot *falltime data*; Y-axis: ADF)

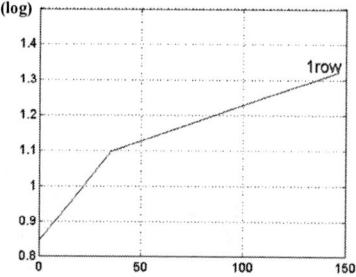

Fig. 5: Failing time with t_{RP} degradation with one row experimental data up to 500 hours with elevated V_{CCP} voltage stress. (X-axis: t_{RP} degradation in percentage; Y-axis: time to take the degradation)

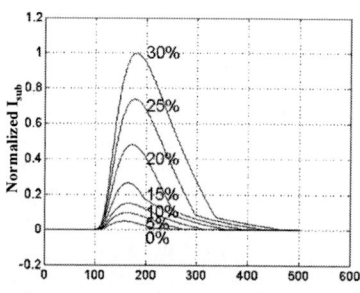

Fig. 6: I_{sub} reduction at PDWD with the PUWD drainage from 5% to 30% before PDWD turns on.

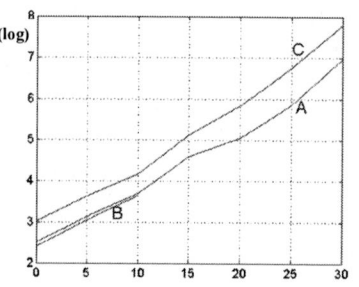

Fig. 7: ADF changes with PUWD drainage technique. (X-axis: reduced voltage percentage through PUWD; Y-axis: ADF values)

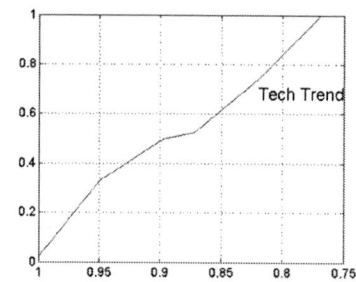

Fig. 8: Transistor I_{DSAT} 10% degradation DC lifetime with V_{CCP} voltage changes in different technologies. (X-axis: normalized V_{CCP}; Y-axis: normalized DC lifetime)

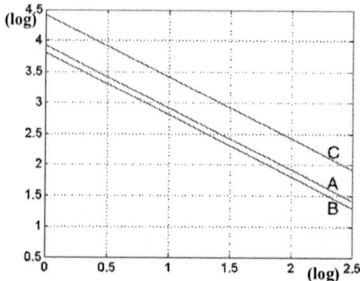

Fig. 9. ADF changes with access rate to a word in MHz.

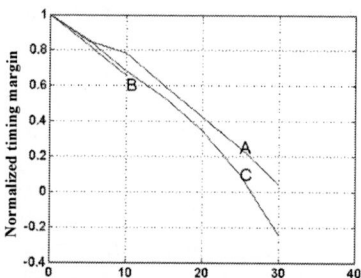

Fig. 10: Timing margin changes with PDWD degradations. (X-axis: PDWD degradation in percentage; Y-axis: normalized timing margin)

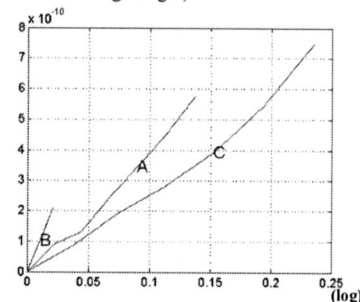

Fig. 11: Timing margin shift in seconds over the changes in ADF due to PDWD degradations.

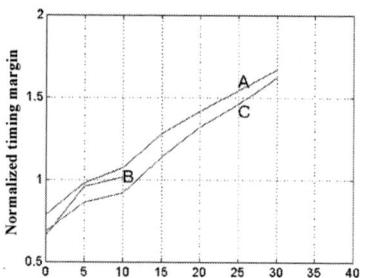

Fig. 12: Timing margin enhancement with PUWD in Fig. 1 when the I_{DSAT} degradation of PDWD is 10%. (X-axis: PUWD drainage in percentage; Y-axis: normalized timing margin)

STI stress-induced degradation of data retention time in DRAM and a new characterizing method for mechanical stress

Tae-Su Jang, Kyung-do Kim, Min-Soo Yoo, Yong-Taik Kim, Seon-Yong Cha, Jae-Goan Jeong, and Sung-Joo Hong

R&D Division, Hynix Semiconductor Inc.,
San 136-1 Ami-ri, Bubal-eub, Ichon-si, Kyoungki-do, 467-701, Korea
Tel) +82-31-639-0814, Fax) +82-31-639-0734, E-mail) taesu.jang@hynix.com

Abstract—The effect of mechanical stress induced by shallow trench isolation (STI) slope on the data retention characteristics of DRAM is investigated and a new electrical parameter for monitoring the mechanical stress is proposed. To maintain high and uniform retention time for the reliable operation of DRAM, the STI slope should not be vertical and should be kept below 86-degree. The new electrical parameter measures the current gain of the parasitic BJT in DRAM cell and shows a strong correlation with the retention time induced by the mechanical stress.

Keywords- Mechanical stress, shallow trench isolation (STI), Data retention time, DRAM

I. INTRODUCTION

Long and uniform data retention time within a wafer should be satisfied at the same time to guarantee low power consumption and reliable chip operation in DRAM. With the shrink of feature size, the retention time tends to degrade while the mechanical stress induced by shallow trench isolation (STI) increases. Therefore, it is necessary to evaluate the effect of STI on the retention time and then apply this learning to the development of next generation DRAM.

II. EXPERIMENTAL

1Gbit DRAM devices were fabricated using a 60nm technology featured with 8F2 and capacitor over bit line (COB) cell structure. The access transistor of DRAM is a recessed gate (RG) type device with an n+ poly/W stack. The formation of STI was processed as follows. After defining the active region using the Si_3N_4 hard mask, a Si trench region was formed with a 300nm depth. Then, an 8nm thick oxide was grown on the side wall of the Si trench and a 5nm thick Si_3N_4 liner was deposited sequentially. After that, the trench was filled with spin-on glass (SOG) and an etch-back process was followed to the depth of 180nm from the bottom Si. Finally, the trench was filled with HDP oxide and planarized with a CMP process. Data retention time was measured in the main chip array while electrical parameters of the cell transistor were measured using test patterns formed in the scribe lane region.

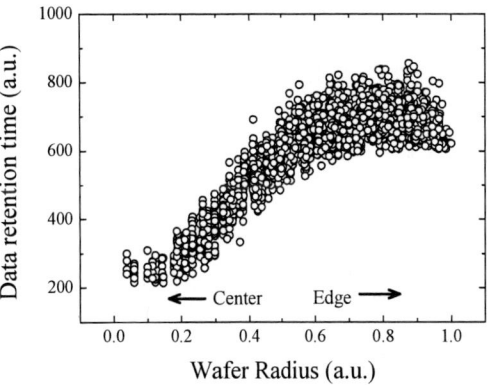

Figure 1. Data retention time of 1Gb DRAM as a function of chip position within a wafer.

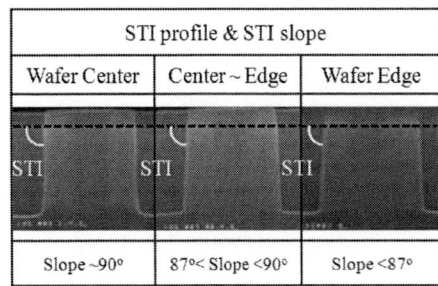

STI profile & STI slope		
Wafer Center	Center ~ Edge	Wafer Edge
STI	STI	STI
Slope ~90°	87°< Slope <90°	Slope <87°

Figure 2. Cross-sectional TEM images of the DRAM cell region after the STI process.

III. RESULTS AND DISCUSSION

Fig. 1 shows the data retention time as a function of chip position within a wafer. The retention characteristics between the center and the edge of the wafer are quite different and the retention time degrades as the chip moves toward the center of the wafer.

Fig. 2 shows the TEM images of devices after the STI process. The STI slope is the steepest at the center and decreases as the device moves toward the edge. The steeper slope of STI induces a larger mechanical stress in the active region [1]. However, the mechanism must be understood and

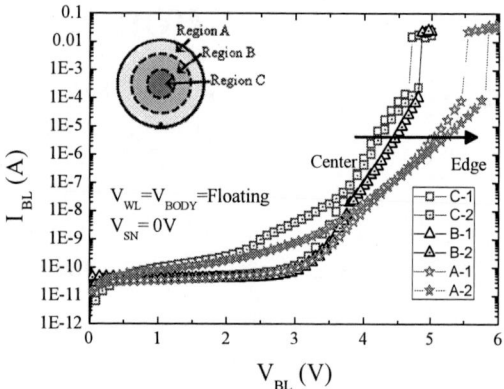

Figure 3. I_{BL}-V_{BL} curves for the devices from three different regions when the WL and Body are floating.

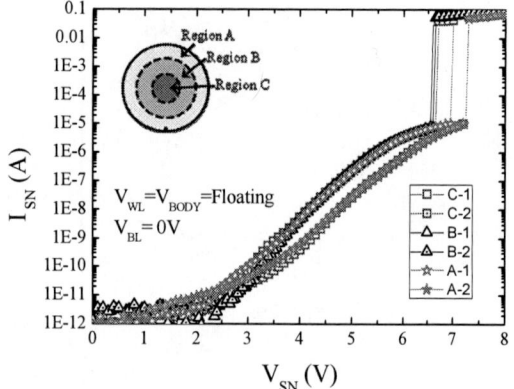

Figure 4. I_{SN}-V_{SN} curves for the devices from three different regions when the WL and Body are floating.

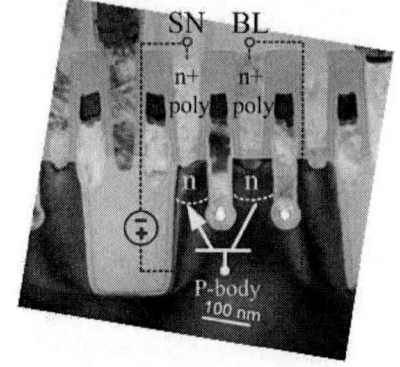

Figure 5. Test configuration for measuring Gummel characteristics of the parasitic BJT of a DRAM cell.

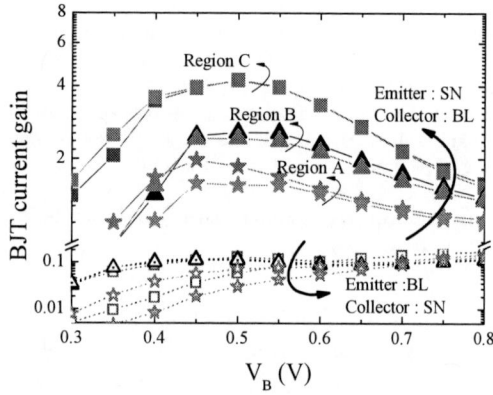

Figure 6. Current gain as a function of body voltage (V_B) for the cases of the SN as an emitter (solid) and the BL as an emitter (open).

an electrical parameter must be found which correlates well with the stress.

The wafer in Fig. 1 was classified into three regions according to the radius as shown in the inset of Fig. 3. Two different devices in each region were chosen to be investigated further in terms of electrical characteristics. Devices A-1 and A-2 were chosen in region A, B-1 and B-2 were chosen in region B, while C-1 and C-2 were chosen in region C.

Fig. 3 shows the I_{BL}-V_{BL} for the devices when the word line (WL) and body are floating electrically. It is seen that the parasitic BJT turns on as the V_{BL} increases from the abrupt increase of I_{BL} and the voltage for requiring the BJT turn-on decreases as the device moves toward the center from the edge of the wafer. However, when the storage node (SN) is swept instead of the BL, the difference of BJT turn-on voltages according to the radius is not found as depicted in Fig. 4.

To further investigate the parasitic BJT in the DRAM cell, the Gummel characteristics were evaluated using the test configuration shown in Fig. 5. For the case of the SN as an emitter and the BL as a collector, the body and the BL are tied together. Then, the body current and the collector current (I_{BL})

are monitored while positive biases are being applied to the body and the BL. As for the other case, where the BL is an emitter and the SN is a collector, the body and the SN are tied together. Then, the body current and the collector current (I_{SN}) are monitored while positive biases are being applied to the body and the SN.

Fig. 6 shows the current gain of the parasitic BJT as a function of the body voltage (V_B) for the cases of the SN as an emitter (solid symbol) and the BL as an emitter (open symbol). The BJT current gain is the highest for the devices at the center of the wafer and decreases as they move toward the edge for the case of the SN as an emitter while there is no noticeable difference of the current gain for the case of the BL as an emitter. Thus, it shows a strong correlation ($R^2 = 0.93$) between the maximum current gain and the retention time for the case of SN as an emitter while it shows a weak correlation ($R2 = -0.1$) between the maximum current gain and the retention time for the case of the BL as an emitter as shown in Fig. 7.

In order to investigate a mechanism of a strong correlation between the current gain and the retention time for the case of the SN as an emitter, the behaviors of the collector current (I_{BL})

978-1-4244-9113-1/11 $26.00 © 2011 IEEE

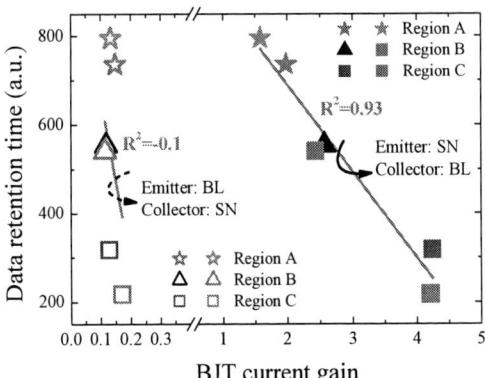

Figure 7. Data retention time versus the current gain of the parasitic BJT for the SN as an emitter (solid) and the BL as an emitter (open).

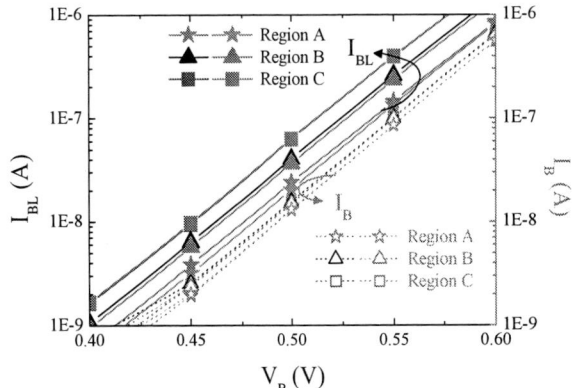

Figure 8. The I_{BL} and the body current (I_B) as a function of V_B for the case of SN as an emitter.

and the body current (I_B) as a function of V_B are examined and shown in Fig. 8. The I_{BL}'s increase as the devices move toward the center of a wafer while I_B's are nearly the same regardless of the position in a wafer. This suggests that the difference in the current gain (I_{BL}/I_B) originates from the diffusion current in the SN-body junction because the emitter current (I_{SN}) is equal to the sum of the collector current (I_{BL}) and the base current (I_B). The effect of mechanical stress on the diffusion current was investigated quantitatively. The equation of the forward diffusion current for an n+-p junction is given as

$$I = qA(\frac{D_p}{L_p}p_n + \frac{D_n}{L_n}n_p)e^{qV/kT} \approx qA\frac{D_n}{L_n}n_p e^{qV/kT} \quad (1)$$

where q, A, D_n, L_n, V, k, and T are electronic charge, area of the junction, the diffusion coefficient of electron, the electron diffusion length, forward bias, Boltzman's constant, and absolute temperature. n_p, the electron concentration in p-type Si, is given by

$$n_p = N_C e^{-(E_{CP}-E_{FP})/kT} \quad (2)$$

where Nc, E_{CP}, and E_{FP} are are the effective density of state in the conduction band, the conduction band level of P-type Si, and the Fermi level of P-type Si, respectively. The built-in potential V_{bi} for the n+-p junction is given by [2]

$$V_{bi} \approx \frac{E_{CP} - E_{FP}}{q} = \frac{E_g}{2q} + \frac{kT}{q}\ln\frac{N_A}{n_i} \quad (3)$$

where E_g, N_A, and n_i are the energy band gap, the doping concentration of p-type Si, and the intrinsic carrier concentration. Combining the above three equations, the following formula is obtained.

$$I = qA\frac{D_n}{L_n}N_C e^{q(V-V_{bi})/kT} \quad (4)$$

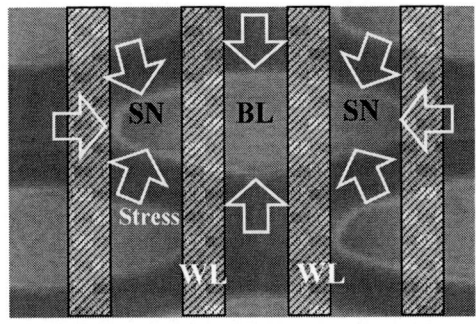

Figure 9. The plane view of an active area and STI of the DRAM cell.

It is known that a mechanical stress induces band gap lowering [3]. This results in a decrease in the built-in potential as can be expected in equation (3). Therefore, it is concluded that the mechanical stress enhanced by the steeper STI slope increases the diffusion current of the SN junction by lowering the built-in potential which increases the current gain of the parasitic BJT in the DRAM cell.

For the case of the BL junction, however, the active area surrounded by STI is larger than the SN area and mechanical stress is only applied to two sides of the active area while it is applied to the three sides for the SN junction as illustrated in Fig. 9. Thus, the stress for the BL junction is smaller than that for the SN junction and this results in different behaviors as discussed in Fig. 4, 6, and 7.

To confirm the effect of STI slope on the retention time, three wafers having different STI profiles were fabricated as shown in Fig. 10. While wafer A has the same profile as in Fig. 2, wafer B has a relatively gentle STI slope and wafer C has a vertical STI slope regardless of position within the wafer.

Data retention characteristics for the three wafers are plotted as a function of chip position in Fig. 11. Wafer B which has a gentle STI slope below than 86-degree in any position within the wafer exhibited high and uniform retention time as

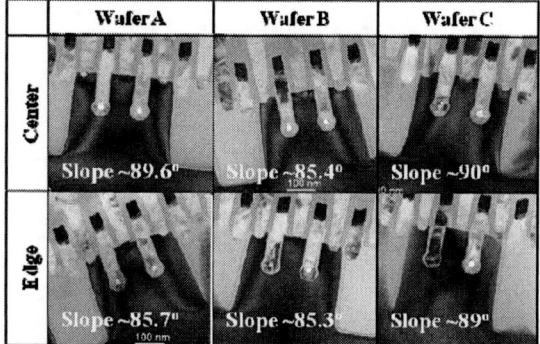

Figure 10. The cross-sectional TEM images for three wafers having different STI profiles.

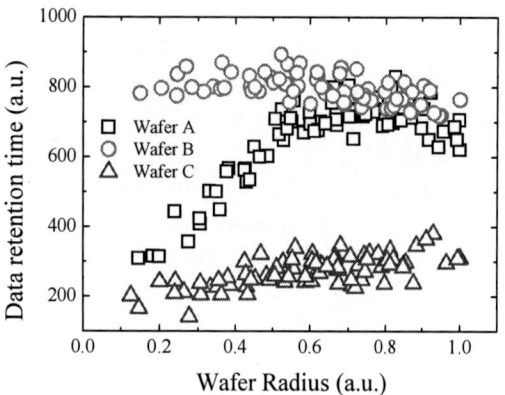

Figure 11. Data retention time as a function of chip position for three wafers having different STI profiles.

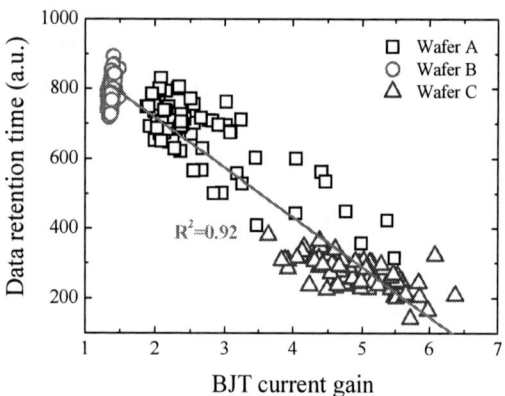

Figure 12. Data retention time versus the current gain for three wafers having different STI profiles.

expected. Wafer C showed the lowest retention time regardless position in the wafer because it has a steep and uniform slope everywhere in the wafer.

Fig. 12 shows the data retention time as a function of the maximum current gain of the parasitic BJT for the three wafers. The current gain was extracted from the Gummel

characteristics of the parasitic BJT using the same test configuration shown in Fig. 5. It has a strong correlation ($R^2=0.92$) between the data retention time and the maximum current gain. Therefore, the current gain is a good electrical parameter to monitor a mechanical stress induced by the STI process.

IV. CONCLUSION

The data retention time of DRAM is strongly dependent on the STI slope and it degrades as the slope goes vertical. To maintain high and uniform retention time for the reliable operation, the STI slope should not be vertical and should be kept below 86-degree. The current gain of the parasitic BJT shows a strong correlation with the retention time induced by the mechanical stress. The higher the mechanical stress, the greater the current gain. Thus, the current gain of the parasitic BJT is a useful parameter for measuring mechanical stress.

REFERENCES

[1] Rui Li et al., "A comprehensive study of reducing the STI mechanical stress effect on channel-width-dependent Idsat," Semicond. Sci. Technol. 22, 2007, pp. 1292-1297.

[2] V. P. Gopinath et al., "STI Stress-Induced Increase in Reverse Bias Junction Capacitance," IEEE Electron Deivce Letters, Vol. 23, No. 6, June 2002. pp. 312-314.

[3] P. Smeys et al, "Influence of process induced stress on device characteristics and its impact on device scaling," IEEE Trans. Electron Devices, vol. 46, pp. 1245-1252, June 1999.

Hot Hole Induced Damage in 1T-FBRAM on Bulk FinFET

M. Aoulaiche, N. Collaert, A. Mercha, M. Rakowski, B. De Wachter, G. Groeseneken, L. Altimime and M. Jurczak

Imec, Kapeldreef, 75, 3001
Leuven, Belgium
Phone: +32 16 28 87 96, e-mail address: *Marc.Aoulaiche@imec.be*

Z. Lu
University of Florida
Gainesville, FL 32611-6130 USA

Abstract— The reliability of a one Transistor Floating Body Random Access Memory (1T-FBRAM) bulk FinFET cell using Bipolar Junction Transistor (BJT) programming is investigated. It is shown that hot holes generated by impact ionization create interface defects close to the drain and positively charged oxide traps, especially at high transverse electric field. These created defects degrade the cell endurance. Moreover, this degradation is enhanced for shorter channel devices and narrower fin widths, which would be a limitation for the scaling of floating body RAM.

Keywords-Floating body cell; RAM; endurance; cycling; BJT; FinFET.

I. INTRODUCTION

One transistor capacitor-less random access memory (1T-RAM) is considered as a candidate to replace the conventional one transistor and one capacitor 1T/1C DRAM, which suffers from the scaling challenges related to the capacitor integration [1,2]. Various device architectures are considered: bulk, Silicon On Insulators (SOI), double gate, surrounded gate etc [3, 4, 5] and different biasing schemes are studied. With regard to the state-1 programming method, three groups can be distinguished: impact ionization, bipolar junction transistor and band-to-band tunneling [6]. One the other hand, with regard to the read method two groups can be noticed in the first group (Gen1), the floating body charge induces a threshold voltage shift, which changes the MOSFET current. The second group (Gen2) [4] uses the bipolar transistor present in the MOS structure. The read is performed by sensing the BJT current. However, to be viable the 1T-RAM candidate has to satisfy conditions such as: high scalability, low intrinsic variations, high programming speed, high sense margin, long retention time and good endurance. The BJT programming mode was reported to improve the retention, giving a high sense margin, which provides fast read and better scalability [1]. Endurance and reliability are still not reported. Therefore, this paper focuses on the reliability study of the bipolar programming method on bulk FinFET devices, which have the advantage to be highly scalable [7].

This paper is organized as follow: in the second section, the device fabrication and the experimental setup are described. In sec. III, the BJT programming method is discussed. In sec. IV, the 1T-RAM reliability is discussed taking into consideration the cycling failure, hot carrier stress and charge pumping current measurements. Finally, also the impact of the gate length (L) and the fin width (W_{Fin}) are investigated.

II. DEVICE FABRICATION

The devices are fabricated on bulk Si-substrate with doped ground plane. Fin widths down to 10nm and a fin height H_{FIN}=60nm were made using 193nm lithography. The gate electrode consists of a 5nm SiO_2 capped with 5nm PE-ALD TiN and 100nm poly. After gate patterning, the extensions were implanted and the nitride spacer was formed. No selective epitaxial growth (SEG) was done on the source/drain areas. A NiPt-based salicide process was used after the deep S/D implants and spike anneal. Finally, a standard Cu back-end-of-line process was used to finish the devices [3].

The measurements of the Floating Body Random Access Memory (FBRAM) were performed by applying short pulses at the drain and gate terminals, while keeping the bulk contact floating. The source current is measured during the read pulse using a current amplifier. All the devices measured consist of five fins.

III. OPERATING CONDITIONS

The floating body memory effect is observed in the bistable BJT state, as shown in Fig.1. During the gate bias (V_G) forward sweep at high drain bias (V_D) and when the gate bias is close to the transistor threshold voltage (V_{tf}), holes are generated by impact ionization near the drain. These holes are injected into the substrate and raise the body potential, and then the parasitic BJT is turned on. During the V_G sweep-back the holes injected by impact ionization keep the BJT current on (state-1) until the positive feedback loop between the impact ionization current and the source-bulk junction forward bias cannot be sustained. Hence, below $V_G = V_{tb}$ the BJT current turns off (state-0).

978-1-4244-9113-1/11 $26.00 © 2011 IEEE

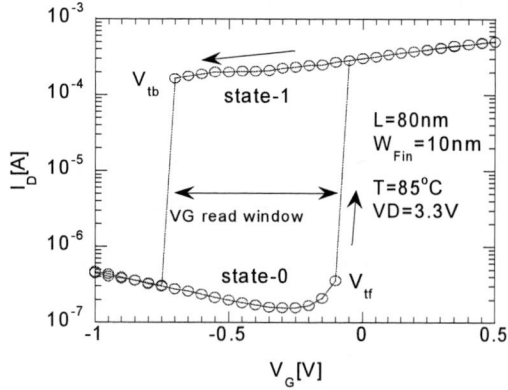

Fig.1. Double sweep I_D-V_G measured on FinFET with L=110nm, W_{Fin}=20nm and at T=25°C, showing the BJT current off (state-0) and on (state-1).

The operating biases are shown in Fig.2. To read, V_D is set high to trigger the parasitic BJT and V_G is defined within the hysteresis window (V_{tf}-V_{tb}). The read drain current follows the expression [7],

$$I_{D,read} = M(I_{ch} + I_{BJT}) = \frac{M}{1-\beta(M-1)} \; I_{ch} \qquad (1)$$

where M is the impact ionization factor, β is the current gain of the BJT, I_{ch} is the channel current and I_{BJT} is the BJT current.

To write state-1 (write-1), the substrate is charged by holes generated by impact ionization using a high V_D and V_G close to the threshold voltage, satisfying the condition turning on the BJT

$$\beta(M-1)\sim1 \qquad (2)$$

To write state-0 (write-0), the holes are removed from the channel by forward biasing the drain-substrate junction.

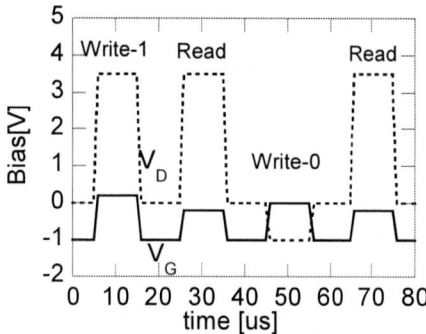

Fig.2. Schematic of the operating biases applied to the cell during write, read and cycling. V_D is the voltage applied to the drain and V_G to the gate. The substrate is left floating.

Fig.3 shows the state-0 and 1 as a function of the retention time measured using the programming biases shown in Fig.2.

As expected from the BJT programming, a high sense margin and long retention time are observed [4].

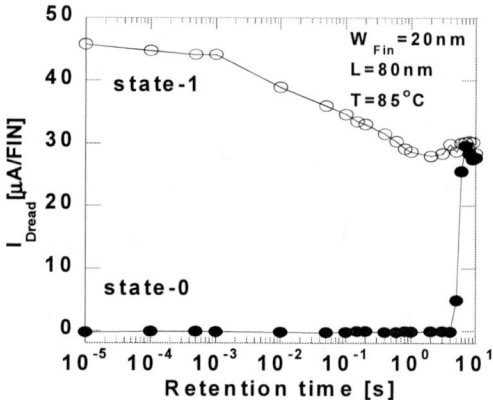

Fig.3. State -0 and 1 as function of the retention time showing high sense margin and long retention time.

IV. ENDURANCE

Endurance measurement are performed by applying a repetitive cycle to the transistor and recurrently after a certain number of cycles the read current for the state-0 and 1 is measured. In this study, one cycle corresponds to a sequence write-1/read-1 and write-0/read-0, where read-0 and read-1 are the BJT currents measured for the state-0 and 1, respectively. The cycling failure is extracted when either the state-0 or 1 shifts by 50% ΔI_S, with ΔI_S is the measured current difference between the state-0 and 1 before cycling.

1) Gate length effect

The write-1 and read bias use the parasitic n-p-n (source-substrate-drain) bipolar transistor current, which is triggered when the expression (2) is satisfied, as mentioned previously. Fig.4 shows the V_D applied during write-1 and subsequent read as a function of the gate length. It is decreased as L is decreased. This is consistent with larger β and M in shorter gate lengths [8]. The decrease of the V_D write bias with the gate length follows the same trend as the common-emitter break down voltage with open base (BV_{CEO}), as shown in Fig.4 by the continuous line [9]. On the other hand, the BJT operation is limited by the short channel effect for shorter L and by the break down for longer L [9]. For very short channels, the source-drain punch-through occurs and the channel cannot be controlled. In this case, the current always flows between the source and drain, even at low V_D. For long channel devices over 130nm, to reach the maximum lateral field needed to induce the impact ionization, a high V_D is required (> 4V) as seen in Fig.4. Therefore, the high transverse field between the drain and gate causes the device breakdown.

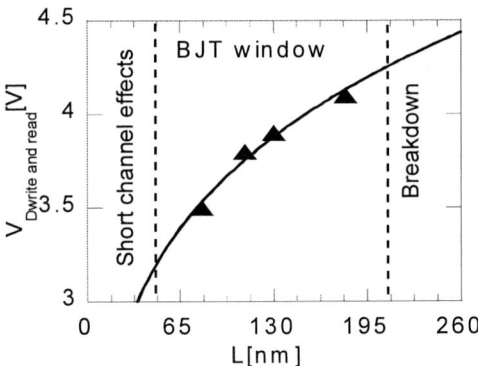

Fig.4. Gate length dependence, of the V_D bias used for the write and read with BJT programming.

Fig.5 shows the cycling failure behavior as a function of the gate lengths, for a fixed fin width of 20nm. Three different regimes are observed as a function of L. In the first regime, the number of cycles to failure increases with L till an optimum, here L=130nm for W_{Fin} of 20nm. In the second regime, the number of cycles decreases. In the third regime, the device breaks after a few cycles.

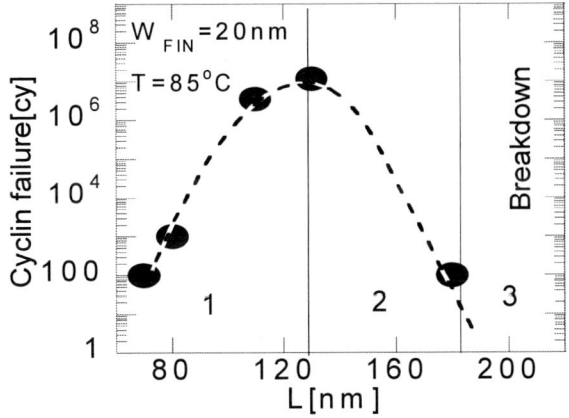

Fig.5. Cycling failure extracted a 50% ΔI_S shift and for different gate lengths. 1 cycle corresponds to the scheme in Fig.2.

In Fig.6, 7 and 8, three different cycling failure types are also observed. The cycling failure can be caused by the state-0 degradation, as shown in Fig.6. This failure is due to the BJT off-current or to the increase of the subthreshold current and it occurs for short channel devices, typically below L=90nm. For L higher than 90nm and below 130nm, the cycling failure is due to the state-1 failure, as shown in Fig.7. In this case, the degradation causes the BJT on-current decrease. For L higher than 130nm both states 0 and 1 fail, as shown in Fig.8.

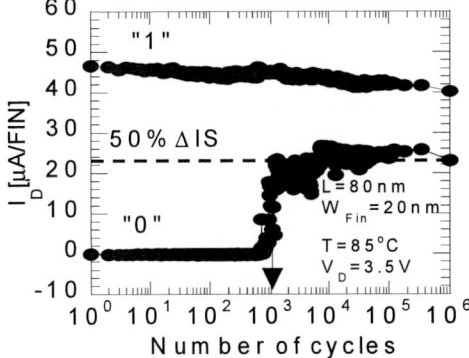

Fig.6. State-1 and 0 shifts as a function of the number of cycles, showing the state-0 degradation.

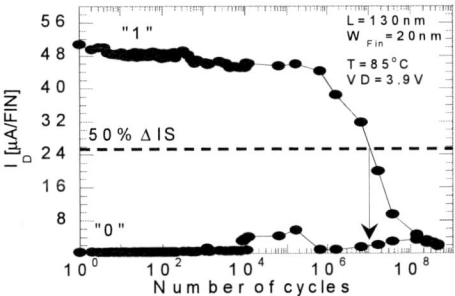

Fig.7. State-1 and 0 shifts as a function of the number of cycles, showing the state-1 degradation

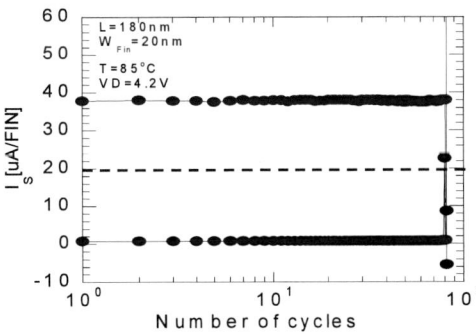

Fig.8. State-1 and 0 measured as a function of the number of cycles, showing the state-0 and 1 failure.

To gain insight into the cycling mechanisms, constant voltage stress and charge pumping currents are used. In order to identify the degrading condition the device is exposed to a constant voltage stress at (V_G=0V, V_D=-2V) and at (V_G=0v, V_D=3.2V) corresponding to the write-0 and write-1 condition, respectively. No significant impact of the stress under the write-0 is shown in Fig.9. I_D-V_G characteristic measured during stress showed a small shift towards more positive V_G indicating a negative charge generated during stress. Conversely, a large shift in I_D-V_G characteristics to more negative V_G is observed under the stress in the write-1 condition, as shown in Fig.9,b. This reveals that the dominant degradation is generated during the write-1 condition, where impact ionization is used. The shift of the I_D-V_G characteristics to more negative V_G indicates positive charge generation. Most probably this related to hot holes induced damage [10].

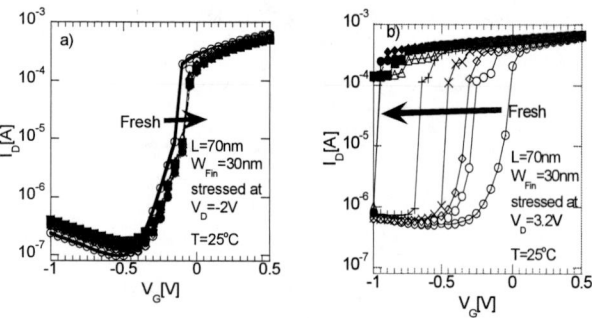

Fig.9. I_D-V_G characteristics measured at V_D=3.2V and at different stress times, a) stressed in the write-0 condition (V_G=0V, V_D=-2V) and b) stressed in the write-1 condition (V_G=0V, V_D=3.2V)

To assess the Si/SiO$_2$ interface degradation during write-1, the charge pumping current is measured before and after stress. In addition, during measurements after the stress the source or drain is disconnected in order to identify where the defects are generated, either close to the drain or source [11]. Interface states generation is observed as shown by the increase of the charge pumping current in Fig.10. Moreover, when the drain contact is disconnected during measurements after the stress all the defects close the drain did not contribute to the measured current (Fig.10). However, when the source contact is disconnected no difference is observed (Fig.10). Accordingly, the interface defects generated are close to the drain, which is consistent with impact ionization occurring in the depletion region close to the drain [12].

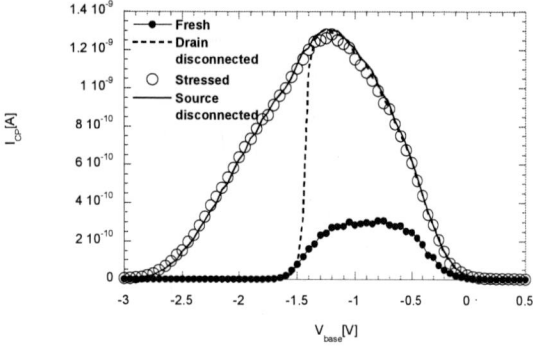

Fig.10. Base level sweep charge pumping current measured before and after stress under high drain bias of a device with 65nm height, W_{Fin}=0.25μm and L= 1μm at 25°C. Disconnecting the drain after stress shows that interface defects are likely generated near the drain.

Furthermore, the normalized charge pumping current before and after stress showed a slight shift in the CP curve to more

negative base level voltage, which is likely related to positively charged oxide traps, either filled or generated by hot holes injection. Indeed, an increase in the current of holes tunneling to the gate can be observed during the hysteresis measurements, as shown in Fig.11.

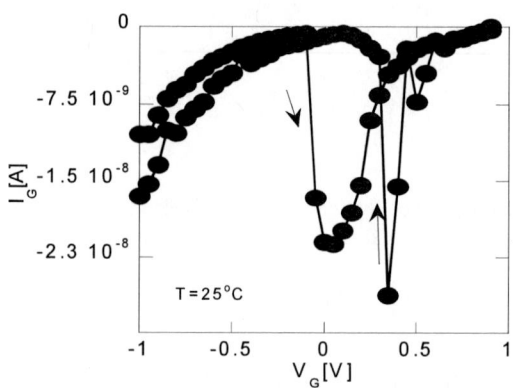

Fig.11. Gate current measured as a function of V_G during the hysteresis measurement as shown in Fig.1, showing hole tunneling to the gate

Fig.12 illustrates the defects generated during cycling and their location. During write-1, it is considered that hot holes generated by impact ionization under high lateral field cause the interface defects close to the drain. For a high transverse electric field between the drain and the gate, most probably some hot holes are injected into the dielectric and generate the positively charged oxide traps.

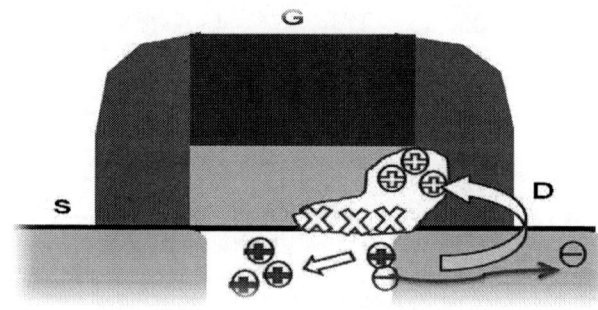

Fig.12. schematic of the defects generated during cycling

For the reason that the generated defects are located close to the drain, they further increase the drain-induced barrier lowering (DIBL) in the short channel devices. This results in a much larger thereshold voltage (V_{th}) decrease in short channel devices, as shown in Fig.13. As the read is done at a fixed negative V_G (Fig.2), the subthreshold current measured during read-0 increases and induces the cycling failure shown in Fig.6. When the gate length increases, the impact on DIBL decreases exponentially and a lower V_{th} shift is observed. However, the devices break earlier during stress for longer L, as shown in Fig.13. In the long channel devices, the V_{th} shift is mainly related to the variation induced by the positively charge defects on the gate bias. As a result, the variation of the read-0 current is negligible (Fig.7).

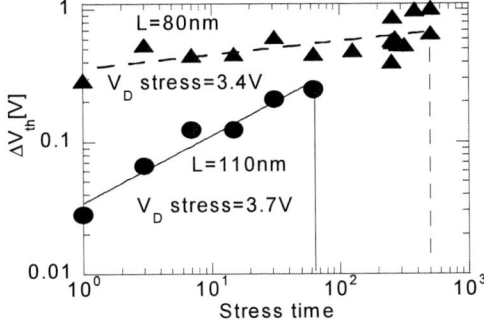

Fig.13. Threshold voltage shift as a function of the stress time measured at room temperature and extracted at high V_D stress. The degradation is enhanced for shorter devices.

Instead, for an increased number of cycles the reduction of the state-1 current (the BJT current) induces the failure. This is due to the degradation of the BJT current gain β and it is caused by hot holes, as illustrated previously in Fig.12 [13, 14]. The decrease of the number of cycles in the second regime for increasing L can be explained by the high transverse electric field between the drain and the gate. For increased V_D to program, holes generated by impact ionization close to the drain gain more energy to cross the SiO_2 potential barrier and generate oxide defects. Consequently, the cycling number is reduced with the increase of L. The change of the cycling behavior, which is seen in Fig.8 for L=130nm or longer, indicates that hot holes tunneling from the drain to the gate is dominant.

2) Fin Width effect

It has been reported that the reduced floating body effects in devices with narrow channel is related to the dopant out diffusion resulting in the carrier lifetime reduction along the channel edges [15]. In the device investigated here, no impact of the fin width is observed. The V_D at which the BJT current is triggered, is similar for the different fin widths shown in Fig.14.

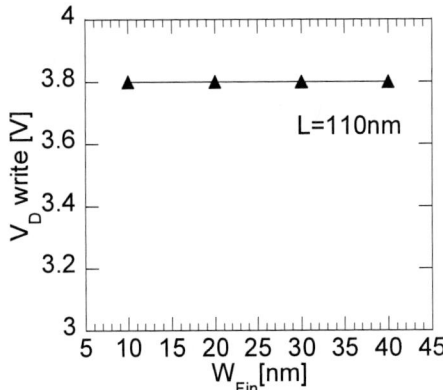

Fig.14. V_D write bias as a function of the fin width and for fixed L=110nm.

Cycling is measured at 85°C for L fixed to 110nm and W_{Fin} 20, 30 and 40nm, as shown in Fig.15. As a comparison devices with W_{Fin}=90nm and L=90nm are used. The number of cycles to failure is increased with the increase of W_{Fin}. However, a saturation trend is expected from Fig.15. The cycling behavior is consistent with the V_{th} shift measured under hot carrier stress (see Fig.16). A larger V_{th} shift is extracted for narrower fins during stress at the same bias conditions. The main reasons for the increased degradation in narrower fins are the corner effect and the stress induced by the shallow trench isolation.

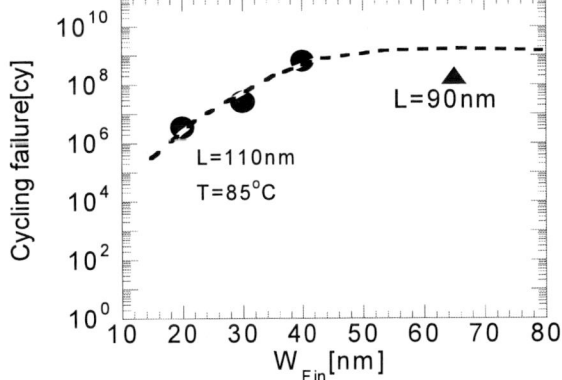

Fig.15. Cycling failure extracted a 50% ΔI_S shift as a function of the fin width.

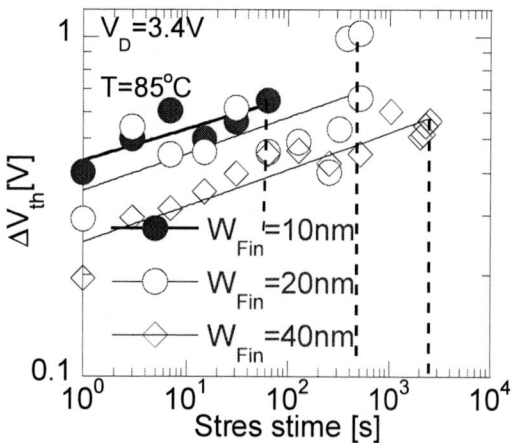

Fig.16. Threshold voltage shift as a function of the stress time measured at room temperature and extracted at high V_D stress, showing higher degradation for narrower fins.

V. CONCLUSIONS

The cycling reliability of one transistor floating body cells on bulk FinFET is studied. It is found that the hot holes generated by impact ionization to write the state-1 create interface defects and positive oxide charges close to the drain. These defects degrade either the MOFET current (state-0 current) or the BJT current (state-1 current) depending on the gate length. It is also shown that the cycling failure is more severe for narrower fin widths. This degradation is inherent to the programming mode and can thus be a possible show stopper for the cell scalability.

ACKNOWLEDGMENT

This work is supported by imec's partners and core partners on the emerging devices research program: Intel, Micron, Panasonic, Samsung, TSMC, ELPIDA, Hynix, PSC, NXP, Sony, Qualcomm, ST.

REFERENCES

[1] S. Okhonin, M. Nagoga, J. M. Sallese, and P. Fazan," A capacitor-less 1T-DRAM cell", Elec. Dev. Let., 02, Vol. 23, No. 2, pp. 85-87.

[2] R. Ranica, A. Villaret, P. Malinge, G. Gasiot, P. Mazoyer, P. Roche, P. Candelier, F. Jacquet, P. Masson, R. Bouchakour, R. Fournel, J. P. Schoellkopf, and T. Skotnicki," Scaled 1T-Bulk devices built with CMOS 90nm technology for low-cost eDRAM applications", in VLSI, 2005, pp. 38–39.

[3] N. Collaert, M. Aoulaiche, B. De Wachter, M. Rakowski, A. Redolfi, S. Brus, A. De Keersgieter, N. Horiguchi, L. Altimime and M. Jurczak," A low-voltage biasing scheme for aggressively scaled bulk FinFET 1T-DRAM featuring 10s retention at 85°C" in VLSI, 2010, p-p. 161 – 162.

[4] S. Okhonin, M. Nagoga, E. Carman, R. Beffa, E. Faraoni," New Generation of Z-RAM", in IEDM, 2007, pp. 925-926.

[5] N. Z. Butt, M. A. Alam ,"Scaling Limits of Double-Gate and Surround-Gate Z-RAM Cells", TED, 2007, Vol. 54, NO. 9, p-p. 2255-2262

[6] M. Bawedina, S. Cristoloveanub, D. Flandrec, F. Udrea," Floating-Body SOI Memory: Concepts, Physics and Challenges", ECS Trans. 2009, 19 (4) 243-256.

[7] N. Collaert, M. Aoulaiche, M. Rakowski, A. Redolfi, B. De Wachter, J. Van Houdt, and M. Jurczak," Optimizing the Readout Bias for the Capacitorless 1T Bulk FinFET RAM Cell", EDL, 2009, Vol. 30, No. 12, pp. 1377-1379.

[8] J-Y Choi, J. Fossum,"Analysis and control of floating-body bipolar effects in fully depleted submicrometer SOI MOSFET's", TED, 1991, Vol.38, N.6,p-p1384-1391.

[9] S.M. Sze, 2nd Edition, New York, Wiley & sons, 1981, pp.484-485.

[10] K.R. Mistry, D.B. Krakauer, and B.S. Doyle,"Impact of snapback-induced hole injection on gate oxide reliability of N-MOSFETs", EFL. 1990, Vol11, No.10, p-p.460-462.

[11] P. Hermans, J. Witters, G. Groeseneken, H. Maes," Analysis of the charge pumping technique and its application for the evaluation of MOSFET degradation", TED. 1989, Vol. 36, NO. 7.p-p.1318-1335.

[12] M. Aoulaiche, N. Collaert, R. Degraeve, Z. Lu, B. De Wachter,G. Groeseneken, M. Jurczak, L. Altimime," BJT-Mode Endurance on a 1T-RAMBulk FinFET Device" EDL, VOL. 31, N. 12, 2010, p-p.1380-1382

[13] J.D. Burnett, and Ch. Hu, TED.,88,Vol. 35,No12,p-p.2238-22

[14] A. Neugroschel, C-T. Sah, M. S. Carroll," Degradation of bipolar transistor current gain by hot holes during reverse emitter-base bias stress",TED, VOL. 43, N. 8, 1996, p-p. 1286-1290.

[15] J. Pretet, N. Subba, D. Ioannou, S. Cristoloveanu, W. Maszara, C. Raynaud,"Explaining the reduced floating body effects in narrow channel SO1 MOSFETs" 2001, SOI Conference pp.25-26.

Effects of BTI during AHTOL on SRAM V_{MIN}

Sun-Me Lim[1], Heebum Hong[1], Sunil Yu[1], Ming Zhang[2], Jongwoo Park[3], Yongshik Kim[1]

Technology Development[1]/Media SOC[2]/Technology Reliability[3], System LSI division, Samsung Electronics
San #24 Nongseo-Dong Giheung-Gu, Yongin-City, Gyeonggi-Do, Korea 446-711
phone: 82-31-209-2496, e-mail: sunme.lim@samsung.com

Abstract—We present an optimal method to characterize and mitigate SRAM Vmin shift during Accelerated High Temperature Operating Life (AHTOL) stress test while taking PG BTI effect into account. Prior work has reported that changes in SRAM V_{MIN} during AHTOL stress test have a strong dependency on Bias Temperature Instability (BTI) degradation on PD and PU transistors [1, 2]. In this work, we have found that PG transistor BTI at a practical duty cycle can appreciably alter SRAM V_{MIN} shift during AHTOL. We have expanded SRAM V_{MIN} shift model on the basis of PG BTI as well as PD and PU BTIs. Statistical SRAM cell design with Response Surface Methodology (RSM)/Response Optimize (RO), Z-score method, is used to extract Optimal Z-score Ridge (OZR). OZR provides the optimal SRAM V_{MIN} at both T0 and EOL. Our work has shown that BTI vector with PD, PG, and PU BTI components should be placed on the OZR in V_T domain to mitigate V_{MIN} shift during AHTOL. The proposed theory has been confirmed with model-hardware correlation of individual disturbance-/writability-limited V_{MIN} degradations. With a consideration of OZR and BTI vector, we can achieve almost zero V_{MIN} shift of SRAM macro during AHTOL in gate-first high-k metal gate 32nm process.

Keywords-component; SRAM, AHTOL, V_{MIN} shift, BTI, OZR

I. INTRODUCTION

Negative and Positive BTIs (NBTI on PMOS and PBTI on NMOS) increase the threshold voltages of PMOS and NMOS during AHTOL stress test, which results in V_{MIN} shift as well as read and write instability[1]. Unlike pSiON gate, PBTI degradation in high-k metal-gate (HKMG) process has been observed in addition to NBTI [2], [3]. For cases in which no read/write is performed during stress, NBTI degradation was considered to be the main contributing factor and analyzed in detail [4-6], and NBTI in pull-up (PU) and PBTI in pull-down (PD) transistors for HKMG were also reviewed in prior works [1-2]. The word line (WL) on-time, during each SRAM read and write operation, is typically a few nanoseconds. As a result most published papers so far have neglected impact of PBTI to pass-gate (PG) due to the much shorter duration of stress time on PG as compared to that on PU and PD. Also, the PU and PD BTI effects have been simplified by using constant degradation [1-6].

However, in a realistic AHTOL test, a chip undergoes alternating current (AC) stress. The characteristics of HKMG BTI degradation have been studied for various stress waveforms in a prior work [7]. In this paper, the AC BTI V_T shifts are calculated by using normalized degradation value from [7], while accounting for our AHTOL stress conditions.

By properly targeting V_T values based on the predictive model, we have achieved close to zero V_{MIN} shift for SRAM macros during AHTOL in the 32nm HKMG technology. The V_{MIN} shift from silicon measurement is also well correlated to calculation by reflecting AC BTI V_T shifts, and we have used Z-score to calculate V_{MIN} by using hspice monte-carlo simulation.

II. AHTOL STRESS AND VT SHIFT

Our AHTOL stress tests are conducted at 140°C with 1.4xVdd for 1000 hours. SRAM cells undergo dynamic read and write operation driven by a 10MHz clock. For the case of static stress, SRAM PU, PG, and PD BTI degradation model followed our previous study [8]. By using normalized degradation value from reference [7] to our static BTI degradation, we have obtained AC BTI degradation values for PU, PG, and PD transistors.

A. AHTOL stress condition

During AHTOL stress, write and read operation are repeated like Figure 1 with 10MHz clock. All SRAM transistors undergo BTI stress periodically. The voltages of n1 and n2 are changing their value at 2.5MHz with 0.5 duty cycle. The WL on-time is a few nanoseconds for the case of read operation, and 50nsec for the write operation. By neglecting read WL on-time, the PG frequency has to be 5MHz having 0.25 duty cycle. Given this non-trivial magnitude of stress duty cycle on PG, we cannot neglect PG PBTI degradation in SRAM cell like prior work [1-6].

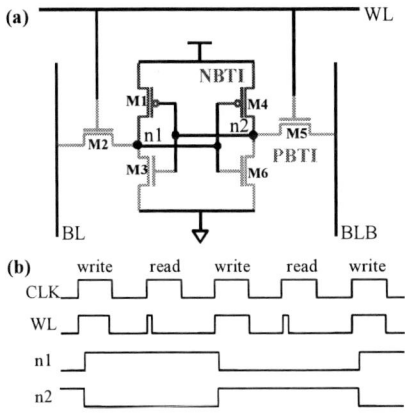

Figure 1. SRAM bit-cell circuit and time diagram. (a) NBTI – M1, M4 (PU), PBTI – M3, M6 (PD), M2, M5 (PG) (b) Timing diagram during AHTOL.

978-1-4244-9113-1/11 $26.00 © 2011 IEEE

B. Static and dynamic V_T shift during AHTOL stress

Static SRAM BTI V_T shift model can be represented by the following equation [8]:

$$\Delta V_T = A_0 \cdot \exp(-Ea_1/kT) \cdot V^\alpha \cdot t^n \quad (1)$$

Where A_0, Ea_1, α, and n are fitting parameters and have been obtained from our BTI stress experiments [8]. PU, PG, and PD static V_T shifts can be calculated as shown in Figure 2(a). PBTI degradation in our 32nm HKMG process is about 40% of NBTI degradation. For the case of AC AHTOL stress, frequency and duty cycle must be considered. According to the reference [7], BTI degradation is insensitive to frequency, and NMOS and PMOS degradation due to duty cycle behave quite differently. PU and PD degradations at 0.5 duty cycle are about 38% and 55% of static stress respectively, and PG degradation at 0.25 duty cycle is 40% of static stress. Figure 2 (a) is calculated V_T shift as a function of AHTOL stress time with constant gate bias stress. Figure 2 (b) is V_T shift for dynamic AHTOL stress having write and read operation. PU V_T shift is the largest, followed by PD and PG V_T shift.

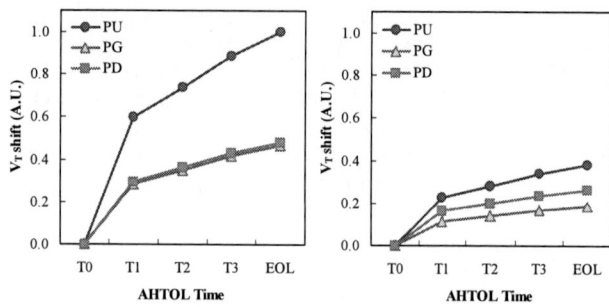

Figure 2. Calculated V_T shift as a function of AHTOL stress time. BTI model is used [4]. T0 is time zero, and EOL is end of lifetime. (a) static, (b) dynamic.

III. SRAM CELL STABILITY AND BTI EFFECTS

The static noise margin (SNM) of SRAM cell degradation due to PU and PD V_T shift is already reported in many papers [1, 2, 4-6]. In this paper we have analyzed SRAM stability parameters as function of cell ratios.

A. SRAM cell ratio and stability

SRAM stability curve can be obtained by node voltage sweep like Figure 3 [10]. Bias condition of BL and BLB determines stability curve shape. SNMR curves are from high BL and BLB, SNMH curves are from low BL and BLB, and SNMW curves are from high BL and low BLB [9]. SNMR, SNMH, and SNMW are the length of diagonal in figure 3, and each means SRAM disturb, retention, and write margin.

SRAM cell ratios are defined as:

$$\alpha = I_{PU}/I_{PD}, \ \beta = I_{PD}/I_{PG}, \ \gamma = I_{PG}/I_{PU} \quad (2)$$

Alpha ratio is originated from inverter voltage transfer curve, and that is determined from PU and PD current ratio. The slope of voltage transfer curve is affected by alpha. SNMH is monotonically depends on the amount of alpha ratio.

$$SNMH \sim \alpha \quad (3)$$

For the case of SNMR, node voltage increased due to PG current. The voltage of node is determined from the current ratio of PD and PG. SNMR is obtained from SNMH – node voltage, thereby SNMR is function of $\alpha - 1/\beta$.

$$SNMR \sim \alpha - 1/\beta \quad (4)$$

Larger beta means smaller node voltage during read, so SNMR is increased.

Finally the green SNMW curve on the lower-left corner of figure 3 is determined from two different BL, BLB biases. In the case of low BLB, the current path is from M4 to M5. Then n2 voltage is changed from high to zero. PG to PU current ratio is defined as gamma, and a high gamma means the node voltage could be easily pulled down to zero. In the case of high BL voltage, current path is formed from M2 to M3. High node voltage means larger SNMW. So SNMW is proportion to $1/\beta + \gamma$.

$$SNMW \sim 1/\beta + \gamma \quad (5)$$

High alpha means larger retention margin, high beta means larger disturb margin, and high gamma means larger write margin. If all cell ratios could be high, then SRAM have no stability issue, but the product of all cell ratios is 1. That means a strong write margin always entails a low retention and/or disturb margin.

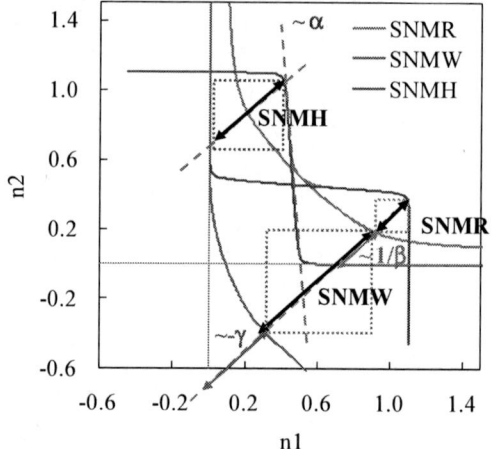

$$SNMH \propto \alpha, \ SNMR \propto \alpha - 1/\beta, \ SNMW \propto 1/\beta + \gamma$$

Figure 3. SRAM cell ratio and stability curve. SNMR, SNMW, SNMH are static noise margins at read, write, and hold operation. Cell ratio definitions are $\alpha = I_{PU}/I_{PD}$, $\beta = I_{PD}/I_{PG}$, $\gamma = I_{PG}/I_{PU}$ which implies $\alpha\beta\gamma = 1$. SNMR represents read disturb or write disturb margin, and SNMW is for writability margin.

In the context of BTI stress, three cases could be categorized. First case is BTI on PU only, and second case is PU and PD BTI V_T shifts, and third case is PU, PD, and PG BTI V_T shifts. The PU V_T shift causes smaller alpha and higher gamma, thereby retention and disturb margin are decreased. The PU and PD V_T shifts bring about smaller beta and higher gamma, so disturb margin is severer than Case 1. Finally in Case 3, PU, PD and PG V_T shifts change every cell ratio, which would entail disturb margin is not any worse than case 2 given the degradation in PG V_T.

B. V_T shift effect on SRAM stability parameters

In view of V_T shift caused by BTI stress, we have defined BTI vectors as illustrated in Figure 4 (a). Figure 4 (b) is SNMR and SNMW variation to various V_T shifts including case 1, 2, 3. The amount of PU, PG, and PD V_T shifts is assumed equal.

PG V_T shift leads to biggest change in SNMR and SNMW, then PU V_T shift, and followed by PD V_T shift. Unlike PU and PD V_T shift, PG V_T shift affects differently on SNMR and SNMW. That means PG V_T shift could help minimize SRAM V_{MIN} shift.

Cases 1 and 2 have smaller SNMR and larger SNMW. On the contrary, Case 3 has larger SNMR and smaller SNMW due to PG V_T shift.

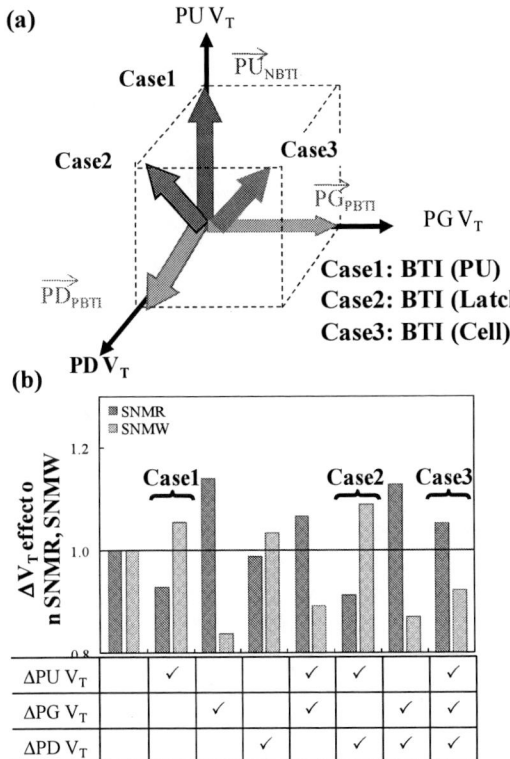

Figure 4. (a) Three BTI vectors at T0 V_T: Case1, 2, 3. The BTI vectors represent cell transistor degradation in cell V_T space. (b) ΔV_T effects on SNMR and SNMW. ΔV_T, $\Delta PUV_T=\Delta PGV_T=\Delta PDV_T$, is assumed.

C. BTI effect on SRAM stability parameter

In this section, we consider the three cases to BTI stress amount reflecting PBTI amount which is about 50 % of NBTI degradation. In Case 1 and 2, SNMR is steadily decreased to BTI stress amount, and Case 2 is worse than Case 1. However Case 3 has little variation of SNMR and SNMW. That means V_{MIN} shift could be almost zero.

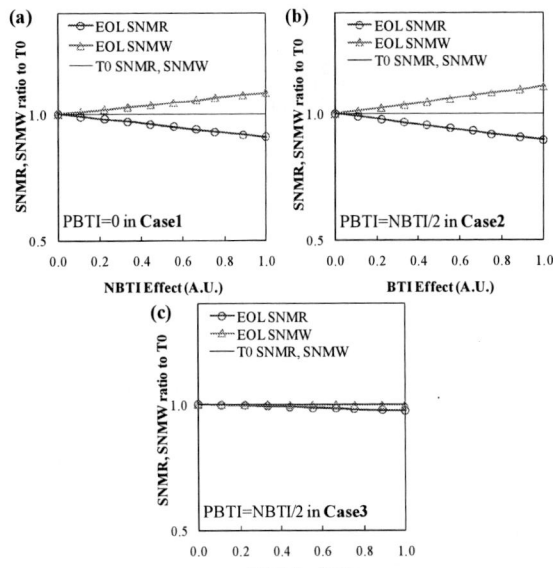

Figure 5. The effect of BTI on SNMR and SNMW. (a) Case1: PBTI=0 (b) Case2: PBTI=1/2 NBTI only in PD. (c) Case3: PU NBTI and PG, PD PBTI with half magnitude of NBTI. The much smaller change in cell stability is seen at Case3 than (a) and (b).

To consider real BTI stress degradation in Figure 2 (b), SNMR and SNMW functions of PU, PG, and PD V_T are needed. To obtain SNMR and SNMW function, we have defined Z-score first and used Z-score for expecting yield and for calculating V_{MIN}.

IV. Z-SCORE AND RSM/RO

A. What is the Z-score?

Z-score or sigma score originated from statistics. A tool of Six Sigma to obtain zero defects and Z-score are used in the evaluation process. The tails of SNMR or SNMW following a Gaussian distribution are regarded as defects. We have defined Z-score as:

$$Z= (L-\mu)/\sigma \qquad (6)$$

L is the lower limit of SRAM parameters. Lower limit is a boundary which separates two kind of bitcell. For example, if SNMR parameters are above L, SRAM bitcells keep their original data during disturb. In Figure 6, blue region below L means defects per opportunity (DPO). For the case of SRAM array, opportunity is defined as SRAM density.

Defects or weak bitcells in a SRAM array are distributed randomly on a chip or wafer, thereby Poisson yield model could be used in like this parametric defects. Then we can calculate Z-score satisfying target yield, for example Z-score is 6.04 at 95% yield for a 64Mb SRAM array.

978-1-4244-9113-1/11 $26.00 © 2011 IEEE

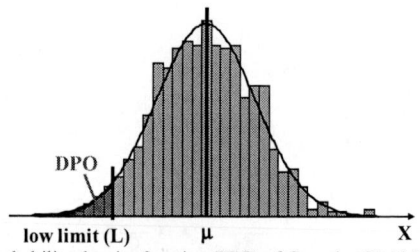

Figure 6. Probability density function (PDF) of Gaussian SRAM parameter X. The low limit (L) is fail criteria. X can be SNMR and SNMW. Z(X) describes Z-score of X parameter. Z-score implies how many sigma is between L and μ, and defect per opportunity (DPO). Yield can be predicted by DPO and Poisson statistics.

By using Monte Carlo simulation, we can obtain the distributions of SRAM parameters such as SNMR and SNMW.

B. Z-score RSM and Optimal Z-score Ridge

Response Surface Methodology (RSM) is a combination of regression modeling and optimization. The optimization is usually needed to design SRAM transistors size or to find out optimum condition to obtain high yield. SRAM consists of 3 kinds of transistor, so 3 factors for Central Composite Design (CCD) are used for RSM to optimize T0 V_T targeting condition. The CCD has center, axial, and corner points as design points, so expectation at edge is more accurate than other designs. Z(SNMR) and Z(SNMW) of 3 factor CCD DOE are obtained from hspice monte-carlo simulation, which reflects transistor performance and variation. 2nd order polynomials are obtained from RSM about Z(SNMR) and Z(SNMW) then Figure 7 (a) is obtained. Figure 7(b) is Z-score contour line. Z-score is defined as minimum of Z(SNMR) and Z(SNMW).

From Figure 7 (a), the Z(SNMR) and Z(SNMW) sensitivities to PU, PG, and PD V_T can be figured out. The PG V_T shift is most sensitivity to Z(SNMR) and Z(SNMW), followed by PU and PD V_T shift. The Z(SNMR) and Z(SNMW) are in inversely proportional to each other.

Optimal Z-score Ridge (OZR) which is defined as Z(SNMR) = Z(SNMW), is represented in Figure 7 (b). Left top area has low Z(SNMR) and right bottom is deficient in Z(SNMW). Converting Z-score contour to yield plot, Milkyway plot [11] or Yamaoka plot [12] could be obtained like [8]. To minimize V_{MIN} shift during AHTOL, cell transistor centering at T0 is to be on OZR. It is supposed that BTI ratio of NMOS to PMOS is around 2 based on Z-score method.

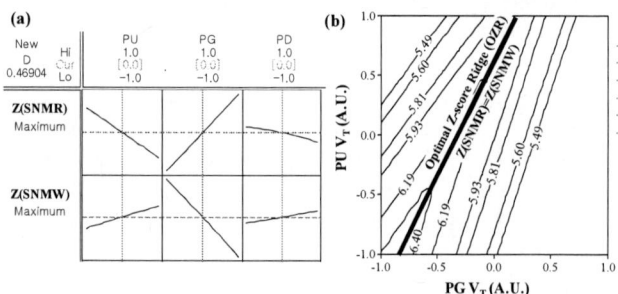

Figure 7. Z-score optimization with RSM/RO. 3 factor central composite design is used for response surface method. (a) Response optimize chart, (b)

minimum of the two Z-score contour between Z(SNMR) and Z(SNMW). The thick solid line means Optimal Z-score Ridge (OZR). OZR is the projection of Z-score satisfying Z(SNMR)=Z(SNMW), indicating the best yield or V_{MIN} at a certain PG V_T or PU V_T.

C. OZR vusus 3 BTI vectors

The V_T shifts due to BTI stress can be expressed by a vector in PU-PG and PU-PD V_T domain. OZR contour is plotted on Figure 8 (a) and (b). In Figure 8 (a), PG V_T is constant, in this case BTI vectors of Case 1 or 2 are across the OZR contour, thereby disturb margin is reduced. Figure 8 (b) is obtained from same change of PG and PD V_T, in that case the BTI vector of Case3 could be on OZR contour.

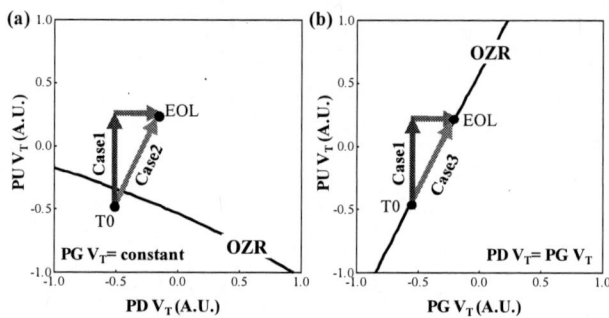

Figure 8. OZR and BTI vectors in (a) PU-PD VT space and (b) PU-PG VT space. Case3 BTI vector is right on OZR, so that maximized Z-score could be obtained not only at T0 but also at EOL.

As evident in Figure 8, PG V_T shift during AHTOL stress in addition to PU and PD V_T shifts could in effect move BTI vector onto the OZR contour line at EOL, which entails an optimal V_{MIN} at both T0 and EOL.

D. V_{MIN} shift calculation of Case1, 2, and 3 BTI vectors

V_{MIN} is obtained from Z-score to Vcc plot as illustrated in Figure 9, by finding when Z-score meets the target Z-score. The Z(SNMR) has convex curvature to Vcc and the Z(SNMW) has concave curvature to Vcc.

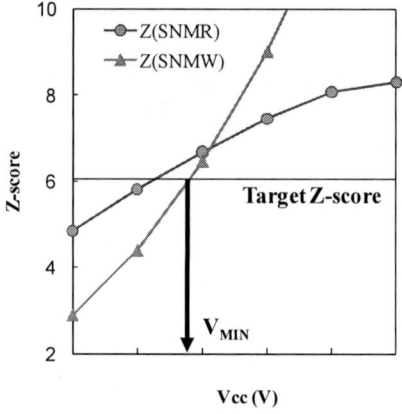

Figure 9. Z-score to Vcc plot. V_{MIN} is the cross point of Vcc when Z-scrore is equals to target Z-score

V_{MIN} shifts of SNMR and SNMW are also obtained from hspice simulation by changing V_T. Figure 10 shows V_{MIN}

shifts of Case 1, 2, and 3. All cases have increases in disturb V_{MIN}, and Case 3 only has increase in write V_{MIN} having small increase of disturb V_{MIN}.

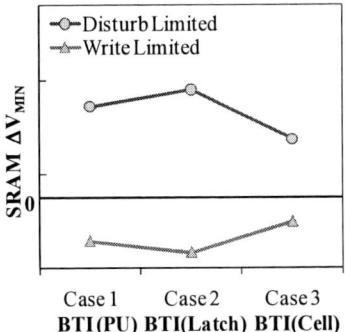

Figure 10. Median SRAM ΔV_{MIN} calculation at Case1, 2, and 3 BTI vectors. Maximum of the disturb and write limited V_{MIN} defines degradation of SRAM V_{MIN} during AHTOL. The Case3 covering 3 BTI components for PU, PG, and PD shows the smallest disturb V_{MIN} shift.

E. Experimental model and hardware correlation of V_{MIN} shift during AHTOL stress test

In our 32nm HKMG process, SRAM V_{MIN} shifts during AHTOL are represented on Figure 11 as a box plot. Disturb V_{MIN} is increased and no degradation of write V_{MIN} is observed. By reflecting BTI shifts in Figure 2 (b), the calculation in Figure 11 is obtained from hspice simulation. The calculated V_{MIN} shift is well correlated to hardware.

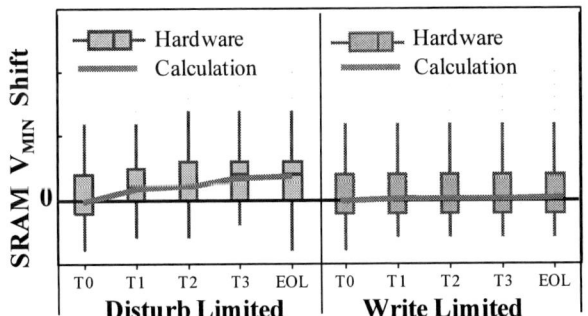

Figure 11. Disturb and write V_{MIN} shift during AHTOL stress. Model describes the median V_{MIN} shift.

F. SRAM V_{MIN} distribution and calculation

Effective V_{MIN} is determined from the worst case of disturb and write. In Figure 12, the distributions of hardware V_{MIN} and calculation V_{MIN} at T0 and EOL are plotted. The calculated V_{MIN} distributions are in agreement with hardware distributions. The PG V_T shift is small during AHTOL stress, so our BTI vector is steeper than OZR contour in Figure 8 (b). By targeting V_T on OZR contour, we have obtained almost zero V_{MIN} shift in hardware.

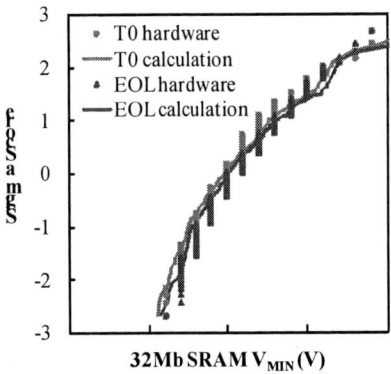

Figure 12. Model and hardware V_{MIN} distributions at T0 and EOL of AHTOL. Model comes from RSM and RO of Z-score. V_{MIN} is maximum of disturb limited V_{MIN} and write limited V_{MIN}.

CONCLUSION

We have found that PG transistor BTI in addition to PU and PG BTI at a practical duty cycle can appreciably alter SRAM V_{MIN} shift during AHTOL. BTI affects SRAM stability parameter, such as SNMR and SNMW. Unlike PU and PD, PG BTI helps to prevent decrease of disturb margin. By using hspice monte-carlo simulation which reflects performance and variation of SRAM transistors, we have obtained the distribution of SNMR and SNMW, and V_{MIN} expectation is done by Z-score. RSM and RO are used to obtain OZR, which provides the optimal SRAM V_{MIN} at both T0 and EOL. By properly targeting V_T values on OZR, we have achieved close to zero V_{MIN} shift for SRAM macros during AHTOL in the 32nm HKMG technology. The V_{MIN} shift from silicon measurement is also well correlated to calculation by reflecting AC BTI V_T shifts.

REFERENCES

[1] Aditya Bansal, Rahul Rao, Jae-Joon Kim, Sufi Zafar, James H. Stathis, "Impact of NBTI and PBTI in SRAM Bit-cells: Relative Sensitivities and Guidelines for Application-Specific Target Stability/Performance," IRPS, pp. 745-749, 2009.

[2] J.C. Lin, A.S. Oates and C.H. Yu, "Time Dependent V_{ccmin} Degradation of SRAM Fabricated with High-k Gate Dielectrics," IRPS, pp. 439-444, 2007.

[3] S. Pae et al, "BTI Reliability of 45nm High-K +Metal-Gate Process Technology,", IRPS, pp. 352-357, 2008.

[4] A.T. Krishnan et al, "SRAM Cell Static Noise Margin and V_{MIN} Sensitivity to Transistor Degradation," IEDM, pp. 1-4, 2006 .

[5] Sanjay V. Kumar, Chris H. Kim, and Sachin S. Sapatnekar, " Impact of NBTI on SRAM Read Stability and Design for Reliability," ISQED, 2006.

[6] J.C. Lin et al, " Predition of Control of NBTI-Induced SRAM V_{ccmin} Drift," IEDM, pp. 1-4, 2006.

[7] Stephen Ramey et al., "Frequency and Recovery Effects in High-k BTI Degradation," IRPS, pp. 1023-1027, 2009

[8] Jongwoo Park et al, "Mature Processability and Manufacturability by Characterizing V_T and V_{MIN} Behaviors Induced by NBTI and AHTOL Test," IRPS, pp. 1023-1027, 2009

[9] Jiajing Wang, Satyananed Nalam, and Benton H. Calhoun, "Analyzing Static and Dynamic Write Margin for Namometer SRAMs," ISLPED, pp. 129-134, 2008.

978-1-4244-9113-1/11 $26.00 © 2011 IEEE

[10] Yongshik Kim, " CMOS Logic – Technology Challenges for the Transition from 32nm to 22nm, Embedded Memory: SRAM," VLSI short course, 2008

[11] Masnanao Yamaoka et al, "Low Poser SRAM Menu for SOC Application Using Yin-Yang-Feedback Memory Cell Technology," VLSI, pp. 288-291, 2004

[12] Yasuhiro Morita et al, "A 0.3V Operating, Vth-Variation-Tolerant SRAM under DVS Environment for Memory-Rish SoC in 90-nm Technology Era and Beyond," IEICE, pp. 3634-3641, 2006

Flexible Biomedical Devices for Mapping Cardiac and Neural Electrophysiology

Dae-Hyeong Kim and John A Rogers
Department of Materials Science and Engineering
University of Illinois at Urbana-Champaign
Urbana, Illinois 61801 USA

Jonathan Viventi and Brian Litt
Department of Bioengineering
University of Pennsylvania
Philadelphia, PA 19104 USA

Abstract— Clinical in-vivo electrophysiological measurements using standard technologies provide spatial resolution limited by the number of electrodes. Recent strategies demonstrate that high resolution is possible by use of active multiplexing silicon electronics, in flexible forms capable of integration with soft, curvilinear tissues of the body. *In vivo* cardiac mapping experiments with such technology illustrate *in-situ* mapping of the spread of electrocardiogram (ECG) waveforms from natural and paced beats. In other examples, circuits in ultrathin mesh formats on sheets of bioresorbable substrates of silk fibroin improve the ability of these systems to conformally wrap the tissue. Neural mapping experiments on feline animal models illustrate the utility of this approach. These concepts provide capabilities for implantable diagnostic and therapeutic systems, which cannot be realized with conventional, wafer-scale device designs.

Keywords - flexible; cardiac; neural; mapping; electrophysiology; silk

I. INTRODUCTION

Flexible electronic devices fabricated with light-weight and low cost plastic substrates have great potential in unconventional electronic application areas, such as wearable computers, flexible displays and personal health monitoring devices [1]. The most common approaches to obtain that flexibility involve organic semiconducting and conducting materials. Recently, however, ultrathin inorganic semiconductor materials, such as silicon, in optimized mechanical layouts have been developed for high speed flexible devices with robust operation, even under harsh environments and conditions, such as complete immersion in bio-fluids [1,2]. These characteristics expand the application opportunities for flexible devices into systems for implantable biomedical instruments for electrophysiological mapping and others [3].

Electronics that must make conformal contacts with curvilinear surfaces, such as brain tissues demand extreme levels of deformability. In particular, the devices must conformally wrap around complex curvilinear biological tissues in ways that are impossible using devices that offer only simple bendability. To realize these outcomes, we explored use of circuis in ultrathin mesh layouts on bio-resorbable thin film substrates[4]. The value of this approach is demonstrated in *in-vivo* experiments for electrocorticography.

The result is a biocompatible electronics platform with the ability to address important problems in human health, not only through the EP systems but also via devices for neural prostheses, electrical stimulation therapies and others. The broad, interdisciplinary nature of this research spans materials science, electrical engineering, clinical medicine and bioengineering, and establishes completely new classes of implantable medical devices, surgical tools and monitors.

II. FLEXILBLE CARDIAC SENSOR

Passive electrodes, consisting of metal pads and interconnecting wires connected to external data acquisition system (DAQ), have limited functionality generally because of the low resolution that they can provide. The number of sensor elements is defined, in these conventional designs, by the number of connecting wires between electrodes and the DAQ. In contrast, the integration of active circuitry for multiplexing permits a large number of sensor elements with associated on-chip amplifiers and filters. Furthermore, flexible active circuits with appropriate designs can dramatically enhance the signal to noise ratio due to improved conformal contact with the tissue, and associated low impedance interfaces, and to local signal amplification. Our recent work demonstrates these capabilities in systems that provide 288 electrodes on thin plastic substrates that can conform to the epicardial surface for high resolution EP recorded with only 36 connections to the external control unit [3].

Figure 1 and its inset show the device and a magnified view of several unit cells, respectively (left frame). Wrapping this circuit on the epicardial surface of a porcine animal model yield high quality EP data, as shown from a single unit cell in the right top voltage trace and the two dimensional maps in the right bottom color frames. With further development of this technology, a new generation of implantable cardiac devices offering a wide variety of high-impact clinical applications can be developed, for safe and efficient surgical procedures in the future.

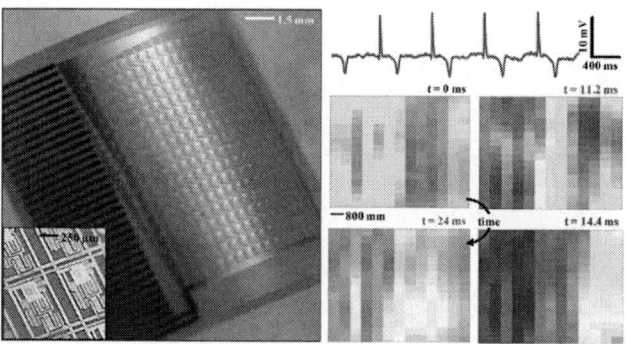

Figure 1. Photograph of a completed sensor array (left frame) and a magnified view of a unit cells (inset); Single voltage trace (right top) and voltage plots from a sensor array at four different times showing propagation of normal cardiac waveform through epicardial surface (right bottom). (adapted from ref [3])

III. DEFORMABLE NEURAL SENSOR

The capability for conformal contact between the sensor electrode and the surface of the monitoring organs, such as the brain or the heart, is one of the most important factors that should be considered in the design of bio-integrated electronic devices. This conformality is required for high quality sensing with low impedance and high signal to noise ratio and for minimal mechanical damage to soft biological tissues. We have developed a series of materials and device strategies for ultrathin, mesh-type biomedical electronic devices and sensors that can establish extremely conformal contact even onto curvilinear surfaces with tight radii of curvature [4]. In one example, we use circuits in ultrathin, mesh designs mounted on bio-resorbable substrates made of silk fibroin. Dissolving the silk drives a spontaneous wrapping process, driven by capillary interactions.

First, we demonstrated how biocompatible and bioresorbable silk can be used to help in fabricating and implanting ultrathin and mesh type passive electrode arrays. Figure 2 shows electrode arrays with various thicknesses and shapes (either sheet type or mesh type) after silk substrate dissolves away. Since the electrode array with conformal contact capability is too floppy to handle, the silk substrate can serve as a temporary, water-soluble carrier to facilitate integration with the tissue. This type of ultrathin mesh electrode array provides much better signal quality than otherwise similar thin film devices, as shown in the right frame of Fig 2. These advances will enable new classes of deformable implantable devices capable of functions that are not possible using technologies that are currently in use.

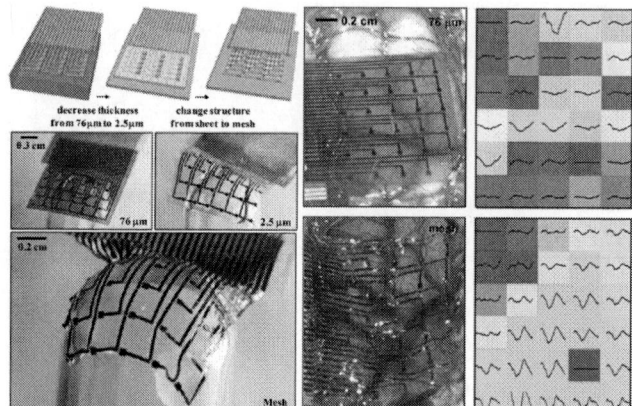

Figure 2. Schematic cartoons and images of electrode arrays wrapped onto a glass hemisphere (76 μm sheet, 2.5 μm sheet and 2.5 μm mesh; left frame). Image of an electrode array on a feline brain and the average evoked response recorded from each electrode (right frame). (adapted from ref [4])

ACKNOWLEDGMENT

This material is based upon work supported by a National Security Science and Engineering Faculty Fellowship and the U.S. Department of Energy, Division of Materials Sciences under Award No. DEFG02-91ER45439, through the Frederick Seitz MRL and Center for Microanalysis of Materials at the University of Illinois at Urbana-Champaign. The aspects of the work relating to silk are supported by the U.S. Army Research Laboratory and the U.S. Army Research Office under contract number W911 NF-07-1-0618 and by the DARPA-DSO and the NIH P41 Tissue Engineering Resource Center (P41 EB002520). Work at the University of Pennsylvania is supported by the National Institutes of Health Grants (NINDS RO1-NS041811-04, R01 NS 48598-04), and the Klingenstein Foundation. One of the authors (JAR) is supported by a National Science Security and Engineering Faculty Fellowship.

REFERENCES

[1] D.-H. Kim, J. Xiao, J. Song, Y. Huang, and J. A. Rogers, "Stretchable, Curvilinear Electronics Based On Inorganic Materials," Adv. Mater., vol. 22, pp. 2108 (2010).

[2] D.-H. Kim, J. Song, W.M. Choi, H.-S. Kim, R.-H. Kim, Z. Liu, Y.Y. Huang, K.-C. Hwang, Y. Zhang, and J.A. Rogers, "Materials and noncoplanar mesh designs for integrated circuits with linear elastic responses to extreme mechanical deformations" Proc. Natl. Acad. Sci. USA, vol. 105, pp. 18675 (2008).

[3] J. Viventi, D.-H. Kim, J. D. Moss, Y.-S. Kim, J. A. Blanco, N. Annetta, A. Hicks, J. Xiao, Y. Huang, D. J. Callans, J. A. Rogers, and B. Litt, "A Conformal, Bio-interfaced Class of Silicon Electronics for Mapping Cardiac Electrophysiology," Science Translational Medicine, vol. 2, pp. 24ra22 (2010)

[4] D.-H. Kim, J. Viventi, J. J. Amsden, J. Xiao, L. Vigeland, Y.-S. Kim, J. A. Blanco, B. Panilaitis, E. S. Frechette, D. Contreras, D. L. Kaplan, F. G. Omenetto, Y. Huang, K.-C Hwang, M. R. Zakin, B. Litt, and J. A. Rogers, "Dissolvable Films of Silk Fibroin for Ultrathin, Conformal Bio-Integrated Electronics," Nature Materials, vol. 9, pp. 511 (2010).

Low-Energy UV Effects on Organic Thin-Film-Transistors

N. Wrachien, A. Cester, D. Bari, G. Meneghesso
Department of Information Engineering
University of Padova
Padova, Italy
+39-049-8277625, wrachien@dei.unipd.it

J. Kovac, J. Jakabovic, M. Sokolsky, D. Donoval, J.Cirak
Slovak University of Technology
Bratislava, Slovakia

Abstract—We subjected Organic Thin-Film-Transistors with different gate dielectrics interface treatment to visible light and low energy UV irradiation. Devices with silicon nanoparticles only feature a remarkable temporary charge trapping, regardless the irradiation wavelength. On the contrary, devices without silicon nanoparticles feature temporary trapped charge under visible light, and permanent mobility degradation if they are irradiated with a wavelength shorter than 420nm.

Keywords-Thin-Film-Transistors, Organic Semiconductors, Irradiation, Reliability, Silicon nanoparticles;

I. INTRODUCTION

Extensive research is currently being performed on organic electronics. In particular, recent achievements in terms of performances made organic thin-film-transistors (OTFT) a very interesting candidate as a possible amorphous silicon replacement for low cost applications, such as liquid crystal displays or active-matrix organic led display (AMOLED). In fact, several works are currently being published on ever increasing hole mobility [1]-[6] with values equaling or exceeding those of amorphous silicon TFTs, at a fraction of their cost. Indeed, one of the main advantages of OTFT with respect to their inorganic counterpart, is the availability of very inexpensive production processes, such as spin-coating, etc [6]-[9], mainly due to the very low temperature required for the organic semiconductor deposition, compared to those of a-Si-TFT. Another advantage is the compatibility of organic semiconductors with plastic materials, which are lighter, more robust and much less expensive than glass. Some plastic substrates can be also bent without damage, opening the doors to a broad range of new applications, such as flexible displays, lab-on-a-chip, smart textiles, ultra low cost RF-ID, etc. [9]-[13].

Nonetheless, in spite of the very large amount of papers published every year, few works investigated the reliability of OTFTs. Among them, most of the works deal with the organic semiconductor intrinsic degradation and stability, or the stress- or light-induced reversible charge trapping [14]-[25]. Even if some works [23]-[25] analyzed the effects of deep ultraviolet light (DUV) on organic semiconductors and devices, the deep ultraviolet exposure is very unlikely to occur at the sea-level, and the exposure mostly occurs only during the manufacturing process.

In the works mentioned above, it has been showed that DUV light permanently affects the device and material characteristics. For instance, from the electrical viewpoint, an increased subthreshold swing and a decreased hole mobility has been observed after DUV irradiation, at relatively low dose levels. Still, the impact of near ultraviolet irradiation has not been fully assessed yet. Low-energy ultraviolet irradiation is much more likely to occur even on a finished device. In fact, excluding peculiar cases (such as the accidental exposure to UV artificial lights such as UV LEDs and Wood/black light/tanning tubes etc), the most likely cause of near UV irradiation is the exposure to sunlight. Indeed, the sunlight spectrum has a remarkable near-UV component [26]-[27], which can be as high as several mW/cm^2. Among the several devices, which could experience direct sunlight exposure, we may cite AMOLED and LCD TFT displays. Those devices, in fact, are built over a transparent substrate, which, at least in principle, does not offer a good low-energy UV shielding (while they usually offer a good DUV protection).

In this work, we show the effects of low energy ultra violet light on organic thin film transistor manufactured, which employ materials and structures widely used [28]-[32]. For the first time, we show that very low energy UV irradiation can permanently compromise the device performances if the TFTs employ some conventionally used materials or processes.

This work is organized as follows: in Section II we describe the devices and the experimental setup; in Section III we show the results; we discuss the data in Section IV; finally, in Section V we draw our conclusions.

978-1-4244-9113-1/11 $26.00 © 2011 IEEE

II. EXPERIMENTAL AND DEVICES

We analyzed two sets of top-contact bottom-gate pentacene-based OTFTs (see Fig. 1). In both cases, n^{++} silicon wafers were used with a 40 nm thick thermally grown SiO_2 insulating layer which was treated in Hexamethyldisilazane (HMDS) vapor for $\approx$50 hours. HMDS treatment is widely used to create a self-assembled monolayer, which improves pentacene layer mobility and stability [28]-[32]. For the first set of devices, a monolayer of 5nm silicon nanoparticles (SiNPs) stabilized with DBSA (Meliorum Technologies) was deposited over the treated SiO_2 surface by Langmuir-Blodgett method. This set of devices is referred as SiNP devices. The layer of silicon nanoparticles is currently being investigated [33]-[34] as it allows controlling the threshold voltage, mobility and other properties of the OTFT. The second set of devices did not contain SiNPs and hereafter is referred as HMDS devices. Both sets feature a 40-nm pentacene organic semiconductor layer. The pentacene layer has been deposited by thermal evaporation, keeping the substrate at a constant 30-°C temperature, which is a good trade-off between stability and hole mobility [35]-[36]. After the pentacene deposition, the 40-nm gold source and drain contacts were deposited by evaporation. All the devices feature a 2-mm channel width and a channel length of 20 μm and 50 μm.

The setup consists of 3 distinct phases, which are shown in Fig. 2. The first phase (P1) consists of an 11-cycles of characterization-irradiation loop. The characterization consists of an I_D-V_{GS} measurement taken in dark with V_{DS}=-8V and V_{GS} swept from -12 to -4V. The bias conditions were empirically chosen to achieve a large enough drain current, while minimizing the perturbation on the I-V induced by the measurement itself. The devices were measured in air, at room temperature and inside a shielded box which ensures shielding from external noise.

The irradiation is performed with grounded terminals and its duration is T_{irr}. At the end of each cycle T_{irr} is doubled. The initial T_{irr} value is 4s. The irradiation intensity was ~9 mW/cm^2. The setup has been performed with different irradiation peak wavelengths (525nm, 470nm, 420nm, 400nm and 385nm) and also in dark. In the latter case, the "irradiation" step consisted only of an "idle" step in which the devices were kept with grounded terminals in dark for T_{irr} time. In this way, we can distinguish between the effects of air exposure or measurement and UV irradiation on the electrical characteristics. In the irradiation phase we also evaluated the effect of visible light, especially at the shorter wavelengths

HMDS device

source contact   drain contact

Au   Au

pentacene   40 nm

HMDS treated SiO_2 40 nm

n^{++} Substrate (gate contact)

(a)

SiNP device

source contact   drain contact

Au   Au

pentacene   40 nm

silicon nanoparticles   40 nm   HMDS treated SiO_2

n^{++} Substrate (gate contact)

(b)

Fig. 1. Cross-section of the devices with HMDS-treated SiO_2 (a) and with silicon nanoparticles (b).

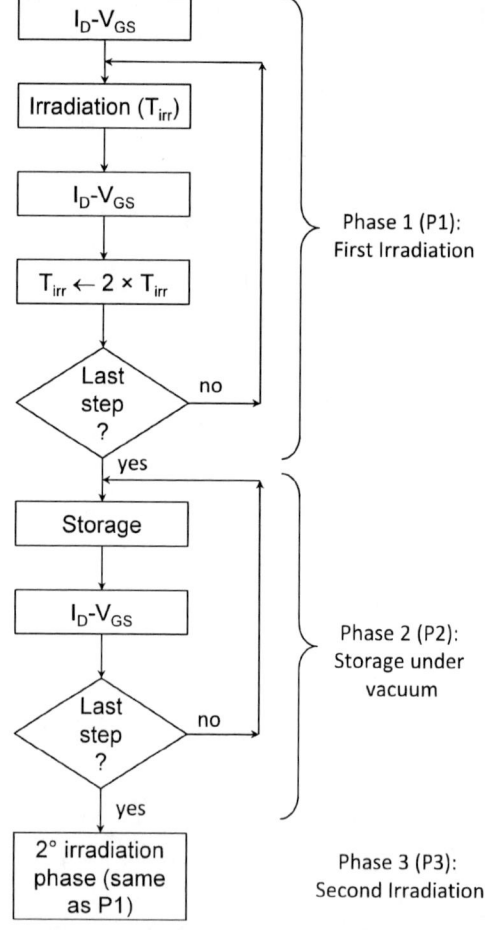

Fig. 2. Flowchart of the experimental setup used in this work.

(470nm and 420nm) to assess whether there is a threshold energy to induce degradation and if the effects of UV and visible light are different.

In the second phase (P2), the devices were stored under vacuum and they were periodically measured (in air) to assess if any recovery or further degradation after irradiation occurs. The third phase (P3) is a new irradiation performed on the same devices already irradiated in P1 and its setup is the same of P1.

To quantitatively assess the impact of the irradiation, we defined and calculated, from the I-V curves measured in P1, P2 and P3, the following parameters:

1. The V* variation (ΔV^*), where V* is defined as the V_{GS} required to achieve a drain-current gate-length product ($I_D \cdot L$) equal to 225 µA·µm and 50µA·µm for SiNP and HMDS devices, respectively.

2. The drain current ($I_D^{(M)}$) at V_{DS}=-8V and V_{GS}=-12V.

3. The transconductance (g_m) at $I_D \cdot L$ equal to 225 µA·µm and 50µA·µm for SiNP and HMDS devices, respectively.

III. RESULTS

A. Irradiation effects on fresh devices

Fig. 3 shows the I_D-V_{GS} taken after each step of P1, for different devices and at different irradiation conditions (dark, visible and UV light). For sake of clarity we plotted only two families of curves for SiNP devices (dark and UV) in Fig. 3a, and three families in Fig. 3b (dark, UV and visible light) for HMDS devices. The irradiation effects are very different, depending on both the wavelength and the device structure. In fact, when no irradiation is performed (i.e. when the devices are just kept at zero bias for the whole time and periodically measured), the I_D-V_{GS} measured after each step feature only some minor changes, both for HMDS and SiNP devices (see solid curves in Fig. 3). These changes arise from the measurement itself and/or from the intrinsic degradation due to air exposure.

A very different behavior is observed if UV light is employed during the irradiation phases of P1 (see dashed curves in Fig. 3). First of all, the variations on the measured I-

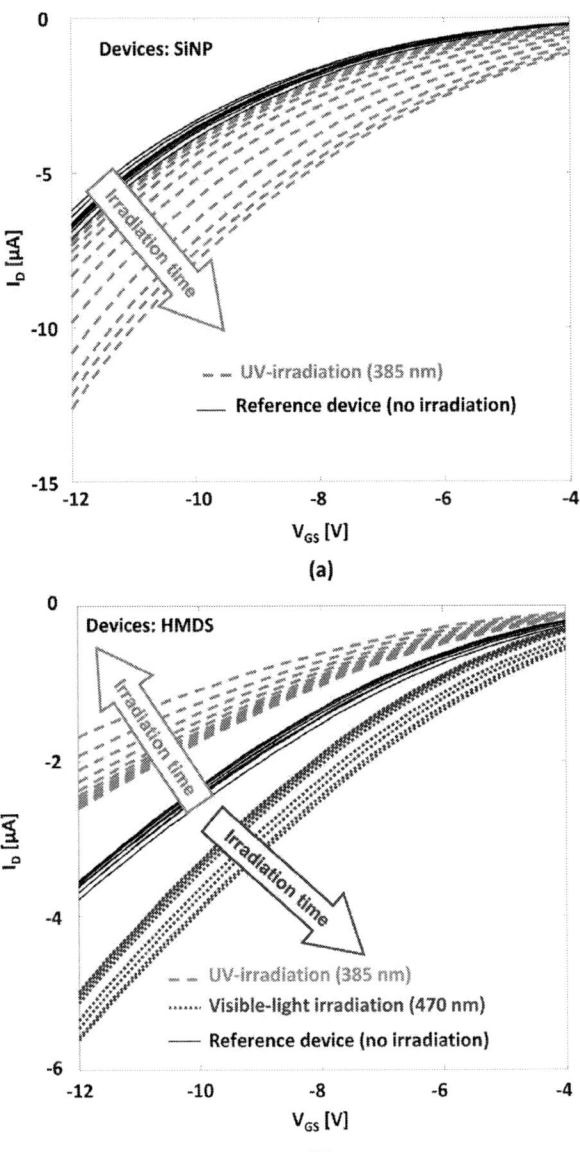

Fig. 3. I_D-V_{GS} curves measured after each step of P1 for SiNP devices (a) and HMDS devices (b), for different irradiation conditions: no irradiation (solid curve), UV irradiation (dashed curves) and visible light irradiation (dotted curves).

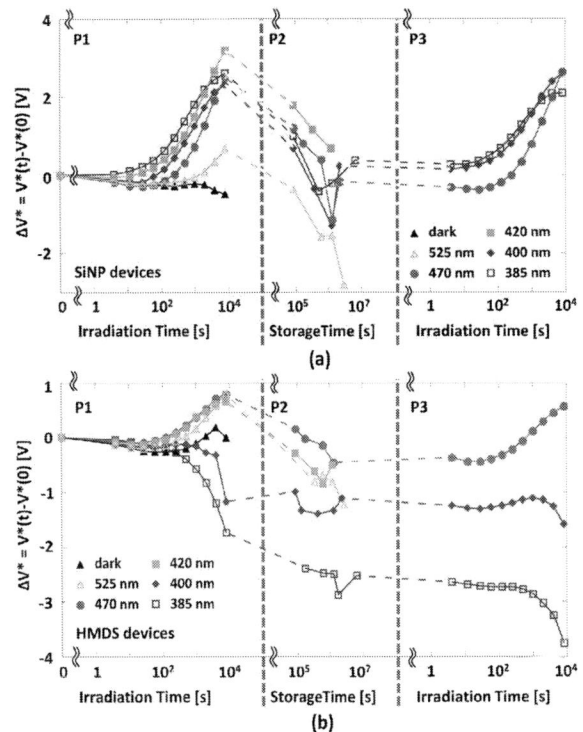

Fig. 4. Evolution of ΔV^* for SiNP (a) and HMDS (b) devices during irradiation (P1 and P3) and storage (P2) at different conditions.

V curves are much more noticeable, with respect to the same measurement performed under dark (compare solid with dashed curves in Fig. 3). Furthermore, a completely different kinetics is observed between SiNP (Fig. 3a) and HMDS (Fig. 3b) devices. Indeed, the HMDS devices (see dashed curves in

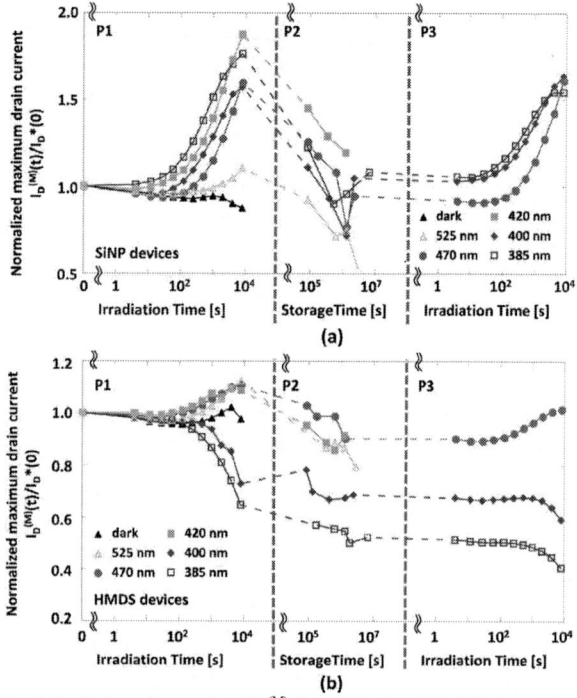

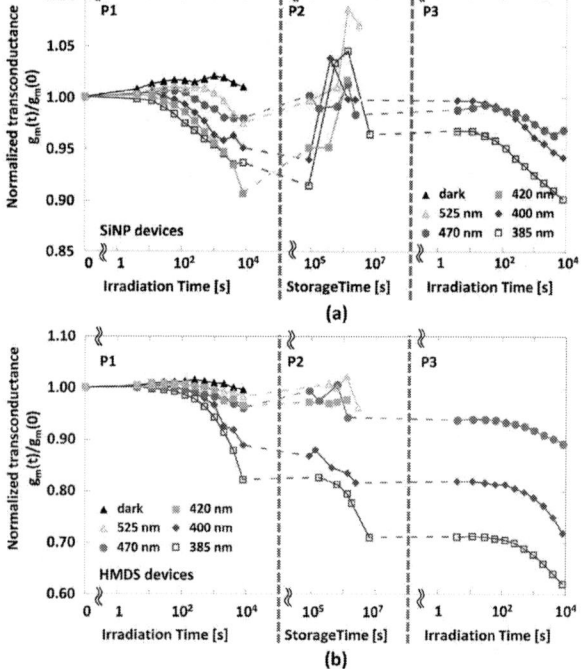

Fig. 5. Evolution of normalized $I_D^{(M)}$ for SiNP (a) and HMDS (b) devices during irradiation (P1 and P3) and storage (P2) at different conditions.

Fig. 6. Evolution of normalized transconductance g_m for SiNP (a) and HMDS (b) devices during irradiation (P1 and P3) and storage (P2) at different conditions.

Fig. 3b) show a reduced maximum drain current, a leftward shift of the I-V curves, and an I-V stretch-out.

If UV irradiation is performed on SiNP devices (see Fig. 3a), the I-V stretch-out still occurs, but the curves are shifted rightward and the maximum drain current is larger. Remarkably, the same phenomenon also occurs on HMDS devices, if they are irradiated with visible light (see dotted curves in Fig. 3b), while the qualitative behavior of SiNP devices is almost independent of the irradiation wavelength at least within the wavelength range 385-525 nm (not shown for sake of conciseness) .

In Figs. 4-6 we plot the evolutions of ΔV^*, $I_D^{(M)}$ and g_m calculated from the I-V measured in P1, P2, and P3 for different devices and irradiation conditions (dark, 525nm, 470nm, 420nm, 400, 385nm). $I_D^{(M)}$ and g_m are plotted normalized to their initial values.

The evolution of ΔV^* (shown in Fig. 4) highlights that, during P1, the I-V shift is negligible if no irradiation is performed (only 0.5 V and 10mV after 8188s for SiNP and HMDS devices, respectively), and very noticeable under the other cases (more than 2.5V and 0.8V for SiNP and HMDS devices, respectively). Furthermore, the ΔV^* is positive in all the cases, except when the HMDS devices are irradiated with UV light (400nm and 385 nm).

The evolution of the $I_D^{(M)}$ (see Fig. 5) mostly reflects the ΔV^* kinetics observed in Fig 5. Indeed, assuming a small impact of the g_m variation, if the I-V is shifted rightward (positive ΔV^*), the $I_D^{(M)}$ increases. Conversely, with a leftward shift (negative ΔV^*) $I_D^{(M)}$ increases.

The g_m evolution, shown in Fig. 6, is always decreasing in P1 (except in the dark conditions), regardless the irradiation conditions and devices.

B. Stability of the irradiation-induced modifications

To assess the irradiation damage stability we stored the devices under vacuum and we periodically measured the I-V (phase P2). The evolutions of ΔV^*, $I_D^{(M)}$ and g_m calculated from the I-V measured during P2 are shown in Figs 4-6.

The results indicate that the irradiation-induced modifications on SiNP devices (Figs. 4a, 5a, and 6a) are unstable and recover over time. Eventually, as the time elapses, the intrinsic degradation dominates.

Instead, HMDS devices (Fig. 4b, 5b and 6b) feature different behaviors, depending on the irradiation wavelength. In fact, those devices irradiated with longer wavelength feature some sort of recovery. However, when the irradiation wavelength falls below 420nm (i.e. 400nm and 385nm in our case), the irradiation damage is stable, on all the considered parameters. As before, the intrinsic device degradation dominates over time and we observed a further degradation (especially in g_m).

C. Cumulative irradiation effects on already irradiated devices

To assess if and how a previous irradiation impacts on a further irradiation procedure and to evaluate the repeatability of the setup and kinetics, we subjected the devices already irradiated in P1 to a further irradiation phase (P3).

Figs 4-6 show also the evolutions of ΔV^*, $I_D^{(M)}$ and g_m calculated from the I-V measured during P3. The degradation kinetics in P3 are very similar, at least qualitatively, to those observed in P1, indicating that the degradation kinetic is reproducible even after the first irradiation. No saturation in the kinetics is observed.

IV. DISCUSSION

The results of Figs. 3-6 show that irradiation induces different modifications on the electrical characteristics, depending on the device type and wavelength:

1. A temporary or permanent shift on the $I_D V_{GS}$ characteristics, which mostly impact on ΔV^* and consequently $I_D^{(M)}$.

2. A temporary or permanent I-V stretch-out, which impact mostly on the g_m and, in a lesser extent, on ΔV^* and $I_D^{(M)}$.

The responsible for the I_D-V_{GS} shift is the charge trapping. Indeed, when positive or negative charge is trapped, the I_D-V_{GS} characteristics exhibit a leftward or rightward shift, respectively.

In addition to the I_D-V_{GS} shift, charge trapping has also a much less obvious impact on the characteristics, inducing either a permanent or temporal stretch-out on the I-V. The I-V stretch-out due to charge trapping might arise from two contributions:

1. Charged defects near the gate dielectric to semiconductor interface might act as scattering centers, inducing mobility degradation. When the charge is neutralized the mobility value is, at least partially, restored.

2. The trapped charge neutralization during the I-V measurement. In fact, if during the I-V measurement some charge is neutralized, the flatband voltage (and consequently, the threshold voltage) varies during the measurement itself, inducing the stretch out of the I_D-V_{GS} curve. In particular, in our case the V_{GS} sweep is in the positive direction (forward sweep from -12V to -4V). If the sign of the trapped charge is positive, any neutralization during the I-V measurement will induce a smaller value of g_m, because the threshold voltage shifts toward more positive values. Of course this apparent reduction is only temporary. As soon as all the trapped charge is neutralized, no more threshold voltage variation can occur during the I-V measurement, resulting in a restored g_m value, which is higher than that observed just after each irradiation step.

Beside the two mechanisms described above, the I_D-V_{GS} stretch-out might also arise from the degradation of the semiconductor layer or semiconductor/gate dielectric interface, with obvious effects on the carrier mobility. In fact, irradiation can, at least in principle, induce photochemical reactions in the pentacene layer or near the pentacene/gate dielectric interface. The photochemical products, in turn, might have a much lower hole mobility than pentacene and/or they might induce or act as new traps, which further reduce the semiconductor mobility.

The data shown in Figs 4a, 5a and 6a, highlight that SiNP devices only feature temporary variations on all the analyzed parameters, i.e. we observe both a temporary I-V stretch-out and a temporary I-V shift. This is consistent with the temporary charge trapping. Indeed, when negative charge is trapped, a threshold voltage increase occurs, and, being our devices PMOS-like, the drain current increases in magnitude. A very similar explanation can be given when HMDS devices are irradiated by visible light. The I-V stretch out (and consequently g_m reduction) can be ascribed to the already discussed temporary impact on the mobility due to charged defects acting as scattering centers but it cannot be ascribed to trapped charge neutralization during the I-V measurement. In fact, the sign of the trapped charge is negative, due to the increased magnitude of the drain current (positive V^* variation). Any neutralization during the forward-sweep I_D-V_{GS} would result in an increased I-V slope, thus an increased g_m. In fact, as the V_{GS} increases from -12V to -4V, the neutralization of the negative trapped charge would decrease the flatband voltage, giving a current which is smaller and smaller with respect to the neutral case.

The negative trapped charge arises from the well know photogeneration, which occurs when a semiconductor is illuminated by a light source with a photon energy larger than the semiconductor energy gap. In particular the smallest photon energy is 2.36eV, while the pentacene energy gap is 1.8-2.2 eV [37]-[38]. We also argue that many of the photogenerated electron-hole pairs quickly recombine. However, some of them, surviving the recombination, might be separated by the presence of the zero-bias electric field (due to the non zero flatband voltage), or simply diffuse away. We expect that the hole mobility is much higher than the electron mobility in pentacene as shown in [39]. In this way, electron could be more easily trapped in bulk pentacene traps or even in the pentacene/gate dielectric interface, while holes can quickly escape the pentacene layer without being trapped, at least at zero bias. These traps might have very a broad range of energy depth as in [22]-[40], hence the carrier release time might span several orders of magnitude, still the majority of the trapped charge is neutralized within 15 days (see Fig. 4). From the comparison of the data in P1 and P2 of Fig 4, we estimate that more than half of the trapped charge is neutralized within the first day, because V^* recovers more than half of its variation.

A completely different kinetics is observed when HMDS devices are irradiated with UV light (i.e. 400nm and 385nm in our case). The two main differences between the previously discussed cases are:

1. The opposite variation on V^* and $I_D^{(M)}$. In particular, Figs 4b-5b show that both V^* and $I_D^{(M)}$ decrease.

2. The stability of the modifications, observed in P2 whereas, if the devices are irradiated with visible light, the variations are temporary.

978-1-4244-9113-1/11 $26.00 © 2011 IEEE

The stability of the variations suggests that UV light induces other degradation mechanisms on HMDS devices, which do not occur on SiNP devices (or HMDS devices if they are irradiated with visible light). Indeed, it is clear that UV induced some sort of permanent degradation on HMDS devices. Noticeably, HMDS and SiNP devices differ only in the additional SiNP layer. At this point, some considerations are worth to be draw. The SiNP layer is obviously the responsible for the improved UV-tolerance of our devices. What are the mechanisms underlying this improvement? We already mentioned about possible UV-assisted photochemical reactions, which might occur in the pentacene layer or at the pentacene-dielectric interface. The most obvious explanation for the improved stability is that photons with enough energy, such as UV, can induce some reactions only between the HMDS treated SiO_2 and pentacene. Instead, the reactions cannot occur between SiNP and pentacene.

However, other mechanisms could come into play. In fact, in [41] it has been observed that UV-induced direct decomposition of pentacene is unlikely to occur in vacuum, without other reactants, at least at photon energies up to 25 eV. However, in the same and other works [41],[42] it has been shown that, in presence of oxygen, UV irradiation could enhance the reaction with pentacene. The layer of SiNP might help in reducing the oxygen diffusion in the pentacene layer. In fact, it has been shown that the pentacene morphology strongly impacts the combined effect of UV and oxygen and that some crystal structures of pentacene films do not allow for oxygen diffusion (see [42] and references cited therein). Incidentally, our pentacene layer feature a much better crystalline quality if it is grown over the SiNP layer, as confirmed both by AFM data on the same structures [34] and also by the higher mobility value. We calculated that the hole mobility of SiNP devices is at least 3 times higher than that of HMDS devices.

We expect also that, in SiNP devices, the degradation (if any) of the HMDS-treated SiO_2 to the SiNP interface plays only a minor role on the hole mobility. Indeed, the conduction takes place at the SiNP to pentacene interface, which is at least 5-nm above the pentacene to SiNP interface.

At this point, one might wonder if the variations (in particular V^* and $I_D^{(M)}$) could be ascribed to positive charge trapping. There are several evidences that allow us to exclude this hypothesis:

- The variations are stable. In particular, g_m does not recover, indicating that there was no or negligible charge neutralization during the I-V measurement. Trapped charge near the interface might still act as scattering centers, but we expect that this would neutralize very quickly, whereas we noticed that, for instance, V^* does not recover even well after 15 days.

- The sign of the V^* variation is consistent with a positive trapped charge. Positive charge might be trapped in the oxide. However, even with the shortest wavelength (385nm, corresponding to a photon energy well below 3.5eV) the direct photogeneration of electron/holes pair in the SiO_2 layer cannot occur, because the SiO_2 has an energy gap larger than 9eV. Positive charge might be photogenerated in the

pentacene but, it cannot be injected in the SiO_2 layer, because of the hole barrier height, which is about 2 eV higher than our maximum photon energy. Still, positive charge could be trapped in the pentacene layer under certain conditions as confirmed by some works [17],[21]-[22]. However, the same should happen in SiNP devices under the same conditions, so we should see the same variations on those devices too. Finally, charge might be trapped at the HMDS-treated SiO_2 to pentacene interface. However, as we have previously discussed, we expect that this charge would neutralize in a timescale much shorter than that considered in P2 (see Figs. 4), where we observed no recovery on UV-irradiated HMDS devices, at least within 2 weeks.

The comparison of the kinetics of HMDS devices irradiated with 420nm and 400nm highlight that a sharp threshold exists between these two wavelengths, corresponding to a threshold photon energy in a very small range, between 2.95 eV and 3.10 eV. When the photon energy is below this threshold value, the degradation is negligible and the photogeneration and subsequent trapping of negative charge dominates. Conversely, if the photon energy is above this threshold value, permanent degradation dominates. This consideration also points out that even SiNP devices might feature a similar degradation kinetic, if a short enough irradiation wavelength is employed.

Finally, we observed a degradation (especially on g_m) at the end of P2 both in SiNP (which featured an initial recovery) and HMDS devices. This behavior might be ascribed to the intrinsic pentacene degradation over time, due to air exposure during measurements

V. CONCLUSIONS

In this work we showed that even low-energy UV light might induce permanent degradation on the electrical characteristics on some commonly employed device structures.

Devices with silicon nanoparticles at the HMDS-treated SiO_2 to pentacene interface featured a very noticeable but just temporary charge trapping, regardless the irradiation wavelength. On the contrary, devices without silicon nanoparticles (i.e. with direct contact between the pentacene and the HMDS-treated SiO_2) showed a strong wavelength dependence on the electrical characteristics modifications. Below the 3.0-eV threshold photon energy, the main degradation mechanism is the temporary charge trapping. With higher photon energies, a strong and permanent degradation occurs. Therefore, if pentacene is directly grown over the HMDS-treated SiO_2 the device suffers from poor stability due to its sensitivity to UV light, and the additional layer of SiNPs helps in reducing the UV-light permanent effects on the hole mobility. However, an adequate visible light shielding is required too, because the effects of visible light on the electrical characteristics, even if just temporary, are very pronounced. This could be a strong issue especially on those devices, which are grown on transparent substrates, such as displays.

ACKNOWLEDGMENT

Part of this work was done in Center of Excellence CENAMOST (Slovak Research and Development Agency Contract. No. VVCE-0049-07) with support of projects APVV-0290-06 and VEGA-1/0689/09.

This work was partially supported by Progetto di Ateneo 2009 – Università di Padova, Italy (Project Number CPDA083941).

REFERENCES

[1] U. Zschieschang, T. Yamamoto, K. Takimiya, H. Kuwabara, M. Ikeda, T. Sekitani, T. Someya, H. Klauk, "Low-voltage, high-mobility organic thin-film transistors with improved stability," *Device Research Conference* (DRC) 2010, pp.177-178, 21-23 June 2010.

[2] Y.Y. Lin, D.J. Gundlach, S.F. Nelson, T.N. Jackson, "High-mobility pentacene-based organic thin film transistors," *5th Device Research Conference Digest, 1997*, pp.60-61, 23-25 Jun 1997.

[3] B. Stadlober, M. Zirkl, M. Beutl, G. Leising, S. Bauer-Gogonea, S. Bauer, "High-mobility pentacene organic field-effect transistors with a high-dielectric-constant fluorinated polymer film gate dielectric," *Applied Physics Letters*, vol.86, no.24, pp.242902-242902-3, Jun 2005.

[4] C.-Y. Wei, S.-H. Kuo, Y.-M. Hung, W.-C. Huang, F. Adriyanto, Y.-H. Wang, "High-Mobility Pentacene-Based Thin-Film Transistors With a Solution-Processed Barium Titanate Insulator," IEEE *Electron Device Letters*, vol.32, no.1, pp.90-92, Jan. 2011.

[5] D.K. Hwang, K. Lee, J.H. Kim, S. Im, C.S. Kim, H.K. Baik, J. Park, H. Ji, E. Kim, "Low-voltage high-mobility pentacene thin-film transistors with polymer/high-k oxide double gate dielectrics," *Applied Physics Letters*, vol.88, no.24, pp.243513-243513-3, Jun 2006.

[6] K. P. Sung, K. Chung-Chen, J.E. Anthony, T.N. Jackson, "High mobility solution-processed OTFTs," IEEE International Electron Devices Meeting, 2005. pp.4 pp.-108, 5-5 Dec. 2005.

[7] S.E. Molesa, S.K. Volkman, D.R Redinger, Ad.F Vornbrock, V. Subramanian, "A high-performance all-inkjetted organic transistor technology," IEEE International Electron Devices Meeting, 2004, pp. 1072- 1074, 13-15 Dec. 2004.

[8] V. Subramanian, P.C. Chang, J.B. Lee, S.E. Molesa, S.K. Volkman, S.K., "Printed organic transistors for ultra-low-cost RFID applications," IEEE Transactions on Components and Packaging Technologies, vol.28, no.4, pp. 742- 747, Dec. 2005.

[9] T. Trigaud, I. El Jazairi, S.Y. Kwon, V. Bernical, J.P. Moliton, "Transparent all organic TFT fabrication by low cost process without clean room," IEEE IECON 2006 - 32nd Annual Conference on Industrial Electronics, pp.4876-4881, 6-10 Nov. 2006.

[10] Yun Shuai, A. Banerjee, D. Klotzkin, I. Papautsky, "On-chip fluorescence detection with organic thin film devices for disposable lab-on-a-chip sensors," 2008 IEEE Sensors, pp.122-125, 26-29 Oct. 2008.

[11] I. Manunza, E. Orgiu, A. Caboni, M. Barbaro, A. Bonfiglio, "Producing Smart Sensing Films by Means of Organic Field Effect Transistors," EMBS '06, 28th Annual International Conference of the IEEE Engineering in Medicine and Biology Society, 2006, pp.4344-4346, Aug. 30 2006-Sept. 3 2006.

[12] P. van Lieshout, E. van Veenendaal, L. Schrijnemakers, G. Gelinck, F. Touwslager, E. Huitema, "A flexible 240×320-pixel display with integrated row drivers manufactured in organic electronics," ISSCC. 2005 IEEE International Solid-State Circuits Conference, pp.578-618 Vol. 1, 10-10 Feb. 2005.

[13] M. Takamiya, T. Sekitani, Y. Kato, H. Kawaguchi, T. Someya, T. Sakurai, "An Organic FET SRAM With Back Gate to Increase Static Noise Margin and Its Application to Braille Sheet Display," IEEE Journal of Solid-State Circuits, vol.42, no.1, pp.93-100, Jan. 2007.

[14] K. Diallo, M. Erouel, J. Tardy, E. Andre, J.-L. Garden, "Stability of pentacene top gated thin film transistors," Applied Physics Letters, vol.91, no.18, pp.183508-183508-3, Oct 2007.

[15] S. J. Zilker, C. Detcheverry, E. Cantatore, D.M. de Leeuw, "Bias stress in organic thin-film transistors and logic gates," Applied Physics Letters, vol.79, no.8, pp.1124-1126, Aug 2001.

[16] Ching-Lin Fan, Tsung-Hsien Yang, Chin-Yuan Chiang, "Performance Degradation of Pentacene-Based Organic Thin-Film Transistors Under Positive Drain Bias Stress in the Atmosphere," Electron Device Letters, IEEE , vol.31, no.8, pp.887-889, Aug. 2010.

[17] K.K. Ryu, I. Nausieda, D.D. He, A.I. Akinwande, V. Bulovic, C.G. Sodini, "Bias-Stress Effect in Pentacene Organic Thin-Film Transistors," Electron Devices, IEEE Transactions on , vol.57, no.5, pp.1003-1008, May 2010.

[18] Tse Nga Ng, M.L. Chabinyc, R.A. Street, A. Salleo, "Bias Stress Effects in Organic Thin Film Transistors," 45th Annual IEEE International Reliability Physics Symposium, 2007, pp.243-247, 15-19 April 2007.

[19] Y. Qiu, Yuanchuan Hu, Guifang Dong, Liduo Wang, Junfeng Xie, and Yaning Ma, "H_2O effect on the stability of organic thin-film field-effect transistors," Appl. Phys. Letters, (83), 8,2003, pp. 1644-1646.

[20] C. Pannemann, T. Diekmann, and U. Hilleringmann, "On the degradation of organic field-effect transistors," Microelectronics, 2004. ICM 2004.

[21] M. Debucquoy, S. Verlaak, S. Steudel, K. Myny, J. Genoe, and P. Heremans, "Correlation between bias stress instability and phototransistor operation of pentacene thin-film transistors," Appl. Phys. Lett. 91, 103508 (2007).

[22] N. Wrachien, A. Cester, N. Bellaio, A. Pinato, M. Meneghini, A. Tazzoli, G. Meneghesso, K. Myny, S. Smout, J. Genoe, "Light, Bias, And Temperature Effects On Organic TFTs", IEEE International Reliability Physics Symposium 2010, May 2-6, 2010.

[23] Jeong-M. Choi, D. K. Hwang, Jung Min Hwang, Jae Hoon Kim, and Seongil Im, "Ultraviolet-enhanced device properties in pentacene-based thin-film transistors", Appl. Phys. Lett. 90, 113515 (2007).

[24] Jae Bon Koo, Seong Yeol Kang, In Kyu You, and Kyung Soo Suh "Hysteresis Effect Induced by UV/ozone Treatment in Pentacene Organic Thin Film Transistor", 213th ECS Meeting. Phoenix, Arizona, May 18-22, 2008.

[25] Woo Jin Kim, Won Hoe Koo, Sung Jin Jo, Chang Su Kim, Hong Koo Baik, D. K. Hwang, Kimoon Lee, Jae Hoon Kim, and Seongil Im, "Ultraviolet-Enduring Performance of Flexible Pentacene TFTs with SnO_2 Encapsulation Films" Electrochem. Solid-State Lett. 9, G251 (2006).

[26] A. V. Parisi and J. Turner, "Variations in the short wavelength cut-off of the solar UV spectra", Photochem. Photobiol. Sci., 2006, 5, 331–335.

[27] "Standard Tables for Reference Solar Spectral Irradiances: Direct Normal and Hemispherical on 37° Tilted Surface", ASTM G173-03 2008.

[28] Sang Chu Lim, Seong Hyun Kim, Jung Hun Lee, Mi Kyung Kim, Do Jin Kim, Taehyoung Zyung, "Surface-treatment effects on organic thin-film transistors", Synthetic Metals, Vol, 148, 3 January 2005, Pages 75-79.

[29] Jae Bon Koo, Seong Hyun Kim, Jung Hun Lee, Chan Hoe Ku, Sang Chul Lim, Taehyoung Zyung, "The effects of surface treatment on device performance in pentacene-based thin film transistor", Synthetic Metals, Volume 156, 1 February 2006, Pages 99-103.

[30] Yong-Hoon Kim, Jae-Hoon Lee, Min-Koo Han and Jeong-In Han, "The Improvement of Electrical Characteristic of Solution Processed Organic Thin-Film Transistors with 6,13-bis (triisopropylsilylethynyl) Pentacene Films Employing HMDS Treatment," ECS 210th Meeting, Cancun, Mexico,October 29-November 3, 2006.

[31] Y. Wang, R. Kumashiro, R. Nouchi, N. Komatsu, K. Tanigaki, "Influence of interface modifications on carrier mobilities in rubrene single crystal ambipolar field-effect transistors," Journal of Applied Physics , vol.105, no.12, pp.124912-124912-5, Jun 2009.

[32] Jeong, Yeon Taek, Dodabalapur, Ananth, "Pentacene-based low voltage organic field-effect transistors with anodized Ta_2O_5 gate dielectric," Applied Physics Letters , vol.91, no.19, pp.193509-193509-3, Nov 2007.

[33] Martin Weis, Jack Lin, Takaaki Manaka, and Mitsumasa Iwamoto, "Trapping centers engineering by including of nanoparticles into organic semiconductors," Journal Of Applied Physics 104, 114502, 2008.

[34] J.Jakabovic, J. Kovac, R. Srnanek, M. Weis, M. Sokolsky, A. Hinderhofer, K. Broch, F. Schreiber, D. Donoval, and J. Cirak, "Pentacene-gate dielectric interface modification with silicon nanoparticles for OTFTs", IVC-18, Beijin, 25-27 August, 2010.

[35] J. Kovac, J. Jakabovic, R. Srnanek, J. Kovac jr, D. Donoval, N. Wrachien, A. Cester, G. Meneghesso, "Growth morphologies and electrical properties of pentacene organic TFT with SiO2/parylene dielectric layer", WOCSDICE 2009, 17-20 May 2009.

[36] N. Wrachien, A. Cester, A. Pinato, M. Meneghini, A. Tazzoli, G. Meneghesso J. Kovac, J. Jakabovic, D. Donoval, "Organic TFT with SiO2-Parylene Gate Dielectric Stack and Optimized Pentacene Growth Temperature," 39th European Solid-State device research Conference, ESSDERC 2009, 14 - 18 September 2009, Athens..

[37] S.S. Kim, S.P. Park, J.H. Kim, and S. Im, "Photoelectric and optical properties of pentacene films deposited on n-Si by thermal evaporation", Thin Solid Films 420 –421 (2002) 19–22.

[38] Li T., Balk J. W., Ruden P. P., Campbell I. H. and Smith D. L., 2002, *J. Appl. Phys.* **91** 4312.

[39] Satoko Takebayashi, Shigeomi Abe, Koichiro Saiki, and Keiji Ueno "Origin of the ambipolar operation of a pentacene field-effect transistor fabricated on a polyvinylalcohol-coated Ta_2O_5 gate dielectric with Au source/drain electrodes," Applied Physics Letters 94, 083305, 2009.

[40] A. R. Volkel, R. A. Street, and D. Knipp, "Carrier transport and density of state distributions in pentacene transistors," Physical Review B, 66, 195336, 2002.

[41] A. Vollmer et al., Surface Science 600 (2006), 4004–4007.

[42] Ashok Maliakal et al., Chem. Mater. 2004, 16, 4980-4986.

A new method for predicting the lifetime of highly stable amorphous-silicon thin-film transistors from accelerated tests

T. Liu, S. Wagner and J. C. Sturm

Department of Electrical Engineering and
the Princeton Institute for the Science and Technology of Materials (PRISM), Princeton University
Princeton, New Jersey 08544, USA
e-mail: tingliu@princeton.edu

Abstract— We present a new method for predicting the lifetime of highly stable amorphous-silicon thin-film transistors (a-Si TFTs) from accelerated tests at elevated temperatures. The rate of DC saturation current drop can be accelerated by a factor of ~10^4 when the test temperature is raised to 160°C. This ability is particularly significant for predicting the stability and lifetime of a-Si TFTs as analog drivers in active-matrix organic light emitting diode (AMOLED) displays.

Keywords- Accelerated lifetime tests, a-Si TFTs, stability, current degradation, stretched hyperbola

I. Introduction

Because conventionally-fabricated a-Si TFTs are unstable, they are usually used only as digital switches at low duty cycle in active-matrix liquid crystal displays (AMLCDs). Ultra-stable a-Si TFTs have been recently reported and proposed for analog drivers of the OLEDs in AMOLED displays [1,2]. This application requires a reliable method for evaluating their stability and predicting their lifetime. a-Si TFTs in AMOLEDs are driven at low gate voltage, where defect creation in a-Si dominates the degradation. The concurrent threshold voltage shift causes a reduction of the TFT's drain current [1,3]. In this work, we obtain the distribution of defect creation energies and the "attempt-to-break" frequency in the a-Si from drain current vs. time data taken at elevated temperatures. We relate these parameters, which characterize the stability of a-Si, to the acceleration factor for TFT current degradation. This work then enables one to predict the room temperature lifetime of devices from accelerated tests at elevated temperatures using a physically-based model.

II. Theoretical Framework

A. Defect Creation

The threshold voltage shift of a-Si TFTs under gate bias leads to reduced drain current and thereby reduced OLED brightness in AMOLED displays. At the low gate fields appropriate for driving OLEDs, defect creation by bond-breaking in a-Si is the mechanism [1,3] that we need to accelerate to predict the lifetime of a-Si TFTs. The degradation of TFT drain current is accelerated at elevated temperatures, where strained bonds are thermally activated and more easily broken. The amorphous nature of a-Si results in a range of Si-Si bond energies. The distribution of electron energies E on strained (weak) bonds is modeled with an exponential function $n(E) \propto e^{E/kT_0}$, where the characteristic temperature T_0 reflects the disorder broadening of the a-Si structure, and a high value of T_0 reflects a broad distribution of strained bonds [4]. Therefore, the conventional Arrhenius model for thermally activated failure of electronic devices cannot be applied directly to a-Si TFTs. Wehrspohn et al. proposed a stretched hyperbola model [5-8] for the defect creation process in the form of

$$\Delta N_{DB}(t) = N_{BT}^0 \left\{ 1 - \left[1 + \left(\frac{t}{t_0} \right)^{T/T_0} \right]^{-1/(\alpha-1)} \right\} \quad (1)$$

$\Delta N_{DB}(t)$ is the number of created defects, N_{BT}^0 is the initial number of carriers in the band-tail at $t = 0$, the fitting parameter α takes into account a superlinear bias dependence, ranging from 1 to 2, and $t_0 = \nu^{-1} \exp(E_{act}/kT)$. ν is the "attempt to break" frequency in the model for the bond-breaking frequency $1/t_0$, and the effective activation energy E_{act} is related to the weighted average energy barrier for defect creation under gate bias [5,8]. E_{act} differs from the conventional activation energy in an Arrhenius model, in that it is a weighted average energy for the distribution of electron energies E on strained (weak) bonds in a-Si, instead of a single energy as in crystalline silicon.

B. Acceleration of Drain Current Degradation

At low gate fields, defect creation in a-Si leads to a threshold voltage shift of a-Si TFTs, with

$$\Delta V_T(t) = q\Delta N_{DB}(t)/C_{ins} \quad (2)$$

where C_{ins} is the capacitance of the gate insulator.

Since

$$N_{BT}^0 = (V_G - V_{T0}) \times C_{ins}/q \quad (3)$$

978-1-4244-9113-1/11 $26.00 © 2011 IEEE

by substituting (2) and (3) into (1), the threshold voltage shift as a function of time under bias can be expressed as [5]

$$\Delta V_T(t) = (V_G - V_{T0}) \left\{ 1 - \left[1 + \left(\frac{t}{t_0} \right)^{T/T_0} \right]^{-1/(\alpha-1)} \right\} \quad (4)$$

In this work we combine the stretched hyperbola model for the defect creation with the current-voltage relation for a-Si TFTs in the saturation regime

$$I_D(t) = \frac{1}{2} \mu_n C_{ins} \frac{W}{L} [V_G - (V_{T0} + \Delta V_T(t))]^2 \quad (5)$$

Since the changes of mobility are much less significant to the drain current degradation than the changes of threshold voltage (shown in section IV. A.), we neglect the mobility changes in our analysis. By substituting (5) into (4), the reduction with time of the normalized drain current $I_{D,nor}(t) \equiv I_D(t)/I_D(t=0)$ caused by threshold voltage shift can be expressed as

$$I_{D,nor}(t) \equiv \frac{I_D(t)}{I_D(t=0)} = \left[1 + \left(\frac{t}{t_0} \right)^{T/T_0} \right]^{-2/(\alpha-1)} \quad (6)$$

Equation (6) shows that the normalized drain current degradation is temperature dependent, with temperature T appearing in parameter $t_0 = \nu^{-1} \exp(E_{act}/kT)$ and the exponential term T/T_0. Thus, accelerated lifetime tests can be carried out by simply monitoring the drain current degradation at elevated temperatures.

Note that the normalized drain current degradation in the saturation regime can be directly used to characterize the stability of a-Si TFTs, without having to be converted to a threshold voltage shift ΔV_T as in the conventional stretched-hyperbola method [5-8]. Also note that while the term $(V_G - V_{T0})$ in (4) can vary from transistor to transistor, the current degradation method based on (6) has the advantage that it is independent of $(V_G - V_{T0})$.

III. EXPERIMENTAL PROCEDURE

A. Sample Preparation

The samples in our measurements are back-channel passivated TFT structures (Fig. 1). A standard bottom-gate non-self-aligned process was used.

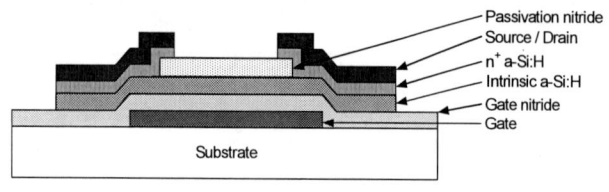

Figure 1. Schematic cross-section of back-channel passivated TFT structure.

The silicon nitride and amorphous silicon were grown in a standard plasma-enhanced chemical vapor deposition (PECVD) system. 300-nm gate nitride, 200-nm intrinsic hydrogenated a-Si and 300-nm passivation nitride were deposited sequentially at 350°C, 250°C and 250°C, respectively. The silicon nitride and a-Si were patterned with dry etching. Before measurements, all samples were annealed at 200°C to remove the damage induced by the dry-etching plasma.

B. Accelerated Lifetime Testing

For accelerated lifetime testing, the TFTs were biased in the saturation regime with a constant gate voltage of 5V (a gate field of ~1.5×10^5 V/cm) and a constant drain voltage of 7.5V. We raised the substrate temperature from 20°C to 160°C in steps of 20°C. At each temperature we biased a fresh TFT without any prior stress in the saturation regime and measured the drain current as a function of time. This measurement was done without any interruption, for example to measure I_D-V_G characteristics at intermediate points. At the end of the continuous bias stress period, a gate-bias voltage sweep was applied to measure the I_D-V_G characteristics of the TFTs. To minimize structural rearrangement and hydrogen diffusion in the a-Si, in the testing we did not exceed 160°C, which is much lower than the a-Si silicon deposition temperature at 250°C and the 200°C annealing temperature.

IV. RESULTS AND DISCUSSION

A. Shifts of Transfer Characteristics after Stress

For our 150μm wide and 15μm long channel TFTs, at 20°C and in the saturation regime, the initial drain current $I_D(t=0)$ was 2.12±0.09μA, with the initial threshold voltage 0.44±0.22V and the field-effect mobility 0.99±0.12cm²/V·s. Higher extracted mobilities were correlated with higher extracted threshold voltages to give the tight current distribution.

The transfer characteristics for samples before and after stress were recorded at 20°C and 160°C. To avoid defect creation during the gate-bias voltage sweep, at 20°C we took a fast sweep, with V_G swept from 5V to -3V in 3 sec. V_D was kept at 10V in order to stress the TFT in the saturation regime, which is consistent with the bias stress condition in the accelerated lifetime tests. The $\sqrt{I_D}$-V_G plots before and after a 5×10^4 sec bias stress at 20°C are shown in Fig. 2(a), from which we can see that the bias stress leads to a 12% drop of the drain current when V_G=5V. The slope of $\sqrt{I_D}$-V_G is almost constant before and after the bias stress, with the extracted field effect mobility changing from 0.88cm²/V·s before the bias stress to 0.89cm²/V·s after the bias stress (~1% change). The threshold voltage increased from 0.30V to 0.62V with ΔV_T=0.32V.

Similar $\sqrt{I_D}$-V_G plots before and after a 1.0×10^4 sec bias stress at 160°C in Fig. 2(b) show that the bias stress leads to a 86% drop of the drain current when V_G=5V. The two I_D-V_G sweeps were also taken at 160°C. From the slope of $\sqrt{I_D}$-V_G, the extracted field effect mobility changing from 3.46cm²/V·s before the bias stress to 3.57cm²/V·s after the bias stress (~9% change). The threshold voltage increased from 0.48V to 3.12V with ΔV_T=2.64V.

At both temperature extremes (20°C and 160°C) in our accelerated lifetime testing, the effect of the threshold voltage shifts was the dominant one on the TFT current degradation, validating our modeling approach to derive (6).

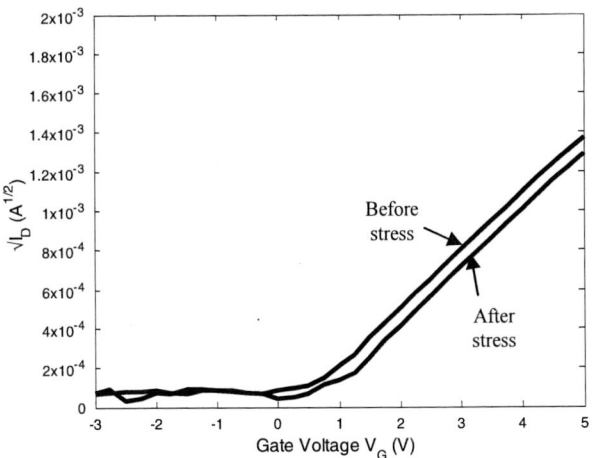

Figure 2(a). $\sqrt{I_D}$-V_G plots before and after a 5×10^4 sec bias stress at 20°C.

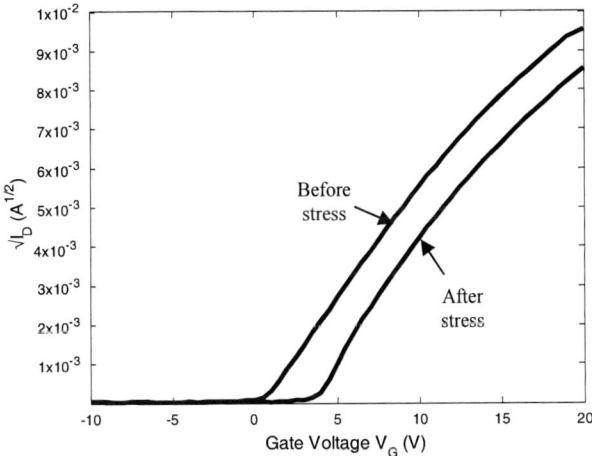

Figure 2(b). $\sqrt{I_D}$-V_G plots before and after a 1×10^4 sec bias stress at 160°C.

B. Dependence of Drain Current Degradation on Temperature

The experimental data of normalized drain current $I_{D,nor}$ as a function of time at different temperatures are shown with open squares in Fig. 3.

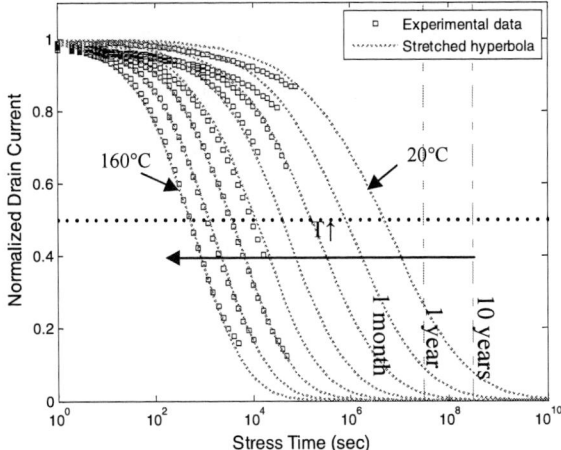

Figure 3. Normalized drain current degradation data (open squares) of the a-Si TFTs at temperature stepped by 20°C from 20°C to 160°C. The dotted lines are stretched hyperbola fits described in Fig. 4 and (7).

We can represent all sets of current vs. time data with a single curve by replacing the time axis with the "thermalization energy" [5-8], which is defined as $E_{th} = kTln(vt)$. That all experimental data (open squares in Fig. 4) cluster so closely on a single curve with $v = 4\times10^5$Hz gives us confidence in this method to predict the drain current degradation at low temperatures with high temperature data. From (6), the relation for $I_{D,nor}$ with E_{th} can be reformulated as

$$I_{D,nor}(E_{th}) = \{1 + \exp[(E_{th} - E_{act})/kT_0]\}^{-\frac{2}{\alpha-1}} \quad (7)$$

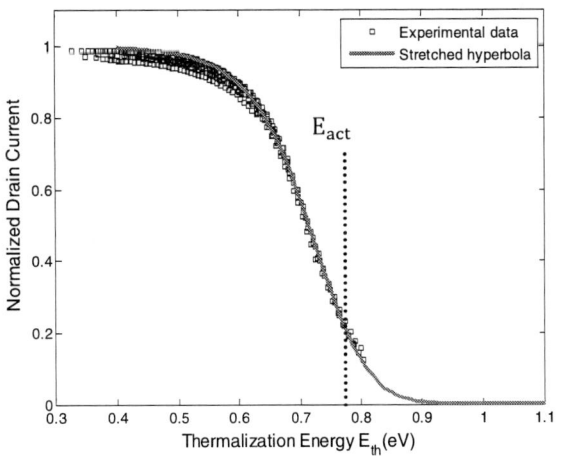

Figure 4. Normalized drain current unified in terms of the thermalization energy $E_{th} = kTln(vt)$, where $v = 4\times10^5$ Hz. Stretched-hyperbola fit (solid line) with $\alpha = 1.9$.

Equation (7) provides an excellent stretched-hyperbola fit with $\alpha = 1.9$, shown as the solid curve in Fig. 4. We find $E_{act} = 0.78$eV and $T_0 = 680$K; both are typical values for a-Si and are independent of temperature. The relation of $E_{th} = kT\ln(vt)$ between thermalization energy and time enables the stretched hyperbola fit in a direct function of time to the data in Fig. 3 (dotted lines in Fig. 3), using the parameters α, v, E_{act} and T_0 (Table I). The fits agree well with the experimental data over the entire temperature range.

TABLE I. FITTING PARAMETERS

α	v (Hz)	E_{act} (eV)	T_0 (K)
1.9	4×10^5	0.78	680

In previous current degradation work [10], the exponent of t/τ (our t/t_0) was referred to as β. Physically, β should equal to T/T_0, where T_0 is the characteristic temperature reflecting the disorder broadening of the a-Si structure and should not change under low temperature stress. In [10], β was used as an independent fitting parameter at each temperature, and the T_0 caculated from T/β was temperature dependent, which is physically implausible. Although the data could be fitted at each temperature, there was no physical basis for using high temperature results to predict drain current degradation at room temperature.

In our work, we used a single fitting parameter T_0 for all stress temperatures, which makes a physically plausible accelerated lifetime test possible.

C. Acceleration Factors

We define the acceleration factor (AF) for drain current reduction as the ratio of the DC saturation current half-life [1] at room temperature (20°C) $t_{RT,50\%}$ to that at the stress temperatures $t_{ST,50\%}$. From (6), AF can be written as

$$AF = \frac{t_{RT,50\%}}{t_{ST,50\%}} = \frac{(50\%^{\frac{\alpha-1}{2}}-1)^{\frac{T_0}{T_{RT}}}v^{-1}\exp\left(E_{act}/kT_{RT}\right)}{(50\%^{\frac{\alpha-1}{2}}-1)^{\frac{T_0}{T_{ST}}}v^{-1}\exp\left(E_{act}/kT_{ST}\right)} \quad (8)$$

With the fitting parameters α, v, E_{act} and T_0, AFs at elevated temperatures are calculated from (8) and listed in Table II. A 160°C AF is 9.5×10^3, which means that accelerated lifetime tests with our method can be used to drastically reduce the duration of a-Si TFT stability tests.

TABLE II. HALF-LIVES AND ACCELERATION FACTORS AT STRESS TEMPERATURES FROM 20°C TO 160°C RELATIVE TO ROOM TEMPERATURE (20°C)

T (°C)	20	40	80	120	160
Half-life (s)	5.0×10^6	8.1×10^5	4.0×10^4	3.7×10^3	5.2×10^2
AF	1.0	6.1	1.2×10^2	1.4×10^3	9.5×10^3

D. Accelerated Threshold Voltage Shifts

With the current-voltage relation for a-Si TFTs in the saturation regime expressed as (5), the drain current degradation can be converted to the threshold voltage shift. The converted threshold voltage shifts as a function time at temperatures from 20°C to 160°C are shown with the open squares in Fig. 5. The threshold voltage shifts are faster at elevated temperatures. Stretched hyperbola fits with (4) and the fitting parameters listed in Table I are shown with the dotted lines in Fig. 5. Our results for the threshold voltage shift of TFTs biased in saturation regime at different temperatures agree well with the stretched hyperbola model and are similar to the threshold voltage shift analysis in [5-8].

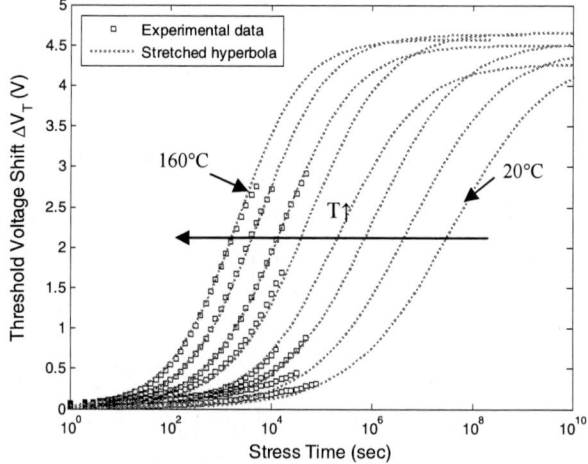

Figure 5. Threshold voltage shift data (open squares) and stretched hyperbola fits (dotted lines) of the a-Si TFTs at temperature stepped by 20°C from 20°C to 160°C.

Note that to convert the threshold voltage shift from drain current data and plot the stretched hyperbola fits require the knowledge of V_{T0} at each temperature. The difference of $(V_G - V_{T0})$ at each temperature leads to the different asymptotic values of ΔV_T at long time.

The analysis of the normalized drain current degradation with (6) and that of the threshold voltage shift with (4) are fundamentally similar. However, the normalized drain current degradation analysis we present here does not require the measurement of transfer characteristics to calculate V_{T0} of each transistor.

E. Discussion

For our benchmark of DC saturation current half-life, the 50% drain current degradation for an initial threshold voltage $V_{T0} = 0.44$V is equivalent to a threshold voltage shift of $\Delta V_T = 1.3$V, which corresponds to a 2D charged defect density of $\Delta N_{DB} = 1.7\times10^{11}$cm^{-2} created during the low gate field stress.

We have explained our approach in terms of weak bonds with the exponential barrier distribution model [12] and additional weakening of the occupied conduction-bandtail states [5]. However, we note that the stretched hyperbola model [5-8], the basis of our method, is a semi-empirical model, which can also be used to model the defect creation by a hydrogen diffusion model [11] with a carrier-dependent hydrogen-diffusion constant [5].

For the 160°C bias stress, subthreshold plots of I_D-V_G are shown in Fig. 6. The subthreshold slopes were ~750mV/dec both before and after the bias stress, with any change difficult to determine within the experimental resolution, which agrees with the observations in [9,10]. It is not straightforward to quantitatively relate the number of created defects as inferred by the threshold voltage shift data to the defects created in the a-Si or at the interface between the a-Si and the silicon nitride from the subthreshold slope data. This is because the threshold voltage shift characterizes the number of defects created deep in the band gap, while the subthreshold slope depends heavily on the localized band tail states as well [13].

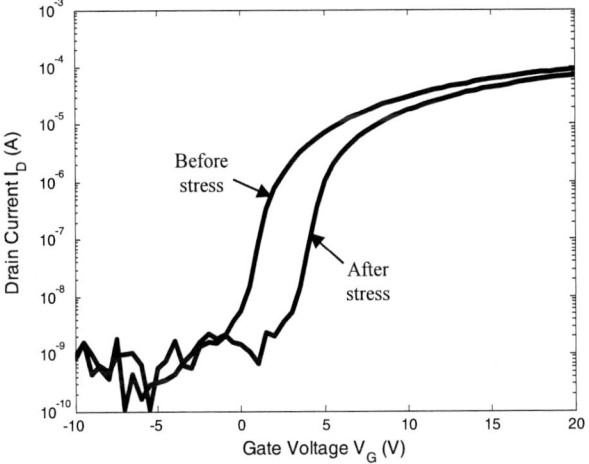

Figure 6. Subthreshold plots before and after a 1×10^4 sec bias stress at 160°C.

Highly stable a-Si TFTs reported recently have extrapolated DC saturation current half-lives that range from 100 to 1000 years, based on only months of testing at room temperature [1,2]. With our method for accelerated lifetime testing by monitoring the DC saturation current degradation at elevated temperatures, an acceleration factor of ~ 10^4 at 160°C is achieved, which enables us to establish the half-life of such devices with much higher confidence than tests conducted at room temperature alone.

V. SUMMARY

a-Si TFTs can be fabricated to have extremely long operating lifetimes under DC gate bias. The time at room temperature for the drain current to drop to 50% of its initial value can reach 100 to 1,000 years. To date, these lifetimes have been calculated by extrapolating tests conducted for several months at room temperature. Now we have shown that the rate of current drop can be accelerated by a factor of nearly 10,000 when the test temperature is raised to 160°C. Thus, a

month-long degradation test conducted at 160°C would yield the same information as a 10,000-month (~800 years) long test conducted at 20°C. This ability is particularly significant for predicting the stability and lifetime of a-Si TFTs in AMOLED displays.

ACKNOWLEDGMENT

This work was sponsored in part by the Industrial Technology Research Institute, Hsinchu, Taiwan.

REFERENCES

[1] B. Hekmatshoar, K. H. Cherenack, S. Wagner, and J. C. Sturm, "Amorphous silicon thin-film transistors with DC saturation current half-life of more than 100 years", Tech. Dig. Int. Electron Devices Meet., pp. 89-92, 2008

[2] B. Hekmatshoar, S. Wagner, and J. C. Sturm, "Tradeoff regimes of lifetime in amorphous silicon thin-film transistors and a universal lifetime comparison framework", Appl. Phys. Lett., Vol. 95, 143504, 2009

[3] M. J. Powell, C. van Berkel and J. R. Hughes, "Time and temperature dependence of instability mechanisms in amorphous silicon thin-film transistors", Appl. Phys. Lett., Vol. 54, No. 14, pp. 1323-1325, 1989

[4] R. A. Street, "Hydrogenated amorphous silicon", Cambridge Solid State Science Series, Cambridge, UK, pp. 88-91, 1991

[5] S. C. Deane, R. B. Wehrspohn, and M. J. Powell, "Unification of the time and temperature dependence of dangling-bond-defect creation and removal in amorphous-silicon thin-film transistors", Phys. Rev. B, Vol. 58, No. 19, pp. 12625-12628, 1998

[6] R. B. Wehrspohn, R. Brüggemann, S. C. Deane, I. D. French, I. G. Gale and M. J. Powell, "Urbach energy dependence of the stability in amorphous silicon thin-film transistors", Appl. Phys. Lett., Vol. 74, No. 22, pp. 3374-3376, 1999

[7] R. B. Wehrspohn, M. J. Powell and S. C. Deane, "Kinetics of defect creation in amorphous silicon thin film transistors", J. Appl. Phys., Vol. 93, No. 9, pp. 5780-5788, 2003

[8] R. B. Wehrspohn, J. R S. C. Deane, I.D. French, I. Gale, J. Hewett, and M. J. Powell, "Relative importance of the Si-Si bond and Si-H bond for the stability of amorphous silicon thin film transistors", J. Appl. Phys., Vol. 81, No. 1, pp. 144-154, 2000

[9] F. R. Libsch and J. Kanicki, "Bias-stress-induced stretched-exponential time dependence of charge injection and trapping in amorphous thin-film transistors", Appl. Phys. Lett., Vol. 62, No. 11, pp. 1286-1288, 1993

[10] C. C. Shih, Y. S. Lee, K. L. Fang, C. H. Chen and F. Y. Gan, "A current estimation method for bias-temperature stress of a-Si TFT device", IEEE Trans. Device and Materials Reliability, Vol. 7, No. 2, 2007

[11] J. Kakolios, R. A. Street, and W. B. Jackson, "Stretched-exponential relaxation arising from dispersive diffusion of hydrogen in amorphous silicon", Phys. Rev. Lett., Vol. 59, pp. 1037, 1987

[12] R. S. Crandall, "Defect relaxation in amorphous silicon: stretched exponentials, the Meyer-Neldel rule, and the Staebler-Wronski effect", Phys. Rev. B, Vol. 43, pp. 4057, 1991

[13] G. W. Neudeck and A. K. Malhotra, "An amorphous silicon thin film transistors: theory and experiment", Solid Stat. Electron., Vol. 19, pp. 721, 1976

Investigation of Ultra Thin Polycrystalline Silicon Channel for Vertical NAND Flash

Bio KIM, Seung-Hyun LIM, Dong Woo KIM, Toshiro NAKANISHI, Sangryol YANG, Jae-Young AHN,
HanMei CHOI, Kihyun HWANG, Yongsun KO, Chang-Jin KANG
Process Development Team, Memory R&D Center, Samsung Electronics Co., Ltd.
San #16 Banwol-Dong
Hwasung-City, Gyeonggi-Do, Korea
82-31-208-0275, bioman.kim@samsung.com

Abstract— We have investigated thin film transistors (TFTs) with ultra-thin polycrystalline silicon (poly-Si) of 77 Å - 185 Å. The TFT charge transfer characteristics such as ON current and effective mobility are dominated not by the thickness itself but by the grain size of poly-Si channel. When the poly-Si channel thickness is decreased with the same grain size, the sub-threshold TFT characteristics are improved without degradation of ON current and reliability properties. These results give us appropriate criteria to establish an excellent poly-Si channel in vertical NAND flash memory.

Keywords - ultra-thin polycrystalline silicon; vertical NAND flash; thin film transistor; grain boundary

I. INTRODUCTION

The increases of mobile alliances market, such as smart phones and high-performance memory cards, require NAND flash memory with large storage capacity and faster performance. To satisfy market requirement, increasing bit density and reducing bit cost of NAND flash memory are indispensable. Several manufactures have already started production with 20nm-class NAND flash memories in 2010, but faced scaling limit of floating gate (FG) NAND flash memories. As an alternative of FG NAND flash memory to overcome scaling limit and to reduce bit cost, 3D stacked vertical NAND flash memories such as BiCS [1] and TCAT [2] have caught great attentions. These vertical NAND flash memories use polycrystalline silicon (poly-Si) as a transistor channel material instead of single crystal silicon in planar FG NAND flash memory. Vertical NAND especially prefers thinner poly-Si channel to improve controllability of vertical transistor [3]. In this paper, we will show the electrical properties of ultra-thin poly-Si thin film transistor (TFT) with a thickness of less than 200 Å for vertical NAND flash application for the first time.

II. DEVICE FABRICATION AND STRUCTURAL CHARACTERIZATION

Fig. 1 shows the plan view and cross-sectional view of poly-Si TFT. First, 500Å amorphous silicon (a-Si) was deposited by low pressure chemical vapor deposition (LPCVD) on 1000 Å thick wet oxide and contact pads were patterned to reduce contact resistance of thin poly-Si channel ('pad poly' in Fig. 1(b)). Secondly, undoped a-Si was deposited as channel layer by LPCVD and followed by solid phase crystallization (SPC). After the active regions were defined, 80Å SiO2 layer was deposited as gate dielectric. Then, another a-Si was deposited and patterned to be the gate of the device. After defining the gate, ion implantations were added to source, drain and body of TFT.

The grain structure of poly-Si channel was analyzed using electron backscattered diffraction (EBSD) as shown in Fig. 2. If the deposition and the annealing conditions are same, the average grain size of poly-Si depends on the thickness of as-deposited a-Si. Therefore, active channel of thin poly-Si contains more grains as shown in Fig. 2. Fig. 3 shows that the grain size decreases with the decrease of poly-Si thickness as reported in thick poly-Si films [4].

To distinguish grain size effect and thickness effect in TFT,

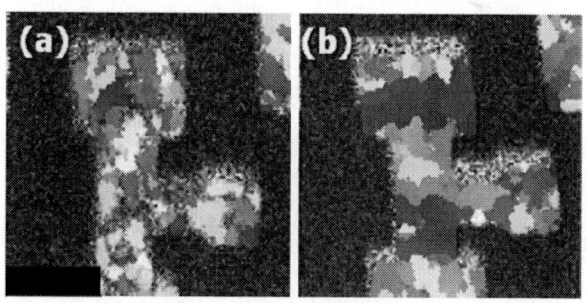

Figure 2. EBSD images of active patterned poly-Si channel with source, drain and body contact. The active area is 1 μm x 1 μn. Poly-Si thicknesses of (a) is ~170 Å, and thicknesses of (b) is ~500 Å.

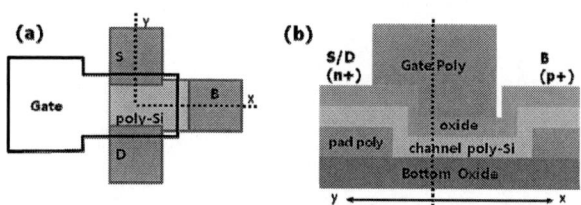

Figure 1. (a) Plan view and (b) Cross sectional view of ultra-thin poly-Si TFT architecture. Gate and pad poly-Si are patterned to minimized external resistance of poly-Si channel.

978-1-4244-9113-1/11 $26.00 © 2011 IEEE

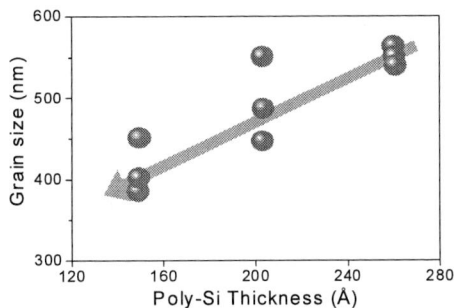

Figure 3. Poly-Si thickness dependence of grain size. Grain size is evaluated from EBSD and the poly-Si thickness is as-deposited thickness.

Sample Number	Poly-Si thickness (Å)	Grain size
1	185	Large
2	88 ~ 138	Medium (variable)
3	87 ~ 130	Large
4	77	Small

TABLE I. POLY-SI CHANNEL SAMPLES. VARIOUS THICKNESS POLY-SI TFTS WERE MADE USING SLOPE ETCH OF SAMPLE 2, 3 AND THE GRAIN SIZE WERE CONTROLLED BY CHANGING THE AS-DEPOSITED A-SI THICKNESS.

we prepared 4 samples as shown in Table 1. The grain size was controlled by changing initial a-Si thickness and the thickness was controlled by etching with a slope. The grain size of sample 3 is same to sample 1 and independent of poly-Si thickness. On the other hand, the grain size of sample 2 is smaller than sample 1 and the grain size is proportional to the thickness of poly-Si. Sample 4 has the smallest grain size and thickness.

III. ELECTRICAL MEASUREMENT RESULTS

The Id-Vg characteristics of samples 1 and 4 are shown in Fig. 4. Due to small grain size of sample 4, we estimated 1µm × 1µm size transistor of sample 4 contains more grains than sample 1 which contains a few grains or single grain without grain boundary. Therefore, Id-Vg characteristics of sample 1 can be strongly affected by the grain boundary condition in the channel. Simulation results of the role of trap in poly-Si channel (Fig. 5) support this behavior. According to the simulation result, drain current is mainly influenced not by the trap location but by the trap density in TFT channel. This explains the Id-Vg curve distribution difference between samples 1 and 4.

Fig. 6 and Fig. 7 show poly-Si thickness dependences of ON current and effective mobility, respectively. Poly-Si thickness of samples 2 and 3 is divided into 3 regions. These graphs indicate that ON current and effective mobility of poly-Si decrease with decrease of grain size of poly-Si channel (samples 1, 2 and 4). On the other hand, these values are independent of poly-Si thickness down to ~88Å (samples 1 and

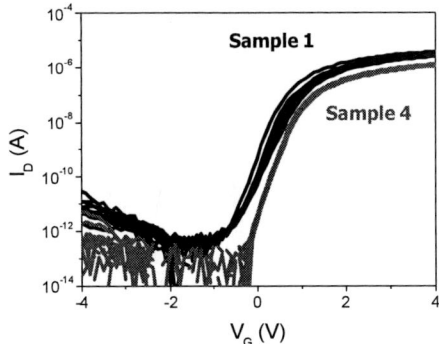

Figure 4. Id-Vg characteristics of sample 1 and sample 4. Transistor size is 1µm × 1µm.

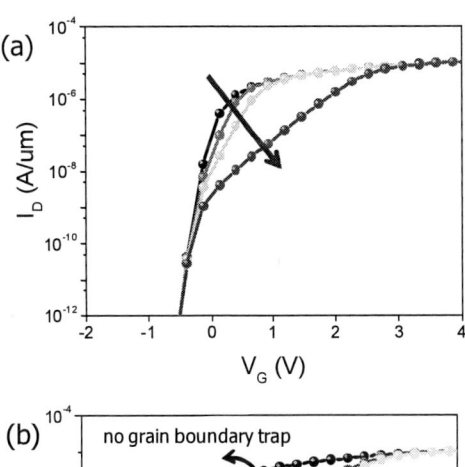

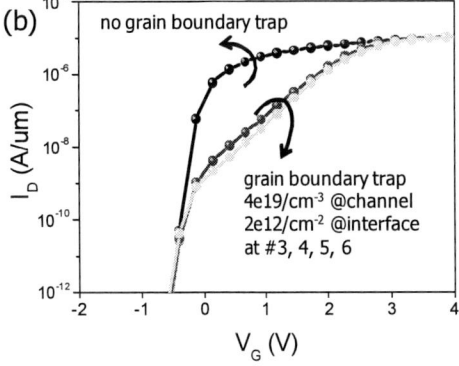

Figure 5. Simulation result of grain boundary effect to drain current. (a) GB is located at #4 in the inset of figure. The trap density in poly-Si channel is changed from $2e18/cm^3$ to $4e19/cm^3$, and interface trap density is changed from $1e11/cm^2$ to $2e12/cm^2$. (b) An effect of grain boundary location in TFT channel. As far as the trap sites are located in the channle region, drain current redution is independent of grain boundary location.

3) if the grain sizes are same. These results indicate that the charge transport in inversion layer is disturbed by the trap sites in grain boundary of poly-Si channel. Figures in Fig. 6 roughly show poly-Si thickness and grain boundary properties.

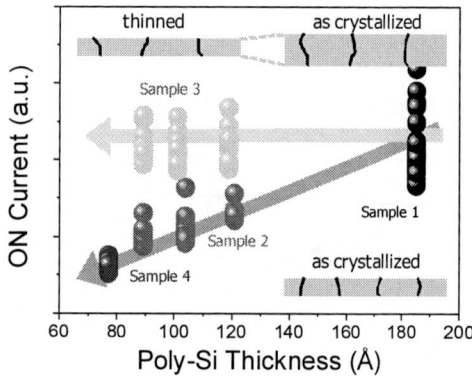

Figure 6. Poly-Si thickness dependence of ON current. Sample 3 shows large current.

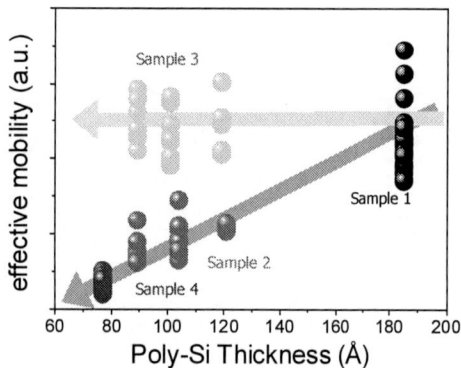

Figure 7. Poly-Si thickness dependence of effective mobility.

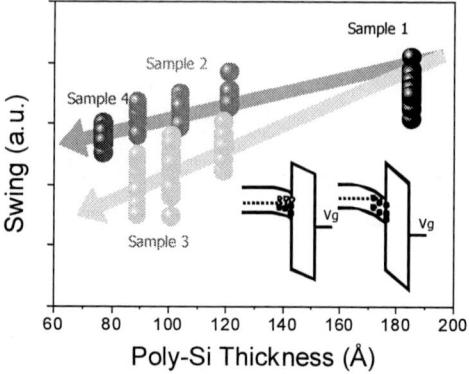

Figure 8. Sub-threshold swing as a function of poly-Si thickness.

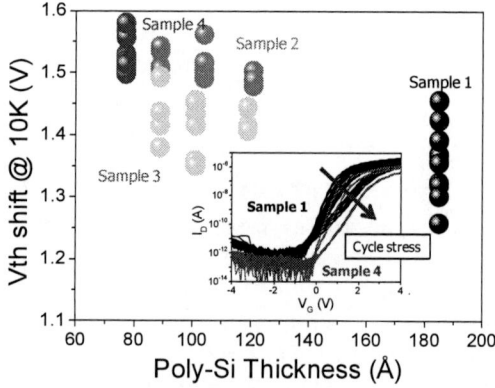

Figure 9. Vth shift after 10K cycle stress as a function of poly-Si thickness. (inset) Id-Vg characteristics of sample 1 and 4 at initial state and after 1K, 10K cycle stresses.

Sub-threshold swing property of TFT is shown in Fig. 8. Due to the thin body of TFT structure which suppresses short channel effect, the thinner poly-Si TFT is expected to have the lower swing value. Although the swing decreases with the decrease of poly-Si thickness regardless of grain size, swing of small grain size sample is larger than that of large grain size sample. In poly-Si TFT, increasing gate voltage fills not only the trap sites in poly-Si and gate oxide interface, but also the trap sites in poly-Si channel. Therefore the increment of grain boundary in sample 2 and 4 compensates the thin body effect and the decrease of swing is less prominent than sample 3 (inset of Fig. 8).

The reliability of poly-Si TFT is measured by applying bipolar pulse stress to gate. Id-Vg characteristics sample 1 and 4 are shown in inset of Fig. 9. As shown in Fig. 4, the curve distribution of sample 4 is better than sample 1 and does not affected by the cyclic pulse stress. Swing degradation is primary factor of poly-Si reliability under cycle stress. Vth shifts after 10K cycle stress are shown in Fig. 9. The poly-Si TFT with large grain size have large resistance to cycle stress regardless of poly-Si thickness.

The trap density of poly-Si channel is compared using grain barrier height model [5] and charge pumping technique. In grain barrier height model, current flows by thermionic emission of electron over energy barrier by trapped electrons at poly-Si grain boundaries (inset of Fig. 11). Grain boundary barrier height (Φ_{GB}) is proportional to the square of grain boundary trap density (Nst). Nst values in inversion layer are obtained by measuring temperature dependence of Id-Vg characteristics and the results are shown in Fig. 10. Nst of sample 2 and 4 is inverse proportional to the grain size as expected. On the other hand, Nst of sample 3 is independent of poly-Si thickness.

In normal Si single crystal transistor, charge pumping current (Icp) is proportional to the frequency using constant rising and falling time. In poly-Si TFT, however, Icp is not proportional to the frequency, i.e. gate pulse frequency dependence of recombined charge per pulse period is not independent of the frequency as shown in Fig. 11(a). This behavior may come from the high resistance of poly-Si channel [6] and long time is necessary to supply enough electrons to channel. In poly channel TFT structure, some free holes (electrons) cannot go back to the body (source/drain) during the trapping (detrapping) process and they are recombined with electrons (holes) from the source/drain (body). It contributes additional charge pumping current (geometric current). This geometric current can be suppressed by increasing rising

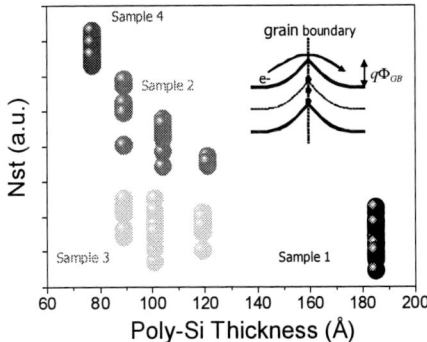

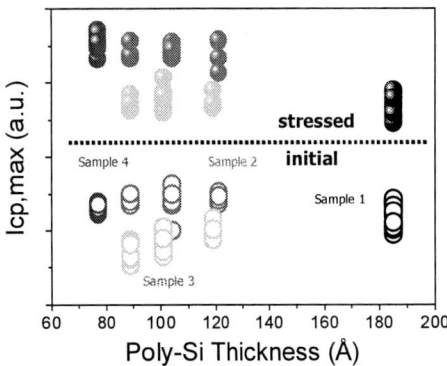

Figure 10.   Grain boundary trap density as a function of poly-Si thickness. (inset) Energy band diagram near grain boundary trap.

Figure 12.   Charge pumping current (Icp) measurement results at initial state (open circle) and after 10K cycle stress (filled cicle).

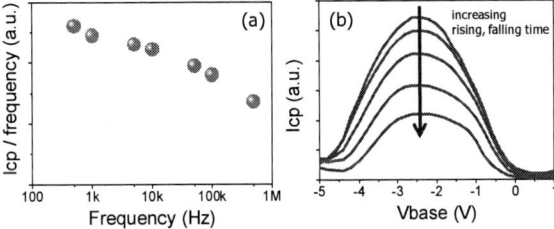

Figure 11.   (a) Frequency dependence of charge pumping current at fixed rsing falling time. (b) Rising and falling time dependence of charge pumping current at 1kHz. Rising and falling time varies from 100 ns to 400 μs.

grain size not by the poly-Si thickness. Increasing grain size is the key for large ON current and better reliability of poly-Si TFT.

(falling) time of gate pulse which gives enough time for holes (electrons) to flow back to the body (source/drain). Fig. 11(b) shows rising and falling time dependence of Icp. By increasing rising and falling time Icp decreases and regular shape charge pumping curve appears [7]. Therefore, low frequency and long rising/falling time gate pulse is used to measure charge pumping current in poly-Si TFT. The charge pumping measurement result before and after 10K cycle stress is shown in Fig. 12. It is similar to previous measurement results. The thinned poly-Si with large grain size (sample 3) shows the least trap sites and the best reliability.

IV.   CONCLUSION

The electrical properties of ultra-thin poly-Si TFT (down to 77Å) are systematically investigated. The measurement results indicate that the electrical property is dominantly affected by

[1]   H. Tanaka, M. Kido, K. Yahashi, M. Oomura, R. Katsumata, M. Kito, Y. Fukuzumi, M. Sato, Y. Nagata, Y. Matsuoka, Y. Iwata, H. Aochi and A. Nitayama, "Bit cost scalable technology with punch and plug process for ultra high density flash memory," VLSI Symp. Tech. Dig. 2007, pp. 14-15.

[2]   J. Jang, H. S. Kim, W. Cho, H. Cho, J. Kim, S. I. Shim, Y. Jang, J. H. Jeong, B. K. Son, D. W. Kim, K. Kim, J. J. Shim, J. S. Lim, K. H. Kim, S. Y. Yi, J. Y. Lim, D. Chung, H. C. Moon, S. Hwang, J. W. Lee, Y. H. Son, U. I. Chung and W. S. Lee, "Vertical cell array using TCAT (terabit cell array transistor) technology for ultra high density NAND flash memory," VLSI Symp. Tech. Dig. 2009, pp. 192-193.

[3]   Y. Fukuzumi, R. Katsumata, M. Kito, M. Kido, M. Sato, H. Tanaka, Y. Nagata, Y. Matsuoka, Y. Iwata, H. Aochi and A. Nitayama, "Optimal integration and characteristics of vertical array devices for ultra-high density, bit-cost scalable flash memory," IEDM Tech. Dig. 2007, pp. 449-452.

[4]   M. K. Hatalis and D. W. Greve., "Large grain polycrystalline silicon by low-temperature annealing of low-pressure chemical vapor deposited amorphous silicon films," J. Appl. Phys. vol. 63, pp. 2260-2266, 1988.

[5]   H. Chen and Ching-Yuan Wu, "An analytical grain-barrier height model and its characterization for intrinsic poly-Si thin-film transistor," IEEE Trans. Electron Device, vol. 45, pp. 2245-2247, 1998.

[6]   N. S. Saks, S. Batrab and M. Manning, "Charge pumping in thin film transistors," Microelectron. Eng. vol. 28, pp. 379-382, 1995.

[7]   L. Lu, M. Wang and M. Wong, "Geometric effect elimination and reliable trap state density extraction in charge pumping of polysilicon thin-film transistors," IEEE Electron Device Lett. vol. 30, pp. 517-519, 2009.

Electrical Reliabilities of Porous Silica Low-k Films

Takamaro Kikkawa[1], Yasuhisa Kayaba[1], Kazuo Kohmura[2] and Shinichi Chikaki[3]

[1] Hiroshima University, Higashi-Hiroshima, Hiroshima, 739-8527 Japan

Phone: (81)-(82)-424-7879, e-mail address: kikkawat@hiroshima-u.ac.jp

[2] Mitsui Chemicals Inc., Chiba, 299-0265 Japan

[3] Renesas Electronics Corp., Kanagawa, 252-5298 Japan

Abstract—Electrical reliability of self-assembled porous silica films was investigated. Vapor phase silylation by use of 1,3,5,7-tetramethylcyclotetrasiloxane (TMCTS) was developed to reduce silanol groups and enhance siloxane cross-linkage, resulting in achieving lower dielectric constant and higher elastic modulus. To promote siloxane cross-linkage, Cs ion was doped to its precursor solution. The self-assembled porous silica low-k film was integrated in Cu damascene interconnects with ultraviolet (UV) irradiation and TMCTS vapor treatment, resulting in the highest elastic modulus of 9 GPa with the dielectric constant of 2.1. Sidewall protection layer was formed in the trench for Cu interconnects to improve time-dependent dielectric breakdown (TDDB) lifetime of more than 10 years at the electric field of 2.3 MV/cm.

Keywords-porous silica; low-k; self-assembly; silylation; cesium, TDDB; Cu damascene

I. INTRODUCTION

Highly reliable ultra-low dielectric constant (low-k) films are needed for advanced ultra-large scale integrated circuits (ULSI) to reduce parasitic resistance-capacitance (RC) time constants of interconnects [1, 2].

Sol–gel derived porous silica (PoSiO) films were developed because of easy control of dielectric constants of the films [3-6]. However, the degree of cross-linkage of the PoSiO film becomes lower when the fabrication temperature decreases down to 350°C to maintain the reliability of copper (Cu) interconnects [7, 8] .

To solve the issue of mechanical degradation of the porous film, a self-assembly technology was developed to control pore distribution of the PoSiO film so that random agglomeration of porogen could be eliminated [9-12]. Another issue of PoSiO low-k films is to lower leakage current of the PoSiO low-k film [13-15]. Since the leakage current is originated from residual silanol groups, reductions of the silanol groups and water adsorption are the most important task to guarantee the reliability of the porous film.

In order to terminate silanol groups, hydrophobic groups such as methyl–silicon and hydrogen–silicon have been used as silylation agents, e.g., hexamethyldisilazane (HMDS) and 1,3,5,7-tetramethylcyclotetrasiloxane (TMCTS) [12, 13]

One of the most important reliability issues for low-k films is time dependent dielectric breakdown (TDDB). The mechanism of TDDB for porous silica is more complex than that of thermal oxide because of its non-uniformity of internal electric field [16]. The degradation of TDDB lifetime is initiated from a weak chemical bond such as silanol groups so that porous silica films have more weak bonds when they are fabricated at lower temperature.

In this paper, reliability of the integration of the self-assembled porous silica low-k film in Cu damascene interconnects is discussed [17, 18]. The effect of silylation with 1,3,5,7-tetramethylcyclotetra siloxane (TMCTS) on cross-linkage of siloxane is discussed. The effects of cesium (Cs) doping on the electrical reliabilities as well as elastic modulus are investigated. The TDDB lifetimes of the low-k films integrated in Cu damascene are also discussed.

II. EXPERIMENTAL

A. Fabrication of self-assembled porou silica

A precursor solution of porous silica was synthesized from an acidic silica sol derived from tetraethylorthosilicate (TEOS) in ethanol diluted with water and a nonionic surfactant polyoxyethylene-20-stearylether (Brij 78). To increase the degree of siloxane crosslinkage of the PoSiO, cesium nitrate ($CsNO_3$) was added to the precursor solution. The doping concentrations of Cs were 0, 5, 15, and 30 wt ppm. Each solution was spin-coated on a 300 mm diameter silicon (Si) wafer and cured under nitrogen (N_2) ambient at 150°C. After being annealed at 350°C in TMCTS vapor for 90 min, the surfactant was removed with the aid of UV irradiation with heating a Si substrate at 350°C, and a few nanometer-scale air pores were formed in the silica skeleton as shown in Fig. 1. The wafers were cut into 3.5 cm × 3.5 cm dies to measure the film properties.

B. Integration of Low-k/Cu damascene

PoSiO in low-*k*/Cu damascene interconnects was deposited sequentially between an etch-stop film and a cap film. The stacked etch-stop films of SiCN and SiC were formed by plasma-enhanced chemical vapor deposition (PECVD) with methylsilane/NH_3 gases and with a precursor of organosiloxane without oxidant, respectively. The cap SiOC film was deposited by PECVD using ethoxymethylsiloxane. The porous SiOC film was deposited by PECVD using ethoxymethylsiloxane and an organic monomer. The SiO_2 film was deposited by PECVD using SiH_4 and O_2 gases. The Cu/Low-k damascene interconnects were formed as shown in Fig. 2. The stacked dielectric films were patterned by plasma etching with CF4 and Ar gases with an ArF photo resist mask. Furnace annealing was carried out at 350°C with a mixture gas

978-1-4244-9113-1/11 $26.00 © 2011 IEEE

of N_2 and TMCTS vapor. A sidewall (SW) protection process was carried out. A SiC film was deposited by PECVD with the same chemicals as the etch-stop layer, followed by anisotropic plasma etching of the SiC film and wet cleaning. After deposition of a barrier of TaN and Ta (Ta/TaN metal), Cu sputtering, Cu plating, and chemical mechanical polishing (CMP) of Cu/Ta/TaN metals were carried out. The SiC/SiCN stack film, SiO_2 film, and PECVD-SiN film were sequentially formed as a passivation layer.

III. RESULTS AND DISCUSSION

A. Effect of Cs doping on reliability

The porosity (P) was determined from the dielectric constant k at RH 0% by using the Rayleigh equation for PoSiO, which has a cylindrical pore and silica skeleton

$$P = \frac{k^2_{silica} - kk_{silica} - k + k_{silica}}{k^2_{silica} + k_{silica}k - k - k_{silica}} \qquad (1)$$

where k_{silica} (=3.9) is the dielectric constant of the silica skeleton. The dielectric constants of Cs-doped PoSiO with 0, 5, 15, and 30 wtppm were 1.89, 1.92, 1.98, and 2.06, respectively. Chemical bonding structures were measured by transmission Fourier transform infrared spectroscopy (FTIR) under N_2 ambient as shown in Fig. 3. Broad and intense peaks at 1000–1250 cm^{-1} correspond to the asymmetric vibration mode of the Si–O–Si network bond. The peak at 1060 cm^{-1} is a ladder-like siloxane network. A peak at 1265 cm^{-1} is the C–H bending mode of H (CH_3)SiO_2 (D-type) structure, which is the non-bridged moiety of the chemisorbed TMCTS. A peak at 1280 cm^{-1} is the C–H bending mode of (CH_3)SiO_3 (Si–Me T-type) structure, which is a bridged structure between TMCTS and pore surface silica or next neighboring TMCTS. The peak intensity increased as 0.08, 0.18, 0.29 and 0.29 for Cs 0, 5, 15, and 30 wt ppm PoSiO. Therefore, the degree of siloxane cross-linkage of TMCTS via Si–Me T-type structure was promoted by Cs doping. A peak at 3740 cm^{-1} corresponds to the O–H stretching mode of an isolated silanol group and a broad peak between 3200 and 3720 cm^{-1} corresponds to the same mode of a hydrogen-bonded silanol group. The peak at 2970 cm^{-1} is the C–H stretching mode of methyl (CH_3) group of TMCTS. The peak at 2920 cm^{-1} and the shoulder at 2880 cm^{-1} is the same mode of the CH_2 group. There are two peaks around 2200 cm^{-1} due to the H–Si stretching mode of TMCTS. The peak at 2180 cm^{-1} corresponds to the Si–H stretching modes of the D-type structure of TMCTS. The Si–H stretching mode of the H–Si–O_3 (H–Si T-type) structure of TMCTS was found between 2250 and 2280 cm^{-1}.

Figure 4 shows the elastic modulus as a function of porosity. From a finite-element analysis, the elastic modulus (E) is described by

$$E = E_W (1 - P)^2 \qquad (2)$$

Here, E_w is the elastic modulus of the silica skeleton [3]. By comparing this equation with the experimental results, E_w were obtained as 10, 17, 31 and 31 GPa for Cs 0, 5, 15 and 30 ppm, respectively. Figure 5 shows elastic modulus versus dielectric constant for PoSiO with UV irradiation process as well as TMCTS treatment. The elastic modulus of 40GPa was

calculated for skeletal silica. The highest elastic modulus (9 GPa) with an ultralow-k value (2.1) was achieved by the new process of UV irradiation and TMCTS silylation.

The leakage current of the Cs doped porous silica low-k film decreased with increasing Cs doping concentration. The leakage current of Cs 0 ppm PoSiO reached a compliance value after bias voltage was applied. Relative humidity (RH) dependence of E-J characteristics was investigated. The E-J characteristic at RH 2% (under N_2 ambient) corresponds to a relaxation current of proton conduction. The E-J characteristic, which was saturated at high electric field strength over approximately 0.5 MV/cm, is the resistive current of proton conducting. Significant improvement was achieved for Cs 30 ppm PoSiO. The leakage current did not depend on RH up to 60% probably because the proton conducting paths through adsorbed water were closed due to the highly cross-linked TMCTS layer on the pore surface.

The electric field strength of time zero dielectric breakdown (TZDB) was measured under N_2 ambient with heating Si substrate at 200°C. Figure 6 shows a Weibull distribution of the TZDB field strength (E_{DB}). In Cs 0 and 5 ppm PoSiO the probability density of the first DB mode was 30 and 20%, respectively. However, Cs 15 ppm PoSiO and Cs 30 ppm PoSiO did not have the first DB mode. The E_{DB} value of 50% probability was −3.4, −4.7, −6.2 and −6.7 MV/cm for Cs 0, 5, 15 and 30 ppm PoSiO, respectively. The enhancement of E_{DB} by Cs doping could be attributed to the replacement of electrically weak bonds such as SiOH with strong bonds such as Si–C.

The TDDB lifetime (t_{DB}) under N_2 ambient is expressed as follows.

$$t_{DB} = A \exp(\frac{\Delta H_0}{k_B T}) \exp[\ \gamma(T)(E_{DB} - E)] \qquad (3)$$

Here, t_{DB} is TDDB lifetime, A is a constant, ΔH_0 is change in enthalpy, k_B is the Boltzmann constant, T is temperature, $\gamma(T)$ is a function of temperature, E_{DB} is TZDB field strength, and E is the applied field strength. Figure 7 shows the t_{DB} of 50% probability [$t_{DB}(50\%)$] at 200°C for various PoSiO with different Cs doping concentrations. For Cs 0 ppm PoSiO and Cs 5 ppm PoSiO the bias stress was −3.5 MV/cm and -4.7 MV/cm. For Cs 15 and 30 ppm PoSiO E was improved to −6.3 and −6.7 MV/cm, respectively. It was found that $t_{DB}(50\%)$ was improved by Cs doping. According to the E model this improvement is attributed to the enhancement of E_{DB} as shown in Fig. 6. Then, the projected TDDB lifetime satisfied 10 years even at 220°C with an $|E| = 1$ MV/cm stress condition.

B. Low-k/Cu damascene integration

Figure 8 shows cross-sectional HAADF-STEM micrographs of PoSiO/Cu damascene interconnects. Bright areas indicate heavy metal lines (Cu). The sample treated with an optimized TMCTS silylation after trench etching before metal deposition showed no dark areas as shown in Fig. 8(a). On the other hand, serious deformation of large voids and Cu precipitation or intrusion occurred in the low-k film without TMCTS treatment as shown in Fig. 8(b). Consequently, the optimized TMCTS silylation could prevent voids and Cu precipitation in the porous low-k film.

Side-wall (SW) of the Cu damascene trench was protected by a low-k film with high step coverage and no damage. Figure 9 shows the cumulative probability of sheet resistance of dense serpentine wirings of the PoSiO/Cu damascene interconnect with and without SW protection. Although sharp distribution was obtained for the straight lines, the distribution for the serpentine patterns spreads to higher resistances without SW protection due to Cu diffusion. The SW protection with the thicknesses of 4.5 nm and 7.0 nm improved the sheet resistance distribution and yield. The median resistance was reduced by 30%. From plan-view TEM, a large Cu void was observed and only barrier metal layers remained at both sides of the Cu line. A Cu void was formed at the side of the barrier metal layer of the folded pattern. The large Cu void was formed by the migration and agglomeration of voids. The TDDB of PoSiO/Cu interconnects having a 100 nm wide and 100 nm spacing comb pattern was carried out at the high applied voltage of 35 V and the temperature of 150°C. Figure 10 shows the mean TDDB lifetime as a function of electric field for PoSiO/Cu damascene interconnects. The slope of the extrapolation for PoSiO/Cu (dotted lines) was assumed to be the same as that of the reference porous SiOC/Cu. The extrapolation shows the lifetime of 10 years under the stress of 2.3 MV/cm.

IV. CONCLUSION

A highly cross-linked porous silica film silylated with TMCTS was fabricated at a low temperature of 350°C. To promote the degree of siloxane linkage of the porous silica, cesium (Cs) was added to its precursor solution with the concentration of 0, 5, 15, and 30 wtppm. Then, the amount of methyl–silicon–three oxygen (Me–Si T-type) and hydrogen–silicon–three oxygen (H–Si T-type) bridged structures of TMCTS was increased, and the silanol group was decreased markedly. The electrical reliabilities were measured and high reliabilities were obtained for Cs 30 ppm-doped porous silica. The influence of humidity on the increase of dielectric constant was suppressed to only 4% even at the RH condition of 65% compared to that of RH 0%. Time zero DB field strength (E_{DB}) was improved from 3.4 to 6.7 MV/cm. Owing to this improvement the TDDB lifetime was prolonged for Cs 30 ppm porous silica, and its projected 10 year lifetime was satisfied under the stress conditions of 220°C and $|E| = 1$ MV/cm. Integration of self-assembled PoSiO film in Cu damascene interconnects was achieved. The high elastic modulus (9 GPa) with an ultralow-k value (2.1) was achieved by UV irradiation and silylation. TMCTS silylation in the sidewall of damascene trenches after trench etching resulted in the suppression of Cu precipitation and void formation in the PoSiO film. A side wall protection process could improve the reliability of PoSiO/Cu damascene interconnects.

ACKNOWLEDGEMENTS

This work was partially supported by NEDO and Hiroshima Prefecture.

REFERENCES

[1] T. Kikkawa, S. Chikaki, R. Yagi, M. Shimoyama, Y. Shishida, N. Fujii, K. Kohmura, H. Tanaka, T. Nakayama, S. Hishiya, T. Ono, T. Yamanishi, A. Ishikawa, H. Matsuo, Y. Seino, N. Hata, T. Yoshino, S. Takada, J. Kawahara, and K. Kinoshita, IEEE International Electron Devices Meeting (IEDM 2005) Technical Digest, (Washington DC, Dec.5, 2005, IEEE), pp.99-102.

[2] S. Chikaki, K. Kinoshita, T. Nakayama, K. Kohmura, H. Tanaka, M. Hirakawa, E. Soda, Y. Seino, N. Hata, and T. Kikkawa, IEEE International Electron Devices Meeting (IEDM 2007) echnical Digest., 2007, pp.969-972.

[3] H. Miyoshi, H. Matsuo, Y. Oku, H. Tanaka, K. Yamada, N. Mikami, S. Takada, N. Hata and T. Kikkawa, Japanese Journal of Applied Physics 43 (2), pp.498-503, (2004).

[4] H. Miyoshi, H. Matsuo, Y. Oku, H. Tanaka, K. Yamada, S. Takada, N. Hata and T. Kikkawa, Japanese Journal of Applied Physics 44 (3), 1161-1165 (2005).

[5] H. Miyoshi, N. Hata and T. Kikkawa, Japanese Journal of Applied Physics, 44, (3), 1166-1168 (2005).

[6] H. Miyoshi, Y. Yamada, K. Kohmura, N. Fujii, H. Matsuo, H. Tanaka, Y. Oku, Y. Seino, N. Hata and T. Kikkawa, Japanese Journal of Applied Physics, 44, (8), pp.5982-5986 (2005).

[7] T. Syozo, H. Nobuhiro, S. Yutaka, F. Nobutoshi and T. Kikkawa, Journal of Applied Physics, 97 (11), pp.113504-113508 (2005).

[8] T. Syozo, H. Nobuhiro, S. Yutaka, F. Nobutoshi and T. Kikkawa, Journal of Applied Physics, Vol. 100, pp. 123512-1 123512-5, (2006).

[9] K. Yamada, Y. Oku, N. Hata, S. Takada and T. Kikkawa, Japanese Journal of Applied Physics, 42 (4B), April, pp.1840-1842 (2003).

[10] K. Yamada, Y. Oku, N. Hata, Y. Seino, C. Negoro and T. Kikkawa, Journal of The Electrochemical Society,151(10), F248-F251 (2004)

[11] N. Fujii, K. Kohmura, T. Nakayama, H. Tanaka, N. Hata, Y. Seino, and T. Kikkawa, Proc. SPIE Optics East 2005, (Oct. 24, 2005, The International Society for Optical Engineering), p.71.

[12] K. Kohmura, H. Tanaka, S. Oike, M. Murakami, N. Fujii, S. Takada, T. Ono, Y. Seino and T. Kikkawa, Thin Solid Films, 515(12), pp. 5019-5024, (2007).

[13] T. Kikkawa,S. Kuroki, S. Sakamoto, K. Kohmura, H. Tanaka, and N. Hata, Journal of The Electrochemical Society, 152(7), G560-G566, (2005),.

[14] S. Kuroki and T. Kikkawa, Journal of The Electrochemical Society, 153(8), G759-G764, (2006).

[15] Y. Kayaba, K. Kohmura, and T. Kikkawa, Japanese Journal of Applied Physics, Vol. 47, No. 11, pp. 8364-8368, 2009.

[16] Y. Kayaba, and T. Kikkawa, Japanese Journal of Applied Physics, Vol. 47, No. 7, pp. 5314-5319, 2008.

[17] Y. Kayaba, K. Kohmura, H. Tanaka, Y. Seino, T. Ohdaira, F. Nishiyama, K. Kinoshita, S. Chikaki, and T. Kikkawa, Journal of The Electrochemical Society, 155 (11), G258-G264, (2008).

[18] S. Chikaki, K. Kinoshita, K. Kohmura,H. Tanaka,.E. Soda,.T. Suzuki,. Y. Seino, N. Hata, S. Saito, and T. Kikkawa, Journal of The Electrochemical Society, 157 (5) H519-H525, (2010).

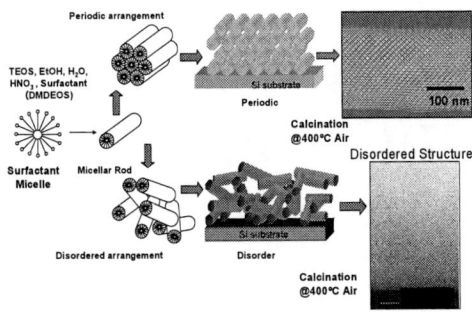

Fig. 1. Fabrication of self-assembled porous silica low-k film.

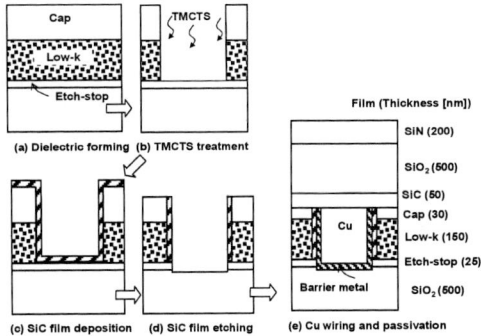

Fig. 2. Integration of porous low-k/Cu damascene.

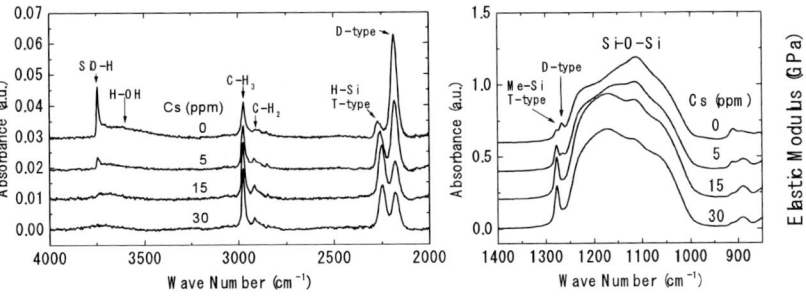

Fig. 3. Fourier transform infrared spectroscopy (FTIR) absorbance spectra of Cs doped porous silica.

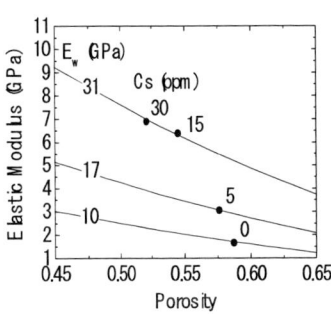

Fig. 4. Elastic modulus of Cs doped porous silica.

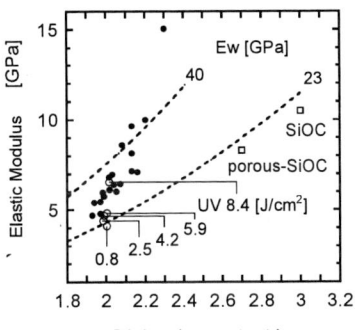

Fig. 5. Elastic modulus versus dielectric constant for PoSiO.

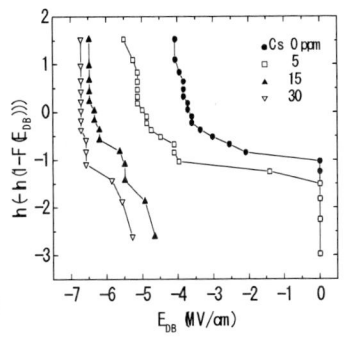

Fig. 6. Weibull distribution of the TZDB field strength for PoSiO with Cs doping.

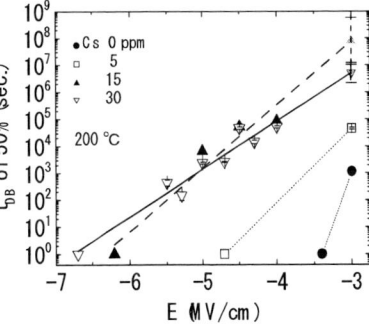

Fig. 7. Effect of Cs doping on TDDB lifetime of 50% probability at 200°C.

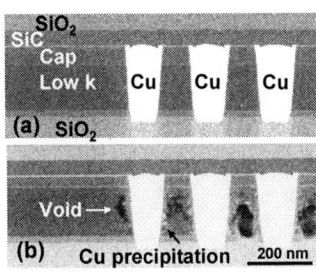

Fig. 8. Cross-sectional HAADF-STEM micrographs of PoSiO/Cu damascene interconnects. (a) With TMCTS. (b) Without TMCTS.

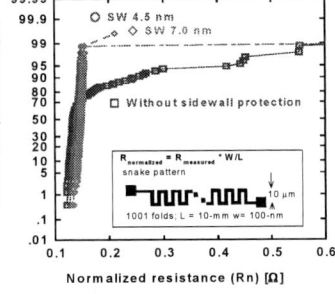

Fig. 9. Cumulative probability of normalized resistance of isolated straight and dense serpentine wirings of the PoSiO/Cu damascene interconnects

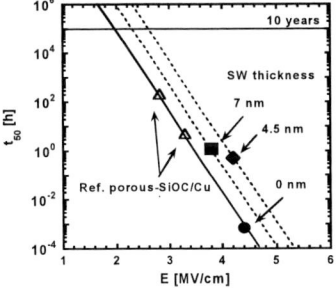

Fig. 10. Mean TDDB lifetime as a function of electric field for PoSiO/Cu damascene interconnects.

978-1-4244-9113-1/11 $26.00 © 2011 IEEE

INVASION PERCOLATION MODEL FOR ABNORMAL TDDB CHARACTERISTIC OF ULK DIELECTRICS WITH CU INTERCONNECT AT ADVANCED TECHNOLOGY NODES

F. Chen, M. Shinosky, B. Li, J. Aitken

IBM Microelectronics, Essex Junction, VT 05452, USA
Phone: 802-769-7917; Fax: 802-769-4139; E-mail: chenfe@us.ibm.com

S. Cohen, G. Bonilla

IBM Thomas J. Watson Research Center, Yorktown Heights, NY 10598, USA

A. Simon, P. McLaughlin, R. Achanta, F. Baumann, C. Parks, M. Angyal

IBM Microelectronics, Hopewell Junction, NY 12533, USA

Abstract — **With the continuing aggressive scaling of interconnect dimensions and introduction of new lower *k* materials, dielectric TDDB reliability margin is greatly reduced. In this paper, a comprehensive investigation on an abnormal low-*k* TDDB characteristic, a systematic degradation of Weibull slopes at lower stress voltages, was conducted for Cu and CuMn alloy interconnect with porous ultra low-*k* (ULK) CVD dielectric (*k*=2.4). Based on extensive electrical and physical evidences, such abnormal TDDB characteristic was attributed to slow metallic diffusion in bulk ULK induced by Mn and Cu segregation. A new TDDB model based on invasion percolation was proposed to successfully model the observed abnormality. CuMn interconnect with robust liner to assure metal free ULK has become important for TDDB reliability.**

Keywords - time-dependent dielectric breakdown, TDDB, low-k reliability, invasion percolationy, CuMn alloy, alloy seed, metal diffusion, TVS, leakage

I. INTRODUCTION

With the continuing aggressive scaling of interconnect dimensions and introduction of new lower *k* materials, the back-end-of-line (BEOL) interconnect reliability margins of TDDB and EM are reduced. Electromigration (EM) is of increasing concern at 32nm because wire cross-section scales by 70% from the 45nm to 32nm node, but circuit voltage and liner thickness do not scale at the same rate. Therefore, even greater current density is imposed for the interconnect wires at the 32nm node. Besides the geometry shrinkage, process induced challenges such as Cu microstructure degradation could further aggravate the EM problem at 32nm. Therefore, there is a great need for EM performance improvement to let EM meet its target at 32nm and beyond. It has been reported that the EM degradation behavior of Cu interconnects could be significantly improved by increasing the bonding strength of the top interface [1-2]. Several approaches have been discussed including metal-cap and local alloying [3-5]. In all cases, the mass transport along the Cu-capping layer interface was reduced. Low-*k* TDDB of Cu interconnects using CoWP metal Cap at 32nm technology node was thoroughly studied by us before [6]. In this paper, we are reporting our latest TDDB evaluation of samples with CuMn alloying seed at 32nm technology node.

Adding dopants to the Cu seed often improves the EM lifetime, with a longer lifetime as dopant concentration increases [7-8]. It is believed that dopant segregation at interfaces and grain boundaries reduces Cu diffusion, therefore resulting in longer EM lifetimes. While providing significant EM reliability improvement by adopting Mn alloy seed compared with pure Cu seed [9-10], there have been very few studies on the effect of Cu alloys on low-*k* TDDB. Kudo et al. [11] reported a 45X increase in TDDB lifetime for Cu interconnects with Mn-doped Cu compared with undoped Cu. The Mn segregates to the Ta and SiCN interface during annealing. Kudo proposed that the Mn segregation at the interfaces acts as an additional diffusion barrier, which inhibits Cu diffusion into the dielectric, thereby increasing the lifetime for dielectric breakdown.

In contrast to Kudo's finding, we have found that adding Mn could exacerbate already problematic low-*k* TDDB reliability if its process was not carefully optimized. Mn could segregate into bulk low-*k* along the sidewall through the defective liner instead of onto the top interface of Cu during the BEOL fabrication process. Such unwanted Mn segregation causes a severe degradation of low-*k* TDDB reliability with poor Weibull slope and small voltage acceleration. Therefore, optimizing the interconnect process to assure metal free ULK with robust liner has become a major task for 32nm CuMn BEOL process development. Finally, we demonstrated that

978-1-4244-9113-1/11 $26.00 © 2011 IEEE

with a careful liner process optimization, comparable leakage and TDDB performance with substantial improvements in EM could be achieved for porous ULK dielectrics using CuMn interconnect at the 32nm technology node.

II. EXPERIMENTAL

Both physical vapor deposition (PVD) pure Cu seedlayer and CuMn alloy seedlayer (1% atomic Mn level) based interconnects within a k=2.4 CVD porous ULK SiCOH dielectric were fabricated using a 32 nm CMOS process on 300mm wafers. Deposition of TaN/Ta bilayer barriers and Mn-doped Cu seed layers was done using ionized PVD. The structures were filled using electroplating, followed by CMP and $SiC_xN_yH_z$ capping. Both line only (serpentine-comb and comb-comb) and via-related structures with various sizes were used in this study. The actual mean spacing between lines and vias ranged from 20nm to 50nm for different test structures on differently processed wafers. The method of multiple time-steps to apply low sensing voltages to monitor stress-induced leakage current (SILC) was used during the constant voltage TDDB stress. Hard breakdown was defined as a sudden leakage current increase during stress, in this case, corresponding to a SILC current above 100nA. 0.1V steps with 1V/s ramp rate and 200ms delay at each step were used for voltage ramp dielectric breakdown (VRDB) test. All VRDB tests were performed at 125 °C and at wafer level. TDDB tests were performed at 125 °C as well but at both wafer and module levels. True linear voltage sweep by HP4140B was used for triangular voltage sweep (TVS) measurement at 300 °C and at wafer level. 0.2V/s slow ramp rate was used to assure mobile ion signals stand out from the capacitor displacement current. Speeding it up could lower TVS signal sensitivity to mobile ion, depending on mobile ion mobility. However, ramp rates slower than 0.2V/s seem to make little or no difference for TVS sensitivity.

III. RESULTS AND DISCUSSIONS

A. Ultra Low-k TDDB Model Corroboration at 32nm

Pure Cu seed samples with k=2.4 SiCOH dielectric at 32nm were used to corroborate the critical TDDB voltage acceleration model. Figure 1 shows the long-term, low field (~1.8–2 MV/cm) module level TDDB stress (> 1.5 year) results combined with the fast, high field wafer level TDDB data from two different wafers at the same 125 °C stress temperature (the worst device junction temperature on a chip) using the same testing structure. In addition to those two wafers, more long term TDDB stresses were performed using 32nm wafers at different IBM reliability labs in different sites. Based on our long-term TDDB results, it is confirmed that the √E dependence is still the best choice for fitting the 32nm ULK TDDB data over a wide range of applied voltages. No obvious deviation from √E model to even more relaxed models such as power-law or 1/E model was observed based on multiple wafers from multiple lots during our study. Croes and Tokei reported that √E model could be too conservative based on their experimental data [12]. However, our obtained longer time, lower field ULK TDDB data so far couldn't support

using a more relaxed model such as power law or 1/E model to replace the square root of E model for low-k TDDB modeling. It seems the difference between square root of E model and power-law model could be small based on available data. Therefore power-law model can't be excluded definitely at this moment. However, one serious concern of using power-law model is that it fails to predict a well-known low-k TDDB characteristic - spacing variation induced distribution change at lower fields. Based on power-law's $(E_1/E_2)^n$ formula, the ratios of various spacing to cause t_{BD} variation are always a constant, independent of applied voltages. Therefore the t_{BD} distribution slopes from power-law model should always be the same at different voltages even with the existence of severe spacing variation across the wafer. However, many studies from different research groups [13-15] already clearly proved that low-k TDDB distributions displayed a strong function of stress field if spacing variation was significant. On the other hand, for the high volume manufacturing semiconductor industry, using a relatively conservative √E model is always a safer approach even if it is proved to be not the most accurate model later. Therefore, in this paper, √E model is still used for all data interpretation. Figures 2 and 3 show the obtained Weibull distributions at various stress voltages for the samples shown in Figures 1 and 2. Well-behaved TDDB characteristics with no systematic Weibull slope change over the entire range of stress bias was obeyed for our optimized Cu samples, confirming the trustworthy plots of those extracted Weibull scale parameters in Figure 1. The Weibull slope variations seen in Figure 2 and 3 are within 90% confidence bounds. Such overall consistent Weibull slopes across a wide range of stress voltage also suggest that spacing variation is well controlled for our samples.

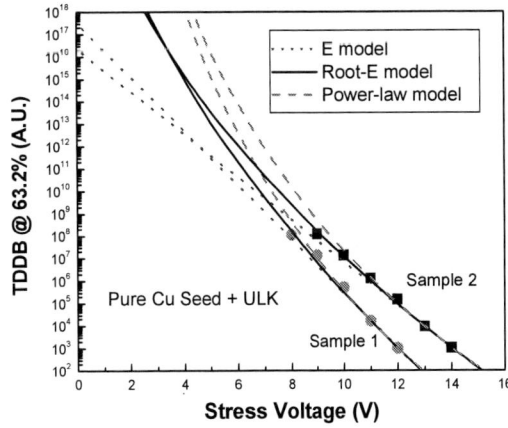

Figure 1: Multiple wafers of ULK with pure Cu seed at 32nm demonstrate the extendibility of using square root of E model

B. Abnormal Low-k TDDB Characteristics

However, in contrast to well-behaved Cu samples, we have found that unoptimized Cu samples could display a systematic degradation of Weibull slopes at lower stress voltages, especially under long-term module level stress condition since the implementation of low-k dielectric at 65nm technology. Similar abnormality was reported by Croes and Tokei recently [12] from their pure Cu seed ULK samples as well. They attributed the degradation of TDDB distributional shape (higher spread) at lower fields to increased dependence of spacing variation across the wafer when less conservative models such as 1/E model became prevalent. However, 1/E model clearly is not the choice for our experimental data as shown in Figure 1. Based on $\sqrt{E}$ model, spacing variation across the wafer would result in improved Weibull slopes at lower electric fields [13,14]. Therefore, any spacing variation related explanations are not reasonable to explain this abnormality here. Further SEM CD measurements proved that the spacing variation was not significant even for the wafers displaying a strong abnormality. Besides, we found that this abnormality could be further worsened by adding Mn alloy into Cu seed at 32nm. As shown in Figure 4, shorter t_{BD}, decreased Weibull slope, and reduced voltage acceleration at lower stress voltages were all observed for our early processed CuMn alloy samples. Therefore, we concluded that there should be a process related root cause instead of a simple geometry variation to explain our observed abnormalities.

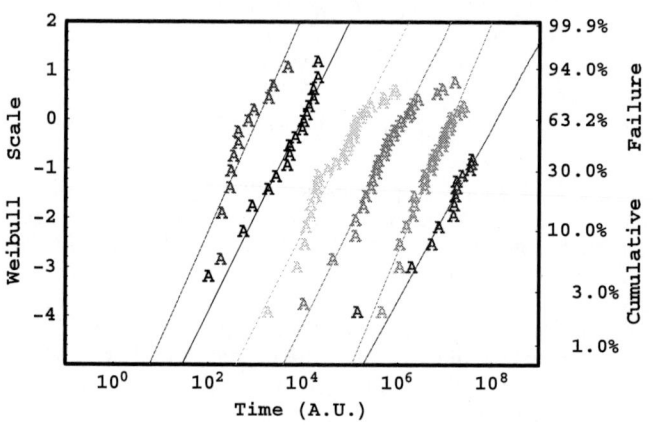

Figure 3: Weibull characteristics at various voltages from sample 2 in Figure 1. Observed Weibull slope variation is within 90% confidence bounds (not shown)

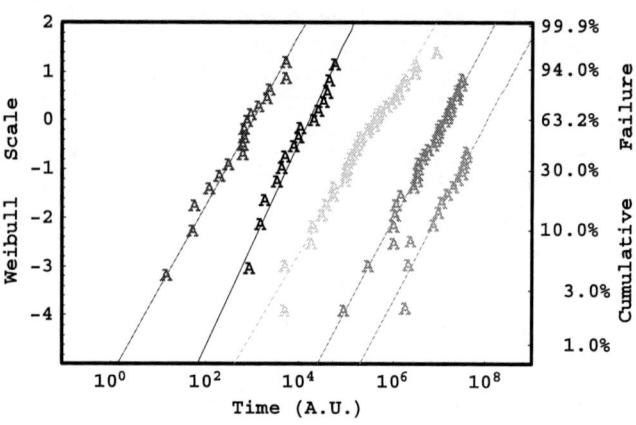

Figure 2: Weibull characteristics at various stress voltages from sample 1 in Figure 1. Observed Weibull slope variation is within 90% confidence bounds (not shown)

Such abnormalities caused a serious concern for the adoption of CuMn with ULK and made it impossible to have an accurate TDDB lifetime projection unless all the stresses were performed at use condition. The Weibull slope degradation, seemingly, was both voltage and temperature driven. Longer passive thermal bake due to stress temperature at lower stress voltages resulted in decreased Weibull slopes. On the other hand, the reduced voltage acceleration at lower stress voltages is obvious regardless of the choice of different TDDB voltage acceleration models, suggesting a change in TDDB kinetics. The degree of such abnormality was found to have a strong wafer-to-wafer variation and a strong dependence on BEOL liner process.

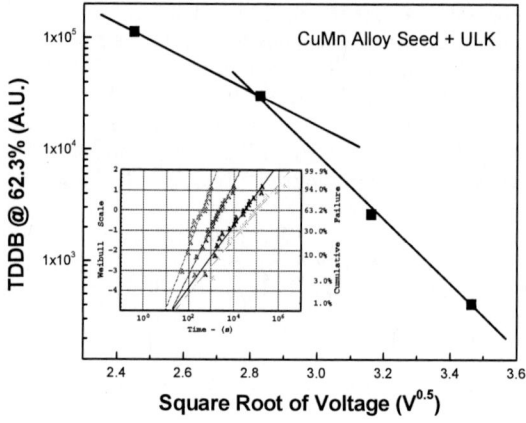

Figure 4: CuMn seed ULK sample showing TDDB abnormality with reduced Weibull slope and voltage acceleration at lower stress voltages

To understand the root cause of the abnormality, both high temperature TVS (triangular voltage sweep) test and precision physical failure analysis were performed on pure Cu seed and CuMn alloy seed samples. True linear voltage sweep by HP4140B was used for TVS measurement at 300 °C. Embedded in the TVS testing program, a bias-temperature (7-10V and 300 °C) pretreatment with time period of 5-30 minutes was performed before the voltage sweep start to activate and drive mobile ions. As shown in Figures 5 and 6, TVS showed obvious metallic peaks in CuMn alloy wafers but such peaks were not significant in the control Cu wafers from the same process except with different Cu seed. Based on the broad peak shape, we hypothesized that such peaks were caused by slow moving, heavy metals like Mn and Cu. Sodium and Hydrogen in general move fast and should give sharp peaks. By testing multiple samples from both pure Cu seed and CuMn alloy seed, we consistently found that TVS of pure Cu overall showed no or little peak compared with CuMn. Our TVS results suggested that a metal source in the dielectric, which caused ion motion detected by TVS, was enhanced significantly by adding Mn into Cu seed. In addition, the TVS peaks also suggested there was not an infinite source of metal in the dielectric as the peaks did not become larger with longer stress time. Additional dielectric breakdown stress at -30 °C on CuMn samples showed that the number of stress fails at -30 °C was reduced, but not eliminated, compared with those at 150 °C. This result indicated that preexisting metal ions also existed in ULK, otherwise the stress fails at -30 °C would disappear. Since the stress fail rate was higher at 150 °C, this further suggests that metallic ion drift, as well as preexisting defects, were the root cause to this overall problem. By performing constant voltage TDDB stresses of CuMn seed and pure Cu seed samples, CuMn seed sample showed shorter breakdown times with an obvious defect tail as compared with CuMn seed sample shown in Figure 7.

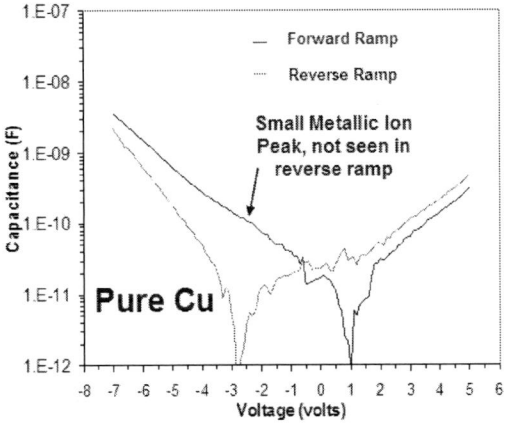

Figure 5: TVS of pure Cu seed sample at 300 °C

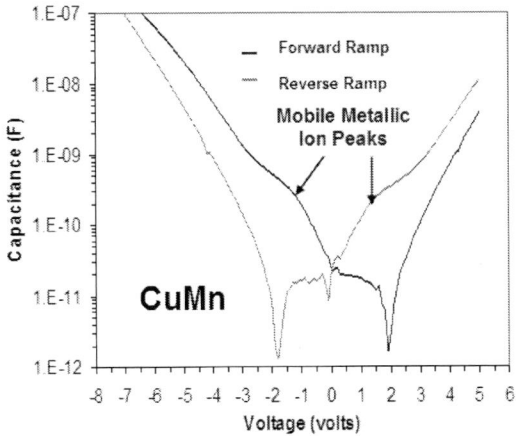

Figure 6: TVS of CuMn seed sample at 300 °C

Intensive TEM analysis with EELS and SIMS analysis were performed on poorly behaved CuMn alloy samples to investigate the physical root cause. SIMS found no Mn segregation on Cu top, but rather along sidewall and at the bottom of the trench as shown in Figure 8. Careful line-scans through the samples at midsidewall provided detail about the oxygen and Mn within the liner layer. Figure 8 shows a strong oxygen signal permeating through the full-width of the TaN/Ta liner layer. A Mn signal is also visible within the barrier layer, suggesting that this condition results in a liner layer with poor hermeticity. TEM analysis confirmed that Mn spread at the outside trench, mainly in the vicinity of TaN/Ta liner, but was also present in the bulk of ULK as well as some Cu in the bulk of ULK (Figure 9). Combining the evidence from electrical analysis and physical analysis, we proposed that both Mn and Cu could diffuse through the leaky liner into bulk ULK if liner process is not optimized. Ta liner could be oxidized (which is often the case when Ta is deposited on porous low-*k* materials) if liner process is not optimized. Simon et. al studied the effect of TaN stoichiometry on liner oxidation and defect density in 32nm Cu/ULK interconnects [16]. They found the liner oxidation could occur in the low nitrogen-content Ta(N) liner, which was absent at the higher stoichiometry case. It is well-known that the driving force for dopant reaction with the dielectric and segregation in the dielectric is the negative free energy of metal oxide formation. It was reported that Mn oxide has a much more negative free energy than the free energy of Cu oxide [17]. On the other hand, the diffusivity of Mn in dielectric should be significantly higher than the diffusivity of Cu based on the results of successful development of Mn self-forming barrier reported from several research groups [9, 10]. As far as self-forming CuMn barrier samples presented no Cu outside MnOx liner, it was direct experimental evidence to support that Mn diffused faster in low-*k* than Cu did, which is one of the core requirements to select Mn as the alloying element. Because of that, the MnOx barrier formation reaction always occurs before there is significant Cu diffusion into the dielectric. Therefore, if liner was oxidized, Mn could segregate out of leaky liner faster than Cu, and form MnOx phase in bulk ULK at time zero stage (time zero defects).

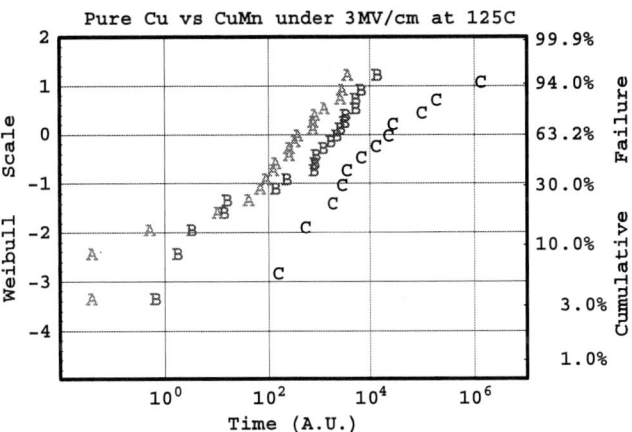

Figure 7: TDDB comparison between CuMn seed samples and pure Cu seed sample under the same electric field. A and B are samples from two CuMn wafers and C is pure Cu.

Our study demonstrated a clear correlation between oxygen permeation of the TaN/Ta liner layers, and the resulted "self-forming barrier" when Mn binds with oxygen to augment the TaN/Ta liner layers. Such unstable MnOx could provide the mobile Mn ions to form bulk metallic diffusion during high temperature and high bias stress. Further, any permeated liner also could cause follow-on Cu diffusion during bias-temperature stress. Therefore, it was likely that both Mn and Cu massive out diffusion happened during the entire TDDB stress period for our poorly processed samples.

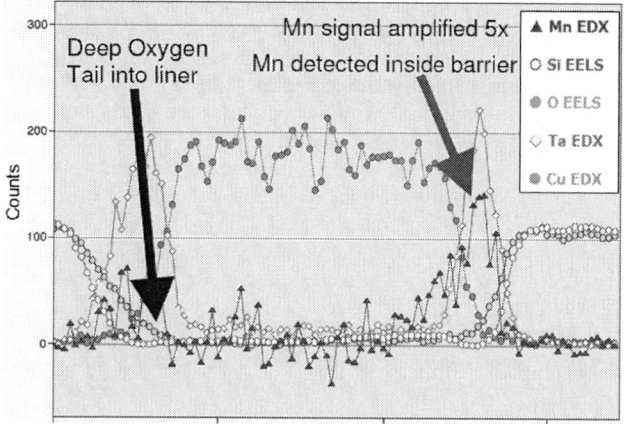

Figure 8: SIMS Linescan across the width of half trench height from a poor CuMn sample. The permeability of the TaN/Ta barrier sidewalls is confirmed by the presence of a deep oxygen tail inside the liner, and Mn inside the liner (arrows)

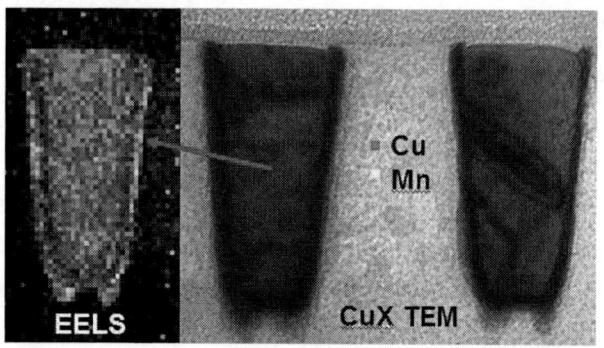

Figure 9: TEM/EELS analysis of one leaky CuMn ULK sample as shown in Figure 8

As we mentioned before, even pure Cu seed samples could show similar abnormalities if liner process was not optimized. Poor liner coverage and quality could cause poor Weibull slope, soft breakdowns, and poor kinetics due to the out diffusion of Cu [18-20]. However, adding Mn demands even higher robustness of liner (dense, full coverage and oxidation free liner) because of Mn's more negative free energy and faster diffusivity in dielectric than Cu. Therefore, optimizing and improving liner process to assure full coverage and oxidation free liner becomes imperative for the adoption of CuMn. Various experiments such as varying magnetron and pedestal powers, gas flows, and top-surface TaN/Ta bilayer

thickness were conducted to achieve a targeted TaN/Ta coverage along the trench sidewalls together with the highest hermeticity. SIMS plot of Figure 10 demonstrates an optimized liner exhibiting neither presence of oxygen nor manganese along the sidewall, which explains the full segregation of the Mn at the top-surface of Cu as shown in Figure 11. No Cu signal was detected in bulk low-k as well. The top-segregation shown in Figure 11 is desirable for retarding Cu diffusion along the capping layer and Cu interface, and therefore improving electromigration lifetime. CuMn ULK samples, with the optimized liner process, also displayed normal TDDB behavior as shown in Figure 12. TaN/Ta liner layer deposition conditions can have a great impact on liner hermeticity, and on the successful use of Mn-dopants in advanced interconnects to achieve improved EM performance while preserving acceptable TDDB reliability.

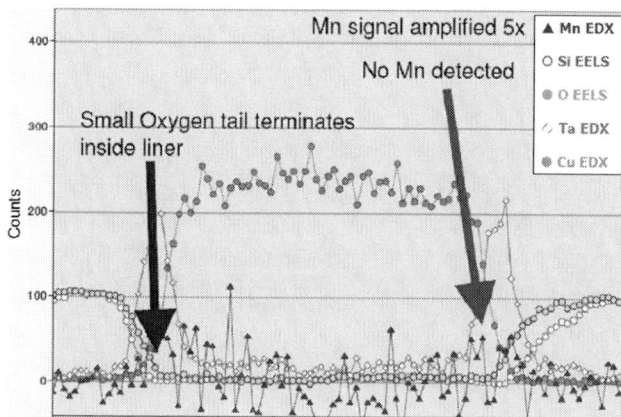

Figure 10: Equivalent SIMS linescan for a CuMn ULK sample with optimized TaN/Ta liner. The oxygen and Mn-signals are absent, illustrating the correlation between barrier integrity and capping-layer interface (top-surface) Mn-segregation.

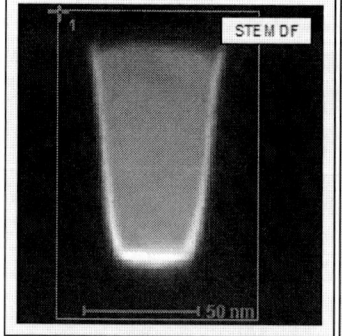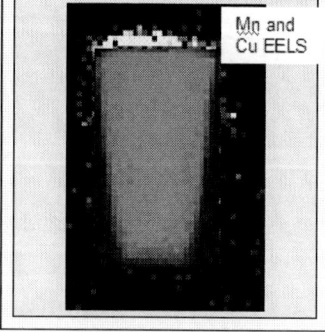

Figure 11: TEM/EELS analysis of one optimized CuMn ULK sample.

C. *Invasion Percolation Modeling of TDDB Abnormality*

Based on experimental observations, a new TDDB breakdown model, invasion percolation (IP) model, is proposed to model ULK breakdown induced by slow bulk metallic diffusion. Dynamic invasion percolation model has been introduced by Wilkinson and Willemsen [21] to describe the evolution of the front between two immiscible liquids in a random medium. Many modifications of the original invasion percolation model were proposed later. They considered, for example, the action of an external gravitational field [22], or the flux with a privileged direction [23]. In all IP models, the cluster spreads follow the path of least resistance. The best-case for invasion percolation is to describe the flow through porous material, and is highly applicable to slow metal diffusion in ULK case. TVS broad peak and SIMS/TEM/EELS physical analysis confirmed slow metal diffusion in ULK under bias and temperature stress.

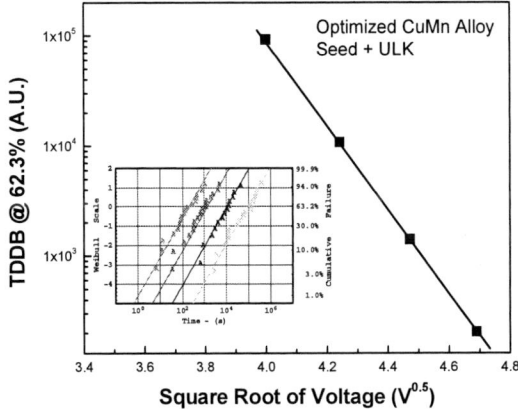

Figure 12: Optimized CuMn seed ULK showing good TDDB behavior

Our simulation theory is based on mapping the invasion resistance of each pore to an occupation likelihood under different forward electric fields. To generate nontrapping IP cluster in 2D, we perform the following steps. (i) We first assign random numbers, r, uniformly distributed in the range [0,1], to each site of an $L \times L$ lattice. Then (ii) we initiate the invasion by putting a seed at (1,1) of the lattice with various forward force, f_a (>1), to mimic different forward electric fields. In step (iii), we define a list of invading sites composed by its first neighbors. The invading flow occupies one growth perimeter site. That site has the smallest p, the product of the random number, r, and the forward force, f_a, under the condition that f_a can only be multiplied to the sites at the forward boundary edge along the force. This list of invading sites is updated at each time step (iv). The last two steps (iii) and (iv) are then repeated to grow larger and larger until the invading cluster reaches the boundary at the opposite side of the lattice. The cluster generated in this way is connected to the critical ordinary site percolation. The number of steps in a connected percolation path can be directly related to the time of breakdown of the invaded medium. During our simulation, invasion without trapping is assumed and no multiple invasion or burst is allowed. Figure 13 illustrates some typical IP pathways simulated by Monte Carlo simulation under one fixed

forward bias starting from one common seed (1,1). The distributions of such pathways under different forward biases are shown in Figure 14. It is instructively noted that the Weibull slopes of each distribution decrease with decreasing forward voltages, which is consistent with our experimental observations from the CuMn samples with unoptimized liner process. The voltage acceleration of low percentile fails (early fails) is smaller than the voltage acceleration of high percentile fails (late fails), which is in contrast to the voltage acceleration trend caused by spacing variation. The reason of degraded Weibull slope at smaller biases can be well explained by IP theory. At higher forward biases, the contribution from f_a to determine p dominates. Therefore cluster grows more straightforward to the opposite side with an overall tighter distribution. While at lower forward biases, the contribution from r to determine p dominates. Parameter r is randomly generated without any directional preference. Therefore, at lower forward biases, higher randomness leads a cluster to grow less directional than grown at higher forward biases with an overall wider distribution. Besides, based on our IP Monte Carlo (MC) simulation, we also found that Weibull slopes exhibit a strong spacing dependence. As shown in Figure 15, under the same forward bias, the smaller the spacing is, the worse the Weibull slope is. This trend is similar to the trend of gate oxide TDDB Weibull slope predicted by single percolation simulation [24]. Such similarity is not a surprise as in two dimension (2D) as well as in three dimension (3D), numerical studies of nontrapping IP and convincing arguments indicate that IP falls into the same universal class as regular single percolation [21].

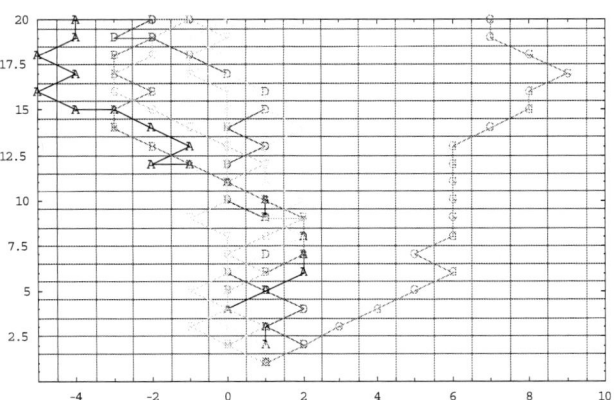

Figure 13: Monte Carlo simulation showing typical invasion percolation pathways under a fixed forward bias starting from site (1,1)

However, the difference between single percolation and invasion percolation does exist. In IP, each site has a different probability to join the cluster and such probabilities are changed under different external forward forces. In single percolation, each site has the same probability to join the cluster based on the complete randomness. Each joined site can be represented as a defect which contributes to final dielectric breakdown. Since this randomness won't be changed under different biases, as a consequence, no bias-dependent TDDB Weibull slope can be produced from single percolation.

978-1-4244-9113-1/11 $26.00 © 2011 IEEE   139

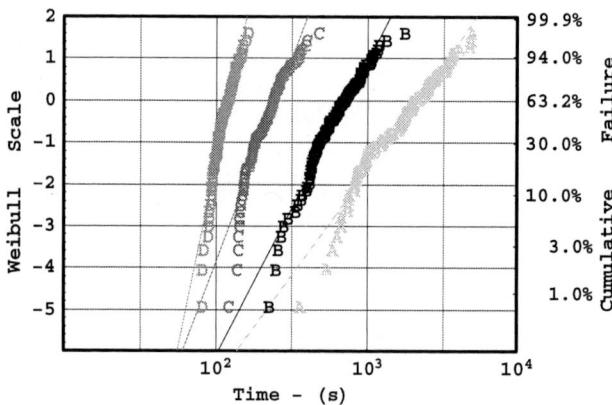

Figure 14: IP simulation showing degraded Weibull slopes with decreasing stress biases. Bias A < B < C < D.

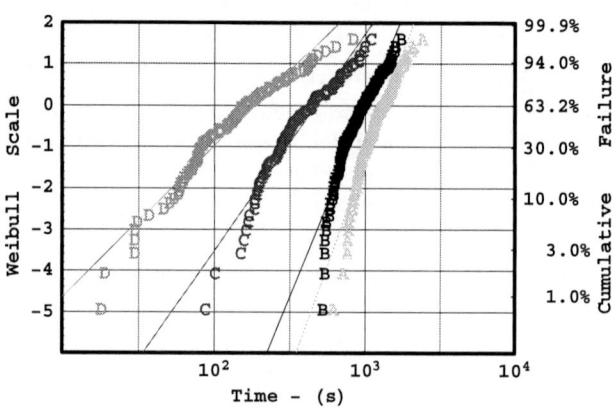

Figure 15: Invasion percolation simulation showing degraded Weibull slopes with decreasing spacing under the same bias. Spacing A > B > C > D.

The real situation for ULK breakdown is more complicated. It is believed that in addition to invasion percolation, conventional single percolation controlled ULK-cap interface breakdown process should still exist, and most likely dominates at high stress voltages. In conventional single percolation low-k interface breakdown model [25], fast diffused Cu ions catalyze the bond breakage reaction under field and temperature by changing the chemical bond or inserting strain field at low-k/Cap interface if their quantity is limited. If both interface and bulk breakdown coexist, at lower applied stress biases, bulk diffusion would take over the leading position to control ULK breakdown due to the fact that interface and bulk breakdown have different voltage accelerations. The bulk leakage has significantly less voltage dependence than interface leakage does from the experimental study. Therefore, bulk breakdown should have lower voltage acceleration than interface ULK

breakdown based on the physics of square root of E model (an electron fluence driven breakdown model). The ultimate dielectric breakdown time is determined by the mechanism that induced the shortest t_{BD}. Therefore, as shown in Figure 16, by considering the coexistence of two competing mechanisms in controlling ULK TDDB, the observed voltage acceleration reduction at lower stress voltages as shown in Figure 4 could be explained as well. Although invasion percolation occurred within bulk ULK for our unoptimized CuMn seed case reported here, please note that IP could occur at low-k/Cap interface as well if massive metallic diffusion occurs along the interface during TDDB stress. It was found that moisture ingression and longer queue-time induced Cu surface oxidation could cause extremely poor low-k TDDB characteristics even for pure Cu seed samples at 65nm technology node with decreased Weibull slope and voltage acceleration [26-28], and all such breakdowns physically were found to occurre at low-k/Cap interface.

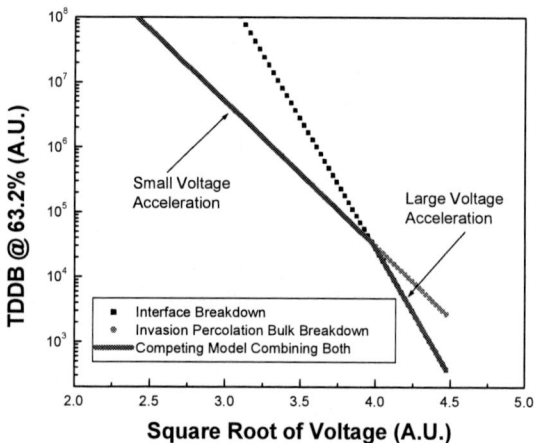

Figure 16: Competing TDDB model including both interface breakdown and invasion percolation bulk breakdown mechanisms

IV. CONCLUSIONS

As a conclusion, impact from CuMn on porous ULK TDDB has been thoroughly studied. TDDB abnormality observed in CuMn samples was attributed to slow metallic diffusion in bulk ULK induced by Mn segregation into bulk ULK. A new TDDB model based on invasion percolation was proposed to successfully model the observed abnormality. Optimizing CuMn process to assure Mn free ULK has become a major task for CuMn implementation at 32nm.

ACKNOWLEDGMENT

The authors wish to acknowledge all the people working within IBM alliance programs for 32nm SOI and Bulk CMOS technology development. This work has been supported by the independent Bulk CMOS and SOI technology development projects at the IBM Microelectronics Division, Semiconductor

Research & Development Center, Hopewell Junction, NY 12533.

REFERENCES

[1] M. W. Lane, E.G. Liniger, and J. R. Lloyd, J. Appl. Phys., vol. 93, pp.1417-1421, 2003

[2] E. Zschech, M.A. Meyer, and E. Langer, Proc. Mater. Res. Soc. Symp, 2004, vol. 812, pp. 361-372

[3] Y. S. Diamand, et al., Proc. Electrochem. Soc., 2000, vol. 99-34, pp.102-105

[4] C.K. Hu, et al., Appl. Phys. Lett., vol. 81, no. 10, pp. 1782-1784, 2002

[5] T. Usui, H. Nasu, S. Takahashi, et al., IEEE Transaction on Electron Devices, vol 53, issue 10, pp. 2492-2499, 2006

[6] F. Chen, et al, Proc. 48th Annu. IEEE Rel. Phys. Symp, 2010, pp. 566-573

[7]T. Vanypre, T. Mourier, N. Jourdan, M. Cordeau, J. Torres, Proc. of Adv. Metallization Conference 2007, MRS, pp. 385-389, 2008

[8] W.F.A. Besling, T. Vanypre, X. Federspiel, T. Mourier, J. Flake, S. Courtas, J. Torres, Proc. of Adv. Metallization Conference 2005, MRS, pp. 735-743, 2006.

[9] S. C. Pan, et al., IITC Proc., 2010, pp. 1-3

[10] T. Watanabe, H. Nasu, G. Minamihaba, N. Kurashima, A Gawase, M. Shimada, Y. Yoshimizu, Y. Uozumi, H. Shibata, IITC Proc., 2007, pp. 7-9

[11] H. Kudo, et al., IITC Proc. 2008 pp. 117-119

[12] K. Croes and Zs. Tőkei, Proc. 48th Annu. IEEE Rel. Phys. Symp, 2010, pp. 543-548

[13] F. Chen, J. R. Lloyd, K. Chanda, R. Achant[2], O. Brav, A. Strong, P. S. McLaughlin, M. Shinosky, S. Sankaran, E. Gebreselasie, A. K. Stamper, and Z.X. He, 46th Annu. IEEE Rel. Phys. Symp, 2008, pp. 132-137

[14] M. Vilmay, D. Roy, C. Monget, F. Volpi, J.-M.Chaix, 47th Annu. IEEE Rel. Phys. Symp, 2009, pp. 606-612

[15] Shou-Chung Lee; A.S. Oates, and Kow-Ming Chang, Device and Materials Reliability, IEEE Transactions on, vol 10, issue 3, pp. 307-316, 2010

[16] A.H. Simon, et al., Proc. of the MRS 2010 Spring Meeting, Symposium N, 1249-F01-02, 2010

[17] http://www.doitpoms.ac.uk/tlplib/ellingham_diagrams/index.php

[18]F. Chen, et al., Proc. 43rd Annu. IEEE Rel. Phys. Symp., 2005, pp. 501–507

[19] F. Chen, et al., Proc. 47th Annu. IEEE Rel. Phys. Symp, 2009, pp. 464-475

[20] Zs. Tőkei, Y-L. Li, and G.P.Beyer, Microelectronics Reliability 45 pp. 1436–1442, 2005

[21] D. Wilkinson and J. F. Willemsen, J. Phys. A: Math. Gen. 16, pp. 3365-3376, 1983

[22] D. Wilkinson, Phys. Rev. B 30, pp. 520-531, 1984

[23] R.N. Onody, Int. J. Mod. Phys. C 6, pp. 77-83, 1995

[24] J. H. Stathis, J. Appl. Phys. **86**, pp. 5757-5756, 1999

[25] F. Chen, et al., Proc. 44th Annu. IEEE Rel. Phys. Symp., 2006, pp. 46-53

[26] J. Noguchi, N. Miura, M. Kubo, T. Tamaru, H. Yamaguchi, N. Hamada, K. Makabe, R. Tsuneda, and K. Takeda, Proc. 41th Annu. IEEE Rel. Phys. Symp., 2003, pp.287-292

[27] F. Chen and M. Shinosky, IEEE Trans. On Electron Devices, vol. 56, pp. 1-12, 2008

[28] Yunlong Li, Ivan Ciofi, Laureen Carbonell, Guido Groeseneken, Karen Maex, and Zsolt Tőkei, 45th Annu. IEEE Rel. Phys. Symp., 2007, pp. 405-409

Comparison between Intrinsic and Integrated Reliability Properties of Low-k materials

K. Croes[1], M. Pantouvaki[1], L. Carbonell[1], L. Zhao[2], G.P. Beyer[1] and Zs. Tőkei[1]

[1]Imec, Kapeldreef 75, B-3001 Leuven, Belgium
[2]Intel assignee at imec
phone: (32)16/281621; fax: (32)16/281576; e-mail: Kristof.croes@imec.be

Abstract - Using a dedicated test vehicle (low-k planar capacitor) for studying the intrinsic properties of low-k materials and using standard single damascene 50 and 90nm ½pitch test vehicles, differences in reliability behavior between intrinsic and integrated SiOCH porous low-k materials were investigated. The studied parameters were leakage current, breakdown field, distributional shape of failure times and TDDB lifetimes. Compared to the intrinsic material, the integrated properties significantly deteriorated and these differences were quantified. The low-k planar capacitor test vehicle was also used to study the intrinsic breakdown behavior at low fields. Using statistical simulations, we found that the proper choice of high field conditions is crucial to be able to discriminate between different lifetime models at low fields and recommendations are given on how to choose these high field conditions. Based on tests of several months, the E- and 1/E-model can be excluded (all test material had a standard TaNTa barrier between copper and dielectric). The √E-model and the power law are statistically not different, but a strong tendency towards the power law is observed. As a corollary, we compared the benefit of choosing less conservative lifetime models for extrapolations to lower fields to the loss that occurs when predicting median failure times to lower percentiles. We show that process optimizations leading to a lower spread in the failure times are more important than choosing less conservative lifetime models.

Keywords - BEOL, copper, low-k, TDDB, low field

I. INTRODUCTION

Time dependent dielectric breakdown (TDDB) of intermetal dielectrics (IMD) in Back-End-Of-Line interconnects [1,2] became a serious concern with the replacement of SiO_2 as IMD by low-k materials as these materials are electrically weaker and break faster. Figure 1 shows a cross-sectional cartoon of integrated single damascene copper interconnects. During integration, dielectrics are damaged when being subjected to different plasma treatments, barrier deposition processes and CMP chemistries, which all influence the dielectric performance. Patterning small dimensions leads to line-edge-roughness (LER), which influences narrow space IMD reliability [3-5]. Also, tapered trenches after etch lead to different dielectric thicknesses along the height of the structure. Finally, it is reported that the interface between the IMD and the dielectric cap layer is a leaky interface which also impacts the overall dielectric performance [2,6]. In summary, the reliability of the

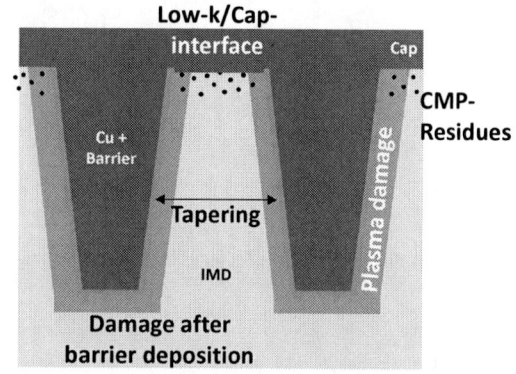

Figure 1.   Cross-sectional cartoon of a damascene structure, where different influences to dielectric reliability are indicated.

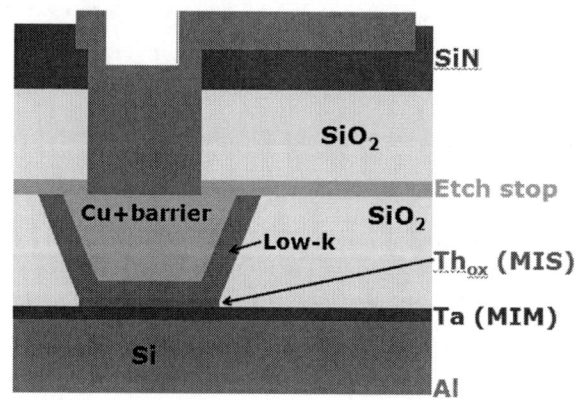

Figure 2.   Cross-sectional cartoon of the p-cap structure.

pristine low-k material is different from the one after full integration and these differences need to be studied.

Recently, imec developed a dedicated test vehicle (low-k planar capacitor or p-cap) to study the pristine properties of low-k/barrier systems [7-8], of which a cross-sectional cartoon is shown in figure 2. Since no plasma's are involved after low-k deposition and the low-k material is not directly exposed to CMP-chemicals, the material is kept pristine. Due to optimized rounded profiles, edge failures do not occur and nice area scaling was demonstrated [9].

In the first part of this paper, we compare the reliability properties of the pristine material to those after full integration in 50 and 90nm ½pitch test vehicles, where leakage current densities, breakdown fields, failure times, distributional shapes and acceleration factors are evaluated.

Besides studying the differences in reliability behavior between intrinsic and integrated dielectrics, knowing the dependence of TDDB times to failure (TTF) on the applied electric field E for pristine low-k materials is useful. Over the past years, a significant amount of studies have been conducted about this field dependence for damascene structures. Here, different authors have put forward the E- (ln(TFF)~E) [10-12] and the √E-model (ln(TTF)~√E) [2,6]. Recently, it was pointed out that these models might be too conservative to explain extremely low field TDDB [13]. Compared to the E and the √E model, the impact damage model (ln(TTF)~A√E+B/E) [14] and the numeric model proposed by G. Haase [15] are less conservative. Besides these two models that have been proposed for BEOL TDDB, also the pure 1/E-model (ln(TTF)~1/E) [16,17] and the power law (TTF~E^n) [18,19] are often considered as less conservative alternatives compared to E and √E as they have been widely studied for the application of gate oxide TDDB.

Since integrating copper lines impacts the dielectric reliability, knowledge of the lifetime model of the pristine dielectric can deliver useful information. Recently, the p-cap structure has been used to demonstrate a clear 1/E-dependence for a system with SiO_2 as dielectric, where copper was directly deposited onto the SiO_2 without the presence of a barrier [20]. The second part of the current paper is dedicated to low field TDDB measurements performed on p-caps using low-k dielectrics in presence of a barrier between copper and low-k.

II. EXPERIMENTAL

The fabrication of p-caps starts with a doped Si-wafer on which PECVD SiO_2 is deposited and patterned with squares of 100x100μm (Figure 2). After this, the low-k material is deposited and cured. For the current study, a 55nm thick SiOCH CVD porous low-k material with 25% porosity (k= 2.5) was deposited. Then, a 3/3nm PVD TaN/Ta layer was deposited followed by copper seed, plating, CMP and a 35nm SiCN, 300nm SiO_2 and 500nm Si_3N_4-stack as passivation. To avoid mechanical damage to the dielectric during probing, contact was made using Al-bondpads which are put some distance away from the structure. The wafer is finished with a thin Al-layer at the back to ensure a good backside contact.

We evaluated the dielectric using both a metal-insulator-semiconductor (MIS) and a metal-insulator-metal (MIM) lay-out. In order to minimize interface effects between the Si substrate and low-k for the MIS lay-out, a 1nm-thin SiO_2 layer is thermally grown onto the Si before low-k deposition. The MIM lay-out was produced by putting a Ta-layer underneath the dielectric. This layer was deposited before the deposition of the PECVD SiO_2.

All breakdown tests were done with a positive voltage applied to the bondpads and with the Al backside contact grounded. Short tests were done at wafer level and long term tests were done at package level. No difference in failure times

was observed between wafer and package level tests at similar voltages. Hence, the packaging process did not influence the TDDB behavior.

For the integrated structures, single damascene copper lines were integrated in the same low-k material with also a PVD TaN/Ta barrier between copper and IMD and full passivation on top of the wafers. Two types of structures have been used: a) a 1cm long meander-fork structure with 90nm ½pitch and b) a 1cm long fork-fork structure with 50nm ½pitch [13]. Both structures were printed using 193nm immersion lithography and patterned by using a metal hardmask approach. The 90nm ½pitch structures were integrated with conventional patterning techniques [21], while the 50nm ½pitch structures were integrated using the double patterning approach [22]. Cross-sectional TEM pointed out that the actual spacing of the 50 and 90nm ½pitch structures was 45 and 105nm, respectively.

III. RESULTS

A. Comparison intrisic versus integrated properties

Figure 3 shows typical current density versus field graphs for the MIS and MIM p-caps and the 50 and 90nm ½pitch structures. The integrated structures show higher leakage currents and lower breakdown fields. For the 90nm ½pitch structure, the reduction in breakdown field is 1.3MV/cm, while this goes up to 2.4MV/cm for the 50nm ½pitch structures. This increase in leakage current and decrease in breakdown field are caused by etch plasma's, CMP, LER, different leakage paths, etc. Also the decrease in breakdown field when scaling the dimensions is significant. The contribution of the damage to the low-k and LER increases when scaling dielectric spaces.

Figure 4 shows a Weibull probability plot of failure times collected at 4MV/cm. Both the shape parameter β and the scale parameter $t_{63.2\%}$ of the underlying Weibull distribution is significantly deteriorated when going from intrinsic to integrated materials and when scaling dimensions.

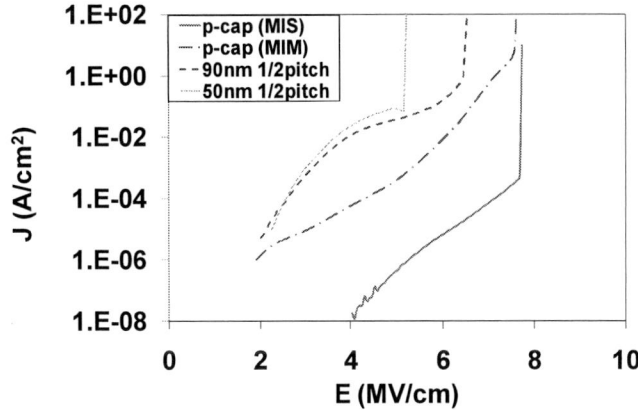

Figure 3. Typical current density versus field curves for the different test vehicles. Higher leakage currents and lower breakdown fields are observed for damascene structures.

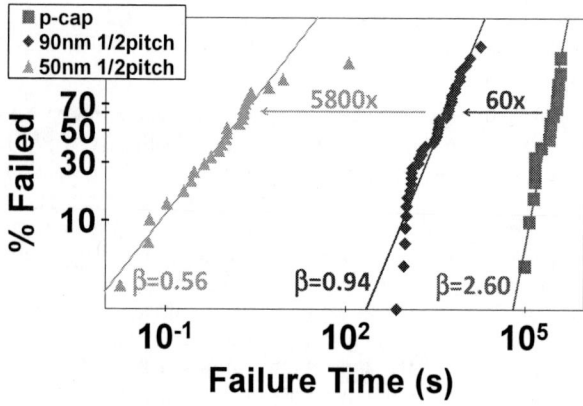

Figure 4. Weibull plot of the failure times obtained at 4MV/CM for the different test vehicles

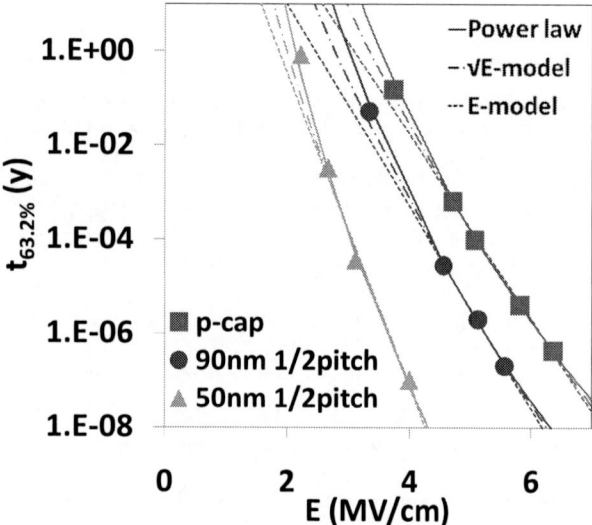

Figure 5. 63.2%-percentile as a function of applied field E for the different test vehicles, together with extraploations to lower fields using the E-model, the √E-model and the power law.

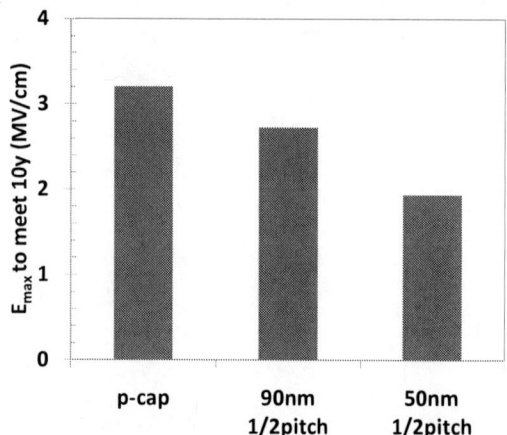

Figure 6. Maximum allowed field E_{max} to meet a 10y lifetime criterion of the 63.2%-precentile for the different test vehicles. E_{max} was taken from the power law fits of Figure 5.

The impact of the reduction of the $t_{63.2\%}$-percentile on the maximum allowed field E_{max} to meet a lifetime specification of 10 years is explored in figure 5 where estimated 63.2%-percentiles at different fields are used to extrapolate to lower fields using the E- and √E-model and the power law. Note that this graph contains data of several months at the lowest fields for each test vehicle. To investigate the differences in extrapolation models, the data points at the lowest field have not been taken into account during the fit of the data. In [13], we showed with statistical significance that the data obtained with the 90 and 50nm ½pitch test vehicles did not follow the E-nor the √E-model. The model dependence of the p-cap test vehicle will be discussed in section B. From Figure 5, the E_{max} to meet a 10y lifetime was determined based on the power law for the different test vehicles (Figure 6). E_{max} goes down by about 0.5MV/cm when going from intrinsic to integrated 90nm ½pitch and further reduces by about 0.8MV/cm when scaling further to 50nm ½pitch. Note that the assumption of different lifetime models did not have a significant impact on this quantification (not shown here).

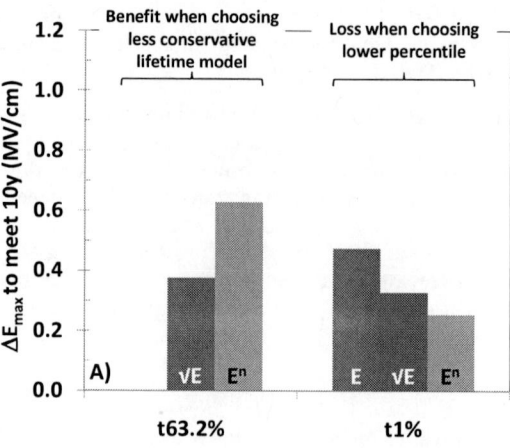

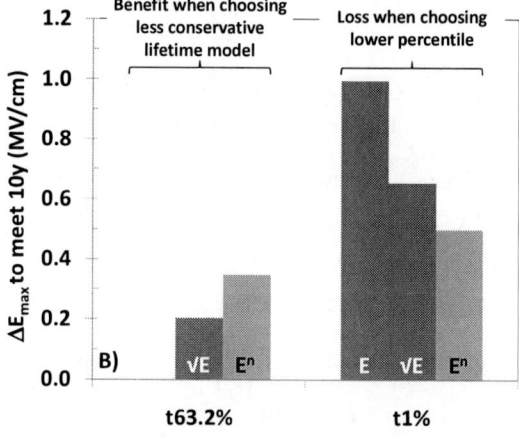

Figure 7. Differences in E_{max} to meet 10y lifetime criterion when choosing different percentiles and lifetime models. The bars indicated with "t63.2%" show the beneifit when choosing the √E-model or the power law (E^n) instead of the E-model. The bars indicated with "t1%" show the loss in E_{max} for each of the models when choosing the 1%-percentile instead of the 63.2%-percentile. Figure A shows these data for the intrinsic low-k material, while figure B considers the 50nm ½pitch integrated test vehicle.

The influence of the choice of the lifetime model compared to the impact of the differences in distributional shape β is quantified in figure 7. The graphs plot differences in the maximum allowed field, E_{max}, to meet a 10 year lifetime criterion when choosing different percentiles and lifetime models. Figure 7A considers these differences for the p-caps, where the bars indicated with "t63.2%" show the differences in E_{max} when choosing the $\sqrt{E}$-model or the power law (E^n) instead of the E-model, while the bars indicated with "t1%" show the differences in E_{max} for each of the models when choosing the 1%-percentile instead of the 63.2%-percentile. For the intrinsic low-k material, the drop in E_{max} when going from large to low percentiles is similar to the benefit of choosing a less conservative lifetime model.

Figure 7B shows the same information as figure 7A, but now for the 50nm ½pitch test vehicle. It is clear that the benefit of choosing less conservative lifetime models is much smaller than the loss of choosing lower percentiles. As such, the choice of the lifetime model has a lower impact than the shape of the distribution of failure times.

This is a surprising result as people tend to believe that, once one can prove that the commonly used E-model is too conservative, the predicted lifetimes using less conservative models would become orders of magnitude higher. With our study, we show that, when looking at maximum allowed fields and when considering actual reliability data, this benefit is much lower than expected and that process optimizations that lead to higher distributional shapes β are at least as beneficial compared to using different lifetime models. Note that the above conclusion relies on the assumption that the distributional shape does not depend on the applied electrical field. We believe this is a valid assumption because the lowest fields at which actual measurements are performed are not too far away from the obtained E_{max}-values. At these low fields, no significant change in β is observed with respect to the b at higher fields (see [13] and section B).

B. Low-field tests on pristine low-k material

Studies have been conducted by different authors to determine the correct lifetime extrapolation model for integrated dielectrics. To differentiate between different lifetime models, low field tests are often performed. In this section, low field tests on the p-cap test vehicle will be presented, where we want to differentiate between different lifetime models for pristine low-k materials. In an earlier paper, we used ~1 year-data to show that the E- and the $\sqrt{E}$-model were too conservative to explain our low field data on 50 and 90nm ½pitch structures [13]. In such experiments, the choice of the high field conditions used to predict failure times at low fields needs to be made in a way that the width of the uncertainty intervals of the estimated failure times at these low fields is as small as possible. In parallel, in order to be able to distinct between valid and non-valid models, the extrapolation error of the non-valid models needs to be as big as possible. Given a fixed highest allowed field E_{high}, both the width of the uncertainty intervals and the extrapolation errors are dominated by the lowest field E_{low} used for the prediction. The lower E_{low}, the narrower the uncertainty intervals become, but the smaller

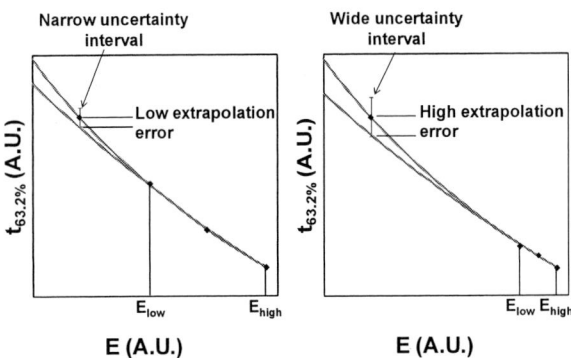

Figure 8. 63.2%-percentile for different fields and extrapolations to lower fields with fixed E_{high} and a) low E_{low} and b) high E_{low}.

the extrapolation errors (Figure 8). Hence, the optimum E_{low} is where a balance between extrapolation errors and width of uncertainty intervals is obtained.

For our p-caps, this optimum E_{low} was determined using Monte-Carlo simulations. The approach was as follows:

1. Consider the $\sqrt{E}$-model and the power law.
2. Set a field to predict to. For the current simulations, 3.5MV/cm was taken, which is below the lowest field shown in figure 5.
3. Determine the highest allowed field for our test material on which TDDB tests can be performed without overstressing the device. For this, the shape parameter β of the underlying Weibull distribution is used as a criterion; this β has to be the same than the one obtained at lower fields. For our test material, this field is determined to be 6.4MV/cm.
4. Simulate experiments with three different fields: E_{high}=6.4MV/cm, E_{low}=Variable and E_{mid}=(E_{high}+E_{low})/2. Input parameters are the number of devices/condition=15, the shape parameter β and the acceleration factors for the different models. The latter two were obtained from the high field data.
5. For each set of simulated experiments, determine the width of the uncertainty intervals and the extrapolation errors and obtain their average values from 1,000 such simulations for each E_{low}.

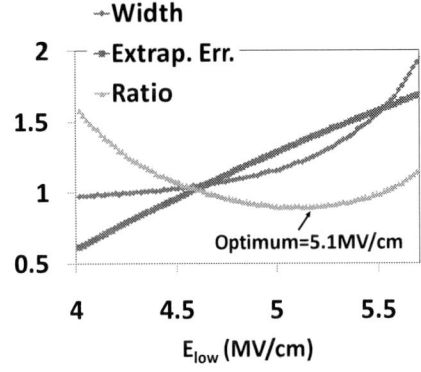

Figure 9. Width of uncertainty interval on predictied failure times, extrapolation error and their ratio as a function of E_{low}

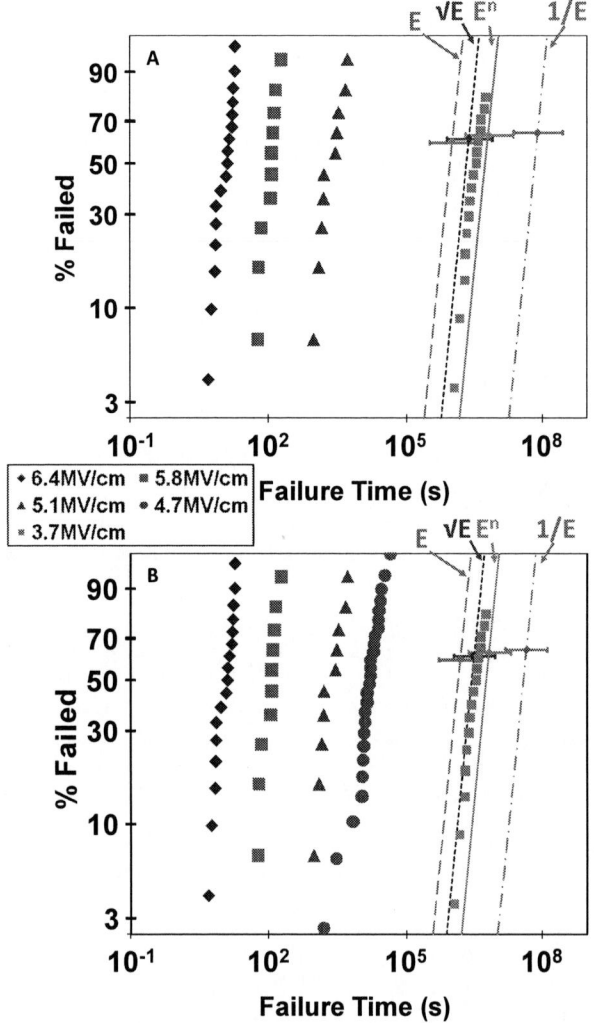

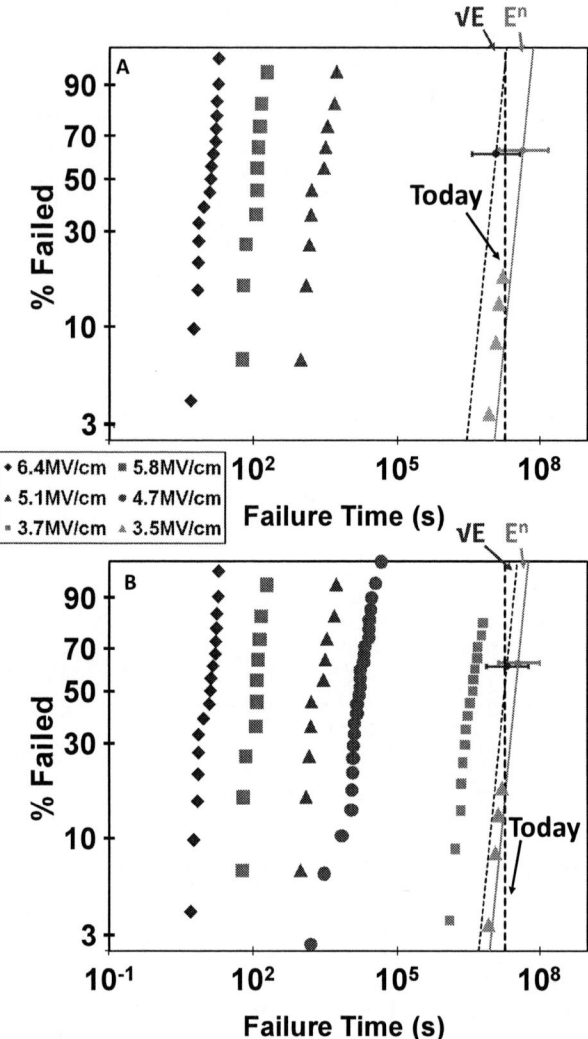

Figure 10. Weibull probability plot of the failure times obtained on the p-cap test vehicle. Extrapolation to 3.7MV/cm using the E-model, the √E-model, the power law and the 1/E-model has been done based on A) the three and B) the four higher fields.

Figure 11. Weibull probability plot of the failure times obtained on the p-cap test vehicle. Extrapolation to 3.5MV/cm using the √E-model and the power law has been done based on A) the three and B) the five higher fields.

The width of the uncertainty interval decreases with decreasing E_{low} but the extrapolation error decreases as well (Figure 9). An optimum value for E_{low} is where the ratio of width and extrapolation error is minimal. For our test material, this is about 5.1MV/cm. As such, we decided to take 6.4, 5.8 and 5.1MV/cm as high fields to extrapolate from to low fields.

Figure 10A shows the Weibull probability plot of low field tests performed at 3.7MV/cm. Using maximum likelihood fitting, the three highest conditions have been fitted with the E-model, the √E-model, the power law (E^n) and the 1/E-model. From this fit, an extrapolation is done to 3.7MV/cm for all percentiles and the 95% prediction interval of the extrapolated $t_{63.2\%}$ is indicated on the graph. Since the failure times from the 2 month test at 3.7MV/cm fall outside of the prediction intervals of both the E- and the 1/E-model, these models are not suited to describe p-cap data with a barrier between copper and

dielectric. If we would have taken a lower E_{low} into account, we would not have been able to reject the E-model (Figure 10B). The fact that taking more data into account lead to less significant conclusions sounds counterintuitive. This is due to the choice of the statistical method, where we use high field data to predict what would happen at lower fields, given a certain lifetime model. As discussed above, this method requires significant extrapolation errors of non-valid lifetime models which does not allow too low fields to predict from. Other statistical methods to differentiate between lifetime models that are based on maximum likelihood fitting are, to the authors knowledge, not existing and need to be developed.

Based on these data and the data presented in [13], we can conclude that the E-model does not describe the TDDB-behavior of both integrated and intrinsic low-k materials. Earlier models to motivate the E-model which were based on the assumption that weak bonds in the dielectric can be broken by applying an external electric field [23] do not hold for low-k dielectrics. Besides the E-model, we can also exclude the 1/E-model as valid lifetime model for pristine low-k material. Adding a barrier between the copper and the dielectric prevents copper drift during TDDB and earlier observations of the 1/E-model in the presence of copper [20] are not valid in a copper/barrier/low-k system.

To further discriminate between the √E-model and the power law, long term tests at 3.5MV/cm have been launched (Figure 11). After 7 months of testing, 4/20 devices have failed. The predicted number of failures based on the √E-model would be about 14-20/20, while the power law nicely predicted 3-4/20 failures. However, given the wide prediction intervals, the √E-model cannot be excluded with statistical significance.

Summarizing our current and our earlier [13] data, we observe that a) the E-model does not hold for both integrated and pristine material, b) the √E-model is too conservative to explain integrated data, but cannot be excluded as valid model for pristine material, c) the power law cannot be excluded for both pristine and integrated materials and d) the 1/E model is too optimistic to explain our data on pristine low-k's. If we assume that the same lifetime model holds for pristine and integrated low-k materials, we can conclude that the power law describes the data best. This implies that the impact of copper on BEOL TDDB [2,6], which was the main motivation for the √E-model, could be lower than assumed. One supporting argument of the power law is the dependence of the distributional shape of failure times on the electric field. Different authors proved that that, when assuming the E or √E-model, this distributional shape would decrease with decreasing fields due to LER-contributions and spacing variations between structures [4,5,13], while the distributional shape is independent of the field in case of the power law [13]. Since our previous [13] and current data over a wide range of fields show no significant change in distributional shape we believe the power law is better suited to describe our data

IV. CONCLUSIONS

Differences in reliability behavior between pristine and integrated low-k materials were studied. The intrinsic properties were studied using a novel test vehicle, the so-called low-k planar capacitor. Compared to the pristine material, the leakage currents of the materials after integration were significantly higher and the breakdown fields dropped by 1.3-2.4MV/cm depending on the pitch after integration. Also the maximum allowed field in order to meet a 10year lifetime criterion, E_{max}, dropped significantly. Depending on the chosen lifetime model, E_{max} goes down by 0.5-0.6MV/cm when going from intrinsic to integrated 90nm ½pitch and further reduces by 0.4-0.8MV/cm when scaling further to 50nm ½pitch. Comparing the benefit of choosing a less conservative lifetime model to the loss of going to lower percentiles showed that the choice of the lifetime model has a lower impact than the shape of the distribution of failure times.

We also used the low-k planar capacitors to study the intrinsic low field TDDB behavior. Using Monte-Carlo simulations, we showed that the choice of the high field conditions used to predict the lifetimes at lower fields is crucial

when it comes down to discriminate between different models. Using data of 2 months, we were able to exclude the E- and the 1/E-model as valid models for systems with TaNTa as barrier between copper and low-k. Longer tests to making the distinction between the √E-model and the power law did statistically not show significant differences, but a strong tendency towards the power law was observed.

ACKNOWLEDGMENT

The authors would like to thank the participants of imec's barrier reliability meeting and also the members of the interconnect and remo group are acknowledged for fruitful and stimulating discussions. In particular, Dr. Michele Stucchi and Dr. Philippe Roussel are acknowledged for the nice discussions.

REFERENCES

[1] R. Tsu, J. W. McPherson and W. R. McKee, "Leakage and breakdown reliability issues associated with low-k dielectrics in a dual-damascene Cu process", IEEE Int. Reliability Physics Symposium (IRPS), p. 348, 2000

[2] F. Chen, O. Bravo, K. Chanda, P. McLaughlin, T. Sullivan, J. Gill1, J. Lloyd, R. Kontra and J. Aitken, "A comprehensive study of low-k SiCOH TDDB phenomena and its reliability lifetime model development", IEEE Int. Reliability Physics Symposium (IRPS), p. 46, 2006

[3] Zs. Tökei, Ph. Roussel, M. Stucchi, J. Versluijs, I. Ciofi, L. Carbonell, G.P. Beyer, A. Cockburn, M. Agustin and K. Shah, "Impact of LER on BEOL dielectric reliability: a quantitative model and experimental validation", IEEE Int. Interconnect Technolody Conference (IITC), p. 228, 2009

[4] M. Vilmay, D. Roy, C. Monget, F. Volpi and J.-M. Chaix, "Copper line topology impact on the SIOCH low-k reliability in sub 45 nm technology node", IEEE Int. Reliability Physics Symposium (IRPS), p. 606, 2009

[5] S.-C. Lee, A.S. Oates and K.-M. Chang, "Geometric Variability of Nanoscale Interconnects and Its Impact on the Time-Dependent Breakdown of Cu/Low-k Dielectrics", IEEE Transactions on Device and Materials Reliability, Vol. 10, p. 307, 2010

[6] N. Suzumura, S. Yamamoto, D. Kodama, K. Makabe, J. Komori, E. Murakami, S. Maegawa and K. Kubota, "A new TDDB degradation model based on Cu ion drift in Cu interconnect dielectrics", IEEE Int. Reliability Physics Symposium (IRPS), p. 484, 2006

[7] L. Zhao, Zs. Tökei, G. G. Gischia, M. Pantouvaki, K. Croes and G. P. Beyer, "A novel test structure to study intrinsic reliability of barrier/low-k", IEEE Int. Reliability Physics Symposium (IRPS), p. 848, 2009

[8] L Zhao, Zs. Tökei, G. G. Gischia, H. Volders and G. P. Beyer, "A New Perspective of Barrier Material Evaluation and Process Optimization", IEEE Int. Interconnect Technolody Conference (IITC), p. 206, 2009

[9] G. G. Gischia, K. Croes, G. Groeseneken, Zs. Tökei, V. Afanas'ev and L. Zhao, "Study of leakage mechanism and trap density in porous low-k materials", IEEE Int. Reliability Physics Symposium (IRPS), p. 549, 2010

[10] W. Wu, X. Duan and J. S. Yuan, "A physical model of time-dependent dielectric breakdown in copper metallization", IEEE Int. Reliability Physics Symposium (IRPS), p. 282, 2003

[11] G. S. Haase, E. T. Ogawa and J. W. McPherson, "Breakdown Characteristics of interconnect dieletrics", IEEE Int. Reliability Physics Symposium (IRPS), p. 466, 2005

[12] J. Kim, E. T. Ogawa and J. W. McPherson, "Time dependent dielectric breakdown characteristics of low-k dielectric (SiOC) over a wide range of test areas and electric fields", IEEE Int. Reliability Physics Symposium (IRPS), p. 399, 2007

[13] K. Croes and Zs. Tőkei, "E- and √E-model too conservative to describe low field time dependent dielectric breakdown", IEEE Int. Reliability Physics Symposium (IRPS), p. 543, 2010

[14] J. R. Lloyd, E. Liniger and T. M. Shaw, "Simple model for time-dependent dielectric breakdown in inter- and intralevel low-k dielectrics", Journal of Applied Physics, Vol. 98, p. 084109, 2005

[15] G. S. Haase, "A model for electric degradation of interconnect low-k dielectrics in microelectronic integrated circuits", Journal of Applied Physics, Vol. 105, p. 044908, 2009

[16] I. C. Chen, S. Holland and C. Hu, "A quantitative model for time-dependent breakdown in SiO2", IEEE Int. Reliability Physics Symposium (IRPS), p. 24, 1985

[17] J. W. McPherson, R. B. Khamankar and A. Shanware, "Complementary model for intrinsic time-dependent dielectric breakdown in SiO2 dielectrics", Journal of Applied Physics, Vol. 88, p. 5351. 2000

[18] J. Suñé and E. Wu, "A New Quantitative Hydrogen-Based Model for Ultra-Thin Oxide Breakdown", Digest of Technical Papers on Symposium on VLSI Technology, p.97, 2001

[19] E. Wu, J. Suñé, W. Lai, A. Vayshenker and D. Harmon, "A Comprehensive Investigation of Gate Oxide Breakdown of P+Poly/PFETs Under Inversion Mode", Technical Digest of the International Electron Devices Meeting (IEDM), p. 339, 2005

[20] L. Zhao, Zs. Tőkei, K. Croes, C.J. Wilson and G.P. Beyer, "Direct observation of the 1/E dependence of time dependent dielectric breakdown in the presence of copper", Applied Physics Letters, in press.

[21] M. Pantouvaki, H. Struyf, D. Hendrickx, N. Heylen, O. Richard and G. P. Beyer, "The impact of ash on TDDB of metal-hard-mask-etched Cu/low-k interconnects", ", Proceedings of the Advanced Metallization Conference (AMC), p. 733, 2008

[22] J. Van Olmen, A. Al-Bayati, G. P. Beyer, P. Boelen, L. Carbonell, C. Zhao, I. Ciofi, M. Claes, A. Cockburn, G. Druais, D. Hendrickx, N. Heylen, E. Kesters, S. Lytle, A. Noori, M. Op de Beeck, H. Struyf, Zs. Tőkei and J. Versluijs, "Integration of 50nm half pitch single damascene copper trenches in BDII by means of double patterning 193nm immersion lithography on metal hardmask", Proceedings of the Advanced Metallization Conference (AMC), p. 355, 2007

[23] J. W. McPherson and H. C. Mogul, "Underlying physics of the thermochemical E model in describing low-field timedependent dielectric breakdown in SiO2 thin films," Journal of Applied Physics, Vol. 84, p. 1513, 1998

Statistics of Breakdown Field and Time-Dependent Dielectric Breakdown in Contact-to-Poly Modules

Shinji Yokogawa, Satoshi Uno, Ichiro Kato, Hideaki Tsuchiya, Tatsuo Shimizu, and Mitsuhiro Sakamoto

Advanced Device Development Department, Device and Analysis Technology Division, Renesas Electronics Corporation
1120 Shimokuzawa, Chuou-ku, Sagamihara, Kanagawa 252-5298, Japan
phone: +81-42-771-4274; fax: +81-42-771-0916; E-mail: shinji.yokogawa.vx@renesas.com

Abstract—**In this paper, we present the results of voltage-ramp dielectric breakdown and time-dependent dielectric breakdown experiments for contact–polysilicon control gate intra-level dielectric stacks. Lifetime distribution and area scaling are discussed statistically with the analysis of global and local space deviations using the electrical method. Optimized process reliability is evaluated by performing a SRAM lifetime test that measures the early life failure rate.**

Key words: *VRDB; TDDB; CA/PC; Area scaling; Global deviation; Local deviation; ELFR*

I. Introduction

The minimum spacing between a contact (CA) and a polysilicon control gate (PC) in advanced integrated circuits is progressively decreasing in a step-wise manner with successive generations of technology. As reported in a study on field failures in flash memory [1], leakage between the word and bit lines are often found or suspected to be caused by a particle- or process-defect-induced breakdown between the second polysilicon (control) gate and the drain contact. This phenomenon represents a major vulnerability of flash cells. Aggressive technology scaling will result in the exacerbation of this problem not only for the flash memory but also for the standard CMOS technology in general.

In addition, the rapid adoption of new materials to contacts and gates may increase the risk of leakages. For example, Cu contact for 22-nm nodes and beyond was considered as a critical challenge [2]. As the contact size shrinks, tungsten contact metallization faces scaling issues owing to poor gap fill. A significant increase and broad distribution will be observed in the contact resistance. To overcome these issues, Cu was identified as a potentially useful material owing to its lower resistance. One of the concerns for the technology is a containment capacity of the barrier liner for Cu diffusion. It is well known that the Cu drift affects inter-metal dielectric reliability. Similarly, metal gates have been put into practical use instead of the traditionally used polysilicon gate. Because film depositions are to be repeated for the gate-forming process, many interfaces exist between a CA and a PC, which may contribute to the metal ion drift by the electric field acceleration with scaling. Therefore, it is important to understand the basic behavior of CA/PC breakdown reliability for the successful adoption of the new materials. However, there are not many previous studies on the basic behavior of CA/PC breakdown reliability.

In this paper, we discuss the results of voltage-ramp dielectric breakdown (VRDB) and time-dependent dielectric breakdown (TDDB) experiments for CA/PC intralevel dielectric (ILD) stacks. Lifetime distribution and area scaling are discussed statistically with the analysis of global and local deviations (Fig. 1) by using a recently proposed electrical method [3]. Optimized process reliability was evaluated by performing a SRAM lifetime test that measures the early life failure rate (ELFR).

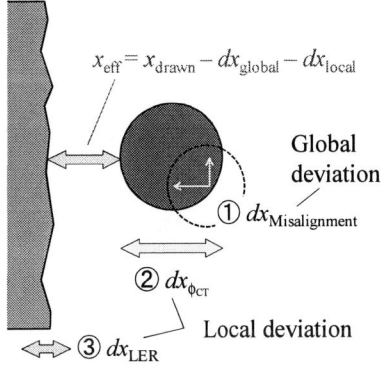

Figure 1. Schematic diagram of process deviations in CA/PC module. They can be categorized into a global deviation (i.e., misalignment) and local deviations (i.e., deviation in CA diameter and line-edge roughness of gate).

II. Experiments

A. Test structure

The spacing between CA and PC, x_{drawn}, is defined by the layout (Fig. 2). Test structures were fabricated using standard 40-nm technology on 300-mm wafers. CA/PC modules were layouted on shallow trench isolation (STI) to avoid a gate dielectric breakdown. The number of CA/PC pairs, N, was set to 1, 100, and 10,000 (10k) to investigate the area scaling.

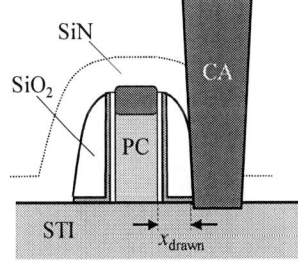

Figure 2. Cross-sectional diagram of CA/PC test structure.

B. Statistical methodology

In the VRDB technique, the breakdown voltage V_{bd} is measured as a function of the as-drawn CA/PC spacing (Fig. 3). The spacing between CA and PC is modulated on the layout in order to simulate the impact of misalignments. Steps of 0.1 V with a ramp rate of 1 V/s and 200 ms delay at each step were used for the VRDB test at room temperature.

The effective nominal breakdown field, E_{bd_eff} under a specific voltage ramp condition for the ILD material between CA and PC is estimated from the slope of the linear regression. The deviation from drawn space, the etch bias, is estimated by the x-intercept [3], which represents a global deviation of the CA/PC space. Residual error of the fitted line represents local deviations of the CA/PC space. Global and local deviations can be analyzed by the measurements for each lithography shot.

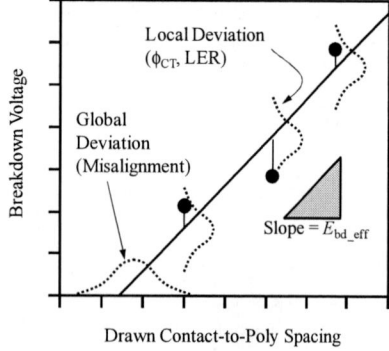

Figure 3. Schematic of electrical space evaluation method used to measure global and local deviations in a lithography shot.

C. TDDB test

The standard constant-voltage stress (CVS) test was performed to evaluate the TDDB lifetime distribution. The TDDB tests were carried out under constant-voltage and 110 °C. The time needed to observe a sudden change in leakage current (hard breakdown) was defined as a lifetime. Lifetime distribution was assumed to adapt to Weibull distribution, and the maximum-likelihood method was used to estimate parameters of the distribution.

A transmission electron microscope (TEM) was used for a cross-sectional analysis of the breakdown point.

D. SRAM ELFR test

To confirm the product level reliability, a highly accelerated ELFR test was performed with 1.5 Mbit that were misaligned purposely, and 32 Mbit SRAM. Dynamic burn-in stress was carried out with 1.8 V and 125 °C for about 4000 packaged SRAMs. The failure rate was evaluated for both pre- and post-optimized process conditions.

III. RESULTS AND DISCUSSIONS

A. Behavior and leakage conduction mechanism of CA/PC

As shown in Fig. 2, CA/PC ILD stacks are formed from layers. Hence, to check the uniformity of the conduction mechanism in the experiments, a current-versus-field relation

was evaluated on the basis of Poole-Frenkel (PF) emission conduction [4][5]. PF and Schottky emission (SE) conductions show a current (I) versus electric field (E) relation as follows:

$$I \propto A \cdot \exp\left(\frac{\beta\sqrt{E} - \varphi}{k_B T}\right), \qquad (1)$$

where A is a pre-factor, β is the field-dependent pre-factor that changes value with the change in the conduction mechanism, φ is the potential barrier, k_B is the Boltzmann's constant, and T is the temperature. The value of β for the PF conduction (β_{PF}) and SE conduction (β_{SE}) are represented as follows:

$$\beta_{PF} = k_{slope} k_B T \sqrt{s} = \left(\frac{e^3}{\pi\varepsilon_0 k}\right)^{1/2}, \qquad (2)$$

$$\beta_{SE} = k_{slope} k_B T \sqrt{s} = \left(\frac{e^3}{4\pi\varepsilon_0 k}\right)^{1/2}, \qquad (3)$$

where k_{slope} is the slope of $\ln(I)$ versus the square root of V (Fig. 4), s is the spacing between CA and PC, e is the charge of an electron, ε_0 the dielectric constant of free space, and k is the relative dielectric constant. In other words, although PF and SE conduction have different physical bases, both of them are represented by the similar current versus field relation shown in eq. (1), and their field-dependent pre-factors differ by a factor of only two.

Based on PF conduction theory, median values of β_{PF} are estimated by the observed median k_{slope} and the dielectric constants of SiO$_2$ ($k = 4.2$) and SiN ($k = 7.0$). As shown in Fig. 4, the observed value of β seems to be constant for each drawn space. This suggests that the conduction mechanism is unique. Moreover, values of β that are estimated by the observed k_{slope} are reasonably consistent with the β_{PF} that are calculated by the dielectric constant of SiO$_2$ and SiN. If the SE conduction is dominant, the observed values of β will correspond with values of β_{SE} that are half the value of β_{PF}. Consequently, these results suggest that the conduction mechanism of CA/PC is similar to that of PF conduction.

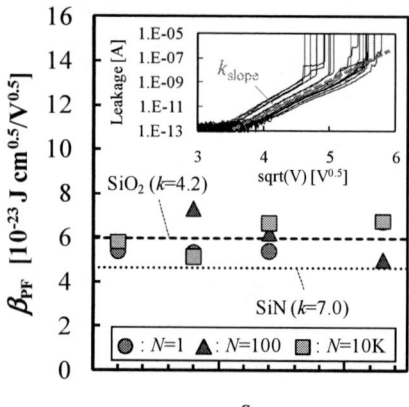

Figure 4. CA/PC space dependence of normalized lnI–sqrt(V) slope at 25 °C.

B. Behavior of CA/PC TDDB

In the TDDB test, initial current leakage and lifetime showed a strong dependence when a designed spacing between CA and PC of 24 nm was used. Figure 5 shows an example of TDDB test results for 24 nm CA/PC designed spacing and N = 10k. As shown by Fig. 5(a), samples that show a small value of an initial leakage current tend to have a long lifetime. Therefore, the Weibull plot of the accumulated oxide charge (Q_{bd}) shows an improved Weibull slope compared to that of lifetime, as shown by Fig. 5(b). These results suggest that lifetime during the CVS test strongly depends on the space between CA and PC.

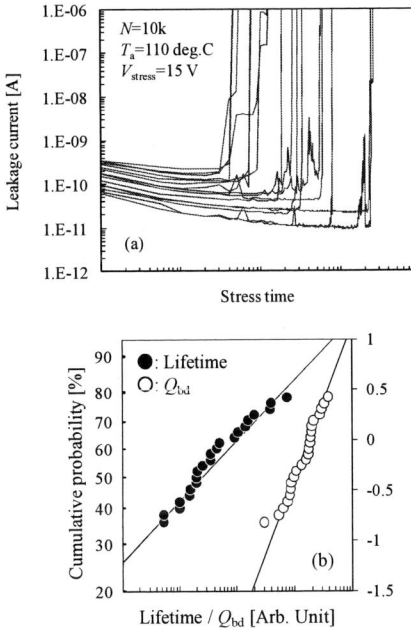

Figure 5. Examples of CA/PC TDDB test results. (a) Leakage vs. time of CA/PC CVS test (N = 10k). (b) Weibull plots of lifetime and Q_{bd}.

Cross-sectional S-TEM analysis was performed to find the breakdown site using a purposely misaligned sample. As shown in Fig. 6, the breakdown failure was located between the top edges of the PC and CA. Hence, breakdown occurs at the location of minimum space, and it agrees with the initial current dependence of the lifetime. In other words, the maximum electric field in the test structure will be dominant for the lifetime. A method to estimate the maximum electric field in the structure is required to evaluate an electric field dependence of the lifetime on global and local deviations of CA/PC spacing.

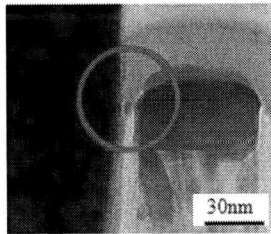

Figure 6. Cross-sectional TEM observation of CA/PC TDDB failure. The red circle indicates the breakdown site that was misaligned purposely.

C. Global and local variations of estimated CA/PC spacing

VRDB tests were performed in order to separate and analyze global and local deviations of CA/PC spacing. The E_{bd_eff} and etch bias were measured for N = 1 structures with several different CA/PC spacings. As shown in Fig. 7, the median V_{bd} shows a highly linear dependence of drawn CA/PC spacing. However, the observed V_{bd} for each lithography shot show significant variation of the x-intercept. In addition, residual error from each trend is also significant. These results suggest that neither global nor local deviations were negligible. In other words, the etch bias estimated for each lithography shot will be a good indicator of the global deviation. The residual error of individual test structures is closely related to the local deviation, and will affect the area scaling, as described below.

Moreover, a similar E_{bd_eff} trend was observed for both negative and positive biases, as shown in Fig. 8. This result suggests that the fast ion drift of a specific species does not contribute to the V_{bd} and E_{bd_eff}. Henceforth, PC was maintained under negative bias in this study.

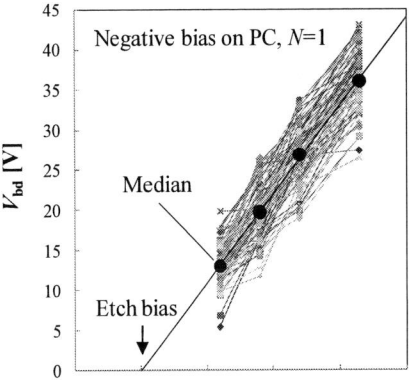

Figure 7. V_{bd} of CA/PC ILD with varying values of spacing, measured at 25 °C for N = 1.

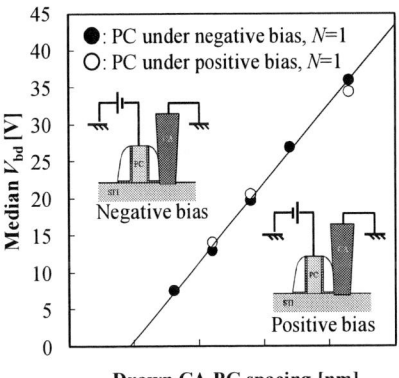

Figure 8. Comparison of median V_{bd} for PC under negative and positive bias at 25 °C.

D. Area scaling of CA/PC spacing

To discuss the deviations separately, the area scaling of the median V_{bd} and E_{bd_eff}, as well as local deviation, were investigated. The median V_{bd} clearly exhibited area scaling, as shown in Fig. 9, although the median E_{bd_eff} was stable. The estimated etch bias, determined by the x-intercept, appeared to increase with the number of CA/PC pairs. On the other hand, the estimated E_{bd_eff} distributions in each lithography shot did not show clear area scaling, as shown by Fig. 10. Namely, the median V_{bd} area scaling should depend on the local deviation of CA/PC spacing.

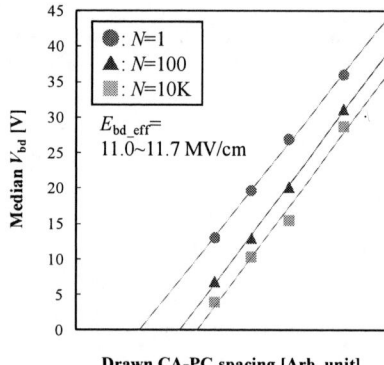

Figure 9.   Area scaling of median V_{bd} for CA/PC ILD at 25 °C.

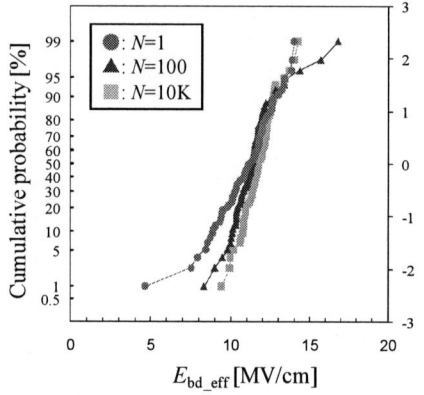

Figure 10. Breakdown field distributions at 25 °C for N = 1, 100, and 10k.

To investigate the area scaling of the local deviation of CA/PC spacing, the global deviation should be subtracted from drawn spaces, x_{drawn}. The estimated etch biases in each lithography shot, dx_{global}, correspond to the global deviation. Hence, $x_{drawn}-dx_{global}$ represents one of the indicators for the local deviation of CA/PC spacing. If V_{bd} depends on the minimum spacing in the structure, the estimated indicators will indicate an area scaling due to the local deviation. Figure 11 indicates the area scaling of $x_{drawn}-dx_{global}$. It assumed that V_{bd} corresponding to the minimum space was obtained, and is converted to the cumulative probability of single CA/PC pairs based on the Poisson law. The cumulative probability of $x_{drawn}-dx_{global}$ for individual CA/PC pairs shows linearity as a whole, indicating that V_{bd} depends on the minimum space and its distribution. Especially, it is speculated that biases of the slope in N = 100 and 10k from the fitted line depend on the residual error of the estimated global variations.

These results suggest that the local deviation of CA/PC spacing will also have large impact on the lifetime distribution and its area scaling. It is important to demonstrate these correlations for exactly estimating the product level reliability.

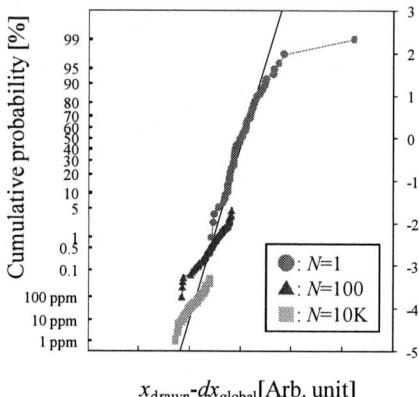

Figure 11. Area scaling of estimated local deviation of CA/PC space.

E. Area scaling of TDDB lifetime

TDDB lifetime distribution under the CVS test is drastically affected by global and local deviations. Figure 12 shows the cumulative probability of the lifetime at the single CA/PC level that is converted to the cumulative probability of single CA/PC pairs based on the Poisson law. As shown in Fig. 12, the plot did not exhibit a linear curve with several values of N (1, 100, and 10k). This result indicates that the Poisson assumption is not valid for the area scaling. In addition, each distribution exhibited a very small shape parameter, β_{life}, due to global and local deviations. On the other hand, the conventional estimation of β_{area} by area scaling of the scale parameter η results in a larger estimate, as shown in Fig. 13. These results do not indicate that the phenomenon of the CA/PC TDDB in this study has a decreasing failure rate, but do indicate that the β_{life} appears because of spatial deviations. Consequently, a method to normalize global and local deviations of CA/PC spacing is required to estimate a product level reliability.

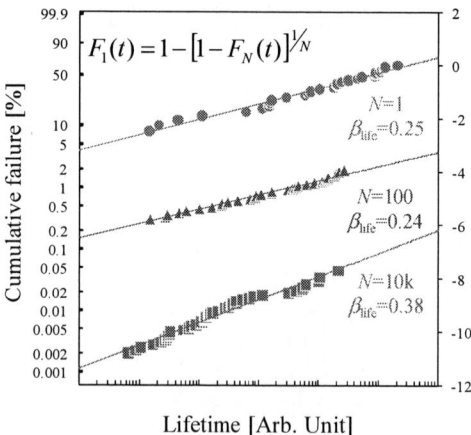

Figure 12. TDDB lifetime distributions of CA/PC CVS test at 110 °C and V_{stress} = 15V. Cumulative failure is estimated on individual CA/PC pair. The drawn space of CA/PC is 33 nm.

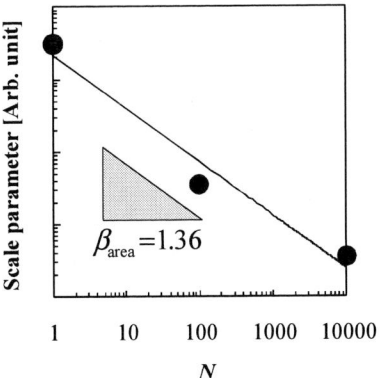

Figure 13. Area scaling of scale parameter of CA/PC TDDB lifetime distributions. CA/PC CVS test was performed at 110 °C and $V_{stress} = 15$ V. The drawn space of CA/PC is 33 nm.

The lifetime distribution, in which a fatal ratio k [6] and voltage acceleration are assumed, was evaluated. The cumulative distribution function for N pair structures is defined based on Weibull distribution as follows:

$$
\begin{aligned}
F_N(t,V,N) &= 1 - \left\{ \exp\left[-\left(\frac{t}{\eta(V)} \right)^{\beta_{\text{life}}} \right] \right\}^{N^k} \\
&= 1 - \exp\left[-N^k \left(\frac{t}{\eta(V)} \right)^{\beta_{\text{life}}} \right] \\
&= 1 - \exp\left[-\left(\frac{t}{\eta(V) / N^{k/\beta_{\text{life}}}} \right)^{\beta_{\text{life}}} \right], \quad (4)
\end{aligned}
$$

and,

$$
\eta(V) = A \exp\left(-\gamma \sqrt{V} \right), \quad (5)
$$

where $\eta(V)$ is a voltage-dependent scale parameter for a pair, A is a constant, and γ is a voltage acceleration factor. The maximum likelihood estimation (MLE) was used to estimate parameters. Observed lifetime distribution and MLE of eq. (4) were indicated in Fig. 14. Solid lines in Fig. 14 represent the best fit to the data. The estimated fatal ratio k of 0.251 indicates that Poisson area scaling cannot be assumed, and suggests that only a portion of the total CA/PC pairs, those which have the smallest CA/PC spacing, will contribute to lifetime. Such critical CA/PC pairs can be defined as TDDB "fatal" pairs. In other words, the maximum electric field that is defined at the fatal pairs will be dominant for the lifetime.

In eq. (4), an empirical power-law relation is assumed for the fatal pair ratio as $N_{\text{actual}} = (N_{\text{drawn}})^k$. Under this assumption, β_{area} is represented as β_{life}/k. For example, β_{area} is estimated to be $\beta_{\text{life}}/k = 1.04$ in Fig. 14. In the case of Fig. 13, $\beta_{\text{area}} = 1.32$ is estimated from $k = 0.22$ and the average $\beta_{\text{life}} = 0.29$, and corresponds with the value in Fig. 13 estimated by conventional area scaling. In other words, the fatal ratio can be inversely estimated as $\beta_{\text{life}}/\beta_{\text{area}}$.

Based on the assumption of the fatal pairs, the maximum electric field dependence of the lifetime is indicated on Fig. 15. The maximum electric field, $E_{\max}$, of the individual sample was estimated by $V/(x_{\text{drawn}} - dx_{\text{global}})$, where V is the stress voltage and $(x_{\text{drawn}} - dx_{\text{global}})$ is the estimated CA/PC spacing, except for the global deviation based on the VRDB. As shown in Fig. 15, the lifetime at $N = 100$ and 10k strongly depends on the maximum electrical field with no area scaling. The slope of the plots is estimated as 21.7 $(\text{cm/V})^{0.5}$. This value is very close to the actual electric field dependence indicated in Fig. 14 (22.4 $(\text{cm/V})^{0.5}$). Namely, the lifetime distribution in Fig. 14 seems to contain the CA/PC spacing variation and its electric field dependence.

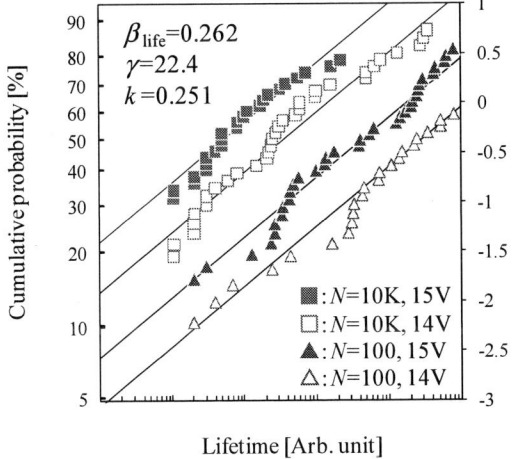

Figure 14. CA/PC TDDB lifetime distributions with various CVS and N. CA/PC CVS test was performed at 110 °C. The drawn space is 24 nm.

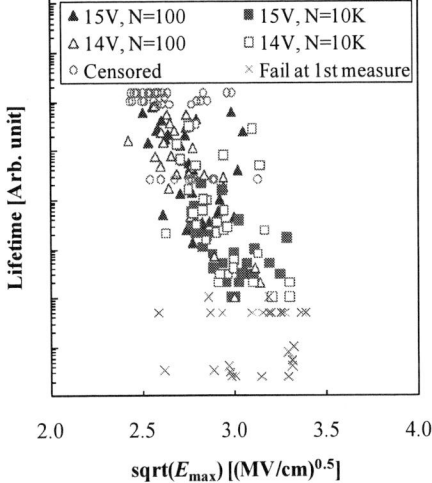

Figure 15. Estimated electric field dependence of lifetime. Maximum field is estimated by using the estimated global deviation on the same shot. CA/PC CVS test was performed at 110 °C. The drawn space is 24 nm.

If the empirical power-law relation is valid for different areas up to the product level, the fatal ratio concept will provide an exact extrapolation of product lifetime. In addition, it is appropriate to use minimum CA/PC spacing due to the global deviation, which is assumed to an integrated process, as a worst-case estimation.

F. SRAM ELFR test

To evaluate the product level reliability, a highly accelerated ELFR test was performed with 1.5 Mbit SRAM. The sample for which misalignment, CA diameter, and line-edge roughness of gate were not optimized was used. The drawn minimum space of CA/PC is about 31 nm. Dynamic burn-in stress was carried out with 1.8 V and 125 °C for approximately 2000 samples. Figure 16 shows the cumulative failure probability. The Weibull shape parameter of the observed cumulative failure probability, β_{SRAM}, was estimated to be 0.264. This value nearly corresponds with the Weibull shape parameter that was evaluated by the CA/PC test structure (Fig. 12). The observed failure mode was confirmed by a physical failure analysis, and was identified as the CA/PC breakdown at very narrow CA/PC spacing pairs. These results suggest that CA/PC TDDB will be apparently observed as a decreasing failure mode for products under use conditions.

After optimization of misalignment, the CA diameter and line-edge roughness of gate, ELFR, were evaluated with 32 Mbit SRAM. In spite of being over 20 times the SRAM size, no failure was observed during the test that was performed with the same conditions with 2000 samples. The cumulative failure probability that was normalized to 1.5 Mbit SRAM is indicated in Fig. 16. If the fatal ratio $k = 0.251$ is valid for the area scaling, 32 Mbit SRAM is regarded as only approximately 2 times the size of 1.5 Mbit SRAM. In this case, a failure should be observed during the ELFR test. Therefore, it is thought that a remarkable improvement of CA/PC TDDB reliability was proved successfully.

Additionally, the voltage dependence of the β_{life} and β_{SRAM} should be investigated as a future work. Based on the root-E model, if global and local deviations are significant, an improvement of distribution at low stress conditions should be observed. For this investigation, an additional development of an efficient test structure and a test method is strongly required.

IV. CONCLUSIONS

The E_{bd_eff} and TDDB characteristics of CA/PC modules were investigated statistically. The VRDB-based electrical measurement provides information on global and local deviations that affect the TDDB lifetime distribution.

Because of the local deviation of the CA/PC spacing, only the few CA/PC pairs that have the smallest CA/PC spacing will contribute the most to the lifetime under stress conditions. The fatal ratio concept was well adapted to the observed lifetime data. The lifetime distribution contains the CA/PC spacing variation and its electric field dependence.

Optimized product reliability was demonstrated by performing a SRAM lifetime test. The optimization of misalignment, CA diameter, and line-edge roughness of the gate is effective for achieving remarkable improvements in the CA/PC TDDB reliability.

ACKNOWLEDGMENT

The authors would like to thank R. Okamura, T. Fukai, K. Umeda, K. Shiba, and Y. Yagami of Renesas Electronics Corporation for their useful discussions. The authors would like to thank N. Nakamura, K. Imai, and I. Akiba of Renesas Electronics Corporation for their research support.

REFERENCES

[1] P. Muroke; "Flash Memory Field Failure Mechanisms", *Proceedings of 2006 International Reliability Physics Symposium*, pp.313-316 (2006).

[2] S.-C. Seo, C.-C. Yang, C.C. Yeh, B Haran, D. Horak, S. Fan, C. Koburger III, D. Canaperi, S.S. Papa Rao, F. Monsieur, A. Knorr, A. Kerber, C.-K. Hu, J. Kelly, T. Vo, J. Cummings, M. Smalley, K. Petrillo, S. Mehta, S. Schmitz, T. Levin, D.-G. Park, J.H. Stathis, T. Spooner, V. Paruchuri, J. Wynne, D. Edelstein, D. McHerron, and B. Doris; "Copper Contact Metallization for 22nm and Beyond", *Proceedings of 2009 International Interconnect Technology Conference*, pp.8-10 (2009).

[3] K.-Y. Yiang, M. Chin, A. Marathe, and O. Aubel; "A Simple Electrical Method for Etch Bias and Process Reliability Determination", *Proceedings of 2010 International Reliability Physics Symposium*, pp.562-565 (2010).

[4] N. Suzumura, S. Yamamoto, D. Kodama, H. Miyazaki, M. Ogasawara, J. Komori, and E. Murakami; "Electric-field and Temperature Dependencies of TDDB Degradation in Cu/Low-k Damascene Structures", *Proceedings of 2008 International Reliability Physics Symposium*, pp.138-143 (2008).

[5] F. Chen, E. Huang, M. Shinosky, M. Angyal, T. Kane, Y. Wang, and A. Kolics; "A Comparative Study of ULK Conduction Mechanisms and TDDB Characteristics for Cu Interconnects with and without CoWP Metal Cap at 32nm Technology", *Proceeding of 2010 International Interconnect Technology Conference*, pp.1-3 (2010).

[6] F. Chen, M. Shinosky, B. Li, J. Gambino, S. Mongeon, P. Pokrinchak, J. Aitken, D. Badami, M. Angyal, R. Achanta, G. Bonilla, G. Yang, P. Liu, K. Li, J. Sudijono, Y. Tan, T.J. Tang, and C. Child; "Critical Ultra Low-k TDDB Reliability Issues for Advanced CMOS Technologies", *Proceedings of 2009 International Reliability Physics Symposium*, pp.464-475 (2009).

[7] S. Yokogawa and H. Tsuchiya; "Statistical Analysis of Lifetime Distribution of Time-dependent Dielectric Breakdown in Cu/low-k Interconnects by Incorporation of Overlay Error Model", *Japanese Journal of Applied Physics*, Vol.49, No.5, 05FE01-1-4 (2010).

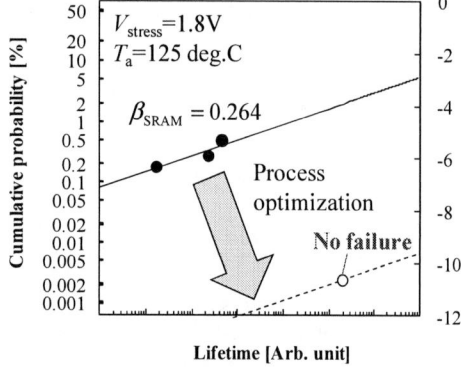

Figure 16. Early life failure rate evaluation by highly accelerated SRAM burn-in test due to CA/PC ILD TDDB.

RELIABILITY LIMITATIONS TO THE SCALING OF POROUS LOW-K DIELECTRICS

Shou-Chung Lee and A. S. Oates

TSMC, 168, Park Ave. 2, Hsinchu Science Park Hsinchu County, Taiwan 30844, email: scleec@tsmc.com

ABSTRACT

We show that processes used to fabricate advanced porous dielectrics can exhibit reliability approaching the intrinsic capability of the material. Combining this with simulations of failure distributions as a function of porosity and line edge roughness we demonstrate that failure times due to electrical breakdown rapidly decrease below k=2.3. The rapid failure time decrease is due to the statistical nature of increasing porosity (decreasing k), which leads to a shortening of the percolation path for dielectric breakdown. Continued scaling will require greater understanding of the breakdown impact on circuits as well as materials innovations to improve robustness.

INTRODUCTION

Much effort has been expended to understand the processing influences and the physical mechanisms of low-k dielectric failure [1-4]. Clearly reliability considerations are important constraints on the reduction of k, but so far it has not been considered whether electrical breakdown will ultimately limit scalability, or if it will be constrained by factors such as mechanical stability or process damage [5-8]. To assess the ultimate scalability of porous dielectrics due to breakdown, first it must be determined whether advanced Cu/low-k technologies are capable of exhibiting intrinsic breakdown behavior. If not, process optimization can be implemented to ensure scalability. Second, since reliability is determined by failure at low percentiles (e.g. 0.1%), failure distributions must be accurately predicted to include the effects of porosity and local line edge roughness (LER), which becomes increasingly significant as scaling proceeds. Here we consider both of these factors to determine how failure times scale with reduction of k.

Percolation theory has been very successful in describing gate oxide breakdown statistics [9]. Ogawa considered the pore as a pre-existing defect and used a percolation model to predict relative breakdown fields and Weibull slope (β) for low-k materials [6]. We further explained the pore acts as a local high electric field region. The pore shortens the percolation path for breakdown but does not change the physics of failure, so the field acceleration factor is unchanged for silica-based dielectrics with the increasing porosity [7]. Theoretically, percolation theory is capable of predicting the dielectric reliability as k is decreased for Cu/low-k interconnects. However, it is well known that the low-k dielectric failure distribution shape is stress voltage dependent because of the LER effect [10]. The intrinsic material properties will dominate the failure distribution at low voltage use conditions while LER will dominate at high voltage acceleration test conditions. This makes the interpretation of reliability data become difficult since now failure distributions are convoluted with material and LER properties.

The damascene structure has been widely used to study the overall quality of the process integration of Cu / low-k interconnects. However, from this structure it is difficult to obtain accurate estimates of the intrinsic reliability of the barrier / low-k because the result may be affected by other aspects of the process such as low-k / cap interface quality of CMP [3], moisture absorption in low-k dielectrics [4], and LER. Zhao reported a novel planar low-k structure to evaluate the intrinsic barrier / low-k reliability with

various barrier and low-k materials and found the intrinsic breakdown field is in the range of 7~8 MV/cm for k=2.5 dielectrics with various barrier materials [11,12]. The planar structure excludes the CMP and geometric variation effects and hence provides the intrinsic material properties of the dielectrics. However, it is still a question whether process optimization of the damascene structure is capable of meeting the intrinsic material capability. In this work, we combined both LER and porosity effects in a percolation model to describe low-k failure distributions in the range of k=2.0~2.8.

PREDICTION OF FAILURE DISTRIBUTIONS

The intrinsic breakdown of a porous dielectric can be determined using percolation theory to have Weibull statistics with distribution parameters given by [6,7,9]:

$$t_{63} = t_0 (L/a_0)^{\frac{-1}{\beta}} f(E_{ILD}) \tag{1}$$

$$\beta = (ms/a_0)[1 - (1 + \alpha)P] \tag{2}$$

where t_o and a_o are the characteristic failure time and dimension of the percolation unit cell respectively, m is a constant, L is the length of the sample, s is the dielectric thickness, and P is the porosity of the dielectric. In the percolation formulation of breakdown, processing artifacts are considered to impact the length of the breakdown path via the parameter α, which relates to field enhancement at defects. Therefore, processing primarily impacts β, which then modifies the failure time, while t_0 is a material parameter that relates to the intrinsic breakdown strength of the dielectric. The field dependence $f(E_{ILD})$ remains controversial despite intense recent scrutiny [5,13-16]. Our recent work on the distortion of failure distributions from Weibull due to LER suggests that the field dependence must be of the E- or √E-model [17]. Therefore, in this study we use a √E field model for the failure distribution modeling study since recent experimental evidence tends to support this functionality [18]. However, we wish to emphasize that the ultimate conclusions regarding scaling capability are independent of the field model, as discussed later.

We have shown previously that the effect of porosity on failure is accurately described by (1) and (2) [6,7,9]. LER impacts failure distributions at high voltage, distorting the failure distribution shape from Weibull as well as introducing systematic errors in the measured field dependence of failure [10]. Moreover, LER effects are a strong function of voltage, and so it is important to model LER effects accurately to predict failure distributions at low operating voltages. LER effects can be calculated by considering that at high voltage, to a good approximation breakdown occurs at the minimum thickness on a test structure, s_{min} [10]. From the extreme value theorem of statistics, the distribution of s_{min} is readily obtained as the extreme value distribution of s (independent of the detailed distribution of s). We show in fig.1 that with a knowledge of s_{min} distributions it is possible to accurately model failure distributions as a function of both porosity and magnitude of LER. In all cases s_{min} was obtained from routine electrical measurements of s performed in the manufacturing environment. Fig.1 compares experimental TDDB failure distributions with Monte-Carlo simulations using eq.(1) for metal combs with dielectric thickness of s=40 nm as a function of k in the range 2.3 to 2.8. Our experimental data is determined using hard breakdown as a failure criterion. Since the k value of these dielectrics is only controlled by the degree of porosity, we varied β to simulate failure distributions because β is a function of porosity as depicted by (2). We keep all other parameters in (1) constant and the LER magnitude was σ=4% of s. The β value is therefore determined to be β=6,4,and 3 for k=2.8,2.5, and 2.3 dielectrics. These β values are in good agreement with the percolation model prediction in (2) (fig.2) and are consistent with that measured from length scaling experiments where the LER effect is diminished (not shown here).

To further investigate the material properties at extreme low k values, we switch to VRDB testing because the lifetime difference between k=2.8 and k=2.0 is about 6 orders of magnitude from the model prediction [7] and it is difficult to collect TDDB failure distributions under the same test voltage conditions. Fig.3 shows experimental VRDB failure distributions are also in agreement with model predictions for metal combs with dielectric thickness of s=60 nm as a function of k in the range 2.0 to 2.5. Since s_{min} is a function of the length of the test structure, fig. 4 illustrates the robustness of the simulation procedure by accurately reproducing experimental data as a function of comb length with a k=2.3 dielectric. To sum up our findings from fig.1 to fig.4, we conclude that the failure distribution of the porous low-k dielectric damascene structure can be well modeled by the percolation theory incorporating LER.

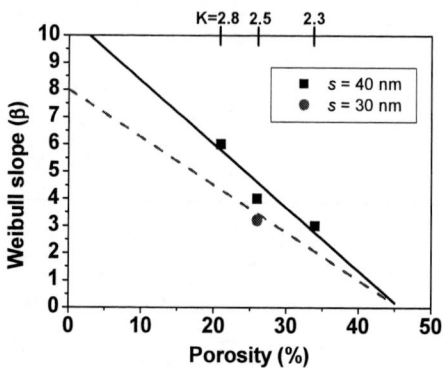

Fig.2 Weibull β dependence on porosity (or dielectric constant k) for thickness of s=40nm and s=30nm dielectrics. Experimental data (symbols) is in good agreement percolation model prediction (equation (2))(lines)

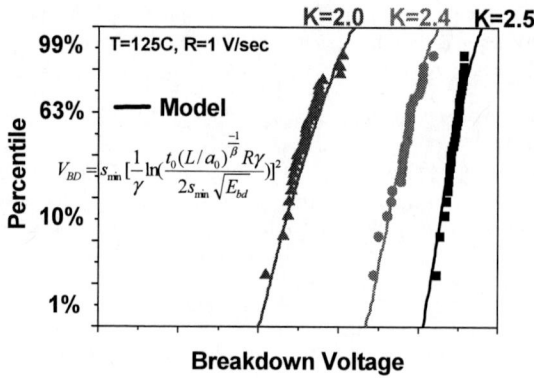

Fig.3 Experimental VRDB failure distribution and model predictions for k=2.0~2.5 dielectrics with s=60nm. Model distributions (lines) were calculated with LER $1\sigma = 0.03s$.

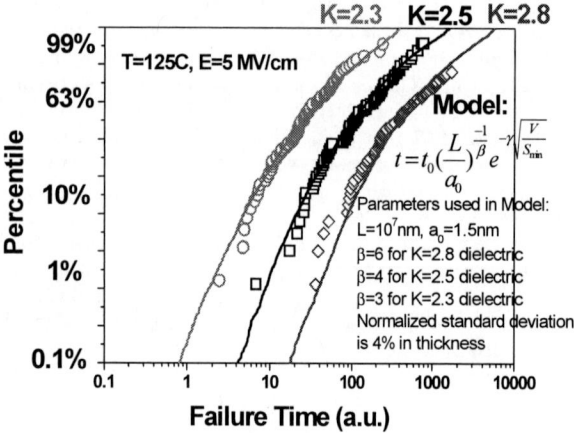

Fig.1 Experimental TDDB failure distribution and model predictions for k=2.3~2.8 dielectrics with s=40nm. Model distributions (lines) were calculated with LER $1\sigma = 0.04s$, which was determined from electrical measurements of the distribution of s on test wafers.

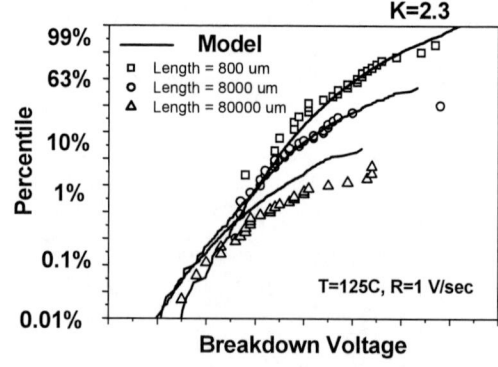

Fig.4 Experimental VRDB failure distributions (symbols) and calculated distributions (lines) for k=2.3 dielectrics of s=40nm for various metal comb lengths between 800 and 80000μm.

INTRINSIC RELIABILITY CAPABILITY OF POROUS LOW-K DIELECTRIC BREAKDOWN

It is well known that low-k breakdown can be strongly affected by processing artifacts (Cu contamination, CMP defects, moisture etc.), which degrade the ultimate reliability of the dielectric [4-7]. We have performed an extensive set of experiments designed to determine if advanced process technologies are capable of exhibiting intrinsic breakdown. Fig. 5 summarizes experiments that determine the impact of the trench barrier process on breakdown with s=30nm for k=2.5 dielectric of metal comb structures. In total 4 different barrier schemes were studied with different barrier materials deposited with PVD, ALD, or self-formed techniques. In all cases the Cu layer was formed by deposition of a seed layer followed by ECP and passivation with a SiN-based dielectric. For comparison we plot in fig. 5 the range of intrinsic lifetime expected for k=2.5 from the same percolation model parameters used in fig.1 and β=3.2 predicted from fig.2 for s=30nm dielectrics, and the s_{min} as determined from electrical measurements. (Calculation details omitted here for the sake of brevity). For this experiment the overlay tolerance of the structure was not characterized directly on the wafers and intrinsic distributions are calculated assuming the manufacturing range. All barrier processes exhibit similar breakdown voltages, and all are consistent with intrinsic characteristics.

To further confirm the intrinsic β of TDDB distribution in fig.5, we performed within-die TDDB experiments in fig.6 for k=2.5 dielectrics to exclude the overlay tolerance impact, where each die has 10 identical metal comb structures with constant overlay tolerance. In this within-die structure, the LER effect is still present so it can be seen that the failure distribution distortion is still present in some dies and we focus on low percentile distributions for intrinsic material properties characterization. We found that the β value of each die is in good agreement with the model prediction in fig.2. Another confirmation of intrinsic breakdown performance is shown using a large sample size experiment (~200), as shown in fig.7, where the line-to-line overlay tolerance will reach its maximum at low percentiles of the distribution and it is possible to unambiguously determine the intrinsic β [10]. The measured β=3.2 is in close agreement with β=3.2 expected from (2). This large sample size experiment is practically useful for intrinsic material properties (β) characterization since only low percentile distribution (<10%) is of interest and there is no need to measure the whole failure distribution.

Two further considerations are important: i) these Cu/low-k damascene structure failure times agree to within an order of magnitude with those of metal combs consisting of W lines with pure SiO₂ dielectric (see fig.8 in next section), and ii) these Cu/low-k damascene failure times are only about a factor 5 lower than those of the planar capacitor structure proposed by Zhao et al [11,12]. The β values of the damascene and planar capacitors are 3.2 and 3.6 respectively and are close to the intrinsic β=3.2 from (2). We conclude that advanced Cu/low-k process technologies are capable of exhibiting breakdown approaching the intrinsic material capability.

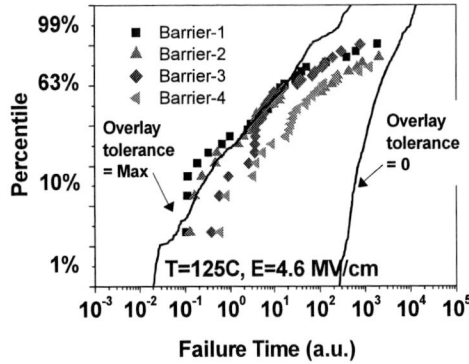

Fig.5 Experimental failure distributions (symbols) of a k=2.5 dielectric with s= 30nm with various trench barrier layers deposited by PVD, ALD, and self-form techniques. The solid lines are calculated distributions for the range of overlay tolerance as specified by the manufacturing process.

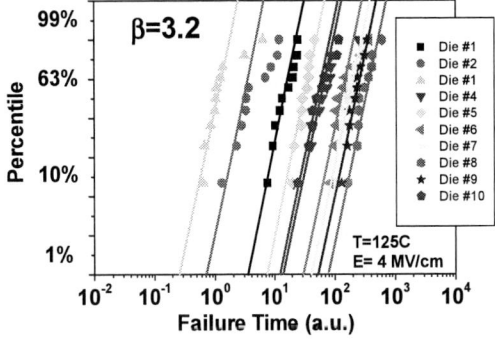

Fig.6 Within-die TDDB distribution for k=2.5 and dielectric thickness of s=30nm. In each die distribution the line-to-line overlay tolerance is constant while the LER effect may still present. Intrinsic breakdown behavior was characterized by low percentile distribution.

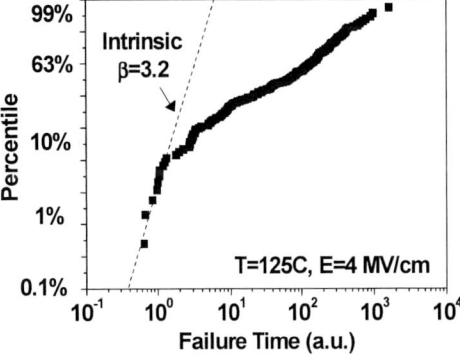

Fig.7 Failure distribution of k=2.5 dielectric with single PVD barrier showing intrinsic Weibull β =3.2.

978-1-4244-9113-1/11 $26.00 © 2011 IEEE

LOW-K RELIABILITY SCALING

Fig. 8 shows calculations and experimental results of the median failure time (MTF) of the dielectric as a function of P and k at E=4 MV/cm, including the limiting case of P=0 for the W/SiO$_2$ and Cu/FSG structures. While MTF is relatively insensitive to k between k=3.9 and k=2.5, there is a precipitous drop in MTF as k is reduced below 2.3. MTF predictions are in good agreement with the experimental data. This rapid decrease in k is independent of the field dependence and is a fundamental feature of breakdown. It occurs due to the decrease in the length of the percolation path with increasing P, which is illustrated by 2 dimensional Monte Carlo simulations in fig.9. When porosity exceeds P=30%, the number of breakdowns required to form a high current path decreases rapidly with increasing P. This result is consistent with Hwang's finding that TDDB failure time will drop significantly when P>30% for MIS structures [19]. Since we use accelerated field data to verify our model predictions, the choice of field dependence is not critical to this verification. However, we note that the functional form of the field dependence does not impact the rapid reduction of MTF for k < 2.3, but only impacts the magnitude of failure times. It is the rapid rate of reduction of MTF with k that implies a practical limitation to reliability scaling of porous dielectrics.

As indicated above based on the W/SiO$_2$ and planar capacitor data, process improvements may be able to increase lifetimes by about an order of magnitude, but this will not change the clear conclusion that lifetimes rapidly drop below k=2.3.The model parameters used for MTF prediction in fig.8 are experimentally determined. To improve the scaling capability of porous dielectrics two possibilities can be considered: (i) increase unit cell characteristic failure time (t_0), or (ii) increase the effective length of the percolation path for breakdown (i.e. increase β). Fig. 10 shows the relative impact of changes in these parameters. Changes to t_0 e.g. by implementing new porous dielectric materials do not affect the precipitous failure time drop when P>30%. However, increasing the effective length of the percolation path for breakdown can prolong k scalability. The effective path length can be increased by pore engineering to reduce the field enhancement associated with pores, or by adding a dielectric with increased bond strength for electrical breakdown into trenches [20].

(a) Porosity = 10%   (b) Porosity = 20%

(c) Porosity = 30%   (d) Porosity = 40%

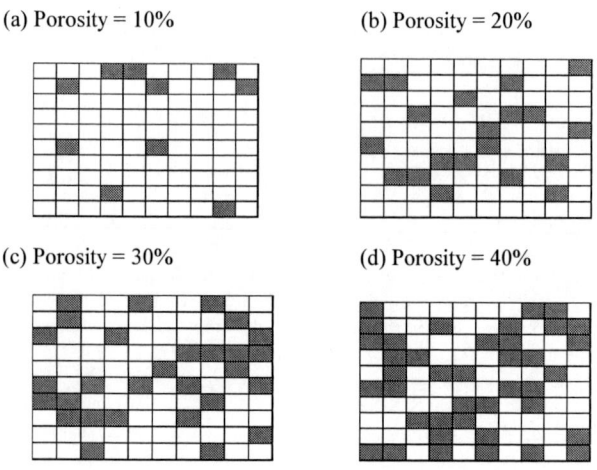

Fig.9 Schematic diagram of porous Low-K dielectric matrix with porosity in the range of P=10%~ 40 %, where solid cell represent pore and open cell represents silica unit cells.

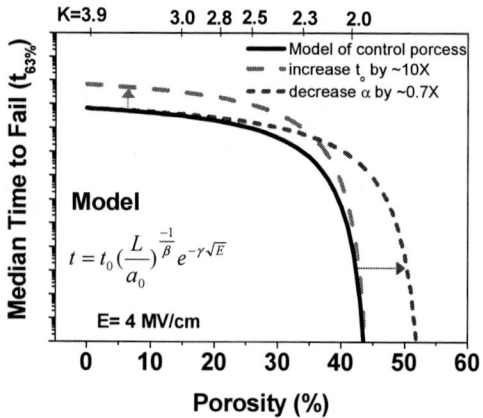

Fig.10 Model prediction for porous silica-based dielectrics as a function of P (or k) with the process optimization effect on t_0 and β. Improving unit cell characteristic failure time (t_0) does not slow down precipitous failure time drop at P>30%, while increasing percolation path for breakdown (β) can give more lifetime margin for k scalability.

Although our work implies that reliability may limit low-k scalability, several other important factors must be considered. In this work we have assumed that breakdown exhibits similar characteristics at stress and operational voltages. If "soft-breakdown" dominates at low voltage, increases in scalability may be achieved since soft-breakdown can be viewed as increasing the effective percolation length for breakdown. Low voltage breakdown behavior needs to be carefully examined to determine the propensity for, and characteristics of soft-breakdown. For the same reason detailed studies of breakdown under AC behavior are important. Further studies of the impact of breakdown on circuit performance will also be valuable to assess the scalability of porous dielectrics.

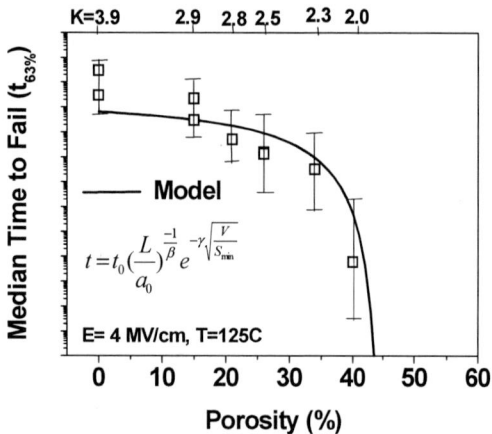

Fig.8: Measured failure times (symbols), and model prediction (lines) for porous silica-based dielectrics as a function of P (or k). The failure time decreases rapidly for k<2.3.

CONCLUSIONS

We show that processes used to fabricate advanced porous dielectrics of Cu damascene interconnect can exhibit reliability approaching the intrinsic capability of the material. Combining intrinsic behavior with simulations of failure distributions as a function of porosity and line edge roughness we demonstrate that failure times due to electrical breakdown rapidly decrease below k=2.3. The rapid failure time decrease is due to the statistical nature of increasing porosity (decreasing k), which leads to a shorting of the percolation path for dielectric breakdown. Our model predictions indicate that process optimization associated with increasing the percolation path for breakdown by advanced pore engineering or Cu diffusion barrier integration can give more lifetime margin for k scalability. Continued scaling will require greater understanding of the circuit impact of breakdown as well as materials innovations to improve robustness.

REFERENCES

[1] Junji Noguchi, Tatsuyuki Saito, Naofumi Ohashi, Hiroshi Ashihara, Hiroyuki Maruyama, Maki Kubo, Hizuru yamaguchi, Daisuke Ryuzaki, Ken-ichi Takeda, and Kenji Hinode," Impact of Low-K Dielectrics and Barrier Metals on TDDB Lifetime of Cu Interconnects", IEEE Int. Reliability Physics Symposium (IRPS), 2001, p. 355

[2] M. N. Chang, Robin. C. J. Wang, C.C. Chiu and Kenneth Wu, " An Efficient Approach to Quantify the Impact of Cu Residue on ELK TDDB ", IEEE Int. Reliability Physics Symposium (IRPS), 2009, p. 619

[3] J. R. Lloyd, X-. H. Liu, G. Bonilla, T.M. Shaw, E. Liniger, and A. Lisi, " On the Contribution of Line-edge Roughness to Intre-level TDDB Lifetime in Low-K Dielectrics ", IEEE Int. Reliability Physics Symposium (IRPS), 2009, p. 602

[4] J. R. Lloyd, C.E. Murray, S. Ponoth, S. Cohen, and E.Liniger, "The Effect of Cu diffusion on the TDDB behavior in a low-k interlevel dielectruics", Microelectronics and Reliability, ,Vol 46, Issue 9-11, Sep.-Nov. 2006 p. 1643

[5] J. Kim, E. T. Ogawa, J. W. McPherson, " Time Dependent Dielectric Breakdown Characteristics of Low-k Dielectrics (SiOC) Over a Wide Range of Test Areas and Electric Fields ", IEEE Int. Reliability Physics Symposium (IRPS) 2007, p.399

[6] E. T. Ogawa, Jinyoung Kim, Gad. S. Haase, Homi C. Mogul, and Joe W. McPherson, " Leakage, Breakdown, and TDDB of Porous Low-k Silica-Based Interconnect Dielectrics ", IEEE Int. Reliability Physics Symposium (IRPS) 2003, p.166

[7] S.C. Lee, A.S. Oates, K.M. Chang, "Fundamental Understanding of Porous Low-k Dielectric Breakdown", IEEE Int. Reliability Physics Symposium (IRPS), 2009, p.481

[8] Ivan Ciofi, Zsolt Tokei, Domenica Visalli and Marleen Van Hove, "Water and Copper Contamination in SiOC:H Damascene: Novel Characterization Methodology based on Triangular Voltage Sweep Measurements" IITC 2006, p.181

[9] Jordi Sune, "New Physics-Based Analytic Approach to the Thin-Oxide Breakdown Statistics", IEEE Electron Device Letters, Vol.22, No.6, June 2001, p.296

[10] S.C. Lee, A.S. Oates, K.M. Chang, "Geometric Variability of Nano-Scale Interconnect and Its Impact on Time Dependent Breakdown of Cu/Low-k Dielectrics " in IEEE Transaction on Devices and Materials Reliability (T-DMR), 2010, Vol.10, No.3, p.307

[11] Larry Zhao, Zsolt Tokei, Gianni Giai Gischia, Henny Volders , and Gerald Beyer, "A New Perspective of Barrier Material Evaluation and Process Optimization ", IEEE Interconnect Technology Conference (IITC), 2009, p.206

[12] Larry Zhao, Gianni Giai Gischia, Kristof Croes, Guido Groeseneken, and Zsolt Tokei, "Study of Leakage Mechanism and Trap Density of Porous Low-k Materials", IEEE Int. Reliability Physics Symposium (IRPS) 2010, p.549

[13] F. Chen, O. Bravo, K. Chanda, P. McLaughlin, T. Sullivan, J. Gill, J. Lloyd, R. Kontra, and J. Aitken, "A Comprehensive Study of Low-k SiCOH TDDB Phenomenon and Its Reliability Lifetime Model Development", IEEE Int. Reliability Physics Symposium (IRPS) 2006, p.46

[14] N. Suzumura, S. Yamamoto, D. Kodama, K. Makabe,, J. Komori, E, Murakami, S. Maegawa, and K.Kubota, IEEE Int. Reliability Physics Symposium (IRPS), 2006, p.484

[15] J. R. Lloyd, S.Ponoth, E. Liniger and S. Cohen, " Role of Cu in TDDB of Low-k Dielectrics " IEEE Int. Reliability Physics Symposium (IRPS) 2007, p.410-411

[16] K. Croes and Zs Tokei, "E- and √E-model too conservative to describe low field time dependent dielectric breakdown", IEEE Int. Reliability Physics Symposium (IRPS), 2010, p.543

[17] S.C. Lee, A.S. Oates, K.M. Chang, "Field Dependence of Porous Low-k Dielectric Breakdown as Revealed by the Effect of Line Edge Roughness on Failure Distributions " in IEEE Transaction on Devices and Materials Reliability (T-DMR), 2011, in press

[18] J. M. Atkin, R. B. Laibowitz and T. F. Heinz, "Effects of Photoinduced Carrier Injection on Time-Dependent Dielectric Breakdown", IEEE Int. Reliability Physics Symposium (IRPS) 2009, p.851

[19] S. S. Hwang, Hee-Chan Lee, Hyun Wook Ro, Do Yeung Yoon, and Young-Chang Joo," Porosity Content Dependence of TDDB Lifetime and Flat Band Voltage Shift by Cu Diffusion in Porous Spin-On Low-k", IEEE Int. Reliability Physics Symposium (IRPS), 2005, pp. 474

[20] S. Chikaki, K. Kinoshita, T. Nakayama, K. Kohmura, H. Tanaka, M. Hirakawa, E. Soda, Y. Seino, N. Hata, T. Kikkawa and S. Saito, "32nm node Ultra-low k (k=2.1)/ Cu Damascene Multilevel Interconnect using High-Porosity (50%) High Modulus (9Gpa) Self-Assembled Porous Silica", IEEE International Electron Device Meeting (IEDM), 2007, pp. 969

A Comprehensive Process Engineering on TDDB for Direct Polishing Ultra-Low K Dielectric Cu Interconnects at 40nm Technology Node and Beyond

W. C. Lin, T. C. Tsai, H. K. Hsu, Jack Lin, W. C. Tsao, Willis Chen, C. M. Cheng,
C. L. Hsu, C. C. Liu, C. M. Hsu, J. F. Lin, C. C. Huang and J. Y. Wu

United Microelectronics Corp., Advanced Technology Development Division,
No 18, Nanke 2nd Rd. Tainan Science Park, Sinshih, Tainan County 741, Taiwan, R.O.C.
Tel: 886-6-505-4888 Ext.86-13257, fax: 886-6-505-0960, e-mail: welch_lin@umc.com

Abstract — The failure ratios of the three typical time-dependent dielectric breakdown (TDDB) failure modes, including top interface, sidewall and bottom corner areas, have been identified for a direct polishing ultra low k (ULK) dielectric Cu back-end-of-line (BEOL) structure at 40nm node. The Cu surface roughness of the metal lines, and the adhesion and thickness of the metal capping layers are strongly correlated to the top interface failure mode. The dielectric constant of the ULK and the concentration of the aluminum-doped Cu (CuAl) seed layer could be related to the sidewall failure mode. The bottom corner failures are induced by inappropriate Cu barrier re-sputter processes. In this study, the TDDB reliability performance can be effectively improved by evaluating a post-Cu chemical mechanical polishing (Cu CMP) cleaning process with smooth Cu surface roughness, developing a better step coverage with multi-layer capping layer, using a slightly higher dielectric constant ULK film, replacing a conventional pure Cu with a CuAl seed layer and optimizing the Cu barrier layer deposition process. The lifetime of the TDDB can be significantly improved over three orders (larger than 10000 years) as implementing an optimized integrated Cu with ULK BEOL structures at 40nm technology node.

Keywords — TDDB; ULK; BEOL; aluminum-doped Cu; seed; barrier; CMP

I. INTRODUCTION

The Cu damascene with direct chemical mechanical polished (CMP) porous type ultra low k (ULK) interconnects has been introduced to realize the high performance of the resistance-capacitance (RC) delay for 45nm technology node and beyond. However, the long-term reliability of the back-end-of-line (BEOL) structures, especially for time-dependent dielectric breakdown (TDDB), for such the fragile and incorporation of ULK dielectric materials in Cu interconnects is becoming one of the critical challenges for product qualification [1~8]. Generally, the dielectric constant and intrinsic breakdown strength simultaneously decrease with increasing the porosity of ULK film. Lower intrinsic breakdown strength with porous properties of ULK could result in the Cu extrusion at the sidewall and bottom areas of Cu lines because of the ULK film damaged post dry etching with in-situ plasma treatment and weaker adhesion between Cu barrier and ULK layer [2, 5]. On the other hand, the interface integrity among the ULK, the Cu metal surface and the diffusion capping layer has also been pointed out to be the key factor to impact the TDDB performance [6, 7]. Previous research articles have discussed the sensitivity and acceleration models of TDDB characteristics to the geometry and the electric field factors for different low k and ULK dielectrics integrated scheme [3, 8, 9]. Inappropriate Cu metal-to-metal lines spacing control across whole wafers could normally induce the varied TDDB reliability performance. However, as for the minimized feature size in 40 nm and beyond, the TDDB behaves not only sensitive to geometries but also highly sensitive to the engineering in the fabrication process. Recently, the physical vapor deposition (PVD) Ti, Al, Mn and chemical vapor deposition (CVD) Co based self-forming barrier (SFB)/liner (seed) approaches were investigated as an alternative to conventional PVD Ta based barrier and pure Cu seed layers for improving BEOL reliability (10~13). In this study, three TDDB failure modes with related key process modules have been identified. The major process engineering schemes to prevent the failures with lifetime extension for TDDB tests are reported as well.

II. EXPERIMENTAL

300mm Cu damascene interconnect structures were constructed by 40nm metal hard mask/ trench first scheme with porous type ULK dielectric film (k value 2.5~2.6), PVD Cu barrier (Ta/TaN) and Cu or Al-doped Cu seed layers, Cu electroplating, Cu CMP and plasma enhanced (PE) CVD SiCN dielectric barrier layer. The Cu surface roughness was characterized by high-resolution atomic force profiler and inspected by top viewed scanning electron microscopy (SEM). The SiCN thickness post-Cu CMP was measured by KLA-F5x thin film measurement system. Breakdown behavior, including voltage rapid dielectric breakdown (VRDB) and TDDB tests, were probed on a combed type test structure comprised of three Cu and one top Al interconnect layers. The TDDB test structure is subjected to a wafer-level constant voltage test at a temperature of 125°C and positive bias voltage ranging from 19V to 25V. The failure location was detected by optical beam induced resistance change (OBIRCH) and then applied transmission electron microscopy (TEM) to do the failure mode analysis.

III. RESULTS and DISCUSSION

A. The failure modes of TDDB:

Figure 1 shows three major TDDB failure modes, including top interface, sidewall and bottom corner areas, could be observed in porous type ULK Cu interconnects. The early, middle and final TDDB failure stages and the TDDB failure ratios of these three major failure modes are indicated in Figure 2a and Figure 2b, respectively. The failure modes of the top interface and bottom corner are dominantly found at the early and middle TDDB failure stages. However, the sidewall failure mode becomes more important than bottom corner failure mode at the final TDDB failure stage.

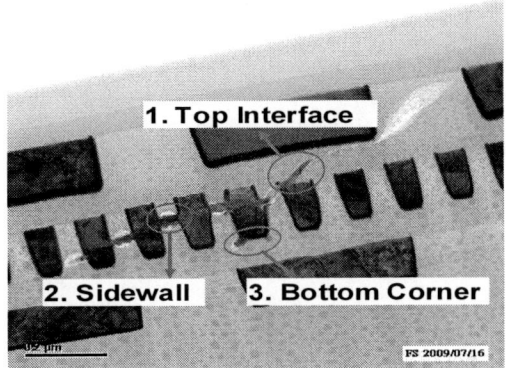

Figure 1- Three major TDDB failure modes were found at 1. the top interface failure areas; 2. the sidewall failure areas; and 3. the bottom corner failure areas.

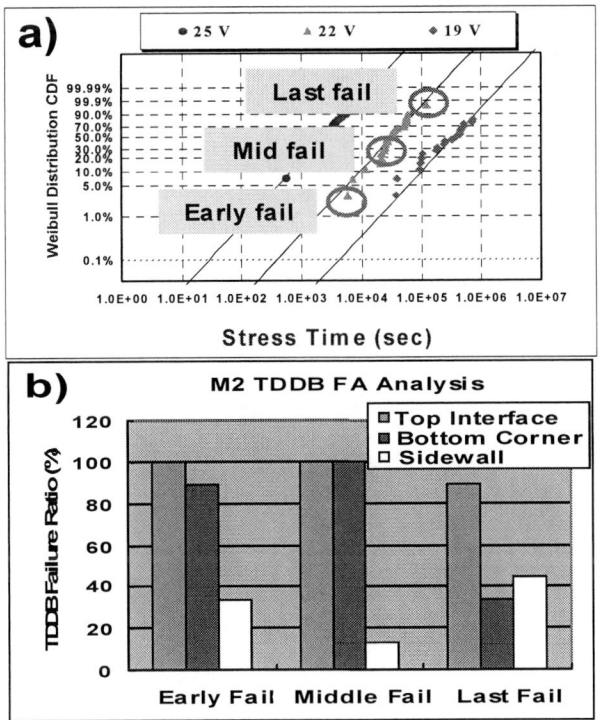

Figure 2- (a) Typical early/ middle/ last TDDB failure points; (b) TDDB failure ratios of these three major failure modes.

B. Process engineering for preventing top interface failure:

Figure 3 shows obvious film cracking of the metal capping layer at the top corner of Cu lines and the Cu penetration phenomenon on the top surface of the ULK film at the TDDB test failure areas. The smooth Cu surface condition derives the better TDDB reliability performance, as indicated in Figure 4. More roughing Cu surface condition could induce the film cracking of the metal capping layer because of the poor step coverage of capping film. These film-cracking areas would become a week point to create the diffusion path of Cu ions and result in the Cu bridge issue. Therefore, how to eliminate the Cu surface roughness and improve the deposition step coverage of metal capping film are crucial to enhance the TDDB performance. Previous study demonstrates optimized post-Cu CMP cleaning chemical dilute ratio and brush gap conditions with smaller Cu surface roughness can obviously extend the lifetime of TDDB (3).

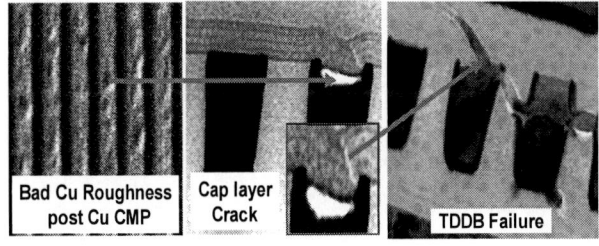

Figure 3- Cu surface roughness induces TDDB failure on top in interface areas.

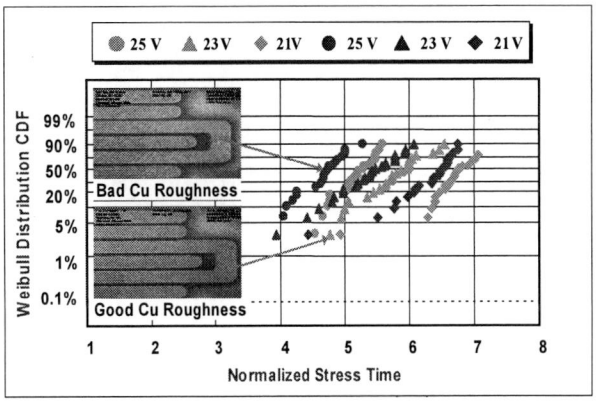

Figure 4- Effects of the Cu surface roughness on TDDB performance.

Figure 5 demonstrates the metal capping film with multi-layers possesses a better VRDB break-down voltage performance than the single-layer capping film. Multi-layer PECVD deposition process can enhance the step coverage of capping film as compared with a single layer deposition process. More conformal step coverage of metal capping layer can effectively retard the formation of film cracking at the top corner interface areas of the Cu metal structures. Furthermore, the VRDB voltage also increases with increasing the deposition thickness of the multi-layer metal

978-1-4244-9113-1/11 $26.00 © 2011 IEEE   161

capping film, as shown in Figure 6. Thicker capping layer could introduce higher compressive strain upon the top of Cu surface and longer film crack propagation path to extend the failure voltage level post VRDB tests [14].

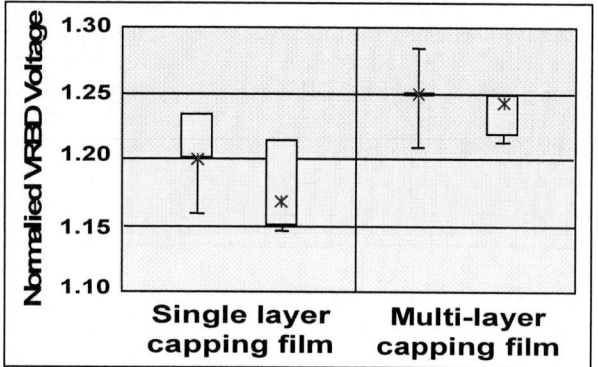

Figure 5- Effects of the metal capping film type on VRDB performance.

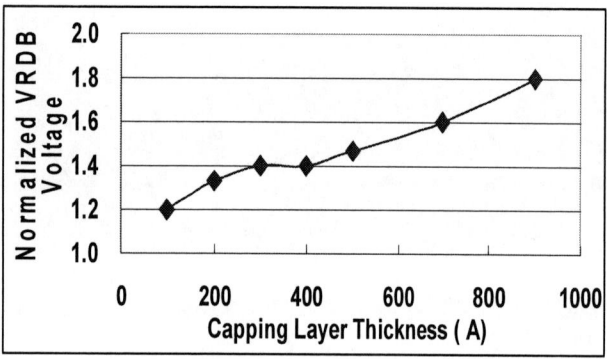

Figure 6- VRDB voltage as a function of the deposition thickness of the multi-layer capping film.

C. Process engineering for restraining sidewall failure:

Figure 7 demonstrates the TDDB lifetime increases with increasing the k value of ULK. Around three orders of TDDB enhancement on M2 Cu dual damascene structures can be demonstrated as the k value of ULK increasing from 2.5 to 2.6.

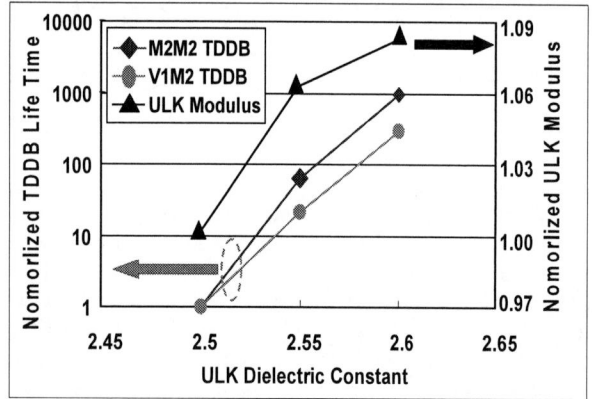

Figure 7- Effects of the ULK dielectric constant on TDDB performance.

The CuAl seed layer has been reported to replace the conventional pure Cu seed structure for improving BEOL electro-migration (EM) [13]. Figure 8 reveals the EDX signal of the Al element can be obviously found at the sidewalls of the Cu metal trench structures as using the CuAl seed layer. Actually, the CuAl seed layer could not only benefit for EM but also better for TDDB. Figure 9 indicates the effects of process engineering for CuAl seed, multi-capping layer, modified ULK film and Cu CMP on TDDB performance. The implementation of the CuAl seed layer can significantly improve around two orders life time on the M2 TDDB test structures as compared with original baseline. Extra-one and two orders life time improvements of the M2 TDDB can be reached as CuAl seed combined with multi-capping layer and modified ULK film, respectively. Slight release the dielectric constant of ULK film is more reliable to improve TDDB the life time for both M1 and M2 Cu interconnects. Overall two orders for M1 and four orders for M2 TDDB life time enhancements can be achieved as simultaneously carrying out the CuAl seed structure, multi-capping layer, modified ULK film and optimized Cu CMP process.

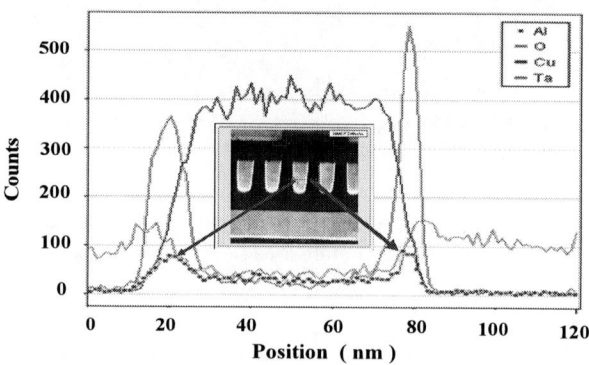

Figure 8- Line-scan profiles of Al, O, Cu and Ta elements measured by energy dispersive X-Ray spectroscopy (EDX).

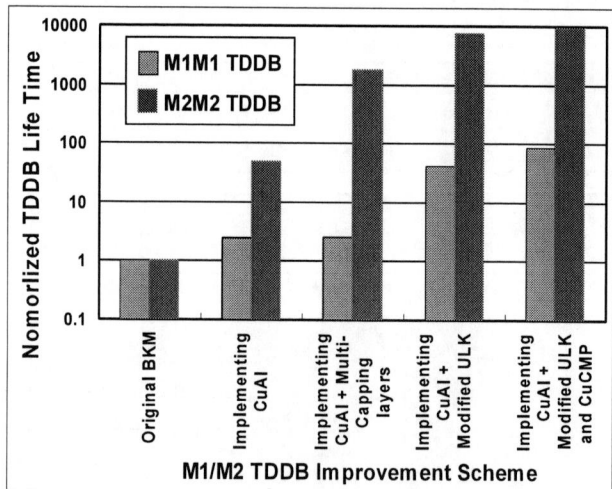

Figure 9- Effects of process engineering for multi-capping layers, CuAl seed, modified ULK and CuCMP on TDDB performance.

978-1-4244-9113-1/11 $26.00 © 2011 IEEE

D. Process engineering for retarding bottom corner failure:

The bottom corner failures of TDDB are strongly related to the Cu barrier process. In Cu PVD barrier deposition process, the re-sputter step post film deposition was widely applied to improve the barrier bottom corner step coverage. However, too aggressive re-sputter amount would damage trench bottom ULK dielectric film and result in the following barrier deposition anchoring into ULK film, as shown in Figure 10(a). Figure 10(b) indicates the Cu extrusion can be found at the trench bottom corner areas due to weakness Cu barrier structures after TDDB test. Table I introduces an appropriate barrier film deposition approach with optimized TaN/Ta deposition thicknesses, Ta re-sputter amount and Ta re-deposition condition for enhancing the step coverage of the metal trench bottom corner areas. The unwilling teeth on the trench bottom can be effectively minimized to obtain the better TDDB performance as implementing an optimized Cu barrier process with better step coverage, as shown in Figure 11 and Figure 12, respectively. Figure 13 demonstrates the time to failure distribution plots for final optimized M1 and M2 structures with over 10000 years TDDB life time.

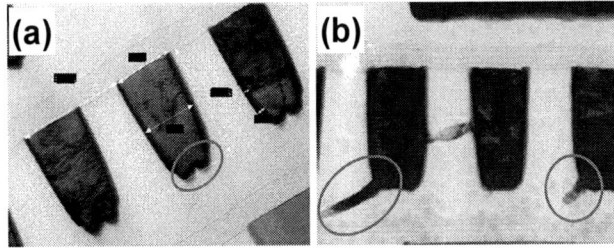

Figure 10- (a) too aggressive re-sputter induces the teeth profiles on the bottom corners of the metal trenches; (b) Cu extrusion from trench bottom corner areas after TDDB tests.

Table I – Optimized TaN/Ta barrier process conditions for the improvement of TDDB performance.

Barrier Deposition Steps	TaN Thickness	Ta Thickness	Ta Re-sputter Amount	Re-deposite Ta Thickness
Original Baseline	Thicker	Thinner	Lower	Thicker
Optimized Condition	Thinner	Thicker	Higher	Thinner

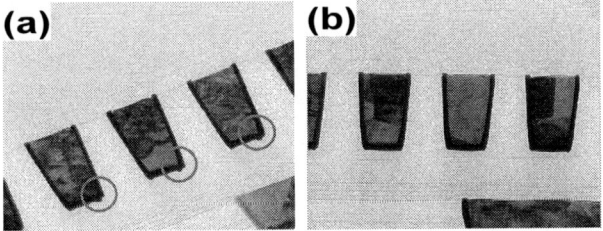

Figure 11- (a) slight and (b) no re-sputter damaged conditions for different Cu barrier deposition processes.

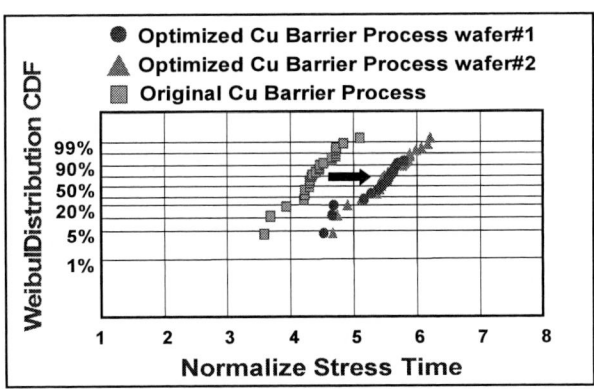

Figure 12- Effects of Cu barrier process on M2 TDDB performance.

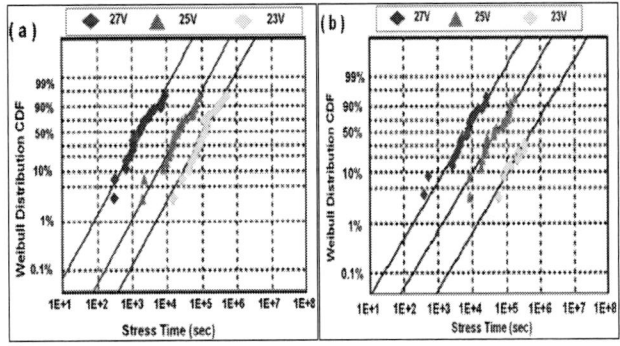

Figure 13- (a) M1 and (b) M2 TDDB time to failure distribution plots.

IV. CONCLUSION

The overall process engineering on the enhancement of the TDDB life time (larger than 10000 years) for direct polishing ULK dielectric Cu interconnects at 40nm node is shown as following:

1. The Cu surface roughness of the metal lines and the metal capping layer on the top of metal trench structures are strongly correlated to the top interface TDDB failure mode. The TDDB performance can be effectively improved as optimizing the post-Cu CMP cleaning process with smooth Cu surface roughness and developing a multi-layer capping film with better step coverage on the top of metal trench areas.

2. The dielectric constant of the ULK and the film structure of the seed layer are corresponded to the metal trench sidewall TDDB failure mode. Introducing a ULK film with around 2.6 dielectric constant and replacing a conventional pure Cu with an Al doped Cu seed layer is benefit to extend the life time of TDDB.

3. The TDDB failures at the bottom corner of metal trench are induced by inappropriate Cu barrier deposition and re-sputter processes. The step coverage of Cu barrier layer can be enhanced to eliminate the unwilling teeth on the trench bottom corner areas and improve TDDB performance by adjusting TaN/Ta deposition, Ta re-sputter and Ta re-deposition conditions.

REFERENCES

[1] L. Arnaud, et al., "Reliability failure modes in interconnects for the 45nm technology node and beyond", *Proceedings of the IITC*, 2009, pp. 179~181.

[2] C. L. Hsu, et al., "The TDDB failure mode and its engineering study for 45nm and beyond in porous low k dielectrics direct polish scheme", *Proceedings of the IRPS*, 2010, pp. 918-921.

[3] W. C. Lin, et al., "Effects of Cu surface roughness on TDDB for direct polishing ultra low k dielectric Cu interconnects at 40nm technology node and beyond", *Proceedings of the AMC*, 2010.

[4] F. Chen, et al., "Comprehensive investigations of CoWP metal-cap impacts on low-k TDDB for 32nm technology application", *Proceedings of the IRPS*, 2010, pp. 566-573.

[5] N. Posseme, et al., " In situ post etching treatment as a solution to improve defect density for porous low-k integration using metallic hard masks", *Proceedings of the IITC*, 2009, pp. 240~242.

[6] M. Vilmay, et al., "Key process steps for high reliable SiOCH low-k dielectrics for the sub 45nm technology nodes", *Proceedings of the IITC*, 2009, pp. 122~124

[7] E. G. Liniger, et al., "Processing and moisture effects on TDDB for Cu/ULK BEOL structures", *Proceedings of the AMC*, 2010.

[8] F. Chen, et al., "Line edge roughness and spacing effect on low-k TDDB characteristics", *Proceedings of the IRPS*, 2008, pp. 132-137.

[9] Gaddi S. Haase, "An alternative model for interconnect low-k dielectric lifetime dependence on voltage", *Proceedings of the IRPS*, 2008, pp. 556-565.

[10] K. Ohmori et al., "A key of self-formed barrier technology for reliability improvement of Cu dual dama-scene interconnects", *Proceedings of the IITC*, 2010.

[11] T. Indukuri et al., "Electrical and reliability charac-terization of CuMn self forming barrier interconnects on low k CDO dielectrics", *Proceedings of the AMC*, 2010.

[12] T. Nogami et al., "CVD Co and its application to Cu damascene interconnections", *Proceedings of the IITC*, 2010.

[13] Y. Hayashi et al., "Robust low oxygen content Cu alloy for scaled-down ULSI interconnects based on metallurgical thermodynamic principles", *IEEE Transactions on Electron Devices, vol. 56, No. 8*, 2009, pp. 1579-1585.

[14] H. Miyazaki et al., " The observation of stress-induced leakage current of damascene interconnects after bias temperature aging", *Proceedings of the IRPS*, 2008, pp. 150-157.

Space Radiation Effects and Reliability Considerations for the Proposed Jupiter Europa Orbiter

Allan Johnston, *Fellow, IEEE*

Jet Propulsion Laboratory, California Institute of Technology
Pasadena, California USA 91109
(1) 818 354-6425 allan.h.johnston@jpl.nasa.gov

Abstract – **The proposed Jupiter Europa Orbiter (JEO) mission to explore the Jovian moon Europa poses a number of challenges. The spacecraft must operate for about seven years during the transit time to the vicinity of Jupiter, and then endure unusually high radiation levels during exploration and orbiting phases. The ability to withstand usually high total dose levels is critical for the mission, along with meeting the high reliability standards for flagship NASA missions. Reliability of new microelectronic components must be sufficiently understood to meet overall mission requirements.**

I. INTRODUCTION

NASA flagship missions, such as the proposed JEO mission to Europa, are expected to operate for long periods of time in the harsh radiation environment of space. Design and operational rules have been established to achieve this goal, as evidenced by previous deep space missions (Galileo and Cassini), as well as in Mars surface exploration missions and the Hubble space telescope.

The unusually high total dose levels of the proposed JEO mission affect part performance as well as reliability. A related problem is device scaling, which introduces new issues for reliability and radiation effects that were not important for the 1970 and 1980 technologies used in older flagship missions with high radiation levels.

This paper discusses some of the underlying issues in selecting suitable components, testing and qualifying them for this unique environment, and incorporating system design methods that can be used for such a mission.

II. JOVIAN RADIATION ENVIRONMENT

A. Total Dose

The Jovian trapped belts are far more intense than the earth's radiation belts, primarily because Jupiter has a magnetic field that is about 20 times higher than the earth. The trajectory selected for the proposed mission avoids the inner proton belts; most of the radiation is due to trapped electrons with energies up to several hundred MeV. Electrons with such energies require much thicker shielding compared to electrons in earth orbits, limiting the effectiveness of shielding (particularly spot shielding). Selection of radiation-tolerant microelectronics is critically important because of the difficulty of adding additional shielding.

The research in this paper was carried out at the Jet Propulsion Laboratory, California Institute of Technology, under contract with the National Aeronautics and Space Administration (NASA).

Figure 1 shows the total dose for a mission with a 105 day orbit as a function of aluminum shield thickness, along with the specification for the previous Galileo mission [1]. The initial Galileo environment is shown, along with the actual environment after several extensions of the original mission. Note that the Europa requirement is slightly above the extended mission environment of Galileo.

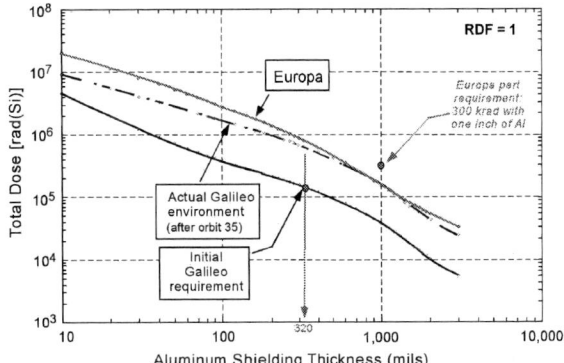

Fig. 1. Total dose vs. shielding thickness for the proposed JEO mission, along with the requirements and actual total dose for Galileo.

B. Galactic Cosmic Rays and Solar Particles

Although the solar particle intensity is lower at Jupiter compared to regions near the earth, the charged particle environment that produces single-event effects (SEE) is not very different from that of other deep space missions. Various SEE (such as upset, functional interrupts, and latchup) remain important issues for the proposed mission, but are similar to other space missions. Therefore single event effects will be discussed only briefly in this paper, emphasizing synergistic effects between SEE sensitivity and total dose damage. SEE effects are discussed in more detail in references 2 through 4.

C. Displacement Damage

Displacement damage is also a concern. There are two possible sources: neutrons from the potential on-board radioisotope power system, and electrons from the natural environment. Contributions from neutrons depend on the location of components relative to the power system, and are expected to be less than 1/3 of the overall displacement damage requirement, except in special circumstances.

Electrons are less effective than protons in producing displacement damage, but that mechanism is still important because of the high energy and high fluence. For the proposed JEO mission, the equivalent neutron damage for a shielding thickness of one inch is 5 x 10^{11} n/cm^2 [1]. Displacement

978-1-4244-9113-1/11 $26.00 © 2011 IEEE

damage is particularly important for detectors, but also affects conventional components using bipolar technology.

Figure 2 shows how gain degradation of a 2N2222 transistor is affected by displacement damage. Below 100 krad(Si) ionization damage dominates, but displacement damage becomes more important at higher radiation levels due to saturation of ionizing radiation damage. One of the reasons for saturation is the buildup of internal electric fields within the oxide as charge accumulates near the silicon-SiO$_2$ interface.

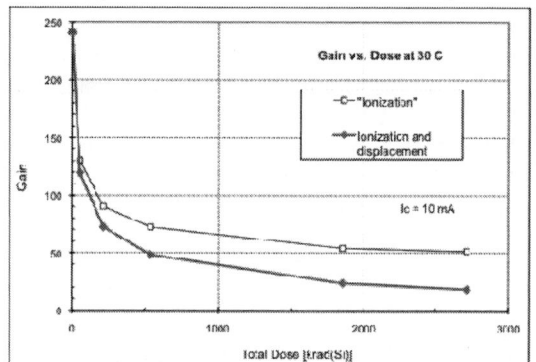

Fig. 2. Effect of electron displacement damage on transistor gain degradation.

Displacement damage is even more important for bipolar linear circuits, which typically contain wide-base pnp transistors. Figure 3 compares shifts in the internal bandgap reference voltage of a voltage regulator when it is irradiated with gamma rays and protons; displacement damage causes much more damage [5].

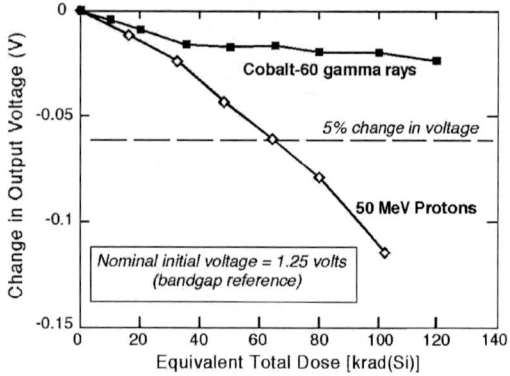

Fig. 3. Proton and gamma ray tests of a voltage regulator. The output voltage scales with changes in the internal bandgap reference.

The key point is to require evaluation of displacement damage effects in addition to total dose damage for most bipolar devices. Certain types of light-emitting diodes (LEDs) are also highly sensitive to displacement damage; for example, the Galileo tape recorder failed at the end of the mission due to LED displacement damage [6].

III. DEALING WITH HIGH RADIATION REQUIREMENTS

A. Basic Issues

The high total dose level is a major issue for the proposed mission. It is not only higher than that of other space missions, but is above the maximum radiation level used to qualify most hardened components. Consequently, little information is available about whether devices will actually function at those levels, imposing additional risk for component selection Other difficulties include:

a. Substantial shielding is planned to reduce the total dose level to lower levels. Consequently, adding additional shielding to further reduce the total dose for problem components is a costly and weight-penalizing option.

b. The actual total dose will be close to the design requirement, with reduced margins compared to typical space missions.

c. Qualification methods used by manufacturers begin with tests at high dose rate, followed by an annealing step which is applicable to missions with very low dose rate, but may not be appropriate for the somewhat higher dose rate conditions of the proposed JEO mission during the orbiting phase.

B. Total Dose and Displacement Damage

General Concerns

For initial design concept purposes, it is convenient to divide active components into basic categories. There is generally less concern about digital devices, first because radiation-hardened components are readily available; and second, because total dose hardness generally increases with scaling [7]. The most critical component families are detectors, analog circuits, and power control devices.

Many analog circuits exhibit more damage when they are exposed at the low dose rates in typical space environments compared to tests at high dose rate; the term "ELDRS" (enhanced low-dose rate sensitivity) is often use to describe this effect [8].

Although the ELDRS phenomenon has been investigated for many devices, most tests are not carried out above approximately 50 krad(SiO$_2$) because of the lengthy time required for irradiation, and the fact that few space missions have requirements above 100 krad(SiO$_2$). Figure 4 compares tests at high and low dose rate for an analog-to-digital converter, used on the Juno program [50 krad(SiO$_2$) requirement]. Although this part was acceptable for that mission, catastrophic failure occurred for unbiased parts when the tests were extended to higher levels in order to evaluate their potential use for JEO. No precursor was observed for the onset of catastrophic failure from the tests at 50 krad(SiO$_2$).

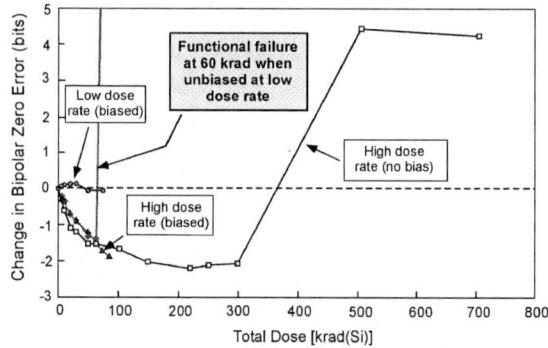

Fig. 4. Evaluation of a radiation-tolerant analog-to-digital converter at low dose rate in the region above 50 krad(SiO$_2$).

This illustrates the importance of evaluating analog bipolar parts under low-dose rate conditions at high radiation levels, despite the long testing times. Fortunately, the dose rate for the JEO orbiting phase is ~ 40 mrad(SiO₂)/s [1], about an order of magnitude higher than that of conventional space missions, which reduces the overall test time.

ELDRS in Scaled CMOS

CMOS devices with feature sizes ≤ 0.25 µm use shallow trench isolation (STI). STI oxides extend laterally for distance of 0.08 to 0.25 µm, a distance that is comparable to the oxide thickness in bipolar oxides. Recent work has shown that an effect similar to ELDRS occurs in STI oxides, with more damage taking place when tests are done at low dose rate [9]. Figure 5 illustrates this, along with test results for 10-keV X-rays, which further overestimates radiation hardness.

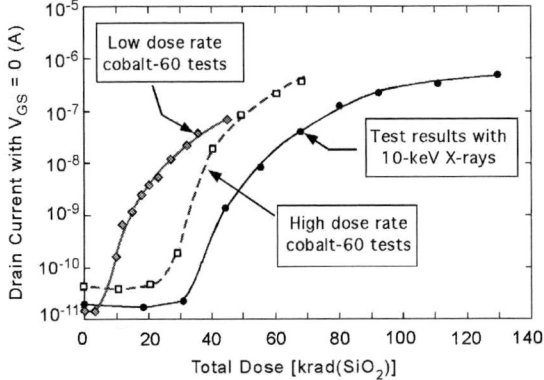

Fig. 5. Increase in drain current vs. total dose for test transistors from a CMOS process with 180 nm feature size [9]. The total dose for inversion is significantly lower at low dose rate compared to test results at high dose rate.

This result was unexpected because leakage current in CMOS devices typically anneals, leading to the conclusion that less damage should occur at low dose rate. The mechanism is related to the long transport time for holes in thick oxides; increased recombination at higher dose rates decreases the charge that is actually transported to the interface region between the STI and the body region of the MOS transistor.

Lot Variability

There can be large differences in the radiation hardness of different production lots [10]. An example is shown in Fig. 6 for a voltage regulator. Two points are important: first, the total dose hardness differs by a factor of 2.2 to 3.5 (depending on the load conditions) for devices produced within about 15 months; and second, short circuit current is a critical parameter for applications of this type of device. If the device load exceeds the current drive capability, large changes in output voltage will take place that usually disable circuits that are powered by the regulator unless the load can be reduced. It is far more difficult to deal with this type of response compared to gradual parametric shifts that often allow circuits to continue functioning with some degradation in performance.

Displacement Damage

Even though the displacement damage requirement for JEO is relatively low, it can still be important. One reason for this is saturation of total dose damage at higher radiation levels (see Fig. 2); displacement damage does not saturate, and can

potentially affect the overall hardness of some types of components.

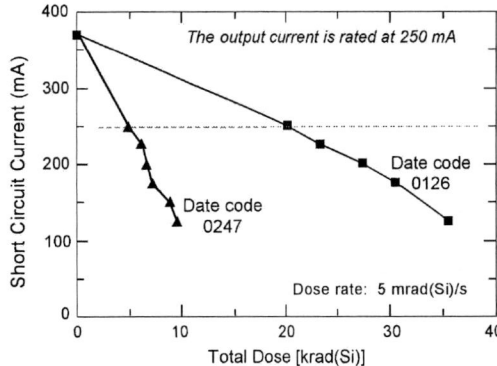

Fig. 6. Lot variability of radiation degradation of short-circuit current of a voltage regulator.

It is usually impractical to perform radiation tests with the exact type of particle and energy range that occurs in the environment. Although total dose and displacement damage are produced simultaneously in the real space environment, the effects are generally considered separately from the standpoint of test and qualification. One approach that can be used when displacement damage effects are expected to be less than total dose damage is to pre-irradiate qualification samples to the expected displacement dose before total dose tests are performed.

Fortunately, the displacement damage fluence is low enough for JEO so that relatively few components are affected. Table 1 lists the device types where displacement damage is potentially important. For detectors and light-emitting diodes, displacement damage is often the dominant source of damage. It can usually be considered as second-order for bipolar linear circuits and discrete transistors.

Table 1. Components Affected by Displacement Damage

Device Type	1-MeV Fluence Damage Threshold (n/cm²)	Comments
Detector	10^9 to 10^{10}	Depends on technology
Light-emitting diode	2×10^{10}	Dominant failure mode
Bipolar linear circuit	4×10^{10}	TID and displacement effects are both important
Discrete transistor	3×10^{11} to 10^{12}	Mainly effects low frequency transistors

IV. RELIABILITY STRATEGIES FOR SPACE MISSIONS

Several steps are necessary in order to obtain the high reliability needed for space missions such as JEO. The very high reliability of earlier flagship missions is frequently cited to show the effectiveness of existing practices for reliability. However, the components used in those missions were designed and fabricated quite differently from present-day devices. New reliability challenges introduced by device scaling and complexity are not necessarily solved by older reliability methods [11,12]. Compound semiconductors impose additional reliability difficulties because of their limited history (other than GaAs MESFETs) and use of new fabrication technologies [13,14].

The specific environment for JEO must be taken into account for reliability evaluation. Some space missions (such as those involved in Mars surface exploration) must endure (Martian) daily temperature cycles; consequently, reliability tests based on thermal cycling are heavily emphasized. The proposed JEO mission is quite different. The spacecraft will undergo extreme vibration during launch, but the thermal environment after launch is relatively benign.

A. Component Testing and Qualification

The first reliability step is thorough electrical testing and burn-in of all components. In the past this has included tracking devices at individual wafer levels, as well as logging parametric data before and after burn-in. Although this approach was highly successful for the less complex components used in earlier missions, there may be limitations in its effectiveness for newer technologies.

CMOS reliability mechanisms are heavily influenced by advances in manufacturing technology [15]. Manufacturers continue to evaluate fundamental mechanisms such as threshold shifts from hot carriers, time-dependent dielectric breakdown, and electromigration. From a user standpoint, the most effective way to deal with these topics is to review the design rules and methods used in manufacturing to verify reliability. Other issues are more difficult. The most important is probably the extremely large number of transistors on a single die. Performance variations are caused by statistical fluctuations in threshold voltage (due to the small number of dopant atoms in each transistor) [16] along with manufacturing defects, including mask misalignment, that may allow individual transistors to function during initial tests, but reduce margins when we consider the very long operational life. Although "outliers" can often be identified by testing completed devices, statistical fluctuations are inherent in the technology and generally cannot be eliminated by testing. New approaches need to be developed to deal with these issues. Note that older methods (such as I_{ddq} testing) have limited effectiveness, particularly for processes with many levels of metallization.

Packaging is another concern. Nearly all components in previous flagship missions have used hermetic packages, and the general approach towards testing and qualification is focused on such package types. Non-hermetic packages (e.g., ball grid or column grid arrays) may be the only option for digital parts with high frequency response and large numbers of pins. For the typical range of activation energies associated with most failure mechanisms, burn-in temperatures for these types of packages are too low to establish reliability thresholds for long life applications.

Reliability of complex packages is likely to be one of the most important issues [17,18]. Much of the work on emerging package technologies has emphasized thermal cycling, which causes cracks to form in BGA and CGA packages [19].

Although appropriate for some missions, the relatively constant temperature of electronics on JEO diminishes the importance of thermal cycling.

An example of ongoing work on package reliability is work by Lall, *et al.*, [20]. They proposed using the growth of intermetallic compounds in BGA test structures during thermal aging as a reliability indicator when limited thermal cycling is expected. Thickness was evaluated by (destructively) sectioning some of the samples from a larger group at periodic intervals, using a SEM to measure intermetallic thickness. As shown in Fig. 7, the data fit a $\sqrt{t}$ dependence, consistence with a diffusion-controlled mechanism.

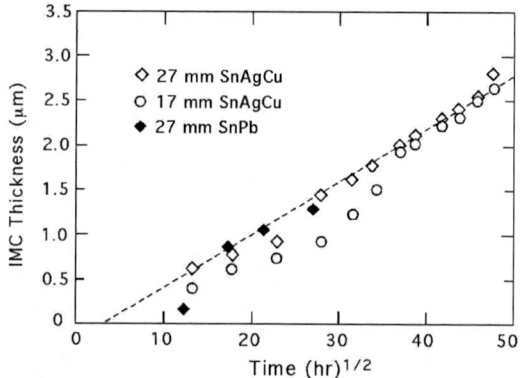

Fig. 7. Time dependence of inter-metallic compounds in ball grid arrays (after Lall, *et al.* [20]).

B. Derating

The second reliability step, which is probably the most important, is component derating. Although individual components are designed and tested to work over an extended temperature range[1], temperature ranges in typical applications are much narrower. Additional derating factors have been used in older flagship missions, where power dissipation, chip temperature, current and voltage are reduced from the maximum values allowed by manufacturers. For example, the maximum application voltage of a power MOSFET is 75% of the manufacturer's rating, with an additional derating factor for single-event gate rupture.

These derating methods provide significant margins in the operating stress of individual components. Although no attempt is made to quantify the improvement in reliability, the approach appears satisfactory for older components and will likely be beneficial for new technologies as well (including the inherent difficulties associated with advanced packaging).

C. System Design

The typical approach used to evaluate components is to combine worst-case values for initial parameters, reliability, temperature, and radiation damage in an additive mode. Although it is somewhat conservative, it can usually be accommodated by systems with less stringent requirements. For JEO, the inherent hardness of some parts is so close to the actual requirement that it becomes very difficult to use such an approach.

- - - - - - - - - - - - - - - -

[1]The military temperature range (-55 to 125 °C) was used in the past, but narrower ranges may be necessary for large-scale devices.

A statistical design approach is being considered for JEO that reduces some of the conservatism by using statistical representations of radiation degradation, reliability, temperature, and initial parameter values. The concept is illustrated in Fig. 8, recognizing that it is not practical to carry out statistical analyses to the point where the actual statistical distribution is characterized. A significant improvement in mission lifetime can be obtained with this approach, as illustrated in the figure.

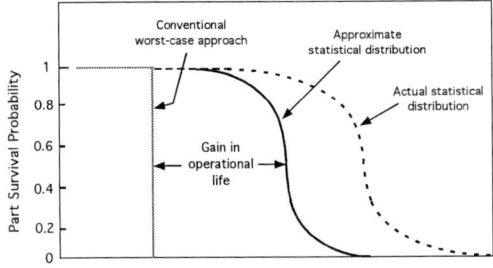

Fig. 8. Relationship between worst-case and statistical approaches for system lifetime.

D. Concerns and Synergistic Effects for JEO

Identification of Key Problem Areas

Reliability is a complex topic. Fundamental reliability mechanisms are usually evaluated using special test structures, and nearly always are taken into account in establishing design rules for complex circuits. For a mission like JEO, we need to know why devices really fail in field applications, and concentrate on ways to decrease the actual failure rate, relying on manufacturers to deal with the more fundamental mechanisms that have to be dealt with at the root manufacturing level. Field failure data from manufacturers may be helpful in identifying the key failure mechanisms, even though the conditions may differ from those used in space.

When we consider the design and thermal derating methods that are likely to be imposed by JEO, the main concerns are probably interconnects and packaging. As discussed earlier, new approaches will have to be identified for those factors in large-scale devices.

Other areas of concern for reliability are new technologies (particularly compound semiconductors), and special device technologies used in instruments, particularly detectors [21,22].

Synergistic Effects

Restricting operating conditions, power dissipation and temperature generally improve overall reliability. However, we have to consider possible interactions between the unusually high radiation environment and reliability mechanisms that can potentially make the overall reliability problem worse [23,34].

One example is the effect of small changes in the internal threshold voltage of SRAMs with small feature size. Although tests of highly scaled CMOS have shown that gate threshold leakage and leakage through the STI isolation regions are negligible [25], that is not the case for devices with narrow channel widths, such as those used in SRAMs; cache memories are designed with very narrow changes, and must be able to function with "5 sigma" parameter valuations at the 90 nm feature size node. Figure 9 shows how small changes in the internal switching point affect the write margin of an SRAM [26]. This particular case is for 90 nm cache cells, which are only affected at radiation levels > 1 Mrad(SiO_2). The effect has not been investigated for larger feature sizes, but it will likely affect devices manufactured at the 130 nm as well. From total dose scaling studies [27], we expect that such effects would become important between 200 and 500 krad(SiO_2) at that node.

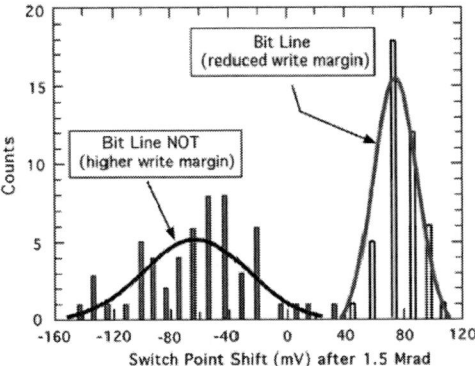

Fig. 9. Influence of small changes in the internal switching margin from total dose degradation on the SEU response of an SRAM with 45 nm feature size [26]. The margin is reduced for the normal bit line (compared to the bit line NOT line), increasing the SEU rate compared to an unirradiated device.

Other synergistic effects are more obvious, including the effect of gradual increases in standby leakage (or overall power consumption for bipolar devices) as the total dose increases during the mission. This not only increases the load on power distribution systems, but may increase the temperature of other components on conduction-cooled circuit boards, affecting reliability as well as performance characteristics.

CONCLUSIONS

The proposed JEO mission to Europa must deal with fundamental reliability issues as well as an usually high level of total ionizing dose, beyond the normal range considered for most radiation-hardened components. Improvements in hardened part technology combined with the very high reliability of older flagship missions show that it is possible to meet the demanding requirements of this mission.

The main areas of concern are those related to new technologies, where older methods of design and reliability do not necessarily apply. Additional work is needed on packaging and interconnect reliability, as well as on synergistic effects between reliability and radiation damage for advanced devices, where the approaches used by device manufacturers to maintain reliability will not necessarily apply.

We also have to be concerned about the special components used in detectors for spacecraft control (including star scanners), as well as those used in instruments. Earlier missions (including the Galileo mission) were able to deal successfully with those components. The challenge for the proposed JEO mission is to ensure that more advanced components of this type could also meet the demanding requirements of the mission.

978-1-4244-9113-1/11 $26.00 © 2011 IEEE

REFERENCES

[1] Jupiter Europa Mission Study 2008: Final Report, Feb. 12, 2009 (available on line through the NASA website).

[2] L. W. Massengill, "Cosmic Ray and Terrestrial Effects in Dynamic Random Access Memories," IEEE Trans. Nucl. Sci., **43**(2), pp. 576-593 (1996).

[3] P. E. Dodd, "Basic Mechanisms and Modeling of Singe-Event Upset in Digital Microelectronics," IEEE Trans. Nucl. Sci., **50**(3), pp. 583-602 (2003).

[4] F. W. Sexton, "Destructive Single-Event Effects in Semiconductor Devices and ICs," IEEE Trans. Nucl. Sci., **50**(3), pp. 603-621 (2003).

[5] B. G. Rax, et al., "Proton Damage Effects in Linear Integrated Circuits," IEEE Trans. Nucl. Sci., **45**(6), pp. 2632-2637 (1998).

[6] G. M. Swift, et al., "In Flight Annealing of Displacement Damage in LEDs: A Galileo Story," IEEE Trans. Nucl. Sci., **50**(6), pp. 1991-1997 (2003).

[7] T. R. Oldham and F. B. McLean, "Total Ionizing Dose Effects in MOS Oxides and Devices," IEEE Trans. Nucl. Sci., **50**(3), pp. 483-499 (2003).

[8] R. L. Pease, R. D. Schrimpf and D. M. Fleetwood, "ELDRS in Bipolar Linear Circuits: A Review", EEE Trans. Nucl. Sci., **56**(4), pp. 1894-1908 (2009).

[9] A. H. Johnston, R. T. Swimm and T. F. Miyahira, "Low Dose Rate Effects in Shallow Trench Isolation Regions," IEEE Trans. Nucl. Sci., **57**(3), pp. 3279-3287 (2010).

[10] D. M. Fleetwood and H. A. Eisen, "Total-Dose Radiation Hardness Assurance," IEEE Trans. Nucl. Sci., **50**(3), pp. 552-564 (2003).

[11] D. A. Southerland, "Qualification Issues and Pitfalls for Advanced Semiconductor Devices in Space," 2009 IEEE Reliability Physics Symposium, pp. 221-246.

[12] A. H. Johnston, "Space Radiation Effects and Reliability Considerations for Micro- and Optoelectronic Devices", IEEE Trans. on Device and Materials Reliability (in press).

[13] P. C. Chao, M. Y. Kao, K. Nordheden and A. W. Swanson, "HEMT Degradation in Hydrogen Gas", IEEE Elect. Dev. Lett., **15**(5), pp. 151–153 (1994).

[14] A. H. Johnston, *Reliability and Radiation Effects in Compound Semiconductors,* World Scientific: New Jersey (2010).

[15] D. L. Crook, "Evolution of VLSI Reliability Engineering," 1990 IEEE Internat. Reliability Physics Symposium, pp. 2-11.

[16] K. Takeuchi, R. Koh and Tohru Mogami, "A Study of the Threshold Voltage Variation for Ultra-Small Bulk and SOI CMOS," IEEE Trans. Elect. Dev., **48**(9), pp. 1995-2000 (2001).

[17] S. Stoyanov, et al., "Lifetime Assessment of Electronic Components for High Reliability Aerospace Applications," 2004 IEEE Electronics Packaging Conference, pp. 324-329.

[18] H. Qi, M Osterman and M. Pecht, "Plastic Ball Grid Array Solder Joint Reliability for Avionics Applications," IEEE Trans. on Components and Packaging Technologies, **30**(2), pp. 242-247 (2007).

[19] R. Ghaffarian, "Thermal Cycle Reliability and Failure Mechanisms of CCGA and PBGA Assemblies with and without Corner Staking", IEEE Trans. Comp. and Packaging Tech. **31**(2), pp. 285-296 (2008).

[20] P. Lall, et al., "Leading Indicators of Failure for Prognostication of Leaded and Lead-Free Electronics in Harsh Environments", IEEE Trans. Comp. and Packaging Tech. **32**(1), pp. 135-144 (2009).

[21] J. C. Pickel, et al., "Radiation Effects on Photonics Imagers – A Historical Perspective," IEEE Trans. Nucl. Sci., **50**(3), pp. 671-688 (2003).

[22] G. R. Hopkinson, et al., "Radiation Effects in InGaAs and Microbolometer Infrared Sensor Arrays for Space Applications," IEEE Trans. Nucl. Sci., **55**(6), pp. 3483-3493 (2008).

[23] D. M. Fleetwood, et al., "Effects of Device Aging on Microelectronics Radiation Response and Reliability," Microelectron. Reliability, **47**, pp. 1075-1085 (2007).

[24] R. D. Schrimpf, et al., "Reliability and Radiation Effects in IC Technologies," 2008 IEEE Reliability Physics Symposium, pp. 97-106.

[25] R. LaCoe, et al., "Improving Integrated Circuit Performance Through the Application of Hardness-by-Design Methodology," IEEE Trans. Nucl. Sci., **55**(4), pp. 1903-1925 (2008).

[26] X. Yao, et al., "The Impact of Total Ionizing Dose on Unhardened SRAM Cell Margins", IEEE Trans. Nucl. Sci., **55**(6), pp. 3280-3287 (2008).

[27] P. E. Dodd, et al., "Current and Future Challenges in Radiation Effects on CMOS Electronics", IEEE Trans. Nucl. Sci., **57**(4), pp. 1747-1763 (2010).

RELIABILITY AND PERFORMANCE CHARACTERIZATION OF A MEMS-BASED NON-VOLATILE SWITCH

Roberto Gaddi, Cor Schepens and Charles Smith
Cavendish Kinetics
's Hertogenbosch, The Netherlands
phone: (31) –(73)-6249110, roberto.gaddi@cavendish-kinetics.com

Cristian Zambelli, Andrea Chimenton and Piero Olivo
Dipartimento di Ingegneria
Università di Ferrara
Ferrara, Italy

Abstract—**In this paper we report data on the reliability and performance characterization of a CMOS-based non-volatile memory (NVM) array, the operating principle of which is based on stiction forces within a MEMS switch. Unlike any other NVM technology, the data retention of this technology improves with increasing temperatures. The switches have been proven to operate over an extremely wide temperature range from -150°C to 300°C, in a 4MRad/s radiation environment and can withstand acceleration forces up to 20,000g. The technology is an ideal candidate for highly reliable non-volatile memory in harsh environmental applications, like auto-motive, defense, space, down-well and geo-thermal. This NVM switch and a tunable RF-MEMS capacitor will be the first products based on this CMOS integrated MEMS platform.**

Keywords-component; MEMS, CMOS, non-volatile memory, NVM, harsh environment.

I. INTRODUCTION

We report for the first time the reliability performance of non-volatile memory arrays based on Nanomech™ MEMS technology [1], integrated in a standard CMOS process. The Nanomech™ non-volatile switch is able to maintain a programmed bit state across harsh environmental conditions, making this technology suitable for applications where traditional non-volatile memories fail.

The Nanomech™ process integrates mechanical switches within the CMOS back end of line metallization. Figure 1 shows a SEM cross section through an array of switches implementing a memory array. The following key advances in such technology compared to traditional MEMS technologies allow harsh environment reliability performance: 1) the cavity is sealed within the back end of line without the interior ever being exposed to the environment, generating pristine surfaces and hermetic vacuum conditions; 2) refractory metals are used to implement the cavity structure and the mechanical switch, providing hermeticity and stability over a wide temperature range. The baseline reliability performances of the

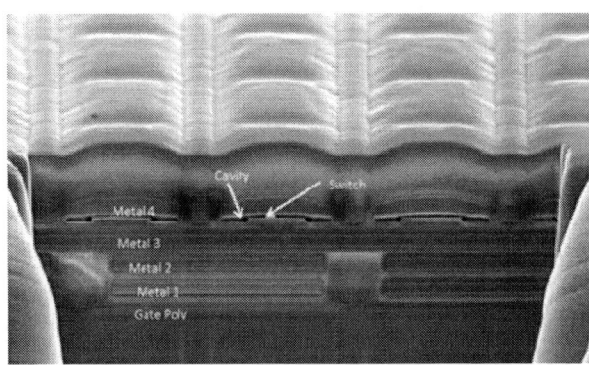

Figure 1. Nanomech™ 3rd generation MEMS: FIB + SEM of non-volatile switches in cavities integrated within standard CMOS.

Nanomech™ technology is confirmed by reporting no significant drift in parametric and device operations after submitting bare dies to thermal shock, temperature cycling from -150°C to 300°C, unbiased autoclave, temperature storage and standard mechanical shock and vibration tests.

II. TEMPERATURE BEHAVIOR OF NVM TECHNOLOGIES

Non-volatile memory (NVM) can be an important feature for many automotive integrated circuits. For under-the-hood applications high operating temperatures are mandatory and the memory needs to continue to work reliably under harsh conditions. Existing non-volatile memory technologies are very vulnerable to high temperatures and this is the bottleneck for many automotive as well as military and aerospace applications.

All mainstream NVM technologies are based on the principle of storing charge on a floating gate in a non-equilibrium energy state, which can leak away. This phenomenon gets accelerated at higher temperatures [2].

978-1-4244-9113-1/11 $26.00 © 2011 IEEE

Other types of NVM technologies have been developed over a number of years, such as ferroelectric, phase change and magnetic based memories. Most of these technologies require special materials, which are not used in CMOS fabs and fab managers are reluctant to include these materials into the process flow. Most of these technologies provide a faster write and erase of the data bits and have a higher endurance. However, also the data retention in these technologies is negatively influenced by applying higher temperatures.

FeRAM or FRAM is based on a ferroelectric material deposited in between two plates of a capacitor. A high-voltage across the ferroelectric material provides a shift to its electric dipoles, causing a change in dielectric constant which is detected as a variation in capacitance. A memory cell consists of one transistor and one capacitor or two transistors plus two capacitors for faster and more reliable differential readout. Unfortunately, ferroelectric materials lose their hysteresis at higher temperatures. As electrons vibration energy increases at higher temperatures, gradually more and more electric dipoles lose their alignment and the ferroelectric behavior fades out. Therefore, the highest temperature reached by FeRAM is equivalent to robust floating gate technologies. The maximum temperature found today for stand-alone FeRAM memories is 125 °C. [3]

MRAM technology is based on a magneto-resistive material in which the magnetic field can be altered. A permanent magnetic material and the ferromagnetic material need to be placed closely together separated by a thin insulating layer. By applying a high current (density) in one direction, the magnetic field orientation in the ferromagnetic material can be altered. Due to the magnetic tunnel effect, the electrical resistance of the cell changes due to the orientation of the fields in the two plates. By measuring the resulting current, the resistance inside any particular cell can be determined, and from this the polarity of the writable plate. Unfortunately, ferromagnetic-resistive materials also lose their magnetic field direction above certain temperatures and, in fact, some power-saving techniques make use of this. MRAM products are sold that are rated up to 125 °C. [4].

Phase Change Memories (PCM, OUM or PRAM) use the unique behavior of chalcogenide (GeSbTe) glass, which can be "switched" between two states, crystalline (low resistance) and amorphous (high resistance), with the application of heat. At 400°C it moves to a crystalline state and at 600°C it becomes amorphous. Although, these temperatures are well above the required automotive specifications, so far available devices are specified up to 85°C. Recently, PCM memories have been reported being rated up to 150°C [5].

RRAM or memristor technology [6] uses a dielectric, which is normally insulating, but can be made conductive through a filament or conduction path formed after the application of a sufficiently high voltage. The conduction path formation can arise from different mechanisms, including defects, metal migration, etc.. These are also very sensitive to temperature and today none have reached 10 year retention at room temperature.

Our stiction-based NVM technology uses moving micro-electromechanical conductive elements, which make contact with a conductive electrode and remain in contact due to

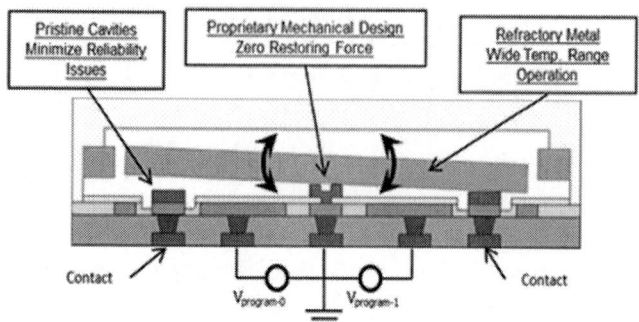

Figure 2. Cross section drawing of a Nanomech™ non-volatile switch integrated in a cavity inside the CMOS back end of line.

stiction. Stiction is created by attractive forces at the interfaces, mainly Metal-to-Metal bonding and van der Waals forces and to a lesser extent Casimir forces [7]. These forces are known to increase with increasing temperatures [8] and therefore data retention further improves with temperature for this operating principle.

III. OPERATING PRINCIPLE

The standard cantilever originally reported in [1] has been replaced by a revolutionary new design, which can best be described as a teeter-totter principle, as depicted in the cross-section of Figure 2. It consists of a mechanically floating plate, which lands on the contact electrodes when the sacrificial material gets released during fabrication. In general, forces acting on the MEMS mechanical element are: elastic deformation restoring force; adhesion forces at contact interfaces; electrostatic forces from charge or actuation voltages when the device is subject to program or erase signals. By designing a mechanical structure of the MEMS switch with zero mechanical restoring force, surface adhesion forces acting on the device at the contact points will dominate and force the acquired position to be maintained indefinitely. The contact points are electrically conductive and a resistance measurement can be made to sense the switch state.

The working principle of a single memory element is as follows. The floating plate is grounded as it makes contact with the two grounded electrodes in the center, which are slightly higher positioned since they overlap with the oxide layer on the pull-electrodes. The stiction forces between these contact electrodes and the floating plate will keep it in position. When applying a voltage on the right pull-in electrode (the program electrode) an electrostatic force will tilt the plate to touch the right contact electrode. The stiction forces at this contact point will keep the floating plate in this position. While in this position, a voltage can be applied on the left pull-in electrode (the erase electrode), resulting in an electrostatic force on the left side of the floating plate. When this electrostatic force reaches a threshold value, it will overcome the stiction forces at the right contact electrode and the plate will tilt to the left and land on the left contact electrode.

This describes how two alternative states of the device can be generated and maintained, which is the basic principle of a memory switch element. In addition, this teeter-totter device

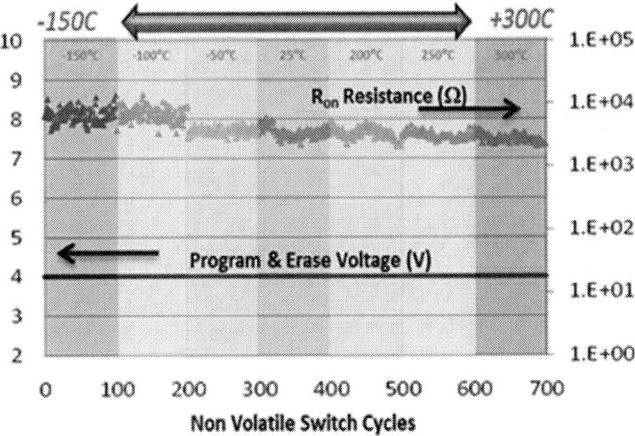

Figure 3. Operation of the Nanomech™ non-volatile switch (back end only implementation) across a wide temperature range showing no change in device parameters.

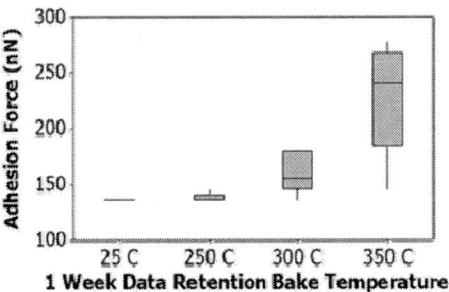

Figure 4. Extracted adhesion force in a Nanomech™ non-volatile switch subjected to 168 hours retention test at increasing temperatures.

is in a low energy state when in either the program or erase state and there are no forces present to move it from one to the other. Thus, contrary to all the previously described NVM technologies, the non-volatile working principle of stiction improves with increasing temperatures and very long lifetimes are ensured making them ideal for long time archiving of data. All NVM technologies rely on some kind of hysteresis model, which can degrade over time due to external events, but will never be restored. For stiction-based principles, external events may cause a minor lateral shift in contact points. Such shift may result in somewhat more or somewhat less intimate contact, depending on the local surface roughness. The next event may cause a slightly different position again, but eventually more intimate contacts will be harder to break and be more stable. So, the stiction force may vary over time, but will eventually never decay.

When an array of switches is implemented to form a memory, performing a single bit read requires sensing a contact resistance and comparing it to a reference value. Changing the state of the switch is performed by applying appropriately engineered waveforms to the program or erase electrodes. The resulting electrostatic forces are designed to overcome adhesion and allow the movement of the switch structure to the desired position, performing a program or erase action of the single bit.

IV. SWITCH-LEVEL RELIABILITY

The single switch device has been thoroughly investigated in order to characterize which forces are temperature-dependent and may influence the stiction behavior. This study included the traditional cantilever type design as it generates useful experimental results for characterizing forces into play. For example, different thermal expansion of the mechanical plate and substrate may result in a lateral force higher than the friction at the contact electrode. When this occurs the cantilever will move over the contact and this will momentarily affect the stiction force at the contact. For anchored cantilevers with a relatively high restoring force, this restoring force will work opposite to the adhesion forces and thus result in a reduction in vertical stiction force. Thus when the temperature-induced

lateral friction results in movement of the contact point, this may result in a momentarily smaller stiction force, which may be less than the restoring force, resulting in loss of contact.

On the contrary, with the new teeter-totter design of Figure 2 there is no restoring force and van der Waals forces will keep the contact in place so that it will be re-established at a slightly different position. The new stiction force will differ because of respective local surface roughness variations. In any case, with the zero restoring force in the current device design shown in Figure 2, such movements will NOT lead to loss of contact, as the stiction force will never go to zero and there is no restoring force.

Thus, thermal expansion may result in repositioning of the contact point and in a new value for the contact resistance, which may become worse or may become better. But as a lower contact resistance relates to a stronger stiction and a stronger friction force, it will require a stronger thermal expansion to re-position again, while a high contact resistance will be more easily repositioned . This process has a tendency to iterate the device contact to a lower resistance value and more stable stiction. It is interesting to note that this process is not dependent on the absolute temperature, but on the change in temperature.

A. Experimental Results

The design of the non-volatile MEMS switch takes advantage of the technology to achieve an operating temperature range of -150°C to 300°C. In order to demonstrate such a wide operating window, single non-volatile switches were fabricated on back end only wafers (no CMOS active circuitry) and operated with external voltage waveforms [9]. Figure 3 shows typical operation of a back end only single switch cycled through program and erase at varying temperatures, covering a range from -150°C to 300°C: No sign of program voltage change or degradation of the contact resistance emerged from such temperature operating tests [9].

The adhesion forces can be extracted from measurements of the MEMS switch threshold voltages and analyzed versus temperature. Figure 4 shows data from retention tests performed by unbiased storage at high temperature for 168 hours. Adhesion is shown to be slightly increasing with temperature, confirming that the memory storage mechanisms is not negatively impacted by temperature and that the

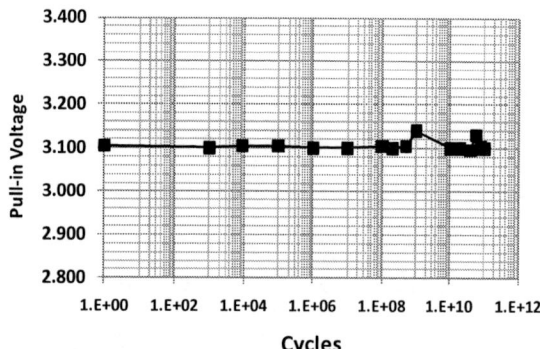

Figure 5. Threshold voltage values of a back end only implemented switch submitted to lifetime cycling test up to 10^{11} cycles at room temperature.

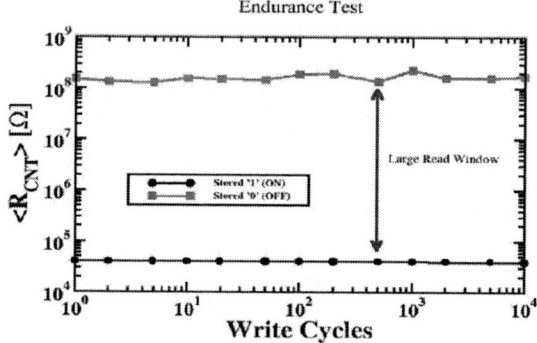

Figure 6. Read window showing average read resistance exhibited by a Nanomech™ array during an Endurance experiment. Red squares indicate OFF-state resistance which is limited by tester.

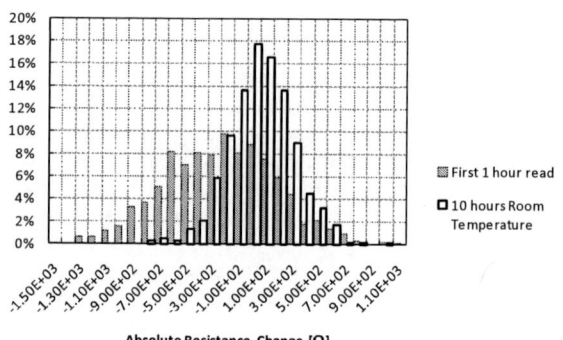

Figure 7. Resistance drift the first hour after a programming event and after the next 10 hours storage at room temperature.

technology is suitable for high temperature applications. Further increase in temperature will start to reach the limit where degradation of the Aluminum interconnects starts to occur.

The lifetime stability of the non-volatile switches has also been proven to withstand 10^{11} consecutive write cycles at room temperature, as shown in Figure 5.

V. ARRAY LEVEL EXPERIMENTAL RELIABILITY RESULTS

The integration of a topological matrix of Nanomech™ switches into array structures allowed the validation of the technology as a non-volatile memory targeted to embedded applications. The stored '1' (ON) state is represented by the switch closure and therefore by a low contact resistance value. The stored '0' (OFF) state is represented by an open contact, and the memory control circuitry can set the threshold at a few MΩ. Preferably, a differential readout circuitry is implemented, which senses the difference between the two contact electrodes.

The advantages of using this approach instead of the traditional voltage/current sense typical for charge-storage memory technologies is clearly seen in Figure 6, where the measured OFF-state resistance is limited by test equipment resolution. The read-window is comfortably large (more than 5 orders of magnitude), allowing safe storage of the information and giving plenty of margin in the design of the sense amplifier for bit state read.

Furthermore, the write and erase speeds of traditional floating gate based technology is known to be slow [10]. NanoMech™ write/erase speed being in the sub-μs range matches that of other emerging technologies, such as PCM and RRAM. The power required for write / erase action is very low as it will be dominated by address logic and write lines in the array, while intrinsic MEMS switch energy required to overcome the stiction holding it in position is much lower. After programming the full array, all bits are read back about 2 msec later. A further read operation after 1 hour at room temperature shows a small average drift, while the next 10 hours show stability as no average drift is visible, as shown in Figure 7. These small short-term small resistance adjustments have no impact in reliability as they are orders of magnitude smaller than the memory read window.

The table in Figure 8 provides a summary of the reliability tests. applied to the memory arrays. Figure 9 shows the average relative resistance-drift for different MEMS switch architecture variations, for all the performed tests. It should be stressed once again that these percentage variations are still orders of magnitude smaller than the memory read window. Therefore, no data retention loss is experienced on any of these tests.

Remarkably, most tests result in a negative drift in resistance, which indicates an increase in the contact area and thus an increase in stiction and an improvement in retention. From the mechanical tests, consisting of a mechanical shock and vibration, a small resistance increase is observed, while a constant acceleration of 20,000g (positioned such that the acceleration counteracts the stiction force), results in a smaller resistance. The other tests clearly show that temperature has a positive effect. In any case, all resistance shifts have magnitudes way below the memory read window, indicating no loss of information during any of the stress tests.

Figure 10 shows the difference in resistance of the measured average bit resistance (RCNT) during Liquid-Liquid stress test. The test is performed on ceramic packaged dies with open lid, over a temperature window from -55°C to 150°C. The total sensed resistance is a measure of the MEMS switch contact resistance combined with 6 pass transistors. By

978-1-4244-9113-1/11 $26.00 © 2011 IEEE   174

TEST	JEDEC	TYPE	CONDITION	DRIFT kOhm
HTSL	JESD22-A103-C	High temp. storage life	6 weeks @ 200°C	-8.6
TC	JESD22-A104	Thermal cycling	500 cy -65°C to +150°C	-3.4
LIQ-LIQ	JESD22-A106	Fast temp. changes 100C/sec	100 cy -55°C to +150°C	-3.5
HAST	JESD22-A118-C	Humidity test	130°C 85% humidity, 2 atm for 96 hours	-3.9
CA	MIL-STD883	Constant acceleration	20,000g / 1min / Y1 axis	-2.4
MS	JESD22-B104	Mechanical shock	10 half sine shocks 1500g @ 0.5ms	1.4
VFV	JESD22-B103	Variable frequency vibration	10 sweeps of 20-20,000Hz @ 50g	0.8
LIQ-LIQ + HAST		Fast temp changes followed by humidity test	see above	0.7
ROOM		simple retention proof	24h @ 23°C	-0.4

Figure 8. Reliability tests executed on NanoMech™ memories. No failing bits.

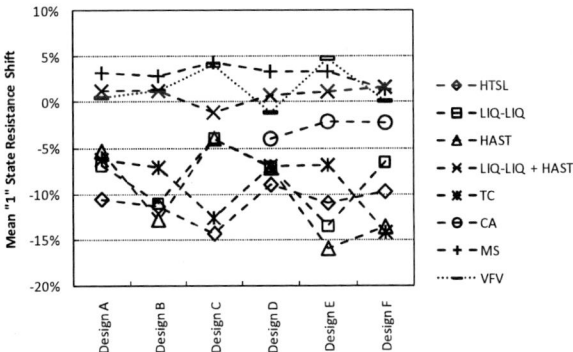

Figure 9. Relative resistance drift, average across full array, for all reliability tests performed on 6 alternative MEMS switch architecture variations.

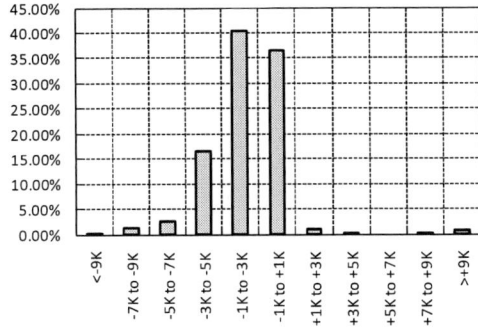

Figure 10. Contact resistance behavior on Liquid-Liquid test with -55C – 150°C temperature range. The distribution shows an improvement of the RCNT after the test

analyzing the distributions of the bit resistances before and after testing one can observe a small improvement of the stored '1' state resistance values. This results in a larger read window, which also indicates stronger stiction and thus improved

retention. This test was perceived to be the harshest test for these non-volatile MEMS switches.

As the liquid-liquid test was supposed to have the highest probability for creating 'cracks' in the passivation layers protecting the MEMS devices' cavities, it was decided to execute a post stress HAST test to see whether hermeticity was broken and humidity could enter the cavities. This test also resulted in no failing bits. Both liquid-liquid and HAST tests were performed with the lids of the ceramic packages removed to allow temperature and humidity to reach the bare die.

The NanoMech™ memories have also undergone a 1 million write-erase endurance test without any failures. The average resistance showed a minor drift indicating there is no wear-out after 1 million programming events.

VI. CONCLUSION

The non-volatile retention characteristic of all existing NVM technologies, whether based on charge, ferroelectric, magnetic or otherwise, all show a negative trend with respect to higher temperature. The NanoMech™ non-volatile switch is based on stiction, which shows an improved retention at higher temperature. This is a unique feature, which is mandatory for reliable memories at very high temperatures.

The executed reliability tests show an improved average resistance, indicating these environmental conditions tend to lead to a stronger stiction force and thus an improved retention.

The reported technology uses standard CMOS materials and CMOS production equipment, adding a few process steps in between two metal interconnect layers, which makes it compatible with any standard process or special processes for high-voltage, analog or RF applications. It is an ideal candidate for highly reliable non-volatile memory in harsh environmental applications, like automotive, defense, space, down-well and geo-thermal markets.

The cavity integrity of the process also enables the creation of a MEMS platform, of which this NVM and a tunable RF-MEMS capacitor will be the first products

ACKNOWLEDGMENT

The research leading to these results has received (partial) funding from the European Community's Seventh Framework Programme under grant agreement n°216436 (project ATHENIS).

REFERENCES

[1] M. Beunder, R. van Kampen, D. Lacey, M. Renault and C. Smith, "A New Embedded NVM Technology for Low-Power, High Temperature, Rad-Hard Applications", NVMTS 2005.

[2] G. Tempel, "Abnormal Charge Loss of Flash Cells at Medium Temperatures," IEEE NVSM Workshop, pp. 105-107, 2000.

[3] T. Mikolajick, C. Dehm, W. Hartner, I. Kasko, M.J. Kastner, N. Nagel, M. Moert, C. Mazure, "FeRAM technology for high density applications", Microelectronics Reliability, vol. 41, pp. 947-950, 2001

[4] J. Akerman, P. Brown, D. Gajewski, M. Griswold, J. Janesky, M. Martin, H. Mekonnen, J.J. Nahas, S. Pietambaram, J.M. Slaughter, S. Tehrani, "Reliability of 4Mb MRAM", in Proceedings 43rd Annual Reliability Physics Symposium, 2005, pp. 163-167, 2005.

[5] Y.H. Shih, J.Y. Wu, B. Rajendran, M.H. Lee, R. Cheek, M. Lamorey, M. Breitwisch, Y. Zhu, E.K. Lai, C.F. Chen, E. Stinzianni, A. Schrott, E. Joseph, R. Dasaka, S. Raoux, H.L. Lung and C. Lam, "Mechanisms of retention loss in Ge2Sb2Te5-based Phase-Change Memory," Electron Devices Meeting, 2008. IEDM 2008. IEEE International , pp. 1-4, 2008

[6] D. Wouters, "Oxide resistive RAM (OxRRAM) for scaled NVM application", in Innovative Mass Storage Technologies – IMST, 2009

[7] C. Smith, R. Kampen, J. Popp, D. Lacy, D. Pinchetti, M. Renault, V. Joshi and M. Beunder, "Nanomechanical cantilever arrays for low-power and low-voltage embedded nonvolatile memory applications," in Proc. SPIE 6464, 2007, p. 646406.

[8] C Brown, O. R. "Temperature dependence of asperity contact and contact resistance in gold RF MEMS switches," JOURNAL OF MICROMECHANICS AND MICROENGINEERING , 1 (2009).

[9] V. Joshi, R. Knipe, R. Kampen, D. Lacey, T. Nagata, D. Yost, and C. Smith, "A non volatile mems switch for harsh environment memory applications," in International Conference and Exhibition on High Temperature Electronics Network (HiTEN), 2009

[10] Peter Desnoyers,"Empirical Evaluation of NAND Flash Memory Performance" in SOSP Workshop on Hot Topics in Storage and File Systems (HotStorage '09), 2009.

978-1-4244-9113-1/11 $26.00 © 2011 IEEE

Microanalysis for Tin Whisker Risk Assessment

Maribeth Mason, Genghmun Eng, Martin Leung, Gary Stupian and Terence Yeoh
Microelectronics Technology Department
The Aerospace Corporation
El Segundo, CA
Maribeth.S.Mason@aero.org

Abstract—We use microanalysis to illustrate several effects of plating microstructure on tin whisker growth, such as grain size, plating composition, and plating thickness. Starting from the diffusion-based theory of tin whisker growth, we explain why large-grained and thin platings grow fewer whiskers, and why Ni diffusion barriers and Sn-3% Pb solder composition may not be protective against tin whisker growth. We also use tin whisker growth rate data to quantify the diffusion model and predict the time evolution of whisker length distributions. The results provide a consistent framework for tin whisker risk assessment from a materials perspective.

I. INTRODUCTION

Tin whiskers, microns in diameter and hundreds of microns long, can grow from pure tin plating. Whisker-induced electrical shorts can develop into sustained plasmas that can carry tens to hundreds of amperes and blow spacecraft fuses. Although pure tin has long been prohibited in space systems, it is still found in space hardware, and tin whiskers have been implicated in several commercial space-system anomalies.

The following case studies use microanalysis to determine the effect of plating microstructure on whisker growth, along with physical modeling to predict the time evolution of whisker length distributions. We interpret these results in the context of the qualitative tin whisker growth model outlined by K.N. Tu [1], which gives three necessary conditions for spontaneous tin whisker growth for the Sn-Cu system: 1. grain boundary diffusion of Cu in Sn; 2. formation of Cu-Sn intermetallic compounds (IMC), which creates compressive stress on Sn grains as the driving force for whisker growth, and 3. breaking of surface tin oxide, through which Sn extrudes and recrystallizes into whiskers. Tin platings are highly variable, and the propensity to grow whiskers can depend on grain size, grain boundary integrity, thickness, tin oxide thickness, impurity content, and other factors.

II. EXPERIMENT

Several tin and tin-lead platings were investigated using scanning electron microscopy (SEM), focused ion beam (FIB) milling and ion imaging, and energy dispersive x-ray spectroscopy (EDX). Examples of the information produced by these techniques are shown in Fig. 1, which shows a FIB cross-section of a tin-lead plating on a capacitor. While SEM is sensitive to surface features, imaging with the ion beam provides more information about grain structure and chemical composition due to ion channeling effects. Finally, EDX analysis provides quantitative information about spatial elemental distributions within the plating.

III. RESULTS

A. Effect of Grain Size

Although all Sn on Cu systems are at risk for whisker growth, anecdotal evidence suggests that commercial tin-plated copper wire or coaxial cable may be less prone to whisker growth. Fig. 2 shows a FIB cross-section of a tin-plated copper coaxial cable on which no whiskers were observed. Although the surface of the cable showed micron-size features, the underlying grains were discovered to be 20-40 um in diameter. Such large grains leave few grain boundaries for IMC invasion that would cause compressive stress and whisker growth. The Cu-Sn IMC was largely confined to the Cu-Sn interface, with very little in the grain boundaries. By contrast, tin platings that have grown whiskers often have small, columnar grains, allowing as much as 100x more grain boundary diffusion of Cu in Sn [2,3].

For the Sn on Cu system, the dominant IMC formed by Cu invading the Sn grain boundaries is Cu_6Sn_5. The activation energy for Cu_6Sn_5 growth is $U_{IMC} = 0.27$ eV [4], which is comparable to that for diffusion along pipe or threading dislocations in Sn and to the recrystallization energy for pure tin of 0.26 eV [5]. Thus, the Sn-Cu interaction results in a competition between Sn recrystallization and IMC nucleation. In a large-grained, very pure polycrystalline Sn plating, the dominant activation energy for Cu migration would be the 0.41-0.47 eV required for grain boundary diffusion [6]. The lower tin recrystallization energy could allow the tin plating to almost fully recrystallize, shutting off nearly all Cu migration pathways. Such a system would then primarily form a Sn/IMC/Cu layered structure, where the growth rate of these layers would be dominated by a bulk diffusion activation energy of 0.97-1.1 eV [7-9]. The remaining pure tin grains would incorporate little lateral stress, and tin whisker growth should be nonexistent. This appears to be the case for the plating observed in Fig. 2.

For a very small-grained polycrystalline pure tin plating, tin recrystallization could result in the coalescence of a small amount of impurities at the grain boundaries. Grain boundary diffusion of Cu and IMC nucleation would then compete with grain growth. Under these conditions, tin grain growth may fully surround and isolate IMC nuclei, but there would be no

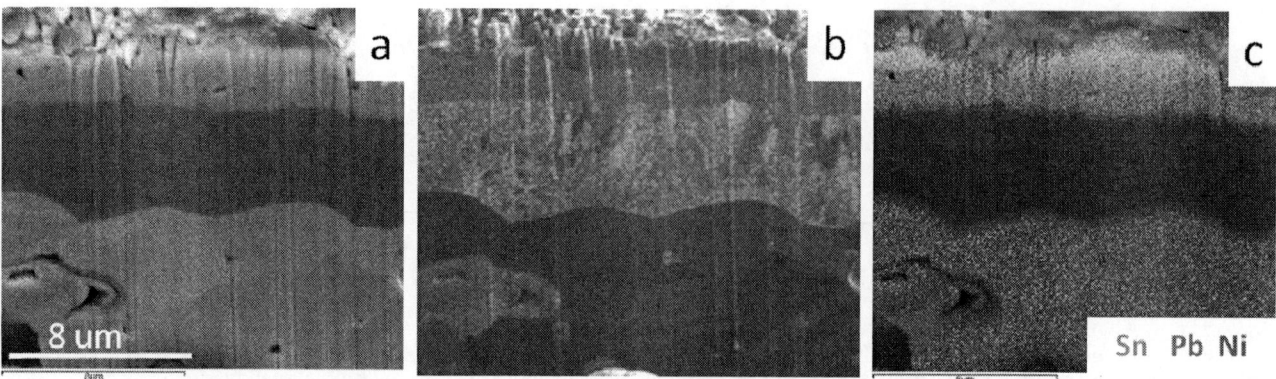

Figure 1. Examples of microanalysis techniques applied to a FIB cross-section of a Sn-Pb on Ni plating. (a) SEM. (b) Ion-electron imaging. (c) EDX.

mechanism for an ongoing lateral stress buildup, and whisker growth should again be nonexistent.

In a columnar tin plating, the columns offer fast migration channels for Cu. Small differences in plating impurity content could then easily tip the balance between Sn recrystallization and IMC formation. Once a grain boundary has a connected Cu diffusion channel, it is likely to support ongoing IMC growth, providing a continuing source of pressure buildup to support whisker formation. Impurity content may thus be a primary factor in IMC formation and whisker growth.

B. Effect of Plating Composition

Short whiskers were observed on capacitor end bells coated with Sn solder over Ni on Ag. The solder had >3% Pb content as verified by x-ray fluorescence (XRF). A uniform bulk Sn 97% Pb 3% solder is generally considered protective against tin whisker growth [10], as is a Ni underlayer [11].

EDX mapping of the plating surface, shown in Fig. 3, showed that Pb had segregated into patches on the tin surface. These Pb patches behaved like weak spots in the tin oxide, allowing whiskers to preferentially extrude from these areas. EDX revealed that each tin whisker had a lead cap. Because XRF is a bulk technique, it cannot distinguish between 3% Pb on the surface and 3% Pb properly alloyed within the bulk.

The Monte Carlo code BEAMnrc [12] was used to simulate the XRF spectra for various Sn-Pb sample configurations. The spectrum of a 35 kV x-ray tube with a silver anode (similar to some commercial handheld XRF units) was modeled independently in BEAMnrc, as illustrated in Fig. 4. The detector is modeled as an annular ring that collects all incident photons, and the sample is a multilayer stack with layers of Sn and/or Pb.

Fig. 5 illustrates how ambiguities can arise in XRF spectra depending on sample geometry. The simulated XRF spectra representing (a) an intimate mixture of 3 wt. % Pb in Sn and (b) a layered sample of 100 nm Pb on pure Sn are very similar. We recommend that plating inspection standards require EDX and FIB cross-sections to confirm proper alloying of Pb in Sn. Extensive Pb segregation as determined by EDX may be a cause for concern.

C. Effect of Plating Thickness

Fig. 6 shows FIB cross-sections of two regions on the same capacitor end-bell, one with whiskers and one without. The area with whiskers had a thick (~10 um) tin plating with voids, while the region without whiskers had a thin (~1 um) tin plating. Thin tin platings with large grains are likely to be uniformly consumed by IMC formation and thus less likely to grow whiskers. By contrast, thick tin platings, especially those with small columnar grains, are more likely to build up compressive stress and grow whiskers due to IMC formation.

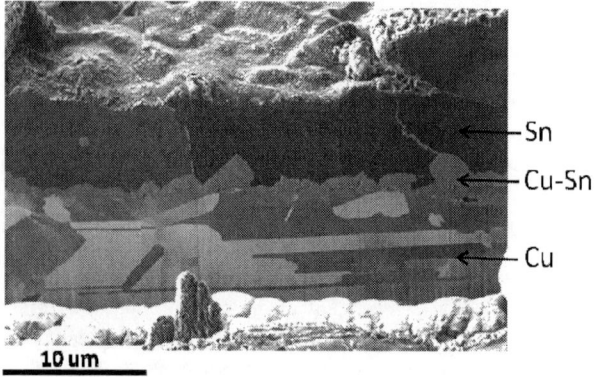

Figure 2. FIB cross-section of tin-plated copper coaxial cable.

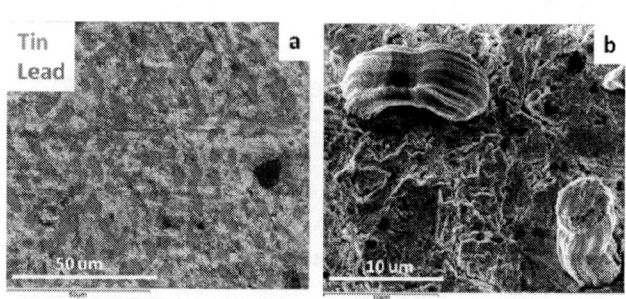

Figure 3. EDX maps of Sn-Pb solder plated capacitors. (a) Lead is segregated into surface patches. (b) Tin whiskers have lead caps.

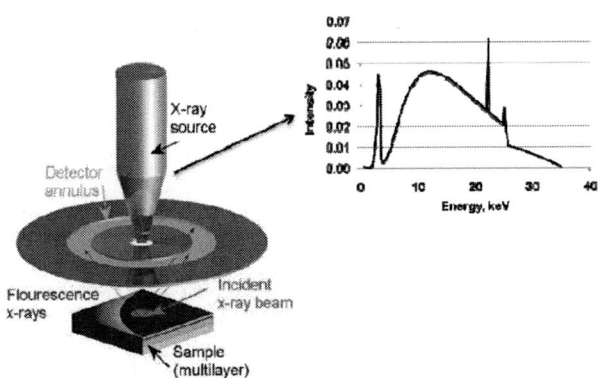

Figure 4. Schematic of simulated XRF spectrometer.

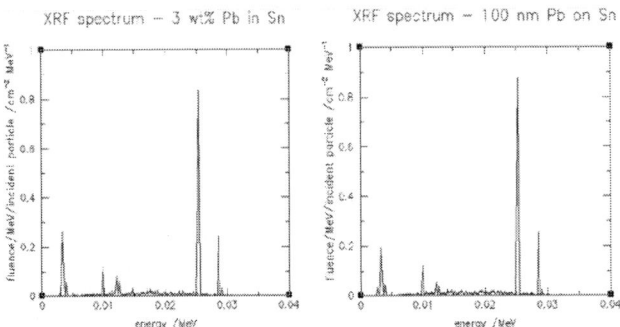

Figure 5. Simulated XRF spectra of (a) uniform mixture of 3 wt. % Pb in Sn and (b) 100 nm Pb layer on Sn..

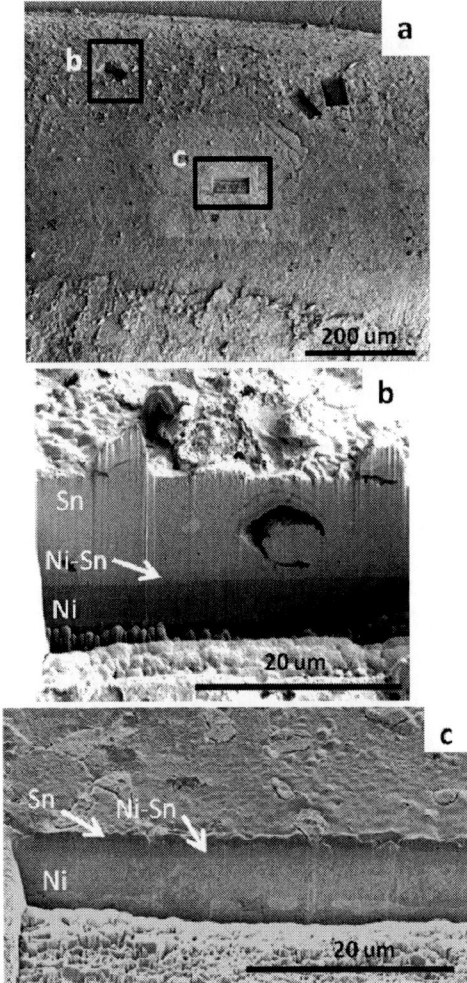

Figure 6. (a) SEM micrograph of capacitor end-bell, showing regions (b) with whiskers and (c) without whiskers. (b) FIB cross-section of region with whiskers, which has a thick tin plating. (c) FIB cross-section of region without whiskers, which has a thin tin plating.

IV. QUANTITATIVE PREDICTION OF WHISKER LENGTH DISTRIBUTIONS

Long whiskers, along with a partially broken whisker, were observed on capacitors with a pure Sn on Ni plating. We expanded on the qualitative model of whisker growth outlined by Tu [1] and combined it with modern whisker growth rate data from Panashchenko [13] for Sn on Ni systems at ambient after thermal cycling. These data represent a worst case limit for constant temperature tin whisker growth without prior cycling, and were used to calibrate the diffusion-based model for a quantitative estimation of a whisker length distribution at later times to simulate mission life. The results also allowed us to estimate the probability of forming longer whiskers, which would be more likely to break, and were factored into reliability calculations and risk assessments.

The model predicts the following qualitative results for whisker growth, which will be discussed in the following section:

- Whiskers grow as the square root of time, with growth stopping when all the Sn is consumed by IMC formation or extruded into whiskers..

:

- Whisker length variations arise from differences in the tin oxide strength. The weakest tin oxide breaks first, and forms short whiskers that grow more slowly, while stronger tin oxide areas break later and form long whiskers that grow faster.
- The longest whiskers likely have an enhanced diffusion supply due to local variations in the tin plating. The standard lognormal distribution is thus unbounded for the longest whiskers, and since these present the majority of the risk, they must be considered separately from the main population.

A. Diffusion-Based Model for Whisker Growth

Worst-case assumptions for the diffusion-based model were used in order to provide an upper bound for whisker growth in most cases. The following primary assumptions were used to model whisker growth in the temperature range from 20-85 C:

1. Impurity effects govern the diffusion coefficient prefactor for IMC formation along grain boundaries, given by $D(T) = D_0\, exp[-U_{IMC}\,/kT]$. D_0 varies by orders of magnitude between a large-grained structure with no IMC invasion and a columnar grain structure with nearly complete IMC invasion. $U_{IMC} = 0.26$ eV for Cu_6Sn_5 IMC formation and 0.47 eV for Ni_3Sn_4 IMC formation [14].

2. D_0 is a critical parameter and a primary unknown quantity in the model. To estimate the maximum possible tin whisker growth rate and determine a numerical scaling factor for D_0, actual whisker data from Ref. 13 for a Sn on Ni system was used.

3. The underlying IMC growth function is approximately uniform across the entire tin plating.

4. For most whiskers (excluding some very long outliers), spatial variation in whisker lengths arises from local variations in oxide strength. For a constant IMC growth rate, a region with a weaker oxide might allow several tin whiskers to grow, while a region with a stronger oxide may only have one whisker. This is equivalent to local mass conservation.

5. The sole effect of lateral pressure gradients in the tin layer is to restrict the ultimate length of any tin whisker to 1 cm independent of the original tin layer thickness.

Consider a Sn plating of thickness L on a Cu substrate. At t=0, the Cu concentration C is zero at the tin surface and a constant C_0 at the Sn-Cu boundary. As Cu diffuses into the Sn, each Cu atom displaces one Sn atom, which is then available to become part of a growing whisker. Some fraction of the total tin is reserved for IMC growth; for example, C_0 could be chosen to be 6/11 of the original tin amount for a Cu_6Sn_5 IMC. The diffusion of Cu through the Sn along the x direction from the Cu substrate ($x = 0$) toward the Sn surface is governed by the following equations:

$$\frac{dC}{dt} = D(T)\frac{d^2 C}{dx^2} \qquad (1)$$

$$\frac{dC}{dx} = 0 \text{ at } x = 0 \qquad (2)$$

The solution to these equations for the given boundary conditions [15] is

$$C(x,t) = C_0 \frac{4}{\pi} \sum_{m=0}^{\infty} \frac{(-1)^m}{2m+1} \exp[-(2m \qquad (3)$$
$$+ 1)^2 Z]\cos\frac{(2m+1)\pi X}{2L}$$

where

$$Z(t) \equiv D(T)t\left(\frac{\pi}{2L}\right)^2. \qquad (4)$$

We can subsequently derive a fraction of Sn $g(t)$ that has been extruded into whiskers due to underlying IMC formation:

$$g(t) = \frac{8}{\pi^2} \sum_{m=0}^{\infty} \frac{1}{(2m+1)^2}\{1 - \exp[(-2m+1)^2 Z(t)]\} \qquad (5)$$

A simple analytic approximation for $g(t)$ is

$$\left(\frac{1}{g(t)}\right)^6 \approx \frac{8}{Z^3(t)} + 1. \qquad (6)$$

The whisker length $X(t)$ for a given whisker is then

$$X(t) = g(t)X_{max} \qquad (7)$$

where X_{max} is the whisker's final length as t approaches infinity. $X(t)$ then forms the progress variable for tin whisker growth. These approximations show that whiskers grow as the square root of time, with growth stopping when all the Sn is either consumed by IMC formation or extruded into whiskers.

For each whisker, X_{max} is set by the local tin oxide strengths, with the ultimate upper bound to X_{max} restricted to 1 cm. The growth of a 1 cm whisker of diameter d growing from a Sn plating with thickness L by the formation of M_kSn_n IMC could draw from an underlying area of radius

$$R = \sqrt{\frac{k+n}{4n}\frac{d^2}{L}} \qquad (8)$$

For a 10,000 μm long, 2 μm diameter whisker growing by Cu_6Sn5 IMC formation in a 10 μm thick Sn plating, R = 47 μm.

While normal whisker growth is associated with an R on the order of the tin grain size, the longest whiskers likely have an enhanced diffusions supply due to a large pressure release after oxide breakage. This can be accompanied by a hydrodynamic tin flow that breaks up the underlying tin grain structure, creating a larger feed zone for that whisker. The standard lognormal distribution that is observed for short and medium length whiskers does not properly account for these longest whiskers, and, since they present the majority of the risk, they must be considered separately from the main population.

B. *Predicting Whisker Length Distributions*

Fig. 7 illustrates how the model described above was used to reanalyze a critical subset of the data in Ref. 13. The distribution at t=t0 reflects the measured whisker length distribution after thermal cycling, and a final whisker length distribution was measured after two years at ambient. The model explains several distinct regimes of whisker growth that are observed:

- Normal whisker growth is associated with a tin supply on the order of the tin grain size. It is characterized by a constant whisker growth rate, and should be associated with relatively thinner tin oxides and the simultaneous growth of many whiskers of short X_{max}.
- Anomalous whisker growth is characterized by a nearly constant but faster growth rate, and is associated with a thicker tin oxide. This regime may be separated from normal whisker growth by a transition zone, which is likely to contain a mixture of oxide strengths.
- Extreme whisker growth results in the longest whiskers, and is characterized by a pressure release resulting in hydrodynamic tin flow which produces a larger tin supply.

Fitting these different regimes to the diffusion-based tin whisker growth model then sets an X_{max} for each initial whisker length, as shown in Fig. 7. This can then be used to predict a worst-case whisker length distribution at mission life for hardware which has not undergone thermal cycling.

V. CONCLUSIONS

We used microanalysis to illustrate several general principles about the effect of microstructure on tin whisker growth:

- Diffusion and IMC formation drive whisker growth. This occurs for Sn-Ni as well as Sn-Cu systems; i.e., a nickel underlayer does not necessarily protect against whisker growth.
- Plating microstructure determines the extent of diffusion and IMC formation. For example, large-grained tin platings and thin tin platings grow fewer whiskers. The plating microstructure can be affected by several factors, such as thermal history.
- Impurity content and alloying are also important. Since XRF cannot necessarily determine whether or not a Sn-Pb solder plating is properly alloyed, we recommend that standards require cross-sectioning and EDX to confirm composition and alloying.

These results can be integrated into a consistent framework for tin whisker risk assessment, along with our understanding of tin plasma risk [16], tin whisker growth modeling, and other hardware specific factors.

REFERENCES

[1] K.N. Tu, Solder Joint Technology: Materials, Properties and Reliability, New York: Springer, 2007.

[2] G.T. Galyon, "Sn-Cu Whisker Nodule," iNEMI Whisker Fundamentals, 2004.

[3] T. Kakeshita, K. Shimizu, R. Kawanaka and T. Hasegawa, "Grain size effect of electroplated tin coatings on whisker growth," J. Mat. Sci. 17 (9), pp. 2560-2566, 1982.

[4] M. Braunovic, "Effect of intermetallic phases on the performance ot tin-plated copper connections and conductors, Proc. 49th IEEE Holm Conf. Elec. Contacts, pp. 124-131, 2003.

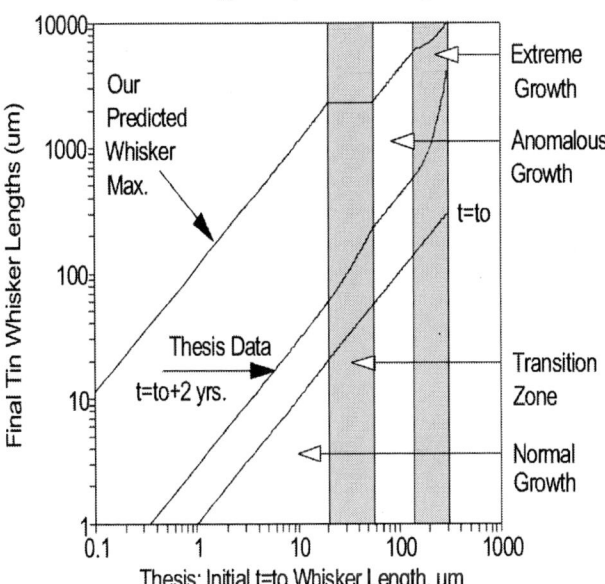

Figure 7. Prediction of time-evolved whisker length distributions from data in Ref. 13.

[5] E.L. Holmes and W.C. Winegard, "Grain growth in zone-refined tin," Acta Metallurgica 7 (6), pp. 411-414, 1959.

[6] K. Jung and H. Conrad, "Microstructure coarsening during static annealing of 60Sn40Pb solder joints," J. Electron. Mater. 30, pp. 1303-1307, 2001.

[7] C. Coston and N.H. Nachtrieb, "Self-diffusion in tin at high pressure," J. Phys. Chem. 68 (8), pp. 2219-2229, 1964.

[8] W. K. Warburton and D. Turnbull, in Diffusion and Solids, ed. A.S. Nowick and J.J. Burton, Academic: New York, 1975.

[9] P. Adeva, G. Caruana, O.A. Ruano, and M. Torralba, "Microstructure and high temperature mechanical properties of tin," Mat. Sci. Eng. A 194, pp. 17-23, 1995.

[10] W. Boettinger, C. Johnson, L. Bendersky, K. Moon, M.Williams, and G. Stafford, "Whisker and hillock formation on Sn, Sn-Cu and Sn-Pb electrodeposits," Acta Materialia 53, pp. 5033-5050, 2005.

[11] iNEMI Report, "iNEMI recommendations on lead-free finishes for components used in high-reliability products," version 4, December 2006, available at http://thor.inemi.org/webdownload/projects/ese/tin_whiskers/Pb-Free_Finishes_v4.pdf.

[12] BEAMnrc, http://irs.inms.nrc.ca/software/beamnrc-V4-2.3.1/ .

[13] L. Panashchenko, "Evaluation of Environmental Tests for Tin Whisker Assessment," Univ. of Maryland thesis, December 2009.

[14] J. W. Yoon and S. B. Jung, "Growth kinetics of Ni₃Sn₄ and Ni₃P layer between Sn-3.5Ag solder and electroless Ni-P substrate, J. Alloys Compounds 375, pp. 105-110, 2004.

[15] J. R. Crank, The mathematics of diffusion, Clarendon, Oxford, p. 24, 1979.

[16] M.S. Mason and G. Eng, "Understanding tin plasmas in vacuum: a new approach to tin whisker risk assessment," J. Vac. Sci. Technol. A 25, pp. 1562-1566 (2007).

ACKNOWLEDGMENTS

This work was sponsored by The Aerospace Corporation's Independent Research and Development Program.

Random Telegraph Noise Reduction in Metal Gate High-κ Stacks by Bipolar Switching and the Performance Boosting Technique

W.H. Liu, K. L. Pey*, N. Raghavan, X. Wu
Microelectronics Center
Nanyang Technological University
50 Nanyang Avenue, Singapore 639798.
Phone: (65) – 6790 6371, eklpey@ntu.edu.sg
* Current address is at Singapore University of Technology and Design.

M. Bosman
A*STAR Institute of Materials Research and Engineering
3 Research Link, Singapore 117602.

T. Kauerauf
imec
Kapeldreef 75, B-3001 Leuven, Belgium

Abstract— **In high-κ (HfSiON and HfLaO) metal gate stacks, the traps leading to gate current I_g random telegraph noise (RTN) are found to be effectively passivated by bipolar switching from negative gate bias, where RTN and threshold voltage variation (ΔV_T) are reduced significantly or even disappear. The reduction of RTN, ΔV_T and I_g in degraded gate dielectrics is modeled by oxygen ion drift from the oxygen gettering metal gate electrode to re-passivate the traps upon negative gate bias. A *performance boosting technique* for transistors during operation is proposed. In this technique, the gate is swept by a small negative voltage to induce a bipolar switching and thus boost up performance after long duration of operation.**

Keywords - Bipolar Switching, High-κ Metal Gate (HK-MG), Performance Boosting Technique, Random Telegraph Noise (RTN), RTN Reduction, Threshold Voltage Shift (ΔV_T), Triggering Voltage (V_{trig}),

I. INTRODUCTION AND EXPERIMENT

Random telegraph noise (RTN) is a low-frequency noise (LFN) component measured in nanoelectronic devices. RTN is an indication of single or multiple active traps in the conduction mechanism of small dimension devices. The capture or emission of a charge carrier (electron or hole) by a trap gives a transient measurable change in the device resistance, which can be measured by fluctuations in the current or voltage from the device terminals. For MOSFETs, RTN is usually measured at the drain or the gate terminal. Under constant voltage bias condition, RTN is observed in the channel current (I_d RTN) [1], [2] or the gate leakage current (I_g RTN) [1], [3]. The capture and emission events in RTN are stochastic processes – the capture and emission of carriers are

continuous and independent of one another and cannot be predicted [1]. Therefore RTN is usually characterized by its average properties, which include the capture time constant (τ_c), the emission time constant (τ_e) and the power spectrum. Taking I_g RTN schematic in Fig. 1 as an example, τ_c is the time it takes for an empty trap to capture an electron, corresponding to the higher level of I_g. On the other hand, τ_e is the time it takes for a filled trap to release an electron, corresponding to the lower level of I_g. I_g is lower at the filled-trap state because an electron trapped inside a dielectric trap can block the current flowing through, due to coulombic repulsion [3].

It is well documented that RTN can lead to threshold voltage shift (ΔV_T) or fluctuations [4] and cause functional instabilities, affecting the performance of logic and memory devices and challenging the future scaling of high-κ gate dielectrics. It is also believed that RTN causes erratic behavior in SRAM [5]. As a result, suppressing RTN to minimize

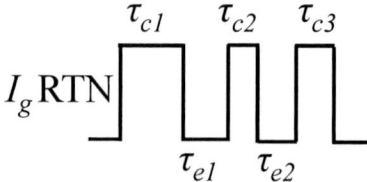

Figure 1. The RTN parameters in I_g, capture time τ_c and emission time τ_e. τ_c corresponds to the time spent at the higher level of I_g for an empty trap to capture an electron. On the other hand, τ_e corresponds to the lower level of I_g and it is the time taken for a filled trap to release an electron [3].

degradation in device performance has become an important field of study.

It has been reported [6] that controlled forming gas annealing (FGA) during fabrication processes can suppress RTN arising from process induced traps (PIT). However, it is important to note that, although FGA can passivate PIT and reduce RTN effectively, it cannot be used to passivate the stress induced traps (SIT) that originate during prolonged device operation [3]. First, it is not practical to detach the chip during operation and perform FGA treatment. Second, the high temperature during FGA may introduce unexpected degradation to other components of the chip.

On the other hand, it was reported [7] that applying substrate bias on a submicron MOSFET has an impact on the channel current (I_d) RTN parameters such as the emission and capture time constants. The proposed explanation from the quantum-mechanical approach was that the substrate bias can cause changes in both the channel inversion layer surface charge density N_s and the distance between the centroid of the inversion layer and the interface. Therefore, with the existence of the substrate bias, the I_d RTN parameters are strongly modulated.

The time constants can be expressed by the Shockley-Read-Hall theory [8]: capture time $\tau_c = \dfrac{1}{\sigma(x_t)v_{th}n}$ and emission

time $\tau_e = \dfrac{1}{g\sigma(x_t)v_{th}N_C \exp[(E_T - E_{CS})/kT]}$, where v_{th} is

the average thermal velocity of the charge carriers, n is the volume concentration of electrons in the inversion layer, $\sigma(x_t)$ is the capture cross section of a trap located in the oxide at a distance x_t from the Si-oxide interface, g is the degeneracy factor, E_T is the trap energy, E_F is the Fermi level, E_{CS} is the conduction band energy at the Si-oxide interface, and N_C is the conduction band effective density of states. With an increasing positive gate bias, n increases and it leads to the decrease in τ_c, while ($E_T - E_{CS}$) decreases and it leads to the increase in τ_e. Zanolla *et. al.* [9] have shown that, under the bias configurations where the gate is switched between ON and OFF states, a strong reduction in RTN is obtained when a forward substrate bias is applied during the device OFF-state. From the RTN emission and capture time constants measured, a reduction in RTN is found to be caused by a significant decrease in the emission time constant, when a low gate voltage and a positive substrate voltage are applied concurrently [9]. The decrease in the emission time constant leads to an imbalance between the capture and emission time and a departure from the condition $\tau_e = \tau_c$, where significant RTN signal exists (The $\tau_e = \tau_c$ condition does not necessarily occur at zero bias condition). As a result, a reduction in RTN is achieved. A proposed explanation for the decrease in the emission time constant under forward substrate bias is the fact that a transient accumulation layer in the silicon at the oxide interface is induced and that leads to a significant recombination rate between the trapped carriers and the accumulated holes during the OFF state [9].

The literatures cited above focused on the modulation of the inversion layer [7] or the transient accumulation layer [9] by substrate bias on MOSFETs to achieve the purpose of RTN parameter modulation. Because there is no actual change of the defects giving rise to I_d RTN here, the reduction in I_d RTN [9] obtained is not permanent and the reduction effect vanishes once the substrate bias is removed.

It is believed that I_g RTN [3] is sensitive to the bulk traps in the dielectric, while the interface traps have an impact on the I_d RTN. In this paper, the I_g RTN was measured and studied in different degraded high-κ metal gate (HK-MG) stacks. The I_g RTN was found to significantly reduce or even disappear after a bipolar switching event, induced by a negative gate bias sweep. In addition, the threshold voltage shift, ΔV_T, measured in the degraded dielectrics is also effectively reduced. The reduction of I_g RTN was found to be accompanied by the reduction of the leakage current I_g. The reduction in the I_g RTN power spectral density, the I_g level and the ΔV_T are found to be permanent and stable to any subsequent bias configuration applied to the MOSFET.

The devices studied were nMOSFETs with NiSi FUSI/HfSiON(25Å)/SiOx(12Å) (*Stack-A*) and TiN/HfLaO (18Å) (*Stack-B*) gate stacks, having a channel width of 0.35-1.3μm and a channel length of 70-200nm. The samples were subjected to an inversion mode positive constant voltage stress (CVS) at the gate to induce a soft breakdown (BD) capped by a gate leakage current compliance, I_{gl} of 0.5-5μA, after which the devices were still functional with a certain degree of degradation. The degradation is evidenced by an increase in the I_g and $\Delta I_g / I_g$ levels and clear existence of Lorentzian I_g RTN. We measure the I_g RTN at low sense gate voltage, V_g, with $V_s = V_d = V_{sub} = 0$V at temperature = 25°C. The measurements were performed using a Keithley 4200 semiconductor characterization system, having a resolution down to femtoampere current level with pre-amplifiers. The highest time resolution the probing system can achieve is ~20ms, and the time resolution in the actual measurement also depends on the current level.

After this, a negative V_g was applied to the gate to induce a switching in I_g [10]. As we used an opposite polarity of the ramping voltage (i.e., negative polarity) with reference to the CVS-induced BD using positive polarity, this is called bipolar switching. The I_g measurement was repeated at the same V_g to examine the effect of bipolar switching on RTN. The experiment procedures are summarized as a flowchart in Fig. 2.

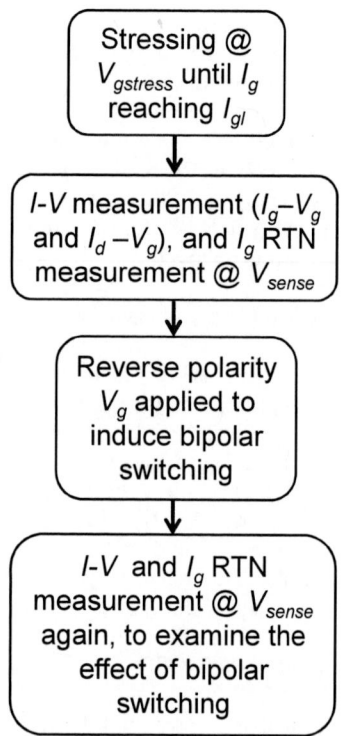

Figure 2. The flowchart summarizing the experiment procedures.

II. REDUCTION OF RTN BY BIPOLAR SWITCHING

A. Full and Partial Recovery from Bipolar Switching in NiSi FUSI/HfSiON stacks

A *Stack-A* nMOSFET of area $1\times0.13\mu m^2$ was subjected to CVS at 3.1V to induce a soft BD capped at I_{gl}=1µA. Fig. 3(a) shows clear 2-level I_g RTN at V_g=0.9V, arising from the traps created in the percolation path. Then, the gate was ramped in the opposite polarity voltage from 0 to -2.5V. The I_g-V_g plot in Fig. 3(b) shows that the negative ramping voltage switched I_g from a low resistance state (LRS) after BD to a high resistance state (HRS) as compared to that of the fresh devices measured before BD. I_g measurement was repeated at V_g=0.9V and RTN disappeared as shown in Fig. 3(c), in addition to a significant reduction of I_g (I_g dropped from nA to pA level). The I_g versus time data from Figs. 3(a) and (c) were transformed to the frequency domain and the power spectral density (PSD) computed using Fast Fourier Transform [11] for better comparison. The corresponding RTN signals exhibit a Lorentzian spectrum with slope of γ=2 [1]. In Fig. 3(d), the I_g PSD directly measured after 1µA BD has a slope of γ=1.8 indicating strong existence of RTN, while the one after switching shows a slope γ~0, which is mainly due to white noise. Corresponding to the reduction in I_g, there is a reduction in the PSD magnitude by ~5 orders after switching.

Using the current separation method developed by imec [12], the 1µA BD location was determined to be S_{BD} = $L\times I_d/(I_s+I_d)$=0.99L (where L=channel length) from the source,

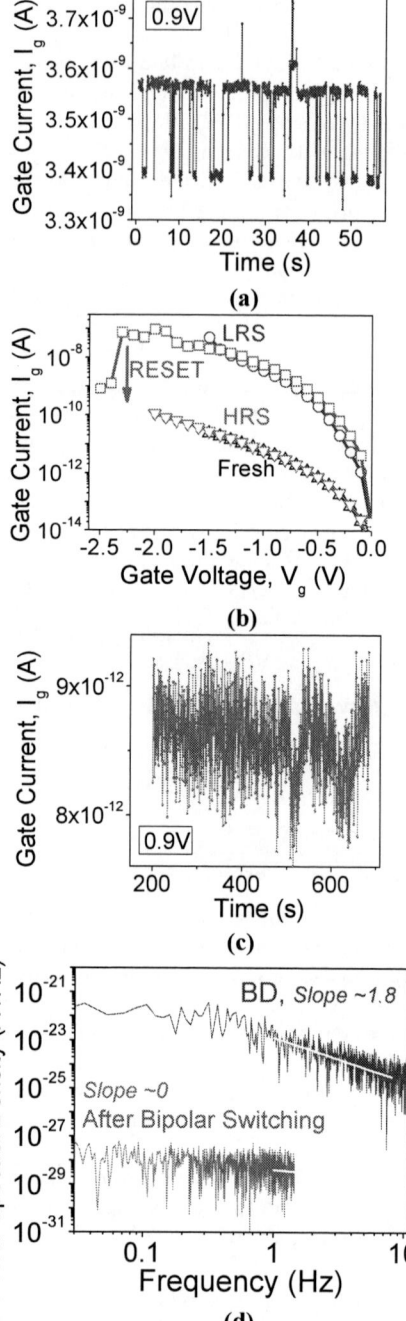

Figure 3. **(a)** Clear 2-level I_g RTN at V_g=0.9V, arising from the traps in the percolation path created from I_g=1µA BD in nMOSFET (NiSi FUSI /HfSiON(25Å)/SiOx(12Å) and W×L=1.0×0.13µm²). This indicates the existence of one active trap in dielectric layer in the percolation path. **(b)** The I_g-V_g trend shows a bipolar switching, after which I_g recovered fully to the fresh device condition. **(c)** At V_g=0.9V after switching, no I_g RTN signal was observed, in addition to a lower I_g level. **(d)** Comparison of I_g PSD between BD (Fig. 3(a)) and after switching (Fig. 3(c)). The PSD measured at BD shows a slope of γ=1.8, indicating the existence of Lorentzian RTN, while the one after switching is mainly attributed to white noise with γ~0. There is also a strong reduction in the PSD magnitude.

which means the percolation path was generated at the drain upon the BD event. However, after switching, the calculated S_{BD} shifted to the center of the channel, which typically suggests a non-BD case. This suggests that the traps in the percolation path close to the drain could be fully passivated by the bipolar switching, and therefore I_g recovers to almost the fresh device condition, leading to the disappearance of RTN in Fig. 3(c).

Fig. 4(a) shows a *partial* recovery of I_g after a bipolar switching in an I_{gl} =0.5μA BD 1×0.18 μm^2 nMOSFET of *Stack-A*. The HRS after switching is in between the BD and the fresh condition. I_g exhibits a 4-level RTN signal immediately after the BD as shown in Fig. 4(b), while RTN is hardly seen in I_g measured after a bipolar switching, initiated at -1.6 V (Fig. 4(c)). This implies that already a partial recovery through bipolar switching can achieve a significant reduction in RTN as well.

B. Partial Recovery from Bipolar Switching in the TiN/HfLaO stack

A *Stack-B* nMOSFET (W×L = 0.5×0.07μm^2) was subjected to CVS to induce an I_{gl}=2μA soft BD, after which the device was still functioning with a degraded performance. Following BD, in the RTN signal, 16 discrete levels were observed at V_g=1.3V as shown in Fig. 5(a). This indicates that there are at least 4 active traps present in the percolation path at V_g=1.3V. It is assumed that the 4 traps are independent of each other in the corresponding capture and emission events and each of the traps gives rise to a different RTN amplitude. Therefore, 2^4=16 levels of RTN were generated. It is easier to observe multi-level RTN in smaller dimension devices (e.g., smaller dimension in Fig. 5 against larger dimensions in Figs. 3 and 4). One possible explanation is that, for devices with smaller dimensions, the capture or emission of a single charge carrier can shows up as a more significant and measurable resistance change. A bipolar switching, as shown in Fig. 5(b), was applied to examine its effect on the multi-level RTN. The HRS achieved after a switching at -1.6V is near the fresh condition, in addition to a ~3 orders of magnitude reduction in I_g at V_g=-1V. For comparison, I_g is measured at V_g=1.3V again in Fig. 5(c), with only a 2-level RTN signal evident after the bipolar switching, instead of 16 levels. There was also a reduction in the magnitude of I_g by more than two orders. This suggests that the proposed bipolar switching effectively passivated 3 out of the 4 traps in the BD path. Moreover, there is no clear RTN measured at $V_g \leq 1.2$V as shown in Fig. 5(d) after switching. The reason is, after switching, the single trap accessed at 1.3V was no longer accessible at 1.2V for channel electrons due to a higher tunneling barrier. This implies that if the operating voltage $V_{op} \leq 1.2$V, there is no I_g-RTN problem after bipolar switching even if all the traps are not fully passivated.

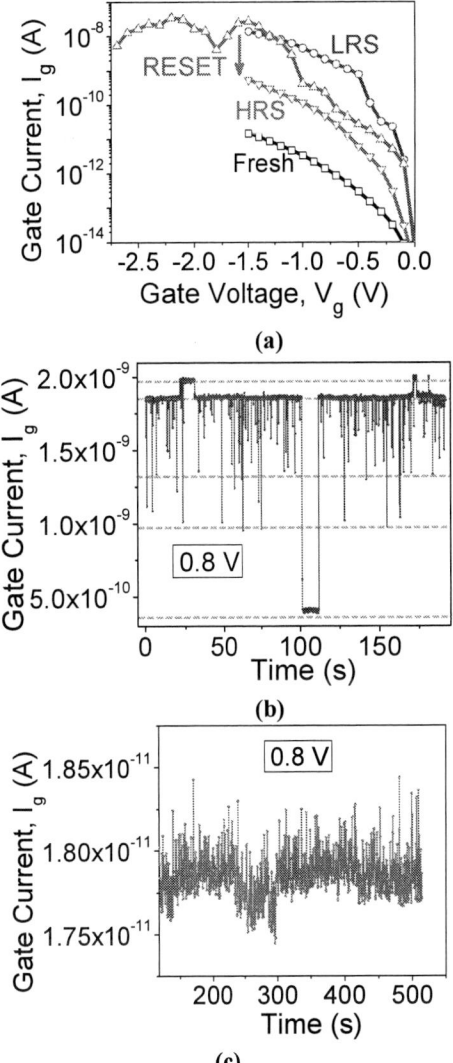

Figure 4. (a) A partial recovery of I_g after a bipolar switching in a I_{gl} =0.5μA BD 1×0.18μm^2 nMOSFET. The HRS of I_g after switching is in between the BD and fresh conditions. (b) A 4-level I_g RTN at V_g=0.8V, arising from traps in the BD path created upon a 0.5μA BD. (c) I_g is measured at V_g=0.8V again after a bipolar switching, and no RTN is observed. This implies that even a partial recovery can effectively reduce RTN.

C. Threshold Voltage Shift ΔV_T Recovered by Bipolar Switching

To study the effect of bipolar switching on the charging effect associated ΔV_T in degraded dielectrics, the transconductance G_m was measured for three stages: before stressing, after a 1μA soft BD, and after a bipolar switching. The G_m results in Fig. 6 show that G_m after a bipolar switching (Fig. 6 inset) recovered to the fresh condition which was confirmed by the negligible ΔV_T shift after switching.

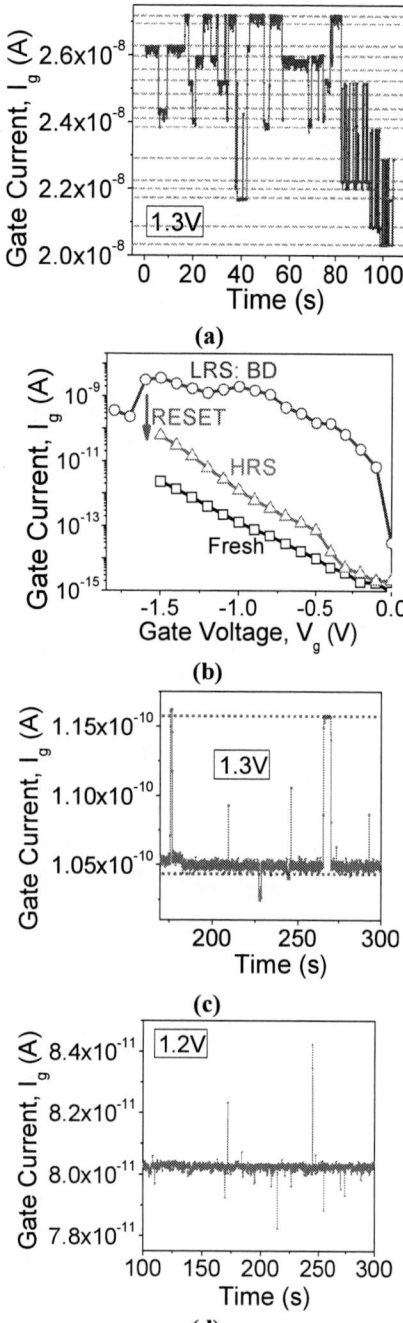

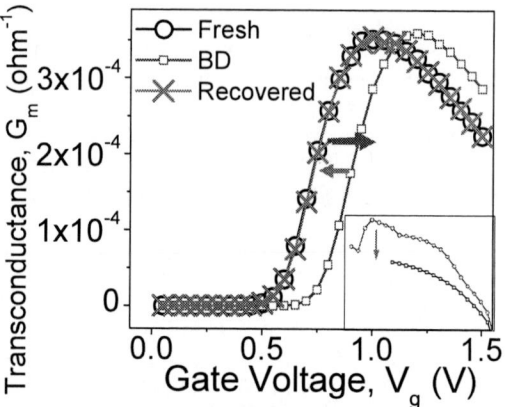

Figure 6. The G_m measured after bipolar switching recovers fully to the fresh condition. Thus, ΔV_T suffering from 1μA soft BD is eliminated. Inset shows the bipolar switching from post-BD high conduction to low conduction state.

D. Faster Traps and Slower Traps

Fig. 7(a) shows the I_g RTN at V_g=0.6V after I_{gl}=2μA BD, with both relatively faster and slower traps (long τ_c, τ_e) existing. After bipolar switching, only RTN from slower traps was observed as shown in Fig. 7(b). As faster traps are located near the oxide-Si or oxide-gate interface, this indicates that traps in the percolation path near to the interfaces of oxide are easier to be passivated by the bipolar switching. It is noted that the faster trap here has a ΔI_g of ~5nA, and this is different from the white noise shown in Figs.3(c) & (d). It is also noted that the magnitude and ΔI_g of the slower trap after bipolar switching are different from the ones measured before bipolar switching. Further study on this is necessary for a better understanding.

E. RTN Triggering Voltage V_{trig}, Improved by the Bipolar Switching

The triggering voltage V_{trig} for RTN is the V_g above which I_g RTN is initiated, which is the minimum voltage level that is required to access the RTN traps [13]. If a device is operated at $V_g < V_{trig}$, the device is immune to RTN noise, even with certain degradation mechanisms already present in the dielectric. It has been observed that as the I_{gl} increases (which means a harder BD), V_{trig} tends to drop to lower values [13]. A lower value of V_{trig} implies that I_g-RTN is observed at a lower V_g.

However, by reduction of RTN, bipolar switching increases the V_{trig}, thereby improving device immunity to the instability problems and performance degradation arising from RTN. The V_{trig} was obtained after BD and after bipolar switching were tested for more than 15 devices and the results plotted in Fig. 8. There is an obvious increase in V_{trig} after the bipolar switching, with at least an additional 0.2-0.3V margin before RTN arises, considering the lower and upper bound of V_{trig} before and after switching.

Figure 5. (a) 16-level I_g RTN at V_g=1.3V, arising from 4 traps (2^4=16) in the percolation path created from a I_{gl}=2μA BD in an nMOSFET (TiN/HfLaO(18Å)/p-Si and W×L=0.5×0.07μm²). (b) The I_g-V_g trend shows a bipolar switching, after which I_g recovers nearly to the fresh device condition. (c) I_g was measured again at V_g=1.3V after switching and only a 2-level RTN is observed. This indicates that 3 out of the 4 traps in the percolation path created by BD were passivated by bipolar switching. (d) I_g is measured at V_g=1.2V after switching, without RTN observed. As no RTN was observed below V_g=1.2V, this means that the remaining traps are no longer accessible for charge carriers at 1.2V. It implies that RTN is not a problem after bipolar switching in this case, if $V_{op} < 1.2V$.

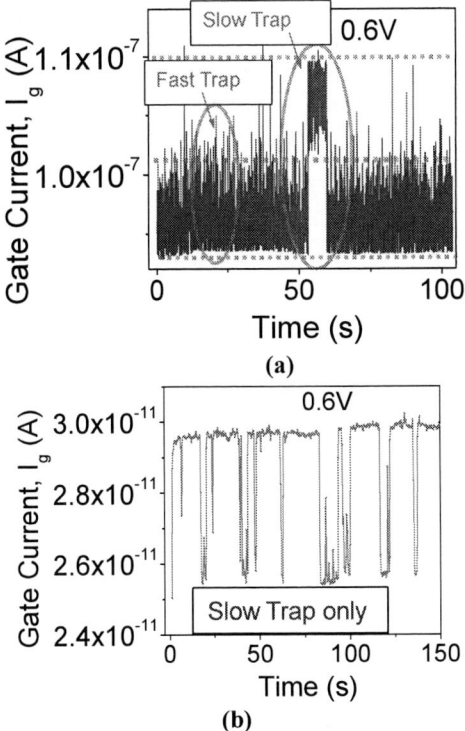

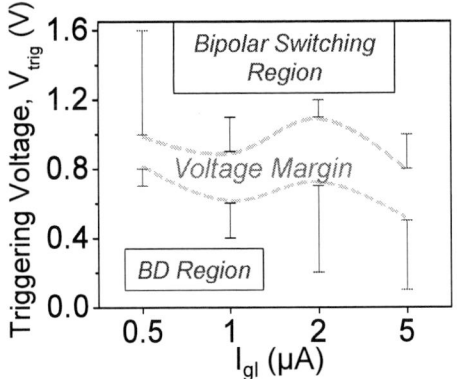

Figure 7. **(a)** I_g RTN at V_g=0.6V, containing both fast traps (time constant τ~25ms) and slow traps (τ~2-20s) comprising the percolation path for I_{gl}=2μA BD. **(b)** After a bipolar switching, only RTN from slow traps is observed, with the fast traps passivated. As the fast traps are located close to the oxide-Si or oxide-gate interface, this indicates that the traps near oxide interfaces are easier to be passivated by the bipolar switching.

Figure 8. The V_{trig} vs I_{gl} trend comparing V_{trig} measured after BD and after bipolar switching in a NiSi FUSI/HfSiON(25Å)/SiO$_x$(-12Å) stack. V_{trig} is the minimum V_g needed to initiate I_g RTN. Bipolar switching provides an additional 0.2-0.3V margin before RTN problem arises.

III. MODEL

Using electron energy loss spectroscopy (EELS) analysis in transmission electron microscope, it has been reported [14] that the traps giving rise to I_g RTN in degraded dielectrics could be oxygen vacancies. This has been supported by the RTN simulation results reported by Padovani *et. al.* [15]. During the

dielectric degradation process, oxygen vacancies are generated from bond breaking, and the detached oxygen ions drift towards the gate electrode consisting of NiSi or TiN which serves as a good oxygen reservoir [10]. With the aid of a reverse polarity gate bias, which is the bipolar switching mentioned above, the oxygen ions "stored" in the gate can drift back to the BD region in the dielectric and re-passivate the traps in the percolation path. Thus, a strong reduction in RTN and ΔV_T is observed after bipolar switching.

The model of moving oxygen ions has been applied to understand the oxide resistance change in resistive switching memory (RRAM) devices [16]. The Joule heating generated from the gate leakage may assist the movement of the oxygen ions [10] as well as in repairing the dangling bonds.

In the schematic shown in Fig. 9(a), 4 traps exist in a degraded high-κ dielectric layer, corresponding to the 16-level RTN in Fig. 5(a). After a bipolar switching, it is believed that oxygen ions move back and re-passivate 3 out of the 4 traps as shown in Fig. 9(b), effectively reducing RTN to only 2 levels shown in Fig. 5(c).

IV. PERFORMANCE BOOSTING TECHNIQUE

Based on the effective reduction of RTN and ΔV_T by bipolar switching in degraded dielectrics, a *performance boosting technique* is proposed. This technique comprises applying a negative polarity voltage sweep (-V_g for nMOSFET) on the gate after certain operation duration, as shown in Fig. 10(b). The $|V_g|$ applied is small and found to be ≤ 2.5V for (25Å)HfSiON + (12Å)IL stack and ≤1.6V for (18Å)HfLaO

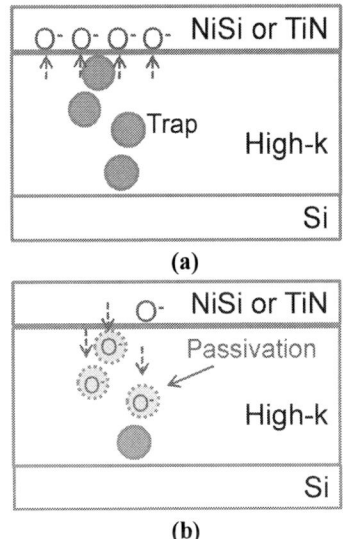

Figure 9. The reduction in RTN is modeled by moving oxygen ions passivating traps in the percolation path. 4 traps in the dielectric layer corresponding to a 16-level RTN in Fig. 5(a) are shown. (b) 3 out of 4 traps in the percolation are assumed to be passivated by moving oxygen ions [7]. As a result, only 2-level RTN was observed due to the remaining trap, in Fig. 5(c), after passivation from bipolar switching.

978-1-4244-9113-1/11 $26.00 © 2011 IEEE   187

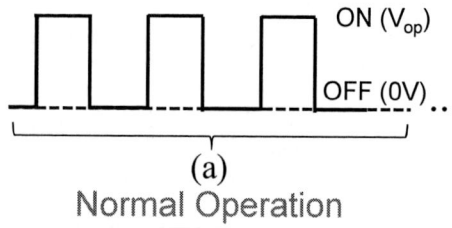

(a)
Normal Operation

Performance Boosting
Technique (b)

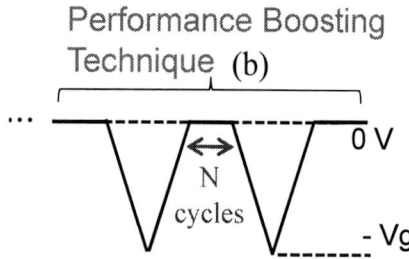

Figure 10. (a) Normal operation of an nMOSFET. (b) The proposed *performance boosting technique* is simply by sweeping the device with $-V_g$ to induce bipolar switching. This technique is inserted after a long duration of the device normal operation (e.g., after N cycles of switching between ON and OFF states as indicated by the double arrow). ($|V_g| \leq 2.5$V for 25Å HfSiON+12Å IL stack and $|V_g| \leq 1.6$V for 18Å HfLaO stack.) Comparing with FGA, this technique is practical and easy to reduce RTN generated after prolonged normal device operation.

stack. Compared to FGA, this technique is much more practical to reduce RTN and ΔV_T by passivating the stress induced traps generated after a certain time of device operation (leading to percolation BD), thereby boosting the performance after certain operational duration.

V. CONCLUSION

I_g-RTN and ΔV_T in degraded high-κ dielectrics are significantly reduced by bipolar switching with a small reverse polarity gate voltage sweep (negative V_g for nMOSFETs), regardless of full or partial recovery. In addition, the triggering voltage V_{trig} for RTN is increased by 0.2-0.3 V after the bipolar switching, giving an additional operating voltage margin before RTN problem arises. The effect of bipolar switching is modeled as a re-passivation of traps in the percolation path by moving oxygen ions in high-κ dielectrics. The oxygen ions are temporarily "stored" in the oxygen soluble gate electrode [17] material (NiSi and TiN in this work).

The I_g RTN reduction achieved in this work is different from the previous research by applying substrate bias, where I_d RTN is reduced through modulating RTN parameters such as capture time constant τ_c or emission time constant τ_e. Through trap passivation from bipolar switching, the reduction in RTN and leakage I_g is permanent and it does not depend on the subsequent bias configuration applied to the device. A performance boosting technique based on bipolar switching is

proposed to reduce RTN and ΔV_T problems arising from degradation during device operation and to boost and prolong device performance. It is possible that this technique be programmed in the circuit for RTN reduction at the circuit level, and this can be one of the tools for design-for-reliability (DFR) initiative in HK-MG stacks.

REFERENCES

[1] M. J. Kirton and M. J. Uren, "Noise in Solid-State Microstructures – A New Perspective on Individual Defects, Interface States and Low-Frequency (1/f) Noise" *Adv. Phys.*, **38**, pp. 367-468 (1989)

[2] S. Lee, H-J. Cho, Y. Son, D. S. Lee, and H. Shin, "Characterization of oxide traps leading to RTN in high-k and metal gate MOSFETs," in 2009 *Proc. IEDM Tech. Digest.*, pp.763-766.

[3] S.S. Chung and C. M. Chang, "The investigation of capture/emission mechanism in high-k gate dielectric soft breakdown by gate current random telegraph noise approach", *App. Phys. Lett* 93, 213502 (2008).

[4] N. Tega, H. Miki, F. Pagette, D. J. Frank, A. Ray, M. J. Rooks, W. Haensch, and K. Torii, "Increasing threshold voltage variation due to random telegraph noise in FETs as gate length scale to 20 nm", in 2009 *Proc. VLSI Tech. Digest.*, pp.50-51.

[5] M. Agostinelli, J. Hicks, J. Xu, B. Woolery, K. Mistry, K. Zhang, S. Jacobs, J. Jopling, W. Yang, B. Lee, T. Raz, M. Mehalel, P. Kolar, Y. Wang, J. Sandford, D. Pivin, C. Peterson, M. DiBattista, S. Pae, M. Jones, S. Johnson, and G. Subramanian, "Erratic fluctuations of SRAM cache V$_{min}$ at the 90 nm process technology node," in *IEDM Tech. Dig.*, 2005, pp. 655–658.

[6] N. Tega, H. Miki, Z. Ren, C. P. D'Emic, Y. Zhu, D. J. Frank, J. Cai, M. A. Guillorn, D.-G. Park, W. Haensch, and K. Torii, "Reduction of Random Telegraph Noise in High-κ / metal-gate Stacks for 22 nm Generation FETs", in 2009 *Proc. IEDM Tech. Digest.*, pp.771-774.

[7] N. B. Lukyanchikova, M. V. Petrichuk, N. P. Garbar, E. Simoen, and C. Claeys, "Influence of the substrate voltage on the random telegraph signal parameters in submicron n-channel metal–oxide–semiconductor field-effect transistors under a constant inversion charge density," *Appl. Phys. A*, vol. 70, no. 3, pp. 345–353, Mar. 2000.

[8] W. Shockley and J. W. T. Read, "Statistics of the recombination of holes and electrons," *Phys. Rev.*, vol. 87, no. 5, pp. 835–842, Sep. 1952.

[9] N. Zanolla, D. Siprak, M. Tiebout, P. Baumgartner, E. Sangiorgi, C. Fiegna, "Reduction of RTS Noise in Small-Area MOSFETs Under Switched Bias Conditions and Forward Substrate Bias", *IEEE Trans. Electron. Dev.*, vol.57, No.5, May 2010.

[10] W. H. Liu, K. L. Pey, X. Li and M. Bosman, "Observation of Switching Behaviors in Post-Breakdown Conduction in NiSi-gated Stacks", in 2009 *Proc. IEDM Tech. Digest.*, pp.123-126.

[11] A. K. Raychaudhuri, "Measurement of 1/f noise and its application in materials science," *Curr. Opinion Solid State Mater. Sci.*, vol. 6, no. 1, pp. 67–85, Feb. 2002.

[12] R. Degraeve, B. Kaczer, A. D. Keersgieter, G. Groeseneken, "Relation between breakdown mode and breakdown location in short channel NMOSFETs and its impact on reliability specifications", in 2001 *Proc. IRPS*, pp.360-366.

[13] W. H. Liu, K. L. Pey, N. Raghavan, X. Wu, and M. Bosman, "Triggering Voltage for Post-Breakdown Random Telegraph Noise in High-κ Metal Gate MOSFETs Due to Tunneling Barriers", *Unpublished*, (2011).

[14] X. Li, C. H. Tung, K. L. Pey, and V. L. Lo, "The physical origin of random telegraph noise after dielectric breakdown", *App. Phys. Lett* 94, 132904 (2009).

[15] A. Padovani, L. Morassi, N. Raghavan, L. Larcher, W.H. Liu, K. L. Pey, and G. Bersuker, "A physical model for post-breakdown digital gate current noise", *IEEE Electron Dev. Lett.*, Vol. 31, No. 9, pp.1032-1034, (2010).

[16] R. Meyer, L. Schloss, J. Brewer, R. Lambertson, W. Kinney, J. Sanchez, and D. Rinerson, " Oxide Dual-Layer memory Element for Scalable Non-Volatile Cross-Point Memory Technology"*Proc. 9th Annual Non-Volatile Memory Tech. Symp.* pp. 54 (2008).

[17] N. Raghavan, K. L. Pey, X. Wu, W. H. Liu, X. Li, M. Bosman, and T. Kauerauf, "Oxygen soluble gate electrodes for prolonged high-κ gate stack reliability", *IEEE Electron Dev. Lett.*, In press.

Correlation of I_d- and I_g-Random Telegraph Noise to Positive Bias Temperature Instability in Scaled High-κ/Metal Gate n-type MOSFETs

Chia-Yu, Chen[+], Qiushi Ran[+], Hyun-Jin Cho[*], Andreas Kerber[**], Yang Liu[+], Ming-Ren Lin[*], and Robert W. Dutton[+]

[+]Department of Electrical Engineering, Stanford University, Stanford, CA, 94305, U.S.A.
Tel: +1-(650) 8623663, Email: yu0528@stanford.edu

[*]Strategic Technology Group, GLOBALFOUNDRIES Inc., Sunnyvale, CA, 94088, U.S.A.

[**]T. J. Watson Research Center, GLOBALFOUNDRIES Inc., Yorktown Heights, NY, 10598, U.S.A.

Abstract—**Random telegraph noise (RTN) in high-κ nMOSFETs is directly linked to Positive Bias Temperature Instability (PBTI). For the first time, the correlation between I_d- and I_g-RTN is clearly observed in high-κ MOSFET. I_g-RTN is directly related to physical trapping or de-trapping and the I_d-RTN reflects sensitivity to charge trapping as determined by gm, which is confirmed by both experiments and TCAD simulations.**

Keywords-Random telegraph noise; high-κ; Metal gate; scaling; MOS transistor; PBTI

I. INTRODUCTION

There is growing concern regarding random telegraph noise (RTN) in small-area MOS devices [1] [2]. Yet a detailed understanding of the physical mechanisms involved in RTN remains elusive, including consideration of devices containing high- (HK) dielectrics. For HK devices, positive bias temperature instability (PBTI) has previously been reported [3] and attributed to negative charge trapping in the HK dielectric, causing positive threshold voltage shift and simultaneously increasing gate current (I_g), in the phenomenon known as SILC (Stress-Induced Leakage Current) [4]. This paper focuses on the mechanism of RTN in HK devices and correlates it to the PBTI behavior, as both phenomena share the same physical origin: charge-trapping in dielectric. For the first time, the correlation of PBTI, I_g-RTN, and I_d-RTN in HK n-type MOSFETs is investigated and the physics is clarified, based on comparisons with measured noise data and TCAD simulation results.

II. EXPERIMENTAL RESULTS

Metal gate devices with Hafnium-based dielectrics fabricated on SOI substrates with size width/gate length=70nm/40nm are studied [5]. Measurements are performed in the linear drain current regime at room temperature.

A. PBTI measurements

PBTI test was done by measuring the device with normal situation at first, then applying a 1.8V voltage at the gate electrode for 1000 seconds as the positive stress, after that a -1.8V bias was added as the negative stress to form a recovery circle. The I_d/I_g-V_g characteristics measured at three conditions (before stress, after positive stress, and after the negative recovery cycle) are shown in Fig. 1 [4].

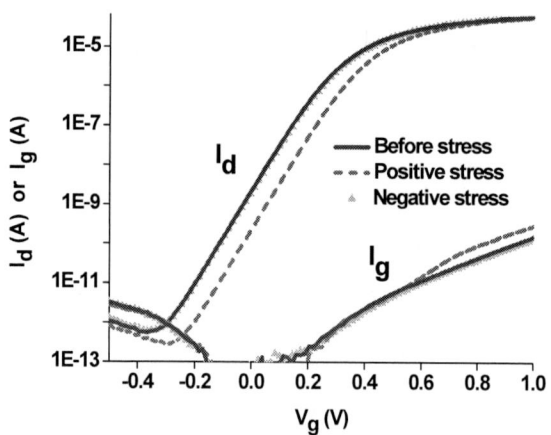

Figure 1. Current-Voltage characteristics of a high-κ nMOSFET (W/L=450nm/30nm) before (solid line) and after stress. Stresses are conducted with positive stress (1.8V, 1000 sec., dashed line) and followed by negative stress (-1.8V, 1000sec., triangle). Notice that large V_{th} shift and increased I_g occur after positive stress. Both V_{th} shift and high gate current recover during negative stress.

Three phenomena are noticed from the figure. First, Stress-Induced Leakage Current (SILC) is clearly observed in the

978-1-4244-9113-1/11 $26.00 © 2011 IEEE

stressed devices. Second, V_{th} is positively shifted after the positive stress (decrease of I_d). Third, both additional I_g and V_{th} shift are recovered after negative stress (triangle mark in Fig. 1). During the stress condition, electrons are trapped in the dielectric, triggering trap-assisted tunneling (TAT) and resulting in higher I_g. The trapped charge influences the gate potential and shifts threshold voltage. When charges are de-trapped (or recovered), both SILC and the threshold voltage shift disappear. The trap location is in the high-κ dielectric stack; therefore mobility degradation due to remote Coulomb scattering is directly related to interfacial layer thickness [6] [7].

B. I_g-RTN results

During the time domain measurement of both I_g and I_d (Fig. 2), I_g shows a two level stochastic behavior, which is a signature of RTN. The high I_g current step corresponds to a trapped state and the low I_g current step reflects a de-trapped state. This is consistent with previous PBTI results, as the trapped charge increases I_g. In addition to I_g–RTN, Fig. 2 also shows that higher I_g corresponds to lower I_d and vice versa. This result confirms that negatively trapped charge increases threshold voltage, resulting in a decrease of I_d, and at the same time increases I_g, which is consistent with the PBTI results.

Notice that many devices have the same gate bias (V_g) dependence: for I_g-RTN, the ratio of average time spent in the upper I_g state (τ_+) to average time spent in the lower I_g state (τ_-) increases with increasing V_g as shown in Fig. 3. We attribute the gate bias dependence to a narrowing of the energy gap between quasi-Fermi level (E_f) in the channel and trap level (E_t) in the oxide as shown in Fig. 4.

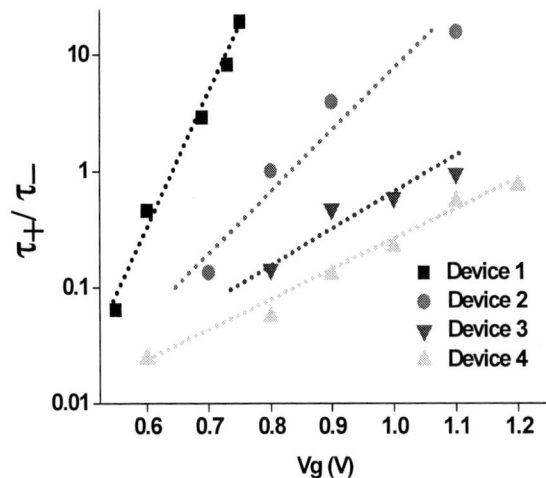

Figure 3. The ratio of average time spent in the upper gate current state to average time spent in the lower I_g state ratio with respect to gate bias for high-κ n-MOSFETs (W/L = 70nm/40nm).

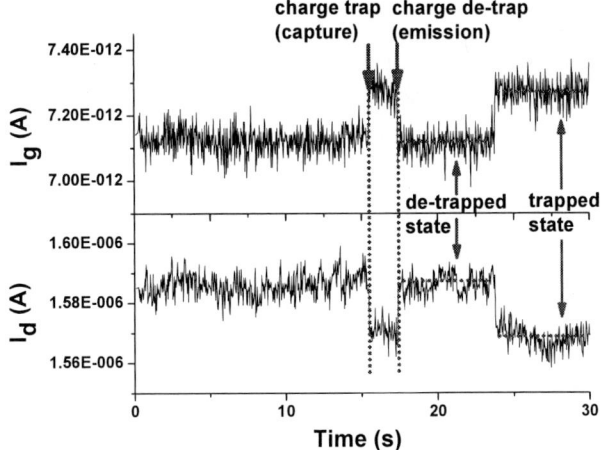

Figure 2. Drain and gate current noise signal. Notice that the drain and gate currents track each other. Electron trapping (capture) increases I_g and decreases I_d. Electron de-trapping (emission) recovers both I_g and I_d (W/L = 70nm/40nm).

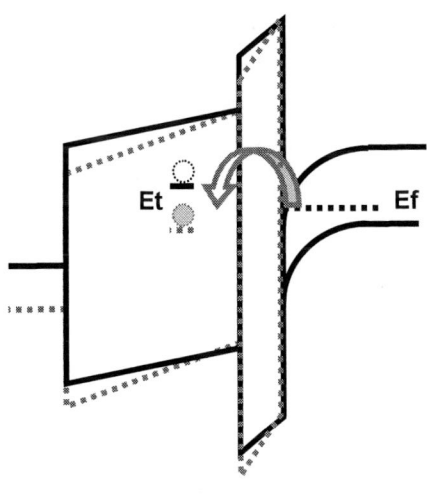

Figure 4. Higher gate bias narrows the energy gap between quasi-Fermi level and trap level, resulting in higher probability of capturing electrons.

C. I_d- and I_g-RTN results

I_d- and I_g-RTN behavior with respect to V_g was analyzed further as shown in Fig. 5. The measurements are done in the same device and different gate bias conditions. From Fig. 5, it appears that I_g-RTN is manifest at most V_g conditions whereas I_d-RTN disappears at either low or high V_g conditions. This can be explained by the maximum sensitivity of the device to trapped charges at peak g_m, which is around half V_{dd}.

978-1-4244-9113-1/11 $26.00 © 2011 IEEE   191

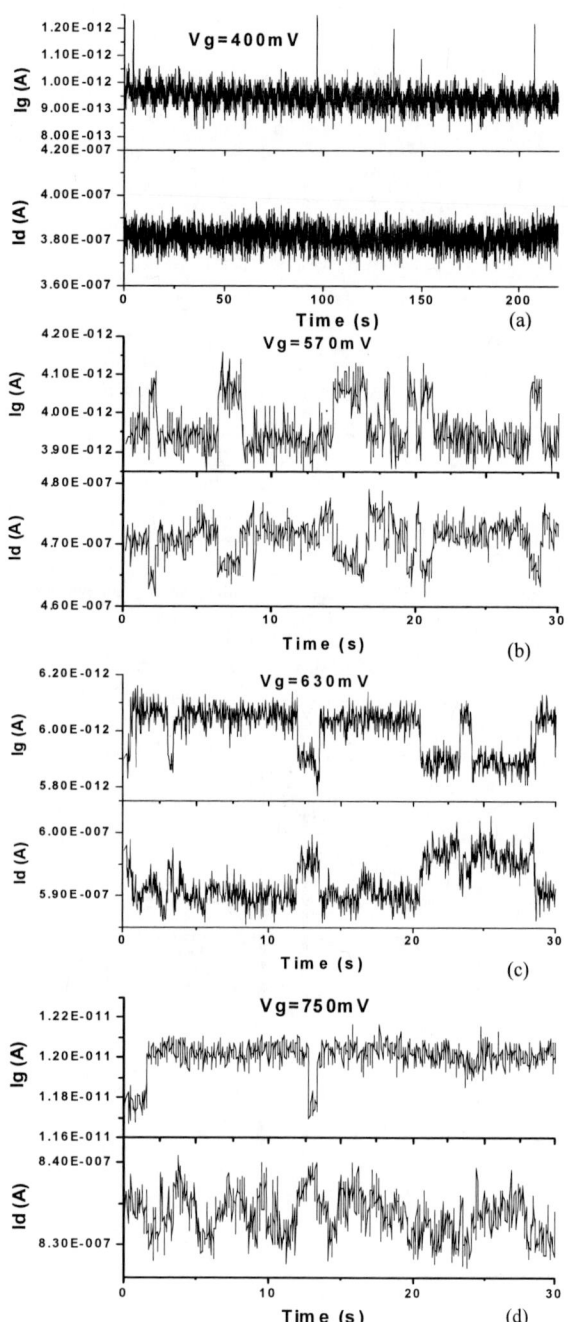

Figure 5. I_d-RTN and I_g-RTN measured in different gate bias conditions (W/L = 70nm/40nm). A high correlation between I_d- and I_g-RTN is observed in (b) and in (c); no correlation in (a) and in (d).

I_g-RTN comes directly from the physical event of charge trap/de-trap phenomena and can be observed at most V_g conditions. On the contrary, I_d-RTN comes from the output characteristics of the single stage transistor amplifier, with I_g-RTN providing the input noise source. In low g_m bias conditions, other noise components become dominant and distort I_d-RTN. Fig. 6 shows the gate bias dependence of I_d-

RTN as well as the noise spectra extracted from the Fourier transform of the time domain signal. A clear I_d-RTN and corresponding Lorentzian curve $(1/f^2)$ appear at V_g=0.57V; both disappear for operation with low g_m.

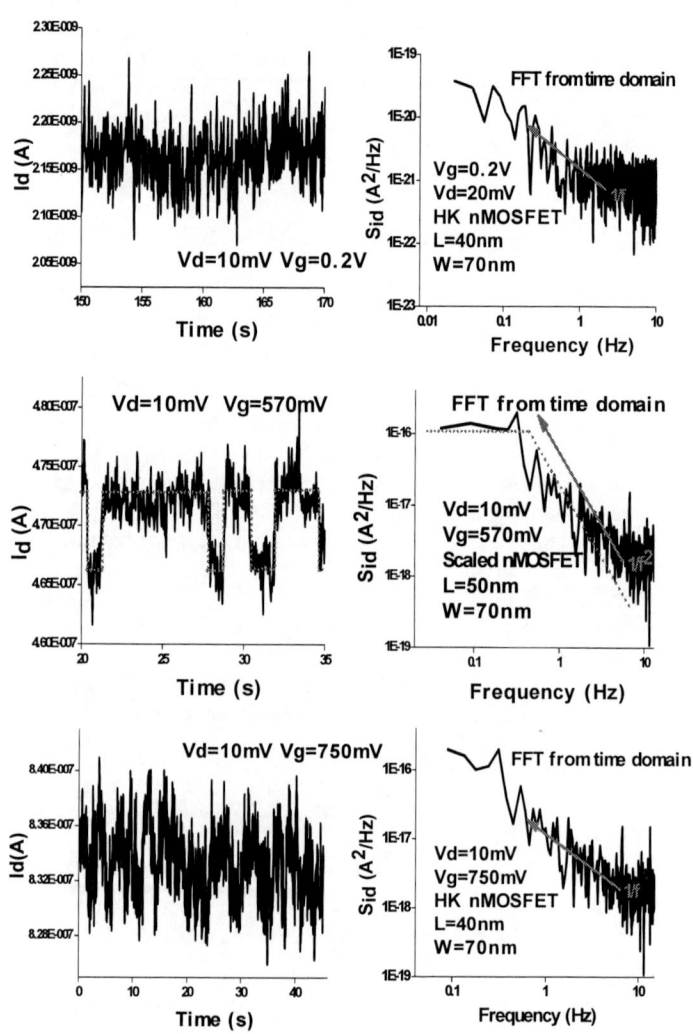

Figure 6. Gate bias dependence of I_d-RTN and corresponding noise spectra extracted from Fourier transform of time domain signals (W/L = 70nm/40nm).

III. TCAD SIMULATIONS

To understand the gate bias dependence of I_d-RTN, TCAD simulations are developed based on the McWhorter theory [8]. The simulation schematic is illustrated in Fig. 7 (a). The noise source is modeled as traps inside the HK layer. Through the impedance field method [9] the trap/de-trap fluctuation is correlated to the drain current noise. Fig. 7 (b) and (c) show the trap in both energy space and real space. Gaussian functions with narrow characteristic length are used to define one trap location in space and in energy. In addition to the trapping/de-trapping fluctuations from the gate oxide, the Hooge mobility fluctuation [10] from bulk phonon scattering was incorporated as shown in Fig. 7 (d).

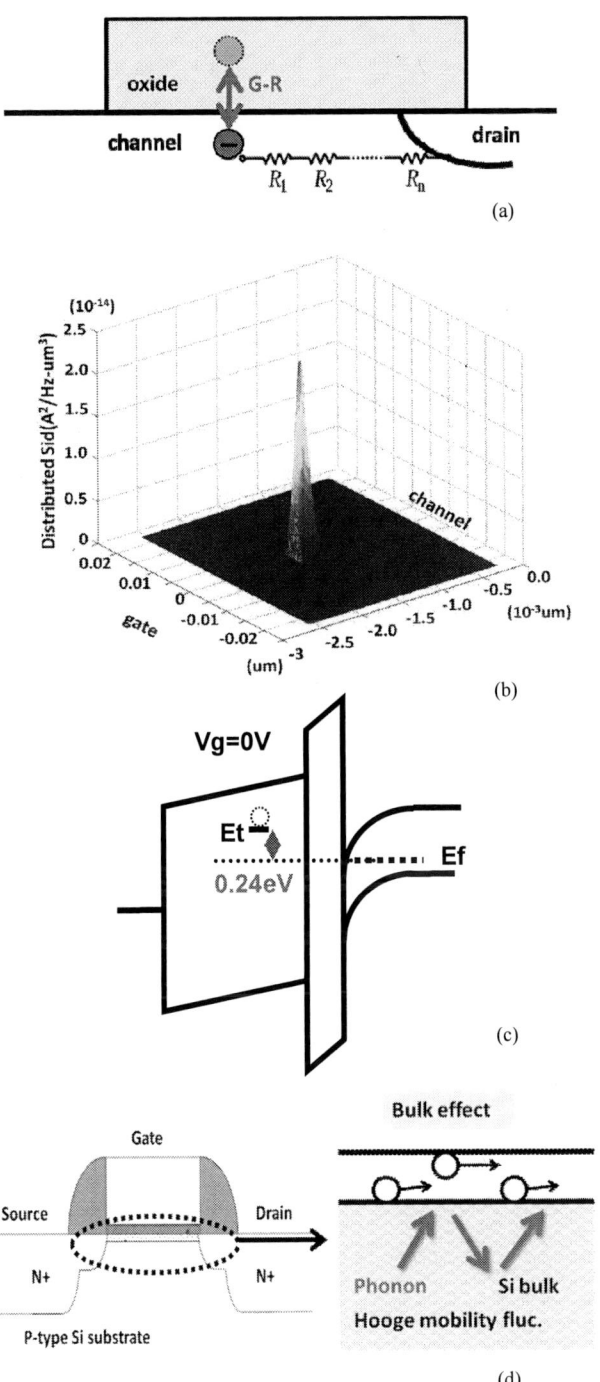

fluctuation is amplified, thus there is a clear I_d-RTN signal at the drain terminal. When the V_g is biased in low g_m regime (0.1V and 1.1V), Hooge mobility fluctuations overcome the RTN effect at the drain. In this condition, instead of RTN, 1/f noise is shown at the drain, which is consistent with the measurement results (Fig. 6).

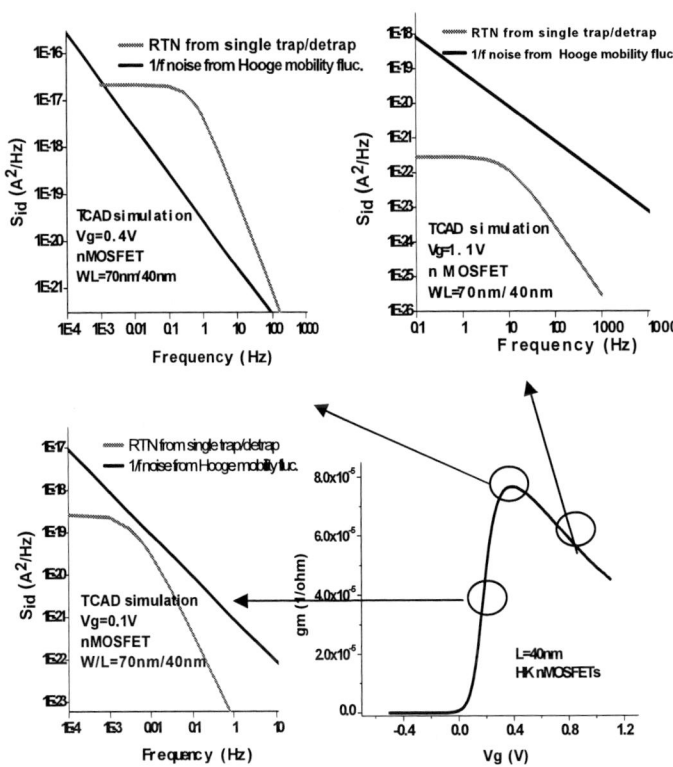

Figure 8. Noise simulation results in different gate bias conditions. The low frequency noise mechanism changes between the Hooge mobility fluctuation (1/f) and RTN (Lorentzian spectrum) depending on the g_m value (W/L=70nm/40nm).

Figure 7. Random telegraph noise simulation: (a) shows the impedance field method used in TCAD simulator, PROPHET. (b) and (c) are the single trap located in energy space and in real space. (d) the schematic of the Hooge mobility fluctuation originating from bulk phonon scattering.

Fig. 8 shows the noise simulation results in different V_g conditions. When the device is biased in the high g_m regime (V_g=0.4V), RTN becomes dominant at the drain. For the high g_m condition, the I_g-RTN caused by oxide trapping/de-trapping

IV. STATISTICAL PROPERTY

To study the statistical properties of RTN, measurements have been made with over 1000 small area devices (W/L=70nm/40nm). Fig. 9 shows four representative cases: 1) no RTN, 2) I_g-RTN only, 3) I_d-RTN only and 4) both I_d-/I_g-RTN. The physical mechanism involved in case 3 (I_d-RTN only) is still an open issue.

Fig. 10 shows the statistical properties of the RTN distribution. In the measurements V_d is 10mV and V_g is 570 mV (the maximum g_m regime). For advanced fabrication processes technology (W/L=70nm/40nm), about 12% of the devices show random telegraph noise with 9% in I_g-RTN, 2% I_d-RTN and 1% both in I_d-/I_g- RTN. Through the statistical data we can observe that RTN already induces device variability in gate and drain current which will become important when considering SRAM yield and other device reliability issues.

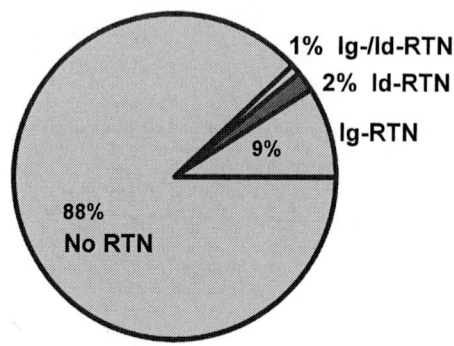

Figure 10. About 12% of n-MOSFETs (W/L=70nm/40nm) show RTN: 9% Ig-RTN, 2% Id-RTN, and 1% Ig-/Id-RTN. The statistical results are from 1000 samples with the same technology and structures.

V. CONCLUSIONS

Firstly, RTN in HK n-MOSFETs is directly linked to PBTI. PBTI and RTN originate from the same physical process: charge trapping in the HK dielectric. For the first time, the correlation between I_d- and I_g-RTN is clearly observed. I_g-RTN is directly related to physical trapping / de-trapping and the I_d-RTN reflects sensitivity to charge trapping as determined by g_m.

REFERENCES

[1] N. Tega, H. Miki, F. Pagette, D. J. Frank, A. Ray, M. J. Rooks, W. Haensch, and K. Torii,"Increasing Threshold Voltage Varaiation due to Random Telegraph Noise in FETs as Gate Lengths Scale to 20 nm," *Symp. VLSI Tech. Dig.*, pp. 50-52, 2009.

[2] C.M. Chang, S.S. Chung, Y.S. Hsieh, L.W. Cheng, C.T. Tsai, G.H. Ma, et al. "The observation of trapping and detrapping effects in high-κ gate dielectric MOSFETs by a new gate current random telegraph noise approach," *IEEE International Electron Devices Meeting*, p. 787, 2008.

[3] A. Kerber, E. Cartier, L. Pantisano, R. Degraeve, T. Kauerauf, Y. Kim, A. Hou, G. Groeseneken, H. E. Maes, and U. Schwalke,"Origin of the Threshold Voltage Instability in SiO₂/HfO₂ Dual Layer Gate Dielectrics," *IEEE Elec. Dev.Lett.*, Vol. 24, No. 2, p87, 2003.

[4] E. Cartier, and Andreas Kerber, "Stress-Induced Leakage Current and Defect Generation in nFETs with HfO₂/TiN Gate Stacks during Positive-Bias Temperature Stress," *Proc. IEEE International Reliability Physics Symposium*, p. 486, 2009.

[5] M. Chudzik, B. Doris, R. Mo, J. Sleight, E. Cartier, C. Dewan, D. Park, H. Bu, W. Natzle, W. Yan, C. Ouyang, K. Henson, D. Boyd, S. Callegari, R. Carter, D. Casarotto, M. Gribelyuk, M. Hargrove, W. He, Y. Kim, B. Linder, N. Moumen, V.K. Paruchuri, J. Stathis, M. Steen, A. Vayshenker, X. Wang, S. Zafar, T. Ando, R. Iijima, M. Takayanagi, V. Narayanan, R. Wise, Y. Zhang, R. Divakaruni, M.Khare, T.C. Chen, "High-Performance High-κ/Metal Gates for 45nm CMOS and Beyond with Gate-First Processing," *IEEE Symp. VLSI Tech.*, pp194-195, 2007.

[6] K. Maitra, M. M. Frank, V. Narayanan, V. Misra, and E. Cartier, "Impact of metal gates on remote phonon scattering in titanium nitride/hafnium dioxide n-channel metal–oxide–semiconductor field effect transistors-low temperature electron mobility study," *J. Appl. Phys.*, vol. 102, no. 11, p. 114507, Dec. 2007.

[7] M. Cassé, L. Thevenod, Bernard Guillaumot, L. Tosti, François Martin, J. Mitard, O. Weber, F. Andrieu, Thomas Ernst, G. Reimbold, T. Billon, M. Mouis, and F. Boulanger, "Carrier Transport in HfO₂/Metal Gate MOSFETs: Physical Insight Into Critical Parameters," *IEEE Trans. Elec. Dev.*, Vol. 53, no. 4, 2006.

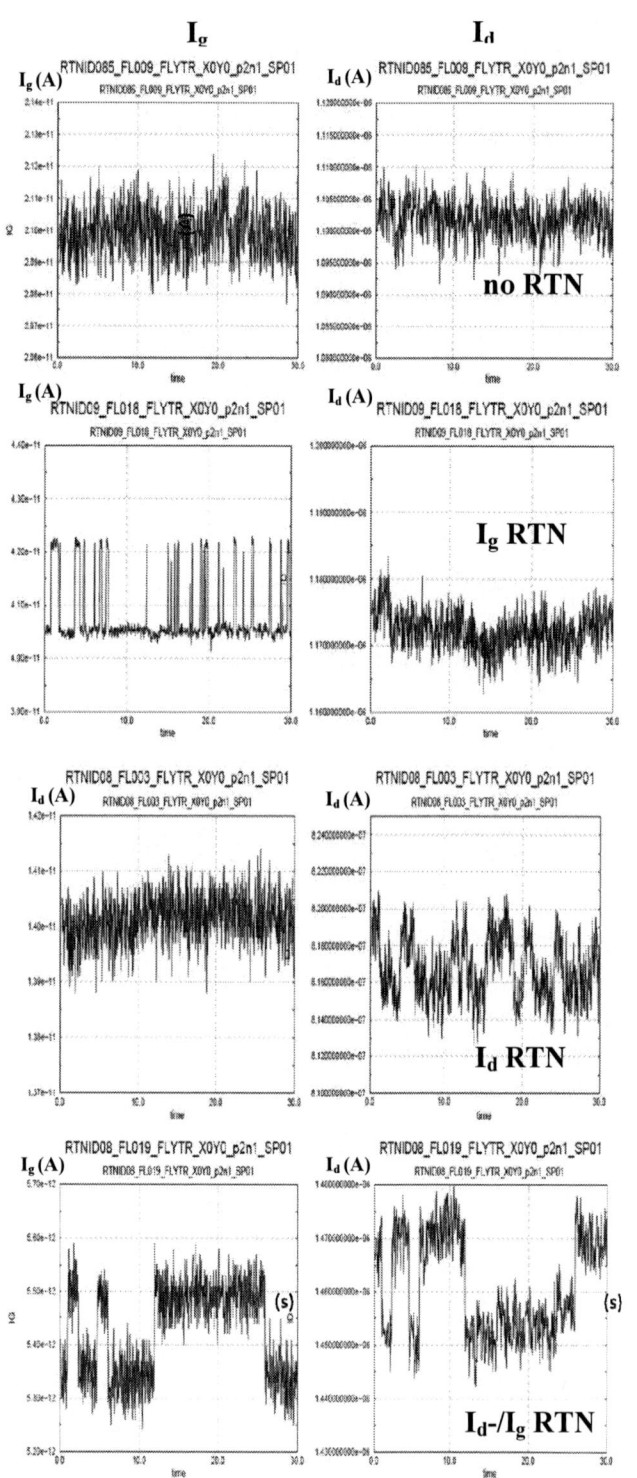

Figure 9. Drain and gate current noise signal. Notice that the drain and gate currents track each other. Electron trapping (capture) increases I_g and decreases I_d. Electron de-trapping (emission) recovers both I_g and I_d (W/L = 70nm/40nm).

[8] A. McWhorter, "1/f noise and germanium surface properties," in Semiconductor Surface Physics. Philadelphia, PA: Univ. Pennsylvania Press, pp. 207–228, 1957.

[9] Y. Liu, S. Cao, and R. W. Dutton, "Numerical investigation of low frequency noise in MOSFETs with high-κ gate stacks," *in Proc. SISPAD*, pp. 99–102, 2006.

[10] C.-Y. Chen, Y. Liu, R. W. Dutton, Junko Sato-Iwanaga, Akira Inoue, and Haruyuki Sorada, "Numerical Study of Flicker Noise in p-Type $Si_{0.7}Ge_{0.3}$/Si Heterostructure MOSFETs*", IEEE Trans. Elec. Dev.*, Vol. 55, No. 7, pp. 1741-1748, 2008.

[11] L. D. Yau and C.-T. Sah, "Theory and experiments of low-frequency generation- recombination noise in MOS transistors," *IEEE Trans. Elec. Dev.*, Vol. 16, no. 2, pp. 170-177, 1969.

SILC-Based Reassignment of Trapping and Trap Generation Regimes of Positive Bias Temperature Instability

J.Q. Yang[1,2], M. Masuduzzman[1], J.F. Kang[2], M.A. Alam[1]

[1]Department of ECE, Purdue University, West Lafayette, IN 47907, USA
[2]Institute of Microelectronics, Peking University, Beijing, 100871, P.R. China
phone: 765-494-5988, fax: 765-494-6441, e-mail: {yang259, mmasuduz, alam}@purdue.edu

Abstract— We report a simple but effective SILC-based methodology to separate and identify trapping and trap generation dominated regimes of positive bias temperature instability (PBTI). We use theoretical model and experiments to demonstrate that the sign for stress induced leakage current (SILC) reverses as PBTI transitions from trapping to trap generation dominated regimes; this is in contrast to threshold voltage shift with no corresponding sign reversal. SILC crossover methodology further verifies that initial and fast saturated trapping is temperature independent while trap generation is voltage and temperature activated. The SILC-based reassignment not only indentifies trapping and trap generation regimes of PBTI, but also suggests a remarkable universality of trap generation in wide variety of High-k samples.

Keywords-positive bias temperature instability (PBTI), stress induced leakage current (SILC), Trapping, Trap Generation

I. INTRODUCTION

With the recent integration of High-k (HK) dielectric in the gate stack, Positive Bias Temperature Instability (PBTI) has reemerged as an important reliability concern for ultra-scaled transistors. Despite industry-wide characterization of various aspects of PBTI phenomena and a general consensus regarding its empirical features, detailed physical interpretation of the degradation kinetics is not fully understood [1-9]. Indeed, although there is a broad consensus that charging of pre-existing traps (*Trapping-TP*) and/or newly generated traps in the bulk oxide (*Trap generation-TG*) are responsible for the PBTI degradation [1-9], the identification of the (time) regimes dominated by the respective processes remains challenging; and this ambiguity leads to different types of PBTI models in the literature. Various version of TP-only theories [1-4,6] suggest a two step process: electrons are first trapped into preexisting bulk traps and they subsequently migrate to other pre-existing traps [5]. Meanwhile, other groups observe steeper increase of PBTI degradation at very long time stress [2] or at shorter time, but at extreme stress conditions [1,3]. This additional contribution is attributed to TG; indeed, CP measurements also confirm generation of new traps within interface layer or both in interface layer and high-k films [7,8] during PBTI. The phenomenology and mechanism of TP and TG are so different that the detailed physical model appropriate for reliability projection and their correlation need further discussions. The above uncertainty

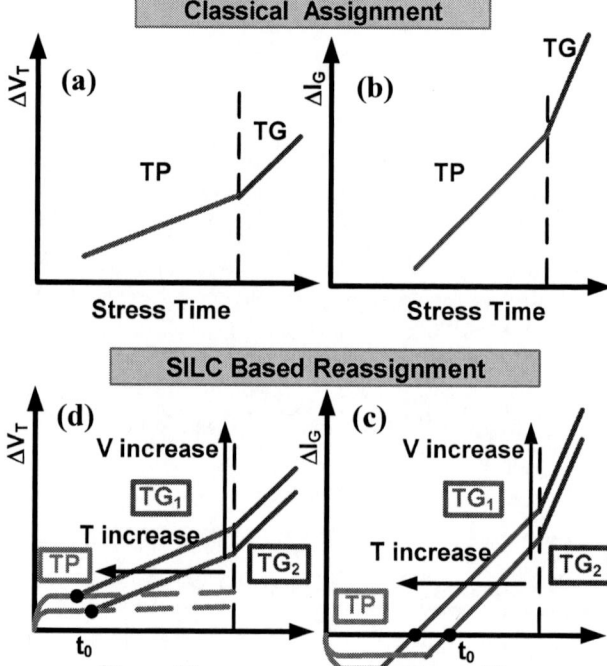

Figure 1. The different degradation rates (different time exponent, n) of (a) ΔV_T and (b) ΔI_G are sometimes interpreted as due to Trapping (TP) and Trap generation (TG). (c) However, although TP and TG have same sign for ΔV_T, they have opposite sign (see Fig. 3, 4) for ΔI_G. This allows us to separate TP from TG. Since TP is characterized by negative SILC, any change in exponent n during the positive SILC must be due to different trap generation mechanisms (TG1, TG2, etc.) as summarized in (d).

arises primarily because PBTI is typically characterized by threshold voltage shift ΔV_T degradation, to which both TP and TG contribute with same sign (increase).

In this paper, we make the following contributions:

(i) Identify that although the *sign* of degradation is *same* for ΔV_T due to TP and TG, it must be *opposite* for stress induced leakage current (SILC) $\Delta I_G/I_G$ experiments (distinguished from the model in [2,3]) (see Fig. 1),

(ii) Establish the TP/TG crossover by physically detailed trapping model (Fig. 3) [10].

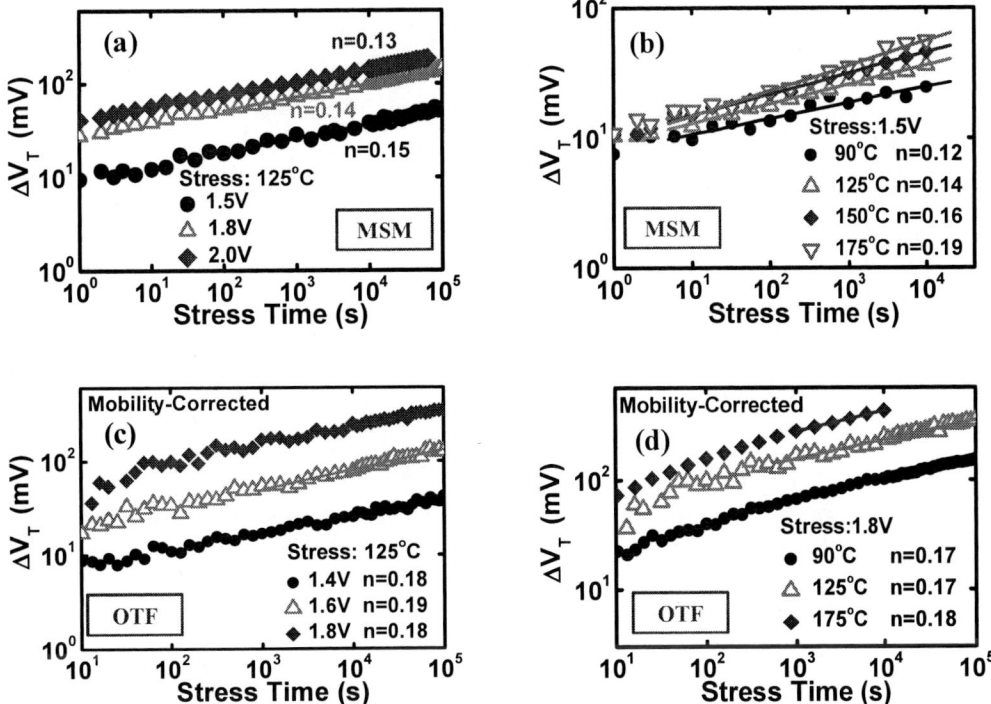

Figure 2. PBTI ΔV_T shift under (a) different voltage and (b) temperature stresses using MSM measurements. Since this methodology is contaminated with trapping and recovery, time exponents are stress dependent and non-universal, consistent with existing literature. (c, d) However, the on-the-fly measurements with mobility correction show uniform time exponent (~0.18) for all stress conditions. The results imply universality in trap generation mechanism.

(iii) Experimentally demonstrate a consistent and unambiguous switchover in the sign of SILC, as initial, relatively fast TP phase is taken over by relatively slow TG phase at longer times (Fig. 4, 5).

(iv) And Show that 1/f noise analysis provides independent validation of the reassignment of TP/TG regimes (Fig. 6).

This reassignment proposed in this paper *reinterprets* broad range of PBTI (ΔV_T) results from literature (Fig. 7), and (once the technology-specific TP component is subtracted) identifies intrinsic, universal TG characteristics of HK dielectrics regardless of the process details.

II. EXPERIMENTS

The nFETs used in this study are hafnium based High-k dielectric on a chemical SiO_2 interface layer (IL) with metal gate. The thickness of SiO_2 IL and HfO_2 bulk dielectric is about 1.0nm and 2.5 nm, respectively. Constant voltage PBT stresses are applied to gate terminals under different voltages and temperatures. Conventional Measure-Stress-Measure (MSM) and recovery-free On-the-fly (OTF) methodologies are applied to monitor the threshold voltage degradation. To capture the characteristics of SILC, all the gate leakage currents are measured at sense voltage of 1V during short intervals of the stress. The flicker noise spectrum measurements are also performed in the linear region before and after PBTI stress to further verify our speculations regarding trapping and trap generation.

III. RESULTS AND DISCUSSION

A. Seperation of Trapping and Trap Generation

Fig. 2 shows the observed ΔV_T shift on a HK sample under different PBT stress conditions and measurement approaches. As can be seen in Fig. 2a, and Fig.2b, the MSM experiments suggest power-law degradation with stress-dependent non-universal exponents with n in the range of 0.1-0.2. Such variation in power-exponent is consistent with broad range of reports in the literature [1,2,4,13,21] and therefore considered typical of PBTI degradation. We attribute this variation to the presence of trapping in pre-existing traps and well known recovery during measurement interval that contaminate the MSM results. However, the intrinsic, universal time-exponent of (n~0.18) *independent of stress conditions* is restored if we use OTF experiment with mobility correction [11], see Fig. 2c, and Fig. 2d. Moreover, it is observed in Fig. 4 that the SILC experiments are described by long term power exponent n~0.5, so that $\Delta I_G \propto (\Delta V_G)^3$, also a well known result [2,3]. These results are broadly consistent with PBTI literature, indicating the robustness/validity of our characterization protocols.

As noted previously, however, that it is difficult to assign the physical mechanism (TP and/or TG) *based on ΔV_T measurements alone*. Indeed, some groups use different time exponents of ΔV_T characterizing intermediate and late phases of degradation to interpret transition from TP to TG [2], while others argue that this change in exponents should be assigned to different TG processes (TG_1, TG_2, etc.), not TP to TG transition [12]. In contrast, *if we complement ΔV_T with SILC*

978-1-4244-9113-1/11 $26.00 © 2011 IEEE

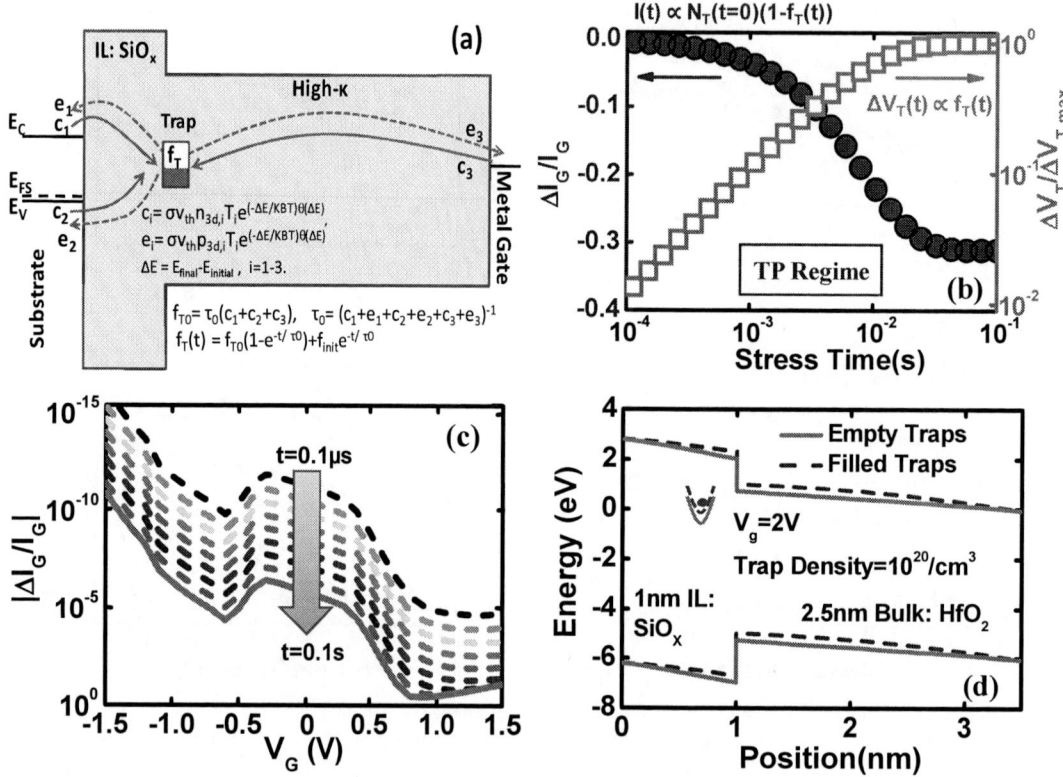

Figure 3. Physics of TP Regime: The simulation of trapping and trap-assisted tunneling for a given trap distribution located within the oxide. (a) The simulation framework is shown with all the relevant fluxes. (b) The transient trap occupancy increases from zero to $f_{T,max}$ and correspondingly ΔV_T increases to $\Delta V_{T,max}$ -a positive shift. On the other hand, the trapping current decreases as f_T increases, and as a result, the sign of SILC is negative. (c) This decrease of SILC is universal for all stress voltages, although the magnitude may vary depending on the voltage and specific trap distribution. Note that this simulation only accounts the trapping part. Once the Trap generation (TG) becomes significant, the SILC increases, eventually makes it positive. This sign reversal allows us to separate the trapping from trap generation regimes. (d) Energy band before and after TP, from self consistency solving of Possion equation, where the midgap of Si substrate is taken as zero point of energy and distribution of trap is uniform over dielectric with density as high as $10^{20}/cm^3$. The change of barrier heighth is so negligible even under the ultra high trap density that it is not necessary to consider the local electric field relaxation due toTP.

measurements, our analysis shows that TP and TG are easily separable and TP/TG transitions are readily identified by just looking at the sign of SILC (see Fig. 1c and 1d) – an intuitive result that appears to have been thus far overlooked in PBTI literature. Briefly, trapping within a fixed pre-existing traps (TP) results in a negative SILC transient (Fig. 3b) while that by newly generated traps (TG) causes positive SILC transient, a simple insight and a key result that we will discuss in more detail in the theory section below. We will also see later that excellent experimental agreement unambiguously confirms this hypothesis.

B. *Theoretical Model*

In order to rigorously establish the above intuitive statement regarding signs of TP-SILC and TG-SILC, we use a detailed theoretical model to simulate the transient ΔV_T as well as SILC for a given pre-existing trap distribution. The simulation [10] is based on appropriately generalized SRH-like formalism with inelastic tunneling and activated capture, see Fig. 3. Different flux components from the substrate and the gate are related by detailed balance, as shown in Fig. 3a. The capture rates accounts for any thermally activated processes as well as possible relaxation of trapping state following capture. The model is self-consistent, i.e., it solves

for the transient flux continuity and the Poisson equations simultaneously. The simulation of self-consistent transient trap occupancy, $f_T(t)$ allows us to calculate corresponding trapping current. Since for a given trap density, threshold voltage shift is proportional to charging, we always see *positive transients* in ΔV_T (Fig. 3b).

However, trapping current (I_T) is maximum when the trap is empty (t=0) and as the trap fills, it is reduced as

$$I_G(t) \propto N_T\left(t=0\right)\left(1-f_T\left(t\right)\right), \qquad (1)$$

and $\Delta I_G(t) < 0$ as $\Delta f_T(t) > 0$, and finally reaches minimum when fluxes are balanced, which is an intuitive classical result [13,14]. Hence, at the initial stage, TP-SILC ($\Delta I_G/I_G$) is *described by a negative transient*. Note that this negative TP-SILC is a general characteristic of all stress voltages (Fig. 3c). Although local electric field relaxation due to electrons trapping may also result in negative SILC evolution, this self-consistent change in barrier height and local electric field relaxation in thin oxides is generally negligible. This is illustrated in Fig. 3d, where we have considered TP into uniformly distributed traps across dielectric with density as high as $10^{20}/cm^3$.

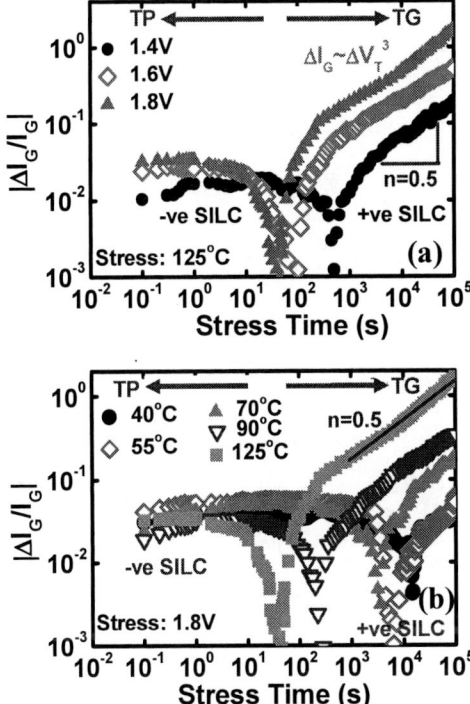

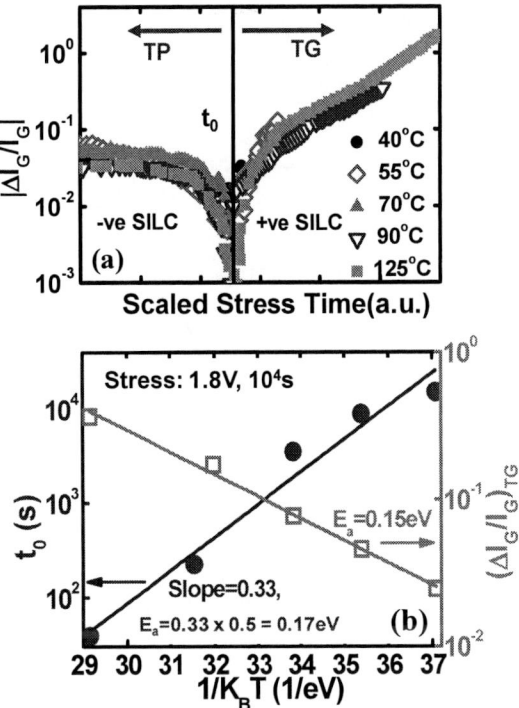

Figure 4. The SILC transients under (a) different voltages and (b) temperatures stress. Initially the SILC is negative (as predicted in Fig. 3) and temperature insensitive confirming the trapping process (TP-SILC). With time as TG becomes dominant (which is V and T dependent), there is a crossover to positive SILC (TG-SILC). The cross point from negative to positive shift decreases with higher temperature and bias, indicating strong V, T dependence of trap generation (TG).

Figure 5. (a) The horizontally scaled SILC transients at different temperatures show universal trend for TP-TG crossover. (b) Activation energies extracted from long time SILC evolution (Fig 4b- vertical) and crossover time of SILC (Fig. 4b-horizoltal) coincides with each other. The systematic crossover further verifies the co-contributions of TP-SILC and TG-SILC.

Once TG becomes dominant at longer stress times,

$$I_G(t) \propto N_T(t)(1 - f_T),\qquad(2)$$

and the increase of $N_T(t)$ makes SILC positive and makes the composite SILC due to TP and TG reverse sign. Therefore, the key to explore TP regime is to observe SILC transient very early in the degradation, a regime that has not been explored in typical SILC experiments.

C. Experimental Confirmation

To prove that our reassignments of TP and TG regimes are not only valid from the theoretical consideration of the TP model, but are also reflected in typical experimental conditions, we have characterized PBTI with very large number of SILC transient and 1/f measurements.

SILC transient experiments shown in Fig. 4 confirm the existence of a negative TP-SILC, as predicted by our preceding analysis. With time, there is a turnaround in the SILC and it becomes positive, reflecting the transition to TG-SILC process – also consistent with theory. We also note that the magnitude of initial negative TP-SILC is temperature independent and weakly voltage dependent, as expected of TP process. The zero-crossing points from TP-SILC to TG-SILC are voltage and temperature dependent, which is also expected. From our reassignment, higher voltage and temperature stress leads to

faster trap generation, consequently results in a shorter crossover time (t_0) to the positive SILC. Finally, to clearly demonstrate different regimes of TP and TG, we scale the SILC transients in Fig. 4b in the time axis and establish the universality of TP-TG crossover, see Fig. 5a.

The overall SILC evolution includes negative TP-SILC and positive TG-SILC as given by

$$\Delta I_G / I_G = (\Delta I_G / I_G)_{TP} + (\Delta I_G / I_G)_{TG}.\qquad(3)$$

Since TP-SILC is temperature independent and fast saturated (within seconds) [15,16], while TG-SILC is temperature activated and power-law time dependent, we can express the sum as

$$\Delta I_G / I_G \sim -A(t) + B(V,...)\cdot exp(-E_a / K_B T) t^n,\qquad(4)$$

where A is the contribution of TP and is a constant after saturation, B is voltage dependent prefactor of TG-SILC. Eq. (4) suggests that we can extract the activation energy E_a in two ways, and E_a extracted from long term SILC (at t=t', vertically) should be related to those obtained from the crossover time $t_0(T)$ (at $\Delta I_G/I_G$=0, horizontally), by the following relationship

$$t_0 = \left(\frac{A(t_0)}{B(V,...)}\right)^{1/n} \cdot exp(E_a / nK_B T),\qquad(5)$$

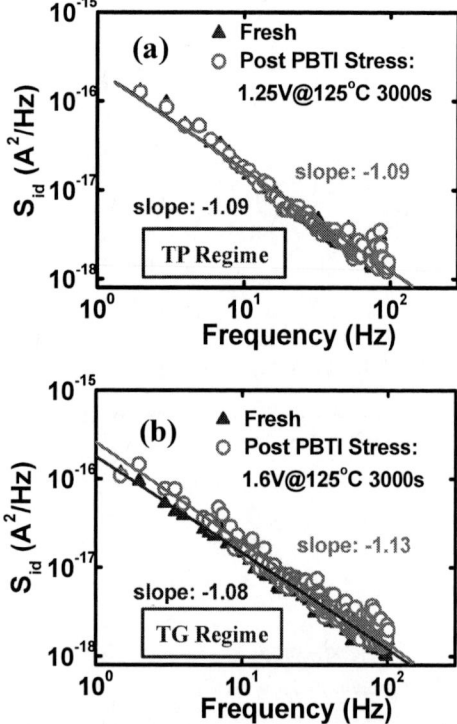

Figure 6. S_{id} before and after PBT stress for HfSiO dielectrics of n channel MOSFETs. V_G-V_T was fixed to keep the constant channel charge density. (a) For low PBT stress (in TP regime, Fig. 1c), S_{id} is nearly constant after stress. (b) For high PBT stress (in TG regime, Fig. 1c), S_{id} increases after PBT stress, suggesting that PBT stress causes the generation of new bulk traps. The increased slope also indicates that the bulk traps are generated far from the interface.

i.e., in order to calculate E_a at constant SILC (t_0) rather than constant time, we need to multiply the slope of Fig 5b with n (=0.5 for SILC). Fig. 5b shows excellent coincidence of E_a extracted from two different methods, which indicates the robustness of t_0 (V, T) between TP and TG.

Further Confirmation by 1/f Noise Spectrum

Our new assignment of TP and TG regimes can also be analyzed by 1/f noise. The 1/f noise spectrum density is sensitive for the detection of bulk traps density N_t in gate dielectrics [17,18]. In unified noise model, noise spectrum magnitude S_{id} is proportional to N_t, which is shown in [18]

$$S_{Id} / I_d^2 = N_t q^2 kT / C_{EOT}^2 WLf g_m^2 (V_g - V_t)^2 \gamma, \qquad (6)$$

where γ is attenuation coefficient, g_m is the transconductance, and C_{EOT} is the gate dielectric inversion capacitance. We follow this model and measure 1/f noise spectrum at a constant gate overdrive before and after PBTI stresses to reconfirm TP and TG regimes.

As shown in Fig. 6, magnitude of noise characteristics remains unchanged in a TP-dominated regime (when TG is comparatively negligible). On the other hand, spectrum magnitude increases significantly beyond the cross-over time t_0, indicating significant TG. Observed TP and TG regimes from spectrum characteristics are consistent to those shown in

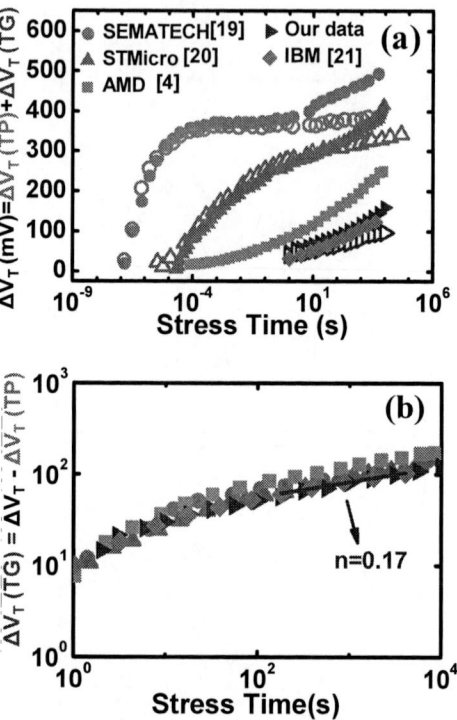

Figure 7. ((a), open symbol: room T and closed symbol: 125°C) Different processing conditions can result in different extrinsic trap concentration and (T,V) insensitive TP contribution. SILC results in Fig. 4, 5 suggest that once the T-insensitive TP is subtracted, (b) the long-time PBTI would indicate TG. We find that the TG is universal, implying similar intrinsic property of the different samples.

Fig. 4, where crossover time from TP to TG is voltage dependent.

D. Reinterpretation of Trap Generation in $\triangle V_T$ Data

After confirming the theoretical model of TP/TG regimes by systematic experiments of SILC and noise, it is important to explore the implications of these new results in reassessing PBTI data from literatures. As demonstrated previously, TP phase is essentially temperature and voltage independent, while TG phase has strong voltage and temperature dependencies, the time-dependent SILC can allow us to separate the trapping from trap generation. Therefore, we can use the same characteristic features to investigate and reinterpret the TP and TG regime in HK ΔV_T data, which is otherwise indistinguishable. In Fig. 7a, we collect different PBTI ΔV_T data (V~2.0V, 125°C) on HfO$_2$ devices from broad range of industrial/academic sources [4,19-21].We re-plot the data in Fig. 7b by subtracting the initial temperature-independent ΔV_T (related to TP regime). We find that although the original PBTI data look very different and appear technology-specific, the post-subtraction TG data are characterized by the same universal features of TG regime, including the same power-law exponent of n~0.17, as expected. This implies that although the different samples have different extrinsic pre-existing traps depending on process conditions and therefore have different TP contributions, theses samples stressed at similar stress conditions have intrinsically similar TG process.

Apart from the identification/isolation of TP/TG regimes, however, we make no comment on the physics of TG during PBTI or the kinetics of the process: this is a topic of future research, with no implications for the empirical confirmation TP/TG assignment discussed in this paper.

IV. CONCLUSIONS

Trapping and trap generation have radically different reliability implications for high-k MOSFET. We have presented a very simple algorithm and methodology to separate and unambiguously identify the crossover from trapping to trap generation regimes in PBTI degradation. We have also confirmed the hypothesis by extensive measurements including the temperature independent TP-SILC at short time-scales, temperature and voltage activated crossover points, and 1/f noise spectrum in respective regimes. Finally, once the regimes are identified and isolated, we have suggested remarkable universality of TG in wide variety of (apparently different) PBTI results reported in the literature, which is a key insight that might have implications for future PBTI optimization of high-k gate stack. The reassignment is not only important to identify degradation mechanism of PBTI, but also helpful to provide universal lifetime prediction of variety of HK dielectric devices and circuits.

ACKNOWLEDGMENT

We acknowledge Birck Nanotechnology Center for experimental facilities and NCN for computational resources. J.Q. Yang has received funding from China Scholarship Council No. 2009601194.

REFERENCES

[1] A. Kerber and E. Cartier, "Reliability challenges for CMOS technology qualifications with Hafnium Oxide/Titanium Nitride gate stacks," *IEEE Trans. Device Mater. Rel.*, vol. 9, no. 2, pp. 147-162, 2009.

[2] D. Ioannou, S. Mittl, and G. La Rosa, "Positive Bias Temperature Instability effects in nMOSFETs with HfO2/TiN gate stacks," *IEEE Trans. Device Mater. Rel.*, vol. 9, no. 2, pp. 128-134, 2009.

[3] E. Cartier and A. Kerber, "Stress-induced leakage current and defect generation in nFETs with HfO2/TiN gate stacks during positive-bias temperature stress," in *Proc. Int. Rel. Phys. Symp.*, 2009, pp. 486-492.

[4] A. Kerber, K. Maitra, A. Majumdar, M. Hargrove, R. Carter, and E. Cartier, "Characterization of fast relaxation during BTI stress in conventional and advanced CMOS devices with HfO2/TiN gate stacks," *IEEE Trans. Electron Devices*, vol. 55, no. 11, pp. 3175-3183, 2008.

[5] G. Bersuker et al., "Mechanism of Electron Trapping and Characteristics of Traps in HfO2 Gate Stacks," *IEEE Trans. Device Mater. Rel.*, vol. 7, no. 1, pp. 138-145, 2007.

[6] K. Zhao, J. Stathis, A. Kerber, and E. Cartier, "PBTI relaxation dynamics after AC vs. DC stress in high-k/metal gate stacks," in *Proc. Int. Rel. Phys. Symp.*, 2010, pp. 50-54.

[7] C. Young et al., "Electron trap generation in high-k gate stacks by constant voltage stress," *IEEE Trans. Device Mater. Rel.*, vol. 6, no. 2, pp. 123-131, 2006.

[8] M. Rafik, G. Ribes, and G. Ghibaudo, "Contributions and limits of charge pumping measurement for addressing trap generation in high-k/SiO2 dielectric stacks," in *Proc. Int. Rel. Phys. Symp.*, 2008, pp. 341-346.

[9] S. Pae et al., " Characterization of SILC and its end-of-line reliability assessment of 45nm high-k and metal-gate technology," in *Proc. Int. Rel. Phys. Symp.*, 2009, pp. 499-504.

[10] M. Masuduzzaman, A. Islam, and M. Alam, "Exploring the capability of multifrequency Charge Pumping in resolving location and energy levels of traps within dielectric," *IEEE Trans. Electron Devices*, vol. 55, no. 12, pp. 3421-3431, 2008.

[11] A. Islam, V. Maheta, H. Das, S. Mahapatra, and M. Alam, "Mobility degradation due to interface traps in plasma oxynitride PMOS devices," in *Proc. Int. Rel. Phys. Symp.*, 2008, pp. 87-96.

[12] S. Sahhaf, R. Degraeve, P. Roussel, B. Kaczer, T. Kauerauf, and G. Groeseneken, "A new TDDB reliability prediction methodology accounting for multiple SBD and wear out," *IEEE Trans. Electron Devices*, vol. 56, no. 7, pp. 1424-1432, 2009.

[13] S. Zafar, A. Kumar, E. Gusev, and E. Cartier, "Threshold voltage instabilities in high-κ gate dielectric stacks," *IEEE Trans. Device Mater. Rel.*, vol. 5, no. 1, pp. 45-64, 2005.

[14] K. Sakakibara, N. Ajika, M. Hatanaka, and H. Miyoshi, "A quantitative analysis of stress induced excess current (SIEC) in SiO2 films," in *Proc. Int. Rel. Phys. Symp.*, 1996, pp. 100-107.

[15] S. Deora, V. Maheta, A. Islam, M. Alam, and S. Mahapatra, "A common framework of NBTI generation and recovery in plasma-nitrided SiON p-MOSFETs," *IEEE Electron Device Lett.*, vol. 30, no. 9, pp. 978-980, 2009.

[16] S. Mahapatra, V. Maheta, A. Islam, and M. Alam, "Isolation of NBTI stress generated interface trap and hole-trapping components in PNO p-MOSFETs," *IEEE Trans. Electron Devices*, vol. 56, no. 2, pp. 236-242, 2009.

[17] Y. Yasuda, T.-J. Liu, and C. Hu, "Flicker-Noise Impact on Scaling of Mixed-Signal CMOS With HfSiON," *IEEE Trans. Electron Devices*, vol. 55, no. 1, pp. 417-422, 2008.

[18] P. Magnone, F. Crupi, L. Pantisano, and C. Pace, "Fermi-level pinning at polycrystalline silicon-HfO[sub 2] interface as a source of drain and gate current 1/f noise," *Applied Physics Letters*, vol. 90, no. 7, p. 073507, 2007.

[19] D. Heh, C. Young, and G. Bersuker, "Experimental evidence of the fast and slow charge trapping/detrapping processes in High- k dielectrics subjected to PBTI stress," *IEEE Electron Device Lett.*, vol. 29, no. 2, pp. 180-182, 2008.

[20] J. Mitard et al., "Large-scale time characterization and analysis of PBTI In HfO2/metal gate stacks," in *Proc. Int. Rel. Phys. Symp.*, 2006, pp. 174-178.

[21] S. Zafar et al., "A comparative study of NBTI and PBTI (Charge Trapping) in SiO2/HfO2 stacks with FUSI, TiN, Re gates," in *VLSI Symp. Tech. Dig.*, 2006, pp. 23-25.

978-1-4244-9113-1/11 $26.00 © 2011 IEEE

A New Interface Defect Spectroscopy Method

J.T. Ryan[1], L.C. Yu[1,2], J.H. Han[1], J.J Kopanski[1], K.P. Cheung[1,*], F. Zhang[3], C. Wang[4,5],
J.P. Campbell[1], J.S. Suehle[1], V. Tilak[2], and J. Fronheiser[2]

[1]National Institute of Standards and Technology, Gaithersburg, MD
[2]GE Global Research, Niskayuna, NY
[3]Michigan State University, East Lansing, MI
[4]Purdue University, West Lafayette, ID
[5]Fudan University, Shanghai, China
*301-975-3093, kin.cheung@nist.gov

Abstract— **A new interface defect spectroscopy method based on variable height charge pumping capable of observing the amphoteric nature of Si/SiO_2 interface states in production quality sub-micron devices is demonstrated. It can help to resolve the long standing debate about the true nature of Si/SiO_2 interface states. Additionally, we show that this is a powerful technique for studying other important material systems.**

Keywords; interface states, P_b centers, charge pumping

I. INTRODUCTION

There is a long standing debate on whether P_b centers account for all electrically observed Si/SiO_2 interface states. The inability to establish the amphoteric nature of Si/SiO_2 interface states in high quality samples is a key obstacle for resolving this issue. In this work, we introduce a new interface defect spectroscopy method based on variable height charge pumping and report the amphoteric signature of Si/SiO_2 interface states in production quality samples for the first time. The new technique is demonstrated on sub-micron size devices, highlighting the exceptional level of sensitivity.

Our new technique is a modified charge-pumping (CP) measurement. When applied to production quality pure SiO_2 MOSFETs with negligible bulk trapping centers, we can be confident that the CP signal is due only to interface states. Fig. 1 shows the key result demonstrating the new technique. The characteristic double peaks (which maintain their shape and increase in magnitude following moderate gate stress) at the expected energy locations clearly show that the interface states are of the P_b center family. Please note the absolute number of defects (left vertical axis) and the defect density (right vertical axis). This is the first ever reported double peak signature on a high quality sub-micron device.

All previously reported successful electrical observations of the amphoteric nature of Si/SiO_2 P_b centers relied on devices with very poor quality interfaces ($D_{it}>10^{11}$ cm^{-2}) and/or very large device areas [1-14]. In some cases, the amphoteric nature was only observable after very harsh irradiation [2, 7, 8, 12, 14]. In addition to a high density of interface states, these samples also exhibit high densities of non-interface bulk trapping centers. As most electrical measurements are not capable of truly excluding non-interface defects, quantitative disagreements are common when compared to P_b center data obtained from electron spin resonance. Such disagreements fueled the controversy [15, 16].

Using CP to perform defect spectroscopy is not new [1-3, 11]. Previous attempts were based on trapped charge emission which had difficulty probing mid-gap and (in some cases) require complicated gate voltage pulse trains that limit access near the band edges (slower pulse rise time (t_r) and fall time (t_f)) [1-3, 11]. Our approach utilizes a simple square wave, relies on charge capture rather than emission, is able to probe defects through mid-gap, and allows the use of faster t_r and t_f to probe closer to the band edges. Additionally, **choosing a frequency low enough to ensure complete trap filling is the key new aspect** in our approach.

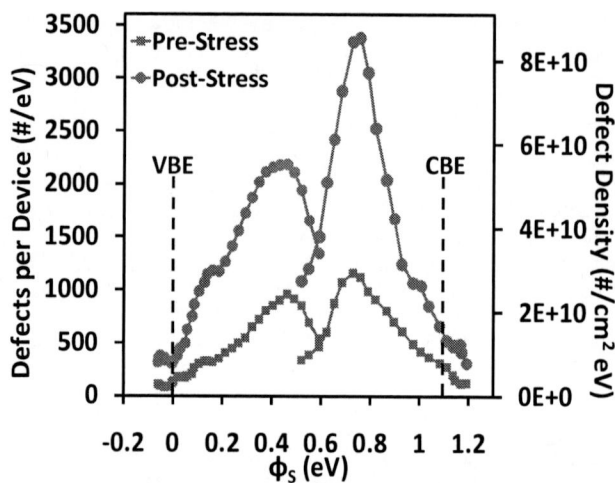

Fig. 1: Amphoteric nature of Si/SiO_2 interface states in a production quality sub-micron device. This is the first ever reported double peak signature on a high quality sub-micron device.

II. EXPERIMENTAL

The devices used are production quality 16.45 x 0.24 μm^2 nMOSFETs with 5.5 nm pure SiO_2 gate dielectrics ($V_{th} \approx 0.65$ V). CP was performed by applying a simple square wave (provided by a commercially available pulse generator) to the gate while shorting the source and drain to ground. The substrate (charge pumping) current (I_{CP}) was measured with an ultra-low noise current preamplifier and recorded with a digital storage oscilloscope for several seconds. The CP current was then averaged offline to further increase the signal to noise ratio. This approach allowed us to accurately measure very small currents with excellent sensitivity. Our CP method is

978-1-4244-9113-1/11 $26.00 © 2011 IEEE

very similar to the variable pulse height method [17]. The upper half of the band gap is measured by setting the pulse low level (V_{GL}) at strong accumulation (V_{GL} = -2 V in these samples) while sequentially stepping the pulse high level (V_{GH}) from strong inversion to depletion. The lower half of the band gap is measured by setting V_{GH} at strong inversion (V_{GH} = 1 V in these samples) while sequentially stepping V_{GL} from strong accumulation to depletion. By varying the pulse height in this fashion, we can obtain CP current as a function of the probed energy window in the band-gap. Fig. 2 schematically illustrates the pulse conditions. To reduce emission loss and push the measurement window closer to the band edges, t_r and t_f are both held at 2.3 ns for all measurements (unless otherwise noted). Based on instrumentation limits, slower t_r and t_f times can be used, but this will result in greater emission loss and will reduce how close to the band edges one can probe. Alternatively, faster t_r and t_f can be used to push the measurement window closer to the band edges, but care must be taken to avoid CP geometric effects based on the device dimensions as well as waveform distortion. Additionally, for more careful analysis of the band edges, the slope of the pulse transition for various pulse heights should be considered.

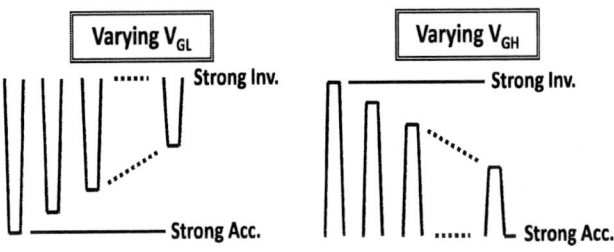

Fig. 2: Schematic illustration of the pulse bias conditions. For the case of fixed strong inversion (left), the pulse low level (V_{GL}) is sequentially stepped from strong accumulation to deep depletion. For the case of fixed strong accumulation (right), the pulse high level (V_{GH}) is sequentially stepped from strong inversion to deep depletion.

For each pulse bias level, the CP current was measured at 1, 2, 3, and 4 kHz. Such low frequency is of key importance and separates our approach from the conventional variable pulse height method, as explained later. Since the CP current should be zero at a CP frequency of 0 Hz, a linear plot of CP current vs. frequency for each pulse bias level allows for leakage current and amplifier offset to be corrected (by subtracting off the linear fit y-intercept value) and the CP current at a 2 kHz to be extracted accurately. Fig. 3 illustrates corrected CP current vs. frequency for two different pulse bias levels along with their respective linear fits. In thinner gate dielectrics, correcting for gate leakage current with this approach becomes difficult as the gate leakage current starts to overwhelm the CP current. Since the ratio of gate leakage current versus CP current can become quite large, this correction looses accuracy and introduces errors in the extracted measurement results. Thus, we estimate that below about 2 nm of pure SiO_2 gate

oxide thickness, the gate leakage current becomes large enough such that this method should be used carefully and the results viewed with care.

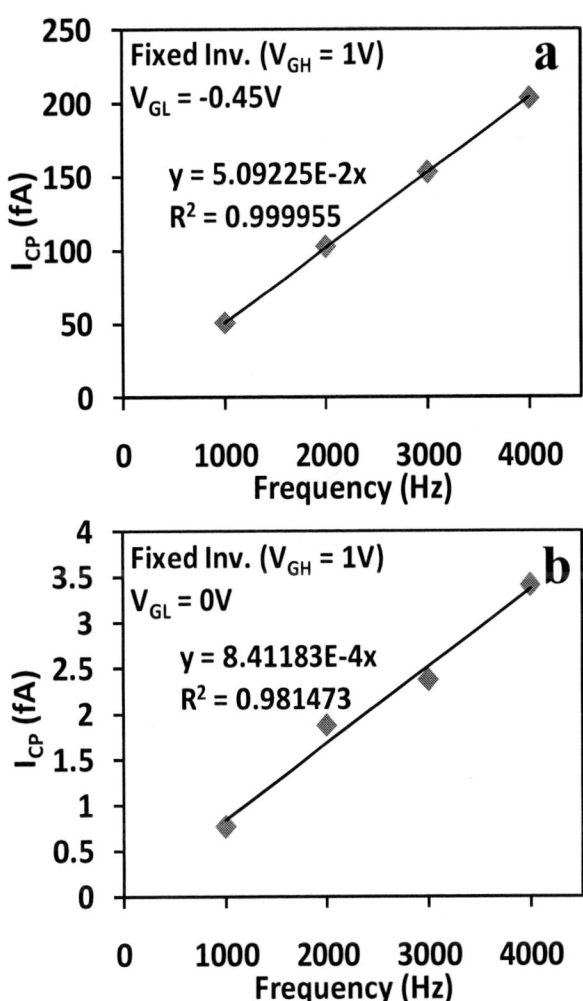

Fig. 3: Linear fits of CP current vs. frequency at fixed inversion for (a) V_{GL} = -0.45 V and (b) V_{GL} = 0 V after correcting for gate leakage current and amplifier offset. The linear fit (R^2) is clearly degraded in (b) indicating the onset of incomplete trap filling, as discussed in the text.

III. DISCUSSION

As mentioned above, choosing a CP frequency low enough to ensure complete trap filling is a vital necessity for an accurate measurement. When we vary the pulse height deep into depletion, the carrier density can become very small. Since the time constant associated with trap filling is inversely proportional to the number of carriers available to fill, the trap filling time becomes longer as the pulse height is pushed deep into depletion. This means that traps expected to participate in the CP process are not filled completely if sufficient time (slow enough frequency) is not given. The result is an apparent loss of measured CP current which detrimentally alters the measurement result. This is the reason we extract our CP results at 2 kHz.

The specific frequency is experimentally determined by measuring the CP current as a function of frequency for various pulse bias conditions as shown in Fig. 4 for the case of fixed accumulation (fixed V_{GL}). It clearly demonstrates that as the frequency is increased and/or V_{GH} is reduced into depletion, charge per cycle decreases; this is caused by insufficient time to ensure complete filling of the traps within the measurement window and **must** be avoided.

Charge per cycle is essentially the number of traps participating in the CP process and is extracted by dividing the measured CP current by the CP frequency and electronic charge (q). When incomplete trap filling is avoided, we equate "charge per cycle" to "defects per device" since all traps within the probed energy window in the device are contributing to the CP current. **Ensuring the measured result is free from incomplete trap-filling effects separates our new method from conventional variable pulse height CP**, which is usually done at higher frequency.

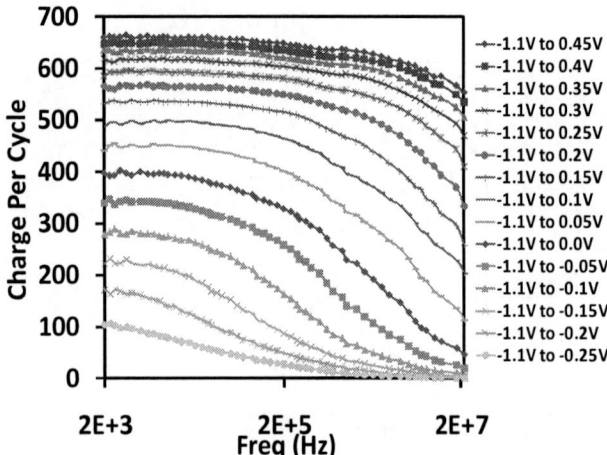

Fig. 4: Charge per cycle vs. frequency at fixed accumulation for a variety of different pulse high levels (V_{GH}). As the CP frequency is increased and/or V_{GH} is pushed into depletion, charge loss due to incomplete trap filling clearly occurs. **This effect must be avoided.**

Fig. 5 shows the measured defect density (#/cm^2) vs. gate voltage (V_G) for both fixed inversion and fixed accumulation cases at a frequency of 2 kHz. For the fixed accumulation case, V_G indicates the stepped pulse high level (V_{GH}) and for the fixed inversion case, V_G indicates the stepped pulse low level (V_{GL}). To obtain the defect density, defects per device (explained above) is extracted at 2 kHz and is then divided by the device area. Superimposed in Fig. 5 is the simulated [18] pre-stress capacitance vs. V_G (CV) curve.

It is clear that the defect density in Fig. 5 refers to the total defects within the probed energy window. As we vary V_{GL} (or V_{GH}) we change the probed energy window. By taking the derivative of the measured defect density with respect to V_G, we can extract the density of states (DOS) of the traps participating in the CP process. This derivative is shown in Fig. 6 (the pre-stress curves of Fig. 1) after converting V_G to silicon surface potential (ϕ_s) using the simulated CV curve

(superimposed). The left hand curve (extracted from fixed inversion) peaks at about 0.45 eV above the valence band edge (VBE) and the right hand curve (extracted from fixed accumulation) peaks at about 0.76 eV above the VBE.

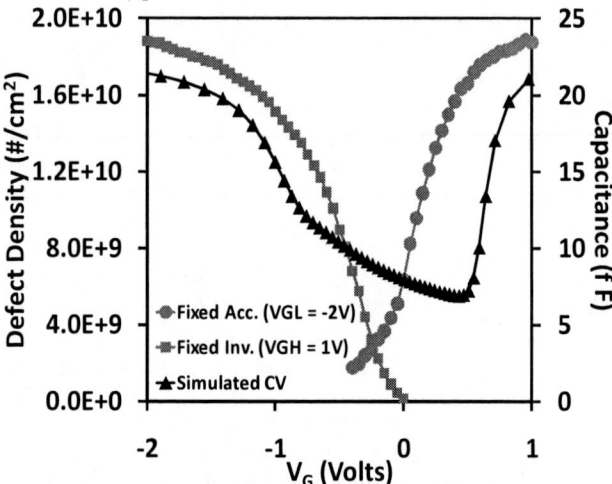

Fig. 5: Defect density (#/cm^2) vs. V_G extracted from the CP current at 2 kHz and a superimposed simulated CV curve. V_G indicates pulse high level (V_{GH}) and low level (V_{GL}) for fixed accumulation and fixed inversion cases, respectively.

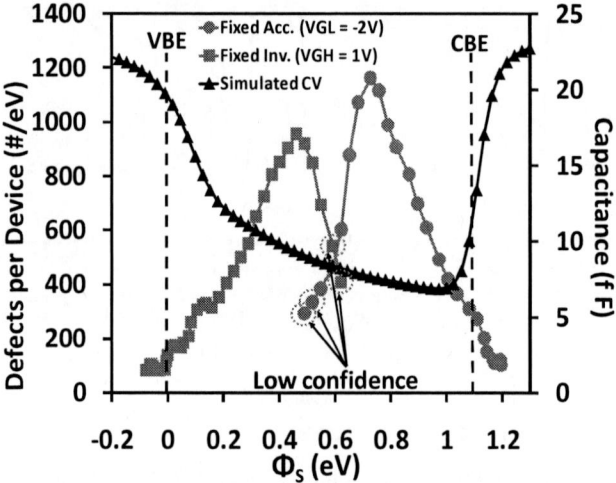

Fig. 6: DOS illustration vs. silicon surface potential (ϕ_S) and a simulated CV curve. The DOS (#/eV) was computed by taking the derivative of the data from Fig. 5 with respect to V_G and ϕ_S was calculated from the simulated CV curve. The indicated data points have low confidence, as explained in the text.

An obvious question is why the two sides of the measured DOS are not in agreement where they overlap? The answer involves the accuracy of the measurement falling off as the pulse height is pushed deep into depletion. As mentioned above, for each pulse bias condition, frequency dependent CP was performed at frequencies of 1, 2, 3, and 4 kHz, and the CP current at 2 kHz was extracted from a linear fit correction. As we push deep into depletion, incomplete trap filling starts to set in (as illustrated in Fig. 4) and the linearity of the frequency dependent CP data will start to degrade (the linear fit R^2 value in Fig. 3b is worse than in Fig. 3a which is in part due to this). Fig. 7a compares the linear fit R^2 values for fixed inversion with V_{GL} = -0.7 V to 0 V. At certain V_{GL} levels (as we push deeper into depletion), the linear fit R^2 value clearly starts to degrade. This is indicative of the on-set of incomplete trap filling (as well as signal to noise degradation). Similarly, the phenomenon is also clearly observed in the fixed V_{GL} case of Fig. 7b. This R^2 approach provides a convenient, objective and stringent method to monitor incomplete trap filling issues.

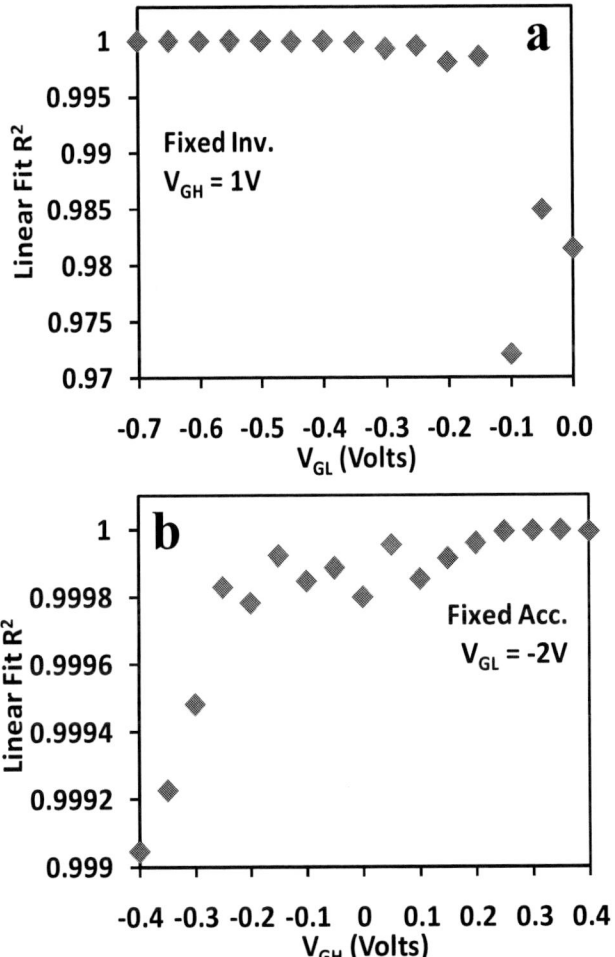

Fig. 7: Liner fit quality (R^2) vs. pulse height for fixed inversion (a) and fixed accumulation (b). As the pulse height is decreased, the linear fit is degraded; the degraded data has low confidence.

Clearly, the reliability of the last three points in Fig. 7a and the first three data points in Fig. 7b should not be trusted. These points are already dropped in Figs. 1 and 6. However, our differentiation routine propagates the influence of these points to the next three points. (We use the Savitzky-Golay method for differentiation, which is formally equivalent to taking a small segment of the curve and fitting it to a second order polynomial, and then differentiating the polynomial [19].) As differentiation tends to amplify small errors, the influence at the first point is significant. The next point is less. From the fourth point on, the influence is no longer there.

Based on this consideration, we circle the data points with a low confidence level in Fig. 6, thereby eliminating the disagreement.

To illustrate the new technique and to show that the peak locations are not measurement artifacts, we apply this technique to a different material system, namely SiC/SiO$_2$. For this system, one would intuitively expect a different DOS distribution than conventional Si/SiO$_2$ as SiC/SiO$_2$ could have silicon dangling bonds as well as carbon dangling bonds each with two DOS peaks. Fig. 8 shows the results of applying our new technique on a SiC nMOSFET (400 x 2 μm^2, t_{ox} = 50 nm) and four peaks are observed. Measured CV curves are superimposed and the VBE and CBE markers are estimates only. In this figure, the DOS was extracted at 10kHz, which was experimentally determined to avoid the interface trap filling time issue discussed above, and t_r and t_f were both held at 50 ns to avoid CP geometric effects. This result illustrates the power of this technique when studying different material systems as well as strongly supporting the validity of our Si/SiO$_2$ DOS peaks.

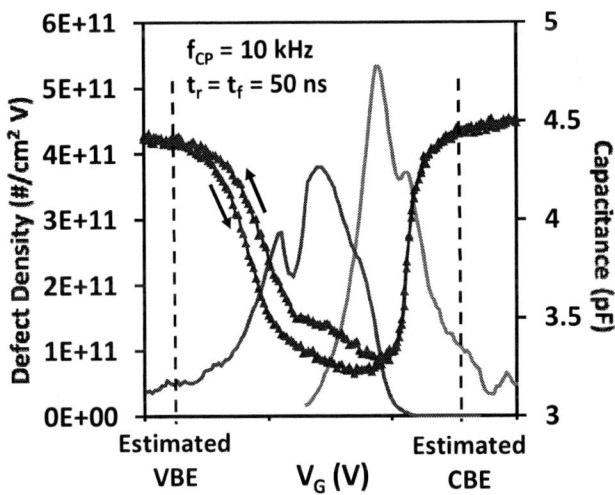

Fig. 8: DOS illustration and a measured CV curve for a SiC nMOSFET. Four peaks clearly appear in the DOS which are intuitively due to silicon dangling bonds and carbon dangling bonds.

IV. CONCLUSIONS

We have shown, for the first time, the amphoteric nature of Si/SiO$_2$ interface states in production quality sub-micron devices with excellent sensitivity and resolution. This work should help resolve the ongoing debate as to whether or not P$_b$ centers are responsible for the "true" electrically measured interface states. Performing the measurements at sufficiently low frequency to ensure complete trap filling separates this method from conventional variable height CP, and is crucial to the accurate extraction of defect density as a function of energy in the band-gap. Furthermore, we have shown that this is a powerful technique for extracting the DOS in other important material systems.

J.T.R. acknowledges funding support by the National Research Council. This work was partially prepared with the support of the U.S. Department of Commerce under Award No. NIST 60NANB10D109. However, any opinions, findings, conclusions or other recommendations expressed herein are those of the author(s) and do not necessarily reflect the views of the U.S. Commerce Dept.

REFERENCES

[1] G. Van den Bosch, G.V. Groeseneken, P. Heremans, and H.E. Maes, "Spectroscopic charge pumping: a new procedure for measureing interface trap distributions on MOS transistors," IEEE Trans. Elec. Dev., vol. 38 (8), pp. 1820-1831 (1991).

[2] J.L. Autran, C. Chabrerie, P. Paillet, O. Flament, J.L. Leray, and J.C. Boudenot, "Radiation-induced interface traps in hardened MOS transistors: an improved charge pumping study," IEEE Trans. Nuc. Sci., vol. 43 (6), pp. 2547-2257 (1996).

[3] J.L. Autran, F. Seigneur, C. Plossu, and B. Balland, "Characterization of Si-SiO$_2$ interface states: comparison between different charge pumping and capacitance techniques," J. Appl. Phys., vol. 74 (6), pp. 3932-3935 (1993).

[4] M.J. Uren, J.H. Stathis, and E. Cartier, "Conductance measurements on P$_b$ centers at the (111) Si/SiO$_2$ interface," J. Appl. Phys., vol. 80 (7), pp. 3915-3922 (1996).

[5] E.H. Poindexter, G.J. Gerardi, M.E. Rueckel, P.J. Caplan, N.M. Johnson, and D.K. Biegelsen, "Electronic traps and P$_b$ centers at the Si/SiO$_2$ interface: band-gap energy distribution," J. Appl. Phys., vol. 56 (10), pp. 2844-2849 (1984).

[6] N.M. Johnson, D.K. Biegelsen, M.D. Moyer, S.T. Chang, E.H Poindexter, and P.J. Caplan, "Characteristic electronic defects at the Si/SiO$_2$ interface," Appl. Phys. Lett., vol. 43 (6), pp. 563-565 (1983).

[7] P.B. Parchinskii, "Density of states at a gamma irradiated Si/SiO$_2$ interface: the effect of ultrasonic treatment," Solid State Dev. Cir., vol. 34 (6), pp. 420-423 (2005).

[8] Y. Nishioka, E.F. da Silva, and T.P. Ma, "Evidence for (100) Si/SiO$_2$ interfacial defect transformation after ionizing radiation," IEEE Trans. Nuc. Sci., vol. 35 (6), pp. 1227-1233 (1988).

[9] Y.G. Fedorenko, L. Truong, V.V. Afanas'ev, and A. Stesmans, "Interface state energy distribution in (100) Si/HfO$_2$," Mat. Sci. Semi. Process., vol. 7, pp. 185-189 (2004).

[10] L.A. Ragnarsson and P. Lundgren, "Electrical characterization of P$_b$ centers in (100) Si/SiO$_2$ structures: the influence of surface potential on passivation during post metallization anneal," J. Appl. Phys., vol. 88 (2), pp. 938-942 (2000).

[11] J.L. Autran, C. Plossu, F. Seigneur, B. Balland, "A comparison of Si/SiO$_2$ interface trap properties in thin film transistors with thermal and plasma nitrided oxides," J. Non-Crys. Solids, vol. 187, pp. 374-379 (1995).

[12] Y. Wang, T.P. Ma, and R.C. Barker, "Orientation dependence of interface trap transformation," IEEE Trans. Nuc. Sci., vol. 36 (6), pp. 1784-1791 (1989).

[13] P.K. Hurley, A. Stesmans, V.V. Afanasev, B.J. O'Sullivan, and E. O'Callaghan, "Analysis of P$_b$ centers at the Si(111)/SiO$_2$ interface following rapid thermal annealing," J. Appl. Phys., vol. 93 (7), pp. 3971-3973 (2003).

[14] N. Haneji, L. Vishnubhotla, and T.P. Ma, "Possible observation of P$_{b0}$ and P$_{b1}$ centers at irradiated (100) Si/SiO$_2$ interface from electrical measurements," Appl. Phys. Lett., vol. 59 (26), pp. 3416-3418 (1991).

[15] P.M. Lenahan and J.F. Conley, "What can electron paramagnetic resonance tell us about the Si/SiO$_2$ system," J. Vac. Sci. Technol. B, vol. 16 (4), pp. 2134-2153 (1998).

[16] E. Cartier and J.H. Stathis, "Atomic hydrogen induced degradation of the Si/SiO$_2$ structure," Microelectron. Eng., vol. 28, pp. 3-10 (1995).

[17] G. Groeseneken, H.E. Maes, N. Beltran, and R.F. De Keersmaecker, "A reliable approach to charge pumping measurements in MOS transistors," IEEE Trans. Elec. Dev., vol. 31 (1), pp. 42-53 (1984).

[18] http://www-device.eecs.berkeley.edu/qmcv/index.shtml, This document includes links to websites that may have information of interest to the reader. NIST does not necessarily endorse the views expressed or the facts presented on these sites. Further, NIST does not endorse any commercial products that may be advertised or available on these sites.

[19] A. Savitzky and M.J.E. Golay, "Smoothing and differentiation of data by simlified least squares procedures," Anal. Chem., vol. 36 (8), pp. 1627-1639 (1964).

Experimental Identification of Unique Oxide Defect Regions by Characteristic Response of Charge Pumping

Muhammad Masuduzzaman[1,*], Ahmad Islam[1], Robin Degraeve[2], Moonju Cho[2], Mohammed Zahid[2], Muhammad Alam[1]

[1]Department of ECE, Purdue University, West Lafayette, IN 47907, USA
[2]IMEC, Kapeldreef 75, B3001 Leuven, Belgium
[*]Phone: (765)-543-3616, Email: masuduzzaman@ieee.org

Abstract – **Although multi-frequency charge pumping (MFCP) is a widely used characterization technique to study oxide defects, the range and type of defects probed by the technique awaits conclusive identification. In this paper, we use characteristic response of variable channel length, as well as variable T_{charge}-$T_{discharge}$ CP experiments to self-consistently isolate and assign specific defect energy regions to CP signals. The results confirm the existence of shallow traps in Al_2O_3 samples and deep traps in HfO_2 bulk oxide, as has been speculated by various research groups in the past. We also provide specific experimental guidelines to identify the most prominent defect region for a given transistor technology. Such identification is essential to correctly interpret CP experiments and decide on optimization schemes for gate stacks.**

[Keywords: Bulk trap, charge pumping, high-k dielectric, trap profiling]

I. INTRODUCTION

As high-k gate dielectrics find mainstream applications in ultra-scaled transistors, memories, and sensors, it has become important to be able to precisely characterize energy-resolved bulk defect density (N_{OT}) within oxide, especially for their implications in NBTI/PBTI characteristics in MOS transistors [1]. The defects (either pre-existing or stress-induced) are often characterized by using charge pumping (CP) experiments [2, 3]. Although classical CP [2] has been used primarily to determine Si-SiO2 interface defect density based on high frequency gate pulses, the technique has recently been generalized for characterizing bulk oxide defects using lower-frequency gate pulse [4] – leading to the concept of Multi-frequency CP (MFCP) experiment.

Today the experimental simplicity of MFCP makes it a versatile technique for studying a wide variety of characterization problems, including the study of defects in flash memory [5], defect dynamics for dielectric breakdown [6], nature of defects in high-κ (HK) oxides of modern MOSFETs [7], etc. Despite its versatility and experimental simplicity, however, the MFCP technique involves complex (but not necessarily complicated) dynamics of charge trapping and electron-hole recombination at various defect sites and energies, making it difficult to correlate the MFCP signal to specific defect region. This complexity arises in part because, as opposed to classical (high frequency) CP that involves carrier transfer exclusively between conduction and valence bands (fluxes d_1, d_2, Fig. 1), the longer time-constants involved in low frequency CP measurement make broader ranges of fluxes (fluxes s_1-s_3,d_1,d_2, Fig. 1) and corresponding detailed flux balance, relevant for interpretation of MFCP signal (explained later) [8]. Despite this complexity, however, we find that with appropriately designed experiments, MFCP signals can be uniquely and self-consistently assigned to specific energy regions (shallow or deep), with important implications for design of gate stacks and other applications.

The identification/assignment of defect bands by MFCP is possible because CP signal responds differently to different defect regions. For example, as speculated theoretically in [8], there could be two different defect regions – defined as regions D and S in Fig. 1b – that, due to their energy location relative to conduction and valence bands, should respond very differently to the CP pulses (the symbols S and D here refer to trap bands and should not be confused with Source/Drain of a transistor). Specifically during a CP cycle, deep traps (D) exchange electrons with conduction and valence bands, while shallow traps (S) exchange electrons only with single (conduction *or* valence) band (as discussed later).

In this paper, we use characteristic response of variable channel length (V_{Ch}-CP) [9], as well as Variable T_{charge}-$T_{discharge}$ CP (VT^2CP) [10] experiments to self-consistently isolate and assign specific defect energy regions to CP signals. We experimentally confirm the characteristic signature as predicted from our simulation for shallow and deep traps (Fig. 6). The results show the existence of shallow traps in the particular Al_2O_3 samples and deep traps/trap relaxation in HfO_2 bulk oxide [11]. Indeed, since defects in regions D and S respond differently for voltage level sweep of MFCP [8, 12], any attempt to interpret the MFCP signal without knowing the associated dominant CP defect region would result in mischaracterization of the trap energy and location.

978-1-4244-9113-1/11 $26.00 © 2011 IEEE

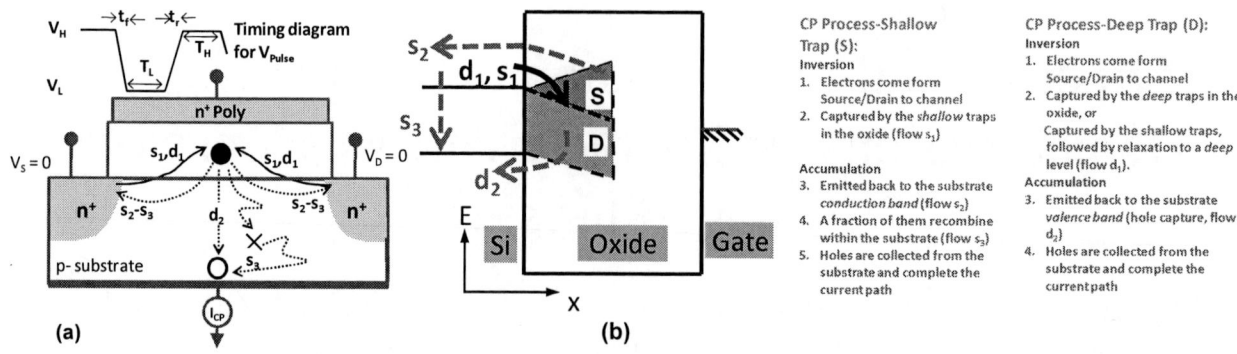

Fig. 1: The two current pathways for low f charge pumping is shown in (a). Defect bands D and S within the dielectric in flat band condition are shown in (b). In either case, during CP, a net flow of electrons occurs from source/drain (via the conduction band) to substrate (via valence band) for NMOS. For region D, the electron flow consists of d_1 (inv), d_2 (acc) and that for S, it consists of s_1 (inv), s_2, s_3 (acc) as summarized on the right. Flow s_3 is L_{CH} dependent.

II. LOW FREQUENCY CHARGE PUMPING

As summarized in Fig. 1, with the application of gate pulse (low f) in a CP experiment, two different current (I_{CP}) pathways are possible from source/drain to substrate. In either case, during the inversion bias-V_H (for NMOS), the channel is inverted and electrons are trapped (through inelastic tunneling) by the oxide defects (both shallow and deep traps, see Fig. 1b). Depending on the oxide, some traps may relax to a deep level [11], provided the relaxation time- τ_R is faster than the inversion time, T_H of gate pulse. Next, during the accumulation bias-V_L, the trapped electrons respond in two different ways based on their energy level: If the 'relaxed' position of the defect is located within region-D, it captures holes from the substrate valence band (V_B). On the other hand, if the relaxed position of the defect is in region-S, it can emit back the electron (due to the lower electron barrier as compared to hole barrier within the oxide) to the substrate conduction band (CB). The electron diffuses as minority carrier within the substrate conduction band with lifetime τ, and a fraction of them (higher for long L_{CH} and lower for shorter L_{CH}) recombine with holes within the bulk of the substrate. Consequently, process-S depends on fractional recombination within the substrate and hence the CP signal scales with L_{CH}. In contrast, process-D depends on electron-hole recombination within the oxide and hence is independent of L_{CH}.

III. L_{CH} DEPENDENT V_{CH}-CP

Let us now consider the implication of defects in region D and S for an identical set of transistors (nominally with the same defect density) with varying L_{CH}, ranging from relatively short (~0.5 μm) to relatively long (4 μm) channel length. We will call this approach [9]- variable channel length CP, or V_{Ch}-CP for short. As discussed in the last paragraph, we expect L_{CH} dependency in the integrated areal defect density D_T ($=I_{CP}/qAf$, in cm^{-2}), only if defects are located in region S. Our transistors were intentionally designed with relatively thick dielectric (0.9nm SiO$_2$+10nm ALD Al$_2$O$_3$) so that I_{CP} is not corrupted by gate leakage. We measure I_{CP} at different frequencies (kHz-MHz) using gate pulses of 50% duty cycle. The rise/fall time of gate pulse is 95ns, which is sufficient [13] to avoid the geometric component of CP for transistors used in this study. When the frequency-averaged D_T is plotted

against L_{CH} (Fig. 2), we find that for L_{CH}>0.5μm, D_T increases monotonically with L_{CH}. This channel length dependency confirms that regardless the details of the capture and emission processes and independent of any theoretical model, there must be shallow defects in Al$_2$O$_3$ that participate in the CP process by recombination in the substrate, confirming the existence of region S in these oxides. Note that region S is inaccessible to classical CP measurements [2]; therefore MFCP *does* broaden the range of traps probed by CP methodology.

The L_{CH}-dependent MFCP signal in Fig. 2 allows us to calculate the density of traps in region S and D in the following manner: First note that the CP contribution through the substrate recombination should diminish as L_{CH} is reduced and essentially vanish (s_3=0, although $s_2 \neq 0$, in Fig. 1b) at short channel length of ~0.5μm, therefore the residual CP signal at very short channel lengths must originate from D traps; we find that the residual trap density belonging to band D being (~5×10^{11} cm^{-2}). At longer channel length of L_{CH}=4μm, the extracted trap density (8×10^{11} cm^{-2} in Fig. 2) depends on the sum of the defects in band D calculated above (*i.e.*, 5×10^{11} cm^{-2}), plus a fraction (=s_3/s_2) of defects located in band S that contributed to CP current (*i.e.* ΔD_T ~3×10^{11} cm^{-2}). Had L_{CH} been sufficiently long (many times the diffusion length) so that s_3=s_2, the integrated D_T would have represented total traps from both D and S bands. To estimate the substrate recombination ratio s_3/s_2 as a function of L_{CH}, we use carrier-

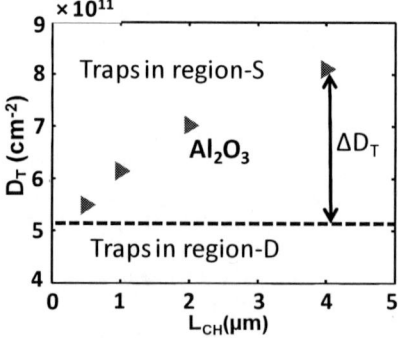

Fig. 2: V_{Ch}-CP experiment shows L_{CH} dependence on the CP response. This implies that the CP process must involve recombination in the substrate, which is possible for shallow traps only.

separation technique and numerical simulation by Medici™ to find a bulk recombination time constant $\tau \sim 1\mu s$ (typical for industry grade sample)[14]. This recombination time suggests a flux ratio $s_3/s_2 \sim 0.075$ for $L_{CH}=4\mu m$. The total defect density in band S in this experiment is thereby extracted to be $\sim 4\times10^{12}$ cm^{-2} ($\sim s_2/s_3 \Delta D_T$ cm^{-2}); an order of magnitude higher than that in band D. In sum, this technology is dominated by defects in region S.

IV. CHARACTERISTIC RESPONSE OF VT²CP

For additional and independent identification of both the conclusions above, *i.e.*, the existence of bands D/S and the magnitude of defect densities assigned to respective bands - let us consider the following variable T_{charge}-$T_{discharge}$ CP (VT²CP) experiment. This experiment requires only a single long channel transistor, rather than multiple transistors used in Fig. 2 – thereby simplifying the characterization process. In VT²CP [10], T_H (charging time) and T_L (discharging time) are varied independently, as opposed to conventional MFCP (50% duty cycle). In this particular experiment, we apply a gate pulse with constant T_H and variable T_L (and vice versa) and measure the charge pumping current on a transistor (L_{CH} = 4 µm) of the same technology (Al$_2$O$_3$) that we previously used for V$_{Ch}$-CP experiment. The responses ($D_T = I_{CP}/qAf$) are shown in Fig 3a and 3b. Interestingly we note that the response is constant with variable T_L (Fig. 3a, triangle), while it increases monotonically for variable T_H pulse sequence (Fig. 3b). These results clearly demonstrate that the probing of traps in Al$_2$O$_3$ is dictated by charging time (T_H) for a given CP pulse. A completely different response is obtained for device with HfO$_2$, which indicates the T_L dominance in CP

experiment for this particular sample (Fig. 3-square). Since we already know from V$_{Ch}$-CP experiment discussed in Sec. III that the specific Al$_2$O$_3$ oxide is dominated by S-traps, we expect that the VT²CP signal should also reflect the dominant contribution from region S for Al$_2$O$_3$.

To see if this is true, we simulate the flux dynamics for charge pumping pulse sequence to find probing distance within the oxide as a function of different VT²CP waveforms (varying T_H, T_L), using the formulation discussed in [8, 14]. The simulation is based on detailed balance of transient trapping-detrapping fluxes (Fig. 4) within an appropriately generalized SRH-like formulation, *e.g.*, the capture rate from conduction band is calculated as $\sigma v_{th} n T_e exp(-(\Delta E/K_B T)\theta(\Delta E))$, where symbols have their usual meanings [8]. The capture cross section $\sigma = \sigma_0 exp(\Delta E_B/k_B T)$ accounts for any thermally activated process influencing the capture process, where ΔE_B has been variously interpreted as multi-phonon emission barrier [15], Coulomb blockade barrier [16], etc. The gate bias dependence of σ has been assumed symmetric for large binary pulses (*e.g.*, inversion and accumulation) characteristic of MFCP experiments. Furthermore, at constant temperature measurements, as is the case for VT²CP experiments under consideration, σ can be replaced by an effective σ_{eff} (10^{-15} cm^2, in this simulation). The probing depth (in x) does depend on the specific value of σ_{eff}, but the key quantities of interest, *i.e.*, the existence of defect regions and the integrated defect densities in respective bands (/cm^2) do not. Indeed, different values of σ_{eff} would only (affine)-scale the probing depth in x, while keeping the relative change of the probing depths invariant as a function of T_H, T_L. Given the uncertainty in σ, we will therefore plot the x-axis in arbitrary units – which will have no implications for our final results.

The contours in Fig. 5a and b shows the probing area for variable T_L and T_H respectively. We note that for variable T_L (1µs to 1ms) with constant T_H (1µs), the contours does not change for region S, while it increases and finally saturates for region D (Fig. 5a). We can explain the behavior as follows: For NMOS, charge pumping for region-D involves flow d_1 (from conduction band) and d_2 (to valence band) in Fig. 1b. Since flow d_2 is more difficult than flow d_1, as flow d_2 involves tunneling through a comparatively higher hole barrier [8], the probing depth in region D is primarily determined by T_L for comparable T_H. As a result, we get increased probing area with higher T_L for defect in region-D.

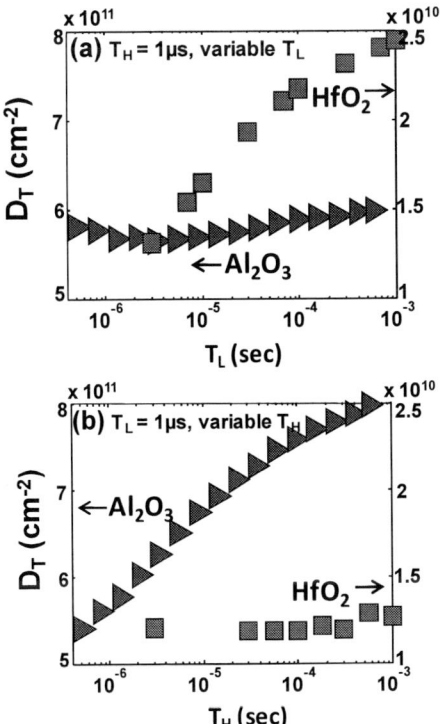

Fig. 3: VT²CP Experiments with (a) variable T_L (constant T_H) and (b) variable T_H (constant T_L). For Al$_2$O$_3$ (triangle) the response is sensitive to T_H only (and not T_L) and vice versa for HfO$_2$ (square).

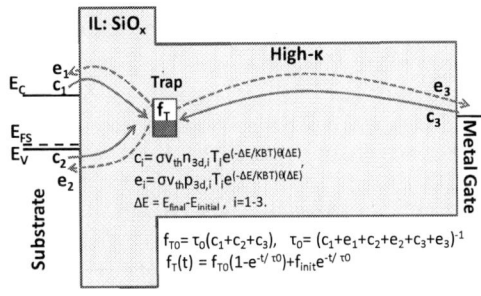

Fig. 4: The CP simulation framework involves calculation of different fluxes with proper detailed balance among them. For any given gate pulse, from the transient and net flow of different fluxes, the CP response can be calculated for a given trap location.

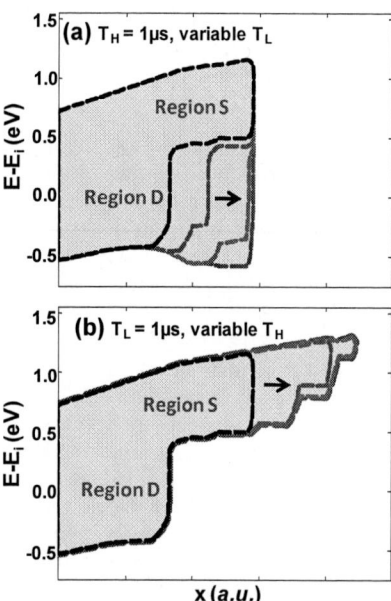

Fig. 5: VT²CP simulation: The simulated contours of the probing area (in x-E space, Fig. 1b) within the dielectric for T_L (a) and T_H (b) sweep as in Fig. 3. T_L (a) or T_H (b) is varied for 1, 10, 100, 1000μs. E_i refers to Si mid-gap energy. The x axis is plotted in *a.u.* as for this study we focus only on the relative variation of the probing area in response to T_L or T_H sweep. The absolute probing depth does depend on the proper choice of σ, however the relative value does not. The simulations suggest that region S (shallow trap) is sensitive to T_H while region D (deep trap) is sensitive to T_L.

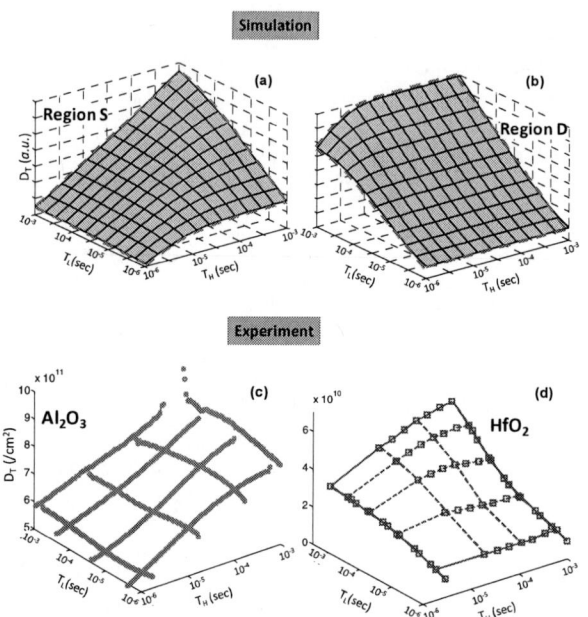

Fig. 6: The simulation for expected CP response (D_T) for all practical T_L - T_H space is shown for both region-S (a) and –D (b). Distinct characteristic responses are resulted for the two cases (*i.e.*, 2D surfaces lying along specific directions and bent at specific corners). The experimental VT²CP responses for different T_L - T_H combinations are shown for Al_2O_3 (c) and HfO_2 (d) respectively. The response for Al_2O_3 matches closely with that for shallow traps (a & c) and for HfO_2 with deep traps (b & d).

However, as T_L increases far beyond T_H, the probing area is limited by T_H and hence saturates (see Fig. 5a). On the other hand, charge pumping for region-S involves flow s_1 and s_2 (in Fig. 1b) to and from the same conduction band. The availability of free- states above conduction band edge makes the rate of s_2 higher than s_1. Hence, the probing area in region S is primarily determined by T_H, for comparable T_L and thus, does not change for fixed T_H (Fig. 5a). This also explains the case when T_L is fixed (1μs) and T_H is varying (Fig. 5b). We see that the contours for region D are fixed in this case (fixed T_L) and region S is increasing with T_H until limited by T_L.

Thus, the VT²CP simulation explains that depending on the trap energy, the CP response can be limited by either charging (region S) or discharging time (region D). Hence, taken together, the results of Fig. 3 and 5 confirm the existence of shallow traps (T_H sensitive) in Al_2O_3 devices and deep traps (T_L sensitive) in HfO_2 devices.

Note that for Al_2O_3 samples, we reached the same conclusion from V_{Ch}-CP experiment discussed in Sec. III. For a quantitative comparison, we note that the range of extracted D_T for this experiment is ~5-8×10¹¹ cm⁻² for the minimum and maximum scanning of region S, respectively (Fig. 3). Once again, this result from VT²CP experiment matches closely with ΔD_T in Fig 2 obtained from V_{Ch}-CP measurement. This cross-corroboration suggests that instead of channel length dependent V_{Ch}-CP experiment, one can equivalently change the T_H/T_L ratio in VT²CP technique to modulate the contribution from shallow band S, and thereby separate contributions from S and D bands. The result is also consistent with the shallow defect region found on transistors of the same technology using a separate characterization technique [17].

We also perform simulation, not only for the two cases of Fig. 3 and 5, but also for all practical combinations of T_L and T_H in VT²CP to find the expected response of traps located in region-S and D (Fig. 6a and b). The two regions provide very distinct characteristic responses (note the different identifiable orientation of the 2D surfaces of CP response). It is remarkable that the corresponding experiments give exactly similar characteristic response with either region-S (Fig. 6c-Al_2O_3) or region D (Fig. 6d- HfO_2). The robust experimental match unambiguously supports our claim about the specific assignment of trap energy to the CP response.

Relaxation of Traps:

There is a conceptual issue regarding the assignment of defects in region D that requires further discussion. For many oxides, the traps are known to relax after capturing electrons. If τ_R is much larger than T_H or T_L (μs to ms), we can safely ignore such effect in CP. But in case of smaller τ_R, we must consider this effect in the simulation. If we assume that the structural relaxation is very fast ($\tau_R \ll T_H, T_L$), it effectively splits the capture and emission energies in the detailed balance equation for the electron capture flux to/from the trap. Since the trap energy is no longer a constant, we define the defect energy level as the final (relaxed) level of the trap after capturing the electron (which is relevant for electrical device operation).

The simulation (Fig. 7) shows that greater the relaxation (so that some electrons could be captured in 'shallow' energy

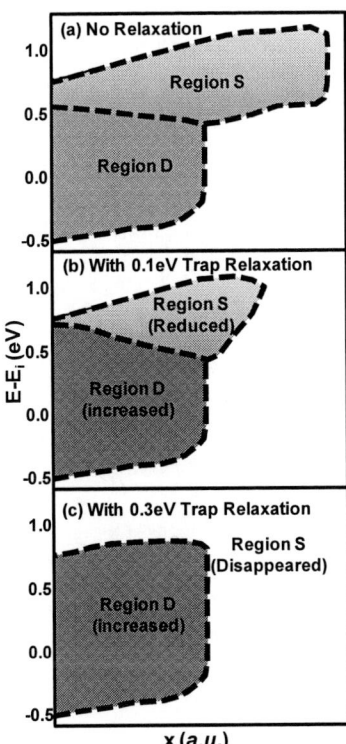

Fig. 7: Effect of trap relaxation is simulated for different amount of relaxation energy, V_R (eV). As V_R increases, it is difficult to emit back to CB in the CP process (Fig. 1b) and hence region-S starts disappearing. However, CP process D becomes easier for the relaxed trap. Note that the characteristic response of the relaxed trap does not change with relaxation, as being in region D, it is still T_L sensitive in the CP signal and the same response of Fig. 6a,b results for the respective regions. Only it is the likelihood of probing deep trap that is increased due to relaxation.

state, but relax to the deep state), more difficult it is to emit back to CB (intuitively) and hence region-S decreases in size. With this redefinition, some traps that are shallow when empty (before the capture of an electron) should be more appropriately interpreted as deep trap (region D) after relaxation in the experimental CP signal. Note that apart from increasing the likelihood of finding deep trap and diminishing that of shallow trap, relaxation does not change the characteristic response of respective regions (*i.e.*, region-D$\leftrightarrow$T$_L$ sensitive$\leftrightarrow$Fig. 6b like response, *etc.*).

Our conclusion about dominant deep traps for the HfO$_2$ devices remains unchanged with this extended definition of trap energy with relaxation. We conclude that the traps in the specific HfO$_2$ devices may be either deep to begin with (without relaxation) or it becomes deep after structural relaxation (the final energy being deep in either case). The possibility of energy relaxation of HfO2 devices has been discussed in [11].

V. CONCLUSION

In this paper, we demonstrate that both L_{CH} dependent V_{Ch}-CP and single L_{CH} VT^{2}CP experiments provide simple, elegant and independent experimental protocols to identify the dominant defect regions (D or S) in a transistor. The robust experimental match with the characteristic response (Fig. 6) for different defect energy regions (also consistent with

independent V_{Ch}-CP exp, Fig. 2) unambiguously supports our respective assignment of defect regions. We conclude that for the specific devices with Al$_2$O$_3$ dielectric, the traps are shallow (with or without structural relaxation) and for HfO$_2$ dielectric, the traps are deep – either pre-existing or are dynamically converted to deep levels following structural relaxation after electron capture. Such identification is essential to correctly interpret CP experiments and decide on optimization schemes for gate stacks.

ACKNOWLEDGEMENT

The authors would like to thank the MIT-MSD program for financial support, Network for Computational and Nanotechnology (NCN) for computational facilities, IMEC for the samples, Birck Nanotechnology Center for experimental facilities, and Maria Toledano-Luque, IMEC for useful discussion.

REFERENCES

[1] G. Ribes, *et al.*, "Review on high-k dielectrics reliability issues," *IEEE Transactions on Device and Materials Reliability*, vol. 5, pp. 5-19, 2005.

[2] G. Groeseneken, *et al.*, "A Reliable Approach To Charge-Pumping Measurements In Mos-Transistors," *IEEE Transactions On Electron Devices*, vol. 31, pp. 42-53, 1984.

[3] X. Garros, *et al.*, "In-depth Analysis of VT Instabilities in HfO2 Technologies by Charge Pumping Measurements and Electrical Modeling," in *IEEE International Reliability Physics Symposium*, 2007, pp. 61-66.

[4] R. E. Paulsen and M. H. White, "Theory And Application Of Charge-Pumping For The Characterization Of Si-SiO$_2$ Interface And Near-Interface Oxide Traps," *IEEE Transactions on Electron Devices*, vol. 41, pp. 1213-1216, Jul 1994.

[5] A. Arreghini, *et al.*, "New Charge Pumping model for the analysis of the spatial trap distribution in the nitride layer of SONOS devices," *Microelectronic Engineering*, vol. 80, pp. 333-336, JUN 2005.

[6] D. Varghese and M. A. Alam, "Charge Pumping as a Monitor of OFF-State TDDB in Asymmetrically Stressed Transistors," *IEEE Electron Device Letters*, vol. 30, pp. 972-974, 2009.

[7] G. Bersuker, *et al.*, "Breakdown in the metal/high-k gate stack: Identifying the "weak link" in the multilayer dielectric," *IEEE International Electron Devices Meeting 2008, Technical Digest*, pp. 791-794, 2008.

[8] M. Masuduzzaman, *et al.*, "Exploring the Capability of Multifrequency Charge Pumping in Resolving Location and Energy Levels of Traps Within Dielectric," *Electron Devices, IEEE Transactions on*, vol. 55, pp. 3421-3431, 2008.

[9] A. Kerber, *et al.*, "Charge trapping in SiO2/HfO2 gate dielectrics: Comparison between charge-pumping and pulsed I-D-V-G," *Microelectronic Engineering*, vol. 72, pp. 267-272, 2004.

[10] M. B. Zahid, *et al.*, "Defects Generation in SiO2/HfO2 Studied With Variable Tcharge-Tdischarge Charge Pumping (VT2CP)," in *IEEE International Reliability Physics Symposium*, 2007, pp. 55-60.

[11] D. Veksler, *et al.*, "Understanding noise measurements in MOSFETs: the role of traps structural relaxation," in *IEEE International Reliability Physics Symposium (IRPS)*, 2010, pp. 73-79.

[12] E. Cartier, *et al.*, "Fundamental Understanding and Optimization of PBTI in nFETs with SiO2/HfO2 Gate Stack," in *International Electron Devices Meeting (IEDM) Technical Digest*, 2006, pp. 317-320.

[13] G. Vandenbosch, *et al.*, "On The Geometric Component Of Charge-Pumping Current In MOSFET," *IEEE Electron Device Letters*, vol. 14, pp. 107-109, 1993.

[14] M. Masuduzzaman, *et al.*, "Physics and Mechanisms of Dielectric Trap Profiling by Multi-Frequency Charge Pumping (MFCP) Method," in *IEEE International Reliability Physics Symposium*, 2009, pp. 13-20.

[15] C. H. Henry and D. V. Lang, "Nonradiative Capture And Recombination By Multiphonon Emission In Gaas And Gap," *Physical Review B*, vol. 15, pp. 989-1016, 1977.

[16] M. Schulz, "Coulomb Energy Of Traps In Semiconductor Space-Charge Regions," *Journal of Applied Physics*, vol. 74, pp. 2649-2657, 1993.

[17] R. Degraeve, *et al.*, "Trap Spectroscopy by Charge Injection and Sensing (TSCIS): a quantitative electrical technique for studying defects in dielectric stacks," *Ieee International Electron Devices Meeting 2008, Technical Digest*, pp. 775-778, 2008.

The Implications of RoHS on Active Implantable Medical Devices

Timothy Scott Savage

Medtronic Tempe Campus
Tempe, Arizona, USA
011-480-303-4749, scott.savage@medtronic.com

Abstract—**This paper discusses the Restriction of Hazardous Substances (RoHS) directive as it relates to active implantable medical devices (AIMD). AIMD have remained out of the scope of RoHS at this time. The changes to large volume commercial electronics targeted by RoHS have resulted in changes to component sourcing that have impacted AIMD components in the availability of non-RoHS legacy components and changes to materials and lead plating. The issues that would need to be addressed if RoHS legislation changes to include AIMD are discussed from a use condition perspective and lessons learned during the commercial change-over to RoHS compliant materials/ systems/ processes.**

Keywords- RoHS; active implantable medical devices; use conditions

I. INTRODUCTION

In July 2006 the Restriction of Hazardous Substances (RoHS) directive 2002/95/EC took effect in the European Union (EU) [1]. This legislation and its companion legislation the Waste Electrical and Electronic Equipment (WEEE) directive 2002/96/EC were created to reduce the amount of hazardous electronic materials that were being created and ending up in landfills [2]. The substances that were restricted are listed in Table I. The main focus of the legislation was consumer electronics and appliances that accounted for the majority of the electronic waste. The prime changes with electronics that had to take place as a result of this were 1) the removal of lead (Pb) from solder interconnects and component finishes and 2) the removal of polybrominated biphenyls (PBB) and polybrominated diephenyl ethers (PBDE) used as flame retardants in plastic encapsulant molding compounds in electronics. The categories of devices that these restrictions applied to are listed in Table II. Categories that are out of scope of RoHS for safety/ criticality reasons are: Military, Equipment designed to be sent into space, Large industrial tools/ installations, Large industrial machines, R&D equipment, photovoltaic cells, Transportation (planes, trains, automobiles), Active Implantable Medical Devices and Non-road machinery (bulldozers, excavators, cranes, lawnmowers, chainsaws).

Of particular interest in this discussion is the broad category of medical devices. In 2006, medical devices were left completely out of the scope of the RoHS legislation. A recent review of the legislation has brought the general category "medical devices" into the scope of RoHS but has left the subcategory active implantable medical devices (AIMD) out of scope. Medical devices now in scope will be phased in as listed in Table III assuming that the legislation is approved by the EU in early 2011.

TABLE I.   RoHS RESTRCITED SUBSTANCES

Substance	Allowed amount (weight in homogeneous material)
Lead (Pb)	<0.1%
Mercury (Hg)	<0.1%
Cadmium (Cd)	<0.01%
Hexavalent Chromium (Cr 6+)	<0.1%
polybrominated biphenyls (PBB)	<0.1%
polybrominated diephenyl ethers (PBDE)	<0.1%

TABLE II.   RoHS CATEGORIES

Category	Description
1	Large household appliances
2	Small household appliances
3	IT and telecommunications equipment
4	Consumer equipment
5	Lighting equipment
6	Electrical and electronic tools
7	Toys, leisure and sports equipment
8	Medical devices
9	Monitoring and control instruments
10	Automatic dispensers
11	Other electrical and electronic equipment not covered by categories 1-10

TABLE III.   MEDICAL DEVICE RoHS PHASE-IN DATES

Category	Phase in date
Medical devices/ monitoring and control instruments	2014
In-vitro medical devices	2016
Active Implantable Medical Devices	Out of scope – no phase in date but may be reviewed for inclusion at a later date

II. COMMON CHANGES DRIVEN BY RoHS

A. Solder Alloy/ Reflow Temperature

Based on the large number of consumer driven electronics that were required to be compliant to RoHS legislation starting in 2006, changes were made to electronics components and assembly processes for integrated circuit boards. The main change that precipitated a number of other changes was to the solder system used in the electronics assembly process. Historically, the eutectic binary alloy Sn/Pb solder (63/37 wt%) was used for its relatively low, narrow melting point (183°C) and compatibility with many different soldering processes (IR reflow, wave soldering, hand soldering), board finishes and component finishes with typical reflow peak temperatures ranging from 220°C to 240°C. The necessity of removing the Pb from the solder resulted in the development of a number of SnAg, SnCu and SnAgCu solder alloys as shown in Table IV [3]. All of these alloys melt at temperatures above ~220°C, considerably higher than SnPb, resulting in reflow temperatures as high as 260°C.

The list of alloys available to tailor the properties of Pb-free solder to a given application has changed every year with the addition of rare earth metals, Ni, Ge, Mn and many others [4]. Each change seems to bring some new discovery on an evolving platform of solders.

The new alloys also have significantly different material properties than the SnPb solder that they were replacing. The SnAg and SnAgCu alloys have higher elastic moduli and yield stresses than SnPb (i.e. they are stiffer) and are less prone to creep. The morphologies of the microstructures are also considerably different. Eutectic SnPb has a fairly homogeneous structure with interspersed nearly pure Sn and Pb phases with no Sn/Pb intermetallic compounds (IMC) or solid solutions. The Sn rich Pb-free alloys however have large Sn rich regions that can result in a small number of grains in each solder joint with large SnAg or SnCu based IMC running across the solder joint [5]. This change from a homogeneous structure to one of large grains, often spanning from the circuit board all the way to the component side of the solder joint, results in significantly different properties. The mechanical properties are dependant on the grain orientation and amount and location of the IMC and can make it hard to predict the behavior of individual solder joints. A large amount of research has gone into better understanding the material properties of these joints but there is still much to be learned.

TABLE IV.   COMMONLY USED SOLDER MATERIALS

Alloy (weight %)	Melting point	Common name
63Sn37Pb	183	SnPb eutectic
96.5Sn3.5Ag	221	
99.3Sn0.7Cu	227	
96.5Sn3.0Ag0.5Cu	217	SAC305
98.35Sn1.15Ag0.5Cu	225	SAC105

B. Re-engineered Components

The increase in the reflow temperatures by up to 40°C required components to be re-engineered to be able to withstand the higher temperatures without excessive delamination, cracking or thermal degradation. The electronic component suppliers spent large amounts of time and money re-engineering components and printed circuit boards between 2003 and 2006 to meet the new requirements. The JEDEC standard JSTD-020D was updated for Moisture Sensitivity Levels to reflect these changes [6]. Component suppliers were successful in creating components that met the new reflow profile requirements. One change that did occur on some components was to lower the amount of time that components can be left out in ambient atmospheres prior to board mounting commonly referred to as moisture sensitivity level (MSL). This reduced margins that existed in the handling of components prior to reflow.

C. Component Plating

A third change that was required for components was the removal of Pb from the plating finishes on metal surfaces of leads. The typical lead plating material prior to RoHS was SnPb with approximately 10% Pb. The Pb was added 60 years ago due to its ability to reduce the possibility of the formation of metal filaments on pure Sn plating [7]. When the concentration of Pb drops below about 3%, the Sn regions are prone to forming filaments or whiskers that can result in shorts between electronic component connections. The industry ended up choosing several different plating materials to replace SnPb. The most popular ended up being matte Sn with less popular solutions being SnCu, NiPd and NiPdAu.

D. Board Finish

The combination of the solder alloy and the board finish can greatly affect the outcome of drop testing and longer term aging characteristics. SnPb solders were fairly insensitive to this combination as along as there was adequate wetting, proper IMC formation (some but not too much) and a lack of plating defects such as black pad. For instance, drop testing with SAC305 shows better results with CuOSP than electroless nickel immersion gold (ENIG) plating [8].

III. REQUIREMENTS FOR AIMD

AIMD range from devices that can be life supporting to devices that improve the quality of life (reduction in pain, brain stimulation for tremors, localized drug delivery, nerve stimulation). They are implanted for durations from months (diagnostic devices) up to 12 years (pacemakers) or longer. In all of these applications, a failure can result in the need to remove and replace the device, harm to the patient or even death. The removal process requires surgery that can result in risks to the patient in terms of infection or other ancillary effects. The critical nature of these devices has resulted in them being left off of the RoHS list along with other critical categories.

Most implantable devices derive their operational current from a battery. As such, the use of current is kept to the bare minimum during operation. Leakage currents due to board insulation resistance and off state circuitry are key to meeting these long life requirements. Additionally, AIMD that are used to defibrillate the heart develop high voltage levels (up to ~1000V) to charge capacitors that will deliver a large energy pulse of 40 to 100J. This high voltage on the circuit board requires sufficient insulation resistance to prevent surface arcing or printed circuit board dielectric leakage or breakdown.

Key to AIMD is high reliability of the components and finished circuits. A typical reliability requirement for an AIMD circuit board would be in the range of 99.9% reliable after 10 years of operation. High reliability is achieved through circuit design, process control, component selection and adequate screening/ testing.

Since implantable devices are exposed to bodily fluids that can be corrosive and conductive, most AIMD are contained in a titanium enclosure that is hermetically sealed to protect the circuits. This titanium enclosure provides both a chemical and mechanical barrier. Figure 1 shows several typical AIMD including a diagnostic monitor, pacemaker, defibrillator and drug pump.

Figure 1. Example of AIMD.

IV. USE CONDITIONS FOR AIMD

The use conditions for AIMD are the best way to examine the risk that can be posed by conversion to typical reflow RoHS compatible systems. Like other electronic components, circuits for AIMD are processed through a circuit assembly process. Devices in semi-finished and finished states are shipped between manufacturing sites and to doctors in the field. The devices are then implanted where they are exposed to the forces of the human body.

A. Assembly Use Conditions

During the assembly process circuits are exposed to thermal (solder reflow operations, bakes, cures), mechanical (singulation, clamping, flex/ bend) and chemical (clean operations) stresses.

1) Thermal

The SnPb reflow profile has a typical peak temperature of 220°C (235°C maximum) with 60-150 seconds above liquidus (183°C) and 20 seconds within 5°C of the peak temperature per J-STD-020D. The Pb-free reflow profile has a typical peak temperature of 245°C (260°C maximum) with 60-150 seconds above seconds above liquidus (217°C) and 30 seconds within 5°C of the peak temperature per J-STD-020D. Reflow

profiles targeting tighter windows near 240 – 245°C are recommended in supplier literature to reduce the possible detrimental effects of 260°C reflows [9]. Maximum ramp and cool down rates are specified to be the same for both profiles (3°C/sec) resulting in longer exposures for Pb-free reflow profiles than SnPb profiles. Also recommended is a minimum cool down rate of 1°C/sec to ensure that the microstructure of solders, such as SAC 305, maintains a fine grain structure and inhibits the growth of IMC in the bulk solder to help with long term reliability. This effectively requires much tighter controls on the reflow process between the melting point and peak temperature.

Other thermal processes typically involve bakes to remove moisture (~125°C for 24 hours) and polymer cure steps that can range from snap cures at 175°C for several seconds to longer duration cure steps at ~125°C to 150°C for several hours. If circuit level burn-in is used as a screening methodology, finished circuits may be exposed to temperatures of ~125°C for up to several hundred hours with the Mil Std 883E being 160 hours for Class B circuits [10]. Storage at 125°C for 160 hours can result in grain structure and IMC layer changes in both SnPb and Pb-free solders that can impact reliability [11].

2) Mechanical

Mechanical stresses in the manufacturing line can be brought about based on any process that has the potential to clamp or bend the circuit board. Circuit boards typically run through the assembly line in a square or rectangular coupon with test contacts. For use in the final application, the square coupon is either cut or sawn into the final shape. This is of particular importance for AIMD as the form factors are typically rounded to make the overall device more comfortable in the patient. The act of singulation and final form factor creation can result in bending or flexing of the board. IPC-9704 was created to provide a standard strain gage based methodology of evaluating the strain levels and rates during assembly or use of a circuit board [12]. Characterization of the strain levels during manufacturing can be useful in creating a design that is manufacturable with low defect levels.

3) Chemical

Chemical cleans are often used to remove flux and other contamination from circuit boards as part of the manufacturing process. Terpene and other chemical based systems are used to remove solder flux from boards and create a clean surface that will pass high surface insulation resistance requirements of low leakage circuits. For SnPb based solders, the flux can readily be removed with Terpene or water based systems. Considerable work has been done to achieve the same levels of clean on Pb-free solders but the flux residues are generally harder to remove. The higher reflow temperatures and higher flux activity needed to boost wetting result in additional side reactions that make the flux residues harder and less soluble. The higher amount of Sn salts that form in the flux can result in more residues [13]. In general, Pb-free flux residue removal requires higher concentrations of chemicals, slower process times, more agitation and faster flow rate.

B. Shipping and Handling Use Conditions

Components, circuits and finished devices are subjected to shipping and handling stresses prior to implant. Typical shipping and handling stresses include vibration, thermal cycling, pressure cycling, mechanical shock and drop. There are both device level and packaged product level requirements that are used to assess the acceptability of the design for production. Table V summarizes the device and package level requirements that exist in either industry or international standards. The methods described in the standards are based on pass/ fail criteria. In order to provide industry-wide guidance, IPC-9703 was created to obtain more detailed understanding of the actual strain or acceleration seen at the circuit level [14]. While the IPC-9703 technique is aimed at shock conditions, it can be used for any condition where strain or acceleration is of concern. Another technique that can be used to understand the risk to solder joint or board failure is strain states. With this technique, the strain data from IPC-9703 can be used to understand how the board is bending locally [15].

TABLE V.   STANDARD THERMAL AND MECHANCIAL TESTS FOR AIMD

Industry standard	Test	Condition
ISO 14708-1	Thermal storage, cycling	-10C to 55°C, 1°C/ min ramp
ISO 14708-2/6	Random vibration	5Hz to 500Hz, $0.7(m/s^2)^2$/Hz, 30 min per axis
ISO 14708-2	Mechanical shock	500g, haversine, pulse duration 1ms, 1 shock per axis per direction (6 total)
ISO 14708-2	Barometric pressure	Ambient to 50kPa, 25 cycles, 3 min ramp (maximum), 3min dwell (minimum) Ambient to 304kPa, 40 cycles, 2 min ramp(maximum), 2 min dwell (minimum)
ISO 14708-1	Barometric pressure	50 kPa, 150kPa, 1 hour
Medtronic labeling	Device Freefall Drop	12 inches
ASTM D4169	Package (Freefall) drop	24 inches, 6 sides, worst orientation 48 inches
ASTM D4169	Loose load vibration	30 minutes bottom, 15 minutes x2 adjacent sides
ASTM D4169	Random vehicle vibration	Distribution cycle, 180 minutes

C. Post-implant Use Conditions

Once the AIMD is implanted into the human body it maintains a fairly steady temperature around 37°C in a sealed hermetic environment. Mechanically, however, it can be subjected to pressure variations (airplane to scuba), cyclical deflection/ bending, mechanical shock (falls, impacts) and low frequency vibration.   Aside from the standards already mentioned for shipment use conditions, there are no additional standards that detail required tests specific to post implant use. ISO 14708 testing for mechanical shock, barometric pressure and shipping vibrations cover the majority of the use conditions once implanted. Cyclical deflection/ bending are not covered in these standards. These forces can be broken down into 1) large forces that happen infrequently (due to the possible infliction of pain/ injury) and 2) much smaller forces that are applied every time a patient flexes muscles or places normal movement forces on the implant. The location and magnitude of these forces is not easy to predict given patient populations that vary in age, size, activity level and location of implant.   The design of the AIMD needs to minimize the transfer of these forces to the circuit assembly based on the expected use condition. Figure 2 shows locations of typical implants that are used for heart, brain and nerve stimulation and drug delivery. These locations include the upper chest, lower abdomen and lower back (posterior).

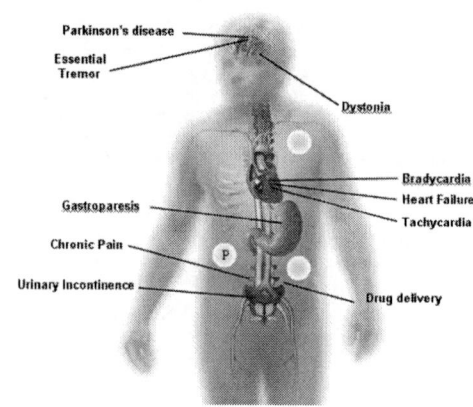

Figure 2.   Example of AIMD implant locations.

V.   AIMD RoHS ISSUES

There are a number of issues that can be gleaned from the discussion of RoHS and AIMD including regulatory oversight, component availability and most importantly long term reliability.

A. RegulatoryOversight/ Approval

The regulatory environment adds considerable complexity to process component and design changes.   Since there are governing bodies around the world that regulate the certification for sale of AIMD, the level of scrutiny and evidence required to release for use is significantly higher than typical consumer or commercial electronics.   For instance, simply changing to a RoHS compliant component in the circuit may require the complete requalification of the component and the AIMD. This could take upwards of a year to complete once samples are built.   Switching the entire assembly over to a Pb-free reflow process would require the process, components, design and finished device to be characterized and qualified. Once this work is complete, the qualification packet would need to be submitted to the regulatory bodies in the countries across Europe and across the world if the entire line is changed over to RoHS compliant. Typical review times for such a submission could easily be 180 days and since process, component and design are involved may require multiple submissions. Changes that are not reported and reviewed correctly can result in the recall of product from the field and the inability to sell new product until the changes are approved. This can be quite costly to the business and its image if done incorrectly.

B. Component Availability/ Change Issues

There are a number of issues that arise in the AIMD component world brought on by the change to RoHS in the commercial world. The first issue is the availability of non-RoHS compliant components. As the consumer electronics industry switched to RoHS compliant components many suppliers switched their entire product portfolio over to RoHS compliant to reduce the costs of maintaining multiple production lines. Lack of availability of non-RoHS compliant components that are already qualified has created supply issues. Once a supplier change notification is received, the purchaser has several options: 1) approve the change and go through a qualification and regulatory approval process, 2) reject the change and try to bargain with the supplier to continue use of the current process/ materials, 3) perform a last time buy of components to satisfy demand for remaining planned builds or 4) look to the open market for components. The longer product cycles typically involved with AIMD requires a strategy to manage these changes.

1) Approval of Change Request

Approval of the change to materials or processes will require the component and AIMD manufacturers to perform reliability testing and associated regulatory submission. Many of the changes are actually beneficial to producing high reliability circuits as the changes in mold compounds and manufacturing processes to improve the performance of components at higher reflow temperatures reduce the likelihood of failures at lower SnPb reflow temperatures. There have been some notable exceptions to this, however, including a switch away from brominated flame retardants to phosphorous. Sumitomo Bakelite, an epoxy mold compound manufacturer, created an epoxy formulation that had field failures related to large red phosphorous particles in the mold compound bridging internal electrical connections [16]. This failure mode was identified and newer epoxy flame retardant materials have been extensively tested and fielded with no issues. The change, however, does illustrate the risks of switching from known materials to newer, relatively un-field tested materials.

2) Rejection of Change Request

Rejection of the change can work if the AIMD manufacturer has significant business with the supplier. However, AIMD sales are typically not high enough volume-wise to maintain a stand alone production line with mainstream electronic component companies. There are exceptions to this

where electronic component suppliers have carved out business models that support high reliability applications but it is not the norm in the industry that drives fast technology innovation for consumer products. Fostering close relationships with the suppliers will be the key to this option.

3) Lifetime Buys

Lifetime or last time buys is a technique that is used to procure one last order of components from the original manufacturer prior to the process/ material or end of life change that makes the component no longer available. To adequately perform a last time purchase, the expected quantity required needs to be accurately forecasted. With AIMD products that can be manufactured for many years on the same design, the forecast may be for several years into the future. This can lead to large over or under-estimation of the actual product need. This can drive costs in two ways: 1) underestimating quantities can result in lost revenue due to an inability to manufacture the product or 2) overestimating quantities can result in excess components that will be scrapped after the product ceases production. Once the last time buy is completed, it is prudent to remove the component from the active component list so that new designs are not created with a component that can no longer be sourced in its current incarnation. Also of consideration with last time buys is the storage of components that may not be used for long time periods. Storage needs to ensure that the components maintain proper temperature, humidity and environmental conditions such that when the components are placed onto the circuit assembly they are reliable and have not degraded.

4) Counterfeit Components

Turning to the open market for components of the correct material/ process can be fraught with issues. The market for non-RoHS components has created a large counterfeit component business. Companies with products that are near the end of their program life typically do not want to invest additional resources in qualifying new components with associated regulatory submissions. One option is to look for components on the open market through component distributors, resellers and brokers. Counterfeit electronic components can be parts that are simply remarked at the distribution level (packaging labels), remarked at the component level, scrapped product, product built to look like the original from other material sources, reworked components (replated SnPb from Sn) or even components produced with original molds, materials and active components by a third party [17]. The end result is that the components may be of unknown origin, quality and reliability resulting in potential regulatory, manufacturing quality and field issues. Supplier control and monitoring is an important piece of understanding change once a component is qualified for medical implantable use. Purchasing components on the open market makes this traceability and control difficult at best, impossible at worst if the parts turn out to not be genuine. The rise of counterfeit components has resulted in symposiums by organizations like the Center for Advanced Life Cycle Engineering (CALCE) at University of Maryland.

C. Long Term Reliability Issues

A number of long term reliability concerns have been raised on the individual discussions for the changes due to RoHS and use conditions. Ones that require more discussion here are latent reflow/ assembly issues, long term mechanical issues and Sn whiskers.

1) Latent Reflow/ Assembly Defects

The increase of the reflow temperature results in increased risks of excessive warping, cracking and delamination of packages and materials. Large plastic packages, ceramic capacitors, flip chip die and printed circuit board materials are particularly at risk for these issues. Delamination of printed circuit board or components can compromise functionality, voltage breakdown and leakage properties. These failures can be latent in nature and hard to detect until a large increase in leakage current or absolute failure is obtained.

The increased reflow temp results in flux residues which are harder to remove and require more aggressive cleaning. This could lead to residual flux on the surfaces and increased surface leakage. Increased leakage could lead to early battery depletion or catastrophic arcing in high voltage situations.

2) Mechanical discussion

The change to new solder alloys with higher stiffness can have an impact on both high strain rate (drop/ shock) and low strain rate (thermal cycling and mechanical bending stresses) events. In high strain rate events, the stiffer solder may result in brittle fracture as compared to SnPb solders that deform rather than fracture. At low strain rate events, the stiffer solder will transmit more stress to interfaces where brittle IMC may exist resulting in brittle failure at the interface. High Sn, Pb-free solders often have cooling rate and aging effects on the microstructure and mechanical properties of the solder joint. Large grains can develop with varying orientation. The crystallographic orientation of the grains determines the strength of the solder joint resulting in some joints that are stronger than others [18]. The ductility and general uniformity of SnPb solder joints is part of the reason that they have been of primary use for many years. Pb-free solders also tend to produce more IMC such as Ni_3SN_4, Cu_6Sn_5 and Cu_3Sn that can contribute to brittle fracture.

3) Sn Whiskers

There is a long history of research into the phenomena that results in Sn whiskers. Going back to 1948, Bell laboratories studied the physics behind Cd, Zn and Sn whiskering [19]. In 1959, it was found the addition of 3-10% Pb greatly reduced the amount and length of whiskers that could be grown on electrodeposited films [20]. Addition of Pb allows for a reduction in the film stress over time while the removal of Pb results in the Sn film having increased compressive stress over time. Sn whisker formation can be influenced by temperature, relative humidity, film stress, method of deposition, plating chemistry, plating thickness, substrate and post-plate treatments. The physics behind Sn whisker growth has not been completely determined and to this day is not predictable.

With the elimination of Pb from the plating solutions, the plating industry focused on managing the grain structure, grain size, impurity levels and post plating treatments to create Sn plated surfaces that would be less prone to whiskering than historical systems. The resulting films are called matte or satin Sn due to the lower reflectivity of the films. The change to matte Sn, however, does still result in the risk of Sn whiskers. JEDEC standard JESD22A121 was created to assess the propensity of films to grow Sn whiskers [21]. Most vendors are testing components to this standard but without a physics based model to predict if, when, how many and how long whiskers will grow, the only sure why to know which solutions work is to perform long-term studies at use conditions to see which plating solutions and mitigation actions are effective.

VI. CONCLUSION

The change in materials and processes for the commercial electronics sector to meet RoHS requirements is now going on 5 years. A number of reliability issues with higher reflow temperatures, material property changes and Sn whisker growth have been identified. AIMD are not tolerant to these types of latent failures given their life critical application and as such should rightfully remain outside the scope of RoHS. Staying with non-RoHS compliant components and processes does pose a supply risk that will need to be managed to ensure un-interrupted production. The added complexity of regulatory oversight will require long review cycles when components or processes are changed.

ACKNOWLEDGMENT

The author would like to thank Peter Tortorici, PhD, George Raiser, PhD, and Jim Neville for inputs on this manuscript.

REFERENCES

[1] Directive 2002/95/EC of the European Parliament and the Council on 27 Jnauary 2003 on the restriction of the use of certain hazardous substances in electrical and electronic equipment (RoHS).

[2] Directive 2002/96/EC of the European Parliament and of the Council on 27 January 2003 on waste electrical and electronic equipment (WEEE).

[3] K.J Puttlitz, K.A. Stalter, Handbook of Lead-Free Solder Technology for Microelectronics Assemblies, Marcel Dekker, Inc., NY, NY 2004.

[4] W. Liu, N. Lee, "Novel SACX solders with superior drop test performance," SMTA Proceedings, Chicago, 2006.

[5] L.P. Lehman, R.K. Kinyanjui, J. Wang, Y Xing, L. Zavalij, P. Borgesen, E. J. Cotts, "Microstructure and damage evolution in Sn-Ag-Cu solder joints," Proc. IEEE Electronic Components Technology Conf., 2005, pp. 674-681.

[6] JSTD-020D.1 Joint IPC/ JEDEC standard for moisture/reflow sensitivity classification for nonhermetic solid state surface-mount devices, JEDEC Solid State Technology Association/ IPC, March 2008.

[7] S.M Arnold, "The growth of metal whiskers on electrical components," Proc. IEEE Electronic Components Technology Conf., 1959, pp. 75-82.

[8] M. Alajoki, L. Nguyen, J. Kivilahti, "Drop test reliability of wafer level chip scale packages," Proc. IEEE Electronic Components Technology Conf., 2005, pp.637-644.

[9] Intel Packaging Databoook, Chapter 9 SMT Board Assembly Process Recommendations, Revised 12-2007, pp.5-10.

[10] MIL-STD-883E, Method 1015.9 Burn-in test, 1 June 1993.

[11] L. Xu, J.H.L. Pang,K.H. Prakash, T.H. Low, "Isothermal and thermal cycling aging on IMC growth rate in lead-free and lead-based solder interface," IEEE Trans on Component and Packaging Technologies, 2005, 28(3), pp.408-414.

[12] IPC/JEDEC-9704 Printed wiring board strain gage test guideline, JEDEC Solid State Technology Association, June 2005.

[13] N.C. Lee, M. Bixenman, "Flux technology for lead-free alloys & its impact on cleaning," Proc. IEEE Electronic Components Technology Conf., 2002, pp. 1484-1490.

[14] IPC/JEDEC-9703 Mechanical shock test guidelines for solder joint reliability, JEDEC Solid State Technology Association/ IPC, March 2009.

[15] F.Z. Liang, R.L Williams, "Board strain states and FCBGA solder joint shock analysis," Proceedings of 2005 IMAPS Device Packaging Conference, Scottsdale, AZ, March 13-16, 2005.

[16] R.R. Hylton, "Techniques for identification of silver migration in plastic encapsulated devices assembled with molding compound containing red phosphorus flame retardant material," Proc. 34th International Symposium for Testing and Failure Analysis, 2008, pp. 112-120.

[17] M. Pecht, S. Tiku, "Bogus: Electronic manufacturing and consumers confront a rising tide of counterfeit electronics," IEEE Spectrum, May 2006, 43(5) pp. 37-46.

[18] J. Gong, C. Liu, P.P. Conway, V.V. Silberschmidt, "Grain features of SnAgCu solder and their effect on mechanical behaviour of micro-joints," Proc. IEEE Electronic Components Technology Conf., 2005, pp. 250-257.

[19] K.G.Compton, A. Mendizza, S.M. Arnold, "Filamentary growths on metal surfaces – Whiskers," Corrosion, 7(10), pp. 327-334, Oct 1951.

[20] S.M Arnold, "The growth of metal whiskers on electrical components," Proc. IEEE Electronic Components Technology Conf., 1959, pp. 75-82.

[21] JEDEC JESD22A121 Measuring Whisker Growth on Tin and Tin Alloy Surface Finishes, JEDEC Solid State Technology Association, May 2005.

Implantable microtechnologies for the brain: challenges and strategies for reliable operation

Jit Muthuswamy, Sindhu Anand, Jemmy Sutanto
School of Biological Health Systems Engineering
Arizona State University
Tempe, AZ, USA
Phone: (001)-(480)-965-1599, e-mail address: jit@asu.edu

Michael Baker and Murat Okandan
MEMS Science and Technology division
Sandia National Laboratories
Albuquerque, NM, USA

Abstract— Implantable microtechnologies and microelectrodes are crucial for the success for the next generation of prosthetic devices for the brain. This paper reviews some of the challenges in developing reliable technologies that will interface with neurons in the brain. We also recount here some of the milestones in our own successful effort to create interfaces with neurons in the brain. We have used a novel approach of developing neural interfaces using MEMS technologies that can be moved within the brain after implantation. The strategy is to enhance reliability by either creating stable neuron-electrode interfaces or by seeking new neuron electrode interfaces in the event of failure or deterioration of function in the interface. We have developed novel microactuators, microelectrodes, packaging and interconnect technologies to achieve the above goals. The success of this technology in recording good quality neuronal signals in chronic experiments, analysis of failure modes, the packaging efforts and reliability analysis to improve the longevity of the implant is addressed below.

Keywords-MEMS, microelectrode, neural recording, flip-chip, packaging

I. INTRODUCTION

Implantable microelectrode arrays play a crucial role in a variety of current and emerging biomedical prosthetic devices such as cochlear implants, visual prostheses, brain-machine interfaces, deep-brain stimulation devices etc. These emerging devices promise to restore function in patients with hearing loss, retinitis pigmentosa, patients with amputated upper-extremities or paralyzed limbs, or Parkinsons' disease. While current deep brain stimulation (DBS) devices are not microscale, but microtechnologies hold promise in improving the efficacy of DBS based therapy for current and a variety of emerging pathologies such as clinical depression, obesity etc. However, many of the above microtechnologies (apart from the cochlear implants) are either currently research devices or are on the cusp of wide-spread translation in the clinic. In all of the above applications, implantable microelectrode arrays are the primary portal for communication with single neurons or a family of neurons in a specific region of interest in the brain. In some of the above prosthetic devices, the implanted microelectrodes are used to electrically stimulate specific neurons while in others such as cortical prosthetic devices and

brain-machine interfaces the microelectrodes are used to record electrical activity from specific neurons of interest. While both classes of applications continue to face significant reliability issues in vivo, this review here will focus only on implantable microelectrodes used for recording from specific neurons in the brain that are absolutely crucial to the success of the emerging, exciting cortical prosthetic devices and other brain-machine interface devices. Arguably, the single, biggest challenge to cortical (brain) prosthetic devices that remains unsolved is the unreliable brain tissue-microelectrode interface. Current microelectrode technologies fail after a few months or years, which is inadequate for brain prostheses that needs to last the lifetime of the patient.

The reasons for failure are not fully understood, particularly in long-term experiments. It is widely hypothesized that the vigorous foreign body reaction mounted by the brain tissue surrounding the microelectrode could be a mode of failure. It has been quite well documented that the foreign body response of the brain tissue dramatically changes the cellular composition of the microenvironment sampled by the microelectrode. Over a period of 4 weeks or more after implantation, a 50-100 μm thick layer of glial sheath covers the microelectrode shank and the recording site, potentially forming a conductive barrier between the recording site and the neuron of interest. A diminishing number of neurons have also sometimes been observed in the immediate vicinity of the microelectrode suggesting neuronal migration away from the microelectrode over the first several weeks of implantation.

Another reason for the failure is hypothesized to be relative micromotion between the surrounding brain tissue and the microelectrode shank. The recording site of the implanted microelectrodes has to be within a few tens of microns from a neuronal cell body (depending on the location and orientation of the neuron in the brain) for optimal recording quality. Any relative micromotion between the surrounding brain tissue and the microelectrodes fixed to the skull can therefore result in the loss of a functional interface. Micromotion is typically caused by pulsations in the vasculature in the visco-elastic brain tissue, and also pressure waves induced in the brain during breathing. There is also significant micromotion of the brain within the skull due to behavior of the animal, the

---

Identify applicable sponsor/s here. *(NIH R01NS055312, R01NS055312-S!, Arizona Biomedical Research Commission)*

magnitude of which is typically higher in bigger species such as non-human primates and human beings (in the order of few millimeters). Micromotion in the brain is even less understood than the brain tissue response as a failure mode in long-term implant experiments.

We have been pursuing a novel strategy of mobile microelectrodes (using MEMS technologies) to overcome some of above failure modes and enhance reliability in long-term experiments. These mobile microelectrodes can now be moved to a new location within the same functional region of the brain in the event of a failure after implantation. Microactuators are used to power individual microelectrodes in an array making the microelectrodes independently positionable. This also increases the yield (number of microelectrodes in an array that actually pick up electrical activity after implantation) of the microelectrode array since each microelectrode can now be optimally positioned to within the recording radius of a neuron of interest. In the event of a failure or loss of electrical interface due to any number of reasons, the microelectrodes can now be moved to a new location to seek a new neuron. This strategy will potentially increase the chances of success of the emerging brain prostheses technology by achieving a long-lasting functional interface between the neuron and the implanted microelectrode.

II. CHALLENGES

The above implantable MEMS technology and associated packaging techniques will have to address other significant challenges to the long-term functionality such as:

1) Biological fluid entry – Biological fluids such as blood and CSF (cerebro-spinal fluid) often enter the active areas of MEMS and other implanted microelectronics and can be potentially catastrophic. The problems get worse when a sensory part of the MEMS device must be extended toward the biological fluid while the active MEMS mechanism must be protected from the fluidic contamination [1]. For this type of MEMS application, MEMS must be sealed semi-hermetically – such a sealing technology must provide the barrier for the biological fluid to enter the MEMS active part while provide access for the MEMS sensory part to be extended in/out the MEMS package through microchannels [1-3].

2) Temperature and humidity – MEMS package is exposed to a normal body temperature of 37° C that increases the chances of failure. Evaporation of biological fluid and some condensation on the die is possible if the MEMS die's temperature is relatively cooler than the air inside the MEMS enclosure. This condensation may impact the reliability of the MEMS structures.

3) Mechanical vibration and shock due to behavior – The active behavior of the animal causes significant mechanical stresses due to constant vibration and often mechanical shock to the MEMS packages and mechanical structures in the MEMS device.

4) Mounting strategy – for MEMS devices that need to be implanted in the brain, mounting is a significant surgical issue. Mounting is highly customized for different applications and devices. For the long-term brain implant, such mounting techniques must accommodate for the animal skull growth, regrowth of skin, and other meningeal layers on the brain etc. Mounting must also robust enough to protect the packaged MEMS devices from the mechanical impacts due to behavior.

5) Reliable Interconnects – Most MEMS package is still using wire bonding technique. Wire bonding is certainly not reliable MEMS interconnects method for chronic application. Furthermore, wire-bonding results in large and heavy packages that interfere with the behavior of the animal and also increases the mechanical stresses.

Over the past 10 years, our successful effort to develop these implantable MEMS devices for the brain have resulted in (a) several generations of novel electrostatic and electrothermal microactuators using the SUMMiT V™ (in collaboration with Sandia National Laboratories, Albuquerque, NM) [4-6] (b) several novel interconnect and packaging technologies for implantable MEMS devices [1, 7] (c) development of associated novel materials and processes used in the packaging for long-term in vivo implantation (d) novel discoveries in structure-property relationships in polycrystalline silicon microstructures [8-9] and (e) development of optimal processes for fabricating the polycrystalline silicon microstructures [9]. We outline below some key milestones in our progress towards developing a reliable interface with neurons in the brain.

III. MEMS MICROELECTRODE ARRAY

The microelectrodes and the associated electro-thermal microactuators are fabricated using a standard MEMS multi-user process SUMMiT V™ at Sandia National Laboratories. It is a five layer polysilicon surface micromachining process. The die size is 3 mm × 600 μm × 6 mm and consists of an array of three microelectrodes, which are polysilicon shanks 50μm in width and 4μm in thickness. The microelectrodes are actuated by a two pairs of Chevron-type bent beam actuators based on the principle of electrothermal actuation. Pair of actuators for locking the electrode in position is called the 'release up' and 'release down' actuators respectively and the 'move up' and 'move down' actuators are responsible for the bi-directional movement. A more detailed description of the actuation mechanism is published here .

A. Long-term rodent experiments

We have reported long term neural recording lasting over 80 days in a cohort of rodent experiments using the MEMS microelectrode devices [6]. The MEMS device was wire-bonded to an external PCB and a non-hermetic mesh encapsulation was used to prevent contamination from biological fluids in the brain like CSF from entering the device active area [1]. The wire bonded package was almost three times the size of the MEMS chip. The overall dimensions of the packaged device were 14 mm × 17 mm × 3 mm.

B. Results of failure modes in chronic experiments

Five of the 8 experimental MEMS devices used in long-term experiments remained functional for 4-13 week durations [6]. Analysis of the failure modes of the implanted devices in the above successful long-term experiments indicated that all 8 experimental devices (with semi-hermetic encapsulation) had surgical mount (based on dental cement) failures. Two of the 8 also had fluid infiltration into the mechanical structures leaving behind solid debris that was catastrophic. One of the eight devices also had other mechanical structural failure resulting in broken springs and actuators. In the control MEMS devices which did not have an encapsulation, all 4 of the control devices had fluid infiltration into the mechanical structures. An example of a device with fluid infiltration after 14 weeks of implantation is shown in Fig. 1.)

Figure 1. Wire bond packaged device with fluid infiltration after 14 weeks of implantation. The solid debris left behind after evaporation of the fluid results in catastrophic failure of the device.

C. Reliability of MMFI interconnects

We have also developed a novel MEMS Microflex Interconnect (**MMFI**) technology to addresses three critical needs in packaging our MEMS based devices that involve movable microelectrodes that extend off the edge of the chip and are implanted in the brain tissue [7]. The first need is for operating space within the package to allow for movement of mechanical parts. The second need is for flexible interconnects that will mechanically isolate the skull-mounted devices against stresses induced while engaging the interconnects. And the third need is for a scalable interconnect and packaging technology that can be expanded later on to multiple microchips simultaneously.

We fabricated a thin polyimide substrate with embedded bond-pads, vias, and conducting traces for the interconnect. A backside dry etch and a sidewall perimeter were created around the perimeter of the device, so the flexible circuit can act as a thin-film cap for the MEMS package. A double gold stud bump that sandwiched the polyimide substrate was used to form electrical connections to the chip. The use of a double stud bump gave sufficient spacing to prevent the flexible substrate from adhering to the actuators. The MMFI approach achieved a chip scale package (CSP) that is lightweight, biocompatible,

having flexible interconnects, and does not require the use of an underfill. Reliability tests were performed on the MMFI demonstrated minimal increases in mean contact resistances of 0.35 mΩ, 0.23 mΩ and 0.15 mΩ under high humidity, thermal cycling, and thermal shock conditions respectively. High temperature tests resulted in an increase in resistance of > 90mΩ when aluminum bond pads were used, but an increase of ~ 4.2mΩ with gold bond pads. We concluded from the results above that the unique MMFI technology provides a feasible and reliable approach for packaging and interconnecting MEMS based brain implants.

IV. FAILURE MECHANISMS IN THE MICROACTUATORS

Significant wear was observed in the thermal strips of the electrothermal actuators. The operating voltage for the actuators was set below the threshold voltage, i.e., the voltage at which the actuators melt. After a million cycles of operation, some of the actuator strips as shown in Fig. 2 had melted and there were debris on the polysilicon surface. The release actuators that lock on to the teeth of the microelectrode and keep it in position were found to be misaligned. There was mechanical failure with a broken joint between the release lock structures and the beams connecting the lock structures and the actuators as shown in Fig. 3. Physical deformations in the electrothermal strips that connect to the move actuator can also be observed in Fig. 3.

A. Mechanical Wear

Significant wear was observed in the thermal strips of the electrothermal actuators. The operating voltage for the actuators was set below the threshold voltage, i.e., the voltage at which the actuators melt. After a million cycles of operation, some of the actuator strips as shown in Fig. 2 had melted and there were debris on the polysilicon surface. The release actuators that lock on to the teeth of the microelectrode and keep it in position were found to be misaligned. There was mechanical failure with a broken joint between the release lock structures and the beams connecting the lock structures and the actuators as shown in Fig. 3. Physical deformations in the electrothermal strips that connect to the move actuator can also be observed in Fig. 3.

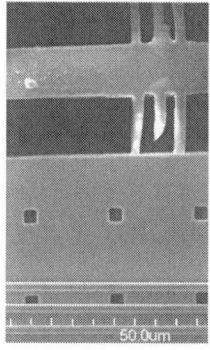

Figure 2. SEM micrograph taken at x1000 resolution and 10kV accelerating voltage. Mechanical wear visible in an actuator strip that has melted after a million cycles of operation.

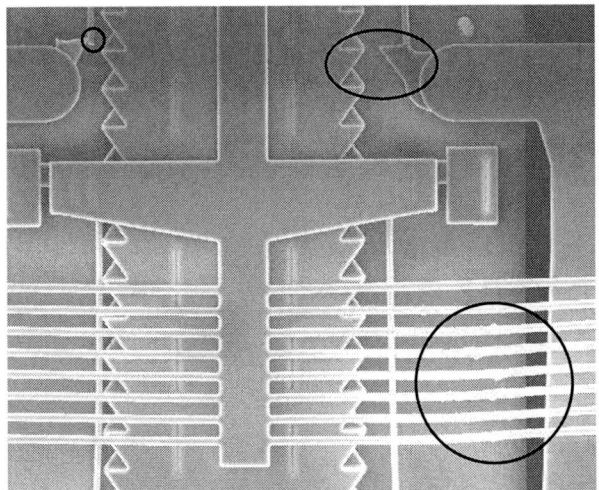

Figure 3. SEM micrograph taken at x600 resolution and 10kV accelerating voltage. The highlighted regions signify the failure mechanisms identified in the device- a broken joint, misalignment of lock structures and wear in the electro-thermal strips (clockwise direction).

V. CONCLUSION

Reliable neural data recording by implantable MEMS sensors/electrodes is dependent on two critical factors: (1) reliable functioning of the microelectrodes and microactuators over the lifetime of the implant (2) reliability of the interface between the microelectrodes and the brain tissue. The first concern can be mitigated by improved design of the MEMS actuators and microelectrode system. The new design should take into account the possible failure modes and try and minimize the interaction of bulky polysilicon moving structures that could lead to catastrophic joint failures. Also, mechanism of movement should be simplified such that the microelectrode can move reliably in both the forward and backward direction. The position of the microelectrode in the brain can then be reliably estimated (without taking into account the viscoelastic properties of the brain tissue). Improved packaging and mounting structures can enhance the longevity of the brain implant by creating a stable interface at the surgical site (between the implant and the craniotomy) and inside the brain tissue.

REFERENCES

[1] Jackson, N., et al., *Nonhermetic Encapsulation Materials for MEMS-Based Movable Microelectrodes for Long-Term Implantation in the Brain.* J Microelectromech Syst, 2010. **18**(6): p. 1234-1245.

[2] Erismis, M.A., et al., *A water-tight packaging of MEMS electrostatic actuators for biomedical applications.* Microsystem Technologies, 2010. **16**(12): p. 2109-2113.

[3] Panchawagh, H.V., et al., *A flip-chip encapsulation method for packaging of MEMS actuators using surface micromachined polysilicon caps for BioMEMS applications.* Sensors and Actuators A:Physical, 2007. **134**(1): p. 11-19.

[4] Muthuswamy, J., et al., *An array of microactuated microelectrodes for monitoring single-neuronal activity in rodents.* IEEE Trans Biomed Eng, 2005. **52**(8): p. 1470-7.

[5] Muthuswamy, J., et al., *Electrostatic microactuators for precise positioning of neural microelectrodes.* IEEE Trans Biomed Eng, 2005. **52**(10): p. 1748-55.

[6] Jackson, N., et al., *Long-Term Neural Recordings Using MEMS Based Movable Microelectrodes in the Brain.* Frontiers in Neuroengineering, 2010. **3**: p. 1-13.

[7] Jackson, N. and J. Muthuswamy, *Flexible Chip Scale Package and Interconnect for Implantable MEMS Movable Microelectrodes for the Brain.* J Microelectromech Syst, 2009. **18**(2): p. 396-404.

[8] Saha, R., et al., *Highly doped polycrystalline silicon microelectrodes reduce noise in neuronal recordings in vivo.* IEEE Trans Neural Sys Rehab Eng, 2010. **18**(5): p. 489-97.

[9] Saha, R. and J. Muthuswamy, *Structure-property relationships in the optimization of polysilicon thin films for electrical recording/stimulation of single neurons.* Biomed Microdevices, 2007. **9**(3): p. 345-60.

A Swallowable diagnostic capsule with a direct access sensor using anisotropic conductive adhesive

Pio Jesudoss, Alan Mathewson, Karen Twomey and Frank Stam

Heterogeneous System Integration Group, Microelectronics Applications Integration

Tyndall National Institute

Cork, Ireland

+353214904441, pio.jesudoss@tyndall.ie

William M.D. Wright

Department of Electrical and Electronic Engineering

University College Cork

Cork, Ireland

Abstract— Technological developments in biomedical microsystems are opening up new opportunities to improve healthcare procedures. Swallowable diagnostic sensing capsules are an example of this. In this paper, a novel direct access sensor (DAS) has been demonstrated which uses Flip Chip (FC) technology to expose the sensor to the liquid medium. An electrochemical study showed that the Anisotropic Conductive Adhesive (ACA) joint provides good connection and does not impair the sensor functionality. The reliability test results showed that most of the samples survived the humidity aging test and that only 2 out of 9 ACA connections of the same electrode failed. For the failed samples, the failure analysis showed that the tensile stress at the chip/epoxy interface caused a delamination at this interface.

Keywords-component; Anisotropic conuctive ahesive (ACA); Flip hip Over Hole (FCOH); Direct Access Sensor (DAS); reliability; swallowable diagnostic sensing capsule.

I. INTRODUCTION

In medicine, Inflammatory Bowel Disease (IBD) is a group of inflammatory conditions that affect the Gastro-Intestinal (GI) tract [1]. Although the IBD can be divided into several categories, the two major forms of IBD are Crohn's disease (CD) and ulcerative colitis (UC) [2]. There has been a rapid growth of IBD in Europe and North America during the second half of the twentieth century and it is becoming more prevalent in the rest of the world as they adopt the western life style [3]. CD and UC are chronic disease which can lead to long-term

and sometime irreversible impairment of the GI tract [1].

One of the conventional methods to investigate any suspected pathology is to use an endoscope which is inserted through patient's mouth, nose or rectum. These procedures provide some information: gastroscopy provides information about the Oesophagus and the stomach while the colonoscopy helps investigate the large intestine. These procedures are not only unpleasant for the patients but are also unable to provide information from the small intestine.

With recent advances in microelectronics, wireless communication and sensor development, the limitations of endoscopy is overcome in the format of a biomedical swallowable capsule [4]. The swallowable capsule is an autonomous system which contains a sensor, the associated electronics for signal conditioning and amplifying and a radio transmitter all encapsulated in a biocompatible material. The swallowable capsule involves a non-invasive technique which can provide information about the whole GI tract. The concept of first radio telemetry ingestible capsule was put forward by R.Stuart Mackay and Bertil Jacobson in 1957 [4, 5].The swallowable capsules can be classified into families of imaging (PillCam, Olympus Optical) [5-7], drug delivery systems [5, 7, 8] and sensing capsules [4, 5, 7-15]. Unlike the imaging and the drug delivery capsules where none of the parts are exposed, the chemical sensing capsules have one or more sensors that measures biochemical variables related to the gut ecosystem through exposed sensors. But in none of the diagnostic sensing capsules, the sensor attachment -the first level packaging of the

978-1-4244-9113-1/11 $26.00 © 2011 IEEE

sensor in a swallowable capsule- is achieved by Flip Chip Over Hole (FCOH) method using anisotropic conductive adhesive. ACA's relatively simple process steps [16] make it suitable for bonding a DAS. In a DAS, ACA not only provides the electrical interconnection but simultaneously seals the interconnect area and the underlying electronics from the sensor area in a capsule application, as shown in Fig. 1. The adhesive is more prone to moisture uptake than chip or metals used in track and pads [17]. The capsule transition through the GI tract takes around 72hrs during which the epoxy matrix will come in direct contact with GI fluids. As the adhesive comes in contact with fluid, one of the most severe tests to be considered is the moisture sensitivity test. As there are no standardarized test procedures that could be found for biomedical humidity aging, the moisture humidity test was considered at 50°C/95%RH to study the reliability of the ACA for a DAS.

This paper describes the DAS achieved through FCOH and its endurance in terms of reliability of the ACA joint during humidity aging. The following section describes the development of the DAS and its electrochemical testing. This was followed by the reliability study of the DAS ACA joint and conclusion.

II. DAS DEVELOPMENT AND ELECTROCHEMICAL TEST

A. Chip

The test chip is a 6x6mm^2 die with a thickness of 0.525mm. It was fabricated using a multi-layer silicon technology and photolithography techniques. Gold and platinum were deposited on the chip sensing area. The sensing chip has 5 pads on the periphery with Input/Output (I/O) pad size of 300 micron square as shown in Fig. 2(a).The microelectronic sensor comprised of four gold working electrodes (WE) of 1 mm diameter and a platinum counter electrode (CE) of 2mm diameter. The distance between centers of counter electrode and the working electrode was set at 0.5mm. The sensor chip electrodes are used to evaluate the liquid medium with cyclic voltammetry.

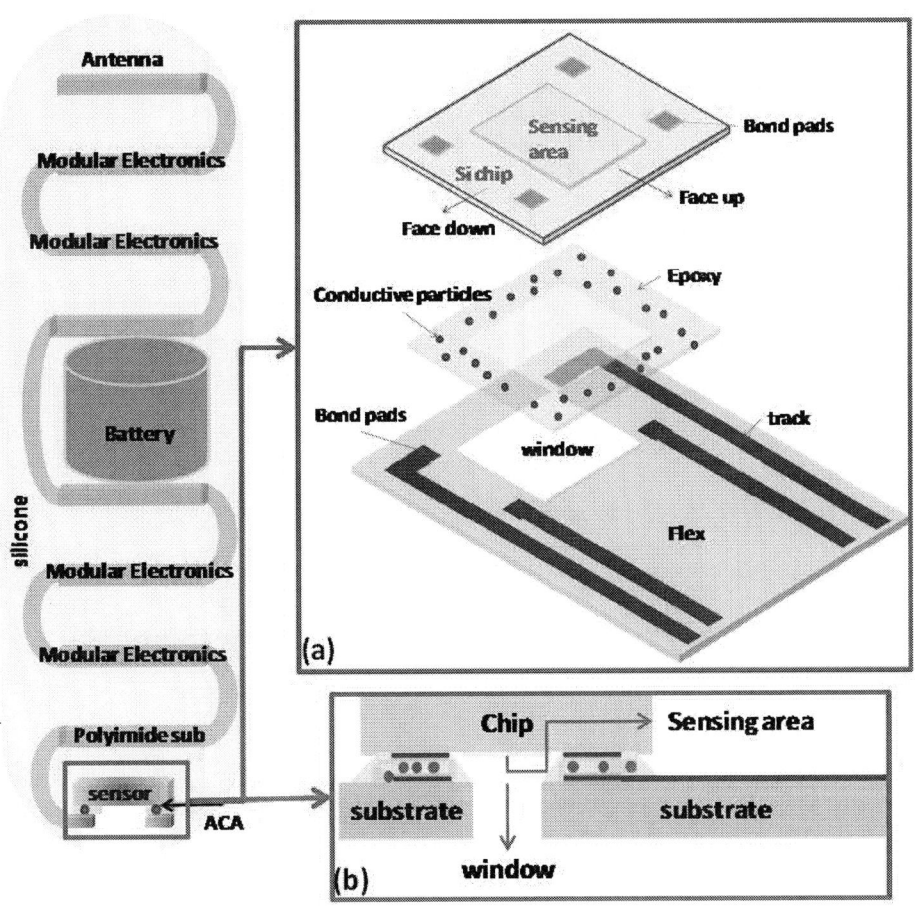

Fig. 1 Schematic of capsule with expanded image of FCOH (a) and (b) Cross sectional view of FCOC/DAS interconnections based on ACA

978-1-4244-9113-1/11 $26.00 © 2011 IEEE

B. Test Substrate

A single layer thin polyimide substrate with thickness 0.025mm was fabricated. The copper track on the flexible substrate was 15μm thick while on the bond pads an additional 5μm Ni and 0.05μm of electroplated flash gold was deposited. The flexible substrate is 77mm long, with a circular part of radius 6 mm at one end, a 3mm wide middle section and a 11.25mm width fanned out region on the other end. A square window of 4.4mm was cut from the centre of the board with a laser cutting tool to expose the sensors to the external environment, see Fig. 2(b).

C. Assembly process

A pre-cleaning procedure was carried out separately on both the chip and the substrate. This involved placing the chips and the substrates into a barrel type chamber of a March Plasmod system and exposing them to an oxygen plasma for 40sec at 150 watts. This was followed by IPA immersion in a bench top ultraware ultrasonic precision cleaner for 5 minutes followed by a DI water rinse. The samples were then dried in a conventional Heraeus vacuum oven at 150°C for an hour.

Gold stud bumps were formed on the die pads using a Kulicke and Soffa ball wedge gold bonder. The bumps had a mean diameter of 103μm and a mean height of 108μm. The gold stud bump was coined at 26N/mm^2 at a coining temperature of 180°C for 20 sec and the coined bump's diameter and the height were around 123μm and 18μm respectively.

ACA material from Loctite was dispensed on the test board using a CAM/ALOT 1414 liquid dispense system. A brown viscous epoxy paste with gold coated nickel filler particles of 7μm was used. The alignment and bonding of the chip/substrate was performed using Finetech Flip-Chip bonder. Bonding was carried out at a ramp rate of 2K/sec with a hold at 180°C for 40 sec and a cool down rate at 3K/sec and with a bonding pressure of 22N/mm^2 for 8 sec. The average bond line thickness of the adhesive taken from 10 samples was approximately 21μm with a standard deviation of 1μm.

D. DAS packaging

The test assemblies were then encapsulated in silicone. The encapsulation process consisted of the following: A region of silicone was dispensed on the perimeter of the window using

CAM/A LOT and cured in the oven at 80°C for 3hrs. The cured silicone acted as a dam around the window. The protection of the sensor was achieved by applying AZ photoresist – Diazonnaphthoquinones (AZ Electronic Materials GmbH) via pendant drop method - 6 drops of AZ on the sensor area - and cured at 80°C for 1 hr. This had a height of around 596.7μm and acted as a plug covering the exposed area of 19.36mm^2. Once the dam and the plug were ready, the assembly was inserted into a gelatin glycerin capsule (33mm*13mm) and secured in place. The fixed assembly was then filled with silicone and cured at room temperature for 24 hrs. This was followed by immersing the capsule in warm water (50°C) for 10-15 minutes to dissolve the glycerin capsule. The sensor was then exposed by dissolving the AZ photoresist in acetone for 5-10 minutes, as shown in Fig. 3.

E. DAS Testing – Cyclic Voltammetry

The electrical connection and the robustness of the packaging as well as the functionality of the sensor were tested using cyclic voltammetry on the three electrode cell comprising of WE, CE and the Pt R.E on the sensing chip. In such an analytical electrochemical setup CE, WE and R.E are placed in an electrolyte, the current is passed between the WE and the CE and the voltage measurement is made between WE and the R.E. Elecrochemical reactions occurring at the interface between the WE and the solution were monitored by a CH instruments 620B computer controlled potentiostat. The fabricated test assemblies were dipped into a solution of 0.5M of H$_2$SO$_4$ and cyclic voltammetry test at a scan rate of 0.2V/sec was applied to the electrode system. The chemical reactions that occurred at the gold WE in this solution are well documented [18] and any change in the performance of ACA or the component will be identified at this stage.

F. Results of DAS Testing – Cyclic Voltammetry

Fig. 4 shows the plot of current versus potential for a forward and reverse sweep as a result of the cyclic voltammetry. A peak was obtained at 1.4V during the positive voltage sweep – oxidation (ox) - from 0 to 1.5V, and a corresponding peak was obtained at 0.9V during the negative voltage sweep, reduction (red). These gold peaks were due to gold oxide formation and reduction and the corresponding gold reactions taking place are shown in equation (1) [19], and they illustrated the correct function of the sensor and interconnect.

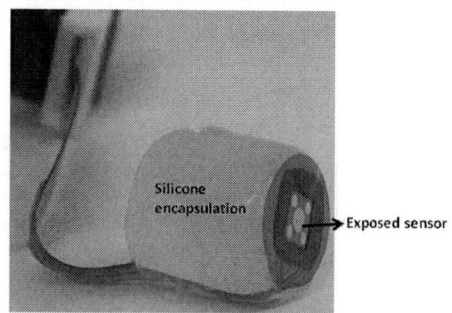

Fig. 3. Package sensor.

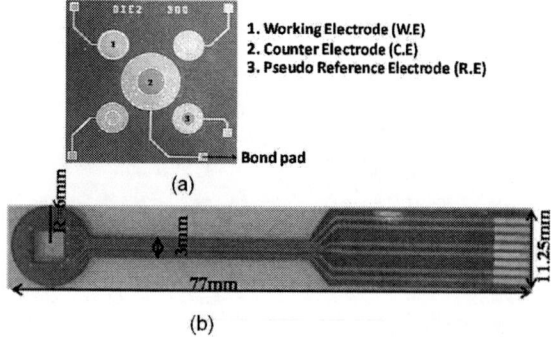

Fig. 2. (a) The direct access sensor chip and (b) Flexible substrate with window opening for sensor access.

$$2\,Au + 3H_2O \rightarrow Au_2O_3\,(hyd) + 6H^+ + 6e^- \quad (ox)$$
$$Au_2O_3\,(hyd) + H_2O \rightarrow 2AuO_2 + 2H^+ + 2e^- \quad (red)$$

(1)

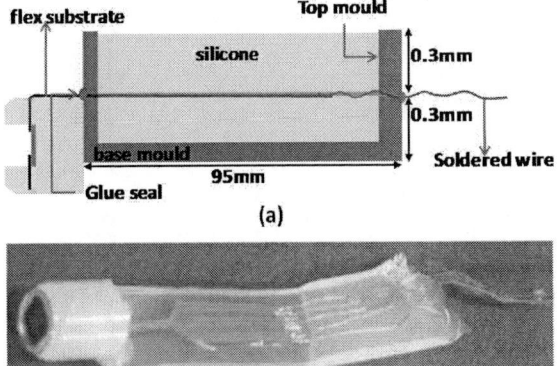

Fig. 5. (a) Cross-sectional schematic of the second encapsulation process and (b) picture of encapsulated sample for reliability testing.

III. RELIABILITY OF THE DAS ACA JOINT - CONSTANT HUMIDITY AGING

Once again, it is important to highlight that the ACA interconnect used in the DAS was destined to be used in the gut environment. It is crucial that the adhesive survive the 72 hour transition through the GI environment. As the adhesive comes in contact with fluid, one of the most severe tests that could be considered was the moisture sensitive test. As no standard test procedures that could be found for biomedical humidity aging, the constant humidity aging was carried out at 50˚C/95%RH to study the reliability of the ACA for a DAS. This part of the work focused on the endurance of the ACA attachment for a DAS.

A. Further Encapsulation process

Wires were soldered on to each flex board connector bond pads. A two part Polypropylene mould of 95mm length, 13.25mm wide and 0.3mm thick was fabricated where the first part had a base while the second part side was hollow. As shown in the cross-sectional Fig. 5(a), the sensor encapsulated sample was sandwiched between the base and the hollow mould and the edges sealed with glue. Silicone was poured and cured for 24 hrs. The cured sample was taken out by pulling the moulds apart and the encapsulated sample is shown in Fig. 5(b).

B. Electrical Measuremnts

As seen in Fig. 2(a), each electrode had its corresponding bond pad on the chip periphery. These bond pads made connection via the ACA joint to the matching band pad on the substrate thus making individual joint connection for each electrode. The DAS structure developed in this research was destined to be tested via electrochemistry and therefore an online four point measurement of the contact resistance was not possible. As a result the contact resistance of the ACA interconnect was measured using a HP 3441A multimeter

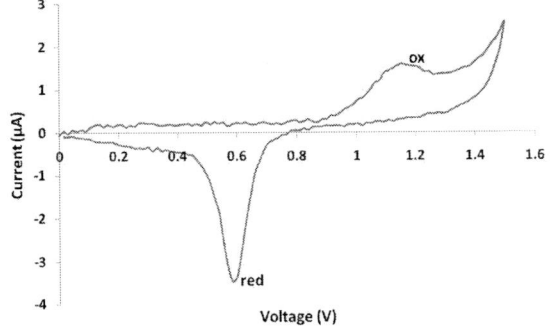

Fig. 4. Cyclic current-voltage curve of the Au W.E.

acting as a two probe ohmmeter. The resolution of the resistance measurement is $1m\Omega$. The error in the repeatability of the measurement was in the order of $8\text{-}10m\Omega$. The measurement was made by placing the probe tip on the electrode surface and at the tip of the wire. The resistance was separately measured for the chip, and the substrate. The same measurements were conducted on the device – chip mounted on the substrate via ACA and encapsulated in silicone. The measurements carried out on the substrate and the chip were subtracted from the contact resistance measurement on the device, consequently measuring the contact resistance of the ACA joint.

C. Results and Discussion

A batch of 9 samples were made. Each sample had 5 sensor connections. Fig. 6 shows the variation of the adhesive joint contact resistance on the 5 different connections on the same sample versus the duration of hygrothermal aging at 50˚C/95%RH. It can be seen that the initial contact resistance of the 5 adhesive connections range from $100m\Omega$ to $600m\Omega$. This discrepancy in the contact resistance could be explained by the number of conductive particles trapped in parallel configuration as shown in Fig. 7.

Fig. 8 shows the variation of ACA contact resistance corresponding to the same electrode connection of all 9 samples. Only 2 out of 9 ACA connections showed a slow increase in contact resistance until an open circuit was formed. It also showed that most of the samples showed no resistance drift during the whole reliability test. It can be seen in Fig. 6 and 8 that as the test went on, the contact resistance measurements showed some ripples that occurred at around 100 hrs of constant humidity aging.

As shown in Fig. 2 (a) only three electrodes noted 1, 2 and 3 were used in the electrochemistry. Only 2 out of 9 samples showed failure after hygrothermal aging for the contact joints representing the three electrodes. The samples that showed contact joint failure were samples 3 and 4. A 100% failure was observed in samples 3 and 4 at around 100hr of aging.

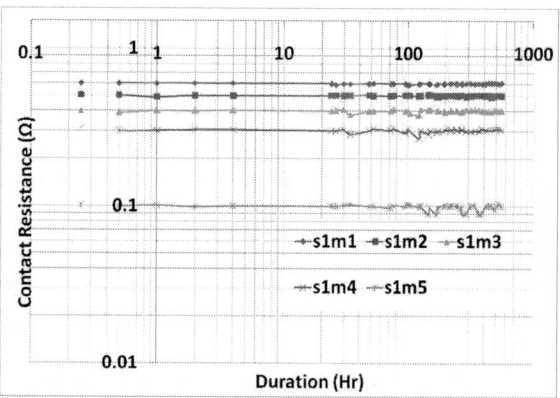

Fig. 6. Plot of ACA contact resistance vs. duration of humidity aging of all the electrodes in one sample.

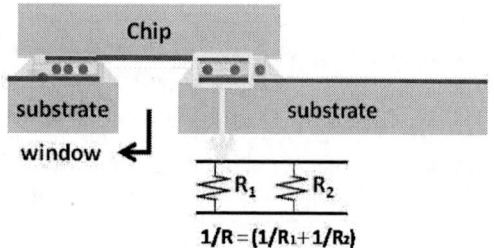

$$1/R = \{1/R_1 + 1/R_2\}$$

Fig. 7. Schematic of ACA interconnect showing the parallel configuration of the trapped conductive particles.

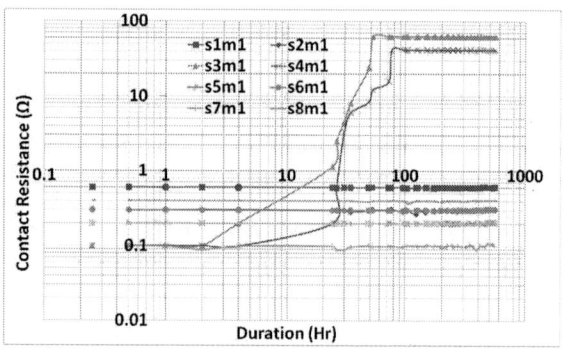

Fig. 8. Plot of ACA contact resistance vs. duration of humidity aging corresponding to the same electrode in all the 9 samples.

Epoxy expands due to moisture ingress while the silicon die and the other metal pads experience no expansion [20]. The loading conditions created a coefficient of thermal expansion (CTE) and coefficient of moisture expansion (CME) mismatch between the chip, substrate and the adhesive. Typical material parameters are shown in table 1.

Both CTE and CME mismatch between materials lead to thermal stress and hygroscopic stress at the interface between epoxy/chip and epoxy/flex substrate system. The diffusion

coefficient is greater for the adhesive. This will force the adhesive to absorb more water than other materials in the humidity chamber. Moreover the CME of the adhesive is higher than other materials. In our package configuration, the epoxy swelling due to moisture absorption produced a perpendicular as well as parallel expansion of the epoxy with respect to the die and the substrate and induced tensile and shear stress at the ACF interface. However former studies have showed that when the epoxy is saturated with moisture, the swelling was uniform and that the induced shear stress was insignificant when compared to the normal stress [17, 20].

For contact resistance measurement the samples were taken out of the humidity chamber through a side door and the measurements were carried out manually at room temperature using HP 3441A multimeter. The samples were out of the chamber for around 5- 10 minutes. One previous study suggested that the momentary transfer of the sample in and out of the humidity chamber caused a temporary concentration gradient which lead to either absorption or desorption of water from the adhesive. This constant absorption and desorption would cause the materials to expand and shrink. This constant movement of the sample in and out of the chamber for testing could have caused microscopic sliding. This slow fatigue like process coupled with slow relaxation of the contact pressure could lead to a steady increase in contact resistance until an open circuit was observed [21]. This could explain the observed slow and steady degradation of the contact resistance shown in Fig. 8. The ripples that are observed could be attributed to the error in contact pressure when placing the probes on the sample during the measurement.

Polymer expansion due to moisture ingress and the mismatch of moisture expansion coefficients of materials may cause the formation of defects like cracks and delaminations [22, 23]. It is more likely that there are some initial microscopical delaminations present at the ACA interface caused by the defects present on the chip or substrate surface or by the process [20]. Moisture has an adverse effect on ACA's interfacial adhesion and may accelerate the delamination process by weakening the polymer interface. Fig. 10 (a) and (b) and Fig. 11 (a) and (b) present examples of the Scanning Acoustic Microscopy (SAM) study that was used to assess the sealing efficacy of the DAS before and after humidity testing. Fig.10 (a) and (b) show the SAM images of one the samples that survived the hygrothermal testing. No delamination was observed in the sample before and after humidity aging. On the contrary, Fig. 11 (a) and (b) show the SAM images taken on the failed samples after hygrothermal aging. It can be seen that

Table.1. Material parameters. Collected from [17, 24-26].

	Silicon chip	Polyimide substrate	Gold bump	ACA
Coefficient of thermal expansion (CTE) (ppm/ºC)	2.7	20	12.9	47
Coefficient of moisture expansion (CME) (mm³/g) (*10²)	0	1	0	4
Diffusion Coefficient (mm²/s) (*10⁻⁶)	1e⁻²⁴	5.0	1e⁻²⁴	9.7
Specific Heat, c (J/Kg-K)	700	1090	800	1000

there are delaminations present, these are indicated by rectangles on the figure, after the humidity aging. The moisture aging has a deleterious effect on the adhesion strength and the weakening of the adhesion is reflected by white patches in the SAM image. The SAM analysis does not show where the failure occurred and in order to get a precise picture of the failure a cross-sectional analysis was performed.

An Olympus BH2-UMA Optical microscope image of the samples that failed during aging were cross-sectioned and presented in Fig. 12. Failure occurred by an interfacial delamination at the epoxy/silicon and epoxy/bump interfaces. Previous studies have shown that the ACA adherence to another polymeric material is higher than that for silicon [27, 28]. In addition, the bump/pad interface was found open, resulting in the loss of conductivity and increase in resistance. The delamination seems to proceed along the chip passivation layer with a higher gap on the left side of the bump/pad towards the edge of the chip than on the right side of the bump/pad. This could imply that the delamination was initiated at the edge of the die. The disruption of the hydrogen bonds by water molecules caused plasticization in the short term by spreading the polymer molecules apart and causing expansion of the adhesive joint [29, 30]. In addition, the temperature gave rise to CTE mismatch between the adhesive and the other metals and contributed to the degradation of the ACA interface. Thus the interface was under tensile and shear stress with a

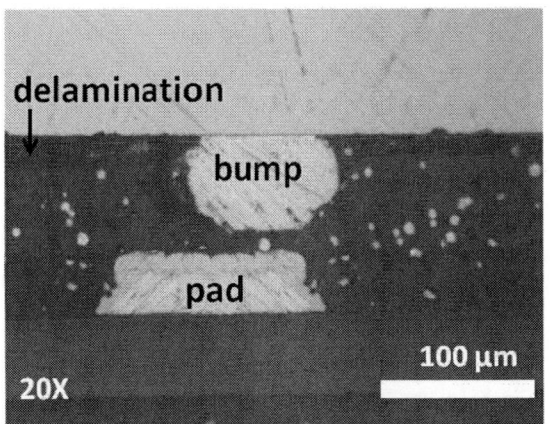

Fig. 12. Cross-section of samples after humidity testing - failure by crack propagation.

dominant tensile stress at the bump and die/epoxy interface. The fact that the delamination was parallel to the chip interface suggests that a pure tensile force was in action during the swelling process.

IV. CONCLUSION

In this study, a novel DAS has been developed with a modified FC technique to expose the sensor to the liquid medium. Furthermore ACA can be used as a suitable material for applications with few relatively large bond pads and particularly in relation to measurements in the fluidic environment when the sensing area needs to be sealed off from the electronics.

The electrochemical study of the ACA joint and the sensor after encapsulation with silicone showed the ACA joint provides good connection and the electronic functionality and chemical sensing performance wasn't compromised.

The reliability of the ACA attachment for a DAS was studied by constant humidity aging. At the end of the reliability test, out of the 9 samples tested, for the same electrode in all 9 samples, only 2 of the ACA connections exhibited a slow increase in resistance until an open circuit was observed. The slow resistance increase of the failed joint was attributed to a fatigue like process induced by removal and replacement of the sample in the humidity chamber. Failure analysis of the failed joint after reliability testing showed that the constant movement of the sample in and out of the chamber for room temperature testing resulted in swelling and shrinking of the adhesive causing a crack to initiate and propagate along the die-epoxy interface.

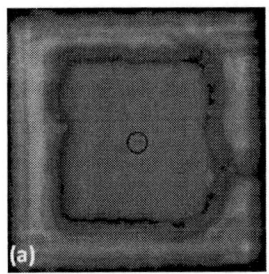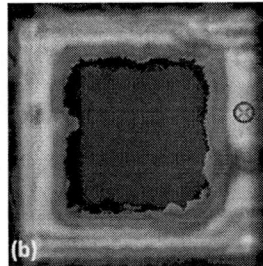

Fig. 10. SAM images (a) before and (b) after hygrothermal testing of one the samples that survived.

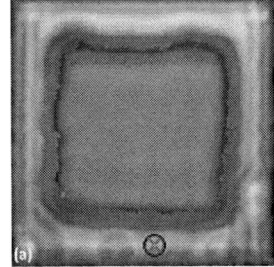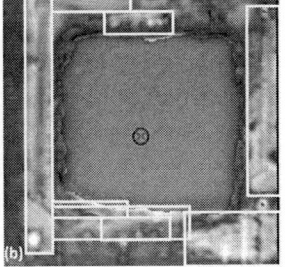

Fig. 11. SAM images (a) before and (b) after hygrothermal testing of one the samples that failed.

ACKNOWLEDGMENT

The author would like to acknowledge Enterprise Ireland CFTD /05 / 122 and HEA PRTLI-IV project NEMBES for providing the opportunity to carry out the work presented in this paper.

REFERENCES

[1] G. Bouma and W. Strober, "The immunological and genetic basis of inflammatory bowel disease", Nature Reviews Immunology, vol. 3, pp. 521-533, 2003.

[2] J. Lennard-Jones, "Classification of inflammatory bowel disease", Scandinavian Journal of Gastroenterology, vol. 24, pp. 2-6, 1989.

[3] R. Sartor, "Mechanisms of disease: pathogenesis of Crohn's disease and ulcerative colitis", Nature Clinical Practice Gastroenterology and Hepatology, vol. 3, p. 390, 2006.

[4] H. Battery, V. Capsule, E. Contract, and G. IDEAS, "Lab-on-a-Pill: The International Context of the Work: The invention of the transistor enabled".

[5] C. McCaffrey, O. Chevalerias, C. O'Mathuna, and K. Twomey, "Swallowable-capsule technology", IEEE Pervasive Computing, vol. 7, pp. 23-29, 2008.

[6] G. Costamagna, S. Shah, M. Riccioni, F. Foschia, M. Mutignani, V. Perri, A. Vecchioli, M. Brizi, A. Picciocchi, and P. Marano, "A prospective trial comparing small bowel radiographs and video capsule endoscopy for suspected small bowel disease", Gastroenterology, vol. 123, pp. 999-1005, 2002.

[7] K. Twomey and J. Marchesi, "Swallowable capsule technology: current perspectives and future directions", Endoscopy, vol. 41, pp. 357–362, 2009.

[8] "Philips Intellicap technology": http://www.research.philips.com/newscenter/backgrounders/081111-ipill.html.

[9] R. Allan, "Smart Pill Goes On A Fantastic Voyage", Electronic Design, vol. 54, pp. 27-66, 2006..

[10] R. Dettmer, "Fantastic voyage [wireless capsule endoscopes]", IEE Review, vol. 51, pp. 28-32, 2004.

[11] L. Wang, E. Johannessen, P. Hammond, L. Cui, S. Reid, J. Cooper, and D. Cumming, "A programmable microsystem using system-on-chip for real-time biotelemetry", IEEE Transactions on Biomedical Engineering, vol. 52, pp. 1251-1260, 2005.

[12] J. Cooper, E. Johannessen, and D. Cumming, "Bridging the gap between micro and nanotechnology: using lab-on-a-chip to enable nanosensors for genomics, proteomics, and diagnostic screening", Network and Parallel Computing, pp. 517-521, 2004.

[13] E. Johannessen, L. Wang, S. Reid, D. Cumming, and J. Cooper, "Implementation of radiotelemetry in a lab-in-a-pill format", Lab on a Chip, vol. 6, pp. 39-45, 2006.

[14] E. A. Johannessen, W. Lei, C. Li, T. Tong Boon, M. Ahmadian, A. Astaras, S. W. J. Reid, P. S. Yam, A. F. Murray, B. W. Flynn, S. P. Beaumont, D. R. S. Cumming, and J. M. Cooper, "Implementation of multichannel sensors for remote biomedical measurements in a microsystems format", Biomedical Engineering, IEEE Transactions on, vol. 51, pp. 525-535, 2004.

[15] T. Tang, E. Johannessen, L. Wang, A. Astaras, M. Ahmadian, L. Cui, A. Murray, J. Cooper, S. Beaumont, and B. Flynn, "IDEAS: a miniature lab-in-a-pill multisensor microsystem", 2002.

[16] R. Briegel, M. Ashauer, H. Ashauer, H. Sandmaier, and W. Lang, "Anisotropic conductive adhesion of microsensors applied in the instance of a low pressure sensor", Sensors and Actuators A: Physical, vol. 97-98, pp. 323-328, 2002.

[17] K. Saarinen and P. Heino, "Moisture effects on reliability of non-conductive adhesive attachments", in 2nd IEEE International Interdisciplinary Conference on Portable Information Devices, and the 7th IEEE Conference on Polymers and Adhesives in Microelectronics and Photonics, 2008, pp. 1-6.

[18] L. D. Burke and B. H. Lee, "An investigation of the electrocatalytic behaviour of gold in aqueous media", Journal of Electroanalytical Chemistry, vol. 330, pp. 637-661, 1992.

[19] E. Á. D. Eulate, "Investigation of an Electronic Tongue Array for Gastro-Intestinal Disease Detection", in Department of Chemistry, Master of Science Cork: University College Cork, 2007, unpublished.

[20] L. L. Mercado, J. White, V. Sarihan, and T. Y. T. Lee, "Failure mechanism study of anisotropic conductive film (ACF) packages", IEEE Transactions on Components and Packaging Technologies, vol. 26, pp. 509-516, 2003.

[21] J. F. J. M. Caers, X. J. Zhao, J. W. C. de Vries, E. H. Wong, G. Kums, and A. R. C. Engelfriet, "HART: A new highly accelerated robustness test for conductive adhesive interconnects", in 58th Electronic Components and Technology Conference, 2008, pp. 1695-1699.

[22] Y. Lin and J. Zhong, "A review of the influencing factors on anisotropic conductive adhesives joining technology in electrical applications", Journal of Materials Science, vol. 43, pp. 3072-3093, 2008.

[23] J. Liu, X.-z. Lu, and L.-q. Cao, "Reliability aspects of electronics packaging technology using anisotropic conductive adhesives", Journal of Shanghai University (English Edition), vol. 11, pp. 1-16, 2007.

[24] L. Teh, M. Teo, E. Anto, C. Wong, S. Mhaisalkar, P. Teo, and E. Wong, "Moisture-induced failures of adhesive flip chip interconnects", IEEE Transactions on Components and Packaging Technologies, vol. 28, pp. 506-516, 2005.

[25] C. Yin, H. Lu, C. Bailey, and Y. Chan, "Macro-micro modelling of moisture induced stresses in an ACF flip chip assembly", Soldering & Surface Mount Technology, vol. 18, pp. 27-32, 2006.

[26] C. Wu, J. Liu, and N. Yeung, "The effects of bump height on the reliability of ACF in flip-chip", Soldering & Surface Mount Technology, vol. 13, pp. 25-30, 2001.

[27] S. Gi-Dong, C. Chang-Kyu, P. Kyung-Wook, and L. Soon-Bok, "Experimental analysis on the mechanism of moisture induced interface weakening in ACF package", European Microelectronics and Packaging Conference, EMPC, pp. 1-7, 2009.

[28] G.-D. Sim, C.-K. Chung, K.-W. Paik, and S.-B. Lee, "Moisture induced interface weakening in ACF package", Materials Science and Engineering: A, vol. 528, pp. 698-705, 2010.

[29] T. Ferguson and J. Qu, "The Effect of Moisture on the Adhesion and Fracture of Interfaces in Microelectronic Packaging", in Micro- and Opto-Electronic Materials and Structures: Physics, Mechanics, Design, Reliability, Packaging, E. Suhir, Y. C. Lee, and C. P. Wong, Eds.: Springer US, 2007, pp. B431-B471.

[30] G. Xian and V. M. Karbhari, "DMTA based investigation of hygrothermal ageing of an epoxy system used in rehabilitation", Journal of Applied Polymer Science, vol. 104, pp. 1084-1094, 2007.

Application Based Reliability Assessment and Qualification Methodology for Medical ICs

Xiaowei Zhu, Karthik Vasanth, Xiaochen Xu, Charles Smyth and Brent Rhoton
Medical/HiRel Business Group, High Performance Analog
Texas Instruments
Dallas, TX, US
phone: (1) –(972)- 836-6778, e-mail address: vivi@ti.com

Abstract—Reliability assessment and qualification system has strong economic implications for both manufacturers and customers. The best system should have a good balance among cost of verification, market timing requirement, and acceptable risk that meets the targeted user's application conditions and requirements. With the increasing use of innovative electronics in the medical applications, it becomes difficult to have a single reliability assessment and qualification approach to serve all applications. In this paper, we review the existing reliability assessment and qualification framework, and discuss their applicability in medical ICs. We will discuss the tradeoff and challenges in defining reliable medical IC products based on the application demands using a couple of medical IC examples.

Keywords-medical IC; ultrasound; qualification; reliability assessment; aging

I. INTRODUCTION

In a very simple view, quality is a measure of the ability of a product to meet its data sheet specifications at time zero, whereas reliability evaluates its ability to continue meeting those specifications throughout the product lifetime. In some life critical medical applications, the quality and reliability of the semiconductor chips can have a direct impact on human lives. Naturally, these chips need to go through a very stringent verification and qualification process. Meanwhile, the ICs in some of the disposable medical devices are expected to last only for several hours in a very benign use condition. It is not economically viable to put these ICs through the same process as those used in the life-critical applications. Therefore, having a tailored approach of reliability assessment and qualification for the medical ICs is critical to the business success in the medical market segment.

Historically, the semiconductor reliability assessment methodologies were developed for military or space applications, where long life time and harsh environment are expected, and for which, it is often very expensive to replace or repair a defective component. The qualification approach is more or less stress test driven [1][2]. As the application types shifted to be more consumer oriented, many of these stress test standards may be overkill for the intended application use condition. Another qualification approach based on Physics of Failure (PoF) has been developed [3]. This approach takes into the consideration of the field use conditions, the expected failure modes and mechanisms, and the acceleration method that covers the known failure modes for the entire product lifetime through testing.

There are pros and cons in both approaches. Having one single reliability assessment and qualification approach to serve all applications is economically appealing, but may not be feasible in the medical IC market segment, as we will show later.

In this paper, we present a tailored approach to perform reliability assessment and qualification for medical ICs based on their application classifications. In section II, we review the general reliability assessment and qualification considerations. In section III, we present an overview on medical ICs and how we classified them based on their application and qualification requirements. In section IV we review the trends, challenges and solutions in defining successful medical IC products using medical ultrasound application as an example. We discuss the challenges in developing the optimal reliability assessment and quality system in the context of the specific application conditions and requirements. We have developed a method to estimate the equivalent lifetime when multiple competing failure mechanisms age through multiple application modes and stress conditions with different pace, such as seen in the medical ultrasound transmitter chip.

II. RELIABILITY ASSESSMENT GENERAL CONSIDERATION

Prior to diving into the specifics of the application and use conditions of medical ICs, we go through the process of reliability assessment and consideration in general. First, we need to understand the product manufacture and test flow, the transportation and storage conditions, and the field use conditions. Usually, component providers do not have direct control over the steps beyond shipment to customer; however, those steps could affect the component reliability. Therefore, it is the key for a successful reliability assessment to gather accurate post-shipment handling information from the customer. The relevant information includes, but not limited to the following:

- wafer level probe test conditions

- assembly conditions

- packaged chip final test conditions including power supply voltages, pin configurations, temperature, duty cycles etc

- shipping conditions (package form, temperature, humidity, vibration and shock)

- storage conditions (temperature and humidity)

- board level assembly and test conditions

- operating conditions (temperature and voltage, duty cycle)

- expected device operating lifetime, mode and power on hours

It is very important to understand the reliability requirement from the end use equipment manufacturers. Typically the specifications are the early life failure rate and cumulative end of life failure percentage.

After gathering the above information, we can then perform a failure mode and effect analysis with a cross functional team. During the analysis, a list of known failure modes and mechanisms will be verified, and the potential failure modes should be explored.

III. MEDICAL IC OVERVIEW

There are four major trends that have become the driving force behind the medical device market. 1) Aging populations: By 2020, there will be more than 1 billion people of age 60+; 2) Rising healthcare costs: US healthcare spending is roughly 16% of GDP in the recent years; 3) Need for access to medical diagnosis and treatment in remote and emerging regions; and 4) Need to manage personal healthcare in our own homes [4]. These trends will continue to fuel and power the innovative use and development of the semiconductor devices in the medical applications for years to come.

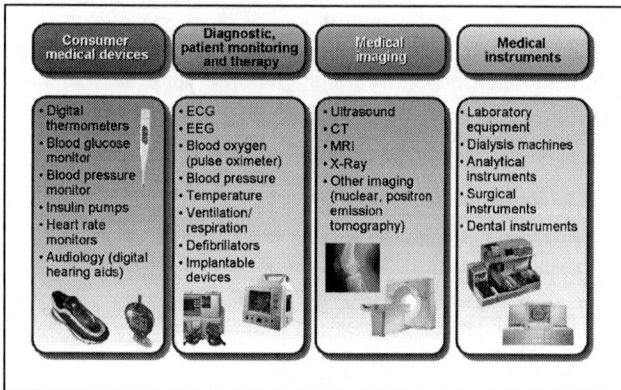

Figure 1. The opportunities for electronics in medical applications are diversified and abundant.

From semiconductor industry perspective, the opportunities for electronics in medical applications are diversified and abundant, from consumer medical devices to patient monitoring and therapy, from medical imaging to large medical instruments, the electronics are ubiquitous.

Despite of the diversified usage of medical applications, many medical electrical systems can be described by a generic block diagram, as shown in Figure 2. The heart of the system is a digital processor, being a microcontroller or a DSP. Analog

chips such as amplifiers, data converters take the real world signals from the sensors and convert them into electric digital signal for the system to process. Necessarily, all these devices require power management, whether the system uses a plug from the wall or a battery. Finally, if there is a need to connect these systems to data collection centers, or patient monitoring devices to IT systems, either through a wired connection or a wireless connection, there are opportunities for products in the portfolio of connectivity solutions.

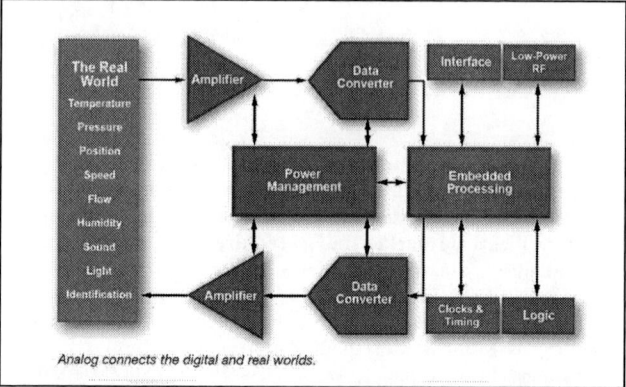

Figure 2. A generic block diagram for a typical medical system

Classified by volumes and application types, medical ICs portfolio could be viewed as a pyramid, shown in Figure 3. The bottom of the pyramid is composed of broad analog and digital catalogue devices; they find their ways into almost every medical electronic application. The middle section of the pyramid identifies application-specific products, such as the analog front end chips targeted for the ultrasound applications, which will be discussed at length in section IV. These products are very specifically designed for targeted medical applications and finely tuned to their signal processing, low power and wireless connectivity requirements. At the top are found niche devices such as implantable pacemakers, defibrillators, bio-sensors and virus diagnostic chips. They represent one of the most creative and fascinating applications of electronics. For example, thanks to the arrival of implanted deep brain stimulator, the life quality of patients with Parkinson's disease has improved greatly. Electrically controlled micro-dose drug delivery system will undoubtedly impact close-loop medical treatment in the future.

The quality and reliability are the cornerstones of building a successful medical electronics portfolio, as it has direct impact to a company's reputation, customer loyalty and satisfaction. However, good quality and reliability for medical ICs imply the ability to serve customers of all sizes with a variety of needs, and translate into quite different requirements for the three tiers of chips in the pyramid. For those catalogue chips at the bottom, their design consideration and qualification methods are very similar to the ones used in other applications. For the ICs used in life critical application, such as the implantable ICs, reliable operation is on the top of the care-about list. The life critical devices need to go through much elaborate verification and validation steps compared to the devices for general use. The component suppliers need to work closely with the medical device manufacturers to ensure reliability goals are

978-1-4244-9113-1/11 $26.00 © 2011 IEEE

met and to adequately assess and mitigate any potential risks to the end user [5]. For the application specific chips in the middle section of the pyramid, the requirements need to be balanced with power, performance and cost demands coming from that specific application. The needs in some cases translate into the enhanced qualification and screening, enhanced package offering, while in other cases, they translate into a more robust process change control notification, and/or long product lifetimes and/or a robust product obsolescence policy to ensure continuity of supply.

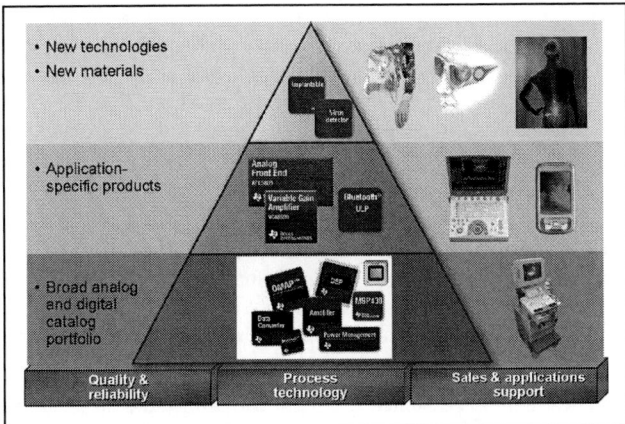

Figure 3. Medical ICs include large quantities of catelog devices, some application specific products and a small volume of the niche devices such as implantable ICs.

In the next section, we use a medical ultrasound system as example to dive into the details of some application specific demands, and illustrate some of the challenges when all these demands are considered together.

IV. TRENDS, CHALLENGES & SOLUTIONS

A. High Integration and Low Power Trend Leads to Improved System Reliability

In all medical market segments, we see high levels of miniaturization and integration in portable equipment. For example, manufacturers of today's advanced portable or handheld ultrasound systems demand highly integrated, scalable solutions. This allows medical professionals to move beyond the lab or office to reach clients in remote settings or emergency situations around the world [6].

Along with the desire for portable equipment comes the demand for low power solution. Making power management decisions early in the design cycle will help define system-level tradeoffs that may be necessary to meet portability and run-time targets. Smaller portable medical products may use disposable batteries, whereas larger systems might leverage various rechargeable battery chemistries and battery pack sizes. Power reduction, in general, is a favorable trend from reliability point of view, as many failure mechanisms accelerate with increasing temperature, and lower power implies lower junction temperatures.

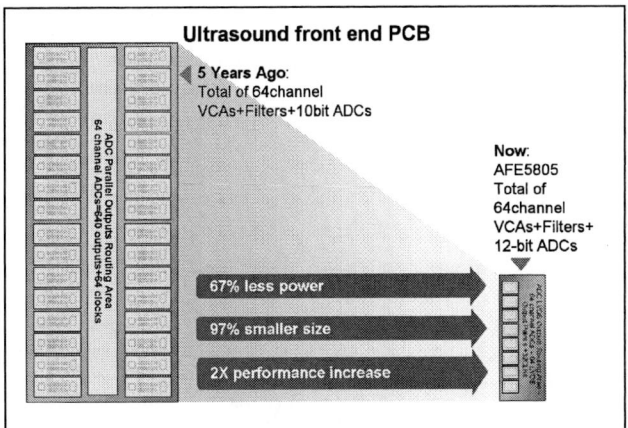

Figure 4. Enabled by high integration level, ultrasound analog front end chips have a greatly reduced footprint on the PCB board.

With the demand for miniaturization and integration, many of the blocks are integrated into one single chip. As an example, shown in Figure 4. to build signal chains on the receiving end for a 64 channel ultrasound system, we only need eight TI's AFE5805 chips as each chip integrates eight channels of voltage-controlled-amplifiers (VCAs), filters and analog-digital-converters (ADCs). Using the components available five years ago, we needed 32 two-channel VCAs and 32 two-channel ADCs, plus 384 discrete passive components for the differential filters. Assuming a defect level of 100 DPPM for each component, the block failure fraction could be as high as 3.8% five years ago, in today's system it is only 0.08%. Therefore, the high degree of integration not only leads to cost reduction by saving the board room and reduces power consumption, but also greatly improved imaging performance and the system level reliability. At the same time, some of the system level requirements and tradeoffs between performance, power dissipation, noise, size and reliability are transferred to the component level.

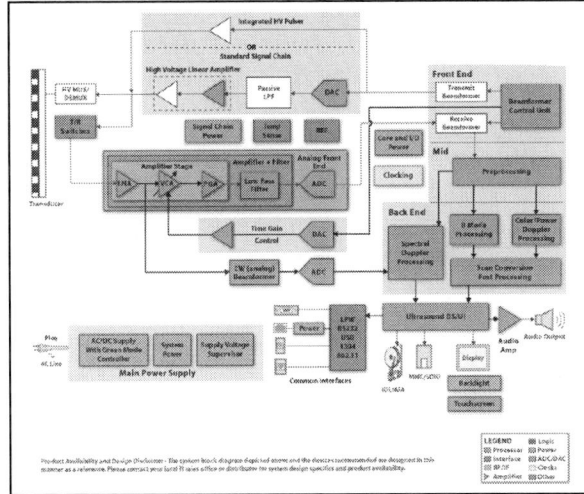

Figure 5. Block diagram of an ultrasound machine.

978-1-4244-9113-1/11 $26.00 © 2011 IEEE

B. Medical Ultrasound Imaging Basics

To understand system requirements imposed on the chip level due to the high integration, and how these application specific requirements translate to the chip level, some basic understanding of how medical ultrasound works is necessary. Because the analog front end chips are chosen for the case studies, we mainly focus on the principle associated with the signal chain of the system.

A block diagram of a medical ultrasound machine is shown in Figure 5. Ultrasound images are obtained by sending a narrow beam of acoustic energy into the body and reconstruction of the signals reflected back by structures within the body. Ultrasound transducers, consisting of multiple piezoelectric elements, are responsible for converting the electric pulses to acoustic energy and back. To produce acoustic waves with sufficient energy to penetrate the body, the transmitter (TX) chip needs to output a high-voltage impulse, which is typically a shaped burst of 1 to 50 cycles at frequencies ranging from 1MHz to 20MHz with amplitudes of up to ±100V. Meanwhile, the returning signal levels range from large input signals (~1V) due to reflections near the surface of the body to less than 10 μV for echoes occurring deep within the body [8][10]. The operation of an ultrasound imaging machine is very similar to that of radar, it sends acoustic waves to the target, and then "listens" to the reflected signals.

When sound travels through a medium, its intensity diminishes with distance. Attenuation coefficient in biological tissues has units of dB/(MHz*cm). This means that although higher frequencies provide better resolution, their penetration depth is limited. The magnitude of the pulse determines the amount of energy beamed into the object. The object to be imaged, energy needed for tissue penetration, and acquisition modes dictate the transmitter pulse amplitude, frequency, pulse cycle and pulse repetition frequency (PRF).

TABLE I.   ULTRASOUND MODES FOR TYPICAL MEDICAL APPLICATIONS

	Mode		
	B-Mode	**PW Doppler**	**CW Doppler**
Voltages (V)	+/-90	+/-75~30	+/-8~4
Frequency (MHz)	1-20	1-10	1.5-6
PRF(kHz)	1-10	1-40	N/A
Number of Cycles	1-3	4-32	infinite
Effective duty cycle at typical application condition	0.2%	1-3%	100%
Operation time (B-Mode)	100%	N/A	N/A
Operation time (B-D Duplex Mode)	70%	30%	N/A
Operation time (B-D-CW Tripple Mode)	50%	30%	20%

TABLE I. lists the operation modes and associated parameters in that mode that the TX chip experiences in typical applications for different ultrasound systems. In the current market, the low end system mainly use only B-mode, which is used for black and white 2D imaging; while mid-end system

has PW Doppler-mode added to it, where PW or color Doppler-mode is used to acquire blood flow information using Doppler principle; More complex systems also include continuous Doppler (CW-mode), which is for measuring ultra-high or ultra-low flow velocities. Very often the color image that indicates blood flow speed and direction is overlaid on the B-mode display of the blood vessel, so there is a need to switch between the B-mode and color modes [9].

We now analyze two case studies of challenges and quality control requirements of the front end analog chips in the medical ultrasound system. In the first case, we use the analog transmitter chip to review some of the challenges encountered for high voltage and high current applications. In the second case, we use a receiving chip specification to illustrate how an application needs translate into a quality control requirement at chip or component level.

C. Challenges presented by application condition

The complicated application profile presents several challenges to the chip verification, validation and qualification for the analog chips used in ultrasound system. Frequency, voltage level and temperature are the three primary parameters to review.

The frequency selected for a specific ultrasound application condition is based on the tissue penetration characteristic and dictated by the central frequency of the piezoelectric transducer. High level of integration, combined with multiple key variables, sometimes make exhaustive characterization covering all permutations unrealistic if not impossible. Usually the typical performance data are characterized at a single frequency by the chip manufacturer and listed in the datasheet. The customer may expect identical level of performance at other frequencies, thus a possible gap in customer expectation may exist. In addition, some of the system performance parameters at different frequencies could be affected by external components such as coupling capacitors. As a result, the system design decisions could greatly impact the chip performance. Very often chip supplier and system designer need to work together to solve the issues. In some cases, a solution for one application may lead to an issue for another, and this is a common challenge faced by a component supplier to support the needs of a broad customer base.

For ultrasound TX chips, the peak drive current of 2A is not an uncommon condition when the high voltage pulses are generated. As a result, the temperature of the transmitter chip can become quite high when operating under long duty cycle condition. Measured data in Figure 7. shows the TX chip temperature increases with the duty cycle. The fact that high voltage high temperature condition exists in the normal application modes makes it quite difficult to define a failure acceleration test for the TX chip.

To verify that a chip can operate reliably through its intended operation life time, accelerated life tests are carried out to quicken the degradation process. For silicon related failure mechanisms, the common acceleration methods are either by elevated voltage or elevated temperature or both. The total acceleration factor AF is the product of voltage acceleration factor AFV and temperature acceleration AFT, as shown in (1):

$$AF = AFV(Vstress, Vop) \cdot AFT(Tstress, Top) \qquad (1)$$

The Arrhenius equation (2) below is used to model acceleration due to elevated temperature test. The Arrhenius equation relates how increased temperature accelerates the age of a product as compared to its normal operating temperature.

$$AFT(Tstress, Top) = e^{\frac{E_a}{k}\left(\frac{1}{Top} - \frac{1}{Tstress}\right)} \qquad (2)$$

AFT = temperature acceleration factor
Ea = activation energy in electron-volts (eV)
k = Boltzmann's constant (k = 8.617 x 10-5eV/Tk)
Top = junction temperature during operation, in degrees Kelvin
Tstress = junction temperature during test, in degrees Kelvin
e = 2.71828 (base of the natural logarithms)

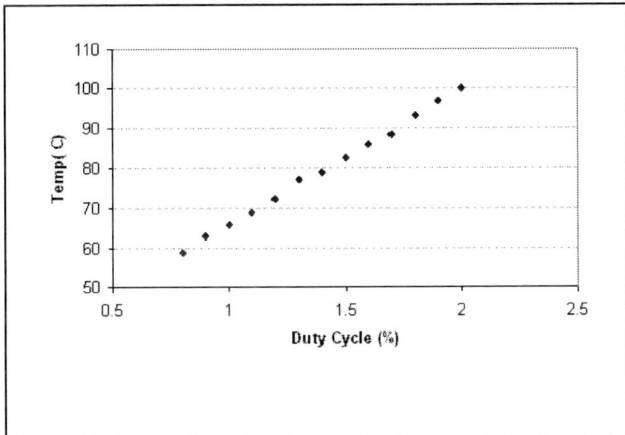

Figure 6. Measured maximum temperature of the TX chip vary signficantly as a function of duty cycle.

While for almost all types of silicon-related failure mechanisms, the temperature dependence of their lifetime could be modeled by Arrhenius relation, the voltage dependence of their life time can not be described by a single model. Some defects follow an exponential model described by (3), while others follow a power law model described by (4).

$$AFV(Vstress, Vop) = e^{N(Vstress-Vop)} \qquad (3)$$

$$AFV(Vstress, Vop) = \left(\frac{Vstress}{Vop}\right)^{N} \qquad (4)$$

AFV = voltage acceleration factor
N = model coefficient

Vop = voltage applied during operation
Vstress = voltage applied during test

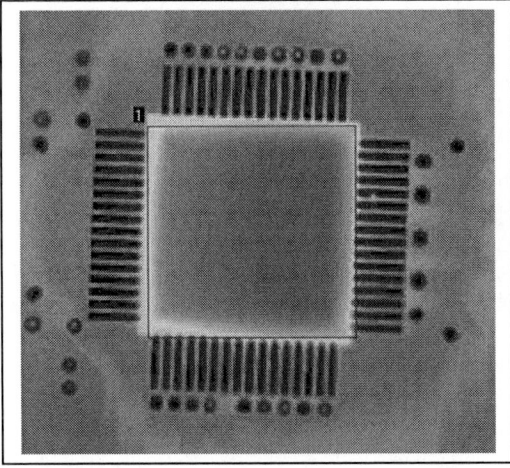

Figure 7. Thermal map of the transimitter chip indicate significant temperature difference, 50C in some cases, exists on the chip under the same application mode.

It is desirable to have a lifetime test with high acceleration factor without altering the failure mechanism. For TX chips, high voltage coexists with high temperature condition; this leaves little room for failure acceleration. In addition, since the chip temperature is dependent of the power and the duty cycle, which is dependent of operation mode, scoping out voltage and temperature conditions during the field operation itself is a challenging task. As shown in TABLE I. depending on which type of ultrasound machine the TX chips go into, the operation modes and then the percentage of its life time the chip is used in that mode can be quite different. Although it seems intuitive to believe that the worst aging case would be to put the chip in B-mode 100% of the time during the test, since the voltage amplitude of the transmission pulse is the highest, we will demonstrate in section E it is important to pay attention to the stress level during the mode switching period. Potentially there could be transient voltage overshoot and/or undershoot during the mode switching period, and they could consume quite large reliability budget, and greatly shorten the equivalent lifetime of the product.

Further analysis of the measured device thermal map data indicates that large temperature gradients exist on the device. Figure 7. shows the thermal map of a TX device. In some application mode, the temperature difference at different location of the die is as large as 50C. To account for this effect, chip area partitioning by the chip thermal map needs to be considered in the reliability model.

The problem is further complicated by the fact that there could be multiple failure mechanisms, each accelerated at different pace at different voltage and temperature conditions [11]. Therefore, there is a need for a method to evaluate and compare the amount of damage under different application scenarios.

978-1-4244-9113-1/11 $26.00 © 2011 IEEE   235

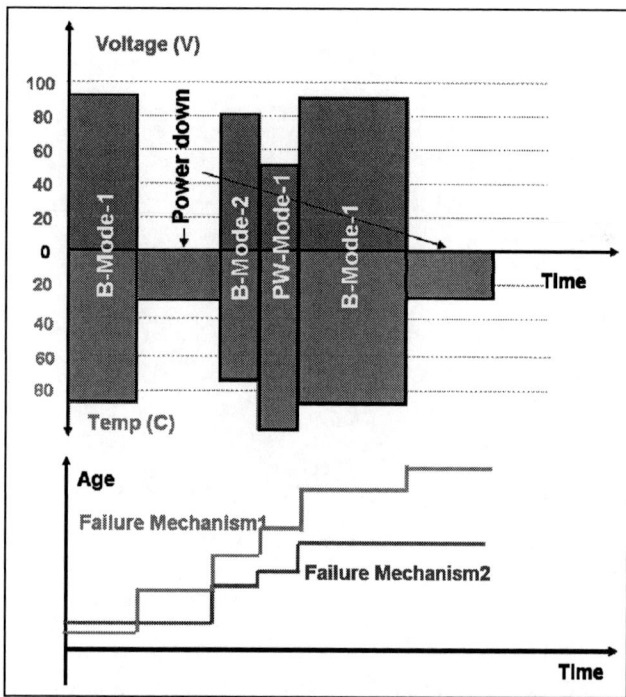

Figure 8. The amount of damage at any stress condition for a period of time can be converted and normalized to add to the cumulative failure fraction, or age, until 100% of the failure fraction is reached.

D. Application based aging model and method

In response of the challenges presented by the complicated application conditions such as seen in the ultrasound case, we have developed a method of reliability assessment. This method can be used to identify the worst case application use profile, and evaluate if a certain design meet the reliability target. It can also be used to map out the equivalent life time proved by a particular accelerated life test condition, which is very useful to define the high temperature operating life (HTOL) test during the chip qualification.

The core idea of this method is to use a normalized age to benchmark and compare the cumulated damage caused by different failure mechanisms. Figure 8. illustrates the aging process. An ultrasound TX chip experiences high voltages and high temperatures when an ultrasound operator sets the system in B-mode and PW-mode. When the machine is powered down, the voltage of the chip becomes zero, and the temperature returns to the ambient temperature. Each stress step adds damage to the chip and advances its aging process. The amount of damage caused by the same stress environment, however, is different for different failure mechanisms. A given voltage may be necessary to advance the aging process, of a failure mechanism and not others. Multiple failure mechanisms therefore, race with each other through the aging process. The first mechanism advancing to its full course leads to the visible failure symptom associated with that failure mechanism. To model this process, we can borrow a concept used in the area of dielectric reliability, and extend it further to account for multiple acceleration factors, multiple aging processes in one application mode.

Berman use a lifetime integral to describe the transformation between the ramp voltage breakdown distributions and TDDB curves [12], and we can write in a more generalized form as in (5).

$$ 1 = \int \frac{1}{t_{life}\left(V_{stress}, Temp\right)} \, dt \tag{5} $$

An implication of the way this life integral is written is that it is a quasi-static approximation, i.e., one can convert the amount of damage at any stress condition for a period of time and add to the cumulative failure fraction, or age, until 100% of the failure fraction is reached.

Equation (5) can be rewritten into a discrete form, and the equivalent lifetime teqlife after a series of voltage and temperature stress is expressed as in (6).

$$ t_{eqlife}\left(\sum_{i=1}^{N} t_i\right) = \frac{1}{\displaystyle\sum_{i=1}^{N}\left(\frac{t_i}{\displaystyle\sum_{i=1}^{N} t_i} \cdot \frac{1}{tlife\left(V_i, T_i\right)}\right)} \tag{6} $$

t_{eqlife} = equivalent life time
t_i = time duration of the stress condition
$t_{life}(Vi,Ti)$ = life time at the stress level of Vi and Ti
Vi = voltage applied during ti
Ti = temperature during ti

E. Apply the aging model to a TX chip reliability assessment case

For the purpose of demonstrating how to use the aging model and method in the previous section, we use a hypothetical case. In this case, we need to evaluate and compare the equivalent lifetime of a TX chip for three application profiles listed in TABLE II. In the application profile 1, the TX chip is used in B-mode only, and in the application profile 2, the TX chip is used in both B-mode and PW-mode. In both B-mode and PW-mode case, the main component under stress in B-mode is a couple of high voltage transistors. In application profile 3, CW-mode is used, and the main component under stress in this mode is a couple of low voltage transistors. An arbitrary lifetime model shown in (7) is chosen and arbitrary model parameters for the two types of transistors are listed in TABLE III. These model parameters are not representative of any particular technology.

$$ tlife(Vstress, Tstress) = t0 \cdot e^{-N \cdot Vstress} \cdot e^{\left[\frac{Ea}{k} \cdot \left(\frac{1}{Tstress+273}\right)\right]} \tag{7} $$

TABLE II. EXAMPLE ULTRASOUND APPLICATION PROFILES

	Application Profile 1		
	Duration (hrs)	Voltage (V)	Temp (C)
B-Mode	200	90	90
Power down	99800	0	27
Teqlife (hrs)	**3.26×10^{13}**		
	Application Profile 2		
	Duration (hrs)	Voltage (V)	Temp (C)
B-Mode	100	90	90
PW-Mode	10	50	100
PW-B-Mode switching	1	100	100
Power down	9989	0	27
Teqlife (hrs)	**2.2×10^{10}**		
	Application Profile 3		
	Duration (hrs)	Voltage (V)	Temp (C)
B-Mode	100	90	90
PW-Mode	10	50	100
PW-B-Mode	1	100	100
CW-Mode	25000	5	90
Power down	7489	0	27
Teqlife$_{HV}$ (hrs)	**2.2×10^{10}**		
Teqlife$_{LV}$ (hrs)	**1.832×10^{10}**		
	HOTL Test Profile		
	Duration (hrs)	Voltage (V)	Temp (C)
B-Mode	10	90	125
PW-Mode	1	50	125
PW-B-Mode Switching	0.2	100	125
Power down	989	0	125
Teqlife (hrs)	**2.8×10^{8}**		

TABLE III. MODEL PARAMETERS

	HV transistor	LV transistor
Ea (Ev)	0.7	0.7
N (V^{-1})	1.2	12
t0 (hrs)	10^{48}	10^{26}

In comparison of the application profile 2 and 3, the low voltage transistors are the component under stress in CW mode, and they age faster compared to the high voltage transistor under the specific application conditions in the application profile 3. Therefore, between the two competing aging mechanisms, the low voltage transistor wins the race. This result emphasizes that the failure mechanism dominates in one application condition does not necessarily dominate in another. Therefore, making sure the condition in the HTOL test covers the dominant fail modes is very important.

Also shown in TABLE II. is the capability to use this method as a tool to design a HTOL test which offers enough acceleration within reasonable test time. In this particular example, we design the HOTL test to toggle between B-Mode and PW-mode through entire HOTL test duration, and make use of the high voltage stress during the mode switching as the main acceleration force. The equivalent lifetime for the HTOL use profile is roughly 100 times of that for the application profile 2. If the parts survived the 1000 hrs of the HOTL test, then we can infer that the parts can survive 10^5 hrs if it were to operate in the field using the application profile 2.

F. Gain Control for Receiving Front End ICs

On the receiving side, the wide range of signals must be amplified to 2 V to drive an analog-to-digital converter. This is accomplished by adjusting the gain of the amplifier to be small for strong echoes in the near field and to be large to enhance the weak signals from deep scans, and enabled by a voltage controlled attenuator and a programmable gain amplifier (PGA). Figure 9. shows the total gain through the device as a function of applied control voltage for several PGA settings. Along with other critical parameters such as noise, linearity and power dissipation, gain error and channel matching are very important specifications for the system performance. A well controlled gain error ensures that the gain of the amplifier is precisely set when a certain control voltage is applied. Gain channel matching requirement is derived from the level of uniformity of the ultrasound image. To ensure ultrasound image consistency, from system stand point, not only does the gain of each channel of a ultrasound machine need to be matched, but also the whole batch of the machine using the same system software version need to have uniform gain match.

The calculated equivalent lifetime for each application profiles is summarized in TABLE II.

In comparison of the application profile 1 and 2, although the voltage stress in the PW mode is less, a transient overshoot voltage exists during the mode switch from PW to B-mode. For a lifetime model that is very sensitive to voltage stress, such as oxide reliability, the transient voltage although brief, could significantly shorten the system lifetime. Therefore, equation (6) provides a method to identify the circuit operating conditions that consume most of the reliability budget, and can serve as the guidance for design change to extend the lifetime of the part after removing these high risk conditions.

978-1-4244-9113-1/11 $26.00 © 2011 IEEE

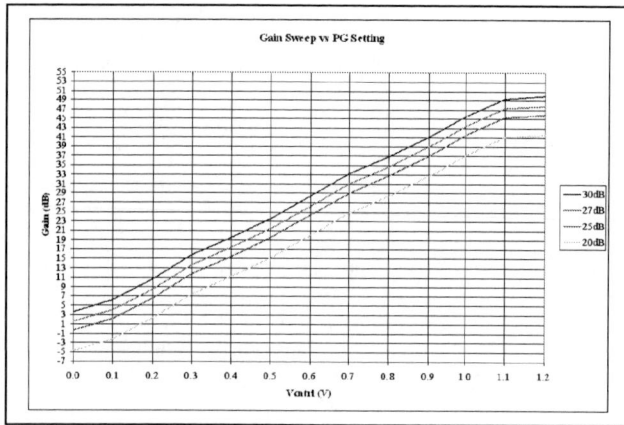

Figure 9. The total gain through the device as a function of applied voltage on the VCA for several PGA settings.

As shown in Figure 10. to meet the need for the ultrasound machines to have consistent gain performance, the chip manufacturers need to develop a quality control system to accomplish a tight distribution of the device specification within the limits, and make sure that the distribution does not drift out of the bounds. This can only be achieved by a combination of good chip design, adequate test coverage and tight process control. The design should have the built-in margins to allow some degree of variations in chip manufacturing process. Test teams always seek an optimal solution which has a good balance between the test coverage and cost. In this case, finer voltage sweep step increases the precision of the gain error, also increase the test time. The benefits and risks of test time reduction should be evaluated together with the process control. As the process is improved and refined, there is less risk to increase the voltage sweep step, and the gain variation can be kept to the minimum. This is a continuous process.

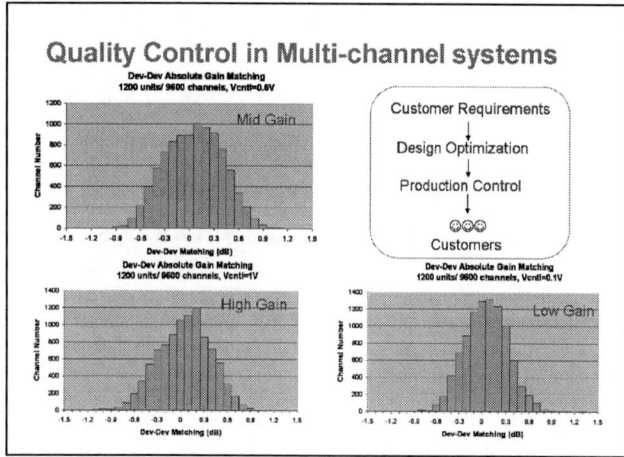

Figure 10. Gain needs to be matched from channel to channel as well as from device to device, which can only be achieved by a combination of a good chip design, an adequate test coverage and a tight process control.

CONCLUSION

Companies that can economically design and market products that meet their customers' reliability expectations have a strong competitive advantage in today's marketplace. It is no exception for the medical IC market. Due to the diversified applications conditions and broad spectrum of IC types that goes into these applications, it is difficult if not impossible to have a single method of the reliability assessment for medical ICs. In this work, we classify the ICs based on their intended use in the medical space, and show how we use different approaches based on their classification. The traditional qualification method still applies for the catalog devices; for application specific products, we need to carefully balance the requirement of performance, power and reliability imposed by the system and application; for niche devices such as implantable ICs used in life critical applications, special attention and care are needed as new material and technologies continue to emerge in this sector.

ACKNOWLEDGMENT

The authors would like to thank Eduardo Bartolome, Claude Cirba, Jim Lampos, Srikanth Krishnan, Colin Martin and John Rodriguez for their invaluable advice and support.

REFERENCES

[1] MIL-STD-883, "Test methods and procedures for microelectronics".

[2] JEDEC JESD47F, "Stress-test-driven qualification of integrated circuits".

[3] JEDEC JEP148A, "Reliability qualification of semiconductor devices based on physics of failure risk and opportunity assessment".

[4] C. Rampell, "U.S. healthcare break from the pack," *New York Times*, 8 July 2009.

[5] M. Porter, P. Gerrish, L. Tyler, S. Murray, R. Mauriello, F. Soto, G. Phetteplace and S. Harehand, "Reliability Considerations for Implantable Medical ICs," *IEEE Proc. Int. Reliability Physics Symp.*, pp 516-523, 2008.

[6] S. Dean, "Current and future trends in medical electronics," *EE Times Asia*, Aug 2009.

[7] R. Sarpeshkar, "Ultra low power bioelectronics: fundamentals, biomedical application, and bio-inspired systems," *Cambridge University Press*, 2010.

[8] Havlice J.F. and Taenzer J.C., "Medical Ultrasonic Imaging: An Overview of Principles and Instrumentation," *Proceedings of the IEEE*, Vol. 67, No. 4, pp. 620-640, April 1979.

[9] X. Xu, S. Baier, H. Venkataraman, "Ultrasound Imaging System Reshaped by AFEs," *EE Times*, Sep 2008.

[10] I. Oguzman, A. Loloee, "State-of-the-Art IC: Transmitter in Ultrasound Devices," *Wireless Design and Development*, 2009.

[11] J. Bernstein, J. Qin, "Semiconductor Device Qualification with Multiple Failure Mechanisms," *the Journal of the reliability information analysis center*, fourth quarter, 2006.

[12] A. Berman, "Time zero dielectric reliability test by a ramp method" in *IEEE Proc. Int. Reliability Physics Symp.*, pp204-209, 1981.

Quantification and Mitigation Strategies of Neutron Induced Soft-Errors in CMOS Devices and Components

-The Past and Future-

Eishi Ibe, Ken-ichi Shimbo, Hitoshi Taniguchi, Tadanobu Toba

Production Engineering Research Laboratory, Hitachi, Ltd.

292 Yoshida, Totsuka, Yokohama

Kanagawa, 244-0817 Japan

Koji Nishii

Telecommunication & Network System Division, Hitachi, Ltd.

Yokohama, Kanagawa, 244-8567 Japan

Yoshio Taniguchi

Corporate Quality Assurance Division, Hitachi, Ltd.

Chiyoda

Tokyo, 101-8608 Japan

Abstract—As semiconductor device scaling is on-going far below 100nm design rule, terrestrial neutron-induced soft-error typically in SRAMs is predicted to be worsen furthermore. Moreover, novel failure modes that may be more serious than those in memory soft-error are recently being reported. Therefore, necessity of implementing mitigation techniques is rapidly growing at the design phase, together with development of advanced detection and quantification techniques. The most advanced such techniques are reviewed and discussed.

Keywords-terrestrial neutron; CMOS; SRAM; soft-error, logic device; MCU; MNT; electronic system; FTF; DOUB; LABIR

I. INTRODUCTION

A. Historical Overview

Scaling down of semiconductor devices to sub-100nm technology encounters a wide variety of technical challenges like V_{th} variation [1], Negative Bias Temperature Instability (NBTI)[2], short-channel effect[3], gate leakage[4] and so on. Terrestrial neutron-induced single event upset (SEU) is one of such key issues that can be a major setback in scaling. As scaling proceeds below 130nm, a number of new error mode are found to be emerging as summarized in Table 1 [5].

Table 1 Error modes of single event effects in semiconductor devices [5]

Category	Mode		Memo
Soft Error (SEU)	SBU/SBE		Single bit upset/error
	MCU/MCE		Multiple-bits upset/error as one event
	MBU/MBE		MCU/MCE in the same word (not correctable by ECC)
	Block error		Multiple bits errors along with BL or WL originally due to errors in peripheral circuits.
	MCBI		Multiple-bits upset due to paracitic bipolar action triggered by snapback in channel. Correctable by re-writing. Low current sometimes associates.
	FBE		Main error mode of SOI. Mitigated by Body Tie.
	SET		Error mode of logic devices such as latch, inverter and
Pseudo Hard Error	SESB		Bipolar action in S/D channel. Impact ionization may affect.
	PCSE	SEL	High current continues to flow due to paracitic cylister. Only power cycle can resume, but sometimes destructive (hard error)
		SEFI	PCSE of logic devices.
		Firm	Error mode of SRAM-based FPGA
Hard Error	SEGR		Destruction of thin oxide layer mainly due to high-energy heavy ion. Power MOSFET, Flash memory.
	SEB		Destructive/explosive error of power MOSFET.

In particular, "multi-cell upsets (MCUs)" which are defined as simultaneous errors in more than one memory cell induced by a single event, have been under close scrutiny and their ratio to the total SEU are drastically increasing [6-10]. The concept of the MCU, therefore, contains both upsets that can be corrected by error detection/correction code (EDAC/ECC) as well as those which cannot. The latter is called "multi-bit upset" (MBU) of memory cells in the same word, and can lead, for example, to hang-ups of computer systems. Though MBUs can be avoided by a combination of ECC and the interleaving technique [10], MCUs that can be corrected by EDAC/ECC can still be problematic in high performance devices such as contents addressable memories (CAMs) [11] or registers used in network processors and routers. In the case of system design, it is therefore very important to evaluate MCUs as well as soft-error rates (SERs) of the device in design phase.

In addition, MCU can be a threat in mission-critical systems with a extreme number of logic devices that are mainly protected by spatial or time redundancies. Typically redundancy circuits such as Triple Module Redundancy (TMR)[12], duplication[13], replication[14] and redundant-nodes latches/FFs [15-19] like the Dual Interlocked storage Cell (DICE)[20] cannot be effective when relevant nodes are corrupted simultaneously by an MCU [21, 22]. Since such redundancy systems in electronic systems are strictly relevant to the international standard IEC61508[23] that defines the functional safety of electrical /electronic/ programmable electric safety related systems, protection technologies against MCUs may have to be consistent with the scope of the standard.

B. Rapidly Emerging Threat

Historically, MCUs are understood as taking place as a result of collection of charges produced by secondary ions from nuclear spallation reaction in a device. As device scaling down proceeds, novel MCU modes are being reported as "charge sharing" among memory storage nodes in the vicinity [9, 24,25] or bipolar effects in p-well [10, 26]. Ibe *et al.* have proposed multi-coupled bipolar interaction (MCBI) for one of the bipolar MCU mechanisms that is regarded as a parasitic thyristor effect triggered by a single event snapback (SES) when an energetic secondary ion passes through a p-n junction on the wall of p-well, not necessarily on the storage node, and causes MCU multiplicity of more than 10 bits [10]. It is also reported that MCU physical address pattern differs depending on written data patterns typically between the groups ALLX (All "1" or All "0") and Checkerboard (CB or its complement CBc).

II. SCALING IMPACTS ON NEUTRON SOFT-ERROR IN CMOS DEVICES

Scaling of CMOS devices has nose-dived into 22nm design rule at academic-level or 28nm even in commercial level. **Figure 1** demonstrates an example of Monte-Carlo simulation results on scaling impacts on terrestrial neutron soft-error in CMOS SRAMs, showing spread of failed physical address along with the bit-line(BL) and word-line(WL). [27, 28] Solid dots correspond to failed bit after about 60,000 nuclear spallation reactions take place in the center 4 bits around the origin due to the high energy terrestrial neutrons at Tokyo sea-level. It is seen that while the failed bits spread over about 100x100bits area for 130nm design rule, they spread over as broad as 1000x1000(=1 million) bits area for 22nm design rule. It is

also found that while the bit multiplicity in MCU (Multi-Cell Upset) is 10 bits at the maximum, it grows over 100bits with the MCU ratio to the total SEU is as high as about 50%.

These predicted trends can be understood the following considerations:

(i) The range of secondary ion created from the nuclear spallation reaction is not changed with the device scaling, while the bit size does shrink.

(ii) As device scaling proceeds, the critical charge Q_{crit} to flip data stored in the memory decreases. This means that light secondary particles like protons and alpha particles which deposit less charge than heavier particles can flip the stored data. These light particles have longer ranges than heavier particles.

As a consequence from (i) and (ii) above, the area of failed bits spreads as device scaling.

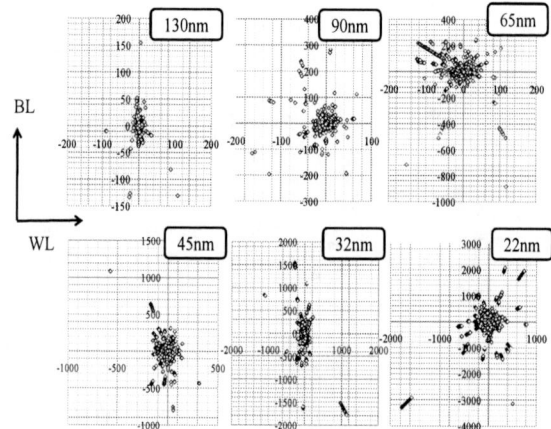

Fig. 1 Simulation results for failed bit map with device scaling after about 60,000 nuclear spallation reactions in the 4bits around the origin with neutron spectrum at the Tokyo sea level [48]

Even the MCU size in one event increase drastically, more than 99% of the MCUs along with single WL are adjacent 2 bits so that MBU (MCU in the same word) can be avoided by combination of ECC (Error Correction Code) and interleaving of bits in the same word. On the other hand, the impact for logic gates should be different from that for memories. The multiple hits on two or more nodes cause multiple SETs (Single Event Transient), or Multi-Node Transient (MNT). If these transients are latched to different flip flops (FFs), they cause simultaneous errors in FFs. If these transients involved in the different inputs of redundancy circuits like DICE or TMR, they may cause malfunction in the component/system so that the impact of MNT is much more fatal that MCU in the memory cells.

Furthermore, SETs from the global control lines like clock line or the Set/Reset lines to FFs reportedly cause multiple upsets in FFs. [29-32]

978-1-4244-9113-1/11 $26.00 © 2011 IEEE

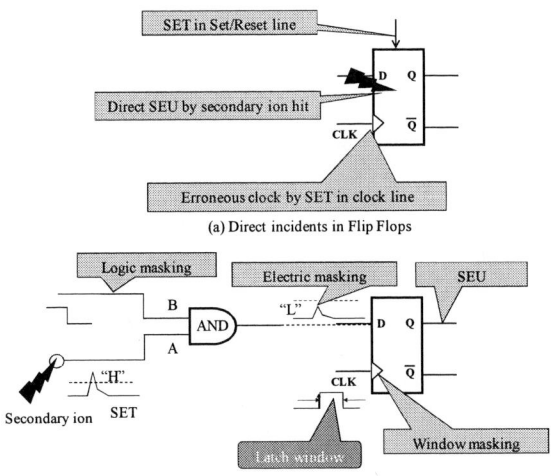

(a) Direct incidents in Flip Flops

(b) Indirect SEU incidence and various masking effects

Fig.2 General description of (a) direct incidents in Flip Flops and (b) Indirect SEU incidence with various masking effects [48]

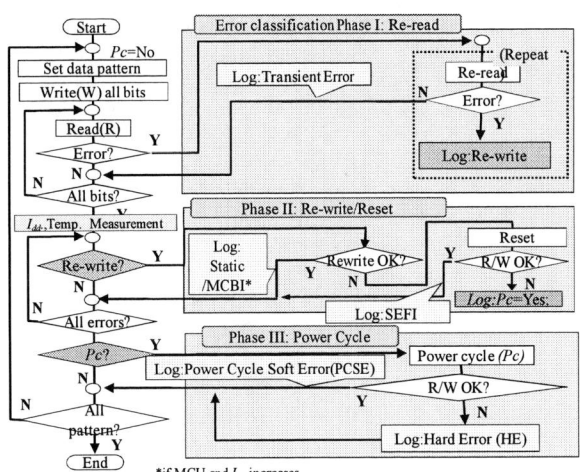

Fig. 3   Sequential classification algorithm of SEE in time domain ([10]© 2006 IEEE).

Establishment of mitigation techniques for logic gates is quite urgent and crucial. Relevant technologies toward that direction are reviewed with some proposals in the present paper.

III. VISUALIZATION OF SER

3.1 Memory Devices

A. SEU Classification Techniques in Time Domain

Figure 3 shows the sequential test algorithm basically to classify nature of SEU [10]. One normal *write/read* cycle takes 8 –9 seconds for all bits in a DUT. Two or more errors in the same sampling interval are basically regarded as MCU.   Once the data in a certain bit is in an error and the error classification algorithm is applied thereafter. If the data is recovered by re-reading, the error is regarded as a transient error. If the error is not a transient error, then compliment data is written to the bit in the phase II. If the bit is re-writable, then the error is regarded as "static-soft-error" part of which may be MCBI as discussed later or SEFI if it can be corrected by resetting. After all the bits are checked and if there are any error bits which are not re-writable or cannot be corrected by resetting, the DUT power is turned off and then turned on to see if the bit is re-writable in the phase III. If the bit is re-writable after power cycle, the errors are categorized as the "power-cycle soft-error" (PCSE)[33]. Non-destructive SEL may be among PCSE mode. If the errors cannot be corrected even by power cycle, then those errors may be classified as hard error (HE). I_{DD} current and device temperature are measured with a certain time interval independently.

B. MCU Classification Techniques in Topological Space Domain

A space-domain topological classification algorithm, which automatically identifies and classifies MCUs within a single sampling time window, is implemented in a specially designed program MUCEAC [7].

Figure 4 outlines the basic algorithms in MUCEAC. Any two errors within a certain distance along both BL and WL directions in the same time interval are regarded as contained in a MCU. If any two errors in different MCUs satisfy these criteria, the two MCUs make a single MCU. This procedure is continued until all SEU/MCUs are isolated.

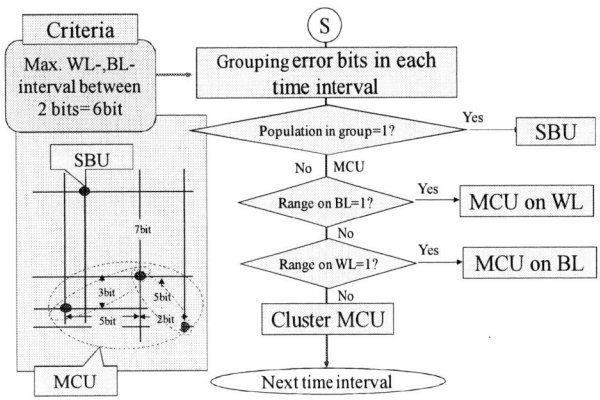

Fig. 4 Topological classification algorithm of MCU in space domain ([10]© 2006 IEEE).

3.2 Logic Devices, Circuits and Components

As for sequential logic like flip-flops, error can be detected and evaluated by utilizing FF scan chain under static mode, in which known data are written in each FFs

978-1-4244-9113-1/11 $26.00 © 2011 IEEE   241

before irradiation and FFs in scan chain with flipped data can be checked after irradiation. But under actual operation mode of chips, this method cannot be apply because there are no method to know what is correct state in each FFs. Considering this setback, statistical approaches are pursued to estimate SER in sequential logic devices by Yoshimura, et al.[34]

Soft errors in logic devices are originated from faults in combinational logic devices (SETs) and fixed as errors after they are latch in sequential logic devices. Those errors may be masked during propagation in the components by logic/window/electrical masking mechanisms so that the final functional abnormality at the outputs of the chip/board is problematic in reality. Primary importance is, therefore, tracing technology of the faults in the circuits.

Evaluation of fatality of failure modes and pin-pointing techniques to identify the vulnerable parts and circuits to be mitigated are quite important steps. Some approaches are introduced as follows:

(i) Fault injection

While operating benchmark application in a chip emulator constructed by using, for example, FPGAs, particular noises are injected randomly into the circuits. Relationship between the failure modes, frequency and particular nodes or circuits can be clarified [35].

(ii) Circuit simulation

Propagation of noise can be analyzed by using circuit simulator like SPICE with a certain number of benchmark circuits.

(iii) Stepwise analysis

With a database for gate-level SET rate SET^i_{gate}, the raw chip-level SET, SET_{Chip} is calculated.

$$SET_{Chip} = \sum_{i=1}^{N} N_i \times SET^i_{gate}. \qquad (1)$$

Starting with this raw chip-level SET, various masking effects of faults are taken into account and necessary corrections are made step by step.[35] Probabilistic approach for SER in combination logic devices with making effects is pursued by Takata, et al [36].

As for experiments, irradiation test of the chains of the logic gates [37-39] seems to be appropriate to obtain each SET^i_{gate}. Pulse expansion through the log gate chains may be an intrinsic problem to be solved in this method.

There are a number of (hundreds of) logic gate types even in a chip, so that all the gate level SETs SET^i_{gate} are very difficult to obtain. Combination with experiments and simulations may be realistic approach.

Current standard test method like in JESD89A [33] deals almost solely single bit soft-error in DRAM and SRAM cells and spallation test method ignores contribution of low energy neutrons with energy of below 10MeV. Looking at the current status of the terrestrial neutron-induced failures, standard test method for SEU in other devices, SET in logic gate, MCU, MNT, and chip/board irradiation test should be examined.[41-43] Furthermore, current standard test method should be modified to include low energy neutron contribution since the soft-error susceptibility will be drastically increasing as

shown in **Fig.5** in low energy range below 10MeV as scaling proceeds further.[27,28,43,44]

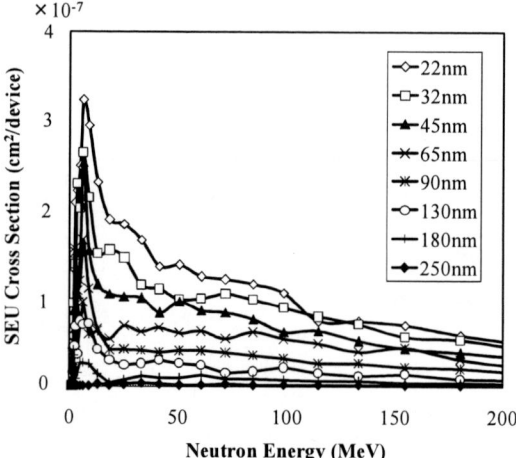

Fig. 5 Prediction Results of Scaling Trend in SEU cross Section [27,48]

(iv) Chip/board level partial/full irradiation test

A part or over all chip or printed circuit board (PCB) is irradiated while operating the benchmark application and failure modes and vulnerable parts are identified.[45-50] An example of test set-up is shown in **Fig.6**. A Part of the board for the router is irradiated by the quasi-monoenergetic neutron beam in CYRIC. The most typical error mode is rebooting of the entire board. By measuring the fluence to failure (rebooting) FTF Φ_i, single event upset cross section σ_{seu} can be calculated through Mean Fluence to Failure MFTF proposed by Shimbo, et al[46,47].

$$MFTF = \frac{1}{n} \sum_{1}^{n} \Phi_i \ . \qquad (2)$$

Then,

$$\sigma_{seu} = \frac{1}{MFTF}, \qquad (3)$$

where,

n: total number of cycles for the relevant chip or board.

This approach can be very effective to evaluate all actual failure modes including consequence of MNT and is necessary to validate the approach (i)-(iii).

As the approaches (i)-(iii) start with each device, circuit data to analyze the chip/board level failure, they can be called as the bottom-up approach. As the approach (iv) starts with chip/board to identify the source of failure, it can be called as the top-down approach. Both approaches, however, have merits and demerits or trade-offs. The followings are such examples:

From the viewpoint of quality assurance, quantitative data of both the average and variation in failure rate of the chip/board is necessary to design the system properly with reasonable margin.

The approaches (i)-(iii) cannot analyze MNT in principle. This means that these approaches may overestimate fatal failure modes.

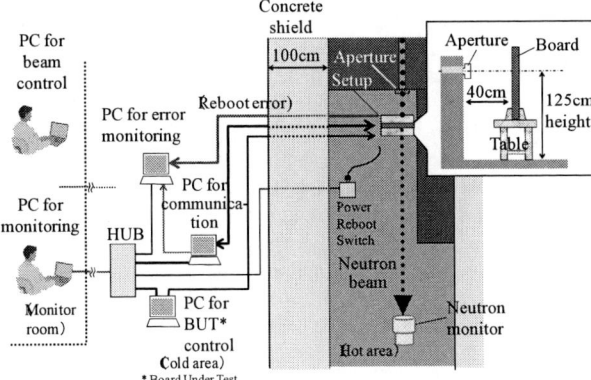

Fig.6 Board setup and conceptual layout of experimental components in

The approaches (i)-(ii) have variations inevitably in analysis due to initial conditions, circuitry, and benchmark applications. It takes tremendous amount of time and cost to evaluate such variations. In addition, the factors such as initial conditions, circuitry, and benchmark applications cannot be incorporated in the approach (iii).

Analysis of variation in the approach (iv) is also very difficult due to the limits of cost and time. In addition, as the most typical failures in approach (iv) are freeze or wild run of the chip or board and it needs relatively long exposure time, statistics in failure data may not be sufficient.

IV. MITIGATION TECHNIQUES

Cloud computing and data center are current major trend in network systems. One of the most crucial priorities in these applications is lowering power consumption to suppress the global warming. Mitigation techniques for SER, therefore, must be power conscious. Trade-offs among power, speed, cost and reliability must be considered to determine the mitigation techniques to be applied.

A. Device Layer Mitigation Techniques

The followings exemplify the device layer mitigation techniques:

(i) Confinement of charge collection volume, or p-well in triple well structure to make amount of charge less. It is noteworthy that this is effective only for charge collection mode of soft-error, but has adverse effect on bipolar mode soft-error such as MCBI (Multi-Coupled Bipolar Interaction)[10].

(ii) Modification of STI structure/layout to limit charge flow.

The techniques (i) and (ii) may associate with process difficulty.

Some other techniques are aiming at increasing Q_{crit}:

(iii) Addition of capacitor to increase Q_{crit}.

(iv) Addition of resistance between two nodes of SRAM

to suppress to flip.

(v) Interleaving bits in the same word to suppress MBU.

These mitigation techniques may cause sophistication of device structure and become more and more difficult with device scaling.

B. Circuit Layer Mitigation Techniques

A number of mitigation techniques are proposed in circuit layer.

ECC is involved in such techniques. It is known that some additional circuits are added in FF or delay circuits in output or input of FFs to detect/mask/resume FF errors as follows:

(i) Space domain redundancy in FF

DICE has two redundant nodes. When one node is about to flip, feedback loop from the other node suppress the flip to keep the initial condition. Seifert, et al. proposed SEUT with an extended concept of DICE [29]. These mechanisms are vulnerable to simultaneous hit of the two redundant nodes as mentioned before, mitigation techniques in DICE to avoid simultaneous hit are proposed by changing layout of the relevant nodes [31,52]. Mitra proposed BISER with a C-element and a weak keeper to prevent propagation of erroneous input to FFs [51]. BISER is, however, known to be vulnerable to direct hit to the C-element as the clock frequency increases. Furuta, et al. propose the dual-modular FFs as the possible extension of BISER to avoid such a erroneous operation of C-element by applying redundant C-elements and cross-coupled weak keepers [53, 54].

(ii) Time domain redundancy in FF

RAZOR [55] has a branch with a additional latch in the input line. The input data is kept with some delay compared to the FF in the additional latch. If SET pulse passes through within the delay length, the outputs of the FF and the latch differ and thus the SET can be detected. In this case, delayed latch output is always correct so that the output of the FF can be resumed to feed the latch data to the input of the FF. This type of detection and resume technique by shifting and branching the phase of input is activated only when the fault is detected so that speed area penalties can be minimum compared with other space redundancy technique. However, this type of technique may become helpless if the SET pulse width is lengthen due to scaling or lowering operational voltage.

C. Module/Processer Layer Mitigation Techniques

System level mitigation techniques for micro processors, servers, and routers in networks are principally based on redundancy in execution flow of instructions. Such techniques can be classified into roughly three types:

(i)TMR (Triple Module Redundancy)

TMR has three modules and one voter. One instruction is executed in the three modules simultaneously and two out of three voting is executed by the voter. The highest reliability seems to be achieved, but both area and power penalties in TMR are x3 and some speed penalty exists in the voter. Moreover, it must be noted that reliability improvement in total is possible only when redundant modules and the voter have a certain level of high reliability. This technique is also helpless when MNT cause errors in

978-1-4244-9113-1/11 $26.00 © 2011 IEEE   243

two redundant modules [56].

(ii)Duplication+Comparison+Checkpointing (DCC or Double Module Redundancy, DMR)

Two redundant modules are used and a certain number of checkpoints are set in instruction flow in DCC. Necessary data to restart the execution from the checkpoint are stored. The same instructions are executed simultaneously in the redundant modules to the checkpoint (duplication), and results are compared (comparison). If the results are different, error is supposed to take place in one of modules, and execution is resumed from the previous checkpoint.

DCC is also vulnerable to MNT and area, power penalties are x2. Some speed penalty exits due to restart from checkpoint and delay in comparator.

(iii)Replication+Rollback (RR)

The same instruction is executed twice in one module (Replication). Initial execution results are stored in the keeper and compared with the second execution results. If the results are different, instruction restarted from the previous step (rollback). This technique has x2 speed and power penalties [14].

In addition to the redundancy approaches explained above, there are many cases to prepare stand-by units for replacement of the failed unit. Care must be paid for these units to see if errors are accumulated in the memory parts. MBUs can take place unless refresh of memory content is done. Regular patrol of the memory parts in the stand-by unit may be necessary.

V. DOUB (DESIGN ON UPPER BOUND) AND LABIR (INTER LAYER BUILT-IN RELIABILITY) [57]

In the Section IV, stack layer (device, circuit, module/processor) level mitigation techniques are reviewed and no single mitigation technique seems to fulfill simultaneously the reliability and performance requirements with minimum penalties and reasonable costs.

The authors, therefore, are working on the different and novel approach named (v) Design on Upper Bound (DOUB) by which the upper-bound failure rate can be estimated explicitly.

The equation (A), for example, gives the maximum upper-bound of chip-level SET because the equation does not include any masking effects. By modifying the maximum upper-bound with various physical limits determined by device structure/layout, circuit complexity, structure of logical layers, the realistic upper-bound failure rate free from the variations may be obtained. If this upper-bound of a chip is low enough, the chip can be ignored for further analysis. If the upper-bound is of concern, mitigation techniques are applied from a simple and low cost method in design phase as shown in like:

(i) Exchange of weak logic gate/memory to robust logic gate/memory. DRAM is currently very robust device and can be substitute of SRAMSs where speed is not critical [46,47].

(ii) Minimization of active memory area

(iii) Limited and local use of space or time redundancy

techniques in the circuit level, and so on.

The authors are proposing a novel method LABIR (Inter LAyer Built-In Reliability) as is illustrated in **Fig.7**. LABIR proposes interactive or communicative mitigation techniques in which a recovery action such as rollback to the checkpoint ignited when a layer finds any error symptom, not necessarily error or fault itself. BIST (Built –In Self Test) [58], Built-In Current (Pulse) Sensor (BICS[59], BIPS) can be used for such kind of technique for symptom detection. The symptom may not appear so often so that power and area penalties can be minimized with minimum additional structure and circuits. By using BIPS, a pulse current propagated from an MCBI (Multi-Coupled Bipolar Interaction) zone in p-well can be detected in I_{dd} line as demonstrated in [10]. By capturing such a symptom by applying a sense amp between adjacent two p-wells, for example, errors or failures can be resumed by the rollback and replication operation in CPU level of the ULSI chip. Other sources of noises like EMI (Electro-Magnetic Interference) [47] propagate in wider area than soft-error over many wells so that they can be eliminated by the differential method between adjacent wells.

CONCLUSIONS

As semiconductor device scaling is on-going far below 100nm design rule, terrestrial neutron-induced soft-error typically in SRAMs is predicted to be worsen furthermore. Moreover, novel failure modes that may be more serious than those in memory soft-error are recently being reported. Therefore, necessity of implementing mitigation techniques with marginal penalties including power dissipation is rapidly growing at the design phase, together with development of advanced detection and quantification techniques. The most advanced such techniques are reviewed and discussed with proposal of novel mitigation strategies of the Design on Upper Bound (DOUB) and the inter LAyer Bulit-In Reliability (LABIR).

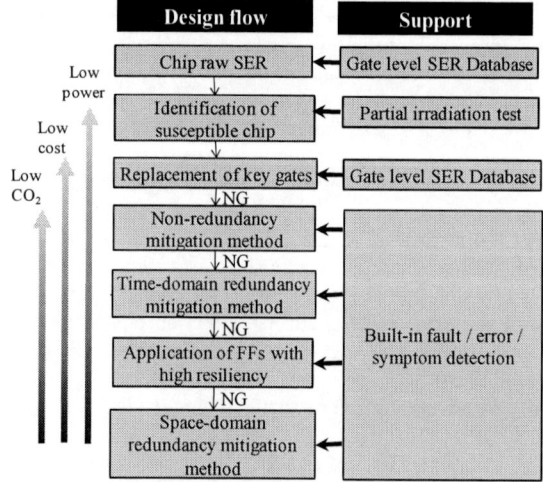

Fig. 7 General design flow of stepwise reduction in SER under the design on upper bound concept . Power consumption, cost, and global warming are key issues.

978-1-4244-9113-1/11 $26.00 © 2011 IEEE   244

ACKNOWLEDGEMENTS

We gratefully acknowledge Profs. Emeritus T. Nakamura, M. Baba and Prof. Y. Sakemi for helpful discussions and support for the database on nuclear reactions and high energy neutron experiments. Communicative discussions with Drs. C. Slayman, R. Baumann, M. Nicolaidis, D. Alexandrescu, S. Rezugui, P. Roche, A. Bougerol, T. Uemura, H. Kobayashi and Y. Zorian are acknowledged. We thank gratefully to Profs. K. Kobayashi, H. Onodera, Drs. M. Yoshimura, and Y. Matsunaga for giving valuable information on SEU tolerant flip flops.

REFERENCES

[1] N. Sugii, R. Tsuchiya, T. Ishigaki, Y. Morita, H. Yoshimoto, K. Torii, and S. Kimura, "Comprehensive Study on Vth Variability in Silicon on Thin BOX (SOTB) CMOS with Small Random-Dopant Fluctuation: Finding a Way to Further Reduce Variation," *IEDM, San Francisco, Dec. 15-17*, pp. 249-253 (2008).

[2] S. Wen, R. Wong, and A. Silburt, "IC Component SEU Impact Analysis," *SELSE4, University of Texas at Austin, March, 26,27* (2008).

[3] D. Villanueva, A. Pouydebasque, E. Robilliart, T. Skotnicki, E. Fuchs, and H. Jaoue, "Impact of the Lateral Source / Drain Abruptness on MOSFET Characteristics and Transport Properties," *2003 IEDM, Washington, DC, December 7 - 10, 2003*, No.9.4 (2003).

[4] L.T. Clark, K.C. Moh, K.E. Holbert, X. Yao, J. Knudsen, and H. Shah, "Optimizing Radiation Hard by Design SRAM Cells," *TNS*, Vol.54, No.6, pp. 2028-2036 (2007).

[5] T. Nakamura, M. Baba, E. Ibe, Y. Yahagi, and H. Kameyama, "Terrestrial Neutron-Induced Sift-Errors in Advanced Memory Devices," New Jersey, World Scientific (2008).

[6] E. Ibe, S. Chung, S. Wen, Y. Yahagi, H. Kameyama, and S. Yamamoto, "Multi-Error Propagation Mechanisms Clarified in CMOSFET SRAM Devices under Quasi-Mono Energetic Neutron Irradiation," *2007 NSREC, Ponte Vedra Beach, Florida, July 17-21, 2006*, No.PC-6 (2006).

[7] E. Ibe, S. Chung, S. Wen, S., Y. Yahagi, H. Kameyama, S. Yamamoto, T. Akioka, and H. Yamaguchi, "Valid and Prompt Track-down Algorithms for Multiple Error Mechanisms in Neutron-Induced Single Event Effects of Memory Devices," *Workshop on Radiation Effects on Component and Systems (RADECS), Athens, Greece, September 27-29, 2006*, No.D-2 (2006).

[8] D. Radaelli, H. Puchner, P. Chia, S. Wong, and S. Daniel, "Investigation of Multi-Bit Upsets in a 150 nm Technology SRAM Device," *2005 NSREC, Seattle, Washington, July 11-15, 2005*, No.F-4 (2005).

[9] N. Seifert, and V. Zia, "Assessing the impact of scaling on the efficacy of spatial redundancy based mitigation schemes for terrestrial applications," *IEEE SELSE3, Austin Texas, April 3, 4* (2007).

[10] E. Ibe, S. Chung, S. Wen, H. Yamaguchi, Y. Yahagi, H. Kameyama, S. Yamamoto, and T. Akioka, "Spreading Diversity in Multi-cell Neutron-Induced Upsets with Device Scaling," *2006 CICC, San Jose, CA., September 10 - 13, 2006*, pp. 437-444 (2006).

[11] K. Pagiamtzis, N. Azizi, and F. Najm, "A Soft-Error Tolerant Content-Addressable Memory (CAM) Using An Error-Correcting-Match Scheme," *idem,*, pp. 301-304 (2006).

[14] R. Noji, S. Fujie, Y. Yoshikawa, H. Ichihara, and T. Inoue, "An FPGA-based Fail-Soft System with Adaptive Reconfiguration," *IOLTS2010, Corfu Island, Greece, July 5-7*, No.6.3, pp. 127-132

(2010).

[15] L.R. Rockett, "An SEU-Hardened CMOS Latch Design", *TNS,* Vol.35, No.6, pp.1682-1687.1988.

[16] M. Li, M. Pradeep,R.S. Sahoo, S. Adve, V. Adve, and Y.Y. Zhou, "SWAT: An Error Resilient System," *SELSE4, University of Texas at Austin, March, 26,27* (2008).

[17] R. Velazco, D. Bessot, S. Duzellier, R. Ecoffet, and R. Koga, "Two CMOS Memory Cells Suitable for the Design of SEU-Tolerant VLSI Circuits," *TNS*, Vol.41, No.6, pp.2229-2234, 1994.

[18] T. Uemura, R. Tanabe, Y. Tosaka, and S. Satoh, "Mitigation Techniques Using Low Pass Filters Against Single Event Transients in 45nm-technology LSIs," *IOLTS, Rhodes, Greece, July 7-9,* No.iolts08-15

[19] K.J. Hass, J.W. Gambles, B. Walker, and M. Zampagione, "Mitigating Single Event Upsets From Combinational Logic," 7th NASA Symposium on VLSI Design 1998.

[20] T. Calin, M. Nicolaidis, and R. Velazco, "Upset Hardened Memory Design for Submicron CMOS Technology," *TNS*, Vol.43, No.6, pp.2874-2878 Dec.1996.

[21] A. Lesea, and K. Castellani-Coulie, "Experimental Study and Analysis of Soft Errors in 90nm Xilinx FPGA and Beyond," 2007 *RADECS, Deauville, France, Sept. 10-14*, No.DWL-16 (2007).

[22] S. Rezgui, J.-J.Wang, E.C. Tung, B. Cronquist, and J. McCollum, "New Methodologies for SET Characterization and Mitigation in Flash-Based FPGAs," *NSREC, Honolulu, Hawaii, July 23-27*, No.J-8 (2007).

[23] IEC61508 Functional Safety of Electrical/Electronic/Programmable Electronic Safety- Related Systems (1998).

[24] N. Seifert, "Soft Error Rates of Hardened Sequentials utilizing Local Redundancy," *IOLTS 2008, Greece, July 6-9, 2008*, No.S1.3, pp. 49-52 (2008).

[25] O.A. Amusan, L.W. Massengill, B.L. Bhuva, P.R. Fleming, and M.L. Alles, "Charge Collection and Sharing in a 130nm CMOS Technology," *NSREC, Ponte Vedra Beach, Florida, July 17-21, 2006*, No.C-3 (2006).

[26] B.D. Olson , D. Ball, K.M. Warren, L.W. Massengill, N.F. Haddad, S.E. McMorrow, and D. Doyle, "Simultaneous Single Event Charge Sharing and Parasitic Bipolar Conduction in a Highly-Scaled SRAM Design," *TNS*, Vol.52, No.6, pp. 2132-2136 (2005).

[27] E. Ibe, H. Taniguchi, Y. Yahagi, K. Shimbo, and T. Toba, "Scaling Effects on Neutron-Induced Soft Error in SRAMs Down to 22nm Process," *Third Workshop on Dependable and Secure Nanocomputing, June 29, 2009, Estoril, Lisbon, Portugal*, No.2.1 (2009).

[28] E. Ibe, H. Taniguchi, Y. Yahagi, K. Shimbo, and T. Toba, "Impact of Scaling on Neutron-Induced Soft Error in SRAMs from a 250nm to a 22nm Design Rule," *IEEE Trans. on Electronic Devices*, Vol.57, No.7, pp. 1527-1538 (2010).

[29] N. Seifert, V. Zia, "Assessing the impact of scaling on the efficacy of spatial redundancy based mitigation schemes for terrestrial applications, " *IEEE Workshop on Silicon Errors in Logic - System Effects 3, Austin Texas, April 3, 4* (2007).

[30] T.D. Loveless, L.W. Massengil, B.L. Bhuva, W.T. Holman, R.A. Reed, D. McMorrow, J.S. Melinger, and P., Jenkins, "A Single-Event-Hardened Phase-Locked Loop Fabricated in 130 nm CMOS," *TNS, Honolulu, Hawaii, July 23-27*, Vol.54, No.6, pp. 2012-2020 (2007).

[31] M. Cabanas-Holmen, E. H. Cannon, A. Kleinosowski, J. Ballast, J. Killens, and J. Socha, "Clock and Reset Transients in a 90 nm RHBD Single-Core Tilera Processor," *TNS*, Vol.56, No.6, pp.3505-3510 (2009).

[32] T. Uemura, Y. Tosaka, H. Matsuyama, K. Shono, C.J. Uchibori, K. Takahisa, M. Fukuda, and K. Hatanaka, "SEILA: Soft Error Immune Latch For Mitigating Multi-Node-SEU and Local-Clock-SET," *Proc. Int'l Reliability Physics Symposium, May 2-6, 2010, Anaheim, CA.*, pp.218-223 (2010).

[33] JEDEC, "Measurement and Reporting of Alpha Particles and Terrestrial Cosmic Ray-Induced Soft Errors in Semiconductor Devices : JESD89A," *JEDEC STANDARD, JEDEC Sold State Technology Association*, No.89, pp. 1-85 (2006).

[34] M. Yoshimura, Y. Akamine, and Y. Matsunaga, "An SER Analysis Method for Sequential Circuits," *SELSE 2011, Champaign, Illinoi, March 29-30*, 2011

[35] P. Roche, "Industrial Impacts of SER on Today's Consumer Electronic Arena," *IOLTS2010, Corfu Island, Greece, July 5-7*, Keynote 1, p. xv (2010).

[36] H. Chapman, E. Landman, A. MargalitIlovich, Y.-P. Fang, A.S. Oates, D. Alexandrescu, and O. Lauzeral, "A Multi-Partner Soft Error Rate Analysis of an InfiniBand Host Channel Adapter," *SELSE6, Stanford University, March 23, 24* (2010).

[37] T. Takata, and Y. Matsunaga, "A Robust Algorithm for Pessimistic Analysis of Logic Masking Effects in Combinational Circuits," *SELSE 2011, Champaign, Illinoi, March 29-30*, 2011

[38] T. Makino, D. Kobayash, K. Hirose, D. Takahashi, S. Ishii, M. Kusano, S. Onoda, and T. Hirao, and T. Ohshima, "Soft-Error Rate in a Logic LSI Estimated From SET Pulse-Width Measurements," *TNS2009*, Vol.56, No.6, pp. 3180-3184 (2009).

[39] H. Nakamura, K. Tanaka, T. Uemura, K. Takeuchi, T. Fukuda, and S. Kumashiro, "Measurement of Neutron-induced Single Event Transient Pulse Width Narrower than 100ps," *IRPS, Anaheim, CA, May 2-6*, pp. 694-697 (2010).

[40] E.H. Cannon, and M. Cabanas-Holmen, "Heavy Ion and High Energy Proton-Induced Single Event Transients in 90 nm Inverter, NAND and nor Gates," *NSREC, Quebac, Canada, July 20-24, 2009,*, No.PI-3 (2009).

[41] E. Ibe, K. Shimbo, T. Toba, Y. Taniguchi and H. Taniguchi, "Novel SER Standards: Backgrounds and Methodologies," *ICICDT2010, Grenoble, France, June, SER Session No.1 (2010)*

[42] E. Ibe, "Novel Features in SER Characteristics toward New Standards Special Session 1-Panel :SER standards: Where we are? What's next?," *IOLTS2010, Corfu Island, Greece, July 5-7* (2010).

[43] D. Alexandrescu, R. Baumann, A. Bougerol, E. Ibe, S. Rezgui, and C. Slayman, "Special Session 1-Panel: SER standards: Where are we? What's next?," *IOLTS2010, July5-7, 2010, Corfu Island, Greece* (2010).

[44] B.D. Sierawski, R.A. Reed, R.D. Schrimpf, R.A. Weller, M.H. Mendenhall, M.A. Xapsos, R.C. Baumann, and X. Deng, "Impact of Low-Energy Proton Induced Upsets on Test Methods and RatePredictions," NSREC, Quebac, Canada, July 20-24, 2009, No.A-8 (2009)

[45] D.F. Heidel, K.P. Rodebell, P.W. Marshall, J.A. Pellish, K.A. LaBel, M.A. Xapsos, S.E. Hakey, M.C. Rauch,J.R. Schwank, P.E. Dodd, M.R. Shaneyfelt, M.D. Berg, M.R. Friendlich, A.D. Phan, and C.M. Seidleck, "Single-Event Upsets and Multiple-Bit Upsets on a 45 nm SOI SRAM," *idem.,*, No.I-9 (2009).

[46] K. Shimbo, T. Toba, "Correlation of Mitigation of Soft-Error Rate of Routers between Neutron Irradiation Test and Field Soft-error Data," IEICE Technical report CPM2009-139, *Kochi, Dec. 2-4*, Vol.109, No.317,318, pp. 51-55 (2009)(In Japanese)

[47] K. Shimbo, T. Toba, K. Nishii, E. Ibe, Y. Taniguchi, and Y. Yahagi, "Quantification & Mitigation Techniques of Soft-Error Rates in Routers Validated in Accelerated Neutron Irradiation Test and Field Test," *SELSE7 , Champaign, Illinoi, March 29-30* (2011).

[48] N. Kanekawa, E. Ibe, T. Suga, and Y. Uematsu, "Dependability in Electronic Systems-Mitigation of Hardware Failures, Soft Errors, and Electro-Magnetic Disturbances-," New York, Springer (2010).

[49] L. Borucki, G. Schindlbeck, and C. Slayman, "Comparison of Accelerated DRAM Soft Error Rates Measured at Component and System Level," *IRPS 2008, Phoenix, Arizona, April 27-May 1, Phoenix Convention Center*, No.5A.4 (2008).

[50] H. Ando, and S. Hatanaka, "Accelerated Testing of a 90nm SPARC64 V Microprocessor for Neutron SER," *IEEE Workshop on Silicon Errors in Logic - System Effects 3, Austin Texas, April 3, 4* (2007).

[51] S. Mitra, M. Zhang, N. Seifert, T. Mak, and K.S. Kim, "Built-In Soft Error Resilience for Robust System Design," *ICICDT2007, Austin, Texas, May 18-20*, 2007, pp. 263-268 (2007).

[52] H.-H. K. Lee, K. Lilja, M. Bounasser, P. Relangi, I.R. Linscott, U.S. Inan, and S. Mitra ,"Leap: Layout Design Through Error-Aware Transistor Positioning for Soft-Error Resilient Sequential Cell Design," *idem.*, pp. 203- 212(2010).

[53] J. Furuta, K. Kobayashi, and H. Onodera, "An Area/Delay Efficient Dual-Modular Flip-Flop with Higher SEU/SET Immunity," *IEICE Trans. on Electronics*, Vol.E93-C, No.3, pp. 340-346 (2010).

[54] J. Furuta, C. Hamanaka, K. Kobayashi, and H. Onodera, "A 65nm Bistable Cross-coupled Dual Modular Redundancy Flip-Flop Capable of Protecting Soft Errors on the C-element," *VLSIC, Honolulu, HI, USA, June 16-18*, pp. 123-124 (2010).

[55] D. Ernst, N.S. Kim, S. Das, S. Pant, R. Rao, T. Pham, C. Ziesler, D. Blaauw, T. Austin, K. Flautner1, and T. Mudge, "Razor: A Low-Power Pipeline Based on Circuit-Level Timing Speculation," *MICRO-36*, 2003

[56] H. Quinn, K. Morgan, P. Graham, J. Krone, M. Caffrey, and K. Lundgren, "Domain Crossing Events: Limitations on Single Device Triple-Modular Redundancy Circuits in Xilinx FPGAs," *NSREC, Honolulu, Hawaii, July 23-27, 2007*, No.C-5 (2007)

[57] E. Ibe, K. Shimbo, T. Toba, H. Taniguchi, and Y. Taniguchi, "LABIR: Inter-LAyer Built-In Reliability for Electronic Components and Systems," *SELSE7, Champaign, Illinoi, March 29-30* (2011).

[58] G. Theodorou, N. Kranitis, A. Paschalis, and D. Gizopoulos, "A Software-Based Self-Test Methodology for In-System Testing of Processor Cache Tag Arrays," *IOLTS2010, Corfu Island, Greece, July 5-7, 2010*, No.7.4, pp. 159-164 (2010).

[59] S.A. Bota, G. Torrens, B. Alorda, J. Verd, and J. Segura, "Cross-BIC Architecture for Single and Multiple SEU Detection Enhancement in SRAM Memories," *IOLTS2010, Corfu Island, Greece, July 5-7, 2010*, No.7.1, pp. 141-146 (2010).

Effects of Scaling on Muon-Induced Soft Errors

Brian D. Sierawski*, Robert A. Reed*, Marcus H. Mendenhall*, Robert A. Weller*, Ronald D. Schrimpf*,
Shi-Jie Wen†, Richard Wong†, Nelson Tam‡, Robert C. Baumann§

*Institute for Space and Defense Electronics
Vanderbilt University, Nashville, TN 37212
Email: brian.sierawski@vanderbilt.edu
†Cisco Systems, Inc.
San Jose, CA 95134
‡Marvell Semiconductor, Inc.
Santa Clara, CA 95054
§Texas Instruments
Dallas, TX 75243

Abstract—**Experimental results are presented that indicate technology scaling increases the sensitivity of microelectronics to soft errors from low-energy muons. Results are presented for 65, 55, 45, and 40 nm bulk CMOS SRAM test arrays. Simulations suggest an increasing role of muons in the soft error rate for smaller technologies.**

I. Introduction

The natural background radiation is well-known as one of the primary sources of soft errors in microelectronics. Among the species of particles that are produced in cosmic ray showers in the Earth's atmosphere, neutrons have been the primary reliability concern as they are able to generate transient currents in devices through recoiling or secondary particles and have historically had the largest effect. While terrestrial neutrons are quite numerous, they rarely interact with nuclei and therefore present a small cross section for interaction and single event effects.

Recently, research has shown that commercial static random access memories (SRAMs) are now so small and sufficiently sensitive that single event upsets (SEUs) may be induced from the electronic stopping of a proton [1]–[4]. This sensitivity appears near the 65 nm technology node as the critical charge to upset a cell is on the order of 1 fC; merely 6,000 electrons are required to cause a change in data state. In the space radiation effects community, low-energy proton sensitivity is a concern for reliability since the proton flux is large for low-Earth orbits and during solar particle events.

In terrestrial applications, the proton flux is effectively shielded by the atmosphere and building materials. However, muons, also singly-charged particles, do not participate in the strong interaction and are therefore not easily stopped by shielding. For this reason they dominate the hard component of the natural radiation environment. Further, the overall flux of muons even exceeds that of neutrons.

Unlike neutrons, muons can be positively or negatively-charged. They have identical electronic stopping power curves as a function of velocity to protons. In very sensitive devices, the continuous energy loss to direct ionization of charge carriers will have a larger upset cross section than nuclear reaction induced events.

Single event effects due to the muon ionization wake were predicted in a number of early works on microelectronic reliability [5]–[7]. Wallmark and Marcus provided a brief investigation of the role of these particles as one of the fundamental physical limits to continued microelectronic scaling [5]. Ziegler and Lanford provided a much expanded investigation of cosmic ray induced error rates and predicted the coming of a dramatic increase in errors with decreased critical charge [6]. The authors predicted, although with limited environmental measurements, that devices with extremely low critical charge values will be susceptible to upset from, and errors dominated by, the muon ionization wake. While the history of process scaling has dramatically altered the device geometries, the crossover where this mechanism was predicted to exceed the neutron contribution was for devices with critical charges below 5 fC in [7].

In experimental investigations Dicello published a series of papers comparing muon and pion error counts and their contribution to the sea level error rate [8]–[11]. In [8], the authors report on error counts measured with a 4K NMOS SRAM at LAMPF. A 109 MeV/c μ^+ beam only produced 3 errors, and no errors were observed for the μ^- beam. The μ^+ errors were attributed to pion contaminants, however. In [9] the authors obtained similar data, but used a lucite degrader in the beam line to moderate the energy. The pion-induced error counts decreased with increasing degrader thickness until the device entered the stopping region in which the π^- error counts increased by a factor of 7. The lack of a similar response for π^+ indicates that this was a result of π^- capture and not direct ionization.

This work was supported in part by the Defense Threat Reduction Agency grants HDTRA1-08-1-003 and HDTRA1-08-1-0034, the Defense Threat Reduction Agency Radiation Hardened Microelectronics Program under IACROs #09-45871 and #10-49771 to NASA, and the NASA Electronic Parts and Packaging Program.

978-1-4244-9113-1/11 $26.00 © 2011 IEEE

Another experiment used a cloud muon beam produced by pion decay with a lucite degrader to obtain error counts for low-energy muons [11]. By placing the device in the μ^- stopping region, the authors were able to measure three errors over the course of 24 hours. No errors were observed with μ^+. Based on these results the stopping muon error rate at sea level was estimated at 2% of the total rate from all species. The authors also state that there are large uncertainties in the relative fluence of the terrestrial species.

In previous work [12], we presented experimental data that show low-energy muons are able to cause single event upsets in CMOS SRAMs through the electronic stopping process. Energy deposition measurements using a surface barrier detector were presented to characterize the kinetic energy spectra produced by the M20B surface muon beam at TRIUMF [13] and Geant4 was used to estimate the energy spectra incident on the memories. The probability of an upset for a given SRAM was shown to increase near the energy region corresponding to muon stopping, indicative of direct ionization effects.

In this work, we present experimental data showing muon-induced upsets as a function of technology node for nominal operating bias conditions. We report on the upset count for 65, 55, 45, and 40 nm bulk CMOS SRAMs. For a constant beam tune, the number of muon-induced upsets is greater for smaller devices. Because muons interact through direct ionization, whereas neutrons only interact through rare nuclear reactions, their contribution to a part's soft error rate is potentially large. We use Monte Carlo simulations to estimate the contribution of low-energy muons to the soft error rate for future technologies.

II. Experimental Data

In this experiment we measured the number of single event upsets for four bulk CMOS SRAMs of different technology nodes. The 65 nm and 45 nm generations were 8 Mb test arrays that were operated at their nominal supply voltage of 1.2 V. The 55 nm and 40 nm generation devices were 1 Mb and 5 Mb arrays, respectively, and the supply voltage for these devices was 1.0 V. The M20B surface muon beamline at TRIUMF was used to direct low-energy muons to the device under test. This facility regularly produces muons for scientific research. At the device location, the beam is a distribution of muon energies that is determined by the momentum selection in the beamline and unavoidable energy straggling through intermediate matter. The kinetic energy characterization of the beam and justification of beam purity are presented in [12]. This beam allows a test with muons that are stopping or near stopping in the active device regions.

A. Effect of muon momentum

The measured upset probabilities versus momenta are presented in the top plot of Fig. 1. The abscissa is related to the kinetic energy by associating the upsets at the mode value of the simulated energy distribution for the experimental momentum selection. An approximate event cross section is indicated based on an estimate of the beam fluence. In this dataset, all SRAMs were operated at a supply voltage of

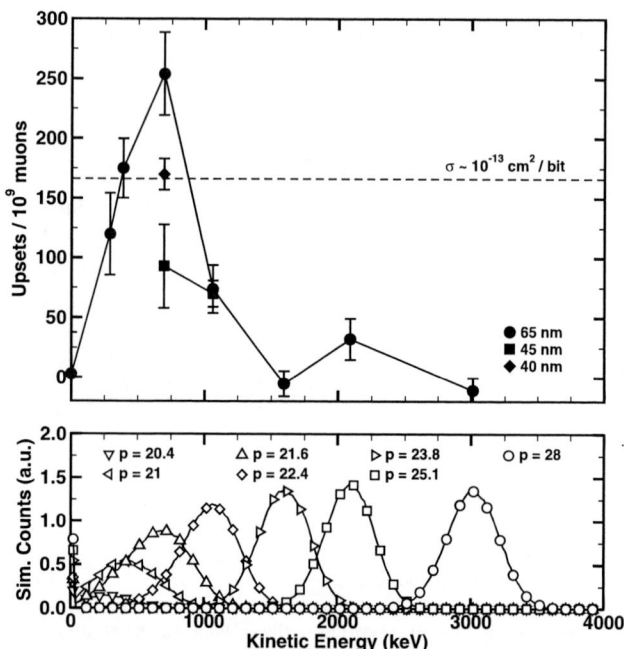

Fig. 1. Simulated muon kinetic energy distributions, as seen at the front of the part, corresponding to experimental momenta including upstream energy losses and straggling (bottom). Experimental error counts for 65 nm, 45 nm, and 40 nm SRAMs versus estimated muon kinetic energy at 1.0 V bias (top). Dashed horizontal line represents an approximate muon-induced SEU cross section for reference.

1.0 V. This bias was chosen because it produced a statistically significant error count.

At the highest energy, 3 MeV, the upset count was indistinguishable from the baseline for the 65 nm device. This was because muons pass through the device without generating sufficient charge to result in an SEU. Additionally, this measurement confirms that reported upsets cannot be attributed to noise sources while the beam is in operation. Near 700 keV the range of the beam through the metallization is such that a large portion of the muons traversing the active silicon are close to the Bragg peak and the collected charge is sufficient to exceed the critical charge.

Similar to low-energy proton testing, the peak SEU sensitivity occurs when the beam is near the end of range and particles deposit the maximum energy in the sensitive volume. The accelerator tune needed to achieve this is therefore dependent on the target's distance from the beam window and the thickness and composition of the metallization and overlayer stacks. A quick characterization of each device indicated that this peak sensitivity was near 21.6 MeV/c. As the mean energy is further decreased, the beam begins to range out and the error counts return to the baseline. The 45 nm and 40 nm devices were spot checked and also exhibited muon sensitivities.

B. Effect of technology

The data presented in Fig. 2 show the probability of upset at nominal bias. All four SRAMs were irradiated using the 21.6 MeV/c μ^+ beam. The probabilities are presented as the

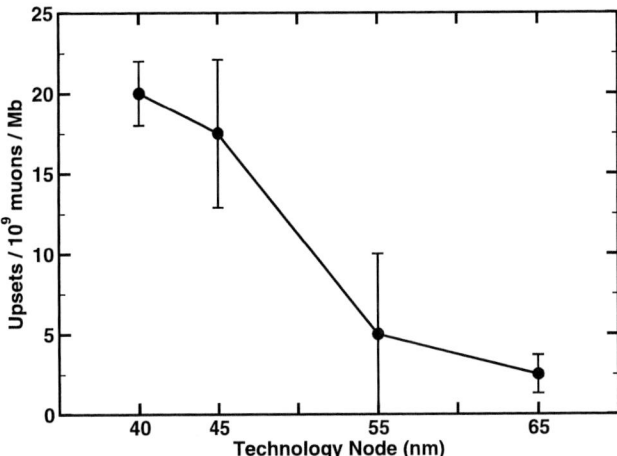

Fig. 2. Experimental muon-induced single event upset probability for each device under test operated at nominal supply voltage. Momentum selection was 21.6 MeV/c.

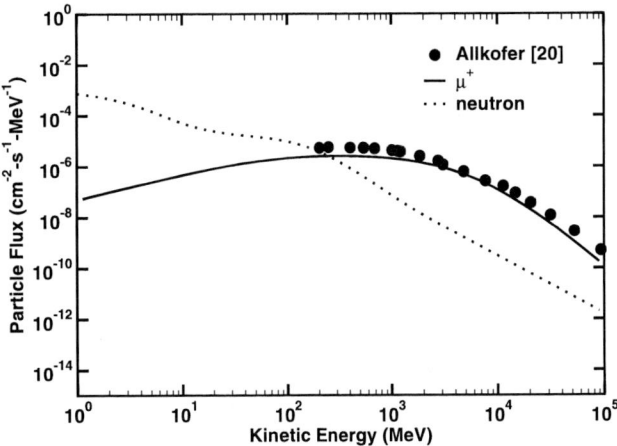

Fig. 3. Muon and neutron differential flux specta for New York City (40° lat., 73° long., 0 m. elevation) generated by EXPACS. The negative muon spectrum follows the positive muon spectrum closely, but is omitted for clarity. Data points represent measured values in [20].

number of upsets per 10^9 muons counted by the scintillator and scaled by the memory capacity. We have refrained from presenting the data as SEU cross sections to emphasize that the beam was not monoenergetic.

These data show a clear increase in the SEU susceptibility and significance of energy deposition by muons for scaled technologies. To first order, the reduction of the device area results in a decrease in the number of particles passing through the cell and capable of producing an upset. This scaling would reduce the probability of a bit being in error in the beam. The increase in upset probability is therefore due to differences in the geometry of the charge collection, an increase in the fluence of energy deposition events exceeding the critical charge, or both.

Further, in these experiments, the incident muons have a distribution of kinetic energies and therefore a distribution of stopping powers. As the technology node decreases, the charge required to upset a single memory cell decreases. The effect of this trend is an increase in the fraction of the distribution that is able to induce an upset. For the 40 nm SRAM, a larger portion of the 21.6 MeV/c beam exceeds the stopping power threshold as compared with the 65 nm SRAM. While the probability of upset is increasing for future devices, it cannot be ascertained from these data whether the trend is linearly or super-linearly increasing.

III. TERRESTRIAL ENVIRONMENT

In the sea-level environment, muons are the most numerous species, with a flux of nearly 60 cm^{-2}hr^{-1} for momenta greater than 0.35 GeV/c [14]. Several works have measured the high momentum portion of the muon spectra. However, few report on the differential flux below 1 GeV/c. As it is postulated that muons of low momentum will induce single event effects, this region of the flux is important. In our analysis we use a larger energy range as produced by a model. We then compare the model's energy spectrum to various data sets. For the purposes of this work, we have selected the spectra produced by the EXPACS program [15]. This model is implemented in a spreadsheet summarizing the results of PHITS Monte Carlo simulations [16], [17]. Given the parameters specifying a sea level location in New York City, differential flux spectra with respect to energy are produced for neutrons and positive and negative muons from 1 MeV to 100 GeV.

The differential flux spectra are shown in Fig. 3. With the exception of thermal neutron capture, neutron-induced single event effects are the result of fast neutrons. It is typically assumed that the threshold energy for neutron reactions leading to SEE is 10 MeV, although, thresholds nearing 1 MeV have been reported [18]. The flux of neutrons in this high energy tail is accepted to be between 13 and 26 cm^{-2}hr^{-1} for the respective thresholds [19].

As previously mentioned, the high energy tail of the muon spectrum is inconsequential for single event effects. In order to provide further validation of the selected model, we have made a comparison to stopping muon rates measured underground. These experiments often used nuclear emulsions placed at varying depths in meters of water equivalent. The EXPACS differential flux spectra with respect to energy are converted to different flux spectra with respect to range using an energy to range conversion. This curve is plotted in Fig. 4 along with experimental measurements. The model agrees well with data and is consistent with the estimated stopping rate of 1.6×10^{-5} cm^{-3}s^{-1} in silicon provided in [6].

IV. SIMULATION RESULTS

To estimate the effect that muon sensitivity has on devices fabricated in various technology nodes, we performed a series of simulations with the Vanderbilt Monte Carlo Radiative Energy Deposition (MRED) code [26]. In this analysis, we examine the implication of advancements in process technology leading to smaller, more densely-packed memory cells, by

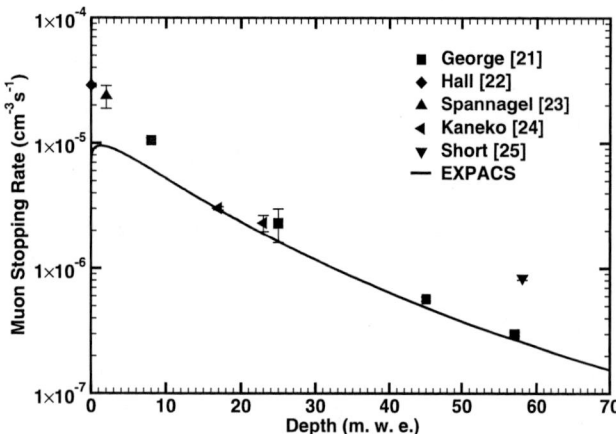

Fig. 4. Comparison of EXPACS muon stopping rate in water to experimental data reported in [21]–[25]. Normalization of stopping rates to material density is unclear.

TABLE I
CRITICAL CHARGE ESTIMATES

Technology (nm)	65	45	32	22	16
Vdd (V)	1.2	1.1	0.97	0.90	0.84
Q_{crit} lower (fC)	0.32	0.21	0.13	0.088	0.056
Q_{crit} upper (fC)	1.3	0.71	0.44	0.36	0.19

the use of a simplified charge collection model. The changes in process technology are represented as changes in the geometry of the volumes as well as the threshold for upset. The analysis is performed for bulk CMOS technologies in general and is not tied to any one technology in particular. Other drivers besides reliability may influence the course of process advancement such that our assumption of planar MOSFETS at the smallest technology nodes studied is not accurate. For the purposes of this study, we will assume that simple assumptions of scaling trends will continue.

A. Critical Charge Estimations

The evaluation of a cell design's sensitivity to ionizing radiation begins with an estimate of the circuit critical charge (Q_{crit}) value. Few publications provide critical charge estimations for recent or future deep-submicron technologies. For existing technologies, this value can be inferred from tests, device simulations, or circuit simulations. Device simulation can be used to provide an estimate of Q_{crit}, but requires at the least basic knowledge of the process. If a chip has been fabricated, accelerated testing can be useful to determine the LET threshold, but the estimation of Q_{crit} still requires the assumption of a charge collection depth and may require at-angle testing and advanced modeling to validate. Other works have attempted to provide models for pulse shaping for circuit simulations and critical charge analysis, but often lack any comparison to other forms of determination.

In the first evaluation of critical charge values, circuit simulations were performed. Predictive models [27], [28] for various nodes were downloaded from the Predictive Technology Model website [29]. A standard 6T SRAM cell based on an existing 65 nm design was constructed to establish transistor width to length ratios. For each smaller technology node, the appropriate MOSFET models were included in the netlist and the transistor dimensions were scaled by the square root of two. The voltage supply was also reduced using projections in the ITRS Roadmap [30]. A current source based

on [31] was added to the netlist to perform transient charge injection. The parameters of the source were varied and the current integrated to obtain the injected charge. The largest quantity of charge that did not upset the circuit was recorded as the critical charge. These values are provided in Table I.

The problems with a critical charge estimation from SPICE are the omission of parasitic capacitance in a predictive circuit netlist and the need for a current source with accurate pulse shaping. Several works have shown that this method may provide overestimates simply due to the current pulse shape [32], [33]. Therefore the SPICE values will be taken as an upper bound of the critical charge.

One could also base estimates of Q_{crit} on the trend in values projected by the ITRS Roadmap. The values are independent of the particular process and can be used to generalize the industry as a whole. In the roadmap, the total gate capacitance for a minimum length device is expected to remain relatively constant at 1 fF/μm over the next several generations. From the voltage and capacitance on an NMOS and PMOS pair, we derive an estimate of the Q_{crit} values for 65 through 16 nm technologies. These values are used as a lower bound for Q_{crit} in our subsequent analysis.

B. Sensitive Volumes

In this work we have chosen to use a single sensitive volume model for each SRAM cell. The sensitive volume reports the amount of energy deposited within its boundaries for each particle event. The simple approximation of the model is that a cell upset occurs if the deposited charge is greater than or equal to the critical charge. In other works, we have discussed use of multiple weighted sensitive volume models [3]. For deep-submircon designs, it is probably the case that more complicated sensitive volume models would provide a better description of the amount of charge seen at the circuit node. However, in the absence of SEU cross section data and the complications of capturing multiple node charge collection and multiple cell upsets across technology nodes, we will forgo this model. Instead, the choice of a single sensitive volume will be used as an indicator for the trend in the soft error rates.

At deep-submicron technology nodes, it is likely that sensitive volume for charge collection does not scale proportionally with the technology feature size. In the original works that developed the concept of a sensitive volume, the dominant mechanism for charge collection was drift in the depletion region around the junction. For large devices, circa 1990, this volume may have scaled well with the junction size. However, diffusion of charge to devices with nanometer dimensions is an important consideration. Thus diffusion transport may establish the dimensions of the sensitive volume.

978-1-4244-9113-1/11 $26.00 © 2011 IEEE

For each technology node investigated, an array of volumes was constructed to represent a small group of SRAM cells. An individual sensitive volume was sized according to the NMOS gate and drain geometry. As mentioned, the correlation of the volume lateral dimensions with the gate length is one possible, but not necessarily the correct, approximation of the effects of technology scaling on charge collection. At these technology nodes, it is likely that charge carriers from a particle strike can diffuse from distances much greater than the gate length. However, for the simplicity of analysis, we will make the previous assumption for all technologies. To scale the model between technology nodes, the lateral dimensions of the model were reduced by the square root of two, as well as the spacing between volumes.

We note that there is little motivation for decreasing the well depth in the same way that scaling drives packing density. Therefore we assume a constant collection depth of 0.5 μm, although a 1.0 μm depth would not be unreasonable. We will discuss the implications of our assumptions later.

C. Soft Error Rate Estimations

Our preliminary estimates indicate that the muon-induced error rates for the 65, 55, 45, and 40 nm technology nodes are insignificant. In all cases, the muon-induced FIT rate is much lower than those commonly quoted from neutron events. Although it has been shown experimentally that these nodes are sensitive to direct ionization from muons, the flux of muons with sufficient stopping power to upset a cell is quite low.

We build on the assumptions laid out for the scaling of sensitive volumes with technology nodes to examine if and when such a muon susceptibility would become a reliability issue. For smaller and more sensitive technologies, the flux of muons that are capable of exceeding a cell's critical charge value is larger. Consequently, the muon-induced soft error rate will increase.

A multilayer planar stack structure was used to model the substrate and back-end-of-line materials for radiation transport. The layers were constructed of several layers of common microelectronics materials such as silicon, SiO_2, and copper. The exact details of the composition have very little effect on the results as the primary energy loss mechanism involved is electronic stopping. The material structure was reused for all technologies and did not change with scaling. The sensitive volume models were placed within the substrate layer.

The particle gun used the μ^+ and μ^- spectra obtained from EXPACS as described in Section III. The spectra include the flux of muons greater than 1 MeV. In order to generate stopping muons in the sensitive volume, an additional 200 μm of silicon was added to the top of the multilayer stack.

The simulations produce histograms of energy deposition events weighted appropriately by the input flux spectrum. The simulation data are integrated and scaled to present plots of soft error rate versus the charge generated in the cell. Therefore the imposition of a critical charge value determines the error rate for the represented memory cell. The curves are necessarily monotonically decreasing, so circuits with larger

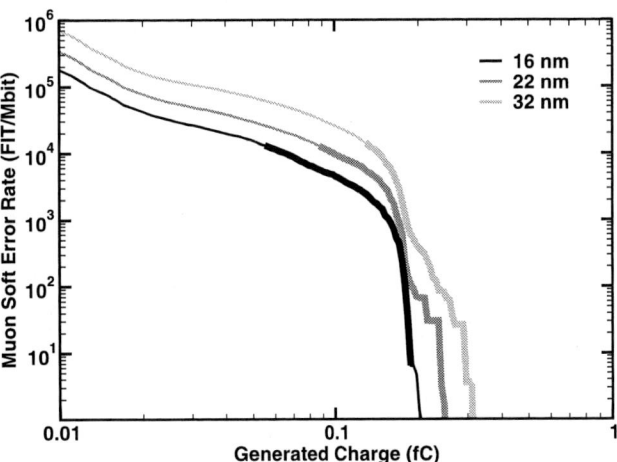

Fig. 5. Estimated NYC muon-induced event rate versus generated charge curves for 32, 22, and 16 nm bulk CMOS representative sensitive volumes. Thick lines indicate the error rate for a technology node based on the range of critical charge values.

critical charge values will have less frequent errors. This presentation of data is also useful to explore the sensitivity to uncertainty in the critical charge.

The estimated sea level error rates are shown in Fig. 5 for 32, 22, and 16 nm representative sensitive volumes. Each curve increases for lower values of generated charge as one would expect a more sensitive device to have a higher error rate. The thick curves represent the range of Q_{crit} values that fall between our established upper and lower bounds for the technology node. Below the 32 nm technology node, the critical charge is decreased enough that this threshold now permits a significant rate of errors to occur.

The error rates predicted by model are not definitive, and depend on the previously stated assumptions. We can be sure that the rate will decrease for a decreasing charge collection volume; however, it will also increase for a decreasing critical charge value.

For each technology node, a single sensitive volume model has been used to capture energy deposition within the active device area. At deep-submicron technologies, this is likely a conservative assumption. In fact charge diffusion will increase the sensitive area beyond the dimensions used in this analysis. Larger diffusion areas would cause the generated charge curves to translate to high error rates. Further, the sensitive volume depth has been assumed to be 0.5 μm. An increase in the sensitive volume depth will increase charge collection for any given particle's linear energy transfer through an increase in pathlength. The energy deposition curves would translate to the right causing the soft error rates for an assumed Q_{crit} to increase.

The results indicate the potential for large variations in the error rate for 16 nm devices. The dramatic increase in the error rate for devices with a threshold below 0.2 fC suggest that even minor design differences may have a large impact on the reliability of the memory. Small changes in

978-1-4244-9113-1/11 $26.00 © 2011 IEEE

charge collection depth, cell area, and even critical charge variability may produce large changes in the error rate. Low-power memories may be especially susceptible to the effects of reduced critical charge.

V. SUMMARY

Accelerated neutron testing and soft error rate predictions are standard practices for terrestrial microelectronics applications with high reliability requirements. The data presented here, however, suggest that the SER of future technologies also may be affected by muons. Whereas neutrons only rarely interact with nuclei, charged particles such as muons are able to generate charge carriers through the electromagnetic force. Therefore, the low-energy muon fluxes have the potential to be significant components of the SER for sensitive devices.

ACKNOWLEDGMENT

The authors would like to thank Ewart Blackmore and Michael Trinczek for their support at the TRIUMF facility and Michael Alles and Jeffrey Kauppila at Vanderbilt University for contributions to critical charge determinations.

REFERENCES

[1] K. Rodbell, D. Heidel, H. Tang, M. Gordon, P. Oldiges, and C. Murray, "Low-energy proton-induced single-event-upsets in 65 nm node, silicon-on-insulator, latches and memory cells," *IEEE Trans. Nucl. Sci.*, vol. 54, no. 6, pp. 2474–2479, Dec. 2007.

[2] D. Heidel, P. Marshall, K. LaBel, J. Schwank, K. Rodbell, M. Hakey, M. Berg, P. Dodd, M. Friendlich, A. Phan, C. Seidleck, M. Shaneyfelt, and M. Xapsos, "Low energy proton single-event-upset test results on 65 nm SOI SRAM," *IEEE Trans. Nucl. Sci.*, vol. 55, no. 6, pp. 3394–3400, Dec. 2008.

[3] B. Sierawski, J. Pellish, R. Reed, R. Schrimpf, K. Warren, R. Weller, M. Mendenhall, J. Black, A. Tipton, M. Xapsos, R. Baumann, X. Deng, M. Campola, M. Friendlich, H. Kim, A. Phan, and C. Seidleck, "Impact of low-energy proton induced upsets on test methods and rate predictions," *IEEE Trans. Nucl. Sci.*, vol. 56, no. 6, pp. 3085–3092, Dec. 2009.

[4] R. Lawrence, J. Ross, N. Haddad, R. Reed, and D. Albrecht, "Soft error sensitivities in 90 nm bulk CMOS SRAMs," in *Radiation Effects Data Workshop, 2009 IEEE*, Jul. 2009, pp. 123–126.

[5] J. T. Wallmark and S. M. Marcus, "Minimum size and maximum packing density of nonredundant semiconductor devices," in *Proc. of the IRE*, Mar. 1962, pp. 286–298.

[6] J. F. Ziegler and W. A. Lanford, "Effect of cosmic rays on computer memories," *Sci.*, vol. 206, no. 4420, pp. 776–788, 1979.

[7] R. Silberberg, C. H. Tsao, and J. R. Letaw, "Neutron generated single-event upsets in the atmosphere," *IEEE Trans. Nucl. Sci.*, vol. 31, pp. 1183–1185, Dec. 1984.

[8] J. F. Dicello, C. W. McCabe, J. D. Doss, and M. Paciotti, "The relative efficiency of soft-error induction in 4k static RAMs by muons and pions," *IEEE Trans. Nucl. Sci.*, vol. 30, no. 6, pp. 4613–4615, Dec. 1983.

[9] J. F. Dicello, M. E. Schillaci, C. W. McCabe, J. D. Doss, M. Paciotti, and P. Berardo, "Meson interactions in NMOS and CMOS static RAMs," *IEEE Trans. Nucl. Sci.*, vol. 32, no. 6, pp. 4201–4205, Dec. 1985.

[10] J. Dicello, "Microelectronics and microdosimetry," *Nucl. Instrum. Methods Phys. Res. B*, vol. 24-25, no. Part 2, pp. 1044–1049, 1987.

[11] J. F. Dicello, M. Paciotti, and M. E. Schillaci, "An estimate of error rates in integrated circuits at aircraft altitudes and at sea level," *Nucl. Instrum. Methods Phys. Res. B*, vol. 40, pp. 1295–1299, Apr. 1989.

[12] B. D. Sierawski, M. H. Mendenhall, R. A. Reed, M. A. Clemens, R. A. Weller, R. D. Schrimpf, E. W. Blackmore, M. Trinczek, B. Hitti, J. A. Pellish, R. C. Baumann, S.-J. Wen, R. Wong, and N. Tam, "Muon-induced single event upsets in deep-submicron technology," *IEEE Trans. Nucl. Sci.*, vol. 57, no. 6, pp. 3273–3278, Dec. 2010.

[13] G. M. Marshall, "Muon beams and facilities at TRIUMF," *Zeitschrift fur Physik C Particles and Fields*, vol. 56, pp. 226–+, Mar. 1992.

[14] P. K. Grieder, "Cosmic rays at sea level," in *Cosmic Rays at Earth*. Amsterdam: Elsevier, 2001, pp. 305–457.

[15] "Expacs homepage," [Online] Available: http://phits.jaea.go.jp/expacs/.

[16] T. Sato and K. Niita, "Analytical functions to predict cosmic-ray neutron spectra in the atmosphere," *Radiation Research*, vol. 166, no. 3, pp. 544–555, Sep. 2006.

[17] T. Sato, H. Yasuda, K. Niita, A. Endo, and L. Sihver, "Development of PARMA: PHITS-based analytical radiation model in the atmosphere," *Radiation Research*, vol. 170, no. 2, pp. 244–259, 2008.

[18] Y. Yahagi, E. Ibe, Y. Takahashi, Y. Saito, A. Eto, M. Sato, H. Kameyama, M. Hidaka, K. Terunuma, T. Nunomiya, and T. Nakamura, "Threshold energy of neutron-induced single event upset as a critical factor," in *Proc. IEEE Int. Physics Rel. Symp.*, April 2004, pp. 669–670.

[19] *JESD89A: Measurement and reporting of alpha particle and terrestrial cosmic ray-induced soft errors in semiconductor devices.* JEDEC Solid State Technology Association, 2006.

[20] O. C. Allkofer, K. Carstensen, and W. D. Dau, "The absolute cosmic ray muon spectrum at sea level," *Physics Lett. B*, vol. 36, no. 4, pp. 425–427, 1971.

[21] E. P. George and J. Evans, "Further observations of cosmic-ray events in nuclear emulsions exposed below ground," *Proc. Physical Soc. A*, vol. 68, no. 9, p. 829, 1955.

[22] H. E. Hall, Jr. and M. E. Richmond, "Stopping rate and energy loss of cosmic ray muons in sand," *Geophysical Research*, vol. 79, pp. 5503–5506, Dec. 1974.

[23] G. Spannagel and E. L. Fireman, "Stopping rate of negative cosmic-ray muons near sea level," *Geophysical Research*, vol. 77, no. 28, pp. 5351–5359, 1972.

[24] S.-I. Kaneko, T. Kubozoe, M. Okazaki, and M. Takahata, "Observations of slow particles and stars in nuclear emulsions exposed at 17 m.w.e. underground," *Journal Physical Soc. of Japan*, vol. 10, pp. 600–+, Aug. 1955.

[25] A. M. Short, "Mesons stopped underground," *Proc. Physical Soc.*, vol. 81, no. 5, p. 841, 1963.

[26] R. Weller, M. Mendenhall, R. Reed, R. Schrimpf, K. Warren, B. Sierawski, and L. Massengill, "Monte carlo simulation of single event effects," *IEEE Trans. Nucl. Sci.*, vol. 57, no. 4, pp. 1726–1746, Aug. 2010.

[27] Y. Cao, T. Sato, M. Orshansky, D. Sylvester, and C. Hu, "New paradigm of predictive mosfet and interconnect modeling for early circuit simulation," in *Proc. IEEE Custom Integrated Circuits Conf.*, 2000, pp. 201–204.

[28] W. Zhao and Y. Cao, "New generation of predictive technology model for sub-45 nm early design exploration," *IEEE Trans. Electron Devices*, vol. 53, no. 11, pp. 2816–2823, Nov. 2006.

[29] "Predictive technology model (ptm)," [Online] Available: http://ptm.asu.edu.

[30] Semiconductor Industry Association (SIA), "International roadmap for semiconductors 2010 update," 2010. [Online]. Available: http://www.itrs.net

[31] J. S. Kauppila, A. L. Sternberg, M. L. Alles, A. M. Francis, J. Holmes, O. A. Amusan, and L. W. Massengill, "A bias-dependent single-event compact model implemented into BSIM4 and a 90 nm CMOS process design kit," *IEEE Trans. Nucl. Sci.*, vol. 56, pp. 3152–3157, Dec. 2009.

[32] R. Naseer, Y. Boulghassoul, J. Draper, S. DasGupta, and A. Witulski, "Critical charge characterization for soft error rate modeling in 90nm SRAM," in *IEEE Symp. Circuits and Systems*, May 2007, pp. 1879–1882.

[33] T. D. Loveless, M. L. Alles, D. R. Ball, K. M. Warren, and L. W. Massengill, "Parametric variability affecting 45 nm SOI SRAM single event upset cross-sections," *IEEE Trans. Nucl. Sci.*, vol. 57, no. 6, pp. 3228–3233, Dec. 2010.

Neutron Induced Single Event Multiple Transients With Voltage Scaling and Body Biasing

Ryo Harada[†*], Yukio Mitsuyama[†*], Masanori Hashimoto[†*], and Takao Onoye[†*]

[†]Dept. Information Systems Engineering, Osaka University, JAPAN [*]JST, CREST

{harada.ryo, mituyama, hasimoto, onoye}@ist.osaka-u.ac.jp

Abstract—This paper presents measurement results of neutron induced SEMT (single event multiple transients). We devise an SEMT measurement circuit and evaluate the dependency of SEMT on supply and body voltages using test chips fabricated in a 65nm CMOS process. Measurement results show that transients can arise simultaneously at adjacent six inverters sharing the same well, and SEMT ratio to all the single event transients reaches 40% at 0.7V with reverse body biasing. We also investigate the correlation between the spatial spreading of SEMT and the distance between sensitive nodes in layout. Furthermore, referring to the occurrence rates of single event single transient (SEST) and single event single upset (SESU), we validate the measured results.

I. INTRODUCTION

Along with the scaling of semiconductor devices, single event multiple faults, which is a kind of soft error that two or more nodes are affected by one radiation particle, is becoming a serious problem [1], [2], [3], [4], [5]. Single event multiple faults is mainly classified into single event multiple(-bit) upsets (SEMU), which causes bit flips in two or more memory elements, and single event multiple transients (SEMT), which induces pulses in two or more combinational logic nodes.

To understand and avoid SEMU, SEMU characterization[1], [2], [3], SEMU mitigation techniques, such as error checking and correction, memory interleaving [1] and layout design through error-aware transistor positioning (LEAP) [6], has been intensively studied. However, no SEMT measurement results have been reported though a few evaluations based on device and circuit simulation are reported [4], [5].

Soft error mitigation techniques often exploit spatial redundancy such as triple modular redundancy and built-in soft error resilience [7], [8]. On the other hand, for instance, even when a circuit is duplicated, the duplicated cells may simultaneously receive pulses induced by SEMT due to their location proximity. Without the consideration of SEMT, the reliability of the spatially redundant circuits can be overestimated. To avoid unexpected reliability degradation, understanding the characteristic of SEMT and deriving a layout guideline for redundancy are crucially important. Moreover, voltage scaling and body biasing are becoming popular for power reduction, and low voltage yet highly reliable devices are demanded for medical applications, such as implantable devices. Therefore, the dependency of SEMT on supply and body voltages should be investigated.

In this work, we propose a measurement circuit for acquiring the spatial spreading of SEMT. Experimental results of neutron radiation tests using fabricated test chips show that six transients can arise simultaneously at 3x3 inverter matrix and decrease in supply voltage and reverse body biasing increases SEMT ratio to all the single event transients. We also show and investigate the occurrence number and spatially-distributed distance of SEMTs for each spatial SEMT pattern. We furthermore confirm that the measured SEMTs are sufficiently isolated from multiple SESTs (single event single transient) and multiple SESUs (single event single upset) in neutron radiation tests whereas the isolation is not sufficient in alpha-particle radiation tests.

The remainder of this paper is organized as follows. Section II explains requirements for measuring SEMT and introduces the proposed SEMT measurement circuit. Section III presents an implementation of the proposed circuit and shows experimental results of accelerated neutron radiation tests. Section IV discusses difference between neutron- and alpha-induced soft errors and verifies the reliability of the measured SEMT results. Finally, the conclusion is given in Section V.

II. SEMT MEASUREMENT CIRCUIT

We first review the requirements for measuring characteristic of SEMT. We then introduce the proposed measurement circuit.

A. Requirements for SEMT measurement circuit

SEMT causes multiple transients in a combinational circuit. In order to distinguish all the transients, they must be delivered to different memory elements (or different channels of a measurement equipment) respectively. Otherwise, some transients diminish and become invisible. This means that logical and electrical masking must be eliminated in the measurement circuit, and hence parallel cell-chains, each of which is terminated with an individual memory element, are desirable for SEMT measurement.

Furthermore, to finely characterize SEMT spatial distribution, the distance between sensitive nodes included in different cell-chains should be small. From such a perspective, inverter-chains are desirable, since the footprint of inverter cells is small compared to other logic cells. By adjacently placing inverter cells that are included in different cell-chains, the spatial spreading of multiple transients is expected to be captured with high spatial resolution.

Another requirement is the isolation from multiple SESUs and SESTs. When the isolation is not enough, the measured SEMTs are no longer reliable. This isolation must be ensured by circuit organization and/or measurement setup.

978-1-4244-9113-1/11 $26.00 © 2011 IEEE

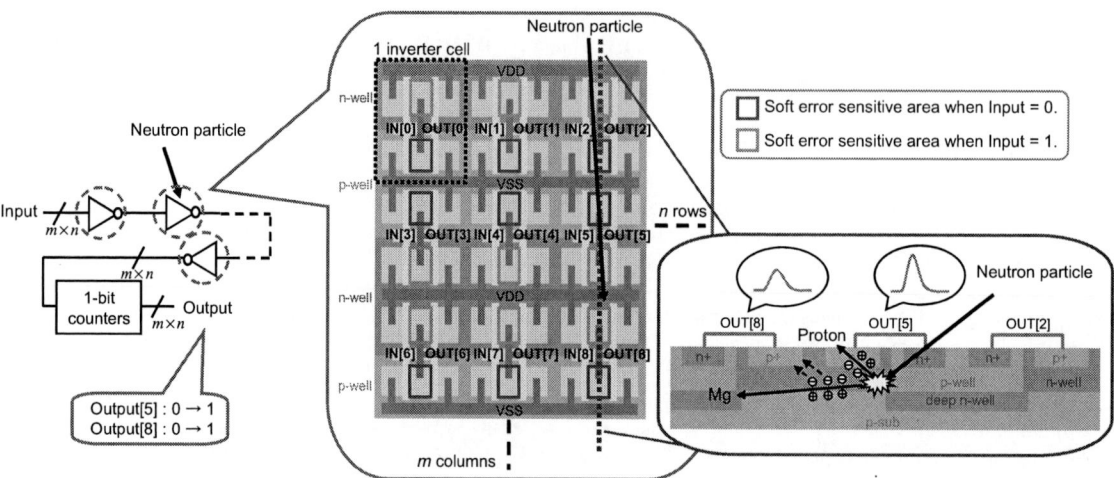

Fig. 1. Structure and operation of SEMT measurement circuit. At each stage, inverters are placed in $m \times n$ array. Each inverter-chain has a 1-bit counter at output, and SEMT across cells can be observed.

B. Proposed measurement circuit

According to the requirements described above, we devised a SEMT measurement circuit. Figure 1 illustrates the proposed measurement circuits composed of $m \times n$ inverter-chains, where each of them has the same stage length and a 1-bit counter at the output. For each stage, inverters are adjacently placed in a $m \times n$ array, and this array is serially connected on the whole.

In this circuit, we can measure SEMTs that spatially spread within $m \times n$ inverters at the maximum. A hit of neutron to silicon substrate could deliver charge to multiple sensitive nodes[1] across cell boundaries due to, for example, diffusion/drift and bipolar action [2]. In the example of Fig. 1, SEMT pulses arise at OUT[5] and OUT[8] and propagate into counters. By reading out the flips of counter values during the short time interval in which multiple SESTs will not happen, we can observe spatial distribution of SEMTs.

III. EXPERIMENT

This section first explains the experimental setup and then shows experimental results of accelerated neutron radiation test.

A. Experimental setup

To measure SEMT using the proposed circuit, a test chip was fabricated in a 65 nm CMOS process. The structure of test circuit is depicted in Fig. 2. 3×3 array using 2X-sized inverters in triple-well structure was selected as SEMT target circuit, and 15,896-stage chains were constructed for obtaining a large number of samples. Supply and body voltages of SEMT target circuit can be controlled separately. With double-back layout policy, power/ground lines are shared by inverters at successive columns as depicted in Fig. 1.

Neutron radiation test was performed at RCNP (Research Center for Nuclear Physics, Osaka University) using accelerated wide spectrum neutron beam [9], [10]. Figure 3 shows

[1]Soft error sensitive area is NMOS drain area when its inverter input is 0 and PMOS drain area when its inverter input is 1.

the test setup. The flux density of neutron beam was $2.41 \times 10^9 \mathrm{cm}^{-2}\mathrm{h}^{-1}$. In the experiment, the beam was irradiated to 30 test chips. The inputs of the inverter-chains were fixed to zero. The counter values were read out in every five seconds. In this implementation, we need to examine how much combinations of SESTs and/or SESUs are mistakenly included in the measured SEMT data. We will discuss it and show that the probability of multiple particle hits is low enough compared to that of SEMTs in Section IV.

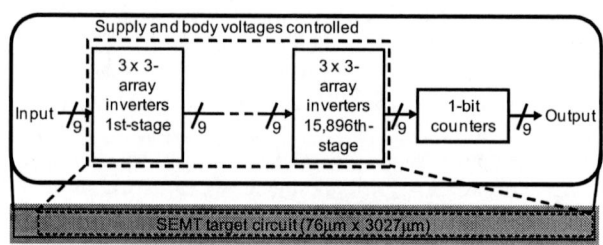

Fig. 2. Test chip fabricated in 65 nm process. SEMT target circuit is 15,896-stage 3×3-array inverter-chains.

Fig. 3. Photograph of test environment. 30 test chips are installed on 8 DUT boards in total which are placed in series in front of beam aperture.

TABLE I

NUMBER OF SESTs AND SEMTs OBSERVED IN NEUTRON RADIATION
TEST AND SEMT RATIO.

Supply voltage [V]	1.2			0.7		
Body bias	ZBB	FBB	RBB	ZBB	FBB	RBB
Measurement time [h]	1.48	1.96	1.65	1.67	1.01	0.81
# of total transients	122	0	169	127	0	50
# of SESTs [/h]	72.9	0	80.5	50.9	0	37.2
# of SEMTs [/h]	9.46	0	21.8	25.1	0	24.8
SEMT ratio [%]	11.5	-	21.3	33.1	-	40.0

SEST: a flip at one 1-bit counter, SEMT: flips at two or more 1-bit counters.

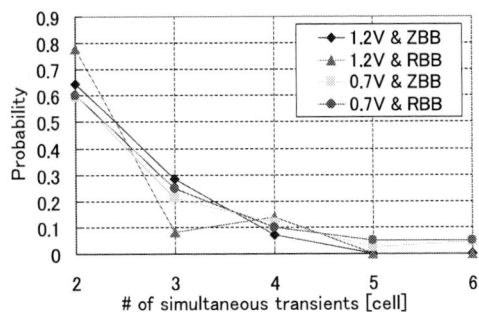

Fig. 4. Probability distribution of number of simultaneous transients, where for each curve the sum of probability is normalized to 1.

B. Results

Table I lists the measured results of experiments at two supply voltages, which are 1.2 V and 0.7 V, with three body-biasing configurations, zero body biasing (ZBB), 0.6 V forward body biasing (FBB), and 1.3 V reverse body biasing (RBB). While in FBB, there was neither SEST nor SEMT in this experiment, in RBB compared to ZBB, the numbers of SESTs and SEMTs increase at 1.2 V and they decrease at 0.7 V. The ratio of SEMTs to the total transients increases as the supply voltage decreases and the body is biased in reverse direction.

Figure 4 shows the probability distribution of the number of simultaneous transients, where for each curve the sum of probability is normalized to 1. We observed six transients at the maximum for a single neutron hit. Five and six transients occurred only in the operation of 0.7 V supply voltage. The occurrence probability decreases as the number of simultaneous transients increases, and this tendency is less dependent on supply and body voltages.

Figure 5 illustrates the layout of sensitive nodes in the 3x3 array and possible horizontal and vertical patterns of two-transient SEMTs. Figure 6 shows the number of occurrence for each SEMT spatial pattern. The horizontal axis is the farthest distance between related sensitive areas in the corresponding SEMT pattern. Here, any symmetric patterns are regarded as a single pattern. We did not observe SEMT across three cells in the vertical direction. This means that SEMTs arise within a single well and no SEMTs across well boundary happen, which is consistent with the SEMU observations [3]. Looking at two-transient cases, the numbers of occurrence are 38 and 35 in horizontal and vertical cases respectively, and they are almost the same, although the distance between sensitive nodes is twice

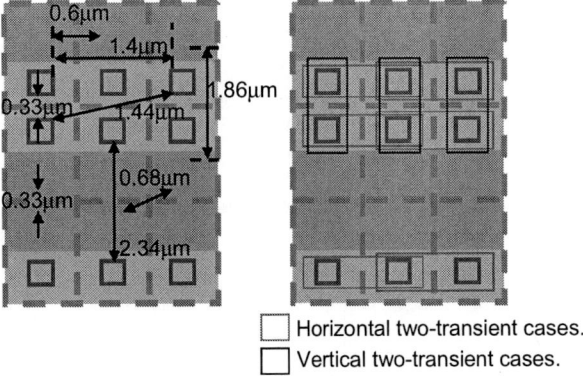

Horizontal two-transient cases.
Vertical two-transient cases.

Fig. 5. The distance between sensitive nodes and possible physical patterns of two-transients cases.

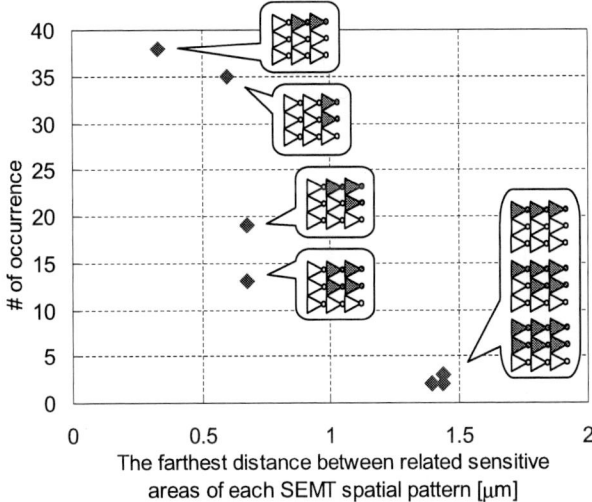

Fig. 6. The farthest distance between two sensitive areas versus the number of occurrence for each SEMT spatial pattern.

larger in horizontal case (0.6μm) than in vertical case (0.33μm). On the other hand, in 3x3 array, the number of possible physical patterns in horizontal case is six whereas it is three in vertical case when considering SEMTs arise only within the same well. The distance between sensitive nodes is canceled by the number of possible patterns, and then the similar numbers of occurrence were observed. In this experiment, we observed six-transient SEMTs, which are the largest measurable SEMTs in the fabricated test chip, and hence it might be possible that SEMTs were more widely spread. Larger array of inverter-chains would be necessary for further investigation.

IV. DISCUSSION ON SEMT ISOLATION FROM MULTIPLE SESTs AND SESUS

As described in Section III-A, by the measurement circuit implemented on the test chip, an true-SEMT cannot be distinguished from pseudo-SEMT, which is multiple SESTs, multiple SESUs and a combination of SEST and SESU. We investigate the misrecognition probability of SEMT.

TABLE IV

RATIO OF ESTIMATED OCCURRENCE PROBABILITIES OF TRUE-SEMT AND PSEUDO-SEMTS TO TOTAL OCCURRENCE PROBABILITY OF ALL SEMTS IN BOTH NEUTRON AND ALPHA-PARTICLE TESTS. EACH OCCURRENCE PROBABILITY IS CALCULATED BY USING THE MAXIMUM λ AMONG SIX SUPPLY AND BODY BIAS CONDITIONS.

| | Neutron | | | | Alpha | | | |
| | Pseudo-SEMT | | | True-SEMT | Pseudo-SEMT | | | True-SEMT |
	2 SESTs	2 SESUs	1 SEST & 1 SESU		2 SESTs	2 SESUs	1 SEST & 1 SESU	
Probability ratio[%]	0.72	0.00	0.00	99.28	6.80	22.09	24.57	46.55

TABLE II

CROSS SECTIONS OF SESTS AND SESUS OBSERVED IN NEUTRON AND ALPHA RADIATION TESTS.

	Neutron	Alpha
Measurement time [h]	8.38	5.47
SEST cross section [μm^2/cell]	4.03×10^{-7}	1.28×10^{-6}
SESU cross section [μm^2/bit]	6.63×10^{-6}	1.28×10^{-2}

TABLE III

ESTIMATED NUMBER OF SESTS AND SESUS PER SECOND IN A 1-BIT COUNTER.

	Neutron	Alpha
#SESTs [/(s·counter)]	2.49×10^{-4}	2.66×10^{-5}
#SESUs [/(s·counter)]	6.63×10^{-8}	4.81×10^{-4}

We first examine the cross-sections of SESU and SEST. In the same test chip, a shift register consisting of 100 D-FFs was implemented. We measured SESUs occurred in these D-FFs at the same time as measuring SESTs and SEMTs in the neutron radiation test. We also measured SESTs, SEMTs, and SESUs in alpha radiation test for clarifying the difference between neutron induced soft errors and alpha-particle induced soft errors. The alpha particle tests were performed using an Americium-241 foil whose flux is $9 \times 10^9 cm^{-2}h^{-1}$. The radiation source was put immediately above on a test chip [10].

Table II shows the experimental results of both the neutron and alpha radiation tests. Compared to the neutron results, alpha-induced SESU has much larger cross section than SEST. We here associate the cross section difference with the energy spectrum difference between neutron and alpha particles as one of possible reasons, though other factors in terms of charge generation and collection are different as well. In the measurement circuit, SESU critical charge for flipping a D-FF is much smaller than SEST critical charge for flipping a 1-bit counter at the end, because the mismatch of rise and fall delay and electrical masking through 15,896 inverters at maximum reduce the SEST pulse width before arriving at the 1-bit counter. Besides, the main peak energy of Americium-241 is 5.49 MeV, while the energy of neutron beam used for the experiments distributes over 100 MeV [9]. If the mean of the charge induced by alpha particles is between the critical charge of SESU and that of SEST, SEST/SESU ratio in alpha radiation test becomes low. On the other hand, the wide spectrum neutron beam includes many particles which can induce the charge that is larger than the SESU and SEST critical charges. This is one of possible reasons that

explain lower SEST/SESU ratio in alpha particle tests.

Next, the obtained SEMT results, which might include pseudo-SEMTs, are verified. First, the number of SESTs and SESUs observed in a 1-bit counter of SEMT measurement circuit are estimated from the total number of SESTs observed in the SEMT measurement circuit and the number of SESUs observed in the 100 D-FFs. Table III lists the estimated number of SESTs and SESUs per second in a 1-bit counter. From now, two-transient SEMTs are validated using the estimated number of SESTs and SESUs. The occurrence probabilities of SEST and SESU that N events are observed in a unit time follows the Poisson distribution and is expressed by

$$P(N) = \frac{e^{-\lambda}\lambda^N}{N!}, \tag{1}$$

where λ is the expected number of occurrences in the time interval. Similarly, the occurrence probabilities of all types of SEMTs that flip two of any nine counters during the time interval of counter read, which consist of pseudo-SEMTs and true-SEMTs, are estimated by Table I and (1). Using these occurrence probabilities, we calculate the occurrence probabilities of pseudo-SEMTs and true-SEMT. Table IV lists the ratio of calculated occurrence probabilities of pseudo-SEMTs and true-SEMT to the total occurrence probability of all SEMTs. In the case of two-transient SEMT, the ratio of true-SEMT occurrence probability in alpha particle tests is estimated to be 46.6% because of higher SESU event rate, which means that true-SEMTs are overwhelmed by pseudo-SEMTs. On the other hand, we confirm that the SEMT results in neutron particle tests were sufficiently isolated since the ratio in neutron particle tests is 99.3%.

V. CONCLUSION

In this paper, we proposed an SEMT measurement circuit and presented the measurement result of neutron-induced SEMTs in 65 nm test chips. The measurement result showed that SEMT ratio increases as decreasing the supply voltage and biasing the body in reverse direction. On the other hand, neither SEST nor SEMT occurred in forward body biasing. We observed that SEMT can induce six transients in adjacent inverter cells and the occurrence probability of SEMT decreases according to increase in the number of simultaneous transients. We also showed that the occurrence tendency of spatial patterns of SEMTs and no SEMT observation across three cells in the vertical direction due to well boundary. Furthermore, we validated the SEMT measurement results of the neutron radiation test on the basis of the occurrence probabilities of SEST and SESU. As future works, we will investigate the reason why soft error did not occur in

FBB case and we will characterize SEMTs using larger array structure.

Acknowledgments

The authors would like to acknowledge technical advices from Dr. Eishi Ibe of Hitachi. The authors also would like to thank VDEC, the University of Tokyo in collaboration with STARC, e-Shuttle, Inc., and Fujitsu Ltd. for their support in getting the chip fabricated. The authors appreciate the support of Professor Kichiji Hatanaka and Assistant Professor Keiji Takahisa of Osaka University for the neutron radiation test at RCNP. This work was partly supported by NEDO.

REFERENCES

[1] N. Seifert, P. Slankard, M. Kirsch, B. Narasimham, V. Zia, C. Brookreson, A. Vo, and S. Mitra, "Radiation-induced soft error rates of advanced CMOS bulk devices," in *Proc. IRPS*, pp. 217–225, 2006.

[2] E. Ibe, S. Chung, S. Wen, H. Yamaguchi, Y. Yahagi, H. Kameyama, S. Yamamoto, and T. Akioka, "Spreading diversity in multi-cell neutron-induced upsets with device scaling," in *Proc. CICC*, pp. 437–444, 2007.

[3] Y. Yahagi, H. Yamaguchi, E. Ibe, H. Kameyama, M. Sato, T. Akioka, and S. Yamamoto, "A novel feature of neutron-induced multi-cell upsets in 130 and 180 nm SRAMs," *IEEE Transactions on Nuclear Science*, vol. 54, no. 4, pp. 1030–1036, 2007.

[4] D. Rossi, M. Omana, F. Toma, and C. Metra, "Multiple transient faults in logic: An issue for next generation ICs?" in *Proc. DFT*, pp. 352–360, 2005.

[5] C. Rusu, A. Bougerol, L. Anghel, C. Weulerse, N. Buard, S. Benhammadi, N. Renaud, G. Hubert, F. Wrobel, T. Carriere *et al.*, "Multiple event transient induced by nuclear reactions in CMOS logic cells," in *Proc. IOLTS*, pp. 137–145, 2007.

[6] H. Kelin, L. Klas, B. Mounaim, R. Prasanthi, I. Linscott, U. Inan, and S. Mitra, "LEAP: Layout Design through Error-Aware Transistor Positioning for soft-error resilient sequential cell design," in *Proc. IRPS*, pp. 203–212, 2010.

[7] S. Mitra, M. Zhang, S. Waqas, N. Seifert, B. Gill, and K. Kim, "Combinational logic soft error correction," in *Proc. ITC*, vol. 2, p. 824, 2006.

[8] J. Furuta, C. Hamanaka, K. Kobayashi, and H. Onodera, "A 65nm Bistable Cross-coupled Dual Modular Redundancy Flip-Flop capable of protecting soft errors on the C-element," in *Proc. VLSIC*, pp. 123–124, 2010.

[9] Y. Tosaka, H. Ehara, M. Igeta, T. Uemura, H. Oka, N. Matsuoka, and K. Hatanaka, "Comprehensive study of soft errors in advanced CMOS circuits with 90/130 nm technology," in *Proc. IEDM*, pp. 941–944, 2005.

[10] JEDEC standard JESD89, "Measurement and reporting of alpha rarticles and terrestrial cosmic ray-induced soft errors in semiconductor devices," 2001.

DOUBLE-PULSE-SINGLE-EVENT TRANSIENTS IN COMBINATIONAL LOGIC

J. R. Ahlbin, T. D. Loveless, D. R. Ball, B. L. Bhuva, A. F. Witulski, L. W. Massengill
Dept. of Elec. Eng. and Comp. Sci.
Vanderbilt University
Nashville, TN USA
phone: (01) - (615) - 343-6705, e-mail: jon.ahlbin@vanderbilt.edu

M. J. Gadlage
NSWC Crane
Crane, IN USA

Abstract — For the first time, double-pulse-single-event transients (DPSETs) are observed during heavy-ion broad beam testing. The transients are generated in a serially connected string of inverters and measured with an autonomous on-chip SET pulse-width measurement circuit. Three-dimensional mixed-mode technology computer aided design (TCAD) simulations show that DPSETs are the result of multiple inverters being upset by a single ion strike.

Keywords - soft error, double-pulse-single-event transients, single event, single-event transient, SET, SER, pulse width, charge sharing, pulse quenching, radiation environment

I. INTRODUCTION

The scaling of integrated circuits to feature sizes below the 100 nm technology node presents a host of challenges for circuit applications for use in a space environment due to a growing susceptibility to soft errors [1-6]. As transistor density increases on chip with each new scaled technology, the probability of a single ion causing a single-event transient (SET) in a circuit or depositing charge on multiple nodes increases [7-8]. SETs are a significant reliability concern, as they can compete with legitimate logic signals and corrupt downstream logic and latches. Additionally, multiple node charge collection from a single ion strike can render hardened storage cells and SET mitigation techniques (such as guard rings, guard drains, additional well contacts, etc.) ineffective. Therefore, it is important to understand the SET characteristics of each new technology node and the SET impact on circuit designs.

In this paper, recent experimental data from a 65 nm bulk CMOS process is discussed and analyzed showing the existence of double-pulse-single-event transients (DPSETs). The existence of DPSETs means that soft-error rates would need to be calculated differently because nominally only one error is associated with an incident ion, whereas DPSET will show two errors for each incident ion. These DPSETs were observed for the first time during a heavy-ion experiment in a

---

This work was supported in part by the DTRA Radiation-Hardened Microelectronics Program and Cisco Systems.

J. R. Ahlbin, T. D. Loveless, D. R. Ball, B. L. Bhuva, A. F. Witulski, and L. W. Massengill are with Vanderbilt University, Nashville, TN 37235 USA (e-mail: jon.ahlbin@vanderbilt.edu).

M. J. Gadlage is with NAVSEA Crane, Crane, IN 47522 USA.

chain of inverters where every pFET was located in its own separate n-well. However, DPSETs were not observed in a second target inverter chain where inverters shared a common n-well.

To investigate these experimental results, three-dimensional mixed-mode technology computer aided design (TCAD) simulations were performed. The two fabricated inverter target chain designs were replicated and the experimental conditions were modeled. The simulations, just like the experimental results, showed that DPSETs were created in the separate n-well design but not for the common n-well design. TCAD simulations also indicate a complex relationship among multiple phenomena such as charge sharing [4,5], pulse quenching [9] and strike location. For DPSETs to occur in the separate n-well design, various mechanisms interact in a way to allow multiple pFETs to collect charge and change state. However, in the common n-well design, pulse quenching is the dominant mechanism preventing DPSETs from being generated.

II. SET MEASUREMENT CIRCUIT DESIGN

The test circuits used during the heavy-ion experiment were fabricated in the IBM 65 nm bulk CMOS technology. The two target circuits (common n-well and isolated n-well) were part of an on-chip SET measurement circuit based on the autonomous pulse-width measurement technique described in [10], which digitizes the SET pulse width following the arrival of an SET. This implementation of the SET measurement circuit was able to measure transient pulses ranging from 75 ps to 2 ns with a 25 ps measurement resolution. Thus measureable DPSETs were restricted to those separated by at least 25 ps in time. Following the generation of DPSETs and the propagation to the measurement circuit, the first pulse of the DPSET triggered the measurement circuit to "freeze" or stop acquisition, as in the nominal case of a single SET. This circuit reaction, however, is delayed in time on the order of hundreds of picoseconds (measurement lag time), thus allowing for a second SET arriving within the measurement lag-time to be captured by the on-chip measurement circuit.

In both target designs, a chain of 1000 inverters was used and designed in 10 rows of 100 inverters, each connected in a serpentine manner. These inverters were designed to be current matched with a pMOS transistor width and length (W/L) of 400 nm / 50 nm and nMOS transistor W/L of 200 nm / 50 nm.

The differences between the two inverter chains were the size of the n-well and the transistor spacing. The first target chain contained inverters with each row having 100 pMOS transistors in the same n-well. These inverters were spaced 0.75 μm from gate center to gate center, as shown in Fig. 1. The purpose of this target chain was to enhance charge sharing among the multiple transistors spaced close together in the common n-well. Additionally, the entire n-well was contacted with a strip well-contact leading to a total n-well-contact-area-to-well-area percentage of 19%.

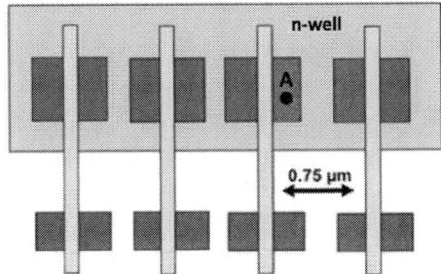

Fig. 1. Layout of the 65 nm target inverter chain where each row of inverters shared a common n-well (common n-well).

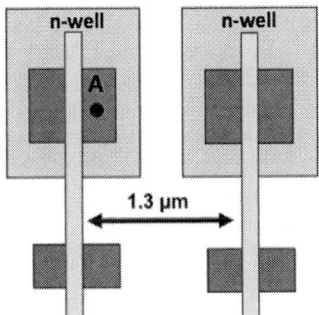

Fig. 2. Layout of the 65 nm target inverter chain where each inverter had its own separate n-well (separate n-well).

In the second inverter target chain, each pFET was placed in an isolated n-well and spaced 1.3 μm from gate center to gate center, as shown in Fig. 2. With a p-well between each n-well and increased spacing, a reduction in charge sharing was expected. The p-well would act as a doped barrier to charge diffusion, and the increased spacing would reduce the amount of charge that could reach adjacent transistors. Last, each inverter had it's own n-well contact such that the contacted-area to well-area percentage would be similar to that for the common n-well design. The total contacted-area to well-area percentage was 14%. Gadlage et al., has shown that the amount of n-well area covered by contacts can significantly influence SET pulse widths so it is important to keep the value similar between the two different target strings [11].

III. DPSET EXPERIMENTAL RESULTS

The circuits were tested at the Lawrence Berkley National Laboratory with ions having LETs of 3.5, 9.7, 21, 30, and 58 MeV-cm^2/mg at a 60° angle of incidence. The test fixture was arranged so that particle penetration was parallel to the power rails and longitudinal to the n-wells in order to promote charge collection among multiple transistors. For an LET of 58 MeV-cm^2/mg the circuit was tested to an effective fluence of 5×10^7 particles/cm^2. Mid-range LETs of 21 and 30 MeV-cm^2/mg were tested to an effective fluence of 1×10^8 particles/cm^2. Effective fluences of 2×10^8 and 5×10^8 particles/cm^2 were used for LETs of 9.7 and 3.5 MeV-cm^2/mg respectively. The effective fluence differed between ion species to improve the statistics of the error counts for the lower LET ion exposures.

During the experiment, each time an ion strike caused a DPSET in the separate n-well target circuit, the on-chip measurement circuit recorded it. Fig. 3 shows the number of DPSETs for each ion LET tested. DPSETs were observed at LETs as low as 9.7 MeV-cm^2/mg. As ion energy increased, the number of DPSETs observed increased, suggesting that the amount of charge deposited in the ion strike radius is a key factor in the generation of DPSETs.

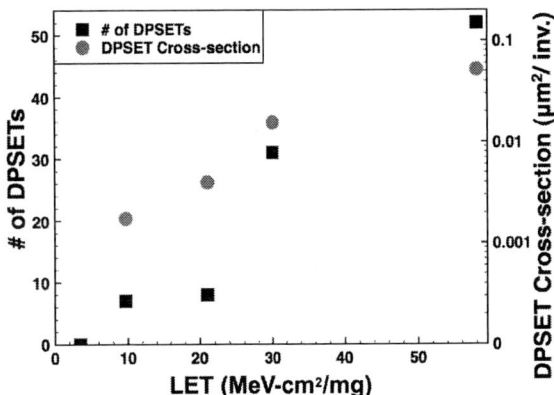

Fig. 3. Number of DPSETs and cross-section for separate n-well inverter target chain with all ions incident at 60° angle longitudinal to the n-well

In addition to counting the number of DPSETs that occurred, characteristics of the DPSET pulse shape were recorded. In conventional SET measurement tests in older technologies, only a single SET is measured and a pulse width for that single SET is calculated, as shown in Fig. 4. However for a DPSET, two pulses were measured along with the time difference between the two of them for one upset event. For example, Fig. 5 shows a measured DPSET at an LET of 58 MeV-cm^2/mg. In this DPSET, the first pulse has a width of 100 ps and the second pulse has a pulse width of 150 ps with the two pulses occurring 700 ps apart.

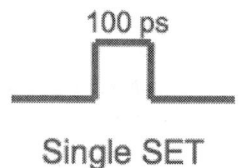

Fig. 4. Single pulse voltage transient.

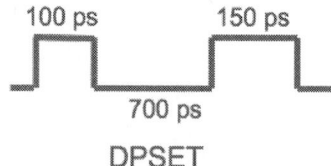

Fig. 5. Double pulse voltage transient

The DPSET described in Fig. 5 is one of 52 DPSETs observed at 58 MeV-cm^2/mg. Moreover, each DPSET tended to have unique characteristics. Out of the 52 events observed at an LET of 58 MeV-cm^2/mg, only two DPSETs were observed where the first and second pulse had the same pulse width. This trend was similarly seen in LETs below 58 MeV-cm^2/mg as well.

The DPSET's can be broken down as two distinct events, each with is own pulse width, and separated by a time constant. The pulse width data for the first and second pulses can be seen in Fig. 6.

Fig. 6 gives insight into the generation of DPSETs by separating out the pulse widths of each event. The first pulse has a typical distribution of pulse widths similar to previously reported SET pulse width distributions for particles at normal incidence [11][12], and would indicate that the ion is passing through the drain region of the pMOS transistor. The first pulse

has a wider distribution spanning from 25 ps to 275 ps, whereas the second pulse, has a narrower distribution of pulse widths from 50 ps to 175 ps. The second pulse would seem to be a result of substrate-related charge collection from the portion of the ion track that is passing below inverters farther down the chain. These mechanisms are investigated in the simulation section of this paper.

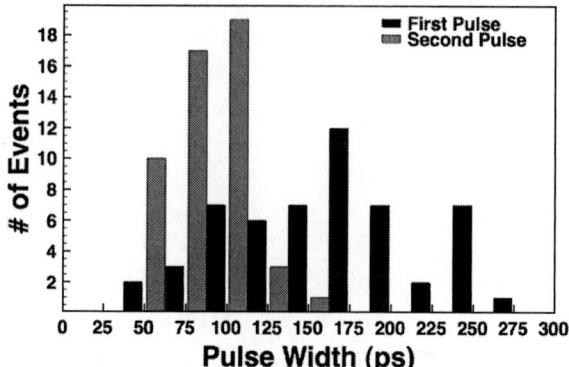

Fig. 6. Plot showing the different pulse widths measured for the 1st and 2nd pulses of the DPSETs that occurred during the LET of 58 MeV-cm^2/mg experimental run.

Fig. 7 shows the time duration distribution between the first and second pulses for all the DPSETs observed during the LET of 58 MeV-cm^2/mg run. The interval between double pulses appears to be uniformly distributed. The probable reason for this distribution of data is because the duration between pulses is directly related to the strike location. During heavy-ion testing, there is an equal chance an ion can strike anywhere in the sensitive area of the circuit to cause an SET. Since the double-pulse mechanism depends on diffusion of charge, and diffusion is a function of distance, the shape of the double pulse

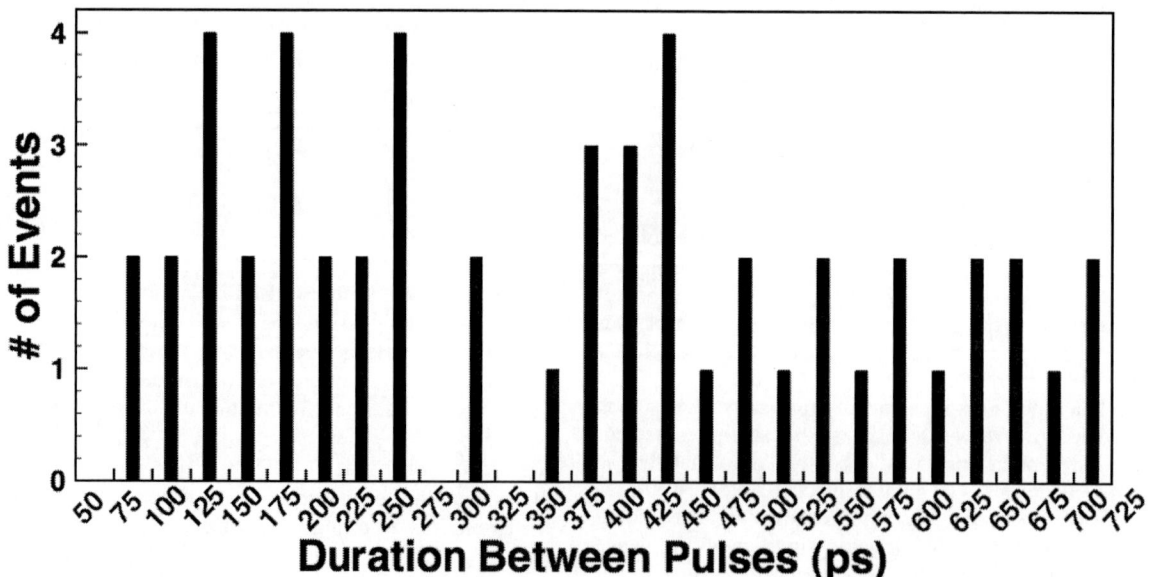

Fig. 7. Plot showing the time durations measured between the 1st and 2nd pulses of the DPSETs that occurred during the LET of 58 MeV-cm^2/mg experimental run.

transient is a strong function of strike location.

IV. DPSET SIMULATION ANALYSIS

Initial analysis of the DPSET data suggest that the likely cause of a DPSET is more than one inverter being upset through charge sharing. Charge sharing was found to be prevalent at 65 nm technology node with these specific test structures [13]. To take into account charge sharing and understand the experimental data, large-scale simulations were carried out using Synopsys SDevice 3D mixed-mode tools. Mixed-mode simulations allow specific transistors in a circuit to be modeled fully in 3D TCAD while the rest of the circuit can be implemented using compact models. Such an approach is very efficient and effective in understanding various mechanisms involved in single-event effects.

Two TCAD models were used in the mixed-mode simulation with each containing eight 65 nm pFETs. One model was based on the separate n-well target circuit layout, and one was based on the common n-well target circuit layout previously discussed. The pFETs in each model were calibrated to match dc and ac electrical characteristics (e.g., I_d-V_d and I_d-V_g curves) based on the IBM CMOS10SF PDK. The remainder of the simulation setup included calibrated compact models for the nMOS transistors in a chain of inverters. This chain matched the sizing of the target inverters for the compact model portion of the simulation.

The TCAD models are pictured in Figs. 8 and 9. Notice the n-diffusion p-body diodes. The actual nFETs are not modeled in TCAD, and are modeled as compact models because of computing limitations. These diodes serve as representations of the drains of the nFETs, matching the active area size and placement to the actual nFETs. The diodes are electrically biased as if they are nFETs but are not electrically connected to the pFETs. They increase the accuracy of the simulation by serving as charge sinks mimicking the actual nFETs in the design.

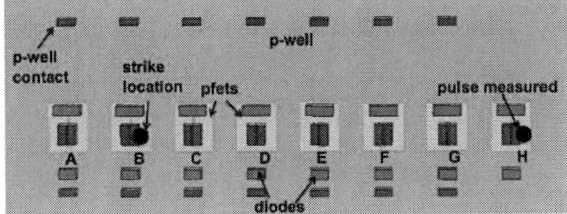

Fig. 8. TCAD model of 65 nm separate n-well inverters with pFETs modeled in TCAD and nFETs as compact models.

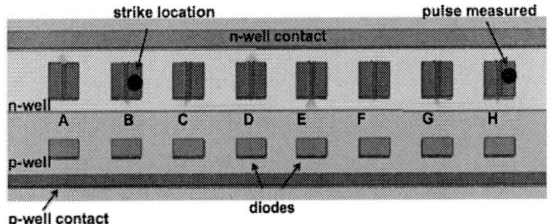

Fig. 9. TCAD model of 65 nm common n-well inverters with pFETs modeled in TCAD and nFETs as compact models.

The eight simulated pFETs in each TCAD model represent the pFETs in the second through final stages of the nine-inverter chain, while the first inverter in the chain is a compact model. All simulations used the following physical models: Fermi-Dirac statistics, SRH and Auger recombination, and the Carrier-Carrier Scattering mobility model. The incident heavy-ions were modeled using a Gaussian radial profile with a characteristic 1/e radius of 50 nm, and a Gaussian temporal profile with a characteristic decay time of 2 ps. Simulations were carried out using the Vanderbilt ACCRE computing cluster [14]. The inverter string was statically biased such that pFETS A, C, E, and G were on and the remaining pFETS were off.

During and following the ion strike, the simulations are setup to monitor the voltage at each node of the inverter chains. Voltage monitoring of each node allows any SETs created to be observed during formation and propagation through the circuit. As a reference to understand the charge transport mechanisms during and after the ion strike, Fig. 10 provides a node-labeled cross section of Fig. 8 along the direction of the ion strike for the separate n-well model.

Matching the experimental setup, each TCAD model is struck with a single 58 MeV-cm^2/mg heavy-ion incident at 60° from the Si surface that penetrates the second pFET (node B). As the ion penetrates the device and the surrounding n-well, charge is deposited and the potential of the well collapses. In the separate n-well simulations, the well collapse is limited because of the use of p-well between the different n-wells. However, the well collapse is larger in the common n-well simulations because the n-well spans the entire model and there is no p-well between n-wells to act as a charge diffusion barrier. In both cases, the pFET B is in the OFF state prior to the ion strike, and charge deposited near the transistor is collected and causes the transistor to be upset. Thus, pFET B turning ON changes the voltage at node B and produces a first SET of about 150 ps in duration for the separate n-well and about 140 ps for the common n-well, as shown in Fig. 11. This SET propagates very rapidly through the inverter chain and slightly attenuates to a pulse width of 133 ps for the separate n-well and about 120 ps for the common n-well. This initial pulse is the first pulse of the DPSET. It requires charge to be deposited directly into the drain node.

At node D (Fig. 12), after the first SET, a second SET is formed in both full TCAD simulations as a result of substrate-related charge collection from the ion passing below the OFF-state pFET at node D. The charge deposited by the ion has diffused from the substrate up into the n-well surrounding node D, debiasing the well, and causing node D to collect charge. The resulting voltage pulse is the second pulse of the DPSET.

In order to separate the two pulses (and their mechanisms) from one another, two additional simulations were performed. Node B was replaced with a compact model in one simulation with all other transistors remaining in TCAD. For the second simulation, node D was replaced with a compact model, again with all remaining transistors still modeled in TCAD. The resulting pulses are shown in Fig. 12. With node D having been replaced with a compact model, only the first pulse of the DPSET, the one that originates from the hit to the drain of node B, can be seen. With node B having been replaced with a compact model, only the second pulse of the DPSET, the one that results as substrate-related charge collection, can be seen.

It is also important to note the presence of pulse quenching [9] because of the charge still present in the area from the angled ion strike. (Transistor E starts off ON, then is electrically turned off by the transient from D on its gate, then it starts collecting charge to turn on again.) At node E (Fig. 13), pulse quenching is significant and reduces the pulse width of the second SET from about 900 ps to 195 ps in the separate n-well simulation. By running an additional simulation with node E replaced with a compact model, we see that pulse quenching no longer occurs. The compact model does not account for the pulse quenching mechanism, instead letting the second pulse in the DPSET propagate through the chain, and reaching pulse widths over 1 ns long. However in the common n-well simulation with the pFET modeled in TCAD at node E, pulse quenching is more significant and eliminates the second pulse, which is why we do not see DPSETs in the common n-well data. For the separate n-well simulation, a DPSET is seen at node F (Fig. 14), and is representative of the double pulse that will propagate through the rest of the inverter chain.

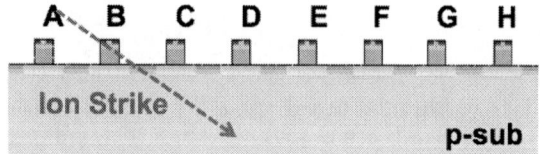

Fig. 10. Cross-section of 65 nm TCAD model showing the 60° ion strike.

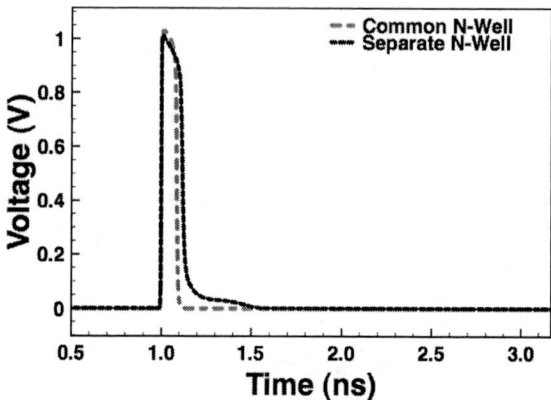

Fig. 11. Plot of the voltage output for node B.

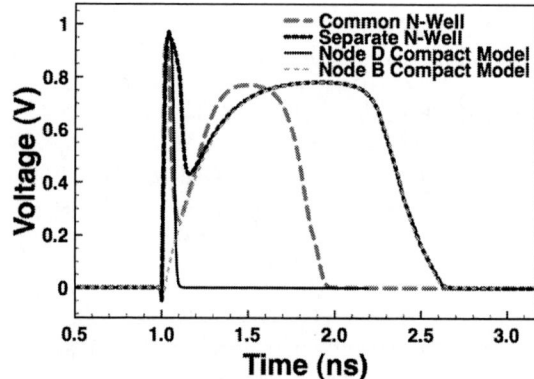

Fig. 12. Plot of the voltage output for node D.

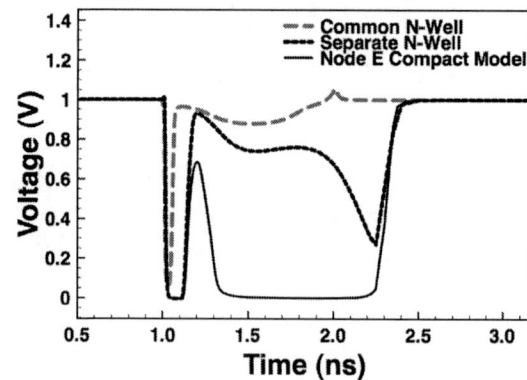

Fig. 13. Plot of the voltage output for node E.

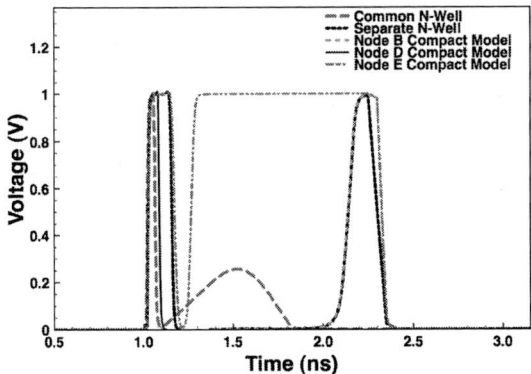

Fig. 14. Plot of the voltage output for node F.

To further understand the wide variety of DPSETs observed in the heavy-ion experiment, simulations with different ion strike locations were performed. A total of forty locations were struck with a 58 MeV-cm^2/mg heavy-ion incident at 60° from the Si surface. From those forty locations, only five DPSETs were observed. Between the 1st and 2nd pulses, the five DPSETs had durations as short as 281 ps and as long as 981 ps. Last, no two DPSETs simulated had the same characteristic pulse widths – similar to the DPSETs experimentally recorded.

Matching the device size and placement of the 65 nm heavy-ion experiments; the TCAD simulations definitively support the existence and location dependence of DPSETs in combinational logic. These simulations show how an angled individual ion strike can cause multiple transistors to collect charge and be upset. The simulations also explain how DPSETs can be observed in the separate n-well target chain but not be observed in the common n-well target chain.

V. CONCLUSIONS

In this paper, measurements and TCAD simulations of digital SETs in 65 nm bulk CMOS inverter strings show that a single angle particle strike can cause double-pulse-single-event transients (DPSETs). Experiments with inverter chains comprised of pFETs in separate n-wells show significant DPSETs. Inverter chains comprised of pFETs within a common well region show no DPSETs. Through experimental and TCAD analyses, we have determined that the DPSET effect is a direct result of cross-boundary charge sharing (in the case of separate n-well isolation of devices) leading to a secondary pulse. In the case of pFETs within a common well, the second pulse is mitigated by the pulse quenching phenomenon. The wide disparity among primary and secondary pulse widths, as well as the inter-pulse spacing, is driven by complex interaction of ion strike location, layout geometry, and electrical transmission characteristics.

ACKNOWLEDGEMENTS

The authors would like to thank the Defense Threat Reduction Agency and Cisco Systems, Inc. for their support of this effort.

REFERENCES

[1] O. Musseau, F. Gardic, P. Roche, T. Corbiere, R. A. Reed, S. Buchner, P. McDonald, J. Melinger, L. Tran, and A. B. Campbell, "Analysis of multiple bit upsets (MBU) in a CMOS SRAM," *IEEE Trans. Nucl. Sci.*, vol. 43, no. 6, pp. 2879–2888, Dec. 1996.

[2] S. P. Buchner and M. P. Baze, "Single-event transients in fast electronic circuits," *NSREC Short Course*, 2001, Sec. V.

[3] D. G. Mavis and P. H. Eaton, "Soft error rate mitigation techniques for modern microcircuits," in *Proc. IRPS*, Apr. 2002, pp. 216–225.

[4] B. D. Olson, D. R. Ball, K. M. Warren, L. W. Massengill, N. F. Haddad, S. E. Doyle, and D. McMorrow, "Simultaneous Single Event Charge Sharing and Parasitic Bipolar Conduction in a Highly-Scaled SRAM Design," *IEEE Trans. Nucl. Sci.*, vol. 52, no. 6, pp. 2132-2136, Dec. 2005.

[5] O.A. Amusan, A. F. Witulski, L. W. Massengill, B. L. Bhuva, P. R. Fleming, M. L. Alles, A. L. Sternberg, J. D. Black, and R. D. Schrimpf, "Charge Collection and Charge Sharing in a 130 nm CMOS Technology," *IEEE Trans. Nucl. Sci.*, vol. 53, no. 6, pp. 3253-3258, Dec. 2006.

[6] V. Ferlet-Cavrois, P. Paillet, D. McMorrow, N. Fel, J. Baggio, S. Girard, O. Duhamel, J. S. Melinger, M. Gaillardin, J. R. Schwank, P. E. Dodd, M. R. Shaneyfelt, and J. A. Felix, "New insights into single event transient propagation in chains of inverters—evidence for propagation-induced pulse broadening," *IEEE Trans on Nucl. Sci.*, vol. 54, no. 6, pp. 2338-2346, Dec. 2007.

[7] B. D. Olson, D. R. Ball, K. M. Warren, L. W. Massengill, N. F. Haddad, S. E. Doyle, and D. McMorrow, "Simultaneous Single Event Charge Sharing and Parasitic Bipolar Conduction in a Highly-Scaled SRAM Design," *IEEE Trans. Nucl. Sci.*, vol. 52, no. 6, pp. 2132-2136, Dec. 2005.

[8] O.A. Amusan, A. F. Witulski, L. W. Massengill, B. L. Bhuva, P. R. Fleming, M. L. Alles, A. L. Sternberg, J. D. Black, and R. D. Schrimpf, "Charge Collection and Charge Sharing in a 130 nm CMOS Technology," IEEE Trans. Nucl. Sci., vol. 53, no. 6, pp. 3253-3258, Dec. 2006.

[9] J. R. Ahlbin, L. W. Massengill, B. L. Bhuva, B. Narasimham, M. J. Gadlage, and P. H. Eaton, "Single event transient pulse quenching in Advanced CMOS logic," *IEEE Trans. Nucl. Sci.*, vol. 56, no. 6, pp. 3050-3056, Dec. 2009.

[10] B. Narasimham, V. Ramachandran, B. L. Bhuva, R. D. Schrimpf, A. F. Witulski, W. T. Holman, L. W. Massengill, J. D. Black, W. H. Robinson, D. McMorrow, "On-chip characterization of single event transient pulse widths", *IEEE Trans. on Dev. and Mat. Rel.*, vol. 6, p. 542-549, 2006.

[11] M. J. Gadlage, J. R. Ahlbin, B. Narasimham, B. L. Bhuva, L. W. Massengill, R. A. Reed, R. D. Schrimpf, and G. Vizkelethy, "Scaling Trends in SET Pulse Widths in Sub-100 nm Bulk CMOS Processes," *IEEE Trans. Nucl. Sci.*, vol. 57, no. 6, pp. 3336-3341, Dec. 2010.

[12] J. R. Ahlbin, M. J. Gadlage, D. R. Ball, A. W. Witulski, B. L. Bhuva, R. A. Reed, G. Vizkelethy, and L. W. Massengill, "The Effect of Layout Topology on Single-event transient pulse quenching in a 65 nm bulk CMOS process," *IEEE Trans. Nucl. Sci.*, vol. 57, no. 6, pp. 3380-3385, Dec. 2010.

[13] ACCRE Computing Cluster. Nashville, TN [Online] Available: http://www.accre.vanderbilt.edu

978-1-4244-9113-1/11 $26.00 © 2011 IEEE

Thermomechanical Reliability of Through-Silicon Vias in 3D Interconnects

Kuan-Hsun Lu[1], Suk-Kyu Ryu[2], Jay Im[1], Rui Huang[2] and Paul S. Ho[1]

[1]Microelectronics Research Center
[2]Department of Aerospace Engineering and Engineering Mechanics
University of Texas at Austin
Austin, TX 78712
paulho@mail.utexas.edu

Abstract—This paper investigates two key aspects of thermomechanical reliability of through-silicon vias (TSV) in 3D interconnects. One is the piezoresistivity effect induced by the near surface stresses on the charge mobility for p- and n- channel MOSFET devices. The other problem concerns the interfacial delamination induced by thermal stresses including the pop-up mechanism of TSV with a 'nail head'. We first analyze the three-dimensional distribution of the thermal stresses near the TSV and the wafer surface. The stress characteristics are inherently 3D in nature with the near-surface stress distributions distinctly different from the 2D solution. The energy release rate for interfacial delamination of TSV is evaluated under both cooling and heating conditions, using an analytical solution for a steady-state crack growth as an upper bound and numerical solutions by finite element analysis (FEA) for more detailed calculations. Based on these results, we examine the piezoresistivity effect induced by the near surface stresses on the charge mobility for p- and n- channel MOSFET devices, including the study of the effect of TSV scaling on the keep-out zone for MOSFET devices. This is followed by analyzing the energy release rate for interfacial delamiantion for a fully filled TSV and the potential mechanisms for TSV pop-up due to interfacial fracture.

Keywords-3D interconnect, TSV, Thermo-mechanical reliability, FEA, Crack driving force

I. INTRODUCTION

Three dimensional (3D) integration with through-silicon-vias (TSVs) has emerged as an effective solution to meet the future interconnect requirements beyond the 32nm technology node [1, 2]. The through-silicon via (TSV) is a critical structural element in the 3D interconnects, which directly connects stacked structures die-to-die. The incorporation of TSV structures can significantly impact the thermomechanical reliability of 3D interconnects. The mismatch in thermal expansion between the TSV and Si can induce large thermal stresses during TSV fabrication to degrade the performance of stress-sensitive devices. It has been reported that 100 MPa of stress can change more than 7% of carrier mobility in MOSFET devices [3]. In addition, the thermal stresses can drive interfacial delamination between the TSV and the silicon matrix [4-8], thus the stress impact on device reliability and

performance has to be considered in the development of 3-D integrated circuits.

In this paper, we first analyze the characteristics of the thermal stress induced by TSVs. Three-dimensional finite element simulation and analytical solutions are applied to characterize the stress distribution in the Si wafer surrounding an isolated TSV. Here the analysis is focused on the near surface region in the Si surrounding the TSV since most of the devices are located very close (about 1 μm) to the surface. Then we examine two key issues of TSV reliability based on the stress characteristics deduced. One is the piezoresistivity effect induced by the near surface stresses on the charge mobility for p- and n- channel MOSFET devices. Here we extend the study to examine the impact of TSV scaling on the keep-out zone for MOSFET devices. The other issue concerns the interfacial delamination and its impact on TSV reliability. We will deduce an analytical solution for the steady state energy release rate as an upper bound for the fracture driving force. The analysis, together with numerical results using FEA, is used to investigate the mechanism of TSV pop-up due to interfacial delamination.

II. THREE DIMENSIONAL STRESS ANALYSIS

Finite element methods have been used to analyze the thermo-mechanical stresses of TSV structures, which were found to depend on the materials, processes and structural designs of the 3D integrated structures [4-8]. To assess the thermo-mechanical reliability, the driving forces for the growth of both cohesive and interfacial cracks were calculated based on the concepts of facture mechanics [8-10]. So far, the previous studies have focused mainly on the TSV structural reliability and little information was available about the stress behavior near the wafer surface. The latter was required to understand the stress effect on device performance since most of the active devices are located very close to the Si surface. In a recent study, a 2-D analytical solution was deduced to analyze the stress interaction in TSV arrays [11]. The 2D solution, however, does not capture the 3D nature of the stress field near the wafer surface around a TSV. For the present study, we analyze the 3D near-surface stress field around a single TSV embedded in an infinite Si wafer. In this analysis, we first assume that all materials are isotropic and linear elastic. Under the assumption of linear elasticity, the stress field in the TSV structure can be obtained by superposition of the two problems sketched in Fig. 1. In Problem A, the system is subjected to a thermal loading ΔT and a uniform stress σ_z on the surface of the via, where the stress field is homogeneous in the via. The exact solution to Problem A in Fig. 1 is identical to the 2D plane-strain solution to the classical Lame problem [12]. The 2D solution does not satisfy the traction-free boundary condition on the surfaces in the original problem (Fig. 1a) because of the presence of the axial stress σ_z^A in the via. To recover the traction-free boundary condition, the normal stress on the surface is removed by superimposing Problem B, in which the via is subjected to a pressure of the same magnitude ($p = \sigma_z$) at both ends, but no thermal load [13]. The stress field due to the surface pressure is typically localized near the ends of the via. Thus, the stress distribution from the 2D solution is an accurate solution far away from the TSV ends, especially for TSVs with a high aspect-ratio (height/diameter, H/D_f) embedded in a thick wafer. However, the correction due to Problem B renders a very different stress distribution near the wafer surface around the TSV. For a relatively thin wafer, the stress in the entire via and its surrounding can be affected and thus comes out different from the 2D solution. This can impact the near-surface thermal stress distribution and the device characteristics surrounding the TSV.

Problem A can be solved analytically, while an approximate solution to Problem B can be obtained semi-analytically. The thermal stress in the via is uniform and tri-axial. The detailed solutions can be found in our previous publication [14].

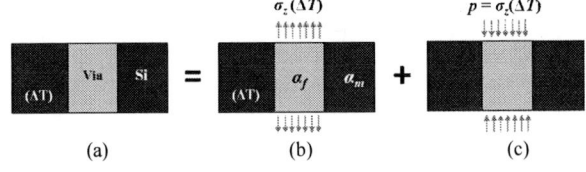

Fig. 1. Illustration of the method of superposition to obtain the semi-analytical solution for the thermal stresses in a TSV structure: (a) the original problem, with a thermal load and traction-free surfaces; (b) Problem A, with a thermal load; (c) Problem B, with surface load only.

Figure 2 shows the distributions of various stresses in the model structure for the negative thermal load, $\Delta T = -250$ °C. Figure 2a shows that the normal stress σ_z is zero on the surface ($z = 0$), as required by the traction-free boundary condition. The normal stress is non-uniform in the via and Si near the surface. Unlike the 2D solution, the shear stress σ_{zr} is not zero near the end of the via. In fact, a concentration of the shear stress is predicted at the junction between the surface ($z = 0$) and via/Si interface ($r = D_f / 2$), which can contribute to the driving force to cause interfacial delamination. The distributions of the radial stress σ_r and the circumferential

stress σ_θ near the end of TSV are also very different from the 2D solution (Figs. 2c and 2d).

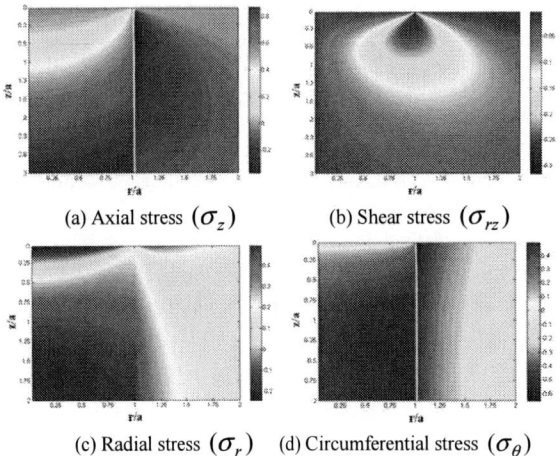

(a) Axial stress (σ_z) (b) Shear stress (σ_{rz})

(c) Radial stress (σ_r) (d) Circumferential stress (σ_θ)

Fig. 2. Near-surface stress distributions predicted by the semi-analytical solution for thermal load, $\Delta T = -250$ °C. The stress magnitudes are normalized by $p = -E\varepsilon_T /(1-v)$.

The fracture behavior depends on the character of the stress components and the sign of the thermal mismatch strain, $\varepsilon_T = \left(\alpha_f - \alpha_m\right)\Delta T$, which can be either tensile or compressive where the subscripts f and m refer to the TSV and Si, respectively. For example, if $\alpha_f > \alpha_m$, $\varepsilon_T > 0$ for heating ($\Delta T > 0$) and $\varepsilon_T < 0$ for cooling ($\Delta T < 0$). Under cooling, the radial stress is tensile along the via/Si interface, which can contribute to the driving force for interfacial delamination. The radial stress is also tensile in Si near the surface, which can induce circumferential cracking of the Si. Under heating, the circumferential stress is tensile in Si, which can induce radial cracks to form in Si. In contrast, the shear stress σ_{rz} along the TSV/Si interface contributes to the driving force for interfacial delamination under both heating and cooling. In the present study we focus on interfacial delamination as the critical failure mode under both heating and cooling conditions.

To verify the semi-analytic solution, finite element analysis (FEA) is performed using the commercial package, ABAQUS (v6.8). Since the thickness of the Si wafer is one of the key design parameters for the TSV structure, the effect of wafer thickness on thermal stress distribution is examined by FEA models with two different thicknesses. The model structure is shown in Fig. 1a, with the TSV diameter $D_f = 30$ μm and the wafer thickness $H=300$ μm and 60 μm. A negative thermal loading (cooling), $\Delta T = -250$ °C, is assumed. The material properties are: $E_f = E_m = 110$ GPa, $v_f = v_m = 0.35$, and $\alpha_f = 17$ ppm/°C and $\alpha_m = 2.3$ ppm/ °C. The model is an approximation to a Cu TSV in Si, neglecting the elastic mismatch between Cu and Si. In practice a thin barrier layer is

978-1-4244-9113-1/11 $26.00 © 2011 IEEE

typically placed between the Cu via and Si, which has minimal effect on the stress distribution and is thus ignored here.

In Figure 3, the results from the FEA are compared with those obtained by the semi-analytical solution. First, the axial stress (σ_z) along the center line of the TSV ($r = 0$) shows a transition from zero stress at the surface ($z = 0$) to a tensile stress away from the surface (Fig. 3a). For the thick wafer ($H/D_f = 10$), the FEA result shows excellent agreement with the analytical solution, both approaching the 2D solution (the dashed line) away from the surface. Similarly, the radial stress (σ_r) at the interface for the thick wafer approaches a finite value at $z = 0$ and the 2D solution far away from the surface (Fig. 3b). In contrast, for the thin wafer, the stresses are not built up high enough to reach the 2D solution; more so for the axial stress.

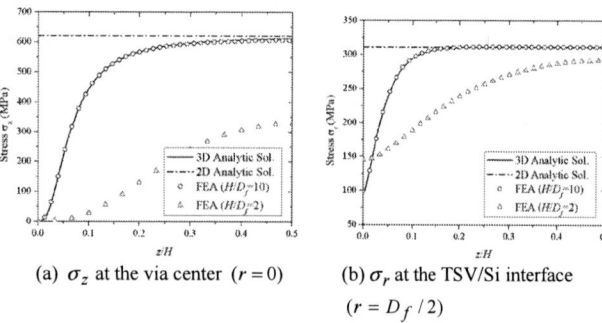

(a) σ_z at the via center ($r = 0$)   (b) σ_r at the TSV/Si interface ($r = D_f/2$)

Fig. 3. Effect of wafer thickness on stress distributions ($D_f = 30$ μm, $\Delta T = -250$ °C).

It is worth noting the 3D nature of the near-surface stresses, which has to be taken into account to determine the keep-out zone around the TSV. This is discussed in the following section. In addition, the near surface stresses can induce channeling cracks at the Si surface near the TSV in either the radial or the circumferential direction, depending on the magnitudes and signs of the stresses.

III. TSV-INDUCED PIEZORESISTIVITY EFFECT ON MOSFETs

A. Piezoresistivity of silicon

Piezoresistivity of Si refers to the effect of stresses on the mobility of the charge carriers of Si. The incorporation of TSV can introduce undesired stresses in the Si matrix to degrade the performance of the adjacent MOSFETs. As illustrated in a previous study, a Cu TSV of 30 μm diameter can induce a thermal stress greater than 100 MPa at a distance 10 μm away from the TSV edge [14]. Stresses of such magnitude can affect the carrier mobility and will have to be taken into account in the design of the keep-out zone (KOZ). To design the KOZ, the piezoresistivity effect in Si induced by stresses around the TSV on the carrier mobility has to be considered. This effect has been reviewed recently [15]. In a Si substrate, the general relations among the electric field E, current density J, and the stress components can be expressed as:

$$
\begin{aligned}
\frac{\Delta E_1}{\rho} &= [\pi_{11}\sigma_1 + \pi_{12}(\sigma_2 + \sigma_3)]J_1 + (\pi_{44}\sigma_6)J_2 + (\pi_{44}\sigma_5)J_3, \\
\frac{\Delta E_2}{\rho} &= [\pi_{11}\sigma_2 + \pi_{12}(\sigma_1 + \sigma_3)]J_2 + (\pi_{44}\sigma_6)J_1 + (\pi_{44}\sigma_4)J_3, \\
\frac{\Delta E_3}{\rho} &= [\pi_{11}\sigma_3 + \pi_{12}(\sigma_1 + \sigma_2)]J_3 + (\pi_{44}\sigma_5)J_1 + (\pi_{44}\sigma_4)J_2,
\end{aligned}
\tag{1}
$$

where ρ is the resistivity of the unstrained Si. π_{11}, π_{12}, and π_{44} are the piezoresistivity coefficients of Si. The subscripts of E and J (1, 2, and 3) designate the components along the three (100) crystal axes, while the six subscripts of σ designate the components of a stress matrix. In the following, we consider first the case where a MOSFET is located on a (001) Si wafer with its channel direction aligned with the [100] direction (Fig. 4a). Assume that the electrical current only flows along the channel direction, i.e. $J_2 = J_3 = 0$, and only the electric field across the channel (ΔE_1) is measured. The average resistivity (or mobility) change along the channel direction can be obtained from Eq. (1):

$$
\frac{\Delta \rho_1}{\rho} = \frac{\Delta \mu_1}{\mu} = \pi_{11}\sigma_1 + \pi_{12}(\sigma_2 + \sigma_3),
\tag{2}
$$

The channel direction of a MOSFET is often aligned along the [110] direction on a (100) wafer (Fig. 4b). In this case, the resistivity or the mobility change in Eq. (2) can be deduced by rotating the coordinate system as [15, 16]:

$$
\begin{aligned}
\frac{\Delta \rho_{1'}}{\rho} &= \frac{\Delta \mu_{1'}}{\mu} = \pi'_l \sigma_{1'} + \pi'_t(\sigma_{2'} + \sigma_{3'}), \\
\pi'_l &= \frac{\pi_{11} + \pi_{12} + \pi_{44}}{2}, \\
\pi'_t &= \frac{\pi_{11} + \pi_{12} - \pi_{44}}{2},
\end{aligned}
\tag{3}
$$

where σ' are the stress components in the new coordinate system, which is illustrated by red dash lines in Fig. 4b. π'_l and π'_t are the longitudinal and transverse piezoresistive coefficients of the new channel direction, respectively. Figure 5 shows the graphic representations of the variations of the longitudinal and transverse piezoresistive coefficients with the channel directions on (001) surface in p- and n-Si [15, 17].

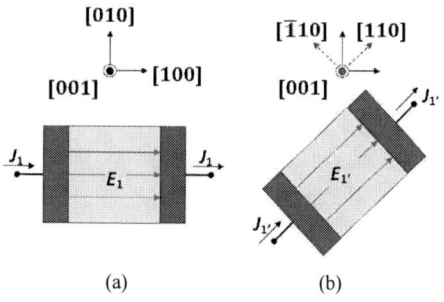

(a)   (b)

Fig. 4. Channel direction of MOSFETs on (001) Si wafer.

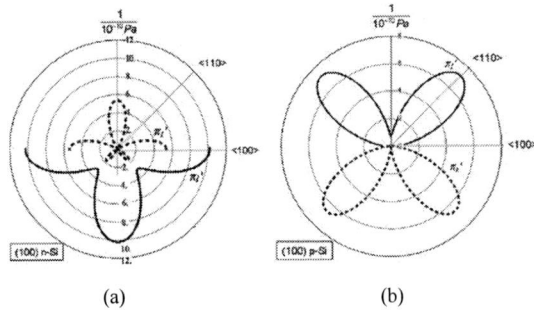

Fig. 5. Longitudinal and transverse piezoresistive coefficients on (001) Si wafer [15, 17].

B. Analysis of mobility change and keep-out zone

Because the MOSFETs are usually fabricated very near the wafer surface, the out-of-plane stress component σ_3, is about two orders smaller than the in-plane component σ_2, and thus is negligible. Additionally, it is clear from Eq. (1) and Eq. (2) that the shear stress components do not contribute to the mobility change. Therefore, the configuration of the KOZ is mainly controlled by the in-plane normal stresses parallel and perpendicular to the channel direction of MOSFETs. To deduce the piezoresistivity effect for Si devices, it is convenient to express the stress distribution in Cartesian coordinates. Accordingly, the distribution of normal stresses on the wafer surface is two-fold rotational symmetric and the distribution of the mobility change (or the KOZ) should have a two-fold rotational symmetry as well.

To calculate the mobility change induced by an isolated TSV, thermal stress analysis is performed using FEA simulation. The model consists of a quarter of a TSV embedded in a Si matrix with a TSV diameter D_f of 20 μm and wafer thickness of 100 μm. The TSV material is taken to be Cu and a negative thermal load, $\Delta T = -180^\circ C$, is applied to calculate the thermal stresses. The distributions of the normal stresses σ_{xx} and σ_{yy} are shown in Fig. 6. Since the devices are fabricated very near the Si surface, the normal stresses (σ_{xx}, σ_{yy}, and σ_{zz}) are simply taken to be on the wafer surface (z = 0) from the simulation results. In the calculation, the channel direction of MOSFETs is assumed to be parallel to the [110] direction, and the piezoresistivity coefficients of bulk n-Si and p-Si (Table 1) are used.

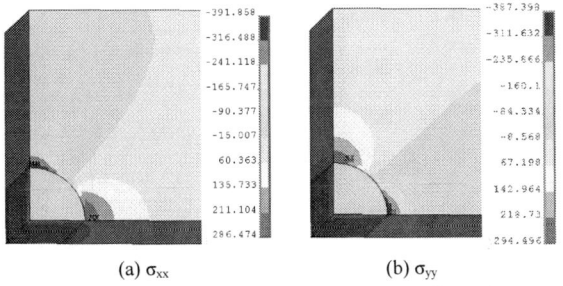

(a) σ_{xx} (b) σ_{yy}

Fig. 6. Distribution of thermal stress component σ_{xx} and σ_{yy} in a quarter Cu TSV model.

Table 1. Piezoresistivity coefficients for bulk Si (10^{-4} MPa^{-1}) [15]

	π_{11}	π_{12}	π_{44}	$\dfrac{\pi_{11} + \pi_{12} + \pi_{44}}{2}$	$\dfrac{\pi_{11} + \pi_{12} - \pi_{44}}{2}$
Direction	$\pi \,/\!/$ [100]	$\pi \perp$ [100]		$\pi \,/\!/$ [110]	$\pi \perp$ [110]
n-Si	-10.22	5.37	-1.36	-3.16	-1.76
p-Si	0.66	-0.11	13.81	7.18	-6.63

The results obtained for the mobility changes along the [110] channel direction are shown in Figure 7 revealing a significant difference of the stress effect on the carrier mobility for the n- and p-MOS devices. With the same TSV geometry and thermal load (-180°C), the maximum mobility change can reach 35% for the p-MOSFET while only 7% for the n-MOSFET. The result can be attributed to the combination of the sign and magnitude of the stresses and the piezoresistivity coefficients for these two types of devices. First, TSV-induced in-plane normal stresses (σ_{xx} and σ_{yy}) are of similar magnitude but the piezoresistivity coefficients parallel and perpendicular to [110] crystal direction are opposite in sign for p-Si (7.18 v.s. -6.63$\times 10^{-4}$ MPa^{-1}), but of the same sign for n-Si (-3.16 v.s. -1.76$\times 10^{-4}$ MPa^{-1}). This results in an addition of the contributions from σ_{xx} and σ_{yy} for the p-MOSFET, but a subtraction for the n-MOSFET.

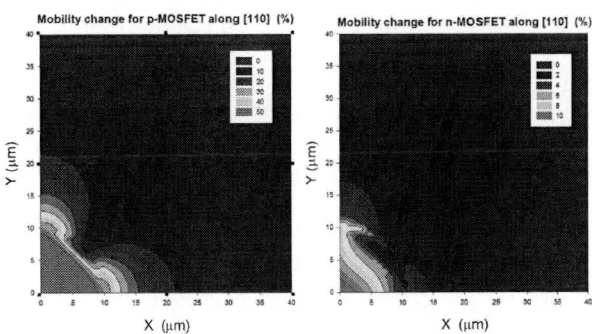

Fig. 7. Distribution of mobility change for p- and n-MOSFETs along [110].

Along the [100] direction, the effect is quite different. Here the piezoresistivity coefficients for n-Si along the parallel and perpendicular directions are opposite in sign (-10.22 v.s. 5.37$\times 10^{-4}$ MPa^{-1}) and also about an order of magnitude larger than that for p-Si (0.66 v.s. -0.11$\times 10^{-4}$ MPa^{-1}). Consequently, n-MOSFETs are more sensitive to the TSV-induced thermal stresses than p-MOSFETs along the [100] direction.

The scaling effect on KOZ is investigated as a function of TSV diameter following a similar procedure. For this purpose, FEA models of Cu TSVs are built with a fixed wafer thickness (100 μm) and varying diameter (D_f = 10 ~ 30 μm). A negative thermal load, $\Delta T = -180^\circ C$, is applied. After calculating the mobility change on the wafer surface for both p-Si and n-Si, an arbitrary criterion for KOZ with 10% change in mobility is applied to calculate the area of KOZ surrounding the TSV. The results are plotted in Fig. 7 for the p-MOSFET devices.

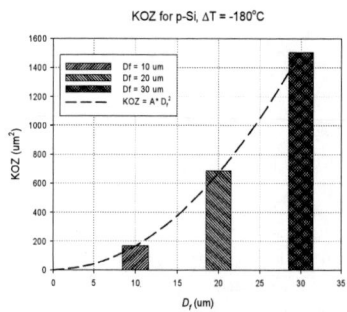

Fig. 8. Effect of TSV diameter on Keep-out Zone.

The results in Fig. 8 indicate a significant effect due to the scaling of the TSV diameter on the area of KOZ for p-MOSFET, increasing approximately with the square of D_f. Overall, the effect yields an area ratio between KOZ and TSV of about 2. It is interesting to note that the TSV-induced stresses cannot generate a mobility change greater than 10% in n-MOSFET under the given thermal load (-180°C), therefore, there is no KOZ for n-MOSFET in this study.

IV. INTERFACIAL DELAMINATION OF TSV

A. Fully filled TSV

The stress analysis in the previous section suggests a potential failure mechanism of the TSV structure due to interfacial delamination. Figure 9 depicts two modes of interfacial delamination for a fully-filled TSV structure. With a negative thermal load ($\Delta T < 0$), the radial stress along the via/Si interface is tensile (assuming $\alpha_f > \alpha_m$). Consequently, the interfacial delamination crack can grow in a mixed mode (peeling and shearing). With a positive thermal load ($\Delta T > 0$), however, the radial stress is compressive which does not contribute to the driving force for delamination. This results in an interfacial crack with a pure shearing mode (mode II). In this case, the two crack faces are in intimate contact and may be subject to friction. For simplicity, we assume a frictionless contact in the present study and develop analytical solutions for the steady-state energy release rates of the interfacial crack, under both cooling and heating conditions. The analytical solutions are then compared to finite element analysis, which is extended to study the effects of crack length and wafer thickness on the fracture driving force.

For a TSV with a relatively high aspect ratio (H/D_f), the energy release rate for interfacial delamination reaches a steady state when the crack length is several times greater than the via diameter. Since the energy release rate is usually lower for shorter cracks, the steady-state value sets an upper bound for the fracture driving force, which may be used as the critical condition for conservative design of reliable TSV structures.

Consider an infinitely long fiber (TSV) in an infinite matrix (Si wafer), with a pre-existing semi-infinite, circumferential crack along the interface and subjected to a thermal load ΔT. The steady-state energy release rate (ERR) for the interfacial crack growth (per unit area) is obtained by comparing the elastic strain energy far ahead of the crack front and that far behind the crack front. While the stress field near the crack front is complicated with singularity and 3D distribution, it merely translates in the steady state as the crack front advances. Far ahead of the crack front, the stress field can be obtained analytically by solving the 2D plane-strain problem (Problem A in Fig. 1). Far behind the crack front, since the TSV is debonded from Si, the stress is relaxed in both the via and Si. For the case of cooling ($\Delta T < 0$), the stress is zero in both TSV and Si. For heating ($\Delta T > 0$), however, the contact between the crack faces induces a stress field similar to Problem A, but the axial stress (σ_z) in the via is zero from the assumption of frictionless contact. If the elastic mismatch is neglected (i.e., $\alpha = 0$ and $v_f = v_m = v$), the steady state energy release rate under cooling can be expressed in a simple form:

$$G_{cooling}^{SS} = \frac{E \varepsilon_T^2 D_f}{4(1-v)}. \qquad (4)$$

Under heating with $\Delta T > 0$, due to the contact of the crack faces (Fig. 9b), the stress state in the TSV far behind the crack front is equibiaxial. As a result, the steady state ERR under heating can be obtained by neglecting the elastic mismatch, namely

$$G_{heating}^{SS} = \frac{1+v}{8(1-v)} E \varepsilon_T^2 D_f. \qquad (5)$$

To verify the steady state ERR solution, a FEA model of the TSV structure is constructed, and the energy release rates are calculated by the J-integral method. As expected, the energy release rate increases with the crack length and approaches the steady-state solution when the crack length is longer than about 2-3 times the via diameter (Fig. 10).

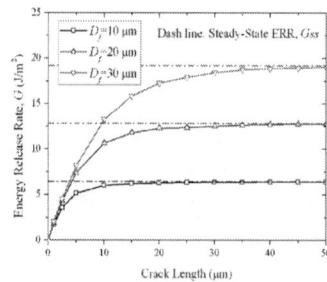

Fig. 10. Effect of crack length on the energy release rate for interfacial delamination of TSVs ($H = 300$ μm and $\Delta T = -250$ °C).

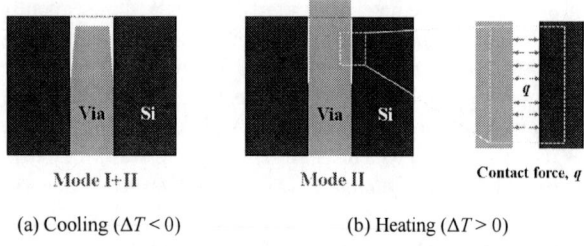

(a) Cooling ($\Delta T < 0$)   (b) Heating ($\Delta T > 0$)

Fig. 9. Schematics of interfacial delamination of TSV under cooling and heating conditions.

978-1-4244-9113-1/11 $26.00 © 2011 IEEE

Several interesting results can be deduced based on the analytical solutions for the steady-state energy release rates. First, the steady-state ERR for interfacial delamination is linearly proportional to the TSV diameter, which may set an upper bound for the via diameter in order to avoid delamination. Second, the ERR is proportional to the square of the thermal mismatch strain, $\varepsilon_T = \left(\alpha_f - \alpha_m\right)\Delta T$. Thus the delamination driving force can be reduced by either using TSV materials with smaller thermal expansion mismatch ($\alpha_f - \alpha_m$) and/or by reducing the thermal loads (ΔT). Third, the energy release rate for interfacial delamination increases with the elastic modulus of the TSV material, however, the effect is less important than the effect of the thermal expansion mismatch. Finally, a comparison between Eq. (4) and (5) indicates that, with the same thermal load ΔT, the driving force for interfacial delamination under cooling is about twice that under heating, a result that can be attributed to the presence of a tensile radial stress (σ_r) across the interface (opening mode) for the case of cooling.

B. The "pop-up" of TSV with a nail head

In practice, a hard mask for etching Cu TSVs in silicon substrate often results in a ledge or overhang called 'nail head' over the TSV. The nail head can also be used on purpose to facilitate connection to the upper die. The presence of the nail head changes the boundary conditions at the crossing point of Cu/Si/Nail head and interfacial end of Si/Nail head, which in turn affects the stress distribution around both the TSV and Si. In particular, under negative thermal loading, the concentration of the shear stress along the TSV/Si will decrease but the shear concentration along nail head/Si will develop due to the nail head. In addition, the opening stress at the nail head perimeter/Si interface possesses singularity. As a result, the stress concentration contributes to interfacial failures. To analyze this problem, we calculate the steady-state energy release rate for the TSV structure with nail head under cooling. Here the dimensions of thickness and diameter for nail head simply are assumed to be sufficiently large. This yields zero radial and circumferential stresses far behind the crack tip while the out-of plane stress exists due to the constraint by the nail head. The ERR obtained under these assumptions is close to that obtained by FEA.

To evaluate the effect of nail head on energy release rate, we compare ERR between fully filled TSV ($D_f = 30\,\mu m$) and TSV with a nail head ($H_n = 0.5D_f$) under cooling ($\Delta T = -250\,°C$). The cooling condition contributes to failure of only the vertical surface, so that we vary the vertical crack length (c_1) with zero horizontal crack ($c_2 = 0$). As crack propagates along the vertical interface, the ERR approaches the steady-state solution. The steady-state ERR for TSV with a

nail head drops about 30% due to the constraint effect of the nail head (Fig.11). Thus the nail head can be helpful to improve the TSV reliability.

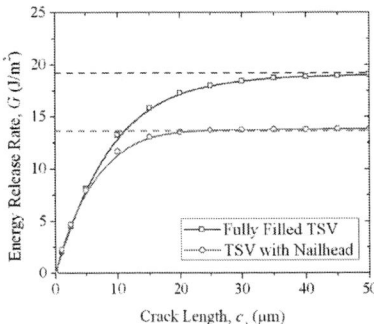

Fig. 11. Comparison of steady-state ERRs between a fully filled TSV and a TSV with nail head ($D_f = 30\,\mu m$, $H_n = 0.5D_f$, $c_2 = 0$, and $\Delta T = -250\,°C$).

TSV pop-up describes a phenomenon of TSV lifting off from the surrounding matrix as schematically depicted in Fig. 12. The interface between the nail head and Si is subjected to shearing near the perimeter of the nail head, which may cause delamination at that location. Therefore, analysis of the interfacial reliability of TSVs with nail head should consider both interfaces, vertical and horizontal. If both interfaces fail during thermal cycling, TSV can be extruded from the Si substrate. Here, we describe two thermal processes that can cause TSV pop-up.

First, by comparing the energy release rates under cooling and heating, we observe that the interface is more prone to delamination under cooling than under heating. Under cooling, a vertical crack (c_1) initiated at the TSV/Si interface can reach a stationary crack through repeated thermal cycles. Then, during the ensuing heating cycling, a horizontal crack (c_2) can be generated at the Si/nail head interface from inside toward the outside of the nail head (Fig. 12a). Finally, the delamination at both interfaces can bring about TSV pop-up. Alternately, after vertical crack failure, horizontal crack could start from outside toward the inside of the nail head. However, the ERR for this failure mode is relatively small, e.g. less than $1\,J/m^2$. Thus, this process can be ruled out from consideration.

Another possibility for TSV pop-up arises from the concentration of opening and shearing stress under cooling (Fig. 12b). The driving force can cause initial crack growth at the free end of the nail head/Si interface, followed by an interfacial crack expanding inward due to the positive opening stress. Then, after fully debonded at the nail head/Si interface, a vertical crack propagates along the TSV/Si interface under an ERR corresponding to that of the fully filled TSV.

978-1-4244-9113-1/11 $26.00 © 2011 IEEE

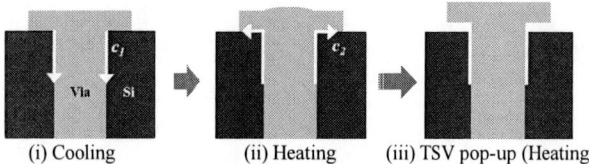

(i) Cooling   (ii) Heating   (iii) TSV pop-up (Heating)

(a) A first scenario for TSV pop-up

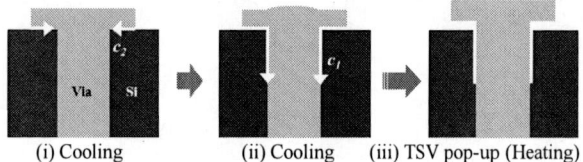

(i) Cooling   (ii) Cooling   (iii) TSV pop-up (Heating)

(b) A second scenario for TSV pop-up

Fig. 12. Two probable thermal processes for TSV pop-up

V. CONCLUSIONS

In this study, the thermo-mechanical reliability of a TSV structure in 3D interconnect is investigated by a semi-analytical approach and FEA calculations. The stress characteristics are inherently 3D in nature with the near-surface stress distributions distinctly different from the analytical solution based on a simple 2D model. The energy release rate for interfacial delamination of TSV is evaluated under both cooling and heating conditions, using an analytical solution for a steady-state crack growth and numerical solutions obtained by FEA for non-steady state crack growth. Based on these results, the interfacial reliability of a fully filled TSV together with the potential mechanisms for TSV pop-up are discussed. In this paper, we also examine the piezoresistivity effect induced by the near surface stresses on the charge mobility for p- and n- channel MOSFET devices and investigate the impact of scaling on the keep-out zone for MOSFET devices. Together, the results of this study provide a basis to explore the potential of using materials and structure optimization for improving TSV reliability for 3D interconnects.

ACKNOWLEDGMENTS

The authors gratefully acknowledge Semiconductor Research Corporation for the financial support of this work.

REFERENCES

[1] International Technology Roadmap for Semiconductors (ITRS), 2009.

[2] M.S. Bakir, J.D. Meindl, Integrated interconnect technologies for 3D nanoelectronic systems. Artech House, Norwood, MA, 2009.

[3] S. Thompson, Guangyu Sun, Youn Sung Choi, and T. Nishida, "Uniaxial-process-induced strained-Si: extending the CMOS roadmap," *Electron Devices, IEEE Transactions on*, vol. 53, 2006, pp. 1010-1020

[4] A.P. Karmarker, X. Xu, and V. Moroz, Performance and reliability analysis of 3D-integration structures employing

through silicon via (TSV), Proc. IEEE 47th Annual International Reliability Physics Symposium, Montreal, 2009, pp. 682-687.

[5] C.S. Selvanayagam, J.H. Lau, X. Zhang, S.K.W. Seah, K. Vaidyanathan, and T.C. Chai, Nonlinear thermal stress/strain analyses of copper filled TSV (Through Silicon Via) and their flip-chip microbumps, Electronic Components and Technology Conference, 2008, pp. 1073-1081.

[6] N. Ranganathan, K. Prasad, N. Balasubramanian, and K. L. Pey, A study of thermo-mechanical stress and its impact on through-silicon vias, J. Micromech. Microeng. 18, 1-13 (2008).

[7] X. Liu, Q. Chen, P. Dixit, R. Chatterjee, R. Tummala, and S. Sitaraman, Failure mechanisms and optimum design for electroplated copper through-silicon vias (TSV), Electronic Components and Technology Conference, 2009, pp. 624-629.

[8] K. Lu, X. Zhang, S. Ryu, J. Im, R. Huang, and P. Ho, Thermo-mechanical reliability of 3-D ICs containing through-silicon-vias, Electronic Components and Technology Conference, 2009, pp. 630-634.

[9] A. Karmarkar, Xiaopeng Xu, and V. Moroz, "Performanace and reliability analysis of 3D-integration structures employing Through Silicon Via (TSV)," *Reliability Physics Symposium, 2009 IEEE International*, 2009, pp. 682-687.

[10] X. Liu, Q. Chen, P. Dixit, R. Chatterjee, R. Tummala, and S. Sitaraman, Failure mechanisms and optimum design for electroplated copper through-silicon vias (TSV), Electronic Components and Technology Conference, 2009, pp. 624-629.

[11] K. Lu, X. Zhang, S. Ryu, R. Huang, and P. Ho, Thermal stresses analysis of 3-D interconnect, 10th International Workshop on Stress-Induced Phenomena in metallization, Proc. American Institute of Physics Conference 1143, 2009, pp. 224-230.

[12] A.E.H. Love, The stress produced in a semi-infinite solid by pressure on part of the boundary, Philos. Trans. A 228, 377-420 (1929).

[13] T.C. Lu, J. Yang, Z. Suo, A.G. Evans, R. Hecht, and R. Mehrabian, Matrix cracking in intermetallic composites caused by thermal expansion mismatch, ACTA Metall. Mater. 39, 1883-1890 (1991).

[14] S. Ryu, K. Lu, X. Zhang, J. Im, P. Ho, and R. Huang, IEEE Trans. TDMR, 2010.DOI:10.1109/TDMR.2010.2068572.

[15] Y. Sun, S. Thompson, and T. Nishida, Strain Effect in Semiconductors: Theory and Device Applications, New York: Springer, 2010.

[16] W.P. Mason and R.N. Thurston, "Use of Piezoresistive Materials in the Measurement of Displacement, Force, and Torque," The Journal of the Acoustical Society of America, vol. 29, Oct. 1957, pp. 1096-1101.

[17] Y. Kanda, "A graphical representation of the piezoresistance coefficients in silicon", *IEEE Transactions on*, vol. 29, 1982, pp. 64-70.

Characterization of Steady and Transient Heating of Interconnects – A Review

Banafsheh Barabadi, Yogendra Joshi, and Satish Kumar

G.W. Woodruff School of Mechanical Engineering, Georgia Institute of Technology,
801 Ferst Dr, Atlanta GA 30306, USA
Phone: +1 (404) 385-2810
Fax: +1 (404) 894-8496
Email: yogendra.joshi@me.gatech.edu

Abstract — Continued scaling of transistors and metal interconnects have resulted in high current densities and significant Joule heating in the metal lines, exacerbating thermally driven reliability issues in microprocessors. It is imperative, therefore, to develop an accurate and rapid predictive thermal characterization capability for on chip interconnect arrays to facilitate chip design. This is a multi-scale problem for which the traditional finite difference and finite element methods are generally inefficient due to their large computational times. Also, the thermophysical properties needed as inputs to the models are size dependent at the scale of interest. In this paper, we provide a review of some of the techniques developed recently for steady state and transient thermal characterization.

Keywords-Joule Heating, Interconnect, Multi-Scale Modeling, Transient Thermal Analysis, Thermal Characterization Techniques

1. INTRODUCTION

The continued scaling of transistors and metal interconnects has resulted in a greatly increased level of integration of interconnect wiring and high current densities through individual wires, resultingina propensity for localized hot spots. These may accelerate failures due to fatigue associated with thermal cycling, open- and short-circuit failures due to morphological changes, such as the formation of hillocks, whiskers and voids, that occur during its high temperature operation and thermal cycling.[1-2]

In this paper, we review metrology and modeling approaches developed recently for characterizing JJoule heating in interconnects under steady state and transient conditions. From the steady state perspective, we mostly focus on compact finite element thermal modeling techniques. For the transient analysis, different approaches for reduced order modeling are described.

2. JOULE HEATING IN INTERCONNECTS

Joule heating in interconnects has been studied by various methods using different levels of approximation. These methods include analytical (2D) models, finite difference (FD) models, and finite element (FE) models [3-6]. A set of long uniformly spaced interconnects is often approximated as two-dimensional thermal spreading from a localized heat source, as is shown in Fig. 1. Bilotti et al. [3] obtained analytical expressions for the heat spreading factor using double Schwarz-Christoffel conformal transformation. Chiang et al. [7] considered finite length interconnects using a fin type equation. Resistance to heat conduction through the dielectric was assumed to be constant in this 2D model, with consideration given to heat conduction along the interconnect itself. This type of fin equation can be solved for any specified temperature condition near the ends. By adding average contributions to temperature rise from other interconnect levels, Chiang et al. [7] were able to include multi-level interconnects in their 2D modeling.

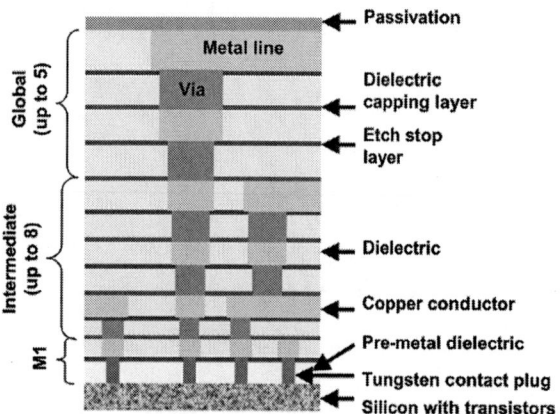

Figure 1- A schematic of interconnect stack with metal lines and vias [7].

FE modeling is exemplified in the work of Chen et al. [6]. Interconnect temperature rise was computed using a commercially available finite element solver for more realistic structures. A combination of analytically and numerically fitted solutions provided expressions for temperature rise in interconnect structures. Banerjee et al. [8] studied the thermal break down of metal interconnects under conditions of short-time high JJoule heating under electrostatic discharge (ESD) and electrical overstress (EOS) conditions. These particular failure types play an important role in interconnect/dielectric reliability. Banerjee, et al. developed a model that included the heating of the layered metal layers and the surrounding oxide. Their model related the maximum allowable current density to the signal pulse width. A methodology to quantify the role of electro-migration (EM) reliability and interconnect performance in determining optimal interconnect design in low-k/Cu systems was presented in [9]. The issue of reliability of interconnects has also been addressed [10-11]. Chiang et al [11] also showed that heating effects can severely degrade reliability and speed performance.

3. IMPORTANCE OF SIZE EFFECTS

Performance and reliability design of future microelectronic and nanoelectronic systems requires knowledge of the thermal and electrical properties of thin films. As the size of a metal interconnects becomes comparable to, or smaller than the electron mean free path (~40 nm in Cu at room temperature), electron transport becomes dominated by scattering at the metal-dielectric interface, and at grain boundaries. This scattering can reduce the electrical and thermal conductivity to less than half that of the bulk value [12-17]. This reduction in conductivity has been explained by the Fuchs-Sondheimer model [12-13], and subsequently more refined models [14-17]. This confirms the need for experimental methods for measurement of thermal properties of thin films materials for interconnects and dielectrics.

4. MEASUREMENTS OF THIN FILM PROPERTIES

Various techniques have been investigated for the measurement of thin film thermal properties, more specifically, thermal conductivity for thin dielectric and metal layers. Some of the well-established measurement techniques are described briefly below. A thorough review on various micro and nanoscale thermal characterization techniques is provided by Christofferson et al. [18].

4.1. 3Ω METHOD

The most conventional technique for the measurement of thin semiconductor films is the 3ω method established by Cahill [19]. This method was first introduced for an isotropic bulk material. In this method, a long metal line is deposited on top of the material of interest and is excited with a sinusoidal current at a frequency of ω. This originats an AC temperature rise at a frequency of 2ω, as well as a DC temperature rise. Considering the temperature dependence of the resistance, a 3ω component appears in the voltage signal that is proportional to the temperature amplitude within the deposited line. Subsequently, using appropriate analytical or numerical solutions, the thermophysical properties of the substrate material can be extracted.

4.2. PUMP-PROBE TRANSIENT THERMOREFLECTANCE (PPTTR) TECHNIQUE

The (PPTTR) technique was proposed by Paddock and Eesley [20]. In contrast to the 3ω method, PPTTR has the ability to differentiate between the thermal conductivity of thin films and their interface thermal resistance [21-22]. It is a time resolved methodology that extends the standard thermo-reflectivity technique [23] to very short time scales by applying optical sampling. Among the advantages of this technique are the fully optical, noncontact, and nondestructive nature of it, together with a high temporal and special resolution. This makes PPTTR a prominent methodology in determining the thermal properties of thin dielectric and metal layers.

4.3. NETWORK IDENTIfiCATION BY DECONVOLUTION (NID) ASSISTED PPTTR

Another approach to analyze the thermal decay of a PPTTR signal, based on RC network theory of linear passive elements, is Network Identification by Deconvolution (NID). NID was originally introduced by Székely and Bien [24]. This technique, which includes structural information obtained using thermal transient measurements, is used to analyze the temperature response of semiconductor device packages. Using this technique, one can separate the different contributions to the total thermal resistance and capacitance of the package being studied. Most of the applications of NID have been for the case of step function thermal transient measurement [25-26]. Recently, Ezzahri and Shakouri

978-1-4244-9113-1/11 $26.00 © 2011 IEEE

considered the case of a delta function excitation applied to the structure [27]. They demonstrated the capability of NID in extracting the thermal conductivity of the thin dielectric layer, as well as the metal layer interface thermal resistance from a single PPTTR signal. Of its strengths is that it does not assume *a priori* (number of layers or interfaces) for the structure of interest.

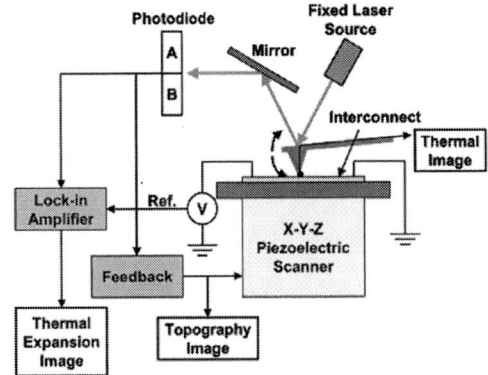

Figure 2- Schematic diagram of the experimental setup for the scanning Joule expansion microscopy (SJEM) adopted from [34].

4.4. SCANNING THERMAL MICROSCOPY (SThM)

Another approach to measure thermal conductivity is by utilizing Scanning Thermal Microscopy (SThM), [28] in which a very small thermocouple is fabricated on the tip of an atomic force microscope (AFM) [29-32]. Such a method can provide a maximum resolution of 50 nm. The advantages of SThM are significant spatial resolution, on smooth objects, and high bandwidth. One of the most important drawbacks of this method is that heat transfer between the probe and sample depends on the tip contact with the sample, which can vary with sample hardness, wear, or contact force [33].

4.5. SCANNING JOULE EXPANSION MICROSCOPY (SJEM)

SJEM is an AFM based technique to extract in-plane thermal conductivity of thin metallic films whose thickness is comparable to the electron mean free path. Fig. 2, adopted from [34], shows the schematic diagram for an experimental setup that is used for Scanning Joule Expansion Microscopy. SJEM measures the periodic thermal expansion amplitude at the sample surface, which corresponds to the periodic temperature at the surface. This technique is capable of mapping the temperature gradient near a constriction between wide and narrow metal lines. Thermal conductivity is extracted by using a numerical fit to the measurements. Extracted thermal conductivities of thin gold films showed good consistency with predictions from Wiedemann-Franz for 43 nm and 131 nm gold films. SJEM is a more robust method

than SThM because SJEM has no dependence on tip-sample heat flow.

Gurrum et al. proposed an approach to extract in-plane thermal conductivity of the 43 nm and 131 nm gold films using SJEM with 10 nm resolution. They used a three-dimensional FE model of the frequency-domain heat transfer problem, to fit the in-plane thermal conductivity to the measured data. Gurrum, et al. determined that for a heating frequency of 100 kHz, the thermal conductivity of a 43 nm film was 82 ± 7.7 W/mK, and for a heating frequency of 90 kHz, the thermal conductivity of a 131 nm film was 162 ± 16.7 W/mK. These values are significantly smaller than the bulk thermal conductivity of 318 W/mK for gold, showing thermal conductivity dependence on size due to electron scattering [35].

A proper heat transfer model for the interconnect structure can be developed once the thermal properties of the thin film material are determined. Different techniques are categorized below for different problems.

5. MULTI SCALE THERMAL MODELING

The problem of heat dissipation in ICs, and specifically interconnects, is a multi-scale problem. To precisely predict device performance and temperature effects that arise from Joule heating, it is necessary to have a detailed temperature profile of the interconnect stack under various operating conditions. Traditional FD and FE methods require large computation time even for a unit cell (micro-models). This has led to the development of reduced, or compact, thermal models, which trade resolution and accuracy for shorter computational time to carry out parametric design studies. Reduced models have been employed to address both steady-state and transient Joule heating in interconnects, although predominantly the former . Reduced or compact thermal models for off-chip or package level analyses have also been investigated ([36-39]). Most of these investigations have been directed towards obtaining an equivalent thermal resistance network.

In review of the different available techniques, we first look at the use of compact steady-state modeling using finite element analysis. We then focus on the transient thermal behavior of the interconnect structures under varying electrical current conditions. Two methods for predicting transient thermal behavior, *Transmission Line Matrix* and *Proper Orthogonal Decomposition* are described.

6. STEADY STATE THERMAL ANALYSIS

6.1. COMPACT THERMAL METHOD

Some of the first compact modeling work was done by Kreuger and Bar-Cohen in 1992 [40]. They modeled a chip package with a simplified resistor network. They showed that with a reduction in mesh size, shorter simulation times can be achieved. Compact thermal models for on-chip interconnect

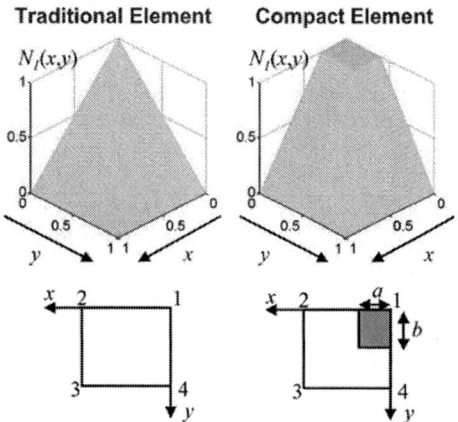

Figure 3- A standard and a compact element are shown at the bottom. The compact element has both the metallic region (shaded) and the dielectric region. The weighting function for Node 1 in both approaches is shown above. In compact element case (top right), this function ignores temperature drop across the metallic region [42].

heating analysis have also been developed [41-42]. The three-dimensional compact finite element modeling of interconnects with vias by Gurrum et al. [42] is briefly explained.

In a standard FE analysis, the spatial domain of interest is discretized into elements such that the region inside an element is homogeneous, although it can have anisotropic thermal conductivity. In the compact approach, investigated by Gurrum et al.[42], a given element is permitted to have both metal and dielectric regions. This will reduce the number of nodes at which temperature needs to be computed. Considering T_i to be the temperature at an arbitrary node i and $N_i(x,y)$ to be the weighting function for node i, the temperature distribution within the element (shown in Fig. 3) is:

$$T(x,y) = N_1(x,y)T_1 + N_2(x,y)T_2 + N_3(x,y)T_3 + N_4(x,y)T_4$$

(1)

The weighting function for node i is expressed in Fig. 3 for both standard and compact finite element models. These functions estimate the metallic region to have infinite thermal conductivity. Setting $a=b=0$ in Eq. 1 results in having conventional bilinear weighting functions instead.

Gurrum et al. [42] applied the compact FE modeling technique to a three dimensional structure consisting of 500-link M1/V1/M2 serpentine test structure, shown in Fig. 4. The length of the interconnects and the spacing between the rows (shown in Fig. 4a) are 7μm. The structure consists of two levels of copper lines (M1 and M2) and a via (V1) that connects M1 to M2. Lines widths are 180 nm and line height is 350 nm. Interconnect temperature rise for this case has been measured by Ramakrishna et al.[43]. The test structure consists of 500 metal lines and 500 vias, and is modeled by properly choosing the domain for simulation. There are

roughly 200,000 elements in the model. The temperature predictions form a detailed finite element model of a unit cell. Modeling of this test structure has also been conducted by Ramakrishna et al.[43]. Fig. 4b shows the temperature map of the top surface of the heat-generating region. The compact FE modeling technique predicts the maximum temperature rise in the metal to within 5–10% of the detailed numerical computations. To do this, only a fraction of elements are needed. The result is a computational time of several seconds for the compact model solution, versus the many hours required for the detailed solutions obtained through successive mesh refinement until grid independence is achieved.

The thermal resistor network approach is applicable to simple interconnect structures. As addressed previously in the literature [44], geometrical complexity complicates the network topography of resistor type compact models. Finite element based compact modeling can be utilized for more complex architectures, with relatively straight forward scaleup. A key challenge associated with any compact modeling approach is the *a priori* estimation of error bounds in predicted temperatures.

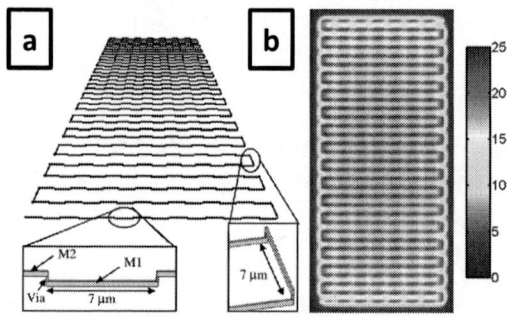

Figure 4- Schematic of 500-link M1/V1/M2 serpentine test structure (a) and the temperature map of the heat generating region for a current density of 23.8 MA/cm^2 (b) [42].

7. TRANSIENT THERMAL ANALYSIS

Although steady state modeling of Joule heating is of great significance, electric current pulses may be encountered during operation, which necessitates studying the effects of transient Joule heating in interconnects.

7.1. TRANSMISSION LINE MATRIX (TLM) METHOD

An established method for solving the transient heat conduction equation is known as Transmission Line Matrix (TLM) [45-46]. The TLM formulation is based on a resistance and capacitance network that represents a thermal system. The TLM method provides an advantage of the traditional FE and FD methods by allowing for temperature-dependent and inhomogeneous material parameters, non-uniform mesh, and non-uniform time stepping. Details of the TLM approach can be found in [47]. The foundation of the TLM method is based

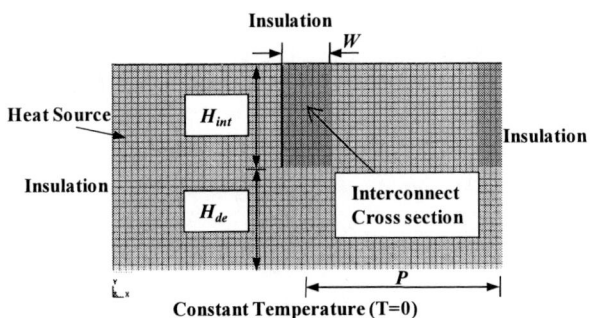

Figure 5- Schematic of the model consisting of a set of W = 180 nm wide interconnects evenly spaced and embedded in the dielectric. $H_{int} = H_{de}$ = 2W and P = 4W [47].

on the physics of traveling potential pulses. The incident pulse at an arbitrary node (x) travels along the transmission lines approaching the node. This pulse is then scattered as it passes the node. The pulse then gets transmitted along the line with coefficient, τ, or it gets reflected back in the same line with the reflection coefficient ρ. These coefficients are calculated based on the resistance and impedance encountered by the pulse in the transmission line. The TLM formulation has two forms: link-resistor (LR), with capacitance centered on the nodes and resistors placed mid-way between nodes, and link-line (LL), in which resistors are placed at the nodes and transmission lines link the adjacent nodes.

Barabadi et al. previously investigated transient Joule heating in 180-nm-wide and 360-nm-high copper interconnects embedded in silicon dioxide dielectric with constant current density of 10 MA/cm^2 using the TLM and FE methods (Fig. 5) [47]. The effect of duration and amplitude of rapid square-wave source current pulses on the transient temperature history were also investigated by the same authors.[48]. Shown in Fig. 6, the TLM method is compared with conventional FE results for a 1 µs square-pulse heat source. The longitudinal spatial variations of the top-edge temperature at six different times for the two models are plotted. Temperature is constant in the interconnects, but decreases in the dielectric along the x-axis, as expected. The figure shows both the increase in temperature during the square-pulse source and then the passive relaxation after the source is turned off at t = 1 µs. The TLM-predicted value for peak temperature in the source (left) interconnect differed by less than 0.5% from the FE result. TLM results during the transient phase, however, generally lag the corresponding FE results. This difference in temperature rise between the two models increases at the beginning of the relaxation period, then shrinks over time and reaches zero at the steady state.

In summary, the mesh refinement issue is of practical importance when solving the Joule heating problem in interconnects. Multi-stack interconnect architectures are characterized by a wide range in the length scales (from 10^{-9} m to 10^{-2} m) and significant material inhomogeneity (thermal conductivity variation of two to three orders of magnitude

difference). It is often desirable to use an inhomogeneous mesh that is more refined in the areas where there are large temperature gradients in order to reduce the total simulation time. The rate of reducing the element size in small distances can be significantly greater than what can be accomplished using conventional FE simulations if the quad-tree mesh method, which can be combined with the TLM method, is used [49]. Otherwise, the increasing complexity of the structure increases the simulation time, making the TLM model inefficient for the purpose of multiscale modeling.

Another important feature of the TLM algorithm is that it is unconditionally stable, despite being an explicit method, so the time step can be increased when needed, which can significantly reduce the simulation time compared to the conventional FD and FE methods [50]. Additionally, it has been shown that the stability of the LL results depend on the time-step of the simulation. Instability in LL results occurs at a particular time-step or smaller, causing unwanted oscillations that initiate near the insulating boundaries [46-47, 50].This indicates that the LR formulation is more accurate for transient thermal analysis.

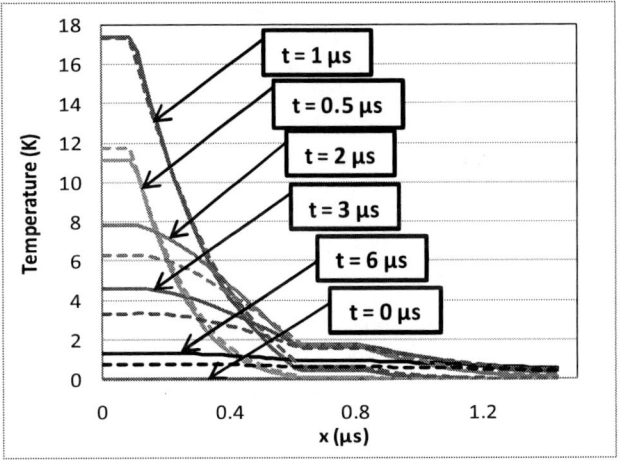

Figure 6- Transient Temperature response for the upper edge of the structure for 1 µs square-pulse heat source between the TLM (solid line) and FE (dashed line) models [48].

7.2. PROPER ORTHOGONAL DECOMPOSITION (POD) METHOD

The POD is a robust and elegant method of data analysis that enables gaining low-dimensional approximate descriptions of a high-dimensional process. It expands a set of data on empirically determined basis functions for modal decomposition and can be used to numerically predict the temperature distribution more rapidly than full-field simulations. The history of POD goes back over 100 years [51], when it was used as a means for processing statistical data. Since that time, it has been applied in many engineering fields, like fluid flow and turbulence [52-54], structural

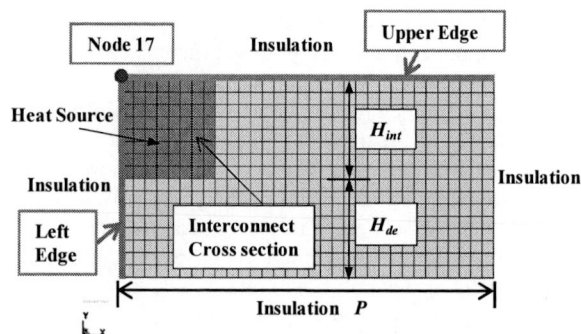

Figure 8- Schematic of the model with the cross sectional area of 1.44 μm 720 nm consisting of a 360 nm 360 nm interconnect embedded in the dielectric. $H_{int} = H_{de} = 360$ nm [65].

vibrations [55-56], control theory [57], and others. More specifically, POD has been used in the field of microelectromechanical systems (MEMS) and electronic packaging [58-59]. More recently, the POD technique has been applied to transient heat conduction problems [60-62]. Bleris and Kothare [63] studied Microsystems using empirical eigenfunctions obtained from the POD technique to address the problem of thermal transient regulation. More recently, a Boundary-Condition-Independent POD together with Galerkin Methodology for one dimensional heat conduction was studied by Raghupathy et al. [64]. Berkooz et al [54] provided a thorough summary for applications of POD in various fields.

The POD-based model reduction technique is a method that provides an optimal set of empirical basis functions (also known as POD modes) from an ensemble of observations obtained either experimentally or from numerical simulation. In this technique, data sets are expanded for modal decomposition on the basis functions which minimize the least squares error between the true solution and the truncated representation of the POD model. The temperature distribution can be determined from the expansion theorem:

$$T(x,y,t) = T_0(x,y) + \sum_{i=1}^{m} b_i(t)\varphi_i(x,y) \qquad (2)$$

where T_0 is the time average of temperature, $\varphi_i(x, y)$ is the i-th POD mode, and $b_i(t)$ is the i-th POD coefficient, explained later. The algorithm is illustrated in Fig. 7. A detailed procedure to generate a POD based reduced order model is provided in [39, 65].

To model transient Joule heating in interconnects using POD, we look at the work of Barabadi et al [65]. They considered a two dimensional (2D) inhomogeneous structure with insulated boundary conditions corresponding to regions of conductor wiring embedded in the bulk dielectric of a microelectronic device, shown in Fig. 8. Using the POD method and the Galerkin Projection Technique,, the effect of different types of current pulses, pulse duration, and pulse amplitude on thermal distributions in the system were investigated in this study. Transient heat generation in the system was simulated using a sinusoidal wave function with one cycle, period of $\Pi_3 = 20$ μs,

and amplitude of $A_3 = 10 \times q''' = 2 \times 10^{14}$ (W/m³). In this case, a single noise signal enters the interconnects. The temporal dependence of temperature rise inside the interconnect region is compared between conventional FE and POD models (Fig 9) for the mentioned case at node 17 (noted in Fig. 8). Based

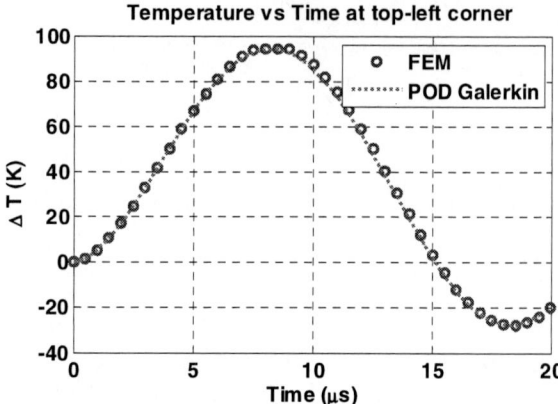

Figure 9- Comparison of Temporal dependence of temperature, at the top edge in the left-most node (node 17) in the interconnect, x=0, between FE (Blue markers) and POD (dotted red line) models for sinusoidal function with one cycle [65]

on the results, this POD model using Galerkin Projection Technique can predict the transient thermal behavior for a sinusoidal heat wave. No new observations are necessary for using the proposed method so long as there is one set of observations availablefor the value of the heat source for one case (for example, a step function for the heat source).

The model predicted the exact transient thermal behavior of the system, using just a few POD modes, for all other cases when using a representative step function as the heat source without generating any new observations. The proposed POD method also predicts the transient temperature distribution in the 2D structure considered, regardless of the temporal dependence of the heat source being applied. Computational cost can be significantly decreased now as a result of being able to use smaller sample sets to predict other cases using POD. One of the most remarkable characteristic of the POD is its optimality. POD offers the most efficient method of capturing the dominant components of an infinite-dimensional process with a finite number of modes [54, 66].

8. RESEARCH CHALLENGES

While there has been extensive research on thermal analysis of Joule heating in interconnect structures, a number of key challenges remain due to its multi-scale nature. Reduced order models are approximations by definition. Observational methods like proper orthogonal decomposition can incur large errors outside the range of observations. However, tools to systematically predict the boundary beyond which this large error is incurred currently do not exist. Another vitally important topic is how to connect nano-scale models with continuum based micro-scale models. In general, these connections are only effective over a limited range of length scales. Development of more generic models that are effective over the continuum of nano- to micro-scale is needed.

ACKNOWLEDGMENTS

The authors acknowledge the support from Semiconductor Research Corporation (SRC) under Task 1883.001.

REFERENCES

[1] J. R. Lloyd, "Materials Reliability in Microelectronics III," in *Materials Research Society*, Pittsburgh, PA, 1993, p. 177.

[2] T. Phan, *et al.*, "Thermomechanical study of AlCu based interconnect under pulsed thermoelectric excitation," *Journal of Applied Physics*, vol. 81, pp. 1157-1157, 1997.

[3] A. A. Bilotti, "Static temperature distribution in IC chips with isothermal heat sources," *IEEE Transactions on Electron Devices*, vol. ED-21, pp. 217-26, 1974.

[4] Y. L. Shen, "Analysis of Joule heating in multilevel interconnects," *Journal of Vacuum Science & Technology B (Microelectronics and Nanometer Structures)*, vol. 17, pp. 2115-21, 1999.

[5] C. C. Teng, *et al.*, "iTEM: A temperature-dependent electromigration reliability diagnosis tool," *Computer-Aided Design of Integrated Circuits and Systems, IEEE Transactions on*, vol. 16, pp. 882-893, 2002.

[6] D. Chen, *et al.*, "Interconnect thermal modeling for accurate simulation of circuit timing and reliability," *Computer-Aided Design of Integrated Circuits and Systems, IEEE Transactions on*, vol. 19, pp. 197-205, 2002.

[7] T. Y. Chiang, *et al.*, "Analytical thermal model for multilevel VLSI interconnects incorporating via effect," *Electron Device Letters, IEEE*, vol. 23, pp. 31-33, 2002.

[8] K. Banerjee, *et al.*, "High-current failure model for VLSI interconnects under short-pulse stress conditions," *IEEE Electron Device Letters*, vol. 18, pp. 405-407, 1997.

[9] K. Banerjee, *et al.*, "Quantitative projections of reliability and performance for low-k/Cu interconnect systems," 2002, pp. 354-358.

[10] S. Im and K. Banerjee, "Full chip thermal analysis of planar (2-D) and vertically integrated (3-D) high performance ICs," 2002, pp. 727-730.

[11] T.-Y. Chiang, *et al.*, "Impact of Joule heating on scaling of deep sub-micron Cu/low-k interconnects," in *2002 Symposium on VLSI Technology Digest of Technical Papers, June 11, 2002 - June 13, 2002*, Honolulu, HI, United states, 2002, pp. 38-39.

[12] K. Fuchs, "The conductivity of thin metallic films according to the electron theory of metals," 1938, pp. 100-108.

[13] E. Sondheimer, "The mean free path of electrons in metals," *Advances in Physics*, vol. 1, pp. 1-42, 1952.

[14] J. M. Ziman and P. W. Levy, "Electrons and phonons," *Physics Today*, vol. 14, p. 64, 1961.

[15] Y. Namba, "Resistivity and temperature coefficient of thin metal films with rough surface," *Japanese Journal op Applied Physics Vol*, vol. 9, 1970.

[16] S. B. Soffer, "Statistical Model for the Size Effect in Electrical Conduction," *Journal of Applied Physics*, vol. 38, pp. 1710-1715, 1967.

[17] S. P. Gurrum, *et al.*, "Numerical simulation of electron transport through constriction in a metallic thin film," *Electron Device Letters, IEEE*, vol. 25, pp. 696-698, 2004.

[18] J. Christofferson, *et al.*, "Microscale and Nanoscale Thermal Characterization Techniques," *Journal of Electronic Packaging*, vol. 130, pp. 041101-6, 2008.

[19] D. G. Cahill, "Thermal conductivity measurement from 30 to 750 K: the 3 method," *Review of Scientific Instruments*, vol. 61, p. 802, 1990.

[20] C. Paddock, "Transient Thermoreflectance From Thin Metal Films," *J. Appl. Phys.*, vol. 60, pp. 285-290, 1986.

[21] D. G. Cahill, *et al.*, "Nanoscale thermal transport," *Journal of Applied Physics*, vol. 93, p. 793, 2003.

[22] D. G. Cahill, *et al.*, "Thermometry and Thermal Transport in Micro/Nanoscale Solid-State Devices and Structures," *Journal of Heat Transfer*, vol. 124, pp. 223-241, 2002.

[23] S. Dilhaire, *et al.*, "Calibration procedure for temperature measurements by thermoreflectance under high magnification conditions," *Applied Physics Letters*, vol. 84, p. 822, 2004.

[24] V. Szekely and T. Van Bien, "Fine structure of heat flow path in semiconductor devices: a measurement and identification method," *Solid-State Electronics*, vol. 31, pp. 1363-1368, 1988.

[25] P. Szabo, *et al.*, "Short time die attach characterization of LEDs for in-line testing application," 2007, pp. 360-366.

[26] V. Székely, in *International Workshop on Thermal Investigations of ICs and Systems (THERMINICS)*, Rome, Italy, September 24–26, 2008 (unpublished).

[27] Y. Ezzahri and A. Shakouri, "Application of network identification by deconvolution method to the thermal analysis of the pump-probe transient thermoreflectance signal," *Review of Scientific Instruments*, vol. 80, p. 074903, 2009.

[28] A. Majumdar, "Scanning thermal microscopy," *Annual review of materials science*, vol. 29, pp. 505-585, 1999.

[29] G. B. M. Fiege, *et al.*, "Quantitative thermal conductivity measurements with nanometre

resolution," *Journal of Physics D: Applied Physics,* vol. 32, p. L13, 1999.

[30] H. Fischer, "Quantitative determination of heat conductivities by scanning thermal microscopy," *Thermochimica Acta,* vol. 425, pp. 69-74, 2005.

[31] V. Gorbunov, *et al.,* "Probing surface microthermal properties by scanning thermal microscopy," *Langmuir,* vol. 15, pp. 8340-8343, 1999.

[32] F. Ruiz, *et al.,* "Determination of the thermal conductivity of diamond-like nanocomposite films using a scanning thermal microscope," *Applied Physics Letters,* vol. 73, p. 1802, 1998.

[33] H. Pollock and A. Hammiche, "Micro-thermal analysis: techniques and applications," *Journal of Physics D: Applied Physics,* vol. 34, p. R23, 2001.

[34] J. Varesi and A. Majumdar, "Scanning Joule expansion microscopy at nanometer scales," *Applied Physics Letters,* vol. 72, p. 37, 1998.

[35] S. P. Gurrum, *et al.,* "Size Effect on the Thermal Conductivity of Thin Metallic Films Investigated by Scanning Joule Expansion Microscopy," *Journal of Heat Transfer,* vol. 130, p. 082403, 2008.

[36] V. Adams, *et al.,* "Application of compact model methodologies to natural convection cooling of an array of electronic packages in a low profile enclosure," *Advances in Electronic Packaging, Pt,* vol. 2, pp. 1967–1974.

[37] A. Bar-Cohen, *et al.,* "&thetas;< sub> jc</sub> characterization of chip packages-justification, limitations, and future," *Components, Hybrids, and Manufacturing Technology, IEEE Transactions on,* vol. 12, pp. 724-731, 2002.

[38] D. S. Boyalakuntla and J. Y. Murthy, "Hierarchical compact models for simulation of electronic chip packages," *Components and Packaging Technologies, IEEE Transactions on,* vol. 25, pp. 192-203, 2002.

[39] H. I. Rosten, *et al.,* "The world of thermal characterization according to DELPHI-Part I: Background to DELPHI," *Components, Packaging, and Manufacturing Technology, Part A, IEEE Transactions on,* vol. 20, pp. 384-391, 2002.

[40] W. Krueger and A. Bar-Cohen, "THERMAL CHARACTERIZATION OF A PLCC-EXPANDED Rjc METHODOLOGY," 1992, p. 263.

[41] M. R. Stan, *et al.,* "HotSpot: a dynamic compact thermal model at the processor-architecture level," *Microelectronics Journal,* vol. 34, pp. 1153-65, 2003.

[42] S. P. Gurrum, *et al.,* "A compact approach to on-chip interconnect heat conduction modeling using the finite element method," *Journal of Electronic Packaging,* vol. 130, pp. 031001-1, 2008.

[43] K. Ramakrishna, *et al.,* "Prediction of maximum allowed rms currents for electromigration design guidelines," 2004, p. 156.

[44] D. Celo, *et al.,* "Hierarchical thermal analysis of large IC modules," *IEEE Transactions on Components and Packaging Technologies,* vol. 28, pp. 207-17, 2005.

[45] C. Christopoulos, "The transmission-line modeling method: TLM," *Antennas and Propagation Magazine, IEEE,* vol. 39, p. 90, 2002.

[46] D. De Cogan, *et al., Transmission line matrix in computational mechanics*: CRC, 2006.

[47] B. Barabadi, *et al.,* "Thermal characterization of planar interconnect architectures under transient currents," in *ASME 2009 International Mechanical Engineering Congress and Exposition, IMECE2009, November 13, 2009 - November 19, 2009,* Lake Buena Vista, FL, United states, 2010, pp. 1381-1389.

[48] B. Barabadi, *et al.,* "Thermal characterization of planar interconnect architectures under different rapid transient currents using the transmission line matrix and finite element methods," in *2010 12th IEEE Intersociety Conference on Thermal and Thermomechanical Phenomena in Electronic Systems (ITherm), 2-5 June 2010,* Piscataway, NJ, USA, 2010, p. 8 pp.

[49] T. Smy, *et al.,* "Transient 3D heat flow analysis for integrated circuit devices using the transmission line matrix method on a quad tree mesh," *Solid-State Electronics,* vol. 45, pp. 1137-1148, 2001.

[50] R. Ait-sadi and P. Naylor, "An investigation of the different TLM configurations used in the modelling of diffusion problems," *International Journal of Numerical Modelling: Electronic Networks, Devices and Fields,* vol. 6, pp. 253-268, 1993.

[51] K. Pearson, "LIII. On lines and planes of closest fit to systems of points in space," *Philosophical Magazine Series 6,* vol. 2, pp. 559-572, 1901.

[52] D. Ahlman, *et al.,* "Proper orthogonal decomposition for time-dependent lid-driven cavity flows," *Numerical Heat Transfer, Part B: Fundamentals,* vol. 42, pp. 285-306, 2002.

[53] G. Berkooz, *et al.,* "The proper orthogonal decomposition in the analysis of turbulent flows," *Annual Review of Fluid Mechanics,* vol. 25, pp. 539-575, 1993.

[54] G. Berkooz, *et al.,* "Turbulence, coherent structures, dynamical systems and symmetry," *Cambridge Monographs on Mechanics, Cambridge University Press,* 1996.

[55] J. Cusumano, *et al.,* "Experimental measurements of dimensionality and spatial coherence in the dynamics of a flexible-beam impact oscillator," *Philosophical Transactions of the Royal Society of London. Series A: Physical and Engineering Sciences,* vol. 347, p. 421, 1994.

[56] B. Feeny and R. Kappagantu, "On the physical interpretation of proper orthogonal modes in vibrations," *Journal of Sound and Vibration,* vol. 211, pp. 607-616, 1998.

[57] J. A. Atwell and B. B. King, "Proper orthogonal decomposition for reduced basis feedback controllers for parabolic equations* 1," *Mathematical and Computer Modelling,* vol. 33, pp. 1-19, 2001.

[58] Y. Liang, *et al.,* "Proper orthogonal decomposition and its applications-part II: Model reduction for MEMS dynamical analysis," *Journal of Sound and Vibration,* vol. 256, pp. 515-532, 2002.

[59] L. Codecasa, *et al.,* "An Arnoldi based thermal network reduction method for electro-thermal analysis," *Components and Packaging Technologies, IEEE Transactions on,* vol. 26, pp. 186-192, 2003.

[60] Bia and R. lstrok, "Proper orthogonal decomposition and modal analysis for acceleration of transient FEM thermal analysis," *International Journal for Numerical Methods in Engineering,* vol. 62, pp. 774-797, 2005.

[61] R. Bia ecki, *et al.,* "Reduction of the dimensionality of transient FEM solutions Using Proper Orthogonal Decomposition," 2003, p. 2003.

[62] A. Fic, *et al.,* "Solving transient nonlinear heat conduction problems by proper orthogonal decomposition and the finite-element method," *Numerical Heat Transfer, Part B: Fundamentals,* vol. 48, pp. 103-124, 2005.

[63] L. G. Bleris and M. V. Kothare, "Reduced order distributed boundary control of thermal transients in microsystems," *Control Systems Technology, IEEE Transactions on,* vol. 13, pp. 853-867, 2005.

[64] A. P. Raghupathy, *et al.,* "Boundary-Condition-Independent Reduced-Order Modeling of Heat Transfer in Complex Objects by POD-Galerkin Methodology: 1D Case Study," *Journal of Heat Transfer,* vol. 132, p. 064502, 2010.

[65] B. Barabadi, et al., "Prediction of Transient Thermal Behavior of Planar Interconnect Architecture Using Proper Orthogonal Decomposition Method," *submitted to InterPACK,* Portland, Oregon, USA, July 6-8, 2011.

[66] A. Chatterjee, "An introduction to the proper orthogonal decomposition," *Current science,* vol. 78, pp. 808-817, 2000.

978-1-4244-9113-1/11 $26.00 © 2011 IEEE

Impact of Scaling on the Performance and Reliability Degradation of Metal-Contacts in NEMS Devices

Hamed F. Dadgour *, Muhammad M. Hussain **, Alan Cassell [†], Navab Singh [††] and Kaustav Banerjee *

* Department of Electrical and Computer Engineering, University of California, Santa Barbara, CA 93106
** King Abdullah University of Science and Technology, Thuwal, Saudi Arabia, 23955-6900
[†] NASA Ames Research Center, Moffett Field, CA 94035
[††] Institute of Microelectronics, A*STAR, Singapore, 117685

ABSTRACT- Nano-electro-mechanical switches (NEMS) offer new possibilities for the design of ultra energy-efficient systems; however, thus far, all the fabricated NEMS devices require high supply voltages that limit their applicability for logic designs. Therefore, research is being conducted to lower the operating voltages by scaling down the physical dimensions of these devices. However, the impact of device scaling on the electrical and mechanical properties of metal contacts in NEMS devices has not been thoroughly investigated in the literature. Such a study is essential because metal contacts play a critical role in determining the overall performance and reliability of NEMS. Therefore, the comprehensive analytical study presented in this paper highlights the performance and reliability degradations of such metal contacts caused by scaling. The proposed modeling environment accurately takes into account the impact of roughness of contact surfaces, elastic/plastic deformation of contacting asperities, and various inter-molecular forces between mating surfaces (such as Van der Waals and capillary forces). The modeling results are validated and calibrated using available measurement data. This scaling analysis indicates that the key contact properties of gold contacts (resistance, stiction and wear-out) deteriorate "exponentially" with scaling. Simulation results demonstrate that reliable (stiction-free) operation of very small contact areas ($\approx$ 6nm $\times$ 6nm) will be a daunting task due to the existence of strong surface forces. Hence, contact degradation is identified as a major problem to the scaling of NEMS transistors.

Keywords- Contact Degradation, Contact Reliability, Contact Resistance, Digital Circuits, Nano-electro-mechanical Switches (NEMS), Process Variations, Scaling Analysis, Stiction, Wear and Fatigue.

I. INTRODUCTION

In the recent years, NEMS transistors have generated a great amount of interest due to their superior subthreshold behavior [1]-[7]. This is primarily because NEMS offer steep switching characteristics between ON and OFF states and hence, they can effectively eliminate the subthreshold leakage issue that has troubled CMOS technology for many years. For instance, it has been shown through experiments that micro-scale electromechanical switches can exhibit an incredibly low subthreshold swing of 2mV/decade [8]. However, wide-spread usage of NEMS transistors in logic applications is hindered by the need for high supply voltages because pull-in voltage (equivalent of the threshold voltage for CMOS devices) for current NEMS devices is about 5~10V [9]-[11].

To overcome this shortcoming, researchers are attempting to reduce the physical dimensions of NEMS in order to reduce the

Authors would like to acknowledge SEMATECH for providing the experimental data and valuable scientific insights.

Nomenclature			
a	Bar's original length	P_{avg}	Average stress
A	Hamaker constant	P_{crit}	Critical stress (the yield strength)
Area	Area of the metal bar	r	Contact area radius
b	Bar's elongation due to stress	r_0	Contact radius under zero-load
D	Distance between two spheres	R	Equivalent radius of curvature
d_0	separation between two surfaces	R_i	Radius of curvature of surface i
E	Elastic modulus of surface	R_M	Maxwell resistance
F	External force	R_S	Sharvin resistance
g	Air gap size	s	The contact area's length
H	Hardness of materials	t	The cantilever beam's thickness
HR	Humidity ratio	T	Temperature
h	Planck's constant	V	Volume
I_{DS}	Drain-source current of NEMS	$V_{pull-in}$	Pull-in voltage of NEMS
k	Boltzmann's constant	V_{GS}	Gate-source voltage of NEMS
K	Archard's wear coefficient	w	Width of NEMS devices
l	Length of the NEMS gate	W_{12}	Surface energy (per unit area)
N	Lifetime of contact (cycles)	z_i	Asperity height of surface i
n_i	Refractive index of medium i		
Greek Letters			
α	Embracing angle	ε_i	Dielectric constant of medium i
β	Scaling factor	λ	Electron mean free path
γ	Liquid surface tension	ρ	Resistivity of contact material
δ	Vertical deformation	σ_R	rms surface roughness
δ_C	δ at the onset of plastic yielding	v_e	UV absorption frequency
θ	Contact angle		

pull-in voltages and consequently, the required supply voltages. However, the scalability of NEMS is predicted to be challenging since the processing techniques for fabricating extremely small NEMS have not been developed yet [12]. More importantly, the impact of scaling on other properties of NEMS devices has not been thoroughly investigated. For instance, one of these characteristics is the performance/reliability of electrical contacts [13]-[15]. Although there have been numerous studies on modeling the properties of metal contacts in NEMS [16]-[22], there has been no systematic study on the impact of scaling on these contacts. Therefore, this study, for the first time ever, highlights the key challenges regarding "metal contacts" that arises due to scaling. This contribution is especially valuable because the electrical/mechanical properties of the contacts have significant effects on the performance/reliability of NEMS devices [23]-[25].

In this study, a comprehensive modeling environment is developed to calculate the important performance and reliability properties of the contacts (**Fig. 1**). This environment takes into account the physical and material properties of NEMS devices and their contacts. Accurate analytical formulas are employed for modeling the rough surfaces where electrical contact occurs exclusively at asperities (a-spots) or local maxima of the films' surfaces [14]. Since the exact profiles of surfaces are random in nature, these

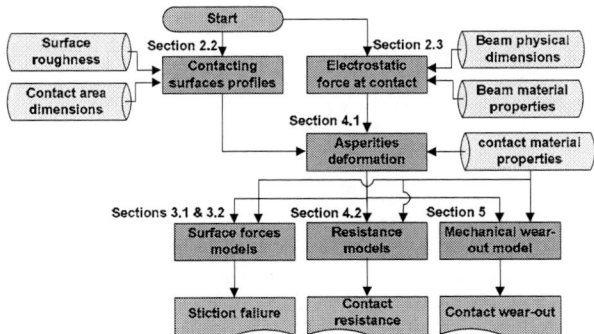

Fig. 1. Schematic diagram of different steps involved in the modeling approach and the corresponding sections that cover each step.

profiles are generated assuming particular surface roughness values [15]-[16]. Moreover, various inter-molecular forces (such as, Van der Waals and capillary) [26], are precisely taken into account. These forces can significantly impact the electrical and mechanical characteristics of scaled NEMS. The accuracy of the modeling approach is verified using existing experimental measurement data. Using the proposed simulation environment, a comprehensive scaling analysis is conducted to study various properties of contact such as, resistance, stiction and wear-out [27]-[29]. This analysis reveals several interesting results:

(1) The contact resistance increases "exponentially" as the device dimensions shrink. This is counter-intuitive because one would expect that the resistance, proportional to the observable contact area, must increase in a quadratic fashion. However, it is shown that the actual contact area decreases faster than a quadratic function because the external (electrostatic) force that presses contact's mating surfaces decreases significantly with scaling. Note that higher contact resistance values increase the delay associated with charging/discharging capacitive nodes (RC delay). Thus, this analysis indicates that the contact resistance will play a more critical role in limiting the performance of the scaled devices.

(2) The simulation results indicate that the surface forces will be a more dominant factor in reducing the reliability of scaled NEMS. In this analysis, the ratio of Van der Waals and capillary forces to that of the elastic force (which attempts to restore the beam to its original position) are evaluated at the contacting surfaces. It is shown that for smaller devices these ratios can be larger than unity. This implies that for very small contacts, once the contacting faces meet, the surface forces will be so strong that they will permanently hold the beam in the contact position.

(3) The contact failure caused by mechanical wear-out of the metal surfaces is also studied in this work. It is shown that various metal elements and alloys differ significantly in terms of mechanical wear-out. However, the contact materials that exhibit superior mechanical stability often perform poorly in terms of electrical conduction and vice versa. Simulation results reveal that gold, which is the dominant contact material for NEMS relays, exhibits considerably smaller time-to-failures as devices scale down [17]. This is another key scaling challenge because gold is the most desirable contact material due to its excellent electrical properties.

The rest of this paper is organized as follows. Section 0 provides an overview of NEMS structure and operation. Section III includes the analytical modeling approach used for considering surface forces (Van der Waals and capillary). Section IV represents the analytical framework, which is used to calculate the contact resistances. Section V studies the mechanical wear-out of contacts and Section VII concludes this paper.

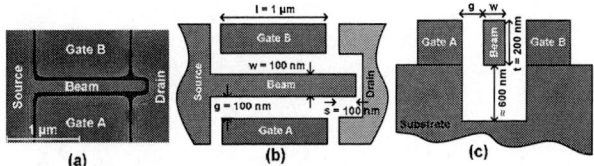

Fig. 2. A laterally-actuated double-gate NEMS device: (a) SEM picture of the fabricated device, (b) schematic of the top-view and (c) the side-view schematic. Typical values of the different dimensions of the structure are also shown here.

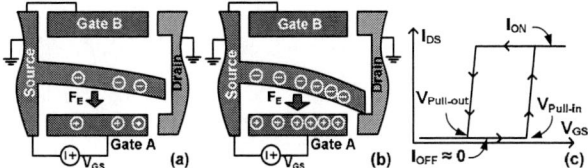

Fig. 3. The basic operation of laterally-actuated NEMS devices: (a) when $V_{GS} < V_{pull-in}$, (b) when $V_{GS} > V_{pull-in}$, and (c) I_{DS}-V_{GS} characteristics of the device.

II. PRELIMINARIES

A. Device Structure and Operation Principle

The SEM picture of a fabricated NEMS transistor along with its schematics is shown in **Fig. 2** [4]. The top-view schematic of the device (**Fig. 2 (b)**) corresponds to the SEM picture shown in **Fig. 2 (a)**. Similarly, the sketch shown in **Fig. 2 (c)** provides a cross-sectional view of the transistor. This device has two gate terminals (Gate A & Gate B in **Fig. 2 (a)**), which are controlled independently. The basic operation of the device is illustrated in **Fig. 3** where, by applying a bias voltage between one of the gates (for example, Gate A) and the source (Gate B is biased at the same voltage as the source), opposite charges appear on the beam and the corresponding gate terminal, generating an electrostatic force. If the gate voltage is smaller than a threshold value ($V_{pull-in}$), as shown in **Fig. 3 (a)**, the beam bends slightly, but does not touch the drain terminal. However, if the bias voltage is larger than $V_{pull-in}$, the beam deflects sufficiently to touch the drain and hence, creates a conduction path from the source to the drain as shown in **Fig. 3 (b)**.

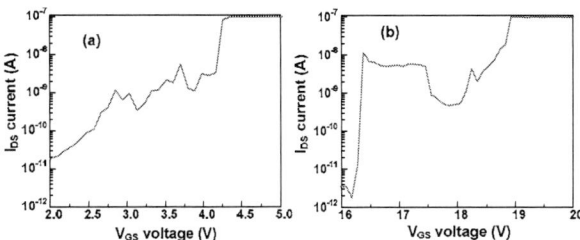

Fig. 4. The switching behavior of fabricated NEMS: (a) I_{DS}-V_{GS} characteristics of a NEMS with a long beam (steady increase in the I_{DS}) and (b) I_{DS}-V_{GS} characteristics of a NEMS with a short beam (erratic increase in the I_{DS}). The switching characteristics of these scaled devices degrade by scaling.

A sketch of the typical I_{DS}-V_{GS} characteristic of such a device is presented in **Fig. 3 (c)** where I_{DS} denotes the source-drain current and V_{GS} refers to the gate-source voltage difference. It can be observed that the device exhibits a hysteresis behavior, which is because of the existence of surface forces that prevent the beam from being restored to its original position once it is pulled down. **Fig. 4** reports the switching characteristics of two NEMS devices with identical physical dimensions except for their beam length. The I_{DS} of the longer device increases gradually as the V_{GS} voltage is ramped up (**Fig. 4 (a)**). However, the I_{DS} curve for the shorter

978-1-4244-9113-1/11 $26.00 © 2011 IEEE

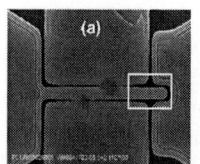 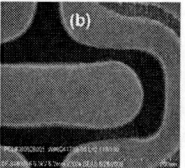 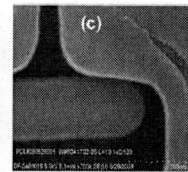

Fig. 5. The SEM pictures of the contact area: (a) the NEMS device (the area inside the box is enlarged and shown in (b) and (c) panels), (b) the contact area when the device is OFF and (c) the contact area when the device is ON.

device reveals a more erratic behavior (**Fig. 4 (b)**). This type of unpredictability, which is observed to deteriorate with further scaling down of the devices, is believed to be due to the contact degradation in shorter devices. Hence, this study aims at investigating the impact of device scaling on the performance and reliability of metal contacts.

B. Contact Area

Another SEM picture of a fabricated device is shown in **Fig. 5 (a)** where the enlarged pictures of its contact area are also presented. Here, **Fig. 5 (b)** and **Fig. 5 (c)** depicts the contact area when the device is OFF and ON, respectively. Although it seems that the beam and the drain terminal form a perfect contact, the effective contact area is dictated by surface irregularities. That is because real metal surfaces, if considered in the nano-scale regime, are not indefinitely smooth, but instead rather rough (**Fig. 6 (a)**). This implies that when two such rough films come into electrical contact (as in a NEMS device), the actual area of conduction is much smaller than the total surface area since electrical contact occurs exclusively at asperities (a-spots), or local maxima of the films' surfaces [30]-[31].

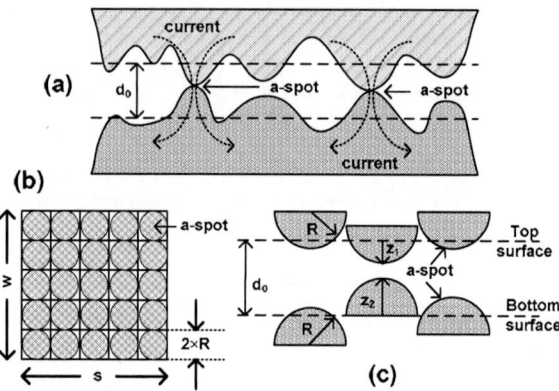

Fig. 6. Contacting surfaces: (a) surface roughness and current conduction (b) top-view of the proposed model for the contact area and (c) side-view of the contact area's model. In this model, the contact asperities are modeled as semi-spheres with radius R. Here, z_1 and z_2 are surface heights, which can be modeled using statistical distributions; d_0 is the mean distance between two surfaces.

A rigorous modeling of the contact area requires an accurate profile of the contact surfaces. However, since the actual profiles of contact areas are random in nature, it is logical to adopt a statistical approach based on some simplified surface models [32]. In this work, it is assumed that surfaces are composed of half-spherical asperities with a certain radius (R) whose heights vary over a range, which is determined by the surface roughness (σ_R). For instance, two typical contacting surfaces are represented in **Fig. 6 (b)**. Here, the contacting asperities are assumed to be semi-spheres with radii R on both surfaces. The average distance between two surfaces is denoted as d_0 and it is assumed that the height of asperities on top and bottom surfaces can be modeled

using random variables z_1 and z_2. These random variables are often assumed to follow a Gaussian distribution. Using such a model, it is possible to statistically determine the effective contact area of two surfaces in terms of the applied force and their material properties as will be discussed later.

C. Scaling Analysis

This paper investigates the implications of scaling NEMS devices on the performance and the reliability of their contacts. In order to achieve this goal, a dimension scaling approach is assumed where the critical device dimensions (l, g, w, t and s) are reduced by a factor of $1/\beta$ ($\beta > 1$) as shown in **Table 1**. The first generation NEMS devices (Node 1) are assumed to have critical dimensions identical to those of the fabricated transistors (**Fig. 2**).

This scaling scenario results in reduced pull-in voltages for NEMS devices and consequently decreases the required supply voltages. Such a trend is shown in **Fig. 7 (a)** where the vertical axis represents the pull-in voltage values. In this figure, the horizontal axis shows the scaling generations (nodes) where each generation is indicated by its corresponding beam length value (l). As it can be observed, the pull-in voltages decrease in a quadratic fashion as the beam length reduces linearly. However, **Fig. 7 (b)** reveals that as NEMS devices scale down, the external electrostatic force that is required to turn the device ON reduces almost exponentially. This is because smaller beams are less stiff and hence, can be deflected with smaller forces. However, this results in higher contact resistances because the mechanical deformations of contacting asperities are a function of the external force as it will be discussed later. Thus, the effective contact area decreases by lowering the applied load on the mating surfaces. The higher contact resistances have significant implications for the performance of NEMS-based circuits as they limit the ON current that can be supplied by NEMS transistors to charge/discharge load capacitances.

Table 1. The scaling scenario: all five device dimensions (l, g, s, t, and w) (defined in Fig. 2) are scaled down by a factor of $1/\beta$ ($\beta = 2$).

	Scaling scenario				
	l (nm)	g (nm)	s (nm)	t (nm)	w (nm)
Node 1	1000	100	100	100	100
Node 2	500	50	50	50	50
Node 3	250	25	25	25	25
Node 4	125	12.5	12.5	12.5	12.5
Node 5	62.5	6.25	6.25	6.25	6.25

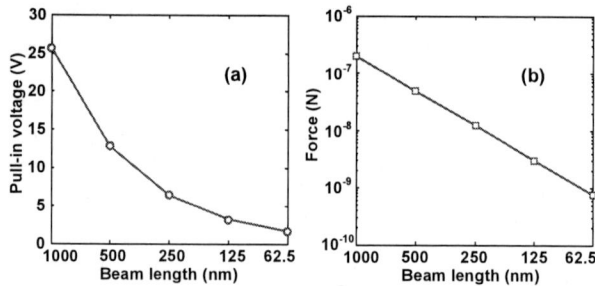

Fig. 7. The impact of device scaling on (a) the pull-in voltage and (b) the force that is required for turning the device ON. In these graphs, the x-axis represents the beam length at each generation (node).

III. SURFACE FORCES AND ADHESION

Surface forces play a detrimental role in increasing the probability of failures in NEMS devices. One of the most common failure modes in NEMS relays is adhesion or 'stiction' where the movable parts of the structures get stuck and are unable to return to their original position [28]-[29]. There are two main causes responsible for the stiction phenomenon: capillary and Van der

978-1-4244-9113-1/11 $26.00 © 2011 IEEE

Waals forces. Capillary forces result in the adhesion between mechanical parts of NEMS in the presence of humidity [33]. Techniques to eliminate capillary forces include coating the structure with a low-surface-energy hydrophobic molecular-monolayer [34]-[36]. However, these techniques are not always practical especially in the case of metal contacts where the surface must be covered with a metal with low resistivity.   The other major causes of stiction are Van der Waals dispersion forces, which cannot be eliminated. Thus, the surface forces are the most dominant threat to the reliability of NEMS structures [37].

A. Capillary Forces

Capillary forces are commonly encountered in nature because of the spontaneous condensation of liquid from surrounding vapor, leading to formation of liquid bridges (**Fig. 8 (a)**). Water molecules are inherently polar molecules and hence, they are essentially permanent dipoles. The strong electrostatic attractions between such dipoles are called "hydrogen bonds" as shown in **Fig. 8 (b)**. Since water molecules adhere to contact surfaces and also attract each other through hydrogen bonds, they can create a strong force (known as capillary force), which attempts to reduce the separation of two surfaces [37].

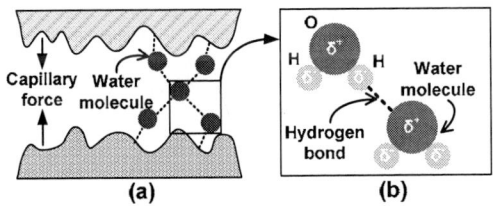

Fig. 8. Capillary forces created between two surfaces due to condensed water drops: (a) water molecules which adhere to the contact surface and each other and (b) a detailed illustration of hydrogen bonds, which are formed among polar water molecules.

In NEMS applications, the formation of capillary forces leads to an increase in the stiction failures both during and after the fabrication process. The final step in the fabrication of NEMS is usually a wet chemical etch. During this step, the sacrificial layer that encapsulates the moveable mechanical structures is removed. Removal of the wafer from the liquid etchant results in a meniscus (liquid-air interface) that often pulls moveable structures into contact via capillary forces [27]. Even if stiction does not occur during fabrication, capillary forces can result in adhesion after the devices are manufactured. This is because when the metal surfaces are in contact, the liquid from the humidity of the surrounding environment can infiltrate the contact area. As a result, capillary force together with other types of surface forces (such as Van der Walls) can prevent the mechanical beam from retrieving to its original position.

Capillary forces significantly increase the attractive force between two surfaces and hence, it can result in permanent device damage or malfunction. Therefore, the prediction and control of the magnitude of capillary forces is necessary for eliminating or minimizing these undesirable events [27]. As mentioned earlier, the contacting asperities are modeled as spheres. Thus, in order to estimate the capillary force between two surfaces, which is composed of a number of asperities, the capillary force between two sphere-shaped asperities should be calculated. Then, the total force between two surfaces can be evaluated by a summation over all the capillary forces between contacting asperities.

An expression for the calculation of the capillary force between two spheres (**Fig. 9 (a)**) is given by (1)-(2) as proposed in [38]. In this equation, γ is the liquid surface tension, θ is the

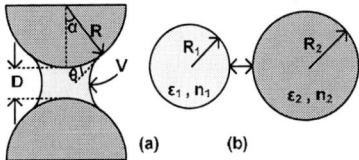

Fig. 9. Schematic illustrating different parameters, which are used in the calculations of (a) capillary and (b) Van der Waals forces.

"contact angle", α is the "embracing angle", and D is the shortest distance between the spheres. This formula assumes that the liquid bridge contains a fixed volume (V). However, this volume itself is a function of humidity (HR) and can be estimated using (3) as shown in [38]-[39].

$$F_{Cap}(D,V) = -\frac{2\pi R\gamma\cos\theta}{1+(D/2d(D,V))} - 2\pi R\gamma\sin\alpha\sin(\alpha+\theta) \qquad (1)$$

$$d(D,V) = (D/2)\times\left[-1+\sqrt{1+2V/(\pi RD^2)}\right] \qquad (2)$$

$$V = \frac{kT}{\gamma}\ln(HR)\left(\frac{1}{\dfrac{1}{R\sin\alpha}-\dfrac{\cos(\theta+\alpha)+\cos\theta}{D+R(1-\cos\alpha)}}\right) \qquad (3)$$

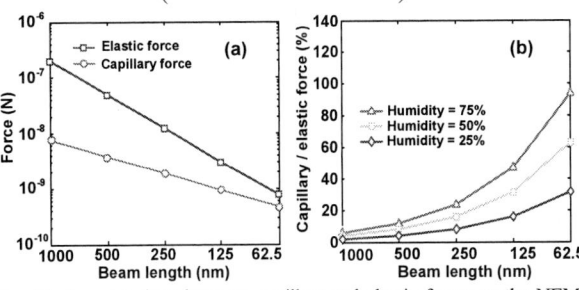

Fig. 10. A comparison between capillary and elastic forces as the NEMS device is scaled down: (a) Although both forces decrease, the elastic force reduces at a faster rate and (b) the ratio of capillary force to the elastic force for various levels of humidity.

Using (1)-(3), the capillary force at the contact area for NEMS devices with various sizes is calculated as shown in **Fig. 10 (a)**. In this graph, the elastic forces of the beams are also included for comparison purposes. Here, the vertical axis represents the force and the horizontal axis is the beam length of the device (according to **Table 1**). While both forces decrease with scaling, the elastic force decreases at a faster rate. This is a critical result because it indicates that by scaling down the devices, the ratio of capillary to elastic force increases and hence, the stiction due to capillary forces is significantly more likely in scaled NEMS. Moreover, the ratios of capillary to elastic forces are also plotted in **Fig. 10 (b)** considering various levels of humidity. Clearly, higher humidity increases the adhesive capillary forces and hence, their ratio to the elastic forces. Note that at l = 62.5nm, the capillary force is nearly as strong as the elastic force at humidity levels higher than 75%. Therefore, it can be concluded that all the devices with l = 62.5 nm will be presumably stuck if the humidity levels rise above 75%.

B. Van der Waals Forces

Van der Waals forces are another type of intermolecular forces, which must be considered in the reliability analysis of NEMS. Van der Waals forces are defined as the attractive or repulsive forces between atoms or molecules other than those due to covalent bonds or due to the electrostatic interaction of ions with one another or with neutral molecules. Van der Waals forces are usually induced by three means [26]: (1) attraction between two permanent dipoles, (2) force between a permanent dipole and a corresponding induced dipole and (3) instantaneously induced dipole-dipole forces (London dispersion force) [40]. These three cases are explained in further detail as shown in **Fig. 11**. In the case of two permanent dipoles, the system reaches equilibrium when two dipoles are aligned such that they attract each other as shown in **Fig. 11 (a)**. On the other hand, the presence of a molecule that is a permanent dipole can temporarily distort the balance of negative and positive charges' distribution in nearby nonpolar molecules, thereby inducing further polarization. Thus, an attractive force results from the interaction of the permanent dipole with the neighboring induced dipole as shown in **Fig. 11 (b)**. Furthermore, it is also possible for two nonpolar molecules to attract each other due to Van der Waals forces. Such a scenario can happen when a temporary dipole is randomly formed in one of the nonpolar molecules. This temporary dipole can in turn induce a dipole in the adjacent molecule and create attraction between two molecules as shown in **Fig. 11 (c)**.

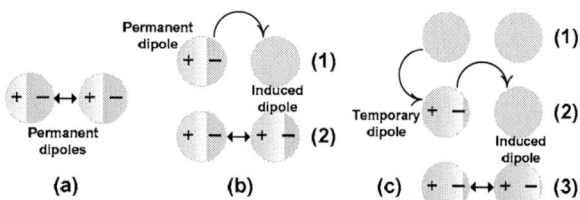

Fig. 11. The inter-molecular Van der Waals forces are generated due to three reasons: (a) permanent dipoles, (b) a permanent dipole that induces another dipole on a nonpolar molecule and (c) a temporary dipole formed in a nonpolar molecule that can induce and create another temporary dipole in the adjacent nonpolar molecule.

Although the Van der Waals forces are inherently inter-molecular forces, they can create attraction between objects. This attraction between two large bodies can be calculated by the summation of attractive forces that exist among the molecules (dipoles) of two bodies. Since, in this work, it is assumed that asperities in the contact area are modeled as half-spheres, it is useful to consider the Van der Waals attraction between two spheres as shown in **Fig. 9 (b)**. Note that Van der Waals forces between dipoles impede rapidly as the distance between dipoles increases. As a result, the Van der Waals attraction between two half-spheres can be estimated with the Van der Waals force that would exist between two spheres. Equation (4) presents an expression for the calculation of the Van der Waals force between two spheres with radii R_1 and R_2, which are separated by a distance D [38]. Here, A is the Hamaker constant which can be estimated using the Lifshitz theory [41] as shown in (5).

$$F_{VdW}(D) = -\frac{A}{6D^2}\frac{R_1 R_2}{R_1 + R_2} \qquad (4)$$

$$A \approx \frac{3}{4}kT\left(\frac{\varepsilon_1 - \varepsilon_3}{\varepsilon_1 + \varepsilon_3}\right)\left(\frac{\varepsilon_2 - \varepsilon_3}{\varepsilon_2 + \varepsilon_3}\right)$$
$$+ \frac{3h\nu_e}{8\sqrt{2}}\frac{\left(n_1^2 - n_3^2\right)\left(n_2^2 - n_3^2\right)}{\left(n_1^2 + n_3^2\right)^{1/2}\left(n_2^2 + n_3^2\right)^{1/2}\left\{\left(n_1^2 + n_3^2\right)^{1/2} + \left(n_2^2 + n_3^2\right)^{1/2}\right\}} \qquad (5)$$

where ε_1, ε_2, and ε_3 are the static dielectric constants of the three media and n_1, n_2, and n_3 are their refractive indexes. In this formula, h is the Planck's constant; k is the Boltzmann's constant; T is the temperature and ν_e is called the plasma frequency of the free electron gas (or UV absorption frequency).

Using (4)-(5), the Van der Waals forces at the contact area of NEMS devices are shown in **Fig. 12 (a)** for various device generations. Here, the beam's elastic force is also plotted to provide a basis for comparison. In this figure, the vertical axis represents the force and the horizontal axis is the beam length of

the device (according to **Table 1**). As shown earlier in **Fig. 10 (a)**, the elastic force of the beam declines exponentially with scaling. However, the effect of Van der Waals force remains approximately unchanged. This is because it is assumed in these simulations that the radii of asperities decrease at the same rate as the device dimensions. Thus, both large and small NEMS have nearly identical numbers of asperities and the strength of Van der Waals forces remain the same. The ratio of Van der Waals to the elastic forces is plotted in **Fig. 12 (b)**. As it can be observed, this ratio is larger than unity for $l = 62.5$ nm, which indicates that NEMS devices with such dimensions will be stuck permanently once it is switched ON. This is a significant conclusion because it suggests that for smaller devices the stiction phenomena will be inevitable.

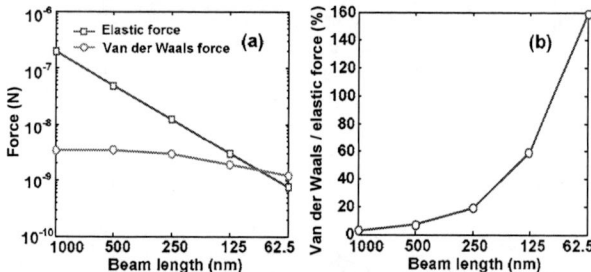

Fig. 12. A comparison between Van der Waals and elastic forces as the device is scaled down: (a) the elastic force reduces at a faster rate and (b) the ratio of Van der Waals force to the elastic force increases significantly.

IV. ELECTRICAL CONTACTS

A. Contact Theory

Due to the surface roughness, when two surfaces are brought together, the contact is made at a finite number of points, where the asperities on both sides touch (**Fig. 6 (a)**). When two asperities are pressed against each other, they deform and form a circular contact area. The amount of this deformation and the radius of the contact area are determined by the applied stress (force) and the elastic properties of the contacting materials [13]-[16]. For small amounts of load, the deformation of the asperities can be modeled with an elastic model. However, for large values of the load, the contact asperities follow a plastic behavior.

A typical relationship between stress (force per area) and strain (longitudinal deformation), called the stress-strain curve, is shown in **Fig. 13 (a)**. The elastic deformation is a change in the shape of a material at low stress, which is recoverable after the stress is removed. This type of deformation involves stretching of the atomic bonds inside the material, but the atoms do not slip past each other. It can be observed that at the low levels of stress (elastic region), the strain increases linearly with stress.

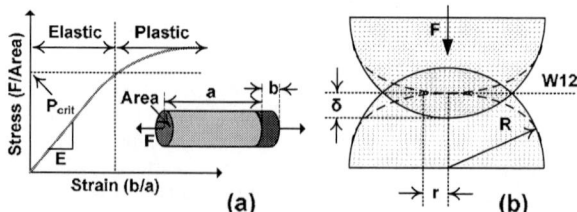

Fig. 13. The deformation of metals under stress: (a) the stress-strain curve for a typical metal where at stress levels below/above a critical stress (P_{crti}), the deformation is elastic/plastic. Here, F, A and E denote the applied force, the cross-sectional area and Young's modulus of the metal bar, respectively; a is the metal bar's original length and b is its elongation due to the applied stress. (b) Deformation of two half spheres under an external force (F).

The slope of the stress-strain curve in this region is equal to the Young's modulus of the material. When the stress is higher than a threshold value, which is known as "yield strength", the metal's deformation is permanent. Such a deformation is called a plastic deformation, which involves the breaking of a limited number of atomic bonds by the movement of dislocations inside the lattice. One can easily model the mechanical behavior of a single asperity using elastic/plastic deformation models. However, in real contacts, there are more than one pair of asperities, which come into contact.

As the applied load on the contacting surfaces is increased, more and more asperities touch until the elastic force created by the deformation of contacting asperities counter-balances the external force. Moreover, the distributions of asperities' height on the mating surfaces follow a Gaussian profile whose standard deviation is determined by the surface roughness. Thus, the number of asperities that come into contact is not always identical even under an equal external force. Therefore, the accurate modeling of contacts' mechanical behavior requires a statistical approach that can take into account the surface roughness and various types of mechanical deformations (elastic or plastic) depending on the applied force.

The electrical properties of contacts (resistance) are dictated by the number of contacting asperities and their radii as shown in **Fig. 6 (a)**. Since the electrical current is constricted to flow through these regions, it is also essential to have accurate models for the scattering of electrons. The statistical models that predict the distributions and radii of contacting asperities combined with the electron scattering models provide a framework to accurately determine the contact resistance values.

B. Contact Resistance

Contact resistance exists because the flow of electrical current is limited at the surface where two metals touch. Therefore, the electrical resistance is determined by the size of those asperities. The force, which is compressing two metal films significantly increases the contact area and hence, it reduces the contact resistance. For small applied forces, only a few metal asperities come into contact when the switch is closed, so the area through which the current flows is initially small. By increasing the applied force, the actual contact area increases, resulting in a decreased contact resistance [14]. Therefore, in order to calculate the contact resistance between two rough surfaces, one can follow a three-step approach proposed in [15]: (1) evaluating the force applied to the junction, (2) calculating the contact radii of the asperities that come into contact and (3) evaluating the resistances of contacting asperities. This method is explained in more detail in the following subsections.

i) Contact Force

The distribution of electrostatic force along the cantilever beam can be calculated using the Euler-Bernoulli's beam equation. The magnitude of the deflecting force is a function of applied voltage, the physical dimensions of the device and the material properties of its beam [28]. Note that according to the Euler-Bernoulli's equation, the distribution of electrostatic force along the length of the beam is not uniform. In this work, the contact force is determined considering such non-uniformities. For instance, **Fig. 7 (b)** represents an estimation of the electrostatic force at the contact for NEMS with different lengths.

ii) Contact Radius for One Asperity

The contact area for two sphere-shape asperities pressed against each other can be calculated using a well-known method

called JKR model which is proposed in [42]. Unlike, the Hertz model, which ignores the existence of surface forces [43], JKR model accounts for those forces. One limitation of JKR method is that it assumes an elastic deformation for both spheres. While this can be true for small levels of stress, it can be shown that nanoscale contacts often operate in the plastic deformation regime [15]. Recently, a model has been proposed in [44], which calculates the deformation of a plastically deformed asperity on the basis of volume conservation. Therefore, in this work, a combined modeling approach is employed where JKR model is used to model the deformation of the asperities until the stress is equal to the yield strength (P_{crit}). After the stress at the asperity exceeds P_{crit}, the deformation follows the volume-conservation-based model proposed in [44]. This approach is explained below in more details.

According to the JKR theory (**Fig. 13 (b)**), two spheres of radii R_1 and R_2, elastic modulus, E, and surface energy, W_{12} per unit area, will flatten when pressed against each other by an external force, F, such that their contact area will have a radius, r, given by (6):

$$r^3 = \frac{R}{E}\left[F + 3\pi R W_{12} + \sqrt{6\pi R W_{12} F + \left(3\pi R W_{12}\right)^2} \right] \quad (6)$$

The vertical deformation of the asperities in this case can be estimated using (7):

$$\delta = \frac{r^2}{R}\left[1 - \frac{2}{3}\left(\frac{r_0}{r}\right)^{3/2} \right] \quad (7)$$

where r_0 is the radius of the contact area under zero-load condition ($F = 0$) which has a non-zero value. Using these formulas, it is possible to calculate the average stress (P_{avg}) applied to the contact area in order to identify when the material's deformation switches from the elastic to the plastic regime. Note that, the onset of plastic deformation is assumed to occur when the stress exceeds the yield strength ($P_{avg} > P_{crit}$). A typical value for the yield strength is 0.6H ($P_{crit} \approx 0.6H$) where H is the hardness of the contact material. In the plastic yielding regime, the average pressure in the contact area remains constant ($P_{avg} = 0.6H$) and the contact radius varies such that it always satisfies this criteria. The vertical deformation of the asperities can then be estimated by using the volume conservation assumption. The formula given in [44] proposes that the contact radius during the plastic deformation is predicted by (8):

$$r = \sqrt{R\delta(2 - \frac{\delta_c}{\delta})} \quad (8)$$

where δ_c denotes the vertical deformation of the spheres at the onset of plastic deformation when $P_{avg} = 0.6H$.

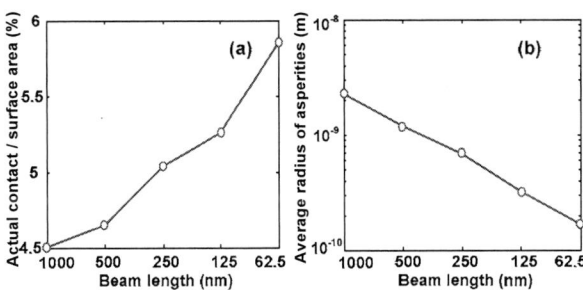

Fig. 14. The contact deformation results: (a) the ratio of contacting asperities' area (actual contact area) to the contact's surface area and (b) the average radius of contacting asperities as NEMS scale down.

Using (6)-(8), the deformation of contact asperities are evaluated at contact areas for NEMS with various sizes. **Fig. 14 (a)** shows the actual contact area (the sum of all contacting asperities' areas) to the surface area of the contact. It can be observed that normally only 5% of the contact area contributes to the electrical conduction. Furthermore, **Fig. 14 (b)** presents the average radius of contacting asperities. Note that these radii decrease rapidly as the NEMS scale down. This results in significant increase in the electron scattering probability at asperities and their boundaries and hence, it significantly raises the contact resistance.

iii) Contact Resistance for One Asperity

Using the contact radius, one can calculate the contact resistance associated with any particular asperity. The flow of current through each asperity is limited by two mechanisms: (1) the resistance due to the lattice scattering mechanism, which is also called Maxwell resistance (R_M) [14]-[51] and (2) the resistance caused by the boundary scattering of electrons, which is also described as the Sharvin resistance (R_S) [48].

The Sharvin resistance is dominant when the contact radius (r) is small compared to the electron mean free path (λ). In this case, electrons can be projected ballistically through the contact area without scattering. Assuming that the resistivity of the contact material is ρ, the Sharvin resistance can be calculated using (9):

$$R_S = \frac{4\rho\lambda}{3\pi \times r^2} \qquad (9)$$

When the contact radius is larger than the electron mean free path, the electrical current is limited by the lattice scattering mechanism (Maxwell resistance). This resistance can be calculated according to (10):

$$R_M = \frac{\rho}{2r} \qquad (10)$$

Both Maxwell and Sharvin resistances contribute to the overall contact resistance (R_C), which can be evaluated using a model proposed in [49]-[50] as described by (11):

$$R_C = f(\frac{\lambda}{r})R_M + R_S \quad where \quad f(\frac{\lambda}{r}) = \frac{1+0.83(\lambda/a)}{1+1.33(\lambda/a)} \qquad (11)$$

In (11), $f(\lambda/r)$ is an interpolation to account for the transition between the lattice and boundary scattering regimes. When the electron mean free path is much smaller than the contact radius ($\lambda/r \approx 0$), both mechanisms will contribute equally to the contact resistance ($f(\lambda/r) = 1$). However, when the electron mean free path is considerably larger than the contact radius ($\lambda/r \to \infty$), Sharivn resistance plays a more important role ($f(\lambda/r) = 0.624$).

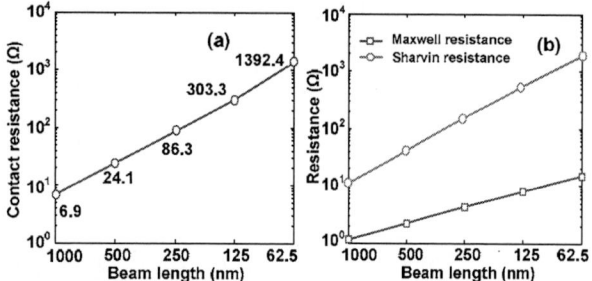

Fig. 15. The impact of scaling on contact resistance: (a) the total contact resistance and (b) contributions of Sharvin and Maxwell resistances to the total resistance.

The impact of scaling on the contact resistance of NEMS is demonstrated graphically in **Fig. 15 (a)** where the vertical and horizontal axes represent the total contact resistance and the beam length of NEMS devices as scaling proceeds. These simulations reveal that the contact resistance increases rapidly. This is an extremely important result because the higher contact resistances can considerably limit the performance of scaled NEMS devices by reducing their ON current. Moreover, **Fig. 15 (b)** illustrates the

contributions of Sharvin and Maxwell resistances to the total resistance. It can be observed that the dominant source of electrical resistance is the boundary scattering (Sharvin resistance). The reason is that the radii of contacting asperities are significantly below the electron mean free path in NEMS devices.

V. CONTACT DEGRADATION DUE TO MECHANICAL WEAR

NEMS devices with impacting contacts are vulnerable to mechanical wear-out induced by continuous impacting during switching cycles. Impact failures are dependent on the stress exerted on the contacting surfaces [27]-[28]. While low stress levels can prolong the life-time of a contact, the final failure of a contact due to mechanical wear-out is inevitable. In reality, the failure rate of NEMS devices due to mechanical wear-outs follows the behavior of a bathtub curve as shown in **Fig. 16 (a)**. The bathtub function consists of three regions. In the initial stage (infant mortality), the failure rate is high due to the latent defects in the device. During the second phase (useful lifetime), the device enters a stable stage with a constant failure rate. Finally, in the wear-out stage, the failure rate increases significantly due to failures induced by mechanical wear-out especially in the contact regions.

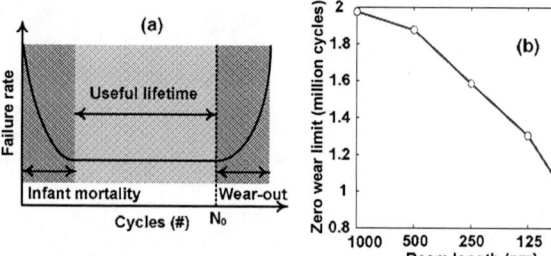

Fig. 16. The mechanical wear-out of contacts in NEMS: (a) the bathtub curve illustrating the failure rate of NEMS over time and (b) the impact of scaling on the zero wear limit of these devices.

In **Fig. 16 (a)**, N_0 represents the 'zero wear' limit of this particular repetitive impact wear process. Zero wear is defined as that number of cycles when the depth of the wear deformity has just reached $\sigma_R/2$, i.e. half the original surface roughness, significantly distinguishing the worn surface from the original one.

The mechanical wear in NEMS structures is caused mainly due to a mechanism called adhesive wear [27]-[29]. This wear mechanism adhesive wear occurs when two interacting surfaces are not sufficiently lubricated, which results in the adhesive transfer or removal of near-surface material. Surface adhesion is dependent on the nature of the contacting surface. The classic expression used to describe quantitatively the value of the adhesive wear debris (V) removed for a given load F, on a material of hardness H, over the sliding distance x, for a dry contact, is given by Archard's equation [55]-[56] as shown (12):

$$V = \frac{K \times x \times F}{3 \times H} \qquad (12)$$

where K is Archard's wear coefficient. This coefficient is considerably different for various materials. For gold contacts, Archard's wear coefficient is estimated to be 5×10^{-4} [57]. Assuming that sliding distance is on average 10% of the radius of asperities, zero wear limits can be determined for NEMS contact with various sizes (**Fig. 16 (b)**). In this graph, the vertical and horizontal axes represent the zero wear limit (N_0 in **Fig. 16 (a)**) and scaling trends, respectively. The zero wear limit decreases by scaling, which implies that the contacts in smaller NEMS devices will age considerably faster. This is another important implication of NEMS scaling indicating that reliability of these devices deteriorate significantly as their physical dimensions shrink.

VI.   IMPLICATIONS FOR CIRCUIT AND PROCESS DESIGN

Increasing contact resistances have significant circuit-level implications, which include higher resistances and hence, higher propagation delays. In this subsection, N-stage ring oscillators are used to investigate such circuit-level implications. An N-stage ring oscillator is composed of N (an odd number) inverters which are connected in a loop configuration. An N-stage oscillator has no stable state and hence, oscillates as long as it is connected to a power supply. The transistor level implementations of NEMS-based inverters are illustrated in [4]-[5]. The circuit simulations are performed using compact NEMS circuit models which are developed in previous studies [4]-[5]. For the sake of completeness, in the following subsection, a brief overview of the NEMS device models is presented.

A.   Compact Circuit Models

These compact models are illustrated in **Fig. 17** where the coupling capacitances between the beam and the Gate A and Gate B are denoted as C_A and C_B, respectively [5]. Furthermore, the voltage difference between Gate A and Gate B are labeled as V_A and V_B, correspondingly. If the device is in the OFF state, the value of C_A and C_B depend on V_A and V_B as indicated in **Fig. 17** (a)-(b) because the curvature of the beam (and hence, the coupling capacitance) is a function of the gate bias. Using the models proposed in [4], one can take into account the dependency of C_A and C_B on V_A and V_B. While both C_A and C_B are functions of the gate bias in the OFF state, these variables are independent of the gate voltage when the device is ON. This is because, in the ON state, the shape of the bended beam is invariant; therefore, the values of C_A and C_B are independent of the gate biases (as indicated in **Fig. 17** (c)).

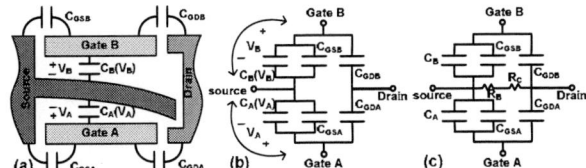

Fig. 17. Compact circuit models for laterally-actuated double-gate NEMS: a) the schematic of the device in the OFF state along with all capacitances, the compact model of the device in b) the OFF state and c) the ON state.

The parasitic capacitances between the gate and the source (C_{GSA} and C_{GSB}) and the gate and the drain (C_{GDA} and C_{GDB}) are also included in these compact models. Since these parasitic capacitances exist between stationary parts of the device, their values can be easily calculated using the parallel plate capacitance model. The source and the drain terminals are electrically isolated when the transistor is OFF (**Fig. 17** (b)), a conduction path is formed once the device turns ON (**Fig. 17** (c)). Note that there will be a contact resistance at the interface of the beam and the drain terminal when they come into contact. Therefore, this electrical connection can be modeled with a series combination of the beam resistance (R_B) and the contact resistance (R_C). The values of R_B and R_C can be calculated using existing models in the literature or extracted from the I_{DS} -V_{GS} characteristics of devices.

B.   Circuit Simulation Results

To illustrate the circuit-level impact of contact degradation, propagation delay of N-stage NEMS oscillators are evaluated here. Note that the delay of a NEMS device consist of two components: (1) mechanical delay: the time that is needed to deflect the beam and form a contact and (2) electrical delay: the time which is required to charge/discharge the output load. The mechanical delay of NEMS devices can also be integrated in the circuit model (though not shown in **Fig. 17**) using available analytical models [4]-[5]. The electrical delay of the NEMS transistors/circuits can be determined using commercial circuit simulators and the model proposed in **Fig. 17**.

Assuming the physical dimensions of the NEMS devices (**Table 1**) and the fact that the cantilever beam is made of metal elements with different resistivity (**Table 2**), the resistances of the beams are calculated. Using the methodology proposed in this work, the contact resistances (R_C) are estimated for each case. Moreover, the gate capacitances, C_A and C_B, as well as, other parasitic capacitances (C_{GSA}, C_{GSB}, C_{GDA} and C_{GDB}) are estimated.

Several N-stage ring oscillators are simulated using the above compact models. For instance, the simulation results for five-stage ring oscillators (N=5) are shown in **Fig. 18** where NEMS transistors are scaled according to the scaling scenario discussed before. These results indicate that the performance (inverse of the total delay) improves with scaling. Note that in NEMS-based circuits the total delay consists of the mechanical delay (needed to physically deflect the beam) and the electrical delay (required to charge/discharge the circuit nodes). Both delay components decrease by scaling, however, the mechanical delay component reduces at a faster rate. The reason is that while the electrical delay (RC delay) reduces due to smaller capacitive loads, the benefits of device scaling are largely lost as a result of the increase in the contact resistance. Therefore, as shown in **Fig. 18**, in scaled NEMS circuits, the performance will be dominated by the electrical delay (due to high contact resistances) rather than mechanical delay.

As discussed above, the electrical and mechanical properties of the metal contacts play a critical role in determining the contact performance. In order to investigate the impact of using different metals, the key electrical and mechanical properties of commonly used metal elements are summarized in **Table 2** from the literature.

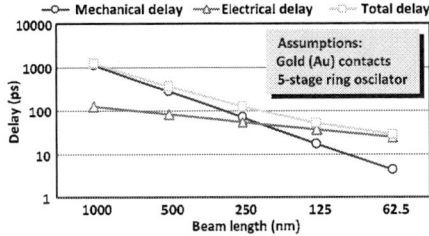

Fig. 18. Delay of a five-stage ring oscillator assuming gold contacts for the NEMS devices.

Table 2. Electrical and mechanical properties of various metals. elements, which can possibly be used to from the electrical contacts.

	Young's Modulus (GPa)	Hardness (GPa)	Resistivity ($\mu\Omega$.cm)	Archard's wear constant
Ag	50	1.5	1.6	2.2×10^{-4}
Al	70	1.2	2.8	4.1×10^{-4}
Au	78	1	2.2	5×10^{-4}
Cu	120	1.3	1.7	1.5×10^{-4}
Pt	184	5.4	10.5	0.2×10^{-4}
Rh	256	9.8	4.3	0.03×10^{-4}
Ru	292	15.3	7.1	0.01×10^{-4}
Ti	110	2.8	42	2.5×10^{-4}
W	405	4.1	5.3	0.4×10^{-4}

Using the proposed modeling framework and the data from **Table 2**, various metal elements are ranked in terms of contact conductance (inverse of resistance), reliability (lower probability of stiction) and useful lifetime (zero-wear limit) as shown in **Table 3**. It can be observed that there is no clear winner (no metal is ranked high on all three columns). Therefore, this study reveals that the choice of the contact-metal will be a key and challenging issue for NEMS device designers.

Table 3. Ranking of various metal elements in terms of electrical conductance (1/resistance), reliability (stiction free operation) and useful lifetime ('zero wear' limit).

Ranking	Contact conductance	Reliability (stiction-free)	Useful lifetime
#1	Ag	Ru	Ru
#2	Au	Rh	Rh
#3	Cu	W	Pt
#4	Al	Pt	W
#5	W	Ti	Ti
#6	Rh	Ag	Cu
#7	Pt	Al	Ag
#8	Ti	Au	Al
#9	Ru	Cu	Au

VII. CONCLUSIONS

This paper reports a comprehensive NEMS scaling analysis, which aims at investigating the impact of reducing NEMS transistors' physical dimensions on the performance and reliability of their metal contacts. An analytical modeling framework is developed, which can accurately take into account the effects of contact surface roughness, the elastic/plastic deformation of contacting asperities, and inter-molecular forces between metal surfaces including Van der Waals and capillary forces. It is shown that the contact resistances as well as the probability of stiction failures increase rapidly with scaling. The rate of increase of the contact resistance is considerably higher than the rate of reduction of contact area, which is predicted with a quadratic function. Moreover, for aggressively scaled NEMS, the surface forces become so strong compared to the restoring elastic forces that the contact adhesion will be inevitable. It is also found that the performance of NEMS devices will be mainly limited by high RC delays due to degraded contact resistances rather than their mechanical delay. More importantly, it is shown that there is no traditional metal that can be considered as "optimal" for simultaneously overcoming all these issues. Hence, contact performance and reliability degradation will be dominant factors limiting the scalability of NEMS devices.

VIII. REFERENCES

[1] H. F. Dadgour and K. Banerjee, "Design and analysis of hybrid NEMS-CMOS circuits for ultra low-power applications," *IEEE/ACM Design Automation Conference (DAC)*, 2007, pp. 306-311.

[2] H. F. Dadgour and K. Banerjee, "Hybrid NEMS-CMOS integrated circuits: A novel strategy for energy-efficient designs," *IET Transactions on Computers and Digital Techniques - Special Issue on Advances in Nanoelectronics Circuits and Systems*, vol. 3, no. 6, pp. 593-608, Nov. 2009.

[3] S. Chong, et al., "Nanoelectromechanical (NEM) relay integrated with CMOS SRAM for improved stability and low leakage," *IEEE International Conference on Computer-Aided Design* (ICCAD), pp. 478-484, 2009.

[4] H. F. Dadgour, M. M. Hussain, C. Smith and K. Banerjee, "Design and analysis of compact ultra low-power logic gates using laterally-actuated double-electrode NEMS," *IEEE/ACM Design Automation Conference (DAC)*, 2010, pp. 893-896.

[5] H.F. Dadgour, M.M. Hussain and K. Banerjee, "A new paradigm in the design of energy-efficient digital circuits using laterally-actuated double-gate NEMS," *IEEE International Symposium on Low Power Electronics and Design (ISLPED)*, 2010, pp. 7-12.

[6] H. Kam et al., "A new nano-electro-mechanical field effect transistor (NEMFET) design for low-power electronics," *IEDM*, 2005, pp. 463- 466.

[7] K. Akarvardar et al., "Design considerations for complementary nanoelectromechanical logic gates," *IEDM*, 2007, pp. 299-302.

[8] N. Abele et al., "Suspended-gate MOSFET: bringing new MEMS functionality into solid-state MOS transistor," *IEDM*, 2005, pp. 479- 481.

[9] M. Li, H. X. Tang and M. L. Roukes, "Ultra-sensitive NEMS-based cantilevers for sensing, scanned probe and very high-frequency applications," *Nature Nanotechnology.*, vol. 2, pp. 114–120, 2007.

[10] W. W. Jang et al., "Fabrication and characterization of a nanoelectromechanical switch with 15-nm-thick suspension air gap", *Appl. Phys. Lett.*, vol. 92, 103110, 2008.

[11] L. Duraffourg, et. al., "Compact and explicit physical model for lateral metal-oxide-semiconductor field-effect transistor with nanoelectromechanical system based resonant gate," *Appl. Phys. Lett.*, vol. 92, 174106, 2008.

[12] The International Technology Roadmap for Semiconductors (ITRS), 2009.

[13] S.T. Patton, and J.S. Zabinski, "Fundamental studies of Au contacts in MEMS RF switches" *Tribology Letters*, vol.18, pp. 215-30, 2005.

[14] R. Holm, *Electric Contacts: Theory and Applications*, 4th edition, Berlin, Springer, 1967.

[15] S. Majumder, N.E. McGruer, G.G. Adams, P.M. Zavracky, R.H. Morrison and J. Krim , "Study of contacts in an electrostatically actuated microswitch," *Sens. Actuators A*, vol. 93, pp. 19–26, 2001.

[16] O. Rezvanian, M. A. Zikry, C. Brown and J. Krim, "Surface roughness, asperity contact and gold RF MEMS switch behavior," *J. Micromech. Microeng.*, vol. 17, pp. 2006–2015, 2007.

[17] H. Kwon, S. -S. Jang, Y. -H. Park, T. -S. Kim, Y. -D. Kim, H. -J. Nam and Y. -C. Joo, "Investigation of the electrical contact behaviors·in Au-to-Au thin-film contacts for RF MEMS switches," *J. Micromech. Microeng.*, vol. 18, pp. 105010-105019, 2008.

[18] C. Brown, A. S. Morris III, A. I. Kingon, and J. Krim, "Cryogenic performance of RF MEMS switch contacts," *Journal of Microelectromechal Systems*, vol. 17, pp. 1460-1467, 2008.

[19] E. J. J. Kruglick, and K. S. J. Pister, "Lateral MEMS microcontact considerations," *Journal of Microelectromechal Systems*, vol. 8, pp. 264-271, 1999.

[20] D. Hyman and M. Mehregany, "Contact physics of gold microcontacts for MEMS switches," *IEEE Transactions on Components and Packaging Technologies*, vol. 22, pp. 357-364, 1999.

[21] N.E. McGruer, G.G. Adams, L. Chen, J. Guo, and Y. Du, "Mechanical, Thermal, and material influences on ohmic-contact-type MEMS switch operation," *IEEE International Conference on MEMS*, 2006, pp. 230-233.

[22] J. Tringe, W. Wilson, and J. Houston, "Conduction properties of microscopic gold contact surfaces," Reliability, Testing, and Characterization of MEMS/MOEMS, *Proceedings of SPIE*, vol. 4558, pp. 151-158, 2001.

[23] D. J. Dickrell and M. T. Dugger, "Electrical contact resistance degradation of a hot-switched simulated metal MEMS contact," *IEEE Transactions on Components and Packaging Technologies*, vol. 30, pp. 75-80, 2007.

[24] S. Majumder, N.E. McGruer, and G.G. Adams, "Adhesion and contact resistance in an electrostatic MEMS microswitch," *18th IEEE International Conference on Micro Electro Mechanical Systems*, MEMS 2005, pp. 215-218.

[25] C. Brown, O. Rezvanian, M. A. Zikry and J. Krim, "Temperature dependence of asperity contact and contact resistance in gold RF MEMS switches," *J. Micromech. Microeng.*, vol. 19, pp. 025006-025014, 2009.

[26] J. Israelachvili, *Intermolecular and Surface Forces*, 2nd ed., Academic, London, 1991.

[27] S. L. Miller, M. S. Rodgers, G. LaVigne, J. J. Sniegowski, P. Clews, D. M. Tanner, and K. A. Peterson, "Failure modes in surface micromachined microelectromechanical actuators", *IEEE International Reliability Physics Symposium Proceedings (IRPS)*, 1998, pp. 17-25.

[28] J. A. Walraven, "Future challenges for MEMS failure analysis," *International Test Conference*, 2003, pp. 805-855.

[29] S. M. Allameh, "An introduction to mechanicalproperties-related issues in MEMS structures," *Journal of Material Science*, vol. 38, pp. 4115–4123, 2003.

[30] D. Tabor, "Surface forces and surface interactions," *Journal of Colloid and Interface Science*, vol. 58, pp. 2-13, 1976.

[31] A. Greenwood and J. P. Williamson, "Contact of nominally flat surfaces," *Proceedings of the Royal Society (London)*, A295, pp. 300-319, 1966.

[32] J.A. Greenwood and J.H. Tripp, The contact of two nominally flat rough surfaces, *Proc. Inst. Mech. Eng.*, vol. 185, pp. 625–633, 1971.

[33] R. Maboudian and C. Carraro, "Surface chemistry and tribology of MEMS," *Annual Review of Chemical Physics*, vol. 55, pp. 35–54, 2004.

[34] U. Srinivasan, M. R. Houston, R. T. Howe, R. Maboudian, "Alkyltrichlorosilane-based self-assembled monolayer films for stiction reduction in silicon micromachines," *Journal of Microelectromechal Systems*, vol. 7, pp. 252–260, 1998.

[35] B. H. Kim, T. D. Chung, C. H. Oh, & K. Chun, "A new organic modifier for anti-stiction," *Journal of Microelectromechal Systems*, vol. 10, pp. 33–40, 2001.

[36] W. R. Ashurst, C. Yau, C. Carraro, C. Lee, G. J. Kluth, R. T. Howe, R. Maboudian, "Alkene based monolayer films as anti-stiction coatings for polysilicon MEMS," *Sensors and Actuators A: Physical*, vol. 91, pp. 239–248, 2001.

[37] F.W. DelRio, M.P. De Boer, J.A. Knapp, E.D. Reedy, P.J. Clews, M.L. Dunn, "The role of van der Waals forces in adhesion of micromachined surfaces," *Nat. Mater.*, vol. 4, pp. 629–634, 2005.

[38] Y. Rabinovich, M. Esayanur, and B. Moudgil, "Capillary forces between two spheres with a fixed volume liquid bridge: theory and experiment," *Langmuir*, vol. 21, pp. 10992–10997, 2005.

[39] X. Xiao and L. Qian, "Investigation of humidity-dependent capillary force," *Langmuir*, vol. 16, pp. 8153–8158, 2000.

[40] F. London, "The general theory of molecular forces," *Trans. Faraday Soc.*, vol. 33, pp. 8–26, 1937.

[41] E. M. Lifshitz, "The theory of molecular attractive forces between solids" *Sov. Phys. JETP 2*, vol. 73, pp. 73-83, 1956.

[42] K. L. Johnson, K. Kendall, and A. D. Roberts, *Proc. R. Soc. London A*, vol. 324, pp. 301-313, 1971.

[43] H. Hertz, "The contact of elastic solids," *J. Reine Angew. Math.*, vol. 92, pp. 156–171, 1881.

[44] W. R. Chang, I. Etsion, and D. B. Bogy, "An elastic-plastic model for the contact of rough surfaces," *Journal of Tribology*, vol. 109, pp. 257-263, 1987.

[45] B. V. Derjaguin, V. M. Muller, and Y. P. Toporov, *J. Colloid Interface Sci.*, vol. 53, pp. 314-326, 1975.

[46] B. N. J. Persson and E. Tosatti, "The effect of surface roughness on the adhesion of elastic solids," *Journal of Chemical Physics*, vol. 115, pp. 5597-5610, 2001.

[47] W. Wang, Y. Wang, H. Bao, B. Xiong, and M. Bao, "Friction and wear properties in MEMS," *Sensors and Actuators A*, vol. 97-98, pp. 486–491, 2002.

[48] Y. V. Sharvin, "A possible method for studying Fermi surfaces," *J. Exp. Theor. Sci.*, vol. 21, pp. 655–656, 1965.

[49] G. Wexler, "The size effect and the non-local Boltzmann transport equation in orifice and disk geometry," *Proc. Phys. Soc.*, vol. 89, pp. 927–941, 1966.

[50] B. Nikolic and P. B. Allen, "Electron transport through a circular constriction," *Phys. Rev. B*, vol. 60, pp. 3963–3969, 1999.

[51] A. G. M. Jansen, A. P. van Gelder and P. Wyder, "Point-contact spectroscopy in metals," *J. Phys. C: Solid State Phys.*, vol. 13, pp. 6073–118, 1980.

[52] D. M. Tanner and M.T. Dugger, "Wear mechanisms in reliability methodology," *Proceedings of the SPIE: The International Society for Optical Engineering*, vol. 4980, pp. 22-40, 2003.

[53] M. Mehregany, S.D. Senturia and J.H. Lang, "Measurement of wear in polysilicon micromotors," *IEEE Transactions on Electron Devices*, vol. 39, pp. 1136-1143, 1992.

[54] A. D. Corwin and M.P. de Boer, "Effect of adhesion on dynamic and static friction in surface micromachining," *Applied Physics Letters*, vol. 84, pp. 2451-2453, 2004.

[55] J. F. Archard, "Contact and rubbing of flat surfaces", *J. Applied Physics*, vol. 24, pp. 981-988, 1953.

[56] P. A. Engel, *Impact Wear of Materials*, Elsevier, 1976.

[57] R. Lewis, "A modeling technique for predicting compound impact wear," *Wear*, vol. 262, pp. 1516–1521, 2007.

The Effect of Temperature on Dielectric Charging of Capacitive MEMS

M. Koutsoureli, L. Michalas and G. Papaioannou

Solid State Physics Section, Physics Dpt.
University of Athens
Athens, Greece
+30-210-7276722, mkoutsoureli@phys.uoa.gr
+30-210-7276722, lmichal@phys.uoa.gr
+30-210-7276817, gpapaioan@phys.uoa.gr

Abstract— **The present paper investigates the effect of temperature on the charging process in dielectric films of MEMS capacitive switches. The investigation includes the assessment of MIM capacitors and MEMS capacitive switches. The data analysis shows that the dielectric charging is thermally activated and the process can be described by a system with a wide distribution of relaxation times that exhibits power-law relaxation. The activation energies obtained from MIM and MEMS are attributed to different charge collection mechanisms.**

Keywords-component; capacitive MEMS; dielectrics; dielectric charging;

I. INTRODUCTION

The effect of dielectric charging constitutes a major reliability issue in MEMS capacitive switches. In spite of the intensive research during the last decade, the effect is still not fully understood. The presence of a high electric field during actuation gives rise to charge injection that in turn shifts the capacitance voltage characteristic [1] and limits the device lifetime, the latter depending significantly on the actuation voltage magnitude [2]. The dielectric films used in MEMS are deposited at low temperatures thus leading to highly disordered materials, which deviate from stoichiometry and thus contain large concentrations of defects and dipoles [3,4]. Since charge trapping and emission as well as oriented dipole randomization are thermally activated, the dielectric charging is expected to be strongly affected by temperature. The presence of thermally activated mechanisms was first reported on the temperature dependence of switch-on capacitance transient [5] and later on the shift of bias for capacitance minimum with temperature [6]. In the case of SiO_2 MIM (Metal-Insulator-Metal) capacitors, the discharge current transient method revealed two charge trapping mechanisms [7] while in the case of Si_3N_4 devices, the trapping mechanisms were found to exhibit wide distributions around certain activation energies [8]. Here it must be pointed out that MIM capacitors, in spite of the persisting disagreements, may behave like MEMS when instead of measuring the short circuit current in the external circuit we measure the potential of the floating electrode [9]. Beyond this, no direct correlation has been made between the thermally activated mechanisms in MIM capacitors and MEMS.

The aim of the present work is to show that the charging in silicon nitride MEMS is thermally activated and to correlate the mechanisms determined in MIMs with those in MEMS in order to provide an easy and efficient tool to predict the charging mechanisms hence the lifetime of MEMS switches. Finally the aim of the present work is to show that the charge redistribution through the hopping process plays a significant role in the dielectric charging.

II. THEORETICAL BACKGROUND

The dielectric charging process in amorphous thin insulating films, such as those used in MEMS capacitive switches constitutes a complex process that involves tunneling of metal contact electrons into traps located in the dielectric film band gap and vice versa as well as the redistribution of the injected and existing charges through the Poole-Frenkel effect as well as hopping through traps in the bulk material.

In such materials the trap distribution in energy and space in amorphous dielectrics is not well determined and depends strongly on the film deposition conditions, e.g., in SiN_x [10]. If for the sake of simplicity we assume a single trap level at an energy E_t below the conduction band the distribution of trapped electrons with distance in the dielectric will depend on the trap distribution in space, the capture cross section of the traps, and the tunneling current density during injection. For low injection levels, which is when the concentration of traps (N_t) is large compared to the number of injected electrons and/or when the traps have a small capture cross-section, the spatial distribution of trapped electrons on a first approach will be given by:

$$n(x) = N_0 \exp\left(-\frac{x}{\lambda}\right) \qquad (1)$$

where λ is the mean free path for trapping in the nitride during injection. N_0 ($< N_t$) depends on the number of injected electrons [11]. For the case of amorphous dielectrics and under the presence of electric field the injected charge will be derived by the integration over the band gap [12]. The integration must take into account the modification caused by

---

The paper has been partially supported by ESA projects AO/1-5288/06/NL/GLC, AO/1-5851/08/NL/NA and ENIAC-2010 NANOCOM.

the trap assisted tunnelling and Poole-Frenkel effect transport [12].

The discharge current primarily takes place through tunneling towards the injecting electrodes and simultaneously through transport/diffusion towards the opposite electrodes. This gives rise to two currents with opposite direction, towards the injecting electrodes and through the film, which diminish the measured current in the external circuit. Moreover, because as a rule no external field is applied during discharge, the process is more complex because it may arise from dipole-dipole interactions, anisotropy of the internal field in which the dipoles are reoriented, the random walking of sequential trapping, and emission of the charges diffusing towards the contacts, etc. [13, 14] Here it must be emphasized that these processes are plausible in any disordered material with a high dipole and/or defect concentration.

A. Thermally Stimulated Depolarization Current (TSDC)

The thermally stimulated depolarization current (TSDC) technique is used as an efficient tool for the investigation of dipolar and dc relaxation phenomena observed by heating a variety of materials over a wide temperature range after polarization by an electrostatic field at a polarization temperature (T_p). The observed phenomena occur due to the orientation sensitivity of bond dipoles and charges (electrons and ions) to the external electrostatic field.

During temperature scan, the current density produced by the progressive decrease in polarization in the course of a TSDC experiment, where time and temperature are simultaneously varied, is approximated by [13]:

$$J_D(T) \approx \frac{P_S(T_p)}{\tau_0} \cdot \exp\left(-\frac{E}{kT}\right) \cdot \exp\left[-\frac{1}{\gamma\tau_0} \cdot \frac{kT^2}{E} \exp\left(-\frac{E_A}{kT}\right)\right] \quad (2)$$

where τ_0^{-1} is the escape attempt frequency related to the lattice, P_S is the steady state polarization, E_A the activation energy of the contributing polarization mechanism, γ the heating rate and

$$\tau(T) = \tau_0 \cdot \exp\left(\frac{E_A}{kT}\right) \quad (3)$$

the thermally activated relaxation time for each contributing mechanism with a corresponding activation energy E_A [15]. The electric charge density Q produced in the external circuit during the depolarization (heating) stage in a TSDC experiment can be calculated by integration over the TSDC spectrum (from T_0 to T_f):

$$Q = \frac{A}{\gamma} \cdot \int_{T_0}^{T_f} J(T) \cdot dt \quad (4)$$

where A is the capacitor area.

B. Kelvin Probe method

The current through the dielectric film has been calculated from the decay of the potential of the top electrode in open circuit mode. This procedure simulates discharge current measurement with blocking electrodes which due to the very low value of discharge current cannot be measured in the conventional way. The use of an ammeter in the femto-Ampere range is limited by noise and in MEMS capacitive switches it was shown to allow the current measurement for a relatively short time interval (t ≤ 200 sec) after bias removal [16].

This limitation is overcome by employing the Kelvin Probe method, which can be used to measure the potential decay of the top electrode of the MIM capacitor [9, 17]. The average discharge current density through the dielectric film, if surface leakage current is negligible, will be given by

$$J(t) = -\frac{\varepsilon_r \varepsilon_0}{d} \cdot \frac{dV(t)}{dt} \quad (5)$$

where ε_r and ε_0 are the relative and vacuum dielectric constants respectively, d the dielectric film thickness and V the applied bias to null the electric field between the MIM capacitor top electrode and the Kelvin probe oscillating electrode. Here it must be pointed out that the applied bias V must also compensate the electrodes work functions, which according to (5) does not affect the discharge current measurement.

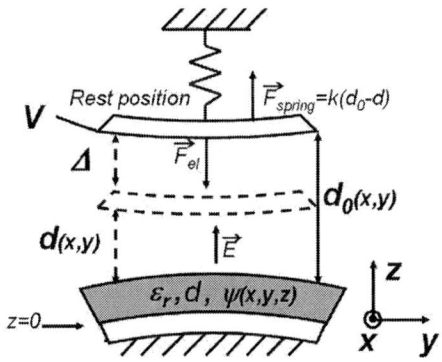

Figure 1. Model of a capacitive switch with non uniform trapped charge and air gap distributions [18]

C. MEMS switches

The determination of the average charge in a MEMS capacitor is based on the model presented by X. Rottenberg et al. in [18]. For this we consider the setup in Fig. 1 that includes a fixed non-flat metal plate of area A that is covered with a dielectric layer of uniform thickness d, dielectric constant ε_r, and volume charge density $\psi(x, y, z)$. Above it a rigid but nonflat movable metal plate is fastened with a linear spring k to a fixed wall above the dielectric layer at a rest position $d_0(x, y)$. A dc bias source of amplitude V is applied to the two plates. Following the procedure analyzed in [18] we find that the electrostatic force F_{el} can be written in a compact form of

$$F_{el}(\Delta) = \frac{A}{2\varepsilon_0}\left[(V\mu_\alpha - \mu_\beta)^2 + V^2\sigma_\alpha^2 + \sigma_\beta^2 - 2V\,\mathrm{cov}(\alpha,\beta)\right] \quad (6)$$

where μ, σ^2, and cov denote the mean, variance, and covariance, respectively, of the $\alpha(x, y, \Delta)$ and charge $\beta(x, y, \Delta)$ distributions:

$$\alpha(x,y,\Delta) = \frac{\varepsilon_0}{(d_0(x,y)-\Delta)-\dfrac{d}{\varepsilon_r}} \quad (7)$$

which is the distribution of capacitance per unit area and

$$\beta(x,y,\Delta) = \frac{d}{\varepsilon_r\varepsilon_0}\cdot\psi_{eq}(x,y)\cdot\alpha(x,y) \quad (8)$$

the distribution of charge density induced at armature area and $\psi_{eq}(x,y)$ and Δ are the equivalent surface charge distribution and the displacement from equilibrium respectively. The equilibrium position of the system is determined by equating the electrostatic and spring forces.

The experimental data that can be exploited from a C-V characteristic are the pull-in and pull-out voltages as well as the bias at which the capacitance, in the up state attains a minimum (V_{Cmin}). In an experiment involving measurements at different temperatures the pull-in and pull-out voltages need to be treated with care since the mechanical properties of the switch may vary significantly with temperature. A parameter that depends only on the average value of the charge at the dielectric surface is the bias for capacitance minimum (V_{Cmin}). The value of V_{Cmin} is obtained from:

$$\frac{d}{dV}\left[(V\mu_\alpha - \mu_\beta)^2 + V^2\sigma_\alpha^2 + \sigma_\beta^2 - 2V\,\mathrm{cov}(\alpha,\beta)\right] = 0 \quad (9)$$

thus

$$V_{Cmin} = \frac{\mu_\alpha\mu_\beta + \mathrm{cov}(\alpha,\beta)}{\mu_\alpha^2 + \sigma_\alpha^2} \quad (10)$$

The variance of both the capacitance and the charge non uniform distributions, in (6) and (10), determine the switch behavior [18]. The variance of the capacitance distribution can be controlled by device design or selection. In contrast the variance due to non uniform charge distribution is unpredictable and may vary with time, as demonstrated with Kelvin Probe Force Microscopy method assessment [19].

For a switch with parallel plates or small capacitance variance Eq.10 reduces to

$$V_{Cmin} = \frac{\mu_\beta}{\mu_\alpha} = \frac{d}{\varepsilon_r\varepsilon_0}\cdot\overline{\psi_{eq}} \quad (11)$$

where $\overline{\psi_{eq}}$ is the dielectric film average equivalent surface charge density.

D. Hopping transport

In all theoretical descriptions of transport in disordered insulators and semiconductors, hopping transitions between localized states in the exponential band tails $\left[g(E) = \dfrac{N_0}{E_0}\exp\left(-\dfrac{E}{E_0}\right)\right]$, where $g(E)$ is the density of states and N_0 the total concentration of tail states, play the decisive role and determine the temperature dependence of transport, provided the thermal energy kT is small comparing to the characteristic energy of the band tail E_0. At higher temperatures, all these approaches converge to the conventional multiple-trapping model in which hopping between localized states is unimportant, the movement of carriers through extended states determining the current. The field non-linearities of the transport are most pronounced at low enough temperatures, when transport is determined by hopping, see [19] and references therein. Under strong electric field B.I. Shklovskii [21] has shown that the electric field plays a role similar to that of temperature, see Fig. 2.

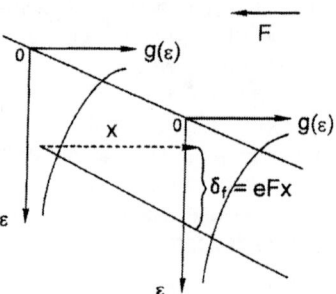

Figure 2. Electron hop over distance x in the presence of high electric field [20].

In order to obtain the field dependence of the conductivity $\sigma(F)$, F is the electric field intensity, at high fields he replaced the temperature T in the well known dependence $\sigma(T)$ for low fields by a function $T_{eff}(F)$ of the form:

$$T_{eff} = \frac{qF\alpha_r}{2k} \quad (12)$$

where α_r is the decay length of the wave function in the tail states and the other parameters have the usual meaning. According to this the hopping rate ν_{ij} from a site i to a site j is governed by

$$\nu_{ij} = \nu_0\exp\left(-\frac{2r_{ij}}{\alpha_r}\right)\exp\left[-\frac{(E_j - E_i) - qF(x_j - x_i)}{kT}\right] \quad (13a)$$

for $E_j - E_i - qF(x_j - x_i) > 0$, and

$$\nu_{ij} = \nu_0\exp\left(-\frac{2r_{ij}}{\alpha_r}\right) \quad (13b)$$

for $E_j - E_i - qF(x_j - x_i) < 0$, where r_{ij} is the distance between the sites involved, x_i and x_j their coordinates, v_0 the attempt-to-escape frequency and $(E_j - E_i)$ the energy difference between these sites.

III. EXPERIMENTAL PROCEDURE

The dielectric charging has been monitored in both MIM capacitors and MEMS capacitive switches. The dielectric film was silicon nitride deposited with PECVD method at 250C. XPS assessment revealed a N/Si ratio of about 0.81. The MIMs have symmetric metallic contacts (Ti/SiNx/Ti) of 1mm diameter and thickness of 200nm. The assessment was performed by charging the dielectric with electric fields ranging from 1 MV/cm to 2MV/cm at 450K for 15 min. Then the devices were cooled to 200K under electric field. The, short circuit, thermally stimulated depolarization current (TSDC) was measured by ramping the temperature range of 200K to 450K with a constant rate of 2.5K/min. The current was measured with a Keithley 6487 voltage source current measure system. The surface potential was measured with a Kelvin probe setup after 10 min charging under electric field of 1MV/cm at room temperature. The resolution of the system was 1mV.

The MEMS switches, with dielectric film of 200nm and 430nm, were assessed in the temperature range of 300K to 450K. The bias rate was about 60mV/sec and bias step 50mV. Each C-V characteristic was swept with bias ascending in the range of -25V to 25V. Finally the pull-in voltages were about ±20V and pull-up voltages of about ±10V. The dependence of the shift of bias for minimum capacitance (V_{Cmin}) as a function of temperature was monitored in the pull-up state and during the sweep ascending from negative actuation to positive. The choice of studying V_{Cmin} was based on the fact that the condition for minimum electric field is not affected by any bridge thermal expansion and/or deformation.

IV. RESULTS AND DISCUSSION

A. MIM capacitors assessment

The TSDC spectra were obtained for both electric field polarities and revealed asymmetric charging in spite of the symmetric electric contacts (Fig. 3a). Furthermore the structure of the spectra shows that below room temperature (~275K) a discrete contribution prevails. Here it must be pointed out that processes occurring below 300K in the TSDC assessment are characterized by short time constants at room temperature [8] and therefore do not contribute significantly to the device lifetime. According to this, at higher temperatures, due to the fact that the charging/discharging mechanisms are thermally activated (3), the very long time constants, which determine the device lifetime decrease [8] and therefore can be monitored within the experiment time window. Above room temperature the TSDC structure arises from a continuous distribution of contributing discharge mechanisms.

A detailed analysis assuming three contributions (C_1, C_2, C_3), based on (2), have been fitted to experimental data and are presented in Fig. 3b.

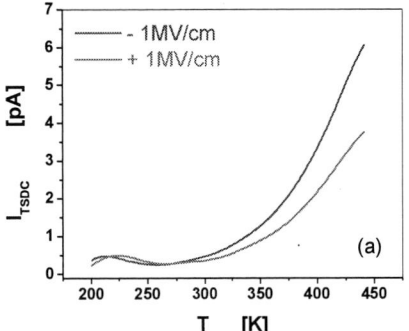

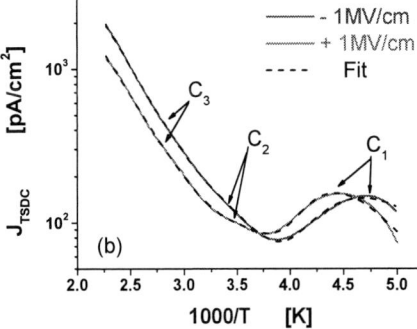

Figure 3. TSDC spectra for 200nm SiN (a) as a function of temperature and (b) as a function of 1/T to determine the corresponding activation energies E_A and time constants τ.

The fitting results, which correspond to activation energies and time constants normalized with (3) to room temperature, are presented in Table 1.

TABLE 1. CHARACTERISTIC OF CHARGING PROCESSES

	Positive	Negative
E_1 (eV)	0.116	0.136
τ_1 (sec)	5.9×10^{-5}	3.9×10^{-3}
E_2 (eV)	0.203	0.155
τ_2 (sec)	0.98	3.8×10^{-2}
E_3 (eV)	0.219	0.235
τ_3 (sec)	-	-

No time constant was determined for mechanism C_3 because no peak was detected and $\tau_{0\infty}$ could not be determined through the Arrhenius plot. This behavior clearly denoted that above room temperature the TSDC structure is composed of self-similar combination of Debye-relaxing sub-processes, which describe a system with a wide distribution of relaxation times that exhibits power-law relaxation, thus revealing the depolarization of a fractal dielectric [22]. In this temperature regime the TSDC current can be described by:

$$I_{TSDC}(T) \approx P(0) \cdot \tau_{0\infty}^{-1} \exp(-E/kT) \qquad (14)$$

where E is the apparent activation energy, E_3 in present case, and τ_0 the time constant for assuming an infinite number of relaxing sub-processes.

The collected charge during the TSDC sweep was obtained by integrating the current in the external circuit over the temperature range, using (4), taking into account that the temperature was linearly swept. The dependence of collected charge on the applied electric field exhibits a rather linear dependence on the charging electric field intensity (Fig. 4).

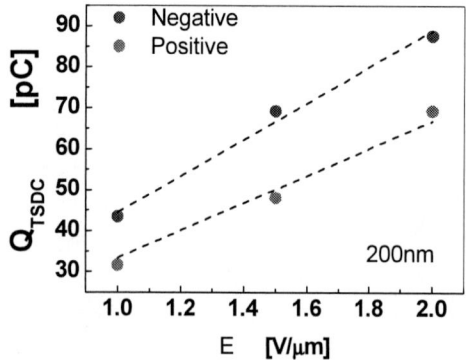

Figure 4. Dependence of charge collected during a TSDC sweep on charging electric field

This result is in good agreement with the Discharge Current Transient (DCT) method results obtained in SiNx deposited under different conditions, Fig. 9 of [23], and KPFM (Kelvin Probe Force Microscopy) assessment performed after charging with similar electric field intensity [24]. This asymmetry is always observed in silicon rich SiN films and SiN MEMS capacitive switches, which degrade even under bipolar cycling.

This is attributed to the presence of a defect band or differences in material quality [25, 26] located near the bottom electrode and introduced during deposition and the favored hole over electron transport because of the much higher density of tail states in the valence band compared with the conduction band [26].

As already mentioned, in TSDC assessment the measured discharge current arises from the algebraic sum of the charge flowing towards the injecting contacts and the charge flowing through the dielectric film and collected by the opposite contacts. In order to determine the contribution of the two currents we employed the Kelvin probe method to measure the top electrode potential decay and calculate the current through the dielectric film using (5). The calculated current density is presented in Fig. 5 and clearly shows that its magnitude is several orders lower than the TSDC one. This leads us to the conclusion that the injected and trapped charges are located close to the injecting electrodes, so that they can be collected by the injecting electrodes and a rather negligible part by the bottom electrode. This result is in good agreement with

preliminary calculations assuming the presence of band tails, similar to amorphous silicon, and current injection through Trap-Assisted-Tunneling and charge redistribution through Poole Frenkel effect. The latter is the dominant mechanism at room temperature and under the presence of high electric fields ($\geq$1MV/cm) as shown by S.P. Lau et al. in [28] and

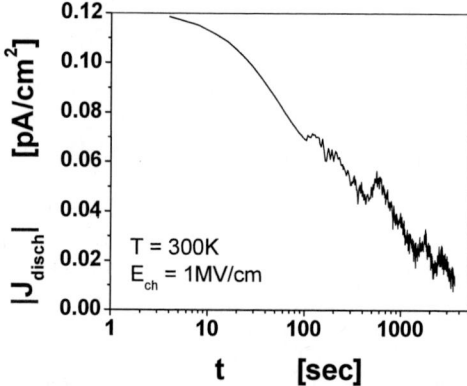

Figure 5. Discharge current transient at room temperature using blocking electrodes

demonstrated in the dependence of leakage current in charging current transient assessment of MIM capacitors [29]. According to hopping transport theory the calculations revealed that the injected charges are distributed close to the injecting electrodes [27].

A comparison between the currents measured in the external circuit during TSDC assessment and those calculated from the Kelvin probe assessment reveals a ratio (J_{TSDC}/J_{disch}) of three orders of magnitude. This leads us to the conclusion that the TSDC assessment is not affected by the MIM capacitor intrinsic leakage, and thus the calculated activation energies in Table 1 are not affected by other processes. Moreover, this suggests that the injected charge must be located close to injecting electrodes because any charge distribution across the film thickness would lead to a much smaller current ratio.

B. MEMS Capacitive Switches

Fig. 6 shows the temperature dependence of the shift of V_{Cmin}, for MEMS capacitive switches with dielectric film thicknesses of 200nm and 430nm is presented in Fig. 6. The experimental data clearly show that the charging during pull-down state is thermally activated process with apparent activation energy of about 0.42eV. Here it must be pointed out that the activation energy was determined in the temperature range (T $\geq$ 370K) where the data show minimum dispersion.

Comparing the temperature evolution of the bias for minimum capacitance we conclude that the responsible mechanism must be similar to the one responsible for the contribution C_3 in MIM capacitors; the C_1 and C_2 are characterized by very short time constants which cannot be observed under the large observation time window of this experimental procedure. The fact that the activation energy is larger suggests a different self-similar combination of Debye-relaxing sub-processes and taking into account the distribution of time constants presented in [8] we may argue that the present behavior can be related to the center/s E_2 (~0.37eV) in [8] which was/were detected after charging with lower electric field, of 0.2MV/cm.

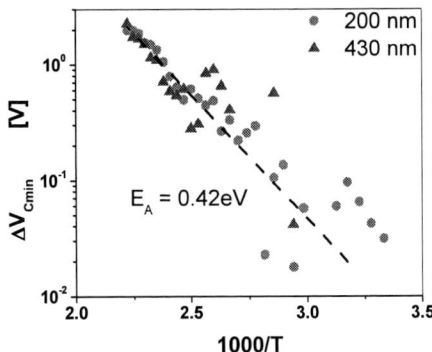

Figure 6. Arrhenius plot of the shift of V_{Cmin} with temperature

In order to obtain a better understanding of the observed different activation energies, it is essential to bear in mind that:
i.   in MEMS the charge injection occurred under high electric field and temperature range of 300K to 450K
ii. the transport towards the bottom electrode takes place under intrinsic electric field, which is lower than the injecting one and decreases continuously with time
iii. during charging, both the temperature and electric field are high so that the hopping effect can be considered negligible leaving the Poole-Frenkel effect to control the charge redistribution [20]
iv. after pull-out and during the C-V sweep until the minimum capacitance is attained, the discharge is performed under low electric field.

Taking all these into account and the fact that the bias for minimum capacitance is determined after a time interval of about 170sec, roughly determined from the value of pull-up voltage and sweep rate, we may assume that the majority of the charges trapped in shallow levels (C_3) are released and redistributed. This process is assisted by the intrinsic electric field barrier lowering through Poole-Frenkel, which is more efficient for shallow levels than deeper ones. Here it must be pointed out that such a process does not lead to charge collection by the bottom electrode but the emitted charges are expected to be trapped in empty sites deeper in the dielectric film and not affected by direct tunneling charge injection form metallic contacts.

Based on the foregoing discussion, we propose that the strong temperature dependence (0.42eV) of the shift of bias for minimum capacitance, which is proportional to the equivalent mean charge density at the dielectric film surface (11), arises from charge redistribution by hopping. This hypothesis is also supported by reported observations on charge trapping instabilities in amorphous silicon-silicon nitride thin-film transistors [30] with 0.5µm thick SiN gate and applied gate field of 0.24MV/cm. In these devices the threshold voltage was found to be thermally activated, with activation energy 0.3eV. There the observation of the threshold voltage shift was attributed to a more favorable electron hop from filled states to empty ones and the temperature dependence was attributed to thermally activated tunneling, i.e., hopping within a band of states at least 0.3eV wide. For a continuum of trap states, which agrees with the present work results, the hopping will be variable rather than nearest neighbor.

The different activation energies observed in MIMs and MEMS may be attributed to the contribution from free surface charging in MEMS, a phenomenon that does not occur in MIM capacitors and where the observed charging arises from the bulk material near the injecting contacts. In MEMS the dielectric material free surface contribution to charging is significant. As already demonstrated, the charging in MEMS is strongly affected by the ambient environment [31,32]. Although the ambient effect has not been intensively investigated, it has to be attributed to surface free bonds. Moreover, the fact that the dielectric material is silicon rich [31] may lead us to the conclusion that the surface charging is significantly affected by the silicon free bonds since in a nitrogen environment the charging decreases significantly [31,32]. Beyond this there is no concrete information available, and the contribution of free surface to charging requires further investigation.

Finally, the fact that charging in MEMS is thermally activated allows us to draw the conclusion that the parameters affecting the device lifetime are thermally activated too. This has a significant impact on the reliability of MEMS since it allows us to detect the contributions responsible for the device aging from their "signature", which is the Arrhenius plot. Moreover, this practically means that it will be possible to derive a temperature accelerated lifetime test procedure, rather in a similar way like in semiconductor devices. Obviously, in order to achieve this several issues need to be addressed, e.g., the significance of surface charging over the bulk charging, the impact of controlled environment of a sealed package, etc.

V.   CONCLUSIONS

In conclusion it has been demonstrated that the dielectric charging in silicon nitride based MEMS is thermally activated. The dielectric material electrical properties studied with the TSDC method revealed that silicon nitride exhibits a self-similar combination of Debye-relaxing sub-processes, which describe a system with a wide distribution of relaxation times that exhibits power-law relaxation revealing a fractal dielectric. Moreover, it was demonstrated that the TSDC assessment is not affected by the charge redistribution in the dielectric film bulk, because the latter gives rise to currents of three orders of magnitude lower than the TSDC one. In MEMS the charging monitored from the temperature dependence of bias for minimum capacitance, which is

978-1-4244-9113-1/11 $26.00 © 2011 IEEE

proportional to the average equivalent surface charge on the top surface of the dielectric, revealed a different and large activation energy. This has been attributed to hopping in a continuum of trap states where the hopping is expected to be variable rather than nearest neighbor. Finally the fact that the charging consists of thermally activated mechanisms allows us to conclude that it is possible to detect and monitor, through Arrhenius plot, the parameters that determined the device lifetime and look forward for a temperature accelerated lifetime test bench.

ACKNOWLEDGMENT

The authors would like to acknowledge the partial support from ESA projects AO/1-5288/06/NL/GLC "High Reliability MEMS Redundancy Switch" and AO/1-5851/08/NL/NA on "Avoidance of dielectric charge trapping" as well as ENIAC-2010 NANOCOM project.

REFERENCES

[1] J. Wibbeler, G. Pfeifer and M. Hietschold, Parasitic charging of dielectric surfaces in capacitive microelectromechanical systems (MEMS), Sensors and Actuators A vol. 71, pp 74-80, 1998

[2] C. Goldsmith, J. Ehmke, A. Malczewski, B. Pillans, S. Eshelman, Z. Yao, J. Brank, and M. Eberly, Lifetime Characterization of Capacitive RF MEMS Switches, IEEE MTT-S Digest pp.227-230, 2001

[3] G. Papaioannou, "The Impact of Dielectric Material and Temperature on Dielectric Charging in RF MEMS Capacitive Switches" in NATO Science for Peace and Security Series – B: Physics and Biophysics, E. Gusev, E. Gurfankel and A. Dideikin Eds., Springer, 2009, pp. 141-153

[4] G. Papaioannou, "Dielectric Charging" in Advanced RF MEMS, S. Lucyszyn Ed., Cambridge University Press, 2010, pp.140-187

[5] G. Papaioannou, M. Exarchos, V. Theonas, G. Wang, and J. Papapolymerou, Temperature Study of the Dielectric Polarization Effects of Capacitive RF MEMS Switches, IEEE Trans. on Microwave Theory and Techniques, vol. 53, pp 3467-3473, 2005

[6] G. Papaioannou, J. Papapolymerou, P. Pons and R. Plana, Dielectric charging in radio frequency microelectromechanical system capacitive switches: A study of material properties and device performance, Applied Physics Letters 90, pp 233507, 2007

[7] X. Yuan, Z. Peng, J. C. M. Hwang, David Forehand and Charles L. Goldsmith, Acceleration of Dielectric Charging in RF MEMS Capacitive Switches, IEEE Trans. on Device and Materials Reliability, 6, pp 556-563, 2006

[8] E. Papandreou, M. Lamhamdi, C.M. Skoulikidou, P. Pons, G. Papaioannou and R. Plana, Structure dependent charging process in RF MEMS capacitive switches, Microelectronics Reliability 47, pp 1812–1817, 2007

[9] U. Zaghloul, M. Koutsoureli, H. Wang, F. Coccetti, G. Papaioannou, P. Pons, R. Plana, Assessment of dielectric charging in electrostatically driven MEMS devices: A comparison of available characterization techniques", Microelectronics Reliability 50, pp.1615–1620, 2010

[10] J. Robertson and M.J. Powell, "Gap states in Silicon Nitride", Applied Physics Letters 44, 415, 1984

[11] L. Lundkvist, I. Lundstrom and C. Svensson, "Discharge of MNOS Structures", Solid-State Electronics, vol. 16, pp. 811-823, 1973

[12] R. Ramprasad, "Phenomenological theory to model leakage currents in metal-insulator-metal capacitor systems", Physica Status Solidi (b) 239, 2003, 59–70.

[13] J. Vandershueren and J. Casiot, P. Braunlich, Ed. Berlin, Germany: Springer-Verlag, 1979, vol. 37, ch. 4.

[14] J. van Turnhout in: G.M. Sessler (Ed.) Topics in Applied Physics: "Electrets", vol. 33, ch. 3, Springer-Verlag, Berlin, 1987, pp. 81-216

[15] J. Ross Macdonald, "Transient and Temperature Response of a Distributed, Thermally Activated System", J. Applied Physics, vol. 34, pp. 538-552, 1963

[16] D. Molinero and L. Castaner, "Dielectric charging characterization on microelectromechanical switches by discharge current transient", Applied Physics Letters, vol. 96, pp. 183503, 2010

[17] http://www.kelvinprobe.info/

[18] X. Rottenberg, I. De Wolf, B. K. J. C. Nauwelaers, W. De Raedt and H. A. C. Tilmans, "Analytical Model of the DC Actuation of Electrostatic MEMS Devices With Distributed Dielectric Charging and Nonplanar Electrodes", J. of Microelectromechanical Systems, vol. 16, No.5, pp 1243-1253, 2007

[19] R.W. Herfst, P.G. Steeneken, J. Schmitz, A.J.G. Mank and M. van Gils, Kelvin probe study of laterally inhomogeneous dielectric charging and charge diffusion in RF MEMS capacitive switches, 46th Annual International Reliability Physics Symposium, Phoenix, pp. 492-495, 2008

[20] B. Cleve, B. Hartenstein, S. D. Baranovskii, M. Scheidler, P. Thomas and H. Baessler, "High-field hopping transport in band tails of disordered semiconductors", Phys. Rev. B, vol. 51, pp. 16705–16713, 1995

[21] B.I. Shklovskii, "Hopping conduction in semiconductors in a strong electric field", Sov. Phys.-Semicond. Vol. 6, pp. 1964, 1973

[22] J. Bisquert, G. Garcia-Belmonte, "Analysis of the power-law response in the fractal dielectric model by thermally stimulated currents and frequency spectroscopy", J Applied Physics 89, pp.5657–62, 2001

[23] U. Zaghloul, G. Papaioannou, F. Coccetti, P. Pons, R. Plana, "Dielectric charging in silicon nitride films for MEMS capacitive switches: Effect of film thickness and deposition conditions", Microelectronics Reliability, vol. 49, pp. 1309–1314, 2009

[24] U. Zaghloul, F. Coccetti, G. Papaioannou, P. Pons, R. Plana, "A novel low cost failure analysis technique for dielectric charging problem in electrostatically actuated MEMS devices", IEEE International Reliability Physics Symposium (IRPS) 2010, pp.237-245, Anaheim , USA

[25] K. J. B. M. Nieuwesteeg, A. A. van der Put, M. T. Johnson, and C. G. C. de Kort, "dc-Bias Stress of Non-Stochiometric Amorphous Silicon Nitride Thin Film Diodes", J. Applied Physics, vol. 79, pp.842-849, 1996

[26] J. M. Shannon and B. A. Morgan, "Hole transport via dangling-bond states in amorphous hydrogenated silicon nitride", J. Applied Physics, vol. 86, pp. 1548-1551, 1999

[27] G. Papaioannou, F. Coccetti and R. Plana, "On the Modeling of Dielectric Charging in RF-MEMS Capacitive Switches", Topical Meeting on Silicon Monolithic Integrated Circuits in RF Systems (SiRF), New Orleans, LA, pp. 108-111, 2010

[28] S.P. Lau, J.M. Shannon, B.J. Sealy, "Changes in the Poole–Frenkel coefficient with current induced defect band conductivity of hydrogenated amorphous silicon nitride", Journal of Non-Crystalline Solids, vol. 277–230, pp. 533–537, 1998

[29] M. Lamhamdi, P. Pons, U. Zaghloul, L. Boudou, F. Coccetti, J. Guastavino, Y. Segui, G. Papaioannou, R. Plana, "Voltage and temperature effect on dielectric charging for RF-MEMS capacitive switches reliability investigation", Microelectronics Reliability, vol. 48, pp. 1248–1252, 2008

[30] M. J. Powell, "Charge trapping instabilities in amorphous silicon-silicon nitride thin-film transistors", Applied Physics Letters, vol. 43, pp. 597-599, 1983

[31] P. Czarnecki, X. Rottenberg, R. Puers, I. De Wolf, Effect of gas pressure on the lifetime of capacitive RF MEMS switches, MEMS 2006, pp 890-893

[32] U Zaghloul, B Bhushan, P Pons, G J Papaioannou, F. Coccetti and R. Plana, "On the influence of environment gases, relative humidity and gas purification on dielectric charging/discharging processes in electrostatically driven MEMS/NEMS devices", Nanotechnology vol. 22, pp. 035705, 2011

ELECTROMIGRATION INDUCED VOID KINETICS IN CU INTERCONNECTS FOR ADVANCED CMOS NODES

L. Arnaud [1,2], P. Lamontagne[2], R. Galand[2], E. Petitprez[2], D. Ney[2], P. Waltz[2]

[1] CEA LETI-MINATEC, 17 rue des martyrs, 38054 Grenoble cedex 09, France
[2] STMicroelectronics, 850 rue Jean Monnet, 38926 Crolles, France
e-mail: lucile.arnaud@st.com

Abstract—**Time evolution of resistance during EM tests is extensively analyzed for various Cu interconnect structures and processes from the 40 nm node technology. Resistance evolution is used to model void nucleation and growth kinetics. We show that adding Al or other impurities in the line is effective to increase electromigration lifetime. This lifetime increase is due, as expected, to Cu drift velocity decrease but also to an increase of the time to void formation. TEM picture shows that Al precipitates are formed at grain boundaries and are most likely responsible for the occurrence of an incubation time**

Resistance saturation is observed for short lines thanks to Blech effect. A resistance model is developed to characterize short length effect in 40 nm node. The model is also used to explain EM lifetime improvement thanks to a pre-stress condition where compressive stress is added at cathode end of long line structures.

Keywords---electromigration, Cu interconnect, voiding, microstructure, stress

I. INTRODUCTION

In semiconductor integrated circuits (ICs), interconnects dimensions are continuously scaled down to follow the increase of device density and overall cost reduction. Meanwhile, the demand for increasing current density needs to carefully consider interconnect reliability. It is important to identify the fundamental reliability limits of interconnects schemes and to correctly describe the active degradation mechanisms. Numerous improvement of both materials and processes, such as Cu alloys (e.g. CuAl [1], CuMn [2]) or novel metal wire capping materials (e.g. CoWP [3], Co [4]), have been proposed to cope with increasing interconnect current density. Metal reliability is generally assessed with electromigration (EM) tests. EM degradation is a mass transport driven by an electron flow and leads to the formation of voids causing electrical opens in metallic wires. While the theoretical description of EM mechanisms is well understood, careful testing and analyzing is very important in order to correctly assess interconnect performance.

The classical approach studies EM with lifetime experiments. Elementary test structures are stressed under accelerated conditions (high current density and temperature)

until degradation occurs. Line failure is detected once the resistance has increased by a certain amount, typically 10%. This approach provides a significant statistic from which interconnect lifetime at operating conditions is deduced. To get experimental data at much lower failure percentage some authors [5, 6] proposed to use novel test structures based on multiple links that enable a significant sample size increase. It is then possible to discriminate between multiple failure modes and their respective characteristics.

In this paper we propose to analyze EM test results with an alternative to the lifetime method. We study EM induced void kinetics through an extensive analysis of resistance evolution of single link interconnect. Many authors [7-9] have already taken advantage of the resistance evolution analysis of interconnect structures during an EM test. It has been shown to provide valuable information about physical mechanisms involved in void nucleation and growth. Quantitative data on EM related parameters, such as Cu electromigration drift velocity (v_d), void length (L_v) or void location can be obtained [10].

In the following sections, we discuss some aspects of Cu alloy materials on Cu drift velocity and void formation time. The influence of interconnect geometry and its relation with microstructure is then analyzed. The final study is related to EM improvement thanks to stress induced effects.

II. RESISTANCE EVOLUTION VERSUS TIME

A. Method and test structure

The principle of the method is straightforward and involves recording resistance variation of a metal wire submitted to electromigration degradation. Usual exploitation considers large sample size as for lifetime exploitation. However, the behavior of each single device can also be analyzed individually.

The EM test structures are built from 40nm node damascene Cu interconnects with ultra low-k SiOCH (k=2.5) as dielectric material. Cu material is encapsulated with TaN/Ta sidewall and bottom barrier and SiCN cap on top, (see Transmission Electron Microscopy (TEM) cross section in Fig. 1). The test samples are 140nm thick, 63nm wide and 225µm long lines. Dual damascene structures have symmetrical design with single via at both ends. Upstream (Fig. 2-top) and downstream (Fig. 2-bottom) structures are both considered.

978-1-4244-9113-1/11 $26.00 © 2011 IEEE

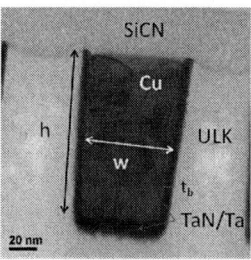

Figure 1. TEM cross section of interconnect line showing the materials and geometry of the 40 nm node samples

Figure 2. Schematic of test structures. Top=upstream , bottom=downstream.

EM tests are carried out at package level with usual accelerated conditions (temperature T of 300°C and current density j ranging from 2 to 3 MA/cm²). A set of at least 30 samples is used for each stress condition.

A typical resistance curve, (see Fig. 3), is composed of three phases. First the resistance remains almost constant which is attributed to the nucleation and growth of the EM induced void until it is electrically detectable. The second stage consists of a significant resistance step which occurs when the void spans over the whole line thickness (see Fig. 4). This creates an additional resistive path as the current flows through the barrier material over the void length (Lv). The final stage shows a continuous resistance increase where the resistance increase rate (Rslope) results from a void front motion along the conductor.

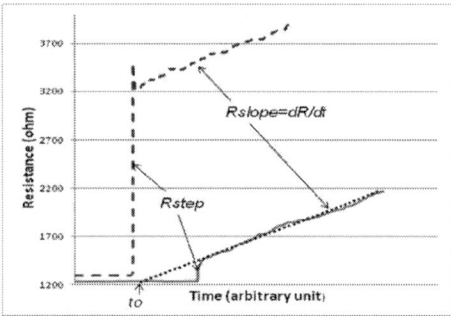

Figure 3. Typical resistance evolution of Cu interconnect upstream or downstream structures. The step of resistance (Rstep) is followed by a constant resistance increase (Rslope). Either large Rstep (> 1500Ω) or low Rstep (~150Ω) may coexist. Extrapolation of resistance slope defines incubation time (to).

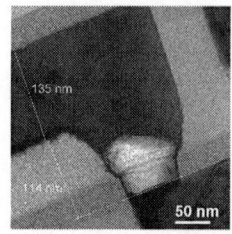 

Figure 4. TEM picture of typical EM cathode void . Left is upstream structure with void at cathode via . Right is void in the line a few nm from the cathode. Test is stopped right after step of resistance is recorded.

Resistance saturation may be possibly observed [11, 12] for test structures with short line length where a stress backflow progressively opposes the void growth. Discussion on short length effect is the scope of section V.

B. EM Cu drift velocity (vd)

For long line lengths, once the EM void has reached the bottom of the line, the rate of resistance increase versus time Rslope is constant and proportional to Cu drift velocity (vd) following :

$$R_{slope} = \rho_b \, v_d / [(2h+w)t_b] \qquad (1)$$

Where ρ_b, t_b, h and w are, respectively, the TaN/Ta barrier resistivity, its thickness, and the copper line cross section dimensions (see Fig. 1).

The experimental observation of a constant Rslope value during void growth at experimental conditions is in good agreement with our understanding of Cu diffusion in Cu interconnects. It has been observed by in situ EM Scanning Electron Microscopy (SEM) experiments that voids mainly grow by front edge displacement [13]. Multiphysics model [14, 15] also provides a constant void front movement when Cu diffusion at top interface is the main diffusion path.

In Fig. 5, Rslope follows a single mode distribution. For the same process and line geometry, up- and downstream structures give the same median value of Rslope, (also named Rslope in the following) and the same standard deviation (σ). This means that void front motion occurs in line for both structures with similar Cu diffusion drift velocity. It is also worth noting that standard deviation values (σ) of Rslope and lifetime distributions compare well (see Table II for instance). It seems that line cross section and interconnect microstructure variations from one device to another have the same influence on lifetime and Cu drift velocity.

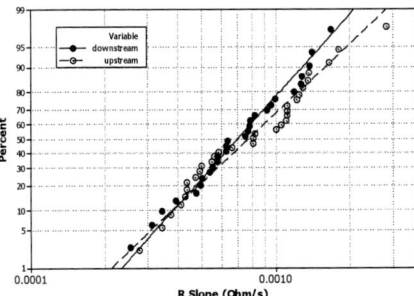

Figure 5. Comparison of *Rslope* distributions for upstream and downstream structures.

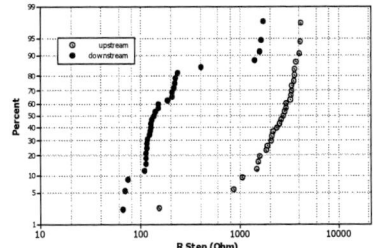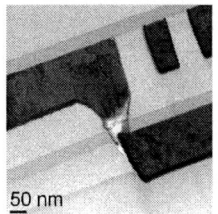

Figure 7. Left: Comparison of *Rstep* for upstream and downstream structures with via misalignment issue. Right: TEM of via voiding .

C. EM Void length (Lv) and void location

Once the EM void reaches the bottom of the trench, the test structure can be modeled with 2 serial resistances. The step of resistance (*Rstep*) then simply reads [16]:

$$R_{step} = \rho_b \, L_v \,/[(2h+w)t_b] \qquad (2)$$

Equation 2 allows to calculate the void length (*Lv*) corresponding to the time to failure (*TTF*). It assumes simple void shape geometry where the void length is fixed in the line cross section. Good correlation has been previously established between calculated void length and void length measured on TEM pictures validating simple void geometric assumption made in (2) [8].

Nevertheless, as shown in Fig. 3, resistance curves may also show very high *Rstep* not matching (2). As discussed in one of our previous studies [10], high *Rstep* value is due to a void located underneath the via in downstream structures or at the via bottom for upstream ones (as shown in Fig. 3-left).

In Fig. 6, most of the samples give similar *Rstep* values for both upstream and downstream structures issued from the same wafer. This indicates that, for a given process and line geometry, void in line is the preferred failure mode. It occurs with the same percentage independently of upstream or downstream current flow. Usually, at least a small proportion of high *Rstep* failures remain in both structures with similar percentage too.

This indicates that void formation at extreme cathode end also occurs independently of upstream or downstream current flow. On the contrary, some process issues, such as via misalignment, give rise to different *Rstep* distributions for down- and upstream structures (see Fig. 7-left). This indicates that a weak failure mode is preferred for upstream configuration only. In this case, TEM pictures (Fig.7-right) show that void in via occurs for all upstream samples.

D. Incubation time (t₀)

It should be recognized that EM failure is composed of stages of void nucleation and void growth [17]. But for each process change, a question mark still remains: is the void growth the dominant failure mechanism or should one also consider additional void nucleation time? First, we look at a possible correlation between *Rstep* and time to failure for trench voiding failure mode.

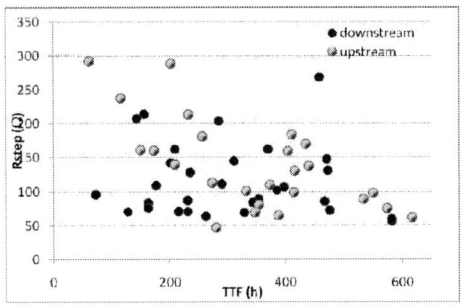

Figure 8. Plot of *Rstep* versus *TTF*

Fig. 7 shows no obvious correlation between *Rstep* and *TTF*, which is a general trend for 40nm node technology. Correlation would mean that the failure time is proportional to the void length. This would be an ideal case where a single void continuously grows with stress time so that the larger the void, the longer the time to failure. In other words, this would mean that void growth is the dominant failure mechanism, assuming reasonably a more or less constant Cu drift velocity for all samples. Our data suggest that this hypothesis is not fulfilled. Although some variation in void shape from the ideal case of (2) may occur and add some jitter to *Rstep*, we believe that the failure time is the sum of a void formation time and a

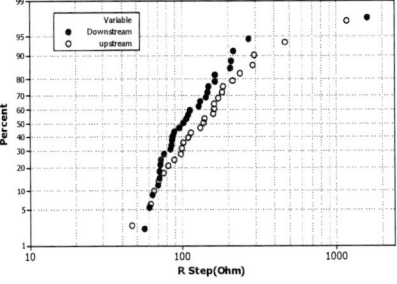

Figure 6. Comparison of *Rstep* distributions for upstream and downstream structures.

void growth time. SEM in situ EM experiments performed by other groups [18, 19], show that voids smaller than the line width move and grow simultaneously before reaching the cathode end of the line. Significant incubation time has also been observed with CoWP capped Cu lines [20]. The authors suggest that higher stress is required for void nucleation.

Considering the resistance versus time characteristic during EM degradation for trench voiding mechanism, we define EM incubation time t_0 as the projection of intersection of *Rslope* with initial resistance plateau (see Fig. 3):

$$t_0 = TF - Rstep / Rslope \qquad (3)$$

t_0 represents an incubation time before final void starts to grow. Equation 3 presumes that t_0 is the time to form an ideal void at the upper interface of interconnect trench. This ideal void is an horizontal slit void with thickness zero and length *Lv*. Equation 3 also assumes that the void will first grow vertically from the upper interface to the bottom of the trench while its length *Lv* remains constant. In a second step the void grows horizontally as its front moves along the line. Both vertical and longitudinal void growth velocities are assumed to be constant and similar so average t_0 can be determined from average *Rstep/Rslope* value. Next section provides a complementary discussion on the evolution of average t_0 (also named t_0 in the following) with Cu alloying process in interconnects.

III. INFLUENCE OF CU DOPING

A. Lifetime comparison of Cu and CuAl process

In this section we compare failure times and void growth kinetics extracted from the resistance curves of test samples with downstream structures. The major process differences are summarized in Table I. CuAl seed layer is investigated because Al atoms deposited within CuAl seed diffuse in the Cu interconnect cross section [1, 21]. Authors reported that the Al atoms mostly segregate at the Cu/SiCN interface slowing down Cu drift and providing lifetime increase.

TABLE I.   SAMPLE DESCRIPTION

Sample	Process summary
A	Cu seed / ref TaN/Ta barrier
B	CuAl seed (1%Al) / ref TaN/Ta barrier
C	CuAl seed (1%Al) / improved TaN/Ta barrier

The comparison of the experimental times to failure is given in Fig. 9. As expected we observe a significant Median Time to Failure (*MTF*) boost when adding Al in the Cu seed (sample B), and some additional increase with an improved TaN/Ta barrier (sample C). It should be noted that all the distributions exhibit a σ of lognormal plot about 0.4 indicating a single failure mode.

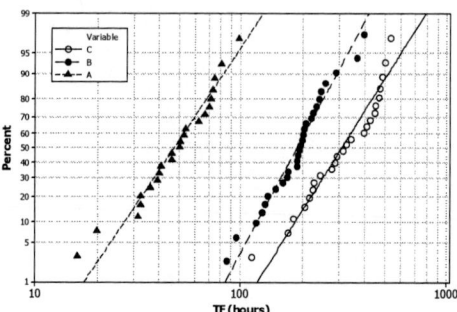

Figure 9.   Cumulative lognormal plot of failure times for samples A, B, C.

B. Drift velocity versus lifetime comparison

In order to provide understanding of lifetime increase, we analyze both *Rstep* and *Rslope* for A, B and C samples following the definition given in Fig. 3.

Bimodal distribution of *Rstep* is clearly seen in Fig. 10-top. As discussed previously, high *Rstep* are due to voids underneath the via while small *Rstep* are due to voids located in line a few nm from cathode end. In this section, we restricted our investigation to the core distribution of *Rstep*, i.e, to trench voiding.

Rstep Median values (*Rstep* in Table II) are very similar for the 3 processes indicating that the significant *MTF* increase in Fig. 9 cannot be attributed to any variation of average void length *Lv* versus process variation.

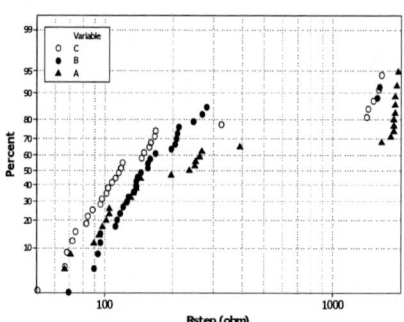

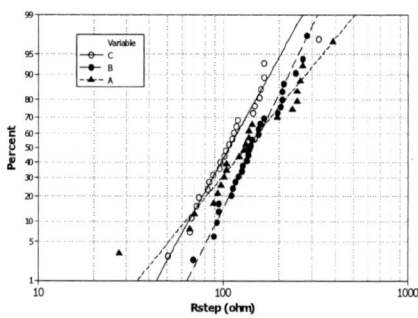

Figure 10. Distribution of resistance step (*Rstep*). Top is bimodal distribution due to different possible EM void location. Bottom gives only *Rstep* distribution for voiding in the line.

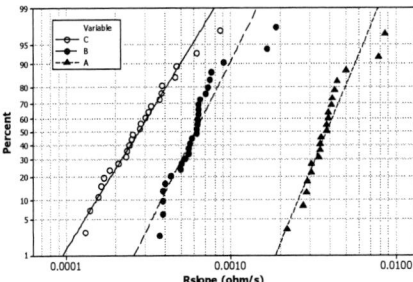

Figure 11. CDF of slope of resistance (*Rslope*) for the A, B, C samples with voiding in line.

On the contrary, the comparison of *Rslope* distributions for samples A, B and C (Fig. 11) shows significant decrease for Cu process with Al doping. This means, as we would expect, that lifetime increase is the consequence of Cu drift velocity (v_d) decrease.

With the data of Table II, and (1) and (2), one can compare Cu drift velocity decrease and lifetime increase for samples with Al doping. We prefer to use *Rlope/Rstep* ratio to figure v_d in order to cancel any geometry variation within wafer. The calculation shows that the decrease of vd in samples B and C with Al doping, versus sample A does not fully compensate the *MTF* increase of both samples. We suspect that the difference is due to different incubation times t_0. In Table II, significant increase of t_0 is also reported for B and C samples. Further discussion on the effect of impurities in Cu interconnects on t_0 is the scope of section D.

TABLE II. SUMMARY OF EM PARAMETERS OF SAMPLES A, B AND C

Sample	MTF (h)	σ	Rstep (Ω)	Rslope (Ω/s)	σ	t0 (h)
A	47	0.43	133	$3.3\ 10^{-3}$	0.31	36
B	188	0.35	145	$6.1\ 10^{-4}$	0.37	122
C	310	0.40	108	$2.7\ 10^{-4}$	0.45	199

In addition no correlation has been found between *TTF* and *Rslope* for each device of all samples showing that for all 3 given samples, the variation of void growth velocity from one device to another is independent upon the device time to failure. This means that time to failure is not controlled by Cu drift velocity possible variation from one void location to another at least for trench voiding mechanism.

C. Drift velocity comparison for Al and Co doping

Additional process solutions have been studied to improve EM lifetime. CVD Co is either deposited as a liner between TaN/Ta barrier and the CuAl seed (Co liner), or deposited as a cap layer between Cu top surface and SiCN cap (Co cap). Process details are summarized in Table III, while Fig. 12 shows *Rslope* distributions extracted from resistance plots. Stress conditions and line geometry were kept the same as for samples A, B and C.

TABLE III. Co SAMPLE DESCRIPTION

Sample	Process summary
D	Cu seed/ CVD Co sidewall liner (~1nm) / ref TaN/Ta
E	CVD Co cap (~1.5nm) / CuAl seed (1%Al) / ref TaN/Ta sidewall barrrier

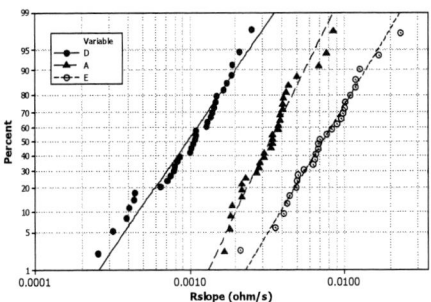

Figure 12. Distribution of slope of resistance (*Rslope*) for the A, D, E samples with voiding in line.

It happens that adding a Co liner (sample D) decreases the Cu drift velocity in comparison to Cu only seed process. This behavior is similar to the decrease of Cu drift velocity when Al is incorporated in the Cu seed (see sample B data). *MTF* increase is also similar between sample D and B (see Table IV).

On the contrary to our expectations, sample E, with the Co cap, displays a larger Cu drift velocity than pure Cu process despite its CuAl seed. The comparison of sample E and B (see Table IV) shows that Cu drift velocity is about one decade larger in Co cap sample. Co mapping with EDX shown in Fig. 13 let us conclude that incomplete Co capping in this sample may be responsible of easy Cu drift at the top interface.

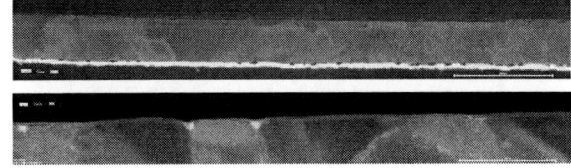

Figure 13. TEM cross section of sample E, together with chemical mappings of Co (blue), Al (green) and Ta (yellow).

However, further comparison of samples A and E shows an increase of *MTF* for sample E in spite of a higher average Cu drift velocity *(Rslope)*. This apparent contradiction may be explained by a difference in incubation time t_0. In table IV we report an increase of t_0 between samples E and A. Al mapping (see Fig. 13-bottom) shows the presence of Al in grain boundaries. This may be the reason of incubation time increase for sample E.

We have previously shown [22, 23] that void nucleation in Cu interconnects preferentially occurs at grain boundaries with high angle misorientation (>15°). Given that Al precipitates are located at the void nucleation sites, this may block void formation and hence provide an increase in incubation time, t_0.

978-1-4244-9113-1/11 $26.00 © 2011 IEEE 301

TABLE IV. SUMMARY OF EM PARAMETERS OF SAMPLES A, D, E

Sample	MTF (h)	σ	Rstep (Ω)	Rslope (Ω/s)	σ	t0 (h)
A	47	0.43	133	$3.3\ 10^{-3}$	0.31	36
D	196	0.5	113	$9.5\ 10^{-4}$	0.37	163
E	57	0.44	341	$7.3\ 10^{-3}$	0.45	44

We believe that the difference of incubation times (t_0) between samples A and E is meaningful. Confidence intervals of both distributions are not overlapping showing that results are statistically different. Error estimates are +/-3h for sample A and +/-4h for sample E.

D. Evolution of EM incubation time versus line resistivity

In order to link experimental EM parameters with EM physics, we compare many samples issued from different lots and processes but with the same line cross-section. Because line resistivity is a good indicator of the amount of impurities added for a given line section, we calculate the line resistivity with Temperature Coefficient of Resistance (TCR). In Fig. 14, a linear increase of t_0 versus resistivity is obtained for the samples A, B, C and D already studied. This result shows that either Al or Co impurities play an important role in the first steps of EM induced void formation. The void formation time (t_0) then depends upon the amount of diffusing impurities.

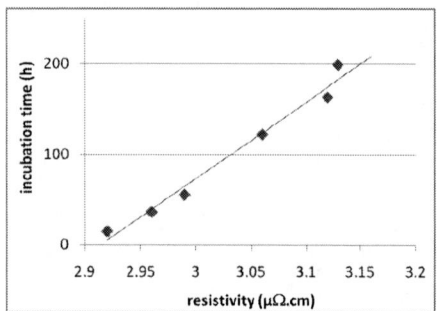

Figure 14. EM incubation time ($t0$) evolution versus resistivity of Cu lines for A, B, C, D process

IV. INFLUENCE OF CU MICROSTRUCTURE

The influence of interconnect microstructure on void formation and kinetics is also an important concern for EM performance. As interconnect downscaling continues, copper microstructure is shown to move from bamboo to polygrain structure. Grain boundary diffusion becomes more and more important and may degrade the EM performance. Zhang et al. [24] recently studied EM lifetime for different average grain sizes where CoWP capping strongly reduced easy interfacial Cu diffusion. They observed that large grain structures have longer lifetime in comparison to small grain ones and concluded that mass transport along grain boundaries is more important in structures with small grains. Similar grain boundary diffusion has also been reported in small linewidth

[25] with polygrain structure where a low EM activation energy value was observed.

Here we study the influence of Cu line microstructure on EM performance by comparing 3 different linewidths. We measure lifetimes and void kinetics parameters with resistance curves on downstream structures. EM results with minimum linewidth (wmin) as studied in previous sections are compared to EM results with smaller (wmin-5nm) and larger linewidths (2wmin).

Detailed analysis of the Cu microstructure versus line width has been carried out with Electron Backscattered Diffraction (EBSD) technique on 45nm node interconnects [26]. It was shown that grain size increases linearly with linewidth and polycrystalline grain structure remains in the line thickness for linewidths up to about 200nm. For larger linewidths columnar microstructure is seen in cross section. We believe that microstructure evolution versus linewidth is the consequence of a change of Cu re-crystallization mechanism in trench with linewidth variation. In large line, large grains from overburden Cu can invade the trench. In narrow line re-crystallization is limited by the trench sidewall.

EM tests carried out on 45 nm node downstream structures with [22] showed that MTF was not sensitive to variation of average grain size. But the presence of a large proportion of high angle misorientation boundaries (>15°) was responsible for the presence of early fails with voids located in the immediate vicinity of the via.

In 40 nm node technology, the comparison of EBSD analysis on wmin wide structures for samples A (pure Cu seed) and B (CuAl seed) shows the same average grain size for both processes (85nm) and some differences between high angle grain boundary ratio (22 to 29 %) [23].

Here, for 40 nm node technology and process B, Fig. 15 displays a significant increase of *TTF* for increasing linewidth. At the same time, the Cu drift velocity (*Rslope/Rstep*) increases with decreasing linewidth (see Table V). This is in agreement with previous published results [9] where grain boundaries diffusion becomes significant versus interface diffusion as soon as interconnect length has an important proportion of polycrystalline segment, e.g more than 80%. This is the usual case for linewidths below 100nm.

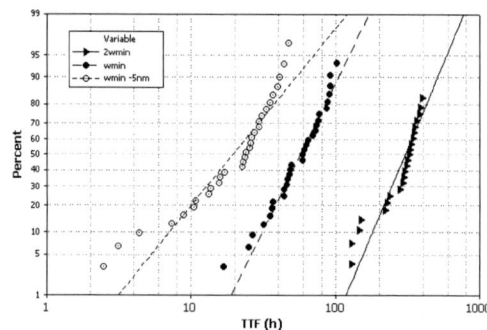

Figure 15. Lifetime distribution versus linewidth for downstream structures of process B. Test is done at same current density and temperature for all.

TABLE V.   EM PARAMETERS VERSUS LINEWIDTH

Sample	MTF (h)	Rstep (Ω)	Rslope (Ω/s)	t0 (h)
wmin -5nm	19	228	8.4 10⁻³	11
wmin	54	190	3.7 10⁻³	39
2wmin	300	114	4.6 10⁻⁴	231

In Table V, we can see that both Cu drift velocity decrease and incubation time (*t0*) increase account for the *MTF* improvement in larger grain microstructure. However it is not possible, at this level, to determine if grain boundary diffusion is decreased in large linewidths (large grains). It might also be possible that a larger amount of Al atoms segregates at the surface for large linewidths and results in an additional decrease of interface diffusion. Because, incubation time is also increased whit increasing linewidth, we make the same hypothesis than before: Al precipitates are more effective in blocking void formation at void nucleation sites and provide more delay for final void formation in large linewidths. This may be due to the number of grain boundaries to consider for void nucleation versus linewidth.

In addition to the decrease of *MTF* whit decreasing linewidth, we generally also observe an increase of the lognormal plot standard deviation (σ) and a clear bimodal *TTF* distribution. Early fails are well correlated with high *Rstep* values or void located underneath via or at via bottom. We believe that such an increase is due to an increase of the proportion of high disoriented grain boundaries for the smaller linewidth (wmin-5nm).

The distribution of *Rstep* plotted in Fig. 16 for the 3 linewidths shows that the proportion of high *Rstep* is larger for the smaller linewidth (wmin-5nm). But we also observe at least 30 % of high Rstep with monomodal lifetime distribution and σ ~ 0.4 for the 2 other linewidths. This means that voids located underneath via may exist even with good lifetime and good σ value.

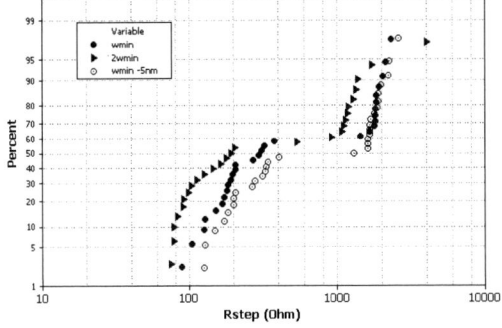

Figure 16. Bimodal Rstep distributions versus linewidth.

V.   STRESS EFFECT ON EM LIFETIME

A.   Modeling Blech length

The demand of increasing current for advanced nodes interconnects pushes one to take advantage of Blech effect for short lines where a stress induced back flow counterbalances the EM induced flux. The resistance evolutions for short length interconnects have been studied with varying length (*L*) and current density (*j*). It was shown for 65 nm node interconnects [12] that as the *jL* product decreases, *TTF* increases and the rate of resistance increase (*Rslope*) also decreases.

The study of the void growth kinetics gives complementary understanding on the interconnect degradation in the presence of Blech effect. The evolution of stress as a function of time is given by the solution of the Korhonen's equation that simulates the presence of a void at the early stage of EM process [27].

Void volume evolution versus time V(t)) is calculated with integration of the stress over the whole line length [28]:

$$V(t) = V_{sat}\left[1 + \frac{32}{\pi^3}\sum_{n=1}^{\infty}\frac{(-1)^n}{(2n-1)^3}\exp\left[-\left(\frac{2n-1}{2}\pi\right)^2\frac{t}{\tau}\right]\right] \quad (4)$$

$$\tau = \frac{L^2 kT}{DB\Omega} \quad (5) \qquad V_{sat} = \frac{e\rho Z^* A_{cu}}{2\Omega B}\cdot jL^2 \quad (6)$$

where D, ρ, Ω and Z* are respectively the copper diffusion coefficient, resistivity, atomic volume and effective charge. A_cu is the line section, B the effective modulus and e is the elementary charge.

Assuming a two phase void growth, with simple void shape assumption, the resistance evolution versus time can be simulated. Perfect vertical void growth is assumed to start with an ideal initial slit void of length *Lv* and is followed by longitudinal void front movement [12].

Good agreement between experiments and simulations were obtained for a set of material parameters D, B and Z* reported in Table VI.

Fig. 17 shows experimental and simulated resistance evolutions for 40 nm nodes interconnect structures with process B and various jL conditions. At low jL condition, EM flux is entirely compensated by a backflow flux. Stationary state is reached and the EM induced void no longer grows. We can notice that stationary state can be reached once EM induced void opens the whole line section. As a consequence, resistance saturation is simulated.

978-1-4244-9113-1/11 $26.00 © 2011 IEEE

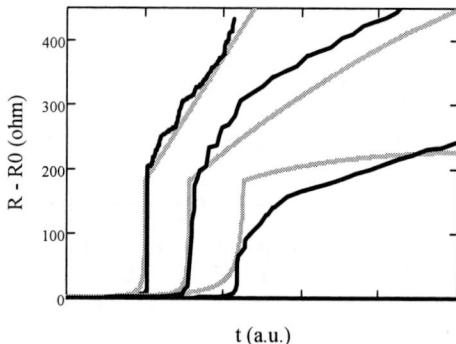

Figure 17. Experimental (black) and simulated (grey) resistance evolutions for L = 225µm (left), 45µm (middle) and 22.5 µm (right) structures stressed at the same current density

TABLE VI. MATERIAL PARAMETER EXTRACTED FROM SIMULATIONS

Technology node	D (m²/s)	B (GPa)	Z*
65 nm	9.10^{-17}	15	0.34
40 nm	$2.2.10^{-17}$	5	0.22

In comparison to the material parameters of the previous technology node, 40 nm node technology provides lower D, B and Z*. Lower diffusion coefficient is explained by integration of CuAl which is aimed to slow down Cu diffusion. The 40 nm node process seems to induce a slightly lower interaction between electrons and local electronic environment of the atoms. Because advanced technology node uses porous intermetal dielectric (IMD) and thinner sidewalls barrier, the whole interconnect system is expected to be more ductile. We also get lower value of the effective modulus B with this method.

Equations 5 and 6 help to compare void growth kinetics for both technologies. Assuming a constant atomic volume, Ω, a ratio of 12 and 0,70 are respectively found for τ and $Vsat$ moving from 65nm technology to 40nm. On one hand, kinetic of void growth is slower thanks to lower D and B. But the introduction of a less stiff IMD is not necessarily an issue for the Blech effect as long as extrusions are avoided. On the other hand, the void volume $Vsat$ needed to stop completely the flux also becomes smaller for advanced technology mostly because of line section scaling. Normalization by the line sections shows a ratio of 2 between $Vsat$ of both technologies. In summary, reduction of Cu diffusion (D) and interaction between electrons and defects (Z*) slow down the degradation kinetics. The improvement of both parameters remains necessary to keep the advantage of the Blech effect in future technologies.

B. Compressive stress effect.

Theory for EM failure [29] has explained the influence of stress state in the Cu line on void formation behavior. Therefore tensile stress will favor void nucleation. In this study we intend to show that compressive stress state will delay EM void formation and increase EM failure time. For this purpose,

we used bidirectional EM stresses [30][15][31] to locally induce compressive stress state at cathode end.

Experiments consist of usual EM tests on upstream structures built from 40 nm node interconnects. A first EM stress condition creates voids in lines. We intentionally maintain the stress to grow voids up to a size of ~1µm [31]. No other degradation of the structures occurs thanks to robust TaN/Ta barriers. Large voids created at one end of the structure let us assume that matter is accumulated along the line length and preferentially at the other side.

A second EM stress condition is applied on all test structures. For this second condition EM experiments are carried out on samples with pre-stress. Resistance evolution versus time is displayed in Fig. 18 for 3 conditions of EM stress. Lifetimes distributions of the 3 conditions are given in Fig. 19.

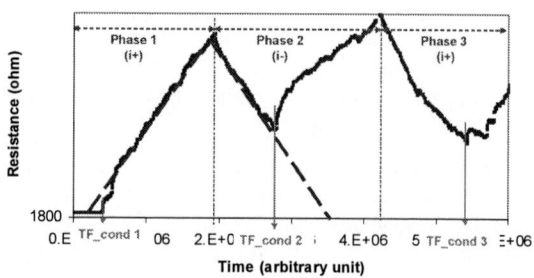

Figure 18. Resistance evolution of a given device for a sequence of 3 phases of bidirectionnal current stress. In phase 1, electron flux is from left to right with cathode end at left side. In phase 2 electron flux is from right to left with cathode at right side. Phase 3 is identical to phase 1.

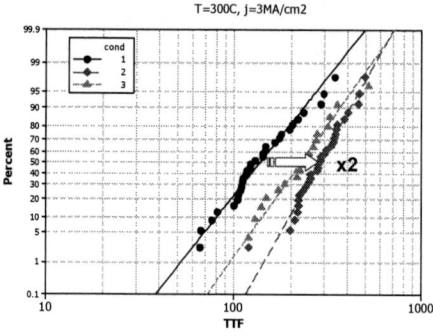

Figure 19. Lifetime distributions of all devices for 3 EM stress conditions

Significant improvement of lifetime (a factor of 2) is obtained after pre-stress. In order to better understand the physics involved in lifetime improvement, we carried out analytical simulation of stress states at the end of pre-stress.

Two solutions of the Korhonen's equation [27] are used in order to calculate i) the compressive EM induced stress at the anode side at the end of condition 1 and ii) the tensile stress needed for void nucleation. Experimental resistance evolutions are used to determine the relevant material parameters (D, Z*, B), following the method introduced in [12][31].

978-1-4244-9113-1/11 $26.00 © 2011 IEEE

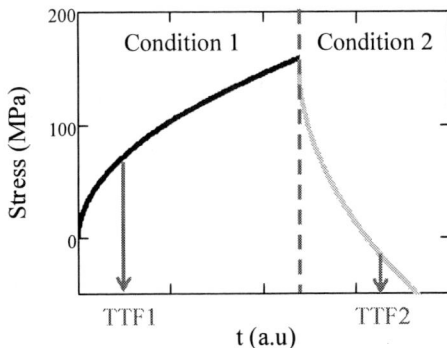

Figure 20. Compressive stress evolution versus time (t) at line end opposite to void nucleation site in condition 1 followed by stress released caused by the reversed current. Maximum stress value is 160MPa at the end of condition 1

We found that applying pre-stress is a way to induce a compressive stress state at anode side of the structure. Reverse current releases compressive stress at new cathode end up to the condition of void nucleation. This provides an additional time for void nucleation or an increase in the incubation time needed for void formation without any change in other EM parameters i.e. Cu drift velocity.

We also found that lifetime improvement is dependent on the duration of pre-stress [31] or the amount of compressive stress to release before void creation.

VI. CONCLUSION

In this paper, we show that resistance evolution analysis provides quantitative data on formation and growth of EM induced voids. The influence of Al doping in Cu interconnects is evidenced through Cu drift velocity decrease, as expected, but also through time to void formation increase. In addition we show that the increase of time to void formation is likely to be due to the amount of Al (or other impurities) available for diffusion in the Cu line.

Usually, trench voiding provides similar EM characteristics in upstream and downstream structures. For some process issues, the analysis of resistance curve provides information on void localization. Indeed, resistance step value allows to discriminate between EM failure due to void located in the immediate vicinity of the via and failure due to void located in the trench.

The significant increase of EM lifetime when linewidth is increased is explained with a decrease of grain boundary diffusion when grain size increases. In Al doped interconnects, time to void formation also increases for large linewidths.

Analytical simulations are used to model the resistance evolution in short line lengths and the stress induced in long lines. The creation of voids with EM tests may block the EM degradation in short line lengths. In long lines large voids created at one line end provide a compressive stress calculated at 160MPa at the other end. This is a pre-stress condition where EM failure is significantly delayed thanks to an additional incubation time.

ACKNOWLEDGMENT

The authors wish to thank all the people involved in process development and integration for the 40nm node technology. Dedicated acknowledgement is made to D. Galpin, S. Chhun, C. Monget, Y. Le Friec, J. Guillan for providing appropriate samples for this study and for valuable discussions about the details of the process involved. The authors are happy to thank the failure analysis team for providing TEM pictures and mostly R. Pantel and L. Clement for their work in EDX impurities mapping.

REFERENCES

[1] S. Yokogawa, H. Tsuchiya, *J. Appl. Phys.*, 101, p.013513 (2007).

[2] C. Christiansen, B. Li, M. Angyal, V. McGahay, S. Yao, In *IEEE International Reliability Physic, Symposium Proceedings*, 2011 in press.

[3] C.K. Hu, et al, *Applied Physics Letters*, 84, p. 4986 (2004)

[4] T. Nogami, et al, *Proceeding of IEE-IITC 2010*

[5] M. Hauschildt, M. Gall, R. Hernandez, *J. Appl. Phys*, 108, p. 013523 (2010)

[6] R.G. Filippi , P.C. Wang, A. Brendler, J.J. Demarest, J.R. Lloyd, *IEEE International Reliability Physic, Symposium Proceedings*, 2009 p. 444

[7] H. Kawasaki, and C.K. Hu, *Symposium on VLSI Technology digest*, p. 196, (1996.)

[8] L. Doyen, E. Petitprez, P. Waltz, X. Federspiel, L. Arnaud, Y. Wouters,, *J. Appl. Phys.*, 104, p. 123521 , (2008).

[9] M.H. Lin, S.C. Lee, 1.S. Oates, *IEEE International Reliability Physics Symposium Proceedings 2010*, p. 705, (2010)

[10] L. Arnaud, et al , *Proceeding of IEE-IITC 2009*, p. 179 (2009)

[11] R.G. Filippi et al, *Applied Physics Letters*, 69, 2350 (1996)

[12] P. Lamontagne, D. Ney, L. Doyen, E. Petitprez, Y. Wouters, *IEEE International Reliability Physic Symposium Proceedings 2010*, p. 922 (2010)

[13] N. Claret, C. Guedj, L. Arnaud, G. Reimbold, *Microelectronic Engeeneering*, 83, p. 2715 (2006).

[14] F.Cacho, V. Fiori, L. Doyen, C. Chappaz, C. Tavernier, H. Jaouen, *Eurosime*, pp. 649-654 (2008)

[15] L.Arnaud, F. Cacho, L. Doyen, F. Terrier, D. Galpin, C. Monget, *Microelectronic Engineering*, 87, pp. 355-360, (2010)

[16] X. Federspiel, L. Doyen, S. Courtas, *IEEE Transactions on device and Materials reliability*, vol 7, p 236, (2007)

[17] J.R. Lloyd, *Microelectronics Reliability*, 47, p. 1468,(2007)

[18] E. Zschech, R. Hubner, O. Aubel, P.S. Ho, *IEEE International Reliability Physics Symposium Proceedings 2010*, p. 574.(2010)

[19] A.V. Vairargar, et al *Appl. Phys. Letters*, 85, p. 2502 (2004)

[20] C. Witt, V. Calero, C.K. Hu, F. Feustel, G. Bonilla, AMC 2010.

[21] K. Maekawa et al, *Microelectronic Engineering*, 85, p. 2137, (2008)

[22] R. Galand K. Haxaire, L. Arnaud, E. Petitprez, L. Clement, P. Waltz, Y. Wouters, *Microelectronics Engineering*, in press (2010)

[23] R. Galand, L. Arnaud, E. Petitprez, G. Brunetti, L. Clement, P. Waltz, Y. Wouters, *Microelectronic Engineering*, in press.

[24] L. Zhang, J.P. Zhou, J. Im, P.S. Ho, O. Aubel, C. Hennestal, E. Zschech, *IEEE International Reliability Physics Symposium Proceedings 2010*, p. 581. (2010).

[25] C.K. Hu, L. Gignac, B. Baker, E. Liniger, R. Yu, P. Flaiz, *Proc. of IEEE -IITC 2007* p. 93, (2007).

[26] R. Galand L. Clement, P. Waltz, Y. Wouters, *AIP Conf Proceedings*, vol 1300, p.47, (2010).

[27] M.A. Korhonen P. Borgesen, K.Tu, C. Li, *J. Appl. Phys.*, 73, p. 3790, (1993).

978-1-4244-9113-1/11 $26.00 © 2011 IEEE

[28] J. He, Z. Suo, T.N. Marieb and J.A. Maiz, *J. Appl. Phys.*, 85, p. 4639 (2004)

[29] J.R. Lloyd, C.E. Murray, T.M. Shaw, M.W. Lane, X.H. Liu, E.G. Liniger, *AIP Conf Proceedings*, 817, p.23 (2005).

[30] L. Doyen, L. Arnaud, X. Federspiel, P. Waltz, Y. Wouters, *IEEE International Reliability Physic, Symposium Proceedings 2008*, p. 681, (2008).

[31] L. Arnaud, P. Lamontagne, E. Petitprez, R. Galand, *IEEE International Reliability Physics, Symposium, Proceedings 2011*, in press (2011)

Formation of Highly Reliable Cu/Low-k Interconnects by Using CVD Co Barrier in Dual Damascene Structures

Hye Kyung Jung, Hyun-Bae Lee, Matsuda Tsukasa, Eunji Jung, Jong-Ho Yun, Jong Myeong Lee, Gil-Heyun Choi, Siyoung Choi and Chilhee Chung

Process Development Team, Semiconductor R&D Center, Samsung Electronics Co., LTD.,
San #16 Banwol-Dong, Hwasung-City, Gyeonggi-Do, Korea, 445-701
phone: 82-31-208-2410; e-mail: hyekyung.jung@samsung.com

Abstract— **CVD Co film was investigated as an alternative barrier layer to the conventional PVD TaN\Ta in V1\M2 structure for 32nm node. We improved via filling performance and upstream (V1→M2) electromigration (EM) lifetime by more than three times. Excellent step coverage of CVD barrier makes it possible to reduce the thickness of the barrier metal by 30% and to increase the volume of Cu in metal lines. RC delay also reduced with decrease in resistance. Since adhesion at the interface between the barrier-Co and Cu also is strong, migration of Cu atoms is dramatically slowed down. EM in the via is finally deterred due to absence of pre-existing voids, consequently lifetime increases. This CVD Co process is expected to be beneficial for the next technology generation beyond 20nm node.**

Keywords-Electromigration (EM), BEOL, Dual Damascene, Cu, CVD, Co

I. INTRODUCTION

While the width of metal lines has been gradually decreased every generation, the concerns about EM reliability as well as Cu filling have been increased. The grain size of Cu becomes smaller and smaller in narrow lines and consequently the diffusion path increases through the grain boundaries. Besides, high current density also contributed to accelerating Cu migration.

BEOL process of the logic devices consists of multiple layers of the dual damascene interconnect with fragile low-k inter-metal-dielectric (IMD), therefore, there are some problems associated with patterning, filling performance and reliability. So, many studies have been continuously conducted to solve them. It has been reported that migration of Cu atoms has been controlled by metal doping [1,2]. Doped species are known to interrupt the diffusion path by segregating at Cu grain boundaries. However, this method gave rise to higher resistance. Recently, CVD Ru and Co barriers have been actively studied as a next candidate, and showed a lot of promising results in filling properties, resistance and reliability [3-5].

EM is the result of the movement of Cu atoms due to an electron wind caused by an externally applied electric field. The rate of this movement varies with how fast Cu atoms migrate on a sub-layer. Stress accumulation during electromigration of Cu atoms causes void nucleation at the cathode end, and the void grows to cause a fatal failure. Void growth is predicted by calculating net atomic flux around the void with a boundary condition that assumes no flux into the void since the void is located at the cathode end [6]. Time to failure is predicted and extrapolated using a model developed by Black [7]. Most voids are commonly generated at the interface between a top of Cu line and an upper capping dielectric due to the lowest activation energy for Cu diffusion (0.9eV). Thus, a number of studies have been published that selective deposition of thin metal on Cu lines can prevent this movement [8-10].

EM is divided into upstream (V1→M2) and downstream (V2→M2) by the direction of current flow in a dual damascene structure, as shown in Fig. 1. It is natural that voids originate at the interface of Cu and capping layer regardless of the current flow (Fig. 1 a,b). If it is difficult to obtain a void-free via due to small pitch, a void firstly occurs in the via, not at the Cu\Capping layer interface (Fig. 1 c). The pre-existing void can readily cause open failure by electromigration while a device is in operation. This is the reason why upstream EM is more critical than downstream in a dual damascene and it is important to optimize the process by minimizing pre-existing voids in the interconnect.

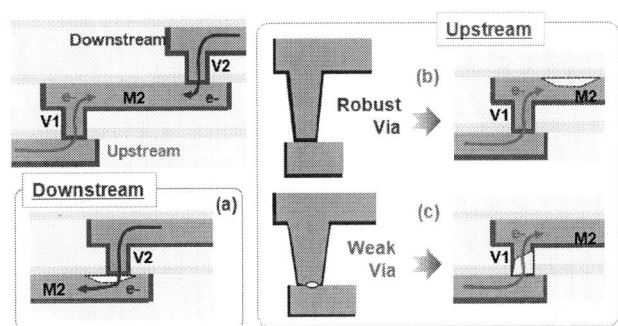

Figure 1. Two types of EM along the current flow was shown.

In this paper, CVD Co, which provides higher activation energy for Cu migration than Ta (2.0eV vs. 1.4eV) and

978-1-4244-9113-1/11 $26.00 © 2011 IEEE

improved gap-fill performance, has been investigated barrier metal at V1\M2 dual damascene structure with 32nm node. The resistance and EM reliability were presented along with possible mechanism.

II. EXPERIMENTAL

CVD Co layer was deposited on PVD TaN by using a di-cobalt hexacarbonyl t-butylacetylene (CCTBA) precursor and hydrogen gases at 150oC. Then, hydrogen plasma treatment followed to remove C impurities which were incorporated during Co deposition. Conventional PVD seed Cu, electroplating, annealing, and CMP processes were carried out.

The split conditions of the samples for different PVD TaN thickness and stack structures are summarized in Table 1. The 32 samples per split were tested with a current density of 1.2MA/cm2 at 300°C. Resistance of each sample was monitored throughout the test. Failure was assumed if the resistance increase exceeded 10% of the initial resistance, and the current is no longer applied to the test line. Finally, the failure analysis was carried out by FIB and TEM.

TABLE I.   THREE BARRIER CONDITIONS USED IN THIS STUDY

Sample group	1 Without CVD Co	2 With 2nm-CVD Co	3 With 2nm-CVD Co
Stack	PVD TaN\Ta	PVD TaN1 \CVD Co	PVD TaN2 \CVD Co
Normalized total thickness	1	0.81	0.73

III. RESULT AND DISCUSSION

A. Step overage of CVD Co barrier

We optimized the process of CVD Co deposition in order to obtain smooth and continuous thin film with the lowest resistance [12]. Continuous thin film of 2nm-thickness could be successfully formed in oxide trenches with the aspect ratio of 4:1. The step coverage and the distribution for Co were analyzed with scanning transmission electron microscopy-energy dispersive spectrometer (STEM-EDS) and conformal step coverage was observed as shown in Fig. 2.

B. Gapfill and Resistance by CVD Co barrier

Optimized stack of TaN\Co was deposited on V1\M2 low-k structure (k~2.7). The use of TaN\Co barrier stack in lieu of TaN\Ta showed a remarkable decrease in the total number of void defects after CMP to 1/15 of TaN\Ta barrier stack as shown in Fig. 3 and also revealed an about 10% decrease in line resistance while showing similar chain via resistance (Fig. 4). The lower resistance is probably attributed to the thinner barrier and the correspondingly increased Cu volume. Fig. 5 shows RC curve for TaN\Ta and TaN\Co stacks and about ~10% decrease in the RC value was also observed when a TaN\Co stack was applied.

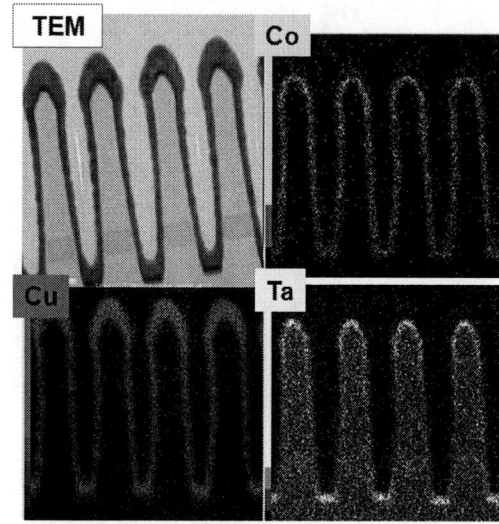

Figure 2.   Almost 100% step coverage of CVD Co could be obtained in cross sectional STEM-EDS images of TaN\Co\Cu seed.

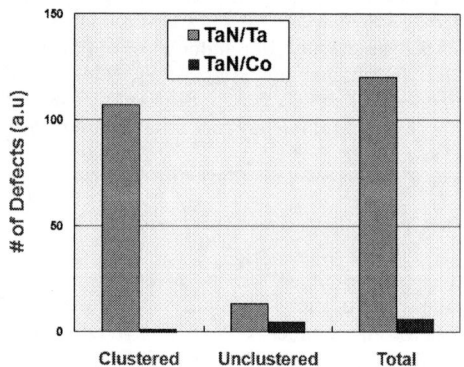

Figure 3.   Classified numbers of defects after CMP was significantly decreased by using TaN\Co barrier.

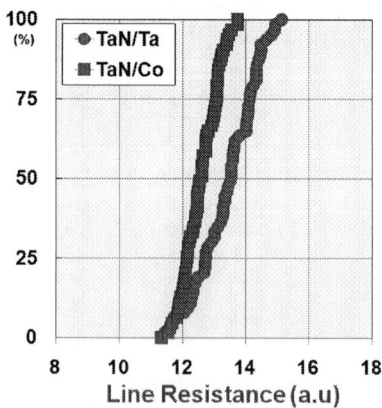

Figure 4.   About 10% decrease in line resistance was seen in TaN\Co samples.

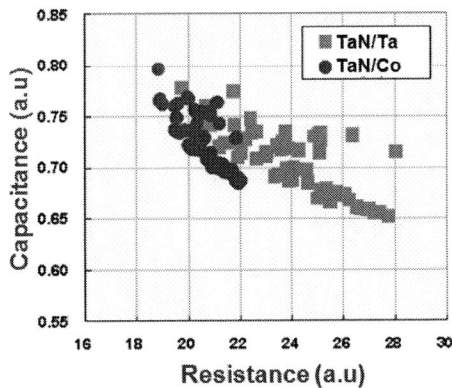

Figure 5. RC value decreased by 10% with the resistance when a TaN\Co stack was applied.

C. Resistace gradation trend during Electromigration

When CVD Co barrier was adopted, two different trends of resistance degradation were observed in upstream EM graph. Firstly, MTTFs (median time to failure) of CVD Co groups were increased by three times of the reference's group as shown in Fig. 6. Secondly, some CVD Co samples show gradual increases in resistance, contrary to the abrupt increase in case of PVD TaN\Ta. The gradual increment of resistance for TaN\Co barrier has been known to result from the growth of voids. Since TaN\Ta barrier has high resistivity, the current flow might be stopped when voids nucleate or grow. However, Co layer can act as a shunting layer even after a void formation due to a relatively low resistivity, and endure until it reached to a final failure size.

The physical limitation of PVD barrier is shown in Fig. 7. We can see that the thickness of PVD TaN\Ta layer depends predominately on the profile of the sub-layer. Very thin or even discontinuous layers (dashed red arrows) were observed along with the normal coverage (blue solid arrows). Due to the intrinsic asymmetric deposition in PVD system and variation of etching\deposition ratio over the wafer, it is difficult to achieve a uniform deposition of barrier metals on the surfaces with negative or 45°-tilted slope. If the coverage of barrier metal and seed Cu is insufficient, multiple micro voids would form during the subsequent electroplating process (Fig. 7 b).

D. EM failure mechanism

Lognormal plot (Fig. 8) helps us see the enhanced lifetime clearly. All time to failures (TTFs) of the TaN\Co group were shifted three times longer than those of the TaN\Ta, regardless of TaN thickness.

In order to understand the role of CVD Co barrier in EM failure mechanism, focused ion beam (FIB) analysis was performed for the samples with the whole range of lifetime that were plotted in Fig. 8. All of voids were observed in the via in the TaN\Ta samples, which might be related with a lack of barrier coverage as mentioned above. This phenomenon is similar to the CVD Co samples. However, other samples (Fig. 8, 9 f,g) showed voids in the M2 line, not in the V1 via. These showed longer TTF and it is believed to be a major factor for increased EM lifetime. It is the prominent evidence that robust

via can be achieved by adopting CVD Co barrier and thus enhance the EM lifetime. The voids are bigger in the samples with Co than those without it. Even the void occurred in the via, it grows towards the direction of M2 line. This means that killer void which cause open failure is larger with Co barrier.

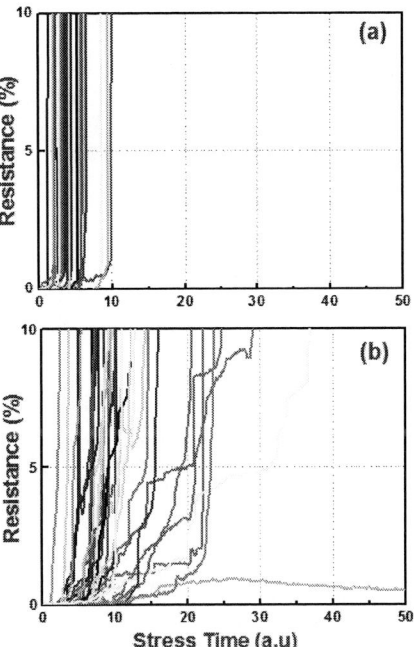

Figure 6. For upstream EM, resistance degradation trends are quite different between two groups ; (a) TaN/Ta, (b) TaN/CVD Co. When CVD Co barrier was adopted, increased MTTFs (median time to failure) and gradual increase in resistance were observed.

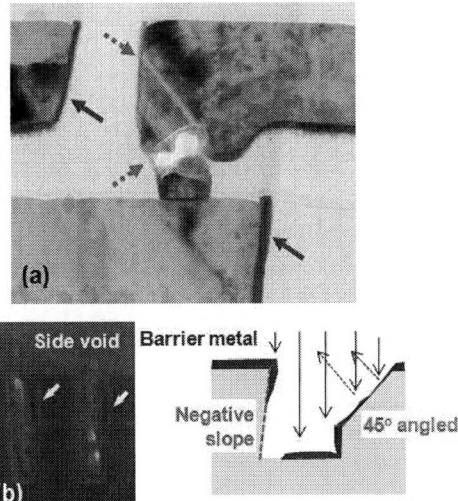

Figure 7. (a) Very thin or discontinuous layers (dashed red arrows) were observed along with the normal coverage (blue solid arrows) in TEM image of PVD TaN\Ta sample after EM test.
(b) Vertical SEM image shows the voids during elecroplating due to insufficient coverage of barrier metal and seed Cu.

978-1-4244-9113-1/11 $26.00 © 2011 IEEE

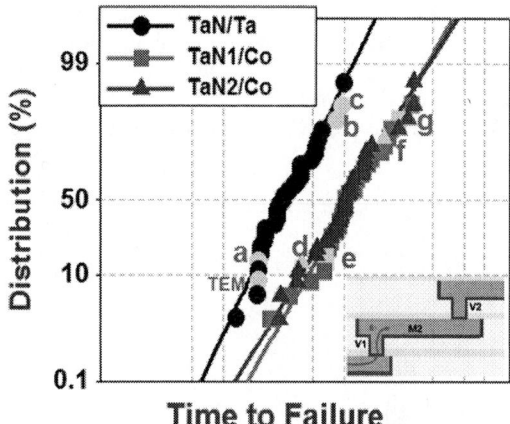

Figure 8. Lognormal probability plot of upstream EM show the enhanced lifetime clearly.
* Green samples were characterized by FIB and TEM in Fig. 7, 9.

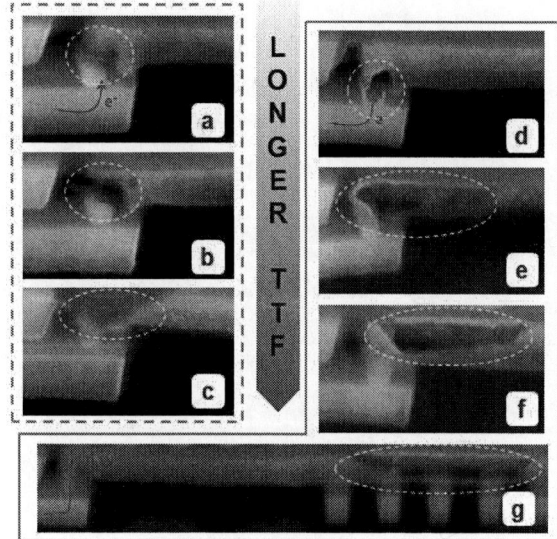

Figure 9. FIB analysis was performed for the samples with the whole range of lifetime that were plotted in Fig. 8; (a)-(c) : TaN/Ta, (d)-(g) : TaN/CVD Co

In case of a downstream (V2→M2) structure, we could not observe the shift of TTF like the upstream. It means the failure mechanism did not change and FIB images also confirmed that all of voids show the same trend for both TaN\Ta and TaN\Co groups (Fig. 11). When a void is generated under the via, it gives rise to the fatal failure in a very short time. Longer lifetime would be achieved if a void nucleates in distance of the via and grows to the opposite direction. Extrapolated lifetime to operation condition based on Black's equation is expected to increase since lognormal distribution became stiff by removal of early failures when CVD Co barrier was involved. This comes from enhanced fill performance and eliminates pre-existing voids prior to EM stressing. The vulnerable point for void nucleation was found to locate at the interface of Cu line contacted under the via, at which good adhesion of CVD Co liner to Cu could hardly affect the migration. Metal capping

process would be much more effective for suppressing the void nucleation.

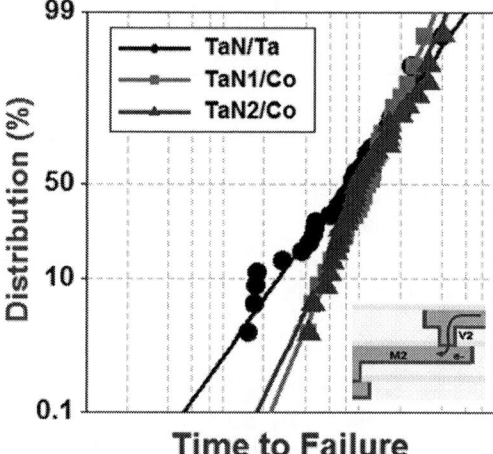

Figure 10. The shift of TTF was not observed in lognormal probability plot of downstream EM like the upstream, but the distribution became stiff by removal of early failures when CVD Co barrier was involved.

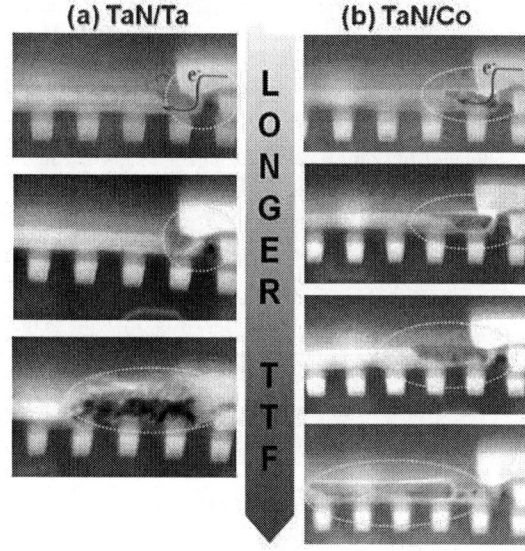

Figure 11. In FIB images, all of voids show the same trend for both barrier groups; (a) : TaN/Ta, (b) : TaN/CVD Co

In conclusion, we demonstrated that highly reliable dual damascene structure could be achieved by using CVD Co barrier. All properties which are required for logic device including void-free, low-resistance and better EM reliability were improved with an optimized condition of CVD Co process. It reduces the thickness of the barrier by 30%, and helps void-free Cu filling due to large opening. Then, EM in the via is deterred due to absence of pre-existing voids, consequently lifetime increases.

The importance of CVD barriers will be focused more and more, because the imperfect filling issue becomes more serious with the advanced technology node. In this study, we found that the use of CVD Co barrier dramatically enhanced EM lifetime by formation of robust via in a dual damascene structure due to its excellent gap-fill performance as well as better adhesion to Cu.

REFERENCES

[1] M. Ueki, M. Hiroi, N. Ikarashi, T. Onodera, N. Furutake, N. Ionue and Y. Harashi, "Effects of Ti Addition on Via Reliability in Cu Dual Damascene Interconnects", IEEE Transl. Elec. Dev., vol. 51, pp. 1883-1891, November 2004

[2] S. Yokogawa, H. Tsuchiya, Y. Kakuhara and K. Kikuta, " Analysis of Al Dopping Effects on Resistivity and Electromigration of Copper Interconnects", , IEEE Transl. Device and Materials Reliability vol. 8, pp 216-221, March 2008

[3] T. Ponnuswamy, S. Park, N. Dubina, T. Mountsier, F. Greer, R. Shaviv, K. Chattopadhyay, A. McKerrow, E. Webb and J. Reid, "Reliability of Cu/Ru/TaN Interconnects", Adv. Metallization Conf., pp15-16, 2008

[4] J. Lu, H.C. Ha, J. Aubuchon, P. Ma, S.H. Yu and M. Narasimhan, "Conformal CVD Co deposition for enhancement of Cu gapfill application", Adv. Metallization Conf., pp21-22, 2008

[5] P. Ma, Q. Luo, A. Sundarrajan, J. Lu, J. Aubuchon, J. Tseng, N. Kumar, M. Okazaki1, Y. Wang, Y. Wang, Y. Chen, M. Naik, I. Emesh and M. Narasimhan, "Optimized Integrated Copper Gap-fill Approaches for 2x Flash Devices", International Interconnect Tech. Conf., pp38-40, 2009

[6] M.A. Korhonen, P. Borgensen, K.N. Tu and C.-Y. Li, "Stress evolution due to electromigration in confined metal lines", J. Appl. Phys., vol. 73, pp. 3790-3799, 1993

[7] J.R Black, "Mass Transport of Aluminum by Momentum Exchange with Conducting Electons", Proc 6th Annual Int Rel Phy Symp, pp148-159, 1967

[8] J. Gambino, J. Wynne, S. Mongeon, D. Meatyard, H. Bamnolker, L. Hall, N. Li, M. Hernandez, P. Little, M. Harmed and I. Ivanov, "Yield and Reliability of Cu Capped with CoWP using a Self-Activated Process", International Interconnect Tech. Conf., pp30-32, 2006

[9] N. Kawahara, M. Tagami, B. Withers, Y. Kakuhara, H. Imura, K. Ohto, T. Taiji, K. Arita, T. Kurokawa, M. Nagase, T. Maruyama, N. Oda, Y. Hayashi, J. Jacobs, M. Sakurai, M. Sekine and K. Ueno, "A Novel CoWP Cap Intergration for Porous Low-k/Cu Interconnects With NH3 Plasma Treatment and Low-k Top (LKT) Dielectric Structure", International Interconnect Tech. Conf., pp.152-154, 2006

[10] C.-C. Yang, D. Edelstein, K. Chanda, P. Wang, C.-K. Hu, E. Liniger, S. Cohen, J.R. Lloyd, B. Li, F. McFeely, R. Wesnieff, I. Ishizaka, F. Cerio, K. Suzuki, J. Rullan, A. Selsely and M. Jomen, "Integraion and Reliability of CVD Ru Cap for Cu/Low-k Development", Interconnect Tech. Conf., pp255-257, 2009

[11] S. Yokogawa, K. Kikuta, H. Tsuchiya, T. Takewaki, M. Suzuki, H. Toyoshima, Y. kakuhara, N. Kawahara, T. Usami, K. Ohto, K. Fujii, Y. Tsuchiya, K. Arita, K. Motoyama, M. Tohara, T. Taiji, T. kurokawa and M. Sekine, "Tradeoff Characteristics Between Resistivity and Reliability for Scaled-Down Cu-Based Interconnects", IEEE Transl. Elec. Dev., vol. 55, pp. 350-357, January 2008

[12] M. H. Lin, Y. L. Lin, K. P. Chan, K. C. Su and T. Wang, "Copper Interconnect Electromigration Behavior in Various Structures and Precise Bimodal Fitting", Japanese Jounal of Appl. Phys. Vol 45, pp700-709 , February 2006

[13] H.B. Lee, T. Matsuda, J.H. Kang, H.K. Jung, J.W. Hong, J.H, Yun, J.S. Park, J.M. Lee, I.S. Park, G.H. Choi, S. Choi and C. Chung, " CVD Cobalt as an enhancement layer to improve Cu/Low-k interconnect performance", Adv. Metallization Conf., 2010

Electromigration-resistance enhancement with CoWP or CuMn for advanced Cu interconnects

Cathryn Christiansen[1], Baozhen Li[1], Matthew Angyal[2], Terence Kane[2], Vincent McGahay[2], Yun Yu Wang[2], Shaoning Yao[2]

IBM Systems and Technology Group

[1]Essex Junction, VT 05452, [2]Hopewell Junction, NY 12533

christia@us.ibm.com

Abstract— **Suppressing Cu diffusion along the Cu/Cap interface has proven to be one of the most effective ways to enhance the electromigration (EM) resistance of advanced Cu interconnects. Two methods, depositing a thin layer of CoWP on the Cu surface and doping the Cu seed layer with Mn, are presented in this paper. While each effectively enhanced the EM performance, they behaved somewhat differently in improving the line-depletion and via-depletion EM performance. CoWP functioned primarily as a Cu surface modifier and did not alter the Cu diffusion behavior below the surface, making Cu interconnects capped with CoWP very sensitive to defects in the via. As a result, the hardware processed with CoWP had greatly increased EM failure times, but also had large variability in failure times and activation energy. On the other hand, the hardware with the CuMn seed layer relied on Mn segregation to the Cu surface to slow down the Cu diffusion, plus Mn also may have diffused to grain boundaries and defective areas of the liner. Although the EM failure times of Cu interconnects with CuMn seed in some cases were not as long as those with CoWP, the variability and sensitivity to process defects was reduced.**

Keywords-electromigration; line; via; CoWP; CuMn

I. INTRODUCTION

EM is one of the main reliability concerns for integrated circuits as technology dimensions shrink, because circuit current density requirements increase while the capability of Cu via and line structures to withstand EM induced damage decreases. Several effects of technology scaling make EM damage occur more rapidly in more advanced technologies. First, smaller dimensions mean a smaller void is needed to cause a circuit failure. In addition, Cu grain growth becomes more difficult in very fine lines, resulting in less bamboo type structures and a larger grain boundary diffusion contribution to EM failures. Furthermore, new dielectric materials introduced to lower capacitance are mechanically weaker with more integration challenges for insuring the good liner and seed coverage necessary to completely fill the lines with Cu. Consequently, the measured median failure times (t_{50}s) for pure Cu interconnects consistently decreased from each technology generation to the next, as shown by the black squares in Fig. 1. At the same time as these challenges are occurring, the current densities required in the metal lines are increasing because the demanded performance and required voltages are not scaling at the same rate as the dimensions, leading to the introduction of new materials to enhance the EM-resistance of Cu.

Under the influence of an electric current, Cu movement proceeds along several different diffusion paths, and different EM enhancement techniques address these diffusion paths differently. The most important of these is interface diffusion

at the Cu/cap interface. Grain boundary diffusion, Cu/liner interface diffusion and bulk diffusion are of secondary importance. In order to meet the high performance, high current density needs of new technologies, the EM performance of metal lines and vias can be improved by strengthening the Cu/cap interface to slow down Cu diffusion along the historically fastest diffusion path. This can be achieved by depositing a material directly on the top Cu surface, such as CoWP [1], or by doping the Cu seed with other elements [2,3,4], such as Mn, which then move to the top surface during a further processing step. This paper discusses the differences in EM-resistance enhancement of two methods, depositing a CoWP cap or using a CuMn dopant.

Both CoWP and CuMn have each been proven to offer EM-resistance enhancement [1-4]. Though the manufacturing processes are very different, the final effect on EM is that both function as Cu/Cap interface modifiers. The CoWP processing involves selectively depositing a thin layer of CoWP on the Cu surface, which is very effective to slow down Cu/cap interface diffusion. Because Co diffusion into Cu is very slow at temperatures below 300C, the CoWP layer, in theory, should have little or no impact below the Cu surface. Thus CoWP cannot slow down Cu diffusion along grain boundaries and the Cu/liner interface. This means it is very important to control Cu grain size and Cu fill quality for CoWP to be effective. In contrast, doping Mn into Cu seed is a PVD process followed by a Mn segregation step to the top Cu surface for Cu diffusion suppression. In addition, some Mn may diffuse to and remain at grain boundaries to help slow Cu diffusion there. Furthermore, Mn may help patch some liner defects when it accumulates at defective spots.

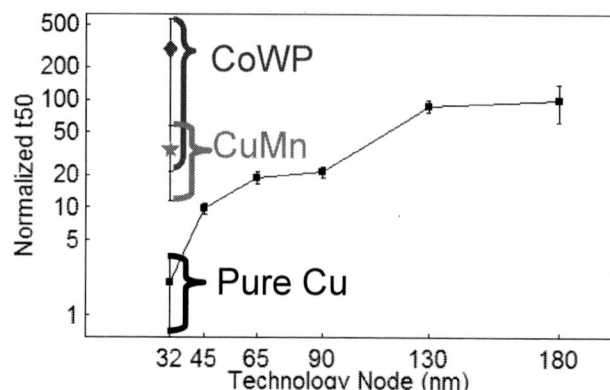

Figure 1. Normalized median failure times versus technology node for 6 technology generations. The symbols are as follows: black square for pure Cu, blue diamond for Cu with CoWP capping layer, red star for Cu with Mn doping.

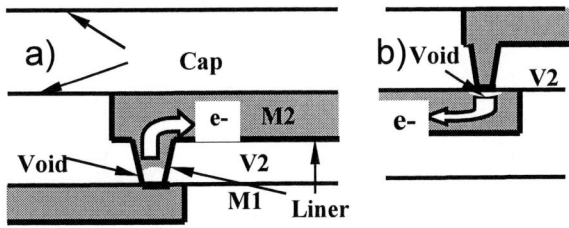

Figure 2. Schematic of dual damascene test structures with typical void locations for two cases: a) via-depletion or upstream stress (V1/M2), and b) line-depletion or downstream stress (V2/M2)

The overall effects on t_{50} of CoWP and CuMn at the 32nm node are shown in Fig. 1 as the blue diamond and red star, respectively. This paper examines the details of the EM failure time distributions and kinetics for hardware processed with CuMn or CoWP, and discusses how the differences in the nature of the CoWP and CuMn processing may have caused differences in the EM performance.

II. EXPERIMENTAL

EM experiments were performed on dual damascene Cu via/line structures with a Ta/TaN liner in a dielectric material with k=2.7 Applied temperatures were 275-345°C, and current densities were up to 35mA/μm^2. A 10% resistance change was the failure criterion. The 200μm long lines had widths of 0.05μm to 0.18μm, and the via diameter was 0.05μm. Both via-depletion mode, where the electrons flow from the via below the line upstream into the line under stress (V1/M2), and line-depletion mode, where the electrons from the via above the line downstream into the line under stress (V2/M2), were studied, as shown schematically in Fig. 2.

III. RESULTS

For either CoWP or CuMn, the effectiveness of the enhancement of EM failure times depended greatly on the health of the base process, in particular the via profile and liner coverage. In both Figs. 3 and 4 A and C are data from an un-optimized process and B and D are data from base processes which are better optimized for EM performance. When CoWP and CuMn processes were each used with varying base processes, smaller vias and/or poor liner coverage in the vias resulted in worse EM failure time distributions. However, there were differences in the effects of the two types of EM enhancers on via-depletion vs. line-depletion mode.

CoWP had a large variation in EM failure times of about 10X-25X for both line-depletion and via-depletion modes with base process changes, as shown by comparing A to B and C to D in Fig 3. The changes were correlated between the two modes; line-depletion and via-depletion both got better or worse together.

On the other hand, CuMn had a moderate variation in failure times of about 5X for via-depletion (A, B of Fig. 4) and small variation in failure times of only 1.5X for line-depletion (C, D of Fig. 4). In other words, changing the base

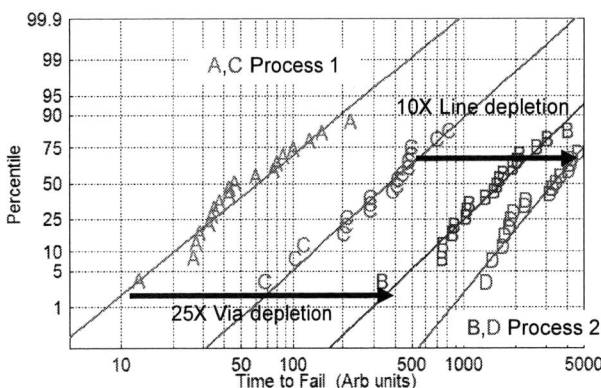

Figure 3. V1→M2 (A, B) and V2→M2 (C, D) failure time distributions for wafers with the same CoWP process and two different base processes. A & C are from process 1; B & D are from process 2. M2 lines are all 0.05μm wide.

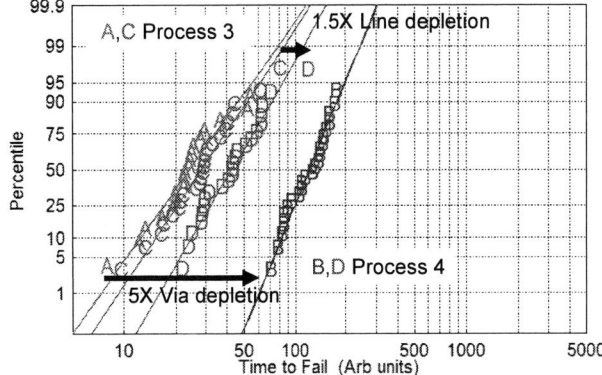

Figure 4. V1→M2 (A, B) and V2→M2 (C , D) failure time distributions for wafers with the same CuMn process and two different base processes. A & C are from process 3; B & D are from process 4. M2 lines are all 0.05μm wide. The stress temperature in this figure was the same as in Figure 3.

process with CuMn modulated via-depletion but did not affect line-depletion very much.

Another difference between CoWP and CuMn was seen by direct comparison of line-depletion and via-depletion mode failure distributions with the same process. For CoWP, no matter what the base process, line-depletion always had longer failure times than via-depletion. This was different from CuMn, where via-depletion had as long or longer failure times than line-depletion. In addition, the spread in the distributions as measured by the lognormal sigma was much smaller for CuMn (0.3-0.5) than CoWP (0.6-0.9), especially for the via-depletion mode.

Wide lines also had different EM behavior in CoWP and CuMn. For CoWP, no EM failures were observed for lines 0.1μm or wider at stress conditions of 345 °C and 35mA/μm^2 in 1000 hours. CuMn also enhanced the EM behavior of wider lines, but not as much. Failures were observed in wider lines with CuMn, with failure times for 0.1μm or wider with CuMn about 5X longer than for 0.05μm, as illustrated in Fig. 5.

978-1-4244-9113-1/11 $26.00 © 2011 IEEE   313

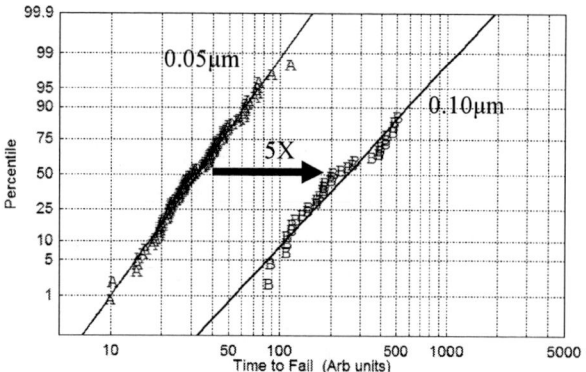

Figure 5. V1→M2 failure time distributions for different width structures on the same wafer processed with CuMn. A is 0.05μm wide and B is 0.10μm wide. The stress temperature in this figure was the same as in Figure 3.

The results of kinetics studies of line-depletion and via-depletion stress modes for CoWP and CuMn are shown in Figs. 6-8 and Table 1. For either CoWP or CuMn, the lognormal sigma was constant with temperature within experimental error, as shown in Figs. 6 and 7. For either CoWP or CuMn, the activation energy, ΔH, was always lower for via-depletion than for line-depletion by 0.07 to 0.20eV, as indicated in Table 1. Although this difference is within the 90% confidence bounds, the fact that via-depletion consistently had lower activation energy in all four cases studied makes it worth noting.

Either CoWP or CuMn had higher ΔH for the better optimized processes with the longer failure times (CoWP process 2 in Fig. 3 Table 1, and CuMn process 4 in Fig. 4 and Table 1). In addition, the activation energy for CuMn was approximately the same as CoWP at 1.05-1.20eV, for the respective optimized processes for EM. However, CoWP had a greater variation in ΔH between base processes and a greater variation in ΔH between line and via-depletion.

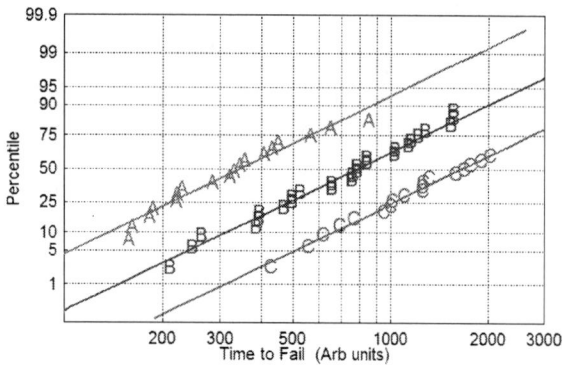

Figure 6. V1→M2 failure time distributions at 3 different temperatures for 0.05μm wide lines from a wafer processed with CoWP. A is 345°C, B is 325°C, and C is 300°C. Solid lines are lognormal fits to the data with a common sigma of 0.72.

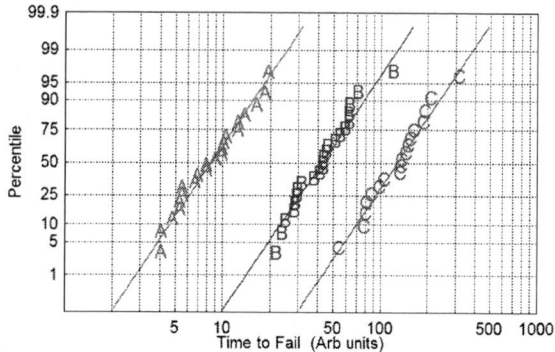

Figure 7. V1→M2 failure time distributions at 3 different temperatures for 0.05μm wide lines from a wafer processed with CuMn seed. A is 345°C, B is 300°C, and C is 275°C. Solid lines are lognormal fits to the data with a common sigma of 0.47.

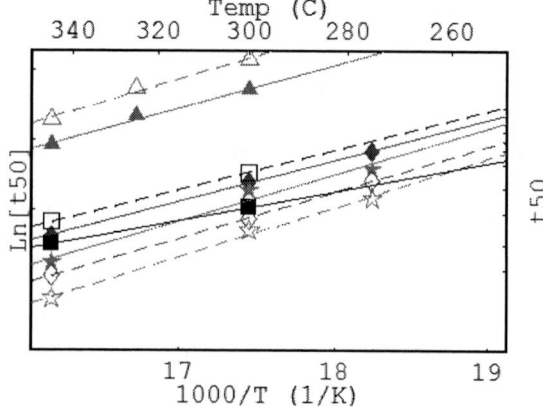

Figure 8. Failure time vs. temperature for 0.05μm M2 lines with 4 different process variations. The symbols are as follows: black square for process 1 CoWP, green triangle for process 2 CoWP, red star for process 3 CuMn, blue diamond for process 4 CuMn. Solid symbols are via-depletion. Open symbols are line-depletion.

TABLE I.   ACTIVATION ENERGIES FOR 0.05UM M2 LINES WITH 4 DIFFERENT PROCESS VARIATIONS.

Process	Via Dep. ΔH (eV)	Line Dep. ΔH (eV)
1 CoWP	0.70	0.96
2 CoWP	1.05±0.11	1.15±0.12
3 CuMn	1.00±0.04	1.13±0.04
4 CuMn	1.13±0.18	1.20±0.09

IV. DISCUSSION

For either CoWP or CuMn, the EM performance enhancement was realized by suppressing the Cu diffusion along the top surface. This made local defects (e.g. pre-existing voids) and grain size distribution very important factors for their effectiveness. As a result, the EM enhancement effectiveness was different between via-depletion and line-depletion, and between narrow and wide lines.

The aggressive scaling of advanced technologies created more challenges for narrow lines than wide lines for liner coverage and Cu filling quality. The narrow lines were more prone to Cu filling defects (voids or high vacancy density) and were harder to grow grains in during anneal. Consequently, the minimum width lines became the worst case for EM reliability.

The dominant failure location in this hardware for via-depletion was in the middle of the via (Fig. 9). Due to un-optimized patterning, the liner in this region was sometimes too thin with small discontinuities. Mn from the adjacent seed layer may have helped to patch these spots if it accumulated there as shown in Fig. 10 [2]; however, CoWP was only on the surface. A local defect, such as a pre-existing void from poor Cu fill due to poor liner coverage, only needed short range Cu diffusion along grain boundaries and/or defect surfaces to cause a fail. Though both CoWP and CuMn can slow Cu surface diffusion, only CuMn may help local defects, which was why CuMn had better improvement of via-depletion EM lifetime projections. Similarly, local defects in the via also may explain the lower activation energy measured for via-depletion as compared to line-depletion.

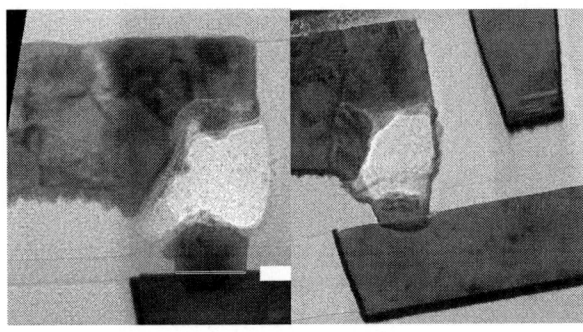

Figure 9. TEM micrographs of V1→M2 failures showing similar void locations in the center of the via. The sample on the left was processed with CoWP; the sample on the right was processed with CuMn.

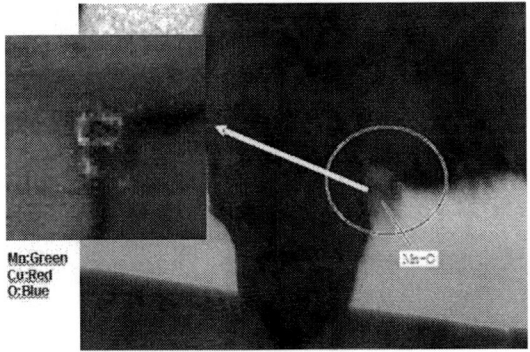

Figure 10. Main figure: TEM micrograph of an unstressed sample showing liner weakness. Inset: EELS mapping of Mn(green), Cu(red), O(blue), showing Mn segregation to the location of the weak liner. [2]

The sensitivity of the measured activation energy to integration process indicates that the measured activation energy was a composite value for different Cu diffusion processes. Higher values than those listed in Table 1 are expected for bamboo grain size structures, as reported in [1]. The lower measured activation energy was believed caused by fast Cu diffusion along grain boundaries of the fine grain lines and along local defect surfaces [5]. Vias are more prone to process defects, such as poor liner coverage along sidewalls and preexisting voids. These defects not only contribute to early fails and shallow failure time distributions (high lognormal sigma), but also cause the lower apparent activation energy.

In addition to the differences in failure times and activation energy, there was a significant difference in the spread of the distributions as measured by sigma. CoWP was more effective at slowing down Cu surface diffusion, but less effective on local defects than CuMn. Thus samples with less local defects got a larger increase in failure times (strong ones got much stronger) from CoWP, while defective ones got less of an increase, leading to a high observed sigma. CuMn may not be as effective as CoWP at slowing down Cu/cap interface diffusion, since its function was not purely limited to the Cu surface; however, it may have helped more on the samples with more local defects such as small grains and weaker liner coverage resulting in a tighter failure time distribution.

To get the maximum EM benefit from CoWP, a good quality via was needed to get good liner and seed coverage. In contrast, CuMn was more tolerant of via quality. In addition, trade-offs such as electrical resistance increases and manufacturing complexity must also be considered when choosing an EM-resistance enhancement method.

V. SUMMARY

While either a CoWP layer or Mn doping in Cu seed could significantly improve EM performance, due to the subtle differences in their functionality, they reacted differently to process defects, design layout features and EM failure modes. CoWP primarily functioned as a Cu surface modifier, thus reducing process defects in the via and in the line were critical to its effectiveness. Mn doping in Cu seed improved the Cu diffusion behavior below the surface, so that it was less sensitive to those process defects. Nevertheless, minimizing process defects is essential to maximizing the effectiveness of either CoWP or CuMn.

ACKNOWLEDGEMENTS

The authors wish to acknowledge Chad Burke, Kevin Lindstam, and Nicholas Hogle, who helped to collect the EM stress data. Management support from Dinesh Badami and Stephan Grunow, and technical discussions within the IBM reliability and process integration teams are also highly appreciated.

REFERENCES

[1] C.-K Hu, et al, "Effect of overlayers on electromigration reliability improvement for Cu Low-k interconnects,", IEEE Inter. Rel. Phys. Symp (IRPS), 2004, pp. 222-228

[2]	T. Nogami, et al, "High reliability 32nm Cu/ULK BEOL based on PVD CuMn seed and its extendibility", IEDM 2010, 33.5.1 – 33.5.

[3]	M. Iguchi, et al, "Optimization of metallization process for 32nm node highly reliable ultralow-k (k=2.4)/Cu multilevel interconnects incorporating a bilayer low barrier cap (k=3.9); IEDM 2009, 36.1.1-36.1.4

[4]	S. C. Pan, et al "Interface effect on Mn-containing self-formed barrier formation with extreme low-k dielectric integration," IEEE Intern. Interconnect Tech. Conf (IITC) 2010, 3.3

[5]	Zhang, L., et al, "Effects of cap layer and grain structure on electromigration reliability of Cu/low-k interconnects for 45 nm technology node", IEEE Inter. Rel. Phys. Symp (IRPS), 2010, pp. 581 - 585

A Study of Via Depletion Electromigration with Very Long Failure Times

Baozhen Li[1], Cathryn Christiansen[1], Kaushik Chanda[2], Matt Angyal[2], and Jennifer Oakley[2]

IBM Systems and Technology Group

1)   Essex Junction, VT 05452; 2) Hopewell Junction, NY 12533

lib@us.ibm.com

Abstract—**Liner coverage in the via plays a critical role on via depletion EM for dual damascene Cu interconnects. Poor liner coverage at the via bottom often results in early EM fails. On the other hand, if the liner at via bottom is permeable to Cu diffusion, thanks to the constant Cu supply into the via from the line below, a very long or even "immortal" EM failure mode can be observed. This paper discusses how to modulate the Cu diffusion through the via bottom liner and its impact on product reliability.**

Keywords-electromigration; liner; via; immortal; Cu diffusion

I.   INTRODUCTION

There are two distinct electromigration (EM) failure modes for dual damascene Cu interconnects due to the blocking nature of Cu diffusion by the via bottom liner [1-5]. The so called line depletion mode (aka down stream) EM failure mode refers to Cu voiding underneath a via resulting from electrons flowing from a via down to the line below. The via depletion (aka up stream) EM failure mode is named for the case of Cu voiding within the via as a consequence of electrons flowing from the line below upwards into the via. For via depletion EM, bimodal failure time distributions are often observed; with early fails (the weak mode) from voiding in the via and late fails (the strong mode) from voiding away from the via weak spot or away from the via in the Cu line. These different failure modes are determined by the liner quality (coverage and thickness) in the via [2-4].

For the weak mode of via depletion EM failures, two preferred void locations have been reported: voiding at the via bottom [2] and voiding at the via chamfering region [4]. The cause for this subtle difference lies in the via process features. Poor liner coverage at the via bottom perimeter due to via RIE or clean undercut along the Cu and cap interface was believed to lead to EM voiding at the via bottom. And poor liner coverage along the via sidewall in the chamfering region due to liner deposition re-sputtering was thought to cause early EM fails from the via chamfering area.

In addition to the commonly discussed two failure modes (weak and strong), another failure mode is sometimes observed for via depletion EM – a very long lasting or even non-failing (immortal) population with a reasonably long stress duration, say a few times longer than the last observed stress fail time [6-8]. This has been attributed to be caused by poor Cu diffusion blocking (or permeable) liner at the via bottom. The theory is that when the liner at via bottom is too thin or not continuous [6,7], the liner becomes permeable to Cu diffusion. The permeable liner allows Cu diffusion from the usually wide feeding line below into the via and effectively eliminates the atomic flux divergence in the via to make the EM failure times unusually long or immortal. A feature of the EM test structure design (the large feeding line) is one of the major contributing factors to these very long failure times. For different design layouts in real product applications, this failure mode may behave very differently. Furthermore, for advanced Cu interconnects, in addition to the aggressive scaling of the liner thickness, the liner engineering has become very complicated and can result in very different coverage features at the via bottom. Attention needs to be paid to these features of the liner at via bottom and their impact on reliability. To address these concerns, this paper first demonstrates how to modulate this "immortal" failure mode by controlling the Cu diffusion in the feeding line and by different liner engineering, and then discusses the potential impact of this mode to chip design.

II.   EXPERIMENTS AND RESULTS

EM structures with 2 levels of Cu were fabricated with a dual damascene Cu process using a K=2.7 low K dielectric, a Ta/TaN liner and a $SiC_xN_yH_z$ cap layer. The lower level feeding line (~2 um wide, 3.5 um long) was connected to an upper level stress line (0.1um wide, 200um long) by a single via (0.1umx0.1um), as shown in Figure 1. EM stresses were conducted at 300 °C and 25mA/um^2 current density based on the stress line Cu cross sectional area. An EM fail was defined as a resistance increase by more than 10% at stress conditions.

Two methods were used to modulate the potential Cu diffusion through the via bottom liner: 1) suppression of Cu diffusion in the feeding line (limiting the source for Cu diffusion into the via); and 2) adjustment of the via bottom liner process to vary the Cu permeability through the via bottom liner. A comparative via depletion EM stress was then conducted to compare the differences in failure time distributions and resistance increase characteristics with time.

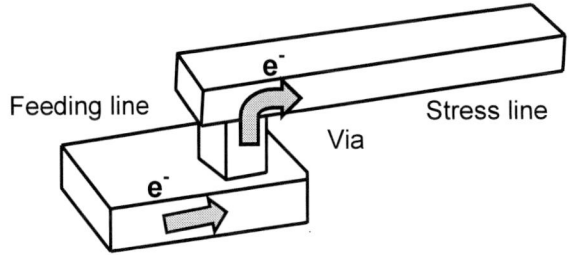

FIGURE 1.   SCHEMATIC OF THE EM STRUCTURE (NOT IN PROPORTION) AND STRESSING ELECTRON CURRENT FLOW (UPSTREAM OR VIA DEPELTION).

A.   *Impact of Cu diffusion in the feeding line*

To modulate the Cu supply to the via during via depletion EM stress, two types of feeding lines were fabricated. One feeding line had the Cu interface diffusivity suppressed by depositing a thin layer of CoWP on the Cu line surface [9, 10];

978-1-4244-9113-1/11 $26.00 © 2011 IEEE

while the other was a normal pure Cu feeding line. The via and the upper level stress line were processed identically with pure Cu for both cases. Figure 2 illustrates the impact of the Cu diffusivity in the feeding line on EM failure distributions. Figure 3 compares the resistance variation characteristics of the two stresses. With the suppression of Cu diffusion in the feeding line, the failure time distribution is clearly bimodal, and no "immortal" population is observed – all samples have failed before 50 hrs of stress. Without suppressing the Cu diffusivity in the feeding line, a high percentage of the population exhibited immortal characteristics -- about half of the samples did not show any resistance increase after 240hrs stress at which time the stress was terminated. The failure time distribution of the non-immortal samples matches well with the strong failure mode of the samples with suppressed Cu diffusivity of the feeding line. The corresponding weak failure mode was not observed in the samples with the pure Cu feeding line.

If sufficient Cu can diffuse from the feeding line through the via bottom liner into the via, the atomic flux divergence at the via bottom liner and Cu interface could be greatly diminished. Therefore, the weak mode failures could change into very long failures or an "immortal" mode due to the weakness of liner coverage in the via. When the Cu diffusion along the feeding line interface was effectively suppressed, Cu diffusion through the permeable liner into the via would be limited by the slow grain boundary or bulk diffusion in the feeding line. With the strong driving force of the high stress current density, though the via bottom liner is still permeable, the EM induced voids due to the divergence at the via bottom (near the liner and Cu interface) could form the early EM failure population. On the other hand, as expected that the strong mode was not impacted by the Cu diffusivity in the feeding line, since these fails were not determined by the via bottom liner weakness, and the EM induced voids were supposed to form away from the via bottom. This result confirms that 1) the permeable liner at the via bottom is the cause for the very long EM failure time population; and 2) this very long failure time population can become the weak mode failure population if the Cu supply into the via is limited.

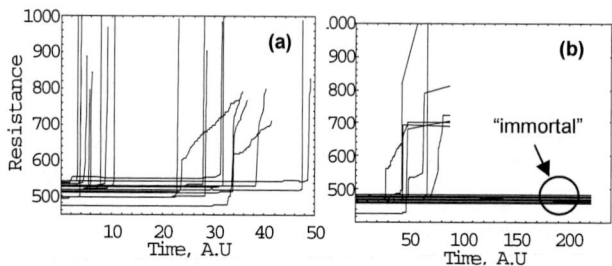

FIGURE 3. RESISTANCE VARIATION WITH STRESS TIME FOR 2 DIFFERENT FEEDING LINES. A – FEEDING LINE WITH CU DIFFUSION SUPPRESSED; B – FEEDING LINE WITH PURE CU

B. Impact of the liner processes

Since the via bottom liner quality is critical for via depletion EM failure distributions [2,4, 6-8], it is important to understand how the liner quality is modulated by the processes. Four different liner processes were evaluated with different thickness and quality at the via bottom. Figure 4 shows EM failure time distributions and Figure 5 shows the resistance variation characteristics with time under via depletion stress for the four different liners. A pure Cu feeding line (no Cu diffusion suppression) was used for all four stresses. Liner A, as a control, used the same process as discussed in the previous section (a known permeable liner to Cu diffusion). While liners B & D might have been effectively non-permeable for Cu diffusion at the via bottom (no immortal population was observed), the significant weak failure mode population indicated some weak liner coverage in the via (either at the via bottom or in the chamfering area). Liner C not only demonstrated non-permeable to Cu diffusion at the via bottom to eliminate the "immortal" population, but also suppressed the weak failure mode. It is the only liner among these 4 processes that showed good current shunting redundancy for all samples stressed -- a gradual resistance increase with time after initial resistance jump as shown in Figure 5 (c). This is another indication that liner C has superior coverage and quality in the via compared to the other three liners.

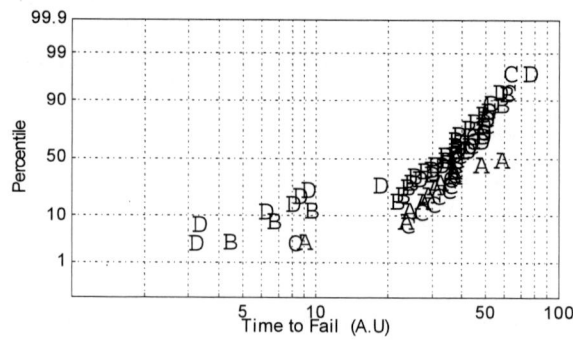

FIGURE 4. VIA DEPLETION EM FAILURE TIME DISTRIBUTION SHOWING IMPACT OF DIFFERENT VIA BOTTOM LINER PROCESSES

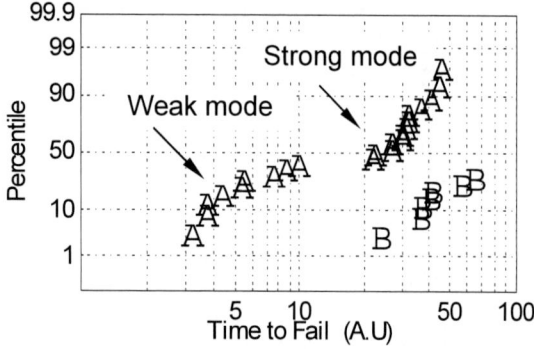

FIGURE 2. VIA DEPLETION EM FAILURE TIME DISTRIBUTION. A – FEEDING LINE CU DIFFUSION SUPPRESSED BY DEPOSITING A THIN LAYER OF COWP ON CU SURFACE; B -- PURE CU FEEDING LINE

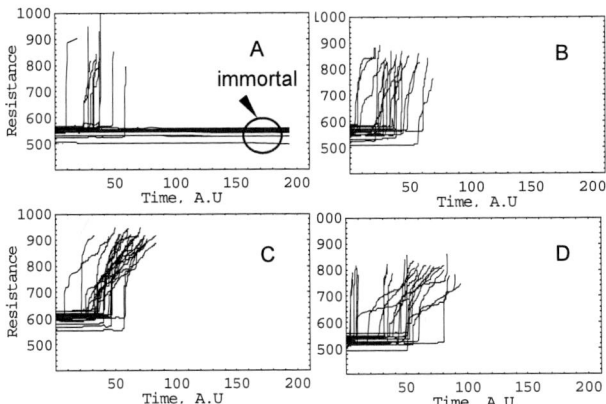

FIGURE 5. RESISTANCE VARIATION WITH STRESS TIME FOR 4 DIFFERENT LINER PROCESSES

C. EM damage evaluation on immortal population

Although the "immortal" population showed no resistance increase during stress, it did not necessarily mean no damage was caused by the EM stress in the structure. The cause for the no resistance increase of the immortal samples was assumed to be no voiding in the critical path of the stressing via and line. It was highly possible that voids caused by the stressing existed in the EM structure. One possibility could be that the voids in the stress line/or via were not large enough yet to cross the entire line or via to cause any observable resistance increase. Another possibility could be that voids were formed somewhere in the feeding line outside the resistance sensing region, as observed by Hu et al [6-8]. While it is important to know if and where the voids have formed, it was found that physically locating these voids through failure analysis was quite challenging and tedious on immortal samples, because a large population needs to be examined on large areas of the stress structure. Due to the size and low current density in the feeding line (a feature of the EM stress structure), it is highly possible that no single large void was formed there. For any other non-critical voids (voids not causing observable resistance increase) in the stress line or via, they may be subsurface. To complement the physical failure analysis difficulties, a quick electrical evaluation by ramping current was performed on a group of immortal samples after 240 hrs stress and on a group of control samples which had never been stressed. The idea was that if significant damage existed on the critical current path, such as voiding in the line or in the via but still sufficient Cu remained along the trench bottom and/or sidewalls with essentially no resistance increase, the current needed to blow the line open should be lower than a line without any damage. If there was no damage or the damage was not in the critical current path, such as minor voiding in the feeding line away from the via, the current needed to blow the structure open should be comparable to the non-stressed samples.

Figure 6 shows the I-R curves of the current ramping. For most of the immortal samples, the breakdown current was very comparable to that of the un-stressed samples. This suggests that the EM stress did not cause significant damage in the critical current path. Two immortal samples show lower

breakdown current, suggesting that damage existed in the critical current path, i.e. in the stress line and/or in the via, though no resistance increase was observed after 240 hrs stress. All post stress samples were electrically verified for their resistance change after cooling down to room temperature before being taken out of the EM stress oven, and all the immortal samples were confirmed having virtually the same resistance values as those when they were initially loaded into the oven. It has not been understood yet why one post stress "immortal" sample showed higher resistance after being loaded to the current ramping tool. The sample resistance change was unlikely to have resulted from handling damage, which usually led to an open circuit. It was speculated that an EM induced void in the via or line might have reshaped itself somewhat during storage to cause the resistance increase.

It was noted that the current ramping may have its own limitations on pre-existing damage (EM induced voids) evaluations. The line or via breakdown during current ramping most likely was caused by joule heating. Due to the layout characteristics of the EM stress structure, a huge temperature gradient could exist within the structure. The large feeding line could be significantly cooler than the narrow stress line, even the via and the ends of the stress line could be much cooler than the bulk portion of the stress line [11]. This brought concerns of the sensitivity of breakdown current to void locations by current ramping. Since the intention of the current ramping was to look for gross EM induced damages in the via or in the stress line, the breakdown current should still offer a good indication of the degree of the EM induced damage in the critical current path.

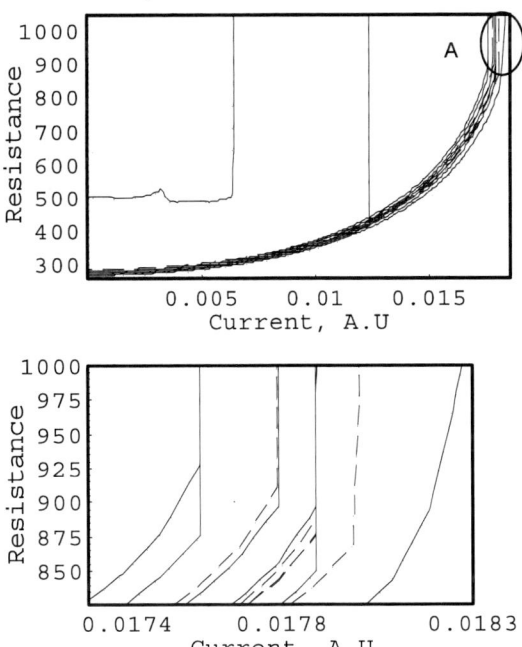

FIGURE 6. CURRENT RAMPING RESULTS OF IMMORTAL SAMPLES (BROKEN LINES) AFTER 240 HRS STRESS AND UN-STRESSED SAMPLES (SOLID LINES). THE BOTTOM PLOT IS THE EXPANSION OF PORTION A FROM THE TOP PLOT

978-1-4244-9113-1/11 $26.00 © 2011 IEEE

III. DISCUSSION

There are two important questions still yet to be clearly answered: 1) why the samples of weak failure mode for non-permeable liners would become immortal, rather than strong mode for permeable liners; 2) what would be the different response (if any) to permeable vs non-permeable liners for the weak EM failure mode due to poor liner coverage at the via bottom and at the via chamfering region. While future studies are clearly needed to answer these questions, we suspect that question 1) might be related to the Blech effect caused void saturation. Detailed discussions on these questions are out of the scope of this paper.

Whether the permeable liner at via bottom is beneficial or detrimental to via depletion EM for a real product depends on the layout and process control. For those layouts similar to the commonly used EM structures with feeding line much larger/wider than the stress line, the immortal EM feature due to the permeable liner at via bottom may help EM reliability. While the constant supply of Cu to the via extends the via EM lifetime, the amount of Cu depleted from the large feeding line does not cause any detrimental voiding or significant resistance increase of the structure. On the other hand, if the feeding line is much smaller than the stress line, the immortal benefit may no longer exist. It could actually be detrimental when voiding in the small feeding line causes early fails. For the case of when the feeding line is comparable to the stress line, it would be very hard to predict if the permeable liner at via bottom would be detrimental or beneficial. Furthermore, the process control of the permeable liner is a major concern. While good liner quality in the via ensures predicable EM behavior, marginal non-permeable liner could translate the "immortal" population to early fail population. Especially if the permeable liner is a result of the variation in liner process, the effect of permeable liner on via depletion EM could become truly unpredictable from wafer to wafer or lot to lot, or from different via density areas on the same wafer. Thorough considerations should be made before deciding if the advantage of permeable liner on via depletion EM can and/or should be taken.

It is a common practice for reliability evaluation to focus on weak mode or early fails. However, the strong mode or late fails should not be simply ignored without understanding the cause and mechanisms. This is especially true for the cases with samples showing unusually long failure times or even immortality. As discussed in this paper, these long lasting failures may not always behave in the same way. Under certain conditions, such as for different design layouts and process variations, this very long lasting population in via depletion EM stress could change to weak mode or early fail population. Understanding and control of these populations could become very important to ensure the final product level reliability.

IV. SUMMARY

The via bottom liner permeability to Cu diffusion can lead to very long lasting or immortal via depletion EM failures. These long lasting or immortal parts could become early fail population if the permeable liner becomes marginally blocking to Cu diffusion. While the early EM stress fails will be the obvious focus for reliability evaluation, it is also important to understand the samples with unusually long failure times. These samples could be deceptive due to the feature of the stress structure and can potentially turn into early fails with design layout and process variations. If not well understood and properly controlled, these very long lasting failures could have important implications to product EM reliability.

ACKNOWLEDGEMENTS

The authors would like to acknowledge the technical discussions with the IBM reliability and process integration teams, especially with Tim Sullivan and C. K. Hu. Chad Burke, Kevin Lindstam and Nick Hogle helped to collect the EM stress data. Management support from Dinesh Badami and Stephan Grunow is also highly appreciated.

REFERENCES

[1] E. Ogawa, et al, "Electromigration reliability issues in dual-damascene Cu interconnections," In *IEEE Trans. Relia.*, 2002, 51(4), pp. 403–419.

[2] J. Gill, et al, "Investigation of via-dominated multi-modal electromigration failure distributions in dual damascene Cu interconnects with a discussion of the statistical implications." In *40th Annual IEEE Int. Relia. Phys. Symp. (IRPS) Proc.*, 2002, pp. 298–304

[3] S. Lee, A. Oates, "Identification and analysis of dominant electromigration failure modes in copper/low-k dual damascene interconnects," In *IEEE Int. Rel. Phys. Symp. (IRPS)*, 2006, pp. 107-114

[4] C. Christiansen, B. Li, J. Gill, "Via depletion electromigration in Cu interconnects," In *IEEE Trans. Devices and Materials Reliability, 6(2006)*, pp. 163-168

[5] A. Oates, M. Lin, "Analysis and modeling of critical current density effects on electromigration failure distribution of Cu dual damascene vias," In *IEEE Int. Rel. Phys. Symp. (IRPS)*, 2008, pp. 385-391

[6] C.-K. Hu, et al, "Mechanisms for very loing electromigration lifetime in dual damascene Cu interconnections," Appl. Phys. Lett., 78 (2001) pp 904-906

[7] E. Liniger, et al, "Effect of liner thickness on electromigration lifetime,"J. Appl. Phys., 93 (2003), pp 9576-9582

[8] O. Aubel, et al, "Investigation of via bottom barrier integration impact on electromigration," In *IEEE Int. Rel. Phys. Symp. (IRPS)*, 2007, pp. 648-649

[9] C.-K Hu, et al, "Effect of overlayers on electromigration reliability improvement for Cu Low-k interconnects," IEEE Inter. Rel. Phys. Symp (IRPS), 2004, pp. 222-228

[10] E. Huang, et al, "CoWP metal caps for reliable 32 nm 1X Cu interconnects in porous ULK (k=2.4)," Proc. of Adv. Metallization Conference, 2009, pp. 33-34

[11] H. Schafft, "Thermal analysis of electromigration test structures," IEEE Trans Electron Device, ED34, 1987, pp664-672

Study of Void Formation Kinetics in Cu Interconnects using Local Sense Structures

K. Croes, M. Lofrano, C.J. Wilson, L. Carbonell, Y.K. Siew, G.P. Beyer and Zs. Tőkei

Imec, Kapeldreef 75, B-3001 Leuven, Belgium
phone: (32)16/281621; fax: (32)16/281576; e-mail: Kristof.croes@imec.be

Abstract - **A test structure that allows the study of void formation kinetics during electromigration is proposed and characterized. Compared to a standard single-via electromigration test structure voltage-senses are placed near the via. This allows monitoring resistance changes before final void formation, while the void formation process is not affected. Part of the samples show single void formation, while for other samples, multiple voids are formed. For the single void case, a model is proposed to calculate void-depth as a function of time. Initially, voids grow faster and this growth slows down towards the end of the void formation process. Estimated velocities during void formation are in the same order of magnitude compared to literature results of drift velocities during void growth. Cases where multiple voids are formed show that voids which initially form further away from the via stop growing upon formation of a void closer to the via.**

Keywords - BEOL, copper, electromigration, void formation, local sense test structure, FEM

I. Introduction

Electromigration (EM) is one of the main metal failure mechanisms of BEOL interconnects. EM is the result of the movement of copper due to an electron wind caused by an electrical current [1]. The result of EM in copper interconnects is voids that span the whole line width and height. Figure 1 shows a typical resistance versus time trace of a copper interconnect which is subjected to an electromigration stress (high temperature and current density). Three phases are observed. In phase I, which is linked to the void formation process, no change in the resistance is seen. Once a void is formed that spans the whole line width and height, the current locally has to flow through the metallic barrier, a phenomenon often referred to as current shunting [2]. Such an event is characterized by a jump in the resistance versus time curve (phase II in figure 1). After this jump, a linear increase of the resistance as a function of time is observed (phase III). This increase is linked to void growth. Quite some literature exists about the drift velocity in metal lines after full void formation [3-5], where the conclusions are based on the resistance behavior after this resistance jump.

As the resistance increase caused by void formation (phase I) is very small, EM systems are not able to detect any resistance change and void formation cannot be electrically characterized. Dedicated test methods have been proposed in literature to study this void formation process, where both

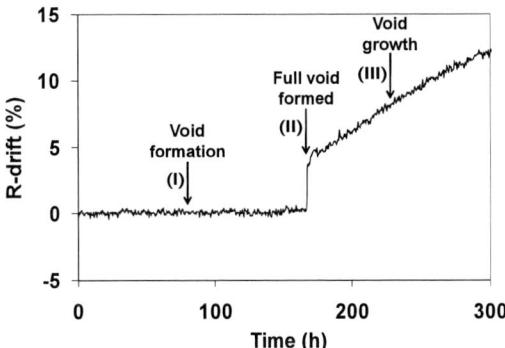

Figure 1   Typical resistance versus time curve of a copper interconnect that is subjected to electromigration stress. Different phases of the degradation process are indicated.

top-down and cross-sectional SEM pictures were made of interconnects during the initial stages of the EM process [6-7]. Such studies are very time consuming and might also influence the void formation process.

This paper summarizes a study where a test structure is proposed which is sensitive enough to electrically measure void formation, i.e. the structure enables measuring resistance changes before the typical jump in the resistance versus time curve.

The layout of the paper is as follows: after discussing the experimental details and introducing the new test structure, typical resistance traces obtained with this structure are shown and discussed. Then, a simple model to derive the void size from these measured resistance traces is put forward. After this, the model fitting is discussed and the experimental range in which the model is valid is argued. Finally, drift velocities during void formation are discussed, both for the single and multiple void case.

II. Experimental

Dual damascene copper structures were integrated on 300mm wafers. Intermetal (IMD) and interlayer dielectric (ILD) was a SiOCH CVD low-k dielectric. The M1-IMD was with k-value ~3.0, porosity ~9% and E-modulus ~15GPa, while via-ILD and M2-IMD was with k-value~2.5, porosity~25% and E-modules ~9GPa. A 5/25nm SiCN/SiCO etch stop layer (E-modulus ~100 GPa) was used between the M1- and via/M2-layer. A 3/3nm TaN/Ta barrier was used between copper and

978-1-4244-9113-1/11 $26.00 © 2011 IEEE

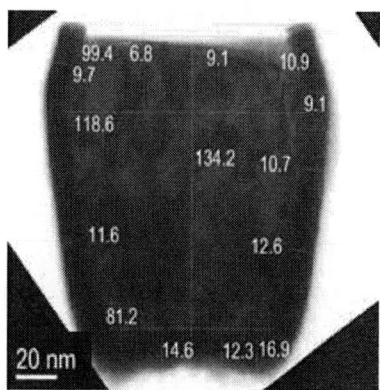

Figure 2   Cross-section of the M2-lines used in this study. Exact copper and barrier dimensios were determined using this TEM.

dielectric. The wafers were passivated with a 35nm SiCN, 300nm SiO$_2$ and 500nm Si$_3$N$_4$-stack.

Dimensions of the lines used for our study were both for M1 and M2 100nm wide and 134nm high, while the via-size was 90nm. About 30 devices were prepared for package level EM at 330°C and 1.1MA/cm^2. During the tests, the electron flow is sent upward into the via (upstream EM). To support our modeling, about 10 samples were inspected after failure using cross-sectional SEM. Our modeling requires exact copper and barrier dimensions as input. These dimensions were obtained using the cross-sectional TEM of an M2-line (Figure 2).

Figure 3A shows a typical EM test structure. For single via-structures, long lines are typically chosen to avoid that the current injectors act as a copper reservoir. The problem with such long structures is that they are not sensitive to small resistance changes. Figure 3B shows the newly proposed local sense test structure where the voltage terminals are put much closer to the via and thus the sensitivity to resistance changes in this region is increased.   Since voids are typically formed within a few microns from the via, we placed these terminals 10µm away from the via.

To better study dominating diffusion interfaces for EM, wafers having a SiC(N)-layer in the M1-trenches were produced. For these wafers, the copper in M1 was fully

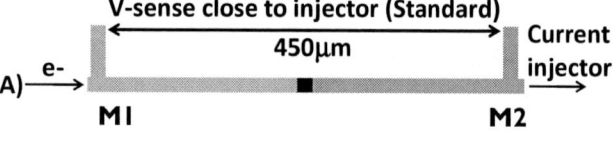

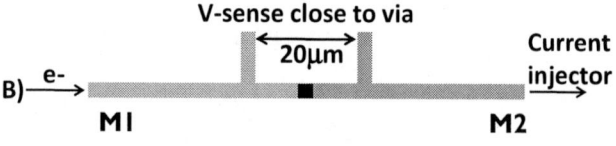

Figure 3   Test structures used A) Standard single via EM-structure with voltage-senses close to the current injector and B) New local sense test structure with voltage-senses closer to via.

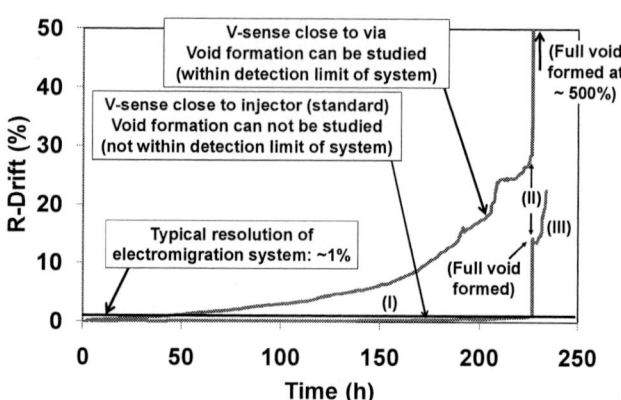

Figure 4   Typical resistance drift versus stress time when sensing at the injector/local senses.

surrounded by SiC(N), which is often reported to be the dominant diffusion path for EM in bamboo or semi-bamboo lines [8-10]. On these wafers downstream electromigration was done at temperatures between 200 and 300°C and 0.75MA/cm^2. Sister wafers with a standard 3/3nm TaN/Ta barrier were used as a reference, where EM tests were performed at temperatures between 270 and 330°C.

III. RESULTS

Figure 4 shows the resistance drift as a function of time when sensing either close to the current injector (standard) or at the local sense probes. Sensing at the standard probes shows the typical behavior as discussed in figure 1: no voltage increase (phase I) followed by a jump that indicates the formation of a full void (phase II) and the void growth phase (phase III). The behavior of the voltage drift in phase I is clearly different when sensing the voltage over the local senses; a clear voltage increase is observed, proving that the newly proposed local sense structure is sensitive to measure resistance shifts in phase I.

To investigate whether adding the local senses have an impact on the void formation process, electromigration tests were conducted on devices from the exact same wafer using either the standard or local sense test structures. Figure 5 shows a lognormal probability plot of the times till the 1st jump (phase II on figure 4) for samples tested at 330°C-1.1MA/cm^2. Both plots fall on top of each other. Hence, moving the voltage senses closer to the via does not change the time to full void formation and consequently also not the void formation process.

The void formation process is further explored in figure 6, where the resistance shift $\Delta R = R - R_{t=0}$ over the local senses of a representative device is plotted as a function of time till full void formation. The first 5-10% (region I) increase happens during almost 100% of the total time to full void formation, while in the last minutes of the void formation process, the remaining 90-95% increase (region II) happens. When copper is left in a forming void, the resistance of the void does not change dramatically as the current can still flow through the

978-1-4244-9113-1/11 $26.00 © 2011 IEEE   322

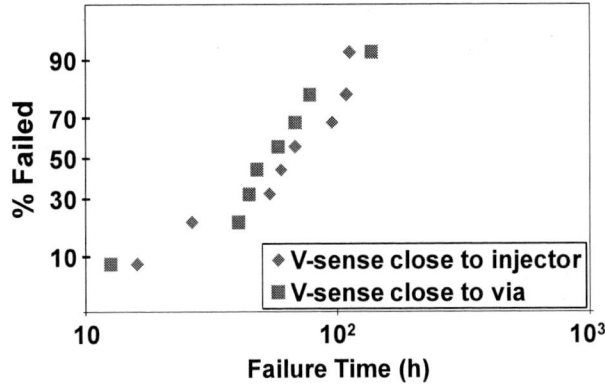

Figure 5 Lognormal plot of times till jump in resistance versus time curves for standard and local sense structures.

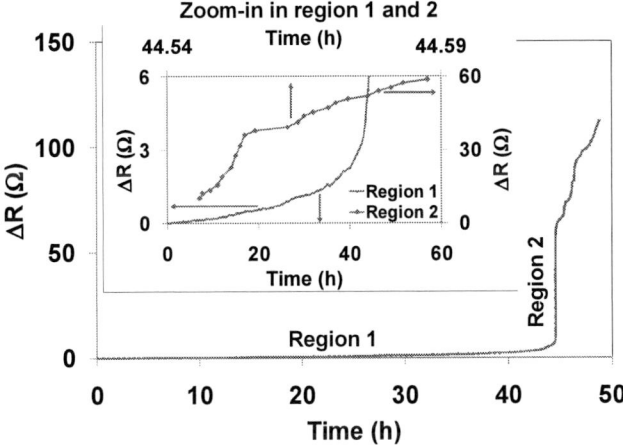

Figure 6 Typical ΔR versus time curve when using the local sense structure with a zoom-in at the different regions during the void formation phase.

remaining copper. Only at the very end of the void formation phase, the resistance will increase dramatically as current has to shunt locally through the metallic barrier. The crucial advantage of the local sense structure is that it allows monitoring the resistance drift even when copper is still left in the line (region I).

During this study, two different shapes of ΔR versus time in region I are observed. A first typical shape is where the voltage-drop increases monotonically at a higher rate with time (Figure 7A), while a second typical shape is a non-monotonic increase of the resistance (Figure 7B). The first case is linked to situations where single voids are formed in the line, while the second case corresponds to samples where multiple voids are formed. Both cases will be treated separately in the coming sections.

IV. SINGLE VOID CASE

To link the measured resistance increase ΔR in figure 7A to the depleted copper volume, we propose the simple model explained in figure 8, where we assume that the length of the full void (L_{void}) is fixed and spans the full line width from the

start and grows into the line with a certain distance h_{void} as a function of time. This model allows considering voids that grow in/above the via or in the line within a small distance away from the via.

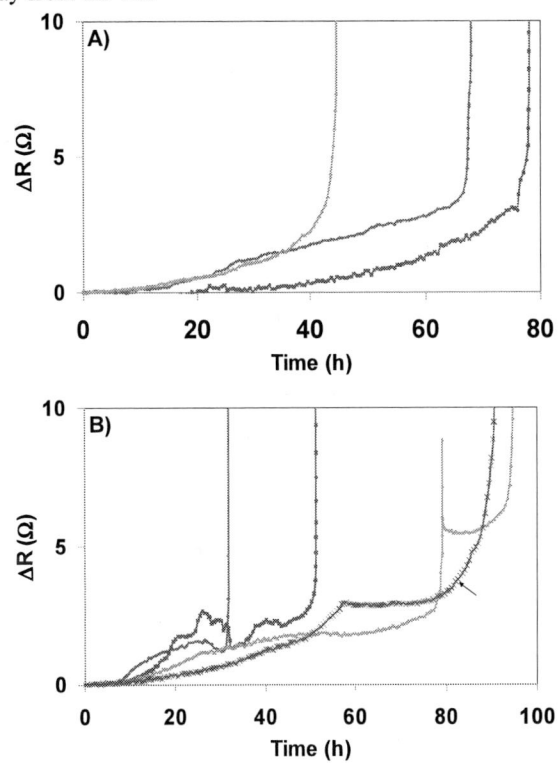

Figure 7 Typical shapes of ΔR versus time observed in this study. A) Monotonic increase at a higher rate with time (single void case). B) Non-monotonic increase (multiple void case).

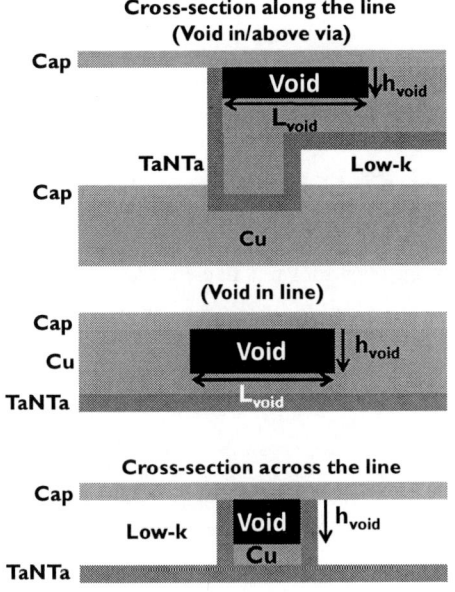

Figure 8 Proposed model to explain single void case. Void length L_{void} is fixed form the beginning and grows into the line (h_{void}) with time.

A. Model fitting

The parameter L_{void} can be estimated from the voltage-shift ΔR after full void-formation [11]:

$$L_{Void} = \frac{\Delta R * Area_{TNT}}{\rho_{TNT}}, \quad (1)$$

where $Area_{TNT}$ and ρ_{TNT} are the cross-sectional area and the resistivity of the TaNTa-barrier, respectively.

Based on a cross-sectional SEM picture after void formation, we calibrated the ρ_{TNT} to be $\sim 250\mu Ohm.cm$ (Figure 9), which is in very good agreement with existing literature data [11] on similar dimensions. Within our experimental range, values for L_{void} are between 80 and 500nm, which are typical sizes of voids as reported by many different authors.

Once L_{void} is estimated, $h_{void}(t)$ can be determined from $\Delta R(t)$ during void formation by considering the formed void as a parallel chain of the available copper height and the TaNTa barrier:

$$\Delta R(t) = \frac{1}{\left(\frac{1}{R_{TNT}(L_{void})} + \frac{1}{R_{Cu}(h_{void}(t), L_{void})} \right)} - \frac{1}{\left(\frac{1}{R_{TNT}(L_{void})} + \frac{1}{R_{Cu}(h_{void} = 0, L_{void})} \right)} , \quad (2)$$

where $R_{Cu}(h_{void}(t), L_{void})$ and $R_{TNT}(L_{void})$ are the resistances at the void location of the copper and the TaNTa barrier for a given h_{void} and L_{void}, respectively:

$$R_{Cu}(h_{void}(t), L_{void}) = \frac{\rho_{Cu}(h_{void}(t)) * L_{void}}{(t - h_{void}(t)) * w} \quad (3)$$

and

$$R_{TNT}(L_{void}) = \frac{\rho_{TNT} * L_{void}}{Area_{TNT}}, \quad (4)$$

where t and w are the copper thickness and width respectively and ρ_{Cu} is the copper resistivity. Note that for fitting this model, the 3 parameters t, w and $Area_{TNT}$ were determined using the cross-sectional TEM shown in figure 2.

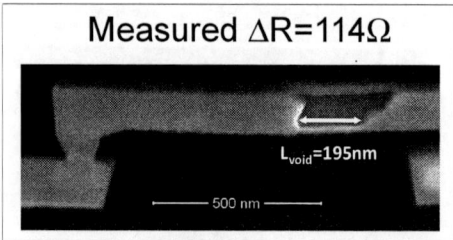

Measured $\Delta R = 114\Omega$

$L_{void} = 195nm$

500 nm

Figure 9   Cross-sectional SEM picture used to calibrate ρ_{TNT} from eq. 1. ΔR of 114Ω was measured after full void formation.

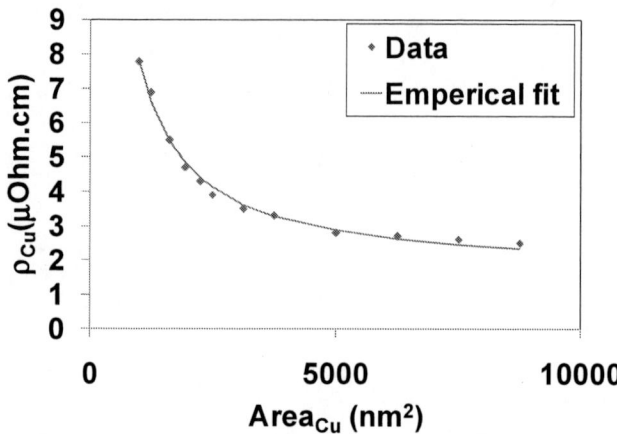

Figure 10   Empirical fit of recently published copper resistivity data at room temperature [12] which allows correcting $\rho_{Cu}(h_{void}(t))$ from eq. 3 for different values of h_{void}.

As the copper resistivity ρ_{Cu} strongly depends on the copper area, we used recently published data [12] to calculate the copper resistivity for a given h_{void} by making use of the empirical fit shown in figure 10:

$$\rho_{Cu}(Area_{Cu}, T) = 1.63 + \frac{6220}{Area_{Cu}} + 6.77 * 10^{-3} * (T - 25), \quad (5)$$

where T is the stress temperature in °C. The first two terms are a direct fit of the data shown in figure 10, while the third term compensates for higher temperatures ($d\rho_{Cu}/dT = 6.77 * 10^{-3}\mu\Omega.cm/K$).

$\rho_{Cu}(h_{void})$ can be calculated from $\rho_{Cu}(Area_{Cu})$ using:

$$Area_{Cu} = (t - h_{void}) * w . \quad (6)$$

For the copper dimensions involved in this study, this empirical fit allows to reliably estimate ρ_{Cu} for values of h_{void} between 0nm and t-10nm.

To determine whether local Joule heating during void formation needs to be accounted for in the model, we performed Finite Element Modeling (FEM), where we estimated the Joule heating of the copper in the region of the void. Figure 11A shows a color map of the Joule heating in a line with a void with $h_{void} = t-10nm$, where the ambient temperature was 304°C and the current density was $0.4MA/cm^2$. Figure 11B shows the maximum observed temperature in the void for current densities between 0 and $2.5MA/cm^2$. Even for such a big void, the Joule heating stays within 10°C and can thus be neglected for the resistivity calculations in Eq. 5 (third term). The low Joule heating is due to the low thermal resistivity of the copper itself, such that the copper line acts as a sink for the heat generated in the void-region.

Figure 12 shows the resulting h_{void} when fitting our model to the data displayed in figure 7A. h_{void} starts to increase from the very beginning of the test. For our long lines, a reported [13-14] incubation time is not observed. As the void formation

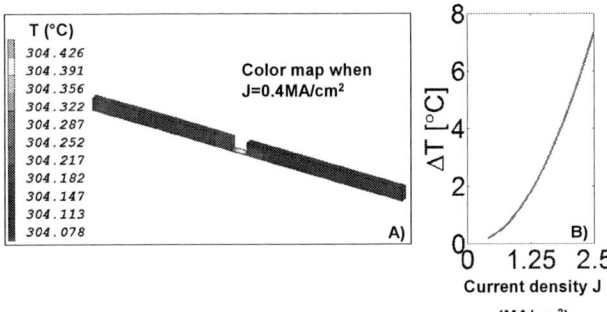

Figure 11 FEM-study of the local Joule heating when a big void of h_{void}=t-10nm is present in the line.

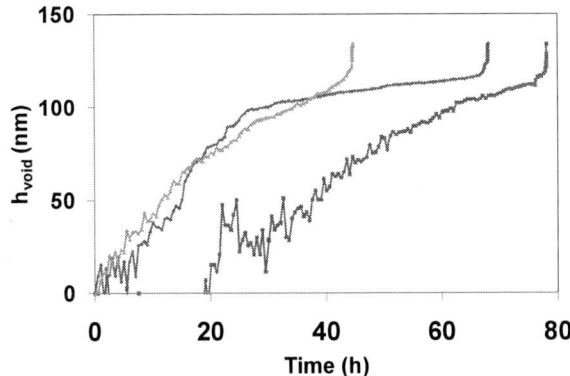

Figure 12 h_{void} as a function of time as estimated from our proposed analitical model for the data displayed in figure 7A as function of time.

progresses, the increase of h_{void} slows down with time. At the end, a rapid increase of h_{void} is observed.

B. Model discussion

We believe, within certain limits, our simple model is able to describe the formation of a single void. The key-assumption in the model is that the voids initially form at the copper/cap interface and not in the via itself, which is a valid assumption for our 90nm-diameter via's, where the barrier deposition is considered conformal and thus the copper/barrier interface in the via is a reliable interface. For such cases, it is well-accepted that void formation starts at the copper/dielectric cap-interface. Initial stages of void growth are modeled [15] and observed experimentally [6-7]. Voids form at interfaces and triple points of copper and these voids agglomerate to form a shallow void that extends typically between grain boundaries. The initial formation at interfaces and triple points is consistent with our observation, where at the initial stages of the test, several small voids of 5-30nm deep are observed at the copper/cap interface (Figure 13).

To further confirm this, we produced samples where we deposited SiC(N) in the M1-trenches and performed downstream EM. This way, the copper is embedded in a

Figure 13 Tiny voids after limited test time

TABLE I   MAIN RELIABILITY PARAMETERS OF DOWNSTREAM EM-TESTS PERFORMED ON DEVICES WITH TaNTa AND SiC(N) AS TRENCH-BARRIER AT M1

Trench-barrier	$t_{50\%}$ at 315°C-0.75MA/cm² (h)	E_a (eV)	σ
TaNTa	230 (+54,-44)	0.80 (+0.26,-0.26)	0.67 (+0.10,-0.13)
SiC(N)	1.7	0.87 (+0.11,-0.11)	0.51 (+0.19,-0.12)

dielectric cap at all interfaces. Table I shows the most important reliability parameters of this test. Compared to a standard 3/3nnm TaNTa–barrier, the SiC(N) barrier shows equal σ and E_a, proving that the failure mechanism is the same. However, the failure time is much shorter and thus the Cu/SiC(N) interface is the faster diffusing interface.

In conclusion, our simple model of void growth into the line with constant void length might not hold during the initial stages of void formation, i.e for low values of h_{void} (say below 30nm). Also at the very end of the voiding process, when the remaining copper thickness becomes very small, copper islands or fractals might be formed, which are not described by our model.

However, we believe our model decently represents the void formation process in the region between $h_{void} = 30nm$ and $h_{void} = t-30nm$. For values of h_{void} between these values, we further checked the validity of our model by making cross-sectional images of the lines after moderate resistance increases before final void formation. As shown in figure 14, the observed void shapes are, within good approximation, well described by our model proposed in figure 8. The main reason to assume a fixed void length L_{void} throughout the void formation process is in the micro-structure of our lines. Based on top-down TEM-pictures, we earlier showed that the grain structure of our lines was semi-bamboo [16]. For such semi-bamboo lines, it is reasonable to assume that, once a void is formed, it grows into one single grain [15].

C. Velocity of void formation

The derivative of h_{void} as a function of time for the samples shown in figure 7A is displayed in figure 15. A decrease of the velocity with time is observed, where velocities at the

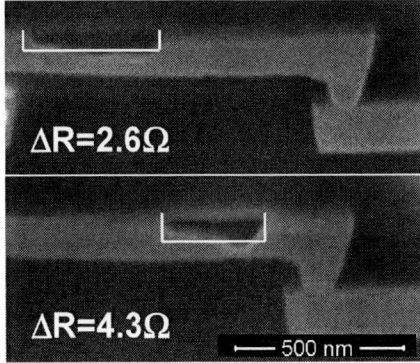

Figure 14 Typical void shapes after moderate resistance changes before final void formation. White shapes indicate how our model approximates the void shapes.

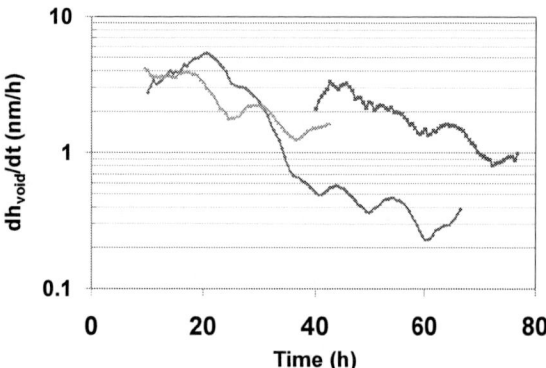

Figure 15 dh_{void}/dt of the estimated h_{void} from figure 12.

earlier stages of void formation are, for all devices evaluated in this study between 2 and 10nm/h, while at later stages this velocity is between 0.1 and 2nm/h.

Literature values of drift velocities after void formation at 330°C for semi-bamboo lines reported by different authors [3-5,10] are between 0.1 and 25nm/h, which is in the same order of magnitude of the velocities of h_{void} we obtained during void growth. As such, we conclude that the diffusion paths of copper leaving the void during void formation are similar compared to those during void growth.

An explanation for the decrease of the velocity of h_{void} during the void formation can possibly be found in the fact that the dominant diffusion paths change during void formation. Different authors report that both copper surface [17,18] and copper grain boundaries [9-10] are faster diffusion paths compared to the copper/dielectric cap interface. Possibly, grain boundary and surface diffusion are more dominant during the earlier phases of void formation.

V. MULTIPLE VOID CASE

In some cases the degradation occurs in two steps, where at first a saturating resistance increase is observed and then the resistance starts to increase again (Figure 7B). A cross-section of the particular sample indicated with the arrow in figure 7B shows one void with a limited depth further away from the via and a full void closer to the via (Figure 16).

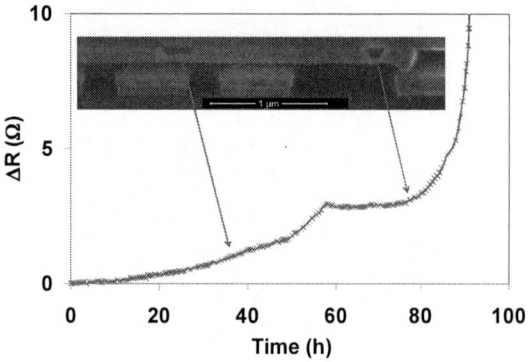

Figure 16 ΔR versus time and FIB-cross section of the sample indicated with an arrow in figure 7B

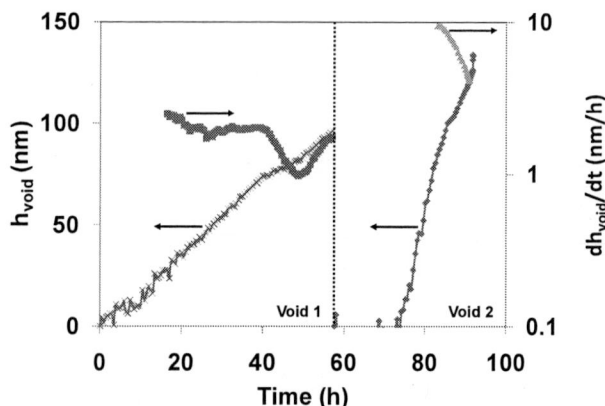

Figure 17 h_{void} and dh_{void}/dt versus time for the sample depicted in figure 16

We used our model to estimate the void height h_{void} and dh_{void}/dt as a function of time for each of the voids (Figure 17). For this fit, L_{void} was estimated from the cross-section shown in figure 16.

We believe that the void further away from the via is initiated first, but stops growing when the void closer to the via is initiated. The tensile stress that is built up by the EM process is higher closer to the via and vacancies are attracted by the void which is closest to this point. Also, the second void grows faster because the first void is partially depleted.

VI. CONCLUSIONS

We characterized a test structure to study void formation using electrical measurements, where void formation is defined as the degradation that takes place before the void spans the whole line width and height. The structure senses the voltage close to the via. This structure is sensitive to study void formation and the failure times are comparable to traditional structures, so the senses do not influence the voiding process. A simple model is proposed where the void grows into the line, with a fixed length. It is argued that the model holds for semi-bamboo lines in stages where the void is not too shallow and not too deep. The void height increases fast in the beginning of the voiding process and this growth slows down towards the end. Since the calculated drift velocities the voiding process are comparable with literature values of drift velocities after full void formation, we hypothesize that the dominating diffusion paths for copper atoms leaving the void change towards the voiding process. In cases where multiple voids are formed, we show that voids that initially grow further away from the via stop growing upon the initiation of a void closer to the via. This second void forms at a faster rate because it partially depletes the first void.

ACKNOWLEDGMENT

The authors would like thank the participants of imec's electromigration meeting and the members of the interconnect and remo group for fruitful and stimulating discussions. Also, Rudy Caluwaerts is acknowledged for the failure analysis work, Veerle Simons, Annelies Vanderheyden and Myriam

Van De Peer for the operational work and Yuichi Miyamori, Sony assignee at imec, for the nice discussions.

Also the personnel of the intrinsic reliability test business unit of Aetrium Inc. is acknowledged for their valuable support during this study.

REFERENCES

[1] C.-K. Hu, L. Gignac and R. Rosenberg, "Electromigration of Cu/low dielectric constant interconnects", Microelectronics Reliability, Vol. 46, p. 213, 2006

[2] J. Michelon, C. Bruynseraede, D. Tio Castro, Ph. Roussel, R. Hoofman and K. Maex, "Electromigration study of sub-100nm Cu-lines", Proceedings of the Advanced Metallization Conference (AMC), p. 253, 2004 (published in 2005)

[3] S. Yokogawa, "Electromigration-Induced Void Growth Kinetics in SiN_x-Passivated Single-Damascene Cu Lines", Japanese Journal of Applied Physics, Vol. 43, p. 5990, 2004

[4] L. Doyen, E. Petitprez, P. Waltz, X. Federspiel, L. Arnaud and Y. Wouters, "Extensive analysis of resistance evolution due to electromigration induced degradation", Journal of Applied Physics, Vol. 104, p. 123521-1, 2008

[5] M. H. Lin, S. C. Lee and A. S. Oates, "Electromigration Mechanisms in Cu Nano-Wires", IEEE Int. Reliability Physics Symposium. (IRPS), p. 705, 2010

[6] E. Zschech, R. Hübner, O. Aubel and P. S. Ho, "EM and SM Induced Degradation Dynamics in Copper Interconnects Studied Using Electron Microscopy and X-Ray Microscopy", IEEE Int. Reliability Physics Symposium. (IRPS), p. 574, 2010

[7] C. Witt, V. Calero, C.-K. Hu, F. Feustel and G. Bonilla, "Copper Microstructure: Effect on Electromigration Void Evolution", Abstract book of the Advanced Metallization Conference (AMC), 2010

[8] C.-K. Hu, R. Rosenberg and K.Y. Lee,"Electromigration path in Cu thin-film lines", Applied Physics Letters, Vol. 74, p. 2945, 1999

[9] L. Arnaud, T. Berger and G. Reimbold, G, "Evidence of grain-boundary versus interface diffusion in electromigration experiments in copper damascene interconnects", Journal of Applied Physics, Vol. 93, p. 192, 2003

[10] C.-K. Hu, L. Gignac, B. Baker, B.; E.G.Liniger, R. Yu and P. Flaitz, "Impact of Cu microstructure on electromigration reliability", IEEE International Interconnect Technology Conference (IITC), p. 93, 2007

[11] X. Federspiel, L. Doyen and S. Courtas, "Use of Resistance-Evolution Dynamics During Electromigration to Determine Activation Energy on Single Samples", IEEE Transactions on Device and Materials Reliability, Vol. 7, p. 236, 2007

[12] L. Carbonell et al., "Metallization options for sub- 30 nm interconnects: comparison of Cu and W metallizations", Abstract book of the Advanced Metallization Conference (AMC), 2010

[13] M.H. Wood, S.C. Bergman and R.S. Hemmert, "Evidence for an incubation time in electromigration phenomena", IEEE Int. Reliability Physics Symposium. (IRPS), p. 70, 1991

[14] S. Yokogawa and H.Takizawa, "Electromigration Induced Incubation, Drift and Threshold in Single-Damascene Copper Interconnects", IEEE International Interconnect Technology Conference (IITC), p. 127, 2002

[15] H. Ceric, R.L. de Orio, J. Cervenka and S. Selberherr, "The effect of microstructure on electromigration induced voids", IEEE International Symposium on the Physical and Failure Analysis of Integrated Circuits (IPFA), p. 694, 2009

[16] C. J. Wilson, K. Croes, C. Zhao, T. H. Metzger, L. Zhao, G. P. Beyer, A. B. Horsfall, A. G. O'Neill and Zs. Tőkei, "Synchrotron measurement of the effect of line-width scaling on stress in advanced Cu/Low-k interconnects", Journal of Applied Physics, Vol. 106, p. 053524, 2009

[17] Z.-S. Choi, R. Monig and C.V. Thompson, "Activation energy and prefactor for surface electromigration and void drift in Cu interconnects", Journal of Applied Physics, Vol. 102, p. 083509-1, 2007

[18] J.V. Jo and R.W. Vook, "In-situ ultra-high vacuum studies of electromigration in copper films", Thin Solid Films, Vol. 262, p. 129, 1995

3D Integration Technology and Reliability

Mitsumasa Koyanagi

New Industry Creation Hatchery Center (NICHe)
Tohoku University
Sendai, 980-8579 Japan
phone: +81-22-795-6906, e-mail address: koyanagi@bmi.niche.tohoku.ac.jp

Abstract—Three-dimensional (3D) integration technologies including a new 3D heterogeneous integration of super-chip are described. In addition, reliability issues in these 3D LSIs such as mechanical stresses induced by through-silicon vias (TSVs) and metal microbumps and Cu contamination in thinned wafers are discussed. Cu TSVs with the diameter of 20μm induced the maximum compressive stress of ~1 GPa at the silicon substrate adjacent to them after annealed at 300℃ for 30 min. Mechanical strain/stress and crystal defects were produced in extremely thin wafers (thickness ~10μm) of 3D LSIs not only during wafer thinning, but also after wafer bonding using fine-pitch, high-density metal microbumps and curing. The influence of Cu contamination at the back surface of the thinned wafer has been evaluated by C–t analysis. C–t curves measured in MOS capacitors without IG layer and EG layer were seriously degraded after annealing even at 200°C whereas the C–t curves exhibited only a little change even after annealing up to 350 min at 300°C. It was revealed that the generation lifetime of minority carrier is significantly reduced by the Cu contamination.

Keywords-3D LSI; TSV; Microbump; Mechanical stress; Cu contamination

I. INTRODUCTION

Recently three-dimensional (3D) integration technology using through-Si vias (TSVs) has attracted much attention since it gives rise to the higher packing density, shorter interconnections, lower power consumption and hetero-geneous device integration [1]-[25]. However, there are many concerns in 3D LSIs with TSVs to be solved before the volume production starts. The most serious concern is the heat accumulation in 3D stacked chips. TSVs and metal micro-bumps act as effective heat conductors among many stacked chip layers. Therefore, heat generated at a hot layer is quickly transferred to cool layers and consequently the average temperature of 3D stacked chip increases. Influences of mechanical stress and strain introduced in thinned Si substrates are another concerns in 3D LSIs with TSVs. TSVs and metal microbumps introduce significant mechanical stress and strain into thinned Si substrates. In addition, the LSI chip with thinned Si substrate is more easily affected by metal impurity contamination and crystal defects. Usually intrinsic gettering layer and extrinsic gettering layer are formed in Si substrates of LSI chips to minimize the influences of metal impurity contamination and crystal defects. These gettering layers might be removed by thinning the Si substrate in 3D LSI fabrication process. In this paper, the influences of mechanical stress and metal impurity contamination in 3D LSIs are discussed.

II. 3D INTEGRATION TECHNOLOGY

The cross-sectional structure of the 3D LSI with TSVs is illustrated in Fig.1 [16], [19]. The thinned upper layers are stacked onto the thick LSI wafer in a 3D integration technology based on a wafer-to-wafer bonding. In these 3D LSIs, a relatively thick Si substrate remains after completing the fabrication process. This remaining Si substrate is useful for reducing the mechanical damage caused to the devices during the 3D fabrication process. The electrical interconnection in the vertical direction is created by the TSVs and the metal micro bumps. TSVs are fabricated before the transistor formation in a via-first process whereas those are fabricated after the transistor formation in a via-middle process as shown in Fig.2. Furthermore, TSVs are fabricated after completing the BEOL of CMOS process in a via-last process. There are two kinds of via-last processes of the front-via process and back-via process. We had initially developed a 3D integration technology with the via-first process using poly-Si TSVs [3]-[5]and fabricated several prototype 3D LSI test chips such as a 3D stacked image sensor chip, 3D s tacked memory, 3D-stacked artificial retina chip and 3D stacked microprocessor chip for the first time [7]-[9],[13]. An SEM cross-sectional view of fabricated 3D stacked microprocessor test chip is shown in Fig.3. Subsequently we developed the tungsten (W) TSV and copper (Cu) TSV techno-logies to reduce the TSV resistance as shown in Fig.4.

LSI wafer with TSVs is bonded onto another wafer in the 3D integration technology. In this case, the electrical connection has to be built between two wafers. Then

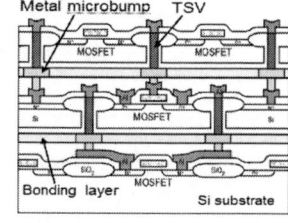

Fig.1 Cross-sectional structure of 3D LSI with TSVs.

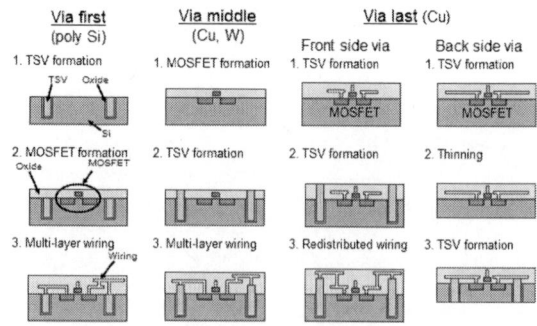

Fig.2 TSV formation processes.

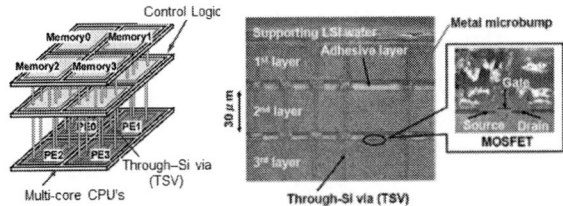

Fig.3 SEM cross-sectional view of fabricated
3D stacked microprocessor test chip.

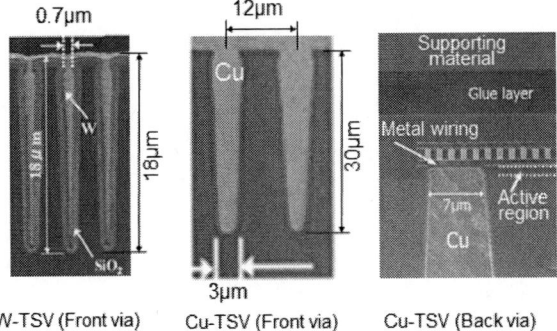

W-TSV (Front via)　Cu-TSV (Front via)　Cu-TSV (Back via)

Fig.4 Cross-sectional micrograph of three kinds of TSVs.

metal micro-bumps are formed onto LSI wafers to build the electrical connection. Therefore two wafers are bonded by both metal microbumps and bonding material. Three kinds of wafer bonding methods have been proposed a s shown in Fig.5. Metal is used for both microbumps and bonding material in the direct metal bonding [6]. Metals such as Cu, CuSn, CuSnAg, Au, and InAu etc. are used in this bonding method. Organic material and silicon oxide are used for the bonding material in the adhesive/metal hybrid bonding and oxide/metal hybrid bonding, respectively. We should choose one of these bonding methods carefully considering the bonding temperature, bonding pressure, temperature tolerance, alignment accuracy, manufacturability, influences on device characteristics, reliability and so on. We employed a new adhesive/metal hybrid bonding with CuSn or InAu in our 3D integration technology where the liquid epoxy adhesive is injected into the gap among microbumps in a vacuum chamber after the temporary bonding of two wafers by InAu or CuSn microbumps [5]. SEM cross-sectional views after bonding with CuSn microbumps and InAu microbumps are shown in Fig.6.

We have also developed a new 3D integration technology using multichip-to-wafer bonding to achieve a 3D super chip as shown in Fig.7 [20]-[25]. A chip-to-wafer bonding is required to stack known good dies (KGDs) or LSI chips with different

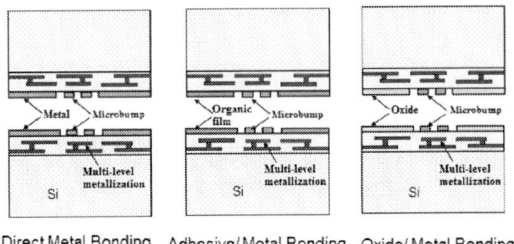

Direct Metal Bonding　Adhesive/Metal Bonding　Oxide/Metal Bonding

Fig.5 Three kinds of wafer bonding methods.

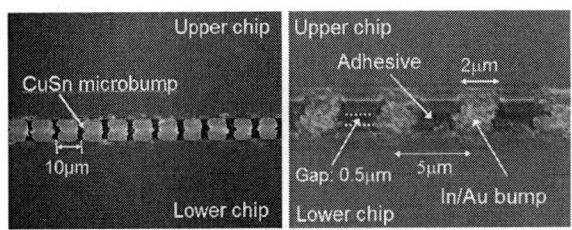

(a) CuSn microbump　　　(b) In Au microbump

Fig.6 SEM cross-sectional view of stacked die
after adhesive/metal hybrid bonding.

sizes and different technologies. However the fabrication throughput is very low in the conventional chip-to-wafer bonding method since chips are sequentially picked up and placed onto a wafer. Then we proposed a new multichip-to-wafer

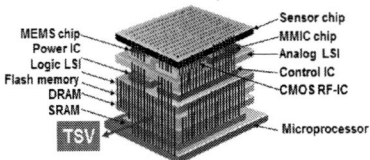

Fig.7 Structure of 3D super-chip.

bonding method as shown in Fig.8 in which many KGDs or chips with different sizes and different technologies more than several hundred or several thousand are simultaneously aligned and bonded onto a wafer. We have developed a new self-assembly technique using the surface tension of liquid for our multichip-to-wafer bonding method. We have developed a self-assembly machine for 8 inch wafer. We simultaneously aligned and bonded more than five hundreds of KGDs onto an 8 inch wafer using this machine as shown in Fig.9. The total alignment time was less than 1 sec. and the average alignment accuracy was 0.5μm. We have fabricated 3D test chips with different sizes using the self-assembly technique. We also succeeded in stacking MEMS devices and optical devices onto LSI chips using the self-assembly technique.

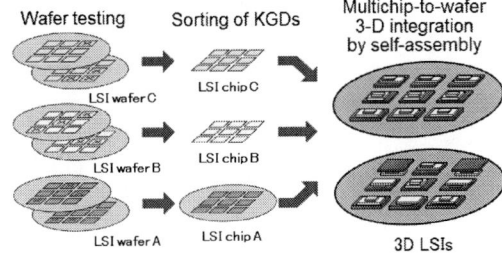

Fig.8 Super-chip fabrication by multichip-to-wafer bonding.

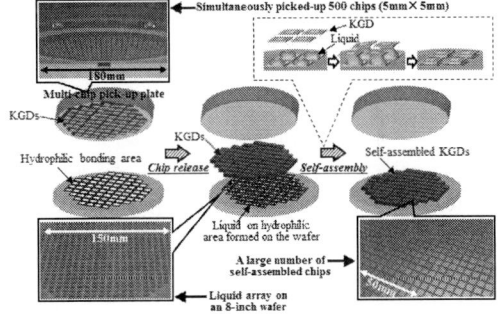

Fig.9 Multichip-to-wafer bonding using 8-inch self-assembly machine.

978-1-4244-9113-1/11 $26.00 © 2011 IEEE　　329

III. MECHANICAL STRESS INDUCED BY TSVs

Big concerns in 3D LSIs with TSVs are influences of mechanical stress and metal contamination on device characteristics as shown in Fig.10 [26]. Large mechanical stresses are induced in the silicon substrate by TSVs and metal microbumps. It is known that Cu TSVs induce both

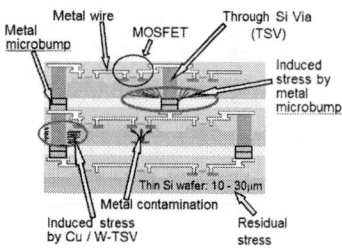

Fig.10 Mechanical stress and metal contamination in 3D LSI with TSVs.

tangential compressive stress and radial tensile strain in the surrounding silicon substrate [27],[28]. The mechanical stress is also induced by thinning the wafer since the wafer thinning process introduces crystal defects into the silicon substrate. We have evaluated these mechanical stresses and crystal defects by a micro-Raman (μR) spectroscopy, X-ray photoelectron spectroscopy (XPS) and TEM [29]. Two-dimensional micro-Raman (2D-μR) mapping images of Cu TSVs are shown in Fig.11 where the annealing temperature after the TSV formation is changed. As is obvious from the upper photographs of Cu TSV, Cu atoms segregate upward at the center of TSV by annealing and eventually popped up after annealing at 400°C. This pop-up of Cu TSV gives rise to serious damages to the multilevel metal wirings. μR images indicate that large compressive stress is induced in the silicon substrate adjacent to the Cu TSV which is surrounded by large tensile strain. These stress and strain induced by the Cu TSV increased as the annealing temperature increased and eventually decreased after annealing at 400°C since Cu pop-up occurs. Fig.12 shows the spatial mechanical stress distribution

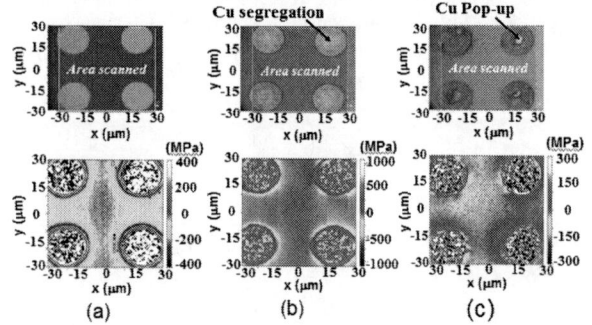

Fig.11 Micro-Raman mapping images of Cu TSVs with the diameter of 20μm.

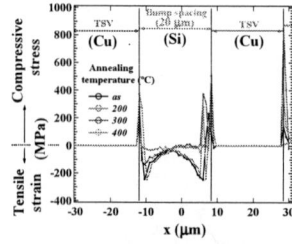

Fig.12 Spatial mechanical stress distribution along the horizontal direction.

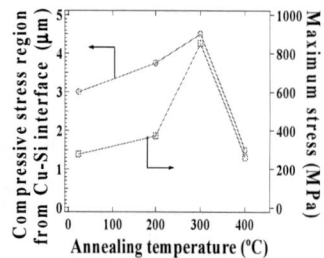

Fig.13 Annealing temperature dependence of mechanical stress induced by TSVs.

across two Cu TSVs along the horizontal direction. It is more clearly observed in the figure that large compressive stress is induced in the silicon substrate adjacent to the Cu TSV which is surrounded by large tensile strain. It is also clear from Fig.13 that the maximum compressive stress and the compressive stress region edge from the Cu-silicon substrate interface increase with increasing annealing temperature and eventually decrease after annealing at 400°C. These mechanical stresses induced by Cu TSVs change depending on their layout pattern and position. The Cu TSV located at the center of TSV array exhibits larger mechanical stress than those of TSVs at the periphery of array since mechanical stresses from other TSVs are accumulated on the TSV at the center of array. In addition, tensile strains induced by two adjacent TSVs become to be overlapped as the TSV spacing is reduced and as a result the silicon substrate in the central region of TSV array is completely occupied by tensile strains.

The mechanical stress induced by TSVs seriously influences on the device characteristics. It is well known that carrier mobility is changed by the mechanical stress. As a result, device characteristics are changed by the mechanical stress. However, the influences of mechanical stress on carrier mobility and device characteristics are very complicate. For example, the mechanical stress gives rise to different effects to electron and hole. Therefore, the effects of mechanical stress induced by TSVs on device characteristics of n-channel MOSFET (nMOSFET) are different from those of p-channel MOSFET (pMOSFET). The hole mobility is more significantly influenced by the mechanical stress than the electron mobility. Furthermore, the influences of mechanical stress on carrier mobility and device characteristics change by the crystallographic orientation. Therefore, the drain current changes depending on the layout of MOSFET. The drain current decreases in nMOSFET and increases in PMOSFET when the current flow of MOSFET is perpendicular to the radial tensile strain and vise-versa when the current flow is parallel to it when (100)/<110> wafer is used as summarized in Fig.14 and Table 1.

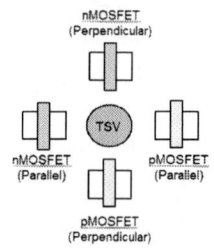

Fig.14 Different effects of mechanical stress on MOSFETs placed around TSVs.

The drain current change by TSVs is larger in pMOSFET. To avoid the influences of TSVs on device characteristics, we have to lay down the area around TSVs where transistors do not exist. This area is defined as the keep-out zone (KOZ).

Table 1 Changes of MOSFET channel current perpendicular or parallel to the tensile strain induced by TSVs.

	nMOSFET	pMOSFET
Perpendicular	Decrease	Increase
Parallel	Increase	Decrease

IV. MECHANICAL STRESS INDUCED BY METAL MICROBUMPS

In order to evaluate the thermo-mechanical stress induced in the Si substrate by CuSn microbumps, we have carried out the cross sectional 2D-μR mapping analysis at every stages of the bonding process, right after bump formation, heat treatment up to 280 °C before bonding, after bonding at

280 ℃, and finally post heat treatment at 300 ℃ for 5 min after bonding [29]. CuSn microbumps consist of 3μm thick Cu and 2μm thick Sn. Before bonding, we mainly observed a compressive stress below the microbump. When the die with these microbumps cools down, the Cu shrinks more than the Si substrate due to the difference in the coefficient of thermal expansion (CTE). Hence, the Cu of the microbump compresses the Si die and generates compressive stress in the die. There exists a maximum compressive stress of 125 MPa for the Si region under the microbump with the size of 20×20 μm^2 (40 μm pitch), and it fades away towards the edges of microbump. We observed an increase in the compressive stress in the Si substrate under the microbump to ~250 MPa immediately after bonding at 280 ℃ for 10 sec. This compressive stress further increased to >300 MPa after post annealing at 300 ℃ for 5 min. as shown in Fig.15. The compressive stress in the Si surface by microbumps decreases exponentially along the cross-sectional direction as shown in Fig.16. In the horizontal direction, the compressive stress induced by microbumps propagates to the microbump space region. Here the propagation length is defined as the distance where the induced stress during bonding becomes negligibly small or completely vanish. The penetration length increases as the compressive stress in the microbump region increases. In addition, the compressive stresses produced by the two adjacent microbumps become overlapped each other in the bump space region as the microbump pitch is reduced. As a result, the Si substrate in the bump space region exhibits the compressive stress before filling the bump space region by the underfill material. However, we observed the maximum tensile strain of -100 MPa in the Si substrate of the bump space region after filling the gap between microbumps by epoxy resin. This is because large volume expansion during heating due to larger CTE of epoxy and slow volume compression during cooling due to the extremely poor thermal conductivity of epoxy induce the tensile strain in the Si substrate of bump space region.

We evaluated the influences of mechanical stress induced by metal microbumps on device characteristics. The 4 point substrate bending technique has been widely employed to deduce the influences of mechanical stress/strain on device characteristics. However we can not use this method to introduce large local stress into the Si substrate. Then we em-

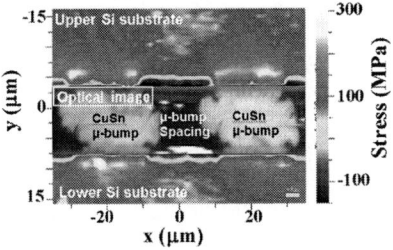

Fig.15 Cross-sectional 2D-μR mapping image of mechanical stress induced by CuSn microbump after bonding and post annealing at 300 ℃ for 5 min.

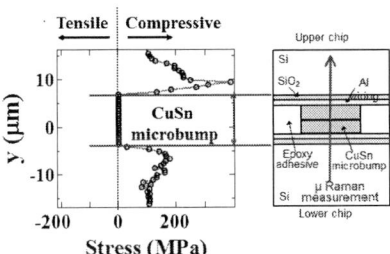

Fig.16 Vertical distribution of compressive stress induced by CuSn microbumps along the cross-sectional direction.

loyed the structure with Si bumps as shown in Fig.17 to introduce large local stress into the Si substrate. We can precisely control the mechanical stress introduced into the Si substrate by changing the Si bump height, size and pitch, the CTE of underfill material and the annealing temperature. As is obvious in the figure, in the bump region, a very large compressive stress is introduced in the bottom surface of Si substrate which is in contact with Si bump and tensile strain in the top surface on it. Meanwhile, in the bump space region, large tensile strain is induced in the bottom surface of Si substrate which is in contact with the underfill material (epoxy resin) and compressive stress in the top surface on it. Using this Si bump array test structure, we have successfully measured local bending on 10 μm thick die. It was revealed that the amount of created local bending varies by the stress relief treatment after wafer thinning as shown in Fig.18. The local bending is ~ 0.23 μm for wafers stress-released by plasma etching (PE), dry polishing (DP) and ultra ploy grinding (UPG) which is nearly twice of that of the wafer treated by chemical-mechanical polishing (CMP). Therefore, PE, DP and UPG methods are not the suitable stress relief process in terms of local bending although they might be suitable in terms of extrinsic gettering effect for metal contamination. The amount of local bending is 5 fold reduced as the bump spacing is reduced from 100 μm to 20 μm as shown in Fig.19.

We performed μR spectroscopy (μRS) analysis to measure two-dimensional stress distribution in the top surface of 10 μm thick die bonded on the Si bump array structure with the bump size of 5μm and the spacing of 100μm. The stress-relief treatment by plasma etching is done in the backside of 10 μm thick die. The result is shown in Fig.20. It is clear in the figure that large tensile strain of ~1.8 GPa is induced at the top

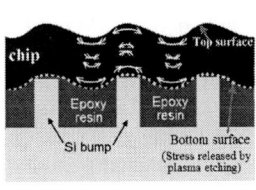

Fig.17 Si bump array test structure employed to introduce the mechanical stress into thinned Si die.

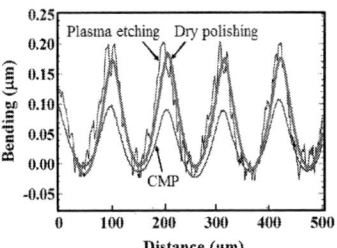

Fig.18 Local bending of 10 μm thick die with various stress-released back-surface introduced by Si bump array structure.

surface of Si substrate in the bump region whereas large compressive stress is induced at the top surface in the bump space region. We fabricated MOSFETs in the top surface of 50 μm thick die bonded on the Si bump array structure with the bump size of 5μm and the spacing of 300μm to evaluate the influences of mechanical

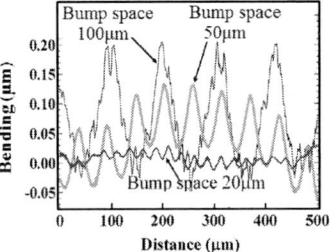

Fig.19 Local bending of 10 μm thick die introduced by Si bumps with the size of 5×5μm^2 and various spacings.

978-1-4244-9113-1/11 $26.00 © 2011 IEEE   331

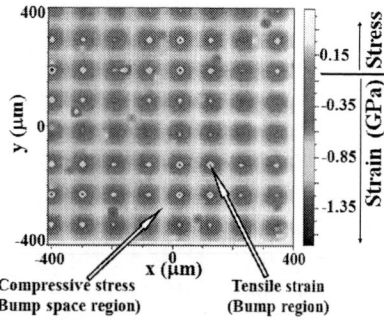

Fig.20 Two-dimensional stress distribution in the top surface of 10 μm thick die bonded on the Si bump array structure with the bump size of 5μm and the spacing of 100μm which was obtained by μRS analysis.

stress on device characteristics. Fig. 21 shows the measured drain current-voltage characteristics of nMOSFET with the gate length of 0.5μm and the gate width of 10μm. As is obvious in the figure, the drain current is reduced by ~10% due to the maximum local stress of -500 MPa induced in the top surface of

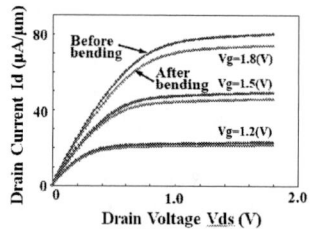

Fig.21 Drain current-voltage characteristic degradation of nMOSFET by the mechanical stress.

Si substrate. These results indicate that we can not neglect the influences of mechanical stress induced by metal microbumps on device characteristics as well as the influences of TSVs.

V. MINORITY CARRIER LIFETIME DEGRADATION BY CU DIFFUSION BROM BACKSIDE SURFACE

To fabricate 3D LSIs, each functional LSI wafer should be thinned to 10-50μm thickness by mechanical grinding and stress-relief polishing methods. However, this may causes severe degradation in device reliability since the intrinsic gettering (IG) region and the extrinsic gettering (EG) region might be removed by wafer thinning and consequently active regions of 3D LSI might be more easily contaminated by metallic impurities as shown in Fig.22. Metallic contamination is caused mainly by Cu atoms originated from Cu TSVs and Cu microbumps. Cu atoms stuck on the back-ground surface during wafer thinning are not completely removed even after the cleaning process. Cu atoms diffuse into both SiO_2 and Si substrate even at low temperature. These Cu atoms generate the deep impurity levels in the Si substrate which reduce the minority carrier lifetime, and hence deteriorate the device characteristics. Both IG layer and EG layer are very effective to prevent Cu atoms from deeply diffusing into the Si substrate.

We have evaluated the influences of Cu diffusion

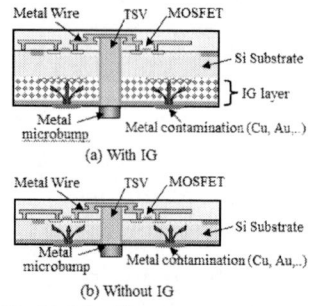

Fig.22 Metallic contamination in thinned Si substrate with and without gettering layer of 3D LSI.

into Si substrate by a transient capacitance measurement using an MOS capacitor which is called a capacitance-time (C-t) method [30]. We can deduce the generation lifetime of minor carriers from the Zerbst plot of C-t curve. The C-t method is far more sensitive for metal contaminations in the Si substrate compared to analytical methods such as SIMS (Secondary ion mass spectrometry) and XPS. MOS capacitors were fabricated on the thinned Si wafers of 50μm thickness without IG layer and with IG layer. Al gate electrodes with the thickness of 800nm and the diameter of 700μm were formed on the 10nm thick gate oxide by evaporation method. The wafer with these MOS capacitors was thinned down to 50μm thickness by mechanical grinding and stress-relief CMP process. A 50nm thick Cu layer was evaporated on the back surface of the thinned substrate as a contamination source. To diffuse Cu atoms into the Si substrate, the wafers were annealed at 200°C and 300°C in N_2 ambient for various annealing time. C-t curves after annealing are plotted in Fig.23 which are measured in MOS capacitors without IG layer. In the C-t measurement, high step voltage is applied to the gate electrode to induce a deep depletion region in the surface of Si substrate. The width of this deep depletion region decreases with time, hence the capacitance increases with time and eventually reaches the constant value C_f as a result that an inversion layer is formed at the surface of Si substrate. The time t_f for the capacitance to reach C_f decreases as the generation lifetime of minority carrier (electron) becomes shorter. It is obvious in Fig.23 that t_f and hence the generation lifetime of electron significantly decreases even after annealing at low temperature of 200°C. This indicates that device characteristics are easily degraded by diffused Cu atoms when the backside surface of Si substrate without IG layer is contaminated by Cu atoms. The generation lifetime of electron deduced from C-t curves is plotted versus the surface concentration of Cu atom in Fig.24. The Cu concentration was measured at the front surface by a total reflection X-ray fluorescence (TRXF) analysis after Cu diffusion from the backside of Si substrate [31]. As is clear in the figure, the generation lifetime significantly decreases as the Cu concentration increases. It is also obvious in the figure that the C-t

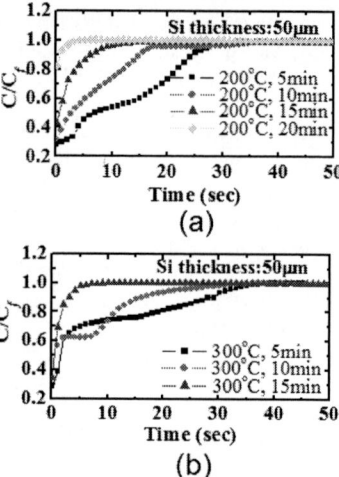

(a)

(b)

Fig.23 C-t curves after Cu diffusion into Si substrate from the backside surface.

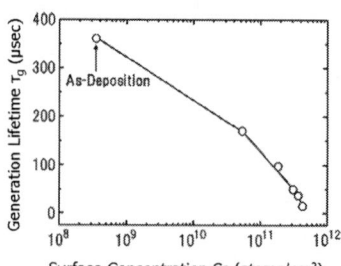

Fig.24 Generation lifetime of minority carrier versus surface concentration of Cu atoms.

978-1-4244-9113-1/11 $26.00 © 2011 IEEE   332

method is very sensitive to the Cu contamination. We had similar results for the Au contamination.

Fig.25 shows the C-t curves of the MOS capacitors formed on the thinned Si substrate with the 30μm thick IG layer and the 20μm

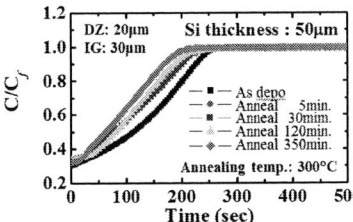

Fig.25 C-t curves after Cu diffusion into Si substrate with intrinsic gettering (IG) layer.

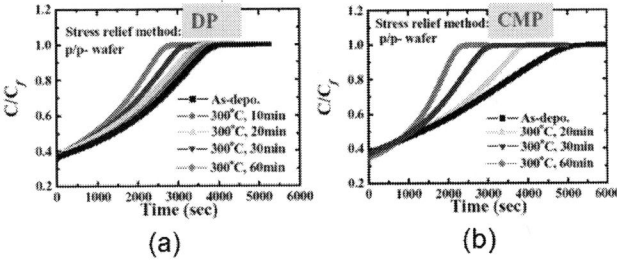

Fig.26 C-t curves after Cu diffusion into Si substrate with stress-released backside surface.

thick denuded zone which were measured after annealing at 300°C with different annealing time. As clearly seen in the figures, C-t curves exhibit only a little change even after annealing up to 350min at 300°C. It means that Cu atoms do not diffuse into the active region in the Si substrate and do not reach to the Si-SiO$_2$ interface as a result of Cu gettering by IG layer and the generation lifetime of minor carriers is not reduced even after at a relatively high temperature and long time annealing. We also confirmed from the C-t curves of MOS capacitors with thicker IG layers that a thicker IG layer more effectively retards the Cu diffusion. We have also evaluated the effects of extrinsic gettering (EG) by the C-t method. Fig.26 shows the C-t curves of the MOS capacitors with the backside surface stress-released by DP and CMP. The DP treatment introduces more mechanical damages at the backside surface of Si substrate than the CMP treatment which act as gettering sites for impurities. As is clear in Fig.26, C-t curves of MOS capacitors with DP stress-released backside surface exhibit less degradation compared to MOS capacitors with CMP treated backside surface since the mechanical damaged region at the DP stress-released backside surface acts as the EG layer.

VI. CONCLUSIONS

Locally induced stress by wafer thinning, bonding and TSVs in the 3D-LSI fabrication process has been investigated. Cu TSVs with the diameter of 20μm induced the maximum compressive stress of ~1 GPa after annealed at 300°C for 30 min and cooled down to room temperature. Mechanical strain/stress and crystal defects were produced in extremely thin wafers (thickness ~10 μm) of 3D-LSIs not only during wafer thinning, but also after wafer bonding using fine-pitch, high-density microbumps and curing. It was revealed that the induced stress by CuSn microbumps penetrated deeper for larger bump size and wider for smaller bump pitch. The maximum local stress of -500 MPa which was introduced

using Si bumps caused a 10% change in the drain current of n-channel MOSFET. The influence of Cu contamination at the back surface of the thinned wafer has been evaluated by C–t analysis. C–t curves measured in MOS capacitors without IG layer and EG layer were seriously degraded even after annealing at relatively low temperature of 200°C whereas the C–t curves exhibited only a little change after annealing up to 350 min at 300°C. It indicates that Cu atoms hardly diffused into the active region in a wafer with the IG layer. It was also shown by the C–t analysis that the EG layer is effective to retard the Cu diffusion. The quantitative relationship between the generation lifetime of minority carrier reduced by Cu atom and surface concentration of Cu atom was evaluated and it was revealed that the generation lifetime is significantly reduced by the Cu contamination.

ACKNOWLEDGMENT

The author would like to thank Prof. T. Tanaka, Assoc. Prof. T. Fukushima, Assoc. Prof. K-W Lee, Dr. J. Bea and Dr. M. Murugesan for their fruitful discussions and collaborations. This work was performed at Micro/Nano-Machining Research and Education Center (MNC) and Jun-ichi Nishizawa Research Center at Tohoku University. We would also like to thank the staff of the centers. A part of this work was entrusted by NEDO, "Development of Functionally Innovative3D-integrated Circuit (Dream Chip) Technology".

REFERENCES

[1] M. Koyanagi, "Roadblocks in Achieving Three-Dimensional LSI," Proc. 8th Symposium on Future Electron Devices, pp.50-60, 1989.

[2] H. Takata, M. Koyanagi et al., "A novel fabrication technology for optically interconnected three-dimensional LSI by wafer aligning and bonding technique," Int. Semiconductor Device Research Symposium, pp.327-330, 1991.

[3] T. Matsumoto, M. Koyanagi et al., "Three-dimensional integration technology based on wafer bonding technique using micro-bumps," Conf. on Solid State Devices and Materials (SSDM), pp.1073- 1074, 1995.

[4] M. Koyanagi, H. Kurino, K-W Lee, K. Sakuma and H. Itani, "Future system-on-silicon LSI chips," IEEE MICRO, 18 (4), pp.17 – 22, 1998.

[5] T. Matsumoto, M. Koyanagi et al., "New three-dimensional wafer bonding technology using the adhesive injection method," Jpn. J. Appl. Phys., 1 (3B), pp.1217 – 1221, 1998.

[6] A. Fan, A. Rahman, and R. Reif, "Copper wafer bonding," Electrochem. Solid State Lett., vol.2, no.10, pp.534-536, 1999.

[7] H. Kurino, M. Koyanagi et al., "Intelligent image sensor chip with three dimensional structure," Int. Electron Devices Meeting (IEDM) Dig., pp.879–882, 1999.

[8] K W. Lee, M. Koyanagi, "Three-dimensional shared memory fabricated using wafer stacking technology," Int. Electron Devices Meeting (IEDM) Dig., pp.165-168, 2000.

[9] M. Koyanagi et al., "Neuromorphic vision chip fabricated using three-dimensional integration technology," Int, Solid State Circuits Conf. (ISSCC) Dig, pp.270-271, 2001.

[10] J. A. Davis et al., "Interconnect limits on gigascale integration (GSI) in the 21st century," Proc. IEEE, vol.89, no.3, pp.305-324, 2001.

[11] P. Ramm et al, "Interchip via technology for vertical system integration," Int. Interconnect Technology Conf. (IITC), pp.160-162, 2001.

[12] J. Burns et al., "Three-dimensional integrated circuits for low-power, high-bandwidth systems on a chip," Int, Solid State Circuits Conf. (ISSCC) Dig., pp.268-269, 2001.

[13] T. Ono, M. Koyanagi et al., "Three-dimensional processor system fabricated by wafer stacking technology," Int. Symp. on Low-Power and High-Speed Chips (COOL Chips), pp.186-193, 2002.

[14] J.-Q. Lu et al., "Evaluation procedures for wafer bonding and thinning of interconnect test structure for 3D ICs," Int. Interconnect Technology Conf. (IITC), pp.74-76, 2003.

[15] B. Swinnen et al., "3D integration by Cu-Cu thermocompression bonding of extremely thinned bulk-Si die containing 10 μm pitch through-Si vias", Int. Electron Devices Meeting (IEDM) Dig., pp. 371-374, 2006.

[16] M. Koyanagi et al., "Three-Dimensional Integration Technology Based on Wafer Bonding With Vertical Buried Interconnections," IEEE Trans. on Electron Devices, vol. 53, No. 11, pp. 2799-2808, 2006.

[17] Jan Van Olmen et al., "3D Stacked IC Demonstration using a Trough Silicon Via First Approach", Int. Electron Devices Meeting (IEDM) Dig., pp.603-606, 2008.

[18] Uksong Kang et al., "8Gb 3D DDR3 DRAM Using Through-Silicon-Via Technology," Proc. IEEE Int, Solid State Circuits Conf. (ISSCC) Dig., pp.130-131, 2009.

[19] M. Koyanagi, T. Fukushima and T. Tanaka, "High-Density Through Silicon Vias for 3-D LSIs," Proceedings of THE IEEE, Vol.97. No.1, pp.49-59, (2009).

[20] T. Fukushima, Y. Yamada, H. Kikuchi and M. Koyanagi, "New Three-Dimensional Integration Technology Using Self-Assembly Technique," Int. Electron Devices Meeting IEDM) Dig., pp.359-362, 2005.

[21] T. Fukushima, M. Koyanagi et al., "New Three-Dimensional Integration Technology Based on Reconfigured Wafer-on-Wafer Bonding Technique," Int. Electron Devices Meeting (IEDM) Dig., pp.985-988, 2007.

[22] T. Fukushima, M. Koyanagi, "Three-Dimensional Integration Technology Based on Reconfigured Wafer-to-Wafer and Multichip-to-Wafer Stacking Using Self-Assembly Method," Int. Electron Devices Meeting (IEDM) Dig., pp.349-352, 2009.

[23] K-W Lee, M. Koyanagi et al., "3D Heterogeneous Opto-Electronic Integration Technology for System-on- Silicon (SOS)," Int. Electron Devices Meeting (IEDM) Dig., pp.531-534, 2009.

[24] T. Fukushima, E. Iwata, K.-W. Lee, T. Tanaka and M. Koyanagi, "Self-Assembly Technology for Reconfigured Wafer-to-Wafer 3D Integration," Electronic Components and Technology Conference (ECTC), pp.1050-1053, 2010.

[25] Kang-Wook Lee, M. Koyanagi et al., "Three-Dimensional Hybrid Integration Technology of CMOS, MEMS, and Photonics Circuits for Optoelectronic Heterogeneous Integrated Systems," IEEE Trans. on Electron Devices,Vol.58, No3, March, 2011.

[26] M. Murugesan, M. Koyanagi et al., "Impact of Remnant Stress/Strain and Metal Contamination in 3D-LSIs with Through-Si Vias Fabricated by wafer Thinning and Bonding," Int. Electron Devices Meeting (IEDM) Dig., pp.361-364, 2009.

[27] C. Okoro et al., "Extraction of the Appropriate Material Property for Realistic Modeling of Through-Silicon-Vias using μ-Raman Spectroscopy," Int. Interconnect Technology Conf. (IITC), pp.16-18, 2008.

[28] A. Mercha et al., "Comprehensive Analysis of the Impact of Single and Arrays of Through Silicon Vias Induced Stress on High-k / Metal Gate CMOS Performance," Int. Electron Devices Meeting (IEDM) Dig., pp.26-29, 2010.

[29] M. Murugesan, M. Koyanagi et al., "Wafer Thinning, Bonding, and Interconnects Induced Local Strain/Stress in 3D-LSIs with Fine-Pitch High-Density Microbumps and Through-Si Vias," Int. Electron Devices Meeting (IEDM),pp.30-33, 2010.

[30] Jichel Bea, K. W. Lee, T. Fukushima, T. Tanaka and M. Koyanagi, "Evaluation of Cu Contamination at Backside Surface of Thinned Wafer in 3-D Integration by Transient-Capacitance Measurement," IEEE Electron Device Lett., Vol.32, No.1, January, pp.66-68, 2011.

[31] K. Hozawa, K. Takeda and K. Torii, Symp. on VLSI Tech. Dig., pp.172-173, 2009.

IMPACT OF AIR-INDUCED POLY-SI /OXYNITRIDE INTERFACE LAYER DEGRADATION ON GATE-EDGE LEAKAGE

Ziyuan Liu, Shuu Ito, Tomoya Saito
Device & Analysis Technology Division, Renesas Electronics Corporation,
Kawasaki, Kanagawa, Japan
phone: (81) – (44)- 435-1438, ziyuan.liu.jc@renesas.com

Soon W. Chang,
Renesas Electronics America, Inc.
Roseville, CA, United States

Arito Ogawa, Sadayoshi Horii,
Semiconductor Equipment System Laboratory, Hitachi Kokusai Electric Inc.,
Toyama-city, Toyama, Japan

Tsuyoshi Horikawa,
Nanodevice Innovation Research Center, National Institute of Advanced Industrial Science and Technologies (AIST),
Tsukuba, Ibaraki, Japan

Markus Wilde, Katsuyuki Fukutani
Institute of Industrial Science, University of Tokyo and CREST-JST,
Tokyo, Japan

Toyohiro Chikyow
Advanced Electric Materials Center, National Institute for Materials Science (NIMS)
Tsukuba, Ibaraki, Japan

Abstract—**The air-sensitivity of the poly-Si interface in MOS transistors and its impact on the electrical properties are studied. It is found that the gate leakage localized near the side of air-exposed edges is possibly caused by air-induced degradation of the poly-Si interface, which supplies mobile NH_3-like species to the gate edge side surface, resulting in the formation of a non-stoichiometric as well as impurity-retaining, hence conductive, SiO_xN_y edge layer. Control of the air-sensitive interfacial oxynitride and its NH_3-related decomposition reaction is considered to be essential for improving the gate-edge leakage.**

Keywords- Gate-edge leakage, poly Si interface, hydrogen

I. INTRODUCTION

Aggressive scaling of tunnel oxide thickness for future Flash memories requires, among other concerns, the assurance that the oxide meets the reliability demands. Among the major reliability concerns, the data loss through tunnel-oxide leakage in a few tail cells after program/erase (P/E) cycling is expected to increase as the oxide thickness is downscaled and the projected number of P/E cycles is increased [1, 2]. This may impose severe limitations on thickness reduction and overall Flash scalability.

An oxide thickness around 4.5 nm as shown in Figure 1 has been indicated to be sufficient for intrinsic retention in FLASH nonvolatile memories [3, 4]. Currently however, reliable tunnel films are remaining at a thickness of 8 nm. Continuous downscaling of the tunnel oxide thickness thus

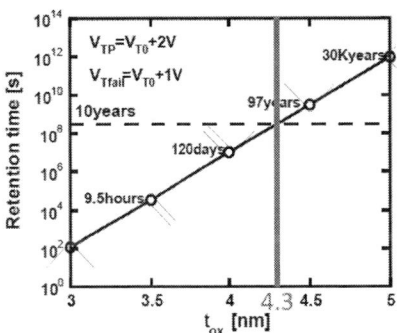

Figure 1. Dependence of the retention time on the tunnel dielectric thickness for FLASH memory devices. An oxide thickness around 4.5 nm is considered to be sufficient for intrinsic data retention (after SPINELLI [3]).

978-1-4244-9113-1/11 $26.00 © 2011 IEEE

becomes a limitation for FLASH memory development [5, 6, 7]. Since gate-edge leakage directly induces a drain of the stored charge, it is one of the causes limiting the scalability of the tunnel dielectric. Clarification of the gate-edge leakage mechanism is therefore urgently required. In practice, sidewall oxidation is an effective process step to improve the gate-edge leakage, but an alternative method that does not increase the tunnel dielectric film thickness in the gate-edge is strongly desired for further downscaling purposes [8, 9].

We have previously identified an ultrathin oxynitride layer that exists in the poly-Si/oxynitride interface as well as in the near surface region of N_2-annealed nitride films as being air-sensitive [10, 11]. This specific oxynitride functions as a potential storage layer for stable hydrogen (H) species, but it readily degrades upon air contact, generating mobile, presumably NH_3-related H species [11]. Since the poly-Si/oxynitride interface is usually exposed to air at the edge-side surface until the sidewall formation during manufacturing, its decomposition may influence the dielectric quality of the gate edge.

In this work, we therefore focus on the gate edge characteristics, especially on the edge surface leakage property, and investigate whether it is related to the air-induced poly-Si interface degradation. We found evidence indeed that the air-contact induced decomposition of the poly-Si interfacial oxynitride layer relates to the gate-edge leakage near the dielectric side surface, which eventually limits the lifetime of the gate dielectric. Control of the air-sensitive interfacial oxynitride and its NH_3-related decomposition reaction is considered to be essential for improving the gate-edge leakage. Our results emphasize that the poly-Si interface has impact not only on the intrinsic but also on the extrinsic life time of MOS devices. We therefore suggest that understanding and refining the poly-Si interface (a part of the MOS stack that has not received much attention yet) is of highest importance.

II. EXPERIMENTAL

Two test nMOS device types were compared in this study. Figure 2 illustrates their architectures, referred to as type A (**without air-exposed gate edges**) and type B (**with air-exposed gate-edges**). The devices were fabricated in the AIST line, which is connected by a load lock system to integrated multi-chamber equipment for MOS stack fabrication. This system allows for *in-situ* analysis of intermediate fabrication stages, e.g., the condition between gate nitridation and the poly-Si deposition step. Gate dielectric films were formed by depositing 10 nm CVD oxides onto Si (100) substrates and subsequent nitridation by N_2O, NO, NH_3 and plasma processes, respectively. As shown in Figure 3, the following poly-Si deposition was split into *in-situ* (no air-contact) and *ex-situ* processes, resulting in two different poly-Si interfacial layers, referred to as **with air-contact** (7 h) and **without air-contact**. Electrical properties such as charge-to-breakdown (Q_{bd}),

interface trap density estimated by charge pumping (CP), and gate leakage (I-V) curves were evaluated.

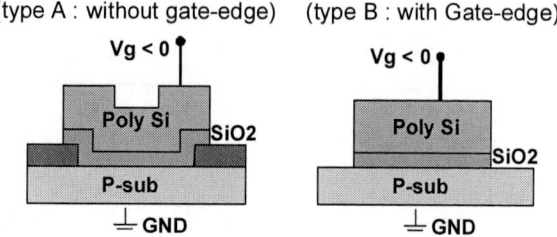

Figure 2. Schematics of the nMOS structure in the test devices. The difference between the MOS stacks is the gate-edge, type A: without air-exposed gate-edge surfaces, type B: with air-exposed gate-edge surfaces. (side wall and pads are not shown in this illustration).

With the same process, poly-Si (5-50 nm) /oxynitride blanket films were fabricated separately for H depth profiling with ^{1}H (^{15}N,$\alpha\gamma$) ^{12}C nuclear reaction analysis (NRA), and for H thermal stability analysis by thermal desorption spectroscopy (TDS). H depth profiling based on NRA has been applied as a probe for emerging applications such as monitoring the diffusivity of H [12] and detecting H-diffusion barriers inside nitride/oxide stacks [13].

In this study, NRA H depth profiling was applied to assess the H impurity-retaining property of the buried poly-Si/dielectric interface, which has been shown to degrade during air-exposure [11]. Thermal desorption spectroscopy (TDS) was applied to the poly-Si/ oxynitride stack. By varying the poly-Si thickness the signal of H desorbed from the buried interface can be distinguished from that desorbed from the surface.

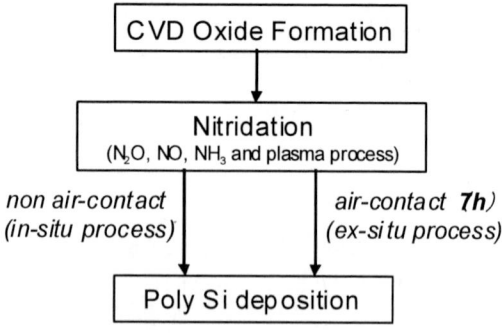

Figure 3. Experimental schemes for the key process steps. No air-contact and 7 h of air-contact were arranged before poly silicon deposition.

978-1-4244-9113-1/11 $26.00 © 2011 IEEE

III. RESULTS AND DISCUSSION

A. The gate-edge leakage phenomenon

Figure 4 shows a comparison of typical Q_{bd} distributions for type A and type B devices, in which the poly-Si was fabricated by *in-situ* and *ex-situ* processes, respectively. In order to emphasize the gate edge effect, the tested devices in Fig. 4 received no gate-edge sidewall oxidation.

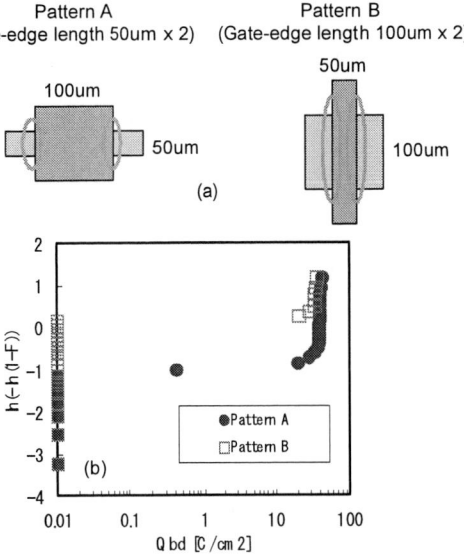

Figure 5. Top view of two test patterns which differ only in their air-exposed gate-edge length (type B device) (a); their Qbd distributions ($V_G>0$) (b). The probability of initial failure (18/25) in pattern B is close to twice that probability (8/25) in pattern A.

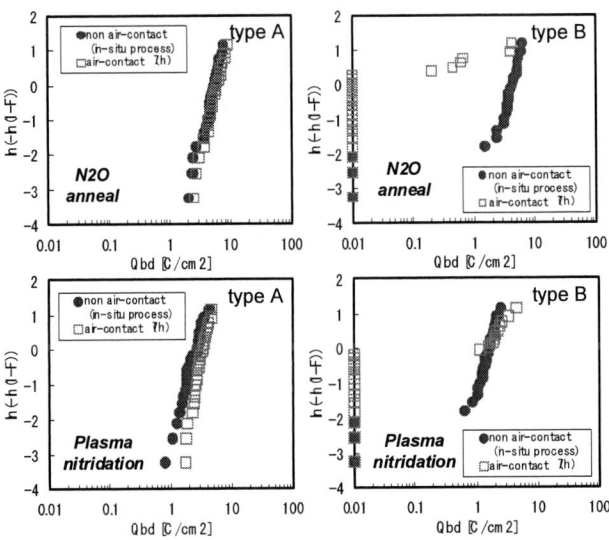

Figure 4. Typical Qbd Weibull plots. Initial failure bits only appear in type B devices (with air-exposed gate-edge), particularly clustering on devices with an air-contacted poly-Si interface. This trend is independent of the nitridation process.

from the normal and the initial failure bits are compared in Fig. 7. The gate leakage currents of initial failure bits shown in Fig. 7(b) are unstable and remarkably higher than those of normal bits (Fig. 7 (a)). In the second V_g scan, the leakage ($V_g>6V$) current decreased in normal bits while it significantly increased in initial failure bits. The leakage currents for both bits became stable afterwards. Thus the origin of the leakage current in initially failing bits appears to be different from that of normal bits.

Two clear trends can be seen in Fig. 4: (i) considerable initial failure bits only occur in type B devices (with air-exposed gate-edge), and (ii) especially those devices are affected whose poly-Si interface had once contact to air. These results imply that the initial failure relates not only to the presence of an exposed gate-edge but also to the air-contact of the poly-Si interface.

Figure 5 compares the Q_{bd} distributions of the patterns A and B, which differ in their air-exposed gate-edge lengths (A/B=1/2) but have the same active area. The probability of initial failure in pattern B is 18/25, which is close to twice as often as in pattern A (8/25). This gate-edge length dependence of the initial failure bits thus indicates that the initial failure probability rises in proportion to the gate-edge length. This demonstrates that the events resulting in initial failure take place near the gate-edge side surface. As shown in Fig. 6, this type of failure can be improved dramatically by gate-edge sidewall oxidation. Typical I-V curves obtained

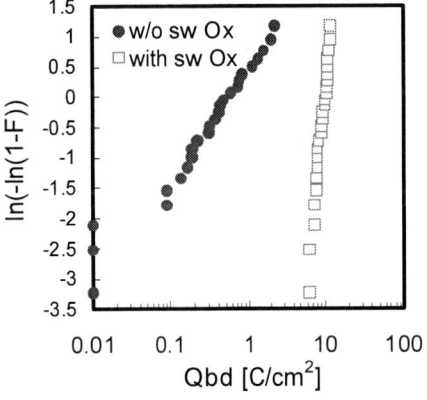

Figure 6. Effect of sidewall oxidation (sw Ox) on the Qbd distribution of type B devices (with gate-edge). Sidewall oxidation dramatically improves the initial failure as well as the Qbd.

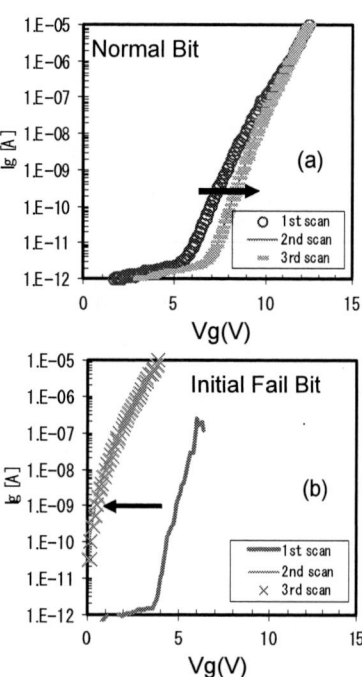

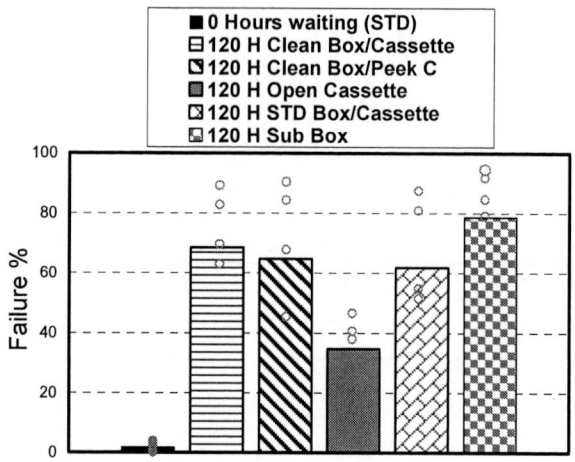

Figure 9. Waiting location split, suggesting that instead of atmospheric contaminations, the wafer themselves apparently provide the edge degradation agent. Each circle represents the failure rate of one wafer.

Figure 7. Typical I-V curves for normal bits (a) and initially failing bits (b), observed in 1st VG scans (virgin states) and 2nd, 3rd VG scans (stressed states).

B. Process influence on gate-edge leakage

In order to understand the mechanism of the gate-edge leakage, a related control experiment was performed as below. The time interval (wait time) between the gate edge acid clean and the sidewall formation steps, during which the gate edge side surface had air-contact, was arranged to be 0 h, 12 h, 24 h, 48 h, 72 h, and 96 h, respectively. Figure 8 shows the wait

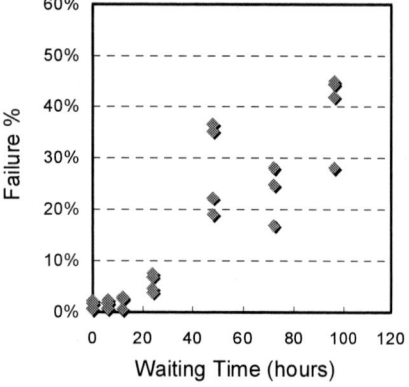

Figure 8. Wait time dependence of the failure rate. Wait time represents the time interval between the gate edge acid cleaning step and the sidewall formation step.

time dependence of the specific failure caused by gate edge leakage. Each point represents the failure rate of one wafer. Failures are definitely noticeable after 24 hours of air-contact, and increase up to 40% when the wait time reaches 96 h. Moreover, in the first 24 hours of air-contact time, a slightly increasing trend exists depending on the air-contact time. The results observed in Fig. 8 thus give the important hint that the gate-edge leakage relates to events occurring during the air exposure of the gate side surface, with a direct proportionality to the wait time.

The air-contact influence on the gate-edge leakage might be attributed to contaminations adsorbing on the gate side surface, which originate either from the atmosphere or from the wafer container. Under the hypothesis that such air-borne contaminations might act as a trigger for gate-edge leakage, the wafer locations during the wait time that exposed the gate edges were varied systematically, testing 5 different conditions: 1) Clean box/cassette (right after normal cleaning), 2) Clean box/Peek C (Peek C is a cassette type of a less outgassing material), 3) Open cassette (the wafer waiting outside of the Box in a Class 1 clean room), 4) STD box/cassette (standard condition: 1~2 month since last cleaning), 5) Sub box (substrate box: air tight condition). This wait location evaluation exhibits interesting results (Fig. 9). Of the various wafer wait locations, contrary to our expectation, the 'sub box' location gives the worst consequence while the open cassette shows the best result. The Peek C cassette and the cleaned cassette do not display any improvement over the standard box condition, although the Peek C cassette should be least outgassing. These results point out that the contamination which causes the gate-edge leakage apparently comes from the wafers themselves rather than from the atmosphere or the material of the wafer container.

978-1-4244-9113-1/11 $26.00 © 2011 IEEE

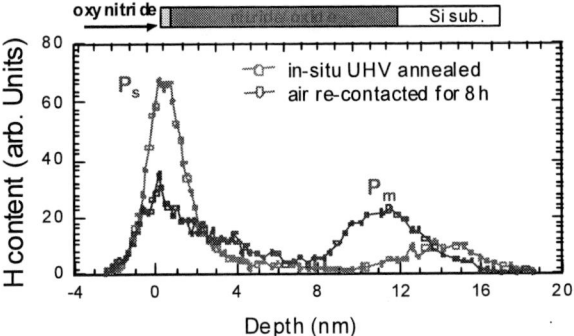

Figure 10. NRA H depth profiles of a N_2-annealed nitride / oxide stack. The specimen conditions in this experiment differ only by the air-contact prior to the NRA measurements.

C. Degradation of the poly Si interface

We have recently discovered an ultrathin oxynitride in the poly-Si/oxynitride interface as well as in the near-surface region of N_2-annealed nitride film, which may function as a potential storage layer for stable H-species, but degrades rapidly in air [10, 11]. Figure 10 shows NRA H-depth profiles of a nitride/oxide stack covered with this H-storing surface layer. The two profiles in Fig. 10 were recorded from the same specimen, which received an UHV-anneal after installation from air into the NRA vacuum chamber. The only difference between the H profiles in Fig. 10 is that one was recorded after re-admitting air (8 h) to the in-situ annealed sample prior to the NRA measurement, while the other was recorded directly in as-annealed condition. The H depth profiles exhibit two distinctly peaked signals, labeled as Pm and Ps, respectively. We have demonstrated that the peak (P_s) centered at a depth <1.0 nm is attributed to stable H species retained in the storage

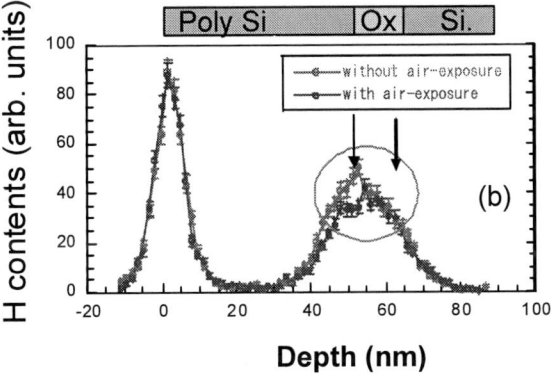

Figure 11. NRA H depth profiles of poly-Si (50nm)/ oxynitride (10nm) stacks, fabricated by in-situ (without air-contact) and ex-situ (with air-contact) processes.

layer near the Si_3N_4 surface, while the peak (P_m) close to the SiO_2/Si interface corresponds to mobile H species, which were initially present in the surface of Si_3N_4 but migrated to the SiO_2/Si interface under the NRA ion irradiation [10]. The P_s/P_m peak ratio in the NRA H depth profiles roughly represents the H-storage ability of the surface layer. Obviously, the H retention in this surface layer decreases dramatically as a consequence of the air admission. This result provides clear evidence that air-contact induces degradation of the H-storage property and results in generation of mobile H-species (P_m). Figure 11 compares NRA H-depth profiles taken from NH_3-nitrided CVD gate films (10 nm) covered by a poly-Si (50 nm) layer, which were formed by in-situ and ex-situ processes, respectively. The two NRA H depth profiles show excellent agreement except near the poly-Si/oxide interface. The H intensity at the poly-Si interface is reduced in the sample that had air-contact, implying degradation of the H-retaining ability of the poly-Si interface during the 7 h in air. This difference between the NRA H depth profiles of specimen with and without air-contact is visible for NH_3-nitrided as well as for plasma-nitrided CVD oxides. The effect of air-contact is not visible in NRA H depth profiles for NO and N_2O nitrided films, because no clear H peak was observable in the oxynitride beneath the poly-Si layer, probably due to their lower H concentration.

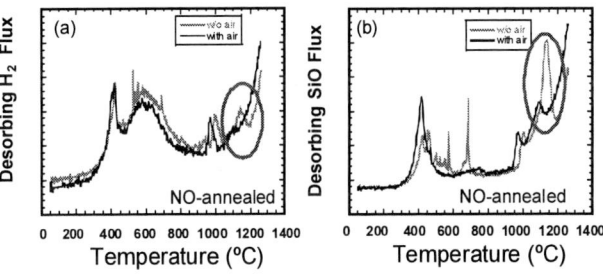

Figure 12. Typical TDS spectra of H_2 (a) and SiO (b) species observed from the NO-annealed oxynitride. The stable H_2 desorption peak (1150 °C), associated with the SiO desorption, almost disappears after 7 hours of contact to air.

Further support for the interface degradation in NO as well as N_2O nitrided condition is given by TDS spectra of H_2 and SiO species obtained from poly-Si (5 nm)/ oxynitride stacks (Fig. 12). It has been demonstrated that the poly-Si/oxynitride interface contains stable H species, which desorb at a temperature above 1100°C [11]. We focus on the desorption peak of these stable H species and compare them for the in-situ and ex-situ fabricated stacks in this study. Clearly, the stable H species desorbing together with SiO around 1150°C reduce strongly due to 7 h of air-contact. The TDS peak at 1150 °C shifts towards higher temperature as the poly-Si layer thickness is increased from 5 nm to 50 nm. It was therefore

978-1-4244-9113-1/11 $26.00 © 2011 IEEE

attributed to the H desorbing from the buried interface underneath the poly-Si layer. The NRA and TDS results thus consistently confirm that air-contact affects the properties of the poly-Si interface, especially the stability of certain H-containing species.

D. Model

Due to their common characteristic air-sensitivity and the pronouncedly Q_{bd} degrading effect of combined air-exposure of the gate-edge and the poly-Si/oxynitride interface, we suggest that the air-induced gate-edge leakage and the H-storage layer degradation are related. We propose a new model shown in Fig. 13 to explain gate-edge leakage and its correlation to the air-sensitive poly-Si interface. Ultra thin N-rich oxynitride (Si_2N_2O-like) forms on the gate oxide surface after the nitridation process. Upon contact with atmospheric moisture, the H-retention property degrades due to local decomposition (Fig.13 (a, b)):

$$Si_2N_2O + 3\,H_2O \rightarrow 2\,SiO_2 + 2\,NH_3 \qquad (1)$$

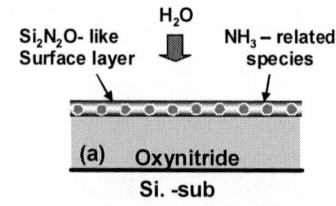

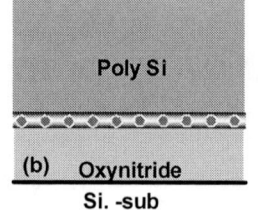

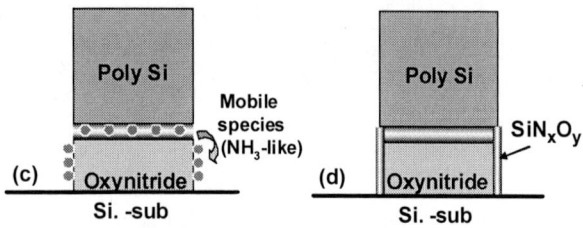

Figure 13. Proposed model to explain gate-edge leakage and its correlation to the air-sensitive poly-Si interface. Illustration of MOS gate stack states in different process steps. (a) gate oxynitride before poly-Si deposition: the oxynitride surface degrades under air-contact. (b) after poly-Si layer deposition, the poly-Si interface contains mobile species. (c) after gate stack patterning and the following gate edge acid cleaning, the mobile species seep through the poly-Si interfaces. (d) The gate stack pretreated by N_2-anneal prior to sidewall formation: oxynitride layers are formed at the side surfaces of the gate stack, which retains H-related species.

The produced NH_3–related species may partially remain beneath the surface layer and at the interface of poly Si/oxynitride (observed as part of the mobile H by NRA), and reveal themselves at the gate edge surface when air is admitted to the side surface (Fig.13 (c)). The abundance of this residual NH_3 then possibly enables the formation (Fig.13 (d)) of a non-stoichiometric as well as impurity-retaining oxynitride (SiNxOy) in the gate side surface during the N_2-anneal prior to sidewall formation (2):

$$SiO_2 + NH_3 \rightarrow SiN_xO_y + H_2O \qquad (2)$$

The leakage is assumed to be due to electrically active defects and impurities retained in this SiN_xO_y surface layer [14, 15]. The H content in SiN_xO_y films is known to be a function of the O/N ratio [16, 17]. Referring to the effect of H-induced semiconductor surface metallization [18, 19, 20], we assume the possibility that the enclosed impurities (e.g., H) contribute to surface conductivity of the air-exposed gate edge (Fig. 13 (d)). We may thus attribute the gate-edge leakage to the formation of SiN_xO_y in the side surface. Further study is necessary to corroborate this mechanism and to clarify the correlation between near surface hydrogen and the surface conductivity of silicon oxynitride. If confirmed, controlling the chemical reactions related to the NH_3-derived species as well as the presence of moisture will likely be keys for suppressing gate-edge leakage.

IV. CONCLUSIONS

A new model based on the air-sensitive hydrogen storage layer in the poly-Si interface is proposed for the origin of air-induced gate-edge leakage. The gate leakage localized near the side of the exposed edge is possibly caused by air-induced degradation of a specific compound in the poly-Si/oxynitride interface, which supplies mobile NH_3-related species to the gate side surface, resulting in the formation of a non-stoichiometric as well as impurity-retaining, hence conductive, SiO_xN_y edge layer. Edge-protecting device architecture combined with control of the air-sensitive interfacial oxynitride and its NH_3-related decomposition reaction during production is considered to be essential for improving the resistance to gate-edge leakage.

ACKNOWLEDGMENTS

The author, Z. Liu, wishes to thank M. Fukuma, N. Nakamura, T. Ishiyama of Renesas Electronics Corporation for their warm support. M. Wilde and K. Fukutani are grateful for assistance in the MALT tandem accelerator operation by H. Matsuzaki and C. Nakano at the University of Tokyo.

This work was supported by New Energy and Industrial Technology Development Organization of Japan (NEDO).

REFERENCES

[1] International technology roadmap for semiconductors, ITRS, 2010, (update).

[2] A. Hoefler, J. M. Higman, T. Harp and P. J. Kuhn, "Statistical modeling of program / erase cycling acceleration of low temperature data retention in Floating gate nonvolatile memories", IEEE Int. Reliability Physics Symposium, pp.21, 2002.

[3] A. S. Spinelli and C. M. Compagnoni, "Flash Memory Reliability 1: Basic and Emerging issues", Tutorial, S-131, IEEE Int. Reliability Physics Symposium, 2009.

[4] D. Ielmini, A. S. Spinelli and A. L.Lasaita, "Recent developments on FLASH memory reliability", Microel., Eng. **80**, pp. 321, 2005.

[5] Jonghan Kim, Jung Dal Choi, Wang Chul Shin, Dong Jun Kim, Hong Soo Kim, Kyong Moo Mang, Sung Tae Ahn, Oh Hyun Kwon, "Scaling down of tunnel oxynitride in NAND Flash memory, oxynitride selection and reliability", IEEE Int. Reliability Physics Symposium, pp. 12, 1997.

[6] H. Watanabe, S. Aritome, G. J. Hemink, T. Maruyama, R. Shirota, "Scaling of tunnel oxide thickness for flash EEPROMs realizing stress-induced leakage current reduction", Symposium on VLSI Technology Digest of Technical Papers, pp. 47, 1994.

[7] K. Naruke; S. Taguchi; M. Wada; "Stress induced leakage current limiting to scale down EEPROM tunnel oxide thickness", Tech. Dig. Int. Electron Device Meeting, pp. 424, 1988.

[8] K. N. Yang, H. T. Huang, M. J. Chen, Y. M. Lin, M. C. Yu, S. M. Jang,. D. C. H. Yu, and M. S. Liang, "Characterization and modeling of edge direct tunneling (EDT) leakage in ultrathin gate oxide MOSFETs", IEEE Trans. on Electron Devices., **48**, pp. 1159, 2001.

[9] M. Park, M. K. Suh, K. Kim, S. Hur, K. Kim, W. Lee, "The effect of trapped charge distribution on data retention characteristics of NAND Flash memory cells", IEEE Electron Device Letters, **28**, pp. 750, 2007.

[10] Z. Liu, S. Ito, M. Wilde, K. Fukutani, I. Hirozawa, and T. Koganezawa, "A hydrogen storage layer on the surface of silicon nitride films", Appl. Phys. Lett., **92**, pp. 192115, 2008.

[11] Z. Liu, S. Ito, S. Koyama, M. Makabe, M. Wilde and K. Fukutani, "Mobile and stable hydrogen species in the interface layer between poly silicon and gate oxynitride", IEEE Int. Reliability Physics Symposium, pp. 417, 2010.

[12] Z. Liu, S. Fujieda, F. Hayashi, M. Shimizu, M. Nakata, H. Ishigaki, M. Wilde, and K. Fukutani, "Influence of hydrogen permeability of liner nitride film on Program/Erase endurance of split-gate type flash EEPROMs", IEEE Int. Reliability Physics Symposium, pp. 190, 2007.

[13] Z. Liu, S. Ito, T. Ide, M. Nakata, H. Ishigaki, M. Makabe, M. Wilde, K. Fukutani, H. Mitoh and Y. Kamigaki, "Indications for an ideal interface structure of oxynitride tunnel dielectrics", IEEE Int. Reliability Physics Symposium, pp. 902, 2009.

[14] J. W. Osenbach and W. R. Knolle, "A model describing the electrical behavior of a - SiN:H alloys", J. Appl. Phys. **60**, pp. 1408, 1986.

[15] J. W. Osenbach and W. R. Knolle "Semi-insulating silicon nitride (SinSiN) as a resistive field shield", IEEE trans. Electron Devices, pp. 1522, 1990.

[16] F. H. P. M. Habraken, R. H. G. Tijhaar, W. F. van der Weg, A. E. T. Kuiper, and M. F. C. Willemsen, "Hydrogen in low –pressure chmical-vaper-deposited oxynitride films", J. Appl. Phys., **59**, pp. 447, 1986.

[17] Z. Liu, M. Wilde and K. Fukutani , to be submitted.

[18] V. Derycke, .P. G. Soukiassian, F. Amy, Y. J. Chabal, M. D. D'angelo, H. B. Enriquez and M. G. Silly, "Nanochemistry at the atomic scale revealed in hydrogen-induced semiconductor surface metallization", Nature materials, **2**, pp. 253, 2003.

[19] Y. Wang, B. Meyer, X. Yin, M. Kunat, D. Langenberg, F. Traeger, A. Birkner, and Ch. Woll, "Hydrogen Induced Metallicity on the ZnO (1010) Surface", Phys. Rev. Lett., **95**, pp. 266104, 2005.

[20] C. Wang, G. Zhou, J. Li, B. Yan, and W. Duan, "Hydrogen-induced metallization of zinc oxide (2110) surface and nanowires: The effect of curvature", Phys. Rev., B **77**, pp. 245303, 2008.

On the Thermal Failure in Nanoscale Devices: Insight towards Heat Transport Including Critical BEOL and Design Guidelines for Robust Thermal Management & EOS/ESD Reliability

Mayank Shrivastava[1], Manish Agrawal[2], Jasmin Aghassi[3], Harald Gossner[3], Wolfgang Molzer[3], Thomas Schulz[3], V. Ramgopal Rao[2]

[1]Intel Mobile Communications, East Fishkill, USA, email: mayank.shrivastava@intel.com; shrivastva.mayank@gmail.com

[2]Center for Nanoelectronics, Department of Electrical Engineering, Indian Institute of Technology-Bombay Mumbai-400076, India, email: rrao@ee.iitb.ac.in

[3]Intel Mobile Communications, Munich, Germany, email: thomas.schulz@intel.com; harald.gossner@intel.com

Abstract— **For the first time we have reported thermal failure of FinFET devices related to fin thickness mismatch, under the normal operating condition. Pre and post failure characteristics are investigated. Furthermore, a detailed physical insight towards heat transport in a complex back-end of line (BEOL) of a logic circuit network is given for FinFET and extreme thin silicon on insulator (ETSOI) devices. Self heating behavior of both the FinFET and ETSOI devices is compared. Moreover, layout, device and technology design guidelines (based on complex 3D TCAD) are given for robust thermal management and electrical overstress / electrostatic discharge (EOS/ESD) reliability.**

Keywords- BEOL Reliability, FinFET, ETSOI, Electrothermal, ESD .

I. INTRODUCTION

A number of non-planar SOI devices such as FinFETs and nano-wire FETs are proposed as technology options for sub 22 nm node gate lengths [1][2]. However, thermal management in these nanoscale devices on SOI wafer is a big concern, which is due to (i) significantly poor thermal conductivity of ultra thin materials and (ii) tight thermal boundary conditions, i.e. cooling conditions, resulting in performance degradation and serious reliability issues. Recently ultra thin body (UTB) SOI MOSFETs were demonstrated as an option for sub 20nm gate lengths, with advantages such as lower 3D parasitics, ease of incorporating strain effects, flexibility in width selection and suppressed random dopant fluctuation [3] [4]. However, its self heating behavior needs further evaluation and clarification. So far, previous work reported in [5] modeled a thermal resistance network for FinFET devices and evaluated the impact of device dimensional parameters on self heating behavior. However, this work does not provide any guidelines for BEOL design, which is due to (i) the nature of approximations for heat generation, i.e. Q=$V_{DD}I_{ON}$ and (ii) inappropriate thermal boundary conditions. Furthermore, the thermal boundary conditions considered in the work done by [6] restrict a proper

understanding of heat flow in real 3D geometries and complex BEOL. A proper understanding/characterization of the heat transport in the BEOL of a logic circuit network is therefore still missing. Currently, during the technology development phase, the device design is driven by front end of the line (FEOL) topics and core circuit performance optimization without taking into account the back end of the line (BEOL) issues in great detail regarding its impact on thermal management or EOS/ESD reliability issues. This imposes serious reliability constraints for BEOL designs. Keeping this objective in mind, for the first time we accounted for complete BEOL/FEOL of FinFET and ETSOI devices, while investigating impact of layout/device/technology parameters on the self-heating effects providing useful FEOL/BEOL design guidelines.

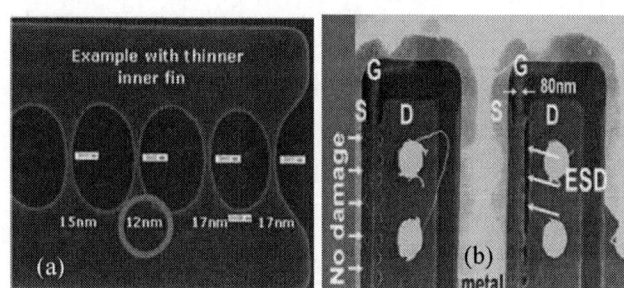

Figure 1. (a) SEM image showing ±3nm deviation in Fin thickness. This leads to weak spots which may cause early damage under thermal stress. (b) SEM image showing typical failed device from high current or thermal stress [9].

II. DEVICE FABRICATION AND EXPERIMENTAL RESULTS

Undoped trigate FinFET devices were realized with a metal mid-gap TiN gate and SiON dielectric (EOT=1.9nm). The fins have a target width of 15nm, length of 70nm and are of 60nm height. **Fig. 1a** shows SEM picture of the realized Fins where ±3nm deviation from the targeted fin thickness is evident. This leads to physically weak spots and may cause early failure

978-1-4244-9113-1/11 $26.00 © 2011 IEEE

under high thermal stress. Moreover, **fig 1b** shows SEM image of a typical FinFET device failed under the ESD stress. It is evident from mitigated impact ionization at higher currents (**fig. 2**) that the device suffers from the significant self heating, which leads to device failure after a few measurements as shown in **Fig. 3** Furthermore, we also found that current density under nominal operating conditions was getting close to as it was under ESD failure conditions [8], which further validates presence of strong heating.

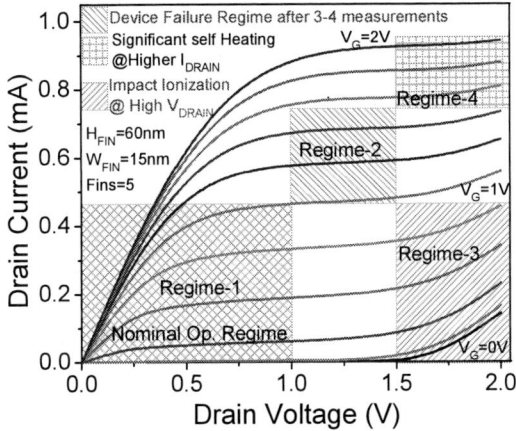

Figure 2. Measured I_D-V_D characteristics of FinFET device. Mitigated impact ionization at higher bias conditions proves the presence of significant self heating.

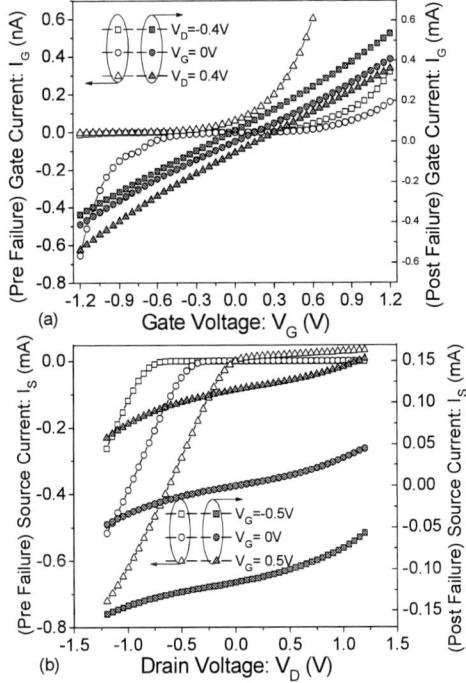

Figure 3 (a) & (b). Post and pre failure currents. Fig. shows significantly high leakage current after 2-3 measurements. Post failure characteristics shows very low resistance path from drain-to-gate and drain-to-source, which in conjunction with self heating behaviour (Fig. 2) validates an early thermal failure.

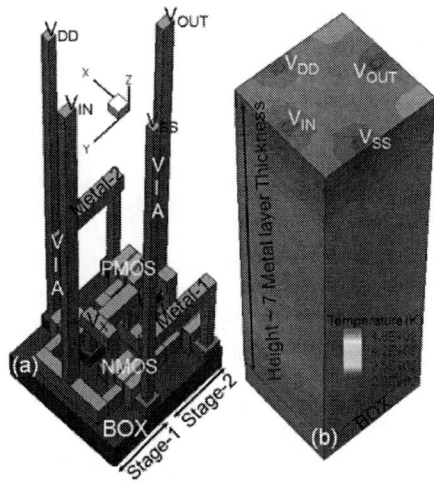

Figure 4 (a): Two stage inverter-driving-inverter with full BEOL definition. (b) Temperature distribution up-to 7 metal layers shows significance of BEOL definition.

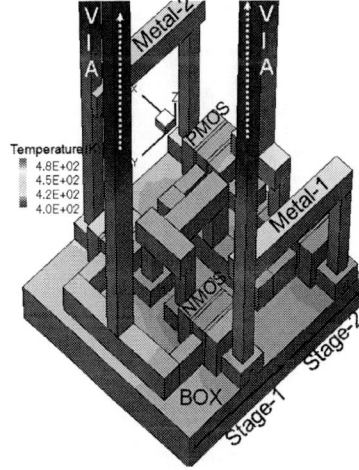

Figure 5. Temperature distribution along the interconnect metals. Figure shows max. Contribution of heat flux through metal interconnects.

III. SIMULATION FRAMEWORK AND DESIGN GUIDELINES

In order to fully capture the heat flow through various interconnect lines and inter layer dielectric (ILD) regions, a two stage inverter-driving-inverter with full BEOL definition is realized for TCAD simulations (**Fig. 4a**). Devices are realized as given in our previous work [4] and their layout is defined by using a predictive technology model [7]. The simulation approach used in this work to extract quasi-static temperature (or the worst case temperature), which should be the cumulative rise in peak temperature after 1000s of pulses, is discussed in our work elsewhere [9]. Thermal properties of various regions are calibrated as per their exact dimensions and region specific material used on silicon. **Fig. 4b and 5 and 6b** show the temperature distribution across BEOL (thickness equivalent up-to 7 metal layers), along the interconnect metals and in the active device region. Fig. 6a shows maximum contribution of heat flux through metal interconnects instead of Si substrate. This is attributed to (i) very high thermal resistance contributed by the BOX and Silicon substrate region,

978-1-4244-9113-1/11 $26.00 © 2011 IEEE

compared to metal interconnects and (ii) significantly lower thermal conductivity of Silicon region at higher temperatures. This eventually leads to most of the heat sink into the overlying back end metallization instead of Silicon substrate. Furthermore, temperature distribution across the interconnect metals and active (Si) region along different planes show: (i) NMOS has a higher temperature rise as compared to PMOS, which is due to NMOS having a relatively higher drive current (ii) Devices close to I/O pads have lower heating as compared to others, which is due to their better thermal coupling. (iii) Maximum heat flux is through the interconnect (metal) lines as compared to BOX and ILD layers. Significance of proper BEOL definition for electrothermal modeling is evident from **Figs. 4, 5 & 6. Fig. 7** shows calibration of TCAD models for 3D drift diffusion transport (considering quantum corrections) with our measurement data.

A. Impact of Device Scaling

Fig. 8 shows impact of (a) SOI/fin thickness and (b) scaling of active width of ETSOI device & fin width of FinFET device on self-heating effects. Moreover, **fig. 9** shows self heating effect due to ETSOI device. It is also evident from **Figs. 8 & 9** that ETSOI device has a more relaxed self heating compared to FinFET device. Rise in lattice temperature depends on (i) volume of power source (active region), (ii) volume of heat sink in the surrounding and (iii) thermal boundary condition in the exteriors- defines the quasi-static temperature rise. Relaxed lattice temperature can be attributed to a smaller power density in a given active Si volume compared to FinFET device, while keeping the BEOL definition same for both the devices (**Fig. 9**). Furthermore, fin height & fin width scaling in FinFET devices and SOI thickness & active area scaling in ETSOI devices reduce the temperature rise due to simultaneous drop in drain current. Moreover, channel length scaling (data not shown here) in both devices leads to a temperature rise, which is due to the slightly higher drive currents and slightly higher thermal resistance, i.e. lower thermal coupling with I/O pads. This gives an indication that technology scaling should not affect the thermal performance significantly if fin width and fin height are scaled simultaneously with channel length scaling.

Figure 6. (a) Heat Flux per unit area from various heat sinkers (or thermal contacts). (b) Temperature distribution across the interconnect metals and active (Si) region along different planes. Fig. shows: (i) NMOS has higher temperature rise as compared to PMOS, (ii) Devices close to I/O pads have lower heating as compared to others, which is due to better cooling condition, (iii) Max. Heat flux is through interconnect metals as compared to BOX and ILD layers.

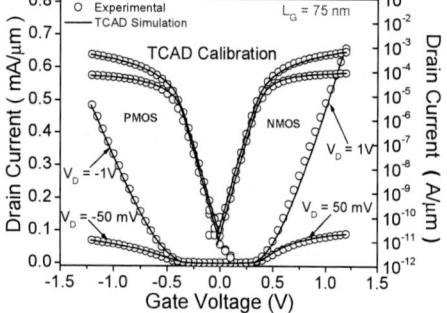

Figure 7. Calibration of TCAD models for 3D drift diffusion transport (considering quantum corrections) with experimental data.

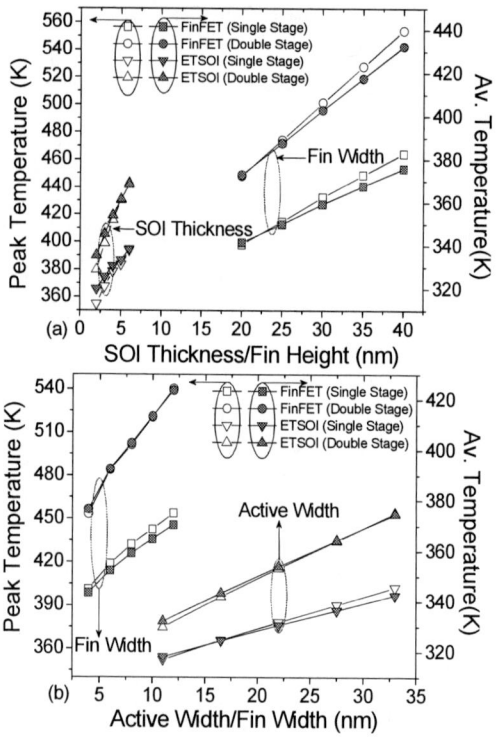

Figure 8. Impact of (a) SOI thickness and Fin thickness and (b) active width of ETSOI device & Fin Width of FinFET device scaling on self heating effects. Fig. shows: (i) ETSOI device has smaller temperature rise as compared to FinFET device, which is due to smaller power density in a given volume compared to FinFET device, (ii) Fin height scaling in FinFET device and active area scaling in ETSOI device relaxes the self heating effects and (iii) SOI thickness scaling in ETSOI and Fin Width scaling in FinFET device reduces the temperature rise due to simultaneous drop in drain current.

978-1-4244-9113-1/11 $26.00 © 2011 IEEE

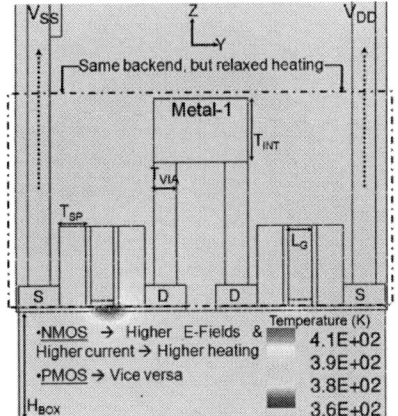

Figure 9. Self heating in ETSOI devices. It shows relaxed self heating in ETSOI device, which is due to lower volume of power source while keeping the BEOL same.

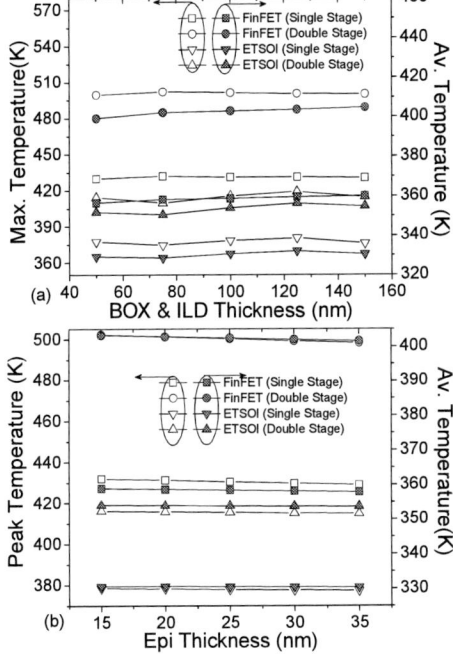

Figure 10. Impact of (a) Buried Oxide, Inter Layer Dielectric thickness (at level M1) and (b) Epitaxial raised S/D thickness on the self heating behaviour. It shows that both the technology parameters have no impact on the self heating in FinFET and ETSOI devices.

B. Impact of Technology Parameters

Fig. 10 show that (a) Buried oxide (BOX); (b) Inter-Layer Dielectric (ILD at Metal-1); and (c) Epitaxial (raised S/D); thickness has no impact on the self heating in both FinFET and ETSOI devices unlike to [6]. BOX thickness has a negligible impact due to the significantly lower thermal coupling between power source (active Si) and heat sink (external boundary) through BOX regions as shown in Fig. 6a. A lower ILD thickness does not impact much on the interconnect distribution at the Metal-1 level, i.e. total thermal resistance from metal lines, which attributes to no change in

the self heating behavior while changing ILD thickness. Note that increasing (decreasing) the thickness of epitaxial region increases (decreases) the volume of heat sink but at the same time it also increases (decreases) thermal resistance and eventually balance the self heating behavior.

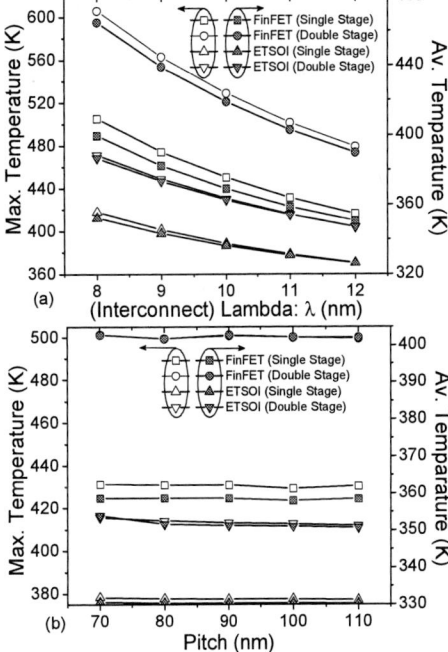

Figure 11. Impact of (a) Interconnect scaling and (b) pitch scaling on the self heating behaviour. It is evident from figure that active area pitch can be scaled without affecting self heating behaviour, whereas interconnect height and interconnect thickness impacts significantly on the self heating behaviour.

C. Impact of Layout Parameters

Fig. 11a shows that interconnect height and thickness ($\sim\lambda$) significantly impact the self heating behavior. Increasing (decreasing) the interconnect dimensions improves (degrades) the thermal performance due to lower (higher) thermal resistance. Moreover, **fig. 11b** shows that active area pitch can be scaled without affecting the self heating behavior. This is attributed to a balance between (i) rise in temperature due to higher coupling between two active regions and (ii) fall in temperature due to improved coupling between heat sink with active region.

D. Impact of Materials Thermal Properties

Fig. 12 shows that (a) interconnect and (b) interlayer dielectric thermal conductivity have a significant impact on the self heating in both FinFET and ETSOI devices. Increasing interconnect material's thermal conductivity reduces the thermal resistance and improves the cooling conditions, which eventually leads to a significantly relaxed temperature rise. Increasing the ILD thermal conductivity improves the thermal coupling of ILD layer with metal interconnects and eventually relaxes the self heating effects. Moreover, it also shows a significantly relaxed lattice temperature when interconnect thermal conductivity was equivalent to carbon nano tube

978-1-4244-9113-1/11 $26.00 © 2011 IEEE

(CNT) like material (~35W/K-cm). On the other hand, (a) gate oxide and (b) BOX thermal conductivity has almost a negligible impact on self heating effects. This is attributed to (i) very small physical thickness of gate-oxide material and (ii) due to a very weak thermal coupling between the heat sink and active region through BOX region, respectively.

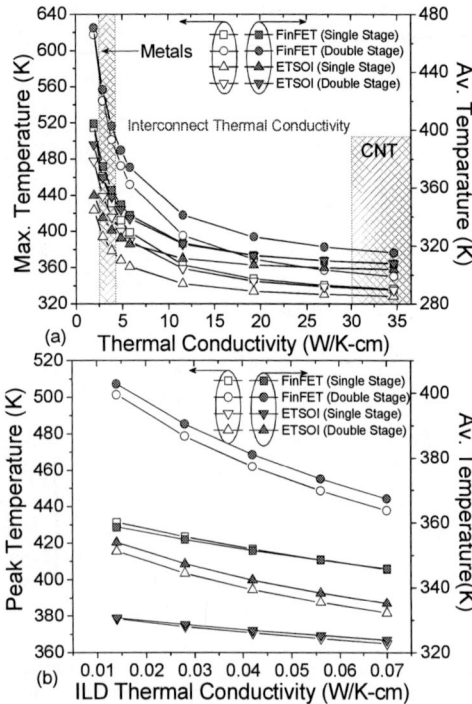

Figure 12. Impact of various material parameters (a) interconnect and (b) interlayer dielectric thermal conductivity on self heating behaviour. It is evident from figure that both the parameters have significant impact on self heating in FinFET and ETSOI devices.

E. Impact of Power Supply Scaling

Fig. 13 shows significantly relaxed lattice temperature with the supply voltage scaling. It is evident from the figure that power supply scaling is also one of the key requirements to mitigate the self heating effects in nanoscale devices.

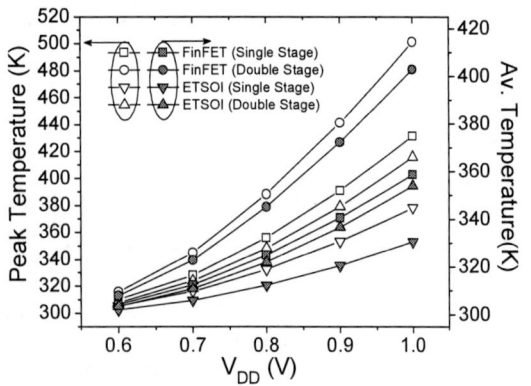

Figure 13. Impact of power supply scaling. Peak temperature (i.e. Joule Heating) proportional to I instead of I^2, which is unlike to standard theories for bulk CMOS devices.

IV. CONCLUSION

We reported thermal failure of FinFET devices, even under nominal operating conditions. A new simulation framework and importance of thermal boundary condition is discussed for accurate electrothermal modeling and for gaining a better physical insight towards self heating behavior in nano-scale SOI devices. Our results demonstrate that in order to build 3D thermal resistance network for FinFET and ETSOI like devices, proper BEOL definition is as important as the FEOL definition. It was found that scaling of all the layout/technology parameters does not have the same impact on the self heating behavior, i.e. only few parameters in BEOL have the maximum impact and need to be considered while building electrothermal models. Based on this study a design guideline is extracted and summarized in **table-I**. Furthermore, either (i) power supply scaling or (ii) use of ETSOI devices in conjunction with CNT like high thermal conductivity materials for interconnects are the key requirements in order to alleviate the thermal issues in these nano-scale CMOS technologies.

TABLE-I: Influence of Technology and design parameters on the self heating behavior. (Markers: - ↑→Rising, •→No significant change and ↓→Falling)

Parameters	Impact
Metal Thermal Cond. (K_M)	$\uparrow K_M \rightarrow \downarrow\downarrow T_{MAX}$
Power Supply (V_{DD})	$\downarrow V_{DD} \rightarrow \downarrow\downarrow T_{MAX}$
Interconnect Lamda (λ)	$\uparrow\lambda \rightarrow \downarrow T_{MAX}$
Fin Height (H_{FIN})	$\downarrow H_{FIN} \rightarrow \downarrow T_{MAX}$
Active Si Width (W_A)	$\downarrow W_A \rightarrow \downarrow T_{MAX}$
BOX Thickness (T_{BOX})	$\downarrow\uparrow T_{BOX} \rightarrow \bullet T_{MAX}$
Epi Thickness (T_{EPI})	$\downarrow\uparrow T_{EPI} \rightarrow \bullet T_{MAX}$
ILD Thickness (T_{ILD})	$\downarrow\uparrow T_{ILD} \rightarrow \bullet T_{MAX}$
Spacer Width (W_{SP})	$\uparrow W_{SP} \rightarrow \downarrow T_{MAX}$
SOI Thickness (T_{Si})	$\downarrow T_{Si} \rightarrow \downarrow T_{MAX}$
G_{ox} Thermal Cond. (K_{GOX})	$\downarrow\uparrow K_{GOX} \rightarrow \bullet T_{MAX}$
BOX Thermal Cond. (K_{BOX})	$\downarrow\uparrow K_{BOX} \rightarrow \bullet T_{MAX}$
Fin Width (W_{FIN})	$\downarrow W_{FIN} \rightarrow \downarrow T_{MAX}$
ILD Thermal Cond. (K_{ILD})	$\uparrow K_{ILD} \rightarrow \downarrow T_{MAX}$
Gate Length (L_G)	$\downarrow L_G \rightarrow \uparrow T_{MAX}$

ACKNOWLEDGMENT

Authors would like to acknowledge Prof. Dinesh Kumar Sharma (EE Department, IIT Bombay) for some insightful discussions.

REFERENCES

[1] H. Kawasaki *et. al.*, IEDM-2008.

[2] Y. Jiang, *et. al.*, IEDM 2008.

[3] O. Weber *et. al.*, IEDM, 2008.

[4] Mayank Shrivastava, *et. al.*, IEDM, 2009.

[5] B. Swahn, *et. al.*, Transactions on VLSI System, July 2008.

[6] S. Kolluri, *et. al.*, IEDM, 2007.

[7] Predictive Technology Models [Online: www.eas.asu.edu/~ptm].

[8] Harald Gossner, *et. al.*, IEDM-2006.

[9] Mayank Shrivastava, *et. al.*, T-ED, June 2010.

Resistance increase due to electromigration induced depletion under TSV

T. Frank[1,3], C. Chappaz[1], P. Leduc[2], L. Arnaud[1,2], F. Lorut[1], S. Moreau[2], A. Thuaire[2], R. El Farhane[2], L. Anghel[3].

(1) STMicroelectronics 850 rue Jean Monnet F-38926 Crolles, France.
(2) CEA-Leti, Minatec, 17 rue des Martyrs F-38054 Grenoble, France.
(3) TIMA, 46 avenue Felix Viallet F-38000, Grenoble, France.
phone: +33 4 38 92 27 30 ; e-mail: thomas.frank@st.com

Abstract - **This paper focuses on the EM induced voiding in a line ended by a TSV, and proposes an analytical model based on the link between the monitored electrical resistance increase and the matter depletion flow.**

Keywords – Through Silicon Via (TSV); electromigration; copper; resistance trace; void; model

I. INTRODUCTION

3D integration using Through Silicon Via (TSV) is becoming an alternative to overcome obstacles of CMOS scaling. As TSV processes reach maturity, reliability investigation becomes critical. Electromigration (EM) is the main reliability concern in BEOL interconnects and has been well studied these last decades. It is a metal mass transport driven by the electron wind and enhanced by temperature. It can result in voids that eventually lead to open circuits. This wear-out mechanism can also be a major reliability issues for TSVs [1-6].

Nowadays CMOS damascene copper interconnects EM wear-out mechanism is characterized by the nucleation of a void and its growth. Maximal current flow divergence occurs close to the via at the cathode of the line and leads to a void nucleation at the copper-dielectric interface (Cu/SiN or Cu/SiCN), which is the weakest interface largely due to its poor adhesion and consequent high diffusion rate [7,8].

EM is well studied through monitoring electrical resistance of a via-ended interconnect structure under accelerated current and temperature stress conditions. Typical package level EM resistance trace for via-ended copper lines is divided in 3 parts [9, 10] (Figure 1): (a) a first flat constant part, during which the void appears and grows but remains electrically undetectable; (b) a sudden step up as the void fully spans the line copper section, and the electrons have to flow through the barrier; (c) a linear increase as the void continues to grow progressively in the direction of electron flow.

For a copper metal line connected to a TSV of about 2 µm diameter, electromigration physics are still the same regarding void nucleation and copper diffusion. But the resistance trace versus time is observed to be logarithmic instead of having a step followed by a linear increase (Figure 4). Failure analyses reveal that a void nucleates in the line under the TSV, and grows with a circular shape. This paper proposes a model of electrical resistance increase based on void growth kinetic, which enables extraction of current exponent and activation energy in Black's law. Comparison is finally made between parameters calculated with the model or determined by the Mean Time to Failure.

II. EXPERIMENTAL

A. Technology

The device, illustrated on Figure 2, consists of a stack of two 200 mm silicon wafers – an upper wafer and a bulk wafer – bonded with SiO_2 using Direct Bonding technology [11]. The first damascene metal level, M_{BOT}, is processed on the upper wafer before flipping and bonding wafers. The copper TSV is processed after bonding the upper wafer, which has a thickness of 15 µm. It is surrounded by 25 nm thick TiN Barrier, and 45 nm thick SiO_2 liner. The TSV has a nearly square section. The second damascene metal level, M_{TOP} is processed after TSV. Both metal levels are processed with 45 nm thick TiN barriers at bottom and sidewalls, and 35 nm thick SiN capping on top. At final stage, top and bottom interfaces of the TSV are the same: copper (TSV) / TiN / copper (metal line).

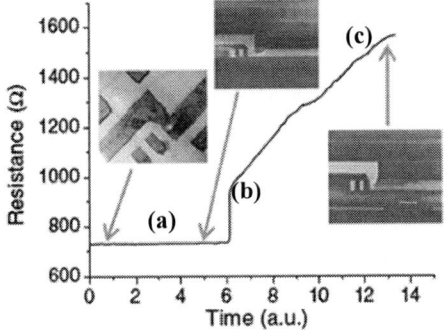

Figure 1. Typical package level EM resistance trace [10]

978-1-4244-9113-1/11 $26.00 © 2011 IEEE

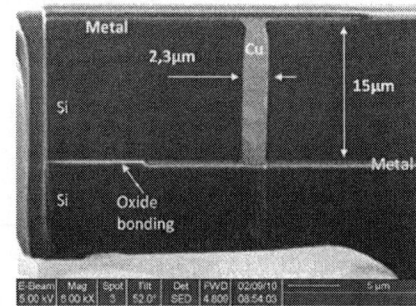

Figure 2. SEM view of TSV and the two metal levels.

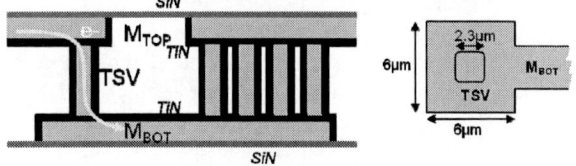

Figure 3. a) schematic cross section view of the test structure; b) schematic top view of the cathode

B. Test structure and stress conditions

EM tests are performed on a 500 μm long, 4 μm wide, and 0.25 μm thick line ended with one TSV of 2.3 μm diameter on a 6 μm square pad at the cathode, and 4 TSVs at anode (Figure 3). This configuration ensures copper depletion at the cathode of the line where the current divergence is maximal. The structure is supplied and monitored at M_{TOP} using 2 currents leads of 10 μm width, and 2 voltage monitors of 4 μm width.

EM tests are done at package level using temperatures of 250 and 300 °C, and currents of 15 and 25 mA. Respective current densities in the 4 μm wide 0.25 μm thick line are 15 and 25 mA/μm², ensuring Joule-Heating less than 5 °C. 128 samples were tested.

C. Resistance curve and failure analyses

Figure 4 shows resistance curves during EM test. They are divided in 2 parts: (1) a latency period; (2) and an increase of resistance. Plots in logarithmic time scale (Figure 4 b) reveal that the increase of part (2) follows a logarithmic law. After the logarithmic increase, resistance increase is observed to end with either a noisy trace or with an abrupt jump. In comparison to typical resistance traces of usual CMOS copper BEOL interconnects (Figure 1), there is no sudden step, and the resistance increase versus time is not linear.

Failure analyses on samples stopped during part (1) of test (Figure 5) reveal void nucleation at the cathode of the structure in the pad under the TSV. Regarding the void shape, nucleation and diffusion interfaces reveal the same capping Cu/SiN interface.

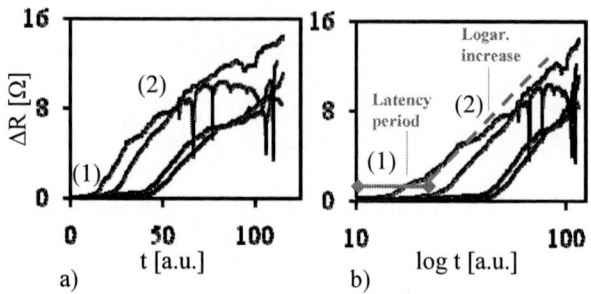

Figure 4. a) resistance traces for TSV-ended structure b) same traces than a) in a logarithmic time scale

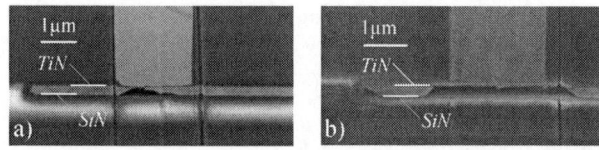

Figure 5. SEM views of void under TSV.
a) test stopped in part (1) of resistance trace; b) test stopped in the logarithmic resistance increase, part (2) of resistance trace

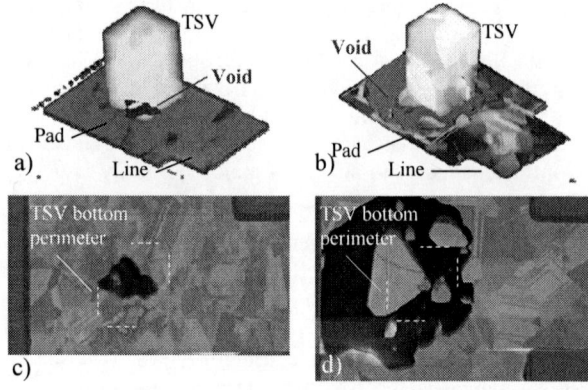

Figure 6. 3D FIB-SEM reconstructed views: a) test stopped in latency period part of resitance curve; b) test stopped after logarithmic increase

During resistance increase, part (2), the void is still located under the TSV, but has a bigger section than the bottom of the TSV. At this stage, the shunt is already effective, and electrons have to flow through the TiN barrier.

Reconstructed views of the cathode by 3D FIB-SEM (Focus Ion Beam – Scanning Electron Microscopy) analyses, confirmed that the void nucleates under the TSV, and grows in a quasi-cylindrical shape as shown in Figure 6.

Figure 6 b) and d) also show that the matter depletion can leave some Cu islands in the middle of the void.

III. MODEL OF RESISTANCE INCREASE

A. Model of electrical resistance curve as function of time

Regarding the nucleation site under the TSV and the quasi-cylindrical void shape during growth, an analytical model is developed to fit both latency period and logarithmic increase. It is based on five hypotheses.

The general expression of EM vacancy flow F along an interconnect is the sum of the flow induced by the electron wind, called the EM driving force F_{EM}; and the backflow induced by the gradient of matter concentration built up by the accumulation of matter at the anode and the depletion at cathode F_B.

$$F = F_{EM} + F_B = N_i \frac{D_{eff}}{k_B \cdot T} eZ^* \rho_{Cu} i - D_{eff} S \frac{\partial c}{\partial x} \quad (1)$$

N_i is the ion concentration, D_{eff} the effective diffusion coefficient, T the line temperature, k_B the Boltzman constant, Z^* the effective charge, ρ_{Cu} the electrical resistivity of copper, i the current, S the interconnect section, and $\partial c/\partial x$ the EM-induced matter concentration gradient between anode and cathode as matter flows towards the anode.

Following the general understanding of electromigration mechanism, backflow may be neglected when the product of current per line length (jL) is far above the critical Blech product $(jL)_C$ [12, 13, 14]. For copper interconnects $(jL)_C$ is assumed to be between 3000 and 4000 A/cm [14, 15]. In our tests (jL) is respectively equal to 75000 A/cm and 125000 A/cm for stresses of 15 mA and 25 mA. Therefore, EM flow is expressed as:

$$F = F_{EM} = N_i \frac{D_{eff}}{k_B \cdot T} (eZ^* \rho_{Cu} i) \quad (2)$$

This is the first hypothesis of this model. H1: flow of vacancy through the line F is constant during a test at constant current and temperature. In other words, this means that at any moment no backflow is built up, which would induce a slowdown and a saturation of void growth. After H1, the volume of matter V_M depleted by electromigration along whole line at the instant t can be expressed as:

$$V_M(t) = \int_0^t F dt = Ft \quad (3)$$

Vacancy migration induces nucleation and growth of voids on sites of flow divergence. Analyses of Figure 5 and Figure 6 reveal one main void located at the cathode under the TSV. But there are also other small voids at the cathode. Each void captures a part of the whole vacancies flow. The second hypothesis, H2, presumes that the ratio of vacancy flow captured by the main void over the whole vacancy flow is constant for given time and stress conditions.

According to H2, void volume is:

$$V_{void} = \alpha F t , \quad \alpha \in]0;1[, \quad d\alpha/dt = 0 \quad (4)$$

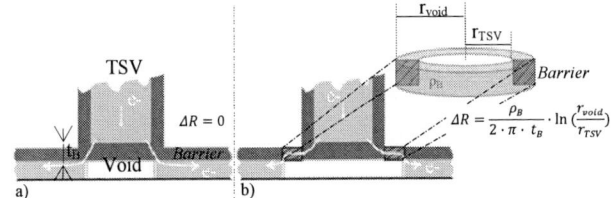

Figure 7. a) Void smaller than TSV section – no resistance increase; b) Void larger than TSV section – logarithmic resistance increase.

Observations of failure analyses of Figure 5 and Figure 6, also lead to the third hypothesis, H3, which claims that the void growth under the TSV is isotropic in the plane, that is to say, equal in all directions. This is in contrast to the usual via-ended line, where void grows toward anode once it spanned the line section. We assumed that the void under the TSV grows with cylinder geometry (Figure 7) with a radius of r_{void} and copper thickness, it induces a measurable resistance increase as soon as the radius of the void becomes larger than the radius of the TSV r_{TSV}. In this case the resistance increase is relative to the radial length of barrier $(r_{TSV} - r_{void})$, through which electrons have to flow as void grows.

Considering H3, the void volume is:

$$V_{void} = \pi r_{void}^2 t_{Cu} \quad (5)$$

where t_{Cu} is the copper thickness of the pad and line.

Fourth hypothesis, H4: the TSV bottom is approximated as a circle, although it is indeed quasi-square. This leads to an error illustrated on Figure 8, but it is of second order regarding the scale of the void shape during resistance increase as illustrated on Figure 6. Last hypothesis, H5, assumes that others voids along the line, which do not overtake the line width, do not increase significantly the resistance of the structure, because they do not force the electrons through the barrier.

Considering the modeling of the TSV bottom illustrated in Figure 8a), the infinitesimal variation of the resistance dR, once the void is larger than TSV section, is therefore:

$$dR = \rho_B \frac{dr_{void}}{2\pi t_B r_{void}} \quad (6)$$

Where dr_{void}, t_B, are respectively the infinitesimal variation of the void radius and the thickness of the barrier.

The integration of (6), from instant void is larger than the TSV section gives the resistance increase as a function of the void radius:

$$R(r_{void}) - R_0 = \int_{r_{TSV}}^{r_{void}} dR = \frac{\rho_B}{2\pi t_B} * \ln\left(\frac{r_{void}}{r_{TSV}}\right) \quad (7)$$

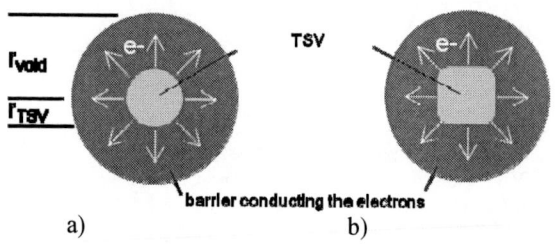

a)   b)

Figure 8: Schematic top view of the barrier conducting the electrons when void size is larger than TSV section: a) the model is based on a circular TSV bottom; b) the real TSV bottom is between a circle and a square as on Figure 6. This detail is not analytically modeled

Combining (4), (5) and (7), the resistance increase can be expressed as a function of time t as:

$$R(t) - R_0 = A \ln\left(\frac{t}{t_0}\right), \quad t > t_0 \qquad (8)$$

with:

$$A = \frac{\rho_B}{4\pi t_B} \quad \text{and} \quad t_0 = \frac{t_{Cu}\pi r_{TSV}^2}{\alpha F} \qquad (9)$$

A represents the slope of the resistance trace in a logarithmic plot, and t_0 is the time when the void becomes larger than TSV section. R_0, ρ_B, t_B, t_{Cu}, r_{TSV} and αF are the initial electrical resistance of the test structure at test temperature, the barrier resistivity, the barrier thickness, the copper thickness, the radius of TSV and the portion of vacancy flow which generates the void under TSV.

IV. FITTING THE MODEL TO THE EXPERIMENTAL TRACE

For each sample, the model is fitted to experimental data by using least squares regression, for the time interval t_{int} on which the experimental trace has a logarithmic increase (see Figure 9). Out of 128 tested samples, 89 fit to the model. Samples which do not fit the model have a resistance trace either completely noisy, or with an abrupt increase. Both cases are linked to failures in the monitoring interconnects at the beginning or during the test.

To evaluate suitability of the model, four points are studied: the time interval t_{int}, which has to be above a given number of hours to ensure a relevant fit; the independence of extracted A parameter regarding test conditions as expected by (9); the values of barrier resistivity ρ_B; and calculation of Black's parameters through the extracted t_0 parameter.

A. Time interval of fitting

The time interval t_{int} is defined between the moment the trace starts to increase logarithmically, and the moment it abruptly becomes noisy or jumps abruptly. The values of t_{int} of all fitted samples are plotted and summarized in Figure 10. In order to have a relevant fit, a comfortable t_{int} value is of at least 5 to 10 hours. Below these values, any curve behavior might be incoherent regarding the model. Among 89 fits, 83 have a t_{int} above 5 hours. Log-normal distribution plots of t_{int} on Figure 10 have a constant slope regarding the stress-conditions. And their values decrease with higher stress conditions as void growth velocity increase (Figure 10, table b)).

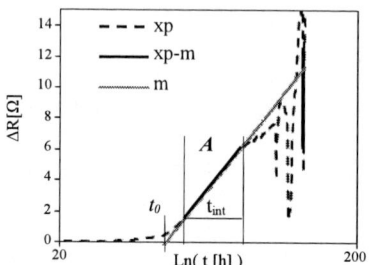

Figure 9.   Resistance traces, logarithmic abscissa.
xp: experimental trace; xp-m: experimental part used to fit model; m: model fitted to experimental trace.

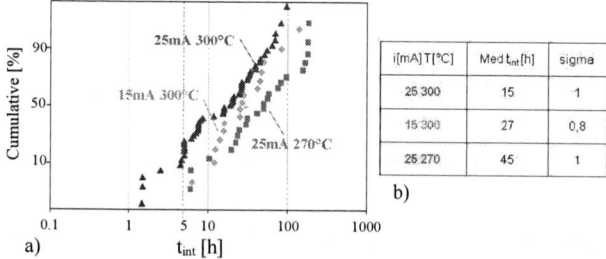

Figure 10. a) Log-normal plot of t_{int} distributions; b) table of median values of t_{int} and its log-normal slope (sigma)

B. Extraction of logarithmic slope A and the deriving barrier resistivity ρ_B

Figure 11 shows the distribution of the extracted logarithmic slope A for all samples in a log-normal plot. The distributions are quasi overlaid, confirming that the logarithmic resistance slope is independent of stress condition as it is expressed in (9). A depends only on barrier resistivity and thickness.

Given a constant barrier thickness of 45nm, the distribution of barrier resistivity extracted via the A parameter of the resistance trace model is compared to values extracted via a Kelvin TSV measurement and using (10).

$$\rho_B = \frac{\pi r_{TSV}^2}{2t_B}\left(R_{KEL,TSV} - \rho_{Cu}\frac{h_{TSV}}{S_{TSV}}\right) \qquad (10)$$

Where $R_{KEL,TSV}$, ρ_{Cu}, h_{TSV}, and S_{TSV} are respectively the electrical measured resistance of the Kelvin TSV, the copper resistivity, the height of the Kelvin TSV, and the section of the Kelvin TSV. The value of the copper resistivity is measured on a long serpentine line structure.

978-1-4244-9113-1/11 $26.00 © 2011 IEEE   350

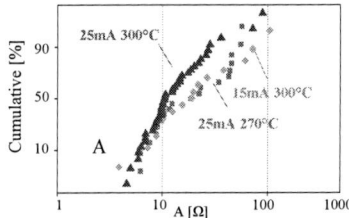

Figure 11. Log-normal plots of extracted A parameter

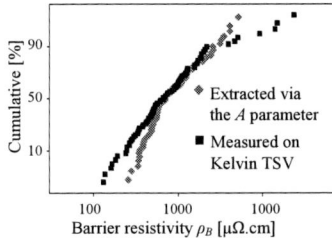

Figure 12. Distributions of barrier resistivity ρ_B values: a) extracted by the model from A parameter; b) extracted from measured Kelvin TSV

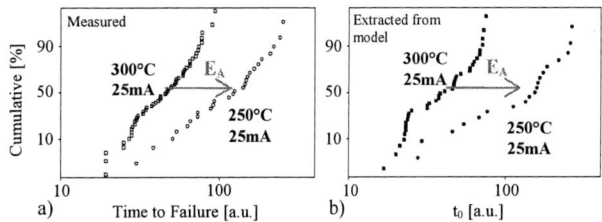

Figure 13. Log-normal plots of: a) measured Time To Failure; b) extracted t_0. Activation energy E_A is calculated for both cases.

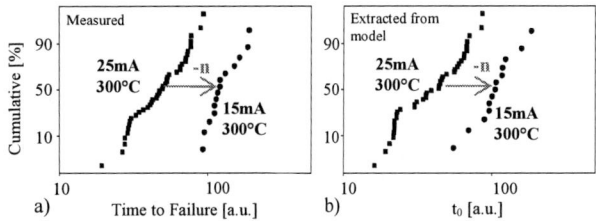

Figure 14. Log-normal plots of: a) measured Time To Failure; b) extracted t_0. Current exponent n is calculated for both cases.

As illustrated on Figure 12, values of the TiN barrier resistivity extracted via the A parameter are similar to values measured via Kelvin TSV. Moreover, the lower tail of the distributions of the barrier resistivity is about 200 to 300 $\mu\Omega$.cm, also reported by others [16]. The high dispersion is related to the imperfect process of the bottom of the TSV as described in a previous paper [6].

C. Extrapolation of activation energy E_A and current exponent n

The effective diffusion coefficient D_{eff}, appearing in (1), follows an Arrhenius law and is expressed as:

$$D_{eff} = D_{eff_0} \exp\left(-\frac{E_A}{k_B T}\right) \quad (11)$$

E_A is the EM activation energy.

Regarding (3), (9) and (11), t_0 is:

$$t_0 = \frac{t_{Cu} \pi r_{TSV}^2 k_B T}{\alpha N_i D_{eff_0} Z^* \rho_{Cu}} i^{-n} \exp\left(-\frac{E_A}{k_B T}\right) \quad (12)$$

n is current exponent in Black's equation, and is theoretically believed to be close to 1 [17]. Hence, considering (12) and the hypothesis H2, activation energy E_A and current exponent n can be extracted with the parameter t_0. The results give $E_A = 0.9 \pm 0.2$ e.V. and $n = 1.8 \pm 0.2$. From Black's equation with experimental Mean Time to Failure (MTF), $E_A = 0.9 \pm 0.1$ eV and $n = 2.0 \pm 0.2$. A failure criterion of 10% resistance increase is used to get the time to failure. These values compare well with the theoretical activation energy and current exponent for copper damascene electromigration of $E_A = 0.9$ e.V. and $n = 1$.

In the present study, the rather high extracted n value might be the consequence of local Joule Heating in the barrier due to high current density after shunting occurs for the case when the void is larger than TSV as previously assumed by some authors [8, 18, 19].

V. DISCUSSION

A. Voiding site always under TSV

Voiding is always observed to be localized in the 6 µm x 6 µm pad under the TSV, and neither in the 4 µm wide 500 µm long line where the current density is maximal, nor in the transition site between the pad and the line, where there is a divergence of the current density. Indeed, voiding should not happen at a current density divergence site, but with divergence of copper flow, as explained by [20]. Electromigration diffusion in copper damascene is assumed to be dominated by the copper/capping layer and secondarily by the grain-boundary path [7]. As long as the grain size is the same, the diffusivity is thus also expected to be the same regardless of line width. In the case of the test structure investigated in this study, with a constant TSV section, the void site should always be under the TSV whatever the dimensions of either pad or line, as long as the microstructure remains the same.

B. Time t_0 depends on TSV section and not on line width

The expression of t_0 in (9), time where resistance trace starts to increase, is proportional to the surface of TSV bottom section and is independent of line or pad width. Hence increasing the TSV bottom section should improve the Time To Failure. Increasing either line or pad width, as it is done in classic BEOL, where electromigration induces no via-related voids, will have no impact.

C. Adding a redundant TSV

Adding one or more redundant TSVs at the cathode would have two consequences. First it should increase the size of the critical section needed to be depleted to provide an open circuit, and secondly it would reduce the divergence of copper flow under the TSVs. Both are presumed to increases Time To Failure. However, the model proposed in this study is only valid for one TSV at the cathode, and has not yet been adapted for a structure with two or more TSVs.

D. Using this model for other technologies

This model is able to cover other TSV approaches such as via-middle, or via-first; or other technology choices including changes in SiN capping or TiN barrier. The general limit of this model is to use TSV with 1 to 10 µm diameter connected to a copper damascene line. With larger TSV diameters the void might be not localized under the center of the TSV. The other conditions to fulfill are (1) the presence of a conductive barrier such as TiN or TaN/Ta between TSV and copper line to ensure the current shunt, (2) the line has to be thin enough in comparison to TSV dimension to ensure a void nucleation at capping layer (i.e. neither at line/TSV interface nor in the bulk TSV).

E. Failure Criterion

The common failure criterion for CMOS copper interconnects lifetime is generally 10% resistance increase. This is based on the 10% RC delay tolerance for signals used in circuit design. For via-ended line structure the sudden step up of the resistance trace – about 100 to 200 Ω for a 65 nm node [10] – is monitored by the 10% criterion.

There is no sudden step up for the case of the TSV-ended line structure. The time t_0 where the resistance starts to continuously increase happens before the resistance reaches 10% increase. To ensure a physical approach, lifetime has to be extrapolated with a failure criterion as close as possible to t_0. This is more relevant regarding a failure definition linked to the open of the copper conductive section.

VI. CONCLUSION

A model is proposed for the resistance evolution versus time monitored during electromigration test on damascene copper line connected at cathode to a TSV. Activation energy and current exponent extracted via the model are close to values calculated from Black's equation using time to failure. Moreover, it appears that the time at which resistance starts to increase depends only on TSV bottom section and is independent of the width of the line. Hence only a larger TSV section or adding redundant TSV will improve lifetime.

ACKNOWLEDGMENT

The author would like to thank STMicroelectronics Physical Characterization team at Crolles for providing SEM and TEM analyses.

REFERENCES

[1] JEP158, "3D Chip Stack with Through-Silicon Vias (TSVs): Identifying, Evaluating and Understanding Reliability Interactions", JEDEC publication, 2009.

[2] A. D. Trigg et al., "Design for Reliability in Via Middle and Via Last 3-D Chipstacks Incorporating TSVs", IEEE Electronics Packaging Technology Conference (EPTC), 2010, pp. 328-332.

[3] A. D. Trigg et al., "Design and Fabrication of a Reliability Test Chip for 3D-TSV", IEEE Electronics Components and Technology Conference (ECTC), 2010, pp. 328-332.

[4] Z. Chen, et al., "Modeling of Electromigration of the Through Silicon Via Interconnects", IEEE International Conference on Electronic Packaging Technology & High Density Packaging (ICEPT-HDP), 2010, pp. 1221-1225.

[5] Y. C. Tan, et al., "Electromigration performance of Through Silicon Via (TSV) - A modeling approach", Microelectronics Reliability, 2010, pp. 1336-1340.

[6] T. Frank, et al., "Reliability approach of high density Through Silicon Via (TSV)", IEEE Electronics Packaging Technology Conference (EPTC), 2010, pp. 321-324.

[7] E. T. Ogawa, et al., "Electromigration reliability issues in dual-damascene Cu interconnections", Transactions on Reliability, 2002, Vol. 51, Iss. 4, pp. 403-419.

[8] C.K. Hu, et al, "Electromigration path in Cu thin-film lines", Appl. Phys. Lett., 74, 1999, pp. 2945-2947.

[9] K. D. Lee, et al., "The impact of partially scaled metal barrier shunting on failure criteria for copper electromigration resistance increase in 65nm technology", IEEE International Reliability Physics Symposium (IRPS), 2005, pp. 31-35.

[10] L. Doyen, et al., "Extensive analysis of resistance evolution due to electromigration induced degradation", Journal of Applied Physics, Vol. 104, Iss. 12, 2008, p. 123521.

[11] P. Leduc, et al., "Enabling technologies for 3D chip stacking", VLSI Technology, Systems and Applications, 2008, pp. 76-78.

[12] I. A. Blech, et al., "Electromigration in thin aluminum films on titanium nitride", Journal of Applied Physics, Vol. 47, Iss. 4, 1976, pp. 1230-1208.

[13] M.A. Korhonen, et al., "Stress evolution due to electromigration in confined metal lines", Jour. App .Phy., 73, 1993, pp. 3790-3799.

[14] D. Ney, et al., "Electromigration threshold in copper interconnects and consequences on lifetime extrapolations", IEEE International Interconnect Technology Conference (IITC), 2005, pp. 105-107.

[15] P.-C. Wang, et al., "Electromigration threshold in copper interconnects", Journal of Applied Physics, Vol. 78, Iss. 23, 1976, pp. 3598-3600.

[16] Fabreguette, et al., "Correlation between the electrical properties and the morphology of LP-MOCVD titanium oxinitride thin films grown at various temperatures", Chem. Vap. Dep. 6, 2000, pp. 109-114.

[17] J. Lloyd, "Black's law revisited – Nucleation and Growth in electromigration failure", Microelectronics Reliability, 2007, pp. 1468-1472.

[18] R. Kirchheim and U. Kaeber, "Atomistic and Computer Modeling of Metallization Failure of Integrated Circuits by Electromigration," Journal of Applied Physics, 70, 1991, pp. 172.

[19] Federspiel X, et al., "Effect of Joule heating on the determination of electromigration parameters", IEEE International Reliability Workshop (IRW), 2003, pp. 139-142.

[20] C. Hau-Riege, et al., "The effect of a width transition on the electromigration reliability of Cu interconnects", IEEE International Reliability Physics Symposium (IRPS), 2008, pp. 377-380.

978-1-4244-9113-1/11 $26.00 © 2011 IEEE

Reliability- and Process-Variation Aware Design of Integrated Circuits – A Broader Perspective

Muhammad A. Alam, Kaushik Roy, and Charles Augustine[1]

[1]Department of ECE, Purdue University, West Lafayette, IN 47907, USA
phone: 765-494-5988, fax: 765-494-6441, e-mail: alam@purdue.edu

Abstract— A broad review the literature for Reliability- and Process-variation aware VLSI design shows a re-emergence of the topic as a core area of active research. Design of reliable circuits with unreliable components has been a challenge since the early days of electro-mechanical switches and have been addressed by elegant coding and redundancy techniques. And radiation hard design principles have been used extensively for systems affected by soft transient errors. Additional modern reliability concerns associated with parametric degradation of NBTI and soft-broken gate dielectrics and proliferation of memory and thin-film technologies add new dimension to reliability-aware design. Taken together, these device, circuit, architectural, and software based fault-tolerant approaches have enabled continued scaling of integrated circuits and is likely to be a part of any reliability qualification protocol for future technology generations.

Keywords-positive reliability physics, circuit design, variability, modeling, lifetime projection

I. INTRODUCTION

Most introductory courses on VLSI design presumes interchangeability and uniformity of components whose properties remain invariant with time and posit that the fundamental challenge of IC design is the trade-off among area, performance, testability, and power dissipation of a circuit. The key feature of such *deterministic* optimization is that it involves analysis of a *static and uniform network* of transistor and interconnect nodes in response to a set of time-dependent test-patterns. Since in practice no two transistors are quite alike even within a single die due to process-variability [1], nor do they retain the same characteristics over time as defects accumulate within the transistor in response to nodal activity and input patterns [2][3][4], the idealized design principles are hardly defensible in practice. Traditional VLSI design bypasses the analysis and optimization of such *nonuniform, dynamic network* by approximating the problem into optimization of *uniform static network with certain guard band*. This allows the optimized static circuit to continue functioning even if the devices have randomly distributed parameters (with some margin), or become faster or slower over time.

Such worst-case guard-band limited VLSI design is widely used in microelectronic integrated circuits in modern CPU, as well as in macroelectronic technologies such as Liquid Crystal Displays (LCD). Despite its routine use, the methodology is inefficient and involves considerable penalty in the area-performance-power-reliability budget. After all, the process-induced variation [5], time-dependent degradation, radiation-induced soft and hard errors make it difficult to design integrated circuits within a conservative guard-bands without unacceptable compromise of power/performance targets of the circuit design. Given the constraints, the following question frames the discussion of this paper: "How does one design and optimize for power/performance/area metrics various VLSI and macroelectronic circuits with components that are occasionally faulty, but more generally have statistically distributed parameters that evolve in time either abruptly or gradually over a period of time?" There is no global optimization principle that addresses all problems of course, but the design and device communities have collaborated over the years to find a series of device/circuit/architecture/software solutions to reliability concerns in specific context. We will summarize some of the key development from the perspective of a reliability engineer.

Although the process and reliability considerations across the electronics industry is multi-faceted and is a vast topic, in general, they can be classified in four groups (see Fig. 1): *The first group* involves transistor-to-transistor, die-to-die, wafer-to-wafer process induced variation that introduces parametric variation among otherwise two nominally identical and functional transistors. *The second group* involves irreversible parametric degradation arising

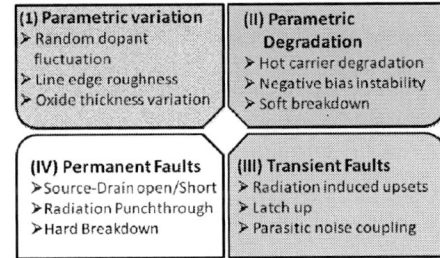

Figure 1: ICs might experience different types of faults during the fabrication and/or operation of the system. And these faults must be managed at the hardware or software levels. In this article, we classify the faults in four categories depending on how they affect system performance and the classes of solution strategies used to address them.

from loss of passivated surfaces (*e.g.*, broken Si-H bonds for NBTI/HCI damages in microelectronic and macroelectronic applications [6], loss of Si-H bonds and increase in dark current in amorphous-Si based solar cells [138], *etc.*), salt penetration through oxides in biosensors [7], etc. The third group of reliability issues involving transient errors (*e.g.*, soft error due to radiation [8][9], transient charge loss in Flash and ZRAM memories [10], etc.). *Finally, the fourth group* involves permanent damage arising from pre-existing or new generated of bulk defects (e.g., broken Si-O bonds) in SiO_2 that leads to gate dielectric breakdown in logic transistors [11][13], anomalous charge loss in Flash transistors [14], loss of resistance ratio in MRAM cells, radiation induced permanent damage in SRAM cells [25], open interconnects due to imperfect processing or electromigration, *etc.*

In each section of the following discussion, we will discuss definition, detection, and solution strategies associated with these four class of defects. We conclude with a brief discussion regarding the status, challenge, and opportunities of CAD tools in implementing process- and reliability-aware design methodologies and provide an outlook for the emerging research directions for VLSI design.

II. PHYSICAL ORIGIN OF VARIABILITY

In this section, we consider the general features of the four defect classes associated with process- and reliability-aware design.

A. Origin/Measure of 'Time-Zero' Variation

Modern VLSI designers are intimately aware of process-related parametric fluctuation arising from (i) random-dopant fluctuation [26][27][28][29], (ii) fluctuation of oxide thickness [30], (iii) statistically distributed channel lengths due to line-edge roughness, etc. These randomness reflects local, submicron-scale processing history of individual transistors and translates to threshold voltage variation, fluctuation in gate leakage, and variability of series resistance.   Regardless of the type of transistor (e.g., FINFET, ultra-thin body SOI, double-gate transistors, etc. [32]), the continued technology scaling of guarantees the susceptibility of these transistors to the processing conditions.

Other types of electronic components are similarly affected. The fluctuation in the number of nanocrystals in NC-Flash memories translates to distribution of threshold voltage and retain time, the number of grains within a TFT channel dictates its ON current [34], and as Nair et al. [7]

has shown that dopant-induced statistical fluctuation of drain-current is so significant that absolute biosensing based on Nanowire biosensors is actually impossible. In short, all areas of modern electronics (*e.g.*, logic/memory in microelectronics, macroelectronics, and bioelectronics) are influenced by randomness of design parameters.

To design an IC with transistors with random parameters, the randomness is first captured in the lowest level of abstraction (*e.g.* process level) and its effects on quantities like threshold voltage, and leakage current can either be determined numerically or by using percolation theory and stochastic geometry. These effective electrical parameters are then propagated to higher levels of abstractions (*e.g.* circuits/systems level) using numerical or analytical approaches [35][36]. Although widely used, the Monte-Carlo based numerical determination of V_T has the disadvantage that limited sample-size makes the calculation of tails of the distribution difficult and probably inaccurate.

Finally, even for nominally identical transistors, the local operating environment (nodal activity, passivation, etc.) or R-C drop due to statistically distributed length of interconnects results in local inhomogeneity in opearating temperature and ultimately to difference in I_{ON}, V_T, etc. Fortunately, given the nodal activity and operating conditions, this aspect of the problem may be predictably defined by self-consistent solution of the electro-thermal design problem [37][38].

B. 'Time-Dependent' Variation: Physical Models of Reliability

The second set of variability concerns involve permanent degradation of individual transistor characteristics that evolve with time as a function of transistor activity [2][3][11][12][53][56]. Unlike 'time-zero' variation discussed above, this variation would arise for two identically processed transistors with the same initial characteristics and that the shift in parametric values are generally permanent and cannot be restored to their pristine value by turning the IC off, see Fig. 2.

Several degradation mechanism, such as, negative bias temperature instability (NBTI), positive bias temperature instability (PBTI), hot carrier degradation (HCI), (soft) dielectric breakdown (S-TDDB), have been extensively characterized by many groups over the years. Moreover, the implications of these degradations on circuit and system performance have also been explored extensively. The key difference between models for degradation for reliability aware design vs. the standard model for qualification is the emphasis on time (t), frequency (f), duty-cycle (d) dependencies of degradation in addition to traditional focus on voltage acceleration (γ_V), temperature acceleration (γ_T) and statistical distribution (β) of failure time. This generalization of reliability models emphasize the fact that it is too conservative to qualify a technology for a given operating voltage and temperature and dictate that all IC design based on the technology reside within the boundary. Instead, the quantification of time-dependence of degradation allows trade-off among voltage, temperature, *and* time based on local nodal activity and specific usage condition of the IC.

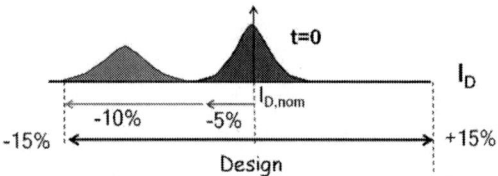

Figure 2 Type 1 faults are associated with distribution of initial parameters due to process variation (blue). When coupled with Type 2 time-dependent parametric degradation (red), the total guard-band necessary to design an IC can become unacceptably large.

Consider the case of NBTI degradation associated with PMOS transistors biased in inversion. The degradation is presumed to involve dissociation of Si-H bonds at the Si/SiO_2 interface followed by either repassivation of the bonds by newly created free hydrogen or diffusion away from the interface following H_2 dimerization [2]. (An introductory lecture, a set of lecture notes, and numerical simulation tool (DevRel) to calculate degradation characteristics are available at www.nanohub.org [15][16][17]. Based on decades of research by many groups across the industry (see Fig. 3), the degradation of threshold voltage can be succinctly described by [18]

$$\Delta V_T = [\frac{d}{1+\sqrt{(1-d)/2}}]^n \ (N_0)^{\frac{2}{3}}$$
$$* e^{-\gamma E} e^{\frac{a}{k_B T}} e^{\frac{-E_D - E_{F0} + E_R}{kT}} \ (D_0 \ t)^n$$

where t is the integrated stress time seen by a transistor, $n \sim 1/6$ is the time exponent, E is the surface field, N_0 is the number of Si-H bonds at the interface, T is the temperature, a is polarization constant, and D_0 is the diffusion coefficient, E_D, E_{F0}, and E_R are activation energies for H_2 diffusion, forward-dissociation, and reverse annealing of Si-H bonds. The net degradation is frequency independent [19] and equals the degradation at 50% duty cycle [20]. Remarkably, PBTI degradation associated with NMOS transistors biased in inversion is also defined by similar formula, except physical meaning of the individual terms are different and the magnitude of degradation is somewhat reduced. NBTI is also a key reliability concern for solar cells and TFTs based on a-Si or poly-crystalline Si and are described by analogous formula [21].

There are corresponding formula for hot carrier degradation based on bond-dissociation model [22][23][24]. The soft-breakdown characteristics after dielectric breakdown can be characterized by [11][52]

$$\Delta I = I_0 \left(\frac{t}{\gamma(V,T)}\right)^\beta$$

where I_0 is jump in gate leakage for each breakdown event, γ is the voltage and temperature dependent acceleration factor, β is the Weibull slope. Similar formula exists for backend degradation arising from electro-migration and stress migration, etc.

Finally, while radiation damage was primarily associated with transient fault (soft error) or gate punch-through (hard error), close analysis of radiation detector show that steady generation of permanent traps and type inversion of donor levels lead to time-dependent increase excess leakage and depletion width. Such parametric degradation has been a serious concern for viability of radiation detectors.

Once these degradation characteristics are characterized, its effect on inverting logic [3], SRAM cells [56][12], pipeline microprocessor [94], etc. can be readily analyzed and the susceptibility of failure due to a degradation mode or a combination thereof can be predicted. Based on these predictive characterizations, we will explore various solution methodologies later in the paper.

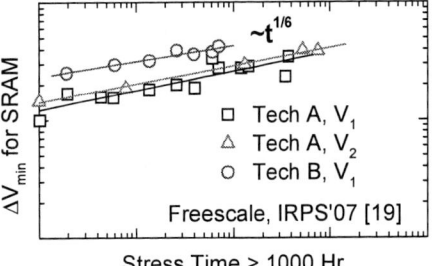

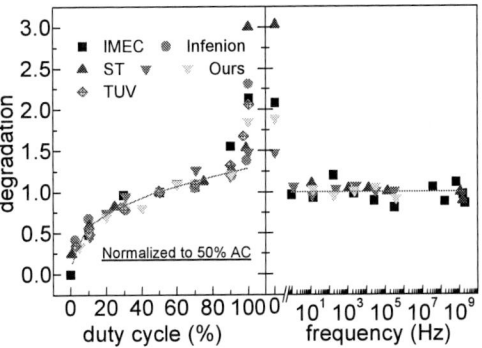

Figure 3. Time evolution of degradation at very long stress time from various published results showing power law dependence with universally observed exponent of n=1/6AC NBTI degradation normalized to 50% AC versus pulse duty cycle and frequency from various published results. R-D solutions (red lines) are shown [2]. Taken from S. Mahapatra et al., Proc. IRPS, 2011.

C. Physical Origin of Transient Faults and Soft Errors

Many researchers attribute the mysterious charge-loss electrometers in 1800s as the first reported occurrence of radiation damage in electronic components and the US Navy also reported radiation related failures of electronic components during tests for atomic bomb during WWII. Radiation-induced transient faults gained increasing prominence in 1960s and 1970s as transistors were miniaturized and transient effects associated with radiation strike and latch-up [9] caused serious reliability concerns. Transient soft errors are run-time computational error that does not lead to permanent degradation of parameters, nor is it related to process variation, but the random nature of the transient upset makes its detection and correction a challenging problem. For example, tests of SGI Altix 4700 computer (32 blades, 35 GB SRAM) shows approximately 4-5 SRAM upsets every week – errors that must be corrected for the proper operation of the machine.

Sources of radiation include alpha particles and high energy neutrons in solar winds and cosmic particles as well as low energy neutrons from packaging materials [39], making design of satellite and space-crafts particularly challenging. Although the single events effects (SEE) is generally related transient errors related to single bit upset (SEU) or multi-bit upset (MBU), permanent faults related to Single event gate rupture or burnout (SEGR/SEB) are also possible and are discussed in the next section [39][40].

Modeling/characterization of soft errors involves three steps [41] (i) understanding the radiation environment of the electronics, which may involve interaction of the atmosphere with radiation fluxes, physics of nuclear fission, etc., (ii) given the relative fluxes of particles, probability that they will interact with a given active volume of a transistors, and finally, (iii) calculation of critical charge necessary to upset a particular type of device. These effects are modeled at various levels of sophistication depending on the degree of precision necessary for a particular application and compared extensively with measurements. For example, simulation show and experiments confirm that the SER rate seems to increase exponentially with reduction in supply voltage – increasing by a factor of 10-50 as supply voltage is scaled from 5V to 1V. In the worst case scenario, a strike can change the state of multiple nodes (MBU), making it difficult to recover the lost data. In memories, techniques are present to detect and correct one error in every column/row, but in the case of multiple errors, present day techniques are not sufficient to tackle such errors and will result in complete system failure [8]. Thus, it is important to explore techniques that can detect (and if possible correct) the errors, without allowing the errors to propagate. In general, transition to SOI technologies reduces the amount of charge that is generated within bulk of the body. Since this increase the LET threshold for upset, the SER probability is reduced by a factor of 2-3 [42]. However, trapping and other hysteresis effects could be a concern for SOI transistors.

Finally, memories like Flash and ZRAM are also susceptible to soft error [10], since the energetic particle may eject the stored charge and change the memory state permanently. We will briefly summarize the well known detection and solution strategies later in the paper.

D. Origin of Permanent Faults

Permanent faults has been a reliability concerns from the early days of electronic industry based on electro-mechanical switches and punch-cards based on optical readers [43]. The unreliability of mechanical switches in 1940 and 1950 were so severe that many great scientists like von Neuman, Shanon, and Hamming spent considerable time working on fundamental principles of designing reliable systems with unreliable components. The techniques include judicious use of combination of techniques involving redundancy, stability analysis, and coding. To this day, when computer scientists discuss reliability of components,

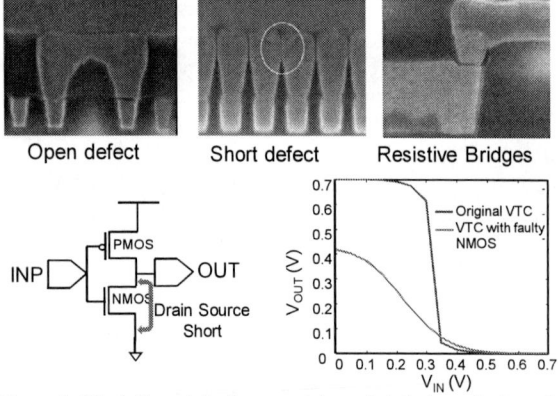

Figure 4. (Top) Type 4 faults present in scaled devices. (bottom, left) Defective NMOS in INV and corresponding (less than perfect) voltage transfer characteristics (VTC) (bottom, right).

the language and terminology ('stuck-at-zero' fault meaning a relay which cannot be closed or 'stuck-at-one' fault meaning a relay permanently connected to the output) can be dated back to early days of computer engineering.

Modern semiconductor processing has vastly improved process-reliability. While it is still possible to have physical defects such as shorts, opens and resistive bridges (see Fig. 4) that can be modeled as "stuck-at-zero' and 'stuck-at-one' or stuck-open, [44]; fortunately, these challenges are likely to be resolved with process debugging or detected at test time using standard test generation techniques. However, once the circuit has been used for a prolonged period of time, 'stuck-at-zero' type faults may occur with a hard-dielectric breakdown at the gate for NMOS (Fig. 4). Moreover, given the reduced oxide thickness, there is increasing likelihood of punch-through of the dielectric film (SEGR/SEB) by energetic particles leading to permanently damaged transistors. These types of hard errors also occur in transistors based on CNT, poly-silicon NW, etc. since excessive heating may lead to burning of the channel region [45]. The classical theories of fault-tolerance therefore can be useful in these contexts, with the implicit assumption that a certain percentage of devices may fail randomly during IC operation. However, since all such redundancy techniques incur extra area/power penalty, ensuring high performance with high reliability requires intelligent trade-off and thoughtful design at architecture, circuit, and transistor levels. We will discuss the detection and solution strategies of such failures later in the next section.

III. CIRCUIT SOLUTIONS STRATEGIES FOR LOGIC CIRCUITS

A. Process Variation and Solution Strategies.

Process variation and reliability consideration pose special design challenges for logic circuits. Historically, the circuit designers first became aware of the challenges of design under variability as they had to reckon with the widening distribution of leakage due to random dopant fluctuation. In response, device physicists proposed channel dopant-free transistor designs [26]. At the circuit design level deterministic 'global' techniques like back-gate bias techniques, and adaptive V_{DD} have been proposed to tighten V_T and leakage current distributions [46]. These approaches are particularly effective for systematic process fluctuation arising from line-edge roughness or oxide thickness, etc.

On the other hand, CRISTA [47] is a circuit/architecture co-design technique for variation-tolerant digital system design, which allows for aggressive voltage over- scaling. The CRISTA design principle (a) isolates and predicts the paths that may become critical under process variations, (b) ensures that they are activated rarely, and (c) avoids possible delay failures in the critical paths by adaptively stretching the clock period to two cycles. This allows the circuit to operate at reduced supply voltage while achieving the required yield with small throughput penalty (due to rare two-cycle operations). A schematic of the approach is shown in Fig. 5.

A recently introduced approach called 'RAZOR' [48] uses a shadow latch to detect the transient failures in pipelined logic, which causes an output to be incorrectly

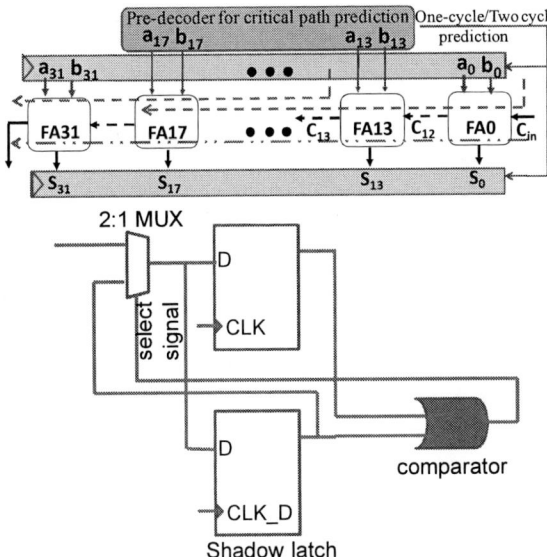

Figure 5. (Top) Fig. 7 32bit Full Adder designed using CRISTA design methodology. (bottom) RAZOR flip flop (CLK_D is the delayed CLK).

latched. Upon the detection of error, the instruction is re-executed by flushing the pipeline and using the correct data from the shadow latch (Fig. 5). Razor is an adaptive tuning technique for correct operation under parametric process variation.

At system level, design techniques like Robust core checker [49], leakage based PVT sensor [Kim04], or on-chip PLL-based sensors [50] have also been explored to generate self-diagnostic signals to dynamically control adaptive body-bias for inter-die variation. An excellent review of the process tolerant design techniques is summarized in [1].

B. Variation Due to Parametric Degradation.

Parametric time-dependent degradation associated with NBTI, HCI, TDDB, and EM degradation are challenging problems – made even more complicated by increasing additional contributions from 'time-zero' process variation. Given the voltage, temperature, and time dependent analytical models, however, statistical design methodologies [51] offers promising solutions in such a scenario. We offer a brief review of the approaches related to NBTI to illustrate the principles of statistical VLSI design in the presence of variability; previously there have been extensive work regarding gate-dielectric breakdown [52] and there are suggestions by many that PBTI and HCI are likely to be analyzed by similar approaches.

To appreciate the role of such parametric degradation on logic circuits, we used 70 nm Predictive Technology Model, (PTM) an NBTI-aware static timing analysis (STA) tool [3] to estimate temporal degradation in circuit delay under NBTI. Both simple logic gates (e.g. NAND, NOR gates), as well as various ISCAS benchmark circuits involving a few hundred to a few thousand transistors were simulated to estimate the degradation. In general, the results indicate that the circuit delays can degrade by more than 10% within 10 years operation time [3]. Such large change in parameters from just one of the degradation mechanisms is a genuine

cause for concern. The techniques proposed to address this degradation are discussed next.

Pre-Silicon Solution Strategies: The solution strategies discussed in the literature roughly fall into two categories: deterministic and statistical. Various global deterministic techniques like dynamic V_{DD} scaling, duty-cycle scaling, *etc.* have been discussed in the literature [53]. The 'V_{DD} scaling' may not always be optimal because in addition to increasing reliability, it also increases delay [54][55]. The proposal of 'reduced duty-cycle' [56] is misleading, because in an inverting logic, reduced duty-cycle at one stage translates to enhanced duty-cycle for the following stage [57].

The statistical technique involves Lagrange-Multiplier (LM) based sizing given the constraints of area, gate delay, and product lifetime. Both cell based sizing [3] and transistor based sizing [57] can be effective. Since the threshold voltage of a gate depends on switching activity, which in turn depends on the sizes of the other transistors, the sizing algorithm therefore is by definition self-consistent [58]. The results show that cell-sizing (where NMOS and PMOS are scaled by the same factor) [3] of several ISCAS benchmark circuits can be made NBTI-tolerant only with modest area overhead (~9%). Even better optimization is desirable and possible if PMOS is scaled, (but NMOS is not), thereby providing – in the best case and as shown in Fig. 6 – a factor of two improvements in area penalty [57] so that a wide variety of ISCAS benchmark circuits at 70nm node can be designed with only 5% area-penalty for NBTI tolerant design (with no delay-penalty). One limitation of the specific work discussed above is that the model does not include the effect of NBTI relaxation – a key feature of NBTI characteristics. More recent work [59] addresses the relaxation issue specifically and arrives at improved (but comparable) estimates of various trade-off.

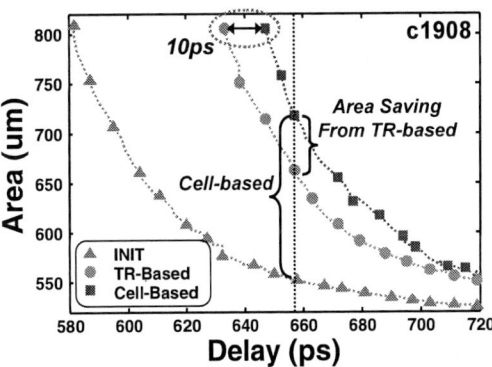

Figure 6. PMOS-only LR optimization of ISCAS benchmark circuit C1908 based on 70 nm PTM offers approximately 45% saving in area overhead (11.7% for cell-based optimization, 6.13% for PMOS-only optimization).

Post-Silicon Silicon' Odometers: Since NBTI degradation is actually specific to local operating environment, all ICs do not degrade equally. An operating condition specific indicator of NBTI degradation would be of great value. A new class of techniques, broadly known as "Silicon

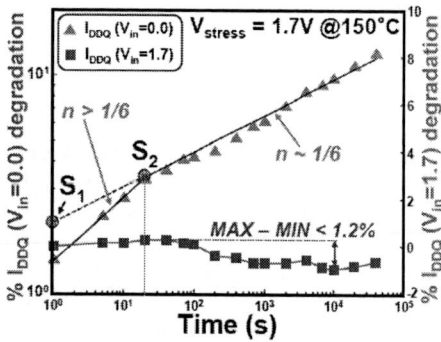

Figure 7: The change in Iddq (red triangle) is measured by temporarily flipping the states of the transistors (Vin=0) so that leakage from degraded PMOS is reflected in Iddq. Note that the power-law exponents of individual devices and that of an IC (here) is identical.

Odometers", have recently been proposed to allow simple and direct estimate of actual usage of an IC.

The first approach suggests the use of quiescent leakage current (IDDQ) to establish on-the-fly time-dependent degradation due to NBTI, see Fig. 7 [60][61]. Over the years, Iddq has been used a process-debugging tool for time-zero short- or open-circuit issues [95]. Since Iddq is also affected by V_T-shift, it can also be an effective and efficient monitor of application- and usage-specific IC degradation as a function of time.

The second approach is based on differential signals from build-in modules or newly designed structures. For example, Ref. [63][64] compare the phase difference between two ring oscillators (DLL) - one is stressed in actual operation and the other is not – to estimate the actual usage history of the IC. Other groups [50] have used variation in PLL signal to arrive at the same signatures.

Finally, Ref. [65] uses a PMOS transistor biased at sub-threshold region as a NBTI sensor for actual usage of the IC and this degradation is then mapped back to frequency through 15-stage NAND-gate ring oscillator. In sum, a wide variety of techniques are now being used to measure – in situ – the actual usage of the IC. Other types of locally embedded sensors have also been used to dynamically monitor the degradation [66][67]. This enables the circuit to be run at its maximum possible frequency, while preventing any failure to occur due to degradation.

These indicators – be it IDDQ or DLL phase shifts – would then allow one to use adaptive design to extend lifetime of the ICs or to indicate incipient failure and preventive maintenance, allowing a powerful post-Si design approach for future ICs. The effectiveness of the Iddq methodology has been demonstrated based on ISCAS benchmark-circuits at 70 nm PTM node – establishing a perfect correlation between NBTI degradation and decrease in Iddq. The technique has also been verified experimentally both for NBTI and TDDB to show ease of implementation of the methodology. Although recent results show encouraging trends; however, the area- and power- overhead of the sensors as well as algorithmic complexity associated with various corrective techniques are limitations of this technique that needs careful consideration.

Solution Based on Degradation-Free Transistors: Classical MOSFETs are undergoing significant changes: the channels have been strained by Si/Ge Source and drain and high-k/metal gate have been introduced in the gate stack. The material changes have led to an intriguing suggestion of the possibility of NBTI *degradation-less* transistor operation for high performance logic circuit [68]. Briefly, the ON-current of a transistor is given by $I_D \sim C_{ox}$ $(V_G - V_T)$ (μ is the mobility, V_T the threshold voltage), so that

$$\Delta I_D / I_{D,0} = \Delta \mu / \mu_0 - \Delta V_T / (V_G - V_{T,0}),$$

where subscript '0' indicates pre-stress parameters t. When V_T degrades (increases) with generation of bulk and interface traps, the E_{eff} at the Si/SiO2 interface is reduced. Since $\mu - E_{eff}$ characteristics is negative, this translates to positive $\Delta \mu$. Take together, the self-compensating effects of $\Delta \mu$ and ΔV_T reduce drain-current degradation, and in some strain-values can make the transistor degradation-free. A combination of circuit techniques as well device engineering may keep NBTI degradation within acceptable limits. There may be similar opportunities for PBTI and HCI hardened design principles.

C. Detection and Solution Techniques for Transient Failures

Similar to degradation-free transistors for NBTI, we have already discussed the intrinsic improvement in soft errors robustness for SOI transistors that has smaller collection volume and therefore higher LET for upset. In addition, there are many Circuit/Architectural/Software detection/correction algorithms for soft errors in logic circuits, as discussed below.

At circuit level, flip-flops have been proposed with built-in soft error tolerance [69]. The flip-flop implementation consists of two parallel flips-flops with four latch elements and a C-element for comparing the output of two flip-flops. Out of the four latch elements, one can be subjected to single event upset (SEU). In that case, output of the two flip-flops differs and error does not propagate to the output through the C-element. In the event of more than one error in four latch elements, the output can go wrong, however, the probability of such an event is very low. Another topology for soft-error tolerant flip-flop is proposed in [70]. Along with immunity towards soft-error, the proposed flip-flop offers enhanced scan capability for delay fault testing.

Finally, in the 'Check point and roll back' [71] technique instructions stop executing once a fault has been detected and the state of the system is restored from the point of error introduction (i.e. rolled back). This technique is associated with micro-architecture Access-Control-Extension (ACE) instruction set [72] for detecting the faults. Although the approach can cause stalling of the processor for few cycles, but it results in correct execution of instructions without the need for complete flushing of the pipeline.

The above-mentioned approaches for soft error tolerance are applicable in general purpose computing systems. On the other hand, for probabilistic applications such as Recognition and Mining and Synthesis (RMS), Error Resilient System Architecture (ERSA) is proposed in [73]. ERSA combines error resilient algorithms and software optimization with asymmetric reliability cores (some cores are more reliable than others due to parameter variations, say). Results show that the architecture is resilient to high error rates on the order of 20,000 errors/cycle per core.

D. Detection and Solution Techniques for Hard Faults.

In the gate level 'stuck-at' fault model, each circuit node can be at stuck-at zero (1/0: good value is ONE and faulty value is ZERO), stuck-at one (0/1), 0/0, 1/1 or x/x value. Not that this model is used in reference to the manifestation of defects at the electrical level and is not a physical defect model. For detecting stuck-at one fault (say), the node is activated to have a logic value of 0 using an input vector such that the effect of the fault is propagated to a primary output.

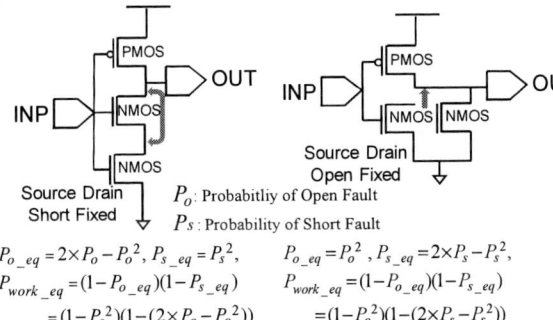

P_o: Probabitliy of Open Fault
P_s: Probability of Short Fault

$$P_{o\_eq} = 2 \times P_o - P_o^2 , P_{s\_eq} = P_s^2, \qquad P_{o\_eq} = P_o^2 , P_{s\_eq} = 2 \times P_s - P_s^2,$$
$$P_{work\_eq} = (1 - P_{o\_eq})(1 - P_{s\_eq}) \qquad P_{work\_eq} = (1 - P_{o\_eq})(1 - P_{s\_eq})$$
$$= (1 - P_s^2)(1 - (2 \times P_o - P_o^2)) \qquad = (1 - P_o^2)(1 - (2 \times P_s - P_s^2))$$

Figure 8. Transistor level redundancy techniques for Type 4 short and open faults.

Although we have already discussed IDDQ (Quiescent current) tests to determine NBTI and HCI degradation, the IDDQ test was originally developed to detect hard faults (such as short between two nodes, pseudo-stuck at faults, stuck-on faults, etc. [74]). The assumption that in CMOS circuits, if a fault, such as a bridging defect is activated, will lead to large increase in quiescent current, worked well in early technology generations to detect manufacturing defects. However, with increased leakage and leakage spread (due to parameter variations) in scaled technologies, it is becoming increasingly difficult to use IDDQ to test manufacturing defects. The sensitivity of IDDQ tests to detect manufacturing under increased leakage and parameter variations has decreased considerably.

Pre-Silicon Solution Strategies: Once the faults have been identified, its effects can be mitigated by various types of redundancy. At the <u>transistor level</u>, one can deploy various types of series and parallel transistor redundancies depending on the prevailing types of defects (as shown in Fig. 8). Redundancies can be distributed at random throughout the circuit. The introduction of transistor level redundancy may decrease the finite impulse response (FIR) filter failure probability by almost 20% [75].

At the <u>system level</u>, triple modular redundancy (TMR) and Triplicated Interwoven Redundancy (TIR) are popular options to address both transient and permanent faults [76]. The final result of the computation in TIR and TMR are decided by majority voting of three identical sets of hardware (see Fig. 9). For a given area overhead, however, TIR seems to provide somewhat improved robustness compared to TMR. Note that, in this analysis the voting circuit is assumed to have zero failures.

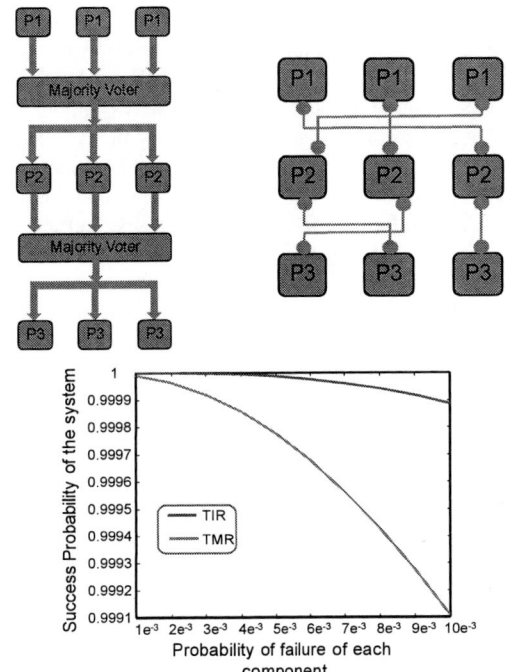

Figure 9. (Top, left) Cascaded TMR and probability for correct system operation . (Top, right) Triplicated Interwoven Redundancy (TIR) and probability for correct system operation. (Bottom) Comparison of TMR and TIR.

The main challenge of dealing with hard failure is that many of the classical techniques proposed above to address soft degradation are not suitable to handle such catastrophic failure: global solutions like adaptive body-bias, *etc.* are not helpful, redundancy could help, but since the BIST may have to monitor the cell failure at exceptionally high rate, the approach may not be practical. Various other versions of reconfigurable logic based on FPGA are being investigated and offer some promise [77][78], but more research is needed for practical, cost-effective solutions. Therefore, the challenges of circuit design under random permanent faults offer new avenues for research in this field.

III. DETECTION AND SOLUTION TECHNIQUES FOR SRAM

The relative footprint of memory in microprocessors is expected to reach 50% by 2008 and 90% by 2011. As such robust SRAM design under process-variation and parameter degradation has been a persistent challenge for many years [27][79].

A. Process Variation and Solution Strategies

Four functional issues define SRAM yield: **ACCESS** failure (AF) defines the unacceptable increase in cell access time, **READ** failure (RF) defines the flipping of the cell while reading, **WRITE** failure (WF) defines the inability to write to a cell, and **HOLD** failure (HF) defines the flipping of the cell state in the standby, especially in low V_{DD} mode [80]. Only those memory blocks that have none of these failures in any of their cells are expected to remain functional. Systematic inter-die variation and random inter-die variation limit the yield of SRAM cells – as expected.

Of the many proposals discussed in the literature to address this issue, we will discuss two representative cases. Similar to logic transistors, the first approach involves using adaptive body bias to tighten distribution of V_T and leakage current. The signal may be self-generated by monitoring the on-chip leakage. It has been shown both theoretically and experimentally that such adaptive body bias can increase SRAM yield significantly [81]. Once again, the methods are effective for inter-die variation, but unsuitable for fine-grained intra-die variations.

The second approach consists of two circuit level approaches: one involving more complex SRAM cell design, especially for low-voltage application [82] and the other involving redundant columns to replace faulty cells. In the 'redundant column' approach, the Built-in Self Test (BIST) circuit periodically (e.g. every time operating condition is changed) checks for the faulty bits and subsequently resizes the memory to prevent the access to the faulty bits [96]. It has been shown that such design can be transparent to the processor with negligible effect on access time or area/energy overhead and can still improve yield by as much as 50%.

B. Reliability Aware SRAM Design

Similar to design of logic transistors, SRAM design with the simultaneous presence of process variation and parameter degradation is significantly more challenging problem. For example, regardless of whether an SRAM cell stores a '0' or '1', if the cell is not accessed frequently, one of the PMOS pull-up transistors could be in NBTI stress for a prolonged period of time. As such the divergence of parameters of the pair of PMOS transistors within a SRAM cell can erode the read and write stability of the cell [12]. Likewise, since gate dielectric breakdown (TDDB) is a statistical process, breakdown in one of the transistors would induce excess gate leakage in one transistor without corresponding effect on the other five transistors [83]. The asymmetric degradation is a key reliability concern for SRAM design and needs to be reflected in IC design process.

An analysis of typical 6T SRAM array shows that READ failure probability increases by a factor of 10-50 due to NBTI, if the initial ΔV_T due to process variation ranges between 30-40 mV (70 nm PTM technology). On the other hand, the "WRITE stability' improves somewhat with NBTI as the cell can be written at smaller voltage. Overall, however, the metrics like static noise margin (SNM) reduces so quickly with NBTI that it emerges as an important reliability concern [12]. Indeed it was found that parametric yield of a large sized SRAM arrays (e.g., 2MB, PTM 70nm) can degrade up to 10% in 3 years time period (Fig. 10).

Solution Strategies: Several groups have proposed and explored potential remedies for NBTI-induced SRAM degradation. These approaches involve improvement in pre-silicon design phase, post-silicon test phase, and post-silicon operation phase, as discussed below.

In device design phase, the most effective route to combating the NBTI related asymmetry is to reduce initial asymmetry as much as possible. Therefore, techniques that reduce process variation are well-suited for reliability-aware design as well. These may include dynamic V_{DD} scaling,

body biasing, or redundant arrays to replace faulty cells, *etc.* [53][56][12].

During the test-phase, a body-bias based burn-in strategy could also be effective [55]. In particular, if the body bias is used such that the SNM is reduced, then NBTI induced

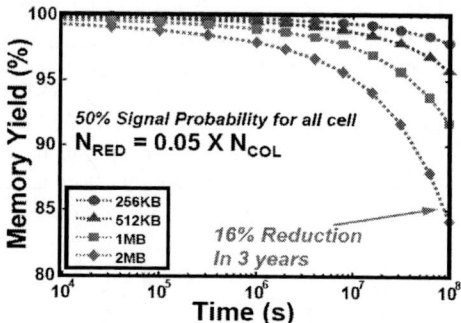

Figure 10. Temporal parametric yield of 6T-SRAM array. The decrease in memory yield for larger memories, especially towards the end of the product cycle, could be addressed by pre-silicon design (e.g. cell flip) or post-silicon monitoring (Fig. 8).

degradation will immediately identify vulnerable cells. With an additional N-well bias, the SNM could be increased for those cells and this increases memory yield by a factor of 3 – a remarkable improvement.

Finally, cell-flipping algorithm proposes to restore symmetry in PMOS transistors by inverting the stored data in the memory every other day, so that any given PMOS transistors are not subjected to prolonged uninterrupted stress. While good results are obtained, the main limitation of the approach involves the algorithmic and hardware complexity/overhead associated with inverted data every other day [56].

C. Soft Errors: SRAM Solutions

For SRAM memory, various types of error detection/ correction schemes are now routinely used for wide variety of IC. These include (i) Single Error Correction (SEC) that requires 6-8 additional error correction bits depending on bit-length, (ii) Single Error Correction, Double Error Detection (SEC-DEC) schemes with 7-9 error correction bits, and (iii) Double error Correction (DEC) schemes with 12-24 bits [84]. These techniques based on coding theory require substantial overhead in area, power, and execution time and requires careful implementation, however in general these techniques are effective and widely used.

D. Permanent Faults: Detection and Solution Strategies.

The possible manufacturing defects (such as shorts, gate oxide defects, opens, etc) can be modeled as circuit nodes being stuck-patterns at zero or one. There exists several test pattern generators that optimizes the number of test vectors while trying to achieve close to 100% fault coverage. We have already discussed use of test patterns and IDDQ for logic circuits, here we discuss approaches that are relevant for faults which occur in SRAMs.

SRAM fault models not only consider standard stuck-at faults but also transition (cell fails to make a 0→1 or 1→ 0 transition), coupling (transition in bit 'j' cause unwanted change in bit 'i') faults. A popular testing methodology

978-1-4244-9113-1/11 $26.00 © 2011 IEEE

known as MARCH test involve writing/reading 0 and 1 in successive memory locations [85]. The test failure data provided by the MARCH test enables identification of fault locations in the memory array. The address of faulty bit-cells can then be remapped to redundant bit-cells in the array (column redundancy) to increase memory yield.

Techniques to address soft errors in SRAM memory relies either on error correction codes (ECC) discussed above, or by raising the critical charge necessary to flip a cell by various circuit techniques. A classic example involves Dual Interlocked Storage Cells where SRAM connected back to back stores the memory state [97]. Regardless the position of the strike, the bit does not flip because the other SRAM prevents racing of the signal necessary to complete the bit-flip. Other rad-hard methods add resistance to the feedback loop to slow the racing of signal and thereby reduce probability of bit flip. Such resistive hardening however may not be scalable in high-performance circuits.

IV. OUTLOOK/CONCLUSIONS

A. Evolution of CAD Tools

One of the primary challenges in process and reliability-aware design is the lack of widely available CAD tools and well-recognized design philosophy that could encapsulate the information about device properties to circuit designers for appropriate corrective action.

Recently, there have been changes across the board. Commercial device simulators like MiniMOS and Medici now incorporate models for NBTI, HCI, and TDDB degradation. And there are university-based efforts on reliability models (e.g. DevRel at www.nanohub.org), which are also being developed. Circuits design techniques are being explored by several groups from Purdue (CRISTA), University of Minnesota, Arizona State, University of Michigan. And industrial groups like TI and IBM are publishing their design methodologies at various stages of design process [86]. Recent simulation software from Arizona group called 'NewAge' integrates self-consistently activity dependent temperature enhancement, transistor degradation along with timing analysis to provide a systematic prediction of the role of parametric degradation for various integrated circuits [87]. Eventually, we anticipate that these issues will propagate to the system level tools.

The multi-scale radiation damage and soft error models have grown increasingly sophisticated over the years. For example, simulators like SEMM-2 from IBM contains particle transport, Nuclear physics, and particle source generator modules as well as detailed layout information for front and backend information to create a realistic simulation environment for prediction of soft errors [41]. These simulators however focuses on charge generation and critical charge, however, if hot carrier information is desired, modeling framework that GEANT4 along with full-band Monte Carlo simulation is appropriate and has been successfully used to study charge ejection from Flash and ZRAM memories [10].

Several Automatic Test Pattern Generator (ATPG) tools are available from vendors like Mentor Graphics, Synopsys and Cadence that can generate test patterns for stuck-at faults, IDDQ along with at-speed testing for path delay faults [88][89][90]. Test failure data coupled with test patterns and gate level design description can be used to identify different defects causing the test failure and the defect locations [93]. However, the cost of external testing using automatic test equipment is increasing as the chips become more complex thereby increasing the total chip cost. Research work is in progress to achieve lower test cost using on-chip testing (Built in Self Test, BIST) instead of relying on expensive external testing methodologies practiced today [91][92].

In sum, in this paper we have reviewed four classes of faults and various fault-tolerant strategies to detect and manage them. We have explored in some depth the various approaches proposed to design with these faults both for logic as well as for memory transistors. The techniques are innovative and often appear promising as they offer saving of power/area over more conservative design. *However, the implication for test, burning, and yield are important questions that needs to be explored.* Eventually, we suggest that the only way for implementing any design philosophy is through the CAD tools. Progress towards that goal is already underway.

ACKNOWLEDGEMENT

We wish to acknowledge contributions from members of our groups, K.-H. Kang, B. Paul, A. E. Islam, D. Varghese and G. Karakonstantis. This work was supported by National Science Foundation, Applied Materials, Texas Instruments, Taiwan Semiconductor Manufacturing Company.

REFERENCES

[1] S. Borker, "Designing Reliable Systems from Unreliable Components: The Challenge of Transistor Variability and Degradation", *IEEE Micro*, p. 10-16, 2005.

[2] M. A. Alam, S. Mahapatra, "A comprehensive model of PMOS NBTI degradation," *Microelectronics Reliability*, 45 (1) p. 71, 2005.

[3] B. C. Paul et al. *EDL-26*, 560-562, 2005. Also see, B. Paul, Proc. of DATE, p. 780-785, 2006.

[4] M.A. Alam, N. Pimparkar, S. Kumar, and for Macroelectronics Applications", MRS Bulletin, 31, 466, 2006.

[5] G. Groeseneken, R. Degraeve, B. P. Roussel, "Challenges in Reliability Assessment of Advanced CMOS Technologies", Keynote Presentation at the Int. Conf. on. Physical and Failure Analysis of Integrated Circuits, p. 1-9, 2007.

[6] H. Kufluoglu and M.A. Alam, "A geometrical unification of the theories of NBTI and HCI time-exponents and its implications for ultra-scaled planar and surround-gate MOSFETs," *IEEE IEDM Technical Digest*, p. 113 (2004).

[7] P. Nair and M. Alam, "Design Considerations of Nanobiosensors", IEEE Trans. Elec. Devices, 2007.

[8] R. Baumann, "The impact of technology scaling on soft error rate performance and limits to the efficacy of error correction," *IEDM Tech. Digest*, pp. 329-332 (2002).

[9] P. E. Dodd, M. R. Shaneyfelt, J.R. Schwank, and G. L. Hash, "Neutron Induced Latchup in SRAM at Ground Level, *IRPS Proceeding*, pp. 51-55, (2003).

978-1-4244-9113-1/11 $26.00 © 2011 IEEE

[10] N. Butt, Ph.D. Thesis, Purdue University, 2008.

[11] M. Alam and K. Smith, "A Phenomenological Theory of Correlated Multiple Soft-Breakdown Evetns in Ultra-thin Gate Dielectrics," *IEEE IRPS Proceedings*, p. 406, (2003).

[12] K. Kang, H. Kufluoglu, K. Roy, and M. A. Alam, "Impact of Negative Bias Temperature Instability in Nano-Scale SRAM Array: Modeling and Analysis," to be published in *IEEE Transactions on CAD*, 2007.

[13] J. H. Stathis, "Reliability Limits for the Gate Insulator in CMOS Technology," *IBM Journal of Research and Development*, vol. 46 (2/3), pp. 265-286, (2002).

[14] R. Degraeve, B. Govoreanu, B. Kaczer, J. Van Houdt, G. Groeseneken, "Measurement and statistical analysis of single trap current-voltage characteristics in ultrathin SiON", *Proc. of IRPS*, 2005. pp. 360- 365.

[15] https://www.nanohub.org/resources/1647/

[16] https://www.nanohub.org/resources/1214/

[17] http://cobweb.ecn.purdue.edu/~ee650/handouts.htm

[18] A.E. Islam, Ph.D. Thesis, Purdue University, 2010.

[19] R. Fernandez, B. Kaczer, *IEDM Tech. Dig.* 2006.

[20] H. Kufluoglu et al., *Proc. of IRPS*, 2007.

[21] D. Allee et al., "Degradation Effects in a a-Si :H Thin Film Transistors and their Impact on Circuit Perfomance," Proc. of IRPS, p. 158, 2008.

[22] E. Takeda, C. Yang, A. Miura-Hamada, "Hot Carrier Effects in MOS Devices," Academic Press (1995).

[23] http://www.eecs.berkeley.edu/Pubs/TechRpts/1990/ ERL-90-4.pdf

[24] D. Varghese, Ph.D. Thesis, Purdue University, 2009.

[25] G. Cellere, L. Larcher, A. Paccagnella. A. Visconti, M. Bonanomi , "Radiation induced leakage current in floating gate memory cells", IEEE transactions on nuclear science 2005, vol. 52 (1), n°6, pp. 2144-2152.

[26] R. W. Keyes, "Physical Limits of Silicon Transistors and Circuits, *Rep. Prog. Phys.* 68, 2701-2746, 2005.

[27] A. Bhavnagarwala, X. Tang, J. Meindl, The impact of Intrinsic Device Fluctuations on CMOS SRAM Cell Reliability", IEEE J. of Solid State Circuits, 36 (4), pp. 658-664, 2001.

[28] V. De, X. Tang, and J. Meindl, "Random MOSFET Parameter Fluctuation Limits to Gigascale Integration (GSI)", 1996 *Sym. Of VLSI Technology*, paper 20.4.

[29] M. Agostinelli, S. Lau, S. Pae, P. Marzolf, H. Muthali, S. Jacobs, "PMOS NBTI-induced circuit mismatch in advanced technologies," *IEEE IRPS Proceedings*, p. 171 (2004).

[30] M. A. Alam, B. Weir, P. Silverman, J. Bude, A. Ghetti, Y. Ma, M. M. Brown, D. Hwang, and A. Hamad , "Physics and Prospects of Sub-2nm Oxides," in *The Physics and Chemistry of SiO2 and the eSi-SiO2 interface-4* H. Z. Massoud, I. J. R. Baumvol, M. Hirose, E. H. Poindexter, Editors, PV 2000-2, p. 365, The Electrochemical Society, Pennington, NJ (2000).

[31] D. G. Flagello, H. V. D. Laan, J. V. Schoot, I Bouchoms, and B. Geh, "Understanding Systematic and Random CD Variations using Predictive Modelling Techniques," *SPIE Conference in Optical Microlithography XII*, pp. 162-175, 1999.

[32] M. Ieong, B. Doris, J. Kedzierski, K. Rim, M. Yang, "Silicon device scaling to the sub-10 nm regime," *Science*, vol. 306 p.2057 (2004).

[33] A. Asenov, "Random dopant induced threshold voltage lowering and fluctuations in sub-0.1 μm MOSFET's: A 3-D "atomistic" simulation study" *IEEE Transactions on Electron Devices* 45(12):pp. 2502-2513, 1998.

[34] Simon W-B. Tam, Yojiro Matsueda, Hiroshi Maeda, Mutsumi Kimura," Improved Polysilicon TFT Drivers for Light Emitting Polymer Displays," Proc Int Disp Workshops, 2000.

[35] R. Gusmeroli, A.S. Spinelli, C. Monzio Compagnoni, D. Ielmini and A.L. Lacaita, "Edge and percolation effects on V_T window in nanocrystal memories", *Microelectronics Engineering*, doi: 10.1016 /j.mee2005.04.066.

[36] N. Pimparkar, J. Guo and M. Alam, "Performance Assessment of Sub-percolating Network Transistors by an Analytical Model," IEEE Trans. of Electron Devices, 54(4), 2007.

[37] J. Choi, A. Bansal, M. Meterelliyoz, J. Murthy, and K. Roy, "Leakage Power Dependent Temperature Estimation to Predict Thermal Runaway in FinFET Circuits," *International Conference on CAD*, pp. 2006.

[38] S. Kumar, J. Murthy, and M. Alam, "Self-Consistent Electro-Thermal Analysis of Nanotube Network Transistors" Journal of Applied Physics, 109, 014315, 2011. Also see, S.-C. Lin and K. Banerjee, "An Electrothermally-aware Full Chip Substrate Temperature Gradient Evaluation Methodology for Leakage Dominant Technologies with Implications for Power Estimation and Hot Spot Management", Proc. ICCAD, pp. 568-574, 2006.

[39] Robert Bauman, Tutorial Notes, International Reliability Physics Symposium, 2004.

[40] D.F. Heidel et al. "Single Event Upsets and Multiple-Bit Upsets on a 45 nm SOI SRAM", IEEE Trans. Nuclear Science. 56(6), pp. 3499-3504, 2009.

[41] H.H. K. Tang, "SEMM-2: A New Generation of Single-Event-Effect Modeling Tool," IBM Journal of Research and Development, 52(3), pp. 233-244, 2008.

[42] P. E. Dodd, F. W. Sexton, M. R. Shaneyfelt, J. R. Schwank, and D. S. Walsh, Sandia National Laboratories "Epi, Thinned and SOI substrates: Cure-Alls or Just Lesser Evils", IRPS Proc. 2007.

[43] R. Lucky, in Silicon Dreams: Information, Man, and Machine, 1989.

[44] E. J. McCluskey, Chao-Wen Tseng, "Stuck-fault tests vs. actual defects", Proc. ITC, pp.336-342, 2000.

[45] S. Shekhar, M. Erementchouk, M. Leuenberger and S. Khondaker, "Correlated breakdown of CNT in aligned arrays", arxiv.org/ftp/arxiv/papers/1101/1101.4040.pdf, 2011.

[46] J. Tschanz, J. Kao, S. Narendra, R. Nair, D. Antoniadis, A. Chandrakasan, V. De, "Adaptive Body Bias for Reducing Impacts of Die-to-Die and Within-Die Parameter Variations on Microprocessor Frequency and Leakage," *International Solid State Circuit Conference*, pp. 477-479, 2002.

[47] S. Ghosh, S. Bhunia, and K. Roy, "CRISTA: A New Paradigm for Low- Power,Variation-Tolerant, and Adaptive Circuit Synthesis Using Critical Path Isolation", Transactions on CAD, 2007.

[48] D. Ernst, N. Kim, et al., "Razor: A Low-Power Pipeline Based on Circuit-Level Timing Speculation," Proc. IEEE/ACM Int. Symposium on Microarchitecture (MICRO), Dec. 2003.

[49] T. Austin, "DIVA: A Reliable Substrate for Deep Submicron Microarchitecture Design," *ACM/IEEE 32nd Annual Symposium on Microarchitecture*, p. 196 November 1999.

[50] K. Kang, K. Kim, and K. Roy, "Variation Resilient Low-Power Circuit Design Methodology using On-Chip Phase Locked Loop", to be published in *ACM/IEEE Design Automation Conference*, 2007.

[51] C. Visweswariah, "Death, Taxes and Failing Chips," *ACM/IEEE Design Automation Conference*, pp. 343-347, 2003.

[52] B. Kaczer, IRPS Tutorial of Gate Dielectric Breakdown, 2009.

[53] R. Vattikonda et al. *Proc. of DAC*, 1047-1052, 2006.

[54] T. Krishnan, V. Reddy, S. Chakravarthi, J. Rodriguez, S. John, S. Krishnan, "NBTI impact on transistor and circuit: models, mechanisms and scaling effects", *Proc. of IEDM*, 2003.

[55] A. Krishnan et al., NBTI Effects of SRAM Circuits, *IEEE IEDM Digest*, 2006.

[56] S. Kumar et al., Proc. of Quality Electronic Design, pp. 210-218, 2006.

[57] K. Kang, H. Kufluoglu, M. A. Alam, and K. Roy, "Efficient Transistor-Level Sizing Technique under Temporal Performance Degradation due to NBTI," *IEEE International Conference on Computer Design*, pp. 216-221, 2006.

[58] C. P. Chen, C. C. N. Chu, and D. F. Wong, "Fast and exact simultaneous gate and wire sizing by Langrangian relaxation," *IEEE Trans. on Computer-Aided Design of Integrated Circuits and Systems*, vol. 19, pp. 1014-1025, (1999).

[59] S. Kumar, C. Kim and S. Sapatnekar, "NBTI Aware Synthesis of Digital Circuits", *Design Automation Conference*, pp. 370-375, San Diego, California, USA, June 2007

[60] K. Kang, K. Kim, A. E. Islam, M. A. Alam, and K. Roy, "Characterization and Estimation of Circuit Reliability

978-1-4244-9113-1/11 $26.00 © 2011 IEEE

Degradation under NBTI using On-line IDDQ Measurement," to be published in *ACM/IEEE Design Automation Conference*, 2007.

[61] K. Kang, M. A. Alam, and K. Roy, "Estimation of NBTI Degradation using On-Chip IDDQ Measurement", *International Reliability Physics Symposium*, pp. 10-16, 2007.

[62] C. H. Kim, K. Roy, S. Hsu, R. Krishnamurthy, S. Borkhar, "An On-Die CMOS Leakage Current Sensor For Measuring Process Variation in Sub-90nm Generations", *VLSI Circuits Symposium*, June 2004

[63] T. Kim, R. Persaud, and C.H. Kim, "Silicon Odometer: An On-Chip Reliability Monitor for Measuring Frequency Degradation of Digital Circuits", *IEEE Journal of Solid-State Circuits*, pp 874-880, Apr. 2008. A

[64] T. Kim, R. Persaud, and C.H. Kim, "Silicon Odometer: An On-Chip Reliability Monitor for Measuring Frequency Degradation of Digital Circuits", VLSI Circuits Symposium, pp122-123, June 2007

[65] E. Karl, P. Singh, D. Blaauw, D. Sylvester "Compact In-Situ Sensor for Monitoring Nagative Bias-Temperature-Instability Effect and Oxide Degradation", *IEEE International Solid State Circuit Conference*, pp. 410-411, California, USA, 2008

[66] M. Agarwal, S. Mitra, B. C. Paul, M. Zhang, "Circuit Failure Prediction and its Application to Transistor Aging," *VLSI Test Symposium*, 2007.

[67] S. Mitra, Proc. of Int. Rel. Phys. Symposium, 2008.

[68] A.E. Islam and M. A. Alam, "On the Possibility of Degradation free Field-Effect Transistors", *Appl. Phys. Lett.* 92, 173504, 2008.

[69] M. Zhang et. al., "Sequential Element Design With Built-In Soft Error Resilience," IEEE Transactions on VLSI Systems, 2006.

[70] A. Goel et. al., "Low-overhead design of soft-error-tolerant scan flip-flops with enhanced-scan capability," ASP-DAC, 2006.

[71] R. Koo and S. Toueg, "Checkpointing and rollback-recovery for distributed systems," IEEE Trans. Software Engineering 13, 1, 23–31, 1987.

[72] K. Constantinides et. al., "Software-Based Online Detection of Hardware Defects: Mechanisms, Architectural Support, and Evaluation", International Symposium on Micro architecture.

[73] L. Leem et. al., "ERSA: Error Resilient System Architecture for Probabilistic Applications", DATE, 2010.

[74] J.M. Soden, C.F. Hawkins, R.K. Gulati, W. Mao, "IDDQ Testing: A Review," Journal of Electronic Testing: Theory and Applications, Vol. 3, No. 4, pp. 291-303, Dec. 1992.

[75] N. Banerjee, C. Augustine, K. Roy, "Fault-Tolerance with Graceful Degradation in Quality: A Design Methodology and its Application to Digital Signal Processing Systems", DFT, 2008.

[76] Jie Han , Jianbo Gao , Yan Qi , Pieter Jonker , Jose A. B. Fortes, Toward Hardware-Redundant, Fault-Tolerant Logic for Nanoelectronics, IEEE Design & Test, v.22 n.4, p.328-339, July 2005

[77] M. Z. Hasan and S. G. Ziavras, "Runtime Partial Reconfiguration for Embedded Vector Processors", International Conference on Information Technology: New Generations", Las Vegas, Nevada, April 204, 2007.

[78] J. R. Heath, P. J. Kuekes, G. Snider, and R. S. Williams, "A Defect-Tolerant Computer Architecture: Opportunities for Nanotechnology", *Science*, 280, pp. 1716-1721, 1998.

[79] S. Mukhopadhyay, H. Mahmoodi, and K. Roy, "Modeling of Failure Probability and Statistical Design of SRAM Array for Yield Enhancement in Nanoscaled CMOS", *IEEE TCAD*, 2004.

[80] S. Mukhopadhyay, A. Raychowdhury, H. Mahmoodi, K. Roy, "Statistical Design and Optimization of SRAM cell for yield improvement," *International Conference on Computer Aided Design*, pp. 10-13, 2004.

[81] S. Mukhopadhyay, K. Kang, H. Mahmoodi, K. Roy, "Reliable and self-repairing SRAM in nano-scale technologies using leakage and delay monitoring," *International Test Conference*, 2005.

[82] , K. Roy, J. Kulkarni, M.-E. Hwang, "Process-Tolerant Ultralow Voltage Digital Subthreshold Design" IEEE Topical Meeting on Silicon Monolithic Integrated Circuits in RF Systems, 2008. SiRF. pp. 42-45, 2008.

[83] R. Rodriguez, J. Stathis, B. Liner, S. Kowalczyk, C. Chung, R. Joshi, G. Northrop, K. Bernstein, A. Bhavnagarwala, and S. Lombardo, "The Impact of Gate Oxide Breakdown on SRAM Stability," IEEE Electron. Dev. Lett. 23, 559 (2002).

[84] Heidergott, NSREC Short Course, 1999.

[85] A.J. van de Goor, "Using March Tests to Test SRAMs," *IEEE Design & Test of Computers*, Vol. 10, No. 1, Mar. 1993, pp. 8-14.

[86] [Goda05] A. Goda, ISQED, 2005.

[87] M. DeBole, Int. J. Parallel Prog. 2009

[88] http://www.mentor.com/products/silicon-yield

[89] http://www.synopsys.com/Tools/Implementation/RTLSynthesis/Pages/TetraMAXATPG.aspx

[90] http://www.cadence.com/products/ld/true_time_test

[91] V. D Agrawal, "A tutorial on built-in self-test. I. Principles", Design and Test of Computers, 1993.

[92] V. D Agrawal, "A tutorial on built-in self-test. II. Applications", Design and Test of Computers, 1993.

[93] W. T Cheng et. al., "Compactor independent direct diagnosis", Proc. Asian Test Symp, 2004.

[94] B. Vaidyanathan, B.; A.S. Oates, X. Yuan Intrinsic NBTI-variability aware statistical pipeline performance assessment and tuning, ICCAD, pp. 164-171, 2009.

[95] [Rajsuman95] R. Rajusman, *Iddq Testing for CMOS VLSI*, Artech House, 1995.

[96] [Agarwal05] A. Agarwal, B. C. Paul, H. Mahmoodi, A. Datta, K. Roy, "A process-tolerant cache architecture for improved yield in nanoscale technologies," *IEEE Transactions on VLSI*, vol. 13, no. 1, pp. 27-38, 2005.

[97] Calin, et al., IEEE Trans. Nucl. Sci. 43, p. 2874, 1996.

Response of a single trap to AC Negative Bias Temperature Stress

M. Toledano-Luque[1,2], B. Kaczer[2], Ph.J. Roussel[2], T. Grasser[3], G.I. Wirth[4], J. Franco[2,5],
C. Vrancken[2], N. Horiguchi[2], G. Groeseneken[2,5]

[1]Universidad Complutense de Madrid, Spain.
Phone: +34 91 3944434, mtluque@fis.ucm.es

[2]imec, Leuven, Belgium.

[3]Technische Universität, Wien, Austria.

[4]Universidade Federal do Rio Grande do Sul, Brazil.

[5]Katholieke Universiteit Leuven, Belgium.

Abstract — **We study the properties of a single gate oxide trap subjected to AC Bias Temperature Instability (BTI) stress conditions by means of Time Dependent Defect Spectroscopy. A theory for predicting the occupancy of a single trap after AC stress is developed based on first order kinetics and verified on experimental data. The developed theory can be used to develop circuit simulators and predict time dependent variability.**

Index Terms — **Negative Bias Temperature Instability, constant voltage stress, AC stress, MOSFET, reliability, variability, SiON**

I. INTRODUCTION

Only a handful of gate oxide defects is expected to be active in future nm-sized CMOS devices [1-3]. Since digital circuits operate with binary AC signals, a good understanding of individual defect properties under AC conditions is compulsory to predict the device reliability [3] and to develop circuit simulators for future device generations [4-7]. Recently, a new methodology has been introduced to study the statistical properties of individual traps after Bias Temperature Instability (BTI) [8]. This new technique called Time Dependent Defect Spectroscopy (TDDS) is based on the quantized V_{TH} relaxation transients observed on nano-scale devices following constant stress [9,10]. It has been demonstrated that the emission and capture of single traps under constant stress are described by first order kinetics [8].

In this work, we study the response of a single gate oxide trap in a deeply-scaled SiON pMOSFET during AC negative bias temperature stress. We demonstrate that the behavior of individual traps as a function of the stress time and duty factor is dictated by their characteristic capture and emission times at DC high and low voltages. The developed AC model can be straightforwardly implemented in circuit simulators [5].

In this paper, a detailed description of the experimental setup is first given. The methodology to determine the emission and capture times of a single trap under DC stress is reviewed [8]. Afterwards, the statistical properties of the traps are studied under different AC stress conditions and temperatures. The theory to explain the experimental data is elaborated based on first principle kinetics and successfully verified.

II. EXPERIMENTAL

The setup used in our experiment is schematically depicted in Fig. 1. The four terminals of *a single selected 70×90 nm² 1.6 nm-SiON pMOSFET* were connected to Keithley 2602 source measurement units (SMUs). The SMU voltages were V_{RELAX} at gate, V_D = -0.1 V at drain, and $V_S = V_B = 0$ V at source and bulk.

The device was gate stressed with an AC signal switching from V_{STRESS} to V_{RELAX} with a frequency f =100 kHz and duty factor DF (= $f \times t_H \times 100$ where t_H is the time that the signal is at high level) for stress time t_{STRESS}. During t_{STRESS} the gate terminal was connected to an HP8110 pulse generator.

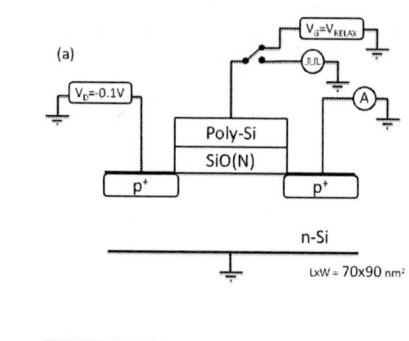

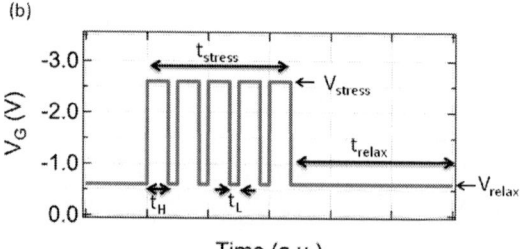

Fig. 1: Measurement procedure for investigating AC trap response. (a) AC signal is applied to the gate of *a selected 70×90 nm² SiON pMOSFET* and the source current I_{SOURCE} is registered as a function of time. (b) The applied AC signal switches between V_{STRESS} and V_{RELAX} for a stress time t_{STRESS}. Afterwards, the relaxation transient is registered at V_{RELAX} for a relaxation time t_{RELAX}.

978-1-4244-9113-1/11 $26.00 © 2011 IEEE

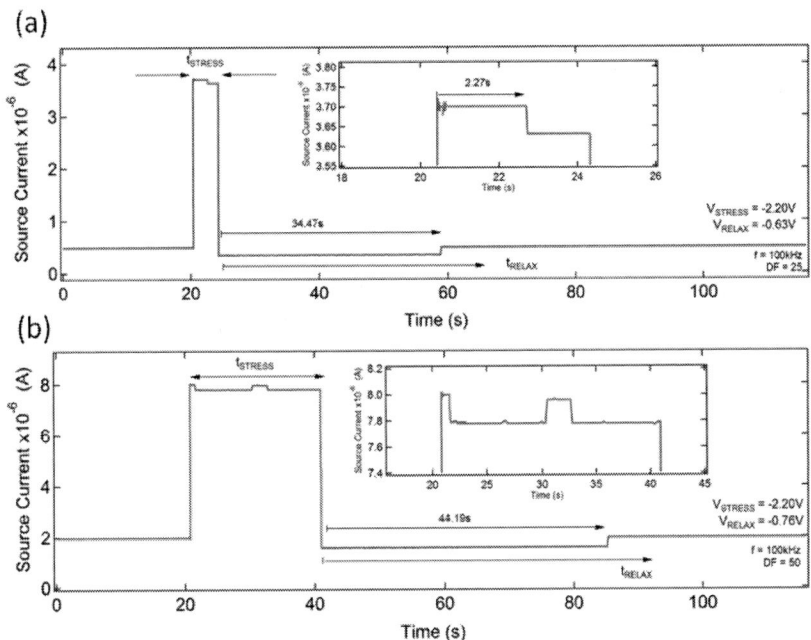

Fig. 2: Typical I_{SOURCE} vs. time curves obtained from *the single selected device* during the experiment. Measured I_{SOURCE} during stress is integrated over many AC periods. In (a) hole capture is observed at t_{STRESS} = 2.27 s, the charge is emitted after 34.74 s at V_{RELAX}. In (b), *capture* and *emission* are observed during stress.

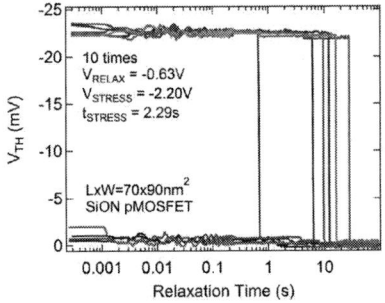

Fig. 3: 10 typical V_{TH} transients after DC stress at 25 °C and -2.2 V for 2.29 s. Note that 6 out of 10 traces show a giant discrete step of ~ 23 mV.

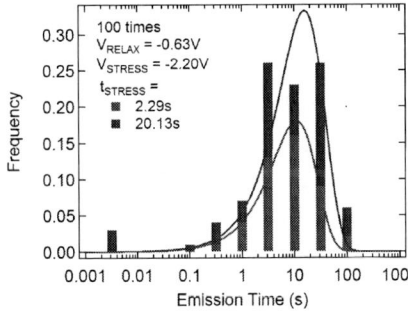

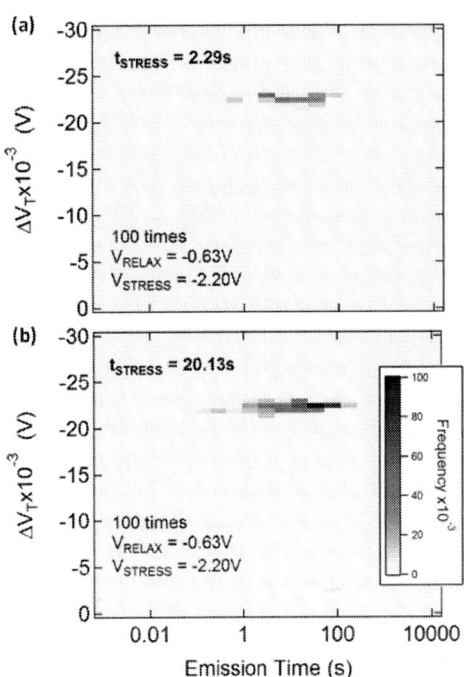

Fig. 5: Histograms f_E of the emission times t_e extracted from 100 V_{TH} relaxation transients under the condition of Fig. 3 for two different stress times 2.29 s and 20.13 s. Note that the emission times are binned on the logarithmic scale. The histograms can be fitted with eq. 2.

Fig. 4: TDDS spectra for two stress times (a) t_{STRESS} = 2.29 s and (b) t_{STRESS} = 20.13 s extracted from 100 recovery traces from the selected device one after the other under the condition of Fig. 3. A homogenous cluster appears at ~14 s and 23 mV for both spectra. Note that the intensity increases with increasing t_{STRESS} indicating that trap occupancy after longer stress increases.

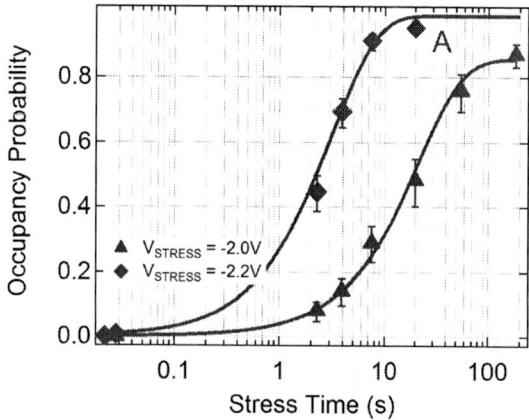

Fig. 6: Trap occupancy probability P_C with $\pm\sigma$ error bars vs. t_{STRESS} for a constant V_{STRESS} of -2.0 and -2.2 V and at 25 °C. For the V_{STRESS} of -2.0 V, the occupancy does not reach 1, indicating that the emission time at V_{STRESS} = -2.0 V is comparable to the capture time. Capture and emission times extracted from the fitting to the data with eq. 3 are shown in Table I.

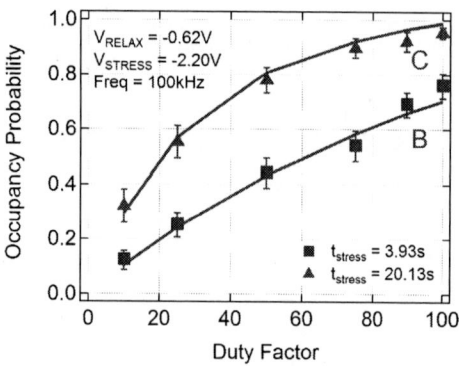

Fig. 7: (Symbol) Trap occupancy probability P_C (number of traces that presents the giant step / number of trials) applying Benard correction with $\pm\sigma$ error bars for V_{STRESS} = -2.2 V. (Lines) Predicted occupancy according to the proposed theory considering the emission and capture times at V_{STRESS} = -2.2 V and V_{RELAX} = -0.62 V (see Fig. 10 line B for t_{STRESS} = 3.93 s and line C for t_{STRESS} = 20.13 s).

T (°C)	V_G (V)	$\tau_{capture}$ (s)	$\tau_{emission}$ (s)
25	-0.62	> 200	14 ± 1
60	-0.62	> 200	0.14 ± 0.01
25	-0.76	> 200	57 ± 3
25	-2.0	25.5 ± 1.6	150 ± 32
60	-2.0	5.9 ± 0.3	5.7 ± 0.5
25	-2.2	3.2 ± 0.2	> 315

Table I. Emission and capture times obtained from the fitting of the experimental data acquired under DC stress with the maximum likelihood method to eq. 2 for emission times and eq. 3 for capture times.

Afterwards, the gate electrode was switched within less that a ms range to the SMU during a relaxation time t_{RELAX} = 200 s. The reed relay between the gate SMU and the pulse generator was controlled by the gate SMU. During this entire process the source current I_{SOURCE} was registered as a function of the time.

Fig. 2 shows typical I_{SOURCE} vs. time curves obtained in the experiment *on the selected device*. Note that measured I_{SOURCE} during stress is integrated over many AC periods due to the slow response of the SMU with respect to the applied signal (f = 100 kHz). In Fig. 2a, an abrupt drop of the source current I_{SOURCE} is observed at t_{STRESS} = 2.27 s, indicating a hole capture event. When the gate voltage switches again to V_{RELAX}, the source current is lower than the current at the start of the measurement (t = 0 s). The current suddenly recovers at t_{RELAX} = 34.75 s when the hole is emitted. Fig. 2b shows a trace where both capture and emission events are detected during stress. During relaxation the hole is emitted at t_{RELAX} = 44 s. The events observed during stress indicate that the occupancy at the end of the stress period is determined by the convolution of emission and capture processes.

III. DETERMINATION OF EMISSION TIME UNDER DC STRESS

Fig. 3 shows the relaxation curves at V_{RELAX} = -0.63 V after DC stress at 25 °C and -2.2 V. The V_{RELAX} is chosen close to the threshold voltage of the device so that ΔI_{SOURCE} varies strongly with respect to the absolute value of I_{SOURCE}. The source current I_{SOURCE} obtained following the procedure of Fig. 1 was transformed into a V_{TH} shift via a reference I_{SOURCE}-V_G curve of the device taken prior to stress [12]. Conversely to the continuous relaxation curves obtained in large devices [11,13], a discrete behavior is observed. Fig. 3 shows a giant V_{TH} shift of ~ 23 mV for 6 out of 10 traces at the start of the relaxation period that drops abruptly to 0 mV at t_{RELAX} ~ 14 s. This step height is significantly larger than the expected threshold voltage shift by the simple charge sheet approximation. This is due to the amplifying effect of the random dopants in the FET channel [2,14]. The step heights from device to device follow an exponential distribution and the number of steps per device is Poisson distributed [15]. The device under study on this paper was selected because it presented a large and single step, simplifying the analysis of the data.

Fig. 4 shows the corresponding TDDS spectra, i.e. two-dimensional histogram of the emission times t_e and the V_{TH} step heights, obtained from 100 traces of the device under study for two stress times: (a) 2.29 s and (b) 20.13 s. A homogenous cluster is observed at about 23 mV (cf. Fig. 3), indicating the presence of a single active trap. Fig. 5 shows the histograms of the emission times t_e (i.e. relaxation time at which the step is detected) obtained at the two stress times 2.29 s and 20.13 s shown in Fig. 4. The emission time t_e follows an exponential distribution as expected from first-order kinetics, i.e., in a Markov-process the transitions between two neighboring states are exponentially distributed [5,8]

$$P_E(t_{RELAX}) = \frac{\tau_c}{\tau_c + \tau_e}\left\{1 - \exp\left[-\left(\frac{1}{\tau_e} + \frac{1}{\tau_c}\right)t_{RELAX}\right]\right\} \qquad (1)$$

The emission times can be fitted with the maximum likelihood method and the histogram f_E when binned on logarithmic scale follows:

$$f_E(t_{RELAX}) = \frac{t_{RELAX}}{\tau_e}\exp\left[-t_{RELAX}\left(\frac{1}{\tau_e} + \frac{1}{\tau_c}\right)\right] \qquad (2)$$

where τ_e and τ_c are the mean emission and capture times at V_{RELAX}, respectively. The fit of the data presented in Fig. 5 provides a characteristic emission time τ_e of about 14 s for both stress times. The characteristic emission time is independent of the stress time [8,11]. On the other hand, the characteristic capture time τ_c cannot be fitted accurately since this time is at least one order of magnitude larger than τ_e according the fit.

When the experiment was repeated at a higher absolute value of the relaxation voltage, V_{RELAX} = -0.76 V, the characteristic emission time increased to 57 s, i.e., the higher the relaxation voltage is, the larger the emission time is. Therefore, the characteristic emission time τ_e depends strongly on V_G (see Table I) [8], and, for this reason, τ_e has to be always referred to a particular voltage. The capture time for low voltages (-0.62 and -0.76 V) was estimated to be larger than 200 s for the modeling purpose presented in section VI.

IV. DETERMINATION OF CAPTURE TIME UNDER DC STRESS

The TDDS spectra of Fig. 4 show that the number of traces which present the giant step increases with t_{STRESS}. Fig. 6 shows the intensity of the cluster, i.e. the cumulative probability of charging the trap P_C (=occupancy probability), as a function of t_{STRESS} for V_{STRESS} -2.0 and -2.2 V. Note that the probability of occupancy saturates at 1 for V_{STRESS} = -2.2 V, and at 0.82 for V_{STRESS} = -2.0 V. As soon as the emission time at stress condition enters the same range as the capture time, the probability of intermediate emission during the stress cannot be neglected (see Fig. 2b). All these features can be described by first order kinetics

$$P_C(t_{STRESS}) = \frac{\tau_e}{\tau_c + \tau_e}\left\{1 - \exp\left[-\left(\frac{1}{\tau_e} + \frac{1}{\tau_c}\right)t_{STRESS}\right]\right\} \qquad (3)$$

where τ_e and τ_c are the mean emission and capture times at V_{STRESS}. If τ_c is much shorter than τ_e, the probability of occupancy reaches 1 as shown in Fig. 6 for V_{STRESS} = -2.2 V. In the case of the occupancy saturating at a lower value, the emission events during stress are not negligible and the characteristic emission and capture times can be determined simultaneously from the fit of the data. The corresponding likelihood function to maximize becomes:

$$\prod_{k=1}^{n_k} P_C(t_{STRESS,k})^{n_k}\left[1 - P_C(t_{STRESS,k})\right]^{p_k - p_k} \qquad (4)$$

where p_k is the number of traces that show the step out of n_k traces observed for each stress time $t_{STRESS,k}$. Table I shows the obtained values for the different conditions applied in this experiment.

Note that the capture time decreases (i.e. the probability of capture increases) and emission time increases with increasing V_{STRESS} [8]. Therefore, at high gate voltages the capture events are dominant, while at low voltages it is the emission events.

V. PROBABILITY OF OCCUPANCY UNDER AC STRESS

Fig. 7 shows the occupancy probability P_C of the trap as a function of the duty factor DF of the AC signal for two values of t_{STRESS}. The AC signal switched between V_{STRESS} = -2.2 V and V_{RELAX} = -0.62 V. For both cases, P_C increases with DF up to the value corresponding to DC (see Fig. 6 for V_{STRESS} = -2.2 V).

For low DF (10, 25, and 50%), the capture and emission processes can be detected from the I_{SOURCE} traces during stress. For larger DF values, the delta of I_{SOURCE} with respect to the total I_{SOURCE} produced at the emission and capture events was too low to be distinguished with a reasonable level of certainty from the noise. Fig. 8 shows four typical I_{SOURCE} traces during stress at DF = 10%. In Fig. 8a, the source current keeps constant, conversely to Fig. 8b where a clear drop of the source current is visible at about 5.5 s, indicating the capture of a hole. In addition to the single capture process shown in Fig. 8b, emission events can also be visible during stress. In Fig. 8c, the hole capture is observed at ~7 s and its emission 2 s afterwards. In Fig. 8d, a second capture process is observed at ~16 s. Therefore the occupancy after a stress time t_{STRESS} is determined by the capture and emission events that take place during that period. Additionally, the number of traces that does not contain any capture event reduces with reducing the DF, i.e. the time that the voltage is at high level.

Consequently, the trap occupancy at the end of the stress time depends on the stress time and the capture and emission processes that take place during the AC stress. Additionally, the capture and emission processes are governed by the characteristic emission time τ_e and capture time τ_c at V_{STRESS} and V_{RELAX}, respectively. In the next section a theory to explain these features is developed.

VI. AC MODEL

In the sections III and IV, it has been demonstrated that the emission and the capture events are described by first-order kinetics and the occupancy probability are given by τ_e and τ_c at certain voltage. Therefore, the occupancy probability for the special case of digital signal switching between two levels (high (H) and low (L) levels) is described by considering the occupancy probabilities at high voltage P_{CH} and at low voltage P_{CL}, the times t_H and t_L that each voltage is applied during each period (see Fig. 1b), and the previous state. For the n-th period we can thus write

$$P_{CH}(n) = \frac{\tau_{eH}}{\tau_{eH} + \tau_{cH}} + \left\{P_{CL}(n-1) - \frac{\tau_{eH}}{\tau_{eH} + \tau_{cH}}\right\}\exp\left[-\left(\frac{1}{\tau_{eH}} + \frac{1}{\tau_{cH}}\right)t_H\right] \qquad (5)$$

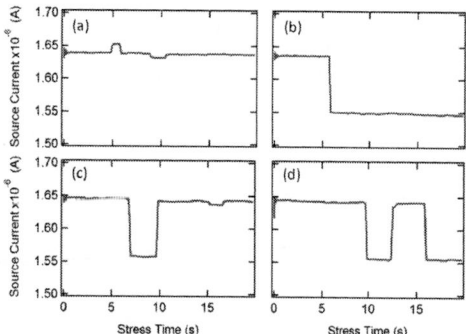

Fig. 8: Four typical I_{SOURCE} integrated over many AC cycles of frequency f = 100kHz vs. stress time t_{STRESS} curves obtained for V_{STRESS} = -2.2 V, V_{RELAX} = -0.62 V, and DF = 10%. During stress time at these conditions, different capture and emission processes are observed.

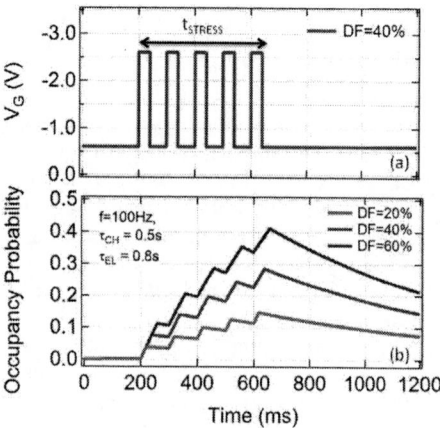

Fig. 9: Calculated occupancy probability P_C as a function of stress time t_{STRESS} (b) for the stress signal shown in (a). Occupancy was calculated according to eqs. (5) and (6) for the emission and capture times shown in (b). P_C after stress time t_{STRESS} depends on capture time τ_c, emission time τ_e at high (H) and low voltages (L), stress time t_{STRESS}, and duty factor DF.

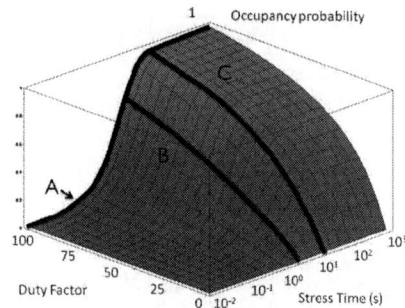

Fig. 10: Occupancy probability P_C as a function of t_{STRESS} and DF considering the conditions of experiment presented in Figs. 6 and 7. Line A traces P_C vs. t_{STRESS} at DC stress (see Fig. 6). Lines B and C trace P_C vs. DF (see Fig. 7). The model can predict correctly P_C for all t_{STRESS} and DFs provided that the emission and capture times are known under DC conditions.

$$P_{CL}(n) = \frac{\tau_{eL}}{\tau_{eL} + \tau_{cL}} + \left\{ P_{CH}(n) - \frac{\tau_{eL}}{\tau_{eL} + \tau_{cL}} \right\} \exp\left[-\left(\frac{1}{\tau_{eL}} + \frac{1}{\tau_{cL}} \right) t_L \right] \quad (6)$$

Fig. 9 shows the occupancy probability evaluated according to these equations. It is observed that the occupancy probability increases during the high level and decreases during the low gate voltage level. The occupancy probability increases up to a saturation level which can be lower than 1. This curve resembles the behavior of the charge and discharge of a capacitor under an AC signal. It is also noted that the higher the duty factor, the higher the occupancy probability P_C. When the stress signal is removed, the system returns to the occupancy probability dictated by eq. 6, i.e., the low voltage occupancy probability P_{CL}.

Expressing the increase in P_{CL} per period during the AC stress period, one can obtain the occupancy probability P_C as a function of the number of applied pulses n ($= f \times t_{STRESS}$):

$$P_C(n) = \frac{b}{a}\left(1 - e^{-an}\right) \quad (7)$$

where a and b are functions of τ_{eH}, τ_{cH}, τ_{eL}, τ_{cL}, DF, and f [5]. Fig. 10 shows P_C as a function of t_{STRESS} and DF calculated using eq. 7 and using the emission and capture times corresponding to the V_{STRESS} = -2.2 V and V_{RELAX} = -0.62 V (see Table I): the experimental conditions in Fig. 6 and Fig. 7. Line A in Fig. 10 corresponds to DF = 100% (DC stress). Line A is also a trace in Fig. 6. Under DC conditions, t_L is equal to 0 and eq. 7 becomes eq. 3.

Lines B and C in Fig. 10 correspond to the occupancy probability as a function of DF at fixed stress time t_{STRESS} of 3.93 s and 20.13 s. These lines are also traced in Fig. 7. It can be noted that the derived model follows correctly the experimental data.

If the characteristic emission and capture times are significantly larger than the period of the stress signal, eq. 7 can be simplified to a more intuitive equation

$$P_C(t_{STRESS}) = \frac{\tau_e^*}{\tau_c^* + \tau_e^*} \left\{ 1 - \exp\left[-\left(\frac{1}{\tau_e^*} + \frac{1}{\tau_c^*} \right) t_{STRESS} \right] \right\} \quad (8)$$

where τ_c^* and τ_e^* are the effective capture and emission times under AC stress which depend on capture and emission times at V_{STRESS} and V_{RELAX}. These effective times are defined as:

$$\frac{1}{\tau_c^*} = \frac{DF}{\tau_c(V_{STRESS})} + \frac{1-DF}{\tau_c(V_{RELAX})} \quad (9)$$

$$\frac{1}{\tau_e^*} = \frac{DF}{\tau_e(V_{STRESS})} + \frac{1-DF}{\tau_e(V_{RELAX})} \quad (10)$$

Eqs. 7 and 8 provide the same occupancy probability since all the characteristic times of the trap under study (see Table I) are significantly larger than the period ($1/f$) of the AC stress signal.

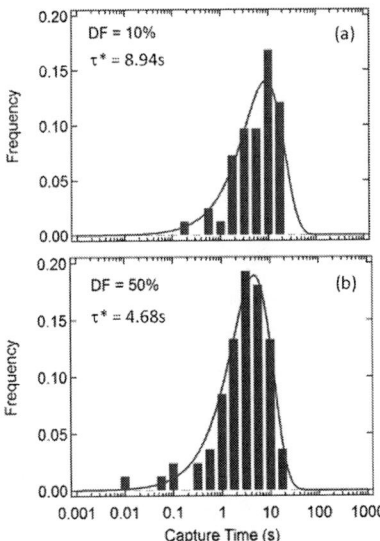

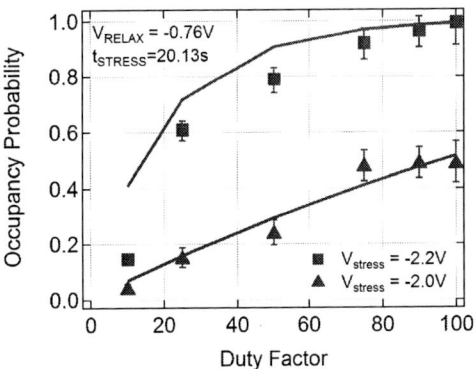

Fig. 11: Histograms of the first capture time t_c extracted from 80 I_{SOURCE} stress transients under V_{STRESS} = -2.2V and V_{RELAX} = -0.62V at (a) DF = 10%, and (b) DF = 50%. The capture times are binned on a logarithmic scale. Note that the effective time τ^* obtained from the fit of the date to eq. 11 is longer than the capture time τ_c obtained at DC stress (Table I). τ^* decreases with increasing the DF.

Fig. 14: (Symbol) Trap occupancy probability P_C with $\pm\sigma$ error bars for V_{RELAX} = -0.76 V. (Lines) Predicted occupancy according to the derived model matches the experimental data.

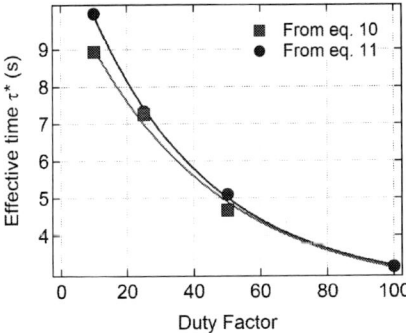

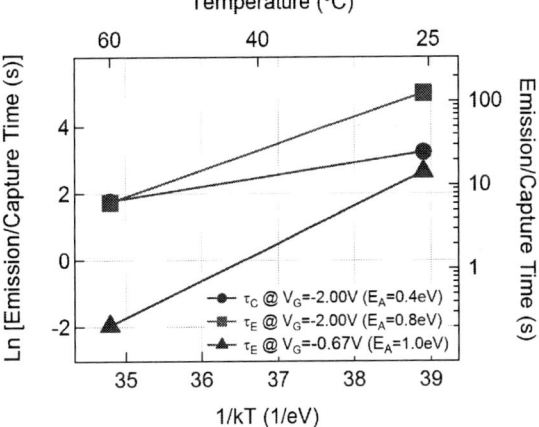

Fig. 12: Effective time τ^* vs. DF obtained (■) from the fit of the capture times t_c registered during stress (see Fig. 11) to eq. 11 and (●) from the τ_e and τ_c obtained under DC conditions and eq. 12.

Fig. 15: Arrhenius plot of τ_e and τ_c. Note the large shift of the values for only a 35 °C increase of temperature, indicating that the capture and emission of charge are thermally activated processes [8,11]. The reduction of E_A for τ_e with increasing V_G is in line with the prediction of ref. [8].

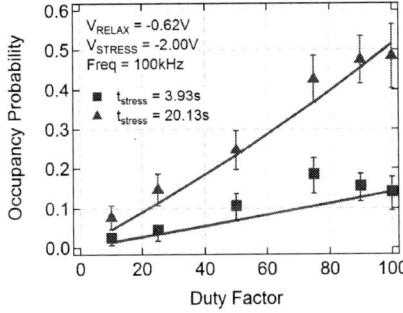

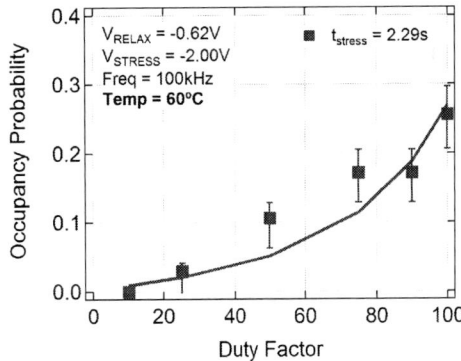

Fig. 13: (Symbols) Trap occupancy probability P_C with $\pm\sigma$ error bars for V_{STRESS} = -2.00 V and V_{RELAX} = -0.62 V as a function of the AC duty factor DF for two different values of t_{STRESS} (3.93 and 20.13 s). (Lines) Predicted occupancy according to eq. 7. Note that the experimental data match the derived model.

Fig. 16: (Symbols) Trap occupancy probability P_C with $\pm\sigma$ error bars as a function of duty factor DF at the temperature of 60 °C. Stress and relaxation voltages are the same as in Fig. 13. (Lines) Calculated occupancy probability obtained by means of eq. 7. Again, the experimental data match the derived model.

VII. FURTHER EXPERIMENTAL VALIDATION OF THE MODEL

Fig. 11 shows the histograms for the first capture time registered from 80 I_{SOURCE} stress transients under $V_{STRESS} = -2.2$ V and $V_{RELAX} = -0.62$ V at two different values of DF. Similarly to the emission times t_e, the capture times t_c follow the exponential distribution shown in eq. 7 and 8. The histogram of the capture times when binned on a logarithmic scale is given by

$$f_C(t_{STRESS}) = \frac{t_{STRESS}}{\tau_c^*} \exp\left(-\frac{t_{STRESS}}{\tau^*}\right) \tag{11}$$

where τ^* is the effective time:

$$\frac{1}{\tau^*} = \frac{1}{\tau_e^*} + \frac{1}{\tau_c^*} \tag{12}$$

Fig. 12 shows (1) the effective times τ^* obtained from the fit of the data in Fig. 11 to eq. 11, and (2) the effective times τ^* obtained by eq. 12 using the characteristic times obtained previously under DC stress (see Table I). Note that Fig. 12 only shows the effective time τ^* obtained by eq. 10 for low DF values, since only for low DF values, the capture/emission processes during stress are detectable (see Fig. 8). Fig. 12 shows a perfect agreement between both sets of data. Note that the average time τ^* decreases with increasing DF down to the value at DC stress, $DF = 1$, $\tau^* = \tau_c^* = \tau_c$. This again shows that the emission and capture event and the stress time are important parameters to determine the occupancy.

To further check the validity of the equations extracted in the last section, different stress and relax conditions were applied (Figs. 13 and 14). Note that the stress and relaxation voltages determine the capture and emission times of the trap (see Table I). Fig. 13 shows the occupancy probability P_C as a function of the duty factor DF for two stress times t_{STRESS}: 3.93 and 20.13 s. In this figure, V_{STRESS} was reduced to -2.0V with respect to the stress condition of Fig. 7, therefore the capture time τ_{cH} increases and the emission time τ_{eH} decreases (see Table I). The characteristics times (τ_{cL} and τ_{eL}) at low voltage are constant with respect to Fig. 7 since V_{RELAX} was not changed. Comparing Figs. 7 and 13, one can observe that the occupancy probability P_C increases with V_{SENSE}. The occupancy P_C according to eq. 7 is also traced in Fig. 13. We can observe that the equations developed in this paper correctly predict the experimental data.

In Fig. 14, the occupancy probability is plotted for two stress voltages V_{STRESS} at fixed $t_{STRESS} = 20.13$s and $V_{RELAX} = -0.76$ V. The V_{RELAX} was increased with respect to Figs. 7 and 13, so the emission time τ_{eL} at low level was increased (see Table I). An increase of the emission time means an increase of the occupancy since the probability of emission at low voltage is reduced. This fact is observed in Fig. 14 when it is compared to Figs. 7 and 13. Once more, the lines represent the predicted occupancy according to eq. 7 using the emission and capture times corresponding to the stress and relax voltages of the experiment. The experimental data and

eq. 7 developed in the last section based on first order kinetics are once again in agreement.

VIII. TEMPERATURE DEPENDENCE

Finally, the capture and emission processes were studied at a higher temperature (60 °C) under the stress and relax conditions of Fig. 13: $V_{STRESS} = -2.0$ V and $V_{RELAX} = -0.62$ V. Remarkably, during this experiment, the step height produced by the trap was reduced to 10mV with respect to the 23mV step height observed at 25°C. This feature has already been observed in Refs. [8] and [11]. The splitting of the cluster was explained by the electrostatic interaction with another charge trap [8].

By means of a DC stress, the emission and capture times at V_{RELAX} and V_{STRESS} were again first determined (see Table I) following the procedure presented in Sections III and IV. Fig. 15 shows the Arrhenius plot for the obtained characteristic emission and capture times at high and low voltages. As expected, these times decrease with increasing temperature. Large activation energies were extracted from the Arrhenius plots (see Fig. 15). Note that the activation energy for emission was about 1eV. These values are close to the activation energies for capture and emission times obtained for pMOSFETs in ref. [8] and for nMOSFETs in ref. [11]. This indicates that without any doubt the capture and emission times are thermally activated processes [16,17].

In Fig. 16, the probability occupancy P_C at 60°C is displayed as a function of DF. The lines show the predicted occupancy according to eq. 7 using the emission and capture times at 60°C. Experiment and theory are in agreement once more.

Summarizing, the model can predict correctly the P_C for all t_{STRESS} and DFs provided that the emission and capture times under DC conditions are known. Furthermore, the model can be used in simulations of the response of CMOS circuits under AC conditions [5], taking into account that (1) the V_{TH} shift produced by a single defect is exponentially distributed, (2) the number of defects per device follows a Poisson distribution [15], and (3) the emission and capture times are distributed inversely with time [10]. A circuit simulator based on the AC theory presented in this paper has been developed in ref. [5].

IX. CONCLUSIONS

In conclusion, we have studied thoroughly the response of a single trap under various DC and AC stress conditions. The theory to explain the obtained experimental data under AC stress has been elaborated based on first-order kinetics. This theory passes all sets of performed experimental tests. The model can thus be transferred to circuit simulators in order to simulate the effect of individual traps under AC workloads.

ACKNOWLEDGMENT

This work was carried out as part of IMEC's Industrial Affiliation Program funded by IMEC's core partners. M.T.L. thanks the IberCaja Postdoctoral grant program and Spanish

MEC (TEC2010-18051) for financial support. The authors would like to thank the imec DRE and NVM groups, and AMSIMEC for helpful discussions and input throughout this work.

REFERENCES

[1] A. Asenov, R. Balasubramaniam, A.R. Brown, J.H. Davies, "RTS amplitudes in decananometer MOSFETs: 3-D simulation study," *IEEE Trans. Electron Devices*, vol.50, no.3, pp.839- 845, March 2003.

[2] M. F. Bukhori, S. Roy, and A. Asenov, "Simulation of Statistical Aspects of Charge Trapping and Related Degradation in Bulk MOSFETs in the Presence of Random Discrete Dopants," *IEEE Trans. Electron Devices*, vol.57, no.4, pp.795-803, April 2010.

[3] V. Huard, "Two independent components modeling for Negative Bias Temperature Instability," in IRPS 2010, pp.33-42.

[4] R. Fernandez, B. Kaczer, A. Nackaerts, S. Demuynck, R. Rodriguez, M. Nafria, G. Groeseneken, "AC NBTI studied in the 1 Hz -- 2 GHz range on dedicated on-chip CMOS circuits," in *IEDM Tech. Dig.*, 2006, pp.1-4.

[5] B. Kaczer, S. Mahato, V. Valduga de Almeida Camargo, M. Toledano Luque, Ph. Roussel, T. Grasser, F. Catthoor, P. Dobrovolny, P. Zuber, G. Wirth, G. Groeseneken, "Atomistic approach to variability of bias-temperature instability in circuit simulations" in *IRPS* 2011 (to be published).

[6] M. Miranda Corbalan, B. Dierickx, P. Zuber, P. Dobrovolny, F. Kutscherauer, Ph. Roussel, and P. Poliakov, "Variability aware modeling of SoCs: from device variations to manufactured system yield." In: *International Symposium on Quality Electronic Design*. 2009, pp. 16-18.

[7] M. Alam, "Reliability and Process Variation Aware Design of Integrated Circuits," in *19th Europena Symposium Reliability of Electron Devices, Failure and Analysis,* 2008.

[8] T. Grasser, H. Reisinger, P. Wagner, F. Schanovsky, W. Goes, B. Kaczer, "The time dependent defect spectroscopy (TDDS) for the characterization of the bias temperature instability," in *IRPS* 2010, pp.16-25.

[9] B. Kaczer, T. Grasser, J. Martin-Martinez, E. Simoen, M. Aoulaiche, Ph. J. Roussel, G. Groeseneken, "NBTI from the perspective of defect states with widely distributed time scales," in *IRPS* 2009, pp. 55-60.

[10] T. Grasser, H. Reisinger, W. Goes, T. Aichinger, P. Hehenberger, P.-J. Wagner, M. Nelhiebel, J. Franco, B. Kaczer, "Switching oxide traps as the missing link between negative bias temperature instability and random telegraph noise," *IEDM Tech. Dig.*, 2009, pp.729-732.

[11] M. Toledano-Luque, B. Kaczer, Ph. Roussel, M.J. Cho, T. Grasser, and G. Groeseneken, "Temperature dependence of the emission and capture times of SiON individual traps after positive bias temperature stress," *J. Vac. Sci. Technol. B* 29, 01AA04, January 2011.

[12] B. Kaczer, T. Grasser, P.J. Roussel, J. Martin-Martinez, R. O'Connor, B.J. O'Sullivan, G. Groeseneken, "Ubiquitous relaxation in BTI stressing—New evaluation and insights," in *IRPS 2008*, pp.20-27.

[13] A. Kerber, K. Maitra, A. Majumdar, M. Hargrove, R.J. Carter, E.A. Cartier, "Characterization of Fast Relaxation During BTI Stress in Conventional and Advanced CMOS Devices With Gate Stacks," *IEEE Trans. Electron Devices*, vol.55, no.11, pp.3175-3183, Nov. 2008

[14] A. Ghetti, C.M. Compagnoni, A.S. Spinelli, and A. Visconti, "Comprehensive Analysis of Random Telegraph Noise Instability and Its Scaling in Deca–Nanometer," *IEEE Trans. Electron Devices*, vol.56, no.8, pp.1746-1752, Aug. 2009.

[15] B. Kaczer, Ph.J. Roussel, T. Grasser, G. Groeseneken, "Statistics of Multiple Trapped Charges in the Gate Oxide of Deeply Scaled MOSFET Devices—Application to NBTI," *IEEE Electron Device Letters*, vol.31, no.5, pp.411-413, May 2010.

[16] M.J. Kirton, M.J. Uren, "Capture and emission kinetics of individual Si:SiO2 interface states," *Applied Physics Letters*, vol.48, no.19, pp.1270-1272, May 1986.

[17] Y. Shi, H.M. Bu, X.L. Yuan, S.L. Gu, B. Shen, P. Han, R. Zhang and Y.D. Zheng, "Switching kinetics of interface states in deep submicrometre SOI n-MOSFETs", *Semicond. Sci. Technol.*, 16, pp. 21-25, 2001.

PBTI UNDER DYNAMIC STRESS: FROM A SINGLE DEFECT POINT OF VIEW

K. Zhao, J. H. Stathis, B. P. Linder and E. Cartier
T.J. Watson Research Center, IBM
1101 Kitchawan Road, Yorktown Heights, NY, 10598, USA
phone: +1 (845) 894 5308, email: kzhao@us.ibm.com

A. Kerber
Technology Research Group, GLOBALFOUNDRIES Inc.
1101 Kitchawan Road, Yorktown Heights, NY, 10598, USA

Abstract— In this paper, fundamental aspects of the Bias Temperature Instability (BTI) in FETs with metal gate/high-k (HKMG) gate stacks are discussed from a single defect point of view. First, Random Telegraph Noise (RTN) measurements are used to show that the capture/emission processes of individual defects in highly scaled HKMG FETs exhibit very similar Poisson statistics and can be fully characterized by a characteristic electron/hole capture, τ_c, and emission time, τ_e, in NFET/PFET. In all cases, capture and emission are found to be thermally activated. These observations suggest that NBTI and PBTI share similar microscopic trapping/de-trapping mechanism, for holes and electrons, respectively. Based on these findings, a simple physical model is introduced which describes the behavior of a distribution of identical defects (characterized by τ_c and τ_e) but provides deep insights into the BTI dynamics under AC stress in general. The occupancy level of identical defects at equilibrium is found to becomes frequency, f, independent for $f >> [1/\tau_c, 1/\tau_e]$, such that the BTI behavior at operation conditions (~GHz) can be measured at relatively low frequencies (in the kHz range). The single defect model was then expanded to predict the macroscopic BTI behaviors in NMOS devices for arbitrary stress conditions. Excellent agreement between model prediction and experimental data is demonstrated, confirming that PBTI in HKMG gate stacks can be understood as a superposition of trapping/de-trapping events from individual defects in the gate stack. The overall dynamics of PBTI is thus largely governed by the distribution of electron capture and emission times of the defects in the gate stack. The challenges for using a capture and emission time based model for product lifetime predictions are addressed.

Index Terms—HKMG, RTN, NBTI, PBTI, Bias Temperature Instability, single defects, frequency dependence, duty cycle dependence.

Manuscript received October 9, 2010. This work was performed by the Research Alliance Teams at various IBM Research and Development Facilities.
K. Zhao is with IBM research division, Hopewell Junction, NY 12533 USA phone: 845-894-5308, fax:845-894-5308, e-mail: kzhao@us.ibm.com.
J. H. Stathis is with IBM research division, Yorktown Heights, NY 10598
E. Cartier is with IBM research division, Yorktown Heights, NY 10598
B. P. Barry is with IBM research division, Yorktown Heights, NY 10598
A. Kerber is with Technology Research Group, GLOBALFOUNDRIES Inc. Yorktown Heights, NY 10598

I. INTRODUCTION

Over the past technology generations, progressive scaling has resulted in much thinner gate dielectrics and higher oxide fields. This exacerbates the bias temperature instability for advanced CMOS devices. Understanding the underlying physical mechanisms on an atomistic level and accurate modeling of the V_t instability associated with the Bias Temperature Instability (BTI) remains a central task in the BTI community. In recent years, substantial progress has been made. With the availability of deeply scaled FETs, the properties of individual defects in gate dielectric have been studied through techniques such as Random Telegraph Noise (RTN) and Deep Level Transient Spectroscopy [1-4], and it has been demonstrated that the macroscopic NBTI dynamics can be reconstructed through summation of the microscopic trapping /de-trapping events of individual defects in PFETs with SiON gate stacks [3], providing strong evidence that trapping/de-trapping of single defects in SiON gate stack is one of the fundamental process causing NBTI in PMOS devices.

In this paper, we will explore the single defect behavior in FETs with HKMG stack using RTN. Based on the measured statistical properties of single defects with simple stress-relax sequences, a simple model is derived to study the behaviors of a distribution of defects (described by characteristic capture and emission times, τ_c and τ_e) under general AC stress condition. The model is then expanded to describe the macroscopic BTI behavior of FETs under general dynamic stress conditions. The BTI behaviors in HKMG NMOS devices under various AC stress conditions can be modeled as the superposition of the stress response of a collection of single defects with characteristic trapping and de-trapping time constants. Practical implications to circuit and system reliability are discussed and the possibilities to use this model approach for BTI lifetime predictions are addressed.

II. RESULTS AND DISCUSSION

A. Single Defects in HKMG FETs

In this section, our study of the trapping and de-trapping

behavior of single defects in FETs with HKMG gate stacks is summarized. The trapping and de-trapping characteristics of single defects in both NFETs and PFETs are investigated with RTN measurements. To isolate individual defects, highly scaled NFETs and PFETs (W x L ~200 nm x 40 nm) with HfO$_2$-based HKMG gate stack were used. The RTN signals were measured by monitoring the drain current fluctuations under constant gate bias. Fig. 1 shows typical RTN signals at various gate bias conditions in a PFET. Similar RTN signals were also measured in scaled NFETs. At a given gate bias, the drain current switches between a high current state and a low current state, corresponding to charge capture or emission events from a single defect, respectively. The magnitude of the current fluctuation is defect specific and depends not only on the electronic structure of the defect but also on its location in the gate stack and in the x-y plane of the transistor. The trapping/de-trapping is bias dependent, as can be easily seen from the current–time traces shown in Fig. 1. 'Defect 1' (with larger amplitude) predominantly switches at high gate bias, V_g, whereas 'Defect 2' (with smaller amplitude) becomes more active at the lower V_g. The physical origin for the difference in behavior can not be easily identified. It may be structural in nature, or a location effect or both.

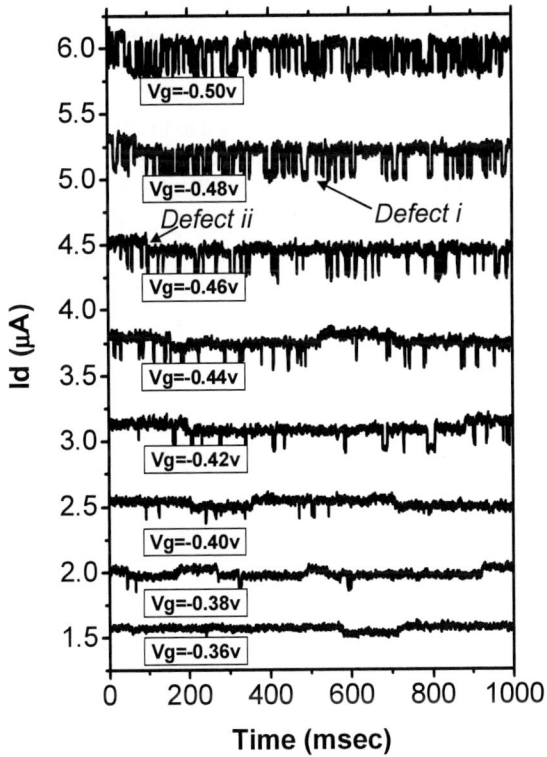

Fig. 1. Random Telegraph Noise (RTN) in the drain current, measured on a 0.2umx0.04um PFET at different gate bias conditions. Note that RTN signals from different defects can be observed at different bias conditions.

To evaluate the statistical distribution of trapping and de-trapping times, the time intervals between switching events are extracted from the current-time traces in Fig. 1, resulting in switching time distribution plots as shown in Fig. 2. In Fig. 2, typical time distribution plots for nFET (PBTI) and pFETs (NBTI) are compared. As can be seen, the capture and emission process for both, NFETs and PFETs, can be described by Poisson statistics and the times can be parameterized in terms of a characteristic capture and emission time constant as in Eq. (1) and (2), respectively.

$$P_c(t) = 1 - \exp\left(-\frac{t}{\tau_c}\right). \qquad (1)$$

$$P_e(t) = 1 - \exp\left(-\frac{t}{\tau_e}\right). \qquad (2)$$

The characteristic capture and emission times, τ_c and τ_e, are gate bias dependent. This aspect can be used to gain insight into the physical nature of the defect [4]. As will be shown later, the characteristic τ_c and τ_e can be used to determine the occupancy of a single defect under AC stress condition.

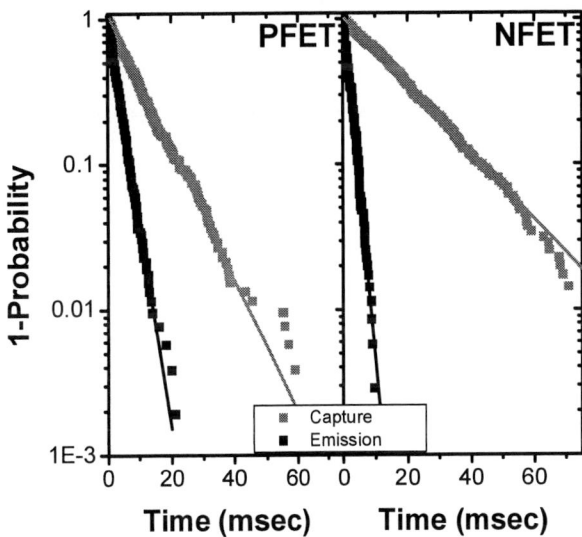

Fig. 2, Typical time distribution plot for NFETs and PFETs, showing similar Poisson distribution for capture and emission process.

In Fig. 3, the bias dependence of the characteristic times, τ_c and τ_e, is shown at various temperatures. The temperature dependence follows Arrhenius temperature activation for both, NFETs and PFETs, as shown in Fig. 4. The dependence of τ_c and τ_e on V_g and temperature can be expressed as,

$$\tau_c = \tau_{c0} \exp\left(-\beta_c \cdot V_g\right) \exp\left(-\frac{E_c}{kT}\right), \qquad (3)$$

and,

$$\tau_e = \tau_{e0} \exp\left(\beta_e \cdot V_g\right) \exp\left(-\frac{E_e}{kT}\right), \qquad 4)$$

where β_c and β_e are the voltage dependence factors, and E_c and E_e are the activation energies for capture and emission, respectively. The strong dependence on temperature suggests that for these MGHK gate stacks, direct tunneling process cannot be used to model the trapping/de-trapping process, consistent with earlier reports for SION [5, 6]. It is interesting

978-1-4244-9113-1/11 $26.00 © 2011 IEEE

to note that there is no fundamental difference between nFETs (PBTI) and pFETs (NBTI). This implies that similar physical processes control electron and hole trapping/de-trapping. Since a strong temperature dependence is observed in both cases, it can be expected that the concept of local atomic structure rearrangement upon electron (hole) capture and emission do occur not only during hole capture/emission (NBTI) [7-10] but also during electron capture/emission (PBTI).

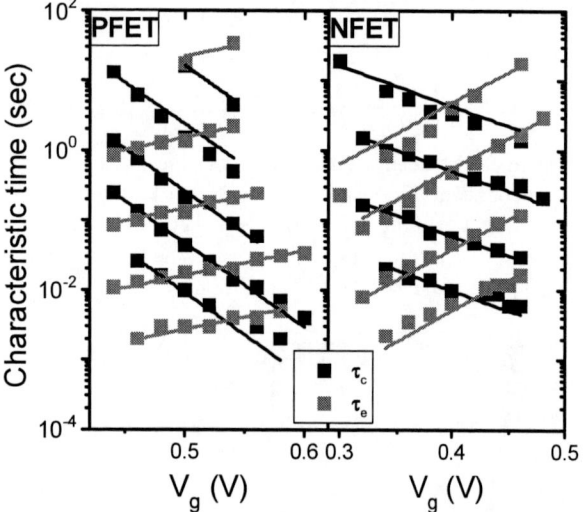

Fig. 3, Comparison of the gate bias dependence of capture times and emission times at different temperatures for for pFETs and nFETs The temperatures for individual data sets can be inferred from the Arrhanius plots of τ_c and τ_e in Fig. 4.

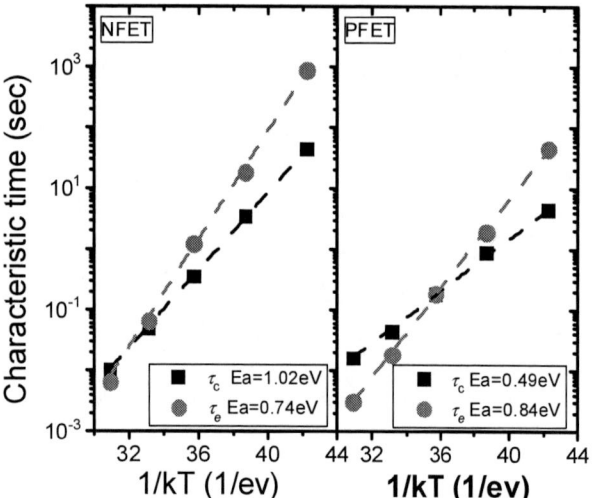

Fig. 4. Temperature dependence of the capture and emission time from derived from data in Fig. 3. An Arrhenius behavior with comparable activation energies is observed for HKMG nFETs and pFETs

B. Modeling of Single Defects in HKMG FETs

The capture and emission process of individual defects follows Poisson statistics, as shown in the previous section. Under this condition, it is possible to model a distribution of single defect as a RC element with characteristic charging and discharging time constant, $R_cC = \tau_c$, and $R_eC = \tau_e$, as previously proposed by Reisingers et al. [3]. This approach greatly

simplifies BTI modeling. Our single trap studies show that the "RC" approach can be applied to both NBTI and PBTI in MGHK FETs.

In the following we use the "RC" approach to model the occupancy of a distribution of defects characterized by time constants, τ_c and τ_e, under AC stress condition.

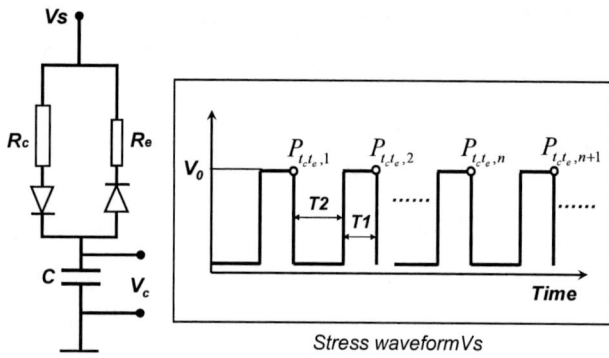

Fig. 5, Schematics, illustrating the single defect modeling using an equivalent "RC" circuit [3]. $R_cC = \tau_c$, and $R_eC = \tau_e$.

In Fig. 5, the threshold voltage shift caused by charging of a distribution of identical defects with characteristic capture and emission times is equivalent to a simple RC circuit with time constants $R_cC = \tau_c$ and $R_eC = \tau_e$ [3]. The V_t-shift is equal to the amplitude V_c across the capacitor, C. Therefore, when all defects are occupied, $V_c = V_s$. Let's assume now that an AC stress signal, V_s, with frequency, $f = 1/(T_1+T_2)$ and duty cycle, $DC = T_1/(T_1+T_2)$, is applied to the input terminal. Because each defect contributes an identical amout to the net V_t shift of the device, we can evaluate the defect occupancy level, $P = V_c/V_s$, at any moment in time simply by calculating the charging state of the capacitor. After the n-th stress pulse (see voltage-time trace in Fig. 5), the occupancy level, $P_{\tau_c\tau_e,n}$ (at the end of a stress half-cycle, corresponding to 'Sense 1') and the occupancy level, $P'_{\tau_c\tau_e,n}$ (at the end of a relaxation half-cycle, corresponding to 'Sense 2') can be derived analytically (see Appendix I), and are given by

$$P_{\tau_c\tau_e,n} = \left(\left(1-\exp\left(-\frac{T_1}{\tau_c}\right)\right)\left(1-\frac{1}{1-\exp\left(-\left(\frac{T_1}{\tau_c}+\frac{T_2}{\tau_e}\right)\right)}\right)\right) \quad (5),$$

$$\times \exp\left(-(n-1)\left(\frac{T_1}{\tau_c}+\frac{T_2}{\tau_e}\right)\right) + \frac{\left(1-\exp\left(-\frac{T_1}{\tau_c}\right)\right)}{1-\exp\left(-\left(\frac{T_1}{\tau_c}+\frac{T_2}{\tau_e}\right)\right)}$$

and,

$$P'_{\tau_c\tau_e,n} = P_{\tau_c\tau_e,n}\exp\left(-T_2/\tau_e\right) \quad (6)$$

For $T_2 \to 0$, Eq. (5) converges to the DC case,

$$P_{\tau_c\tau_e,n}\big|_{T_2=0} = \left(1-\exp\left(-\frac{nT_1}{\tau_c}\right)\right) \quad (7)$$

From Eq. (5) and (6), the occupancy of a distribution of identical defects can be calculated for different stress

978-1-4244-9113-1/11 $26.00 © 2011 IEEE

conditions. Examples for $\tau_c = \tau_e = 1$ sec are shown in Fig. 6. As can be see, under DC stress, the occupancy increases exponentially and approaches 1 for stress times, $T_s \gg \tau_c$. More interestingly, under AC stress at long stress times, the occupancy does not saturate at 1, but reaches an intermediate equilibrium level given by,

$$P_{\tau_c\tau_e,n}\Big|_{T_s\gg\tau_c,\tau_e} = \frac{1-\exp\left(-\dfrac{T_1}{\tau_c}\right)}{1-\exp\left(-\left(\dfrac{T_1}{\tau_c}+\dfrac{T_2}{\tau_e}\right)\right)} \quad (5a)$$

for 'Sense 1', and by,

$$P'_{\tau_c\tau_e,n}\Big|_{Ts\gg\tau_c,\tau_e} = \frac{\exp\left(\dfrac{T_1}{\tau_c}\right)-1}{\exp\left(\dfrac{T_1}{\tau_c}+\dfrac{T_2}{\tau_e}\right)-1} \quad (6a)$$

for Sense 2. At long times, the occupancy level oscillates between the two extreme values at Sense 1' and 'Sense 2', showing that at equilibrium, the capture probability during each stress half-cycle equals the emission probability during the following relaxation half-cycle. The occupancy levels at equilibrium are determined by the defect capture/ emission times, τ_c, τ_e, as well as by the frequency and the duty cycle. Note that, the occupancy level under AC stress is always less than 1, reflecting the observation that AC BTI degradation is always lower than DC BTI degradation. Furthermore, at high frequency, 'Sense 1' and 'Sense 2' merge into a single value.

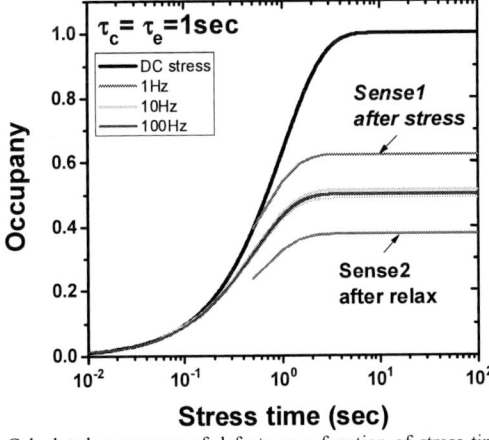

Fig. 6. Calculated occupancy of defects as a function of stress time under different stress modes. 'Sense 1' is the occupany of the defects calculated at the end of a stress half-cycle. 'Sense 2' is the occupancy of the defects calculated at the end of a relaxation half-cycle. Note that an equilibrium state is reached at stress time, $T_s \gg \tau_c$, τ_e.

The frequency dependence of 'Sense 1' and 'Sense 2' is shown in Fig. 7a for the three cases, $\tau_c/\tau_e = 0.1$, 1, and 10, respectively. The duty cycle was set to be 50% in these examples. At low frequency, $f \ll [1/\tau_c, 1/\tau_e]$, the occupancy of the defects at 'sense 1' and 'sense 2' switches between 1 and 0. All defects are filled at the end of the stress half-cycle, and empty at the end of the relaxation half-cycle. As the frequency

increases, the occupancy levels at 'sense 1' and 'sense 2' gradually converge to a common value for $f \gg [1/\tau_c, 1/\tau_e]$ and the occupancy value becomes frequency independent. From Eq. (5), the occupancy at equilibrium can be expressed as,

$$P_o\Big|_{T_s\gg T_c,f\gg[1/\tau_c,1/\tau_e]} = \frac{1}{1+\dfrac{t_c}{t_e}\left(\dfrac{1}{Dutycycle}-1\right)}. \quad (8)$$

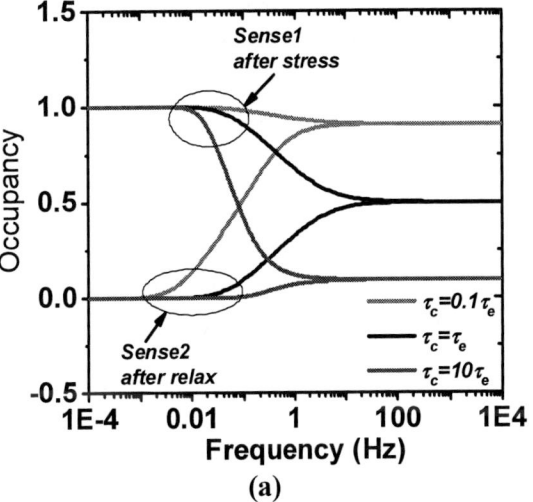

(a)

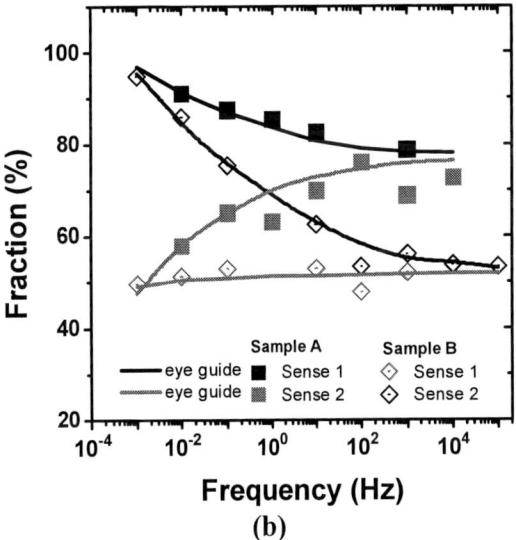

(b)

Fig. 7. Frequency dependence of the PBTI instability. (a) Calculated occupancy of defects with different capture time, t_c, and emission time, t_e. Note that the occupancy becomes frequency independent for $f \gg [1/t_c, 1/t_e]$. (b) Measured frequency dependence for two different HKMG NFETs ('Sample A' and 'Sample B').

From the analysis provided above, the first important conclusion for BTI qualification under AC stress conditions can be drawn. As can be seen in Fig. 7a, for the parameters used, the V_t shifts under AC stress approach the steady state condition already in the kHz frequency range. Therefore, measurements in the kHz range will provide a good approximation for the long time BTI behavior under operation conditions in the GHz

978-1-4244-9113-1/11 $26.00 © 2011 IEEE

frequency range. The experimental data in Fig. 7b confirms that this conclusions also apply to realistic HKMG FETs. In fig. 7b, the measured PBTI frequency dependence for two different types of HKMG FETs are compared, exhibiting significantly different saturation values. However, qualitatively, the overall trend of the measured frequency dependence agrees well with the model calculation. For both samples, the V_t-shifts measured 'Sense 1' merges with the shifts measured at 'Sense 2' around 1-10 kHz and become frequency independent.

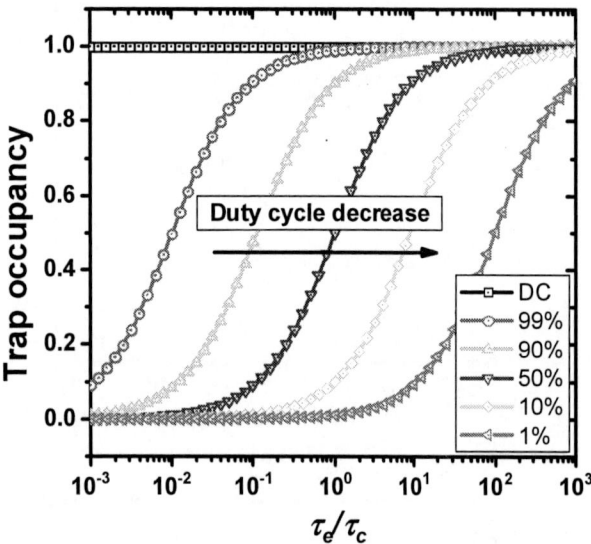

Fig. 8. Calculated AC stress occupancy at equilibrium ($T_s >> t_c, f > [1/t_c, 1/t_e]$) as a function of the t_e/t_c-ratio for different duty cycle values.

In Fig. 8, the calculated trap occupancy at equilibrium is shown as a function of the τ_e/τ_c-ratio for duty-cycle values of 1%, 10%, 50%, 90%, 99% and for the DC case. For defects with very small τ_e/τ_c-ratio, most defects remain empty at equilibrium because the emission process during the relax half-cycles is much faster than the capture process during the stress half-cycles. In the other extreme for defects with large ratio, $\tau_e/\tau_c >> 1$, the capture process during the stress phase becomes much more efficient and most defects remain filled at equilibrium. The transition time between the two extremes is modulated by the value of the duty cycle.

In Fig. 9a, the calculated duty-cycle dependences for a distribution of identical defects with different τ_e/τ_c-ratios are compared. For defects with $\tau_e = \tau_c$, the occupancy at equilibrium shows a linear dependence on duty cycle. For defects with $\tau_e > \tau_c$, the duty cycle dependence is strong at low duty-cycle values, while for defects with $\tau_e < \tau_c$, the variability is shifted to the high duty-cycle region. In practical cases, the duty cycle dependence is more S-shaped. This is caused by multiple defects with different τ_e/τ_c-ratios. To illustrate the impact of multiple defects, simulations were performed for samples with multiple defects (see Fig. 9b). The differrence between the three cases shown is illustrated by the defect maps shown in Fig. 9c. For multiple defects with equal emission and capture times, the duty cycle dependence shows a 'symmetric' "S" shape behavior. For defects with asymmetric time distributions, the

"S" curve moves up (with more defects with $t_e/t_c > 1$) or down (with more defects with $t_e/t_c < 1$), but maintains the S-shape. The shape of the duty-cycle dependence contains information on the trapping/de-trapping time distribution.

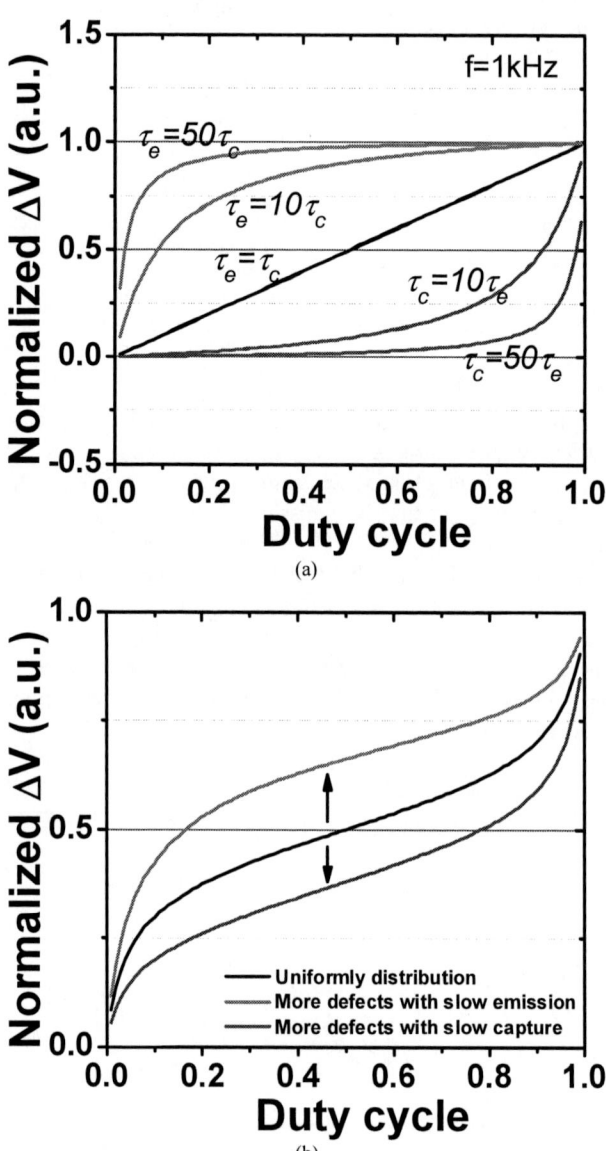

Fig. 9, (a) calculated duty cycle dependence of single sets of defects with different te//tc ratios. (b) calculated duty cycle dependence of defects with different distributions.

All these examples demonstrate that all important aspects of the BTI dynamic during AC stress can be rationalized by modeling the trapping and detrapping of defects with different characteristic trapping and detrapping time constants. The examples show that the occupancy level under AC stress always reaches a constant equilibrium state at which the charge capture process during the stress half-cycle is in balance with the charge emission process during the following relaxation half-cycle. For a single set of defects, the ratio, τ_e/τ_c, is the key parameter that determines the occupancy level at equilibrium. The model also shows that the occupancy level at equilibrium becomes frequency independent at high frequency, $f >> [1/\tau_c, 1/\tau_e]$. This explains the widely reported observation that BTI shifts becomes frequency independent above a critical AC stress frequency [12]. It therefore provides a theoretical base for the assertion that the BTI behavior at operation frequencies (GHz) can be inferred form the frequency behavior in the kHz range, which is experimentally readily accessible.

C. PBTI under AC stress in HKMG FETs

In this section, the PBTI behavior under various AC stress modes are experimentally measured in HKMG FETs and the results are compared with model prediction. To obtain a more realistic model, the single defect picture is expanded to multiple defects with different characteristic emission and capture times. Practical implications of this comparison for circuit and system reliability are also discussed.

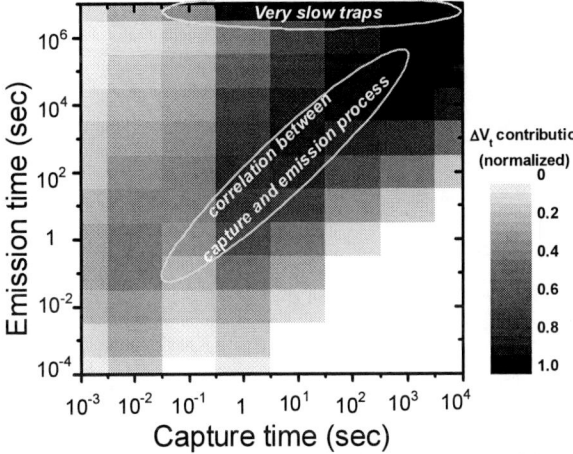

Fig. 10, defect distribution of capture and emission time extracted from DC stress/relax traces [3]. A clear correlation between capture time and emission time is observed. Note the distribution of the traps with very long emission times ($\tau_e > 10^4$sec) are extracted through the universality fitting of the relaxation time traces [13].

Expanding on what was said earlier, the V_t-shift at time t in a large device can be calculated by summing up the ΔV_t contribution of all active defects as,

$$\Delta V_t(t) = \sum_{\tau_c} \sum_{\tau_e} P_{\tau_c \tau_e}(t) \Delta V_{t(\tau_c, \tau_e)}. \qquad (10)$$

$P_{\tau_c \tau_e}(t)$ is the occupancy of defects with capture time, τ_c, and emission time, τ_e, and $\Delta V_{t(\tau_c, \tau_e)}$ is the corresponding V_t-shift.

The characteristic emission and capture time distributions describing the defects in a real device are not known *a priori*. To obtain these distributions, the method discussed for NBTI stress in [3] is applied here. The basic idea of this method is to extract the capture time and emission time from a series of individual DC stress/relax measurements. The data in Fig. 10 shows a representative defect map for the HKMG NFETs used in this work. The distribution of defects with very long emission times ($\tau_e > 10^4$sec) are extracted through the universality fitting of the relaxation time traces [13]. The defect

map is in general bias dependent. The map was extracted from stress/relax sequences with a stress voltage of 1.45 V and a relax voltage of 0.3 V. A clear correlation between capture and emission times is observed. The empirical defect map shown in Fig. 10 can now be used to predict the PBTI behavior of this spesific nFET under various dynamic stress conditions and the predictions can then be verified directly by experiment.

In Fig. 13, modeled V_t-shifts for three cases (DC stress, AC stress at 5 mHz and at 10 Hz) are compared to the measured V_t-shift for the same experimental conditions. Again, two sense measurements, 'Sense 1' and 'Sense 2' were performed. A 1 ms measurement time delay was included in the simulations. As can be seen, the model predictions show excellent agreement with the measured data for all cases, supporting the notion that PBTI in HKMG nFETs is largely caused by the collective response (trapping/de-trapping) of individual defects to the applied stress. The overall dynamics of PBTI is controlled by the capture time and emission time distribution of all the defects in the FET.

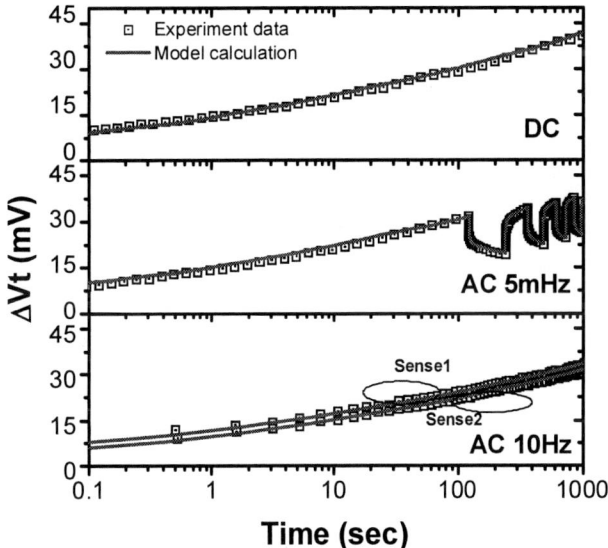

Fig. 11, Comparison of measured (symbols) and calculated (lines) V_t-shifts during PBTI in a MGHK nFET. Three different stress modes (DC, AC 5 mHz, and AC 10 Hz) are shown. The defect map shown in Fig.10 is used for the calculation. Excellent agreement between calculation and experiment is obtained for all three stress modes.

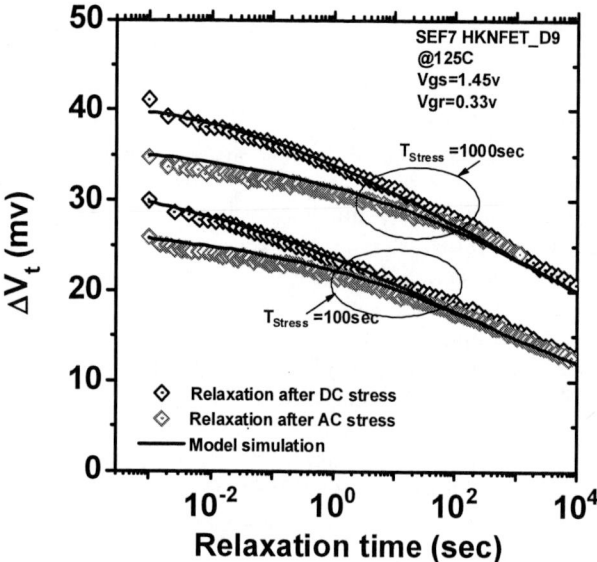

Fig. 12, PBTI relaxation after DC and AC stress. Symbols are experimental data. Solid lines are model prediction.

The difference in the PBTI relaxation dynamics after AC (100 Hz, 50% duty cycle) and DC stress from Ref. [11] is compared to model predictions in Fig. 12. Model predictions are based on the empirical defect distribution shown in Fig. 10. As can be seen, the difference in the relaxation behavior can readily be predicted via Eq. (10) (solid lines in Fig. 12). The AC relaxation rate early in the relaxation phase is small compared to the relaxation rate after DC stress. However, the two relaxation curves gradually merge at longer relaxation times. The "time-to-merge" depends on the stress time and it is also predicted correctly. To gain further insight into the origin of this difference, the occupancy maps after AC and DC stress (with the same net stress time) are compared in Fig. 13. First, it is noticed that the defects with short emission time, $\tau_e < 1$ ms) are mostly unoccupied. This is caused by the measurement delay time (~1 msec) which is considered in the modeling. More importantly, if one compares the occupancy maps for AC and DC stress, it can be seen that 'shallow' defects with $t_e < t_c$ are unoccupied after the AC stress. The absence of these fast defects is the cause for the slow initial relaxation after AC stress – these states relax already during the AC stress. At longer times, the maps are virtually identical and the relaxation behavior becomes indistinguishable from the DC case. The single defect modeling discussed in the context of Fig 7 provides further insight into this stress-type dependence of the relaxation. The occupancy of certain defect at equilibrium under AC stress was shown to be strongly dependent on the ratio of t_e/t_c and on the value of the duty cycle. In the example discussed here, the duty cycle was 50%. Therefore, for the defects with $\tau_e/\tau_c < 1$, the occupancy at equilibrium is small.

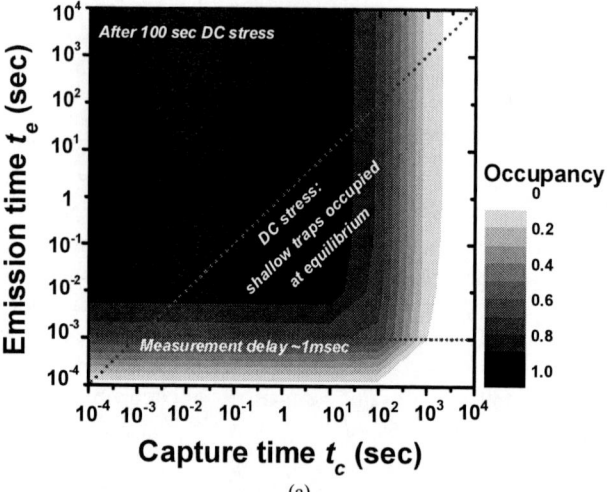

(a)

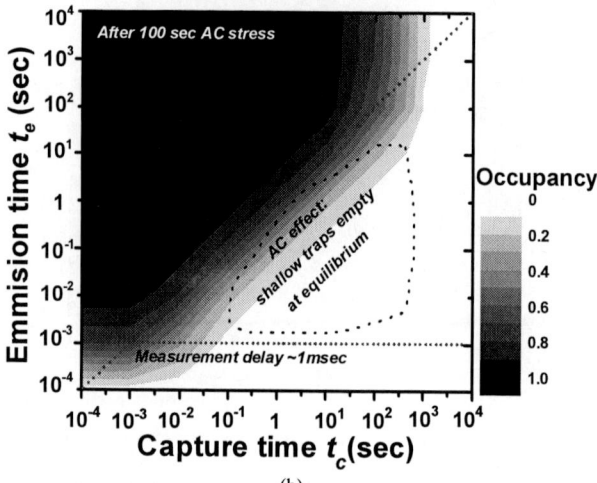

(b)

Fig. 13. Calculated occupancy maps after PBTI DC (top panel) and AC stress (bottom panel). It can be see that the difference in the relaxation behavior reported in Fig. 12 is due to absence of 'shallow' traps at equilibrium. Note that the measurement delay time is taken into account in the model calculation.

For many circuit applications, the stress mode may change over time. The model proposed here can also describe the PBTI degradation for such non-periodic stress conditions. An example is shown in Fig. 14. In this example, the stress mode is changed from AC to DC stress at time, $t = 100$ s and then back from DC to AC stress at $t = 1000$ s. The measured degradation dynamics during a continuous DC stress is shown for comparison. When the stress mode is changed from AC stress to DC stress, the measured V_t-shift quickly merges with the DC degradation curve. Then, when the stress mode is changed from DC stress back to AC stress, the V_t-shift relaxes towards the 'extrapolated' AC degradation curve and follows the AC degradation trend. The solid lines show the calculated PBTI behavior using Eq. (10) and the empirical occupancy map from Fig. 10. Again, good agreement with the experimental data is obtained. These simple examples clearly demonstrate that the

978-1-4244-9113-1/11 $26.00 © 2011 IEEE

occupancy of each individual defect reaches an equilibrium value dictated by the stress mode (DC, AC, frequency, duty cycle). When the stress mode is changed, the occupancies rapidly transit to a new equilibrium condition appropriate for the new stress condition.

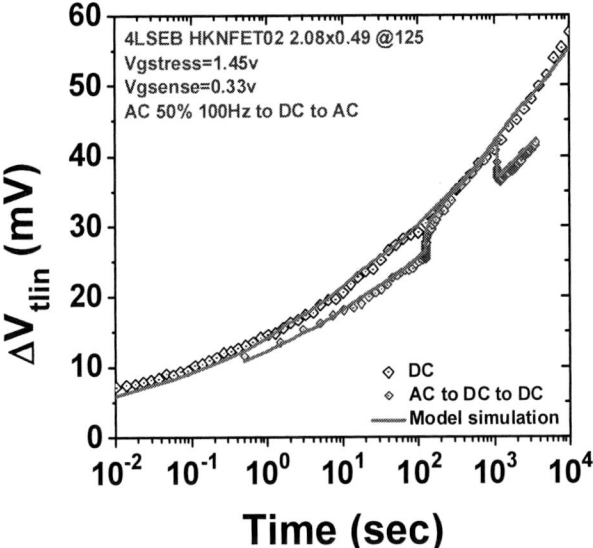

Fig. 14. Comparison of calculated (solid lines) and measured (symbols) V_t-shift for non-periodic stress conditions. The stress mode is changed from DC to AC stress at $t \sim 100$ s and back to AC stress at $t \sim 1000$ s. The AC frequency was 100 Hz and the duty cycle was 50 % .

III. BTI LIFETIME PREDICTIONS

In the BTI modeling approach described here, all the relevant physical properties of the gate stack are assumed to be contained in the capture-emission time distribution maps like those shown in Fig. 10. No pathway has been described to obtain such a map directly from the actual physical properties – like the defect structure of the gate stack. The maps were inferred from BTI experiments using the approach described in [3]. What is demonstrated here is that such emission-capture time maps can be most successfully be used in combination with Eq. (10) to describe the BTI behavior of a transistor under arbitrary stress conditions. It remains important to keep in mind that the time domain used for the predictions in examples shown was of the same order of magnitude as the time domain used to establish the capture-emission time distributions. It is evident, that any attempt to predict the BTI behavior of the FET out side this time domain – like a lifetime prediction – would utterly fail. This is simply due to the fact that the BTI dynamics at long times is controlled by defects with much longer time constants.

These facts generate a dilemma for lifetime predictions and it remains to be seen whether the kind of modeling discussed here can help to overcome the completely empirical methodologies used for lifetime predictions used today and put them on more solid scientific footings.

Finally, it should also be mentioned, that this approach currently does not consider the interface state and bulk trap generation by the high stress field or by hot carriers in the FET.

Since some believe that the long term behavior may be dominated by interface state generation for NBTI [14-16] and by bulk trap generation (possibly oxygen vacancies) in the HfO$_2$ layer, for PBTI [17, 18], much work remains to be done.

IV. CONCLUSIONS

Random Telegraph noise (RTN) measurements were used to show that charge trapping during positive and negative bias temperature stress in nFETs (electron trapping) and pFETs (hole trapping) with ultra thin HKMG gate stacks exhibit similar statistical trapping behavior. The capture and emission processes were shown to follow Poisson statistics for electron trapping in the High-k layer as well as for hole trapping in the interfacial oxide layer. A simple trapping model based on the emission and capture time concept [3] was used to describe the behavior of a distribution of identical defects under arbitrary stress conditions, resulting in a number of important observations. Under AC stress, the occupancy level of the defects is found to be set by the equilibrium condition between charge capture during the stress half-cycle of the AC signel and the charge emission during the relaxation half-cycle of the AC signal. The capture to emission time ratio, t_e/t_c, was found to be the key parameter that determines the occupancy level at equilibrium. The occupancy level at equilibrium becomes frequency independent at high frequencies ($f \gg [1/t_c, 1/t_e]$), such that it becomes possible to predict the high frequency behavior at operation conditions (~GHz) on the bases of BTI measurements at relatively low frequency (~kHz). Under a given stress mode, the occupancy of individual defects reaches an equilibrium state that is controlled by the capture and emission times of the defect, only. For different stress modes, different stress-mode specific "steady-state" values for the V_t-shift are established at long times. Excellent agreement between calculated and measured PBTI shifts under various AC stress conditions was demonstrated for large area FETs, demonstrating clearly that the macroscopic BTI behavior is predominantly governed by the microscopic response of individual defects. The model was shown to accurately describe the PBTI behavior also for non-periodic mode-switching. Finally, the limitations of the described model have been discussed. BTI lifetime predictions are not possible at this time with in the proposed modeling framework.

APPENDIX: DERIVATION OF DEFECT OCCUPANCY UNDER ARBITRARY STRESS MODEL.

As discussed earlier, the occupancy of defects follows Poisson statistics and can be modeled using an equivalent RC circuit as shown in figure 5. The stress signal can be generalized by assuming different "Up" time T1 (at stress) and "Down" time T2 (at relax). When $T_2 = 0$, the stress represents the DC case. When $T_2 \neq 0$, it becomes AC stress case with frequency

$$f = 1/(T_1 + T_2)$$

and

$$dutycycle = 1 - \frac{T_2}{(T_1 + T_2)}$$

At the end of the first stress phase, the occupancy of defect with capture time of τ_c and emission time of τ_e,

$$P_{\tau_c \tau_e, 1} = 1 - \exp\left(-\frac{T_1}{\tau_c}\right) \qquad [11]$$

The occupancy of the defect at the end of the following relax phase,

$$P'_{\tau_c \tau_e, 1} = \left(1 - \exp\left(-\frac{T_1}{\tau_c}\right)\right) \times \exp\left(-\frac{T_2}{\tau_e}\right) \qquad [12]$$

Similarly, at the end of the (n+1)th stress phase,

$$P_{\tau_c \tau_e, n+1} = 1 - \exp\left(-\frac{T_1}{\tau_c}\right) + P_{\tau_c \tau_e, n} \exp\left(-\left(\frac{T_1}{\tau_c} + \frac{T_2}{\tau_e}\right)\right) [13]$$

Equation 13 shows that the occupancy of defect at end of the each stress phase follow geometric progression with a common ratio of $\exp\left(-\left(T_1/\tau_c + T_2/\tau_e\right)\right)$. Combining equation 13 and 11, we have the defect occupancy at the end of nth stress phase as a function of the number of stress cycles (n), the "Up" time T_1 and "Down" time T_2.

$$P_{\tau_c \tau_e, n} = \left(\left(1 - \exp\left(-\frac{T_1}{\tau_c}\right)\right)\left(1 - \frac{1}{1 - \exp\left(-\left(\frac{T_2}{\tau_e} + \frac{T_1}{\tau_c}\right)\right)}\right)\right) \qquad [14]$$

$$\times \exp\left(-\left(\frac{T_2}{\tau_e} + \frac{T_1}{\tau_c}\right)\right)^{n-1} + \frac{\left(1 - \exp\left(-\frac{T_1}{\tau_c}\right)\right)}{1 - \exp\left(-\left(\frac{T_2}{\tau_e} + \frac{T_1}{\tau_c}\right)\right)}$$

and the defect occupancy at the end of the following relax phase,

$$P'_{\tau_c \tau_e, n} = P_{\tau_c \tau_e, n} \exp\left(-T_2/\tau_e\right) \qquad [15]$$

ACKNOWLEDGMENT

This work was performed by the research Alliance Teams at various IBM Research and Development Facilities. Many people contributed to various aspects of this work, and we especially acknowledge Frederic Monsieur, Miaomiao Wang, Sufi Zafar.

REFERENCES

[1] S. Lee, et al., IEDM, (2009)
[2] C. M. Chang, et al., IEDM (2009)
[3] H. Reisinger, et al., IRPS, (2010)
[4] T. Grasser. et al., IRPS, (2010)
[5] J. P. Campbell, et al., IRPS, (2009)
[6] L. B. Freeman, et al., Solid State Electron, 13, 11, (1970)
[7] W. Goes, et al., IEEE TDMR, (2008)
[8] D. Veksler, et al., IRPS, (2010)
[9] P. Vashishta, et al., Phys. Rev. B, Condens. Matter, 41, 17, (1990)
[10] G. Bersuker, et al., IEDM, (2008)
[11] K. Zhao, et al., IRPS, (2010)
[12] M. Li, et al., Jpn. J. Appl. Phys. 43 (2004)
[13] T. Grasser, et al., IEDM (2007)
[14] M. Alam, et al., IEDM (2003)
[15] S. Tan, et al., Appl. Phys. Lett. 82 (2003)
[16] J. H. Stathis, et al., IRPS (2004)
[17] E. Cartier, et al., IEDM (2007)
[18] M. Jo, et al., Microelectronic Engineering 84 (2007)

Understanding of Traps Causing Random Telegraph Noise Based on Experimentally Extracted Time Constants and Amplitude

Kenichi Abe[1], Akinobu Teramoto[2], Shigetoshi Sugawa[1], and Tadahiro Ohmi[2, 3]

1 Graduate School of Engineering
2 New Industry Creation Hatchery Center
3 World Premier International Research Center
Tohoku University
6-6-10, Aza-Aoba, Aramaki, Aoba-ku, Sendai, Japan, 980-8579
Phone: +81-22-795-3977, teramoto@fff.niche.tohoku.ac.jp

Abstract— **We develop a high-speed method to extract time constants and noise amplitude of random telegraph noise (RTN). We investigate distributions of these RTN parameters for more than 270 n- and p-MOSFETs and clarify spectroscopy of traps causing RTN. Most of traps are distributed in an energy range of 220 meV, and mean times to capture/emission are measured in a wide range between 10 μs and 20 ms.**

Keywords-random telegraph noise, trap, time constant, energy distribution

I. INTRODUCTION

Random telegraph noise (RTN) of metal oxide semiconductor field effect transistors (MOSFETs) has been investigated energetically in recent years because it has a singular characteristic which relates to single electron trapping/detrapping physics and it gives us beneficial information on a trap spectroscopy of the gate dielectric film.

Moreover, RTN has a negative impact on reliable operations of flash memory [1] and static random access memory (SRAM) [2, 3]. The traps causing RTN is also suspected as a cause of negative bias temperature instability (NBTI) in MOSFETs with a high-k/Metal Gate (MG)-gate stack [4]. Therefore, the importance of RTN analysis has become greater.

The transition process of a carrier between the channel region and the trap is still unclear. We focus on time constants of RTN, which represent trapping/detrapping probability and we measure them for numerous samples by a newly developed method to describe the trap spectroscopy experimentally. We have reported that the energy distribution of traps in the gate insulator can be evaluated by an extracting only the ratio of time constants in our newly developed test pattern [5]. However, the time constants themselves cannot be extracted by the previous method because of the long data sampling period. In this paper, very short sampling period of 1 μs was realized and this allows us to understand relationships among RTN parameters which include its amplitude, transition rate, and time constants

themselves and physical description of RTN origins. Moreover, samples with different RTN characteristics are investigated by the same way and we clarify the differences precisely.

II. ANALYTICAL APPROACH

We employ an arrayed device under test (DUT) test pattern previously reported by [6] and a newly measurement system to measure numerous transistors (Fig. 1). In this system, RTN waveform appears as source voltage fluctuation at each cell under applied constant drain current.

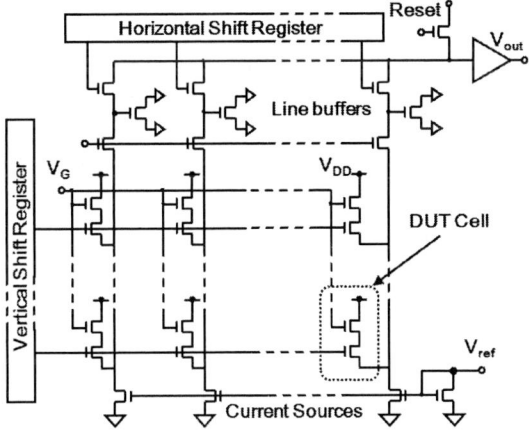

Figure 1. Schematic of the test circuit. The chip includes 131,072 nFETs and 81,920 p-MOSFETs as DUT transistors.

Two-level type RTN is characterized by only three parameters, which are mean time to capture ($<\tau_c>$), mean time to emission ($<\tau_e>$), and amplitude (ΔV_{gs}). The time constants correspond to two physical states of a trap, that is τ_c and τ_e represent spans in low V_{gs} level (carrier trapping state) and those in high V_{gs} level (carrier emission state) (Fig. 2(a)). The RTN amplitude ΔV_{gs} is defined as a difference between two normal distributions in a voltage histogram (Fig. 2(b)).

978-1-4244-9113-1/11 $26.00 © 2011 IEEE

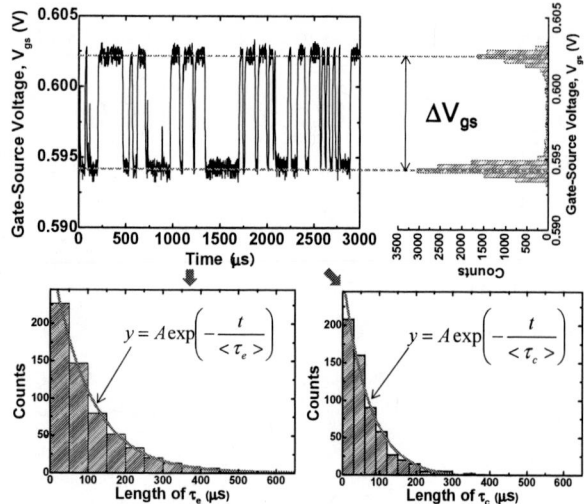

Figure 2. Definitions and extractions of two time constants ($<\tau_c>$, $<\tau_e>$) and amplitude (ΔV_{gs}) from RTN waveform data.

To extract time constants accurately, we select a particular cell showing RTN by pseudo-addressing with two shift registers and digitize the noise signal by an oscilloscope with fast sampling rate and long record length. A determination of RTN existence uses two criteria, which consists of over 1.7 mV of ΔV_{gs} due to limitation of the instruments accuracies and over 80 jumps between high- and low-level. In this work, we measured the test circuit under 1 MHz of the sampling rate and 10^6 points (1 s) of the record length, respectively. We extract the time constants by fitting the distributions of τ_c and τ_e to the exponential distribution ($Ae^{-t/<\tau>}$) because the phenomenon is ruled by Poisson process.

The test chip includes 131,072 n-MOSFETs and 81,920 p-MOSFETs with 0.22 μm of gate length, 0.28 μm of gate width, and 5.7 nm wet thermal gate oxide. It was fabricated by 0.22 μm logic CMOS technology.

III. RESULTS AND DISCUSSIONS

Fig. 3 shows a histogram of E_t-E_{fn} where E_t is the trap energy, E_{fn} is quasi-Fermi energy of electron. We can calculate the trap energy E_t from τ_c and τ_e by a following equation,

$$\frac{\langle \tau_c \rangle}{\langle \tau_e \rangle} = g \cdot \exp\left\{ \left(\frac{E_t - E_{fn}}{kT} \right) \right\} \quad (1)$$

where g is the degeneracy, k is Boltzmann constant, and T is the temperature [7]. In this graph, the distribution has a symmetrical form at the point of $E_t = E_{fn}$ and E_t-E_{fn} spreads to about ±110 meV. We have reported the histogram E_t-E_{fn} for a similar measurement condition (I_d=0.1 μA) and different record period [5]. Fig. 4 shows the histogram E_t-E_{fn} reported in [5]. The record period of 1 s in this experiment is much shorter than that of 4,200 s in [5]. The distributions of the detected trap energy are different between two conditions. Fig. 5 shows the trap energy distributions with the band diagram of the silicon channel region at the source edge for

both record periods. The bottom energy of the conduction band at in the channel at the source edge is 12.3mV, which is simulated by the device simulator (SILVACO, Inc. ATLAS). The sub-band energy E_0 (Lowest energy) and E_1 (second lowest energy) are calculated by following equation [8].

$$E_j = \left\{ \frac{3hq\varepsilon_s}{4\sqrt{2m_x}} \left(j + \frac{3}{4} \right) \right\}^{\frac{2}{3}}, j = 1,2,3 \cdots \quad (2)$$

Where, ε_s is electric field. E_j is j-th's sub-band energy, h and m_x are Planck's constant and effective mass of electrons, respectively. The E_0 and E_1 are 52.3 and 82.8 mV, respectively. The detected traps at only higher energy than E_{fn} increase with the sampling period though those at lower energy than E_{fn} do not increases. These suggest that the trap having a higher energy than E_{fn} is hard to detect in short record period because the high energy carrier density in the channel region is smaller than that near E_{fn}, and the number of traps located at lower energy level than E_{fn} are less than that at higher energy.

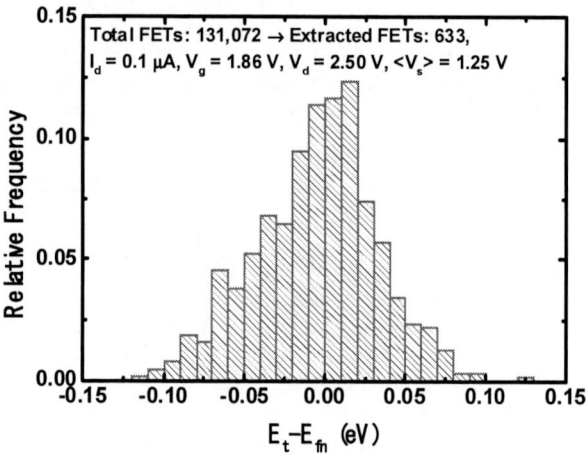

Figure 3. Histogram of E_t-E_{fn} for n-MOSFETs calculated from $<\tau_c>/<\tau_e>$. Sampling cycle=1 μs, Record Period=1 s.

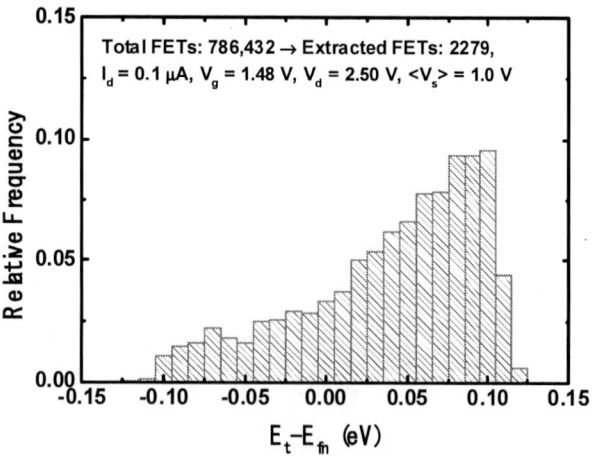

Figure 4. Histogram of E_t-E_{fn} for n-MOSFETs calculated from $<\tau_c>/<\tau_e>$. Sampling cycle=0.7 s, Record Period=4,200 s [5].

978-1-4244-9113-1/11 $26.00 © 2011 IEEE

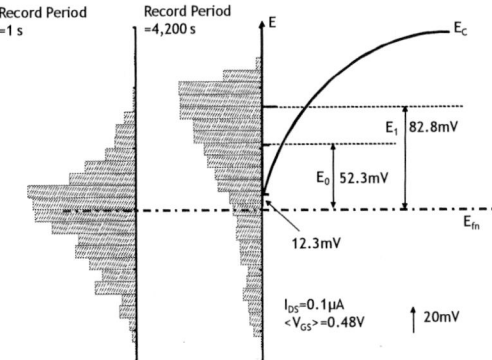

Figure 5. Trap energy distributions with the band diagram of the silicon channel region for the measurement record periods of 1 s and 4,200 s.

Fig. 6 show distributions of $\langle\tau_c\rangle$ and $\langle\tau_e\rangle$ for n-MOSFETs. Both of them distribute based on a power function in the wide range of 10 µs to 20 ms because of the variability of the trap energy and position. In these graph, the vertical axis also corresponds to a trap density $N_t(E_t, x_t)$ where x_t is trap depth from the interface between Si and the gate dielectric. In our experiments, numerous traps where are near the interface and which have small E_t-E_{fn} are observed and they may affect actual circuit operations and reliabilities.

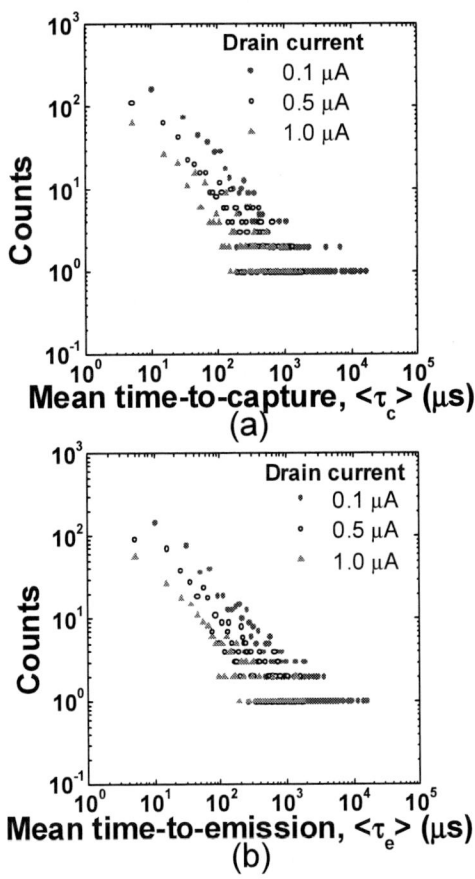

Figure 6. Distributions of time constants of (a) $\langle\tau_c\rangle$ and (b) $\langle\tau_e\rangle$ for three drain currents.

These experimental results can be used for a prediction of occurrence frequency of problematic time constants in an actual circuit design such as SRAM [9]. Furthermore, the distributions move to shorter as drain current increases. It can be explained by a simple relation,

$$\tau_c = \frac{1}{n\sigma v_{th}} \tag{3}$$

where n is the carrier density in the inversion layer, σ is the capture cross section, and v_{th} is the average thermal velocity of the carriers [7]. Increase in the drain current means increase in the carrier density (n) and it results in decrease in τ_c and increase in the transition rate averagely. Therefore, the relationship between τ_c and τ_e shows positive correlation (Fig. 7). This means that the trapping and detrapping process are not independent of each other. In other words, they almost occur between the inversion layer and the trap site, the case of detrapping to the gate electrode rarely occurs.

Fig. 8 show scatter plots between number of jumps N_J and $\langle\tau_c\rangle$, $\langle\tau_e\rangle$. Negative correlation appears because the inverses of time constants represent the transition rate. Fig. 9 shows the relationship between the number of jumps N_J and E_t-E_{fn}. Here, the relation of the number of jumps, the sampling times, and the trap energy level is as follows [5, 7, 10]

$$\frac{\langle\tau_c\rangle}{\langle\tau_e\rangle} = g\cdot\exp\left\{\left(\frac{E_t - E_f}{kT}\right)\right\} = \frac{Count\_L}{Count\_H} \tag{4}$$

Where Count_L and Count_H are the data count number of the low (τ_c period) and high (τ_e period) states, respectively. In this experiment, the total sampling number is 10^6. Then,

$$Count\_L + Count\_H = 10^6 \tag{5}$$

From (3) and (4),

$$Count\_L = \frac{10^6 \times A}{1+A} \tag{6}$$

$$Count\_H = \frac{10^6}{1+A} \tag{7}$$

Where,

$$A = g\cdot\exp\left(\frac{E_t - E_f}{kT}\right) \tag{8}$$

The maximum number of jumps N_{JMAX} is indicated by (9).

$$N_{JMAX} = 2 \times \min(Count\_L, Count\_H) \tag{9}$$

As indicating (6)-(9), N_{JMAX} has to depend on the trap energy level. The shape of maximum value of N_J in Fig.9 is defined by these reasons. N_J is widely distributed in each trap energy level. It is considered that this distribution indicates the distance between traps and channels. The large and small N_J's mean the near and far between traps and channels,

978-1-4244-9113-1/11 $26.00 © 2011 IEEE

respectively. On the other hand, the minimum number of jumps was 80, then, the minimum Count_(L or H) becomes 40. From (3), the detectable E_t-E_f range become about ±200 meV. In this experiment, E_t-E_f distribution was in ±110 meV. This indicates that the E_t-E_f of detected traps in this experiment was ±110 meV.

Fig. 10 show scatter plots between ΔV_{gs} and $<\tau_c>$, $<\tau_e>$. There are not clear correlations. It seems that the trap locating on near the Si/SiO$_2$ interface causes shorter $<\tau_c>$/$<\tau_e>$ and larger ΔV_{gs} simultaneously, however ΔV_{gs} is mainly dominated not by the trap depth from the interface, but by the randomness of the channel percolation and the lateral trap location [11, 12].

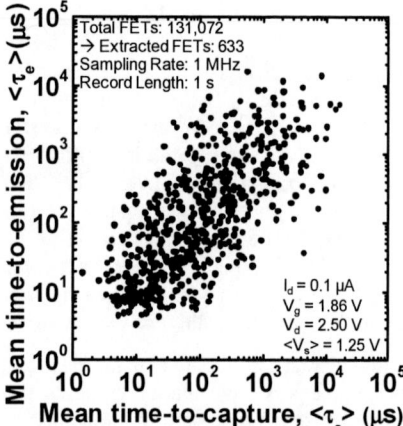

Figure 7. Scatter plot between $<\tau_c>$ and $<\tau_e>$. Positive correlation appears within the range of 10 μs to 10 ms.

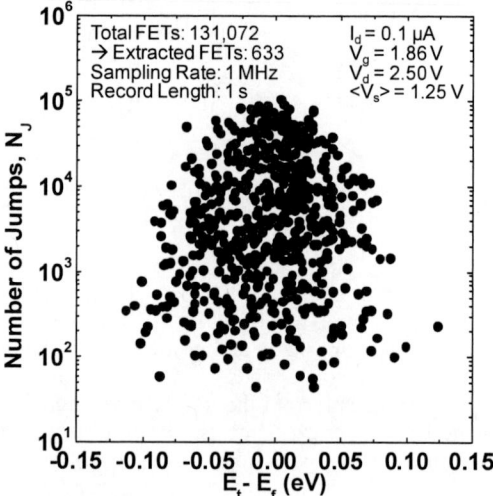

Figure 9. Relationship between the number of jumps and E_t-E_{fn}.

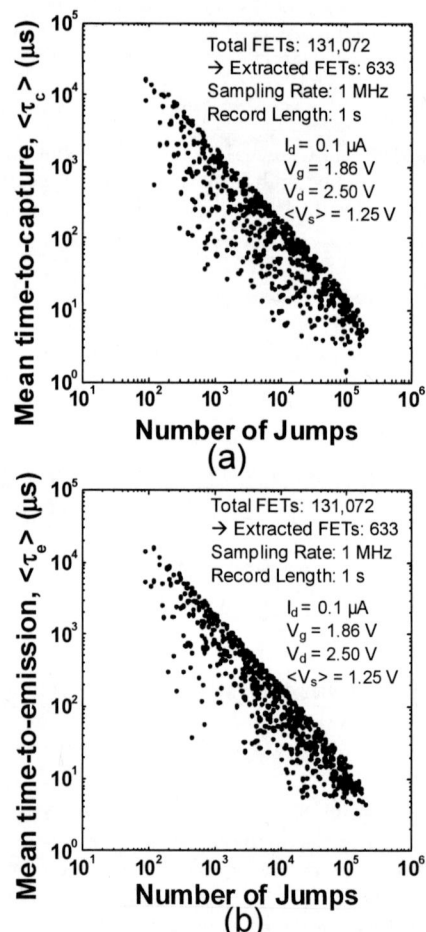

Figure 8. Scatter plots between number of jumps and (a) $<\tau_c>$, (b) $<\tau_e>$.

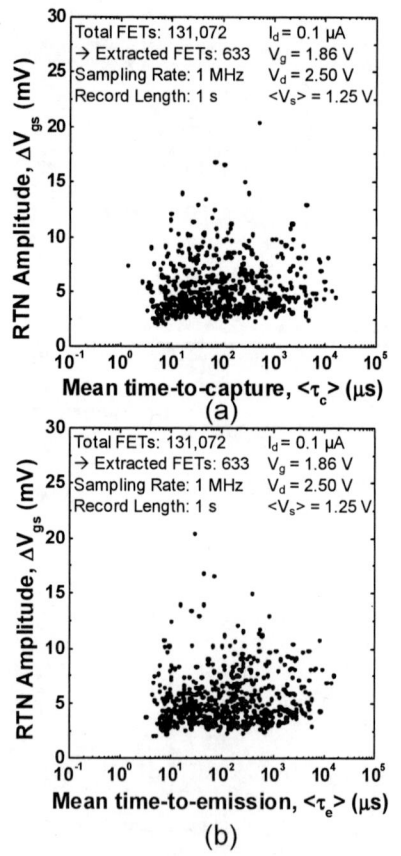

Figure 10. Scatter plots between number of jumps and (a) $<\tau_c>$, (b) $<\tau_e>$.

We can also investigate RTN for p-MOSFETs with the same system. Figure 11 indicate relative frequency distributions of the time constants for n- and p-MOSFETs. Slopes of p-MOSFET in the double logarithmic plots are smaller for $<\tau_c>$ and $<\tau_e>$ than ones of n-MOSFETs. Possible causes are considered such as differences in the trap distribution, the carrier density, and the capture cross section for electron and hole. Fig. 12 shows histograms of E_t-E_f for n- and p-MOSFETs where E_f means quasi-Fermi energy of electron and hole, respectively. A slight difference in the symmetric property can be caused by the differences in the trap distributions for electrons and holes.

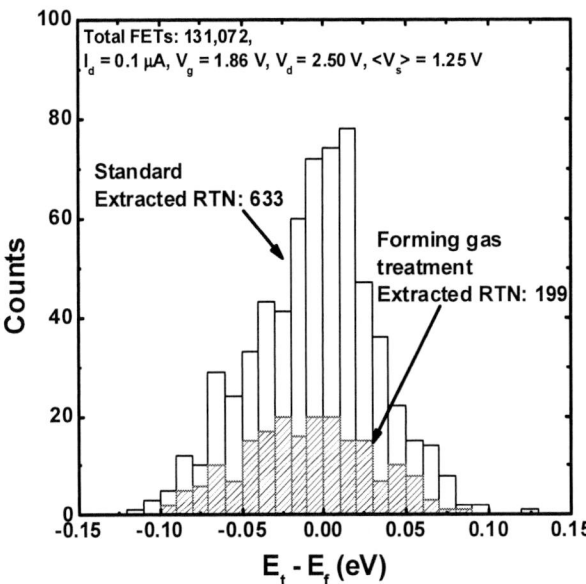

Figure 12. Comparison of histograms of E_t-E_f for n- and p-MOSFETs.

IV. CONCLUSION

We have clarified the trap spectroscopy relating to RTN from the viewpoint of $<\tau_c>$ and $<\tau_e>$ for numerous MOSFETs experimentally. The result of the positive correlation between $<\tau_c>$ and $<\tau_e>$ shows that trapping and detrapping process are same fundamentally. The amplitude ΔV_{gs} has no relation to $<\tau_c>$ and $<\tau_e>$, and ΔV_{gs} is almost ruled by randomness of trap location and channel percolation. RTN time constants of n-MOSFET and p-MOSFET are slightly different but the trap density of p-MOSFET is very small than that of n-MOSFET. With this method, RTN trap spectroscopy will be measured in a short time and we can easily obtain important information on circuit operation margins, reliabilities, and physical mechanism of RTN.

ACKNOWLEDGMENT

We would like to thank Mr. Y. Kamata and Mr. K. Shibusawa for useful discussions and manufacture of the test structure.

REFERENCES

[1] N. Tega, H. Miki, T. Osabe, A. Kotabe, K. Otsuga, H. Kurata, S. Kamohara, K. Tokami, Y. Ikeda, and R. Yamada, "Anomalously Large Threshold Voltage Fluctuation by Complex Random Telegraph Signal in Floating Gate Flash Memory," in *International Electron Devices Meeting*, San Francisco, 2006, pp. 491-494.

[2] S. O. Toh, Y. Tsukamoto, G. Zheng, L. Jones, L. Tsu-Jae King, and B. Nikolic, "Impact of random telegraph signals on V_{min} in 45nm SRAM," in *IEEE International Electron Devices Meeting*, Baltimore, 2009, pp. 767-770.

[3] M. Tanizawa, S. Ohbayashi, T. Okagaki, K. Sonoda, K. Eikyu, Y. Hirano, K. Ishikawa, O. Tsuchiya, and Y. Inoue, "Application of a Statistical Compact Model for Random Telegraph Noise to Scaled-

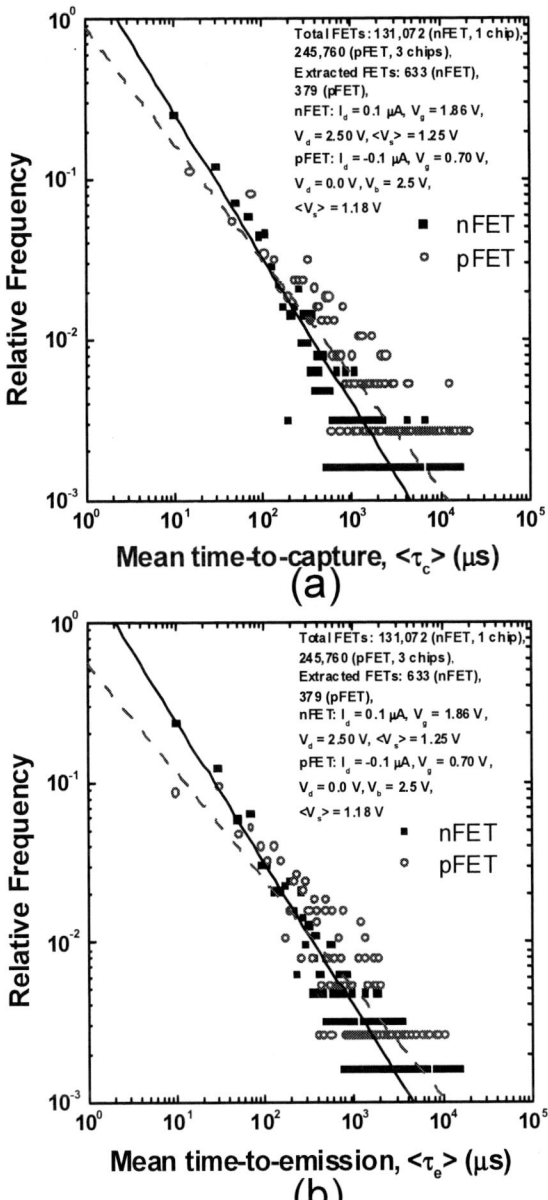

Figure 11. Relative frequency distributions of time constants of (a) $<\tau_c>$ and (b) $<\tau_e>$ for n- and p-MOSFETs.

SRAM Vmin Analysis," in *Digest of Technical Papers Symposium on VLSI Technology*, Honolulu, 2010, pp. 95-96.

[4] H. Reisinger, T. Grasser, W. Gustin, Schlu, x, and C. nder, "The statistical analysis of individual defects constituting NBTI and its implications for modeling DC- and AC-stress," in *2010 IEEE International Reliability Physics Symposium* Anaheim, 2010, pp. 7-15.

[5] A. Teramoto, T. Fujisawa, K. Abe, S. Sugawa, and T. Ohmi, "Statistical evaluation for trap energy level of RTS characteristics," in *Symposium on VLSI Technology* Honolulu, 2010, pp. 99-100.

[6] K. Abe, S. Sugawa, S. Watabe, N. Miyamoto, A. Teramoto, Y. Kamata, K. Shibusawa, M. Toita, and T. Ohmi, "Random Telegraph Signal Statistical Analysis using a Very Large-scale Array TEG with 1M MOSFETs," in *IEEE Symposium on VLSI Technology*, Kyoto, 2007, pp. 210-211.

[7] M. J. Kirton and M. J. Uren, "Noise in solid-state microstructures: A new perspective on individual defects, interface states and low-frequency (1/f) noise," *Advances in Physics*, vol. 38, pp. 367 - 468, 1989.

[8] Y. Taur and T. H. Ning, "Fundamentals of Modern VLSI Devices," 2nd ed Cambridge: Cambridge University Press, 2009, pp. 234-239.

[9] K. Takeuchi, T. Nagumo, K. Takeda, S. Asayama, S. Yokogawa, K. Imai, and Y. Hayashi, "Direct observation of RTN-induced SRAM failure by accelerated testing and its application to product reliability assessment," in *Symposium on VLSI Technology*, 2010, pp. 189-190.

[10] T. Fujisawa, K. Abe, S. Watabe, N. Miyamoto, A. Teramoto, S. Sugawa, and T. Ohmi, "Accurate Time Constant of Random Telegraph Signal Extracted by a Sufficient Long Time Measurement in Very Large-Scale Array TEG," in *IEEE International Conference on Microelectronic Test Structures*, 2009, pp. 19-24.

[11] K. Abe, A. Teramoto, S. Watabe, T. Fujisawa, S. Sugawa, Y. Kamata, K. Shibusawa, and T. Ohmi, "Experimental Investigation of Effect of Channel Doping Concentration on Random Telegraph Signal Noise," *Jpn. J. Appl. Phys.*, vol. 49, p. 04DC07, 2010.

[12] A. Ghetti, C. M. Compagnoni, A. S. Spinelli, and A. Visconti, "Comprehensive Analysis of Random Telegraph Noise Instability and Its Scaling in Deca-Nanometer Flash Memories," *IEEE Trans Electron Devices*, vol. 56, pp. 1746-1752, Aug 2009.

Mechanistic Understanding of Breakdown and Bias Temperature Instability in High-K Metal Devices Using Inline Fast Ramped Bias Test

Siddarth A. Krishnan, Eduard Cartier, James Stathis, Michael Chudzik
Semiconductor Research Development Cooperative (SRDC)
IBM
Hopewell Junction, NY, USA
1-845-892-0042, Siddarth.Krishnan@us.ibm.com

Andreas Kerber
GLOBALFOUNDRIES
Yorktown Heights, NY, USA

Abstract— Reliability Qualification has historically been a time consuming affair, taking up several months in each technology node's development cycle. The recent introduction of High-K/Metal Gates (HKMG) and the additional complexity they bring to the gate stack have placed increased demands on reliability and the reliability feedback for gate stack definition. It is demonstrated that these demands can be met with a Fast Ramped Bias Test. Applying these tests to a large variety of High-K/Metal Gate stacks, it is shown that Breakdown depends almost exclusively on time zero gate leakage. PBTI is found to depend predominantly on the Interface layer (IL) and High-K thickness, while NBTI depends most strongly on the nitrogen content in the IL.

Keywords-Reliability; TDDB; BTI; High-K; Metal Gate, Ramped Bias.

I. INTRODUCTION

With the introduction of HKMG in technology roadmaps [1,2], the complexity of reliability qualification has substantially increased. A typical HKMG gate stack has

A. An interface layer (IL) containing Nitrogen.
B. A High-K Dielectric (with Nitrogen).
C. Capping Layers (for Bandedge Threshold Voltages – V_{TH}).
D. Metal Gates.

The interplay of these disparate elements of the gatestack has a profound impact on reliability and in order to be able to fairly assess and improve the gate stack, traditional constant voltage reliability tests are largely inadequate, due to the amount of time they take to perform and analyze. In this paper, ramped bias tests are shown to be a time efficient alternative to constant voltage stress, providing equivalent information to the conventional constant voltage stress methods under the restriction of constant ramp rate. The results provide insight into physical parameters driving the various reliability parameters.

II. EXPERIMENT

The devices used in this paper are all nominal channel length devices fabricated using a state of the art gate first process[1,2], utilizing HfO2 as the high-K gate dielectric (unless stated otherwise) and TiN metal gate, with capping layers set the work function on n-Channel and p-Channel transistors (nFETs and pFETs). As part of the gate stack evaluation for performance, we changed the interface layer, high-K dielectric, and the metal gate stoichiometry. The devices all see high thermal budgets to activate the source/drain dopants. The evaluated splits range in inversion electrical thickness or Tinv from ~ 12Å to ~16Å. In order to evaluate the large number of splits generated for Time Dependent Dielectric Breakdown and Bias Temperature Instability (BTI), we use a very fast ramp based testing methodology [2,3], with both breakdown ("Ramped-BD") and BTI ("Ramped-BTI") algorithms implemented in our inline testers. The breakdown and BTI measurements were done on areas ranging from $0.01 um^2$ to $1um^2$ on nominal channel length devices. All the measurements are done at room temperature because the inline testers, configured for speed of measurement primarily operate at room temperature.

A. Breakdown:

Kerber et al [3,4] have previously shown that the Ramped-BD test is, in essence, equivalent to the traditional constant voltage test. The test is done at room temperature on 20 chips around a 300mm wafer, on three different areas. A typical Weibull plot of the breakdown voltage (V_{bd}) is shown in Fig. 1. Two tracking parameters are extracted from these distributions. One is the breakdown voltage which is read off this chart at weibit = 0 ($V_{63\%}$). The other is the slope of this distribution. The slope is equal to $\beta*(n+1)$, where β and n are the Weibull slope and voltage acceleration factor obtained from constant voltage stress[4]. In this test, higher $V_{63\%}$ values and a higher $\beta*(n+1)$ product represent longer breakdown times.

978-1-4244-9113-1/11 $26.00 © 2011 IEEE

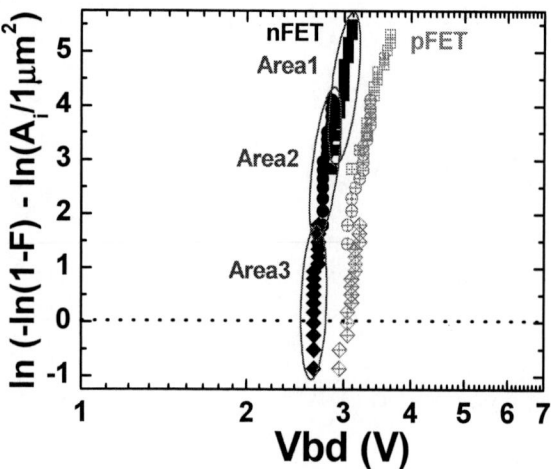

Fig. 1: Weibull plot on the Vbd-scale for nFETs and pFETs. The data from three areas is linearly fit for the slope and the intercept at y-axis = 0 is V63%

B. BTI:

Similar to the ramped-BD test, the Ramped-BTI test [5] tracks the increase of threshold voltage (V_{TH}) when the gate of the transistor is ramped up at a constant rate. The outputs of the test are the gate voltage required to reach a pre-specified shift in V_{TH} and the slope of the ΔV_{TH}-gate voltage plot in the log-log scale (Fig 2).

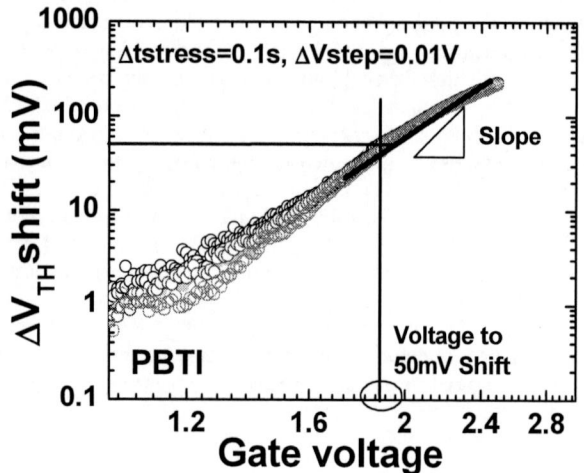

Fig. 2: Typical V_{TH} increase with ramped BTI stress (PBTI in this example). The slope of the curve and the gate voltage to 50mV Shift are outputs – the different symbol colors are different devices around the wafer.

The lifetime of the specified device at an operating voltage of 1V can be calculated from

$$t = (\frac{n}{RR*slope}) * Vg^{slope/n},$$

where n is the time evolution of the threshold voltage shift (we assume this to be 0.17 for all splits, for both NBTI and PBTI, in the following), RR is the ramp rate, slope is the slope of the ΔV_{TH}-gate voltage plot, V_g is the gate voltage required to hit the target shift. Better lifetimes are yielded by higher V_g and higher slopes.

III. RESULTS

A. Breakdown:

The time to breakdown and therefore the breakdown voltage have a very strong dependence on the initial gate leakage at a constant reference voltage[6]. Empirically, we show here that breakdown voltage is *only* dependent on the initial gate leakage. Fig. 3 shows the dependence of $V_{63\%}$ as a function of gate leakage when the interface (IL) thickness, and HfO_2 and HfZrOx (alternate high-K) thicknesses are changed. The interface layer thickness is changed from ~10Å-14Å and the High-K thickness is changed from 14-20Å. Changing the high-K thickness and IL thickness all fall on the same line for HfO_2 and HfZrOx. The gate leakage is normalized to area in Fig. 3 and 4 and spans about 3 orders of magnitude.

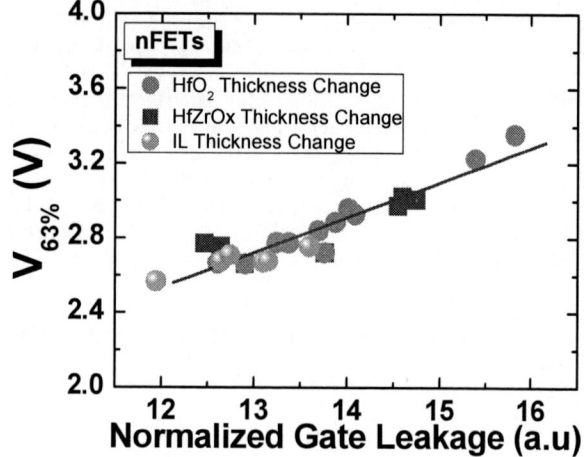

Fig. 3: V63%-Gate leakage: Interface Layer thickness change, HfO2 thickness change and HfZrOx thickness change all fall on the same line. Note: "Normalized Gate Leakage" increases with reduction in gate leakage; Normalized Gate leakage α -log(Leakage/Area).

Indeed, in evaluation of splits ranging from IL thickness, High-K type, High-K thickness, metal gate thickness, Nitrogen content in the dielectric, deposition methods and integration schemes, the dependence of $V_{63\%}$ on gate leakage is inviolate on both nFETs and pFETs (Fig. 4).

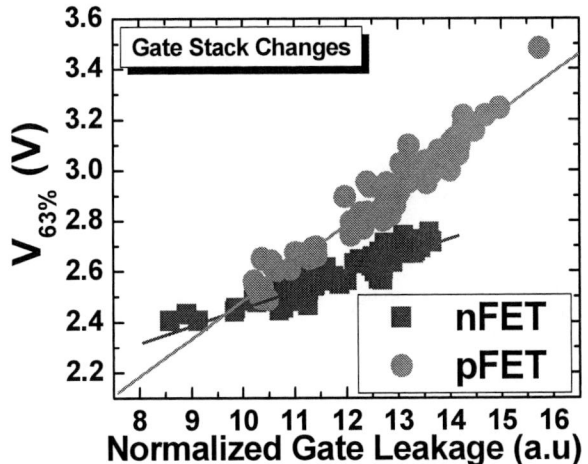

Fig. 4: Breakdown voltage, V63%, versus normalized gate leakage. Data for 114 wafers with over 50 process splits follow a universal trend for nFETs and for pFETs.

pFETs show a significantly larger dependence on gate leakage than nFETs, placing a tighter constraint on process control over the pFETs than nFETs. The inline algorithm provides a powerful tool in predicting changes in the β or n, as well, as shown in Fig. 5, where we compare the slopes of the Vbd-weibull plots on two different dielectrics (High-K 1 and High-K 2). The increase in slope on these pFETs was reflected in TDDB measurements in the voltage acceleration factor which increased from ~34 for High-K 1 to ~ 44 for High-K 2 (Fig. 6).

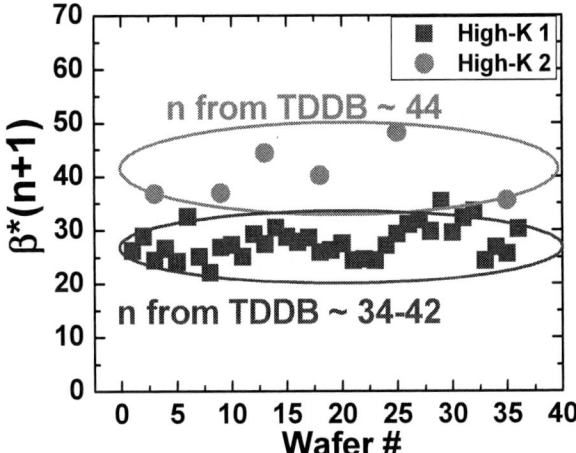

Fig. 5: The product, β*(n+1) from the inline testers was used to compare two different types of Hf containing dielectrics. The slope predicted an increase in n.

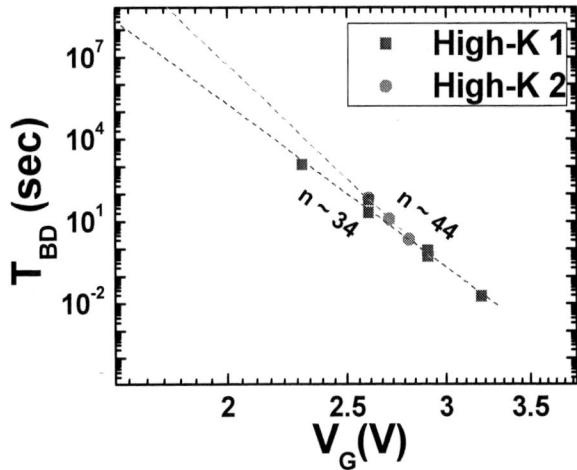

Fig. 6: TDDB measurements for selected wafers from Fig. 5. The measured voltage acceleration factor confirms the prediction from the ramped test.

The Ramped-BD test is also a powerful tool for monitoring the health of the gate oxide, as illustrated in Fig. 7, where we plot $V_{63\%}$ and the slope for close to 60 lots.

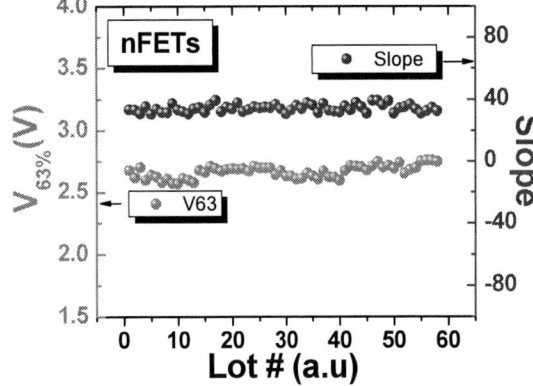

Fig. 7: The breakdown voltage and slope as a function of lot # across many yield waves. The inline algorithm is a powerful predictor of the health of the gate oxide.

B. PBTI:

PBTI: We look at the interface thickness and HfO2 thickness dependence of Positive BTI ("PBTI" on nFETs). As before, the interface thickness is varied from 10-14Å and the High-K thickness is varied from 14-20Å. IL thickness reduction reduces both the gate voltage to 50mV ΔV_{TH} and the slope simultaneously, leading to a lower predicted lifetime at operating voltage. HfO2 thickness reduction on the other hand, leads to a higher voltage to shift, but a lower slope (Fig. 8).

978-1-4244-9113-1/11 $26.00 © 2011 IEEE

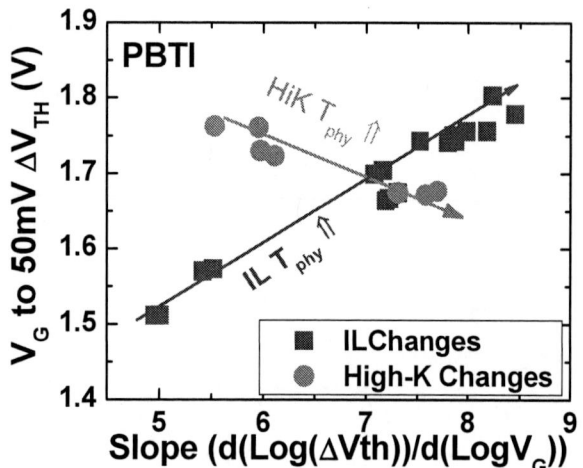

Fig. 8: PBTI: Reducing the interface layer reduces both gate voltage and slope while High-K thickness reduction improves gate voltage, but reduces slope.

Now, the slope is simply a sum of the voltage acceleration factor (n) and the slope (m) of the time evolution of PBTI. The lower slope with high-k thickness reduction means that the voltage acceleration factor reduces with thinner high-k (m<<n). This data is consistent with previous results shown by Iaonnou et al [7]. In final analysis, while the interface thickness reduction rapidly reduces lifetime, with a cliff below 12.5Å Tinv, high-k thickness reduction either changes lifetime very little or actually decreases lifetime (Fig. 9).

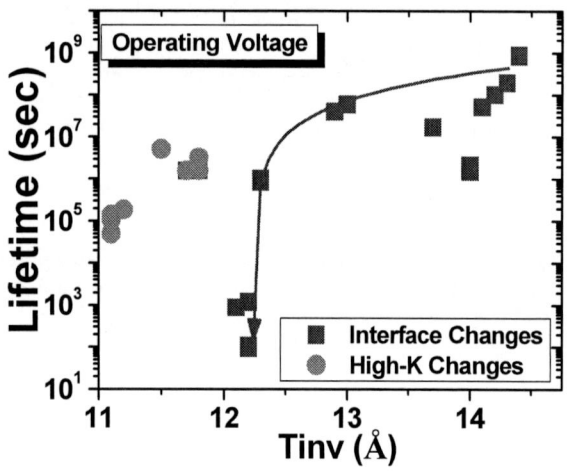

Fig. 9: PBTI: Lifetime reduces rapidly with reduction in interface thickness, whereas the reduction in lifetime is more gentle with High-k thickness change.

C. NBTI:

While High-k Thickness and IL thickness are the most important gate stack parameters for PBTI, the nitrogen in the inferface layer has been shown to be one of the most decisive parameters [8,9] for HKMG stacks. In Fig. 10, the gate voltage to 50mV ΔV_{TH} is plotted as a function of slope in devices where the nitrogen and oxygen contents in the interface are changed by changing the temperature of the thermal nitridation and thermal oxidation of the interface layer. Increasing nitrogen content reduces both the V_g to 50mV shift and the slope (voltage acceleration factor), while, in contrast, increasing oxygen content increases the V_g minimally and reduces the voltage acceleration factor.

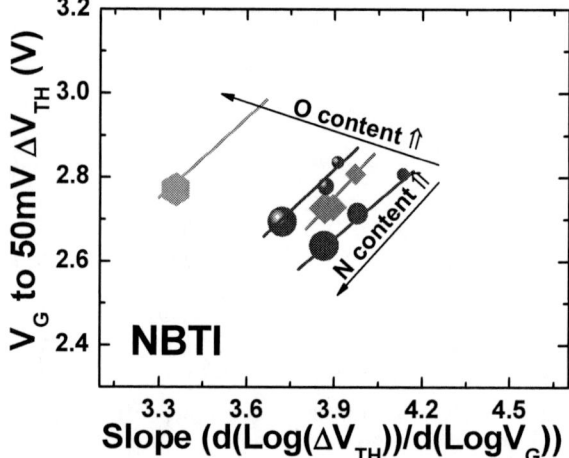

Fig. 10: NBTI, Increasing nitrogen content in the IL reduces slope and gate voltage to 50mV ΔV_{TH} while increasing oxygen content in the IL reduces the slope with a small increase the gate voltage to 50mV ΔV_{TH}.

Lifetime, at operating voltage, therefore, reduces rapidly with increasing nitrogen content and changes little with changing oxygen content (Fig. 11).

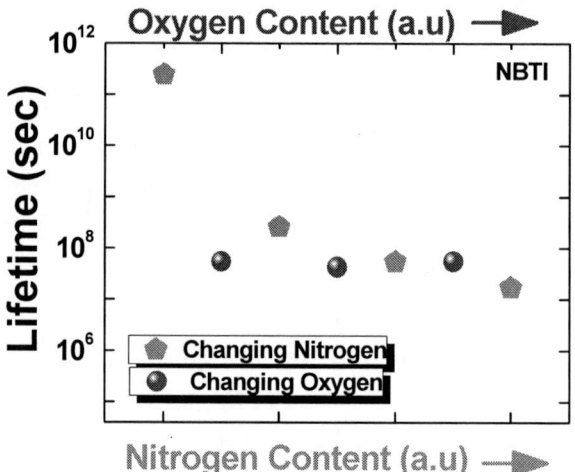

Fig. 11: NBTI: Lifetime reduces rapidly with increasing nitrogen while it does not change substantially with increasing Oxygen content.

IV. RESULTS

Reliability estimates and predictors from fast Inline algorithms have been presented. The ramped breakdown test was utilized to show that breakdown in High-K Metal Gatestacks is almost exclusively dependent on gate leakage. Interface layer thickness and high-k thickness were shown to have subtly different impacts on PBTI, while NBTI was dependent predominantly on Interface layer nitrogen content. Optimizing the gatestack for reliability is a tedious and recursive task requiring subtle engineering of many different parameters, while keeping performance requirements in mind and the ramped bias tests provide an invaluable, quick feedback process for engineers.

ACKNOWLEDGEMENTS

The authors would like to acknowledge Maryjane Brodsky, Yanfeng Wang, Barry Linder, Ernesto Shiling, Unoh Kwon, Edward Maciejewski, Keith Wong, Joseph Shepard and Min Dai of IBM, Jamie Schaeffer of GLOBALFOUNDRIES and Seiji Inumiya of Toshiba for their many contributions to this work.

REFERENCES

[1] M. Chudzik et al, Symposium on VLSI Technology, pp 194-195, 2007

[2] W. Henson et al, IEDM, pg : 1-4, Dec 2008

[3] A. Kerber et al, IEEE EDL, vol 27, # 7, pg 609-611, 2006

[4] A. Kerber et al, WODIM, June 2006

[5] A. Kerber et al, IEEE EDL vol 30, # 12, pg 1347-1349, 2009

[6] A. Kerber et al, IEEE TDMR, vol 9, # 2, pg 147-162, 2009

[7] D. Ioannou et al, IRPS, pg 1044-1048, Apr 2010

[8] J. Yugami et al, IWGI, pg 140-145, Nov. 2003

[9] S. Tsujikawa et al, IEDM, pg 824-827, Dec 2005

DESIGN OF MODIFIED ESD PROTECTION STRUCTURE WITH LOW-TRIGGER AND HIGH-HOLDING VOLTAGE IN EMBEDDED HIGH VOLTAGE CMOS PROCESS

Tai-Hsiang Lai, Lu-An Chen, Tien-Hao Tang, and Kuan-Cheng Su

ESD Engineering Department, Reliability Technology & Assurance Division
United Microelectronics Corp., Hsinchu Science Park, Taiwan

INTRODUCTION

Electrostatic discharge (ESD) protection design in high-voltage (HV) technologies becomes more and more serious reliability issue due to weak ESD robustness, high trigger voltage and low holding voltage of HV NMOS transistor. The tough ESD protection capability is a requirement for automotive or biomedical applications. On the other hand, the holding voltage of ESD device under power supply lines is smaller than the power supply voltage, which could cause a very grave latchup or latchup-like problem in HV integrated circuits (ICs) by system-level ESD test. Therefore, how to enhance the ESD performance for high-level ESD specification of customers' demand and improve the holding voltage for latchup-free solution in HV CMOS process are indeed two rigorous reliability challenges.

The embedded SCR structure within LDMOS transistor has been demonstrated that it is beneficial to improve the inherent ESD weakness of HV NMOS device [1]. However, the holding voltage of SCR structure is too low not to overcome latchup issue. For this reason, the embedded HV SCR device is not suitable for power-rail ESD clamp circuit between VDD and GND supply lines. To increase the holding voltage and decrease the turn-on voltage, the stacked-field-oxide structure with substrate-triggered technique was reported [2]. Nevertheless, a large layout area would be occupied by the ESD detection circuit. For the sake of improving the immunity against transient latchup (TLU) test, a new HV NSCR structure with CR-based ESD detection circuit was proposed [3]. By means of a HV PMOS device embedded in the HV NSCR structure, the holding voltage could be increased over the normal power supply voltage. But, the structure still needs an extra layout space for the necessary detection circuit and HV PMOS transistor.

In this work, we propose a modified structure with low trigger voltage and high holding voltage for efficient component-level and system-level ESD protections. Here, the TLP measurement, the TLU test and the technology computer-aided design (TCAD) simulation for the common and the modified HV N-channel MOSFETs are investigated in detail.

ESD STRUCTURES AND EXPERIMENTAL RESULTS

The 100-ns TLP measured I-V curve of HV gate-grounded NMOS (GGNMOS) device by Barth 4002 TLP system is shown in Fig. 1(a). From the measured result, there are both the double snapback and the soft leakage degradation characteristics in HV GGNMOS. The weakness of HV GGNMOS resulted from the soft leakage degradation problem was discussed in 2009 IRPS [4]. After the first snapback voltage at 44.4V, the device snaps back to 31.4 V and then the device quickly enter the second snapback. At the moment, the final holding voltage drops to only 8.2V. Such low holding voltage is very subject to latchup or latchup-like damage induced by ESD gun test. Moreover, the second breakdown current (It2) by TLP instrument and the HBM ESD level by Thermo Keytek MK2 tester are only 1.1 A and 0.8 kV, respectively. By the scanning electron microscopy (SEM) analysis, the contact spiking was found at the drain side of only one finger in the multi-finger HV GGNMOS with 300-μm device width. It implies that the HV GGNMOS can be

easy to be damaged in the corner of shallow trench isolation (STI) under high-energy ESD bombardment and has the non-uniform turn-on issue in the multi-finger structure.

Since the embedded HV technology is used for the system on chip (SOC) design of TFT LCD driver IC in small panel application, the process must include LV (1.5V) devices for SRAM circuit, MV (5.5V) devices for source driver circuit, and HV (20V) devices for gate driver circuit. In order to search out a dual-purpose HV ESD protection structure with latchup-free and higher ESD robustness, we apply the N-well implant of LV and MV PMOS device region in the drain side of HV NMOS device. However, the N-well implant is inherent in the embedded HV process, so we do not need to increase the additional mask layers or modify the process steps to obtain the revised device. The layout top view of the modified HV ESD structure is exhibited in Fig. 2(a), and the N-well layer can not be drawn beyond the drain diffusion layer in the y-axis direction for preventing the close N-well/N-drift junction from affecting the device breakdown voltage substantially. Fig. 2(b) displays the device cross-sectional view along A-A' line in Fig. 2(a). The larger the implant energy is, the deeper the depth is. This is the reason why the N-well is underneath the N-drift.

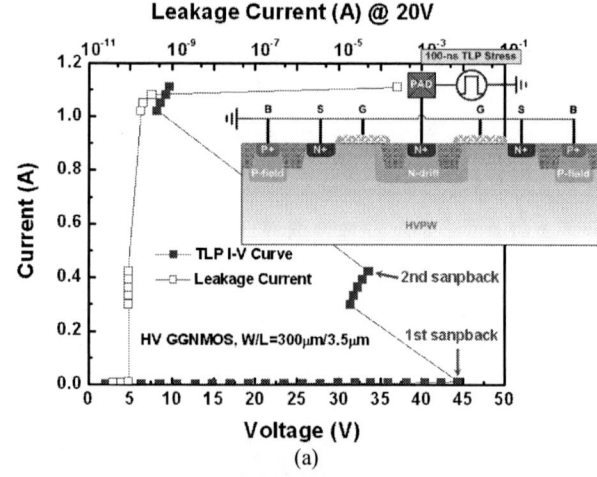

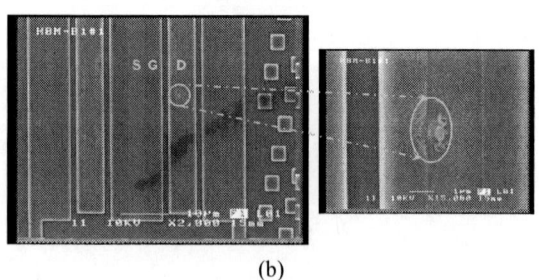

FIGURE 1. (A) THE 100-NS TLP-MEASURED I-V CHARACTERISTIC OF HV GGNMOS DEVICE WITH DEVICE WIDTH OF 300 μ m. (B) THE SEM FAILURE PICTURE OF HV GGNMOS AFTER 0.8-KV HBM ESD TEST.

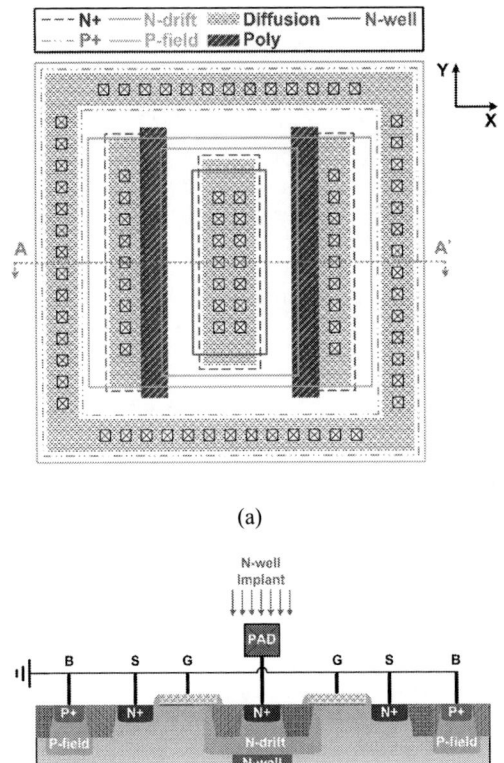

(a)

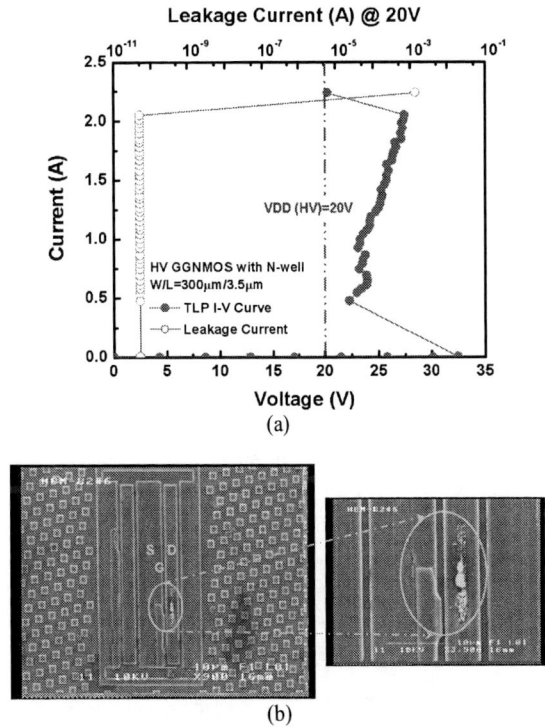

(a)

(b)

FIGURE 2. (A) THE LAYOUT TOP VIEW AND (B) THE DEVICE CROSS-SECTIONAL VIEW ALONG A-A' LINE OF THE MODIFIED HV GGNMOS STRUCTURE WITH N-WELL IN DRAIN REGION.

The 100-ns TLP measured I-V characteristic of modified HV GGNMOS device with N-well in the drain side is exhibited in Fig. 3. The result shows that the double snapback characteristic of HV GGNMOS device is suppressed by the N-well implant. The abrupt increase issue of leakage current after the snapback in HV GGNMOS is also solved completely because of the N-well implant in the modified HV GGNMOS device. Furthermore, the trigger voltage (Vt1) is reduced from 44.4V to 32.5V. This is because the concentration of N-well implant is higher than that of N-drift implant and then the avalanche breakdown formation is shifted from the N-drift/HVPW junction to the N-well/HVPW junction. Besides, the It2 level is increased from 1.1A to 2.1A as a result of the N-well effect. It is more important to see that the device holding voltage (Vh) of 22.3V is safely above the power supply voltage of 20V. Hence, the proposed ESD protection device can have high latchup immunity against system-level test. After 3.2-kV HBM ESD test, there was an obvious damage region in the modified HV GGNMOS structure because the device was burned out from drain to source, as shown in Fig. 3(b). The failure photograph hints that the N-well implant in the drain side of HV GGNMOS will enable the ESD current to pass through the deeper HVPW region and consequently avoid the current crowding at STI edge. From the HBM ESD test results, as shown in Fig. 4, the HBM ESD robustness of HV GGNMOS without N-well is not linearly increased with the device channel width due to non-uniform turn-on issue which exists in the multi-finger HV GGNMOS structure. On the contrary, the HBM ESD level could be successfully multiplied per scaling up the device channel width of revised HV

GGNMOS structure with N-well implant. It is very helpful to achieve the demanding ESD requirement through modulating the device dimension of proposed ESD structure. For example, the device channel width of modified HV GGNMOS may be designed as 750μm for 8kV HBM ESD specification.

(b)

FIGURE 3. (A) THE 100-NS TLP-MEASURED I-V CHARACTERISTIC OF PROPOSED HV GGNMOS STRUCTURE WITH N-WELL IN DRAIN REGION UNDER DEVICE WIDTH OF 300 μ m. (b) THE SEM FAILURE PICTURE OF PROPOSED HV GGNMOS STRUCTURE WITH N-WELL IN DRAIN REGION AFTER 3.2-kV HBM ESD TEST.

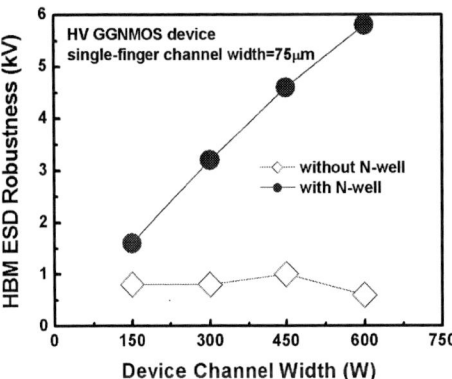

FIGURE 4. THE HBM ESD ROBUSTNESS OF HV GGNMOS WITHOUT AND WITH N-WELL IN DRAIN ZONE UNDER DIFFERENT DEVICE CHANNEL WIDTHS AND THE SAME SINGLE-FINGER CHANNEL WIDTH OF 75 μ m.

DEVICE SIMULATION AND ANALYSIS

By means of the 2-Dimension Synopsys TCAD simulator, the physical mechanism of common and modified HV GGNMOS devices can be analyzed further. Fig. 5(a) shows the simulated distribution of electric field in the HV GGNMOS under the first snapback condition, where the maximum electric-field intensity is located at the N-drift/HVPW junction. However, the maximum electric-field intensity under the second snapback condition shown in Fig. 5(b) would be transferred to the N+/N-drift junction due to the base push-out effect (Kirk effect) [5]. The considerable difference in doping concentration between N+ and N-drift causes a strong snapback occurrence for the secondary avalanche breakdown at N+/N-drift junction, so the holding voltage of HV GGNMOS would become very low. In the proposed HV GGNMOS structure with N-well implant, the avalanche generation is initiated through the N-well/HVPW junction and consequently the maximum electric-field intensity of the snapback state under the low current region is situated in the N-well/HVPW junction, as shown in Fig. 6(a). Since the N-well implant heightens the doping concentration of N-drift area, the difference in doping concentration between N+ and N-drift could be decreased substantially. This is reason why the proposed ESD structure has no double snapback phenomenon and the location of maximum electric-field intensity can remain in the N-well/HVPW junction for the snapback condition under the high current region illustrated in Fig. 6(b). Therefore, the holding voltage is not to be pulled down but to be pulled up by adding the N-well implant into the drain region of HV GGNMOS device. From the TCAD simulation results, we can clearly explain why the proposed ESD structure with N-well implant is capable of effectively enhancing the holding voltage.

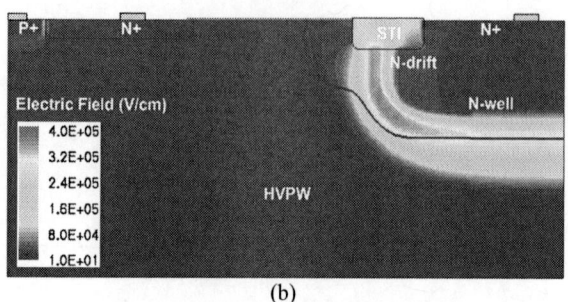

(b)

FIGURE 6. THE SIMULATED ELECTRIC-FILED DISTRIBUTIONS OF PROPOSED HV GGNMOS STRUCTURE FOR THE SNAPBACK CONDITION UNDER (A) LOW CURRENT REGION (B) HIGH CURRENT REGION.

TRANSIENT LATCHUP (TLU) MEASUREMENT

To precisely simulate the system-level ESD-induced noises on the power lines of CMOS ICs during the normal circuit operation condition, a component-level TLU measurement setup [6]-[7[is used in this work. The measurement setup to investigate TLU susceptibility of the device-under-test (DUT) is shown in Fig. 7. Here, the charging voltage with positive and negative polarities stored on 200-pF capacitor can generate the induced noise to trigger the DUT biased at the power supply voltage of 20V into the latch state when the relay is closed. The purpose of the 5-Ω current-limiting resistor (R) is to protect the DUT from electrical-over-stress (EOS) damage under such high-current latchup status. Although the current-blocking diode (D) can be utilized to avoid the damage to the power supply during TLU test, it has been demonstrated that the TLU measurement setup with the current-blocking diode will overestimate the TLU immunity level of the DUT [7]. Hence, the current-blocking diode will be not used in this paper for the accuracy of TLU level. The measured voltage waveforms of common and modified HV NMOS structures at X node under TLU test are shown in Fig.8 and Fig. 9, respectively. From the measured results of Figs. 8(a) and 8(b), the voltage waveform of HV GGNMOS is initially kept at the power supply voltage of 20V and then clamped down at ~10V after the transient trigger with Vcharge of +75V or -51V. The clamped voltage level of HV GGNMOS is close to the holding voltage measured by TLP system. The TLU test has confirmed that the HV GGNMOS is susceptible to transient-induced latchup hazard due to such low clamped or holding voltage which is much smaller the power supply voltage of 20V. In Figs. 9(a) and 9(b), the observed waveforms show that the clamped voltage waveform of the proposed HV GGNMOS with N-well implant can still come back to the power supply voltage of 20V after the transient trigger with Vcharge of +85V or -72V. Therefore, there are not any TLU-induced failure issues in the modified HV NMOS with N-well implant under system-level test.

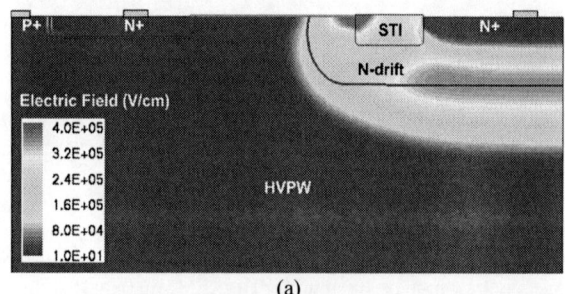

(a)

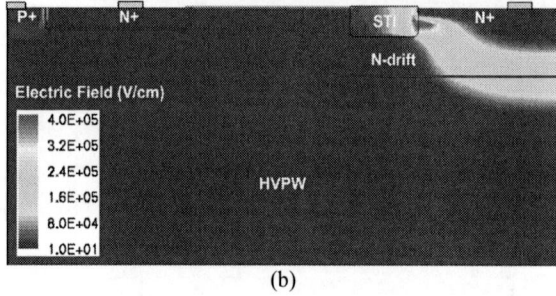

(b)

FIGURE 5. THE SIMULATED ELECTRIC-FILED DISTRIBUTIONS OF HV GGNMOS STRUCTURE UNDER (A) THE FIRST SNAPBACK CONDITION (B) THE SECOND SNAPBACK CONDITION.

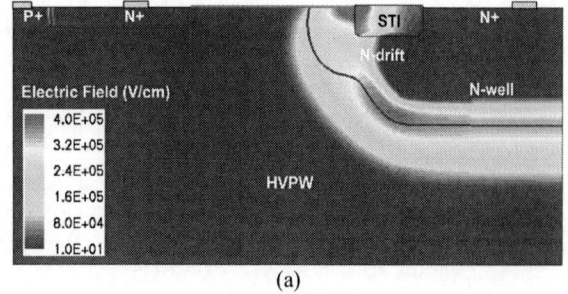

(a)

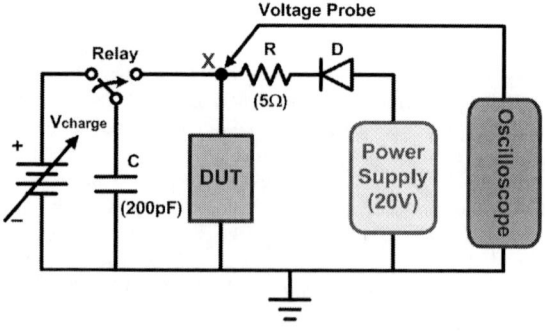

FIGURE 7. THE TRANSIENT LATCHUP MEASUREMENT SETUP FOR COMPONENT LEVEL SUPPLY TRANSIENT STIMULATION.

978-1-4244-9113-1/11 $26.00 © 2011 IEEE

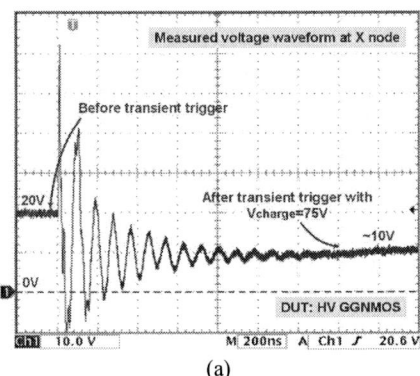

(a)

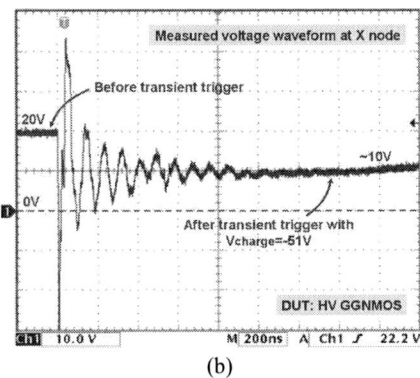

(b)

FIGURE 8. THE MEASURED VOLTAGE WAVEFORMS OF HV GGNMOS UNDER TLU TEST WITH VCHARGE OF (A) +75V (B) -51V.

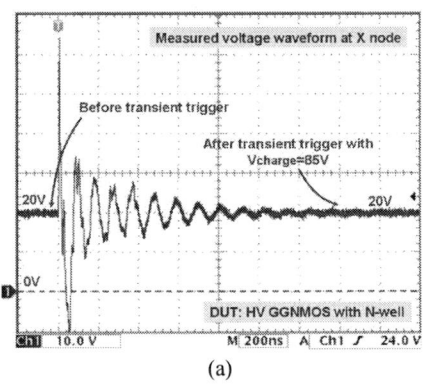

(a)

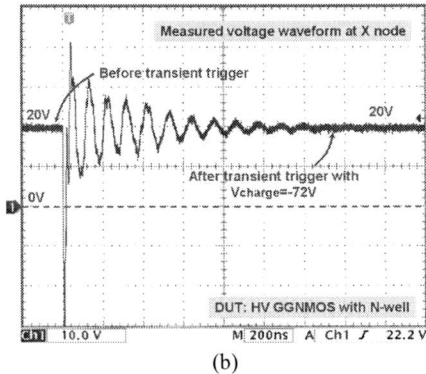

(b)

FIGURE 9. THE MEASURED VOLTAGE WAVEFORMS OF MODIFIED HV GGNMOS WITH N-WELL IMPLANT UNDER TLU TEST WITH VCHARGE OF (A) +85V (B) -72V.

TABLE I

COMPARISON OF TLP DATA, HBM/MM ESD ROBUSTNESS, AND TLU IMMUNITY BETWEEN HV GGNMOS STRUCTURE WITH AND WITHOUT N-WELL IMPLANT IN DRAIN REGION

HV GGNMOS, W=300μm		without N-well	with N-well
TLP Measurement	Vt1 (V)	44.4	32.5
	Vh (V)	8.2	22.3
	It2 (A)	1.1	2.1
ESD Test	HBM (kV)	0.8	3.2
	MM (V)	150	250
TLU Measurement	Positive TLU Level	+75	+85
	Negative TLU Level	-51	-72
Occurrence of TLU		Yes	No

CONCLUSION

The modified ESD protection structure with N-well implant in the drain region has been proposed and investigated in this paper. Table I lists the comparison of TLP, HBM/MM ESD robustness, and TLU immunity between HV GGNMOS structure with and without N-well implant. By the influence of N-well implant, many drawbacks such as double snapback, soft leakage degradation, non-uniform current conduction, low ESD robustness, and weak TLU immunity in the common HV GGNMOS device are overcome efficaciously. Without additional mask or process cost, the proposed ESD device with low trigger voltage and high holding voltage is effectively employed for power clamp protection in HV CMOS ICs without latchup or transient-induced latchup damage.

REFERENCES

[1] J.-H. Lee, J.-R. Shih, C.-S. Tang, K.-C. Liu, Y.-H. Wu, R.-Y. Shiue, T.-C. Ong, Y.-K. Peng and J.-T. Yue, "Novel ESD protection structure with embedded SCR LDMOS for smart power technology," in *Proc. IEEE Int. Reliability Physics Symp.*, 2002, pp. 156–161.

[2] K.-H. Lin and M.-D. Ker, "Design on latchup-free power-rail ESD clamp circuit in high-voltage CMOS ICs," in *Proc .EOS/ESD Symp.*, 2004, pp. 265–272.

[3] M.-D. Ker, C.-L. Hsu, and W.-Y. Chen, "ESD protection circuit for high-voltage CMOS ICs with improved immunity against transient-induced latchup," in *Proc . IEEE Int. Symp. Physical and Failure Analysis of Intergraded Circuit*, 2010, pp. 989–992.

[4] T. Imoto, K. Mawatari, K. Wakiyama, T. Kobayashi, M. Yano, M. Shinohara, T. Kinoshita, and H. Ansai, "A novel ESD protection device structure for HV-MOS Ics," in *Proc. IEEE Int. Reliability Physics Symp.*, 2009, pp. 663–668.

[5] S. M. Sze, *Physics of Semiconductor Devices*, 2nd ed. New York: Wiley, 1981.

[6] M. Kelly, L. Henry, J. Barth, G. Weiss, M. Chaine, H. Gieser, D. Bonfert, T. Meuse, V. Gross, C. Hatchard, and I. Morgan, "Developing a transient induced latch-up standard for testing integrated circuits," in *Proc. EOS/ESD Symp.*, 1999, pp. 178–189.

[7] M.-D. Ker and S.-F. Hsu, "Evaluation on efficient measurement setup for transient-induced latchup with bi-polar trigger," in *Proc. IEEE Int. Reliability Physics Symp.*, 2005, pp. 121–128.

978-1-4244-9113-1/11 $26.00 © 2011 IEEE

An EOS-Free PNP-Enhanced Cascoded NMOSFET Structure for High Voltage Application

Shih-Yu Wang, Yao-Wen Chang, Yan-Yu Chen, Chieh-Wei He, Guan-Wei Wu, Tao-Cheng Lu, Kuang-Chao Chen, and Chih-Yuan Lu

Macronix International Co., Ltd., No.16 Li-Hsin Road, Science Park , Hsin-Chu, Taiwan, R.O.C.
Phone: 886-3-5786688; E-mail: bryanwang@mxic.com.tw

Abstract—**An EOS-free PNP-enhanced cascoded NMOSFET structure for high voltage application is proposed. By controlling the bias of the first gate of cascoded NMOSFET, the EOS-free control circuit can not only raise the holding voltage in normal high voltage (HV) operation to prevent from the threats of EOS damages, but also retain the strong ESD robustness during ESD stress. The improved EOS immunity is successfully verified by a simple EOS-emulating test in this paper.**

Keywords- EOS-Free; PNP-Enhanced; Cascoded; High Voltage Application

I. INTRODUCTION

Cascoded NMOSFET structure has been widely used in mixed voltage IC applications wherein the I/O pad must tolerant voltages in excess of V_{DD} [1-3]. As compared to the single NMOS structure, ESD current carrying capability of the cascode structure usually degrades due to the wider base region of the parasitic lateral NPN for bipolar action during ESD events. To improve the ESD robustness of the cascode structure, several approaches, including additional ESD implantation [4], substrate-triggered technique [3,5], layout optimization [6], and bias control of the couple of gates [7,8], have been proposed.

In this paper, the application of cascode configuration is extended to a HV pin of a 3V parallel flash memory IC, which has to sustain 10.5V operation for accelerated programming function. The cascode configuration can have good ESD robustness, but it is difficult to simultaneously raise the holding voltage (V_h) to be higher than 10.5V, making the HV pin always under EOS risk as the memory circuit is put in a noisy system. The HV pin may be mistakenly triggered by the noise, held at 10.5V with high conducting current, and then burn out as shown in Figure 1, the SEM picture of an EOS-failed sample.

An additional base-emitter diode of the vertical PNP is stacked in series with the cascode configuration in the product, for the purpose to further raise the trigger and holding voltages by the added forward diode, as shown in Figure 2. The TLP characteristics of both cascode structures with and without the stacked diode are depicted in Figure 3. EOS immunity of the cascode structure with stacked diode is indeed improved by the higher trigger and holding voltages, which result from the extra

voltage drop on the forward diode. However, the raised holding voltage cannot exceed 10.5V and EOS risks still exist.

Besides the increases of trigger and holding voltages, it is observed from the inset of Figure 3 that the second breakdown failure current (I_{t2}) is also improved from 3.29A to 3.77A. The ESD robustness enhancement results from the current amplification of PNP bipolar transistor, which helps to share some ESD currents direct to the substrate. The increment of leakage current, from 0.28nA to 1.3nA, which can also be observed in the inset, is the payment of the PNP-enhanced structure. Fortunately, the leakage is still far below the specification of the HV pin even under high-temperature operation.

An EOS-free control circuit for cascoded NMOSFET structure will be proposed. The control circuit can raise V_h to be higher than 10.5V in normal HV operation, and can still retain the good ESD robustness when the protection is under ESD zapping. Since V_h judged by TLP measurement is possible to be too optimistic for EOS risk evaluation in HV applications [9], additional test is usually needed to further confirm the EOS immunity. A simple EOS-emulating test, which will be also introduced in this paper, is implemented to demonstrate that the proposed protection can successfully escape from being latched at 10.5V even if triggered by emulated noise. As compared to transient latch-up test [10], the EOS-emulating test is a simpler implementation.

II. EXPERIMENTAL RESULTS

A. First-gate bias for maximizing the holding voltage

Under ESD conditions, the bias of the first gate in the cascode configuration plays an important role on its ESD performance [2,11,12]. The trigger voltage (V_{tr}) can be raised as the first-gate bias increases [2,12]. It is because high first gate bias reduces E-field under this gate thus reducing impact ionization. Higher drain voltage for avalanche breakdown is needed accordingly. At the same time, its I_{t2} would deteriorate significantly with the increasing first-gate bias [11]. As first-gate bias is high, the first MOSFET device will be turned on during ESD event. Too much dissipated heat may crowd at the device surface causing the degradation of I_{t2}.

978-1-4244-9113-1/11 $26.00 © 2011 IEEE

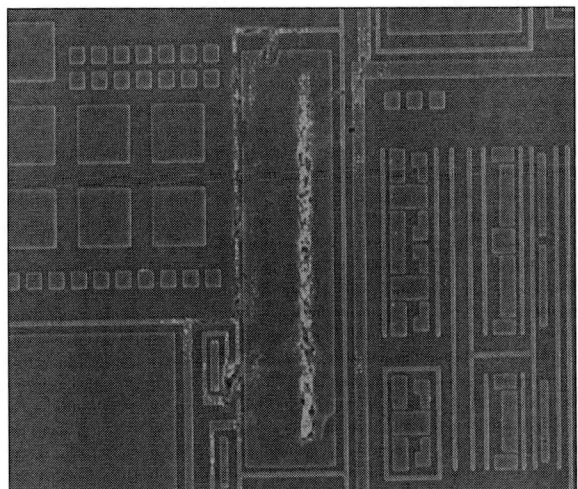

Figure 1.   The SEM picture of a failed sample caused by an EOS event.

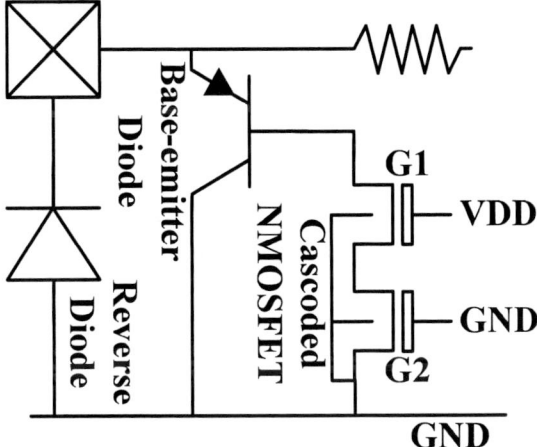

Figure 2.   The protection scheme of PNP-enhanced cascoded NMOSFET.

V_{tr}, V_h and I_{t2} of cascoded NMOSFET structures with varied first-gate bias (VG1) were measured using Barth TLP 4002 and the results are shown in Figure 4 and 5. The structures under test include cascoded NMOSFET and PNP-enhanced cascoded NMOSFET. The total width of both cascoded NMOSFET's is 400μm with channel length of 0.6μm for both gates. The space between the 1st gate and 2nd gate is 0.5μm. All the test devices were fabricated by the non-silicide process. From Figure 4, it can be observed that V_{tr} increases with the increasing VG1, which has the similar trend as the previous works. V_h has the same VG1 dependence as V_{tr}, and it can exceed 10.5V when VG1 is high enough. Unfortunately, as VG1 gets higher than 9V, as shown in Figure 5, I_{t2} of the protections downgrade dramatically.

From the above analysis, it can be derived that the cascode configuration with the first gate coupled to drain will have higher V_{tr} and V_h but very weak ESD capability. Although the holding voltage is higher than 10.5V, which makes the protection free from the risk of EOS, the degraded ESD capability will make it not a qualified protection.

B.   TCAD simulation

TCAD simulation of the cascode structure with the first gate coupled to drain is implemented to study the above characteristics under ESD conditions. Figure 6(a) shows the contours of impact ionization rate at trigger point. It can be observed that the avalanche point transfers to the junction under the second gate, from that under the first gate. Additional voltage drop across the diode-connected first device generates higher V_{tr} and V_h.

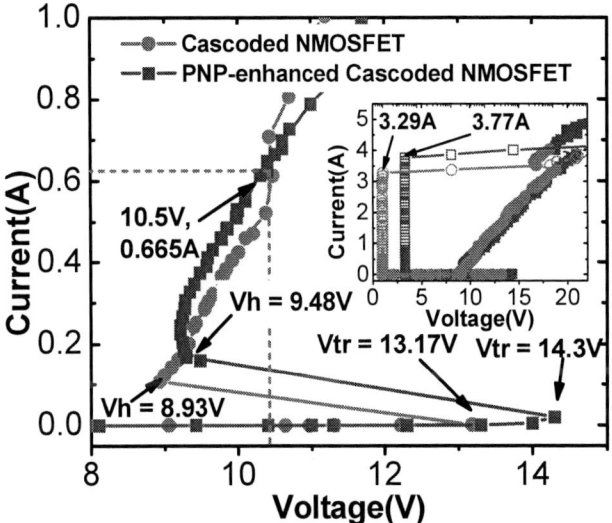

Figure 3.   TLP results of PNP-enhanced cascoded NMOSFET and cascoded NMOSFET.

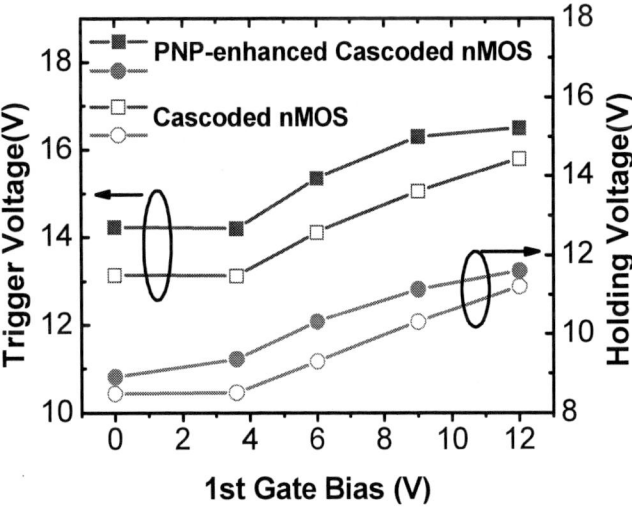

Figure 4. V_{tr} and V_h of cascoded NMOSFET structures vs VG1.

Figure 6(b) shows the current flow-lines after the protection is snapbacked. With the first gate at high bias, most ESD current conduction is through the surface channel of first MOSFET device instead of the sub-surface region by the parasitic NPN under the MOSFET device. It seems that too much dissipated heat, crowding at the surface under the first MOSFET device, significantly downgrades the ESD robustness.

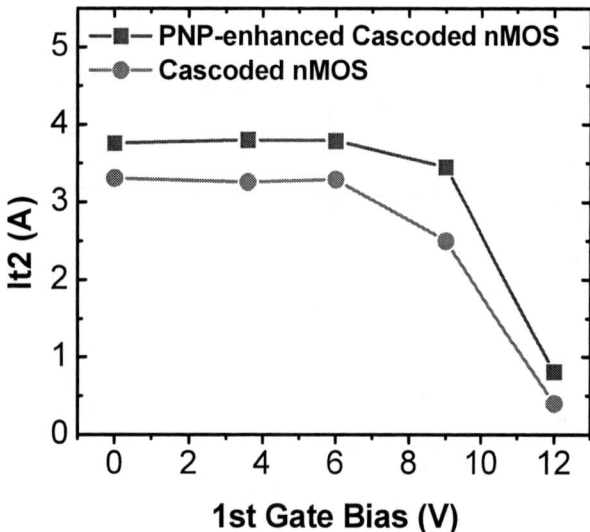

Figure 5. I_{t2} of cascoded NMOSFET structures vs VG1.

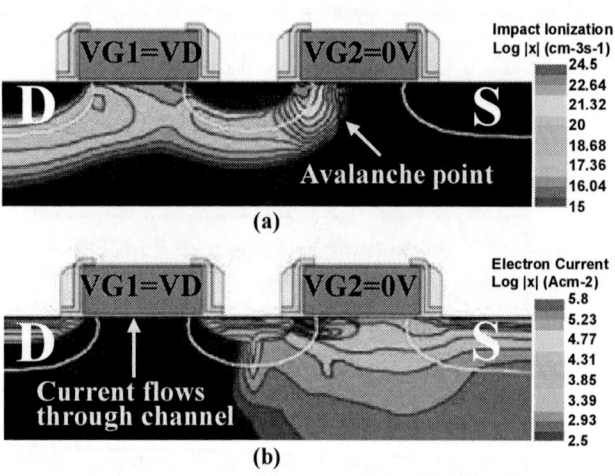

Figure 6. TCAD simulation results of cascoded NMOSFET structure. (a) contours of impact ionization rate at trigger point and (b) the current flowlines after snapbacked.

C. EOS-free control circuit for cascoded NMOSFET structure

To prevent EOS event and retain good ESD robustness as well, an EOS-free control circuit for cascoded NMOSFET structures is proposed, which is shown in Figure 7.

When the ESD protection is under normal operation with V_{DD} biased at 3.6V, the inverter composed of N2 and P2 delivers high voltage to the first gate of cascoded NMOSFET. In this condition, cascoded NMOSFET has high V_h to be free from EOS risks. On the other hand, when the protection is under ESD event, V_{DD} is floating. The gates of N2 and P2 are coupled to high voltage by ESD pulse through C1 and then the inverter will pull down the first gate of cascoded NMOSFET to

ground. NMOSFET N1 can also be replaced by a resistor to avoid the possible damage as it is under high voltage operation.

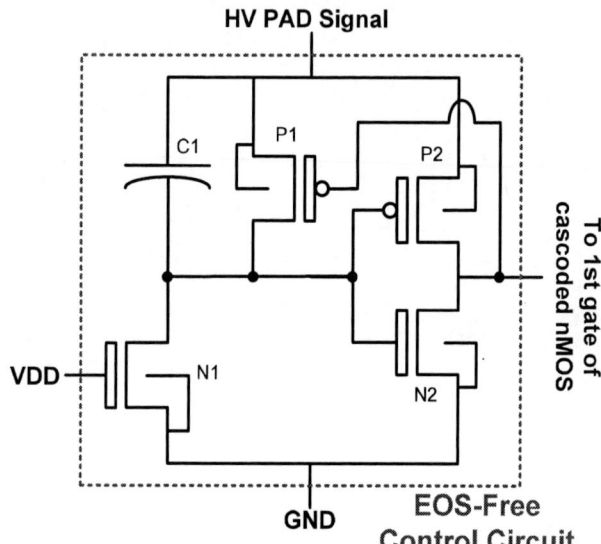

Figure 7. EOS-free control circuit for cascoded NMOSFET structures.

Figure 8 shows the SPICE simulation results of the biases at HV pin and the first gate under ESD zapping. With the first gate pulled down to ground, the cascode structure will maintain the original robust ESD capability. In the control circuit, PMOS P1 can also help to stabilize the bias so that large C1 is not necessary.

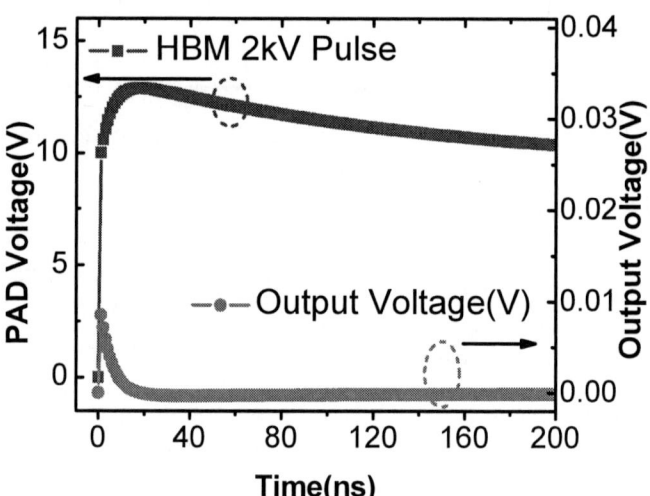

Figure 8. The SPICE simulation results of the biases at HV pin and output voltage of EOS-free circuit to the first gate under ESD zapping.

Figure 9 shows the TLP test results of PNP enhanced cascoded NMOSFET with EOS-free control circuit when V_{DD} is ready or floating. V_{tr} and V_h are high when V_{DD} is ready; and I_{t2} is high when the protection is under ESD zapping with V_{DD} floating.

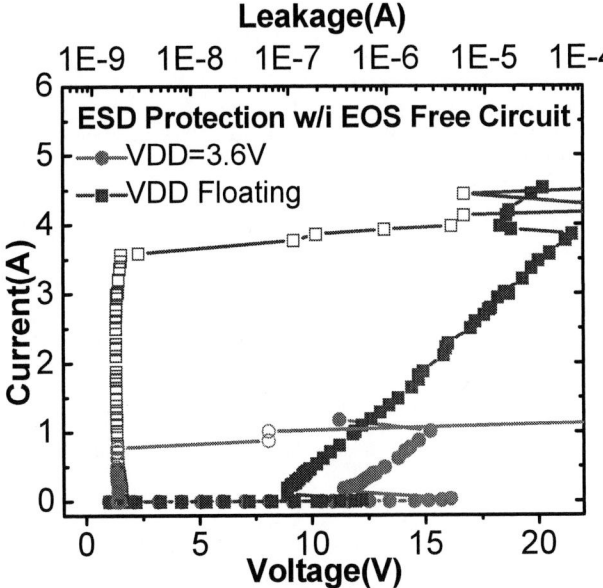

Figure 9. The TLP test results of PNP enhanced cascoded NMOSFET with EOS-free control circuit when V_{DD} is ready or floating.

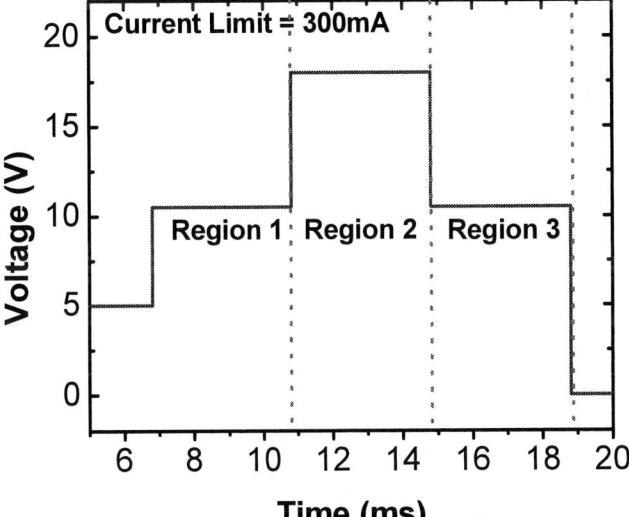

Figure 10. A three-phase pulse generated by Keithley2420 for the EOS-emulating test.

D. EOS-emulating tests

Holding voltage judged by TLP measurement is a key reference for the evaluation of EOS risk, but the extracted V_h may be overestimated, especially in HV applications [9]. Additional verification test, such as transient latch-up test, is necessary to evaluate the real susceptibility of the protection to system noises. In this paper, a simpler EOS-emulating test is implemented to verify whether the protections can be certainly free from EOS risks in normal operation. The test is implemented by Keithley 2420. A three-phase pulse is generated for the EOS-emulating test as shown in Figure 10. In region 1, the pulse rises to 10.5V as in HV operation. In region 2, voltage rises to a higher voltage to emulate system noise

triggering the protection. Referring to the TLP characteristics in Figure 9, the high voltage is set to 18V to trigger the protection with and without V_{DD} applied. The current compliance is set to 300mA to sustain the ESD path to be fully turned on and to avoid the burn-out of the protection during this period. In region 3, the voltage falls back to 10.5V. If V_h of the protection is lower than 10.5V, it will be latched and continuously conduct a high current. EOS immunity can then be easily judged by the measured conducting current in this region.

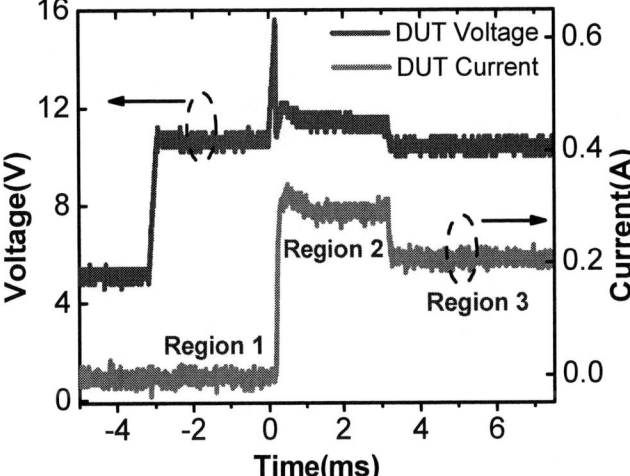

Figure 11. Real-time I-V characteristics of PNP-enhanced cascoded NMOSFET without EOS-free control circuit

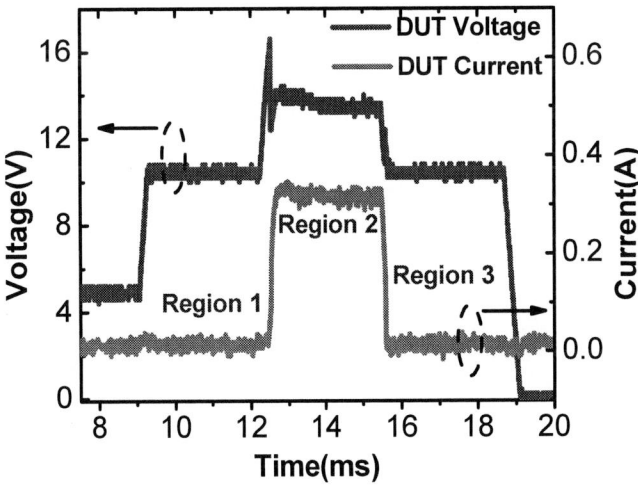

Figure 12. Real-time I-V characteristics of PNP-enhanced cascoded NMOSFET with EOS-free control circuit.

Real-time I-V characteristics of two configurations of PNP-enhanced cascoded NMOSFET with and without EOS-free control circuit can also be monitored during the similar EOS-emulating tests. When the first gate is coupled to V_{DD} without EOS-free circuit, the protection scheme is triggered in the second region and still latched in the third region, as shown in Figure 11. When there is EOS-free circuit, the protection scheme is similarly triggered in the second region but will not be latched, as shown in Figure 12, which validates EOS-free control circuit can indeed prevent the PNP-enhanced cascoded NMOSFET free from EOS risks. The raised holding voltage by the EOS-free circuit is also checked by the measurement of curve tracer, which is considered as a DC measurement. When V_{DD} is ready, the measured V_h is 12V, which is indeed higher than 10.5V. The improved EOS immunity is confirmed again.

III. CONCLUSION

An EOS-free control circuit for PNP-enhanced cascoded NMOSFET structures is proposed. The control circuit retains strong ESD robustness as its conventional configuration and the EOS-free circuit helps to raise V_h of the protection under high-voltage operation to prevent EOS events. The good EOS immunity is successfully verified by a very simple EOS-emulating test.

REFERENCES

[1] Warren R. Anderson and David B. Krakauer, "ESD Protection for Mixed-Voltage I/O Using NMOS Transistors Stacked in a Cascode Configuration", ESD/EOS Symposium Proceedings, p. 54, 1998.

[2] James W. Miller, Michael G. Khazhinsky, James C. Weldon, "Engineering the Cascoded NMOS Output Buffer for Maximum V_{t1}", ESD/EOS Symposium Proceedings, p. 308, 2000.

[3] Agha Jahanzeb, Charvaka Duvvury, Roger Cline, Scott Sterrantino, Siva Kothamasu, Ananth Somayaji, "High Voltage ESD Protection Strategies for USB and PCI Applications for 180nm/130nm/90nm CMOS Technologies", ESD/EOS Symposium Proceedings, p. 222, 2006.

[4] Ming-Dou Ker, Hsin-Chyh Hsu and Jeng-Jie Peng, "Novel Implantation Method to Improve Machine-Model Electrostatic Discharge Robustness of Stacked N-Channel Metal-Oxide Semiconductors (NMOS) in Sub-Quarter-Micron Complementary Metal-Oxide Semiconductors (CMOS) Technology", Jpn. J. Appl. Phys. Vol. 41, pp. L1288, 2002.

[5] Ming-Dou Ker, Kun-Hsien Lin, and Chien-Hui Chuang, "On-Chip ESD Protection Design With Substrate-Triggered Technique for Mixed-Voltage I/O Circuits in Subquarter-Micrometer CMOS Process", TED, p. 1628, 2004.

[6] V. A. Vashchenko, A. Concannon, M. ter Beek, and P. Hopper, "Increasing the ESD Protection Capability of Over-Voltage NMOS Structures by Comb-Ballasting Region Design", Proc. IRPS, p.261, 2003.

[7] Teruo Suzuki, Masayoshi Kojima, Junji Iwahori, Teruo Morita, Nobuyoshi Isomura, Kenji Hashimoto, Noboru Yokota, "A Study of ESD Robustness of Cascoded NMOS Driver", ESD/EOS Symposium Proceedings, p. 7A.4-1, 2007.

[8] Shuqing Cao, Jung-Hoon Chun, Eunji Choi, Stephen Beebe, Warren Anderson, Robert Dutton, "Investigation on Output Driver with Stacked Devices for ESD Design Window Engineering", ESD/EOS Symposium Proceedings, p. 3A.6-1, 2010.

[9] Wen-Yi Chen, Ming-Dou Ker, Yeh-Jen Huang, Yeh-Ning Jou and Geeng-Lih Lin, "Measurement on Snapback Holding Voltage of High-Voltage LDMOS for Latch-up Consideration", APCCAS, p.61, 2008.

[10] Ming-Dou Ker and Sheng-Fu Hsu, "Evaluation on Efficient Measurement Setup for Transient-Induced Latchup with Bi-polar Trigger", Proc. IRPS, p.121, 2005.

[11] Jian-Hsing Lee, J.R. Shih, Y. H. Wu, T.C. Ong, "The Failure Mechanism of High Voltage Tolerance IO Buffer under ESD", Proc. IRPS, p. 269, 2003.

[12] V.Vassilev, M.Lorenzini, Ph.Jansen, V.Vashchenko, J.-J.Yang, A.Concannon, D.Archer, G.Groeseneken, M.I.Natarajan, M.Terbeek, S.Thijs, B.-J. Choi, MSteyaert, H.E.Maes, "Snapback circuit model for cascoded NMOS ESD over-voltage protection structures", ESSDERC, p.561, 2003.

Latch-up Free ESD Protection Design With SCR Structure in Advanced CMOS Technology

Chang-Tzu Wang, Tien-Hao Tang, and Kuan-Cheng Su

ESD Engineering Department, Reliability Technology & Assurance Division, United Microelectronics Corporation
No. 3, Li-Hsin Rd. II, Hsinchu Science Park, Taiwan 300, R.O.C.
phone: +886-3-578-2258; e-mail: isaac_wang@umc.com

Abstract—An electrostatic discharge (ESD) protection circuit with silicon-controlled-rectifier (SCR) device has been designed without latch-up risk. After fabrication in a 0.13-μm CMOS process, the ESD protection circuit with SCR width of 60μm can sustain 6.2kV human-body-model (HBM) and 475V machine model (MM) ESD tests. The latch-up test shows the immunity against 500-mA triggering current under 3.3V supply voltage.

Keywords: electrostatic discharge (ESD), latch-up, silicon-control-rectifier (SCR).

I.   INTRODUCTION

Silicon-controlled-rectifier (SCR) device, which can sustain a high electrostatic discharge (ESD) level within a small silicon area, is commonly used for ESD protection in advanced CMOS processes [1]-[3]. The main concerns of SCR device as the ESD device are the slow turn-on speed and latch-up issue. There are several methods reported to improve the turn-on speed of the SCR device. One concept is using the additional ESD detection circuit to reduce the native SCR snapback trigger voltage during the ESD stress event. Another concept is to introduce other devices into the SCR structure for lower trigger voltage. The low voltage triggered SCR (LVTSCR) and the diode triggered SCR (DTSCR) were reported by introducing a MOS device or diodes into the SCR structure to improve the SCR turn-on efficiency [4], [5].

To solve the latch-up issue, several modified SCR structures have been reported to increase the holding voltage or the holding current [6]-[8]. However, the additional epitaxial or isolation layer in the prior works is not available in all productive CMOS technologies [8]. The high-holding current created by embedded MOS structure is around tens of milli-amperes, which is not higher than the latch-up condition with trigger current of several hundreds of milli-amperes. Therefore, how to design an effective ESD protection circuit with advantages of low cost, area efficiency, and latchup immunity is important in advanced CMOS technologies.

In this work, a new ESD protection design with SCR device and frequency dependent driving circuits is proposed without latch-up danger by inserting a controlled guard wall into the SCR device. This concept can be easily implemented in general CMOS technologies. The proposed ESD protection design has been successfully verified in a 0.13-μm CMOS process.

II.   THE PROPOSED LATCH-UP-FREE ESD PROTECTION DESIGN WITH SCR STRUCTURE

The circuit diagram and the layout top view of the proposed latch-up free ESD protection circuit are shown in Figs. 1(a) and 1(b). The latch-up-free ESD protection structure is composed of the SCR device with low freq. driving circuit connected to the N+_ctrl node. The low freq. driving circuit ties the base of PNP BJT high when the anode is supplied under the power-on condition. The N_well region is biased to the voltage level at Anode via both the N+ tag and the low freq. driving circuit. Under normal operating conditions, the N+_ctrl node is controlled by the low freq. driving circuit and used as the guard wall to block the SCR path.

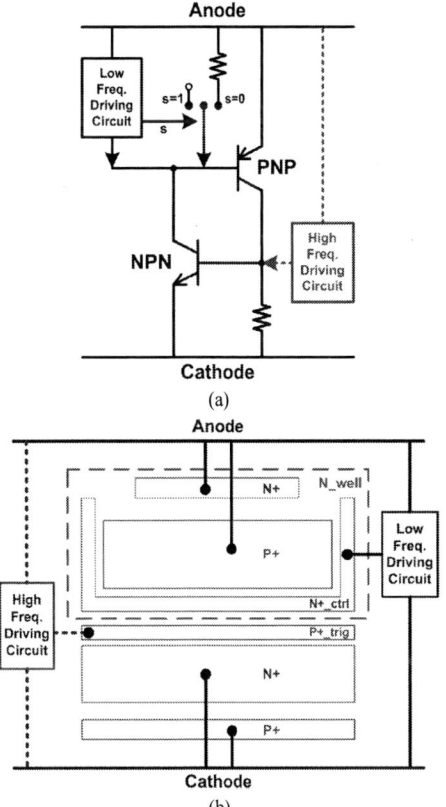

Figure 1. (a) The circuit diagram and (b) the layout top view of the proposed ESD protection structure composed of SCR device and the low freq. driving circuit, where the high freq. driving circuit is used to improve the turn-on speed.

978-1-4244-9113-1/11 $26.00 © 2011 IEEE   401

The circuit implementation of the ESD protection structure with the low freq. driving circuit is shown in Fig. 2(a). The low freq. driving circuit is composed of a MOS capacitor (MC), a resistor (R1), and a pMOSFET (MP) which is connected to the N+_ctrl node. In normal operation of Anode at 3.3V and Cathode at 0V, the gate of MP is biased at 0V through R1. The on-state MP will pull up the N+_ctrl node to 3.3V via a small turn-on resistance (Ron_MP). Since the voltage level of the N_well region is well biased at the same voltage of Anode, the PNP BJT in the SCR structure is hard to be turned on and in turn to be latched under the normal operating condition.

The high freq. driving circuit is used to improve the turn on speed of the SCR device, which is natively triggered on by the N_well to P_well avalanche breakdown. The circuit implementation of the ESD protection structure with the high freq. driving circuit is shown in Fig. 2(b). The RC timer in the low freq. driving circuit can shared for controlling the nMOSFET (MN) in the high freq. driving circuit.

When ESD voltage is applied to the Anode with Cathode relatively grounded, the capacitor MC will couple some ESD transient voltage to the gate of MN. Therefore, MN with gate voltage at high voltage level can be quickly turned on by ESD energy to generate the driving current into the P+_trig node of the SCR device. The gate of MP in the low freq. driving circuit is kept as the same voltage level as that at Anode. The off-state MP will not influence the behavior of the SCR device. The base voltage of the PNP BJT in the SCR structure is biased at Anode initially through the N_well resistance R2. After the driving current is injected into the SCR device, the collector current of NPN BJT causes a voltage difference across R2 to turn on the PNP BJT. Therefore, the SCR structure can be quickly triggered on to discharge ESD current from Anode to Cathode.

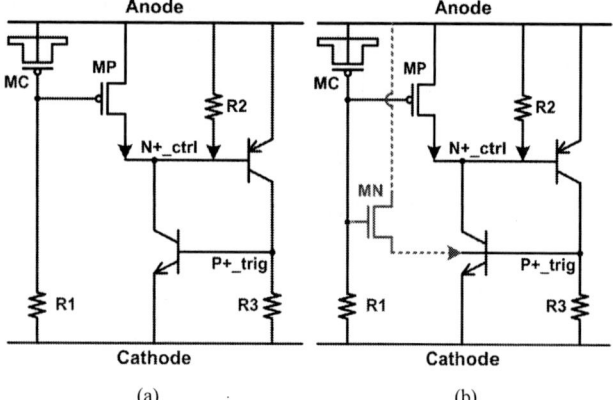

(a)   (b)

Figure 2. The circuit implementations of the ESD protection structure (a) without and (b) with the high freq. driving circuit. The RC time constant is around 300ns

III. EXPERIMENTAL RESULTS

The new proposed ESD protection design has been fabricated in a 0.13-μm CMOS process. The proposed ESD protection circuit with and without the high freq. dependent driving circuit are demonstrated for turn-on speed comparison.

The SCR device is fully-silicided and the widths of the SCR device in ESD clamp device are varied in 30μm, 40μm, and 60μm.

A. DC I-V Characteristic

The DC I-V characteristics of the proposed ESD protection structure and the typical lateral SCR are measured by using Tek370 curve tracer. To avoid the issue of false triggering and latchup, the holding voltage of the SCR device needs to be designed to be greater than the maximum voltage level of Anode (3.3V in 0.13-μm CMOS process) under the normal circuit operating condition. The measured DC I-V characteristics of the typical lateral SCR device and the proposed ESD structure with low freq. driving circuit are compared in Fig. 3. The holding voltage of the typical lateral SCR is only 1.5V, which is much lower than the operating voltage of 3.3V. On the other hand, the holding voltage of the proposed ESD protection structure is higher than 4V up to 400mA. Moreover, the proposed ESD protection structure is hard to be triggered on or be kept at holding state (without a sharp snapback phenomenon) due to the effectiveness of the low freq. driving circuit.

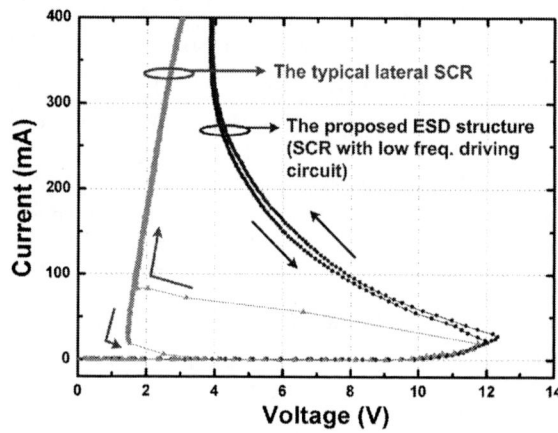

Figure 3. The DC-measured I-V curves of the proposed ESD structure and the typical lateral SCR device.

B. TLP-measured Fast Transient I-V Characteristic

To investigate the ESD capability and the related characteristic of the proposed ESD protection circuits during the ESD stress event, a transmission line pulse (TLP) generator with a pulse width of 100ns and a rise time of 10ns is used to measured the second breakdown current (It2) of the ESD protection circuits. The TLP-measured I-V characteristics of ESD protection structure with and without the high freq. driving circuit are shown in Fig. 4. With the high freq. driving circuit, the trigger voltage of the ESD protection structure can be reduced from 12V to 5.8V.

The TLP-measured I-V characteristics of ESD protection structure with high freq. driving circuit under SCR device of different widths are shown in Fig. 5. The ESD protection circuit with SCR widths of 30μm, 40μm, and 60μm can achieve an It2 of 1.91A, 2.64A, and 3.78A, respectively. The

It2 of the ESD protection circuit is proportional to the width of the SCR device. The clamped voltage of the ESD protection structure is around 4V. Such holding voltage is higher than the voltage level of 3.3V under the normal circuit operating condition. Even if the ESD protection circuit mis-triggers due to the noise disturbance, it will automatically recover to the normal condition after the noise source is removed.

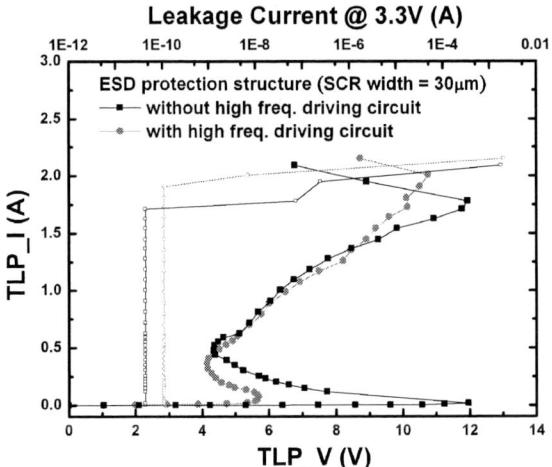

Figure 4. The TLP-measured I-V curves of the proposed ESD structure with and without high freq. driving circuit.

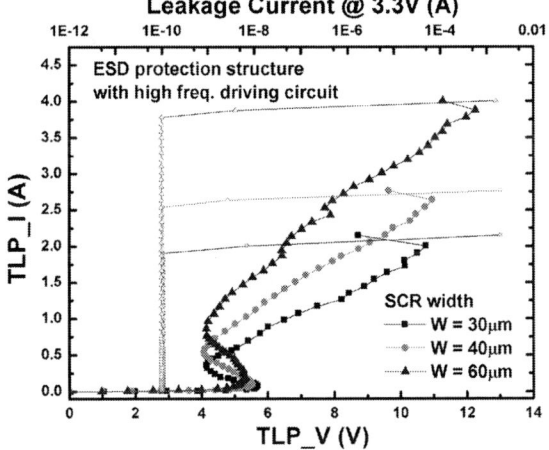

Figure 5. The TLP-measured I-V curves of the proposed ESD structure with high freq. driving circuit under different widths of SCR device.

C. Turn-on Verification

The turn-on behavior of SCR device is one of important factors for ESD protection [9], [10]. To verify the turn-on efficiency of the proposed ESD protection circuit, a square-type voltage pulse, which is generated by the voltage pulse generator, with a rise time of ~10ns and a pulse height of 20V is used to simulate the rising edge of a human-body-model (HBM) ESD pulse. The voltage waveforms clamped by of the proposed ESD protection structure with and without high freq. driving circuit are shown in Fig. 6. With the high freq. driving circuit, the fast-transient voltage pulse can be clamped to a low voltage level within a shorter time.

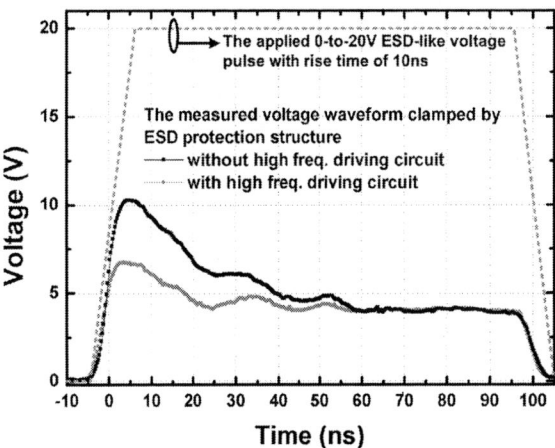

Figure 6. The measured voltage waveforms clamped to verify the turn-on efficiency of the proposed ESD protection structure with and without high freq. driving circuit by applying a 0-to-20V voltage pulse with rise time of 10ns and pulse width of 100ns.

To evaluate the effectiveness of the proposed ESD protection circuit in faster ESD-transient events, the TLP with rise time of 200ps is used to measure the transient behavior. The clamped voltage waveforms by the proposed ESD structure with and without high freq. driving circuit under TLP voltage pulses with pulse heights of 10V, 20V, and 30V are compared in Figs. 7(a) to 7(c). From Fig. 7(a), the ESD protection structure with high freq. driving circuit can be turned on efficiently and the voltage waveform can be clamped to a lower voltage level when a 10V pulse is applied into the anode. The overshooting observed results from the delay time in the TLP tester itself. Without the high freq. driving circuit, the ESD protection structure cannot turned on well even a 20V pulse is applied, as shown in Fig. 7(b). From Fig. 7(c), the turn on time of the ESD protection structure with high freq. driving circuit to clamp the voltage waveform is a little shorter than that without high freq. driving circuit when a 30V pulse is applied to trigger both circuits on.

D. ESD Robustness & Latch-up Immunity

The HBM ESD levels and machine-model (MM) ESD levels of the proposed ESD protection circuit with SCR of different widths are listed in Table I. The corresponding It2 measured by TLP is also listed in Table I. The low enough trigger voltage and high It2/HBM/MM values of the ESD protection structure with high freq. driving circuit can ensure the effective ESD protection capability.

The latch-up results of the ESD protection circuit under different supply voltages and positive/negative currents are listed in Table II. Under 3.3V operation, the proposed ESD protection circuit is free from latch-up issue even a 500-mA current is applied into the circuit. Only when the supply voltage is higher than 5V (also higher than the gate-oxide breakdown voltage), the circuit is going to be latched when a high triggered current is injected. Therefore, the latchup immunity of the proposed ESD protection structure can be guaranteed for normal operations.

978-1-4244-9113-1/11 $26.00 © 2011 IEEE   403

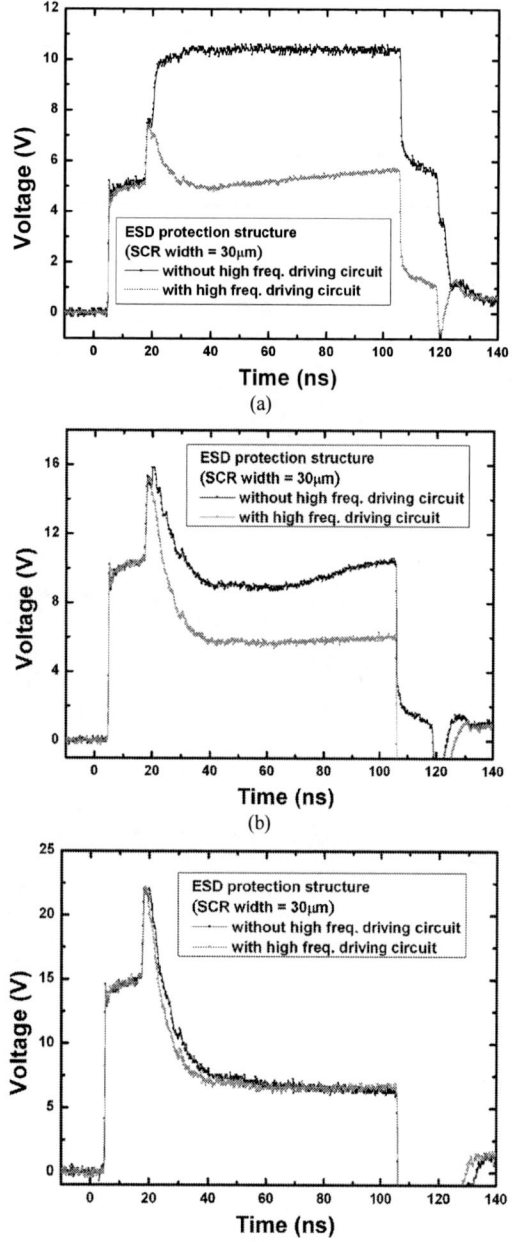

Figure 7. The measured voltage waveforms clamped to verify the turn-on efficiency of the proposed ESD protection structure with and without high freq. driving circuit by applying a TLP voltage pulse with rise time of 200ps, pulse width of 100ns, and voltage level of (a) 10V, (b) 20V, and (c) 30V.

TABLE I
THE ESD ROBUSTNESS OF THE PROPOSED ESD PROTECTION STRUCTURE WITH
HIGH FREQ. DRIVING CIRCUIT

ESD Protection Structure with High Freq. Driving Circuit			
SCR Width	It2	HBM ESD Level	MM ESD Level
30μm	1.91A	3.2kV	300V
40μm	2.64A	3.8kV	375V
60μm	3.78A	6.2kV	475V

Min. value within 4 samples

TABLE II
THE LATCH-UP IMMUNITY OF THE PROPOSED ESD PROTECTION CIRCUIT
UNDER DIFFERENT SUPPLY VOLTAGE

	Temp. = 25°C		Temp. = 125°C	
VDD	+I Test	-I Test	+I Test	-I Test
1.2V	> 500mA	> 500mA	> 500mA	> 500mA
3.3V	> 500mA	> 500mA	> 500mA	> 500mA
5V	440mA	> 500mA	280mA	> 500mA

IV. CONCLUSION

A new proposed ESD protection design has been successfully verified in a 0.13-μm CMOS process. The ESD capability and latchup immunity of the proposed ESD protection design are well demonstrated in this work. With the low freq. driving circuit, the ESD protection structure is free from latch-up damage under normal operating condition. The latch-up-free ESD protection circuit has performed the efficient turn-on speed and robust ESD levels, which is an excellent design solution for protecting I/O buffers or power rails in advanced CMOS technology.

REFERENCES

[1] C. Russ, M. Mergens, J. Armer, P. Jozwiak, G. Kolluri, L. Avery, and K. Verhaege, "GGSCRs: GGNMOS triggered silicon controlled rectifiers for ESD protection in deep submicron CMOS processes," in *Proc. EOS/ESD Symp.*, 2001, pp. 22–31.

[2] M.-D. Ker and K.-C. Hsu, "Overview of on-chip electrostatic discharge protection design with SCR-based devices in CMOS integrated circuits," *IEEE Tran. Device Mater. Reliab.*, vol. 5, no. 2, pp. 235–249, Jun. 2005.

[3] P.-Y. Tan, M. Indrajit, P.-H. Li, and S. H. Voldman, "RC-triggered PNP and NPN simultaneously switched silicon controlled rectifier ESD networks for sub-0.18μm technology," in *Proc. of IEEE Int. Symp. on Physical and Failure Analysis of Integrated Circuits*, 2005, pp. 71–75.

[4] A. Chatterjee and T. Polgreen, "A low-voltage triggering SCR for onchip ESD protection at output and input pads," *IEEE Electron Device Lett.*, vol. 12, no. 1, pp. 21–22, Jan. 1991.

[5] M. Mergens, C. Russ, K. Verhaege, J. Armer, P. Jozwiak, R. Mohn, B. Keppens, and S. Trinh, "Diode-triggered SCR (DTSCR) for RF-ESD protection of BiCMOS SiGe HBTs and CMOS ultrathin gate oxides," in *IEEE Int. Electron Devices Meeting Tech. Dig.*, 2003, pp. 515–518.

[6] G. Notermans, F. Kuper, and J.-M. Luchies, "Using an SCR as ESD protection without latch-up danger," *Microelectronics Reliability*, vol. 37, no. 10/11, pp. 1457–1460, 1997.

[7] M. P. J. Mergens, C. C. Russ, K. G. Verhaege, J. Armer, P. C. Jozwiak, and R. Mohn, "High holding current SCRs (HHI-SCR) for power ESD protection and latch-up immune IC operation," in *Proc. EOS/ESD Symp.*, 2002, pp. 10–17.

[8] G. Meneghesso, A. Tazzoli, F. A. Marino, M. Cordoni, and P. Colombo, "Development of a new high holding voltage SCR-based ESD protection structure," in *Proc. of IEEE Int. Reliability Physics Symp.*, 2008, pp. 3–8.

[9] J. D. Sarro, K. Chatty, R. Gauthier, and E. Rosenbaum, "Evaluation of SCR-based ESD protection devices in 90nm and 65nm CMOS technologies," in *Proc. of IEEE Int. Reliability Physics Symp.*, 2007, pp. 348–358.

[10] G. Wybo, S. Verleye, B. V. Camp, and O. Marchial, "Characterizing the transient device behavior of SCRs by means of VFTLP waveform analysis," in *Proc. EOS/ESD Symp.*, 2007, pp. 366–375.

Transient Latchup in Power Analog Circuits

V.A. Vashchenko, D. LaFonteese, A. Concannon

National Semiconductor Corp.
Santa Clara, CA, USA,
(650) 906-0172, vladisllavv@att.net

Abstract— **Two cases of transient latchup specific to power management analog integrated circuit design are described and analyzed experimentally. The representative case studies include the interaction of a power array and ESD clamp and the interaction of two high voltage ESD clamps**

Keywords- ESD, latchup, TLP, power, analog

I. INTRODUCTION

The challenge of providing ESD protection for a power circuit is not limited to implementing a clamp that nominally triggers within the required ESD protection window. Interaction of ESD protection components with the internal circuit can critically impact the triggering regime of the snapback ESD devices under both ESD stress and operational conditions. The scenario of "accidental" ESD device turn-on under non-ESD time domain transient conditions may result in failure during functional testing. Additionally, circuit board design is important due to the significant effect its reactive load has on pulsed voltage waveforms. In principle, a premature turn-on of the ESD device in snapback mode due to additional injection from the internal circuit components can be reversible and even favorable for ESD circuit operation. However under normal conditions with power supply voltage applied, turn on of the ESD device may cause damage due to excessive power generation. This study presents two cases specific to power management circuits with the purpose to clarify the transient latchup scenario.

II. POWER ARRAY-ESD CLAMP INTERACTION

A simplified circuit diagram with a switch pin (SW) is presented in Fig. 1. The original circuit met all ESD specification requirements and was designed with an absolute maximum drain-source voltage of 45 V for switch (SW) and other control pins. The maximum operation voltage is specified at 40 V. The ESD clamp is based upon an N-type drain-extended silicon controlled rectifier (NDeMOS-SCR). It is designed for a triggering voltage of 50 V, above the lower ESD protection window limit of 45V. In this particular design an identical ESD protection clamp has been used for protection of both the switch power pin and a number of high voltage control protection pins. At the switch pin the topological separation of the clamp from the large power array is ~250μm (Fig. 1). The space is filled with a P-guard ring diffusion.

The observed application problem was related to accidental turn-on of the clamp in certain transient test regimes close to the absolute maximum limits. The transmission line pulsed (TLP) measurements demonstrate a significant change in the triggering voltage observed at the pin in comparison with the

characteristics of the stand alone clamp (Fig. 2a). For the switch to ground pin combination (SW-GND), TLP test results differed depending on the state of the 5 V input voltage pin (VIN), which controls the power array gate ("Gate", Fig. 1); at high voltage the gate is held low. For floating VIN, the triggering voltage was measured at only ~40 V (Fig. 2a). Prior to triggering into snapback a significant difference in pulsed I-V characteristics was observed. A much higher ~3 A current level preceded the snapback mode. This current is representative of the on-state conditions of the array connected in parallel to the pin.

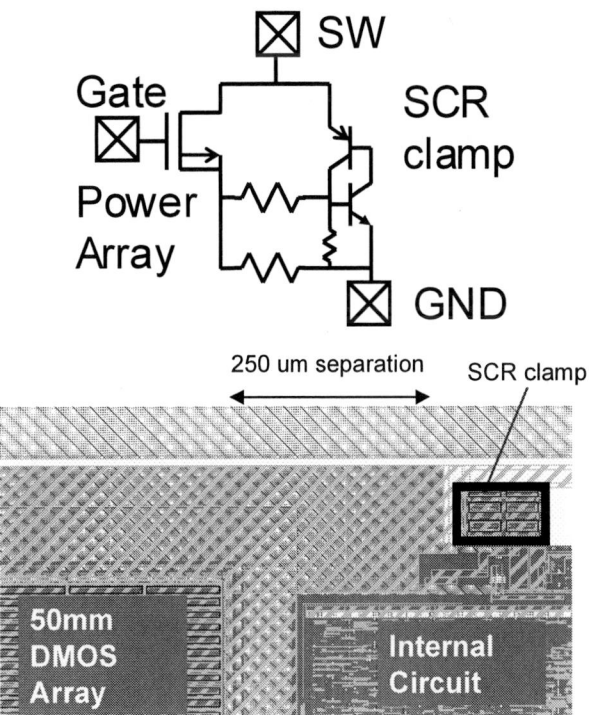

Figure 1. Simplified application diagram for the part of the driver circuit and layout of the power array and ESD clamp

When the VIN pin is biased high the power array is in off-state, and clamp turn-on matches more closely the stand alone conditions. However even in this case it is hard to conclude to what extent the internal avalanche injection conductivity modulation mechanism of the clamp itself is involved. With bias applied to the input pin, TLP characteristics of the SW-GND pin combination show a triggering voltage of 48 V, much closer to the stand-alone clamp with low pre-trigger current.

978-1-4244-9113-1/11 $26.00 © 2011 IEEE   405

However, even in this case the switching DMOS draws over 150 mA prior to entering snapback mode (Fig. 2a), indicating an additional current path. At floating VIN node the array gate to drain coupling of the switch transistor is high. This results in significant drain current (caused by the gate pull-up) and corresponding substrate current. In spite of the large guard-ring protected distance between the power array and ESD device, a 10 V triggering voltage reduction is experimentally observed (Fig. 2a).

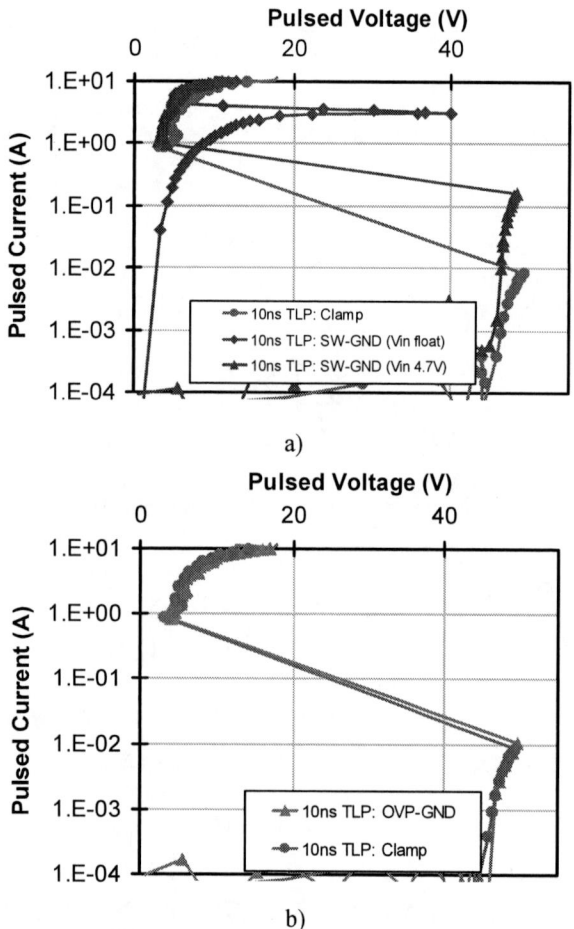

a)

b)

Figure 2. Comparison of the TLP characteristics of the stand-alone ESD clamp with the characteristics at SW SW pin (a) and at OVP pin (b)

Depending on the process, array design, and electrical regime, the array substrate current can reach a level of 10-30% of the drain current. The current path is realized between the array drain and any bulk p+-diffusion in the vicinity of the arrays, including P-guard ring of the ESD clamp. Only a few milliamps of the current injected into the SCR clamp is sufficient to trigger snapback mode (Fig. 2b). At the same time, independent from each triggering scenario the holding voltage at each pin is observed at the level similar to the stand alone clamp. This indicates an identical final current path in the high current snapback mode.

For comparison TLP characteristics have been measured for a control pin - overvoltage protection (OVP) pin. Unlike the

power pin behavior, the combination of OVP to ground demonstrates TLP characteristics similar to the stand alone clamp (Fig. 2b). Thus despite layout isolation of the SCR clamp, the clamp is still likely to be triggered by the parasitic current from the internal circuit. The substrate current in the DeMOS array is acting similarly to the base current of the parasitic NPN structure and thus provides a lower turn-on voltage.

To improve the design for the low impedance switch pins, for large switching LDMOS arrays (total gate width > 30 mm) the circuit should rely on array self-protection. An alternative approach may involve application of non-snapback high holding voltage ESD clamp based on a lateral PNP or NPN device. For small LDMOS arrays a gate clamp should be introduced to ensure off-state condition of the power array during ESD stress to guarantee a high avalanche breakdown voltage, however this measure might be inadequate under fast transient test conditions. Finally, for high impedance control pins an ESD clamp with pulsed triggering voltage significantly above absolute maximum can be used to realize a two-stage protection approach.

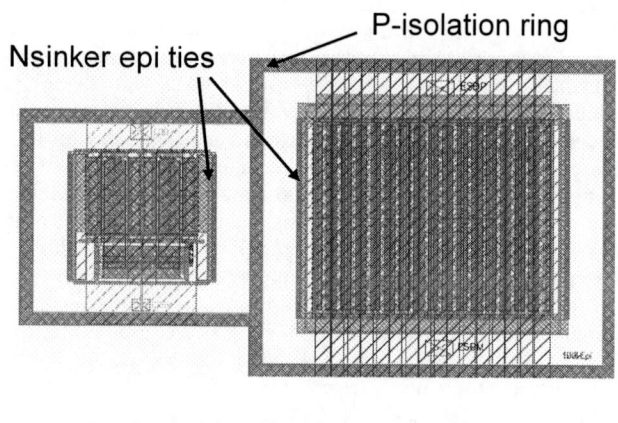

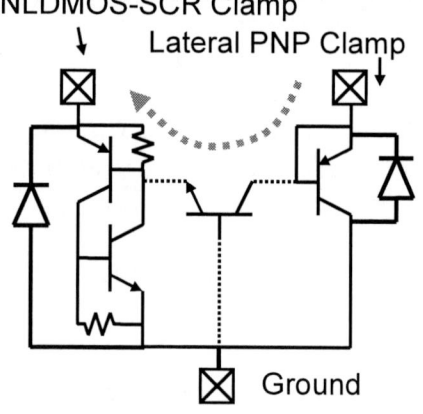

Figure 3. Illustration of the alternative current path formed between two adjacent H pins protected by NLDMOS-SCR and Lateral PNP clamps

III. NEPI-NEPI TRANSIENT LATCHUP SCENARIO

Another case of transient latchup is observed in a 100 V BCD process technology. This latchup current path can be realized either between two adjacent high voltage pins protected by high voltage ESD devices or the ESD device and the internal circuitry. The primary reason for this latchup path formation is snapback of the parasitic Nepi – p-isolation – Nepi NPN BJT. In principle under ESD test conditions this parasitic NPN activates with variable base potential or even current. Thus the critical triggering voltage can be realized in a very broad range.

An example of this parasitic NPN and consequent current path is illustrated in Fig. 3 between a high-voltage control pin protected by NLDMOS-SCR ESD clamp and the power pin protected by a high holding voltage lateral PNP ESD clamp. Both clamps are referenced to the same power ground pin. However during ESD test the power ground node is floating. This in general creates variable conditions at the base of the parasitic NPN formed by Nepi regions of both structures and P-isolation ring.

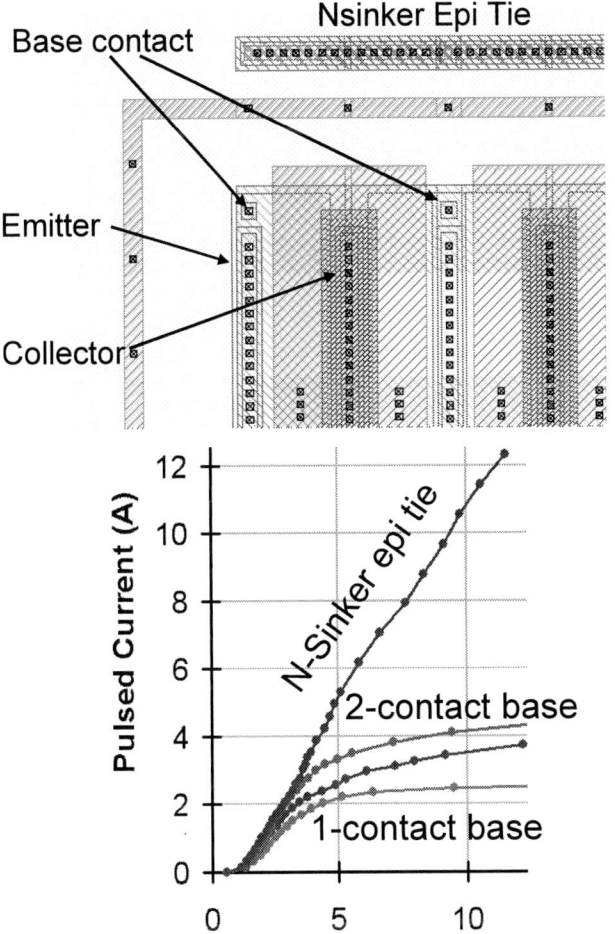

Figure 4. Partial layout view of the Lateral PNP Clamp with the device regions and comparison of the negative TLP pulse characteristics for versions of the clamp with different N-base connection

Originally, the expected current path is through the HV snapback clamp of positively biased pin and the reverse path

diode of the negatively biased pin clamp. However if two clamps are placed with minimum isolation rules an alternative current path is formed through the parasitic NPN structure described above. This alternative current path remains open for ESD current. Therefore the final result depends upon the metallization routing and the internal circuit design. Since the real current path through the internal circuit is hard to generally predict, an irreversible burnout may occur.

To allow the lateral PNP clamp to clamp reverse ESD stress at low voltage, the emitter and base are shorted, allowing the forward-biased collector-base diode to clamp reverse stress. Significant improvement of these diode characteristics [1, 2] in a BCD process technology for reverse ESD protection can be achieved by enlarging n+ base contact area, adding base contacts, or by adding an N-sinker epi tie to take an advantage of low buried N-layer resistance. The effects of these measures are presented in Fig. 4 for a 100 V clamp. All methods allow 1.5 A protection (> 2 kV HBM) at less than 3.5 V voltage drop. The highest current and lowest resistance are achieved with N-sinker as expected. Meanwhile, the main approach to suppression of the parasitic current involves introduction of ESD layout guidelines for spacing limits between two epi-regions connected to the high voltage pins.

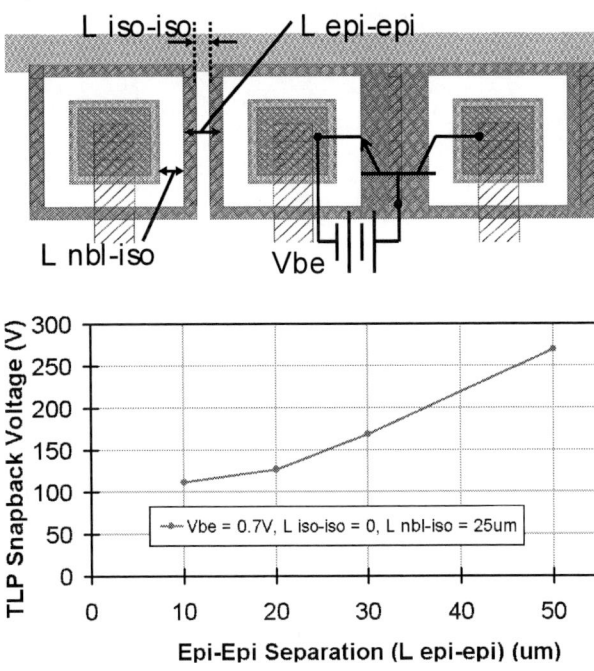

Figure 5. Layout view for experimental structures for Epi-to-Epi isolation rules evaluation and experimental dependence of the critical snapback voltage upon Epi region separation

To determine ESD design rules for integrated circuit design, a simple 3-terminal experimental device structure to study the current path through parasitic NPN can be created and characterized to determine the rules for different high voltage epi region isolation under ESD conditions (Fig.5). According to the experimental data the critical voltage is a

978-1-4244-9113-1/11 $26.00 © 2011 IEEE

function of the isolation region bias and the spacing between the epi-regions that represents the base length (Fig. 5).

Several important conclusions can be derived from this case study. First of all the high voltage reverse path diode design in high voltage clamps is critically important. For the pin protection itself the body diode of the clamp is often sufficient. However for the highlighted HV pin-to-HV pin combination the total voltage may exceed the parasitic NPN turn on voltage especially in fast transient mode due to excessive voltage drop on the reverse path diode.

IV. CONCLUSIONS

In a power analog circuit, multiple current paths can be realized during an ESD event. Two scenarios for such event are demonstrated in this study followed by proposed countermeasures based upon physical analysis. It is demonstrated that development the ESD design guidelines based upon pulsed measurements in the transient latchup conditions is required for successful design of high voltage power analog integrated circuit.

REFERENCES

[1] Vashchenko VA, LaFonteese D "System Level and Hot Plug-in Protection of High Voltage Transient Pins" in Proc. EOS/ESD Symposium 2009

[2] Vashchenko VA, LaFonteese D "Lateral PNP BJT ESD Protection Devices", BCTM 2008 pp53-56.

WCDM2 – Wafer-Level Charged Device Model Testing with High Repeatability

Nathan Jack,[1] Timothy J. Maloney,[2] Bruce Chou,[2] and Elyse Rosenbaum[1]

(1) Department of Electrical and Computer Engineering
University of Illinois at Urbana-Champaign, 1308 W. Main St., Urbana, IL 61801
ndjack2@illinois.edu

(2) Intel Corporation
3601 Juliette Lane, Santa Clara, CA 95054 USA

Abstract—**CDM-like unipolar pulses are generated at the wafer level with excellent repeatability and linearity. Pulse width and rise time resemble that of FICDM testers. In-situ pre- and post-stress curve tracing reveals the current failure threshold for the device under test.**

I. Introduction

Several test methods have been introduced for generating charged device model (CDM) stress at the wafer level [1]-[4]. Wafer-level testing is desirable for multiple reasons, including reliability assessment earlier in the design phase (before packaging), elimination of packaging cost, and elimination of unrealistic package parasitics when stressing small test structures.

Wafer-Level CDM (WCDM) [1] was designed to replicate the air discharge event of field-induced CDM (FICDM)—the package-level industry standard [5]. In WCDM, charge is stored on the capacitance formed between the grounded wafer and an overhead field plate; discharge occurs through the arc formed between the wafer and an approaching probe tip connected to the field plate. While this air discharge mechanism does resemble FICDM testing, the variability of the arc causes the peak current to vary by 30% or more from zap to zap. It was also observed that the arc caused the stressed pad to crater at moderate current levels and spray pad metal across the wafer. Moreover, the process of repetitively contacting the probe tip to the wafer for each discharge accelerates the dulling of the probe tip. Finally, the discharge current was measured through a band-limited transformer, the behavior of which had to be de-embedded from the measurement data to obtain the true time domain waveform [1][6].

Capacitively-coupled TLP (CC-TLP) [2][3] offers a more reproducible method for generating wafer-level CDM-like stress. During CC-TLP, a very fast TLP (VF-TLP) pulse is applied to the device under test (DUT) via a probe that contacts the wafer, with the return path formed by the capacitance C_{DUT} between a large ground plane and the wafer over which it is suspended. Because the VF-TLP pulse is relay-actuated, repeatability can be very good. Displacement current during the pulse rising and falling edges generates a dual-polarity stress with each VF-TLP pulse. However, it is often desirable to generate a single-polarity stress when debugging CDM failures or to avoid reverse-bias damage when testing standalone devices.

In this work, WCDM is modified to be relay-initiated for excellent repeatability. The single-polarity stress generated with the new system closely resembles FICDM waveforms and provides testing flexibility. The new system is named WCDM2 [7].

II. WCDM2 Description

A schematic of the WCDM2 tester is shown in Figure 1. The probe card is a three-layer PCB with field plates constructed on the outer two layers and a 50 Ω stripline trace in the center. The two field plates are shorted together and are connected to the center conductor of a SMA cable. The stripline connects the probe tip to the grounded shield of the SMA cable. A stripline was chosen over a microstrip design for its superior high frequency performance; microstrips suffer from multimodal, non-TEM (transverse electromagnetic) dispersion arising from the boundary discontinuity at the air-dielectric interface [8]. As shown in the photo in Figure 2, the probe card can be mounted to a micromanipulator for positioning. A hole in the center of the probe card allows the probe tip and wafer pads to be viewed through a microscope while positioning the tip. The probe tip lifetime is significantly improved over WCDM, since the tip may remain in contact with the wafer during multiple stresses. The probe tip used in this work has a radius of 10 μm, allowing for probing of small pads.

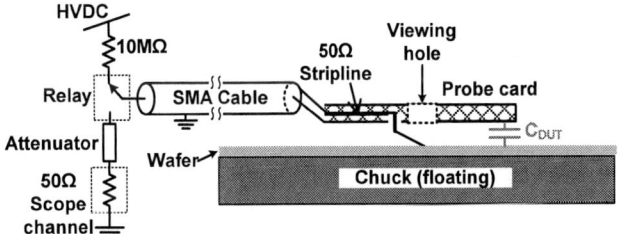

Figure 1: Schematic representation of WCDM2. The probe card is a 3-layer PCB with a 50Ω stripline connected to the probe tip.

978-1-4244-9113-1/11 $26.00 © 2011 IEEE

Figure 2: Photo of the WCDM2 probe card mounted onto a micromanipulator. A viewing hole in the center of the probe card allows the use of the microscope when positioning the probe.

With the field plate at 0 V, the probe tip is brought into contact with the pad to be zapped; this brings the potential of the pad and the wafer to ground. Next, a mercury-wetted relay connects the SMA cable to the high voltage DC source through a 10 MΩ resistor, so that the cable, stripline, and plate-to-wafer capacitance C_{DUT} are slowly precharged to the desired voltage. To initiate the stress, the mercury-wetted relay is actuated to disconnect the high voltage supply and discharge the system through the oscilloscope input channel. Attenuators are used to limit the voltage appearing at the scope input. The rapid discharge of C_{DUT} delivers a CDM-like stress to the DUT.

Non-air-discharge CDM has benefits for package-level testing as well. The WCDM2 charging and discharging mechanisms [7] are similar to that of the package-level tester "CDM2" proposed by Given et al. [9].

III. WAVEFORM ANALYSIS AND CALIBRATION

A calibration board was used to measure the WCDM2 current stress felt by the DUT. The calibration board consists of a 1 ft by 1 ft square copper board with a 1 Ω disk resistor soldered into a hole in the center, as shown in the photo in Figure 3. A SMA connector soldered to the backside of the disk resistor is connected through a cable to an oscilloscope. During a zap, the WCDM2 probe card is suspended over the calibration board, and the probe needle makes contact with the center conductor of the disk resistor. The charge induced on the copper must travel through the disk resistor to reach the grounded center during the zap. The oscilloscope measures the voltage across the resistor, from which the current through the resistor can be extracted. All measurements presented in this work were taken using a 6 GHz oscilloscope.

A. Discharge Capacitance

Figure 4 shows the current waveforms measured while zapping the calibration board with a 100 V precharge. C_{DUT} increases with the area of the field plate on the probe card; C_{DUT} is a decreasing function of the separation between the field plate and the DUT. The measurements shown are for two probe cards differing only in the length of the probe needle, i.e.

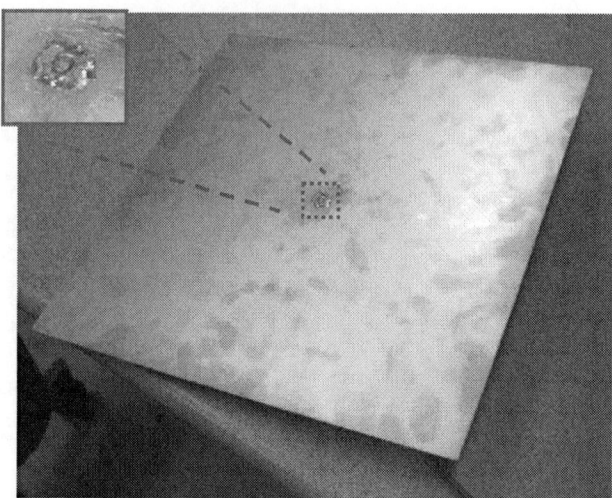

Figure 3: Photo of the calibration board. A 1 Ω disk resistor is soldered into a hole in the center of the 1 ft by 1 ft square copper board. A SMA connector soldered to the backside of the disk resistor is connected by a cable to an oscilloscope. The oscilloscope measures the resistor voltage, and thus current, during a WCDM2 zap to the center conductor of the resistor.

the card with the longer probe needle has a smaller C_{DUT}. The length of the longer probe tip was approximately 2 mm. To extract the capacitance, the total charge injected during each zap was calculated by integrating the current waveform over time, as shown in Figure 5. Dividing the total charge by the 100 V precharge voltage yields C_{DUT}. The total charge for the card with the shorter needle was roughly 1.7 nC, implying a C_{DUT} of 17 pF. The total charge for the other card was 1.2 nC, yielding a C_{DUT} value of 12 pF.

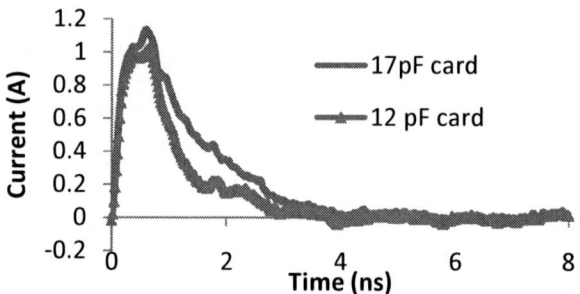

Figure 4: Current measured when zapping a calibration board using a 100 V precharge voltage. Pulse width decreases with C_{DUT}, as does the peak current. C_{DUT} is controlled by the probe tip length and the sizes of the field plate and DUT.

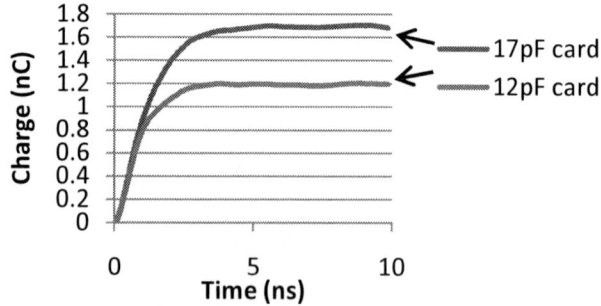

Figure 5: Integrals of the current waveforms from Figure 4. Dividing the total charge by the 100 V precharge voltage yields the capacitance of each card.

The pulse width and peak current both increase with C_{DUT}, as seen in Figure 4. The pulse produced by the 12 pF card discharge is approximately 1 ns wide with an amplitude of 1 A at a precharge of 100 V; the 17 pF card produces a pulse roughly 1.4 ns wide with peak current of approximately 1.1 A at 100 V. These results may be understood by modeling the WCDM2 waveform as the response of a RC network to a step function with a non-zero fall time. This is illustrated schematically in Figure 6. The 1 Ω resistor represents the calibration board disk resistor, and the capacitance represents C_{DUT}, which is precharged to V_{pre}. The 56 Ω resistor represents the 50 Ω input of the oscilloscope plus an estimated 6 Ω for the spark resistance in the mercury-wetted relay. Closing the relay initiates the discharge of C_{DUT} by completing the path to ground. However, the ground loop is not instantaneously established; there is a finite fall time associated with the probe tip inductance and the non-ideal transmission line and relay. This fall time is captured by the finite step function in Figure 6.

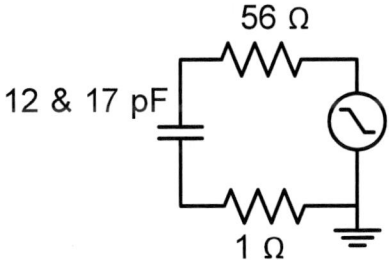

Figure 6: RC model of the WCDM2 discharge path. The 1 Ω resistor represents the disk resistor on the calibration board, the 56 Ω resistor represents the scope impedance plus spark resistance, and the capacitor represents C_{DUT}. The non-zero fall time of the grounding signal represents the non-idealities of the transmission line and relay.

The model in Figure 6 can be used to solve for the transient WCDM2 current $i(t)$. This is can be accomplished easily in the Laplace domain. The step function can be approximated as a decaying exponential with a fall time τ_f, which is represented in the Laplace domain as $V_{pre}/(s + a)$, where $a = 1/\tau_f$. Applying Kirchhoff's Current Law to Figure 6 and combining the resistances into a single resistor R, one obtains

$$\frac{1}{C_{DUT}}\left[\frac{I(s)}{s} - \frac{C_{DUT}V_{pre}}{s}\right] + RI(s) = \frac{-V_{pre}}{s+a}. \quad (1)$$

Rearranging (1) and taking the inverse Laplace transform, one obtains

$$I(s) = \frac{aV_{pre}}{s+a}\left(\frac{C_{DUT}}{sRC_{DUT}+1}\right) \Longleftrightarrow \quad (2)$$

$$i(t) = \frac{V_{pre}}{\frac{1}{aC_{DUT}}-R}\left[\exp(-at) - \exp\left(\frac{-t}{RC_{DUT}}\right)\right]. \quad (3)$$

Increasing C_{DUT} will proportionally increase the total charge stored for a given precharge voltage. This is manifest primarily by the increase of the RC time constant of the discharge path, which appears in the second exponential term in (3). The result is a slower decay time and hence a wider current pulse. Increasing C_{DUT} also increases the coefficient of

the exponential terms, resulting in an increase in the peak current. Figure 7 shows the current predicted by (3) for two different values of C_{DUT}, i.e. two different WCDM2 probe cards, and $V_{pre} = 100$ V. The waveforms obtained analytically closely resemble the measured waveforms in Figure 4.

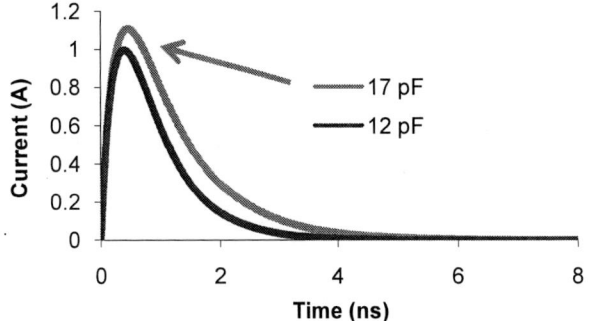

Figure 7: Analytic solution of the WCDM2 current for the two different C_{DUT} values using (3) with $V_{pre} = 100$ V. The waveforms obtained analytically closely resemble the measurements in Figure 4. The larger RC time constant of the 17 pF probe card results in a larger pulse width and slightly higher peak current.

B. System Discharge Waveform

The calibration board measurements show the current waveforms at the 1 Ω DUT (Figure 4). However, the waveform measured at the scope input into which the discharge current flows (see Figure 1) represents the discharge of the entire WCDM2 system. This waveform is shown in Figure 8 for a 100 V zap of the calibration board. Dividing this waveform by the 50 Ω input impedance of the oscilloscope channel yields the system discharge current. In Figure 8, the first 9 ns represent the discharge of the SMA cable. Between 9 ns and 10 ns, a small spike is seen as the probe card stripline discharges. The last spike of the waveform, occurring at about 10 ns, is the result of C_{DUT} discharging.

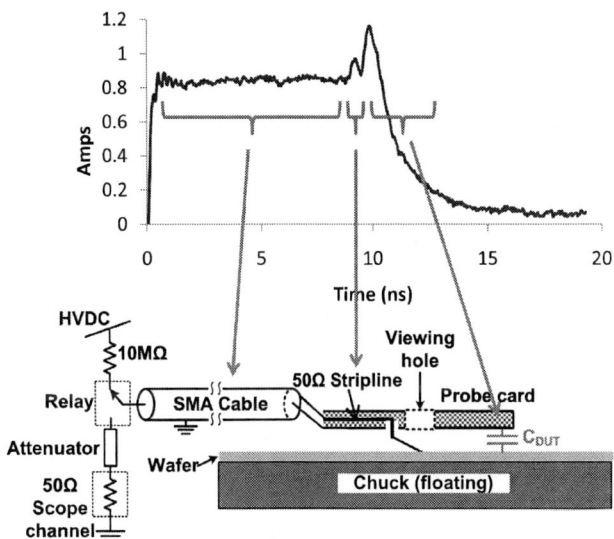

Figure 8: Waveform measured by the scope on the channel into which the discharge occurs. The first 9 ns of the current represent the cable discharging. Next, the stripline discharges. Finally, C_{DUT} is discharged, occurring at the 10 ns point and onward.

The spikes in the system discharge waveform arise from impedance discontinuities along the discharge path, i.e. impedances that are unequal to the 50 Ω characteristic impedance of the SMA cable and oscilloscope input. The reflection coefficient Γ at such points in the system is given by

$$\Gamma = \frac{Z_L - 50\Omega}{Z_L + 50\Omega} \qquad (4)$$

where Z_L is the impedance looking into the system at the point of the impedance discontinuity. The spike arising from the discharge of C_{DUT} is expected since the impedance of this capacitance is not 50 Ω [10]. However, the stripline was designed to have an impedance of 50 Ω, so it causing a spike was unexpected.

To extract the apparent impedance of the stripline, time-domain reflectometry (TDR) was used. The oscilloscope screenshot in Figure 9 compares the reflected waveform for the cable plus WCDM2 board to that of the cable only; in both cases the ends of the transmission lines were open-circuited. The initial negative reflection seen on the waveform from the WCDM2 board indicates that its impedance is less than 50 Ω. Application of (4) yields $Z_L \approx 40$ Ω.

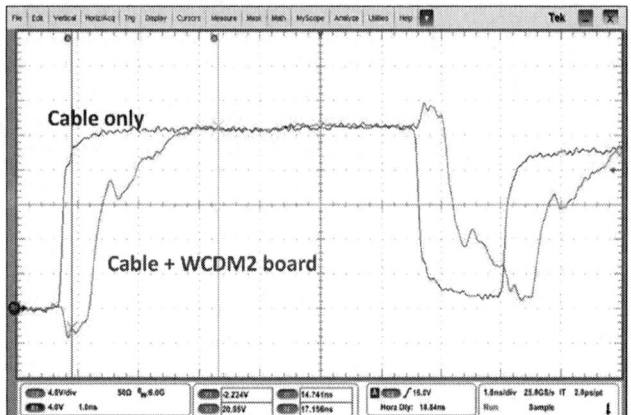

Figure 9: Oscilloscope screenshot showing TDR response of the WCDM2 plus the cable connected to it, compared with the cable alone. The negative reflection at the WCDM2 board interface indicates a 40 Ω stripline impedance, contrary to the designed-for 50Ω impedance. The scales for this plot are 4 V / div. on the vertical axis and 1 ns / div. on the horizontal axis.

To understand why the stripline impedance is not 50 Ω, it is important to note that the transmission line is inverted at the cable-to-PCB interface. As illustrated in Figure 8, the center conductor of the SMA cable is connected to the outer planes of the stripline, while the SMA shield is connected to the stripline center conductor. The characteristic impedance Z_O of an ideal transmission line is given by

$$Z_O = \sqrt{\frac{L}{C}} \qquad (5)$$

where L and C are the inductance and capacitance per unit length, respectively. Conventionally, the outer planes of a stripline are at earth potential, so there is no charge stored on the free-space capacitance from the outer planes to other nearby, grounded objects; C is therefore entirely contained

within the PCB. The stripline was designed using impedance equations derived under this assumption. However, in WCDM2, the outer planes of the stripline are connected to the SMA center conductor and they are charged to a non-zero potential. C in (5) now comprises the PCB stripline capacitance plus the stray capacitance of the planes to ground, such as to the nearby grounded shield of the SMA cable (see Figure 2). Because C in (5) is larger than intended, Z_O is smaller than the designed-for 50 Ω.

The stripline could be connected conventionally (i.e. SMA center to stripline center, SMA shield to stripline shield) to regain the designed-for 50 Ω impedance signal transmission. Indeed, TDR measurements on the same stripline connected in the conventional fashion reveal a characteristic impedance of 50 Ω. In such a configuration, the probe needle, and therefore the wafer and the chuck on which it rests, would be charged to the potential on the high voltage supply. This effectively increases C_{DUT} by charging the stray chuck capacitance with respect to ground. The result is a much longer discharge time constant and therefore a much wider current pulse, which is undesirable for CDM-like testers.

If a 50 Ω stripline is desired, the PCB capacitance could be designed to be smaller, i.e. the stripline could be designed to have an impedance greater than 50 Ω if connected conventionally. Then, when connected in the inverted fashion depicted in Figure 8, the effective capacitance would be reduced and the impedance would be closer to 50 Ω because of the stray capacitance. However, the small mismatch in the present system does not seem to cause significant reflections to the discharge path as indicated by the very sharp, narrow discharge current measured by the calibration board (Figure 4).

C. System Calibration

When zapping something other than the calibration board, a calibration method is required to extract the DUT current from the system discharge waveform. This method is illustrated in Figure 10. First, the calibration board current and the system discharge current are measured simultaneously during a zap at the desired precharge voltage. Next, the calibration board current waveform is aligned such that its rising edge is coincident with the rising edge of the last peak in the system discharge current waveform (Figure 10a); this last peak represents the discharge of C_{DUT} as discussed earlier. Subtraction of the calibration board current from the system discharge current measured yields a calibration current pulse (Figure 10b). For subsequent zaps at the same precharge voltage, the calibration pulse can be subtracted from the measured current (Figure 10c) to produce the DUT current (Figure 10d). The calibration pulse represents the system discharge and is independent of the DUT; therefore, the calibration pulse can be used to calculate the current felt by any on-wafer DUT at the same precharge voltage.

A calibration pulse is needed for every precharge voltage that will be used during product testing. The user could take calibration measurements at each of the desired precharge voltages to extract and store a calibration pulse at each voltage. Or, a single calibration pulse can be appropriately scaled for

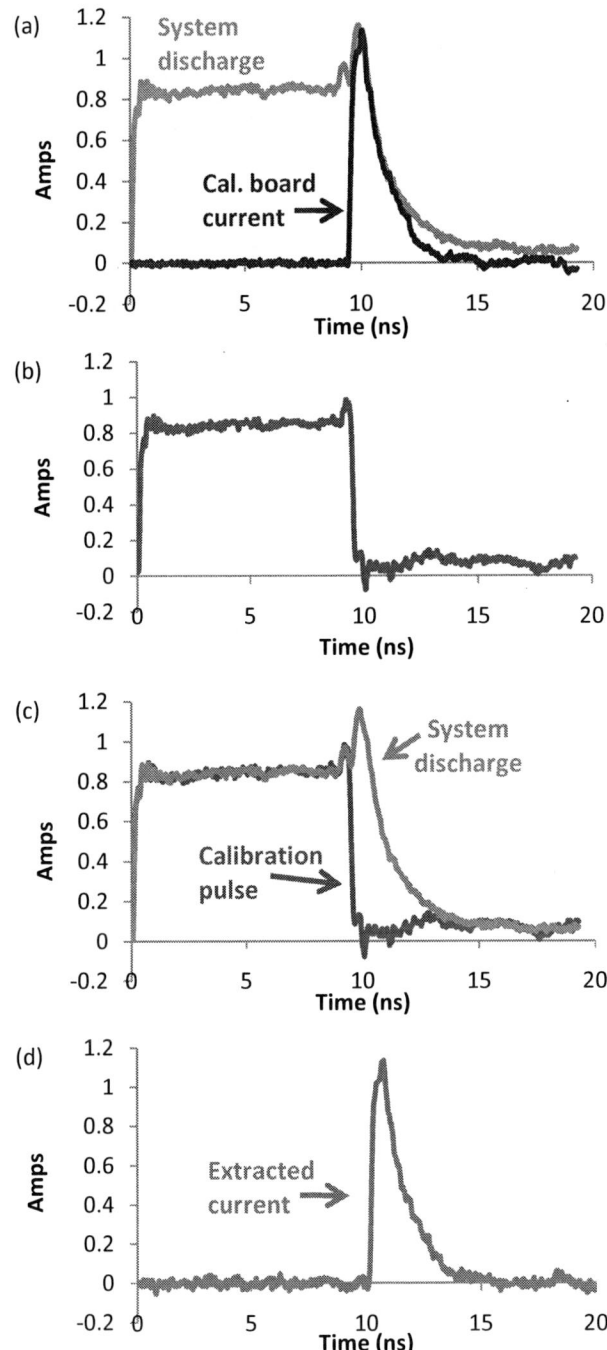

calibration board zap. As expected, the measured and calculated waveforms are nearly identical, as shown in Figure 11a. The 100 V calibration pulse current was then scaled up by a factor of 6 and was used to extract the DUT current from a 600 V zap. As shown in Figure 11b, the calculated and measured currents show close agreement, demonstrating that a single calibration pulse can be scaled and used at multiple precharge voltages with considerable accuracy.

The accuracy of the calculated calibration pulse will depend on proper alignment of the system discharge and calibration board waveforms (Figure 10a). Achieving good alignment is not difficult if a fast-sampling oscilloscope is used to acquire the waveforms; such a scope will provide several data points on the sharp rising edges for more precision in the alignment process. In any case, misalignment by several tens of picoseconds does not significantly alter the peak or the shape of the extracted DUT current.

The calibration procedure can be carried out entirely on an oscilloscope equipped with arithmetic capabilities. The system discharge and disk resistor waveforms can be simultaneously measured on separate channels then subtracted using built-in functions to produce the calibration pulse. The calibration pulse can be stored in memory, and on subsequent zaps it can be subtracted from subsequent system discharge waveforms to produce the DUT current instantaneously. A stored calibration pulse can also be appropriately scaled using built-in math functions and applied to subsequent measurements at multiple precharge voltages.

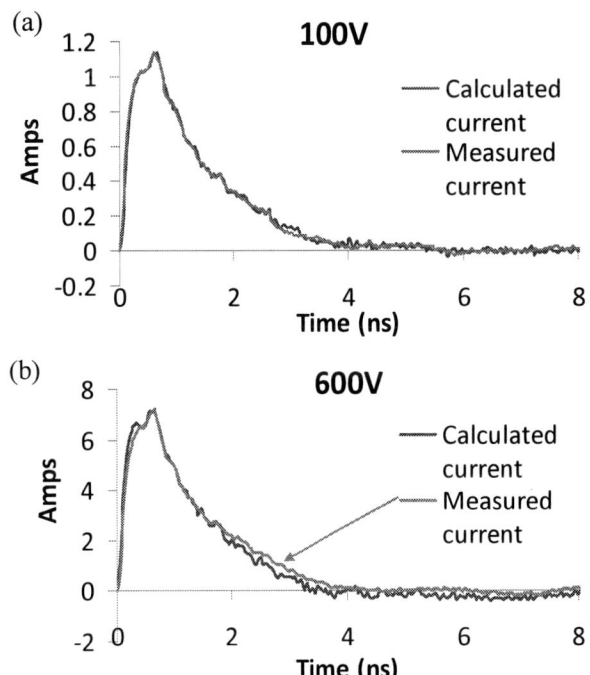

Figure 11: (a) A calibration pulse calculated for a 100 V precharge voltage was used to calculate the DUT current during a subsequent 100 V zap of the calibration board for a nearly identical match. (b) The 100 V calibration pulse was scaled by a factor of 6 and applied to a 600 V measurement; the calculated and measured currents show close agreement, demonstrating that a single calibration pulse can be scaled and used at multiple precharge voltages with considerable accuracy.

Figure 10: (a) Current measured by the scope channel into which the system was discharged, compared with the current measured by the calibration board. (b) The calibration pulse calculated by subtracting the DUT current measured from the system discharge current. (c) On subsequent zaps, subtraction of the calibration waveform from the system discharge yields the DUT current. (d) The extracted DUT current from a subsequent zap.

use at multiple precharge voltages. The results are still quite accurate since the system response is essentially a linear function of the precharge voltage (see Figure 12). This is demonstrated in the following experiment. A calibration pulse was extracted for a 100 V precharge voltage; it was then used to calculate the DUT current during a subsequent 100 V

978-1-4244-9113-1/11 $26.00 © 2011 IEEE   413

D. Discharge Linearity and Repeatability

The resistance of the spark generated during FICDM discharge is a decreasing function of the precharge voltage. The consequence is that the FICDM peak current is a nonlinear function of the precharge voltage [9]. During a WCDM2 discharge, the spark is initiated in an environmentally-controlled mercury-wetted relay, and the spark resistance is essentially independent of precharge voltage. The WCDM2 peak current is plotted as a function of precharge voltage in Figure 12 for the case of a 17 pF card discharging into the calibration board. The peak current is a linear function of the precharge voltage, with roughly 1.2 A per 100 V precharge. This linearity was verified from 100 V to 1000 V.

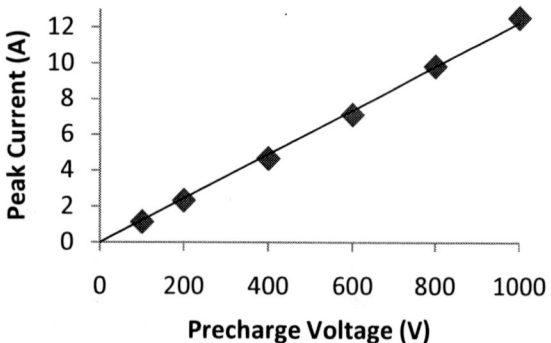

Figure 12: WCDM2 peak current as a function of the precharge voltage using a 17pF probe card to zap the calibration board. Because the spark occurs within the controlled environment of the mercury-wetted relay, the spark resistance is constant, and the current increases linearly with voltage.

The repeatability of WCDM2 is significantly improved over WCDM and FICDM. Figure 13 shows a screen capture of 50 repeated zaps of the calibration board at 100 V. Virtually no variation is seen. In contrast, WCDM showed a variation of about 30% in the peak current at the same precharge voltage [1]. The FICDM measurement tolerance is specified at 20% [5], but this is often exceeded under various test conditions [9].

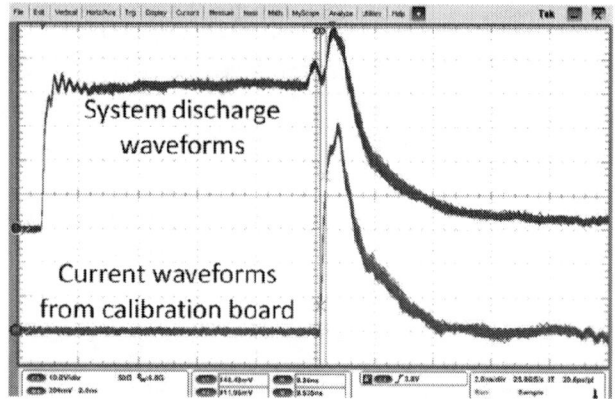

Figure 13: Oscilloscope screen capture showing 50 consecutive zaps of the calibration board at 100 V. Variations are negligible. The peak current is approximately 1 A, and the timescale is 2ns per major division.

IV. COMPARISON WITH FICDM

It is desirable that the current pulse generated by WCDM2 is comparable to that of FICDM since it is an industry-wide standard [5]. A FICDM waveform and a WCDM2 waveform are shown in Figure 14. The FICDM waveform was obtained while zapping the VSS pin of a large, packaged Intel product with an effective discharge capacitance C_{eff} of 16 pF. The WCDM2 waveform is from a zap to a standalone ESD structure on a full wafer with C_{DUT} equal to 13 pF. Both were taken at a 500 V precharge voltage, and the peak currents have been normalized to 1 A in Figure 14. The WCDM2 pulse width and rise time are very similar to those of FICDM pulses. The 10%-90% rise time of the WCDM2 waveform in Figure 14 is approximately 260 ps, while that of the FICDM waveform is approximately 240 ps. The pulse width at the 50% point is approximately 1.08 ns, while for FICDM it is approximately 0.8 ns.

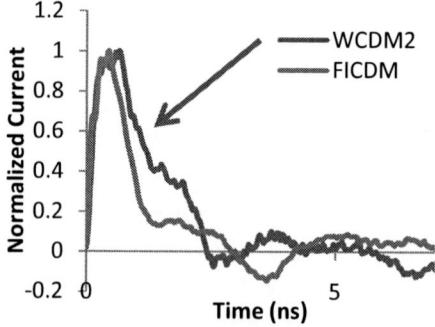

Figure 14: WCDM2 waveform from the zap of a standalone ESD structure on a full wafer compared to the FICDM waveform from the zap of a VSS pad on a packaged Intel product. Both were taken at 500 V precharge, and the peak currents have been normalized to 1 A in both cases. The WCDM2 pulse width and rise time are very similar to FICDM pulses.

An even closer match of the WCDM2 pulse width to that of FICDM could be achieved by more closely matching the RC time constant in the two systems. The resistance in the discharge path of a FICDM tester consists of the 1 Ω disk resistor plus an average spark resistance of about 25 Ω [1]; this is less than half the discharge resistance for WCDM2, which includes the 50 Ω scope input impedance. Therefore a WCDM2 card designed to have C_{DUT} roughly equal to half of typical FICDM C_{eff} values should produce a similar RC time constant and therefore similar pulse widths. The FICDM C_{eff} can range from a few pF for small packages to 10 - 20 pF for large packages [11], so C_{DUT} values of 5 – 10 pF should result in comparable WCDM2 pulse widths.

V. DEVICE TESTING RESULTS

An on-wafer standalone N+/Pwell diode was stressed using WCDM2. The N+ terminal was grounded, and a negative bias was applied to the field plate, thereby inducing a positive charge on the substrate. Upon discharge, the positive substrate charge travels to the grounded N+ terminal, forward biasing the junction. A semiconductor parameter analyzer was connected to the chuck of the system and was in high-impedance mode during the WCDM2 zap. Pre- and post-stress curve traces were obtained by forcing current from the chuck to the grounded terminal. As shown in Figure 15, damage to the junction occurred after a 6 A stress.

978-1-4244-9113-1/11 $26.00 © 2011 IEEE

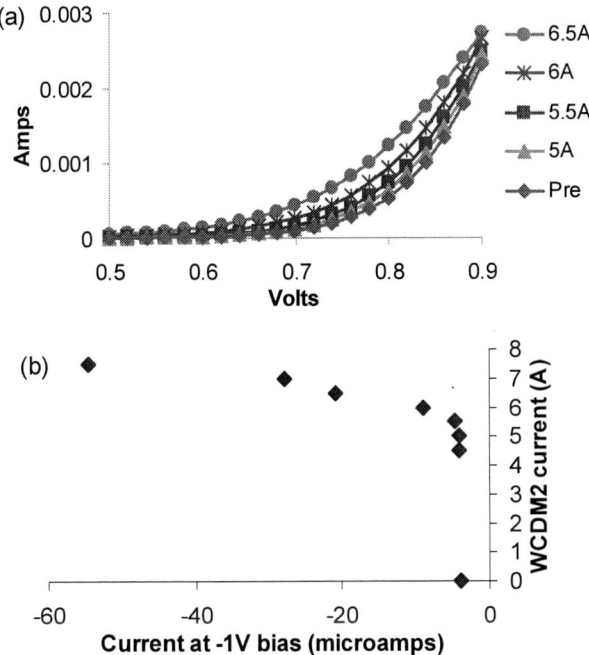

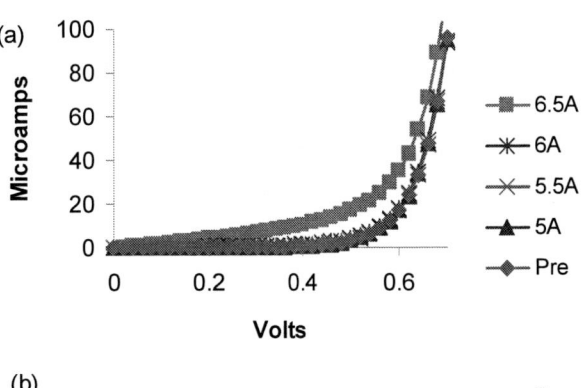

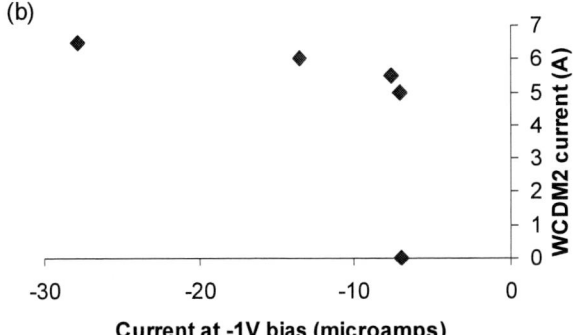

Figure 15: (a) Forward-bias dc I-V curve of an on-wafer standalone N+/Pwell diode before ("Pre") and after WCDM2 stress of increasing current levels. (b) DC current measured with a -1 V bias applied. Both plots indicate the device was damaged after a 6 A stress.

Figure 16: (a) Forward-bias dc I-V curve of a dual-diode-protected receiver before ("Pre") and after WCDM2 stress of increasing current levels. (b) DC reverse leakage current measured with a -1 V bias applied. The circuit was damaged after a 6 A stress. Experiments revealed that the top diode was the damaged device.

A dual-diode-protected receiver circuit with rail clamp was also stressed using WCDM2. Both the top and bottom diodes were sized similarly to the standalone N+/Pwell diode described above. The receiver input pad was grounded, and a positive potential was applied to the WCDM2 field plate, thereby storing negative charge on the substrate. Upon discharge, current flows from the input pad through the top diode and then through the rail clamp to ground. As seen in Figure 16, damage occurred after a 6A stress. It is concluded that the top diode was damaged because, in separate experiments, a standalone rail clamp passed 10 A of WCDM2 testing, and a similar receiver structure with diodes sized twice as large passed 10 A.

VI. CONCLUSIONS

WCDM2 produces repeatable current stresses that resemble those of FICDM testers, though with much better repeatability. Unlike FICDM, the peak current per precharge voltage is highly linear. WCDM2 allows for CDM-like stressing of products and smaller test structures without the need for packaging. The generated stress is unipolar – a benefit for debugging failure modes. A scalable calibration method was presented for calculating the current felt by the DUT. In-situ pre- and post-stress curve tracing may be performed to identify the current failure threshold.

ACKNOLWEDGEMENTS

The authors thank A. K. M. Ahsan of Intel for the FICDM waveform.

REFERENCES

[1] B. Chou, T. Maloney, and T. Chen, "Wafer-level charged device model testing," in *Electrical Overstress/Electrostatic Discharge Symposium Proceedings,* 2008, pp. 115-124.

[2] H. Wolf, H. Gieser, W. Stadler, and W. Wilkening, "Capacitively coupled transmission line pulsing CC-TLP—A traceable and reproducible stress method in the CDM-domain," in *Electrical Overstress/Electrostatic Discharge Symposium Proceedings,* 2003, pp. 338-345.

[3] H. Wolf, H. Gieser, and D. Walter, "Investigating the CDM Susceptibility of IC's at Package and Wafer Level by Capacitive Coupled TLP," in *Electrical Overstress/Electrostatic Discharge Symposium Proceedings,* 2007, pp. 297-303.

[4] A. Gerdemann, E. Rosenbaum, and M. Stockinger, "A novel testing approach for full-chip CDM characterization," in *Electrical Overstress/Electrostatic Discharge Symposium Proceedings,* 2007, pp. 289-296.

[5] ESD Association, "Electrostatic Discharge Sensitivity Testing, Charged Device Model, Component Level," Standard Test Method ANSI/ESD STM 5.3.1-1999, 1999.

[6] T. Maloney, "Evaluating TLP Transients and HBM Waveforms," in *Electrical Overstress/Electrostatic Discharge Symposium Proceedings*, 2009, pp. 143-151.

[7] T. Maloney, "TDT-CDM," Presentation to CDM Standards Committee, ESD Association Meeting, Anaheim, CA, Feb. 2009.

[8] H. Johnson. (2001, Apr.). Strange Microstrip Modes. *EDN: Electronics Design, Strategy, News.* [Online]. P. 32. Available: http://www.edn.com/article/497702-Strange_microstrip_modes.php

[9] R. Given, M. Hernandez, and T. Meuse, "CDM2 – A New CDM Test Method for Improved Test Repeatability and Reproducibility," in *Electrical Overstress/Electrostatic Discharge Symposium Proceedings,* 2010, pp. 359-367.

[10] W. Dally and J. Poulton, *Digital Systems Engineering.* Cambridge, United Kingdom: Cambridge University Press, 2001.

[11] B. Atwood, Y. Zhou, D. Clarke, and T. Weyl, "Effect of Large Device Capacitance on FICDM Peak Current," in *Electrical Overstress/Electrostatic Discharge Symposium Proceedings,* 2007, pp. 273-282.

Reliability of GaN-HEMTs for High-Voltage Switching Applications

(invited paper)

Wataru Saito

Semiconductor Company, Toshiba Corporation
Kawasaki, Japan
+81-44-549-2683, wataru3.saito@toshiba.co.jp

Abstract—**This paper reports that the maximum electric field is a dominant factor for reliability in high-voltage GaN-HEMTs. Four types of the GaN-HEMT with different field plate (FP) structures were tested in continuous switching operation mode to analyze the degradation mechanism and the optimal device design. From the on-resistance degradation dependence on the FP structure, we extract that the gate-edge electric field strongly affects the increase of the dynamic on-resistance. Although the FP-edge field also increased the dynamic on-resistance, its influence was weaker than that of the gate-edge field. The optimal FP structure minimizes the increase of the dynamic on-resistance by reducing the electric field peaks and showed no degradation of power efficiency at the boost converter operation.**

Keywords: GaN, HEMT, High-Voltage Switching

I. INTRODUCTION

The AlGaN/GaN HEMTs can produce high-power-density operation with low power loss in RF and power electronic systems due to high carrier mobility in two-dimensional electron gas (2DEG) and high breakdown voltage due to large critical electric field [1], [2]. Recent demonstrations show that AlGaN/GaN devices can attain breakdown voltage of several hundred volts with ultra-low on-resistance lower than the Si-limit [2]-[4].

Power electronic applications require fast digital switching under high applied voltage and thus low dynamic on-resistance. In high-voltage GaN-HEMTs, dynamic on-resistance increases with voltage due to current collapse. In previous works, it was reported that the increase of dynamic on-resistance in the first minutes of continuous switching operation, and showed that this increase could be controlled and suppressed through field plate (FP) engineering [4]-[6]. In this work, in contrast, we report on the long term reliability of GaN switches.

The reliability of GaN-HEMT in the RF applications has been extensively studied [7]-[9]. Two main degradation mechanisms has been proposed, inverse piezoelectric [7] and hot carrier injection, both of which depend on the electric field distribution in the device. In the previous work, the effect of the FP structure upon RF stress has also been studied [10].

This paper reports on the dynamic on-resistance degradation of GaN-HEMTs during continuous switching operation at 300 V. The influence of the electric field on

reliability is discussed and analyzed through HEMTs with different FP configurations. Finally, we evaluate the performance of a boost converter using the GaN-HEMT with the optimal FP structure.

II. DEGRADATION MECHANISM AND DEVICE DESIGN

Two degradation mechanisms, hot carrier injection and inverse piezoelectric effect has been proposed in GaN HEMTs designed for RF applications. The hot carrier generates defects in the AlGaN barrier and the passivation interface due the injection impact. The inverse piezoelectric effect induces the mechanical stress by the electric field [7] and generates the defect. The electric field is a key factor for either mechanism. Therefore the electric field must be managed to realize high reliability.

The FP structure modulates the electric field distribution due to the shield effect [5]. The FP length and the insulator thickness under the FP electrode are design parameters. A long FP and a thin insulator enhance the shield effect. In those cases, the electric field reduces at the gate edge and increases at the FP edge. Therefore the FP length and the insulator thickness must be optimized to minimize the electric field peaks.

An experiment with four types of FP structures, described in Fig. 1, were used to analyze the influence of the electric field on the long term electrical reliability of GaN-HEMTs in high-voltage switching operation mode. The dual-FP structure is a combination of the G-FP and the S-FP. Since the multi-step FP

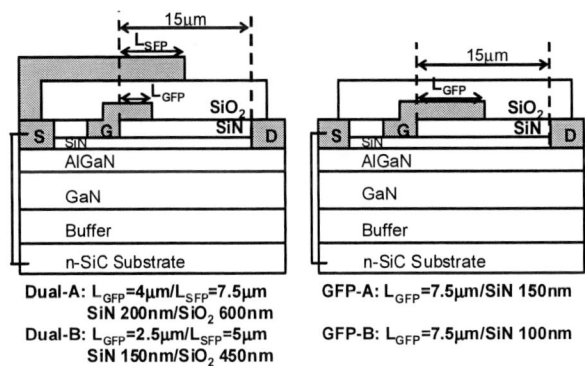

Dual-A: L_{GFP}=4µm/L_{SFP}=7.5µm
SiN 200nm/SiO$_2$ 600nm
Dual-B: L_{GFP}=2.5µm/L_{SFP}=5µm
SiN 150nm/SiO$_2$ 450nm

GFP-A: L_{GFP}=7.5µm/SiN 150nm

GFP-B: L_{GFP}=7.5µm/SiN 100nm

Fig. 1 Cross-sectional structure of fabricated GaN-HEMTs with FPs.

978-1-4244-9113-1/11 $26.00 © 2011 IEEE

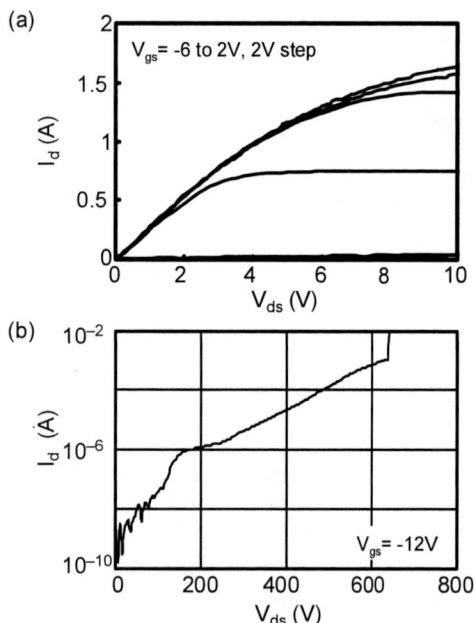

Fig. 2 (a) on-state and (b) off-state I-V curves of the fabricated GaN-HEMTs.

Table I Sample list of continuous switching test.

Device Structure	initial α	Electric Field Peak (Vertical Term) (MV/cm)	
		Gate Edge	FP Edge
Dual-A	15%	1.76 (0.72)	1.45 (0)
Dual-B	23%	**1.81 (0.93)**	1.51 (0)
GFP-A	24%	1.42 (0.56)	**2.42 (0.24)**
GFP-B	37%	1.10 (0.35)	**2.98 (0.51)**

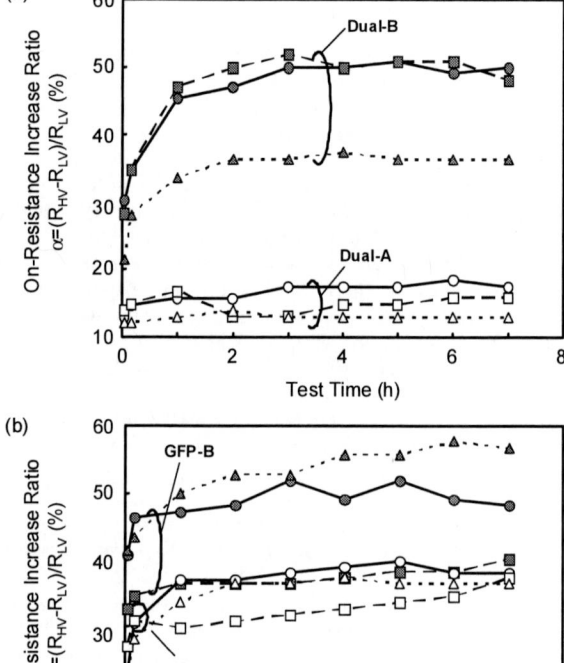

Fig. 3 Change of on-resistance increase ratio α with switching test time.

structure reduces whole peaks by spreading the electric field concentration [11], the dual-FP structure is more effective for high reliability than the G-FP structure. In this work, the substrate was connected to the source for a role of the backside FP electrode.

AlGaN/GaN heterostructures were grown on n-SiC (ρ = 0.02 Ωcm) substrate by MOCVD. First, 180 nm-thick AlN/n-GaN/AlN layers were grown as a buffer layer [12]. Then a 5-μm-thick undoped GaN layer as a channel layer and a 28-nm-thick $Al_{0.25}Ga_{0.75}N$ as a barrier layer were grown. The device processing consisted of conventional HEMT fabrication steps [5]. The MIS gate structure with 10 nm-thick SiN gate insulator film was employed to reduce the gate leakage current. SiN and SiO_2 were deposited by CVD as passivation films, whose thicknesses were shown in Fig. 1. The gate-drain offset length was 15 μm and the gate length was 1 μm. The gate width was 3 mm. The active device area was 0.089 mm², which included source and drain contact regions.

The specific on-resistance of the fabricated GaN-HEMT was as low as 3.2 mΩcm² (10.8 Ωmm), which was estimated from the on-state drain voltage for the drain current of 0.5 A at V_{gs} = 2 V as shown in Fig. 2(a). The threshold voltage was –6 V. The maximum drain current was 1.7 A (570 mA/mm). The breakdown voltage was 640 V as shown in Fig. 2(b). These static characteristics were independent from the FP structure because the 2DEG density was not modulated by the FP structure at low V_{ds} and the breakdown voltage was determined by the vertical electric field in the epitaxial layers between the drain electrode and the substrate.

III.   CONTINUOUS SWITCHING TEST

The dynamic on-resistance after applying high-voltage R_{HV} was measured from the average on-state voltage at continuous switching with a resistive load in 5 μs of time interval starting 20 μs after turn-on signal using the oscilloscope [5]. The switching frequency was 10 kHz, the duty ratio was 25%, and the on-state drain current was 0.4 A. The on-resistance without applying high-voltage R_{LV} was measured using the curve-tracer. The on-resistance degradation is discussed using the on-resistance increase ratio $\alpha = (R_{HV} - R_{LV})/R_{LV}$ in this work.

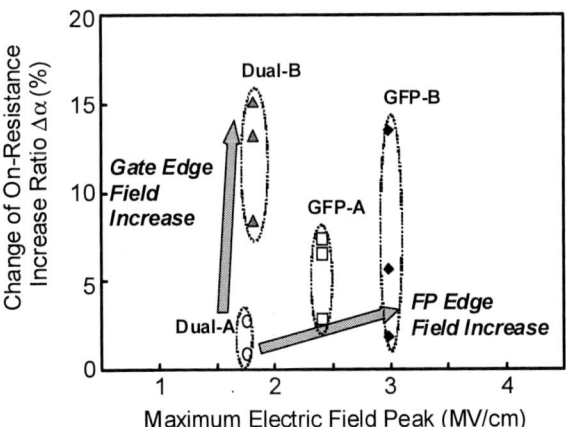

Fig. 4 Relation between the change of on-resistance increase ratio and maximum electric field peak.

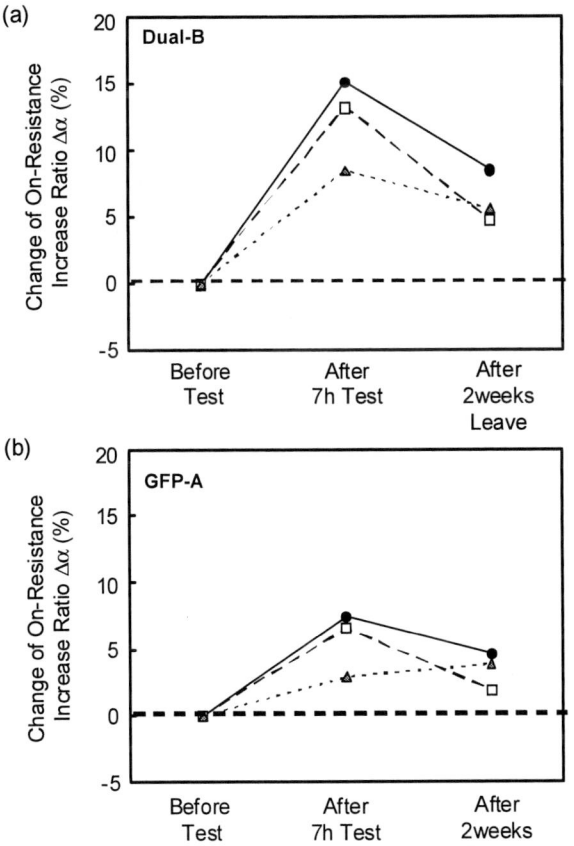

Fig. 5 The change of on-resistance increase ratio α after test and leaving for 2 weeks for (a) Dual-B and (b) GFP-A.

To discuss the relation between the on-resistance degradation and the electric field, the electric field distribution was calculated using the device simulator ISE-Dessis [13]. It was simulated at V_{ds} = 300 V and V_{gs} = –12 V, which were the off-state conditions of the switching test for R_{HV} measurement.

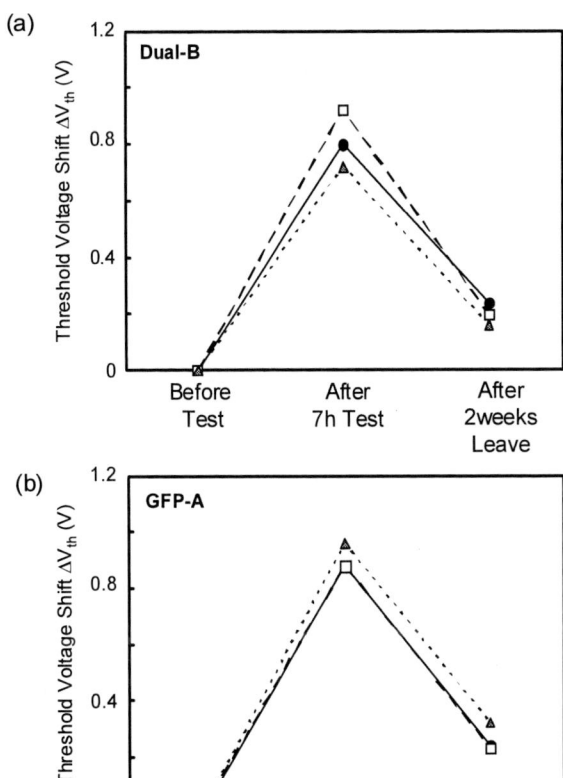

Fig. 6 The change of threshold voltage after the test and leaving for 2 weeks for (a) Dual-B and (b) GFP-A.

In these simulations, the SiC substrate was assumed metallic. To discuss the inverse piezoelectric effect, the vertical terms of the electric field peaks are also shown in the table.

The initial dynamic on-resistances increased by the current collapse phenomena and the simulated electric field peaks are shown in Table I. Dual-A had the smallest initial α in the tested structures due to minimum electric field peaks. Although initial α of Dual-B was almost same as that of GFP-A, the peak point of maximum electric field was different between Dual-B and GFP-A. GFP-B had the largest initial α and electric field peak in the tested structure. The degradation of α was measured during the continuous switching test for 7 hours to discuss the reliability. For each structure, three devices were tested.

The degradation of the dynamic on-resistance depended on the FP structure. The change of α for Dual-A samples was negligibly small as shown in Fig. 3(a). On the other hand, Dual-B samples were drastically degraded. In addition, the change of GFP-A samples was weaker than that of Dual-B even with the same initial α as shown in Fig. 3(b) and Table I. It is verified that the degradation depended on not only the electric field peak but also peak position. The degradation for GFP-B samples was stronger than that for GFP-A.

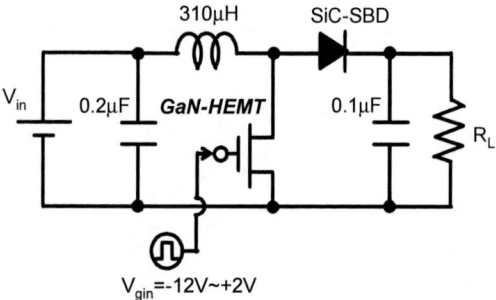

Fig. 7 Demonstrated boost converter circuit, in which the GaN-HEMT was used as a main switch.

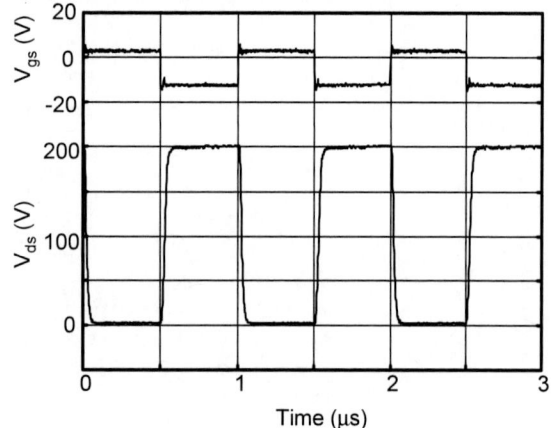

Fig. 8 Operating waveform of the boost converter circuit using dual-FP GaN-HEMT with an output voltage of 200 V and switching frequency of 1 MHz.

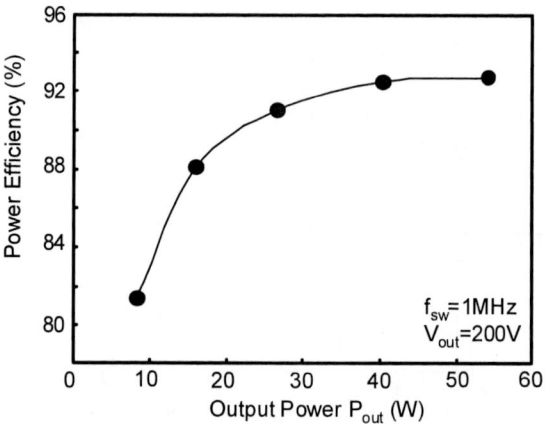

Fig. 9 Power efficiency of demonstrated boost converter with a constant output voltage of 200 V.

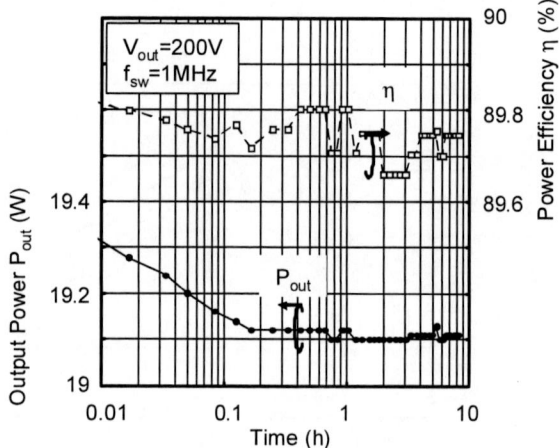

Fig. 10 Continuous operation of the boost converter using dual-FP device with an output voltage of 200 V. The output power fluctuation was as low as 1.2 %.

Degradation and maximum electric field are both plotted in Fig. 4. The gate-edge electric field strongly affected the reliability. The FP-edge electric field also cause degradation, but its influence was much weaker. This behavior is similar to that of the initial on-resistance degradation α [14]. The results confirm that the optimal FP structure not only suppress current collapse (initial α), but also improves its long term reliability due to reduction of peak field.

The degradation mechanism is discussed as follows. The degradation of the dynamic on-resistance did not recover completely even after leaving for 2 weeks and the remained damage depended on the FP structure as shown in Fig. 5. On the other hand, the threshold voltage was almost recovered after 2 weeks for all devices as shown in Fig. 6. The influence of the threshold voltage shift upon the on-resistance was estimated less than 1% even with the shift of 1 V. From these results, the samples were damaged irreversibly during the test, although the temporary electron trapping was also occurred at the SiN/AlGaN interface.

The influence of the electric field depends on the peak position as shown in Fig. 4. In the Dual-B structure, not only the total electric field peak but also the vertical term at the gate edge was larger than that in the Dual-A structure as shown in Table 1. On the other hand, at the GFP structures, although the

total peak was increased, the vertical terms were smaller than the Dual-A structure.

A hot carrier injection depends on the total electric field due to the scattering and the impact ionization. The inverse piezoelectric effect is determined only by the vertical electric field in the AlGaN barrier [7]. From these discussions, it is shown that the gate-edge peak causes the dynamic on-resistance degradation due to the hot carrier injection and the inverse piezoelectric effect. On the other hand, the hot carrier injection also occurs at the FP-edge, but the degradation by the inverse piezoelectric effect is not appeared due to small vertical field. Therefore the gate edge peak strongly affects on the reliability comparing the FP edge peak.

IV. BOOST CONVERTER OPERATION

A boost converter circuit was operated using the device with the optimal FP structure, which was almost the same as the Dual-A [5]. The test circuit is shown in Fig. 7. Since the boost converter circuit is a major dc/dc converter circuit and employed to the PFC circuit in the ac/dc converter for PC power supplies, the power electronics application of high-voltage GaN-HEMTs can be verified using the demonstrated circuit.

The input and output powers were measured using digital meters. The pulse-wise gate-source voltage swinging from 2 to −12 V was supplied using a pulse generator and the duty ratio was 50%. As a recovery diode, a SiC-SBD was used to minimize the recovery loss.

The fabricated device operated at a dc output voltage of 200 V and a switching frequency of 1 MHz as shown in Fig. 8. High power efficiency of more than 90% was achieved as shown in Fig. 9. The maximum output power was 54 W with a power efficiency of 92.7%.

The continuous operation was tested with an output voltage of 200 V. The power efficiency did not degrade after 8 hours of the continuous operation, as shown in Fig. 10. The output power decrease on the first 10 minutes was due to self heating effects on the load resistance. Therefore the power efficiency was not changed. The output power fluctuation was as low as 1.2% even with the influence of the load change. It is shown that the optimal FP design realizes stable circuit operation under high applied voltage.

V. CONCLUSIONS

The degradation of the dynamic on-resistance in high-voltage GaN-HEMTs was measured by continuous switching test to discuss the reliability in power electronic applications. From the FP structure dependence, the gate-edge electric field strongly affects the increase of the dynamic on-resistance. Although the FP edge field also increased the dynamic on-resistance, its influence was weaker than that of the gate-edge field. The degradation was not recovered even after the leaving for 2 weeks. From these results, it is verified that irreversible damage was generated by hot carrier injection at the switching. The optimal FP structure realizes no degradation in the circuit operation under high applied voltage due to minimizing the electric field peak.

ACKNOWLEDGMENT

The author wishes to thank K. Morizuka, T. Sugiyama, Y. Saito, T. Noda, H. Fujimoto, T. Nitta, Y. Kakiuchi, A. Yoshioka and T. Ohno for their support and fruitful discussion of this work.

REFERENCES

[1] U. K. Mishra, P. Parikh and Y.-F Wu, "AlGaN/GaN HEMTs – an overview of device operation and application," Proceedings of IEEE, vol. 90, pp. 1022-1031. 2002.

[2] N. -Q. Zhang, B. Moran, S. P. DenBaars, U. K. Mishra, X. W. Wang and T. P. Ma, "Effects of surface traps on breakdown voltage and switching speed of GaN power switching HEMTs," in IEDM'01 Tech. Digest, pp. 589-592, 2001.

[3] N. Kaneko, O. Machida, M. Yanagihara, S. Iwakami, R. Baba, H. Goto and A. Iwabuchi, "Normally-off AlGaN/GaN HFET using NiOx gate with recess," in Proceedings of ISPSD'09, pp. 25-28, 2009.

[4] N. Ikeda, S. Kaya, J. Li, T. Kokawa, M. Masuda and S. Katoh, "High-power AlGaN/GaN MIS-HFETs with field-plates on Si substrates," in Proceedings of ISPSD'09, pp. 251-254, 2009.

[5] W. Saito, M. Kuraguchi, Y. Takada, K. Tsuda, Y. Saito, I. Omura and M. Yamaguchi, "Current collapseless high-voltage GaN-HEMT and its 50-W boost converter operation," in Technical Digest of IEDM'07, pp. 869-872, 2007.

[6] N. Tipirneni, V. Adivarahan, G. Simin and A. Khan, "Silicon dioxide-encapsulated high-voltage AlGaN/GaN HFETs for power-switching applications," IEEE Electron Device Lett., vol. 28, pp.784-786, 2007.

[7] J. Joh and J. A. del Alamo, "Mechanixms for electrical degradation of GaN high-electron mobility transistors," in IEDM'06 Tech. Digest, 2006.

[8] G. Meneghesso, G. Verzellesi, F. Danesin, F. Rampazzo, F. Zanon, A. Tazzoli, M. Meneghini and E. Zanoni, "Reliability of GaN high-electron-mobility transistors: state of the art and prespectives," IEEE Trans. Device and Reliability, vol. 8, pp..332-343, 2008.

[9] M. Faqir, G. Verzellesi, G. Meneghesso, E. Zanoni and F. Fantini, "Investigation of high-electric-field degradation effects in AlGaN/GaN HEMTs," IEEE Trans. Electron Devices, vol. 55, pp. 1592-1602, 2008.

[10] C. Lee, H. Tserg, L. Witkowski, P. Saunier, S. Guo, B. Albert, R. Birkhahn and G. Munns, "Effects of RF stress on power and plused IV characteristics of AlGaN/GaN HEMTs with field –plate gates," IEE Electronics Letters, vol. 40, online no. 20046921, 2004.

[11] W. Feiler, E. Falck and W. Gerlach, " Multistep field plates for high-voltage planar p-n Junctions," IEEE Trans. Electron Devices, vol. 39, pp. 1514-1520, 1992.

[12] W. Saito, T. Noda, M .Kuraguchi, Y. Takada, Y. Saito, I. Omura and M. Yamaguchi, "Effect of buffer layer structure on drain leakage current and current collapse ohenomena in high-voltage GaN-HEMTs," IEEE Trans. Electron Devices, vol. 56, pp. 1371-1376, 2009.

[13] ISE TCAD Manuals, (ISE Integrated Systems Engineering AG, Zurich, 2002) part 11, release 8.

[14] W. Saito, T. Nitta, Y. Kakiuchi, Y. Saito, T. Noda, H. Fujimoto, A. Yoshioka and T. Ohno, "Influence of Electric Field upon Current Collapse Phenomena and Reliability in High Voltage GaN-HEMTs," in Proceedings of ISPSD'10, pp. 339-342, 2010.

Time Evolution of Electrical Degradation under High-Voltage Stress in GaN High Electron Mobility Transistors

Jungwoo Joh and Jesús A. del Alamo
Microsystems Technology Laboratories
Massachusetts Institute of Technology
Cambridge, MA, USA
+1-617-258-5752, Jungwoo@mit.edu

Abstract—**In this work, we investigate the time evolution of electrical degradation of GaN high electron mobility transistors under high voltage stress in the OFF state. We found that the gate current starts to degrade first, followed by degradation in current collapse and eventually permanent degradation in I_D. We also found that the time evolution of gate current degradation is unaffected by temperature, while drain current degradation is thermally accelerated.**

Keywords-GaN, HEMT, reliability, degradation, time evolution

I. INTRODUCTION

In GaN high electron mobility transistors (HEMTs), electrical reliability remains a key concern. To address reliability issues, it is critical to develop detailed physical understanding of the mechanisms behind device degradation. High-voltage stress has been found to result in device degradation with a critical voltage behavior [1-2] and characterized by trap formation and trapping [3-4]. To date, few studies have investigated the time evolution of degradation under high-voltage stress conditions [3, 5-8]. Some of these studies focused on the evolution of the gate/drain leakage current [6, 8]. Although there have been efforts to understand the time evolution of the drain current [3], its detailed nature is still largely unknown. This is important in order to develop degradation models that can predict device lifetime under given operational conditions.

In this work, we investigate the time evolution of degradation in the gate and drain currents and other figures of merit under high voltage stress with voltages in excess of the critical voltage. In particular, we separately monitor permanent and trapping-related degradation of the drain current, as well as degradation in the gate current [9]. We have found that there is an *incubation time* during which degradation does not take place and that this incubation time is different for different device figure of merits. The temperature dependence of this incubation time is also different. The incubation time for gate current degradation was found to be almost temperature independent while that of the drain current degradation was thermally activated.

II. EXPERIMENTAL

We have studied 0.25 μm GaN HEMTs on SiC [10]. The device width is 2x25 μm. We performed OFF-state stress at V_{GS}=-7 V and V_{DS}=40 V. This voltage was much larger than the critical voltage at which sudden gate current degradation takes place (typical V_{DGcrit} for these devices is about 20 V at 150 °C) [1]. These experiments were performed at various base-plate temperatures between 75 and 150 °C. Throughout the experiment, at regular intervals, we performed detailed device characterization at 30°C, including comprehensive trapping analysis [4]. The effect of stress times as short as 10 ms and as long as several days was studied.

Among various figures of merit that we monitored at 30°C, we focused on reverse-bias gate leakage current I_{Goff} (I_G at V_{GS}=-5 V, V_{DS}=0.1 V), permanent degradation in I_{Dmax} (I_D at V_{GS}=2 V, V_{DS}=5 V) and I_{Dlin} (I_D at V_{GS}=1 V and V_{DS}=0.5 V), and drain current collapse (a measure of trapping related degradation). We have also monitored linear threshold voltage (defined as V_{GS} at I_D=1 mA/mm for V_{DS}=0.1 V). Permanent degradation was defined as the decrease in uncollapsed I_{Dmax} (or I_{Dlin}) that is measured after a detrapping step where the device was illuminated with microscope light for 30 s. We use the term "permanent" because this degradation is completely irreversible after light illumination, heating up to 150 °C, or waiting for 1 month. This is unlike drain current collapse which can be reversed by electron detrapping. Our definition of uncollapsed or permanent I_D reflects device behavior with most of the traps empty after a detrapping step [4].

In order to measure drain current collapse, we performed I_D transient measurements where trapping was produced by applying a 1 s voltage pulse with V_{GS}=-10 V and V_{DS}=0 V [4]. The collapsed value of I_{Dlin} was measured 10 ms after the removal of the pulse. We used I_D in the linear regime for the current collapse measurements in order to suppress self-heating and measurement-induced trapping [4]. Current collapse was defined as the relative change in I_{Dlin} before and after the trapping pulse.

These experiments were performed on devices that are located side by side in a small chip (~10 mm^2) and thus have closely matched characteristics. This allowed us to compare different stress conditions in a consistent manner.

978-1-4244-9113-1/11 $26.00 © 2011 IEEE

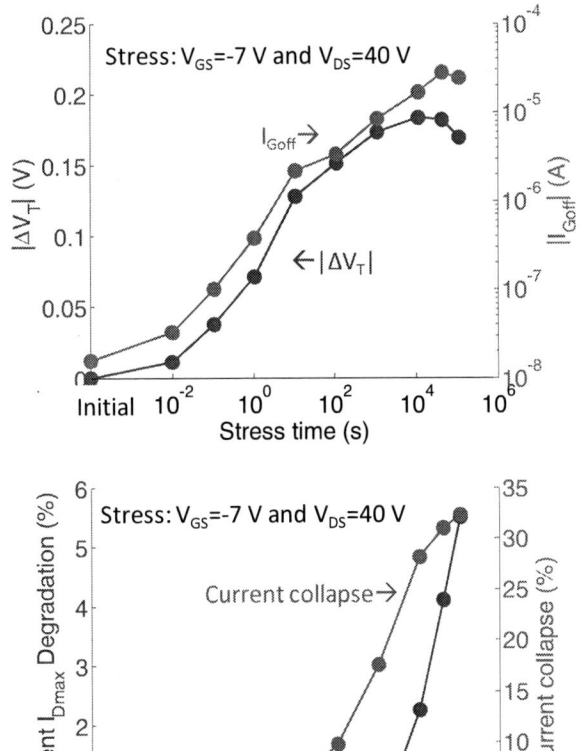

Fig. 1. Time evolution of I_{Goff}, V_T (actual change is negative. See Fig. 4), current collapse, and permanent I_{Dmax} degradation for a device stressed in the OFF state (V_{GS}=-7 V, V_{DS}=40 V) at 125 °C. Device characterization was performed at 30 °C.

III. TIME EVOLUTION OF DEGRADATION

Fig. 1 shows the time evolution of V_T, I_{Goff}, I_{Dlin} current collapse, and permanent I_{Dmax} degradation for a device stressed at T_{base}=125 °C. The gate current is seen to increase even after just 10 ms of stressing. It eventually tends to saturate after a 2-3 orders of magnitude increase, although this takes as long as ~10^4 s. The threshold voltage changes in a similar way to I_{Goff}. On the other hand, degradation of the drain current, permanent or trapping related is a significantly slower process. It can be seen that there is an incubation time for which negligible degradation occurs in current collapse and permanent I_{Dmax} degradation (uncollapsed I_{Dlin} behaves in a similar way). Current collapse starts to increase at around 10 s of stress time and tends to saturate at ~10^4 s. Permanent degradation in drain current (non-trapping related) starts even later – it took more than 1000 seconds to see noticeable permanent degradation – and it does not seem to saturate.

Detrapping analyses at various times during the experiment reveal that the increase in current collapse originates from two different sources. As shown in the time constant spectrum (described in more detail in [4]) in Fig. 2, a detrapping component with a well defined time constant (marked as DP1)

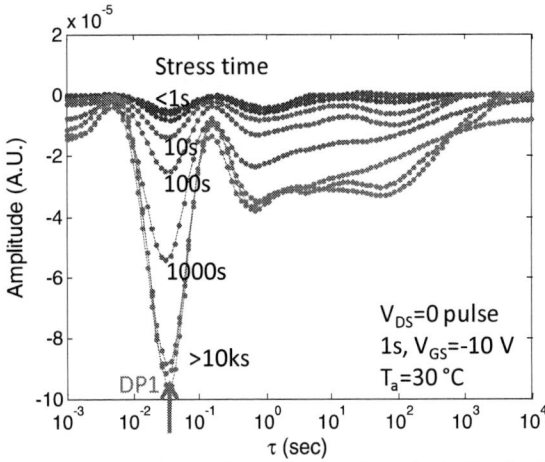

Fig. 2. Time constant spectra of detrapping transients after 1s V_{DS}=0 pulse with V_{GS}=-10 V measured at 30 °C for the same device stressed in Fig. 1 at 125 °C.

abruptly increases for stress times beyond 10 s and saturates after a few hours. A trap with a similar time constant was found in earlier work to sharply increase for stress beyond the critical voltage and to have an energy level of ~0.56 eV [11]. In addition, a broad spectrum of traps with longer detrapping time constants (detrapping time of 1-100 s) emerges with a similar stress time dependency (Fig. 2). The time dependence for the growth of these trapping peaks matches well the time evolution of current collapse in Fig. 5 suggesting that these are different manifestations of the same physics which is prominent trap generation over a time window of 10-10^4 s.

IV. TEMPERATURE DEPENDENCE

The time evolution of gate and drain current degradation exhibits different behavior at different stress temperatures. As shown in Fig. 3, gate current degradation depends little on temperature. At all stress temperatures between 75 and 150 °C, the gate current starts to increase even after 10 ms of stressing. Interestingly, the threshold voltage exhibits a similar behavior to I_{Goff}. As shown in Fig. 4, in an initial stage (t_{stress}<10 s), the

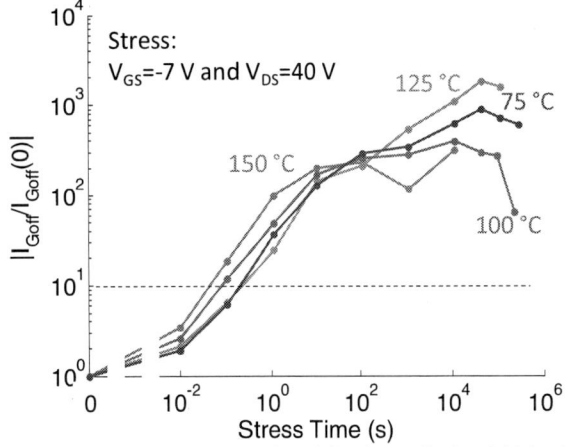

Fig. 3. Evolution of gate current degradation (normalized to initial value) at various stress temperatures. Four different devices were used. The stress condition was V_{GS}=-7 V and V_{DS}=40 V. Device characterization performed at 30 °C.

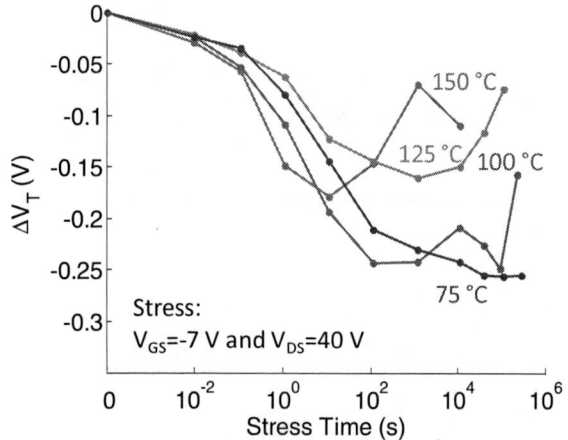

Fig. 4. Evolution of threshold voltage (relative to its initial value) in the experiment of Fig. 3. Device characterization performed at 30 °C.

change in V_T is largely unaffected by temperature. For longer times, ΔV_T tends to saturate. This similarity in the behavior of I_{Goff} and V_T is reasonable because both are mainly affected by degradation in the intrinsic gate region.

In contrast, drain current degradation, both permanent and trapping related, show clear stress temperature dependence (Fig. 5Fig. 6). Degradation in I_{Dlin} current collapse and uncollapsed I_{Dmax} (and I_{Dlin}) start earlier at higher temperatures and the final value also increases as the stress temperature increases. This later observation is consistent with earlier studies [12]. The fact that permanent I_{Dlin} evolves in a similar way as permanent I_{Dmax} (Fig. 6) suggests that the changes that we are observing are taking place in the extrinsic portion of the device, more concretely the drain since that is where the high field region appears during stress.

Our experiments reveal that there is an incubation time for device degradation and that the incubation time is different and evolves in a different way with temperature for the different figures of merit. We have extracted the incubation time for these modes of degradation with the following definitions. For I_{Goff} degradation, the incubation time is defined as the time for I_{Goff} to increase by a factor of 10. For permanent I_{Dmax}

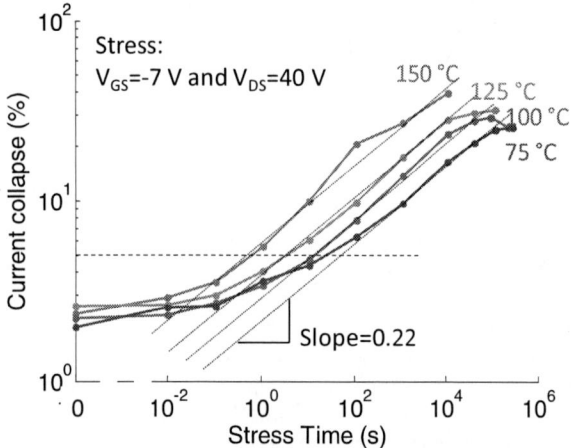

Fig. 5. Evolution of I_{Dlin} current collapse in the experiment of Fig. 3. Current collapse is defined as the relative change in I_{Dmax} after applying 1s $V_{DS}=0$ pulse with $V_{GS}=-10$ V at 30 °C.

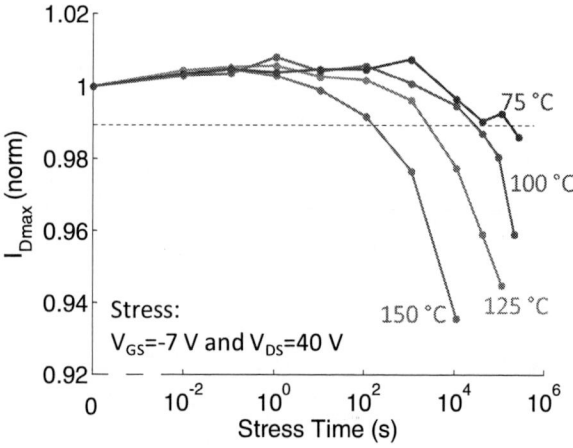

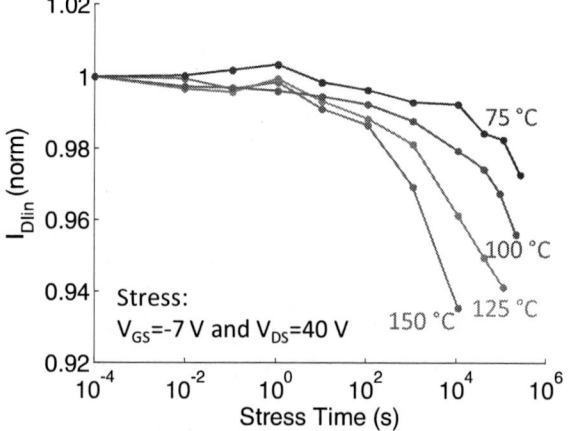

Fig. 6. Evolution of permanent I_{Dmax} and I_{Dlin} degradation in the experiment of Fig. 3. The values are normalized to their unstressed levels. Device characterization performed at 30 °C.

degradation, it is defined when the uncollapsed I_{Dmax} decreases by 1%. For trapping-related (current collapse) degradation, we monitor the time for current collapse to reach 5%.

As shown in Fig. 7, the incubation time for I_{Goff} is almost unaffected by temperature ($E_a \sim 0.17$ eV). This is consistent with the previous observation in [8]. In contrast, the incubation times for drain current degradation are thermally activated but with different activation energies (1.12 and 0.59 eV, respectively), suggesting that different physical processes are involved. The activation energy for permanent I_{Dmax} degradation is very close to that of I_{Dlin} and that extracted from long-term life test data in similar devices ($E_a=1.05$ eV) [13]. Also, a similar activation energy was reported for degradation in the DC value of drain current ($E_a=1.13$ eV) under the stress condition of $V_{DS}=40$ V and $I_D=200$ mA/mm in different devices by a separate team [3].

V. DISCUSSION

It is interesting to compare the electrical results obtained in this work with the structural analysis performed in GaN HEMTs after high voltage stress as performed in [7]. In that earlier study, we investigated the time evolution of structural defect formation under high voltage electrical stress with

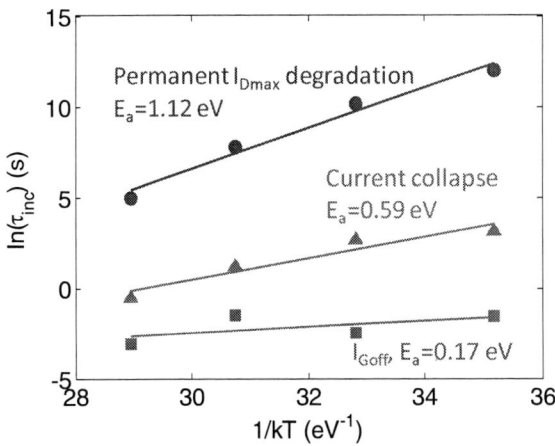

Fig. 7. Arrhenius plot of incubation time for I_{Goff}, permanent I_{Dmax} degradation and current collapse degradation as a function of stress temperature. The incubation time is defined as dashed lines in Fig. 3-Fig. 6.

$V_{DS}=0$ in GaN HEMTs as observed by planar techniques after the passivation and metals had been removed from stressed devices. In [7], we observed relatively fast formation (less than 10 s) of a continuous groove on the GaN cap running along the edges of the gate. On a longer time scale, we also observed the formation of prominent pits at the source and drain edges of the gate. The average pit cross-sectional area increased as $t^{0.25}$. We found a close correlation between the size and density of the pits and permanent current degradation as a function of stress time. Current collapse degradation also correlated well except for a similar saturating pattern as observed here.

A comparison between the study in [7] with the present work, allows us to postulate hypotheses for the evolution of degradation observed in the present work. The early phase of degradation that is characterized by temperature independent gate leakage current degradation and a corresponding shift in V_T could be associated with the formation of the groove. The groove has been observed to be formed even for stress voltages below V_{crit} [7, 14]. Also, degradation in I_{Goff} has been reported for stress voltage below V_{crit} if the device is stressed for long enough time [8].

The later phase of stress in which both current collapse and permanent drain degradation increase seems associated with the formation and growth of the pits. Interestingly, the time evolution for the increase in current collapse in Fig. 5 evolves as $t^{0.22}$ (slope in Fig. 5) regardless of temperature. This is very close to the observations made in [7] for the time evolution of the growth of the pits.

This temperature-independence degradation rate seems to agree with [7] where pit growth rate exhibited a very weak temperature dependence (E_a=0.11 eV). In addition, it was found that it takes about 10-100 s for the pit formation to be initiated. This appears to correspond to the incubation time for current collapse degradation. Developing a detailed understanding of the physics behind these findings will require structural degradation studies similar to those performed in [14-15].

VI. CONCLUSION

In summary, we have investigated the time evolution of electrical degradation of GaN HEMTs under high-voltage stress beyond the critical voltage in the OFF state. We found that the gate current starts to degrade first, followed by current collapse and eventually permanent degradation in I_{Dmax}. We also found that while the time evolution of gate current degradation is largely unaffected by temperature, drain current degradation is thermally accelerated. Our findings will be instrumental in developing predictive models for operational lifetime of GaN HEMTs

ACKNOWLEDGMENT

This work was funded by a DARPA program under ARL contract #W911QX-05-C-0087 (Dr. Alfred Hung) and by a DRIFT MURI program under ONR Grant #N00014-08-1-0655 (Dr. Paul Maki). We also acknowledge collaboration with TriQuint Semiconductor (Dr. Jose Jimenez).

REFERENCES

[1] J. Joh and J. A. del Alamo, "Critical voltage for electrical degradation of GaN high-electron mobility transistors," *IEEE Electron Dev. Lett.*, vol. 29, pp. 287-289, 2008.

[2] J. A. del Alamo and J. Joh, "GaN HEMT reliability," *Microelectronics Reliability*, vol. 49, pp. 1200-1206, 2009.

[3] G. Meneghesso, et al., "Reliability issues of Gallium Nitride High Electron Mobility Transistors," *International Journal of Microwave and Wireless Technologies*, vol. 2, pp. 39-50, 2010.

[4] J. Joh and J. A. del Alamo, "A Current-Transient Methodology for Trap Analysis for GaN High Electron Mobility Transistors," *Electron Devices, IEEE Transactions on*, vol. 58, pp. 132-140, 2011.

[5] D. W. Gotthold, S. P. Guo, R. Birkhahn, B. Albert, D. Florescu, and B. Peres, "Time-dependent degradation of AlGaN/GaN heterostructures grown on silicon carbide," *J. Elec. Mat.*, vol. 33, pp. 408-411, 2004.

[6] T. Ohki, et al., "Reliability of GaN HEMTs: current status and future technology," *IRPS*, pp. 61-70, 2009.

[7] J. Joh, P. Makaram, C. V. Thompson, and J. A. del Alamo, "Planar view of structural degradation in GaN high electron mobility transistors: time and temperature dependence," presented at IWN, 2010.

[8] D. Marcon, et al., "A Comprehensive Reliability Investigation of the Voltage-, Temperature- and Device Geometry-Dependence of the Gate Degradation on state-of-the-art GaN-on-Si HEMTs," *IEEE IEDM*, pp. 472-475, 2010.

[9] J. Joh and J. A. del Alamo, "Trapping vs. permanent degradation in GaN high electron mobility transistors," *ICNS Proc.*, pp. 947-948, 2009.

[10] J. L. Jimenez and U. Chowdhury, "X-band GaN FET reliability," *IEEE IRPS Proc.*, pp. 429-435, 2008.

[11] J. Joh and J. A. del Alamo, "Impact of electrical degradation on trapping characteristics of GaN high electron mobility transistors," *IEEE IEDM Tech. Digest*, pp. 461-464, 2008.

[12] J. Joh and J. A. del Alamo, "Effects of temperature on electrical degradation of GaN high electron mobility transistors," *ICNS*, p. 39, 2007.

[13] P. Saunier, et al., "Progress in GaN performances and reliability," *IEEE DRC Conference Digest*, pp. 35-36, 2007.

[14] J. Joh, J. A. d. Alamo, K. Langworthy, S. Xie, and T. Zheleva, "Correlation between electrical and material degradation in GaN HEMTs stressed beyond the critical voltage," *ROCS Proc.*, pp. 103-107, 2010.

[15] P. Makaram, J. Joh, J. A. del Alamo, T. Palacios, and C. V. Thompson, "Evolution of structural defects associated with electrical degradation in AlGaN/GaN high electron mobility transistors," *Applied Physics Letters*, vol. 96, pp. 233509-3, 2010.

Reliability-limiting defects in AlGaN/GaN HEMTs

Tania Roy, En Xia Zhang, Daniel M. Fleetwood, Ronald D. Schrimpf
Department of Electrical Engineering and Computer Science
Vanderbilt University
Nashville, TN USA
phone: (001) –(615)- 343-6705, email: tania.roy@vanderbilt.edu

Yevgeniy S. Puzyrev and Sokrates T. Pantelides
Department of Physics and Astronomy
Vanderbilt University
Nashville, TN USA

Abstract— Low-frequency noise measurements and density functional theory calculations are combined to show that N-anti-site and C impurity defects can lead to changes in the low frequency noise of GaN/AlGaN HEMTS fabricated with three typical process conditions. Implications for device reliability are discussed.

Keywords- high electron mobility transistors, electrical stress, 1/f noise, carbon substitutional impurity, nitrogen antisite.

I. INTRODUCTION

The extent to which defects in AlGaN/GaN HEMTs may affect operating device lifetimes is presently unknown. Hydrogenated Ga vacancies and N anti-site defects contribute to hot-carrier-stress induced degradation in GaN HEMTs grown under Ga-rich, N-rich, and NH_3-rich conditions [1]. Here we show results from $1/f$ noise measurements and first-principles density functional theory calculations that identify additional point defects that can lead either to an increase, or surprisingly, a decrease in the low-frequency noise in AlGaN/GaN high electron mobility transistors. The increase in $1/f$ noise and hot-carrier degradation appears to be caused by dehydrogenation of hydrogenated nitrogen anti-site defects, while a decrease in noise with stress can be caused by dehydrogenation of carbon impurities at a nitrogen site in AlGaN. This reinforces the strong role of N anti-site defects in limiting the reliability of GaN HEMTs.

II. EXPERIMENTS

For this study, surface-passivated GaN/AlGaN HEMTs were fabricated under three different growth conditions – Ga-rich, N-rich, and NH_3-rich. Fig. 1 shows the structures of these devices. Both GaN and AlGaN layers are grown under Ga-rich condition for Ga-rich process, and under ammonia-rich condition for the ammonia-rich process. The devices do not have field plates. The devices were electrically stressed at a drain voltage of 20 V and a gate voltage close to $V_{pinch-off}$ to keep a high electric field under the gate. Low frequency $1/f$ noise was measured before and after stress, at gate voltages varying from $V_{pinch-off} + 0.1$ V to $V_{pinch-off} + 1$ V, ensuring that the noise is coming from the gated portion of the channel [2].

The noise was measured before and after stress, as a function of temperature. The temperature was varied from 85 K to 450 K. The drain voltage during the noise measurement was fixed at 0.05 V and the gate voltage was fixed at $V_{pinch-off} + 0.2$ V.

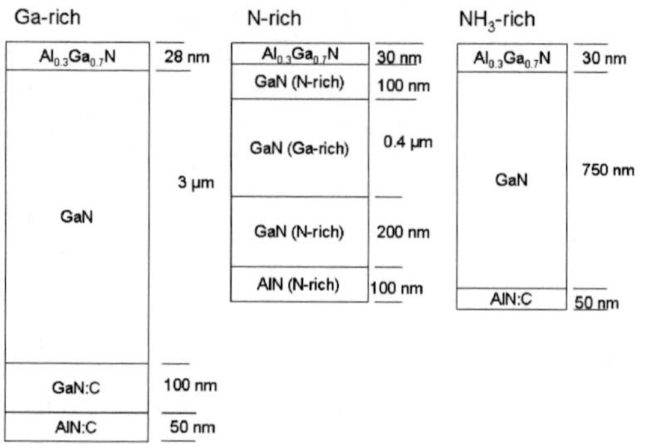

Fig. 1. Epitaxial structures of the GaN/AlGaN HEMTs grown on SiC substrates. The gate length of these devices is 0.7 μm; $L_{GD} = 1.2$ μm and $L_{GS} = 0.7$ μm. The devices are 150 μm wide.

III. RESULTS

Electrical stress of GaN HEMTs causes $V_{pinch-off}$ to shift positively for devices grown under Ga-rich and N-rich conditions, and negatively for devices grown under NH_3-rich conditions [1]. For devices fabricated using NH_3-rich conditions, hot carrier stress causes hydrogen to be released from N-anti-sites, shifting $V_{pinch-off}$ negatively after stress. Dehydrogenation of N-anti-sites results in defects with a less negative charge state than the initial charge state, causing negative shift in pinch-off. Similar dehydrogenation of Ga-vacancies after stress causes $V_{pinch-off}$ to shift positively in devices grown under Ga-rich and N-rich conditions, since the dehydrogenated Ga-vacancies have a more negative charge state compared to the charge state of the hydrogenated defect [1].

Fig. 2 shows the post-stress vs. pre-stress leakage current of the devices. After stress, the leakage current increases by more than an order of magnitude for N-rich devices. The increase in leakage with stress is consistent with defect formation in the AlGaN barrier due to electrical stress [3].

Fig. 3 compares the noise magnitudes before and after hot-carrier stress. The noise increases for some devices after stress and decreases for others, with no clear trend based on the growth conditions [4].

Figs. 4 and 5 show the normalized low-frequency noise magnitude at a frequency of 10 Hz as a function of temperature, before and after stress, for Ga-rich and ammonia-rich devices. The excess drain voltage power spectral density of the $1/f$ noise, S_{vd}, is proportional to $(V_G - V_{off})^{-3}$ [4]. The selected normalization facilitates comparison among different devices and over a range of temperatures. A peak in the noise spectrum is observed below ~ 100 K, corresponding to a trap energy of ~ 0.2 eV, as we discuss below.

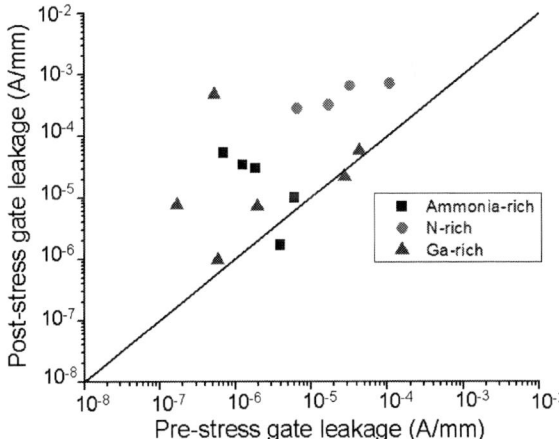

Fig. 2. Post-stress vs. pre-stress leakage current for Ga-rich, N-rich and NH₃-rich devices.

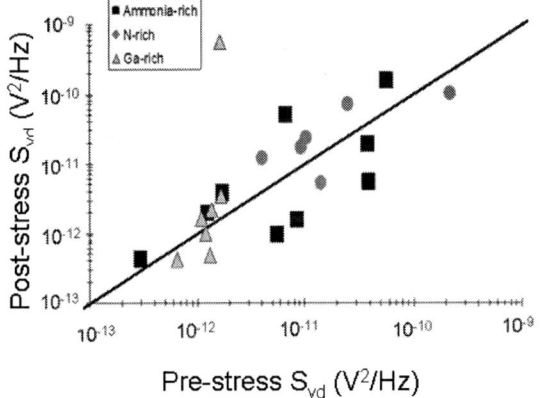

Fig. 3. Post-stress vs. pre-stress noise for Ga-rich, N-rich, and NH₃-rich devices; $V_G - V_{pinch-off} = 0.2$ V, $V_D = 0.02$ V; frequency = 10 Hz. After [4].

IV. PHYSICAL MECHANISMS

The position of the Fermi level in the GaN bulk and AlGaN barrier is 0.5 eV below E_c in the GaN and ~ 1 eV below E_c in the AlGaN during stress conditions (V_G close to $V_{pinch-off}$), as shown in Fig. 6. Hydrogen forms defect complexes with impurities and native defects in AlGaN [5-7]. Carbon is a

common impurity that is typically incorporated during the MBE process. Figs. 7-9 show the formation energies of hydrogenated carbon in N-sites in $Al_{0.3}Ga_{0.7}N$ and hydrogenated N-anti-sites in $Al_{0.3}Ga_{0.7}N$ and GaN, as calculated by density functional theory. A change in slope signifies a change in charge state of the defect, which is equivalent to an electron trap level in the band gap.

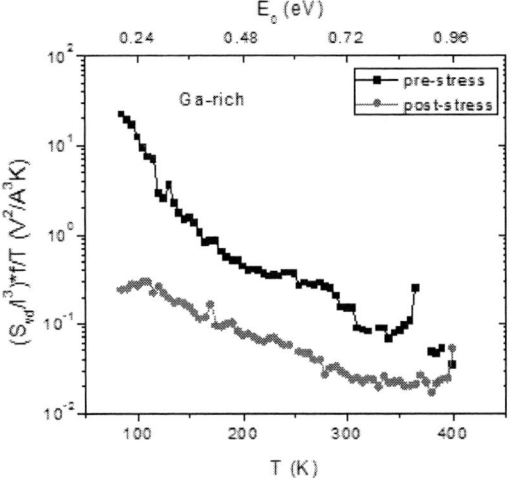

Fig. 4. $S_{vd}/I^3 f/T$ as a function of temperature and E_0 for a Ga-rich device, showing a decrease in noise after stress. Noise was measured at $V_G = V_{pinch-off} + 0.2$ V, ensuring that noise comes from the channel of the device; $f = 10$ Hz.

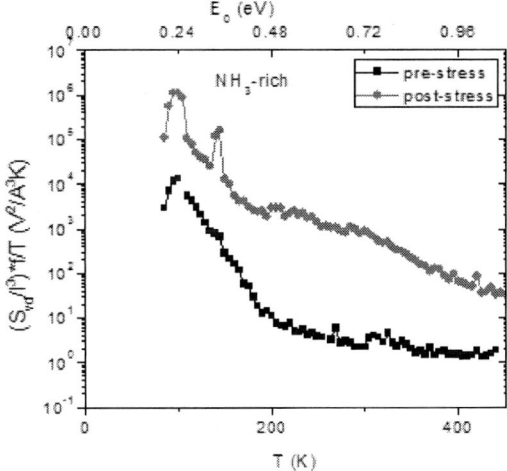

Fig. 5. $S_{vd}/I^3 f/T$ as a function of temperature and E_0 for a NH₃-rich device showing an increase in noise after stress. Noise was measured at $V_G = V_{pinch-off} + 0.2$ V, ensuring that noise comes from the gated portion of the channel of the device; $f = 10$ Hz.

For the capture and release of electrons that can contribute to $1/f$ noise, there must be a change in the charge state of a defect near the position of E_f during the noise measurement. Hydrogenated C on an N site shows such a change in charge state ~ 1.0 eV below the conduction band, as shown by the change of slope of the defect formation energy curve in Fig. 7. During the noise measurements, the gate voltage is close to $V_{pinch-off}$, and E_f is ~ 1.0 eV below E_c. After hydrogen is removed

978-1-4244-9113-1/11 $26.00 © 2011 IEEE

by hot electrons during electrical stress, the substitutional C no longer has a defect level near the position of E_f during the measurement. This can explain the decrease in noise observed in some devices after stress. Hence, these C impurity defects do not appear to play a significant role in the degradation of the devices. In contrast, the hydrogenated N-anti-site (Figs. 8-9) does not have a transition level close to E_f before electrical stress. Anti-sites that are dehydrogenated during stress have defect levels close to E_f during the noise measurement. This level captures and releases electrons, causing the $1/f$ noise to increase after stress-induced dehydrogenation in devices dominated by N-anti-site defects. Thus, the anti-site defect not only causes a shift in pinch-off voltage after stress, but also causes the post-stress noise to increase.

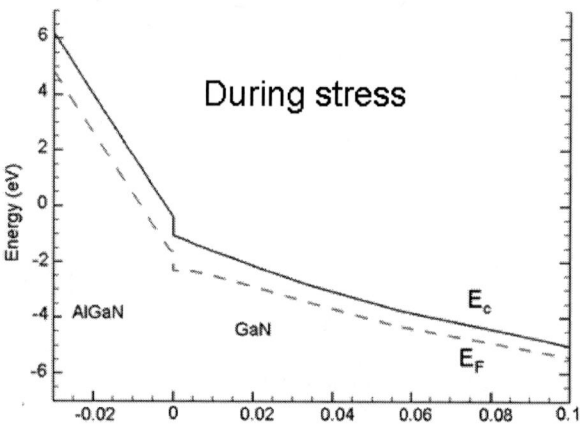

Fig. 6. Position of Fermi level with respect to conduction band minima at $V_G = V_{pinch-off}$. The GaN bulk and AlGaN barrier are simulated for a density of 10^{16} cm^{-3} donor traps, with trap levels 0.1 eV below E_c, corresponding to the unintentional doping of GaN. During stress, $E_c - E_f = \sim 1$ eV in AlGaN, and $\sim$ 0.5 eV in GaN.

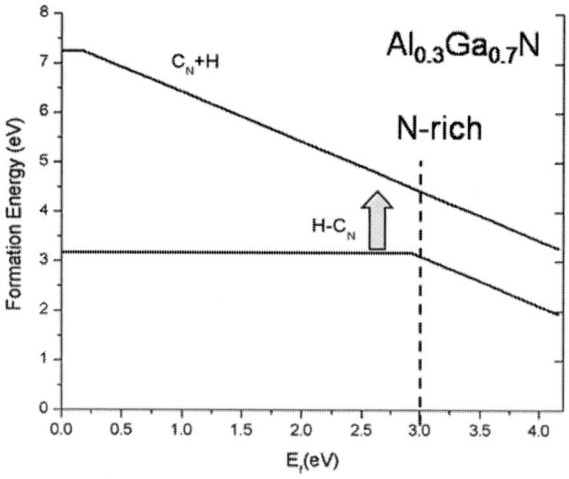

Fig. 7. Formation energy of substitutional carbon in N-site as a function of Fermi energy in AlGaN.

In Figs. 4 and 5, the temperature dependence of the noise is shown from 85 K to 300 K as a function of stress for devices grown under Ga-rich and NH$_3$-rich conditions, respectively.

The defect energy can be parameterized as a function of temperature, through the equation $E_0 = -k_B T ln(\omega \tau_0)$ [7,8], where $\tau_0 = 10^{-14}$ s, corresponding to a typical inverse phonon frequency in GaN [9]. For the Ga-rich devices, the post-stress noise is consistently lower than the pre-stress noise at all temperatures. Fig. 5 shows the noise characteristics of a representative NH$_3$-rich device in which the post-stress noise is higher than the pre-stress noise. In this case the noise is higher at all temperatures. These results confirm a strong link between defects, hot carrier stress, and low-frequency noise, across a broad range of device operating conditions and temperatures. Also, the noise shows a peak at $\sim$ 100 K in both Ga-rich and NH$_3$-rich devices, corresponding to a trap energy level of ~ 0.2 eV. The microstructure of this defect and its potential role in GaN/AlGaN HEMT reliability are presently unknown, but are clearly topics of interest for future study.

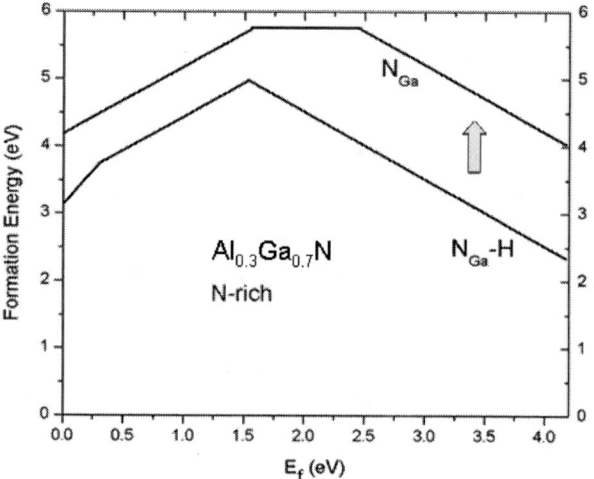

Fig. 8. Formation energy of hydrogenated N-antisite as a function of Fermi energy in AlGaN.

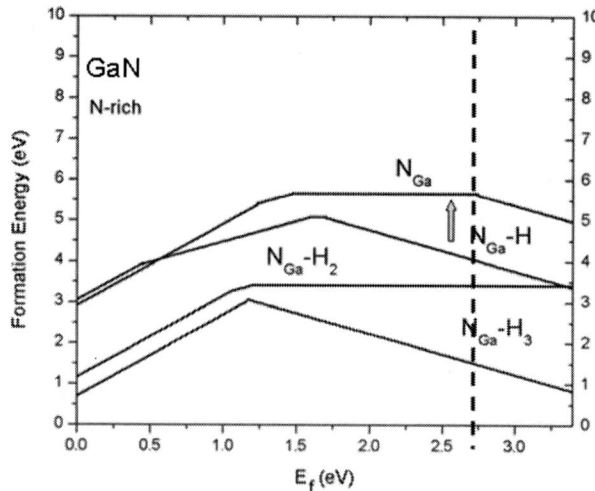

Fig. 9. Formation energy of hydrogenated N-anti-site as a function of Fermi energy in GaN.

V. CONCLUSION

Dehydrogenation of N-anti-site defects increase the low-frequency-noise of GaN/AlGaN HEMTS and lead to pinch-off voltage shifts. A reduction in the density of these defects should significantly enhance the reliability of GaN/AlGaN HEMTs. The noise spectrum as a function of temperature reveals peaks at ~ 100 K, corresponding to a trap energy of 0.2 eV. The potential role of this low-energy trap in GaN/AlGaN HEMT reliability is unknown.

ACKNOWLEDGMENTS

We would like to thank Profs. U. K. Mishra and J. S. Speck of the University of California, Santa Barbara, for providing us with the GaN HEMTs grown under Ga-rich, N-rich, and NH_3-rich conditions.

REFERENCES

[1] T. Roy, Y. S. Puzyrev, B. R. Tuttle, D. M. Fleetwood, R. D. Schrimpf, D. F. Brown, U. K. Mishra, and S. T. Pantelides, "Electrical-stress-induced degradation in AlGaN/GaN high electron mobility transistors grown under gallium-rich, nitrogen-rich, and ammonia-rich conditions", *Appl. Phys. Lett.*, vol. 96, no. 13, pp. 133503:1-3, 2010.

[2] A. Balandin, "Gate-voltage dependence of low-frequency noise in GaN/AlGaN heterostructure field-effect transistors", *Electron. Lett.*, vol. 36, no. 10, pp. 912-913, 2000.

[3] A. V. Vertiatchikh and L. F. Eastman, "Effect of the surface and barrier defects on the AlGaN/GaN HEMT low-frequency noise performance", *IEEE Electron Dev. Lett.*, vol. 24, no. 9, pp. 535-539, 2003.

[4] T. Roy, E. X. Zhang, S. DasGupta, S. A. Francis, D. M. Fleetwood, and R. D. Schrimpf, "1/f noise in GaN HEMTs grown under Ga-rich, N-rich, and NH_3-rich conditions," *2010 IEEE Reliability of Compound Semiconductors Workshop*, Portland, OR, May 17, 2010.

[5] Y. S. Puzyrev, B. R. Tuttle, R. D. Schrimpf, D. M. Fleetwood, and S. T. Pantelides, "Theory of hot-carrier-induced phenomena in GaN high-electron-mobility transistors", *Appl. Phys. Lett.*, vol. 96, no. 5, pp. 053505:1-3, 2010.

[6] C. G. Van de Walle and J. Neugebauer, "First-principles calculations for defects and impurites: Application to III-nitrides", *J. Appl. Phys.*, vol. 95, no. 8, pp. 3851-3879, 2004.

[7] P. Dutta and P. M. Horn, "Low-frequency fluctuations in solids: 1/f noise", *Rev. Mod. Phys.*, vol. 53, pp. 497–516, 1981.

[8] D. M. Fleetwood, H. D. Xiong, Z. Y. Lu, C. J. Nicklaw, J. A. Felix, R. D. Schrimpf, and S. T. Pantelides, "Unified model of hole trapping, 1/f noise, and thermally stimulated current in MOS devices, " *IEEE Trans. Nucl. Sci.*, vol. 49, no. 6, pp. 2674-2683, 2002.

[9] H. Siegle, G. Kaczmarczyk, L. Fillippidis, A. P. Litvinchuk, A. Hoffmann, and C. Thomsen, "Zone-boundary phonons in hexagonal and cubic GaN", *Phys. Rev. B*, vol. 55, no. 11, pp. 7000-7004, 1997.

250 GHz HETEROJUNCTION BIPOLAR TRANSISTOR:
FROM DC TO AC RELIABILITY

Malick DIOP, Salim Ighilahriz, Florian Cacho, Vincent Huard
STMicroelectronics, 850 rue Jean Monnet, 38926 Crolles Cedex, France
phone: + 33 438 922 498; fax: + 33 438 922 953; e-mail: malik.diop@st.com

ABSTRACT

Continued node shrinks have given rise to the emergence of ultra-fast *SiGe BiCMOS* opportunities. This class of technology integrates high performance *HBTs* with state of art *CMOS* technology and targets.millimeter wave applications such as *77 GHz* automotive radars and non invasive imaging for airport security. In order to reach this performance, *HBTs* are designed to operate at high current density and often in some cases at voltages beyond the collector – emitter (C/E) breakdown voltage with an open circuit base (BV_{CEO}). Regarding these severe conditions which are susceptible to create serious damage and limit the device performance, reliability emerges as an important challenge. In this work, *LF* noise, *DC* current measurements and *DCIV* method [1] are used to evaluate the interface trap density induced in Si/SiO_2 interfaces during forward stress and systematically extract reliability model parameters. We propose a global time to failure (*TTF*) model for the reliability qualification of high speed *SiGe HBTs*. Finally, the application of the degradation model to RF circuit is proposed and discussed.

[*Keywords: BiCMOS, Bipolar transistor, SiGe:C, high speed, reliability, modelling*].

INTRODUCTION

Recent development in bipolar devices in terms of reduction of dimensions and the increase in doping levels, have led to an increase in electric field strength around the base – emitter (*B/E*) junction. Moreover, significant self heating is induced in these devices due to strong current density in the forward mode (around $10mA/\mu m^2$)

Following pioneering research, it is now widely accepted that degradation of bipolar transistors is caused by:

- emitter resistance R_E decrease in the high injection regime ($V_{BE} = 0.8V$) [2] [3]

- base current I_B decrease (or increase) and consequent β gain increase (or decrease) in mid – injection V_{BE} range ($0.5 < V_{BE} < 0.8$) [2] [3] [4]

The second characteristic is widely observed in our devices. Base current drift is attributed to hot carriers generated in the *B/E* junction space charge region leading to the creation of interface traps in the oxide spacer [5]. Moreover, in highly doped junctions, which have a narrow depletion layer, the relatively short tunneling distances facilitates carrier trapping by trap assisted (*TAT*) or band to band (*BTT*) tunneling [6]. Then carriers can later recombine via Shockley – Read - Hall process [7].[8].

In this paper we report a new base current I_B degradation model for high performance bipolar transistors. The transistors were stressed under high currents through the emitter with high voltages V_{CE}

around and beyond BV_{CEO}. LFN measurements are done using an *EGG 5182 I/V* converter for fluctuations amplification and *FFT* analyzer for power spectral density *(PSD)* calculated at $I_B = 50\mu A$ and $V_{CE} = 1V$ from *10Hz to 100Hz*. Electrical stress were periodically interrupted for Gummel plot and/or *LFN* measurements.

Degradation of the base current drift was modeled and parameters systematically extracted using *DCIV* approach [1]. Finally, using *HiCUM* spice model and Golden Gate simulator, Beta gain and I_{DB} compression point after 10 years of use is provided for the first time for a *77 GHz LNA* test case.

EXPERIMENTAL DETAILS

The architecture of this *HBT* is a fully self-aligned (*FSA*) double-polysilicon structure using Selective Epitaxial Growth (*SEG*) of the *SiGe:C* base. The process uses a standard collector module (n+ buried layer / collector sinker), a self-aligned selective implanted collector (*SIC*), a boron doped *SEG SiGe:C* base, and an arsenic in-situ doped mono emitter. An inside spacer module is employed to obtain emitter widths $W_E < 0.13\mu m$. The inside spacer module used is a D-shaped spacer formed by a thin oxide and a thick nitride. The *HBT* front-end fabrication is finalized with a spike activation annealing, cobalt silicidation and contacts formation [9]

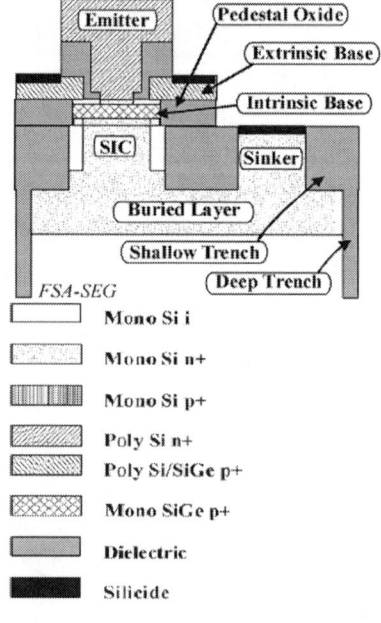

FIGURE 1: SCHEMATIC CROSS SECTION OF FULLY SELF ALIGNED HBT

The technology is based on a *0.13-μm CMOS* core process with a copper *BEOL*. The tested device, with emitter size of *0.13x4.85 μm²*, exhibits peak dc current gain of around *2000* and breakdown volt-

978-1-4244-9113-1/11 $26.00 © 2011 IEEE

ages BV_{CBO} of *5V* and BV_{CEO} of *1.6 V*. The cut-off frequency f_T, which represents the high-speed performance of the device, is around *250 GHz* at the Nominal Current (*N.C.*) density around *10 mA/μm²* and $V_{CE} \approx BV_{CEO}$

LOW FREQUENCY NOISE MEASUREMENTS

One of the most relevant figures of merit of *SiGe HBTs* with regard to CMOS is their superior low *1/f* noise at low frequencies (*LF*) [10]. In several RF applications, it is important to pay attention to the degradation of the base and / or the collector *1/f* noise. Indeed degradation of *1/f* noise affects the spectral purity of non linear radiofrequency (*RF*) circuits [11].

It has been established experimentally that only the base current has *1/f* noise in bipolar transistor. Its physical origin is the mobility or number fluctuation (or both) of the carriers due to defects [12]. Degradation of *B/E* junction will impact *1/f* noise.

Thus, this paper focuses on 1/f noise measurement on the base. This measurement could be effective to discriminate the physical degradation mechanisms (generation-recombination, trap assisted tunneling, etc) caused by current and/or voltage stress. Fluctuations on I_B are represented by the noise power spectral density (*PSD*) S_{IB}:

$$S_{I_B} = K_F \frac{I_B^{\alpha}}{f} + 2qI_B$$

where α is a constant and K_F correspond to model parameter used in SPICE.

Noise *PSD* has been characterized under the same conditions before and after a stress in the high forward mode ($J_E = 10$ mA/μm² and $V_{CB} = 2V$, beyond BV_{CEO}) in order to create avalanche current. These bias conditions are generally used in *RF* and mixed-signal applications to achieve maximum circuit performance.

For $I_{Bmeas.} = 0.5$ μA we plot in fig. 2 *PSD* which exhibits a strong increase of excess noise level with a Lorentzian spectrum. This shows that serious damage has been introduced by hot energetic carriers due to high electric field.

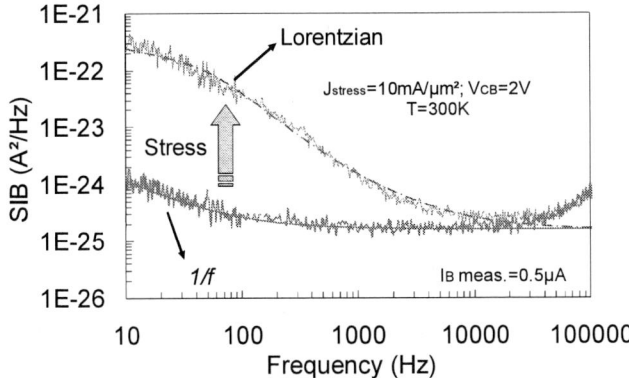

FIGURE 2: S_{IB} SPECTRA BEFORE AND AFTER MIXED − MODE STRESS

These hot carriers generate interfacial and/or bulk oxide traps in the *E/B* junction depletion region [13]. This Lorentzian noise signature is generally attributed to trapping/detrapping via mid-gap traps through SRH processes. The total *1/f* noise after stress follows the empirical model:

$$S_{I_B} \cong \sum_i^{N_T} \frac{\tau_i}{1 + (2\pi f \tau_i)^2} I_B^{\alpha} + 2qI_B$$

where N_T is the total number of traps and τ_i is the characteristic time constant of the ith trap which will contribute to excess *1/f* noise. In order to evaluate trap density induced under stress and build a new *DC* time to fail model for our high performance *SiGe HBTs*, we introduce the *DCIV* methodology [1]

THEORY, RESULTS AND DISCUSSIONS

A. DC PARAMETERS INVESTIGATION

DCIV method is demonstrated to be particularly suitable for determining the mid-gap interface trap density representing the major defect distribution in *HBTs* junctions [1] Based on "degradation bundle" ($\Delta I_B(t)$) which allows us to follow base current I_B (or current gain β) degradation during stress [14], we introduce a new parameter A which is directly linked to interface trap states N_{eff} with the mid − gap energy which can be expressed as [1]:

$$A(t) = \frac{\Delta I_B(t)}{\exp\left(\frac{qV_{BE}}{nkT} - 1\right)} = \frac{Bqc_0 n_i}{2} \Delta N_{eff}(t)$$

Where c_0 represent the average trap capture, $n_i = 1.5 \times 10^{-10} cm^{-3}$ the intrinsic electron density of *Si*, n the ideality factor of the *PN* junction current and B a constant. The excess base current is attributed to trap generation in the depletion region. Then it is important to note that ΔN_{eff} should generate a change in base current of ΔI_B.

First of all, electrical stress was applied and periodically interrupted to perform Gummel plot measurements using a parameter analyzer. This allow us to follow I_B (or β) drift which we use here as figures of merit for lifetime at 10% I_B increase (or β decrease).

We start by studying the influence of V_{CE} on degradation by forcing a constant emitter current $I_{Estress}$ that yields the peak f_T. Figure 3 represents the I_B drift as a function of stress time extracted at $V_{BE} = 0.8V$ and shows a power law dependence ($t^{0.25}$). We can notice that the degradation of base current increases significantly with V_{CE}.

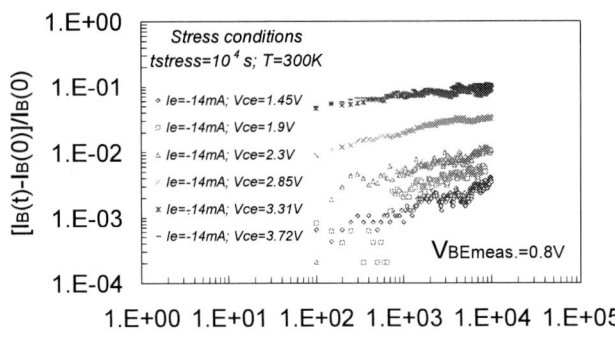

FIGURE 3: I_B DEGRADATION DYNAMICS WITH V_{CE} VERSUS STRESS

No additional degradation is observed between $V_{CE} = 3.3V$ and $V_{CE} = 3.7V$. This phenomenon of saturation indicates to the first order that the effective density of created traps and/or activated traps has reached a final value towards *3.3V*. In order to extract the acceleration factor (*acc. fact.*) of I_B degradation with voltage stress, we define a function *Age*. *Age* variable is directly related to the degradation and previously depicted in [15]:

$$Age = t \times V_{CE}^{acc.fact}.$$

The variable *Age* reflects the amount of degradation that takes place in the device during the stress. Figure 4 shows the plot of *A(t)* versus *Age(t)* which allow us to normalized the degradation and then extract a strong power life time dependence on V_{CE} found as *acc. fact.* $\alpha = 5.5$ $\pm 10\%$ and stress time factor $\rho = 0.25$

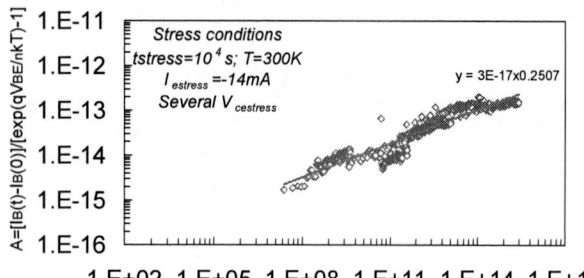

FIGURE 4: I_B NORMALIZED DEGRADATION

Next we investigate the I_E dependence at constant $V_{CE}=2.8V$. Figure 5 (arbitrary units) shows an increase of traps generation (N_{eff}) with $I_{Estress}$ until 7 mA following by a decrease whenever $I_{Estress} > 7mA$. Following the same approach, we define two age function:

$$Age_1 = t \times I_E^{\mu_1}$$

$$Age_2 = t \times I_E^{\mu_2}$$

we extract a power life time dependence $\mu_1 = 2$ for low currents stress and $\mu_2 = -5.5$ for high current stress. We have developed a function in *ORIGIN* as

$$P = cte \cdot \left(\frac{1}{\frac{1}{\lambda_1 I_{Estress}^{2}} + \frac{1}{\lambda_1 I_{Estress}^{'-5.5}}} \right)^{0.25}$$

allowing us to fit the experimental N_{eff} as shown in Figure 5. This behaviour is synonym of two degradation mechanisms previously depicted in [16]; beyond 7 mA, it was shown that strong local heating occurs in the E/B junction. This can lead to the passivation of the current induced interface traps resulting in N_{eff} decrease.

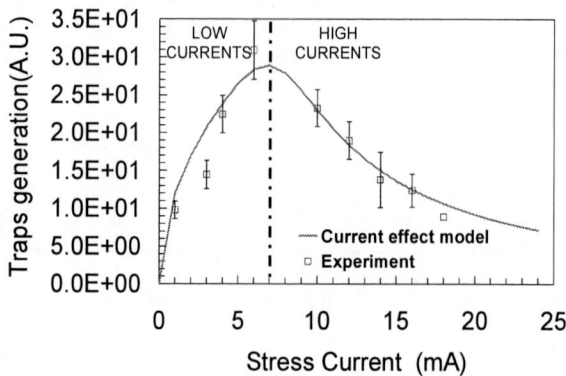

FIGURE 5: I_B TRAPS GENERATION VERSUS STRESS CURRENT

Finally, we also investigate the temperature dependence by stressing devices at seven different temperatures ranging from -40°C to 125°C. Base current drift under $I_E=10mA$ $V_{CE}=2.8V$ shown in Figure 6 (a) with extraction of $E_a=0.25 eV$ (Figure 6 (b)).

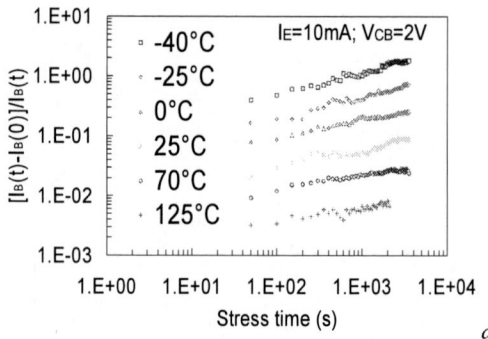

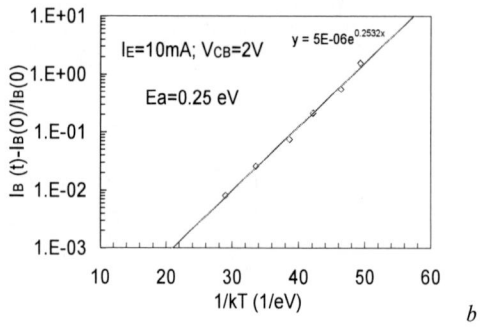

FIGURE 6: I_B DRIFT WITH T°C WITH EA EXTRACTION

This gives our final model which is continuous on V_{CE}, I_E, and $T°C$ and which allow us to evaluate the "failure criterion" for all reliability evaluations (- 10% degradation in I_B from its initial value):

$$\Delta I_B(t) = cte \left(\frac{1}{\frac{1}{\lambda_1 I_{stress}^{\mu_1}} + \frac{1}{\lambda_1 I_{stress}^{'\mu_2}}} \right) x V_{CEstress}^{\alpha} x \exp(\frac{qV_{BEsens}}{nkT}) x \exp(\frac{E_a}{kT_e}) \times t_{stress}^{\rho}$$

Our reliability model is consistent with experiments and fits the entire range of data (additional plots not shown here). Thus, we propose now to investigate *AC* parameters.

B. *AC* PARAMETERS INVESTIGATION

With the static model of degradation having been achieved, *AC* parameter degradation will now be investigated. *AC* parameters require considerable attention in analog *RF* circuits. In order to investigate their degradation, *AC* parameters measurements are performed on several devices before and after *DC* stress. We summarize in *Table 1* the different measured devices, the stress times and stress conditions.

	3 structures	W=0.27 (μm)	L=3, 10, 15 (μm)
DC stress / AC measurement	6 stress conditions 2 dies/stress	V_{BE} (V), V_{CE} (V) stress Stress time = {100, 1000, 3000} sec	V_{BE}=0.9 / V_{CE}=2.5 V_{BE}=0.95 / V_{CE}=2.5 V_{BE}=0.95 / V_{CE}=3 V_{BE}=1 / V_{CE}=2 V_{BE}=1 / V_{CE}=2.5 V_{BE}=1.2 / V_{CE}=2

TABLE 1: MEASUREMENT CONDITIONS SUMMARY FOR AC PARAMETERS DEGRADATION STUDY

Junction capacitances (*B/E* C_{tbe}, *B/C* C_{tbc} and *C/Bulk* C_{tcs}) and resistances (Emitter resistance R_{em} and Base intrinsic-extrinsic $R_{bb'}$) are measured through S parameters up to *27 GHz* while figures of merit like cut-off frequency f_T and maximum oscillation frequency f_{max} are obtained from extrapolation from measurements up to *110 GHz*. All the measurement conditions are summarize in *Table 2*.

Temperature = 300 K	Base-Emitter Voltage (V)	Collector-Base Voltage (V)	Measurements conditions
AC parameters (f_T, f_{max})	Sweep from 0.78 to 0.97	0.5	s parameters up to 110 GHz, extrapolated beyond
C_{tbe} and C_{tbc} (V_{BC}= 0V)	Sweep from -1 to 0.5	Sweep from -1 to 0.5	s parameters up to 27 GHz
C_{tcs} (V_{CE}= 1.5V)	Sweep from 0 to 1	1.5	
R_{em} (bias independant)	0.84	independent	
$R_{bb'}$ (V_{CE}= 0V)	Sweep from 0.84 to 0.87	Sweep from -0.84 to -0.87	

TABLE 2: AC PARAMETERS MEASUREMENTS CONDITIONS

The following results are related to the nominal structure (*W=0.27μm*, *L=5μm*). Whatever the applied stress conditions, no significant degradation occurs on these *AC* parameters. Moreover, concerning the C_{tbe} parameter, the small *3%* drift is due to the inaccuracy of the measurements, as shown in figure 7. Additionally, a cut-off frequency drop is represented in figure 8; around *10 GHz*, occurs for the more aggressive stress condition {V_{BE}=0.95V; V_{CE}=3V}. Note that a contribution of the measurement inaccuracy must be taken into account in that drift.

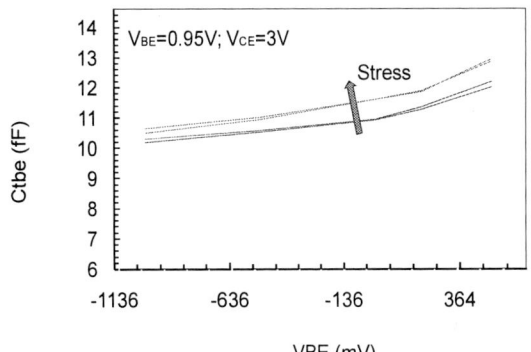

FIGURE 7: B/E TRANSITION CAPACITANCE C_{TBE} FUNCTION OF V_{BE} BEFORE (BLUE) AND AFTER (RED) STRESS

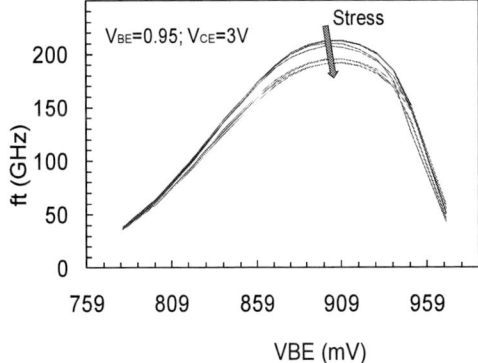

FIGURE 8: CUT-OFF FREQUENCY F_T FUNCTION OF V_{BE} (B) BEFORE (BLUE) AND AFTER (RED) STRESS

In spite of no significant degradation observed on *AC* parameters after *DC* stress, more measurements are necessary to confirm these results.

APPLICATION

The time evolution of bipolar transistor *DC* and *AC* parameters was studied under high forward stress configurations in the previous section. Here, we propose to investigate how the changes in these parameters may affect circuit performance. For this, simulations on a single stage common-emitter Low Noise Amplifier (*LNA*) have been performed.

To reduce losses in the lines (adding noise in a receiver chain), the *LNA* block is generally the first one in a receiver. Its role is to amplify the received signal while generating the minimum noise.

A first step consists in a calculation of the base current degradation using the equations of our new model described above. The base current drift is integrated along a period (*130 ns*) according to the realistic waveform of I_E and V_{CE}. For an operating time of *10* years, a ΔIb= 2.10^{-7}A value for V_{BE}=0.5V is found Then, in a second step, this degradation is implemented in *HICUM* spice model. For the sake of illustration, the Gummel plots of a virgin and a degraded device are presented in Figure 9.

978-1-4244-9113-1/11 $26.00 © 2011 IEEE

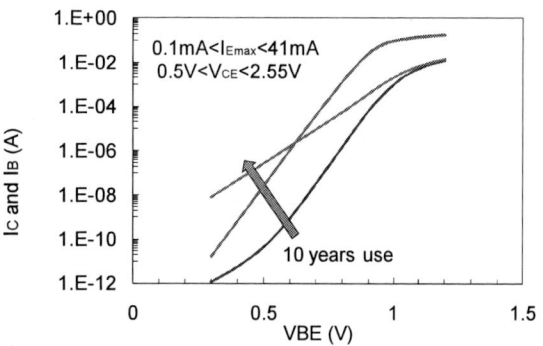

FIGURE 9: GUMMEL PLOT FOR VIRGIN (BLUE) AND AGED (RED) HBT

An amplifier maintains a constant gain for low-level input signals. However, at higher input levels, the amplifier goes into saturation and its gain decreases. The *1 dB* compression point (*P1dB*) indicates the power level that causes the gain to drop by 1 dB from its small signal value. *Figure 10* shows the *1dB* compression point simulation before and after 10 years of use. We can observe that there is no significant degradation on this characteristic after stress.

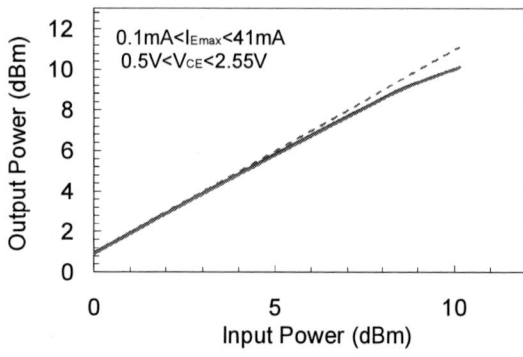

FIGURE 10: *1DB* COMPRESSION POINT BEFORE AND AFTER 10 YEARS USAGE

The compression point is not impacted by the ageing of the *HBT*. Indeed, as the collector current remains the same, there is no modification of this dynamic gain. At the circuit level, no significant drift is observed on these parameters even though the HBT has suffered significant degradation of the base current.

CONCLUSIONS

High speed *SiGe HBTs* degradation under high forward stress has been investigated using *LFN* and modeled using *DCIV* measurement method. The introduction of an *Age* function allowed us to determine how *HBT* device damage was impacted by voltage stress, current stress and temperature effects. This was simulated through an acceleration factor and E_a extraction. Our reliability model was tested over a broad parametric space covering multiple orders of magnitude in I_{Eref} (and f_T peak) and V_{CE} (including beyond BV_{CEO}). *AC* measurements demonstrate no significant degradation after *DC* stress. Finally, *HBT* degradation equivalent to *10* years of operation is introduced in a spice model and *LNA* degradation simulation is performed. No degradation of compression point and *NF* parameter is observed

nevertheless. Further *RF* circuits will be assessed using this methodology.

REFERENCES

[1] B.B.Jie, M.F. Li, and K.F. Lo "Energy dependence of interface trap density investigated by *DCIV* method Proc. of 7th IPFA pp. 206 209 1999.

[2] D.D. Tang, E. Hackbarth, and T.-C. Chen, "Degradation of very high – current degradations on Si n – p – n transistors" *IEEE Trans. Electron Devices, vol.37, pp. 1698 – 1706, July 1990*

[3] M.S. Carroll, A. Neugrochel, and C.-T. Sah, "Degradation of silicon bipolar junctiontransistors at high forward current densities " *IEEE Trans. Electron Devices, vol.44, pp. 110 – 117, Jan 1997*

[4] T.-C. Chen, C. Kaya, M.B. Ketchen, and T.H. Ning, "Reliability analysis of self – aligned bipolar under forward active current stress", *in IEDM Tech. Dig., 1986, pp. 650 – 653*

[5] R.A. Wachnick, T.J. Bucelot, and G P Li. "Degradation of bipolar transistors under high current stress" at 300K. *J.Appl. Phys., 63 (9): 4734-4740, 1988.*

[6] G.M. Hurkx, D.B.M. Klaassen, M.P.G. Knuvers "A new recombination model for device simulation including tunneling" *IEEE transaction on electron devices, vol 39, n°2, Fevr. 1992*

[7] J.A. del Alamo, Richard M. Swanson "Forward bias tunneling: A limitation to bipolar device scaling" *IEEE Electron Device Letters, vol EDL-7, n°. 11, November 1986*

[8] Johannes M.C.Stork, Randall D. Isaac " Tunneling in base – emitter junctions" *IEEE Transactions on electron devices, vol. ED-30, n° 11, November 1983*

[9] P. Chevalier et al., *IEEE JSSC*, vol. 40, no. 10, pp.2025-2034, October 2005.

[10] L.S.Vempati et al., "Low-frequency noise in UHD/CVD epitaxial Si and SiGe bipolar transistors," *IEEE J. Solid State Circuits, vol. 31, n°10, pp. 1458-1467, 1996.*

[11] Van Damme LKJ. "Noise as diagnostic tool for quality and reliability of electronic devices" *IEEE Trans. Electron Dev. 1994; vol 41 n°11, pp. 2176-2187.*

[12] G. Niu, Zhenrong Jin, Jonh D. Cressler, Rao Rapeta, Alvin J Joseph, and David Harame "Transistor noise in SiGe HBT RF Technology" *IEEE Journal of Solid State circuits, vol 36, no 9, 2001*

[13] J.D.Burnett and C. Hu, "Modeling hot-carriers effects in poly-silicon emitter bipolar transistors", *IEEE Transactions on Electron devices, Vol. 35, No. 12, pp. 2238-2244, December 1988.*

[14] M.Diop, N Revil, M.Marin, F. Monsieur, G. Ghibaudo "Coupled approach for reliability study of FSA 250 GHz HBTs, *In proc. of IRW pp. 77 october 2008*

[15] P.M. Lee M.M. Kuo, K. Seki, P.K.Lo, C. Hu, , *IEEE IEDM*, pp 134-137, *1988*

[16] M. Diop, M. Marin, N. Revil, F. Pourchon, C. Leyris P. Chevalier, G. Ghibaudo Reliability review of 250 GHz HBTs for millimeter wave applications *IEEE. IRPS, pp. 76 – 82 April 2009*

Intrinsic Reliability of RF Power LDMOS FETs

David C. Burdeaux, Member, IEEE, david.burdeaux@freescale.com
Wayne R. Burger, Ph.D., Member, IEEE, wayne.burger@freescale.com
Freescale Semiconductor, RF Division: RF, Analog, and Sensors Group, 2100 E. Elliot Rd., Tempe, AZ 85284

Abstract -- **RF-LDMOS is the dominant RF power device technology in the cellular infrastructure market, having successfully displaced vertical MOSFETs and silicon bipolar transistors in the 1990s. A similar technology shift towards RF-LDMOS is occurring today in adjacent RF power markets such as UHF Broadcast, VHF Broadcast, L-Band and S-Band Radar, and the Industrial/Scientific/Medical markets (MRI, CO2 Laser, synchrotron, etc.). This increasing adoption of RF-LDMOS into these other RF power applications is the direct consequence of continuing progress at improving the intrinsic reliability and application-specific customization of LDMOS device structures.**

RF power applications, whether cellular infrastructure or the adjacent non-cellular markets, present unique and challenging thermal and electrical environments for the RF power transistor. While the design and architecture of the power amplifier is critically important in defining the stress environment, this presentation will focus on improvements of the intrinsic reliability of RF-LDMOS FETs. The most important of these intrinsic reliability characteristics are Hot Carrier Injection (HCI), Electromigration (EM), and device ruggedness. The stress environment presented to the RF power transistor will be described in detail, including the linkage between the RF stress and these intrinsic reliability metrics. Detailed models have been created to simulate these stresses, and the results of various device design strategies to mitigate these stresses will be presented.

I. INTRODUCTION

LDMOS devices for RF PAs were commercially introduced during the 1990s and quickly displaced bipolar transistors and vertical MOSFETs as the dominant technology used for cellular infrastructure [1, 2]. The underlying reasons for this technology transition are two-fold: the ability to easily manipulate the FET characteristics (surface dominated device) and the ability to leverage the low cost silicon manufacturing technologies developed for CMOS. While not to minimize the commercial importance of the cost structure, it is the control over the FET characteristics that is the most important element to the dominance of this technology for RF PAs. LDMOS is a surface dominated MOSFET structure which gives the manufacturing process a high degree of control over the device characteristics and by design results in low parasitic (capacitance and inductance) behaviors important to achieving excellent RF performance [3]. Intelligent re-use of proven CMOS manufacturing technologies continues to enable improved intrinsic device structures, pushing LDMOS devices toward ideal levels of performance for RF PAs. As these improved designs demonstrate improved reliability, particularly for Hot Carrier Injection (HCI), Electromigration (EM), and ruggedness, they can be successfully deployed into more advanced infrastructure applications and into new RF applications where the performance and reliability requirements are more aggressive [4].

II. LDMOS STRUCTURE

Figure 1 shows a cross sectional diagram of a typical high power n-channel LDMOS FET finger from source to drain (left to right). The essential features that distinguish this MOSFET from a typical

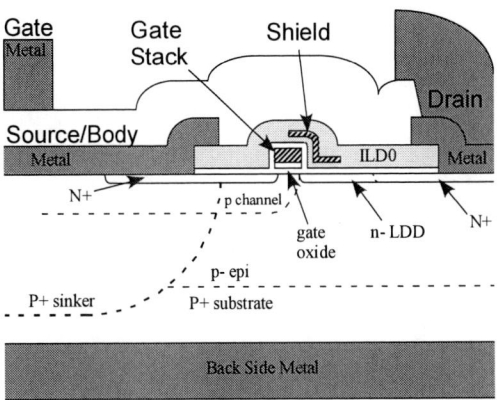

Figure 1: LDMOS FET Cross Section

CMOS FET mostly involve those features which allow the device to operate under high drain supply potentials (to allow the PA to achieve high power output). These essential features include the drain extension region or Lightly Doped Drain (LDD), and a thick lightly-doped epitaxial deposited layer on the starting substrate, both of which support high drain terminal potentials, and a laterally diffused channel doping (the LD in LDMOS) to control gain and threshold along with other channel and gate structures. Other features of note, particularly as they relate to device reliability, include the shield or field plate (connected to ground, and helps control the E-fields near the drain edge of the gate, among other functions) and the drain metallization and cross section (which determine the electromigration resistance). The die backside contact metallization and die attach influence heat transfer and thus the device temperature, which directly contributes to operational reliability. These features influence the junction current and voltage waveforms which along with the device temperature are the key drivers for device reliability.

III. RF PAs & DEVICE RELIABILITY

An RF PA offers unique challenges to a MOSFET for both performance and reliability. It is a large signal analog application where the PA circuit design, the loads being driven, and the FET characteristics combine and interact to define the critical environmental factors that impact reliability – in particular the voltage and current waveforms and the resulting power dissipation within the FET (thermal environment).

Figure 2 shows the final stage of a typical 100W CW RF PA with details around the LDMOS FET (this model PA will be referenced throughout the remainder of this paper). The simulated RF behavior at 100W output power shows time varying potential, current, and dissipated power waveforms around the final FET shown in Figure 3. Internally the device has a distribution of charge carriers related to the terminal conditions illustrated in Figure 4 by the 2D electrostatic potential simulation using TCAD. These simulation tools are critical in developing an understanding of the stress environment experienced by a FET in an RF PA application and the models associated with understanding intrinsic reliability.

978-1-4244-9113-1/11 $26.00 © 2011 IEEE

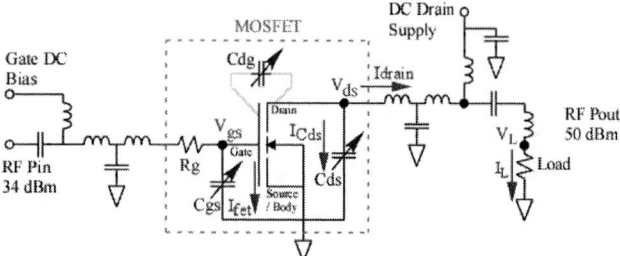

Figure 2: Model 100W RF PA with FET Details

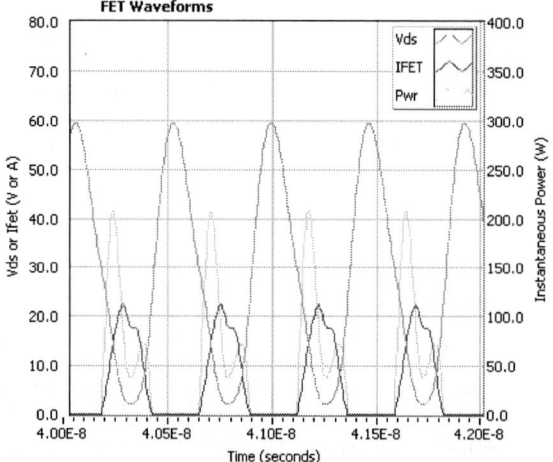

Figure 3: Model PA FET Waveforms at 100W

Figure 3 illustrates the reliability challenges for LDMOS (or any FET) in an RF PA application: large steady-state potential and current waveforms that result in significant power dissipation within the device during normal operation which creates a very challenging thermal environment. A more meaningful way to examine this data is illustrated in Figures 5a and 5b, which show two FET Load Line views of our model PA. The dynamic load lines provide useful insights that assist in understanding device behavior, reliability and PA performance. Figure 5a shows the relationship of between the FET channel current (Ifet) and drain to source potential (Vds) overlaid with a set of I-V curves for various gate voltages (Vgs). Note that the FET channel current <u>does not</u> include the current associated with charging and discharging the intrinsic drain non-linear capacitance (Cds) and thus represents only the current transiting across the channel of the device. Figure 5b shows the drain terminal current (the combination of the FET channel current and the current associated with Cds) mapped onto the same IV curves. These two figures will be of interest in the analysis and understanding of the LDMOS key reliability characteristics.

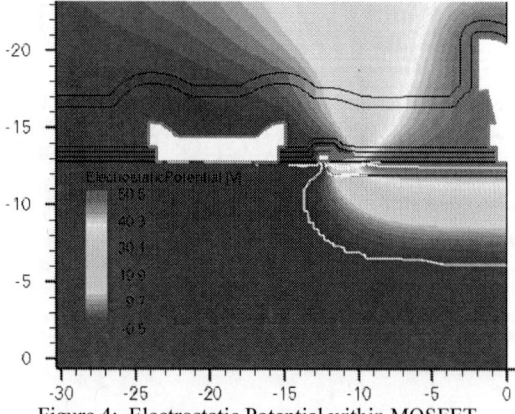

Figure 4: Electrostatic Potential within MOSFET

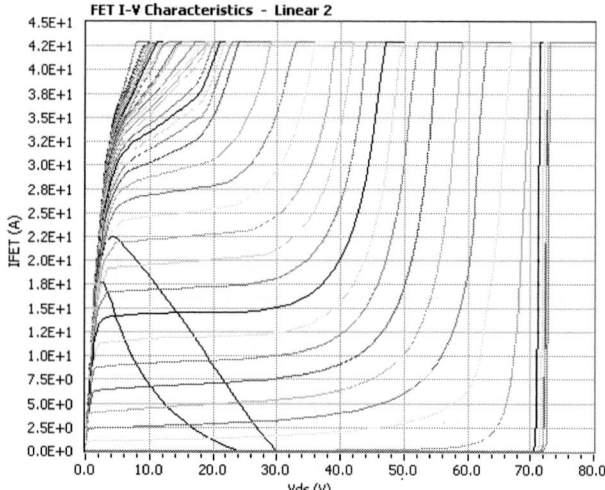

Figure 5a: MOSFET Channel Current vs Drain Potential

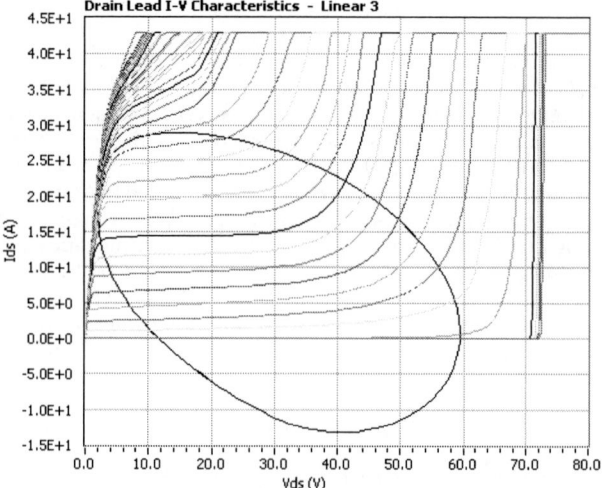

Figure 5b: MOSFET Drain Terminal Current vs Potential

The load line shown in Figures 5a and 5b can be used to illustrate several important concepts. First, the peak current reached on the drain terminal (5b) correlates to the peak current (and thus power) delivered to the Load by the PA (this peak current correlates very strongly with the peak FET channel current from 5a). Second, the 'trajectory' of the "load line" in Figure 5a for a given device structure is primarily determined by the output circuit of the PA. Small adjustments to this circuit to maximize a particularly set of performance characteristics is referred to as 'RF tuning'. For Figure 5a this has the effect of changing the effective 'slope' and shape of the Load Line at the FET. Figure 5c shows an identical load line 5a but now overlaid with the instantaneous FET power dissipation contours which is the product of the channel current (Ifet) and the drain to source potential (Vds) from Figure 5a. The average power dissipation in the FET is the time averaged integration of the curve shown in Figure 5c. Due to the density of these constant power contours, anything which moves the FET load line can quickly change the power dissipated in the FET which directly impacts the operating temperature of the device.

So far only steady-state un-modulated RF waveforms have been illustrated – but real life RF PA applications use either modulated signals (e.g. cellular infrastructure, other communication networks), pulsed signals (e.g. radar, MRI), or large load modulations (e.g. laser, plasma). Load lines analogous to Figure 5a are shown for a pulsed CW transient with an undesirable turn-on phase in Figure 5d and a

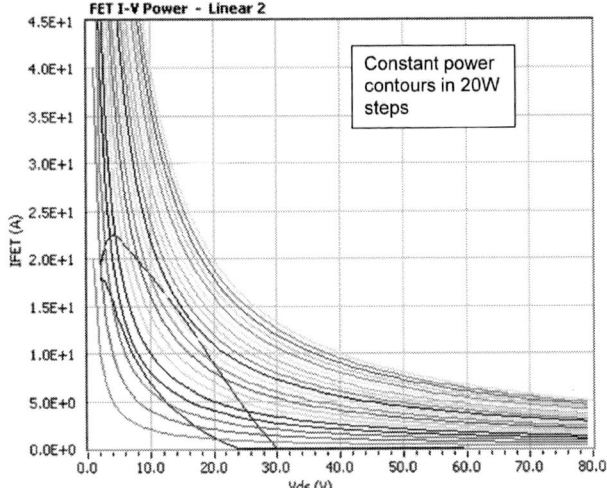

Figure 5c: MOSFET Load Line with Power Dissipation Contours

complex modulated CDMA transient (amplitude and phase change) in Figure 5e. These transient behaviors deviate significantly from the steady-state RF conditions experienced by the FET and are particularly important for the intrinsic ruggedness of the device.

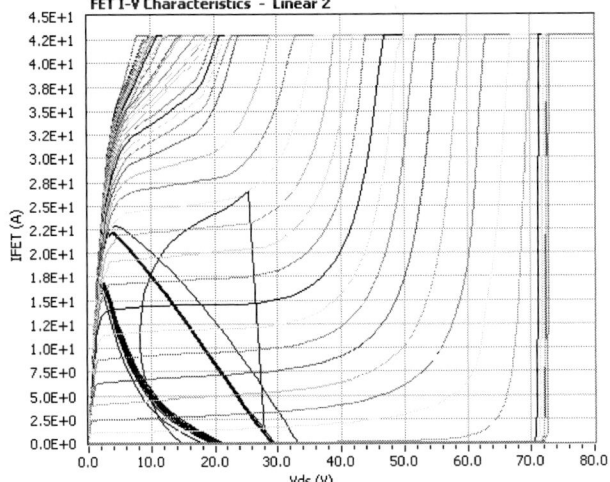

Figure 5d: Pulsed CW Load Line with Undesirable Turn-on Phase

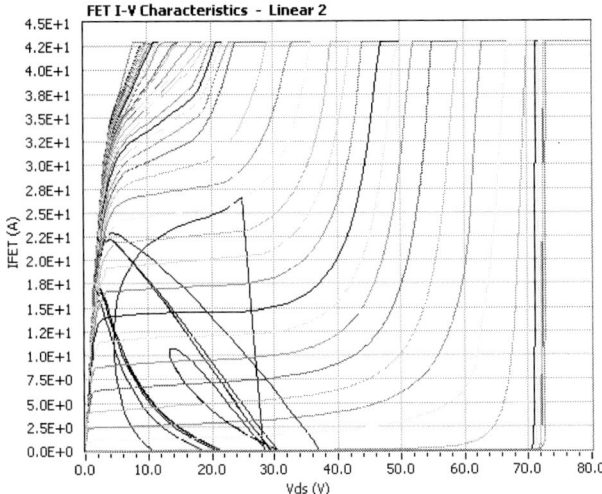

Figure 5e: WCDMA Load Line with Phase/Amplitude Transient

Some outstanding papers have already been presented and published on LDMOS reliability [5], particularly for cellular infrastructure applications. The purpose of this paper is to not repeat these discussions on reliability but to expand upon them. A thorough understanding of the three key reliability characteristics (Hot Carrier Injection, Electromigration, and Ruggedness) has resulted not only in improved reliability for cellular infrastructure applications but has allowed for recent commercial deployment of LDMOS devices into adjacent RF markets (e.g. UHF Broadcast, Industrial Applications) that have more demanding ruggedness and operating temperature requirements.

IV. HOT CARRIER INJECTION

Hot Carrier Injection (HCI) exists in all MOSFET devices and is the trapping of charge carriers in the oxides (gate or other dielectric layers) above the active regions of the FET which then impact the performance of the device because of the resulting distortion of the local E fields. In LDMOS there are two HCI effects that are important for the RF PA which can be easily monitored by DC behaviors: an increase in threshold voltage (V_t) and an increase in on-resistance (Rdson). Typically a change in V_t behavior can be compensated by a bias change for the RF PA, but the change in Rdson can lead to loss of PA peak power capability and is generally the more important issue. Loss of RF PA peak power capability can be compensated for by allowing for a larger tolerance in design but as LDMOS devices are deployed into more advanced infrastructure or adjacent RF market applications, PA peak power output levels (and the resulting MOSFET stress) are increasing, which makes the cost of incorporating these design tolerances prohibitive. Therefore, improved resistance to HCI is required in order to successfully deploy into these applications.

HCI is a cumulative effect of charge trapping and de-trapping mechanisms occurring within the device – but for most MOSFETs (including LDMOS) the dominant effect is a charge trapping mechanism which is driven by acceleration of the charge carriers in high internal Electric (E) fields. For a typical CMOS FET, HCI is driven primarily by the gate to source potential driving very high vertical E fields within the device [6] and trapping charge carriers in the gate oxide above the channel causing V_t shifts only. Due to the design of LDMOS RF power devices there are both vertical and lateral E fields which are important for HCI effects – but they are driven primarily by high drain to source terminal potentials. Figures 6a and 6b show 2D internal device simulations which illustrate the lateral and vertical E fields that can occur in an LDMOS device.

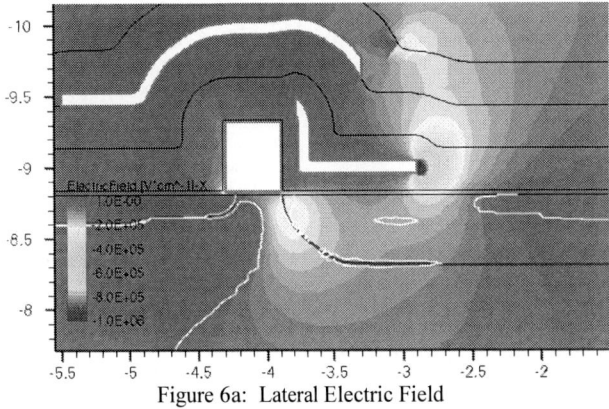

Figure 6a: Lateral Electric Field

This results in charge carriers (electrons) in LDMOS scattering and trapping in the gate oxide over the channel and the insulators (generally also oxides) in the dielectric layer over the lightly doped drain region. Figure 6c illustrates a 2D device simulation showing impact ionization rates. Impact ionization is driven by a combination of E field strength and carrier availability, from which it is concluded

978-1-4244-9113-1/11 $26.00 © 2011 IEEE

that two regions in the device dominate impact ionization – adjacent to the gate, and at the end of the shield. To first order, changes in V_t and Rdson can be attributed to the charge trapping above these two regions, in the channel region gate oxide and the adjacent dielectric layer over the LDD region, respectively.

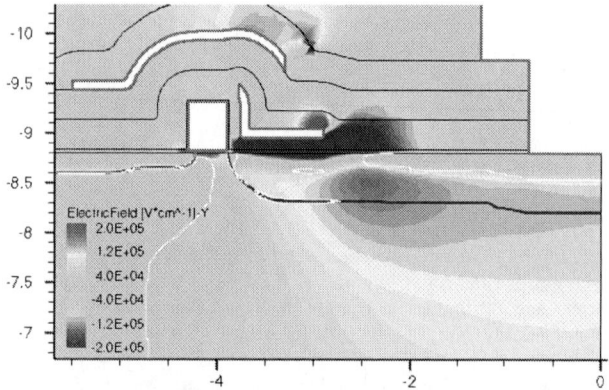

Figure 6b: Vertical Electric Field

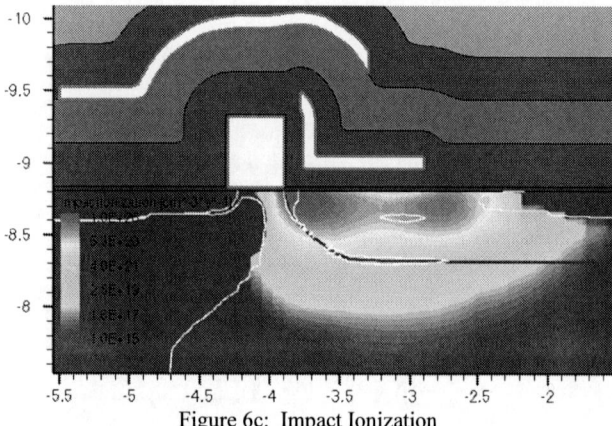

Figure 6c: Impact Ionization

Regardless of the source of the traps, they can be characterized by two parameters illustrated in Figure 7: a classical view of a barrier energy level (E_b) and a trap energy level (E_t). For a given device structure there are a distribution of E_b and E_t levels that when combined with the energy distribution of the charge carriers and trapping/de-trapping probability coefficients allows the net charge trapping rate to be estimated. This ends up being a very complicated analysis and it is much more expedient to use the fundamental mechanistic model (similar to that used for chemical and nuclear reactions) and expect a classic exponential behavior with time for a fixed set of stimuli. A first order exponential behavior is observed as

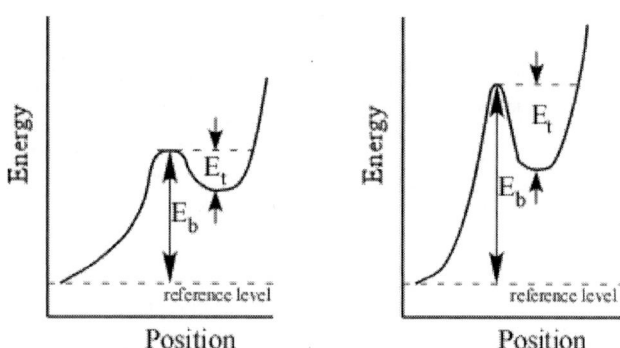

Figure 7: Classic Energy View of Charge Traps

shown in Figures 8a and 8b, which show drain current (I_d) and Rdson vs time for a fixed DC stress condition indicating a primary dominant mechanism (the charge trapping behavior in this case). LDMOS device characterization for HCI behavior is typically conducted under DC conditions where the stresses driving these behaviors (internal E fields, charge carrier distributions, and device temperature) are constant. Table 1 shows examples of DC HCI measurement data showing Rdson and I_d change for wide range of device design features. This underlying mechanistic model suggests how to improve LDMOS HCI behavior: reduce the stress applied to the critical regions of the FET (drain and channel E fields), and change the quantity and energy levels of the charge traps.

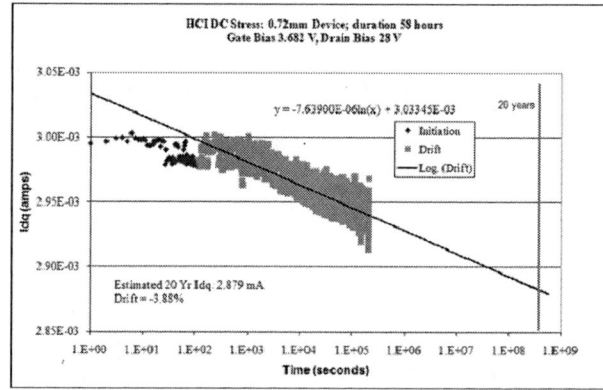

Figure 8a: Drain Current vs Time for DC HCI Stress

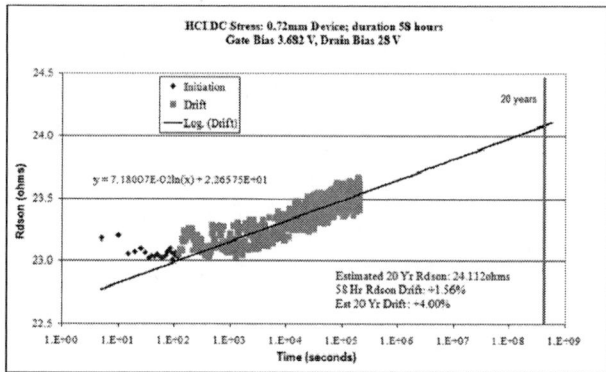

Figure 8b: Rdson vs Time for DC HCI Stress

Table 1: DC HCI Behavior vs MOSFET Design Features

NHV Profile	Gate Length	Shield Design	20Yr Rdson	20Yr Idq
A	H	G	7.2%	-3.2%
A	L	G	7.4%	-3.3%
B	L	B	15.6%	-6.3%
B	L	G	7.2%	-2.9%
B	N	G	3.3%	-1.3%
C	H	F	5.1%	-4.3%
C	L	B	8.6%	-2.1%
C	L	G	4.9%	-3.0%
C	N	G	3.3%	-1.8%
D	H	F	14.1%	-9.6%
D	L	F	20.6%	-9.6%
D	L	G	10.5%	-4.8%
D	N	G	1.3%	-1.3%

The steady-state and transient behavior in an RF circuit results in these stress conditions continuously changing with time. While it is possible to simulate/calculate or measure [7] this behavior, it is more expedient to select a DC condition that represents a reasonable compromise to reflect the stresses encountered under RF conditions. Figure 9 shows the steady-state simulated load lines for our model

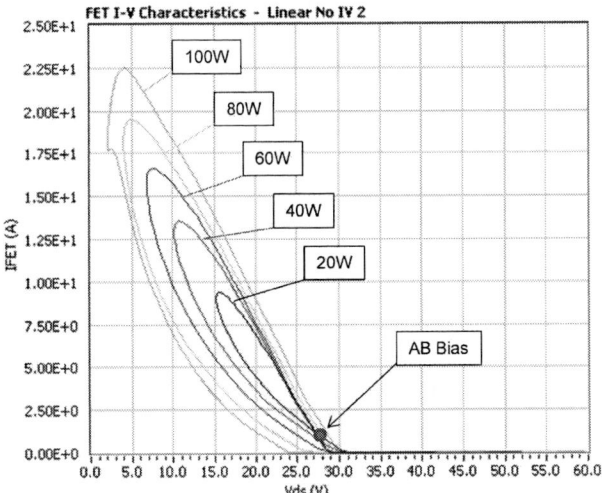

Figure9: Steady-State RF Load Line vs Output Power

PA between 20W and 100W output power and a typical class AB bias condition. This class AB bias condition is a reasonably good DC stress point for evaluating HCI performance because the charge trapping rate is proportional to the charge carriers available but exponentially related to their energy (E field / terminal potential). This results in the DC point of maximum HCI stress (along a specific load line) being weighted toward high potentials where there are a moderate concentration of carriers. The typical class AB bias point for an infrastructure RF PA ends up being a reasonably good representative DC HCI stress condition to reflect RF HCI related changes. Table 2 shows the measurement and simulation results for DC HCI combined with results from an RF life test on a commercial LDMOS part over a 1000 hour period which shows reasonable agreement, with simulation under predicting the actual DC and RF measurements.

Table 2: DC, Simulation and RF Life Test for LDMOS Part

Parameter	DC HCI	HCI Sim	RF Life
20 Yr Idq	-4.30%*	-4.02%	-
16 Hr Rdson	1.52%	1.07%	-
1000 Hr Rdson	3.35%	2.36%	4.10%
20 Yr Rdson	5.6%*	3.9%*	6.9%*
1000 Hr P1dB	-	-	-3.90%
20 Yr P1dB	-	-	-6.5%*

* - estimated from model behavior

Temperature effects are also very important for HCI measurements as they alter charge carrier distributions and the resulting E fields and carrier mean free path (which both alter the carrier energy distributions). Once again the complexity of the time dependent variation in the stress factors drives the need for a mechanistic model that can be used to understand and provide appropriate correlation to real world HCI behavior in order to make practical choices in device design to optimize HCI performance for a specific application.

The important HCI effect for an RF PA is the increase in Rdson because of how this impacts the peak power that can be delivered by the PA to the load. Figure 10a shows an illustration that exaggerates the change in Rdson (shown in red) on the FET "DC" load line (shown in blue). This figure highlights two essential impacts to the RF PA resulting from this HCI effect, power loss and efficiency loss. First, the peak currents that can be obtained from the FET decrease (assuming the Load Line remains constant) as this current is truncated earlier due to the changes the trapped charges have on the MOSFET IV characteristics – which translates to less power available at the load. If the load line is altered in order to obtain the FET peak current level (due to a 'tuning' change) more power is

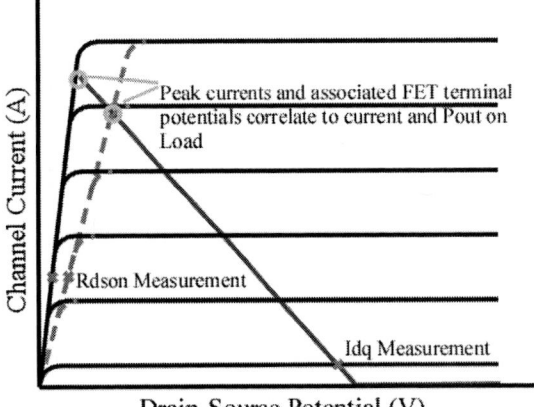

Figure10a: Illustration of Rdson Impact to RF Load Line

dissipated in the FET to obtain this peak current and therefore the PA drain efficiency will be reduced. In real RF PAs or simulated RF PAs the practical observation is that both power and efficiency loss are observed. Figure 10b shows a simulation of Load power for our 100W model PA with a 10% and 20% increase in Rdson. Under these conditions output power is degraded by ~2% and efficiency is degraded by 0.7 points. This simulation under-predicts the observations from similar measurement data shown in Table 2 which emphasizes the need for measurements to validate RF PA lifetime performance. A robust design process for an RF PA will include a lifetime performance analysis that accounts for such HCI related power output degradation.

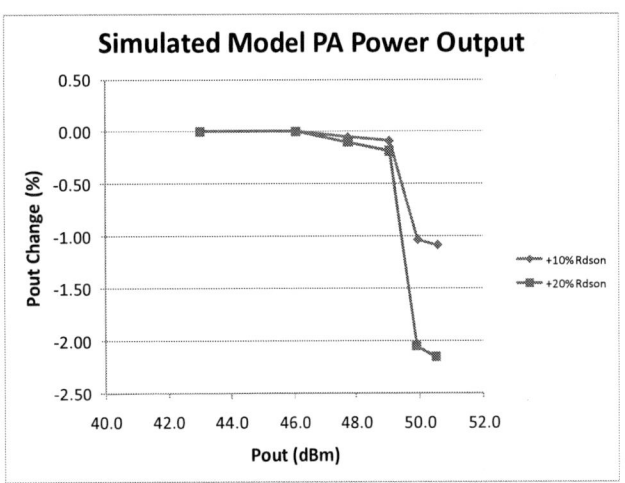

Figure10b: Simulated Output Power vs Rdson Change

In terms of design-for-reliability the operating temperature for any LDMOS device is only partially influenced by the design of the MOSFET. The PA circuit and signal modulation determine how much power is dissipated within the MOSFET and the effective heat transfer determines the operating temperature. Therefore, effective design strategies for improving the HCI behavior of LDMOS have focused on reducing the internal E fields that are driving the charge trapping regardless of the device temperature or PA circuit. The primary features within the LDMOS structure (Figure 1) which influence the E-fields in and around the drain extension region include the field plate, gate length, and the drain extension doping level and dimension. Figure 11 shows 2D TCAD simulations under identical terminal conditions which change the impact ionization generation rate as a function of some of these variables. Many device design variables must be optimized for the best trade off for reliability and performance to meet the needs of a given application.

978-1-4244-9113-1/11 $26.00 © 2011 IEEE

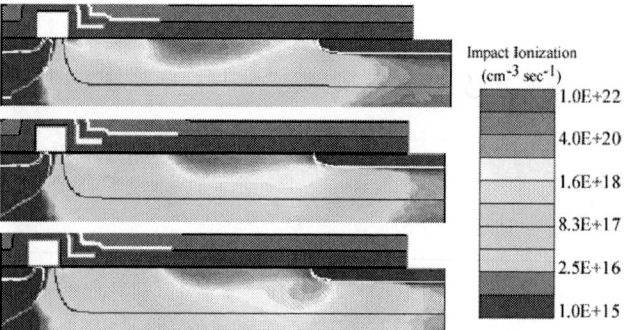

Figure11: Impact Ionization vs Drain Design

V. ELECTROMIGRATION

Electromigration (EM) is the movement of metal atoms in a conductor due to momentum transfer from electrons impacting the metal atoms in the crystal lattice. This phenomenon was first described by Black et al. [8] and has since been studied extensively by many others [9-15]. Black established a correlation model that relates an EM failure to the current density (J) and temperature (T) as shown in Equation 1.

$$MTTF = \frac{A}{J^n} \exp\left[E_a/kT\right] \quad \text{Eqn. 1}$$

The model coefficients A and E_a (activation energy) are determined experimentally under well controlled DC conditions and the exponent n is generally assumed to be 2. The Median Time To Failure (MTTF) criteria is arbitrarily defined but a typical failure criteria is a 10% increase in the resistance of the specific interconnect structure being examined (this can be anything from a single layer metal transmission line to a multilevel series of transmission line sections interconnected through vias). Due to this phenomenon the metal interconnect in MOSFETs in general and LDMOS in particular tends to be the weak link from a catastrophic reliability failure viewpoint.

The grain boundary structure of the metallic conductor is important to the observed EM performance since atoms at the grain boundary are more easily displaced than those within the interior of a crystal lattice [10]. This makes the grain boundary and transmission line dimensions important for the observed EM MTTF. Figure 12a and 12b illustrate grain structure within two lines with dimension larger than and smaller than the mean metal grain size, respectively. The structure illustrated in Figure 12b leads to the so called 'bamboo' metal line. The experimentally determined Black's Equation coefficient (A) will be much larger for the transmission line structure from Figure 12b than from 12a. The reason for this improved EM performance is that nearly all the grain boundaries are perpendicular to the direction of current flow and easy atomic displacement along the grain boundary (in the direction of current flow) is nearly shut down as a displacement mechanism. Al and Au based metal interconnect systems for semiconductors have been studied with reasonably wide ranging values reported for A and E_a coefficients (and n when it has not been assumed as 2) [11-14]. E_a has a large influence on the lifetime calculations and in general for interconnect systems relevant to RF-LMDOS devices have typically been reported in the range of 0.6-0.7eV for Al and 0.8-1.0eV for Au systems. This difference between Al and Au has been attributed to a combination of lattice binding energy and the mass of the nucleus (amount of momentum transfer required to achieve a displacement).

Figure 5b shows the RF steady-state load line for the FET drain terminals (metal interconnect) of the model PA which has a fundamental frequency of 2.14GHz . Studies of EM behavior vs frequency [15] have shown that when the frequency is $> 10^5$ Hz the EM performance is no longer correlated to the peak currents observed

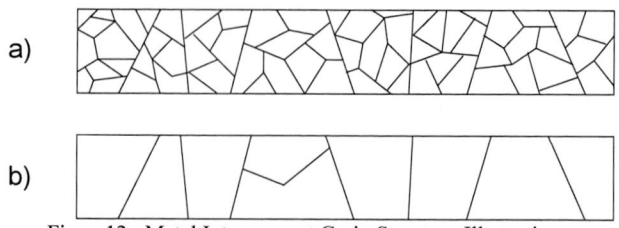

Figure12: Metal Interconnect Grain Structure Illustration

on the interconnect but to the average or net DC current flow. This makes estimates of the EM performance of a specific PA operating above this frequency simple because all that is required is the average drain DC supply current flow (Figure 2b), the temperature of the device, the Black's equation coefficients (A, E_a), and the device interconnect design, from which the MTTF can be calculated. A typical minimum criterion for a cellular infrastructure LDMOS is >20 year MTTF at 200 °C under nominal rated output power. Figure 13 shows the model PA MTTF vs temperature under both a rated and low power output condition based upon such calculations. Under nominal output power conditions this 100W PA has a device (junction) temperature in its reference circuit of 129 °C resulting in an estimated MTTF of 450 years.

From a MOSFET design standpoint EM can be improved by increasing the drain interconnect metal cross sectional area (reduces J), using materials with higher E_a (higher mass nuclei metals), increasing the interconnect grain sizes to significantly larger than transmission line dimension (increases A), by modification of the grain boundaries to impede atomic displacements, and by developing MOSFETs that can achieve higher drain efficiencies (reduces J). From a PA design viewpoint EM can be improved by operating the devices at maximum efficiency operating points (reduces J and T), by advanced device packaging materials to extract heat more efficiency (reduces T), and by improved PA design heat transfer from the MOSFET (reduces T).

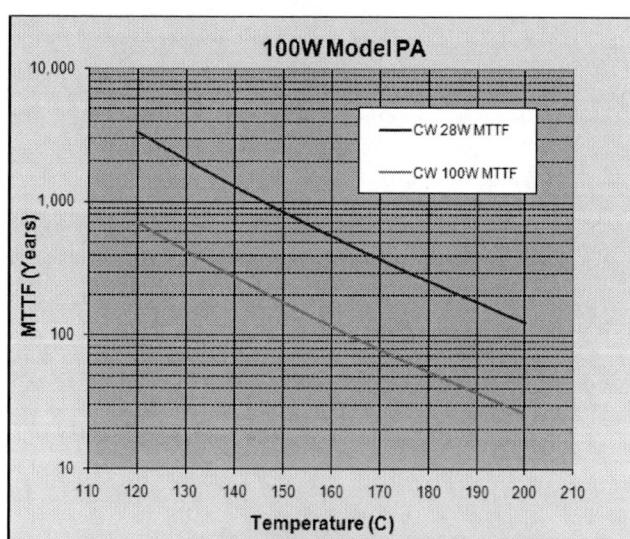

Figure13: EM MTTF for Model PA vs Temperature

VI. RUGGEDNESS

Ruggedness is probably the least well defined reliability characteristic for MOSFETs (or any semiconducting device) because it is extremely application and PA design dependent. A reasonable definition for ruggedness is the ability of a device to withstand unusual electrical conditions on the input or output section of a PA without degrading the RF performance when conditions return to normal. While this definition is reasonable, it is very non-specific, dependent upon the particular PA design, and most importantly

978-1-4244-9113-1/11 $26.00 © 2011 IEEE

'normal' conditions which vary widely by application, frequency range, and signal modulation. Given this background it should be no surprise that there is significant divergence of opinion when it comes to the topic of ruggedness for MOSFETs. Ruggedness-induced failures are a major failure mode in PAs, and must be considered both during the design of the RF transistor and for the application environment surrounding the transistor.

The most common ruggedness failure mechanism for RF-LDMOS is a catastrophic failure resulting from the snapback of a parasitic Bipolar Junction Transistor (BJT) formed by the drain (collector), channel (base), and source/body (emitter) of the MOSFET. Figure 14a shows a cross sectional diagram of a typical LDMOS FET overlaid with the circuit schematic for the parasitic BJT. Figure 14b shows a schematic of the LDMOS FET, parasitic BJT, and other parasitic elements of this circuit that are relevant for the discussion of this failure mechanism which will be the focus for LDMOS ruggedness.

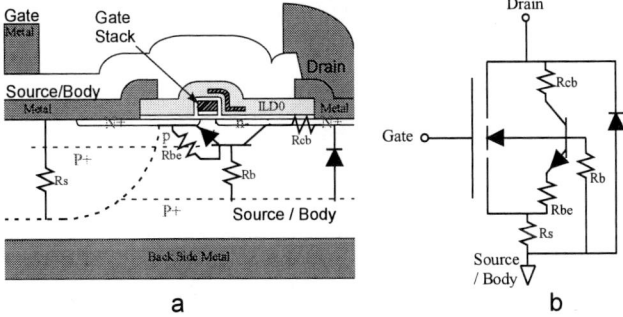

Figure14: LDMOS FET with Parasitic BJT Overlay

The parasitic BJT formed as a result of the LDMOS construction is actually a series of parallel BJTs with differing characteristics. Therefore the collective behavior of this 'device' is not simply dependent upon the near surface silicon characteristics but also upon the internal starting material characteristics. For simplicity we will discuss this device as a monolithic BJT. The ruggedness failure of the LDMOS device occurs because the BJT can be turned on and reaches a thermal "runaway" condition due to its negative temperature coefficient for device on-resistance. This effectively creates for the LDMOS FET a very low impedance path in very small discrete locations within the FET between the drain and source/body. The resulting current flow and thermal dissipation that occurs when this event is triggered rapidly and dramatically escalate, and essentially vaporizes the localized silicon and associated metal interconnect which then cascades in a series of other failures resulting in an end-state catastrophic failure like the image in Figure 15.

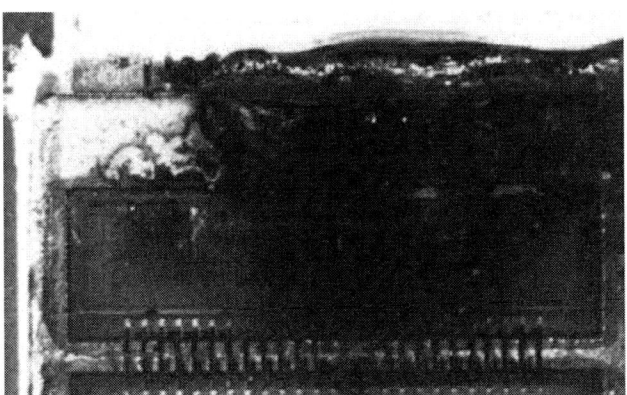

Figure15: LDMOS Catastrophic Ruggedness Failure

In order for the parasitic BJT to reach this state two necessary conditions must be met. First, in order to forward bias the base-emitter junction the drain-source of the LDMOS FET must reach an

impact ionization breakdown condition; this is the only sufficient source of injected charge carriers (holes) into the base to sustain a forward bias condition in the parasitic bipolar transistor. The second condition that must be met is for the parasitic BJT common-emitter current gain (β_F) to be >1. At this point the BJT is unstable and reaches a point of thermal runaway; depending upon the energy available in the circuit, different levels of catastrophic damage to the LDMOS device can occur. This is commonly referred to as a snapback in BJTs, but the description of turning on and the subsequent thermal runaway of a common emitter BJT is more complete and better in terms of understanding how to resolve this problem for LDMOS devices. Previously referenced 2D device simulation Figure 6c shows typical impact ionization locations within the MOSFET, while Figures 16a and 16b show electron and hole current densities under similar bias conditions.

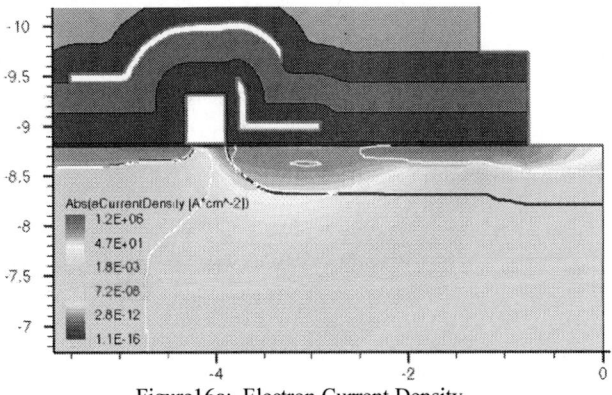

Figure16a: Electron Current Density

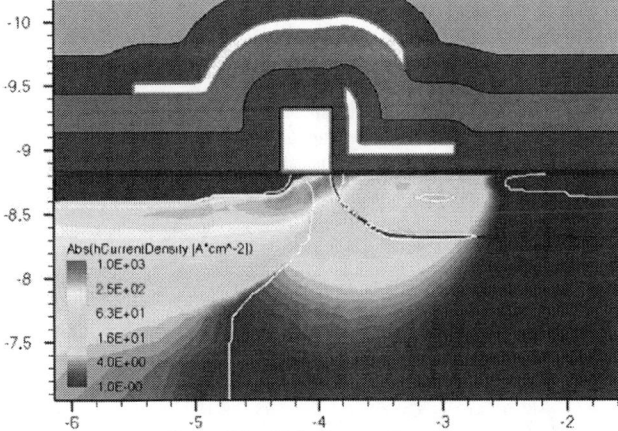

Figure16b: Hole Current Density

The parasitic BJT resistances (Rbe and Rcb) shown in Figure 14b are relevant to snapback behavior because for a given level of injected charge into the base (primarily impact ionization) these resistances determine the effective bias voltage seen on the base which controls β_F. Therefore there are two effective strategies for reducing the BJT "snapback" ruggedness failure mechanism; mitigation of the impact ionization events to prevent carriers from being generated that can forward bias the emitter-base junction, and reduction of the BJT common emitter current gain. Elimination of the impact ionization breakdown condition by increasing BVdss has been a method for improving ruggedness for decades in MOSFETs. However, this didn't always improve the ruggedness observed on the devices and in general had negative consequences on other performance parameters. This is the case for two reasons: impact ionization and subsequent triggering of the BJT can happen in on-state as well as off-state conditions, and the common emitter current gain can be increased by certain changes which increase BVdss. In

978-1-4244-9113-1/11 $26.00 © 2011 IEEE   441

addition it can be the transient behaviors, as illustrated in Figures 5d and 5e, that is often the triggering event for ruggedness failures.

There are several strategies for reducing the common emitter current gain of the parasitic BJT but the most effective are reducing the base-emitter resistance (Rbe) and increasing the effective base-collector resistance (Rbc). Power vertical MOSFETs change the effective Rbc by using a zener diode to shunt much of the impact ionization current away from the parasitic BJT, thereby increasing Rbc for the parasitic BJT. Reducing Rbe is also an easy way to reduce the gain of the parasitic BJT because a higher level of charge injection current is required to reach the same forward base-emitter voltage (Vbe) with lower Rbe. Typical LDMOS device design strategies focus on careful tailoring of the p-type dopant in the channel region to lower Rbe, and controlling the location of the impact ionization events to help shunt current away from the base of the BJT.

Making only minor changes to the fundamental LDMOS FET structure can yield ruggedness improvement with minimal changes in RF performance as demonstrated in Table 3 which shows that the MRFE6S9160 transistor has significantly enhanced ruggedness, with virtually no change in RF performance.

Table 3: Ruggedness Comparison of MRF6S9160 and MRFE6S9160

Parameter	MRF6S9160	MRFE6S9160
Vdd (V)	28	32
VSWR	10:1	10:1
Input (dB)	0	3

Both part have 160W rate Pout for
the same devcie size.

With more aggressive design strategies, LDMOS devices can now be created that essentially eliminate BJT snapback as a failure mechanism as demonstrated by the ultra-high ruggedness MRFE6VP1K25H. Figure 17a shows a PA using the MRFE6VP1K25H operating at 130 MHz at 1250W output where the transmission line is being alternatively presented a short and open circuit condition (sparking) in an attempt to cause damage to the MOSFET; the transistor was not damaged!

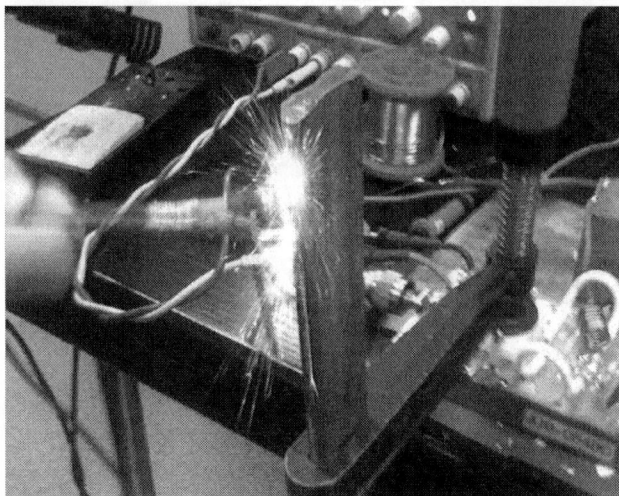

Figure17a: MRFE6VP1K25H with Shorted Load

Figure 17b shows the output power of the same part as a function of drain supply potential up to 90V where the load line for the MOSFET reaches impact ionization conditions for much of the RF cycle with no failures occurring in matched or mis-matched load conditions. Clearly such a demonstration at low frequency (230 MHz) without failure under extreme PA input and output conditions indicates that the BJT transistor failure mechanism is now well understood and

appropriately accounted for in the device design. LDMOS RF applications do not have significant secondary mechanisms outside of BJT snapback which are of relevance for device ruggedness. It is simply a matter of design, processing, and performance tradeoff to create a device, part, and circuit to meet specific RF PA performance and ruggedness behavior.

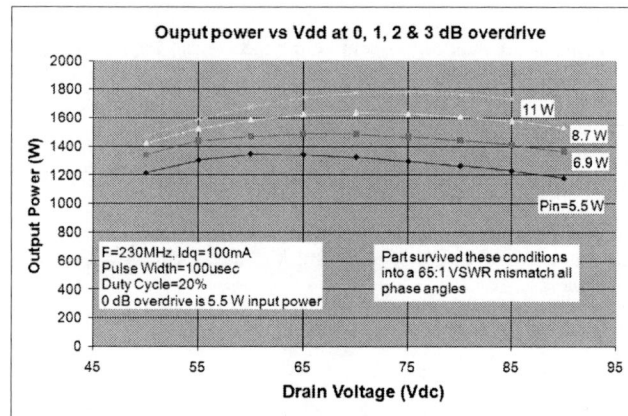

Figure17b: MRFE6VP1K25H as a Function of Drain Supply Vdd

VII. SUMMARY

Improvements in LDMOS reliability as they relate to HCI, EM and Ruggedness have opened up applications in adjacent channel RF markets, such as UHF Broadcast, Industrial Lasers, or S-Band Radar, that have significantly more challenging reliability requirements than cellular infrastructure applications. This is driven by the resulting steady-state and transient voltage and current waveforms around the LDMOS FET, and the thermal environment under 'normal' operational conditions imposed by the PA designs and the applications in these areas. This paper shows how three critical LDMOS FET reliability characteristics, Hot Carrier Injection (HCI), Electromigration (EM), and ruggedness, are related to fundamental FET terminal voltage and current waveforms and the thermal environment experienced in the RF PA design. Strategies to mitigate and design reliability into the LDMOS transistor for these failure modes have been discussed. Improvements in LDMOS behavior in these areas have allowed RF-LDMOS devices to successfully transition to RF applications adjacent to cellular infrastructure, including UHF Broadcast, Industrial Lasers, Pulsed Radar, etc. Addressing these fundamental reliability failure mechanisms has allowed successful deployment of LDMOS devices into applications that used to be considered too harsh for reliable operation.

ACKNOWLEDGEMENTS

The authors would like to gratefully acknowledge the contributions of many Freescale colleagues for the fundamental development of the failure mechanism models, device design improvements, and the assembly of this paper. They include, but are not limited to, Hernan Rueda, Xiaowei Ren, Bob Davidson, Warren Brakensiek, Chris Dragon, Bob Pryor, Agni Mitra, and Basim Noori.

REFERENCES

[1] A. Wood, C. Dragon, and W. Burger, "High Performance Silicon LDMOS Technology for 2GHz RF Power Amplifier Application", in *IEDM Tech. Dig.*, 1996, pp. 87-90.

[2] H. Brech, W. Brakensiek, D. Burdeaux, W. Burger, C. Dragon, G. Formicone, B. Pryor, and D. Rice, "Record Efficiency and Gain at 2.1GHz of High Power RF Transistors for Cellular and 3G Base Stations", in *IEDM Tech. Digest*, 2003, pp. 15.1.1-15.1.4.

[3] M Trivedi, and K. Shenai, "Comparison of RF performance of vertical and lateral DMOSFET", in *Proc. 11th International Symposium on Power Semiconducting Devices*, May 1999, pp. 245-248.

[4] G. Formicone, F. Boueri, J. Burger, W. Cheng, Y. Kim, and J. Titizain, "Analysis of Bias Effects on VSWR Ruggedness in RF-LDMOS for Avionics Applications", in *Microwave Integrated Circuit Conference EuMIC 2008*, pp. 28-32.

[5] M. De Souza, G. Cao, D. Hinchley, "Design for Reliability: The RF Power LDMOSFET", in *IEEE Transactions on Device and Materials Reliability*, Vol. 7, NO. 1, March 2007, pp. 162-174.

[6] E. Takeda et al, *Hot-Carrier Effects in MOS Devices*, (San Diego: Academic Press, 1995), Chapter 2.

[7] G.T. Sasse, F.G. Kuper, and J.Schmitz, "MOSFET Degradation Under RF Stress", in *IEEE Transactions on Electron Devices, Vol. 55, No. 11*, November 2008.

[8] J.R. Black, "Electromigration failure Models in Aluminum Metallization for Semiconductor Devices", in *IEEE Transactions on Electron Devices, Vol. 57, No. 9*, September 1969, pp. 1587-1594.

[9] K.N.Tu, "Recent advances on electromigration in very-large-scale-integration of interconnects", in *Journal of Applied Physics, Vol. 94, Issue 9*.

[10] J. Böhm, C.A. Volkert, R. Mönig, T.J. Balk, and E. Arzt, "Electromigration-induced damage in bamboo Al interconnects", in *Journal of Electronic Materials, Vol. 32, No.1*, pp. 45-49.

[11] B.J. Klein, "Electromigration in thin gold films", in *Journal of Physics F: Metal Physics, Vol. 3, No. 4*, April 1973.

[12] H.U. Schreiber, "Activation energies for the different electromigration mechanism in aluminum", in *Solid State Electronics*, Issue 6, June 1981, pp.583-589.

[13] M.J. Attardo, R. Rosenberg, "Electromigration Damage in Aluminum Film Conductors", in *Journal of Applied Physics, Vol. 41, Issue 6*, May 1970.

[14] I.A. Blech and E. Kinsbron, "Electromigration in thin gold films on molybdenum surfaces", in *Thin Solid Films, Vol. 25, Issue 2*, February 1975, pp. 327-334.

[15] B.K. Liew, N.W. Cheng, C. Hu, "Electromigration Interconnect Lifetime Under AC and Pulse DC Stress", in *Proc. 27th International Reliability Physics Symposium*, 1989, pp. 215-219.

Investigation of Multistage Linear Region Drain Current Degradation and Gate-Oxide Breakdown Under Hot-Carrier Stress in BCD HV PMOS

Yu-Hui Huang, J.R. Shih, C.C. Liu, Y.-H. Lee, R. Ranjan, Puo-Yu Chiang, Dah-Chuen Ho, and Kenneth Wu

TQRD, Taiwan Semiconductor Manufacturing Company, 121, Park Ave. 3, Hsinchu Science Park, Taiwan 300

e-mail: yhhuangx@tsmc.com

Abstract-**Hot-carrier injection (HCI) at maximum gate current (I_G) stress condition for BCD HVPMOS has been studied. It is found that HCI not only causes linear region drain current degradation and minimizes the operation window, but also degrades the gate oxide (GOX) and may result in GOX breakdown. A multistage I_{Dlin} degradation behavior has been observed during HCI stress, which is associated with two competing mechanisms, i.e., interface-state (N_{it}) generation and electron trapping caused by hot electrons originated from impact ionization. HCI leads to the gate oxide breakdown even at very low e-field of ~1.5MV/cm across the GOX. TCAD simulation results by placing N_{it} and negative charges at different location of the device also support a multistage I_{Dlin} degradation. It is found that both initial I_G and bulk current (I_B) are well correlated with GOX time-dependent-dielectric-breakdown (TDDB). In addition, better TDDB has been observed at higher temperature compared to lower temperature, which verifies that GOX breakdown is associated with HCI.**

Keywords- (BCD HVPMOS, Hot-Carrier Injection (HCI), Gate Oxide (GOX), interface-state (N_{it}), Electron Trapping, Time-Dependent-Diekectric-Breakdown (TDDB))

INTRODUCTION

Hot-carrier injection induced device degradation is of increasing concern for BCD HVPMOS because of its high voltage application [1]. Most of the studies in the area focused on the drain current degradation due to the impact-ionization induced surface-state generation or charge trapping near the Si/SiO_2 interface and/or Si/STI [2], while there has been limited or no report of the GOX deterioration. GOX degradation under low-V_G/high-V_D bias in BCD HVPMOS should be taken into consideration because of the significant impact-ionization along with the downward electric field in drift-region (Fig. 1), which is very prone to electron injection into GOX. In this study, we demonstrated that HCI not only induces the multistage I_{Dlin} degradation behavior but also causes the localized GOX leakage under max-I_G HCI stress. TCAD simulation is also performed to mimic the post-stress behavior of I_{Dlin} degradation. Moreover, temperature dependent GOX TDDB under max-I_G stress with different V_D-bias and different initial I_G current levels are also studied to clarify the origin of oxide damage.

EXPERIMENTAL

The HVPMOS device used for this study is fabricated using 0.18um technology with 130Å GOX thickness. STI is introduced in the drift-region HVPW (High-Voltage-P-Well), in order to sustain gate-to-drain voltage drop. The stress bias is performed at an elevated drain voltage to accelerate the device HCI degradation, i.e., $V_D=Vcc*1.375 = -55V$ under the max-I_G condition shown in Fig. 1, ($V_G= -1.5V$ and $V_D= -55V$). Under this stress condition, the impact-ionization is mainly located within the drift-region (Fig. 2). For GOX TDDB evaluation, the HCI stress is also performed under high voltages with various temperatures between 5°C and 125°C.

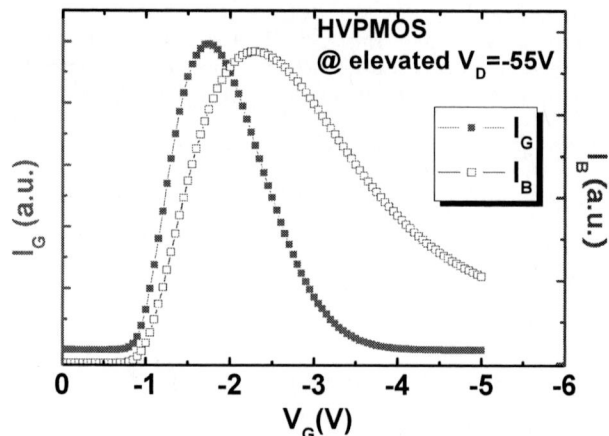

Figure 1. The max I_G of the HV PMOS biased at elevated V_D=-55V is at V_G=-1.5V. Such low-V_G /high -V_D bias leads to noticeable impact-ionization along with a vertical downward electric field in the drift-region (HVPW), which attracts electron injection upward.

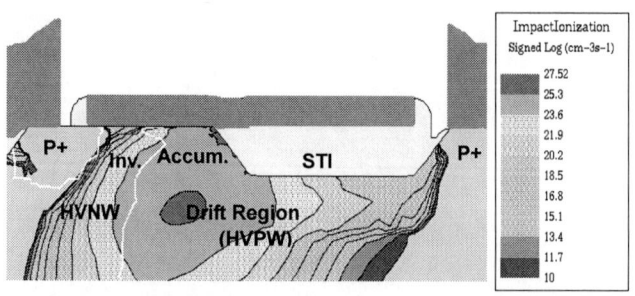

Figure 2. TCAD simulation shows that impact-ionization under the HCI max-I_G stress is located within the drift-region (HVPW) in the HV PMOS.

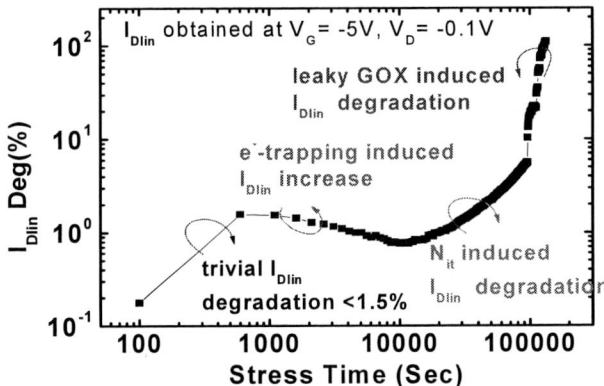

Figure 3. Multi-stage behavior of I_{Dlin} degradation under max-I_G stress (V_G=-5V and V_D=-0.1V) is attributed to two competing mechanisms, i.e., N_{it} formation and electron trapping. The trivial I_{Dlin} degradation in the earlier stage (<600sec) is attributed to N_{it} formation dominance. After that, the I_{Dlin} increases as a result of electron trapping dominance (600~10Ksec) and decreases again during the following stress due to again N_{it} formation dominance over electron trapping (10K~100Ksec). In the ultimate stage (>100Ksec), I_{Dlin} undergoes a different mechanism of degradation as a result of gate leakage.

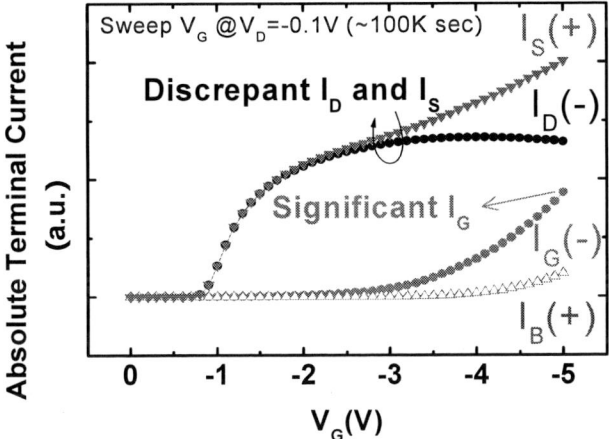

Figure 4. The terminal current characteristics of HVPMOS after HCI stress ~100Ksec (in the ultimate stage of I_{Dlin} degradation) show discrepancy between I_D and I_S and significant I_G current increase. The I_{Dlin} (V_G=-5V, V_D=-0.1V) degradation is due to the significant leakage electron current from gate-terminal, partially compensating the I_{Dlin} hole current.

RESULT AND DISCUSSION

Fig. 3 shows the multistage I_{Dlin} degradation (@V_G= -5V and V_D= -0.1V) behavior during maximum I_G HCI stress. It may be due to two simultaneous competing mechanisms, i.e., electron trapping and N_{it} formation during the stress. In the very initial stage of stress, the trivial I_{Dlin} degradation <1.5% is attributed to N_{it} formation dominance over electron trapping. After that, the I_{Dlin} showed increment (i.e. slow I_{Dlin} degradation recovery), which is due to electron trapping overpowers the N_{it} generation, and then followed by two different stages of I_{Dlin} degradation, i.e. gradual and abrupt I_{Dlin} degradation. The gradual I_{Dlin} degradation is attributed to again dominance of N_{it} generation on electron trapping. As can be seen in Fig. 4, the abrupt I_{Dlin} degradation in the ultimate stage of the stress is due

to the leaky electron current through GOX (I_G) from gate-terminal, partially compensating the hole current collected by drain-terminal and accounting for the difference between drain current (I_D) and source current (I_S). In spite of the GOX breakdown, HVPMOS still holds its device characteristic such as gate-controlled turn-on and -off ability, except for the significant I_{Dlin} degradation as a result of GOX leakage.

A. Prior to GOX Breakdown

Fig. 5 shows the typical linear I_D-V_G (@V_D= -0.1V) plot of HVPMOS after the long stress (~50Ksec) prior to GOX breakdown. It can be seen that the post-stress I_D behavior is different at low ($|V_G|$: 1V ~ 3V) and high V_G ($|V_G| > 4V$) compared to pre-stress I_D, one increasing, and the other decreasing. The post-stress I_D degradation at high V_G prior to GOX BD is caused by the hot-electron induced N_{it} generation along the STI, which dominates over the electron trapping in the accumulation region, rather than by GOX leakage (I_G), as supported by Fig. 6. From Fig. 6, it can be seen that I_{Dlin} (V_G=-5V, V_D=-0.1V) gradually degrades with stress time prior to GOX breakdown and abruptly dropping by abrupt increase of I_G due to GOX breakdown in the ultimate stage. On the other hand, the increase of the post-stress I_D at low-V_G (i.e., $|V_G| < 3V$) (Fig. 5) is associated with a possible negative charge trapping dominant phenomena in the accumulation region. This phenomenon is also reflected from the maximum G_{mlin} increase in Fig. 7. The maximum G_{mlin}, which reflects the low-V_G behavior, initially increases due to the negative charge trapping along the accumulation region and after long stress time (50Ksec), it starts to decrease. It indicates that N_{it} generation dominates over negative charge trapping along the accumulation region after long stress time.

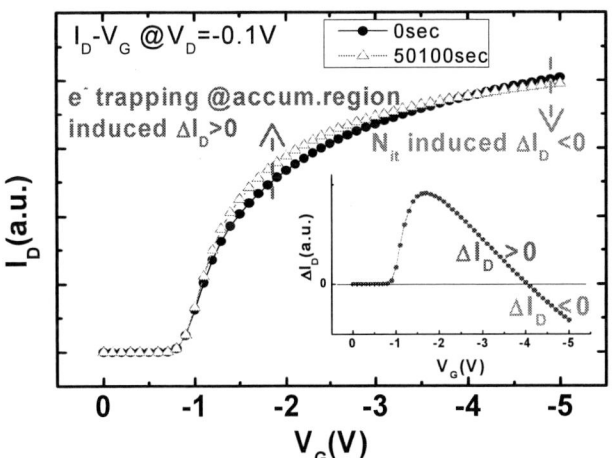

Figure 5. Typical I_D-V_G (@V_D= -0.1V) curves before and after ~50Ksec HCI stress at max-Ig condition. The I_{Dlin} (V_G/V_D= -5V/-0.1V) degrades due to N_{it} generated along the STI. Moreover, lower-V_G ($|V_G|$<3V) current increases due to the electron trapping along the accumulation region.

978-1-4244-9113-1/11 $26.00 © 2011 IEEE   445

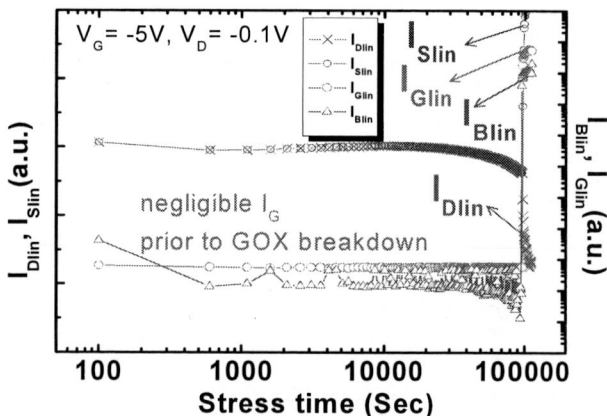

Figure 6. The terminal current of HVPMOS changes as a function of the HCI stress time. The linear-region (V_G= -5V, V_D= -0.1V) current evolution shows negligible I_G change prior to GOX breakdown, indicating the I_{Dlin} degradation prior to GOX breakdown is due to interface-state formation.

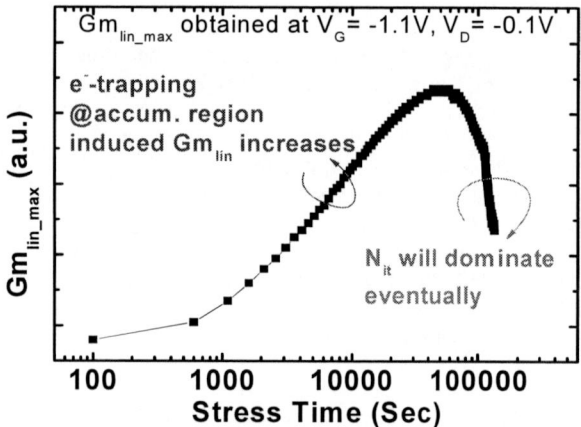

Figure 7. The max G_{mlin} increases at first, indicating the accumulation region behavior is dominated by electron-trapping. Eventually, it changed to interface-states formation dominating along the accumulation region.

TCAD simulation is carried out by placing negative charges and N_{it} in the different regions of the HVPMOS device and compared the I_D and V_G characteristics against the experimental data (Fig. 5) to further understand the physical mechanism behind the multistage I_{Dlin} degradation behavior. The results shows that negative charges placed along the ac-cumulation region (Fig. 8) increase the linear I_D more signifi-cantly at low V_G, while N_{it} placed along the STI (Fig. 9) de-grades the linear I_D more noticeably at high V_G. At lower V_G, the accumulation region is under weak gate-control and be-comes more resistor-like. Hence, the trapping species in the resistor-like region can change the effective resistance and significantly affect the linear I_D at low V_G, which can be ex-pressed by the following equation:

$$I_D = V_D/(R_{Accum}+R_{STI}) \quad \text{(at low-}V_G\text{, }V_D\text{= -0.1V)} \quad (1)$$

But at higher V_G, the accumulation region is under strong gate-control and the STI becomes the only resistive-like re-gion. The resistor term of R_{Accum} in Equation (1) is therefore neglected, and the linear I_D at high V_G can be expressed as:

$$I_D = V_D/R_{STI} \quad \text{(at high-}V_G\text{, }V_D\text{= -0.1V)} \quad (2)$$

Hence, the trapping species along STI can significantly af-fect linear I_D at high V_G. As both effects are considered (Fig. 10) in the TCAD simulation, the obtained I_D-V_G curve in Fig. 10 can completely mimic the 50Ksec I_D-V_G experimental data in Fig. 5 and it can also explain the maximum G_{mlin} (@V_G= -1.1V, V_D= -0.1V) increase in Fig. 7. Therefore, it can be con-cluded that HCI stress generated hot electrons and created N_{it} initially. Hot-electrons are trapped along the accumulation region and N_{it} are piling up in the STI simultaneously. These two mechanisms are competing with each other and resulting in multistage I_{Dlin} degradation at high V_G and maximum G_{mlin} increase followed by G_{mlin} decrease at lower V_G.

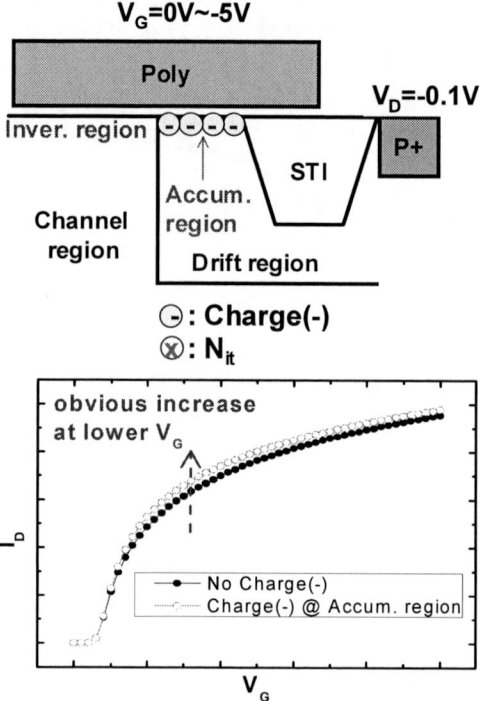

Figure 8. Comparison of I_D-V_G characteristics for HVPMOS with and without charge placement in the accumulation region. Negative charges placed along the accumulation region (upper figure) shows obvious current increase at lower V_G (lower figure).

978-1-4244-9113-1/11 $26.00 © 2011 IEEE

446

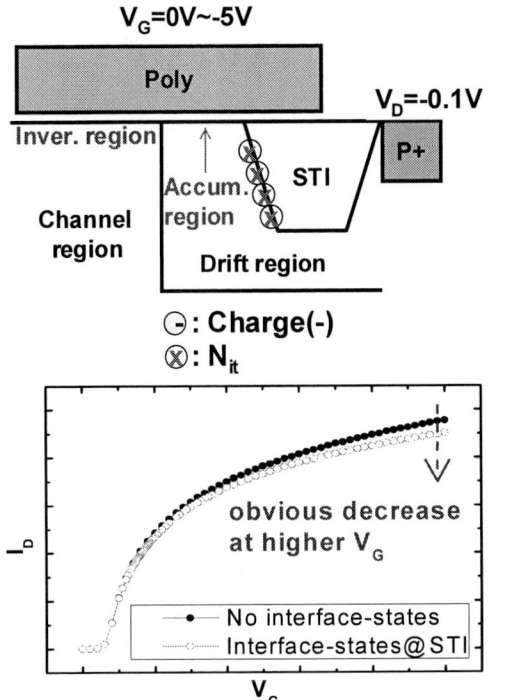

Figure 9. Interface-states placed along the STI (upper figure) shows I_{Dlin} ($V_G= -5V$, $V_D= -0.1V$) degradation (lower figure).

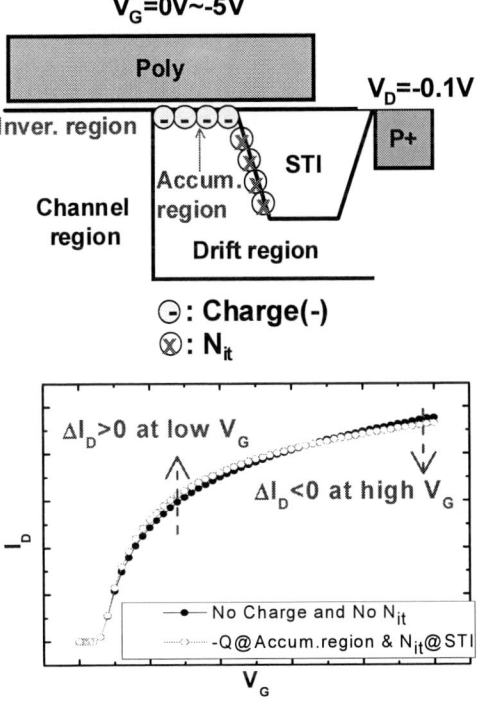

Figure 10. As both negative charges (–Q) at accumulation-region and N_{it} at STI are considered (upper figure), the TCAD result of I_D-V_G plot (lower figure) can mimic the post-stress I_D-V_G behavior prior to GOX breakdown as shown in Fig. 5.

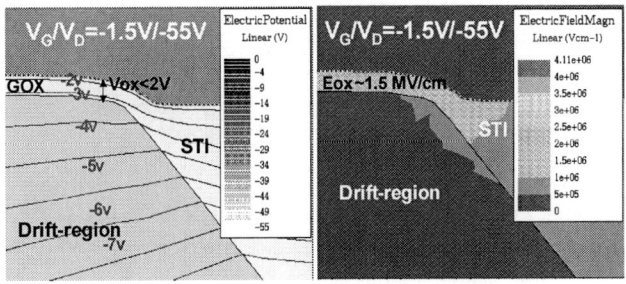

Figure 11. Simulation of e-field in HV PMOS under max-I_G stress at $V_G/V_D= -1.5V/-55V$. Only <2V voltage-drop and 1.5MV/cm downward e-field across the 130A GOX under max-I_G stress. Hence, a low GOX TDDB concern due to electric field induced oxide damage.

B. After GOX Breakdown

Although V_D of 55V during HCI stress is quite high, a significant portion of it is expected to cross the drift region underneath the STI and resulted in a much lower e-field cross the GOX. TCAD simulation is carried out to quantify the actual voltage drop on the GOX. Simulation result showed that only 1.5MV/cm of e-field (Fig. 11) exists within the GOX under the max-I_G HCI stress condition. Hence, it can be concluded that GOX degradation during HCI stress is very unlikely to be induced by such a low e-field on the GOX.

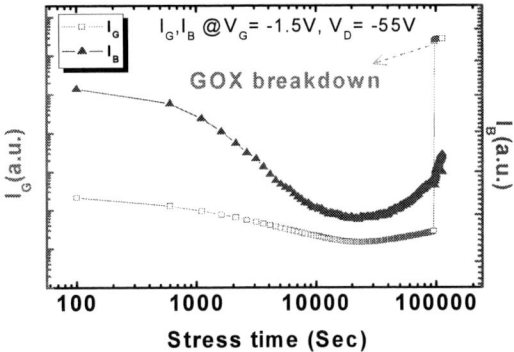

Figure 12. The I_G time-evolution shows similar trend with I_B because I_G is the electron component of impact-ionization while I_B is the hole component.

On the other hand, it is noted that the I_G-time evolution during HCI stress follows the similar trend as substrate current I_B (Fig. 12) and TDDB performance is found to correlate well with the I_B current (Fig. 13). These two characteristics support the argument that the GOX breakdown is affected by HCI induced impact-ionization. To further clarify if the GOX TDDB model played a role, HCI stress with the same V_D and V_G bias under various temperatures have been carried out. Data showed that T_{BD} (Time-to-Breakdown) at the same HCI bias condition is worse at low temperature than at high temperature (Fig. 14), while HCI stress under the same I_G had a much worse T_{BD} at higher temperature (Fig. 15). These two results also convincingly support the HCI impact ionization as the primary cause for the GOX breakdown. The critical location is along the interface of GOX/Accum. region and near STI top corner where there is serious impact-ionization in Fig. 2. And the I_G(e-) direction is upward due to the downward electric-field at the stress bias as shown in Fig. 11. The continuous electron-injection continuously and latently damage GOX until it breakdown.

978-1-4244-9113-1/11 $26.00 © 2011 IEEE   447

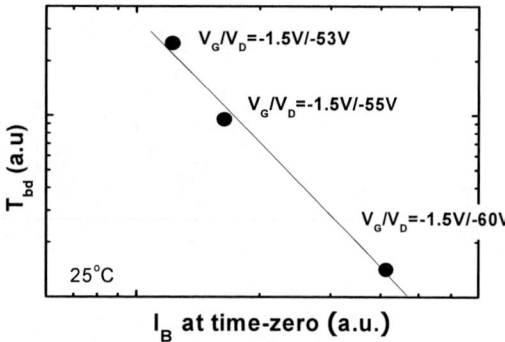

Figure 13. GOX TDDB is correlated with I_B current, indicating that the GOX breakdown mechanism is affected by HCI induced impact-ionization.

Therefore, it can be concluded that the GOX breakdown mechanism is associated with HCI impact-ionization induced hot electrons injection into the GOX. The leakage path within the GOX is localized only in the gate-to-accumulation overlapped region, rather than in channel. Hence, it explains why this mechanism will not destruct the low-V_G device characteristic even with the serious gate leakage (Fig. 16). In Fig. 16, the HVPMOS still holds its turn-on ability and low-V_G characteristic after a > 100Ksec stress, which means the channel (inversion region) is as normal as before despite the obvious current degradation at higher V_G resulting from the GOX leakage path localized within accumulation region.

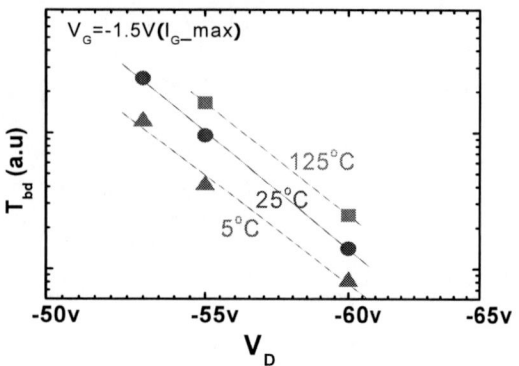

Figure 14. Higher temperature under the same stress condition gets the better TDDB because higher temperature moderates HCI effect.

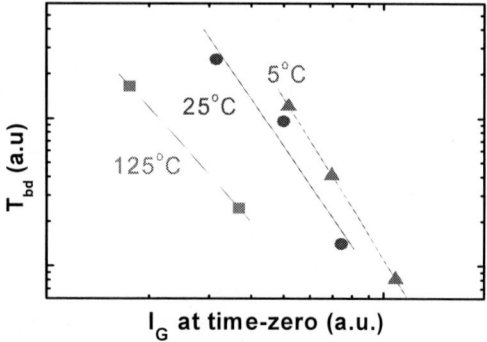

Figure 15. Higher temperature under the same injected I_G gets worse TDDB because GOX insulation ability degrades with temperature increase.

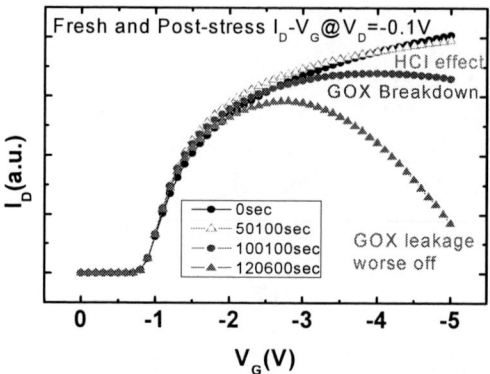

Figure 16. After GOX breakdown, the PMOS can preserve the device characteristic at lower V_G, but becomes abnormal at higher V_G. It indicates the GOX leakage path is localized only in the gate-to-accumulation overlapped region and not in the channel.

CONCLUSION

A multistage I_{Dlin} degradation behavior associated with simultaneous N_{it} generation and electron trapping has been observed during max-I_G HCI stress prior to GOX breakdown. TCAD simulation results correlate well with the experimental data and supports the simultaneous electron trapping and N_{it} generation during HCI stress prior to GOX breakdown. Eventually, the electron-injection during HCI stress deteriorates the localized GOX insulating ability in the drift-region, leading to the significant gate leakage current and responsible for the ultimate stage of abrupt I_{Dlin} degradation. Such GOX failure is more critical for BCD HVPMOS because electrons can overcome the GOX/Si barrier more easily than holes and becomes the major injected component for HV PMOS under low-V_G/high-V_D HCI stress.

ACKNOWLEDGMENTS

The authors would like to express their gratitude to Dr. N.S. Tsai for the managerial support and to TSMC R&D team for the Silicon manufacturing.

REFERENCES

[1] E. Riedlberger, R. Keller, H. Reisinger, W. Gustin, A. Spitzer, and M. Stecher, "Modeling the lifetime of a lateral DMOS transistor in repetitive clamping mode." IEEE IRPS, p.175, 2010.

[2] Y. Huang, J. Shih, Y. Lee, S. Hsieh, C. Liu, K. Wu, and H. Chou, "Investigation of monotonous increase in saturation-region drain current during hot carrier stress in N-type lateral diffused MOSFET with STI." IEEE IRPS, p.170, 2010.

978-1-4244-9113-1/11 $26.00 © 2011 IEEE

New Investigation of Hot Carrier Degradation of RF Small-Signal Parameters in High-k/Metal Gate nMOSFETs

[1,3,4]Hyun Chul Sagong, [3]Chang Yong Kang, [1]Chang-Woo Sohn, [1]Min Sang Park, [1]Do-Young Choi, [1]Eui-Young Jeong, [4]Jack C. Lee, and [1,2]Yoon-Ha Jeong

[1]Dept. of Electronic and Electrical Engineering, Pohang University of Science and Technology (POSTECH), Pohang, Korea
[2]National Center for Nanomaterials Technology (NCNT), Pohang, Korea,
[3]SEMATECH, 2706 Montopolis Drive, Austin, TX 78741, USA,
[4]Microelectronics Research Center, University of Texas at Austin, Austin, TX 78758, USA
(82) –(54)- 279-2897, hcsagong@postech.ac.kr, yhjeong@postech.ac.kr

Abstract—Hot carrier effects on RF small-signal parameters in high-k/metal gate nMOSFETs are characterized by DC and RF measurements. To explain the novel hot carrier-induced degradation, we suggest a modified surface channel resistance model that can be applied to both conventional SiO_2 and high-k nMOSFETs.

Keywords-high-k; hot carriers; MOSFET; resistance; RF

I. INTRODUCTION

Si devices are attracting greater attention in the radio-frequency (RF) arena because of their improved performance after physical scaling (Fig. 1); however, their increased lateral field can result in hot carrier (HC)-induced device degradation [1-5]. This HC effect has been studied thoroughly in the DC and low frequency regions, but few studies have investigated HC-induced device degradation in the RF region of high-k MOSFETs. In this work, we report results of RF small-signal parameter degradation in high-k/metal gate MOSFETs and SiO_2/poly-Si gate devices under HC stress. It is found that a new analytical model is needed to explain the degradation mechanism in high-k/metal gates. Therefore, we propose a new simple model to explain HC-induced device degradation in both conventional devices and high-k/metal gate MOSFETs.

II. EXPERIMENTAL

High-k dielectric gate stacks were processed by atomic layer deposition (ALD) of 3-nm of HfO_2 on a 1-nm interfacial layer (IL) followed by an ALD TiN electrode. Their equivalent oxide thickness (EOT) was 1.3 nm. For comparison, a control n-poly Si/SiO_2 device with an EOT of 2.37 nm was fabricated. For RF characterization, two-finger nMOSFETs with a finger width of 20-μm were prepared with Ground-Signal-Ground (GSG) probing pads. For specific DC measurements, 150-nm long nMOSFETs with 10-μm single finger width and four terminal probing pads were used. S-parameters were measured to analyze the RF characteristics by an Anritsu 37397C vector network analyzer (VNA). Pad parasitic components were removed by a two-step de-embedding technique. An Agilent

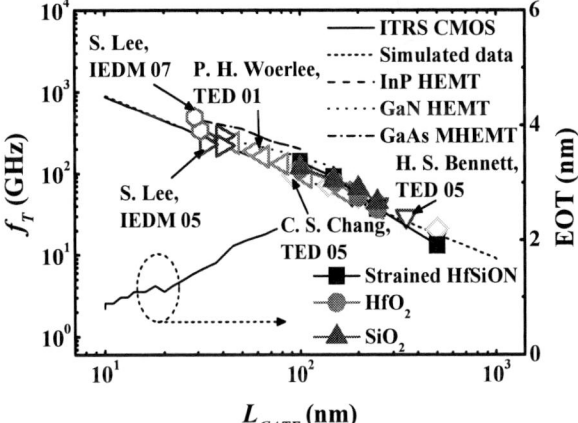

Figure 1. f_T and EOT versus L_{gate}. Trend based on 2007 ITRS roadmap is expressed by lines. Solid symbols indicate the results of this work.

IC-CAP program was used to extract DC and RF small-signal parameters.

III. EQUIVALENT CIRCUIT MODEL & DEVICE PARAMETERS

To explain HC-induced capacitance degradation in SiON/poly-Si gate electrode devices, a surface channel resistance model was introduced as shown in Fig. 2 [6-8]. In this model, the area under the voltage line (v_{ac}) represents the gate-related capacitances as shown in (1a)-(1b).

$$C_{gs} = -\frac{\partial Q_g}{\partial V_s} = \frac{WC_{ox}}{V_m}\int_{x=0}^{x=L} v_{ac}(x)dx \qquad (1a)$$

$$C_{gd} = -\frac{\partial Q_g}{\partial V_d} = \frac{WC_{ox}}{V_m}\int_{x=0}^{x=L} v_{ac}(x)dx \qquad (1b)$$

where L and W is the length and width of the device, respectively; C_{ox} is the gate oxide capacitance per unit area; $v_{ac}(x)$ is the applied AC voltage; and V_m is the amplitude of the AC voltage.

978-1-4244-9113-1/11 $26.00 © 2011 IEEE

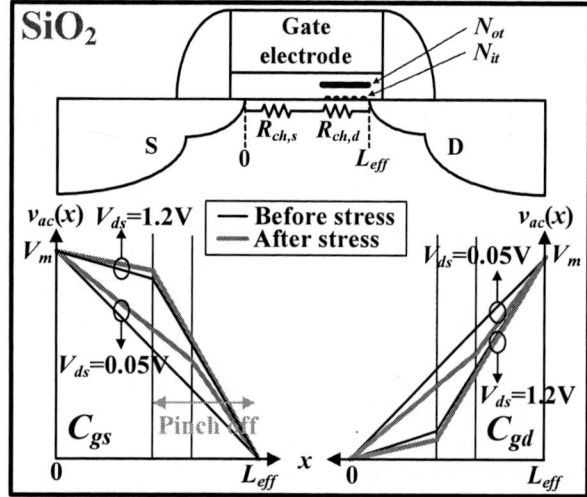

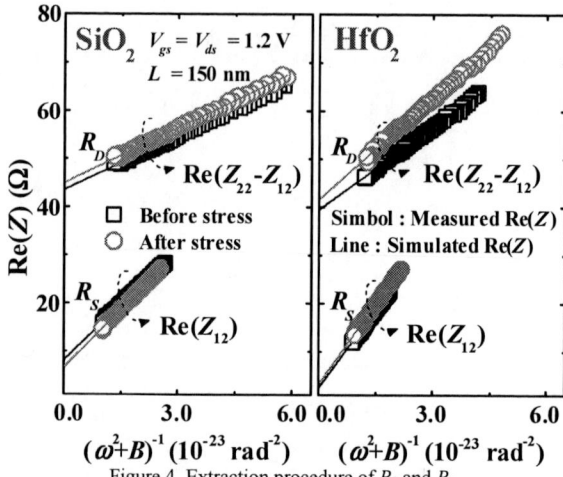

Figure 4. Extraction procedure of R_S and R_D.

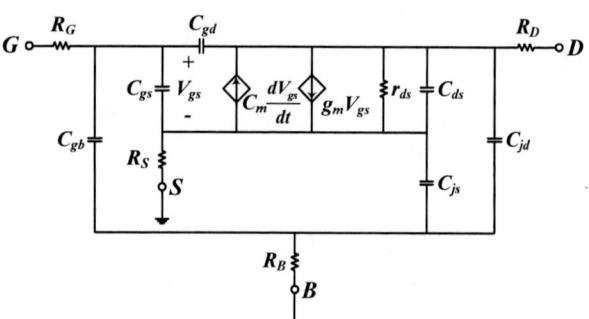

Figure 2. Conventional surface channel resistance model. Resistance in channel region is divided into two parts, such as $R_{ch,s}$ and $R_{ch,d}$. AC voltage (v_{ac}) is applied from source or drain terminal. In lower graph, voltage line of the v_{ac} through the two resistances is plotted.

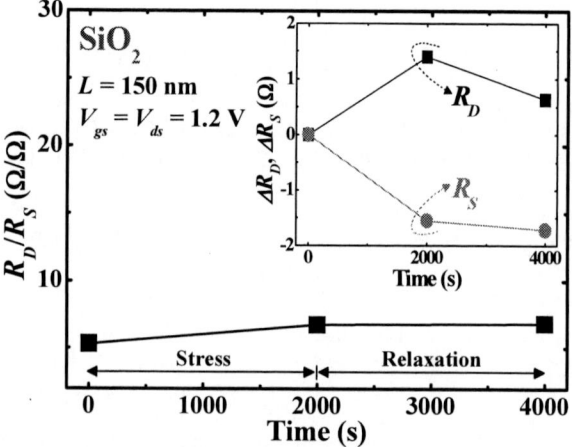

Figure 3. Small-signal equivalent circuit for RF MOSFET.

Figure 5. Variation of ratio (R_D/R_S) by 2000 s stress and 2000 s relaxation. In inset, terminal resistance shift (ΔR) is plotted to help understanding for variation of the ratio.

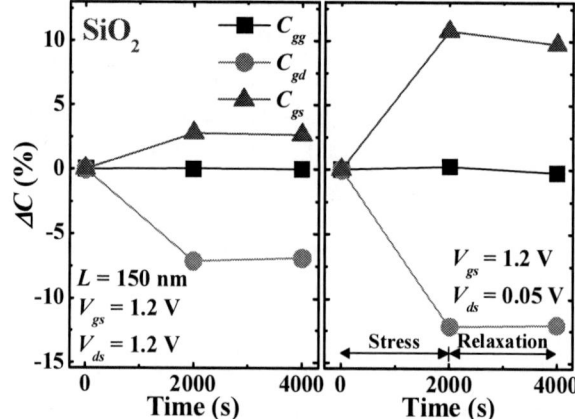

Figure 6. Capacitance degradation in SiO_2 nMOSFET by HC stress.

Before applying this model to high-k/metal gate MOSFETs, we verified it using the SiO_2 control devices with the small-signal equivalent circuit shown in Fig. 3. At low frequencies, the influence of the substrate is usually ignored. At high frequencies (>GHz), however, the substrate-related parameters become important because the signal at the drain couples to the source and bulk terminals through source/drain junction capacitance and substrate resistance, respectively [9]. Therefore, this equivalent circuit should take into account substrate-related parameters such as source junction capacitance (C_{js}), drain junction capacitance (C_{jd}), and bulk resistance (R_B). Y-parameters were obtained by analyzing the equivalent circuit [9-10]. The Y-parameters are expressed as

$$Y_{11} \approx \omega^2 C_{gg}^2 R_G + j\omega C_{gg} \tag{2a}$$

$$Y_{12} \approx -\omega^2 C_{gg} C_{gd} R_G - j\omega C_{gd} \tag{2b}$$

$$Y_{21} \approx g_m - \omega^2 C_{gg} C_{dg} R_G - j\omega\left(C_{dg} + g_m R_G C_{gg}\right) \tag{2c}$$

$$Y_{22} \approx g_{ds} + \omega^2\left[\frac{C_{jd}^2 R_B}{1+\omega^2 C_{jd}^2 R_B^2} + C_{gd}\left(C_{dg} R_G + g_m R_G^2 C_{gg}\right)\right]$$
$$+ j\omega\left(\frac{C_{jd}}{1+\omega^2 C_{jd}^2 R_B^2} + C_{ds} + C_{gd} + g_m R_G C_{gd} - \omega^2 C_{gg} C_{gd} C_{dg} R_G^2\right) \tag{2d}$$

The device parameter equations given by (3a)-(3e) are determined from the Y-parameters (2a)-(2d). Equations (4a)-(4d) are also given for extracting resistance [11]. The resistances were extracted using Raskin's method as shown in Fig. 4 [12]-[13].

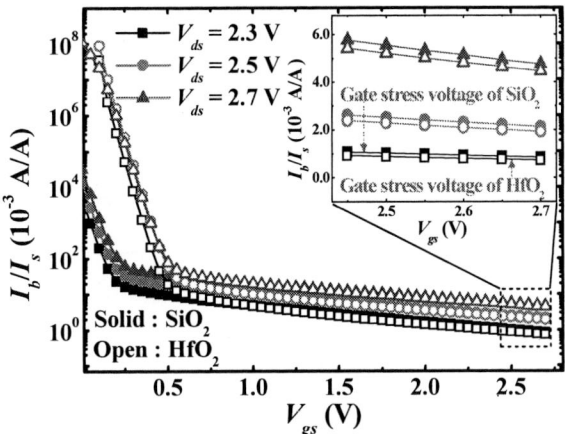

Figure 7. Current ratio (I_b/I_s) as impact ionization index. The current ratio of HfO$_2$ nMOSFET and SiO$_2$ nMOSFET is compared to determine stress voltage condition.

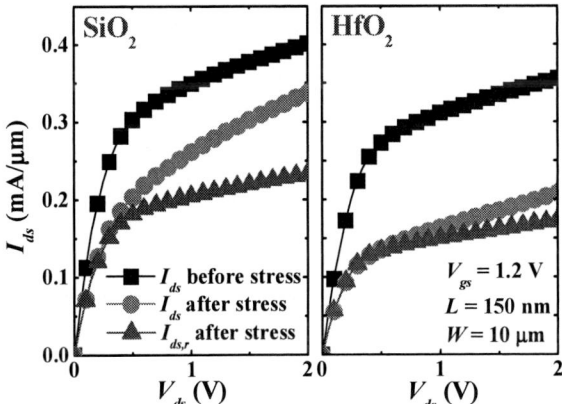

Figure 8. Normalized I_{ds} as a function of V_{ds} before and after stress. For comparison, $I_{ds,r}$ is measured at source/drain reverse connection after 2000 s stress.

$$C_{gg} = \frac{\text{Im}(Y_{11})}{\omega} \quad (3a)$$

$$C_{gd} = -\frac{\text{Im}(Y_{12})}{\omega} \quad (3b)$$

$$C_{gs} = \frac{\text{Im}(Y_{11} + Y_{12})}{\omega} \quad (3c)$$

$$g_m = \text{Re}(Y_{21})\big|_{\omega=0} \quad (3d)$$

$$g_{ds} = \text{Re}(Y_{22})\big|_{\omega=0} \quad (3e)$$

$$\text{Re}(Z_{11} - Z_{12}) = R_G + \frac{A_g}{\omega^2 + B} \quad (4a)$$

$$\text{Re}(Z_{22} - Z_{12}) = R_D + \frac{A_d}{\omega^2 + B} \quad (4b)$$

$$\text{Re}(Z_{12}) = R_S + \frac{A_s}{\omega^2 + B} \quad (4c)$$

$$B = \left[\frac{g_m C_{gd} + g_{ds}(C_{gs} + C_{gd})}{C_{gs} C_{ds} + C_{gs} C_{gd} + C_{gd} C_{ds}}\right]^2 \quad (4d)$$

IV. RESULTS AND DISCUSSION

After HC stress, bulk trap density (N_{ot}) and interface trap density (N_{it}) in Fig. 2 increased near the drain region, which in

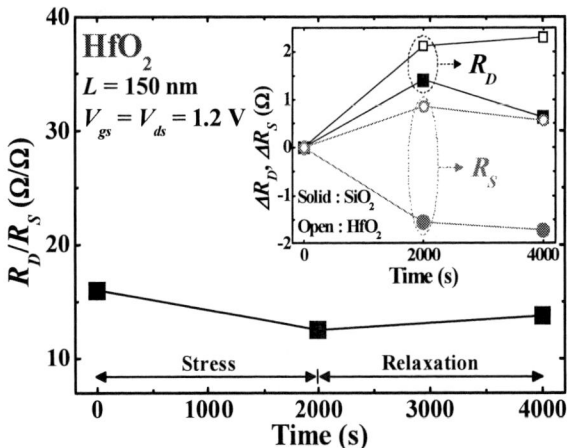

Figure 9. Ratio (R_D/R_S) in high-k nMOSFET. In inset, resistance shift (ΔR) of SiO$_2$ nMOSFET and HfO$_2$ nMOSFET is compared.

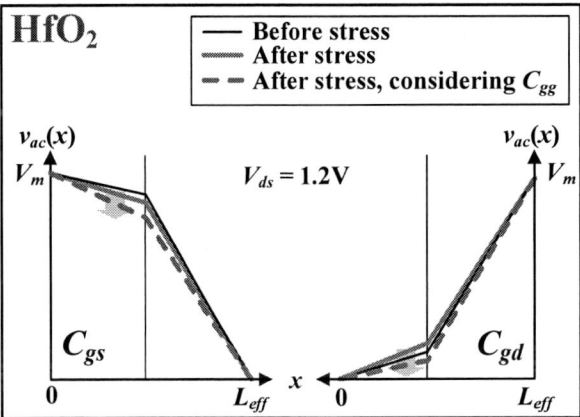

Figure 10. Modified surface channel resistance model.

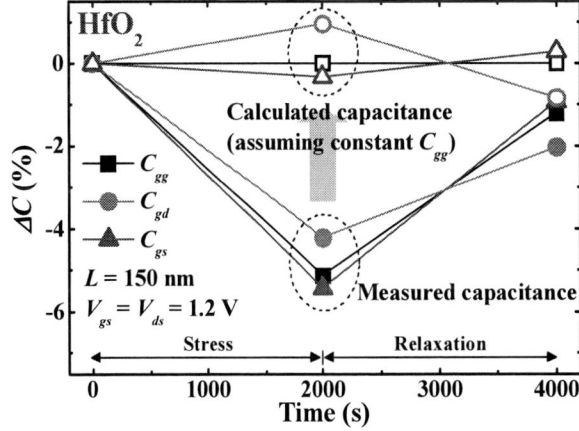

Figure 11. Capacitance degradation by HC stress in HfO$_2$ nMOSFET. Measured capacitances are expressed by solid symbols. Calculated capacitances assuming constant C_{gg} are exhibited by open symbols.

turn increased $R_{ch,d}$ and decreased $R_{ch,s}$. The variation of the resistances was confirmed by the extracted terminal resistances (R_D & R_S) as shown in the inset of Fig. 5. These extracted terminal resistances can include some part of the channel resistance damage near each terminal. The damage near the

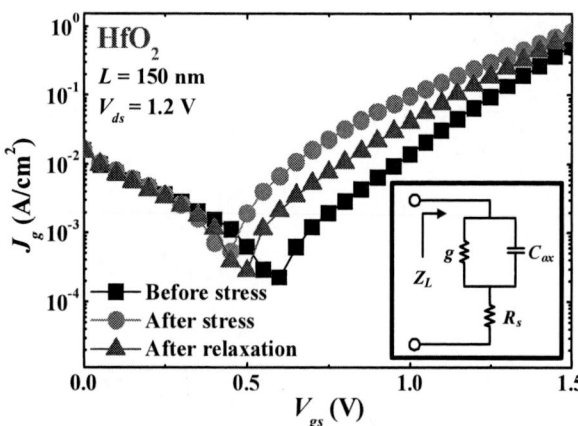

Figure 12. Gate current density (J_g) as a function of V_{gs}. Inset is approximated three-element circuit model of MOSFET. Gate capacitance (C_{gg}) is expressed as (5) from the approximated three-element circuit model.

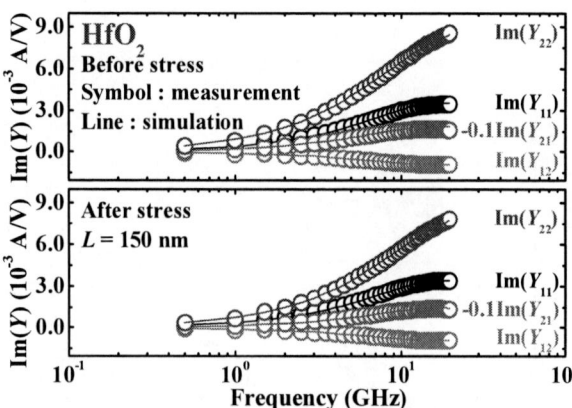

Figure 13. Measured Y-parameter and simulated Y-parameter before and after HC stress for HfO$_2$ nMOSFET biased at $V_{gs} = V_{ds} = 1.2$V.

drain region increased the terminal resistance R_D while the R_S decreased due to the increase in the source-side effective potential shown in the inset of Fig. 5 [14-15]. To explain this change in resistance, the slope of the channel voltage line should be varied as shown in Fig. 2. From (1a)-(1b), in turn, the C_{gd} decreases while the C_{gs} increases after HC stress. These effective voltage (capacitance) changes are confirmed by the results of RF measurements (Fig. 6).

In the same manner, we applied this model to high-k devices. For an exact comparison with the SiO$_2$ nMOSFETs, the stress voltage condition was $V_{gs} = V_{th} + 2$ V and $V_{ds} = 2.3$ V, which ensured almost the same quantity of hot carriers generated by impact ionization as shown in Fig. 7. From the asymmetric current-voltage (I-V) characteristics after HC stress, HC-induced damage in the control SiO$_2$ device occurs near the drain region (Fig. 8). In the high-k nMOSFET, however, a near symmetrical IV was observed; both R_D and R_S values in the HfO$_2$ nMOSFET increased after HC stress (Fig. 9) resulting in the thick solid lines in Fig. 10. This symmetric IV was caused by charge trapping near the source in conjunction with HC injection in the drain [16]. However, the calculated effective capacitance (C_{gd} & C_{gs}) was significantly different from the measured results (Fig. 11), which suggests that a conventional

TABLE I. COMPARISON BETWEEN HC-INDUCED VARIATION OF MEASURED RF PARAMETERS AND SIMULATED RF PARAMETERS FOR 150-NM NMOSFET BIASED AT $V_{GS} = V_{DS} = 1.2$ V. THE PARAMETERS OF HFO$_2$ NMOSFET IS ALSO COMPARED WITH THAT OF SIO$_2$ NMOSFET.

RF parameter	SiO$_2$ nMOSFET		HfO$_2$ nMOSFET	
	Measured	Simulated	Measured	Simulated
g_m (%)	-4.6	-4.5	-22.4	-21.3
r_o (%)	-18.3	-19.1	23.5	23.6
C_{gg} (%)	0.0	0.0	-5.1	-5.1
C_{gd} (%)	-7.1	-7.1	-4.2	-4.2
C_{gs} (%)	2.8	2.9	-5.4	-5.5
R_D (%)	7.0	7.0	7.3	7.3
R_S (%)	-7.8	-7.9	63.8	63.6

surface channel resistance model cannot be used to explain the degradation in capacitance in high-k nMOSFETs.

Previous reports did not take C_{gg} into account as an important factor to determine capacitances such as C_{gd} and C_{gs} because variations in C_{gg} are almost constant in conventional SiO$_2$ nMOSFETs as shown in Fig. 6 [6-8]. In high-k nMOSFETs, however, variations in C_{gg} should be considered a separate fitting parameter to explain HC-induced device degradation.

An approximated three-element circuit model of a MOSFET is shown in the inset in Fig. 12. In this model, the gate capacitance C_{gg} extracted from Y_{11} is expressed as

$$C_{gg} = \frac{C_{ox}}{(1 + g \cdot R_s)^2} \tag{5}$$

where R_s is the series resistance including the effective gate, channel, and substrate resistances, and g is the effective conductance related to the gate direct tunneling leakage, which is also the reciprocal of the effective resistance (R_p) [17].

After stress, J_g in the HfO$_2$ device increased due to trap-assisted tunneling as shown in Fig. 12 [18], which is directly related to the increase of the effective conductance (g) in (5), resulting in a reduction of C_{gg}. We also identified that the reduction in C_{gg} is reasonable compared to the measured C_{gg} in Fig. 11. To explain the capacitance variation in high-k nMOSFETs, we suggest a modified surface channel resistance model that considers the C_{gg} variation. When considering the reduction in C_{gg}, the thick solid voltage lines (Fig. 10) should be pushed down to the thick dashed voltage lines. By pushing down the voltage lines, the measured data in Fig. 11 can be explained.

Based on this approach, a simulation verified the validity of the parameter variation. Fig. 13 shows that the measured Y-parameters before and after HC stress match well with the simulated Y-parameters. Additionally, the variations in the parameters extracted from the simulation are in close agreement with the variation of the measured parameters in TABLE I.

V. CONCLUSION

Hot carrier degradation of the RF small-signal parameter was analyzed in high-k devices. To explain the measurement results, we suggest a modified surface channel resistance model to analyze the HC-induced degradation of the RF parameter in both conventional SiO_2 nMOSFETs and high-k/metal gate nMOSFETs. The modified model's ability to estimate the RF parameter variation will enhance the accuracy of RF simulation, which enables us to estimate HC-induced RF performance.

ACKNOWLEDGMENT

This research was supported by World Class University (WCU) program through the National Research Foundation of Korea funded by the Ministry of Education, Science and Technology (R31-2008-000-10100-0). This work was partially supported by the BK21 program and the National Center for Nanomaterials Technology (NCNT) in Korea.

REFERENCES

[1] C. S. Chang, C. P. Chao, J. G. J. Chern, and J. Y. -C. Sun, "Advanced CMOS technology portfolio for RF IC applications," *IEEE Trans. Electron Devices*, vol. 52, no. 7, pp. 1324–1334, Jul. 2005.

[2] H. S. Bennett, R. Brederlow, J. C. Costa, P. E. Cottrell, W. M. Huang, A. A. Immorlica, J. -E. Mueller, M. Racanelli, H. Shichijo, C. E. Weitzel, and B. Zhao, "Device and technology evolution for Si-based RF integrated circuits," *IEEE Trans. Electron Devices*, vol. 52, no. 7, pp. 1235–1258, Jul. 2005.

[3] S. Lee, B. Jagannathan, S. Narasimha, A. Chou, N. Zamdmer, J. Johnson, R. Williams, L. Wagner, J. Kim, J.-O. Plouchart, J. Pekarik, S. Springer, and G. Freeman, "Record RF performance of 45 nm SOI CMOS technology," in *IEEE IEDM*, pp. 255–258, Dec. 2007.

[4] S. Lee, L. Wagner, B. Jagannathan, S. Csutak, J. Pekarik, M. Breitwisch, R. Ramachandran, and G. Freeman, "Record RF performance of sub-46 nm L_{gate} NFETs in microprocessor SOI CMOS technologies," in *IEEE IEDM Tech. Dig.*, pp. 241–244, Dec. 2005.

[5] P. H. Woerlee, M. J. Knitel, R. van Langevelde, D. B. M. Klaassen, L. F. Tiemeijer, A. J. Scholten, and A. T. A. Zegers-van Duijnhoven, "RF-CMOS performance trends," *IEEE Trans. Electron Devices*, vol. 48, no. 8, pp. 1776–1782, Aug. 2001.

[6] C. H. Ling, S. E. Tan, D. S. Ang, "A Study of Hot Carrier Degradation in NMOSFET's by Gate Capacitance and Charge Pumping Current," *IEEE Trans. Electron Devices*, vol. 42, no. 7, pp. 1321–1328, Jul. 1995.

[7] M. J. Deen and T. A. Fjeldly, "CMOS RF Modeling, Characterization and Applications," Singapore: World Scientific, 2002.

[8] M. -C. Tang, Y. -K. Fang, W. -S. Liao, D. C. Chen, C. -S. Yeh, S. -C. Chien, "Investigation and Modeling of hot carrier effects on performance of 45- and 55- NMOSFETs With RF Automatic Measurement," *IEEE Trans. Electron Devices*, vol. 55, no. 6, pp. 1541-1546, Jun. 2008.

[9] Y. Cheng, M. J. Deen, and C. -H. Chen, "MOSFET Modeling for RF IC Design," *IEEE Trans. Electron Devices*, vol. 52, no. 7, pp. 1286–1303, Jul. 2005.

[10] I. Kwon, M. Je, K. Lee, and H. Shin, "A Simple and Analytical Parameter-Extraction Method of a Microwave MOSFET," *IEEE Trans. Microwave Theory Tech.*, vol. 50, no. 6, pp. 1503–1509, Jun. 2002.

[11] S. Lee, H. K. Yu, C. S. Kim, J. G. Koo, and K. S. Nam, "A Novel Approach to Extracting Small-Signal Model Parameters of Silicon MOSFET's," *IEEE Microwave and Guided Wave Lett.*, vol. 7, no. 3, pp. 75–77, Mar. 1997.

[12] J. P. Raskin, G. Dambrine, and R. Gillon, "Direct extraction of the series equivalent circuit parameters for the small-signal model of SOI MOSFET's," *IEEE Microwave and Guided Wave Lett.*, vol. 7, no. 12, Dec. 1997.

[13] G. -B. Choi, S. -H. Hong, S. -W. Jung and Y. H. Jeong, "On the RF series resistance extraction of nanoscale MOSFETs," *IEEE Microwave and Wireless Components Lett.*, vol. 18, no. 10, pp. 689-691, Oct. 2008.

[14] C. -L. Lou, W. -K. Chim, D. Chan, and Y. Pan, "A Novel Single-Device DC Method for Extraction of the Effective Mobility and Source-Drain Resistances of Fresh and Hot-Carrier Degraded Drain-Engineered MOSFETs," *IEEE Trans. Electron Devices*, vol. 45, no. 6, pp. 1317–1323, Jun. 1998.

[15] C. Yu and J. S. Yuan, "MOS RF Reliability Subject to Dynamic Voltage Stress-Modeling and Analysis," *IEEE Trans. Electron Devices*, vol. 52, no. 8, pp. 1751–1758, Aug. 2005.

[16] K. T. Lee, C. Y. Kang, O. S. Yoo, R. Choi, B. H. Lee, J. C. Lee, H. D. Lee, and Y. H. Jeong, "PBTI Associated High-Temperature Hot Carrier Degradation of nMOSFETs With Metal-Gate/High-k Dielectrics," *IEEE Electron Device Lett.*, vol. 29, no. 4, pp. 389–391, Apr. 2008.

[17] G. -B. Choi, S. -H. Hong, S. -W. Jung, H. -S. Kang, and Y. -H. Jeong, "RF Capacitance Extraction Utilizing a Series Resistance Deembedding Scheme for Ultraleaky MOS Devices," *IEEE Electron Device Lett.*, vol. 29, no. 3, pp. 238–241, Mar. 2008.

[18] D. Han, J. Kang, C. Lin, and R. Han, "Reliability characteristics of high-k gate dielectrics HfO2 in metal-oxide semiconductor capacitors," *Microelectron. Eng.*, vol. 66, pp. 643–647, 2003.

DESIGN-IN RELIABILITY APPROACH FOR HOT CARRIER INJECTION MODELING IN THE CONTEXT OF AMS/RF APPLICATIONS

Vincent Huard, Thomas Quemerais, Florian Cacho, Laurence Moquillon, Sebastien Haendler and Xavier Federspiel

STMicroelectronics

850 rue jean Monnet 38926 Crolles, France

phone: + 33(0)438922907 ; fax : + 33(0)438923227 ; email : vincent.huard@st.com

Abstract— **Both AMS/RF applications and High-Speed digital designs present voltage transitions much faster than the conventional quasi-static experiments used to build up Hot Carrier models. This work combined reliability simulations and experimental DC/RF stresses to demonstrate that our framework of HCI reliability simulations is suitable to re-produce electrical ageing up to 60GHz.**

Keywords-HCI; reliability; RF; Power Amplifier

I. INTRODUCTION

Reliability concerns are becoming increasingly difficult to handle in advanced nodes. Design-in Reliability (DiR) seeks to provide a quantitative assessment of reliability – CMOS device reliability in this case – at early design stage, so to enable designers to optimize their designs for reliability prior to silicon validation.

In this work, we provide details of DiR methodology developed to address the effects of the Hot Carrier Injection (HCI) mechanism. In most published works, the focus is set on digital applications which allow model simplifications by the nature of the electrical stimuli. On the contrary, we aimed to develop HCI models based on physics foundations which in turn would allow supporting not only digital designs but also AMS/RF applications.

First, the modeling aspects of HCI degradation modes will be thoroughly described, emphasizing AMS/RF specific parameters. In a second time, implemented CMOS device reliability models are used to predict RF parameters degradation after both DC and RF stress conditions.

II. TRANSISTOR RELIABILITY MODELING

The transistor reliability models will be described here below as well as the global approach (including required methodologies) to extract model parameters. The parameter ΔD chosen to represent the degradation forms the bridge between the physical degradation and the evolution of MOS parameter. The models are made linear with respect to time for being suitable for integration during circuit simulation. The boundaries of the integrals represent the window during which the stress assessment is made (during circuit simulation).

A. Hot Carrier Injection Modeling

Hot-Carrier Injection degradation presents a renewed interest in the more recent nodes where high level of device reliability is difficult to achieve at high temperature as a function of supply voltage V_{DD}. This point is mainly explained by a continuous increase in lateral electric field since 120nm node. Both digital and AMS/RF applications require HCI modeling over the whole V_{gs}/V_{ds} design space. Recent experimental HCI analysis at transistor level allowed separating the contributions of three independent modes. The first mode is related to carriers bringing individually enough energy to break the Si-H bond. The second mode is related to moderate carrier energy range for which Electron-Electron Scattering allows Si-H bond breaking [1]. Altogether, these Channel Hot Carrier (CHC) modes are related to carrier energy. Finally, a third mode in low energy range was recently attributed to Multiple Vibrational Excitation (MVE) [2-3]. In this configuration, the degradation is lead mostly by the number of carriers "hitting" the bond and should be considered as Channel Cold Carrier (CCC) mode since the degradation is no longer driven by carrier energy. This description in three modes allows modeling the HCI degradation in the whole V_{gs}/V_{ds} design space (cf. fig. 1).

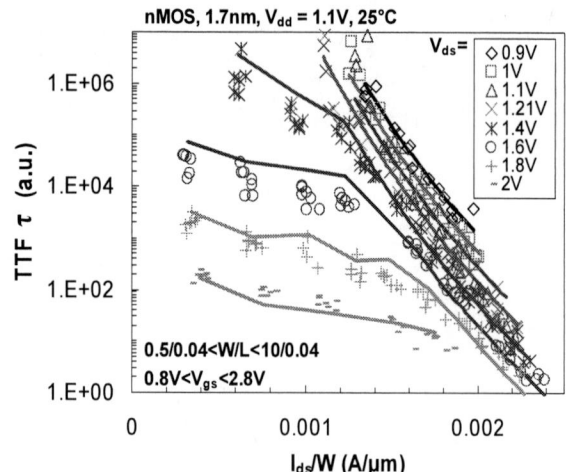

Fig. 1: HCI modelling (lines) at RT compared to experimental dataset covering the whole V_{gs}/V_{ds} design space.

Understanding accurately the underlying physics of each mode as extensively described in [4] allows also describing the HCI degradation behavior over a wide range of temperature at the cost of no additional fit parameters (cf. fig. 2).

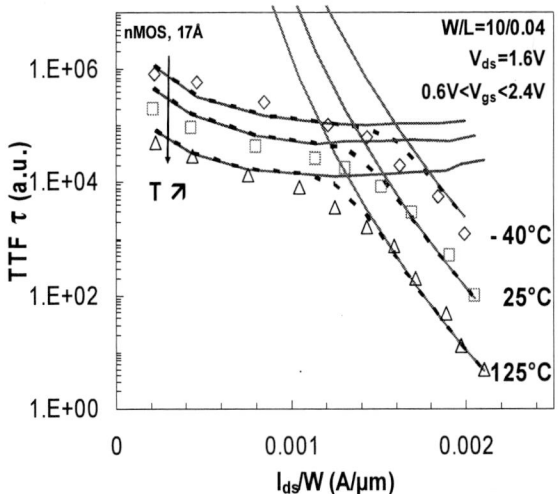

Fig. 2: HCI modelling (blue line: high-energy modes; green line: low-energy modes; dotted line: full model) at different temperatures compared to experimental dataset.

The amount of HCI degradation (Age) impacting the device can thus be normalized whatever the stress conditions as (see also [3] for more details):

Age	$Age = \dfrac{t}{\tau} = t \cdot \left[C_1 \cdot \left(\dfrac{I_{ds}}{W}\right)^{a_1} \cdot \left(\dfrac{I_{bs}}{I_{ds}}\right)^{m} + C_2 \cdot \left(\dfrac{I_{ds}}{W}\right)^{a_1} \cdot \left(\dfrac{I_{bs}}{I_{ds}}\right)^{m} + C_3 \cdot V_{ds}^{a_2/2} \cdot \left(\dfrac{I_{ds}}{W}\right)^{a_1} \cdot \exp\left(\dfrac{-E_{emi}}{k_B T}\right) \right]$

In addition to the conventional electrical parameters degradations, AMS/RF circuits might be impacted by electrical effects which have negligible impact in digital circuits. We report here for the first time the HCI-related degradation of the Low-Frequency Noise (LFN). LFN degradation is accelerated with stress time (cf. Fig. 3) and with increased drain voltage (cf. Fig. 4).

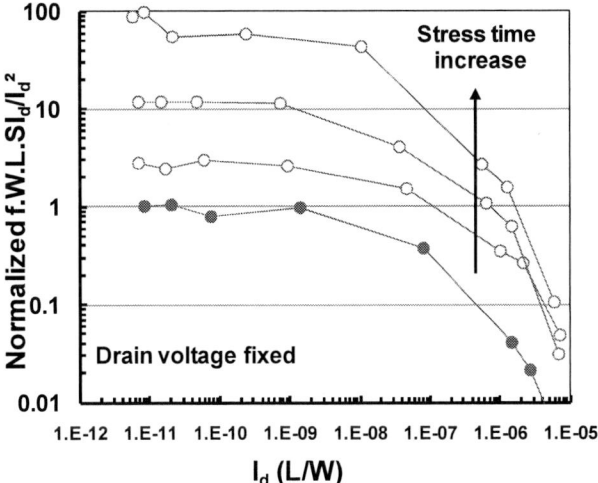

Fig. 3: Low-Frequency Noise (LFN) increase due to HCI stress for various stress times at a given drain voltage.

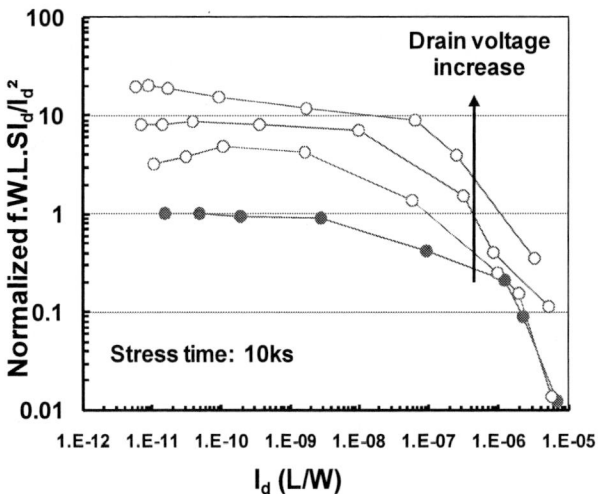

Fig. 4: Low-Frequency Noise (LFN) increase due to HCI stress for various drain voltages at a given stress time. Stress gate voltage is fixed.

This LFN degradation increase is not observed after NBTI stress [4] (cf. Fig. 5).

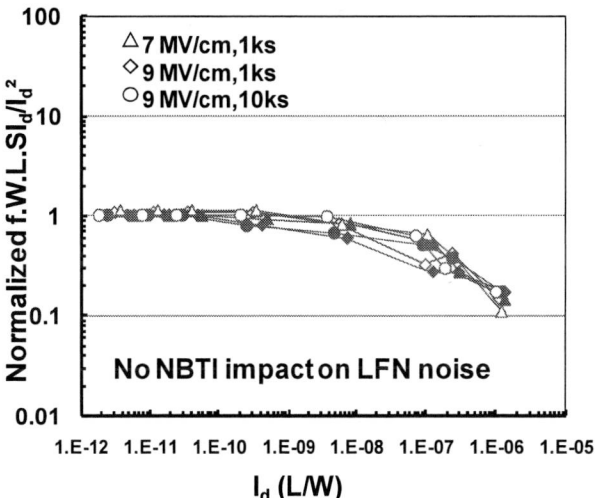

Fig. 5: Low-Frequency Noise (LFN) remains unchanged after NBTI stress for various gate voltages and stress times.

The Hot-Carrier situation is often described as the interaction of a high-energy solitary carrier with either the silicon substrate (impact ionization (ii)) or an interface defect (interface states creation (it)). The probability of interaction is mainly dependent of the energy of the carrier and is described by a scattering rate function $S_i(E)$ (respectively $S_{ii}(E)$ for the impact ionization and $S_{it}(E)$ for interface traps generation). The hot carriers are distributed in energy according to the Electron Energy Distribution Function (EEDF) f(E). Finally, the total probability of interaction is described as the integral of f(E).$S_i(E)$ over the whole energy range:

$$R_i = \int f(E) . S_i(E) d(E) \qquad \text{Eq. 1}$$

As shown in Fig. 7, the rate R_i peaks at one preferential energy referred as the "dominant energy" (E_{dom}). It describes

978-1-4244-9113-1/11 $26.00 © 2011 IEEE

the carrier population with carriers combining both large population and high energy. Using a Keldysh-like impact ionization scattering rate, the impact ionization rate can thus be described as [5]:

$$R_{ii} = \int f(E).S_{ii}(E)dE \propto \int f(E).(E-\phi_{ii})^{p_{ii}} dE \qquad \text{Eq. 2}$$

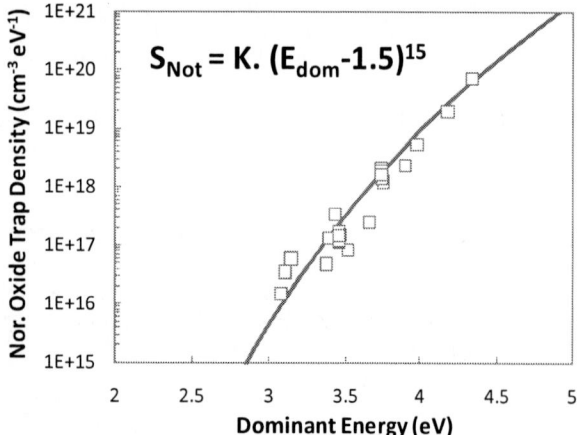

Fig. 6: Schematic illustration of the EEDF f(E) and the interaction function Si(E): (a) in the classical LEM picture assuming Keldysh scattering rate function and high energy EEDF, (b) in the energy-driven picture assuming soften Keldysh scattering rate function and TCAD-based EEDF with a knee at a given energy. The shaded area is the schematic illustration of the impact ionization/interface traps creation rate. It might be described by a dominant energy E_{dom}.

Similarly to interface traps creation, we can assume the existence of a scattering rate function S_{ot} relative to oxide traps defects creation. By mixing several HCI stress conditions, and using the same AGE function as described above, we can for the first time extract experimentally such scattering rate function (cf. Fig. 7). The dominant energy E_{dom} is extracted as in [5] and further calibrated on Monte-Carlo TCAD simulations. The threshold energy is about 1.5eV, pretty much in line with the experimental observations on interface traps. The power law exponent is in this case more important than for impact ionisation (~4.2) and interface defects (~11). It is also worth noticing that N_{ot} creation occurs for $E_{dom}<3.5eV$.

Fig. 7: Experimental Scattering rate function Sot relative to the generation of oxide traps in the energy-driven context of HCI.

Similarly, it is possible using long-term experiments to show that the time dynamics of oxide traps defects follows closely a power law with a time exponent ~0.5 (cf. Fig. 8).

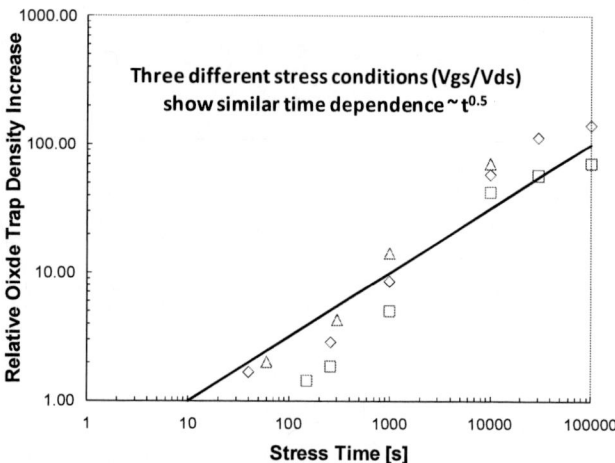

Fig. 8: Low-Frequency Noise (LFN) increase due to HCI stress for various stress times.

B. Degradation mapping to compact models

ELDO simulator has been enhanced to communicate with a proprietary Application Programming Interface named UDRM (fig. 9), in which two set of equations/parameters are encoded. The first set named 'Defect creation' is related to stress models which aim to calculate the number of defects generated by the stimuli coming from the simulation. This set of equations is generally based on quasi-static modeling of HCI degradation without any experimental validation of the validity in AMS/RF range.

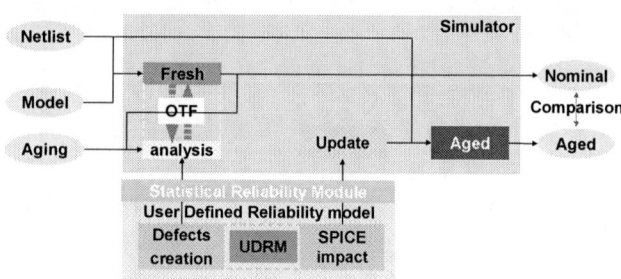

Fig. 9: Schematics of the reliability simulation flow. Standard BSIM and PSP equations are equally understood [4].

The second set named 'SPICE impact' is related to the translation of the generated defects into SPICE parameters variations which finally turn into device performances degradation. Fig. 10 shows an example on how accurate is our HCI modelling approach to describe transistor DC characteristics.

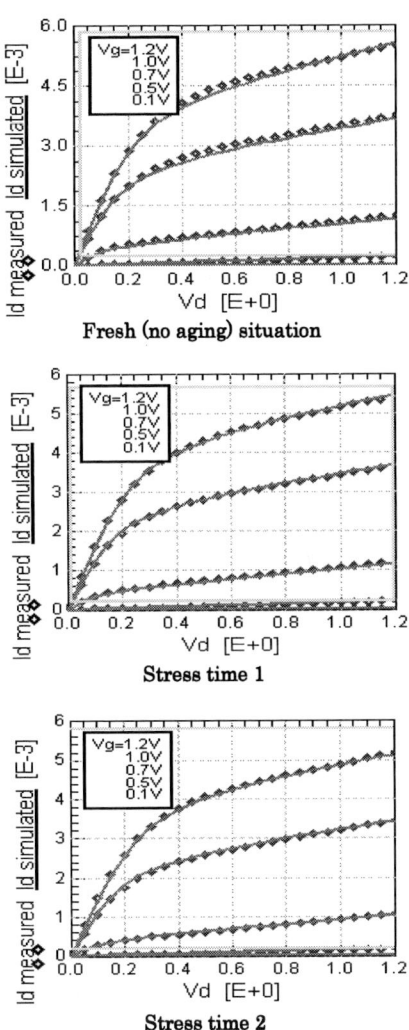

Fig. 10: HCI modeling results (red lines) versus experimental DC characteristics for a nMOS 65nm transistor without aging (top) and for different increasing stress times (middle and bottom).

III. HCI MODELING VALIDITY IN AMS/RF CONTEXT

Recent CMOS technologies are enabling integration of very high frequency applications like HDMI, WLAN or WPAN communications in the range of 60GHz. In parallel, high frequencies are also commonly observed in high-speed digital designs. Both AMS/RF applications and High-Speed digital designs present voltage transitions much faster than the conventional quasi-static experiments used to build up Hot Carrier models. Until now, it remains a challenge to demonstrate that a given HCI modeling approach might yield to accurate predictions in such aggressive contexts.

A. Experimental procedure

Based on STMicroelectronics 65nm process capability, we have designed a dedicated test structure to embed a single nMOS in common source configuration into a network of passive elements to adapt to 50 ohms external impedances at 60GHz (cf. Fig. 11).

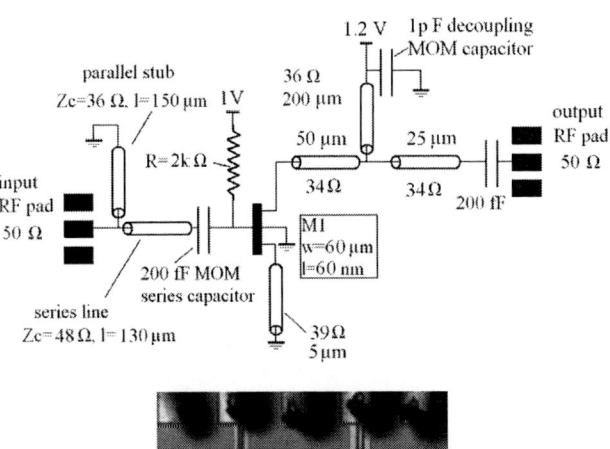

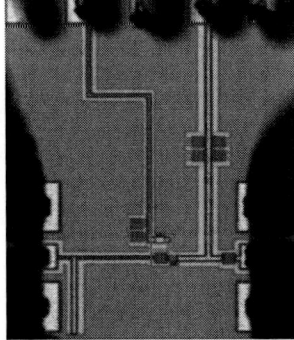

Fig. 11: Schematics of the test structure containing a single nMOS in common source configuration within its 50 ohms impedance matching network, and its corresponding microphotography.

This test structure acts like a millimeter wave one stage power amplifier (PA) with the following specification (cf. Fig. 12).

Test Structure	P_{sat} (dBm)	P_{1dB} (dBm)	G_p (dB)	Power (mW)	PAE (%)
1 stage PA	9.2	6.4	4.5	20.5	26

Fig. 12: Test structure (1 stage PA) performances.

This test structure opens the way to realize both DC and RF (60GHz) stresses on the same transistor so to validate the two steps approach of HCI modeling in a regime favoring non quasi static behaviors. DC supply voltages for drain and gate can be used to accelerate the degradation. Experimentally, large 65nm MOSFETs width (W>>10 µm) avoid statistical approach of the degradation. Indeed, the degradation of the presented PAs can be measured on few devices. To study the impact of HCI degradation, the transistor is first characterized in line with DC methodologies using I-V curves. In a second time, both small-signal and large-signal RF parameters are monitored among which the power gain, the input and output matching (S_{11} and S_{22}), the output saturated power (P_{sat}) and the output 1dB compression point (OCP1dB). The transistor is degraded by applying accelerated stress conditions on gate and drain voltage pads in a very similar way than product operating lifetests. Three stress voltages and several stress times (up to 50 hours) were used.

B. DC stress

As a first step, only a DC stress was applied in a way very similar to conventional Wafer Level Reliability. DC I-V curves were monitored for the transistor under stress prior and after the DC stress. In parallel, SPICE simulations were run to reproduce the same test sequences. Fig. 13 shows that a very good agreement is obtained between silicon measurement (red lines) and our HCI modeling approach (dashed black lines). Though this step is similar to the one used to develop the HCI SPICE models (cf. Fig 10), it was important to check that the passive network present for impedance matching does not introduce measurements issues compared to simple isolated transistor.

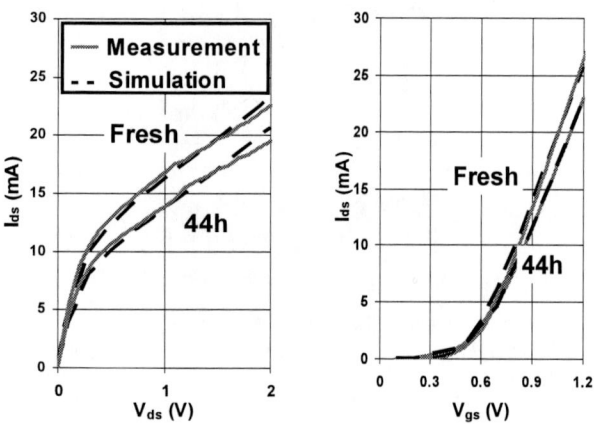

Fig. 13: Comparison between I·V characteristic curves in between silicon (in red) and SPICE (in black) for fresh devices (no stress) and after DC stress.

In a second time, we wanted to check the accuracy of the set of equations related to SPICE parameters variations to reproduce RF parameters drift under DC stress. This approach allows referring to DC stress which are similar to our quasi-static approach modelling approach and only focus on validating our SPICE parameters strategy.

Fig. 14 and 15 shows experimental drifts observed on several small-signal S parameters for various stress times and drain voltages.

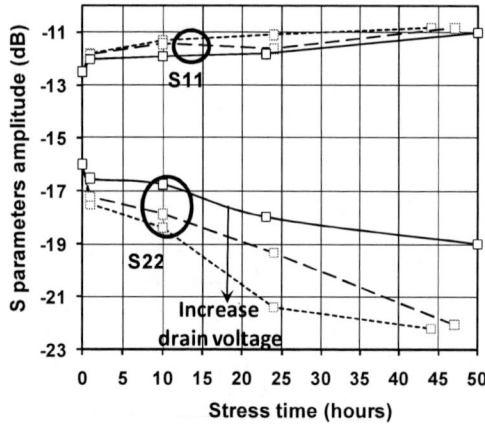

Fig. 14: Experimental S parameters (S_{11} and S_{22}) drifts observed for various stress times and drain voltages, measured at 60GHz.

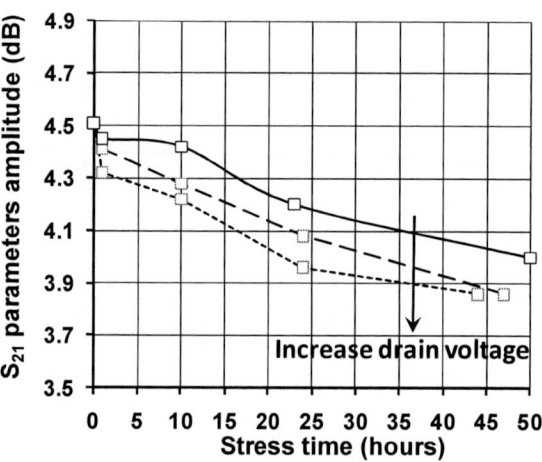

Fig. 15: Experimental PA gain (S_{21}) drifts observed for various stress times and drain voltages, as measured at 60GHz.

Reliability SPICE simulations were run to reproduce all experimental set of conditions. Fig. 16 shows one example of good agreement between reliability modeling and experimental results for small-signal parameters.

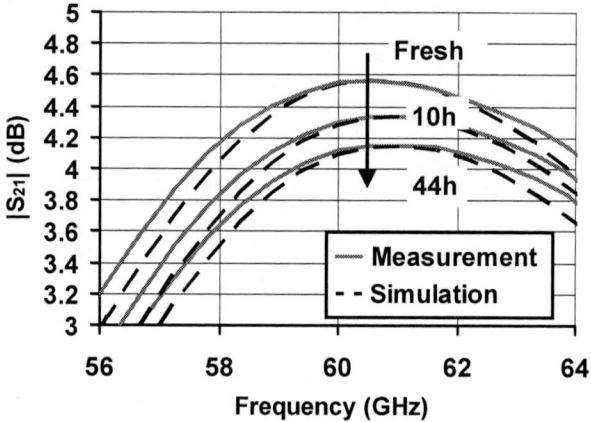

Fig. 16: Comparison between measured and simulated 1 stage PA gain S_{21} as a function of stress time for HCI DC stress.

Similarly, large-signal measurements can also be modeled using our reliability tool. Once again, it is found that our reliability predictions are very well aligned with silicon measurements, in spite of the fact that measurements are led into millimeter wave domain (~60GHz).

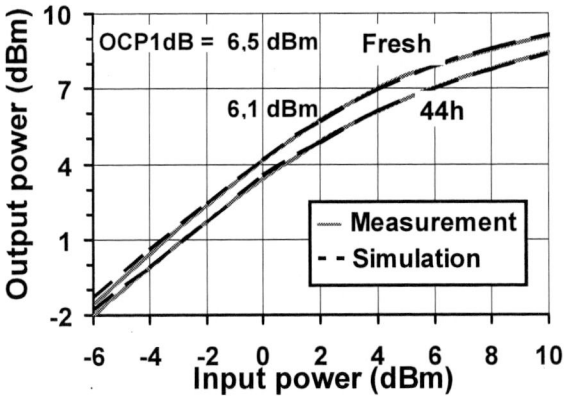

Fig. 17: Comparison between measured and simulated 1 stage PA output saturated power and the output 1dB compression point (OCP1dB) versus input power at 60GHz for HCI DC stress.

This first step of validation allows concluding that, for a HCI DC stress, our reliability SPICE simulation tool provides predictions well aligned to silicon for characteristics as different as DC ones (I-V curves), small-signal parameters (S parameters) and large-signal parameters (like ouput power). Overall, we can conclude at this point that the set of equations related to SPICE parameters impact is adequate for a large range of signals.

C. RF stress

The last step is related to the validation of quasi-static stress models (i.e. related to the equations driving defect generation) up to 60GHz input frequency. RF stresses are applied by setting the input frequency to 60GHz so to insure that most of input RF power is transferred to the transistor with minimum attenuation. Two input RF powers have been used as stress conditions: 0dBm (corresponding to OCP1dB) and -10dBm.

We first validated that experimental DC I-V curves degradation monitored prior and after RF stress are in line with our modeling prediction. Fig. 18 shows that once again our reliability predictions are very accurate when compared to silicon.

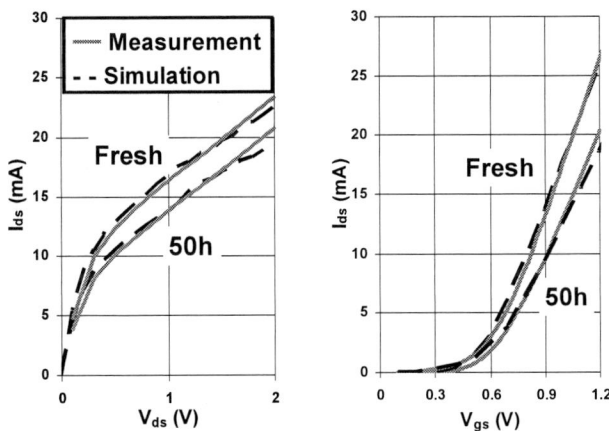

Fig. 18: Comparison between I-V characteristic curves in between silicon (red) and SPICE (black) for fresh devices and after RF stress at 60GHz with Pin=0dBm.

It allows validating that our quasi-static-based approach for defect generation rate is still valid for high frequency domain up to 60GHz, which allows covering both our AMS/RF applications but also the fast voltage transitions occurring in high-speed digital designs.

Similarly to the case of DC stress, we had also compared the results of our reliability simulations to small-signal (Fig. 19) and large-signal (Fig. 20) parameters drifts.

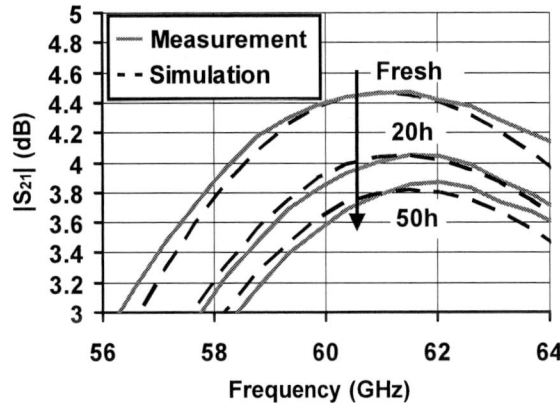

Fig. 19: Comparison between measured and simulated 1 stage PA gain S_{21} as a function of stress time for RF stress at 60 GHz with Pin=0dBm.

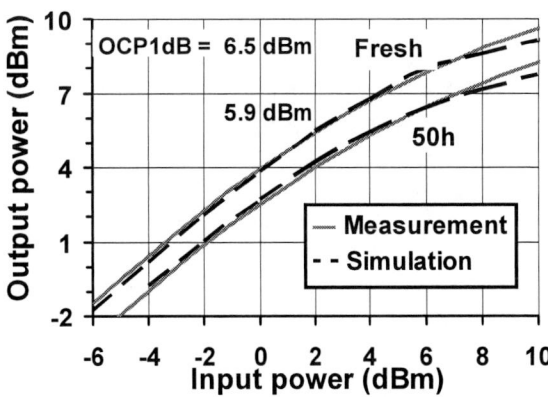

Fig. 20: Comparison between measured and simulated 1 stage PA output saturated power and the output 1dB compression point (OCP1dB) versus input power at 60GHz for RF stress at 60 GHz with Pin=0dBm.

Once again, very good agreement is achieved for various parameters, stress times and stress conditions. Overall, we have demonstrated that the quasi-static approach used to generate HCI reliability simulations is suitable to reproduce electrical ageing phenomena up to 60GHz. These important results were mandatory to demonstrate that AMS/RF applications can be supported by our reliability simulations but also High-Speed digital designs with fast transitions in range of tens of picoseconds.

IV. APPLICATION TO A REAL TESTCASE

Finally, similar RF stress experiments have been led on a more realistic (on a product sense) 4 stages Power Amplifier (PA) [6], designed to work at 60GHz.

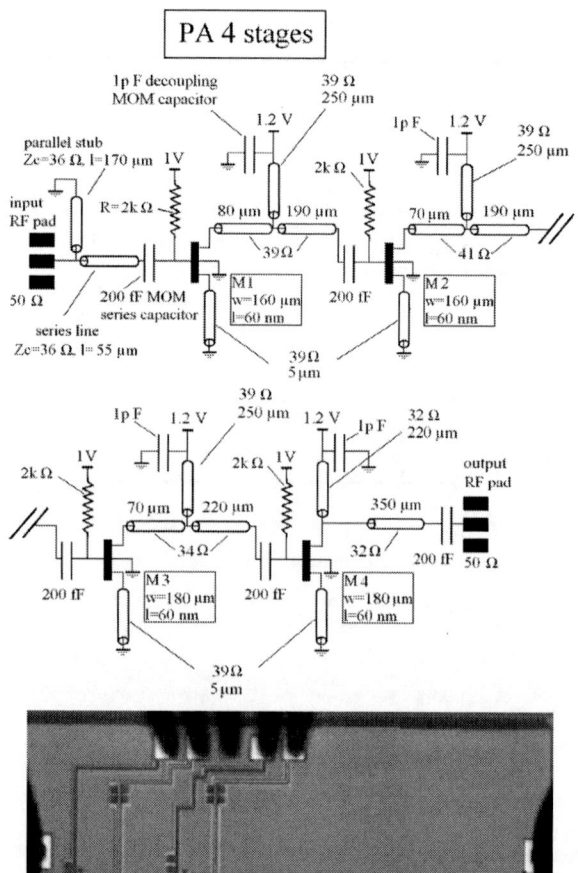

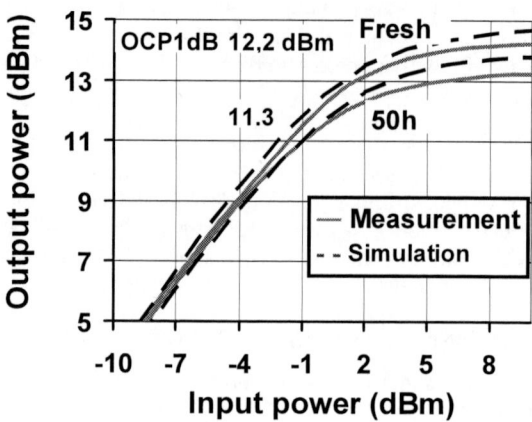

Fig. 21: Schematics of the 4-stages Power Amplifier designed for 60GHz use and its corresponding microphotography.

This 4-stages PA presents state-of-the-art performances for a 60GHz use.

Test Structure	P_{sat} (dBm)	P_{1dB} (dBm)	G_p (dB)	Power (mW)	PAE (%)
4 stage PA	14.2	12.2	13.7	300	8.4

Fig. 22: 60GHz 4-stages PA performances.

Similarly to simpler 1-stage testcase, reliability simulations are well aligned to silicon measurements prior and after both DC and RF stress conditions.

Fig. 23 shows the silicon-CAD correlation for small-signal parameters after a DC stress as an example.

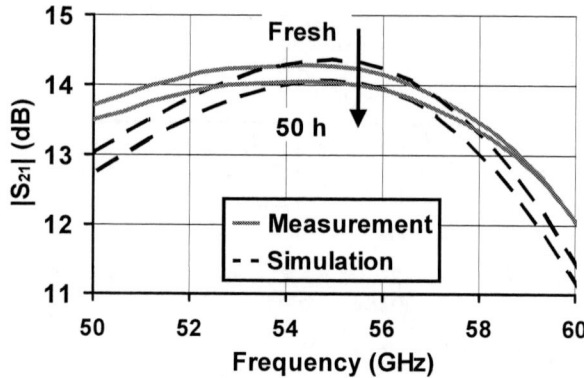

Fig. 23: Comparison between measured and simulated 4 stage PA gain S_{21} as a function of stress time for DC stress.

Fig. 24 shows the silicon-CAD correlation for large-signal parameters after a RF stress as an example.

Fig. 24: Comparison between measured and simulated 4 stage PA output saturated power and the output 1dB compression point (OCP1dB) versus input power at 60GHz for RF stress at 60 GHz with Pin=0dBm.

V. CONCLUSIONS

We have demonstrated that the quasi-static approach used to generate HCI reliability simulations is suitable to reproduce electrical ageing phenomena up to 60GHz using both simple 1-stage PA testcase and a more realistic, state-of-the-art 4-stages PA. These important results were mandatory to demonstrate that AMS/RF applications can be supported by our reliability simulations but also High-Speed digital designs with fast transitions in range of tens of picoseconds.

REFERENCES

[1] S. E. Rauch, TDMR, vol. 1, (2001).

[2] C. Guerin et al., IEEE IRPS (2008).

[3] A. Bravaix et al., IEEE IRPS (2009).

[4] V. Huard et al., IEEE IRPS (2010).

[5] C. Guerin et al., J. Appl. Phys., vol 105 (2009).

[6] T. Quemerais et al., IEEE RFIC (2010).

Impact of Source/Drain Contact and Gate Finger Spacing on the RF Reliability of 45-nm RF nMOSFETs

Rajan Arora, Sachin Seth, John Chung Hang Poh, John D. Cressler

School of Electrical and Computer Engineering, Georgia Institute of Technology,
777 Atlantic Drive NW, Atlanta, GA 30332-0250 USA
phone: (678) 641-3634, e-mail: arora@ece.gatech.edu

Akil K. Sutton, Hasan M. Nayfeh, Giuseppe L. Rosa, and Greg Freeman
IBM Semiconductor Research and Development Center,
Hopewell Junction, NY 12533 USA

Abstract— We report the radio frequency (RF) stress reliability response of 45-nm SOI RF nMOSFETs. The dependence of gate oxide degradation and off-state leakage (due to RF stress) on the contact spacing of the Source/Drain (S/D) terminals and the gate finger-to-gate finger spacing is investigated. The RF device performance trade-offs vs. RF stress reliability that result are investigated. Devices with "tight" S/D contact spacing have improved RF performance but worse RF reliability than devices with "loose" S/D contact spacing. Devices with "loose" gate-finger to gate-finger spacing have better RF performance and also better RF reliability. The net result of this investigation is that fundamental tradeoffs between RF performance and reliability exist at these advanced scaling nodes.

Keywords-: RF, RFCMOS, f_T, P1dB, reliability

I. INTRODUCTION

Scaling of CMOS technology has resulted in significant improvements in RF performance. Cut-off frequency (f_T), and maximum oscillation frequency (f_{max}) above 300 GHz at room temperature have been reported [1]. The performance of state-of-the-art MOSFETs is attractive for RF circuit design for system-on-a-chip applications in which digital baseband, mixed-signal, and RF transceiver blocks are integrated on a single die. CMOS-based monolithic transmit-receive (T/R) modules for wireless applications already exist [2]. One major challenge for CMOS technology in RF applications is the reliable design of power amplifiers (PAs), due to the low breakdown voltages inherent in short channel length MOSFETs. To achieve maximum efficiency, the PA is typically operated close to compression conditions which happens at high input power [3]. Applying high input power to nanoscale MOSFETs creates strong electric fields across the ultrathin gate oxide and shallow p-n junctions and can result in reliability concerns. Another significant concern in monolithic CMOS solutions is the leakage of high power RF signals from the transmit path to the receive path. CMOS single-pole

double throw (SPDT) switches are used to switch between transmit and receive paths. The switch's isolation can degrade due to the high RF power output (e.g., ~ 1 W) from the power amplifier, thus leaking undesired power into the receive path. CMOS technology has been previously studied for the effects of dc hot carrier and RF stress on RF properties [4]-[5]. In addition, the effects of dc and RF stress on RF circuits has been studied [6]-[7]. In the present paper, we investigate the RF reliability of 45-nm SOI CMOS devices and discuss ways to improve the RF reliability by simple device-level modifications. Significant improvements in RF performance can be achieved (especially for sub-50 nm channel length FETs) using only layout modifications [8]. Here we investigate the effects of device layout optimization on gate oxide degradation and off-state leakage (due to RF stress) for the first time. Device degradation due to the following three stress conditions is considered:

A. Hot Carrier Stress

DC bias in a circuit can cause shifts in device parameters. To investigate the effect of hot carrier stress on device parameters, the device is stressed for constant gate and drain voltage ($V_g = 1.0$ V, $V_d = 1.5$V).

B. Constant RF power stress for varying time

To investigate the effect of a constant RF power stress, for instance, in a LNA circuit, the device is stressed in on-state ($V_g = 0.6$ V, $V_d = 0.9$ V) for constant RF power stress ($P_{in} = 0$ dBm) for varying stress time. The device bias condition ($V_g = 0.6$ V, $V_d = 0.9$ V) also causes some degradation due to hot carrier stress.

C. Varying RF power stress for constant time

To investigate the effect of varying power stress, for instance, in a SPDT switch, the device is stressed in on-state ($V_g = 0.6$

V, V_d = 0.9 V) for varying RF power stress for constant stress time (t = 100 s).

II. EXPERIMENTAL DETAILS

This 45-nm SOI RF CMOS technology features 1.16 nm gate oxide, dual stress liners (DSL), embedded SiGe in the S/D regions of the pFETs, advanced activation annealing and stress memorization techniques [9]. The devices investigated here are n- and p-MOSFETs with varying source-drain contact pitch (CA-CA spacing), gate finger-to-gate finger (PC-PC) spacing variations, and S/D contact to gate finger (CA-PC) variations (Fig. 1 and Table I) [1]. Both floating-body and body-contact device options are available. Here we focus on the floating-body devices. Devices with symmetric S/D halo doping are investigated. All devices have 20-fingers (L=45 nm, W = 2μm) and are wired for RF measurements using G-S-G co-planar probes. The measurements are performed on a custom integrated S-parameter and load-pull system, which allows both small- and large-signal measurements with a single probe contact (Fig. 2). The device source and load impedances are matched for maximum gain. The RF stressing is performed in the impedance matched configuration with the device biased at peak f_T conditions (V_{gs} = 0.6 V, V_{ds} = 0.9 V). A two-step de-embedding technique based on the open/short calibration test structures is used. For the RF parameter extraction (C_{gd}, R_g, R_d), a physics-based RF model is assumed and fitted to the measured S-parameters [10]. All measurements are performed at room temperature. Three samples of each device were tested and the results are repeatable.

TABLE I. CONTACT SPACING FOR DIFFERENT DEVICES USED

Device	W/L (μm)/(μm)	Number of Fingers	PC-PC spacing (μm)	CA-PC spacing (μm)	S/D CA-CA spacing (μm)
A	2/0.045	20	0.150	0.045	0.084
B	2/0.045	20	0.340	0.140	0.084
C	2/0.045	20	0.150	0.045	2.000
D	2/0.045	20	0.340	0.140	2.000

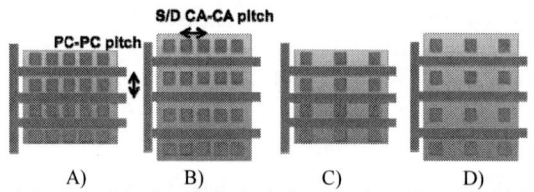

Fig. 1. Layout of the floating-body devices. Device with A) close PC-PC and Source/Drain (S/D) CA-CA spacing. B) Larger PC-PC and close S/D CA-CA spacing. C) Close PC-PC and larger S/D CA-CA spacing D) Large PC-PC and large S/D CA-CA spacing.

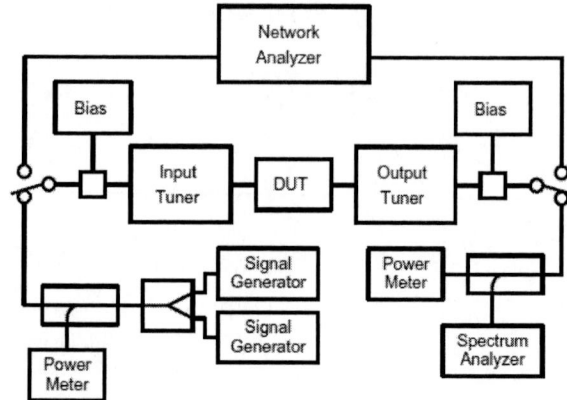

Fig. 2. Block Diagram of the load-pull measurement setup used.

III. RF PERFORMANCE & STRESS RESULTS

Devices with tighter S/D CA-CA spacing (devices A, B) have improved RF performance: higher cut-off frequency (f_T), higher maximum oscillation frequency (f_{max}), higher $P1dB$ (1 dB gain compression), higher power gain, higher output third-order inter-modulation distortion point (OIP_3), than devices with looser CA-CA spacing (devices C, D) (Figs. 3, 4, 5). The above measurements are made at a CW frequency of 9 GHz. In addition, devices with looser CA-gate spacing (devices B, D) show better RF performance due to the increased volume of the stress liner and hence higher net channel strain [1]. Lower gate to source parasitic capacitance (C_{gs}) also results in higher RF performance for the devices B and D. Thus, having tight S/D contact spacing and relaxed poly gate pitch is a very promising way of designing devices used in applications requiring the highest speed. Here, it is shown for 45 nm RF-CMOS technology, but it will apply in general to short-channel length MOSFETs where the parasitic start dominating device performance.

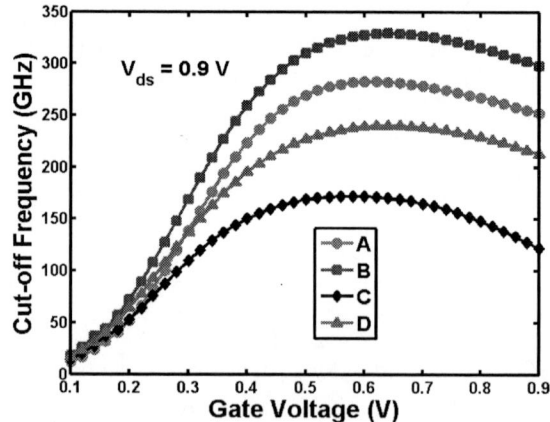

Fig. 3. Cut-off frequency for n-MOSFETs with the same dimensions (L=45 nm, W= 2μm), but varying contact spacing and gate pitch.

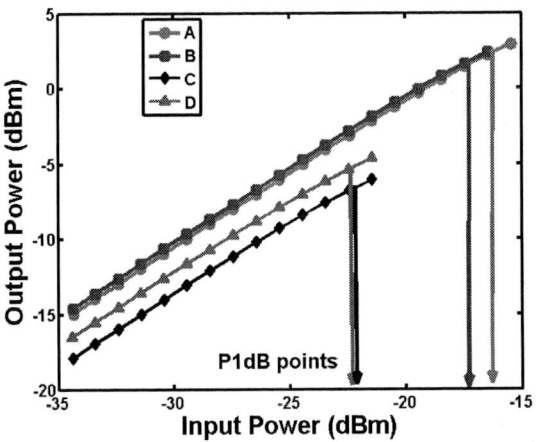

Fig. 4. Output power versus input RF power of nFETs with varying contact spacing and gate pitch.

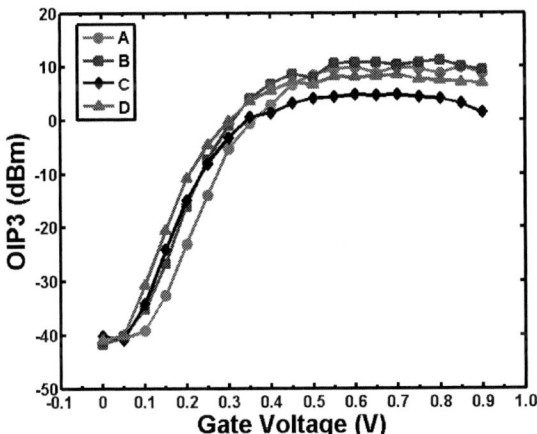

Fig. 5. Output third order intercept point (OIP3) of nFETs with varying contact spacing and gate pitch.

The ESD and DC hot carrier reliability dependence on contact spacing has been reported previously [11], [12]. NMOS devices with tight drain contact spacing had more ESD damage, which was attributed to a non-uniform current flow in this case [11]. The devices with tighter S/D spacing were shown to be more susceptible to hot carrier damage [12]. Device A and B show larger threshold voltage shifts in the hot carrier stress condition than devices C and D (Fig. 6) [12]. The pFETs show similar trends as the nFETs [12]. Devices with tighter S/D contact (multiple contacts) spacing have larger electric field at the drain than devices with looser spacing (single contact), due to the decreased silicide diffusion resistance (Fig. 7) [12]. Degradation mechanisms for hot carrier stress were shown to be due to higher electric fields in the drain and channel region [12].

Here we describe the effects of contact spacing and gate finger-to-gate finger width on RF stress results. The devices are stressed at constant RF input power (0 dBm) (CW frequency = 9.0 GHz) for varying stress times and also

stressed with varying RF power for constant time (t = 100 s) at the peak f_T bias condition (V_g = 0.6 V, V_d = 0.9 V).

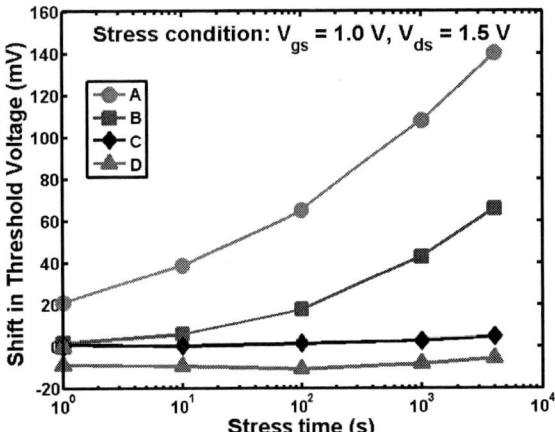

Fig. 6. Shift in threshold voltage with DC hot carrier stress (V_g =1 V, V_d = 1.5 V) for nFETs with varying contact spacing and gate pitch [12].

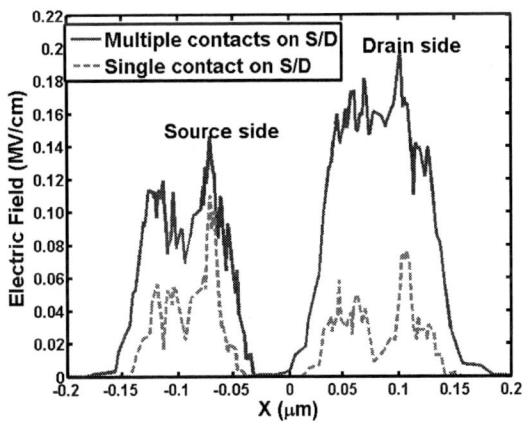

Fig. 7. Comparison of TCAD simulated electric field at drain for single contact and multiple S/D contact device [12].

Consistent degradation in DC and RF parameters is observed with increasing stress time and power. We observe a positive shift in threshold voltage (indicating electron traps in gate oxide) and a decrease in drive current (mobility degradation) (Fig. 8). The threshold voltage shifts in the RF stress condition (Fig. 8) are mostly due to the DC hot carrier degradation. The magnitudes of threshold voltage shifts in Fig. 8 are much smaller than Fig. 6 because of lower DC voltages applied for peak f_T bias condition. The devices are stressed up to 3000 s under constant RF power stress (RF power = 0 dBm) (Fig. 9a). The device with wider S/D contact spacing (devices C, D) is able to operate for longer times and shows smaller threshold voltage shifts than devices with tight contact spacings (devices A, B). The devices are stressed at the same RF input power (although they have different P1dB). The P1dB power level defines the maximum power level where the device provides gain. It does not limit the maximum input power that can be applied to the device. Therefore, for a fair evaluation of the impact of layout on device RF stress reliability, the same input power levels are applied. The devices are stressed at up to 13.6 dBm of RF power for constant stress time (time=100 s) (Fig. 9b). This high power would be a representation of the worst case scenario for a device used in SPDT switch. The RF parameters of interest, small-signal gain (S21), isolation (S12), cut-off frequency (f_T), maximum oscillation frequency (f_{MAX}) and unilateral power gain all degrade with applied stress (Fig. 9). As shown in Fig. 9a, Device C shows much smaller degradation in RF parameters for varying stress time than device A. In addition, device C (with wider Source/Drain CA-CA spacing) can operate up to higher RF power levels (Fig. 9b).

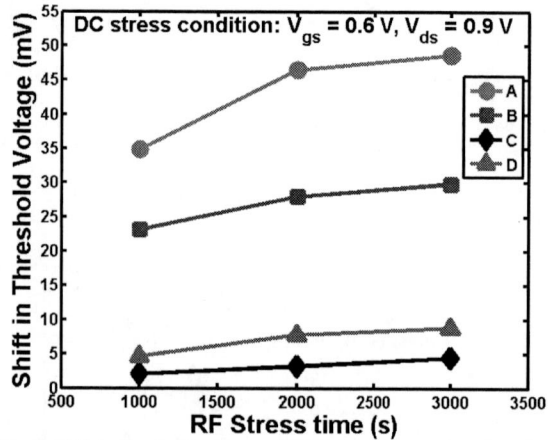

Fig. 8. Shift in threshold voltage with RF stress for varying time (V_g = 0.6 V, V_d = 0.9 V) for nFETs with single contact and multiple S/D contact device (Frequency = 9GHz, Pin = 0dBm).

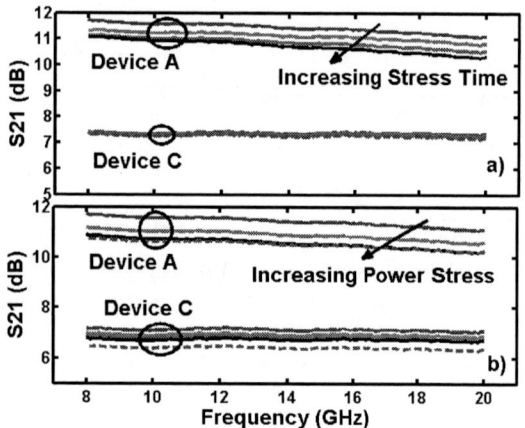

Fig. 9. Comparison of S21 degradation with increasing: a) RF stress time for 1000s, 2000s, and 3000s stress (for constant power = 0dBm) b) input RF power of +5, +6, +7 dBm (for constant stress time = 100s) of nFETs (at peak f_T bias).

IV. MECHANISMS

For the three stress conditions (hot carrier, constant power stress for varying time, and constant time stress for varying power) the physical mechanisms causing the degradation are not the same:

A. Hot Carrier Stress

With hot carrier stress as a function of increasing stress time only threshold voltage shift and mobility degradation is observed. This is due to the hot electron trapping in the gate oxide [12]. These DC parametric shifts further show up in RF parameter degradation [12].

B. Constant RF power stress for varying time

With constant RF power stress for varying time threshold voltage shift and mobility degradation is observed at moderately high input power levels (P_{in} = 0 dBm). The shifts are mostly due to the hot carrier stress by DC bias on the device. The device input is looking at an impedance of 49 + j119 in the source matched condition. That corresponds to around 0.36 V at input (for P_{in} = 0 dBm). These DC parametric shifts further show up in RF parameter degradation. Thus under normal device operating conditions of circuits like low-power LNA/switches the device is expected to operate without much degradation for long time.

C. Varying RF power stress for constant time

In the catastrophic instance of leakage of RF power to an unintended port (e.g power transmission from the power amplifier to the receiver on the same chip), it is important to study parameter degradation as a function of varying power level. For varying RF power stress (constant time) the off-state leakage current continuously increases along with positive threshold-shifts and mobility degradation (Fig. 10). The gate current increases continuously with increasing power stress, initially soft oxide breakdown is observed (upto P_{in} ~ 10 dBm) followed by hard oxide breakdown (Fig. 11). Device A, B show soft gate-oxide breakdown at much lower power levels than device C, D. This increased off-state leakage can be attributed to the Fowler-Nordheim (FN) tunneling of electrons from the gate to the drain above certain power levels [13]. Further higher RF power results in gate-oxide breakdown. High RF power stress (above 0 dBm) results in large electric field in the drain region under the gate-drain overlap. In a 50 Ohm system 10 dBm corresponds to 0.7 V. The device input is looking at an impedance of 49 + j119 in the source matched condition. That corresponds to around 1.2 V at input. This voltage stress due to RF signal when added to the DC bias causes significant damage as shown in Fig. 9, 10. The high frequency RF signal can also pass through the parasitic gate-drain overlap capacitance (C_{gd}) and also gate-source overlap capacitance (C_{gs}) to directly appear at the drain/source terminal of the FET. This can form a positive feedback loop and hence causing enhanced degradation [15]. This fact is supported by the observation that device A, B degrade at lower power levels compared to device C, D (Fig. 9). The reason being that device A, B have greater parasitic capacitance (C_{gs}, C_{gd}) than device C, D due to greater interaction between S/D contacts (Fig. 12). Also Device A, C degrade at lower power levels compared to device B, D again due to greater parasitic capacitance (C_{gs}, C_{gd}).

The devices with closely spaced Source/Drain CA-CA spacing (devices A, B) exhibit greater degradation of drive current than devices with looser Source/Drain CA-CA spacing (device C, D) (Fig. 13, 14). This may be attributed to the higher electric field in the device with tightly spaced contacts, which in turn results in greater damage of the gate oxide.

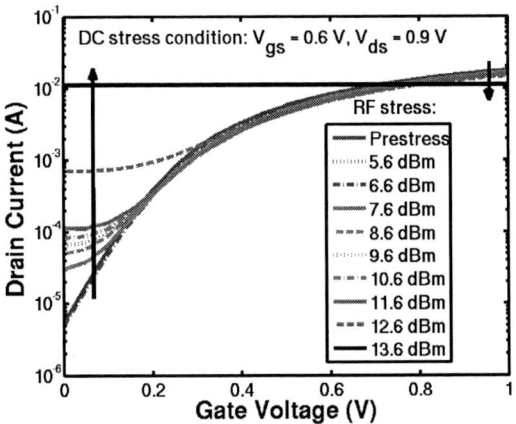

Fig. 10. Transfer characteristics of nFET (device C) with varying RF power stress (V_g =0.6 V, V_d = 0.9 V). Stressed for 100s at each power step (frequency = 9 GHz).

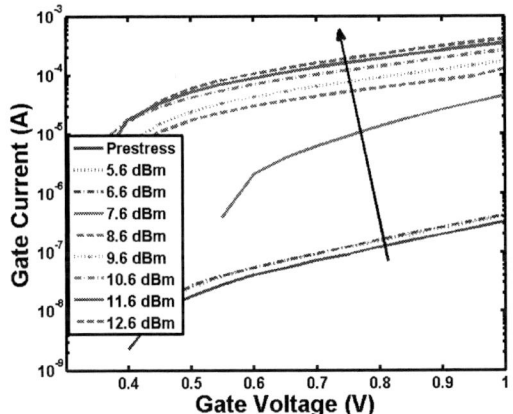

Fig. 11. Gate-current characteristics of nFET (device C) with varying RF power stress (V_g =0.6 V, V_d = 0.9 V).

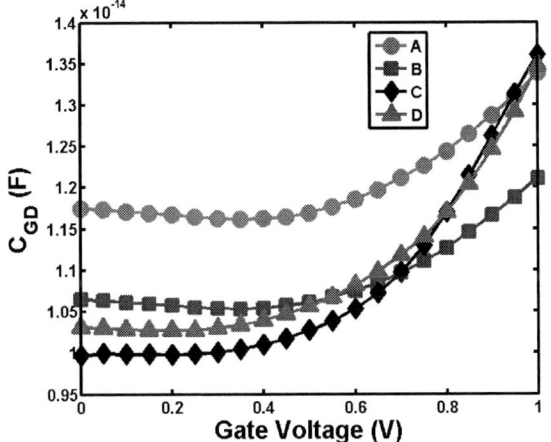

Fig. 12. Comparison of gate-to-drain capacitance with gate voltage of nFETs with varying contact spacing.

The gate-to-drain capacitance (C_{gd}) decreased after stress, indicating an increase in drain junction depletion region width (eventually leading to punch-through). We see the same trade-offs in hot carrier stress as devices with wider S/D contact spacing can work under higher RF input power levels without catastrophic failure. The device on-state drain current and gate-to-drain capacitance decreases for all four device variants with increasing power (Fig. 14). Devices C and D are able to operate at higher power levels without catastrophic failure (despite the fact that they have a lower P1dB).

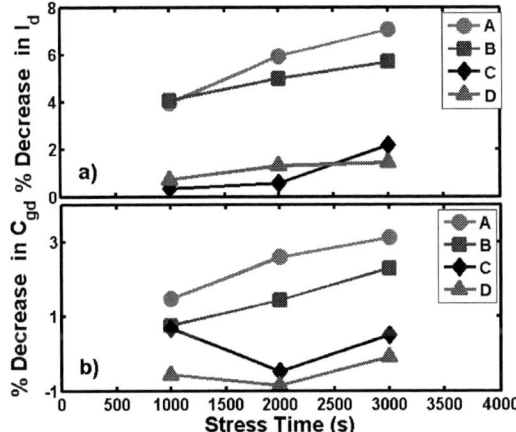

Fig. 13. Comparison of a) I_d degradation (percentage) as a function of RF stress time (at peak f_T bias) b) C_{gd} degradation (percentage) as a function of RF stress time of nFETs (at peak f_T bias).

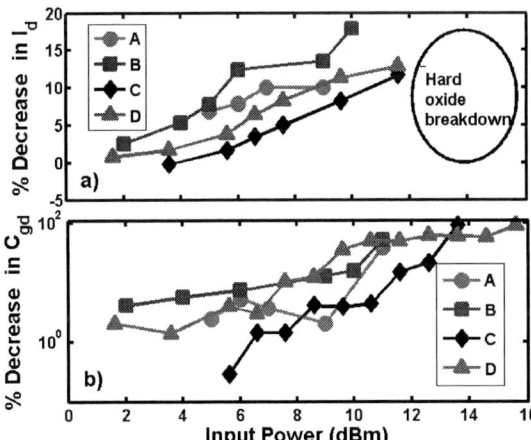

Fig. 14. Comparison of a) I_d degradation (percentage) with input RF power of nFETs (at peak f_T bias) b) C_{gd} degradation (percentage) with input RF power of nFETs (at peak f_T bias).

V. SUMMARY

We have examined the impact of source/drain contact spacing and gate finger-to-gate finger spacing on the RF stress reliability of 45-nm SOI RF CMOS technology. FETs with looser S/D contact spacing and wider gate finger-to-finger spacing are able to operate at higher input power levels and also for longer stress times. The role of parasitics in RF reliability has been investigated. For circuit designs in ultra short channel length FETs, device layout optimization can result in improved RF performance and acceptable RF stress reliability.

ACKNOWLEDGMENT

The authors would like to acknowledge the support provided by members of the SiGe devices and circuits team at Georgia Tech and the CMOS team at IBM.

REFERENCES

[1] S. Lee, B. Jagannathan, S. Narasimha, A. Chou, N. Zamdmer, J. Johnson, R. Williams, L. Wagner, J. Kim, J. –O. Plouchart, J. Pekarik, S. Springer, and G. Freeman, "Record RF performance of 45-nm SOI CMOS technology", *IEEE International Electron Devices Meeting (IEDM)*, pp. 255-258, 2007.

[2] E.–H. Kim, J.–K. Choi, S.–O. Yun, J. Ko, and K. Lee, "A highly efficient 5.8 GHz CMOS transmitter IC with robustness over PVT variations", *IEEE Radio Frequency Integrated Circuits Symposium (RFIC)*, pp. 403-406, 2010.

[3] T. H. Lee, The design of CMOS radio-frequency integrated circuits, Cambridge University Press, 2004.

[4] C.–H. Liu, R.–L. Wang, Y.–K. Su, C.–H. Tu, and Y.–Z. Juang, "DC and RF degradation induced by high RF power stresses in 0.18-μm nMOSFETs", *IEEE Transactions on Device and Materials Relaibility*, vol. 10, pp. 317-323, 2010.

[5] G. T. Sasse, F. G. Kuper, and J. Schmitz, "MOSFET degradation under RF stress", *IEEE Transactions on Electron Devices*, vol. 55, pp. 3167-3174, 2008.

[6] T. Quemerais, L. Moqoillon, V. Huard, J. –M. Fournier, P. Benech, N. Corrao, and X. Mescot, "Hot-carrier stress effect on a CMOS 65-nm 60-GHz one-stage power amplifier", *IEEE Electron Device Letters*, vol. 31, pp. 927-929, 2010.

[7] E. Xiao, and J. S. Yuan, "Hot carrier and soft breakdown effects on VCO performance", *IEEE Radio Frequency Integrated Circuits Symposium (RFIC), pp. 459-462, 2002.*

[8] J. W. Sleight, L. Lauer, O. Dokumaci, D. M. Fried, D. Guo, B. Haran, S. Narasimha, C. Sheraw, D. Singh, M. Steigerwalt, X. Wang, P. Oldiges, D. Sadana, C. Y. Sung, W. Haensch, and M. Khare, "Challenges and opportunities for high performance 32 nm CMOS technology", *IEEE International Electron Devices Meeting (IEDM)*, pp. 1-4, 2006.

[9] S. Narasimha, K. Onishi, H. M . Nayfeh, A. Waite, M. Weybright, et al. "High performance 45-nm SOI technology with enhanced strain, porous low-k BEOL, and immersion lithography", *IEEE Internation Electron Devices Meeting (IEDM)*, pp. 1-4, 2006.

[10] Y. Cheng, M. J. Deen, and C.–H. Chen, "MOSFET modeling for RF IC design", *IEEE Transactions on Electron Devices*, vol. 52, pp. 1286-1303, 2005.

[11] J.–H. Lee, Y.–H. Wu, C.–H. Tang, T.–C. Peng, S–H. Chen, and A. Oates, "A simple and useful layout scheme to achieve uniform current distribution for multi-finger silicided grounded-gate NMOS", *IEEE International Reliability Physics Symposium (IRPS)* , pp. 588-589, 2007.

[12] R. Arora, K. A. Moen, A. Madan, J. D. Cressler, E. Zhang, D. M. Fleetwood, R. D. Schrimpf, A. K. Sutton and H. M. Nayfeh, "Impact of body-tie and source/drain contact spacing on the hot carrier reliability of 45-nm RF-CMOS", *IEEE International Integrated Reliability Workshop (IIRW)*, 2010 (in press).

[13] K. Roy, S. Mukhopadhyay, H. M-. Meimand, "Leakage current mechanisms and leakage reduction techniques in deep sub-micrometer CMOS circuits",Proceedings of the IEEE, vol. 91, pp. 305-323, 2003.

[14] R. Arora, K. A. Moen, E. X. Zhang, R. D. Schrimpf, D. M. Fleetwood, J. D. Cressler, A. K. Sutton and H. M. Nayfeh, "Impact of body-contacting and source/drain contact spacing on the total dose reliability of 45 nm RF-CMOS technology", (to be submitted) *IEEE Transactions on Nuclear Science*, 2011.

[15] B. Razavi, "Design of Analog CMOS Integrated Circuits", McGraw-Hill Science/Engineering/Math, 2000.

Soft Error Testing at Advanced Technology Nodes

B. Bhuva[1], B. Narasimham[2], A. Oates[3], K. Patterson[4], N. Tam[5], M. Vilchis[6], S.-J. Wen[7], R. Wong[7] and Y. Xu[8]

[1] Vanderbilt University, Nashville, TN, USA
[2] Broadcom Corporation, Irvine, CA, USA
[3] TSMC, Taiwan
[4] Avago Technologies, Fort Collins, CO, USA
[5] Marvell Semiconductor, San Jose, CA, USA
[6] LSI Corporation, Milpitas, CA, USA
[7] Cisco Systems, Inc., San Jose, CA, USA
[8] Altera Corporation, San Jose, CA, USA

Abstract— Abstract— A test vehicle concept for soft error testing of flip-flop designs has been developed and verified at 40 nm technology node. Key contribution of this test vehicle has been identification of any new unpredictable failure mechanisms, training of engineers for soft error testing and mitigation, establishment of knowledge database, and to build a start-to-finish (designers-foundry-system) link for a complete solution for soft error related issues. Major issues faced by the designers and test results are discussed for 40 nm technology node.

Keywords-component; Soft errors, testing, neutrons, alpha particles, IC design

I. INTRODUCTION

For reliability engineers, Failure-in-time (FIT) rates are most indicative of the potential impact of different failure mechanisms on product reliability in the field. With latest advances in semiconductor manufacturing, novel failure mechanisms may emerge that will dominate (or significantly affect) overall FIT rates for a given product. The biggest challenge faced by individual semiconductor companies is the process, and associated cost, of evaluating a circuit design for such novel failures when they are not fully characterized and/or known. Among all other reliability problems at advanced technology nodes, soft errors (SE) in electronic circuits are expected to contribute significantly towards the overall FIT rates for a given product. The SE FIT rates are dependent on the fabrication process, circuit design, and system design. Since, the mechanisms associated with SE errors are evolving with technology, it is hard to carry out predictive analysis of circuits and systems. Novel approaches are necessary that can identify SE related failure mechanisms for a technology node, apply these mechanisms to circuit-level designs, and predict system-level failure rates. Since, TCAD simulations alone may not be sufficient to identify novel failure mechanisms, experimental results are required for each technology node. This means semiconductor companies must test circuit designs, IC designs, and system designs at great expense. This paper presents a novel approach to reduce cost and test complexity associated with soft error testing at advanced technology nodes to provide designers with requisite data for predictive analysis.

The root cause requiring testing of parts to identify failure mechanisms and failure modes (and FIT rates) for SE is the complexity of the physical mechanisms associated with a soft error. The primary physical cause of the SE is an incident particle, such as Alpha particles, neutrons, or heavy-ions, on Silicon. When such a particle is incident on an Integrated Circuit (IC), it creates electron-hole pairs in Silicon substrate through coulombic interactions with Si lattice either directly (Alpha, heavy-ions, etc.) or through secondary reactions (neutrons, high-energy protons, etc.). Most of these charges recombine and does not affect the circuit operation. However, some of the charges may reach a circuit node through drift and/or diffusion processes to create a voltage perturbation at that node. These perturbations may alter the data stored at the circuit nodes, resulting in circuit malfunction. The problem of understanding the effects of SE arises because one needs full understanding at the physical level (transport of electrons and holes in a Si substrate), at the circuit level (effects of voltage perturbation a circuit node), and at the system level (incorrect data in a flip-flop affecting the system-level operation) to fully characterize the error. With billions of transistors on an IC, any miscalculation at any of the steps in the SE characterization will result in large errors in FIT rates. As a result, it is imperative for engineers to have accurate models for predictive simulations.

All of the processes involved for SE (charge generation, charge collection, charge transport, node voltage perturbations, transient propagation, error latching, system-level error propagation, etc.) are very well understood at this point. However, FIT rate estimation and SE mitigation is still elusive for most engineers because the advances in fabrication technologies add novel failure mechanisms in the soft error generation. For example, for advanced technologies, due to close proximity of transistors, multiple circuit nodes will collect charge resulting in multiple voltage perturbations within a circuit. Another mechanism, pulse-quenching

978-1-4244-9113-1/11 $26.00 © 2011 IEEE

mechanism, shortens the voltage perturbation due to charge collection by electrically connected nodes. Process-parameter variations add another dimension to the problem by making all variables non-deterministic. In addition, the temporal and spatial characteristics of SE transient pulses strongly influence the FIT rates. These SE transients, in turn, are dictated by the fabrication process parameters, layout, and circuit design. The problem engineers face when trying to estimate SE related FIT rates is that not all of these factors are known *a priory*. In fact, most times reliability engineers are required to characterize a fabrication process that is not yet commercially available. With the rapid pace of advancement of technologies, designers are constantly designing circuits for technologies that will be available in the future. Designers need to know the FIT rates due to soft errors for the flip-flops being used in the design. But without a test IC, reliability engineers can not estimate accurate FIT rates due to evolving failure mechanisms. And by the time ICs are available for testing, it is too late unless designers are willing to redesign the whole IC.

Another problem faced by reliability engineers is that fabrication houses are reluctant to grant access to a process that is not yet production-ready. Until the fabrication process is released in the open market, fabrication houses restrict access to the fabrication line. This results in the familiar scenario of designers not knowing what FIT rates to expend during design phase. As a result, they end up using data from a previous generation of technology (for example, designers are using data from 65 nm process for estimating FIT rates for their 28 nm designs). Since failure mechanisms for soft errors are evolving, data based on 65 nm technologies may not be valid for 28 nm technologies. For example, a DICE FF design was shown to be 100X better than conventional DFF design at 65 nm technologies. However, at 28 nm technology most DICE FF designs are only 5X better than conventional DFF designs. Such knowledge is required by the designers during early stages of the design to achieve early understanding of SE failure mechanisms. The only way this is possible is through collaborative efforts with fabrication companies to develop failure models, and subsequent mitigation techniques.

Cisco-TSMC-Vanderbilt team along with Altera, Avago, Certichip, Bayview, Broadcom, LSI Logic, and Marvell started projects at 40 nm and 28 nm technology nodes with the primary goal of characterizing soft-error failure modes. Secondary goals were to estimate FIT rates for specific flip-flop designs for advanced technologies so designers can use them for predictive analysis and to develop mitigation strategies. The 40 nm test vehicle was fabricated in 2010 and tested with neutrons, alpha particles, thermal neutrons, heavy ions, protons, muons, and laser beams. The 28 nm test vehicle uses two separate die and will be fabricated during the first quarter of 2011. The following sections detail the legal and logistical hurdles to collaboration between such a large and diverse team. The collaborative effort was successful at the 40 nm technology node in characterizing SE effects and determining the FIT rates for a set of flip-flop designs for neutrons, alpha particles, thermal neutrons, and heavy-ions.

Details regarding efforts to streamline the design and data collection processes are also discussed. Major contributions of this project are the development of a community vehicle for addressing issues related to soft errors, such as identification of any new unpredictable failure mechanisms, training of engineers for soft error testing and mitigation, establishment of knowledge database, and to build a start-to-finish (designers-foundry-system) link for a complete solution for soft error related issues.

II. COLLABORATIVE APPROACH FOR SER MODELING

For success of any approach to characterize an advanced fabrication process, both the fabrication engineers and design engineers need to collaborate closely. The fabrication house will need to grant access to an advanced or a premature fabrication process, while the benefit they receive will be in the form of necessary data related to soft errors for making adjustments in their fabrication process. Design engineers will need to spend time to develop and design a test IC. Information available from the test IC will help designers understand effects of novel mechanisms on their circuit designs, provide enough details to estimate FIT rates of their products, and allow for development of mitigation strategies.

The biggest concern for such a collaborative approach is to convince fabrication engineers and design engineers to expend time for process and design efforts. Additionally, there is the matter of legal issues related to protection of IP and proprietary information. The fabrication house will need to reveal some details of the fabrication process so reliability engineers can understand and interpret test results. Designers will have to reveal details of their design as data interpretation without design details will be meaningless. Both parties, design engineers and fabrication engineers, must coordinate their efforts so that protection of proprietary information is guaranteed.

Since it will be very expensive for a single industry partner to carry out such a project, multiple industry partners are necessary to reduce the cost and the risk associated with such a project. The cost for the project must be equally shared by all partners. With multiple partners, the cost is reduced but the legal and logistical issues arise that need attention even before the project starts. Industry partners will be reluctant to share their designs with a competitor. So a neutral entity is needed that can handle the proprietary data and manage expectations. Since most industry partners will be competing against each other in one business area or another, an outsider is necessary as a neutral entity.

A University research team is just such an entity that can provide neutral management of the project. The university team must treat all industry partners as equal and keep everything transparent to ensure project success. The university team must have capability to handle proprietary information and be able to restrict the distribution of information among graduate students (who may end up working for a competitor after graduation). The university partner must also have enough capability to handle a large design, test, and analysis effort. Vanderbilt University is ideally poised to address all of these issues. At Vanderbilt, Institute for Space and Defense Electronics (ISDE) has capability to handle classified and ITAR data in a secure facility. Involvement of graduate students is limited through limited access to information and data for projects associated with ISDE. Professors at Vanderbilt, with a combined experience of over 200 years in radiation effects on semiconductor and materials can actively contribute towards this project. Vanderbilt professors also have experience with designing and testing ICs in technologies with minimum feature sizes ranging from 1.5 μm to 40 nm. Additional engineering support, for design, test, and data analysis, is provided by 16 full time engineers and support staff. As a result, Vanderbilt was chosen to manage this project with multiple partners and characterize SE effects in advanced technologies.

III. PROJECT DETAILS

The first and foremost task for the project was to address legal issues related to IP protection and IP sharing. Vanderbilt legal team developed a document that allows for complete transparency between all partners regarding the details of the project. This document also specified the handling of all IP provided by the partners. All test data and characterization of IP was owned by the respective IP owners. This information was not shared with others. The Vanderbilt design team provided baseline test structures that can be used to compare IP designs by different partners. Vanderbilt was responsible for handling all IP matters in a confidential, responsible and appropriate manner.

Once the IP issues were settled, details of the test IC design were finalized. Test structure design should be carried out such that answers required by the fabrication engineers and designers are provided. Fabrication engineers want to understand the effects of fabrication process parameters on SER modeling. These parameters are usually charge collection efficiency, charge-sharing, and other factors related to charge generation, charge transport, and charge collection at circuit nodes. Designers are usually interested in the performance of their designs, namely SER for flip-flop designs used in their ICs and the factors affecting the SER. This means test structures must be able to measure charge collection processes and evaluate flip-flop designs from partners. Since each partner will have their own flip-flop designs that are very different from all others, careful design of test structure is necessary for timely completion of the project. The test IC must also provide information from baseline public domain flip-flop designs that is open to all

partners. The test IC can be used to compare partner's hardened designs against a common baseline structure. This will enable partners to compare their designs against other partner's designs, if they agree to do so.

The target fabrication process was determined to be the 40 nm bulk CMOS fabrication process at TSMC. Multiple partners provided their flip-flop cell designs for soft error evaluation. Usually, single-event testing of storage cells is carried out by arraying basic cells on an IC. This may result in a two dimensional array or a one dimensional array. Two dimensional arrays are used for basic storage cells, such as SRAM cells. For flip-flop designs, one dimensional arrays configured as shift registers are preferred for ease of I/O requirements. Usually, in the shift register configuration, one input and one output are required with a known input pattern (all 0's, all 1's, 0101 pattern, or any other similar pattern). The biggest problem with such a configuration is that the clock speed is limited by what can be input through I/O pads, usually much slower than the internal clocks used on ICs. As the error rates are a strong function of operating frequencies for sequential circuits, it is important to measure error rates for flip-flops at operating frequencies experienced in the field. In order to operate flip-flops at higher clock frequencies, the Circuit for Radiation Effects Self Test (CREST) technique was developed by NASA personnel [1]. The basic circuit for the CREST approach is shown in Fig. 1. In this circuit, the flip-flops are connected in the conventional shift-register fashion. However, it differs in the manner in which input and outputs are connected. Instead of providing an input signal from outside the IC, an on-chip random pattern generator (RPG) circuit is used. The data pattern generator is designed such that it repeats the data after (2^n-1) cycles. The number of flip-flop stages was also 2^n-1, allowing for output of the RPG and the shift register to match in the absence of an error. This technique will allow for on-chip testing of flip-flop designs at very high frequencies. A PLL design and random pattern generator were also designed for the test IC. Additional designs for characterizing SET pulse widths [2] and charge collection mechanisms [3] were also designed and placed on the test IC.

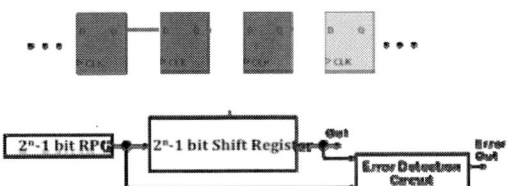

Fig. 1. CREST circuit for testing flip-flops using a random pattern generator.

978-1-4244-9113-1/11 $26.00 © 2011 IEEE

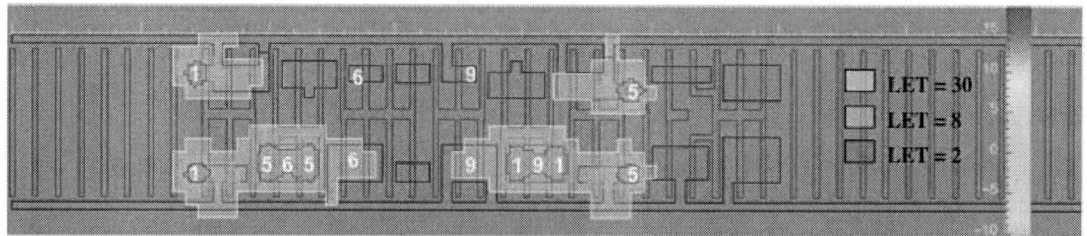

Fig. 2. Sensitive areas as a function of particle LET for the DFF design.

The test IC contained (8K-1) stage CREST designs for each of the 14 flip-flop designs. This IC was exposed to neutrons, alpha particles, thermal neutrons, and heavy-ions after fabrication and FIT rates for each flip-flop were estimated. Additionally, 3D TCAD simulations were carried out to identify sensitive areas for each flip-flop and mitigation strategies developed by individual designers. Table 1 shows the normalized FIT rates for various flip-flop designs for neutron and alpha exposures [4]. Fig. 2 shows the predictive analysis carried out with 3D TCAD software to show the most vulnerable parts of a flip-flop design as a function of the LET of incident particle. Such data and analysis allowed designers to fully understand the limitations of their flip-flop designs and to develop techniques to improve the SE performance of their designs. More testing on this IC is still continuing (heavy-ions, alpha particles, protons, thermal neutrons, and laser) and future results will show the effects of charge sharing on multiple upsets and pulse quenching effects.

Table 1: 14 flip flop designs with relative FIT rates due to neutron and alpha particles

Flip Flop design	Relative FIT wrt DFF	
	Neutron	Alpha
Standard DFF	1	1
High speed	0.91	0.95
High speed	0.87	0.98
High speed	1	1.31
Low power	0.68	0.23
Low power	0.79	0.82
Low power	1.04	0.94
Inverter, TG	1.08	0.87
High speed	1.04	0.75
Hardness by Redundancy	0.68	0.1
Hardness by Redundancy	0.23	0.01
Hardness by Redundancy	0.28	0.08
Hardness by TMR	0.3	0.09
Capacitive hardness	1	0.95

IV. CONCLUSION

The SE errors are expected to contribute significantly towards the overall FIT rates of complex electronic systems of the future. The complexities involved in modeling and predicting the SE response of deep-sub-micron technologies has frustrated design and fabrication engineers alike. Without proper planning, designers end up using FIT rates from older technologies for predictive analysis. But experimental results have shown newer failure mechanisms for advanced technology nodes, making such predictive analysis overly optimistic. Test ICs designed at the most advanced technology nodes are necessary to fully understand SE failure characteristics and to develop mitigation strategies. Vanderbilt University, along with multiple industry partners, has developed test ICs at the 40 nm technology node to support design and fabrication engineers to characterize SE failures. The framework in place has allowed launching projects for SE characterization of 28 and 20 nm technology nodes. For the 28 nm project, two separate die are being designed with tape out planned for February 22nd, 2011. With an increasing number of industry partners willing to participate, the 20 nm test vehicle is expected to launch with three separate die during the third quarter of 2011. Such projects will increase the awareness of SE related issues and instigate research for mitigation approaches needed at the advanced technology nodes in the future.

REFERENCES

[1] P. Marshall, M. Carts, S. Currie, R. Reed, B. Randall, K. Fritz, K. Kennedy, M. Berg, R. Krithivasan, C. Siedleck, R. Ladbury, C. Marshall, J. Cressler, Niu Guofu, K. LaBel, and B. Gilbert. "Autonomous bit error rate testing at multi-gbit/s rates implemented in a 5AM SiGe circuit for radiation effects self test (CREST)," *IEEE Trans. Nucl. Sci.*, vol. 52, no. 6, pp. 2446-2454, Dec. 2005.

[2] B. Narasimham, V. Ramachandran, B. L. Bhuva, R. D. Schrimpf, A. F. Witulski, W. T. Holman, L. W. Massengill, J. D. Black, W. H. Robinson, D. McMorrow, "On-chip characterization of single event transient pulse widths", *IEEE Trans. on Dev. and Mat. Rel.*, vol. 6, p. 542-549, 2006.

[3] O. A. Amusan, M. C. Casey, B. L. Bhuva, D. McMorrow, M. J. Gadlage, J. S. Melinger, L. W. Massengill, "Laser verification of charge sharing in a 90 nm bulk CMOS process," IEEE Trans. On Nucl. Sci., Vol. 56, No. 6, pp. 3065-3070, December 2009.

[4] S. Jagannathan, T. D. Loveless, Z. Diggins, B. L. Bhuva, S.-J. Wen, R. Wong, L. W. Massengill, "Neutron and alpha particles induced soft error rates for flip-flops at a 40 nm technology node," Accepted for publications at 2011 IRPS Conference.

Measurement of Neutron-induced SET Pulse Width Using Propagation-induced Pulse Shrinking

Jun Furuta∗, Chikara Hamanaka†, Kazutoshi Kobayashi† and Hidetoshi Onodera∗‡
∗Graduate School of Informatics, Kyoto Univesity, ‡JST, CREST
†Graduate School of Science & Technology, Kyoto Institute of Technology

Abstract— We propose a single event transient (SET) pulse width measurement circuit using propagation-induced pulse shrinking on a clock buffer chain. It achieves the resolution of less than 1ps since the target circuit of the buffer chain is directly connected to the pulse capture FFs. Experimental results using the spallation neutron beam accelerated test show SET pulse widths are exponentially-distributed and number of SETs longer than 350 ps are reduced to 9% by inserting tap-cells closely. The SET rate on the clock buffer is 23x smaller than SEU rate on FFs.

I. Introduction

According to process scaling and increase in the clock frequency, not only SEU (Single Event Upset) but also SET (Single Event Transient) become more significant issues[1]. They are collectively called soft error. SEU flip a stored value on SRAMs or flip-flops (FFs), while SET generates temporal pulse on logic gates. To protect FFs from SET pulse, delay element and low-pass filter are commonly used. To optimize LSI reliability and performance, it is important to measure distribution of SET pulse widths. Previous works proposed several SET pulse width measurement circuits and reported its measurement results[2][3][4]. But, there are few measurement results of SET pulse distributions by neutron irradiation[5][6]. Previous work[5] measured neutron-induced SET pulse widths on an inverter chain using a D-latch chain. However, an SET pulse width linearly expands or shrinks as it propagates through the inverter chain[7]. SET pulse widths depend on where a pulse is injected.

In this paper, we propose a SET pulse width measurement circuit according to similar ideas to the process variability monitor[8] and the time-to-digital converter[9]. To measure SET pulse widths, it uses the propagation-induced pulse shrinking on a buffer chain and achieves the resolution of less than 1ps with no dependency where a pulse is injected. We also show accelerated test results of SET pulse widths on a 65 nm process using the proposed circuit. Accelerated tests were carried out by the spallation neutron beam at RNCP(Research Center for Nuclear Physics, Osaka University). The rest of this paper is organized as follows. Section II explains the proposed circuit design and propagation-induced pulse shrinking in detail. Section III shows our neutron-beam experimental setup in RCNP, followed by Section IV which discusses experimental results. Section V concludes this paper.

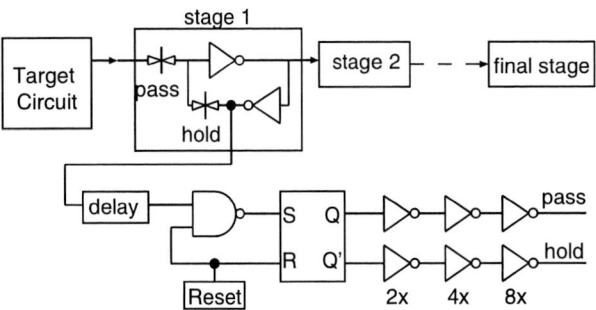

Fig. 1. Conventional SET pulse width measurement circuit[5].

II. SET Pulse Width Measurement Circuit

A. Conventional SET Pulse Width Measurement Circuit

Fig. 1 shows conventional SET pulse width measurement circuit proposed by Narasimham[5]. The target circuit consists of an inverter chain and the measurement circuit consists of a D-latch chain and an SR-latch. If a particle hits on the target circuit and an SET is injected, it propagates to the D-latch chain. The SR-latch detects the SET pulse and then D-latch chain turns to be in the hold state. Therefore, the SET pulse width can be computed by the number of flipped D-latches and delay time of them.

However, an SET pulse width linearly expands or shrinks as it propagates through an inverter chain, since measurement circuit is connected to target circuit in series[7]. Measurement results of the SET pulse width depend on where it is injected. Thus conventional circuit structures cannot measure pulse-width distributions accurately. It is difficult to eliminate the propagation-induced effect, since it is composed of target and capture circuits placed separately. Therefore, we propose a SET pulse width measurement circuit which explicitly utilizes the propagation-induced effect (pulse shrinking) to measure SET pulse width[9]. First, we explain propagation-induced pulse shrinking in the following section.

B. Propagation-induced Pulse Shrinking on a Buffer Chain

As shown in Fig. 2, propagation-induced pulse shrinking is caused by the difference between the fall propagation delay, d_{fall} and the rise propagation delay, d_{rise} of each inverter along the buffer chain. They can be approximated

978-1-4244-9113-1/11 $26.00 © 2011 IEEE

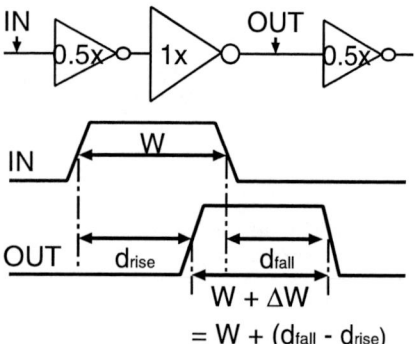

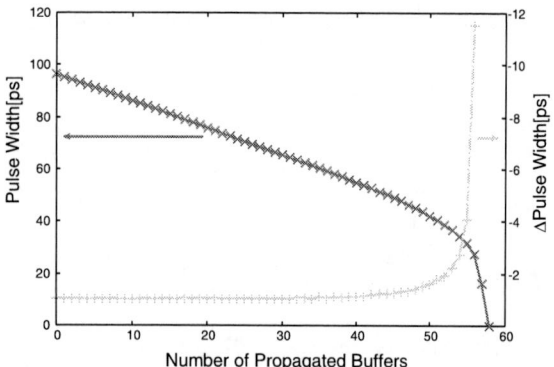

Fig. 2. Concept of propagation-induced pulse shrinking on a buffer chain.

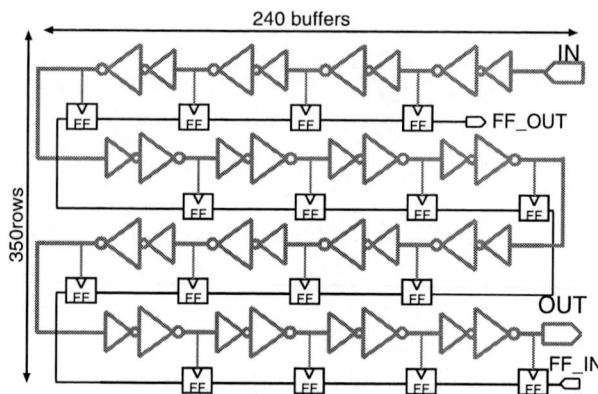

Fig. 4. Proposed SET pulse width measurement circuit.

TABLE I

COMPARISON OF CONVENTIONAL AND PROPOSED CIRCUIT.

	conventional	proposed
connection to target cir.	in series	in parallel
influence of propagation (accuracy)	high (low)	low (high)
measurement method (resolution)	propagation delay (>10 ps)	pulse shrinking (< 1 ps)
measurement of SEU	no	yes
circuit size	small	big

Fig. 3. Simulation result of propagation-induced pulse shrinking.

simplistically as follows[9].

$$d_{\mathrm{fall}} \approx k \left(\frac{C_{\mathrm{g}}}{0.5 g_{\mathrm{mp}}} + \frac{0.5 C_{\mathrm{g}}}{g_{\mathrm{mn}}} \right) \quad (1)$$

$$d_{\mathrm{rise}} \approx k \left(\frac{C_{\mathrm{g}}}{0.5 g_{\mathrm{mn}}} + \frac{0.5 C_{\mathrm{g}}}{g_{\mathrm{mp}}} \right) \quad (2)$$

where g_{mp}, g_{mn} are the transconductance parameters of the 1x inverter, C_{g} is its gate capacitance and k is a factor of proportionality. Then, amount of pulse width variation through a buffer, ΔW can be calculated as follow.

$$\Delta W = d_{\mathrm{fall}} - d_{\mathrm{rise}} = \frac{3}{2} k C_{\mathrm{g}} \left(\frac{1}{g_{\mathrm{mp}}} - \frac{1}{g_{\mathrm{mn}}} \right) \quad (3)$$

Therefore, ΔW can be controlled by PMOS and NMOS sizes of each inverter. If g_{mp} is bigger than g_{mn}, ΔW is negative and pulse width is linearly shrunk as propagating through the inverter chain.

Fig. 3 shows a simulation result of pulse shrinking on a buffer chain. A 100 ps input pulse is linearly shrunk up to 40 ps. Then it vanishes immediately after becoming less than 40 ps. Therefore, if we can retrieve the location where SET is injected and vanishes on a buffer chain, the pulse width can be measured by pulse shrinking with the resolution of ΔW.

C. Proposed SET Pulse Width Measurement Circuit

Fig. 4 shows the proposed SET pulse width measurement circuit. To accurately measure SET pulse widths, each buffer output is directly connected to a clock input of an FF. All FFs

construct a shift register to retrieve stored values of FFs. In order to prevent malfunction of the shift register due to hold time violations, FFs are connected in the reverse direction of the buffer chain. In this structure, to detect where an SET is injected and vanishes on a buffer chain, all FFs are initialize to the stripe pattern as shown in the upper part of Fig. 5. If an SET is injected on the buffer chain, stored values of FFs are shifted from the injected point until SET vanishes as shown in the lower part of Fig. 5. The stripe pattern can detect where SET is injected and vanishes on a buffer chain and measure how many buffers SET propagated through. Therefore, SET pulse widths can be measured by the propagation-induced pulse shrinking using the proposed circuit.

Table I shows comparison of the conventional and proposed circuits. The proposed circuit is less affected by the propagation and achieves high accuracy since the measurement circuit (FFs) is directly connected to the target circuit (buffers). It also achieves high resolution of less than 1ps while that of the conventional circuit is more than 10 ps. It is because the resolution of the conventional circuit is eliminated by inverter propagation delay, while that of the proposed circuit can be changed by transistor sizes as shown Eq. (3). However, it has the drawback of bigger area than the conventional circuit. We can reduce the circuit area by changing insertion interval of FFs from every buffer to two or more buffers.

III. Experimental Setup

Fig. 6 shows a chip micrograph fabricated in 65 nm bulk CMOS process. In order to measure neutron-induced SET pulse widths, the proposed circuit including 161,000 buffers and FFs is implemented in a 1.1×1.3 mm^2 region on a 2×4 mm^2 die. Tap-cells to stabilize potentials of well or bulk are

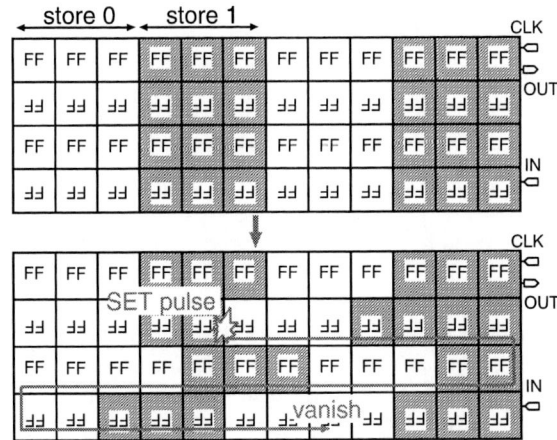

Fig. 5. Initial stripe pattern to detect an SET pulse (3bit stripe version).

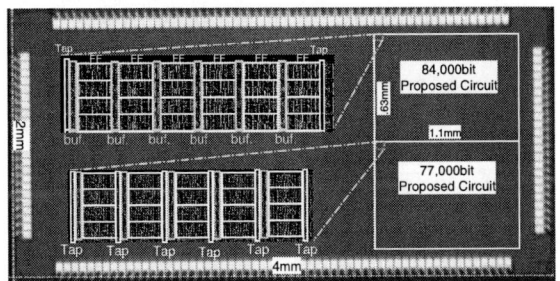

Fig. 6. Chip micrograph and partial layout structure.

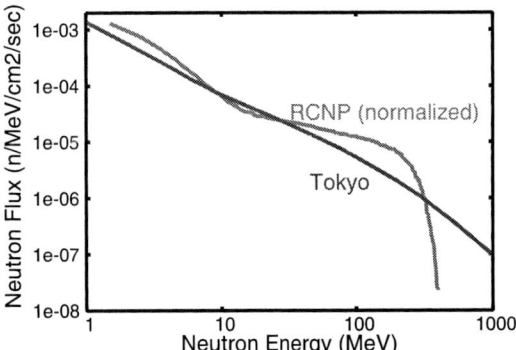

Fig. 7. Neutron spectrum at RCNP.

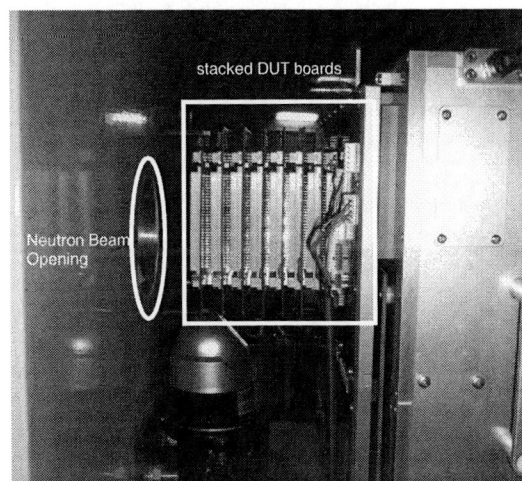

Fig. 8. Neutron-beam opening and stacked DUT boards.

inserted every 28 μm in the first 84,000 bit of the proposed circuit. On the other hand, they are inserted every 5μm during the latter 77,000 bit.

Fig. 7 shows the neutron beam spectrum compared with the terrestrial neutron spectrum at the ground level of Tokyo. The average accelerated factor is 3.8×10^8 in this measurement. We used 5-stacked DUT boards to increase error counts. Each DUT board has four segments, each of which is equipped with a single DUT. Note that input signals are common for every segment, while output signals are independent for 5 DUTs to minimize time for the shift operation. In this measurement, 18 DUTs out of 20 are fully functional. Fig. 8 depicts the neutron-beam opening and the stacked DUT boards. An engineering LSI tester is used to control DUTs and collect shifted error data.

During irradiation, input of the buffer chain is fixed to 1. Since the proposed circuit has no self-trigger circuit to detect SET, we must frequently retrieve stored values of FFs. In this measurement, stored values of all FFs were retrieved every 5 minutes. After finishing retrieving (shifting), all FFs are restored to the initial state of the stripe pattern. Note that we initialize FFs to 20 bit stripe pattern in this measurement.

IV. Experimental Results and Discussions

A. Effective Resolution of the Proposed Circuit

In order to accurately measure SET pulse widths, we must measure the effective resolution (ΔW as shown eq. (3)) of the proposed circuit. To calibrate effective resolution of the proposed circuit, rectangular pulses are injected into the buffer chain. Fig. 9 shows the relationship between the input pulse width and the number of shifted FFs from 3 DUTs. These measurement results are average results of 20 times measurement and each point has about $\pm 5\%$ margin of error. All DUTs have linear dependence between input pulse width and number of FFs. In this result, slopes are different by the tap-cell interval. We assume that it is caused by difference of well resistance.

Fig. 10 shows effective resolutions calculated by the slopes. Effective resolutions are different for each DUT. We expect that it is caused by die-to-die variations. Table II shows simulation results of effective resolutions in corner models. There are big difference between each corner especially SF and FS. It is because that effective resolution, ΔW is determined by difference in transconductances of PMOS and NMOS as shown Eq. (3).

TABLE II

SIMULATION RESULTS OF EFFECTIVE RESOLUTION IN A CORNER MODEL.

corner model	SS	TT	FF	SF	FS
resolution	−1.2 ps	−0.60 ps	−0.89 ps	−3.0 ps	1.1 ps

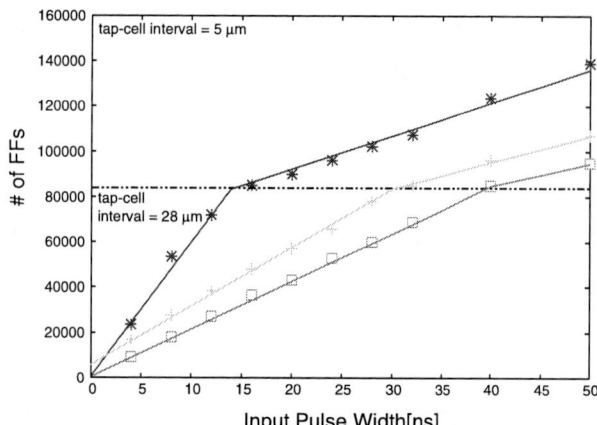

Fig. 9.   Calibration results of 3 DUTs.

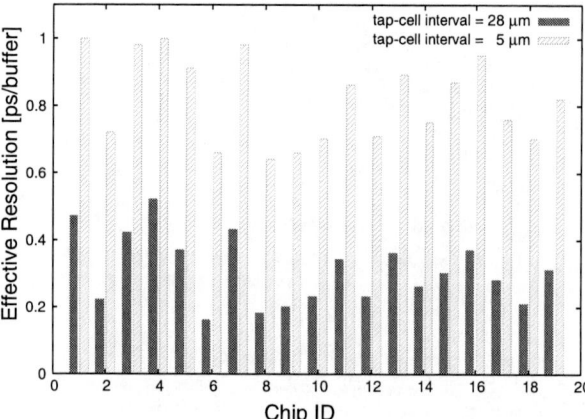

Fig. 10.   Effective resolutions of 18 DUTs

B. Measurement Result of SET Pulse Widths

Fig. 11 shows distribution of SET pulse widths and Table III shows total number of SETs and average pulse widths. Fig. 12 shows a stripe pattern affected by an SET pulse and SEUs. Note that SET pulse vanishes immediately after becoming less than 40 ps in the simulation as shown in Fig. 3. However, this measurement results do not consider this immediate disappearance. SET pulse widths are just calculated by number of FFs shifted by SET and the effective resolutions as shown Fig. 10. In this measurement, shorter SETs have higher probability, which is similar to [6]. SET pulse widths are reduced by inserting tap-cell closely and number of SETs longer than 350 ps is reduced to 9%. This result is similar to [10] and suggests that SET pulse widths are increased by the parasitic bipolar effect. Fig. 13 shows shows distribution of SET pulse widths in a log scale. SET pulse widths on the buffer chain inserted tap-cell every 5 μm are exponentially-distributed, and SET pulse widths up to 350 ps on the buffer chain inserted tap-cell every 28 μm are also exponentially-distributed. We assume parasitic bipolar effects prolong the pulse width over 350 ps in the case of the every-28 μm tap-cell insertion.

In this measurement, multiple SETs (MSETs) can be

TABLE III
TOTAL NUMBER OF SET AND AVERAGE SET PULSE WIDTHS.

tap-cell interval	# of SET[n/Mbuffer/h]	Avg. width[ps]
28μm	18.7	180
5μm	17.2	130

TABLE IV
TOTAL NUMBER OF SET, MSET AND SEMT.

tap-cell interval	SET	MSET	SEMT
28μm	123	4	7
5μm	110	6	12

observed in different regions because the proposed circuit has 161,000 buffers and FFs to capture SETs. In addition, we also observed SEMTs(Single Event Multiple Transient) which indicate two or more SETs are caused by a single particle hit. Table IV shows total number of SET, MSET and SEMT. We regard two SETs as an SEMT when they started from the vertically-adjacent buffers as shown in Fig. 14. In this result, number of MSET is only 4 and SETs of MSET were surely injected on sufficiently separated buffers.

As shown in Fig. 12, the proposed circuit can simultaneously measure SEU on the FF. Table V shows SET rate on a buffer and SEU rate on a FF. This result shows SET rate on a buffer is 23x smaller than SEU rate on a FF. For SEU hardened circuit such as DICE latch[11], SET cannot be ignored, especially SET injected on the clock buffer.

V. Conclusion

We propose an SET pulse width measurement circuit using propagation-induced pulse shrinking on a clock buffer chain on a shift register. Pulse shrinking reduces an SET linearly as it propagates through the clock buffer chain. Pulse shrinking is a critical issue to measure SET pulse width with accuracy in the conventional circuit. On the contrary, the proposed circuit utilizes the pulse shrinking to measure the SET pulse.

TABLE V
NUMBER OF SEUs AND SETs BY THE NEUTRON IRRADIATION.

SEU[n/Mbit/h]	SET[n/Mbuffer/h]	SET/SEU[%]
412	18.0	4.4

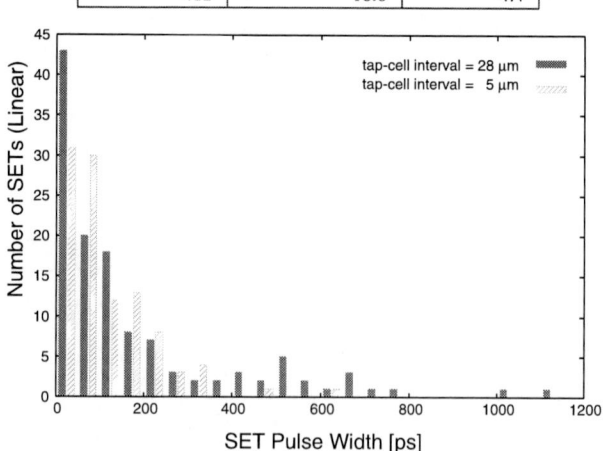

Fig. 11.   Distribution of SET pulse width by the neutron irradiation.

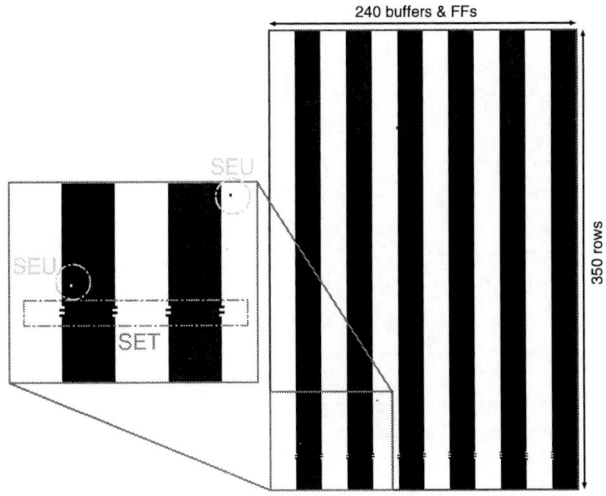

Fig. 12. Measured stripe pattern affected by an SET pulse and two SEUs.

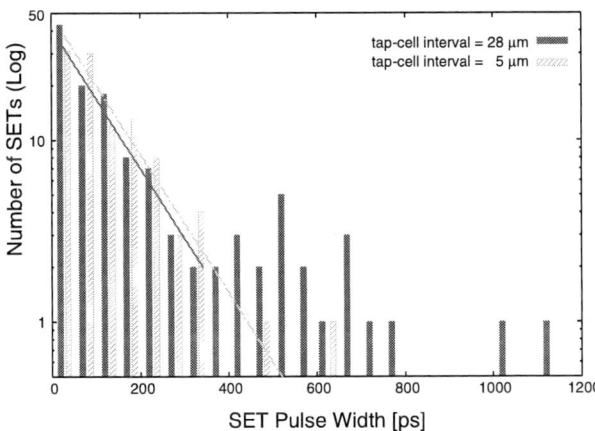

Fig. 13. Distribution of SET pulse width by the neutron irradiation (Number of SETs in log scale).

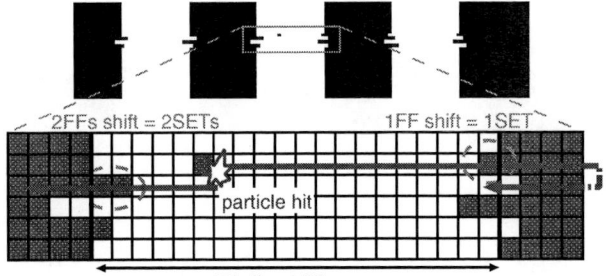

Fig. 14. a measurement result of a Single Event Multiple Transient (SEMT).

fabricated in the chip fabrication program of VLSI Design and Education Center(VDEC), the University of Tokyo in collaboration with STARC, e-Shuttle, Inc., and Fujitsu Ltd.

REFERENCES

[1] P. Shivakumar, M. Kistler, S. Keckler, D. Burger, and L. Alvisi, "Modeling the Effect of Technology Trends on the Soft Error Rate of Combinational Logic," in *Dependable Systems and Networks, 2002. DSN 2002. Proceedings. International Conference on*, 2002, pp. 389 – 398.

[2] P. Eaton, J. Benedetto, D. Mavis, K. Avery, M. Sibley, M. Gadlage, and T. Turflinger, "Single Event Transient Pulsewidth Measurements Using a Variable Temporal Latch Technique," *Nuclear Science, IEEE Transactions on*, vol. 51, no. 6, pp. 3365 – 3368, Dec. 2004.

[3] Y. Yanagawa, K. Hirose, H. Saito, D. Kobayashi, S. Fukuda, S. Ishii, D. Takahashi, K. Yamamoto, and Y. Kuroda, "Direct Measurement of SET Pulse Widths in $0.2\mu m$ SOI Logic Cells Irradiated by Heavy Ions," *Nuclear Science, IEEE Transactions on*, vol. 53, no. 6, pp. 3575 –3578, Dec. 2006.

[4] S. Jagannathan, M. J. Gadlage, B. L. Bhuva, R. D. Schrimpf, B. Narasimham, J. Chetia, J. R. Ahlbin, and L. W. Massengill, "Independent Measurement of SET Pulse Widths from N-Hits and P-Hits in 65-nm CMOS," *Nuclear Science, IEEE Transactions on*, 2010.

[5] B. Narasimham, M. Gadlage, B. Bhuva, R. Schrimpf, L. Massengill, W. Holman, A. Witulski, X. Zhu, A. Balasubramanian, and S. Wender, "Neutron and Alpha Particle-induced Transients in 90 nm Technology," in *Reliability Physics Symposium (IRPS), 2010 IEEE International*, May 2008, pp. 478 –481.

[6] H. Nakamura, K. Tanaka, T. Uemura, K. Takeuchi, T. Fukuda, and S. Kumashiro, "Measurement of Neutron-induced Single Event Transient Pulse Width Narrower than 100ps," in *Reliability Physics Symposium (IRPS), 2010 IEEE International*, May 2010, pp. 694 –697.

[7] M. Gadlage, J. Ahlbin, B. Narasimham, V. Ramachandran, C. Dinkins, N. Pate, B. Bhuva, R. Schrimpf, L. Massengill, R. Shuler, and D. Mc-Morrow, "Increased Single-Event Transient Pulsewidths in a 90-nm Bulk CMOS Technology Operating at Elevated Temperatures," *Device and Materials Reliability, IEEE Transactions on*, vol. 10, no. 1, pp. 157 –163, March 2010.

[8] T. Iizuka, T. Nakura, and K. Asada, "Buffer-Ring-Based All-Digital On-Chip Monitor for PMOS and NMOS Process Variability and Aging Effects," in *Design and Diagnostics of Electronic Circuits and Systems (DDECS), 2010 IEEE 13th International Symposium on*, April 2010, pp. 167 –172.

[9] P. Chen, S.-L. Liu, and J. Wu, "A CMOS Pulse-Shrinking Delay Element For Time Interval Measurement," *Circuits and Systems II: Analog and Digital Signal Processing, IEEE Transactions on*, vol. 47, no. 9, pp. 954 –958, Sep. 2000.

[10] B. Narasimham, B. Bhuva, R. Schrimpf, L. Massengill, M. Gadlage, T. Holman, A. Witulski, W. Robinson, J. Black, J. Benedetto, and P. Eaton, "Effects of Guard Bands and Well Contacts in Mitigating Long SETs in Advanced CMOS Processes," *Nuclear Science, IEEE Transactions on*, vol. 55, no. 3, pp. 1708 –1713, june 2008.

[11] P. Hazucha, T. Karnik, S. Walstra, B. Bloechel, J. Tschanz, J. Maiz, K. Soumyanath, G. Dermer, S. Narendra, V. De, and S. Borkar, "Measurements and Analysis of SER-tolerant Latch in a 90-nm Dual-VT CMOS Process," *Solid-State Circuits, IEEE Journal of*, vol. 39, no. 9, pp. 1536 – 1543, Sep. 2004.

It achieves much higher accuracy of 1 ps resolution. It is because that a pulse is shrunk by 1 ps thorough one stage of the clock buffer, while the conventional circuit utilizes the delay time of logic gates which is usually over 10 ps.

Experimental results by the neutron irradiation show that SET pulse widths are exponentially-distributed and they are reduced by inserting tap-cell closely. Comparing the pulse width distributions of the closely-placed (every 5 μm) tap-cell with that of the sparsely-placed (ever 28 μm) one, the average pulse width is reduced from 180 ps to 130 ps and rate of SET pulses longer than 350 ps is reduced to 9% by it. Tap-cell density is very important thing to mitigate SET. SET rate on a buffer is 23x smaller than SEU rate on a FF. This results shows that if SEU hardened circuit has non-redundant clock buffer, its soft error resilience is only 23x better than that of the conventional FF.

ACKNOWLEDGMENT

The authors would like to thank to Prof. K. Hatanaka at RCNP and all the other RCNP staffs for our neutron-beam experiments. The VLSI chip in this study has been

Multicenter comparison of alpha particle measurements and methods typical of semiconductor processing

Jeffrey D. Wilkinson, Brett M. Clark, Richard Wong, Charles Slayman, Barry Carroll, Michael Gordon, Yi He,
Olivier Lauzeral, Keith Lepla, Jennifer Marckmann, Brendan McNally, Philippe Roche, Mike Tucker, Tommy Wu

Abstract— Alpha counting measurement methods have been widely used in the semiconductor industry for many years to assess the suitability of materials for semiconductor production and packaging applications. Although a number of published articles describe aspects of this counting, a multicenter, comparative trial has not been carried out to assess the methodological accuracy of current methods. This paper reports on experience with a 9 center, international, round-robin style trial using a shared set of samples to quantify variability in alpha emission measurements. Four samples representing low and ultralow alpha materials were counted by each participating lab in a blinded trial. The consensus mean emissivity for low alpha material was estimated as 30.9 khr^{-1}-cm^{-2} with a range from 20.2 to 45.5, less than half of which can be attributed to counting uncertainty or other known sources of error. A strong correlation for replicate measurements within a lab was also observed supporting the conclusion that there are systematic variations in equipment or calibration among labs. Eleven of 23 measurements of ultralow alpha materials were within 1 standard deviation of the consensus mean and 7 were at or below background. The high level of counting uncertainty for these measurements is thought to be sufficient to mask any systematic variation similar to the low alpha observations. Comparison of the reported values with a standard calculation demonstrates that there are also differences in the interpretation of the values reported for emissivity and error, underscoring the need for careful interpretation of results.

Keywords-component; soft errors, alpha particles, semiconductor device reliability, device packaging

I. INTRODUCTION

Radioactive impurities in integrated circuit (IC) packaging materials are an important contributor to a circuit's soft error

J.D. Wilkinson, Medtronic, CRDM Device Technology, Mounds View, MN, USA, jeff.wilkinson@medtronic.com, 763-526-0483. Inquiries related to the alpha consortium may be directed to Jeff Wilkinson, Brett Clark, Richard Wong or any of the other authors.
B.M. Clark, Honeywell, Spokane, WA, Brett.Clark@Honeywell.com, 509-252-8716 – R. Wong, Cisco Systems – C. Slayman, Ops A La Carte – B. Carroll, Freescale Semiconductor Inc. – M. Gordon, IBM TJ Watson Research Center, Yorktown Heights, NY USA, gordonm@us.ibm.com. – Y. He, Intel Corporation, CH5-157, ATTD, Chandler, AZ 85226 USA. – O. Lauzeral, iRoC Technologies – K. Lepla, Teck Metals Ltd. – J. Marckmann, Medtronic Tempe Campus, Tempe, AZ – B. McNally, XIA LLC – P. Roche, STMicroelectronics – M. Tucker, Alpha Sciences, Inc. – T. Wu, SGS Taiwan Ltd, Material and Engineering Lab, Polymer Lab, Kauhsiung, Taiwan, tommy.wu@sgs.com, +886-7-3012121.
This work was sponsored by funding from Cisco Systems, Medtronic, and Honeywell, along with donations of counter time from all of the participant companies.

rate (SER), particularly as the process geometry and operating voltage are scaled to smaller feature sizes. Multiple reports have stressed the importance of ultralow alpha materials and the potential for the SER to be dominated by alpha emissions even when these materials are utilized [1, 2].

In the semiconductor industry, alpha particle emission levels are most commonly measured using sensitive, large area gas proportional detectors. Through careful sample preparation and control of detector parameters, it is thought to be possible to measure alpha emission rates of materials down to 1 khr^{-1}-cm^{-2} (1 emitted particle per square centimeter per 1000 hours). For many alpha emitting isotopes this corresponds to concentrations in parts per trillion (PPT) or less [1, 2]. For short half-life isotopes, such as ^{210}Po the corresponding concentration is as low as a few parts in 10^{18}. At these levels it is generally impractical to identify the particular isotope or combination of isotopes responsible for the alpha emission by chemical analysis. Clark [3] reports on an attempt to match uranium and thorium concentrations measured using glow discharge mass spectroscopic (GDMS) analysis with calculated and measured values of alpha emission. In all cases, the calculated emission rate based on the GDMS analysis underestimates the measured alpha emission flux by a factor between 4.2×10^2 and 1.5×10^5. Although assumptions of the particular alpha emitters and their secular equilibrium status have been made in previous work (see, for example [2, 4]) it is the authors' opinion that this has not been adequately demonstrated for these materials and would be particularly challenging due to the extremely low decay rates. This position is consistent with the guidance provided in the soft error test standard JESD89A ([5] page 76). To further complicate matters, there is no accepted flux standard for a large area alpha particle source with an emission rate within 1000X of the expected activity level. Likewise, there is no published standard for low-level alpha measurements, sample handling or laboratory quality control procedures, although a JEDEC standard is in preparation [6]. To better understand and quantify alpha counting measurements a loose consortium has been formed to coordinate joint projects. This paper reports on the results of the consortium's first project which is an effort to quantify the range of measurements made by different labs. This study did not attempt to separate different potential sources of variation, preferring to let "the chips fall where they may."

Nine participants representing Alpha Sciences; Freescale Semiconductor; Honeywell Electronic Materials; IBM TJ Watson Research Center; Intel; iRoC Technologies; Medtronic, Inc.; SGS Taiwan Ltd; Teck Metals and XIA LLC (iRoC Technologies and Alpha Sciences participated jointly) made measurements for this study. Each participant agreed to use their normal laboratory procedures to make at least 4 measurements of samples representing a low alpha (LA) and an ultralow alpha (ULA) material. Although one previous publication has presented counter efficiency measurements made at multiple sites [7], this is the first study to directly compare alpha counting measurements at multiple sites at emission levels pertinent to the semiconductor industry.

All of the counters used in this study detected ionization from alpha particles as they passed through a gas-filled counting chamber. For the proportional mode counters (Alpha Science and Ordela) the gas was a combination of purified argon with a small fraction of methane used to quench the ionization multiplication effect. The ionization mode counter (XIA) used purified argon only. A large electric field is applied across the chamber so that charge carriers (Ar+ and e-) created by the passage of an ionizing particle drift to the high voltage electrodes in the chamber to be collected, amplified and counted as discrete events. The counter records the number of detected events (N) and the length of a counting period (T, in hours). Due to the stochastic nature of radioactive decay, the count rate, N/T, has a Poisson distribution with a mean value of N/T. When N is greater than 20, which is true for all measurements in this study, the Poisson distribution is accurately approximated by the normal distribution with mean value N/T and standard deviation equal to $\sqrt{N}/T$. This random variation in count values is referred to as "counting uncertainty" to differentiate it from other systematic errors such as equipment calibration error or sample area measurement.

All alpha counters have a "background" rate caused by alpha emissions from the counting chamber walls, alpha emissions from the sample tray, electronic noise in the detector and other sources. During sample counting, with the sample in the counting chamber, both the sample emission and the background events are present so the resulting count values of N and T are subscripted with "S+B", reflecting the contribution both from the sample and the background. Depending on the counter design the background events may either be discriminated from sample emissions and rejected during counting, or they may be estimated by measuring the count rate when no sample is present and subtracting this value from the count rate when sample and background are both present. Counting values made without a sample in the chamber are subscripted with "B" referring to the background. The corresponding count rates for these cases are simply N_{S+B}/T_{S+B} and N_B/T_B, respectively. All counters also have an associated efficiency (ϵ), empirically determined during calibration and primarily related to the counter geometry, which expresses the fraction of counts recorded vs. those emitted from the sample. The reader is referred to [8] for a thorough discussion of counters and counting statistics.

II. METHODS

Due to the sensitive nature of these measurements, the study was designed to protect the identity of each participant and only publish values identified by anonymous laboratory codes. It was decided that this lack of transparency does not undermine the primary objective of the study which is to quantify the variability of alpha emissivity measurements as typically employed by manufacturers at all levels of the semiconductor industry. Author C.S. and his company, Ops A La Carte, were contracted to execute confidential disclosure agreements with all participating labs and manage all data reports. All counting reports and the results of the final survey were gathered by C.S. during the experiment. Identifying information was removed, coded identifiers were assigned to each participant and the complete, disguised dataset was provided for analysis at the conclusion of the data collection phase. Some particular patterns that could be inferred from the full data set necessitated that the coded identifiers be assigned independently for the LA samples and for each ULA sample to protect confidentiality. For similar reasons the survey results are only reported in summary rather than being attributed to specific lab codes. The records of the original data are maintained by C.S. for reference. Each participant's lab codes and results were sent to participants individually at the conclusion of the data collection.

Counting samples for this study were prepared representing an ultralow alpha (ULA) material with an emissivity of less than 2 khr^{-1}-cm^{-2} and low alpha (LA) material with an emissivity from 2 to 50 khr^{-1}-cm^{-2}. Due to the limited time available for collecting data and the large number of participants, 2 matched samples of the ULA and LA material were prepared. These samples are referred to as ULA-1, ULA-2, LA-1 and LA-2. Low alpha (LA) samples were prepared by melting 135 g of commercial grade aluminum 6061 alloy (170 khr^{-1}-cm^{-2}) and 435 g of 99.9995% pure aluminum (0.3 khr^{-1}-cm^{-2}) at 700 C in a nitrogen atmosphere. After solidification, all surfaces were machined to remove oxidation or scale and the ingots were rolled to a thickness of 1 to 2 mm, and then sheared into appropriately sized sections to fit the counting chambers. All pieces were cleaned with caustic and acid solutions followed by deionized water rinse and high purity isopropanol wipe. The resulting samples were stored in bags backfilled with nitrogen. Each LA sample was provided as a set of 4 pieces totaling more than 1000 cm^2.

Ultra low alpha (ULA) samples were prepared from 99.995% pure titanium bars that were rolled to 0.5 mm, cut, cleaned and then stored similar to the LA samples. The ULA samples were narrower and required 8 pieces to provide adequate area of more than 1000 cm^2.

To confirm emission stability and a reasonable match between replicates, all 4 samples were counted multiple times over a period of several weeks prior to the start of data collection and again once at the end by B.M.C using a common protocol and equipment. Pre- and post- measurement counting results are provided in table 1. Values for emissivity are in percent, normalized to the mean values of the LA and ULA measurements. This choice was made since there is no available counting standard to provide an absolute calibration

978-1-4244-9113-1/11 $26.00 © 2011 IEEE

TABLE I: Pre- and Post- data collection measurements to assess the stability and matching of sample emissivity.

| Sample | Relative Emissivity | | Time (days) |
	Pre-	Post-	
LA-1	92 ± 4	93 ± 6	224
LA-2	115 ± 7	101 ± 4	207
ULA-1	150 ± 60	100 ± 80	238
ULA-2	100 ± 60	40 ± 60	239

and these measurements are only to check the stability and match among the samples. The "Time" column records the number of days between the pre- and post- emissivity measurement. Although the values in table 1 might suggest some emissivity mismatch or instability, particularly in the LA samples, the reader is referred to the results reported for all paired LA samples below.

Each sample was labeled with a coded identifier that was changed prior to being sent to each lab for counting, blinding the participant to the particular sample's identity. The label also instructed the participant to count the reverse side of the sample to ensure that only 1 surface was used. Samples were enclosed and heat sealed in a commercial polyethylene bag material before being shipped for counting. Participants were told if the sample they received was LA or ULA so that they could adjust their procedures, if desired (it is common industry practice to adjust the counting time or sample area based on the expected level of activity). No other specification was made with regard to procedure so that the full range of methodological variation would be sampled. At the conclusion of a measurement the participant filled out a data collection form and returned it to C.S. where identifying information was removed. At the end of the data collection phase the full, disguised data set was provided for analysis. A smaller and more fully disguised subset of data was provided to J.D.W. early in the collection phase to confirm the procedures.

By design, the order of sample counting was not randomized, although it was not disclosed to the participants. The objective of the counting order was to have samples LA-1 and ULA-1 sent repeatedly to a subset of participants, while

samples LA-2 and ULA-2 were sent to each participant once. Due to time constraints it was not possible to completely meet this schedule. The participant code assignment was a result of patterns discernible in the partial data set and the requirement to protect the participant's confidential information.

Following data collection a voluntary survey was sent to each participant with the instructions that no proprietary information was to be disclosed. Each question included an option for "no response" that was used to indicate these purposeful omissions. Surveys were gathered by the C.S., identifying information was removed and the remaining responses were summarized.

III. RESULTS AND OBSERVATIONS

Table 2 summarizes the number of participants measuring the different samples and their reported values. The JEDEC calculated values will be discussed in detail below but are included here for comparison. Each sample was sent to a participant for counting one or more times and in several cases more than one count was performed by a participant on a particular round. The number of measurement rounds for each sample and the total number of measurements are both detailed.

Table 3 details all values from the data collection in the "Reported Values" columns. Each participant's readings have been assigned a code in the format "SS.PPNa" to identify the sample; participant, order, for paired readings; and participant order, when multiple readings were made by a participant. Sample codes are L1, L2, U1 or U2 for samples LA-1, LA-2, ULA-1 and ULA-2, respectively. The identifying code for LA samples is a single character while ULA samples use a number-letter or letter-number code for ULA-1 and ULA-2, respectively. For participants that made measurements of the same sample at different times during the study, the participant code is followed by "1" or "2," to denote the order that the readings were made. Finally, for participants that provided multiple measurements when they received a sample a lower case alphabetic character is assigned to differentiate them. The order of these multiple readings, other than those for U1.4G1a and U1.4G1b, was not available from the collected data so temporal order should not be assumed.

Data presented in table 3 have been appropriately censored, as defined by each participant's counting procedure, to remove unrepresentative data at the beginning of a counting cycle when contaminants are being flushed from the counting chamber. The "Reported Values" columns are directly as reported on the data forms, with the following exceptions:

- All emissivity and error values have been uniformly reported in units of khr^{-1}-cm^{-2}, although other units may have been used on the data collection forms. The same number of significant digits and any other notations have been retained.

- The reporting of background information was not uniform due to variations in participant procedures. Although the data collection form requested values for N_B and T_B, some reports instead provided N_B/T_B and T_B. In these cases, the value of N_B was inferred from the data. For participants

TABLE II: Experimental summary of the collected data. Emissivity values are in khr^{-1}-cm^{-2}.

| | LA | | ULA | |
	1	2	1	2
Number of participants counting this sample	6	9	6	9
Samples submitted to participant	10	9	10	8
Data forms submitted	10	9	13	10
Reported Values				
Mean / median reported mean emissivity	31.2	30.6	0.8	0.7
	29.15	29.8	0.8	0.6
Minimum / maximum reported mean emissivity	20.2	20.3	0.22	0.0
	45.3	45.7	1.6	0.812
Minimum / maximum reported upper bound	—	—	0.19	<0.3
			0.8	1.4
JEDEC Calculated Values				
Consensus mean / median α	31.0	30.6	0.7	0.8
	29.2	29.8	0.7	0.1
Minimum / maximum α	20.2	20.2	0.0	0.0
	45.1	45.5	1.3	0.8
Minimum / maximum counting uncertainty (1 σ)	1.8%	1.8%	20%	56%
	10. %	13. %	67%	210%

978-1-4244-9113-1/11 $26.00 © 2011 IEEE

reporting N_B and T_B the calculated value of N_B/T_B is reported.

- Participants were requested to report emissivity as an estimated value or upper bound, at their discretion. These

are denoted with an upper case 'E' or 'B' when reported.

- All inferred or calculated values are shown in red italics.

The precise definition of the reported emissivity and error values are self-defined by the participant, as is the choice of

TABLE III: SUMMARY OF ALL DATA REPORTED.

		REPORTED VALUES										**JEDEC**			
Lab	Reading	A (cm2)	ε	Ns+b	Ts+b (hrs)	Nb	Tb (hrs)	Nb/Tb	Est/UB	Emiss-ivity	Error	α	σ	LOD90%	Notes
	**Sample LA-1 (α_{mean}=31.0 khr^{-1} cm^{-2}, 20.2 < α < 45.1, 1.8% < σ < 10%)**
A	L1.A	804.1	88%	2120	60	548	159	3.45	E	45.3	1.1	45.1	1.1	1.8	
B	L1.B1	1000	84%	901	48	86	48	1.8	E	20.2	0.8	20.2	0.8	1.3	
B	L1.B2	1000	84%	1024	48	149	48	3.1	E	21.7	0.8	21.7	0.8	1.4	
J	L1.J1	706.86	93.96%	597	20.5			0	E	44.945	1.783	43.8	1.8	2.9	
J	L1.J2	706.86	93.84%	157	6			0	E	39.449	3.05	39.4	3.1	5.2	
N	L1.N1	848	90%	1401	48	173	48	3.60	e	33.5	2.2	33.5	1.1	1.8	
N	L1.N2	848	90%	1413	48	183	48	3.81	e	33.6	2.2	33.6	1.1	1.8	
Q	L1.Q	866	84%	1947	92	320	92	3.48	U	24.4	0.3	24.3	0.7	1.2	
Y	L1.Y1	848	90%	2324	100	431	100	4.31	E	24.8	0.7	24.8	0.7	1.1	
Y	L1.Y2	840	90%	3322	150	602	150	4.01	E	24	0.55	24.0	0.6	0.9	
	**Sample LA-2 (α_{mean}=30.6 khr^{-1} cm^{-2}, 20.2 < α < 45.5, 1.8% < σ < 13%)**
A	L2.A	546.48	88%	4186	166	403	120	3.36	E	45.7	0.9	45.5	0.9	1.4	
B	L2.B	1000	84%	1122	48	149	48	3.1	E	24.1	0.9	24.1	0.9	1.5	
D	L2.D	754	84%	1715	87	375	87	4.31	E	24	0.8	24.3	0.8	1.4	
J	L2.J	706.86	94.24%	110	4			0	E	41.28	3.82	41.3	3.9	6.5	
N	L2.N	852	90%	1452	48	183	48	3.81	e	34.5	2.2	34.5	1.1	1.8	
Q	L2.Q	548	84%	1199	92	344	92	3.74	U	20.3	0.3	20.2	0.9	1.5	
V	L2.V	1065	84.20%	2902	95	1925	500	3.85	e	29.8	0.8	29.8	0.6	1.0	
Y	L2.Y	848	90%	2248	100	602	150	4.01	E	24.2	0.7	24.2	0.7	1.1	
Z	L2.Z	950	85%	4691	164	462	164	2.82	e	31.9	0.9	31.9	0.5	0.9	
	**Sample ULA-1 (α_{mean}=0.7 khr^{-1} cm^{-2}, -0.5 < α < 1.3, 20% < σ < 67%)**
1A	U1.1A1	940	84%	427	87	369	87	4.24	E	0.8	0.4	0.8	0.4	0.7	
1A	U1.1A2	940	84%	369	87	341	87	3.92	E	0.4	0.4	0.4	0.4	0.6	
2E	U1.2E	900	85.60%	444	79	469	100	4.69	e	1.6	0.4	1.2	0.4	0.7	
4G	U1.4G1a	799.8	90%	417	100	431	100	4.31	U	0.19	0.38	-0.2	0.4	0.7	1
4G	U1.4G1b	799.8	90%	417	100	602	150	4.01	E	0.22	0.38	0.2	0.4	0.6	1
4G	U1.4G2	757.8	90%	328	150	123	67	1.84	E	0.514	0.24	0.5	0.3	0.5	
5H	U1.5H1	706.86	89%	30	66			0	e	0.722	-0.131/-0.275	0.7	0.1	0.2	
5H	U1.5H2	706.86	90.64%	21	40			0	E	0.819	+.170/-0.253	0.8	0.2	0.3	
8U	U1.8U1a	505.8	81%	446	138	948	322	2.94	E	0.6	0.4	0.7	0.4	0.7	2
8U	U1.8U1b	528	90%	396	138	803	326	2.46	E	1	0.4	0.9	0.4	0.6	2
8U	U1.8U1c	649.8	81%	491	162	803	326	2.46	E	1.1	0.3	1.1	0.3	0.5	2
8U	U1.8U2	633	81%	335	108	565	234	2.41	E	1.3	0.4	1.3	0.4	0.6	
9F	U1.9F	806.2	84%	363	116	401	116	3.46	U	0.8		-0.5	0.4	0.6	
	**Sample ULA-2 (α_{mean}=0.3 khr^{-1} cm^{-2}, -0.6 < α < 0.8, 56% < σ < 210%)**
A5	U2.A5	825	84%	370	87	409	87	4.7	U	< 1	0.4	-0.6	0.5	0.8	
C6	U2.C6	1000	84%	164	46	156	52	3	E	0.7	0.4	0.7	0.4	0.7	
H2	U2.H2	1000	85%	932	332	1006	332	3.03	U	< 0.3	0.3	-0.3	0.2	0.3	
L0	U2.L0	818	90%	214	48	223	48	4.65	u	1.4	1.1	-0.3	0.6	1.0	
M9	U2.M9	819.5	90%	162	70	123	67	1.84	E	0.65	0.33	0.6	0.3	0.5	
N4	U2.N4	706.86	92.70%	25	48			0	E	0.812	+.156/-.395	0.8	0.2	0.3	
P8	U2.P8a	537.4	90%	378	113	743	240	3.10	E	0.6	0.4	0.5	0.4	0.7	2
P8	U2.P8b	538.6	81%	326	113	692	240	2.88	E	0.0	0.4	0.0	0.4	0.7	2
P8	U2.P8c	583.2	90%	404	120	872	265	3.29	E	0.1	0.3	0.1	0.4	0.6	2
Q1	U2.Q1	828	84%	276	92	281	92	3.05	U	0.8		-0.1	0.4	0.6	

Notes

1. Reading U1.4G1a is reported as an upper bound and includes a note that the background data was taken before sample counting for 100 hours. Reading U1.4G1b is reported as an estimated value and background data was taken after counting for 150 hours. The count data is identical for these measurements.
2. Notes on the data collection form indicate that various combinations of sample pieces and counters were used for these measurements.

reporting an estimated value or an upper bound. Three columns have been added to right hand side of table 3 for comparison with these data. The columns labeled "α" and "σ" are calculated from the counting data (N_{S+B}, N_B) and parameters (T_{S+B}, T_B, A and ε) according to the definition in the proposed JEDEC counting standard [6]. These values are the expected value for the emissivity, in khr^{-1}-cm^{-2}, and the counting uncertainty expressed as the standard deviation in the same units. The "$LOD_{90\%}$" column, also from the proposed JEDEC standard, is the "limit of detection" expressed at the 90% confidence level. This value is the minimum emissivity required from a sample so that it can be discriminated from the background at the 90% confidence level. The $LOD_{90\%}$ is determined by the choice of counting parameters and is proportional to the counting uncertainty. Specifically, the definitions from the JEDEC proposal are:

$$\alpha = \frac{\dfrac{N_{S+B}}{T_{S+B}} - \dfrac{N_B}{T_B}}{A\,\varepsilon} \times 1000, \tag{1}$$

$$\sigma = \frac{\sqrt{\dfrac{N_{S+B}}{T_{S+B}^{2}} + \dfrac{N_B}{T_B^{2}}}}{A\,\varepsilon} \times 1000, \quad \text{and} \tag{2}$$

$$LOD_{90\%} = 1.64\,\sigma. \tag{3}$$

A. Low alpha measurements

Figure 1 presents the emissivity measurements for the LA samples, grouped by participant.

It is immediately apparent that each lab provides consistent results for its own measurements, independent of the sample or measurement order. None of the 19 measurements are within 1 σ of the consensus mean value of 30.9 khr^{-1}-cm^{-2}, indicated by the purple dashed line, and only 2 (L2.V and L2.Z) are within 2 σ.

Participants B, J, N and Y each measured sample LA-1 twice with an average of 69 days between readings. When considered pair-wise for the first vs. the second measurement (e.g., L1.B1 vs. L1.B2) the average deviation between samples is -0.2 σ, with a range of -1.8 to +1.2 σ, supporting the assumption that sample LA-1's emissivity is stable over time these measurements were made. These readings may also be combined with the pre- and post- data collection readings made by B.M.C (table 1) for 5 paired measurements. In this case the resulting average deviation is 0.0 σ and the range remains the same, again supporting an assumption of stability.

In a similar analysis participants B, J, Q, N and Y made measurements of both LA-1 and LA-2. By averaging the multiple measurements of LA-1 made by B, J, N and Y a set of 5 pair-wise comparisons may be made for LA-1 vs. LA-2. The

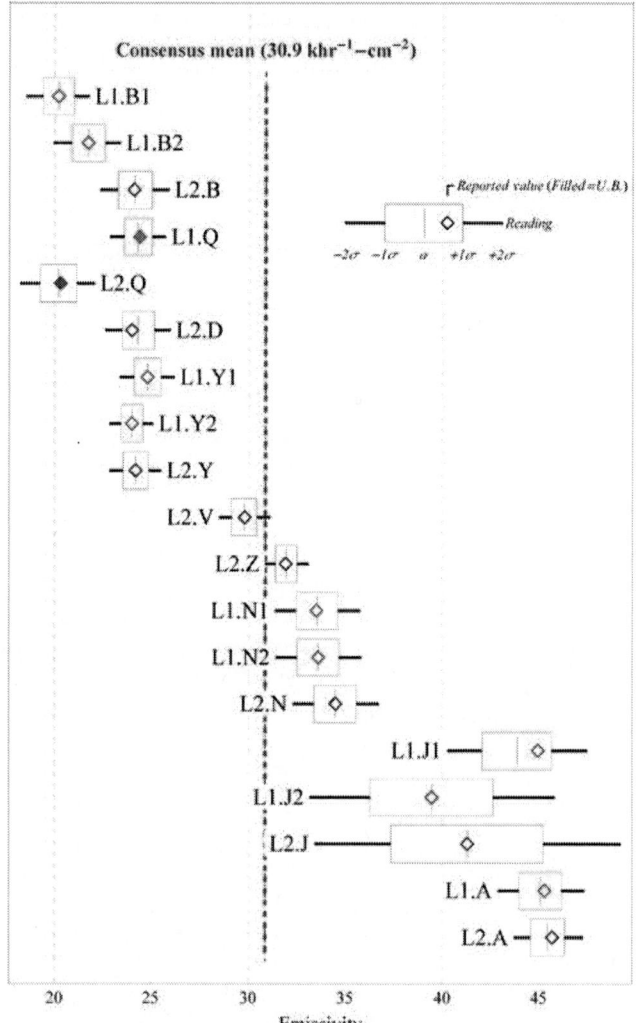

Figure 1. Emissivity measurements for LA-1 and LA-2 showing more than a 2X range in the estimated values. The key for interpreting each measurement is shown in the figure. Filled diamonds represent values reported as upper bounds and empty diamonds are reports as estimated values.

average deviation for these readings is 0.0 σ with a range from -4.4 σ to +3.6 σ, supporting the assumption that sample LA-1 and LA-2 are matched. These measurements may also be augmented with the readings from table 1 resulting in an average deviation of 0.1 σ and no change in the range.

B. Ultralow alpha measurements

The large counting uncertainty and mixture of upper bounds vs. estimated values in these reported results makes the reported emissivities very difficult to compare directly. All of the values are available in table 3 for the interested reader and some commentary is provided in the discussion in this regard. The remainder of the quantitative discussion therefore focuses on the calculated values for α and σ determined from the more objective reports of counts and time.

Figure 2 presents the 13 measured values for emissivity (α) from 6 participants for ULA-1 and Fig. 3 presents 10 measurements from 8 participants for ULA-2. The purple dashed line indicates the consensus mean values of 0.6 and 0.2 khr^{-1}-cm^{-2} for ULA-1 and ULA-2, respectively. Paired measurements of sample ULA-1 were made by participants 1A, 4G, 5H and 8U, separated by an average of 93 days. Participants 2E and 9F each contributed a single reading for ULA-1. Eight of the 9 participants were able to provide measurements for ULA-2.

Six of 13 measurements for ULA-1 and 5 of 10 for ULA-2 include the consensus mean within the 1σ range of their counting uncertainty. This distribution of measurements does not demonstrate the range of emissivity values seen in the LA data, where the counting uncertainty is so much smaller than the range of emissivity measurements that it is obvious that they are distinctly different values. It may be possible that similar lab dependence exists in these measurements but is masked by the counting uncertainty. Unfortunately, the large uncertainties lead to qualitatively weak statistical conclusions that there is no evidence of a difference.

Participants 4G, 5H, 1A and 8U each measured sample ULA-1 twice with an average of 93 days between readings. Considered pair-wise, the average deviation between samples is +.4 σ, with a range from -1.1 σ to 1.2 σ. This supports the assumption that sample ULA-1's emissivity is stable over time.

C. Participant survey

The results from the post-data collection survey are summarized in table 4. One participant did not reveal the counter that was used, 1 used an Ordela 8600A-LB, 1 used an XIA UltraLo-1800 and the remaining 6 were Alpha Sciences Model 1950. Six of 8 labs followed a written procedure for counting, and only 1 of 9 reports taking any "extra" care with this these measurements compared to others that are routinely made. Determination of a background level was done in several ways combining measurements made immediately before or following a sample count with historical data. Procedural choices, such as pre-cleaning of samples prior to counting, handling precautions, and packaging for return shipment varied widely. The energy detection threshold was also a variable, ranging from 1 to 3 MeV for reported values. The impact of this particular variation is discussed in more depth below. It is hoped that the survey results will be helpful for formulating hypotheses for future studies in this area.

IV. DISCUSSION

A. Variability in LA data

The most striking observation that can be made from these data is the variation of approximately 2X in the reported emissivity values for the LA samples, correlated with the lab making the measurement. From the survey results it is possible to estimate the magnitude of the impact of errors in the measurement of the sample area, counter efficiency and background determination, relative to the expected emissivity, as being well under 10% each. The resulting combined error is therefore expected to be less than 20% from these random factors. Eight of 9 counters were calibrated within 24 months making this an unlikely source of additional error. Although there are many unexplored sources for potential error from

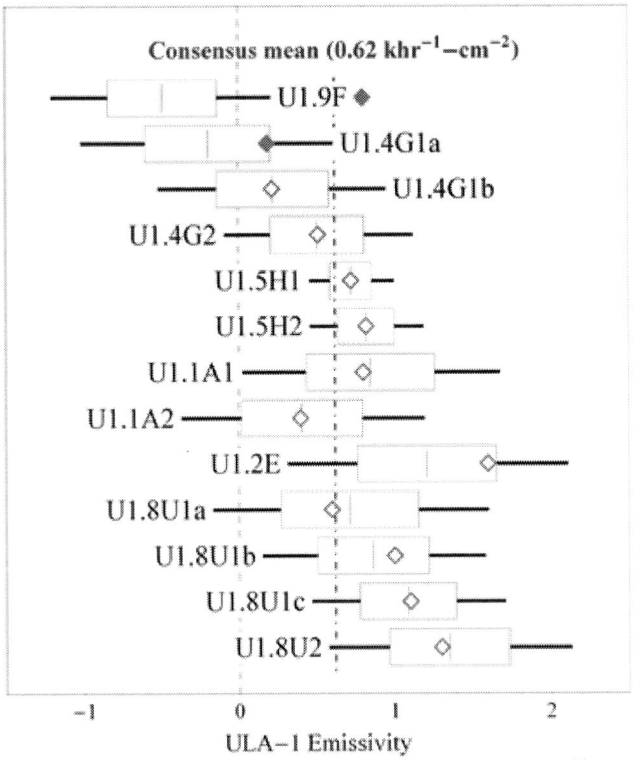

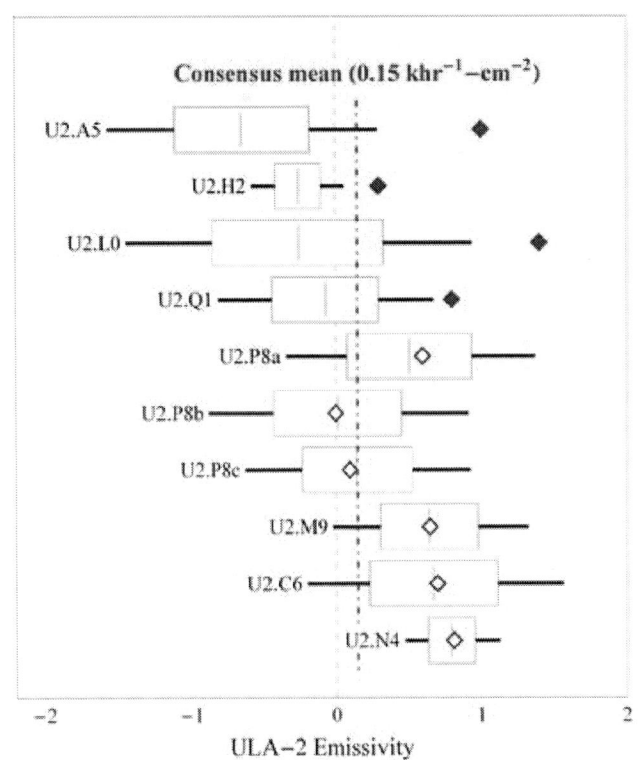

Figure 2. Results for measurement of ULA-1. The key is the same as Fig. 1

Figure 3. Results for measurement of ULA-2.

procedural differences, such as sample contamination, improper procedures, and misreporting it is the authors' opinion that these would not explain the pattern seen in these data.

It is likely that most of this measurement difference is not the result of counting uncertainty or measurement process capability. Given the strong correlation of the variation with a particular lab that persists when the JEDEC values are calculated using a single procedure, it is quite likely that the variation is linked to the particular counter used at each site.

TABLE IV. SUMMARIZED RESULTS FROM PARTICIPANT SURVEY.

	Question	Responses
1	Was a <u>written</u> procedure followed for this testing?	*NR - 2, Yes - 6, No - 1*
2	Was the testing for this experiment conducted by personnel that are generally responsible for counting operations?	*Yes - 9, No - 0*
3	Was any "extra" care taken with this experiment? For example, were additional tests run that are not normally required, extra quality checks conducted, counters used other than those normally used for production?	*Yes - 1, No - 8*
4	What brand(s) and model(s) of counter were used for LA samples?	*NR - 1, Ordela 8600A-LB - 1, Alpha Sciences 1950 - 6, XIA Ultralo 1800 - 1*
5	What brand(s) and model(s) of counter were used for ULA samples?	*NR - 1, Ordela 8600A-LB - 1, Alpha Sciences 1950 - 6, XIA Ultralo 1800 - 1*
6	Was the counter for this experiment built in-house or modified in-house?	*NR - 5, Built in-house - 3, Modified in-house - 1*
7	What is the lower detection threshold for the counter used for LA samples?	*NR - 3, 1 MeV - 2, 1.1 MeV (Est.) - 1, 3 MeV - 1, Other [no description] - 1 , Other - LLD of 0.9 khr-1 cm-2*
8	What is the lower detection threshold for the counter used for ULA samples?	*NR - 3, 1 MeV - 2, 1.1 MeV (Est.) - 1, 3 MeV - 1, Other [no description] - 1, Other - LLD of 0.8 khr-1 cm-2*
9	How was the background determined for LA samples?	*NR - 1, Measured and combined with historical data - 1, Measured after counting - 0, Measured before counting - 4, Historical data only - 1 , Measured every month before counting - 1, Background measured earlier in the month - 1 (notes 2, 3, 4)*
10	How was the background determined for ULA samples?	*NR - 1, Measured and combined with historical data - 1, Measured after counting - 0, Measured before counting - 4, Historical data only - 1, Measured every month before counting - 1, Background measured before and after sample] - 1 (notes 2, 3 & 4)*
11	How was the length of the background count determined for LA samples (for either per-sample or historical measurements)?	*Same as the length of the sample count - 4, Standard length of time based on sample type - 1, Standard length of time, independent of sample type - 2, No background - 1 (note 5)*
12	How was the length of the background count determined for ULA samples (for either per-sample or historical measurements)?	*Same as the length of the sample count - 4, Standard length of time based on sample type - 0, Standard length of time, independent of sample type - 2, No background - 1 (note 5), Other (note 5)*
13	What is in the counting chamber during background determination?	*NR - 1, Only the sample tray - 3, Known emissivity standard - 0, Zero or near-zero emissivity standard - 1, Other (note 5), "Empty" - 1*
14	Was a correction made to the background to account for the portion of the sample tray that would be covered by the counting sample?	*NR - 2, Yes - 0, No - 7*
15	Does the counter efficiency contribute to the background determination?	*NR - 2, Yes - 7, No - 0*
16	Were samples handled according to your normal procedure or work flow?	*Yes - 9, No - 0*
17	Was the sample cleaned or otherwise prepared for counting? If so, how?	*Yes (please describe) - 4 (notes 7, 8, 9), No - 5*
18	Was the sample packaged for return according to a standard procedure? If so, what packaging was used?	*NR - 2, Yes (packaging in which sample was received) - 4, Yes ("standard bag used for shipping low alpha material") - 1, No - 2*
19	When was the counter last calibrated?	*NR - 1, < 6 months - 3, < 12 months - 2, < 24 months - 2, Other (please describe) - 1 (note10)*
20	Who performs counter calibration?	*Manufacturer - 4, Independent laboratory - 0, In house - 5*
21	How was counter efficiency determined?	*Manufacturer - 7, Independent laboratory - 0, In house - 2*
22	What criteria were used for determining the counting time?	*NR - 1, Single time used for all samples - 0, Based on early count results - 0, Based on expected sample activity - 5, Other (please describe) - 3 (notes 11, 12 & 13)*
23	What criteria were used for determining the area of the sample to be used for counting?	*Maximized area able to fit in counter - 7, Standard area for all samples - 1, Other (please describe) - 1 (note 14)*
24	What is your estimated error of the sample area measurement?	*NR - 2, < 1% - 1, < 2% - 2, < 5% - 1, < 10% - 1, Other (please describe) - 2 (note 15)*

Responses in italic type are from the survey multiple choices. Non-italic responses are provided by the respondent. Longer, non-standard, responses are provided as notes. NR indicates "No Response"
2. One respondent that measured background before counting noted that it may alternatively be measured following counting if the schedule is tight.
3. "Sample areas are sufficient to cover region which produces background from the sample tray. Other sources of background are included in the count and can't be measured independently yet."
4. Only one respondent had different background determination procedures for LA vs. ULA samples.
5. One response (same respondent as note 3) said that there is 0 background due to the counter design.
6. "Other, Time to achieve reasonable relative Standard deviation but maximize sample throughput."
7. "Yes, ..." Solid samples are cleaned with ethanol wipe and then vacuumed. Powder and polymer samples are spread on an acrylic plate, covered with Mylar film and stored in flowing P-10 for at least 36 hours.
8. "Yes, Sample was not cleaned, but arranged on tray in a N2-filled glove box."
9. "Yes, Wiped with high purity isopropanol."
10. "Other, Thorough calibration 13 months ago, calibration check 3 months ago."
11. "Other, Comb[ination] of expected activity and early results, rounded to nearest convenient time."
12. "Other, Sample emissivity relative to background."
13. "Other, For solid sample: 48hrs. For colloid or powder sample: 72hrs."
14. "Other, Since the counting area of the sample is set by the electrode size, and the electrode size is a manufactured part, the error in the electrode size is very small." (same respondent as notes 3 and 5)
15. "Other, Hard to estimate the error with one number. Some samples were irregularly shaped, some were not."

One mechanism that could explain the variation is hinted at by the response to survey questions Q7 and Q8, reporting a range of energy thresholds used by each participant. All of the alpha counters in this study use some sort of energy threshold to differentiate between alpha and beta particle emission, and to prevent electronic noise from introducing spurious counts. Alpha particles entering the count chamber at low energies are therefore eliminated from the counting.

For solid materials the maximum range of an alpha particle is less than 100 μm. Materials with alpha emissions spread to a depth of more than a few 10's of micrometers are colloquially referred to as "thick" while materials with the emissions concentrated very near the surface are "thin." Thick samples are most commonly encountered and are representative of emissions from bulk source materials. Thin samples are less common and generally associated with surface contaminant deposition. This study used thick samples for both the LA and ULA material.

For these thick samples it is assumed that the alpha particle energies entering the sample chamber are distributed from 0 up to several MeV. The reader is referred to [4] where a simulation of a thick sample's alpha emission energies are shown in figure 2. The counter's threshold setting determines the fraction of these alpha particles that are rejected during counting. As the particular isotopes and their relative ratios are unknown for these samples it is not possible to make a quantitative estimate of this effect. It is also not possible to compare the energy threshold of each site with the reported values using this data set due to restrictions related to confidentiality. All that can be said is that this mechanism would be expected to create a variation between counts made using different equipment, while demonstrating a strong repeatability for measurements made on the same equipment. This observation is consistent with the data. A more rigorous test of this hypothesis awaits the availability of a different data set.

Although a precise statistical test has not been developed yet, it is believed that the weak trends in the ULA data are not inconsistent with the more robust findings for the LA data.

B. Inconsistent reporting conventions

Comparing the reported emission values from each lab with the mean and standard deviation calculated according to the JEDEC method reveals differences, particularly in the ULA data. It is not expected that the participant's reported value would necessarily match the JEDEC value, due to different assumptions regarding the counting process capability, interpretation of the reported values and their intended application. Nevertheless, the JEDEC values of α (sample mean) and σ (standard deviation) are useful metrics to highlight the choices made in this reporting.

As an example, measurements of ULA materials are commonly reported as an upper bound due to the high level of uncertainty associated with acceptable choices for counting parameters. For these reports, it is reasonable to assume that the emissivity and the portion of the error from counting uncertainty would be reported as

$$Emissivity_{U.B.} = \alpha + N\sigma$$
$$Error_{U.B.} = \sigma \qquad (4)$$

The value of N would be chosen to reflect the desired confidence level of the reported result and would be expected to range from 1 to 3, corresponding to confidence levels of 84% and 99.9%, respectively. A variety of choices are reported in table 3 with values of N ranging from 1 to 2.3. Reading U1.9F reports an emissivity equal to 2σ and an unspecified error. Two LA readings, L1.Q and L2.Q, are reported as upper bounds although the emissivity is equal to α in both cases.

There is no "correct" method regarding this reporting. The observed variability in these values emphasizes the need for a standard reporting method, a clear definition of the meaning of the numerical values being reported, or both. The JEDEC proposed standard requires that the mean emissivity and the standard deviation of the emissivity both be reported, along with many other parameters that can be helpful in interpreting the report.

C. Sample quality

Referring to the paired comparisons presented earlier, these data do not demonstrate a mismatch in emissivity between LA-1 and LA-2, or instability in emissivity for LA-1 or ULA-1. By association, it is reasonable to assume both that LA-2 and ULA-2 are also stable given their common starting source materials and subsequent processing (i.e. LA-2 matched with LA-1 and ULA-2 matched with ULA-1). The only data available to compare the matching of ULA-1 and ULA-2 is the single paired measurement made by B.M.C. before and after the data collection phase (see table 1). Pooling the before and after data, under the assumption of stability, the difference in the mean value of these measurements is 0.5 σ, consistent with an assumption that the samples are matched. Given the large uncertainty in the sample measurements this conclusion is necessarily weak statistically. It is not unlikely that the ULA and LA samples are both matched and stable. Given the apparent match between samples it is very likely that the activities of the individual pieces making up each sample are also well matched. There is a possibility, that is not explored here, that the emissivity of the individual pieces could vary due to differences in surface contamination but this is considered unlikely.

D. Role of Counting Uncertainty and Background Determination

When making measurements of emissivity very near the limit of the instrument's capability, the accurate determination of the sample+background and background counting rates is extremely important. From the measurements shown in table 3 it is evident that the range of N_B/T_B (1.84 to 4.70 hr-1) substantially overlaps the range of N_{S+B}/T_{S+B} (0.45 to 5.62 hr-1) values. Three example ULA measurements are shown in fig. 4. For each measurement the values for N_{S+B}/T_{S+B} and N_B/T_B are shown as individual points accompanied by the distribution of the counting uncertainty. The distance between these distributions, shown by the dotted lines, is the net count rate, in counts per hour, from which the emissivity is calculated.

Measurement U1.8U2 demonstrates adequate separation between the rates, while the differences for measurements U2.C6 and U2.L0 are so small, compared to the counting uncertainties that the net difference is overwhelmed by the uncertainty. The lower right corner of Fig. 4 simulates the effect of increasing the count times by 4X for measurement U2.C6, reducing the uncertainty to a level where the difference can be adequately determined.

Small errors in the background determination, either due to counting uncertainty or systematic error, can easily result in a substantial error in the measured emissivity. Likewise, inadequate counting time results in large uncertainty that will overwhelm the small difference in rates that are trying to be discerned. The proper choice of counting parameters and the careful control of the background rate are obviously vitally important when making measurements for ULA samples.

V. Conclusions

These preliminary data demonstrate that there is a wide variation for LA emissivity measured by various centers that cannot be attributed to normal statistical variations inherent in the measurement that are typically referred to as "counting uncertainty." Although the ULA data match within the expected uncertainty of the measurements this is more a reflection of magnitude of the uncertainty than confidence in the results. The generally good match for paired measurements made within a lab, and the much greater variability between labs, tends to support this interpretation.

These data should not necessarily be interpreted as a serious problem for the semiconductor industry. Monitoring of a manufacturing process is generally an ongoing operation with all measurements made by a single lab following its own procedure and with a consistent relative calibration method. When interpreted for this application these data support the view that monitoring methods are well behaved, at least for the period of time studied.

Estimations of soft error rates commonly make use of

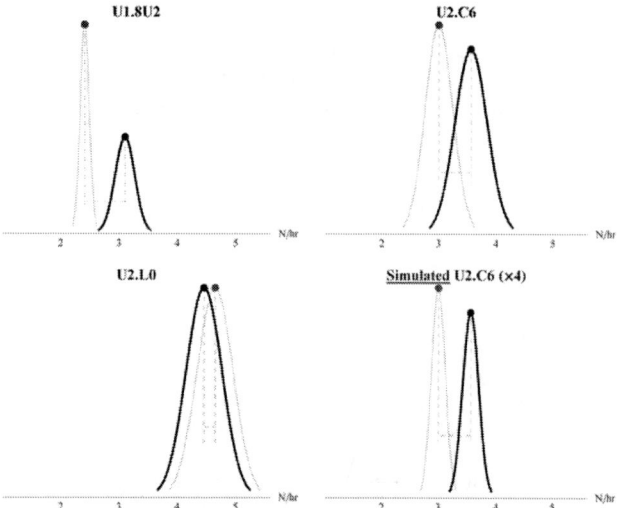

Figure 4. Sample+background and background rates shown with the range of uncertainty. A simulation of the effect of increasing the values of T_{S+B} and T_B by 4X each for measurement U2.C6 is shown in the lower right corner.

accelerated data taken using a high emissivity source and then scaled by the ratio of the high emissivity source and the LA or ULA target material. Since this ratio is generally greater than 10^8 the determination of each emissivity must necessarily be made using different methods. There are generally a multitude of other uncertainties and assumptions that also have to be quantified during this estimation process. These data help to quantify the uncertainty in this portion of the analysis when scaling from accelerated testing to "as built" devices. Since common practice is for this scaling to be done conservatively it is not likely that the uncertainty will generally invalidate prior analyses. Still, it is prudent to consider these findings and assess their impact on soft error estimates.

The lack of an accepted emissivity calibration standard in the measurement range of interest may be a significant contributing factor to the LA mismatch. The authors would like to suggest that the development and acceptance of a standard would be a valuable contribution to both the science and application of alpha counting.

After discussing these results the authors are planning additional projects to understand the sources of the variability. We are planning a follow on experiment in which more of the experimental variables are fixed. For example we may work with a single sample, standardize our handling procedures, or require a specific measurement period. It is our hope that further data will quantify the impact of these variables and improve the certainty of these measurements.

Acknowledgements

A project of this scope requires help from a large team of people and the authors would like to express their particular gratitude to Xiaoting Niu, Cisco Systems, who provided early project management and logistical support; Molly Lu, Advanced Semiconductor Engineering, Shi-Jie Wen, Cisco Systems, and Norbert Seifert, Intel, for coordination and encouragement; Roger Kuo, Advanced Semiconductor Engineering, for technical support; Professor J.L. Autran, University of Provence, France, for facilitating measurements in 2 laboratories in the French Alps; Mike Silverman, Ops A La Carte, for managing the confidentiality arrangements; and William K. Warburton, XIA LLC for early guidance and development of the alpha consortium concept. All of the authors also gratefully acknowledge the generous donation of over 9000 hours of valuable counter time by the counting participants, without which this study would not have been feasible. Finally, J.D.W. would like to particularly thank C.S. for going above and beyond his contracted responsibilities by facilitating confidential communications with participants in order to resolve errors, providing thoughtful review and input on this paper, and for identifying every obscure case that could compromise our participant's confidentiality. All of the authors are deeply indebted to him for the success of this project.

References

[1] Baumann, R. C. and D. Radaelli (2007). "Determination of Geometry and Absorption Effects and Their Impact on the Accuracy of Alpha

Particle Soft Error Rate Extrapolations." Nuclear Science, IEEE Transactions on 54(6): 2141-2148.

[2] Kobayashi, H., N. Kawamoto, et al. (2009). Alpha Particle and Neutron-Induced Soft Error Rates and Scaling Trends in Sram. Reliability Physics Symposium, 2009 IEEE International.

[3] Clark, B. M., M. W. Weiser, et al. (2004). "Alpha Radiation Sources in Low Alpha Materials and Implications for Low Alpha Materials Refinement." Thin Solid Films: 384-386.

[4] Wrobel, F., F. Saigne, et al. (2009). "Radioactive Nuclei Induced Soft Errors at Ground Level." Nuclear Science, IEEE Transactions on 56(6): 3437-3441.

[5] Measurement and Reporting of Alpha Particle and Terrestrial Cosmic Ray-Induced Soft Errors in Semiconductor Devices. JEDEC Standard. Arlington, VA, JEDEC: 94.

[6] Clark, B. M. (2010). Personal communication.

[7] Mistry, A., S. Lee, et al. (2000). Characterization of Low Alpha Emissivity System on Electroplated Solder Bumps. 2000 Electronic Components and Technology Conference.

[8] Knoll, G. F. (2010). Radiation Detection and Measurement, John Wiley & Sons, Inc.

The Impact of New Technology on Soft Error Rates

Anand Dixit[1] and Alan Wood[2]

[1] Systems Group, [2] Oracle Labs

Oracle Corporation

Santa Clara, CA USA

[1] 408-276-6335 anand.x.dixit@oracle.com, [2] alan.wood@oracle.com

Abstract—This paper presents the impact of new microprocessor technology on microprocessor soft error rate (SER). The results are based on Oracle's (formerly Sun Microsystems) neutron beam testing over the past several years. We describe how the tests were conducted and how the test results are used to influence microprocessor design. As microprocessor feature sizes decreased from 180nm to 65nm, memory error rates per bit decreased, but our data indicates a reversal of this trend at 40nm. Flop error rates still appear to be decreasing, even at a 28nm feature size We measure SER as a function of power supply voltage (Vdd) over a range of 1.2V down to 0.5V, and the data shows SER significantly increases as Vdd decreases. This result implies that dynamic voltage frequency scaling (DVFS), a commonly used microprocessor energy reduction technique, could cause a significant decrease in microprocessor reliability. The data also show that more energy-efficient transistors using back bias technique do not appear to significantly impact microprocessor reliability.

Keywords-soft error; single-event upset; neutron beam testing; bit error rates

I. INTRODUCTION

Reliability has always been one of the key requirements in microprocessor designs at Oracle (formerly Sun Microsystems). With shrinking cell geometries, neutron-induced soft errors have become a greater concern. We conduct accelerated tests at the LANSCE test facility in Los Alamos [1] to characterize cell upsets, usually called single-event upsets (SEUs), in our microprocessors. We measure the SEU rate of unprotected memory cells and logic blocks of flops and/or latches in a neutron beam to determine the raw SEU rate of SRAM cells and flip flops on microprocessors. These data are used to help define appropriate error detection and correction in microprocessor designs.

This paper presents some of the trends and lessons learned from our accelerated test experiments over the past ten years, ranging from 250nm through 28nm technology nodes. In particular, we describe how the test results are used to influence microprocessor design. Our results show that:

- The SEU rate for individual SRAM cells increased at a 40nm feature size compared to a 65nm feature size. This reverses a long-term trend of SEU rate decline with technology scaling that had been true from the 250nm node to the 65nm node. In the past, the reduction in critical charge caused by declining power supply voltage has been more than offset by cell area reduction and technology improvement, but this no

longer appears to be true for SRAMs. However, there is still a reduction in latch/flop SEU when moving from 65nm to 40nm to 28nm, which may be caused by flops being larger than SRAMS and/or by flop design having a big impact on SEU sensitivity.

- Flop SEU rate increases significantly as Vdd decreases due to the reduction in critical charge of the storage nodes. The SEU rate approximately doubles when dropping from 1.25V to 0.7V and doubles again when dropping from 0.7V to 0.5V. This result implies that dynamic voltage frequency scaling (DVFS), a commonly used microprocessor energy reduction technique, could cause a significant decrease in microprocessor reliability.

- Flop SEU rates do not seem to vary much with new circuit designs that trade off device speed with leakage current, hence improving energy-efficiency. The new designs usually involve back-bias transistors, in which a bias voltage is applied to the "back gate" or "body" terminal of a transistor - thus changing the effective threshold voltage of the device. However, forward bias does seem to decrease SEU rates at the 28nm node.

- Shrinking feature sizes have made multi-cell upsets more prominent, requiring that ECC be supplemented by appropriate spacing between bits belonging to the same logical word to minimize the system level effect of soft errors in microprocessor memory cells.

- Flop SEU rates are approximately the same as SRAM cell SEU rates for recent technology nodes. Flop SEU rates are expected to play a greater role in defining the system level soft error rates for future technologies.

- As feature sizes decrease, the apparent neutron beam attenuation due to the beam passing through a device and package material increases, perhaps because of greater sensitivity to lower energy neutrons. This limits the number of devices that can be simultaneously tested.

II. TEST SETUP

Accelerated testing is necessary to provide data for microprocessor designers because cell upset events due to neutrons are relatively rare events in a terrestrial environment. Although neutron flux is higher at higher altitudes and we have performed testing at high altitudes with hundreds of units in the past, we have found that a neutron beam is the most practical

978-1-4244-9113-1/11 $26.00 © 2011 IEEE

way to get sufficient neutron flux rates for accelerated testing. The LANSCE facility at Los Alamos [1] provides an energy distribution very close to the cosmic neutron flux at the earth's surface at approximately 10^8 times the intensity, and we have performed neutron beam testing at that facility once or twice every year for the past ten years.

Our test methodology is fairly standard for beam testing. We begin by initializing memory and logic to a known state, turn on the beam, and monitor errors while the beam is on. We generally initialize to all zeros for symmetric memory cells but have also experimented with checkerboard patterns. For SRAM cells, we sweep through memory while the beam is on and record the location and time (to within one sweep) of bit flips. For flops we only check for bit flips at the end of the test because the number of flops on a microprocessor is small compared to the number of memory cells, meaning that the total number of logic cell errors is relatively small compared to the total number of memory cell errors.

After the test, we analyze the data to determine the number of single and multi-cell upset events. We use time and spatial correlation to determine if a single neutron has upset multiple

SRAM cells. If, during a single sweep through memory, we record bit flips for cells within a few microns of each other, we consider those events as multi-cell upset events. Otherwise, we consider them as single cell upset events.

Figure 1 shows our test setup. The beam direction is shown by the arrow as it travels through the neutron counter and two microprocessor boards. We use a portable tester [12], not shown in Fig. 1, for neutron beam testing (see [2] for the benefits of using this approach instead of testing complete systems). We test both microprocessors and specially designed test chips using the arrangement shown in Fig. 1. Using test chips allows us to test and compare various types of flops, as well as to vary voltages and other parameters. An example of a board with 20 test chips is shown in Fig. 2. The beam diameter and distance to the beam are carefully selected so that the transmitted beam flux is approximately the same for all the test chips on the board. Each test chip for these latest experiments contains approximately 55,000 flops, so there are more than a millions flops on the test board. At the current time, we have 28nm results only for flop test chips.

As can be seen in Fig. 1, multiple microprocessors are placed in the neutron beam path. This allows us to gather more data in a single experiment as long as the neutron beam attenuation, caused by neutron absorption and the spreading of the beam cross-section with distance, is reasonable. We can measure the attenuation by analyzing the total number of upset events for each microprocessor position in the beam path. It has been observed that the attenuation has increased from about 10% per microprocessor position in 180nm to 40% per microprocessor position in 65nm technology. This has direct impact on the number of boards that can give meaningful results when placed in series in the neutron beam. In 180nm technology, we would place 4-5 boards with microprocessors in the beam path. In 65nm and smaller technology nodes, we only place 2 microprocessors or test boards in the beam path.

To further investigate the beam attenuation, the neutron energy spectrum was captured before and after the neutron beam passed through a single board. The beam radius was collimated to approximately 1 inch in diameter to ensure that the entire beam passed through the board material, package lid, and silicon die. Figure 3 shows the beam transmitted flux as a

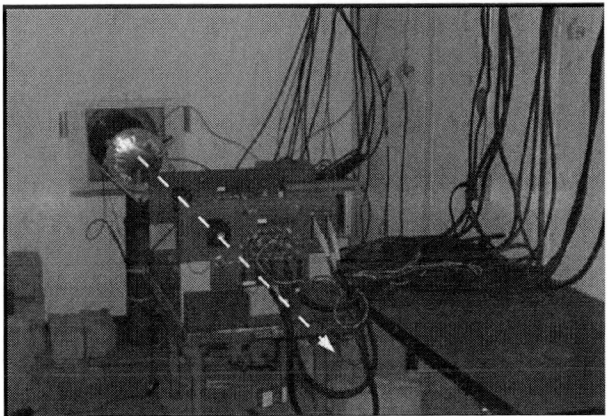

Figure 1.   Test setup using a special tester [12]

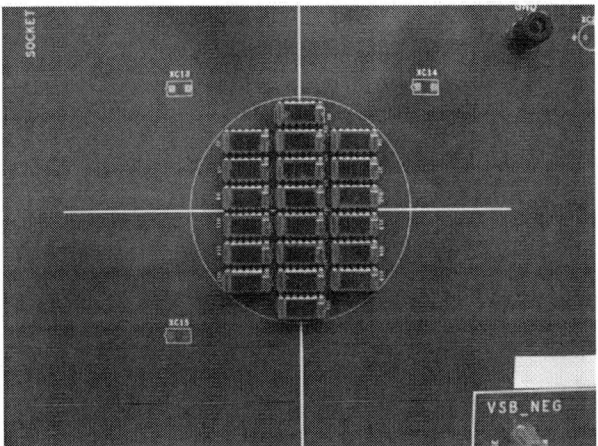

Figure 2.   Test board containing 40nm test chips

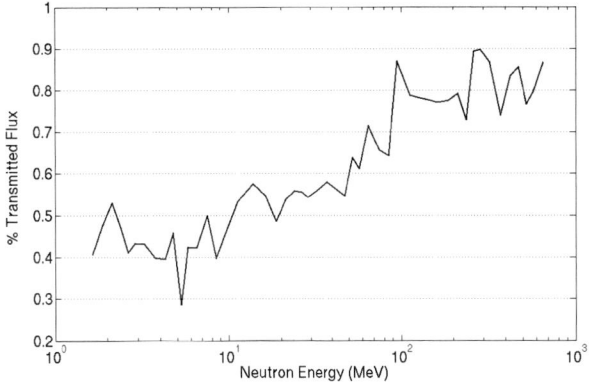

Figure 3.   Neutron beam transmission spectrum through a circuit board and microprocessor

function of the neutron beam energy. Beam attenuation is the inverse of beam transmitted flux (attenuation = 1 − normalized transmitted flux). The apparent attenuation in Fig. 3 is due to absorption of neutrons by the circuit board and beam spreading. The data in Fig. 3 is noisy due to the narrow beam diameter required for this test and the limited test time that we were able to devote to it. However, it does show the correct trend for transmitted flux as a function of neutron energy. Lower energy neutrons are more attenuated (around 50%) than the higher energy neutrons (0-20%). The increase in attenuation as technology has scaled mentioned in the previous paragraph, together with the change in beam attenuation observed in the measurements, could be indicative of memories being more sensitive to lower energy neutrons as the technology has scaled. Due to the attenuation, it is important to compare the test results from each board independently to see if there are differences.

In all beam testing, it is important to avoid contaminating the data with secondary particles originating from neutron interactions on the first board and propagating to the second board. For all our beam testing, the spacing between the boards was 12 to 18 inches to allow the secondary particles to decay and disperse.

III. TECHNOLOGY TREND TEST RESULTS

Figure 4 shows the microprocessor SRAM single-event upset (SEU) rate and voltage as a function of the technology node, normalized to a value of 1 at the 90nm technology node for ease of comparison and to protect proprietary data. (SEU rate is reported in FITs/kbit, equivalent to cell upset events per bit per million hours.) The nominal power supply voltage (Vdd) has been slowly decreasing, which decreases the critical charge necessary for a cell upset event, thus making the cells more vulnerable to bit flips. However, cell size reduction and the corresponding sensitive area reduction with technology scaling has led to a reduction in SRAM cell SEU, even as the voltage has decreased, until the 40nm technology node. The large reduction in SRAM cell SEU rate from 130nm technology to 90nm technology was most likely the result of an

SRAM design change. The layout of the SRAM cell changed from the traditional one to a lithographically friendly one with uni-directional poly orientation [3]. As a result, 90nm SRAMs have a significantly reduced amount of NMOS active area - over and above what just technology shrink would indicate, thus significantly reducing the sensitive area most contributing to SEUs.

Because soft error susceptibility increases exponentially as voltage decreases and decreases linearly as area decrease (quadratically as feature size decreases), it has long been expected that the voltage reduction that has accompanied feature size reduction would eventually cause SEU rates to increase. Our data for 40nm SRAMs appear to be the first evidence of this affect. The 40nm SRAM SEU rate is 30% higher than the 65nm SRAM and nearly the same as the 90nm SRAM. One sigma error bars for SRAM FIT rate (not shown in Fig. 4) are less than 3%, so the results are not due to uncertainty. However, there was a foundry change between 65nm and 40nm that may have impacted the results. When we have data for 28nm SRAM SEU rates, we will be able to determine the relative contributions of the technology node and foundry changes.

Kobayashi et al [5] have recently published an analysis on how various particles generated in the nuclear reactions of neutron and silicon nuclei contribute to the final SEU rate of a 45nm SRAM. Their analysis indicates a sharp increase in SEU rate as a function of critical charge when the cell critical charge drops below than ~0.6fC. For cell critical charge less than 0.6fC, the contribution from protons (with average energy of ~5MeV for all interactions) becomes the dominant term. Since these relatively low energy particles are much more abundant than heavier ions like He, Al and Mg, this may explain the sharp increase in SEU rate. This sharp increase in SEU rate as a function of critical charge has been noted by others as well [6] [10]. Another recent result [7] adds more evidence to the SRAM cells being sensitive to low energy particles. In Section II, we reported on the change in apparent beam transmission while stacking and testing multiple boards in the neutron beam. Our results indicate that SRAM cells are becoming more sensitive to reaction products from low energy neutrons as they are being scaled in each technology node. However, SEU rate depends on the balance of critical charge and critical volume, and other authors [11] predict that SRAM cell SEU rate will continue to decline to the 22nm technology node, so our results may not be generic.

Figure 4 also shows how flop SEU rates compare to SRAM SEU rates as a function of technology. Flop data presented here is averaged over all the different drive strengths and different flop design families in the product. Flop data also has higher error bars because the total number of flops is smaller compared to the total number of memory cells on the same die. As shown in Fig. 4, the SEU rate for flops in 130nm product is similar to flops in 90nm product, while flops in 65nm product show a reduction in SEU rate.

Unlike SRAMs, flops in 40nm technology show a reduction in SEU rate compared to 65nm, although not as big a reduction as predicted by our simulations [4]. Flops in 28nm technology show a significant reduction in SEU rate compared to 40nm.

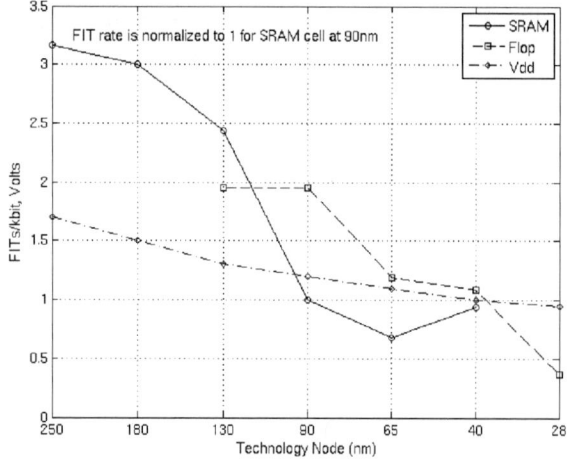

Figure 4.   SRAM and Flop SEUs as a function of technology node

One reason that the trend may not yet have appeared in the flop data is that flops are bigger than SRAMs, so the sensitive area reduction may have more relative importance. However, it should be noted that due to the special treatment by fabs, SRAM cells get close to the expected area shrink entitled by the feature size reduction. Flops don't enjoy the same treatment, and hence, the sensitive area reduction for flops is more modest. Another contributing factor to the flop trend is that flop design style plays a big role in determining the flop SEU rate. Due to the manufacturing and design dependencies, some difference can be expected in the SEU rate versus technology trend for memory cells and flops. This will be especially true when comparing trend results from different vendors.

In the past, designers have not been greatly concerned about soft errors in microprocessor logic because the number of flops/latches on a microprocessor is much fewer than the number of SRAM cells, and flop SEU rates were lower than SRAM SEU rates. In 90nm, 65nm, and 40nm technology, flop SEU rates are larger than SRAM SEU rates. Because flop protection mechanisms such as state machine encoding and invariant checking are more difficult to implement than simple parity and ECC, flops are quickly becoming the major contributor to system soft error rate as technology scales to smaller feature sizes.

While SEU rates per memory cell have been decreasing, the amount of memory on a microprocessor has been increasing. Table 1 shows the raw SEU rate per microprocessor as a function of the technology node for representative microprocessors built in that technology. The example microprocessor in 180nm was a simple die shrink of the 250nm microprocessor, so the amount of memory did not change. However, for later technology generations, the amount of on-chip microprocessor SRAM has been significantly increased to create larger caches and improve performance. Due to the larger caches, the total uncorrected SEU rate per microprocessor has increased with technology generation even though the SEU rate per bit has decreased. The uncorrected rates in Table 1 are input into error correction code (ECC) design to determine the appropriate protection for each memory structure. With ECC, the corrected memory array failure rate on a microprocessor is orders of magnitude smaller than the numbers in Table 1.

Table 1 also shows how the microprocessor designs have evolved over the years. Early designs were done in a way to ease production in the next technology generation via optical scaling when moving from the 250nm node to the 180nm node. This was abandoned in favor of using additional silicon real estate on the scaled chip for multiple copies of the device when moving from the 180nm to the 130nm node. Technological complexity also dictated that porting a design to the next node was no longer a relatively simple optical process but required significant engineering resources. Larger cache sizes were deemed necessary to bridge the widening speed gap between microprocessor and memory I/O. This explains the almost 10x increase in memory capacity when moving from 130nm to 90nm node. The next node (65nm) was the first multi-core architecture with high-speed serial I/O links and shared level-2 caches and hence shows only a modest increase in the total

TABLE I. RAW SEU RATE PER MICROPROCESSOR

Tech. (nm)	Relative SEU rate in FITs/kbit	Approx. Mbits per microprocessor	Relative uncorrected SEU rate per microprocessor (kFIT)
250	3.2	1.52	5.0
180	3.0	1.52	4.3
130	2.4	3.28	7.9
90	1.0	33.6	33.6
65	0.7	44.3	30.5
40	0.94	71.0	67.0

number of bits. The current node (40nm) continues the trend in multi-core architecture with extra chip real estate being used to increase on-chip memory by close to 50% while doubling the number of cores on the die.

IV. VOLTAGE TEST RESULTS

As shown in Fig. 5 for the 40nm technology node, the average flop SEU rate increases significantly as Vdd decreases. This is expected because the critical charge decreases linearly as Vdd decreases and SEU rate has an exponential dependence on critical charge. The SEU rate increases by approximately 30% per 0.1V as the voltage decreases from 1.25V to 0.5V as shown by the exponential fit in Fig. 5. The error bars in this paper indicate +/- 1 standard deviation range around the average value. The SEU rate approximately doubles when the voltage decreases from 0.7V to 0.5V. The 28nm technology node is also shown in Fig. 5 and indicates the same trend, albeit with fewer data points. Note that the data in figures 5, 6, and 7 is normalized to a SEU rate of 1 for the 40nm node data at 0.95V for ease of trend comparison and to protect proprietary data.

Because the dynamic power consumption of transistors is proportional to V^2F, where V is the voltage (usually Vdd) and F is the frequency, microprocessor designers would like to reduce Vdd as much as possible to make the microprocessor more energy-efficient. These test results provide an indication of the SEU rate impact of that Vdd reduction.

Dynamic voltage frequency scaling (DVFS) is the (usually simultaneous) reduction of both voltage and frequency to reduce microprocessor power consumption. Many modern microprocessors use this technique to reduce power consumption during periods of light utilization or during periods when external conditions require that the system reduce its power consumption. If the microprocessor is idle while the voltage is reduced, the increase in flop SEU rate may not matter, but if the microprocessor is operating at a reduced rate, the increase in flop SEU rate may negatively impact reliability based on our test results.

Figure 6 shows the trend of SEU rate increasing as voltage decreases for four different flop design topologies: master slave, dynamic [8] and two different flavors of sense amplifier based flops [9]. As shown in the figure, all flops show similar trends for SER as a function of supply voltage. While this is true for these particular flop designs tested, it should not be taken for granted for all designs. A counterexample is shown in Fig. 7, where a different master slave design on our 40nm test chip shows a very different slope for SER as a function of

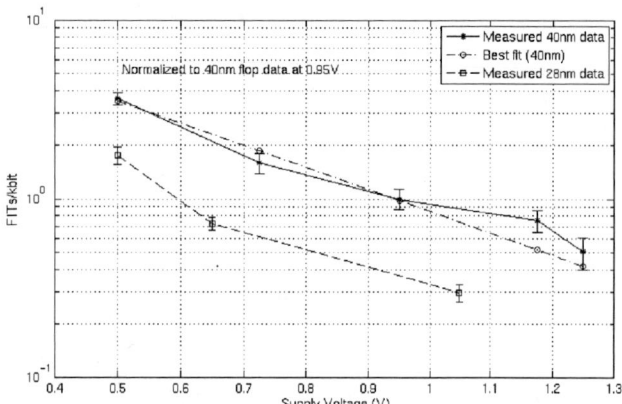

Figure 5.   Flop SEU rates as a function of Vdd

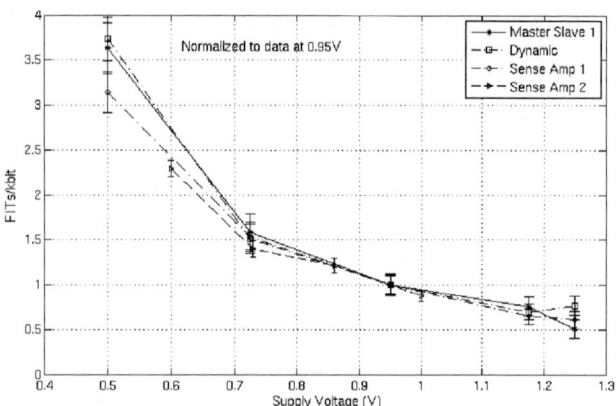

Figure 6.   SEU rates as a function of VDD for various flops
Note that all flop families show similar trends.

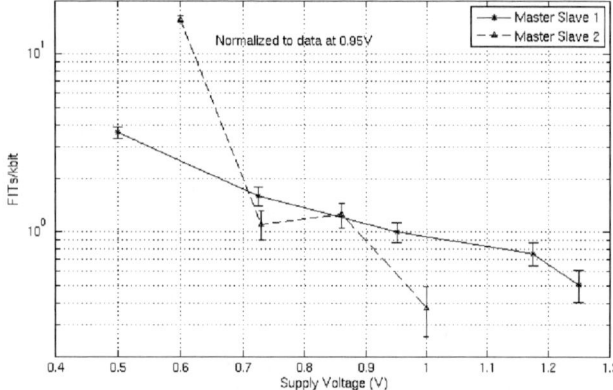

Figure 7.   SEU rates for two different master slave flops on 40nm test chip
Note the very different slopes for the two designs.

supply voltage. The second master slave flop shows 2.3x more sensitivity to voltage compared to first – its SEU rate increases ~3x for each 100mV drop in supply voltage. Please note that our data for the second master slave flop has large error bars at higher voltages but the difference in slope and its higher SEU rate at lower voltages is clearly visible. Its higher SEU rate at 0.86V compared to 0.73V is believed to be an artifact of these large error bars.

Clearly, for designs using DVFS, comprehending such differences in SEU sensitivity is of great importance. It would be highly desirable to exclude extremely sensitive flops from the design. The master slave flops in Fig. 7 are different circuit designs and follow different layout styles. Master Slave 1 follows all DFM (Design For Manufacturing) guidelines with only straight poly while Master Slave 2 uses the minimum design rules and allows for poly bends in order to reduce the circuit area as much as possible. For SEU sensitivity reduction in this application, the Master Slave 1 design approach would be preferred.

To reduce microprocessor power consumption, designers are creating more energy-efficient circuits. One example is the use of back-bias transistors, in which a bias voltage is applied to the "back gate" or the "body" terminal of a transistor. This can be used to increase the effective threshold voltage of an extremely leaky device, thus reducing leakage current and making the device usable in a system that would have otherwise rejected it due to power consumption. Back bias can also be used to reduce the threshold voltages of the devices. A reduction in threshold voltage increases the operating speed of the device and allows it to be used in a system for which it originally would have been too slow at the cost of added leakage current as noted above. SEU rate implications as a function of both forward bias and reverse bias for the complementary NMOS and PMOS transistors need to be understood before this scheme can be qualified for field usage.

For an NMOS transistor, the substrate voltage (Vsb) or the p-well voltage can be varied, and for a PMOS transistor, the n-well voltage (Vnw) can be varied. Changing the back bias (Vsb or Vnw) affects both the charge collection volume and the feedback dynamics of the storage nodes. Forward bias (+Vsb, -Vnw) reduces the charge collection area due to lower applied voltage across the p-n junction area and increases feedback speed, while reverse bias has the opposite effect. Table II shows the results of varying the substrate voltage (Vsb) and the n-well voltage (Vnw) for a flop on a 65nm test chip. Typical operating ranges of interest are +/-300mV. No statistically significant trend in SEU rate as a function of bias is evident even when testing up to +/-500mV, well beyond the expected range. All measured results are within 10% of nominal. This indicates that SEU rate is insensitive to this transistor energy-reduction technique in these circuits. The circuits tested are twin well circuits manufactured in a triple-well capable process.

TABLE II. NORMALIZED FLOP SEU RATE FOR DIFFERENT TRANSISTOR BIAS VOLTAGE FOR 65NM PARTS

Vnw	Vsb = +0.5V	Vsb = 0V	Vsb = -0.5V
	Part A, 65nm, Vdd = 1.0V		
1.5	1.50	1.35	1.36
1	1.47	1.37	1.43
0.7	1.42	1.35	1.25
	Part B, 65nm, Vdd = 1.2V		
1.7	1.11	1.04	1.15
1.2	1.15	1.06	1.1
0.9	1.13	1.1	1.14

Values are FITs/kbits, normalized to a value of 1 for 90nm technology and 1.2V Vdd.

Vsb is the substrate voltage, and Vnw is the n-well voltage. Forward/reverse bias is +/-500mV for Vsb and -300/+500mV for Vnw.

TABLE III. NORMALIZED FLOP SEU RATE FOR DIFFERENT TRANSISTOR BIAS VOLTAGE FOR 40NM AND 28NM PARTS

Vnw	Vsb = +0.3V	Vsb = 0V	Vsb = -0.3V
	Part C, 40nm, Vdd = 1.0V		
1.3	1.10	-	1.23
1	-	1.09	-
0.7	0.96	-	1.22
	Part D, 28nm, Vdd = 0.65V		
0.95	0.96	-	1.02
0.65	-	0.80	-
0.35	0.77	-	0.95

Values are FITs/kbits, normalized to a value of 1 for 90nm technology and 1.2V Vdd.

Vsb is the substrate voltage, and Vnw is the n-well voltage. Forward/reverse bias is +/-300mV for Vsb and -/+300mV for Vnw.

Table III shows the results of varying Vsb and Vnw for flops on a 40nm test chip and a 28nm test chip. Due to lack of test time, data was only gathered for the corner cases in the table. Although some evidence of SEU rate reduction with forward bias (+Vsb, -Vnw) appears, the trend is minimal and not statistically significant. Note that testing was done with a nominal Vdd of 0.65V for the 28nm test chip in an attempt to accelerate the results.

V. MULTI-CELL UPSETS – DESIGN IMPLICATIONS

As shown in Fig. 8, SRAM multi-cell upsets are much more common in newer technology nodes, and microprocessor designs need to protect against them. Standard ECC can detect up to two bit errors per word and correct one, but the prevalence of single neutron events that upset 3 or more bits means that designers need to appropriately physically separate bits that appear in the same protected word.

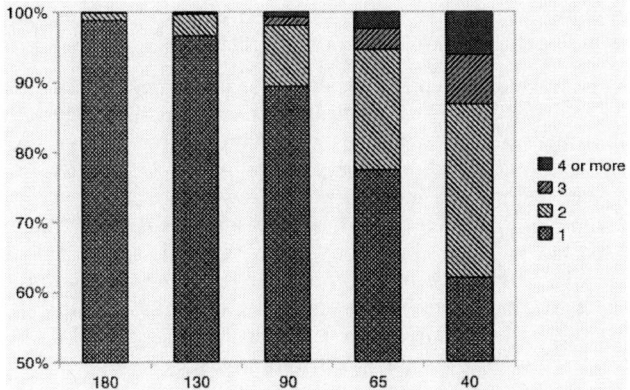

Figure 8. Multi-cell error percentages by technology node

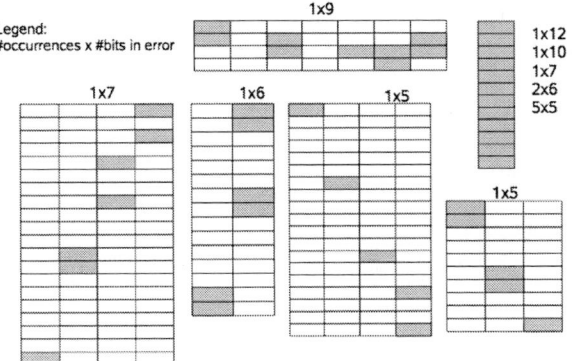

Figure 9. Multi-cell upset patterns

Multi-cell upset events that impact different rows of an array are not a concern because a single word only uses cells in the same row. The concern is a multi-cell upset event that flips multiple bits in the same row. Many of the double, triple, and quadruple cell upset events are exactly of this nature. From our study of these event patterns, bits in the same word must be separated by a number of cells to avoid multi-cell upsets in the same word.

The cell upset patterns for the events that upset 5 or more cells are a bit more complex and problematic as shown in Fig. 9. These events come from a recent experiment that obtained over 36,000 SEUs from a 90nm part. Fig. 9 shows the number of occurrences for each pattern and the number of bits in the pattern. The "1x9" pattern at the top middle of Fig. 9 means that there was one event that upset 9 cells. It was spread over 4 rows and 7 columns as shown in the figure. Similarly, the "1x7" pattern on the left was one event that upset 7 cells, spread over 4 rows and 20 columns. The "1x7" event was about the maximum cell separation we would consider as a single event; events with larger separation would be considered as separate upsets (recall that an event is recorded during a single sweep through memory, so the upsets are correlated in time as well as space). There were a number of events indicated on the top right of Fig. 9 that impacted several rows in a column, but these are not an obvious concern for a design because each upset would be in a different word.

However, cell upsets events are spatially random. In the "1x7" pattern, it appears as though a charged particle traveled from the top right corner to the bottom left (or vice versa), leaving a trail of cell upsets in its wake. In the "1x5" pattern, it appears that the trail was upper left to lower right or vice versa. In either of these events, there is no reason the charged particle could not have travelled along a row rather than diagonally. The SRAM cell size in Fig. 9 is 2.1 microns by 0.7 microns. The maximum distance between cells upset in a multi-cell upset event is 16 microns from corner to corner in the "1x7" event. Thus, if a designer wanted to protect against the worst possible event, they would protect against events that were 16 microns or 8 cells apart in the same row. For example, if simple parity is used for protection, bits in the same word should be separated by at least 8 cells in order to guard against such errors. In other words, there needs to be at least a 9-word interleave in each row or a design rule that states that cells in the same word must be on at least an 18 micron center-to-center pitch.

The approach described in the previous paragraph is very conservative. It assumes the worst case multi-cell upset event at the worst case geometry. A more reasonable approach is to compare the protection mechanisms against the upset data to determine the probability of an event that would defeat the protection. For example, if double error detection, single error correction ECC is used with a 2-word interleave, there would have to be 3 upsets within 5 cells in a row to cause an undetectable error. This occurs once in 36,000 events - in the third row of the "1x9" configuration. With an event rate of around 30 kFITs per Table 1, this is an undetected error rate of 1 FIT, something that the microprocessor designers may decide is acceptable. It is also possible to create a probability vs. distance distribution using the test data and do a more conservative calculation by assuming worst-case geometry for the multi-cell upset events.

VI. CONCLUSION

The Oracle design teams have found the accelerated test data from our experimentation at LANSCE to be very valuable. We have used them to guide error detection and correction in our microprocessor designs. Our data shows trends that will make system error protection even more important:

- SEU rates per SRAM cell have reversed a long-term trend and show an increase at the 40nm technology nodes. Data from the 28nm node is needed to confirm the trend.

- SEU rates per flop continue to show an improvement trend, although a trend reversal is expected in the next few technology nodes. Flop design and manufacturing appears to be more important factors for SEU rate than charge collection volume reduction and increased voltage sensitivity.

- Multi-cell upsets have become much more frequent due to shrinking feature sizes. Designers of memory

structures on microprocessors need to consider multi-cell upset protection techniques.

- Microprocessor energy reduction techniques can negatively impact SEU rate, e.g., reducing Vdd to reduce energy consumption.

- The use of back-bias transistors to improve energy-efficiency does not seem to significantly impact SEU rate.

We intend to continue our accelerated test program to track these trends and guide our microprocessor design.

ACKNOWLEDGMENT

The authors would like to thank the entire soft error team at Oracle for all the experimental work at Los Alamos.

REFERENCES

[1] B. Takala and S. A. Wender, "Accelerated neutron testing of semiconductor devices". http://wnr.lanl.gov/see/poster.pdf

[2] A. Dixit, R. Heald, and A. Wood, "Trends from Ten Years of Soft Error Experimentation", SELSE5. http://selse5.selse.org/program.html

[3] M. Ishida, T. Kawakami, A. Tsuji, N. Kawamoto, M. Motoyoshi, and N. Ouchi, "A novel 6T-SRAM cell technology designed with rectangular patterns scalable beyond 0.18um generation and desirable for ultra high speed operation," IEDM Tech. Dig., pp.201-204, Dec. 1998.

[4] A. Dixit and R. Heald, "Soft Error Estimates for Fabless Companies", ICICDT 2009, pp. 125-127, May 2009.

[5] H. Kobayashi, N. Kawamotom, J. Kase, and K. Shiraish, "Alpha particle and neutron-induced soft error rates and scaling trends in SRAM", IRPS 2009, pp. 206-211, April 2009.

[6] B. Narasimham, M. J. Gadlage, B. L. Bhuva, R. D. Schrimpf, L. W. Massengill, W. T. Holman, A. F. Witulski, Z. Xiaowei, A. Balasubramanian, and S. A. Wender, "Neutron and alpha particle-induced transients in 90nm technology", IRPS 2009, pp. 478-481, April 2009.

[7] D. F. Heidel, P. W. Marshall, J. A. Pellish, K. P. Rodbell, K. A. LaBel, J. R. Schwank, S. E. Rauch, M. C. Hakey, M. D. Berg, C. M. Castaneda, P. E. Dodd, M. R. Friendlich, A. D. Phan, C. M. Seidleck, M. R. Shaneyfelt, and M. A. Xapsos, "Single-Event Upsets and Multiple-Bit Upsets on a 45 nm SOI SRAM," IEEE Trans. Nucl. Sci., vol. 56, no. 6, pp. 3499-3504, Dec. 2009.

[8] F. Klass. C. Amir, A. Das, K. Aingaran, C. Truong, R. Wang, A. Mehta, R. Heald, and G. Yee, "A new family of semidynamic and dynamic flip-flops with embedded logic for high-performance processors," IEEE J. Solid-State Circuits, vol. 34, pp. 712–716, May 1999.

[9] M. Matsui, H. Hara, Y. Uetani, K. Lee-Sup, T. Nagamatsu, Y. Watanabe, A. Chiba, K. Matsuda, and T. Sakurai, " A 200 MHz 13 mm 2-D DCT macrocell using sense-amplifying pipeline flip-flop scheme," IEEE J. Solid-State Circuits, vol. 29, pp. 1482–1490, Dec. 1994.

[10] E. Cannon, D. Reinhardt, M. Gordon, and P. Makowenskyj, "SRAM SER in 90, 130 and 180 nm bulk and SOI technologies", IRPS 2004, pp. 300-304, April 2004.

[11] E. Ibe, H. Taniguchi, Y., Yahagi, K. Shimbo., and T. Toba, "Impact of Scaling on Neutron-Induced Soft Error in SRAMs from a 250nm to a 22nm Design Rule," IEEE Trans. on Electron Devices, Vol.57, No.7, pp. 1527-1538 (2010).

[12] Verigy tester, http://www1.verigy.com/cntrprod/idcplg?IdcService= GET_FILE&dID=1471&Rendition=Primary.

978-1-4244-9113-1/11 $26.00 © 2011 IEEE

Quantitative, nanoscale free-carrier concentration mapping using terahertz near-field nanoscopy

J. Wittborn and R. Weiland

Infineon Technologies AG
Munich, Germany
+49 89 234 29111, Jesper.Wittborn@infineon.com

A. J. Huber

Neaspec GmbH
Martinsried, Germany

F. Keilmann

Max Planck Institut of Quantum Optics and Center for NanoScience
Garching, Germany

R. Hillenbrand

CIC nanoGUNE Consolider, Donostia – San Sebastian, Spain
Ikerbasque, Basque Foundation for Science, Bilbao, Spain

Abstract— We use ultra-resolving terahertz (THz) near-field microscopy based on THz scattering at atomic force microscope tips to analyze 65-nm technology node transistors. Nanoscale resolution is achieved by THz field confinement at the very tip apex to within 30 nm. Images of semiconductor transistors provide evidence of 40 nm (λ/3000) spatial resolution at 2.54 THz (wavelength $\lambda = 118\mu m$) and demonstrate the simultaneous THz recognition of materials and mobile carriers in a single nanodevice. The mobile carrier contrast can be clearly related to near-field excitation of THz-plasmons in the semiconductor regions. The extraordinary high sensitivity of our microscope provides THz near-field contrasts from less than 100 mobile electrons in the probed volume.

Keywords-component; atomic force microscope; terahertz near-field; carrier concentration mapping; microscopy

I. INTRODUCTION

A. Background

Measurement of carrier- or doping-concentration of nanostructured devices still remains a challenge for the semiconductor industry. Secondary ion mass spectroscopy (SIMS) and spreading resistance profiling (SRP) are useful methods for measuring dopant concentration and carrier concentration, respectively, but are limited to 1-dimensional depth profiles. In addition, both these methods require relatively large, laterally homogenous sample areas. Atom probe microscopy faces the opposite problem; it yields 3-dimensional measurements of the nanoscale dopant concentration [1] but the maximum sample size limits its use for many failure analysis applications. Scanning probe microscopy (SPM) based methods [2] such as scanning capacitance microscopy (SCM) [3, 4, 5, 6, 7], scanning spreading resistance microscopy (SSRM) [8, 9] and scanning microwave microscopy (SMM) utilizes different tip-sample interaction mechanisms to yield 2-dimensional doping maps. Spatial resolution and detectable doping range varies between the methods.

Here we demonstrate the benefits of near-field nanoscopy. We use laser light at 2.54 THz (wavelength $\lambda = 118\mu m$) scattering at metalized atomic force microscope tips to achieve 2-dimensional, quantitative mapping of carrier concentration in the range 10^{16} to 10^{19} carriers/cm^3 at a spatial resolution of 40 nm (λ/3000). Nanoscale resolution is achieved by THz field confinement at the very tip apex to within 30 nm. Images of 65-nm technology node CMOS transistors demonstrate the simultaneous THz recognition of materials and mobile carriers in a single nanodevice.

Electromagnetic radiation at THz frequencies addresses a rich variety of light-matter interactions because photons in this low energy range can excite molecular vibrations and phonons, as well as plasmons and electrons of non-metallic conductors [10, 11, 12, 13]. Consequently, THz radiation offers intriguing possibilities for material characterization currently motivating major efforts in the development of THz imaging systems [14, 15]. Diffraction unfortunately limits the spatial resolution to

978-1-4244-9113-1/11 $26.00 © 2011 IEEE

about half the wavelength which is in the order of 100 μm. For this reason, THz mapping of micro- or nanoelectronic devices could not be attained. A promising route to break the diffraction barrier and to enable sub-wavelength scale imaging is based on fine-focusing of THz radiation by millimeter-long tapered metal wires [16, 17, 18, 19]. Acting as antennas, the wires capture incident THz waves and convert them into strongly confined near fields at the wire tip apex [20]. When this confined field becomes modified by a close-by scanned sample, the scattered radiation carries information on the local dielectric properties of the sample [18, 21, 22]. THz images can be obtained by recording the scattered radiation by a distant THz receiver. Attempts of realizing such THz-scattering near-field optical microscopy (THz-SNOM), however, suffered from extremely weak signals and faint material contrasts owing to strong background scattering. Novel probes for THz focusing are thus a subject of current interest [23, 24]. In 2008, Huber et al. [25, 26] introduced THz near-field microscopy achieving unprecedented resolution of about 40 nm, paired with extraordinarily high image contrast and acquisition speed. This is enabled by interferometric detection of THz radiation scattered from cantilevered atomic-force-microscope (AFM) tips. Building on this pioneering work we demonstrate THz mapping of mobile carriers within a single nano-device and its relevance for semiconductor technology.

B. Theory

The optical near-field scattering, depending on refractive index and absorption of the sample, can be enhanced by phonon polaritons in polar dielectrics or plasmon polaritons in metals and doped semiconductors [27]. Using a wavelength near the polariton resonance of the material of interest thus causes sharp contrast between this and other materials.

A dipole model for the field scattered by the tip that has proven to agree well with experimental results has been developed and is thoroughly described in a paper by Keilmann and Hillenbrand [21].

The essential ingredient to nanoscale resolved THz near-field microscopy is a strongly confined THz near field for generating highly localized scattering. Nanoscale near-field confinement can be achieved by plane-wave illumination of a conical metal tip, similar to visible [28] and infrared [29] frequencies. Strong confinement as well as an about 25-fold field enhancement (compared to the incident field) are essentially caused by the lightning-rod effect [30]. We note that because of the short tip length (L << λ), geometrical antenna resonances [16, 23] can be neglected. Also, dielectric plasmon resonances of a metal tip are absent at THz frequencies.

C. The experimental set-up

Our experimental set-up basically consists of a home-built AFM combined with a THz laser light source and interferometric detection of the scattered light.

We use AFM-tips (L ≈ 20 μm) with a Pt metallization of about 20 nm thickness (Mod. CSC37/Ti-Pt, Mikromasch).

Financially supported by BMBF within the NanoFutur program, grant no. 03N8705, Deutsche Forschungsgemeinschaft Clusters of Excellence "Nanosystems Initiative Munich (NIM)" and "Munich-Centre for Advanced Photonics (MAP)", and Etortek Nanotron project from the Government of the Basque Country.

While the sample is scanned in conventional AFM mode, both the tip (with curvature radius of about 30 nm) and the cantilever are illuminated with a focused laser beam at a frequency of 2.54 THz from a continuous-wave CH_3OH gas laser (Mod. SIFIR-50, Coherent), at a power of about 5 mW (Fig. 1). THz imaging is performed by collecting the back-scattered radiation with a parabolic mirror. For detection we use a Michelson interferometer featuring a 23 μm thick polyethylene beam splitter at 45° incidence. Interferometric detection offers the advantage of signal amplification which is crucial for measuring the back-scattered THz radiation with a high signal-to-noise ratio at reasonable integration times less than 100 ms per pixel. As THz detector we use a hot-electron bolometer (Mod. RS1-5T, Scontel). In order to eliminate background scattering the AFM is operated in dynamic mode where the cantilever oscillates at its mechanical resonance frequency Ω, here at 35 kHz, with amplitude of about 100 nm_{pp}. The bolometer signal is subsequently demodulated at harmonic frequencies $n\Omega$ (with n = 2 or 3) yielding a background-free THz signal amplitude s_n [21, 31].

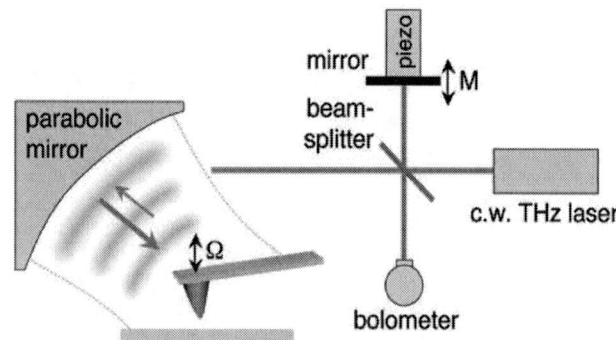

Figure 1. Scheme of our experimental setup based on an AFM: A laser emitting a monochromatic beam at 2.54 THz is used for illuminating a cantilevered AFM tip and interferometric detection is used for recording the backscattered THz radiation simultaneously with the AFM topography.

II. EXPERIMENTAL RESULTS

A. Sample preparation

Samples were prepared for imaging by cleaving and mechanical polishing to the cross-section of interest. No further sample preparation such as etching or metal-coating was necessary [32].

B. Terahertz near-field nanoscopy

By imaging a polished cross section through a 65-nm technology node CMOS transistor we demonstrate the simultaneous recognition of materials and free carriers by THz near-field microscopy. While the AFM topography (Fig. 2a) only shows some depressions where metal contacts (Cu, W) were differently polished than SiO_2, the THz image (Fig. 2b) clearly recognizes seven transistors manufactured in Si, with polycrystalline Si gates, and with SiO_2 as insulating material. The THz contrast can be clearly related to the different materials composing the transistors as we verify by scanning electron microscopy (SEM) of a similar sample where decoration etching was employed to highlight the different materials (Fig. 2d). We find that regions with metals or highly conductive semiconductors give the highest signal in the THz image, the lowly doped semiconductors less, and the low-refractive-index oxides the lowest signal, similar to earlier findings of scattering near-field microscopy at optical and infrared frequencies [21, 33]. This near-field contrast can be explained by dipolar near-field coupling between tip and sample which predicts higher signals s_2 for materials with higher dielectric values.

Near-field microscopy at THz frequencies particularly enables us to recognize mobile carriers and their distributions, in a concentration range centrally important for semiconductor science and technology ($n = 10^{16}$ to 10^{19} carriers/cm^3) where visible and infrared methods lack sensitivity. This sensitivity can be clearly seen from the strong THz signal variations within the Si substrate of the device structure shown in Fig. 2a. We observe a decreasing THz signal just below the transistors and a local maximum at 500 nm depth. After passing a second minimum at about 900 nm depth, the THz signal reaches a constant level on intrinsically doped Si. A comparison with the nominal mobile carrier concentrations (the numbers in Fig. 2b and c are from device simulations) provides clear evidence that the THz contrast maps the mobile carrier distribution with nanoscale resolution. For comparison we also show an infrared near-field image taken with a CO_2 laser at $\lambda \approx 11$ μm (Fig. 2c). It does not exhibit signal variations below the transistors.

To explain the THz contrast in the theoretical framework mentioned above, we calculate the frequency-dependent THz signal s_2 (Fig. 2e), between a metallic sphere and an extended Si surface where the mobile carrier response is described by a Drude term which depends on the mobile carrier concentration n [34, 35]. The calculated spectra have highest THz signals at low frequencies and a shape which can be assigned to the near-field coupling between the tip and the mobile carrier plasmons in the Si sample.

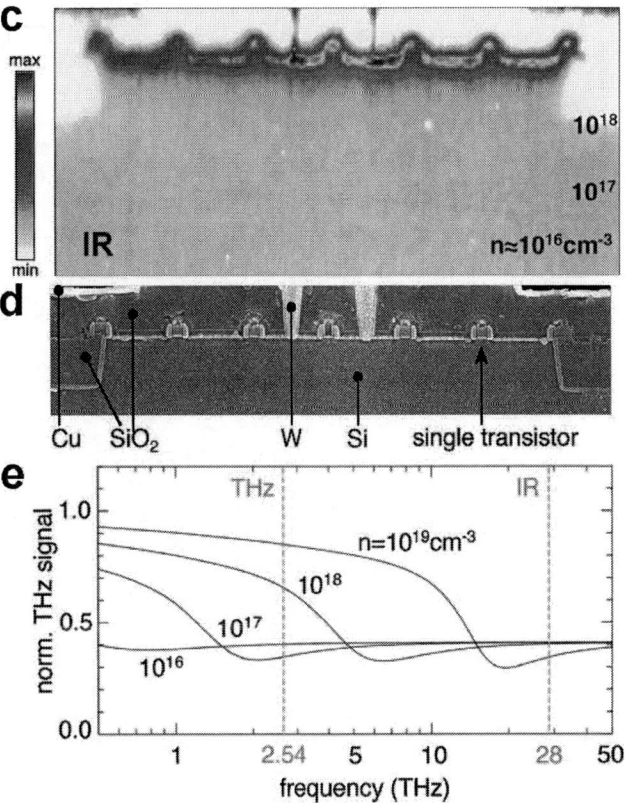

Figure 2. THz near-field microscopy of a polished cut through a multiple-transistor device structure. AFM topography (a) and simultaneously acquired THz near-field image (b). The varying THz signal within the Si substrate reveals the different mobile carrier concentrations n indicated by numbers obtained from device simulations. The infrared near-field image (c) ($\lambda \approx 11$ μm; taken for comparison) clearly demonstrates that only with THz illumination the varying free-carrier concentration can be recognized. The SEM image (d) of a similar but decoration-etched sample validates that the THz image distinguishes different materials and the single transistors. The rectangle in 2a and b marks the zoom-in area depicted in Fig. 4a. In 2e the THz signal amplitude s_2 calculated for n-doped Si as a function of the illumination frequency is shown.

With increasing n, the spectral signature shifts to higher frequencies. Interestingly, a minimum near the plasma frequency is predicted, which occurs at 2.54 THz (red line in Fig. 2e) for $n \approx 2 \cdot 10^{17}$ cm^{-3}. A minimum is indeed observed in the THz image (Fig. 2b), at a depth of 900 nm below the transistors where the designed concentration gradient assumes a value of $n \approx 2 \cdot 10^{17}$ cm^{-3}.

In Fig. 3 a comparison between a simulation of the transistors net doping and a terahertz near-field nanoscopy image is shown, as well as a line profile through the center of the gate (along the white, dashed line) showing the normalized THz signal. It can be noted that the image and line profile both correspond well to the carrier concentrations given by the simulation. This good agreement between design, experiment and theory enables immediate applications of THz near-field microscopy in semiconductor science and technology. Owing to the distinct relation between THz-plasmons and carrier concentration, our results open the door to quantitative mobile carrier profiling at the nanometer scale. The capability of quantifying relevant mobile-carrier distributions with nanoscale lateral resolution could be augmented by improving the model and by employing broadband THz spectroscopy [36].

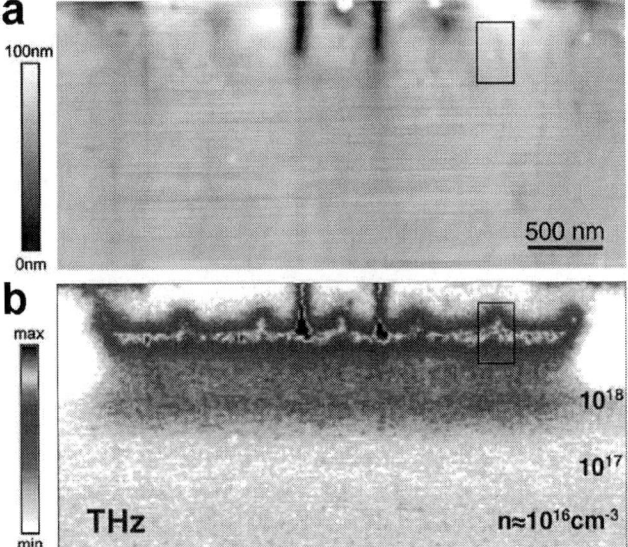

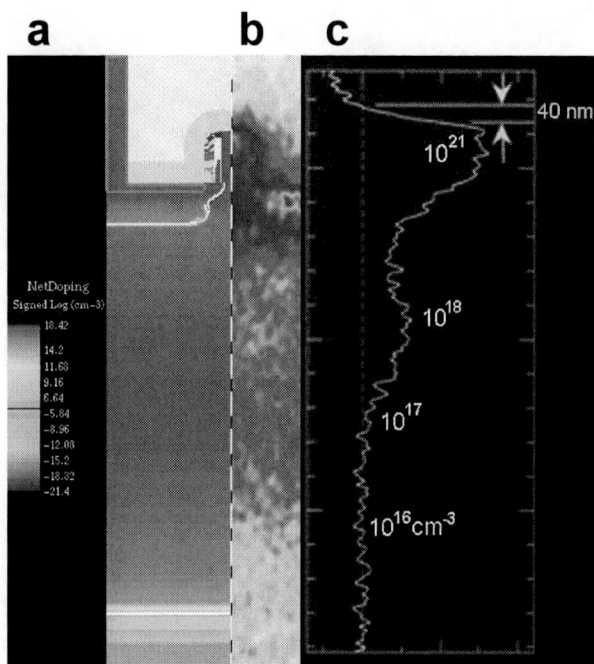

Figure 3. Comparison of the THz near-field nanoscopy image with simulation. Please note that the carrier concentrations indicated by the simulation (a) can be clearly resolved in both the THz image (b) and in the line profile (c) through the center of the gate, as marked by the dashed white line.

Further evidence of the values of THz near-field microscopy is provided by scanning a single transistor (marked in Fig. 2a and b) at reduced pixel size. Fig. 4 shows a 65-nm technology node transistor. In (a) a terahertz near-field nanoscopy image is shown. Note that the poly-Si gates, NiSi metallization, as well as the source- and drain-diffusions can be clearly out-lined! In (b) a line profile along the dashed, black line in (a) through the center of the gate shows a sharp signal increase from the low-refractive-index material SiO$_2$ to the metallic NiSi gate contact. Due to the smooth sample surface (see Fig. 2a) we can exclude any topography-related artifact and also prove the pure dielectric origin of the THz contrast. This allows us to unambiguously determine a near-field optical resolution of about 40 nm (λ/3000), defined as the distance between 20% and 80% of the increase from minimum to maximum THz signal. In (c) a transmission electron microscopy (TEM) image of a transistor from the same device is shown for comparison. The component material and an out-line of the source- and drain-diffusions are indicated in the image. The comparison between TEM and high-resolution THz image clearly verifies the capability of THz near-field microscopy to map the basic entities of the transistor: source, drain and gate. The metallic NiSi is more or less opaque to the electron beam, and thus appears as black areas in this high contrast TEM image.

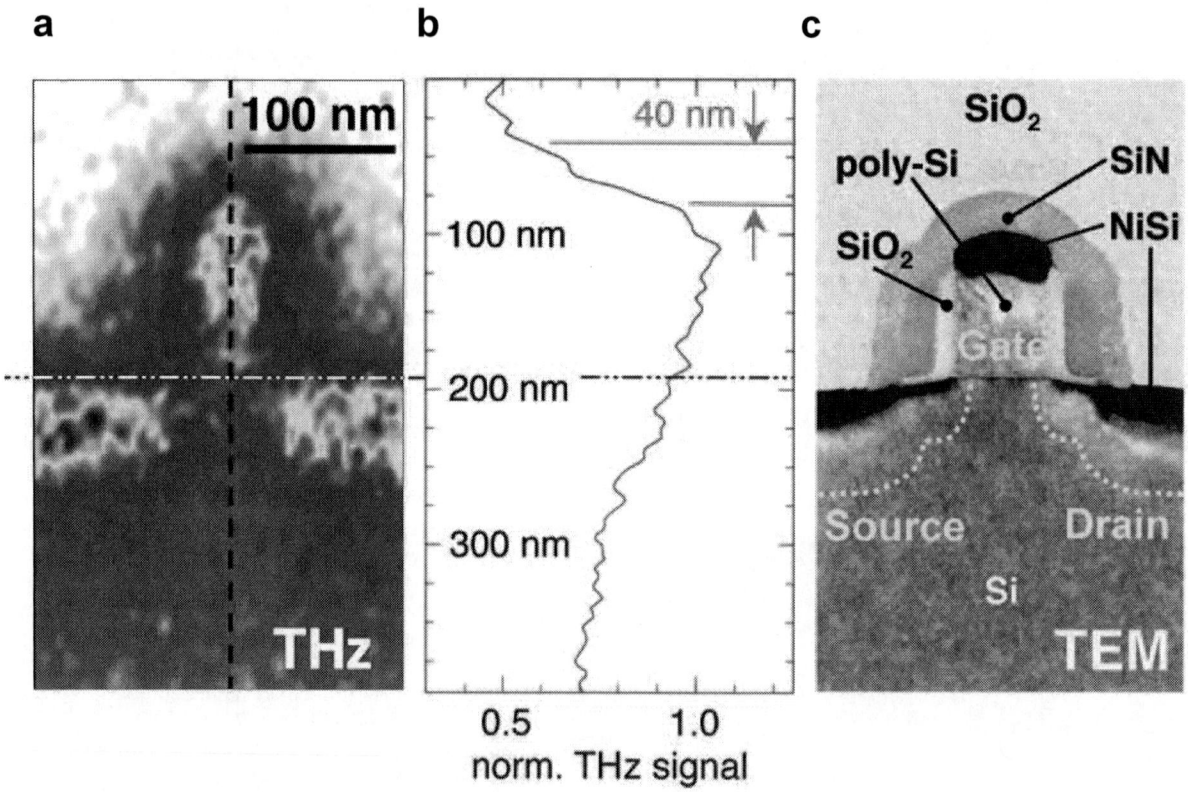

Figure 4. 65-nm technology node CMOS transistor: (a) THz near-field nanoscopy image. (b) Line profile demonstrating the 40 nm resolution, defined as the 20% to 80% signal increase distance. (c) TEM image of a transistor from the same device as shown in the THz near-field nanoscopy image.

In the THz image on the other hand, a gradient in signal within the NiSi layer can be observed. This is probably caused by the interaction depth of the THz nanoscopy of about 40 nm which averages the signal from the relatively rough Si-NiSi interface. Furthermore, we find that between the NiSi source and drain contacts (black parts in the TEM image) the THz signal level reaches an intermediate value. According to our calculation in Fig. 2e, this indicates a mobile carrier concentration in the order of 10^{18} cm^{-3} which is in good agreement with results from device simulations. Obviously, THz near-field microscopy allows for quantitative probing of mobile carriers in the 65 nm wide region between source and drain. In combination with future spectroscopic extensions [17, 18, 19, 22, 36, 37, 38], this possibility could open the door to even measure the carrier mobility in this most important part of nanoscale semiconductor devices.

C. Comparison with SCM and SSRM

For comparison, the same sample was imaged using SSRM and SCM, the two most widely used, SPM based methods for 2-dimensional carrier concentration mapping. In Fig. 5a a SSRM image is shown, demonstrating the excellent spatial resolution achievable [9]. However, it also shows the lack of sensitivity at lower carrier concentration, as well as its completely missing material contrast. Moreover, even though SSRM in principle can give quantitative results, it is for real failure analysis samples seldom possible to achieve this due to sample preparation complications. The SCM image in Fig. 5b demonstrates the capability of SCM of differentiation between p- and n-doped areas, and even though its spatial resolution is limited compared to SSRM it is comparable to that achieved using THz or IR s-SNOM. Moreover, sub-10 nm resolution as well as quantitative measurement using SCM has been reported [7].

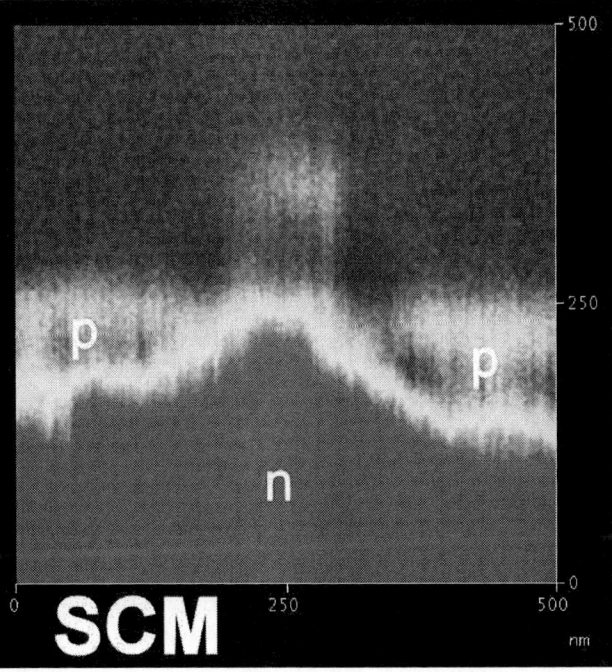

Figure 5. In (a) an SSRM image is shown, demonstrating its high spatial resolution. In (b) an SCM image is shown – demonstrating its p- and n-doping differentiation.

III. DISCUSSION

An intriguing consequence follows from the sensitivity of our setup to mobile carrier concentrations in the range 10^{17} - 10^{18} cm^{-3}, evident from the THz contrast shown in Fig. 4a. Since the spatial resolution of 40 nm infers [30] that the volume probed by the THz near-field is about (40 nm) [12], we conclude that an average of less than 100 electrons in the probed volume suffices to evoke significant THz contrast. This opens the fascinating perspective that straightforward improvements of the current setup could master THz studies of single electrons - and in conjunction with ultrafast techniques even their dynamics. Moreover, THz near-field microscopy seems predestined to study also other charged particles and quasiparticles in condensed matter—for example in superconductors, low-dimensional electron systems or conducting biopolymers—which possess intrinsic excitations at THz quantum energies and thus should exhibit resonantly enhanced THz contrast. Also biological molecules and cellular entities may be possible to investigate using THz nanoscopy.

IV. CONCLUSIONS

We succeeded in quantitative mapping areas with various materials, as well as free carrier concentrations between 10^{16} and 10^{19} carriers/cm^3. The capacity of THz near-field nanoscopy to perform quantitative studies of local carrier concentration and mobility at the nanometer scale gives it the potential to find many applications within semiconductor industry and research.

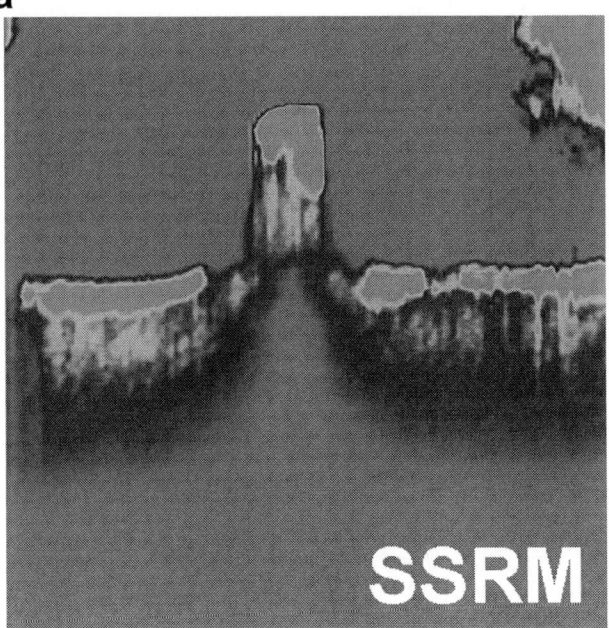

ACKNOWLEDGMENT

Angela Collantes (Infineon) is acknowledged for TEM imaging, and Stephan Schömann (Infineon) is acknowledged for SSRM imaging.

REFERENCES

[1] Alvis, R. and Kelly, T. F., *Microscopy Today*, September 2008, pp. 6-11.

[2] Schweinböck, T., Schömann, S., Alvarez, D., Buzzo, M., Frammelsberger, W., Breitschopf, P., and Benstetter, G. "New trends in the application of Scanning Probe Techniques in Failure Analysis", *Microelectronic reliability*, Vol. 44, (2004), pp. 1541-1546.

[3] Born, A. and Wiesendanger, R., "Scanning capacitance microscope as a tool for the characterization of integrated circuits", *Applied Physics A*, Vol. 66, (1998), pp. 421-426.

[4] Anand, S., "Another Dimension in Device Characterization – Scanning Capacitance Microscopy of InP-Based Laser Structures", *Circuits & Devices,* March, (2000), pp. 12-18.

[5] Zimmerman, G., Born, A., Ebernsberger, B., and Boit, C., "Application of SCM for the microcharacterization of semiconductor devices", *Applied Physics A*, Vol. 76, (2003), pp. 885-888.

[6] Benstetter, G., Breitschopf, P., Frammelsberger, W., Ranzinger, H, Reislhuber, P., and Schweinböck, T., "AFM-based scanning capacitance techniques for deep sub-micron semiconductor failure analysis", *Microelectronic reliability*, Vol. 44, (2004) pp. 1615-1619.

[7] Kikuchi, Y., Kubo, T., and Kase, M., "Quantitative Ultra Shallow Dopant Profile Measurement by Scanning Capacitance Microscope", FUJITSU Sci. Tech. J., Vol. 38, No. 1, (2002) pp. 75-81.

[8] De Wolf, P., Clarysse, T., Vandervorst, W., Hellemans, L., Niedermann, Ph., and Hänni, W., "Cross-sectional nano-spreading resistance profiling", *J. Vac. Sci. Technol. B,* Vol. 16(1), (1998), pp. 355-361.

[9] Schömann, S. and Alvarez, D., "Doping mapping by SSRM – reaching maturity and sub-10nm resolution", *Proceedings from the 30th International Symposium for Testing and Failure Analysis,* (2004), pp. 346-349.

[10] Kuzmany, H., *Solid-State Spectroscopy.* Springer: Berlin, 1998

[11] Ferguson, B.; Zhang, X. C. *Nature Materials* (2002), 1, pp. 26-33.

[12] Mittleman, D., *Sensing with Terahertz Radiation.* Springer: Berlin, 2003.

[13] Tonouchi, M. *Nature Photonics* (2007), 1, pp. 97-105.

[14] Chan, W. L.; Deibel, J.; Mittleman, D. M. *Reports on Progress in Physics* (2007), 70, 1325-1379.

[15] Withayachumnankul, W.; Png, G. M.; Yin, X. X.; Atakaramians, S.; Jones, I.; Lin, H. Y.; Ung, B. S. Y.; Balakrishnan, J.; Ng, B. W. H.; Ferguson, B.; Mickan, S. P.; Fischer, B. M.; Abbott, D. *Proceedings of the IEEE* (2007), 95, pp. 1528-1558.

[16] Matarrese, L. M.; Evenson, K. M. *Applied Physics Letters* **1970,** 17, pp. 8-9.

[17] van der Valk, N. C. J.; Planken, P. C. M. *Applied Physics Letters* (2002), 81, pp. 1558-1560.

[18] Chen, H. T.; Kersting, R.; Cho, G. C. *Applied Physics Letters* (2003), 83, pp 3009-3011.

[19] Chen, H. T.; Kraatz, S.; Cho, G. C.; Kersting, R. *Physical Review Letters* (2004), 93, 267401.

[20] Planken, P. C. M.; van der Valk, N. C. J. *Optics Letters* (2004), 29, pp. 2306-2308.

[21] Keilmann, F. and Hillenbrand, R., "Near-field microscopy by elastic light scattering from a tip" *Philosophical Transactions of the Royal Society London. A,* Vol. 362 (2004), pp. 787–805.

[22] Buersgens, F.; Kersting, R.; Chen, H. T. *Applied Physics Letters* (2006), 88, 112115.

[23] Wang, K. L.; Mittleman, D. M.; van der Valk, N. C. J.; Planken, P. C. M. *Applied Physics Letters* (2004), 85, pp. 2715-2717.

[24] Maier, S. A.; Andrews, S. R.; Martin-Moreno, L.; Garcia-Vidal, F. J. *Physical Review Letters* (2006), 97, 176805.

[25] Huber, A. J., Keilmann, F., Wittborn, J., Aizpurua, J. and Hillenbrand, R., *Nano Letters*, (2008), 8 (11), pp. 3766-3770.

[26] Highlighted by "A terahertz nanoscope", *Nature* 456, (2008), pp. 454-455.

[27] Ocelic, N., Hillenbrand, R., "Subwavelength-scale tailoring of surface phonon polaritons by focused ion-beam implantation", *Nature Materials,* Vol. 3, (2004), pp. 606-609.

[28] Novotny, L.; Sanchez, E. J.; Xie, X. S. *Ultramicroscopy* (1998), 71, pp. 21-29.

[29] Cvitkovic, A.; Ocelic, N.; Aizpurua, J.; Guckenberger, R.; Hillenbrand, R. *Physical Review Letters* (2006), 97, 60801.

[30] Novotny, L.; Hecht, B., *Principles of Nano-Optics.* Cambridge University Press (Cambridge, 2007).

[31] Ocelic, N.; Huber, A.; Hillenbrand, R. *Applied Physics Letters* (2006), 89, 101124.

[32] Wittborn, J., Weiland, R., Kazantsev, D., Huber, A., Keilmann, F. and Hillenbrand, R., "Material and doping contrast in semiconductor devices at nanoscale resolution using scattering-type scanning near-field optical microscopy", *Proceedings of the 32nd International Symposium for Testing and Failure Analysis*, Austin, Texas, USA, November, (2006), pp. 98-101.

[33] Taubner, T.; Hillenbrand, R.; Keilmann, F. *Journal of Microscopy-Oxford* (2003), 210, pp. 311-314.

[34] Huber, A. J.; Kazantsev, D.; Keilmann, F.; Wittborn, J.; Hillenbrand, R. *Advanced Materials* (2007), 19, pp. 2209-2212.

[35] Knoll, B.; Keilmann, F. *Applied Physics Letters* (2000), 77, pp. 3980-3982.

[36] von Ribbeck, H. G.; Brehm, M.; van der Weide, D. W.; Winnerl, S.; Drachenko, O.; Helm, M.; Keilmann, F. *Opt. Express* (2008), 16, pp. 3430-3438.

[37] Zhan, H.; Astley, V.; Hvasta, M.; Deibel, J. A.; Mittleman, D. M.; Lim, Y. S. *Applied Physics Letters* (2007), 91, 162110.

[38] Naftaly, M.; Miles, R. E. *Proceedings of the IEEE* (2007), 95, pp. 1658-1665.

Rapid and Automated Grain Orientation and Grain Boundary Analysis in Nanoscale Copper Interconnects

K. J. Ganesh[1], S. Rajasekhara, D. Bultreys[2], P. J. Ferreira[1]

[1]*Materials Science and Engineering Program, The University of Texas – Austin, Austin, Texas – 78712, USA*
[2]*NanoMEGAS Inc. SPRL Blvd. Edmond Machtens 79, B-1080, Brussels, Belgium*

Abstract - **A combination of diffraction scanning transmission electron microscopy (D-STEM) and automated precession microscopy is used to obtain orientation information from 108 copper grains in 120 nm wide copper interconnect lines. Grain boundary analysis based on this orientation data reveals that $\sum 3^n$ (n = 1, 2) boundaries are predominant in these lines. Finite element analysis reveals regions of high and low stresses within the copper microstructure.**

I. INTRODUCTION

The downscaling of feature sizes in complex CMOS devices require that the interconnect wiring needed for these devices also shrink in dimensions [1-2]. The current generation of copper interconnect (CI) lines, which are widely used for interconnect wiring, have linewidths of approximately 90-120 nm and were expected to decrease to approximately 45 nm by 2010 [3, 4]. As these CIs are getting narrower, stress-induced voiding (SIV) becomes one of the reliability issues of concern. SIV phenomenon is thought to occur when residual quasi-hydrostatic stresses ($\sigma_x \neq \sigma_y \neq \sigma_z$ and σ_z > 0) present in narrow CI lines relax during thermal cycling [4-6]. These residual stresses are present due to the difference in the coefficients of thermal expansion between the CI lines and the silicon substrate [1-6]. Recently, direct evidence of this relationship between SIV formation and local stresses in 180 nm CI lines was demonstrated by *in-situ* transmission electron microscopy heating experiments [7]. Previously, the stress analysis of CIs with different aspect ratios using Finite Element Method (FEM) simulations [6, 8] have shown that large hydrostatic stress gradients at the CI line/dielectric interface may provide a driving force for void formation. However, these studies assumed the CI lines to be isotropic, thus neglecting the anisotropic elastic behavior of copper. Nucci et al [9, 10] recognized this limitation and suggested that local variations in the CI microstructure, such as local texture, misorientation between grain boundaries, and the presence of triple junctions, should contribute to stress induced void formation in narrow CI lines [9, 10]. However, these studies were performed on Cu thin films and wide CI lines (~ 1μm). Until recently, no analysis was available for CI

lines with narrow linewidths (~ 120 nm) primarily because of the difficulties in obtaining crystal orientation information from these lines.

In a recent article [11, 12], we demonstrated that a novel diffraction scanning transmission electron microscopy (D-STEM) technique may be used to obtain local orientation information from nano/submicron grains present in CI lines. However, for grains oriented off-axis with respect to the beam direction and regions with a high density of defects, the orientation analysis is limited. As a consequence, the overall statistical analysis is affected. In the current work, we have advanced this technique further and combine, for the first time, D-STEM with precession microscopy to obtain local orientation information from 108 grains in multiple 120 nm CI lines. The combination of these two techniques (D-STEM and precession microscopy) allows for rapid (at least ten times faster than the original D-STEM technique) collection of orientation information from nanoscale grains, allowing us to make a statistically significant analysis of texture and the relation between grain boundaries and local stresses in 120 nm CI lines.

II. EXPERIMENTAL PROCEDURE

The D-STEM configuration is combined with full-precession using the ASTAR™ system from NanoMEGAS Inc. The precession system allows for deflection of the beam prior to incidence on the specimen to form an oblique illumination condition. After interacting with the specimen, the diffracted beams are re-deflected using a complimentary deflection system called de-scan to produce a stationary diffraction pattern which has the geometry of a conventional pattern [13]. A separate scan generator installed on the microscope is used to raster the precessed beam on the sample plane to obtain diffraction patterns on a pixel by pixel basis. Under the above mentioned conditions, diffraction intensities are primarily acquired from off-zone axis reflections and are integrated through the Bragg condition, thereby generating pseudo-kinematical patterns [14]. Such near-kinematical patterns are directly interpretable and thus, allow us to rapidly obtain orientation information from nanoscale grains, even those oriented off-axis with respected to the beam direction and located in regions with a high density of defects.

To conduct this work, CI lines 120 nm wide, 200 nm deep, and spaced approximately 120 nm apart were provided by Freescale[TM] Semiconductor Inc. Planar electron-transparent CI lines for TEM analysis were prepared by conventional grinding, dimpling and ion-milling techniques.

II. RESULTS AND DISCUSSION

Figure 1a and 1b show bright-field STEM images of two sets of 120 nm CI lines taken under the D-STEM conditions, and the corresponding orientation maps generated by precession microscopy are shown in Figs. 1c and 1d, respectively. The various colors in Figs. 1c and 1d represent grains with different orientations. The grains within the CI lines are predominantly bamboo-type with regions of non-bamboo type nanoscale grains and annealing twins. No dominant texture is present in these CI lines. Furthermore, these results confirm the observation made in an earlier work that narrow CI lines comprise of few {111} type grains [11].

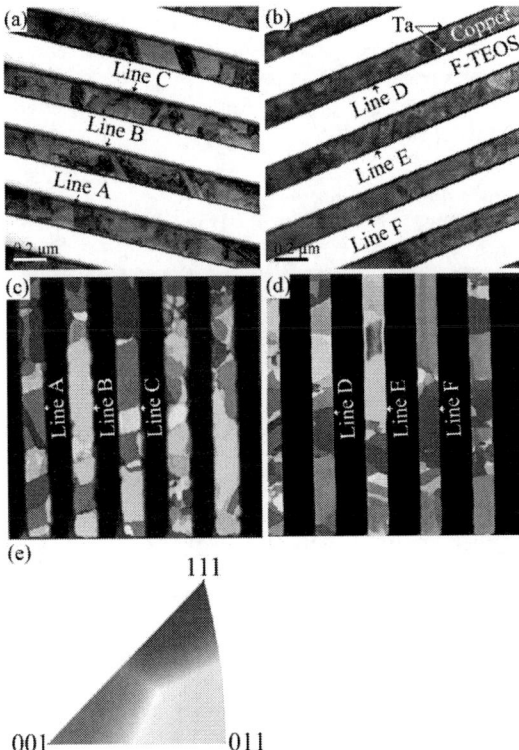

Fig.1. (a) and (b) Bright-field STEM images of two sets of CI lines, (c) and (d) orientation map of the CI lines shown in (a) and (b), which were obtained by a combination of D-STEM and precession microscopy. The precessed beam was sequentially scanned along the length of the CI lines, and independent orientation images were reconstructed as shown in (c) and (d); the rotation of the lines in Figs. (c) and (d) with respect to Figs. (a) and (b) is a result of the image-reconstruction procedure, (e) orientation index for images (c) and (d).

The orientations from each arbitrary grain were recorded with respect to a reference coordinate system in the form of Euler angles (φ_1, ϕ and φ_2), where φ_1 is the angle of rotation of the grain of interest about its z-axis in the crystal reference frame, ϕ the angle of rotation of the grain of interest about its transformed x-axis in the crystal coordinate system, and φ_2 the angle of rotation of the grain of interest about its transformed z-axis. These angles may be used to determine the misorientation matrices M'. A detailed procedure is outlined in Ref. 15. The misorientation matrix with the largest trace (Tr) and the elements of this matrix m'_{ij} may be used to determine the minimum misorientation angle and the common misorientation axis <UVW> between two grains, given by [15]:

$$\Theta = \cos^{-1}\left[\frac{Tr-1}{2}\right] \qquad (1a)$$

$$U:V:W = m'_{32} - m'_{23} : m'_{13} - m'_{31} : m'_{21} - m'_{12} \qquad (1b)$$

On the basis of the Euler angles of the 108 grains present in the two sets of CI lines shown in Figs 1c and 1d, the misorientation angles and misorientation axes between adjacent grains have been determined.

Based on the abovementioned calculations, we have found a significant fraction of grain misorientation angles with $\Theta \approx 39°$ and $\Theta \approx 60°$ (Fig. 2). The misorientation axes for $\Theta \approx 39°$ and $\Theta \approx 60°$ are of the type <110> and <111>, which correspond to the $\sum 9$ and $\sum 3$ Coincident Site Lattice (CSL) type boundaries, respectively [15]. The presence of such CSL boundaries is expected in annealed polycrystalline *fcc* metals with low stacking fault energies [16, 17].

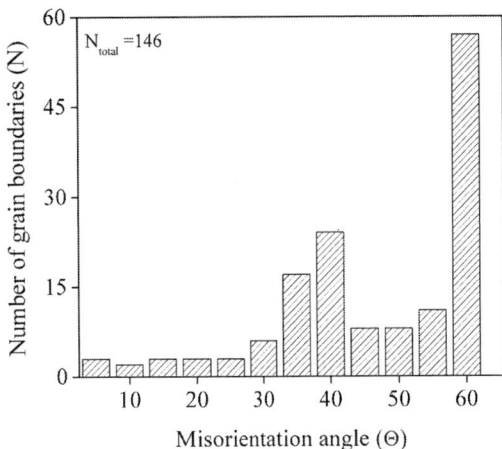

Fig. 2. Misorientation angles between adjacent grains obtained from 144 grain boundaries in the two sets of CI lines shown in Fig. 1.

978-1-4244-9113-1/11 $26.00 © 2011 IEEE

Upon knowing the misorientations between grains, local stress distribution within the CI lines was computed using Finite Element Method (FEM) on lines A, B and C from Fig. 1a and lines D, E and F from Fig. 1b. The simulations were performed with the OOF2 program, a FEM package that takes into account the local grain misorientations present in a 2D microstructure, while solving for a stress solution. The details of the procedure are available in a previous article [11]. The computer simulations were performed on 2D microstructures, using x and y displacements as boundary conditions. Although the OOF2 program only accepts 2D boundary conditions, more complex simulations using a 3D model were not performed because (i) the grains would need to be assumed columnar, which is not applicable for a heavily twinned material like copper, and (ii) automated grain orientation would need to be determined in 3D, a technique which is currently unavailable. Nevertheless, the OOF2 2D simulations provide a correlation between the relative stress distribution and the 2D microstructure of the CI lines, which can offer a critical insight into the fundamental mechanisms of stress-induced void formation [7]. The results provided by the FEM simulations show that the hydrostatic stresses present within the interior of grains are in the range 500 – 550 MPa (Figs. 3b and 3d). This finding agrees well with previously obtained results on a single CI line [11], as well as FEM studies on isotropic CI lines [6, 8].

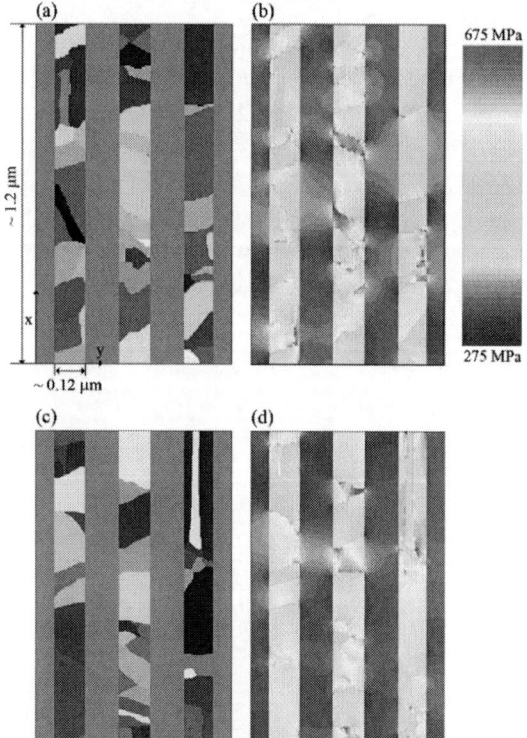

Fig. 3. (a) A color-coded replica of lines A, B, and C from Fig.1b that orientation information for each grain (b) the corresponding FEM stress solution, (c) a color-coded replica of lines D, E, and F from Fig. 1d that that contains orientation information for each grain, (d) the corresponding FEM stress solution.

In summary, the combination of D-STEM with precession electron microscopy provides a rapid way to analyze local texture in downscaling nano CI lines in an automated manner with a spatial resolution of 1-2 nm. Misorientation information in the CI lines can be used for microstructural FEM analysis using OOF2 to locally identify regions of high and low stresses in these CI lines This would be critical to investigate the influence of grain orientations and grain boundaries on SIV

ACKNOWLEDGMENT

The authors acknowledge the financial support from Semiconductor Research Corporation (SRC), contract 2010-KJ-2072. The authors also thank Professor Katyun Barmak and Amith Darbal, from Carnegie Mellon University, for valuable discussions regarding grain boundaries.

REFERENCES

1. T. D. Sullivan, Annual Reviews in Materials Science, **26** (1996) 333.
2. B. Li, T. D. Sullivan, T. C. Lee, D. Badami, Microelectronics Reliability **44** (2004) 365.
3. International Technology Roadmap for Semiconductors – Interconnects (2007) pp. 7-8.
4. J. H. An, P. J. Ferreira, Applied Physics Letters **89** (2006) 151919.
5. J. Zhang, J. Y. Zhang, G. Liu, Y. Zhao, J. Sun, Thin Solid Films **517** (2009) 2936.
6. S-H. Rhee, Y. Du, P. S. Ho, Journal of Applied Physics, **93** (2003) 3926.
7. J. H. An, *Thermal Stress Induced Voids in Nanoscale Cu Interconnect by in-situ TEM heating,* The University of Texas – Austin, Ph.D. Dissertation, (2007).
8. D. Ang, C. C. Wong, R. V. Ramanujan, Thin Solid Films **515** (2007) 3246.
9. J. A. Nucci, R. R. Keller, J. E. Sanchez Jr., Y. S-Diamand, Applied Physics Letters **69** (1996) 4017.
10. J. A. Nucci, R. R. Keller, D. P. Field, Y. S-Diamand, Applied Physics Letters **70** (1997) 1242.
11. K. J. Ganesh, S. Rajasekhara, J. P. Zhou, P. J. Ferreira, Scripta Materialia **62** (2010) 843.
12. K. J. Ganesh, M. Kawasaki, J. P. Zhou, P. J. Ferreira, Microscopy and Microanalysis, Vol.16, pp. 614-621 (2010)
13. T. A. White, A. S. Eggeman, Midgley, Ultramicroscopy (2009) doi: 10.16/j.ultramic. 2009.10.013.
14. C. S. Owen, L. D. Marks, Review of Scientific Instruments **76** (2005) 033703.
15. V. Randle, *The Measurement of Grain Boundary Geometry*, IOP Publishing 1st Ed. (1993) pp. 17-45.
16. V. Randle, Acta Metallurgica **47** (1999) 4186.
17. V. Randle, Materials Charaterization **47** (2001) 411.

High Reliable Strain Measurement For Power Devices Using STEM-CBED Method

N. Nakanishi, H. Arie, H. Maeda, Y. Hirose, N. Hattori, T. Koyama and E. Murakami

Renesas Electronics Corp., Devices & Analysis Technology Division,
4-1, Mizuhara, Itami, Hyogo 664-0005, Japan
phone: +81-72-787-2440, e-mail address: nobuto.nakanishi.jz@renesas.com

Abstract—Scanning transmission electron microscopy convergent beam electron diffraction (STEM-CBED) was applied to strain analysis of deep trench electrodes for devices such as power devices. Source-drain current leak, which was one of the crucial failures of this kind of structure, depends on boron concentration in boron doped poly-Si (BP) layers. TEM/STEM and diffraction analysis showed that the BP layers consist of epitaxial phase and poly-Si phase, and the proportion of these phases depends on the boron concentration in the BP layer. Clear strain distribution around the BP layers was obtained with STEM-CBED. This revealed that the origin of the strain is volume shrinkage of the epitaxial phase in the BP layer, and the poly-Si phase acts as buffer against this strain. Relationship between Si phases and boron concentration in the BP layers was examined with STEM and scanning capacitance microscopy. These analyses suggested that boron segregation occurred in samples having a higher boron concentration, and prevented epitaxial growth in the BP layers. As a result, the core of the BP layer remains as poly-Si or amorphous Si and acts as strain buffer. Our analysis concluded that boron concentration in the BP layer is one of the most important factors enabling high production yield for the structure.

Keywords-Strain measurement, Convergent beam electron diffraction, Scanning transmission electron microscopy, Power devices, Boron doped poly-Si

I. INTRODUCTION

Trench type electrode structures penetrating into P−epitaxial layers are employed for devices that need vertical electrical connections such as power devices. Since strain induces crystal defects which cause current leak in these devices, it has to be evaluated precisely in order to improve production yield [1]. Transmission electron microscopy (TEM) is one of the most suitable tools for strain measurement at the transistor level. Among the different TEM techniques, nano-beam diffraction (NBD) [1, 2] and convergent beam electron diffraction (CBED) [3-6] are used for quantitative strain measurement. These techniques have some advantages. NBD can be applied near interfaces or surfaces. Therefore NBD seems to be suitable for strain measurement of advanced CMOS devices. CBED is the most sensitive to strain among the TEM techniques [6]. Obtaining strain distribution is desirable for detailed analysis especially for power devices which have vertical structure such as trench type electrodes or isolations. Using NBD and CBED in TEM mode to obtain strain distribution at the nano-meter level is not convenient and

practical because the data is acquired manually and the quality of the data depends to a great extent on the operator's skill. Furthermore, it takes a long time to acquire the patterns because the operation is based on point measurements (For example: $0.5 - 1$ min/point). A combination of scanning TEM (STEM), one of the various TEM imaging techniques [5], and CBED appears quite promising for precise strain measurement because both use the same optical configuration. Obtaining the strain distribution helps us to understand the failure mechanism. In this paper, we introduce a STEM-CBED method to analyze the strain and the failure mechanisms of trench type electrode structures.

II. EXPERIMENT

A. Samples

Figure 1 shows a schematic drawing of the test structure used in this study. The source electrode is the backside of a wafer in this structure. A pair of trench type electrodes is used to connect the diffusion layer to the P++ substrate (source). This structure is formed by deposition of boron doped amorphous silicon in the trench and etching back to remove excess silicon from the substrate. Due to a series of thermal treatments after deposition, the amorphous boron doped silicon changes into a poly crystalline state. This electrode is known as a boron doped poly-Si (BP) layer. In this study, samples with different boron concentrations in the BP layer were prepared.

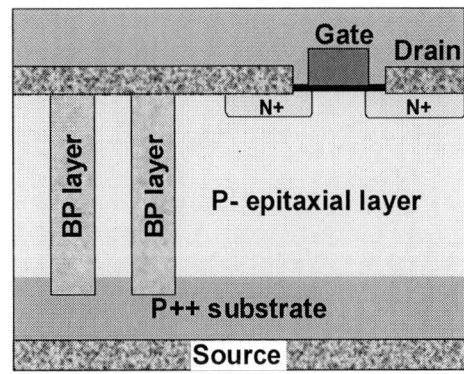

FIGURE 1 Schematic drawing of the test structure of this study.

B. STEM-CBED measurement

Figure 2 shows a schematic drawing of the STEM-CBED system. A finely focused convergent electron beam is used as probe and the detector or CCD is set on the back-focal plane (i.e. diffraction plane) for both STEM and CBED data acquisition. Therefore, the STEM image and CBED pattern can be obtained without obstruction using an annular dark field (ADF) detector as shown in Fig. 2. This enables us to obtain ADF-STEM image and CBED patterns simultaneously. Since a PC controlled system ensures precise position measurement, highly reliable strain measurements can be made. Furthermore, this system reduces measurement time drastically ($0.5 - 1$ sec/point) and allows us to obtain line distributions and two-dimensional mappings of strain.

TEM samples were prepared with focused ion beam (FIB). The target thickness for TEM samples for strain measurement is about 200 nm, compared to the normal thickness of 100 nm, in order to minimize stress relaxation due to sample thinning [7]. TEM/STEM observations and STEM-CBED measurements were performed with an FEI Tecnai G2 F20 field-emission electron microscope, using an accelerating voltage of 200 kV. CBED patterns were collected using Gatan Model 863 GIF Tridiem. In order to obtain clear higher-order Laue zone (HOLZ) lines in the CBED patterns, the patterns were acquired in energy-filtering mode which eliminates the effect of inelastic scattering in thick TEM samples [8]. The data were processed using the ASAC software [9] which extracts the HOLZ lines from the experimental CBED pattern and determines the lattice parameters. Strain is calculated as the relative lattice parameter variation $\Delta d/d$.

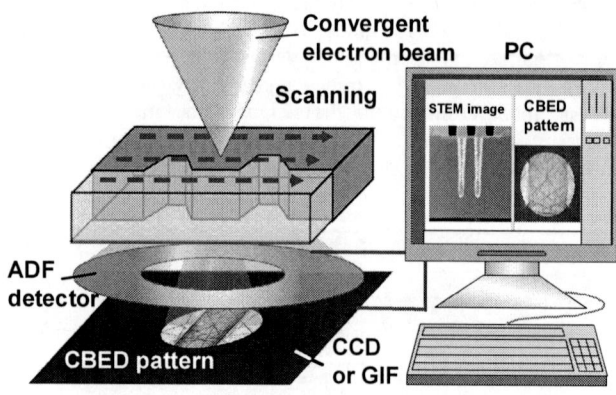

FIGURE 2 Schematic drawing of the STEM-CBED system.

III. RESULTS AND DISCUSSION

A. Electric Characteristics

The relationship between the boron concentration of the BP layer and source-drain current (Idss) leak is shown in Fig. 3. Idss leak increases as boron concentration of the BP-layers decreases. The Idss leak is caused by crystal defects such as dislocation from the BP layers. Samples used in this strain measurement are identified as samples A-C as indicated in Fig. 3.

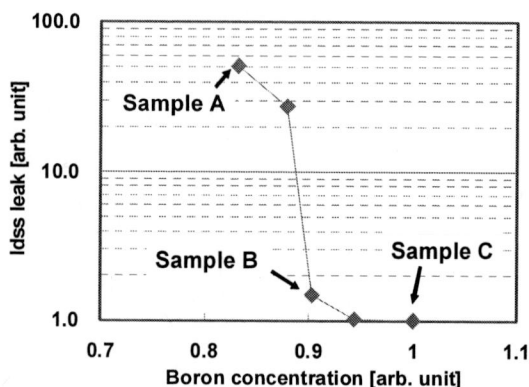

FIGURE 3 Relationship between boron concentration and Idss leak.

B. Crystal Structure Analysis

Bright field TEM images and diffraction patterns were taken from each sample (Fig. 4). The BP layers of sample A show only dark regions while those of samples B and C show dark and bright regions. In order to determine the crystal micro structure, selected area diffraction (SAD) patterns were taken from the BP layers as indicated by the circle in Fig. 4(a). In sample A, extra spots appear in addition to the fundamental spots from the Si substrate (Fig. 4(d)). These extra spots are identified as 1/3(111) and 2/3(111) spots that come from twins [10]. This result shows that Si in the BP layers consists of epitaxial Si and twins. Figure 5 shows a high-resolution TEM (HRTEM) image obtained at the P− epi./BP layer interface of sample A. There are several lamellar twins in the BP layer and they are epitaxially grown on substrate Si (or P− epi. layer). This phase having epitaxial Si and twins in the BP layer is denominated as the epitaxial phase. In the SAD patterns taken from samples B and C (Fig. 4(e) and (f)), the ring pattern characteristic of poly-Si [5] appears in addition to the spots from the epitaxial phase. This phase is identified as the poly-Si phase. Thus, the dark regions in the BP layer correspond to the epitaxial phase and the bright regions correspond to the poly-Si phase in bright field TEM images. This TEM analysis shows that there are two phases of silicon state in the BP layers having a higher concentration of boron, and the proportion of the epitaxial and poly-Si phase in the BP layer depends on the boron concentration in the BP layer. Figure 6 shows ADF-STEM images of sample C. Figure 6(a) shows a high angle ADF-STEM (HAADF-STEM) image which shows strong atomic number contrast [11]. The poly-Si phase appears slightly darker than the epitaxial phase and substrate Si while contrast of the epitaxial phase is almost the same as substrate Si. This suggests that the poly-Si phase consists of not only Si but also elements lighter than Si. Low-angle ADF-STEM (LAADF-STEM) images reflect contrast from differences in the atomic number and strain [12]. Therefore, LAADF-STEM is useful for analysis of strain related problems such as crystal defects [1, 2]. In the LAADF-STEM image in Fig. 6(b), the twins in the epitaxial phase appear to have increased contrast. LAADF-STEM is suitable for observing the BP layers in STEM mode.

978-1-4244-9113-1/11 $26.00 © 2011 IEEE   504

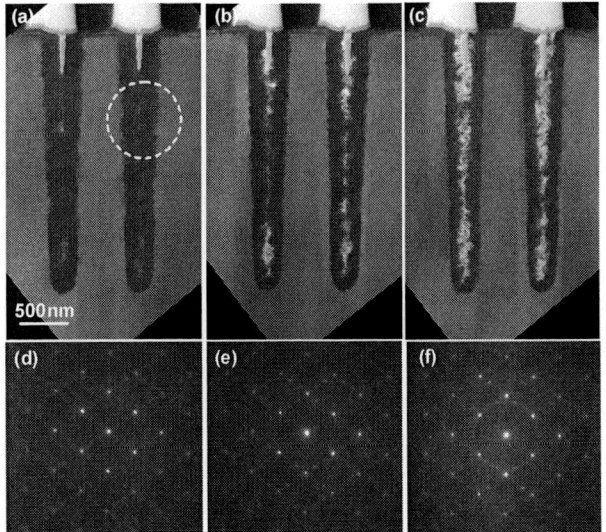

FIGURE 4 (a)-(c) Bright field TEM images of the BP layer of samples A-C, respectively. (d)-(f) SAD patterns taken from the BP layers of samples A-C, respectively.

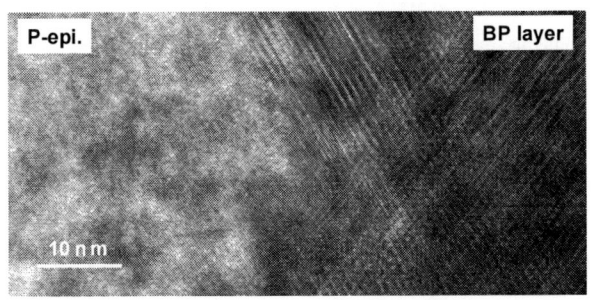

FIGURE 5 HRTEM image taken at P–epi./ BP layer interface of sample A.

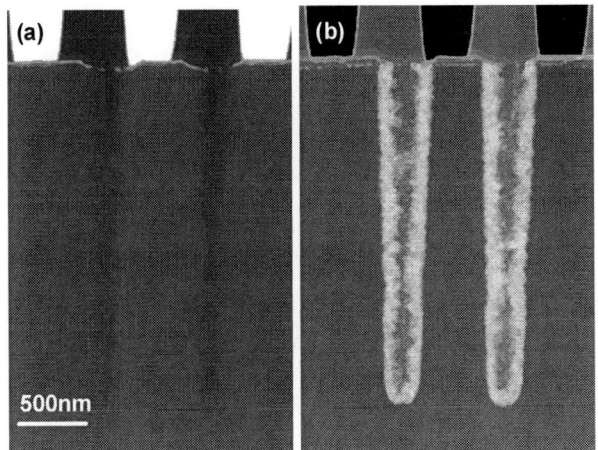

FIGURE 6 (a) HAADF-STEM and (b) LAADF-STEM image of sample C.

C. Strain Measurement using STEM-CBED

STEM-CBED measurement was carried out to examine strain fields around the BP layers. Figure 7 shows the LAADF-STEM image (a) and strain measurement results of sample A with STEM-CBED ((c) and (d)). STEM-CBED measurements were carried out at the outside (line A) and middle (line B) of the BP layer as indicated in Fig. 7(a) (by 100 nm steps). The direction of strain is defined as x (horizontal) and z (vertical) as shown in Fig. 7(a). In the graphs of strain distribution, plus/minus signs in the vertical axis corresponds to tensile/compressive strain. It is found that a clear strain distribution corresponding to device structure (Fig. 7(b)) can be obtained with STEM-CBED. Both lines A and B show tensile strain in the x-direction and compressive strain in the z-direction at the top and bottom of the trench. This suggests that the volume of the BP layers has shrunk due to crystallization.

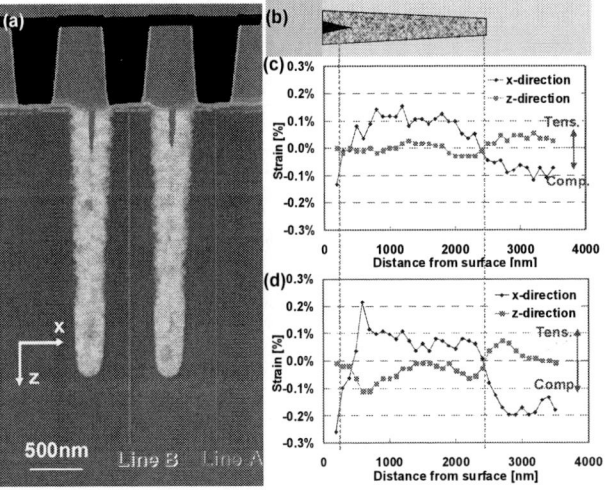

FIGURE 7 (a) LAADF-STEM image of sample A. (b) Schematic drawing of the BP layer. (c) and (d) Strain distributions of lines A and B shown in (a), respectively.

The strain measurement results at the middle of the BP layers of samples A-C are shown in Fig. 8. Strain fields due to the volume shrinkage of the BP layers are observed in every sample. Figure 8 shows that sample A shows the strongest strain and sample C shows the weakest. These results suggest that the epitaxial phase, which has a simpler grain structure than poly-Si, causes more volume shrinkage than the poly-Si phase. Samples with more epitaxial phase (such as sample A) show more strain due to volume shrinkage of the BP layers. It should be noted that strain around the BP layer of sample C is almost zero even though the epitaxial phase is present. Furthermore, little Idss leak is detected in samples B and C, in which the poly-Si phase is present in larger amounts in the BP layer (Fig. 3). These facts suggest that the poly-Si phase in the BP layer plays a role as a buffer against strain. A strain generation model is constructed from these results (Fig 9). Before the annealing process, the BP layers are formed by deposition of amorphous Si (Fig. 9(a)). Amorphous Si changes into the epitaxial and the poly-Si phases during the annealing process. When the concentration of boron in the BP layer is low, the BP layers consist of the epitaxial phase only. Volume

978-1-4244-9113-1/11 $26.00 © 2011 IEEE   505

shrinkage due to formation of the epitaxial phase generates strong strain (Fig. 9(b)). On the other hand, the BP layer changes into the epitaxial phase and poly-Si phase when the boron concentration is high. The poly-Si phase reduces the strain because volume shrinkage of the poly-Si is smaller than that of the epitaxial phase. Therefore, the existence of poly-Si phase in the BP layers results in strain relaxation.

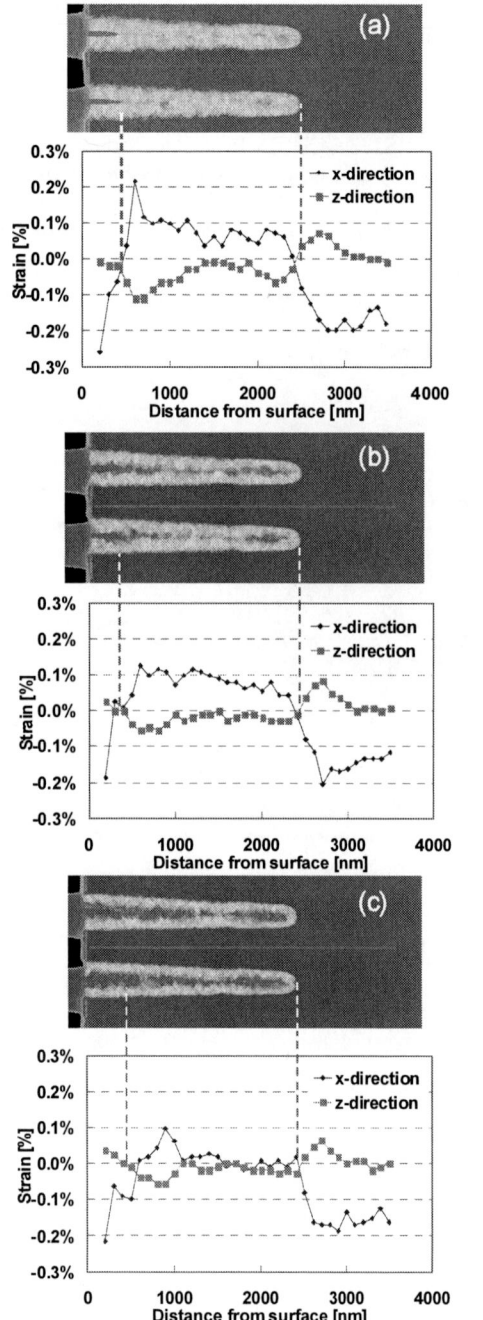

FIGURE 8 (a)-(c) Strain distribution at the middle of the BP layers of samples A-C, respectively. Upper images show LAADF-STEM images of each sample and red lines show measurement position.

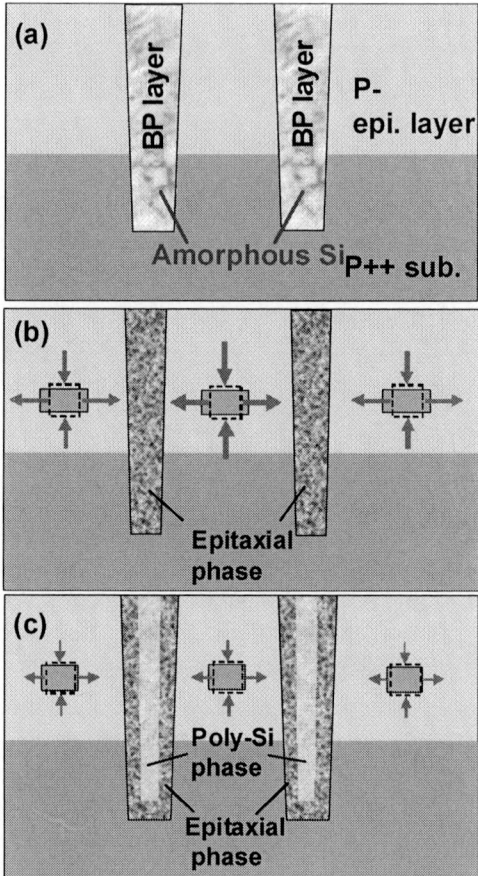

FIGURE 9 Schematic drawing of strain generation model. (a) BP layers after deposition of amorphous Si. (b) and (c) After annealing and crystallization in case of low and high boron concentration of the BP layers, respectively.

D. Disscussion

The relationship between boron concentration and Si phases in the BP layer is discussed in this section. First, we focused on the HAADF-STEM image as shown in the previous section (Fig. 6(a)). It is remarkable that the poly-Si phase in the BP layers is darker than that of the epitaxial phase and substrate Si in HAADF-STEM image. This HAADF-STEM image indicates that the poly-Si phase has a higher concentration of boron (lighter element than Si), and suggests that boron segregation occurs in the poly-Si phase during crystallization (i.e. annealing process). On the other hand, boron diffusion into P− epitaxial layer and P++ substrate should coincide with the segregation. Boron diffusion was examined with scanning capacitance microscopy (SCM) whose intensity corresponds to carrier concentration [13]. Figure 10(a) shows SCM image of sample A. It is found that P+ region in the P− epitaxial layer is broader than the width of the BP layer. This shows that boron atoms diffuse into the P− epitaxial layer from the BP layers. The intensity profile of the SCM image was used to measure the diffusion width. Figure 10(b) shows SCM intensity profile of sample A taken along the broken line drawn in Fig. 10(a).

978-1-4244-9113-1/11 $26.00 © 2011 IEEE

The boron diffusion width is defined at the full width at half maximum. Figure 11 shows diffusion width of boron in samples A-C. The boron diffusion width of sample A whose boron concentration in the BP layers is the lowest is shorter than that of other samples. It should be noted that the boron diffusion widths of samples B and C are almost the same while the boron concentration in the BP layers is different. This indicates that the amount of boron atoms diffusing into P–epitaxial layer is the same in samples B and C. Therefore, more boron atoms should remain in the poly-Si phase in sample C. TEM analysis supports this assumption. The volume of the poly-Si phase in sample C is larger than that in sample B (Fig. 4). Therefore, it is considered that boron segregation occurs in the poly-Si phase of the BP layers having a higher boron concentration. Boron segregation might prevent epitaxial growth in the BP layers. As a result, the core of the BP layer consists of the poly-Si phase when boron concentration is high and results in strain relaxation. Controlling the boron concentration in the BP layer affects strain and is one of the key factors for improvement of production yield.

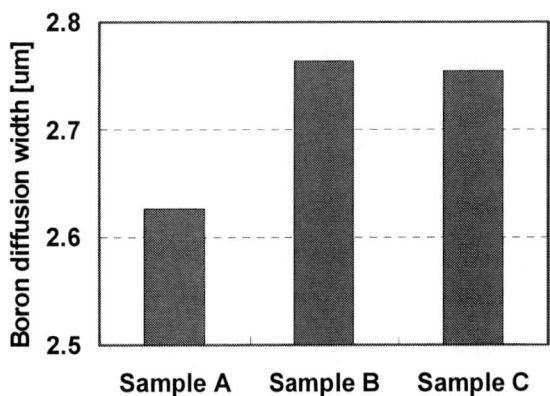

FIGURE 11 Boron diffusion widths of samples A-C measured from SCM.

IV. CONCLUSIONS

Our systematic analysis based on strain measurement using STEM-CBED revealed the mechanism of strain generation in Si substrate with deep trench electrodes. TEM analysis revealed that the BP layers consist of epitaxial and poly-Si phase, and that the proportion of these phases depends on the boron concentration in the BP layer. Strain analysis using STEM-CBED revealed that the magnitude of the strain increases as the boron concentration in the BP layer decreases. Furthermore, it was found that the origin of the strain is volume shrinkage of the epitaxial phase in the BP layer, and that the poly-Si phase in the BP layer acts as a strain buffer. As a result, an increase in the boron concentration in the BP layer leads to an increase in the amount of the poly-Si phase in the BP layer and a subsequent decrease in strain. The relationship between Si phases and boron concentration in the BP layers was examined. SCM and HAADF-STEM analysis showed that boron segregation prevented epitaxial growth in the BP layers having a higher boron concentration. As a result, the core of the BP layer remains as poly-Si or amorphous Si and plays a role as strain buffer. We conclude that a strain distribution measurement using STEM-CBED contributes to our understanding of strain related problems in devices having deep trench structure such as power devices.

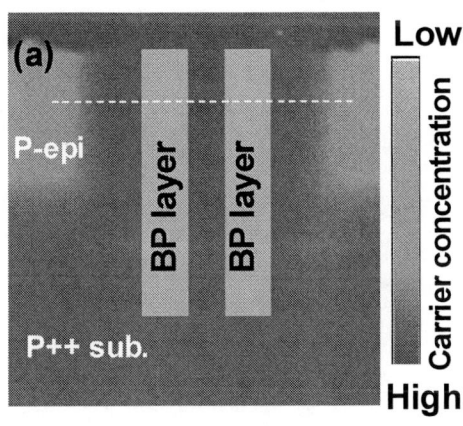

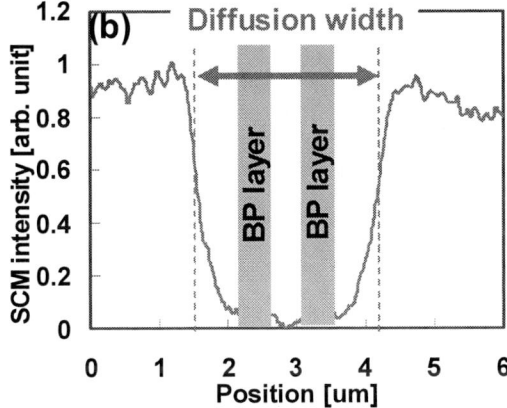

FIGURE 10 (a) SCM image of sample A. (b) SCM intensity profile taken along broken line shown in (a).

REFERENCES

[1] S. Kudo, N. Nakanishi, Y. Hirose, K. Sato, T. Yamashita, H. Oda, K. Kashihara, N. Murata, T. Katayama, K. Asayama, J. Komori and E. Murakami, "Three-dimensional visualization technique for crystal defects in high performance p-channel MOSFET with embedded SiGe-source/drain" Jpn. J. Appl. Phys, Vol. 49, pp.04DA22, 2010.

[2] N. Nakanishi, S. Kudo, M. Kawakami, T. Hayashi, H. Oda, T. Uchida, Y. Miyagawa, K. Asai, K. Ohnishi, N. Hattori, Y. Hirose, T. Koyama, K. Asayama, and E. Murakami, "Strain mapping technique for performance improvement of strained MOSFETs with scanning transmission electron microscopy," IEDM 2008, pp 431-434.

[3] J.C.H. Spence, and J.M. Zuo, "Electron microdiffraction," Plenum Press, New York, 1992.

[4] M. Ishibashi, K. Horita, M. Sawada, M. Kitazawa, M. Igarashi, T. Kuroi, T. Eimori, K. Kobayashi, M. Inuishi and Y. Ohji, "Novel shallow trench isolation process from viewpoint of total strain process design for 45 nm node devices and beyond", Jpn. J. Appl. Phys., 44, pp.2152, 2005.

[5] D.B. Williams, and C.B. Carter, "Transmission electron microscopy," Plenum Press, New York, 1996.

[6] T. Yamazaki, T. Isaka, K. Kuramochi, I. Hashimoto, and K. Watanabe, "Presice measurement of local strain fields with energy-unfiltered convergent electron diffraction," Acta Cryst., Vol. A62, pp. 201-207, 2006.

[7] L. Clement, R. Pantel, L.F.Tz. Kwakman, J.L. Rouviere, "Strain measurements by convergent- beam electron diffraction: The importance of stress relaxation in lamella preparations", Appl. Phys. Lett., Vol. 85 pp. 651-653, 2004.

[8] N. Hashikawa, K. Fukumoto, T. Kuroi, M. Ikeno, Y. Mashiko, "Direct observation of local strain field for ULSI devices," Microelectronics Reliability, Vol. 38, pp. 913-917, 1998.

[9] A. Armigliato, R. Balboni, G.P. Carnevale, G. Pavia, D. Piccolo, S. Frabboni, A. Benedetti and A.G. Cullis, "Application of convergent beam electron diffraction to two-dimensional strain mapping in silicon devices," Appl. Phys. Lett., Vol. 82, pp.2172-2174, 2003.

[10] K. Watanabe, Y. Anzai, N. Nakanishi, T. Yamazaki, K. Kuramochi, K. Mitsuishi, K. Furuya, and I. Hashimoto, "Microstructures formed in recrystallized Si," Appl. Phys. Lett., Vol. 84, pp. 4520-4521, 2004.

[11] T. Yamazaki, N. Nakanishi, A. Recnik, M. Kawasaki, K. Watanabe, M. Ceh and M. Shiojiri, "Quantitative high-resolution HAADF-STEM analysis of inversion boundaries in Sb_2O_3-doped zinc oxide," Ultramicroscopy, Vol. 98, pp.305-316, 2004.

[12] Z. Yu, D.A. Muller and J. Silcox, "Study of strain fields at a-Si/c-Si interface," J. Appl. Phys., Vol. 95, pp.3362-3371, 2004.

[13] R. Stephenson, P.D. Wolf, T. Trenkler, T. Hantschel, T. Clarysse, P. Jansen, and W. Vandervorst, "Practicalities and limitations of scanning capacitance microscopy for routine integrated circuit characterization" J. Vac. Sci. Technol. B, Vol. 18(1), pp. 555-560, 2000.

AUTHOR INDEX

Abbas, Syed Mohsin.....................863
Abe, Kenichi............................381
Achanta, R.........................1, 134
Achanta, Ravi..........................665
Aghassi, Jasmin........................342
Agostinelli, M.........................533
Agrawal, Manish........................342
Ahlbin, J. R...........................258
Ahn, Jae-Young.........................126
Aichinger, Th..........................605
Aitken, J.........................134, 696
Alaeddine, A...........................674
Alam, M. A. 196, 207, 353, 557, 614
Allee, David R.........................904
Altimime, L.............................99
Amashita, Takuro.......................876
Anand, Sindhu..........................220
Ang, D. S.........................935, 943
Anghel, L..............................347
Angyal, M. 134, 312, 317
Aoulaiche, M............................99
Arie, H................................503
Aritome, Seiichi.......................641
Arnaud, L.........................297, 347, 746
Arora, Rajan...........................461
Athikulwongse, Krit.....................65
Augustine, Charles.....................353
Aymerich, X............................920
Backes, Benjamin.......................527
Badami, D..............................696
Bae, Ilchan............................852
Bae, Kidan........................17, 852
Baeg, Sanghyeon....................91, 863
Baek, Dong-Cheon.......................734
Bagatin, M.............................751
Bai, P.................................533
Bai, Shawn.............................706
Baik, Seung Jae........................650
Baker, Michael.........................220
Balasubramanian, S......................41
Ball, D. R.............................258
Banerjee, Kaustav......................280
Banerjee, Sanjay K.....................539
Bansal, Aditya.............47, 696, 700
Barabadi, Banafsheh....................271
Bari, Daniele.....................113, 566
Bashir, Muhammad........................65
Baumann, F.............................134
Baumann, Robert C......................247
Beltrami, S............................751
Bersuker, Gennadi.................802, 807
Berthold, J. M..........................56
Bertuccio, M...........................751
Beyer, G. P.......................142, 321
Bhuva, B. L.......................258, 467, 886

Bhuva, Bharat..........................891
Bieler, Thomas R.......................573
Black, Dolores A.......................897
Black, Jeffrey D.......................897
Blair, Lauren..........................573
Bloom, Ilan............................819
Boit, Christian...................514, 774
Bolam, R...............................696
Bonilla, G.............................134
Boo, A. A.........................935, 943
Boret, S...............................811
Bosman, M.........................182, 786
Brain, Ruth.............................61
Bravaix, A. 704, 798, 926
Brown, Thomas M........................566
Bultreys, D............................500
Burdeaux, David C......................435
Burger, Wayne R........................435
Byun, Jun Young........................656
Cacho, Florian....................430, 454
Camargo, V. Valduga De Almeida.........915
Campbell, J. P.........................202
Cao, Yu.................................36
Carbonell, L......................142, 321
Carroll, Barry.........................476
Cartier, E. 41, 372, 387, 696, 933
Carulli, John...........................36
Cassell, Alan..........................280
Castillo, James........................868
Catthoor, F............................915
Cester, A..............................113
Cester, Andrea.........................566
Cha, Seon-Yong..........................95
Chanda, Kaushik........................317
Chang, Soon W..........................335
Chang, Tang-Long.......................717
Chang, Yao-Wen.........................396
Chappaz, C.............................347
Cheffah, Saad..........................704
Chen, An...............................843
Chen, Chia-Yu..........................190
Chen, F................................134
Chen, Frederick T......................847
Chen, K. C.............................645
Chen, K. F.............................645
Chen, Kuang-Chao.......................396
Chen, Lu-An.......................392, 714
Chen, M. S.............................645
Chen, Min...............................36
Chen, Pang-Shiu........................847
Chen, Wei-Su...........................847
Chen, Willis...........................160
Chen, Yan-Yu...........................396
Chen, Ying-Je..........................824
Chen, Yu-Sheng.........................847

AUTHOR INDEX

Chen, Z. 913
Cheng, C. H. 645
Cheng, C. M. 160
Cheng, Chun-Min 661
Cherman, Vladimir..................................592
Cheung, K. P. 202
Chevallier, Remy....................................704
Chia, Pierre 91
Chiang, Puo-Yu 444
Chikaki, Shinichi....................................130
Chikyow, Toyohiro 335
Chimenton, Andrea........................171, 828
Chin, H. W. 636
Cho, Hyun-Jin 190
Cho, M. 624
Cho, Moonju 207
Choi, Do-Young.............................449, 908
Choi, Gil-Heyun 307
Choi, Hanmei...126
Choi, Seung-Man 734
Choi, Siyoung 307
Choi, Wonsup 650
Chong, L. H. 645
Chou, Bruce...409
Christiansen, Bradley D. 680
Christiansen, Cathryn 312, 317
Chu, W. T. 636
Chu, Y. S. 636
Chua, C. M. 780
Chuang, Ching-Te 47
Chudzik, Michael 387
Chung, Chilhee.......................................307
Chung, Steve S.......................................941
Ciappa, M. 28
Cirak, J. 113
Claeys, C. 882
Clark, Brett M..476
Classe, F. C. 685
Cohen, S. 134
Collaert, N. 99
Compagnoni, Christian Monzio.................833
Concannon, A. 405
Coutu, Ronald A. 680
Cressler, John D......................................461
Cristoloveanu, S. 882
Croes, K. 142, 321, 592
Crupi, F..624
Curello, G. 533
Czeppel, L. T. 751
Dadgour, Hamed F..................................280
Dalal, Vikram L......................................546
Daoud, K. 674
Darbandi, Payam 573
Das, K..696
De Wachter, B. 99

Degraeve, Robin 7, 207, 815
Del Alamo, Jesús A..................................422
D'Emic, Christopher P..............................630
Deora, S. 614
Deshpande, H. 533
Dey, Aritra 904
Di Carlo, Aldo 566
Dick, Kevin D. 897
Diggins, Z. 886
Diop, Malick...430
Dixit, Anand 486
Do, Jinho 852
Dobrovolny, P. 915
Dongaonkar, Sourabh 557
Donoval, D. 113
Dosseul, Franck 592
Dunn, S. 51
Dutton, Robert W. 190
Eaton, Paul 868
El Farhane, R. 347
Elkins, Don 868
El-Mamouni, F. 882
Eneman, G. 624
Eng, Genghmun 177
Ermisch, Karsten 597
Falk, John 75
Fang, Y.-P. 891
Federspiel, Xavier.............................454, 798
Ferreira, P. J. 500
Fichtner, W. 28
Fleetwood, Daniel M. 426
Floyd, Rich 868
Foran, Brendan 509
Franco, J. 364, 605, 624
Frank, David J. 630
Frank, T. 347
Freeman, Greg..461
Frei, Michel 557
Fronheiser, J. 202
Fujii, Shosuke 839
Fujitsuka, Ryota 839
Fukutani, Katsuyuki 335
Furuta, Jun 471
Gaddamraja, Sesha 685
Gaddi, Roberto 171
Gadlage, M. J. 258
Galand, R. 297, 746
Galloway, K. F. 882
Ganesh, K. J. 500
Gao, Y...935
Geer, Robert E..527
Gerardin, S. 751
Ghibaudo, G. 811
Gloria, D. 811
Gordon, Michael.....................................476

AUTHOR INDEX

Gossner, Harald 342
Grasser, T. 364, 597, 605, 624, 915
Green, Ronald 756
Groeseneken, G. 7, 99, 364, 624, 915, 920
Gu, Pei-Yi 847
Guillorn, Michael A. 630
Gustin, W. 56, 597
Gyun, Byung Gu 656
Habersat, Daniel 756
Haendler, Sebastien 454, 798
Haensch, Wilfried 630
Hafez, W. 533
Hamanaka, Chikara 471
Han, J. H. 202
Han, Jeong-Uk 582
Han, T. T. 645
Harada, Ryo 253
Harm, Gregory J. 740
Hashimoto, Masanori 253
Hattori, N. 503
Hau-Riege, Christine 588
Hauser, M. 696
He, Chieh-Wei 396
He, Jun 61
He, Yi 476
Hekmatshoar, Bahman 562
Heller, Eric R. 680
Heo, Jinchul 852
Herfst, Rodolf W. 929
Hicks, Jeff 61
Higgins, R. 51
Hillenbrand, R. 493
Hirose, Y. 503
Ho, Dah-Chuen 444
Ho, Paul S. 264
Hoffman, M. 56
Hoffmann, T. Y. 624
Hong, Heebum 105
Hong, Sung-Joo 95, 641
Hopstaken, Marinus 562
Horiguchi, N. 364
Horii, Sadayoshi 335
Horikawa, Tsuyoshi 335
Hsieh, E. R. 941
Hsu, C. L. 160
Hsu, C. M. 160
Hsu, Chia-Lin 670
Hsu, Chi-Mao 661
Hsu, H. K. 160
Hsu, T. H. 824
Hsu, Yen-Ya 847
Hu, Y. Z. 913
Huang, C. C. 160, 661
Huang, Clement 731
Huang, Climbing 670

Huang, Hsin-Fu 661
Huang, I. J. 645
Huang, J. S. 645
Huang, R. M. 941
Huang, Ru 847
Huang, Rui 264
Huang, Y. C. 22
Huang, Yu-Hui 444, 690
Huard, Vincent 430, 454, 704, 798, 926
Huber, A. J. 493
Hussain, Muhammad M. 280
Hwang, Kihyun 126
Hwang, Lira 17, 852
Ibe, Eishi 239
Ighilahriz, Salim 430
Im, Jay 264
Ioannou, D. P. 696
Islam, A. E. 614
Islam, Ahmad 207
Ito, Kyosuke 710
Ito, Shuu 335
Iwai, H. 786
Jack, Nathan 409
Jagannathan, Srikanth 886
Jain, A. 614
Jakabovic, J. 113
Jan, C.-H. 533
Janai, Meir 819
Jang, Tae-Su 95
Jenkins, Keith A. 47
Jeon, Chulhee 852
Jeong, Eui-Young 449
Jeong, Jae-Goan 95
Jeong, Tae-Young 734
Jeong, Yoon-Ha 449, 908
Jesudoss, Pio 224
Jha, N. K. 22
Jin, Minjung 17, 852
Joh, Jungwoo 422
Johnston, Allan 165
Joo, Moon Sig 656
Joshi, K. 614
Joshi, Yogendra 271
Juan, Alex 731
Jung, Eui-Young 908
Jung, Eunji 307
Jung, Hye Kyung 307
Jurczak, M. 99
Kaczer, B. 364, 605, 624, 915, 920
Kadi, M. 674
Kakushima, K. 786
Kane, Terence 312
Kang, A. C. 636
Kang, Chang Yong 449, 908
Kang, Chang-Jin 126

AUTHOR INDEX

Kang, J. F. 196
Karthik, Y. 557
Kato, Ichiro 149
Kauerauf, Thomas 7, 182
Kawaguchi, Hiroshi 876
Kayaba, Yasuhisa 130
Keilmann, F. 493
Keller, Robert R. 740
Kendig, Dustin 725
Ker, Ming-Dou 717
Kerber, Andreas 41, 190, 372, 387
Kikkawa, Takamaro 130
Kim, Andrew T. 734
Kim, Bio 126
Kim, Chris H. 47
Kim, Dae Hyun 65
Kim, Dae-Hyong 111
Kim, Dong Woo 126
Kim, Hyungseok 641
Kim, Hyunjin 17
Kim, Jae-Joon 47, 696, 700
Kim, Jung Nam 656
Kim, Jungin 17
Kim, Kyoung-Hwan 582
Kim, Kyung-Do 95
Kim, Sook Joo 656
Kim, Taeh Wan 656
Kim, Tae-Hyoung 47
Kim, Wan Gee 656
Kim, Won 656
Kim, Yong Tae 582
Kim, Yongshik 105
Kim, Yong-Taik 95
Kindereit, Ulrike 514
King, Everett E. 81
Ko, Yongsun 126
Kobayashi, Kazutoshi 471, 710
Koh, L. S. 780
Kohmura, Kazuo 130
Komeyli, K. 533
Kopanski, J. J. 202
Kouda, M. 786
Koudymov, A. 802
Koutsoureli, M. 290
Kovac, J. 113
Koyama, T. 503
Koyanagi, Mitsumasa 328
Krishnan, A. T. 51
Krishnan, S. 36, 51, 387
Ku, S. H. 645
Kumar, Satish 271
Kumari, Sangita 740
Kurtz, Sarah 551
Kushvaha, S. S. 786
Kwasnick, Robert 75

La Rosa, G. 696
Labie, Riet 592
Lacaita, Andrea L. 833
Lafonteese, D. 405
Lai, Tai-Hsiang 392, 714
Lamontagne, P. 297, 746
Larcher, Luca 807
Lauzeral, Olivier 476
Leduc, P. 347
Lee, C. H. 645
Lee, Chienying 645
Lee, Heng-Yuan 847
Lee, Hyun-Bae 307
Lee, Jack C. 449
Lee, Jeong-Soo 908
Lee, Jian-Hsing 690
Lee, Jong Myeong 307
Lee, Jongho 852
Lee, Kyong Taek 17
Lee, Miji 734
Lee, Seokkiu 641
Lee, Shou-Chung 155
Lee, Tackhwi 539
Lee, Tae-Kyu 573
Lee, Y. H. 22, 444, 636
Lelis, Aivars 756
Leong, K. C. 943
Lepla, Keith 476
Leung, Martin 177
Li, Baozhen 1, 134, 312, 317
Li, Jinglong 765
Li, Zhao 546
Liang, C. W. 941
Liang, James W. 731
Liao, P. J. 22
Liehr, M. 802
Lim, Hajin 852
Lim, Koeng Su 650
Lim, Seung-Hyun 126
Lim, Sung Kyu 65
Lim, Sun-Me 105
Limbrick, Daniel B. 897
Lin, C. P. 690
Lin, Chun-Yu 717
Lin, Hung Sung 520, 770
Lin, J. F. 160
Lin, Jack 160
Lin, Jin-Fu 661
Lin, Kun-Hsien 661
Lin, Ming-Ren 190, 843
Lin, Mingte 731
Lin, W. C. 160, 661, 670
Linder, Barry P. 1, 47, 372, 696
Litt, Brian 111
Liu, C. C. 160, 444

AUTHOR INDEX

Liu, Cheng-Jye 824
Liu, Kuo-Chuan 573
Liu, T. 121
Liu, W. H. 182, 847
Liu, Yan-Chun 661
Liu, Yang 190
Liu, Ziyuan 335
Lo, Chun-Yuan 824
Lofrano, M. 321
Lorut, F. 347
Loveless, T. D. 258, 886
Lowrie, Anthony 61
Lu, Chih-Yuan 396, 645
Lu, Kuan-Hsun 264
Lu, Pong-Fei 47
Lu, Tao-Cheng 396
Lu, W. P. 645
Lu, Z. 99
Machavolu, K. S. 913
Maeda, H. 503
Mahapatra, Souvik 557, 614
Mahatme, Nihaar 891
Mahato, S. 915
Maheta, V. D. 614
Maize, Kerry 725
Malandruccolo, V. 28
Maloney, Timothy J. 409
Marckmann, Jennifer 476
Martin-Martinez, J. 920
Mason, Maribeth 177, 509
Massengill, L. W. 258, 886
Massey, G. 696
Masuduzzaman, Muhammad 196, 207
Mathewson, Alan 224
Matsumoto, Takashi 710
Mavis, David 868
McDonough, Colin 527
McGahay, Vincent 312
McLaughlin, Paul 134, 665
McMahon, W. 41
McMorrow, D. 882
McNally, Brendan 476
Mendenhall, Marcus H. 247
Meneghesso, Gaudenzio 113, 566
Mercha, A. 99
Miccoli, Carmine 833
Michalas, L. 290
Miki, Hiroshi 630
Milor, Linda 65
Min, Hong Kook 582
Misra, Durgamadhab 792
Mitani, Yuichiro 857
Mitard, J. 624
Mitsuyama, Yukio 253
Mittl, S. 1, 696

Molzer, Wolfgang 342
Monsieur, Frederic 933
Moquillon, Laurence 454
Moreau, S. 347
Moss, Steven C. 509
Motohiko, Masuda 765
Mukhopadhyay, Saibal 47
Murakami, E. 503
Muthuswamy, Jit 220
Mutihac, Oana-Mihaela 514
Nafria, M. 920
Nakanishi, N. 503
Nakanishi, Toshiro 126
Nam, Hyeowoo 91
Nam, Jongik 17
Narasimham, B. 467
Nayfeh, Hasan M. 461
Negre, L. 811
Nelhiebel, M. 605
Ney, D. 297
Nielen, Heiko 597
Nigam, T. 41
Nishii, Koji 239
Nishizawa, Shinichi 710
Oakley, Jennifer 317
Oates, A. S. 155, 467, 706, 891
O'Connor, Robert 7
Ogawa, Arito 335
Ohmi, Tadahiro 381
Okandan, Murat 220
Okumura, Shunsuke 876
Olivo, Piero 171, 828
Olson, Nicholas 719
Ong, T. C. 636
Onodera, Hidetoshi 471, 710
Onoye, Takao 253
Osborn, Jon V. 81
O'Shea, S. J. 786
Paccagnella, A. 751
Padovani, Andrea 807
Pagano, Carlo 774
Pantelides, Sokrates T. 426
Pantuovaki, M. 142
Papaioannou, G. 290
Papathanasiou, Athanasios E. 75
Park, Dae-Gyu 630
Park, Jongwoo 17, 105, 533, 734, 852
Park, Junekyun 17, 852
Park, Min Sang 449, 908
Park, Se Yeoul 582
Park, So Ra 582
Park, Sung Ki 656
Park, Sungju 863
Park, Sungkye 641
Parks, C. 134

AUTHOR INDEX

Patterson, K. 467
Pavan, Paolo 807
Penna, Stefano 566
Petitprez, E. 297, 746
Pey, K. L. 182, 786
Phang, J. C. H. 780
Phoa, K. 533
Pimparkar, N. 41
Pobegen, G. 605
Poh, John Chung Hang 461
Poling, Brian S. 680
Pourboghrat, Farhang 573
Prabhumirashi, Prad 61
Presser, Nathan 509
Puzyrev, Yevgeniy S. 426
Quemerais, Thomas 454
Rafik, M. 798, 926
Raghavan, N. 182, 786
Ragnarsson, Lars-åke 7
Rahim, Nilufa 792
Rahman, A. 533
Rajasekhara, S. 500
Rakowski, M. 99
Ramaswami, Sesh 524
Ran, Qiushi 190
Randriamihaja, Y. Mamy 798, 926
Ranjan, R. 22, 444, 636
Rao, R. 696
Rao, Rahul M. 47, 700
Rao, V. Ramgopal 342
Rashid, Al 75
Read, David T. 740
Reale, Andrea 566
Reddy, Vijay 36, 51
Reed, Robert A. 247, 882, 897
Reilly, Matthew 75
Reisinger, H. 56, 597, 605
Ren, Zhibin 630
Rentala, Vijay 36
Rhoton, Brent 231
Rideau, D. 798, 926
Robinson, William H. 897
Roche, Philippe 476
Rodriguez, R. 920
Rogers, John A. 111
Roh, Jae Sung 656
Rosa, Giuseppe L. 461
Rosenbaum, Elyse 409, 719
Rothleitner, H. 28
Roussel, Ph. J. 7, 364, 624, 915
Roy, D. 798, 811, 926
Roy, Kaushik 353
Roy, Tania 426
Ryan, J. T. 202
Ryu, Suk-Kyu 264

Sadana, Devendra 562
Sagong, Hyun Chul 449, 908
Sahhaf, Sahar 7
Saito, Tomoya 335
Saito, Wataru 417
Sakamoto, Mitsuhiro 149
Savage, Timothy Scott 213
Scarpulla, John 81
Scheer, P. 811
Schepens, Cor 171
Schluender, Christian 597
Schlünder, C. 56
Schmitz, Anthony 61
Schmitz, Jurriaan 929
Scholten, Andries J. 929
Schrimpf, Ronald D. 247, 426, 882
Schulz, Thomas 342
Sekine, Katsuyuki 839
Seo, Soonok 641
Seth, Sachin 461
Shahrjerdi, Davood 562
Shakouri, Ali 725
Shaviv, Roey 740
Shealy, Jeffrey B. 680
Shih, J. R. 22, 444, 690
Shim, Byung Sup 582
Shimbo, Ken-Ichi 239
Shimizu, Tatsuo 149
Shin, Dongseok 852
Shin, Hyunjung 650
Shinosky, M. 134
Shrivastava, Mayank 342
Shubhakar, K. 786
Shukla, Vrashank 719
Shur, Yael 819
Shusterman, Yuriy 61
Sibley, Mike 868
Sierawski, Brian D. 247
Siew, Y. K. 321
Simoen, E. 882
Simon, A. 134
Sin, Yongkun 509
Singh, Navab 280
Slayman, Charles 476
Smith, Charles 171, 231
Sohn, Chang Woo 449, 908
Sokolsky, M. 113
Song, Grace 765
Spinelli, Alessandro S. 833
Srinivasan, Venkatesh 36
Stam, Frank 224
Stathis, James 372, 387, 696, 933
Stupian, Gary 177
Sturm, J. C. 121
Su, David 690

AUTHOR INDEX

Su, Kuan-Cheng 392, 401, 714, 731
Suehle, J. S. 202
Sugawa, Shigetoshi 381
Sun, Wein-Town 824
Sunagawa, Hiroki 710
Suñé, J. 1
Sung, Min Gyu 656
Sutanto, Jemmy 220
Sutton, Akil K. 461
Tagliaferro, Roberto 566
Tam, Nelson 247, 467
Tang, Tien-Hao 392, 401, 714
Taniguchi, Hitoshi 239
Taniguchi, Yoshio 239
Taylor, B. 802
Tega, Naoki 630
Teo, J. K. J. 780
Teo, Z. Q. 935, 943
Teramoto, Akinobu 381
Thuaire, A. 347
Tian, Li 765
Tilak, V. 202
Tillack, Bernd 514
Ting, Yun-Jen 824
Toba, Tadanobu 239
Tökei, Zs. 142, 321
Toledano-Luque, M. 364, 815, 915, 920
Torii, Kazuyoshi 630
Toriumi, Akira 857
Tsai, C. H. 533, 847, 941
Tsai, C. T. 941
Tsai, Ming-Jinn 847
Tsai, T. C. 160, 670
Tsao, W. C. 160
Tsuchiya, Hideaki 149
Tsukasa, Matsuda 307
Tucker, Mike 476
Twomey, Karen 224
Uno, Satoshi 149
Vaidyanathan, Balaji 706
Van Houdt, J. 815
Vandelli, Luca 807
Vandevelde, Bart 592
Varghese, D. 51
Vasanth, Karthik 231
Vashchenko, Vladislav 405, 725
Veksler, D. 802
Vetury, Rama 680
Via, G. David 680
Vilchis, M. 467
Visconti, A. 751
Viventi, Jonathan 111
Vrancken, C. 364
Wagner, P.-J. 605
Wagner, S. 121

Waltz, P. 297
Wang, C. 202, 636
Wang, Chang-Tzu 401, 714
Wang, Dapeng 557
Wang, Shih-Yu 396
Wang, Shun-Min 847
Wang, Tahui 645
Wang, W. 22, 527, 765
Wang, Y. H. 636
Wang, Yun Yu 312
Wang, Z. R. 786
Webers, Tomas 592
Weigmann, J.-M. 56
Weiland, R. 493
Weller, Robert A. 247
Wen, S. J. 91, 247, 467, 886
Wilde, Markus 335
Wilkinson, Jeffrey D. 476
Wilson, C. J. 321
Winters, Christophe 592
Wirth, G. 364, 915
Wittborn, J. 493
Witters, L. 624
Witulski, A. F. 258
Wohlgemuth, John H. 551
Wong, Richard 91, 247, 467, 476, 886
Wood, Alan 476
Wrachien, Nicola 113, 566
Wright, William M. D. 224
Wu, Chunlei 765
Wu, E. 1, 792
Wu, Guan-Wei 396
Wu, J. Y. 160, 661, 670
Wu, Kenneth 22, 444, 636, 690
Wu, Miao 765
Wu, Mong Sheng 520
Wu, Tommy 476
Wu, X. 182
Xia, Feng 61
Xiong, W. 882
Xu, J. 533
Xu, Xiaochen 231
Xu, Y. 467
Yamaguchi, Kosuke 876
Yang, I. C. 645
Yang, J. Q. 196
Yang, Sangryol 126
Yang, Seung Jin 582
Yao, Shaoning 312
Yasuda, Naoki 839
Yau, You-Wen 588
Yeh, J.-Y. 533
Yeo, Myung-Soo 734
Yeoh, Terence 177
Yew, T. Y. 22

AUTHOR INDEX

Yokogawa, Shinji 149
Yokoyama, Yoshiyuki..774
Yoo, Hyunjun 650
Yoo, Jong Hee..656
Yoo, Min-Soo..95
Yoshimoto, Masahiko 876
Yoshimoto, Shusuke 876
Young, C. D. 802
Yu, H. Y. 786
Yu, Joe 765
Yu, L. C 202
Yu, Nick 588
Yu, Sunil 105
Yum, Jung 807
Yun, Jong-Ho 307
Zahid, M. B...207, 815
Zaka, A. 798, 926
Zaknoon, Bashir...75
Zambelli, Cristian 171, 828
Zamiri, Amir 573
Zhang, E. X. 426, 882
Zhang, F. 202
Zhang, Lijie 847
Zhang, Ming 105
Zhao, K. 372, 696, 700
Zhao, L..142
Zhou, Bite 573
Zhou, X. 913
Zhu, Xiaowei 231
Zhu, Yu 630
Zous, N. K...645
Zuber, P. 915

CURRAN ASSOCIATES INC.
proceedings
.com

9781424491131

2011 IEEE International Reliability Physics Symposium (IRPS 2011)

Monterey, California, USA
10-14 April 2011

IEEE Catalog Number: CFP11RPS-POD
ISBN: 978-1-42449-113-1

2011 IEEE International Reliability Physics Symposium

(IRPS 2011)

Monterey, California, USA
10 - 14 April 2011

Pages 509 - 947

IEEE Catalog Number:	CFP11RPS-PRT
ISBN:	978-1-4244-9113-1

Copyright © 2011 by the Institute of Electrical and Electronic Engineers, Inc
All Rights Reserved

Copyright and Reprint Permissions: Abstracting is permitted with credit to the source. Libraries are permitted to photocopy beyond the limit of U.S. copyright law for private use of patrons those articles in this volume that carry a code at the bottom of the first page, provided the per-copy fee indicated in the code is paid through Copyright Clearance Center, 222 Rosewood Drive, Danvers, MA 01923.

For other copying, reprint or republication permission, write to IEEE Copyrights Manager, IEEE Service Center, 445 Hoes Lane, Piscataway, NJ 08854. All rights reserved.

This publication is a representation of what appears in the IEEE Digital Libraries. Some format issues inherent in the e-media version may also appear in this print version.

IEEE Catalog Number:	CFP11RPS-PRT
ISBN 13:	978-1-4244-9113-1
ISSN:	1541-7026

Additional Copies of This Publication Are Available From:

Curran Associates, Inc
57 Morehouse Lane
Red Hook, NY 12571 USA
Phone: (845) 758-0400
Fax: (845) 758-2633
E-mail: curran@proceedings.com
Web: www.proceedings.com

TABLE OF CONTENTS

SESSION 2A: GATE DIELECTRICS:

Post-Breakdown Statistics and Acceleration Characteristics in High-K Dielectric Stacks 1
E. Wu, J. Suñé, B. Linder, R. Achanta, B. Li, S. Mittl

Methodologies for Sub-1nm EOT TDDB Evaluation 7
Thomas Kauerauf, Robin Degraeve, Lars-Åke Ragnarsson, Philippe Roussel, Sahar Sahhaf, Guido Groeseneken, Robert O'Connor

Frequency Dependent TDDB Behaviors and Its Reliability Qualification in 32nm High-k/Metal Gate CMOSFETs 17
Kyong Taek Lee, Jongik Nam, Minjung Jin, Kidan Bae, Junekyun Park, Lira Hwang, Jungin Kim, Hyunjin Kim, Jongwoo Park

Re-Investigation of Gate Oxide Breakdown on Logic Circuit Reliability 22
Y. C. Huang, T. Y. Yew, W. Wang, Y.-H. Lee, R. Ranjan, N. K. Jha, P. J. Liao, J. R. Shih, K. Wu

SESSION 2B: CIRCUIT RELIABILITY:

In Situ Screening Techniques for Defective Oxides in Devices for Automotive Applications 28
V. Malandruccolo, M. Ciappa, W. Fichtner, H. Rothleitner

A TDC-Based Test Platform for Dynamic Circuit Aging Characterization 36
Min Chen, Vijay Reddy, John Carulli, Srikanth Krishnan, Vijay Rentala, Venkatesh Srinivasan, Yu Cao

Fast Characterization of the Static Noise Margin Degradation of Cross-Coupled Inverters and Correlation to BTI Instabilities in MG/HK Devices 41
A. Kerber, N. Pimparkar, S. Balasubramanian, T. Nigam, W. McMahon, E. Cartier

Reliability Monitoring Ring Oscillator Structures for Isolated/Combined NBTI and PBTI Measurement in High-K Metal Gate Technologies 47
Jae-Joon Kim, Barry P. Linder, Rahul M. Rao, Tae-Hyoung Kim, Pong-Fei Lu, Keith A. Jenkins, Chris H. Kim, Aditya Bansal, Saibal Mukhopadhyay, Ching-Te Chuang

Negative Bias Temperature Instability "Multi-Mode" Compact Model Based on Threshold Voltage and Mobility Degradation 51
D. Varghese, R. Higgins, S. Dunn, A. T. Krishnan, V. Reddy, S. Krishnan

A New Smart Device Array Structure for Statistical Investigations of BTI Degradation and Recovery 56
C. Schlünder, J. M. Berthold, M. Hoffman, J.-M. Weigmann, W. Gustin, H. Reisinger

SESSION 2C: FABLESS AND PRODUCTION RELIABILITY:

Characterization and Challenge of TDDB Reliability in Cu/Low K Dielectric Interconnect 61
Feng Xia, Jun He, Prad Prabhumirashi, Anthony Schmitz, Anthony Lowrie, Jeff Hicks, Yuriy Shusterman, Ruth Brain

Backend Low-k TDDB Chip Reliability Simulator 65
Muhammad Bashir, Dae Hyun Kim, Krit Athikulwongse, Sung Kyu Lim, Linda Milor

Determination of CPU Use Conditions 75
Robert Kwasnick, Athanasios E. Papathanasiou, Matthew Reilly, Al Rashid, Bashir Zaknoon, John Falk

Si3N4 Extrinsic Defects and Capacitor Reliability 81
John Scarpulla, Everett E. King, Jon V. Osborn

SESSION 2D: MEMORY:

AC-DC Factor Sensitivity for DRAM Components Lifetime under Hot-Carrier Injection 91
Sanghyeon Baeg, Hyeowoo Nam, Pierre Chia, ShiJie Wen, Richard Wong

STI Stress-Induced Degradation of Data Retention Time in DRAM and a New Characterizing Method for Mechanical Stress 95
Tae-Su Jang, Kyung-Do Kim, Min-Soo Yoo, Yong-Taik Kim, Seon-Yong Cha, Jae-Goan Jeong, Sung-Joo Hong

Hot Hole Induced Damage in 1T-FBRAM on Bulk FinFET..........99
M. Aoulaiche, N. Collaert, A. Mercha, M. Rakowski, B. De Wachter, G. Groeseneken, L. Altimime, M. Jurczak, Z. Lu

Effects of BTI during AHTOL on SRAM VMIN..........105
Sun-Me Lim, Heebum Hong, Sunil Yu, Ming Zhang, Jongwoo Park, Yongshik Kim

SESSION 2E: THIN FILM TRANSISTOR:

Flexible Biomedical Devices for Mapping Cardiac and Neural Electrophysiology..........111
Dae-Hyong Kim, John A. Rogers, Jonathan Viventi, Brian Litt

Low-Energy UV Effects on Organic Thin-Film-Transistors..........113
N. Wrachien, A. Cester, D. Bari, G. Meneghesso, J. Kovac, J. Jakabovic, M. Sokolsky, D. Donoval, J. Cirak

A New Method for Predicting the Lifetime of Highly Stable Amorphous-Silicon Thin-Film Transistors from Accelerated Tests..........121
T. Liu, S. Wagner, J. C. Sturm

Investigation of Ultra Thin Polycrystalline Silicon Channel for Vertical NAND Flash..........126
Bio Kim, Seung-Hyun Lim, Dong Woo Kim, Toshiro Nakanishi, Sangryol Yang, Jae-Young Ahn, HanMei Choi, Kihyun Hwang, Yongsun Ko, Chang-Jin Kang

SESSION 2F: BEOL DIELECTRICS:

Electrical Reliabilities of Porous Silica Low-k Films..........130
Takamaro Kikkawa, Yasuhisa Kayaba, Kazuo Kohmura, Shinichi Chikaki

Invasion Percolation Model for Abnormal TDDB Characteristic of ULK Dielectrics with Cu Interconnect at Advanced Technology Nodes..........134
F. Chen, M. Shinosky, B. Li, J. Aitken, S. Cohen, G. Bonilla, A. Simon, P. McLaughlin, R. Achanta, F. Baumann, C. Parks, M. Angyal

Comparison between Intrinsic and Integrated Reliability Properties of Low-k Materials..........142
K. Croes, M. Pantouvaki, L. Carbonell, L. Zhao, G. P. Beyer, Zs. Tőkei

Statistics of Breakdown Field and Time-Dependent Dielectric Breakdown in Contact-to-Poly Modules..........149
Shinji Yokogawa, Satoshi Uno, Ichiro Kato, Hideaki Tsuchiya, Tatsuo Shimizu, Mitsuhiro Sakamoto

Reliability Limitations to the Scaling of Porous Low-K Dielectrics..........155
Shou-Chung Lee, A. S. Oates

A Comprehensive Process Engineering on TDDB for Direct Polishing Ultra-Low K Dielectric Cu Interconnects at 40nm Technology Node and Beyond..........160
W. C. Lin, T. C. Tsai, H. K. Hsu, Jack Lin, W. C. Tsao, Willis Chen, C. M. Cheng, C. L. Hsu, C. C. Liu, C. M. Hsu, J. F. Lin, C. C. Huang, J. Y. Wu

SESSION 2G: EXTREME ENVIRONMENTS:

Space Radiation Effects and Reliability Considerations for the Proposed Jupiter Europa Orbiter..........165
Allan Johnston

Reliability and Performance Characterization of a MEMS-Based Non-Volatile Switch..........171
Roberto Gaddi, Cor Schepens, Charles Smith, Cristian Zambelli, Andrea Chimenton, Piero Olivo

Microanalysis for Tin Whisker Risk Assessment..........177
Maribeth Mason, Genghmun Eng, Martin Leung, Gary Stupian, Terence Yeoh

SESSION 3A: GATE DIELECTRICS:

Random Telegraph Noise Reduction in Metal Gate High-kappa Stacks by Bipolar Switching and the Performance Boosting Technique..........182
W. H. Liu, K. L. Pey, N. Raghavan, X. Wu, M. Bosman, T. Kauerauf

Correlation of I_d- and I_g-Random Telegraph Noise to Positive Bias Temperature Instability in Scaled High-kappa/Metal Gate n-Type MOSFETs..........190
Chia-Yu Chen, Qiushi Ran, Hyun-Jin Cho, Andreas Kerber, Yang Liu, Ming-Ren Lin, Robert W. Dutton

SILC-Based Reassignment of Trapping and Trap Generation Regimes of Positive Bias Temperature Instability..........196
J. Q. Yang, M. Masuduzzaman, J. F. Kang, M. A. Alam

A New Interface Defect Spectroscopy Method 202
  J. T. Ryan, L. C. Yu, J. H. Han, J. J. Kopanski, K. P. Cheung, F. Zhang, C. Wang, J. P. Campbell, J. S. Suehle, V. Tilak, J. Fronheiser

Experimental Identification of Unique Oxide Defect Regions by Characteristic Response of Charge Pumping 207
  Muhammad Masuduzzaman, Ahmad Islam, Robin Degraeve, Moonju Cho, Mohammed Zahid, Muhammad Alam

SESSION 3B: MEDICAL ELECTRONICS:

The Implications of RoHS on Active Implantable Medical Devices 213
  Timothy Scott Savage

Implantable Microtechnologies for the Brain: Challenges and Strategies for Reliable Operation 220
  Jit Muthuswamy, Sindhu Anand, Jemmy Sutanto, Michael Baker, Murat Okandan

A Swallowable Diagnostic Capsule with a Direct Access Sensor Using Anisotropic Conductive Adhesive 224
  Pio Jesudoss, Alan Mathewson, Karen Twomey, Frank Stam, William M. D. Wright

Application Based Reliability Assessment and Qualification Methodology for Medical ICs 231
  Xiaowei Zhu, Karthik Vasanth, Xiaochen Xu, Charles Smyth, Brent Rhoton

SESSION 3C: SOFT ERRORS:

Quantification and Mitigation Strategies of Neutron Induced Soft-Errors in CMOS Devices and Components - The Past and Future 239
  Eishi Ibe, Ken-ichi Shimbo, Hitoshi Taniguchi, Tadanobu Toba, Koji Nishii, Yoshio Taniguchi

Effects of Scaling on Muon-Induced Soft Errors 247
  Brian D. Sierawski, Robert A. Reed, Marcus H. Mendenhall, Robert A. Weller, Ronald D. Schrimpf, Shi-Jie Wen, Richard Wong, Nelson Tam, Robert C. Baumann

Neutron Induced Single Event Multiple Transients with Voltage Scaling and Body Biasing 253
  Ryo Harada, Yukio Mitsuyama, Masanori Hashimoto, Takao Onoye

Double-Pulse-Single-Event Transients in Combinational Logic 258
  J. R. Ahlbin, T. D. Loveless, D. R. Ball, B. L. Bhuva, A. F. Witulski, L. W. Massengill, M. J. Gadlage

SESSION 3D: THERMO-MECHANICAL AND MEMS:

Thermomechanical Reliability of Through-Silicon Vias in 3D Interconnects 264
  Kuan-Hsun Lu, Suk-Kyu Ryu, Jay Im, Rui Huang, Paul S. Ho

Characterization of Steady and Transient Heating of Interconnects - A Review 271
  Banafsheh Barabadi, Yogendra Joshi, Satish Kumar

Impact of Scaling on the Performance and Reliability Degradation of Metal-Contacts in NEMS Devices 280
  Hamed F. Dadgour, Muhammad M. Hussain, Alan Cassell, Navab Singh, Kaustav Banerjee

The Effect of Temperature on Dielectric Charging of Capacitive MEMS 290
  M. Koutsoureli, L. Michalas, G. Papaioannou

SESSION 3E: ELECTROMIGRATION/VOIDING:

Electromigration Induced Void Kinetics in Cu Interconnects for Advanced CMOS Nodes 297
  L. Arnaud, P. Lamontagne, R. Galand, E. Petitprez, D. Ney, P. Waltz

Formation of Highly Reliable Cu/Low-k Interconnects by Using CVD Co Barrier in Dual Damascene Structures 307
  Hye Kyung Jung, Hyun-Bae Lee, Matsuda Tsukasa, Eunji Jung, Jong-Ho Yun, Jong Myeong Lee, Gil-Heyun Choi, Siyoung Choi, Chilhee Chung

Electromigration-Resistance Enhancement with CoWP or CuMn for Advanced Cu Interconnects 312
  Cathryn Christiansen, Baozhen Li, Matthew Angyal, Terence Kane, Vincent McGahay, Yun Yu Wang, Shaoning Yao

A Study of Via Depletion Electromigration with Very Long Failure Times 317
  Baozhen Li, Cathryn Christiansen, Kaushik Chanda, Matt Angyal, Jennifer Oakley

Study of Void Formation Kinetics in Cu Interconnects Using Local Sense Structures 321
K. Croes, M. Lofrano, C. J. Wilson, L. Carbonell, Y. K. Siew, G. P. Beyer, Zs. Tökei

SESSION 3F: PROCESS INTEGRATION AND 3D/TSV:

3D Integration Technology and Reliability 328
Mitsumasa Koyanagi

Impact of Air-Induced Poly-SI/Oxynitride Interface Layer Degradation on Gate-Edge Leakage 335
Ziyuan Liu, Shuu Ito, Tomoya Saito, Soon W. Chang, Arito Ogawa, Sadayoshi Horii, Tsuyoshi Horikawa, Markus Wilde, Katsuyuki Fukutani, Toyohiro Chikyow

On the Thermal Failure in Nanoscale Devices: Insight Towards Heat Transport Including Critical BEOL and Design Guidelines for Robust Thermal Management & EOS/ESD Reliability 342
Mayank Shrivastava, Manish Agrawal, Jasmin Aghassi, Harald Gossner, Wolfgang Molzer, Thomas Schulz, V. Ramgopal Rao

Resistance Increase Due to Electromigration Induced Depletion under TSV 347
T. Frank, C. Chappaz, P. Leduc, L. Arnaud, F. Lorut, S. Moreau, A. Thuaire, R. El Farhane, L. Anghel

SESSION 4A: TRANSISTORS/ORGANICS TFTS:

Reliability- and Process-Variation Aware Design of Integrated Circuits - A Broader Perspective 353
Muhammad A. Alam, Kaushik Roy, Charles Augustine

Response of a Single Trap to AC Negative Bias Temperature Stress 364
M. Toledano-Luque, B. Kaczer, Ph. J. Roussel, T. Grasser, G. I. Wirth, J. Franco, C. Vrancken, N. Horiguchi, G. Groeseneken

PBTI under Dynamic Stress: From a Single Defect Point of View 372
K. Zhao, J. H. Stathis, B. P. Linder, E. Cartier, A. Kerber

Understanding of Traps Causing Random Telegraph Noise Based on Experimentally Extracted Time Constants and Amplitude 381
Kenichi Abe, Akinobu Teramoto, Shigetoshi Sugawa, Tadahiro Ohmi

Mechanistic Understanding of Breakdown and Bias Temperature Instability in High-K Metal Devices Using Inline Fast Ramped Bias Test 387
Siddarth A. Krishnan, Eduard Cartier, James Stathis, Michael Chudzik, Andreas Kerber

SESSION 4C: ESD AND LATCH-UP:

Design of Modified ESD Protection Structure with Low-Trigger and High-Holding Voltage in Embedded High Voltage CMOS Process 392
Tai-Hsiang Lai, Lu-An Chen, Tien-Hao Tang, Kuan-Cheng Su

An EOS-Free PNP-Enhanced Cascoded NMOSFET Structure for High Voltage Application 396
Shih-Yu Wang, Yao-Wen Chang, Yan-Yu Chen, Chieh-Wei He, Guan-Wei Wu, Tao-Cheng Lu, Kuang-Chao Chen, Chih-Yuan Lu

Latch-Up Free ESD Protection Design with SCR Structure in Advanced CMOS Technology 401
Chang-Tzu Wang, Tien-Hao Tang, Kuan-Cheng Su

Transient Latchup in Power Analog Circuits 405
V. A. Vashchenko, D. LaFonteese, A. Concannon

WCDM2 - Wafer-Level Charged Device Model Testing with High Repeatability 409
Nathan Jack, Timothy J. Maloney, Bruce Chou, Elyse Rosenbaum

SESSION 4E: COMPOUND OPTO-ELECTRONICS:

Reliability of GaN-HEMTs for High-Voltage Switching Applications 417
Wataru Saito

Time Evolution of Electrical Degradation under High-Voltage Stress in GaN High Electron Mobility Transistors 422
Jungwoo Joh, Jesús A. del Alamo

Reliability-Limiting Defects in AlGaN/GaN HEMTs 426
Tania Roy, En Xia Zhang, Daniel M. Fleetwood, Ronald D. Schrimpf, Yevgeniy S. Puzyrev, Sokrates T. Pantelides

250 GHz Heterojunction Bipolar Transistor: From DC to AC Reliability 430
Malick Diop, Salim Ighilahriz, Florian Cacho, Vincent Huard

SESSION 5A: HIGH VOLTAGE/RF:

Intrinsic Reliability of RF Power LMDOS FETs 435
David C. Burdeaux, Wayne R. Burger

Investigation of Multistage Linear Region Drain Current Degradation and Gate-Oxide Breakdown under Hot-Carrier Stress in BCD HV PMOS 444
Yu-Hui Huang, J. R. Shih, C. C. Liu, Y.-H. Lee, R. Ranjan, Puo-Yu Chiang, Dah-Chuen Ho, Kenneth Wu

New Investigation of Hot Carrier Degradation of RF Small-Signal Parameters in High-k/Metal Gate nMOSFETs 449
Hyun Chul Sagong, Chang Yong Kang, Chang-Woo Sohn, Min Sang Park, Do-Young Choi, Eui-Young Jeong, Jack C. Lee, Yoon-Ha Jeong

Design-in Reliability Approach for Hot Carrier Injection Modeling in the Context of AMS/RF Applications 454
Vincent Huard, Thomas Quemerais, Florian Cacho, Laurence Moquillon, Sebastien Haendler, Xavier Federspiel

Impact of Source/Drain Contact and Gate Finger Spacing on the RF Reliability of 45-nm RF nMOSFETs 461
Rajan Arora, Sachin Seth, John Chung Hang Poh, John D. Cressler, Akil K. Sutton, Hasan M. Nayfeh, Giuseppe L. Rosa, Greg Freeman

SESSION 5B: SOFT ERRORS:

Soft-Error Testing at Advanced Technology Nodes 467
B. Bhuva, B. Narasimham, A. Oates, K. Patterson, N. Tam, M. Vilchis, S.-J. Wen, R. Wong, Y. Xu

Measurement of Neutron-Induced SET Pulse Width Using Propagation-Induced Pulse Shrinking 471
Jun Furuta, Chikara Hamanaka, Kazutoshi Kobayashi, Hidetoshi Onodera

Multicenter Comparison of Alpha Particle Measurements and Methods Typical of Semiconductor Processing 476
Jeffrey D. Wilkinson, Brett M. Clark, Richard Wong, Charles Slayman, Barry Carroll, Michael Gordon, Yi He, Olivier Lauzeral, Keith Lepla, Jennifer Marckmann, Brendan McNally, Philippe Roche, Mike Tucker, Tommy Wu

The Impact of New Technology on Soft Error Rates 486
Anand Dixit, Alan Wood

SESSION 5C: FAILURE ANALYSIS:

Quantitative, Nanoscale Free-Carrier Concentration Mapping Using Terahertz Near-Field Nanoscopy 493
J. Wittborn, R. Weiland, A. J. Huber, F. Keilmann, R. Hillenbrand

Rapid and Automated Grain Orientation and Grain Boundary Analysis in Nanoscale Copper Interconnects 500
K. J. Ganesh, S. Rajasekhara, D. Bultreys, P. J. Ferreira

High Reliable Strain Measurement for Power Devices Using STEM-CBED Method 503
N. Nakanishi, H. Arie, H. Maeda, Y. Hirose, N. Hattori, T. Koyama, E. Murakami

Electron Beam Induced Current Characterization of Dark Line Defects in Failed and Degraded High Power Quantum Well Laser Diodes 509
Maribeth Mason, Nathan Presser, Yongkun Sin, Brendan Foran, Steven C. Moss

Spectral Resolution of Photon Emission from SiGe:C Heterojunction Bipolar Transistors (HBTs) 514
Ulrike Kindereit, Oana-Mihaela Mutihac, Christian Boit, Bernd Tillack

Isolating Light-Sensitive Defects Using C-AFM 520
Hung Sung Lin, Mong Sheng Wu

SESSION 5D: PROCESS INTEGRATION AND 3D/TSV:

A Holistic Approach to Process Co-Optimization for Through-Silicon Via 524
Sesh Ramaswami

Thermal and Spatial Profiling of TSV-Induced Stress in 3DICs 527
Colin McDonough, Benjamin Backes, Wei Wang, Robert E. Geer

Reliability Studies of a 32nm System-on-Chip (SoC) Platform Technology with 2nd Generation High-K/Metal Gate Transistors 533
A. Rahman, M. Agostinelli, P. Bai, G. Curello, H. Deshpande, W. Hafez, C.-H. Jan, K. Komeyli, J. Park, K. Phoa, C. Tsai, J.-Y. Yeh, J. Xu

VTH Shift Mechanism in Dysprosium (Dy) Incorporated HfO$_2$ Gate nMOS Devices 539
Tackhwi Lee, Sanjay K. Banerjee

SESSION 5E: PHOTOVOLTAIC DEVICES:

Physics of Instability of Thin Film Si and (Si,Ge) Alloy Solar Cells 546
Vikram L. Dalal, Zhao Li

Reliability Testing beyond Qualification as a Key Component in Photovoltaic's Progress toward Grid Parity 551
John H. Wohlgemuth, Sarah Kurtz

Identification, Characterization and Implications of Shadow Degradation in Thin Film Solar Cells 557
Sourabh Dongaonkar, Muhammad A. Alam, Karthik Y., Souvik Mahapatra, Dapeng Wang, Michel Frei

Metastability of Hydrogenated Amorphous Silicon Passivation on Crystalline Silicon and Implication to Photovoltaic Devices 562
Bahman Hekmatshoar, Davood Shahrjerdi, Marinus Hopstaken, Devendra Sadana

Optical Stress and Reliability Study of Ruthenium-Based Dye-Sensitized Solar Cells (DSSC) 566
Daniele Bari, Nicola Wrachien, Andrea Cester, Gaudenzio Meneghesso, Roberto Tagliaferro, Stefano Penna, Thomas M. Brown, Andrea Reale, Aldo Di Carlo

SESSION 5F: CPI/ELECTROMIGRATION/VOIDING:

The Role of Elastic and Plastic Anisotropy of Sn on Microstructure and Damage Evolution in Lead-Free Solder Joints 573
Thomas R. Bieler, Bite Zhou, Lauren Blair, Amir Zamiri, Payam Darbandi, Farhang Pourboghrat, Tae-Kyu Lee, Kuo-Chuan Liu

Robust Pad Layout to Improve Wire Bonding Reliability 582
Kyoung-Hwan Kim, Hong Kook Min, Se Yeoul Park, So Ra Park, Seung Jin Yang, Byung Sup Shim, Yong Tae Kim, Jeong-Uk Han

Electromigration Characterization of Lead-Free Flip-Chip Bumps for 45nm Technology Node 588
Christine Hau-Riege, You-Wen Yau, Nick Yu

Electromigration Failure Mechanisms for Different Flip Chip Bump Configurations 592
Riet Labie, Tomas Webers, Christophe Winters, Vladimir Cherman, Kristof Croes, Bart Vandevelde, Franck Dosseul

SESSION 6A: TRANSISTORS/ORGANICS TFTS:

Understanding and Modeling AC BTI 597
Hans Reisinger, Tibor Grasser, Karsten Ermisch, Heiko Nielen, Wolfgang Gustin, Christian Schluender

The 'Permanent' Component of NBTI: Composition and Annealing 605
T. Grasser, Th. Aichinger, G. Pobegen, H. Reisinger, P.-J. Wagner, J. Franco, M. Nelhiebel, B. Kaczer

A Critical Re-Evaluation of the Usefulness of R-D Framework in Predicting NBTI Stress and Recovery 614
S. Mahapatra, A. E. Islam, S. Deora, V. D. Maheta, K. Joshi, A. Jain, M. A. Alam

On the Recoverable and Permanent Components of Hot Carrier and NBTI in Si pMOSFETs and Their Implications in Si$_{0.45}$Ge$_{0.55}$ pMOSFETs 624
J. Franco, B. Kaczer, G. Eneman, Ph. J. Roussel, M. Cho, J. Mitard, L. Witters, T. Y. Hoffmann, G. Groeseneken, F. Crupi, T. Grasser

Impact of HK / MG Stacks and Future Device Scaling on RTN 630
Naoki Tega, Hiroshi Miki, Zhibin Ren, Christopher P. D'Emic, Yu Zhu, David J. Frank, Michael A. Guillorn, Dae-Gyu Park, Wilfried Haensch, Kazuyoshi Torii

SESSION 6B: MEMORY:

Split-Gate Flash Memory for Automotive Embedded Applications..636
  Y. S. Chu, Y. H. Wang, C. Y. Wang, Y. H. Lee, A. C. Kang, R. Ranjan, W. T. Chu, T. C. Ong, H. W. Chin, K. Wu

Novel Negative Vt Shift Program Disturb Phenomena in 2X~3X nm NAND Flash Memory Cells............641
  Soonok Seo, Hyungseok Kim, Sungkye Park, Seokkiu Lee, Seiichi Aritome, Sungjoo Hong

Junction Optimization for Reliability Issues in Floating Gate NAND Flash Cells................................645
  C. H. Lee, I. C. Yang, Chienying Lee, C. H. Cheng, L. H. Chong, K. F. Chen, J. S. Huang, S. H. Ku, N. K. Zous, I.
  J. Huang, T. T. Han, M. S. Chen, W. P. Lu, K. C. Chen, Tahui Wang, Chih-Yuan Lu

Charge Diffusion in Silicon Nitrides: Scalability Assessment of Nitride Based Flash Memory................650
  Seung Jae Baik, Koeng Su Lim, Wonsup Choi, Hyunjun Yoo, Hyunjung Shin

The Effect of Crystallinity of HfO2 on the Resistive Memory Switching Reliability................................656
  Min Gyu Sung, Wan Gee Kim, Jong Hee Yoo, Sook Joo Kim, Jung Nam Kim, Byung Gu Gyun, Jun Young Byun,
  Taeh Wan Kim, Won Kim, Moon Sig Joo, Jae Sung Roh, Sung Ki Park

BEOL DIELECTRICS:

**A Novel Pre-Clean Process of BEOL Barrier-Seed Process to Enhance Reliability Performance of
Advanced 40nm Node**..661
  Chun-Min Cheng, Chi-Mao Hsu, W. C. Lin, Hsin-Fu Huang, Yan-Chun Liu, Kun-Hsien Lin, Jin-Fu Lin, C. C.
  Huang, J. Y. Wu

A Charge Transport Based Acceleration Model for Interlevel Dielectric Breakdown................................665
  Ravi Achanta, Paul McLaughlin

A Model for Post-CMP Cleaning Effect on TDDB..670
  Chia-Lin Hsu, Wen-Chin Lin, Teng-Chun Tsai, Climbing Huang, J.-Y. Wu

COMPOUND OPTO-ELECTRONICS:

Performance and Structure Degradations of SiGe HBT after Electromagnetic Field Stress................674
  A. Alaeddine, M. Kadi, K. Daoud

Reliability Testing of AlGaN/GaN HEMTs under Multiple Stressors..680
  Bradley D. Christiansen, Ronald A. Coutu Jr., Eric R. Heller, Brian S. Poling, G. David Via, Rama Vetury,
  Jeffrey B. Shealy

CPI / ELECTROMIGRATION / VOIDING:

Long Term Isothermal Reliability of Copper Wire Bonded to Thin 6.5 μm Aluminum................................685
  F. C. Classe, Sesha Gaddamraja

**A New ESD Model Induced Yield Loss during Chip-On-Film Package Process and It's Failure
Mechanism**..690
  Jian-Hsing Lee, J. R. Shih, Yu-Hui Huang, C. P. Lin, David Su, Kenneth Wu

CIRCUIT RELIABILITY:

**A Robust Reliability Methodology for Accurately Predicting Bias Temperature Instability Induced
Circuit Performance Degradation in HKMG CMOS**..696
  D. P. Ioannou, K. Zhao, A. Bansal, B. Linder, R. Bolam, E. Cartier, J.-J. Kim, R. Rao, G. La Rosa, G. Massey, M.
  Hauser, K. Das, J. H. Stathis, J. Aitken, D. Badami, S. Mittl

Bias Temperature Instability Model for Digital Circuits – Predicting Instantaneous FET Response............700
  Aditya Bansal, Kai Zhao, Jae-Joon Kim, Rahul Rao

Soft Oxide Breakdown Impact on the Functionality of a 40 nm SRAM Memory................................704
  Saad Cheffah, Vincent Huard, Remy Chevallier, Alain Bravaix

**The Relationship between Transistor-Based and Circuit-Based Reliability Assessment for Digital
Circuits**..706
  Balaji Vaidyanathan, Shawn Bai, Anthony S. Oates

The Impact of RTN on Performance Fluctuation in CMOS Logic Circuits..710
  Kyosuke Ito, Takashi Matsumoto, Shinichi Nishizawa, Hiroki Sunagawa, Kazutoshi Kobayashi, Hidetoshi Onodera

ESD AND LATCH-UP:

The Modified P+ Electrode Layout Schemes to Enhance ESD Robustness of SCR Structure for PMIC Applications 714
Lu-An Chen, Chang-Tzu Wang, Tai-Hsiang Lai, Tien-Hao Tang, Kuan-Cheng Su

Impact of Shielding Line on CDM ESD Robustness of Core Circuits in a 65-nm CMOS Process 717
Ming-Dou Ker, Chun-Yu Lin, Tang-Long Chang

Test Chip Design for Study of CDM Related Failures in SoC Designs 719
Nicholas Olson, Vrashank Shukla, Elyse Rosenbaum

Nanosecond Transient Thermoreflectance Imaging of Snapback in Semiconductor Controlled Rectifiers 725
Kerry Maize, Dustin Kendig, Ali Shakouri, Vladislav Vashchenko

ELECTROMIGRATION/VOIDING:

Degradation and Failure Analysis of Polysilicon Resistor Connecting with Tungsten Contact and Copper Line 731
Clement Huang, Mingte Lin, James W. Liang, Alex Juan, K. C. Su

A Practical Modeling for Transient Thermal Characteristics of Multilevel Interconnects 734
Seung-Man Choi, Dong-Cheon Baek, Tae-Young Jeong, Myung-Soo Yeo, Miji Lee, Andrew T. Kim, Jongwoo Park

Electromigration of Cu Interconnects under AC, Pulsed-DC and DC Test Conditions: Ramifications on Accelerated Testing 740
Roey Shaviv, Gregory J. Harm, Sangita Kumari, Robert R. Keller, David T. Read

Improving Lifetime of Cu Interconnects with Adding Compressive Stress at Cathode End 746
L. Arnaud, P. Lamontagne, E. Petitprez, R. Galand

EXTREME ENVIRONMENTS:

A Study on the Short- and Long-Term Effects of X-Ray Exposure on NAND Flash Memories 751
S. Gerardin, M. Bagatin, A. Paccagnella, A. Visconti, S. Beltrami, M. Bertuccio, L. T. Czeppel

Application of Reliability Test Standards to SiC Power MOSFETs 756
Ronald Green, Aivars Lelis, Daniel Habersat

FAILURE ANALYSIS:

A Novel and Low-Cost Method to Detect Delay Variation by Dynamic Thermal Laser Stimulation 765
Chunlei Wu, Masuda Motohiko, Winter Wang, Grace Song, Jinglong Li, Joe Yu, Li Tian, Miao Wu

A Study of the Influence of High Voltage Device Characteristics by Electron Beam Irradiation during Nanoprobing 770
Hung Sung Lin

Detecting Laser Beam Reflectance Modulated by Electronic Device Operation with a Simple Setup 774
Carlo Pagano, Christian Boit, Yoshiyuki Yokoyama

Backside Reflectance Modulation of Microscale Metal Interconnects 780
J. K. J. Teo, C. M. Chua, L. S. Koh, J. C. H. Phang

GATE DIELECTRICS:

Nanoscale Electrical and Physical Study of Polycrystalline High-kappa Dielectrics and Proposed Enhancement Techniques 786
K. Shubhakar, K. L. Pey, S. S. Kushvaha, M. Bosman, S. J. O'Shea, N. Raghavan, M. Kouda, K. Kakushima, Z. R. Wang, H. Y. Yu, H. Iwai

Investigation of Progressive Breakdown and Non-Weibull Failure Distribution of High-k and SiO_2 Dielectric by Ramp Voltage Stress 792
Nilufa Rahim, Ernest Y. Wu, Durgamadhab Misra

Oxide Defects Generation Modeling and Impact on BD Understanding 798
Y. Mamy Randriamihaja, V. Huard, A. Zaka, S. Haendler, X. Federspiel, M. Rafik, D. Rideau, D. Roy, A. Bravaix

Comprehensive Analysis of Charge Pumping Data for Trap Identification 802
D. Veksler, G. Bersuker, A. Koudymov, C. D. Young, M. Liehr, B. Taylor

A Physics-Based Model of the Dielectric Breakdown in HfO$_2$ for Statistical Reliability Prediction 807
Luca Vandelli, Andrea Padovani, Gennadi Bersuker, Jung Yum, Paolo Pavan, Luca Larcher

HIGH VOLTAGE / RF:

**Advanced 45nm MOSFET Small-Signal Equivalent Circuit Aging under DC and
RF Hot Carrier Stress** 811
L. Negre, D. Roy, S. Boret, P. Scheer, D. Gloria, G. Ghibaudo

MEMORY:

Characterization of Hexagonal Rare-Earth Aluminates for Application in Flash Memories 815
M. B. Zahid, R. Degraeve, M. Toledano-Luque, J. Van Houdt
Charge Gain, NBTI Recovery and Random Telegraph Noise in Localized-Trapping NVM Devices 819
Meir Janai, Ilan Bloom, Yael Shur
A Highly Reliable Embedded P-Channel SONOS Memory Using Dynamic Programming Method 824
Ying-Je Chen, Cheng-Jye Liu, Chun-Yuan Lo, Yun-Jen Ting, T. H. Hsu, Wein-Town Sun
Analysis of Edge Wordline Disturb in Multimegabit Charge Trapping Flash NAND Arrays 828
Cristian Zambelli, Andrea Chimenton, Piero Olivo
**Investigation of the Programming Accuracy of a Double-Verify ISPP Algorithm for Nanoscale NAND
Flash Memories** 833
Carmine Miccoli, Christian Monzio Compagnoni, Alessandro S. Spinelli, Andrea L. Lacaita
**Precise Understanding of Data Retention Mechanisms for MONOS Memories: Toward Simultaneous
Improvement of Retention and Endurance Performances by SiN Engineering** 839
Shosuke Fujii, Ryota Fujitsuka, Katsuyuki Sekine, Naoki Yasuda
Variability of Resistive Switching Memories and Its Impact on Crossbar Array Performance 843
An Chen, Ming-Ren Lin
Statistical Analysis of Retention Behavior and Lifetime Prediction of HfO$_x$-Based RRAM 847
*Lijie Zhang, Ru Huang, Yen-Ya Hsu, Frederick T. Chen, Heng-Yuan Lee, Yu-Sheng Chen, Wei-Su Chen, Pei-Yi Gu,
Wen-Hsing Liu, Shun-Min Wang, Chen-Han Tsai, Ming-Jinn Tsai, Pang-Shiu Chen*

PROCESS INTEGRATION AND 3D / TSV:

**Behaviors and Physical Degradation of HfSiON MOSFET Linked to Strained CESL
Performance Booster** 852
*Kidan Bae, Minjung Jin, Hajin Lim, Lira Hwang, Dongseok Shin, Junekyun Park, Jinchul Heo, Jongho Lee, Jinho
Do, Ilchan Bae, Chulhee Jeon, Jongwoo Park*
Experimental Study on Origin of V$_{TH}$ Variability under NBT Stress 857
Yuichiro Mitani, Akira Toriumi
Multiple Cell Upsets Tolerant Content-Addressable Memory 863
Syed Mohsin Abbas, Sanghyeon Baeg, Sungju Park
An Automated Approach to Isolate Dominant SER Susceptibilities in Microcircuits 868
James Castillo, David Mavis, Paul Eaton, Mike Sibley, Don Elkins, Rich Floyd
Bit Error and Soft Error Hardenable 7T/14T SRAM with 150-nm FD-SOI Process 876
*Shusuke Yoshimoto, Takuro Amashita, Shunsuke Okumura, Kosuke Yamaguchi, Masahiko Yoshimoto,
Hiroshi Kawaguchi*
Pulsed Laser-Induced Transient Currents in Bulk and Silicon-On-Insulator FINFETs 882
*F. El-Mamouni, E. X. Zhang, R. D. Schrimpf, R. A. Reed, K. F. Galloway, D. McMorrow, E. Simoen, C. Claeys, S.
Cristoloveanu, W. Xiong*
Neutron- and Alpha-Particle Induced Soft-Error Rates for Flip Flops at a 40 nm Technology Node 886
Srikanth Jagannathan, T. D. Loveless, Z. Diggins, B. L. Bhuva, S.-J. Wen, R. Wong, L. W. Massengill
Analysis of Multiple Cell Upsets Due to Neutrons in SRAMs for a Deep-N-Well Process 891
Nihaar Mahatme, Bharat Bhuva, Y.-P. Fang, Anthony Oates
Impact of Ion-Induced Transients on High-Speed Dual-Complementary Flip-Flop Designs 897
Dolores A. Black, Robert A. Reed, William H. Robinson, Jeffrey D. Black, Daniel B. Limbrick, Kevin D. Dick

THIN FILMS:

Stability Improvement of a-ZIO TFT Circuits Using Low Temperature Anneal 904
Aritra Dey, David R. Allee

TRANSISTORS:

Low-Frequency Noise Behavior of La-Doped HfSiON/Metal Gate nMOSFETs 908
Do-Young Choi, Min Sang Park, Chang Woo Sohn, Hyun Chul Sagong, Eui-Young Jung, Jeong-Soo Lee, Yoon-Ha Jeong, Chang Yong Kang

Neutral Interface Traps for Negative Bias Temperature Instability 913
Z. Chen, X. Zhou, Y. Z. Hu, K. S. Machavolu

Atomistic Approach to Variability of Bias-Temperature Instability in Circuit Simulations 915
B. Kaczer, S. Mahato, V. Valduga de Almeida Camargo, M. Toledano-Luque, Ph. J. Roussel, T. Grasser, F. Catthoor, P. Dobrovolny, P. Zuber, G. Wirth, G. Groeseneken

Probabilistic Defect Occupancy Model for NBTI 920
J. Martin-Martinez, B. Kaczer, M. Toledano-Luque, R. Rodriguez, M. Nafria, X. Aymerich, G. Groeseneken

MOSFET's Hot Carrier Degradation Characterization and Modeling at a Microscopic Scale 926
Y. Mamy Randriamihaja, A. Zaka, V. Huard, M. Rafik, D. Rideau, D. Roy, A. Bravaix

Simultaneous Extraction of Threshold Voltage and Mobility Degradation from On-The-Fly NBTI Measurements 929
Rodolf W. Herfst, Jurriaan Schmitz, Andries J. Scholten

Analysis of Recoverable and Non-Recoverable NBTI and PBTI Using AC and DC Stresses 933
Frederic Monsieur, Eduard Cartier, James Stathis

On the Evolution of the Recoverable Component of the SiON, HfSiON and HfO₂ P-MOSFETs under Dynamic NBTI 935
Y. Gao, A. A. Boo, Z. Q. Teo, D. S. Ang

New Observations on the Physical Mechanism of Vth-Variation in Nanoscale CMOS Devices after Long Term Stress 941
E. R. Hsieh, Steve S. Chung, C. H. Tsai, R. M. Huang, C. T. Tsai, C. W. Liang

On the Cyclic Threshold Voltage Shift of Dynamic Negative-Bias Temperature Instability 943
Z. Q. Teo, A. A. Boo, D. S. Ang, K. C. Leong

Author Index

Electron Beam Induced Current Characterization of Dark Line Defects in Failed and Degraded High Power Quantum Well Laser Diodes

Maribeth Mason, Nathan Presser, Yongkun Sin, Brendan Foran, and Steven C. Moss
Microelectronics Technology Department
The Aerospace Corporation
El Segundo, CA
Maribeth.S.Mason@aero.org

Abstract—We investigate the dependence of electron beam induced current (EBIC) contrast from dark-line defects (DLDs) on temperature and voltage bias in failed and degraded high power quantum well laser diodes (HPLDs). Voltage bias induced contrast variations in EBIC allowed us to make the first observation of what may be the DLD initiation point in a degraded, but not failed, HPLD. Wavelet analysis of temperature and voltage dependent EBIC contrast reveals three distinct regions with different defect properties within the DLD. These results can be correlated to destructive physical analysis and other defect characterization techniques, such as cathodoluminescence and deep-level transient spectroscopy, to provide insight into defect types and failure mechanisms in these devices.

Keywords-EBIC, laser diode, dark-line defect

I. INTRODUCTION

Multimode InGaAs/AlGaAs strained quantum well single emitter laser diodes at 920-980 nm have become indispensable as pump lasers for various applications [1-2]. Most reports have focused on performance characteristics, with limited reports focused on understanding the physics of failure critical to the development of high reliability applications. In this paper, we characterize the temperature and voltage dependence of electron beam induced current (EBIC) contrast of failed and degraded HPLDs, and correlated these results to destructive physical analysis and other techniques, such as cathodoluminescence (CL) and deep-level transient spectroscopy (DLTS) to provide insight into defect types and failure mechanisms in these devices.

The Aerospace Corporation's EBIC capability is integrated with a dual-beam system combining a scanning electron microscope (SEM) with a focused ion beam (FIB). Defects not visible with SEM can be observed with EBIC and targeted for destructive physical analysis using the FIB [3]. Defect depth profiles are obtained by varying the electron accelerating voltage [4] or by applying an external bias to modulate the depletion width [5]. Other defect properties can be extracted from temperature-dependent EBIC contrast [6,7]. By correlating temperature and voltage dependent EBIC to

destructive physical analysis and other defect analysis techniques, we are constructing a database of defect types in failed and degraded HPLDs.

II. EXPERIMENT

We studied 975 nm multimode InGaAs/AlGaAs HPLDs with a strained single quantum well grown by MOCVD. The laser cavity is 2.5 mm long by 0.1 mm wide. Fig. 1 shows a device cross-section obtained by high-angle annular dark field scanning transmission electron microscopy (HAADF-STEM). The quantum well consists of a 10 nm layer of InGaAs surrounded by AlGaAs cladding.

Accelerated life testing was used to generate HPLD failures. In some cases, life testing was stopped after the output power was observed to have degraded, but before complete failure. EBIC was then used to locate and characterize defects in the failed and degraded HPLDs. Although the life testing is performed in a p-side down configuration, EBIC measurements require the devices to be p-side up in order for a 30kV electron beam to penetrate to the p-n junction. The failure signature observable by EBIC is known as a dark-line defect (DLD). EBIC was used to target DLDs for FIB cross-sectioning and analysis by transmission electron microscopy (TEM).

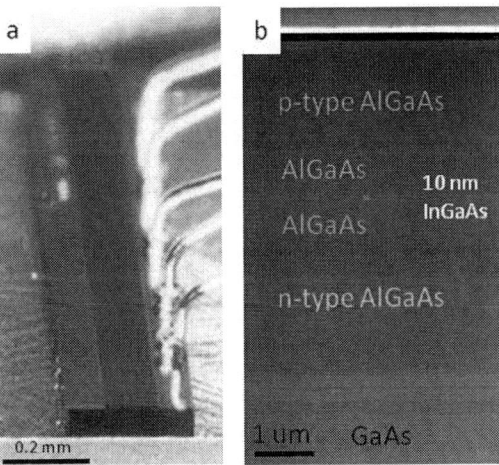

Figure 1. (a) Top-down optical micrograph and (b) HAADF-STEM cross-section of HPLDs.

EBIC has been used to investigate DLDs in HPLDs with both unpassivated and passivated front facets. The presence of facet passivation can affect the failure signature of the device, as shown in Fig. 2. Typically, DLDs within the waveguide region of unpassivated HPLDs intersect the front facet and are coincident with catastrophic optical mirror damage (COMD), while DLDs in passivated HPLDs do not typically intersect the front facet. It is unclear whether the DLDs in either type of device propagate to or from the area near the front facet. The front facets of the HPLDs discussed in the following sections are passivated.

III. RESULTS

A. Correlation between EBIC and TEM analysis

Fig. 3 shows a top-down EBIC image of a branched DLD within the waveguide region of a degraded HPLD. The laser output power at the front facet had decreased by 60% during a 5 A constant current life test with a junction temperature of 65°C. The temperature-dependent EBIC contrast along the line shown in Fig. 2, through the three branches of the defect, was characterized as a function of temperature from 94—417K. The results are shown in Fig. 3. The EBIC contrast is defined as $1 - (I_{defect}/I_{bg})$, where I_{defect} and I_{bg} are the measured EBIC currents for the DLD and background, respectively. The EBIC contrast arises from differences in the recombination rates of electron hole pairs produced by the injection of high energy electrons by the SEM.

The increased carrier lifetime at low temperature leads to decreased EBIC contrast, since more incident electrons can escape the defect region. Fig. 4 shows that the contrast in branches 1 and 3 of the DLD saturates at low temperature, while the contrast in branch 2 continues to decrease. This contrast behavior can be explained by the defect structures observed by TEM.

Fig. 5 shows cross-sectional ADF-STEM images of each branch of the DLD. ADF-STEM images strain contrast from dislocations within the failure region, which appear bright in these images. The bright horizontal line across each image represents the location of the 10 nm quantum well, which is

disturbed in all three branches of the DLD. This disturbance has been shown to be associated with In migration from the InGaAs QW into the AlGaAs cladding layers [8]. However, while the disturbance in branch 2 of the DLD is confined to a narrow area surrounding the quantum well, the disturbance in branches 1 and 3 is much more extreme, with dislocations extending well beyond the quantum well. The temperature dependent EBIC contrast in branches 1 and 3 is then likely dominated by these dislocations, while the EBIC contrast in branch 2 is dominated by defects and dislocations within the quantum well region.

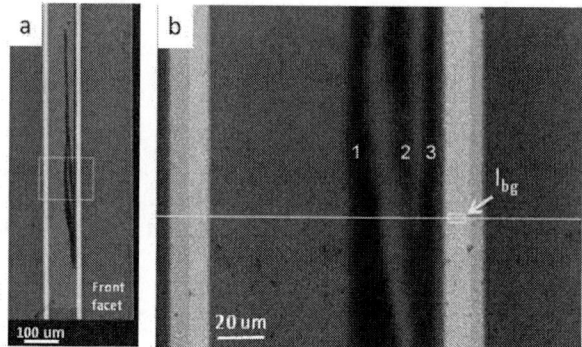

Figure 3. (a) Top-down EBIC image of a branched DLD in 60% degraded HPLD at 94K. (b)EBIC image of area outlined in (a), illustrating line along which temperature-dependent EBIC contrast was measured.

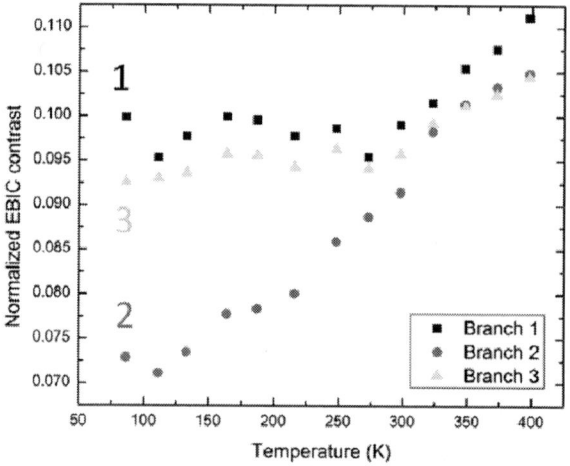

Figure 4. Temperature-dependent EBIC contrast of a branched DLD structure as measured along the line in Fig. 3.

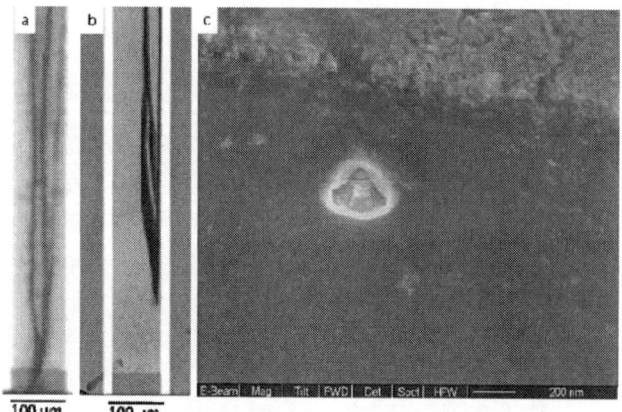

Figure 2. EBIC images of DLDs in (a) unpassivated and (b) passivated HPLDs. (c) Catastrophic optical mirror damage at front facet associated with failure in (a), where the DLD intersects the front facet.

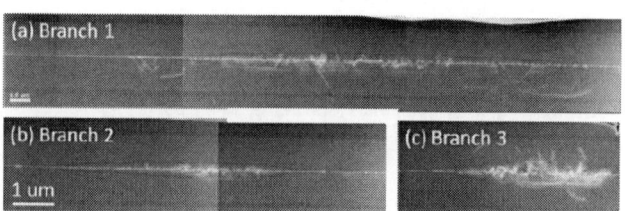

Figure 5. ADF-STEM images of DLD branches indicated in Fig. 3.

978-1-4244-9113-1/11 $26.00 © 2011 IEEE

B. Temperature and bias dependent EBIC contrast

Figs. 6 and 7 show the temperature and voltage bias dependent EBIC contrast for a passivated HPLD with a 10% loss in output power. The dark spot near the rear facet of the HPLD may correspond to a DLD initiation point. Fig. 6 shows that, at 298K and 0V, a dark spot in the EBIC contrast is visible. At high temperature and high reverse bias, an extended defect structure is observed surrounding the central spot. Under reverse bias, the EBIC contrast of this region increases due to expansion of the depletion width of the p-n junction, but the contrast changes little under forward bias. The EBIC contrast for defects far from the central spot is also enhanced at high temperature due to a decrease in the minority carrier diffusion length. Based on prior observations [8], the central spot may correspond to an area where the quantum well is severely disturbed, while the surrounding region may have undergone a lesser disturbance, with some intermixing between the InGaAs quantum well and the AlGaAs cladding layers.

C. Wavelet analysis of EBIC contrast

The experiments described above result in a large, 5-dimensional data set, in which each location (x,y) in the defect region has an associated contrast that depends on temperature and voltage. Fig. 8 illustrates two of these locations with different temperature and voltage dependent EBIC contrast. Our goal was to capture this temperature and voltage dependent contrast behavior in each location in a small number of transform coefficients, and to map these coefficients to identify defect regions with similar properties.

Wavelets are orthogonal functions that provide a compact, representation of discrete data. In contrast to Fourier sine and cosine functions, wavelets are localized in space as well as frequency, and the wavelet transform matrix acts to sample the dataset to which it is applied at both coarse and fine scale resolution. The simplest wavelet is the Haar wavelet [9], whose mother wavelet has the form

$$\psi(t) = \begin{cases} 1 & 0 \le t < \frac{1}{2} \\ -1 & \frac{1}{2} \le t < 1 \\ 0 & otherwise \end{cases} \qquad (1)$$

with scaling function

$$\varphi(t) = \begin{cases} 1 & 0 \le t < 1 \\ 0 & otherwise \end{cases} \qquad (2)$$

We applied two-dimensional Haar wavelet transforms, *not* to the EBIC images themselves, but to the matrix of temperature and voltage dependent EBIC contrast at each location (x,y), such that t in the above equations represents a step in either temperature or voltage [10]. The results are shown in Fig. 9.

With appropriate thresholding, the temperature and voltage dependent EBIC contrast at every point can be captured in two significant wavelet coefficients, which are plotted in Fig. 9. Mapping of the second order wavelet coefficient reveals three distinct subregions within the defect, which can be correlated to the defect structure observed by transmission electron microscopy and other defect analysis techniques.

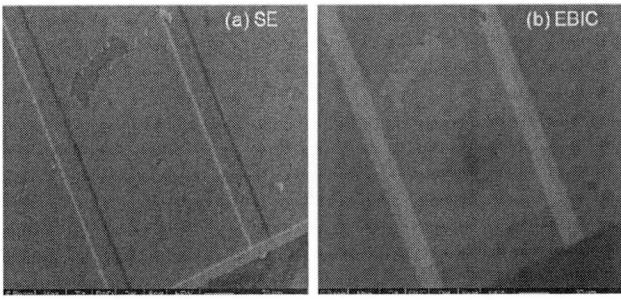

Figure 6. (a) SE and (b) EBIC images of possible DLD initiation point in 10% degraded HPLD at 0V bias and 298 K.

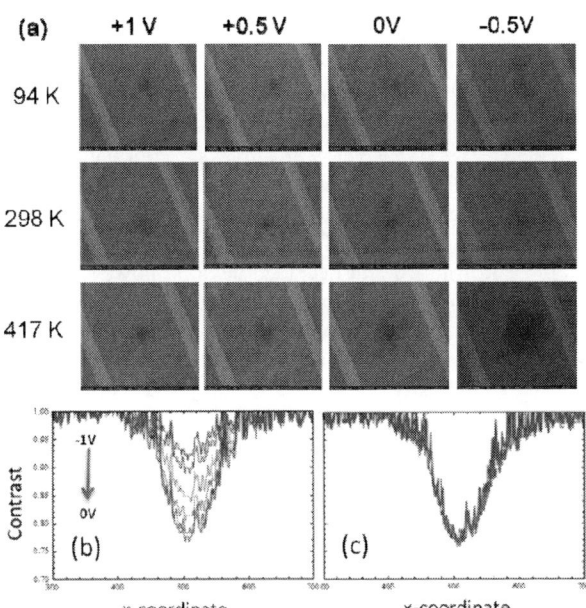

Figure 7. (a) Temperature and voltage dependent EBIC contrast of possible DLD initiation point in 10% degraded HPLD. (b) Voltage dependent EBIC contrast at 94 K under reverse bias. (c) Voltage dependent EBIC contrast at 94 K under forward bias.

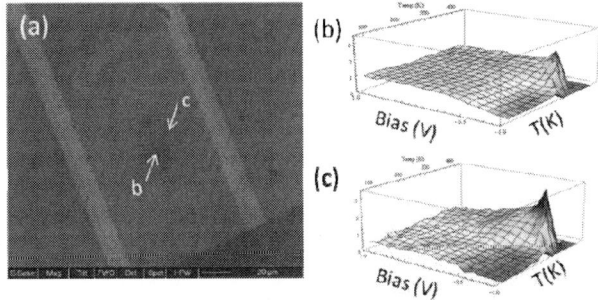

Figure 8. (a) EBIC image of HPLD illustrating regions for temperature and voltage dependent EBIC contrast analysis. (b) and (c) Temperature and voltage dependent EBIC contrast of regions labeled in (a). Contrast in (b) and (c) is presented on the same linear and color scales.

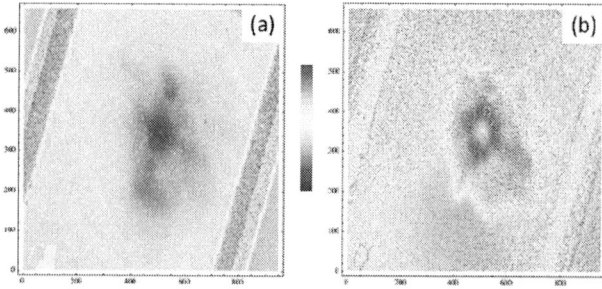

Figure 9. Maps of (a) first-order and (b) second-order wavelet coefficients representing temperature and voltage dependent EBIC contrast. Three distinct defect subregions are visible in (b).

D. Optical and electrical recombination centers

Temperature and voltage dependent EBIC analysis provides a spatial map of electrical recombination centers within the HPLD. However, EBIC gives no information about specific types of defects, or about whether these defects act as optical as well as electrical recombination centers. Deep-level transient spectroscopy (DLTS) was used to identify defect types present in failed and degraded devices; however, DLTS provides no information about the spatial location of the defects it identifies. Cathodoluminescence (CL) can provide a spatial map of optical emission from defects, and the results can be correlated to DLTS and EBIC to provide information about the spatial location of specific traps within the DLD.

Conventional capacitance DLTS measurements from the 10% degraded HPLD reveal two major types of defects: EL2 traps (E_C-E_T = 0.65 eV) and DX centers (E_C-E_T = 0.37 eV). The DX centers have been attributed to lower n-doped AlGaAs cladding layers and the EL2 traps to the InGaAs QW and GaAs layers [11].

Fig. 10 directly compares a single EBIC image at 94K and 0.5 V reverse bias to the results of the temperature and voltage dependent EBIC contrast analysis and monochromatic CL maps taken at room temperature and 30 kV. CL emission was observed at 675 nm, corresponding to the AlGaAs cladding layers, and at 940 nm, corresponding to the InGaAs QW. The dark areas in Fig. 10 (c) and (d) correspond to the absence of CL emission.

The small region with no CL emission at 675 nm, shown in Fig. 10 (c), directly corresponds to the dark orange region in Fig. 10 (a) and the center dark spot in Fig. 10 (b). This region shows no CL emission at 675 or 940 nm. The larger area with no CL emission at 940 nm, shown in Fig. 10 (d), directly corresponds to the larger dark region in Fig. 10 (b) and the light orange region in Fig. 10 (a). This extended defect region shows CL emission at 675 nm but not 940 nm. These results predict a defect structure similar to that shown in Fig. 5, in which a large QW disturbance is surrounded by an area of intermixing between the InGaAs QW and the AlGaAs cladding layers. We predict that TEM analysis will confirm this type of defect structure, and may possibly reveal the origins of the center light orange region in Fig. 10 (a).

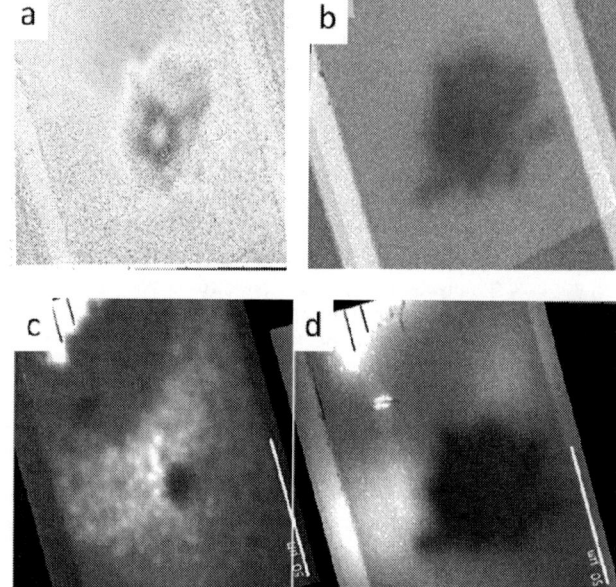

Figure 10. DLD in 10% degraded HPLD. All images show the defect with the same coordinate axes and the same scale. (a) Results of temperature and voltage dependent EBIC contrast analysis. (b) EBIC image at 94K and -0.5V. (c) Map of CL emission at 675 nm. (d) Map of CL emission at 940 nm.

The CL measurements give some indication as to the location of the defect types observed with DLTS within the DLD, since DX centers do not participate in non-radiative recombination processes, but EL2 traps are efficient non-radiative recombination centers [12]. To further isolate the location of each type of trap within the DLD, we plan to experiment with a scanning DLTS technique to be performed in conjunction with EBIC characterization. Where conventional DLTS analyzes temperature and rate-window dependent capacitance transients, scanning DLTS analyzes the EBIC current transients produced by a pulsed electron beam at a single location. This technique can non-destructively provide spatial maps of defect concentrations and activation energies to complement EBIC characterization.

IV. CONCLUSIONS

We have developed a set of complementary failure analysis techniques to investigate dark-line defects in HPLDs. Integration of the EBIC technique with the dual-beam FIB allows defect regions to be easily targeted for destructive physical analysis, and temperature-dependent EBIC contrast has been shown to correlate to defect structures observed byTEM. Wavelet analysis of temperature and voltage dependent EBIC contrast can reduce large EBIC datasets to a limited set of maps which identify distinct defect regions for further analysis. Finally, cathodoluminescence as well as conventional and scanning DLTS can non-destructively correlate EBIC contrast to defect types and defect-induced failure modes in HPLDs.

REFERENCES

[1] Y. Sin, N. Presser, B. Foran, N. Ives, and S.C. Moss, "Catastrophic facet and bulk degradation in high power multimode InGaAs strained quantum well single emitters," Proc. SPIE, Vol. 7198, 719818, 2009.

[2] V. Rossin, E. Zucker, M. Peters, M. Everett, and B. Acklin, "Highly reliable high-power broad area laser diodes," Proc. SPIE, Vol. 6104, 610407, 2006.

[3] N. Presser, B. Foran, M. Mason, Y. Sin, and S.C. Moss, "EBIC targeting for dual beam FIB-Based TEM sample preparation," Microsc. Microanal., vol. 14 (suppl. 2), pp. 392-393, 2008.

[4] T. E. Everhart and P.H. Hoff, "Determination of kilovolt electron energy dissipation vs. penetration distance in solid materials," J. Appl. Phys. 42 (13), pp. 5837-5846, 1971.

[5] N. C. MacDonald and T.E. Everhart, "Direct measurement of the depletion layer width variation vs. applied bias for a p-n junction," Appl. Phys. Lett. 7, p. 267, 1965.

[6] J. D. Zook, "Theory of beam-induced currents in semiconductors," Appl. Phys. Lett. 42 (7), pp. 602-604, 1983.

[7] M. Kittler and W. Seifert, "On the origin of EBIC defect contrast in silicon: a reflection on injection and temperature dependent investigations," Phys. Stat. Sol. A 138, pp. 687-693, 1993.

[8] B. Foran, N. Presser, Y. Sin, M. Mason, and S.C. Moss, "Two distinct types of dark line defects in a failed InGaAs/AlGaAs strained quantum well laser diode," in Conference on Lasers and Electro-Optics/International Quantum Electronics Conference, OSA Technical Digest (CD) (Optical Society of America), paper CWF5, 2009.

[9] R.S. Stankovic and B.J. Falkowski, "The Haar wavelet transform: its status and achievements," Comp. Elec. Eng. 29, pp. 25-44, 2003.

[10] I. Popanov and R.J. Miller, "Similarity search over time-series data using wavelets," Proc. 18th Intl. Conf. Data Eng., pp. 212-221, 2002.

[11] Y. Sin, N. Presser, N. Ives, and S.C. Moss, "A study of degradation in high power multimode InGaAs-AlGaAs strained quantum well lasers as pump lasers," Mater. Res. Soc. Symp. Proc. Volume 1195, Warrendale, PA, 2010, 1195-B01-06.

[12] J.C. Bourgoin and N. DeAngelis, "The defect responsible for non-radiative recombination in GaAs materials," Semicond. Sci. Tech. 16 (6), 2001.

ACKNOWLEDGMENTS

The authors would like to thank Mr. Mark Cornish of the Materials Research Laboratory at the University of California, Santa Barbara and Dr. David Stowe of Gatan UK for assistance with cathodoluminescence measurements. This work was supported under The Aerospace Corporation's Independent Research and Development Program.

Spectral Resolution of Photon Emission from SiGe:C Heterojunction Bipolar Transistors (HBTs)

Dr. Ulrike Kindereit
Circuit Test and Diagnostics Technology
IBM T.J. Watson Research Center
Yorktown Heights, NY, USA
phone: +1-914-945-1469, e-mail: ukinder@us.ibm.com

Dr. Oana-Mihaela Mutihac and Prof. Christian Boit
Semiconductor Devices
TUB – Berlin Institute of Technology
Berlin, Germany

Prof. Bernd Tillack
Technology
IHP GmbH – Innovations for High Performance Microelectronics
Frankfurt (Oder), Germany

Abstract—This publication presents photon emission measurements of SiGe:C-HBTs acquired with Si-CCD and InGaAs detector, proving the InGaAs-camera capability for this application. The emission characteristic helps distinguish operating modes like saturation, active or avalanche. Spectral response shows a local maximum at 1300 nm, representing the decreased bandgap due to additional germanium.

Photon emission, heterojunction bipolar transistor, HBT, SiGe, Si-CCD-camera, InGaAs-camera, spectral analysis.

I. INTRODUCTION

Static photon emission of hetero-junction transistors (HBTs) has been investigated and correlated to the device activity ([1]-[4]). The results indicate that the photon emission sources are distinct for the three different device states: saturation, forward active mode and avalanche. Polonsky et al. used a Si(silicon)-CCD(charge coupled device)-camera with high quantum efficiency at 700 nm and a rapid roll-off towards the near infrared (NIR; quantum efficiency < 1 % at 1100 nm wavelength). Wavelengths higher than 1100 nm can not be detected with such a camera. The wavelength of the radiant recombination for Si is expected to be around 1107 nm, corresponding to the 1.12 eV bandgap energy of silicon at 300 K. For the detection of the radiant recombination emission, the authors of the cited work used a RG780 filter (low-pass filter; cut-off wavelength 780 nm). To verify the strong visible (VIS) part of the radiation in the avalanche region, a BG39 filter (bandpass filter; transmission window between 300 and 700 nm) was used.

The application of SiGe(germanium):C(carbon)-HBTs gives rise to the question, which wavelength the emitted photons consist of and if the Si-CCD-camera is the proper tool to detect the light. Especially the germanium but also the carbon in the SiGe:C-HBTs alter the bandgap of the device, such that the resulting bandgap is narrower than the bandgap of pure silicon, with the according increase of the optical wavelengths.

Up to date, there has been no investigation on the spectral response of the HBT photon emission for wavelengths higher than 1100 nm with an InGaAs(indium-gallium-arsenide)-camera.

For this work, the photon emission results of the standard Si-CCD-camera of the system used (Hamamatsu Phemos-1000), which has a spectral sensitivity in the range from 400 nm up to 1100 nm, will be compared to those acquired with a cooled (-70 °C) InGaAs-camera. The spectral sensitivity of the InGaAs-camera reaches from 950 nm up to 1550 nm, enabling photon emission measurements from hot carriers (wavelength lower than bandgap wavelength), but also for radiant recombination from a narrower bandgap than that of pure silicon (higher wavelengths). For the spectral resolution, various filters were used.

II. PHOTON EMISSION FROM HBTs

For an npn-HBT in common-base configuration in saturation, the emitter-base junction is forward biased, hence the electrons flow from the emitter to the base, where they become minority carriers and recombine. The resulting photon emission is caused by this radiant recombination in the NIR region. The base-collector junction is forward biased, as well, in this operating state, but for higher collector currents the forward bias is decreasing, which could result in a reduced photon emission level.

In the non-saturation region (i.e. forward active mode), the emitter-base junction is still forward biased, but the base-collector junction is in reverse bias, which becomes stronger for higher base-collector voltages. Polonsky et al. found no dependency on the base-collector voltage, which means that the emission does not depend on the scattering of the hot electrons, but they found a linear dependency on the emitter current. The photon emission was weaker here than in saturation, because the number of electrons recombining in the base was low compared to the combination of recombination in the base and the collector (saturation).

A very high collector-base voltage causes a strong electric field, which leads to impact ionization. The generated hot electrons contribute to the collector current (increase in I_C), whereas the hot holes move into the base and decrease the base current (until it becomes negative). The intensity of the photon emission in the avalanche region was found to be the strongest in [1] and it increased with the collector-base voltage. The wavelengths of the emitted photons are in the VIS range (hot electron radiation with higher energies and lower wavelengths). Nonetheless, simultaneously, the radiant photon emission, caused by the forward bias of the emitter-base junction, is still present even if it is much weaker. For large emitter currents (I_E > 1 mA), base widening occurs and reduces the maximum electric field, which in turn decreases the photon emission intensity (see [1] pp. 346). The proportion of the emission levels in all three operating conditions strongly depends on the device structure and its electrical characteristics and hence can vary for each device.

For easier comparison of the results discussed in this work to the results of the cited work, the following table summarizes the measurements and expectations of Polonsky et al. [1].

Operating mode	Relative mission intensity levels	Dominating emission spectrum
Saturation	Strong	NIR
Forward active	Weaker	NIR
Avalanche	Strongest	VIS

Table 1: Summary of measurements and expectations of the cited work [1], compare to results of this work in Figures 9-11.

III. DEVICE UNDER TEST

The structure investigated here is a large SiGe:C-HBT with a 100 x 50 µm² emitter area built in IHP's H1 0.25 µm process technology. The layout of the structure is shown in Figure 1 (emitter area indicated by the dashed orange line). The emitter (E) pad is located at the bottom, the collector (C) pad is at the top and another metal line at the bottom of the emitter area leads to the base (B) pad. The according schematic diagram is shown in Figure 2 and the Gummel and output characteristic for the common-emitter configuration can be found in Figure 3 and Figure 4. The crosses in Figure 4 indicate the operating points, in which the measurements, shown in the next paragraph, were performed: a) saturation, b) forward active mode and c) + d) avalanche.

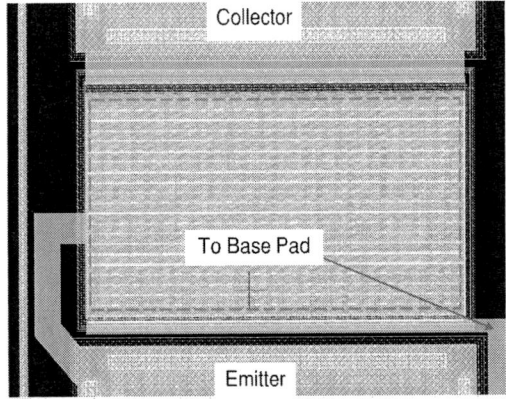

Figure 1: Layout of the HBT used for the investigations of this work; emitter area 100 x 50 µm², indicated by the dashed orange line.

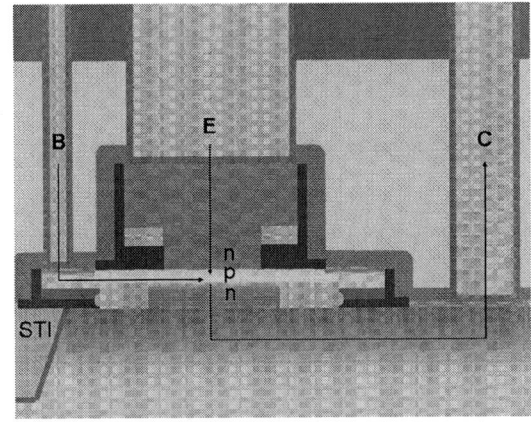

Figure 2: Schematic diagram of the npn-HBT used for this work; arrows indicate electron flux in active forward mode; B: base, E: emitter, C: collector.

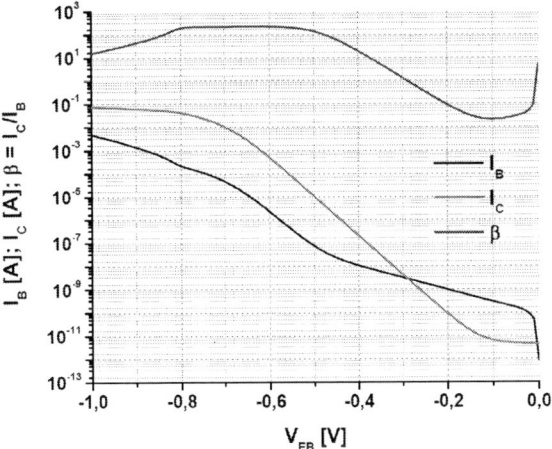

Figure 3: Gummel characteristic: base and collector current and current gain versus emitter-base voltage V_{EB} at $V_{CB} = 0$ V.

978-1-4244-9113-1/11 $26.00 © 2011 IEEE   515

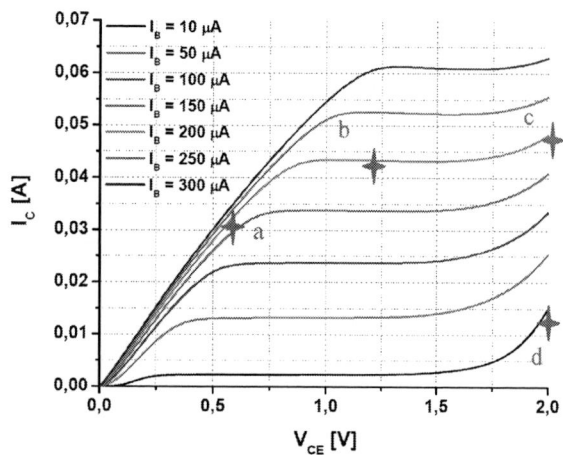

Figure 4: Output characteristic; crosses indicate the operating points for the measurements: a) saturation, b) forward active mode and c) + d) avalanche.

Figure 5 shows a typical graded SiGe:C-HBT profile of the HBTs produced in the IHP. According to People and Bean [5] the resulting bandgap can be calculated by the following equation.

$$E_{g,SiGe} = E_{g,Si} - x \cdot 0.74 \text{ (eV)} \qquad (1)$$

Here $E_{g,Si}$ is the silicon bandgap, x is the germanium content in the silicon and $E_{g,SiGe}$ is the resulting bandgap of the alloy. Since the germanium content in the base is around 20 %, the according bandgap of the SiGe base is 0.97 eV, which corresponds to wavelengths of around 1300 nm. Photons of such long wavelengths can not be detected with the Si-CCD-camera, but the InGaAs-camera is well-suited for this scope. The carbon content (about 0.1-0.2 %) has only minor influence on the bandgap (according to [6] the bandgap increase is around 5 meV).

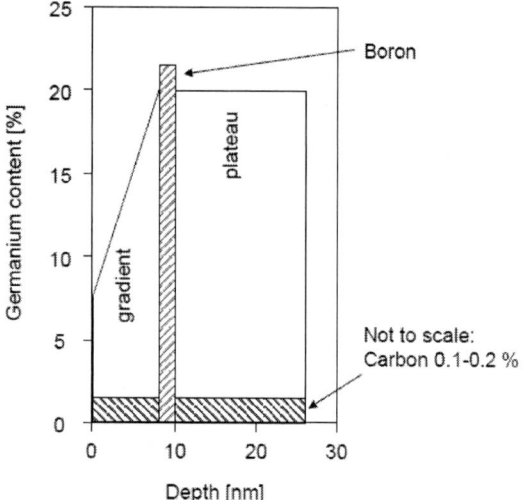

Figure 5: Typical graded SiGe:C-HBT profile.

IV. MEASUREMENTS AND RESULTS

All measurements were performed at room temperature (about 300 K). Since the emission intensities varied strongly for the different operating points and the two different cameras, it was necessary to vary the exposure time (exposure time: time the camera records the emission; varied between 2 and 200 s) in various measurements. The linear dependency between the exposure time and the relative emission intensity has been proven. In order to enable direct comparison between the measurement results in the following graphs, the relative emission intensity was divided by the exposure time, hence resulting in the absolute emission intensity (in arbitrary units [a.u.]). The single data points shown in the graphs were determined by acquiring a photon emission intensity map of the structure, taking the average of the signal level and subtracting the background noise. This method allows basic correlations of signal levels with the applied voltages or currents; nonetheless it is important to note that the signal levels in the maps can be much lower or higher than the absolute values in the graphs in different places, as will be shown in Figure 9 and Figure 10.

A. Amplitude measurements

For the following investigations the HBT operated in the common-emitter configuration. The emission intensity was recorded for various V_{CE} values and different I_B steps, which means that the graph contains the results for the HBT in saturation, forward active mode and avalanche (compare Figure 6 to the output characteristic in Figure 4). The results for the Si-CCD-camera and the InGaAs-camera are shown in the same graph for comparison. For identical operating points, the signal levels detected with the Si-CCD-camera are one to two orders of magnitude lower than those recorded by the InGaAs-camera. The signal levels for the operating points in saturation and forward active mode – depending on the base current, but basically for $V_{CE} < 1.4$ V – are much lower than those acquired in avalanche. However, for the Si-CCD-camera the behavior in avalanche and in non-avalanche operating areas is distinct: there is a steep increase in the signal level for the avalanche region. The signal levels detected with the InGaAs-camera seem to follow a linear dependency for the whole V_{CE}-range. In Figure 7 the same results are shown with respect to the base current. Neglecting the results for $I_B = 10$ µA (signal levels detectable and, depending on the V_{CE} level, strong but noisy), the following characteristic can be extracted. For the Si-CCD-camera, the signal levels for $V_{CE} \geq 1.7$ V decrease clearly with increasing base current. In saturation and forward active mode ($V_{CE} < 1.7$ V) this trend is reversed. The behavior in avalanche and forward active mode is comparable to what was found by Polonsky et al. (compare to [1]), but the authors also detected strong photon emission levels in saturation for a common-base configuration, whereas here the signal levels in saturation are lower than those in forward active mode.

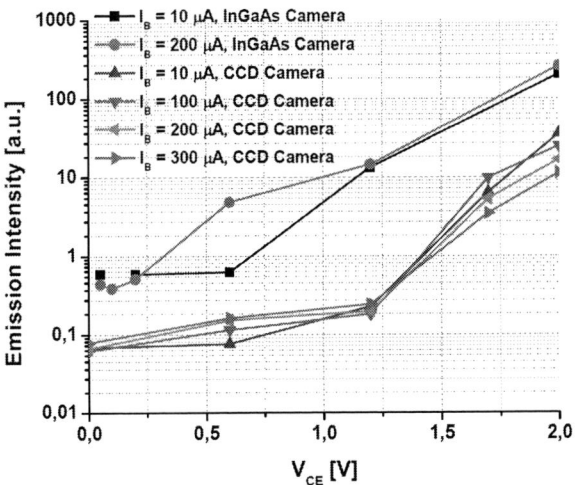

Figure 6: Si-CCD and InGaAs-camera; emission intensity versus V_{CE} (saturation, active forward mode and avalanche).

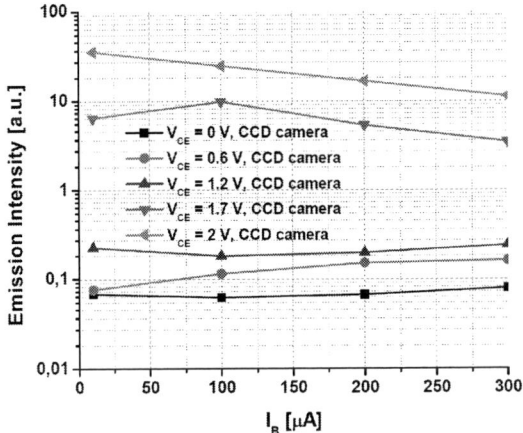

Figure 7: Si-CCD-camera; emission intensity versus I_B.

The signal levels in saturation are lower than those in forward active mode, which is different of what Polonsky et al. detected (strong photon emission levels in saturation for a common-base configuration). The devices used for this investigation are much larger than the devices employed by Polonsky et al. (IHP: 100 x 50 µm² / IBM: 0.5 x 2.5 µm²). The ratio of surface to edge is 16.67 (IHP) compared to 0.21 (IBM), which could be one reason for the increased recombination reported in [1]. Another reason may be the electrical characteristic of the device: in addition to the recombination paths, the profile of the radiant recombination depends on the density of free carriers, which will differ for the two devices. To assure the accuracy of the detected signals, the emitter-base diode in forward bias has been investigated separately for both cameras to check the emission intensity only caused by the diode itself ($V_{CB} = 0$ V); see Figure 8. For a base current of 200 µA, an emitter-base voltage of around 0.75 V can be assumed (see Gummel characteristic in Figure 3); the emission intensity for such an emitter-base voltage in Figure 8 is about 0.25 a.u. for the Si-CCD-camera and 1.75 a.u. for the InGaAs-camera. Taking into account that applying a collector-emitter voltage will reduce the number of carriers that can contribute to the

recombination process, the saturation results are understandable.

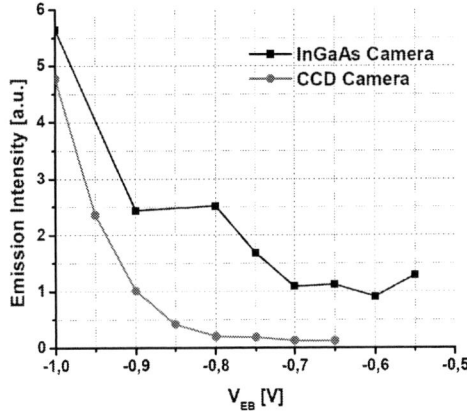

Figure 8: Si-CCD and InGaAs-camera; emission intensity of the EB-diode in forward bias ($V_{CB} = 0$ V).

The following figures show the photon emission results for the three operating points labeled with crosses in the output characteristic – a), b) and c) – for the Si-CCD-camera and the InGaAs-camera. Most of the maps were scaled individually (no direct comparison between the colors of the maps) due to the large span of signal levels. The maps of the HBT in saturation reveal stronger emission close to the contact that leads to the base pad – see layout and schematic diagram in Figure 1 and Figure 2. The emission source in saturation is the forward bias of the EB-junction (radiant recombination) and since the resistance decreases towards the contact, the emission becomes stronger in this area. This signal is present in all operating points and hence appears in all images. The second signal source becomes apparent for increasing collector-emitter voltages (forward active mode and avalanche). This signal is caused by the increase of the electric field strength in the BC-junction and the generated hot carriers herein and therefore located close to the collector contact, as can be found in the images for the HBT in forward active mode and avalanche (see Figure 9 and Figure 10). The closer the device is operating to the avalanche region the stronger the emission intensity close to the collector contact becomes.

978-1-4244-9113-1/11 $26.00 © 2011 IEEE   517

Figure 9: Si-CCD-camera; emission intensity of the HBT for three different operating points for $I_B = 200$ μA.

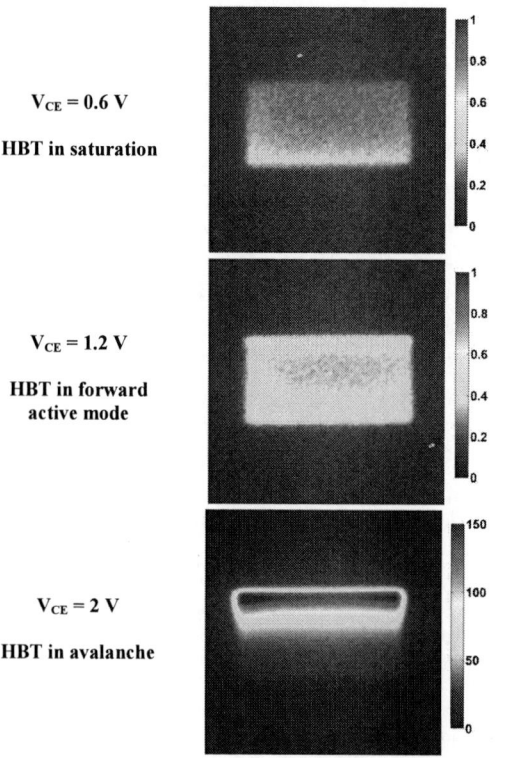

Figure 10: InGaAs-CCD-camera; emission intensity of the HBT for three different operating points for $I_B = 200$ μA.

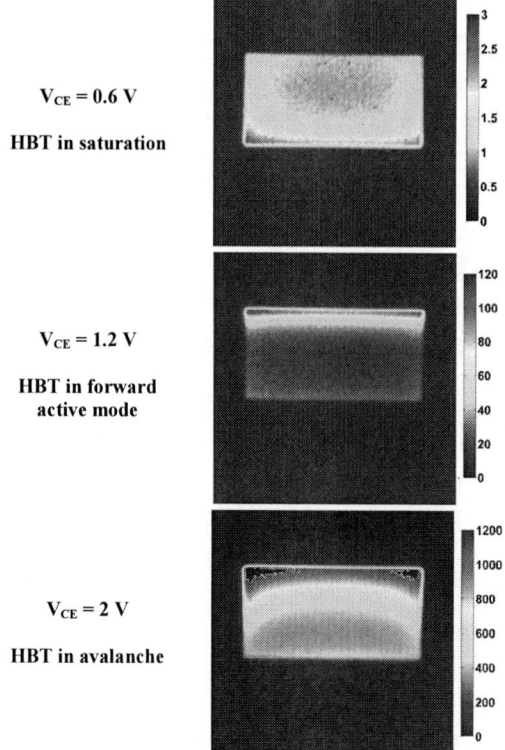

B. Spectral resolution

To resolve the spectrum of the photon emission, filters from 950 nm up to 1550 nm (spectral sensitivity of the InGaAs-camera) in 50 nm steps were used. The spectral sensitivity of each filter has been determined by using a calibrated light source (halogen lamp). The results of the spectral measurements in the following have been weighted with the achieved values. Figure 11 shows the result of the spectral measurement of the InGaAs-camera for wavelengths within the range of its spectral sensitivity. Except in avalanche ($V_{CE} = 2$ V: compare to crosses c) and d) in Figure 4) the HBT generates basically no signal over the whole range of wavelengths. This might be caused by the strong absorption due to free carriers in the collector: since the collector is highly doped, especially photons of higher wavelengths, such as 1300 nm for the radiant recombination in saturation and forward active mode, can be absorbed by the free carriers in the path (absorption increases with the wavelength squared). At lower wavelengths, below 1100 nm, this effect is not as strong, which explains the increase of the signal levels in this range. The spectrum of the photon emission in avalanche reveals two distinct maxima: one larger peak around 1100 nm and a smaller local maximum at 1300 nm. Even for wavelengths higher than 1300 nm relatively high signal levels have been detected. Photons with wavelengths above 1100 nm can not be detected with the Si-CCD-camera (sensitivity: 400 – 1100 nm), which means that a wide range of the spectrum is suppressed for the analysis and in turn explains the stronger signal levels for the InGaAs-camera.

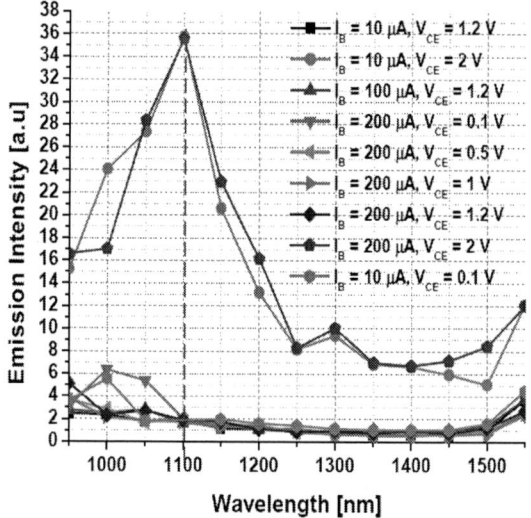

Figure 11: InGaAs-camera; emission intensity versus wavelength for various operating points. For comparison the upper detection limit for the Si-CCD-camera is indicated by a dashed line.

V. CONCLUSION

For both cameras photon emission has been detected in all three cases of operation – saturation, forward active mode and avalanche. In general, the signal level increases for increasing V_{CE}, such that the strongest signal levels were detected in avalanche. The photon emission intensity was found to be low in saturation, which is not in accord with the cited literature. This was led back to the fact that here structures of much larger surface to edge ratios (due to the large devices size) were used.

Especially for the HBT in avalanche (high field photon emission), the emission intensity levels for the InGaAs-camera are up to two orders of magnitude higher that those detected with the Si-CCD-camera. But also for generally lower signal levels, in saturation and forward active mode, the signal levels detected with the InGaAs-camera are at least a factor of 4 higher than the Si-CCD-camera data. This means that the InGaAs-camera is the detector of choice for photon emission measurements of an HBT independent of the operating point.

The emission intensity for the HBT in saturation and active forward mode were very low over the whole range of wavelengths, but in avalanche high emission levels were detected at the wavelength corresponding to the Si bandgap (1107 nm), and at 1300 nm, the wavelength of the SiGe alloy in the base. This means that the recombination takes place mostly in the silicon, but also – at least – partly in the SiGe base. In between these two wavelengths, high emission intensities have been recorded, as well, showing the wide range of energies of the emitted photons. In addition, as expected, relatively high emission levels were recorded for wavelengths lower than 1107 nm, since hot electron radiation takes place for the HBT in avalanche.

ACKNOWLEDGMENT

The authors would like to thank IHP for providing the test chip and specifically Dr. Heinemann, Dr. Knoll, Dr. Fursenko, Dr. Yamamoto and M.Sc. Lischke for fruitful discussions. The samples were prepared for the emission measurements by A. Eckert (Berlin Institute of Technology) using an UltraTec tool. We appreciated being able to use the Phemos-1000 with the Si-CCD-camera and the InGaAs-camera and hence we would like to thank Hamamatsu for this possibility, Dr. Yokoyama for setting up the camera and Dr. Glowacki for his help with the evaluation of the InGaAs-camera data.

REFERENCES

[1] S. Polonsky, A. Talalaevskii and M. McManus, "Characterization of light emission from SiGe heterojunction bipolar transistor for photon emission microscopy applications", IEEE International Reliability Physics Symposium (IRPS), pp. 344-346, 2003.

[2] Christian Boit, "Photoemission microscopy – advanced/theory of operation", Microelectronics Failure Analysis (4th edition), Desk Reference, pp. 213-229, 1999.

[3] N. Akil, S. E. Kerns, D. V. Kerns, A. Hoffmann and J-P. Charles, "Photon generation by silicon diodes in avalanche breakdown", Applied Physics Letters, pp. 871-872, 1998.

[4] J.C.H. Phang et al., "A review of near infrared photon emission microscopy and spectroscopy", Proceedings of the 12th IPFA, pp. 275-281, 2005.

[5] R. People and J. C. Bean, "Band alignments of coherently strained Ge_xSi_{1-x}/Si heterostructures on ⟨001⟩ Ge_ySi_{1-y} substrates", Applied Physics Letters 48 (8), pp. 538-540, 1986.

[6] P. Boucaud et al., "Band-edge and deep level photoluminescence of pseudomorphic $Si_{1-x-y}Ge_xC_y$ alloys", Applied Physics Letters 64 (7), pp. 875-877, 1994.

Isolating Light-sensitive Defects Using C-AFM

Hung Sung Lin, Mong Sheng Wu

United Microelectronics Corporation, Ltd.
No. 3, Li-Hsin Rd. II, Hsinchu Science Park, Taiwan 300, R.O.C.
Tel: 886-3-5782258 ext. 33231; Fax: 886-3-563-6722; Email: giant_lin@umc.com

Abstract—A soft failure, which is recoverable and sensitive to certain stresses, such as voltage, temperature and light, is defined as a failure, fault, defect, or error that results in a shift in the operating margin of a device. Several studies have been conducted into voltage or temperature dependent failures [1-3]. Research into light sensitive failures, however, has seldom been reported, as it is more difficult to isolate defects which are sensitive to light. Although certain global fault isolation techniques, such as photoelectric laser stimulation (PLS) have been developed to localize a wide range of potentially light sensitive defects, by means of the perturbation of the integrated circuit (IC) properties through carrier generation in the silicon, PLS cannot perform an exact failure localization on a single transistor or junction because of limited spatial resolution. This paper describes the use of a conductive atomic force microscope (C-AFM) within the failure analysis (FA) flow as a local fault isolation method in order to generate a more reliable failure hypothesis and successful physical root cause visualization for light-sensitive defects, and, using this technique, such failures, which pose potential reliability issues for devices as the affected circuit degrades over time or under stress, can be easily screened before any quality assurance test.

Keywords-PLS, C-AFM, soft, light, sensitive, perturbation

I. BACKGROUND

The interaction of light and semiconductors has been studied for many years, including the photoelectric effect, which proved that light acted as a particle, known as a photon, in many cases. Photons can also interact with impurity atoms, either as a donor or as acceptors, or they can interact with defects within the semiconductor. When a semiconductor is exposed to light, the photons may be absorbed or they may propagate through the semiconductor, depending on the photon energy $h\nu$ and on the bandgap energy E_g. When $h\nu > E_g$, an electron-hole pair is created. Based on this principle, photoelectric laser stimulation (PLS) techniques, with a wavelength laser where the photon energy is higher than the silicon bandgap, have been developed to detect areas sensitive to laser perturbation by generating electron-hole pairs, which are dissociated at the pn junctions in the silicon active area, thus creating photocurrents. PLS, however, cannot perform an exact failure localization on a single transistor or junction because of limited spatial resolution. As a result, a local fault isolation method, which can pinpoint light-sensitive defects, is needed. C-AFM is a precision probing system based on a direct on-axis optical microscope system. The optical path from the sample to the camera is a straight line. In addition, in order for C-AFM to work, a deflection sensor that functions by reflecting a laser beam off the back of the cantilever onto a photon detector is usually used to measure the extremely small movements of the cantilever. This technique is also known as beam bounce detection. The light exposure, sourced from a halogen bulb inside the optical microscope and a laser beam used for the beam bounce technique, can cause several possible semiconductor photon interaction mechanisms [4-6]. This paper will present an example of C-AFM application that helps isolate defects sensitive to light by taking advantage of the photoelectric effect induced by the C-AFM. The study was done using a C-AFM system called Digital Instruments CP-II manufactured by Veeco. With the guidance of a topography image, a doped tip coated with diamond of 50nm apex was used to probe tungsten plugs 125nm in diameter. The DC voltage was supplied from the bulk, which formed a closed circuit path with a grounded probe tip. The laser diode was powered by a low voltage supply with a maximum output of 0.2mW in the wavelength range of 600 to 700nm.

II. EXPERIMENTAL

In this study, the failure mechanism that causes light-induced leakage failure at room temperature (25 degrees celsius) was analyzed. Test results show that a leakage current can be detected in a 4k SRAM when the SRAM was accepting the ambient incident light. In order to measure the optical characteristics, a light source from a halogen bulb inside an optical microscope (OM) was used, as shown in Fig. 1.

Figure 1. Optical microscope and microprobing system. The sample was placed on the stage both without photon exposure (a), and with photon exposure (b).

The measurement results, as shown in Fig. 2, indicate that the leakage current during light exposure of the failed sample was 8.6E-07A@2V and was more than one order of magnitude higher than that of the normal sample where the leakage current was 2.8E-08A@2V.

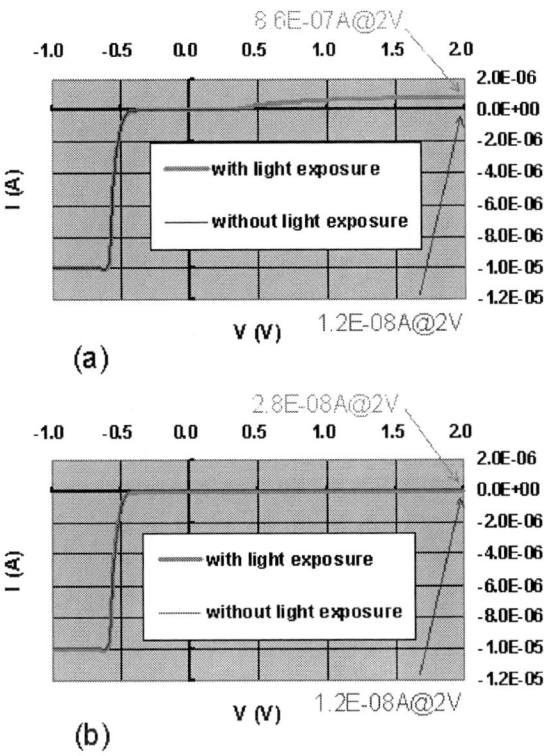

Figure 2. IV characteristics both with and without light exposure of a failed sample (a), and a good sample (b).

III. RESULTS AND DISCUSSION

A global fault isolation technique using IR-OBIRCH was performed from the front of the chip, as shown in Fig. 3. As expected, the thermal laser stimulation (TLS) technique failed to isolate any defects sensitive to light, as the energy of the scanning laser beam is lower than the bandgap energy of silicon, meaning that thermal stimulation will dominate. The absorption of light in a material causes the effect of heating the material lattice without losing the energy for the generation of free carriers, which explains the weaker widespread TLS signals observed in the SRAM, as shown in Fig. 3(c), as the junction reverse saturation current is heavily dependent on temperature. The increase in the reverse saturation current caused by the rise in temperature produces the bright widespread TLS signals when the device is locally heated using the IR laser. As C-AFM can provide light exposure sourced from its halogen bulb and laser beam, as shown in Fig 4, which produces a similar optical testing environment to that under an OM, electrical characterization using C-AFM was performed on the contact level in order to accurately pinpoint the location of defects sensitive to light by generating carriers in the silicon. A negative DC bias was applied to the sample stage during scanning, and the topography and current mapping were recorded simultaneously, as shown in Fig. 5. Current mapping uses different colors to indicate the different current levels in the contact. The darker the contact, the higher the current. NCT denotes a normal n diffusion contact, and DCT denotes a defective n diffusion contact.

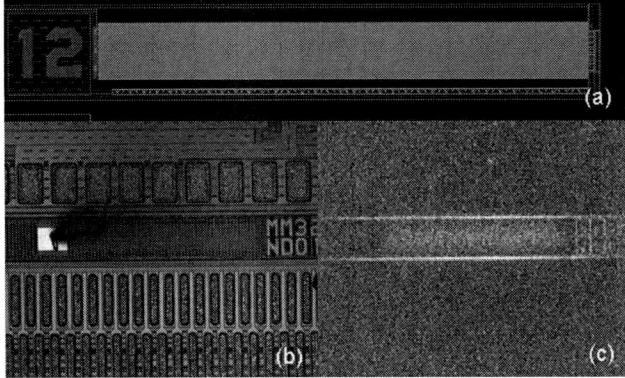

Figure 3. A 4k SRAM layout image (a), front reflected light image overlaid with a TLS image (b), and front TLS image(c).

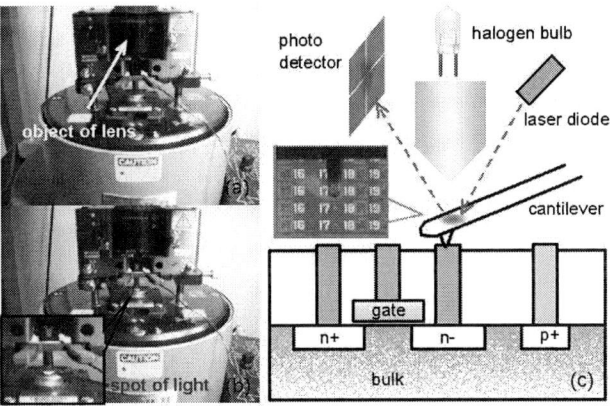

Figure 4. A C-AFM system. The sample was placed on the stage without photon exposure (a), with photon exposure (b), and a schematic diagram of the device cross-section during measurement (c).

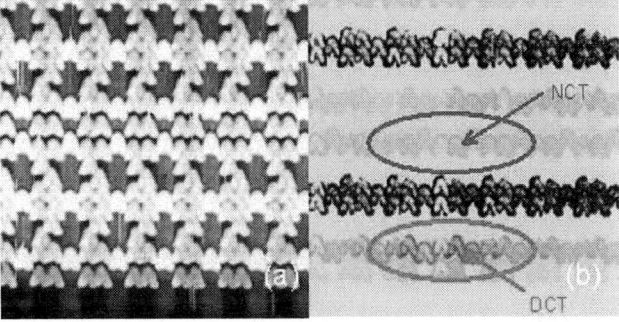

Figure 5. Corresponding topography image (a), and current mapping (b) for C-AFM on the contact level.

Fig. 6 shows the IV curves for both the DCT and the NCT extracted using the C-AFM, indicating that the I-V characteristics of the DCT are different from that of the NCT. These differences in the electrical characteristics of the 3rd and

4th quadrants will be individually discussed in the following sections. In the 3rd quadrant, the n+/p-well junction can be regarded as a photodiode, operated with an applied reverse-bias voltage. The excess carriers generated within the space charge region are swept out of the depletion region very quickly by the electric field. The DCT shows a higher reverse saturation current of about –2.8E-9A@-2V, which is higher than that of the NCT about –5.2E-10A@-2V, and the short circuit current of the DCT is also higher than that of the NCT. The current of the NCT is, however, higher than that of the DCT when the supply voltage is lower than -3.29V. In the 4th quadrant, it can be observed that the cut-in voltage of the DCT (0.37V) is lower than that of the NCT (0.42V).

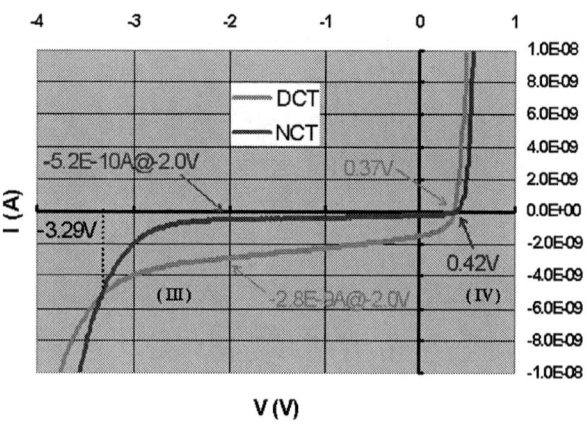

Figure 6.   The IV curves measured on the contact level using C-AFM.

Based on the results observed in the 3rd and 4th quadrant, a proposed physical failure hypothesis is that failures sensitive to light are probably caused by an insufficient n+ junction implant. As the photon-generated current from the space charge region is proportional to the space charge width given by the relationship according to equation (1), an insufficient n+ junction implant can be used to account for the higher reverse saturation current or the short circuit current observed in the DCT.

$$J = eG_L WA \qquad (1)$$

J: photon-generated current

e: electronic charge (magnitude)

G_L: the generation rate of excess carriers

W: the space charge width

A: the area of the pn junction

Because of the existence of the electric field at the junction, a sufficiently strong force may be exerted on a bound electron by the field to expel it from its covalent bond. As a result, new electron-hole pairs will be created, which increases the reverse current. As the field intensity increases, the impurity concentration also increases. For a fixed applied voltage, an insufficient n+ junction implant could explain why the current of the NCT is higher than that of the DCT when the supply

voltage is lower than -3.29V. Compared to the NCT, the DCT has a lower potential barrier due to an insufficient n+ junction implant that could account for why the cut-in voltage of the DCT is lower than that of the NCT. Fig. 7 shows a cross section of the failed contact. The arrow marker indicates the defect, identified as severed salicide. Poor salicide formation was revealed to be an indirect inference. This was induced by a prior poor n+ implant process. An Insufficient n+ junction implant should be caused by a particle which blocks the n+ junction implant, and the particle will also prevent the formation of the salicide in the diffusion area thereafter. The failure mechanism of the severed salicide may pose potential reliability issues for devices as the affected circuit degrades over time or under stress.

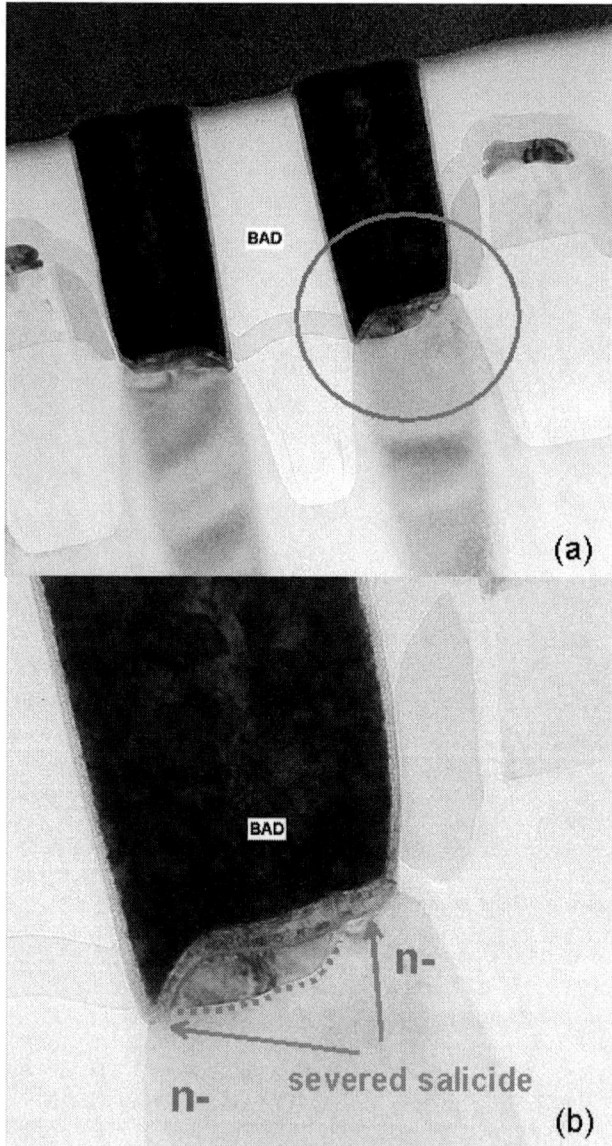

Figure 7.   Cross-sectional TEM images. Low magnification (a), and high magnification (b).

IV. CONCLUSION

In this study, the use of a C-AFM technique that takes advantage of the photoelectric effects induced by the halogen bulb and the laser beam was successfully demonstrated when analyzing a failure caused by light-sensitive defects, such as particle-induced junction n+ doping shadowing. Using this technique, a more reliable failure hypothesis and successful physical root cause visualization can be generated for light-sensitive defects, and, such failures, which pose potential reliability issues for devices as the affected circuit degrades over time or under stress, can be easily screened before any quality assurance test.

ACKNOWLEDGMENT

The author would like to acknowledge the contribution of the SRAM&Logic FA Engineering group in making device level IV measurements. We would also like to thank the TEM group for obtaining cross-sectional TEM images.

REFERENCES

[1] Fubin Zhang, Corey Lewis, Tim Duryea "Case Study of High Temperature Failure Analyses Using an On-Chip Heater," 34th ISTFA Proceedings, Nov. 2008, pp.273-276.

[2] Hung Sung Lin, Mong Sheng Wu, "A Case Study of High Temperature Pass Analysis Using Thermal Laser Stimulation Technique," 48th IRPS Proceedings, Apr. 2010, pp. 801-803.

[3] Hung Sung Lin, Vincent Huang, "Investigation of Thermal Budget Inpact on Core CMOS SRAM Device in an Embedded Flash Technology," 16th IPFA Proceedings, Jul. 2009, pp. 54-58.

[4] Hung Sung Lin, Mong Sheng Wu, " Isolating Marginally Defective Gates Using Photoperturbation Induced via a C-AFM Laser Beam," 48th IRPS Proceedings, Apr. 2010, pp. 801-803.

[5] Hung Sung Lin, Mong Sheng Wu, " A Study of Bipolar Phototransistor Action Existing in CMOS Process Triggered by a Laser Beam Used in a C-AFM System," 47th IRPS Proceedings, Apr. 2009, pp. 801-803.

[6] Hung Sung Lin, Mong Sheng Wu, "A Study of the Photoelectric Effect Caused by a Laser Beam Used in a Beam Bounce Technique in a C-AFM System," 34th ISTFA Proceedings, Nov. 2008, pp.256-259.

A Holistic Approach to Process Co-optimization for Through-Silicon Via

Sesh Ramaswami

Silicon Systems Group, Applied Materials Inc., 974 East Arques Avenue, MS 81151, Sunnyvale, CA 94085 , USA
1-408-584-2789, Sesh_Ramaswami@amat.com

(Invited Paper)

Abstract— As through-silicon via (TSV) technology transitions from development to production, several opportunities exist to co-optimize processes to ensure a wide process window while meeting cost targets and manufacturing robustness. Trade-offs in the via middle, via reveal, and via last integration schemes involving etch, CVD, PVD, ECD, CMP, and wafer support systems (carrier wafers) are addressed.

Keywords—Chemical vapor deposition (CVD), chemical–mechanical planarization (CMP), electrochemical deposition (ECD), etch, physical vapor deposition (PVD), through-silicon via (TSV)

Introduction

THROUGH-SILICON VIA (TSV) has emerged as a mainstream approach for stacking integrated circuits with a high number of interconnects. Vertical micron-sale interconnects passing through the die reduce inter-die interconnect lengths and result in improved performance and compact form factor. DRAMs can be stacked to boost memory capacity per unit board area/volume. When placed adjacent to multi-core CPUs, this method reduces latency and increases bandwidth. A good example for heterogeneous integration of dissimilar die is in the mobile space, with logic + memory stacked structure. TSVs through an interposer are used to achieve high-bandwidth interconnection of several smaller die. Conversely, large die can be partitioned into functional blocks as well using TSV. TSV interconnects are formed by etching deep vias in the silicon wafer, lining them with dielectric and metal-barrier films, and then filling them with metal (typically copper). The two TSV methods commonly used are the "via-middle" and "via-last" process flows. The first method is paired with a "via reveal" flow that exposes the TSV created so that it may be contacted from the back of the wafer. The term "via last" in this paper describes the scheme where the TSV creation occurs on the backside of the wafer after thinning. This paper provides a holistic view and explores trade-offs between key TSV process technologies and the integration flow as the industry prepares a transition from development to production. Risk is mitigated not only through the wide-ranging collaboration between adjacent technologies but through joint prototype testing of process flows. The three overriding objectives that guide the integration work are: wide process window, capability to withstand processing on fragile

bonded/thinned wafers, and overall cost of ownership of the fabrication flow.

I. TSV PROCESSES - OVERVIEW

A. Via Middle Creation:

In the via-middle flow, TSVs are created from the device side of a full-thickness wafer during processing in a wafer fab immediately after transistor and contact formation, but before the formation of BEOL damascene interconnects. In etch, optimization between etch rate, profile and related parameters are well understood, with excellent performance demonstrated on aspect ratios typically ranging from 4:1 to 12:1. A process for dielectric liner deposition (from 0.25 um to 1.2 um on the sidewall) has demonstrated >60% step coverage on such vias. Physical vapor depositions (PVD) of titanium or tantalum barriers and copper seed layers has been co-optimized with electro-chemical deposition (ECD) to ensure void-free metal fill. ECD overburden and PVD barrier/seed on the top surface must subsequently be removed by chemical-mechanical polishing (CMP). Recent work has enabled optimal bulk removal rates with the requisite end-point and process controls to accurately transition between layers and preserve surface topography.

B. Wafer Backside Preparation:

Wafer backside processing encompasses Wafer backside processing encompasses the following three steps: :
(a) The device wafer is temporarily bonded face down with adhesive to a glass or silicon carrier or wafer support system (WSS). While both carriers have displayed the requisite mechanical handling repeatability, silicon carriers are preferred as they are manufactured to industry-standard dimensional specifications, are less prone to breakage (and therefore easier to recycle), and can couple thermally and electrically with the wafer. Some key challenges include bonding tolerances, carrier thickness and flatness, thickness and post-cure properties of the adhesive. The adhesive material must resist material decomposition and interaction during processes such as etch, deposition, wet clean, ECD and CMP.
(b) Thinning the wafer to the desired thickness requires controlling across-wafer total thickness variation (TTV), surface smoothness, contamination, and residual stress.

Inadequate control of these parameters leads to a dramatic reduction in the window for backside processing.

(c) A low-temperature CVD Silicon Nitride (SiN) or silicon oxide is deposited to seal the backside or stress-compensate the wafer.

C. Via Last Creation:

For via-last processes, thermal budget is a significant concern, since device wafers are processed while being temporarily bonded to carriers. TSVs are created from the wafer backside so that the via lands on an underlying metal layer in the device wafer interconnect, thereby creating a back-to-front connection. While via-middle TSVs are subjected to ten or more thermal heat cycles in a typical logic damascene flow, the only heat cycles the via-last TSVs are exposed to are the ones associated with final in-fab anneal with solder-reflow or Cu-Cu bonding during final packaging. All processing conducted on temporarily bonded wafers needs to be at low temperature (e.g., below 200°C), compatible with the adhesive used to temporarily bond the wafer to the carrier.

Hence, via post-etch cleans, dielectric CVD, and metal PVD steps need to be optimized to achieve desired mechanical and electrical film qualities and process performance at these lower temperatures.

D. Backside Via Reveal:

Backside via reveal involves contacting the front-side vias from the wafer backside, so that a back-to-front TSV connection can be made.

E. Backside re-distribution Layers:

Creating re-distribution layers (RDL) involves dielectric (oxide, nitride, and polymer) deposition, sputter deposition, electrochemical deposition, etch, and wet cleans.

F. De-bonding from carrier for singulation:

After wafer processing is completed, the device wafer is separated (de-bonded) from the carrier. When the device wafer is de-bonded, the micro-bump and wafer surfaces be free of any residual adhesive.

G. Some Considerations That Drive Via Dimension:

TSV diameter, depth (desired product wafer/ die thickness), pitch, and pattern density are the key areas where end-product design, process integration, process window; unit-process robustness and cost intersect. Since TSVs extends through the thickness of the die, the TSV depth is driven by two primary factors. The first is the lowest die thickness that can be successfully handled after the wafer is diced (singulated). Wafers used for advanced logic are typically 50-100µm thick in order to maintain structural integrity (i.e., to resist warpage), minimize cracking (especially with low-k materials), and assist with thermal management (spreading hot spots and overall heat dissipation). Wafers for memory die such as NAND and DRAM can be thinner, on the order of 30-50µm so that they may be stacked 8-16 high and still fit within the vertical constraints (0.7mm for die and < 1mm for overall package) imposed by memory cards and other form factors.

It is desirable to have a low TSV capacitance and also minimize the keep-out-area (KOA) between the TSV and the active transistor. Since TSV diameter influences capacitance, vias with smaller radii and lower depth are preferred. Various end-product designs require a high number of connections between the die. To meet this requirement, the roadmap points in the direction of a large number of smaller and shorter vias.

Considering various factors, the 'sweet spot' for TSV diameters is between 5 and 15µm, with roadmaps extending down to 2µm. This puts the via aspect ratio in the range of 5;1 to 12:1.

II. CO-OPTIMIZATION ACROSS UNIT PROCESSES

Typically, the vias are 5-10µm in diameter and 50-100µm deep. Maintaining an aspect ratio less than 10 allows a wider process window. The maximum wafer temperature is in the same range as that of the BEOL films, typically 350-400ºC.

A. PVD - ECD Co-optimization in Via Creation:

A barrier layer (typically Ti, Ta, or their nitrides) is deposited by PVD. A TSV-specific PVD product and deposition process has been developed that enables adequate barrier step coverage in high aspect ratio vias. A continuous and adequately thick copper seed layer is then deposited by PVD, enabling an efficient subsequent copper ECD process. More recent ECD chemistry and processes exhibit enhanced bottom-up fill, with a wide process window and low over-burden. Additionally, this process enables reduced thickness requirements, and hence cost, for Cu seed layers.

B. ECD - CMP Co-optimization in Via Creation:

Over-burden is defined as the thickness of the copper on the field region after the via is filled. Early in the process development (typically associated with conformal electroplating deposition), the overburden could be equal to 0.5-0.75 of the TSV diameter, or up to several microns in thickness for large vias. This added thickness added stress to the wafer and caused excessive wafer bow, which prevented the wafer from being processed in post-ECD steps (especially CMP) and/or induced breakage. A thick copper over-burden also increases the cost of copper CMP. Enhanced bottom-up ECD fill processes outlined above have reduced overburden significantly. Low over-burden ECD lowers the cost of the subsequent CMP process that is needed to remove the copper and barrier material from the field region. The selection of the CMP process and slurry is governed by: (a) Clean copper and barrier removal, (b) No attack or corrosion of the copper in the via or the barrier metal on the sidewall, (c) No divot or attack of the oxide liner on the inner circumference of the via. These films are then annealed to stabilize microstructure and stress. Good post-CMP surface planarity combined with a stabilized via filled with copper, allow interconnect wires to be routed above them.

978-1-4244-9113-1/11 $26.00 © 2011 IEEE

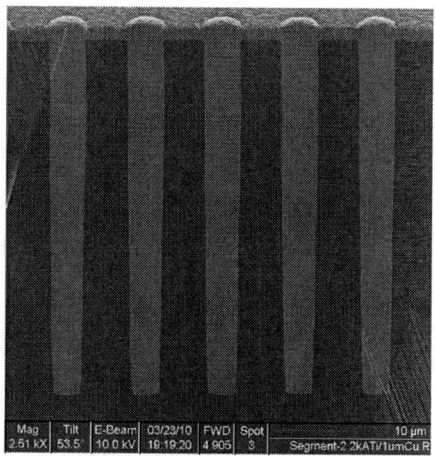

Figure 1: Enhanced bottom-up fill process of 10:1 aspect ratio structure with low copper over-burden in the field region.

C. Backside Via Reveal:

Vias created in the middle of line need to be exposed from the backside of the wafer so that connections can be made. This process, known as 'tip reveal', 'via reveal,' or 'backside contact,' is the companion process to via-middle. Wafer grinding can introduce significant wafer center to wafer edge thickness variations that makes this integration very challenging. There could be similar problems associated with surface roughness, defects and water marks. Hence, silicon CMP after grind is used to achieve a uniform thickness profile and smooth surface. CMP polish heads with zone-tuning capability play a pivotal role in reducing TTV across the wafer.

Total variation for the backside variation includes the following component variations: (a) carrier thickness, (b) adhesive thickness, (c) via-middle etch depth uniformity, and (d) grind thickness uniformity. Hence, successful backside via integration requires that the process chosen accounts for these variations.

After CMP, the silicon is etched back (blanket etch) using plasma etch to reveal the vias. The etch chemistry is set up such that the silicon can be removed rapidly, while at the same time maintaining high selectivity to the silicon oxide that is encasing the via being revealed. Subsequent CVD dielectric deposition on the back of the wafer followed by CMP is then used to reveal the plug completely while simultaneously isolating the revealed vias from one another.

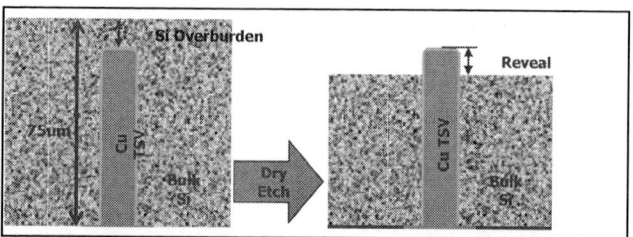

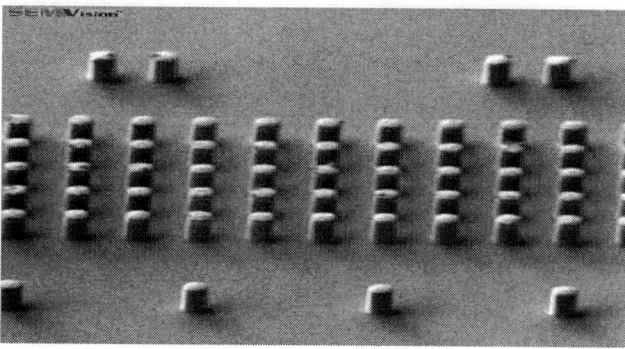

Figure 2: Schematic and actual representations of via reveal etch. The copper pillars represent via-middle TSVs exposed from the wafer backside after thinning.

III. CONCLUSION

Unit processes are optimized and integration is well advanced, with "best known methods" for typical interposer, via-middle, and via-last flows ready. Co-optimization between unit processes has led to a better understanding of process trade-offs early in the technology life-cycle. Technology choices drive costs. The total cost to achieve TSV is split between wafer-level TSV creation (etch, dielectric liner, barrier/seed, ECD fill, and CMP), wafer-level thinning (bonding, thinning, and de-bonding), and die-level processing (dicing, stacking, assembly, and testing). This paper provides visibility to process choices and ties them to a qualitative explanation of the relationship between wafer-level processes and cost, thereby laying a path for transitioning TSV from development to production.

ACKNOWLEDGMENT

The author wishes to acknowledge the contributions of the entire TSV team and the facilities of the Maydan Technology Center at Applied Materials Inc. For specific contributions towards data represented in this paper, the author wishes to acknowledge John Dukovic, Brad Eaton, Max Gage, Aashika Jain, Chris Lazick, Tom Ritzdorf, Jennifer Tseng, Yuchun Wang, and Rao Yallamanchall.

Thermal and Spatial Profiling of TSV-induced Stress in 3DICs

Colin McDonough, Benjamin Backes, Wei Wang, and Robert E. Geer
College of Nanoscale Science and Engineering
University at Albany, SUNY
Albany, NY, USA
01-518-956-7003, rgeer@uamail.albany.edu

Abstract—The thermal and spatial variation of Cu through silicon via (TSV)-induced stress in 300mm Si wafers has been investigated for both isolated TSVs and TSV arrays using top-down and cross-sectional spectral microRaman imaging. The TSV-induced stress in Si results from plastic yield of the Cu, is compressive in the immediate vicinity of the TSV, and transitions to a tensile state at larger separations – in quantitative agreement with finite element modeling (FEM). TSV arrays (linear and square) lead to substantial tensile stress enhancement within the array. Moreover, thermal annealing showed that the intra-array Si stress field became more compressive with increased post-CMP thermal annealing while the Si stress-field external to the arrays exhibited little change. This may open potential avenues for reduction of TSV-induced Si stress in 3DICs.

Keywords-3D integrated circuit; through-silicon via; stress;

I. INTRODUCTION

The development of 3D interconnection methods for future generations of integrated circuits is important for meeting further device scaling demands, as interconnect delay is fast becoming a performance-limiting factor (1-2). In addition, 3D die stacking can significantly improve the areal efficiency and functionality of future chips (3). The use of copper through-silicon vias (Cu TSVs) is a highly promising avenue for 3D integration, but the differences in thermo-mechanical properties between copper (Cu) and silicon (Si) can lead to substantial TSV-induced stress profiles in the surrounding Si. The coupling between this stress and the carrier mobility in Si devices has raised concerns about the impact of TSVs on the performance of nearby devices (4-10). For comprehensive electrical testing of Si-based devices in 3D integrated circuits it is necessary to take into the account the effects of local stress in the active regions induced by TSVs.

One important source of TSV-induced stress in Si is the plastic yield of Cu at elevated processing temperatures. The large difference between the coefficients of thermal expansion in Cu and Si can result in the generation of thermally-induced stress at high temperatures in excess of the Cu yield stress. Upon cooling, the residual strain in Cu associated with high-temperature plastic yield results in a residual stress profile in the nearby Si. Since the elasto-plastic properties of electrochemically deposited Cu used in typical TSV process flows can vary as a function of process conditions, it is important to understand how those changes will affect TSV-

induced stress profiles as inputs to Si device simulators for comparison with electrical testing.

II. EXPERIMENTAL

The experimental structures used for investigation of TSV-induced stress in Si for the work reported here consisted of round isolated Cu TSVs (5 μm diameter, 25 μm depth) and 1×4 linear Cu TSV arrays (10 μm pitch) fabricated in 300 mm Si wafers. Cu TSVs were etched in the Si using reactive ion etching (RIE). Following etch, liner/barrier and Cu seed layers were deposited. Two separate TSV wafers were investigated. For wafer 'A' the TSV was filled using copper electroplating followed by a 150 °C furnace anneal. For wafer 'B' the TSV was filled using copper electroplating followed by a 350 °C furnace anneal. For both wafers the electroplated Cu overburden was removed by chemical mechanical planarization. The top surface of the Cu TSVs (following planarization) was flush with a 1μm SiO_2 cap layer on the Si as noted above. For wafer 'B' only, subsequent processing resulted in the exposure of the Cu TSV structures to a temperature of 400 °C for 1 minute. The wafers were then diced and characterized using scanning micro-Raman spectroscopy to extract the local stress Si profiles (11).

Scanning Raman microscopy was used to measure stress in the Si regions near Cu TSVs. As noted in Ref. 11 Raman scattering provides a measure of the energy of optical or acoustic phonons in a Raman-active crystalline material. Under the application of stress, the deformation of the crystalline lattice results in a modification of the phonon dispersion relation which is reflected in a change in the spectral position of a measured Raman band. If the phonon deformation potentials and the elastic constants of a material are known, the stress state can be extracted from the shifts in the measured Raman bands. In unstrained crystalline Si the 64 meV optical phonon mode leads to a prominent Raman peak at $\omega_0 = 521$ cm^{-1}. This peak is azimuthally symmetric for an excitation photon incident along the Si (100) axis. The shift of this band is a complex function of the crystalline and experimental geometries for arbitrary strain, necessitating a solution of the full secular equation to extract the local stress sate (11). For the Cu TSV geometry under consideration here, the stress state in the Si near the SiO_2/Si interface can be treated as being approximately biaxial. This permits a straightforward analytical solution (5), with the Si stress, σ

978-1-4244-9113-1/11 $26.00 © 2011 IEEE

[MPa] $\approx -434*\Delta\omega$ [cm^{-1}], where $\Delta\omega$ is the difference between the measured Raman peak position and the unstrained value. Raman measurements were carried out in a commercial micro-Raman spectrometer with an excitation wavelength of 532 nm. Relative Raman shifts were referenced to data acquired from uniform regions of Si far (> 20 μm) away from the TSV of interest.

Simulations of stress induced in Si due to cylindrical Cu TSV-based systems were also carried out, and employed the Structural Mechanics module of COMSOL Multiphysics software. The TSV diameter was 5 μm and the TSV depth was 25 μm, corresponding to the experimental geometry. For both isolated Cu TSVs and 1×4 linear TSV arrays, quarter-symmetric 3D models were used, exploiting the symmetry in the systems. In all models the TSVs were surrounded by anisotropic (crystalline <100>) Si with a 1.0 μm SiO$_2$ cap layer. The top of the Cu TSV was flush with the top surface of the SiO$_2$ cap layer – again, in accordance with the experimental structure. For modeling of linear arrays the TSV linear pitch was 10 μm and the array axis was parallel to the <110> crystal orientation of the surrounding Si. For comparison with experiment, values of Si stress from finite element simulations were taken 50nm below the Si/SiO$_2$ interface. This position was chosen as a balance between the depth of channels in nearby transistors and the depth at which the Raman signal is collected for isolated TSVs. The mesh is densest near the edges of the TSVs, where the stress is expected to change most rapidly, and it is kept dense along the Raman scan direction (<110>, along the x-axis) to optimize comparison with experimental data and minimize mesh-based artifacts in the simulation.

The Si is treated as anisotropic, having a rotated stiffness matrix as in (12) which determines its mechanical properties (C11 = 166 GPa, C12 = 64 GPa, and C44 = 80 GPa). The Cu is treated as an aggregate elasto-plastic material with linear isotropic hardening behavior defined by a yield stress (σ_{ys}) and tangent modulus (h). σ_{ys} [MPa] defines a threshold stress beyond which Cu deformation is no longer elastic, and h [MPa] is a measure of the deformation behavior as the strain increases in that plastic regime. A summary of the relevant properties for the materials used in the system is given in Table 1. Each simulation consisted of a two-step annealing process.

TABLE I.   MATERIAL PROPERTIES USED IN ISOLATED AND LINEAR TSV ARRAY SIMULATIONS

Material	Young's Modulus (GPa)	Poisson's Ratio	Density (g/cm^3)	CTE (ppm/°C)
Silicon (anisotropic)	*Direction-dependent*	*Direction-dependent*	2.330	2.3
SiO$_2$ (isotropic)	75	0.17	2.200	0.5
Copper (elasto-plastic)	117	0.3	8.960	16.7

Prior to simulated 'heating' the structure was assumed to be in a stress-free state. The thermal-induced displacement, stress, and strain fields were calculated for the structure at the anneal temperature. These values were used as inputs for the cooling stage, for which the structure was returned to room temperature. In this way a full anneal cycle was simulated.

III.   RESULTS

Figure 1 illustrates a Raman map of TSV-induced stress in Si surrounding a 5×25 μm isolated Cu TSV (wafer B). A biaxially symmetric stress is assumed. Si regions within 1-2 μm of the TSV exhibit a positive Raman shift (compressive stress) relative to unstressed Si. A transition to negative Si Raman shifts (tensile stress) is evident at greater distances (> 2 μm). The four-fold symmetric pattern evident in the Si Raman shift map is attributed to the crystalline anisotropy of the Si. This has been qualitatively confirmed by finite element simulation by the authors (13). That work is not presented here.

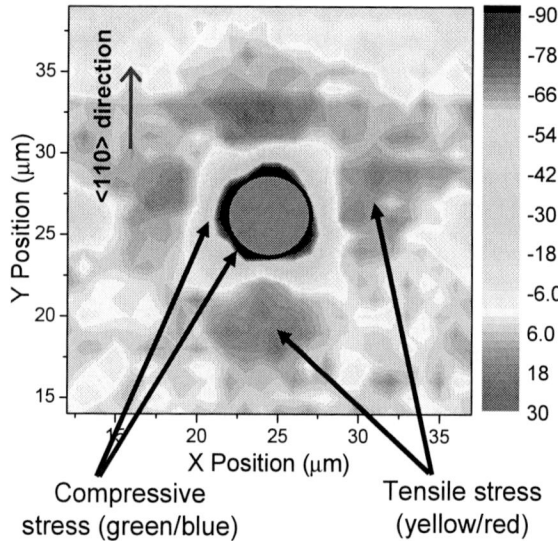

Figure 1.2D Si Raman stress map near an isolated 5 x 25 μm TSV assuming biaxial symmetric stress. Note transition from tensile stress in Si far from the TSV to compressive stress near the TSV.

The symmetric nature of the TSV and the evident four-fold symmetry of the TSV-induced stress field in Si in Fig. 1 support the assumption of biaxial symmetry and the linear conversion from Raman shift to stress (11). The largest measured compressive stresses approached 90 MPa. The largest measured tensile stresses approached 30 MPa.

Linear Si Raman shift profiles were acquired from various die across both wafers A and wafer B. Assuming biaxiality, average linear stress profiles from multiple die on wafer B are shown in Fig. 2. The transition from compressive to tensile stress is clearly evident. Two sets of experimental data are shown in Fig. 2 – one set representing averages over two die and one set representing averages over 5 die. The intent of this comparison is to underscore the die-die variation in the TSV-induced stress profiles in Si. Although the overall shape of the stress profile is invariant, the magnitude of the stress variation shows significant die-to-die fluctuation. The solid line in Fig. 2 corresponds to a finite element model of a TSV undergoing a single 350 °C thermal anneal cycle.

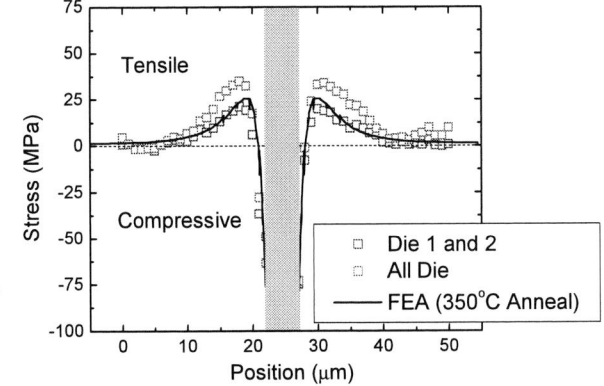

Figure 2. Average linear profile of (biaxial) stress variation in Si near an isolated 5×25 μm TSV (wafer A). Blue squares represent profiles averaged over 2 selected die (see Fig. 5 for die map). Red squares represent averages of 5 die across entire wafer. The solid black line is a simulated stress profile from a finite element model.

The agreement between experimental data for the 2-die average data and the simulation is very good. This agreement confirms that plastic deformation of the Cu within the TSV at elevated temperatures drives the resultant ambient stress profile in the Si. Briefly, the CTE mismatch between Si and Cu results in stresses within the Cu TSV that exceed the Cu yield stress resulting in plastic (irreversible) radial deformation (compression) (5, 10). Upon cooling the 'yielded' Cu further contracts and results in a net tensile stress in the neighboring Si. Geometrically, the circular shape of the TSV induces a transition from tensile to compressive stress in the Si regions directly adjacent to the TSV. This can be understood via a simple Poisson's ratio argument. Tensile deformation of the Si very close to the TSV results in 'azimuthal' Si compression. As confirmed by FE modeling (Fig. 2) the net effect of these coupled deformations is an overall compressive biaxial stress.

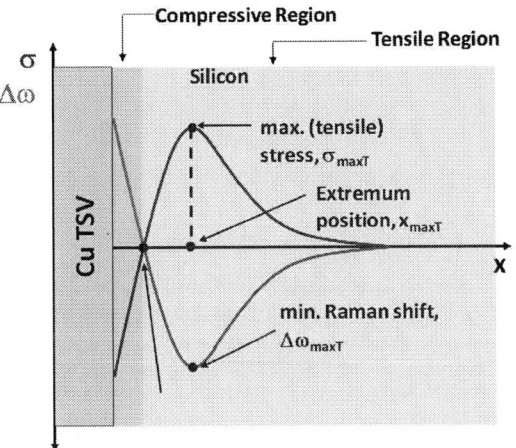

Figure 3. Schematic illustration of points of interest in Si Raman shift profiles near a TSV. σ denotes stress. Δω denotes Si Raman shift.

The typical Si stress profile (or alternately Si Raman shift profile) induced by Cu plastic deformation in a TSV is illustrated in Fig. 3. This figure defines three key metrics used to characterize TSV-induced stress for isolated TSV structures. Firstly, a crossover point is defined characterizing the distance from the TSV at which the Si stress changes from compressive to tensile. Secondly, the maximum tensile stress is defined as the largest tensile stress (σ_{maxT}) measured for distances greater than the crossover point. Alternately, this value is represented by the maximum (negative or tensile) Si Raman shift ($\Delta\omega_{maxT}$). The distances associated with this value (x_{maxT}) and with the point at which the max tensile shift has decreased to 50% of its value are also used as metrics.

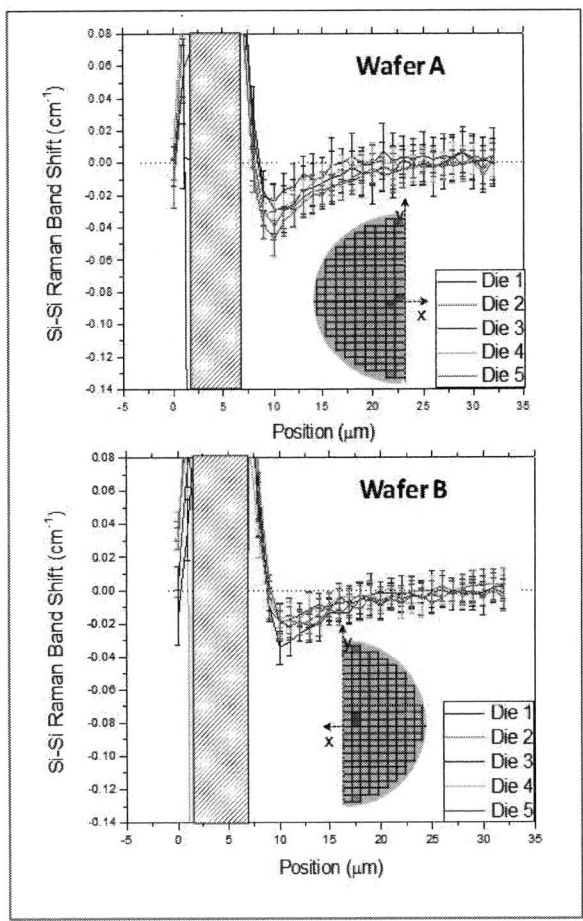

Figure 4. Si Raman shift profiles near isolated TSVs. Top: Profiles from wafer A (Cu Plate, 150 °C Anneal, CMP). Die position noted in inset. Bottom: Profiles from wafer B (Cu Plate, CMP, 350 °C Furnace anneal, CMP, M1 oxide deposition (400 °C)). Die position noted in inset.

Si Raman shift profiles from isolated TSVs taken from 5 die on wafer B and 5 die on wafer A are shown in Fig. 4. Relative die positions are shown in the inset of each panel in Fig. 4. These profiles are characterized by the crossover point, the point of max Si tensile stress (x_{maxT}) and the 50% tensile stress reduction point in Fig. 5 for each wafer. Firstly, high-temperature processing increases the die-to-die variation with respect to these characteristic distances. Secondly, the higher-temperature processing acted to increase the characteristic distances, on average. The average crossover distance increased from 3.2 μm (wafer A) to 4.3 μm (wafer B). x_{maxT} increased from 1.5 μm (wafer A) to 2.2 μm (wafer B). The 50% tensile stress reduction point increased from 6.5 μm (wafer A) to 8.3 μm (wafer B). Note that the standard deviation associated with die-to-die variation exceeded 1 μm for each quantity. This data implies that process temperatures may play a large role in modification of TSV-induced stress profiles, although other parameters (plating chemistry, liner, barrier, seed, etc…) should be considered.

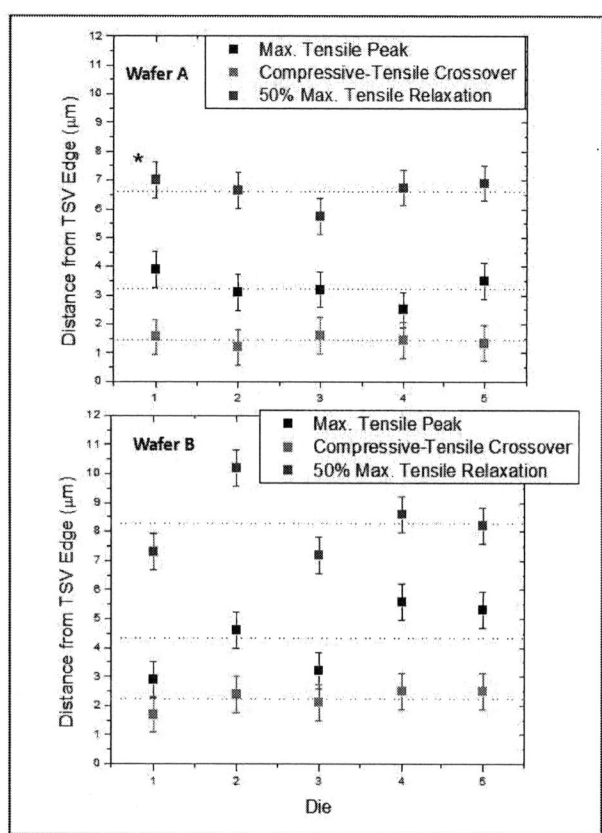

Figure 5. Top: Summary data from Si Raman shift profiles from wafer A. Bottom: Summary data from Si Raman shift profiles from wafer B. The silicon near the isolated TSVs on the wafer exposed to higher-T processing (post-CMP) exhibited larger variations in local stress as compared to the wafer not exposed to higher-T processing.

The Si Raman shift profiles associated with TSV arrays were likewise characterized for 5 die from each wafer. The behavior of the Si Raman shift profiles outside the arrays mirrored that for isolated TSVs. Interior to the array, however, substantial stress field superposition effects were noted. This is illustrated in Fig. 6 which plots the Si Raman shifts from 1×4 TSV arrays. Note the maximum tensile Si Raman shift between TSVs, in some cases, is twice that measured outside the arrays for both A and B wafers. This is summarized concisely in Fig. 7 which plots the maximum tensile Si Raman shift across 1×4 TSV arrays from 5 die from the A wafer and 5 die from the B wafer. Firstly, within the array x_{maxT} is typically midway between TSVs. This is expected based on symmetry. Secondly, analogous to the results shown in Figs. 4 and 5, the

data in Figs. 6 and 7 imply that higher temperature processing contributed to larger die-to-die variations of the TSV-induced stress profiles in Si – at least for the wafer process flow employed here (although other parameters such as plating chemistry, liner, barrier, and seed layers should be considered). This tendency is arguably more dramatic for the 1×4 TSV arrays as compared to the isolated TSVs. This is illustrated clearly by the lower panel of Fig. 7 corresponding to wafer B. Taking the x_{maxT} data for wafer A as a baseline, exposure to higher temperatures results in relaxation or enhancement of tensile stress depending on the die examined. From a device perspective this could lead to some concern with respect to process-induced variations of the size of the keep-out-zone associated with isolated TSVs or TSV arrays.

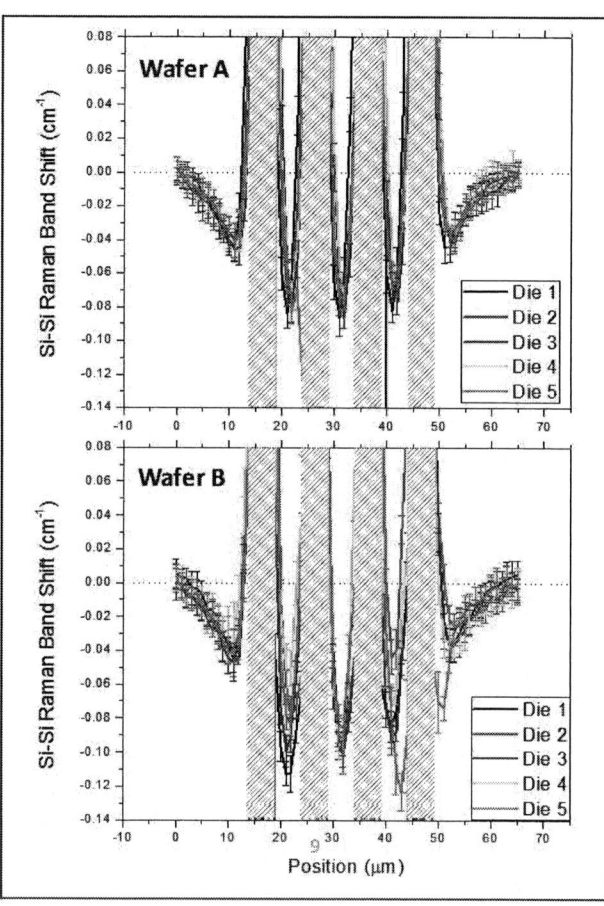

Figure 6. Si Raman shift profiles across 1x4 TSV arrays. Top: Profiles from wafer A. Raman scan geometry shown in inset. Bottom: Profiles from wafer B.

To specifically investigate effects of thermal annealing on the stress in Cu TSVs individual die were annealed in an *in situ* heating state on the Raman spectrometer. A dry nitrogen gas flow was maintained through the stage to prevent oxidation of Cu at elevated temperatures. Si Raman shift profiles were measured at elevated temperatures to investigate the introduction of stress. A representative result is shown in the upper panel of Fig. 8. The red circles denote a measured Si Raman shift profile from an isolated TSV from the B wafer at 250 °C.

978-1-4244-9113-1/11 $26.00 © 2011 IEEE

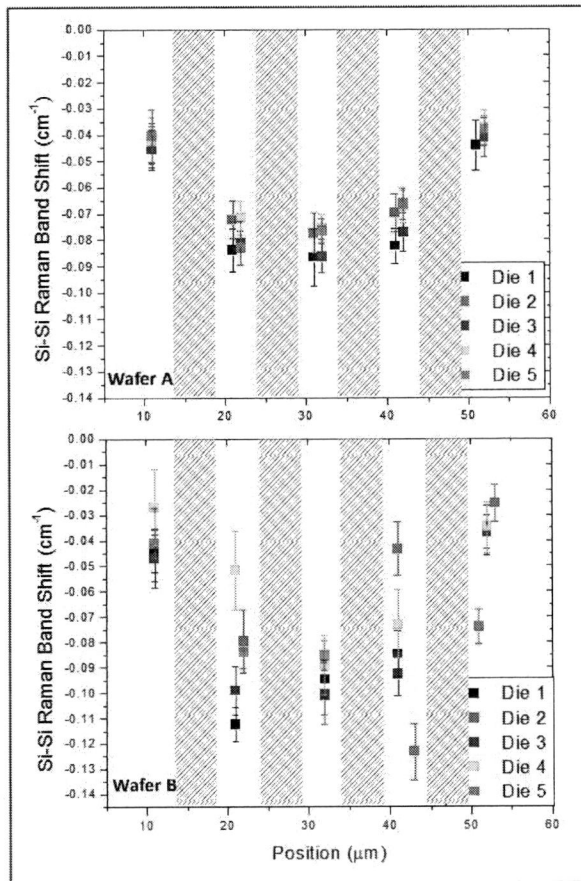

Figure 7. Maximum tensile stress in Si exterior to and interior to 1x4 TSV arrays. Top: Data from wafer A. Bottom: Data from wafer B. Analogous to isolated TSV arrays, the wafer exposed to higher processing temperatures (post-CMP) exhibited substantially larger variations in Si stress within the array.

The blue triangles denote the measured Si Raman shift profile from an isolated TSV (wafer B) at 250 °C. Firstly, it would be assumed that at high temperatures the profile in the Si would be intrinsically compressive which contradicts the at-temperature data in the top panel of Fig. 8. However, it should be emphasized that the intrinsic (pre-heating) TSV-induced stress profile is best described by a 350 °C anneal (Fig. 2). As a result, heating to 250 °C, would be expected to reduce the tensile Raman shift but not induce compressive stress. To the contrary, a 250 °C ramp would be expected to better resemble a cooling ramp from an unstrained Cu state due to the presence of residual Cu strain from 'self-annealing' typically observed in electroplated Cu (14). This is qualitatively consistent with finite element modeling (lower panel of Fig. 8) for a 50 °C cooling ramp (red line).

A more systematic set data illustrating the effect of annealing temperature on measured TSV-induced stress in Si is shown in Fig. 9 for a 1×4 TSV array (wafer A). The maximum tensile stress at points outside and within the 1×4 array is plotted for anneal temperatures varying from 150 °C to 400 °C. Measureable upward shifts in the Raman data were evident at

200 °C. The shifts were more pronounced at 250 °C and 300 °C. The most significant shifts in the Raman profiles were seen following the 350 °C and 400 °C anneals. Note that each set of temperature data were taken from distinct die. However, the trend is clear – on average, increased annealing temperature resulted in only modest changes in Si Raman shifts outside the array. This is roughly consistent with the isolated TSV data shown in Figs. 4 and 5. In contrast, the Si Raman shift data interior to the arrays exhibited upward (compressive) shifts that scaled roughly with temperature. Variations were observed. For example, the Si Raman shift measured after the 300 °C anneal was shifted, nominally, less than the corresponding data measured after the 250 °C anneal.

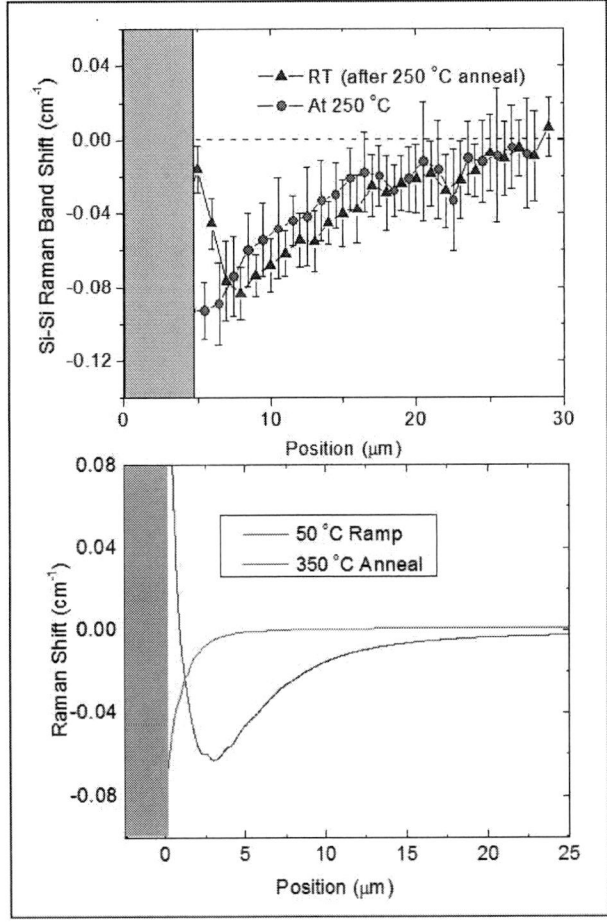

Figure 8. (Top) Stress profile near isolated TSV (B) at 250 °C (red circles) and room temperature following 250 °C anneal (blue triangles). (Bottom) Simulation of TSV-induced Si Raman shift due to 50 °C temperature ramp (red line) as opposed to 350 °C anneal (blue line). These data are for qualitative comparison.

A clarification is warranted regarding the thermal measurements of TSV-induced stress. Firstly, the temperature dependent data in Fig. 9 was acquired following annealing of individual die on the Raman stage weeks after the wafer was originally processed. This is substantially different from the distinct thermal processing histories associated with wafer A and wafer B. Thus, comparing the effect of post-plating anneal temperatures between wafers A and B with thermal annealing of individual die from wafer A on the Raman stage is nontrivial.

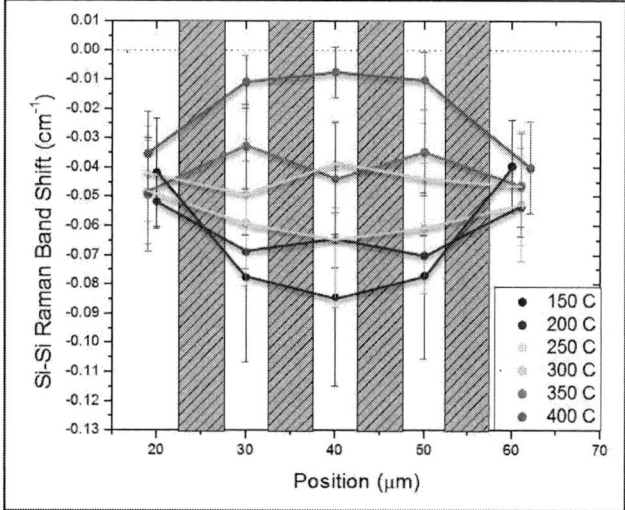

Figure 9. Maximum (tensile) Si Raman shifts exterior and interior to a 1x4 TSV array as a function of anneal temperature (wafer A). The reduction of tensile stress in the Si regions within the array is most dramatic at higher annealing temperatures.

IV. DISCUSSION

The experimental and simulation data presented above clearly supports plastic deformation in Cu as the source of TSV-induced stress in Si. However, the details of the origination and evolution of Cu TSV-induced stress is not as clear. Firstly, finite element simulations (Fig. 2 and lower panel of Fig. 8) imply that the measured spatial profile of TSV-induced stress in Si results from a thermal 'annealing' process. This is sensible in the case of wafer B which saw elevated process temperatures (350-400 °C). However, wafer A, for the most part, exhibited larger stress extrema than wafer B yet saw substantially lower process temperatures (150 °C as compared to 400 °C). This contradiction could be reconciled by the inclusion of a 'self-annealing' step where microstructural evolution in the electroplated Cu leads to an intrinsic residual strain field in the Si equivalent to plastic deformation. In this scenario, the Cu TSV is not in a 'zero-stress' state after deposition. Such a scenario would help explain the observed TSV-induced stress profile in Si for a wafer subjected to 150 °C post-plate anneal which would not be expected to induce substantial Cu yield. If subsequent post-plate annealing occurred at sufficiently high temperatures then further plastic deformation of the Cu in the TSV could easily lead to overall reduction in the stress field. This was observed for the TSV-induced stress in Si within the 1×4 TSV arrays (Fig. 9) and is consistent with the speculation offered above that the TSV-induced Si stress profile observed for wafer A results in large part to a type of microstructural evolution akin to 'self-annealing'.

V. CONCLUSION

In conclusion, the thermal and spatial variation of Cu TSV-induced stress in 300mm Si wafers has been investigated for both isolated TSVs and TSV arrays. The TSV-induced stress in neighboring Si results primarily from plastic yield of the Cu. The measured spatial profile of TSV-induced stress in Si is in quantitative agreement with finite element modeling (FEM). Moreover, thermal annealing measurements showed that the intra-array Si stress field became more compressive with increased post-CMP thermal annealing while the Si stress-field external to the arrays exhibited little change. This may indicate a large 'self-annealing' component to the amount of TSV-induced stress in Si.

ACKNOWLEDGMENT

The authors gratefully acknowledge funding support of SEMATECH, the Semiconductor Research Corporation, and DARPA (through the SRC-FCRP) program.

REFERENCES

[1] International Technology Roadmap for Semiconductors (ITRS). http://public.itrs.net accessed 10 December 2010

[2] Kuhn S, Kleiner M, Ramm P, Weber W (1995). Interconnect capacitances, crosstalk and signal delay in vertically integrated circuits. Proc IEDM 1995. doi: 10.1109/IEDM.1995.499189

[3] E. Beyne (2006). The rise of the 3rd dimension for system integration. Proc IITC 2006. doi: 10.1109/IITC.2006.1648629

[4] Lu K, Zhang X, Ryu S-K, Huang R, Ho P (2009). Thermal stresses Analysis of 3-D interconnect. Proc AIP Conf 2009. doi: 10.1063/1.3169263

[5] Okoro C, Yang Y, Vandevelde B, Swinnen B, Vandepitte D, Verlinden B, De Wolf I (2008). Extraction of the Appropriate Material Property for Realistic Modeling of Through-Silicon-Vias using μ-Raman Spectroscopy. Proc. IITC 2008. doi: 10.1109/IITC.2008.4546912

[6] Barnat S, Fremont H, Gracia A, Cadalen E, Bunel C, Neuilly F, Tenailleau J-R (2010). Design for reliability: Thermo-mechanical analyses of stress in through silicon via. Proc EuroSimE 2010. doi: 10.1109/ESIME.2010.5464559

[7] Ranganathan N, Prasad K, Balasubramanian N, Pey KL (2008). A study of thermo-mechanical stress and its impact on through-silicon vias. J Micromech Microeng. doi: 10.1088/0960-1317/18/7/075018

[8] Chen Z, Song X, Liu S (2009). Thermo-mechanical characterization of copper filled and polymer filled TSVs considering nonlinear material behaviors. Proc ECTC 2009. doi: 10.1109/ECTC.2009.5074192

[9] Vandevelde B , Jansen R, Bouwstra S, Pham N, Majeed B, Limaye P, Beyne E, Tilmans HAC (2010). Thermo-mechanical design of a generic 0-level MEMS Package using Chip Capping and Through Silicon Via's. Proc EuroSimE (2010). doi: 10.1109/ESIME.2010.5464539

[10] Karmarkar AP, Xu X, Ramaswami S, Dukovic J, Sapre K, Bhatnagar A (2010). Material, Process and Geometry Effects on Through-Silicon Via Reliability and Isolation," Proc. Mater Res Soc Symp 2010. doi: 10.1557/PROC-1249-F09-08

[11] I. De Wolf (1996). Micro-Raman spectroscopy to study local mechanical stress in silicon integrated circuits. Semi Sci Tech. doi: 10.1088/0268-1242/11/2/001

[12] De Wolf I, Maes HE, Jones SK (1996), Stress measurements in silicon devices through Raman spectroscopy: Bridging the gap between theory and experiment. J Appl Phys. doi: 10.1063/1.361485

[13] C. McDonough and R. E. Geer, unpublished results.

[14] M. Stangl, V. Dittel, J. Acker, V. Hoffmann, W. Gruner, S. Strehle, K. Wetzig, Applied Surface Science 252 (2005) 158–161.

Reliability Studies of a 32nm System-on-Chip (SoC) Platform Technology with 2nd Generation High-K/Metal Gate Transistors

[1]A. Rahman, M. [1]Agostinelli, [2]P. Bai, [2]G. Curello, [2]H. Deshpande, [2]W. Hafez, [2]C. –H. Jan, [2]K. Komeyli, [2]J. Park, [2]K. Phoa, [2]C. Tsai, [2]J.-Y. Yeh, [1]J. Xu

[1]Logic Technology Development Quality & Reliability, [2]Logic Technology Development
Intel Corporation, 5200 N. E. Elam Young Pkwy
Hillsboro, OR 97124, USA
Primary Author Contact: anisur.rahman@intel.com, 503-840-2647 (phone)/503-613-1068 (fax)

Abstract— Extensive reliability characterization of a state of the art 32nm strained HK/MG SoC technology with triple transistor architecture is presented here. BTI, HCI and TDDB degradation modes on the Logic and I/O (1.2V, 1.8V and 3.3V tolerant) transistors are studied and excellent reliability is demonstrated. Importance of process optimizations to integrate robust I/O transistors without degrading performance and reliability of Logic transistors emphasized. Finally, Intrinsic and defect reliability monitoring for HVM are addressed.

Keywords-SoC, Reliability; High-K dielectric; breakdown, metal gate, SILC; TDDB; BTI; CMOS

I. INTRODUCTION AND BACKGROUND

Highly integrated System-on-Chip (SoC) with numerous functional circuit blocks has recently experienced strong growth within the mainstream IC manufacturing trend. Following Moore's Law for CMOS scaling, an industry leading 32nm technology with 2nd generation HK/MG and 4th generation strain technology was reported in [1-2]. This was subsequently optimized with triple transistor architecture for SoC platform applications (including radio frequency, RF, applications) spanning wide range of power, performance and feature space [3-5]. There were numerous reliability challenges to overcome in order to achieve this highly modular SoC technology, e.g., ensuring high performance, low standby power, integrating conventional and legacy and I/Os and stable high-volume manufacturing trends. The reliability characterizations for the 32nm CPU technology and for 45nm dual gate SoC technology were already reported in [6] and [7], respectively. This paper focuses on the robust transistor reliability achieved for this 32nm SoC technology.

II. DEVICE FABRICATION AND MEASUREMENTS

In this 32nm SoC technology, three transistor families are simultaneously offered with mix-and-match option—logic (HP or SP), low power (LP) and I/O (1.2V LV and 1.8/3.3V HV). The transistor options and design rules are elaborated in table I. This triple transistor architecture ensures independent optimization of device performances to meet the requirements for different SoC circuit blocks. The extremely low gate leakage of the HK dielectric allowed development of a

relatively simple fabrication process flow by allowing logic, LP and LV I/Os to share the same optimized HK dielectric layer. The HV I/O transistors utilize a composite bi-layer gate dielectric stack with a pre-patterned thermal oxide layer underneath the HK layer allowing them to tolerate higher gate voltages. On a single chip the HV I/O can be either 1.8V type or 3.3V type (not both). Reliability results for 3.3V tolerant transistor with composite dielectric are reported for the first time. All transistor families have unique source and drain extension implants. Fourth generation strained silicon technologies including NMOS tensile contact strain, compressive metal gate fill, PMOS embedded high Ge SiGe

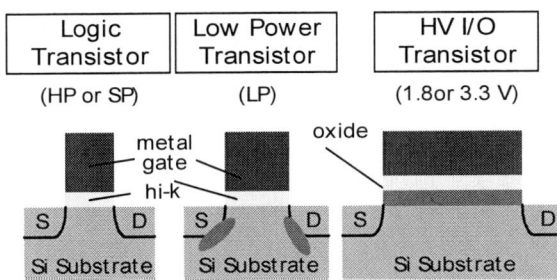

Figure 1. Three transistor families simultaneously offered in 32nm SoC process flow with an option for mix and match. Logic, LP and 1.2V LV (not shown) share the same HK. In addition to longer L, junction engineering through tip/halo doping (highlighted) optimization helped LP to achieve low standby leakage. A composite dielectric stack with a pre-patterned thermal oxide layer beneath the HK layer allows tolerance to higher, 1.8/3.3, voltages.

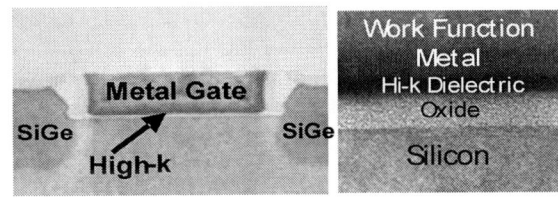

Figure 2. TEM images of the HV I/O devices. Thick SiO2 layer pre-patterned beneath HK layer. Thickness of oxide layer modulates its voltage tolerance, 1.8V or 3.3V. Oxide thickness uniformity is a key aspect of reliability of HV devices. Non trivial process optimization involved to eliminate defects at S/D edge and thickness variability in the STI edge of HV composite gate dielectric.

978-1-4244-9113-1/11 $26.00 © 2011 IEEE

TABLE I. SUMMARY OF TRANSISTOR OPTIONS AND DESIGN RULES

Transistor Type	Logic	Low Power	LV I/O	HV I/O	
	HP/SP	LP	1.2V	1.8V	3.3V
EOT (nm)	0.95	0.95	0.95	~4	~7
Vdd (V)	0.75/1	0.75/1	0.75/1.2	1.5/1.8	3.3
Pitch (nm)	112.5	126	225	338	≥450
Lgate (nm)	30/34	46	≥88	>140	≥300

and reduced proximity raised S/D are employed in a Gate-Last (Replacement Metal Gate) process flow [1-5]. Bias temperature instability (BTI) degradation measurements were performed under quasi-DC conditions with fixed delay between stress and measurement to account for recovery. Hot electron degradation was characterized under peak substrate current condition to ensure maximum impact ionization near the drain end. Constant voltage stress (CVS) was used for time dependent dielectric breakdown (TDDB) characterization of gate dielectric in a stress-measure-stress (SMS) mode and only hard breakdown was considered. No stress to measurement delay was used for hot electron or TDDB characterizations.

III. LOGIC AND LV I/O RELIABILITY

State of the art SoC products usually employ a number of low voltage (LV) I/O blocks, e.g. LP-DDR2, GPIO, MIPI. These circuits are optimally designed only by 1.2V tolerant devices. Here we have demonstrated a LV device sharing the same HK dielectric as logic/LP, but by utilizing process optimization and junction engineering, can tolerate 1.2V. This LV device did ensure about 30% I/O area scaling over alternative options to implement such circuits, e.g. stacked-partial swing design. Relevant degradation mechanism for logic and LV devices is BTI which has been extensively reported in the published literature [6-8]. In HK-MG technologies, PMOS BTI degradation mechanism is generally accepted to be similar to the SiO_2-like interface trap generation

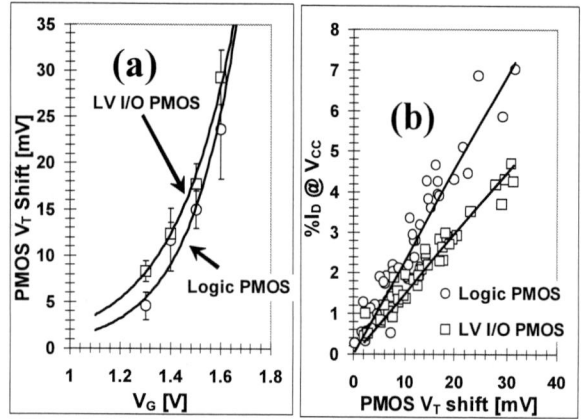

Figure 3. PMOS BTI characteristics for Logic and LV I/O (a) V_T shift vs. gate stress voltage (with 1-sigma error bars). (b) %I_D (at respective operation voltages) vs. V_T shift. Both devices share the same HK layer but LV device is optimized for 1.2V tolerance. Lower %I_D at given V_T shift is from overdrive.

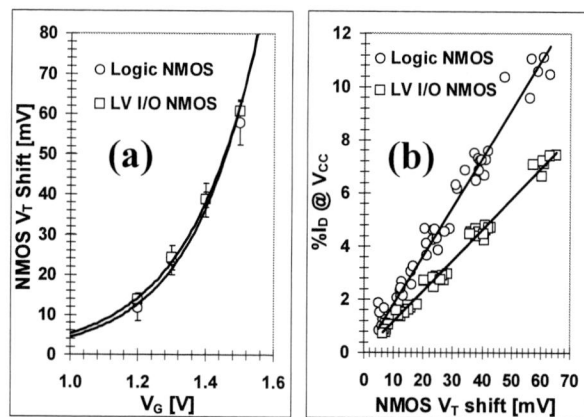

Figure 4. Logic and LV I/O NMOS BTI results (a) V_T shift vs. gate stress voltage (with 1-sigma error bars) (b) %I_D (at respective operation voltages) vs. V_T shift. Same HK for both devices causes same V_T shifts due to HK bulk nature of NMOS BTI. For same V_T shift, LV has less %I_D due to 1.2V operation.

while, NMOS BTI is primarily due to bulk HK trapped charges. The BTI results at 90C, 1000sec for logic and LV devices are shown in Figs. 3 and 4. Drive degradations, %I_D, are measured at 0.95V for logic and at 1.2V for LV devices, respectively.

A. Device Degradation Results

Figure 3 (a) compares the PMOS V_T shifts for logic and LV I/O devices in inversion. Although V_T shift for LV PMOS is on the higher side, the additional gate overdrive in LV softens the V_T shift offset and show lower drive current degradations, as shown in Fig. 3(b). Corresponding NMOS BTI results are presented in figure 4 (a-b). Here due to bulk trapping nature of HK's NMOS BTI, their V_T shifts at a give stress voltage are exactly matched (Fig. 4a). Figure 4(b) shows that %I_D at given V_T shift is lower for the LV I/O transistors compared to logic, which again is from higher operating voltage in the former makes it less sensitive to V_T shifts than the logic device.

The NMOS vs. PMOS V_T shifts sigma difference can be explained from the nature and location of traps involved. Positively charged interface traps generated in PMOS during BTI are more effective in shifting V_T due to their proximity to the hole inversion layer. The discrete nature of such traps results in larger spread in V_T in highly scaled devices. On the other hand the bulk nature of negatively charged NMOS traps in HK acts as a relatively distant charge sheet for channel electrons and therefore, its granular nature is averaged out.

B. Dielectric Reliability Results

In HK-MG technology an asymmetry exists for dielectric band offset between conduction and valence bands. This results in not only unequal gate leakage for PMOS vs. NMOS, but also significantly alters their oxide reliability characteristics. The HK band alignment phenomena and its consequence on TDDB characteristics for NMOS vs. PMOS have been extensively reported in the literature [6], [10-11], [13]. The net outcome of this is in inversion mode of operation, PMOS demonstrates more robust dielectric reliability than NMOS. The gap, however, shrinks as the HK thickness scales.

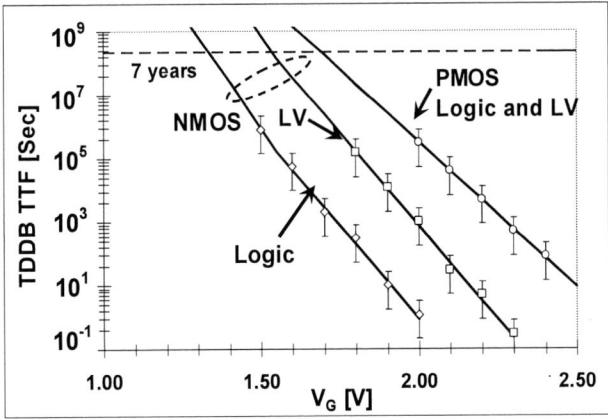

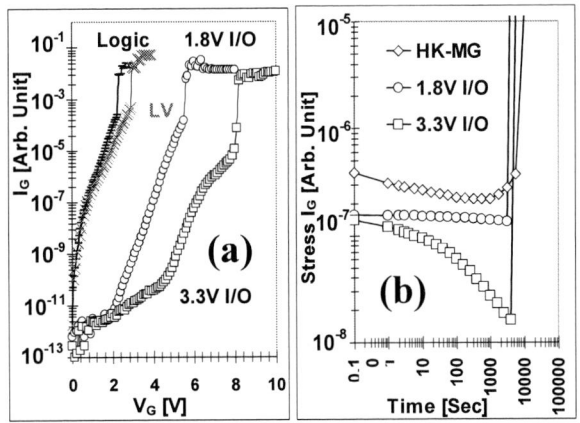

Figure 5. Logic and LV I/O TDDB TTF vs. voltage plot. Thirty devices stressed at 90C for each voltage and the TTF is fitted using Weibull distribution with maximum likelihood. Error bars are from fitted distribution covering 5%-95% TTF. HK/MG PMOS show better TDDB than NMOS, so reliability improvement of NMOS is crucial. Logic and LV are roughly matched for PMOS, but LV NMOS shows better TDDB than Logic enabling their usage in 1.2V circuits.

Figure 5 compares the NMOS/PMOS CVS TDDB results for logic and LV I/O device families, all employing identical HK dielectric thickness. The test structures consist of 2400 identical devices (width 0.4um each) connected in parallel to ensure test area of the order of tens of micron square. The first thing to observe is a clear dielectric reliability margin for PMOS over NMOS. Additionally, the LV NMOS transistor does demonstrate more robust dielectric reliability than logic which is achieved by process optimization. The LV PMOS transistor reliability however is matched to that of its logic counterpart.

NMOS having poorer reliability than PMOS, becomes the limiter for determining the maximum tolerable operational voltage. The improved dielectric reliability demonstrated by LV NMOS in Fig. 5 is key to the usage of LV family as native 1.2V tolerant devices in I/O circuits. Indeed, a detailed TDDB modeling with Weibull TTF distribution, E-filed scaling for V_G, Arrhenius relationship for temperature, and Poisson area scaling confirms that at product like use condition, e.g. area, temperature, lifetime, the LV devices can tolerate >200mV higher operating voltage over the logic device, therefore, can be directly used in special 1.2V I/O circuits required by SoC products. At product operating condition, the full chip p-fail (i.e., cumulative fail probability) budget allocated for core logic and LV I/O blocks is primarily consumed by NMOS, while p-fail from PMOS is insignificant due to its more robust dielectric reliability.

IV. HV I/O RELIABILITY

As mentioned in section II, the gate dielectric stack of HV I/O consists of a pre patterned thick SiO_2 layer beneath the HK. Although the same HK layer is shared by all device families—logic, LP, LV I/O and HV I/O, the 1.8V or 3.3V gate voltage tolerance for the HV devices is a result of this thick SiO_2 layer. Thicker oxide layer is used for 3.3V device compared for 1.8V device to achieve additional voltage margin. The final optimized EOT for these devices can be seen in Table I.

Figure 6. (a) Gate leakage vs. V_G sweep up to oxide breakdown. Progressively increasing breakdown voltages and relative gate leakage for various NMOS. Either 1.8V or 3.3V tolerant HV devices can be fabricated on a given chip, not both. (b) Evolution of stress I_G during TDDB. Electron trapped in dielectric near gate reduces electric filed lowering F-N tunneling.

The relative gate leakages and dielectric breakdown events for NMOS are presented in Fig. 6(a). As gate voltage is ramped at a fast rate from zero, the gate leakage increases until there is a catastrophic and irreversible breakdown for the dielectric. The breakdown can be identified as a discontinuity in the gate leakage. After the breakdown the dielectric loses its insulation property. The progressively increasing voltage tolerance for logic, 1.2V LV I/O, 1.8V and 3.3V HV I/O is demonstrated. Such voltage ramped breakdown characterization of gate dielectric is a valuable tool to identify defects and optimize the thickness during the early state of process development. In Fig. 6(b) stress gate leakages for Logic, and two HV I/Os highlight interesting reliability physics for NMOS TDDB. Here a constant voltage is applied to the gate until the dielectric shows hard breakdown while periodically the leakage at stress is measured. Devices with different thickness undergo different stresses to achieve similar time to breakdown. The gradual decrease in stress

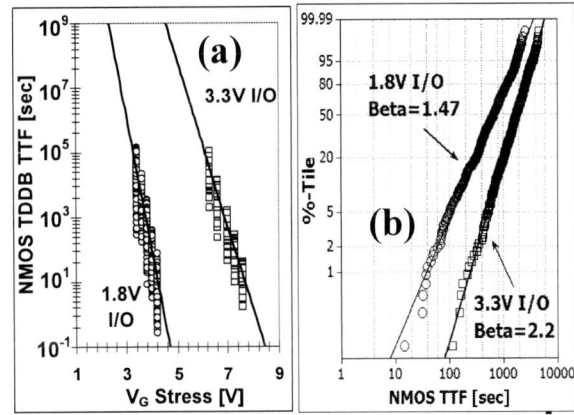

Figure 7. 1.8V and 3.3V tolerant HV I/O NMOS TDDB. (a) Thicker SiO_2 under-layer for 3.3V ensures higher voltage tolerance. PMOS shows additional margin. 50 devices with width 1200um (2000legsX0.6um) each are stressed at 90C for each V_G. (b) Weibull shape scales with oxide thickness, as expected. High value of shape indicates defect free oxide.

978-1-4244-9113-1/11 $26.00 © 2011 IEEE   535

leakage is understood as electron trapping in dielectric adjacent to gate resulting in lower E-field. Electric field being strong modulator for Fowler-Nordheim (FN) tunneling indicates both HK logic and 3.3V HV are stressed in the F-N tunneling regime. However, 1.8V oxide does show flat time evolution of stress leakage indicating that it is necessarily from direct tunneling. When TDDB characterization for a dielectric is done at a number of voltages to extract the voltage acceleration parameter, the oxide can be stressed in FN condition at higher voltage while as direct tunneling at lower voltage. A change in acceleration behavior for time to fail (TTF) with voltage is observed for both HK and oxides technologies when stress is changed from F-N to direct tunneling [7], [14, 15].

A. HV Transistor Dielectric Reliability Results

The oxide reliability of 1.8V and 3.3V tolerant NMOS from HV I/O family have been characterized for TDDB and their voltage acceleration behaviors are compared in Fig. 7(a). The PMOS transistors (not shown here) demonstrate >1V additional margin and hence NMOS reliability becomes the limiter. The composite gate dielectric stack of HV transistors contains a thick SiO_2 layer whose thickness is scaled according to operating voltage. The breakdown from TDDB is a weakest link problem as modeled by percolation theory and hence the time to fail distribution follows Weibull distribution. The Weibull shape factor is a key parameter related to the spread of the distribution. It is crucial to achieve high value of shape (>1) for a technology to minimize the oxide failures during early product lifetime. Early failures are related to poor oxide quality and defects and lowers Weibull shape tow <1. Healthy HV I/O oxide reliability is demonstrated here with excellent Weibull shape factors as seen in Fig 7(b). The Weibull shape also scales with the thickness as expected from the percolation model.

B. HV Transistor Degradation Results

BTI and hot electron are two mechanisms causing gradual degradation of drive current during the normal operation of transistors. For HV transistors the thick SiO_2 layer beneath HK in gate dielectric stack is the source of device degradation. As a result, similar to SiO_2 technologies, the HV PMOS demonstrates BTI degradation while HV NMOS suffers degradation from hot electron effects. The hot electron effects for PMOS and BTI in NMOS in HV devices are insignificant compared to their respective dominant degradation mechanisms. These two degradation results are presented in this section.

Fig. 8(a) describes the 1.8V and 3.3V HV I/O PMOS BTI results. A constant voltage stress is applied at inversion and periodically the stress is removed to measure the drive current degradation. Here TTF is defined as time required to reach certain saturation current, $I_{D,sat}@V_{cc}$, degradation. As in oxide technologies, PMOS BTI is a field dependent phenomenon, therefore, the thicker oxide in 3.3V show longer lifetime compared to 1.8V device. Additionally, their voltage acceleration factors are inversely proportional to oxide thicknesses, confirming same field acceleration parameter for both devices. The overdrive benefit in 3.3V PMOS is highlighted in Fig. 8(b). Here the %$I_{D,sat}$ at their respective V_{cc}

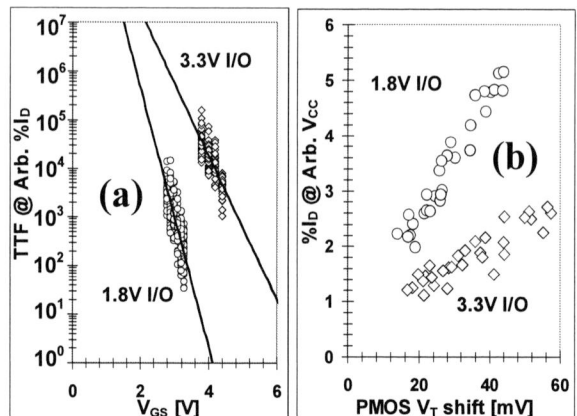

Figure 8. 1.8V and 3.3V HV PMOS BTI. (a) Time to fail is defined as time for %I_D to reach a pre-specified value. Thick oxide under layer in gate stack has same BTI properties as SiO_2 technology. (b) Increased gate overdrive in 3.3V device ensures lower %I_D loss at a given V_T shift

is compared between 1.8V and 3.3V devices as a function of threshold voltage, V_T, shift. A given V_T shift corresponds to a smaller fraction of gate overdrive, (V_G-V_T), loss in 3.3V transistor as seen from the %I_D comparisons. PMOS hot electron effect is insignificant compared to BTI for both devices.

Hot electron (HE) is the most serious degradation mechanism for HV NMOS and can severely degrade the drive current leading to circuit's functional failure. Electrons travelling from source to drain accelerate and gain kinetic energy from the channel directed field at high V_{DS}. Upon reaching near the drain end a fraction of carriers acquire sufficient energy to cause impact ionization from collision with the lattice generating electron hole pairs. Defects are created in the oxide due to these energetic carriers which subsequently lowers the drive current for NMOS. For hot carrier generation a fundamental requirement is simultaneous presence of $V_{GS}>V_T$ and high V_{DS}, which is CMOS circuit is only satisfied during the switching event. Either when the output node is low, $V_{DS}\sim0$, or the output is high then $V_{GS}\sim0$, there is no hot

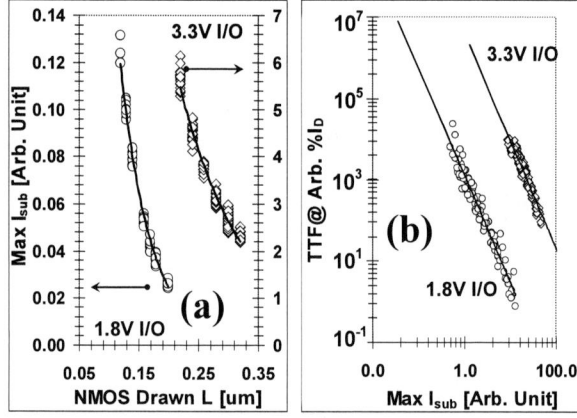

Figure 9. HV NMOS hot electron (a) Impact ionization near drain end gets drastically worse at higher voltages but improves with increasing L (b) Intrinsic hot electron characteristics presented as TTF (time for a pre-specified %I_D) vs. I_{sub}. Longer L, thicker oxide and junction engineering improved hot-E degradation for 3.3V.

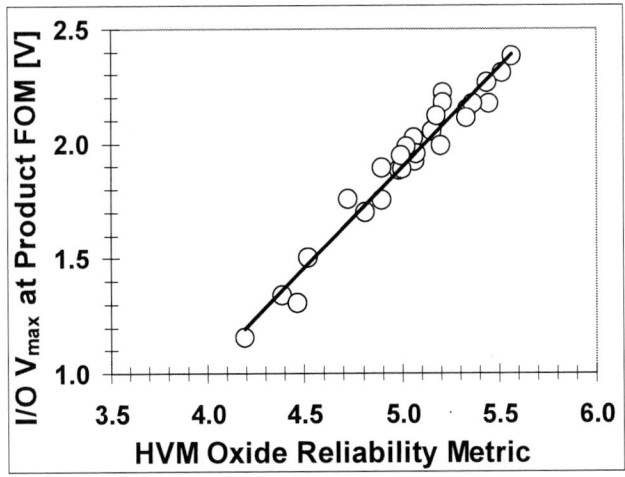

Figure 10. Wafer level High Volume Manufacturing (HVM) Oxide reliability metric shows excelent correlation with the Vmax characterized through elaborate reliability modeling. This metric allows quick monitoring process health.

electron degradation. As degradation happens from energetic carriers generated from impact ionization, the substrate current I_{sub}, a current directly proportional to number of electron-hole pair generated is a direct monitor for the rate at which NMOS undergo degradation during dynamic switching in CMOS. Fig. 9(a) compares the peak substrate currents for 1.8V vs. 3.3V devices as function of channel lengths. At fixed V_{DS}, longer L reduces the channel directed electric field causing less impact ionization, hence, I_{sub} is lowered. Similarly, at a given L, higher voltage drastically increases impact ionization from increased field. Finally, the peak field at drain junction is strongly modulated by doping gradients and a gradual change in doping profile significantly lowers the electric filed making device less susceptible to degradation. Significant process optimization was involved to lower the HE degradation for HV devices, especially for 3.3V devices. Figure 9(b) compares the intrinsic HE characteristics for the two HV NMOS where at a given hot electron generation rate (I_{sub}) 3.3V devices show more robust HE immunity (~100X lifetime).

V. HVM RELIABILITY MONITORING METRIC

Rigorous TDDB characterization for gate dielectric has been performed in order to develop extensive reliability models accounting for voltage, area and temperature scaling for each family of transistors. Such detailed model allows full chip p-fail (cumulative fail probability) prediction at their nominal use condition and to determine maximum voltages that can be applied without exceeding reliability risk. However, it takes substantial time to perform constant voltage stress TDDB, therefore, is not practical for monitoring gate oxide health during their high volume manufacturing (HVM) in Fabs. To address this, an HVM reliability metric was developed to monitor and trend the dielectric health during manufacturing. This metric is much faster to characterize and as shown in Fig. 10, it also is strongly correlated to the maximum tolerable voltage for transistors (V_{max}) calculated from detailed long time TDDB tests. This fast and efficient approach is invaluable to ensure stable manufacturing and identifying excursions.

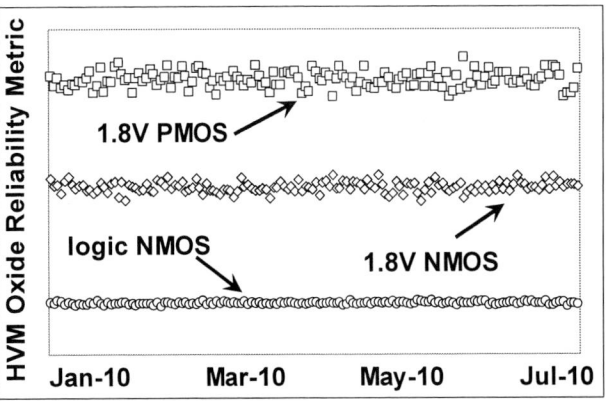

Figure 11. HVM Oxide Reliability Metric shows stable high volume manufacturing trends over a period of several months. The stability highlights the robust and controlled dielectric reliability for the 32nm SoC process flow.

Fig. 11 demonstrates the stable factory trends achieved for Intel's 32nm SoC technology over six months of production.

VI. DISCUSSION

Efficient, stable and cost effective integration of 1.8/3.3V tolerant HV I/O devices to the 32nm CPU process flow has been a non trivial development challenge for the 32nm SoC technology. Cost constraint for high volume manufacturing required most process/fabrication steps be shared between HK only logic and composite dielectric HV I/O transistors. However, in this scheme logic device performance became strongly and adversely coupled to the HV transistor reliability. In other words, performance improvement for logic transistors becomes difficult without sacrificing HV oxide reliability and vice versa. This made achieving an optimal process condition for all transistors families in the 32nm SoC technology challenging. The key to successful HV I/O integration was to ensure quality, uniformity, and geometric aspects of the composite dielectric stack without sacrificing logic/LP performance or reliability. This challenge has been successfully met as evidenced by the 1.5-2.0 Weibull shape achieved for the HV I/O devices (Fig. 7(b) and the stable baseline for oxide reliability metric (Fig. 11).

In addition to overcoming integration challenges for different gate dielectrics on the same chip, minimizing the device performance degradations from BTI and hot electron is also crucial. Hot electron degradation in HV I/O NMOS makes scaling of L practically impossible at fixed supply voltage. Abrupt doping profile at drain to channel junction increases maximum electric filed and worsens hot electron degradation. Although longer implant annealing time is the best approach for a graded junction, the login, LP and LV I/O devices limits the maximum anneal time and temperature for the process. As a result, a number of junction optimization approaches have been implemented for HV devices to achieve a graded junction including tip/halo implant angle, dose and energy modulations. Fabrication of a 3.3V tolerant NMOS was especially challenging where most of the junction grading engineering were implemented.

978-1-4244-9113-1/11 $26.00 © 2011 IEEE

VII. SUMMARY AND CONCLUSIONS

Extensive reliability studies performed for a leading edge, 32nm HK-MG platform technology with triple transistor architecture optimized for ultra low power and high performance SoC products. Robust transistor aging and dielectric reliability, matching or exceeding the 45nm SoC technology, achieved as well as stable high volume manufacturing demonstrated. Simultaneous process optimizations to achieve high performance core logic and robust high voltage tolerant I/O devices emphasized.

ACKNOWLEDGMENT

Authors gratefully acknowledge Chetan Prasad, Sangwoo Pae, and Steve Ramey for valuable technical inputs/discussions and Jeff Hicks, Bruce Woolery and Thomas Marieb for technical reviews and management support (LTD Q&R).

REFERENCES

[1] S. Natarajan et al., "A 32nm logic technology featuring 2nd-generation high-k + metal-gate transistors, enhanced channel strain and 0.171μm² SRAM cell size in a 291Mb array," IEDM Tech. Dig., 2008, pp. 941-943.

[2] P. Packan et al., "High performance 32nm logic technology featuring 2nd generation high-k + metal gate transistors," IEDM Tech. Dig., 2009, pp. 659-662.

[3] C.-H. Jan et al., "A 32nm SoC platform technology with 2nd generation high-k/metal gate transistors optimized for ultra low power, high performance, and high density product applications," IEDM Tech. Dig., 2009, pp. 647-650.

[4] P. VanDerVoorn et al., "A 32nm low power RF CMOS SOC technology featuring high-k/metal gate," VLSI Symp. Proc., 2010, pp. 137-138.

[5] C.-H. Jan et al., "RF CMOS technology scaling in high-k/metal gate era for RF SoC (System-on-Chip) applications," IEDM Tech. Dig., 2010, pp. 604-607.

[6] S. Pae et al., "Reliability characterization of 32nm high-K and Metal-Gate logic transistor technology," IRPS Sym Proc., 2010, pp. 287-292.

[7] C. Prasad et al., "Reliability studies on a 45nm low power system-on-chip (SoC) dual gate oxide high-k/metal gate (DG HK+MG) technology," IRPS Sym Proc., 2010, pp. 293-298.

[8] K. Mistry et al., "A 45nm logic technology with high-k+metal gate transistors, strained silicon, 9 Cu interconnect layers, 193nm dry patterning, and 100% Pb-free packaging," IEDM Tech Dig., 2007, pp. 247-250.

[9] C. Auth et al., "45nm High-k + metal gate strain-enhanced transistors," VLSI Symp. Proc., 2008, pp. 128-129.

[10] S. Pae et al., "BTI reliability of 45 nm high-K+metal-gate process technology," IRPS Sym Proc., 2008, pp. 352-357.

[11] J. Hicks et al, Intel Tech Journal, vol. 12, iss. 2, 2008

[12] S. Ramey et al., "Frequency and recovery effects in high-κ BTI degradation," IRPS Sym Proc., 2009, pp. 1023-1027.

[13] C. Prasad et al., "Dielectric breakdown in a 45 nm high-k/metal gate process technology," IRPS Sym Proc., 2008, pp 667-668.

[14] Chenming Hu and Qiang Lu, "A unified gate oxide reliability model," IRPS Sym Proc., 1999, pp 47-51.

[15] J. W. McPherson, J. Kim, A. Shanware, H. Mogul, and J. Rodriguez, "Trends in the ultimate breakdown strength of high dielectric-constant materials," IEEE Trans. On Elec. Dev., vol 50, p 1771-8, 2003.

V_{TH} shift Mechanism in Dysprosium (Dy) incorporated HfO_2 gate nMOS devices

Tackhwi Lee and Sanjay K. Banerjee

Microelectronics Research Center, R9950, The University of Texas at Austin,

10100 Burnet Road, Austin, Texas 78758 Tel: (512) 471-1627, Fax: (512) 471-5652, Email: tackhwi.lee@gmail.com

Abstract

We discuss temperature-dependent Dy diffusion and the diffusion-driven Dy-silicate formation process in Dy incorporated HfO_2. The Dy-induced dipoles are closely related to the Dy-silicate formation at the high-k/SiO_2 interfaces since the V_{FB} shift in Dy_2O_3 is caused by the dipole and coincides with the Dy-silicate formation. Dipole formation is a thermally activated process, and more dipoles are formed at a higher temperature with a given Dy content. The Dy-silicate related bonding structure at the interface is associated with the strength of the Dy dipole moment, and becomes dominant in controlling the V_{FB}/V_{TH} shift during the high temperature annealing in the Dy-Hf-O/SiO_2 gate oxide system. Dy-induced dipole reduces the degradation of the electron mobility. Charge trapping characteristics in relation to the stress-induced flatband shift and SILC are discussed with a band diagram. The higher effective barrier height of Dy_2O_3, which is around 2.32 eV, calculated from the F-N plot, accounts for the reduced leakage current in Dy incorporated HfO_2 nMOS devices. The lower trap generation rate by the reduced hole trap density and the reduced hole tunneling of the Dy-doped HfO_2 dielectric demonstrate the high dielectric breakdown strength by weakening the charge trapping and defect generation during the stress.

[*Keywords* : Dy_2O_3, Dy incorporated HfO_2, Dy-induced dipoles, Dy-silicate formation, Dy diffusion, mechanism of V_{TH}/V_{FB} shift, effective work function, stress-induced flatband shift, SILC]

Introduction

Controlling the V_{TH} of high-k metal gate (HK-MG) stack has been one of the critical issues for CMOS application. Even though it is still challenging to control the V_{TH} by the work function modulation of metal for the gate first process, the effective work function (EWF) can be modulated to obtain the targeting V_{TH} by introducing the dipoles or charges in the gate oxide. Dysprosium (Dy) incorporation, which is one of the possible candidates in CMOS application, shows the large V_{TH} shift and promising transistor characteristics [1-2]. However, the mechanism of V_{TH} shift by Dysprosium (Dy) incorporation has not been reported yet. Although the high-k/SiO_2 interface is known to be a determining factor in controlling the threshold voltage, experiment data should be analyzed in the context of comparing the V_{TH} shifts of different high-k materials with respect to the structural effects and the diffusion of dopant material corresponding to the different annealing temperatures. The V_{TH} shift mechanism of Dy incorporated HfO_2 will be investigated in relation to the

Dy-induced dipoles formation with a support of material analysis in this report.

In addition, we compared the characteristics of Dy incorporated HfO_2 (DyO/HfO) and HfO_2 gated n-MOSCAP by measuring the gate leakage currents and capacitances with various voltage stresses and by performing the time-zero dielectric breakdown (TZDB) and time-dependent dielectric breakdown (TDDB) with a various constant voltage stress. To clarify the reason for the reduced leakage current and charge trapping characteristics of the Dy-doped HfO_2, the effective barrier heights of Dy_2O_3 and DyO/HfO oxide were extracted from the Fowler Nordheim (FN) tunneling current. Stress-induced leakage current (SILC) and stress-induced flatband shift was analyzed for the charge trapping characteristics and interpreted with a proposed band diagram. TZDB and TDDB were performed with small area capacitors to ensure intrinsic breakdown, and the Weibull slope was obtained.

Experiments

A co-sputtered Hf and Dy and each layer of Hf and Dy were deposited by DC Magnetron sputtering (30mTorr, N_2 ambient), and re-oxidized during PDA (500 °C, 5min) for mixed HfDyO and bi-layered DyO/HfO capacitors. After defining TaN gate electrode, PMA (900 °C, 1min) were performed for S/D activation. Forming gas annealing (600 °C, 300min) was followed by the Al metal S/D contact and backside contact for nMOSFETs.

Results and Discussion

I. The Dy-induced dipole in Dy_2O_3/SiO_2 gate oxide

The EOT-V_{FB} relationship is usually used to clarify the dipole formation or the interface charges. From reference [3], the equation of V_{FB} and EOT can be written for accurate calculation.

$$V_{FB} = \varphi_{ms} + \Delta D - Q_f \frac{EOT}{\varepsilon_{OX}} - \rho_b \frac{EOT^2 - EOT_h^2}{2*\varepsilon_{OX}} \qquad (1)$$

where φ_{ms}, ΔD, Q_f, ρ_b, EOT and EOT_h are the work function difference between the metal gate and the Fermi energy of the Si substrate, dipole term, interface fixed charges, bulk charge density in the high-k layer, EOT of the total gate stack, and EOT of the high-k layer, respectively. According to Eq. 1, the dipole is independent of EOT, so that the V_{FB} will remain unchanged with respect to the EOT increase. Meanwhile, the V_{FB} will be linearly dependent on the EOT change if the fixed charges are formed at the bottom SiO_2/Si interface. If the

978-1-4244-9113-1/11 $26.00 © 2011 IEEE

contribution of bulk charges is significant, then a quadratic dependence of V_{FB} on EOT is expected to be observed.

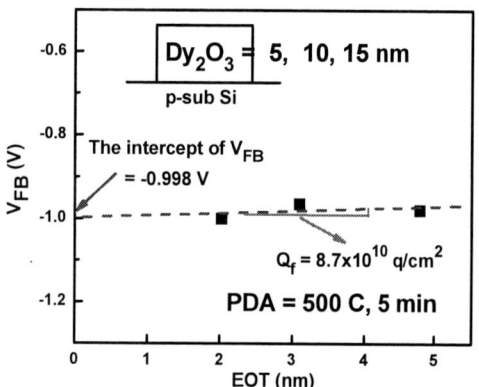

Figure 1 V_{FB} is independent of EOT for the Dy_2O_3 oxide MOSCAPs having different physical thicknesses. This is evidence of Dy-induced dipoles formations rather than interface fixed charges controlling the V_{FB} shift of Dy_2O_3 oxide samples.

In Fig. 1, the dependency of the V_{FB} changes is plotted against EOT. It is clear that the slope of the V_{FB} change is very small, and the fixed charge extracted from the regression is 8.7×10^{10} q/cm² , which is negligibly small, which indicates that the main mechanism of V_{FB} shift is the formation of Dy dipoles at the DyO_x/SiO_2 interface rather than fixed charges in the SiO_2/Si interface or bulk charges in the Dy_2O_3 film. Neither a quadratic EOT dependence can be observed, indicating that the contribution of bulk charge density is also ignorable.

The V_{FB}-axis interceptions of the linear regression lines in EOT vs. V_{FB} plot is now defined by the sum of dipole and φ_{ms}, and the value of -0.998 V is obtained from Fig. 1. Since the work function (φ_m) of TaN is 4.2 eV and the φ_s of the Si substrate used in this experiment with a 3×10^{15} cm^{-3} doping is 4.9 eV, the work function difference ($\varphi_{ms} = \varphi_m - \varphi_s$) is only -0.7 eV. Therefore, the rest of the V_{FB} shift can be attributed to a constant term of 0.3 eV from these set of samples. The V_{FB} and EOT relationship indicates that the main mechanism of V_{FB} shift is the formation of Dy dipoles at the DyO_x/SiO_2 interface is rather than fixed charges in the SiO_2/Si interface or bulk charges in the Dy_2O_3 film. According to the proposed model for the physical origin of the dipole formed at the high-k/SiO_2 interface, the difference in an areal density of oxygen atoms at high-k/SiO_2 interface causes the oxygen movement, and decides the direction and strength of the interface dipole [4]. The number of oxygen atoms per unit area (σ) is approximately $V_u^{-2/3}$, where V_u is defined by the volume of unit structure containing a single oxygen atom, and can be

calculated from the formula weight and the density of oxide. The extracted V_u of Dy_2O_3 is 28.7 since the molecular weight and the density of Dy_2O_3 is 373.99 (g/mol) and 7.81 (g/cm³), respectively. Then, the extracted areal density difference (σ/σ_{SiO2}) of Dy_2O_3 is 0.85, where V_u of SiO_2 is 22.7 and the value of σ_{SiO2} is 1, which is summarized in Table 1.

Oxide	Dy_2O_3	La_2O_3	HfO_2	Al_2O_3
EN (Pauling)	1.22	1.1	1.3	1.61
EN (Sanderson)	2.27	2.18	2.49	2.54
Density (g/cm³)	7.81	6.51	9.6	3.99
Molecular Weight (g/mol)				
	373.99	325.82	210.49	101.96
V_u (Å³)	28.7	27.7	22.7	14.2
σ/σ_{SiO2}	0.85	0.88	1.20	1.37

Table 1. ENs and normalized areal oxygen density (σ/σ_{SiO2}) is summarized in accordance to the structural parameters of Dy_2O_3. The data for La_2O_3, HfO_2 and Al_2O_3 are quoted from reference [4] to compare with the result of Dy_2O_3.

This predicts that the direction of the dipole moment is from the high-k layer to the interfacial SiO_2, causing the band bending, which results in a negative V_{FB} shift.

Dy is one of the lanthanide materials and the atomic behavior seems to be very similar to La. Electronegativity (EN) and normalized areal oxygen density of Dy_2O_3 shown in Table 1 is also similar to La_2O_3, suggesting that Dy would show similar tendency as La in terms of V_{FB} shift. La in SiO_2 was reported to form a silicate layer resulting in the compound $La_2Si_2O_7$. In addition, the La-induced dipole has been explained by the La-silicate formation at the interface. We will investigate the bond structure of Dy-O and Si-O in the interface with the XPS analysis to correlate with the silicate formation. The mechanism of V_{FB}/V_{TH} shift of the Dy-Hf-O/SiO_2 network will be discussed in conjunction with the temperature-dependent Dy-silicate formation at the interface with the material analysis data such as XPS and STEM EDX in the following sections.

In Fig. 2, V_{FB} shift of the co-sputtered DyHfO gate oxide n-MOSCAPs are proportional to the percentage of the Dy atom and their values are between the V_{FB} of Dy_2O_3 and HfO_2. After 900 °C PMA, the larger amount of flatbands was shifted. This implies that the areal density of the net Dy-induced dipole at the interface can be modulated by both Dy concentration and annealing temperature.

II. Material characterization with XPS and STEM EDX data

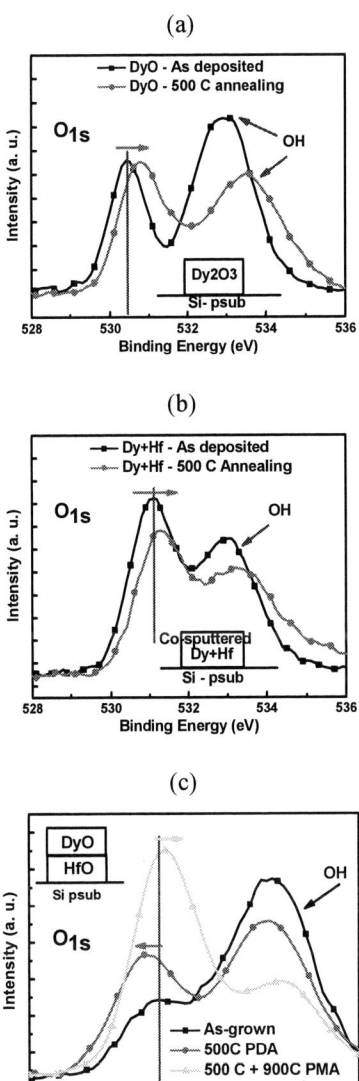

(a)

(b)

(c)

Figure 4. The binding energy (BE) shift of O1s core-level XPS data of (a) Dy_2O_3 and (b) co-sputtered Dy+Hf sample were compared before and after 500 °C PDA. The BE shift toward a higher energy indicates the formation of Dy silicate at the high-k/SiO_2 interfaces. However, in the DyO/HfO case, Dy silicate was formed after 900 °C PMA, but Dy silicate was not formed at the high-k/SiO_2 interfaces at 500 °C annealing.

Figure 2 V_{FB} shift of the co-sputtered DyHfO gate oxide n-MOSCAPs are proportional to the percentage of the Dy atom and their values are between the V_{FB} of Dy_2O_3 and HfO_2. After 900 °C PMA, the larger amount of flatbands was shifted.

To understand the mechanism of V_{FB}/V_{TH} modulation with respect to Dy diffusion in DyO capped HfO_2 samples corresponding the annealing temperature, V_{FB} of DyO/HfO (=22Å/22Å), HfO/DyO (=17Å/26Å), and HfO_2 (=44 Å), are compared after 500 °C PDA and 900 °C PMA, respectively, as shown in Fig. 3. After 500 °C PDA, the V_{FB} shift of DyO/HfO is the same as V_{FB} of HfO_2, and the V_{FB} shift of HfO/DyO corresponds to the V_{FB} of the single Dy_2O_3 oxide, which indicates that the bottom high-k layer deposited on Si substrate directly control the V_{FB} shift.

The binding energy (BE) shift of O1s core-level XPS data of Dy_2O_3 (Fig. 4(a)) and co-sputtered DyHfO (Fig. 4(b)) sample were compared before and after 500 °C PDA to correlate with the Dy-induced dipole. The BE shift to a higher energy indicates the formation of Dy silicate at the high-k/SiO_2 interfaces [5]. According to the BE of DyO/HfO bi-layer sample before and after 900 °C PMA (Fig. 4(c)), Dy-silicate was formed after 900 °C PMA, but Dy-silicate was not

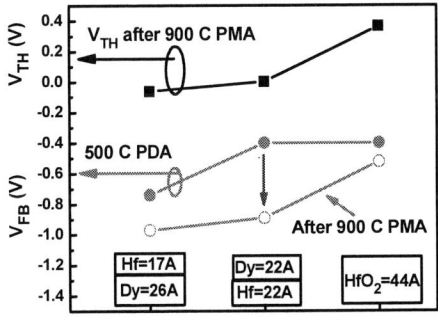

Figure 3 V_{FB} of DyO/HfO (=22Å/22Å), HfO/DyO (=17Å/26Å), and HfO_2 (=44 Å), are compared after 500 °C PDA and 900 °C PMA, respectively.

formed at the high-k/SiO$_2$ interfaces at 500 °C annealing. From STEM EDX analysis of DyO/HfO/Si p-sub sample annealed at 500 °C and 900 °C, Dy atoms and Hf atoms are partially intermixed at the 500 °C PDA, but Dy intermixing with SiO$_2$ is not noticeable in the interfacial layer (Fig. 6 (a)). However, a gradual Dy depth profile is clearly extending into the interfacial layer after 900 °C PMA (Fig. 6 (b)), which confirms the formation of Dy-silicate at the interfacial layer.

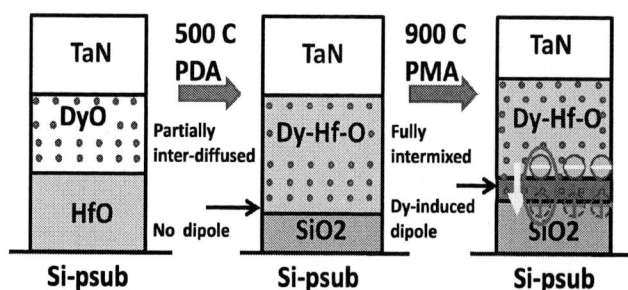

Figure 7 The temperature-dependent Dy diffusion and diffusion-driven Dy dipoles formation process is illustrated in this schematic. DyO layer and HfO layer was partially intermixed, but did not form Dy dipoles at the high-k/SiO$_2$ interfaces for 500 °C annealing. After 900 °C PMA, DyO layer and HfO layer was fully intermixed and Dy dipoles were formed at the high-k/SiO$_2$ interfaces, resulting in V$_{FB}$/V$_{th}$ shift.

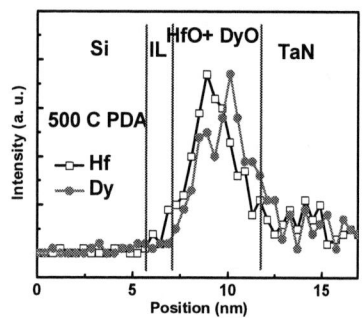

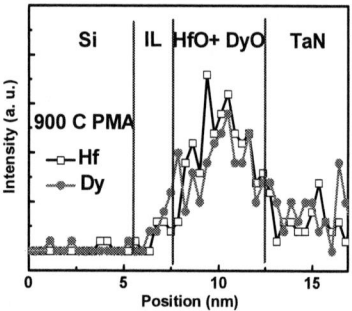

Figure 6 STEM EDX analysis of DyO/HfO system annealed at 500 °C and 900 °C. After 900 °C PMA, Dy atoms are clearly seen at the interfacial layer. This indicates the formation of Dy-silicate at the interfacial layer, which contributes to the Dy-induced dipole, resulting in the V$_{TH}$/V$_{FB}$ shift.

III. Electrical characteristics of DyO/HfO MOS devices

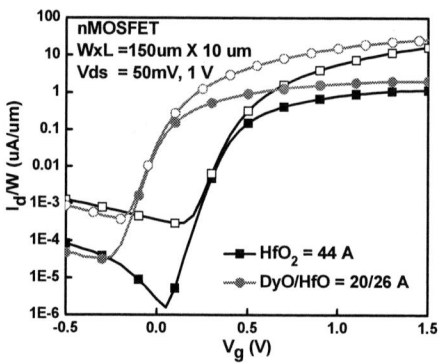

Figure 8 The comparable over-drive currents, subthreshold swing (~70 mV/dec) and lower V$_{TH}$ (~1 mV) were obtained from the DyO/HfO=20Å/26Åsample with EOT 1.38nm.

The temperature-dependent Dy diffusion and diffusion-driven Dy-induced dipoles formation process of the DyO/HfO bi-layer sample are illustrated in this schematic (Fig. 7). DyO layer and HfO layer were partially intermixed, but did not form Dy-silicate at the high-k/SiO$_2$ interfaces for 500 °C annealing. After 900 °C PMA, DyO layer and HfO layer were fully intermixed and Dy-silicate was formed at the high-k/SiO$_2$ interfaces. This implies that Dy-silicate related bonding is associated with the dipole formation, which modulates the V$_{FB}$/V$_{TH}$ shift of the Dy-Hf-O/SiO$_2$ system.

In Fig. 8, the comparable over-drive currents, subthreshold swing (~70 mV/dec) and lower V$_{TH}$ (~1 mV) were obtained from the DyO/HfO=20Å/26Å sample. EOT of this sample is around 1.38nm and the equivalent gate oxide thickness (Tox_gl) is around 1.72nm. The over-drive current, normalized by width and measured at 1V over-drive voltage, is improved.

978-1-4244-9113-1/11 $26.00 © 2011 IEEE   542

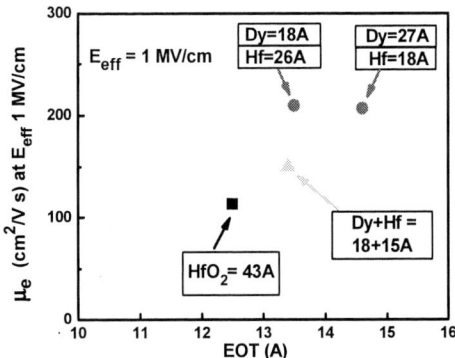

Figure 9 The electron mobility with respect to EOT was plotted to check the mobility degradation in terms of the Dy-induced dipole, where the electron mobility was chosen at 1 MV/cm effective field.

The electron mobility with respect to EOT was plotted to check the mobility degradation in terms of the Dy-induced dipole, where the electron mobility was chosen at 1 MV/cm effective field (Fig. 9). The effective electron mobility is enhanced by Dy incorporation. According to Ref. 6, the mobility degradation at low fields comes from the dipole layer formed at the HfO_2/SiO_2 interface rather than from remote phonon scattering. This dipole layer is also known to be responsible for the anomalous V_{FB} and V_{TH} shift in HfO_2 MOSFETs. However, based on the result, Dy-induced dipole does not seem to degrade the electron mobility similarly to the La-induced dipole case where mobility degradation was interpreted by the thinning of the interfacial layer (IL) rather than the La dipole [7]. This means that the Dy-induced dipole helps reduce the degradation of the effective electron mobility in HfO_2. Data suggest that we may keep scaling down the EOT without degrading mobility, while maintaining the lower leakage current and oxide wear-out by Dy incorporation. Further research is being conducted to check the mobility degradation in terms of the interfacial layer (IL) scaling down (or EOT scaling down).

IV. Stress-induced charge trapping characteristics

To examine the charge trapping characteristics, various stresses are applied to the HfO_2 and DyO/HfO sample having 1nm EOT. Stress-induced flatband voltage and stress-induced leakage current (SILC) are measured with the same electric field to do a fair comparison. According to the stress-induced flatband voltage data shown in Fig. 10(a), hole trapping in oxide are observed for both HfO_2 and DyO/HfO sample, since all C-V curves are shifted in a negative way when the negative voltage stresses applied on the gate. When the hole currents are tunneling through the oxide from the accumulated hole in the Si-substrate, some amounts of holes are trapped in oxide from the hole tunneling currents during the stress.

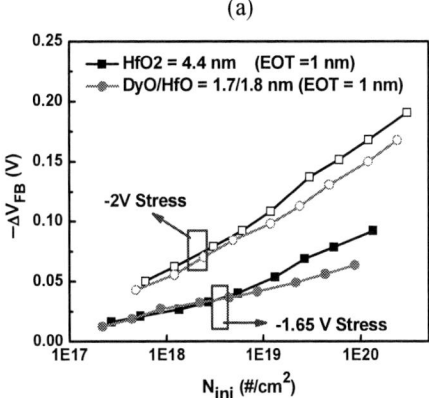

(a)

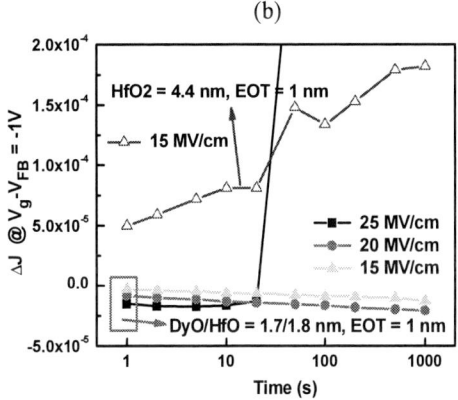

(b)

Figure 10 (a) Stress-induced flatband shift shows a hole trapping in oxide and (b) stress-induced leakage current (SILC) characteristics in HfO_2 and DyO/HfO show a different mode for the oxides wear-out and breakdown.

To compare the charge trapping, we plot the shifted V_{FB} verse the injected charge density, N_{inj}. Initially, at lower level charge injection, the charge trapping is not severe for both samples, but the trap generation rate increases differently at the high level injection. HfO_2 sample shows a higher trap generation rate than DyO/HfO. This decreased hole trapping can be explained by the reduced hole tunneling current and reduced hole trap density. The hole capture cross-section is also believed to be reduced by Dy incorporation, resulting in lower trap generation. This trap generation rate is dependent on the stress voltage, and is also related to the oxide wear-out.

978-1-4244-9113-1/11 $26.00 © 2011 IEEE

Stress-induced leakage current (SILC) characteristics in HfO_2 and DyO/HfO show a different mode for the oxides wear-out and breakdown (Fig. 10(b)). SILC in HfO_2 sample is dominated by the increase of the electron tunneling currents, leading to wear out the oxide gradually into a soft breakdown mode, which can be described with a band alignment as shown in Fig. 12.

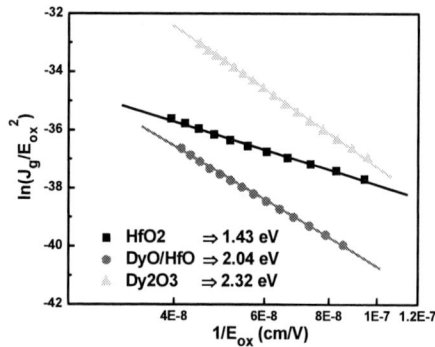

Figure 11. The effective potential barrier height is calculated for Dy_2O_3 and HfO_2 samples from the F-N plot to understand the reduction of the leakage current. The estimated effective barrier heights for Dy_2O_3 and HfO_2 are 2.32 eV and 1.43 eV, respectively. The extracted effective barrier height of the DyO/HfO=1.7/1.8 nm sample is 2.04 eV.

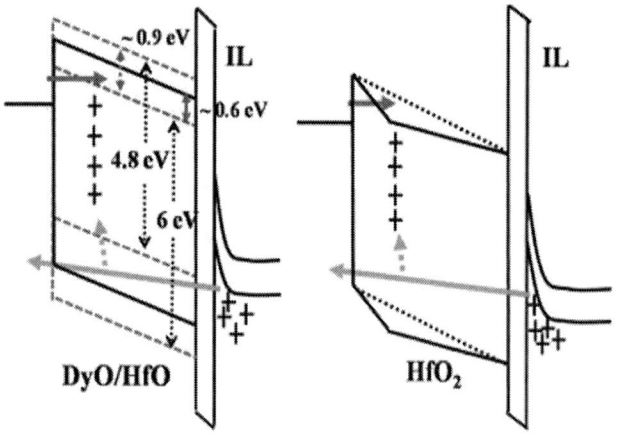

Figure 12 The mechanism of charge trapping for both samples can be explained with the band diagram in relation to the electron tunneling current and hole trapping.

The higher effective barrier height of Dy_2O_3, which is around 2.32 eV calculated from the F-N plot as shown in Fig. 11, accounts for the reduced leakage current in Dy incorporated HfO_2 nMOS devices. The lower barrier height of

HfO_2 characterizes the increasing electron tunneling currents enhanced by the buildup of hole charges trapped in oxide, which causes a severe increase of stress-induced leakage current (SILC), leading to oxide breakdown. However, the increased barrier height in Dy incorporated HfO_2 inhibits a further increase of the electron tunneling from the TaN gate, and trapped holes lessen the hole tunneling currents, resulting in a negligible SILC. The lower trap generation rate by the reduced hole trap density and the reduced hole tunneling of the Dy-doped HfO_2 dielectric demonstrate the high dielectric breakdown strength by weakening the charge trapping and defect generation during the stress.

V. Dielectric breakdown

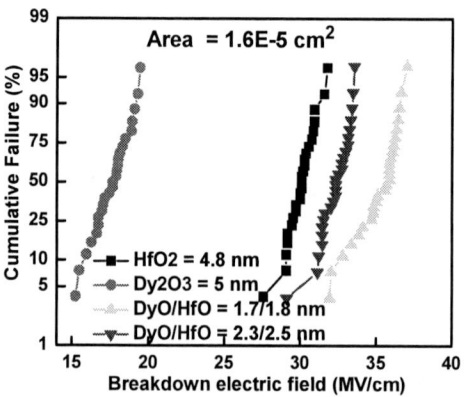

Figure 13 The time-zero dielectric breakdown (TZDB) and time-dependent dielectric breakdown (TDDB) with a various constant voltage stress were performed. Breakdown field distribution for Dy_2O_3, HfO_2 and Dy incorporated HfO_2 samples are compared .

In Fig. 13, we compare the breakdown field of the HfO_2, Dy_2O_3 and Dy incorporated HfO_2 samples. Even though HfO_2 and Dy_2O_3 samples have the lower breakdown field, the increased breakdown field is attained for DyO/HfO = 1.7/1.8 nm and DyO/HfO = 2.3/2.5 nm samples. Defects created during the fabrication processes might be one of the reasons for the lower breakdown field of Dy_2O_3 and HfO_2 samples. The thinner DyO/HfO sample has a highest breakdown field, which implies that defects can be controlled by the oxide physical thicknesses. The lessened trap generation rate and charge trapping are responsible for reduced leakage current in Dy-doped HfO_2 dielectric, and result in the high dielectric breakdown strength. The lower barrier height of HfO_2 characterizes the electron tunneling currents enhanced by the

978-1-4244-9113-1/11 $26.00 © 2011 IEEE

buildup of hole charges trapped in oxide, which contribute to the increase of SILC, leading to a severe increase of leakage current up to an oxide breakdown, distinct from a hard breakdown mode.

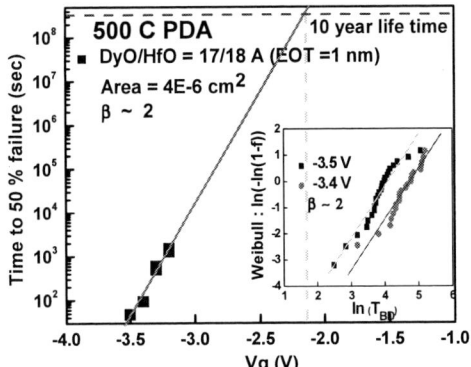

Figure 14 For the DyO/HfO sample with EOT 1 nm, the Weibull slope is 2. A -2.1 V operating voltage is possible for 10 year lifetime.

Breakdown voltage for DyO/HfO=1.7/1.8 nm having 1nm EOT is 4.1 V, the Weibull slope is 2 and -2.1 V operating voltage is achievable for a 10-year lifetime (Fig. 14). Considering the fact that the interfacial layer limits the dielectric reliability and results in low Weibull slopes for the gate injection case [8-9], the interfacial layer created by incorporation of Dy into the HfO_2 might determine the Weibull slope. Therefore, the improved charge trapping and SILC characteristics by the increased effective barrier height of DyO/HfO sample enable a longer lifetime of the dielectric at the same operating voltage.

Conclusion

We investigated the mechanism of V_{FB}/V_{TH} shift in terms of temperature-dependent Dy diffusion and diffusion-driven Dy dipole formation process in Dy-incorporated HfO_2. Improved charge trapping characteristics and oxide reliability was discussed with an explanation of the band diagram

ACKNOWLEDGMENT

This work was partially supported by Micron Foundation and DARPA.

REFERENCES

[1] H. Y. Yu, Chang, S.Z. A. Veloso, A. Lauwers, C. Adelmann, B. Onsia, S. Van Elshocht, R. Singanamalla, M. Demand, R. Vos, T. Kauerauf, S. Brus, X. Shi, S. Kubicek, C. Vrancken, R. Mitsuhashi, P. Lehnen, J. Kittl, M. Niwa, K. M. Yin, T. Hoffmann, S. Degendt, M. Jurczak, P. Absil, S. Biesemans, "Low Vt Ni-FUSI CMOS Technology using a DyO cap layer with either single or dual Ni-phases," *Tech. Dig. VLSI Symp.*, pp 18-19, 2007.

[2] T. Lee, S. Rhee, C. Y. Kang, F. Zhu, C. Choi, I. Ok, M. Zhang, S. Krishnan, G. Thareja, J. C. Lee, "Strutural advantage for the EOT scaling and improved electron channel mobility by incorporating dysprosium oxide (Dy_2O_3) into HfO_2 n-MOSFETs, " *IEEE Electron Device Letter*, vol. 27, pp. 640-643, 2006.

[3] K. Choi, H-C Wen, G. Bersuker, R. Harris and B. Lee, "Mechanism of flatband voltage roll-off studied with Al_2O_3 film deposited on terraced oxide," *Applied Physics Letters*, 93, 133506, 2008.

[4] K. Kita and A. Toriumi, "Origin of electric dipoles formed at the high-k/SiO_2 interface," *Applied Physics Letters*, 94, 132902, 2009.

[5] G. He, M. Liu, L.Q. Zhu, M. Chang, Q. Fang and L.D. Zhang, "Effect of post-deposition annealing on the thermal stability and structural characteristics of sputtered HfO_2 films on Si (1 0 0)," *Surface Science*, vol. 576, pp. 67-75, 2005.

[6] H. Ota, A. Hirano, Y. Watanabe, N. Yasuda, K. Iwamoto, K. Akiyama, K. Okada, S. Migita, T. Nabatame and A. Toriumi, "Intrinsic Origin of Electron Mobility Reduction in High-k MOSFETs - From Remote Phonon to Bottom Interface Dipole Scattering," *IEDM tech. Dig.*, pp. 65-68, 2007.

[7] T. Ando, M. Copel, J. Bruley, M. M. Frank, H. Wantanabe and V. Narayanan,"Physical origins of mobility degradation in extremely scaled siO_2/HfO_2 gate stacks with La and Al induced dipoles," *Applied Physics Letters*, 96, 132904, 2010.

[8] A. S. Oates, "Reliability Issues for High-K Dielectrics," *IEDM Tech. Dig.*, pp. 923-926, 2003.

[9] T. Kauerauf, R. Degraeve, E. Cartier, C. Soens, G. Groeseneken, "Low Weibull Slope of Breakdown Distributions in High-k Layers," *IEEE Electron Device Letters*, vol. 23, no. 4, pp. 215-217, April 2002.

Physics of Instability of Thin Film Si and (Si,Ge) Alloy Solar cells

Authors Name: Vikram L. Dalal
Dept. of Electrical and computer Engineering
Iowa State University
Ames, Iowa, USA
Phone: (515)294-1077, e-mail: vdalal@iastate.edu

Second Author: Zhao Li
Dept. of Electrical and Computer Engineering
Iowa State University
Ames, Iowa, USA
Phone: (515)294-7732

Abstract— The physics of instability of thin film Si solar cells is strongly dependent upon the nature of the Si, whether it is amorphous or nano(also called micro)crystalline. The amorphous phase is much more unstable than the nanocrystalline phase. The instability of the amorphous Si , which is really an alloy of Si and H, and H, is primarily due to the poor microstructure of the material. The amorphous material is not homogeneous, but is rather composed of an inhomogeneous mixture of randomly distributed Si-H bonds, and localized voids which are full of Si-H double bonds, or H atoms in close proximity to each other. The instability depends critically upon the presence of these voids and clustered H. In contrast, nanocrystalline Si, which also has H at the grain boundaries, is much more stable. However, some structural forms of nanocrystalline Si are unstable because of the penetration of impurities such as oxygen into the structure, or poorly bonded Si-H bonds at large-grain grain boundaries. The techniques for making both amorphous and nanocrystalline Si based solar cell materials more stable are discussed in this paper. In general, the alloys such as amorphous (Si,Ge) are more unstable because their structure is more complicated and more inhomogeneous than that of Si.

Keywords: Silicon, Silicon-Germanium, Solar Cells, Instability, Nanocrystalline Silicon, Amorphous Silicon

I. INTRODUCTION

Thin film Si solar cells are an important technology for low-cost solar energy conversion into electricity. Currently, about 10% of the total production of solar cells is based on thin film Si. Thin film Si materials used for solar energy conversion can be divided into two groups: Amorphous Si (a-Si:H) and nano or micro-crystalline Si (nc-Si:H). There are fundamental differences between the two forms of Si. The a-Si:H material is really an alloy of Si and H. The amorphous structure of the material creates broken Si-Si bonds, and H is used to form bonds with the unbonded electron in Si created by the broken bond. Since Si:H is a covalent bond, the presence of such bonds significantly reduces the defect density which would otherwise be present in such amorphous materials. In good materials, the defect density reduces from almost $10^{20}/cm^3$ to

$10^{15}/cm^3$[1]. Unfortunately, a-Si:H is not a homogenous material, where there are few broken Si-Si bonds which are satisfied by a Si-H bond. It also has a significant internal holes or voids [2]. These voids are created during growth, and their internal surfaces are coated with H. Since the void dimensions are small(of the order of a nm or less), the H bonded there is in close proximity to each other. In addition, there may also be Si-H_2 type bonds at the void surfaces.

In contrast, nc-Si:H consists of small grains of Si which are coated with H [3]. The grain size is very small, of the order of 15-20 nm and H is present mainly at the grain boundaries. The presence of H at the grain boundaries passivates the recombination centers at the grain boundaries, thereby allowing for remarkably large minority carrier lifetimes, of the order of a microsecond [4]. The electron and hole mobilities are also reasonable, of the order of a few cm^2/V-s. The combination of reasonable mobilities and lifetimes leads to diffusion lengths which are of the order of several micrometers, in contrast to the case for a-Si:H, where the minority carrier (hole) diffusion length is less than a micrometer. Thus, this material IS fundamentally different from a-Si:H.

In addition to just Si, one can also make alloys such as a-(Si,Ge):H and nc-(Si,Ge):H. Adding Ge to Si reduces the bandgap of the materials, and therefore, such alloys can be very useful for making tandem or multiple-junction solar cells [5].

Note that the very different properties of amorphous and nanocrystalline materials leads to a very different device structure that is needed to make an efficient solar cell. In a-Si:H, because of the very small diffusion length, one needs to have a p-i-n structure, so that the electric field of the p-n junction extends across the entire width of the intrinsic i layer. Then the holes are transported with a field-assist, where the determining parameter for quantum efficiency is range R, given by $R = \mu\tau F$, where μ is the mobility of holes, and τ is the lifetime and F is the electric field. For efficient collection, R>>thickness of the i layer. This forces the i layer to be very thin (0.3-0.4 micrometer) , so that the field F can be large.

978-1-4244-9113-1/11 $26.00 © 2011 IEEE

In contrast, in nanocrystalline Si, the transport is mainly by diffusion and not drift. Thus, the structure is the usual crystalline-type p+nn+ structure, where the n layer may have a doping of a few $10^{14}/cm^3$. Since the diffusion length is of the order of a few micrometer, one can tolerate thicker n layers, about 2 micrometer.

II. PHYSICS OF INSTABILITY IN AMORPHOUS MATERIALS AND SOLAR CELLS

The voids are created in a-Si:H, and in alloys such as a-(Si,Ge):H during the growth of the material. The amterial is usually deposited using plasma deposition (PECVD) using silane and hydrogen mixtures. The plasma decomposes silane into its components, such as SiH_3, SiH_2 and SiH. While it is generally believed that SiH_3 is the primary radical being generated in the plasma, there are always some other radicals also present. Consider the following scenario[6]. Once a SiH_3 radical inserts itself into an open Si bond, it cannot accept any more radicals, since all its bonding electrons (4) are bonded. In contrast, if a SiH_2 radical were to insert and form a layer, its top surface would be open to bonding with another radical. Therefore, it is very likely that a peak and valley type structure arises during growth. The valleys are what forms voids. Since H is bonded at the internal surface of voids, and many of these bonds are likely to be distorted, this bonding energy is significantly lower than the bonding energy for isolated Si-H bonds. Therefore, during an excitation by light, electrons and holes created near this void, when they recombine, release an energy of ~1-1.5 eV. This energy is enough to break the weak bonds. Numerical simulations show that this indeed the case, namely that materials with H atoms which are close to each other are much more unstable than materials which have a smaller density of clustered Si-H bonds [7]. When the bonds break, two defect states are created, leading to excess recombination. This is the primary mechanism for the well known Staebler-Wronski effect in a-Si:H[8], though other factors such as impurities can also play a role[9].

The additional defects have a deleterious effect on the performance of a-Si:H solar cells. First, they lead to a shrinkage of the depletion layer in the intrinsic i layer of the p-i-n device, since the defects also need to change their charge state by the junction field. Secondly, they reduce the minority carrier lifetime, and therefore, the range R. The reduction in range R is catastrophic for the collection of photo-generated holes, particularly when the solar cell is under normal operating conditions, i.e. under forward bias. Under such conditions, the voltage across the junction is reduced, and a reduction in R due to excessive recombination leads to a significant decrease in current at the operating point, i.e. to a loss in fill factor of the cell. The increased recombination also leads to a loss in open-circuit voltage, since the reverse saturation current increases. Typically , a single junction a-Si:H solar cell suffers about a 20% degradation in performance upon exposure to light.

This effect is reversible, because the multi-atom bonds formed by unbonded H break at higher temperatures (>150 C) , freeing up H atoms which can then rebond with Si atoms at the voids.

Whatever is true for a-Si:H is even more true for a-(Si,Ge):H. That is because a-(Si,Ge):H is deposited from two materials, germane and silane. They have very different molecular weights, and therefore, very different radical mobilities on the surface. Germane radicals such as $GeH3$ are much less likely to move on the surface during deposition and find an empty bond, than silane radicals such as SiH_3 [6]. Therefore, such a material should have more clustering of elements, and more voids. This is indeed what is observed experimentally [2]. Thus, it is not a surprise that a-(Si,Ge) solar cells are far more unstable than a-Si:H.

III. REDUCING THE INSTABILITY OF AMORPHOUS SI SOLAR CELLS

From the above discussion of the physics of instability, it is clear that to reduce the instability, one must reduce the deleterious void density. A signature of the void density is the infrared signature of SiH_2 bond, or a Si bond which is in close proximity to significant H. This infrared absorption signature is at ~2100 cm^{-1}, in contrast to the isolated Si-H bond whose IR signature is at 2000 cm^{-1}.

There are two ways to reduce the void density. One is to do a very slow growth in a high hydrogen diluted plasma. The high dilution with hydrogen assures that all Si surface bonds are saturated with H. That means that the surface is uniform, independently of which radical formed the localized Si layer. H ions in the plasma will remove surface H, since H-H bond is stronger than Si-H bond. A second necessary condition is that the SiH3 radical must be able to diffuse to an open bond(statistically, some bonds will be open on the surface) before a second SiH3 radical attaches itself to the first one. This requires conditions (such as low power or a triode geometry) which lead to a slow growth. Work done by Sonobe et al in Japan followed this reasoning, and achieved improved significantly better stability in a-Si solar cells [10].

An alternative approach is to subject a thin film of a-Si (about 1-3 nm) to ion bombardment after growth. This is called chemical annealing The concept is based on pioneering work by Hirose's group in Japan which showed that H ion bombardment immediately following the growth of a thin layer of a-Si restructured the a-Si film, making it more ordered [11]. Our group used this approach, subjecting a thin film of a-Si, to ion bombardment using either He or H, and using repeated cycles of growth/anneal to grow the desired final thickness of the film or the device[12]. The technique is schematically shown in Fig. 1, which shows how chemical annealing is done, by switching silane flow on and off in a cyclical fashion. As expected, we obtained a significant reduction in clustered H bonds by chemical annealing, as shown by the infrared absorption spectrum in Fig. 2. The film prepared using chemical annealing has a significantly lower ratio of SiH2/SiH bonds than the film which was not subjected to chemical annealing.

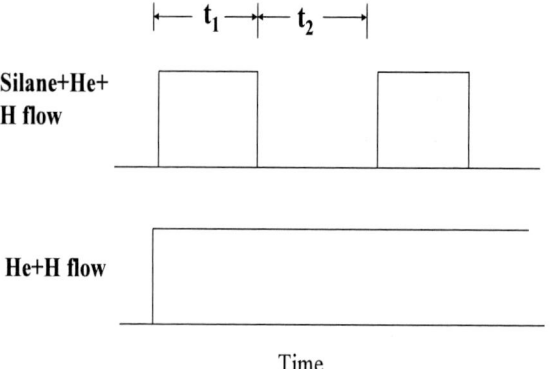

Fig. 1 Schematic diagram of chemical annealing technique

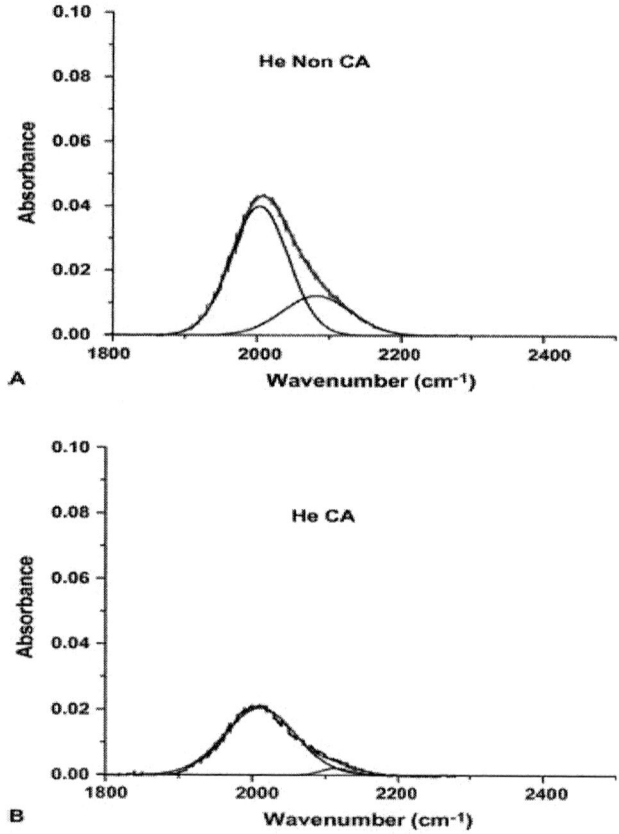

Fig. 2 IR spectra of normal and chemically annealed (CA) films [12]

We then proceeded to make standard p-i-n devices on stainless steel substrates without back reflector. The devices were coated with a top ITO contact. Two sets of devices were made. One was a standard device, and the other, a device with chemical annealing. Both devices had the same thickness of the intrinsic layer, and both techniques produced devices which were nearly identical in all respects. Typical I-V curves for

both devices are shown in Fig. 3 [12], showing very similar I-V data. Then the devices were tested for stability, using a 2xsun intensity. Since fill factor is the solar cell parameter that changes the most, in Fig. 4, we show the degradation in fill factor as a function of time. Note that the solar cell prepared using chemical annealing had a significantly lower degradation in fill factor (~4-5%) than the degradation of fill factor in the cell prepared using continuous growth (~20%). Also, the cell with chemical annealing seems to saturate relatively quickly, whereas the other cell continues to degrade.

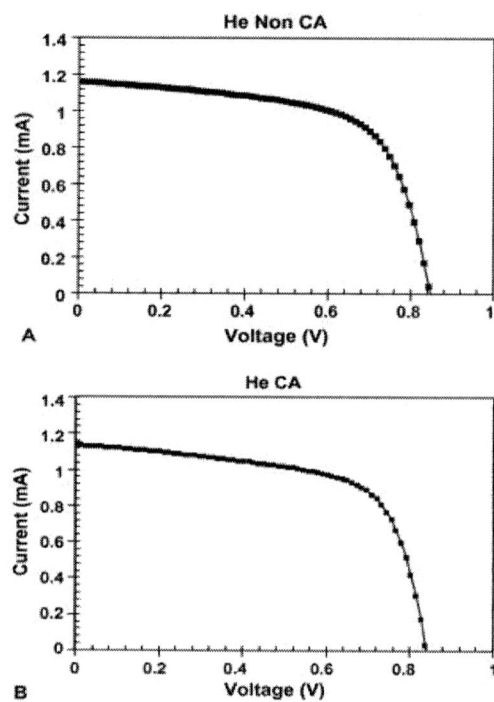

Fig. 3 I(V) curve of two nearly identical cells[12]

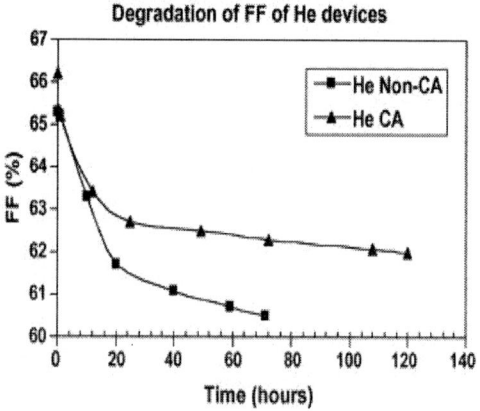

Fig. 4. Degardation in fill factor of chemically annealed vs. non-annealed cells [12]

IV. REDUCING THE ISNTABILITY OF AMORPHOUS SILICON GERMANIUM SOLAR CELLS

We next proceeded to pursue chemical annealing to improve the properties and stability of a-(Si,Ge):H solar cells. As indicated earlier, amorphous (Si,Ge) suffers from many more voids, and a non-homogeneous structure, and as a result, much worse electronic properties and stability than a-Si:H. A systematic set of experiments was carried out to grow a-(Si,Ge) films and devices with varying Si:Ge ratios. Films and devices were made using significant hydrogen dilution for both chemically annealed samples, and ones grown without chemical annealing, so as to obtain meaningful comparisons between the two sets of films and devices. All films were grown on a Corning 7059 glass substrate, and devices on polished stainless steel. All devices were of the p-i-n type, with light incident on the p layer through an ITO contact. In the devices, we measured mobility-lifetime ($\mu\tau$) products of holes for various chemical annealing times. One expects that as the time for annealing part of the growth/anneal cycle increases, the structure will become more homogeneous, and the fundamental material properties, such as the mobility-lifetime products will improve. The $\mu\tau$ products were estimated using the technique of measuring quantum efficiency vs. voltage [13].

Fig. 5 shows the influence of varying chemical annealing time on the $\mu\tau$ product for holes. The figure clearly shows that for all Tauc bandgaps considered in this study, the hole $\mu\tau$ product improves with increasing time for chemical annealing. The improvement is more for lower gap materials, i.e. materials with a higher Ge content, because such materials are likely to have a higher void density [2].

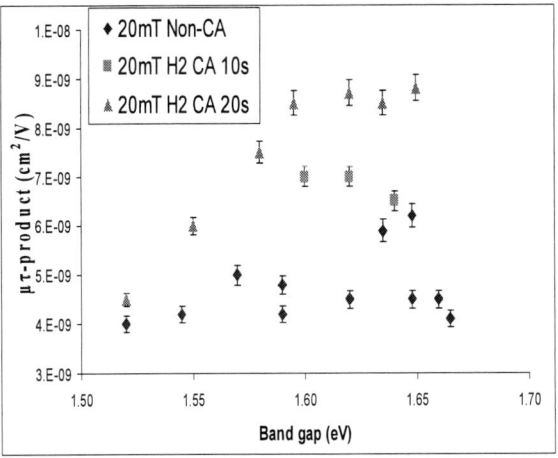

Fig. 5 Influence of chemical annealing at various times on the hole mobility-lifetime product of a-(Si,Ge) devices with varying Tauc gaps

Next, we studied the light-induced degradation of a-(Si,Ge) cells subjected to chemical annealing(CA) and compared this degradation with the degradation of a cell which was made without chemical annealing(Non-CA). The I-V curve for the cell is shown in Fig. 6, showing a good fill factor (0.65). In Fig. 7, we show the results of the study on the degradation of fill factor of the two cells when they were subjected to 2xsun intensity light. The figure clearly shows that the degradation due to light soaking is much less in the chemically annealed cell compared to the degradation for a standard, continuously grown cell.

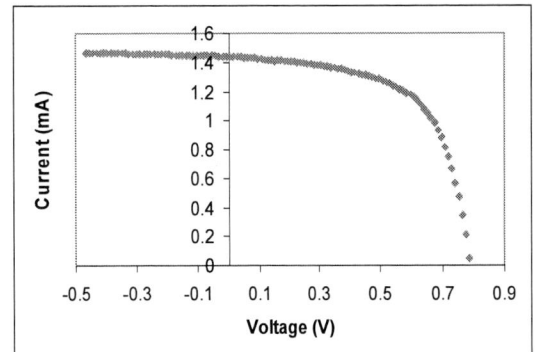

Fig. 6 I(V) curve of an a-(Si,Ge) cell prepared using chemical annealing

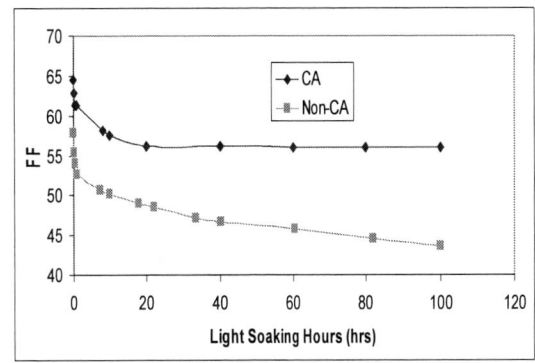

Fig. 7 Comparison of degradation in fill factors of a-(Si,Ge) cells prepared using CA and non-CA.

The above experiments very clearly validate our fundamental physical model, namely that the Staebler-Wronski effect is primarily caused by structural and H bonding inhomogeneities in the material. This understanding should be useful in further improvements in the material properties.

V. STABILITY OF NANOCRYSTALLINE SILICON SOLAR CELLS

The stability of thin nanocrystalline Si solar cells is a much more complicated subject than that of amrophopus Si cells. This reasoning arises from the fact that there is no "one" standard nanocrystalline Si. Nanocrystalline Si is fundamentally a small grained Si, with a thin amorphous Si film coating the surface of the grain. The ratio of amorphous to crystalline phases is strongly dependent on the degree of hydrogen dilution, power used during deposition, and degree of ion bombardment, and also varies with thickness[3,14]. Thicker films become highly crystalline, with a higher ratio of crystalline to amorphous phase, as indicated by the Raman signature of the lattice, with a cauliflower type structure and large grain boundaries[15]. Such boundaries are not usually well passivated with H. As a result, such very highly crystalline films have poorer solar cell properties[16], and surprisingly, degrade more under illumination. In contrast, films which are less crystalline, with only the smaller grains and the absence of the cauliflower type structure, are well passivated with H and tend to give rise to solar cells with better properties.

The stability depends critically upon the structure of these films. The films with large grain boundaries tend to degrade more, whereas the films with a higher amorphous/crystalline ratio, particularly near the critical p-n interface in a p-n junction device, tend to degrade much less[16]. There may be two reasons for this result. First, preventing a cauliflower type structure with large grain boundaries leads to excessive diffusion of impurities such as oxygen into the material. Oxygen at the grain boundary can act as a donor state, leading to a trap which is filed with electrons, and therefore, an effective recombination center for holes. it is known that having significant oxygen present in the nanocrystalline material during growth severely degrades the properties of the solar cells[17]. Preventing the development of the cauliflower structure prevents oxygen ingress at the grain boundaries. Having significant amount of a-Si phase present at the p-n interface also prevents the diffusion of oxygen, since the a-Si structure is pretty dense with no open grain boundaries. A second reason why having a significant a-Si presence at the interface is beneficial ahs to do with the fact that there is always interfacial recombination at the p-n interface in both amorphous and nanocrystalline Si solar cells. Amorphous Si passivates these interfacial states, and therefore reduces recombination. One can even design a graded bandgap profile at the interface (see Fig. 8) to reduce interfacial recombination[]. Such structures can be easily made using an a-Si layer at the interface.

This area is still subject of intensive R&D. Suffice it to say that most device groups do not find a significant degradation in nanocrystalline Si solar cells when they are designed right[ ]. Typical degradation is in 2-3% range.

VI. DISCUSSION AND CONCLUSIONS

We have shown how the basic structures, and localized H bonding environments, in amorphous materials can significantly impact the electronic properties and stability of solar cells made from these materials. In general, making more homogeneous materials with fewer voids, whose signature can be a SiH_2 type bond in infrared spectra, can lead to significant improvements in the stability of solar cells, in both a-Si:H and a-(Si,Ge):H. The case for understanding the stability of nanocrystalline solar cells is more complicated. We still do not know exactly what causes the instability in this material. What is known is that controlling the ratio of amorphous to crystalline pahses during growth, and having a material with a greater degree of amorphous phase at the p-n junction interface, leads to a more stable, and better, solar cell.

ACKNOWLEDGEMENTS

This work was partially supported by grants from NSF, Iowa Energy Center and Iowa Power Fund. We thank Max Noack for his experimental contributions.

REFERENCES

1. R. A. Street, "Hydrogenated Amorphous Silicon", Cambridge University Press (1991)

2. D. L. Williamson , "Microstructure of amorphous and microcrystalline Si and SiGe alloys using X-rays and neutrons", Solar Energy Materials and Solar Cells, 78, 41(2003)

3. A. V. Shah, J. Meier, E. Vallat-Sauvain, N. Wyrsch, U. Kroll, C.Droz and U. Graf, "Material and solar cell research in microcrystalline silicon" Solar Energy Materials and Solar cells, 78, 469 (2003)

4. S. Saripalli and V. L. Dalal , "Transport properties of nanocrystalline Si", to be published in J. Non-Cryst. Solids (2008)

5. V. L. Dalal and E. Fagen, " Design principles for monolithic, multiple gap, a-(Si,Ge) solar cells", Proc. of 14th. IEEE Photovolt. Spec. Conf.,1066(1980)

6. V. L. Dalal, "Fundamental Considerations governing the growth of a-Si and a-(Si,Ge)"(Invited paper), Current Opinions in Solid State Materials, 6, 455 (2002)

7. R. Biswas and B. C. Pan, "Mechanisms of metastability in hydrogenated amorphous silicon" , Solar Energy Materials and Solar cells, 78, 447 (2003)

8. D. Staebler and C. Wronski, "Reversible photo-conductivity in amorphous Si", Appl. Phys. Lett., 31, 292 (1977)

9. V. L. Dalal, P. Sharma and A. Ahmed, "Evidence for trap controlled instability in a-Si", Proc. Of MRS 762,33(2003)

10. H. Sonobe , A. Sato , T. Fujibayashi , A. Matsuda, M. Kondo, "Iproving stabilityof a-Si soalr cells by triode deposition", Proc. Mater. Res. Soc., 862,551(2005)

11. S. Miyozaki, N. Fukuhara and M. Hirose, J. Non-Cryst. Solids, 266-269 (2000), p. 54.

12. Nanlin Wang and Vikram Dalal "Improving stability of amorphous Si using chemical annealing with helium", J. Non-Cryst. Solids, 352, 1937(2006)

13. V. L. Dalal, M. Leonard, J. F. Booker and A. Vaseashta, "Quantum efficiency of a-Si solar cells", Proc. of 18th. IEEE Photovolt. Spec. Conf., 847(1985)

14. Vikram L. Dalal, J. Graves and J. Leib, " Influence of pressure and ion bombardment on the growth and properties of nanocrystalline Si materials", Appl. Phys. Lett., 85, 1413 (2004)

15. J. Kocka, A. Fejfar, H. Stuchlíková, J. Stuchlík, P. Fojtík, T. Mates, B. Rezek, K. Luterová, V. Svrcek and I. Pelant, "Basic features of transport in microcrystalline silicon", Solar Energy Materials and Solar cells, 78, 493(2003)

16. G. Yue, B. Yan, G. Ganguly, J. Yang and S. Guha, "Material structure and metastability of hdyrogenated nanocrystalline Si soalr cells",Appl. Phys. Lett., 88,263507 (2006)

17. P. G. Hugger, J. David Cohen, Baojie Yan, Guozhen Yue, Jeffrey Yang and Subhendu Guha, "Relationship of deep defects to oxygen and hydrogen in nanocrystalline Si", Appl. Phys. Lett., 97,252103 (2010)

Reliability Testing beyond Qualification as a Key Component in Photovoltaic's Progress toward Grid Parity

John H. Wohlgemuth and Sarah Kurtz

National Renewable Energy Laboratory
Golden, Colorado
phone: (1) –(303)- 384-7982, john.wohlgemuth@nrel.gov

Abstract— **This paper discusses why it is necessary for new lower cost PV modules to be tested using a reliability test sequence that goes beyond the Qualification test sequence now utilized for modules. Today most PV modules are warranted for 25 years, but the Qualification Test Sequence does not test for 25-year life. There is no accepted test protocol to validate a 25-year lifetime. This paper recommends the use of long term accelerated testing to compare now designs directly with older designs that have achieved long lifetimes in outdoor exposure. If the new designs do as well or better than the older ones, then it is likely that they will survive an equivalent length of time in the field.**

Keywords-Photovoltaic module reliability, Reliability Testing, Qualification Testing, Levelized Cost of Electricity (LCOE), Grid Parity

I. INTRODUCTION

The photovoltaics market has been growing rapidly. Fig. 1 shows the annual growth rate in worldwide shipments of PV modules from 2004 to 2009 as reported by Navigant [1] and EPIA [2]. Over this timeframe (which includes the financial crisis of 2008) worldwide PV module shipments grew by an average of 50% per year. In late 2010 both SolarBuzz [3] and Renewable Energy World [4] are predicting that the total shipments for 2010 will be approximately 16 GW, which represents more than 100% growth over the 2009 volume.

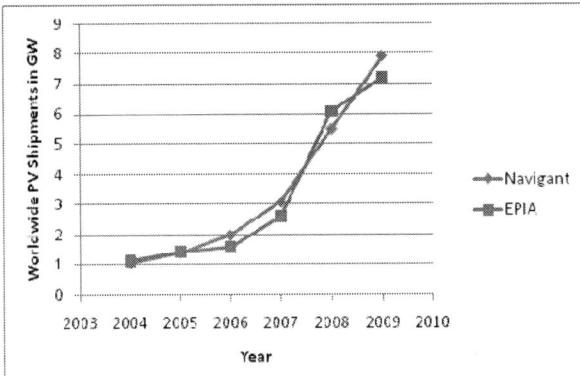

Figure 1: Annual Worldwide Shipments of PV Modules

Since the end of the poly-Si shortage in 2007-2008, the selling price of PV modules has decreased appreciably. SolarBuzz [3] estimated that the retail price of PV modules for residential applications dropped by 27% from late 2007 to late 2010. SEIA and a Lawrence Berkeley 2010 study of the PV industry gave similar estimates for PV module price reductions over this time period. So progress is being made in reducing the cost of PV modules.

Whether you look at the SolarBuzz module retail selling price of $3.50/Watt peak or at the SEIA average module selling price for large systems of $2.21/Wp, the cost of PV generated electricity is still higher than the utility costs in most parts of the US and the world. Using the System Advisor Model (SAM) [5] with a retail module cost of $3.50/Wp from SolarBuzz, an overall systems cost of $7.00/Wp and a module lifetime of 30 years installed in Phoenix, AZ yields a Levelized Cost of Electricity (LCOE) of $0.17/kWh with the 30% tax credit and $0.24/kWh without the tax credit. Neither of these costs can match the local utilities' residential electric rate. EIA reports that average retail electricity prices have been running about $0.12/kWh. [6] The rapid growth in PV shipments is being driven by incentive programs, particularly the feed-in-tariff programs in Germany and other European countries.

Lower module prices and lower system costs are required in order to reach grid parity and therefore stimulate dramatic growth in PV shipments without the need for incentive programs. Lower module manufacturing costs and selling prices must be achieved without adversely affecting the PV module reliability both in terms of overall module lifetime and in terms of continual performance degradation. Increases in the annual degradation rate will have a negative impact on the LCOE while an increase in premature module failures can potentially damage PV's reputation as a reliability electrical source.

So how can PV module manufacturers determine that the changes they make in design, materials and/or processes do not adversely affect the module reliability? They cannot wait 20 or 25 years to see what happens to the lower cost modules when they are deployed in the field. So they must utilize accelerated reliability tests to evaluate the potential for these lower cost modules to survive without increased degradation of output power. However there are no accelerated stress tests that we know of today that can show whether a module type will survive 25 years anywhere it is deployed.

This paper proposes a process that utilizes the past experience of a module type using both field and accelerated stress test results as a baseline for testing of a new module type. The new module type can then be evaluated against the older design using accelerated stress tests that go beyond those utilized in the Qualification Test Sequence. Comparison of the performance between the old and the new module types can give an excellent indication of how the new design is likely to perform outdoors compared with the previous design.

II. QUALIFICATION TESTING

Qualification tests are a set of well-defined accelerated stress tests developed out of a reliability testing program. Qualification tests incorporate strict pass/fail criteria. Such tests are used by customers to qualify modules for purchase and by manufacturers as a means of demonstrating a degree of product reliability. Good examples of these tests are IEC 61215 [7] and IEC 61646 [8] for performance qualification and IEC 61730 – 1 and 2 [9] for safety qualification. These test sequences were developed based on the identification of field failure mechanisms.

A. Failure Modes

Identification of field failure modes has been ongoing since the JPL Block Procurement program in the 1970's and 1980's. [10] A list of major failure mechanisms for crystalline silicon modules is given in Table 1. [11, 12]

TABLE 1
Common Failure Modes for Crystalline Silicon Modules

Failure Modes
Broken Interconnects
Broken Cells
Corrosion
Delamination of Encapsulant
Encapsulant loss of Elasticity or Adhesion
Encapsulant Discoloration
Solder Bond Failure
Broken Glass
Hot Spots
Ground Faults
Junction Box Failures
Connection Failures
Structural Failures
Bypass Diode Failures
Open circuits leading to arcing

Several additional failure modes have been identified for thin film PV modules. These are given in Table 2.

TABLE 2
Common Failure Modes for Thin Film Modules

Technology	Failure Mode
Glass Superstrate Designs	Electrochemical corrosion of TCO
Integrated Modules	Shunts at the scribe lines
Any thin film	Loss of Interlayer Adhesion
Thin Films on Glass	Inadequate edge deletion

The next important step was the identification of accelerated stress tests that would duplicate these failure modes in a reasonable short amount of time. The initial steps in this work were undertaken by JPL in the Block Program

[10, 13]. Table 3 provides a brief summary of the stress tests developed to address the identified failure modes. [14] The list in Table 3 was utilized to establish the initial qualification tests that grew into IEC 61215. [14]

TABLE 3
Accelerated Stress Tests for PV

Accelerated Stress	Failure Mode
Thermal Cycle	Broken Interconnect
	Broken Cell
	Solder Bond Failures
	Junction Box Adhesion
	Module Connection Open Circuits
	Open Circuits leading to Arcing
Damp Heat Exposure	Corrosion
	Delamination of Encapsulant
	Encapsulant loss of adhesion & elasticity
	Junction Box Adhesion
	Electrochemical corrosion of TCO
	Inadequate edge deletion
Humidity Freeze	Delamination of Encapsulant
	Junction Box Adhesion
	Inadequate edge deletion
UV Test	Delamination of Encapsulant
	Encapsulant loss of adhesion & elasticity
	Encapsulant Discoloration
	Ground Fault due to backsheet degradation
Mechanical Load	Broken Interconnect
	Broken Cell
	Solder Bond Failures
	Broken Glass
	Structural Failures
Dry and Wet Insulation Resistance	Delamination of Encapsulant
	Ground Faults
	Electrochemical corrosion of TCO
	Inadequate edge deletion
Hot Spot Test	Hot Spots
	Shunts at the scribe lines
Hail Test	Broken Cells
	Broken Glass
Bypass Diode Thermal Test	Bypass Diode Failures

B. IEC 61215 and IEC 61646 Qualification Tests

IEC 61215 and IEC 61646 include the following stress tests:

- 200 Thermal cycles from -40°C to +85°C with peak power current flow above room temperature.
- Damp heat exposure at 85°C and 85% relative humidity for 1000 hours.
- A combined leg of UV Preconditioning (15 kwhm^{-2}), 50 thermal cycles from -40°C to +85°C, and 10 humidity freeze cycles from +85°C, 85 % RH to -40°C.
- Wet leakage current test at the rated system voltage.
- Mechanical load test of 3 cycles of 2,400 Pa uniform load, applied for 1 hour to front and back surfaces in turn.
- Hail test with 25 mm diameter ice ball at 23 m☐s -1, directed at 11 impact locations.
- A bypass diode thermal test, with one hour at short circuit current and 75 °C and one hour at 1.25 times short circuit current and 75 °C.

- Hot spot test with 3 lowest shunt resistance cells subjected to 1 hour exposure to 1000 $W \cdot m^{-2}$ irradiance in worst-case hot-spot condition and highest shunt cell subjected to 5 hours exposure to 1000 $W \cdot m^{-2}$ irradiance in worst-case hot-spot condition

While Qualification Tests like IEC 61215 and IEC 61646 are important and valuable, they have limitations because the stress levels are by design limited and the goal is to have most commercially available products capable of passing the test sequence. So passing the qualification test means the product has met a specific set of requirements but doesn't say anything about which product is better for long term performance. Most of today's commercial modules pass the qualification sequence with minimum change, meaning that they suffer almost no degradation in power output from the test sequence. This means that the Qualification test itself is not a good tool for determining whether a change in materials, processing or design is likely to reduce the module's lifetime or increase the annual degradation rate. However, the Qualification Test sequence can be utilized as a starting point for developing a methodology to evaluate the impact of new lower cost designs, materials and processes on the modules long term reliability.

III. RELIABILITY TESTING BEYOND QUALIFICATION

The testing required for evaluation of the impact of changes in module construction on the reliability, lifetime and degradation rate must address the observed failure modes and must cause degradation of the product.

A. Establishing Reliability Tests

Since the accelerated stress tests from the Qualification Tests are designed to address the identified field failure modes, these are a good starting point for developing reliability tests for evaluating the impact of changes to the product. How can the Qualification Tests be turned into reliability tests? The following methods may all contribute to the final test plan:

- Increase the test duration, for example do more thermal cycles or expose the modules to damp heat for a longer time.
- Use higher stress levels, but making sure that the higher stress levels don't cause failures that are not seen in the field.
- Combine stress, for example applying voltage to the module during damp heat.
- Utilize step stresses, where the initial stress starts with the stress level from the Qualification Test and increases until failures are seen. Once again care must be taken to insure that the failures seen are the same failure modes identified from field exposure.
- Evaluate new methods to accelerate the failure modes identified in the field, for example using the Dynamic

Mechanical Load Test to accelerate cell breakage caused by wind-induced vibrations. [15]
- Use material or coupon tests in situations where it would be too expensive to test full modules. For example long term UV testing at high temperature to evaluate material discoloration and degradation is better performed on small coupons of the same cross-sectional construction as the module.

B. Measurement Tools

In the Qualification Tests pass/fail measurements include peak power at Standard Test Conditions (STC) and the dry and wet leakage currents. There are additional measurement tools that can be utilized to observe problems before they impact the power or leakage currents. The following measurement tools can be valuable tools to use for identifying failure modes before they are serious enough to cause measureable power loss.

- Visual inspection can be used to observe discoloration of encapsulant, corrosion of metals and delaminations.
- Infrared cameras show heat dissipation, so can be used to determine areas where collection of current has been disrupted (higher series resistance) or where current is flowing where it shouldn't (shunting). An IR camera can also be used to show whether the current path through a string of cells is intact and whether the module bypass diodes are carrying current during normal operation and during shaded operation.
- Electroluminescence looks at the Near IR light generated by carriers transitioning across the cell p-n junction. So electroluminescence can be utilized to see discontinuities in the junction such as cracks in the cell or breaks in the junction itself.
- Testing for adhesion of the package layers, the junction box and the frame.
- Dark I-V curves that can identify small changes in series and shunt resistance before they are large enough to change the light I-V curve.

C. Developing specific reliability tests for specific changes

The first step in this process is to understand which failure modes the proposed change is likely to impact. A guideline for this has been established by Working Group 2 – Modules of IEC Technical Committee TC-82 on PV. These guidelines have been published as an IECEE Decision Sheet [16]. It is now planned for them to be incoporated directly into the third editions of IEC 61215 and IEC 61646. Table 4 provides a summary of these guidelines. These guidelines are utilized to define what retests are necessary in order to maintain product certification under IEC 61215 or IEC 61646.

These guidelines for retest should be used as a starting point for the analysis described in Subsection A. So for example if the change involves the encapsulation system, the test sequences should be based on:

- UV/50 thermal cycles,/10 humidity freeze cycles,
- Damp heat, 1000 hours at 85°C/85% RH

- Hail impact, if not tempered glass, and
- Hot spot, if material composition changes.

So the question becomes "Do these tests need to be modified from the qualification test to assess the ability of the change to survive 25 years without impacting failure rate or degradation rate?" The answer to this question depends upon which test is under discussion. Finally we must decide whether additional tests could be helpful in assessing the ability of the new product to perform as well as the old product throughout the warranted lifetime.

TABLE 4

Guidelines for Retest Requirements for IEC 61215 and IEC 61646

Modifications To	Tests to Repeat
Cell Technology	200 Thermal Cycles
	1000 hours of Damp Heat
	Hot Spot
	Mechanical Load (for reduction in cell thickness)
Encapsulation System	UV/TC 50/HF 10
	1000 hours of Damp Heat
	Hail Impact (if not tempered glass superstrate)
	Hot Spot
Superstrate	UV/TC 50/HF 10
	Mechanical Load
	Hail Impact
	1000 hours of Damp Heat (if non-glass)
	Hot Spot (if non-glass)
	Outdoor Exposure
Increase in Module Size (> 20%)	200 Thermal Cycles
	Mechanical Load
	Hail Impact
Backsheet	UV/TC 50/HF 10
	Robustness of Termination
	1000 hours of Damp Heat
	Hail Impact (if substrate design)
	Mechanical Load (if mounting depends on backsheet)
Frame or Mounting Structure	Mechanical Load
	Outdoor Exposure (if plastic)
	UV/TC 50/HF 10 (if plastic)
	1000 hours of Damp Heat (if adhesive used)
	200 Thermal Cycles (if adhesive used)
Junction Box/Electrical Termination	TC 50/HF 10
	Robustness of Termination
	1000 hours of Damp Heat
	By-pass Diode Thermal Test (if diodes are in J-box)
Interconnection between Cells	200 Thermal Cycles
	1000 hours of Damp Heat
	Hot Spot
Electrical Circuit	Hot Spot (if more cells per diode)
	By-pass Diode Thermal Test (if current level increases)
	200 Thermal Cycles (if internal conductors behind cells)
Higher or lower output (by > 10%)	Hot Spot
	By-pass Diode Thermal Test (if higher)
By-pass Diode	By-pass Diode Thermal Test

D. Establish a Baseline

Since we do not have accelerated stress tests that can show that a module type will survive 25 years outdoors where ever it is deployed, a baseline must be established for the proposed reliability tests. This baseline can be established using modules with similar construction that have proven long term service life. These baseline modules must be of the same PV technology (i.e. crystalline silicon, CdTe, CIGS, etc.) and preferably have similar packaging (i.e. glass superstrate design or glass-glass construction). Baseline modules should then be tested through all of the accelerated stress tests proposed in Section E along with the new module type.

E. Proposed changes to Qualification Tests to assess long term reliability after a change in product design

In this section each of the accelerated stress tests from IEC 61215 /61646 will be discussed and recommendations made as to whether it is necessary to modify them in order to use them for assessing long term reliability of lower cost PV modules.

Thermal cycling: Two hundred (200) thermal cycles have been equated to 10 to 11 years of outdoor exposure via comparison to field data [17 and 18] and via modeling of weather data [19]. Therefore more thermal cycling stress is required to assess a 25 year lifetime. The thermal cycle stress can be increased by cycling faster, using a wider temperature cycle or using more cycles. Cycling faster is limited both by the test chamber capability and the need to avoid thermal shock. Cycling at the fastest rate allowed by IEC 61215 is probably the best compromise. Expanding the temperature range is possible but once again is limited by equipment capability and the potential to cause failures not seen in the field (phase changes at low temperature and polymer damage at higher temperatures. If the test chamber can achieve 90°C a small degree of acceleration can be achieved.

This leaves increasing the number of cycles as the best approach. If 200 cycles equals 10 years of field exposure then 500 cycles would represent 25 years [17 and 18]. If after 500 thermal cycles the control construction and the new, lower cost modules have similar power loss and do not exhibit detrimental changes (i.e. broken interconnects) then the two constructions should have similar field performance for failure modes caused by thermal cycling.

Damp Heat: The 85°C/85% relative humidity exposure is as accelerated as necessary. These conditions probably never happen in the real world as the modules tend to dry out at their highest temperatures, but absorb moisture at lower temperatures. It is difficult to judge what outdoor exposure the 1000 hour exposure at 85/85 represents. In a recent experiment 10 crystalline silicon modules qualified to IEC 61215 were exposed to 1250 hours of 85/85. Only 2 of the 10 types successfully passed the extended test. [18] On the other hand some glass-glass encapsulated modules can easily endure more than 2000 hours of damp heat exposure. So rather than specify a particular length of time, it seems appropriate to test the control technology and the new technology through enough hours of damp heat that both begin to lose some power (say to 90% of the original) in order to verify that the new

978-1-4244-9113-1/11 $26.00 © 2011 IEEE

technology is no worse and has no additional failure modes than the old module technology it will replace.

UV/TC50/HF10: This sequence of tests is mainly a test for the package. If the module fails this test it indicates inadequate adhesion between layers or inadequate cure level in the encapsulant. It is not a lifetime test so typically does not need to be enhanced for reliability testing. So the recommendation is to test both the new module construction and the old module construction to the sequence defined in IEC 61215/61646, but with the addition of the Dynamic Mechanical Load Test discussed below.

Mechanical Load: In the test a specified (wind) load is applied to the front and the back of the modules 3 times. If the module is to be used in a snowy location the load is increased during the last front cycle. The wind load (2400 Pa) and snow load (5400 Pa) are average values from around the world. If modules are to be used in windy or snowy locations higher values should be tested. In addition, mechanical loading can cause cells (especially thin ones) to crack. Modules measured immediately after wind or snow loading may not have degraded power, but if these modules are thermal cycled (say 25 to 50 cycles) significant cell breakage will then cause power loss.

Hail Test: The hail test is only required for changes in non-tempered glass superstrate modules. In this case the test should be run as specified in IEC 61215/61646.

Bypass Diode Thermal Test: No change in this test is recommended for assessing long-term reliability. The bypass diodes will be stressed by the other accelerated tests (thermal cycle, damp heat, etc.) so it is extremely important to ensure that each diode is working correctly after completing the reliability test procedures. In addition, it is important to set up a production line test to ensure that each bypass diode has been installed correctly and is operational before the module is shipped to the customer.

Hot Spot Test: The Hot Spot test in IEC 61215 edition 2 is not a particularly good test. It will be modified in edition 3. In the meantime use the ASTM E 2481-06 Hot Spot Test. [20]

Other tests to consider in the assessment of new products:

Dynamic Mechanical Load Test: The only mechanical test in IEC 61215 is a static mechanical load test that is performed after the accelerated stress tests. A Dynamic Mechanical Load test followed by 50TC/10HF does a much better job of identifying modules with cells that are prone to breakage and would cause subsequent power loss. [15] There is an available DIN Standard (EN12211) that can be utilized for this test. [21] Ultimately it is likely that a similar dynamic mechanical load test (DML) will be incorporated into IEC 61215 as part of the sequence UV/DML/50TC/10HF.

Transportation Testing: PV modules are usually shipped to the installation site. If improperly packaged significant damage and power loss can occur during the shipment. Changes in module construction and/or changes in packaging design can influence this result. Use of a standard transportation test such as ISTA Procedure 3 [22] is recommended until an IEC PV transportation test can be completed.

UV Material Test: While IEC 61215 contains a UV test, this is only meant as a pre-screening test to address UV sensitive bonding issues. This test is not long enough to assess whether the polymeric materials utilized in a module are capable of surviving the UV exposure expected during the lifetime of the module. Long term UV exposure of full sized modules is difficult and expensive. Therefore most long-term UV exposures have been made on coupons with the same cross sectional construction as the modules to be evaluated. STR developed a long term UV exposure protocol during their work evaluating the causes of EVA yellowing. [23] BP Solar reported the use of a similar UV exposure protocol for 26 weeks to verify a 25-year lifetime. [24] A similar UV testing protocol should be used to evaluate any new polymeric material for use in a PV module. The material should be exposed to the UV within the standard package in which it will be used. Since there is no agreement between UV dose and years in the field, it would be best to perform the test with the new material side-by-side with the material it is to replace. The test should proceed for at least the proposed 26 weeks or until one or both of the materials begin to discolor or degrade. At that point a comparison between old and new material will indicate whether the new material will perform as well as that which it is to replace.

IV CONCLUSIONS

A method for assessing the long term reliability and durability of new lower cost PV modules has been presented. Because the new approach compares the results of the accelerated testing with modules that have a known long lifetime, a new module type qualified through this procedure has a high likelihood of also surviving in the field. The recommendations given in this paper can serve as a guideline for the establishment of a specific program of reliability testing for each major cost reduction proposed. Use of this methodology will reduce the risk that a change made to reduce cost will have a major, negative impact on module lifetime or degradation rate.

ACKNOWLEDGMENT

This work was supported by the U.S. Department of Energy under Contract No. DE-AC36-08-GO28308 with the National Renewable Energy Laboratory.

REFERENCES

[1] Navigant Consulting
[2] EPIA "Global Market Outlook for Phtovoltoacis until 2014".
[3] SolarBuzz. http://solarbuzz.com/our-research/recent-findings/global-solar-photovoltaic-demand-forecast-reach-204-gw-2 011-growth-rate
[4] Renewable Energy World, Nov/Dec 2010 edition.
[5] System Advisor Model.
[6] http://www.eia.doe.gov/cneaf/electricity/epm/table5_3.html

[7] IEC 61215 "Crystalline silicon terrestrial photovoltaic (PV) modules – Design qualification and type approval".

[8] IEC 61646 "Thin-Film terrestrial photovoltaic (PV) modules – Design qualification and type approval"

[9] IEC 61730-1 and 2 "Photovoltaic Module Safety Qualification"

[10] M.I. Smokler, D.H. Otth and R.C Ross Jr., "The Block Approach to Photovoltaic Module Development", *Proceedings of the 18th IEEE Photovoltaic Specialist Conference*, 1985, p. 1150

[11] John Wohlgemuth, Daniel W. Cunningham, Andy Nguyen, George Kelly and Dinesh Amin, "Failure Modes of Crystalline Silicon Modules" Proceedings of PV Module Reliability Workshop", http://www1.eere.energy.gov/solar/pv_module_reliability_workshop_2010.html , 2010.

[12] Nick Bosco "Reliability Concerns Associated with PV Technologies" http://www.nrel.gov/pv/performance_reliability/pdfs/failure_references.pdf , 2010.

[13] Otth and Ross "Assessing Photovoltaic Module Degradation and Lifetime from Long Term Environmental Tests", Proceediings of 29th Institute of Environmental Science Annual Meeting, 1983.

[14] John Wohlgemuth, "Overview of Failure Mechanisms and PV Qualification Tests", *Proceedings of PV Module Reliability Workshop*, http://www1.eere.energy.gov/solar/pv_module_reliability_workshop_2010.html, 2010.

[15] John H. Wohlgemuth, Daniel W. Cunningham, Neil V. Placer, George J. Kelly and Andy M. Nguyen, "The Effect of Cell Thickness on Module Reliability" *Proceedings of the 33rd IEEE Photovoltaic Specialist Conference*, 2008.

[16] IECEE CTL Decision Sheet – Retest Guidelines for IEC 61215 and IEC 61646 http://www.iecee.org/ctl/sheet/pdf/PDSH0647A.pdf

[17] J.H. Wohlgemuth, D. W. Cunningham, A.M. Nguyen and J. Miller, "Long Term Reliability of PV Modules", *Proceedings of 20th EUPVSEC*, 2005.

[18] J. Wohlgemuth, D. Cunningham, D. Amin, J. Shaner, Z. Xia and J. Miller, "Using Accelerated Tests and Field Data to Predict Module Reliability and Lifetime", *Proceedings of 23rd EUPVSEC*, 2008.

[19] Nick Bosco and Sarah Kurtz, "Quantifying the Weather: an analysis for thermal fatigue", *Proceedings of PV Module Reliability Workshop*, http://www1.eere.energy.gov/solar/pv_module_reliability_workshop_2010.html, 2010.

[20] ASTM E 2481-06 "Standard Test Method for Hot Spot Protection Testing of Photovoltaic Modules".

[21] DIN EN12211 "Windows and doors resistance to wind load – test method"

[22] ISTA Proceudre 3E "Procedure 3E: Unitized Loads of Same Product"

[23] William H. Holley, Jr, Susan Agro, James P. Galica, Lynne A. Thoma, Robert S. Yorgensen, "Investigation into the causes of browning in EVA encapsulated flat plate PV modules", *Proceedings of the 24th IEEE Photovoltaic Specialist Conference*, 1994.

[24] John H. Wohlgemuth, Daniel W. Cunningham, Paul Monus, Jay Miller, and Andy Nguyen "Long Term Reliability of Photovoltaic Modules" , *Proceedings of the 4th World Conference on Photovoltaics*, 2006.

Identification, Characterization, and Implications of Shadow Degradation in Thin Film Solar Cells

Sourabh Dongaonkar, Muhammad A Alam*
School of Electrical & Computer Engineering
Purdue University, West Lafayette, IN, USA
*alam@purdue.edu

Karthik Y, Souvik Mahapatra
Department of Electrical Engineering
Indian Institute of Technology Bombay, Mumbai, India

Dapeng Wang, Michel Frei
Applied Materials, Santa Clara, CA, USA

Abstract— **We describe a comprehensive study of intrinsic reliability issue arising from partial shadowing of photovoltaic panels (e.g., a leaf fallen on it, a nearby tree casting a shadow, etc.). This can cause the shaded cells to be reverse biased, causing dark current degradation. In this paper, (1) we calculate the statistical distribution of reverse bias stress arising from various shading configurations, (2) identify the components of dark current, and provide a scheme to isolate them, (3) characterize the effect of reverse stress on the dark current of a-Si:H p-i-n cells, and (4) finally, combine these features of degradation process with shadowing statistics, to project 'shadow-degradation' (SD) over the operating lifetime of solar cells. Our results establish shadow degradation as an important intrinsic reliability concern for thin film solar cell.**

Keywords – Thin film solar cells, voltage stress, performance degradation\, reliability

I. INTRODUCTION

Reliability has always been crucial to the economic viability of photovoltaic (PV) technologies [1]. Consequently, the safety issues like hot-spot breakdown [2], and parametric degradation issues like light induced degradation have been studied extensively [3]. One such problem arises due to partial shadowing of the panel, which results in a reverse bias appearing across the shaded cells [4]. The problem of shadowing of solar panels has been studied for quite some time; however, the work has focused primarily on the drop in system energy-yield due to partial shadows[5]. On the device level, the reverse bias caused by shading can also result in long term performance degradation of the shaded cells. In the worst case, the shaded cells may also undergo catastrophic reverse breakdown.

In case of crystalline cells, the problem is avoided by using a bypass diode in parallel to the cell [6]. However, the large area deposition processes, typically used in thin film PV technology, make insertion of such bypass diodes impractical. Consequently, the panel is susceptible to parametric failure due

to reverse stresses induced by shadowing. Therefore, a thorough understanding of the effects of reverse bias stress, especially on long term stability and performance of thin film solar cells becomes very important. We note that the SD is a generic reliability concern for all thin film PV technologies, however, in this paper we illustrate the issue with reference to a-Si:H cells.

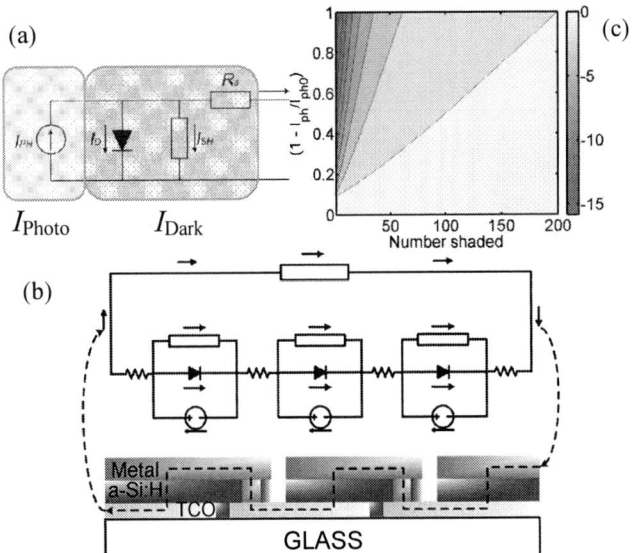

Figure 1: (a) Typical equivalent circuit representation of a solar cell, showing the dark and light current components. (b) The series connection in a solar panel shown in the equivalent circuit picture as well as the physical layout of cells, made by successive depositions of semiconductor and contact layers and scribing. (c) Contour plot showing the operating (reverse) voltage of a cell as a function of photocurrent degradation due to degree of shading, and the number of shaded cells.

II. EFFECT OF SHADOWING

The equivalent circuit shown in Fig 1a identifies the main features of a solar cell. The two current components of a solar

I-V characteristics, namely dark (I_{Dark}) and light (I_{Photo}) currents, are shown shaded. The dark current is a combination of an exponential diode current (I_D) in parallel with a parasitic shunt current (I_{SH}). The photocurrent component is typically represented by a voltage-independent current source (I_{PH}) in parallel with the diode and shunt resistance. In order to obtain high panel output voltage, approximately 100-200 nominally identical cells are connected in series. In case of thin-film cells, this series connection is accomplished in following steps; patterning of TCO on glass by laser scribing, deposition of the semiconductor layer/s (PECVD for a-Si:H, sputtering for CIGS/CdTe), another scribing step to pattern the semiconductor layer, deposition of back contact, and final scribing and isolate, result in a structure shown in Fig. 1b.

This series connection is responsible for the reverse voltage stress due to shadowing. Under normal operating conditions the panel is biased at the maximum power point voltage (MPPV). For a-Si:H cells, this translates to $V_{cell} \sim 1V$. Now, if some of the cells are accidentally shadowed, the photocurrent of the affected cells is suppressed ($I_{PH,shade} < I_{PH0}$), and the new operating point will shift to $V_{cell,shade} < 1V$. Depending on the number of cells shaded and the loss in photocurrent, $V_{cell,shade}$ could be negative depending on the shading conditions [5, 7].

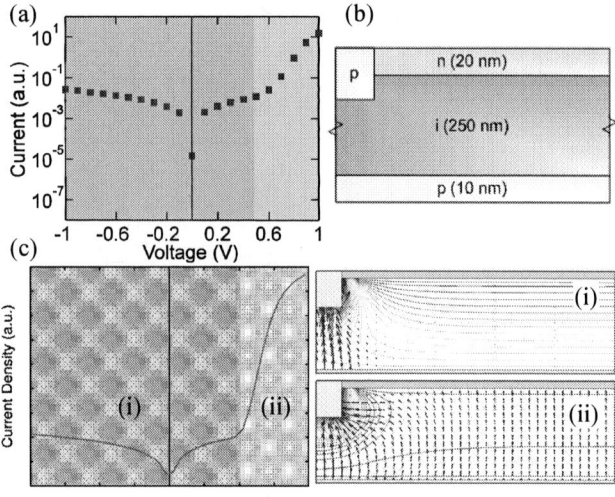

Figure 2: (a) Measured dark IV of typical a-Si:H solar cell (squares), with the shunt current (I_{SH}) dominated region (V < ~0.5V), and diode current (I_D) dominated region (V > ~0.5V) shown shaded. (b) Schematic showing the p-i-n solar cell structure with a p-i-p shunt in the left formed due to Al diffusion and counter doping. Simulated dark IV of the structure shown in schematic (b) reproduces the qualitative features of I_{Dark}, along with the shunt and diode dominated regimes (shown shaded). (i) The potential (contours) and current (quiver) distribution in shunt dominated regime shows current crowding in the p-i-p shunt region; (ii) at higher biases the diode current through p-i-n region takes over and current flow is more uniform.

To determine the new operating point, we simulate a panel of 200 identical cells connected to a load resistance of R_L. Initially, the cells are assumed to be operating at the MPPV. For an accurate estimate of the change in operating condition under shading through the simulations, we use a spline-fit of the measured dark IV characteristics and a constant current source for I_{Photo}, effectively mimicking the equivalent circuit in Fig. 1a. Fig. 1c shows the contour plot of the reverse voltage developed across each of the partially shaded cells, as a

function of the degree of shading and number of cells shaded. This plot of the reverse voltage demonstrates that in majority of cases, when multiple cells are shaded partially ($0 < I_{PH,shade} < I_{PH0}$ & $N_{shade} > 1$), the shaded cells are stressed only at low to moderate reverse bias (< ~ –5V). Extreme reverse biases (~ -15V) causing hot-spot breakdown can only occur if a one or a few cells are shaded completely ($I_{PH,shade} = 0$ & $N_{shade} = 1$).

Fortunately, the form factor of the cells ensures that the probability of worst case shading is statistically rare and catastrophic hot-spot failure is unlikely for typical operating conditions. However, partial shadowing is more probable, which would result in moderate reverse stress on the shaded cells for the duration of shading. We find that such reverse stress leads to a parametric degradation in cell efficiency due to increase in dark current (e.g., a -5V stress for 10^4 sec can reduce the efficiency by 10% in the worst case). Since, it is impossible to control the random shading events, it becomes important to understand its mechanisms and effects clearly to predict the effect of shadow degradation on panel output over its operating lifetime.

III. DARK CURRENT

A. Features of the dark current

In order to precisely understand the effect of shadow degradation on IV characteristics, it is important to understand the components of solar IV characteristics. The dark current (I_{Dark}) of a solar cell is known to consist of two components; an exponential diode current (I_D), and a parallel symmetric, non-ohmic shunt current (I_{SH}), so that $I_{Dark} = I_D + I_{SH}$ [8]. At lower biases the non-ohmic shunt current dominates, while at higher biases the exponential diode current takes over, as shown by shaded regions in Fig. 2a. The shunt current is parasitic component and its magnitude varies from cell to cell even when they are manufactured under nominally identical conditions. The non-ohmic shunts have been observed in all types of solar cells [8, 9]. In case of thin film cells, this shunt current mechanism has been identified as space-charge-limited (SCL) current [10] through parasitic paths formed at different locations on the cell surface.

In a-Si:H p-i-n solar cells, these parasitic shunt paths are consistent with the hypothesis of aluminum incursion in the a-Si:H from the top ZnO:Al of the n contact [11]. This Al can counter-dope the material to p type, and result in a p-i-p parasitic structure, as opposed to the normal p-i-n device (see Fig. 2b) [10]. This parasitic p-i-p path will result in a SCL shunt current in parallel to the exponential diode current through the p-i-n device. Self consistent simulation of the 2D structure shown in Fig. 2b reproduces the measured dark IV with the shunt (marked (i)) and diode (marked (ii)) current dominated regimes (Fig. 2c). The contour and quiver plots in Fig. 2c show the current and potential distribution in different regimes of operation. In reverse and low forward biases (region (i)), the p-i-p shunt path provides the lowest potential barrier resulting in the localized SCL shunt current (arrows). At higher biases the diode current through the bulk p-i-n device becomes large enough and overtakes the shunt component (region (ii)). This results in uniform current flow through the device (see quiver plot in Fig 2c (ii)).

978-1-4244-9113-1/11 $26.00 © 2011 IEEE   558

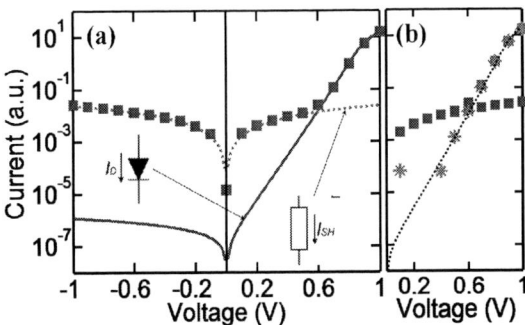

Figure 3: (a) The measured I_{Dark} of a-Si:H p-i-n solar cell (squares) showing I_D (solid line) and I_{SH} (dotted line) components. (b) Plot of $|I|$ vs. $|V|$ (squares) and the 'clean' forward current (stars), obtained by subtracting out I_{SH}.

B. 'Cleaning' the forward dark IV

This analysis of the parasitic shunt current and bulk diode current components of the dark IV (Fig. 3a) allows us to identify the diode (I_D) and shunt (I_{SH}) current components of the dark current (I_{Dark}). Moreover, this picture can help us isolate the $I_{D,fwd}$ from the measured I_{Dark}, by using the symmetry of I_{SH} about $V = 0$ point. From Fig. 3a we see that in reverse bias $I_D \ll I_{SH}$, this means that $I_{Dark,rev} \sim I_{SH}$ in reverse bias. Therefore for finding the forward diode current we can subtract the reverse shunt current (i.e., $I_{D,fwd} = I_{Dark,fwd} - |I_{Dark,rev}|$, see Fig. 3b). This simple subtraction scheme provides a powerfulmethod to isolate the two components of dark IV by 'cleaning' forward current. This means that we can now study the shadow-induced degradation of shunt current (I_{SH}) and diode current (I_D) components individually.

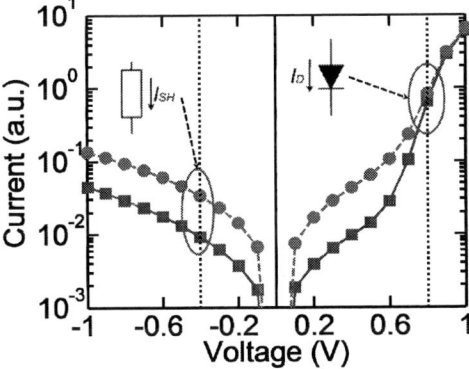

Figure 4: Initial (squares) and post stress (circles) dark IV, after a -5V stress has been applied for 10^4 s. I_{Dark} increases with stress, both in I_{SH} dominated (at -0.4V) as well as in the I_D dominated (at 0.8V) regimes.

IV. REVERSE DEGRADATION MEASUREMENTS

Fig. 4 shows the pre (squares) and post (circles) stress dark IV of a cell (-5V stress for 10^4 s), showing a significant increase in dark current I_{Dark} after stress. It is apparent that both I_D and I_{SH} increase significantly due to the stress. The degradation behavior of I_D and I_{SH} can be monitored separately by observing I_{Dark} at -0.4V (as $I_{Dark} \sim I_{SH}$ here) and 'cleaned' I_{Dark} at 0.8V (as $I_{Dark} \sim I_D$ here). This allows us to

identify and characterize the degradation of these current components separately, as discussed next.

A. Diode current degradation – $I_D(V_R, t)$

In order to identify the features of the degradation mechanism, 22 devices were subjected to stresses ranging from -3V to -7V for a duration of 10^5s each. Fig. 5a shows that time dependence of diode current degradation $\Delta I_D(t)$ for various reverse biases is described by robust *power-law* ($\Delta I_D(t) \sim t^n$). The power exponent n of the power law is $\sim 0.2\text{-}0.26$, across 22 devices measured. The voltage dependence of the ΔI_D can be obtained from the pre-factor ($K(V_R)$) of the power-law fit ($\Delta I_D = K(V_R)t^n$) of the time dependence data. We find that the dependence of ΔI_D stress voltage (V_R) is rather weak, but higher degradation can be observed with higher stress (Fig 5b). Interestingly we find that the power exponent n is *independent of stress voltage* V_R (Fig. 5b inset).

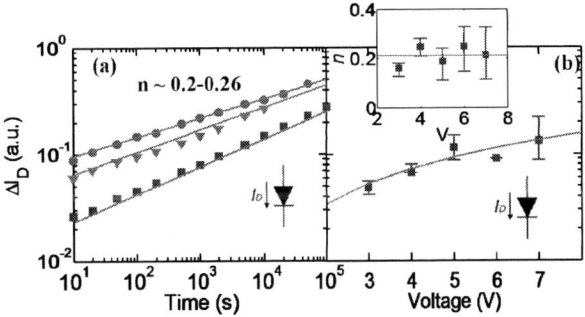

Figure 5: (a) The 'cleaned' $\Delta I_D(t)$ (at 0.8V) for 3 devices showing robust power law behavior with time (t^n with $n \sim 0.2\text{-}0.26$). (b) $\Delta I_D(V_R)$ shows weak linear dependence, inset shows that n is independent of V_R (showing average of 22 devices).

Another feature of the I_{Dark} degradation is the distinct relaxation behaviors of diode and shunt current components. While diode current I_D relaxes to its original value after about a week of relaxation in a dark chamber. The shunt current degradation (ΔI_{SH}) on the other hand appears to be permanent (Fig. 6a). The relaxation kinetics for ΔI_D also shows power-law time dependence, similar to degradation kinetics (Fig. 6b).

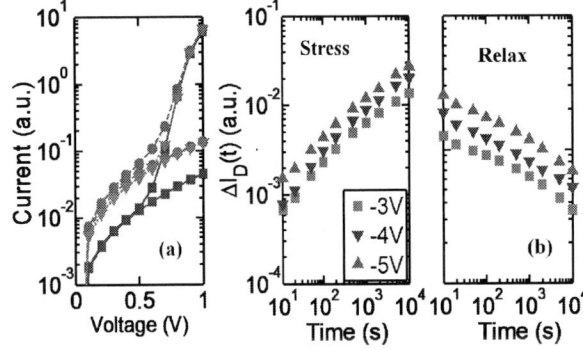

Figure 6: (a) Pre-stress (squares), post-stress (circles) and relaxed (triangles) dark IV, showing that the diode current relaxes with time, but the increase in shunt current is permanent. (b) The stress and relaxation time dependencies of dark current show similar power law features with power exponent independent of stress voltage.

978-1-4244-9113-1/11 $26.00 © 2011 IEEE

Although the physics of time-exponents and weak voltage dependence are not explicity known, the general features of time and voltage dependencies of degradation and relaxation phenomena $I_D(t, V_R)$ suggest metastable defect generation in a-Si:H, due to reverse stress. The most likely cause is metastable defect creation near the midgap due to breaking of Si-H bonds. [12]. In order to explore this hypothesis for diode current degradation, we measured 14 small area devices, which exhibited no shunt current ($I_{SH} = 0$). This allowed us to probe the diode current in the voltage range (-1V to 1V), without any contamination from I_{SH}. For these small area devices, we observe an asymmetrically high increase in I_D at low biases (Fig. 7a). The current increase at low biases is much faster compared to higher biases. This is reflected clearly in the power exponent of $\Delta I_D(t)$ at different values measurement voltages from -0.8V to 0.8V. The exponent decreases monotonically from ~0.95 at -0.8V to ~0.25 at 0.8V measurement voltage (Fig. 7b). Note that the power exponent value of 0.25 at measurement voltage of 0.8V, obtained from these devices with $I_{SH} = 0$, coincides well with the power exponent obtained from the large area devices through 'cleaning' the shunt current (Fig. 5b inset). This shows the effectiveness and usefulness of the cleaning technique in reliability characterization experiments.

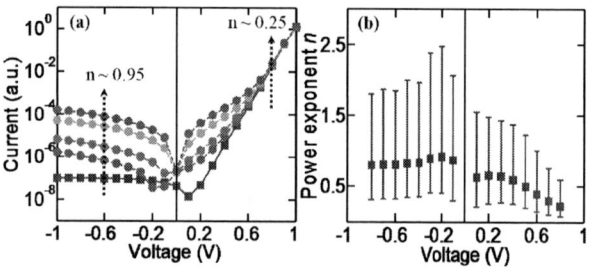

Figure 7: (a) Time evolution of Dark IV at reverse stress of -5V for 10^4s, for a small area device with $I_{SH} = 0$, shows disproportionately high increase in the dark current at low biases. (b) The power exponent n of time dependence ($I_D \sim t^n$) shows a monotonic dependence on measurement voltage (showing and average of 14 devices measured). The value of n at V = 0.8V also coincides well with the power exponent obtained from 'cleaned' diode current of large area devices.

Note that in case of a-Si:H cells, the diode current I_D is dominated by recombination in the i-layer. This means that the ideality factor for a-Si:H p-i-n diodes is ~2. Therefore, an increase in midgap defects leading to higher recombination would imply an equal increase in current at all voltages between -1V and 1V. However, as apparent from Fig 7a, the current increase at reverse and low biases is much larger than the high bias regime. This asymmetry in the degraded diode current cannot be explained through increased recombination in the i-layer alone. Further experimental studies are required to clarify this degradation mechanism further.

B. Shunt current degradation – $I_{SH}(V_R, t)$

The shunt current I_{SH} also increases due to stress (as seen in Fig 4, moreover, this increase in I_{SH} is permanent. However, unlike $\Delta I_D(t)$, the time dependence of I_{SH} does not show a clean monotonic trend (Fig. 8a). The voltage dependence of shunt current ($\Delta I_{SH}(V_R)$) on the other hand shows weak increasing trend (Fig. 8b), for the 22 devices measured.

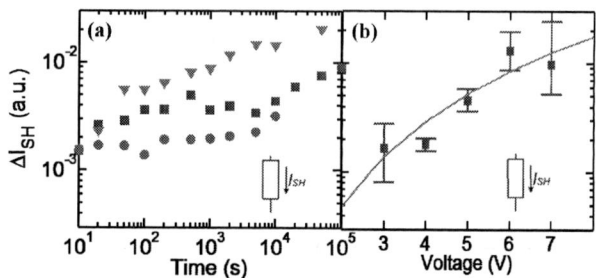

Figure 8: (a) $\Delta I_{SH}(t)$ for the same 3 devices from Fig. 5a, shows more noisy time dependence behavior; and (b) the voltage acceleration for ΔI_{SH} remains weak and positive (average of 22 devices).

We can see that the degradation characteristics of I_{SH} do not show as clear trends as the I_D component. However, based on the understanding of p-i-p shunt picture, we can postulate the possible mechanism. The most likely cause of enhancement in shunt current is due to decrease in shunt length associated with stress-induced diffusion of Al further into the i-layer (creating a larger p region in the process). Since the SCL current is inversely proportional to length ($I_{SCL} \sim L^{-(2\gamma+1)}$), a decrease shunt length will result in an increase in shunt current. There are two possible reasons for this Al incursion during stress. One possibility is that due to current crowding in the shunt region could lead to local temperature rise. This would result in enhanced Al diffusion leading to lower shunt length. Another possibility is that due to the high localized current densities in the shunt region, the Al ions can move due to hopping ionic transport in a-Si:H matrix. The exact mechanism of this Al motion however is not fully established, and further work is needed to identify the physics of this degradation kinetics.

The empirical results on diode and shunt current degradation described above provide a good starting point for exploring the stability of a-Si:H cells. Moreover, these observations when combined with circuit simulations and shading statistics also allow us to estimate the long term effects of this degradation phenomenon.

V. PROJECTION OF EFFICIENCY DEGRADATION

Over the operating lifetime of a panel, a cell will be subjected to repeated shadow stresses of different magnitudes for random lengths of time. This would result in steady parametric increase in dark current due to increase in both I_{SH} and I_D (shown schematically in Fig. 9a). In order to estimate the increase in I_D during a particular shading event; we need to know the reduction in photocurrent of the shaded cells, the number of shaded cells, and the duration of shading. We can use the shading fraction and number of shaded cells, and use the simulation scheme discussed in Sec. II to obtain the reverse stress voltage VR caused by the shadow. We can then use this stress voltage value and stress duration in the empirical relation of $\Delta I_D(t, V_R)$ shown in Fig. 5, to estimate the shadow-induced output degradation for each shading event.

We have three random variables in this statistical shadow stress problem, namely reduction in photocurrent ($X_{shade} = (1 - I_{PH}/I_{PH0})$), number of shaded cells (N_{shade}), and shading duration (t_{shade}). For X_{shade} we can reasonably assume an exponential distribution, to mimic the fact that worst case shading is least likely. For number of shaded cells N_{shade}, and the shading time t_{shade}, we assume a uniformly random distribution. For assessing the long term effect of shadow stresses, we generate random samples for the three random variables as defined above, then use the simulation to estimate V_R, and finally apply the empirical degradation relations to get the distribution of ΔI_D values. Finally, the cumulative ΔI_D is obtained for a particular shading distribution. This calculation process is illustrated in the flowchart in Fig. 9b.

We then use the time and voltage dependencies of Over the operating lifetime of the cell these successive stress induced effects will accumulate to reduce the efficiency of the panel. Fig. 9c shows the mean degradation expected from this analysis, for different shading conditions (resulting in different cumulative shading time τ). For different assumed shading statistics the simulation algorithm provided here can predict the mean degradation over the panel lifetime. While accurate shading statistics need to be obtained for calculating the exact SD value, the method proposed is quite general and will be useful in predicting long term panel performance.

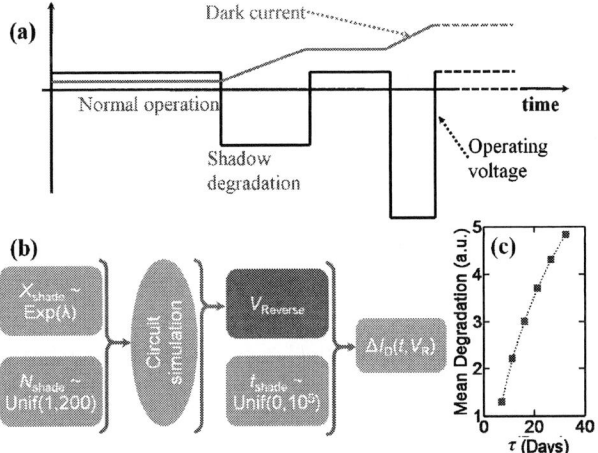

Figure 9: (a) Schematic representation of repeated shadow stresses during operation resulting in I_{Dark} degradation. (b) Flowchart showing the simulation method used for estimating the statistical degradation mechanism. Varying λ results in different shading conditions, and cumulating shading times (τ). (c) Projected degradation over operating lifetime of 30 years, vs. cumulative shading time during the lifetime (τ) in days.

Note that while this first order model of shadow degradation is based on a few reasonable (but untested) assumptions about shadow statistics, the estimates obtained are sufficient for putting a ballpark number for the shadow degradation effect over panel operating lifetime. In this calculation, we have not accounted for the partial relaxation in I_D, making the lifetime estimate somewhat conservative. However this is compensated by the fact that we also did not include the increase in I_{SH}, which tends to underestimate panel lifetime. On balance this preliminary calculation should give a a decent estimate of the total degradation. It must be stressed that the circuit simulation methodology itself is flexible enough to incorporate these features as well, as they are better understood.

VI. CONCLUSION

We have demonstrated that for a panel operating under typical conditions, parametric shadow degradation is more likely than hot-spot breakdown. Consequently, the parametric degradation of the dark current over the operating lifetime, due to reverse stress, is an important intrinsic reliability consideration at par with other reliability issues of light-induced degradation and contact diffusion. We characterized the SD degradation phenomenon in a-Si:H p-i-n solar cells as a function of voltage and temperature, where the diode current enhancement and relaxation $\Delta I_D(t)$ is described by power-law time dependence, and weak voltage dependence. The increase in shunt current however, is non-systematic but permanent. A circuit simulation methodology to include the statistics of shading was developed. This allows a quantitative prediction of panel performance degradation due to shadow degradation.

ACKNOWLEDGMENT

The authors would like to thank Solar Business Group, Applied Materials®, for providing the samples, and for helpful discussions.

REFERENCES

[1] R. Gaston, R. Feist, S. Yeung, M. Hus, M. Bernius, M. Langlois, S. Bury, J. Granata, M. Quintana, C. Carlson, G. Sarakakis, D. Ogden, and A. Mettas, "Product reliability and thin-film photovoltaics," in *Reliability of Photovoltaic Cells, Modules, Components, and Systems II*, San Diego, CA, USA, 2009, pp. 74120N-15.

[2] O. Breitenstein, J. M. Wagner, B. Lim, A. Lotnyk, J. Schmidt, J. Bauer, and H. Blumtritt, "Hot spots in multicrystalline silicon solar cells: avalanche breakdown due to etch pits," *Physica Status Solidi (RRL) - Rapid Research Letters*, vol. 3, pp. 40-2, 2009.

[3] L. K. Wagner and J. C. Grossman, "Microscopic Description of Light Induced Defects in Amorphous Silicon Solar Cells," *Physical Review Letters*, vol. 101, p. 265501, 2008.

[4] A. Johansson, R. Gottschalg, and D. G. Infield, "Modelling shading on amorphous silicon single and double junction modules," Japan, Japan, 2003, pp. 1934-7.

[5] A. Woyte, J. Nijs, and R. Belmans, "Partial shadowing of photovoltaic arrays with different system configurations: literature review and field test results," *Solar Energy*, vol. 74, pp. 217-233, 2003.

[6] N. Dzung and B. Lehman, "A reconfigurable solar photovoltaic array under shadow conditions," in *Applied Power Electronics Conference and Exposition (APEC '08)*, Piscataway, NJ, USA, 2008, pp. 980-6.

[7] M. A. Alam, S. Dongaonkar, Y. Karthik, S. Mahapatra, D. Wang, and M. Frei, "Intrinisic reliability of amorphous silicon thin film solar cells," in *Reliability Physics Symposium (IRPS), 2010 IEEE International*, pp. 312-317.

[8] K. R. Lord, M. R. Walters, and J. R. Woodyard, "Investigation Of Shunt Resistances In Single-Junction A-Sih Alloy Solar Cells," in *Amorphous Silicon Technology-1994*. vol. 336, Pittsburgh: Materials Research Soc, 1994, pp. 729-734.

[9] O. Breitenstein, J. P. Rakotoniaina, M. H. A. Rifai, and M. Werner, "Shunt types in crystalline silicon solar cells," *Progress in Photovoltaics: Research and Applications*, vol. 12, pp. 529-538, 2004.

[10] S. Dongaonkar, J. D. Servaites, G. M. Ford, S. Loser, J. E. Moore, R. M. Gelfand, H. Mohseni, H. W. Hillhouse, R. Agrawal, M. A. Ratner, T. J. Marks, M. S. Lundstrom, and M. A. Alam, "Universality of non-Ohmic shunt leakage in thin-film solar cells," *Journal of Applied Physics*, 2010.

[11] S. Dongaonkar, K. Y, D. Wang, M. Frei, S. Mahapatra, and M. A. Alam, "On the Nature of Shunt Leakage in Amorphous Silicon p-i-n Solar Cells," *Electron Device Letters, IEEE*, vol. 31, pp. 1266-1268.

[12] R. A. Street, "Long-time transient conduction in a-Si:H p-i-n devices," *Philosophical Magazine Part B*, vol. 63, pp. 1343 - 1363, 1991.

Metastability of Hydrogenated Amorphous Silicon Passivation on Crystalline Silicon and Implication to Photovoltaic Devices

Bahman Hekmatshoar, Davood Shahrjerdi, Marinus Hopstaken and Devendra Sadana
IBM Thomas J. Watson Research Center
Yorktown Heights, NY
Email: hekmat@us.ibm.com

Abstract—We present experimental evidence that the surface passivation of crystalline silicon (*c*-Si) by hydrogenated amorphous silicon (*a*-Si:H) is metastable. Photo-conductance decay measurements of the effective minority carrier lifetime in *c*-Si show that the surface recombination velocity of the carriers at the *c*-Si/*a*-Si:H interface is reduced by annealing at temperatures up to ~ 350°C, but relaxed to higher values after further thermal treatment. The relaxation is thermally activated and faster at higher temperatures. We attribute this phenomenon to the thermal equilibration of charged defects in *a*-Si:H. Our finding suggests that *a*-Si:H passivation may require thermal stabilization for realizing reliable photovoltaic devices.

Keywords- hydrogenated amorphous silicon; metastability; surface passivation; carrier lifetime; thermal equilibrium

I. INTRODUCTION

Hydrogenated amorphous silicon (*a*-Si:H) is an appealing alternative to thermally grown oxide for surface passivation of crystalline silicon (*c*-Si) for photovoltaic applications. Plasma-enhanced chemical vapor deposition (PECVD) is the standard technique for depositing *a*-Si:H and may be used for the deposition of high-quality passivation layers at temperatures as low as 200°C. Passivation with *a*-Si:H therefore offers the advantage of a low thermal budget as well as preserving the bulk carrier lifetime of low-cost Si wafers throughout the growth.

Various groups have reported that the quality of *a*-Si:H passivation on *c*-Si improves by annealing as inferred from the improvement of the effective minority carrier lifetime in *c*-Si. The improvement of the quality of surface passivation is also reflected in the increased open circuit voltage of the *c*-Si solar cells passivated with *a*-Si:H [1-3]. However, little work has been done to investigate the mechanisms responsible for the improvement of the passivation quality. Moreover, the passivation has been generally assumed to be stable over time and a systematic study of the passivation reliability is missing.

In this paper, we investigate the mechanism responsible for the improvement of *a*-Si:H passivation on *c*-Si upon annealing and propose that reduction in the density of charged defects in *a*-Si:H during annealing is the responsible mechanism. Also, we provide experimental evidence that the *a*-Si:H passivation is

metastable and that the reduction in the surface recombination velocity of carriers at the *a*-Si:H/*c*-Si interface is reversible. We attribute the metastability of the passivation to the equilibration of charged defects in *a*-Si:H. To the best of our knowledge, this is the first report on the metastability of surface passivation in *a*-Si:H/*c*-Si interfaces.

II. EXPERIMENTS

The samples were prepared by depositing ~120nm-thick undoped *a*-Si:H layers on both sides of ~300μm-thick ~2.5 Ω-cm *n*-type and *p*-type (100) float-zone silicon wafers. After standard cleaning, the wafers were dipped in dilute hydrofluoric acid and loaded immediately into the PECVD machine for *a*-Si:H deposition. The *a*-Si:H films were grown using a mixture of silane (SiH_4) and hydrogen at a substrate temperature of 200°C.

The samples were annealed at temperatures in the range of 200-400°C and the *a*-Si:H films were characterized by secondary ion mass spectroscopy (SIMS) and Fourier transform infrared (FTIR) spectroscopy before and after annealing. In addition, the minority carrier lifetime in *c*-Si was measured using a microwave photoconductance decay (μ-PCD) tool for these samples. Subsequently, the thermal stability of the *a*-Si:H passivation was investigated by thermal treatment of the above-mentioned samples in the range 100-200°C for accelerated testing and measuring the effective minority carrier lifetime in *c*-Si versus treatment time using the μ-PCD tool.

The stability of the samples was also investigated under illumination with an intensity of ~5 suns for accelerated testing. The samples were mounted on a heat sink connected to a radiator for liquid cooling and the temperature of the samples was monitored to assure the samples remain at room temperature during light illumination.

III. RESULTS AND DISCUSSION

A. Annealing Experiments

The effective minority carrier lifetime of the samples annealed at temperatures in the range of 200-350°C for 30 minutes was measured by μ-PCD and plotted in Fig.1. The measured carrier lifetime is seen to rise with increasing the

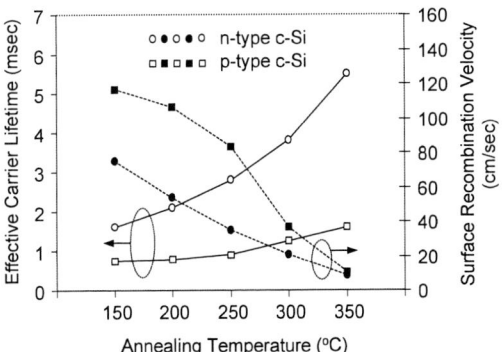

Figure 1. Effective carrier lifetime measured by μ-PCD and calculated surface recombination velocity vs. annealing temperature for n-type and p-type c-Si wafers passivated with a-Si:H.

anneal temperature for both n-type and p-type Si wafers. The improved effective carrier lifetime is a result of the reduced surface recombination velocity at the a-Si:H/c-Si interface. The effective carrier lifetime measured by μ-PCD depends on carrier recombination both in the bulk of c-Si (which remains unchanged during annealing) and at the a-Si:H/c-Si interface and may be expressed as

$$\frac{1}{\tau_{eff}} = \frac{1}{\tau_{bulk}} + \frac{2S}{W} \cdot \qquad (1)$$

where τ_{eff} is the effective carrier lifetime, τ_{bulk} the carrier lifetime in the bulk, S the surface recombination velocity of carriers, and W the thickness of the Si wafers. Knowing the bulk lifetime of the Si wafers (~8ms for n-type and ~1.8ms for p-type wafers), the surface recombination velocity can be calculated (Fig.1). Such an improvement in a-Si:H passivation by annealing is consistent with previous reports [1-3].

To investigate the improvement in a-Si:H passivation, the depth profile of hydrogen was measured before and after annealing at 350°C using SIMS analysis (Fig.2). This analysis shows a negligible change in hydrogen concentration as a result of annealing. The FTIR absorption spectra of the a-Si:H films before and after annealing do not indicate any change in the hydrogen bonding coordination of Si atoms in a-Si:H (Fig.3).

Annealing the samples at 400°C degrades the effective carrier lifetime of the samples (to ~2.4ms for n-type and

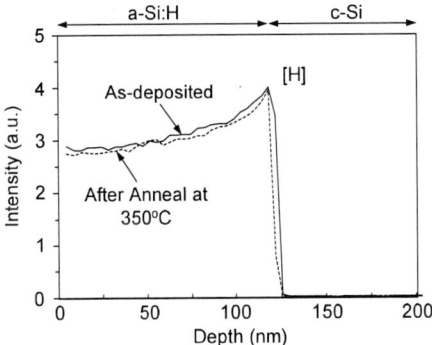

Figure 2. Hydrogen SIMS profile of n-type c-Si wafers passivated with a-Si:H before and after annealing at 350°C.

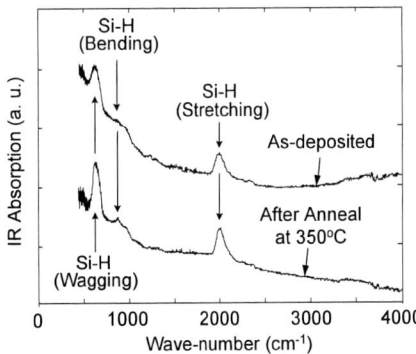

Figure 3. Fourier transform infra-red absorption spectra of a-Si:H passivation on n-type c-Si before and after annealing at 350°C.

~850µs for p-type substrates). This degradation may be attributed to the loss of hydrogen and/or partial crystallization of the a-Si:H film and is consistent with earlier reports [1-3]. Such changes in a-Si:H are irreversible and as such are not attributable to metastability. We show in the next subsection; however, that the improvement in a-Si:H passivation at lower temperatures is reversible.

B. Thermal Relaxation Experiments

The improved carrier lifetime of the samples annealed at temperatures in the range of 200-350°C is relaxed to lower values upon subsequent thermal relaxation treatment over the

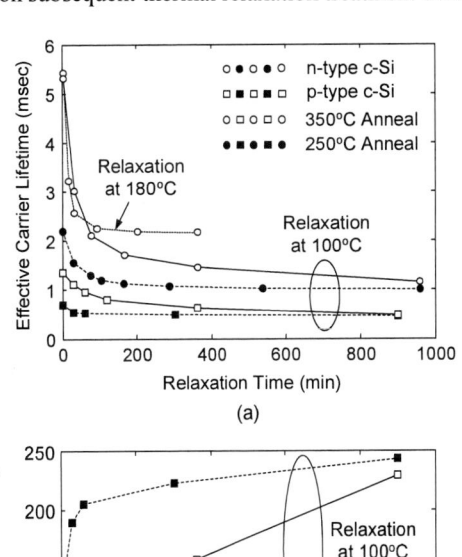

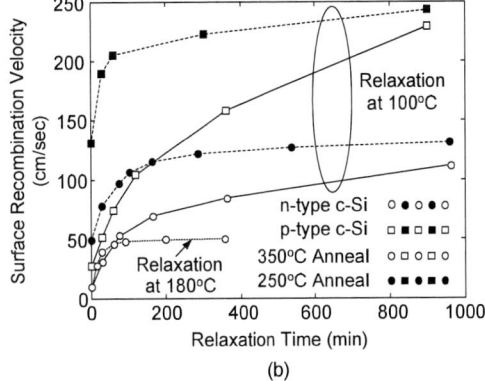

Figure 4. (a) Effective carrier lifetime measured by μ-PCD and (b) calculated surface recombination velocity vs. time at an annealing temperature of 100°C for n-type and p-type samples previously annealed at 250°C and 350°C.

978-1-4244-9113-1/11 $26.00 © 2011 IEEE   563

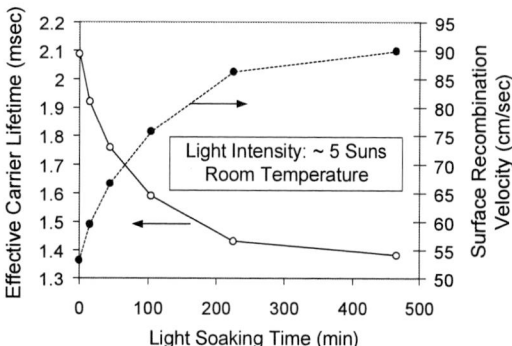

Figure 5. Effective carrier lifetime measured by μ-PCD and calculated surface recombination velocity as functions of light-soaking time under an intensity of ~5 suns at room temperature for *n*-type samples previously annealed at 200°C.

temperature range 100-180°C (Fig.4). The relaxation is faster at higher temperatures, suggesting a thermally activated process. The thermal relaxation process is reversible and the initial lifetime of the samples is recovered by annealing the samples at the corresponding initial annealing temperatures in the range of 200-350°C (data not shown in the figure).

C. Light Soaking Experiments

Illumination degrades the effective carrier lifetime as shown for the samples annealed at 200°C after deposition and then light-soaked under ~5 suns at room temperature (Fig. 5). This degradation is reversible and the lifetime of the sample is recovered by subsequent thermal treatment at ~200°C (not shown in the figure). The metastability of *a*-Si:H passivation under illumination has been reported in the literature; but the rearrangement of Si-H bonding configuration close to the interface has been speculated to be the responsible mechanism rather than the equilibration of charged defects [4].

D. Proposed Mechanism

We explain the experimental results based on the equilibration of charged defects (dangling bonds) in *a*-Si:H. In *a*-Si:H, dangling bonds (defects) are created from weak (strained) Si–Si bonds in a reversible process, with the charge state of the defects depending on the position of the Fermi level [5-8]. If the Fermi level is above the intrinsic level ($E_f > E_i$), i.e. in electron accumulation, negatively charged dangling bonds have the lowest formation energy and are the dominant defects in thermal equilibrium (Fig. 6(a)). This is because the formation energy of a dangling bond is lowered by the energy released from the trapping of a free electron into a dangling bond. Similarly, if ($E_f < E_i$), i.e. in hole accumulation, positively charged dangling bonds have the lowest formation energy, and are the dominant defects (Fig. 6(b)). If E_f is raised above or lowered below E_i by ΔE_f, the density of the negatively charged or positively charged defects is increased by a factor of $\exp(\Delta E_f/kT)$, respectively. Therefore the density of defects in thermal equilibrium depends on the E_f position.

The temperature above which the weak bonds and dangling bonds (defects) are in thermal equilibrium is known as the "freeze-in" temperature and is approximately 200°C for *a*-Si:H. At temperatures below the freeze-in temperature, the rearrangement of the density of states of the charged defects

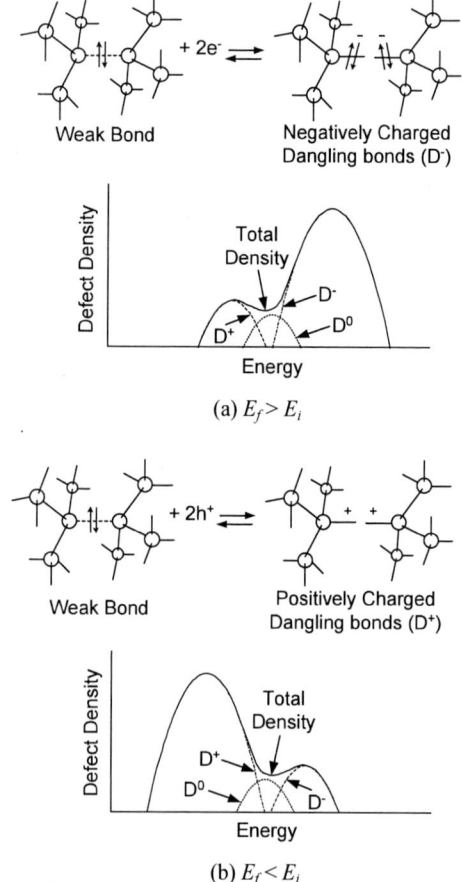

(a) $E_f > E_i$

(b) $E_f < E_i$

Figure 6. Schematic illustration of the formation of charged defects in *a*-Si:H and the resulting defect density under (a) electron and (b) hole accumulation.

towards the equilibrium value does not take place immediately upon a change in the position of the Fermi level. The time required to reach equilibrium is of the order of one year at room temperature but reduced exponentially by increasing the temperature. This is because the creation and annihilation of dangling bonds is a structural change and therefore thermally activated.

The proposed mechanism is illustrated schematically in Fig. 7 (a)-(c) for *n*-type *c*-Si. In thermal equilibrium at room temperature (Fig. 7(a)), the *n*-type *c*-Si is partially depleted of electrons while there is electron accumulation in *a*-Si:H close to the *c*-Si/*a*-Si:H interface. Therefore the negatively charged dangling bonds are the dominant defects in *a*-Si:H. At higher temperatures (during annealing), the Fermi level in *c*-Si moves towards the intrinsic level due the thermal generation of carriers (2.5 Ω-cm *n*-type *c*-Si becomes nearly intrinsic at 300°C) and the band-bending at the junction and therefore the density of negatively charged dangling bonds in *a*-Si:H is reduced in thermal equilibrium (Fig. 7(b)).

Upon cooling down to room-temperature, the Fermi level moves back to its original position (over a short time scale, of the order of the dielectric relaxation time in *c*-Si), while the relaxation of the density of charged defects to its original distribution takes place over a much longer time scale (of the

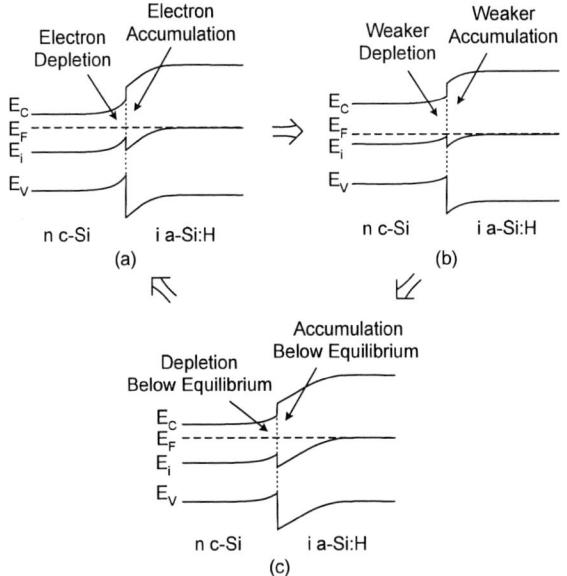

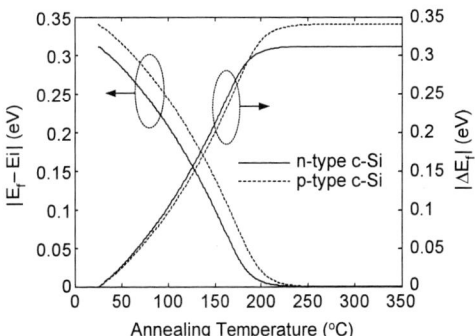

Figure 7. Energy band diagram of the heterojunction formed by *a*-Si:H passivation on *n*-type *c*-Si, (a) in thermal equilibrium, at room temperature, (b) in thermal equilibrium at higher annealing temperatures, and (c) at off-equilibrium conditions upon cool-down to room-temperature after annealing.

Figure 8. The calculated position of the Fermi level (E_f) with respect to the intrinsic level (E_i) and the E_f shift upon annealing, versus annealing temperature in 2.5 Ω-cm *n* and *p*-type *c*-Si.

order of one year). Therefore, due to the lower density of negatively charged defects in *a*-Si:H at off-equilibrium conditions after annealing, the electric field at the junction (which attracts the photogenerated holes to the *c*-Si/*a*-Si:H interface) and the density of the depletion charge in *c*-Si are both reduced compared to the equilibrium conditions prior to annealing. Thus, carrier recombination in *c*-Si close to the interface is reduced and a higher lifetime is measured by μ-PCD. The measured lifetime is reduced with the relaxation of the charged defect density over time (which may be accelerated at higher temperatures), as observed experimentally.

Note that the creation of dangling bonds from weak (strained) Si-Si bonds is a structural change and therefore thermally activated. As a result, the density of dangling bonds is increased by increasing the temperature, by a factor of $\exp(-U/kT)$, where U is the formation energy of neutral (uncharged) dangling bonds, averaging to 0.15-0.2eV for *a*-Si:H [5-7]. However, the change in the position of the Fermi level in *c*-Si is higher than this energy at temperatures above ~150°C as calculated for 2.5 Ω-cm *c*-Si substrates (Fig. 8), and therefore the density of negatively charged dangling bonds in *a*-Si:H is reduced at these annealing temperatures, resulting in a higher carrier lifetime.

The degradation of *a*-Si:H passivation under illumination may be explained by electron trapping (note that $E_f > E_i$ in *a*-Si:H close to the interface) into the dangling bonds (defects) created by the energy released from the recombination of photogenerated electron-hole pairs. The increased density of negatively charged defects induces stronger depletion in *c*-Si and reduces the carrier lifetime over time.

Similar explanations apply to the case of *a*-Si:H passivation on *p*-type *c*-Si, with *a*-Si:H being under hole accumulation close to the junction and positively charged dangling bonds being the dominant defects in *a*-Si:H.

E. Summary and Conclusion

In Summary, we have shown that the improvement of *a*-Si:H passivation on *c*-Si by thermal annealing is reversible and attributed this phenomenon to the equilibration of charged defects in *a*-Si:H. The relaxation of the charged defect density is thermally activated and therefore accelerated by increasing the temperature. Hence annealing at temperatures higher than room temperature may be used for thermal stabilization of *a*-Si:H passivation. This may be a necessary step depending on the reliability requirements of the photovoltaic application employing *a*-Si:H as a passivation layer.

ACKNOWLEDGMENT

The authors are grateful to Prof. Sigurd Wagner of Princeton University for stimulating scientific discussions and allowing usage of his PECVD facility for this work. The authors also gratefully acknowledge technical guidance and encouragement of Dr. Ghavam Shahidi.

REFERENCES

[1] M. Hofmann, C. Schmidt, N. Kohn, J. Rentsch, S. W. Glunz and R. Preu, "Stack System of PECVD Amorphous Silicon and PECVD Silicon Oxide for Silicon Solar Cell Rear Side Passivation", Prog. Photovolt: Res. Appl. vol. 16, pp. 509-518, June 2008.

[2] U. Rau, V. X. Nguyen, J. Mattheis, M. Rakhlin, and J. H. Werner, "Recombination at a-Si:H/c-Si Heterointerfaces and in a-Si:H/c-Si Heterojunction solar cells", Proc. World Conf. Photovoltaic Energ. Conver. Pp. 1124-1127, May 2003.

[3] S. Dauwe, J. Schmidt, and R. Hezel, "Very low surface recombination velocities on p- and n-type silicon wafers passivated with hydrogenated amorphous silicon films", Conf. Rec. IEEE Photovoltaic Spec. Conf. pp. 1246-1249, May 2002.

[4] H. Plagwitz, B. Terheiden, and R. Brendel, "Staebler–Wronski-like formation of defects at the amorphous-silicon–crystalline silicon interface during illumination", J. Appl. Phys. vol. 103, pp. 094506-1-4, May 2008.

[5] M. Stutzmann, "Weak bond-dangling bond conversion in amorphous silicon", Philos. Mag. B, vol. 56, pp. 63-70, January 1987.

[6] Z. E. Smith and S. Wagner, "Band tails, entropy, and equilibrium defects in hydrogenated amorphous silicon", Phys. Rev. Lett. vol. 59, pp. 688-691, August 1987.

[7] R. A. Street, "The origin of metastable states in a-Si:H", Solar Cells, vol. 24, pp. 211-21, July-August 1988.

[8] M. J. Powell, I. D. French and J. R. Hughes, "Evidence for the defect pool concept for Si dangling bond states in a-Si:H from experiments with thin film transistors", J. Non-Crystalline Solids, v 114, pp. 642-644, August 1989.

Optical Stress and Reliability Study of Ruthenium-based Dye-Sensitized Solar Cells (DSSC)

Daniele Bari, Nicola Wrachien, Andrea Cester, Gaudenzio Meneghesso
DEI, Department of Information Engineering
University of Padova
Padova, Italy
phone: +39 –049- 827-7625, daniele.bari@dei.unipd.it

Roberto Tagliaferro, Stefano Penna, Thomas M. Brown, Andrea Reale, Aldo Di Carlo
CHOSE – Department Electronics Engineering
University of Rome "Tor Vergata"
Rome, Italy

Abstract—**In this work, we study the reliability of sensitized solar cells. The reliability study is carried out using accelerated optical and pure thermal stresses with the purpose to extract a degradation law as a function of accelerated parameter and to understand which is the mechanisms involved in the degradation. The responsible of the degradation during thermal or illumination stress is the formation of defects and chemical species at TiO_2/sensitizer/electrolyte interface which reduces the charge transfer at interface and the ion migration across electrolyte.**

Dye-sensitized solar cells; ruthenium-based dye; reliability of solar cell; thermal stress.

I. INTRODUCTION

Since Michael Gratzel advanced the concept of sensitized materials and nanoporous semiconductor, Dye-Sensitized Solar Cells (DSSC) have been studied a long to date [1,2]. Due to the low production and material costs, and the suitability to many applications, this type of solar cell, is very appealing and it could detain a remarkable share in the renewable-energy market in the future. Moreover, DSSC can be fabricated on many different types of substrates such as polymers, flexible materials, glass etc. This makes the DSSC concept a promising alternative to conventional photovoltaic devices especially for low cost applications [3-5]. Up to date this type of solar cells has reached global AM 1.5 power conversion efficiencies up to 11% [6].

The general structure of these devices is shown Fig. 1a and consists of: a transparent conductive oxide (TCO), which acts as photo-anode; a transparent semiconductor oxide; a dye-sensitizer; an electrolyte solution dissolved in an organic solvent or a p-type semiconductor; and a second TCO, which acts as photo-cathode. The choice of the material and the structure of each layer is still a challenge in the fabrication of dye-sensitized solar cells, because the material variations can have a strong impact on the general efficiency of the solar cells.

Different materials were evaluated as transparent semiconductor, the most important are: TiO_2, ZnO, SnO_2, Nb_2O_5. The first is the most used, due to its good properties. Furthermore, TiO_2 is widely available on earth, and it is also biocompatible, and nontoxic. TiO_2 can be deposited on various substrates like as conductive substrates (conductive glasses, metallic films and flexible polymeric sheets). Doctor blading and screen printing techniques are the most used to deposit TiO_2 over different substrates [7]. These techniques are currently used in other production process (such as manufacturing) and they are well-suited for DSSCs manufacture. In the development of the semiconductor, it was observed that a multilayer structure is the best trade-off between large active area and reduction of the internal recombination during the carrier transport in the semiconductor. A typical solution is structure made of three layers with a growing porosity from the photo-anode to the dye-sensitizer interface.

The sensitizer material absorbs and convert all the incident sunlight into photocurrent [8]. A perfect sensitizer should absorb all light with wavelength shorter than 920nm, have a great thermal stability, and it should also strongly bind to the semiconductor oxide. Moreover it should be able to inject electrons in the conduction band of the semiconductor with 100% efficiency, i.e., for each electron photogenerated, an electron is injected in the conduction band of the semiconductor. Finally, it should sustain at least 10^8 redox cycles, which is equal to about 20 years of sunlight illumination exposure [9]. The sensitizers can be organic or inorganic. The former are easier to design and less expensive than the latter. Moreover the latter feature a stronger thermal stability. Organic dyes include natural and synthetic organic dyes. Hemicyanine dyes [10], thienothiophene and thiophene-based dyes [11], phenyl-conjugated oligothiophene dye [12, 13] and coumarin-based dye [14, 15] are some examples of organic dyes. Some examples of inorganic sensitizers are:

978-1-4244-9113-1/11 $26.00 © 2011 IEEE

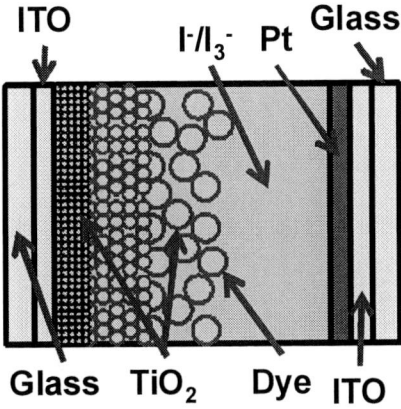

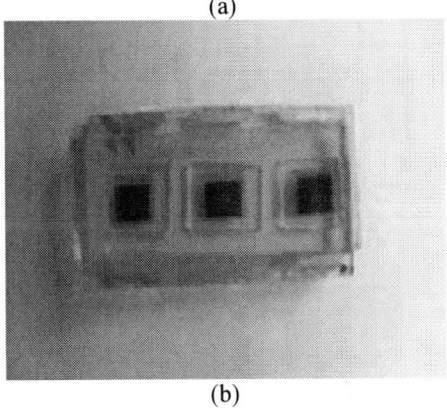

Figure 1. a) Schematic drawing of the cross-section of a DSSC structure (not to scale). b) Picture of a DSSC sample used through this work. Each samples consists of three independent cells.

metal complexes, such as polypyridyl complexes of ruthenium and osmium, metal porphyrin, phthalocyanine and inorganic quantum dots. In particular, ruthenium-based sensitizer was widely investigated because, among all the sensitizers, ruthenium-based dyes show high stability, remarkable redox properties and large absorption on IR and visible sunlight spectrum [4].

The electrolyte is another key parameter to obtain high efficiency solar cells. It transports the positive charge from the dye to the counter electrode. This allows the chemical reduction, i.e. to return to its stable condition. The electrolytes phase can be liquid, quasi-solid and solid. The liquid electrolyte can be divided into organic solvent electrolyte and ionic liquid electrolyte. An ideal electrolyte should have a low viscosity, fast ion diffusion, high efficiency, and it should be easy to design. The electrolyte consists of a solvent, a redox couple and an additive. Examples of solvents are: acetronitrile, valeronitrile, 3-methoxypropionitrile and esters such as ethylene carbonate, propylene carbonate, and gamma-butyrolactone. Some types of used redox couples are: I_3^-/I^-, Br^-/Br_2, $SCN^-/(SCN)_2$, $SeCN^-/(SeCN)_2$, bipyridyl cobalt(III/II). To date the couple I_3^-/I^- has shown the best results as concerned the efficiency of the DSSCs. The presence of additives in electrolytes can improve the efficiency of the solar cells. The additives can suppress the dark current and improve

the conversion efficiency. N-methylbenzimidazole and 4-tert-butylpyridine are the most used additives in the realization of dye-sensitizers solar cells. The efficiency improvement is mainly due to the reduction of the recombination between the electrons in the conduction band of the semiconductor and the electron acceptors in the electrolyte [16]. The encapsulation of the solar cells is a main challenge in the design of solar cell that uses liquid electrolyte. In fact a disadvantage of this kind of solar cells is the evaporation of the electrolyte. To avoid this problem, the liquid electrolyte is substituted with a p-type semiconductor, but the conversion efficiency is much smaller than the liquid electrolytes [17].

The DSSC reliability study is a critical issue. To date, most of the works on DSSCs present in literature deal only with the research of new materials and new structures with the intent to obtain more efficient solar cells, without focusing on the reliability. In this work, we studied the reliability of DSSCs using accelerated optical and thermal stresses by illuminating the devices with an AM 1.5 xenon lamp solar simulator and keeping the solar cells in climatic chambers at different temperatures.

II. DEVICES AND EXPERIMENTAL SETUP

In this work we considered ruthenium-based DSSCs. The core of each cell is the nanoporous semiconductor TiO_2 at which is anchored the sensitized material. The TiO_2 is deposited over one of the conductive TCO-glass substrate by sintering nanoparticles of TiO_2 at high temperature. The ruthenium-based dye is impregnated in the nanoporous semiconductor. The resulting active area, which harvests the sunlight, is $25mm^2$. Using a thermoplastic gasket, the working electrode is hot-sealed with the second conductive Platinum-coated TCO-glass substrate which works as the counter electrode. Finally, the Iodine/Iodide-based liquid electrolyte is injected in the structure through the counter electrode by vacuum back filling technique. The pinhole is finally sealed by the same thermoplastic sealant used previously [18]. Each sample contains three electrically isolated solar cells as shown in Fig. 1b.

To obtain good and reproducible measurements we created an "ad hoc" measurement setup. This setup guarantees good and stable electrical contacts between the devices and the instruments.

The experimental procedure begins with the preliminary characterization of the fresh device. The characterization is performed inside a metallic box, which ensures the electrostatic shielding and it prevents any disturbs from the environmental illumination. After a warm up of the xenon lamp, the device under test is illuminated with an optical intensity of 1Sun ($100mW/cm^2$). Before the electrical characterization we hold the sample under illumination at 1 Sun and at open-circuit condition for five minutes to allow the device to reach the thermal equilibrium. These hold time has been experimentally estimated by continuously monitoring the open-circuit voltage and waiting for it reaches the steady-state value. The characterization consist of: open-circuit voltage (V_{OC}) and short-circuit current (I_{SC}) measure, the current density-voltage curves (J-V), and electrochemical impedance spectroscopy analysis (EIS). Later on the devices were illuminated at a given

978-1-4244-9113-1/11 $26.00 © 2011 IEEE

optical intensity ranging from 8 to 15 Sun. During illumination, each solar cell under stress is loaded with a 12-ohm resistor, which keeps the cell close to the short-circuit operating point. The choice of the load value is arbitrary, but it ensures that the cell works away from its point of maximum efficiency close to short circuit condition. To avoid the light exposure we masked the neighboring cells. The stresses were periodically stopped to allow electrical characterization with the same procedure described above at room temperature.

During optical stresses we measured an increase of the temperature of the sample, especially at the higher optical intensities and long stress steps. To assess the impact (if any) of the "passive thermal stress", we have also monitored the cells that were not under irradiation. The impact of the temperature has also been evaluated performing pure thermal stress at different temperatures. The devices were stressed in dark and at open circuit condition at different temperatures (60°C, 70°C, and 85°C). The thermal stress was periodically stopped to allow the electrical characterization, which has the same setup employed for the optical stress described above and it has been done at room temperature.

III. RESULTS

Figs. 2-4 summarize the results of the illumination stress. Figs 2a and 2b show the degradation of V_{OC} and J_{SC} of a cell stressed at 15Sun. For comparison, we show the evolution of V_{OC} and J_{SC} of one of the two neighboring cells, which is subjected to a parasitic thermal stress due to the glass heating

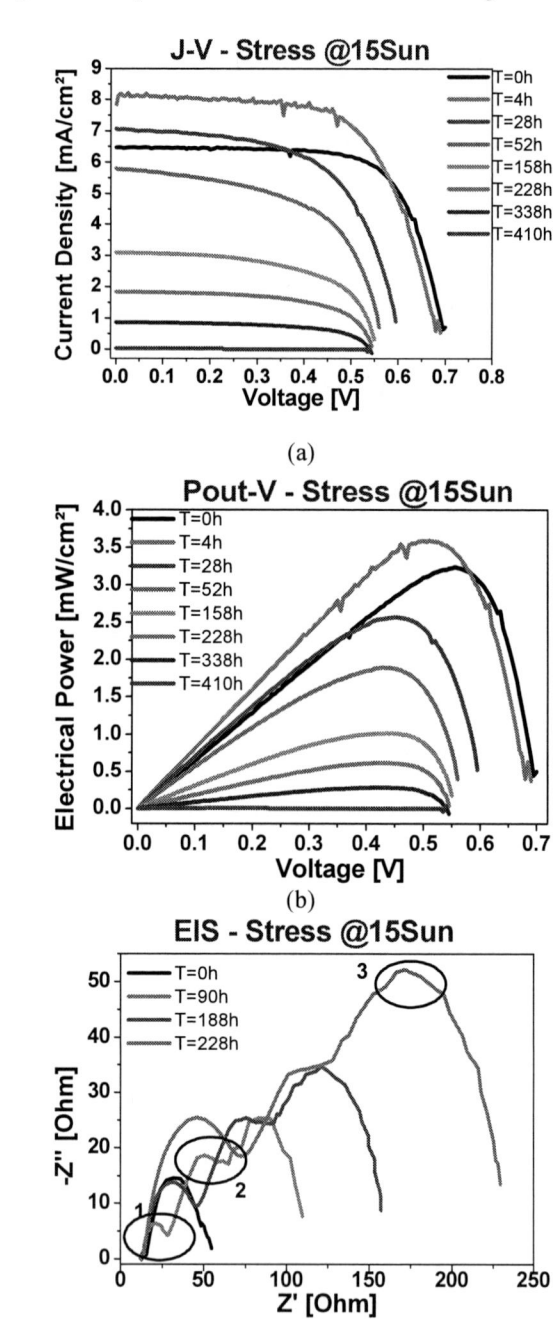

(a)

(b)

(c)

Figure 3. Degradation of the DC and AC characteristics of a DSSC illuminated at 15Sun during stress. Measure have been taken under illumination at 1Sun: a) current density-voltage; b) ouput power-voltage; c) electrochemical impedance spectroscopic analysis (EIS) stress time for four cells illuminated at 8, 10, 12, and 15 Sun.

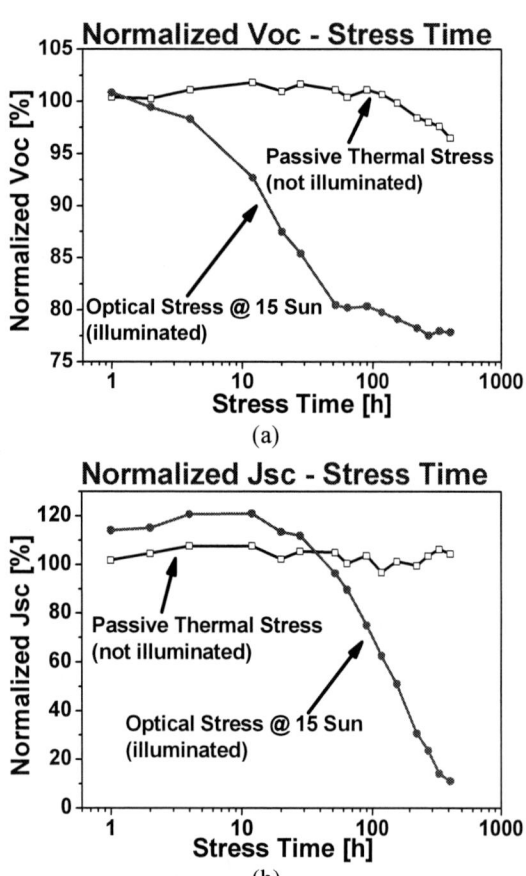

(a)

(b)

Figure 2. Trends of the open circuit voltage (a) and the short circuit current (b), of two cells, one subjected to direct optical stress and the other subjected to passive therml stress. The graph of the open circuit voltage of the passive cell does not change abruptly during the optical stress of the other cell.

(see curve marked as "Passive Thermal Stress" in Fig. 2). The shaded cells show very moderate degradation indicating that the evolution of the illuminated cell is dominated by the effects of light exposure and to electrical conduction.

Fig. 3 shows the current density-voltage (J-V), the power-voltage (P-V) (which corresponds to the efficiency

characteristic) and the EIS characteristics during stress. The reported characteristics refer to the device stressed at 15 Sun. The devices stressed at 8, 10, and 12 Sun have the same behavior.

Figs. 4a, 4b, and 4c show, the evolution of V_{OC}, J_{SC}, and the efficiency as a function of the stress time for different

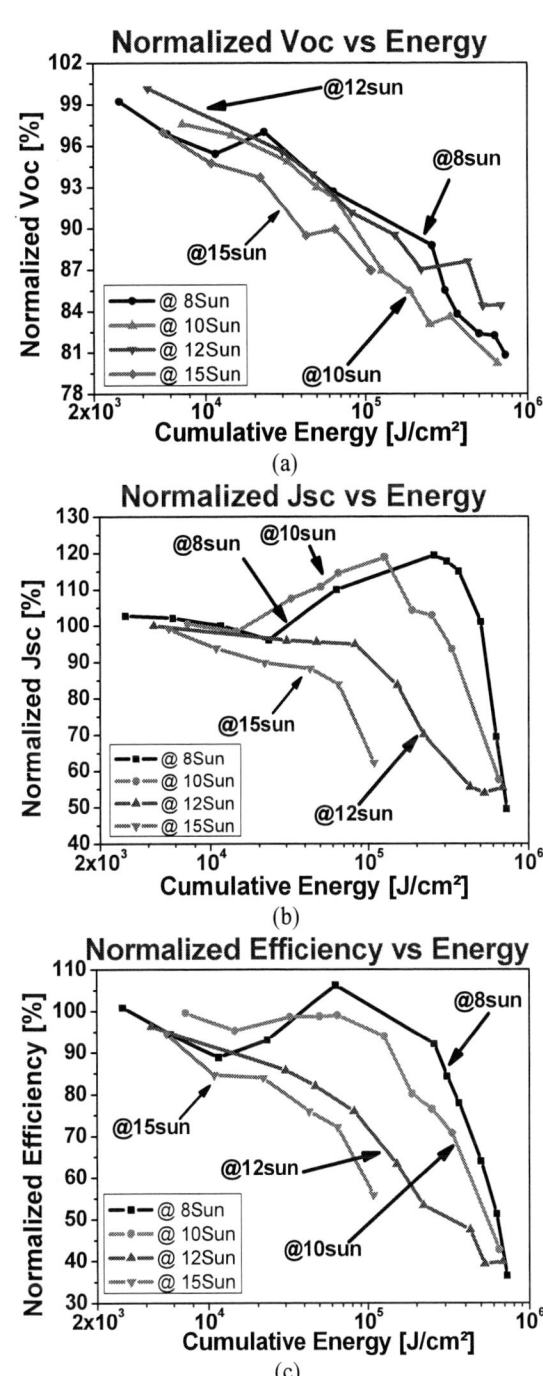

Figure 4. Evolution of V_{OC} (a), J_{SC} (b), and efficiency (c) as a function of stress time for four cells illuminated at 8, 10, 12, and 15Sun.

Figure 5. Evolution of V_{OC} (a), J_{SC} (b), and efficiency (c) as a function of the total cumulative energy irradiated during stress for four cells illuminated at 8, 10, 12, and 15Sun.

978-1-4244-9113-1/11 $26.00 © 2011 IEEE

illumination level (8, 10, 12, and 15 Sun), respectively. Similarly, Figs. 5a, 5b, and 5c show the same three parameters as a function of the cumulative irradiated energy.

Finally, we subjected some cell to a pure thermal stress, by storing the device at 60°C, 70°C and 85°C in a climatic

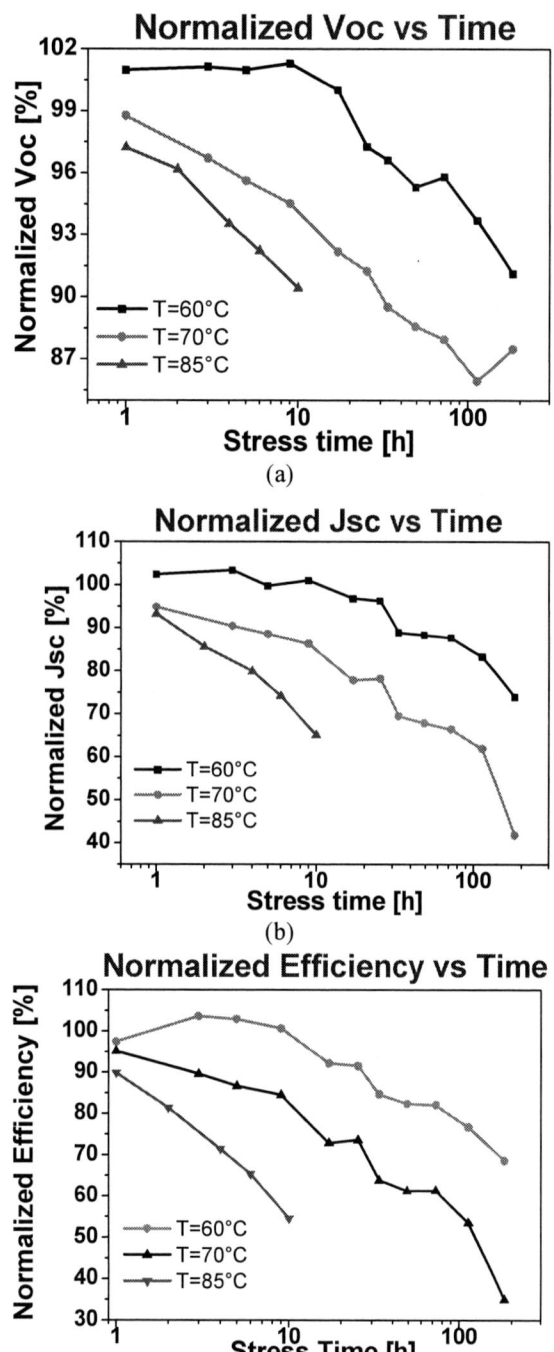

Figure 6. Evolution of V_{OC} (a), J_{OC} (b), and efficiency (c) as a function of storage time during the thermal stress at temperature of 60°C, 70°C and 85°C. The devices has been kept floating during storage.

chamber at open circuit condition. The most relevant results are shown in Fig. 6. Efficiency and short circuit current quickly drop below the 30% of their initial value in less than 30-hours storage at 85°C.

IV. DISCUSSIONS

Our results clearly show that the cells subjected to direct illumination stresses feature: 1) a small decrease of the open-circuit voltage; 2) a very strong reduction of the short-circuit current; 3) a strong reduction of the maximum output power; and 4) an enlargement of the EIS curve.

First of all the strong degradation of the cell characteristics is a consequence of light exposure. In fact, by comparing the open and filled symbol in figures 2a and 2b, we observe a 20% degradation of V_{OC} and almost complete collapse of J_{SC} after 400-hours stress at 15 Sun. In contrast the not illuminated devices show only the 4% variation of V_{OC} and only marginal variation of J_{SC}. This degradation may be due to either parasitic illumination or thermal stress. As described above, the active area of the solar cell is sandwiched between two glasses. In principle, even if the not-illuminated solar cells are masked, the glass can act as light guide and some of the incident light can reach the masked cells. To quantify how much light travels inside the glass and reaches the active area, we have performed characterization measurements at 1Sun, illuminating one cell. and measuring the shaded ones. We have observed that the value of V_{OC} is about three times lower than the one measured illuminating directly the cell, while the value of J_{SC} is about one thousand time lower. Hence we concluded that this parasitic optical stress only marginally impact on the degradation of the shaded cells. This is confirmed by the measurements performed on the shaded cells, as shown in Fig. 2.

The following discussion is split in three parts. In section IV.A we discuss the degradation of DC parameters under illumination. In section IV.B we draw some hypotheses on the nature of damage by analyzing the shape of EIS during illumination stress. Finally, in section IV.C we compare the thermal and illumination stresses.

A. Degradation of the cell characteristics

From the DC measurements we can draw some considerations. The V_{OC} and J_{SC} degradation kinetics are very different: V_{OC} is monotonically and almost linearly decreasing in the logarithmic-scale time of Fig. 4a independently by the optical intensity. Conversely, J_{SC} kinetics features a different behavior: at the beginning of the stress, J_{SC} has an almost constant or even increasing evolution, but after some hours it features a turnover and it starts quickly decreasing. The duration of the initial constant/increasing trend depends on the applied illumination intensity. In fact, the turnover is more pronounced at illumination intensities of 8-10 Sun than for 12-15 Sun as shown in Fig. 4b. We can figure that the turnover is always present independent on the illumination intensity, but it runs out within the first steps of stresses at 12-15 Sun.

From the V_{OC}-Time (Fig. 4a), J_{SC}-Time (Fig. 4b) and Efficiency-Time curves (Fig. 4c), we note that the degradation rate increases as the optical power increases. For instance, the device illuminated at 8 Sun exhibits a 20% efficiency drop after

150-h illumination. Conversely, if the illumination level is increased at 15 Sun (i.e. almost the double), the 20% degradation is reached only after a 6-h stress, i.e., 25 times earlier. This acceleration of the efficiency degradation mainly derives from the reduction of the J_{SC} current. In fact, Fig. 4a shows that V_{OC} is much less affected by the illumination condition than J_{SC}.

If we plot the degradation of V_{OC}, J_{SC}, and efficiency as a function of the total cumulative irradiated energy, as shown in Fig. 5, the large gap between the degradation kinetics at 8 Sun and 15 Sun reduces, but it is still present. We arbitrarily defined a critical value of the cumulative irradiated energy E_C as the energy required to achieve a 20% reduction of a given cell parameters (V_{OC}, J_{SC}, and efficiency). In Fig. 7 we plot E_C as a function of the optical irradiating power during stress. For all parameters, E_C strongly reduces with increasing irradiation intensity, from 8 Sun to 15 Sun. Assuming an exponential relation between E_C and the illumination intensity, we estimated $E_C = 4 \cdot 10^6$ J/cm^2 at 1Sun, corresponding to 46 days. Such exponential relation suggests that at very high illumination levels some additional degradation phenomena occur. This accelerating factor might be strongly correlated to the increase of the interface temperature induced by the high illumination power. Of course this extrapolation has been done by assuming that the degradation rate maintains the same trend also at low illumination intensity.

B. Analysis of cell degradation by EIS measurements

In principle, the reduction of the J_{SC} could be a signature of an artificial sintering of the TiO$_2$ during the illumination exposition, which is more effective at higher illumination intensities (i.e. 12-15 Sun). In fact, the artificial sintering of the TiO$_2$ should reduce the effective surface area and, in turn, the short circuit current. However, the TiO$_2$ sintering requires very high temperature around 500°C. At 15 Sun we measured the glass temperature of about 50°C. Taking in account the typical thermal resistance of glass, this temperature level is not enough to sinter the TiO$_2$ particles. Moreover, at such high temperature the sealing degradation the electrolyte evaporation are much more likely to occur.

Another hypothesis is the formation of defects at the TiO$_2$\sensitizer\electrolyte interface. Such defects may come from the incomplete regeneration of dye molecules by the electrolyte or the formation of weak/dangling/strained bonds between the sensitizer and the electrolyte. The formation of new species is likely activated by temperature, and then, by illumination intensity. In both cases the dye could not be able to generate electron\hole couple when hit by the sunlight and then reducing the short circuit current.

The EIS measurements in Fig. 3c give us further information on the most likely mechanism involved in the cell degradation. There is a clear enlargement of the EIS plot during stress, even though the Nyquist diagram keeps the same shape from the begging to the end of the stress, featuring the three semicircles typical of DSSC [19]. In fact, the three semicircles are always present, but they feature a very different evolution. The second semicircle, which is correlated

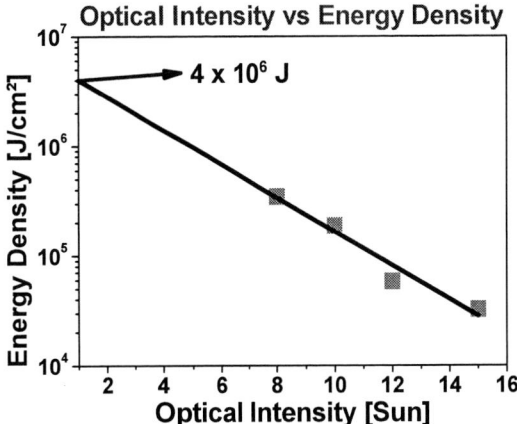

Figure 7. Evolution of E_C as a function of the illumination intensity.

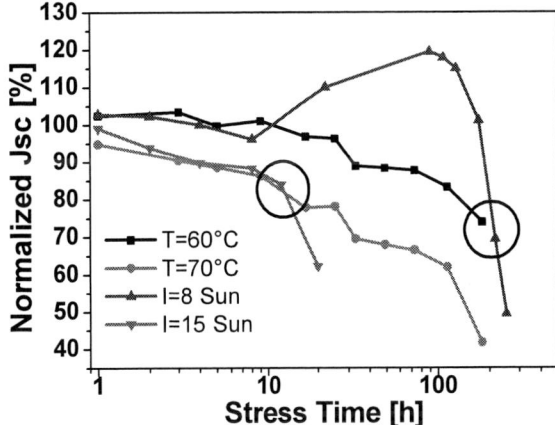

Figure 8. Comaprison between the J_{SC} evolution as a function of stress time during two illumination stress at 8 and 10 Sun, and two thermal stress at 60°C and 70°C.

to the electron transfer at the TiO$_2$/electrolyte interface [19], shows an appreciable variation. This confirms the presence of defects at the TiO$_2$/sensitizer/electrolyte, as discussed above. However, the third semicircle (marked with 3) shows the biggest enlargement. This is correlated with the Nerst diffusion in the electrolyte [19]. This indicates that any chemical reaction due to thermal or illumination stress generates not only defects at the TiO$_2$/electrolyte interface, but also some chemical species within the electrolyte, which may hamper the ion diffusion across the electrolyte.

C. Thermal storage vs. Illumination stress

It is worth to compare the degradation kinetics measured during thermal and illumination stresses (see Fig. 8). The sample illuminated at 15 Sun reaches a 20% degradation of J_{SC} after 10-hours stress. The same degradation is observed after 10-hours thermal stress at 70°C. Similarly, the device illuminated at 8 Sun exhibits a 20% drop after 200-hours which is almost the same drop observed after 200-hours at 60°C (see Fig. 8). These values are in agreement with the interface temperature during illumination stress. At first glance, we can argue that the self heating plays an important role during optical stress and that during illumination the interface temperature may range between 60-70°C, in agreement with the glass thermal resistance. However, the comparison of thermal and optical stress highlights a different shape of the curves plotted in Fig. 8, especially for the stress at 8 Sun. This indicates that besides heating, the degradation is driven by other mechanisms, which are worth of further investigations. Such mechanisms are likely correlated with the current flow. In fact, some preliminary illumination stresses performed at open circuit, i.e. with no current flow, have shown slower cell degradation.

V. CONCLUSIONS

In this work we studied the degradation of DSSC cell subjected to thermal and illumination accelerated stress with different illumination levels. During optical stress the kinetics of degradation is different for the open circuit voltage (V_{OC}) and for the short circuit current (J_{SC}): the former monotonically decreases while the latter increases at the beginning of the stress and then decreases for longer stress time. The duration of the turnover phase is strongly dependent on the illumination intensity used for the accelerated stress: the higher the illumination level, the shorter the turnover phase. Not illuminated cells do not show appreciable degradation of open-circuit voltage and short-circuit current at room temperature as well. The device features faster degradation kinetics with higher illumination levels likely due increase of the interface temperature, as also confirmed by pure thermal stress. The responsible of the degradation during thermal or illumination stress is the formation of defects and chemical species at TiO_2/sensitizer/electrolyte interface which reduces the charge transfer at interface and the ion migration across electrolyte.

These are preliminary results on reliability DSSC cells and many issues remain still open. For instance, what is the nature of these defects? What is the most suitable degradation model? Is this degradation transient or permanent? Can be the damage annealed at higher temperature? How does the degradation kinetics depend on the operating condition of the cell? What is the real acceleration factor at low illumination intensity? All these issues are worth of further investigation.

ACKNOWLEDGMENT

This work was partially supported by Progetto di Ateneo 2009 – Università di Padova, Italy (Project Number CPDA083941), Progetto PRIN 2008 and Polo Solare Organico – Regione Lazio.

REFERENCES

[1] M. Gratzel, "Dye-Sensitized Solar Cell", Journal of Photochemistry and Photobiology A, Vol. 4 p. 145-153, 2003.

[2] B. O'Regan and M. Gratzel, "A low-cost, high-efficiency solar cell based on dye-sensitized colloidal TiO2 films," Nature, vol. 353, p. 737–740, 1991.

[3] M. Gratzel, Nature 2001, 414, 338.

[4] M. K. Nazeeruddin, A. Kay, I. Rodicio, R. Humphry-Baker, E. Mueller, P. Liska, N. Vlachopoulos, M. Graetzel, Journal of the American Chemical Society, vol. 115, p.6382–6390, 1993.

[5] A. Hagfeldt, M. Gratzel Chemical Reviews 1995, 95, 49.

[6] M. Gratzel Journal Photochemical Photobiology A 2004, 164, 3.

[7] L. Zhang, A.J. Xie, Y.H. Shen, S.K. Li "Preparation of TiO2 films by layer-by-layer assembly and their application in solar cell", Journal of Alloys and Compounds Vol. 505 p. 579–583, 2010.

[8] M. Gratzel, "Conversion of sunlight to electric power by nanocrystalline dye-sensitized solar cells" Journal of Photochemistry and Photobiology A, Vol. 164, p. 3–14, 2004.

[9] A. Hagfeldt and M. Gratzel "Molecular photovoltaics", Accounts of Chemical Research, vol. 33, no. 5, pp.269-277, 2000.

[10] Z.-S. Wang, F.-Y. Li, and C.-H. Huang, "Photocurrent enhancement of hemicyanine dyes containing RSO−3 group through treating TiO2 films with hydrochloric acid," Journal of Physical Chemistry B, vol. 105, no. 38, pp. 9210–9217, 2001.

[11] S.-L. Li, K.-J. Jiang, K.-F. Shao, and L.-M. Yang, "Novel organic dyes for efficient dye-sensitized solar cells," Chemical Communications, no. 26, pp. 2792–2794, 2006.

[12] T. Kitamura, M. Ikeda, K. Shigaki, et al., "Phenyl-conjugated oligoene sensitizers for TiO2 solar cells," Chemistry of Materials, vol. 16, no. 9, pp. 1806–1812, 2004.

[13] K. Hara, M. Kurashige, S. Ito, et al., "Novel polyene dyes for highly efficient dye-sensitized solar cells," Chemical Communications, no. 2, pp. 252–253, 2003.

[14] K. Hara, K. Sayama, Y. Ohga, A. Shinpo, S. Suga, and H. Arakawa, "A coumarin-derivative dye sensitized nanocrystalline TiO2 solar cell having a high solar-energy conversion efficiency up to 5.6%," Chemical Communications, no. 6, pp. 569–570, 2001.

[15] K. Hara, Z.-S. Wang, T. Sato, et al., "Oligothiophenecontaining coumarin dyes for efficient dye-sensitized solar cells," Journal of Physical Chemistry B, vol. 109, no. 32, pp. 15476–15482, 2005.

[16] Fan-Tai Kong, Song-Yuan Dai, and Kong-JiaWang, "Review of Recent Progress in Dye-Sensitized Solar Cells", Advances in OptoElectronics, Vol. 2007, 2007.

[17] G. R. A. Kumara, S. Kaneko, M. Okuya, and K. Tennakone, "Fabrication of dye-sensitized solar cells using triethylamine hydrothiocyanate as a CuI crystal growth inhibitor," Langmuir, vol. 18, no. 26, pp. 10493–10495, 2002.

[18] M. Liberatore, F. Decker, L. Burtone, V. Zardetto, T.M. Brown, A. Reale, A. Di Carlo, J Appl Electrochem Vol. 39, p. 2291–2295, 2009. M. Liberatore, F. Decker, L. Burtone, V. Zardetto, T.M. Brown, A. Reale, A. Di Carlo, J Appl Electrochem Vol. 39, p. 2291–2295, 2009.

[19] W. Qing, Moser J., Gratzel M., "Electrochemical Impedance Spectroscopic Analysis of Dye-Sensitized Solar Cells", J. Phys. Chem. B, 109, 14945-14953, 2005.

The Role of Elastic and Plastic Anisotropy of Sn on Microstructure and Damage Evolution in Lead-Free Solder Joints

Thomas R. Bieler, Bite Zhou, Lauren Blair
Chemical Engineering and Materials Science
Michigan State University
East Lansing MI 48824

Amir Zamiri, Payam Darbandi, Farhang Pourboghrat
Mechanical Engineering
Michigan State University
East Lansing MI 48824

Tae-Kyu Lee, Kuo-Chuan Liu
Component Quality and Technology Group
Cisco Systems, Inc., San Jose, CA 95134

Abstract— The elastic, thermal expansion, and plastic anisotropy of Sn is examined to assess how anisotropy affects the microstructural evolution and damage nucleation processes in SAC305 solder joints. Examination of all joints in a package indicates that upon solidification, crystal orientations are nearly randomly distributed. Initial studies of cracked joints after thermal cycling showed that orientations with the c-axis parallel to the joint interface (red orientations) are more likely to crack arising from tensile stresses during the hot part of the cycle. Subsequent studies show that package design has a large influence on how the microstructure evolves; higher strain designs stimulate recrystallization at earlier times. Recrystallization appears to be strongly correlated with crack nucleation and propagation processes, as red orientations often develop and lead to crack nucleation and propagation. The details of the recrystallization process depend strongly on the plastic slip and recovery processes arising from the specific crystal orientation / temperature / strain history that makes microstructural evolution of each joint unique. The unique history for each joint implies that worst case scenarios need to be identified and models developed that can predict microstructural evolution that leads to worst case scenarios.

Keywords- Sn; microstructure; anisotropy; thermal cycling; slip systems; thermal expansion; damage

I. MOTIVATION

In the past decade, Pb-free solder joints have been shown to behave quite differently from Sn-Pb joints, arising largely from the lack of the soft and isotropic properties caused by Pb, which has masked the dramatically anisotropic properties of Sn. Failures in Sn-based solder joints have occurred everywhere in the package, not just predominantly at corners, where extrinsic cyclic CTE strains are largest [1]. Furthermore, Sn-based solder joints exhibit single crystal orientations, tri-crystal orientations, or oligocrystal microstructures [2-8], which cause inhomogeneous deformation behavior. However, design engineers seek simple homogeneous and isotropic descriptions of Sn-based solder deformation behavior for computational models used in package design (e.g. [9]) because more accurate models for the deformation of Sn are not convenient, or even known. With this approach it is impossible to address reliability prediction of Sn-based solder joints in a physically satisfying way, because the deformation behavior of Sn is far from isotropic, and it evolves in complex, but understandable ways.

The effect of crystal orientation on joint performance was dramatically illustrated in a study of one interior row of a package exposed to 2500 thermal cycles, where 4 of 11 joints exhibited cracks. Figure 1 shows the pattern of the cracks observed, and two examples of an optical micrograph and its associated c-axis orientation map. The orientation maps were obtained using automated electron backscattered pattern (EBSP) indexing that can be used to image microstructures and crystal orientations simultaneously (also known as Orientation Imaging Microscopy[TM]). This tool was first used in 2002 to show that lead-free solder joints are not polycrystals like Sn-Pb [10,11], and that heterogeneous deformation of Sn-based solder can be attributed to varying orientations of the Sn phase. Investigations of the evolution and performance of the Sn phase is lacking, especially in contrast to studies of microstructural evolution of interfacial intermetallic (IMC)

978-1-4244-9113-1/11 $26.00 © 2011 IEEE

phases, which dominate the solder literature. (The properties and evolution of the IMC phase is important, especially in high strain rate conditions, where Sn becomes stronger than the IMC, but in the normal operation of electronic equipment that is not dropped, the evolution of the Sn phase microstructure has much greater impact on reliability, because cracks normally grow through the Sn phase.)

This paper focuses on the anisotropic properties of Sn, followed by an examination of how the package design affects the thermal strain history, and consequently microstructural evolution and damage nucleation/propagation. As deformation in metals involves dislocation movement, and elevated temperature causes recovery and recrystallization, the interplay between these phenomena drives the evolution of the microstructure and properties of solder joints, an assessment of how the interplay between activated slip systems arising from thermal strains, recovery and recrystallization lead to damage nucleation and crack growth, and hence, reliability of a given joint.

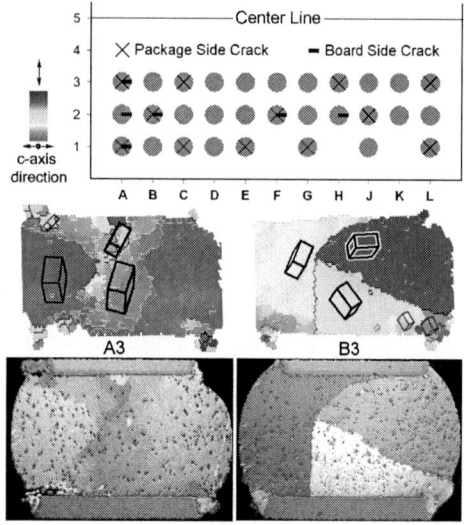

Figure 1. Location of fractured joints after 2500 0-100°C thermal cycles in a sequentially sectioned 10x12 mm package with 0.42mm SAC305 solder balls with a pitch of 800 μm on a 0.32mm NSMD pad. The 8x10 mm die was wire bonded to a thin BT substrate and filled with epoxy [1]. The Sn c-axis was parallel to the substrate (red) in all 4 fractured joints in row 3.

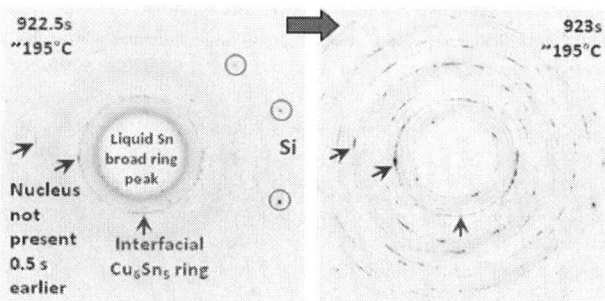

Figure 2. Transition between liquid and solid in diffraction patterns taken 1s apart in a SAC 305 solder joint residing on a Si chip with Au/Ni pads.

II. SOLIDIFICATION MICROSTRUCTURE

The evolution of the microstructure of a Sn-based solder joint begins with the moment of solidification. Because of the small volume of the joints, nucleation is delayed until significant undercooling occurs (this can be altered with minor alloying additions [12]). Figure 2 shows two sequential diffraction patterns taken from synchrotron radiation during in-situ solidification of a Sn-3.0Ag-0.5Cu (wt%) (SAC305) solder ball on a Si substrate with Cu(OSP) interface (Si diffraction spots are also visible). The solidification process takes about 1 s to make the transition from liquid to solid, and there is no evidence of multiple nuclei in the sequential diffraction patterns (contrary to evidence in similar experiments on a different alloy [13]). The diffraction patterns can be indexed (Figure 3), and they generally show the existence of either 1 orientation (sometimes with well defined low angle boundaries) or three orientations present in the entire volume of the joint [13,14]. Preliminary in-situ studies of the solidification process suggest that the as-solidified orientations are retained in joints as they cool.

Studies of solidification in different alloys shows that Cu favors formation of single crystals or tricrystals having solidification twin relationships with ~60° rotations about a common [100] axis [3]. In some joints, this twin orientation relationship leads to a microstructure that resembles a beachball or an orange with 6 segments (3 orientation, shown in Figure 4. More typically, joints do not have grains shaped like an orange slice, such as joint B3 in Figure 1, but even so it has the same beachball orientation relationship. In other alloys and metallization environments, microstructures can differ from the trends observed in SAC305 [6-8].

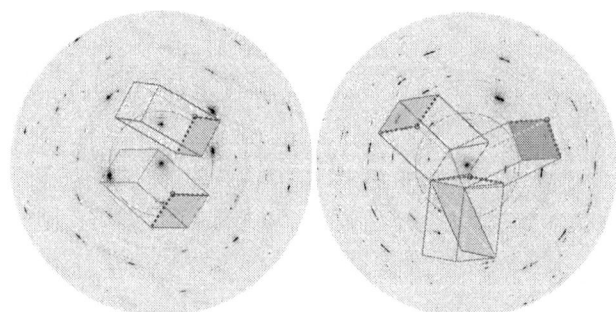

Figure 3. Transmission diffraction patterns indexed to show crystal orientations present in solder joints in a large FCBGA package (Cu/OSP).

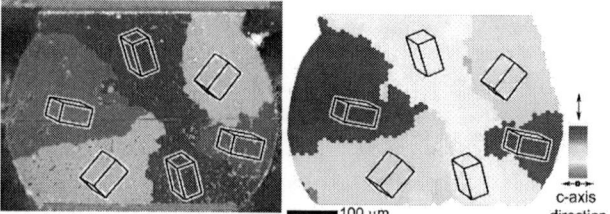

Figure 4. Polarized light image and corresponding c-axis orientation map showing a beach ball microstructure in a SAC305 solder joint.

III. ELASTIC AND CTE ANISOTROPY

As the joint cools, and as joints in packages enter service, internal strains develop due to the differential thermal expansion constraint of the package on the joint and due to the intrinsic anisotropic expansion behavior of Sn itself. With subsequent thermal cycling, internal strains oscillate, leading to stresses sufficient to induce dislocation motion.

The elastic and thermal expansion anisotropy of the Sn unit cell is illustrated in Figure 5 [15-16]. The tetragonal Sn unit cell has an atomic arrangement that is the same as a squashed diamond cubic structure (which is not surprising, given that Sn is in column IV). While it is counter intuitive, the expansion coefficient is greatest in the c-axis direction, which is the direction that has the greatest stiffness. This combination implies that Sn is internally conflicted at grain boundaries, particularly large angle grain boundaries. An effect of

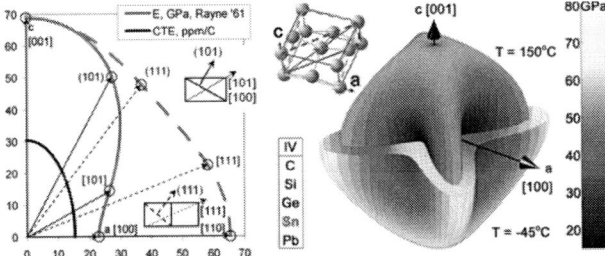

Figure 5. Variation of Young's Modulus on (100) plane (solid red) and (110) plane (dashed red), compared with CTE (solid black) [15,16]; unit cell (c/a = 0.5456), and Young's modulus surface at two temperatures.

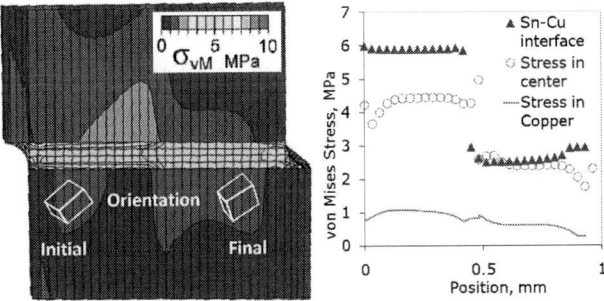

Figure 6. Elastic anisotropic FEM simulation of 115°C temperature change with two twin related orientations; the initial (dominant) orientation has a higher strain energy than the final (minority) orientation [17].

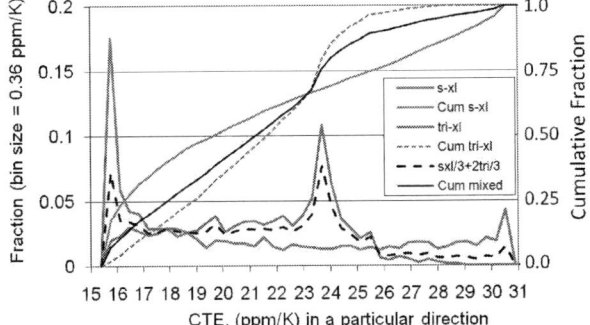

Figure 7. Histogram of fraction of a random population of solder joints with particular values of the CTE in a direction perpendicular to the board for

differential expansion is evident in the bulged grain beneath the grain boundary on the right side of joint B3 in Figure 1. Because of the anisotropic expansion, the stress evolution in a joint is complex, and unique to the microstructure present in the joint, and the position of the ball in the array. Hence, if there are only a few grains in a joint, then every joint is a special case, with a different detailed deformation history.

Because elastic stresses develop heterogeneously during thermal cycling, internal strain energy can provide a driving force for grain boundary migration [17]. In a single shear lap joint which initially had a majority orientation and a minority orientation, the volume fraction switched after about 500 thermal cycles. A temperature change simulation with both orientations in Figure 6 showed that the elastic strain energy resulting from a temperature change was smaller in the final orientation than the initial orientation, indicating that a reduction in internal energy would result from the initial orientation being consumed by the initial minority orientation. Thus, a stress assisted incremental grain boundary migration process caused a recrystallization process that is not commonly observed in cubic metals.

A temperature change will cause not only a shear stresses arising from the differential CTE in the package vs. the board, but axial stresses perpendicular to the board will also develop due to the intrinsic thermal expansion differences in the a and c directions. Figure 7 shows a histogram of the component of the CTE in the axial direction for a random distribution of single crystal orientations, and for twinned tri-crystal orientations. Due to the tetragonal symmetry of the unit cell, the first quartile of single crystal orientations (red line) have a low expansion coefficient in the axial direction. This effect is eliminated with tricrystals (blue line).

Assuming that a package has a random distribution of crystal orientations, then the average (or median) expansion of all of the joints will govern the axial displacement of the package with respect to the board. In joints that have the c-axis parallel to the board, the expansion coefficient is at the minimum (red orientations), so at high temperatures, these joints will be in tension. The observed fractures of this orientation in Figure 1 indicate that tension at high temperature is correlated with damaging conditions, in contrast to blue/purple orientations which are in tension at cold temperatures, as illustrated qualitatively in Figure 8. Orientations with the c-axis inclined to the board will have axial expansion values closer to the average (or median), and these joints should be less likely to develop axial stresses that lead to fracture. The influence of crystal orientation on thermal stresses will be examined later, after considering the role of plastic deformation mechanisms.

IV. PLASTIC ANISOTROPY

Because of the large CTE variation, internal strains arising from both extrinsic (package-board) and intrinsic (anisotropic) CTE mismatches lead to strains on the order of 0.1% or more during the temperature change arising from thermal cycling. This clearly activates plastic deformation at strain rates as large as 10^{-4} /s in accelerated thermal cycling tests common in the industry (e.g. ramp time of about 10 min from low to high T) [18,19]. The plastic properties of Sn are not well understood, as highlighted in a recent review [20]. For example, the critical resolved shear stress (CRSS) for activating slip on different slip systems does not reveal consistent critical resolved shear stress (CRSS) values, though there are some similarities in trends among investigations from several researchers, such as a greater thermal sensitivity for slip on {211) planes [21-23]. From published literature for Sn, the CRSS values of the several slip systems do not seem to be much different, and greater distinctions have been made regarding the rate of hardening (dislocation cutting of forest dislocations) [24]. Consequently, incorporating evolving latent hardening mechanisms and recovery processes are quite important for accurate modeling of slip in Sn, and these features are more difficult to implement than different CRSS values on different slip systems. Table I shows an estimate of the relative ease of activating the 10 relevant slip systems from

TABLE I PLAUSIBLE SLIP SYSTEMS IN Sn

Mode	Slip system	# in family	
1	{100) <001]	2	Most Active
2	{110) <001]	2	
3*	{100) <010]	2	
4	{110) <1-11]/2	4	(like BCC)
5	{110) <1-10]	2	
6*	{100) <011]	4	
7	{001) <010]	2	
8	{001) <1-10)	2	
9*	{011) <01-1]	4	
10	{211) <01-1]	8	Less Likely

{hkl) <uvw] mixed braces recognizes tetragonal crystal symmetry.
Colors are correlated with unit cell in Figure 5.
* Not highly activated in SAC305 room T shear tests [25].

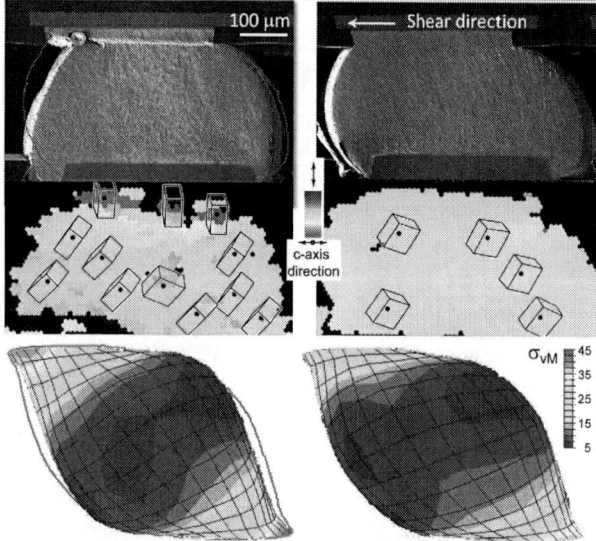

Figure 8. Comparison of two joints with similar c-axis orientation from the board, but different with respect to the direction of shear. In the left joint, the crystal rotates locally to a soft (red) orientation, and develops a shear instability, whereas the orientation in the right joint deformed uniformly. A simple CPFE model was able to simulate this difference qualitatively.

literature data, which were compared with some experimental observations on SAC 305 made recently by Zhou et al. [25].

As the c-axis is the closest packed direction, slip with a Burgers vector in the c-direction is most easily activated. Figure 8 shows an example of how a shear experiment on a section of a package led to intense shear localization facilitated by localized rotation that brings this most easily activated slip system to be optimally aligned with the macroscopic shear direction. In contrast, when the c-axis or its projection is not well aligned with the direction of shear, localized deformation is not facilitated, and the deformation is remarkably uniform, despite the fact that the stress is much lower in the middle than near the interface.

To date, Sn joints are most commonly modeled in finite element models using scalar creep based models developed from data obtained from creep of single shear lap joint specimens consisting of many joints, so that an average creep behavior is extracted. Some modeling efforts use continuum type models with enhancements for anisotropic properties and crack formation [26,27]. There are only two attempts to model deformation in a Sn based solder joint using crystal plasticity finite element (CPFE) modeling [28,29]. Using a very simple crystal plasticity model, where all slip systems have the same flow stress, and slip systems 1,2,4,5 are given a lower hardening rate than the other slip systems, we have been able to obtain qualitative agreement with the two contrasting crystal orientations illustrated in Figure 8 (the shape of the joint that deformed more uniformly is outlined in red with the joint that deformed with a shear instability).

V. LOW STRESS PBGA PACKAGE DESIGN

The preliminary discovery of the susceptibility of red orientations to cracking was investigated in a systematic manner for a plastic ball grid array (PBGA) package having a 14x14 array of 1 mm pitch solder joints with a 5x5 mm silicon die attached. The package side substrates had an electrolytic NiAu surface finish. Assembled parts were attached to a 2.4 mm thick high-T_g FR4 printed circuit board with an organic surface preservative (OSP) surface finish. The solder balls and paste used for the reflow were SAC305, and a typical peak temperature of 240°C with 60 seconds above the liquidus temperature was used for board assembly. Some samples were pre-aged at 100 or 150°C for 500 or 1000 h in ambient air to cause coarsening of precipitates that normally occurs with longer term service, to generate effects of microstructural evolution that would occur in real-time thermal cycling. The thermal cycle was 0-100°C with a ramp rate of 10°C/min and 10 min dwell times.

Single slices were extracted from these PBGA packages and also a large 44x44 (1 mm pitch) flip chip (FCBGA) package (also a low-stress design) for in-situ diffraction with synchrotron x-rays to characterize the strain history using a 70 minute thermal cycle [14]. The advantage of synchrotron x-rays is that the entire ball volume can be non-destructively monitored during thermal cycling. Figure 9 shows the lattice strain history for ball 1 in the FCBGA, which has a near-red orientation (where (110) planes are nearly aligned with the axial direction). Because this position is far from the die, this edge joint has no appreciable extrinsic shear strain, so this result confirms that a red orientation develops a high tensile stress in the hot part of the cycle. In this joint, the magnitude of the strains increased after 50 cycles, but comparing the diffraction patterns revealed no discernable difference in grain orientation or emergence of a new orientation. Similar studies in PBGA samples showed discernable but not dramatic changes in the diffraction pattern in nearly every ball of a 14 joint slice after 450 cycles, confirming that gradual changes in the grain orientation and microstructure occur during thermal cycling.

A. Statistical study of effect of aging and number of cycles

A statistical study of the evolution of joint orientations and microstructure in the PBGA package was made by sectioning packages exposed to several different combinations of aging and thermal cycling parallel to the board in a plan view to reveal all of the joints in one section. Figure 10 shows a c-axis map of one of these packages, showing that many joints appear to be single crystals, while others have 2 or three orientations. With very few exceptions, multicrystal joints showed the solidification twinning orientation relationship illustrated in Figures 3 and 4, though only about a quarter of them gave the appearance of a sectioned beach ball. With the known crystal orientations and area fractions, an average expansion coefficient normal to the board was calculated for each ball, and plotted using the same color scale as the c-axis maps. For example, a triple grain with two red orientations and one green orientation would have an average expansion coefficient that could be represented as a yellow-orange orientation. Figure 11 shows how several of the experimentally measured histograms compare with the theoretical prediction of a random distribution having 1/3 single crystals and 2/3 tricrystals in Figure 7. This shows that the distribution of crystal orientations is not quite random, and

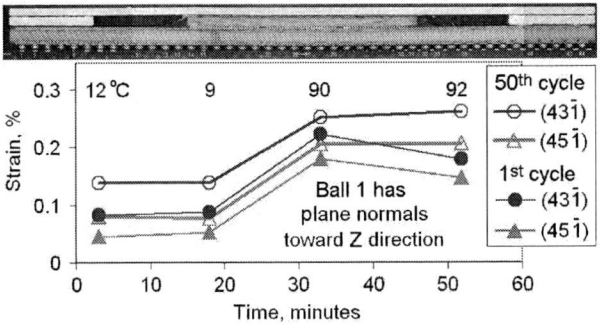

Figure 9. Changes in lattice strains in the full volume of the first joint of a flip chip ball grid array (FCBGA) were assessed non-destructively using synchrotron x-ray diffraction while undergoing thermal cycling [14].

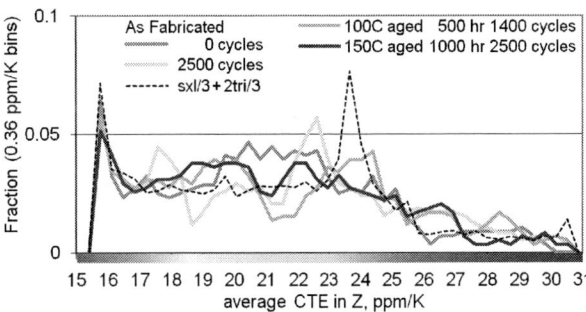

Figure 11. Histogram of crystal orientations for several experimental conditions compared with theoretical 1/3 single crystals and 2/3 tricrystals.

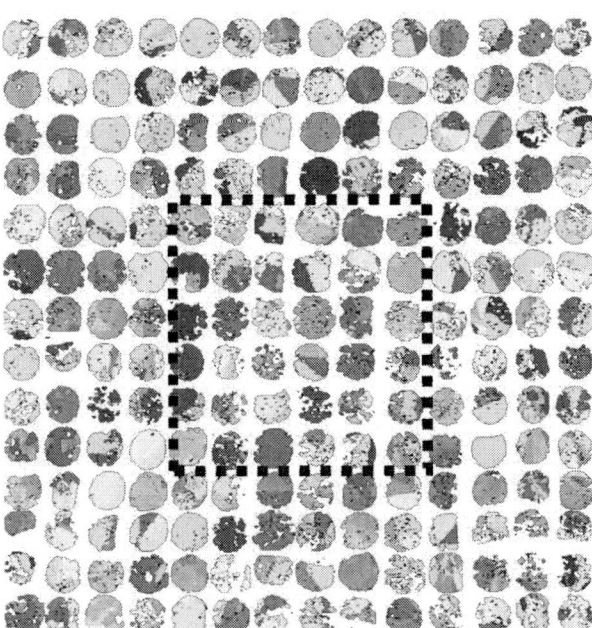

Figure 10. The c-axis orientation map (same color key as above) for a cross section of a PBGA package aged 150°C 1000hr and cycled 2500 times. The dotted line indicates the edge of the silicon die.

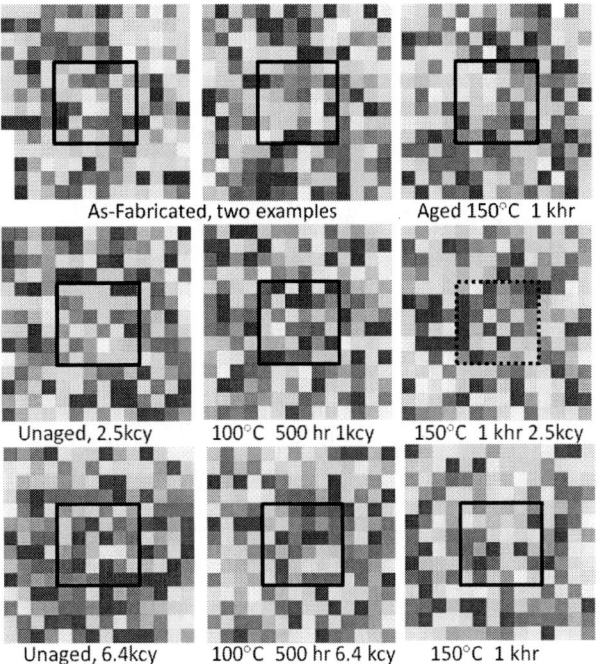

Figure 12. CTE maps of PBGA packages with different processing histories (same color key as Figure 11) The solid line indicates the edge of the silicon die.

that it changes with processing history. Figure 12 shows 9 examples of this outcome expressed spatially, which makes it clear that changes in the crystal orientations occur with thermal cycling. The first two measurements (upper left) are from two separate packages of the as-fabricated condition showing a measure of statistical differences between samples with the same condition. There are subtle differences related to different histories, and the influence of the larger CTE mismatch near the die, discussed elsewhere [11,30]. Figure 13 summarizes how the fraction of joints with different average normal direction CTE values changes with different processing histories. In general, the fraction of blue/purple orientations is constant until large numbers of cycles, when they increase in the unaged sample. There is also some fluctuation of yellow and orange orientations, but with later time, the green fraction becomes smaller as the population favors yellow and red in the package with the greatest time at high temperature. The fraction of single crystals generally decreases with aging and thermal cycling. Gradual grain

boundary migration to minimize elastic strain energy described above can account for such changes, as the presence of multiple orientations when one orientation consumes another will average to green.

B. Crack Nucleation at recrsytallized locations

One shortcoming of characterizing the plan view sections of joints is that fracture does not occur there. To assess damage, either die and pry methods or cross sectioning one row at a time are needed [31]. After 2500 cycles, cross sections of the PBGA packages showed some cracking, with the most cracks observed in samples aged at 100°C. Cracking was much more extensive in samples themally cycled 6400 times, and like the 2500 thermal cycle samples, the 100°C aged specimens showed the greatest amount of damage. Figure 14 illustrates two examples of a center row of joints and two enlarged interfacial regions. Just above the polarized light images of each joint, the length of dark blue indicates the fraction of the joint that is visblty cracked (using higher

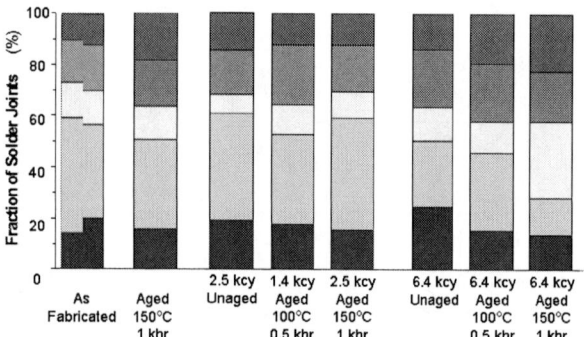

Figure 13. Fraction of c-axis orientations of solder joints in the PBGA package with different processing histories.

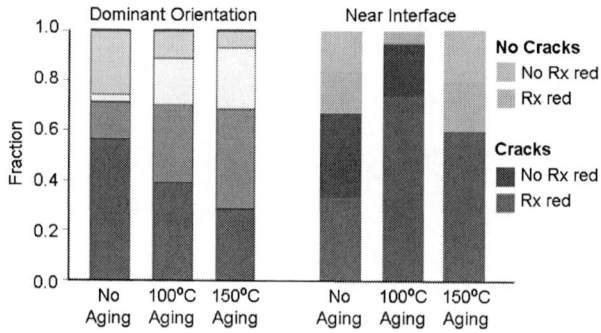

Figure 15. Fraction of cracked joints with different dominant orientations, and the effect of (red) recrystallized grains on cracking in the interface.

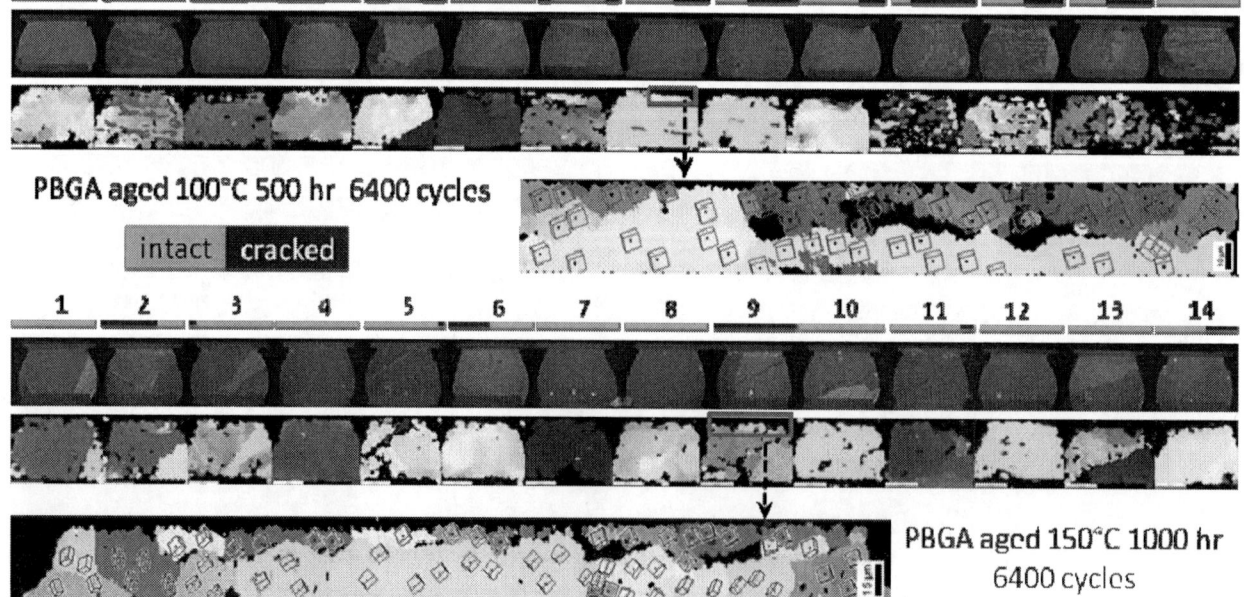

Figure 14. c-axis orientation maps of PBGA joints after 6400 thermal cycles showing how cracks are correlated with recrystallized red orientations. In the 100°C aged joint, only one of 11 joints with cracks showed a recrystallized red orientation and no crack [31].

magnification images). The c-axis maps below show that cracking is not limited to red orienations, but is common among many orientations other than blue/purple, and there are even red joints that show no cracks (note joint 1 in the 150°C aged package). However, Figure 15 shows that the greatest cumulative length of cracks are correlated with dominant red orientations. Upon closer inspection of higher magnification orientation maps of cracked regions in Figure 14, recrystalization causes new red orientations to develop, and in many cases, cracks are associated with these minority red orientations, especially for the 100°C aging condition [31]; this is quantified in the right half of Figure 15. In this low stress package design, which can sustain a very large number of thermal cycles, cracking is facilitated as much by recrystallized orientations as the dominant crystal orientation.

VI. HIGH STRESS FINE PITCH BGA PACKAGE

Systematic studies are under way to investigate microstructural evolution in SAC305 solder joints in a high stress package design. This package is 13mm square with a 10.5 mm die and 0.4 mm pitch BGA with 300 micron diameter solder balls around the perimeter in 4 rows of joints with Ni/Au surface finish. The package is assembled to a 2.4 mm circuit board with an OSP finish. In this package, failures occurred with less than 2000 thermal cycles [32]. Figures 16 and 17 show two joints from an unaged package in high stress locations (in the middle of an edge and at the outer corner) where diagonal cracks not observed in prior studies of PBGA are evident. Even joints with blue orientations exhibit cracks, though they are less well developed than those found in joints having different c-axis orientations in similar locations. Also, Figure 18 shows a parent red orientation in which yellow, green and blue orientations evolve by gradual continuous lattice rotation during recrystallization. This recrystallization resulting in non-red orientations near the interfaces was not observed in low-stress PBGA package, where recrystallized red orientations developed from dominant yellow or green orientations during thermal cycling. In this high stress package design, recrystallization occurred much earlier than in low stress packages, and this process facilitated crack nucleation and growth. These preliminary observations also show that there is a design threshold for cyclic strain where even blue orientations cannot resist damage accumulation.

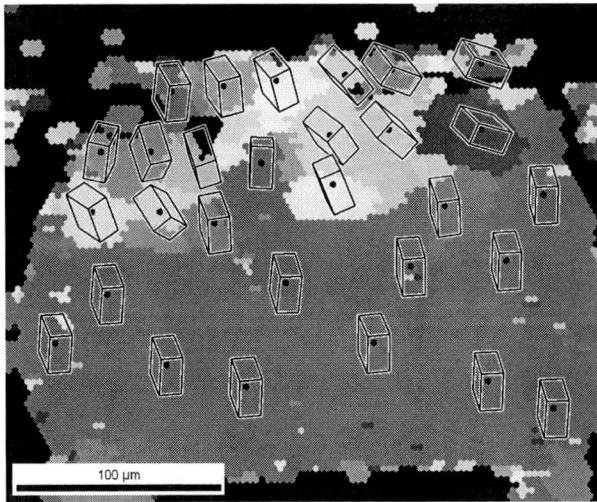

Figure 18. Orientation evolution is apparent in this joint, where the lesser deformed center has the probably original orientation that gradually changes toward the package interface
(aged 150°C thermally cycled to failure interior edge location).

Figure 16. Optical, polarized light, and c-axis orientation map for SAC 305 ball 1 at the corner of an unaged and thermally cycled 13 mm package exhibited recrystallization along the diagonal cracking path.

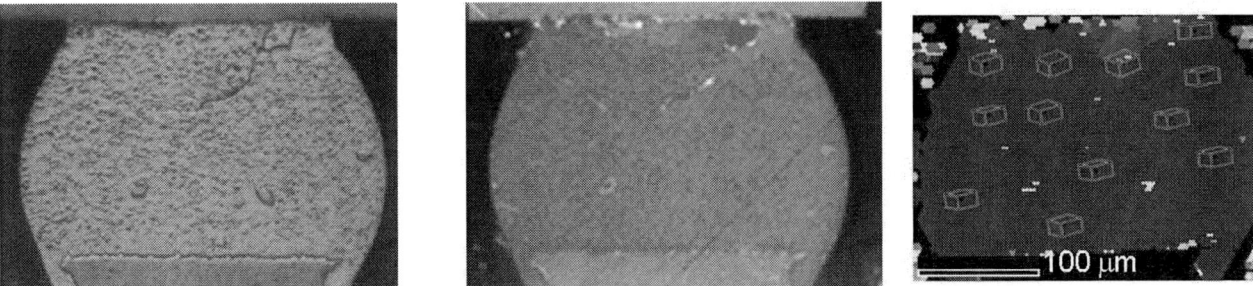

Figure 17. Optical, polarized light, and c-axis orientation map for SAC 305 ball 7 near an edge at of an unaged and thermally cycled 13 mm package exhibited cracked blue joints related to geometrical recrystallization (gradual orientation changes). The crack followed a recrystallized low angle boundary.

VII. TRENDS IN RECRYSTALLIZATION AND CRACK NUCLEATION

As the strains per cycle are small, and because the hysteresis loop for thermal strain is not symmetric [18,28, 33], the gradual accumulation of strains leads to small rotations. This process is not easily discerned from initial glances at orientation maps or diffraction patterns, as small changes in orientation are not evident as a noticeable color change, nor are they visible in diffraction peaks, because the individual peaks are always streaked, even just after solidification. A streaked peak implies that a given crystal orientation has many low angle boundaries or orientation gradients within it. Sometimes this process can be visualized in orientation maps, such as in Figure 18. As the middle of a solder ball has twice the load carrying area as the interface, there are stress (and strain) gradients, so that the evolution of recrystallization processes can sometimes be observed as a stochastic snapshot in a cross section. This joint shows a relatively systematic trend in rotations of the unit cell about a similar crystallographic axis (such as [110], which is commonly observed [34], and evident in Figure 18) from the center to the interface. More smoothly varying orientation gradients are found in some joints in lower stress packages such as the high magnification regions in Figure 14. Some boundaries are low angle (subtle changes in color), and others are high angle (large enough to cause a noticeable change in color in an orientation map). As strain accumulates within a given grain, dislocations are absorbed (recover) in low angle boundaries, leading to gradual increases in the subgrain misorientation by a process known as continuous recrystallization [35]; once the misorientation exceeds about 15°, they are no longer considered subgrains.

Because dislocation slip is the direct cause of continuous recrystallization, knowledge of activated slip can be used to predict subgrain boundary formation. Clarification of what slip systems are activated in thermal cycling conditions is necessary. Furthermore, attention to the relationship between the c-axis of the Sn crystal and the direction of predominant shear will be important, so as insightful as the c-axis maps are, they are insufficient for understanding the mechanisms behind microstructural evolution. The importance of the angle between the c-axis (or its projection) and the direction of primary shear is evident in comparing the two joints in Figure 8, which show monotonic large strain shear. There is remarkable similarity in the sense of shear in the monotonically sheared left joint in Figure 8 and the thermal cycled joint in Figure 18, which will provide insights on dislocation activity in very different modes of deformation in future studies.

In contrast, discontinuous recrystallization arises from highly localized strained regions that develop large local rotations which then become the nucleus of a new grain orientation that can grow into the surrounding lesser deformed material. Such small orientations are commonly found in solder joints, often having the twin relationship. If a small island twin deforms very differently from the surrounding material, this can lead to formation of a discontinuous recrystallization nucleus with a random misorientation; perhaps the blue orientation in Figure 18 developed from a minority orientation, as it does not have a common rotation axis with most of the other grains. Both continuous and discontinuous recrystallization are observed in solder joints, but the majority of boundaries show the continuous process, with systematic and smoothly varying changes in crystal orientation in the same direction across several boundaries traced in sequence.

Finally, Sn has a very strong energetic preference for particular grain boundary misorientations that minimize interfacial energy (such boundaries are highly preferred in orientation misorientation histograms [36]). As deformation pumps dislocations into boundaries, the boundary energy of the low energy boundaries increases dramatically, transforming it into a 'random' boundary. Cracks are commonly observed between grains with a 'random' boundary misorientation (high energy interfaces which require much less energy to separate).

VIII. RESEARCH NEEDS AND OPPORTUNITIES TO SUPPORT PHYSICALLY BASED RELIBILITY PREDICTION

Clearly, the interrelationships between the magnitude of plastic strain per cycle and the processes of recrystallization needs to be addressed and understood quantitatively before prediction of joint lifetimes can be accomplished on the basis of physically understood and modeled mechanisms. This is clearly possible, as control of microstructural evolution is a common outcome of focused metallurgical research for production of high volume products. Indeed one of the most tightly controlled processes is for the aluminum sheet metal used in the beverage can industry, where failures less than 1 ppm are necessary to have a profit margin. In normal polycrystals, heterogeneous deformation in a small microstructural patch of a larger system can often be accommodated by neighboring regions, but there are no neighboring regions to provide constraint on a particular solder joint in an array, which makes them more susceptible to damage nucleation. Current modeling strategies for plasticity and recrystallization consider representative volume elements consisting of many grain orientations, and then homogenize the outcome so that measurable processing parameters such as stress, strain, and crystallographic texture can be used to validate deformation models. Such average properties are useful for basic design, but with regard to reliability prediction, it is most important to understand the deformation and recrystallization process that govern the outliers.

Deformation in Sn differs from common metallurgical processes in that it is done in hot deformation conditions, where strain, recrystallization and recovery all take place concurrently. The most advanced modeling strategies for face centered cubic materials have well characterized and understood slip behavior followed sequentially by a recrystallization anneal, which requires different kinds of simulation models. In solder joints, dislocation and recrystallization phenomena occur concurrently and interdependently, requiring more sophisticated and integrated modeling strategies.

ACKNOWLEDGEMENTS

This research is supported by NSF-GOALI contract 1006656 and Cisco Systems Inc., San Jose, CA. We are grateful to Stefan Zaefferer and Guilin Wu of Max-Planck-Institut für Eisenforschung, Düsseldorf, Germany for indexing x-ray diffraction patterns. The authors thank Douglas S. Robinson, beamline scientist, for help in configuring the experiments in beam line 6-ID-D. Use of the Advanced Photon Source was supported by the US Department of Energy, Office of Science, Office of Basic Energy Sciences, under Contract No. W-31-109-Eng-38.

REFERENCES

[1] T.R. Bieler, H. Jiang, L.P. Lehman, T. Kirkpatrick, E.J. Cotts, B. Nandagopal, IEEE Transactions on Components and Packaging Technologies (CPMT) 31(2), 370-381, (2008).

[2] L.P. Lehman, S.N. Atavale, T.Z. Fullem, A.C. Giamis, R.K. Kinyanjui, M. Lowenstein, K. Mather, R. Patel, D. Rae, J. Wang, Y. Xing, L. Zavalij, P. Borgesen and E.J. Cotts, J. Electronic Materials 33(12), 1581-1588 (2004).

[3] L.P. Lehman, Y. Xing, T. R. Bieler, E. J. Cotts, Acta Materialia, 58(10), 3546-3556 (2010).

[4] K.S. Kim, S.H. Huh, K. Suganuma, Mat Sci Eng A333, 106, (2002).

[5] R. Kinyanjui, L.P. Lehman, L. Zavalij, E. Cotts, J. Mater. Res. 20(11), (2005).

[6] S. Terashima, T. Kobayashi, M. Tanaka, Science and Technology of Welding and Joining 13(8) 732-738 (2008).

[7] Sun-Kyoung Seo, Sung K. Kang, Moon Gi Cho, Da-Yuan Shih, Hyuck Mo Lee, J. Electronic Materials 38(12) 2461-69, 2009.

[8] D.W. Henderson, J.J Woods, T.A. Gosselin, J. Bartelo, D.E. King, T.M. Korhonen, M.A. Korhonen, L.P. Lehman, E.J. Cotts, Sung K. Kang, P. Lauro, Da-Yuan Shih, C. Goldsmith, K.P. Puttlitz, J. of Materials Research, Vol. 19 (6), pp 1608-1612, (2004).

[9] T.T. Nguyen, D. Yu, S.B. Park, J. Electronic Materials, in press, DOI: 10.1007/s11664-011-1534-z, 2011.

[10] A.U. Telang, T.R. Bieler, S. Choi, K.N. Subramanian, J. Materials Research, 17(9), 2294-2306, 2002.

[11] Tae-Kyu Lee, Bite Zhou, L. Blair, Kuo-Chuan Liu, T.R. Bieler, J. Electronic Materials 39(12), 2588-2597, 2010.

[12] I.E. Anderson, J.W. Walleser, J.L. Harringa, F. Laabs, A. Kracher, J. Electronic Materials 38(12), 2770-2779, 2009.

[13] G.J. Jackson, H. Lu, R. Durairaj, N. Hoo, C. Bailey, N.N. Ekere, J. Wright, J. Electronic Materials, 33(12) 1524-1529, (2004).

[14] T.R. Bieler, T.-K. Lee, K.C. Liu, J. Electronic Materials 38(12) 2712-19, (2009).

[15] D.G. House, E.V. Vernon, Brit J Appl Phys, 11, 254-9 (1960).

[16] V.T. Deshpande, D.B. Sirdeshmukh, Acta Cryst. 15, 294-295 (1962).

[17] A.U. Telang, T.R. Bieler, A. Zamiri, F. Pourboghrat, Acta Materialia 55(7), 2265-2277, (2007).

[18] R. Darveaux, K. Banerji, IEEE Trans Components, Hybrids and Manuf Tech, 15, 1013-1024, (1992).

[19] S. Choi, J. Lee, F. Guo, T.R. Bieler, K.N. Subramanian, and J.P. Lucas, JOM 53(6), pp. 22-27 (June 2001).

[20] F. Yang, J. C. M. Li, J. Mater Sci: Mater Electron 18, 191–210 (2007).

[21] B. Düzgün, A.E. Ekinci, I. Karaman, N. Ucar, Journal of the Mechanical Behavior of Materials, vol 10(3), pp 187-203, (1999).

[22] A.E. Ekinci, N. Ucar, G. Cankaya, B. Düzgün, Indian J. Eng. Mater. Sci. 10, 416 (2003).

[23] Y. Kouhashi, Koenronbunshu (Transactions of the Japanese Society for Strength and Fracture of Materials) 12, 15-18, (2000).

[24] M. Fujiwara, T. Hirokawa, Journal of the Japan Institute of Metals, vol 51(9), pp 830-838, (1987).

[25] Bite Zhou, T.R. Bieler, Tae-Kyu Lee, Kuo-Chuan Liu, J. Electronic Materials 38(12), 2702-11, 2009.

[26] R.L.J.M. Ubachs, P.J.G. Schreurs, M.G.D. Geers, Mechanics of Materials 39, 685–701, 2007.

[27] M. Erinc P.J.G. Schreurs, M.G.D. Geers, Mechanics of Materials 40(10), 780-791 (2008).

[28] J. Gong, C. Liu, P.P. Conway, V.V. Silberschmidt, Computational Materials Science 43(1), 199-211, (2008).

[29] A. Zamiri, T.R. Bieler, F. Pourboghrat, J. Electronic Materials 38(2), 231-40, (2009).

[30] T.-K. Lee, K.-C. Liu T.R. Bieler, J. Electronic Materials, 38(12) 2685-93, 2009.

[31] Bite Zhou, T.R. Bieler, Tae-Kyu Lee, Kuo-Chuan Liu, J. Electronic Materials 39(12), 2669-79, 2010.

[32] Tae-Kyu Lee, Hongtao Ma, Kuo-Chuan Liu, Jie Xue, J. Electronic Materials 39(12), 2564-2573, 2010.

[33] C.H. Raeder, R.W. Messler, L.F. Coffin, J. Electronic Materials 28(9) 1045-54, 1999.

[34] A.U.Telang, T.R.Bieler, and M.A. Crimp, Materials Science and Engineering, A421 (1-2) 22-34 (2006).

[35] R.D. Doherty, D.A. Hughes, F.J. Humphreys, J.J. Jonas, D.J. Jensen, M.E. Kassner, W.E. King, T.R. McNelley, H.J. McQueen, A.D. Rollett, Mater. Sci. & Eng. A, 238219, (1997).

[36] A.U. Telang, T.R. Bieler, JOM, 57(6), 44-49, (2005).

Robust Pad Layout to Improve Wire Bonding Reliability

Kyoung-Hwan Kim , Hong Kook Min , Se Yeoul Park , So Ra Park , Seung Jin Yang , Byung Sup Shim , Yong Tae Kim , and Jeong-Uk Han

System LSI Division Samsung Electronics

LSI TD2, TD Team

San #24 Nongseo-Dong, Giheung-Gu, Yongin-City, Gyeonggi-Do, Korea 446-711

phone: 82-10-9047-8689; e-mail: slayerr@samsung.com

Abstract— **Different from the conventional study to improve the pad reliability against the peel-off, this study focuses on the probability that the peel-off could be originated from the perpendicular pushing down mechanical stress (PPMS) during the ball mounting process. Suggested in this paper are a new model which causes the peel-off and a new pad structure that overcomes the pad peel-off without any special procedures or changes in material or dimension. Three sets of layout patterns have been designed and fabricated in a 0.13 *μm* CMOS process. To assess the wire bonding quality, wire pulling tests (WPT) and evaluation of bonding power dependencies by means of wedge wire bonding are conducted. Additionally, FAMMOS simulator is adopted to verify the newly proposed pad structure.**

I. INTRODUCTION

As the dimension of the transistor continues to be scaled down to sub-half micron to achieve more chips in same size of wafer and better performance, the low-k dielectric should be chosen as an IMD (Inter-Metal Dielectric) and the film's thickness is decreased due to the limitation of lithography.[1] Due to this, thinner IMD is easier to be cracked than the original.[2][3] To cut down the costs, moreover, wedge bonding process which is known as having severe mechanical stress compared to ball bonding process is adopted to package process. It is needed to avoid or face these troubles effectively. Thus, it is considered how to secure the wire bonding reliability when the dimension scaled down.

WPT is generally conducted to assess the wire bonding quality.[4]. The typical configuration of this WPT, where the wire is pulled by a hook, is depicted in Figure 1. A wire is hooked and pulled by the hook as in Figure 1(a). The pulling force applied on the wire is perpendicular to the silicon substrate with a force sensor at the top of the hook. When the applied force on the wire is increased continuously, the pulling force to break the wire can be measured from the force sensor. The wire and the bond pad structures should endure a certain force level before failure of the wire or the bonds. As a result of WPT, in general, two kinds of typical failures can be observed, which are peel-off of bond pad as depicted in Figure 1(b) and the neck break of bond wires as depicted in Figure 1(c). The former is considered to be related to the silicon processing, whereas the latter results from the weakness of package process.

In this paper, it is studied in terms of silicon processing enhancement to overcome peel-off failure.

To improve the bonding reliability as like this condition, mainly two kinds of methods can be chosen as known well.

One is to change the materials such as inter-metal dielectric. As the dimension scaled down, new materials are being introduced to undesirable effects, which follow from further miniaturization. Instead of silicon dioxide typically used as insulator, low-k dielectric materials are used. The device cross talk and delay are still dependent on the capacitance associated with it. In order to reduce these impacts, the interlayer dielectric constant has to be reduced. To achieve this feature, low-k material is introduced. It is also known to us that low-k dielectric material has a much lower stiffness and worse adhesion properties than silicon dioxide. All of these changes have weakened the bonding reliability. Therefore, for better bonding reliability, it is needed to change these materials into another one.[5]

The other is to change the pattern layout of metals and holes to improve the bond wire stiffness. Many major semiconductor chip maker have suggested several kinds of pad structure to be strengthened in terms of peel-off failure. [9]

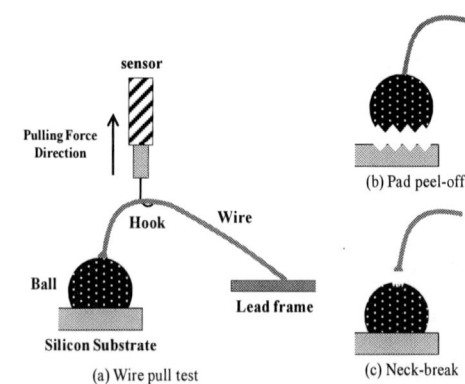

Figure 1. WPT and its types of failure

Up till now, conventional studies for the peel-off failure have been mainly done based on the supposition that the crack occurs during the silicon processing or wire pulling action.[9]

978-1-4244-9113-1/11 $26.00 © 2011 IEEE

However, this paper has emphasized on the probability that the PPMS during the ball mounting process could result in a crack which is the seed of peel-off failure. From this point of view, three sets of pad structure are prepared to evaluate this possibility. With these structures, WPT and power dependencies of wedge wire bonding are conducted for each structure. In addition, it is performed FAMMOS simulation which is known as physical stress simulation tool to verify the structural stiffness. Therefore, this study encompasses four main areas (a) Test Vehicle fabrication, (b) WPT (c) wedge wire bonding power dependencies (d) FAMMOS simulation.

II. PEEL-OFF PHENOMENON

The peel-off failure occurred during the wire bonding process is identified in Figure 2. The top of the metal layers was torn off by ball lifting, thus the pad exposed the inner layer as shown in Scanning Electron Microscope (SEM) image Figure 2(a). The cross sectional view with Focused Ion Beam (FIB) reveals the fact that the top of the metal peels off from the IMD in Figure 2(b). The Microscopic view right after the peel-off failure in Figure 2(c) and chemical reverse etch analysis after ball mount image Figure 2(d) show that the peel-off comes with the crack. With these results, it is noticed that the peel-off failure is accompanied by an IMD crack. In other words, if an IMD crack could be restrained, the peel-off failure would not be happened. Thus, from this point of view, test structures to improve the peel-off failure have been designed.

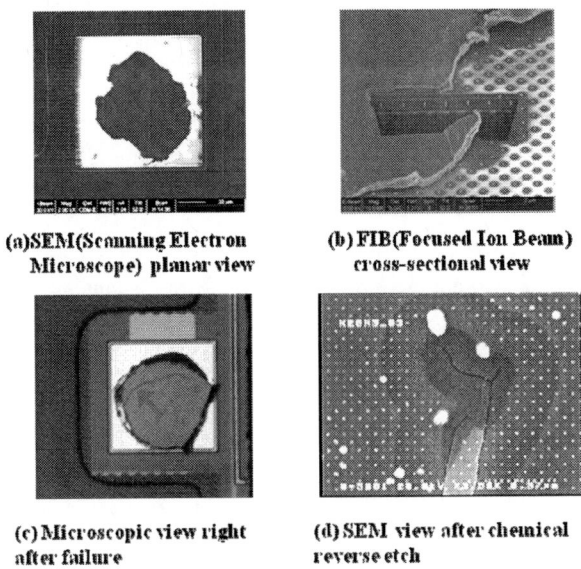

(a)SEM(Scanning Electron Microscope) planar view

(b) FIB(Focused Ion Beam) cross-sectional view

(c) Microscopic view right after failure

(d) SEM view after chemical reverse etch

Figure 2.   Identification of pad peel-off failure

III. EXPERIMENTAL DETAILS

Three kinds of test pad structures were prepared in order to investigate the probability that the crack is caused by PPMS as shown in Figure 3. The samples were fabricated using 0.13um CMOS process. Its backend films are consisted of 4 layers of metal and low-k inter-metal dielectric. The pads are Ti/TiN, Aluminum and TiN sandwich structure.

The conventional bond pad structure is shown in Figure 3(a), and type B is selected as an alternative in Figure 3(b) which could have a good stiffness against the crack, since the crack has been originated from the interface between heterogeneous films in general, Via plug and IMD oxide in this case. [9] Type C is newly proposed pad layout in Figure 3(c) which is expected to disperse the PPMS effectively. When the stress is applied, it is expected that perpendicular vector stress is refracted and weakened as illustrated in Figure 4.

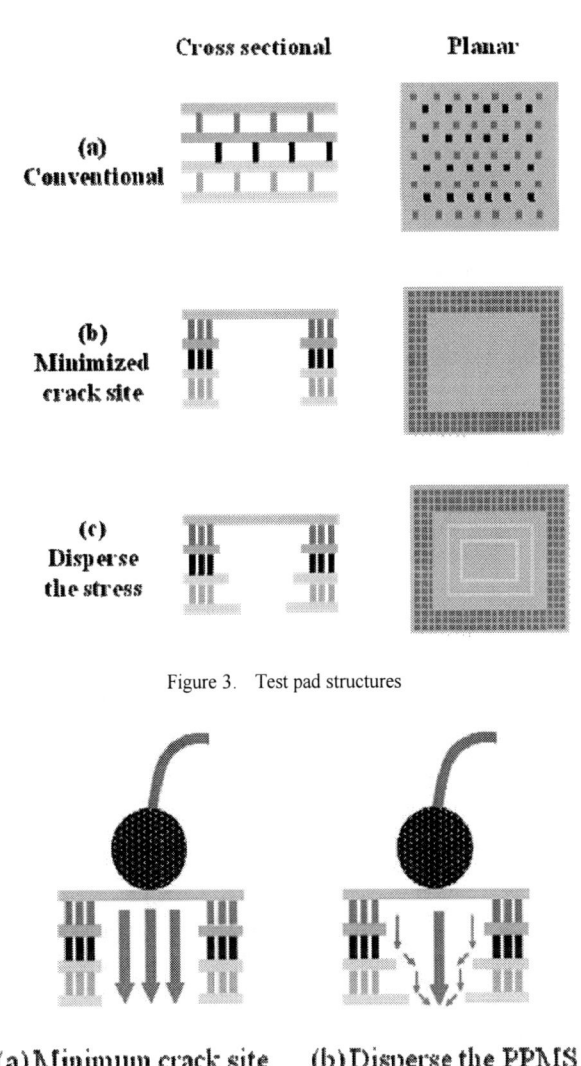

Figure 3.   Test pad structures

(a) Minimum crack site   (b) Disperse the PPMS

Figure 4.   How to disperse the PPMS

With these structures, WPT is conducted to assess the bonding reliability under the same wire bonding condition using the number of 140 pieces of sample, and the wedge wire bonding power dependencies are observed using the number of 20 pieces of sample. For cost reduction in low-priced MCU products, wedge wire bonding has been widely used; otherwise, it is known that wedge wire bonding generates much more damage on pad rather than ball bonding process. [6].

In figure 5 and 6, it is shown that the differences between ball bonding and wedge bonding. Bonding process sequence has arranged in alphabetical order, (a) → (b) → (c). In the ball bonding process, gold wire is melted and come to a ball at the end of the capillary as depicted in Figure 5(a). The ball is mounted on the pad by pushing down the capillary vertical direction to the pad, and ultrasonic vibrations are applied to achieve better adhesive feature. When the capillary is moving to the lead frame, pulling force is applied to the pad, thus the peel-off failure is visualized in appearance in Figure 5(c).

Wedge wire bonding is similar to ball bonding except the absence of ball as shown in Figure 6. Due to this, stronger mechanical stress than that of the ball bonding process could be applied to the pad as mentioned above. Aluminum wire is adhered to the pad directly using the wedge in place of molded golden ball. It is the reason why wedge wire bonding is preferred as low-cost MCU products as mentioned in the introduction.

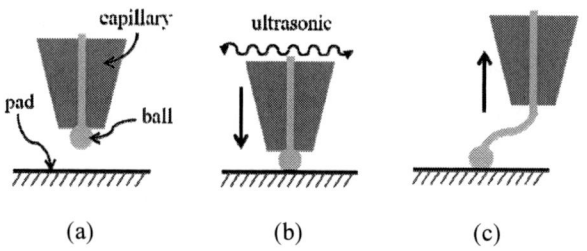

(a)   (b)   (c)

Figure 5. Process sequence of ball bonding

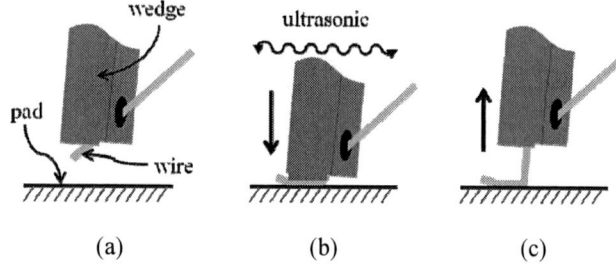

(a)   (b)   (c)

Figure 6. Process sequence of wedge bonding

IV. EXPERIMENTAL RESULTS

The measurement results of the WPT applied on each pad is shown in the form of Weibull distribution in Figure 7. All of tests were done in the same condition including power, force, temperature, time and so on. Type A, the conventional pad layout, shows the most vulnerable feature with wide-range variation. Moreover, it has been proven to fall short of the required value to conduct wire bonding, which is known at least 2.5[g]. [8] In the point of crack source generation, type A

is a richest structure due to so many interfacial region composed of via and metal. Many via holes can create weakly adhering points as well as concrete connection between upper and lower metal.

On the contrary of that, type B is consisted of limited crack generation points compared to type A. Nevertheless, tails still exist in the distribution and it has even a failed point according to the criteria mentioned above. Type C shows very notable results. It has a robust WPT feature stemming from stress dispersion idea as shown in Figure 4.

In view of the results so far achieved, it is considered that the stress dispersion structure has an effect on immunity from peel-off failure. In other words, if we could disperse the PPMS effectively, the peel-off failure would be restrained as needed.

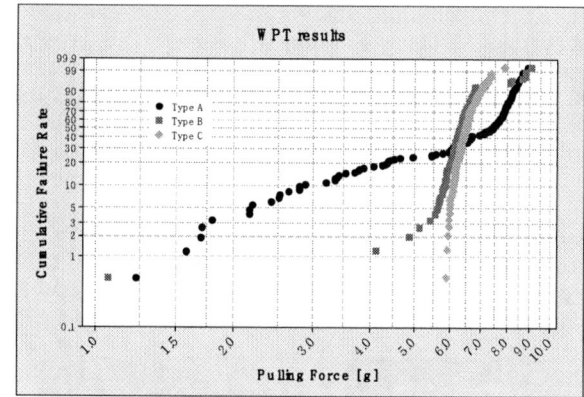

Figure 7. WPT results Weibull Distribution

In addition to this, it is performed wedge wire bonding evaluation to confirm that the immunity from PPMS of type C is still effective in much more severe condition. Bonding power, which is one of parameters about package process, is represented as ultrasonic generator's electric current [mA]. In this case, appropriate power for wedge wire bonding is 60mA, and maximum power not to break down the bond is 80mA. Power dependencies of wedge wire bonding are observed in type B and type C except type A, which is too weak to be carried out the wedge wire bonding process.

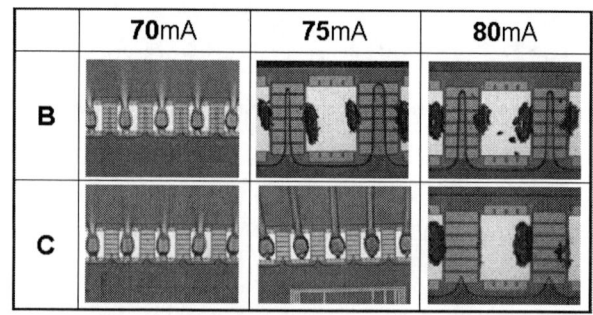

Figure 8. Wedge Bonding Power Split Results

Bonding Power [mA]	WPT [g]	Type B	Type C
70	AVG	5.05	4.62
	Max	5.52	5.05
	Min	4.60	4.26
75	AVG	Peel-off	3.67
	Max		4.63
	Min		2.84
80	AVG	Peel-off	Peel-off
	Max		
	Min		

Figure 9. WPT Values According to the Bonding Power

As a result of each 70, 75, 80mA power test, type B and C is broken down at 75mA and 80mA respectively as shown in Figure 8. It means type C is stronger than type B in terms of ability to withstand the peel-off failure. WPT value measured after wedge wire bonding is arranged in Figure 9. It shows that type B's values could not be measured due to the peel-off failure when 75[mA] of bonding power was applied, otherwise, type C's could be seen as clear even though it's value is lower than that of 70[mA]. With these results, type C also has better immunity from peel-off failure than type B in much more severe condition such as wedge wire bonding process.

FAMMOS simulation is conducted to assess the difference of stress relief according to the pad layout. FAMMOS simulator is to perform 2D and 3D backend process simulation, stress analysis, and failure evaluation for interconnect structures. It can also be used to address layout related stress variation in strained-silicon CMOS technologies. The main feature of FAMMOS is, [7]

- Modeling of the fabrication process to generate 3D interconnect and strained-silicon structures with direct usage of design layout in GDSII.
- Simulation of the thermal and mechanical stress evolutions during the process steps.
- Evaluation of failures related to interface debonding, cracking, and stress voiding.
- Computation of channel mobility variation.

2D is selected as a simulation mode to simplify the structure, thus, make it easy to analyze the simulation results. Process parameters, such as temperature, material, thickness, are given just as same of actual process condition. Definition of the material properties used in the FAMMOS simulation is as below in Table 1. The coefficients of thermal expansion (CTE) refer to measured values at room temperature assumed to be 20°C.[7]

Material	Young's Modulus [GPa]	Poisson's Ratio [unitless]	Coefficients of Thermal Expansion [um/m/°C]
Aluminum	86.0	0.30	24.4
Nitride	162.0	0.28	3.05
Oxide	71.7	0.16	0.51
Titanium	105.1	0.32	24.4
Silicon	162.0	0.28	3.05
Tungsten	366.0	0.296	4.6

Table 1. Mechanical Property Values used in FAMMOS

It is assumed that the perpendicular pushing-down force is applied as amount of 500MPa in each type as same, then the stress distribution in the structure is observed as shown in Figure 10. All the stress is applied uniformly on the aluminum pad regardless of x-coordinates. No stress is applied if there is no aluminum over the structure. In the cross-sectional view of A-A' direction, the perpendicular directional stress is appended as deeper inside in type B along the same x-coordinates. Otherwise, when the width of rim is increased as deeper inside, (in other words, the x-coordinates at the end of rim is toward in the direction of the center of pad,) it is dispersed at each end of metal rim in type C as shown in Figure 11.

It is shown that accumulated stress under the pad is different in accordance with pad structure in Figure 12. In type B shown in blue line with rectangle, applied power (PPMS in other words) is added as deeper inside along the same x-coordinate. As a result of this, the PPMS is maximized at the edge of deepest metal rim, that is, local area around the rim is far more fragile than others.

On the other hand, x-coordinates of each metal rim in type C are different from others. Consequently, it restricts the stress gathering in the same local area such like type B. The key thing we have to focus on in this phenomenon is, the width of metal rim should be longer as deeper inside in the direction of the center of pad. If it were not, most of all amount of stress would be focused on the edge of upper metal rim. For this reason, stress dispersion effect could be lessened when the position of the edge of metal rim is irregular.

In the Figure 13, it is shown that the relationship between the factor, width of rim in type C and the applied peak stress, compressive stress in this case. As the width of rim is increased from zero, the applied stress at the edge of rim is also decreased. When the width of rim reaches 3um, peak stress is saturated at about 110 [Mpa]. In other words, stress dispersion effect by the rim is saturated when it reaches 3um and over. From this result, it is clear that the most effective width of rim is 3um in this experiment.

978-1-4244-9113-1/11 $26.00 © 2011 IEEE

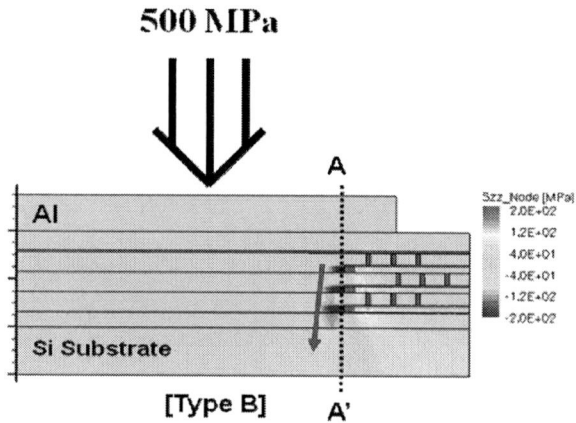

Figure 10. FAMMOS Simulation Results of type B

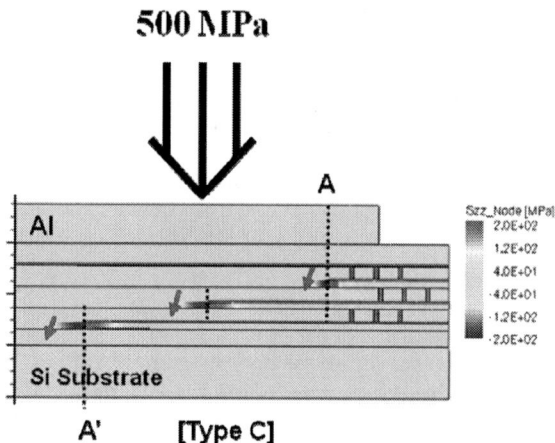

Figure 11. FAMMOS Simulation Results of type C

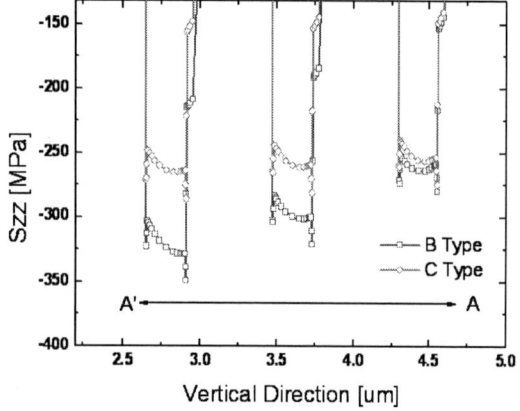

Figure 12. Concentrated Stress Comparison as a function of pad structure

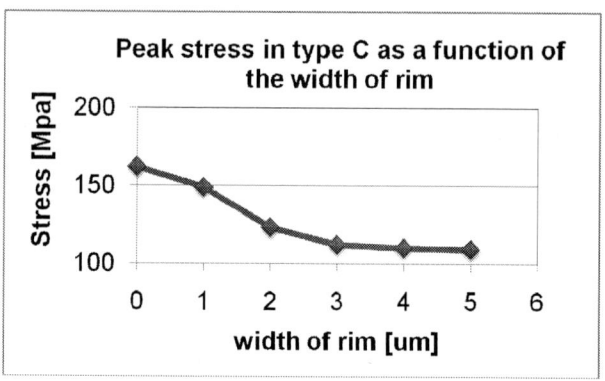

Figure 13 Compressive stress agnitude in according to the rim width

V. CONCLUSION

As devices continue to be scaled down dramatically, we have to be confronted with thinner thickness of IMD and needs for severe cost reduction than before. Consequently, It results in wire bonding vulnerability and wide usage of wedge wire bonding respectively. Most typical troubles due to wire bonding vulnerability is the peel-off failure. To overcome these obstacles, it is needed to investigate the solutions to respond actively. From this point of view, stress dispersion structure is a good solution for a peel-off failure.

It is suggested that the new bond pad structure to disperse the PPMS effectively, reverse-pyramid structure. It is proven that reverse-pyramid structure is more robust against the peel-off failure during a ball bonding process, even a wedge wire bonding process which has much more severe condition in terms of PPMS. The valid reason about its robustness is verified by means of FAMMOS simulation tool.

We have newly investigated the peel-off problem in the point of PPMS while the ball mounting process. The root cause of crack occurring which is the seed of peel-off is not the silicon processing or wire pulling force, but the external PPMS during the wire bonding process. It is proved that reverse-pyramid pad structure is stronger than the original one using WPT, force power split, and FAMMOS simulation.

ACKNOWLEDGMENT

The support and assistance of all the members of this project, in particular our colleagues So Ra Park who is in charge of FAMMOS simulation and Se Yeoul Park who is in charge of package process, is greatly appreciated.

REFERENCES

[1] Vaildyanathan Krispesh, Mohandass Sivakumar, Loon Aik Lim, Rakesh Kumar and Mahadevan K. Iyer, "Wire Bonding Process Impact on Low-K dielectric Material in Damascene Copper Integrated Circuits" in Electronic Components and Technology Conference p873~p880 IEEE 2002.

[2] R Kregting, R.B.R. van Silfhout, O. van der Sluis, R.A.B Engelen, W.D. van Driel, G.Q. Zhang, "An Experimental-Numerical study of Metal Peel off in Cu/low-k Back-End Structures", in 7th International Conference on Electronics Packaging Technology IEEE 2006

[3] X.Gu, Joze Antol, Y.F.Yao, K.H.Chua, "A Reliable Wire Bonding On 130um Cu/low-k Device", in Electronics Packaging Technology Conference p 707~711 IEEE 2003

[4] MIL-STD 883E, Method 2011.7, 1989

[5] Jeng-Jie Peng, Ming-Dou Ker, Nien-Ming Wang, and Hsin-Chin Jiang ,"Layout Design on Bond Pads to Imporve the Firmness of Bond Wire in Packaged IC Products" in VLSI Technology, Systems, and Applications, 1999. International Symposium on 8-10 June 1999 Page(s):147 – 150

[6] Hongjun Ji, Mingyu Li, Chunqing Wang, "Interfacial Characterization and Bonding Mechanism of UnltrasonicWedge Bonding", in Electronic Packaging Technology, 2006. ICEPT '06. 7th International Conference on 26-29 Aug. 2006 Page(s):1 – 5

[7] Synopsys, "Fammox TX Getting Started Guide", page 2, page 21-22, Version A-2008.09, September 2008

[8] David L. O'Meara, "Wirebond pull test failures at RTP TiN/BPSG interface", in SPIE Vol. 2335 p115~p129

[9] Ming-Dou Ker, Senior Member, IEEE, and Jeng-Jie Peng, Member, IEEE, "Fully Process-Compatible Layout Design on Bond Pad to Improve Wire Bond Reliability in CMOS ICs", in IEEE transactions on components and packaging technologies, VOL 25,NO 2, JUNE 2002

Electromigration Characterization of Lead-Free Flip-Chip Bumps for 45nm Technology Node

Christine Hau-Riege
Quality & Reliability Engineering
Qualcomm
Santa Clara, California USA
408-533-9647, chaurieg@qualcomm.com

You-Wen Yau
Quality & Reliability Engineering
Qualcomm
San Diego, California USA

Nick Yu
Process Technology Engineering
Qualcomm
San Diego, California USA

Abstract— **We have conducted electromigration experiments on lead-free SnAg flip-chip bump interconnection for 45nm technology node. We report lifetime distributions, kinetic parameters and intermetallic compound formation. Further, we discuss the impact of Ag-concentration as well as current direction on the electromigration reliability of these flip-chip bumps. Based on these analyses, we conclude that lead-free bumps lead to significantly more robust electromigration reliability than their SnPb counterparts, which render lead-free bumps a suitable replacement for the present and future technology nodes in terms of their current-carrying capability.**

Keywords- electromigration; lead-free; flip-chip bumps

I. INTRODUCTION

As the semiconductor industry moves toward the replacement of lead-based interconnects, many studies are being conducted on the reliability issues associated with the materials and integration changes for lead-free interconnects in the far backend of line (fBEOL). We cite some published works in the areas of mechanical [1, 2] and electrical reliability [3-8]. It can also be noted, however, that the electromigration reliability of fBEOL interconnects as a whole have been much less studied than that of on-die interconnects, even though the ever-increasing demands for current impacts all interconnects, including those used at the package-levels. This study addresses fundamental electromigration characterization of lead-free flip-chip bumps, including kinetic parameters, intermetallic compound formation and the effect of current-direction as well as the effect of process variation in Ag-concentration. We make some general comparisons to on-die Cu wire electromigration behavior where relevant.

II. EXPERIMENTAL SET-UP

Our studies are based on SnAg bumps connected to Si die and organic multi-level substrate, which had Cu metallization. The under bump metallurgy (UBM) on the die-side was 1kÅTi/5kÅCu/3umNi nominal with an 84um diameter and 30um resist opening. The UBM on the substrate-side was 30um thick SAC305 defined by an 80um resist opening (Figure 1). We tested samples with the Ag-concentrations of 1.3, 1.8, 2.4, and 3 wt%. Sn2.4%Ag samples were used for kinetics studies.

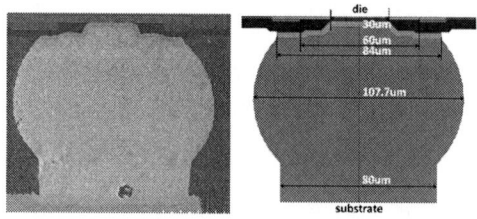

Figure 1. An SEM cross-section and schematic showing the as-processed SnAg bump, including pertinent dimensions.

Our test structure is based on a four-point Kelvin connection, in which a single bump is designated for EM failure (Figure 2). All except one experiment in our study had the cathode (or electron-source) at the die-side. Our tests were conducted at the temperatures of 160 and 180°C (which includes the 10 - 15 degrees of Joule heating) and the currents of 600 and 700 mA/bump, which correspond to the current densities of 1.08E+04 and 1.26E+04 A/cm², respectively, when considering the UBM diameter. The substrate-side test

978-1-4244-9113-1/11 $26.00 © 2011 IEEE

was conducted at a common test condition of 160C and 600mA, and will be referred to as the "reverse current" test. Sample size was 20 in each case. Failure was designated as a 50% increase in resistance, which is about 50 mOhm at test conditions for our structures. Failures were routinely analyzed through SEM, while EDX was used for composition analysis.

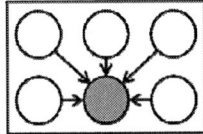

Figure 2. Plan-view schematic of EM test structure, in which circles represent bumps. The tested bump is solid.

III. EXPERIMENTAL RESULTS

The electromigration experiments conducted in this study have been compiled below (Table 1). All distributions were found to be monomodal, which will be shown in the following sections. Also, the sigma values are on the lower-side of the data available on flip-chip bumps [4- 10], but somewhat larger than those reported for on-die Cu wires (i.e., about 0.3). An explanation will be given why the reverse current test (#7) has the largest sigma in the group.

TABLE I. SUMMARY OF ELECTROMIGRATION TEST CONDITIONS AND RESULTS

#	Test Condition	Ag (at%)	e- source	MTF [AU]	σ
1	160C, 600mA	2.4	die	1635	0.35
2	160C, 700mA	2.4	die	992	0.41
3	180C, 600mA	2.4	die	473	0.59
4	160C, 600mA	1.3	die	2293	0.56
5	160C, 600mA	1.8	die	1822	0.26
6	160C, 600mA	3	die	184	0.50
7	160C, 600mA	3	substrate	2682	0.84

A. Kinetics studies

The first three electromigration results shown in Table 1 were used for kinetic studies, and are reported as cumulative lognormal plots in Figure 3. Using Black's law, we have calculated Ea and n to be 1.08 eV and 3.42, respectively (Figure 4). Both parameters are within the (wide) range of values published on PbSn [3] and Pb-free bumps [4-7]. The n-value is noted to be higher than 2, which may be explained by the presence of local Joule heating [8, 9] at the UBM/bump interface.

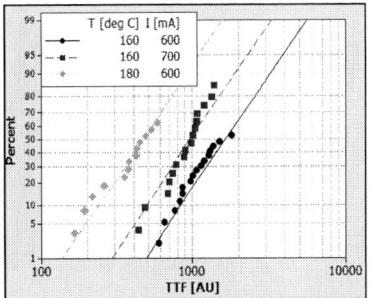

Figure 3. EM lifetime distributions of Sn2.4%Ag bump test at different test temperatures and currents.

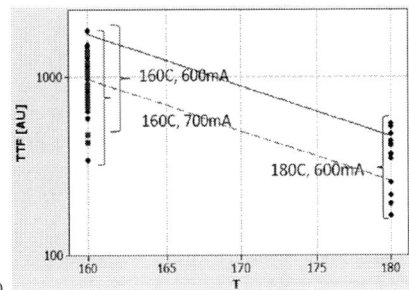

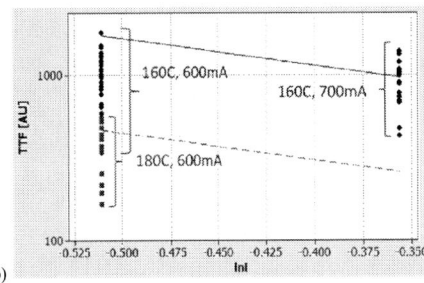

Figure 4. The three EM lifetime distributions from Figure 3 are plotted both as a function of (a) temperature and (b) natural-log of current for the derivation of Ea and n, respectively.

Failure analysis verified that failure occurred by EM-induced voiding at the die-side of the bump at the interface between the intermetallic compound (IMC) and solder. A representative image is shown in Figure 5a. EDX analysis confirmed the IMC to be η-Cu_6Sn_5 (see point 2 in Figure 5b), which is consistent with previous publications, and also shows that Ni is not a perfect diffusion barrier to Cu [2, 4].

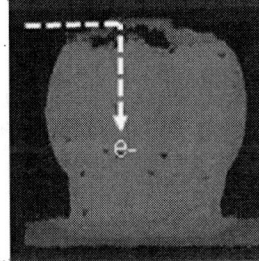

a.

978-1-4244-9113-1/11 $26.00 © 2011 IEEE

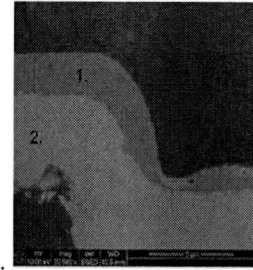

b.

Figure 5. Failure analysis of a tested EM bump, where a) EM voids were observed near the UBM/bump interface at the cathode-side of the bump, while b) EDX results indicate η-Cu_6Sn_5 intermetallic formation at point 2.

The authors note that flip-chip bump EM tests are typically more accelerated by current than temperature, as constrained by the solder Eutectic temperature. Therefore, a stronger emphasis is placed on the current density exponent (n) than activation energy (Ea) for accurate lifetime extrapolation to use conditions. This is a different mindset than typical on-die Cu wire EM tests, which are accelerated more through temperature then current.

B. Effect of Ag-concentration

We report the lifetime results of multiple Ag-concentrations (1.3, 1.8, 2.4 and 3 wt%), which corresponds to experiment #1, 4, 5 and 6 in Table 1. While some variation surely exists, it can be shown using a student's t-test that all distributions are statistically similar within a 90% confidence interval. Because of this, we can combine the distributions into one population, and have determined the sigma to be 0.4 (Figure 6).

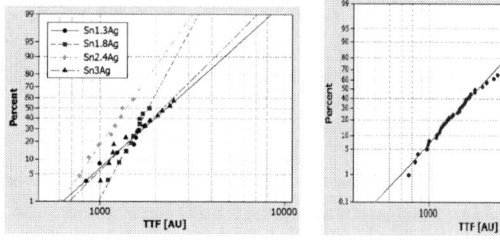

Figure 6. EM lifetime distributions are shown for different Ag-concentrations can be combined into one population.

While Ag_3Sn particulates may be present in the bump, we have shown that the range of Ag-concentrations in our studies did not affect the lifetime distributions. This is consistent with previous studies, which has shown Ag-concentration to affect EM at concentrations lower than 1 wt%, but that the network of Ag-based intermetallic compound remains largely unchanged at concentrations greater than 1% [10]. Interestingly, we also note that the range of Ag-concentrations in this study lead to less than 3% change in melting temperature, as determined through a binary phase diagram [11]. Therefore, we may use these results for setting a process tolerance for Ag-concentration, in order to maintain consistent EM reliability during high volume production.

C. Effect of current direction

We report the lifetime results for a reverse current direction (i.e., #7 in Table 1), in which the cathode is at the substrate-side (Figure 7). We note that the lifetime ranges are similar between the two current directions, but that the spread of the lifetime distribution (or σ) for the substrate-side experiments is relatively large, which is reflected in Table 1. Failure analysis (Figure 8) shows that the failure site for the reverse current tests was at the substrate-side, but that two failure types existed: 1) through the diameter of the bump near the substrate, as defined by the resist, and the 2) at the bump/trace interface. This variation is probably the origin of the larger sigma.

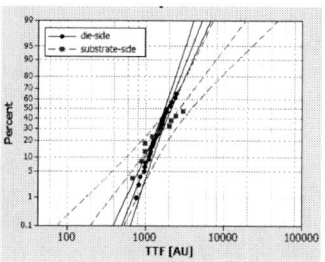

Figure 7. EM lifetime distributions for different current directions.

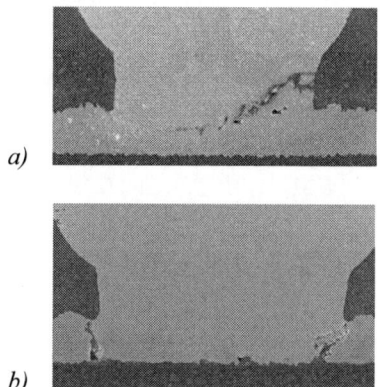

Figure 8. The failure void were found at the substrate-side for reverse-current tests. However, two different types of fails were observed: a) at the bottom of the bump in the resist region, and b) at the bump/trace interface.

It is noteworthy that the ranges of failure times between die- and subsrate-side tests are similar (Figure 7). This may be attributed to the similarities in relevant cross-sectional areas, and therefore current densities. Clearly, the cross-sectional areas of the die-side UBM and substrate side are close (i.e., the diameter of the UBM is 84um and the diameter of the resist is 80um.) Since the Cu trace in the substrate is 18um thick, we have calculated the area of the bump/trace interface at the substrate (which is relevant to the fail shown in Figure 8b) to be 90% of the area relevant to the substrate-side fail shown in Figure 8a, and 82% of the area relevant to die-side fail. The authors concede, however, that die- and substrate-side lifetimes are not necessarily always similar, but depend on the details of the structure, as outlined in this paragraph.

As a general statement, it can be noted that switching the electron flow direction leads to a change in failure site. This "mirror image" of failure sites of the flip-chip bump is quite different from what is observed for typical on-die Cu wire systems, in which voids nucleate under the Cu-cap and grow downwards [12], regardless of current direction. Therefore, different current polarities lead to different electromigration lifetimes for Cu wires [13] but not necessarily for flip-chip bumps.

IV. CONCLUSION

We have provided a summary of electromigration studies on lead-free flip-chip bumps in terms of kinetic parameters (Ea and n), failure mode, Ag-concentration, and current-direction, and have contrasted this behavior with EM behavior of on-die Cu wires where applicable. Based on our studies, we observe that lead-free bumps lead to significant Imax increase over the analogous Eutectic solders of similar geometry [3]. This provides additional margin for EM compliance in package designs that need to be converted to Pb-free, as well as accommodation for designs using increased currents. We also show that a Ag-concentration range of 1.3 – 3 wt% does not change the EM-induced diffusion mechanism of solder depletion at the IMC interface, and leads to statistically consistent EM performance. In addition, we have assessed a similar EM lifetimes for both polarities through EM test, which can be attributed to similar geometries at both sides of the bump as well as a common (and aforementioned) diffusion path found at both sides of the bump. We also report a larger sigma in the reverse flow case, which can be attributed to the greater variation in failure sites at the substrate-side.

REFERENCES

[1] Intel Tech. & Research, Vol. 12, Iss. 01.

[2] K.C. Chan, Z. W. Zhong, K. W. Ong, Soldering & Surface Mount Technology, 15/2 [2003] 46-52

[3] D.H. Eaton, J.D. Rowatt, W.J. Dauksher, 47th Annual International Reliability Physics Symposium 2009.

[4] M. Ding. G. Want, B. Chao, and P.S. Ho, International Reliability Physics Symposium 2005.

[5] J.-H. Lee, Y.-K. Lee, Y.-Bae Park, S.-T. Yang, M.-S. Suh, Q.-H. Chung, K.-Y. Byun, Electronic Components and Technology Conference 2007.

[6] M. Lu and D.-Y. Shih, International Reliability Physics Symposium 2009.

[7] L. Nicholls, R. Darveaux, A. Syed, S. Loo, T. Y. Tee, T. A. Wassick, B. Batchelor, Electronic Components and Technology Conference 2009.

[8] S.-H. Chae, J. Im, T. Uehling, and P.S. Ho, Electronic Components and Technology Conference 2008.

[9] M. Shatzkes and J. R. Lloyd, *J. Appl. Phys.*, vol. 59, pp. 3890 1986.

[10] M. Lu, D.-Y. Shih, P. Lauro, S. Kang, C. Goldsmith, S.-Y. Seo, Electronic Components and Technology Conference 2009.

[11] http://www.metallurgy.nist.gov/phase/solder/agsn-w.jpg

[12] C.-K. Hu and S. Reynolds, Electrochem. Soc. Proc. (1997).

[13] C.L. Gan, C.V. Thompson, K.L Pey, W.K. Choi, H.L. Tay and M.K. Radhakrishnan, Appl. Phys. Letters 79, 4592 (2001).

Electromigration failure mechanisms for different flip chip bump configurations

Riet Labie, Tomas Webers, Christophe Winters, Vladimir Cherman, Kristof Croes and Bart Vandevelde
Imec
Leuven, Belgium
0032 16 281 237, Riet.Labie@imec.be

Franck Dosseul
ST Microelectronics
Tours, France

Abstract— **Different flip chip bump configurations are investigated in terms of their electromigration behavior. Standard SAC (SnAgCu) solder bumps with a Ni/Au finish on the chip side are compared with Cu pillar bumps soldered with a thin layer of SnAg alloy. The substrate finish is identical for both cases and consists of a 17µm thick Cu layer. Depending on the current direction, different interfaces are stressed what results in variable degradation mechanisms. Both the 17µm thick Cu UBM and the Cu pillar bumps outperform the Ni/Au chip finish due to the fast formation of an intermetallic phase which covers the full solder stand-off height. The excessive intermetallic growth indicates significant Cu dissolution but void formation couldn't be detected. When the electrons are forced from the Ni/Au finish to the solder bump, micro-structural degradation and an according bump resistance increase can be clearly monitored for different test conditions. The electromigration parameters of Black's acceleration model are defined for the Ni/Au UBM.**

A TaN temperature sensor is incorporated in the test chip which allows in-situ measurements of the actual device temperature. In this way, the generated Joule heating can be clearly monitored.

Keywords - Electromigration; Flip Chip Solder Joints; Microstructural Evolution; Failure Analysis

I. INTRODUCTION

The ban on Pb forced people to search for reliable alternative solder materials. In addition to the change in solder material, flip chip bumps are also decreasing in size. This could induce an even bigger challenge. The use of smaller bumps not only gives rise to harsher mechanical constrains it may also create additional problems like electromigration (EM). EM failures were formerly seldom incorporated in the solder reliability and lifetime analysis due to the larger bump dimensions and small applied currents. However, with technology scaling, smaller bumps and lower IC voltages while maintaining on-chip power consumption further increases the current density flowing through the solder joint. The international technology roadmap for semiconductors predicts currents as high as 0.2 and 0.4A for each interconnect [1]. Because of this, electromigration effects in solder bumps receive recently more and more attention [1-6].

Due to the ball-shape of standard solder bumps, decreasing the pitch and bump diameter inevitably reduces the stand-off height as well. Since smaller bump heights result in higher shear stresses, which are generated by the CTE (Coefficient of Thermal Expansion) mismatch of the connecting substrates, the thermo-mechanical lifetime of smaller solder joints decreases. Pillar bumps which consist of a thick Cu pillar and a smaller portion of solder have the ability to increase the stand-off height without compromising the pitch. Furthermore, a larger stand-off height facilitates the application of underfill materials. On top of that, finite element simulations have shown a reduction of the current crowding effect when a thick UBM (Under Bump Metallization) is used which limits local current density peaks [7]. For these reasons, pillar bumps are recently introduced for fine pitch flip chip joints [8].

In this work, standard SAC solder bumps and pillar bumps are compared in terms of their electromigration performance. The standard bumps contain the most commonly used UBM layers being Ni/Au on the chip side and Cu on the package side. Ni/Au is often preferred due to the slower reaction kinetics with solder, smaller intermetallic growth and a reduced UBM consumption [9] with thermal ageing.

II. EXPERIMENTAL

A. Sample Preparation

The test structures are fabricated on glass wafers. Electroplated Cu is used for routing the electrical signals. The dielectric material used for isolating the different metal layers is BCB (Benzo-Cyclo-Butadiene). A patterned TaN layer is positioned underneath the current stressed bump and is used as temperature sensor. By calibrating the TCR (Coefficient of Thermal Resistance) of this resistive material, the according temperature can be derived from the resistance value. The TCR calibration is performed during the ramp-up of the oven to the testing temperature (before any current is applied). Any temperature variations throughout the experiment will be notified by a TaN resistance change which is continuously monitored.

978-1-4244-9113-1/11 $26.00 © 2011 IEEE

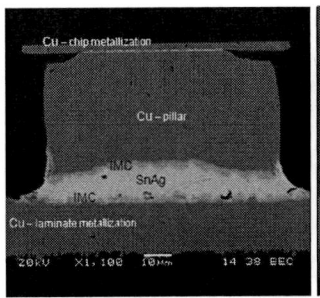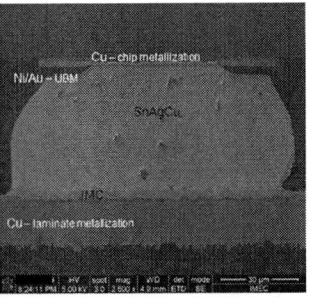

Figure 1. Cross-sectional SEM inspection of a pillar bump (left) and solder bump (right), after flip chip assembly and further packaging steps

The Cu metal of the chip is finished with either a thin Ni/Au layer and a SAC solder bump (with a diameter and height of 90 and 50μm respectively) or the chip is finished with a 40μm thick Cu pillar and 20μm SnAg finish (with a diameter and height of 80 and 60μm). The different test chips are flip chipped on a laminate interposer board which on its turn is connected with a temperature stable PCB (Printed Circuit Board) test board by 300μm solder balls. A cross-section of both flip chip bump configurations after packaging is shown in Figure 1.

B. Test Structure

The used EM test structure is a single bump that is monitored by a four-point measurement which allows accurate monitoring of bump resistance changes. The current is forced in one direction while the reverse current is divided over four bumps. In this way, the highest current densities and electromigration damage are generated in the monitored test bump. A cross-sectional overview is shown in Figure 2. The current direction can be chosen depending on the interface that has to be investigated. Previous work has shown that most damage is generated at the interface where the electrons enter the solder joint (cathode). Micro-structural changes do occur at the anode as well but have limited effect on the bump lifetime and resistance changes [10]. Apart from the electrical analysis, cross-sectional inspection is performed by the aid of SEM (Secondary Electrons Imaging), BSE (Back-scattered Electrons) and EDX (Energy Diffraction Spectroscopy).

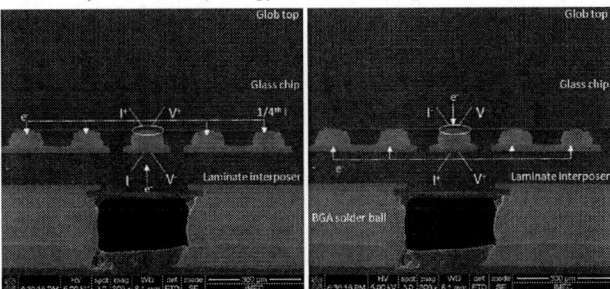

Figure 2. Schematic overview of single bump Kelvin test structure with the electrons entrance (cathode) at the bottom (left) or the top (right) bump interface. The interface where electrons enter is the most critical.

C. Test Conditions

The experiments are performed at 100, 150 or 170°C and the applied currents are either 300mA, 500mA or 1000mA. The generated current densities vary from 0.05 to 0.2mA/μm² and are on the low side compared to some literature references which go up to 0.4mA/μm² [4,5]. The advantage of these low current densities is first of all the reduction of the Joule heating. Fuse experiments, in which the current is gradually ramped up to 3A, show a temperature increase of 80°C at the highest current. Secondly, lower test currents are closest to user conditions.

For each experiment, eight flip chip - BGA packages are tested. For each package, one EM single bump is stressed by a single power source and its resistance is measured continuously. The failure criterion is arbitrarily chosen at a relative bump resistance increase of 150 percent.

III. DISCUSSION

A. Pillar Bumps

The original microstructure after flip chip bonding and further packaging of the pillar bumps is shown in Figure 1 (left). An EM experiment is run at a test temperature of 170°C and with a current of 300mA. An accompanying temperature increase of less than one degree is monitored by the local TaN resistor. Resistance variations of the EM test bumps could not be observed during testing more than 1000 hours and cross-sectional inspection did not reveal any EM induced damage. For all connections, the solder phase is completely transformed into the intermetallic (IMC) phase. There is no difference when comparing the stressed and thermal reference bumps (dummy bumps without electrical current) in terms of Cu consumption or IMC growth as can be seen in Figure 3. It is believed that once the interconnection is fully transformed into an intermetallic, electromigration will no longer be induced with the used current density. Other work shows that electromigration tests of full intermetallic Cu-Sn micro-bumps do not reveal any degradation after more than 2000h of testing at 200°C and for current densities as high as 1.1mA/μm² [11]. For the used dimensions of the pillar bumps, it would require a current higher than 6.5A. This is not only far from the generally in-the-field applied user currents but it would also create excessive Joule heating. Such high testing temperature is not compatible with most packaging materials. It is therefore concluded that further EM testing of the pillar bumps is not relevant since thermal degradation would occur much before EM damage in the bumps.

Although it looks as if the electrical current did not generate any degradation inside the bump and both the EM and thermal reference bumps look similar at the end of the experiment (when all solder is transformed into an intermetallic phase), it is possible that at the beginning of the test, some differences do occur and the electrical current does have an effect on the migration fluxes. It is therefore important to compare the bumps at a point before the solder is completely transformed into an intermetallic. In order to increase the chance for observing a difference between the electrically stressed and thermal reference bump, one additional experiment is executed at a lower temperature of 100°C (in order to reduce the thermal driving force for IMC formation) but with a high testing current of 1A (in order to increase the risk for inducing EM damage). This experiment is only continued for 150h. Again, the resistance of the EM pillar

978-1-4244-9113-1/11 $26.00 © 2011 IEEE   593

bumps remains constant throughout the test. However, cross-sectional inspection now reveals a small difference in the micro-structural evolution when comparing the current stressed and thermal reference bumps (Figure 4). At first sight, one can conclude that more IMC is formed in the EM bump. This could be explained by the fact that these bumps suffer from Joule heating (~ 12°C) due to the higher testing current. However, having a closer look, one can also see that the amount of IMC formation at the top and bottom interface is not the same, which is not the case for the thermal reference bumps that do show a symmetrical amount of IMC growth at both interfaces. The asymmetry is typically induced by electromigration since migration (metal loss) will occur at the entrance of the electrons (cathode) and pile-up will occur where the electrons leave the bump (anode). This leads to less IMC formation at the cathode and more at the anode [10] which is confirmed by the cross-sectional inspection of the EM pillar bump (Figure 4, right). Although the presence of an electrical current results in different rates of IMC formation (higher rate at anode and lower rate for cathode), there is no evidence of any degradation and voids are not detected. Since the intermetallic phases on both sides of the interconnection are approaching and already contacting at some points, one cannot expect further EM damage. It seems that for 20μm thick solder interconnections there is insufficient time to induce micro-structural degradation like void formation before full intermetallic joints are formed and further electromigration is blocked due to reasons mentioned before.

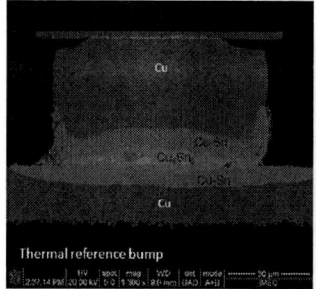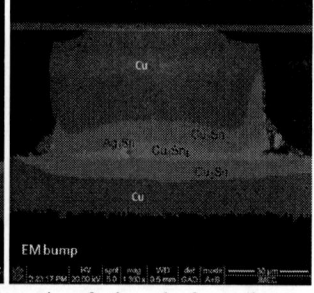

Figure 3. Cross-sectional BSE inspection of a thermal reference bump (left) and the EM bump (right) stressed with 300mA at 170°C. A similar amount and type of intermetallic phases are formed for both bumps.

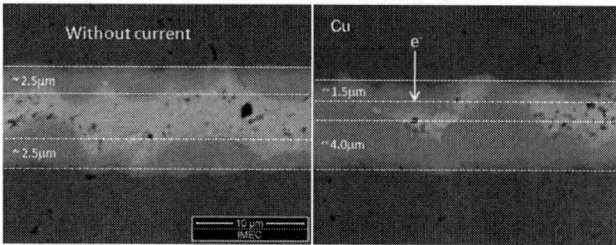

Figure 4. Cross-sectional BSE inspection of a thermal reference bump (left) and the EM bump (right) stressed with 1000mA at 100°C. A similar amount and type of intermetallic phases are formed for both bumps.

B. Standard Solder Bumps

The original microstructure after flip chip bonding and further packaging of the standard SAC bumps is shown in Figure 1 (right). Three different experiments are performed with test conditions of 170°C and 300mA, 150°C and 500mA

(2.3°C Joule heating) and 170°C and 500mA (3.4°C Joule heating).

In case of a stressed Ni/Au UBM (cathode), the monitoring of the EM bump resistance shows an unstable, continuous increase which differs strongly in slope: some bumps show a relatively constant resistance until a sudden increase occurs while other bumps show a more gradual increase. The experiments were stopped when more than 50% of the samples reached a resistance increase of more than 150%. Open failures were not detected within the testing time. The cross-sectional inspection could not show any correlation between amount of micro-structural degradation and varying electrical (bump resistance) behavior. For the Cu UBM stressed bumps (with electrons entering the bump through the bottom Cu UBM) a minor and smooth resistance increase is monitored at the beginning of the experiment which evolves towards a stable resistance with time. The resistance increase remains marginal and 'failures' are not detected.

Cross-sectional inspections are made after testing and the EM bumps, the reverse-current bumps as well as the non-stressed thermal reference bumps are analyzed. The reverse current bumps are those with only 1/4th of the test current and are relevant for observation in order to check for micro-structural damage indications at lower stress conditions and compare failure mechanisms. Identical failure mechanisms for test conditions and real-life conditions are needed in order to allow lifetime extrapolation.

For the Ni/Au UBM stressed interface, void formation in between the Ni-intermetallics and the solder can be observed for all applied test conditions (Figure 5). For all cases, a clear separation between the intermetallic layer and the solder bulk material can be seen. This failure mechanism is in agreement with Kwon et al. [5] (for a testing temperature of 120°C (hot plate) and a current density of 0.4mA/μm^2) but differs from the failure analysis of Ding et al. [4] that claim Ni dissolution eventually leading to interface delamination (for bump testing temperatures of 110, 122 and 155°C and a current density of about 0.4mA/μm^2).

The exact stoichiometry of the formed intermetallic phase could not be determined since EDX measurements indicated atomic ratios in between $(Cu,Ni,Au)_3Sn_4$ and $(Cu,Ni,Au)_6Sn_5$. A comparison with a thermal reference bump is made in Figure 6 (for the experiment with 500mA and at 150°C). Both bumps show a clear difference in micro-structural evolution: for the current stressed bump, the amount of intermetallic formation at the Ni-UBM chip side is much smaller. Although more Ni seems to be consumed, less intermetallics are formed which can be explained by metal migration away from the interface. This conclusion can be confirmed by an EDX analysis of the intermetallic phase formed at the opposing side being the Cu laminate interface which shows the presence of Ni for the stressed bump and no Ni for the thermal reference bump. A larger amount of spalled intermetallics can also be observed for the stressed bump.

A lognormal distribution of the failure data, based on a 150% resistance increase criterion, resulted in MTTF (Mean Time to Failure) values that were used for fitting Black's law.

978-1-4244-9113-1/11 $26.00 © 2011 IEEE

An activation energy of 0.90eV and a current density exponent of 1.69 could be derived. The activation energy is comparable with observations of Chae et al. [6] who found a value of 0.86eV while their current density exponent is slightly larger being 2.1 indicating a higher sensitivity to the applied current.

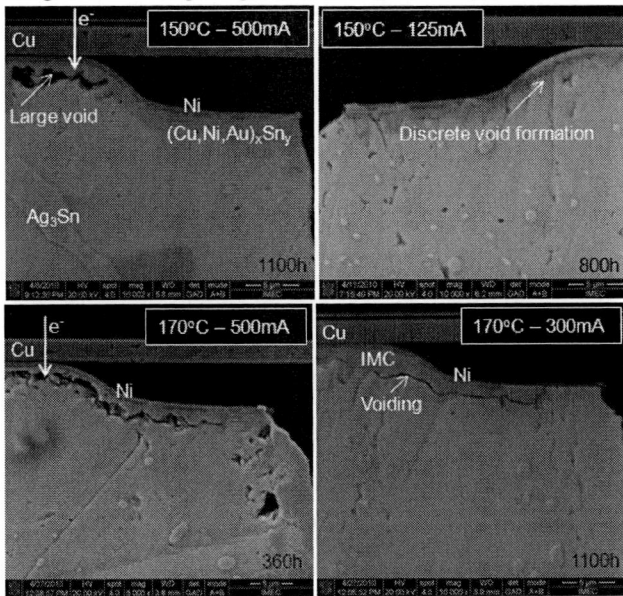

Figure 5. Cross-sectional BSE inspection of the cathodic Ni/Au interface stressed at various conditions. The time after which the experiment is stopped and the image is taken, is indicated. Voiding is observed in between the intermetallics and the solder.

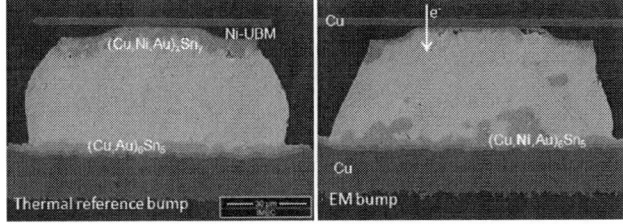

Figure 6. Cross-sectional BSE inspection of a thermal reference bump (left) and the EM bump (right) stressed with 500mA at 150°C. Although a faster consumption rate of the Ni-UBM is detected, a smaller amount of intermetallics is formed for the current stressed Ni/Au interface. A large amount of spalled intermetallics and the presence of Ni is detected at the Cu UBM for the EM bump.

For the Cu stressed interface, similar mechanisms as observed for the Cu pillar bumps, occur for tests at 170°C even for currents as low as 75mA. An increased Cu consumption, leading to excessive intermetallic formation is seen in Figure 7. As soon as the solder material is fully transformed into intermetallic phases, electromigration degradation is no longer expected. Despite the excessive amount of Cu dissolution, voids are not detected which is likely due to the large reservoir of Cu present. For experiments at 150°C, the amount of Cu migration is strongly reduced and even for a current of 500mA, the amount of Cu dissolution and IMC formation is comparable for both current stressed and thermal reference bumps after testing more than 1100h. Since electromigration damage is not observed for such long testing times, the experiments are stopped and the failure time could not be calculated.

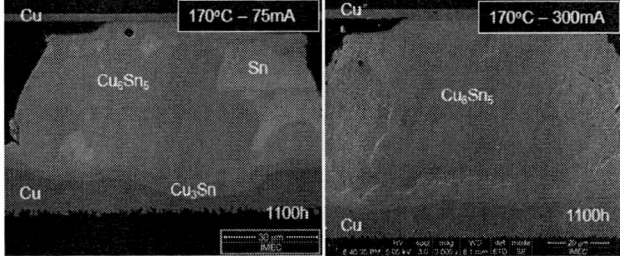

Figure 7. Cross-sectional BSE inspection with electrons flowing from bottom Cu UBM to the solder bump. The time after which the experiment is stopped and the image is taken, is indicated. Excessive IMC formation is observed indicating enhanced Cu dissolution due to the presence of an electrical current.

IV. CONCLUSION

A clear difference in electromigration behavior is observed when comparing different chip finishes for solder flip chip applications. Standard SAC solder bumps with a Ni/Au UBM show a constant failure mechanism of micro-structural degradation through void formation at the interface of the solder and the intermetallics. This occurs for all test conditions used (150-170°C and 300-500mA). The electromigration parameters could be derived by fitting Black's law which gives an activation energy of 0.90eV and a current density exponent of 1.69. Pillar bumps and standard solder bumps with the electrons flowing from the thick Cu UBM into the solder bump, did not show any electrical, nor micro-structural degradation. The rapid formation of a full intermetallic phase is believed to be the main cause of the outstanding electromigration performance of a thick Cu layer. Earlier work has shown an excellent resistance to electromigration of full intermetallic micro-bumps [11]. As soon as all solder material is transformed into the intermetallic phase, the joint is no longer expected to fail at reasonable testing conditions which are compatible with packaging materials.

The used test chip has an incorporated temperature sensor which is located underneath the current stressed test bump. This allows an accurate temperature measurement which has to be taken into account when fitting Black's law for different test conditions. For this work only limited Joule heating is observed for the applied test currents up to 0.5A with a maximum of less than 4°C for tests at 170°C oven temperature. The single experiment in which 1A was used, showed a temperature increase of 12°C.

REFERENCES

[1] ITRS Roadmap

[2] T.Y. Lee, K.N. Tu, S.M. Ku, D.R. Frear, "Electromigration of eutectic SnPb solder interconnects for flip chip technology," Journal of Applied Physics, Vol. 89, pp. 3189-94, 2003.

[3] G.A. Rinne, "Issues in accelerated electromigration of solder bumps," Journal of Microelectronics Reliability, Vol 43, pp. 1975-80, 2003.

[4] M. Ding, Wang G, Chao B, Ho PS., "A study of electro-migration failure in Pb-free solder joints," Proceedings IEEE 43rd Annual International Reliability Physics Symposium, pp. 518-23, 2005.

[5] Y.M. Kwon, K.W. Paik, "Electromigration of Pb-free solder flip chip using electroless Ni-P/Au UBM," Proceedings of ECTC, pp. 1472-76, 2007.

978-1-4244-9113-1/11 $26.00 © 2011 IEEE

[6] S.H. Chae, X. Zhang, K.H. Lu, H.L Chao, P.S. Ho, M. Ding, P. Su, T. Uehling, L.N. Ramanathan, "Electromigration statistics and damage evolution for Pb-free solder joints with Cu and Ni UBM in plastic flip chip packages," Journal of Material Science: Mater Electron, Vol 18, pp. 247-58, 2007.

[7] J.W. Nah, K. Chen, J.O. Suh, K.N. Tu, "Electromigration Study in Flip Chip Solder Joints," Proceedings of ECTC, pp.1450-55, 2007.

[8] B. Ebernsberger, C. Lee, "Cu pillar bumps as a Pb-free drop-in replacement for solder bumped flip chip interconnects", Proceedings of ECTC , 2008, pp 1-7.

[9] R.Labie, W. Ruythooren, J. Van Humbeeck, "Solid state diffusion of Cu-Sn and Ni-Sn diffusion couples with flip chip scale dimensions, " Intermetallics, Vol 15 (3), pp. 396-403, 2007.

[10] R. Labie, T. Webers, E. Beyne, R. P. Mertens, J. Van Humbeeck, "A modified electro-migration test structure for flip chip interconnections, " IEEE Transactions of CPMT, Vol 29, pp. 508-11, 2006.

[11] R. Labie, P. Limaye, K.W. Lee, C.J. Berry, E. Beyne, I. De Wolf, "Reliability testing of Cu-Sn intermetallic micro-bump interconnections for 3D-device stacking", Proceedings of ESTC, 2010, pp. .

Understanding and Modeling AC BTI

Hans Reisinger, Tibor Grasser*, Karsten Ermisch, Heiko Nielen, Wolfgang Gustin, and Christian Schlünder

Infineon Technologies, 85579 Neubiberg, am Campeon 1-12; Germany, phone:+49892349210;
e-mail: Hans.Reisinger@infineon.com
* Institute for Microelectronics, Technical University of Vienna, A-1040 Vienna, Austria

Abstract—We present a model for AC NBTI which is based on capture and emission of charges in and out of oxide border traps. Capture and emission time constants of these traps are widely distributed from <µs to >10⁵s and have been experimentally determined. The model gives a good quantitative understanding of experimental data from alternating stress / recovery sequences. It also provides a physical understanding of all the special features seen in AC NBTI independently of technology parameters.

Keywords- AC-stress, NBTI, recovery

I INTRODUCTION

Recently it has been shown that BTI threshold shift and recovery can be understood and modeled from charging and discharging of defects in the oxide [1-4]. These defects are characterized by their - field- and temperature dependent - time constants for charge capture and emission τ_c and τ_e. The nature of these defects is not thoroughly understood yet. However, for given electrical stress and recovery fields there is a very simple mathematical description of each defect, which is just an asymmetric RC-element (see Fig. 1c). A full description of a macroscopic MOSFET (containing thousands of defects) then is equivalent to a 2d spectral density map (see Fig. 1b). We want to stress that in principle this mathematical model is independent of the physical model behind it. This study is exclusively devoted to AC-BTI stress. The importance of AC-NBTI has been recognized some time ago [5, 6, 7]. AC-NBTI is the dominating degradation mechanism for computational CMOS logic [8]. An AC- instead of DC-assessment will lower the degradation by more than a factor 2 and increase lifetimes by a factor in the order of 100 [9]. To exploit these benefits, however, clearly demands a solid understanding of AC BTI. We will show that all the special features seen in AC-BTI, i.e. the dependence on duty-factor, the experimentally absent frequency dependence and the reduced recovery-rate - which have not been understood so far - can be well understood now, independently of the technology and also for Hi-K gate stacks.

This paper contains three more sections: Section II provides a brief description of samples and our AC measurement technique and parameters. In section III we outline the properties of our model. Properties of the single defects are not discussed in this work. Temperature and field dependencies of τ_c and τ_e are treated in [3, 4, 10] . The ΔVT step heights produced by single charged defects and the variability of ΔVT are discussed in [11, 12, 13] . In the main

section IV we will give explanations for all the special AC-NBTI features mentioned above.

II SAMPLES AND EXPERIMENT

The data analyzed in this study are production-quality pFETs with plasma nitrided oxides (PNO) around 2nm thick and pFETs with 25nm PNO. Also analyzed have been data from Hi-K FETs taken from the literature [14]. Nitrided gate oxides with a thickness of 2nm have been the topic of numerous studies (see [15] and references therein). An inter-technology comparison of some PNO oxides is shown in [4]. It proves that the properties of PNO's with moderate nitridation are comparable w/o major differences.

In our experiments the threshold voltage VT has been measured using a measure-stress-measure sequence and our ultrafast direct VT-measurement [15]. It is important for the discussion of data that measuring is synchronized with the AC stress signal and starts right after the completion of a stress cycle with a 1 µs measuring delay. Most of our measurements have been done at a frequency of 100 kHz. At this frequency we have been able to sweep the duty factor from 1% to 99% w/o requiring pulses shorter than 100 ns. Like for DC NBTI the measuring delay is a parameter with has strong influence on the measured data, especially on frequency- and duty factor dependence. When comparing different data from the literature, differences in measuring parameters may be much more important than differences in the sample parameters.

III NBTI MODEL

A. Model principles

As pointed out in [2-4, 10] charging and discharging of defects is ruled by stochastic capture and emission processes. When we average over many defects with the same capture and emission time constants τ_c and τ_e then the average occupancy P after "applying" stress or recovery, after a stress time t_S or recovery time t_r , respectively, will be described by the following equations:

$$(1) \qquad P(t_s) = P(t_s = 0) \times \left[1 - \exp(-t_s / \tau_c) \right]$$

$$(2) \qquad P(t_r) = P(t_r = 0) \times \exp(-t_r / \tau_e)$$

For the sake of simplicity the Fermi distribution function is neglected and assumed to be 1 during stress and 0 during recovery. Note that a given, single defect has only two states, occupied or empty.

978-1-4244-9113-1/11 $26.00 © 2011 IEEE

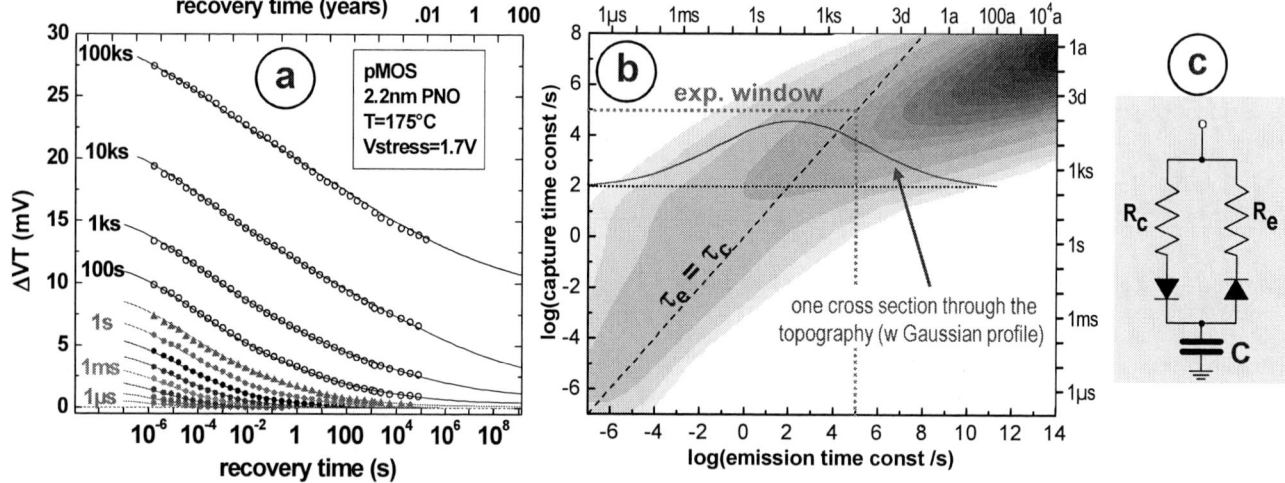

Fig. 1 – a: Measured data (symbols) and simulated data (lines) using the spectral density map of Fig. b; labels are denoting the stress times applied consecutively to the sample. **b:** The spectral defect density which has been varied until the best fit in Fig. a has been achieved. The spectral density distribution is parameterized and consists of 10 parameters in total, determining the topography. The spectral density inside the red rectangle is purely experimental, all values covering times $>10^5$ s are extrapolations. **c:** Equivalent circuit of a single defect. A single RC-element representing one given defect with capture and emission time constants $\tau_C = CR_C$ and $\tau_e = CR_e$. The RC-element is a 1:1 equivalent to the eq. (1) and (2). The voltage on C corresponds to the ΔVT produced by this defect. A real FET contains a large number of these defects, the individual VTs have to be summed up. Information in Fig. a (time domain) is equivalent to Fig. b (spectral domain).

B. Model Verification

A real FET with a WxL around 0.1×0.5 μm² will have a number of ≈500 defects contributing to NBTI degradation (see table 1 in [4]. Clearly to determine the properties of 500 individual defects experimentally is too time consuming to be feasible. Thus a spectral defect density map, taken from the measurement of a very large FET, has to be used instead. In a large FET several thousand defects are active simultaneously. The principles of extracting such a map, of gathering all the information needed, have been outlined in [4]. The extraction method in [4] contains some obvious approximations and is tedious and prone to spurious ΔVT fluctuations. A correct and more straightforward determination of the spectral defect density map is outlined in Fig. 1. Recovery traces are measured covering a regime as wide as possible in stress time as well as in recovery time. Noise and fluctuations in the experimental data should be below a 0.1mV level. Then a defect density map is set up like the one in Fig. 1b. With the complete experimental stress-recovery sequence that has been used for the traces in Fig. 1a this defect density map is then used to re-calculate these traces, resulting in the calculated lines in Fig. 1a. Then, using a nonlinear equation solver, the local density of the defect density map is iteratively varied until a best fit of the calculation (lines) to the experiment (symbols) is achieved. The transformation from the defect density map to the calculated ΔVT curves just employs eq. 1 and 2 for each of the defect classes characterized by their τ_C and τ_e. Each defect class corresponds to a little square in the τ_C / τ_e plane in Fig. 1b. Thus the transformation is trivial, in principle. To avoid any ripple in the calculated data the spacing of the single RC-elements has been chosen to be 0.5 decades. Thus the total number of RC-elements involved in

the calculation is in the order of 1000. It is obvious that the information density from the measurement is not high enough to allow the unique determination of 1000 parameters. As a consequence the defect density map in Fig. 1b has been parameterized. Actually a number of 10 parameters is sufficient to describe the topography of the defect density map in Fig. 1b well.

As seen the fit experiment / simulation in Fig. 1a is perfect. This is not a surprise, however. Our model and the way of parameter extraction actually very much resembles a Fourier transformation: A set of data in the time-domain, (example Fig. 1a) is just converted into a set of coefficients in the frequency domain (example Fig. 1b), i.e. the set of "amplitudes" in the 2-dimensional spectral plot. Fig. 1a is quasi a re-conversion into the time-domain. If done correctly, this re-conversion, like a Fourier transformation, exactly reproduces the experimental data as demonstrated in Fig. 1. Of course the model also reproduces the NBTI degradation power-law $\Delta VT(t_s) \propto t^{0.15}$ correctly (not shown). As a further verification and to prove the claim that the model is able to predict recovery and degradation for any *arbitrary* stress-recovery signal sequence, another model verification is shown in Fig. 15 of ref [4]. The agreement between experiment and model is - to our knowledge for the first time - very satisfactory. This proves that the model actually captures all the relevant (capture- and emission-) time constants, from μs to 100ks, and uses them correctly and thus is also valid for AC stress. It is worth mentioning that the model automatically includes any contributions which might be due to a non-recoverable, permanent contribution to NBTI. They are showing up in the right hand side, $\tau_e = 10^{10}$ s to 10^{14} s, of the spectral landscape Fig. 1b and are practically permanent.

978-1-4244-9113-1/11 $26.00 © 2011 IEEE

IV. AC-NBTI PHENOMENA

In this chapter we will show that the study of the AC-NBTI phenomena will give new insights into some of the NBTI problems. AC-data are easily gained and provide information beyond the DC-NBTI insights. They are more sensitive to extreme tails in the distribution Fig. 1b than DC-NBTI data. We want to stress that any model or theory aiming to explain NBTI (or BTI in general) must be able to explain *all* of the observed AC-BTI phenomena. If a "candidate" for a BTI-theory fails to explain one of these phenomena it just fails to explain BTI at all. We will show that our understanding of AC-NBTI will be able to explain all the special AC-NBTI features. These are the weak frequency dependence, the S-shaped curve of the duty factor dependence, and the AC recovery rate which is initially less than the DC-recovery for short recovery times but has been found to merge the DC recovery curve after long recovery times [14].

Fig. 2 gives an example of the charging of asymmetric RC elements (Fig. 1c), having various τ_c and τ_e values, with a rectangular AC stress signal applied. As seen in Fig. 2 the charging behavior is characterized by a saturation value depending on the duty factor, an envelope charging time constant $\overline{\tau}$ and a value for ripple. A derivation of the charging behavior is straightforward but - to our knowledge - not given in textbooks. The complete derivation will be given in a future publication. We only show the result:

$$(3) \quad V_L(t_{AC}) = \left[1 - \exp(-t_{AC}/\overline{\tau}) \times \frac{d - du}{1 - du}\right.$$

$$V_H(t_{AC} - \Delta T \times (1 - \beta)) = \left[1 - \exp(-t_{AC}/\overline{\tau})\right] \times \frac{1 - u}{1 - du}$$

$$with \quad 1/\overline{\tau} = \frac{\beta}{\tau_c} + \frac{1 - \beta}{\tau_e}$$

where u and d are abbreviations:

$$u = \exp(-\Delta T \beta / \tau_c) \quad ; \quad d = \exp(-\Delta T(1 - \beta)/\tau_e)$$

ΔT is the AC period and t_{AC} is the AC stress time. V_L and V_H are the low- and high levels of the saw tooth function in Fig. 2. $\overline{\tau}$ is the charging time constant seen in Fig. 2, determined by τ_c and τ_e and the duty factor β. Eq.(3) does not contain any approximation. The total computation time for calculating a $V(t_{AC})$ due to AC stress corresponds just to evaluating eq. (3). Done for roughly 1000 defect classes it takes a couple of µs only in total. Thus even for a 10 year stress time at 1 GHz (10^{17} cycles) the calculation is fast. In contrast, calculations based on the reaction / diffusion theory have been reported [16] to consume minutes of computing time only for some 100 stress cycles.

A qualitative and semi-quantitative understanding of all the AC-NBTI features can be obtained from Figs. 3 and 4. These Figs. show the filling state of defects under AC stress, calculated using eq. (3). They explain how the defect density map is sequentially filled during AC stress and emptied again during a recovery phase. Fig. 3 shows an example for the features of the filling state of the defects after applying AC stress with a frequency $f = 1$Hz and duty factor $\beta = 99\%$, for an AC stress time $t_{AC} = 100$ ks and a subsequent recovery time (= measuring delay) of 1 µs. Fig. 3 also serves as a legend for Fig. 4 and uses the same color code. We will see that it is not necessary to consider a special spectral defect density - like the one in Fig. 1b - in order to understand the characteristic AC NBTI features. It

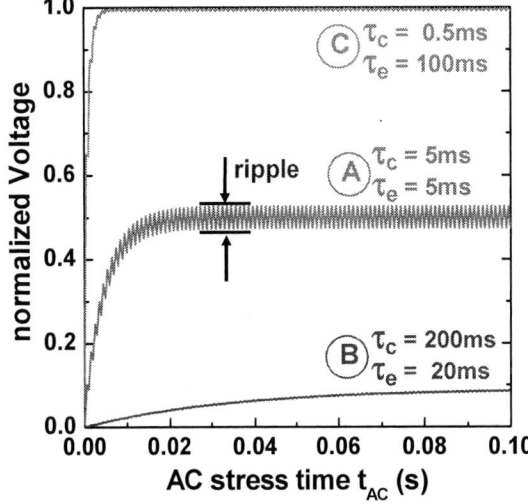

Figure 2: Some examples of charging curves of RC-elements (=defects) with different charging time constants $\overline{\tau}$ (eq. 3) and different ratios τ_e/τ_c, B=0.1, A=1, C=20. Duty factor is 50% for all curves.

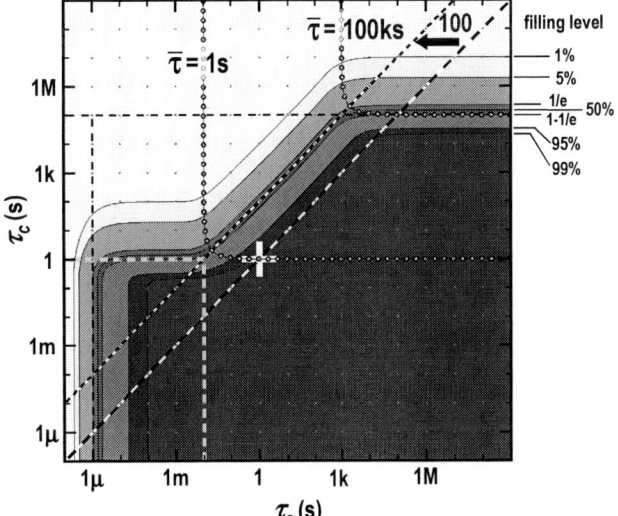

Figure 3: Occupancy level of defects after AC stress with frequency 1Hz, duty factor 99% (ON/OFF ratio = 100) for a stress time $t_{AC} = 100$ ks; the status is taken after a full charging cycle with a 1 µs measuring (=recovery) delay. The diagonal line marks $\tau_c = \tau_e$, the shifted diagonal is shifted by a factor 100 corresponding to the ON/OFF ratio. The white cross marks the AC period-time; the green dashed rectangle marks the charging and discharging periods of the AC signal. The dotted green trajectories mark lines of const. charging times $\overline{\tau}$ 1 s and 100 ks. Right hand labels denote the filling state corresponding to the color code. The black horizontal and vertical dash/dotted lines denote the net stress time and the recovery time (= measuring delay), respectively.

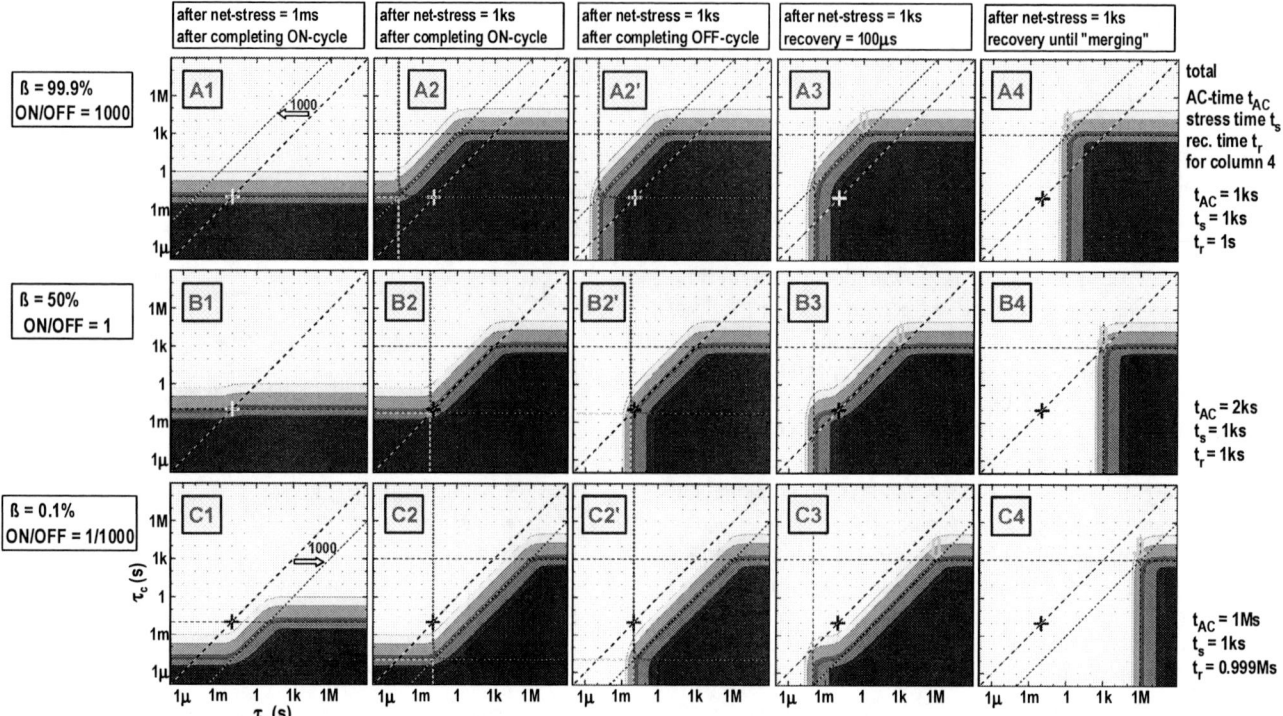

Figure 4.   Shows the occupancy levels of RC elements in 3 rows A, B, C with duty factors β from 99.9% to 0.1%. The columns 1 to 4 correspond to different "readout" stress - and recovery times. Columns 2 and 2' show the status just after completing an ON-cycle of the AC signal (w *zero* measuring delay) and after completing an OFF-cycle, respectively. The green arrows mark the intersection of the horizontal τ_c= net-stress-time line and the diagonal $\tau_c = \tau_e * \beta$ which is the point where "merging" of DC and AC recovery occurs. The other symbols and marks are explained in the caption of Fig. 3.

is sufficient to just consider the filling level of the defects in Figs. 3 and 4, which is the same as to assume a "flat" spectral defect density over the τ_e-τ_c plane. Figs. 3 and 4 are, together with eq. (3), almost self explaining. So we give just a description of Figs. 3 and 4 in brief:

• The total area of filled defects in the τ_e-τ_c -plane corresponds to the total charge, i. e. to the value of ΔVT.

• *Under AC stress* the defects fill from bottom to top up to τ_c-levels which can be divided in 3 main regimes:

(I) In regime I the discharging time constant τ_e is *shorter* than the off-time of the AC stress signal; thus all the defects in this regime are completely "reset" during each of the discharging periods of the AC signal; in Fig. 3 this is the regime left of the vertical, green, dashed line marking a τ_e of 10 ms (that is the discharging time of the 1 Hz AC signal with 99% duty factor).

(II) In regime II, located right of regime I, the occupancy level is continuously rising with stress time; only the levels *right* of a diagonal are filled; the filling state is a *linear* function of the ON/OFF ratio and of the ratio τ_e/τ_c ; a filling state of 50% is obtained at the condition $\tau_c/\tau_e <$ ON/OFF (with ON/OFF = β / (1-β))

(III) Regime III is inside the trajectory having a charging time constant $\overline{\tau}$ (eq. 3) equal to the *net stress time* $t_{AC}*\beta$. Defects inside this trajectory (above $\tau_c = t_{AC}*\beta$) remain uncharged or only partly charged.

• *Upon interruption of stress* the charge is "etched" away from left to right to the τ_e-position equal to the actual recovery time t_r. This etching, which is of course independent of the duty factor β, results in a vertical, denuded front (see Figs. 3, 4).

• A *pure DC-stress* without any off periods would fill *full* rectangles instead of a slanted rectangle. Only $\beta = 99.9\%$ is shown in Fig. 4, see Fig. 7 for 100% .

• *Recovery:* The 3 rows in Fig. 4 have different duty factors but receive all the same net-stress-time $t_{AC}*\beta$. Recovery after stress "etches" the bevel anisotropically away from left to right, and the charged shape will eventually become a rectangle. After a certain recovery time t_r -merge the "AC-rectangle" and the "DC-rectangle" have equal shapes in good approximation as seen in Fig. 4; this is the point were AC- and DC-recovery traces merge (see Figs. 9, 10). The merging point is marked by green arrows in Fig. 4 and is given by the crossing of the dashed lines. As easily seen this crossing corresponds to the condition $t_r = t_s*(1-\beta)$ (given in labels in the right hand side of Fig. 4).

A. Frequency dependence

It has been shown that over 9 decades in frequency (1Hz to 2 GHz) there is no frequency dependence of the AC-degradation [17]. Fig. 5 shows the experimental curves ΔVT vs. frequency from ref. [17] in comparison with calculated

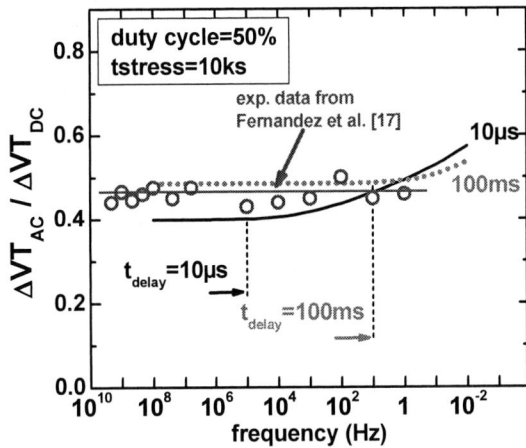

Figure 5: Frequency dependence, calculated with the model parameters from Fig. 1b in comparison with experimental data from ref.[17]. The experimental data were taken with a measuring delay $t_D = 10$ s; the calculations have been done with $t_D = 10$ us and 100 ms (see right hand labels). The frequency dependence vanishes for frequencies $f > 1/t_D$.

curves for two different measuring delays 10 μs and 100 ms. For the calculated curves a realistic spectral defect density (from Fig. 1b) has been used. It can be seen in Fig. 5 that in the calculated curves the frequency dependence vanishes for frequencies *higher* than the inverse measuring delay $1/t_D$. The reason immediately becomes clear when looking at Fig. 3 : The complete regime *right* of the vertical, dashed green line is "ripple-free". In this regime the AC period $1/f$ ($f=1$Hz for the case of Fig. 3) is either inside the $\bar{\tau}=1$s trajectory. This means that $1/f$ is shorter than the characteristic charging time $\bar{\tau}$, thus the defect behaves like a low pass filter. Or, in the regime below the $\bar{\tau}=1$s trajectory the occupancy is 100%, which also means zero ripple. Zero ripple is equivalent with zero frequency dependence, so the population in this regime does not change when the frequency changes. The only frequency dependence is found *left* of the dashed, green line. Now, we see in column 3 and 4 of Fig. 4 that the measuring delay t_D *completely erases* the charges left of the green line, and thus erases any frequency dependence for the case that t_D is larger than $1/f$. From Fig. 4 it is obvious that the "charged" area decreases when t_D increases. In contrast in Fig. 5 ΔVT at high frequency the curve for $t_D = 100$ ms lies significantly higher than the $t_D = 10$ μs curve. This is just a matter of normalization to the DC degradation. When the measuring delay t_D is kept constant and the frequency f is decreased below $1 / t_D$ then ΔVT is increasing, as seen in Fig. 5. This behavior can be most easily understood looking at Fig. 4 / B3: lowering the frequency moves the black cross upward along the diagonal, thus increasing the area of the rectangular base which is at the bottom of the rising triangle. As a result of this chapter we can say that the missing frequency dependence of NBTI is just due to a measuring artifact. A further obvious result is that the general findings about the frequency dependence are just a matter of the wide distribution of capture and

emission time constants and are not dependent on the spectral density. They will be generally valid for any (N)BTI degradation in any technology, also for high-k gate stacks. It should be noted that the apparent frequency-independence does not contain any information about the spectral defect density but is more or less a trivial property.

B. Duty factor dependence

All existing experimental studies about AC-NBTI, e.g. [17, 9] are showing a relation of ΔVT vs. duty cycle having always a typical shape like a lying S. In the vicinity of $\beta = 50\%$ the curve is almost flat. The most prominent and generally observed features are sharp rises in ΔVT between duty factors $\beta=0$ and 5% and between 95% and 100%, that is a derivative $\partial \Delta V_T / \partial \beta$ which is very high at the edges of the curve and is close to zero in the middle. The reason for this shape has not yet been convincingly explained. In [4] we could show that a perfect fit of the "S-curve" can be obtained just by summing up the responses of 3 classes of defects, each contributing about an equal amount of ΔVT, and having ratios τ_e / τ_c in the order of 0.001, 1 and 1000. The fast capturing and slowly emitting class with $\tau_e / \tau_c = 0.001$ is responsible for the sharp rise between $\beta=0$ and 5%. The fast emitting and slowly capturing class with $\tau_e / \tau_c = 1000$ is responsible for the sharp rise between $\beta=95\%$ and 100%. The good fit is due to the fact that the shape of the S-curve is mainly determined by the contributions of the defect classes with the *extreme* (low and high) ratios τ_e/τ_c. The assumption of 3 discrete classes contributing is unrealistic, however. So we will do a more general approach in this paper. Fig. 6 shows examples of measured AC-stress data for 3 different PNO thicknesses. Also shown in Fig. 6 are 2 different calculations. One calculation is done by calculating the AC response of a measured DC spectral

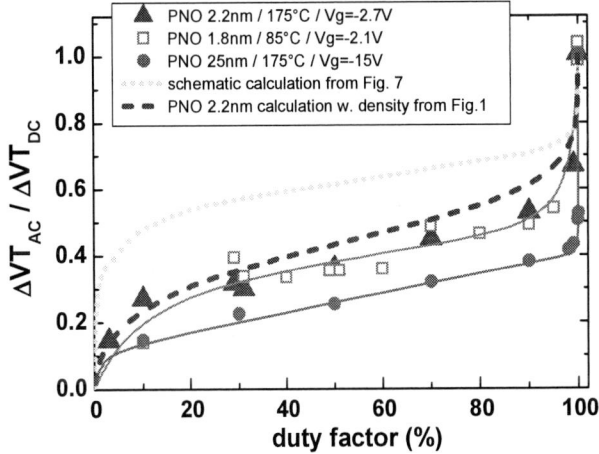

Figure 6. Experimental and calculated curves ΔVT vs duty factor. The calculated curve for the PNO is from the model Fig. 1b and is in fair agreement to experiment. The "schematic" calculation is from an integration over the occupied areas in Fig. 7. All data are for $f = 100$ kHz and a measuring delay of 1 μs after completion of a stress phase.

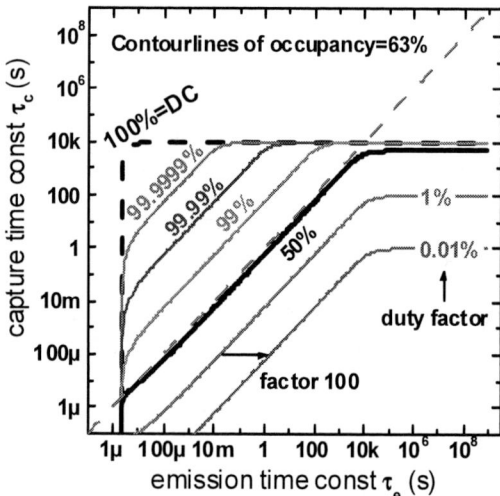

Figure 7. Calculated contour lines of an (1-e) occupancy level in the τ_e-τ_c plane after AC-stress with the duty factor as parameter. Stress frequency is 100 kHz, stress time is 10 ks and measuring delay is 1 µs. For a "flat" spectral defect density the ΔVT corresponds to the area enclosed by the contour lines.

defect density distribution like the one in Fig. 1b, but with stress parameters corresponding to Fig. 6. As seen the fit between calculation and experiment is quite reasonable, though the parameters are coming from a DC measurement. On the other hand even only small contributions of defects having very small or very large ratios τ_e/τ_c, that is defects which are far off the diagonal in Fig. 1b, have a significant influence on the S-curve. The resolution of a DC stress measurements like in Fig. 1 is not sensitive enough to capture these small contributions, though we pushed our measurements to a 0.1mV noise level. Thus a real AC measurement is able to reveal a closer insight into the tails of spectral distribution of defects. Anyway, the fit calculated from the measured DC data is reasonable and can be taken as a proof that our NBTI model is correctly capturing the NBTI physics. In contrast, simulations which have been done using the reaction/diffusion theory, have the characteristic feature that the derivative $\partial \Delta V_T / \partial \beta$ is monotonously decreasing with β [18], which is not in agreement with experiment.

A general understanding of this special derivative $\partial \Delta V_T / \partial \beta$ of AC-NBTI, which is independent of special sample or technology properties, can be drawn from Fig. 7. It shows the calculated contour lines of the charged regimes in the τ_e-τ_c plane, analogously to Fig. 4. The same frequency, measuring delay and stress parameters as for our experimental data have been chosen. The total ΔVT for a duty factor corresponds to the area under the associated contour line in Fig. 7. It becomes obvious from Fig. 7 that this area changes always by roughly the same amount between each of the duty factor steps, for example between 99.9999% and 99.999% or between 1% and 0.1%. Thus it is clear that the derivative $\partial \Delta V_T / \partial \beta$ *must be* diverging close to $\beta = 0\%$ and 100%.

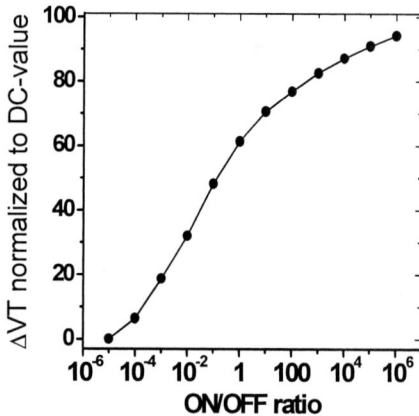

Figure 8. Calculated ΔVT calculated just from integrating over the occupied area in Fig. 7. This curve is identical to the "schematic" curve in Fig. 6. To avoid a divergent integral the area in Fig. 7 has been clipped for values $\tau_e > \tau_e *10^5$. The transformation from duty factor to ON/OFF ratio converts any S-shaped curves into an almost linear curve and vice versa.

As a consequence it makes sense to *not* plot data against the duty factor but against the ON/OFF ratio $\beta/(1-\beta)$. This transformation from β to the ON/OFF ratio is done in Fig. 8. Fig. 8 shows the area under the contour lines in Fig. 7. The infinite derivatives of the S curve are eliminated, resulting in nearly straight lines. Vice versa, any line being roughly straight when plotted vs. the ON/OFF ratio is transformed into an S shaped curve when plotted against the duty factor. So the shape of the experimental curve can be generally understood w/o even taking special technology or measurement / stress parameters into account.

At this point we want to stress that any conclusions based on all the published curves ΔVT vs. duty factor [9, 17] have to be done with care: The values of the ratios $\Delta VT(AC) / \Delta VT(DC)$ are strongly influenced by the measuring delay. If the measuring delay is shorter than $1/f$ then the measured ΔVT is a function of the synchronization of the measurement with the phase of the AC signal. Both these facts have not been considered in any of the "AC-publications" so far.

C. AC recovery

The effect of measuring delay on the measured AC-ΔVT has already been seen qualitatively in Fig. 4. It has been observed experimentally that, compared to the DC value, the AC-ΔVT (cmp. Figs. 4, 6, 7) is reduced even when the duty-factor is only slightly less than 100%, e.g. 99.9%. This is a matter of the "missing triangle" and is self explaining when looking at Figs. 4, 7. It has also been observed experimentally - but not yet been understood - [14] that the AC recovery rate $\partial \Delta V_T / \partial \log(t_r)$ (i) is less than the DC recovery rate, and (ii) that AC and DC recovery traces with the same net stress time as the DC "merge" after a certain recovery time. At this merging point (iii) the initially small AC recovery rate *increases* again. The explanation of the merging of AC- and DC-recovery at a given recovery time has already been explained in Fig. 4. The "merging" occurs

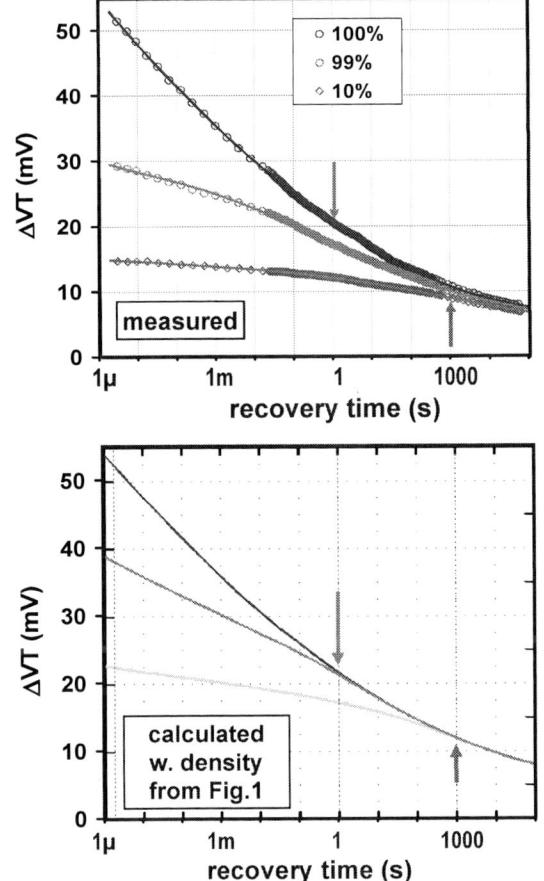

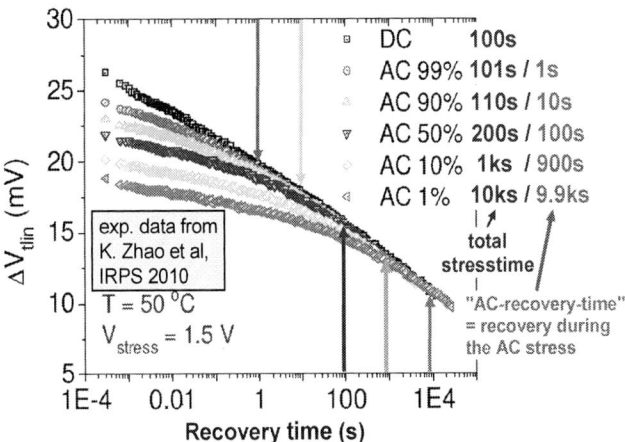

Figure 10. Recovery curves all taken after a net stress time of 100 s with various duty factors. The arrows mark the position were the AC-curves and the DC-curve (100%) are expected to merge.

Figure 9 Measured and calculated recovery after stress with duty cycle 100% / 99% / 10%. The corresponding stress times t_{AC} were 100s /101s / 1000s so that the *net stress times* $t_{AC} \times \beta$ are the same for all 3 curves. Arrows mark the points where the AC recovery curves are supposed to merge with the DC curve (cmp. green arrows Fig. 4).

when the AC occupied area in Fig. 4 has "etched" the slanted rectangles from the left down to rectangles. Seen from another point of view, there is almost no short term recovery after AC-stress because this short term recovery has already been anticipated during the stress phase. For example, for the 10% / 1 ks curve in Fig. 9 there have already been 900 s of recovery during the stress phase, prior to the real stress phase. So after a post-stress recovery phase of 1000 s both the 100%-case and the 10% case have "seen" a net stress time of 100 s and a total recovery time of 1000 s and 1900 s, respectively. Note that 1000 s and 1900 s are almost equal on a log time scale. So 1000 s is the point where both recovery curves merge. Fig. 9 and 10 show AC recovery for different duty factors and the same net stress time for our PNO samples and for Hi-K samples from ref [14]. The merging points in Fig. 9 and 10 are marked with arrows. These arrows correspond to the green arrows in Fig. 4. It is clear now that the difference for example between β=100% and β=99% is due to the cut-off edge which is best

seen in Fig. 7. The cut-off edge is parallel to the diagonal $\tau_e = \tau_c$. It is clear that, no matter how close β is to 100%, even for 99.99%, there will be an effect. It is noteworthy that - like for the duty factor and frequency dependence -, all considerations about the AC-recovery are independent of the exact spectral distribution of defects and thus are generally valid independent of technology or parameters.

V CONCLUSIONS

The AC NBTI model in this work is based on the analysis of single defects. The conclusions we draw are based on the experimental observations only and do not require further assumptions or models. The new understanding of NBTI leads to a straightforward modeling, especially of AC-NBTI, based on a simple equivalent circuit like in Fig. 1c and a set of empiric parameters. The AC-BTI model provides a physical understanding of all the special AC-NBTI features. This understanding is independent of the technology properties and also valid for high-K and p- and n-MOSFETs. Modeling for practical purposes for any arbitrary stress sequence (DC+AC) requires only short computation times and thus allows implementation into commercial aging tools like Cadence□s RelXpert™ [19].

REFERENCES

[1] B. Kaczer, T. Grasser, J. Martin-Martinez, E. Simoen, M. Aoulaiche, Ph. J. Roussel, G. Groeseneken, "NBTI from the perspective of defect states with widely distributed time scales", Proc. IRPS 2009, pp. 55-60.

[2] T. Grasser, H. Reisinger, W. Goes, T. Aichinger, P. Hehenberger, P.-J. Wagner, M. Nelhiebel, J. Franco, B. Kaczer; "Switching oxide traps as the missing link between negative bias temperature instability and random telegraph noise", IEDM technical digest 2009, pp. 729-732.

[3] T. Grasser, H. Reisinger, P.-J. Wagner, F. Schanovsky, W. Goes, B. Kaczer, "The Time Dependent Defect Spectroscopy (TDDS) for the Characterization of the Bias Temperature Instability", proc. IRPS 2010, pp. 26-32.

[4] H. Reisinger, T. Grasser, W. Gustin, and C. Schluender, "The statistical analysis of individual defects constituting NBTI and its implications for modeling DC- and AC-stress", Proc. IRPS 2010, pp. 7-15.

[5] M. A. Alam, "A critical examination of the mechanics of dynamic NBTI for PMOSFETs", IEDM technical digest 2003, pp. 14.4.1-4.

[6] S. Mahapatra P. B. Kumar, and M. A. Alam, "Investigation and Modeling of Interface and Bulk Trap Generation During Negative Bias Temperature Instability of p-MOSFETs", IEEE TED, Vol. 51, No. 9, pp.1371 1379 (2004).

[7] V. Huard, M. Denais, and C. Parthasarathy, "NBTI degradation: From physical mechanisms to modelling", Microelectronics and Reliability, Vol. 46, No. 1, pp. 1-23 (2006).

[8] K. Hofmann, H. Reisinger, K. Ermisch, C. Schluender, W. Gustin, T. Pompl, G. Georgakos, K. v. Arnim, J. Hatsch, T. Kodytek, T. Baumann, .C Pacha, "Highly accurate product-level aging monitoring in 40nm CMOS", Symp. on VLSI Techn. 2010, pp. 27-28.

[9] H. Reisinger, T. Grasser, K. Hofmann, W. Gustin and C. Schluender, "The impact of recovery on BTI reliability assessments", 2010 IIRW final report, pp. 12-16.

[10] T. Grasser, H. Reisinger, P-J. Wagner and B. Kaczer, "Time-dependent defect spectroscopy for characterization of border traps in metal-oxide-semiconductor transistors", Phys. Rev. B, Vol. 82, No. 24, pp. 5318 5327 (2010).

[11] B. Kaczer, Ph. J. Roussel, J. Franco, R. Degraeve, L.-A. Ragnarsson, E. Simoen, G. Groeseneken, T. Grasser, H. Reisinger, "Origin of NBTI variability in deeply scaled pFETs", proc. IRPS 2010. pp. 26-32.

[12] M. F. Bukhori, T. Grasser, B. Kaczer, H. Reisinger, A. Asenov, "'Atomistic' simulation of RTS amplitudes due to single and multiple charged defect states and their Interactions", 2010 IIRW final report, pp.76-79.

[13] M. F. Bukhori, S. Roy, and A Asenov "Simulation of Statistical Aspects of Charge Trapping and Related Degradation in Bulk MOSFETs in the Presence of Random Discrete Dopants", IEEE TED Vol. 57, No. 4, pp. 795-803 (2010).

[14] K. Zhao, J. H. Stathis, A. Kerber and E. Cartier "PBTI Relaxation Dynamics after AC vs. DC Stress in High-K/Metal Gate Stacks" Proc. IRPS 2010, pp. 50-54.

[15] H. Reisinger, O. Blank, W. Heinrigs, W. Gustin, and Ch. Schluender, "A comparison of very fast to very slow components in degradation and recovery due to NBTI and bulk hole trapping to existing physical models", IEEE TDMR, Vol. 7, No. 1, pp. 119-129 (2007).

[16] H. Kufluoglu, V. Reddy, A. Marshall, J. Krick, T. Ragheb, C. Cirba, A. Krishnan, C. Chancellor, "An Extensive and Improved Circuit Simulation Methodology For NBTI Recovery", proc. IRPS 2010, pp. 670-675.

[17] R. Fernandez, B. Kaczer, A. Nackeaerts, R. Rodriguez, M. Nafria, G. Groeseneken, "AC NBTI studied in the 1 Hz - 2 GHz range on dedicated on-chip CMOS circuits", IEDM technical digest 2006, pp.337-340.

[18] A. Jain, A. E. Islam, M. A. Alam, "A Theoretical Study of Negative Bias Temperature Instability in p-Type NEMFET", Proc. Micro/Nano Symposium (UGIM), 2010 , pp. 1-3.

[19] "Reliability Simulation in Integrated Circuit Design - a white paper."[Online Available]: http://www.cadence.com

The 'Permanent' Component of NBTI: Composition and Annealing

T. Grasser*, Th. Aichinger[†,‡], G. Pobegen[†], H. Reisinger[•],
P.-J. Wagner*, J. Franco°, M. Nelhiebel[□], and B. Kaczer°

* Christian Doppler Laboratory for TCAD at the Institute for Microelectronics, TU Wien, Austria
[†] KAI, Villach, Austria   [‡] Now at Penn State University, USA   [•] Infineon, Munich, Germany
° imec, Leuven, Belgium   [□] Infineon, Villach, Austria

Abstract— A number of recent publications explain NBTI to consist of a recoverable and a more permanent component. While a lot of information has been gathered on the recoverable component, the permanent component has been somewhat elusive. We demonstrate that oxide defects commonly linked to the recoverable component also form an important contribution to the permanent component of NBTI. As such, they can contribute to both the threshold voltage shift as well as to the charge pumping current. Under favorable conditions, particularly when subjected to continuous charge-pumping measurements, the permanent component can show recovery rates comparable to that of the recoverable component. We argue that this enhanced recovery is due to a recombination enhanced defect reaction mechanism. We introduce a simple extension to our switching trap model to also capture the impact of charge pumping measurements on the transition rates between the defect states.

I. INTRODUCTION

Recent research indicates that two components dominantly contribute to the negative bias temperature instability (NBTI) [1–6]: while one component dominates the recovery (R) the other one has been suspected to be more or less permanent (P). It has been recently shown that the complete NBTI induced degradation can be annealed at higher temperatures [6–8], implying that P is recoverable as well, albeit at larger time-scales compared to R. The most important aspect regarding P is that it might dominate device degradation at long times and could thus be the crucial degradation mechanism eventually determining the lifetime [6]. Unfortunately, the extraction of P is challenging as within conventional measurement windows (1 µs – 100 ks) it is normally overshadowed by R. As such, our understanding of P is somewhat vague, also regarding its constituents, be it interface and/or oxide defects [3, 6], or fixed positive charges [6]. We show that considerable precautions have to be taken for accurate extraction of P, as it suffers from similar issues than those typically related to the extraction of R, such as measurement *delay*, measurement *duration*, as well as stress/recovery *artifacts introduced by the measurement procedure* itself. Contrary to the work of Huard [6], who links P to interface states and an equal amount of fixed positive charge, *our analysis demonstrates that a significant fraction of P is due to switching oxide traps*, which contribute to both the threshold voltage shift ΔV_{th} and to the frequency dependent fraction of the charge pumping current.

II. ERRONEOUS EXTRACTION OF P

The most straight-forward approach for the extraction of P would be to wait until the recovery of ΔV_{th} has leveled at a plateau, thus directly exposing P. However, the fundamental problem here is the large timescales involved in the recovery of R, as even a short stress of $t_{\text{s}} = 1\,\mu s$ can lead to recovery transients of up to 1 ks, not to mention the recovery of P itself. On the other hand, P is created at a slower rate than R, making it difficult to locate plateaus within reasonable measurement times (< 1 week). As a consequence, plateaus in the recovery are rarely reported in literature [6]. (The plateau reported in [9] was later found to be not reproducible.)

Using different test technologies, from thick SiO_2 to SiON and high-κ gate stacks, we investigated a number of possible extraction methods. Most unfortunately, the extraction of P turned out to be much more complicated than expected. In particular, despite the fact that some attempts are incorrect altogether, it seemed almost as if P was trying to evade our characterization attempts. Fig. 1 summarizes some potential mistakes related to the extraction of P:

M1: The recovery of ΔV_{th} has to be plotted on a *relative logarithmic* scale following the end of stress, otherwise a spurious plateau appears. Such 'plateaus' are commonly found in literature but are completely irrelevant and simply a consequence of the inadequate presentation of the data.

M2: Switching to a lower temperature temporarily freezes recovery, resulting in a spurious plateau [10]. While the example temperature switch from 80 °C to 40 °C given in Fig. 1 may appear pathological, a typical real-world case appears to be given in [6]: The recovery of the devices was monitored at a high temperature on a probe-station for a day. Then, the devices were taken off the probe-station and stored at room temperature to be re-measured after some time. The plateaus obtained from this method are completely arbitrary.

M3: Application of a short positive bias partially removes oxide charges, temporarily accelerating recovery. This is because the emission time constant of switching traps depends strongly on the gate bias [13, 15]. Back at the original recovery voltage, these defects have already been annealed, resulting in a spurious plateau until the original recovery continues.

M4: In order to minimize the recovery, short stress times and low stress voltages can be chosen. This leads to relatively weak stresses and relatively short recovery times. However, particularly in thin oxides, the difference between stress and recovery voltage can be small, leading to notable degradation at the recovery voltage, interfering with the actual recovery. As a consequence, spurious plateaus can appear.

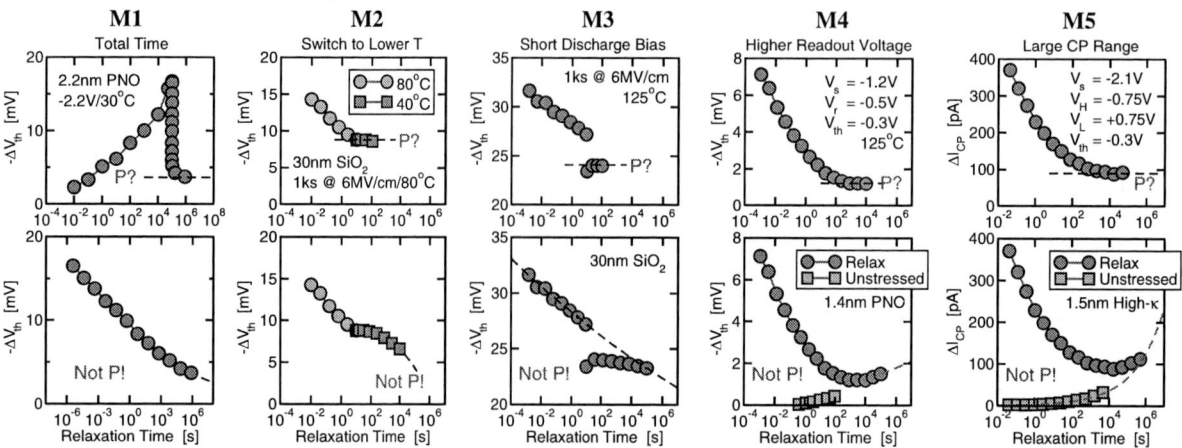

Fig. 1. Potential mistakes encountered while trying to locate plateaus in ΔV_{th} recovery traces. The top figures show apparent plateaus, which have nothing to do with permanent degradation. The reasons for the occurrence of these plateaus are illustrated in the bottom figures. From left to right: (**M1**) The recovery of ΔV_{th} is plotted as a function of the total time, rather than the recovery time t_r. Even if the recovery perfectly follows $\log(t_r)$, a spurious plateau will appear if the data is plotted this way, which has nothing whatsoever to do with P. Similar considerations relate to plotting the data on a linear scale, where the spurious plateau depends solely on the measurement time. (**M2**) Due to the large recovery time required, the device is only kept at stress temperature for a short amount of time [6] and, to ease measurement, recovery is continued at a lower temperature. This, however, is pointless, as a switch to a lower temperature freezes the recovery [10], which results in a spurious plateau. (**M3**) In order to remove R, which is due to trapped holes in the oxide [6, 11–13], a positive bias could be applied [14]. However, since the trap sites are switching traps, this has basically the same effect as a temperature switch, because such a bias switch only removes a few decades from the recovery trace, which continues after that. (**M4**) Relaxation gate voltages only slightly larger than the threshold voltage can already lead to degradation, in this example $V_{\text{relax}} = -0.5\,\text{V}$, with $V_{\text{th}} = -0.3\,\text{V}$. As a result, degradation overlaps with the 'normal' recovery, resulting in a spurious plateau for a certain amount of time. The signature of this plateau is that it disappears when either stress or relaxation voltages are changed. (**M5**) If the charge pumping amplitude is chosen too large, for this 1.5 nm high-κ device for example from $\pm 0.75\,\text{V}$, degradation is observed during the CP measurement, again resulting in a spurious plateau as in (M4). Note the strong relaxation of ΔI_{CP}, which is *anything but constant*.

M5: Similarly to M4, charge pumping (CP) measurements can lead to degradation of ΔI_{CP} when the charge pumping amplitude is chosen too large. Balancing the recovery of ΔI_{CP}, this can lead to spurious plateaus as well, just like M4. M5 already highlights an important issue [16]: ΔI_{CP} is not constant, even within conventional measurement windows, contradicting claims that ΔI_{CP} is nearly constant and equal to P [1, 6]. In particular, the resemblance between the recovery of ΔV_{th} in M4 and ΔI_{CP} in M5 is indeed striking.

III. ATTEMPTS AT EXTRACTING P

We proceed by analyzing ΔV_{th} recovery traces recorded after carefully selected stress/recovery voltages, stress/recovery times, and temperature. A typical plateau at the end of the recovery is shown in Fig. 2. According to Huard [6], this plateau is due to semi-permanent interface states ΔN_{it} and fixed oxide charges. Interface states are fast and can quickly follow changes in the bias ($< 1\,\text{ms}$). Thus, a change of the interfacial Fermi-level would result in a rapid change of the charge stored in these interface states, $\Delta Q_{\text{it}}(E_F)$, according to their density-of-states. In particular, after a temporary bias change, the same ΔV_{th} would be expected back at the original bias. This is clearly not the case. In fact, ΔV_{th} only slowly goes back to its original value, an apparent degradation during the recovery phase [14]. We call this phenomenon *reverse recovery*, which thus indicates that a significant part of P is due to slow oxide defects, ΔN_{ot}, such as those observed previously [11, 13, 17]. The explanation of the reverse recovery effect is as follows: during stress, defects are created inside the oxide. These defects have an energy level in the silicon

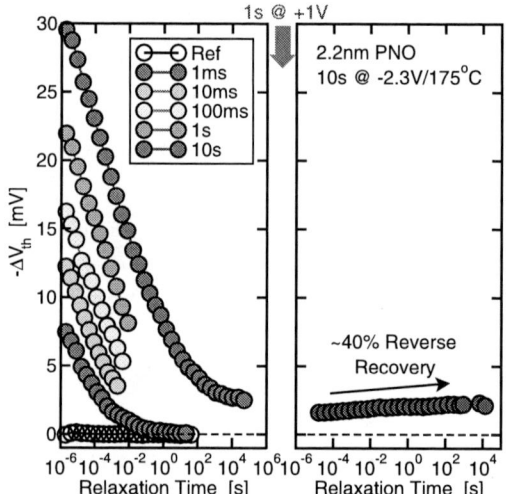

Fig. 2. Typical plateau observed under medium stress conditions. After the plateau has been reached, a positive bias was applied for a short time. For $P \sim \Delta N_{\text{it}}$, one would expect ΔV_{th} to rapidly follow bias changes (within a 1 ms). In fact, a pronounced reverse recovery is observed with time constants as large as 10 ks, indicating that ΔN_{ot} contributes to P.

bandgap and their occupancy depends on the position of the Fermi-level. During application of a positive bias, the defects are discharged. This does not mean that the defects are annealed, the discharging step just makes them electrically neutral and thus invisible in ΔV_{th}. Once in this metastable neutral state, the defects can either completely anneal or they can be charged again when the Fermi-level is moved back to the threshold voltage. However, as the time constants responsible for charging and discharging can be considerably

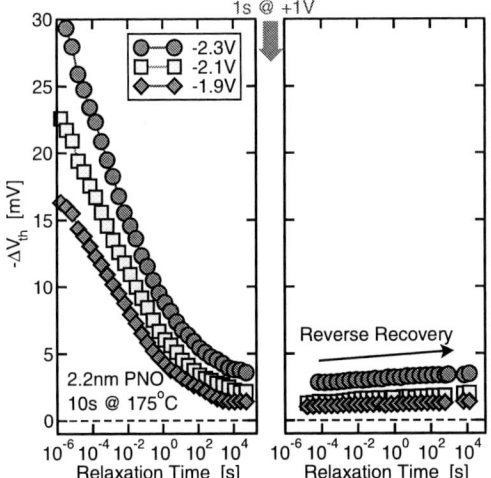

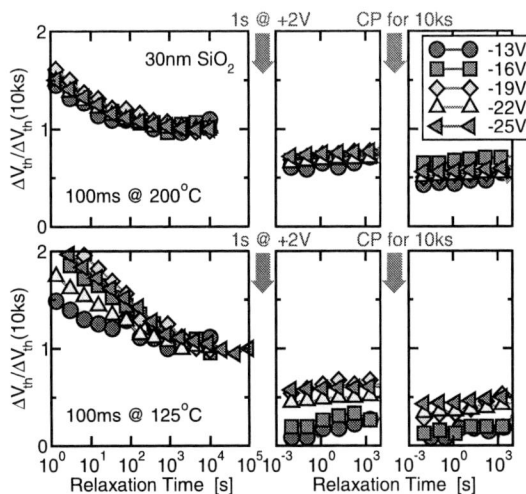

Fig. 3. **Left**: Plateaus in the recovery of ΔV_{th} for three different stress voltages. **Right**: After short application of a positive bias pulse, again considerable reverse recovery is observed, which indicates a significant contribution of slow switching oxide traps to the plateau.

Fig. 5. The same effect as in Fig. 2 is observed on thick SiO_2 devices. The reverse recovery time constants are either somewhat larger or the application of positive bias anneals a fraction of the oxide defects, this being more pronounced at lower T. Continuous CP for 10 ks removes a further fraction, cf. Fig. 9.

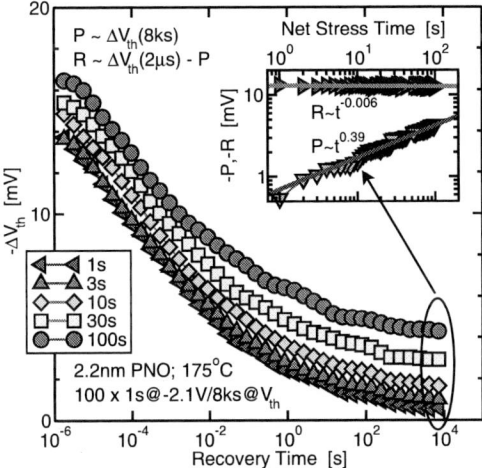

Fig. 4. Low duty factor experiment (1/8000) with 1 s stress followed by 8 ks recovery repeated 100 times. ΔV_{th}(8 ks) recorded at the end of every trace increases as shown in the inset, indicating the build-up of slowly recoverable damage. At the same time, R is slightly reduced, albeit by a much smaller amount than the creation of P (inset).

larger than those typically associated with interface states, their charging is visible as a reverse recovery transient [3]. Note that this is markedly different from the conventional picture of hole trapping, where holes are simply trapped in the oxide. Such trapped holes would not react that sensitively to changes of the Fermi-level and when discharged, they are already fully annealed [3]. Moving back to the threshold voltage would not result in a reverse recovery as these defects can only be charged again by application of a large negative stress pulse.

Fig. 3 shows the bias dependence of these plateaus, demonstrating that the recovery settles at a higher level when larger stress voltages are employed. Fig. 3 also demonstrates the

fundamental dilemma regarding the characterization of the plateaus, namely that even after a stress time of only 10 s, the plateaus may only become gradually visible after a recovery of 10^5 s (about a day), which is already close to the maximum experimentally feasible recovery time.

One possibility to stimulate the build-up of P without excessive creation of R is by repeated stress and recovery experiments with a very low duty factor. Such an experiment results in a slow additive component P to the otherwise unchanged recoverable component R. Unfortunately, due to the intricate dynamic nature of the experiment, the analysis of the data is also much more involved and not possible without assumptions taken from a sensible model. An example of such an experiment with a duty factor of $1/8000$ is shown in Fig. 4. While P increases with a relatively large power-law exponent of about 0.4, the reduction of R is only weak. Still, this reduction in R might be indicative of a coupling between R and P [4, 6].

Fig. 5 documents our search for plateaus on thick SiO_2 devices. In order to make the plateaus clearly visible so that their bias and temperature dependence can be studied, we kept the stress short (10 ms). At 200 °C, where both degradation as well as recovery are strongest, the recovery levels to a plateau after about 10^4 s. Again, the large disparity between stress and recovery times, which differ by six orders in magnitude, is hard to miss. After the plateau had been reached, the devices were driven into accumulation for 1 s. Back at the original read-out voltage ($V_G = V_{th}$), ΔV_{th} was found to be reduced by 40%, the same percentage as observed for the thin SiON devices. Again, reverse recovery was observed, resulting in ΔV_{th} to slowly *increase* following this bias switch. However, even after 10^4 s the original degradation of the plateaus level was not reached, contrary to the thin SiON devices. Also, the effect appears to be about the same for all stress voltages used. Following this reverse recovery phase, a continuous

978-1-4244-9113-1/11 $26.00 © 2011 IEEE

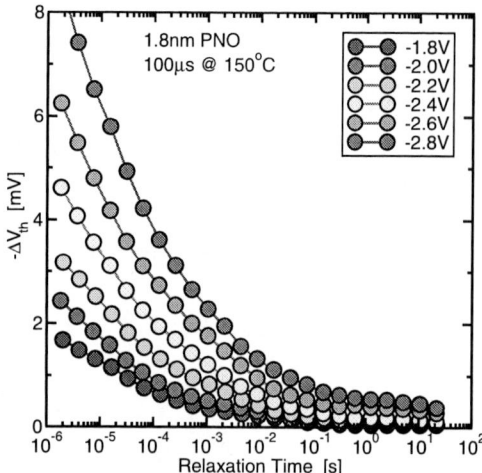

Fig. 6. Plateaus are occasionally also observed after ultra-short stress times, shown for a 1.8 nm PNO device. The plateaus are not permanent and strongly depend on the stress bias. Particularly after such weak stresses, it is important that the degradation at $V_{\mathrm{G}} = V_{\mathrm{th}}$ is negligible in order to avoid mistake M4 of Fig. 1.

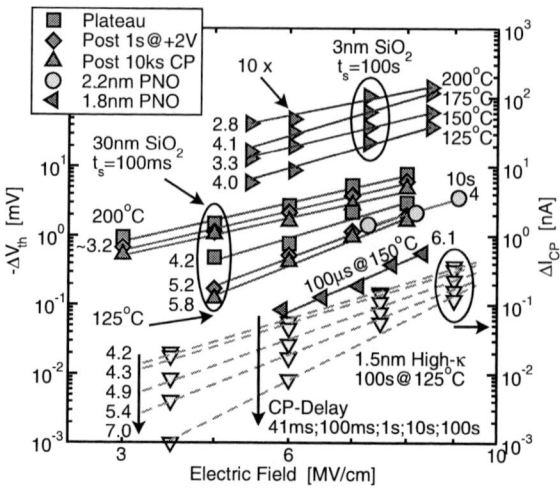

Fig. 7. The field dependence of the 'permanent' part. For the 1.8 nm PNO device from Fig. 6 we take $P \sim \Delta V_{\mathrm{th}}(t_{\mathrm{r}} = 1\,\mathrm{s})$, for the ones of Figs. 3 and 5 $P \sim \Delta V_{\mathrm{th}}(t_{\mathrm{r}} = 10\,\mathrm{ks})$ was chosen. In addition, for the thick $\mathrm{SiO_2}$ devices, the last value at the end of the second and third relaxation cycle are shown (10 ks after the bias switch and 10 ks after the 10 ks CP measurement cycles). Also shown is the field dependence of the CP current, which appears to behave in a similar manner. In any case, P can be fitted by a power-law $\sim E_{\mathrm{ox}}^{\gamma}$.

CP measurement lasting $10^4\,\mathrm{s}$ was performed. At the end of this CP measurement, the degradation level was reduced by about 60% relative to the original plateau value. Again, even after such a long CP measurement, reverse recovery was visible. The same experiment was repeated at 125 °C where the plateaus were only reached after a considerably longer time. Now, the response of the devices to the short accumulation pulse and the long charge pumping measurement depended on the stress voltage used.

IV. BIAS DEPENDENCE OF P

For lifetime back-extrapolation, the bias dependence of P is crucial. This is mostly due to the fact that at higher stress voltages and temperatures, the contribution of P relative to R apparently increases, which has to be corrected for when extrapolating back to operating conditions. Huard [6] observed $P \sim E_{\mathrm{ox}}^{\gamma}$ with a technology-independent $\gamma = 4$, without giving details of the extraction scheme for P.

Another example showing experimentally observable plateaus which appear already after a $t_{\mathrm{s}} = 100\,\mu\mathrm{s}$ stress is given in Fig. 6. The E_{ox} dependence of these plateaus is shown in Fig. 7, together with the plateaus of Figs. 3–5 and related experiments. Contrary to the universal exponent of 4 given by Huard, a wider range is observed, with values smaller and larger than 4. Also, the bias dependence of the plateaus in the thick $\mathrm{SiO_2}$ devices (Fig. 5) shows exponents around 3.2 at 200 °C and in the range 4.2–5.6 at 125 °C. The latter is insofar interesting as the initial plateau has $\gamma = 4.2$, which increases to 5.2 after application of $+2\,\mathrm{V}$ for 1 s, and even to 5.6 after continuous CP measurements for 10 ks. This again demonstrates that P, whatever it is and by whatever means it is extracted, is not really permanent.

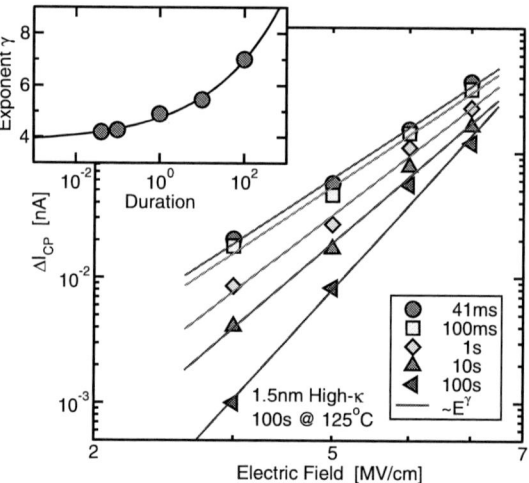

Fig. 8. The permanent component has been suggested to correlate with the change in the charge pumping current, $P \sim \Delta I_{\mathrm{CP}}$. Just like in ΔV_{th} measurements, the CP current is very sensitive to the measurement time in all our investigated technologies. This has a profound impact on the extracted bias dependence of P. Only for short CP measurements an E_{ox}^4 dependence as in [6] is obtained.

V. CORRELATION WITH CHARGE-PUMPING DATA

It has been occasionally suggested [6, 19] that P is correlated to the CP current, ΔI_{CP}. In that context, ΔI_{CP} has been interpreted as being proportional to the number of interface states. Particularly at lower frequencies it has been observed, however, that I_{CP} also contains considerable contributions from oxide traps [20]. This issue has been commonly neglected in the context of NBTI [21].

978-1-4244-9113-1/11 $26.00 © 2011 IEEE

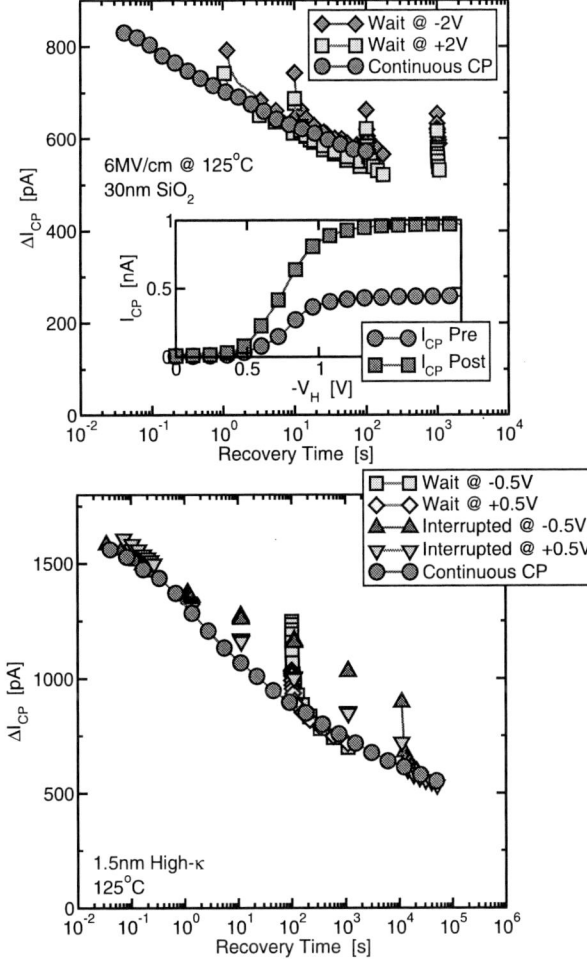

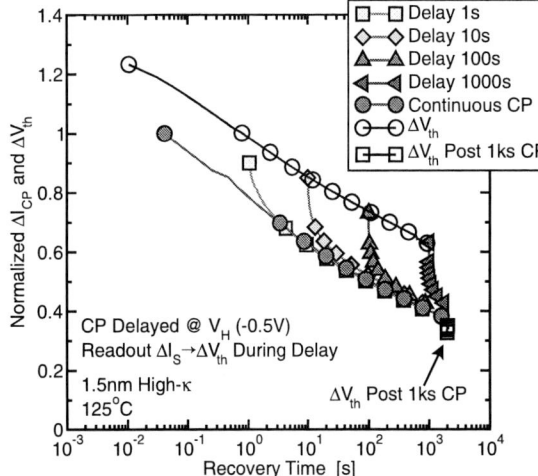

Fig. 10. Extension of the experiment in Fig. 9: the change in the linear source current ΔI_S at the CP high-level V_H is read during the wait phase and converted to ΔV_{th} following [18]. This clearly demonstrates that the onset of the accelerated CP recovery follows the recovery of ΔV_{th}. CP induced accelerated recovery in ΔI_{CP} is paralleled by a recovery in ΔV_{th} by the same fraction.

Fig. 9. Recovery of the CP current at a frequency of 1 MHz. Contrary to other observations [6], the CP current is never constant in any of our technologies, consistent with [16]. Most importantly, continuous CP accelerates recovery. When the beginning of the CP measurement is delayed for variable amounts of time, with V_G being either the CP high- or low-level, rapid recovery is observed at the beginning of the delayed continuous CP measurement. Application of the CP high- and low-levels during the wait phase proves that it is not the voltage but rather the charge pumping itself that causes the rapid recovery. **Top**: For a thick 30 nm SiO_2 device. The inset shows a constant-base-level CP sweep before and after stress. **Bottom**: For a 1.5 nm high-κ gate stack device. The same effect is observed in both dramatically different technologies.

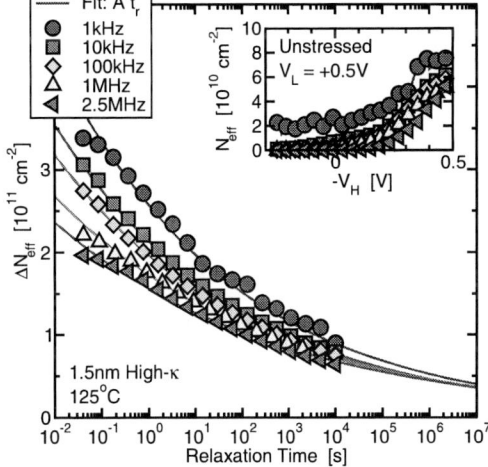

Fig. 11. Frequency dependence of the CP recovery. With lower frequency, a larger amount of oxide defects contributes to ΔN_{eff}. With increasing CP time, however, the oxide defects are 'pumped-away', giving a frequency independent ΔN_{eff} as expected from a pure ΔN_{it} contribution.

The bias dependence of ΔI_{CP} is compared in Fig. 7 to the bias dependence of the ΔV_{th} plateaus, which particularly for fast CP measurements seems to agree well, indicating a correlation between the defects visible in these two experiments. Such a correlation has led previous studies to conclude that ΔV_{th} as well as ΔI_{CP} are dominated by interface states [22].

Interestingly, the bias dependence of ΔI_{CP} is very sensitive to the measurement duration as shown in Fig. 8. While for fast CP experiments (41 ms) we obtain $\gamma = 4$, we observe a strong dependence of the extracted exponent on the duration of the CP measurement. This is reminiscent to ΔV_{th} measurements [23], where the exponent also increases with the measurement duration. The reason for this behavior is that ΔI_{CP} shows

similar recovery rates as ΔV_{th} which is contrary to Huard's work, but consistent with the observation of Rangan *et al.* [16].

Remarkably, recovery is accelerated by the CP measurement, see Fig. 9. The effect is quite similar for thick SiO_2 and thin high-κ devices. At first glance one might relate this to the bias-dependence of the defect annealing rates, since the transistor is continuously pulsed between accumulation and inversion. As it has already been shown for ΔV_{th} recovery, recovery can be accelerated when the transistor is driven towards accumulation [3, 23]. This effect, so undoubtedly present, does not provide the full answer, though. This can be seen in Fig. 9 which also shows reference CP measurements which were interrupted by constant bias phases at the low-

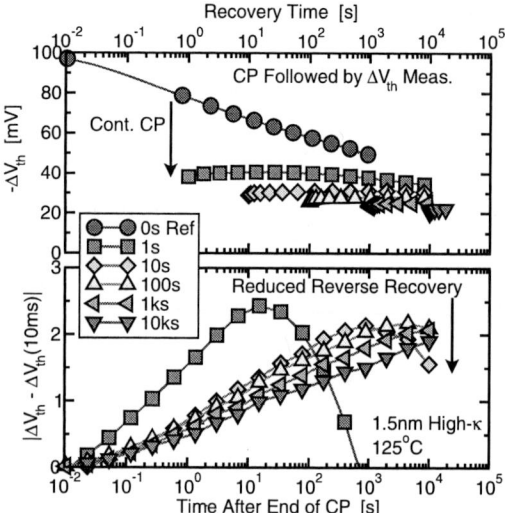

Fig. 12. Impact of continuous CP measurements with variable duration on ΔV_{th}. A considerable recovery in ΔV_{th} is observed. Also, with increasing CP duration, the reverse recovery is reduced, indicating that CP anneals slow oxide defects, ΔN_{ot}.

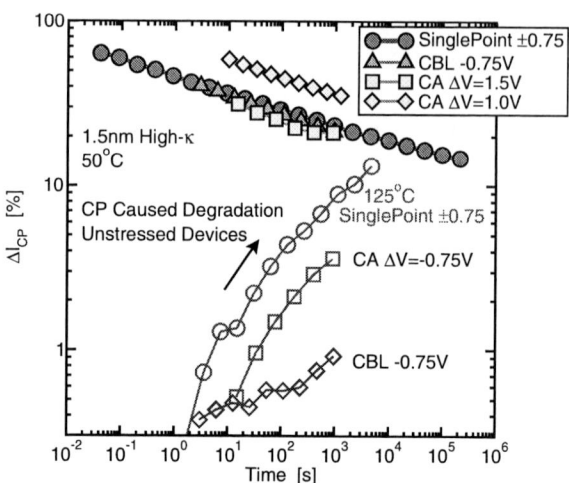

Fig. 13. All CP methods see the same recovery rate (ΔI_{CP}/decade), plotted relative to $I_{\mathrm{CP,0}}$. The single-point CP method is the fastest and gives the same ΔI_{CP} as the full constant-base-level sweep (CBL). Comparable full constant-amplitude (CA) sweeps with proper ΔV also see the same ΔI_{CP}. Particularly at higher temperatures, CP measurements can lead to degradation.

and high-level of the charge pumping pulse, which according to the previous argument should contain the most effective recovery case. However, the ΔI_{CP} recovery in the continuous CP measurement is even larger, implying that it is *accelerated by the pulsing event itself.*

Fig. 10 shows that ΔI_{CP} at the beginning of delayed CP measurements follows the recovery of ΔV_{th}. Also, the amount of recovery induced by the CP measurement is mirrored in the recovery of ΔV_{th}. This data strongly suggests that both ΔV_{th} and ΔI_{CP} are at least partially related to the same microscopic defect, namely switching oxide traps [13, 24–26]. This conclusion is also confirmed by the frequency dependence of the ΔI_{CP} recovery shown in Fig. 11 which gradually becomes smaller after long CP times, indicating that it is oxide defects which can be 'pumped-away'.

Further confirmation that defects visible in ΔV_{th} react to CP measurements is given in Fig. 12: following a CP cycle, ΔV_{th} shows reverse recovery due to slow oxide defects reaching their equilibrium occupancy after long times. Since this reverse recovery becomes smaller and smaller with increasing recovery time, it must be concluded that the defects responsible for reverse recovery can recover as well.

In [27] it was argued that the recovery of I_{CP} was not due to the actual recovery of interface states, but rather due to a reduction of the swing of the surface potential. Although we fail to see why the 'width of the CP hat' should be related to the density of interface states, we performed full constant amplitude and constant-base-level CP measurements in addition to the single-point CP experiments shown in the previous figures. The result is shown in Fig. 13 and quite reassuringly, the same level of degradation and recovery is observed in all measurements, except for the 1 V constant amplitude CP measurements, which use an amplitude too small to cover the whole bandgap. Fig. 13 also shows the detrimental impact of CP measurements which can lead to degradation and thus to artificial plateaus, cf. (M5), particularly at higher temperatures.

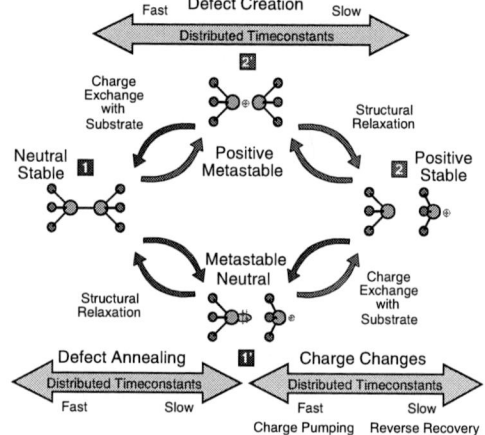

Fig. 14. The switching oxide trap model developed recently [13] is consistent with the experimentally observed behavior. Created defects can be switching traps and exchange charge with the substrate ($1' \leftrightarrow 2$), some of them faster, contributing to ΔI_{CP} in a frequency-dependent manner, some of them very slow, thus causing reverse recovery or transient RTN [13], thereby contributing to ΔV_{th} only (if in state 2).

VI. THE DEFECT MODEL

Except for the newly discovered effect of CP-induced recovery, which will be discussed separately below, all features observed so far are consistent with the detailed defect properties identified using our recent time-dependent defect spectroscopy (TDDS) measurements [13, 15]. The microscopic model we use for the description of the defects is an extension of the switching trap model proposed by Lelis *et al.* [24] and shown in Fig. 14:

- The defects are switching traps, that is, have an energy-level in the Si bandgap. Prior to stress, the defect is in the neutral state 1, while stress transfers it into the positive state 2. Depending on the defect properties, the defect

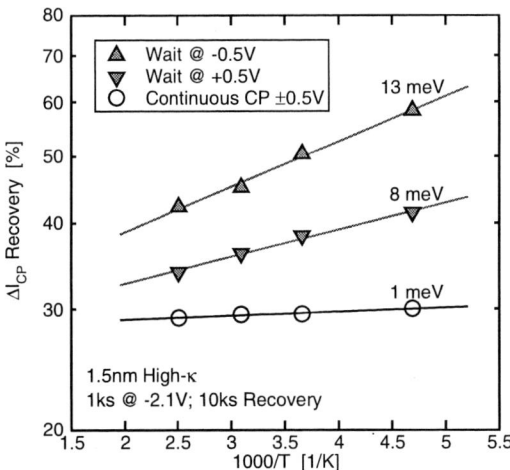

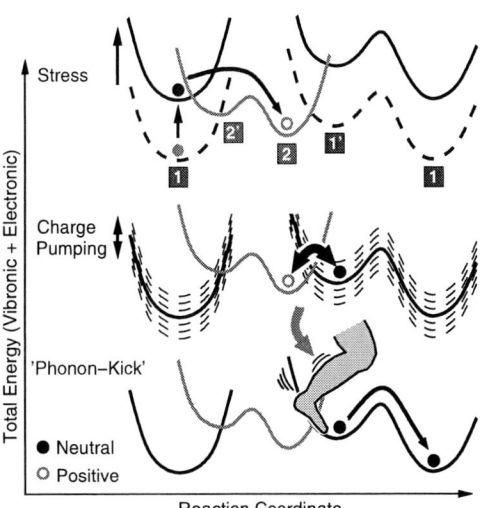

Fig. 15. Temperature dependence of the CP recovery: Continuous CP recovers ΔI_{CP} down to 30% of the post-stress ΔI_{CP}, nearly independently of temperature. Remaining at $-0.5\,\mathrm{V}$ maintains 60% at $-60\,^\circ\mathrm{C}$. Mere application of $+0.5\,\mathrm{V}$, the worst-case in a bias-driven interpretation, reduces ΔI_{CP} only down to 40% at $-60\,^\circ\mathrm{C}$.

Fig. 16. Schematic illustration of the REDR effect: **Top**: During stress, defects become positively charged. **Middle**: During CP, the neutral level is rapidly moved up and down, causing frequent transitions. **Bottom**: The excess energy of the recombination events is deposited into the accepting mode, which leads to a reduction of the activation energy, known as the 'phonon-kick' or REDR effect [30].

may have a neutral metastable state $1'$, which provides the aforementioned energy-level in the bandgap. Transitions from 2 to $1'$ are particularly likely during switches toward accumulation or during CP measurements. While in $1'$, the defect is uncharged and thus not visible in ΔV_{th}.

- The defects may contribute to the CP signal in two ways. First, on an unstressed device, transitions between 1 and $2'$, provided they are sufficiently fast, can create recombination events. Contrary to interface states, whose contribution is temperature-independent, these switching traps will provide a larger contribution at higher temperatures. In particular, they will form the temperature-dependent tail of constant-base-level CP measurements. In addition, as shown in Fig. 13, CP can lead to degradation, which corresponds to a transition to state 2. Once in state 2, transitions between 2 and $1'$ can also contribute to I_{CP}, resulting in a temperature- and bias-dependent hysteresis of the CP curve [21].

- During NBTI stress, the defects move from the neutral state 1 to stable state 2. There, again, transitions between 2 and $1'$ can also contribute to I_{CP}, resulting in a temperature- and frequency-dependent contribution. This appears to be a significant contribution to ΔI_{CP} following NBTI stress.

- Since the time-constants are widely distributed, only the faster transitions between 2 and $1'$ can contributed to I_{CP}. As seen in Figs. 2, 3, 5, and 12, the slower states constitute the reverse recovery effect.

- The reason why these switching traps can contribute to both ΔV_{th} and ΔI_{CP} is simply because once created, these defects can be either positive (state 2) or neutral (state $1'$), depending on the Fermi-level, the former contributing to ΔV_{th}. This Fermi-level dependent defect occupancy also causes the change in the sub-threshold slope reported after NBTI stress [28, 29].

VII. UNDERSTANDING CP-INDUCED RECOVERY

In order to understand the CP-induced recovery, we first studied its temperature dependence which is shown in Fig. 15. At all temperatures, wait phases at $-0.5\,\mathrm{V}$ show relatively weak recovery but a relatively strong temperature dependence. Compared to wait phases at $-0.5\,\mathrm{V}$, wait phases at $+0.5\,\mathrm{V}$ result in a stronger recovery but have a weaker temperature dependence. Finally, continuous CP measurements without an intermediate wait phases accelerate recovery down to 30% of the stress level after 10 ks, nearly independent of temperature.

A possible explanation for this behavior is as follows: the CP measurement at $I_{CP}^{\max}$ is designed to maximize the number of recombination events. Each event releases an energy of the order of the silicon bandgap. With 10^6 cycles per second, this accumulates to an *enormous amount of energy* which has to be *dissipated via phonons*. In due course, reactions near the defect site can be dramatically enhanced, a phenomenon known as the 'phonon-kick', or more recently as *recombination enhanced defect reaction (REDR)* [30, 31]. This is schematically shown in Fig. 16, where the REDR accelerates the transition from the neutral metastable state to the neutral equilibrium state. Following the arguments of Weeks *et al.* [31], the thermal transition rate from state 1 to state $1'$ of our switching trap model [13],

$$k_{1'1} = \nu\, \mathrm{e}^{-\beta \varepsilon_{1'1}} \qquad (1)$$

with $\varepsilon_{1'1}$ as the thermal barrier separating the states $1'$ and $1'$, is replaced by $k_{1'1} + k_{1'1}^*$ with the enhanced rate

$$k_{1'1}^* = \nu^*\, \mathrm{e}^{-\beta(\varepsilon_{1'1} - \varepsilon^*)}, \qquad (2)$$

with $\beta^{-1} = k_{\mathrm{B}}T$. From our experimental data we extract $\varepsilon^* \approx 60\,\mathrm{meV}$ and $\nu^* = 2.5 \times 10^{14}\,\mathrm{s}^{-1}$, with the convincing calibration result shown in Fig. 17. We remark that, as noted

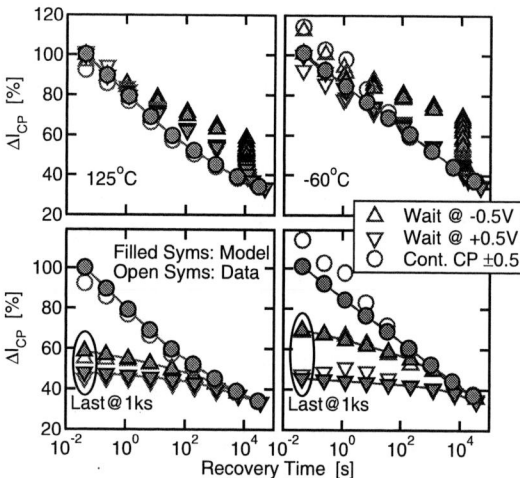

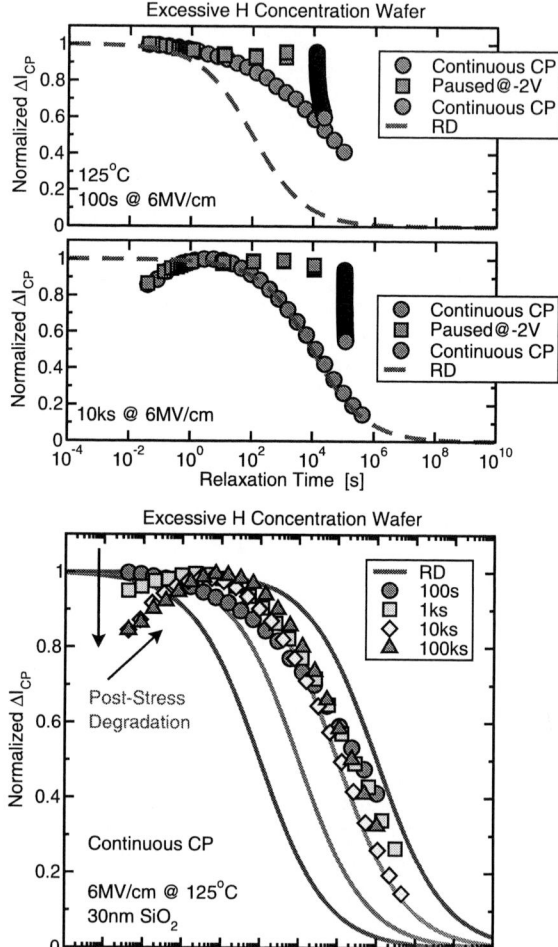

Fig. 17. Consideration of REDR in our switching trap model [3, 13] allows to reproduce the experimental data very well. **Top**: Data from Fig. 15 at 125 °C and −60 °C. **Bottom**: The last long recovery trace during continuous CP at 125 °C and −60 °C. The scatter particularly in the initial data (up to 20%) makes the model calibration challenging.

Fig. 18. **Top**: In an extremely H-rich wafer, recovery in ΔI_{CP} is basically absent provided only occasional CP measurements are made. A diffusion-limited recovery behavior seems to dominate for continuous CP measurements. **Bottom**: The recovery rate appears to be roughly independent of the stress time, meaning that the hydrogen profile is only weakly disturbed during stress. With increasing stress, though, the initial degradation during the recovery phase can last for up to 10 s and amount to 20%. This is a degradation of ΔI_{CP} which was found in this hydrogen-rich wafer only and must not be confused with the reverse recovery visible in ΔV_{th} only, cf. Fig. 12.

previously for ΔV_{th} recovery [11], temperature-activated microscopic defect time constants again result in an apparent temperature-independent macroscopic behavior.

VIII. SPECIAL CASE: HYDROGEN-RICH WAFER

The log-like recovery of ΔI_{CP} as for instance shown in Figs. 9 and 13 is clearly incompatible with the recovery predicted by the reaction-diffusion (RD) model [32], $(1 + \sqrt{t_s/t_r})^{-1}$, which does not depend on bias or temperature [33]. A peculiar exception has been observed on a hydrogen-rich 30 nm SiO_2 split-wafer. Measuring ΔI_{CP} only once per decade results in $\Delta I_{CP} \sim$ const. By contrast, a continuous CP measurement produces recovery traces which bear a striking resemblance to the RD prediction, particularly for $t_s = 10$ ks, see Fig. 18. After studying different stress times, however, we found the measured recovery to be practically independent of the stress-time, not scaling universally over t_s/t_r as expected from RD theory [33]. Still, under continuous CP conditions, recovery could be a diffusion-limited process in this particular wafer. An intriguing feature is that after longer stress times the devices continue to degrade after the end of stress. This is consistent with the idea that hydrogen is released during stress which then depassivates interface states and creates oxide defects [34–37]. Otherwise, degradation after termination of the stress would not be possible. We remark that this is the standard model of irradiation damage [38, 39].

IX. CONCLUSIONS

We have demonstrated that the plateaus occasionally observed in carefully tuned stress/recovery experiments consist of contributions from interface states as well as slower donor-like switching oxide traps. These plateaus are not permanent and normally not too well developed, making a precise definition and extraction difficult. In particular, the plateaus can

be annealed by applying short positive bias pulses or, more effectively, by continuous CP measurements. Particularly the latter provides an efficient means for annealing NBTI degradation, likely due to a recombination enhanced defect reaction mechanism. Under normal recovery conditions, the recovery of ΔV_{th} determines the starting level of ΔI_{CP}, which starts recovering quickly once CP measurements are performed. The latter demonstrates that oxide defects contribute to both ΔV_{th} and ΔI_{CP}. Overall, considering P as permanent will lead to serious errors, even within conventional measurement windows. Finally, we have suggested a simple extension of our switching trap model to also account for recombination enhanced defect reaction (REDR) effects.

ACKNOWLEDGMENT

This work has received funding from the EC's FP7 grant agreement n°216436 (ATHENIS) and from the ENIAC MODERN project n°820379.

REFERENCES

[1] V. Huard, M. Denais, and C. Parthasarathy, "NBTI Degradation: From Physical Mechanisms to Modelling," *Microelectronics Reliability*, vol. 46, no. 1, pp. 1–23, 2006.

[2] T. Grasser, B. Kaczer, P. Hehenberger, W. Goes, R. O'Connor, H. Reisinger, W. Gustin, and C. Schlünder, "Simultaneous Extraction of Recoverable and Permanent Components Contributing to Bias-Temperature Instability," in *Proc. Intl.Electron Devices Meeting (IEDM)*, 2007, pp. 801–804.

[3] T. Grasser, B. Kaczer, W. Goes, T. Aichinger, P. Hehenberger, and M. Nelhiebel, "A Two-Stage Model for Negative Bias Temperature Instability," in *Proc. Intl.Rel.Phys.Symp. (IRPS)*, 2009, pp. 33–44.

[4] T. Grasser and B. Kaczer, "Evidence that Two Tightly Coupled Mechanism are Responsible for Negative Bias Temperature Instability in Oxynitride MOSFETs," *IEEE Trans.Electron Devices*, vol. 56, no. 5, pp. 1056–1062, 2009.

[5] T. Aichinger, M. Nelhiebel, and T. Grasser, "A Combined Study of p- and n-Channel MOS Devices to Investigate the Energetic Distribution of Oxide Traps after NBTI," *IEEE Trans.Electron Devices*, vol. 56, no. 12, pp. 3018–3026, 2009.

[6] V. Huard, "Two Independent Components Modeling for Negative Bias Temperature Instability," in *Proc. Intl.Rel.Phys.Symp. (IRPS)*, 2010, pp. 33–42.

[7] A. Katsetos, "Negative Bias Temperature Instability (NBTI) Recovery with Bake," *Microelectronics Reliability*, vol. 48, no. 10, pp. 1655–1659, 2008.

[8] C. Benard, G. Math, P. Fornara, J. Ogier, and D. Goguenheim, "Influence of Various Process Steps on the Reliability of PMOSFETs Submitted to Negative Bias Temperature Instabilities," *Microelectronics Reliability*, vol. 49, pp. 1008–1012, 2009.

[9] H. Reisinger, O. Blank, W. Heinrigs, W. Gustin, and C. Schlünder, "A Comparison of Very Fast to Very Slow Components in Degradation and Recovery Due to NBTI and Bulk Hole Trapping to Existing Physical Models," *IEEE Trans.Dev.Mat.Rel.*, vol. 7, no. 1, pp. 119–129, 2007.

[10] T. Aichinger, M. Nelhiebel, and T. Grasser, "Unambiguous Identification of the NBTI Recovery Mechanism using Ultra-Fast Temperature Changes," in *Proc. Intl.Rel.Phys.Symp. (IRPS)*, 2009, pp. 2–7.

[11] H. Reisinger, T. Grasser, W. Gustin, and C. Schlünder, "The Statistical Analysis of Individual Defects Constituting NBTI and its Implications for Modeling DC- and AC-Stress," in *Proc. Intl.Rel.Phys.Symp. (IRPS)*, 2010, pp. 7–15.

[12] B. Kaczer, T. Grasser, P. Roussel, J. Franco, R. Degraeve, L. Ragnarsson, E. Simoen, G. Groeseneken, and H. Reisinger, "Origin of NBTI Variability in Deeply Scaled PFETs," in *Proc. Intl.Rel.Phys.Symp. (IRPS)*, 2010, pp. 26–32.

[13] T. Grasser, H. Reisinger, P.-J. Wagner, W. Goes, F. Schanovsky, and B. Kaczer, "The Time Dependent Defect Spectroscopy (TDDS) Technique for the Bias Temperature Instability," in *Proc. Intl.Rel.Phys.Symp. (IRPS)*, May 2010, pp. 16–25.

[14] T. Grasser, B. Kaczer, and W. Goes, "An Energy-Level Perspective of Bias Temperature Instability," in *Proc. Intl.Rel.Phys.Symp. (IRPS)*, 2008, pp. 28–38.

[15] T. Grasser, H. Reisinger, P.-J. Wagner, and B. Kaczer, "The Time Dependent Defect Spectroscopy for the Characterization of Border Traps in Metal-Oxide-Semiconductor Transistors," *Physical Review B*, vol. 82, no. 24, p. 245318, 2010.

[16] S. Rangan, N. Mielke, and E. Yeh, "Universal Recovery Behavior of Negative Bias Temperature Instability," in *Proc. Intl.Electron Devices Meeting (IEDM)*, 2003, pp. 341–344.

[17] T. Grasser, H. Reisinger, W. Goes, T. Aichinger, P. Hehenberger, P. Wagner, M. Nelhiebel, J. Franco, and B. Kaczer, "Switching Oxide Traps as the Missing Link between Negative Bias Temperature Instability and Random Telegraph Noise," in *Proc. Intl.Electron Devices Meeting (IEDM)*, 2009, pp. 729–732.

[18] B. Kaczer, T. Grasser, P. Roussel, J. Martin-Martinez, R. O'Connor, B. O'Sullivan, and G. Groeseneken, "Ubiquitous Relaxation in BTI Stressing-New Evaluation and Insights," in *Proc. Intl.Rel.Phys.Symp. (IRPS)*, 2008, pp. 20–27.

[19] T. Aichinger, S. Puchner, M. Nelhiebel, T. Grasser, and H. Hutter, "Impact of Hydrogen on Recoverable and Permanent Damage following Negative Bias Temperature Stress," in *Proc. Intl.Rel.Phys.Symp. (IRPS)*, 2010, pp. 1063–1068.

[20] R. Paulsen and M. White, "Theory and Application of Charge-Pumping for the Characterization of Si-SiO$_2$ Interface and Near-Interface Oxide Traps," *IEEE Trans.Electron Devices*, vol. 41, no. 7, pp. 1213–1216, 1994.

[21] P. Hehenberger, T. Aichinger, T. Grasser, W. Goes, O. Triebl, B. Kaczer, and M. Nelhiebel, "Do NBTI-Induced Interface States Show Fast Recovery? A Study Using a Corrected On-The-Fly Charge-Pumping Measurement Technique," in *Proc. Intl.Rel.Phys.Symp. (IRPS)*, 2009.

[22] S. Mahapatra, K. Ahmed, D. Varghese, A. E. Islam, G. Gupta, L. Madhav, D. Saha, and M. A. Alam, "On the Physical Mechanism of NBTI in Silicon Oxynitride p-MOSFETs: Can Differences in Insulator Processing Conditions Resolve the Interface Trap Generation versus Hole Trapping Controversy?" in *Proc. Intl.Rel.Phys.Symp. (IRPS)*, 2007, pp. 1–9.

[23] B. Kaczer, V. Arkhipov, R. Degraeve, N. Collaert, G. Groeseneken, and M. Goodwin, "Disorder-Controlled-Kinetics Model for Negative Bias Temperature Instability and its Experimental Verification," in *Proc. Intl.Rel.Phys.Symp. (IRPS)*, 2005, pp. 381–387.

[24] A. Lelis and T. Oldham, "Time Dependence of Switching Oxide Traps," *IEEE Trans.Nucl.Sci.*, vol. 41, no. 6, pp. 1835–1843, Dec 1994.

[25] J. Conley Jr., P. Lenahan, A. Lelis, and T. Oldham, "Electron Spin Resonance Evidence for the Structure of a Switching Oxide Trap: Long Term Structural Change at Silicon Dangling Bond Sites in SiO$_2$," *Appl.Phys.Lett.*, vol. 67, no. 15, pp. 2179–2181, 1995.

[26] J. Ryan, P. Lenahan, T. Grasser, and H. Enichlmair, "Recovery-Free Electron Spin Resonance Observations of NBTI Degradation," in *Proc. Intl.Rel.Phys.Symp. (IRPS)*, 2010, pp. 43–49.

[27] M. Denais, V. Huard, C. Parthasarathy, G. Ribes, F. Perrier, D. Roy, and A. Bravaix, "Perspectives on NBTI in Advanced Technologies: Modelling & Characterization," in *Proc. ESSDERC*, 2005, pp. 399–402.

[28] C. Schlunder, M. Hoffmann, R.-P. Vollertsen, G. Schindler, W. Heinrigs, W. Gustin, and H. Reisinger, "A Novel Multi-Point NBTI Characterization Methodology Using Smart Intermediate Stress (SIS)," in *Proc. Intl.Rel.Phys.Symp. (IRPS)*, May 2008, pp. 79–86.

[29] D. Brisbin and P. Chaparala, "The Effect of the Subthreshold Slope Degradation on NBTI Device Characterization," in *Proc. Intl.Integrated Reliability Workshop*, 2008, pp. 96–99.

[30] H. Sumi, "Dynamic Defect Reactions Induced by Multiphonon Nonradiative Recombination of Injected Carriers at Deep Levels in Semiconductors," *Physical Review B*, vol. 29, no. 8, pp. 4616–4630, 1984.

[31] J. Weeks, J. Tully, and L. Kimerling, "Theory of Recombination-Enhanced Defect Reactions in Semiconductors," *Physical Review B*, vol. 12, no. 8, pp. 3286–3292, 1975.

[32] M. Alam, "A Critical Examination of the Mechanics of Dynamic NBTI for pMOSFETs," in *Proc. Intl.Electron Devices Meeting (IEDM)*, 2003, pp. 345–348.

[33] T. Grasser, W. Goes, V. Sverdlov, and B. Kaczer, "The Universality of NBTI Relaxation and its Implications for Modeling and Characterization," in *Proc. Intl.Rel.Phys.Symp. (IRPS)*, 2007, pp. 268–280.

[34] L. Tsetseris, X. Zhou, D. Fleetwood, R. Schrimpf, and S. Pantelides, "Physical Mechanisms of Negative-Bias Temperature Instability," *Appl.Phys.Lett.*, vol. 86, no. 14, pp. 1–3, 2005.

[35] S. Volkos, E. Efthymiou, S. Bernardini, I. Hawkins, A. Peaker, and G. Petkos, "The Impact of Negative-Bias-Temperature-Instability on the Carrier Generation Lifetime of Metal-Oxynitride-Silicon Capacitors," *J.Appl.Phys.*, vol. 100, no. 12, pp. 124 103–1–124 103–9, 2006.

[36] M. Houssa, V. Afanas'ev, A. Stesmans, M. Aoulaiche, G. Groeseneken, and M. Heyns, "Insights on the Physical Mechanism behind Negative Bias Temperature Instabilities," *Appl.Phys.Lett.*, vol. 90, no. 4, p. 043505, 2007.

[37] D. Dankovic, I. Manic, V. Davidovic, S. Djoric-Veljkovic, S. Golubovic, and N. Stojadinovic, "Negative Bias Temperature Instability in n-Channel Power VDMOSFETs," *Microelectronics Reliability*, vol. 48, pp. 1313–1317, 2008.

[38] F. McLean, "A Framework for Understanding Radiation-Induced Interface States in SiO$_2$ Structures," *IEEE Trans.Nucl.Sci.*, vol. 27, no. 6, pp. 1651–1657, Dec 1980.

[39] D. Brown and N. Saks, "Time Dependence of Radiation-Induced Trap Formation in Metal-Oxide-Semiconductor Devices as a Function of Oxide Thickness and Applied Field," *J.Appl.Phys.*, vol. 70, no. 7, pp. 3734–3747, 1991.

978-1-4244-9113-1/11 $26.00 © 2011 IEEE

A Critical Re-evaluation of the Usefulness of R-D Framework in Predicting NBTI Stress and Recovery

S. Mahapatra[1*], A. E. Islam[2,3], S. Deora[1], V. D. Maheta[1], K. Joshi[1], A. Jain[2] and M. A. Alam[2]

[1]Department of Electrical Engineering, Indian Institute of Technology Bombay, Mumbai 400076, India
[2]School of Electrical Engineering and Computer Science, Purdue University, W. Lafayette, IN 47906, USA
[3]Department of Materials Science and Engineering, University of Illinois, Urbana, IL 61801, USA
*Phone: +91-222-572-0408, Fax: +91-222-572-3707, Email: souvik@ee.iitb.ac.in

Abstract – **Reaction-Diffusion (R-D) framework for interface trap generation along with hole trapping in pre-existing and generated bulk oxide traps are used to model Negative Bias Temperature Instability (NBTI) in differently processed SiON p-MOSFETs. Time, temperature and bias dependent degradation and recovery transients are predicted. Long-time power law exponent of DC degradation and uniquely renormalized duty cycle and frequency dependent AC degradation data from a wide range of sources are shown to have universal features and a broad consensus across industry/academia. These universal features can also be predicted using the classical R-D framework.**

Keywords: NBTI, interface traps, hole trapping, bulk traps, R-D model, SiON p-MOSFETs

I. INTRODUCTION

Negative Bias Temperature Instability (NBTI) is a well-known p-MOSFET reliability issue [1] with broad implications for digital and analog CMOS integrated circuits. Like all other reliability effects, NBTI is measured under accelerated stress conditions for short time and then projected to end-of-life under operating condition. Unlike most reliability effects, however, NBTI degradation recovers after the removal of stress [2], and hence requires fast [2] or ultra-fast [3] measurement methods to avoid experimental artifacts associated with common measure-stress-measure approach [4], as shown in [5]. NBTI recovery also implies reduced degradation under AC stress compared to DC [6],[7]. Estimation of NBTI lifetime requires extrapolations in bias and time for DC qualification, and additional estimation of AC pulse frequency and duty cycle dependence for AC qualification. Hence, the underlying NBTI physical mechanism must be understood in order to make meaningful projection of device/circuit lifetime under DC or AC operating conditions.

In spite of numerous experimental and modeling attempts, a consensus regarding the correct physical interpretation of NBTI is still elusive. Contrary to popular belief (as shown later in this paper), the empirical signatures of NBTI degradation are remarkably consistent across different industry and academic sources. It is the theoretical understanding of the degradation phenomena necessary for lifetime projection that remains a topic of active debate. The Reaction-Diffusion (R-D) model of interface trap generation (ΔN_{IT}), originally proposed more than 30 years ago [8] and later interpreted for modern contexts [9]-[12] has been invoked to explain DC stress [13] and AC frequency independence [11]. However, it was suggested that R-D cannot explain NBTI recovery transient and the shape of AC duty cycle dependence [14]. This apparent *failure* of R-D

framework has inspired a plethora of new "well-based" (three well [14], four well/2-stage [15],[16] and two well [17]) models. However, none of the existing models have been used so far to consistently explain 3 practical features useful for qualification of NBTI: (a) long time degradation under DC stress, (b) AC degradation versus pulse frequency, and (c) AC degradation versus pulse duty cycle. In addition, any model should also be able to successfully explain (d) strong gate insulator process dependence of NBTI [1],[18] and (e) NBTI recovery transients. This work re-evaluates the R-D framework in capturing (a)-(e), and compares key results from other models.

II. PREDICTABILITY OF KEY NBTI FEATURES

A. Long Time DC NBTI Degradation:

Figure 1 shows measured degradation in *industrial grade* devices over very long stress time from different reports [19]-[21]. Note, these data are usually obtained with typically ~ms measurement delay, which however has insignificant effect as the stress time is very long [12]. In spite of disparate sources, all devices unequivocally show power law time dependence with exponent $n \sim 1/6$, a feature of DC NBTI that is now widely accepted in the industry, and is readily predicted by H_2 based R-D framework [10]-[12]. It is important to realize that *the prediction of such power law time exponent is inherent to H_2 based R-D framework and is obtained with zero adjustable parameters.* At present, there is no other theory of NBTI that can replicate this claim.

It is interesting to note that in spite of their *perceived* success in explaining NBTI [14]-[16], the predictability of long time degradation by three and four well models have never been tested or reported. Although the detailed comparison of R-D and well based models is beyond the scope of this paper (see [22] for details), Figure 2 shows the time evolution of NBTI and extracted time exponent as obtained by using three and four well models for various model parameters (including those reported in [14],[15]). It is interesting to note that three-well model cannot predict power law dependence and long time $n \sim 1/6$ unless completely unphysical and unrealistically large parameter (σ) governing dispersion in well energy barrier is used [15]. Though four well model was proposed to circumvent this issue [15], it is worthwhile to note that both models predict saturation in NBTI degradation and therefore anticipate $n \sim 0$ at short to moderately long stress time for physically realistic (for three well) and consistent (for four well, see prediction of AC degradation later in Figure 4) choice of model parameters.

978-1-4244-9113-1/11 $26.00 © 2011 IEEE

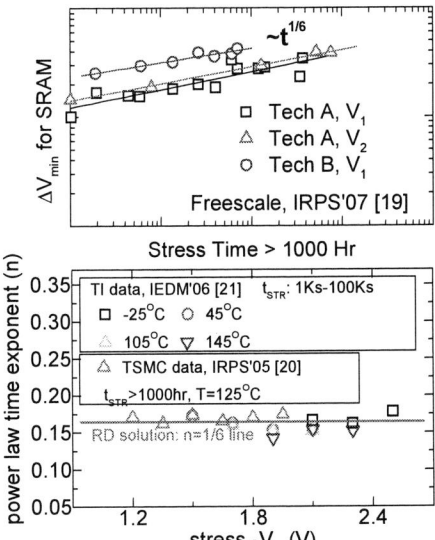

Figure 1. Time evolution of degradation at very long stress time from various published results showing power law dependence with universally observed exponent of n=1/6, as predicted by R-D solution [10]-[12].

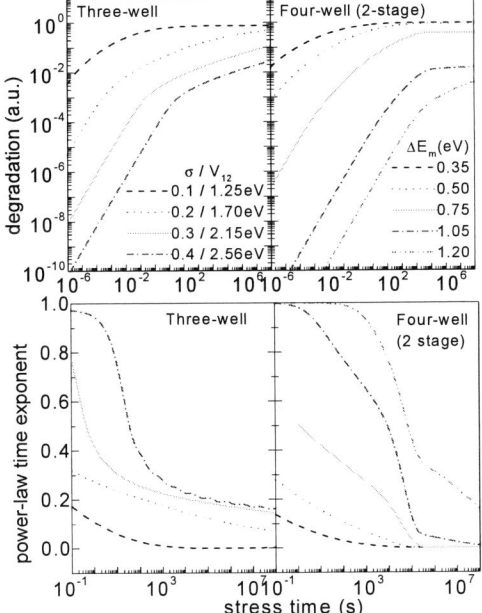

Figure 2. Time evolution of degradation (top panel) and extracted power-law time exponent (bottom panel) obtained using three and four well models [14]-[16], simulated for different parameters governing the barrier between wells 1 and 2 (see [22] for details).

B. Duty Cycle/Frequency Dependence of AC Degradation:

Figure 3 shows comprehensive summary of duty cycle and frequency dependencies from different published reports [6],[7], [14],[17],[23]-[25]. The "shape" of duty cycle dependent data shows very large spread when AC data are normalized to DC. Moreover, although the frequency independence is generally observed (some data show slight drop at higher frequencies and has been demonstrated to be a measurement artifact [6]), the AC/DC ratio as reported by various groups shows huge spread in the range of 0.8-0.3 at 50% duty. Indeed, drastically different shapes and magnitudes of duty cycle dependent data makes modeling a challenge. While both R-D [11] and two well [17] models predict frequency independence, they predict very different AC/DC ratio (~0.8 and ~0.3 respectively), as shown.

The simulated AC/DC ratio values at 50% duty cycle using three [14] and four [15] well models for different parameters (as Figure 2) are shown in Figure 4. Three well model predicts a ratio of ~0.35 for unphysical and large σ, and the ratio is reduced to ~0 (as a consequence of n~0 as shown in Figure 2) for realistic value of σ. Four well model shows very strong parametric impact of AC/DC ratio, which varies from 1 to 0 for small changes in model parameters. Note that AC/DC ratio of ~0.35 is observed for a particular parameter value, which however predicts n~0 for DC stress (see Figure 2). Therefore, it is obvious that both the three and four well models fail to predict key experimental signatures required for NBTI lifetime prediction, and is inconsistent with their *perceived* success in explaining the "physics" of NBTI degradation. Such limitations of the alternate models have not been appreciated thus far.

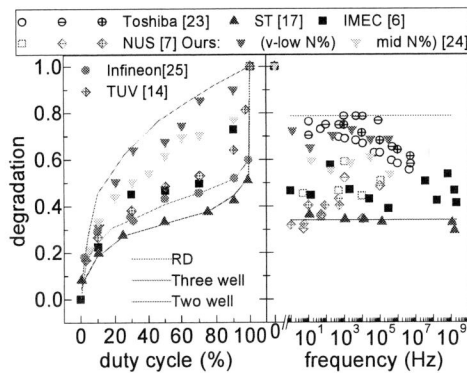

Figure 3. AC degradation (normalized to DC) versus pulse duty cycle and frequency from various reports. R-D, three well [14] and two well [17] solutions are shown.

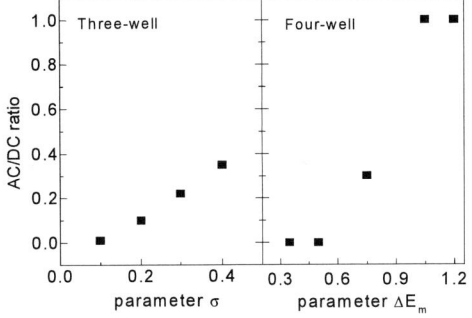

Figure 4. Simulated AC/DC ratio from three and four well models for parameters corresponding to Fig.2.

It is important to note that in addition to ΔN_{IT}, DC stress causes hole trapping in pre-existing bulk oxide traps (ΔN_{hole}) [12],[13] and at relatively higher stress bias ($V_{G,STR}$), additional

978-1-4244-9113-1/11 $26.00 © 2011 IEEE

hole trapping in newly generated bulk oxide traps (ΔN_{OT}) [26]-[28]. The spread of AC data when normalized to DC can be attributed to uncertainties in DC measurement, especially the magnitude of hole trapping that depends on: (i) gate insulator process that governs ΔN_{hole} [1],[3],[13],[18], (ii) value of $V_{G,STR}$ that governs ΔN_{OT} [26]-[28], and (iii) measurement type (OTF or SMS) and associated delay [13],[29]. Role of trapped holes is patently obvious as under identical stress and measurement conditions, gate dielectric with higher %N shows lower AC/DC ratio [24], as shown in Figure 3. This could be attributed to higher hole trapping in heavily nitrided films [18] and therefore a relatively high DC NBTI shift [13]. It is important to resolve uncertainty associated with this usual practice of normalizing AC degradation to DC data before making modeling attempts.

C. Renormalization and Universality of AC Degradation:

Under the assumptions of fast [13],[29] and temporally symmetric [30] hole trapping and detrapping and therefore AC stress to predominantly reflect ΔN_{IT}, AC data at all duty cycle is normalized to 50% duty cycle data to capture the intrinsic ΔN_{IT} behavior as shown in Figure 5. The scaling confirms a robust *universality* of the duty-cycle data (for duty cycle up to ~85%) from *disparate* sources and the *same* scaling factors also makes frequency independent data universal. *Such remarkable universality of AC degradation, which remained hidden under the uncertainty of DC data, has not been appreciated before.* After renormalization, the duty cycle and frequency dependent trends of AC degradation can be explained as have long been anticipated by R-D solution [11],[22] and are shown in Figure 5. The remaining spread in duty cycle dependent data for values close to DC is due to different contribution from ΔN_{hole} and ΔN_{OT}, and any difference in measurement condition or delay between DC and AC measurements.

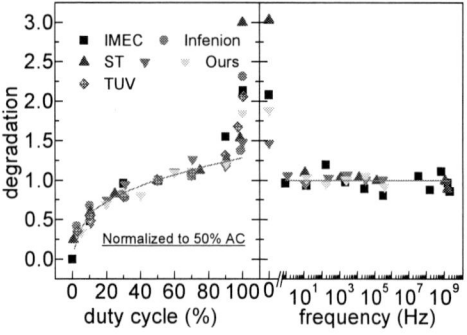

Figure 5. AC NBTI degradation normalized to 50% AC versus pulse duty cycle and frequency from various published results. R-D solutions (red lines) are shown.

Therefore contrary to recent perception, R-D framework not only can predict long time DC stress exponent, but also can successfully predict ΔN_{IT} stress and recovery dynamics as a function of pulse duty cycle and frequency. In the following sections, DC NBTI degradation will be analyzed to determine its underlying ΔN_{IT}, ΔN_{hole} and ΔN_{OT} components that govern the spread in data close to 100% duty cycle. These isolated

components will also be used to explain recovery transients measured following DC stress.

III. EXPERIMENTAL RESULTS ON DC STRESS

A. Device and Measurement details:

It is now well-known that NBTI is strongly influenced by gate insulator processes [1],[3],[13],[18],[23] which needs to be understood for proper modeling of DC stress. To elucidate this effect, DC stress experiments were performed on p-MOSFETs having Plasma Nitrided Oxide (PNO) with and without proper Post Nitridation Anneal (PNA), Rapid Thermal Nitrided Oxide (RTNO) and RTNO+PNO gate insulators [13],[18]. All PNO with proper PNA having low to moderately high N content are classified as Type-A devices, all others as Type-B devices. The following devices (EOT, N%) have been extensively used in this analysis: Type-A (D1: 2.35nm, 16.7%; D2: 1.4nm, 22.6%; D3: 1.56nm, 34.6%; and D4: 2nm, 34.9%) and Type-B (D5: 2.2nm, 5.8%, RTNO) [18]. Threshold voltage shift (ΔV_T) as a function of stress time, Temperature (T) and gate oxide electric field (E_{OX}) was obtained by using Ultra-Fast On-The-Fly (UF-OTF) I_{DLIN} [3] followed by mobility correction [31].

B. Gate Insulator Process Dependence of NBTI Degradation:

Figure 6 shows that the time evolution of ΔV_T for D1 and D5 devices has distinctly different features. D1 shows strong T activation and negligible ultra-short time (<1ms) degradation. D5 shows higher degradation with negligible T dependence at ultra-short time, and very high overall degradation with weak T dependence at long stress time. As illustrated before [13],[18], all Type-A and Type-B devices show similar behavior, and in particular, PNO *without* proper PNA devices also fall under the Type-B category, irrespective of their N content, a fact that has been often missed in previous publications [29],[32].

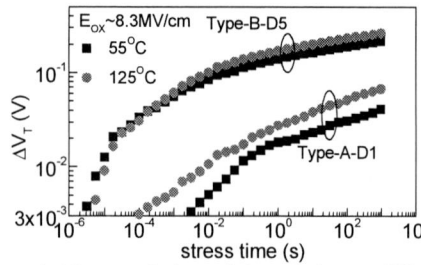

Figure 6. Time evolution of degradation at different stress temperature from ultra-short to moderately long stress time, measured on different devices using ultra-fast technique with 1μs delay.

Figure 7 shows measured ΔV_T as a function of T and E_{OX} for different Type-A and Type-B devices. Type-A devices show considerably lower ΔV_T compared to Type-B device for all T and E_{OX}. All Type-A devices show similar T activation (E_A) and E_{OX} acceleration (Γ), which are higher compared to Type-B device. Such E_{OX} and T dependent measurements were

978-1-4244-9113-1/11 $26.00 © 2011 IEEE

performed on a wide variety of devices (see [18] for details), and time, T and E_{OX} dependent parameters have been extracted. Figure 8 shows extracted n, E_A and Γ for all the different devices used in this work [13],[18] as a function of N%, which illustrate the strong impact of SiON gate insulator processes on measured DC NBTI degradation.

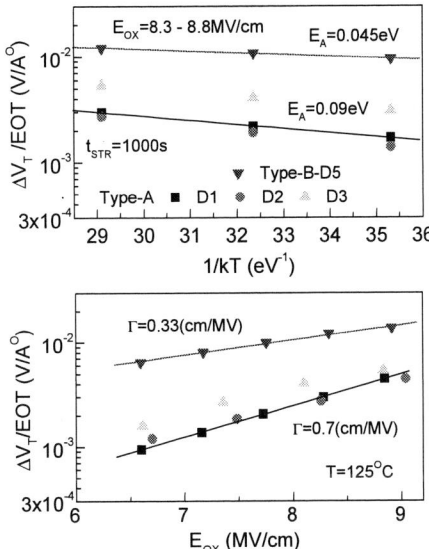

Figure 7. Temperature activation (top panel) and oxide electric field dependence (bottom panel) of degradation for different devices. Note, E_{OX} for T dependent data across different devices are similar (not same). Lines are fit to Type-A-D1 and Type-B-D5 device data.

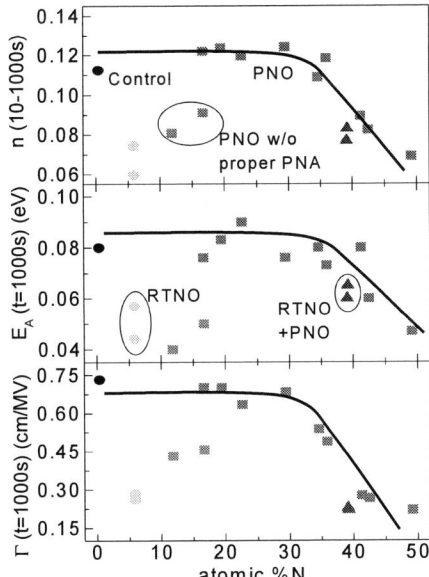

Figure 8. Time exponent (n), temperature activation (E_A) and field acceleration (Γ) of degradation as a function of atomic N content in oxide for different devices. Lines are guide to the eye for PNO with proper PNA devices.

It is evident that Type-A devices (PNO with proper PNA with N up to ~30%) show highest n, E_A and Γ, while all other

devices categorized as Type-B show lower values, *irrespective* of the total N content of the gate insulator. As explained before, the N distribution profile in the gate insulator, in particular, N density close to the Si/SiON interface affects NBTI; overall N content or N density close to the SiON/poly-Si interface has little impact [1],[18]. Although Type-A devices show highest n, the value measured by UF-OTF for moderately long stress time is somewhat lower than $n\sim1/6$ obtained using slightly delayed method at very long stress time as shown in Figure 1, and will be explained in next section. Higher E_A at longer stress time is consistent with stronger T activation at shorter stress time for type-A devices. Longer time E_A reduces and shorter time ΔV_T shows weak T dependence for type-B devices as shown in Figure 6. *SiON process variations resulting in gradual increase in Si/SiON interfacial N density gradually reduces n and E_A at longer stress time; while ΔV_T magnitude at both shorter and longer stress time becomes gradually higher and T activation of ΔV_T at short stress time becomes gradually weaker* [13],[18]. Such strong process dependence of the time and T dependence of measured ΔV_T is due to relative dominance of its underlying ΔN_{IT}, ΔN_{hole} and ΔN_{OT} components (discussed further in the next section) and is not a measurement artifact as claimed in [25]. Finally, note that Γ for mobility corrected ΔV_T as shown in this paper is higher than that for ΔI_{DLIN} [18], and has been explained in [31]. Variation of Γ is consistent with that of n and E_A as SiON processes are varied, as shown in Figure 8.

C. Other (direct) measurement methods:

Since measured ΔV_T captures ΔN_{IT}, ΔN_{hole} and ΔN_{OT}, it is often tempting to measure contribution from only ΔN_{IT} using DCIV [33] or that from ΔN_{IT} and ΔN_{OT} using charge pumping (CP) [34] and compare against ΔV_T, similar to that done in [35]. Such a simple comparison can in principle be used to obtain ΔN_{IT}, ΔN_{hole} and ΔN_{OT} components of overall NBTI. However, note that such direct comparisons are challenging as both DCIV and conventional CP are implemented in the measure-stress-measure mode and suffer from measurement delay. Ultra-fast CP [36],[37] does not suffer from delay, but uses very high bias pulses and can generate very high ΔN_{OT}, which can completely overwhelm ΔN_{IT} due to stronger voltage acceleration of the former (more on this in next section), a crucial fact that was not appreciated in [36],[37]. Both DCIV and CP scan a smaller portion of the energy band gap than ΔV_T [1] and would fail to capture additional (or enhanced) trap generation close to the valence band edge [38]. DCIV and CP also cannot be directly compared due to differences in recovery during measurement delay (for identical delay, CP recovers more as gate is pulsed positive) and portion of band gap scanned. In spite of such shortcomings, these methods can nevertheless be used to make relative estimation of ΔN_{IT} and ΔN_{OT} across different devices to access interface and bulk trap generation.

Figure 9 shows generation of interface traps as measured using DCIV and that of total (interface + oxide bulk) traps as measured using CP for Type-A and Type-B devices. Note that non-negligible trap generation is observed for both DCIV and CP measurements, and Type-A shows lower degradation and higher voltage acceleration when compared to Type-B devices. Non-negligible DCIV signature implies non-negligible ΔN_{IT}, a fact that has been often ignored in recent models [14]-[16]. The relative increase in trap generation magnitude for Type-B over Type-A, as obtained using DCIV and CP, is much lower than the increase in overall NBTI as measured using ΔV_T as shown in Figure 7. It is indeed possible that this is due to much larger interface trap generation near the valence band edge for Type-B device, which is not captured either by DCIV or CP. However, a more plausible explanation is enhanced ΔN_{hole} contribution for Type-B device, since it shows much smaller T activation than Type-A (see Figures 6 and 7). It has been discussed before that Type-B devices having higher N content near the Si/SiON interface show larger hole trapping [1],[13]. This is consistent with flicker noise measurements done in pre-stress on devices having different N content [39], as shown in Figure 10. Higher S_{VG} under identical inversion charge density for devices having higher N content (similar to Type-B) clearly indicates larger N related pre-existing hole traps in these devices, an observation consistent to that reported elsewhere [17]. It will be shown in the next section that ΔN_{hole} due to pre-existing N related traps is largely responsible behind strong NBTI process dependence.

ΔN_{OT} (=$B*t^n$, n=0.3 in the time range used for prediction of measured data [26]) needs to be accounted for in the existing framework of uncoupled ΔN_{IT} and ΔN_h for accurate prediction of stress data over wider stress bias ($V_{G,STR}$) range used in the present set of experiments. Note that the importance of ΔN_{OT} has been emphasized before [28], and though ΔN_{OT} is reduced with proper choice of stress bias, its contribution cannot be ignored for high precision modeling.

Figure 11 shows time evolution of measured ΔV_T, overall model prediction and underlying ΔV_{IT} (=$q*\Delta N_{IT}/CET$), ΔV_{hole} (=$q*\Delta N_{hole}/CET$) and ΔV_{OT} (=$q*\Delta N_{OT}/CET$) components for Type-A and Type-B devices. Note that large degradation at ultra-short (<1ms) time for Type-B device is primarily due to significant increase in ΔV_{hole} contribution, which is consistent with higher magnitude of pre-existing bulk hole trap density for such devices, as verified by flicker noise measurements as shown in Figure 10. ΔV_{IT} also increases for Type-B device, but the increase in ΔV_{hole} is much larger, and ΔV_{OT} remains similar for both devices. This is also consistent with observations made in Figures 7 and 9. Such predictions have been done at different stress T and E_{OX}, and extracted T and E_{OX} dependence of ΔV_{IT}, ΔV_{hole} and ΔV_{OT} components are obtained for different Type-A and Type-B devices, as shown in Figures 12 and 13.

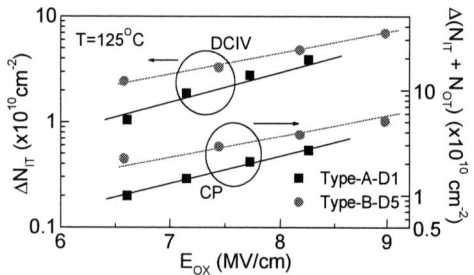

Figure 9. Oxide field dependence of generated interface traps using DCIV (LHS) and generated total (interface and oxide bulk) traps using CP (RHS) for different devices.

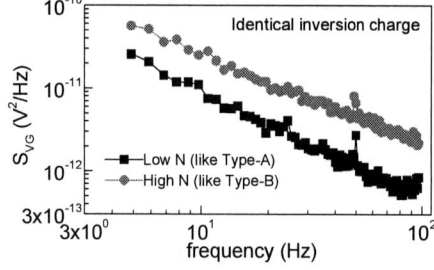

Figure 10. Pre-stress flicker noise measurements as a function of frequency on different devices.

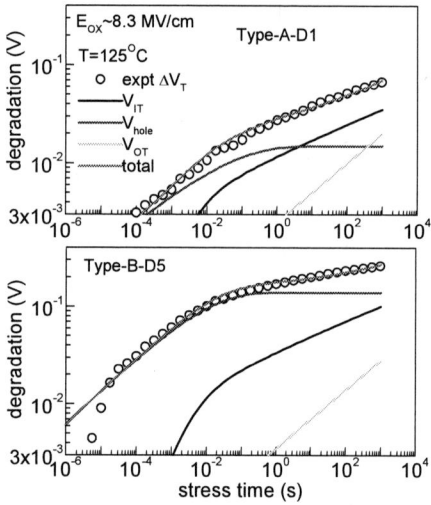

Figure 11. Time evolution of degradation for different devices and model prediction by R-D solution for interface traps (ΔV_{IT}) along with hole trapping in pre-existing (ΔV_{hole}) and in stress generated (ΔV_{OT}) bulk oxide traps.

IV. INTERPRETATION OF DC NBTI STRESS

Time evolution of degradation has been predicted [13] for differently processed SiON p-MOSFETs by using H-H₂ R-D framework for ΔN_{IT} [12], together with an analytical expression for ΔN_{hole} (=$A*(1- \exp(-t/\tau)^\beta)$) that accounts for fast, quickly saturating (<1s) hole trapping in pre-existing bulk oxide defects. Since NBTI is also TDDB stress [26]-[28], contribution from

Figure 12 shows that largest T activation is seen for ΔV_{OT} and smallest for ΔV_{hole}. The *uncoupled* nature of ΔV_{IT}, ΔV_{hole} and ΔV_{OT} is manifested as differences in obtained T activation for these components. All Type-A devices have similar E_A for ΔV_{IT}, ΔV_{hole} and ΔV_{OT}. Type-B device shows similar E_A for ΔV_{hole} and ΔV_{OT} but smaller E_A for ΔV_{IT} when compared to Type-A devices. As per R-D solution [9],[11], E_A for ΔV_{IT} at longer stress time is given by $E_A = 2/3*(E_A (k_F) - E_A (k_R)) + n*E_D$, where $E_A (k_F)$ and $E_A (k_R)$ are the activation of Si-H bond dissociation and annealing and E_D is activation of H₂ diffusion.

All Type-A devices show E_A (k_F) ~ E_A (k_R) [13] and so E_A ~ $n*E_D$ ~0.1 is similar for all devices (E_D~0.6 for H_2 diffusion [40] and n~1/6) [1]. Type-B device shows lower E_A (k_F) [13] possibly due to enhanced reaction favored by larger N density close to the Si/SiON interface [41], which results in lower E_A. In spite of largest n and E_A of ΔV_{OT}, total hole trapping (ΔV_{hole} + ΔV_{OT}) would still show lower n and E_A compared to ΔV_{IT} due to larger ΔV_{hole} contribution, especially for Type-B devices. Since ΔV_{hole} disproportionately increases for Type-B devices due to large pre-exiting bulk hole traps, this in turn results in reduction of n and E_A of overall ΔV_T as shown in Figure 8.

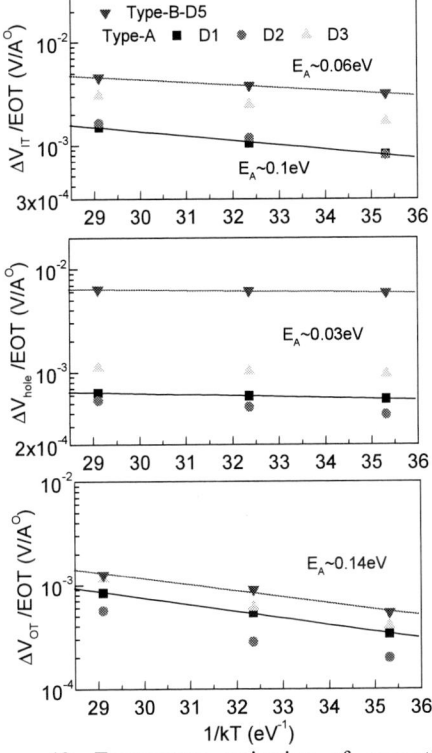

Figure 12. Temperature activation of generation of interface traps (RD solution), hole trapping in pre-existing and generated bulk oxide traps for different devices. Note, E_{OX} for T dependent data across different devices are similar (not same). Lines are fit to Type-A-D1 and Type-B-D5 device data.

Figure 13 shows that ΔV_{IT} and ΔV_{hole} have similar Γ while ΔV_{OT} has much larger Γ for all Type-A and Type-B devices. This makes Γ for overall ΔV_T, as shown in Figure 7, slightly higher than that of ΔV_{IT} and ΔV_{hole}. All Type-A devices have similar Γ's for each of ΔV_{IT}, ΔV_{hole} and ΔV_{OT} components whose values are larger than those of Type-B device. As an additional validation, time to a given ΔV_{OT} that is similar to TDDB voltage acceleration for different Type-A and Type-B devices is shown in Figure 14. Similar slope of power-law $V_{G,STR}$ acceleration is observed across different Type-A devices that is higher than that of Type-B. Note that being p-MOSFETs, the extracted $V_{G,STR}$ acceleration values are lower than that of n-MOSFETs, and is consistent with published reports [26],[27].

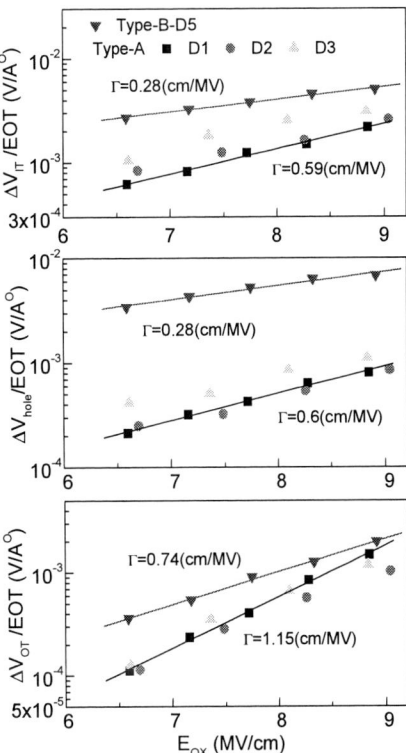

Figure 13. Oxide field dependence of generation of interface traps (RD solution), hole trapping in pre-existing and generated bulk oxide traps for different devices. Lines are fit to Type-A-D1 and Type-B-D5 device data.

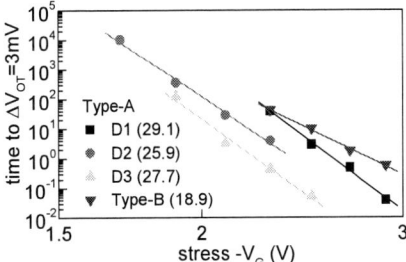

Figure 14. Voltage acceleration of time to a given bulk oxide trap generation for different devices.

Figure 15 shows measured and modeled ΔV_T at different $V_{G,STR}$ for different time-zero (t_0) delay [3]. H-H_2 R-D model [12],[13] solutions for ΔV_{IT} are also shown. Measurement with larger t_0 delay in general captures lower ΔV_{hole} [13], however, ΔV_T at different t_0 delay remains close to each other as ΔV_{hole} contribution is less for this Type-A device. Note that ΔV_{OT} is a slower process and not strongly t_0 dependent. Due to stronger $V_{G,STR}$ acceleration of ΔV_{OT}, total hole trapping (ΔV_{hole} + ΔV_{OT}) is higher and hence ΔV_{IT} is relatively smaller as compared to overall ΔV_T at larger $V_{G,STR}$. However, the difference between ΔV_T and ΔV_{IT} reduces at lower $V_{G,STR}$ and extrapolated long time exponent yields n~1/6 as shown in Figure 13. This is because ΔV_{OT} saturates to n~0.1 at longer stress time [27] and becomes negligible at lower $V_{G,STR}$. Note that the predicted

978-1-4244-9113-1/11 $26.00 © 2011 IEEE

long time power-law exponent of $n \sim 1/6$ is fully consistent with published results shown in Figure 1. It is important to mention that non-optimized (non industrial grade) devices with large pre-existing hole trap density would still show lower value of n at operating condition. This is due to large ΔV_{hole} contribution as ΔV_{hole} and ΔV_{IT} have similar E_{OX} acceleration as shown in Figure 13, the reduction of ΔV_{OT} at lower biases would not have much impact for such (Type-B like) devices.

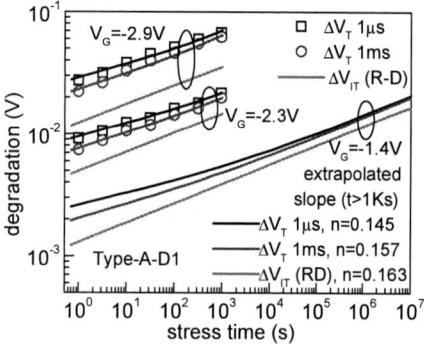

Figure 15. Measured degradation at different time-zero delay at different stress bias, and extrapolated to use condition. R-D solutions are also shown.

V. INTERPRETATION OF NBTI RECOVERY

Original criticisms of R-D model arose from its perceived inability to predict post-stress bias ($V_{G,REC}$) dependence of recovery and the long tail of recovery transients, features that inspired well based models [14]-[16]. It is important to note that these criticisms *predates* the appreciation that experimental data must be cleaned by accounting for hole detrapping from pre-existing and generated bulk traps before comparison with R-D theory is made [24], and that mobility correction [31] is also important for analysis of NBTI relaxation.

Figure 16 shows recovery transients obtained using one-spot I_{DLIN} measured at $V_{G,REC}$ (following stress at $V_{G,STR}$) and converted to mobility uncorrected ΔV_T ($= -\Delta I_{DLIN}/I_{DLIN0} * V_{GT0}$) for Type-A device that shows relatively smaller hole trapping. ΔV_T recovery transients obtained by proper mobility correction [31] are also shown. Mobility uncorrected ΔV_T shows $V_{G,REC}$ dependence while ΔV_T obtained by proper mobility correction shows negligible $V_{G,REC}$ dependence, as long as $|V_{G,REC}| > |V_T|$ and therefore electron capture into interface traps is negligible. This suggests that one of the objections to R-D model is a mere artifact of measurement.

Figure 17 compares measured fractional ΔV_T recovery and R-D model solution for a Type-A device, as is often done to illustrate the *failure* and criticize R-D framework [14],[42]. Note that the start time of recovery as predicted by R-D is ~ 6 orders of magnitude slower than that measured, and is similar to that mentioned in such criticisms. However *as NBTI stress results in ΔV_{hole} and ΔV_{OT} (no matter how small) in addition to ΔV_{IT}, attempting to predict the entire measured ΔV_T recovery transient using R-D solution for ΔV_{IT} recovery is inappropriate and naïve, and therefore requires careful re-evaluation.*

Figure 18 shows measured ΔV_T recovery and its prediction using R-D model solution for ΔV_{IT} recovery and an analytical expression for total hole detrapping ($\Delta V_{hole} + \Delta V_{OT}$) = $C * \exp(-t/\tau)^{\beta}$), that accounts for fast detrapping of trapped holes from pre-existing and generated bulk oxide traps (here one has to note that reduction in ΔV_{OT} is coming from hole detrapping from ΔN_{OT}, and not from the recovery of ΔN_{OT} itself). Once the early part of ΔV_T recovery is accounted for by fast hole detrapping (trapped hole magnitude obtained from analysis of DC stress data in section IV), the onset of ΔV_T recovery as governed by ΔV_{IT} recovery and predicted by R-D solution matches well with experiment.

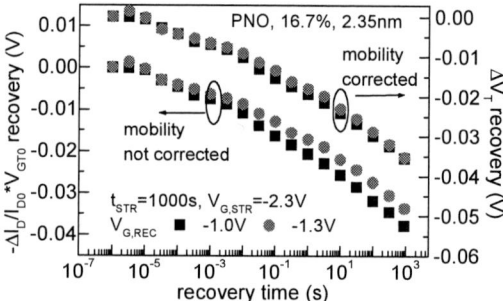

Figure 16. Time evolution of recovery at different recovery bias ($|V_{REC}| > |V_{T0}|$) measured using one spot I_{DLIN} and mobility corrected V_T shift.

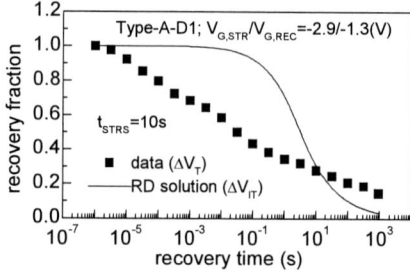

Figure 17. Time evolution of recovery fraction from ultra-short to moderately long time following relatively shorter stress time. Improper comparison of measurement data for ΔV_T and R-D solution for ΔV_{IT} are shown, as are often done to highlight failure of R-D model [14],[42].

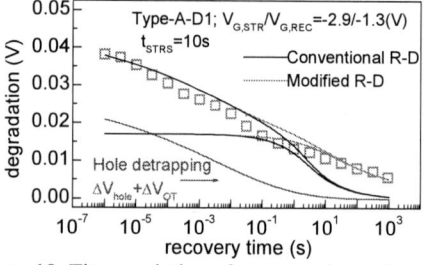

Figure 18. Time evolution of recovery from ultra-short to moderately long time following relatively shorter stress time. Model prediction using hole detrapping and interface trap passivation using conventional R-D and modified R-D solutions.

978-1-4244-9113-1/11 $26.00 © 2011 IEEE

Figure 18 also suggests that although the start of ΔV_{IT} recovery is accurately predicted by the classical R-D theory having symmetric H_2 diffusion during stress and recovery [24], the predicted rate of recovery is somewhat faster than measured data. This failure reflects the inability of simple 1D diffusion formulation to capture the details of an essentially 3D diffusion problem. The distribution of 3D diffusion path-lengths [22] can be accounted for in a simple 1D model by invoking an effective distribution of diffusion coefficient (see Figure 20 for physical mechanism details). This simple and physically meaningful modification of the R-D framework consistently captures the longer term ΔV_T recovery, as governed by the recovery of ΔV_{IT}.

Figure 19 shows measured ΔV_T recovery transients across different devices following stress at different $V_{G,STR}$ and T, and prediction by the proposed model consisting of fast detrapping of holes (ΔV_{hole} and ΔV_{OT}) and slower ΔV_{IT} passivation based on modified R-D framework. The magnitude of trapped holes ($\Delta V_{hole} + \Delta V_{OT}$) and ΔV_{IT} at the onset of recovery is consistent with analysis of DC stress data. Note that the analytic solution for hole detrapping can predict early part of ΔV_T recovery. In addition, the start time of ΔV_{IT} recovery is also accurately predicted, which in-turn re-verifies the consistency of ΔV_{IT}, ΔV_{hole}, and ΔV_{OT} predictions over wide range of $V_{G,STR}$ and T and for device having different N% and EOT. Moreover, the rate of longer term ΔV_T recovery governed by recovery of ΔV_{IT} is also reasonably captured by modified R-D framework for different devices, justifying the proposed hypothesis. Therefore, *contrary to common perception, R-D framework can properly predict portions of ΔV_T recovery that is governed by recovery of ΔV_{IT} (at longer time), provided the shorter time contribution due to recovery of ΔV_{hole} and ΔV_{OT} are properly isolated.*

It is important to justify asymmetric H_2 diffusion during stress and recovery as invoked in the modified R-D framework. As illustrated in Figure 20, for a particular path from Si/SiON interface into oxide bulk as traveled by H_2 during stress (shown by dotted line), there are multiple possible paths for H_2 to come back from oxide bulk to Si/SiON interface during recovery (solid, colored lines). For short recovery time following longer stress, the returning H_2 would see many broken Si-H bonds ("vacant" positions) and can re-passivate (not necessarily at the same location from where it originated during stress) the available ones quite easily. Hence the time required for H_2 to move away up to a certain distance after getting broken during stress and move back to interface and re-passivating would be similar, and in 1D implementation, can be simulated using symmetric H_2 diffusion, as implemented in conventional R-D framework. However, for long recovery time (most Si- bonds are passivated) or for recovery following short stress (few Si-bonds are broken), the returning H_2 would "hover" near the Si/SiON interface until it *finds a "vacant" spot* to re-passivate. Therefore, H_2 would take asymmetrically long time to return back and re-passivate during recovery, and for identical recovery time (as stress), fewer H_2 would be able to find a vacant spot to re-passivate. Though this phenomenon should be ideally modeled using 3D stochastic process, a simpler 1D implementation can be done using lower H_2 diffusivity during recovery. Figure 20 also shows recovery transients simulated using R-D framework by having different (lower than stress)

H_2 diffusivity. The amount of recovery for a given recovery /stress ratio reduces as diffusivity is reduced and is expected. In practice, most H_2 would return with symmetric diffusivity and a few would return with *effective* reduced diffusivity, and the net solution can be obtained by taking a *weighted average* of different solutions, and is illustrated in Figure 18. Note that the present implementation is analogous to the concept of [43], but uses much smaller and physically justifiable dispersion. This is due to proper isolation of contribution due to trapped holes that govern early recovery from that by interface traps that govern long time recovery, as shown in Figures 18 and 19. In [43], the entire recovery was modeled using R-D solution for ΔN_{IT} that of course needed much larger dispersion.

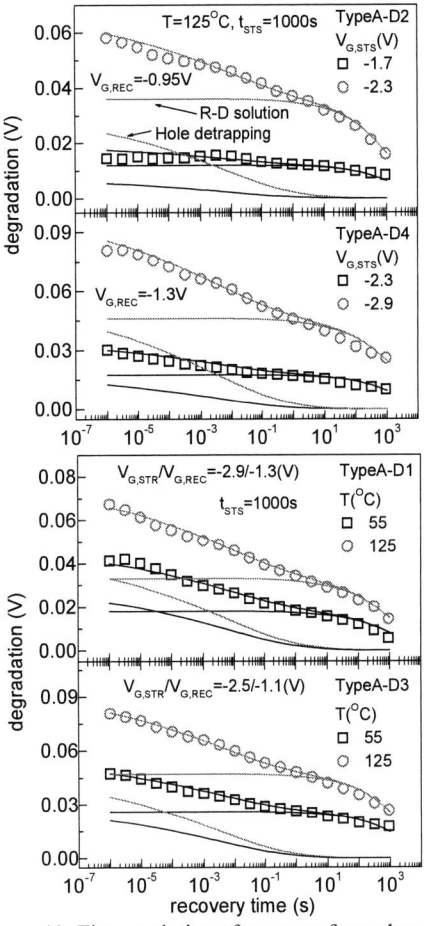

Figure 19. Time evolution of recovery from ultra-short to moderately long time on different devices (top 2 panels) for different stress bias and (bottom 2 panels) for different temperature. Model prediction using hole detrapping and interface trap passivation using modified R-D solutions.

Finally, it is important to re-verify if the modified R-D model violates AC frequency independence. Figure 21 shows AC/DC ratio simulated using different H_2 diffusivity during recovery. Lower diffusivity indeed results in lower recovery and higher AC/DC ratio at lower frequency. However, all ratios saturate to a final value (~0.8) for f>100Hz. Nevertheless, the weighted average (same weights that are used to predict the recovery transients) of AC/DC ratio is frequency independent as shown. Therefore, the proposed generalization of R-D framework can indeed be used to predict all features of NBTI recovery.

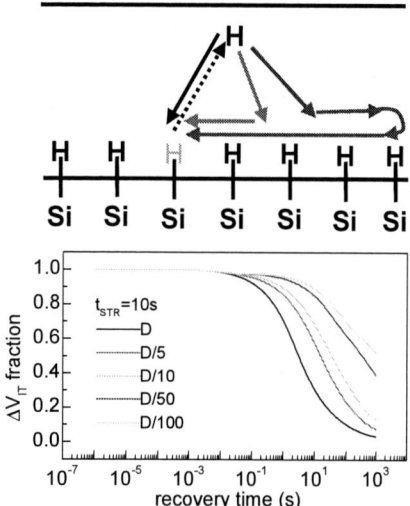

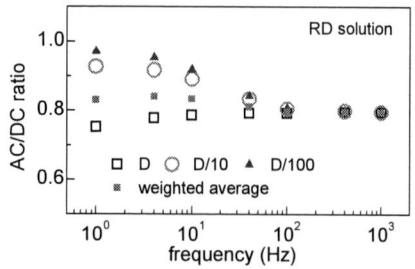

Figure 20. Schematic of H_2 diffusion during stress and recovery showing multiple diffusive pathways possible during recovery (top panel). Simulation of recovery using RD with different (similar or lower than stress) H_2 diffusivity, to "mimic" 3D multiple diffusive pathways in 1D formalism (bottom panel).

Figure 21. Simulated AC/DC ratio versus frequency for different D during recovery. Weighted average (same as used for recovery transients) is shown.

VI. MODEL BENCHMARKING

Given the success of proposed framework in interpreting all the five key features of NBTI, it is important to ask how it relates to other recent models [16],[17]. From our perspective, NBTI involves a fast hole trapping phase in pre-existing traps followed by generation of interface traps, and at higher stress voltages and/or sufficiently long time bulk trap generation and trapping. We have unequivocally shown that all these three components are *not correlated* to each other and their relative magnitude depends on SiON processes and stress conditions. Bulk trap generation is dominant at large biases and is also found by spectroscopic studies (E' centers) [44]. We have previously suggested that the effects of bulk trap generation and trapping can be reduced by reducing the stress bias [28].

Similarly, the reduction of trapping in pre-existing traps is possible by SiON process optimization (e.g., PNO with proper PNA) [18]. However, it is not possible to completely eliminate hole trapping contribution. Similar uncorrelated components were invoked in [17], however, interface trap generation was modeled by using a dispersive reaction framework with no diffusion of H_2.

Recent work [16] relating NBTI to RTN, detailed kinetics of emission constants, and logarithmic time decay in very early phase of relaxation is indeed complementary to our proposed framework in the following sense. We model the effect of hole trapping by a simple phenomenological approach (i.e. stretched exponential) and no attempt has been made to investigate the detailed physics of hole trapping. Instead, [16] focuses on very short time stress (< few seconds, when hole trapping is likely to be dominant) and explores the capture and emission dynamics of holes as well as their connection to RTN. The results of [16] can be easily accommodated in our present framework as an improved version of the hole trapping dynamics. As we have seen, however, that *when this physics of hole trapping is used to create alternate models based on multi-well formulations to replace R-D formulation, none of the five key features of NBTI can be explained* (refer to [22] for details).

Finally, note that the traditional R-D model is based on the breaking of Si_3-Si-H bonds (creation of Pb centers) as reaction followed by H_2 diffusion [8]. However the diffusion process actually controls the time dynamics in moderate to longer stress duration [9]-[12]. In principle, R-D formulation is consistent even if N_3-Si-H bonds (only or additionally to Si_3-Si-H) are broken as suggested in SiON films [45], as long as the broken bonds remain close to Si/SiON interface, and broken H diffuses out eventually as H_2. Therefore, contrary to recent perception, our proposed R-D model based framework is also consistent with recent spectroscopic studies [44],[45].

VII. CONCLUSIONS

Once hole trapping is accurately predicted and isolated, the five broad features concerning NBTI stress, recovery and AC experiments reflect interface trap controlled signatures that can be accurately predicted by using Reaction Diffusion (R-D) framework. This is verified on differently processed devices and under a wide range of stress/recovery conditions. Contrary to recent perception, R-D framework can be readily used to determine NBTI lifetime for process/circuit qualification.

ACKNOWLEDGEMENT

Applied Materials and Renesas Electronics for funding and devices; DIT, Govt. of India for funding; CEN, IIT Bombay for laboratory usage. Khaled Ahmed, Chris Olsen (Applied), Aono Hideki, Eiichi Murakami (Renesas) for useful discussions.

REFERENCES

[1] S. Mahapatra, K. Ahmed, D. Varghese, A. E. Islam, G. Gupta, L. Madhav, D. Saha and M. A. Alam, "On the physical mechanism of NBTI in Silicon oxynitride p-MOSFETs: Can difference in insulator processing conditions

resolve the interface trap generation versus hole trapping controversy?", in *Proc. Int. Rel. Phys. Symp.*, pp. 1-9, 2007.

[2] S. Rangan, N, Mielke and E. C. C. Yeg, "Universal recovery behaviour of negative bias temperature instability", in *IEDM Tech. Dig.*, pp. 331-334, 2003.

[3] E. N. Kumar, V. D. Maheta, S. Purawat, A. E. Islam, C. Olsen, K. Ahmed, M. Alam and S. Mahapatra, "Material dependence of NBTI physical mechanism in silicon oxynitride (SiON) p-MOSFETs: A comprehensive study by ultra-fast On-the-fly (UF-OTF) I_{DLIN} technique", in *IEDM Tech. Dig.*, pp.809-812, 2007.

[4] B. Kaczer, V. Arkhipov, R. Degraeve, N. Collaert, G. Groeseneken and M. Goodwin, "Disorder controlled kinetics model for negative bias temperature instability and its experimental verfication", in *Proc. Int. Rel. Phys. Symp.*, pp. 381-387, 2005.

[5] D. Varghese, D. Saha, S. Mahapatra, K. Ahmed, F. Nouri and M. Alam, "On the dispersive versus Arrhenius temperature activation of NBTI time evolution in plasma nitrided gate oxide: measurements, theory and implication", in *IEDM Tech. Dig.*, pp. 684-687, 2005.

[6] R. Fernández, B. Kaczer, A. Nackaerts, S. Demuynck, R. Rodríguez, M. Nafría, and G. Groeseneken, "AC NBTI studied in the 1 Hz – 2 GHz range on dedicated on-chip CMOS circuits", in *IEDM Tech. Dig.*, pp. 337-340, 2006.

[7] G. Chen, M. F. Li, M, C. H. Ang, J. Z. Zheng, D. L. Kwong, "Dynamic NBTI of p-MOS transistors and its impact on MOSFET scaling", *IEEE Elec. Dev. Lett.*, vol. 23, no. 12, pp. 734-736, Dec. 2002.

[8] K. O. Jeppson and C. M. Svensson, "Negative bias stress of MOS devices at high electric fields and degradation of MOS devices", *J. Appl. Phys.*, vol. 48, no. 5, pp. 2004-2014, May 1977.

[9] M. A. Alam and S. Mahapatra, "A comprehensive model of PMOS NBTI degradation", *Microelectronics Reliability*, vol. 45, pp. 71-81, Jan. 2005.

[10] S. Chakravarthi, A. Krishnan, V. Reddy, C. F. Machala, and S. Krishnan, "A comprehensive framework for predictive modeling of negative bias temperature instability", in *Proc. Int. Rel. Phys. Symp.*, pp. 273-282, 2004 .

[11] M. A. Alam, H. Kufluoglu, D. Varghese, and S. Mahapatra, "A comprehensive model for PMOS NBTI degradation: Recent progress", *Microelectronics Reliability*, vol. 47, pp. 853-862, June 2007.

[12] A. E. Islam, H. Kufluoglu, D. Varghese, S. Mahapatra and M. A. Alam, "Recent issues in negative bias tempertaure instability: Initial degradation, field dependence of interface trap generation, hole trapping effects and relaxation", *IEEE Trans. Electron Devices*, vol. 54, no. 9, pp. 2143-2154, Sep. 2007.

[13] S. Deora, V. D. Maheta and S. Mahapatra, "NBTI Lifetime Prediction in SiON p-MOSFETs by H/H2 Reaction-Diffusion (RD) and Dispersive Hole Trapping Model", in *Proc. Int. Rel. Phys. Symp*, pp. 1105, 2010.

[14] T. Grasser, B. Kaczer and W. Goes, "An energy level perspective of bias-tempertaure instability", in *Proc. Int. Rel. Phys. Symp.*, p. 28, 2008.

[15] T. Grasser, B. Kaczer, W. Goes, T. Aichinger, P. Hehenberger and M. Nelhiebel, "A two-stage model for negative bias temperature instability", in *Proc. Int. Rel. Phys. Symp.*, p. 33, 2009.

[16] T. Grasser, H. Reisinger, P. Wagner, F. Schanovsky, W. Goes, B. Kaczer, "The time dependent defect spectroscopy (TDDS) for the characterization of the bias temperature instability", in *Proc. Int. Rel. Phys. Symp.*, pp.16-25, 2010.

[17] V. Huard, "Two independent components modeling for Negative Bias Temperature Instability", in *Proc. Int. Rel. Phys. Symp.*, pp.33-42, 2010.

[18] V. D. Maheta, C. Olsen, K. Ahmed and S. Mahapatra, "The impact of nitrogen engineering in silicon oxynitride gate dielectric on negative bias temperature instability of p-MOSFETs", *IEEE Trans. Electron Devices*, vol. 55, no. 7, pp. 1630-1638, July 2008.

[19] A. Haggag, G. Anderson, S. Parohar, D. Burnett, G. Abeln, J. Higman and M. Moosa, "Understanding SRAM high temperature operating life NBTI: statistics and permanent vs recoverable damage", in *Proc. Int. Rel. Phys. Symp.*, pp. 452-456, 2007.

[20] C. L. Chen, Y. M. Lin, C. J. Wang, and K. Wu, "A new finding on NBTI lifetime model and an investigation on NBTI degradation characteristic for 1.2nm ultra thin oxide", in *Proc. Int. Rel. Phys. Symp.*, pp. 704 – 705, 2005.

[21] A. E. Islam, G. Gupta, S. Mahapatra, A. Krishnan, K. Ahmed, F. Nouri, A. Oates, and M. A. Alam, "Gate leakage vs. NBTI in Plasma Nitrided Oxides: Characterization, Physical Principles, and Optimization", in *IEDM Tech. Dig.*, pp. 329-332, 2006.

[22] M. A. Alam, S. Mahapatra, A. E. Islam and A. Jain, "On the universality of NBTI degradation", in *Int. Integrated Reliability Workshop*, 2010.

[23] Y. Mitani, H. Satake, and A. Toriumi, "Influence of nitrogen on negative bias temperature instability in ultrathin SiON", *IEEE Trans. on Dev. and Mat. Reliability*, vol. 8, pp. 6-13, Mar 2008.

[24] A. E. Islam, S. Mahapatra, S. Deora, V. D. Maheta and M. A. Alam, "On the differecnes between ultra-fast NBTI experiments and reaction-diffusion theroy", in *IEDM Tech. Dig.*, pp. 733-736, 2009.

[25] H. Reisinger, T. Grasser, W. Gustin and C. Schlunder, "The statistical analysis of individual defects constituting NBTI and its implications for modeling DC- and AC-stress", in *Proc. Int. Rel. Phys. Symp.*, pp.7-15, 2010.

[26] P. Nicollian, "Physics of trap generation and electrical breakdown in ultra-thin SiO2 and SiON gate dielectric material", University of Twente, PhD thesis 2007.

[27] D. Varghese, "Multi-probe experimental and bottom up computational analysis of correlated defect generation in modern nanoscale transistors", Purdue University, PhD thesis, 2009.

[28] S. Mahapatra, P. P. Kumar and M. A. Alam, "Investigation and modeling of interface and bulk trap generation during negative bias temperature instability", *IEEE Trans. Elec. Dev.* , vol. 51, no. 9, pp. 1371-1379, Sep. 2004.

[29] C. Shen, M. F. Li, C. E. Foo, T. Yang, D. M. Huang, A.Yap, G. S. Samudra, and Y. C. Yeo, "Characterization and physical origin of fast Vth transient in NBTI of pMOSFETs with SiON dielectric", in *IEDM Tech. Dig.*, pp. 333-336, 2006.

[30] H. Miki, N. Tega, Z. Ren, C. P. D"Emic, Y. Zhu, D. J. Frank, M. A. Guillorn, D. G. Park, W. Haensch, and K. Torii, "Hysteric drain current behavior due to random telegraph noise in scaled down FETs with high-k/metal gate stacks", in *IEDM Tech. Dig.*, pp. 620-623, 2010.

[31] A. E. Islam, V. D. Maheta, H. Das, S. Mahapatra and M. A. Alam, "Mobility degradation due to interface traps in plasma oxinitride PMOS devices", in *Proc. Int. Rel. Phys. Symp.*, p. 87, 2008.

[32] D. S. Ang, S. Wang, G. A. Du, and Y. Z. Hu, "A consistent deep-level hole trapping model for negative bias temperature instability", *IEEE Trans. on Dev. and Mat. Reliability*, vol. 8, pp. 22-34, Mar 2008.

[33] A. Neugroschel, Chih-Tang Sah; K. M. Han, M. S. Carroll, T. Nishida, J. T. Kavalieros and Yi Lu, "Direct-current measurements of oxide and interface traps on oxidized silicon", *IEEE Tran. Elec. Dev.*, vol. 42, no. 9, pp. 1657-1662, Sep. 1995.

[34] M. Masuduzzaman, A. E. Islam, And M. A. Alam, "A Multi-Probe Correlated Bulk Defect Characterization Scheme For Ultra-Thin High-K Dielectric", in *Proc. Int. Rel. Phys. Symp.*, p. 1069, 2010.

[35] V. Huard and M. Denais, "Hole trapping effect on methodology for DC and AC negative bias temperature instability measurements in PMOS transistors", in *Proc. Int. Rel. Phys. Symp.*, pp. 40-45, 2004.

[36] W. J. Liu, Z. Y. Liu, Darning Huang, C. C. Liao, L. F. Zhang, Z. H. Gan, Waisum Wong; C. Shen and Ming-Fu Li, "On-The-Fly Interface Trap Measurement and Its Impact on the Understanding of NBTI Mechanism for p-MOSFETs with SiON Gate Dielectric", in *IEDM Tech. Dig.*, pp.813-816, 2007.

[37] Z. Q. Teo, D. S. Ang, K. S. See, "Can the reaction-diffusion model explain generation and recovery of interface states contributing to NBTI?", in *IEDM Tech. Dig.*, pp.737-740, 2009.

[38] A. T. Krishnan, C. Chancellor, S. Chakravarthi, P. E. Nicollian, V. Reddy, A. Varghese, R. B. Khamankar and S. Krishnan, "Material dependence of hydrogen diffusion: implications for NBTI degradation", in *IEDM Tech. Dig.*, pp. 705-708, 2005.

[39] G. Kapila, N. Goyal, V. D Maheta, C. Olsen, K. Ahmed, S. Mahapatra, "A comprehensive study of flicker noise in plasma nitrided SiON p-MOSFETs: process dependence of pre-existing and NBTI stress generated trap distribution profiles", in *IEDM Tech. Dig.*, pp.103-106, 2008.

[40] M. L. Reed and J. D. Plummer, "Chemistry of Si-SiO2 interface trap annealing", *App. Phys. Lett.* , vol. 57, no. 2, pp. 162-164, Feb. 1987.

[41] S. S. Tan, T. P. Chen, J. M. Soon, K. P. Loh, C. H. Ang, W. Y. Teo, and L. Chan, "Neighboring effect in nitrogen-enhanced negative bias temperature instability", in *Proc.Solid State Devices and Mater. (SSDM)*, pp. 70-71, 2003.

[42] H. Reisinger, O. Blank, W. Heinrigs, A. Muhlhoff, W. Gustin and C. Schlunder, "Analysis of NBTI degradation- and recovery- behaviour based on ultra fast VT measaurement", in *Proc. Int. Rel. Phys. Symp.*, pp. 448-453, 2006.

[43] H. Kufluoglu, V. Reddy, A. Marshall, J. Krick, T. Ragheb, C. Cirba, A. Krishnan, C. Chancellor, "An extensive and improved circuit simulation methodology for NBTI recovery", in *Proc. Int. Rel. Phys. Symp*, pp.670-675, 2010.

[44] J. T. Ryan, P. M. Lenahan, T. Grasser and H. Enichlmair, "Recovery-free electron spin resonance observations of NBTI degradation", in *Proc. Int. Rel. Phys. Symp.*, pp.43-49, May 2010.

[45] J. P. Campbell, P. M. Lenahan, A. T. Krishnan and S. Krishnan, "Location, structure and density of states of NBTI induced defects in plasma nitrided PMOSFETs", in *Proc. Int. Rel. Phys. Symp.*, pp. 503-510, 2007.

On the Recoverable and Permanent Components of Hot Carrier and NBTI in Si pMOSFETs and their Implications in Si$_{0.45}$Ge$_{0.55}$ pMOSFETs

J. Franco[1], B. Kaczer, G. Eneman[1,2], Ph. J. Roussel, M. Cho, J. Mitard, L. Witters, T. Y. Hoffmann, G. Groeseneken[1]

imec

Leuven, Belgium

Phone: +32 16 28 10 85, e-mail: Jacopo.Franco@imec.be

[1] also at ESAT, K.U. Leuven, Belgium

[2] also FWO-Vlaanderen, Belgium

F. Crupi

DEIS Dept., University of Calabria

Rende, Italy

T. Grasser

Christian Doppler Laboratory for TCAD

Institute for Microelectronics, T.U. Wien

Wien, Austria

Abstract— **The introduction of SiGe channel pMOSFETs for high mobility devices is expected to enhance the impact ionization phenomenon, making it necessary to study Hot Carrier (HC) degradation also for the p-channel MOSFET reliability. The study of pure HC effects on pMOSFETs is complicated due to the mixing with Negative Bias Temperature Instability (NBTI). In the first part of this work the interaction of the two degradation mechanisms is studied thoroughly on Si devices with the extended measure-stress-measure (eMSM) technique which is capable of capturing both the charge trapping and the interface state creation components of the degradation. HC degradation is shown to enhance interface state creation, while eventually reducing the charge trapping w.r.t. standard NBTI. These experimental results are supported by MEDICI simulations. The second part of the paper focuses on the HC reliability of Si$_{0.45}$Ge$_{0.55}$ pMOSFETs. These devices show enhanced degradation w.r.t. their Si counterparts, confirming the importance of studying HC effects for the reliability of this technology. Nevertheless, the SiGe device reliability can be enhanced when reducing the thickness of the Si cap.**

Keywords: Hot Carrier, Negative Bias Temperature Instability, pMOSFETs, SiON, poly-Si, high-k, metal gate, SiGe, Si cap.

I. INTRODUCTION

Hot Carrier (HC) degradation is generally considered a major reliability issue for nMOSFET [1], while the pMOSFET reliability is jeopardized by Negative Bias Temperature Instability (NBTI) [2]. Incorporation of Ge into Si substrate for high-mobility devices, which is considered as one of the most promising options to further improve CMOS performance [3], has been shown to improve NBTI reliability [4-6], while enhancing HC effects due to higher impact ionization caused

by the smaller bandgap of Ge [7-9]. In this scenario, both HC and NBTI need to be considered for the pMOSFET reliability.

Furthermore, a typical HC test implies the application of high bias on both the drain and the gate terminals (Channel HC, CHC) [1]: in such a case, while the oxide at the drain side of the channel experiences the effects of HC caused by the high lateral electric field ('*hot hole*' *injection*), the source side still experiences only the high oxide electric field (E$_{ox}$) due to the applied gate voltage ('*cold holes*' *injection*), resulting in a typical NBTI stress condition [2]. It is hence necessary to study the interplay of these two degradation mechanisms to properly assess the pMOS reliability.

In this work, this study is performed firstly on Si/SiON/poly-Si devices (section III). The experimental results are then validated on a more recent high-k/metal gate technology (section IV) and then further applied to the study of novel Si$_{0.45}$Ge$_{0.55}$ pMOSFETs (section V).

II. EXPERIMENTAL

Three different set of samples are used in this work. Firstly, the interaction between HC and NBTI is studied on Si pMOSFETs with a SiON dielectric (~1.65nm), a poly-Si gate, and a channel length L≈150nm. To validate the results on a more recent technology, Si high-k/metal gate devices are also used. For these samples the gate stack consists of a SiO$_2$ interfacial layer (~0.8nm) and a HfO$_2$ (~2nm) high-k layer. Channel length is L≈70nm in this case. Finally, SiGe pMOSFETs with a buried channel architecture are studied. Ge fraction in the channel is x=0.55, while the gate stack consists of a Si cap of varying thickness (0.65~2nm), a SiO$_2$ interfacial

978-1-4244-9113-1/11 $26.00 © 2011 IEEE

layer (~0.8nm) and a HfO_2 (~1.8nm) high-k layer. Channel length is L≈70nm also in this case. More details on this process can be found in [3].

To compare the effect of NBTI and CHC on pMOS, stress experiments were performed by applying the stress voltage only to the gate for NBTI and to both the gate and the drain for CHC. Since HC effects are only weakly dependent on the temperature (T) [1], while NBTI is strongly T-activated [2], experiments were performed at 125°C (unless otherwise stated). The extended Measure-Stress-Measure technique (eMSM) [10] is used to capture both the so-called 'recoverable' (R) component of the threshold voltage shift (ΔV_{th}), ascribed to the charging of pre-existing oxide defects during stress (ΔN_{ot}), and the 'permanent' (P) or 'slowly-relaxing' one, typically associated with the creation of new interface states (ΔN_{it}) [10,11].

A typical set of relaxation curves obtained with the eMSM technique for NBTI and CHC stress conditions on Si/SiON samples is shown in Fig. 1.

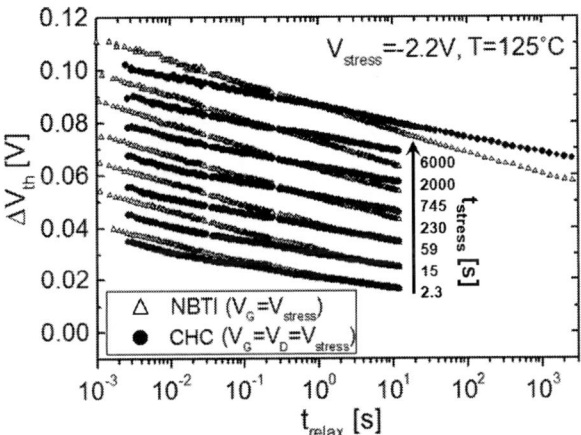

Figure 1. Si/SiON: Relaxation curves for NBTI ($V_G=V_{stress}$) and HC ($V_G=V_D=V_{stress}$) stresses with V_{stress}=-2.2V, for increasing stress times $t_{stress,i}$. For each pair of relaxation curves pertaining to the same $t_{stress,i}$, the HC damage dominates at longer relaxation times, although in the beginning the ΔV_{th} related to NBTI stress is higher. This observation already suggests HC stress is causing a less recoverable degradation.

The eMSM data analysis technique developed previously for NBTI [10,11] is employed in this work also for the CHC stress conditions. Using this technique, based on the previously observed *universality of NBTI relaxation*, the measured ΔV_{th} ($t_{stress,i}$, t_{relax}) are empirically separated into the R and P components, i.e.,

$$\Delta V_{th} (t_{stress,i}, t_{relax}) \approx R (t_{stress,i}, t_{relax}) + P(t_{stress,i}) . \quad (1)$$

Here, $t_{stress,i}$ represents the total stress time after the i-th stress phase, while t_{relax} stands for the time elapsed from the beginning of the last relaxation phase. According to the previous observations [10,11], all the relaxation data obtained

at different stress times fall on the same curve, given by the universal relaxation function r(ξ), where $\xi = t_{relax} / t_{stress,i}$ is the universal relaxation time. Very good fits to experimental data have been obtained in literature using the empirical relation:

$$r(\xi) = \frac{1}{1 + B\xi^\beta} \quad [11] . \quad (2)$$

Therefore the relaxation of the recoverable part of the damage can be described as:

$$R (t_{stress,i}, t_{relax}) = R (t_{stress,i}, t_{relax}= 0) \cdot r(\xi) , \quad (3)$$

where $R (t_{stress,i}, t_{relax}= 0)$ represents the *"full"* R component extrapolated to t_{relax}=0, as if it were measured with zero delay after stress removal. Thus it is possible to estimate the total R from standard delayed measurements.

Conversely, the P component (i.e., $P(t_{stress,i})$ in eq. (1)) is defined as the damage which would be ideally still measured after an infinite time from stress removal, since $R (t_{stress,i}, t_{relax}= \infty) = 0$ (see eqs. (2) and (3), [11]).

Interestingly, the universality of the relaxation is observed to be valid also for CHC stress on pMOSFETs, as shown in Fig. 2 for a stress voltage of, e.g., -2.2V. The same observation can be made for each of the considered stress voltages.

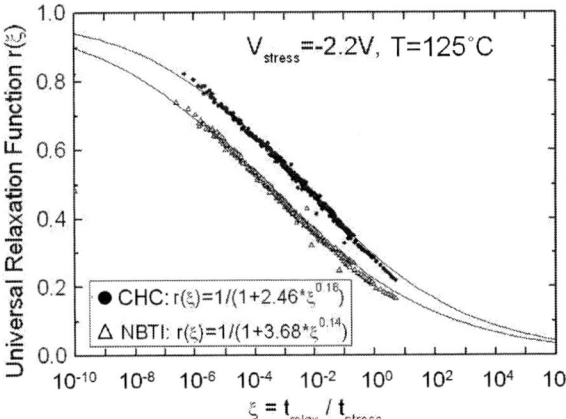

Figure 2. Si/SiON: After extracting the P component, the NBTI relaxation data can be mapped onto the universal relaxation curve r(ξ)=1/(1+Bξ^b) where ξ=t_r/$t_{stress,i}$ (see eqs. (1), (2), and (3), [11]). This is also valid for the relaxation curves after CHC ($V_G=V_D=V_{stress}$), already suggesting a similarity between NBTI and CHC on pMOSFETs. However, the universal relaxation function assumes higher values for the CHC dataset w.r.t. NBTI, again suggesting a less recoverable nature of the HC degradation.

III.   RESULTS AND DISCUSSION ON SI/SION/POLY-SI

As shown in Fig. 3, the CHC stress condition resulted in a reduced R and enhanced P w.r.t. NBTI. While the reduction of R is constant for a wide range of V_{stress} (~1.5x), the P enhancement is observed only at high V_{stress} (Fig. 4). These two experimental observations are discussed and interpreted individually in the following paragraphs.

978-1-4244-9113-1/11 $26.00 © 2011 IEEE

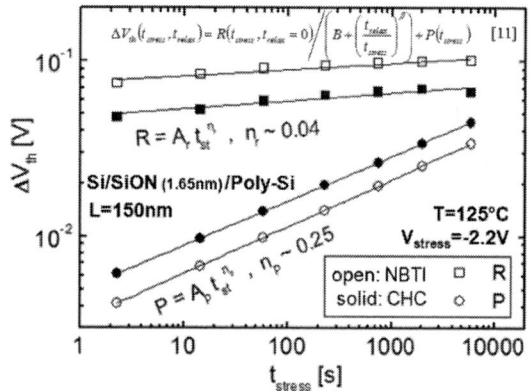

Figure 3. Si/SiON: CHC is observed to cause reduced R and enhanced P w.r.t NBTI. These two observations are attributed to a reduced E_{ox} at the drain side of the channel, and to an enhanced ΔN_{it}, respectively.

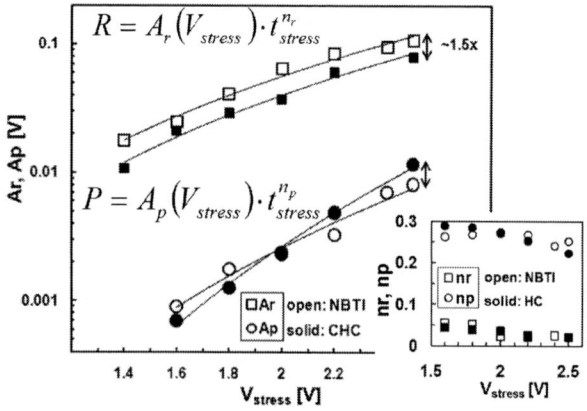

Figure 4. Power law pre-factors for CHC and NBTI: while the reduction of R (~1.5x) is constant for a wide range of V_{stress}, the P enhancement is observed only at high V_{stress}. The inset reports the extracted power law exponents.

A. Recoverable Component

The reduction of R can be related to the E_{ox} reduction at the drain side of the channel due to the high $|V_D|$, i.e. reduction of the *residual* NBTI effects ('*cold holes*') in that region. To support this hypothesis, the E_{ox} profile along the channel was simulated with MEDICI for both the CHC and NBTI stress conditions as reported in Fig. 5a. Using the experimental dependence of R on E_{ox} (inset of Fig. 5a, replotted from the data in Fig. 4), the E_{ox} profile can be converted into the local ΔV_{th} expected to be caused solely by the R component (ΔV_{th_R}) after a fixed stress time, and therefore into local trapped charge, ΔN_{ot} (Fig. 5b). While for the NBTI stress the ΔN_{ot} profile along the channel is constant, the profile for the HC case is decreasing almost linearly toward the drain. The ΔN_{ot} charge profiles are then inserted into the simulated Si/SiON/poly-Si device structures and I_DV_G curves are calculated (Fig. 6). In agreement with the experimental observation, the charge trapping (R) component during a NBTI stress is confirmed to cause increased ΔV_{th} w.r.t. a CHC stress, with the ratio of the two shifts well matching the experimentally observed ratio of the R components after NBTI

and CHC (1.87x vs. 1.5x in Fig. 4). We can therefore conclude that the reduction of the charge trapping component during CHC stress can be indeed explained by the reduction of the oxide electric field along the channel caused by the application of the drain stress bias.

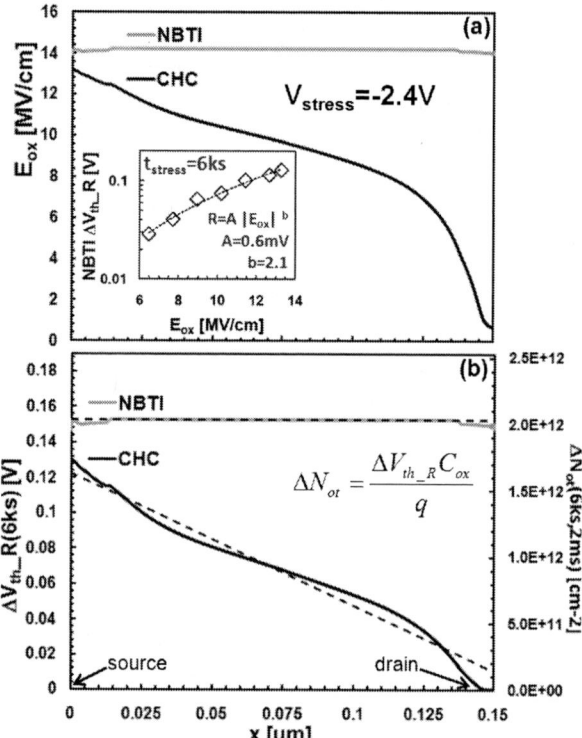

Figure 5. (a) E_{ox} profile along the channel from MEDICI simulations for the CHC and NBTI stress conditions (E_{ox} magnitude is slightly overestimated by neglecting poly depletion and quantum mechanical effects). Using the experimental dependence of R on E_{ox} (inset), the E_{ox} profile can be converted into (b) the local ΔV_{th} expected to be caused solely by the R component (ΔV_{th_R}), and therefore into the local ΔN_{ot} to be inserted into the simulated structure (see Fig. 6).

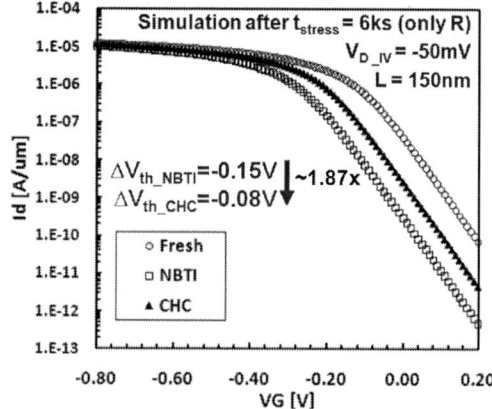

Figure 6. I_DV_G curves simulated in MEDICI before and after inserting the ΔN_{ot} charge profiles from Fig. 5b: in agreement with the experimental observation, CHC stress causes reduced ΔV_{th_R} w.r.t. NBTI.

978-1-4244-9113-1/11 $26.00 © 2011 IEEE

B. Permanent Component

The enhancement of the P component can be interpreted as enhanced ΔN_{it} due to *pure* HC effects (*'hot holes'*). To support this hypothesis, a substrate hot hole injection experiment [12] was performed: a diode next to the device was used to inject hot holes into the channel, while their energy was provided by applying a positive bias to the substrate of the pFET (Fig. 7). In this experiment, HC degradation is distributed uniformly over the whole channel length and, since no drain voltage is applied, E_{ox} is kept constant along the channel. Results in Fig. 8 for different injection current levels confirm that *pure* HC stress causes only *P* enhancement, while *R* is unaffected and caused by *residual* NBTI due to E_{ox}, as seen in the previous paragraph.

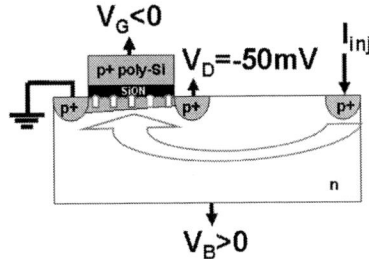

Figure 7. Setup used for the substrate hot carrier injection experiment. Using a p+ injector, hot holes are injected in the n-type bulk; a positive V_B provides potential to push holes toward the gate oxide. $V_D = 0V$ in this HC experiment. V_G stress is chosen as low as –0.8V.

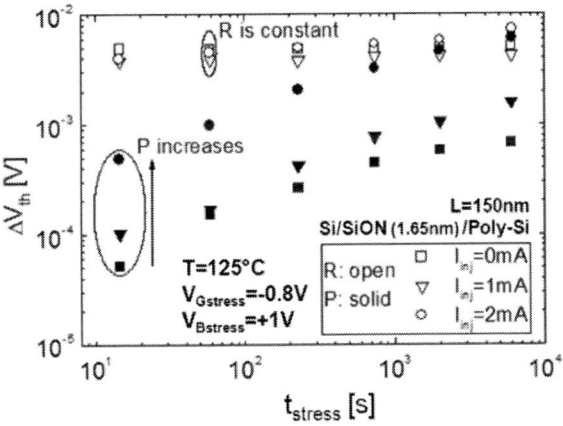

Figure 8. Si/SiON: substrate hot hole injection experiment confirms the increase of the *P* component to be due to *pure* HC effect.

The *P* enhancement during standard CHC stress can be attributed to increased N_{it} creation near the drain due to the peak in the lateral electric field, shown in Fig. 9 as from MEDICI simulations. To show this, lateral N_{it} profiling was performed with the charge pumping (CP) method of [13], where CP current (I_{CP}) contribution from the interface regions above the junctions is controlled by applying a reverse bias to source and drain. Results in Fig. 10 show that while NBTI stress creates N_{it} uniformly, CHC stress enhances N_{it} especially near the drain (higher slope of I_{CP} vs. reverse junction bias).

From the results discussed above we can conclude that the *residual* NBTI at the source side of the channel (*'cold holes'*)

strongly contributes to the total CHC degradation on pMOSFETs, while the *pure* HC effects (*'hot holes'*) are mainly responsible for the interface state creation at the drain side, as for nMOSFETs [1].

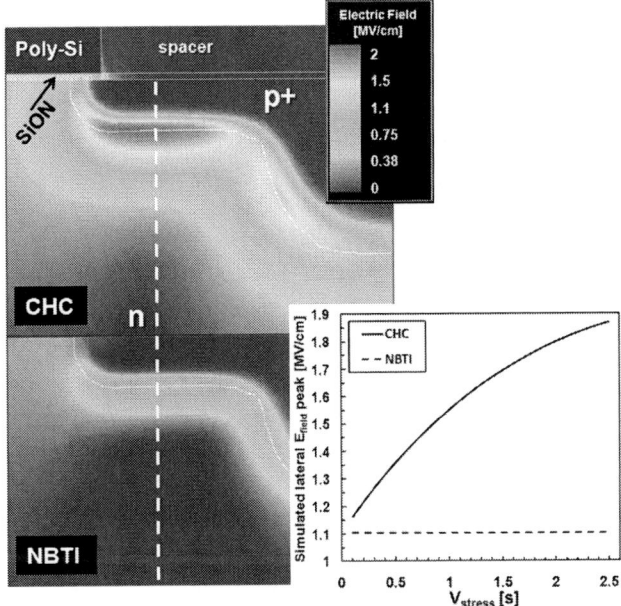

Figure 9. Electric field contour plot near the drain from MEDICI simulations for the NBTI and CHC stress conditions. The lateral field is responsible for the ΔN_{it} enhancement in the CHC case. Inset: the maximum of the lateral field as a function of V_{stress}.

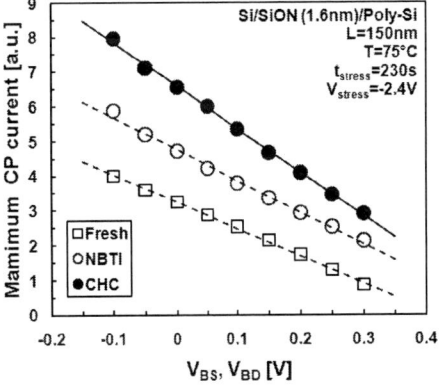

Figure 10. Lateral N_{it} profiling using the CP method proposed in [13]. Reverse junction bias reduces N_{it} contribution from interface regions above the junctions. CHC stress is shown to cause enhanced and more localized N_{it} (higher slope of I_{CP} vs. reverse junction bias)

IV. RESULTS ON SI/SIO₂/HIGH-K/MG

With *R* being initially the largest part of the total ΔV_{th} (as from Figs. 3 and 4), these results predict that for Si pMOSFETs CHC stress can eventually cause a reduced total ΔV_{th} w.r.t. NBTI for typical stress test durations, thanks to the discussed *R* reduction. Moreover *pure* HC degradation (*'channel hot holes'*) causes only moderate extra ΔN_{it} in Si pMOSFETs thanks to the large energy bandgap (E_G) of Si

978-1-4244-9113-1/11 $26.00 © 2011 IEEE

minimizing impact ionization and thanks to the high valence band offset between Si and SiO_2 [1]. This is shown in Fig. 11 also for a Si/SiO_2/high-k/metal gate (MG) device, confirming that CHC do not jeopardize the Si pMOSFET reliability, which is mainly limited by NBTI [14-16].

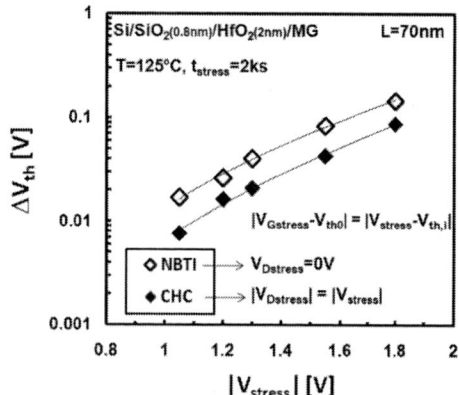

Figure 11. Si/high-k/MG: application of $V_{Dstress}$ on top of NBTI stress reduces total ΔV_{th} thanks to the reduction of R.

V. RESULTS ON $SI_{0.45}GE_{0.55}/SI/SIO_2$/HIGH-K/MG

Higher impact ionization in novel SiGe channel pFETs [7] can change this scenario. Fig. 12a shows the same comparison between NBTI and CHC on a $Si_{0.45}Ge_{0.55}$ device. For low stress voltages, a reduction of the total ΔV_{th} was still observed for the CHC case. Relying on the discussion in the previous sections, this observation can be interpreted as follows: in a regime for which R dominates over P (i.e., low stress voltages and/or short stress time), the R reduction obtained when applying also a drain stress bias directly reflects into a reduced total ΔV_{th}. On the other hand, for higher stress voltages, enhanced ΔN_{it} related to *pure* HC effects (*'hot holes'*), causes additional ΔV_{th} and therefore CHC degradation matches and eventually dominates over NBTI (triangles in Fig. 12a).

A comparison between CHC stress on SiGe and on Si devices with identical high-k/MG gate stacks is also undertaken. It should be noted that due to the known Si cap penalty [3], the SiGe device shows higher capacitance equivalent thickness in inversion (T_{inv}) w.r.t. its Si counterpart. Moreover incorporation of Ge reduces the initial V_{th0} [3]. For a fair comparison it is therefore necessary to *'match'* the stress conditions, artificially *'equalizing'* the V_{th0} to a target value of -0.3V ($V_{th,i} \approx V_{DD}/3$) and to rescale ΔV_{th} to equivalent total charge density ($\Delta N_{eff} = \Delta N_{ot} + \Delta N_{it} = \Delta V_{th} * C_{ox}/q$) in order to account for different C_{ox}. Similarly to Fig. 12a, results in Fig. 12b show that, for low stress voltages where R dominates, SiGe devices experience a reduced ΔV_{th} (confirming to suffer reduced charge trapping w.r.t. Si devices, [5-6]), while for high stress voltages enhanced ΔN_{it} from *pure* HC effects causes higher ΔV_{th} on SiGe.

We have recently shown that NBTI reliability of SiGe pFETs can be optimized by reducing the thickness of the Si

cap [5-6]. It is therefore interesting to study the impact of the Si cap thickness also on the CHC reliability. It is important to highlight once again that, with the Si cap thickness affecting both the T_{inv} and the V_{th0} of the SiGe devices, for a fair comparison it is necessary to carefully *'equalize'* the stress conditions as described above. Results in Fig. 13 show that a reduced Si cap thickness is also extremely beneficial against CHC degradation, remarkably reducing ΔN_{it}.

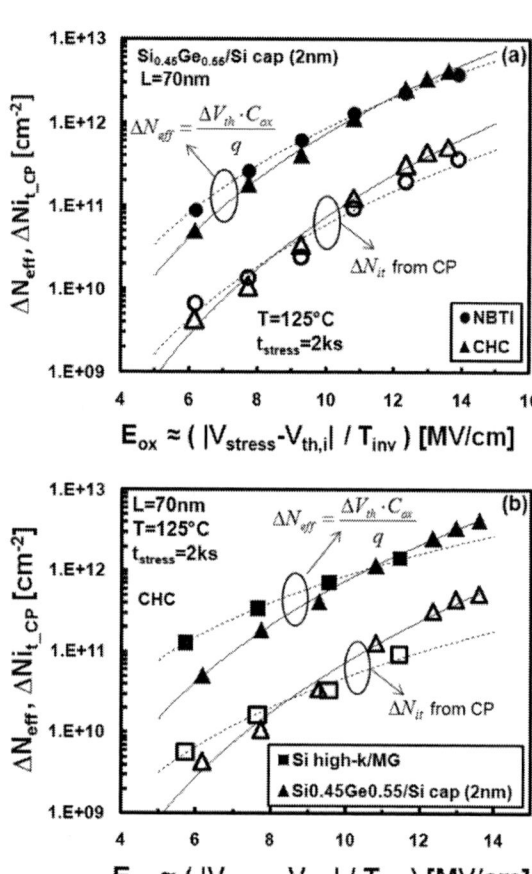

Figure 12. SiGe channel pFETs show higher ΔN_{it} during CHC stress w.r.t. (a) NBTI and w.r.t. (b) CHC stress on Si channel devices with same high-k/MG gate stack. Contrary to Si pFETs (see. Fig. 11), for SiGe devices, the total degradation caused by CHC matches and eventually dominates NBTI. This proves that HC degradation represents the most relevant concern for the reliability of this technology.

As for NBTI, a possible explanation is related to the higher Ge segregation observed with Secondary Ion Mass Spectroscopy (SIMS) at the Si cap/SiO_2 interface for thin Si cap [17,18]. As observed with Electron Spin Resonance (ESR) [19,20], a higher Ge content at the interface can lower the H-passivated Si dangling bonds density which are commonly considered as the interface state precursor defects [6]. Moreover, as one can notice in Fig. 13, a thin Si cap reduced also the ratio $\Delta N_{eff}/\Delta N_{it}$, revealing also a reduction of R. This reduced ΔN_{ot} can be related to a favorable alignment shift of the Fermi level in the SiGe channel w.r.t. the valence band edge of the Si cap at the oxide interface which can reduce the carrier interaction with N_{ot} as we discussed in [6].

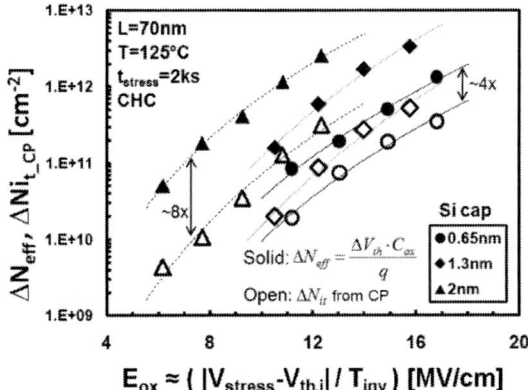

Figure 13. A thin Si cap on a SiGe pFET reduces both ΔN_{it} and ΔN_{ot} (reduced $\Delta N_{eff}/\Delta N_{it}$). This can be explained with reduced N_{it} precursor defect density due to high Ge segregation and reduced interaction with N_{ot} thanks to energy decoupling [6].

VI. CONCLUSIONS

The introduction of SiGe channel devices makes it necessary to study Hot Carrier (HC) degradation also for the pMOSFET reliability. A channel HC (CHC) degradation study on Si pMOSFETs was presented and compared to standard NBTI. The CHC stress condition was shown to reduce the charge trapping component of the degradation (the 'recoverable' component, ΔN_{ot}) due to reduced oxide electric field at the drain side of the channel (i.e., reduced 'cold hole injection'). On the other hand, similarly to nMOSFETs, CHC is shown to enhance the interface state creation (the 'permanent' component, ΔN_{it}) at the drain side of the channel due to the high lateral electric field (i.e., 'hot hole injection').

These experimental results were well supported by MEDICI simulations. Since ΔN_{ot} is the main contribution to the total degradation, and since ΔN_{ot} is significantly reduced for a CHC stress w.r.t. NBTI, these results confirm that CHC do not limit the Si pMOSFET reliability.

However, $Si_{0.45}Ge_{0.55}$ pMOSFETs show enhanced CHC degradation w.r.t. their Si counterparts, confirming the importance of studying HC effects for the reliability of this technology. Nevertheless, the SiGe device reliability can be enhanced when reducing the thickness of the Si cap. Such optimization is shown to reduce both ΔN_{it} and ΔN_{ot}.

ACKNOWLEDGMENT

The imec core partners, the imec pilot line, and Amsimec are acknowledged for their support. We also gratefully acknowledge Profs. A. Stesmans and V. Afanas'ev (Physics and Astronomy Dept., University of Leuven) for useful discussions.

REFERENCES

[1] D. Vuillaume, "Hot carrier injections in SIO_2 and related instabilities in submicrometer mosfets", in *Instabilities in Silicon Devices*", edited by

G. Barbottin and A. Vapaille, Elsevier, Amsterdam, The Netherlands, 1999, Vol. 3, pp. 265–339.

[2] V. Huard, M. Denais, C. Parthasarathy, "NBTI degradation: from physical mechanism to modeling", in Micr. Rel., Vol. 46, No. 1, pp. 1-23, Jan. 2006.

[3] L. Witters et al., "8Å Tinv gate-first dual channel technology achieving low-V_t high performance CMOS", in *Proc.* Symp. on VLSI Technology, pp. 181-182, 2010.

[4] B. Kaczer, J. Franco, J. Mitard, Ph. J. Roussel, A. Veloso and G. Groeseneken, "Improvement in NBTI reliability of Si-passivated Ge/high-k/metal-gate pFETs", in Micr. Eng., Vol. 86, No. 7-9, pp.1582-1584, July-Sept. 2009 (*INFOS 2009*).

[5] J. Franco, B. Kaczer, M. Cho, G. Eneman, T. Grasser and G. Groeseneken, "Improvements of NBTI reliability in SiGe p-FETs", in *Proc.* IRPS, pp. 1082-1085, 2010.

[6] J. Franco et al., "6Å EOT $Si_{0.45}Ge_{0.55}$ pMOSFET with Optimized Reliability (V_{DD}=1V): Meeting the NBTI Lifetime Target at Ultra-Thin EOT", in *Proc.* IEDM, pp. 70-73 , 2010.

[7] D. Maji et al., "Understanding and Optimization of Hot-Carrier Reliability in Germanium-on-Silicon pMOSFETs", IEEE Trans. Electron Devices, Vol. 56, No. 5, pp. 1063-1069, May 2009.

[8] W.-Y Loh et al., "The effects of Ge composition and Si cap thickness on hot carrier reliability of $Si/Si_{1-x}Ge_x/Si$ p-MOSFETs with high-K/metal gate", in *Proc.* Symp. on VLSI Technology, pp. 56-57, 2008.

[9] J. Franco, G. Eneman, B. Kaczer, J. Mitard, B. De Jaeger, and G. Groeseneken, "Impact of halo implant on the hot carrier reliability of germanium p-channel metal-oxide-semiconductor field-effect transistors", J. Vac. Sci. Technol. B Vol. 29, No. 01A804, pp. 1-4, Jan. 2011 (WoDiM 2010).

[10] B. Kaczer et al., "Ubiquitous relaxation in BTI stressing – new evaluation and insights", in *Proc.* IRPS, pp. 20-27, 2008.

[11] T. Grasser et al., "Simultaneous extraction of recoverable and permanent components contributing to Bias-Temperature Instability", in *Proc.* IEDM, pp. 801-804, 2007.

[12] A. Teramoto, R. Kuroda and T. Ohmi, "NBTI Mechanism Based on Hole-Injection for Accurate Lifetime Prediction", in ECS Transactions, Vol. 6, No. 3, pp. 229-243, 2007.

[13] M. G. Ancona, N. S. Saks, D. McCarthy, "Lateral distribution of hot-carrier-induced interface traps in MOSFETs", IEEE Trans. Electron Devices Vol. 35, No. 12, pp. 2221-2228, Dec. 1988.

[14] R. Mishra, S. Mitra, R. Gauthier, D. E. Ioannou, D. Kontos, K. Chatty, C. Seguin, R. Halbach, "On the interaction of ESD, NBTI and HCI in 65nm Technology", in *Proc.* IRPS, pp. 17-22, 2007.

[15] C. Guerin, V. Huard, A. Bravaix, M. Denais, J. M. Roux, F. Perrier, W. Baks, "Combined effect of NBTI and Channel Hot Carrier effects in pMOSFETs", in *Proc.* IIRW, pp. 10-16, 2005.

[16] C.-H. Jeon, S.-Y. Kim, C.-B. Rim, "The impact of NBTI and HCI on deep sub-micron pMOSFETs' lifetime", in *Proc.* IIRW, pp. 130-132, 2002.

[17] M. Caymax et al., "The influence of the epitaxial growth process parameters on layer characteristics and device performance in Si-passivated Ge pMOSFETs", J. Electrochem. Soc., Vol. 156, No. 12, pp. H979-H985, 2009.

[18] B. Vincent, W. Vandervorst, M. Caymax, and R. Loo, "Influence of Si precursor on Ge segregation during ultrathin Si reduced pressure chemical vapor deposition on Ge", Appl. Phys. Lett., Vol. 95, No. 26, Dec. 2009.

[19] A. Stesmans and V. Afanas'ev, "ESR of interfaces and nanolayers in semiconductor heterostructures", in "Characterization of Semiconductor Heterostructures and Nanostructures", edited by C. Lamberti, Elsevier, pp.435-482, Jun. 2008.

[20] J. Franco et al., "Impact of Si-passivation thickness and processing on NBTI reliability of Ge and SiGe pMOSFETs", as discussed at IEEE SISC, Washington - DC, December 2009.

Impact of HK / MG stacks and Future Device Scaling on RTN

Naoki Tega, Hiroshi Miki*, Zhibin Ren**, Christoper P. D'Emic**, Yu Zhu**, David J. Frank**,
Michael A. Guillorn**, Dae-Gyu Park**, Wilfried Haensch**, and Kazuyoshi Torii

Central Research Laboratory, Hitachi Ltd.
1-280, Higashi-koigakubo Kokubunji-shi, Tokyo, 185-8601, Japan
*Semiconductor Innovation Research Project, Hitachi America, Ltd.
**T. J. Watson Research Center, IBM
Phone: +81-42-323-1111, E-mail address: naoki.tega.ub@hitachi.com

Abstract—**This work demonstrates the close relationship between device scaling and the threshold voltage variation (ΔV_{th}) of random telegraph noise (RTN) in high-κ and metal gate (HK / MG) stacks. Statistical analysis clarifies that high temperature forming gas annealing can suppress the RTN ΔV_{th}. And properly annealed HK FETs have smaller RTN ΔV_{th} than SiON FETs, due mostly to fewer traps and partly to thinner inversion thickness in HK / MG stacks. Consequently, the influence of RTN on HK / MG gate stacks is less than that of random dopant fluctuation in the 22 nm generation. However, RTN may pose a difficult challenge for the 15 nm generation. In addition to the scaling dependence, we also find that characterizing hysteretic RTN behaviors due to RTN dependence on bias is essential to determine whether the observed RTN has an impact on SRAM operation or not.**

Keywords; random telegraph noise; RTN; noise; variability; variation; MOSFET; scaling; high-κ / metal-gate stacks; HK / MG

I. INTRODUCTION

Variation due to random telegraph noise (RTN) is becoming a new threat to the future devices because threshold voltage variation (ΔV_{th}) of RTN rapidly rises with scaling, just as variation due to random dopant fluctuation (RDF) [1–3]. The distinctive features of RTN are size dependence and statistical distribution of RTN ΔV_{th}. The size dependence of RTN is relatively stronger than RDF. Moreover RTN has a long-tailed non-Gaussian distribution as log normal, whereas RDF has a Gaussian distribution. Therefore RTN ΔV_{th} could exceed that of RDF in the tail of the distribution, in especially advanced devices. RTN is one of the newly envisioned challenges to SRAM stability at 15 nm and beyond.

High-κ / metal-gate (HK / MG) stacks will be incorporated in these generations as standard technique because of their more optimal power and performance compared to SiON / Poly-Si gate stack [4,5]. Study on the reliability of HK / MG gate stacks has made progress in recent years. However, some challenges in HK interface structure still remain, e. g. bias temperature instability (BTI) and time-dependent-dielectric breakdown [6, 7]. The relationship between RTN on BTI has recently received attention [8, 9].

Because RTN behaviors depend on time and bias [10], the specific RTN with large amplitude does not always cause error in device operation, while the specific device with large shift from optimal V_{th} due to RDF always becomes an error bit. However, there are few reports that investigate the interaction of device transients and RTN [11-13], despite its significance.

This work evaluates the RTN impact on HK / MG FETs and future device scaling from a statistical viewpoint and demonstrates the interaction of device transients and RTN, which is caused by RTN dependence on gate bias (V_g).

II. DEVICE FABRICATION AND MEASUREMENT

A. Device information

In this work, mixed e-beam / optical processing has been used to fabricate small devices (nFETs) with gate length (L_g) from 20 to 90 nm and width (W_g) from 25 to 180 nm. And these small devices incorporate HfO$_2$-based HK / MG stacks. For a comparison, devices with SiO$_2$ and SiON / Poly-Si gate stacks were also fabricated. The HK FETs have smaller T_{inv} and larger transconductance (G_m) than the pure-SiO$_2$ and SiON FETs, as shown in Table 1. To evaluate the effect of hydrogen passivation of interface traps [14], high temperature forming gas annealing (HTFGA) was performed on some HK / MG FETs. The annealing temperature of HTFGA is 475 °C. HTFGA is expected to passivate RTN traps and suppress RTN ΔV_{th}. Because RTN has a long-tailed non-Gaussian distribution, statistical analysis of many FETs is vital. As shown in Fig. 1, test array structures were designed for easy measurement of large numbers of devices (27000 / die) to enable statistical analysis of RTN ΔV_{th} variation.

TABLE I. FEATURES OF nFETs.

Gate dielectric	SiON	High-κ
Gate	Poly Si	Metal Gate
Split N conditions	With or W/O	
Split HTFGA		With or W/O
T_{inv} (nm)	1.83(SiON)	1.49
G_m	Small	Large

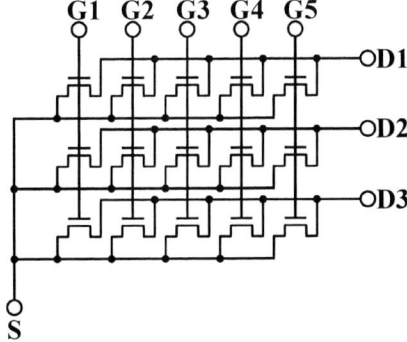

(a) Circuit schematic of array.

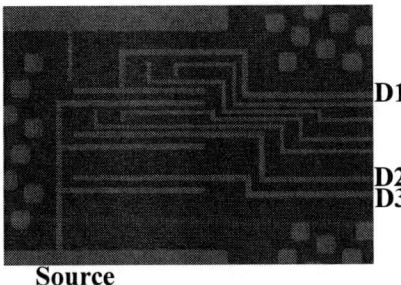

(b) SEM top view

Figure 1. Terminal of array: A matrix has 15 same-size FETs

B. Measurement and analysis methods

A fast measurement unit (Agilent 1530A) which enables a wide band width of up to 1 M / s was used to measure the RTN signals. Figure 2 shows the typical RTN dependence on time obtained by high sampling rate 1 M / s. 10^5 sampling points were used for the measurement. A sampling interval was 1 μs. Drain current (I_d) fluctuation was measured under 50 mV of drain voltage. This RTN behavior has two obvious states due to trapping and detrapping of a carrier at a single trap in HK. Figure 3 illustrates the histogram separation method. RTN I_d variation (ΔI_d) is defined as the peak-to-peak value in this figure. Other noise components are removed from the RTN component by this extraction. RTN variation is transformed from ΔI_d to the input-referred RTN voltage noise (ΔV_{th}) using G_m, in addition to the separation.

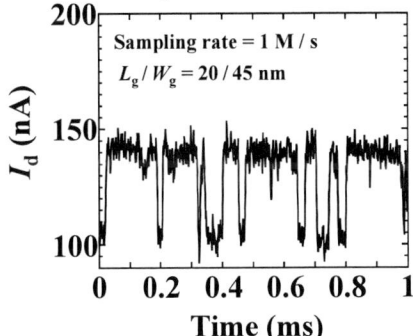

Figure 2. Typical RTN dependence on time.

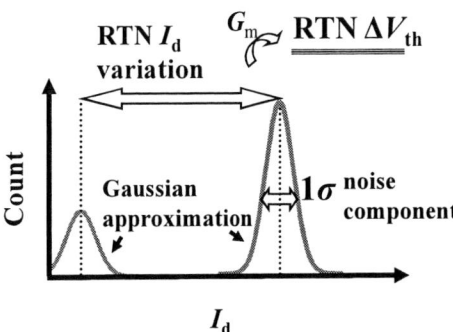

Figure 3. Separation between RTN and other noise components

III. INHIBITORY EFFECT OF HK / MG TECHNIQUES ON RTN

HTFGA is expected to passivate RTN traps and suppress RTN ΔV_{th}. Figure 4 demonstrates the effect of HTFGA on the distributions of RTN ΔV_{th} in 25 nm FETs. As expected, RTN is well suppressed by HTFGA. For instance, Figure 5 shows the samples with and without HTFGA at the 2σ level. The RTN magnitude at the 2σ level for these two cases differs by ~2 times because hydrogen is capable of passivating traps in HK.

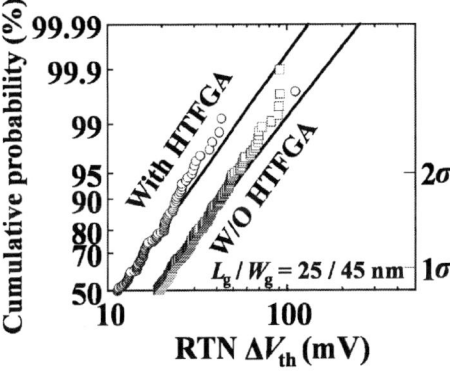

Figure 4. Comparison of RTN in 25 nm FETs with or without HTFGA.

Figure 6 compares RTN ΔV_{th} between HTFGA HK, pure-SiO₂ and SiON FETs. RTN ΔV_{th} of HK and pure-SiO₂ FETs are quite comparable. On the other hand, both the median value and the variation of HK FETs are obviously smaller than those of SiON FETs. The main reason is that trap density in our optimized HK FETs becomes lower because of HTFGA. A secondary reason is that the T_{inv} of HK FETs is 25% smaller than that of SiON FETs. Optimizing HK / MG gate stacks is useful for not only the RDF variation but also the RTN variation in the advanced FETs.

978-1-4244-9113-1/11 $26.00 © 2011 IEEE

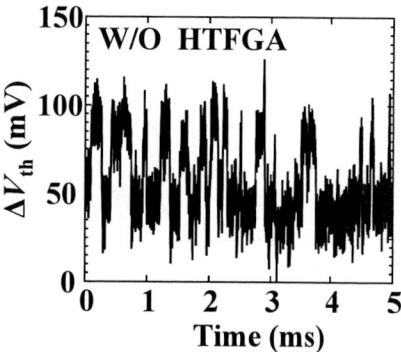

(a) Without HTFGA

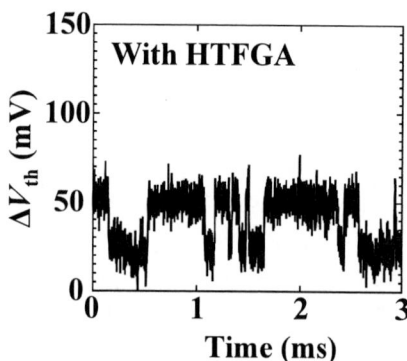

(b) With HTFGA

Figure 5. RTNs at 2σ level in 25 nm FET.

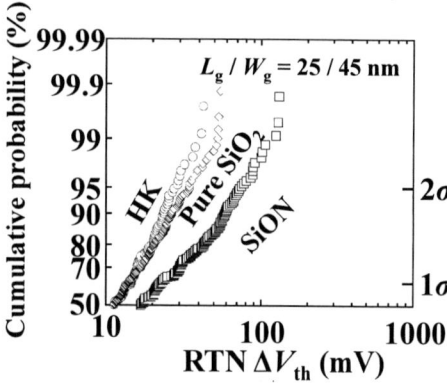

Figure 6. Comparison of RTN between HK with HTFGA, pure-SiO$_2$, and SiON FETs.

IV. IMPACT OF DEVICE SCALING ON RTN

Figure 7 indicates the cumulative distribution of RTN ΔV_{th} of HK / MG FETs with strong device-size dependence. All measured devices were fabricated on same wafer, so process variation is suppressed as much as possible to evaluate only the

impact of scaling on RTN. The RTN distributions clearly show the non-Gaussian like log normal. The RTN amplitudes at high cumulative probability are of greatest importance. The FET with L_g and W_g = 90 / 180 nm has small ΔV_{th} of approximately 4mV at 2 σ. ΔV_{th} of the FET with 45 / 90 nm is more than 20 mV. And the smallest device with 25 / 45 nm has approximately 40 mV of ΔV_{th}. Thus, the variation increases with as the device scales down.

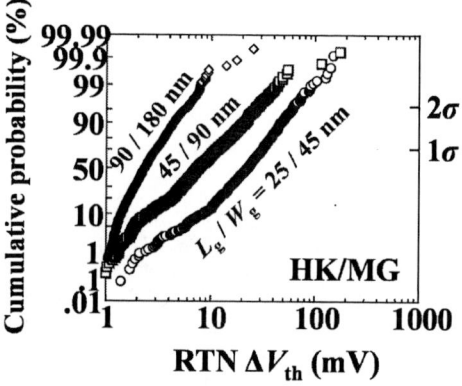

Figure 7. RTN ΔV_{th} distributions.

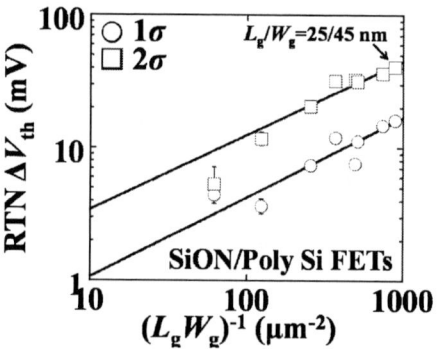

(a) SiON / poly-Si FETs

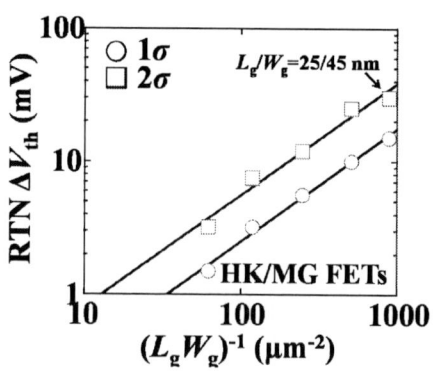

(b) HK / MG FETs

Figure 8. Device-size dependences of RTN ΔV_{th}.

The comparison of the device-size dependence of ΔV_{th} between SiON and HK FETs is seen in Figs. 8. RTN ΔV_{th} steeply increases with device size in both HK FETs and SiON FETs. However, the RTN V_{th} variations of HK FETs are smaller than these of SiON FETs anywhere in the generations. In case of HK FETs, the power law exponent of 0.8 is less than conventional RTN theory (~1). Other factors, e. g. current percolation path, may be associated with the small power law exponent [15]. On the other hand, the RTN power law exponent of 0.8 is much larger than the 0.5 that is generally observed for RDF [16].

RTN has a log-normal distribution, while RDF has been shown to be Gaussian out to at least 5 σ [17]. This work compares RTN with RDF in HK FET. 22 nm generation RTN V_{th} variations exceed RDF V_{th} variations at the 3 sigma level in the SiON FET, while the cross point can be extended to more than 5 sigma level in the case of the HK FET. We therefore conclude that the influence of RTN is less than that of RDF in the 22 nm generation. Considering the size dependence of RTN as shown in Fig. 9, however, RTN may pose a difficult challenge for the 15 nm generation.

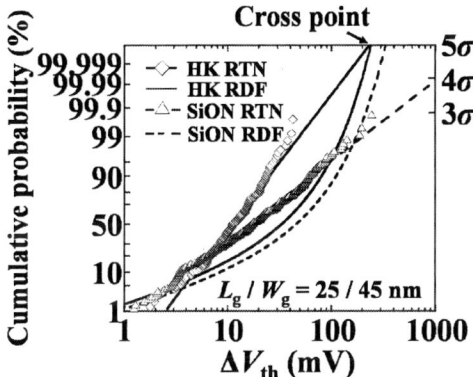

Figure 9. Comparison between RTN and RDF impacts. * RDF data are based on T. Tunomura, et al., VLSI, P. 156, 2008.

V. RTN DEPENDENCE ON V_G AND HYSTERETIC BEHAVIORS

One of the curious RTN behaviors is a time-constant dependence on bias. Our system measured time series of I_d around V_{th} under the constant gate-voltage (V_g) step of 10 mV continuously. Figure 10 shows an example of the V_g dependence of RTN and includes 10 time-series data at $V_g - V_{th}$ from -80 to 10 mV. Figures 11 are extracted time-series data from Fig. 10. The upper state indicates duration time to capture (τ_c), and likewise the lower state indicates duration time to emission (τ_e) as shown in Fig. 11–(a). Figure 11–(a) is the first time series of I_d at $V_g - V_{th} = -80$ mV and has shorter average duration time to emission $<\tau_e>$ than average duration time to capture $<\tau_c>$ clearly. The ratio of duration times $<\tau_e> / <\tau_c>$ is 2.6×10^{-1}. These values change with V_g. Figure 11–(b) is the third time series of I_d at $V_g - V_{th} = -60$ mV. $<\tau_c>$ almost becomes nearly equal to $<\tau_e>$. $<\tau_e> / <\tau_c>$ becomes 7.0×10^{-1}. Furthermore, the gap of the duration times expands with increase of V_g. Figure 11–(c) is the tenth time series of I_d at $V_g - V_{th} = 10$ mV. $<\tau_e> / <\tau_c>$ becomes 2.6×10^{1}.

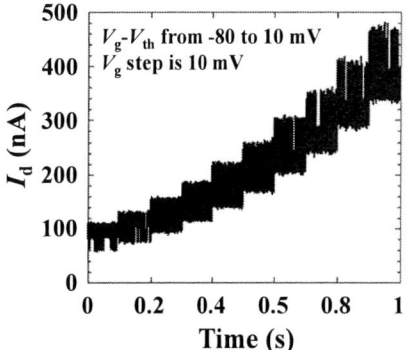

Figure 10. V_g dependence of RTN.

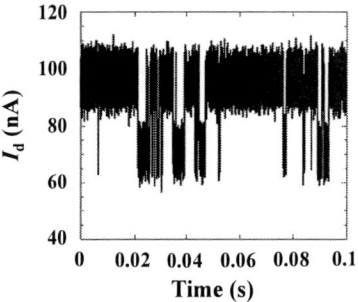

(a) First time series @ $V_g - V_{th} = -80$mV. $<\tau_e>$ is 1.1×10^{-3} s, $<\tau_c>$ is 4.1×10^{-3} s, and $<\tau_e> / <\tau_c>$ is 2.6×10^{-1}

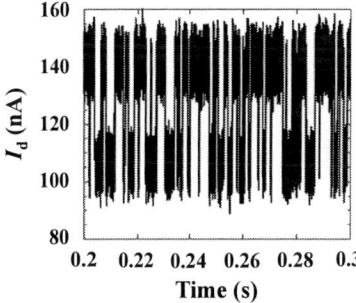

(b) Third time series @ $V_g - V_{th} = -60$mV. $<\tau_e>$ is 1.1×10^{-3} s, $<\tau_c>$ is 1.5×10^{-3} s, and $<\tau_e> / <\tau_c>$ is 7.0×10^{-1}.

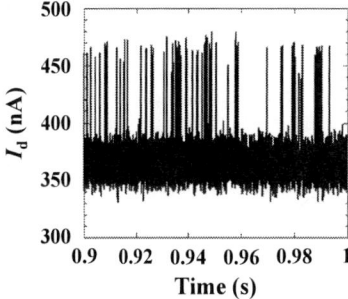

(c) Tenth time series @ $V_g - V_{th} = 10$mV. $<\tau_e>$ is 1.6×10^{-3} s, $<\tau_c>$ is 6.3×10^{-5} s, and $<\tau_e> / <\tau_c>$ is 2.6×10^{1}.

Figure. 11 Each time series from Fig. 10.

The above demonstrations indicate that RTN frequently has a strong V_g dependence. Figure 12 demonstrates the clear V_g dependence of $<\tau_e> / <\tau_c>$ extracted from time series data of Fig. 10. The slope of ln ($<\tau_e> / <\tau_c>$), so called bias sensitivity of RTN, is 1.35 in this case. This is larger than expected and the physical mechanism causing larger bias sensitivity is not yet clear [10]. Figure 13 is a cumulative probability of bias sensitivity. The figure obviously shows that RTN behaviors often depend on bias from the statistical view.

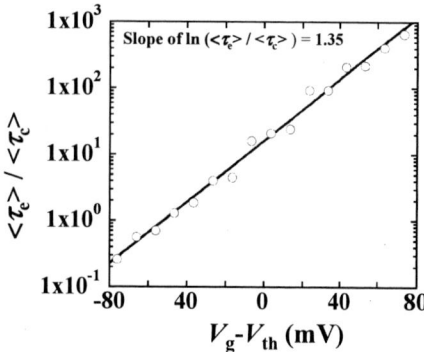

Figure 12. Strong V_g dependence.

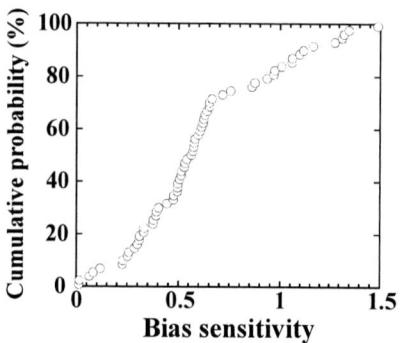

Figure 13. Distribution of V_g sensitivity.

From the above discussion on RTN dependence on V_g, RTN is expected to show transient behavior during device operation [18]. Figure 14 illustrates a pulse sequence to evaluate hysteretic behavior of RTN. The bias is repeatedly applied to one FET. First, the FET is biased Off (V_{set}, 0 V) for time t_{set}, then it is biased On (V_{msr}, 0.7 V) for time t_{on}, the period during which I_d is sampled. Because the transients are stochastic, the Off / On sequence is repeated 256 times for statistics. The measurement result is shown in Fig. 15 using t_{set} = 10 ms. The upper state is observed immediately after the turn-on transient and sustained for several ms in all series in Fig. 15. The hysteretic behavior is a consequence of capture of a single electron at an RTN trap. The hysteresis effect disappears when the initial settling is too short to empty the trap. As a result, some RTN behaviors are extremely sensitive

to change in bias, and probably the hysteric behaviors occur during SRAM operation.

This effect is missing from recent RTN compact modeling, but it is needed to properly assess RTN impact on SRAM operation.

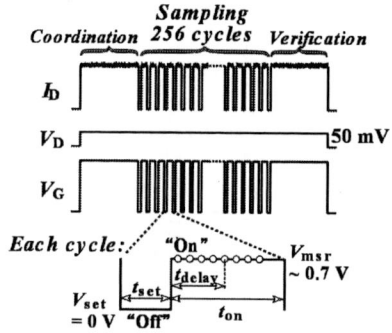

Figure 14. Pulse sequence to evaluate hysteretic behavior.

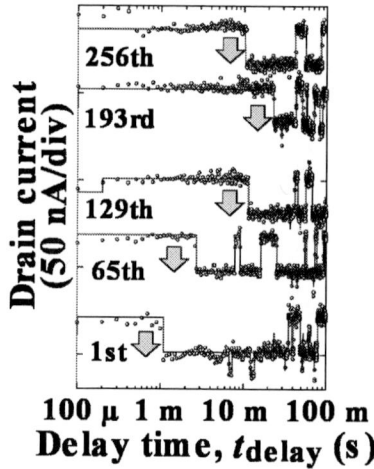

Figure 15. Typical waveforms of turn-on transient of one nFET. Waveforms are offset for clarity.

VI. CONCLUSIONS

The RTN V_{th} variation in HK FET can be suppressed by suitable annealing, such as HTFGA, and by thin T_{inv} and large G_m. As a consequence, properly annealed HK FETs can have smaller RTN variation than SiON FETs. The RTN ΔV_{th} dependence on scaling was demonstrated using both SiON FETs and HK FETs. RTN impact may, however, become severe in 15 nm generation and beyond because of the rapid dependence of RTN on size and because even though HK appears to offer lower RTN, the dimensions will be so small that RTN will become large. We should pay much attention to the RTN hysteretic behavior because the RTN behaviors with the bias dependence are inseparably related to SRAM operation.

ACKNOWLEDGMENT

We gratefully acknowledge the efforts of the staff of the IBM MRL facility, where the samples were fabricated.

REFERENCES

[1] N. Tega, H. Miki, F. Pagette, D. J. Frank, A. Ray, M. J. Rooks, W. Haensch, and K. Torii "Increasing Threshold Votage Variation due to Random Telegraph Noise in FETs as Gate Lengths Scale to 20 nm," VLSI Tech., 2009, pp.51–52.

[2] K. Takeuchi, T. Nagumo, S. Yokogawa, K. Imai, and Y. Hayashi "Single-Charge-Based Modeling of Transistor Characteristics Fluctuations Based on Statistical Measurement of RTN Amplitude," VLSI Tech., 2009, pp.54–55.

[3] N. Tega, H. Miki, Z. Ren, C. P. D'Emic, Y. Zhu, D. J. Frank, J. Cai, M. A. Guillorn, D.-G. Park, W. Haensch, and K. Torii "Reduction of Random Telegraph Noise in High-κ / Metal-gate Stacks for 22 nmGeneration FETs," IEDM Tech. Dig., 2009, pp. 771–774.

[4] C. H. Diaz, K. Goto, H.T. Huang, Y. Yasuda, C.P. Tsao, T.T. Chu, W.T. Lu, Vincent Chang, Y.T. Hou, Y.S. Chao, P.F. Hsu, C.L. Chen, K.C. Lin, J.A. Ng, W.C. Yang, C.H. Chen, Y.H. Peng, C.J. Chen(Ryan), C.C. Chen, M..H. Yu, L.Y. Yeh, K.S. You, K.S. Chen, K.B. Thei, C.H. Lee, S.H. Yang, J.Y. Cheng, K.T. Huang, J.J. Liaw, Y. Ku, S.M. Jang, H. Chuang, and M.S. Liang "32nm Gate-First High-k/Metal-Gate Technology for High Performance Low Power Applications" IEDM Tech. Dig., 2008, pp. 629–632.

[5] C.-H. Jan, M. Agostinelli, H. Deshpande, M. A. El-Tanani, W. Hafez, U. Jalan, L. Janbay, M. Kang, H. Lakdawala*, J. Lin, Y-L Lu, S. Mudanai, J. Park, A. Rahman, J. Rizk, W.-K. Shin, K. Soumyanath*, H. Tashiro, C. Tsai, P. VanDerVoorn, J.-Y. Yeh, and P. Bai "RF CMOS Technology Scaling in High-k/Metal Gate Era for RF SoC (System-on-Chip) Applications, " IEDM Tech. Dig., 2010, pp. 604–607.

[6] K. Zhao, J. H. Stathis, A. Kerber, and E. Cartier, "PBTI RELAXATION DYNAMICS AFTER AC VS. DC STRESS IN HIGH-K/METAL GATE STACKS, " IRPS Proc. , 2010, pp. 50–54.

[7] K.L. Pey, N. Raghavan, X. Li, W.H. Li, K. Shubhakar, X. Wu, and M. Bosman "New Insight into the TDDB and Post Breakdown Reliability of Novel High-κ Gate Dielectric Stacks," IRPS Proc. , 2010, pp. 354–363.

[8] H. Aono, E. Murakami, K. Shiga, F. Fujita, S. Yamamoto, M. Ogasawara, Y. Yamaguchi, K. Yanagisawa, and K. Kubota "A STUDYOF SRAM NBTI BY OTF MEASUREMENT," IRPS Proc., 2008, pp. 67–71.

[9] T. Grasser, B. Kaczer, W. Goes, H. Reisinger, Th. Aichinger, Ph. Hehenberger, P.-J. Wagner, F. Schanovsky, J. Franco, Ph. Roussel, and M. Nelhiebel "Recent Advances in Understanding the Bias Temperature Instability," IEDM Tech. Dig., 2010, pp. 82–85.

[10] N. Tega, H. Miki, Z. Ren, C. P. D'Emic, Y. Zhu, D. J. Frank, M. A. Guillorn, D.-G. Park, W. Haensch, and K. Torii "On the need for a new model: Inconsistencies between observations and physical model for random telegraph noise in HKMG MOSFET" IWDTF proc., 2011, pp. 153–154.

[11] N. Zanolla, D. Siparak, M. Tiebout, P. Baumgartner, E. Sangiorgi, and C. Fiegna, "Reduction of RTS noise in small-area MOSFETs under switched bias conditions and forward substrate bias," IEEE Trans. Electron Devices, vol. 57, no. 5, 2010, pp. 1119–1128.

[12] J. S. Kolhatkar, E. Hoekstra, C. Salm, A. P. van der Wel, E. A. M. Klumperink, J. Schmitz, and H. Wallinga, "Modeling of RTS noise in MOSFETs under steady-state and large signal excitation," IEDM Tech. Dig., 2004, p. 759–762.

[13] B. Dierickx and E. Simoen, "The decrease of "random telegraph signal" noise in meta-oxide-semiconductor field-effect transistors when cycled from inversion to accumulation," J. Appl. Phys., vol. 71, no. 4, 2009, pp. 2028–2029.

[14] K. Onishi, C. S. Kang, R. Choi, H.-J. Cho, S. Gopalan, R. Nieh, S. Krishnan, and J. C. Lee "Effects of High-Temperature Forming Gas Anneal on HfO₂ MOSFET Performance," VLSI Tech., 2002, p.23–24.

[15] A. Ghetti, C. Monzio Compagnoni, F. Biancardi, A. L. Lacaita, S. Beltrami, L. Chiavarone, A. S. Spinelli, A. Visconti "Scaling trends for random telegraph noise in deca-nanometer Flash memories," IEDM Tech. Dig., 2008, pp. 835–838.

[16] H. S. Yang, R. Wong, R. Hasumi, Y. Gao, N. S. Kim, D. H. Lee, S. Badrudduza, D. Nair, M. Ostermayr, H. Kang, H. Zhuang, J. Li, L. Kang, X. Chen, A. Thean, F. Arnaud, L. Zhuang, C. Schiller, D. P. Sun, Y. W. Teh, J. Wallner, Y. Takasu, K. Stein, S. Samavedam, D. Jaeger, C. V. Baiocco, M. Sherony, M. Khare, C.Lage, J. Pape, J.Sudijono, A. L. Steegen, and S. Stiffler "Scaling of 32nm Low Power SRAM with High-K Metal Gate," IEDM Tech. Dig., 2008, pp. 233–236.

[17] T. Tsunomura, A. Nishida, F. Yano, A. T. Putra, K. Takeuchi, S. Inaba, S. Kamohara, K. Terada, T. Hiramoto, and T. Mogami "Analyses of 5σ V th Fluctuation in 65nm-MOSFETs Using Takeuchi Plot," VLSI Tech., 2008, pp.156–157.

[18] H. Miki, N. Tega, Z. Ren, C. P. D'Emic, Y. Zhu, D. J. Frank, J. Cai, M. A. Guillorn, D.-G. Park, W. Haensch, and K. Torii "Hysteretic Drain-Current Behavior Due To Random Telegraph Noise in Scaled-down FETs with High-κ/Metal-gate Stacks, " IEDM Tech. Dig., 2010, pp. 620–623.

Split-Gate Flash Memory for Automotive Embedded Applications

Y.S. Chu, Y.H. Wang, C.Y. Wang, Y.H. Lee, A.C. Kang, R. Ranjan, W.T. Chu, T.C. Ong, H.W. Chin, K. Wu
Technology Quality and Reliability Division
Taiwan Semiconductor Manufacturing Company
Hsinchu, Taiwan 30077, R.O.C.
TEL: 886-3-5636688 ext7022045, e-mail: cywangzh@tsmc.com

Abstract—An embedded split-gate flash memory based on 65nm logic process technology has been developed. The design rules for split-gate flash macro's testability and reliability are discussed. An automotive grade flash memory with 100K endurance, 10 years, 125°C data retention, and 1-ppm requirement has been demonstrated with a comprehensive dielectric screen methodology. Both erase time push out and data retention dominant mechanisms are thoroughly studied with intrinsic lifetime and large sample certification. An automotive embedded split-gate flash solution in 65nm technology is ready for commercialization.

Keywords- split-gate memory; endurance; erase time push out; data retention;

I. INTRODUCTION

A split-gate flash memory, which utilizes source-side injection (SSI) program and poly-poly Fowler-Nordheim (FN) erase with lower power consumption, higher programming efficiency, and immunity of over program and over erase [1], has been processed with compatible CMOS logic technology for embedded or automotive solutions across several generations [2]. Here a new 65nm split-gate flash is developed through structure modification with multiple floating gate (FG) dielectrics [3], which is different from sub-micron source coupling split-gate flash [1]. In comparison with the stack gate flash, which has scaling limitation due to gate control oxide acting as tunnel oxide and drain turn-on program disturb [4], the benefits of this split-gate flash are good scalability with erase tunnel oxide decoupled from FG gate oxide and good program disturb immunity with the select gate (SG) to shut off the channel of unselected cells [3]. In addition, stack gate flash has boosted word line (WL) voltage during read access, which results in higher power consumption and access latency.

In this paper, we discuss the dominant mechanisms of endurance and data retention due to the trap generation inside dielectrics. The traps are created by program/erase cycling, which results in erase time push out (ETPO). By plotting the relationship between cycling and ETPO, flash endurance lifetime can be well predicted. Rigorous experiments are also carried out to demonstrate the dominant data loss mechanisms. Since some dielectrics of this split gate cell are applied by high voltage stress, dielectric screen methodology is utilized to meet stringent reliability requirement of 100K program/erase cycles and 125°C 10 years data storage.

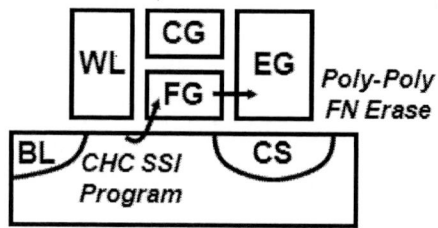

Fig. 1. Cross-sectional view of the split-gate flash cell, which employs source side injection (SSI) program and poly-poly Fowler-Nordheim (FN) erase

II. CELL STRUCTURE AND OPERATION FEATURES

Fig. 1 depicts the cross-sectional view and operation nodes of the split-gate flash cell. After the formation of FG self-aligned to control gate (CG), the SG (also the WL) and erase gate (EG) along with logic gate poly are created, and then standard logic process flow follows. In this study, the data flash macro (256Kb) and code flash macro (16Mb) have been utilized to demonstrate the 65nm automotive grade embedded flash manufacturability and compatibility with 65nm CMOS technology by 7 additional masks. Table 1 lists the typical operation voltages of this split-gate flash cell.

The WL voltage of the split-gate flash employs IO voltage without boost for read operation. It provides low voltage operation, low power consumption, and high cell current with fast read access speed, which is controlled by the SG transistor. Program and erase of the split-gate flash memory employ adaptive algorithms, consisting of a sequence of identical pulses followed by verify to reduce cycling-induced degradation.

TABLE I TYPICAL OPERATION VOLTAGES

	CG	EG	WL (SG)	CS	BL
PGM	11	5	1.3	5	0.2
ERS	0	13	0	0	0
Read	1.8	0	3.3	0	1.3

III. FLASH MACRO CHARACTERISTICS

A. Design for Testability

In order to enhance chip observability, high-voltage external forcing and simultaneously monitoring critical paths with switching on-chip charge pump were designed through specific test modes. Array-level SG, CG or EG stress and drain (also the BL, Bit Line)-inhibit stress were carrier out for dielectric screen and program disturb-verify, which provide the benefits of test time reduction. Tunable erase voltage (Ve) with tighter margin-level-verify was used to reserve guard band for cycling-induced erase time push out (ETPO). Changeable read WL bias was used to raise the cell current in order for subtler investigation of program state such as data retention charge gain study.

B. Design Constraint for Reliability

Because of no program time push out with cycling evolution due to the counter effects of decreased hot carrier injection efficiency and increased FG threshold (Vt) (electron trapping on the FG injection point), the program voltage (Vp) design target was set as close to Vpmin distribution (minimum program voltage to program whole macro with limited pulses, solid circle markers as shown in Fig. 2) to reduce cycling-induced degradation with σVp reserved for natural process variation. On the other hand, Vpmax distribution (maximum program voltage without program disturb, hollow circle markers as shown in Fig. 2) apart from target Vp represents the disturb immunity window. Hence, the cumulative plot of Vpmin and Vpmax distribution as shown in Fig. 2 defines the optimized program voltage target and program window.

Vemin distribution (minimum erase voltage to erase whole macro with limited pulses, solid circle markers for data flash macro, and solid square markers for code flash macro) shift right or ETPO with cycling evolution due to both of decreased erase efficiency (electron trapping on erase tunnel oxide) and increased FG Vt. The Ve design target was set as close to Vemin with ΔVe reserved for cycling degradation (product endurance SPEC-dependent) and σVe for production margin as shown in Fig. 3.

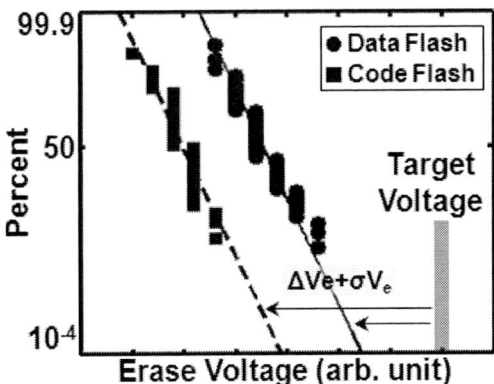

Fig. 3. Minimum erase voltage cumulative plot of erase code flash (circle) and data flash (square). (Target voltage distributes a range)

IV. SCREEN FOR RELIABILITY

Outlier Detection and Dielectric Screening

Multiple dielectrics around FG of the split-gate cell become new reliability concern even though a split-gate structure can achieve low voltage operation and disturb immunity advantages through SG and EG employment. Fig. 4 shows the typical defect bit movement behavior along with tailing bits of main population under a few seconds stress of specific dielectric stress Defect bits showed notable movement after high spec time stress, in the meanwhile some bits at low boundary of distribution became tailing bits. The stress condition was optimized as to push out defective bits as possible prior to dragging tailing away from main population. (Fig. 5)

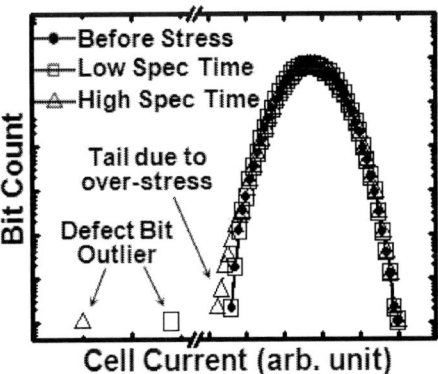

Fig. 4. Bit cell current distribution of before stress (solid) and after stress (hollow) with various stress time.

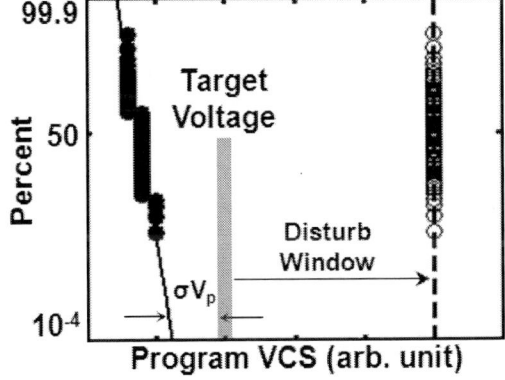

Fig. 2. Minimum program voltage (solid) and maximum program voltage without program disturb (hollow) cumulative plot. (Target voltage distributes a range)

978-1-4244-9113-1/11 $26.00 © 2011 IEEE

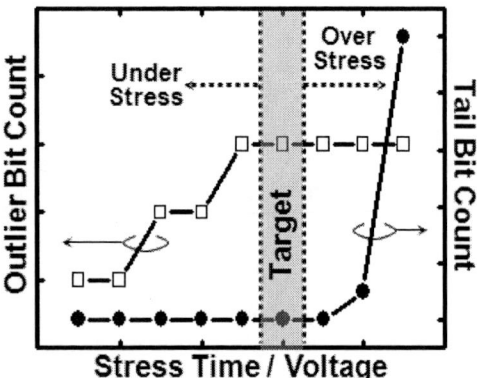

Fig. 5. Outlier bit count and tailing bit count of main population versus stress condition (voltage or time). Target stress time or voltage is set to push out outlier bits with minimum tail bits

V. ENDURANCE

A. Erase Time Push out and Endurance Life Time

Fig. 6 shows logarithmical increase of erase time with a square root of cycling dependence which coincides with bulk oxide trapping behavior of stack gate channel erase [5] with tunnel oxide thickness corner verification at -40^{0}C. The trap-up rate or ETPO relation with cycling was utilized to predict endurance lifetime of various product erase timeout (ETO) spec or process corner. Fig. 7 shows the intrinsic endurance lifetime temperature dependency of data flash macro at -40^{0}C to 125^{0}C to meet 100K cycling at 1ppm automotive requirement. The lower the cycling temperature, the worse the endurance lifetime mainly due to more hot carrier stress (more FG Vt increase) and worse erase efficiency (easier ETO).

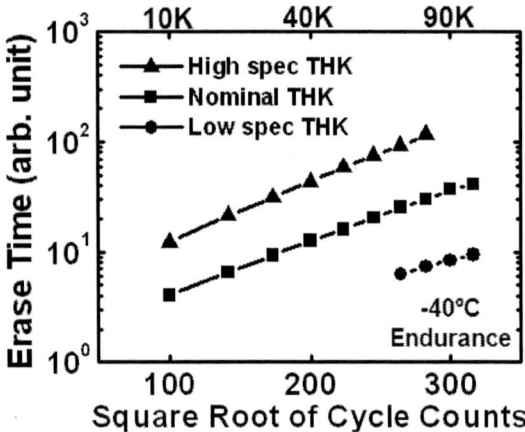

Fig. 6. Erase time versus square root of cycling counts with various erase tunnel oxide thickness at -40^{0}C. Y-axis on the top is cycle counts and on the bottom is square root of cycle counts. Triangle markers are high spec thickness, square makers are nominal spec thickness and circle markers are low spec thickness.

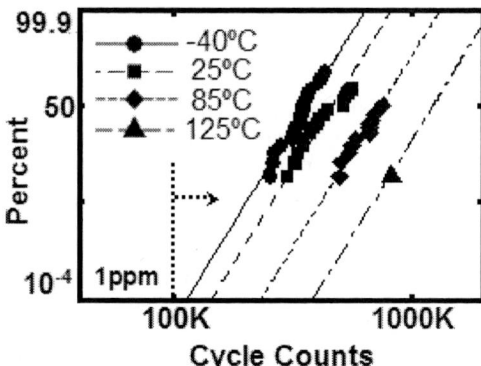

Fig. 7. Endurance lifetime distribution plot with various cycling temperatures between -40^{0}C to 125^{0}C. The endurance lifetime at -40^{0}C (circle markers) is the worst case for endurance, which can meet 1ppm, 100K

B. Large Sample Certification

With data flash (100K) and code flash (10K) cycling of around 10K units for automotive grade readiness, all early cycling failures are marked as dielectric screen bin (Fig. 8). These early cycling failures are attributed to post-cycling program disturb. Not only the dielectric screen effectiveness was established but also the defective dielectric percentage leading to cycling fails could be further validated for process improvement. The inset of Fig. 8 shows 100K cycling endurance with various ETO's and demonstrates 100K, 1ppm endurance prediction at product ETO spec based on Fig. 6.

VI. DATA RETENTION

A. Data Loss Behavior and Data Retention Life Time

The retention characteristics of cycled units are shown in Figs. 9 and 10 with 100K pre-cycle, 1000hrs, 150^{0}C baking. It is worth mentioning that there should be no retention concern on non-cycled units due to good inherent retention dominated by thermionic emission with activation energy (Ea) of around 2.25eV [6]. On the other hand, the extrapolated Ea (from Fig.

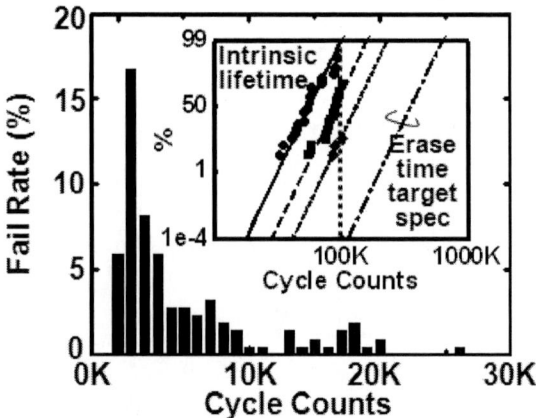

Fig. 8. Early cycling failure rate distribution of dielectric screen bin-out units. Inlet shows intrinsic endurance lifetime with various erase time out conditions. 100K at 1ppm failure rate can be achieved by implanting dielectric screen in chip probing stage

11) is in the range of 0.8eV ~1.1eV which approaches to 0.8eV with more pre-cycling. It indicates that FG oxide detrapping [7] and carrier hopping conduction [6] are the dominant data loss mechanisms. Both mechanisms contribute cell current increase and/or Vt decrease for program state. The net effect of detrapping (which increases cell current) and hopping conduction (which decreases cell current) leads to converged cell current distribution after baking for erase state (Figs. 9 and 10). Fig. 11 shows the intrinsic data retention lifetime with temperature and pre-cycling dependency of data flash macro to meet 10 years at 125^0C, 1ppm automotive requirement.

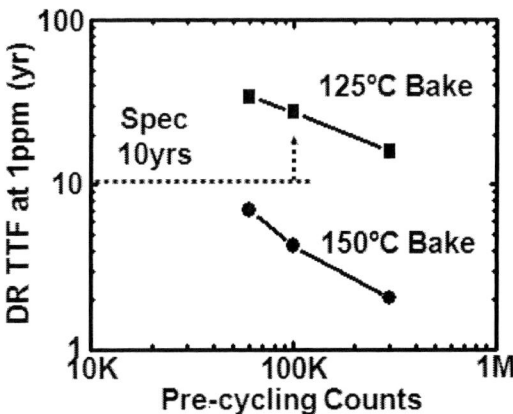

Fig. 11. Data retention lifetime at 1ppm versus pre-cycling counts with various baking temperatures. Intrinsic data retention lifetime with temperature and pre-cycling dependency of data flash macro can meet 10 years at 125^0C, 1ppm automotive requirement.

B. Large Sample Certification

With data flash and code flash 1000hr data retention baking of around 10K units including non-cycled and cycled for automotive readiness, the multiple sensing levels were employed to monitor current shift with uA resolution and all early data loss units are marked with dielectric screen bin (Fig. 12). Dielectric screen bin-out of data retention fails are attributed to FG surrounding dielectrics, not just FG or tunnel oxide with program and erase carriers passing through.

VII. CONCLUSION

A novel 65nm automotive grade embedded split-gate flash solution has been proven here with comprehensive dielectric screen methodology. The memory macro design window and bias setting guidelines have been clarified. The erase time push out has been studied to establish an endurance lifetime prediction method through various temperatures and process corners. Data loss behavior and retention lifetime are summarized here to analyze detrapping and charge hopping conduction phenomenon.

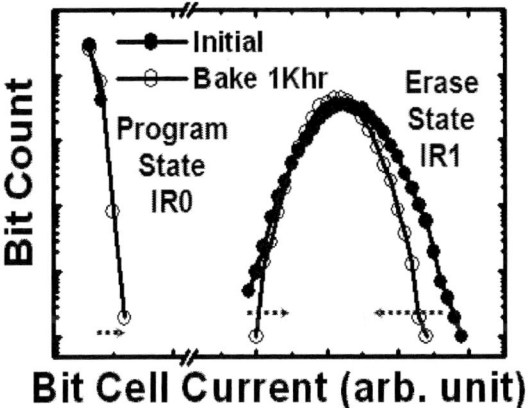

Fig. 9. Bit cell current distribution of before and after 1000 hours baking at 150^0C. Detrapping and carrier hopping conduction lead to cell current movement after backing.

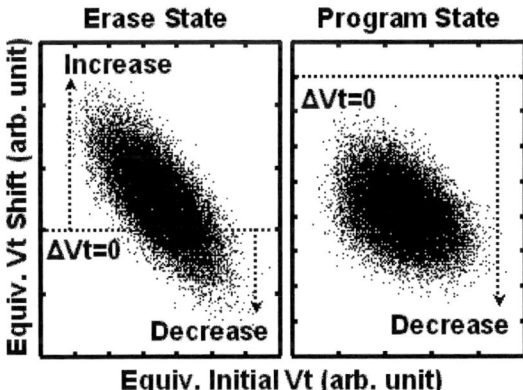

Fig. 10. Equivalent threshold shift after 1000hr, 150^0C baking versus initial threshold for erase state and program state. The graphic on the left shows Vt shift at erase state, while the graphic on the right shows program state. Each data point stands for Vt value of one single bit in micro array

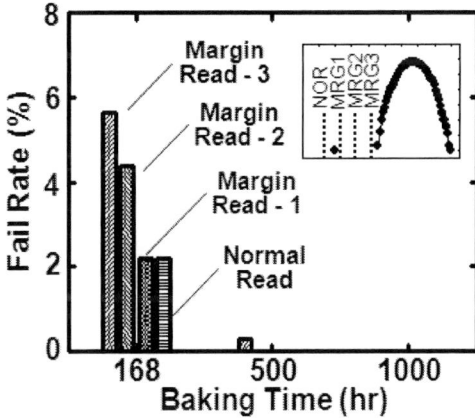

Fig. 12. Early retention failure rate distribution of dielectric screen bin-out units. Inlet illustrates multiple sensing levels to monitor cell current shift.

978-1-4244-9113-1/11 $26.00 © 2011 IEEE   639

Combining with effective dielectric screen, automotive flash Spec with 100K endurance and 150^{0}C, 10 years data retention reliability with 1ppm failure rate are demonstrated.

ACKNOWLEDGMENT

The authors would like to thank R&D and Fab of Taiwan Semiconductor Manufacturing Company for technology and material support.

REFERENCES

[1] Y. H. Wang, *et al.,* "An Analytical Programming Model for the Drain-Coupling Source-Side Injection Split Gate Flash EEPROM," IEEE T-ED, Vol. 52, 2005, p. 385

[2] Embedded non-volatile memory, TSMC's EmbFlash. [Online] http://www.tsmc.com/english/b_technology/b01_platform/b010302_nvm.htm

[3] S. K. Saha, *et al.,* "Design Considerations for Sub-90-nm Split-Gate Flash-Memory Cells," IEEE T-ED, Vol. 54, 2007, p. 3049

[4] S. Lai, "Tunnel Oxide and ETOX^{TM} Flash Scaling Limitation" Nonvolatile Memory Technology Conference, 1998, p. 6

[5] N. Mielke, *et al.,* "Flash EEPROM Threshold Instabilities due to ChargeTrapping During Program/Erase Cycling," IEEE, T-DMR, Vol. 4, 2004, p. 335

[6] H. Kameyama, *et al.,* "A New Data Retention Mechanism after Endurance Stress on Flash Memory," IEEE/IRPS, 2000, p. 194

[7] N. Mielke, *et al.,* "Recovery Effects in the Distributed Cycling of Flash Memories," IEEE/IRPS, 2006, p. 29

Novel Negative Vt Shift Program Disturb Phenomena in 2X~3X nm NAND Flash Memory Cells

Soonok Seo, Hyungseok Kim, Sungkye Park, Seokkiu Lee, Seiichi Aritome and Sungjoo Hong
R&D Division, Hynix Semiconductor Inc.,
San 136-1 Ami-ri, Bubal-eub, Ichon-si, Gyeonggi-do, 467-701, Korea
Tel) +82-31-639-5887, Fax) +82-31-639-0734, E-mail) soonok.seo@hynix.com

Abstract

A novel program disturb phenomena of "negative" cell-Vt shift has been investigated for the first time in 2X~3X nm Self-Aligned STI cell[1,2] of NAND flash memory. The negative Vt shift occurs on an inhibited cell adjacent to a cell being programmed in the WL direction. The magnitude of the shift becomes larger when the programming voltage (V_{Pgm}) is higher, thinner field oxide and slower program speed of the adjacent cell. The mechanism of negative Vt shift is attributed to hot holes that are generated by FN electrons, injected from channel / junction to the control gate (CG) along the isolation. This phenomenon will become worse with scaling since hot hole generation is increased by increasing electron injection due to narrower FG space. Therefore, this negative Vt shift phenomenon is one of the new NAND flash memory cell scaling limiter, that needs to be managed for 2bits and 3bits/cell in 2X nm and beyond.

Introduction

As scaling of NAND flash memory cells, there are several reported obstacles, such as FG-FG coupling interference [3], RTN [4], electron injection spread [5] there are becoming worse with scaling for 2bits/cell and 3bits/cell. This paper introduces a new potential obstacle of "negative" cell-Vt shift program disturb phenomena for 3X and 2X nm NAND Flash cells. Due to this negative Vt shift, read window margin will be further degraded for scaled 2Xnm and 1X nm NAND Flash.

Experimental procedure

Fig.1 shows the cell arrangement for program disturb test. Attack ABL1&2 cells are represented WL direction adjacent cells, and Attack AWL1&2 cells are shown BL direction adjacent cells. The inhibited victim cells Vt shifts when the attack cell is programmed. 3Xnm & 2Xnm NAND flash memory cell and a cell-structured capacitor are used in this experiment.

Negative Vt shift

Fig.2 shows the victim cell Vt shift caused by the attack cell programming. Along the word-line side, the Vt of the victim cell monotonously increases with Attack AWL2 cell due to conventional FG-FG interference. However, along the bit-line side, the Vt of the victim cell initially increases and then decreases as the Attack ABL1 and 2 is programmed. This negative Vt shift of victim cell is presented for the first time in this paper. Fig.3 shows the dependence of this negative Vt shift with Height of the STI Field oxide (FH). Thinner the STI field oxide, larger the Vt shift of the victim cell.

Fig.4 shows victim cell Vt depended on Attack cell (a)Program and (b)Inhibit condition. In case of Attack cell program, large negative Vt shift of victim cell is observed at small FH when program Vt is high. However negative shift is not observed in inhibit mode condition. Difference between Attack cell (a)Program and (b)Inhibit are only the channel Attack cell voltages, 0V for Program or $V_{Boosting}(\sim7V)$ for inhibit. Thus 0V channel voltage is needed to affect the Vt of the victim cells.

Cell array results

This negative Vt shift phenomena is confirmed in test chip as shown in Fig.5. The three different attack cell programming status are used at the three different FH condition, L3 (ABL1)/ L3(ALB2) means that attack cells in both side direction are L3 programming. In case of both attack L3 (attack LSB→L3 ; higher Vt), victim cell delta Vt is much lower than the case of both side attack L1 (Erase→L1 ; lower Vt). This means negative Vt shift is worse in case of L3 programming attack cell. In addition to the program level, negative Vt shift also depends on FH. If we compare two cases: Case 1- both attack cells are programmed to L3 and Case2- one of the attack cell programmed to L3, the victim delta Vt shift is roughly the same for the low FH skew and not for the high FH skew. For the high FH skew, the two sides L3 programming shows the expected higher Vt shift for the victim due to FG-FG interference.

Fig.6 shows the impact of program speed on this negative Vt shift on the same chip, along with the FH dependence. In case of attack cell Vt=1, which is corresponding to attack cell programming to L1, when the attack cell is slow to program, the victim cell shifts more than a fast programming attack cell, programmed to a higher Vt level. On the other hand, in case of attack cell Vt=5, which is corresponding attack cell L3, when the attack cell is slow to program, the victim cell shifts more "negative" than a fast programming attack cell. From this result, victim cell delta Vt of slow attack cell is more positively large for L1 attack cell and negative for L3 attacked cell. The FH dependence holds even this case. Given that the shift depends on program speed along with FH, the variability needs to be character. Fig.7 is a plot of victim cell delta Vt versus attack cell program speed for 16Kbits cells. The distribution of each cell Vt of 16Kbits, X-axes of 1-pulse PGM Vt means after fixed program pulse (Arb.Un. 13V), in other word, cells in left side axes are slow cell and cells in right side axes are faster cell.   In case of (a) Attack cell ABL1=L1 & ABL2 =L1, Victim cell delta Vt shift shows a normal dependence of attack cell delta Vt, due to FG-FG interference. However in case of (b) Attack cell ABL1 and ALB2=L3, the victim shows large negative delta Vt shifts for slow programming attack cell. The observed dependence is due to large number of programming pulse for the slow cell, which increases both time and voltage. This result of Fig.7 is good correlation with that of Fig.6. Consequently this new mechanism results in wider placement distribution, in addition the existing components of variation. In case of (a) Attack cell ABL1=L1 & ABL2 =L1 Victim cell delta Vt distribution is 0.36V, however in case of (b) Attack cell ABL1 and ALB2=L3 Victim cell delta Vt distribution is 0.48V and also, we can see victim cell delta Vt have negative shift of around 0.1V.

Summarizing the above observations, negative Vt shifts predominantly occurs in the bit line direction and is exacerbated by FH, Vt to which the attack cell is placed and program speed. High field effect at the STI is suspected to be the root cause for the observed negative Vt shift. In order to verify this model programming is measured in cell-structured capacitor by using carrier separation method. In Fig.8, hole current (I_{well}) generated by I_{CG} (electron current injected from active/junction to CG) has been observed for the first time. This hole current increases with V_{CG} and smaller FH due increased I_{CG} electron current. These generated holes can be easily to be injected to victim cell FG because of high electric field between CG and FG of victim cell, resulting in the observed negative Vt shift. In Fig.9, for I_{CG}, direct electron injection from substrate to CG is observed at $V_{FG} > 1$ (Arb.Un). I_{CG} is strongly enhanced by the floating gate voltage V_{FG}, even if constant V_{CG} is applied.

Model

From the results of current flow in cell structured capacitor, the mechanism of negative Vt shift is considered as illustrated in Fig.10; hot electrons from channel/junction (0V) to CG (V_{Pgm}) generates hot hole at CG[4,5]. Some of these generated hot holes are injected into to the FG of victim cell through field dielectric or SiN of inter poly dielectric (IPD). Consequently, the Vt of victim cell shifts negatively. This phenomenon is exacerbated with scaling since I_{CG} increases due to narrow FG-FG space field effect, as shown in Fig.9.

Then the negative Vt shift phenomena is one of new scaling limitation factor to manage Vt window of 2bit and 3 bit/cell in 2X nm and beyond NAND Flash memory cell.

Conclusion

Negative Vt shift phenomenon in an inhibited cell has been presented for the fist time. This magnitude of the shift depends on FH, attack cell Vt and attack cell program voltage. This mechanism has been explained by hot hole injection from the control gate generated by hot electrons tunneling from the active STI edge. We believe that the negative Vt shift phenomenon is a new limiter that needs to be managed for 2 and 3bit/cell in 2Xnm NAND and beyond.

References

[1] S. Aritome, S. Satoh, T. Maruyama, H. Watanabe, S. Shuto, G. J. Hemink, R. Shirota, S. Watanabe, and F. Masuoka, "A 0.67 um self-aligned shallow trench isolation cell (SA-STI cell) for 3 V-only 256 Mbit NAND EEPROMs," IEDM 1994, pp 61-64.

[2] S. Aritome, "Advanced Flash Memory Technology and Trends for File Storage Application," IEDM 2000, pp. 763-766.

[3] Jae-Duk Lee, Sung-Hoi Hur, and Jung-Dal Choi, "Effects of Floating-Gate Interference on NAND Flash Memory Cell Operation," Electron Devices Letters, Vol. 23, No5, 2002. pp 264-266

[4] H. Kurata, K. Otsuga, A. Kotabe, S. Kajiyama, T. Osabe, Y. Sasago, S. Narumi, K. Tokami, S. Kamohara, and O. Tsuchiya, "The impact of random telegraph signals on the scaling of multilevel Flash memories," VLSI Symp.Tech. Dig 2006, pp140-141

[5] C. Monzio Compagnoni, A. S. Spinelli, R. Gusmeroli, A. L. Lacaita,S. Beltrami, A. Ghetti and A. Visconti, "First evidence for injection statistics accuracy limitations in NAND Flash Constant-current Fowler-Nordheim programming," IEDM 2007, p 165-168

[6] I. Chen, S. Holland, and C. Hu, "Oxide breakdown dependence on thickness and hole current-enhanced reliability of ultrathin oxide," IEDM 1986, pp660-663.

[7] Z. A. Weinberg, M. V. Fischetti, and Y. Nissan-Cohen, "SiO2-induced substrate current and its relation to positive charge in field-effect transistors", JAP, Vol.59, No3, 1986, pp824-832

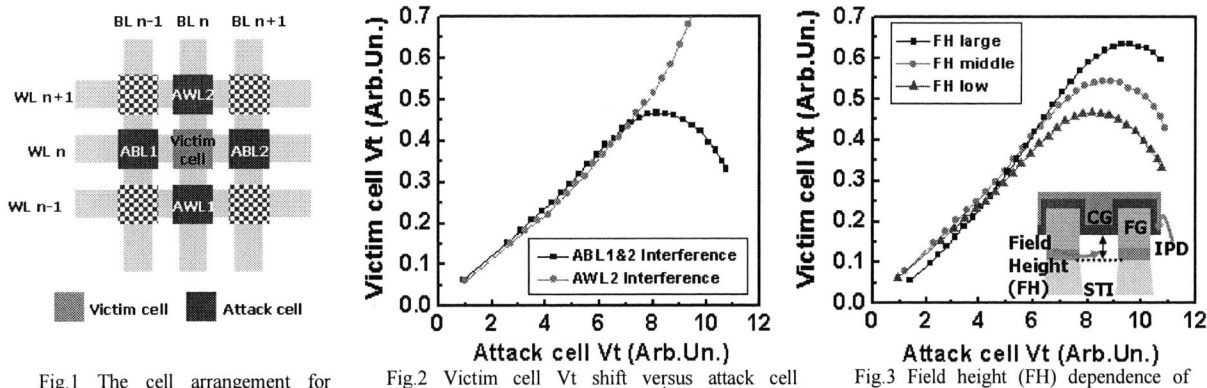

Fig.1 The cell arrangement for disturb test.

Fig.2 Victim cell Vt shift versus attack cell programmed Vt shift. Negative Vt shift phenomena is observed in case of attack ABL1 and ABL2.

Fig.3 Field height (FH) dependence of "negative" Vt. Small FH has the large "negative" shift.

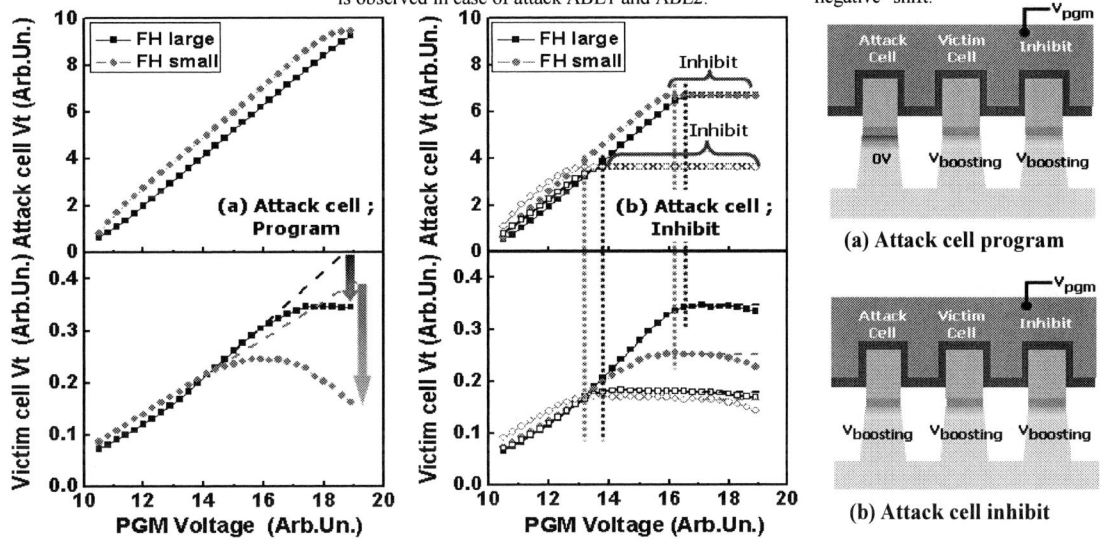

Fig.4 Victim cell Vt shift versus Attack cell (a) Program and (b) Inhibit condition. In case of Attack cell program, large negative shift of Victim cell is observed, however in case of Attack cell inhibit, negative Vt shift is hardly observed.

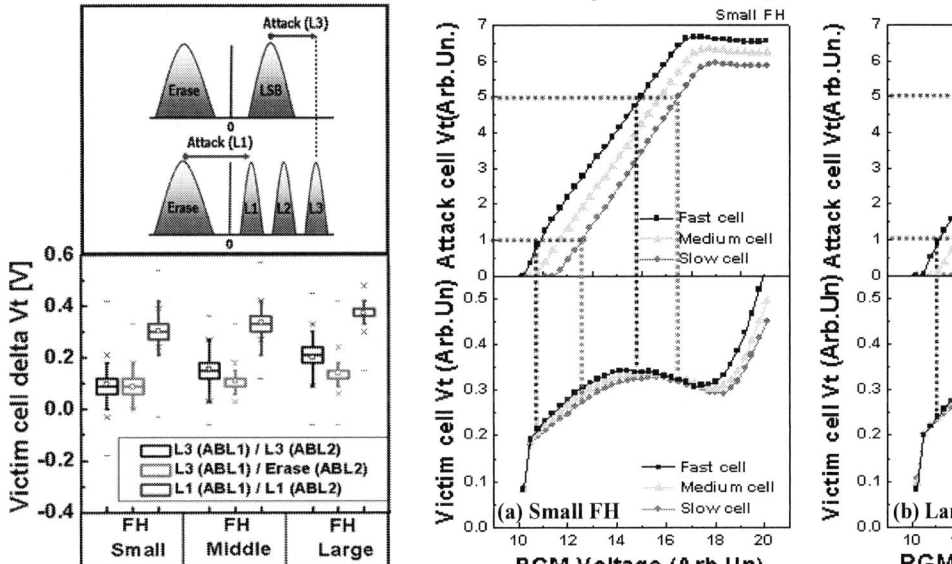

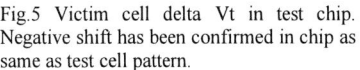

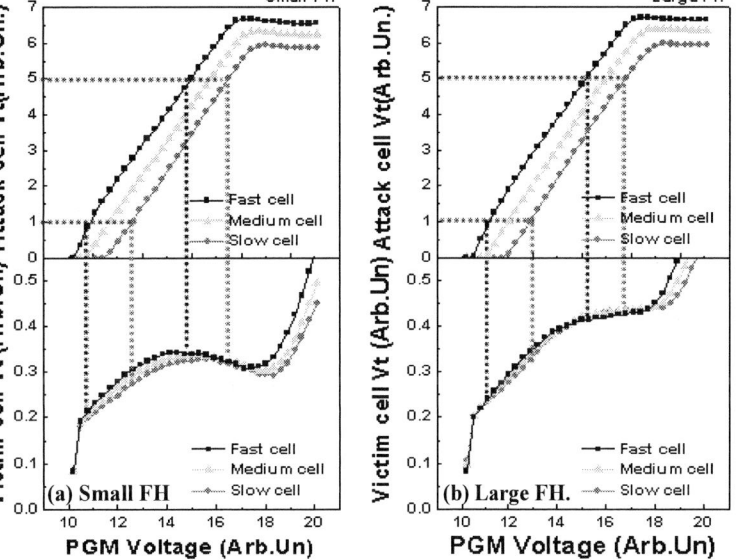

Fig.5 Victim cell delta Vt in test chip. Negative shift has been confirmed in chip as same as test cell pattern.

Fig.6. Victim cell Vt versus divide Attack cell speed (fast / medium/ Slow) at the same chip. (a) Small FH, (b) Large FH.

978-1-4244-9113-1/11 $26.00 © 2011 IEEE   643

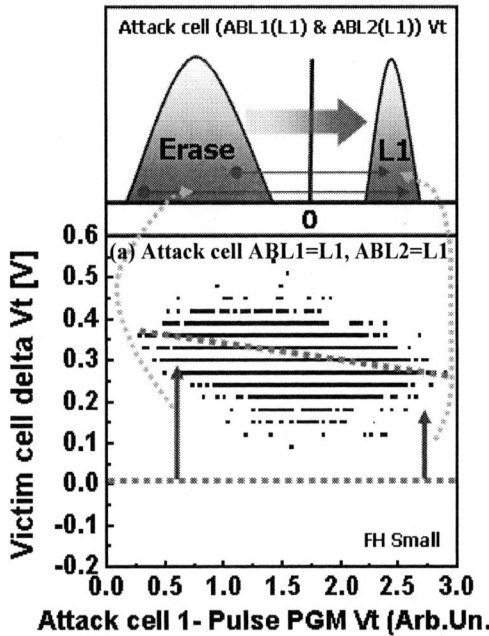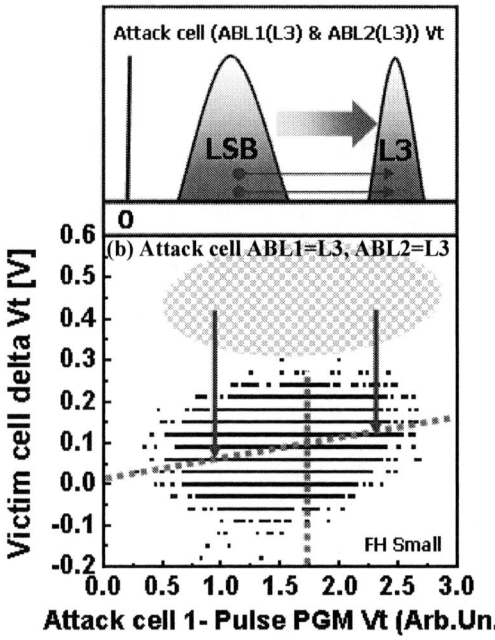

Fig.7 Victim cell delta Vt distribution (16Kbits) versus Attack cell program speed (Attack cell 1-Pulse PGM Vt) in test chip. In case of (a) Attack cell ABL1=L1 & ABL2 =L1, Victim cell delta Vt shift shows normal dependence of attack cell delta Vt due to FG-FG interference. However in case of (b) Attack cell ABL1=L3 & ALB2=L3, Victim cell delta Vt shows large negative Vt shift for slow attack cell program speed.

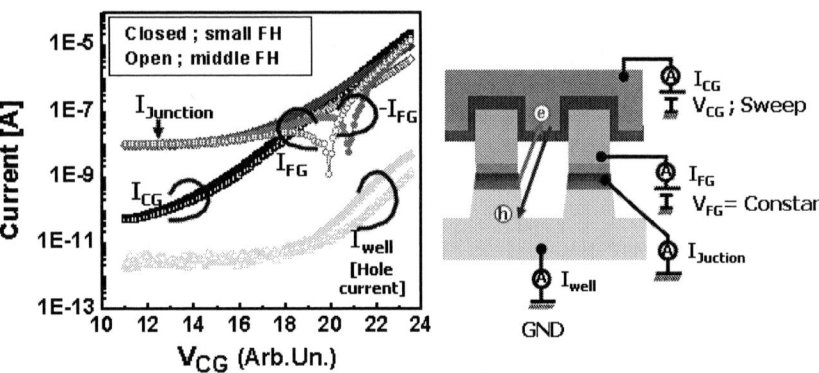

Fig.8 Current analysis of cell structured capacitance (carrier separation). Hole current (I_{well}), which is generated by direct electron injection from Junction to CG, is increased as VCG is increased. And large hole current is generated in case of small FH (Field Height) due to large I_{CG}.

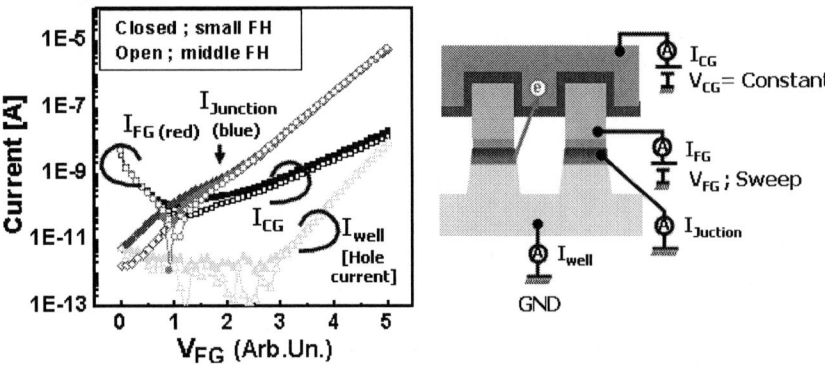

Fig.9 Current analysis of cell structured capacitance. Even if V_{CG} is constant I_{CG} (direct electron injection from substrate to CG) is increased as V_{FG} is increased. This means I_{CG} is enhanced by FG potential at STI Field Height area.

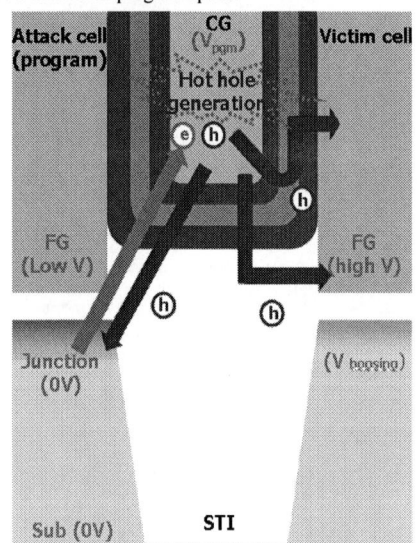

Fig.10 Mechanism of negative Vt shift in victim cell. Hot electron from Junction (0V) to CG (V_{Pgm}) could generate hot hole at CG, and parts of hot holes are injected to FG of victim cell through field dielectric and SiN of inter poly dielectric (IPD) due to electric field. This hot hole injection cause "negative Vt Shift."

This phenomenon ("Negative" shift) is worse on memory cell scaling, because amount of hot hole is increased by increased ICG which is enhanced of narrower FG-FG space.

978-1-4244-9113-1/11 $26.00 © 2011 IEEE

Junction Optimization for Reliability Issues in Floating Gate NAND Flash Cells

C. H. Lee, I. C. Yang, Chienying Lee, C. H. Cheng, L. H. Chong, K. F. Chen, J. S. Huang, S. H. Ku, N. K. Zous, I. J. Huang,
T. T. Han, M. S. Chen, W. P. Lu, K. C. Chen, Tahui Wang, and Chih-Yuan Lu

Macronix International Company Ltd., No. 16, Li-Hsin Road, Science Park, Hsin-Chu, Taiwan, R. O. C
Phone: +886-3-5786688-78062 E-mail: nkzou@mxic.com.tw

Abstract— Source/Drain Reliability issues in a floating gate (FG) NAND string with different junction dosages are investigated. A lighter junction dosage gets better sub-threshold swing (SS) and helps the shrinkage of device channel length. However, some drawbacks, such as worse current fluctuation and abnormal self-boosting (SB), can be observed as the side effects. Charge pumping technique is applied to identify their impact, and then, the noise contribution along the channel can be portrayed by random telegraph noise (RTN) profiling. Second, program/erase (P/E) cycling effect is examined. The degradations of cell performance due to stress-induced oxide charges near gate edges are studied. The correlation between SB behavior and junction profile is established. Contrary to conventional global SB (GSB), the local SB (LSB) is more effective in sustaining sufficient channel potential, which enhances local junction or band-to-band field with modest tunneling induced disturbance. It is observed that an abnormal hot carrier injection results in the tail feature at the upper half of erased-V_T distribution. Furthermore, the program disturbance of a junction-free structure is also reviewed. The optimized window of dosage regarding disturbance is given.

Keywords-component: dopant concentration, erase speed degradation, sub-threshold swing (SS) increase, program/erase (P/E) cycling, on-state current (Ion) reduction, program disturbance

I. Introduction

Recently, demand for high-density storage accelerates the scaling down of NAND flash rapidly beyond 2X nm, in which short channel effect is found to be the most critical bottleneck. To overcome this kind of challenge, lighter S/D implantations accompanied with a reduced I_{ON} become necessary [1,2]. Even more, a junction-free NAND structure, of which the junctions are built by the fringing fields of neighboring gates, is proposed [3,4]. As shown in Fig. 1, cells with and without junctions are simulated at different technology nodes. Since the thickness of inversion layers is only tens of Angstrom, a better SS is achieved. Because the total implanted dosages vary with CD, in Fig. 2, the V_T of traditional cells exhibits

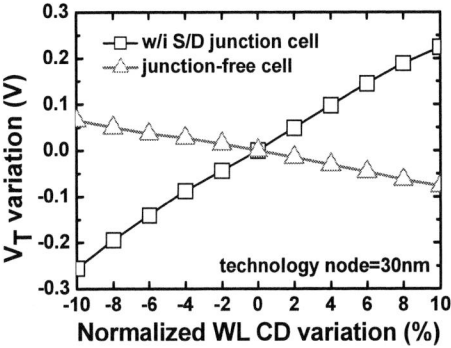

Fig.2 V_T variation for the two splits against the gate length variation when the gate pitch is fixed. Here, the pitch = space + length.

large variation under the same level of CD variation. However, decreasing the dosage of S/D junctions may weaken the potential shielding effect near gate edges, which deteriorates the endurance according to [5,6]. The current fluctuation, which is highlighted as a potential barricade to device shrinkage [8-10], is also affected by channel potential (V_{ch}) near gate edges [11,12]. Moreover, it is known that the ability to boost V_{ch} strongly depends on junction dosages and reduction of S/D dosages will easily transform into a LSB behavior [13]. Because the coupling path between neighbor cells is diminished, the V_{ch} of LSB is dominated by the program gate bias ($V_{g,pgm}$) in a more efficient way as compared to GSB [14]. Higher local V_{ch} enhances the immunity against disturbance from vertical tunneling field but increases the occurrence of hot carriers induced by lateral field as described in [14]. In this work, cell performances with different junction dosages are characterized, and the dilemma of device scaling will be discussed.

II. Experimental results and Discussions

Floating-gate cells with different S/D dosage are fabricated by 70nm technology. A NAND string is composed of 32 cells connected serially. The V_T of a cell is defined as the required gate voltage to reach a specific read current. As for the program disturbance of a junction free structure, both BE-SONOS cells [15] in 30nm technology and a TCAD simulator are employed to explore the underlying mechanisms.

A. Effects of S/D Dosage v.s. Fresh Cell Performances

As shown in Fig.3, a lightly doped junction has a similar initial V_T distribution to a heavily doped one, in which B1 and B2 represent the lower and upper boundaries of V_T distribution. In Fig. 4(a), the fluctuation of read current is compared in time domain between lightly- and heavily-doped cells. The cumulative probability of

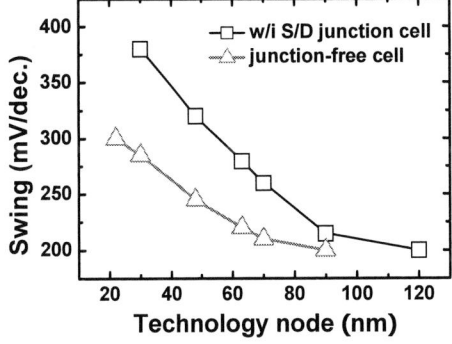

Fig.1 Subthreshold slope (SS) for cells with junction and junction-free as a function of technology node.

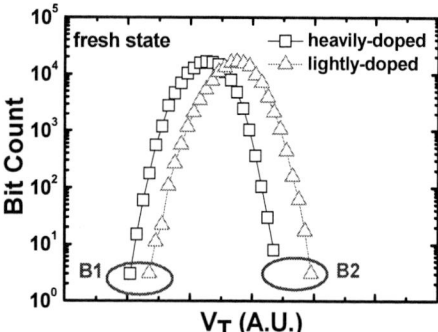

Fig.3 V_T distribution of fresh state for heavily- and lightly-doped splits. B1 and B2 mean the low and high V_T boundary.

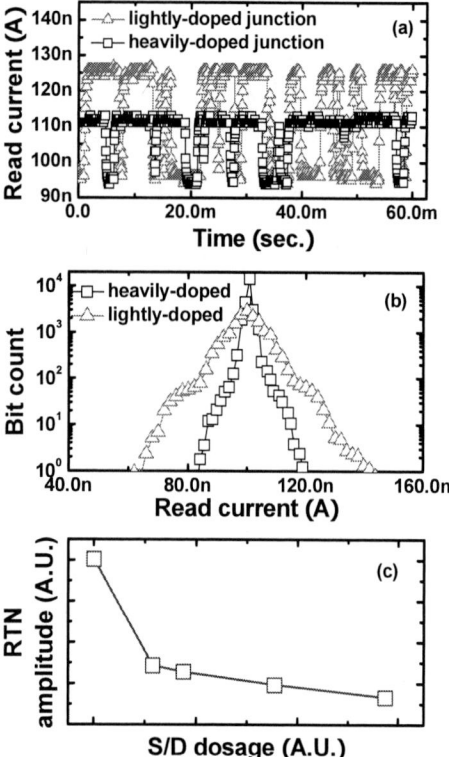

Fig.4 (a) The read current fluctuations are measured between these two splits in time domain. (b) Cumulative distribution of read current fluctuation. (c) S/D dosage effect on RTN amplitude.

collected noise amplitude is plotted in Fig. 4(b). Fig. 4(c) shows RTN is enlarged by lightening the junction dosage. As depicted in Fig. 5(a), distortion of V_T distribution at program verified level might lead to a reduced read margin. To clarify the origin of increased noise, the charge pumping current (I_{CP}) [16] is extracted for the two dosage conditions in Fig. 5(b). The I_{CP} curves converge at high current region but separate from each other as the current decreases.. This implies that their local V_T is not the same around junctions. By using the RTN profiling technique [12], more than 60% of noise contributed from gate edges is found on a lightly doped cell, as shown in Fig. 5(c).

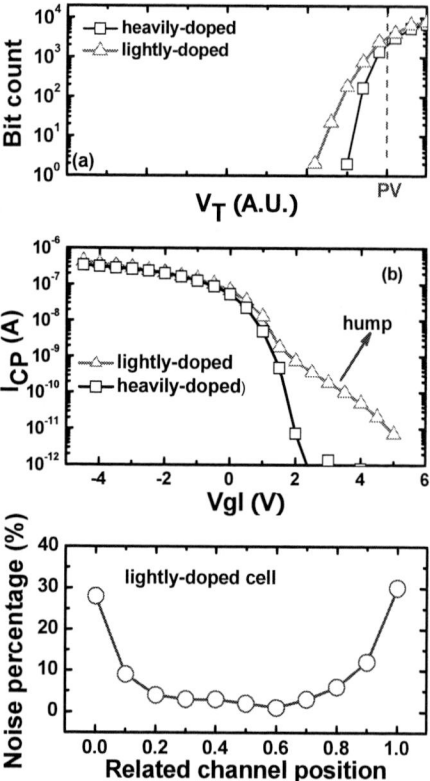

Fig.5 (a) V_T distribution when PV is set. (b) I_{CP} measurement between the two splits. (c) Noise distribution along the channel is extracted. Channel position "0" is the position of source.

B. P/E Cycling Effects on Cell Performances

Cycling performances of cells with lightly and heavily doped junctions are compared in Fig. 6 and Fig. 7. Reduction of S/D dosage

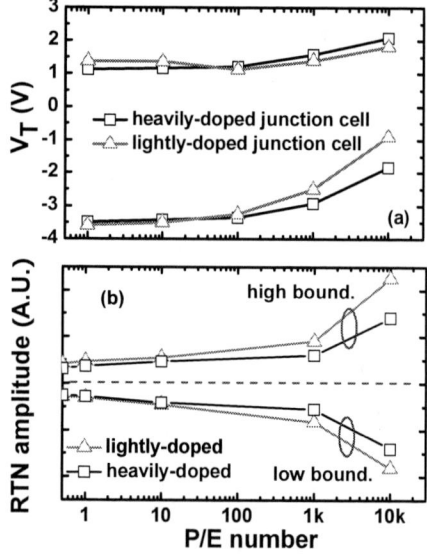

Fig.6 (a) Endurance behavior of single WL. (b) Variation of RTN amplitude against cycle number.

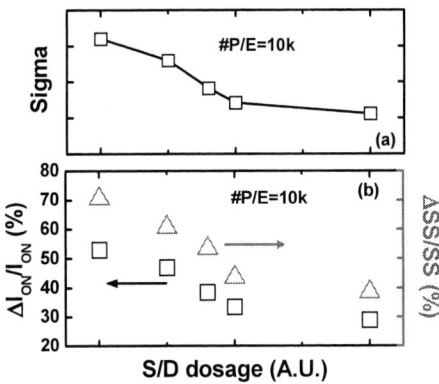

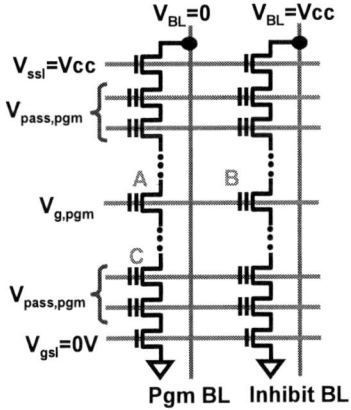

Fig.7 S/D dosage effect on the σ of program V_T distribution (a) and I_{ON} and SS degradation (b).

gets more degradation no matter in endurance, RTN amplitude, width of program distribution, Ion, and SS. It is consistent with previous research [14], which suggests that the S/D overlap region is a key factor to immunize the potential variations beneath gate edges.

C.   Program Disturbance by Abnormal Hot Carrier Injection

The NAND array architecture for evaluation of disturbance is schematically depicted in Fig. 8. Channel Fowler-Nordheim (FN)

Fig.8 Schematic structure of a NAND array.
Bit A: programmed bit.
Bit B: disturbed bit by program bias ($V_{g,pgm}$).
Bit C: disturbed bit by pass gate bias ($V_{pass,pgm}$).

injection is utilized for programming Bit-A, but in the meantime, Bit-B and Bit-C suffer disturbance from $V_{g,pgm}$ and pass gate biases ($V_{pass,pgm}$) respectively. Usually, the SB property is utilized for program inhibition of Bit-B [17,18], of which the V_T distribution is measured under several $V_{g,pgm}$ to evaluate program disturbance for lightly-doped cells. At low $V_{pass,pgm}$, for example 3V in Fig. 9(a), the whole distribution moves upward as $V_{g,pgm}$ increases due to tunneling injection because of insufficient boosting V_{ch}. In contrast, at high $V_{pass,pgm}$ of 14V as the case of Fig. 9(b), only the boundary shifts continuously with increase of $V_{g,pgm}$ while the majority keeps unchanged. In Fig. 9(c), this tail behavior becomes worse especially when the number of program pulse (NOP) increases. The formation

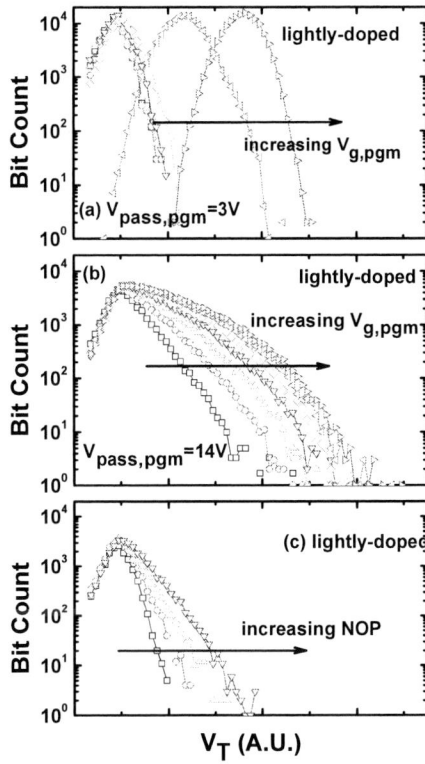

Fig.9 Gate bias effect on program disturbance in lightly-doped cell during (a) low $V_{pass,pgm}$ and (b) high $V_{pass,pgm}$ bias. (c) The tail behavior is enhanced with increasing NOP.

of these tail behaviors can be elucidated as follows. As the dosage of most cells is high enough to generate sufficient GSB potential, a few cells suffer the statistical variation of fabrication processes resulting in more severe hot-carrier disturbance [14]. This explains why the peak of V_T distribution stays and the disturbed ones are randomly scattered over our test chip. The cell count of Bit-B with significant V_T shift is compared in Fig. 10(a) under various $V_{pass,pgm}$ for two different S/D dosages. For the lightly doped one, disturbance with hot carrier mechanism is observed when $V_{pass,pgm}$ is higher than 11V. The distance from peak to B2, i.e. the upper half width of V_T distribution, is plotted against $V_{g,pgm}$ in Fig. 10(b). In the lightly doped case, its B2 moves far away from peak and tends to saturate at higher biases, which is consistent to the feature of hot-carriers [19].On the contrary, both the peak and B2 move together in the heavily doped one. Furthermore, the disturbed behaviors between lightly doped and junction-free BE-SONOS cells are compared as shown in Fig. 10(c). No matter it's FG or BE-SONOS, similar tail behaviors are observed on lighter S/D dosages. Interestingly, in the meantime, nearly all junction free cells are disturbed to high V_T and the tail behavior disappears, which will be illustrated later. After stress, the hot carrier induced disturbance becomes worse, and failure bits appear even at operational $V_{pass,pgm}$ of 10V, as shown in Fig. 11(a). Further, the normal distribution movement starts to worsen the disturbance after P/E number of 10k. The contribution from hot carrier induced tail becomes weak with increasing P/E number, as shown in Fig. 11(b).

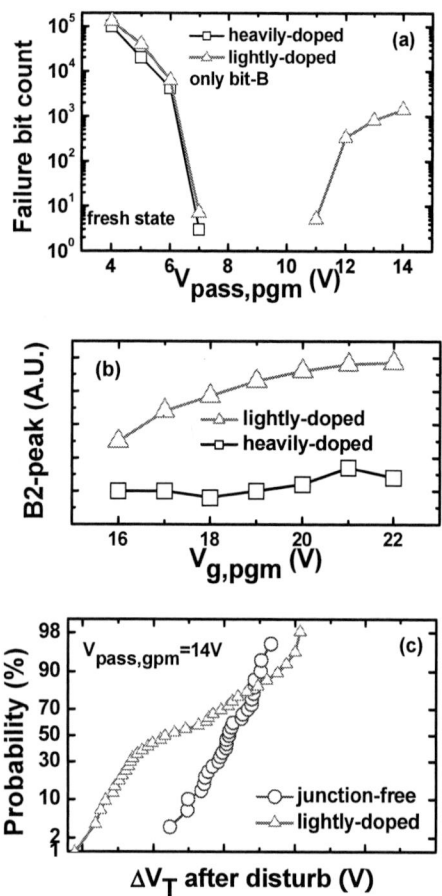

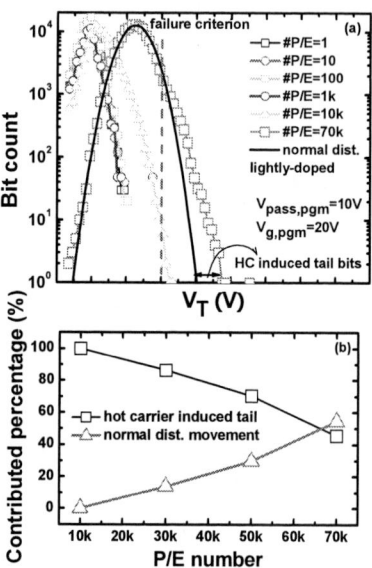

Fig.11 (a) V_T distribution of bit-B after P/E cycle. Red dash line means the failure criterion. The V_T difference between purple symbol and black line (normal distribution) is caused by hot carrier induced tail. (b) Contributions of hot carrier induced tail and normal distribution movement caused failure against P/E cycle number. The normal distribution shift becomes important with increasing cycling number.

Fig.10 (a) Fail bit count of bit-B in a FG NAND array against $V_{pass,pgm}$. (b) V_T difference between B2 and peak for lightly-doped and heavily-doped FG cells. (c) Program disturbance between junction-free and lightly-doped BE-SONOS cells.

As claimed in [14], the source of disturbance in a lightly doped cell comes from excess electrons introduced by junction breakdown. Though V_{ch} is extremely high in a junction-free case, no formation of junction means that the excess carriers do not come from junction breakdown. In Fig. 12(a), a TCAD simulator for a junction-free NAND array is utilized. Band-to-band generation rate is much higher below the cell applied with $V_{g,pgm}$. Fig. 12(b) illustrates the band diagram for the band-to-band induced substrate hot electron injection (BBISHE) [20]. When the valence-band electron (labeled I) tunnels into the conduction band, the free hole is left in the valence band, gains enough energy from the bending of substrate potential, and then, creates electron-hole pairs through impact ionization (labeled II). The generated electrons will drift back to the sub/ox interface. Part of them with energy high enough can overcome the oxide barrier and inject into FG (labeled III). In Fig. 12(c), under a fixed Ion, the program disturbance according to various mechanisms is plotted separately against junction dosage. Unlike FN injection, a peak is observed for hot-carrier type disturbance. This disturbance relates to both injected electric field and electron source. In the junction-free structure, the injected source is ionized by BBISHE. The disturbance

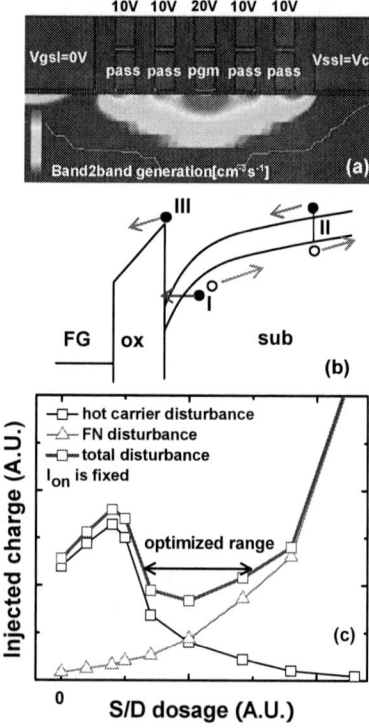

Fig.12 (a) Simulated band-to-band generation rate during disturbance in a junction-free NAND array. (b) Mechanism of the band-to-band induced substrate hot electron injection (BBISHE). (c) Dosage effect on the two disturbance induced by hot carrier and FN injection. Optimized dosage range is observed.

978-1-4244-9113-1/11 $26.00 © 2011 IEEE

increases in a lightly-doped case because the source is transferred to generate by junction breakdown, causing more electrons [14]. As the dosage increases further, however, the injected field decreases owing to the drop of V_{ch} .Thus, a peak is observed.

III. Conclusions

Reliability issues with different junction dosages are investigated. Though the structure with lighter S/D junction is good to control short channel effect, cycling induced degradations, RTN amplitude, and abnormal tail induced by hot carrier injection are enhanced. The tail deteriorates the disturbed window as the product is stressed. Moreover, the mechanism of disturbance is proposed in the junction-free NAND array. Band-to-band induced substrate hot electron injection is the root cause. Finally, an optimized range of junction dosage is given.

References

[1] M. Mizukami, K. Nishihara, H. Ishida, F. Aiso, T. Iguchi, D. Ichinose, A. Fukumoto, N. Aoki, M. Kondo, T. Izumida, H. Tanimoto, T. Enda, T. Suzuki, I. Mizushima, and F. Arai, "Depletion-type Cell-Transistor of 23nm Cell Size on Partial SOI Substrate for NAND Flash Memory," in *Proc. Int. Conf. Solid-State Devices and Materials*, pp. 865-866, 2009.

[2] R. Kuchibhatla, EEtimes, 2010.

[3] H.-S. Oh, S.-C. Lee, C.-S. Lee, D.-Y. Oh, T.-K. Kim, J.-H. Song, K.-H. Lee, Y.-K. Park, J.-H. Choi, and J.-T. Kong, "3-dimensional analysis on the cell string current of NAND Flash memory," *IEEE International Memory Workshop*, pp. 137-139, 2005.

[4] Yohwan Koh, "NAND Flash Scaling beyond 20nm," *IEEE International Memory Workshop*, pp. 3-5, 2009.

[5] C. H. Lee, J. Choi, Y. Park, C. Kang, B. I. Choi, H. Kim, H. Oh, and W. S. Lee, "Highly Scalable NAND Flash Memory with Robust Immunity to Program Disturbance Using Symmetric Inversion-Type Source and Drain Structure", *Symp. VLSI Tech.*, pp. 118-119, 2008.

[6] K. T. Park, J. Choi, J. Sel, V. Kim, C. Kang, Y. Shin, U. Roh, J. Park, J. S. Lee, J. Sim, S. Jeon, C. Lee, and K. Kim, "A 64-Cell NAND Flash Memory with Asymmetric S/D Structure for Sub-40nm Technology and Beyond", *Symp. VLSI Tech.*, pp. 19-20, 2006.

[7] A. Fayrushin, K. S. Seol, J. H. Na, S. H. Hur, J. D. Choi, and K. Kim, "The New Program/Erase Cycling Degradation Mechanism of NAND Flash Memory Devices," *IEDM Tech. Dig.*, pp. 823-826, 2009.

[8] A. Ghetti, C. Monzio Compagnoni, F. Biancardi, A. L. Lacaita, S. Beltrami, L. Chiavarone, A. S. Spinelli, A. Visconti, "Scaling trends for random telegraph noise in deca-nanometer Flash memories," *IEDM Tech. Dig.*, pp. 1-4, 2008.

[9] K. Fukuda, Y. Shimizu, K. Amemiya, M. Kamoshida, and C. Hu, "Random Telegraph Noise in Flash Memories-Model and Technology Scaling," *IEDM Tech. Dig.*, pp. 169-172, 2007.

[10] S. H. Bae, J. H. Lee, H. I. Kwon, J. R. Ahn, J. C. Om, C. H. Park, and J. H. Lee, "The 1/f Noise and Random Telegraph Noise Characteristics in Floating-Gate NAND Flash Memories," *IEEE Trans. Electron Devices*, vol.56, no.8, pp. 1624-1630, 2009.

[11] J. W. Wu, C. C. Cheng, K. L. Chiu, J. C. Guo, W. Y. Lien, C. S. Chang, G. E. Huang, and Tahui Wang, "Pocket Implantation Effect on Drain Current Flicker Noise in Analog nMOSFET Devices," *IEEE Trans. Electron Devices*, vol.51, no.8, pp. 1262-1266, 2004.

[12] Y. L. Chou, J. P. Chiu, H. C. Ma, Tahui Wang, Y. P. Chao, K. C. Chen, and C. Y. Lu, "Use of Random Telegraph Signal as Internal Probe to Study Program/Erase Charge Lateral Spread in a SONOS Flash Memory" *Proc. Int. Reliability Phys. Symp.*, pp. 960-963, 2010.

[13] Dongyean Oh, Changsub Lee, Seungchul Lee, Tae-Kyung Kim, Jaihyuk Song, and Jeonghyuk Choi, "A New Self-Boosting Phenomenon by Source/Drain Depletion Cut-off in NAND Flash Memory" *NVSMW*, pp. 39-41, 2007.

[14] Y. J. Chen, L. H. Chong, S. W. Lin, T. H. Yeh, K. F. Chen, K. F. Chen, J. S. Huang, C. H. Cheng, S. H. Ku, N. K. Zous, I. J. Huang, T. T. Han, T. H. Hsu, H. T. Lue, M. S. Chen, W. P. Lu, K. C. Chen, and C. Y. Lu, "Source/Drain Dopant Concentration induced Reliability issues in Charge Trapping NAND Flash Cells" *Proc. Int. Reliability Phys. Symp.*, pp. 634-637, 2010.

[15] H. T. Lue, E. K. Lai, Y. H. Hsiao, S. P. Hong, M. T. Wu, F. H. Hsu, N. Z. Lien, S. Y. Wang, L. W. Yang, T. Yang, K. C. Chen, K. Y. Hsieh, Rich Liu, and Chih-Yuan Lu, "A Novel Junction-Free BE-SONOS NAND Flash", *Symp. VLSI Tech.*, pp. 140-141, 2008.

[16] S. H. Ku, Tahui Wang, W. P. Lu, Wenchi Ting, Y. H. Joseph Ku, and C. Y. Lu, "Characterization of Programmed Charge Lateral Distribution in a Two-Bit Storage Nitride Flash Memory Cell by Using a Charge-Pumping Technique," *IEEE Trans. Electron Devices*, vol.53, no.1, pp. 103-108, 2006.

[17] K. D. Suh, B. H. Suh, Y. H. Lim, J. K. Kim, Y. J. Choi, Y. N. Koh, S. S. Lee, S. C. Kwon, B. S. Choi, J. S. Yum, J. H. Choi, J. R. Kim, and H. K. Lim, "A 3.3V 32Mb NAND Flash Memory with Incremental Step Pulse Programming Scheme" *IEEE JOURNAL OF SOLID-STATE CIRCUITS*, vol.30, no.11, pp. 1149-1156, 1995.

[18] Taehee Cho, Y. T. Lee, E. C. Kim, J. W. Lee, Sunmi Choi, Seungjae Lee, D. H. Kim, W. G. Han, Y. H. Lim, J. D. Lee, J. D. Choi, and K. D. Suh, "A Dual-Mode NAND Flash Memory: 1-Gb Multilevel and High-Performance 512-Mb Single-Level Modes" *IEEE JOURNAL OF SOLID-STATE CIRCUITS*, vol.36, no.11, pp. 1700-1706, 2001

[19] E. Li, E. Rosenbaum, J. Tao, G. C-F Yeap, M-R. Lin, and P. Fang, "Hot Carrier Effects in nMOSFETs in 0.1um CMOS Technology" *Proc. Int. Reliability Phys. Symp.*, pp. 253-258, 1999.

[20] Ih-Chin Chen, and Clarence W. Teng, "A Quantitative Physical Model for the Band-to-Band Tunneling-Induced Substrate Hot Electron Injection in MOS Devices," *IEEE Trans. Electron Devices*, vol.39, no.7, pp. 1646-1651, 1992.

Charge diffusion in silicon nitrides: scalability assessment of nitride based flash memory

Seung Jae Baik[*] and Koeng Su Lim
Department of Electrical Engineering
Korea Advanced Institute of Science and Technology
Yuseong-gu, Daejeon 305-701, Korea
[*]solar100@kaist.ac.kr

Wonsup Choi, Hyunjun Yoo, and Hyunjung Shin[+]
School of Advanced Materials Engineering
Kookmin University
Seoul, 136-702, Korea
[+]hjshin@kookmin.ac.kr

Abstract—**Electron and hole diffusion coefficients of stoichiometric silicon nitride, silicon rich nitride, and silicon oxynitride were evaluated from variable temperature electrostatic force microscopy (EFM) analysis. Among them, stoichiometric silicon nitride is shown to have smallest diffusion coefficient although silicon oxynitride has the higher temperature activation energy. Scaling charge trap flash towards sub-20nm regime should be accompanied by hole dispersion management, minimization of internal electric field, and adjustment of retention specification.**

Keywords-EFM;nitride;charge trap;flash memory;scalability

I. INTRODUCTION

It has been pointed out that the growth of information triples every two years [1]. This implies that memory density in our information world still has the momentum of rapid growth, and thus development of high density memory for data storage is highly required. NAND flash memories, whose bit density has doubled every year during recent 10 years [2], confront scaling challenges such as electrostatic interference [3] or number of electron fluctuation [4]. Accordingly, nitride trap based flash memories have been widely investigated for their potential scalability gain compared to conventional floating gate flash [5]. However, nitride based flash as a stand-alone solution for a high density data storage does not seem to be probable [6, 7], but its three dimensional implementation would be highly probable due to its potential reduction of bit cost [8, 9].

Performance measure of nitride trap material is usually represented by energy depth of trap levels [10]. Although some variation of stoichiometry in nitride materials [11] or other metal oxide materials [12, 13] have been proposed to show improved memory reliability, long term charge retention properties better than that of stoichiometric silicon nitride have never been demonstrated with small device dimensions. In nitride based flash, charge loss in vertical direction through tunnel oxide or blocking layer does not fully represent the reliability properties of the trap material in small devices. It was already pointed out that the lateral charge migration in nitrides strongly affects the reliability of the nitride based flash memories [14-17]. Hopping mobilities determined by band tails may impact lateral charge dispersion, which could be assessed by transient electrical characterization of test patterns [16, 18].

Charge transport in insulators such as silicon nitrides can be described by Poole-Frenkel emission, hopping conduction through tail states, and trap-to-trap tunneling. Among them, Poole-Frenkel emission and trap-to-trap tunneling mainly represent long term charge transport, i.e., charge retention; and the former has a dominant effect at high temperatures due to its strong temperature dependence. In this work, to eliminate complication due to electric field induced barrier lowering, we investigate transport properties of charged pockets having tiny electric field profiles using a variable temperature EFM to quantify diffusion coefficients, which represents intrinsic transport property of nitrides irrespective of surrounding electric field profiles. Further consideration of enhanced charge transport with the presence of considerable electric fields is performed based on Poole-Frenkel model, which would provide insights into the fundamental assessment for sub-20nm scaling of charge trap flash memories.

II. EXPERIMENTAL: VARIABLE TEMPERATURE EFM

Combined analysis of variable temperature EFM visualization and quantitative modeling of charge retention was also reported focusing on vertical charge de-trapping by Tzeng and Gwo [19]. They have used nitride/oxide/Si (NOS) structures with a sufficiently thin oxide layer (~2 nm) to observe the vertical tunneling loss while suppressing the lateral charge migration. Spatial resolutions of electric potential measurements by scanned probes, which rely on the size of the features, its surroundings, and the probe geometry, are typically a few tens of nanometer; however parameterization of diffusion characteristics in large scale at high temperature could lead to predict nanoscale diffusion. We have visualized both the trapped electrons and holes by a modified and temperature variable EFM under high vacuum. Lateral redistribution of the trapped charges was analyzed by a simple electrostatic modeling with the suppression of vertical charge loss by thick oxide layers (~7 nm) in NOS structures. Sample structure and EFM analysis methods are summarized in Figure 1. Characterized nitride samples are stoichiometric nitride (Si_3N_4), silicon rich nitride (SRN), and silicon oxynitride (SiON). Their compositions were analyzed by X-ray photoelectron spectroscopy as shown in Table I. Measurement temperatures of EFM were varied from $150^{\circ}C$ to $450^{\circ}C$.

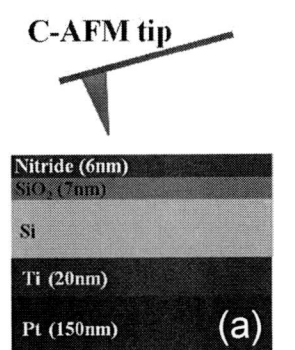
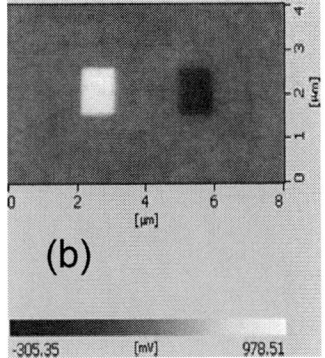

Figure 1. Analysis method of variable temperature EFM. (a) Sample structure and probing schematic. (b) Top view of EFM image. Bright rectangle is positively charged region, and the dark rectangle is negatively charged region.

Trapped electrons and holes in NOS have been visualized by an environmental controlled, variable-temperature scanning probe microscopy (E-sweep, Seiko Instruments, Japan). Probes with Pt-coated tips with the typical resonance frequency of 5 – 25 kHz were used. Injection of the electrons and holes was done by scans (scan size of 1 x 1 μm and speed of 1 Hz) in the contact mode between the tips and the surface of the NOS with the contact forces of 0.002 – 0.08 nN in high vacuum ($10^{-7} \sim 10^{-8}$ torr). Applied voltages for the injection of electrons and holes were ± 2.5 V, and the duration of injection time was set large enough to attain saturated charge density. For imaging, both ac and dc voltages ($V_{app.} = V_{dc} + V_{ac}\sin\omega t$) were applied to the substrate while the conductive tip is grounded. All the images of the injected charges were taken in the mode of non-contact with the scan speed of 0.3 Hz. The surface potential of the tip is equal to the applied external dc voltages obtained from the feedback loop modulation to minimize the electrostatic force between the tip and the sample. The applied external dc voltages were main output for the mapping of surface potentials. This modified EFM mode is also known and called Kelvin probe force microscopy (KPFM). The detailed experimental set up and its principles can be found elsewhere [20].

TABLE I.   ESTIMATED COMPOSITION OF SAMPLES USED IN THIS EXPERIMENT.

Nitrides	Atomic % (XPS)		
	Si	O	N
Si_3N_4	42.84	6.04	52.13
SRN	46.13	5.01	48.86
SiON	42.06	22.78	33.70

III.   PARAMETER EXTRACTION

Charge decay profiles for different nitride materials are shown in Figure 2, where initial charge profiles were formed at the same temperature of decay measurement. Line profiles obtained from the rectangular charged patterns shown in Figure

1(b) are drawn for various elapsed times. At the same measurement temperature, Si_3N_4 shows the smallest decay, and SiON shows the largest decay. The difference of electron and hole injection efficiency results in the different maximum potential for electron and hole charged area. The charge decay originates from the vertical charge loss through tunnel oxide and the lateral diffusion. Vertical charge loss at elevated temperature is attributed to the thermal emission of trapped charges to the band edge of nitrides in combination with direct tunneling through tunnel oxide. The lateral diffusion may depends on the lateral electric field that could contribute barrier lowering of trap energies, but the lateral electric field is small enough in our experiment to alleviate electric field induced diffusion enhancement. Figure 3 shows the variation of charge decay at different temperatures. Enhancement of charge decay at higher temperatures can be attributed to enhanced vertical loss and lateral diffusion, because both are strongly dependent on thermal activation.

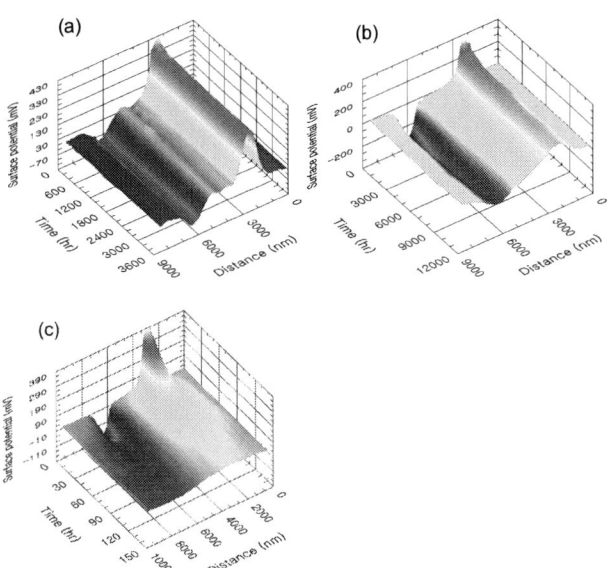

Figure 2.   Charge decay profile measured at 250°C. (a) Si_3N_4 (b) SRN (c) SiON.

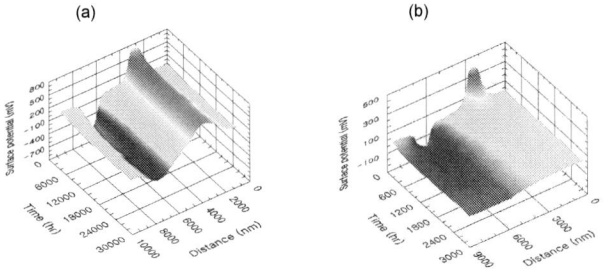

Figure 3.   Charge decay profile measured of SiON measured at different temperatures. (a) 150°C (b) 200°C

Line profiles of charge decay that are used for the parameter extraction are shown in Figure 4(a). The maximum potentials are decreasing and the widths of Gaussian line profiles are increasing as time elapses. To de-convolute the contribution of vertical charge loss and lateral diffusion, integrated values of potential line profiles are used as an approximate measure of the total charge density. That is, we have approximated the Gaussian as a rectangular charge profile with a constant charge density to estimate total charge density. As shown in Figure 4(b), time evolution of integrated line profile corresponds to the total amount of vertical charge loss. Flat region in Figure 4(b) shows the region where vertical charge loss is not pronounced. As indicated in the figure by "diffusion only", line profiles in this time region can be used to characterize diffusion coefficient of nitrides. Integrated charge densities for various nitrides are shown in Figure 5, where Si_3N_4 shows diffusion only decay, while SiON shows dominant vertical charge loss for all the measurement time range. To attain line profiles for diffusion analysis, we have used data from different measurement temperatures: for Si_3N_4, 250°C ~ 450°C, for SRN, 250°C ~ 300°C, and for SiON, 150°C ~ 200°C.

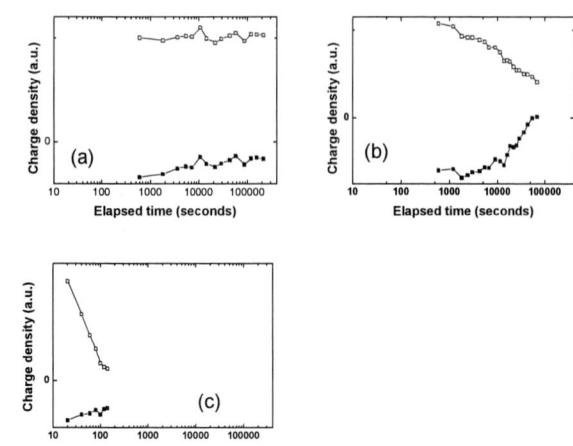

Figure 5. Charge density evaluated from the integration of potential line profiles for (a) Si_3N_4 (b) SRN (c) SiON. All measurements were performed at 250°C.

Selected potential line profiles are fitted by the following equation to quantify diffusion coefficient.

$$p = \frac{U}{2\sqrt{\pi D_p (t + t_0)}} \exp\left(-\frac{x^2}{4 D_p (t + t_0)}\right) \qquad (1)$$

where p, U, D, t and x is charge density, constant related to initial charge density, diffusion coefficient, elapsed time, and lateral dimension, respectively. Typical fitting results are illustrated in Figure 6, where both curves have the same integrated value, i.e., the same charge density. To prove the validity of diffusion assumption on lateral charge transport, extracted diffusion coefficients were plotted against lateral electric fields as shown in Figure 7, which were evaluated from differentiated surface potential profiles. Diffusion coefficients do not vary significantly as the time elapses (represented as decreases in the square root of electric field), which implies that lateral electric field in our experimental setup is small enough to neglect electric field induced barrier lowering (Poole-Frenkel effect) for trapped charges in lateral direction.

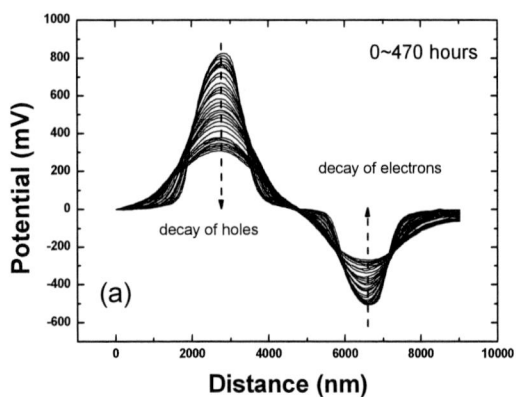

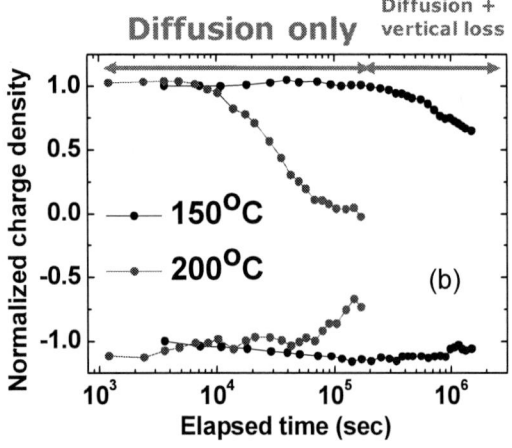

Figure 4. (a) Charge decay line profile of SiON measured at 150°C. (b) Integrated value of line profiles are normalised and plotted versus elaped time for SiON at 150°C and 200°C.

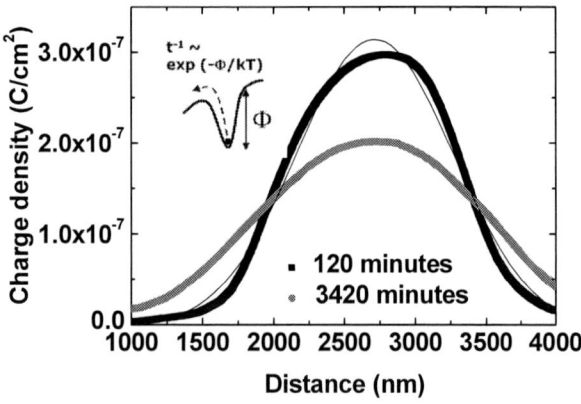

Figure 6. Charge density line profiles (thick lines) and their fitted curve using equation (1) (thin lines) for SiON sample.

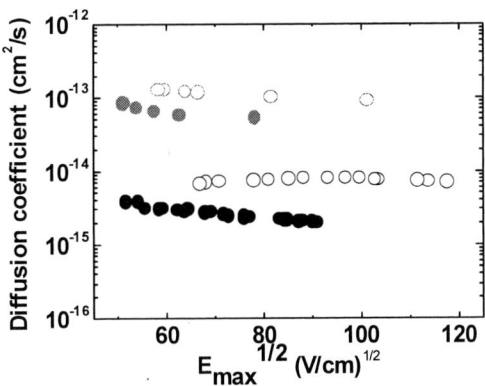

Figure 7. Extracted diffusion coefficients for each line profiles of SiON sample. Filled symbols are for electrons and open symbols are for holes. Black symbols are for data measured at 150°C, and red symbols are for data measured at 200°C. E_{max} represents maximum electric field calculated from potential line profile.

Extracted diffusion coefficients were fitted for different temperatures according to the following equation:

$$D = D_0 \exp\left(-\frac{\Phi_a}{kT}\right) \qquad (2)$$

where Φ_a is the activation energy for diffusion implying the average energy depth of traps, D_0 is the proportionality constant implying diffusion coefficient at infinite temperature, and kT is the thermal energy. Extracted parameters are summarized in Table II. Φ_a for hole is smaller than electron for all cases, and Φ_a of SRN is the smallest and that of SiON is the largest. This is consistent with previous reports on the relatively shallower and deeper nature of trap levels in SRN and SiON [20, 21], which is also consistent with bandgap variation of nitrides.

The extracted diffusion coefficient D is the smallest for Si_3N_4, even though SiON has the largest trap energy. This implies that the lateral diffusion is not determined only by trap energy. That is, in addition to the thermal emission from traps, hopping through tail states and trap-to-trap tunneling should also be considered for the transport between traps. It was empirically demonstrated that Urbach energy, representing the inverse of the slope of tail state density, has a proportional relation with the abundance ratio (atomic ratio: $\{(1/2)O+(3/4)N\}/Si$) [21]. From Table I, the abundance ratio of Si_3N_4, SRN, and SiON corresponds to 0.98, 0.85, and 0.87, respectively. That is, SRN and SiON have a broader tail state distribution than Si_3N_4, which implies lower conductivity of hopping transport. Accordingly, diffusion enhancement of SRN and SiON is not due to the enhancement of hopping conduction, but thus due to the enhancement of tunneling transport, because the broadening of tail states should also be accompanied by the level broadening of deep traps, which would lead to the enhancement of trap to trap tunneling. In addition, SRN is expected to have increased trap density compared to Si_3N_4, which also contributes to the enhanced tunneling transport. However, considerable trap-to-trap tunneling implies that significant overestimation has occurred in the extraction of diffusion coefficients, where we have assumed only diffusion

transport. Those errors reside only in D_0 and not in trap energies.

Although there may have occurred some overestimation of D_0's at high temperatures for SRN and SiON, we believe that the D_0 of Si_3N_4 should be free from those errors. As pointed out in the introduction section, the dominant transport mechanism should be diffusion in the temperature range of our experiment (above 150°C), which has been confirmed in the literature many times [10, 16] for stoichiometric silicon nitrides.

TABLE II. EXTRACTED DIFFUSION PARAMETERS FOR NITRIDE SPLITS. SUBSCRIPTS E DENOTES ELECTRONS, AND H DENOTES HOLES.

Nitrides	Extracted diffusion parameters			
	D_{0_e} (cm²/s)	D_{0_h} (cm²/s)	Φ_{a_e} (eV)	Φ_{a_h} (eV)
Si_3N_4	2.13E-5	4.85E-7	1.14	0.93
SRN	4.47E-3	1.13E-7	1.07	0.59
SiON	2.53	0.284	1.24	1.15

IV. SCALABILITY ASSESSMENT OF NITRIDE BASED FLASH

Lateral dispersion of charges in nitride based flash would result in data loss originating from threshold voltage fluctuation. The fluctuation becomes more critical as the device dimension shrinks, thus the definition of lateral dispersion length for nitrides would give a helpful guide to assess the scalability of nitride based flash. Figure 8 illustrates the definition of dispersion lengths for trap nitrides. The intrinsic dispersion length signifies the shortest dimension of lateral charge migration, which is on the order of average trapping distance ($1/N_t\sigma$, N_t: trap density, σ: capture cross section) that is around or smaller than a few nanometers [22]. This is the minimum charge profile broadening occuring during charge injection (tunneling program/erase). Therefore, a few nanometer would be the minimum length scale of nitride based flash, when we do not consider retention loss.

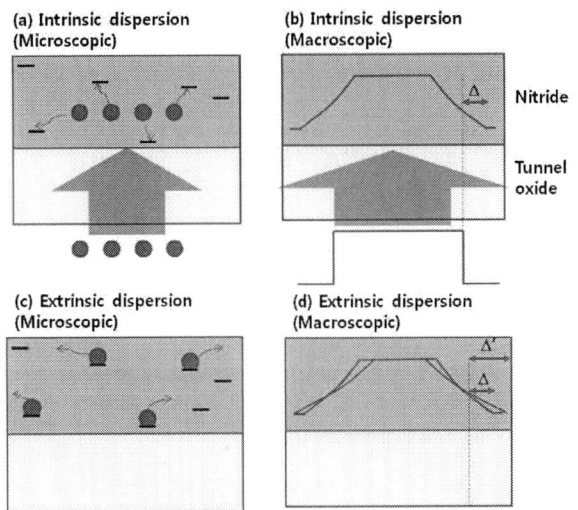

Figure 8. Illustration of lateral charge dispersion. Δ is intrinsic dispersion length occuring instantaneouly after programming, and Δ' represents extrinsic dispersion length causing retention loss.

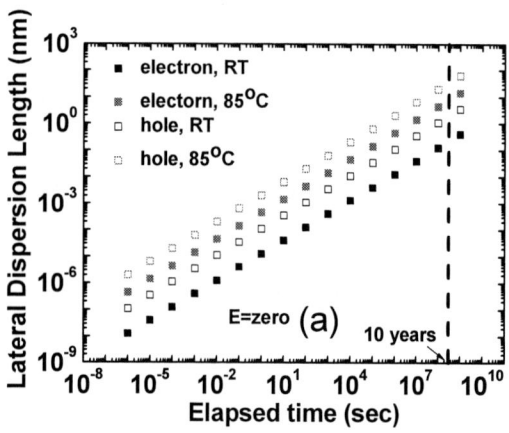

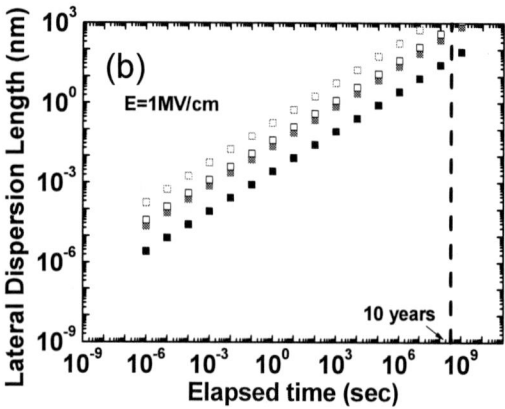

Figure 9. Calculated values of (extrinsic) lateral dispersion length versus elapsed time after programming for stoichiometric silicon nitride. (a) Lateral electric field is neglected. (b) Moderate electric field (1MV/cm) is assumed.

In addition to the intrinsic dispersion length, further broadening of charge profile during retention regime is termed extrinsic dispersion length in Figure 8. In this regime, lateral charge diffusion in realistic devices should also be affected by electric field induced diffusion enhancement (Poole-Frenkel emission). The enhancement of diffusion caused by lateral electric field can be represented as follows:

$$D = D_0 \exp\left(-\frac{\Phi_a - \sqrt{\frac{qE}{\pi\varepsilon}}}{kT}\right) \qquad (3)$$

Figure 9 shows the projected lateral dispersion length $(D\tau)^{1/2}$ (τ: elapsed time) with and without considering barrier lowering parameter E (lateral electric field) in equation (3) for Si_3N_4. When electric field induced barrier lowering is neglected, 10 year retention of lateral dispersion within 10nm is possible both for electrons and holes at room temperature, but hole dispersion is larger than 10nm at high temperature of 85°C. Moreover, when moderate electric field is considered, only electron dispersion at room temperature meet 10nm

requirement, and hole dispersion could be even higher than 100nm at room temperature. Even at this moderate electric field, 10nm lateral dispersion could be met with elapsed time smaller than 10^6. In this projection, only diffusion transport was considered, therefore considering possible contribution of trap-to-trap tunneling, this estimation provides lower bounds of lateral dispersion lengths. That is, these values could become even higher when the contribution of trap-to-trap tunneling exceeds diffusion transport.

Vianello, et al, recently reported a work on charge dynamics in silicon nitride based on similar experimental setup, where they focused on electron diffusion dynamics and additional quantification was performed based on numerical simulation [23]. In this work, we provide a more complete description of charge dynamics attainable by EFM, and present analyses on both carriers (electrons and holes) at various temperatures, which was supported by quantification of temperature dependent diffusion coefficients based on a simple analytic model.

V. CONCLUSION

Scaling charge trap flash memories towards sub-20nm regimes requires (1) hole dispersion management, (2) minimization of internal electric field, and (3) adjustment of retention specification when conventionally available stoichiometric silicon nitride is used for charge trapping layer.

ACKNOWLEDGEMENT

The authors acknowledge the financial support from the NRL Program (2007-0057024), the Nano R&D Program (2009-80 0082717), and the CMPS (R11-2005-048-00000-0) of NRF. This work was in part supported by the 2009 research program of Kookmin University.

REFERENCES

[1] J. F. Gantz, "The expanding digital universe", IDC white paper, March 2007

[2] K. Kim, "From The Future Si Technology Perspective: Challenges and Opportunities", Technical Digest of IEEE International Electron Device Meeting 2010, pp. 1-4, 2010

[3] J.-D. Lee, S.-H. Hur, and J.-D. Choi, "Effects of floating-gate interference on NAND flash memory cell operation", IEEE Electron Device Letters, Vol. 23, Issue 5, pp. 264-266, 2002

[4] G. Molas, D. Deleruyelle, B. DeSalvo, G. Ghibaudo, M. Gely, S. Jacob, D. Lafond, and S. Deleonibus, "Impact of few electron phenomena on floating-gate memory reliability", Technical Digest of IEEE International Electron Devices Meeting 2004, pp.877-880, 2004

[5] C.-H. Lee, J. Choi, Y. Park, C. Kang, B.-I Choi, H. Kim, H. Oh, and W.-S. Lee, "Highly Scalable NAND Flash Memory with Robust Immunity to Program Disturbance Using Symmetric Inversion-Type Source and Drain Structure", 2008 Symposium on VLSI Technology Digest of Technical papers, pp. 118-119, 2008

[6] C.-H. Lee, S.-K. Sung, D. Jang, S. Lee, S. Choi, J. Kim, S. Park, M. Song, H.-C. Baek, E. Ahn, J. Shin, K. Shin, K. Min, S.-S. Cho, C.-J. Kang, J. Choi, K. Kim, J.-H. Choi, K.-D. Suh, and T.-S. Jung, "A Highly Manufacturable Integration Technology for 27nm 2 and 3 bit/cell NAND flash memory", Technical Digest of IEEE International Electron Device Meeting 2010, pp. 1-4, 2010

[7] K. Prall and K. Parat, "25nm 64Gb MLC NAND Technology and Scaling Challenges", Technical Digest of IEEE International Electron Device Meeting 2010, pp. 1-4, 2010

[8] J. Jang, H.-S. Kim, W. Cho, H. Cho, J. Kim, S. I. Shim, Y. Jang, J.-H. Jeong, B.-H. Son, D. W. Kim, J.-J. Shim, J. S. Lim, K.-H. Kim, S. Y. Yu, J.-Y. Lim, D. Chung, H.-C. Moon, S. Hwang, J.-W. Lee, Y.-H. Son, U.-I. Chung, and W.-S. Lee, "Vertical cell array using TCAT (Terabit Cell Array Transistor) technology for ultra high density NAND flash memory", Symposium on VLSI Technology 2009, pp. 192-193, 2009

[9] Y. Komori, M. Kido, M. Kito, R. Katsumata, Y. Fukuzumi, H. Tanaka, Y. Nagata, M. Ishiduki, H. Aochi, and A. Nitayama, "Disturbless Flash Memory due to High Boost Efficiency on BiCs Structure and Optimal Memory Film Stack for Ultra High Density Storage Device", Technical Digest of IEEE International Electron Device Meeting 2008, pp. 1-4, 2008

[10] T. H. Kim, J. S. Sim, J. D. Lee, H. C. Shin, and B. –G. Park, "Charge decay characteristics of silicon-oxide-nitride-oxide-silicon structure at elevated temperatures and extraction of the nitride trap density distribution", Applied Physics Letters, Vol. 85, No. 4, pp. 660-662, 2004

[11] K. H. Wu, H. –C. Chien, C. –C. Chan, T. –S. Chen, and C. –H. Kao, "SONOS Device With Tapered Bandgap Nitride Layer", IEEE Transactions on Electron Devices, Vol. 52, No. 5, 2005

[12] Z. L. Huo, J. K. Yang, S. H. Lim, S. J. Baik, J. Lee, J. H. Han, I.-S. Yeo, U.-I. Chung, J. T. Moon, and B. –L. Ryu, "Band Engineered Charge Trap Layer for Highly Reliable MLC Flash Memory", 2007 Symposium on VLSI Technology Digest of Technical Papers, pp. 138-139, 2007

[13] C. Y. Tsai, T. H. Lee, A. Chin, H. Wang, C. H. Cheng, and F. S. Yeh, "Highly-Scaled 3.6nm ENT Trapping Layer MONOS Device with Good Retention and Endurance", Technical Digest of IEEE International Electron Device Meeting 2010, pp. 1-4, 2010

[14] B. Eitan, P. Pavan, I. Bloom, E. Aloni, A. Frommer, and D. Finzi, "NROM: A novel localized trapping, 2-bit nonvolatile memoroy cell," IEEE Electron Device Letters, Vol. 21, pp. 543–545, 2000.

[15] C. Kang, J Choi, J. Sim, C. Lee, Y. Shin, J. Park, J. Sel, S. Jeon, Y. Park, K. Kim, "Effects of lateral charge spreading on the reliability of TANOS

(TaN/AlO/SiN/Oxide/Si) NAND flash memory", 45th Annual International Reliability Physics Symposium, pp.167-169, 2007.

[16] S. Choi, S. J. Baik, and J.-T. Moon, "Band Engineered Charge Trap NAND Flash with sub-40nm Process Technologies", Technical Digest of IEEE International Electron Device Meeting 2008, pp. 925-928, 2008

[17] W. Sakamoto, T. Yaegashi, T. Okamura, T. Toba, K. Komira, K. Sakuma, Y. Matsunaga, Y. Ishibashi, H. Nagashima, M. Sugi, N. Kawada, M. Umemura, M. Kondo, T. Izumida, N. Aoki, and T. Watanabe, "Reliability Improvement in Planar MONOS Cell for 20nm-node Multi-Level NAND Flash Memory and Beyond", Technical Digest of IEEE International Electron Device Meeting 2009, pp. 1-4, 2009

[18] B. Kim, S. J. Baik, S. Kim, J.–G. Lee, B. Koo, S. Choi, and J.-T. Moon, "Characterization of Threshold Voltage Instability After Program in Charge Trap Flash Memory", 47th Annual International Reliability Physics Symposium, pp.284-287, 2009

[19] S.-D. Tzeng and S. Gwo, "Charge trapping properties at silicon nitride/silicon oxide interface studied by variable-temperature electrostatic force microscopy", Journal of Applied Physics, Vol. 100, 023711, 2006

[20] K. Rue, S. Kim, Y. Choi, H. Shin, C.-H. Kim, J.B. Yun, and B. Lee, "Some Examples of Local Electrical Characterization by Scanning Probe Microscopy", Electronic Materials Letters, Vol. 3, pp.v127-135, 2007

[21] H. Kato, N. Kashio, Y. Ohki, K. S. Soel, and T. Noma, "Band-tail photoluminsecence in hydrogenated amorphous oxynitride and silicon nitride films", Journal of Applied Physics, Vol. 93, pp.239-244, 2003

[22] P. C. Arnett and B. H. Yun, "Silicon nitride trap properties as revealed by charge-centroid measurements on MNOS devices," Applied Physics Letters, Vol. 26, pp. 94-96, 1975.

[23] E. Vianello, E. Nowak, D. Mariolle, N. Chevalier, L. Perniola, G. Molas, J. P. Colonna, F. Driussi, and L. Selmi, "Direct probing of trapped charge dynamics in SiN by Kelvin force microscopy", 2010 IEEE International Conference on Microelectronic Test Structures, pp. 94-97, 2010

978-1-4244-9113-1/11 $26.00 © 2011 IEEE

The effect of Crystallinity of HfO₂ on the Resistive Memory Switching Reliability

Min Gyu Sung, Wan Gee Kim*, Jong Hee Yoo, Sook Joo Kim, Jung Nam Kim, Byung Gu Gyun, Jun Young Byun, Taeh Wan Kim, Won Kim, Moon Sig Joo, Jae Sung Roh, and Sung Ki Park

R&D Division, Hynix Semiconductor Inc.

San 136-1 Ami-ri, Bubal-eub, Ichon-si, Kyoungki-do, 467-701, Korea

Tel) +82-31-639-0929, Fax) +82-31-645-8139, *E-mail) wangee.sung@hynix.com

Abstract - We measured direct physical evidence that can explain different switching behaviors for HfO₂ based ReRAM structure for the first time. The switching behavior depends on the degree of crystallinity of HfO₂. We observed crystallized HfO₂ which is grown under higher temperature with High Resolution Scanning Transmission Electron Microscopy (STEM) Electron Energy Loss Spectroscopy (EELS), which prevents creating enough oxygen vacancies to start electric switching under low forming voltage. However, non-stoichiometric structure region of HfO₂ which has amorphous phase provides enough oxygen vacancies which are the electrical paths. Therefore, non-stoichiometric structure of amorphous HfO₂ during fabricating process is crucial to obtain reliable switching characteristics.

I. Introduction

Various resistance-based nonvolatile memory devices such as ReRAM, MRAM and PRAM have been proposed due to the scaling limit of conventional charge-based flash memory device. Resistance Random Access Memory (ReRAM) is one of the best candidates because it has simple structure, fast switching speed and high scalability [1]. Recently, superior performance of HfO₂ switching using Ti/TiN electrode was reported [2~4]. It showed lower operating voltage and current and satisfactory endurance properties than other transition Metal Oxide (TMO). However, the important thing is that the switching mechanisms of ReRAM devices have not been fully understood yet. Nowadays, it has been shown that the possible resistive switching mechanisms of binary oxides involve a migration of oxygen ions or oxygen vacancies in binary oxides. Many investigators are trying to explain the switching mechanisms of binary oxides based ReRAM devices with only electrical data because it is hard to confirm the behavior of oxygen ions or oxygen vacancies directly.

In this paper, we measured direct physical evidence that can explain oxygen vacancy switching mechanism of HfO₂ based ReRAM. Switching behaviors depend on the degree of crystallinity of HfO₂. We observed that there were different oxygen behaviors between stoichiometric structure with crystal phase and non-stoichiometric structure with amorphous phase of HfO₂. The structure properties of HfO₂ were analyzed via measurement of atomic scale Electron Energy Loss Spectroscopy (EELS) with a spherical aberration corrected Transmission Electron Microscope (Cs-Corrected TEM, FEI).

II. Experimental procedures

The experimented memory structure is a TiN/HfO₂/Ti/TiN fabricated using sub-50 nm process technology with 300 mm wafers. The HfO₂ (~30Å) was deposited by atomic layer deposition method. Ozone was used as the oxidizing source for the HfO₂ deposition and tetrakis-ethylmethylamido hafnium (TEMAHf) was used as a precursor. The bottom electrode (TiN) and top electrode (Ti (100Å) / TiN) were deposited by physical vapor deposition (PVD) method. Schematic process flow of the fabricated ReRAM structure is shown in Fig. 1. Electric I-V measurements were conducted by applying external voltage bias between the top TiN electrode and bottom tungsten layer. Here, the tungsten layer was grounded during the electrical measurement and the voltage ramp speed is 1.06 V/sec. Detailed electrical switching behaviors are presented elsewhere [5]. The Ti is inserted to create oxygen vacancies in the HfO_2, which is believed to be a crucial component in filament ReRAM switching [2].

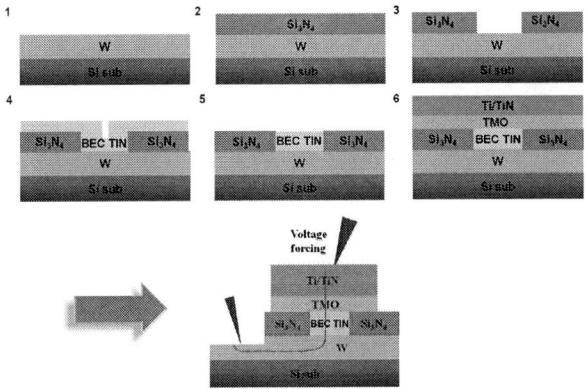

Figure 1. Schematic process flow of fabricated ReRAM structure.

III. Results and discussion

Fig. 2 shows bipolar switching characteristics for the fabricated $TiN/HfO_2/Ti/TiN$ structures. Upper one is for the HfO_2 deposited under 300°C, and lower one is for the 325°C, respectively. Lower deposition temperature shows repeating bipolar switching behavior, whereas higher temperature shows initial hard breakdown caused by local melting in bottom electrode contact (BEC) region during electrical forming. It seems that the lack of oxygen vacancies in the HfO_2 makes it difficult to create conductive filament under low forming voltage.

Fig. 3 shows the sheet resistance (Rs) of non-patterned $TiN/Ti/HfO_2/TiN$ structures depending on the deposition temperature of the HfO_2 film. Rs of the 325°C HfO_2 has higher value that of the 300°C HfO_2. From this result, we can expected that the difference in the Rs depending on the deposition temperature of the HfO_2 film may affect the resistive switching properties at the nano-scale device.

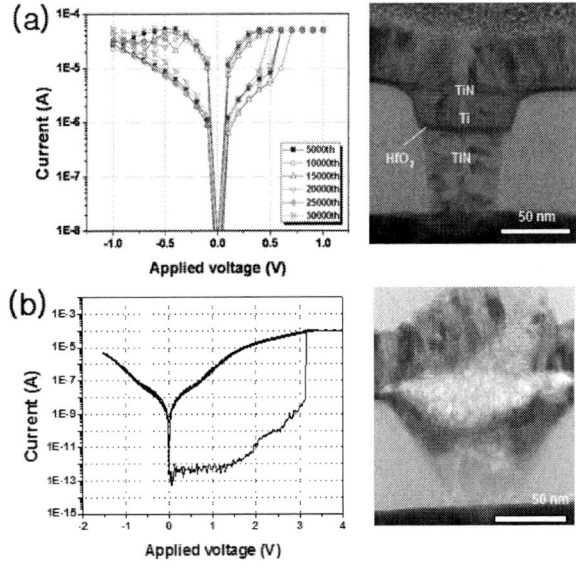

Figure 2. TEM Images and Bipolar switching characteristics for TiN/HfO_2/Ti/TiN structure: (a) is for HfO_2 deposited under 300°C, which shows repeating bipolar switching behavior. (b) is for HfO_2 deposited at 325°C. It shows initial hard breakdown caused by local melting in bottom electrode contact region during electric forming.

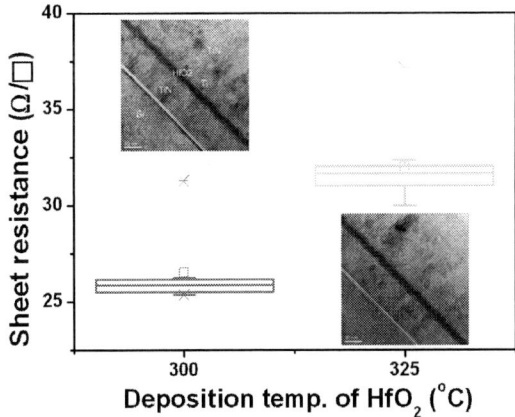

Figure 3. Sheet resistance (Rs) of non-patterned TiN/Ti/HfO_2/TiN structure depending on the deposition temperature of the HfO_2 film.

To explain this different switching behavior physically, we conducted atomic scale electron-energy-loss spectroscopy (EELS) with a spherical aberration corrected scanning transmission electron microscope (Cs-corrected STEM, FEI). Fig. 4 shows the STEM high-angle annular dark-field (HAADF) image and EELS elemental profile for the $TiN/HfO_2/Ti/TiN$ layers deposited at 300°C and 325°C. In the upper Ti/TiN electrode region, nitrogen is found to be diffused

978-1-4244-9113-1/11 $26.00 © 2011 IEEE   657

into the Ti. The nitrogen is provided during reactive sputtering of TiN which is done under a N_2 ambient. However, the observation of change in oxygen at the interface between Ti and HfO_2 can not be made clearly under this magnification. As shown by the nitrogen piled up at the interface between the HfO_2 deposited under 325°C and Ti, the amount of nitrogen at the interface is higher than that under 300°C. It means that the HfO_2 deposited at 325°C is different from that of 300°C physically or chemically and the diffused nitrogen nitrides Ti as well.

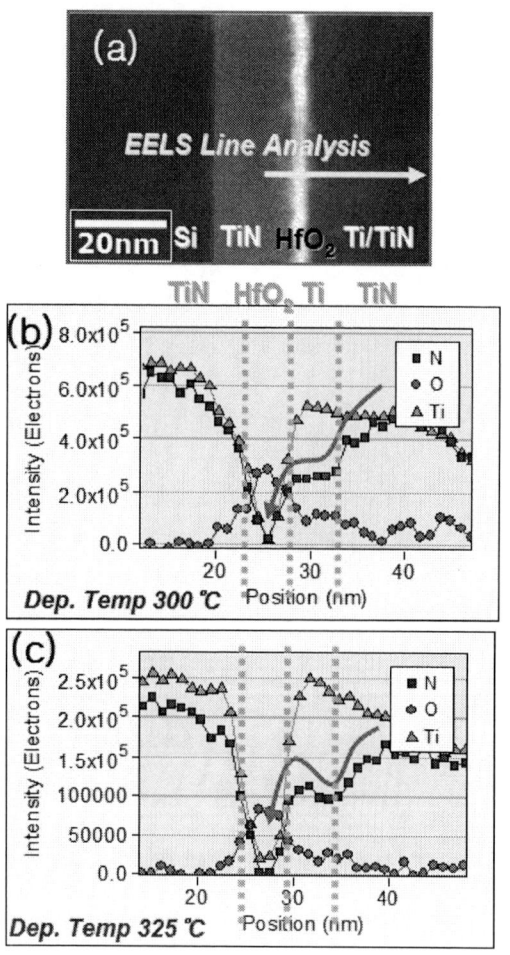

Figure 4. (a) STEM HAADF Image. (b) EELS elemental line profile of the TiN/ HfO_2/Ti/TiN structure whose HfO_2 was deposited under 300°C and (c) 325°C.

We try to observe the change of oxygen at the interface between HfO_2 and Ti, and in the HfO_2 with High Resolution STEM-EELS. At first, we are focusing on the interface. There are STEM HAADF Image and Oxygen EELS Spectra at the interface of HfO_2/Ti in Fig. 5. As shown in Fig. 5, the oxygen EELS Spectrum at the interface of HfO_2/Ti is different from that in the HfO_2. The oxygen EELS Spectrum extracted from center of HfO_2 is typical oxygen EELS Spectrum of HfO_2, whereas the interface region shows Ti-O compound characteristics, which is created by a reaction between the Ti and dissociated oxygen from the HfO_2. The shape of oxygen EELS spectrum at the interface between the HfO_2 deposited at 300°C and Ti is sharper than that under 325°C. It means that the strength of Ti-O bonding in 300°C is stronger than that of 325°C due to nitridation of Ti, which is a consistent result of Fig. 4. Secondly, we try to observe the change of oxygen behavior in the HfO_2. Fig. 6 shows High Resolution STEM HAADF images taken under atomic scale.

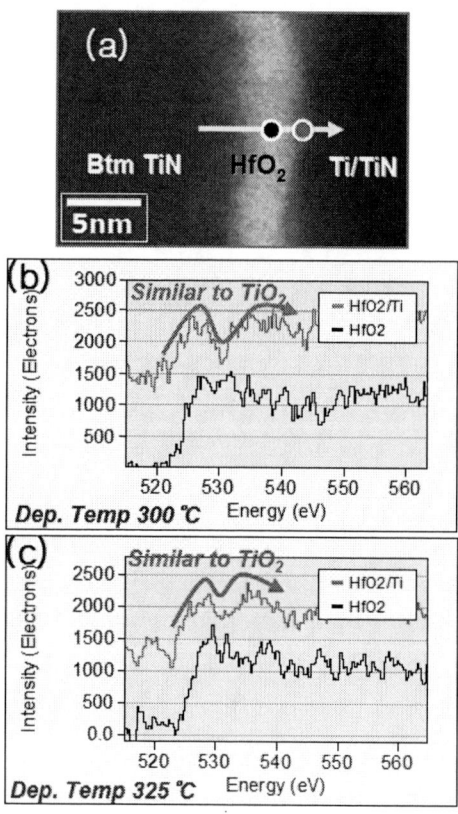

Figure 5. (a) STEM HAADF Image. (b) and (c) Oxygen EELS Spectra of HfO_2 and HfO_2/Ti Interface region. Center of the HfO_2 deposited under 300°C and 325°C shows typical O EELS spectrum of HfO_2, whereas that of Interface region shows Ti-O compound characteristics.

978-1-4244-9113-1/11 $26.00 © 2011 IEEE

As shown in Fig. 7, HAADF intensity profile is not exactly same as oxygen intensity profile. The EELS elemental line profile for the HfO_2 region deposited in 300°C under atomic scale shows the intensity of oxygen in upper part of HfO_2 is smaller than that of lower part, which means oxygen vacancies are created in the upper part of HfO_2. Also, the HfO_2 shows amorphous phase. On the other hand, the amount of oxygen in the upper HfO_2 deposited in 325°C does not decrease. By comparing the STEM images of Fig. 6 we can notice the higher degree of crystallinity of the HfO_2 deposited under 325°C. It seems that enough oxygen vacancies, which are necessary for the electric forming, are not created in the crystallized HfO_2.

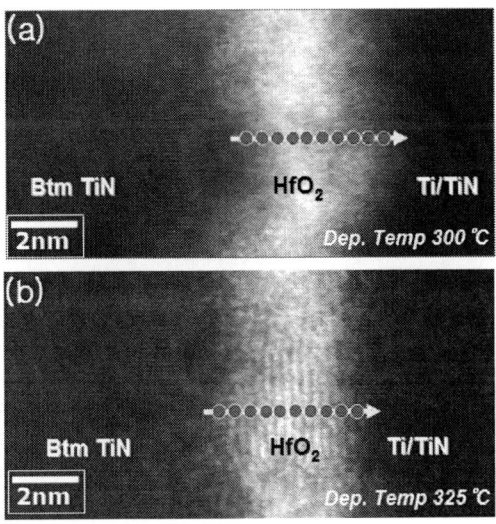

Figure 6. HR STEM HAADF Images. (a) HfO_2 deposited at 300°C (b) deposited at 325°C. The HfO_2 grown under higher temperature shows crystallized phase.

To explain the existence of created oxygen vacancies in the HfO_2, we analyzed non-stoichiometric structure which is attributed to dangling bond of HfO_2 with oxygen pre-peak [6]. Fig. 8 shows the comparison of oxygen EELS spectra, which shows non-stoichiometric structure is observed form the peak shoulder near 527 eV for the HfO_2 deposited under 300°C, which can explain the existence of created oxygen vacancies in the HfO_2 having amorphous phase [7].

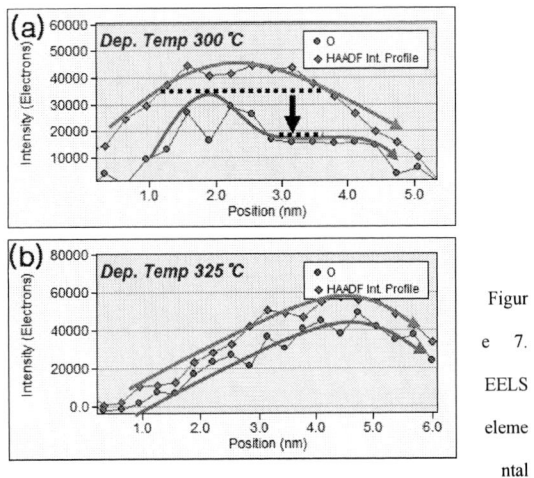

Figure 7. EELS elemental line profile for the HfO_2 region under atomic scale. The HfO_2 grown under lower temperature shows the intensity of oxygen in the upper part of HfO_2 is smaller than that of lower part.

During the deposition of Ti, most of Ti does not react with the HfO_2 (Some of Ti react with the HfO_2, as shown in Fig. 6). Therefore, the unreacted Ti reacts with the nitrogen during the reactive sputtering of TiN, which can explain the observed nitrogen piled up at the interface. From these results, it is believed that the oxygen vacancies are created during the deposition of Ti.

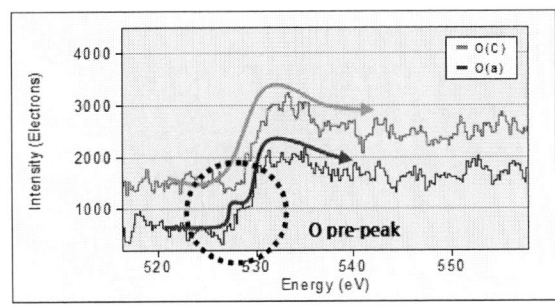

Figure 8. Non-stoichiometric structure is observed from the peak shoulder near 527eV for the HfO_2 grown under 300°C.

Accordingly, additional post anneal process is not necessary to create oxygen vacancies. Post annealing process under the temperature over 400°C leads to the local crystallization of HfO_2, which degrades the switching yield characteristics, which is shown in Fig. 10. Figure 11 shows switching endurance characteristics dependent on the crystallinity of HfO_2. The case without additional metal alloy anneal dose not show any degradation.

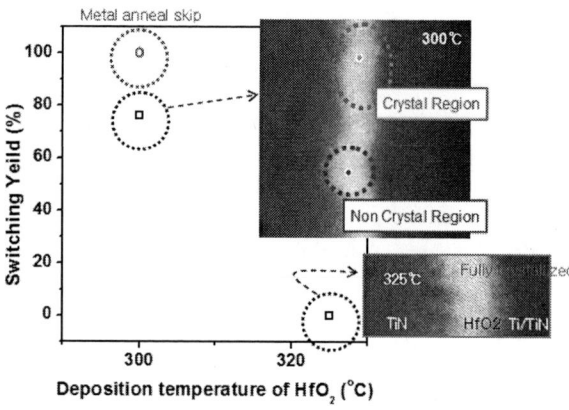

Figure 9. Switching yield dependence on the cystallinity of HfO₂. The case without an additional metal alloy anneal shows the best yield.

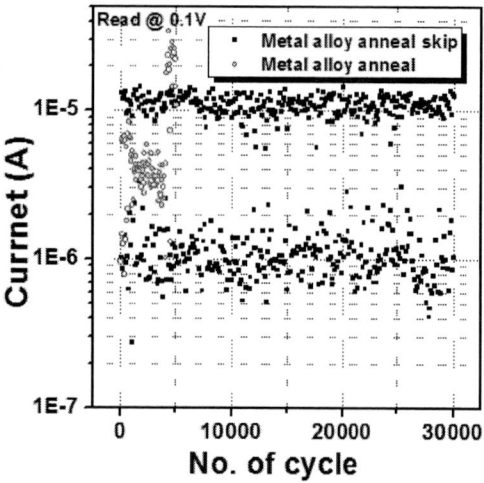

Figure 10. Endurance characteristics dependent on the cystallinity of HfO₂. The case without additional metal alloy anneal does not show any degradation.

IV. Conclusion

We conducted High Resolution STEM-EELS analysis to investigate the effect of non-stoichiometric structure of HfO_2 on switching behavior of ReRAM. As a result, oxygen EELS spectrum had a peak shoulder which could explain existence of oxygen vacancies with non-stoichiometric structure near 527eV for the HfO_2 deposited at 300℃. It was found that high amorphous phase of the HfO_2 is necessary to create enough oxygen vacancies for the better switching characteristics without additional post annealing process.

V. References

[1] S. Q. Liu, et al., "Electric-pulse-induced reversible resistance change effect in magnetoresistive films", Appl. Phys. Lett., Vol. 76, pp. 2749-2751, 2000.

[2] H. Y. Lee, at al, "Low Power and High Speed Bipolar Switching with A Thin Reactive Ti Buffer Layer in Robust HfO₂ Based RRAM", IEDM, 2008, pp. 1-4.

[3] Y. S. Chen, et al, IEDM, 2009, "Highly Scalable Hafnium Oxide Memory with Improvements of Resistive Distribution and Read Disturb Immunity", pp. 105-107.

[4] S.S. Sheu, et. al., "A 5ns Fast Write Multi-Level Non-Volatile 1 K bits RRAM Memory with Advance Write Scheme", VLSI, 2009, pp. 82-83.

[5] W. G. Kim, et al, "Dependence of the Switching Characteristics of Resistance Random Access Memory on the Type of Transition Metal Oxide", ESSDERC, pp. 400-403, 2010.

[6] D. A. Muller, et al., "The electronic structure at the atomic scale of ultrathin gate oxides", Nature, Vol. 399, pp. 758-760, 1999.

[7] H. S. Baik, et al, "Interface structure and non-stoichiometry in HfO₂ dielectrics", Appl. Phys. Lett., Vol. 85, pp. 672-674, 2004.

A Novel Pre-clean Process of BEOL Barrier-seed Process to Enhance Reliability Performance of Advanced 40nm Node

Chun-Min Cheng, Chi-Mao Hsu, W. C. Lin, Hsin-Fu Huang, Yan-Chun Liu, Kun-Hsien Lin, Jin-Fu Lin, C. C. Huang, JY Wu

United Microelectronics Corp., No 18, Nanke 2nd Rd. Tainan Science Park, Sinshih, Tainan County 741, Taiwan, R.O.C.
Tel: 886-6-505-4888 Ext.86-12512, fax: 886-6-505-0960, e-mail: chun_min_cheng@umc.com

ABSTRACT

With scaling down of device geometry and keeping improvement of the chip resistance capacitance (RC) delay, it is necessary to reduce k value. A porous ultra low k-value (ULK) dielectric film is integrated into Cu interconnects of advanced 40 nm. There are several papers discussing about the interface effect between ULK film and barrier on reliability performance [1][2]. This paper will discuss the effect of pre-clean process on reliability performance before barrier and Cu-seed layer deposition that will strong affect the interface properties. Also, the early failure mode of each pre-clean process will be discussed as well to clarify the proposed mechanism.[*Keywords:* time-dependent dielectric breakdown, TDDB, pre-clean, Failure mode, low k]

INTRODUCTION

With geometry scaling down to enhance device performance, it becomes more struggled to pass reliability test under 40nm node integration. Moreover, ULK film is also implemented for RC reduction of 40nm node that is much sensitive to pre-clean process due to its porous film structure. A novel pre-clean process which adopts remote plasma (RPPC) to generate Hydrogen radical for Cu reduction replaces traditional physical bombardment by Ar plasma (named Ar-PC hereafter) This new RPPC technique significantly improves interface properties between barrier and ULK film for reliability enhancement[3]. In order to estimate pre-clean performance on clean capability, we need to develop new methodology and procedure to quickly screen the experimental result .Electron beam inspection (EBI) test on dual damascene structure at post Cu chemical mechanical polish (CMP) can effectively distinguish out process health. Simulated thermal cycling test is also powerful to identify the interface quality between Cu, ULK and Barrier metal.

EXPERIMENTAL

40nm structure wafers with Metal-1 single damascene and Metal-2/Via-1 dual damascene were prepared to check clean process performance. Two kinds of pre-clean processes including physical bombardment (Ar-PC) and reactive clean (RPPC) were split in these structure wafers followed by barrier, electro Cu plating (ECP) and Cu CMP. The EBI inspection was tested at two stages. First EBI was performed right after Cu CMP to check the clean efficiency of pre-clean conditions. The intrinsic interface quality of barrier metal, Cu and ULK film are identified as reference. Then 4 times thermal

cycle with 350degC / 5mins was performed on the same samples to simulate the thermal budget of following BEOL process steps. This thermal cycling test can stress the weak point of interface and screen out the optimized process condition quickly.

For EM (Electron Migration) evaluation, we adopt special test-key layout which more sensitive to via bottom cleanness to estimate clean capability in different pre-clean processes (shown in Fig.1).

TDDB (Time Dependent Dielectric Breakdown) test processed on a general comb-type test-key composed of three Cu interconnect layers and Al layer to check the reliability performance of Ar-PC & RPPC.

The failure sites of reliability samples are detected by optical beam induced resistance change (OBIRCH) and then applied transmission electro microscope (TEM) to identify the failure mode and correlation to different pre-clean processes.

RESULTS AND DISCUSSIONS

1. EBI Inspection Results

The mechanisms of these two processes are shown in Fig. 2. Traditional Ar-PC uses Ar+ plasma bombardment to remove CuO layer and residues of previous etching process under via and trench bottom. However, not only it is high risky to damage the ULK dielectric film and worsen the adhesion between barrier and dielectric film, but also it changes the profile shape by ion bombardment. Meanwhile, some Cu atoms may be sputtered into dielectric film surface to form an Cu diffusion path and source around via and trench sidewall to induce TDDB failed [4]. On the other hand, RPPC generates Hydrogen radical to reduce Copper oxide to pure Cu and to remove etch by-product residues at via bottom at the same time. Thus RRPC is harmless to ULK dielectric film due to no physical bombardment and the profile of trench and via could be kept well.

Fig. 3 shows the EBI results of RPPC & Ar-PC. From pre-thermal EBI results, RPPC shows Dark Voltage Contrast (DVC) free that indicates residues fully removed by RPPC or no discontinuous Cu at via bottom. Compared to RPPC, Ar-PC shows 3.5% DVC failure rate and possibly means some polymer residues or Cu loss at via bottom. After thermal cycle performed, Ar-PC shows tremendous failure rate increment to over 50% but RPPC only shows 4.5% failure rate increased. It

978-1-4244-9113-1/11 $26.00 © 2011 IEEE

indicates that Cu depletion become worse and worse after thermal cycle if some weak points still stay inside of Via/Trench structure in pre-clean process. As a pre-clean process, RPPC shows better cleanness and thermal stability than Ar-PC.

2. EM Performance

Fig. 4 and Fig. 5 show the EM lifetime and CDF chart of two pre-clean processes. RPPC show more than one order improvement of lifetime compared to Ar-PC that mainly contributed by 3X t50 and good CDF distribution. Also, the Ar-PC shows more early failure sites than RPPC that might come from poor clean capability and uniformity [5]. The TEM analysis of early EM failure sites show in Fig. 6 and reveal the different failure mode in both samples. As shown, the early failure site of Ar-PC is at the interface of via bottom and totally different from the early failure site of RPPC at Trench bottom. In theory Ar-PC only removes surface Copper oxide by ion bombardment and cannot fully remove Oxygen at deeper Cu under via bottom. It's trade-off to control Ar-PC removal amount. Less Ar-PC amount will gets high Oxygen remained in Cu and degrade EM performance. More Ar-PC will induce Cu atom re-sputtered into via sidewall to degrade barrier isolation and breakdown performance. From production point of view, RPPC should get more healthy and uniform via interface than AR-PC for EM performance enhancement.

3. TDDB failure analysis and performance

Refer to the CDF chart of TDDB test in Fig. 7, RPPC shows over 100% MTTF improvement and comparable distribution performance. The early failure analysis of TDDB is shown in Fig. 8. Ar-PC failure mode mostly comes from sidewall or bottom Cu diffuse-out which possibly induced by either ULK film damaged or un-expected Cu by ion bombardment. In the other hand, the failure mode of RPPC comes from interface Cu penetration instead of trench sidewall Cu penetration. RPPC shows more typical interface failure mode compared to Ar-PC due to its damage free characteristics.

Refer to physical comparison by TEM in Fig.9, RPPC shows not only damage free advantage but also benefit to keep the spacing of Cu wires. Ar-PC enlarges Cu line CD around 5nm more than RPPC due to Ar ion bombardment [6]. Basically RPPC does almost not consume ULK and HM material and keep Litho/Etch performance as expected CD target. Fig. 10 also shows the correlation of T63.2% of TDDB test with via to via spacing. As shown, you can see that T63.2% is strongly correlated to wire line spacing. The smaller line spacing is, the much worse T63.2% performance is. The MTTF degradation of Ar-PC sample is fully matched to spacing trend.

CONCLUSIONS

ULK has been implemented into production from 45/40nm node. Ar-PC is very challenged for process window in such aggressive geometry and porous material application. A new RPPC technology has been proven better defect and reliability performance than AR-PC in BEoL process. RPPC shows benefit in ULK damage free and no Cu bombardment to get Electron Migration and TDDB performance around one order improvement. Defect and thermal stability are also proven big improvement by new EBI methodology and quick learning procedure.

REFERENCES

[1] C. K. Hu R. Rosenberg and K, Y, Lee, "Electromigration path in thin-film lines," *Appl. Phys. Lett.*, vol.74, no.20, pp. 2945-2947, May 1999.

[2] L. Chen, Net al., "Effect of surface impurity on the Cu/Ta interface" *Thin Solid film*, vol.376, no. 1/2, pp. 115-123, Nov. 2000.

[3] X. Fu, et al., "Advanced Pre-clean for Integration of PECVD Si-COH(k=2.5) Dielectrics with Copper Metallization Beyond 45nm Technology," *IITC*, 2006, pp. 51-53.

[4] Zs. Tokei, et al., "Reliability of Copper Dual Damascene Influenced by Pre-Clean" *IPFA*, 2002, pp. 118-123.

[5] C. C. Yang, et al., "Extendibility of PVD Barrier/Seed for BEOL Cu Metallization," *IITC*, 2005, pp. 135-137.

[6] G. B.Alers, et al., "Barrier-First Integration for Improved Reliability in Copper Dual Damascene Interconnects" *Interconnect Technology Conference*, 2003, pp. 27-29.

FIGURE. 1. Sketch map of EM test-key for estimation of pre-clean process capability. The test-key is more sensitive to via bottom cleanness.

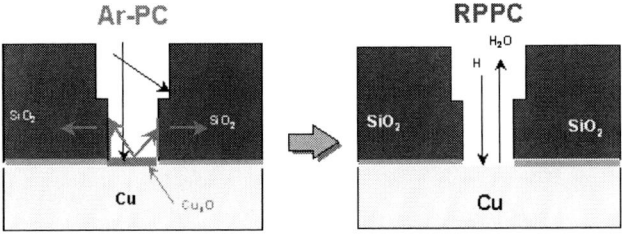

FIGURE. 2 The Pre-clean mechanisms of Ar-PC & RPPC. Traditional Ar-PC bombardment have higher risk to damage ultra-low K film & re-sputter Cu atoms into dielectric film to cause extra leakage path of Cu migration.

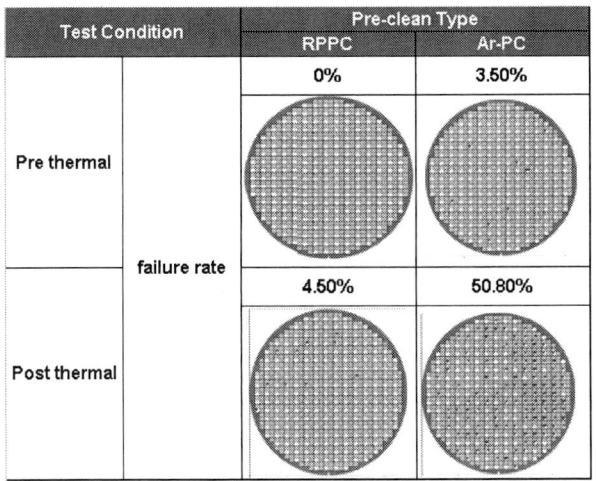

FIGURE 3. EBI inspection results of RPPC and Ar-PC. RPPC shows great thermal stability from EBI inspection

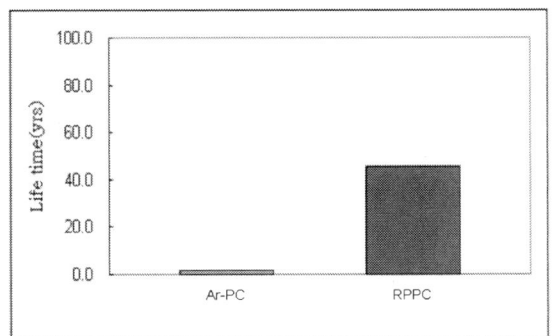

FIGURE. 4. EM lifetime performance of Ar-PC and RPPC . PRPC shows 22times EM lifetime than Ar-PC.

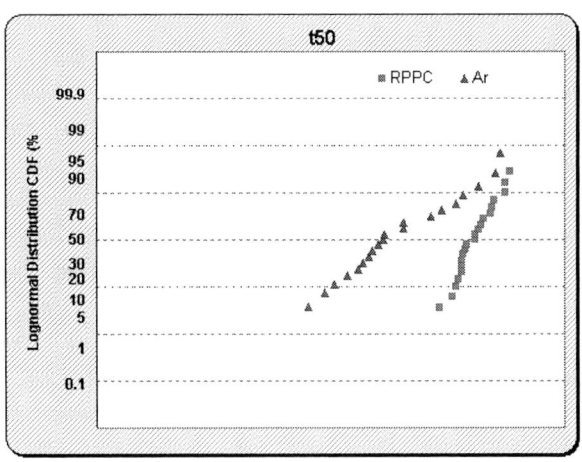

FIGURE 5. CDF chart of EM test. RPPC show better distribution & over 3 times increase of t50 than Ar-PC.

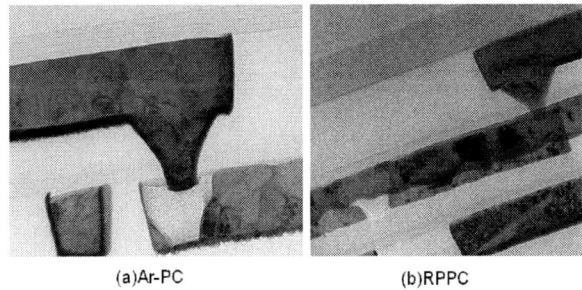

FIGURE. 6 Early failure modes of EM test by TEM analysis. (a)early failure site of Ar-PC ; (b)early failure site of RPPC.

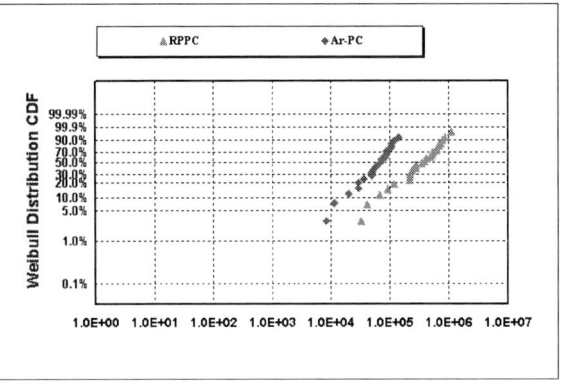

FIGURE. 7. CDF chart of MTTF of TDDB test. RPPC shows over 100% MTTF improvement of TDDB that strong correlated with spacing effect.

978-1-4244-9113-1/11 $26.00 © 2011 IEEE

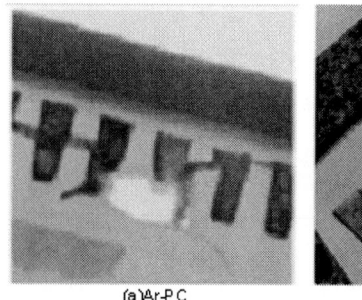 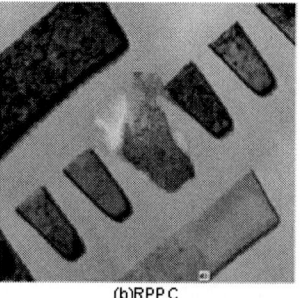

(a)Ar-PC   (b)RPPC

FIGURE. 8 Early failure modes of TDDB test by TEM analysis. (a)Early failure site of Ar-PC comes from trench side-wall or bottom strongly co-related to ULK damaged by Ar Ion bombardment. (b)early failure site of RPPC, no trench side wall failure mode found.

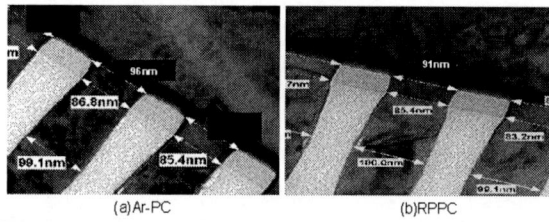

(a)Ar-PC   (b)RPPC

FIGURE. 9 Cu wire space at 90nm diameter via. RPPC show CD in target and Ar-PC enlarge Cu wire CD by 5nm more.

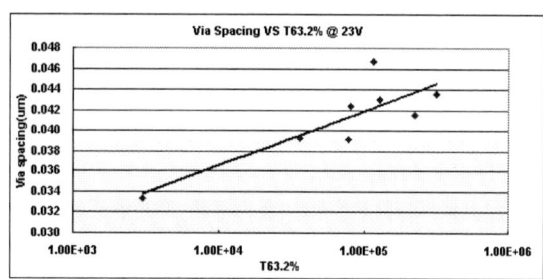

FIGURE. 10 Via spacing V.S. T63.2% of TDDB. It shows strong correlation of spacing on T63.2% performance

A charge transport based acceleration model for interlevel dielectric breakdown

Ravi Achanta

IBM Systems&Technology Group
2070 Rt 52, Hopewell Jn.,NY,USA
rachanta@us.ibm.com

Paul McLaughlin

IBM Systems&Technology Group
2070 Rt 52,Hopewell Jn.,NY,USA

Abstract—**A charge transport model has been applied to predict interlevel dielectric breakdown in copper interconnects. The model very accurately predicts the lifetimes of dense and porous dielectrics at long times. The predictions of the model vary significantly from currently available methodologies which fail to adequately account for copper charge transport. The results have important implications for the reliability of advanced microprocessors with copper interconnects.**

Keywords-copper interconnects, dielectric breakdown, predictive model, charge transport.

I.Introduction

Interlevel dielectrics (ILD) in copper interconnects are known to have reduced lifetime due to copper ion injection [1, 2]. The acceleration models currently used to project the useful life from the higher voltages of reliability testing do not accurately account for the presence or motion of copper ions [3, 4]. The currently used models do not take into account the drift/diffusion of copper ions, the solubility of copper ions in the dielectric and the boundary conditions for the copper ions at the anode (the electrode with the positive applied bias) and the cathode (the grounded electrode, (Fig 1)). Thus, it is unknown if the currently used acceleration models are sufficiently accurate to ensure reliable operation of advanced microprocessors over long periods of time.

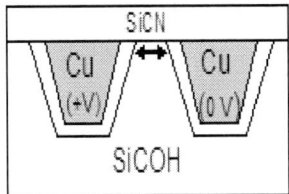

Fig 1: Copper interconnects, anode (+V) and cathode

(Grounded)

In Fig 2 we show long term failure data (~ upto 1.5 years) on comb-comb interconnect test structures, fabricated at the lower metallization levels, using a dual damascene process.

The dielectric is porous SiCOH with a κ=2.4. Each data point is the median fail time for a set of ~10-15 test samples. In this figure we also show fits to the data based on some of the currently used models. Fig 2a is fits to the data based on the 3 highest voltages and Fig 2b are fits based on the 4 highest voltages. It is interesting to note that a) none of the models fit the data accurately over long times b) the final use lifetimes, and the failure projections based on it, changes significantly depending on the number of data points (3 or 4) used for fitting. This suggests that the physics of ILD failure, in these porous materials, is not being accurately captured by any or all of the other models at the practically important lower use voltages.

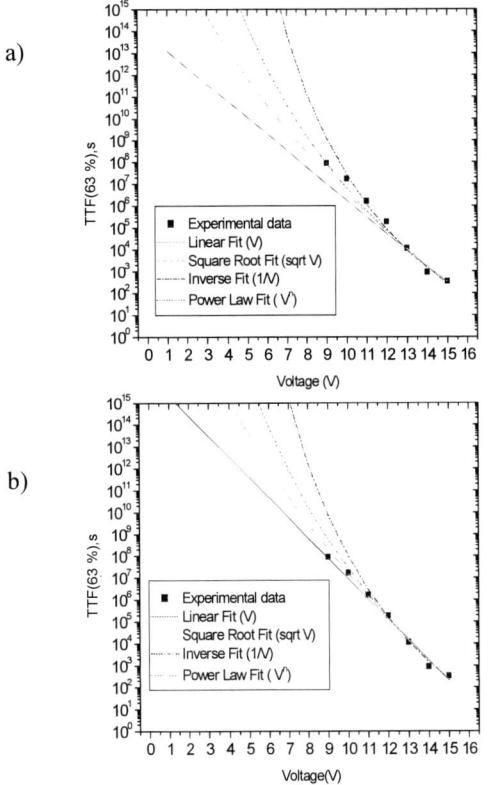

Fig 2:a) Curve fits to data using 3 highest voltages

b) Fits to data using 4 highest voltage data points

II. Model Formulation

A copper ion drift/diffusion based dielectric breakdown model has previously been successfully used to predict breakdown in planar M-I-S type structures with a copper gate (with and without an intervening metallic barrier) on dense dielectrics [5, 6].

The time to fail (TTF) of the dielectric is given by:

$$TTF(s) = A \exp\left(\frac{E_a - \gamma E_{app}^2}{k_b T}\right) . t_{mass} \qquad (1)$$

Here the exponential term describes failure in the presence of copper ions due to intrinsic thermochemical (wear out) effects alone, while t_{mass} accounts for failure due to mass transfer induced internal field build up exceeding the breakdown field of the dielectric at the cathode end. E_a is the activation energy for bond breakage, assumed to be 1.15eV for silica based dielectrics, γ is the 'field acceleration' factor which determines the 'slope' of the prediction while A is a pre-exponential material dependent term. The exponential term ostensibly accounts for the 'bond-breakage' that is supposed to be a contributory factor leading to dielectric breakdown. 'Bond-breakage' during dielectric breakdown has never been experimentally proven. We believe that the exponential term (with the approximate E^2 dependence on the electric field) accounts for the transport of electrons, holes and oxidizing species through the dielectric. Since we do not have the transport properties for any of these species right now, we have not explicitly accounted for them in our mass transfer model and hence they are approximated by the exponential term in equation (1). This form of the exponential term was previously chosen based on its better match to experimental data for dielectric breakdown in planar SiO$_2$ based capacitors [5]. It was subsequently found to apply even to dielectric films with an interposed metallic barrier layer [6]. Here we seek to apply this equation to predict dielectric breakdown in actual interconnects. Much more experimental and theoretical work in determining the transport properties of the various species which move inside the dielectric is needed before we can get rid of the exponential term describing wear-out and replace it with a purely numerical solution to the transport equations.

t_{mass} in our equation is the time needed for sufficient copper ions to accumulate at the cathode to increase the internal space charge above the breakdown strength E_{bd} of the dielectric and thus cause dielectric breakdown. This mass transfer based degradation component becomes important at the practically important lower use voltages whereas at the higher reliability testing voltages the predictions of equation (1) do not differ significantly from the currently available models [7]. t_{mass} is obtained by solving the coupled non-linear continuity/Poisson PDE's (equations 2-6) , with the appropriate boundary conditions, that describe the charge transport of copper ions to that point in time where the internal electric field (E(L)) due to the accumulated copper ions equals the breakdown dielectric field,E_{bd}. The flux equation, as written in equation (3) contains the 'elastic drift' term to account for the concentration dependent effects on the diffusivity of copper ions and the mechanical resistance of the dielectric material to being doped with metallic copper ions.

We solve the transport equations below to get t_{mass}:

Continuity Equation (for Cu+ charge transport)

$$\frac{\partial C}{\partial t} = -\frac{\partial J}{\partial x} \qquad (2)$$

Flux term

$$J(t,x) = -D(1 + \frac{\alpha C}{k_b T})\frac{\partial C}{\partial x} - \mu C \frac{\partial V}{\partial x} \qquad (3)$$

Poisson Equation

$$\frac{\partial^2 V}{\partial x^2} = -\frac{qC}{\kappa \varepsilon_0} \qquad (4)$$

Boundary Conditions for the system

$t = 0$	$C = 0$	$V = 0$
$x = 0$	$C = C_e$	$V = V_0$
$x = L$	$J = -D(1+\frac{\alpha C}{k_b T})\frac{\partial C}{\partial x} - \mu C \frac{\partial V}{\partial x} = 0$	$V = 0$

$$(5)$$

Here we assume a 'blocking' boundary condition at the cathode by which the flux, J, of the copper ions equals zero. We had previously used this boundary condition with planar capacitors based on experimental data which showed an accumulation of copper at the cathode [5]. The choice of the boundary condition is very important and sometimes other boundary conditions have been used in the literature [8.9]. Due to the limited sensitivity of SIMS and other microscopic techniques to detect lower levels of copper ions, it is easy to be confused on the appropriate boundary condition to be used. Data taken on samples exposed to copper drift for long times are more accurate in determining whether copper has been accumulating at the boundaries. One drawback of techniques such as SIMS is they do not differentiate between ionized copper and neutral copper. It is hard to imagine ionized copper being neutralized permanently (as is necessary for the C=0 boundary condition to be used instead of the J=0 condition) while the applied field is still on, which it is during testing or actual device operation.

The Einstein relation relates the diffusivity and mobility through the thermal voltage V_e

$$\frac{D}{\mu} = V_e \qquad V_e = \frac{k_b T}{q} \qquad (6)$$

The nomenclature used in the above equations is explained in the table below:

Symbol	Meaning
C	Concentration of copper ions in dielectric
C_e	Solubility of copper ions in dielectric
D	Diffusivity of copper ions in dielectric
μ	Mobility of copper ions in dielectric
κ	Dielectric Constant
ε_0	Standard permittivity of vacuum
q	Elementary charge
α	Elastic drift coefficient
J	Flux of copper ions
T	Temperature
k_b	Boltzmann Constant
L	Spacing between anode and cathode
V	Voltage
V_o	Applied Voltage
E_{app}	Applied Field, V_o/L
E_{bd}	Breakdown Field of dielectric
t_{mass}	Time for $E(L)=E_{bd}$

III. Results and discussion

In Fig 3 we show the results of our predictive model applied to the comb-comb test structure TTF data. The parameters in our model, namely A and γ, were optimized based on the 3 highest voltage experimental points. The model matches the experimental data excellently at all available voltages, across ~ 7 orders in the time scale. In Figure 4, it can be seen that this model has much greater predictive power in matching the data at the lower experimental voltages than the other models(even though the other models are 'optimized' using the 4 highest voltages) strongly suggesting that charge transport of copper ions is a significant factor in porous ILD breakdown in real interconnects. It is pertinent to note that the projections based off the charge transport model are essentially unchanged once A and γ have been determined (with the help of 2-3 experimental data points) whereas the projections from the other models change significantly based on the number of available experimental data points as illustrated in figures 2-4 . Due to practical reasons many a times it is impossible to obtain more that 3 data points during reliability evaluation and a model proven to give the same final projections regardless of the number of data points utilized to 'fit' its parameters is more useful than one which is not. The charge transport model predicts lifetimes at lower voltages that bend downward compared to the linear E-model whereas the power law or the

1/E models project lifetimes bending upward. This behavior of the charge transport model is due to the increasing dependence of failure on the mass transport of copper ions at lower voltages. Lloyd et al. predicted similar ILD breakdown behavior from their impact damage model based on energy considerations [1,10].

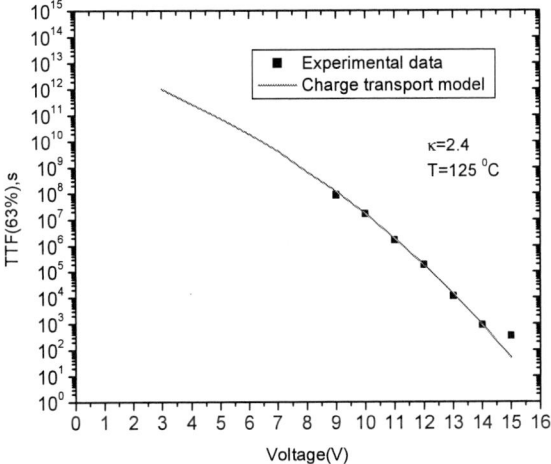

Fig 3: Charge transport model shows excellent match to experimental data.

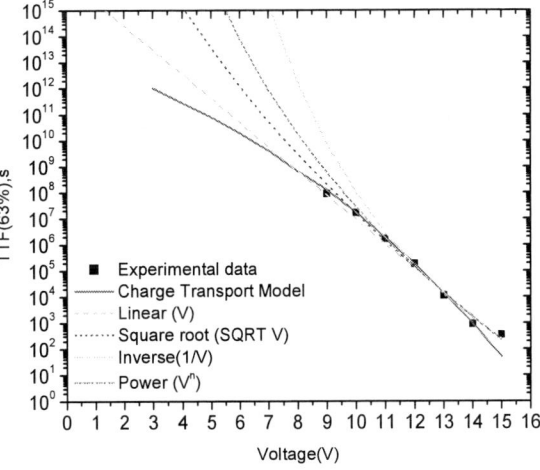

Fig 4: Comparison of data to different models. The best match is with the charge transport model. Significant differences in lifetimes between the charge transport model and currently used models need to be taken into account when assessing risk to the reliability of advanced interconnects.

The predictions of the charge transport model diverge significantly from the currently used models at lower voltages which is a cause for concern as the operating field keeps creeping up for high performance applications since the drop in operating voltage is not making up for the reduction in line-to-line spacing. The differences between the experimental data (at the voltages of 8,9,10 V) and the model predictions are shown in Figure 5. The charge transport model shows the least deviation from the experimental data and hence is more accurate. Obtaining more accurate transport properties in these porous materials could improve the accuracy of the predictions at lower voltages.

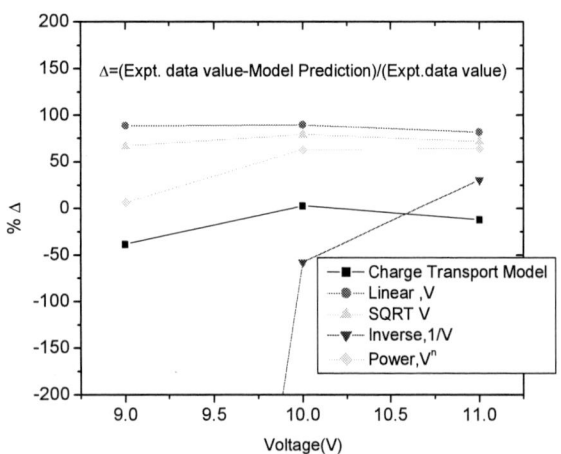

Fig 5: Closeness of fit to the experimental data using various models. Charge transport model shows best accuracy.

The charge transport model predicts a 'threshold' or cutoff applied voltage below which the reliability improves significantly. This occurs due to concentration dependent diffusivity effects which become significant at lower voltages retarding the pile-up induced electric field build-up at the cathode. A steady state internal electric field is ultimately reached, and based on the applied voltage, can be higher or lower than the breakdown voltage of the dielectric, Ebd. In Figure 6 we have calculated the steady state internal electric field reached at the cathode (x=L) at different applied voltages. Based on this analysis, an applied voltage of ~1.5V or lower seems to be a safe operating voltage for this dielectric at these dimensions.

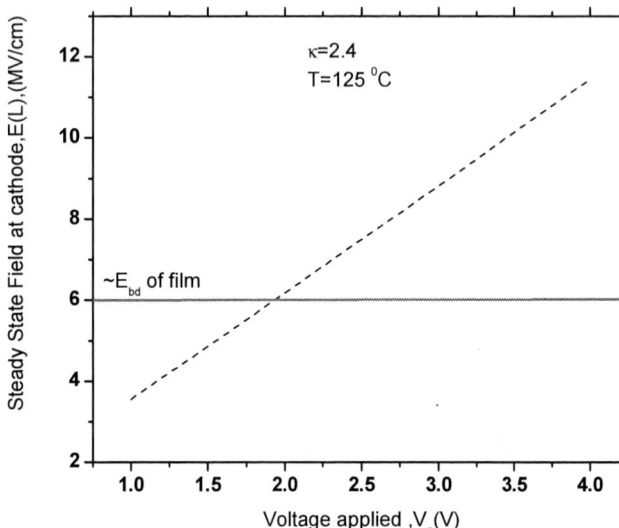

Fig 6: Steady State electric field reached at the cathode as a function of the applied voltage. An operating voltage leading to a steady state field E (L) ~0.5 -0.75 E_{bd} can be considered 'safe'.

IV. Conclusion

We have modeled porous ILD breakdown with a 1-D charge transport model in actual interconnect structures. The model accounts for the transport of copper ions and the internal field build-up due to the accumulation of copper ions at the cathode. This model proves to be more accurate than the other models in predicting and matching our experimental data at longer times. The fact that there is such a clear match with a simple transport model (previously only applied to planar geometries) suggests that mass transfer is a first order effect in the failure of the dielectric and the nature of the failure could be similar regardless of structure (dual damascene structure vs. planar geometry). The predictions of the model differ significantly from the currently used models towards lower voltages due to inclusion of copper ion drift effects and offer a new perspective on the risk assessment for the reliability of advanced high performance microprocessors. We have calculated a 'safe' operating voltage, based on mass transport calculations, below which the breakdown processes are considerably slowed predicting higher device reliability.

Additional longer term data, along with new innovative mechanism experiments to understand the motions of charged species in dielectrics, will help in a more detailed understanding of dielectric stability at low fields and identify and confirm the right acceleration model for these materials.

Acknowledgements

The authors acknowledge IBM SRDC for supporting this work.

References

[1]Lloyd J.R.,Murray,C.E.,Ponoth,S.,Cohen,S., and Liniger,E.G.,et al,Microelectron.Reliab.,46,1643 (2006).

[2]Z.Tokei, K.Croes and G.P.Beyer, Microelectron.Eng. 87,3,348,(2010).

[3]F.Chen et al,IRPS,2006.

[4]N.Suzumura et al,IRPS,2006.

[5] R. Achanta, W.N.Gill, J.L.Plawsky, J.Appl.Phys, 103, 014907, (2008).

[6] R. Achanta, W.N.Gill, J.L.Plawsky, J.Appl.Phys, 106, 074906, (2009).

[7]R.Achanta,W.N.Gill,J.L.Plawsky, Appl.PhysLett.,91,234106, (2007).

[8]Hwang,S-S.,Jung,S-Y.,and Joo,Y-C.,J.Appl..Phys 101,074501,(2007).

[9]Willis,B.G., and Lang,D.V.,Thin Solid Films,467,284 ,(2004).

[10]J.R.Lloyd,IRPS 2010.

A MODEL FOR POST-CMP CLEANING EFFECT ON TDDB

Chia-Lin Hsu, Wen-Chin Lin, Teng-Chun Tsai, Climbing Huang, and J.-Y. Wu

United Microelectronics Corp., No 18, Nanke 2nd Rd. Tainan Science Park, Sinshih, Tainan County 741, Taiwan, R.O.C.
Tel: 886-6-505-4888 Ext.87-12513, fax: 886-6-505-0960, e-mail: chia_lin_hsu@umc.com

ABSTRACT

For 45 nm and beyond, direct polished porous type ultra low-K film (ULK) is integrated in Cu interconnects. Post-cleaning of Cu CMP effect on Time dependent dielectric breakdown (TDDB) was investigated. Cu ions remaining on dielectrics and Cu roughness are found as two dominate factors at different clean time region. High Cu roughness induced capping layer seam results in the degradation of TDDB. A statistical model, said weak element model, was proposed to illustrate the correlation of Cu roughness on TDDB as well.

INTRODUCTION

In order to benefit to chip resistance capacitance delay improvement, advanced VLSI circuits are inevitable continuous scaling. And Copper (Cu) integrated with porous type ultra-low-k dielectric materials (ULK) are widely applied to form interconnects from the 45nm technology node and beyond. Long-term reliability, such as time dependent dielectric breakdown (TDDB), is becoming the most critical challenges for technology qualification because of the small geometry feature size and the ULK film properties, porous, hydrophobic and low modulus.

Metal line geometry is an important factor effecting on TDDB. Lots of papers have discussed the correlations and models of TDDB to the metal line spacing [1], [2]. For example, metal line edge roughness, the cross-sectional shape of copper line, the spacing uniformity within the wafer and from wafer to wafer, and inter-metal dielectrics properties, such as hardness and porosity. Dielectric capping layer properties are also recognized as an important factor for TDDB lifetime. However, the metal line contributing to the TDDB does seldom being addressed [3]. In the processes, Cu CMP is the major one to manage the surface roughness of Cu line, especially the cleaning process. In the past, since metal line width and trench depth much larger than the scale of Cu surface roughness, the contribution seems negligible. However, from 45nm generation, the line-width shrinkage leads that Cu roughness cannot be neglected anymore. It was noted that the Cu roughness performed an important role for yield and reliability. Since Cu surface roughness is highly dependent on Cu CMP cleaning process, in this article, breakdown lifetime for various cleaning time were reported. Two mechanisms in different cleaning time range was proposed and discussed to address the TDDB relationship to post-CMP clean time. A sta-tistical model was purposed to explain the correlation of TDDB to cu roughness.

EXPERIMENTAL

L45 patterned wafers were utilized in all experiments. The structure wafers were prepared with the following steps. The nano-porogen pre-mixed low-k dielectric film was deposited on the wafers by plasma-enhanced chemical vapor deposition (PECVD). The porogens were then removed by UV lamp curing to form porous low-k film (k value ~2.55). The pattern formation of interconnects were based on a dual damascene process (metal and via to form metal conductor concurrently) with the trench 1st metal hard mask scheme. The patterned wafers were then processing PVD type barrier and seed, Cu electroplating and Cu chemical mechanical polish (Cu CMP) sequentially to finish the Cu interconnects. After CMP, the samples received a N-doped silicon carbide (SiCN) dielectric barrier layer and then repeated the above sequence for the next interconnect layer formation.

In Cu CMP process, 3-platen rotary type polisher integrated with cleaning module and dryer was applied. Two colloidal-silica-based slurries were utilized in both Cu polish and barrier polish steps, respectively. Alkali based post-clean solution was applied in brush clean steps to clean the polished wafers. Cu roughness was varied with various cleaning time length with clean solutions. The wafers were finally dehydrated in an IPA dryer.

Roughness was measured with a high-resolution atomic force profiler (AFM) on 10x10um2 Cu pad area of patterned wafers right after polishing. Time of flight secondary ion mass spectrometry (TOF-SIMS) was applied for the Cu concentration detection. A metal line array testkey with line width 15um and spacing 3um was used to collect Cu (ion) signals. Because of the TOFSIMS resolution limitation, Cu concentration would not properly reflect the concentration differences among splits. In this study, Cu concentration was calculated by the signals on dielectric spacing where 3um away from the metal line, as indicated in Figure 1. Breakdown behavior tests were probed on a comb-comb testkey on the structured wafers comprised of 3 Cu interconnect layers and one Al layer. The failure location was detected by optical beam induced resistance change (OBIRCH) and then applied transmission electro microscope (TEM) to do the failure mode analysis.

978-1-4244-9113-1/11 $26.00 © 2011 IEEE

RESULTS AND DISCUSSIONS

The TDDB cumulative failure time was showed in Figure 2 for different post-CMP clean time. If ignored early failure points, the slopes are all similar to each other except the clean time 2x one. The difference of Weibull slope of 2X implies the failure mechanism of 2x one is different from other clean time.

1. Cu Roughness vs Cu Concentration Mechanism

Cu concentrations on line spacing were measured by TOF-SIMS for varied post-CMP clean time and were plotted with TDDB characteristic failure time (t63.2%) in Figure 3. It shows that for short clean time, the high remaining of Cu ions on spacing decreases the effective capacitor spacing and results in the degradation of lifetime to breakdown [4], [5]. On the other side, fairy good agreement between t63.2% and average roughness from 4 to 5A indicates the Cu roughness is a major factor to effect the dielectric breakdown lifetime in this region.

From failure analysis, the leakage occurred at the top corner of the metal line (Not shown). In Figure 5, TEM illustrated if Cu depth deep enough in small-width metal lines, the following dielectric barrier deposition is difficult to deposit into this deep recess and results in a seam formation in the capping layer. This seam degrades the metal diffusion barrier capability of dielectric barrier and becomes the weak point when applying the stress voltage and results in the ruin of the breakdown lifetime.

2. Weak Element Model

A statistical analytical model has been proposed to account for the TDDB dependency to Cu surface roughness. Weak element is defined as the unit location where break down occurs easily. A TDDB comb-comb structure could be considered as an assembly of N small unit capacitors. The breakdown lifetime is dependent on the 1st failure of unit capacitors. For those small units with high possibility tend to fail is defined as weak elements. In Cu roughness case, as presented in Figure 5, when a small unit capacitor includes deep Cu depth to form a capping layer seam, that small unit capacitors is a weak element.

The failure probability increases when the same structure carried with more weak elements.

Let $F_Z(t)$ be the cumulative failure probability of a device with Z weak elements after stress time t. For a device with single weak element, its survival probability can be described as $[1- F_1(t)]$. For the device with Z weak elements, the survival probability of this device decreases by a factor of Z. The correlation could be described as the following relation:

$$[1 - F_z(t)] = [1 - F_1(t)]^Z \qquad (1)$$

A two-parameter Weibull distribution function, as shown in equation 2, is commonly used as the solution of equation 1 [6]:

$$F(t) = 1 - \exp\left[-\left(\frac{t}{\alpha}\right)^\beta\right] \qquad (2)$$

Where α is characteristic lifetime t63.2% and β represents the Weibull shape parameter and approximates to 1 in this study.

To find the lifetime of any capacitor with Z weak elements, one can substitute Eq. (2) back into Eq. (1), take a natural logarithm of both sides and result in the expression:

$$\left(\frac{t}{\alpha_Z}\right)^\beta = Z\left(\frac{t}{\alpha_1}\right)^\beta \qquad (3)$$

α_Z presents the t63.2% of a device with Z weak elements and can be written as a function of α_1:

$$\alpha_Z = \alpha_1 Z^{-1/\beta} \qquad (4)$$

Cu surface depth variation at trench side wall is difficult to be measured directly. If let Cu surface depth at the trench sidewall be regarded as a one-dimensional Cu roughness. The cumulative distribution of Cu surface variation for metal lines in a comb-comb structure could be viewed as the distribution of roughness measured on a 2-dimensional Cu pad and could be simply treated as a normal distribution.

As illustrated in Figure 6, Y is defined as the Cu depth and Function of Y is the distribution of Cu depth (as in the equation 5). Assumed that the capping layer seam would form when Cu depths deeper than "A", the possibility to form weak elements could be described as the equation 6.

$$f(Y) = \frac{1}{\sigma\sqrt{2\pi}} e^{-\frac{1}{2}Y^2/\sigma^2} \qquad (5)$$

Where σ = standard deviation

$$\Pr(Y > A) = \frac{1}{\sigma\sqrt{2\pi}} \int_{Y=A}^{\infty} e^{-\frac{1}{2}Y^2/\sigma^2} dY \qquad (6)$$

The amount of weak elements Z could be then obtained as the product of Pr (Y>A) and N. Substitute into equation 4,

$$\alpha_Z = \alpha_1 \cdot (N \cdot \Pr(Y > A))^{-1/\beta} \qquad (7)$$

When "A" is large enough, this possibility could simply approximate to the area of the triangle as presented in Figure 7. With this approximation, a solution for the probability of $Y > A$ could be expressed as the function of σ (equation 8).

$$\Pr(Y > a) = \frac{1}{2}b \cdot f(a) = \frac{1}{2}b \cdot \left(\frac{1}{\sigma\sqrt{2\pi}} e^{-\frac{1}{2}a^2/\sigma^2}\right) \qquad (8)$$

In this study, Weibull slope β approximates to 1. The standard deviation of Cu roughness is linear correlated with average Cu roughness, shown in Figure 8. As the experimental result presented in Figure 4, α is inversely proportional to the average Cu roughness in this roughness region. It implies that $Pr(Y > A)$ should be proportional to σ as well. By applying numerical method, the probability versus the standard deviation of Cu roughness (σ) for various Y was plotted in Figure 9. It shows the possibility to form weak elements is proportional to σ of Cu depth with good linear agreement when Y is around 7.

CONCLUSIONS

For L45 generation Cu interconnects, the correlation of TDDB lifetime to post-CMP clean time was explored. Two factors, Cu concentration on spacing and Cu surface roughness, dominate TDDB lifetime, respectively, in different clean time region. A statistical weak element model is proposed to explain the correlation of TDDB to Cu roughness. Suggest the weak element would form when Cu depth (Y) larger than A. In this study, when Weibull slope equals to 1, the A approximates to 7 is predicted from numerical calculation.

REFERENCES

[1] J. Kim, et al., *in Proc. IRPS*, pp. 399-404 (2007).
[2] F. Chen, et al., *IEEE Trans. Electr. Dev.*, vol. 56, pp. 2-12 (2009).
[3] C.L. Hsu, *in Proc. IRPS*, pp. 918-921 (2010).
[4] G. S. Haase, *in Proc. IRPS*, pp. 556-565 (2008).
[5] M. Vilmay, *IEEE Trans. Electr. Dev. and Mater. Reliab.*, vol. 9, pp. 120-127 (2009).
[6] W. Weibull, *J. Applied Mechanics*, vol. 18, p. 293, 1951.

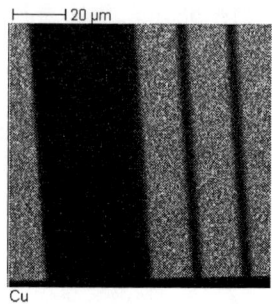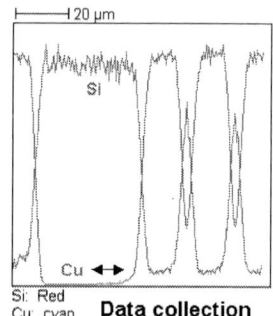

FIGURE 1. CU CONCENTRATION MEASUREMENT BY TOF-SIMS

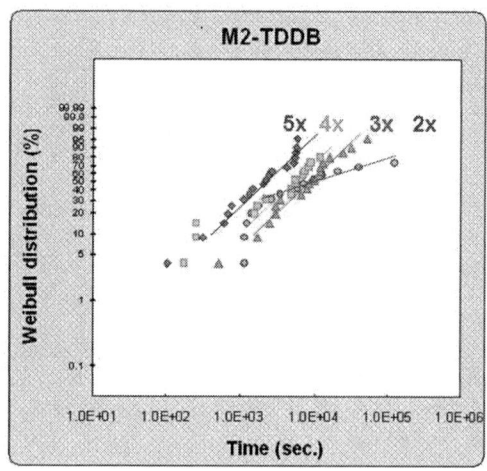

FIGURE 2. WEIBULL DISTRIBUTION OF TDDB FAILURES FOR DIFFERENT POST-CLEAN TIME

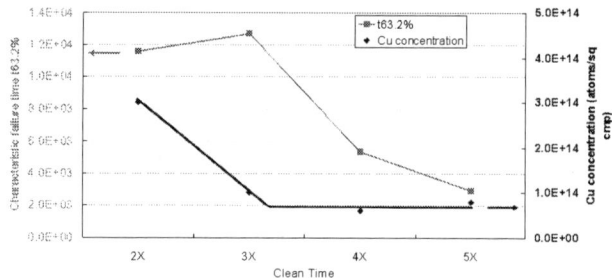

FIGURE 3. THE CHARACTERISTIC FAILURE TIME OF TDDB VS CU DIFFERENT CLEAN TIME AND ROUGHNESS

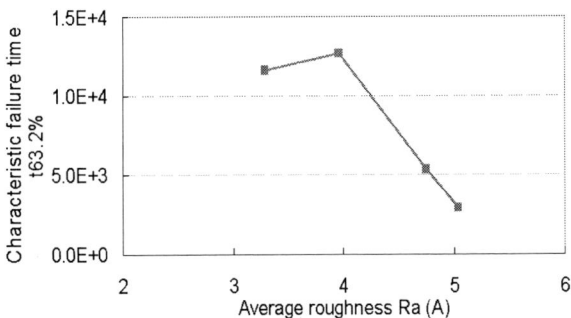

FIGURE 4. THE CHARACTERISTIC FAILURE TIME OF TDDB VS DIFFERENT CU ROUGHNESS

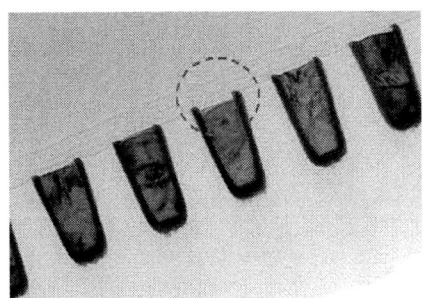

FIGURE 5. TEM OF METAL LINE ARRAY REVEALS THE SEVERE CU RECESS LEADS TO THE SEAM FORMATION IN THE FOLLOWING CAPPING LAYER

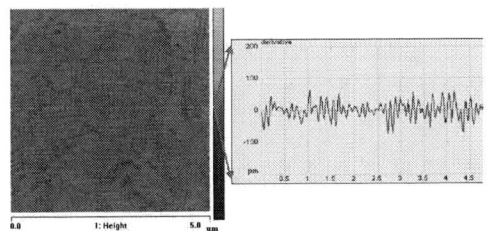

FIGURE 6. THE CUMULATIVE DISTRIBUTION OF CU SURFACE DEPTH FOR LONG CU LINES IS EQUAL TO THE DISTRIBUTION OF ROUGHNESS MEASURED ON A 2-DIMENSIONAL CU PAD

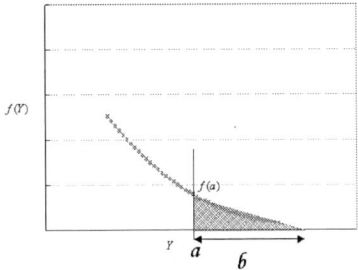

FIGURE 7. THE POSSIBILITY OF $Y > a$ IS SIMPLY APPROXIMATED TO A TRIANGLE AREA.

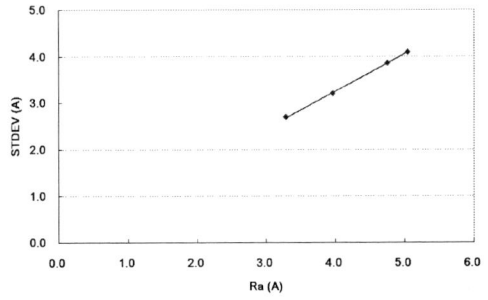

FIGURE 8. THE CORRELATION OF PROBABILITY TO CU ROUGHNESS STANDARD DEVIATION FOR VARIOUS α

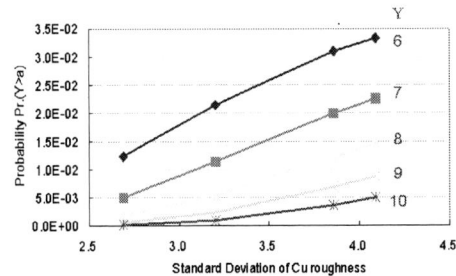

FIGURE 9. THE CORRELATION OF PROBABILITY TO CU ROUGHNESS STANDARD VARIATION FOR VARIOUS α

Performance and Structure Degradations of SiGe HBT after Electromagnetic Field Stress

Alaeddine A.[1,2], Kadi M.[2], Daoud K.[1]

[1] GPM, UMR 6634 CNRS, University of Rouen
[2] IRSEEM, ESIGELEC
Saint Etienne du Rouvray, France
ali.alaeddine@esigelec.fr

Abstract— **This paper addresses failure analysis of electromagnetic field stress effects on SiGe HBTs reliability issues, examining the relation ship between the stress-induced current and device structure degradations. The origin of leakage currents in failed transistors has been studied by complementary failure analysis techniques. Characterization of the structure after aging was performed by Transmission Electron Microscopy (TEM) and Energy Dispersive Spectroscopy (EDS). We found clearly dislocations and interface deformation of the Titanium thin film (Ti) of all contacts. Based on the coupling of high current density and thermal effects due to Joule heating, device failures are explained. These disorders may explain the origin the large shifting of the dynamic characteristics of failed transistors.**

Keywords-SiGe HBT; reliability; EMC; TEM

I. INTRODUCTION

Several companies have produced the first generation SiGe HBT's with transition frequency f_T and cut-off frequency f_{MAX} around 50 GHz, second generation technologies are already at an advanced development stage with f_T and f_{MAX} around 100 GHz [1], while the state of the art indicate a possibility of f_T around 640 GHz [2]. A vast range of RF and mixed-signal circuits are possible with this technology which has demonstrated very attractive capabilities in term of mobile phones, WLNA, satellite communications and Radar applications. As electronic systems are integrating more and more functionalities in a confined volume, some devices can be the source of numerous electromagnetic disturbances which make other components in the vicinity more and more susceptible, In parallel, the immunity of these components has decreased at the same pace, due to a steady reduction in power supply voltage and consequently, noise margin [3]. Many papers have been published on the SiGe HBT's reliability for radiation, thermal and electrical stresses [1] but none of them to our knowledge were carried out on electromagnetic field stress. This paper is organized as follows: the new stress methodology generated by the near-field bench and its effects on the DC and capacitive characteristics are presented in Section II. In Section III, the stress effect on the cut-off frequency at low and high injection levels is studied. In section V, structure identification

before and after stress is presented, using cross section TEM observations and EDS analysis, for providing comprehensive discussions.

II. DC DEGRADATION

As shown in Figure 1, the near-field disturbance method is based on the use of a miniature near-field probe localized above the device under test (DUT) at a given height "H" to produce a strong localized electromagnetic field. It includes automatic near-field mapping system developed by the Research Institute for Electronic Embedded Systems (IRSEEM) [4]. This probe consists of a small loop and it is made up of the inner conductor to produce an electromagnetic field where the magnetic field is dominant [5]. This magnetic probe was fed by a RF generator (0 dBm) at 1 GHz and a 40 dB power amplifier through a directional coupler which allows the measurement of the incident power through a power meter.

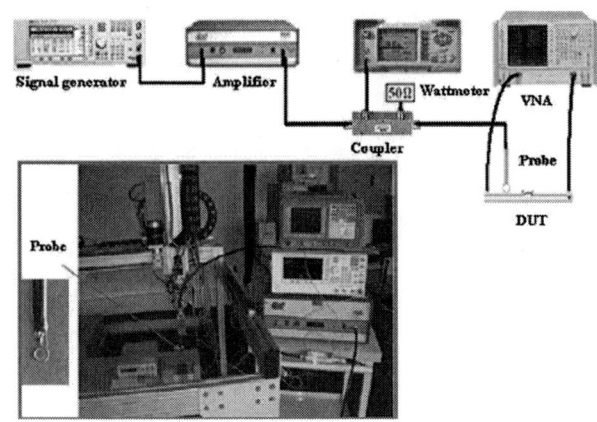

Figure 1. Near-field bench stress system

The tested devices used in this study are SiGe HBTs packaged in SOT-343 with a DC current gain of 300 and a usable cut-off frequency up to 10 GHz. They are mounted on a custom Printed Circuit Board (PCB) like a common emitter amplifier. The magnetic probe is located at 1 mm above the package for the first time and it was fed by 40 dBm at 1 GHz. No degradation has been observed after a given stress duration. The same conditions are applied to the probe located above the microstrip line connecting the collector or the emitter; no important degradation can be seen in these conditions. Finally the transistor was stressed on the microstrip line connecting the base with the same conditions when the probe is located at 3 mm in front of the base input.

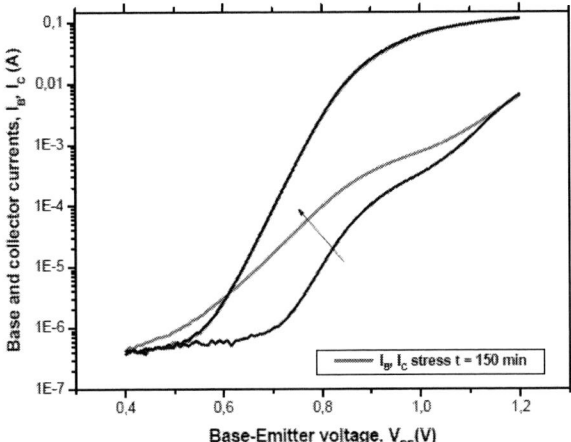

Figure 2. Forward Gummel plots before and after stress.

The typical forward Gummel plots of the HBT measured after 2h30 of stress is shown in figure 2. We can observe that the collector current remains unchanged during the stress while a large degradation of the base current is occurring. We have compared the electromagnetic stress effects with the well-known and previously reported stress effects on I_B degradation from electrical, thermal, or irradiation stresses [1], [6], [7]. The commonly associated mechanism responsible for this shift in base current for reverse bias or radiation stresses is the generation of a damage region at the sidewall-spacer oxide and silicon interface. This damage induces interface traps (Si/SiO_2) in the E-B spacer oxide, due to a hot carrier (HC) injection. These stresses induce Generation/Recombination trap centers and lead to an increase in the recombination component of the base currents which is confirmed by Gummel plots after stress. In addition, the value of the leakage base current (I_B) increases after stress in the reverse Gummel plots while the emitter current remains unchanged, as shown in figure 3. This observed inverse-mode I_B degradation can be created by traps induced at the shallow trench oxide edge between the base and the collector [1], [7], [8]. The reverse current gain degrades with an increase of base current after stress as indicated in the inset of Figure 3. These results show that the electromagnetic field stress induces traps not only in the emitter–base spacer's oxide, but also in the collector–base spacer's oxide. We see that this stress and the mixed mode stress [8] produce similar degradation modes, suggesting similar damage locations.

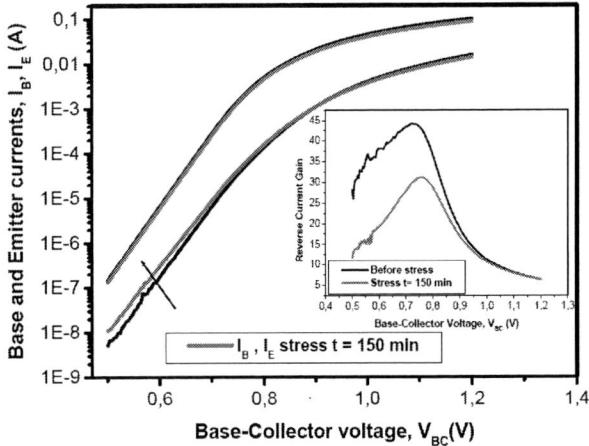

Figure 3. The reverse Gummel plots before and after applying stress. Inset shows the reverse current gain degradation versus Base-Collector voltage.

III. CAPACITANCE MEASUREMENTS

By using the Multi frequency Capacitance Measurement Unit (CMU) added to the Agilent B1500A and to examine the response of the SiGe HBTs under test, capacitance–voltage measurements were carried out at a DC voltage that was swept from -2.5 to 0.5 V. AC voltage was superposed with a frequency of 1 MHz and a signal amplitude of 10 mV. The CMU is equipped with the error correction function used to minimize the effects of the error elements in the extension cables and the DUT interface. Figure 4 plots the Base-Emitter capacitance before and after electromagnetic field stress when the collector is left open during this measurement. Noting that this illustration includes in practice some parasitic components like package or pad capacitances.

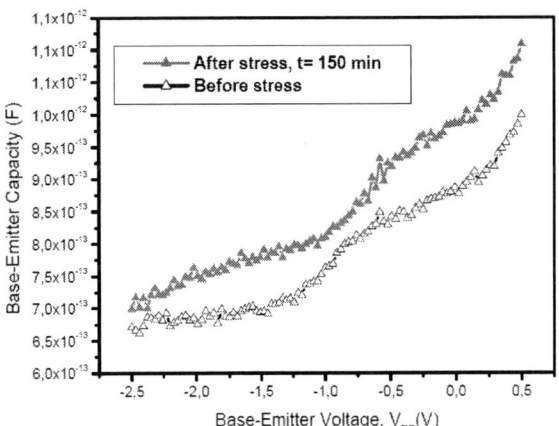

Figure 4. Reverse C-V characteristics of the Base-Emitter junction measured with the collector node open.

The plot shows that there is a considerable increase in the B-E capacitance characteristics after stress. This could be attributed to an increase of the carrier concentration in the B-E junction [9] which is in agreement with the increase of the nonideal base current in forward Gummel plots. The kink in the Base-Emitter capacitance is due to the transition from vertical to horizontal operation. This kink appears when the B–E voltage is equal to the effective vertical punch voltage which is emitter doping dependant. Since the internal part of the base–emitter region can be split up into two parts: one part for the vertical mode and one part for the horizontal one [10].Concerning the Base-Collector capacitance, and in agreement with the increase after stress of the nonideal base current in the reverse Gummel plots, the large number of carriers present in the B-C junction after stress modifies the B-C capacitance causing it to increase (see "Fig.5"). Forward and reverse Gummel plots with capacitance characterizations indicate that the electromagnetic field stress induces traps not only at the Emitter- Base spacer's oxide, but also at the Collector-Base spacer's oxide.

the base-emitter and the base-collector capacitances at low injection levels [11].

$$\frac{1}{2\pi . f_T} = \tau_{EC} = \tau_f + \frac{W_{BC}}{2\nu_S} + C_{BC}(R_C + R_E) + \frac{C_{EB} + C_{BC}}{g_m}$$

Where τ_f is the forward transit time and τ_{EC} is the total transit time. R_C and R_E are the collector and emitter series resistance, respectively. C_{BC} and C_{EB} are the base–collector and the emitter–base capacitance, respectively. The forward transit time τ_f is a sum of the minority carrier base transit time τ_B and the polysilicon transit time τ_E. At high injection levels the forward transit time is dominant in this expression, which represents the minority carrier transit time in the base and the emitter [12]. The transition frequency degradation after stress at high injection levels is due to the increase of the transit time induced by the generation/recombination mechanisms. The deviation of this frequency at low current is consistent with the increase of both capacitances after stress. Noting that, an eventual degradation of the collector and emitter series resistance can be inferred.

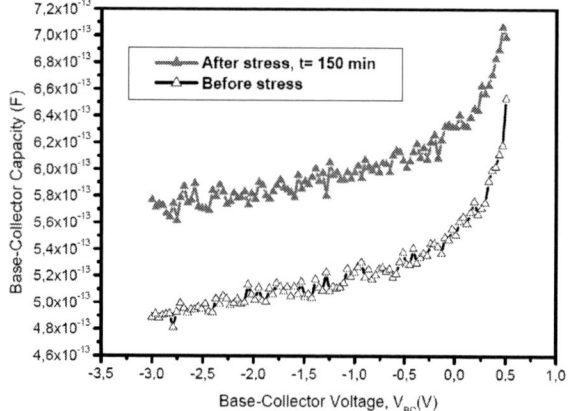

Figure 5. Reverse C-V characteristics of the Base-Collector junction measured with the emitter node open.

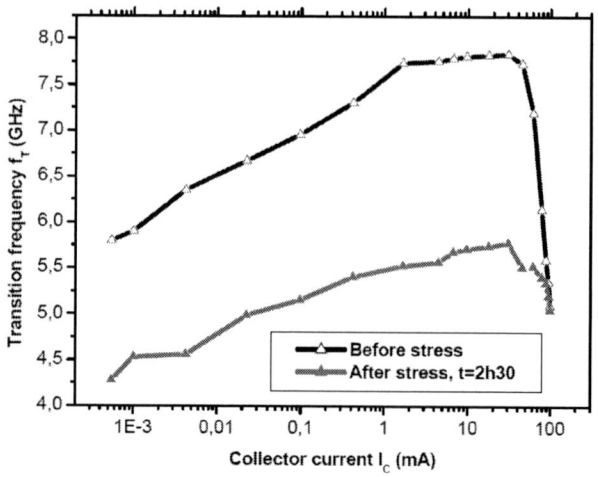

Figure 6. Cut-off frequency versus collector current before and after stress of the packaged HBTs.

IV. AC DEGRADATION

As well as the DC characteristics, the high-frequency characteristics were also affected by electromagnetic field stress. It was found that the amplitude of the transmission parameter (S_{21}) and the input parameter (S_{11}) were degraded after stress. Hence this indicates significant degradation in forward power gain and changes of the input impedance, respectively. This is consistent with the decrease of the output power versus the small and high input power regime. The cut-off frequency which is the most common AC figures-of-merit for RF transistors was extracted using the measured S-parameters. Figure 6 shows this cut-off frequency (f_T) as a function of collector current at V_{CE}= 2 V. It is worth noting that, the stress-induced degradation of (f_T) is significant at low and high currents. As expressed in the well-known equation (1) of (f_T), the cut-off frequency is approximately proportional to

V. FAILURE ANALYSIS

The SiGe HBT's used in this study are Surface Mounted Components (SMC) designed in SOT-343 footprint. Then, to extract the internal die of this component, it was de-packaged by using a chemical procedure. Cross sectioning is often done by Focused Ion Beam (FIB), and then Scanning Electron Microscopy (SEM) can be useful, but simple inspection by SEM is rarely fruitful for microstructural anomalies [13]. Then, cross sections of specimens, about 100 nm in thickness for Transmission Electron Microscopy (TEM) analysis, were made by focused ion beam (FIB).To protect the sample surface from high energy ion beam during sample preparation, ion beam enhanced Platinum (Pt) deposition were applied. TEM observations were carried out on a JEOL 2000FXII microscope operating at 200 kV. Elemental compositions were analyzed by energy dispersive spectroscopy (EDS) with a convergent beam and around 30 nm probe size. Figure 7 displays TEM cross section images of SiGe HBT samples selected from devices before and after 2h30 of electromagnetic field stress. As presented in our previous work, we have to note that the current gain of defective devices is degraded up to (-80%) which exceeds the failure criterion of the reliability evaluations after 2h30 of stress [14]. In the image of stressed sample, there is disorder under the metallization layers of base, emitter and collector contacts. It includes extended defects such as dislocations and interface deformation which may result from some interaction between the metal and the isolation layers. In order to better understand the metal-insulator interaction phenomenon in the stressed device, a higher resolution of the damage area is needed. Figure 8 shows details from figure 7 of the emitter metal contact. The failure signature of (Ti) defects is clearly visible, especially at the interface of Si_3N_4 isolation layers. On the image, we can observe the important degradations of the titanium (Ti) metallization layers situated around the emitter. The same phenomenon is observed at the

titanium layers of the base, and the collector fingers and both in the right and the left side. These transformations are clearly visible at the interface Ti/Si_3N_4 isolation layers and stop at the Ti/SiO_2 interface for the base, the emitter and the collector. This degradation affected all the sharp corners of (Ti) contacts, where the stress is supposed to be higher. It is worth to note that corner area can greatly impact the device reliability performance because some mechanisms give the anomalous device behavior related to the sharp corners. The importance of the electrical behavior of peripheral regions near the area edge is becoming more and more relevant because these areas involve high current density and high local electric field which reduces the oxide reliability [15]. We can note that the Ti layer thickness is reduced by around 50 nm, and it is composed of two different areas which present different morphologies and compositions. EDS profiles have been carried out to determine the composition evolution of the disturbed area. Figure 9 shows the EDS profile of the disturbed Ti layer located at the Si_3N_4 interface.

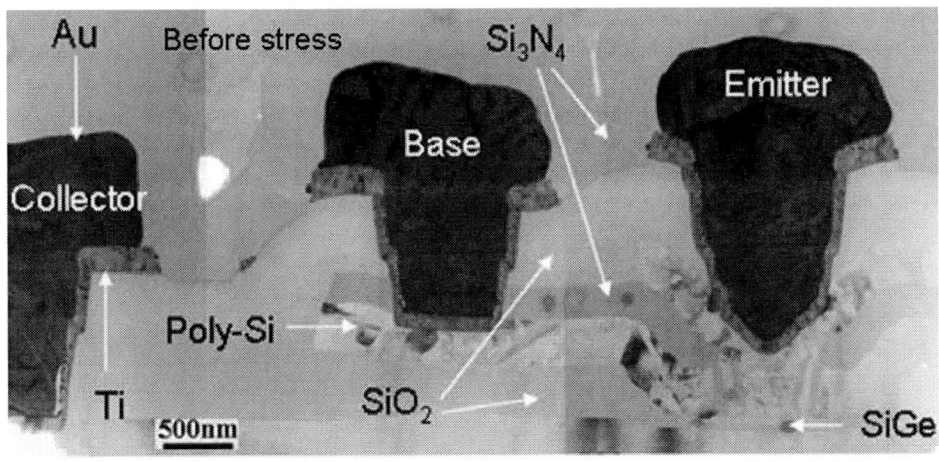

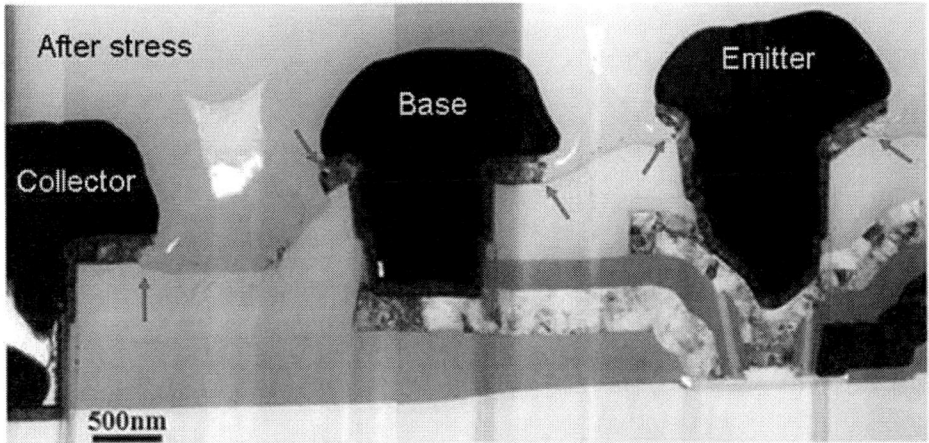

Figure 7. TEM images of both samples before and after 2h30 of electromagnetic field stress. Arrows point to abnormal areas.

At this interface, the initial Ti thin film has evolved into a layer composed of small grains sharply separated. These grains are crystalline and round with a diameter of 5-20 nm. Also, the EDS analysis shows that the dark biggest grains are gold (see "Fig.9"). The importance of the electrical and thermal states of peripheral regions of the metallic layers, near the Au–Ti–Si3N4 interfaces has to be pointed out in the present study because these areas involve high current densities and high local electric field which can affect the device reliability. In fact, the electromagnetic field stress causes high current densities at the base input of the HBT (probe-DUT coupling phenomenon) [14] which can increase the local heating effects [15], [16]. This is confirmed by EDS analyses suggesting that the local high current density and Joule heating induce localized reactive diffusion of Au into the Ti layer to form probably Ti–Au intermetallic compounds [17].

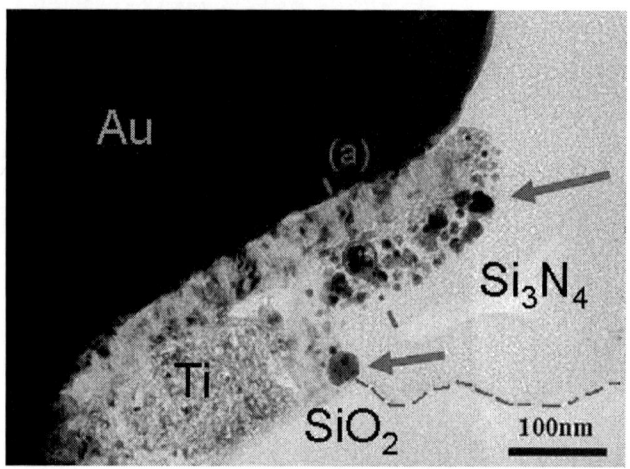

Figure 8. Bright field TEM image of the stressed sample, showing the degraded area of the Ti layer around an emitter (arrows). The line (a) show the EDS profile scans, and the circle points out a gold grain. The dotted line is used to separate silicon dioxide and silicon nitride layers.

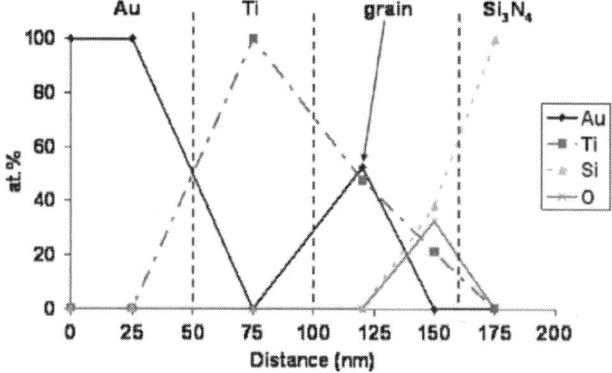

Figure 9. EDS profile carried out through the disturbed Ti layer located at the Ti/Si3N4 interface of the stressed sample.

Some of these Au–Ti reactions are known to increase the resistivity of the conducting layers which directly affects the HBTs' dynamic performances. As presented in our previous work [4], parameter deviations like S_{11} could be then attributed in part to the rise of the metallic resistances. On the other hand, some of our EDS analyses detected silicon in the (Ti) layer, but it was not possible to confirm if this (Si) is dissociated or not from the Si_3N_4 because the analyzed areas being very close to the interface.

CONCLUSION

We have presented a new reliability damage study in SiGe HBTs by the application of electromagnetic field. Important stress effects have been discussed as we have identified a base current leakage. The DC characteristics degradations appear due to hot carrier introducing generation/recombination trap centers at the emitter–base and collector-base spacers' oxide. Interaction phenomena after stress in Au-Ti-Si_3N_4 interfaces have been studied using TEM and EDS analysis. It was shown that the electromagnetic field stress induced the creation of high current densities leading to the Au migration into Ti and Ti-Si_3N_4 interface deformations. Au–Ti reactions may increase the resistivity of the conducting layers and affect the HBTs' dynamic performances.

REFERENCES

[1] J. D. Cressler, Silicon Heterostructure Handbook: Circuits and Applications of SiGe and Si Strained-Layer Epitaxy, USA, First ed, 2006, pp.421-538.

[2] N. Zerounian, E. Ramirez Garcia, F. Aniel, P. Chevalier et al. "SiGe HBT featuring fT>600GHz at cryogenic temperature". Proceedings of the international SiGe & Ge: materials, processing, and device symposium of the joint international meeting of the 214th meeting of ECS, ECS Transactions, vol. 16, pp. 1069–77, 2008.

[3] L. Bouchelouk, Z. Riah, D. Baudry, M. Kadi, A. Louis, B. Mazari. "Characterization of electromagnetic fields close to microwave devices using electric dipole probes". International Journal of RF and Microwave Computer-Aided Engineering, vol 18, pp. 146-156, 2008.

[4] A. Alaeddine, M. kadi, K. Daoud, H. Maanane, Ph. Eudeline, "Study of electromagnetic field stress impact on SiGe heterojunction bipolar transistor performance," International Journal of Microwave and Wireless Technologies, vol. 1, pp. 475-482, 2009.

[5] D. Baudry, C. Arcambal, et al. ''Applications of the Near-Field Techniques in EMC Investigations''. IEEE transaction on EMC, vol 49, pp. 485-493, 2007.

[6] S. Zhang, G. Niu, J.D. Cressler, H-J Osten, D. Knoll. "The Effects of Proton Irradiation on SiGe : C HBTs", IEEE Transactions on nuclear scinece, vol 48, pp. 2233-2237, 2001.

[7] S-Y. Huang, K-M. Chen et al. "Electrical stress effect on RF power characteristics of SiGe hetero-junction bipolar transistors", Microelectronics Reliability, vol 48, pp. 193-199, 2008.

[8] G. Zhang, JD. Cressler, et al. "A new mixed mode reliability degradation mechanism in advanced Si and SiGe bipolar transistors". IEEE Trans ElectronDev, vol 49, pp 2151–2156, 2002.

[9] K.V. Madhu, R. Kumar, M. Ravindra, R. Damle, "Investigation of deep level defects in copper irradiated bipolar junction transistor,", Solid-State Electronics, vol.52, pp. 1237-1243, 2008.

[10] S. Frégonèse, G. Avenier, et al. "A compact model for SiGe HBT on thin-film SOI". IEEE Transactions on Electron Devices, vol 53, pp.296-303, 2006.

[11] R. Liu, W. Qian and T. Wei. "Analytical modelling of current gain and frequency characteristics under high injection levels in Si/SiGe heterojunction bipolar transistors at 77 and 300 K". Microelectronics Journal, vol 30, pp. 1195-1206, 1999.

[12] R. Liu, W. Qian, T. Wei, "Analytical modelling of current gain and frequency characteristics under high injection levels in Si/SiGe heterojunction bipolar transistors at 77 and 300K," Microelectronics Journal, vol. 30, pp. 1195-1206, 1999.

[13] A.E.M. De Veirman. "3-Dimensional TEM silicon-device analysis by combining plan-view and FIB sample preparation''. Materials Science and Engineering, vol. 102, pp. 63-69, 2003.

[14] A. Alaeddine, C. Genevois, M. Kadi, F. Cuvilly, K. Daoud "Degradation of Au-Ti contacts of SiGe HBT's during electromagnetic field stress", Semiconductor Science and technology journal, vol 26 025003 (6pp). doi: 10.1088/0268-1242/26/2/025003, 2011.

[15] K. Banerjee, A. Mehrota, "Coupled analysis of electromigration reliability and performance in ULSI signal nets," Proceedings of the IEEE/ACM international conference on Computer-aided design, vol. 1, pp. 158–164, 2001.

[16] G. Xiang, W. James Haslett, K. Steven Dew. ''Simulation of Temperature Cycling Effects on Electromigration Behavior Under Pulsed Current Stress''. IEEE transactions on electron devices, vol. 45, pp. 380-386, 1998.

[17] W. E. Martinez, G. Gregori, T. Mates, "Titanium diffusion in gold thin films". Thin Solid Films, vol 518, pp. 2585–2591, 2010.

Reliability Testing of AlGaN/GaN HEMTs under Multiple Stressors

Bradley D. Christiansen and Ronald A. Coutu, Jr.
Department of Electrical and Computer Engineering
Air Force Institute of Technology
Wright-Patterson Air Force Base, Ohio, USA
1-937-255-3636 x7230, Ronald.Coutu@afit.edu

Eric R. Heller[1], Brian S. Poling[2], and G. David Via[2]
[1]Materials and Manufacturing and [2]Sensors Directorates
Air Force Research Laboratory
Wright-Patterson Air Force Base, Ohio, USA

Rama Vetury and Jeffrey B. Shealy
Defense and Power
RF Micro Devices, Inc. (RFMD®)
Charlotte, North Carolina, USA

Abstract—We performed an experiment on AlGaN/GaN HEMTs with high voltage and high power as stressors. We found that devices tested under high power generally degraded more than those tested under high voltage. In particular, the high-voltage-tested devices did not degrade significantly as suggested by some papers in the literature. The same papers in the literature also suggest that high voltages cause cracks and pits. However, the high-voltage-tested devices in this study do not exhibit cracks or pits in TEM images, while the high-power-tested devices exhibit pits.

Keywords—reliability, failure mechanisms, GaN, high electron mobility transistor (HEMT).

I. INTRODUCTION

GaN high electron mobility transistors (HEMT) are attractive to the United States Department of Defense for application in communications and sensing systems due to their ability to operate at high frequencies, high voltages, high temperatures, and high power. Interest in this technology is demonstrated by the Defense Advanced Research Projects Agency's Wide Bandgap Semiconductor initiative and by the Multidisciplinary University Research Initiatives funded by the Office of Naval Research and Air Force Office of Scientific Research. Despite the advantages of GaN HEMTs, there is concern that they do not have sufficiently long lifetimes for military systems. This concern has hampered their widespread acceptance and use.

Various stressors are claimed in the literature to cause degradation in GaN HEMTs. Stressor examples include high electric fields, high temperature with electrical stimulus, current with high electric field, and high drain bias with large rf drive. These stressors may result in various degradation mechanisms identified by signatures such as drain current degradation (itself a result of other signatures such as a decrease in transconductance, shifted threshold voltage, or increased on-resistance), an increase in gate leakage current, and/or reduced rf power output.

Two failure mechanisms of concern for GaN HEMTs are identified in [1]. Traps are formed by high electric fields, "hot" electrons (accelerated by high electric fields to energies much greater than the thermal-equilibrium value), and high temperatures within the devices. HEMT performance is degraded since charge collects in the traps and is not available for conduction. The second mechanism, structural damage (called lattice disruptions, pits, or cracks), occurs due to high temperatures in combination with electrical stimulus. The authors propose a current and contaminant interaction that creates the lattice disruptions by an etching process. Device performance is degraded in this case due to a conduction path created in the material beneath the gate.

Another prominent theory of crack formation has been presented. In [2], [3], [4] a critical voltage V_{DG}, inducing the inverse piezoelectric effect, is claimed to cause the pits and cracks in the AlGaN barrier layer of a GaN HEMT. The theory is that the high electric field on the drain side of the gate causes increased mechanical strain in the piezoelectric materials of the HEMT. As the electric field is increased in this region, the mechanical stress causes the lattice to crack at a critical voltage. Once this defect is formed, electrons tunnel from the gate to the conduction channel, which degrades the drain current. Drain current degradation, as measured by a decrease in maximum drain current I_{Dmax}, reportedly occurs in high-power state, OFF state, and, most severely, $V_{DS} = 0$ state tests. Degradation occurs in minutes as the stress voltage V_{DG} is applied in steps of 1 V per minute. I_{Dmax} is measured between steps.

Hot electron degradation is highlighted in [5]. Decreases in saturated drain-source current I_{DSS} and transconductance g_m were caused by hot electrons created by simultaneous high current and high electric field, and not by electric field alone. GaN HEMTs tested in semi-ON-state conditions ($V_{DS} = 20$ V, $V_{GS} = -5.5$ V) experienced a 15% decrease in maximum g_m, while the maximum g_m of devices stressed in ON-state conditions ($V_{DS} = 20$ V, $V_{GS} = 0$ V) and OFF-state conditions ($V_{DS} = 20$ V, $V_{GS} = -7.7$ V) decreased less than 5%. In addition,

---

This research was funded by the Air Force Research Laboratory (AFRL), Sensors Directorate, Aerospace Components and Subsystems Division.

The views expressed in this article are those of the authors and do not reflect the official policy or position of the United States Air Force, Department of Defense, or the U.S. Government.

978-1-4244-9113-1/11 $26.00 © 2011 IEEE

devices tested in ON-state conditions exhibited threshold voltage shifts while the same type of devices tested in OFF-state conditions did not.

Gate leakage current due to tunneling electrons as a dominant failure mechanism in GaN HEMTs is emphasized in [6]. When a HEMT is under a high drain voltage and driven by a large rf signal, the electric field at the drain side of the gate is sufficient to cause electrons to quantum mechanically tunnel from the gate electrode. These electrons can accumulate on the semiconductor surface, and thus be unavailable for conduction. They can also travel over the surface to the drain or through the AlGaN layer beneath the gate. Conduction from the gate to the drain along the surface is the dominant leakage path. The secondary path is through the AlGaN layer to the channel. Field plates can be used to reduce rf power degradation by decreasing the electric field at the gate. However, their use is detrimental to X-band and Ka-band devices due to the feedback capacitance the plates create. Surface passivation is a method to reduce the dominate leakage path over the surface.

With so many proposed stressors, degradation mechanisms, and degradation signatures, it is important to differentiate which stressors cause which effects. Due to this variety of stressors, mechanisms, and signatures, we have begun testing GaN HEMTs under multiple stressors to discover the relevant stressor or stressors, degradation mechanisms, and signatures. Knowing the limitations of a component in terms of potential parameter degradation is important to a circuit designer.

Two objectives of this study were to investigate the effects of different stressors and, specifically, to investigate whether high electric fields alone cause significant degradation.

II. EXPERIMENT DESCRIPTION

The devices used in this study were pulled from two wafers from the same lot. The AlGaN/GaN HEMT structure (from a commercial foundry) consists of a SiC substrate, a gate-integrated field plate, and a source-connected field plate. Gate length is 0.5 µm and periphery is 2×50 µm. See Fig. 1 for a schematic diagram of the tested devices. Additional structure details can be found in [7] and [8].

Two different sets of test conditions were used: one was high voltage ($V_{DS} = 60$ V and 100 V) and low current with the gate pinched off ($V_{GS} = -10$V) and the other was high dc power (≥ 11 W/mm). In all cases, testing was conducted in the dark under dry nitrogen in an Accel-RF dc test station. The base-plate temperatures (T_{bp}) of the power test Conditions 1, 2, and 3 were selected so that the devices had similar estimated peak channel temperatures (based on device modeling). The high-voltage test Conditions 4 and 5 also had similar estimated peak

channel temperatures. Fifteen devices were placed on test, with three devices at each of the five conditions listed in Table I.

For the upper set of power test conditions, the drain voltage V_{DS} was set and the gate voltage V_{GS} was adjusted until the target drain current I_D was reached (within the capabilities of the test station). After the initial setting of V_{GS}, V_{GS} was maintained for the duration of the test. The expected values of V_{GS} for the upper test conditions were based on previous testing and were not anticipated to cause forward gate current based on previous testing at $V_{GS} = 2$ V. For the lower set of high-voltage test conditions, both V_{DS} and V_{GS} were set, and the expected I_D was based on values seen during testing in a probe station.

The intended test sequence for the power test conditions was an initial characterization, followed by stress until I_D degraded to a pre-determined failure criterion, and ending with a post-failure characterization. However, test station measurement was not sufficiently precise and drift was too great to track I_D during stress, and the test was ended at 300 hours to conduct a characterization. The test station has since been upgraded to measure I_D with more precision and less drift.

The devices in the high-voltage set were characterized before stress and after each 100 hours of stress until reaching 300 total hours of stress. Their degradation was tracked with I_{DSS} and I_{Dmax}.

The characterization consisted of I-V and transfer curves conducted at $T_{bp} = 70$ °C. The I-V curves swept V_{GS} from −5 to 1 V in 1-V steps and V_{DS} from 0 to 10 V in 19 steps. The transfer curve was conducted at $V_{DS} = 10$ V with V_{GS} being swept from −5 to 1 V in 0.333-V steps. The characterization was shown to be benign in on-wafer testing. I_{DSS} was measured at $V_{DS} = 10$ V and $V_{GS} = 0$ V. I_{Dmax} was measured at $V_{DS} = 10$ V and $V_{GS} = 1$ V. On resistance R_{on} was calculated with V_{DS}/I_{DS} at $V_{DS} = 0.556$ V and $V_{GS} = 0$ V.

To investigate whether the changes seen after 300 hours of testing would recover with rest, an additional period of testing was begun after more than 48 hours of rest at room temperature in the dark under dry nitrogen. Most devices did not complete the intended additional period of testing for various reasons. The main reason was system glitches that appear to have been caused by building power fluctuations, which also knocked offline a chiller for the cleanroom in the same building.

After testing, four devices were selected for analysis by thermal and photoemission imaging to find apparent weak spots. Then, those select devices were reviewed by scanning electron microscope prior to being imaged by tunneling electron microscope to reveal physical degradation.

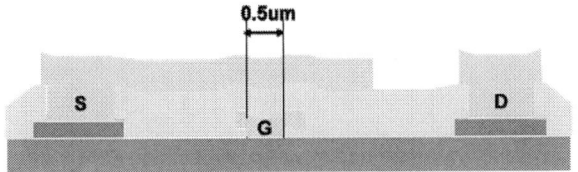

Figure 1.   Schematic diagram of tested devices. [9]

TABLE I.   TEST CONDITIONS FOR PRELIMINARY STUDIES

Condition	T_{bp} (°C)	V_{DS} (V)	Target I_D (mA/mm)	P_{diss} (W/mm)	Expected V_{GS} (V)
1	245	20.0	550	11.0	2
2	133	40.0	550	22.0	< 2
3	130	60.0	367	22.0	< 0.5
			Approx. I_D (mA/mm)		Set V_{GS} (V)
4	245	60.0	0.03	0.0018	-10
5	245	100.0	2	0.2	-10

III. Results and Discussion

The results of thirteen of the fifteen devices placed on test are compared. Of the two devices that are not included in the comparison, one device tested at Condition 4 apparently suffered infant mortality before 100 hours. Another device tested at Condition 3 reached the pre-determined failure criteria for I_D at 133 hours. Two devices that are included in the comparison did not achieve 300 hours. One device tested at Condition 3 reached the pre-determined failure criteria for I_D during stress at 253 hours due to the test station's I_D measurement drift; this device is included in the comparison because other devices (not included in this study) tested at similar conditions showed no significant changes in transfer curves at 200 at 400 hours. The other device was tested at Condition 5 and reached only 263 hours also due to the test station's I_D measurement drift; this device is included because it showed less than 2% change in transfer curves from 263 to 1017 hours in subsequent testing.

A summary of the results of the thirteen devices is in Table II. The percentages are average absolute changes from the pre-stress to the post-stress characterizations since two high-voltage-tested devices were exceptions to the general trends in changes to the selected parameters. One device at Condition 4 and one device at Condition 5 exhibited increases in I_{DSS} and negative threshold voltage V_T shifts. The same device at Condition 5 also exhibited an increase in I_{Dmax} (see Fig. 3). All devices experienced decreases in peak transconductance g_{mp} and increases in R_{on}. The other general trends were positive threshold voltage shifts and decreases in I_{Dmax} and I_{DSS}.

Fig. 2 shows the transfer and transconductance curves of a typical (meaning, following the general trends in I_{Dmax}, I_{DSS}, and V_T) high-voltage-tested device. Fig. 3 shows the transfer and transconductance curves of one of the two exceptional high-voltage-tested devices. Figs. 2 and 3 illustrate the variability in the performance of the tested HEMTs. Although discovering the cause of the negative threshold voltage shift in Fig. 3 was not an objective of this study, a possible explanation is a trapping phenomenon near the gate [10], [11]. The transfer and transconductance curves of Fig. 4 are representative of the power-tested devices. Figs. 3 and 4 contain the transfer and transconductance curves from testing subsequent to the initial 300 hours; these curves show little change after 300 hours.

Comparing the two sets—high voltage and high power—since estimated peak channel temperatures were similar for the respective sets, the devices tested at high power changed more significantly than the devices tested at high voltages and low current, except in g_{mp}. The high-power-tested devices changed more in V_T (13.0%), I_{Dmax} (11.3%), I_{DSS} (14.8%), and R_{on} (11.5%) than the devices tested at high voltage (6.01%, 4.89%, 6.28%, and 4.64%, respectively). Unlike [5], threshold voltage shifts were seen from both ON- and OFF-state conditions. Peak transconductance g_{mp} changed more for the devices tested at high voltage (4.10%) than for the devices tested at high power (2.25%). Similar to [5], the decrease in g_{mp} of devices tested in ON-state and OFF-state was less than 5%. The different degradation signatures in the two sets indicate different degradation mechanisms.

TABLE II. Average Absolute Percentage Changes in Parameters after 300 Hours

Condition	g_{mp}	V_T	I_{Dmax}	I_{DSS}	R_{on}
1	1.68%	10.8%	8.80%	11.8%	8.06%
2	1.98%	10.9%	10.9%	13.2%	12.23%
3	3.53%	19.4%	15.7%	21.5%	15.43%
4	3.45%	5.95%	4.51%	5.09%	3.78%
5	4.53%	6.05%	5.15%	7.08%	5.21%

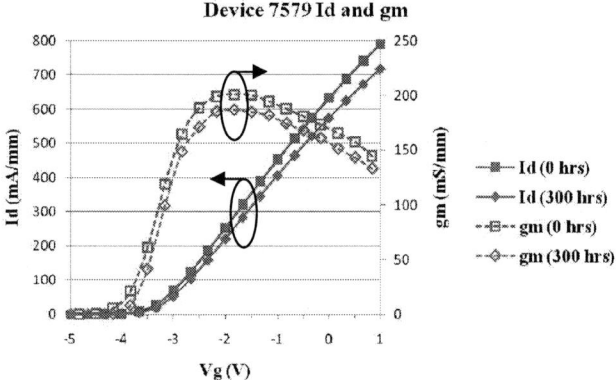

Figure 2. Transfer and transconductance curves at 0 and 300 hours of typical high-voltage-tested device. Device 7579 was tested at Condition 5.

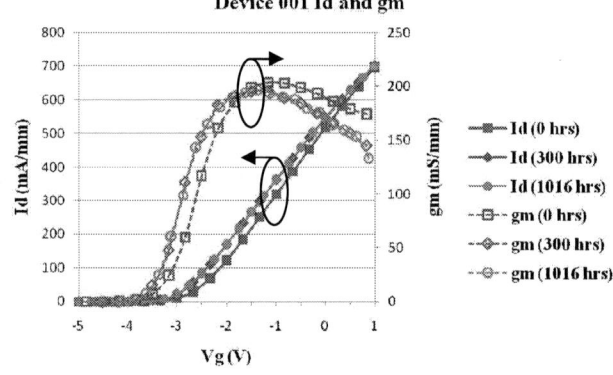

Figure 3. Transfer and transconductance curves at 0, 300, and 1016 hours of exceptional high-voltage-tested device. Device 001 was tested at Condition 5.

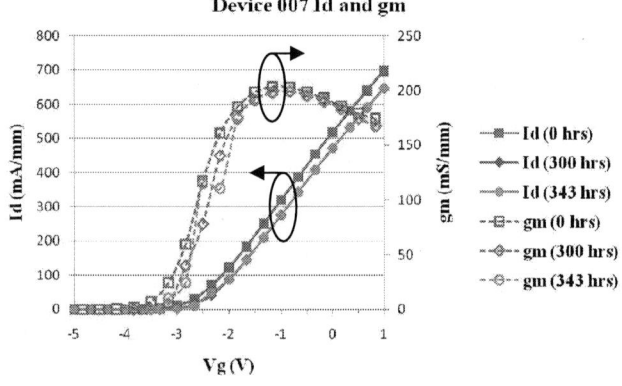

Figure 4. Representative transfer and transconductance curves at 0, 300, and 343 hours of high-power-tested device. Device 007 was tested at Condition 1.

There appears to be a correlation between higher drain biases and greater degradation. Although the estimated peak channel temperatures were similar for Conditions 1, 2, and 3 and separately for Conditions 4 and 5, the average absolute change for the four parameters increased with drain voltage (see Table II). Despite that apparent drain bias and degradation correlation, significant drain current degradation (> 10%) caused by high biases alone was not seen. Comparing Conditions 4 and 5 with other published OFF-state conditions [2], [3], the biases of Conditions 4 and 5 were at least 10 V higher. Yet, the significant degradation seen at the lower voltages after minutes of stress in the other studies was not seen at the higher voltages after hours of stress in this study.

The changes that occurred in the devices during stress seem to be unrecoverable with rest. The average change in I_{Dmax} between the value measured at 300 hours of stress and the value measured after rest was 0.01% with a maximum of 2.3% and a minimum of −2.37%. For I_{DSS}, the average change was 0.12% with a maximum of 2.26% and a minimum of −2.29%. Both maximums were measured on one device and both minimums were measured on another. All other percent changes were less than 1% in absolute value. In addition to the changes in I_{Dmax} and I_{DSS} before and after rest, four of six power-tested devices that began the additional testing period required greater gate voltages to attain the target drain current—an indication of permanent degradation.

After stress testing, we investigated the four select devices by thermal and photoemission imaging in a Quantum Focus Instruments InfraScope™. In three of the four devices, hot spots corresponded with bright spots. We delivered these four devices to NanoTEM for transmission electron microscope (TEM) imaging at the hot and bright spots. Although Device 001 is exceptional (with a negative V_T shift and increases in I_{Dmax} and I_{DSS}), its transfer and transconductance curves and its infrared (IR), photoemission (PE), and TEM images are shown since it had notable IR and PE images, whereas the other imaged high-voltage-tested device did not. Fig. 5 contains IR (radiance) and PE images for the high-voltage-tested Device 001 whose transfer and transconductance curves are in Fig. 3. (An insufficient number of samples were imaged by IR and PE to determine whether the features of Fig. 5 correlate to the negative threshold voltage shift of Fig. 3.) The IR (radiance) and PE images in Fig. 6 are those of the power-tested Device 007 whose transfer and transconductance curves are in Fig. 4.

The TEM images of Devices 001 and 007 are in Fig. 7. As in the TEM image of Device 001 (stressed at Condition 5), the other high-voltage-tested part that was imaged by TEM, Device 7632 (stressed at Condition 4), does not exhibit a crack or pit at the drain edge of the gate. This is contrary to the findings reported in [4]. However, in the TEM images of the two high-power-tested devices, Devices 007 and 008 (not shown), that were both stressed at Condition 1, small pits have formed at the

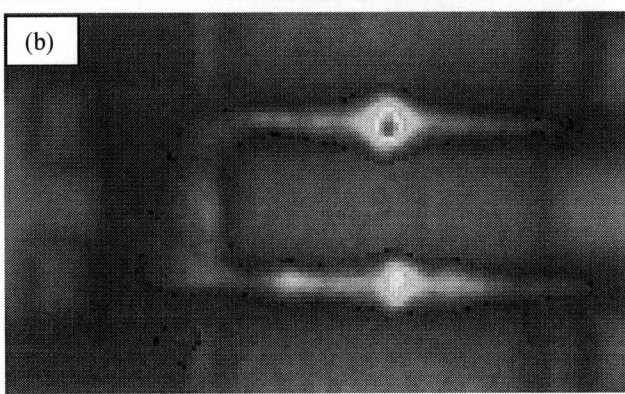

Figure 5. Device 001 (high-voltage-tested) at a baseplate of 85 °C. The upper middle spot was targeted for TEM imaging. (a) IR (radiance) image at 15X magnification. $V_{DS} = 40$ V, $I_D = 10$ mA, $V_{GS} = -2.42$ V, $I_G = -5$ μA. (b) PE image at 20X magnification. $V_{DS} = 100$ V, $I_D = 11$ μA, $V_{GS} = -10$ V, $I_G = -12$ μA.

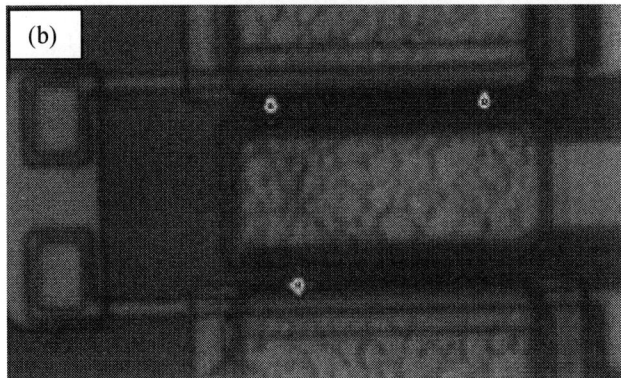

Figure 6. Device 007 (high-power-tested) at a baseplate of 85 °C. The lower left spot was targeted for TEM imaging. (a) IR (radiance) image at 15X magnification. $V_{DS} = 28$ V, $I_D = 10$ mA, $V_{GS} = -1.69$ V, $I_G = -3.3$ μA. (b) PE image at 50X magnification. $V_{DS} = 10$ V, $I_D = 3.2$ mA, $V_{GS} = -1$ V, I_G in nA range.

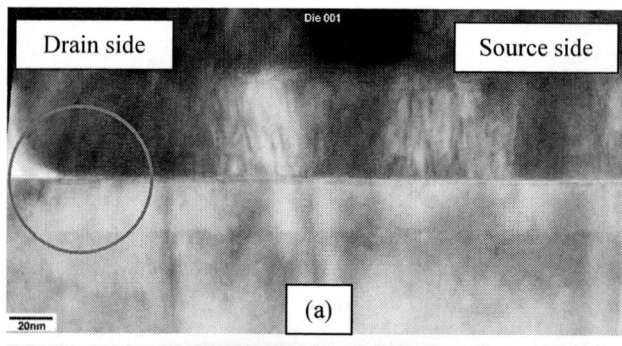

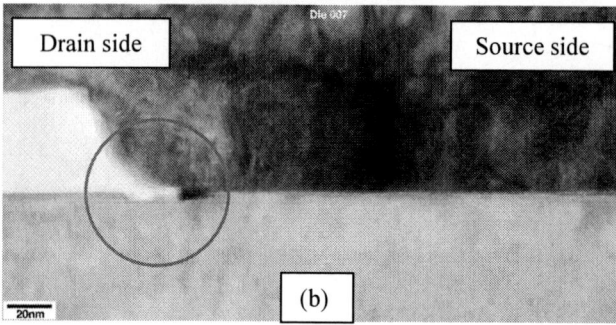

Figure 7. TEM images of (a) Device 001 (high-voltage-tested) and (b) Device 007 (high-power-tested). Notice absence of a pit or crack in Device 001 and the presence of a pit in Device 007.

drain edge of the gate. Thus, current appears necessary to create the pits in the AlGaN layer.

IV. CONCLUSION

We have studied the degradation of AlGaN/GaN HEMTs subjected to the conditions of high dc power and high voltage with the gate pinched off. More degradation was generally observed due to the high-power conditions than to the high-voltage conditions. The degradation seen appears to be unrecoverable with rest. Severe drain current degradation due to high drain biases was not observed as has been reported elsewhere. Pits in the AlGaN layer on the drain side of the gate were observed in the high-power-tested devices. However, pits or cracks were not seen in the high-voltage-tested devices, which is contrary to published reports. Thus, electric field alone does not appear to cause significant degradation, and current in conjunction with high electric fields seems to be required for pit or crack formation. Possible reasons for the differences between our and others' observations include

material quality, fabrication processes, device structure, and bias conditions. The AlGaN/GaN HEMT structure studied herein is robust to high drain biases.

ACKNOWLEDGMENT

B.D.C. thanks the Air Force Research Laboratory High-Reliability Electronics Virtual Center team for fruitful discussions, Steve Tetlak for TEM sample preparation, and Fred Shaapur for TEM imaging.

REFERENCES

[1] K. V. Smith, S. Brierley, R. McAnulty, C. Tilas, D. Zarkh, M. Benedek, P. Phalon, and A. Hooven, "GaN HEMT reliability through the decade," *ECS Trans.*, vol. 19, no. 3, pp. 113-121, May 2009.

[2] J. A. del Alamo and J. Joh, "GaN HEMT reliability," *Microelectron. Reliab.*, vol. 49, nos. 9-11, pp. 1200-1206, Sep. -Nov. 2009.

[3] J. Joh and J. A. del Alamo, "Critical Voltage for Electrical Degradation of GaN High-Electron Mobility Transistors," *IEEE Electron Device Lett.*, vol. 29, no. 4, pp. 287-289, Apr. 2008.

[4] P. Makaram, J. Joh, J. A. del Alamo, T. Palacios, and C. V. Thompson, "Evolution of structural defects associated with electrical degradation in AlGaN/GaN high electron mobility transistors," *Appl. Phys. Lett.*, vol. 96, no. 23, p. 233509, 2010.

[5] G. Meneghesso, G. Verzellesi, F. Danesin, F. Rampazzo, F. Zanon, A. Tazzoli, M. Meneghini, and E. Zanoni, "Reliability of GaN high-electron-mobility transistors: state of the art and perspectives," *IEEE Trans. Dev. Mat. Rel.*, vol. 8, no. 2, June 2008.

[6] R. J. Trew, D. S. Green, and J. B. Shealy, "AlGaN/GaN HFET reliability," *IEEE Microwave*, vol. 10, no. 4, pp. 116-127, June 2009.

[7] S. Lee, R. Vetury, J. D. Brown, S. R. Gibb, W. Z. Cai, J. Sun, D. S. Green, and J. Shealy, "Reliability assessment of AlGaN/GaN HEMT technology on SiC for 48V applications," in *2008 IEEE Int. Rel. Phys. Symp.*, Phoenix, AZ, pp. 446-449.

[8] J. D. Brown, D. S. Green, S. R. Gibb, J. B. Shealy, J. McKenna, M. Poulton, S. Lee, K. Gratzer, B. Hosse, T. Mercier, Y. Yang, M. G. Young, and R. Vetury "Performance, Reliability, and Manufacturability of AlGaN/GaN High Electron Mobility Transistors on Silicon Carbide Substrates", *ECS Trans.*, vol. 3 no. 5, 2006.

[9] RFMD, "RFMD GaN Foundry Services," http://www.rfmd.com/pdf/gan_slick_banner.pdf, accessed Aug. 2010.

[10] H. Rao and G. Bosman, "Device reliability study of high gate electric field effects in AlGaN/GaN high electron mobility transistors using low frequency noise spectroscopy," *J. Appl. Physics*, vol. 108, p. 053707, Sep. 2010.

[11] Z.-Q. Fang, G. C. Farlow, B. Claflin, D. C. Look, and D. S. Green, "Effects of electron-irradiation on electrical properties of AlGaN/GaN Schottky barrier diodes," *J. Appl. Physics*, vol. 105, p. 123704, Jun. 2009.

978-1-4244-9113-1/11 $26.00 © 2011 IEEE

Long Term Isothermal Reliability of Copper Wire Bonded to Thin 6.5 μm Aluminum

Classe, F.C.
Gaddamraja, Sesha
Spansion, Inc, Sunnyvale, CA
915 De Guigne Ave
Sunnyvale, CA
francis.classe@spansion.com

Abstract - **In long term reliability evaluations of Spansion memory products built using copper (Cu) wire bonding in lieu of gold (Au) wire for package-to-die interconnection, results indicated acceptable reliability performance of the copper-aluminum (Cu-Al) bond. Some differences, however, were observed when compared to gold-aluminum (Au-Al) bonds used as a control. In order to determine if these differences represented a true reliability concern, a series of experiments were run on a variety of process technologies (from 200nm to 65nm) to determine wear-out failure mechanisms of these Cu bonds and their associated apparent activation energies. Isothermal reliability tests at three temperatures (150 °C, 175 °C, and 200 °C) were performed using uncoated 0.9 mil Cu wire (with 0.9 mil Au wire as a control) bonded to functional flash die from a variety of process technologies. All bonding was done to 6.5 μm thick Aluminum-Copper (Al-0.5% Cu) bond pads. Bond shear and wire pull values were measured at each readpoint and the experiments were continued through extended long term readpoints to insure that products were stressed until failure. The primary failure mechanism identified was interfacial cracking between the copper bond and the intermetallic layer, starting at the rim of the bond. The apparent activation energy computed for the Cu-Al bond interfacial cracking was 0.70 eV. Subsequent calculations of expected product lifetime in various usage models using this Ea show that the Cu wire bonding provides more than adequate reliability lifetime for all expected product usage scenarios.**

I. INTRODUCTION

Wirebonding remains one of the most popular methods of creating electrical connections from an integrated circuit to leadframe-based or substrate-based packages. By and large, the gold-aluminum (Au-Al) metallurgy is the most common, in which gold wire of varying purities (usually 2N – 4N) is bonded to bond pads consisting mainly of Al with a small percentage of some other material (usually Cu and/or Si, in the 0.5-1% range). With the costs of gold ever increasing along with the need for performance, industry is increasingly moving to copper wire as a higher performance and less expensive alternative bonding wire for semiconductor products.

Considering the vast array of data available for gold-aluminum (Au-Al) interfacial reliability [1-6], the main failure modes for Au-Al bond interfaces are well known (purple

plague, white plague, Kirkendall voiding, etc.). Although a good deal of research has been performed on Cu-Al intermetallic formation and growth [7-10], the data available are less conclusive with respect to actual reliability. Specifically, the "playing field" does not seem to be level for all studies and experiments, as differences in bonding parameters, bond pad thickness and metallurgy, wire diameter, free-air-ball (FAB) formation, and wire coatings all seem to impact the subsequent reliability of the bond, more so than with gold wire [11-17]. Additionally, most bonding studies currently available typically used thicker bond wire (up to 2 mil) [11, 12, 15], and/or thicker bond pads (8 μm – 10 μm) [12, 13, 16, 17]. In this work, 0.9 mil copper was used, and the bond pad metal is only 6.5 μm thick; this is significantly thinner than most of the industry and thinner than what is normally recommended by Cu wire suppliers (8 - 10μm), though evaluations examining the effect of bond pad thickness on reliability are ongoing at Spansion. Further, the majority of available studies examined unmolded parts; a comparison of molded parts to unmolded ones is not a fair one, as the mold compound used in the assembly process plays a significant role in the subsequent reliability of the product [17], not to mention that packaged product shipped to the customer must necessarily be encapsulated.

II. EXPERIMENTAL

A. Description of Assembly and Testing

Electrical performance, bond shear, and wire pull data were obtained for reliability tests performed at 150 °C, 175 °C, and 200 °C, in an attempt to assess the long term reliability. Significant effort was made during the manufacturability assessment to ensure that a uniform thickness of Al bond pad remained after wirebonding to the 6.5 μm thick bond pad (Al-0.5% Cu). 2N 0.9 mil copper wire was thermosonically bonded and the die were packaged into TSOP (Thin Small Outline Packages) and molded with green, low halogen mold compound for reliability test. Forming gas (95%N_2 + 5%H_2) was also used to ensure good bondability. Figure 1 shows a cross section of a Cu bond on a zero-hour, fully assembled unit. Note that at 150 °C, bias was applied

978-1-4244-9113-1/11 $26.00 © 2011 IEEE

during testing whereas at 175 °C and 200 °C, no bias was applied.

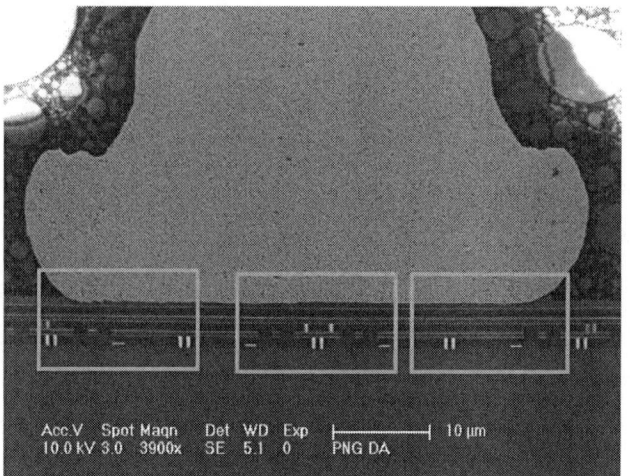

Figure 1 – Cross Section of Typical Cu Bond Post Assembly.

The average remaining bond pad thicknesses measured at the three locations from left to right were 0.451 µm, 0.475 µm, and 0.499 µm respectively. No visible intermetallic was observed, consistent with the slow rate of growth in the Cu-Al metallurgy system. Bond shear and wire pull (described below) were all within Au specifications.

Table 1 shows the matrix of devices used to collect the data. A variety of feature sizes was used across two different device architectures for both Cu and Au wires in order to understand if these factors play a role in the reliability of Cu wire. Note that legs 1c – 6c are the corresponding Au control lots for the Cu lots in 1-6.

TABLE I. DEVICE TEST MATRIX

Leg	Feature	Technology	Wire	150C	175C	200C
1	65 nm	MirrorBit™	0.9 mil Cu	X	-	-
2	90 nm	MirrorBit™	0.9 mil Cu	X	-	-
3	110 nm	MirrorBit™	0.9 mil Cu	X	-	-
4	200 nm	MirrorBit™	0.9 mil Cu	X	-	-
5	130 nm	Floating Gate	0.9 mil Cu	X	-	-
6	200 nm	Floating Gate	0.9 mil Cu	X	X	X
1c	65 nm	MirrorBit™	0.9 mil Au	X	-	-
2c	90 nm	MirrorBit™	0.9 mil Au	X	-	-
3c	110 nm	MirrorBit™	0.9 mil Au	X	-	-
4c	200 nm	MirrorBit™	0.9 mil Au	X	-	-
5c	130 nm	Floating Gate	0.9 mil Au	X	-	-
6c	200 nm	Floating Gate	0.9 mil Au	X	X	X

B. Wire Pull and Bond Shear Data

Wire pull and bond shear data, post decapsulation, were plotted as a function of readpoint for all three temperatures. Because wire pull and bond shear lower limits have not yet

been established for copper wire, values for gold were substituted as a reference; wire pull limits were obtained from JESD22-B116 (8.1 g lower limit for 2.2 mil bond diameter), and bond shear guidelines were obtained from MIL-883 Method 2011 (2.16 g lower limit for 0.9 mil wire post-mold).

Figure 2 shows the average wire pull and bond shear values for 0.9 mil gold wire bonded to 200 nm process silicon (Leg 6c) over time at 150 °C and 200 °C with a dashed line indicating the lower specification limit.

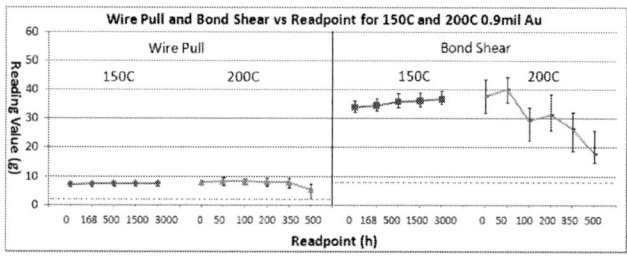

Figure 2 – 0.9mil Au Wire Pull and Bond Shear at 150 °C and 200 °C.

The data indicates that gold wire performance is relatively consistent at 150 °C, but begins to decrease over time at 200 °C.

Figure 3 shows the average wire pull and bond shear values for 0.9 mil copper wire bonded to 200 nm process silicon (Leg 6) over time at various temperatures with a dashed line indicating the lower specification limit.

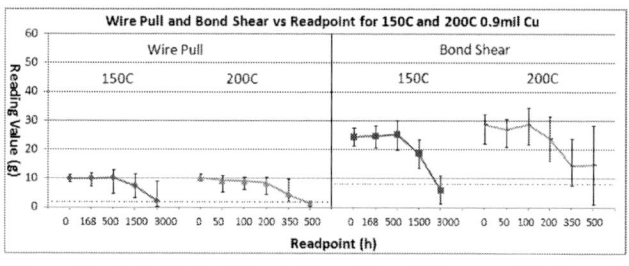

Figure 3 – 0.9mil Cu Wire Pull and Bond Shear at 150 °C and 200 °C

Examining all the process technology wire pull and bond shear data for both Au and Cu wire, the values were remarkably similar across both flash technologies and feature sizes; as such, the figure above should be representative of the performance of the overall Cu wire bonding process.

C. Reliability Data

The data presented in this section are a summary of the isothermal data collected on the devices listed in Table 1. Failure in this case is defined by the part's inability to meet parametric or functional datasheet parameters, verified to have been caused by failure at the bond.

The failure mode observed for copper wires was interfacial cracking, as shown in Figure 4. As noted in [10] and [12], cracking seems to initiate at the outside rim of the bond and then propagate inward. The IMC can also be observed, and appears mostly at the edges of the bond (in a light gray color, lighter than the darker gray of the Al bond pad), consistent with the references cited above. The reported electrical failure mode was "OPEN," indicating significantly increased resistance based on the production electrical test program. The cross section below demonstrates a failing unit from 3000h at 150 °C.

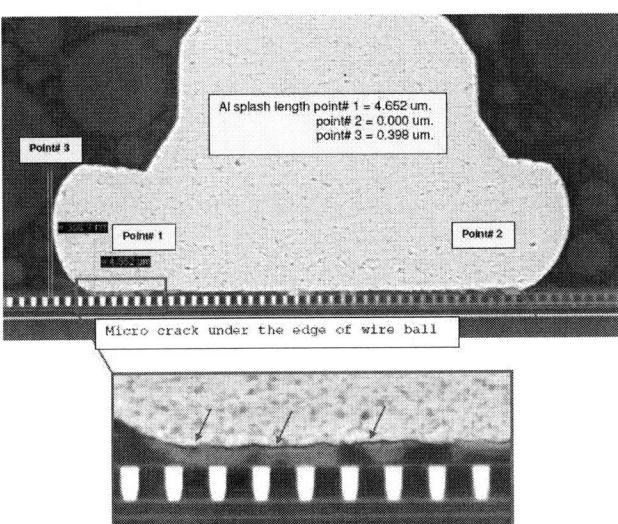

Figure 4 – Typical Cross Section (Non-Decapsulated) of Failing Cu Bond Showing Crack Initiation under SEM.

Figure 5 shows a cross section of a gold bond demonstrating the presence of Kirkendall voiding at 3000h at 150 °C. Note that although Kirkendall voiding was observed, Au wire electrical failures were not observed during any reliability test leg.

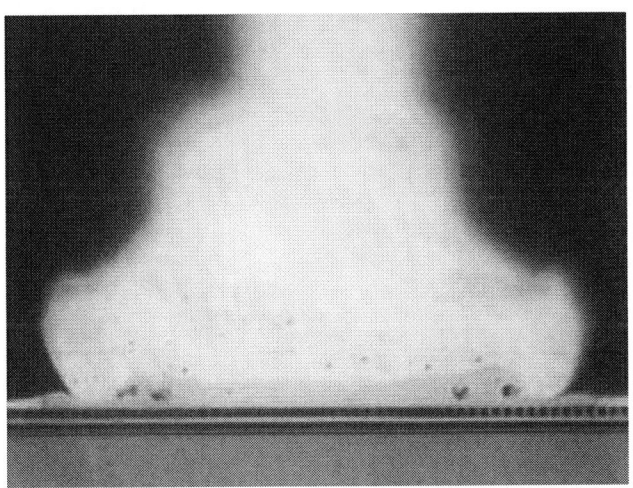

Figure 5 – Typical Cross Section (Non-Decapsulated) of Kirkendall void initiation under optical microscope.

D. Apparent Activation Energy Computation

Because some units were pulled out at interim readpoints to perform wire pull and bond shear, Kaplan-Meier estimation was used to adjust the cumulative failure percentage to account for the units that did not proceed all the way to the end of the test. Figure 6 shows the Lognormal failure plots for 150 °C, 175 °C, and 200 °C. In this plot, CDF (Cumulative Failure Fraction) and the corresponding Cumulative Failure Percentage are plotted on the x-axis, while ln(time) is plotted on the y-axis.

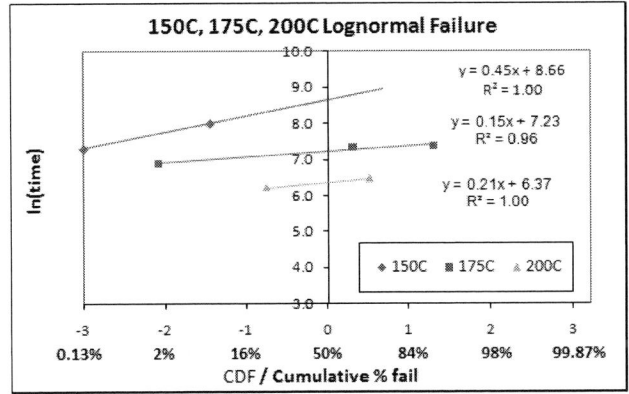

Figure 6 – Lognormal Failure Plot at 150 °C, 175 °C, and 200 °C.

The slopes from all three runs (0.44, 0.15, and 0.21) are similar enough to compute an average sigma value of 0.26 for all runs, indicating that there is likely no defective subpopulation and a single wearout mechanism is probably at work. It should be noted that the final readpoints obtained at 150 °C were at 1500 hours and 3000 hours; since no failures were observed at the 1500 hours readpoint, an estimate of failure was made based on a binomial distribution. At 50% confidence, and based on overall the sample size obtained at the 1500 hours readpoint, the percent defective comes to approximately 0.13%, so this value was used to determine the CDF at 1500 hours to plot the curve for 150 °C. From this data, both the ln(t50) and ln(t16) parameters were extracted from the lognormal plot and used to construct Figure 7.

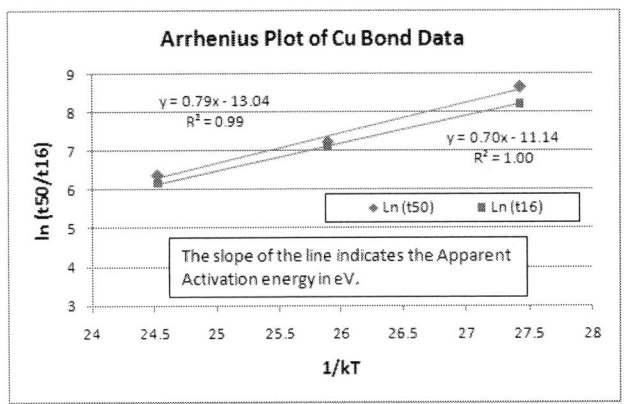

Figure 7 – Arrhenius Plot for ln (t50) and ln (t16).

The slope of the Arrhenius Plot indicates the apparent activation energy for the failure mode. The slope using the natural logarithm of 50% cumulative failure, ln(t50), indicates an activation energy of 0.79 eV, but because of the lack of data at later readpoints in the 150 °C leg, this probably overestimates the activation energy with a larger ln(t50) at 150 °C (a value of 27.5 on the x-axis). A more conservative number is found by utilizing the ln(t16) values, as the first actual datapoint for 150 °C was collected quite close to this failing percentage. This value is 0.70 eV.

III. DISCUSSION

The results of bond shear and wire pull were consistent with the electrical test response ("OPEN" failure signatures), as the drop in performance correlated to the observed failures in electrical test. It is also of interest to note that Au/Al bond shear values at the higher temperatures (175 °C and 200 °C) dropped off at approximately the same time as Cu. There was, however no significant drop in bond shear or wire pull for Au wire at 150 °C; this may be explained by considering the apparent activation energy of gold bond failure, normally reported for a variety of failure modes as 0.70 eV to 1.26 eV, and up to 2 eV (depending on doping) for 2 N wire [1-6]. It should also be noted that the decapsulation process for Cu wire is inherently difficult to perform, and as the parts are exposed to high temperatures over the long term, the process grows even more difficult as more time exposed to the decapsulation chemicals is required to etch away the mold compound. Cross sections revealed a thin crust on the exterior of the mold compound which had a higher percentage of carbon than the mold compound directly underneath it. This crust may be related to further cross linking of the epoxy resin of the mold compound over high temperature, and might be responsible for the wider distribution of bond shear and wire pull data observed at later readpoints.

The fact that copper bond interfacial cracking has a smaller apparent activation energy than those generally accepted for gold wire is not immediately alarming, nor is the fact that copper bonds fail earlier in testing than those of gold. In order to assess the long term reliability requirements for any product, one must take into account the actual field life required for each specific application for the product, and then compute the respective lifetime prediction from the actual accelerated stress test data based upon the apparent activation energy (Ea) and the corresponding acceleration factor calculated from the reliability model being used. Table 2 shows a lifetime estimate for Cu wire bonding, derived from the experimental data herein, for a number of different market segments and usage models. The computations are based upon a 0.1% failure rate at 175 °C, which is approximately 877 hours from the data collected and shown in the lognormal plot (Figure 6).

TABLE II. LIFETIME ESTIMATIONS FOR VARIOUS MARKET SEGMENTS.

MARKET SEGMENT	TYP. LIFE yr	OPERATING CONDITIONS			STORAGE CONDITIONS			EQUIV. TIME @ 175C h	# OF LIVES
		TIME h	TEMP °C	AF	TIME h	TEMP °C	AF		
CONSUMER DESKTOP	5	13000	30	6224.2	30800	30	6224.2	7.0	124.69
HIGH END SERVER	11	94000	30	6224.2	2360	30	6224.2	15.5	56.68
AVIONICS ELECTRONICS	23	150000	50	1170.1	51480	30	6224.2	136.5	6.43
TELECOM HANDHELD	5	43800	40	2627.7	0	30	6224.2	16.7	52.64
TELECOM CONTROLLED	15	131000	70	267.3	400	30	6224.2	490.1	1.79
AUTOMOTIVE UNDERDASH	15	8200	45	1742.4	123200	30	6224.2	24.5	35.82
AUTOMOTIVE UNDERHOOD	15	8200	125	9.9	123200	30	6224.2	847.3	1.04

The operating conditions as well as typical life in the above computations were taken from JESD94A, excepting "Automotive Underdash," which was added to distinguish it from "Automotive Underhood" type applications. The operating temperatures were obtained from the maximum ambient temperature condition shown in the JESD94A. Storage conditions were chosen to be 30 °C uniformly across all applications. The data indicates that for all applications listed, there is more than sufficient margin to meet all listed reliability requirements.

IV. CONCLUSIONS

Flash devices from variety of technologies and architectures were submitted for reliability test, and bond shear and wire pull values were obtained from each readpoint. The values of bond shear and wire pull obtained across various process technologies indicate that the under-pad buildup process has little to no impact on the bond performance on a well controlled bonding process using Cu wire on Al-0.5% Cu bond pads. Based on the electrical data obtained at each readpoint, lognormal and Arrhenius plots were created in order to obtain the apparent activation energy (0.70 V) for the primary failure mode, which was found to be interfacial cracking between the bulk Cu and the IMC formed in the Al bond pad. Based upon this data, acceleration factors were computed for a variety of applications and the corresponding lifetimes were computed. The data indicate that there is more than ample life within the Cu bonds to meet a variety of applications in the field, even some of the more stringent applications such as automotive and avionics electronics. This data was collected on devices using 6.5 µm thick bond metal, which is considered fairly thin by bond wire suppliers (8 - 10 µm is typically recommended as a minimum for Cu bonding), serves to show that a controlled, robust, and reliable process can be achieved even with thin bond pad metal.

It should be noted that, much as the industry has become intimately familiar with Au wire bonding over the decades, the same advances in knowledge will occur with Cu wire bonding as more and more companies study and adopt it. A variety of issues plagued early users of Au bonding on Al bond pads, and through research and innovation, these roadblocks were eliminated and the reliability of the process improved dramatically. With the knowledge gained from Au, improvements and advances in Cu bonding should occur even more quickly, and its adoption in industry even more rapid.

V. FUTURE WORK

The data set analyzed and discussed above was obtained from early evaluations of 0.9 mil Cu wire used in Spansion factories. Additional studies are currently ongoing using 0.8 mil Cu wire with advanced capillaries and optimized bond parameters. Preliminary data (Figure 8) shows significant improvements in as-assembled bond shear and wire pull as well as shear and pull values obtained in subsequent reliability test. Further, no electrical failures were observed at any point in the data shown in Figure 8. This indicates that a significant performance improvement has already been achieved, and that the improved process significantly outperforms the early evaluations presented here.

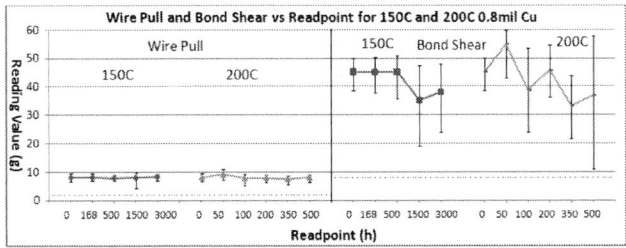

Figure 8 – 0.8mil Cu Wire Pull and Bond Shear at 150 °C and 200 °C

ACKNOWLEDGMENTS

This work would not have been possible without the support and assistance of many Spansion employees, namely: NC Lai, Don Bottarini, Sally, Foong, Pak Wong, Tony Reyes, Richard Blish, and both the Kuala Lumpur and Penang Reliability laboratories, as well as their Device and Construction Analysis laboratories.

REFERENCES

[1] Adams, C.N., "Bonding-Wire Failure Mode In Plastic Encapsulated Integrated Circuits", *11th Annual proceedings, IEEE Reliability Physics Symposium*, pp. 41-44 (1973)

[2] Gale, R.J., "Epoxy Degradation Induced Au-Al Intermetallic Void Formation in Plastic Encapsulated MOS Memories," *22nd Annual proceedings, IEEE Reliability Physics Symposium*, p. 37, (1984)

[3] Khan, M., Fatemi, H., Romero, J. and Delenia, E., "Effect of High Thermal Stability Mold Material on the Gold-Aluminum Bond Reliability in Epoxy Encapsulated VLSI Devices," *26th Annual proceedings, IEEE Reliability Physics Symposium*, p. 40 (1988)

[4] Uno, T., Tatsumi, K. and 0hno, Y.: "Void Formation and Reliability in Gold-Aluminum Bonding", *Proceedings of the Joint ASME/JSME 'Advances in Electronic Packaging'* p. 2 (1992)

[5] Noolu, N., Klossner, M., Ely, K. Baeslack, W. and Lippold, J., "Elevated Temperature Failure Mechanisms in Au-Al Ball Bonds," *International Symposium on Microelectronics*, pp. 478-482 (2002)

[6] Blish, R.C.; Li, S.; Kinoshita, H.; Morgan, S.; Myers, A. "Gold-Aluminum Intermetallic Formation Kinetics," *Device and Materials Reliability, IEEE Transactions*, Volume 7, Issue 1, pp. 51–63 (2007)

[7] Wulff, F. W., Breach, C. D., et al. "Further Characterization of Intermetallic Growth in Copper and Gold Ball Bonds on Aluminum Metallization," *SEMICON*, Singapore, pp. 1-10 (2005)

[8] Braech, C. D., Wulff, F. W., "Intermetallic Growth in Gold Ball Bonds Aged at 175C: Comparison between Two 4N Wires of Different Chemistry," *Gold Bulletin*, Volume 42, No. 2, pp. 92-105 (2009)

[9] Seng, Yeoh Lai. "Characterization of Intermetallic Growth for Gold Bonding and Copper Bonding on Aluminum Metallization in Power Transistors." *Electronics Packaging Technology Conference, 9th*, pp 731-736 (2007)

[10] Kim HJ, Lee JY, Paik KW, et al. "Effects of Cu/Al intermetallic compound (IMC) on copper wire and aluminum pad bondability." *IEEE Transactions on Components and Packaging Technologies*, Volume 26, No. 2, pp. 367-374 (2003)

[11] Passagrilli, C.; Vitali, B.; Tiziani, R.; Azzopardi, C.; "Cu wire bonding: Reliability improvement for high temperature in plastic packages," *Microelectronics and Packaging Conference*, EMPC, pp. 1-4 (2009)

[12] Hang, C.J. et al. "Growth behavior of Cu/Al intermetallic compounds and cracks in copper ball bonds during isothermal aging." *Microelectronics Reliability 48*, pp. 416-424 (2008)

[13] Vath, C.J.; Gunasekaran, M.; Malliah, R.; "Factors affecting the long term stability of Cu / Al ball bonds subjected to standard and extended HTS." *11th Electronics Packaging Technology Conference*, pp. 374–380 (2009)

[14] England, L.; Jiang, T.; "Reliability of Cu Wire Bonding to Al Metallization." *57th Electronic Components and Technology Conference Proceedings*, pp. 1604-1613 (2007)

[15] Thomas, Sven and Dexter Reynoso. "Reliability of Cu Wire Bonding on Active Area for Automotive Applications." *11th Electronics Packaging Technology Conference*, pp. 363-368 (2009)

[16] Saraswati; Ei Phyu et al. "High temperature storage (HTS) performance of copper ball bonding wires." *7th Proceedings, Electronic Packaging Technology Conference Proceedings*, Volume: 2, pp1-6 (2005)

[17] Chylak, B. "Developments in fine pitch copper wire bonding production." *11th Electronics Packaging Technology Conference*, pp. 1-6 (2009)

A New ESD Model Induced Yield Loss during Chip-On-Film Package Process and It's Failure Mechanism

Jian-Hsing Lee, J. R. Shih, Yu-Hui Huang, C.P. Lin, David Su, and Kenneth Wu

Technology Quality and Reliability Division, Taiwan Semiconductor Manufacturing Company

Hsin-Chu, Taiwan, 886–3-5672088 ext. 7022088, jrshih@tsmc.com

Abstract— **A new model of electrostatic discharge (ESD) event is found in ICs during chip-on-film (COF) package. The behavior of this new kind of ESD is different from the human-body mode (HBM), machine model (MM) and charge device model (CDM) model. We call it the charge tape model (CTM). It often damages the gate oxides of the input circuit and output circuit in IC to result in the yield loss. The mechanism of COF package induced yield loss has been identified. Two factors dominate the yield loss. One is the ESD generation on the tape surface during COF tape reeled out process. The other one is the required high temperature for the inner lead bonding, which lowers the breakdown voltage of the gate oxide. As a result, an IC might be damaged to induce the yield loss.**

Keywords-ESD; HBM; MM; CDM

I. INTRODUCTION

Currently, COF package has been widely used for LCD driver IC package since it could implement the electric products with light weight, small volume, flexibility and low cost. The package yield is not only affected by the bond-ability [1] but also affected by the bond-reliability. The thermo-mechanical stress [2] has been attributed as an important bond-reliability issue for an IC chip during COF package. The foil tension transferred from the COF tape to the bump of an IC chip could lead to the chip crack and damage the circuit below the bump. In this paper, another kind of bond-reliability issue for an IC chip during COF package is found. It induces the damages to the input and output circuit, but does not result in any crack on the chip surface. Excluding the thermal-mechanism stress, what kind of damage source can induce such damage on the input and output circuit is investigated. In this paper, the process used to fabricate this LCD driver IC is with a 0.35um 18V double-diffusion drain (DDD) CMOS process.

II. FAILURE ANALYSIS (FA)

One LCD driver product occasionally suffers the yield loss due to malfunction from the voltage waveform check as shown in Fig. 1. It will become a distortion waveform from the repeated bell-shape waveform if a chip is damaged after COF package. From the emission-microscope (EMMI) photograph observation, the hot spots are located at the input circuit and output circuit (Fig. 2). In order to identify what components of the input circuit and output circuit were damaged, the contact voltage contrast (VC) and SEM are used to observe the real damage sites in the chip. From Fig. 3 and by checking with the layout of the input circuit, we can find that the abnormal bright

spot is on the poly gate, which connects to the input pad. Fig. 3b shows the partial layout of the input circuit, where the components connected to the pad are all marked with the green color by the layout tracing tool. After de-layered the chip to the P-substrate, it always could be observed the oxide pin holes on the poly gates (Fig. 4 and Fig. 5) of the input transistor or output transistor.

Fig. 6 shows the schematic diagram of the input circuit and output circuit. The ESD protection scheme, dual diodes (D1 and D2) as well as RC power clamp device, used for this chip is the typical ESD protection scheme for LCD driver IC. Incorporating with the RC power clamp device, the dual diodes can provide the discharge current paths for ESD once the power pad or the ground pad is grounded. In this chip, there are two different power domains. One is for ESD protection devices, input circuit and output circuit and another one is for internal circuit. Based on above results, the damage sites are always at the nodes used to connect the circuits of the two different power domains. Since these nodes connect to the external environment, they are susceptible to the ESD stress. Thus, the damages are suspected as the ESD damages during the COF package.

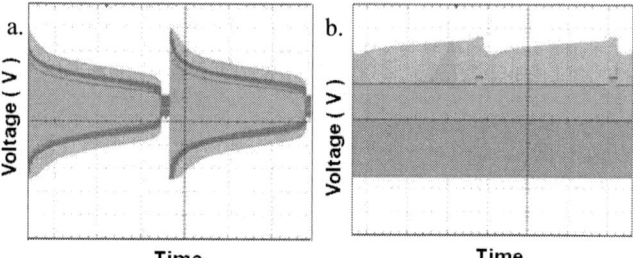

Fig. 1 Waveforms of a chip a. before COF package, b. after COF package.

Fig. 2 EMMI photograph of a failed chip caused by COF package.

978-1-4244-9113-1/11 $26.00 © 2011 IEEE

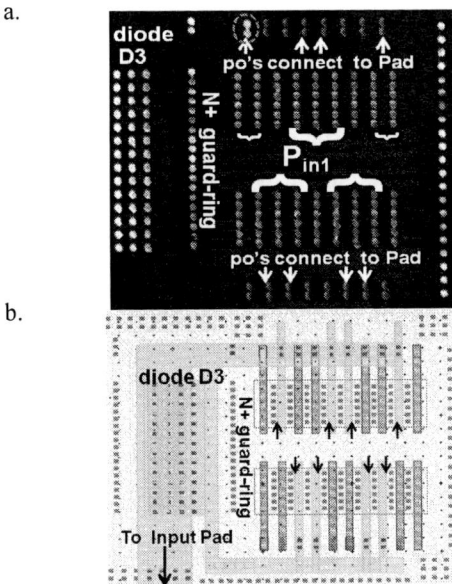

a.

b.

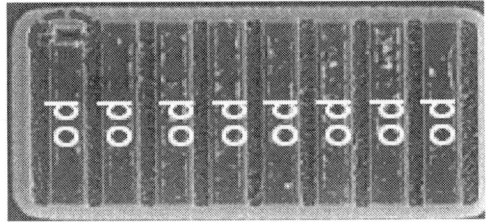

Fig. 3 a. Contact VC of input circuit, b. Layout of the input circuit with the tracing path to the input pad.

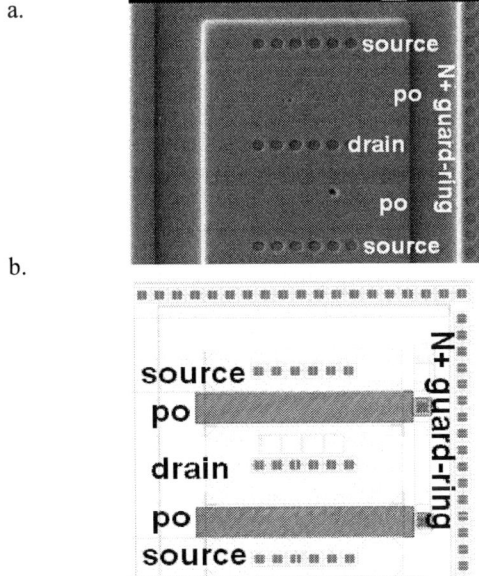

Fig. 4 SEM picture shows the oxide pin hole on poly gate with the abnormal brightness in Fig. 3.

a.

b.

Fig. 5 a. SEM picture shows the oxide pin holes on the poly gates of the output transistor, b. Layout of the output transistor.

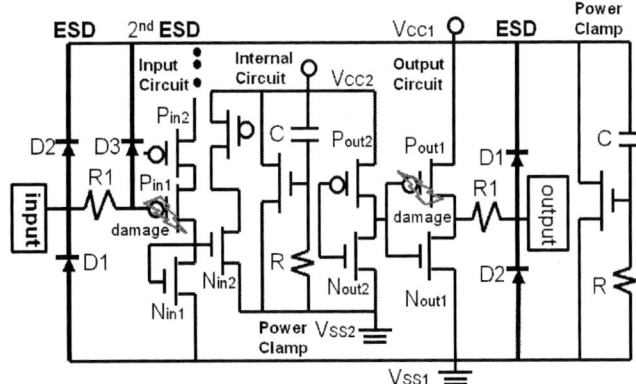

Fig. 6 Schematic diagram of the input circuit, output circuit and power clamp.

III. EXPERIMENT, DISCUSSION AND FAILURE MECHANISM

In order to find out the root cause of the yield loss caused by COF package, different ESD test modes are performed to identify what kind of ESD test could reproduce the same damage on the same location as the failed COF-package chip.

A. Experiments

Fig. 7a and Fig. 7b shows the failure locations of a good packaged chip after CDM test and a good bare-die after wafer level (WL) MM test, respectively. No matter CDM test or WL MM test, all failure sites are at the ESD protection devices, not at the input and output circuits. It implies that the yield loss shouldn't be caused by any step after the bonding since the CDM event occurs at the packaged chip [3], [4]. Moreover, the yield loss is also not caused by the room temperature steps of COF package since the WL MM test does not lead to the damage at the same failure location.

Fig. 7c shows the failure locations of a good bare-die after the high temperature WL (HTWL) MM test. The damage sites are at the output and input circuits, which match well with the failure locations of the failed COF-package chip. From the contact VC in Fig. 8, the abnormal bright spots of the good chip after the HTWL MM test are also at the gates, which connect to the input pad. This result also matches well with that of the failed COF-package chip in Fig. 3. In addition, the gate oxide damages at the output transistor and input transistor can be observed from the SEM pictures in Fig. 9. This implies that the yield loss of COF package might be caused by the process step during the COF package procedure at high temperature, because that only the HTWL MM test can lead to same damages on the same failure locations as the failed COF-package chip.

a.

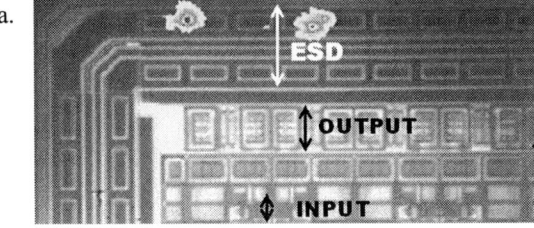

978-1-4244-9113-1/11 $26.00 © 2011 IEEE   691

b.

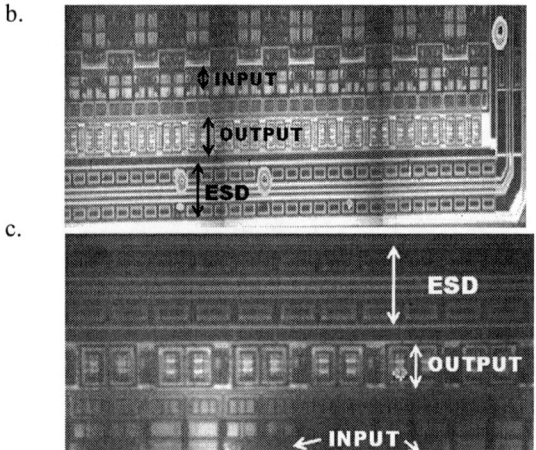

c.

Fig. 7 EMMI photographs for a. a good chip after CDM, b. a good chip after WL MM, c. a good chip after WL high temperature MM

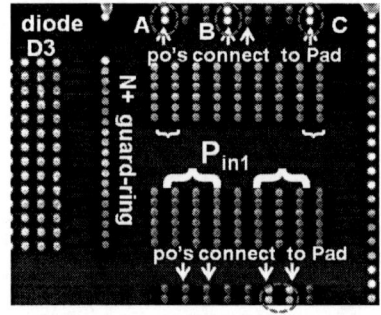

Fig. 8 Contact VC for a good chip after WL high temperature MM.

a.

b.

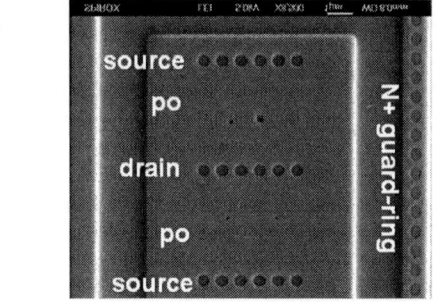

Fig. 9 Oxide pin holes on a. input transistor P_{in1} in Fig. 6, b. output transistor P_{out1} in Fig. 6 for a good chip after HTWL MM.

B. Discussion

The early literature [5] had reported that the package tape could affect the CDM behavior of a packaged chip significantly. From the experiment result, the CDM test does not lead to the same damage as the failed COF-package chip. This verifies that the yield loss should be not caused by the step after the bonding based on the process flow of COF package [6]. Moreover, it has identified a good bare-die after HTWL MM test that can lead to the same damages on the same failure locations as the failed COF-package chip. This implies that the yield loss should occur at the step with high temperature process after the dicing. In addition, the charges should come from the outside to the bare-die, not from the bare-die to the outside. From the process flow of COF package [6], there is only a bonding step between the dicing step and under fill step. Since there are only a few processes for a bare-die during bonding step, it is very easy to find out what process can generate electrostatic charges and what process is operated at the high temperature.

A COF tape is composed of two layers. One is the conductor copper (Cu) film and another is the insulator polyimide (PI) film [1], [2]. When the tape is reeled out from a roll of COF tape, a lot of electrostatic charges are generated and stored on the surfaces of the insulator (Fig. 10). It is because of the triboelectrification after a tape surface contacted and separated with another tape surface. So, the tape after reeling out can be depicted as the equivalent circuit of Fig. 11. The insulator between any Cu inner lead and the tape surface can be treated as a capacitor (C_{A1} or C_{B1}). Every Cu wire is connected in series with the capacitor and connected to the voltage source V_B formed by the stored charges on the tape surface. Obviously, the electrostatic charges are stored outside the bare-die now, which is similar to the bare-die at the transient before the MM test.

The inner lead bonding (ILB) is a process of the bonding step to make the gold bumps jointed to the Cu inner leads of a COF tape (Fig. 12). This is a thermo-mechanical process with the reaction temperature close to 400°C [1]. Since the tape cannot tolerate such high temperature, the heat must be applied to the IC chip first and then transmitted to the inner lead for ILB bonding. Then, the bonding head as well as the tape moves downward with a constant speed to compress an IC chip, resulting in Sn molten and reacting with Au metal to make the Au bumps of the IC chip jointing the inner leads of a tape. Moreover, this action also can induce the charges shored on the tape surface flowing into the bare-die like the MM switch closing to begin discharging its charges to the die. Unlike the bare-die during the MM test, however, there is not any pad of the bare-die that is grounded to sink the charges coming from another pad since all pads of the die connect to the tape. Thus, the only grounded point of a bare-die now is its backside since the stage of the ILB bonding machine is without any charge and can be treated as a virtual ground. If the stage of the ILB bonding machine was not the grounded point, all charges would be stored on the tape surface to induce the CDM event in the following steps as one pad of the bare-die with the tape is grounded. However, this hypothesis cannot match the CDM test result since it cannot generate the same damage as the failed COF-package chip.

978-1-4244-9113-1/11 $26.00 © 2011 IEEE

Based on above, there is a common event for the bare-die during ILB bonding and MM test, which is the electrostatic charges coming from the outside of the bare-die. However, it also has a difference between the bare-die during ILB bonding and MM test. The bare-die during the ILB bonding does not have any grounded pad during the ILB bonding and uses its back side as the ground, but during MM test it has a grounded pad. Thus, it is obvious that the ESD induced by the ILB bonding is different from other kinds of ESD models. Since the charges are stored on the COF tape surface, we call this new kind of ESD the charge tape model (CTM).

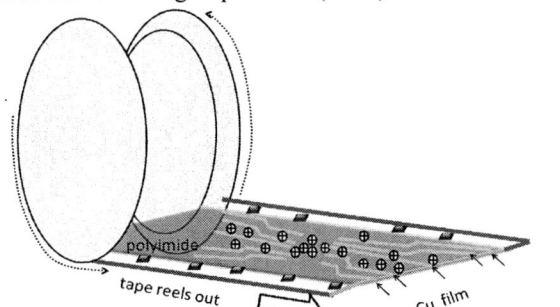

Fig. 10 Tape reels out from a roll of tape at the COF assembly production line.

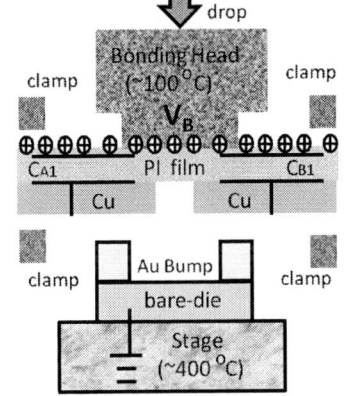

Fig. 11 Schematic for COF package before ILB process.

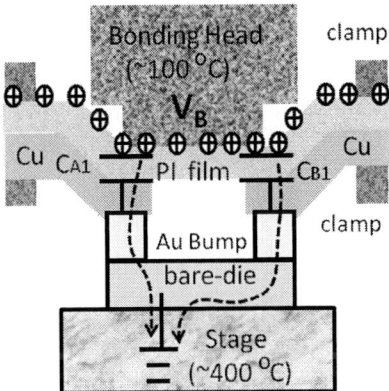

Fig. 12 Schematic for COF package during ILB process.

C. Failure Mechanism

LCD driver IC is usually a high-pin-count IC, and its total pad number might be over one thousand. Fig. 13 shows the partial layout of an IC, where pads are plotted with the blue. Most of the pads are the output pads or input pads and share one power and ground pads. Regardless of the output pad or the input pad, it always connects to the dual diodes except the output transistor and input transistor (Fig. 6). Based on Fig. 12, the equivalent circuit of the single dual diodes for a bare-die during ILB bonding can be depicted as Fig. 14. Any pad (V_{CC} pad, V_{SS} pad, input pad and output pad) is connected in series with a tape capacitor to a voltage source and in series with the junction capacitor or a substrate resistor to the ground. For V_{CC} pad, it is connected in series with a junction capacitor C_{NW} and a substrate resistor R_{sub} to the ground except the charged tape capacitor. For input pad or output pad, it is connected in series with two junction capacitors C_N and C_{NW} and a substrate resistor R_{sub} to the ground except the charged tape capacitor. For V_{SS} pad, it is connected in series with a substrate resistor R_{sub} to the ground except the charged tape capacitor.

Based on above information, the input and output circuits for a chip during ILB bonding can be illustrated as Fig. 16. The main difference between the internal circuit and the input or output circuit is the number of their connecting charged tape capacitors. For internal circuit, there are only two pads (V_{CC} pad and V_{SS} pad) in it. As a result, it only connects to two charged tape capacitors C_{VCCi} and C_{VSSi}. For input and output circuits, there are many pads in it, including all input pads, output pads, one V_{CC} pad and one V_{SS} pad. As a result, it connects with many charged tape capacitors. For a typical IC, the total junction area of the internal circuit is often larger than that of the input circuit and output circuit (Fig. 13). Compared to a single input circuit or output circuit, the internal circuit is with the huge capacitance by capacitor C_{NWi} and connected in series with a tiny resistance by substrate resistor R_{subi}. Moreover, it only connects to two charged tape capacitors. One (C_{VSSi}) is connected in series with the substrate resistor R_{subi} to the ground directly and another one (C_{VCCi}) is connected in series with the capacitor C_{NWi} and the substrate resistor R_{subi} to the ground. Thus, all charges stored on the two charged tape capacitors will be sunk to the stage of the bonding machine instantaneously to cause the voltages of all nodes in the internal circuit almost keeping at zero voltage during the bonding period.

Since the input and output circuit is composed of many repeated cells, it can be simplified to the two parallel capacitors as that of the single dual diodes in Fig. 14. For output circuit, the junction capacitances of the output transistors need to be counted into the C_{NW} and C_N from Fig. 15. For input circuit, the capacitances of the diode D3 and input gate need also be counted into the C_{NW} and C_N from Fig. 15. Thus, each charged capacitor C_{IO} only needs to charge up the two capacitors C_{NW} and C_N. Apparently, the total area of the single dual diodes and the output transistor or input transistor is much smaller that of the internal circuit from Fig. 13. Compared to the internal circuit, the input and output circuit is with two capacitors of tiny capacitance and connected in series with a substrate resistor of huge resistance. Eventually, the charged capacitor C_{IO} can charge up the potential of the two capacitors C_{NW} and

C_N instantaneously and can keep the potentials for a long time due to large RC time constant. This induces the large potential differences between the gates and drains of the output transistors (V_{O1} and V_{O2}) and input transistor (V_{i1} and V_{i2}) as shown in Fig. 15.

Fig. 16 shows the bias condition for each node of the output transistors. Apparently, the stress condition for the gate of the PMOS output transistor is different from that of the NMOS output transistor. For PMOS output transistor, it is the gate stress since the drain, source and N-Well all connect to the high voltage with respect to the grounded gate. For NMOS output transistor, it is the drain-side stress since the high voltage only applies to the drain with respect to the grounded gate. However, the oxide breakdown voltage of the transistor for gate stress had been reported smaller than that for the drain-side stress [7]. Moreover, the gate oxide breakdown voltage decreases with the temperature increase (Fig. 17). For gate stress, the gate oxide breakdown voltage of the PMOS is higher than 45V at room temperature, but it decreases to 42V at 300°C. For drain-side stress, the gate oxide breakdown voltage of the NMOS is higher than 48V at room temperature, but it decreases to 46V at 300°C. Thus, the oxides of the output transistors at the high temperature are more susceptive to the ESD stress than at the room temperature. This is why room temperature MM cannot damage the oxides of the output transistor or input transistor, and only damages the ESD protection device. While, both the ILB bonding of COF package and high temperature MM can damage the gate oxides of the input and output transistors. Due to the smaller oxide breakdown voltage, the PMOS output transistor is more vulnerable to the ESD stress than the NMOS output transistor. Thus, it always can find the damages at the PMOS output transistor, not at the output transistor.

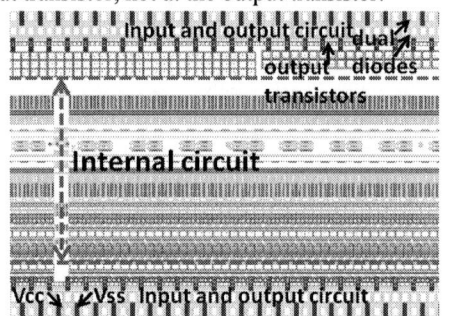

Fig. 13 Layout placement of the pads, input and output circuits, and internal circuit of an IC.

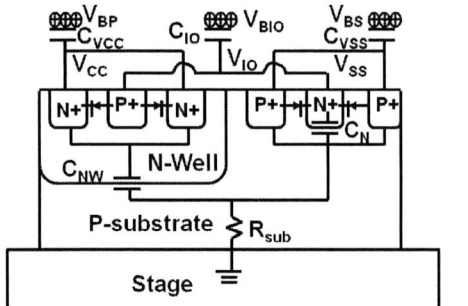

Fig. 14 Schematic diagram of dual diodes during ILB bonding.

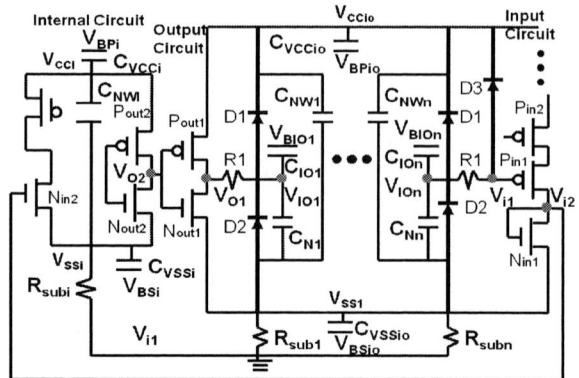

Fig. 15 Schematic diagram of a bare-die during ILB bonding.

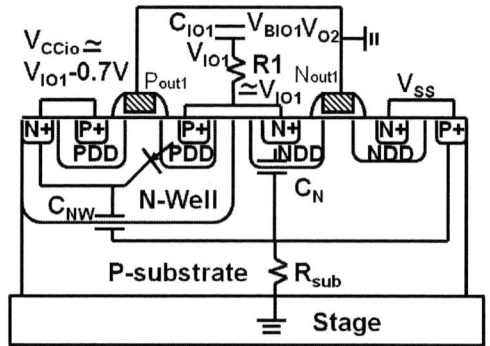

Fig. 16 PMOS is the gate stress and NMOS is the drain-side stress for a bare-die during ILB bonding.

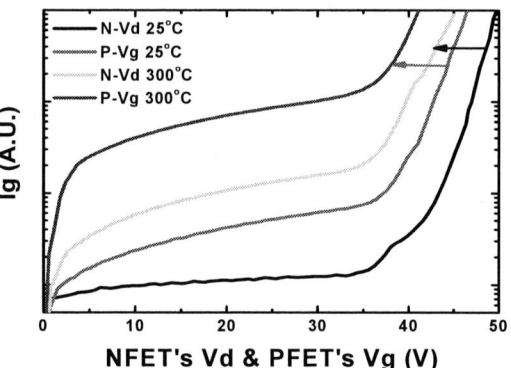

Fig. 17 The DC Ig characteristics of HV N-/P-MOSFET measured at different temperature, where Vg=Vs=Vb=0V with Vd sweep for HV NMOSFET and Vd=Vs=Vb=0V with Vg sweep for HV PMOSFET

D. CDM and CTM Comparison

From the failure mechanism, discussion and early literatures [3]-[4], it can find two commons and some differences between the CDM and the CTM. The first common is that the currents of the CDM and CTM all can flow through the back side, substrate, pads, and the internal circuits of the die. However, the ground for CTM is the back-side of the die, while the ground for CDM is one pad of the die. The second common is that most electrostatic charges are stored on the package.

However, the package is at the outside of the die during the CTM event, while the package is connected with the die during the CDM event. In addition, there is no charge stored in the die before the CTM event occurring, while some electrostatic charges are stored on the bus lines and internal circuits of the die during the CDM event [4].

CONCLUSION

Reeling out the tape induced the electrostatic charge generation and the high temperature process used to joint Au bump of a chip and inner led of a tape are inevitable for a chip during COF package. High temperature will make a chip more vulnerable to the ESD stress since it decreases the gate oxide breakdown voltage. The ESD test at room temperature cannot account for the ESD phenomenon caused by COF package. If a product would like to prevent the yield loss caused by COF package, the designer needs to consider the high temperature ESD event. Otherwise, the product might suffer the yield loss caused by COF package occasionally.

The ESD event occurring at an IC chip during COF package is a new kind of ESD model, which is different from the HBM, MM and CDM. We call it the charge tape model (CTM). Same as HBM and MM, the charges of the CTM come from the external environment to the chip. Different from the HBM, MM and CDM, the charges of the CTM flow through all pads, junctions or gates connected to the pads and the substrate to the backside of the chip. For HBM, MM and CDM, the discharge cannot occur if no pad of an IC chip is grounded. For CTM, however, the discharge can occur even without any pad of an IC chip which is grounded.

REFERENCES

[1] Ching-Yu Ni, Chi-Min Chang, Shao-Chiun Wu, and De-Shin Liu, "Bondability Study of Chip-on-Film (COF) Inner Lead Bonding (ILB) Using Conventional Gang Bonder," IEEE Trans. Electron. Package. Manuf., vol. 31, pp. 285–290, 2008.

[2] Hamit Duran, and Isak Venter, "Die Optimization For COF Assembling," 11th Annual International KGD Packaging and Test Workshop, 2004.

[3] W. Greason, "Analysis of the charge transfer of models for electrostatic discharge (ESD) and semiconductor devices," *IEEE Trans. Ind. Applicat.*, vol. 32, pp. 726–734, 1996.

[4] Jian-Hsing Lee, J. R. Shih, Shawn Guo, Dao-Hong Yang, Jone F. Chen, David Su, Kenneth Wu, "The study Of Sensitive Circuit and Layout For CDM Improvement," in *Pro.16th IPFA* symposium, p. 228-231, 2009.

[5] Alan W. Righter, Javier A. Salcedo, Andrew H. Olney, and Torsten Weyl, "CDM ESD current characterization-package variability effects and comparison to die-level CDM," in *EOS/ESD* symposium, pp. 2B.7.1-2B.7.9, 2009.

[6] Mitsuru Chino, "Development of fine pitch flip chip bonding technology," 9th Annual International KGD Packaging and Test Workshop, 2002.

[7] Nels A. Dumin, Kaiping Liu, and Shyh-Horng Yang, "Gate oxide reliability of drain-side compared to gate stresses," *IEEE Proc. International Reliability Physics Symposium*, pp. 73–78, 2002.

A robust reliability methodology for accurately predicting Bias Temperature Instability Induced Circuit Performance Degradation in HKMG CMOS

D.P. Ioannou, [2]K. Zhao, [2]A. Bansal, [2]B. Linder, R. Bolam, [2]E. Cartier, [2]J.-J. Kim, [2]R. Rao, [3]G. La Rosa, G. Massey, M. Hauser, [2]K. Das, [2]J.H. Stathis, J. Aitken, D. Badami and S. Mittl

IBM Microelectronics, Semiconductor R&D Center, Essex Junction, VT, USA [2] IBM T.J. Watson Research Center, Yorktown Heights, NY, USA, [3] IBM Microelectronics, Semiconductor R&D Center, Hopewell Junction, NY USA

INTRODUCTION

A robust reliability characterization / modeling approach for accurately predicting Bias Temperature Instability (BTI) induced circuit performance degradation in High-k Metal Gate (HKMG) CMOS is presented. A series of device level stress experiments employing both AC and DC stress/relax BTI measurements are undertaken to characterize FET's threshold voltage instability response to a dynamic (inverter type) operation. Results from the AC stress experiments demonstrate that V_T instability is frequency independent, an observation that suggests that V_T degradation under AC stress can be equivalently measured through the simpler DC stress/relax sequence. An AC BTI model is developed that accurately captures the critical BTI relaxation effect through the DC stress/relax predictions on duty cycle dependence. A Ring Oscillator (RO) circuit is used as a model verification vehicle. Excellent agreement is demonstrated between the frequency degradation measurements obtained with a newly developed Ultra-Fast On-The-Fly (OTF) measurement technique optimized for BTI and the AC BTI model based RO simulations.

Keywords: High-k Metal Gate, PBTI, NBTI, Ring Oscillator

Device level AC and DC stress/relax experiments

In the AC BTI measurements, an "inverter" type of AC stress is introduced to simulate the bias condition and evaluate the V_T degradation that FETs would experience in a RO circuit. Fig. 1 shows the typical waveforms used for NFETs, where both V_{gs} and V_{ds} switch quickly during stress. The waveforms are programmed so that V_{ds} is at low when V_{gs} is switched to V_{stress} and V_{ds} is at high when V_{gs} is switched to 0. The timing of switching is also controlled so that Hot carrier Effects are minimized. V_T shift is monitored at both sense1 (right after stress) and sense2 (after the following relaxation phase).

Typical PBTI degradation under inverter AC stress and DC stress is shown in Fig. 2. As can be seen, AC stress results in less V_T shift but very similar time-power exponent n (~0.17) comparing to the DC stress, suggesting that the V_T degradation under AC stress can be expressed as a fraction of the V_T shift under DC stress:

$$\Delta V_{TAC} = C_{FF}\Delta V_{TDC} \qquad \text{(eq. 1)}$$

where ΔV_{TDC} is parameterized with the commonly used power law for the voltage and time dependencies [1] and C_{FF} is the fraction factor function and its value is always less than 1. The fraction factor C_{FF} is the key modeling

parameter and the understanding of its dependence on the AC stress characteristics, such as frequency, duty cycle and

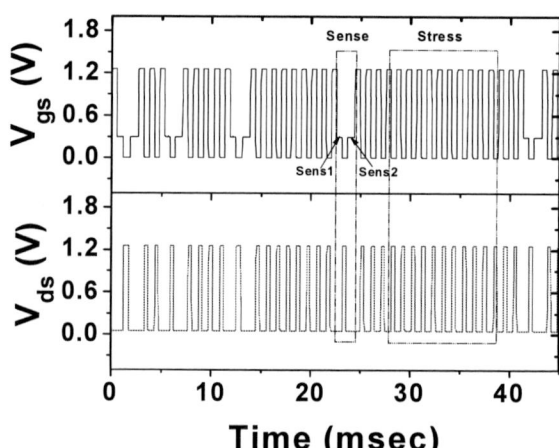

Fig. 1: Typical waveforms of "inverter" type AC stress with both gate bias and drain bias switching during stress. V_T shift is sensed at sense1 (after stress) and sense2 (after the following relax).

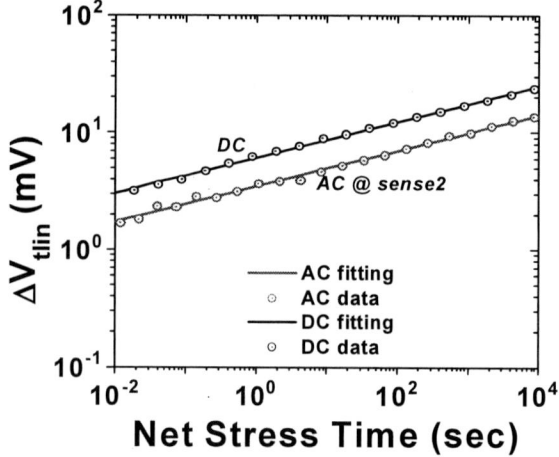

Fig. 2: Typical V_T degradation under AC (1kHz, 50% duty cycle) and DC stress.

operating bias conditions is essentially important. Fig. 3 shows the fraction factor C_{FF} scaling with frequency over a wide range from 1 mHz to 100KHz. It is found that for both NBTI and PBTI, sense1 shows clear frequency dependence, while sense2 shows nearly no frequency dependence. This observation suggests that V_T degradation under

978-1-4244-9113-1/11 $26.00 © 2011 IEEE

AC stress can be well approximated with a simpler DC stress/relax sequence measurement [2]. This is shown in Fig. 4, where the AC PBTI stress measurement results on the Fraction Factor C_{FF} scaling with duty cycle are contrasted with those obtained from the DC stress/relax sequence. The measurement details for the DC-stress relax sequence are similar to the AC stress (shown in Fig. 1) with the only difference that here there is only one cycle involved (a stress phase followed by a relaxation phase).

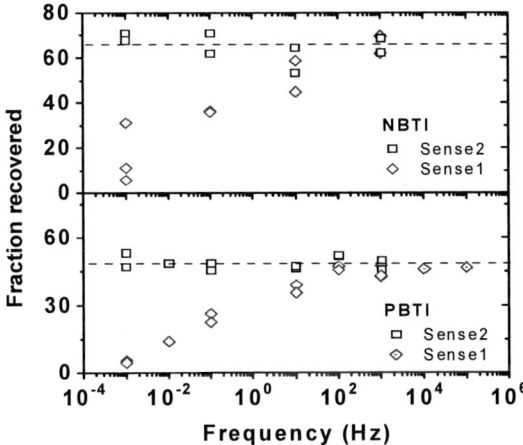

Fig. 3: Frequency dependence for both NBTI and PBTI at 50% duty cycle. Sense 1 shows clear frequency dependence while sense 2 shows nearly no frequency dependence.

Very good agreement is observed between the two techniques (Fig. 4) with regard to the Fraction Factor decrease rate with duty cycle (typical S-shape, [3]) as well as the measured magnitudes.

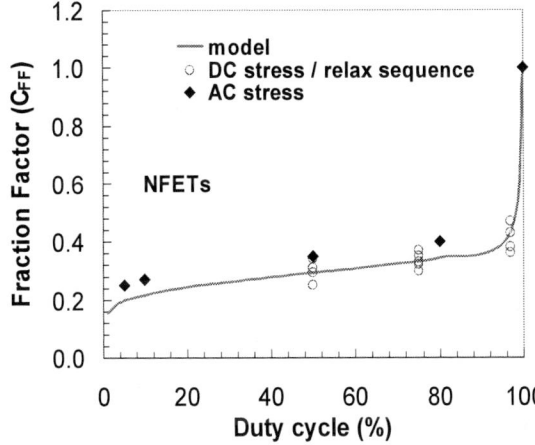

Fig. 4: Fraction Factor scaling with duty cycle characteristics as measured with the AC and DC stress/relax sequence techniques.

Fig. 5 illustrates the measured Fraction Factor at 50% duty cycle vs. stress bias level (V_{dd}) characteristics. The progressive reduction of the Fraction Factor (increase of relaxation) with increasing V_{dd} can be understood if one considers that a higher bias during the "off" phase of an AC signal (relaxation phase) energetically favors charge detrapping from the deeper (in energy) trap levels in addition to that associated with the shallower traps at lower bias. The results shown in Fig. 4 and Fig. 5 are summarized in a semi-empirical form for the Fraction Factor, C_{FF} as:

$$C_{FF} = \left[1 + [a \times \exp(\beta \times V_{dd})] \left(\frac{t_{relax}}{t_{stress}} \right)^n \right]^{-1} \qquad \text{(eq.2)}$$

where α and β are fitting coefficients and n is the time-power exponent for the V_T degradation (Fig. 2).

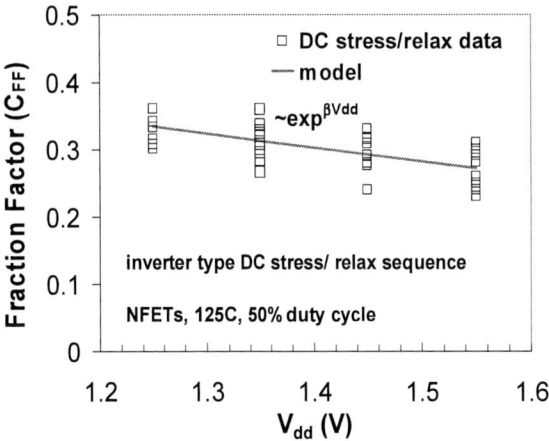

Fig. 5: Fraction Factor scaling with stress bias level measured with the DC stress/relax sequence.

Ultra-Fast On-The-Fly Ring Oscillator frequency measurements

Ultra-Fast On-The-Fly (OTF) Ring Oscillator frequency measurement is our newly developed technique optimized to capture the BTI effect on RO circuit performance degradation. Its main advantages are: 1) enables an ultra-fast (~us) bias transition to the stress bias level ensuring an accurate measurement of the time-zero frequency (f_0) measurement (at stress level). We find that alleviating this initial value issue, previously only recognized as a potential shortcoming for device level OTF measurements [4], is also critical for accurately extracting the frequency degradation time dependency. 2) being an OTF (in-situ) measurement avoids any artificial (measurement induced) relaxation and only captures the intrinsic to BTI relaxation effect due to AC circuit operation.

Fig. 6 shows RO frequency degradation vs. time characteristics measured with the conventional ("slow") and Ultra-Fast OTF RO techniques. The result shown in Fig. 6 is part of our initial methodology calibration study using ROs structures built on an oxynitride based 45 nm CMOS technology. We will report on this study and discuss the implementation details for this new technique elsewhere. Here, the discussion is focused on the key finding with respect to BTI modeling. It is seen from Fig. 6 that the frequency degradation measured with the Ultra-Fast OTF technique follows a perfect time-power law with an exponent value of ~ 0.17 which is in very good agreement with the NBTI induced V_T shift time dependency. By contrast, a time-power exponent of high value (~0.25) is measured

978-1-4244-9113-1/11 $26.00 © 2011 IEEE

with the conventional method resulting in a significant overestimation of the frequency degradation at End-Of-Life and misleadingly alluding to hot carrier related contributions on RO degradation.

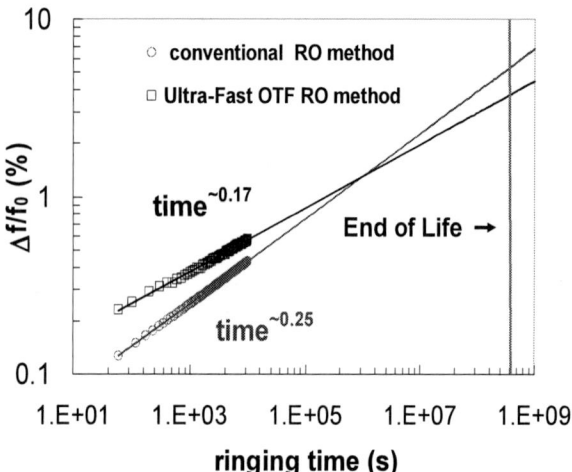

Fig. 6: % RO frequency degradation vs. time characteristics as measured with the conventional and the newly developed Ultra-Fast OTF RO techniques.

Model verification by Ring Oscillator Simulations

Fig. 7 shows the simulation methodology that was adopted for the AC BTI model (eq. 1-2) verification. The FET SPICE models for NFET and PFET were first calibrated to match the measured hardware and subsequently the time-zero RO frequency vs. V_{dd} characteristics were simulated including extracted parasitic resistances and capacitances. The very good agreement obtained between simulations and measurements (Fig. 8) provides additional evidence that the initial value issue is minimized with the Ultra-Fast OTF technique.

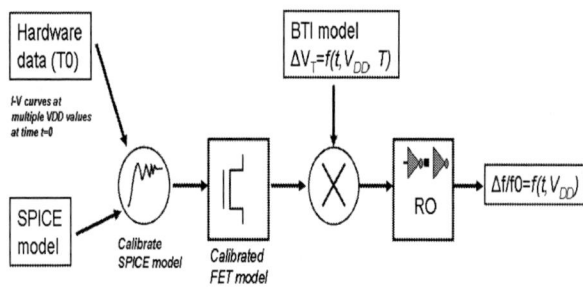

Fig. 7: Simulation methodology flowchart

With regard to the time-dependent frequency simulations, the AC BTI model was coded into the simulation engine as an "add on" voltage source connected to the gate of the FETs. Fig. 9 shows the simulated and measured RO frequency degradation due to BTI. It is seen that the AC BTI model based simulation results match closely with the measured data over a wide range of stress bias conditions.

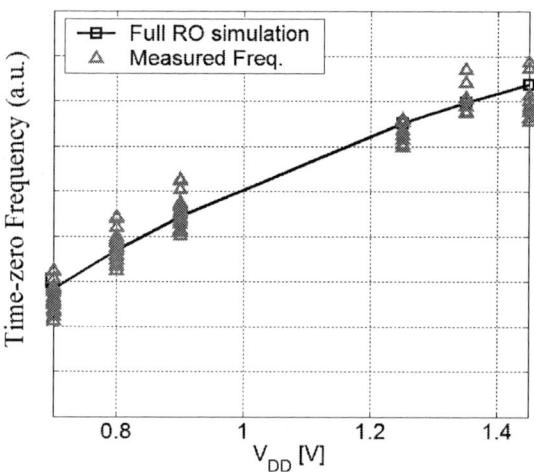

Fig. 8: Time-zero Frequency vs. V_{dd} characteristics (simulation vs. measurements)

It is worth pointing out that accurate simulations were achieved by only properly incorporating a gate voltage offset (through the ΔV_T based BTI model) the to the device SPICE FET models. This is consistent with the negligible transconductance (g_m) device degradation observed during BTI measurements. Finally, although this methodology is exercised here on RO circuits, it can be easily adapted for other logic circuit styles.

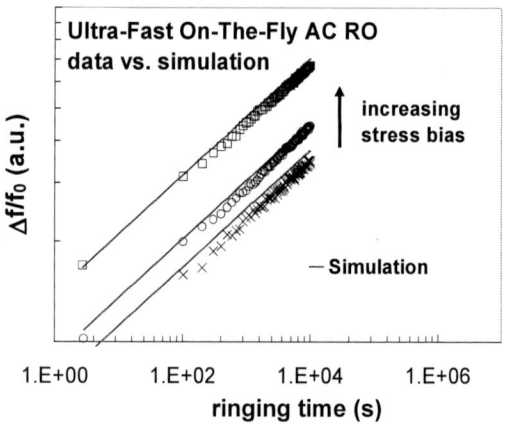

Fig. 9: Measured vs. simulated % RO frequency degradation vs. time characteristics.

ACKNOWLEDGMENT

This work was supported by the research alliance teams at various IBM research and development facilities.

REFERENCES

[1] A. Kerber, and E. Cartier, "Reliability Challenges for CMOS technology qualifications with Hafnium Oxide/Titanium Nitride Gate Stacks," IEEE Trans. Dev. Mat., vol. 9, no. 2, pp. 147-162, Jun. 2009.

[2] D.P. Ioannou, E. Cartier, Y. Wang, and S. Mittl, "PBTI response to interfacial layer thickness variation in Hf-based HKMG nFETs," International Reliability Physics Symposium, pp. 1044-1048, 2010.

[3] V. Huard, C. Parthasarathy, N. Rallet, C. Guerin, M. Mammase, D. Barge, C. Ouvrard, "New Characterization and modeling approach for NBTI degradation from transistror to product level, " International Reliability Physics Symposium, pp. 797-800, 2007.

[4] R. Heisinger, U. Brunner, W. Heinrigs, W. Gustin, C. Schlunder, "A comparison of Fast Methods for measuring NBTI degradation," IEEE Trans. Dev. Mat., vol. 7, no. 4, pp. 531-539, Dec. 2007.

Bias Temperature Instability Model for Digital Circuits – Predicting Instantaneous FET Response

Aditya Bansal, Kai Zhao, Jae-Joon Kim and Rahul Rao
IBM Thomas J. Watson Research Center, Yorktown Heights, NY 10598
Email: bansal@us.ibm.com

Abstract—**We propose a semi-empirically enhanced BTI model to predict the instantaneous shift in V_T due to both NBTI (in PFETs) and PBTI (in NFETs). Our proposed model uses same technology parameters as in existing model, and applied for both NBTI and PBTI. At every step of model generation, we demonstrate the correlation between our model and measured hardware. Further, we discuss the necessary steps to integrate our model with existing digital circuit simulators.**

INTRODUCTION

Negative (or Positive) Bias Temperature Instability increases the threshold voltage of the PFETs (or NFETs) with time under electrical and thermal stresses [1-2]. Change in FET characteristics translates to changes in circuit response. Recently there have been several research papers discussing the impact of temporal degradation in NFETs and PFETs on critical circuit components such as memory cells [3-5], logic gates [6] etc. The NBTI and PBTI induced threshold-voltage (V_T) used in these papers has semi-empirically fitted stress condition and universal recovery model [7-8], and can be expressed as [9]

$$\Delta V_T(t) = \frac{A V_{DD}^a T^b t_s^n}{1 + m(t_r/t_s)^n} \quad (1)$$

where V_{DD} is the stress voltage, T is the temperature, t_s is the net stress time and t_r is the net relaxation time. The constants A, a, b, n and m are technology dependent and fitted using hardware measurements. The above model is separately calibrated for both negative-BTI (NBTI in PFETs) and positive-BTI (PBTI in NFETs) using single cycle stress-relaxation measurements (Fig. 1). This model reasonably predicts the degradation in a FET after long duration of time by accounting for the net stress and relaxation times. However, it is not suitable to determine the *instantaneous BTI induced V_T shift* in a FET. In Fig. 2, we show three stress-relax cases for a fresh device at $t=0$, where at the end of time $t=t_1$, the net t_s and t_r are same in all. Above model (eq. (1)) will predict same degradation in all the cases at time t_1. However, it can be seen that at $t=t_1$, $\Delta V_{T1} > \Delta V_{T2}$, whereas ΔV_{T3} will depend on relative stress and recovery effects. To the best of our knowledge, the existing models that can be practically used for circuit and system reliability prediction can not differentiate the above cases. The ability of a model to closely track instantaneous V_T shift is essential due to the complexity of a FET's operation and the impact of its failure in a real circuit. We'll discuss some examples in the last section on results and discussions.

In this paper, we present our enhanced BTI model which can successfully predict instantaneous V_T shifts due to BTI. Our model is semi-empirical; however, we do not use any new technology constants to keep the analysis simple without losing accuracy. Using our model, one can accurately predict

- Instantaneous PBTI (or NBTI) induced V_T shift in an NFET (or PFET).
- NFET/PFET and circuit response under varied usage conditions – from continuous to intermittent usage while switching between power-on and standby modes.

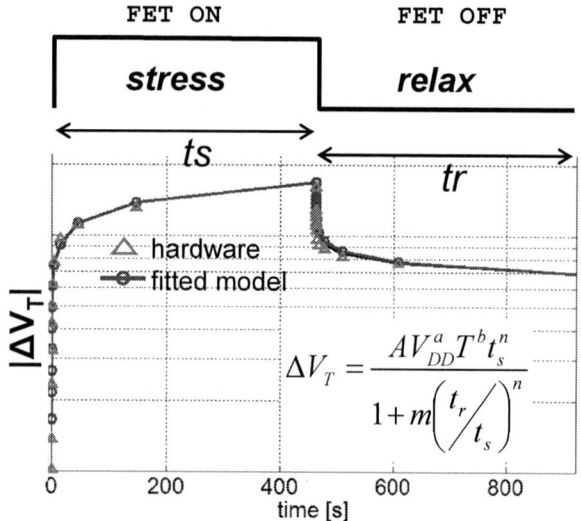

Fig. 1. Building the existing model based on universal relaxation by fitting measurements from stress-relax cycle separately for NBTI and PBTI. *Note: above stress-relax cycle does not represent the gate voltage at a FET.*

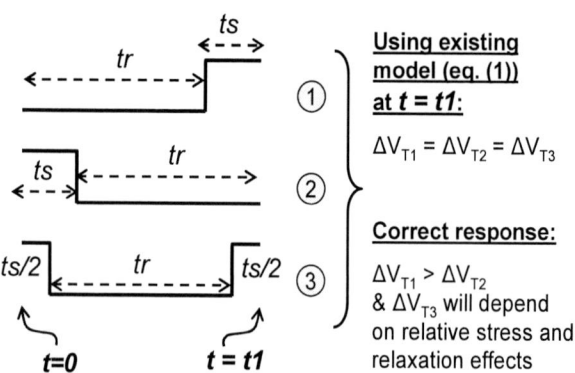

Fig. 2. For a fresh device at t=0: three different cases such that all have same net stress and relax times. Existing model wrongly predicts the same V_T degradation in all three cases as it's based on net stress & relax times seen by a FET.

978-1-4244-9113-1/11 $26.00 © 2011 IEEE

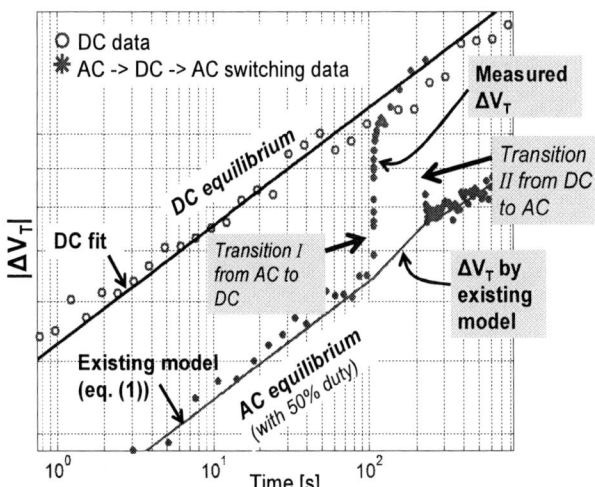

Fig. 3. Existing model against measured data for switching from AC -> DC -> AC. Existing model is not able to predict the transitions.

- Vulnerability of a circuit path to timing faults due to instantaneous large V_T shifts in NFETs and PFETs.
- The effectiveness of recovery in mitigating the NBTI and PBTI induced V_T shift.

PROPOSED MODEL

Our proposed model is based on measured PBTI and NBTI in high-k metal gate hardware. Both NBTI and PBTI were observed to experience similar stress and relax behaviors with different amplitudes. Hence, the V_T shift in both NFETs and PFETs can be expressed using eq. (1) with separately calibrated fitting parameters. In the remaining paper, we'll use the acronym FET for both NFET and PFET, as well as, BTI for both PBTI and NBTI. All the data shown in the figures correspond to PBTI. NBTI data is very similar and omitted without loosing any details of the proposed model.

We observed two key behaviors in the measured hardware. Fig. 3 shows the V_T of a FET while switching between two different stress conditions: (i) Alternating or AC stress where gate of a FET is switching continuously with 50% duty cycle and V_T is measured at the end of each stress-relax cycle, and (ii) Static or DC stress where a FET is continuously stressed without relaxation. Existing model (eq. (1)) reasonably predicts the degradation due to AC stress and also DC stress ($t_r=0$ => only numerator term in eq. (1)). When the stress switches from AC to DC (transition I in Fig. 3) all traps start contributing to overall V_T shift thereby approaching DC equilibrium. Existing model fails to predict the transition from AC to DC because it depends on the net t_r and t_s. On the other hand when the stress switches from DC to AC (transition II in Fig. 3), shallow traps become empty and only deep traps contribute to overall V_T shift thereby re-establishing the AC equilibrium. This behavior has been observed in both NFETs and PFETs.

Fig. 4 shows the instantaneous V_T shift when a FET goes through several stress-relax cycles. It can be seen that the existing model closely predicts the degradation in first cycle

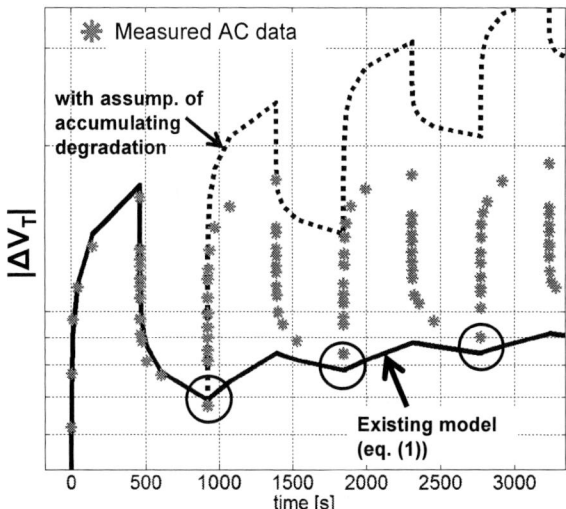

Fig. 4. Existing model accurately predicts instantaneous V_T shift only during first cycle as it's empirically fitted to first cycle. Eq. (1) also closely predicts the degradation *after* every stress-relax cycle (circled).

only. The instantaneous degradation during following cycles can not be predicted, however, the net degradation at *the end* of every stress-relax cycle can still be predicted by existing model (shown as circled in Fig. 4). Note that after 1st stress-relax cycle, device is no longer virgin, hence, existing model can not be applied for subsequent stress-relax cycles. For example, in Fig. 4, 2nd stress starts at $t=900s$. Using $t_r=0$ in existing model to compute degradation during 2nd stress will wrongly give us huge degradation (dotted plot in Fig. 4).

For our proposed model, we start with existing BTI model (eq. (1)) and modify it during stress and relax conditions. In the existing model, at any given point in time, we compute the

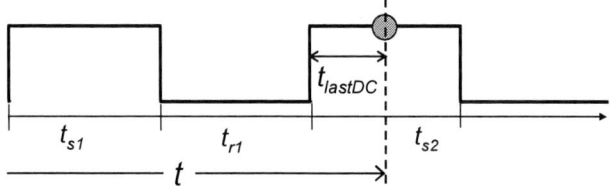

net stress time, $t_s = t_{s1} + t_{lastDC}$

net relax time, $t_r = t_{r1}$

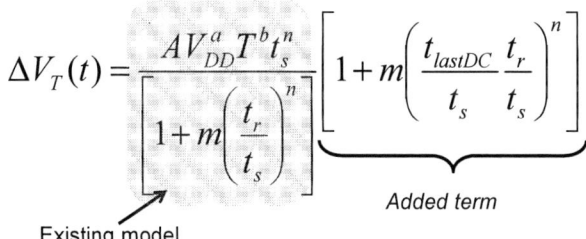

Fig. 5. Our enhanced model for instantaneous V_T shift when a FET is ON or being stressed. For large values of t_{lastDC}, i.e., $t_{lastDC} \sim t_s$ => only the numerator of existing model remains representing transition from AC to DC.

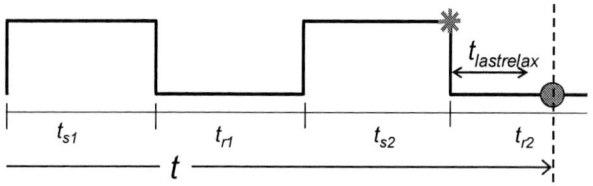

net stress time, $t_s = t_{s1} + t_{s2}$
net relax time, $t_r = t_{r1} + t_{lastrelax}$

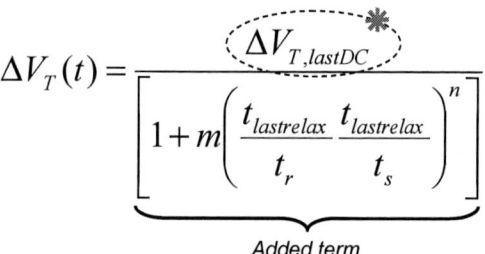

$$\Delta V_T(t) = \dfrac{\boxed{\Delta V_{T,lastDC}}}{\underbrace{\left[1 + m\left(\dfrac{t_{lastrelax}}{t_r}\dfrac{t_{lastrelax}}{t_s}\right)^n\right]}_{\text{Added term}}}$$

Fig. 6. Our modified model for instantaneous V_T shift when a FET is OFF or being relaxed. For large values of $t_{lastrelax}$ i.e., $t_{lastrelax} \sim t_r$, above equation becomes same as existing model (eq. 1) representing transition from DC to AC.

net relaxation times (t_r) and net stress times (t_s). In our proposed model, we consider the last duration of DC (t_{lastDC}), last duration of relax ($t_{lastrelax}$) along with the net t_s and t_r. First we modify the existing model to compute instantaneous V_T shift while a FET is being stressed. Fig. 5 shows the stress-relax durations for a FET. For computing the V_T shift at time t, we note the t_{lastDC}, t_s and t_r. The modified model is also shown in the Fig. 5. The basis for added empirical factor are:

- If $t_{lastDC} \sim t_s$ (net stress time), all traps contribute to the V_T shift making ΔV_T approach DC equilibrium.
- If the net relax time (t_r) is larger than the net stress time (t_s), longer t_{lastDC} is required to reach DC equilibrium.

Next we modify the existing model to compute instantaneous BTI shift while a FET is being relaxed. Fig. 6 shows the model during relax. For computing the V_T shift at time t, we need to compute the V_T shift due to last DC ($\Delta V_{T,lastDC}$) and then apply the relaxation during the last

relaxation time ($t_{lastrelax}$). The relaxation behavior is modified from the universal recovery model. The basis for added empirical factor are:

- If it's the first time FET is relaxing, then $t_r = t_{lastrelax}$, and we'll get the same equation as universal recovery model.
- Assume AC stress where $t_s \approx t_r$, $t_{lastDC} \ll t_s$ and $t_{lastrelax} \ll t_r$. Then the model in Fig. 6 will reduce to numerator only i.e., $\Delta V_{T,lastDC}$ which is computed using the model in Fig. 5. Since $t_{lastDC} \ll t_s$, we'll get the V_T shift due to existing universal recovery model (eq. (1)) thereby predicting AC equilibrium.

RESULTS & DISCUSSIONS

Fig. 7 shows the model to hardware correlation while switching from AC to DC to AC. Our model can reliably predict the V_T shift due to transitions. Prediction of such behavior is crucial under the circuit usage condition where some NFETs and PFETs see intermittent load conditions. For example, certain circuits are designed to work only when data backup is required in a system which can happen infrequently. Also in multi-core processors, based on application, some cores can only be used to handle emergency conditions. These emergency cores may always be powered up but not used. Another example is hardware interrupt circuits. These circuits are typically always powered on and this DC stress condition can cause faults in these paths.

Another example where accurate prediction is required is when a circuit switches from the standby mode to power-on mode. Circuits are normally power-gated to reduce the power dissipation. This implicitly helps in recovering the BTI induced V_T shift. Our model and hardware show that although turning-off a circuit recovers the BTI induced V_T shift, seeing DC stress can again take FETs to DC equilibrium.

Fig. 8 shows the model-to-hardware correlation in computing instantaneous BTI induced V_T shift. Our model can accurately predict the cycle-by-cycle V_T shift in a FET. It is required when observing a FET at finer granularity, for example, while running a benchmark to predict the maximum V_T shift in a FET. Note that our model is applicable for

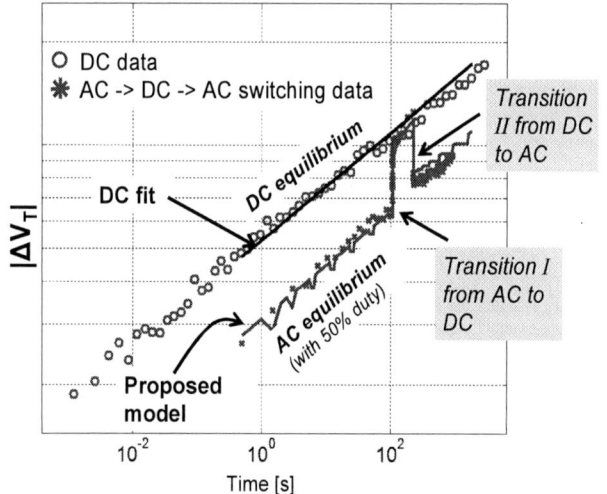

Fig. 7. Model-to-hardware correlation for AC->DC->AC switching. Our model closely matches with measured data.

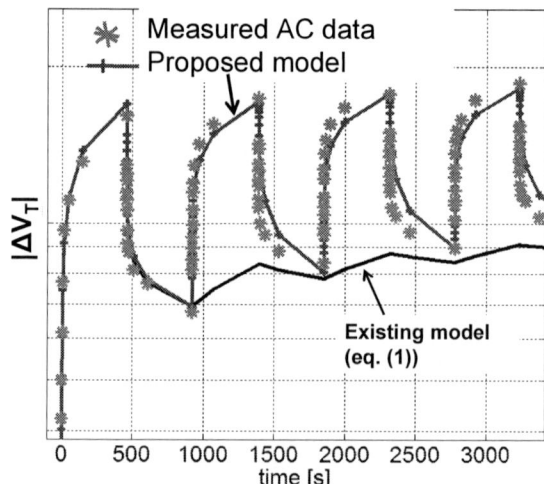

Fig. 8. Model-to-hardware correlation for instantaneous V_T shift.

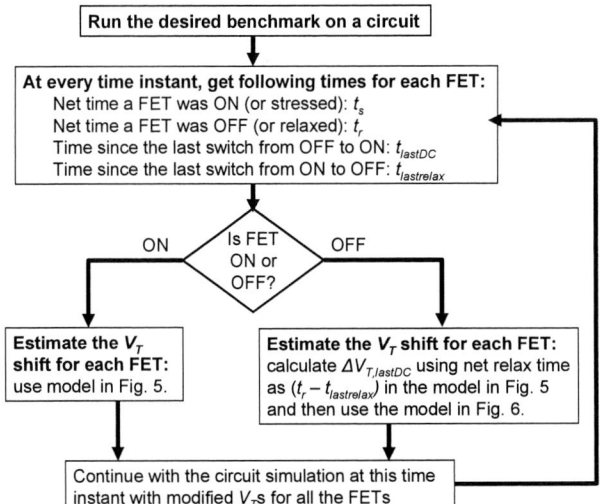

Fig. 9. Methodology for estimating instantaneous circuit performance using our BTI shift model.

varying temporal stress conditions at a given voltage and temperature stresses. In case of multiple voltages and temperatures, the DC and AC equilibriums will shift and any transition will switch between corresponding equilibrium states.

Our model can be integrated in existing digital circuit simulators. We assume that input rise and fall times of a FET are negligibly small, hence, a simulator treats each signal as either high or low. Fig. 9 shows the steps for estimating circuit performance due to BTI shift using our model.

CONCLUSIONS

We demonstrate a semi-empirically enhanced BTI model to predict the instantaneous shift in V_T due to BTI. Our proposed model is simple, does not use any extra technology

parameter beyond what's already used in existing model, and can be applied for both NBTI and PBTI. We accurately predict the instantaneous V_T shift in a FET under stress as well as the transitions from alternating to static stress and vice-versa. At every step, we demonstrate the correlation between our model and measured hardware. Further, we discuss the necessary steps to integrate our model with existing digital circuit simulators.

ACKNOWLEDGMENT

This work was performed by Research Alliance Teams at various IBM Research and Development Facilities.

REFERENCES

[1] M. Alam et al., "A comprehensive model of PMOS NBTI degradation", *Microelectronics Reliability*, vol. 45, no. 1, pp. 71-81, Jan. 2005.

[2] S. Zafar et al., "Charge trapping related threshold voltage instabilities in high permittivity gate dielectric stacks," *JAP* 2003.

[3] J. C. Lin et al., "Time Dependent Vccmin Degradation of SRAM Fabricated with High-k Gate Dielectris", *IRPS,* pp. 439-444, 2007.

[4] K. Kang et al. "Impact of Negative-Bias Temperature Instability in Nanoscale SRAM Array: Modeling and Analysis" *TCAD* 2007, pp. 1770-1781.

[5] A. Bansal. et. al., "Impacts of NBTI and PBTI on SRAM static/dynamic noise margins and cell failure probability," Journal Micro. Rel., 2009, pp. 642-649.

[6] Sanjay V. Kumar. et. al., "NBTI-Aware Synthesis of Digital Circuits," DAC, 2007, pp. 370-375.

[7] S. Rangan. et. al., "Universal Recovery Behavior of Negative Bias Temperature Instability", *Proc. IEEE* IEDM 2003, pp. 341-344.

[8] H. Reisinger. et. al., "Analysis of NBTI degradation- and recovery-behavior based on ultra fast VT-measurements", Proc. IEEE IRPS 2006, pp. 448-453.

[9] S. Bhunia. et. al., *Low-Power Variation-Tolerant Design in Nanometer Silicon*, Springer, 2009.

Soft Oxide Breakdown Impact on the Functionality of a 40 nm SRAM Memory

Saad Cheffah[1,2], Vincent Huard[1], Remy Chevallier[1], Alain Bravaix[2].
[1] STMicroelectronics – 850 rue Jean Monnet 38926 Crolles, France.
[2] IM2NP, CNRS UMR 6242, ISEN-Toulon, Maison des technologies, Place G. Pompidou, 83000 Toulon, France.
Phone: +33(0)476926325, e-mail: saad.cheffah@st.com.

Abstract-As CMOS technology continues to downscale to a deep submicron level (40 nm and beyond), Soft Oxide Breakdown (SBD) is becoming a real problem that could lead to a serious degradation in the performances and the functional operations of SoC. In this paper we study the SBD, using two models, and quantify its impact on the functionality of a 40 nm SRAM memory.

Keywords- Soft Oxide Breakdown (SBD); Transistor SBD models, SRAM functional failure; Delay drift.

I. SRAM LIBRARY OVERVIEW

The impact of intrinsic reliability can have significant impact on Static Random Access Memories (SRAM) libraries as they are more susceptible to functional failure [1]. Besides, SRAM libraries are performance bottlenecks in high-performance VLSI circuits, and occupy a majority of on-chip silicon area while requiring good tolerance throughout the life of usage.

A SRAM library is typically divided in 4 main blocks. The operation in the clock cycle is computed in the control block. When a read or a write cycle is performed, the address is chosen by the decoder block before selecting the right word inside the memory array. In parallel, the input/output block is either collecting the data from the memory array for a read cycle, or collecting from outside of the memory the data which will be saved in the memory array for a write cycle (Fig. 1).

In order to reach the highest level of performance, specific design techniques are implemented in SRAM libraries: dynamic logic for control and decoder, sense-amplifiers for reading the data in the bit-cell. However, this strategy yields the generation of enabling signals like internal clock circuitry which mimics the longest timing path.

II. SBD IMPACT ON SRAM LIBRARY

A. Design sensitivity analysis

In a complex design as an SRAM (more than 10^5 MOS), the defect cannot be simulated on each transistor. As described in fig.1, the main critical parts of the design have been selected to simulate the SBD impact.

Firstly, the SBD effect is statistical: the structure which is covering the biggest area must be tested. In the SRAM context, the majority of the area is covered by the memory bit cell (memcell). As a consequence, the bit cell has been selected.

Secondly, this SRAM is using a complex regulation loop. If this loop is broken, the functionality of the memory is lost. Several critical parts in this loop have been selected.

Finally, the design structures itself has been studied. The sense-amplifier has been selected because it uses a fully symmetric structure in order to detect voltage differences between two nets during read cycles. If this functionality is not fulfilled, the data read is corrupted.

B. SBD electrical modeling strategy

To represent the SBD, we have used the two electrical models used in the literature to emphasis a potential difference between them (Fig. 2):

-The first one is the traditional Ohmic model, which consists of connecting a resistance R_{SBD} between the Gate and the diffusion (Drain or Source) to model a current leakage path (see Fig. 2(i)).

-The second one, which can be obtained via transistor-level characterization [2], is the voltage controlled current source model or Power law model (see Fig. 2(ii)). For the parameters we have chosen $a=5$ and K was calculated at $V=V_{dd}$ for each R_{SBD} so that the SBD from the two models can be compared (R_{SBD} vs. I_{SBD}).

C. SRAM functionality issue due to SBD

In our SRAM functionality issue study, we have focused our analysis on the signals at the output of the SRAM memory ($Q<0>$, ..., $Q<n>$). For each SRAM block studied (e. g. memcell in Fig. 3), we have observed the data loss (reading "0" instead of "1" or the inverse) in the output signal, after the introduction of one SBD that corresponds to a particular value of R_{SBD} or I_{SBD}.

This study allows us, using both Ohmic and power law models, to compare the sensitivity to SBD of the different SRAM blocks (Fig. 4). One can observe that the equivalent breakdown resistances calculated from the more "realistic" power law model (I_{BD}), are lower than those calculated from the Ohmic model (R_{BD}), which suggests that the BD resistances calculated with this model are overestimated. In addition, the Sense Amplifier (SA) appears to be the most sensitive part to SBD, due to its analogical aspect, and its important role in SRAM functionality, and especially in the Read operation. Fig. 5 shows an example of SRAM functional failure (data loss) at the output $Q<0>$ due to SBD in the SA.

An important impact on SRAM functional failure occurs when a SBD defect is introduced in the nMOS Passgate of the memory point (PG in Fig. 3). In fact, we have observed that the error (reading "1" instead of "0" for example) in the output signal related to the defected memcell, propagate through the Word Line (WL) during the Write operation, to reach all the adjacent memcells in the memory array row. One of the consequences of this functional failure is the impossibility to correct it by means of numerical methods such as redundancy.

In the continuity of other previous works [3], Fig. 6 shows that, like the others studied SRAM blocks; the Sense Amplifier (SA) becomes more and more sensitive to SBD as the supply voltage decreases. This could limit the use of such SRAM memories in some low power dedicated applications, in sub-32 nm technology.

D. SRAM performances impact of SBD

Prior to functionality issue, SBD might start to impact SRAM timing performance (or to improve it in some cases) which might yield to SoC failure related to timing violations (Fig. 7).

III. CONCLUSIONS

This paper studies the impact of Soft Oxide Breakdown (SBD) on some sensitive blocks that compose a SRAM 40 nm library, as found in a product. In this study, we have found that the sensitivity of SRAM library versus SBD degradation depends on the location of the SBD defect. Furthermore, we evidenced that different results can be expected if the SBD defect is modeled using either Ohmic model or voltage controlled current source model. In addition, no correlation between NBTI phenomena and SBD has been found in the SRAM memory cell. Finally, we have shown that SRAM sensitivity to SBD defect is enhanced if one wants to use the library at lower voltages.

REFERENCES

[1] V. Huard et al, IEEE IRPS (2010)
[2] R. Rodriguez et al, IEEE IRPS (2003)
[3] V. Ramadurai et al, IEEE IRPS (2006)

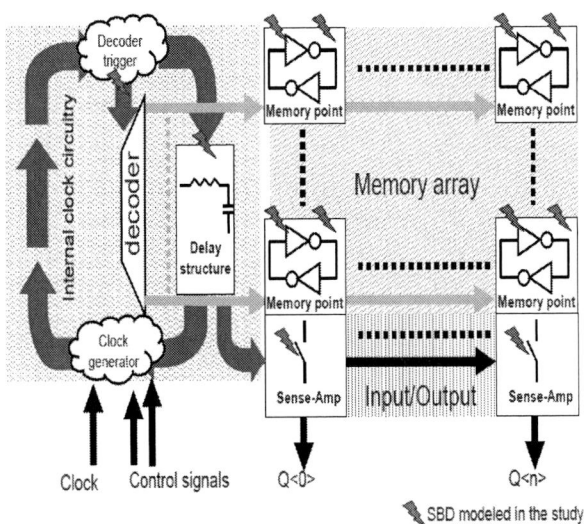

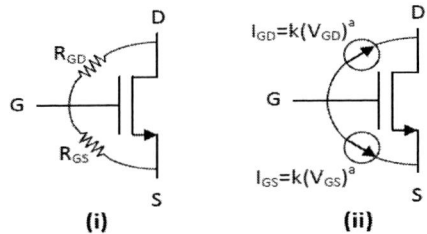

Figure 2. Circuit models used for SBD defect locations in transistors.
(i) Ohmic model. (ii) Voltage controlled current source model.

Figure 1. Bloc diagram of the SRAM memory with the locations of the SBD defects introduced in the study.

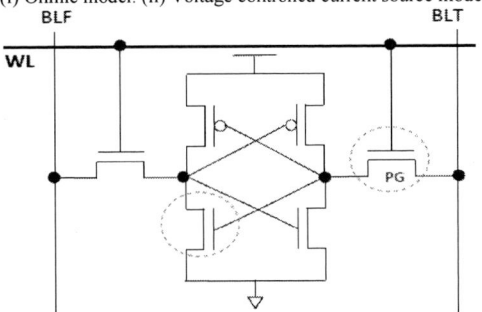

Figure 3. Possible locations for SBD in the memory cell.

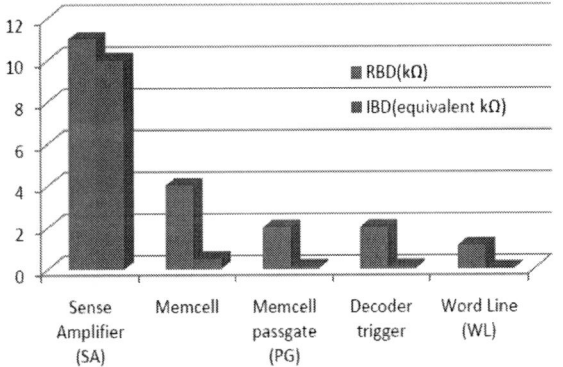

Figure 4. Sensitivity of different parts of the SRAM to Breakdown (BD), modeled by Ohmic model R_{BD} (blue) and power law model I_{BD} (red).

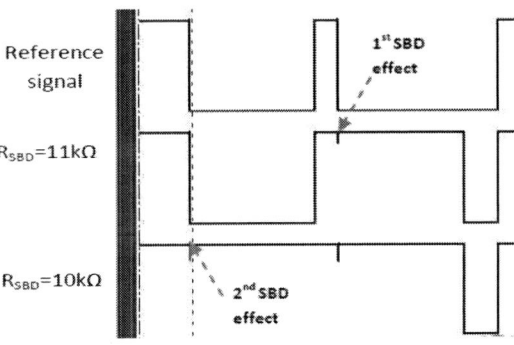

Figure 5. SRAM output signal (Q<0>) degradation due to different values of R_{SBD} introduced in the Sense Amplifier (SA) compared to a free SBD reference signal.

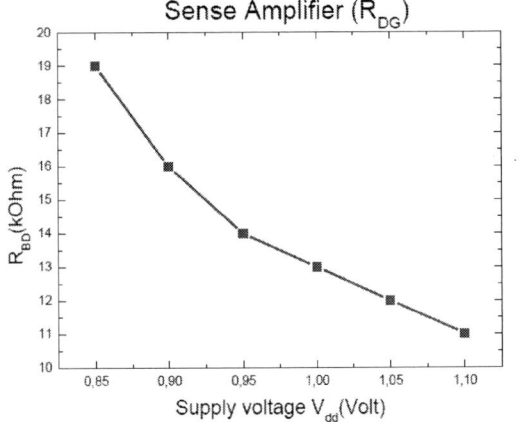

Figure 6. Resistance at Breakdown occurrence in the Sense Amplifier transistor (R_{DG}) as a function of voltage supply.

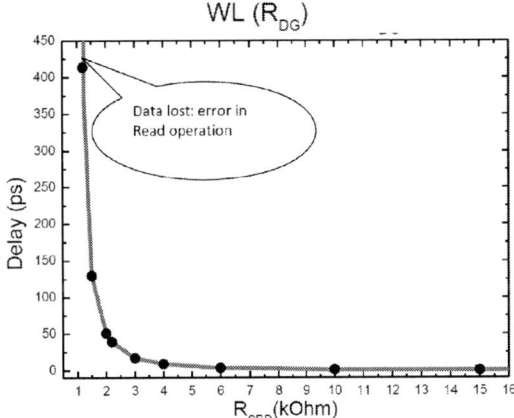

Figure 7. Delay in picoseconds (ps) between output signal reference and the output signal correspondent to R_{SBD} introduced between D and G in the Word Line (WL).

978-1-4244-9113-1/11 $26.00 © 2011 IEEE 705

The Relationship Between Transistor-Based and Circuit-Based Reliability Assessment for Digital Circuits

Balaji Vaidyanathan, Shawn Bai, Anthony S. Oates
Technology Reliability Physics Dept., R&D
Taiwan Semiconductor Manufacturing Company, Hsin-Chu, Taiwan 300-77, R.O.C.
Phone: 886-3-5636688 Ext:7032242; E-mail: {vbalaji@tsmc.com}

Abstract—**Logic, analog, RF, SRAM, and DRAM circuits respond differently to NBTI and HCI induced time dependent parametric shifts. We analyze digital logic susceptibility to these transistor degradation mechanisms and identify the benefits of simulation based aging-induced reliability assurance at the product level.**

I. Introduction

Time-dependent performance degradation due to Negative Bias Temperature Instability (NBTI) and Hot-Carrier Injection (HCI) are major hurdle for product reliability assurance in current and future technologies. As we reach the sub-40nm regime, transistor reliability engineering is approaching its intrinsic limits making performance versus reliability trade-offs cost ineffective to be addressed at the device level [1]. At the technology level all reliability effects are considered equally important and one tries to achieve similar lifetimes. However not all of them show up in equal intensities in products. Depending on the product type one or more of the reliability effects dominate and the rest gets hidden. Further, the aging mechanisms get mostly interlinked with process variation [2] in addition to getting de-rated due to AC operating conditions [3], [4] that the circuits undergo. As a result, low-device lifetime issues can be overcome by circuit level reliability and process variation awareness. In other-words, design awareness can help mask the transistor reliability issues. This has been achieved in the case of SRAM stability degradation due to NBTI [2]. Hence engineering device reliability without product knowledge may prove overly conservative.

A clear understanding of the relation between transistor and product aging is necessary to address reliability at circuit level. There have been relatively few systematic studies linking aging of digital circuits to transistor lifetime. Microprocessor [5] and ring oscillator (RO) [6] aging has been successfully modeled including NBTI and HCI effects. However whether these two circuits are sufficiently representative of the larger range of digital circuits is not entirely clear. In this work, we analyze several digital logic circuits to characterize the relationship between transistor and circuit aging. In the case of traditional transistor based reliability assurance, one achieves a lifetime target for devices without knowledge of various de-rating factors stemming from circuit operation condition, its

structure, and inherent process variation. We demonstrate the reduced pessimism in following circuit simulation based reliability assurance approach compared to the traditional transistor approach (where one achieves a lifetime target for devices without knowledge of various de-rating factors stemming from circuit operation condition, its structure, and inherent process variation).

II. Experimental Setup

This study was performed using an advanced CMOS technology, which utilizes a SiON based gate dielectric. Physically based, silicon-verified NBTI and HCI models were developed to predict the device aging as a function of voltage and temperature Additionally, process variation in circuits is modeled combining both the local and global components.

NBTI Model

NBTI behavior has been widely understood through a Reaction-Diffusion (RD) theory based models [7] and more recently proposed switching trap based models [8]. We follow RD based NBTI stress model in our work. NBTI induced PMOS threshold voltage (V_t) degradation is considered to be a combination of slow interface trapped charges and fast-hole-trapped charges (1). The slow NBTI induced PMOS V_t degradation is modeled as a power law (2) in accordance with the reaction diffusion theory, while the fast stress behavior that is attributed to the hole-trapping/de-trapping mechanism [7] saturates within 100 milliseconds and is found to be negligible compared to interface trapped charges at device operating field less than 4MV/cm [7], [9] at long-term.

$$\Delta V_t = \Delta V_{t\_it} + \Delta V_{t\_h} \tag{1}$$

$$\Delta V_{t\_it} = \Delta V_{t0} * e^{A*E_{Ox}} * e^{-E_a/K_B T} * t^n \tag{2}$$

$$\Delta V_{t\_AC} = R * r(\xi) + P \tag{3}$$

Where fitting parameters ΔV_{t0}, A, and activation energy (E_a) are used in modeling the NBTI behavior due to slow interface traps, and E_{Ox} is the electric field across the gate oxide. Figure 1(a) shows the model parameter extraction for the DC NBTI power-law model using data. One has to incorporate also NBTI recovery to understand the AC behavior

978-1-4244-9113-1/11 $26.00 © 2011 IEEE

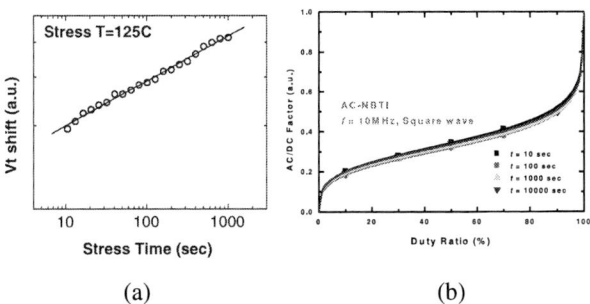

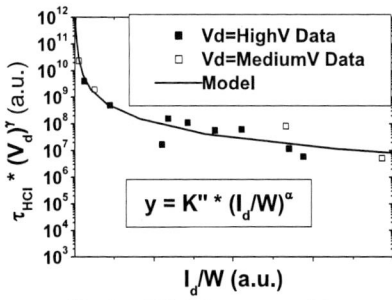

Fig. 2.   HCI data versus model

Fig. 1.   (a) DC NBTI and (b) NBTI AC/DC factor data (symbols) versus model (lines)

of V_t shifts in PMOS transistors. The Universal recovery model is used in our analysis, as proposed by Kaczer et al. [3] as shown in (3).

The total NBTI ΔV_{t_AC} is considered to be a summation of permanent (P, permanent interface traps) and recoverable (R, recoverable interface traps) component (3), ξ is a function of the stress duty factor, while the $r(\xi)$ describes the stress signal duty factor dependence of the recovery [3]. Figure 1(b) shows the AC NBTI model fitting with the AC data. NBTI AC/DC factors derived from the above stress/recovery models are fed into the simulator for circuit lifetime extraction (Figure 4).

HCI Model

Ultra-scaled devices operates mostly in a high drain current mode (high V_g, and low (V_g-V_d)) where the carrier energy is low enough to break a bond by a single direct excitation. However in this low energy mode, electron density close to the surface increases making the interaction between electrons and bonds more probable [10]. Carrier density (I_d) and its energy decided by the drain field (V_d) are required to model HCI lifetime (τ_{HCI}) as shown in (4) and Figure 2. Based on the data, the model parameters α, γ, and pre-factor K'' values are extracted to define the complete HCI lifetime model for an NMOS in a particular technology as given in (4). HCI lifetime (τ_{HCI}) is used in calculating the HCI induced threshold voltage shift (ΔV_{t_HCI}) as shown in (5) where ΔV_{t_LT} is the magnitude of V_t shift leading to 10% drop in Idsat, and n is the power-law time exponent. HCI model is verified with NMOS stressed IV data as shown in Figure 3.

$$\tau_{HCI} = K'' * V_d^{-\gamma} * (I_d/W)^{-\alpha} \qquad (4)$$

$$\Delta V_{t\_HCI} = \Delta V_{t\_LT} * (t/\tau_{HCI})^n \qquad (5)$$

Figure 4 outlines the integration of circuit aging simulation into HSPICE and subsequent verification with ring oscillator data in Figure 5. We used a NAND and NOR based RO (having few hundred stages with large gate-length devices) running below 100MHz frequency when stressed at high voltage (1.4 Vcc) at 125C. HCI induced aging is negligible for such large gate length devices and hence the RO aging is totally described by NBTI alone. The RO frequency was measured within few

hundred milliseconds of time after removal of stress and hence includes recoverable NBTI component on top of the permanent component whose models were used in the verification shown in Figure 5.

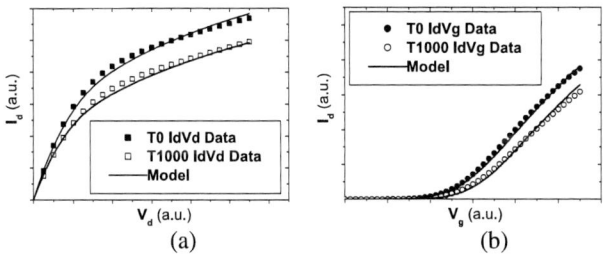

Fig. 3.   (a) IdVd and (b) linear IdVg data versus simulation for with and without HCI-stressed NMOS device

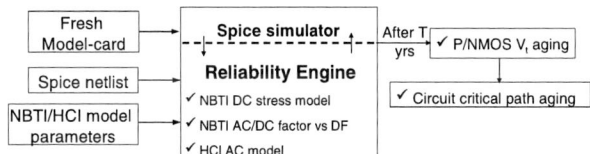

Fig. 4.   Circuit aging simulation setup

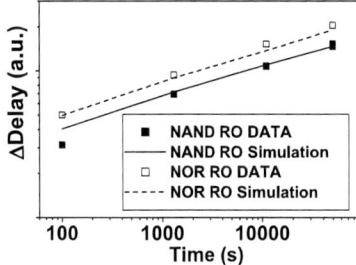

Fig. 5.   NAND and NOR based RO data and HSPICE simulation with aging model

III. RESULTS AND DISCUSSION

For this study we investigated 3 digital logic circuits consisting of a Ring Oscillator (RO), an ARM11 microprocessor, and a logic test circuit (JEDEC). These circuits had operational frequencies close to 1 GHz. The RO is a 59-stage inverter-based ring oscillator; ARM11 is the latest of the widely used ARM processor; and JDC is a JEDEC standard benchmark circuit [11] representative of a processor with basic ALU components. Simulation versus hardware correlation of fresh circuits were achieved. HTOL stressing was carried out with

978-1-4244-9113-1/11 $26.00 © 2011 IEEE   707

functional self-test patterns for 1000hrs at high voltage (1.4 Vcc) at 125C. Only register-to-register access was considered for monitoring the ARM11 critical path delay. Circuit measurements were carried out several hours after being removed from the HTOL test system, and this delay was included in the simulation of circuit level results. The frequency shifts of the circuits due to aging are shown in Figure 6 respectively. Circuit aging simulation was carried out using the aging model outlined in the previous section. A Si-verified process variation model was also included in the aging simulations.

Random Logic Aging

The simulation shows, in general for these high-speed circuits at HTOL and operating voltages, NBTI dominates the frequency aging. Only 20% of RO delay aging is due to NMOS HCI (PMOS HCI lifetimes are few orders of magnitude higher in comparison with NMOS HCI in the considered technology and hence neglected) compared with PMOS NBTI. The simulation predicts RO aging closely, while indicating close to nil aging for ARM11 and JDC circuits (Figure 6). RO has 50% stress duty factor at all its PMOS transistors in addition to having 50% PMOS in its critical path (note that the Inverter RO used in this experiment has almost an equal delay contribution from PMOS and NMOS). For the ARM11 and JDC circuits there is lower combined probability of having higher PMOS contribution to critical path delay as well as to its aging due to NBTI compared to the RO. Though the existence of an aging critical stress test pattern cannot be ruled out, its active period in the pipeline will be relatively less masking the test pattern dependency on the HTOL output to a large extent. Finally the probability of NMOS with larger fan-out (larger fan-out/load NMOS are highly susceptible to HCI) is low in a speed oriented design and if so, it being in the critical path is low when compared with RO. Hence ARM11 and JDC shows negligible aging in HTOL data (Figure 6(b,c)).

Even though RO exhibits the worst-case aging for the random logic category, the observed aging is only 1.5%. The spread in the delay due to process variation is larger than this, so that from a practical perspective the aging induced delay shift if any in ARM11 or JDC is masked to a large extent by the process spread (Figure 6(b,c)). Simulations of product aging at the end of 10yrs under nominal operational voltage at 100C were carried out to understand the on-field aging (Figure 6). HTOL measurements were done with complete NBTI recovery while on-field aging does not recover completely leading to comparatively higher aging entirely due to NBTI as HCI impact decreases rapidly with operational voltage [6]. However, we observe the long-term aging to be still less (1.5%-2.5%). Hence one can conclude that the effect of transistor aging is largely masked in the random logic due to AC/DC de-rating factors that for NBTI and HCI, nominal fan-outs leading to less HCI compared to NBTI, and lesser combined probability of having high PMOS contribution to the critical path delay as well as to its aging.

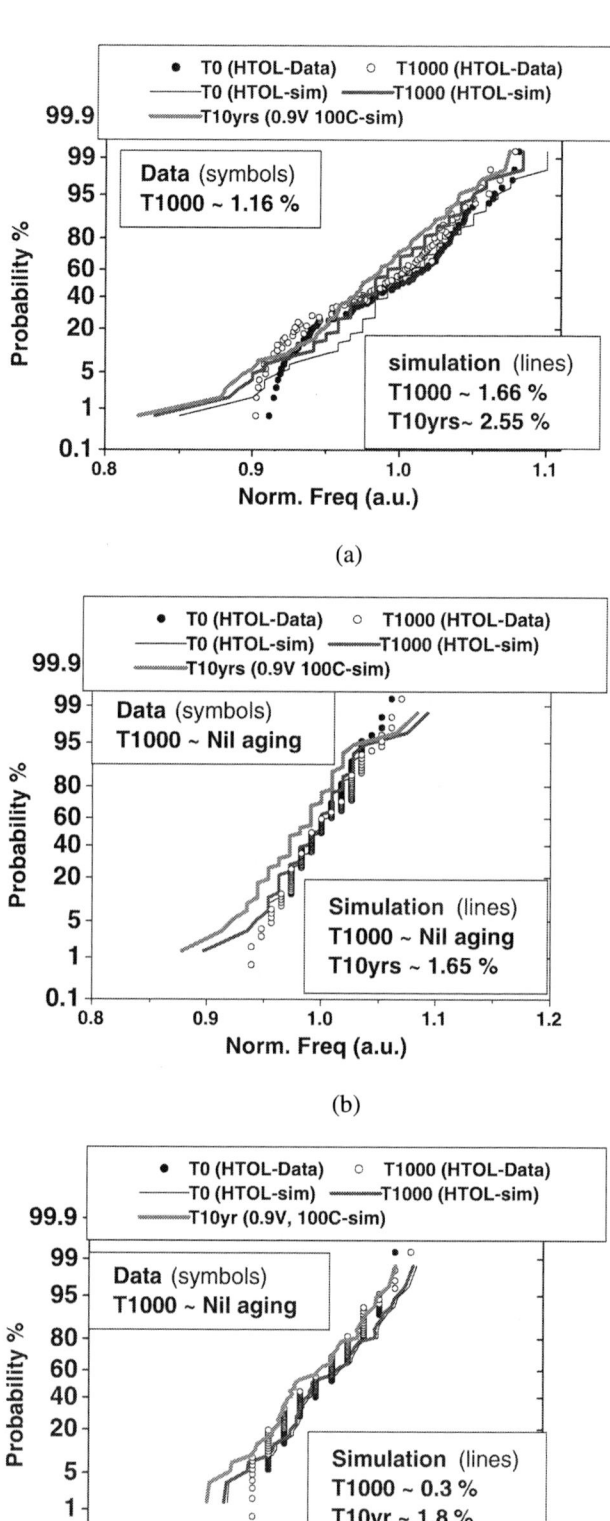

Fig. 6.   (a) RO, (b) ARM, and (c) JDC aging versus simulation

IV. CONCLUSION

A systematic aging study across wide product categories has been conducted to understand the link between transistor and circuit reliability. Random logic delay aging is upper bounded by RO, whose shifts are shown to be negligible due to large DC-to-AC aging de-rating factors in addition to getting buried under process variation induced delay spread. Such studies as we have shown here should help circuit designers to propose circuit solutions to reliability concerns with minimal impact on area and performance.

REFERENCES

[1] M. A. Alam, "Reliability- and Process-Variation Aware Design of Integrated Circuits," *Microelectronics Reliability*, vol. 48, pp. 1114–1122, 2008.

[2] G. L. Rosa, W. L. Ng, and S. Rauch, "Impact of NBTI Induced Statistical Variation to SRAM Cell Stability," in *IEEE Int'l Rel. Phys. Symp.*, 2006, pp. 274–282.

[3] B. Kaczer, T. Grasser, P. J. Roussel, J. Martin-Martinez, R. O'Connor, B. O'Sullivan, and G. Groeseneken, "Ubiquitous Relaxation in BTI Stressing-New Evaluation and Insights," in *IEEE Int'l Rel. Phys. Symp.*, 2008, pp. 20–27.

[4] C. Guerin, V. Huard, C. Parthasarathy, J. Roux, A. Bravaix, and E. Vincent, "Novel Hot-Carrier AC-DC Design Guidelines for Advanced CMOS Nodes," in *IEEE Int'l Rel. Phys. Symp.*, 2008, pp. 741–742.

[5] A. Haggag, M. Lemanski, G. Anderson, P. Abramowitz, and M. Moosa, "Realistic Projections of Product Fmax Shift and Statistics due to HCI and NBTI," in *IEEE Int'l Rel. Phys. Symp.*, 2007, pp. 93–96.

[6] T. Nigam, B. Parameshwaran, and G. Krause, "Accurate Product Lifetime Predictions Based On Device-Level Measurement," in *IEEE Int'l Rel. Phys. Symp.*, 2009, pp. 634–639.

[7] A. E. Islam, H. Kufluoglu, D. Varghese, S. Mahapatra, and M. A. Alam, "Recent Issues in Negative-Bias Temperature Instability: Initial Degradation, Field Dependence of Interface Trap Generation, Hole Trapping Effects, and Relaxation," *IEEE Trans. Electron Devices*, vol. 54, no. 9, pp. 2143–2154, Sept 2007.

[8] T. Grasser, H. Reisinger, P. Wagner, F. Schanovsky, and B. Kaczer, "The Time Dependent Defect Spectroscopy (TDDS) for the Characterization of the Bias Temperature Instability," in *IEEE Int'l Rel. Phys. Symp.*, 2010, pp. 16–25.

[9] J. H. Lee and A. S. Oates, "Characterization of NBTI-Induced Interface State and Hole Trapping in SiON Gate Dielectrics of p-MOSFETs," *IEEE Trans. Device Mater. Rel.*, vol. 10, no. 2, pp. 174–181, June 2010.

[10] C. Guerin, V. Huard, and A. Bravaix, "The Energy-Driven Hot-Carrier Degradation Modes of nMOSFETs," *IEEE Trans. Device Mater. Rel.*, vol. 7, no. 2, pp. 225–235, Jun 2007.

[11] "Standard For Semicustom Integrated Circuits," *JEDEC Standard*, no. 12, Jun 1985.

The Impact of RTN on Performance Fluctuation in CMOS Logic Circuits

Kyosuke Ito∗, Takashi Matsumoto∗, Shinichi Nishizawa∗, Hiroki Sunagawa∗,
Kazutoshi Kobayashi† and Hidetoshi Onodera∗‡
∗Kyoto University, †Kyoto Institute of Technology, ‡JST, CREST, Japan

Abstract— **In this paper, the impact of Random Telegraph Noise (RTN) on CMOS logic circuits observed in a Circuit Matrix Array is reported. We discuss the behavior of RTN under circuit operation, and reveal that the impact of RTN, which is much smaller than that of within-die variation in a 65nm process, can have a severe effect on the performance of a sequential logic gate under low voltage operation.**

I. Introduction

With recent aggressive technology scaling of LSIs for power reduction and die shrink, designing reliable systems becomes more and more challenging. Besides conventional problems such as transistor leakage, degradation and variation of transistor performance have a severe impact on the dependability of VLSI systems [1], [2], [3]. Random Telegraph Noise (RTN) has attracted much attention as a temporal variation caused by the capture and emission of mobile charge carriers by defects inside the dielectric [4], [5]. It has been reported that RTN will emerge as a serious reliability and variability issue in accordance with scaling [6], [7], and even in 90nm process node RTN already has a severe impact on CMOS image sensors [8], Flash memories [9], SRAMs [10]. However, the impact of RTN on CMOS logic circuits has not been well addressed yet. Therefore, we focus on the impact of RTN on circuit delay and have measured RTN-induced delay fluctuation using a circuit matrix array fabricated in a 65nm process.

The remainder of this paper is organized as follows. In Section 2, we will show the test structure fabricated in a 65nm CMOS technology for RTN-induced delay fluctuation measurement. In Section 3, from the measurement data, we discuss RTN characteristics under circuit operation , and evaluate the impact of RTN on combinational circuits and sequential circuits quantitatively. Finally, Section 4 summarizes this paper.

II. Test Structure for RTN-induced Delay Fluctuation Measurement

Fig. 1 shows a test structure fabricated in a 65nm CMOS technology for RTN-induced delay fluctuation measurement. The circuit matrix array contains a 20 × 15 array sections and each section is constructed by a various types of ring oscillators (ROs) including a 7-stage inverter RO. We therefore have 300 identical 7-stage ROs on a chip, which we consider as representatives of combinational circuits. By measuring the fluctuation of oscillation frequencies, we can evaluate the fluctuation of delays in a chain of inverters. ROs are

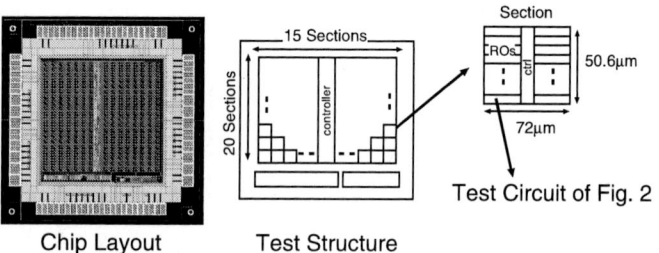

Fig. 1. A test structure for process variation and RTN-induced delay fluctuation measurement in 65nm CMOS technology

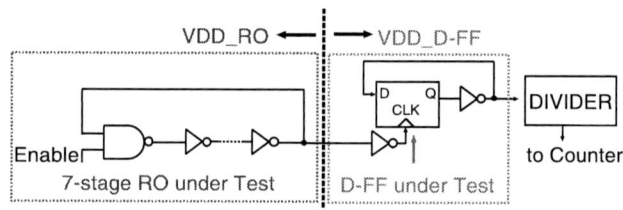

Fig. 2. A test circuit that consists of a ring oscillator (RO) and a divider with a D-FF, for evaluating the impact of process variation and RTN in a combinational circuit(RO) and a sequential circuit(D-FF).

connected to dividers as shown in Fig. 2. The frequency of an RO can be tuned by changing its supply voltage VDD_{RO}, and the output of the RO is connected to the clock input of a D-FF through one inverter so that the circuit acts as a frequency divider as shown in Fig. 2. The power supply of the D-FF (VDD_{D-FF}) can also be varied. In this experiment, we have measured the fluctuation of oscillation frequencies for a period of 250 s by counting the number of oscillation over 20 ms at room temperature. Supply voltages VDD_{RO} and VDD_{D-FF} are set to 0.8 V and 1.0 V, respectively

III. Experimental Results and Discussion

A. Behavior of RTN under circuit operation

Considering the source of $1/f$ noise is superposition of multiple RTN [4], the power spectrum density (PSD) of the oscillation frequency fluctuation of a 7-stage ring oscillator, which consists of 18 MOSFETs, is found to be well represented by an $1/f^s$ response where the power s is between 1 and 2.

978-1-4244-9113-1/11 $26.00 © 2011 IEEE

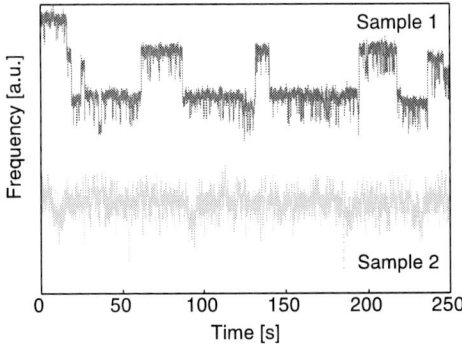

Fig. 3. Two examples of measured RTN-induced frequency fluctuations.

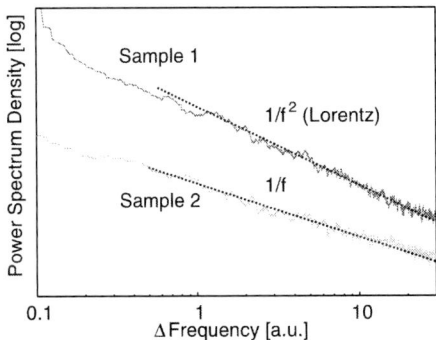

Fig. 4. PSD of RTN-induced frequency fluctuations shown in Fig. 3.

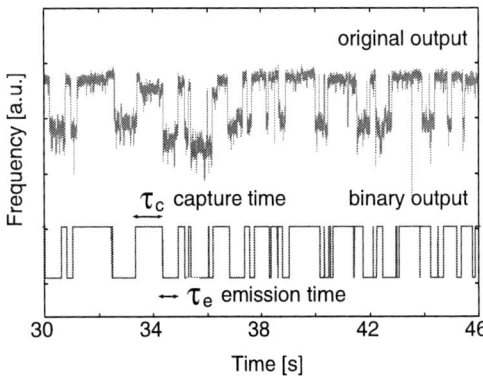

Fig. 5. A measured 2-level frequency fluctuation and its idealized RTN waveform from the noisy measurement data.

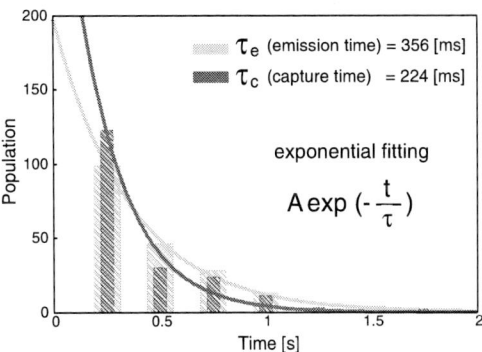

Fig. 6. Under circuit operation (AC bias), τ_e (emission time) and τ_c (capture time) of RTN can be uniquely extracted.

Fig. 3 shows two examples of measured frequency fluctuation observed in 300 samples. Sample 1 has a discrete fluctuation which is a characteristic of RTN while Sample 2 does not. Fig. 4 is the PSD of these frequency fluctuations. Since the PSD of Sample 1 is $1/f^2$-shaped and that of Sample 2 is $1/f$-shaped, they are both believed to be originated from RTN. The increase of VDD_{D-FF} has no influence on the frequency fluctuation. This confirms that the frequency fluctuation is caused by the phase-noise of the ring oscillator due to RTN, not by any functional fluctuation of the D-FF.

The emission time τ_e (averaged time of high V_{th} state) and the capture time τ_c (averaged time of low V_{th} state) are also important factors of RTN. They are known to depend on trap position, trap energy and bias condition strongly (Shockley-Read-Hall statistics [4]). Considering Fermi energy level varies widely under circuit operation, there is a possibility that RTN under circuit operation behaves differently from RTN under DC bias.

Fig. 5 shows an example of measured 2-level frequency fluctuation and its idealized 2-level waveform. From the idealized 2-level fluctuation, we have measured emission times and capture times. Statistical characteristics of those intervals are listed in Fig. 6 where we can extract the time constants of τ_e and τ_c by exponential fitting. This suggests that RTN under circuit operation obeys similar mechanism to that under DC condition.

B. Impact of RTN on delay of combinational circuits

We define the variation of averaged frequencies over measurement time of 250 s for 300 ROs on a chip as WID variation, whereas the maximum frequency shift for each RO in 250s is caused by RTN. Fig. 7 shows the distribution of frequency variation by the WID variation and the frequency fluctuation by RTN measured at $VDD_{RO} = 0.8V$ and $VDD_{D-FF} = 1.0V$.

As shown in Fig. 7, the impact of RTN on delay is as small as 4% of that of WID variation at 3σ. This is because RTN-induced ΔV_{th} is much smaller than that caused by WID variation in a 65nm process for normal logic cells.

Fig. 8 shows the scatter plots of measured RTN-induced frequency fluctuation and WID variation-induced frequency variation. Frequency fluctuation by RTN appears to be uncorrelated to WID frequency variation. This indicates that RTN-induced ΔV_{th} does not depend on the amount of V_{th}.

The distribution of RTN-induced frequency fluctuations at various VDD_{RO} is shown in Fig. 9. We have observed that the ratio of RTN-induced frequency fluctuation to WID variation-induced frequency variation increases as VDD_{RO} increases, as shown in Fig. 10. It has been reported that RTN-induced ΔV_{th} or detected number of traps has V_{gs} dependence [10]. The V_{gs} dependence of the ratio in Fig. 10 confirms the V_{gs}

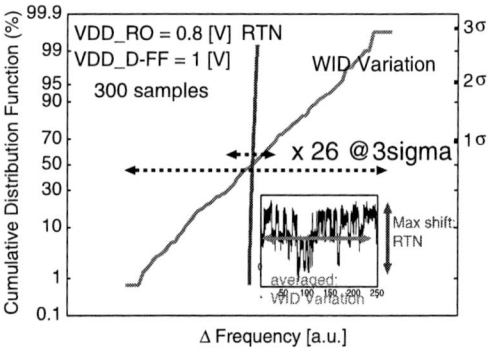

Fig. 7. Distribution of frequency variation by Within die (WID) variation and frequency fluctuation by RTN. In a 65nm logic circuit, the impact of RTN is much smaller than that of WID variation.

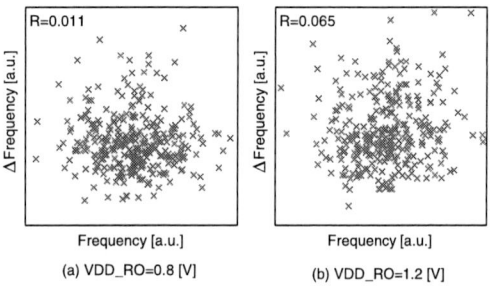

(a) VDD_RO=0.8 [V]   (b) VDD_RO=1.2 [V]

Fig. 8. Scatter plots of measured RTN-induced frequency fluctuation ("Δ Frequency") and WID variation-induced frequency variation ("Frequency"). "Δ Frequency" appears to be uncorrelated with "Frequency".

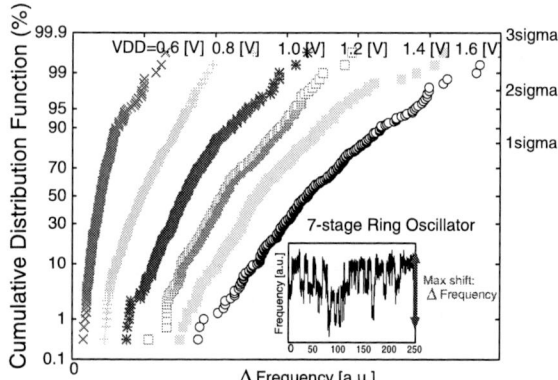

Fig. 9. Distribution of RTN-induced frequency fluctuations under various VDD_{RO}.

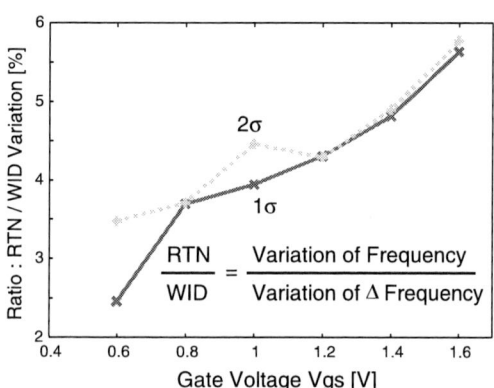

Fig. 10. The ratio of RTN-induced frequency fluctuation to WID variation-induced frequency variation as a function of VDD_{RO}.

dependence of ΔV_{th} under circuit operation.

Fig. 11 shows the estimated ratio of RTN-induced delay fluctuation to WID variation-induced delay variation which appears in a normal logic gate in future technology nodes. The ratio is derived assuming the WL dependence of WID variation ($\sigma/\mu \propto 1/\sqrt{WL}$) known as Pelgrom Plot [11] and that of RTN ($\sigma/\mu \propto 1/WL$ [7]) using the measured data at $VDD_{RO} = 0.8V$. This shows that in more scaled technology RTN is expected to have a non-negligible impact on the delay of combinational circuit, and RTN-aware design margin must be considered.

C. Impact of RTN on sequential circuits

As explained in section II, a ring oscillator in the test circuit is connected to a divider. Fig. 12 shows the simulation result of the output frequency of the divider as a function of the input frequency. There exists a maximum frequency for correct operation (MOF: Maximum Operating Frequency). The MOF corresponds to a D-FF delay with an additional delay of an inverter. Under low voltage operation of the D-FF, we have observed a large frequency fluctuation at the output when the input frequency is close to the MOF, which is shown in Fig. 13. This large fluctuation comes from the failure operation of the D-FF due to the fluctuation of the MOF by RTN. The vulnerability of D-FF timing behavior[12] is triggered by RTN.

Fig. 14 shows the comparison between the distribution of the oscillation frequency fluctuation measured at $VDD_{RO} = 0.8$ V and $VDD_{D-FF} = 0.75$ V, and the one at $VDD_{RO} = 0.8$ V and $VDD_{D-FF} = 1.0$ V. All D-FFs in 300 samples at $VDD_{D-FF} = 1.0$ V can divide the input frequency correctly. When VDD_{D-FF} is reduced to 0.7V, the number of D-FFs that work as a divider correctly decrease to 112 samples because 188 D-FFs stop working due to performance degradation caused by WID variability. Out of 112 working samples, however, 25 samples have exhibit a very large frequency fluctuation originated from RTN in D-FFs. This observation indicates that under low voltage operation, where PVT variation have a large impact, RTN can also have a severe impact on the functionality of a sequential logic such as a D-FF.

IV. Conclusion

We have measured RTN-induced performance fluctuation in CMOS logic circuits using a circuit matrix array test structure. It is revealed that RTN under circuit operation has statistical properties similar to those under DC bias. The impact of RTN is much smaller than that of WID variation on combinational circuits in a 65nm process. However, we

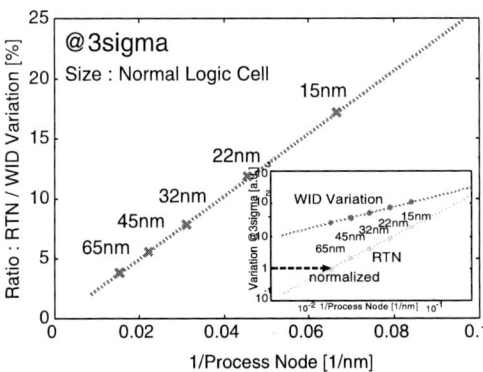

Fig. 11. Predicted comparison between RTN-induced delay fluctuation and WID variation-induced delay variation in a CMOS logic circuit. In more scaled process node, the impact of RTN will become visible on the circuit delay.

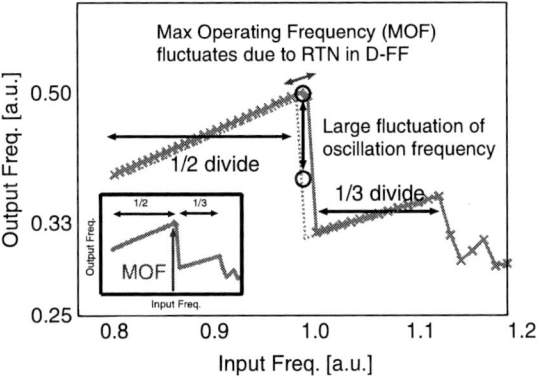

Fig. 12. Simulation result of the output frequency of the divider as a function of input frequency. Large frequency fluctuation results from Max Operating Frequency (MOF) fluctuation induced by RTN in the D-FF.

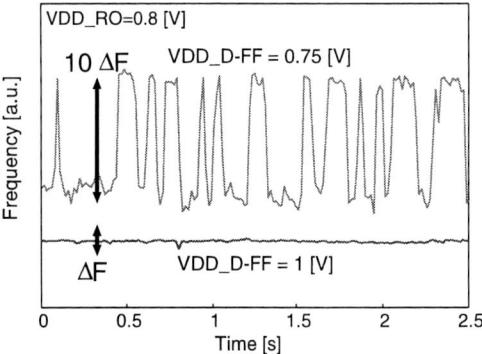

Fig. 13. Oscillation frequency fluctuation due to RTN. Fluctuation increases largely when lowering voltage of D-FF to certain point.

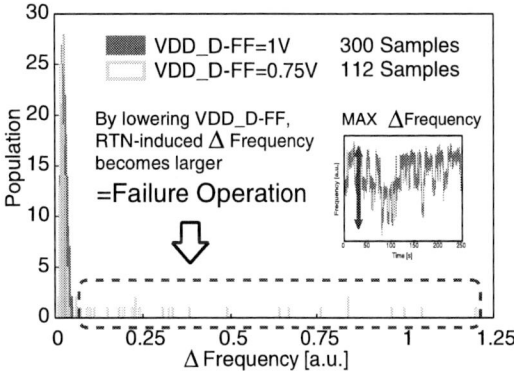

Fig. 14. Distribution of the frequency fluctuation under low VDD_{D-FF} (0.75V). Large frequency fluctuation results from MOF fluctuation due to RTN in the D-FF, as explained in Fig. 12. In low voltage, even RTN produces visible performance degradation in a sequential circuit such as a D-FF.

clarify that the impact of RTN on the circuit delay is expected to increase in more scaled technologies, and RTN has a severe impact on sequential circuits under low voltage operation.

Acknowledgment: The VLSI chip in this study has been fabricated in the chip fabrication program of VDEC, the University of Tokyo in collaboration with STARC, e-Shuttle, Inc., and Fujitsu Ltd.

REFERENCES

[1] S. Borkar. Designing reliable systems from unreliable components: the challenges of transistor variability and degradation. *Micro, IEEE*, Vol. 25, No. 6, pp. 10 – 16, nov. 2005.

[2] S. Mahapatra and M.A. Alam. Defect generation in p-mosfets under negative-bias stress: An experimental perspective. *Device and Materials Reliability, IEEE Transactions on*, Vol. 8, No. 1, pp. 35 –46, mar. 2008.

[3] H. Onodera. Variability modeling and impact on design. *Electron Devices Meeting, 2008. IEDM 2008. IEEE International*, pp. 1 –4, dec. 2008.

[4] M. J. Kirton and M. J. Uren. Noise in solid-state microstructures: a new perspective on individual defects, interface states, and low-frequency noise. *Advances in Physics*, Vol. 38, No. 4, pp. 367–468, jul. 1989.

[5] N. Tega, H. Miki, T. Osabe, A. Kotabe, K. Otsuga, H. Kurata, S. Kamohara, K. Tokami, Y. Ikeda, and R. Yamada. Anomalously large threshold voltage fluctuation by complex random telegraph signal in floating gate flash memory. In *Electron Devices Meeting, 2006. IEDM '06. International*, pp. 1 –4, 11-13 2006.

[6] N. Tega, H. Miki, M. Yamaoka, H. Kume, Toshiyuki Mine, T. Ishida, Yuki Mori, R. Yamada, and Kazuyoshi Torii. Impact of threshold voltage fluctuation due to random telegraph noise on scaled-down sram. In *Reliability Physics Symposium, 2008. IRPS 2008. IEEE International*, pp. 541 –546, apr. 2008.

[7] N. Tega, H. Miki, Z. Ren, C.P. D'Emic, Y. Zhu, D.J. Frank, J. Cai, M.A. Guillorn, D.-G. Park, W. Haensch, and K. Torii. Reduction of random telegraph noise in high-k / metal-gate stacks for 22 nm generation fets. In *Electron Devices Meeting (IEDM), 2009 IEEE International*, pp. 1 –4, 2009.

[8] Jun-Myung Woo, Hong-Hyun Park, Hong Shick Min, Young June Park, Sung-Min Hong, and Chan Hyeong Park. Statistical analysis of random telegraph noise in cmos image sensors. In *Simulation of Semiconductor Processes and Devices, 2008. SISPAD 2008. International Conference on*, pp. 77 –80, 9-11 2008.

[9] H. Kurata, K. Otsuga, A. Kotabe, S. Kajiyama, T. Osabe, Y. Sasago, S. Narumi, K. Tokami, S. Kamohara, and O. Tsuchiya. Random telegraph signal in flash memory: Its impact on scaling of multilevel flash memory beyond the 90-nm node. *Solid-State Circuits, IEEE Journal of*, Vol. 42, No. 6, pp. 1362–1369, jun. 2007.

[10] M. Tanizawa, S. Ohbayashi, T. Okagaki, K. Sonoda, K. Eikyu, Y. Hirano, K. Ishikawa, O. Tsuchiya, and Y. Inoue. Application of a statistical compact model for random telegraph noise to scaled-sram vmin analysis. In *VLSI Technology, 2010 Symposium on*, pp. 95 –96, 16-18 2010.

[11] M.J.M. Pelgrom, A.C.J. Duinmaijer, and A.P.G. Welbers. Matching properties of mos transistors. *Solid-State Circuits, IEEE Journal of*, Vol. 24, No. 5, pp. 1433 – 1439, October 1989.

[12] Hiroki Sunagawa. and Hidetoshi Onodera. Variation-tolerant design of d flipflops. In *SOC Conference, 2010. SOCC 2010. IEEE International*, pp. 147 –151, 2010.

THE MODIFIED P+ ELECTRODE LAYOUT SCHEMES TO ENHANCE ESD ROBUSTNESS OF SCR STRUCTURE FOR PMIC APPLICATIONS

Lu-An Chen, Chang-Tzu Wang, Tai-Hsiang Lai, Tien-Hao Tang, and Kuan-Cheng Su

ESD Engineering Department, Reliability Technology & Assurance Division
United Microelectronics Corp., Hsinchu Science Park, Taiwan

INTRODUCTION

In power management integrated circuit (PMIC) technology, open-drain circuit has been extensively used for output driver design. The output driver with large device dimension to achieve a high driving capability is usually required. In ESD protection design of open-drain circuit, the output driver device can be self-protected from ESD damage by a large efficiency channel width [1]. The simplified diagram of output driver on customer product is shown in Fig. 1(a). We measured the I-V characteristic and leakage current with a Transmission-Line-Pulsing (TLP) System. The TLP-measured I-V curve characteristic of the output driver without ESD clamp is shown in Fig. 1(b). By the TLP-measured result, the efficiency channel width of output driver is not enough to self-discharge ESD current. In order to avoid ESD damage on output driver device, we propose a novel high voltage PMOS with embedded SCR (PSCR) for ESD protection. With regard to the high robustness of PSCR device, it has been presented in other HV process [2].

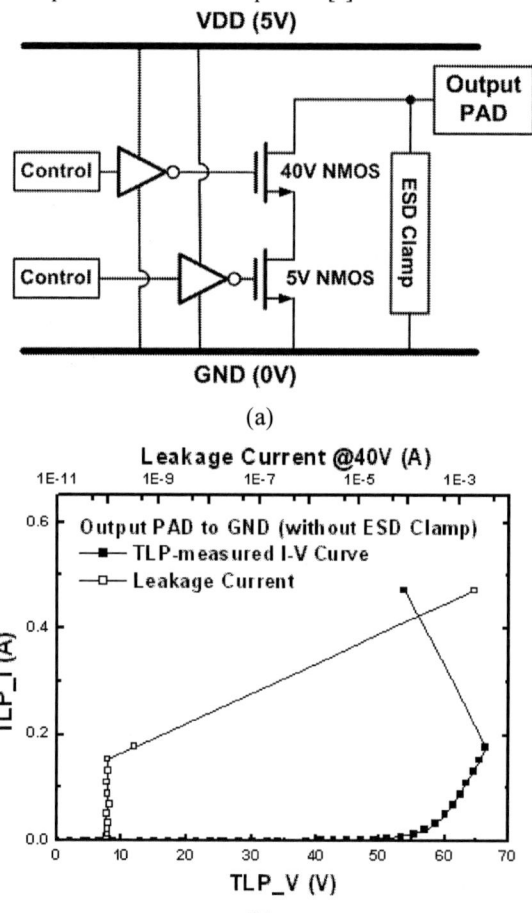

(a)

(b)

Figure 1. (a) The simplified diagram of output driver circuit (B) The TLP-Measured I-V CURVES of the output driver without ESD clamp circuit.

This paper introduces a modified PSCR (MPSCR) structure for high-level ESD protection. This device obtained the low triggering voltage and high second breakdown current by inserting an N+ layer and adjusting the location of P+ electrode in the cathode region of FDPMOS device. The MPSCR device has been verified in 0.35-um 40-V CDMOS technology. The ESD performance of MPSCR under the device width of 300μm can sustain up to 7.2kV for Human Body Model (HBM) and 360V for Machine Model (MM). In addition, the electrical characteristics of the device have been also investigated through TLP measurement and TCAD simulation. The results show that the proposed MPSCR device has high ESD robustness by modifying the vertical n-p-n BJT structure in PSCR device.

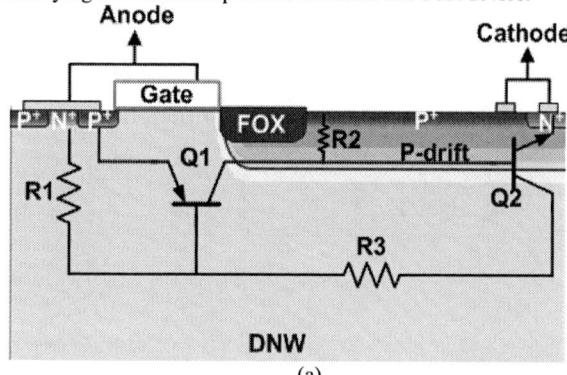

(a)

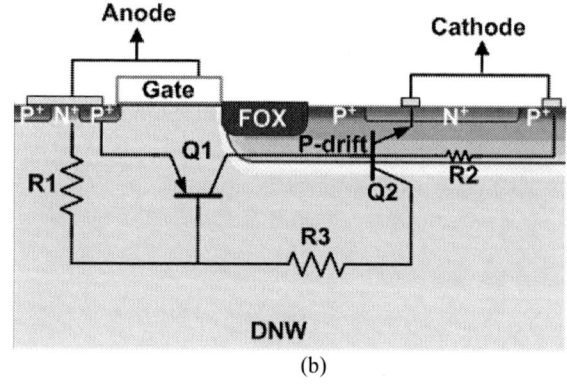

(b)

FIGURE 2. THE CROSS-SECTIONAL VIEWS AND EQUIVALENT CIRCUITS OF (A) PSCR AND (B) MPSCR.

DEVICE STRUCTURES AND EXPERIMENTAL RESULT

The device cross-sectional view and equivalent circuit of traditional PSCR in a 0.35-um CDMOS technology is shown in Fig. 2(a). This device is embedded a SCR path by adding an N+ layer in the cathode region. The device characteristics of the PSCR structure under ESD stress condition were measured by TLP measurement. As shown in Fig. 3(a), the results show that the PSCR device failed after snapback. Therefore, the poor performance of PSCR for ESD protection can not pass ESD spec (HBM< 2kV, MM< 200V). In order to improve the ESD performance, we modified the structure of

978-1-4244-9113-1/11 $26.00 © 2011 IEEE

cathode region, as shown in Fig. 2(b). To modify the place of N+ region at cathode side is same as to adjust the position of SCR path in the FDPMOS device. By swapping the N+ and P+, the NPN base resistance is increased and triggering is facilitated. The P+ electrode of cathode zone distant from filed oxide region (FOX) can increase the turn-on efficiency of SCR path and avoid current crowding at P-drift/FOX junction. The TLP-measured I-V characteristics of the modified PSCR structure are shown in Fig. 3(b). The result shows that the MPSCR device has great HBM/MM ESD performance, and the second breakdown current (It2) can be improved to 10-A, and the trigger voltage (Vt1) can be reduced to 55-V.

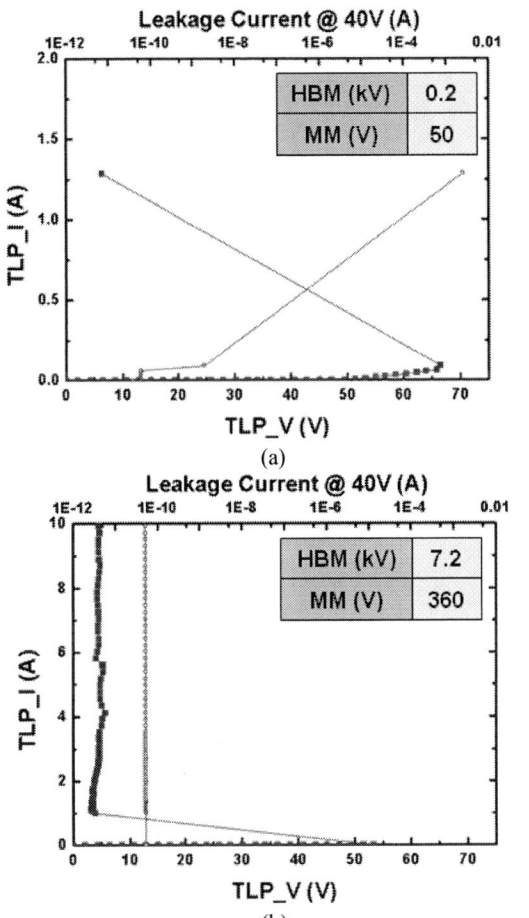

FIGURE 3. THE TLP MEASURED I-V CHARACTERISTICS OF (A) PSCR (B) MPSCR.

DEVICE SIMULATION AND DISCUSSION

We use the 2-Dimension Synopsys TCAD simulator to analyze electrical characteristics of PSCR and MPSCR devices. We defined the device failure point at the maximum lattice temperature up to 1500-K. The simulated DC I-V characteristics of the PSCR and MPSCR devices are shown in Fig. 4(a). The simulation results show that the PSCR device did not failed after snapback point, and this is different from the TLP-measured data. In order to understand the difference, we compared the current-temperature relation of PSCR with MPSCR device. The results show that the PSCR device has a poorer current endurance than the MPSCR device, as shown in Fig. 4(b). The maximum lattice temperature of PSCR is up to 1400K at snapback point (Anode current=1-mA/μm), and this is the reason that PSCR structure failed after the device snapback in the TLP-measured

data. In the other hand, the simulated results show that there is an effective turn-on SCR path in the MPSCR device, and furthermore it has the low triggering voltage (Vt1=50-V) and the large second breakdown current (It2=27-mA/um), as shown in Fig. 4(a). Although the simulated DC I-V data cannot mach the TLP-measured data completely, we can prove that the MPSCR structure has better ESD current endurance than the PSCR structure by TCAD simulation.

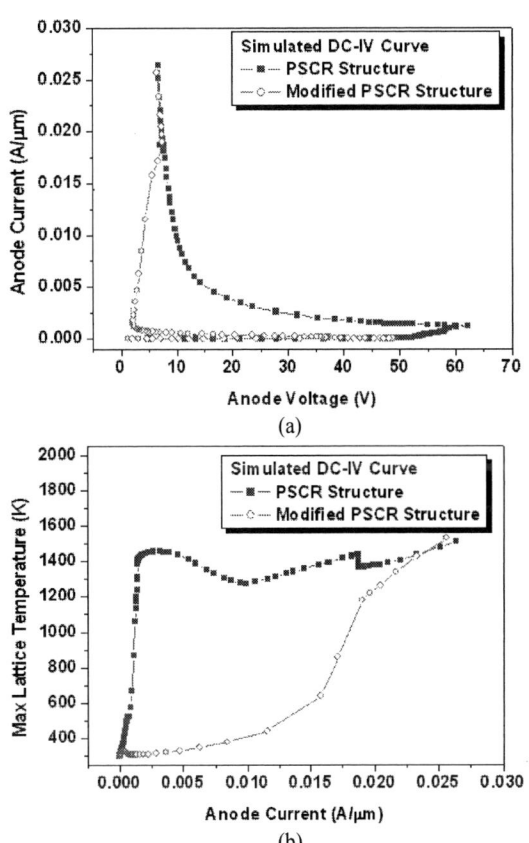

FIGURE 4. THE 2-D SIMULATION DC I-V CHARACTERISTICS OF (A) PSCR (B) MPSCR.

PHYSICAL PHENOMENON SIMULATION

In here, the SCR path mechanism in PSCR and MPSCR devices would be researched by TCAD simulation. The equivalent circuits of PSCR and MPCR devices are shown in Fig. 2(a) and 2(b). Since the anode voltage is increased under ESD event, there is a leakage current path of reverse diode from anode to cathode. As the leakage current passes through R1, there is a cross voltage on R1. The parasitic p-n-p BJT Q1 can not turn on until the cross voltage of R1 is bigger than the emitter-to-base junction voltage of parasitic BJT Q1. The turn-on current of Q1 flows into P-drift resistor R2, and it generates an enough cross voltage on R2 to let BJT Q2 turn on. The parasitic SCR structure will be formed because the BJT Q1 and BJT Q2 are sequentially triggered on. In MPSCR device, the modified P+ electrode location at cathode side can increase the resistance of R2, which can drive BJT Q2 to turn on more effectively. Therefore, the BJT Q2 of PSCR device is difficult to be activated due to the low resistance of R2. The turn-on current of BJT Q1 causes a current crowding effect on the junction of FOX and P-drift before the BJT Q2 is switched on, which makes a local hot spot to result in device damage.

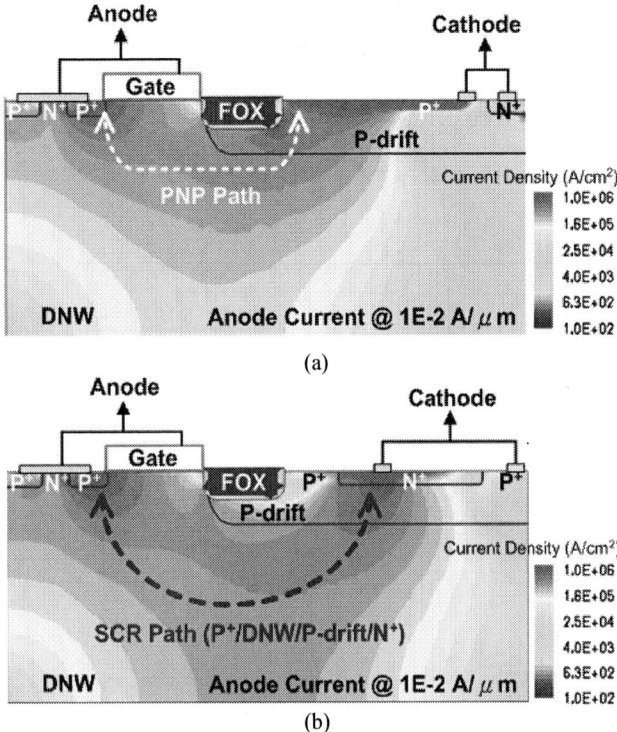

(a)

(b)

Figure 5. The current density plots of (a) PSCR and (b) MPSCR at anode current of 1E-2A/μm.

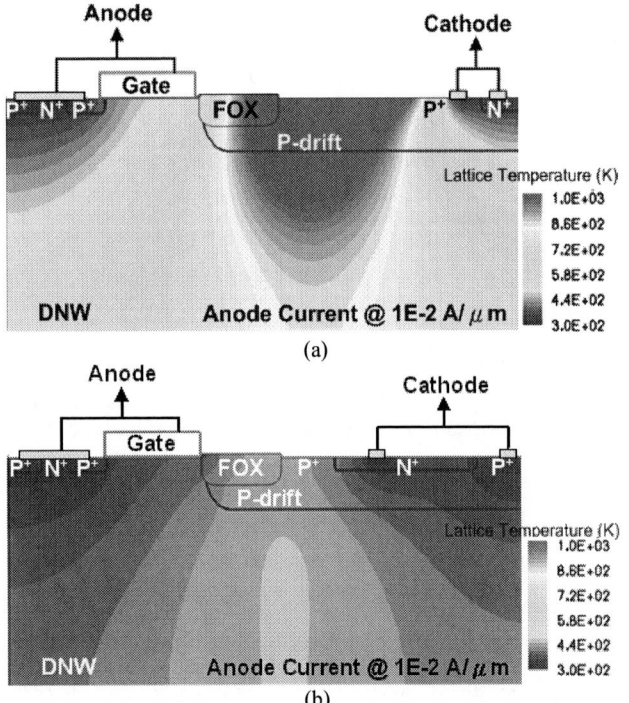

(a)

(b)

Figure 6. The lattice temperature plots of (a) PSCR and (b) MPSCR at anode current of 1E-2A/μm.

The total current density plots for PSCR and MPSCR devices at anode current is 10-mA/μm are shown in Fig. 5. As shown in Fig. 5(a), there is a main current path (PNP path) through the PSCR

device. Obviously, the PNP path causes a surface current to crowd at P-drift/FOX junction. The overcrowding current at P-drift/FOX junction could cause the self-heating effect in the PSCR device [3]. Owing to the current crowding effect, the lattice temperature is up to 1000-K when the anode current is 10-mA/um, as shown in Fig. 6(a). Consequently, the local hot spot makes device to fail easily under ESD stress condition. On the other hand, the MPSCR device has a deep current flow by SCR path (P+/DNW/P-drift/N+), and it avoids the current crowding at FOX/P-drift junction, as shown in Fig. 5(b). Therefore, the lattice temperature is only about 300K as the anode current is 10-mA/um, shown in Fig. 6(b). From the current-temperature relation data of PSCR and MPSCR devices by TCAD simulation, we can know that the MPSCR device has a superior current endurance by altering the site of P+ electrode at cathode side.

CONCLUSION

In this work, the MPSCR structure have been verified in a 0.35-um 40-V CDMOS technology. The MPSCR structure with high ESD robustness has been clearly investigated by TLP instrument and TCAD simulator. By the simulation results, the modified P+ electrode layout of cathode side can enhance the turn-on efficiency of embedded SCR path, and avoid the current crowding effect on the surface of device. The proposed device only need to sweep N+ and P+ regions in drain side, and do not need to increase the additional mask layer. For area reduction, the MPSCR device does not need to increase the layout area and it can sustain up to 7.2kV for HBM and 360V for MM under device width of 300µm. Besides, the proposed MPSCR device has a low trigger voltage (Vt1=54-V) and a high second breakdown current (It2=10-A), which can be extensively applied for ESD protection design of PMIC applications.

REFERENCES

[1] W.-Y. Chen *et al.*, *IEEE EDL*, pp. 159, 2010.

[2] T.-H. Lai *et al.*, *USA Patent 7,368,761, May 6, 2008.*

[3] David Alvarez *et al.*, *in Proc. EOS/ESD Symp.*, pp. 1A.4-1, 2007.

Impact of Shielding Line on CDM ESD Robustness of Core Circuits in a 65-nm CMOS Process

Ming-Dou Ker[1,2], Chun-Yu Lin[1], and Tang-Long Chang[1]

[1] Institute of Electronics, National Chiao-Tung University, Hsinchu, Taiwan
[2] Department of Electronic Engineering, I-Shou University, Kaohsiung, Taiwan

Abstract — **The charged-device-model (CDM) ESD robustness of core circuit with/without the shielding line was studied in a 65-nm CMOS process. Verified in silicon chip, the CDM ESD robustness of core circuit with the shielding line was degraded. The damage mechanism and failure location of the test circuits were investigated in this work.**

Keywords - Charged-device model (CDM), ESD, shielding line.

I. INTRODUCTION

Adding the shielding lines near to the signal lines of high-speed circuits is an efficient method to limit the inductive coupling and to reduce the crosstalk noise between signal lines [1], [2]. The signal lines are generally shielded by the shielding lines which are biased at V_{DD} or V_{SS}. However, the coupling effect between the shielding and signal lines during charged-device-model (CDM) ESD event could induce large transient current to damage the core circuits. Among the chip-level ESD test standards (HBM, MM, and CDM), the CDM ESD events play major roles to cause failures in today's manufacturing and packaging environments [3], [4]. Therefore, several ESD protection designs against CDM ESD events have been reported to protect the input/output (I/O) buffers which are directly connected to the external pins [5]. Besides the I/O buffers, the core circuits would also suffer the dangers when the CDM events happened at the I/O buffers. As shown in Fig. 1, the coupling effect between I/O line and core line during CDM ESD event can induce large transient current to damage the core circuit, even if the shielding line is inserted. In this paper, the impact of shielding line on CDM ESD robustness of core circuit is investigated in a 65-nm CMOS process.

II. TEST CIRCUITS

One set of test circuits is shown in Fig. 2. The I/O buffer realized with inverter is connected to the I/O pad through the I/O line with ESD clamp at the I/O pad. The ESD clamp is realized with the gate-grounded NMOS (GGNMOS) of 360μm/0.12μm (W/L). The core circuits without (with) inserting the shielding line between I/O line and core circuit 1 (core circuit 2) are arranged in parallel to the I/O buffer. The spacing between I/O line and core line is labeled as S, and the length of each line in layout with metal layer is labeled as L. The spacing between I/O line and the shielding line, the width of shielding metal, and the spacing between the shielding line and the core line are labeled as S1, S2, and S3, respectively. The split conditions of the test circuits with different L, S, S1, S2, and S3 are listed in Table I. The layout top view of one test circuit with 100-μm L and 0.78-μm S is shown in Fig. 3. All these circuits have been fabricated in a 65-nm CMOS process with the thin-gate oxide of ~20 Å.

III. MEASUREMENTS RESULTS AND FAILURE ANALYSIS

The test circuits with a die size of ~2x1.5 mm² have been assembled in DIP-40-pin packages. The CDM ESD stresses are applied by the field-induced CDM tester. The failure criterion is 30% shift of the leakage current under 1-V V_{DD} bias from its original level. CDM ESD robustness among these test circuits are listed in Table I. The negative CDM ESD robustness of all core circuits exceeded -600 V. The positive CDM ESD robustness of some core circuits without inserting the shielding line exceeded 600 V. Even if the metal length (L) of core circuits without inserting shielding is reduced, which leads to the lower impedance and higher overshoot voltage, the positive CDM ESD robustness can still achieve 400 V. However, the positive CDM ESD robustness of core circuits with inserting the shielding line are seriously degraded to only 100 V.

As shown in Fig. 4, there are two discharging paths (path1 and path2) from the core circuits to the grounded I/O pad during the CDM ESD stress. The CDM charges located around the p-substrate of core circuits can be conducted through the P+ pickup of core circuits into the V_{SS} line, and then discharged through the ESD clamp device at the I/O pad to external ground, as the dashed line of path1 shown in Fig. 4. The CDM ESD event typically has a fast rise time of only ~0.2 ns. Such very fast transient ESD pulse will be inherently conducted by the displacement current of capacitor, $i = C \times (dv/dt)$, to the external ground through the path with capacitive structures. When the parasitic capacitance becomes larger due to inserting the shielding line between I/O line and core line in chip layout, some of CDM charges will be discharged by the displacement current, as the dashed line of path2 shown in Fig. 4. Since the positive charges located around the PMOS of core circuit can not be efficiently discharged through the P+ pickup and ESD clamp, the positive charges located around the PMOS can only be discharged through the gate oxide of PMOS (path2), which is a capacitive path with low impedance under fast transient. The ESD current discharged through the thin gate oxide of PMOS in the core circuit will cause damage on it.

The scanning-electron-microscope (SEM) photograph of the test circuit with 100-μm L and 0.78-μm S after 200-V CDM ESD test is shown in Fig. 5. The failure point is located only at the poly gate of PMOS transistor in core circuit 2, which fully agrees with aforementioned explanation in Fig. 4.

When the CDM charges stored in the P-substrate is negative, the high reverse breakdown voltage of the parasitic P-substrate/N-well diode will limit the CDM current being discharged through the path2 to external ground. Therefore, the negative CDM ESD robustness was not degraded by inserting the shielding line in chip layout, as compared in Table I.

IV. CONCLUSION

After inserting the shielding line between I/O line and core signal line, the coupling effect during CDM ESD event has been found to induce large transient current to damage the core circuits. This mechanism has been practically verified in silicon chip. The experimental results of this work can help foundries or IC design houses to optimize their layouts for better CDM ESD protection.

978-1-4244-9113-1/11 $26.00 © 2011 IEEE

REFERENCES

[1] Y. Ogasahara, M. Hashimoto, and T. Onoye, "Measurement and analysis of inductive coupling noise in 90 nm global interconnects," *IEEE J. Solid-State Circuits*, vol. 43, no. 3, pp. 718-728, Mar. 2008.

[2] T. Chen, C. Ito, W. Loh, W. Wang, K. Doddapaneni, S. Mitra, and R. Dutton, "Design methodology and protection strategy for ESD-CDM robust digital system design in 90-nm and 130-nm technologies," *IEEE Trans. Electron Devices*, vol. 56, no. 2, pp. 275-283, Feb. 2009.

[3] J. Lee, J. Shih, S. Guo, D. Yang, J. Chen, D. Su, and K. Wu, "The study of sensitive circuit and layout for CDM improvement," in *Proc. IEEE International Physical and Failure Analysis of Integrated Circuits Symp.*, 2009, pp. 228-232.

[4] A. Righter, J. Salcedo, A. Olney, and T. Weyl, "CDM ESD current characterization - package variability effects and comparison to die-level CDM," in *Proc. EOS/ESD Symp.*, 2009, pp. 173-182.

[5] C.-Y. Lin and M.-D. Ker, "CDM ESD protection design with initial-on concept in nanoscale CMOS process," in *Proc. IEEE International Physical and Failure Analysis of Integrated Circuits Symp.*, 2010, pp. 193-196.

ACKNOWLEDGEMENT

This work was supported by National Science Council, Taiwan, under Contract NSC 98-2221-E-009-113-MY2, and by Ministry of Economic Affairs, Taiwan, under Grant 99-EC-17-A-01-S1-104. The authors would also like to thank TSMC University Shuttle Program for providing chip fabrication.

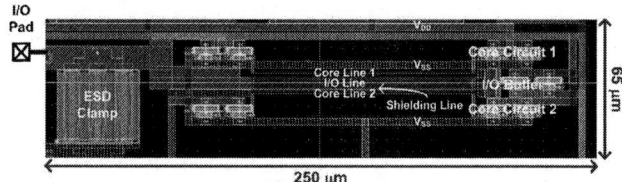

Figure 3. Layout top view of one test circuit for fabrication in a 65-nm CMOS process with 1-V devices.

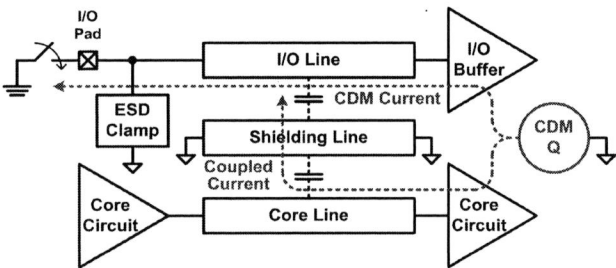

Figure 1. CDM ESD issues at I/O buffer and core circuit.

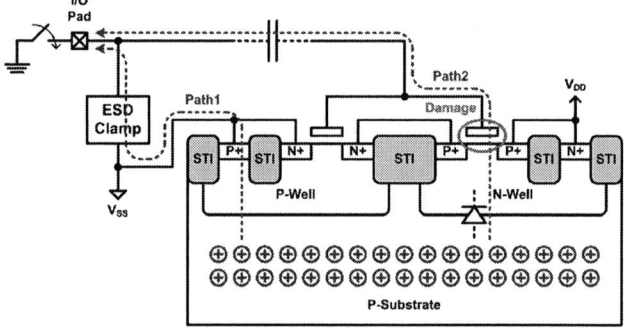

Figure 4. PMOS of core circuit suffers serious CDM damage during CDM ESD event with positive charges.

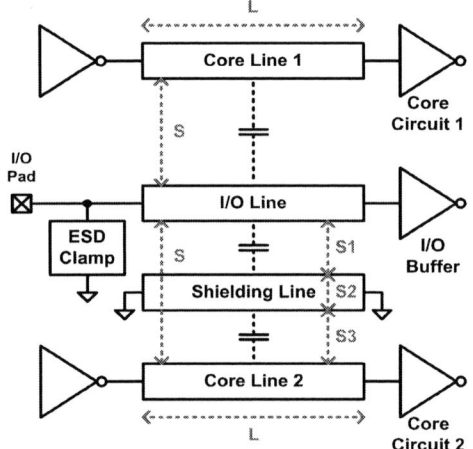

Figure 2. Test circuit without (with) inserting shielding between I/O line and core circuit 1 (core circuit 2).

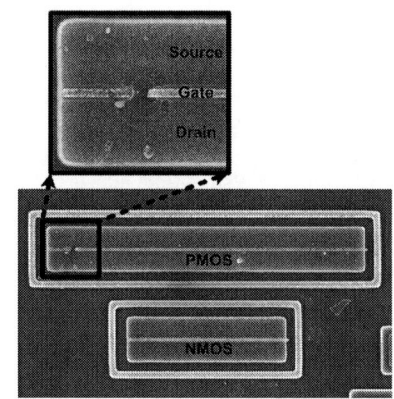

Figure 5. SEM photo of failure point in core circuit 2 after 200-V CDM test.

Table I. CDM ESD robustness of test circuits with/without inserting the shielding line between I/O line and core line.

Core Circuit 1 (Without Shielding)							Core Circuit 2 (With Shielding)						
L (μm)	*S* (μm)	*S1* (μm)	*S2* (μm)	*S3* (μm)	*Positive CDM ESD Robustness (V)*	*Negative CDM ESD Robustness (V)*	*L* (μm)	*S* (μm)	*S1* (μm)	*S2* (μm)	*S3* (μm)	*Positive CDM ESD Robustness (V)*	*Negative CDM ESD Robustness (V)*
20	0.78	N/A	N/A	N/A	400	< -600	20	0.78	0.3	0.18	0.3	**100**	< -600
20	1.98	N/A	N/A	N/A	> 600	< -600	20	1.98	0.3	0.18	1.5	**100**	< -600
50	0.78	N/A	N/A	N/A	400	< -600	50	0.78	0.3	0.18	0.3	**100**	< -600
50	1.98	N/A	N/A	N/A	> 600	< -600	50	1.98	0.3	0.18	1.5	**100**	< -600
100	0.78	N/A	N/A	N/A	> 600	< -600	100	0.78	0.3	0.18	0.3	**100**	< -600
100	1.98	N/A	N/A	N/A	> 600	< -600	100	1.98	0.3	0.18	1.5	**100**	< -600

* The CDM test voltage was increased in 100-V step.

Test Chip Design for Study of CDM Related Failures in SoC Designs

Nicholas Olson, Vrashank Shukla, and Elyse Rosenbaum
Department of Electrical and Computer Engineering
University of Illinois at Urbana-Champaign
Urbana, IL
naolson@illinois.edu

Abstract—**During CDM-ESD testing, SoC (System on a Chip) designs may fail either in the pad ring or in the core circuitry, particularly at the power domain crossings. A specially designed test chip allows one to locate the sites at which ESD-induced damage occurs and also to investigate the efficacy of different CDM protection strategies.**

Keywords- System on a chip; ESD; CDM

I. INTRODUCTION

The traditional approach to CDM protection is to test stand-alone protection devices and then place well performing protection devices at possible failure locations inside the chip. Any failures that still occur are analyzed on a case by case basis. A more general analysis of the relation between chip-level failures and the full chip design is warranted, and test chips are needed to validate the results of such analysis. Test chips for study of CDM failures in the pad ring have been reported [1][2]. However, in SoC (system on a chip) designs, failures are observed to occur at power domain crossing circuits [3][4][5], in addition to the external inputs and outputs. This work reports on the design of a test chip whose purpose is to detect CDM induced failures both in the pad ring and at the domain crossings.

II. METHODOLOGY

Consider a datapath that goes from an external input of the chip to an external output, passing through one or more power domain crossings along the way. A simple example is shown in Figure 1, with the three most likely sites for CDM-induced damage marked with X's. In Figure 1, the I/O domain to core domain crossing circuits are not marked as being susceptible to CDM-induced damage; it is assumed that the level shifters at these interfaces are designed using thick oxide transistors and are thus not likely to fail during CDM testing. Using electrical testing only, it may be difficult to pinpoint which of the three CDM susceptible circuits has failed. Therefore, a CDM test chip must have redundant signal paths added around the expected damage sites to facilitate identifying the failure locations.

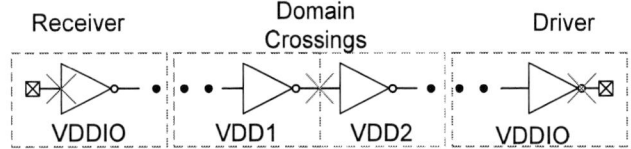

Figure 1. Simple data path. Possible failure locations indicated with red X's. In this design, it is hard to pinpoint the faiture sites using electrical testing only.

The proposed CDM test chip contains an XOR gate on the main datapath, as shown in Figure 2. At any given time, it is intended that only one input will be driven by a data stream, i.e. a time-varying signal, and the other inputs will be tied either low or high. The XOR gate will select the one switching input and will route the signal to all of the cross domain circuits. This architecture provides two benefits. First, the number of external inputs and the number of cross domain circuits have been decoupled. Therefore, the number of cross domain circuits that can be tested for ESD robustness may be much larger than the number of chip inputs. Second, the cross domain circuits may be accessed from any of the external inputs. This provides the ESD engineer with more freedom to investigate I/O protection circuits, even seemingly risky protection circuits.

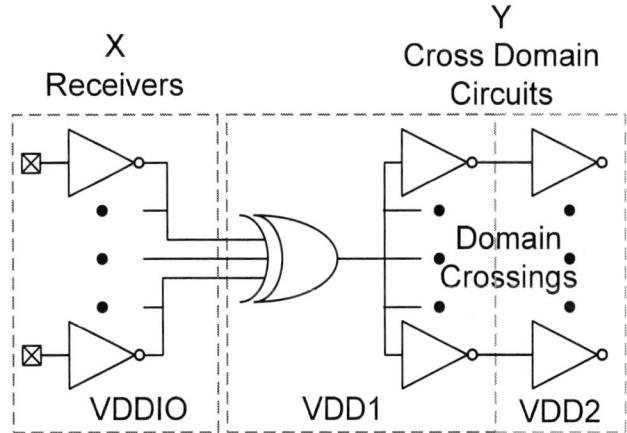

Figure 2. The XOR gate provides redundancy at the input and decouples the values of X and Y where X is the number of receivers and Y is the number of cross domain circuits.

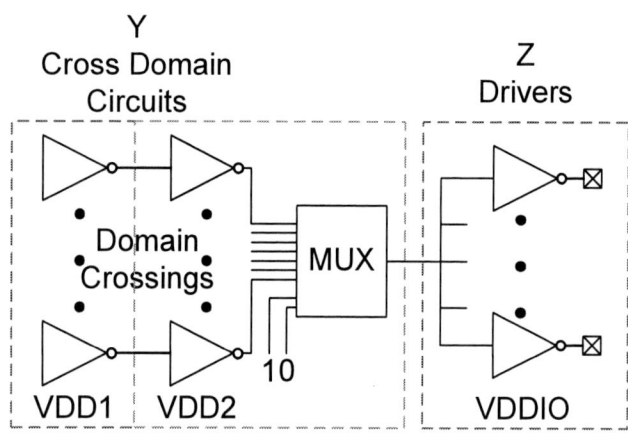

Figure 3. The multiplexer selects one of the cross domain circuits. Y is the number of cross domain circuits and Z is the number of drivers. Y and Z may be different values.

The outputs of the cross domain circuits are connected to a multiplexer, as shown in Figure 3. In turn, the multiplexer output is routed to all of the external outputs. This provides a decoupling between the number of cross domain circuits and the required number of outputs. The combined effect of the multiplexer and the previously described XOR gate is to provide full access to the cross-domain circuits if at least one input and one output circuit are operational. To ensure that the cross domain circuits can be tested, a single I/O pin may be implemented with very robust ESD protection and specified as a no stress pin. As noted later in section IV, the I/O circuits only showed damage or failure if they were directly stressed.

The proposed CDM test chip contains an extra signal path that bypasses the domain crossing circuits (see Figure 4). This signal path never leaves the I/O power domain. If the signal transmission along this path is disturbed after CDM testing, then either the input or output has failed, and this particular input-output circuit pair should not be used to access the domain crossing circuits. Whether the failure has occurred at the input or the output can be surmised from additional testing. The previously described multiplexer has two additional inputs; one is set to logic 1 and the other to logic 0. This allows the outputs themselves to be isolated and tested. Input leakage current measurements provide additional information about the integrity of the input circuit.

This description of a CDM test chip has focused on the main blocks that enable testing. Additional buffers or logic may be inserted between these blocks without affecting the testability. All of the features described here have been included in a test chip whose design and results are described in the next sections.

III. DESIGN

The test chip was fabricated in a 90 nm CMOS process. It contains two 2.5 V I/O domains, two 1.0 V core logic domains, and one 1.0 V analog domain. Inverter chains cross between the two core domains. An overview of the on-chip logic is shown in Figure 4. The data path indicated by the green arrow is used to test all the input and output circuits without the

signals leaving the I/O domain. The I/O test data identifies the working inputs and outputs. As long as at least one (of 7) input circuits and one (of 7) output circuits are functional after CDM stressing, the domain crossing circuits from VDD1 to VDD2 can be accessed through the data path indicated by the blue arrow in Figure 4. The multiplexer was controlled by a counter, reducing the required number of control signals. The control logic was intentionally made uncomplicated to reduce the likelihood that a design error or an unexpected CDM failure would prevent all or part of the chip from being tested. To achieve density representative of a real product design, dummy logic was inserted in the extra silicon area not used by the control logic.

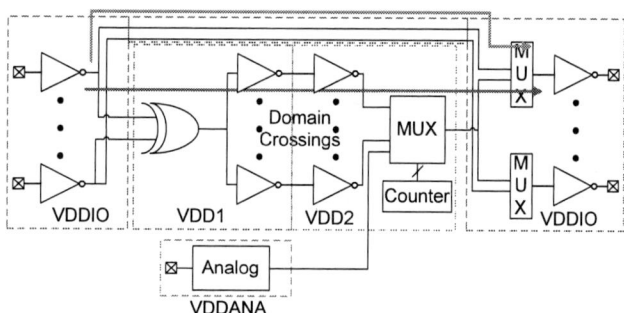

Figure 4. Test chip logic. The data path shown by the green arrow is used to identify working inputs and outputs The data path shown by the blue arrow accesses the domain crossing circuits. The number of logic gates along the datapath is greater than shown here; this simplified logic diagram illustrates the essence of the design.

The domain crossing circuits are varied in regards to the transistor sizes in the transmitter (TX) and the receiver (RX), the length of the interconnect between the TX and RX, and whether or not ESD clamps have been deployed. Some of the domain crossing circuits have GCNMOS [6] at the RX input, some have a small dual-diode circuit at the RX input, or, in some cases, small anti-parallel diodes (APD) are inserted between the local ground busses for the TX and RX (to augment the larger APD near the pad ring). Two identical banks of domain crossing circuits were placed on the test chip, with one bank "A" being accessed from the I/O pads in the VDDIO2 domain and the other "B" being accessible from I/O pads in the VDDIO1 domain. Although the two banks are accessed through different I/O domains, the domain crossing circuits lie between the two core domains and thus the two banks had been expected to respond similarly to CDM events. The full chip signal flow is indicated in Figure 5.

The I/O pads are protected by a variety of different ESD devices, including poly-bound dual diodes, STI-bound dual diodes, DTSCR, GGNMOS, and GCSCR. This will allow us to compare the efficacy of several protection devices within the context of a single chip.

A diagram of the power and ground bus routing in the chip periphery is shown in Figure 6. Collaboration with the foundry was undertaken to ensure that the I/O cell designs and pad ring are representative of commercial products. The test chips were packaged into a 100 pin QFP. The QFP was 20 mm by 16 mm and was 2.85 mm thick.

978-1-4244-9113-1/11 $26.00 © 2011 IEEE

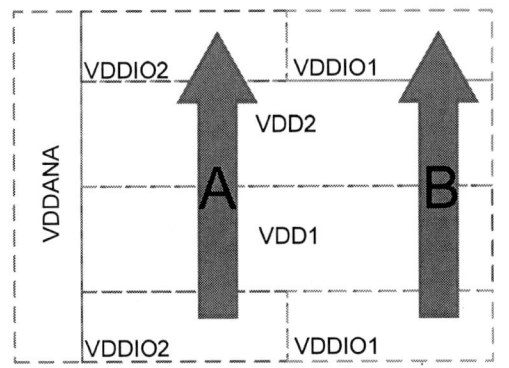

Figure 5. Signal flow of the test chip. Banks A and B contain identical sets of domain crossing circuits.

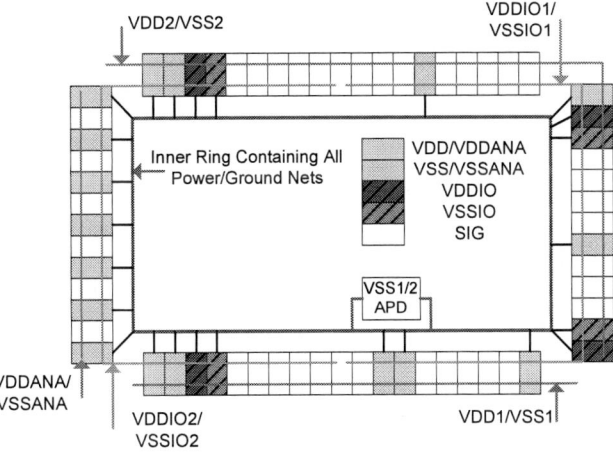

Figure 6. Test chip pad ring and power/ground busses. The busses in the pad ring have lower resistance than those in the inner ring. Rectangles represent I/O blocks with ESD protection built into the cell. The only ESD element not built into the pad cells is the APD between VSS1 and VSS2. This is depicted separately by the block labeled "VSS1/2 APD". This APD was placed outside the pad ring since VSS1 and VSS2 outer busses do not overlap in the pad ring.

TABLE I. CHIPS ARE DIVIDED INTO SEVEN GROUPS. FOR EACH GROUP, ONLY THE PINS NAMED BELOW ARE STRESSED. THREE CHIPS IN EACH GROUP WERE STRESSED AT EACH PRECHARGE LEVEL.

Group Number	Pins Stressed
1	VDD1
2	VSS1
3	VDD2
4	VSS2
5	External I/O
6	VDDIO1, VDDIO2, VDDANA
7	VSSIO1, VSSIO2, VSSANA

I. RESULTS

The electrical tests performed after CDM stressing will detect three types of failures: increased input leakage, stuck at one or zero faults, and reduced maximum frequency. All chips were stressed with a FICDM tester. The chips were divided

into seven groups. Chips within one group had only a subset of their pins subjected to CDM stress, as indicated in Table I. Three different CDM precharge voltages were used: 250V, 500V, and 1000V. Three chips in each group were stressed at each of the precharge voltages.

A. Cross-Domain Circuits

Table II describes the 12 different cross domain circuits found in each bank. The first is the control case, upon which all the other cross-domain circuits are variations. Domain crossing circuit number 1, i.e. the control case, has no local ESD protection; the PMOS in its driver has channel width W_{Ptx} and the receiver PMOS has channel width W_{Prx}. These PMOS devices are sized differently in cross-domain circuits 2 through 5. The small PMOS have channel widths of $0.4W_P$ and the large PMOS have channel widths of $4W_P$. One extra level of inversion was added to the datapath leading to the driver of domain crossing circuit number 6. Domain crossing circuits 7 and 8 have local clamps at the receiver inputs; these protection devices are sized much smaller than those at the external I/O, having a total width of just 8 µm. GCNMOS was used instead of GGNMOS as a local clamp to ensure the trigger voltage would be low enough to protect the core transistors. In this design, the GCNMOS trigger voltage was still much larger than the supply voltage for the core domain, so GCNMOS could be safely deployed at switching nodes of the circuit. A relatively small APD is placed between the VSS1 and VSS2 busses in close proximity to domain crossing circuit 9. The interconnect linking the driver and receiver was lengthened for domain crossing circuit 10. A serpentine shaped interconnect was used to keep the relative locations of the driver and receiver invariant. However, in circuits 11 and 12, the relative positions of the driver and receiver are changed and the signal line length changes accordingly.

TABLE II. LIST OF DOMAIN CROSSING CIRCUITS.

Number	Description
1	Control Case
2	Small PMOS Receiver
3	Large PMOS Receiver
4	Small PMOS Driver
5	Large PMOS Driver
6	Extra Inverter in Logic Path
7	GCNMOS/GCPMOS (Operating as dual diode)
8	GCNMOS
9	Local APD
10	Long Interconnect (Driver & Receiver unmoved)
11	Short Interconnect
12	Long Interconnect

No failures were seen at the cross-domain circuits after CDM stressing at 250 and 500 V. This is due to the small size of the test chip (die, package, and number of pins); a very large precharge voltage is needed to achieve a high peak current during the CDM stress. Based on simulations, this package will produce a peak current of approximately 8.2 A for a 500 V stress and 16.4 A for a 1000 V stress. Cross-domain failures were observed after CDM stressing at 1000 V and thus all measurement data presented in this section is for a 1000 V precharge level.

More CDM induced failures occurred in the B bank of domain crossing circuits than in bank A. In fact, bank A showed no CDM induced failures. The failure data for bank B are summarized in Figure 7. It is seen that the VSS2 stress caused the most damage to the cross domain circuits. It can also be seen that the three circuits that had some form of local ESD protection never failed.

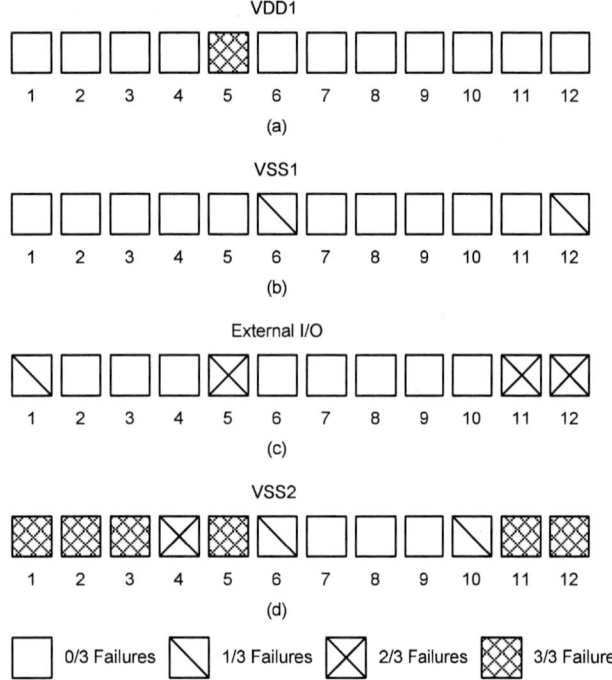

Figure 7. CDM failures in the B bank of cross domain circuits. 1000V precharge voltage. Domain crossing circuits are labeled 1-12 following the naming conventions of Table II. No cross domain failures were caused by stressing any of the VDD2, VDDIO, or VSSIO pins.

The data show that the cross-domain circuits with a large PMOS in the driver were especially vulnerable to CDM stressing. The large driver tends to increase the voltage stress on the receiver. The data also show that having an additional inverter on the datapath did not noticeably change the failure results. This suggests that the propagation delay through the inverter chain exceeds the duration of the CDM event, resulting in logic levels at the driver and receiver that are unaffected by the extra inverter. A similar observation was reported in [7]. If any logic design modifications were made to fix the state of the

driver, these would need to be done as close to the driver as possible.

To aid interpretation of the failure data, CDM simulations were performed using the methodology of [8]. Simulations whose results are shown here were performed using a precharge voltage of 500 V since, based on the experimental data, that is the highest level at which we can be certain that the circuit does not undergo failure and thus the circuit operation can be accurately simulated. Simulations suggest that cross-domain failures occur for a CDM precharge voltage of about 850 V.

Circuit simulations indicate that VSS2 pin zaps should induce the worst case stress at the domain crossing circuits. This prediction is borne out by the measurement data, which show more failures for VSS2 pin zaps than for any other stress condition. Simulations also show that, for the case of a VSS2 pin zap, the stress voltage is higher in bank B than in bank A. This can explain why the incidence of failure was higher in bank B. Figure 8 shows the simulated voltage across the gate oxide of the receiver NMOS in domain crossing circuit number 5, i.e., the circuit with the large PMOS in the driver. The peak voltage is higher for the circuit in bank B. The schematic in Figure 9 highlights the important CDM current paths during a VSS2 pin zap; it is drawn for the case that the VSS2 pin in the upper right corner of Figure 6 is being zapped (i.e. grounded). For domain crossing circuits in bank A, the receiver ground connects to the VSS2 bus at node E in the inner bus ring, and V_{GS} of the receiver NMOS is given by

$$V_{GS}^{A} \approx V_A - V_E = (V_A - V_B) + I1 * R_{VSS1b} + I2 * R_{VSS1a} + V_{APD} + I4 * R_{VSS2a} - (I5 * R_{VSS2b}). \quad (1)$$

In bank B, the receiver grounds connect to the VSS2 bus at node D in the inner bus ring, and the receiver NMOS V_{GS} is given by

$$V_{GS}^{B} \approx V_A - V_D = (V_A - V_B) + I1 * R_{VSS1b} + I2 * R_{VSS1a} + V_{APD} + I4 * R_{VSS2a}. \quad (2)$$

Eqs. (1) and (2) indicate that the receivers in bank B experience stress that is larger by the amount I5*R_{VSS2b}. The value of R_{VSS2b} is layout and floorplan dependent; it depends on the power/ground bus routing from the core circuitry to the inner bus ring and the distance from that connection point to the pad being zapped. In this particular case, R_{VSS2b} is approximately 0.5 Ω. The stress for both banks of domain crossing circuits can be lowered by changing the floorplan to reduce R_{VSS1a} and R_{VSS2a}; this is achieved by extending the core ground busses within the outer ring so that the "VSS1/2 APD" can be placed there, reducing the resistance between points C and D. Furthermore, the VDD1 power clamp should be placed such that the ground bus resistance between points B and C is minimized. This is achieved by identifying the VSS1 and VDD1 pads nearest to where the power and ground busses of the domain crossing circuits tap into the corresponding busses in the I/O ring, moving those two pads closer to each other in the pad ring, and moving a VSS2 pad closer to this pair of VDD1 and VSS1 pads. These beneficial changes for the VSS2 stress should not compromise ESD reliability for non-VSS2 pin zaps.

Three domain crossing circuits did not show any failures, even in bank B. Two of these have ESD protection at the receiver input and the third has local APD. Circuit simulations, Figure 10, predicted that the local APD would be effective.

Although VSS2 pin zaps cause the most failures in the core logic, some failures were induced by VDD1, VSS1, and I/O pin zaps. More failures always occur in bank B than in bank A.

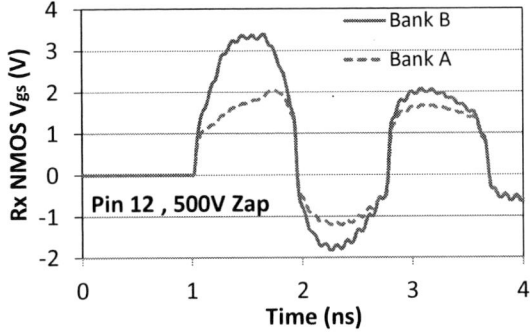

Figure 8. Cross-domain circuit #5; simulated V_{GS} at the receiver NMOS for a 500V VSS2 pin zap. The bank B domain crossing circuit sees a higher voltage stress than the one in bank A.

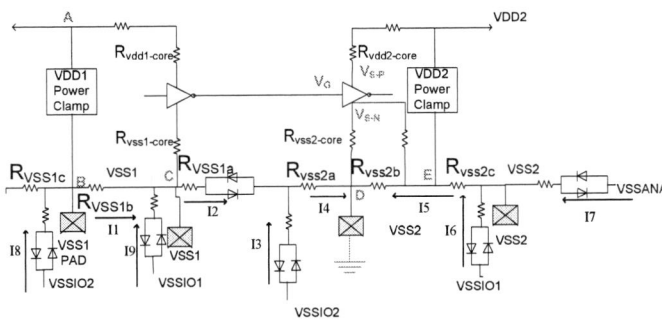

Figure 9. Schematic representation of a domain crossing circuit and important bus elements. Current flow arrows correspond to a VSS2 pin zap.

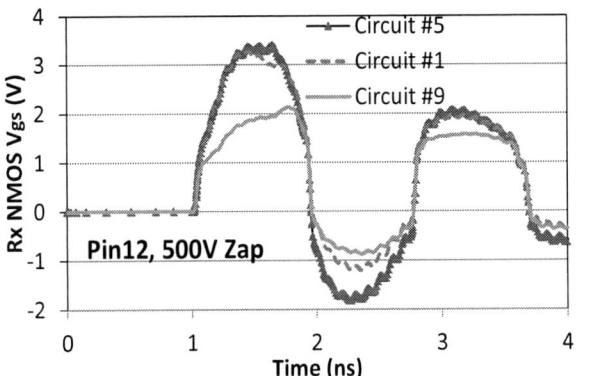

Figure 10. Simulated V_{GS} at the receiver NMOS for three different cross domain circuits for a 500V VSS2 pin zap. All circuits are in bank B. The local APD (in circuit 9) substantially reduce the voltage stress.

An alternate version of the test chip was also fabricated and evaluated. This will be identified as the alternate test chip. The alternate test chip primarily differs from the one presented previously in three ways: the types of rail clamps used in the I/O domains were changed, the amount of decoupling capacitance in VDDIO2 was reduced, and several circuits in VDD1 and VDD2 were removed. Specifically, some dummy logic and some domain crossing circuits were removed from the core. The twelve domain crossing circuits of Table II are present on the alternate test chip; those that were removed are part of a separate study. None of the design changes were expected to change the failure incidence for domain crossing circuits 1-12; however, they did. The results are shown in Figure 11 and indicate that the alternate test chip was more susceptible to CDM induced failures at the domain crossing circuits.

This result is attributed to the reduced density of the core circuitry. Approximately 35% of the transistors in the core were removed. This reduces the parasitic capacitance between VDD1 and VSS1 and also reduces the leakage current between those two busses; both effects are expected to increase the voltage drop between VDD1 and VSS1 during a CDM event. Indeed, circuit simulations confirm that in the alternate test chip, the voltage drop across the VDD1 power clamp was slightly higher. Simulations also confirmed removing some of the cross domain circuits leads to a higher current flowing through the APD. All of these factors lead to a higher voltage stress on the receiver gate oxides.

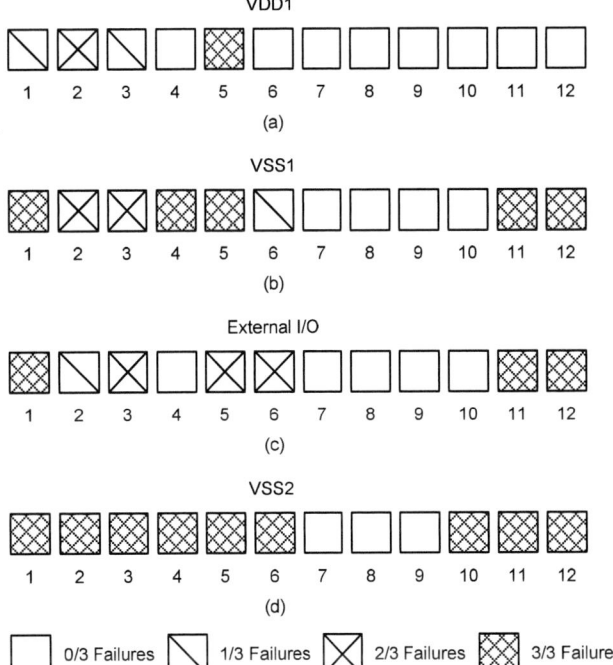

Figure 11. Alternate test chip. CDM failures in the B bank of cross domain circuits. 1000V precharge voltage. Domain crossing circuits are labeled 1-12 following the naming conventions of Table II. There is an increased rate of failure relative to the first test chip (see Figure 7).

978-1-4244-9113-1/11 $26.00 © 2011 IEEE

B. Digital Inputs and Outputs

None of the external inputs or outputs in the VDDIO1 or VDDIO2 domains showed hard failure based on logic tests. This finding is attributed to the I/Os having been well protected. Although no hard failures were observed, there was, on average, a slight increase in the input leakage currents for the I/Os in the VDDIO2 domain, following 1000 V CDM stress. The leakage current increased only for CDM zaps in which the external I/Os were stressed directly. The inputs in the VDDIO2 domain are protected by snapback devices. Diode based protection was used in the VDDIO1 and those inputs did not show increased leakage.

C. Analog I/O

The analog input circuits contain thin oxide transistors and are protected with either a dual diode circuit or a DTSCR. A 60μm diode is placed in parallel with the DTSCR for reverse conduction. There is no secondary CDM protection on any of the analog inputs. Failure is detected by an increase of the input leakage; the results are summarized in Table III. The failure voltage is highly sensitive to the protection device sizing, with the dual-diodes tending to outperform the DTSCR. This is attributed to the larger voltage overshoot inherent to the SCR structure [9].

TABLE III.   ANALOG INPUT FAILURE LEVELS

Protection Device	Failure Level
Dual Diode (60μm)	250V
DTSCR (25μm)	250V
Dual Diode (120μm)	1000V
DTSCR (50μm)	500V

II. CONCLUSIONS

A CDM test chip design methodology has been presented; its objective is to allow for identification of all circuits failing due to CDM ESD. The test chip measurement results support the use of ESD protection at the cross domain interfaces, and demonstrate the efficacy of local APD, a protection circuit located off the signal path. The results further show that certain cross domain interfaces can survive CDM without having ESD protection devices; however, executing a chip design without cross domain protection is risky and should be avoided unless detailed CDM simulations and analysis are performed.

ACKNOWLEDGMENT

The test chip was fabricated by UMC. Rex Lee and Kelvin Hsueh contributed to the design; Howard Tang, Aven Wu, and Isaac Wang facilitated fabrication and ESD testing. All are thanked for their assistance and many useful discussions.

REFERENCES

[1] T. Brodbeck, K. Esmark, and W. Stadler, "CDM Tests on Interface Test Chips for the Verification of ESD Protection Concepts," in *Electrical Overstress/Electrostatic Discharge Symposium*, 2007, pp. 1-8.

[2] C. Chu, A. Gallerano, J. Watt, T. Hoang, T. Tran, D. Chan, W. Wong, J. Barth, and M. Johnson, "Using VFTLP Data to Design for CDM Robustness," in *Electrical Overstress/Electrostatic Discharge Symposium*, 2009, pp. 1-6.

[3] K. Watanabe, T. Hiraoka, K. Sato, T. Sei, and K. Numata, "New protection techniques and test chip design for achieving high CDM robustness," in *Electrical Overstress/Electrostatic Discharge Symposium*, 2008, pp. 332-338.

[4] Y. Huh, P. Bendix, K. Min, J. Chen, R. Narayan, L. Johnson, and S. Voldman, "ESD-induced Internal Core Device Failure: New Failure Modes in System-on-Chip (SOC) Designs," in *International Database Enginerring & Application Symposium*, 2005, pp. 47 – 53.

[5] E. R. Worley, "Distributed Gate ESD Network Architecture for Inter-Power Domain Signals," in *ElectricalOverstress/Electrostatic Discharge Symposium*, 2004, pp. 238-247.

[6] C. Duvvury and C. Diaz, "Dynamic Gate Coupling of NMOS for Efficient Output ESD Protection," in *IEEE International Reliability Physics Symposium*, 1992, pp. 141-150.

[7] X. Fan, and M. Chaine, "ESD Protection Circuit Schemes for DDR3 DQ Drivers," in *Electrical Overstress/Electrostatic Discharge Symposium*, 20010, pp. 375-380.

[8] V. Shukla, N. Jack, and E. Rosenbaum, "Predictive simulation of CDM events to study effects of package, substrate resistivity and placement of ESD protection circuits on reliability of integrated circuits," in *IEEE International Reliability Physics Symposium*, 2010, pp. 485-493.

[9] R. Gauthier, M. Abou-Khalil, K. Chatty, S. Mitra, and J. Li, "Investigation of Voltage Overshoots in Diode Triggered Silicon Controlled Rectifiers (DTSCRs) Under Very Fast Transmission Line Pulsing (VFTLP)," in *IEEE International Reliability Physics Symposium*, 2009, pp. 334-343.

Nanosecond Transient Thermoreflectance Imaging of Snapback in Semiconductor Controlled Rectifiers

Kerry Maize*, Dustin Kendig, and Ali Shakouri
Baskin School of Engineering
University of California, Santa Cruz
Santa Cruz, CA, USA
*kerry@soe.ucsc.edu

Vladislav Vashchenko
National Semiconductor Corp.
Santa Clara, CA, USA

Abstract—Transient thermoreflectance imaging method has been applied for the first time to reveal current distribution in ESD protection devices through the surface temperature change due to self heating. Experimentally calibrated temperature images are obtained of a multiple finger, 80 square micron 100V NLDMOS-SCR device in snapback operation regimes for different current levels (1.15-1.47A) and at different times ranging between 100 nanoseconds to one millisecond after the ESD pulse. The novel applied methodology demonstrates a practical and straightforward way to characterize non-uniform temperature and current distribution in ESD structures, revealing effects of non-simultaneous triggering of individual fingers on the multiple finger SCR device.

Keywords-electrostatic discharge; SCR; thermal imaging; thermoreflectance; snapback.

I. INTRODUCTION

This On-chip local ESD protection is typically implemented using ESD clamps engineered to provide ESD current level according to package or system level standards. Typical clamp design includes distributed multifinger device that provides operation in conductivity modulation mode [1]. Many designs attempt to increase current uniformity, robustness, and area efficiency through multiple finger layouts. It is helpful to characterize such multiple finger designs during operation to improve efficiency or identify undesired behavior. For example, one possible problem is that individual fingers can turn on (trigger into snapback) prematurely, resulting in current crowding and thermal failure in a single finger or other local hot spot [2].

Optimization of ESD protection designs can be assisted with visual characterization tools. Over recent years the method of backside laser transient interferometry (TIM) [3] has been used to inspect current distribution problems in ESD devices during snapback [4,5]. The purpose of this study is to solve the same problem using the simpler alternative method of topside thermoreflectance thermography, which exploits the change in material reflectivity with temperature [6,7]. Thermoreflectance methods that use CCD cameras [8] capture megapixel thermal

images much faster than earlier scanned laser thermoreflectance methods and at lower external illumination levels. Temperature resolution of 10mK has been demonstrated [9]. Recent pulsed illumination techniques have improved CCD thermoreflectance temporal resolution to 100 nanoseconds [10] and even sub-nanosecond, allowing for the first time thermoreflectance CCD imaging of fast transient effects such as snapback in ESD devices. Both thermoreflectance and backside interferometry can measure heating through the sample's substrate using near infrared illumination. However, thermoreflectance can also obtain thermal images directly from the topside of the device using visible wavelength illumination. Topside thermoreflectance with visible light offers better diffraction limited spatial resolution (250nm) than backside infrared methods (1.5-3μm.) The thermoreflectance method has a comparatively simple experiment setup, requiring only a standard optical microscope, LED illumination, CCD camera, timing instruments, and sample biasing instruments. No interferometer is required and the sample does not require special preparation such as backside polishing, thinning, or application of infrared antireflection coating. The main drawback is that, since the thermoreflectance coefficient is small, one cannot do single shot measurements, and averaging (from seconds to minutes) is required. One should note that typically the thermoreflectance coefficient is calibrated for small temperature variations. When temperature rise is several hundreds of degrees, the nonlinear dependence of the reflection coefficient as a function of ambient temperature should be taken into account. Thus more accurate calibrations are necessary in order to quantify the exact temperature rise. However the measured transient reflectance map should provide some information about the current non-uniformity in the device.

This paper presents transient thermoreflectance CCD images of the surface temperature on a multiple finger n-type lateral diffused MOS based silicon controlled rectifier (NLDMOS-SCR) implemented for 100V node protection using 0.5μm biCMOS-DMOS process technology. Thermal images have been captured to analyze non-destructive current localization over the first 300 ns in response ESD like pulses.

This work was supported in part by a grant from UC Discovery and National Semiconductor.

(a)

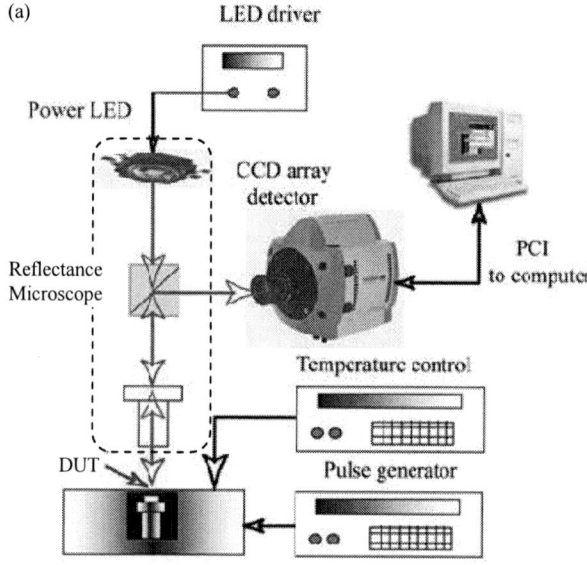

(b)

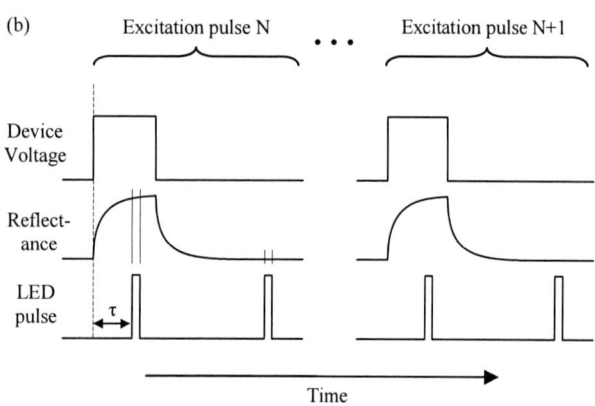

Figure 1. (a) Thermoreflectance CCD imaging and device pulsing configuration. (b) Simplified timing diagram showing relationship between device voltage pulse, optical thremoreflectance, and LED exposure pulse.

Submicron transient thermal images show isolated single finger triggering in the SCR with hot spots that move to different locations on the device in 100's of nanosecond time scale. In addition, dominant heating is shown to alternate between active fingers at different snapback current levels.

II. EXPERIMENT

A. Transient Thermoreflectance CCD Imaging

Fig. 1a shows a diagram of the thermoreflectance CCD imaging setup used to acquire transient thermal images of the MOS-SCR under ESD like electrical pulses. The topside of the SCR chip is illuminated under a reflectance microscope using a narrow-band LED light source. The light reflected from the SCR surface is recorded on a variable frame rate, 12-bit scientific grade Dalsa CCD camera with 1024x1024 pixel resolution. Camera operation and device excitation timing is

controlled by a unified labview program and custom designed hardware trigger board. Repeated square wave voltage pulses are applied to the SCR at 1% duty cycle. For this study, the width of the excitation pulse was varied between 300 nanoseconds to one millisecond to inspect SCR heating under different duration pulses. For each excitation cycle, the pulse causes a temperature rise in the SCR, which in turn induces a change in the optical reflectance (R) at the SCR surface. Images of the SCR surface reflectance during the excited (on) and unexcited (off) states are recorded in the CCD camera. The change in reflectance amplitude ($\Delta R/R$) produces a two-dimension thermoreflectance map across the SCR surface. Thermoreflectance image maps are then converted to temperature maps by applying a thermoreflectance coefficient (C_{TH}), which describes a material's change in optical reflectivity in response to a temperature change. The thermoreflectance change for most materials is very small, on the order of 10^{-4}. However, by averaging over many device excitation cycles signal to noise ratios can be sufficient to yield temperature resolution down to the aforementioned 10mK. Typical averaging time for the images in this study varied between seconds to several minutes.

Calibration of the SCR thermoreflectance coefficient for the silicon substrate and aluminum interconnects was performed in a separate experiment step. The calibration procedure involved heating the entire chip uniformly using an external thermoelectric stage. The thermoreflectance change across the full sample is recorded by the CCD while the temperature is measured simultaneously using a type E microthermocouple. The calibration image and thermocouple measurements are correlated to produce C_{TH} values for each material region on the chip. Material thermoreflectance depends on the wavelength of external illumination [9]. For this reason a blue 470nm LED was used with the SCR sample due to its relatively high C_{TH} for both aluminum and silicon.

Thermoreflectance imaging can be performed using time based or frequency based analysis. Time based methods are easily adapted to measure transient thermal effects. In this approach the "on" and "off" images are subtracted directly. Furthermore, by pulsing the external LED synchronously with device excitation, individual time windows in the device thermal transient can be extracted [11]. The temporal resolution is determined by the minimum width of the LED pulse, which for this experiment was 50 nanoseconds. Fig. 1b shows the simplified timing diagram for time domain transient thermoreflectance CCD imaging. The relationship is shown between device excitation, thermoreflectance change, and LED exposure pulse. Good thermoreflectance signal to noise is achieved by averaging over multiple excitation cycles. Different time windows in the thermal transient are obtained by adjusting the delay (τ) of the LED pulse relative to the rising edge of the device excitation pulse.

B. Device Pulsing Method and I-V Characteristics

An optical image of the NLDMOS-SCR ESD protection cell is shown in Fig. 3a. The 80 micron square layout consists of three anode fingers and two cathode fingers. The electrodes are aluminum. The region surrounding the featured SCR is silicon. Electrical probing and thermal imaging were performed

978-1-4244-9113-1/11 $26.00 © 2011 IEEE

(a)

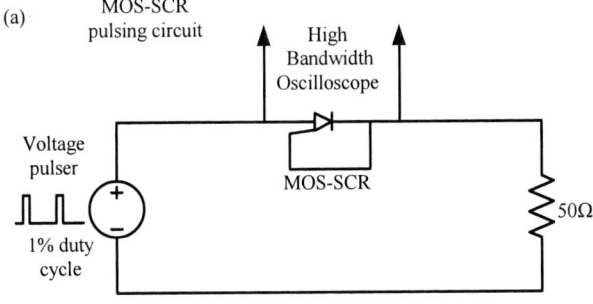

MOS-SCR
pulsing circuit

(b) MOS-SCR electrical transient for 300 nanosecond pulse

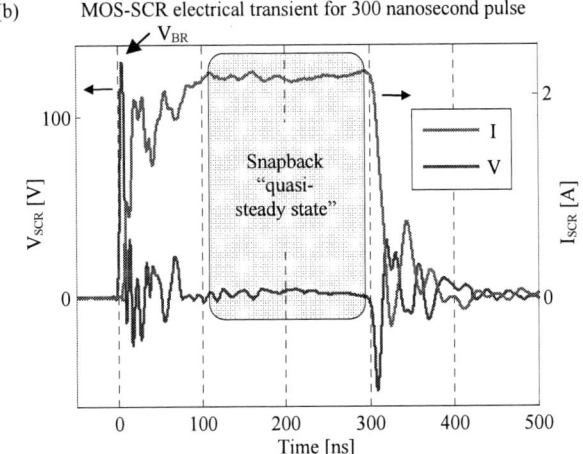

(c) MOS-SCR pulsed I-V and DC gate leakage

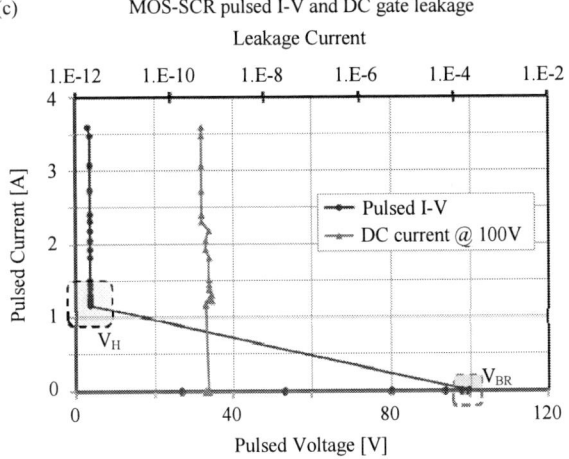

Figure 2. (a) MOS-SCR pulsing circuit. (b) Electrical waveform showing SCR voltage and current for a 300 nanosecond ESD pulse. Thermal images and snapback holding voltage and current were acquired during the "quasi-steady state" region of the waveform. (c) SCR pulsed IV both below breakdown voltage and at snapback holding voltage. DC gate leakage current was measured after each pulsed measurement.

in-situ on a wafer sample. Two high speed, high current probes were used to force the ESD-like pulse at anode and cathode

contact pads near the device (not shown). The other two probes simultaneously measured the voltage waveform across the SCR, which was recorded on a high bandwidth 3.5GHz LeCroy oscilloscope. Fig. 2a shows a diagram of the pulsing circuit with voltage measurement probes indicated. Device excitation was supplied by a commercial Berkeley Nucleonics voltage pulser rated to 300V and five amperes into a 50 ohm load. Pulse rise times during excitation were observed to be less than 5 nanoseconds. Pulse current was measured from the transient voltage across the 50 ohm terminator. These specifications enabled loading of the SCR with pulse waveforms comparable to those produced by transmission line pulsers in the context of human body model and machine model ESD testing [12], or pulses on the order of 3A for 100 nanoseconds with rise time less than 5 nanoseconds.

Fig. 2b shows a typical SCR electrical waveform for a 300 nanosecond long pulse using the described test setup. When supplied with a voltage pulse of sufficient fast rise time and amplitude in excess of the breakdown voltage (V_{BR},) drain-gate capacitive coupling triggers the SCR into snapback operating mode. The structure enters breakdown (V_{BR}=106V) approximately 10 nanoseconds into the ESD pulse, undergoes an unstable period (t=10-100 nanoseconds) of changing device internal resistance, then stabilizes at a holding voltage (V_H) and current after 100 nanoseconds. All thermal images in this study showing SCR self heating in snapback mode were acquired during this "quasi steady-state" part of the electrical waveform after 100 nanoseconds.

The pulsed I-V curve for the SCR is shown in fig. 2c. Values are included to show both the open circuit behavior of the SCR for pulses below the breakdown voltage and also the current sourced while in snapback at the holding voltage. The snapback currents were measured from the quasi-steady state portion of the pulse waveform. Also shown in fig. 2c is gate leakage current, which was checked using 100V direct current after every pulse to ensure the SCR was not damaged. For thermoreflectance imaging, SCR snapback current levels were chosen between 1.15 and 1.47 amperes and pulse widths were chosen between 100 nanoseconds to one millisecond. The influence of photogenerated carriers on SCR device triggering is assumed negligible due to the low intensity of the external LED illumination. (Unlike scanning imaging methods, which focus the illumination to a high intensity spot, camera based thermoreflectance methods distribute the illumination over the full device.) This assumption is supported by the observation during experiment that neither the SCR breakdown voltage nor the electrical waveform showed any dependence on whether the external LED was enabled or disabled.

An optical image of the ESD NLDMOS-SCR is shown in fig. 3a. The 80 micron square layout has three aluminum anode fingers and two cathode fingers. In pulsed mode the snapback device is triggered by drain-gate capacitive coupling in response to ESD pulses with sufficient amplitude and rise time. The NLDMOS-SCR structure enters breakdown (V_{BR}=106V) approximately 10 nanoseconds into the ESD pulse, undergoes an unstable period (t=10-100 nanoseconds) of changing device internal resistance, then stabilizes at holding voltage and current after 100 nanoseconds. All thermal images in this study showing SCR self heating in snapback were acquired during

978-1-4244-9113-1/11 $26.00 © 2011 IEEE   727

this "quasi steady-state" part of the electrical waveform after 100 nanoseconds.

Thermal images have been captured to analyze non-destructive current localization over the first 300 ns in response ESD like pulses. Submicron transient thermal images show isolated single finger triggering in the SCR with hot spots that move to different locations on the device in 100's of nanosecond time scale. In addition, dominant heating is shown to alternate between active fingers at different snapback current levels.

III. RESULTS

A. SCR snapback heating at different current levels

Temperature rise distribution in the SCR was studied for increasing current levels while in snapback mode. In snapback, the SCR maintains a constant reference or "holding" voltage that is effectively independent of the current through the diode. Fig. 3b shows transient thermoreflectance CCD images of the MOS-SCR for snapback current levels I=1.22 and 1.33 amperes. The images were acquired at time = 300 nanoseconds after the rising edge of the excitation pulse. Snapback holding voltage is constant (V_H=3.4V.) The thermal images at 300 nanoseconds revealed dominant heating confined to a 20 micron region of the cathode finger near the cathode pad metal. Fig. 3c shows thermoreflectance temperature change profiles perpendicular to the SCR fingers (profile a-a') for snapback current levels I=1.15 through 1.47A. Between I=1.15A (onset of snapback) and I=1.28A, hot spots occur only in the left cathode finger and the maximum temperature change is 17K. At I=1.33A, dominant heating switches to the right cathode finger, and temperature increases by two degrees. At the next current level, I=1.41A, no hot spot is seen and measured self heating decreases almost to the level of the background thermoreflectance noise level. One possible explanation for the apparent "turn-off" of heating at I=1.41A is that the current filamentation threshold has been exceeded. At lower current, power and heating is concentrated in a single cathode finger. If higher currents produce a more uniform power distribution, for example across both cathode fingers, then localized self heating is reduced.

B. Time dependence of SCR snapback self heating

Thermoreflectance images of heating in the SCR during snapback were obtained at several different times in the thermal transient ranging between 300 nanoseconds and one millisecond. A clear time dependence was observed in both magnitude and spatial distribution of the temperature change across the SCR. Bias conditions were constant for all images in the thermal transient, with snapback current = 1.22A, and holding voltage = 3.4V. Three characteristic time regimes became apparent in the SCR self heating transient. During the first 300-1000 nanoseconds (for a 300 nanosecond square pulse) temperature change was less than 20K and heating was confined to a single cathode finger (fig. 3b.) Fig. 4 shows the thermoreflectance images and temperature change profiles at 30 and 170 microseconds. In this time regime maximum self heating increases to ΔT=400K and dominant heating appears to switch from left cathode finger to right cathode finger. This

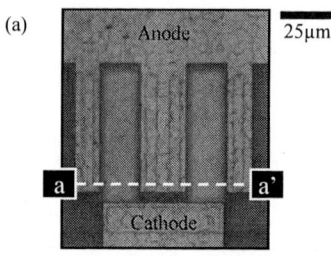

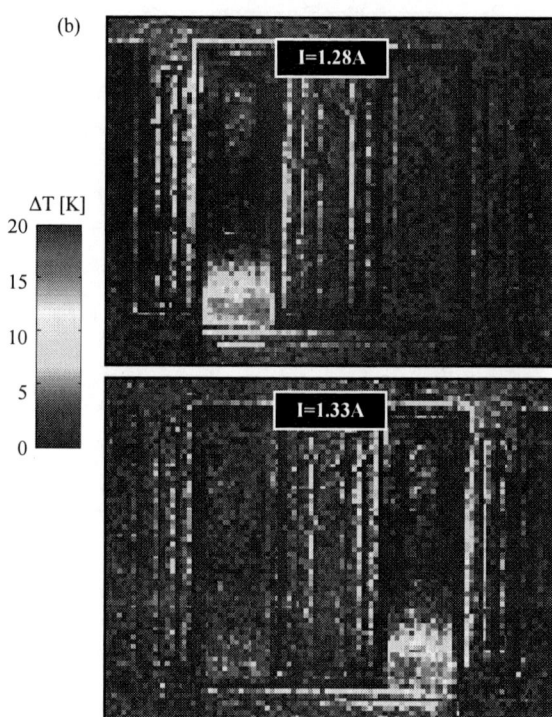

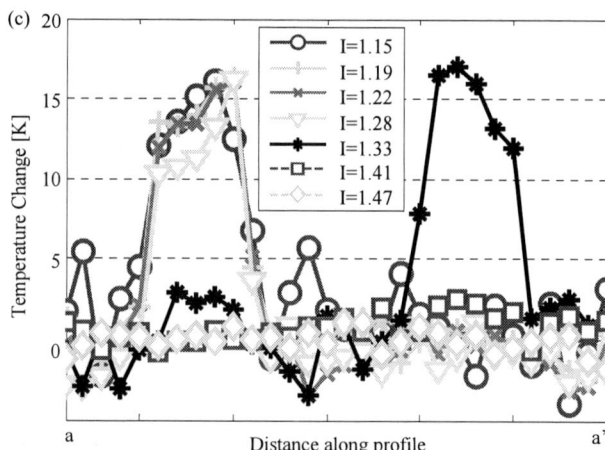

Figure 3. (a) Optical image of MOS-SCR ESD protection device. (b) Thermoreflectance CCD images of SCR for snapback currents 1.28 and 1.33A. Time = 300 nanoseconds. (c) Thermoreflectance surface temperature profiles for the indicated line a-a' across the three anode fingers and two cathode fingers for snapback currents 1.15-1.47A.

978-1-4244-9113-1/11 $26.00 © 2011 IEEE

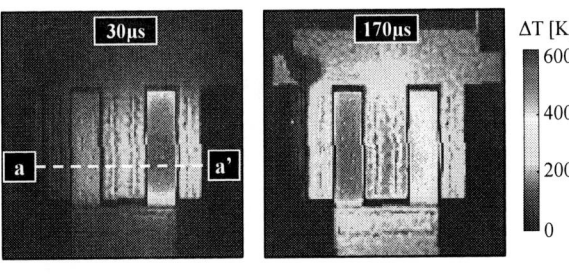

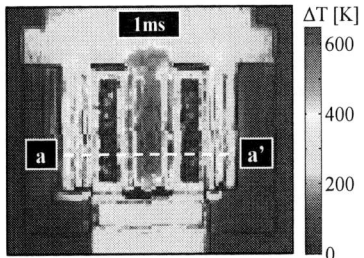

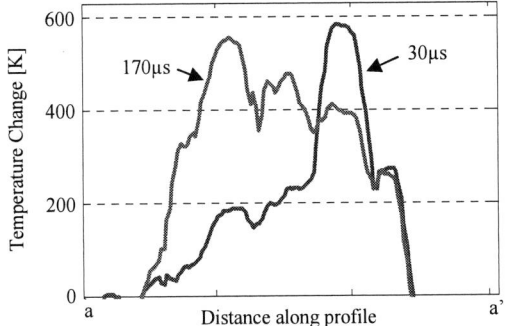

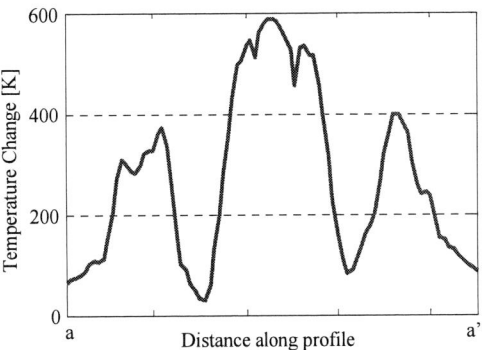

Figure 4. SCR thermoreflectance images and temperature profiles at 30 and 170 microseconds. Snapback current =1.22A for both images.

Figure 5. SCR thermoreflectance image and temperature profiles at one millisecond.Snapback current =1.22A.

phenomenon might be explained by high speed current localization or non-simultaneous triggering of different finger junctions. Fig. 5 shows SCR heating one millisecond into the transient, when device heating appears to reach thermal steady state. Heating is observed to spread to the anode fingers and become symmetric with respect to the device fingers, favoring neither the left nor right side of the SCR.

Current localization and thermal induced changes in the material and device properties should be the cause of heat switching fingers during the microsecond portion of the thermal transient for the SCR in snapback. One should note that typically thermoreflectance coefficient is calibrated for small temperature variations. When temperature rise is several hundreds of degrees, the nonlinear dependence of the reflection coefficient as a function of ambient temperature should be taken into account. Thus more accurate calibrations are necessary in order to quantify the exact temperature rise. However the measured transient reflectance map should provide some information about the current non-uniformity in the device.

IV. CONCLUSION

A new methodology has been developed and applied toward indirect measurement of the current distribution in ESD devices during pulsed operation. The methodology includes transient thermoreflectance images capturing [1] and provides informative characteristics for the transient ESD current distribution during ESD event.

Experimental validation of the new approach is presented for a multiple finger NLDMOS-SCR ESD protection device. The experimental data have demonstrated both non-simultaneous, nonuniform triggering of individual device

fingers as a function of snapback current levels and at different times in the thermal transient.

SCR temperature change during the first 300 nanoseconds of device excitation has been measured in excess of 15K. This demonstrates transient CCD thermoreflectance may be useful in analyzing device current distribution during ESD-like events. In the microsecond time domain temperature increase up to 400K has been observed.

REFERENCES

[1] V.A. Vashchenko, A.A. Shibkov, ESD Design for Analog Circuits, Springer, 2010, and companion website www.analogesd.com.

[2] G. Notermans, Proceedings Electrical Overstress/ Electrostatic Discharge Symposium, pp. 221-229, 1997.

[3] M. Goldstein, G. Sölkner, and E. Gornik, Review of Scientific Instruments, 64(10), 3009, 1993, doi: 10.1063/1.1144348.

[4] C. Furbock, D. Pogany, M. Litzenberger, E. Gornik, N. Seliger, H. Gobner, et al., Electrical Overstress/Electrostatic Discharge Symposium Proceedings 1999, 49, 241-250, 2000.

[5] S. Bychikhin, V. Dubec, M. Litzenberger, D. Pogany, E. Gornik, G. Groos, et al., 2002 EOS/ESD Symposium, Journal of Electrostatics, 59 (3-4), pp. 241-255, 2003.

[6] R. Rosei, and D. Lynch, Physical Review B, 5(10), 3883-3894, 1972.

[7] A. Rosencwaig, J. Opsal, W. Smith, and D. Willenborg, Applied Phys. Letters, 46(11), 1013-1015, 1985.

[8] G. Tessier, S. Holé, and D. Fournier, Applied Physics Letters, 78(16), 2267, 2001.

[9] D. Luerssen, J. A. Hudgings, P. M. Mayer, and R. J. Ram, IEEE Semiconductor Thermal Measurement and Management Symposium, 2005. p. 253-258.

[10] Y. Ezzahri, J. Christofferson, G. Zeng, and A. Shakouri, Journal of Applied Physics 106(11): 114503, 2009.

[11] K. Maize, J. Christofferson, and A. Shakouri. 24th IEEE Semionductor Thermal Measurement and Management Symposium, pp. 55-58, 2008.

978-1-4244-9113-1/11 $26.00 © 2011 IEEE

[12] H. Geiser, "Methods for the characterization of integrated circuits employing very fast high current impulses," in Dissertation Technische Universitaet Muenchen TUM, Shaker-Verlag, Aachen, Germany, 1999.

Degradation and Failure Analysis of Polysilicon Resistor connecting with Tungsten contact and Copper line

Clement Huang, Mingte Lin, James W. Liang, Alex Juan and K. C. Su

Reliability Technology & Assurance Division, UMC Inc.
No.3, Li-Hsin Rd. II, Hsinchu Science Park, Taiwan 300, ROC
Phone: +886-935336621; Fax: +886-3-5776889; e-mail: clement_huang@umc.com

Abstract—**The failure mechanism was studied on Polysilicon Resistor - Tungsten contact - Copper line structures. Silicide resistor could fail at high resistive interface of poly/silicide/barrier/metal because thermal mismatching for varied materials. In the case of Silicide_Block resistor, damage nearby the contact proximity (especially at Cu region) was observed, which originated from local Joule heating at the interface. Finite element analysis (FEA) was demonstrated that failure was dependent on current density and Joule heat generation.**

Keywords: Polysilicon resistor, W, Cu, electromigration, Joule heating, Finite element analysis

I. INTRODUCTION

Resistors provide specific and controlled amounts of electrical resistance. They are useful in a variety of application, ranging from current limiting to voltage division [1]. Polysilicon resistors are available in CMOS and BiCMOS process. The resistivity of polysilicon resistor depends not only on doping but also on grain structure and subsequent stacked interconncet. The grain boundaries and interface of multi-layer interfere with the orderly flow of carriers and raise the resistivity of the stacked material. Silicided poly can't provide enough resistance for most applications, so some processes provide a means for blocking silicide to get higher resistivity. For the advanced technology, designer is concerned about the reliability issue of polysilicon resistor, such as maximum usage current and Joule heating generated from high current [2]. In Cu process, the electromigration (EM) of "Silicide-W plug-Cu line" structure is discussed rarely.

Since the driving force is electric current and polysilicon resistor has complex geometry, we applied FEA to study the current density profiles on these structures. Failure analyses were performed to clarify the failure modes of these structures. Thermal budget increases were also simulated with different resistors. EM performances of polysilicon resistors are explained with these FEA and failure analysis results.

II. EXPERIMENTAL

Polysilicon resistor samples were fabricated with 65 nm Cu/low-k dual damascene process. The test structures are poly line, W contacts and Cu line connecting to probe pad. The width/length (W/L) of poly is 10/40um. The polysilicon resistor structures include silicide and silicide_block (SB) layouts (Fig. 1), both have the same widths of line and contact. Stress current density (I_{stress} / (poly line width * thickenss)) is 45mA/um^2 for silicide-poly and 15mA/um^2 for SB-poly. Stress temperature region is 185~225 °C. EM Failure criterion is 10% resistance increase.

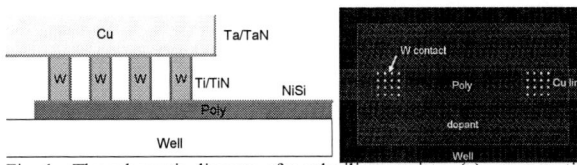

Fig. 1. The schematic diagram of a polysilicon resistor (a) corss- section (b) top-view (Note: SB resistor has no NiSi formation for L=40um, exclude contact region)

III. RESULTS AND DISCUSSION

EM test results are summarized in Table 1. The relative resistance changes with time and EM time to failure (TTF) CDF charts are shown in Fig.2 and Fig.3. The major discrepancy between two types resistor was the ramping rate of resistance. The failure mode of silicide resistor was sudden raise at one specific time for each sample, but SB resistor was gradual. Besides, silicide resistor has longer TTF than SB. The sigma of lognormal TTF distribution for silicide resistor was comparable with SB.

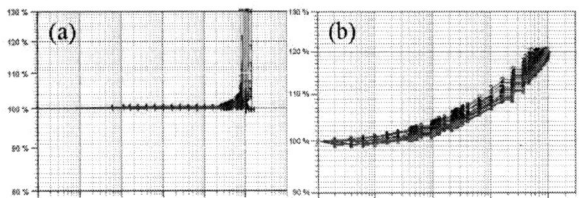

Fig. 2. The resistance change% v.s. normalized stress time of (a) Silicide resistor and (b) SB resistor.

Table 1. EM results of polysilicon resistor.

Resistor type	MTF (A.U.)	Sigma
Silicide	73.0	0.159
Silicide_Block	11.6	0.142

The EM time to failure (TTF) CDF charts are shown in Fig.3. Though the stress current of silicide resistor is much higher than SB resistor (45 v.s. 15 mA/um^2), it is still 6 times the TTF of SB resistor. Thus the current density is not the only one major factor causing failure. Since SB resistor provides higher resistance, joule heat generating during stress process is one reasonable failure mechanism.

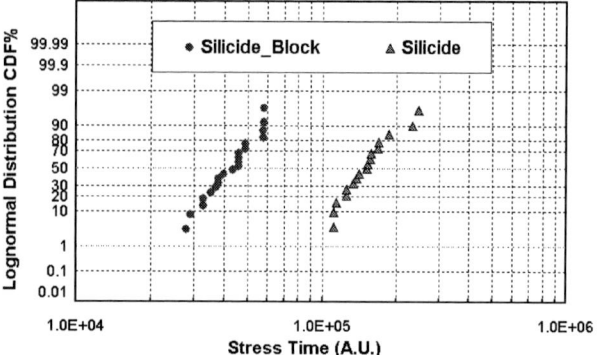

Fig. 3. EM time to Fail CDF charts of polysilicon resistor.

From FIB/TEM failure analyses on stressed sample, considerable damage was observed for silicide resistor (Fig. 4). Not only poly line but W contact and Cu metal line wore out at the nearby contact region. Since current density is larger around NiSi layer upon poly, burst of poly surface lead to distortion of W contact and Cu line. Besides that, the resistor-contact-metal line structure has varied materials (Table 2), the interface of poly/silicide/barrier/metal may induce thermal mismatching. On the contrary, SB resistor showed more robust poly line and W contact since slowly cumulative resistance (Fig. 5). There are several failure modes for SB resistor, (a) void in metal line end (b) slit void in the surface of metal line (c) small void in central metal line. Although there are numerous void inside metal line, none of them caused the Rc-open failure like silicide resistor. Creating such kind of ramping resistance requires considerable force like Joule heating. Since no voids in the W contact, significant heat generation from the bottom of the contact is expected [3].

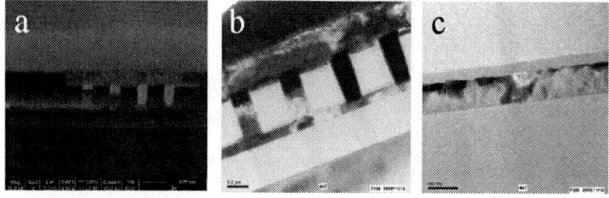

Fig. 4. Failure mode of silicide resistor. (a) SEM (b) W contact region TEM (c) Polysilicon region TEM.

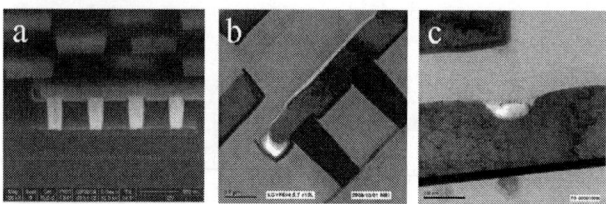

Fig. 5. Failure mode of SB resistor. (a) SEM (b) W contact region TEM (c) Cu line region TEM.

The current density profiles from FEA were shown in Fig. 6. Since silicide resistor has NiSi layer on the poly surface, the current density is much higher than SB. Furthermore, the first row of W contact has the highest current density (5 times than the second row), and 99% current path use the first 2 rows (Fig. 7). This result was verified by failure analysis. The thermal distribution due to Joule heat was simulated (Fig. 8). The Joule heating of SB resistor is significant larger than silicide. Because SB resistor has lower current density but higher Joule heating, the resistance increases gradually. The highest Joule heat generation is located at first row of W contact region nearby Cu metal line end (Fig. 9). Higher temperature of SB resistor could provide more energy for Cu atom migration. The Joule heating is very small in metal area, but EM failure happened in Cu line. Regarding to the activation energy (Ea) of Cu is smaller than W, the momentum for Cu atom movement is provided by Joule heating generated from Poly & W contact, and therefore the void is formed at Cu line but not Poly or W. The metal line end serves as Cu atom reservoir, so the void will extend from this extra Cu volume.

Table 2. Bulk resistivity of materials for Back-end processing.

No.	Material	Resistivity (1E-14 T-ohm*um)
1	Cu	1.7
2	W	5.6
3	NiSi	15
4	Ti	54
5	Ta	70
6	TiN	130
7	TaN	252
8	Poly	3.96E+03

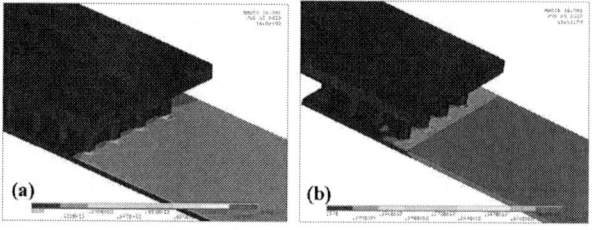

Fig. 6. Contour plot of current density with real stress current value (a) Silicide resistor (b)SB resistor.

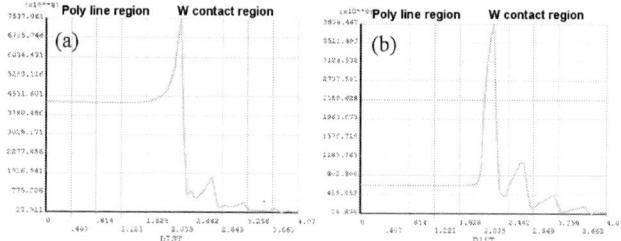

Fig. 7. Current density on (a) Silicide resistor (b)SB resistor. X-axis: distance from poly line Y-axis: current density (A.U.)

978-1-4244-9113-1/11 $26.00 © 2011 IEEE

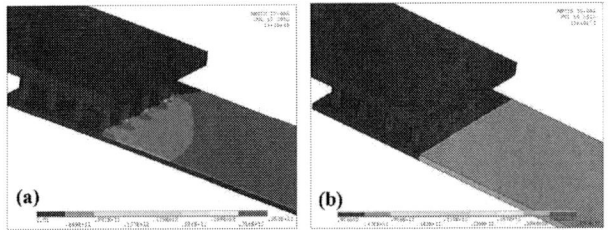

Fig. 8. Contour plot of Joule heat generation (a) Silicide resistor (b) SB resistor.

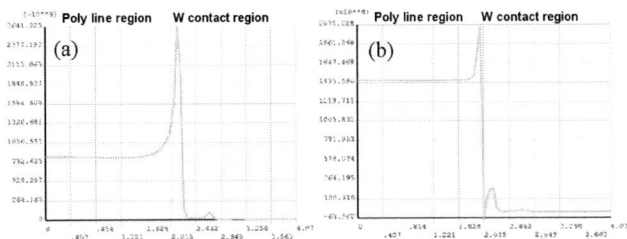

Fig. 9. Joule heat generation on (a) Silicide resistor (b) SB resistor; X-axis: distance from poly line, Y-axis: Joule heating (A.U.)

Fig. 10 shows the Joule heating generation for two type resistors. For silicide resistor, joule heating raises gradually with increasing stress current. When stress current increase from 1 to 2 A.U., the joule heating of silicide resistor increase about 20℃ (2 times). Regarding to SB resistor, the ramping rate of joule heating is considerable higher than silicide. If stress current increase from 0.5 to 1 A.U., the joule heating of SB resistor will raise to 120℃ (11 times). The restriction of operation current for polysilicon resistor is quite significant, or else the integrity of interconnect could be compromised.

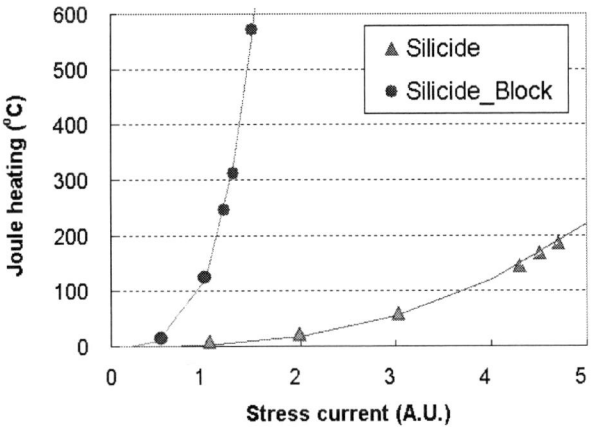

Fig. 10. Joule heat generation of polysilicon resistor.

CONCLUSION

EM of polysilicon resistor was studied in detail with experiments, current density simulation and failure analysis. The major discrepancy between two types resistor was the ramping rate of resistance. Considerable damage is found at poly, W contact and Cu line for silicide resistor because of high current density. As for SB resistor, the path of current is not compromised badly but resistance increase rapidly. Though the current density of silicide resistor is much higher than SB, it still has higher TTF. Resistance simulation showed the Joule heating of SB resistor is significant larger than silicide.

REFERENCES

[1] A. Hastings, "The Art of Analog Layout", 2nd Edition, pp.156-193.

[2] J. Im, B. Ang, S. Tumakha and S. Paak, "Characterization of silicided polysilicon fuse implemented in 65nm Logic CMOS technology", NVMTS, 2006, pp. 55-57.

[3] T. Kauerauf, G. Butera, K. Croes, S. Demuynck, C.J. Wilson, P. Roussel, C. Drijbooms, H. Bender, M. Lofrano, B. Vandevelde, Z. Tokei & G. Groeseneken, "Degradation and failure analysis of copper and tungsten contacts under high fluence stress", IRPS, 2010, pp. 712-716.

A Practical Modeling for Transient Thermal Characteristics of Multilevel Interconnects

Seung-Man Choi*, Dong-Cheon Baek, Tae-Young Jeong, Myung-Soo Yeo, Miji Lee, Andrew T. Kim and Jongwoo Park
Technology Quality and Reliability Group, Q & R Team
System LSI, Semiconductor Business, Samsung Electronics Co., Ltd.
San#24, Nongseo-Dong, Giheung-Gu, Yongin-City, Gyeonggi-Do, 446-711, KOREA
Phone: +82-31-209-4342, e-mail: rirachoi@samsung.com*

Abstract— In this study, intuitive is given on time-dependent thermal characteristics in multilevel interconnects subjected to carry either DC or pulsed-DC. FEM simulation is employed to model the propensity of temperature profile with respect to the variety of interconnects having different geometrical features in terms of metal width, metal height and distance between metal and Si substrate. Accordingly, a practical model that enables to prognosis temperature increase resulting from current-driven metal interconnects and temperature decrease after current carried along metal line stops is developed. It is found that a proposed model precisely predicts thermal transient arisen from metal interconnect, regardless of geometrical factors of metal dimension and location. In addition, transient thermal behavior of metal interconnects carrying pulsed DC with various frequencies is investigated. A circuit designer is required to adjust the maximum allowable current carried along metal interconnects according to the frequency of pulsed DC as well as geometrical dimensions of metal interconnects. Hence, robustness in circuit design even in the earlier stage of development phase can be accomplished for metal interconnects by suppressing electromigration and rupture caused by thermal transient.

Keywords – Transient; Thermal; Joule Heating; Interconnects

I. INTRODUCTION

With the continuing scaling down in the metal pitch and increasing requirement of multilevel interconnects for high performance integrated circuits, interconnects are becoming one of the most important factors determining system performance and power dissipation.[1] Operating current density increases with scaling and moreover the low-k dielectric materials which are widely used for the reduction of RC-type signal time delay have poor thermal conductivity due to higher degree of porosity.[2] As a result, temperature of metal interconnects has become one of the major reliability concerns in integrated circuit and design perspectives. No doubt, Joule heating induced by high current density in metal lines triggers temperature rise and results in the degradation of integrated circuit system due to temperature enhanced failure mechanisms such as electromigration and stress migration [3]. In addition, an overshooting current can cause an instantaneous temperature rise to the melting point of metal resulting in rupture of interconnects, even though current flows for a very short period of time, as shown in Fig.1.

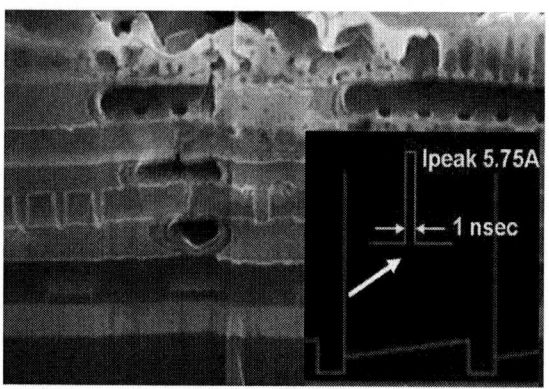

Figure 1. Rupture of metal interconnects due to sudden temperature rise caused by an instantaneous high current flow.

A typical way to investigate temperature profile in metal interconnect is to use FEM (Finite Elements Method) simulation. However, it requires in-depth of knowledge and time consumption to prepare an appropriate model of interconnects and to solve a problem. Harmon et al. [4] proposed an analytical model for predicting metal temperature at steady state. Such modeling is very useful for expecting temperature rise of metal interconnects with various dimensions and agrees well with FEM simulation results. Li et al. [5] and Yokogawa et al. [6] investigated transient thermal characteristics of the specific interconnect structures. However, a universal model for transient thermal characteristics of interconnects having various metal dimensions has not been found yet.

In this paper, we present a universal model for transient thermal characteristics driven by current-carrying interconnects regardless of geometrical features of interconnects, including dimension and location. FEM simulations are performed, in order to gather data for finding thermal time constant τ, and verify the proposed model. With a practical model, temperature profile of interconnects can be precisely modeled and robustness in circuit design is accomplished from reliability perspectives.

978-1-4244-9113-1/11 $26.00 © 2011 IEEE

II. PRACTICAL MODEL OF TRANSIENT THERMAL CHARACTERISTICS

A. Analytical model for thermal characteristics

The analytical model proposed by Harmon [4] for predicting thermal characteristics in steady state can be expressed with (1).

$$\Delta T_{max} = \frac{i^2 \cdot \rho}{W \cdot H \cdot Gp} \tag{1}$$

where i is the current carried along metal line, ρ is the resistivity of metal, W is metal width, H is metal height, and Gp is the thermal conductance per unit length. Gp in (1) can be easily calculated from the following equations.

$$Gp = Gt + Gs + Gb \tag{2}$$

$$Gt = k_i \frac{2}{\pi} Log\left[\frac{\pi W}{2(H+Di)} + 1 + Log\left[1 + \frac{\pi W}{2(H+Di)}\right]\right] \tag{3}$$

$$Gb = k_i \frac{2}{\pi} Log\left[\frac{1}{3} + \frac{2}{3}Cosh\left[1 + 3^{-3/2}\pi + \frac{\pi W}{2Di}\right]\right] \tag{4}$$

$$Gs = k_i \frac{2}{\pi} Log\left[-1 + 2\left(1 + \frac{H}{Di}\right)^2 + 2\left(1 + \frac{H}{Di}\right)\sqrt{\left(1 + \frac{H}{Di}\right)^2 - 1}\right] \tag{5}$$

where k_i is the thermal conductivity and Di is the distance from the bottom of metal to Si substrate. Harmon's model predicts the temperature rise by Joule heating very well after a

TABLE I. INTERCONNECT GEOMETRIES USED FOR FEM SIMULATION.

Parameters	Range
Metal Width, W	0.05 ~ 0.45 um
Metal Height, H	0.1 ~ 0.5 um
Distance (Metal to Si), Di	0.1 ~ 1.7 um

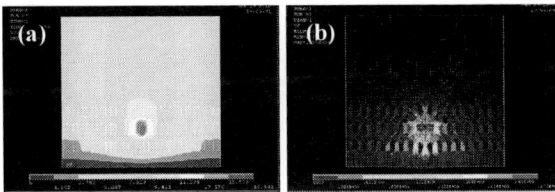

Figure 2. Temperature profile (a) and thermal flux (b) of metal interconnects solved by using FEM simulation.

steady state is reached. However, transient temperature profiles cannot be found by using it.

B. Transient modeling for Joule heating

Generally, the equation (6) is used for depicting time-dependent behavior.

$$F = 1 - Exp\left(-\frac{t}{\tau}\right) \tag{6}$$

where τ is a parameter for time-dependency. Transient temperature behavior by Joule heating can be described with (7) by combining (1) with (6).

$$\Delta T(t) = \frac{i^2 * \rho}{W * H * Gp} * \left[1 - Exp\left(-\frac{t}{\tau}\right)\right] \tag{7}$$

The maximum temperature rise depending on geometry of metal such as dimensions and location can be obtained from the front part of (7) because Gp contains its dependency on interconnect geometry. If we know the dependency of τ on interconnect geometry in the heel of (7), transient thermal characteristics of metal interconnects can be easily obtained from (7). FEM simulations were conducted to investigate time-dependent temperature profile depending on interconnect geometry. The ranges of interconnect dimensions used for FEM simulation are listed in Table 1.

An example of temperature profile and thermal flux of metal interconnects solved by using FEM simulation is shown in Fig.2 (a) and (b), respectively. In comparison to FEM simulation results, it is found that the temperature ramping rate and the maximum temperature driven by Joule heating are dependent on both geometrical features of interconnects and applying current density, as shown in Fig.3. However, if temperature rise is normalized with respect to its respective maximum temperature after reaching the steady state, transient thermal characteristics can be easily obtained by using (6). The

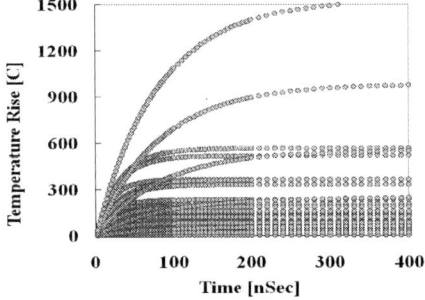

Figure 3. Temperature profiles of metal interconnects with various geometries of such as metal line width, height and distance between metal line and Si substrate.

978-1-4244-9113-1/11 $26.00 © 2011 IEEE

normalized temperature rises are shown in Fig.4. Using the normalized temperature profiles, we can find the dependency of τ on geometrical features of interconnects in terms of metal width, metal height, and distance between metal and Si substrate, as shown in Fig.5. Each equation shown in Fig.5 can be combined together then simplified as (8). As such it provides the total dependency of τ on interconnect geometries. Hence, temperature profiles for interconnects due mainly to Joule heating can be estimated by using (7) and (8).

$$\tau = a(b * e^{-W} + 1)(c * H + 1)(d\sqrt{Di} + e * Di + 1) \qquad (8)$$

Fig.6 shows the comparison of the fitted τ which are calculated by using (8) and the simulated τ by FEM simulation. It is legitimate that the analytical model of (8) agrees well with the results of FEM simulation regardless of geometrical features of interconnects. In Fig.7, the solid line in red represents temperature profiles calculated by using (7) and (8),

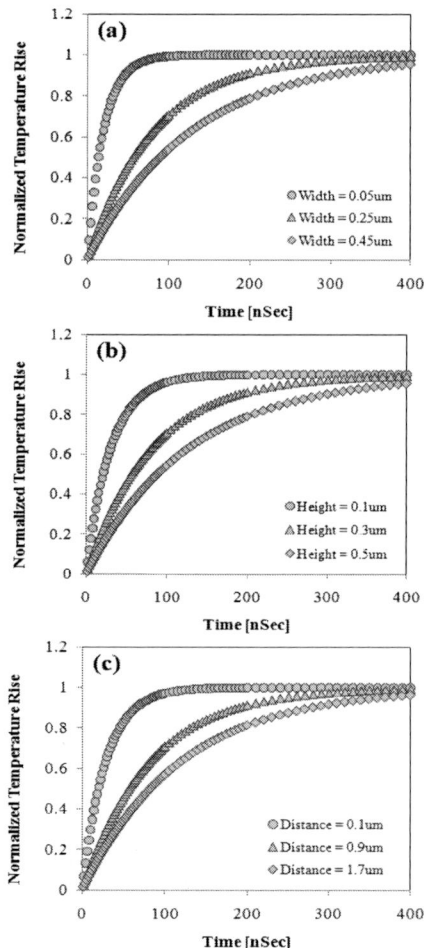

Figure 4. Normalized temperature profiles of metal interconnects with various geometries of such as (a) metal line width, (b) metal line height and (c) distance between metal line and Si substrate.

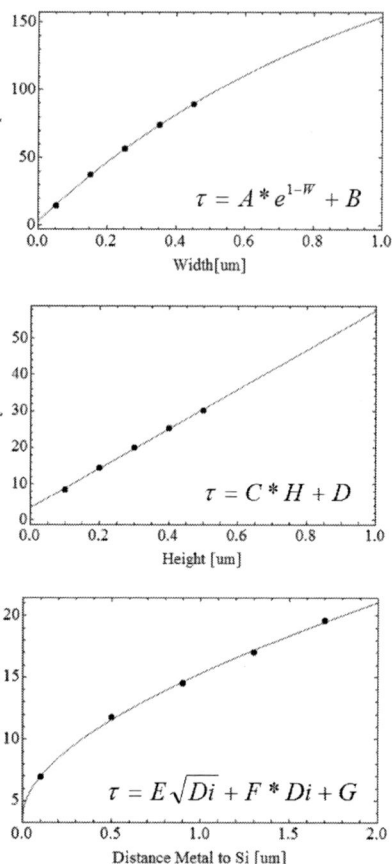

Figure 5. Dependency of τ on metal line width, height, and distance between metal and Si substrate.

while the line with blue dots shows FEM simulated temperature profiles of a metal line with specific dimensions. Although the discrepancy exists after temperature levels off, temperature rising profile of a metal interconnect is well agreed each other as a function of time.

C. Transient modeling for cooling down

If the current carried along metal line stops, the temperature elevated by Joule heating starts to decrease. Time-dependent decreasing behavior can be depicted by (9).

$$F = Exp\left(-\frac{t}{\tau}\right) \qquad (9)$$

Transient temperature behavior of interconnects by cooling can be described with (10) by combining (1) with (9), which is the same way as (7).

$$\Delta T(t) = \frac{i^2 * \rho}{W * H * Gp} * Exp\left(-\frac{t}{\tau}\right) \qquad (10)$$

978-1-4244-9113-1/11 $26.00 © 2011 IEEE

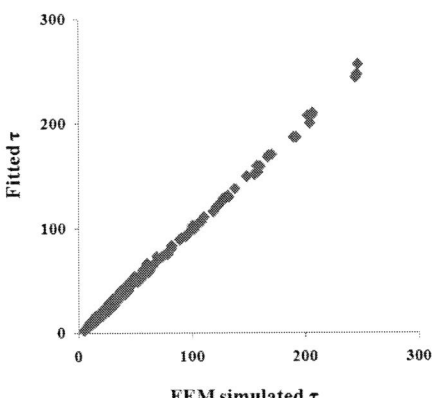

Figure 6. Compatibility between the fitted model with (8) and the simulated by FEM simulation

where τ has the exactly same values as that of Joule heating because it depends on interconnect geometry only. Thus, (8) can also be used for the calculation of τ in transient thermal characteristics of cooling stage. The maximum temperature calculated by the front part of (10) is the maximum temperature rise by Joule heating after the steady state is reached. If the current stops before the steady state is reached, time in (10) needs to be offset. Fig.8 shows the transient temperature profiles of cooling stage of a metal line having the same geometrical dimension as shown in Fig.7. The temperature profile calculated by using (10) and (8) represented by the solid line in red is in accord very well with FEM simulation results represented by the line with blue dots.

D. Thermal characteristics of Periodic Pulsed DC

As shown in Fig.7, when the current is switched on, temperature of a metal line is saturated within a few hundreds nano-seconds even though the maximum temperature rise is as high as several hundred degrees Celsius, which means the metal interconnects can be melt down within a very short period of time if a very high current is applied. As such, it is important to know how long time it takes for metal interconnects to get to a critical temperature. If a periodic pulsed DC is applied to metal interconnects, it gives some time to cool down the temperature of metal interconnects for a period of off-current stage. Fig.9 shows a temperature rise profile calculated by using (7) and (10) with (8) for two periods of pulsed DC. The temperature of metal interconnects returns to its original state if the time of off-current stage is long enough, in other words the frequency of pulsed DC is very low. On the contrary, if pulsed DC has high frequency, there is not enough time for not only cooling down, but also heating up. As a result, the maximum temperature rise decreases as the frequency of pulsed DC increases. Fig.10 shows the temperature rise profiles of metal interconnects having different metal widths with the various frequencies of pulsed DC. The applying current density and the geometrical dimensions except metal width are identical to each other. In

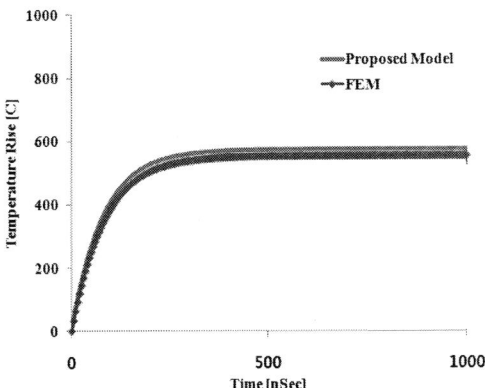

Figure 7. Compatibility between the fitted model (7) with (8) and the simulated by FEM simulation for Joule heating (Metal dimensions: Width = 0.25um, Height = 0.3um and Distance to Si = 0.9um, Applying current density = 60MA/cm^2)

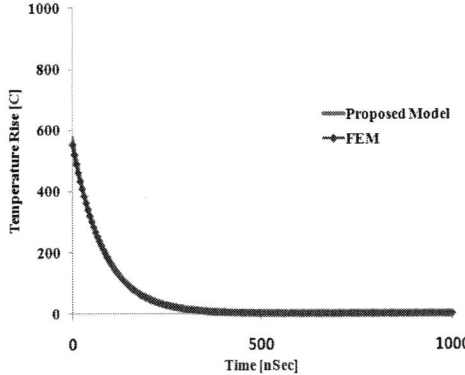

Figure 8. Compatibility between the fitted model (10) with (8) and the simulated by FEM simulation for cooling down. (Metal dimensions: Width = 0.25um, Height = 0.3um and Distance to Si = 0.9um)

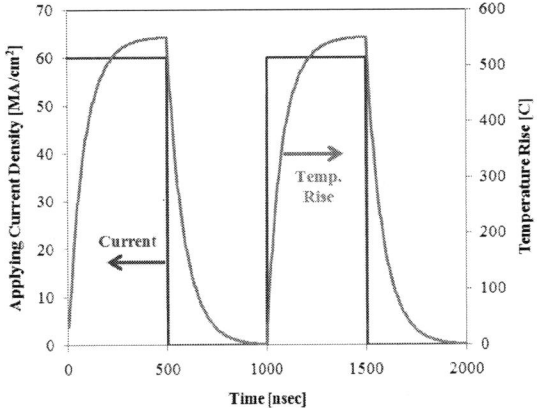

Figure 9. Temperature change by periodic pulsed DC applied to metal interconnects (Metal dimensions: Width = 0.25um, Height = 0.3um and Distance to Si = 0.9um)

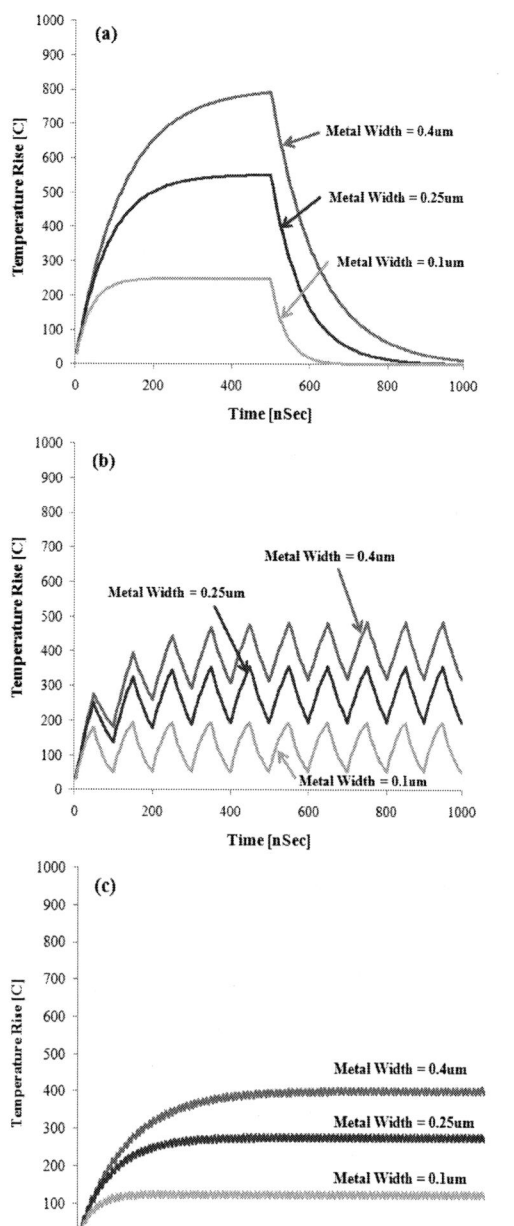

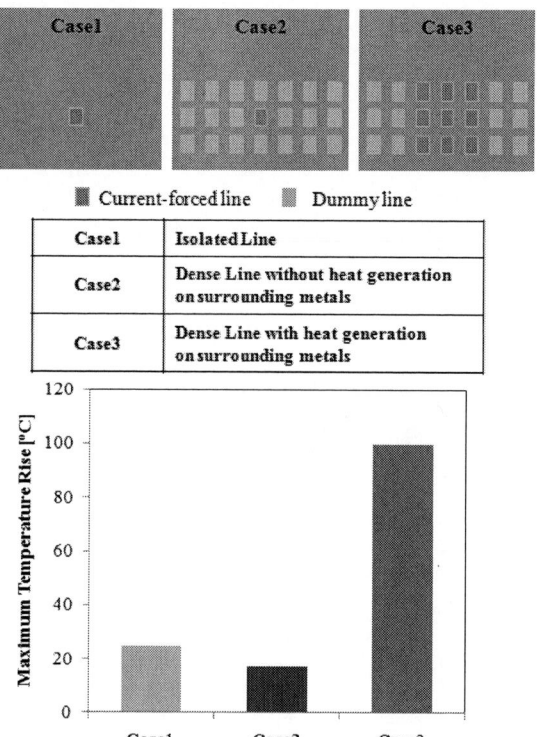

Figure 11. Maximum temperature rises for three different cases solved by using FEM simulation (Metal dimensions: Width = 0.05um, Height = 0.115um and Distance to Si = 0.18um. Applying current density = 43MA/cm²)

starts to be cooled down before it reaches the maximum temperature and it is heated up again before it returns to its original state. Thus, the higher the frequency is, the narrower the amplitude of temperature cycles becomes and the smaller the maximum temperature rise becomes. The maximum temperature rise increases as metal width increases because power dissipation is proportional to the square of the current while the thermal conduction area of metal facing to Si substrate is proportional to metal width. As a natural consequence, if the duty cycle of a pulsed DC is changed, the temperature rise profile and the maximum temperature rise would be changed. These results imply that the maximum allowable current applied to metal interconnects has to be adjusted according to the frequency of pulsed DC as well as geometrical dimensions of metal interconnects.

III. DISCUSSION

Metal lines in real circuits have the neighboring metal lines that may be carrying the current. Fig.11 shows the maximum temperature rises of the metal interconnects in three different cases solved by using FEM simulation. Case 1 is the isolated metal line. Case 2 is the dense metal lines, but the surrounding metal lines don't carry the current. In Case 2, the effective thermal conductivity becomes larger because the thermal conductivity of metal is usually much larger than that of

Figure 10. Temperature rise profiles with the different frequencies of periodic pulsed DC (a) 1MHz pulsed DC (b) 10MHz pulsed DC (c) 100MHz pulsed DC (Metal dimensions: Width = 0.1~0.4um, Height = 0.3um and Distance to Si = 0.9um. Applying current density = 60MA/cm²)

case of pulsed DC with 1MHz frequency shown in Fig.10 (a), the temperature rise starts to cool down after it reaches almost the maximum temperature because the period is long enough for metal interconnects to get to the maximum temperature. However, in other cases of pulsed DC with the higher frequencies shown in Fig.10 (b) and (c), the temperature rise dielectric material. As a result, the maximum temperature rise of Case 2 at the steady state is lower than that of Case 1. Effective thermal conductivity of metal interconnect structures can be measured [7] and the model proposed in this paper can be used for both of Case1 and Case2 if the effective thermal conductivity is properly measured. However, if some or all of the surrounding metal lines carry the current, the maximum temperature rise increases drastically like Case 3 in Fig.11. In real circuits, the transient thermal characteristics as well as the maximum temperature rise would be more complex due to the various geometrical features and the various applying currents of the neighboring metal lines. Thus, it may require the modification of the equations (7), (8) and (10) in order to include the effect of the neighboring metal lines carrying current on the transient thermal characteristics.

IV. CONCLUSION

In order for transient thermal characteristics of multilevel interconnects to be taken into account for circuit design, a universally analytical model is proposed with respect to geometrical factors of metal interconnects in terms of metal width, metal height and distance between metal and Si-substrate. In fact, such heuristic is beneficial especially for circuit designers to predict transient temperature of metal interconnects in the earlier stage of development from design perspective. Simplicity and fast turnaround simulation without using a complex FEM modeling are certainly time and cost effective tool that can be adopted in advanced technology development. In consequence, robustness in circuit design can be attained without compromising the back-end-of-line reliability in connection with electromigration and rupture as well.

ACKNOWLEDGMENT

The authors would like to thank Young-Joon Park for his kind and sincere advices.

REFERENCES

[1] M.T. Bohr, "Interconnect scaling-The real limiter to high performance ULSI," *Tech.Dig. International Electron Device Meeting*, pp.241-244 (1995)

[2] J.W. McPherson, "Reliability challenges for 45nm and beyond," *Proc. Of 2006 DAC*, pp.176-181 (2006)

[3] K. Banerjee and A. Mehrotra, "Global (Interconnect) warning," *IEEE Circuits and Device Magazine*, pp.16-32 (September 2001)

[4] D. Harmon, J. Gill, and T. Sullivan, "Thermal conductance of IC interconnects embedded in dielectrics," *Integrated Reliability Workshop Final Report, IEEE International*, pp.1-9 (1998)

[5] Z. Li, G. Wu, Y. Wang, Z. Li, and Y. Sun, "Numerical calculation of electromigration under pulse current with Joule Heating," *IEEE Transactions on Electron Devices, vol. 46, issue 1*, pp. 70-77 (1999)

[6] S. Yokogawa, H. Tsuchiya, and Y. Kakuhara, "Effective thermal characteristics to suppress joule heating impacts on electromigration in Cu/low-k interconnects," *Proc. of 2010 International Reliability Physics Symposium*, pp.717-723 (2010)

[7] F. Chen, J. Gill, D. Harmon, T.Sullivan, B. Li, A. Strong, H.Rathore, D. Edelstein, C-C. Yang, A.Cowley, and L.Clevenger, "Measurements of effective thermal conductivity for advanced interconnect structures with various composite low-k dielectrics," *42nd Annual International Reliability Physics Symposium*, pp.68-73 (2004)

Electromigration of Cu Interconnects Under AC, Pulsed-DC and DC Test Conditions

Ramifications on Accelerated Testing

Roey Shaviv, Gregory J. Harm, Sangita Kumari*
Novellus Systems Inc., 4000 North First Street,
San Jose CA 95134, USA
(1) - (408)-570-2848. Roey.Shaviv@Novellus.com
*Present address: Department of Material Science and Engineering, University of Arizona,
Tucson AZ 85721-0012 USA

Robert R. Keller, David T. Read
National Institute of Standards and Technology,
Materials Reliability Division,
325 Broadway, Boulder, CO 80305, USA

Abstract—— Electromigration (EM) of a dual damascene, single-via fed test vehicle was measured using DC, AC followed by DC, and three rectangular-wave DC stressing conditions at 598 K. In some of the experiments samples were allowed to cool between stress cycles.

Void formation and migration inside Cu interconnects were filmed. We show void formation by AC and void migration and consolidation by DC.

We found that neither AC stressing, followed by DC stressing to failure, nor DC coupled with thermal cycling had an effect on the net, DC only, EM performance. All tests, regardless of thermal history and current cycling conditions, resulted in similar DC lifetimes and their distributions.

We conclude that for this test structure, only net DC testing time has a significant effect on time to failure. AC stress, thermal history (including cooling to room temperature) and pulsed DC stressing have no effect on electromigration lifetime under DC conditions. This suggests that EM in this test structures depends not on metal microstructure, but only on interface structure.

We further conclude that the standard test methodologies, using accelerated DC stress conditions at elevated temperatures, are adequate. However, since only the net DC stress time has any measurable effect on EM lifetime, accelerated testing provides a good predictor for lifetime expectations only for the DC component of continuous operating conditions, but underestimates lifetime expectation under pulsed DC, and variable usage operating conditions by 50 % or more, depending on duty cycle. We also conclude that neither AC stress nor DC cycling, and changes to grain structure that may result, vaccinate Cu interconnects against EM failures and thus, do not provide any measurable benefit to net, DC only, EM lifetime.

Keywords-Electromigration; accelerated testing; AC; DC

I. INTRODUCTION

Electromigration (EM) is a leading cause of reliability failures in copper metallization of interconnects. EM measurements are typically conducted under accelerated conditions in accordance with JEDEC standards [1], using constant direct current (DC) at elevated temperatures. However, device operation rarely employs pure DC conditions. Interconnect lines are often at an operating temperature of $\sim$ 100° C while running pulsed current. Idle time between pulses can vary between a fraction of a nanosecond, during peak usage, and hours or even days, during low usage periods. Since Cu recrystallization is favored at $T \geq \sim 100°$ C [2], [3] Cu may back-diffuse or recrystallize during the idle time to reverse the effect of electromigration drift, affecting time to fail.

A similar, and possibly enhanced, phenomenon is expected with alternating current (AC). In principle, electromigration should cease under AC conditions because the net momentum transfer is zero. Instead, fatigue can emerge as a key failure mechanism [4]. Grain growth in Al interconnects has been observed and reported during high amplitude AC stressing due to controlled cyclic thermal straining [5]. A combination of AC and DC could therefore affect grain growth and void formation during the AC cycle, and influence electromigration during DC stressing.

None of these effects are typically taken into consideration during the standard EM test. Thus, the projected time to failure under operating conditions does not account for idle time, relaxation, recrystallization of Cu, grain growth and other phenomena that occur only during AC or pulsed-DC conditions. Moreover, even during high usage periods the actual net DC stressing time is only half that of the total time, due to the clock cycle. Consequently, traditional EM testing methodologies may either overestimate or underestimate time to fail since they account for DC only effects.

In a recent communication we reported that only the net DC stressing time had a measurable effect on EM lifetime [6]. That conclusion reaffirms the validity of DC only test methodologies and the reliability of lifetime projections based on those. Here we further report on AC and pulsed DC treatments of Cu interconnect lines, for the purpose of determining whether EM damage depends on metal microstructure or interface structure. We show formation and migration of void during EM testing as well as their effects of these on EM lifetime.

II. EXPERIMENTAL

All experiments were conducted using a two-metal level dual damascene Cu / low-k test-vehicle. Minimum critical dimensions for the EM test structure were 90 nm for the line width and 110 nm for the spaces between adjacent lines in the array. The length of the test line was 500 μm. A single via connected the test line to the lead wires for both upstream and downstream EM testing. An illustration of the test structure is given in Figure 1.

Current densities of 2.5 MA/cm^2 DC and AC (RMS) were employed for package level upstream and downstream EM testing. The EM testing temperature range was 548 K to 598 K (275° C to 325° C). DC-only stress-to-failure conditions were applied at 598 K, to a first batch of test specimens, as benchmark and control. In the second round of tests, AC stressing at 100 Hz for one week at 548 K was applied, followed by DC stress to failure at 598 K. In the third round of tests high current density AC conditions were applied for a duration of 1022 s, for the purpose of altering metal microstructure. The RMS current densities were 76.5 MA/cm^2 and 67.3 MA/cm^2 for downstream and upstream EM testing, respectively. This results in significant Joule heating with power dissipation of ~ 3.6 W or ~ 80 kW/cm^2. The resulting local test temperatures were ~ 630 K and 605 K for downstream and upstream EM testing, respectively. No failures were detected during the AC stress period. Whereas a special test setup was constructed for the AC tests, all DC tests were conducted in commercially available EM test systems.

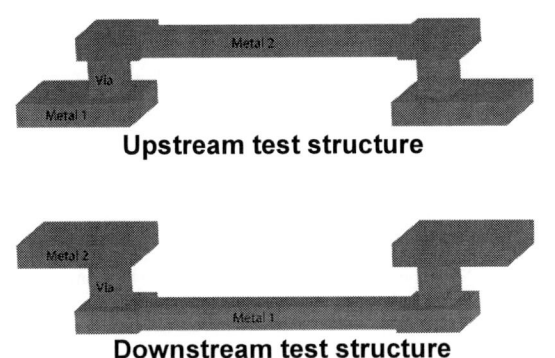

Upstream test structure

Downstream test structure

Figure 1. An illustration of the EM test structures, as prescribed by JEDEC. The width of the lead wires is 500 nm.

In a fourth series of experiments, three variants of pulsed DC stressing were used relative to a constant DC stress to failure at 598 K, as control. The current density in this experiment was 2.5 MA/cm^2. Figure 2a shows the constant DC control condition. The second condition (Figure 2b) used a rectangular DC wave where the samples were allowed to cool between pulses. In each 24 h cycle the samples were soaked at the test temperature of 598 K for 3 h, DC-stressed for about 9 h and then cooled to room temperature for about 12 h. The third variant (Figure 2c) used a similar, but shorter, rectangular wave DC stress. In this experiment the samples were not allowed to cool between pulses. Thus, in each 24 h cycle, samples were heated to the test temperature of 598 K, soaked at that temperature for 2 h and DC-stressed for 1 h. Then, the current was turned off for 2 h, with temperature held at 598 K. The samples were then stressed again for 1 h at 598 K. The samples were cooled to T < 378 K over night to close the cycle. In the fourth experiment (Figure 2d) the samples were not allowed to cool between pulses. The cycle in this test included soak at 598 K for about 3 h, and DC-stress at 598 K for about 3 h. In order to avoid overnight cooling, the samples were continually stressed during that time. Since the test cycles were not programmed, but were rather executed manually, the choice of the test sequences is based on the human ability to operate the test systems.

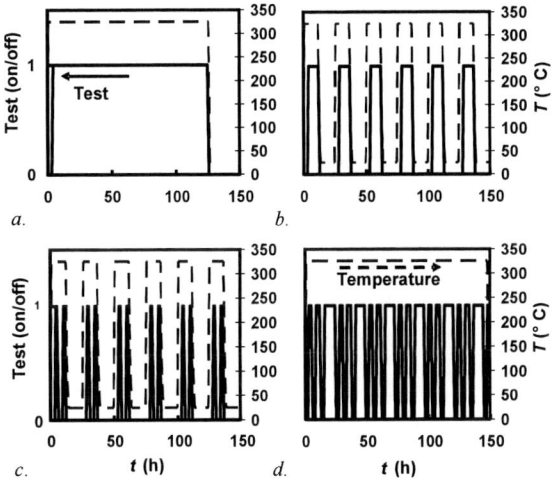

Figure 2. Schematic, temperature (dashed line, right ordinate), and DC stress status (solid line, left ordinate) as a function of time (abscissa) for the four conditions used in the pulsed DC experiment.

The evolution of void migration during high current density AC stressing was also studied, to determine the dominant EM diffusion path. Due to the difficulty of observation of the effects of high currents on 90 nm wide lines, supplementary observations were carried out using a set of 3 μm wide lines of similar metallization surrounded by SiO$_2$ dielectric. Voids were formed by high density AC with peak amplitude of 14.7 MA/cm^2 and a peak temperature of 593 K produced by the Joule heating. Time-lapse photography was then used to film void migration and consolidation, under subsequent DC stressing.

978-1-4244-9113-1/11 $26.00 © 2011 IEEE

III. RESULTS AND DISCUSSION

A. Electromigration Under AC and DC Stress Conditions

Electromigration under AC conditions has been studied extensively in the past [7], [8]. Liew et al reported that lifetime of Al interconnects under AC conditions is about three orders of magnitude longer than EM lifetime under DC conditions and proposed a vacancy relaxation model to explain that observation [9]. While bulk and grain boundary diffusion dominate EM behavior for Al interconnects, interface diffusion is often observed as the dominant failure mechanism for Cu interconnects.

As dimensions scale with technology generations, Cu grains no longer span the entire cross section of the conductor [10], [11]. This leads to increased line resistance, due to electron scattering from grain boundaries, [12]. If grain boundaries in such conductors are parallel to the direction of the current, then EM may take place along them if this diffusion path is easier than that for interface diffusion. Therefore, if AC cycling can lead to a bamboo grain structure in Cu, then AC enhanced grain growth may be expected to slow or even halt grain boundary diffusion and therefore increase EM lifetime. Alternatively, if no change in EM lifetime occurs, despite metal microstructure changes, then interface diffusion can be considered the operative path.

DC pulses present a somewhat different picture. Here the stress profile resembles real operating conditions as discussed above. If grain structure is changed by the cycling, due to thermal fatigue, then EM performance may be dependent on the thermal profile. Since normal operating conditions of most devices consist of intermittent operation at $T > 375$ K, and prolonged idle times at lower temperatures, these experiments provide insights into the relevance of accelerated DC EM testing at elevated temperatures compared to real life operation, and to the validity of extrapolating lifetime expectancy from high temperature, continuous testing, to intermittent lower temperature operations.

The heating and cooling cycles used in this experiment (Figure 2) are significantly more aggressive than those corresponding to normal operating conditions. In particular, during cycling between 598 K and > 373 K (325° C and > 100° C), the device passes through the critical 425 K to 525 K (152° C to 252° C) zone where there is high risk for thermal stress migration in Cu. Single vias connecting to wide lead wires, as prescribed for electromigration test structure design by JEDEC1 and illustrated in Figure 1, are especially susceptible to void formation due to stress migration (SM) in that temperature range. This may lead to a higher probability of early failures as the EM and SM failure mechanisms may be coupled.

The AC experiment included one additional thermal cycle because the transfer of specimens from the AC test system to the DC test system necessitated cooling to room temperature. Consequently, an SM component is expected in the DC electromigration experiment, leading to a somewhat higher probability of early failures.

Results of the first AC / DC electromigration experiment are presented in Figure 3. On the same figure we also show the accumulated time to fail when AC and DC test times are added up. Since the AC test was conducted at 548 K and the DC test was conducted at 598 K a time adjustment was used when merging the test times, using the current acceleration coefficient, n, and the activation energy, E_a, published elsewhere [13].

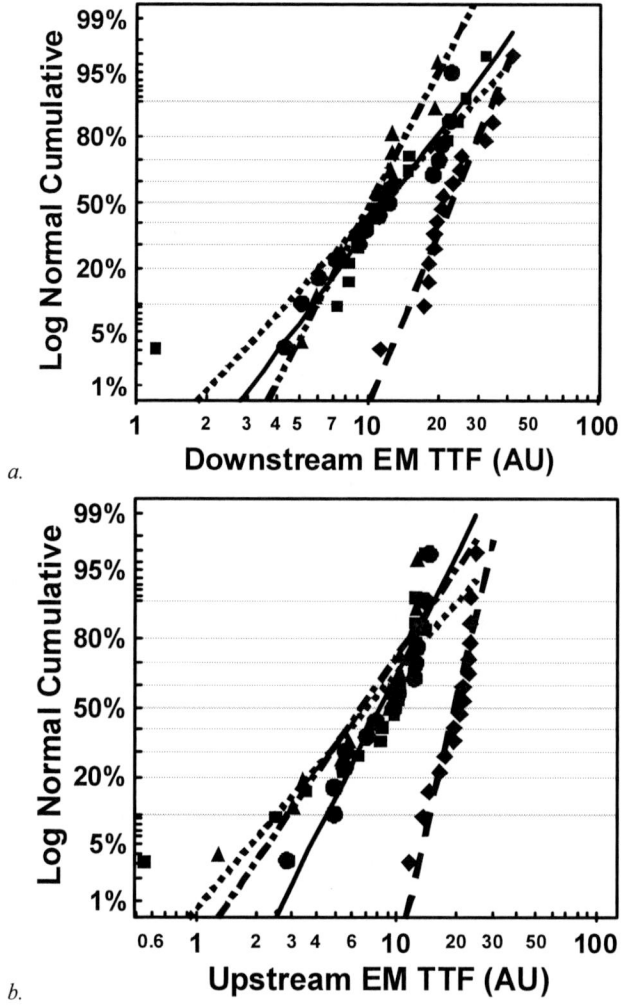

a.

b.

Figure 3. Electromigration time to fail for *a.* downstream, and *b.* upstream DC testing: (----, ■) Samples stressed for a week by 2.5 MA/cm² AC, followed by DC stress to fail, (– – –, ▲) samples stressed for 1002 s at high current density AC, followed by DC stress to fail, and (——, ●) control samples stressed under DC conditions only. The accumulated test time, including both AC and DC stress, is also shown (– –, ◆).

We found that AC stressing had no significant effect on lifetime under DC stress conditions, regardless of AC current density and the Joule heating associated with high current density stressing. The TTF populations resulting from the DC tests, with or without prior AC stressing, are statistically equivalent to the DC control, with $X^2 \sim 0.7$ and ~ 0.4 for downstream and upstream EM respectively. A single early

978-1-4244-9113-1/11 $26.00 © 2011 IEEE

failure was measured in the post AC / DC stress experiment for both upstream and downstream EM. These early fails are from 2 different sites and are not related. The early fails are attributed to SM as discussed above.

When we take into account the entire time in test, including both AC and DC stress times, we find that both t_{50} and σ improve dramatically. Since the AC time at test had no measurable effect on DC time to failure, such an improvement is arbitrary. In this experiment AC stress time was limited to 1 week and as a result we find that t_{50} increased by about a factor of 2 whereas $t_{0.1\%}$ at operating conditions increases by about a factor of 3.

Results of the pulsed DC experiments are presented in Figure 4. Only time under DC stress is presented. Since idle times are arbitrary we did not add them to the final TTF tally. We found that neither the thermal cycling nor the DC pulses had any effect on TTF with X^2 values > 0.2 for all conditions, relative to control, with the exception of square wave downstream test where $C^2 < 0.05$. A single early failure in the downstream testing of the control causes a wider distribution for that condition relative to the other tests. Several early fails for the square wave downstream test cause a lower t_{50} and a wider distribution for this condition. We find no correlation between the early fails and the specific location of the test sites on the wafer. The difference between the t_{50} of that condition and the control is almost a factor of 3. This difference is both statistically and experimentally significant. Moreover, this difference cannot be attributed to experimental equipment because other tests that are equivalent to control were run using the same test equipment and in parallel to this test. We note that packaging yield for this set was also lower than that of other sets, a fact that may provide a possible explanation to the lower time to fail. The two other downstream conditions are statistically equivalent and are well within experimental error. No early fails are measured in these two tests. All are statistically equivalent to the control. For upstream EM testing we observe a small increase of about 25 % in t_{50}. This result is not significant and we conclude that all conditions are equivalent to control in upstream EM testing.

B. Observations on Wide Lines

Optical micrographs from a wide Cu line that had undergone AC stressing for 100 s at 12 MA/cm^2 to induce voids were recorded during a subsequent DC EM test at 15 MA/cm^2, and are shown in Figure 5; contrast in the 92 collected micrographs was enhanced somewhat to improve void visibility. The AC-induced voids were initially evenly distributed throughout the line. During DC stressing about one-third of the voids migrated towards the right along straight paths (inconsistent with influence from microstructure variations), at least one new void formed, and some eventually consolidated. The average speed of the faster-moving voids was approximately $6.7 \times 10^{-2} \mu$m/s. Failure eventually occurred where a moving void intersected a stationary void.

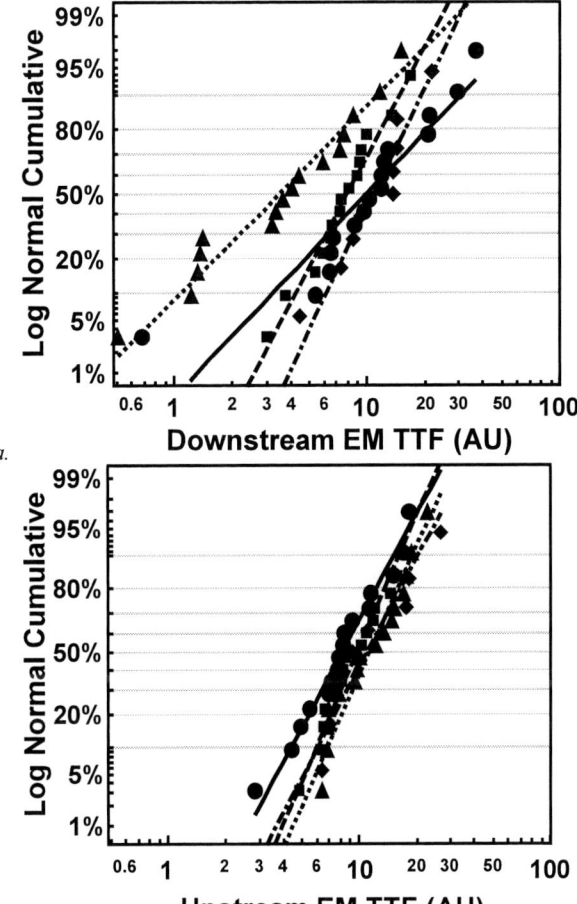

a.

b.

Figure 4. Electromigration time to fail for *a.* downstream and *b.* upstream pulsed DC testing. Continuous DC testing control (— , ●), square wave temperature and DC pulses (- - -, ▲), square wave temperature cycling and shorter DC pulses (– –, ■), and constant test temperature with DC cycling (– - –, ◆).

The classic formula for void electromigration [14] can be written as:

$$v_m = \frac{D_s \delta_s \, e \, z^*_s}{kT} \left(\frac{2E_\infty}{R} \right) , \qquad (1)$$

where v_m is the migration speed, z^*_s is the effective charge for surface electromigration, a measure of the strength of electron-ion scattering, D_s is the surface self diffusion coefficient, δ_s is the thickness of the diffusing layer, k is Boltzmann's constant, T is absolute temperature, R is void radius, and E_∞ is the remote electric field. Using this formula with the present experimental parameters, including $R = 0.5$ μm, produced $D_s \delta_s^* z^*_s = 4.2 \cdot 10^{-6}$ cm^2/s, where $\delta_s^* = \delta_s / r_{nn}$ is the thickness of the diffusing layer, normalized to the equilibrium crystallographic nearest neighbor distance. $D_{s,Cu}$, the surface self diffusion coefficient of copper, had not been measured recently. For the temperature range of interest, < 650 K, $D_{s,Cu}$ values of 5.5×10^{-7} cm^2/s [15] and 2.0×10^{-11} cm^2/s [16] were obtained by interpolating and extrapolating plotted data. These values are much different. Both papers note that the exact condition of the surface may be important. Considering the Butrymowicz et al.[15] value, because the Bowden and Ballufi [16] value is too far away from the present result, we find $z^*_s\delta_s^* = 7.7$. Although this value is as generally expected, this must be considered a coincidence, considering the orders-of-magnitude uncertainty in the surface self diffusion coefficient $D_{s,Cu}$. If δ_s^* is taken as 1, then $z^*_s = 7.7$. Ho's formula [14] for a spherical void would have given $z^*_s = 5.1$. From these supplemental observations, we conclude that the effective surface self diffusion rate in these glass-covered copper lines was much closer to the value of Butrymowicz et al.[15] than to that of Bowden and Ballufi [16], meaning, much closer to the faster of the two surface self diffusion values for copper from the literature. These experiments may be a way to measure parameters governing electromigration.

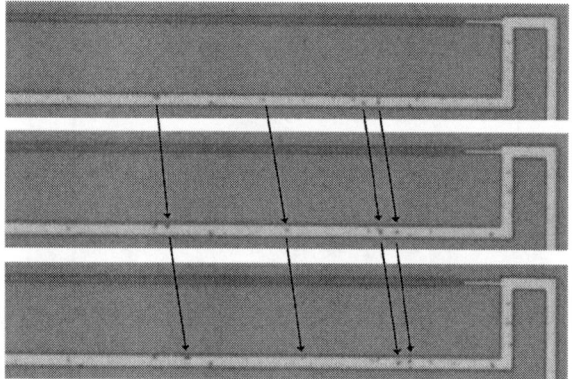

Figure 5. Optical micrographs from a time-lapse EM series on 3 μm wide lines showing voids formed under AC stressing (top), and migrating under DC stressing (middle and bottom)

IV. SUMMARY AND CONCLUSION

Electromigration of a dual damascene, single-via fed, test vehicle was measured using DC, AC followed by DC, and three rectangular-wave DC stressing conditions at 598 K. In some of the experiments samples were allowed to cool between stress cycles.

We find that AC stressing prior to DC stressing to EM failure has no effect on the net, DC only, EM performance. A similar result was obtained with various DC and thermal cycling conditions. All tests, regardless of thermal history and current cycling conditions resulted in similar DC times to fail and distributions of those times.

We conclude that only net DC time at test has a significant effect on time to fail. AC stress has no effect on electromigration performance under DC conditions. Thermal history of the EM test samples, and in particular cooling to room temperature, has no effect of EM lifetime. Pulsed DC stressing also has no effect on EM lifetime. The absence of EM sensitivity to AC stressing, in spite of void formation, suggests that EM in this geometry depends not on metal microstructure, but only on interface voids and structure.

These results have significant implications. Since only the net DC stressing time affects EM lifetime, effective lifetime, when idle and zero-current periods are taken into account, is significantly longer than expected from extrapolations from accelerated tests. In real operating conditions, where zero-current time is at least 50 % of operating time due to the clock cycle, t_{50} is expected to be at least twice as long than the predicated value, based on continuous, DC only, measurements. Subsequently, significant increases in J_{max} and in projected lifetime at operating conditions are obtained. Lifetime is projected to be even longer when idle time and low usage times are taken into account, in particular in light of the conclusion that temperature cycles between operating and idle conditions have no measurable effect on net, DC only, EM lifetime.

We further conclude that the standard test methodologies, using accelerated DC stress conditions at elevated temperatures, are adequate. However, since only the net DC stress time has any measurable effect on EM lifetime, accelerated testing provides a good predictor for lifetime expectations only for the DC component of continuous operating conditions, and underestimates lifetime expectation under pulsed DC, and variable usage operating conditions by 50 % or more. We also conclude that neither AC stress nor DC cycling, and changes to grain structure that may result, vaccinate Cu interconnects against EM failures and thus, do not provide any measurable benefit to EM lifetime.

V. ACKNOWLEDGEMENTS

The authors thank the NIST Office of Microelectronics Programs for support. This manuscript is a partial contribution of the U.S. Department of Commerce, and is not subject to copyright in the United States. We also acknowledge the contribution Jingyan Wang of Novellus, in conducting some of the EM testing, Tom Mountsier and Girish Dixit, also of Novellus, for useful discussions and for their support of this project. We also acknowledge the personnel of the Novellus Customer Integration Center (CIC), and in particular Stefanie Melchor, Nam Huynh, and Surinder Grewal, for preparing the test samples.

REFERENCES

[1] Isothermal Electromigration Test Procedure: JESD61A.01; Method For Characterizing The Electromigration Failure Time Distribution Of Interconnects Under Constant-Current And Temperature Stress: JESD202; Standard Test Structure For Reliability Assessment Of AlCu Metallization With Barrier Materials: JESD87. http://www.jedec.org/standards-documents/results/electromigration. (references)

[2] B.-L. Park, S.-R. Hah, C.-G. Park, D.-K. Jeong, H.-S. Son, H.-S. Oh, J.-H. Chung, J.-L. Nam, K.-M. Park, and J.-D. Byun, "Mechanisms of stress-induced voids in multilevel Cu interconnects," in Proceedings of the International Interconnect Technology Conference, 2002, pp. 130–132..

[3] C. Witt, R. Rosenberg, G. Bonilla, T. Shaw, "Stress relaxation in Cu Thin Films" in Stress Induced Phenomena in Metallization: 10th Workshop, P.S. Ho, S. Ogawa, E. Zschech editors, AIP 2009, p 156.

[4] R.R. Keller, N. Barbosa III, R.H. Geiss, and D.T. Read, "An Electrical Method for Measuring Fatigue and Tensile Properties of Thin Films on Substrates", Key Engineering Materials Vols. 345-346 (2007) pp 1115-1120, (online at http://www.scientific.net).

[5] R.R. Keller, R.H. Geiss, N. Barbosa III, A.J. Slifka, and D.T. Read, "Strain-Induced Grain Growth during Rapid Thermal Cycling of Aluminum Interconnects", Metallurgical and Materials Transactions A, Vol. 38A, (2007) pp 2263.

[6] Roey Shaviv, Gregory J. Harm, Sangita Kumari, Robert R. Keller, David T. Read, "Advanced Metallization Conference 2010", in press.

[7] J. Tao, J.F. Chen, N.W. Cheung, C Hu, Modeling and characterization of electromigration failures under bidirectional current stress, IEEE Transactions on Electron Devices, Vol. 43, No.5, (1996) pp 800.

[8] Z. Li, G. Wu, Y. Wang, Z. Li, Y. Sun, "Numerical calculation of electromigration under pulse current with joule heating", IEEE Transactions on Electron Devices, vol. 46, No. 1, (1999) pp 70.

[9] B.K. Liew, N.W. Cheung, C.Hu, "Electromigration interconnect lifetime under AC and DC pulsed stress", IRPS 1989, pp. 215.

[10] S. Maitrejean, et al. "Cu grain Growth in Damascene Narrow Trenches". Stress Induced Phenomena in Metallization: 10th Workshop, P.S. Ho, S. Ogawa, E. Zschech editors, AIP 2009, pp 135.

[11] L. Zhang et al. "Line Scaling Effect on Grain Structure for Cu Interconnects". Stress Induced Phenomena in Metallization: 10th Workshop, P.S. Ho, S. Ogawa, E. Zschech editors, AIP 2009, pp 151.

[12] W. Zhang,, S.H. Brongersma, O. Richard, B. Brijs, R. Palmans, L. Froyen, K. Maex, Microelectronic Engineering 76 (2004) pp 146–152.

[13] R. Shaviv, H-J. Wu, M. Sriram, W. Wu, A. Pradhan, J. O'Loughlin, K. Chattopadhyay, T. Mountsier, G. Dixit, Advanced Metallization Conference 2008, M. Naik, R. Shaviv, T. Yoda, K. Ueno editors, MRS press (2009) p. 681.

[14] Paul Ho, "Motion of Inclusion Induced by a Direct Current and a Temperature Gradient", Journal of Applied Physics 41 (1), 1970, pp. 64-68.

[15] D. B. Butrymowicz, J. R. Manning, and M. E. Read, "Diffusion in Copper and Copper Alloys Part I. Volume and Surface Self-Diffusion in Copper", Journal of Physical and Chemical Reference Data, 2 (3), 1973, pp. 643-655.

[16] H.G. Bowden and R.W. Ballufi, Measurements of Self-diffusion Coefficients in Copper from the Annealing of Voids, Philosophical Magazine 19 (161), 1969, pp. 1001-1014.

Improving lifetime of Cu interconnects with adding compressive stress at cathode end

L.Arnaud [1,2], P. Lamontagne[2], E. Petitprez[2], R. Galand[2]

1 CEA LETI-MINATEC, 17 rue des martyrs, 38054 Grenoble cedex09, France
2 STMicroelectronics, 850 rue Jean Monnet, 38926 Crolles, France
e-mail: lucile.arnaud@st.com

Abstract— **We show that Cu interconnect lifetime is increased, at least by a factor of 2, with an electromigration (EM) pre-stress. Stress modeling shows that EM induced void at one line end creates compressive stress at the other end. High compressive stress induces an incubation time before EM starts. Incubation time is proportional to the amount of compressive stress induced.**

Keywords- electromigration, Cu interconnect, stress modeling

I. INTRODUCTION

With the continuous reduction of interconnect dimensions in integrated circuits, metal lines are required to carry increasing current densities. To meet electromigration (EM) robustness, major technological breakthroughs, such as Cu alloying [1] or CoWP cap [2] have been introduced. Moreover, it has been proven [3] that Blech length effect provides additional room for current increase in short length interconnect. However, performance improvement is ever required at each node change.

In this work we present our investigations on EM lifetime improvement based on successive EM stressing steps. Our previous studies [4],[5] have already used sequences of bidirectional current stress. Each phase consists of a usual EM test where a constant DC current is applied during enough time to induce an electrically measurable EM degradation in the whole set of tested samples. At each phase change, only the direction of electron flow is reversed by simply using alternatively positive (i+) and negative (i-) current stressing. This stress sequence provides successive void nucleation at both ends [4]. Void healing also occurs with matter transport all over the line length. The present work addresses the lifetime improvement obtained on process with single EM failure mode. We model the stress variation at line end to explain the physics of lifetime improvement.

II. EXPERIMENTAL

The EM test structures are built from 40nm node damascene Cu interconnects with Cu line surrounded by low-k SiOCH dielectric (k=2.5) and capped by a SiCN liner. The test samples have a line height of 130nm and a line width of 63nm with a length of 225µm. Single damascene structures have symmetrical design at both ends (Fig. 1).

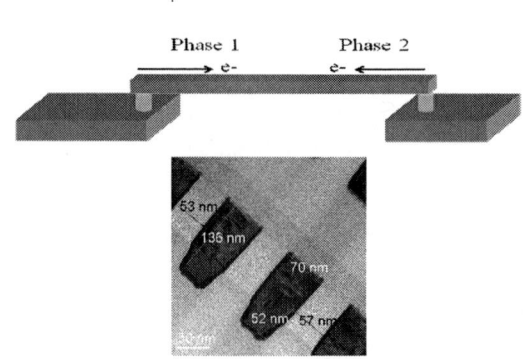

Figure 1. Top : EM Cu test structure with line at metal 1 and a single tungsten contact at each end. Bottom: TEM cross section of Cu M1 line.

The EM tests were carried out at package level with temperature of 300°C and current density of 3 MA/cm². Sample resistance was continuously monitored during 3 phases test (Fig. 2). The trace of resistance is used to study EM induced void kinetics [6].

Resistance evolution is characterized by an initial step (*Rstep*) proportional to the volume of a trench void with length *Lv* [7].

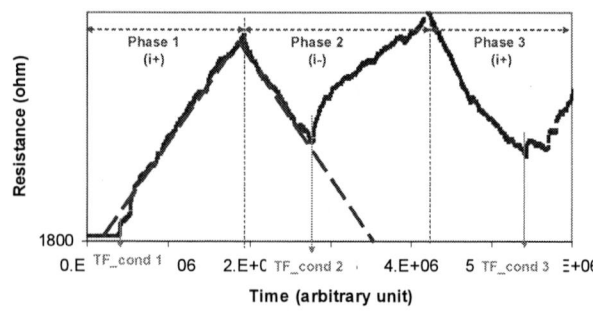

Figure 2. Resistance evolution of a given device for a sequence of 3 phases of bidirectionnal current stress. In phase 1, electron flux is from left to right with cathode end at left side. In phase 2 electron flux is from right to left with cathode at right side. Phase 3 is identical to phase 1.

978-1-4244-9113-1/11 $26.00 © 2011 IEEE

$$R_{step} = \rho_b \, L_v \, /[(2h+w)t_b] \qquad (1)$$

ρ_b and t_b are respectively the TaN/Ta barrier resistivity and thickness and h and w are the copper cross section dimensions (see Fig. 1 bottom).

Here the resistance step is associated to the interconnect time to failure (TTF) of each device for the 3 phases as shown by red arrows in the Fig. 2.

The rate of resistance increase (*Rslope*) is due to the motion of a void front along the conductor [8]. For long lines with length far above Blech length, linear rate of resistance increase is seen and *Rslope* is proportional to Cu drift velocity v_d [7].

$$R_{slope} = \rho_b \, v_d/[(2h+w)t_b] \qquad (2)$$

III. RESULTS

Fig. 3 shows TTF distributions obtained for a set of 30 samples for the 3 phases of current stressing shown in Fig. 2.

The first lifetime distribution (condition 1) has no pre-stress. This is usual experimental condition with constant temperature and current stressing. The second lifetime distribution (condition 2) has a pre-stress. As illustrated in Fig. 2, all the samples have been submitted to 300C and 3MA/cm^2 stress during a given time of 650 hours. Using (1), one can calculated an average void size created during pre-stress of condition 2 with void length ~ 1100nm. The test duration is arbitrary set to a value significantly higher than MTF to induce large void size. During pre-stress of condition 2 voids are created, say, on the left end of the structure. During condition 2 voids are likely to nucleate at the other side, on the right.

Finally the third lifetime distribution (condition 3) has also an additional pre-stress. At right side of the structure, constant stress of 300C and 3 MA/cm^2 during 600 hours created an average void size of about 700nm.

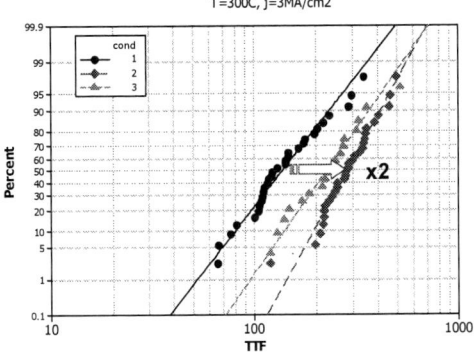

Figure 3. Lifetime distributions of all devices for the 3 phase stress sequence of Fig. 2.

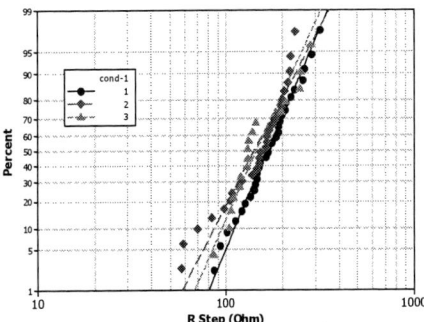

Figure 4. Rstep distributions of all devices and the 3 stress conditions

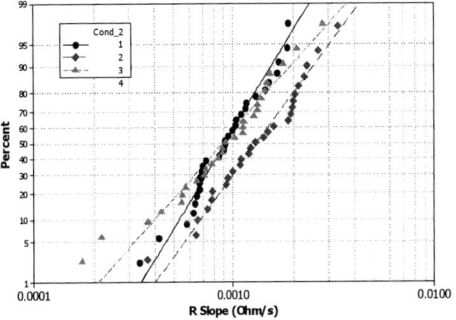

Figure 5. Rslope distributions of all devices and the 3 stress conditions.

A clear lifetime improvement, a factor of 2, is seen between condition 1 and 2 without significant change of distribution slope. It should be added that lifetime improvement is also seen for condition 3, as expected, because similar pre-stress duration is set for this third experiment. To explain the physics of lifetime improvement *Rstep* and *Rslope* distributions for the 3 conditions are plotted in Fig. 4 and Fig. 5.

Very similar *Rstep* and *Rslope* are obtained for all 3 conditions indicating that lifetime increase of conditions 2 and 3 is not due to any change of void size nor Cu drift velocity. Then we believe that lifetime increase is due to an increase in incubation time or time for void formation.

Our hypothesis is that Cu atoms drifted during stress of condition 1, where a large void volume is created at one end of the line, are accumulated at the other side. At the end of condition 1, compressive stress state is expected at line extremity opposite to void nucleation site. The compressive stress state created during pre-stress of condition 2 increases the time to void nucleation before void growth starts. We observe an increase in MTF of condition 2. During pre-stress of condition 3 compressive stress state is also created. Because void volume created during pre-stress of condition 3 is below void volume created during pre-stress of condition 2, low compressive stress state is then expected. MTF increase for condition 3 is then less than MTF increase for condition 2 because smaller time for void formation is needed.

IV. MODELING

In order to explain the experimental results obtained with pre-stress we have used analytical modeling. First we have modeled the stress state at the anode end for first pre-stress condition. Then we have calculated the influence of stress on incubation time at the beginning of condition 2.

For the pre-stress analysis, two solutions of the Korhonen's equation [9] are used in order to calculate i) the compressive EM induced stress at the anode side at the end of condition 1 and ii) the tensile stress needed for void nucleation.

i) compressive stress is obtained with the solution of equation with a vanishing stress at one side and a flux blocked at the other side.

ii) nucleation stress is obtained with the solution where the flux is blocked at both sides.

A. Void growth modeling

Void growth kinetic is given by void volume evolution versus time t, V(t), following [10] :

$$V(t) = Vo\left[1 + \frac{32}{\pi^3}\sum_{n=1}^{\infty}\frac{(-1)^n}{(2n-1)^3}\exp-\left(\frac{2n-1}{2}\pi\right)^2\frac{t}{\tau}\right] \quad (3)$$

$$\tau = \frac{L^2 kT}{DB\Omega} \quad (4)$$

$$Vo = \frac{e\rho Z^* wh}{2\Omega B}jL^2 \quad (5)$$

where D, ρ, Ω, Z^* are respectively the copper diffusion coefficient, resistivity, atomic volume and effective charge. L is the length of interconnect, and B is the effective elastic modulus of interconnect.

From the void growth kinetic, (3), resistance evolution can be modeled at experimental j, L, T conditions. For this purpose, we consider simple void shape where vertical void growth is followed by longitudinal void front movement, (see Fig. 6 [3]). Vertical void growth starts with an initial slit void of length Lv.

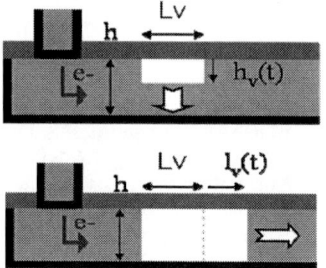

Figure 6. Void growth steps used in model. Top is vertical void growth, bottom is longitudinal void front movement.

The simple picture of Fig. 6 gives reasonable agreement of simulated and experimental resistance evolutions, specially resistance steps and slopes are reproduced. Then experimental resistance evolutions are used to determine the relevant material parameters (D, Z*, B) following the method introduced in [3]. A good agreement between simulated and experimental resistance evolutions is found for values of D = $4.5.10^{-17}$ m/s, Z* = 0.22 and B = 5 GPa. These values are in a reasonable order of magnitude in comparison to the values found in the literature with similar interconnect process [11][12].

B. Stress evolution

Stress evolution versus line length for condition 1 is given in Fig. 7 using the material parameters previously extracted. An evaluation of the tensile stress needed for void nucleation (black curve) and the compression induced at the end of the condition 1 (grey curve) are plotted. We consider that the residual stress in the line is close to zero for our test conditions. This assumption is an approximation consistent with previous experimental stress measurements versus temperature made in Cu interconnects with Synchrotron X Rays [13].

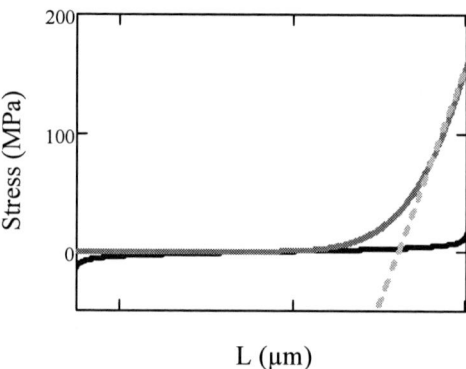

Figure 7. Stress profile along the line length (L).dark grey line: compression induced at the end of condition 1, black line: tensile stress before void nucleation of condition 1, light grey dashed line: linear approximation of the compression used for the simulation of the condition 2;

As expected, for these experimental conditions, almost no change of line stress state over all the line length explains that experimental Cu drift velocity remains constant for all 3 conditions. Stress gradient seems not induce drift velocity variation too. The compressive stress induced at the end of condition 1 is found a decade higher compared with the tensile stress needed for void nucleation. This difference may explain the larger MTF observed in condition 2: compressive stress must first vanish before the cathode is in appropriate tensile stress state for void nucleation.

In the first phase, during condition 1, the stress evolution versus time at anode line end, as shown in Fig. 8, exhibits a rather slow saturation effect. This may explain the origin of the difference of incubation time seen in condition 3 in comparison with condition 2. A lower compressive stress state at the end of condition 2 may induce smaller incubation time of condition 3.

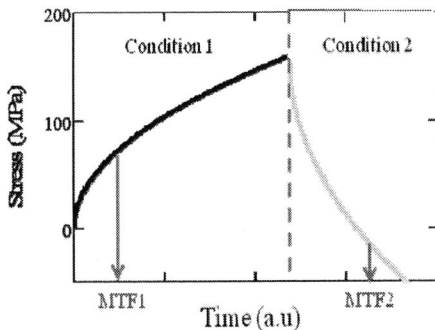

Figure 8. Compressive stress evolution versus time (t) at line end opposite to void nucleation site in condition 1 followed by stress released caused by the reversed current. Maximum stress value is 160MPa at the end of condition 1.

C. Incubation time with initial pre-stress

Additional simulations are made using the Korhonen model to evaluate the effect of the compression at the cathode end at the beginning of the condition 2. Assuming that the vacancy equilibrium is obtained and D is constant along the line length, the continuity equation reads [9]:

$$\frac{\partial \sigma(x,t)}{\partial t} = \frac{\partial}{\partial x}\left[\frac{DB}{kT}\Omega\frac{\partial \sigma(x,t)}{\partial x} + eZ^*\rho j\right] \quad (6)$$

Because the line is confined, the vacancy flux (J) at both ends of the line is equal to zero. J as a function of time (t) and position along the line (x) reads [9]:

$$J(x,t) = -\frac{Dc}{kT}\left(\nabla\mu + eZ^*\rho j\right) \quad (7)$$

where c is the atomic concentration and μ is a chemical potential function which depends on the stress with $\mu = \mu_0 - \Omega\sigma$.

We solve (6) at the cathode end for condition 2 as a function of time. For the simplicity of resolution we assume that the evolution of the stress at one end of the line has no impact on the evolution at the other side. Moreover a linear stress profile $\sigma(x, 0) = ax + \sigma_0$ is used to approximate the stress profile induced at the end of the condition 1 (see dashed line in Fig. 7) where σ_0 is the maximum stress of 160MPa. The solution of (6) finally reads:

$$\sigma(0,t) = -\sqrt{\frac{DB\Omega}{kT}}\left(a + \frac{e\rho Z^* j}{\Omega}\right)\cdot\sqrt{\frac{4t}{\pi}} + \sigma_0 \quad (8)$$

$$\alpha = Z^* e\rho L \quad (9)$$

More details about the resolution method can also be found in [14].

Fig. 8 displays the evolution of the stress during the first and second phase at the right end side (respectively anode during condition 1 then cathode end during condition 2). We can notice that the compressive stress is completely released near the MTF of condition 2 (MTF2). The difference seen on time for compressive stress release and the MTF2 can be explained by the approximation made for the resolution of (6), especially the use of a linear stress profile.

In conclusion, our simulation shows that the compressive stress induced by the EM degradation during condition 1 may explain the MTF gain observed in condition 2.

Because simulation of Fig. 8 shows no saturation of compressive induced stress at the end of our experimental pre-stress conditions, one could expect even more room for lifetime improvement for larger compressive stress. In order to open the validity of interconnect lifetime improvement with adding compressive stress, bidirectional current stress experiments have also been carried out on a few other lots of 40 nm node technology with the same line geometry but with some process differences and different EM lifetimes. There is a general trend of MTF improvement after an initial pre-stress. Fig. 9 shows that the ratio of MTF of condition 1 (no pre-stress) over MTF of condition 2 (initial pre-stress) is increased as a function of pre-stress duration. The pre-stress duration is normalized versus MTF of condition 1 to ignore possible Cu diffusivity differences between wafers.

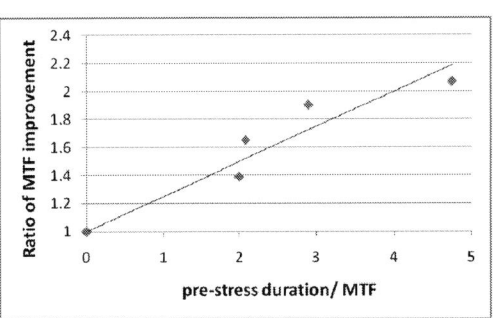

Figure 9. Trend of MTF improvement for EM experiments with an initial pre-stress.

V. CONCLUSION

We have shown that an EM induced void at one line end induced a compressive stress at the other end. Starting EM test with an initial pre-stress where the cathode is in compression induced an incubation time before growth of EM induced void. Lifetime improvement is shown to be proportional to pre-stress duration, i.e. of compressive stress induced. A compressive stress state as high as 160MPa provides a factor of 2 in lifetime in comparison to as grown interconnects.

REFERENCES

[1] M. A. Meyer et al, *Microelectronics Engineering*, vol. 64, p. 375, (2002)

[2] L. Zhang, J.P. Zhou, J. Im, P.S. Ho, O. Aubel, C. Hennesthal, *IEEE International Reliability Physics Symposium Proceedings* 2010, p. 581, (2010)

[3] P.Lamontagne , D. Ney, L. Doyen, Y. Wouters, *IEEE International Reliability Physics Symposium Proceedings* 2010, p. 922, (2010)

[4] L. Doyen, L. Arnaud, X. Federspiel, P. Waltz, Y. Wouters, *IEEE International Reliability Physics Symposium Proceedings* 2008, p. 681, (2008)

[5] L.Arnaud F. Cacho, L. Doyen, F. Terrier, D. Galpin, C. Monget, *Microelectronic Engineering,* vol 87, pp. 355-360, (2010)

[6] L. Arnaud, P. Lamontagne, R. Galand, E. Petitprez, D. Ney, P. Waltz, *IEEE International Reliability Physics Symposium Proceedings* 2011, in Press

[7] L. Doyen, E. Petitprez, P. Waltz, X. Federspiel, L. Arnaud, Y. Wouters,, *J. Appl. Phys.* , vol 104, p. 123521, (2008)

[8] A Oates , *J. Appl. Phy.*, vol 70, p. 5369, (1991)

[9] M.A. Korhonen P. Borgesen, K.Tu, C. Li, *J. Appl. Phys.*, vol 73, p. 3790, (1993).

[10] J.He , Z. Suo, T.N. Marieb, J.A. Maiz, *Appl. Phys. Lett.*, vol 85, p. 4639, (2004).

[11] F. Wei et al., *J. Appl. Phys.*, vol 103, p. 84513 (2008)

[12] A.S. Oates and M.H. Lin, *IEEE International Reliability Physics Symposium Proceedings* 2009, p. 452 (2009)

[13] G. Reimbold, O. Sicardy, L. Arnaud, F. Fillot, J. Torrès, *Proceedings of IEEE-IEDM 2002,* p. 745 (2002)

[14] S.P. Hau-Riege and C.V. Thompson, *J. Appl. Phys.*, vol 89 , p. 601 (2001)

A Study on the Short- and Long-Term Effects of X-Ray Exposure on NAND Flash Memories

S. Gerardin, M. Bagatin, A. Paccagnella

Department of Information Engineering - University of Padova, Padova, Italy,
via Gradenigo 6B, phone: +39-049-8277786; fax: +39-049-8277699; e-mail: simone.gerardin@dei.unipd.it

A. Visconti, S. Beltrami, M. Bertuccio, L.T. Czeppel

Numonyx R&D - Technology Development, Agrate Brianza, Italy

Abstract— **We investigate the effects of X-ray exposure in 41-nm single level NAND Flash memories at small doses, comparable to those used in printed circuit board inspections. We analyze both short-term effects, such as cell threshold voltage shifts during irradiation, and retention and endurance performance of devices exposed to x rays. For doses smaller than 1krad(Si), no effect is observed. At higher doses, charge loss is observed after the exposure and a modest read margin degradation is seen during high-temperature retention tests.**

Keywords- Flash Memories; Floating-gate Cells; X-rays; Radiation Effects;

I. INTRODUCTION

X rays are customarily used to perform inspections during the assembly phase of PCB boards, for instance to assess the quality of solder connections in Ball Grid Arrays (BGA) [1], or during security checks at airports. The doses used by x-ray inspection equipment may vary greatly, from less than 1 to several hundred rads(Si) [1]. In a series of experiments presented in [1], including carry-on and checked baggage and cargo being shipped through several airports, the maximum measured dose was 50 mrads(Si).

In addition, x rays are also used on the ground as a fast tool to emulate total dose effects in the space environment [2]. The doses in space missions, due to particles trapped in planets' magnetospheres and solar activity, are much higher than those on the ground and may reach also multiple Mrad(Si) in a planned mission to the moons of Jupiter [3].

X rays, and ionizing radiation in general, are known to affect dielectric layers in electronic components, causing trapped charge build-up and interface state generation that may lead, for instance, to threshold voltage shifts in MOS-FETs and gain degradation in bipolar devices. There is a vast literature dealing with the effects of radiation. The typical doses to failure vary greatly, depending on the type of component and on the manufacturing technology, ranging from a few krads(Si) to Mrads(Si) in specially designed rad-hard components.

Technology scaling is increasing the tolerance of Commercial Off The Shelf (COTS) components to total dose effects, due to the thinning of the gate oxide, and the consequent reduction in threshold voltage shift. However some issues still remain even in low-voltage devices, such as edge leakage and inter-device leakage.

Relatively less attention has been devoted to possible synergistic interactions between radiation-induced degradation and wear-out. Indeed, wear-out is very important in Flash technologies, where high-voltages are used for program and erase operations, and limits the number of program/erase cycles that can be carried out on a memory.

Some studies have been performed to understand the impact of x-ray inspections in electronic components. In DRAMs, degradation in refresh characteristics was observed even at low doses (tens of rad(Si)) [4] and was attributed to junction leakage increase. It was also demonstrated that after baking at elevated temperature, the refresh time returned to pre-irradiation levels.

Several works have been carried out on the effects of x rays on floating gate cells for use in the harsh space environment [5,6]. The observed phenomena include: failure to program and erase due to charge pump degradation, corruption of memory bits in the floating gate array, and failures in building blocks such as row and column decoders. A reduction in time of the number of x-ray induced errors was also observed [6].

However, quite often, the doses used in these studies are significantly higher than those required by ground inspections and little attention is paid to possible long-term phenomena affecting device reliability. [7] shows that the endurance of Flash NAND memories is practically unaffected after total dose delivered with a Co-60 source. The tested parts were irradiated to doses larger than 50krad(SiO$_2$). [8,9] reported retention issues after heavy-ion irradiation, due to Radiation Induced Leakage Current (RILC), which creates a permanent leakage path in the tunnel oxide of cells struck by heavy ions.

In this contribution we address the impact of x-ray exposure at low doses on the cell threshold voltage and on the retention and endurance characteristics. The paper is organized as follows: Section II describes the devices and the experiments carried out; Section III presents the experimental results, focusing on the effects visible during or immediately after the irradiation and on retention and endurance experiments; finally, Section IV discusses the physical mechanisms underlying the observed phenomena.

II. EXPERIMENTS AND DEVICES

We used commercial 41-nm Single Level Cell (SLC) NAND Flash memories manufactured by Numonyx with a capacity of 8 Gbit. Reserved test mode routines were used to measure the V_{th} distributions of the FG cells. Fig. 1 shows a cartoon of the threshold voltage (V_{th}) distributions in these devices. The program and erase algorithms ensure that the threshold voltage of programmed cells is larger than the Program Verify (PV) level, and that of erased cells is below the

978-1-4244-9113-1/11 $26.00 © 2011 IEEE

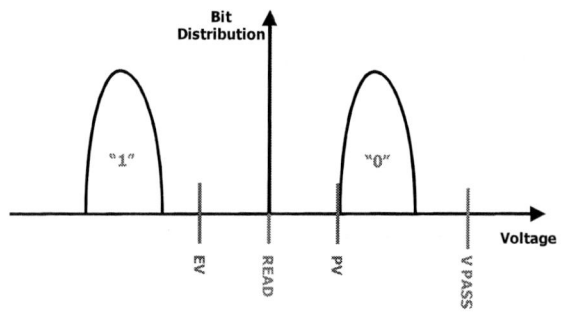

Figure 1. V_{th} distributions and reference levels in NAND arrays. PV = Program Verify; EV = Erase Verify.

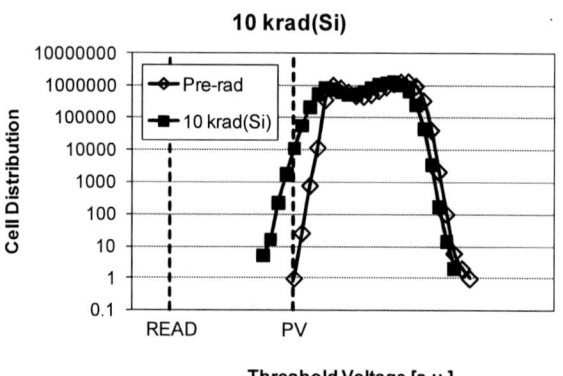

Figure 2. V_{th} distributions for programmed cells before and after a 10-krad(Si) exposure to x rays.

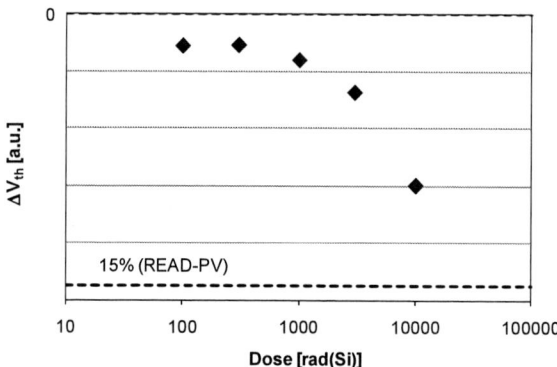

Figure 3. Average V_{th} shift for programmed cells as a function of dose after x-ray exposure.

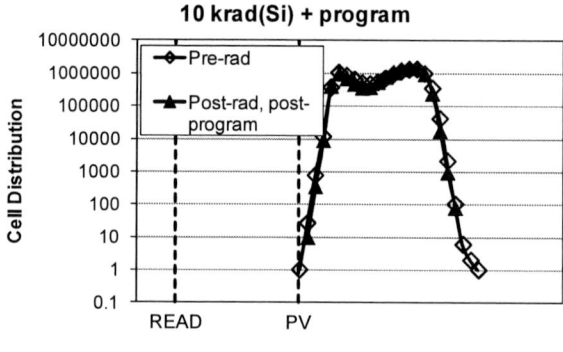

Figure 4. V_{th} distributions for programmed cells before irradiation and after reprogram following a 10-krad(Si) exposure to x rays.

Erase Verify (EV) level. To discriminate between the programmed and the erased states, the cells to be read are biased at 0V (READ). The cells that must let the signal through in the NAND strings are biased at a voltage, V_{PASS}, significantly higher than the threshold voltage of the programmed cells.

In the following we will use the difference between PV and READ (PV-READ) as a metric to evaluate the magnitude of the observed effects in the programmed cells.

Irradiation was performed at wafer level using a 10-keV x-ray probe station at the Legnaro National Laboratories (LNL) in Padova, Italy at different doses: 100, 300, 1k, 3k, 10k. Three dose rates were used: 14 rad(Si)/s for exposures up to 300 rad(Si), 140 rad(Si)/s for exposures to 1 krad(Si) and 3 krad(Si), and 350 rad(Si)/s for exposures to 10 krad(Si). Different sets of devices were used for each dose level. The chips in the wafer were irradiated individually, using a motorized x ray tube stage. A thick aluminum shield connected to the stage and placed just above the wafer, with an opening of the same size as the chip, was used to avoid irradiating neighboring devices.

The devices were left unbiased with floating terminals during the exposure. Before and after the irradiation the threshold voltage distributions of the memories were measured.

After irradiation the following tests were performed:
1. High-temperature retention test: bake at high temperature, up to 1000 hours at 150°C.
2. Room-temperature retention test: bake at room temperature, up to 1000 hours.
3. Endurance: room-temperature Program/Erase (P/E) cycling up to 100 kCycles

The retention tests were performed in steps, each concluding with a measurement of the threshold voltage distribution. Reference chips were measured alongside irradiated devices. Three 300-mm wafers, corresponding to ~1 Tbit, were used for this study.

III. RESULTS

In the following we will focus on programmed cells, since they are expected to be more sensitive to ionizing radiation [5,6], as will be explained in Section IV. Physical checkerboard pattern has been used for all retention tests, allowing the detection of both "0" and "1" fails.

A. Prompt effects

Figs. 2-4 illustrate the short-term effects of x-rays. The exposure causes a threshold voltage reduction in the programmed floating gate cells.

Fig. 2 shows the effects of a 10-krad(Si) x-ray exposure on the threshold voltage distribution: a shift of all the cells towards lower voltages is visible. The shape of the distribution is almost unchanged, even though a certain amount of broadening occurs. In fact, the left part of the distribution shifts more than the right part. The exposed cells with the lowest threshold voltage are still very far from the read voltage. As a result, no digital errors are detected, even though some of the cell V_{th} are below the PV level.

Fig. 3 shows the average threshold voltage shift as a function of dose. The reduction in net stored charge is visible at doses larger than or equal to 1 krad(Si) and is always within

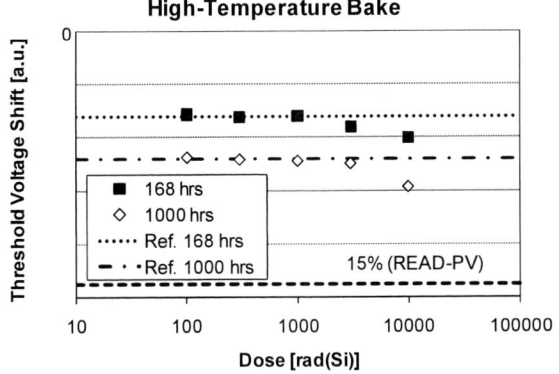

Figure 5. Average V_{th} shift during bake at 150°C on devices reprogrammed after exposure to x rays.

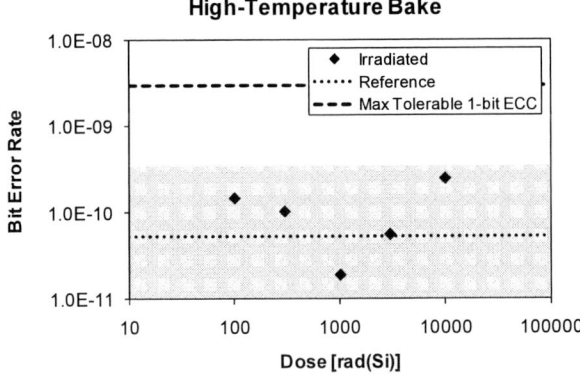

Figure 6. Bit error rate after 500 hours at 150°C on devices reprogrammed after exposure to x rays. The gray area is the reference error bar.

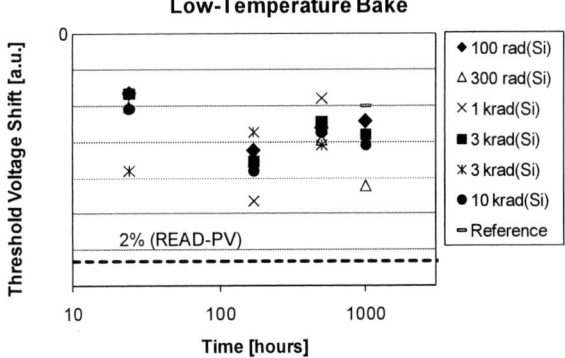

Figure 7. Average V_{th} shift during bake at room temperature in devices reprogrammed after exposure to x rays.

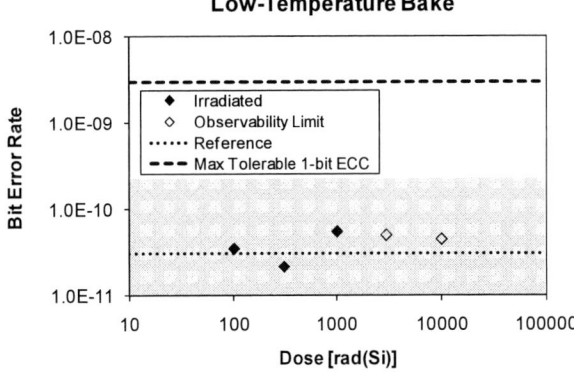

Figure 8. Bit error rate after 500 hours at room temperature on devices reprogrammed after exposure to x rays. The gray area is the reference error bar.

15% of (PV-READ) in the dose range considered here. For this dose range, the threshold voltage shift has an exponential dependence on the received dose.

As displayed in Fig. 4, a program operation on floating gate cells that have previously received 10 krad(Si) successfully restores the initial threshold voltage distribution. In fact, there is no observable difference in the cell distribution before irradiation and after program.

B. Retention tests

Figs. 5-6 show the results of high-temperature retention tests performed at 150°C on devices that have been erased and programmed again after the exposure to x rays. Fig. 5 shows the behavior of the average cell threshold voltage at 168 hrs and 1000 hrs during the retention test, whereas Fig. 6 illustrates the bit error rate after 500 hrs.

As shown in Fig. 5, the cells experience an average threshold voltage decrease with time when submitted to high temperatures, whose magnitude increases with the amount of dose received prior to bake. Our data show that this effect is experimentally observable only for doses equal or larger than 3 krad(Si). No observable radiation-induced tails appear during the high-temperature tests, as shown by the bit error rate (BER) in Fig. 6. BER is determined by those cells whose threshold voltage shifts enough to go below the read voltage. BER in irradiated is in line with reference parts and shows no correlation with the absorbed dose.

Room temperature retention tests are reported in Figs. 7-8. As for high-temperature measurements, we show both the average threshold voltage (Fig. 7) and the bit error rate (Fig. 8). No intrinsic global threshold shift due to the x-ray exposure is visible during these tests (Fig. 7). The data are scattered and there is no correlation with the dose received by the component. The magnitude of the shift is in all cases in line with the reference samples and well within 2% of (READ-PV). In addition, no observable radiation-induced tails appear during room temperature tests, as shown by the BER in Fig. 8.

Similar retention results have been obtained on parts not reprogrammed after irradiation.

C. Endurance tests

Fig. 9 plots the erase time as a function of program/erase cycles for devices irradiated to different doses. No correlation is found between dose and erase time (or program time, not shown). For all the parts shown in the plot, the behavior is in line with the reference unirradiated device. All the devices can be erased in less than the maximum specified time up to 10^5 cycles. After 10^5 cycles, which is the typical spec for single level Flash memories, there is still significant margin.

IV. DISCUSSION

Three mechanisms are responsible for the shifts in the threshold voltage after exposure to x rays [6]:

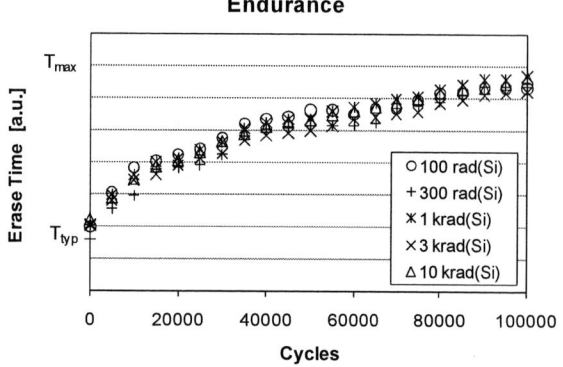

Figure 9. Erase time as a function of dose in devices cycled after exposure to x rays.

i) injection of radiation-induced carriers from the tunnel oxide and the Oxide Nitride Oxide (ONO) layer into the floating gate;

ii) photoemission of carriers stored in the floating gate over the tunnel and ONO barriers;

iii) positive charge trapping in the insulating layers, in particular the tunnel oxide.

The first mechanism causes a reduction in the net stored charge. In fact, when electrons are in the floating gate, the electric field is such that radiation-induced holes are injected from the insulating layers into the floating gate, whereas radiation-induced electrons are swept away from the floating gate. The opposite is true with holes.

The second mechanism again reduces the amount of stored charge: whatever is in the floating gate, either holes or electrons, is removed.

The third mechanism (positive charge trapping) always tends to shift the threshold voltage towards lower values, independently of the sign of the charge stored in the floating gate.

In the case of programmed cells, when electrons are stored in the floating gate, i)-iii) all cause the cell threshold voltage to decrease, which may cause corruption in the data if the threshold voltage goes below the read voltage. This explains the V_{th} shift shown in Fig. 2 for programmed cells. However, as we observed before, the post-rad distribution is broader than the pre-rad distribution. In fact, cells having the same threshold voltage do not necessarily have the same amount of stored electrons; this is because, after each program operation, the so-called compaction algorithm is applied to tighten the V_{th} distributions. As a result, cells that have been irradiated but not reprogrammed are more spread. For erased cells, i) and ii) tend to increase the threshold voltage, whereas iii) tends to decrease it, balancing at least in part the effects of i) and ii).

The net effect of i) and ii) is to move the cell towards the intrinsic state, i.e. with no net stored charge. i) and ii) have been shown to be dominant with respect to iii), because of the relatively small tunnel oxide thickness. It is well known that in a MOSFET, the amount of threshold voltage shift decreases with decreasing oxide thickness, with a $1/t_{ox}^2$ dependence or even faster for ultra-thin gate oxides [10].

Due to the effects of i) and ii), the position of the intrinsic distribution is of paramount importance to determine if the FG information can be corrupted by radiation. For our devices,

the intrinsic distribution is mostly placed below 0V, meaning that as a result of charge loss, only programmed cells can become corrupted (actually, a small tail of the intrinsic distribution is above 0V, so a small fraction of erased cells could become corrupted as well).

In [11], the amount of threshold voltage shift due to charge loss is shown to be linearly dependent on the tunnel oxide electric field, which is determined by the charge stored in the floating gate. From total dose experiments performed on NMOSFETs [12], we can also conclude that the amount of charge trapping is largest when the electric field across the tunnel oxide is maximum. However, as shown in [6] for Multi-Level Cell (MLC) devices, errors reduction due to annealing is more pronounced for the lowest V_{th} program levels than for the highest levels.

If we perform an erase-program operation on the irradiated device and we measure the V_{th} distributions, the situation is exactly the same as before irradiation. This confirms that iii) does not have a large impact, and that, as expected, i) and ii) are reversible.

Quantitatively, our experimental data are consistent with previous reports on less scaled cells [5]. The dose necessary to observe errors is significantly larger than 10 krad(Si). We would like to mention that, when comparing with published data on protons or gamma irradiation, one has to consider that dose enhancement effects take place with x rays (which interact with the target material through the photoelectric effects), in the presence of interfaces between high-Z and low-Z materials, i.e. word lines and floating gates. This means that the dose actually received by the floating gate may be larger than the nominal one.

A. Retention

Let us now focus on the effects of radiation observed during high-temperature tests, where an acceleration with dose was observed. Previous work [13] on SiO_2 layers showed that, for high-enough doses, leakage currents can arise across thin oxide films, due to Trap-Assisted Tunneling (TAT). This Radiation Induced Leakage Current (RILC) was found to significantly anneal only at temperatures well above 150°C [14], i.e. larger than our high-temperature retention tests. Similar to Stress Induced Leakage Current (SILC), RILC is not accelerated by temperature, and therefore should be visible both at high and low temperature, which is not the case (Figs. 5 and 7). Further, with RILC, tails in the V_{th} distributions should appear, because of TAT in clusters of defects, which again is not observed (Figs. 6 and 8).

A possible mechanism for the more pronounced shifts during high-temperature tests in irradiated devices is thermally activated detrapping of electrons from the tunnel oxide. Indeed, the retention measurements were performed a few weeks after the irradiation. In [6] a reduction of x-ray induced errors with time (active up to a few hundreds of hours after the irradiation, and therefore over at the time of our retention tests) was observed after radiation exposure and was attributed to an increase in V_{th} due to the removal or compensation of radiation-induced *positive* charge in the tunnel oxide by tunneling electrons. A stronger removal of these 'compensating' electrons with respect to the radiation-induced trapped holes at high-temperature could be the reason of the dose-dependent shifts in Fig. 5.

B. Endurance

Generally, endurance degradation is related to trapped electron build-up and interface-state generation in the tunnel oxide. The first one reduces tunneling and therefore slows down P/E operations, in addition to increasing V_{th}. The latter only increases V_{th}. Since ionizing radiation typically causes the trapping of positive charge in oxide layers, one would expect the trapped holes to compensate the electrons trapped during P/E cycles. However, this radiation-induced trapped charge component is quite small in thin oxides and, as we saw, it does not have an intrinsic impact on the endurance of the irradiated samples, at least for the dose range considered here. The same can be said for radiation-induced interface traps.

V. Conclusions

X rays have been shown to cause no measurable short- and long-term effects in 41-nm NAND Flash memories for doses smaller than 1 krad(Si). This level is about an order of magnitude larger than that used during typical x-ray inspections, which, therefore, do not pose a significant threat to the information and the reliability of floating gate memories.

Above the 1-krad mark, observed phenomena include charge loss and charge trapping after the exposure and a modest acceleration of V_{th} shift during high-temperature retention tests. Both user-mode retention and endurance are practically unaffected in devices irradiated up to 10 krad(Si).

VI. Acknowledgement

The authors are greatly indebted to Serena Mattiazzo, Devis Pantano, and Mario Tessaro, INFN Padova for their invaluable help with the x-ray probe station, Marco Vezzoli (Numonyx R&D) for helping in data analysis and Caterina Olivieri (Numonyx NAND QA group) for supporting measurements.

References

[1] R.C. Blish, S. Li, D. Lehtonen, "Filter optimization for X-ray inspection of surface-mounted ICs," Device and Materials Reliability, IEEE Transactions on, vol. 2, pp. 102 - 106, dec. 2002

[2] D. M. Fleetwood, R. W. Beegle, F. W. Sexton, P. S. Winokur, S. L. Miller, R. K. Treece, J. R. Schwank, R. V. Jones, P. J. McWhorter, "Using a 10-keV X-Ray Source for Hardness Assurance," Nuclear Science, IEEE Transactions on, vol. 33, pp. 1330 -1336, dec. 1986

[3] H. Schöne, "Trading Margin with Knowledge," Keynote 2, 2010 IEEE International Reliability Physics Symposium

[4] A. Ditali, M. Ma, M. Johnston, "X-Ray Inspection-Induced Latent Damage in DRAM," Reliability Physics Symposium Proceedings, 2006. 44th Annual., IEEE International, vol. , pp. 266 -269, Mar. 2006

[5] M. Bagatin, G. Cellere, S. Gerardin, A. Paccagnella, A. Visconti, S. Beltrami, "TID Sensitivity of NAND Flash Memory Building Blocks," IEEE Transactions on Nuclear Science, vol. 56, pp. 1909 - 1913, Aug. 2009

[6] M. Bagatin, S. Gerardin, G. Cellere, A. Paccagnella, A. Visconti, M. Bonanomi, S. Beltrami, "Error Instability in Floating Gate Flash Memories Exposed to TID," IEEE Transactions on Nuclear Science, vol. 56, pp. 3267 -3273, Dec. 2009

[7] T. Oldham, M. Friendlich, M. Carts, C. Seidleck, K. LaBel, "Effect of Radiation Exposure on the Endurance of Commercial NAND Flash Memory," IEEE Transactions on Nuclear Science, vol. 56, pp. 3280 -3284, Dec. 2009

[8] L. Larcher, G. Cellere, A. Paccagnella, A. Chimenton, A. Candelori, A. Modelli, "Data retention after heavy ion exposure of Floating Gate memories: Analysis and simulation," *IEEE Trans. Nucl. Sci.*, vol. 50, no. 6, pp. 2176–2183, Dec. 2003

[9] G. Cellere, L. Larcher, A. Paccagnella, A. Visconti, M. Bonanomi, "Radiation Induced Leakage Current in Floating Gate Memory Cells," *IEEE Trans. Nucl. Sci.*, vol. 52, no. 6, pp. 2144–2152, Dec. 2005

[10] J. M. McGarrity, "Considerations for hardening MOS devices and circuits for low radiation doses," IEEE Trans. Nucl. Sci., vol. NS-27, p. 1739, 1980

[11] M. Bagatin, S. Gerardin, A. Paccagnella, G. Cellere, A. Visconti, M. Bonanomi, "Increase in the Heavy-ion Upset Cross Section of Floating Gate Cells Previously Exposed to TID," Nuclear Science, IEEE Transactions on, vol. 57, pp. 3407 -3413, December 2010

[12] L. Gonella, F. Faccio, M. Silvestri, S. Gerardin, D. Pantano, V. Re, M. Manghisoni, L. Ratti, A. Ranieri, "Total Ionizing Dose effects in 130-nm commercial CMOS technologies for HEP experiments," Nuclear Instruments and Methods in Physics Research Section A: Accelerators, Spectrometers, Detectors and Associated Equipment Volume 582, Issue 3, 1 December 2007, Pages 750-754

[13] M. Ceschia, A. Paccagnella, S. Sandrin, G. Ghidini, J. Wyss, M. Lavale, O. Flament, "Low field leakage current and soft breakdown in ultra-thin gate oxides after heavy ions, electron or X-ray irradiation," Nuclear Science, IEEE Transactions on, vol. 47, pp. 566 -573, June 2000

[14] C.-H. Ang, C.-H. Ling, Z.-Y. Cheng, S.-J. Kim, B.-J. Cho, "Bias and thermal annealings of radiation-induced leakage currents in thin-gate oxides," Nuclear Science, IEEE Transactions on, vol. 47, pp. 2758 - 2764, December 2000.

Application of Reliability Test Standards to SiC Power MOSFETs

Ronald Green, Aivars Lelis, and Daniel Habersat
Power Components Branch
U.S. Army Research Laboratory, ARL
Adelphi, Maryland 20783, USA
phone: 1 –(301)- 394-5431, email: ronald.greenjr@us.army.mil

Abstract—The application of existing reliability test standards, based on Si technology, to SiC power MOSFET reliability qualification can in some cases result in ambiguous test results. Depending on the exact measurement procedure, a given device stress tested under identical conditions may either pass or fail. The large variations observed in I_D-V_{GS} characteristics, and accompanying shift in threshold voltage (V_T) and change in leakage current, are likely due to the complex time, temperature, and bias dependent nature of the charging and discharging of significant numbers of near-interfacial oxide traps (and possibly mitigated by the movement of mobile ions) which are not present in Si power devices. The variation in V_T following a high temperature gate-bias (HTGB) stress is shown to be dependent on the measurement delay time, sweep direction, and temperature. Negative gate-bias temperature stress results show that device reliability may be limited due to increased drain leakage current in the OFF-state, which is caused by large shifts in V_T depending on the gate-bias stress time, bias magnitude, and stress temperature. In addition, positive gate-bias stressing at elevated temperature may increase power dissipation in the ON-state.

Keywords-component; Power MOSFETs, SiC, V_T instability, oxide traps, HTGB, BTS

I. INTRODUCTION

Silicon Carbide (SiC) power devices provide enhanced temperature operation with higher breakdown field capability, enabling the development of power systems with higher power density, lower losses, and the promise of improved reliability. Advances in material and device technology have resulted in the development of power SiC metal-oxide-semiconductor field-effect transistors (MOSFETs) with higher blocking voltage and lower specific on-resistance in comparison to their silicon (Si) counterparts [1, 2]. The feasibility of SiC MOSFET devices for power electronics applications has been recently demonstrated with the development of several power modules [1-2]. However, device reliability issues, including threshold voltage (V_T) instability [3, 4], must first be resolved.

In this paper, we investigate the effects of high temperature gate bias (HTGB) stress on the reliability of 4H-SiC power MOSFET devices within the guidelines of accepted industrial and military standards for stress test qualification of semiconductor devices. Our findings reveal that the application of existing standards (e.g. JEDEC JESD22-A108C [5], MIL-STD-750E [6], and AEC-Q101 [7]), which are based on Si device technologies, may result in inconsistent pass/fail results when applied to SiC MOSFET devices due to a large threshold voltage variation that strongly depends on measurement conditions.

For instance, the Joint Electron Devices Engineering Council (JEDEC) standard [5] requires that post burn-in electrical measurements be completed as soon as possible, but no longer than 96 hours after removal of the bias. This 96-hour window for electrical testing appears to be inappropriate in qualifying SiC MOSFET devices because of the complex time, temperature, and bias dependent nature of the observed V_T instability, which is likely due to the charging and discharging of near-interfacial oxide traps through a direct

978-1-4244-9113-1/11 $26.00 © 2011 IEEE

tunneling mechanism [3, 8]. This complex dependence causes the drain current versus gate-to-source voltage (I_D-V_{GS}) characteristics to be sensitive to both the measurement sweep speed and direction, which are not addressed at all by the present standards.

Our stress results also show the large variability in the threshold-voltage shift due to variations in stress bias, time, and temperature. The different existing standards, with their varying stress conditions, may also result in inconsistent device evaluations. For example, the Automotive Electronics Council (AEC) standard [7] calls for a 1,000 hour HTRB and HTGB stress, whereas the Department of Defense (DoD) Mil Std 750-E [6] only requires a 48 hour stress, at 80% of the maximum rated gate bias.

This variability in the threshold-voltage shift makes it critical to carefully assess the applicability of existing Si-based testing standards as they begin to be used to qualify SiC-based power electronics. A proper reliability test should be able to quickly and non-destructively evaluate a device's suitability for long-term, reliable operation and separate bad parts from good ones. Before existing reliability standards are applied to SiC power devices, we must determine if they are adequate for evaluating the reliability of SiC-based technology.

The present standards also differ in relation to electrical measurement procedures. For example, JESD22-A108C [5] allows for electrical testing at elevated temperature but only after post-stress room temperature measurements have been performed. AEC-Q101 [7] requires that pre- and post-stress electrical measurements occur at room temperature, but does not explicitly state whether device characteristics could be made at elevated temperatures.

There is little consistency between the standards as it relates to device failure criteria as well. The AEC standard, for example, explicitly requires that specific device parameters remain within electrical test limits of the specification and within $\pm$ 20 percent of their pre-stress values. If not, these devices are deemed as failing. Leakage currents, however, cannot exceed five times their initial value.

The applicable JEDEC and DoD standards are not so clear in regard to device failure criteria.

II. EXPERIMENTAL PROCEDURE

We have examined the effects of HTGB stressing on the I_D-V_{GS} characteristics of large area (0.56 cm^2) SiC power MOSFET devices. These devices are research samples with voltage and current ratings of 1200 V and 67 A, respectively, and fairly representative of the state-of-the-art in SiC MOSFET technology. Two populations of power MOSFETs were evaluated. Previous, but unpublished negative bias temperature stress (NBTS) revealed significant V_T drift in devices from Group A. Devices from Group B were redesigned to mitigate V_T drift issues associated with previously fabricated Group A devices. Both device groups have a thermally-grown gate oxide which received a standard nitric oxide (NO) post-oxidation anneal.

During a stress cycle, all devices were stressed at a temperature of 150 °C with either V_{GS} = +15 V or –15 V depending on whether a positive bias temperature stress (PBTS) or negative bias temperature stress (NBTS) was applied, unless otherwise specified. For the duration of the temperature ramp, V_{GS} was maintained at $\pm$ 15 V with V_{DS} = 0 V. The device temperature generally stabilized in approximately 90 seconds, after which, the high-temperature bias stress was initiated. At the end of the stress, the device was rapidly cooled to room temperature in about 90 seconds while maintaining the gate-bias stress.

I_D-V_{GS} measurements were typically made immediately following an HTGB stress cycle with V_{DS} = 50 mV once the device reached room temperature and the stress bias was removed. The stress and measurement sequences were made using an Agilent 4155C Parameter Analyzer. Following a PBTS cycle, an immediate sweep down of V_{GS} from +15 to −5 V was made to measure the effect of the stress on the device I_D-V_{GS} characteristics. After a NBTS cycle, an immediate sweep up of V_{GS} from −5 to +15 V was made to measure the effect of the stress on the device I_D-V_{GS} characteristics. The SiC MOSFET devices were characterized by measuring shifts in the linear extrapolated V_T and low current (1×10^{-6} A) V_{GS}, both pre- and post-stress. The change in threshold voltage (ΔV_T) and gate bias (ΔV_{GS}) characterizes the effect of the stress and was determined by taking the differences between post- and pre-stress values for V_T and V_{GS}, respectively.

III. RESULTS AND DISCUSSION

In the following sections we examine the effects of HTGB stress testing and variations in the subsequent measurement conditions on the I_D-V_{GS} characteristics of relative state-of-the-art SiC power MOSFET devices and demonstrate the shortcomings in existing reliability and qualification standards when applied to SiC power MOSFETs.

A. Previous Results

It has been previously observed on various SiC MOSFET devices from different manufactures that gate-bias stressing results in instability of the I_D-V_{GS} characteristics, with positive-bias stress causing a positive shift and negative-bias stress causing a negative shift. The resulting V_T shifts are not permanent and may be reversed by a reversal of the applied gate bias. This instability is likely due to the charging and discharging of near-interfacial oxide traps by electrons tunneling to and from the SiC [8]. Recent high-temperature bias-stress testing of power SiC MOSFETs, including self-heating caused by ON-state current stressing, reveals a significant increase in this V_T instability in some devices, which may be due to the activation of additional oxide traps, related to an oxygen vacancy defect referred to as an E-prime center [3]. A smaller variation in V_T instability in other devices with increasing temperature may either be due to improved gate oxide processing or to the presence of mobile ions, which will cause an opposite shift to that caused by charge trapping [9]. A negative V_T shift can give rise to increased leakage current in the OFF-state, especially at elevated temperature where an increased stretch-out of the subthreshold slope has been observed [3]. Therefore, high-temperature gate-bias testing of these devices is necessary for complete reliability monitoring and device qualification.

B. Measurement Parameters

Previous work on SiC power MOSFETs [3,4,8-10] have shown how measurement parameters and conditions can affect the data. Here, we investigate the effects of three measurement-specific parameters and their implications for robust reliability testing: delay time, sweep direction, and temperature.

1) Measurement Delay Time

In this section, we examine the effects of measurement delay time on the transfer characteristics of SiC power MOSFETs following both positive and negative static HTGB stresses. Measurement delay is defined as the time between removing the gate-bias stress and initiating an I_D-V_{GS} measurement sweep. This delay time should not be confused with the time associated with making an I-V measurement (i.e. gate sweep speed). Variation in measurement speed has been previously shown to produce a considerable variation in the I_D-V_{GS} characteristics and corresponding V_T of SiC MOSFETs [8, 10] due to the sensitivity of the oxide trap charging process to the bias applied during the measurement. The JEDEC standard allows for a 96 hour window for electrical testing after removal of the bias stress. However, the data presented below indicates that this large time window may not be appropriate when testing SiC power MOSFETs due to the complex nature of the charge trapping that occurs.

Positive bias temperature stressing generally produces a positive shift in V_T. Fig. 1 illustrates the large variation in I_D-V_{GS} characteristics of a representative Group A SiC power MOSFET following a one-hour gate-bias stress at 150 °C with V_{GS} = +15 V, depending on whether the

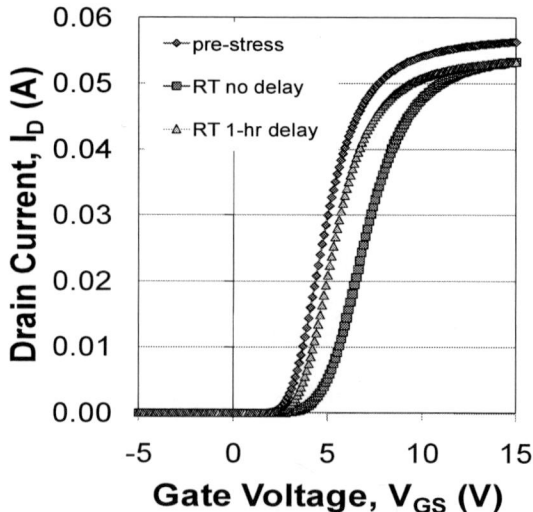

Figure 1. Effect of measurement delay on the I_D-V_{GS} characteristics of a Group A MOSFET following PBTS.

measurement is made immediately after rapidly cooling back to room temperature, or after a delay of one hour. To further dramatize the difference, the immediate measurement was made by sweeping down in gate bias, from +15 V to −5 V, whereas the later measurement was made by sweeping up in gate bias in the conventional manner. None of the present standards address sweep direction. We examine the effect of sweep direction further in Section III.B.2 on the device transfer characteristics following PBTS.

The linearly extracted V_T from the PBTS data shown in Fig. 2 indicates that the immediate post-HTGB measurement, which revealed a significant positive shift in V_T, would result in device failure based on the AEC standard [7] which requires that post burn-in measurement parameters remain within ± 20 percent of the pre-stress value, whereas the later measurement would not. This result clearly illustrates the time-dependent nature of the charge trapping.

Devices from Group B exhibited a similar response to positive HTGB stressing, although the magnitude of these shifts was smaller under identical PBTS conditions due to improved gate oxide processing. The range in ΔV_T with no delay time for devices from Group B was 0.2 V to 0.4 V, compared to a ΔV_T range of 0.9 V to 2.9 V for devices from Group A. It should be noted that although some of these devices may have remained within the 20 percent variation standard based on these test results,

the one-hour bias stress time used in this case is much shorter than the standard stress times. Longer bias stress times would have led to greater V_T shifts—see Section III.C.3.

Conversely, NBTS typically induces a negative shift in measured V_T. Fig. 3 shows the variation in response due to measurement delay time following a one-hour gate-bias stress at 150 °C with $V_{GS} = -15$ V for a typical device from Group A. Although the variation in V_T between the immediate and one-hour delayed measurements is not as great as in the PBTS case, the immediate measurement does result in an increase of the *OFF*-state leakage current due to the larger negative shift in V_T. Similarly, Fig. 4 shows

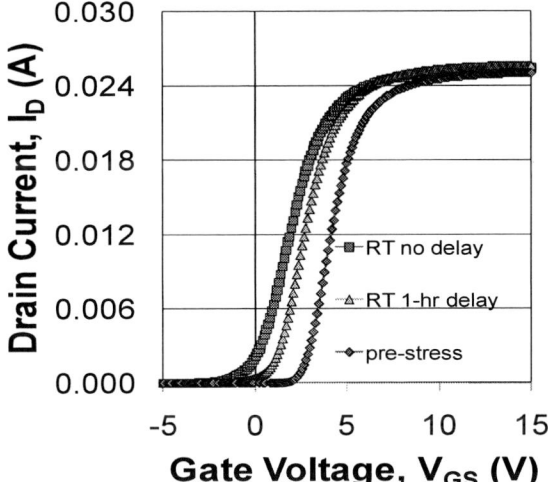

Figure 3. Effect of measurement delay on the I_D-V_{GS} characteristics of a Group A MOSFET following NBTS.

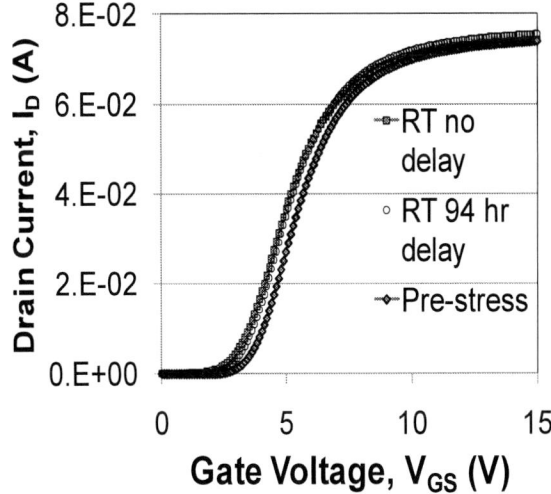

Figure 4. Effect of measurement delay time on the I_D-V_{GS} characteristics on a Group B MOSFET following NBTS.

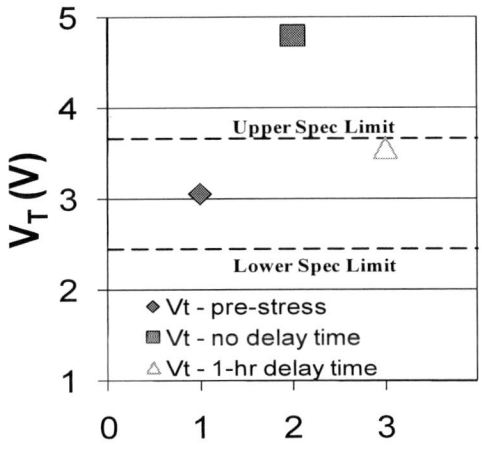

Figure 2. Linearly extracted V_T variation following PBTS within AEC standard (±20%)

the variation in response following a one-hour gate-bias stress at 150 °C with $V_{GS} = -15$ V for a representative device from Group B. The immediate and delayed measurements show a much smaller variation than devices from Group A, even when the delay in the second measurement is much greater—in this case 92 hours later. The initial V_T shifts to the left are reduced for devices from Group B, which results in some improvement in *OFF*-state drain leakage current. This is discussed further in Section III.C.2.

2) Measurement Sweep Direction

Next, we examine the effect of sweep direction on the transfer characteristics of SiC MOSFET devices following PBTS. As before, the stress bias and temperature were +15 V and 150 °C, respectively, although the bias stress time in this case was only 1,800 s. Pre-stress device characterization of the six devices from Group B for this part of the study yielded an average V_T of 3.59 V with a standard deviation of 0.07 V. In all of the pre-stress measurements, the gate was swept up from $V_{GS} = 0$ V to $V_{GS} = +15$ V.

The effect of sweep direction was determined by measuring ΔV_T and comparing the difference between sweeping up and sweeping down during the post-stress measurement. On average, we observed a ten percent larger shift for devices swept down in gate bias following the PBTS. This difference has been found to be much larger when the gate bias is swept up beginning at a negative value. The sensitivity of these results show again the importance of the bias applied during the measurement and how long that bias is applied, i.e., the sweep time. In comparing these results with those of Section III.B.1, it appears that sweep delay time is a more important variable than sweep direction.

3) Measurement Temperature

SiC MOSFETs are expected to operate reliably at junction temperatures approaching 150 °C. It becomes critically important to understand the effects of bias-temperature stressing on the high-temperature I_D-V_{GS} characteristics. In this section we examine the effects of a negative HTGB stress on both the room temperature and high temperature transfer characteristics of several Group A and

Group B SiC MOSFETs. Fig. 5 and Fig. 6 show representative results for devices from Group A and Group B, respectively. Pre-stress characteristic curves were taken at both room temperature and 150 °C. The stress conditions for the NBTS test were $V_{GS} = -15$ V and T = 150 °C. Immediately following the stress, a sweep up of the gate was made in order to measure the change in the high-temperature transfer characteristics. The device was then rapidly cooled to room temperature under bias and the gate was again swept up to measure the device I_D-V_{GS} characteristics.

The high-temperature values for V_T are consistently about 1.5 V more negative than the corresponding room temperature values for all the

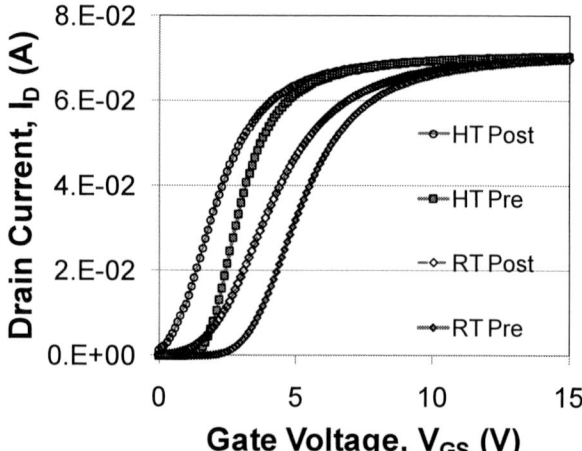

Figure 5. Comparison of transfer characteristics pre- and post-stress at high and room temperature – Group A.

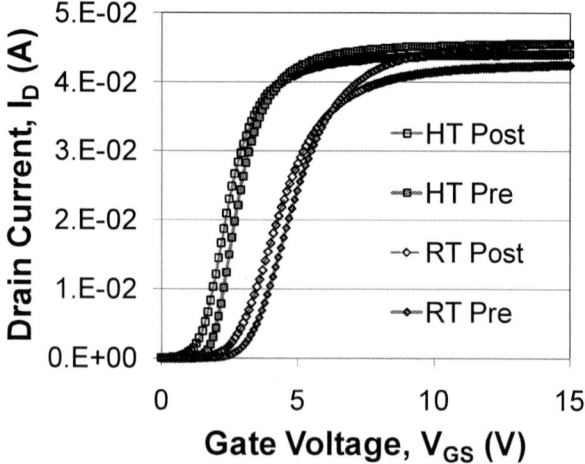

Figure 6. Comparison of transfer characteristics pre- and post-stress at high and room temperature – Group B.

devices tested, although slightly less following the NBTS. Comparing the pre- and post-NBTS results show an additional negative V_T shift of about 1.4 V for Group A devices and about 0.3 V shift for Group B devices, consistent with the PBTS results discussed in Section III.B.1.

As expected, the post-stress high-temperature curves show the largest negative shift of the I_D-V_{GS} characteristic as well as the largest increase in the *OFF*-state leakage current. Fig. 7 provides a summary of the room and high temperature leakage current results. The subthreshold characteristics for Group B devices are only moderately improved over devices from Group A even though ΔV_{GS} (for $I_D = 1 \times 10^{-6}$ A) is significantly less: a 1.5 V negative shift versus a 2.8 V shift, respectively. The subthreshold-voltage instabilities are typically larger than the linear V_T instabilities due to an increased stretch-out of the subthreshold slope under a negative bias stress [3]. The measured high temperature post-stress drain current at $V_{GS} = 0$ V was five orders of magnitude higher than the room temperature pre-stress drain current at $V_{GS} = 0$ V for devices from Group A, whereas devices from Group B exhibited an increase of approximately three to four orders of magnitude in drain-leakage current. The decrease in the pre-stress drain current between room temperature and 150 °C for Group B devices was due to the disappearance of an initial high edge leakage characteristic and not to any voltage shift.

C. Stressing Conditions

Section III.B discussed how variations in the measurement conditions affect the stress results. This section examines the effect of variations in the stress conditions themselves, specifically gate bias, temperature, and time.

1) Bias Stress Dependence

Fig. 8 is a plot of a series of I_D-V_{GS} curves measured on different devices having positive gate voltage stresses of 10, 12, and 15 V, along with a representative pre-stress curve for devices from Group A. The stress temperature and time for the PBTS are 150 °C and 1 hour, respectively. The post-stress curves shift further to the right with increasing bias-stress magnitude. A plot of the increase in ΔV_T with stress bias is shown in Fig. 9.

A similar study looking at variations in NBTS with a constant stress temperature and time of 150 °C and 1 hour, respectively, for negative gate voltage stresses of −5, −10, and −15 V were performed on devices from Group B. The negative shift of the I_D-V_{GS} characteristics increases, as expected, with larger negative stress biases. The measured V_T shift was only 0.02 V for $V_{GS} = −5$ V, whereas the V_T shift was 0.23 V for $V_{GS} = −15$ V. As previously discussed in Section III.B.1, the magnitude of the V_T shifts under NBTS for devices from Group B are less than similarly stressed devices from Group A.

2) Temperature Stress Dependence

The temperature dependence of the gate-bias stress is illustrated in Fig. 10, which shows a plot of ΔV_T as a function of stress temperature. ΔV_T was

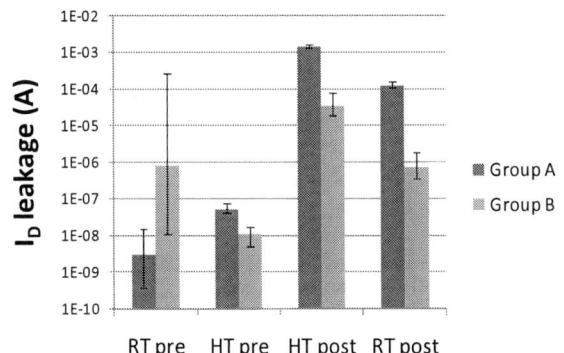

Figure 7. Comparison of pre- and post-stress room and high temperature drain leakage currents at $V_{GS} = 0$ V following NBTS ($V_{GS} = −15$ V; T = 150 °C) for devices from Groups A and B.

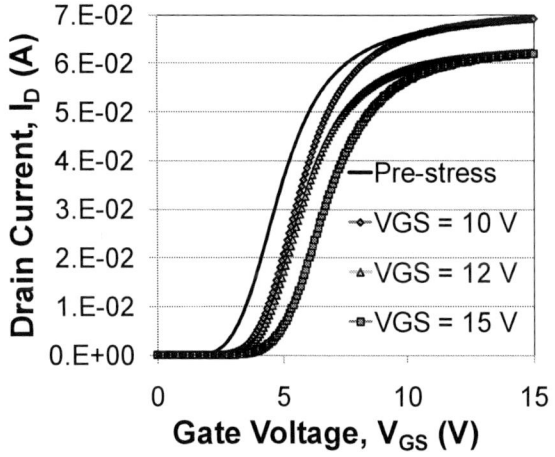

Figure 8. Bias dependence effect on the I_D-V_{GS} characteristics of Group A devices.

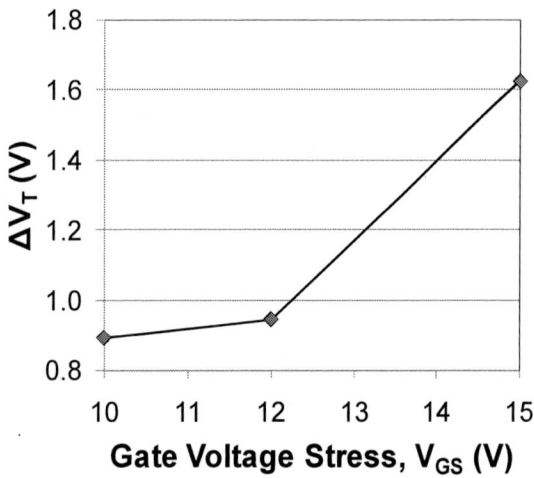

Figure 9. Dependence of V_T shift on stress bias for Group A devices with T = 150 °C and t = 1 hour.

calculated by taking the difference of the post- and pre-stress V_T values extracted from the measured I_D-V_{GS} curves, with a different device from Group B used for each stress temperature—which varied from 25 °C to 150 °C for a one hour NBTS with a constant gate-bias stress of –15 V. Clearly, the greater the stress temperature the larger the negative V_T shift, with a sharp increase observed above 100 °C. These results for NBTS are consistent with a report of previous results for a positive bias temperature stress [3], which attributed an increase in V_T instability to an increase in the number of active near-interfacial oxide traps (see Section III.A).

3) Stress Time Dependence

The time dependence of the bias-stress is illustrated in Fig. 11 for a 150 °C, –15 V negative bias temperature stress of a device from Group B. The individual bias stress times at temperature were 320 s, 1,000 s, 3,200 s, and 10,000 s. At the end of each NBTS, the device was cooled to room temperature under bias and an immediate I_D-V_{GS} measurement was made. Fig. 11 plots the drain current on a log scale to show the effect of bias stress time on the subthreshold device characteristics. Longer stress times result in larger negative shifts. The *OFF*-state leakage current taken at $V_{GS} = 0$ V increased from its pre-stress value of 3×10^{-8} A to 6×10^{-6} A after a cumulative stress time of approximately four hours.

Fig. 12 shows the corresponding smaller shift of the linear V_T value versus the log of the individual stress times. This underscores the importance of monitoring the subthreshold current under NBTS, which can increase more dramatically due to the stretch-out of the subthreshold slope and lead to significant increases in leakage current [3].

These results emphasize the importance of the bias stress time, which varies among the different existing standards. For example, the AEC standard [7] calls for a 1,000 hour HTRB and HTGB stress, whereas the Mil Std 750-E [6] only requires a 48 hour stress, at 80% of the maximum rated gate bias.

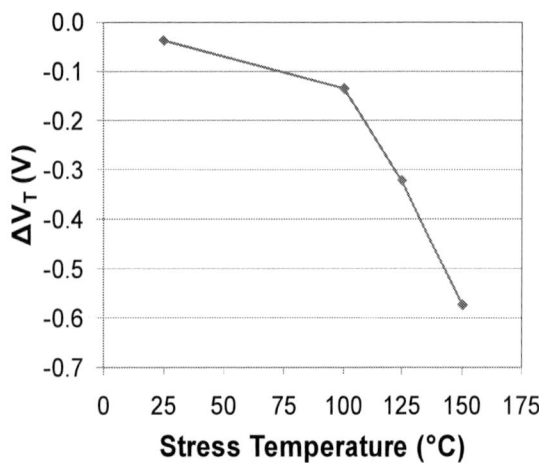

Figure 10. Dependence of V_T shift on stress temperature for Group B devices with $V_{GS} = -15$ V, t = 1 hour.

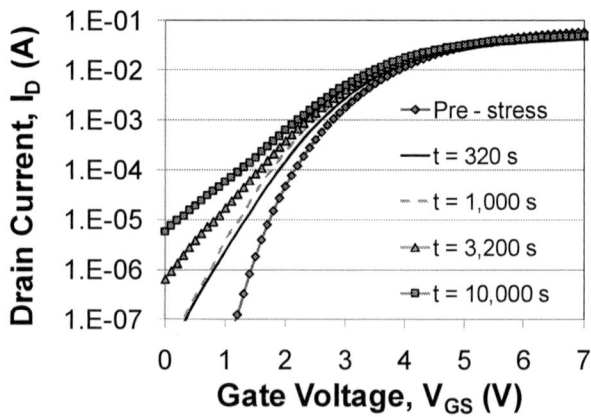

Figure 11. Time dependence of the subthreshold I_D-V_{GS} characteristics for a representative Group B device with $V_{GS} = -15$ V and T = 150 °C.

978-1-4244-9113-1/11 $26.00 © 2011 IEEE

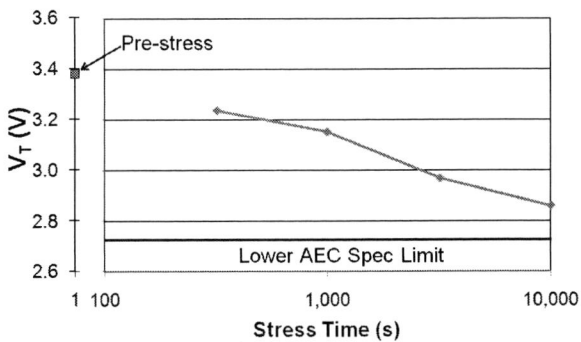

Figure 12. Dependence of V_T shift on stress time for a representative Group B device with $V_{GS} = -15$ V and T = 150 °C.

4) Effect of PBTS on the ON-state Characteristics

Finally, we examined the effect of PBTS testing on the ON-state characteristics (I_D-V_{DS}) for devices from Group B by applying a positive bias temperature stress ($V_{GS} = +15$ V, T = 150 °C). However, because of the large currents involved, a single room-temperature I_D-V_{DS} trace was made using a 371B Tektronix curve tracer by sweeping up in drain bias with a constant $V_{GS} = +15$ V applied during the temperature stress and cool down using an external DC power supply. Fig. 13 shows the effect of the PBTS on the ON-state characteristics for stress times of 24 and 92 hours. These results show an increase in the forward voltage drop (V_{DS-ON}) for increasing stress duration, especially at higher drain currents. If these devices are operated

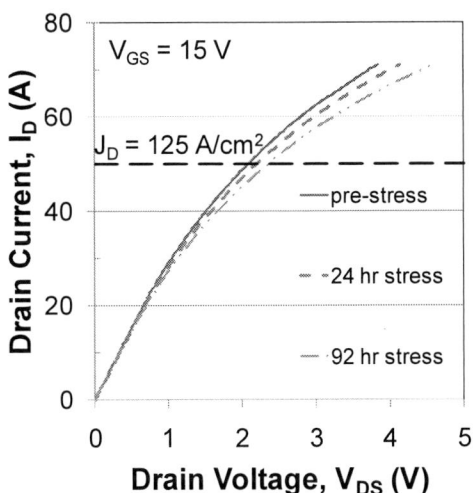

Figure 13. Effect of stress time on the I_D-V_{DS} characteristics of a representative Group B device with $V_{GS} = +15$ V and T = 150 °C.

at $I_D = 50$ A ($J_D = 125$ A/cm^2), then the PBTS results in a 12 percent increase in V_{DS-ON} for a 92-hour stress. This will result in greater ON-state power dissipation and an increase in the junction temperature, which may impact not only performance, but long-term device reliability as well.

IV. CONCLUSIONS

This work has shown the large variation in the I-V and drain-leakage characteristics of SiC power MOSFETs due to variations in both measurement and stress conditions. Specifically, we have shown that the shift in the post high-temperature stress I_D-V_{GS} characteristics is strongly dependent on the measurement delay time, as well as the gate-sweep direction. Previous results have shown that the measurement speed is also of critical importance, with faster measurements revealing larger actual V_T shifts. Although the standards call for pre- and post-BTS measurements to be performed at room temperature, it is important to measure at the stress temperature as well since SiC power MOSFETs are expected to operate at these temperatures and the leakage current under negative bias stress is worse at elevated temperatures.

Similarly, we have shown that these shifts are highly dependent on the bias stress conditions, including bias magnitude, stress time, and temperature. Although all the standards call for testing at elevated temperature, there is a significant variation in the bias-stress time requirements, and some variation in bias magnitude requirements. Both variables have been demonstrated to be significant. In particular, a sharp increase in the magnitude of the V_T shift is observed with increasing stress time at temperatures above 100 °C.

We have shown that, as currently written, power device reliability testing standards such as those from JEDEC, AEC, and DoD do not place enough emphasis on constraining the conditions of measurement when evaluating a device parameter such as V_T. Seemingly minor alterations in measurement procedures, which can be in full compliance with a standard's specifications, may in fact produce vastly different results on a given device. In certain situations, the choice in measurement parameters can make the difference between success and failure. It is therefore important that any standards used be mindful of the unique issues associated with SiC MOSFETs, especially as the technology matures.

For example, the 96 hour window between stress and measurement allowed by all three standards considered here appears to be unsuitable. More specifically, it is important that the existing standards be modified to require faster, and more immediate post-stress measurements with the gate-bias swept down following a positive bias stress, and that elevated temperature measurements be required as well. Otherwise, inconsistent pass/fail results may occur when applied to SiC power MOSFETs.

It is also critical that the bias-temperature-stress times be long enough to allow for the activation of all the elevated temperature mechanisms which may occur under actual operational conditions. Charge trapping effects will get worse, although for some devices this effect may be countered by mobile ion drift. It is also important to determine appropriate accelerated test conditions so that non-operational failure mechanisms are not introduced.

REFERENCES

[1] S.-H. Ryu, B. Hull, S. Dhar, L. Cheng, Q. Zhang, J. Richmond, M. Das, A. Agarwal, J. Palmour, A. Lelis, B. Geil, and C. Scozzie, "Performance, Reliabilty, and Robustness of 4H-SiC Power DMOSFETs," Mat. Sci. Forum, vols 645-648, pp. 969-974, 2010.

[2] K. Matocha, P. Losee, A. Gowda, E. Delgado, G. Dunne, R. Beaupre, and L. Stevanovic, "Performance and Reliability of SiC MOSFETs for High-Current Power Modules," Mat. Sci. Forum, vols 645-648, pp. 1123-1126, 2010.

[3] A. Lelis, R. Green, and D. Habersat, "High Temperature Reliability of SiC Power MOSFETs," to be published Mat. Sci. Forum, 2011.

[4] M. Treu, R. Rupp, and G. Sölkner, "Reliabilty of SiC Power Devices and its Influence on their Commercialization – Review, Status, and Remaining Issues," IEEE IRPS, pp. 156-161, 2010.

[5] "Temperature, Bias, and Operating Life Standard," JESD22-A108C, 2005.

[6] "Test Methods for Semiconductor Devices," MIL-STD-750E, 2006.

[7] "Stress Test Qualification for Automotive Grade Discrete Semiconductors," AEC-Q101-Rev-C, 2005.

[8] A. Lelis, D. Habersat, R. Green, A. Ogunniyi, M. Gurfinkel, J. Suehle, and N. Goldsman, "Time Dependence of Bias-Stress-Induced SiC MOSFET Threshold-Voltage Instability Measurements," IEEE Trans. Electron Devices, vol 55, no. 8, pp. 1835-1840, 2008.

[9] A. Lelis, D. Habersat, R. Green, and N. Goldsman, "Temperature-Dependence of SiC MOSFET Thershold-Voltage Instability," Mater. Sci. Fourm, vols. 600-603, pp. 807-810, 2009.

[10] M. Gurfinkel, J. Suehle, J. B. Bernstein, Y. Shapira, and A. J. Lelis, D. Habersat, N. Goldsman, "Ultra-Fast Characterization of Transient Gate Oxide Trapping in SiC MOSFETs," IEEE IRPS, pp. 462-466, 2007.

A Novel and Low-cost Method to Detect Delay Variation by Dynamic Thermal Laser Stimulation

Chunlei Wu, Masuda Motohiko, Winter Wang, Grace Song, Jinglong Li, Joe Yu, Li Tian, Miao Wu
Product Analysis Laboratory
Freescale Semiconductor (China) Limited
Tianjin, China
b09059@freescale.com

Abstract—Delay variation can be very difficult to localize in function failure analysis. In this paper we combine Delay Variation Mapping (DVM) and Soft Defect Localization (SDL) to develop a novel and low-cost method that can detect delay variation effectively. It just uses Static Thermal Laser Stimulation (S-TLS), Oscilloscope and Function Generator to compose a dynamic thermal laser stimulation (D-TLS) system. The methodology, system configuration and experimental results of this method are presented.

Keywords-delay variation; Dynamic Thermal Laser Stimulation; Optical Beam Induced Resistance Change

I. INTRODUCTION

Thermal Laser stimulation (TLS) techniques are now commonly used in failure analysis for front-side as well as backside defect localization. TLS techniques use a laser with a wavelength around 1340nm to locally heat the device because it is not energetic enough to create electron-hole pairs. Among techniques using local heat, the most commonly used are Optical Beam Induced Resistance Change (OBIRCH) [1], Thermally Induced Voltage Alteration (TIVA) [2] and Seebeck Effect Imaging (SEI) [3] techniques. Defects such as metallic and poly-silicon shorts, voids in vias, Electro Static Discharge (ESD) induced molten silicon spikes and gate oxide shorts are localized with TLS techniques [4]. These techniques are called Static Thermal Laser Stimulation (S-TLS) because current or voltage change is measured in the device which is in a static mode. Static techniques are still commonly used in failure analysis, but there are inefficient for soft defects localization. Soft defects are difficult to localize because they occur when the IC is functional running and could be intermittent. Moreover they are very sensitive to environmental conditions, such as temperature, power supply voltage, frequency and so on. The TLS techniques had to be adapted to be applicable when the IC is dynamically activated.

Dynamic Thermal Laser Stimulation (D-TLS) techniques include Resistive Interconnect Localization (RIL) [5], Soft Defect Localization (SDL) [6] and Delay Variation Mapping (DVM) [4] et al. SDL technique introduces the concept of scanning a laser across a die such that the localized heat can change the pass/fail state of the failing node to identify its physical X, Y location on the die. DVM technique employs laser heating to alter the timing behavior of a functional running device. By scanning the laser beam an image is obtained showing the timing (delay) variation value (which can be positive and negative) for each X, Y laser positions.

Generally, the system configurations of SDL and DVM are complicated. They need synchronize the tester platform with Laser Scanning Module (LSM). And sometimes software program is needed to convert Pass/Fail state or delay variation to voltage or current variation, if they could not be identified directly by system. So in this paper, we combine DVM and SDL to develop a novel and low-cost method. It just uses OBIRCH, an Oscilloscope and 2 Function Generators to compose a D-TLS system that can detect delay variation effectively.

II. METHODOLOGY AND SYSTEM CONFIGURATION

For each functional test sequence, Function Generator 1 generates trigger input signal for the device, and then the device output signal. The time delay between the observed output signal and the trigger input signal compose the delay measurement. OBIRCH employs laser to scan across a die and locally heats each pixel location. The delay time variation is induced by laser heating due to a temperature sensitive soft defect. And then the test result is changed from either a failing state to a passing state or vice versa by the delay time variation.

In each functional test sequence the tester platform need synchronize with LSM of OBIRCH. But OBIRCH has not synchronization function with the tester platform. So we employ a dynamic synchronization method to synchronize the tester platform with the LSM of OBIRCH. The principle of this method is that the device runs test sequence at least once during laser heating a pixel, namely, each pixel dwell time (PDT) of laser is greater than once test sequence duration [7].

For detecting delay variation by OBIRCH, firstly, delay variation is converted to test result variation (voltage variation) by an Oscilloscope and Function Generator 2. Function Generator 2 outputs 5V pulse represents a failing state of test result and 0V pulse represents a passing one or vice versa. For synchronizing LSM of OBIRCH to the tester platform, the pulse duration equals to or less than the test sequence duration. Secondly, the test result variation (voltage variation) is converted to a current variation. The output pulse (Pass/Fail) of the Function Generator 2 sends to the positive side of OBIRCH amplifier that outputs 5V DC voltage via a resistor. The current decrease in the resistor, if Function Generator 2 outputs 5V

978-1-4244-9113-1/11 $26.00 © 2011 IEEE

pulse; the current increase in the resistor, if Function Generator 2 outputs 0V pulse; OBIRCH could detects the current variation in this resistor. So the test result variation (voltage variation) is converted to current variation by this way. Fig. 1 shows the system configuration.

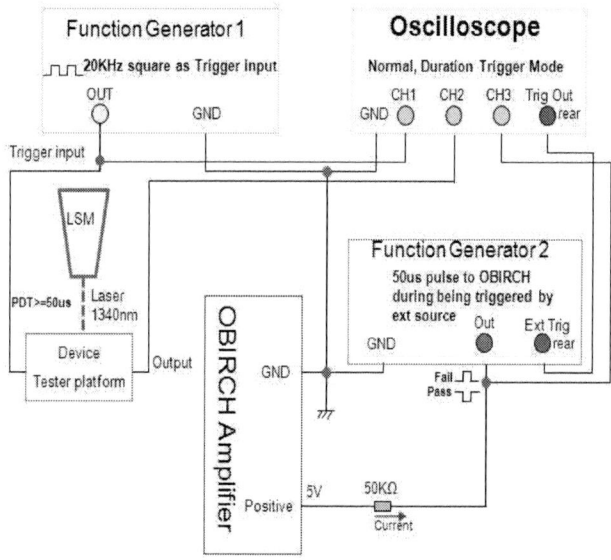

Figure 1.   Dynamic thermal laser stimulation system configuration.

The key of realizing delay variation to test result variation conversion (voltage variation) is Duration Trigger mode function in Oscilloscope which can detect the delay variation and then output a variation voltage (from 5V Fail to 0V Pass or vice versa) via Trigger Out Port in rear panel of Oscilloscope under a right trigger threshold. Fig. 2 shows an example for detecting the delay variation by the Duration Trigger mode. The rising edge delay time between Input signal and Output signal is about 9μs under room temperature, but it is about 5μs under high temperature. So the time (Input=H & Output=H) increases from 16.1μs to 20.1μs as the rising temperature, which is greater than the trigger threshold (19.6μs). So the duration trigger incident happens and Trigger Out Port outputs from 5V to 0V. For synchronizing LSM of OBIRCH to the tester platform, this voltage variation (test result variation) signal is sent to Ext Trig Port in rear panel of Function Generator 2, which could trigger Function Generator 2 to output a pulse (Pass/Fail) whose duration equals to or less than PDT of laser. At last this pulse is sent to the positive side of OBIRCH amplifier.

III.   EXPERIMENTAL RESULTS

A.   Experiment 1

In this paper, we employed a dynamic synchronization method to synchronize the tester platform with the LSM of OBIRCH, because LSM could not synchronize with the tester platform directly. The principle of this method was showed previously. But the accuracy of this method was decided by the relationship between the PDT of laser and Pass/Fail pulse duration (test sequence duration). So in experiment 1, we

researched how to affect the accuracy of this D-TLS technique by PDT of laser and Pass/Fail pulse duration (test sequence duration).

Firstly, we researched how to affect the accuracy of this D-TLS technique by different PDT of laser. Our Pass/Fail pulse duration was 16μs as a constant, so PDT must equal or greater than 16μs for synchronizing the tester platform with LSM. We chose PDT was 16μs and 72μs respectively to get delay variation spot. The black line signified pass test results when laser heating was far away a temperature sensitive place. The white spot presented a fail test result while laser was heating the temperature sensitive place (Fig. 3). So the white spot located the delay variation place. We found the accuracy of delay variation spot under 72μs PDT was better than that under 16μs PDT, because 72μs PDT had more time to heat every spot of die and to process data in the computer of OBIRCH than 16μs PDT.

Secondly, we researched how to affect the accuracy of this D-TLS technique by different Pass/Fail pulse duration. Our PDT of laser was 72μs as a constant, so Pass/Fail pulse duration must equal or less than 72μs for synchronizing the tester platform with LSM. We chose Pass/Fail pulse duration was 16μs and 70μs respectively to get delay variation spot. The black line signified pass test results when laser heating was far away a temperature sensitive place. The white spot presented a fail test result while laser was heating the temperature sensitive place (Fig. 4). So the white spot located the delay variation place. We found the accuracy of delay variation spot under 16μs Pass/Fail pulse duration was better than that under 70μs Pass/Fail pulse duration. Because 72μs PDT had more time to heat every spot of die and to process data in the computer of OBIRCH, when Pass/Fail pulse duration was 16μs.

In conclusion, the longer PDT of laser and the shorter Pass/Fail pulse duration, the more accurate delay variation spot we got. And in experiment 2 and 3, two real function failure cases were presented to show how to locate delay variation spot in detail.

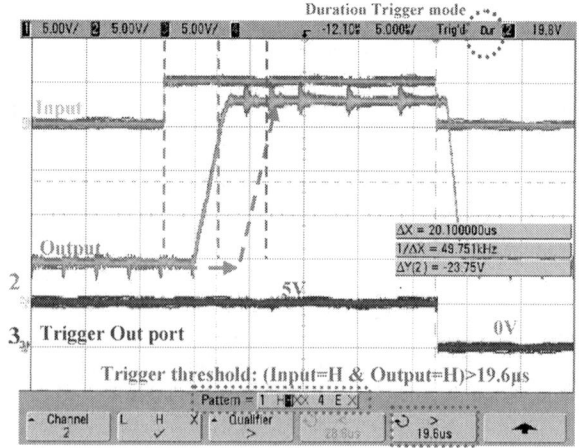

Figure 2.   An example for detecting delay variation by the Duration Trigger mode.

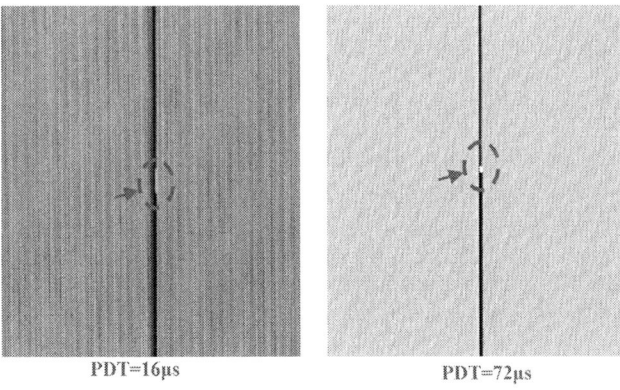

PDT=16μs   PDT=72μs

Figure 3.   Delay variation spot images under different PDT of laser.

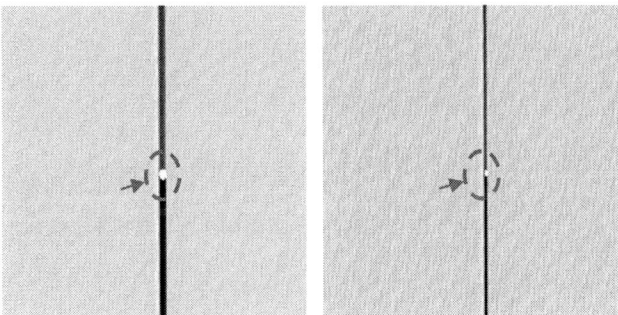

Pass/Fail pulse duration=70μs   Pass/Fail pulse duration=16μs

Figure 4.   Delay variation spot images under different Pass/Fail pulse duration.

B.   Experiment 2

In experiment 2, Function Generator 1 output 100μs square wave as trigger input signal to the device, and the device output a square wave with same frequency. The falling edge delay time between input signal and output signal was about 13μs for the reference device under both room temperature and high temperature, and for the failing device it was about 13μs under room temperature but 113μs under high temperature (Fig. 5). So the rising temperature could change the falling edge delay time and then change the test result from a passing state to a failing state for the failing device. D-TLS technique was employed to detect the falling edge delay variation.

As in Fig. 5, the falling edge delay time between input signal and output signal was about 13μs under room temperature. So the time (Input="1" & Output="1") was about 33μs that was less than the trigger threshold (34.3μs). The duration trigger incident did not happen and Function Generator 2 output 5V pulse (Pass) continuously; But the falling edge delay time increased to 113μs after rising temperature, so the time (Input="1" & Output="1") was about 50μs that was greater than the trigger threshold (34.3μs). The duration trigger incident happened and Function Generator 2 output a 0V pulse (Fail) whose time duration was 25μs. And then the temperature returned back to room temperature, the time (Input="1" & Output="1") changed from 50μs to 33μs. So

the duration trigger incident was gone, and Function Generator 2 output 5V pulse (Pass) continuously again.

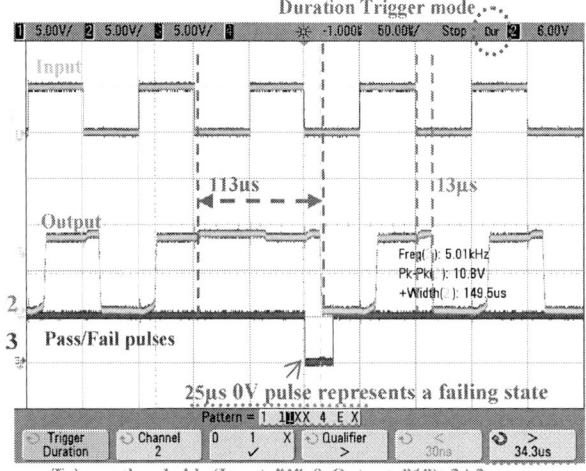

Figure 5.   Detecting the falling edge delay variation by the Duration Trigger mode.

Function Generator 2 output 25μs pulse to signify a passing test result (5V pulse) or a failing test result (0V pulse). So 72μs PDT of laser was chosen for synchronizing LSM of OBIRCH to the tester platform and enough heating. Fig. 6 presents the experimental result with 5x magnification lens. Some delay variation mapping spots were detected in the die of the failing device. Left image was delay variation mapping image, and right image was its overlay image. The black (green) lines signified the falling edge delay time was 33μs (Pass) when laser heating was far away those temperature sensitive areas. The white (red) spots presented the falling edge delay time increased to 50μs (Fail) while laser was heating these temperature sensitive areas. Fig. 7 presents the experimental result with 20x magnification lens. In this experiment, we could not get the delay variation mapping spots with 100x magnification lens, although PDT of laser was increased to 572μs, because the laser heating did not heat the die enough to change the falling edge delay time and trigger a duration trigger incident under 100x magnification lens.

At last, a resistive via was found in the third delay variation spot place by Focus Ion Beam (FIB) cross section (Fig. 8).

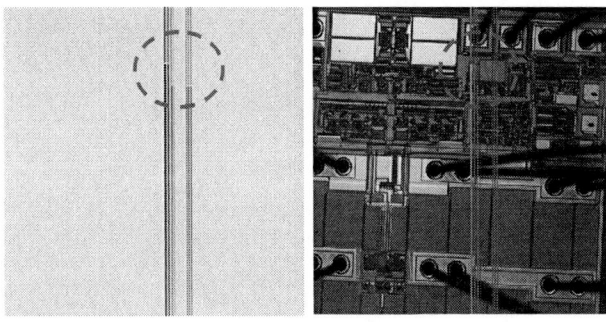

Figure 6.   Delay variation mapping images with 5x magnification lens.

978-1-4244-9113-1/11 $26.00 © 2011 IEEE   767

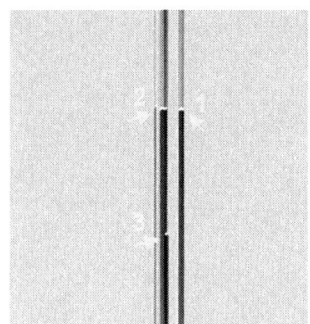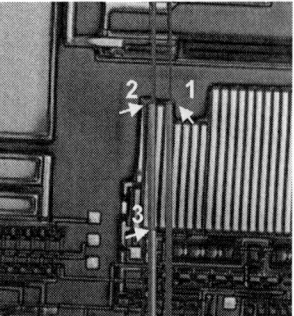

Figure 7. Delay variation mapping images with 20x magnification lens.

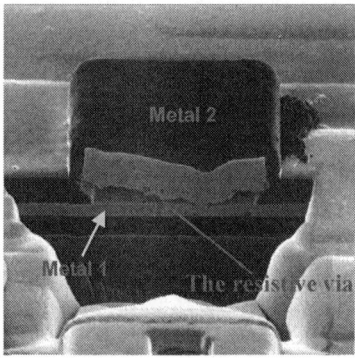

Figure 8. FIB cross section image of the resistive via defect at the third delay variation spot place.

C. Experiment 3

In experiment 3, the rising edge delay time between input signal and output signal of reference device was 1.54µs, but it was 4.36µs (Fail) for the failing device under room temperature (Fig. 9), and it decreased to about 1.54µs (Pass) after rising temperature. So temperature could change the rising edge delay time and then change the test result from a failing state to a passing state for the failing device. D-TLS technique was employed again to detect the delay variation.

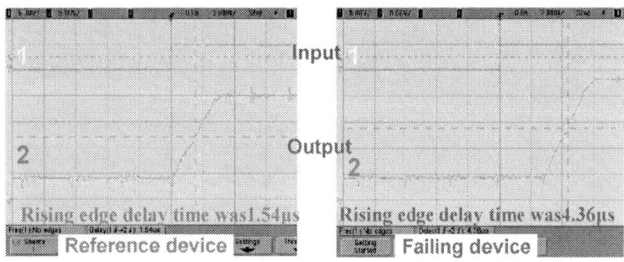

Figure 9. Rising edge delay time images for the reference and failing devices.

Function Generator 1 output 20000 Hz square as trigger input signal to the failing device. Function Generator 2 output 50µs pulse to signify a passing test result (0V pulse) or a failing test result (5V pulse). So 143µs PDT of laser was chosen for synchronizing LSM of OBIRCH to the tester platform and enough heating. Fig. 10 presents the experimental result with 5x magnification lens. Some delay variation mapping spots were detected in the die of the failing device. Left image was

delay variation mapping image, and right image was its overlay image. The black (green) lines signified delay time was 4.36µs (Fail) when laser heating was far away those temperature sensitive areas. The white (red) spots presented delay time decrease to 1.54µs (Pass) while laser was heating these temperature sensitive areas. Fig. 11 presents the experimental result with 20x magnification lens.

The layout of these 5 red delay variation spots areas was examined, and we found all of them related to a signal named DSIFI. So active microprobe was performed and we found an abnormal slower transition from 0V to 5V in DSIFI signal (Fig. 12). At last, a resistive via was found in the fourth delay variation spot place by FIB cross section (Fig. 13).

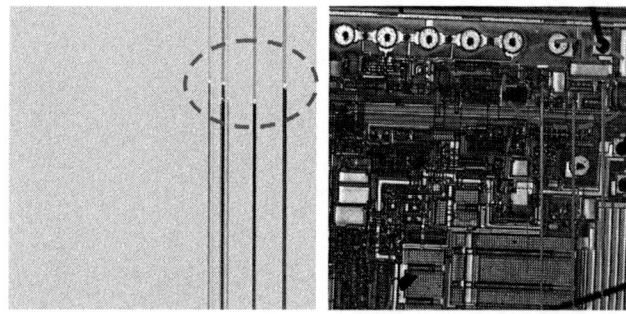

Figure 10. Delay variation mapping images with 5x magnification lens.

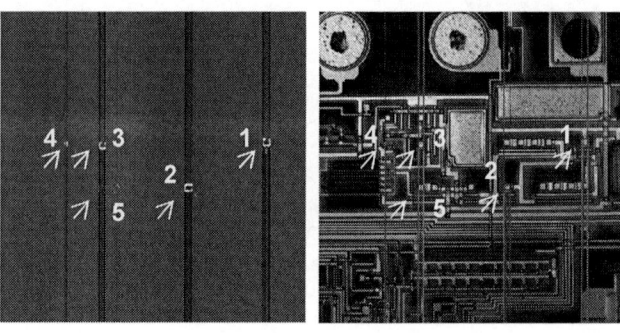

Figure 11. Delay variation mapping images with 20x magnification lens.

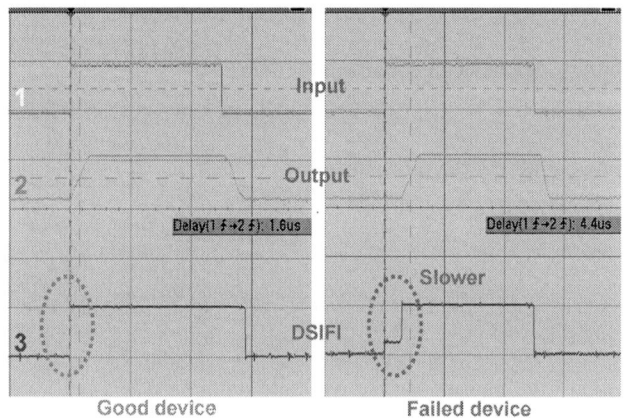

Figure 12. Active microprobe for DSIFI signal.

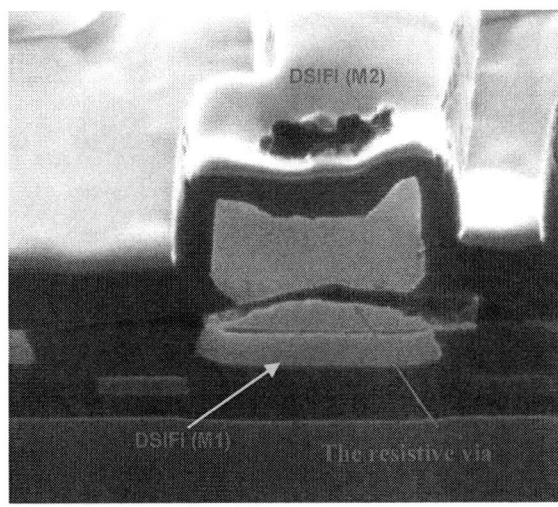

Figure 13. FIB cross section image of the resistive via defect in DSIFI signal.

IV. CONCLUSION

In this paper, we presented a novel and low-cost method to detect delay variation by D-TLS. The proposed methodology extended the application field of OBIRCH to detect delay variation in dynamically functional failed ICs. It just used some common equipment in Failure Analysis (FA) Laboratory such as OBIRCH, Oscilloscope and Function Generator to combine a very effective D-TLS system without additional software program.

For synchronizing the tester platform with the LSM of OBIRCH we employed a dynamic synchronization method to realize it. The principle of this method was shown in detail by some experiments.

Thanks to Duration Trigger mode function in Oscilloscope which could detect the delay variation and trigger a voltage variation via Trigger Out Port in rear panel of Oscilloscope under a right trigger threshold.

In experiment 1, we researched how to affect the accuracy of this D-TLS technique by PDT of laser and Pass/Fail pulse duration (test sequence duration). Some experimental results were presented. And then we found the longer PDT of laser and the shorter Pass/Fail pulse duration, the more accurate delay variation spot we got.

In experiment 2 and 3, two real function failure cases were presented to show how to locate delay variation spot in detail. Experimental results showed this D-TLS system could detect delay variation effectively and found a defect at the delay variation spot place. However, we could not get the delay variation mapping spots with 100x magnification lens, although PDT of laser was greatly longer than Pass/Fail pulse duration, because the laser heating did not heat the die enough to change the delay time and trigger a duration trigger incident under 100x magnification lens. So it is a problem that need us solve in the future.

ACKNOWLEDGMENT

Authors would like to thank Steven Che, Xuezhu Wang, Xiaocui Li, Changyan Qi and Quande Zhang of Product Analysis Lab in Freescale Semiconductor (China) Limited. They gave us great support in Physical Failure Analysis.

REFERENCES

[1] K. Nikawa, S. Tozaki, "Novel OBIC observation method for detecting defects in Al stripes under current stressing," ISTFA 1993, p. 303–310.

[2] E.I. Cole Jr., P. Tangyunyong, D.L. Barton, "Backside localization of open and shorted IC interconnections," IRPS 1998, p. 129–136.

[3] E.I. Cole Jr., P. Tangyunyong, D.A. Benson, D.L. Barton, "TIVA and SEI developments for enhanced front and backside interconnection failure analysis ," *Microelectronic Reliability*, 1999, vol.39, pp. 991–996.

[4] Kevin Sanchez, Romain Deplats, Felix Beaudoin, philippe Perdu, Dean Lewis, Praveen Vedagarbha et al., "Delay variation mapping induced by dynamic laser stimulation," IRPS 2005, p. 305–311.

[5] E.I. Cole Jr., Paiboon Tangyunyong, Charles F. Hawkins, Michael R. Bruce, Victoria J. Bruce, Rosalinda M. Ring et al., "Resistive Interconnection Localization," ISTFA 2001, p. 43–57.

[6] M.R. Bruce, V.J. Bruce, D.H. Eppes, J. Wilcox, "Soft defect localization on ICs," ISTFA 2002, p. 21–27.

[7] Wu Chunlei, Linda Zhai,Masuda Motohiko, Horse Ma,Winter Wang, John Liu et al., "How do power supply voltage, pixel dwell time and test program duration affect the accuracy of soft defect localization technique," IPFA 2010, p. 237–240.

A Study of the Influence of High Voltage Device Characteristics by Electron Beam Irradiation during Nanoprobing

Hung Sung Lin
United Microelectronics Corporation, Ltd.
No. 3, Li-Hsin Rd. II, Hsinchu Science Park, Taiwan 300, R.O.C.
Tel: 886-3-5782258 ext. 33231; Fax: 886-3-563-6722; Email: giant_lin@umc.com

Abstract—**It has been widely reported that floating gate irradiation using a charged beam can shift device parameters with the scaling of the devices [1-4]. For high voltage (HV) devices, the effects of electron beam (EB) induced damage, however, have not been reported. This paper describes how charge damage during EB exposure should also be considered for high voltage (HV) devices when scanning electron microscope (SEM) is employed for probe guidance. In this study, the effects of EB cathode potential on CMOS transistor threshold voltage and off-state current are investigated using HV, middle voltage (MV), and low voltage (LV) devices. The experimental results show that, to avoid damage, the acceleration voltage of EB should be lower.**

Keywords-EB, SEM, nanoprober, HV, charge, damage

I. BACKGROUND

In recent years, using nanoprobing techniques based on a scanning electron microscope (SEM) based to accomplish parametric data extraction has been widely reported as a method of failure analysis. However, this electron beam (EB) technique is also known to induce device degradation for scaled devices [1]. For high voltage (HV) devices, the effects of EB induced damage, however, have not been reported. In this study, HV, middle voltage (MV), and low voltage (LV) devices were prepared in order to implement device parametric measurements using an SEM based nanoprober. Compared with the data extracted from measurement taken in the dark box, degradation of electrical parameters, such as threshold voltage and off-state current, can be observed. EB penetration into the dielectrics, resulting in the modification of the physical properties of the dielectric layer, could be a concern, not only for scaled devices but also for HV devices. Net positive charge buildup in the oxide after EB exposure could increase the off-state current observed in the HV NMOS transistor, or cause the HV PMOS transistor to remain in enhancement mode. To avoid damage, the acceleration voltage of the EB should be lower.

II. EXPERIMENTAL

In this study, six kinds of transistors, whose critical dimensions are tabulated in Table I, were evaluated using an SEM-based nanoprobing system (with electron irradiation). In order to simulate the practical environment of nanoprobing for actual devices of a chip, the backend wiring layers of areas of the interest were mechanically polished to expose isolated tungsten plugs to the bulk, source, drain and the gate. The nanoprobing was performed using a Philips-XL30 SEM subjected to an EB of 2keV/32pA for probe guidance. In order to quantitatively evaluate device degradation caused by electron beam irradiation, before implementing nanoprobing measurements, these six transistors had been evaluated using an HP4156 semiconductor analyzer in the dark box during WAT tests (without electron irradiation). The changes in threshold voltage and off-state current characteristics can be evaluated by comparing the WAT-based data (without electron irradiation) and the nanoprobing data (with electron irradiation).

TABLE I.   CRITICAL DIMENSIONS OF EVALUATED TRANSISTORS

	L (um)	W (um)	GOX (a)
HVNMOS	2.5	10	430
HVPMOS	2.3	10	430
MVNMOS	1.1	10	430
MVPMOS	1.1	10	430
LVNMOS	0.34	10	65
LVPMOS	0.34	10	65

III. RESULTS AND DISCUSSION

The passive voltage contrast (PVC) image stems from the difference in secondary electron emission, which is a function of local surface potential. Different surface potentials caused by electron accumulation will produce different amounts of secondary electron emissions resulting in different contrasts. Fig. 1(a) shows an SEM photograph of an HV device on the contact level subjected to EB irradiation of 2keV/32pA. Under these conditions, the bright contrast of the poly contact can be observed in the both HVNMOS and the HVPMOS, which means that a net negative charge builds up on the surface of the

978-1-4244-9113-1/11 $26.00 © 2011 IEEE

poly contacts. As a result, more secondary electrons will be emitted. Thus, the poly contacts will appear bright in the PVC image. In order to further endorse our belief that a net negative charge builds up on the surface of the poly contacts, the IV characteristics were measured using nanoprobing. A voltage sweeping from a negative bias to a positive bias was applied to the drain, with the source and bulk connected to ground, and with the gate floating. The results of the IV measurements, as shown in Fig. 1(b), indicate that the channel is conducting for the HVPMOS, but the transistor does not conduct appreciably for HVNMOS. In this experiment, it was confirmed that the sample surface subjected to 2keV/32pA EB irradiation during nanoprobing will be biased negatively.

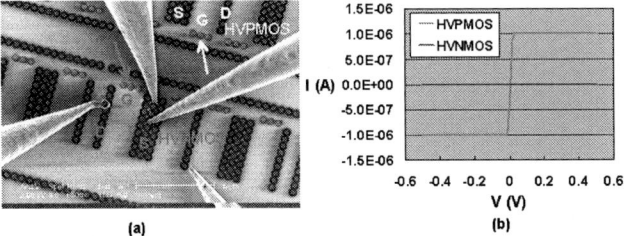

Figure 1. SEM photograph of HV devices on the contact level subjected to EB irradiation of 2keV/32pA (a), and its IV characteristics (b).

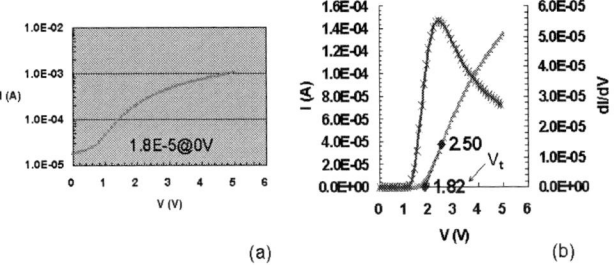

Figure 2. Transfer characteristics of HVNMOS (a), and HVPMOS (b).

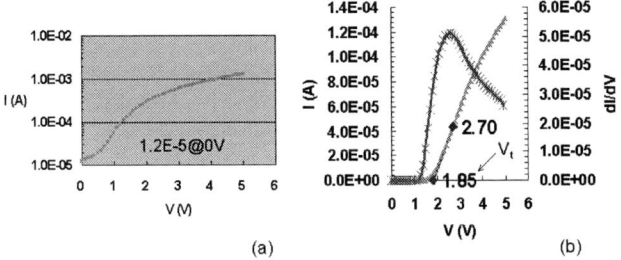

Figure 3. Transfer characteristics of MVNMOS (a), and MVPMOS (b).

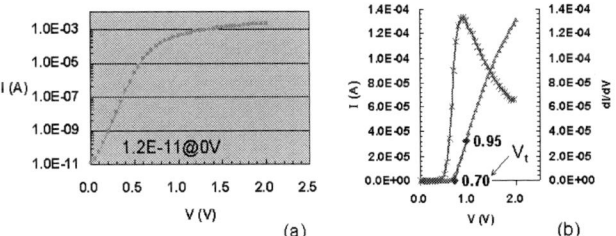

Figure 4. Transfer characteristics of LVNMOS (a), and LVPMOS (b).

Fig. 2 shows the transfer characteristics of the HVNMOS and the HVPMOS. As shown in Fig. 2(a), the off-state current of the HVNMOS approximately equal to E-5 is found to degrade significantly after EB irradiation, compared to that of the off-state current approximately equal to E-12 from the WAT-based data prior to EB irradiation, and the threshold voltage of the HVPMOS equal to 1.82V is also found to degrade significantly after EB irradiation, compared to that of the threshold voltage equal to 1.26V from the WAT-based data prior to EB irradiation, as shown in Fig. 2(b). Fig. 3 shows the transfer characteristics of the MVNMOS and the MVPMOS. As shown in Fig. 3(a), the off-state current approximately equal to E-5 of the MVNMOS is found to degrade significantly after EB irradiation, compared to that of the off-state current approximately equal to E-12 from the WAT-based data prior to EB irradiation, and the threshold voltage of the MVPMOS equal to 1.85V is also found to degrade significantly after EB irradiation, compared to that of the threshold voltage equal to 1.35V from the WAT-based data prior to EB irradiation, as shown in Fig. 3(b). Fig. 4 shows the transfer characteristics of the LVNMOS and the LVPMOS. As shown in Fig. 4(a), the off-state current approximately equal to E-11 of the LVNMOS is not found to significantly degrade after EB irradiation, compared to that of the off-state current approximately equal to E-12 of from the WAT-based data prior to EB irradiation, and the threshold voltage of the LVPMOS equal to 0.7V is also not found to degrade significantly after EB irradiation, compared to that of the threshold voltage equal to 0.68V from the WAT-based data prior to EB irradiation, as shown in Fig. 4(b). The threshold voltage shifts and the off-state current degradation for both the LVNMOS and LVPMOS are negligible. The experimental data for the six transistors is tabulated in Table II.

TABLE II. EXPERIMENTAL RESULTS

	GOX (a)	Vt (V) withou EB	Vt (V) with EB	Ioff (A) without EB	Ioff (A) with EB
HVNMOS	430	1.01		E-12	E-5
HVPMOS	430	1.26	1.82	E-11	E-11
MVNMOS	430	1.25		E-12	E-5
MVPMOS	430	1.35	1.85	E-11	E-11
LVNMOS	65	0.59	0.58	E-12	E-11
LVPMOS	65	0.68	0.70	E-12	E-12

EB:Electron Beam 2keV/32pA

The above observations for the HVNMOS, HVPMOS, MVNMOS, MVPMOS, LVNMOS, and LVPMOS could be attributed to the charge injection into the gate dielectric. The tendencies of the threshold voltage shifts for both the HVPMOS and MVPMOS, and the trends of the off-state current degradation for both the HVNMOS and MVNMOS are consistent with the presence of a positive charge in the gate dielectric after EB irradiation, as the interface trapped charge will change the flat band voltage, which is described as the relationship according to (1).

$$V_{FB} = \phi_{MS} - \frac{Q_{SS}}{C_{OX}} \qquad (1)$$

V_{FB}: flat band voltage

Φ_{MS}: metal semiconductor work function difference

Q_{SS}: effective oxide charges located at the Si/SiO$_2$ interface

978-1-4244-9113-1/11 $26.00 © 2011 IEEE   771

During EB exposure, the gate of the transistor is biased negatively with respect to the body due to the accumulation of electrons on the gate, as shown in Fig. 5, which could result in substrate hot hole (SHH) injection [5]. Fig. 6 shows the energy band diagrams for the PMOSFET and the NMOSFET associated with negatively biased gate conditions. The bands in the semiconductor warp upwards, causing an accumulation of holes in the semiconductor at the oxide interface. The generated holes are injected back into the oxide and cause interface traps at the Si/SiO₂ interface. The higher electric field in the silicon substrate results in greater hot concentration, which can then be injected into the gate dielectric and could be the most important reason for interface trap generation.

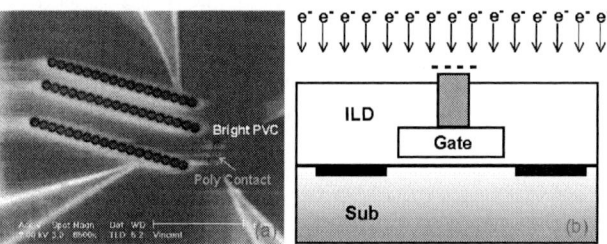

Figure 5. SEM photograph of an HV devices on the contact level subjected to EB irradiatiof 2keV/32pA (a), and a schematic diagram of the device cross-section during EB irradiation (b).

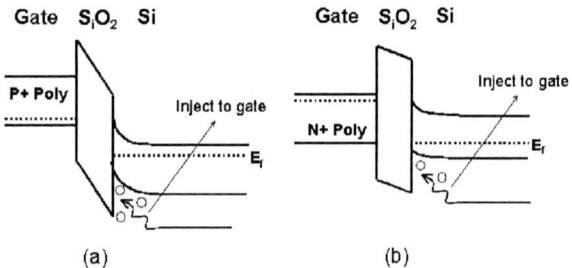

Figure 6. Band diagram of PMOSFET (a), and NMOSFET(b) during EB irradiation (negative bias on gate).

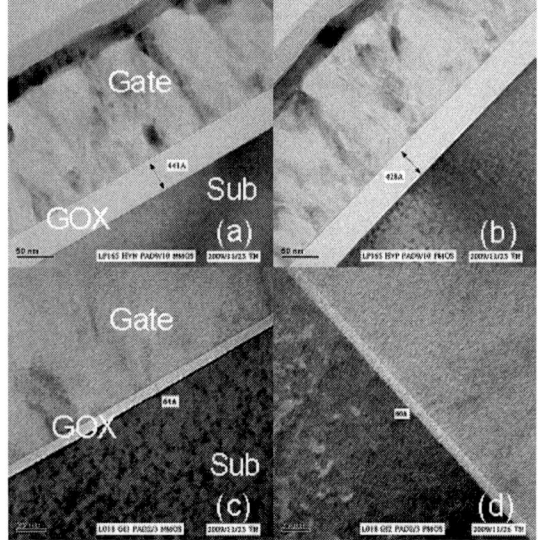

Figure 7. Cross-sectional TEM images for the HVNMOS (a), HVPMOS (b), LVNMOS (c), and LVPMOS (d).

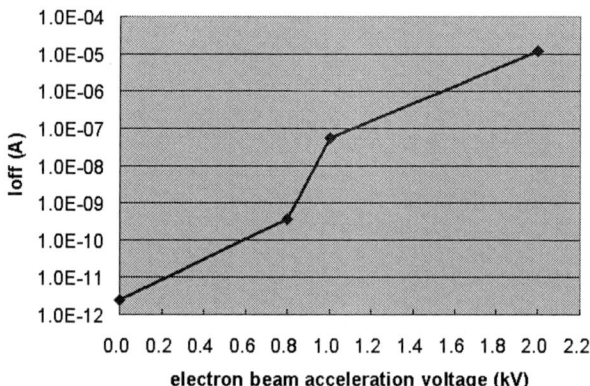

Figure 8. The dependence of the HVNMOS off-state current on the electron beam acceleration voltage.

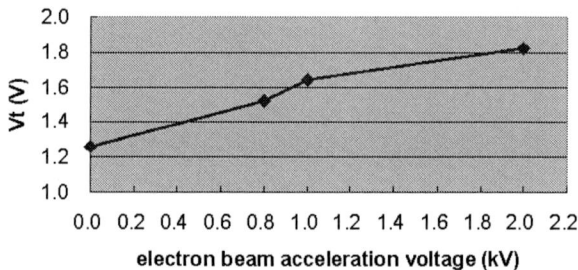

Figure 9. The dependence of the HVPMOS threshold voltage on the electron beam acceleration voltage.

The threshold voltage shifts are an important indicator of the creation of interface states, trapped charges or fixed charges in the gate dielectric. In this experiment, the threshold voltage shifts for both the LVNMOS and LVPMOS are negligible, as shown in Table II, indicating that no significant interface trap creation results from EB irradiation. This could be attributed to the reduced oxide thickness of the LV transistors (65 angstrom), compared with that of the HV and MV transistors (430 angstrom), as shown in Fig. 7. The thinner the gate dielectric is, the higher the gate leakage current could be. As a result, the LV devices could have a faster charge dissipation and lower stress voltages than both the HV and MV devices and, hence, no significant threshold voltage shifts or off-state current degradation were observed for the LV devices.

IV.   CONCLUSION

The influence on HV device characteristics by the EB exposure during nanoprobing observations was studied. Charge damage by EB irradiation resulting in threshold voltage shifts or off-state current degradation could occur on HV or MV thick gage dielectric (430 angstrom) devices while the impact on LV thin gate dielectric (65 angstrom) devices was negligible in the study. As shown in Figs. 8-9, to avoid damage, the acceleration voltage should be lower during device inspections in failure analysis procedures. The effect of the probe current on the degradation of electrical parameters of HV devices, not included in this paper, but which also need to be carefully considered, are currently under investigation.

ACKNOWLEDGMENT

The author would like to acknowledge the contribution of the Logic&MM FA Engineering group in making device level IV measurements. We would also like to thank the TEM group for obtaining cross-sectional TEM images.

REFERENCES

[1] Takayuki Mizuno, Miho Takahashi, Yoshie Azuma, Hiroshi Yanagita, Kyoichiro Asayama, Koji Nakamae, " Maximum Permissible EB Acceleration Voltage for SEM-Based Inspection before Electrical Characterization of Advanced MOS," 45th IRPS Proceedings, Apr. 2007, pp. 618-619.

[2] E. Hendarto, S.L. Toh, P.K. Tan, Y.W. Goh, J.L. Cai, Y.Z. Ma, Z.H. Mai, J. Lam, J. Sudijono, "Investigation on Focused Ion Beam Induced Damage on Nanoscale SRAM by Nanoprobing," 34th ISTFA Proceedings, Nov. 2008, pp.445-448.

[3] T. X. Tong, A. N. Erickson, "Current Image Atomic Force Microscopy (CI-AFM) combined with Atomic Force Probing (AFP) for location and characterization of advanced technology node", 30th ISTFA Proceedings, Nov. 2004, pp. 42-46.

[4] Yasuhiro Mitsui, Takeshi Sunaoshi, Jon C. Lee, "A Study of Electrical Characteristic Changes in MOSFET by Electron Beam Irradiation", Microelectronics Reliability 49, July. 2009, pp. 1182-1187.

[5] Jao-Hsian Shiue, Joseph Ya-min Lee, Tien-Sheng Chao, " A Study of Interface Trap Generation by Fowler-Nordheim and Substrate-Hot-Carrier Stresses for 4-nm Thick Gate Oxides," *IEEE Trans. Electron Devices*, vol. 46, no. 8, pp. 1705-1710, 1999.

Detecting Laser Beam Reflectance Modulated by Electronic Device Operation with a Simple Setup

Carlo Pagano*, Christian Boit
Department of Semiconductor Devices
Berlin University of Technology
Berlin, Germany
*corresponding author: phone: (+49) (0)30- 314-29954, carlo.pagano@tu-berlin.de

Yoshiyuki Yokoyama
Hamamatsu Photonics Deutschland GmbH
Herrsching, Germany

Abstract — **Modulation of reflected light intensity by electronic device operation is well established as a backside contactless probing technique [1]. In this paper, we present a simple setup for mapping reflected light signals across semiconductor devices. In the experiment, metal-oxide-semiconductor field-effect transistors (MOSFET) were used in different operating conditions.**

Keywords-component; Failure Analysis; Backside techniques; Laser Voltage Probing; Contactless measurement; Space charge region; p-n junction; Absorption coefficient; Reflectance.

I. INTRODUCTION

Probing integrated circuits (ICs) to obtain signal waveforms at various nodes is one of the aims of the modern failure analysis (FA). Mechanical probing is the most widely employed way for these measurements, but it presents several limits such as the access to specific locations inside the device. As a result, contactless measurement techniques have become more and more common [2]. Moreover, because of the increasing number of metallization layers, such measurements are often performed from the backside of the chip.

In this paper, we introduce a new and easy laser-based setup for backside mapping of modulated reflected light signals across semiconductor devices.

II. SETUP AND DEVICES USED

The analysis was performed using a Hamamatsu PHEMOS-1000 in combination with a dual-phase wide bandwidth DSP lock-in amplifier, having a reference frequency range of 0.5 Hz to 2 MHz, and an external InGaAs Photodiode unit (PD). The devices under test were MOS transistors (see Table I) manufactured in 120-nm-technology by Infineon Technologies AG. The samples were bonded in SBGA packages in order to have full control over all terminals of the transistors, and then prepared from the backside by mechanical thinning down to 80-100 μm of remaining silicon thickness.

The basic idea for the measurements was to drive the devices with a pulsed signal, scan the transistor area from the backside with the laser and detect the reflected light at each pixel position to create a map of laser beam reflectance modulation intensity (RMI).

Precisely applying a pulsed signal to the transistor affects the space charge region (SCR) and the charge carrier density. Therefore, the device optical properties (absorption coefficient and refractive index) will also change [3]. As a result, the reflected light will be modulated. The intensity of this modulation is not perceptible in the normal DC-reflected image. For this reason, we employed the lock-in amplifier (Fig. 1) to detect desired signals. The PD converted the reflected light into an electrical signal and amplified it. The magnitude of the AC part of the PD-output signal was then extracted using the lock-in amplifier. The lock-in-output was connected to the Phemos DALS-I/O-port in order to obtain the RMI map.

The transistors were operated in three different modes using an arbitrary waveform generator:

- Gate pulsed; drain, source and bulk grounded;
- Gate and drain pulsed; source and bulk grounded;
- Drain pulsed; source, gate and bulk grounded.

TABLE I. Transistors properties

Properties	Transistors			
	nFET-S	pFET-S	nFET-L	pFET-L
L [μm]	0.12	0.12	30	30
W [μm]	10	10	30	30
Vth [V]	0.209	-0.144	0.119	-0.212

For driving the devices we chose a square wave with 50 % duty cycle and a frequency of 2 MHz. The frequency was set to the lock-in cutoff frequency with the aim of reducing the laser dwell-time on each pixel. We were able to set the duration for image acquisition to around 200 s. The choice for the voltage levels was done in order to avoid electroluminescence from the drain and source diffusions. Consequently, we drove the nFETs with a square wave $0\ V/V_{pulse}$, where V_{pulse} was negative for the pFETs and positive for the nFETs. In this way the drain-body and the source-body diodes were biased always in reverse.

The analysis was done using a 1064-nm-laser, having a maximum output power of 200 mW, and with an ULWD 50x lens, having a numerical aperture (NA) of 0.76. The laser power (LP) and the laser power density (LPD) delivered to the sample are represented in Table II. Considering the Abbe diffraction limit [4] of equation (1), we obtain, for our measurement, a minimum resolution of 700 nm.

$$a_{min} = \frac{\lambda}{2NA} \quad (1)$$

TABLE II.   LP and LPD delivered to the sample

%LP$_{out}$	LP [mW]	LPD=LP/($\pi\omega_0^2$) [mW/μm^2]
0	0	0
20	4.74	2.07
40	9.58	4.18
60	15.01	6.55
80	20.09	8.77
100	25.11	10.96

The optical resolution is one of the most important parameters, but it is not the only one influencing our measurement. Two other significant parameters are the laser spot size and the pixel resolution. The first one could be calculated using the equation (2).

$$2\varpi_0 = \frac{1.22\lambda}{NA} \quad (2)$$

The term ω_0 is the radius of the laser spot [5]. For our instrumentation, we achieve a laser spot size ($2\omega_0$) of 1.7 μm.

The pixel resolution is a mechanical limit of the tool resulting from the step movement of the mirrors responsible for scanning the laser beam. For the objective used in this analysis, the pixel resolution is 0.5 μm.

In the fourth section, we will describe in more detail the importance of these parameters for this kind of analysis.

III.   MEASUREMENTS ON LARGE TRANSISTORS

The first analysis was done on large transistors (pFET-L and nFET-L) in order to evaluate the feasibility of our setup.

In Fig. 2 one can see the layout and the DC-reflected-light-image for both transistor pFET-L (a,c) and nFET-L (b,d).

The measurements were done considering a scanning area of 256x256 pixels, i.e. the image scan rate was around 3 ms/pixel.

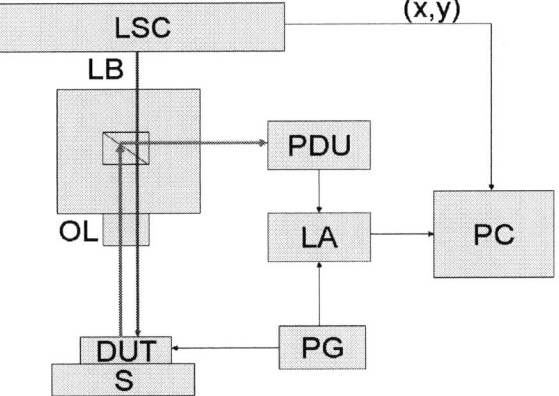

Figure 1.   Block diagram of the setup used. LSC=Laser Scanning Controller; LB=Laser Beam; OL=Objective Lens; DUT=Device Under Test; S=Stage; PDU=Photodiode Unit; LA=Lock-in Amplifier; PG=Pulse Generator.

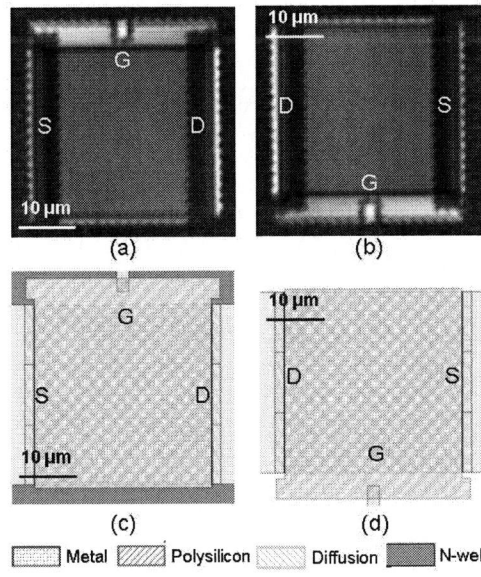

Figure 2.   DC reflected light images of pFET-L (a) and nFET-L (b) respectively. Layouts of pFET-L (c) and nFET-L (d).

A.   Gate pulsed; drain, source and bulk grounded

If we pulse only the gate, we can regard the device as a MOS-varactor in inversion. In Fig. 3, one can see the RMI profiles and the RMI maps for both transistors. In the case of pFET-L (Fig. 3 a, b) the signal is strong and homogenous over the gate, while for the nFET-L (Fig 3. c, d) the signal is not perfectly smooth approaching the drain. The SCR and the charge density in the inversion channel underneath the gate are modulated by the gate voltage and both phenomena affect the reflected light.

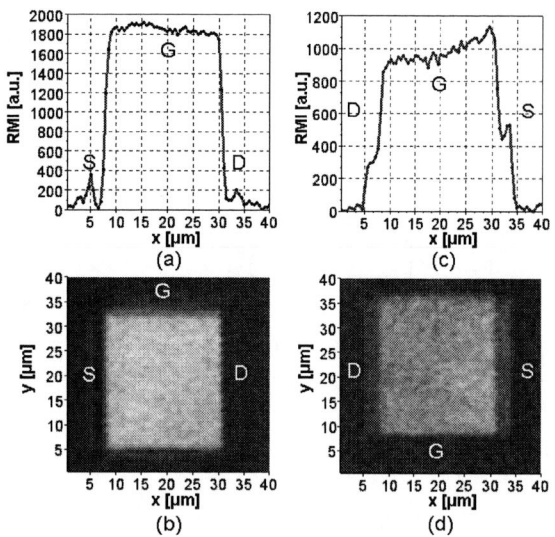

Figure 3. RMI profiles (a, c) and RMI maps (b, d) obtained by pulsing the gate of pFET-L and nFET-L respectively. The voltage levels (V_pulse) used are -1.2 V for pFET-L and 1.2V for nFET-L.

The correlations between the average RMI on the gate and V_{pulse} for both transistors are plotted in Figs. 4 and 5. These graphs were obtained calculating the average value of the RMI on the gate area for different voltages. For the pFET-L (nFET-L) the correlation is almost linear for voltage levels lower than -0.4 V (higher than 0.4 V) and up (below) this value the correlation is weakly dependent on the voltage with a local maximum at -0.3 V (0.2 V).

Now, we take the RMI on the gate area of the pFET-L as an example. Decreasing the voltage will let the SCR underneath the gate deplete more and more influencing the RMI until the energy bands are bent to strong inversion (region I in Fig. 4). This happens when the gate voltage reaches the threshold (V_{th}). Once the inversion channel is created, the contribution on the RMI due to the increasing free carrier density in this layer will dominate on the one due to the SCR [6] (region II in Fig. 4). As a consequence, we obtained the graph of Fig. 4. The same is also true for the nFET-L.

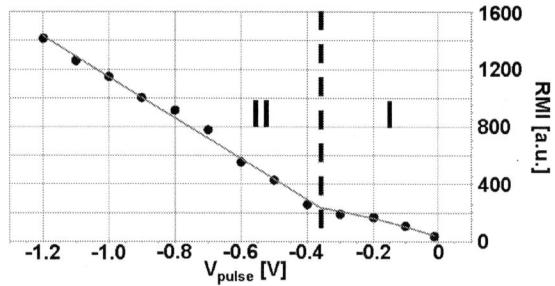

Figure 4. Correlation between average RMI on the gate and V_{pulse} for the pFET-L. The red curve is the fitting curve. In the region I, the curve follows a second degree polinomial. In the region II, the curve is linear.

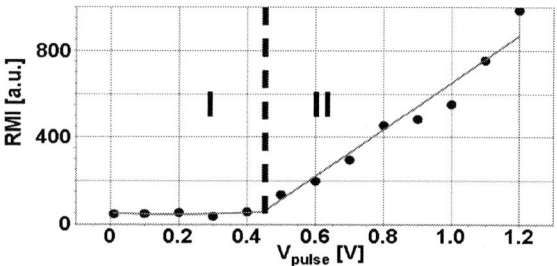

Figure 5. Correlation between average RMI on the gate and V_{pulse} for the nFET-L. The red curve is the fitting curve. In the region I, the curve follows a second degree polinomial. In the region II, the curve is linear.

B. Gate and drain pulsed; source and bulk grounded

If we pulse both drain and gate and keep the other terminals grounded, the transistors will switch between the off and on state (saturation). In the off-state there is no inversion of the channel underneath the gate. When the transistors go in saturation, the channel is formed and pinched off. In Fig. 6, one can compare the RMI profiles and the RMI maps for pFET-L (a, b) and the nFET-L (c, d). For both transistors we can observe the pinch-off region near the drain which is characterized by a very weakly modulated reflection signal. The modulation is stronger on the source side of the channel, where both SCR and the charge density in the inversion layer are contributing to the net modulation of the reflected light. Furthermore, it is possible to distinguish the modulation on the drain region, because the pulse applied on the drain modulates the SCR underneath the drain region. For the pFET-L, the signal from the drain region is confined to a narrow area, while for the nFET-L the same signal spreads more.

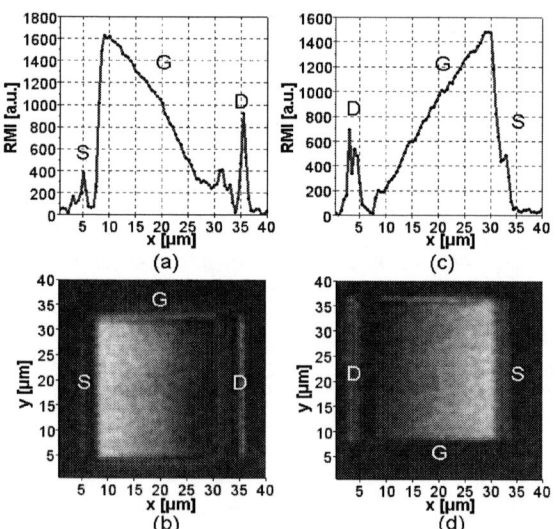

Figure 6. RMI profiles (a, c) and RMI maps (b, d) obtained by pulsing both gate and drain of pFET-L and nFET-L respectively. The voltage levels (V_pulse) used are -1.2 V for pFET-L and 1.2V for nFET-L.

C. Drain pulsed; gate, source and bulk grounded

If we pulse just the drain, we can regard the device as a reverse-biased p-n junction. In Fig. 7 we display the RMI profiles and the RMI maps for pFET-L (a, b) and the nFET-L (c, d). In both cases, the modulation of the reflected light signal is present only over the drain region because only the SCR underneath the drain is modulated.

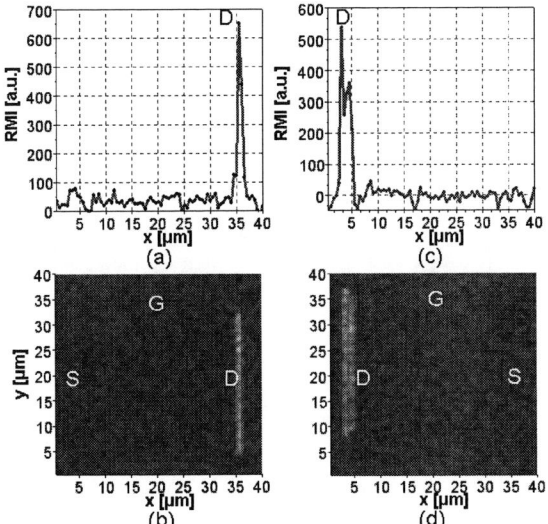

(a)

(c)

(b)

(d)

Figure 7. RMI profiles (a,c) and RMI maps (b,d) obtained by pulsing the drain of pFET-L and nFET-L respectively. The voltage levels (V_pulse) used are -1.2 V for pFET-B and 1.2V for nFET-B.

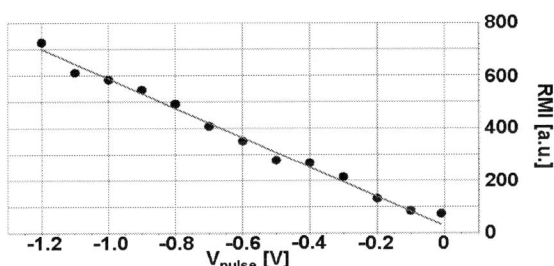

Figure 8. Correlation between average RMI on the drain region and V_pulse for the pFET-L. The red curve is the linear fit.

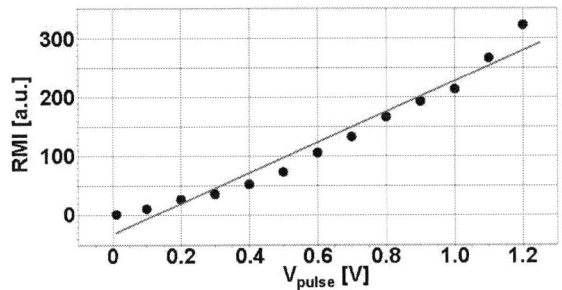

Figure 9. Correlation between average RMI on the drain region and V_pulse for nFET-L. The red curve is the linear fit.

In Figs.8 and 9, the correlation between the average value of the RMI on the drain region and V_pulse for both transistors is plotted. The behavior of these curves is reasonably linear for every voltage level.

IV. MEASUREMENTS ON SMALLER TRANSISTORS

In this section we present the results obtained on transistors with shorter channels (pFET-S and nFET-S). For these devices the dimensions are smaller than the minimum resolution of the system. The area scanned during these measurements was 64x64 pixels, i.e., considering an image scanning time of 200 s, we obtain an image scanning rate of around 49 ms/pixel.

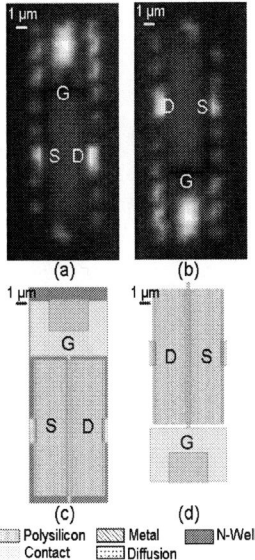

(a)   (b)

(c)   (d)

Polysilicon   Metal   N-Well
Contact   Diffusion

Figure 10. DC reflected light images of pFET-S (a) and nFET-S (b) respectively. Layout of pFET-S (c) and nFET-S (d).

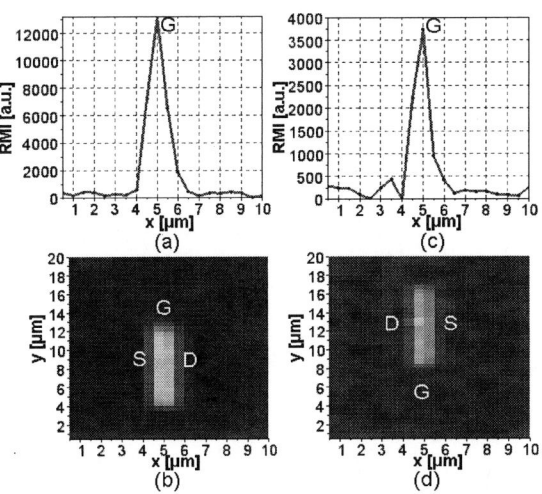

(a)   (c)

(b)   (d)

Figure 11. RMI profiles (a,c) and RMI maps (b,d) obtained by pulsing the gate of pFET-S and nFET-S respectively. The voltage levels (V_pulse) used are -1.2 V for pFET-S and 1.2V for nFET-S

A. Gate pulsed; drain, source and bulk grounded

If we pulse the gate only, we should see in the RMI map a homogenous signal on the gate. However, in this case the channel length is shorter than the pixel resolution; therefore we obtain a signal over an area wider than the real active area as shown in Fig. 11.

978-1-4244-9113-1/11 $26.00 © 2011 IEEE

In Fig. 12 and 13, we have plotted the correlation between the average RMI on the gate and V_{pulse} for both transistors.

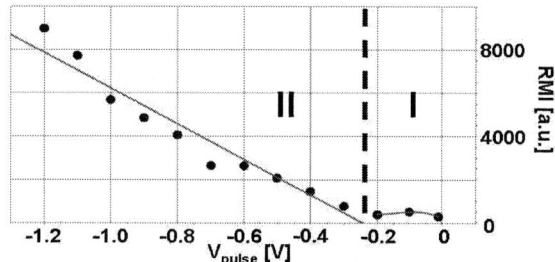

Figure 12. Correlation between average RMI on the gate and V_{pulse} for the pFET-S. The red curve is the fitting curve. In the region I, the curve follows a second degree polinomial. In the region II, the curve is linear.

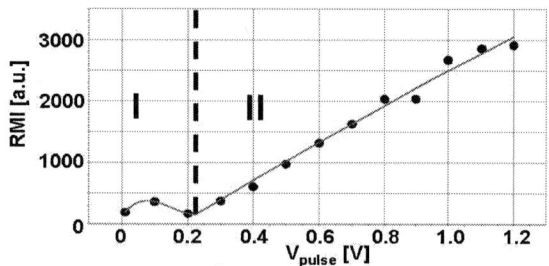

Figure 13. Correlation between average RMI on the gate and V_{pulse} for the nFET-S. The red curve is the fitting curve. In the region I, the curve follows a second degree polinomial. In the region II, the curve is linear.

As for the large transistors, the characteristic for the pFET-S (nFET-S) is partially linear below the voltage of -0.3 V (above the voltage of 0.3 V). In addition, one can see a local maximum at -0.1 V (0.1 V).

B. Gate and drain pulsed; source and bulk grounded

In this case, pulsing gate and drain simultaneously, we obtained a strong signal on the transistors and it becomes quite challenging to distinguish between the signals coming from the active area and the drain region. To overcome this problem, we used less laser power (40 % of the maximum power) and switched off the amplifier present in the PD unit.

In Fig. 14, the RMI profiles and the RMI maps for pFET-S (a, b) and the nFET-S (c, d) are shown.

For the nFET-S, we can recognize the drain region, where the signal is stronger, and it is also possible to see the signal near the source, where both SCR and the charge density in the inversion layer are contributing to the net modulation of the reflected light. Hence, the smaller n-channel transistor behaves in a similar way as the bigger one. However, the detection is limited by the pixel resolution.

For the pFET-S it is more difficult to distinguish particular areas because the intensity is almost the same. Moreover, for this transistor, we can see some artifacts coming from the polysilicon interconnection used to contact the gate (the closed ring in Fig. 14 b) and from the drain metallization (the dashed ring in Fig. 14 b).

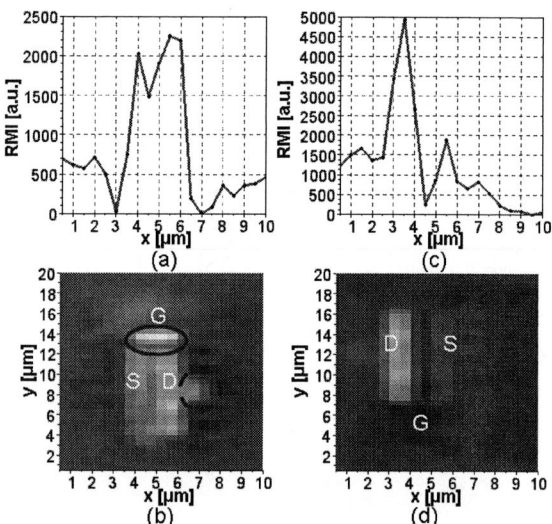

Figure 14. RMI profiles (a, c) and RMI maps (b, d) obtained by pulsing both gate and drain of pFET-S and nFET-S respectively. The voltage levels (V_{pulse}) used are -1.2 V for pFET-S and 1.2V for nFET-S.

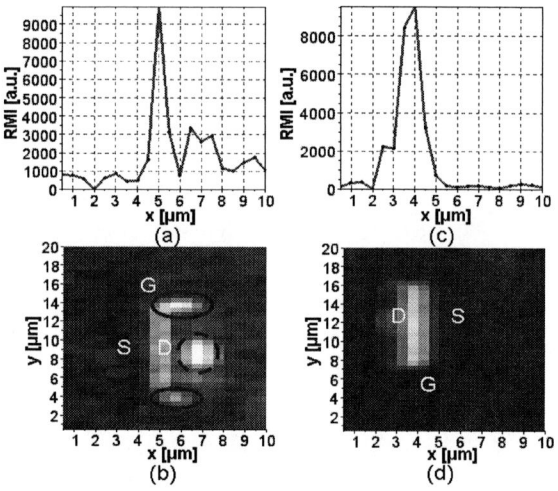

Figure 15. RMI profiles (a,c) and RMI maps (b,d) obtained by pulsing the drain of pFET–S and nFET-S respectively. The voltage levels (V_{pulse}) used are -1.2 V for pFET-S and 1.2V for nFET-S.

C. Drain pulsed; gate, drain and bulk grounded

If we pulse only the drain, we detect the reflectance modulation signal on the drain region as shown in Fig. 15.

For the nFET-S, a strong signal is localized on the drain region, while for the pFET-S, we observe also the signal coming from the metallization (the dashed ring in Fig. 15 b) and, as in the previous case, the artifacts coming from the polysilicon interconnection used to contact the gate (the closed rings in Fig. 15 b).

In Figs. 16 and 17, we have plotted the correlation between the average RMI on the drain and V_{pulse} for both transistors.

978-1-4244-9113-1/11 $26.00 © 2011 IEEE

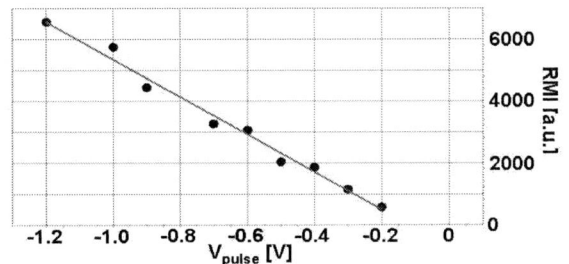

Figure 16. Correlation between average RMI on the drain and V_{pulse} for the pFET-S. The red curve is the linear fit.

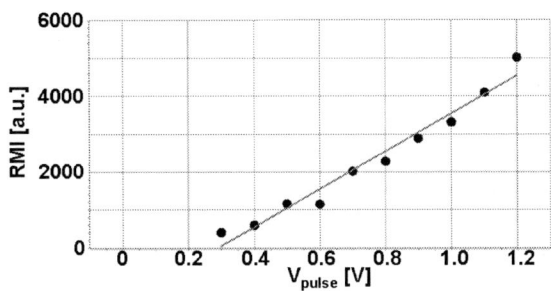

Figure 17. Correlation between average RMI on the drain and V_{pulse} for the nFET-S. The red curve is the linear fit.

From these graphs, we conclude, that for the short-channel transistors the correlation between RMI and voltage is quite linear.

V. CONCLUSIONS

In this paper we presented an easy setup for the RMI mapping without using complicated and high performance instrumentation. We have shown the results for two transistor topologies, long channel and short channel. For the long channel transistors, the analyses ran without problems making it possible to study and monitor the activity of the device. On the other hand, if we use transistors with dimensions smaller than the pixel resolution and the laser spot size, the analysis of the device activity becomes more challenging. However, it is still possible to detect the RMI and the intensity measured is even higher than that for the larger transistors. Moreover, this setup delivers RMI signals also at low voltages ($|V_{pulse}|>0.2$ V).

The problem related to the pixel resolution and the laser spot size could be modeled as shown in Fig. 18.

As already mentioned in section II, the laser is not continuously scanned on the surface, but raster scanned with a certain step that determines the pixel size. The length of each step is the pixel resolution. For every pixel the system acquires the reflected light intensity. If we pulse only the gate, the reflectance should be modulated only on the gate area. Since the spot size is greater than the gate area, the laser acquires signals from the gate for several pixels. In this way, we will detect the signal on an area larger than the gate itself, as shown in Fig. 11. The analysis becomes more challenging if we pulse simultaneously the drain and gate, because for some pixels the reflected light will deliver the information about the

modulation from both the gate and drain areas as already shown in Fig.14.

Despite these resolution limits, we were able to measure the modulation of the reflected light coming from small areas of devices. Moreover, from the Figs. 12, 13, 16 and 17, one can see that the correlations between RMI and V_{pulse} have the same trends as for the large transistors.

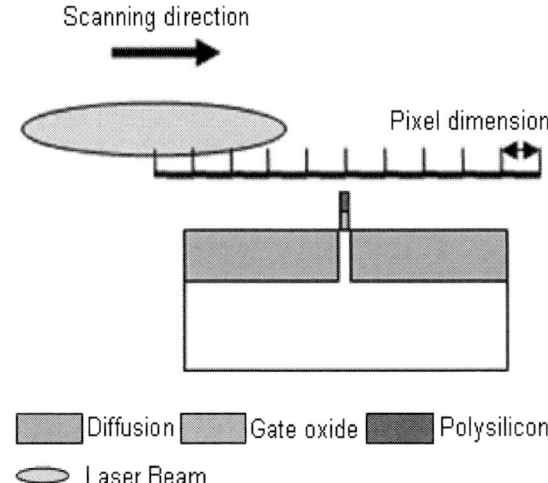

Figure 18. Model for the laser scanning on smaller transistors considering the limit of the pixel resolution and of the spot size.

ACKNOWLEDGMENT

The authors would like to thank R. Hartmann and A. Eckert for the samples preparation and A. Glowacki (all Berlin Univ. of Techn.) for the support and helpful discussions about the laser techniques. Special thanks go to E. Cole of Sandia Labs for reviewing this paper and for useful suggestions.

REFERENCES

[1] U. Kindereit, G. Woods, J. Tian, U. Kerst, R. Leihkauf and C. Boit, "Quantitative Investigation of Laser Beam Modulation in Electrically Active Devices as Used in Laser Voltage Probing," IEEE Tramsactions on Device and Materials Reliability, vol. 7, NO. 1, March 2007.

[2] S. Kolachina, "Introduction to Laser Voltage Probing (LVP) of Integrated Circuits," *Microelectronics Failure Analysis*, 5th ed. EDFAS Desk Reference Commitee, pp. 426–430, October 2004.

[3] R. A. Soref, B. R. Bennett, "Electrooptical effects in Silicon", *IEEE Journal of Quantum Electronics*, pp. 123-129, 1st Januar 1987.

[4] M.V Klein and T. E. Furtak, *Optics*, 2nd edition, New York, Willey, ch. 7, pp 407-500, 1986.

[5] A. E. Siegman, *Lasers*, Mill Valley CA, University Science Books, pp 663-743, 1986.

[6] U. Kindereit, G. Woods, J. Tian, U. Kerst, C. Boit, "Investigation of Laser Voltage Probing Signals in CMOS Transistors", *IEEE proceedings of 45th IRPS Phoenix*, pp. 526-533, 2007.

Backside Reflectance Modulation of Microscale Metal Interconnects

J.K.J.Teo[1], C.M.Chua[2], L.S.Koh[2] and J.C.H.Phang[1,2]

[1]Centre for Integrated Circuit Failure Analysis and Reliability (CICFAR), National University of Singapore
Tel: +65-6516-2244, Fax: +65-6779-1103, E-mail: jasonteo@nus.edu.sg
[2]SEMICAPS Pte Ltd, Singapore

Abstract— The variation of backside reflectance modulation effects on metal line samples at different electrical bias and silicon backside thicknesses is investigated. Negative reflected intensity modulation is observed with temperature increase which is one to two orders of magnitude higher than published results. A backside reflectance model is developed to explain the experimental results.

Keywords-laser reflectance; metal interconnect; substrate thickness;

I. INTRODUCTION

When a laser beam is incident on a material or device, the optical reflectivity may be changed due to temperature [1] or electro-optical effects [2] resulting in a modulation of the reflected laser beam. Laser reflectance modulation has been used to measure the temperature changes of microscale metal thin films [1], silicon (Si) integrated circuits [3] and to acquire timing signals from Si devices [4].

Typical reflectance measurement systems are camera-based or laser-based. Camera-based reflectance systems consist of broadband light source, optics and CCD detector [5,6]. Laser-based reflectance systems consist of monochromatic laser beam, optics and photon detectors [1,3]. The optical setup of laser-based systems is generally more varied and complex.

Most published reflectance modulation results are measured from the frontside using visible wavelengths. These include interferometric detection [1], oblique angle detection [3], CCD imaging [6] and scanning laser [7]. Frontside techniques require accurate measurement of the passivation thickness [3] or calibrations to be performed [8,9] before absolute temperature can be derived.

Backside techniques use near infrared (NIR) wavelengths to penetrate through Si substrates and directly probe the underside metal layers and active layers. Devices with ball grid array (BGA) and flip-chip packages can be characterized with backside probing. Thermoreflectance measurements using NIR light source and InGaAs CCD camera have been reported [5].

This paper reports the variation of backside reflectance modulation effects of microscale metal line samples at different electrical bias and Si backside thicknesses using non-coherent 1340 nm illumination on a scanning optical microscope system. The reflected intensity is found to modulate negatively

with increasing temperature and is one to two orders of magnitude higher than published results. A backside reflectance model is developed to explain the experimental results.

II. THEORY

The laser beam propagation path for backside technique is modeled as shown in Fig. 1. The incident laser intensity is I_0. It is assumed the intensity loss in air is negligible. It is further assumed the incident ray and all the reflected and transmitted rays are at normal incidence.

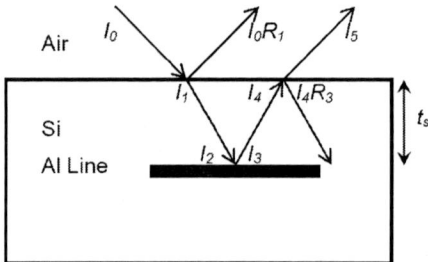

Figure 1. Backside Laser Beam Propagation Model

At the Air-Si interface, the transmitted light intensity I_1 is given by:

$$I_1 = I_0 T_1 = I_0 (1 - R_1) \tag{1}$$

where T_1 and R_1 are the transmittance and reflectance at the air-Si interface. The reflected light intensity is $I_0 R_1$.

For normal incidence, the reflectance R and transmittance T at an interface are given by Fresnel equations [10] as follows:

$$R = \left(\frac{n_t - n_i}{n_t + n_i} \right)^2 \tag{2}$$

$$T = \left(\frac{4.n_t.n_i}{n_t + n_i} \right)^2 \tag{3}$$

978-1-4244-9113-1/11 $26.00 © 2011 IEEE

where n_t is the refractive index of the transmitting medium and n_i is the refractive index of the incident medium.

When the metal line is biased, the temperature of the metal line increases. In the model, the Si substrate is assumed to be at the same temperature as the metal line. The transmitted light propagates through the Si substrate of thickness t_s. For the Si substrate at a constant temperature T, the intensity attenuation follows Lambert's Law [11]. At the Si-Al interface, the light intensity I_2 is given by:

$$I_2 = I_0 . (1 - R_1) . e^{-\alpha t_s} \tag{4}$$

where α is the absorption coefficient of Si.

The light intensity of reflected light at the Si-Al interface I_3 is given by:

$$I_3 = I_0 . (1 - R_1) . R_2 . e^{-\alpha t_s} \tag{5}$$

where R_2 is the reflectance of the Si-Al interface and can be derived from eqn (2).

The reflected light propagates again through the Si substrate of thickness t_s and the resulting intensity I_4 at the Si-Air interface is:

$$I_4 = I_0 . (1 - R_1) . R_2 . e^{-2\alpha t_s} \tag{6}$$

The light intensity of the light transmitted at the Si-Air interface I_5 is:

$$I_5 = I_0 . (1 - R_1) . (1 - R_3) . R_2 . e^{-2\alpha t_s} \tag{7}$$

where R_3 is the reflectance at the Si-Air interface. From eqn. (2), R_3 is the same as R_1 at the Air-Si interface for normal incidence.

The total light intensity I_{tb} measured by the photon detector is the sum of $I_0 R_1$ and I_5 given by:

$$I_{tb} = I_0 . R_1 + I_0 . (1 - R_1)^2 . R_2 . e^{-2\alpha t_s} \tag{8}$$

Eqn. (8) is valid for most backside reflectance systems, with the exception of laser-scanning confocal microscopes [12]. In the latter case, the surface reflectance can be neglected as the objective lenses are corrected telecentrically and the depth of focus is a few micrometers for an objective lens with high magnification and numerical aperture (NA). The total light intensity I_{te} measured by a telecentric system is:

$$I_{te} = I_0 . (1 - R_1)^2 . R_2 . e^{-2\alpha t_s} \tag{9}$$

The detected reflected intensity is dependent on R_1, R_2, α and t_s. From eqn. (2), both R_1 and R_2 are dependent on the refractive index of Si which varies with temperature due to density change from thermal expansion and temperature-dependent polarizability [13]. The variation of α for 1.32 μm photons is dominated by the two photon process [14]. Fig. 2 shows the variation of R_1, R_2 and α over the temperature range of 300K to 400K generated from the models by [13] and [14].

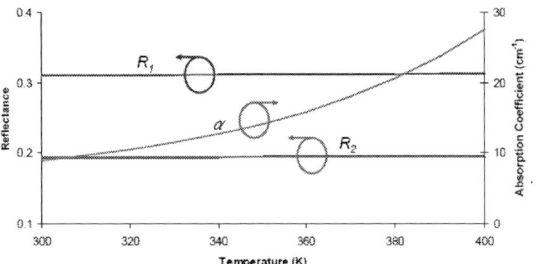

Figure 2. Variation of R_1, R_2 and α with temperature

It is observed α varies significantly with temperature while R_1 and R_2 varies by less than 1.2% for the temperature range from 300K to 400K. Hence, the model predicts that the reflected intensity modulation is due mainly to the temperature variation of the Si absorption coefficient and thickness.

The reflectance coefficient r is defined as:

$$r = \frac{\Delta I}{I} \tag{10}$$

where I is the reflected intensity of the unbiased sample and ΔI is the difference in the reflected intensities between the biased and unbiased sample.

If the physical phenomenon is due mainly to temperature, then the thermoreflectance coefficient k is defined as:

$$k = \frac{\Delta I}{I . \Delta T} \tag{11}$$

where ΔT is the temperature difference between the biased and unbiased condition [8].

III. EXPERIMENTAL PROCEDURE

A. Experimental Setup

Typical backside laser reflectance systems have to address the issue of strong reflection from the upper Si surface, with the exception of laser-scanning confocal microscopy systems. In addition, non-coherent NIR illumination solves the interference issues experienced by coherent NIR illumination methods [5].

Fig. 3 shows the experimental setup of scanning optical microscope using a non-coherent 1340 nm laser for the reflectance measurements in this paper.

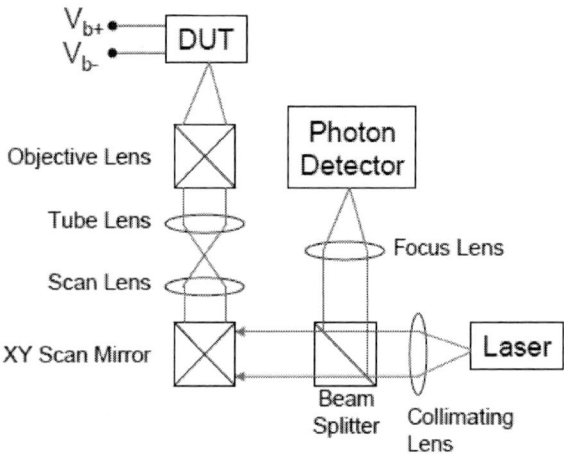

Figure 3.   Experimental Setup

B.   Samples

Two aluminium (Al) resistive structures are backside prepared. Sample #1 is a meandering resistive structure from a commercial 0.18 μm technology process with metal line width of 4 μm and spacing of 4 μm and backside prepared with substrate thickness of 240 μm.

Sample #2 is another meandering resistive structure with line width of 2 μm and spacing of 4 μm. The structure is designed to be temperature calibrated using four-point measurements of the resistance change with temperature. Sample #2 is temperature calibrated and provides the temperature of the metal line when biased. The sample is backside prepared and initially polished to 500 μm. The sample is then polished to reduce the substrate thickness until a series of reflectance measurements are obtained for different substrate thicknesses. Two such samples are used to ensure consistency of the results. The absorption coefficient of Sample #2 at room temperature is experimentally determined to be between 8 to 10 cm^{-1}. This value is used in the model described in Section 2.

IV.   RESULTS AND DISCUSSIONS

A.   Variation of Reflectance Intensity with Electrial Bias

Fig. 4a shows the unbiased reflected image of Sample #1 with a backside substrate thickness of 240 μm. Electrical bias is applied across the pads marked V_{b+} and V_{b-}. Fig. 4b and 4c show the difference image of the biased and unbiased reflectance images at 4 mW and 16 mW respectively of Area D in Fig. 4a transformed in rainbow pseudo-colour where the red areas represent the largest reflectance change and the blue areas represent the least change.

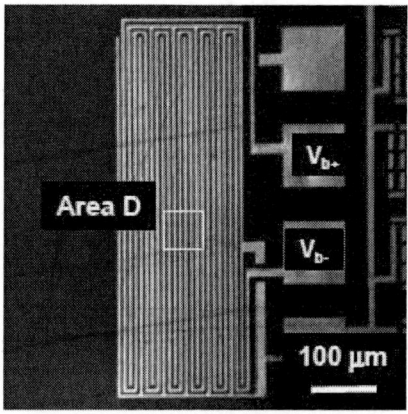

(a) Unbiased

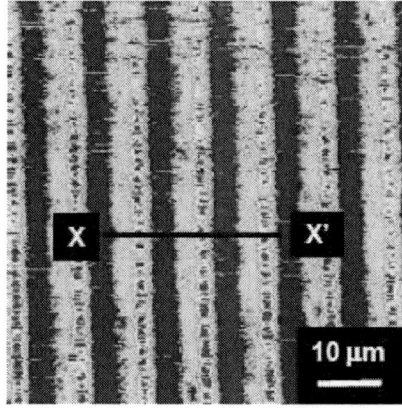

(b) Area D, 0 – 4 mW bias

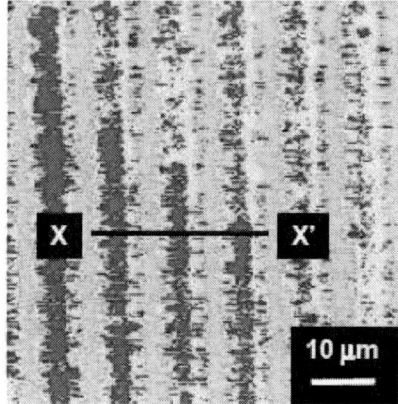

(C) Area D, 0 – 16 mW bias

Figure 4.   (a) Reflected image of Sample #1 and reflectance intensity difference image of Area D (b) For 0 – 4 mW and (c) 0 – 16 mW bias

Fig. 5 shows the reflectance intensity variation across the line profile XX' in Fig. 4b and 4c. The decrease in the reflectance intensity modulation due to an increase in electrical bias is clearly observed. The reflectance intensity in the spacing area is also modulated due to heating from the metal lines. At electrical bias of 4 mW and 16 mW, the temperature change of the metal line is 5.5 K and 11.6 K respectively.

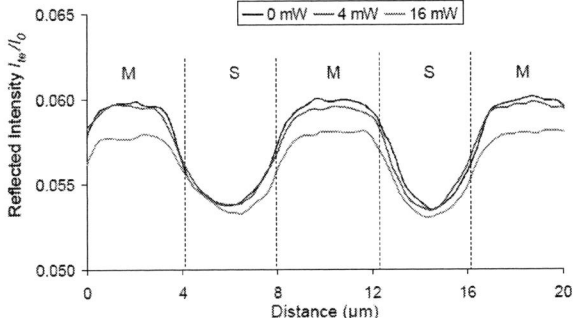

Figure 5. Line Profile XX' of Reflectance Intensity for Sample #1 across metal (M) lines and spacing (S)

Fig. 6 shows the results from Sample #2 with a silicon backside thickness of 340 μm. In this case, as the relative spacing between the metal lines increase, the reflectance intensity in the spacing area is flatter to show that indirect heating from the smaller metal lines is not as severe as in Sample #1. At electrical bias of 6 mW, 14 mW, 25 mW and 40 mW, the temperature change of the metal line are 3.4 K, 10.0K, 19.0K and 25.8K respectively.

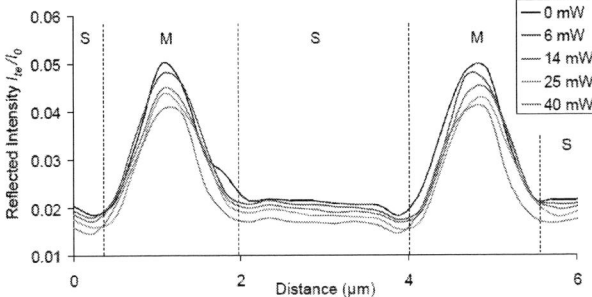

Figure 6. Line Profile XX' of Reflectance Intensity for Sample #2 across metal (M) lines and spacing (S)

The peak reflectance intensity modulation of the metal lines are used to derive r and k. Table I tabulates the coefficients together with calculated values using the model for laser-scanning confocal microscope system and published values for some frontside and backside techniques. The derived r and k values are based on metal line temperature change of 11.6 K for Sample #1 and 10.0 K for Sample #2.

TABLE I.   REFLECTANCE COEFFICIENTS

	r	k (K^{-1})	Remarks
Sample #1, Experiment	-3.34 x10^{-2}	-2.88 x10^{-3}	Backside
Sample #1, Model	-3.83 x10^{-2}	-3.30 x10^{-3}	Backside
Sample #2, Experiment	-7.33 x10^{-2}	-7.33 x10^{-3}	Backside
Sample #2, Model	-4.61 x10^{-2}	-4.61 x10^{-3}	Backside

	r	k (K^{-1})	Remarks
Shimizu [1]	2 x10^{-3}	6.67 x10^{-5}	Frontside
Quintard [3]	1.5 x10^{-3}	5.68 x10^{-4}	Frontside
	3.1 x10^{-3}	1.20 x10^{-3}	Frontside
Tessier [5]	3.80x10^{-3}	1.17x10^{-4}	Backside

The experimental r and k values are negative and agree well with the values generated by the model described in Section II. However, the published r and k values are one to two orders of magnitude lower. This is because for the frontside measurements, the absorption of the passivation layers is low and positive reflected intensity modulation is due mainly to the temperature variation of reflectance. This phenomenon has been verified by published frontside reflectance models [3,7]. For the backside values by [5], undoped Si substrate was used. In that case, the absorption coefficient is negligibly small at the NIR wavelengths used, and the reflected intensity modulation is again due primarily to the temperature variation of reflectance.

The model is modified to neglect the absorption coefficient in the case of undoped Si, so that the backside model only accounts for the temperature variation of reflectance. The calculated values of r and k are compared to Tessier's published results [5] for a substrate thickness of 500 μm. The r and k values are tabulated in Table II. The temperature change on the metal line is 32.5 K in [5] and in the model.

TABLE II.   REFLECTANCE COEFFICIENTS FOR UNDOPED SUBSTRATE

	r	k (K^{-1})	Remarks
Tessier [5]	3.80x10^{-3}	1.17x10^{-4}	ΔT=32.5K
Model	1.45 x10^{-3}	4.45 x10^{-5}	ΔT=32.5K

In this case, the values of r and k obtained are positive when the reflected intensity modulation is primarily due to the temperature variation of reflectance and are close to the values obtained by Tessier [5].

B. *Variation of Reflectance Intensity with Substrate Thickness*

Fig. 7 shows the reflected intensity at different substrate thickness for Sample #2 without electrical bias, i.e. at room temperature. The solid line is generated by the model described in Section II. The experimental points are in good agreement with the model.

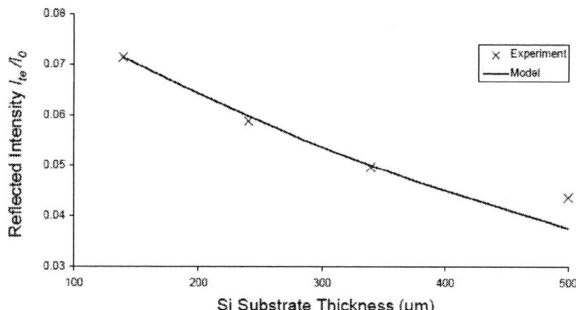

Figure 7. Reflected intensities at different substrate thicknesses and room temperature

In another experiment, the same set of electrical bias is applied to each substrate thickness of Sample #2 to measure the reflected intensities for different substrate thickness. The backside reflectance model is used to determine the reflected intensities at metal line temperatures of 300K, 310K, 320K, 330K and 340K for substrate thickness of 140 μm, 240 μm, 340 μm and 500 μm.

The measured reflected intensities are dependent on instrument gain while the reflected intensities calculated from the model are fully corrected for no gain. As such, a constant scaling factor is used on all the measured reflected intensities so that the measured reflected intensity at a substrate thickness of 140 μm is equivalent to the model. Fig. 8 shows the experimental and calculated reflected intensities.

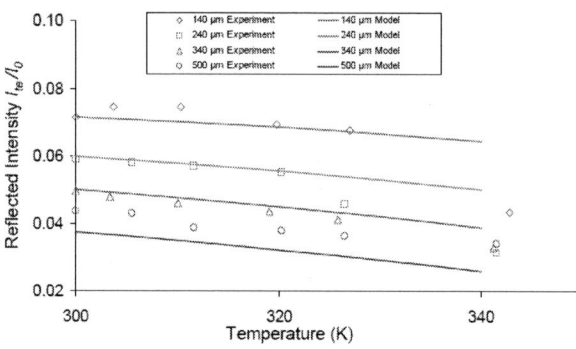

Figure 8. Reflected intensities at different thicknesses and temperature

It is observed the reflected intensities decrease as the temperature on the metal line increases. The experimental data agree well with the model, except at high temperature. An offset is also observed when the substrate thickness is 500 μm. At high substrate thickness, a temperature gradient exists across the substrate. The effective absorption coefficient is therefore lower than the uniform value used in the model.

As before, the peak reflectance intensity modulation of the experimental and model values for the metal lines are used to determine r and k values at different substrate thickness. The derived r and k values are based on metal line temperature

change of 20 K. The coefficients are tabulated in Table III. Experimental and model values are observed to agree well. The r and k values increase with increasing substrate thickness due to the decrease in unbiased reflected intensity and the increase in absorption.

TABLE III. REFLECTANCE COEFFICIENTS AT DIFFERENT SUBSTRATE THICKNESS

Substrate Thickness (μm)	Experiment	Model	Experiment	Model
	r	r	k (K^{-1})	k (K^{-1})
500	-13.14 x10^{-2}	-14.13 x10^{-2}	-6.51 x10^{-3}	-7.07 x10^{-3}
340	-12.15 x10^{-2}	-9.82 x10^{-2}	-6.39 x10^{-3}	-4.91 x10^{-3}
240	-6.32 x10^{-2}	-7.01 x10^{-2}	-3.13 x10^{-3}	-3.50 x10^{-3}
140	-3.01 x10^{-2}	-4.11 x10^{-2}	-1.53 x10^{-3}	-2.06 x10^{-3}

V. CONCLUSIONS

The reflectance intensities are observed to modulate negatively with electrical bias for the case when the absorption coefficient of the backside Si substrate varies significantly with temperature compared to the reflectance. In this case, the reflectance and thermoreflectance coefficients are one to two orders of magnitude larger than the values for frontside measurements or when the Si substrate is undoped. A backside reflectance model is developed and is found to agree well with experimental data. The reflected intensity modulation with electrical bias depends on the reflectance, absorption coefficient and substrate thickness. Positive or negative reflected intensity modulation depends on whether the modulation is due primarily to the temperature variation of the absorption coefficient or the reflectance.

ACKNOWLEDGMENT

JKJ Teo is supported by a National University of Singapore research scholarship.

REFERENCES

[1] Shimizu Y, Ishii J, Baba T, "Reflectance Thermometry for Microscale Metal Thin Films", Jpn J Appl Phys, Vol 46, No 5A, pg 3117-3119, 2007

[2] Soref RA, Bennett BR, "Electrooptical Effects in Silicon", IEEE J Quantum Elec, Vol QE-23, No 1, pg 123-129, 1987

[3] Quintard V, Deboy G, Dilhaire S, Lewis D, Phan T, Claeys W, "Laser Beam Thermography of Circuits in the Particular Case of Passivated Semiconductors", Microelec Eng, Vol 31, pg 291-298, 1996

[4] Paniccia M, Rao RM, Yee WM, "Optical Probing of Flip-Chip Packaged Microprocessor", J Vac Sci Technol B, Vol 16, No 6, pg 3625-3630, 1998

[5] Tessier G, Bardoux M, Boue C, Filloy C, Fournier D, "Back Side Thermal Imaging of Integrated Circuits at High Spatial Resolution", Appl Phys Lett, Vol 90, 171112, 2007

[6] Christofferson J, Vashaee D, Shakouri A, Melese P, Xiaofeng F, Gehong Z, Labounty C, Bowers JE, Croke ET III, "Thermoreflectance Imaging of Superlattice Micro Refrigerators", 17th Ann IEEE Semiconductor Thermal Measurement and Management Symposium, pg 58-62, 2001

[7] Ju YS, Goodson KE, "Short-Time-Scale Thermal Mapping of Microdevices Using a Scanning Thermoreflectance Technique", Trans ASME, Vol 120, May 98, pg 306-313, 1998.

[8] Dilhaire S, Grauby S, Claeys W, "Calibration Procedure for Temperature Measurements by Thermoreflectance Under High Magnification Conditions", Appl Phys Lett, Vol 84, No 5, pg 822-824, 2004

[9] Tessier G, Polignano M-L, Pavageau S, Filloy C, Fournier D, Cerutti F and Mica I, "Thermoreflectance Temperature Imaging of Integrated Circuits: Calibration Technique and Quantitative Comparison with Integrated Sensors and Simulations", J Phys D: Appl Phys, Vol 39, pg 4159-4166, 2006

[10] Hecht E, "Optics", 4th Ed, Chapter 4, Addison Wesley, 2002

[11] Freeman MH, Hull CC, "Optics", 11th Ed, Chapter 11 pg 386, Elsevier, 2003

[12] Pawley JB, "Handbook of Biological Confocal Microscopy" 3rd Ed, Chapter 9, Springer, 2006

[13] Jellison GE, Modine FA, "Optical Functions of Silicon at Elevated Temperatures", J Appl Phys, Vol 76, No 6, pg 3758-3761, 1994

[14] Chernek PJ, Orson JA, "A Simple Thermal Response Model for a p-doped Silicon Substrate Irradiated by 1.06 and 1.32 Micron Lasers", Proc SPIE, Vol 4679, pg 186-197, 2002

Nanoscale Electrical and Physical Study of Polycrystalline High-κ Dielectrics and Proposed Reliability Enhancement Techniques

K. Shubhakar[1,2], K. L. Pey[1, #], S. S. Kushvaha[2], M. Bosman[2], S. J. O'Shea[2], N. Raghavan[1], M. Kouda[3], K. Kakushima[3], Z. R. Wang[1], H. Y. Yu[1] and H. Iwai[3]

[1]Division of Microelectronics, School of EEE, Nanyang Technological University (NTU), Singapore 639798.
[2]Institute of Materials Research and Engineering (IMRE), A*STAR, 3 Research Link, 117602, Singapore.
[3]Tokyo Institute of Technology (TIT), 4259 Nagatsuta, Midoriku, Yokohama 227-8502, Japan.
[#]Ph: (+65) 6790 6371, Fax: (+65) 6792 0415, E-mail: eklpey@ntu.edu.sg

Abstract— Grain boundaries (GBs) in polycrystalline high-κ (HK) dielectric materials affect the electrical performance and reliability of advanced HK-based metal-oxide-semiconductor (MOS) devices. In this work, we present a localized study comparing the electrical conduction through grains and GBs for CeO_2 and HfO_2-based HK dielectrics using scanning tunneling microscopy (STM) and transmission electron microscopy (TEM) at the nanometer scale, in conjunction with macroscopic MOS capacitor device level analysis. Nanoscale STM conduction analysis clearly reveals faster degradation at GB sites and their vulnerability to early percolation. Multi-layer HK dielectric stacks (capping of La_2O_3 on CeO_2 and dual-layer ZrO_2/HfO_2) are proposed as an effective technique to significantly enhance the time-dependent dielectric breakdown (TDDB) robustness of advanced HK metal gate (MG) stacks.

Keywords-component; high-κ, grain boundaries, scanning tunneling microscopy, dual-layer, reliability

I. INTRODUCTION

Reliability of high-κ (HK) dielectrics in advanced complementary metal-oxide-semiconductor (CMOS) technology node is still a major concern, due to the polycrystalline microstructure of post-annealed HK thin films, in which grain boundaries (GBs), with high densities of localized process induced traps (PIT), show complex current conduction mechanisms and serve as high leakage paths. Considering the need to study the role of GBs which have a width of <1 nm [1], we employ scanning tunneling microscopy (STM) and transmission electron microscopy (TEM) as tools for nanometer resolution failure analysis, combining physical and electrical characterization. While Hf-based oxides are touted to be the most suitable candidate to replace $SiO_2/SiON$, CeO_2 and La_2O_3/CeO_2, are considered highly suitable candidates for direct deposition on Si substrate [2] (thereby eliminating the need for interfacial SiO_x layer (IL) and enabling aggressive equivalent oxide thickness (EOT) scaling).

Lanthanum (La) incorporation into HK gate dielectrics has attracted much attention due to its beneficial effects, such as increased crystallization temperature, reduced leakage current and increased dielectric constant 'κ' value [2-4]. HK dielectric structures with two or multiple layers with different metal oxides have also been explored for the application of high-performance transistors [5].

II. EXPERIMENT DETAILS

We used blanket samples of CeO_2 (~4 nm)/n-Si and HfO_2 (~4 nm)/SiO_x (~1 nm)/n-Si for studying the effect of GBs on the performance of polycrystalline HK dielectrics. A layer of ~ 4 nm CeO_2 was deposited on a Si substrate in ultra-high vacuum (UHV) using electron-beam evaporation with a post-deposition annealing (PDA) at 500 °C for 30 minutes. CeO_2 reduces to its sub-oxide, Ce_2O_3, near the Si interface [6] and a very thin layer of SiO_x (x < 2) is formed at the interface of Ce_2O_3 and Si substrate [7]. This Ce_2O_3-SiO_x (x<2) structure results in Ce-silicate [7] formation at the interface of CeO_2 and Si substrate. On the other hand, HfO_2 (~4 nm) gate dielectric film was deposited on n-type Si substrate using atomic layer deposition (ALD) method.

To reduce the effect of GBs in HK thin films and hence the impact on time-dependent dielectric breakdown (TDDB) reliability, the following approaches were adopted. For the La_2O_3/CeO_2/n-Si stack, it was transformed into a LaCe-silicate (~4 nm) layer upon annealing at 800 °C, while a layer of ~2 nm HfO_2 followed by a ~2 nm layer of ZrO_2 was deposited on Si substrate using ALD with PDA at 600 °C after the first and second HK layer deposition. Metal-gate deposited MOS capacitors of 100x100 μm^2 and 50x50 μm^2 area, were also fabricated for macroscopic *I-V* characterization and comparison. The STM experimental setup used in this work is shown in Fig. 1(a), together with a TEM micrograph of the metal/CeO_2/Ce-silictae/Si structure in Fig. 1(b). STM imaging was achieved using a platinum-iridium (Pt-Ir) tip in ultra-high vacuum (~10^{-10} Torr) conditions. The tip was always grounded while a bias was applied to the sample. Since the vacuum has " ~ 1'', a significant portion of the applied sample bias (V_{bias}) drops across the vacuum and hence the voltage drop across the HK layer is only around 40-50% of V_{bias} [8]. By acquiring the *I-V* curves at every pixel in an image while scanning the surface, a topographical image and spatially resolved current imaging tunneling spectroscopy (CITS) data maps [9] can be simultaneously obtained. CITS images the spatial dielectric leakage current over a given area of the dielectric stack and helps in deconvoluting the electrical properties from the morphology [8]. Note that *I-V* characteristics were obtained with the feedback circuit temporarily disabled (i.e. constant

978-1-4244-9113-1/11 $26.00 © 2011 IEEE

spacing between metal tip and HK sample) as defined by the constant-current imaging bias conditions.

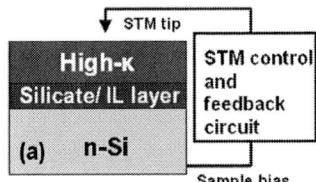

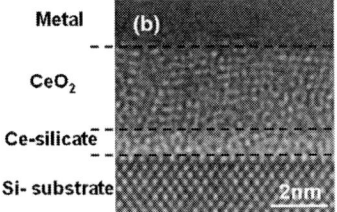

Figure 1. Schematic of (a) STM experimental setup, (b) cross-sectional TEM micrograph of ~4 nm metal/CeO₂/Ce-silicate on n-Si.

III. RESULTS AND DISCUSSION

Grain boundaries, which arise from inter-grain orientational mismatch, serve as a source of defects, which take the functional form of oxygen vacancies, as verified by electron energy loss spectroscopy (EELS) studies [10]. These GB fault lines bridging the gate and substrate have lower effective barrier height to electron transport from the gate to the substrate [1].

A. Grain boundary effects in CeO₂ and effect of La₂O₃ capping

In polycrystalline CeO₂ HK dielectric, GBs contain oxygen vacancies and trivalent Ce^{3+} as reported by Patsalas et al [6]. Fig. 2 depicts the topographical image and the corresponding CITS map, clearly showing the grains and their GB contours with diameter of granular structures varying from 15 nm to 25 nm [11]. A higher gate tunneling leakage current (I_t) is observed at the GBs in the localized I-V characteristics as shown in Fig. 3, attributed to enhanced trap assisted tunneling (TAT) through the dense chain of PITs at the GB.

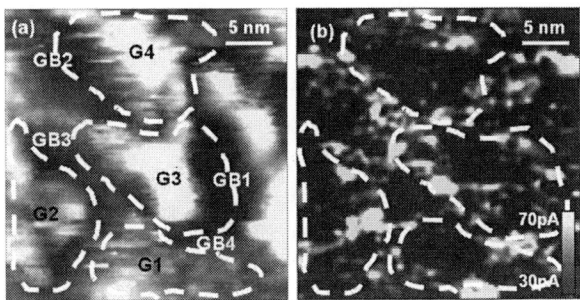

Figure 2. STM (a) topography (bias conditions, +3 V, 30 pA) and (b) corresponding CITS image at +3.5 V of CeO₂, indicating grain boundary contours (bright regions with higher leakage current) [10].

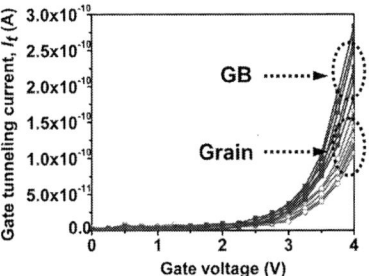

Figure 3. Localized I-V characteristics at grain and GB locations of CeO₂.

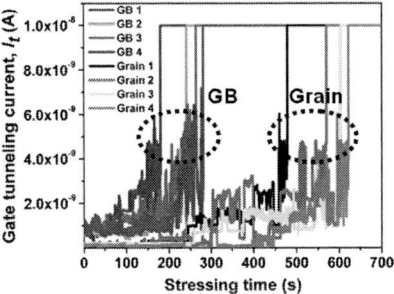

Figure 4. Gate leakage current evolution with stressing time under CVS, +4.5 V, at different locations in grain and GB of CeO₂ [10].

I_t evolution with stressing time under constant voltage stressing (CVS) of +4.5 V, at different locations in the grain and GB, is shown in Fig. 4. In polycrystalline CeO₂ HK dielectric, trap generation is non-uniform and expected to be higher around GBs [11]. In Fig. 4, GBs, on average, show faster degradation compared to grain regions supporting the hypothesis of GB assisted degradation and breakdown (BD) of HK dielectrics [11,12]. The GB accumulates more traps upon stressing, leading to the formation of a percolation path, eventually resulting in a BD spot.

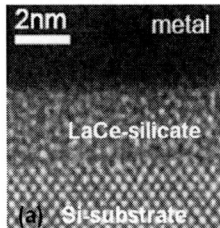

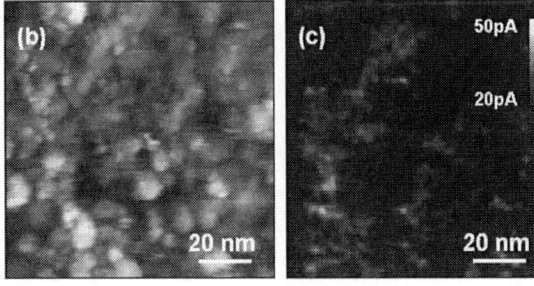

Figure 5. (a) Cross-sectional TEM micrograph of LaCe-silicate on silicon substrate. A direct high-k/Si structure is confirmed. STM (a) topography (bias conditions, +3 V, 30 pA), (b) CITS map (at +3.5 V) of LaCe-silicate dielectric.

Annealing the La_2O_3/CeO_2 stack transforms it into a LaCe-silicate layer directly on Si without any nano-crystallization as shown in Fig. 5(a) [2]. The silicate layer formation occurs with the penetration of Si atoms from the substrate into the gate dielectric [13]. Combination of La_2O_3 and CeO_2 reduces the fixed charges present in La_2O_3 with CeO_2 acting as an oxygen reservoir [14]. La atoms suppress the neutral oxygen vacancies due to the strong ionic character of the La-oxygen bond [15]. And also intermixing of La and Ce atoms takes place during the high temperature PDA [14]. La-based oxides have higher conduction band offsets than CeO_2 and hence reduce the leakage current [14]. STM topography and CITS map of the LaCe-silicate dielectric on n-Si are shown in Fig 5(b) and (c) respectively, indicating better uniformity than the CeO_2 dielectric.

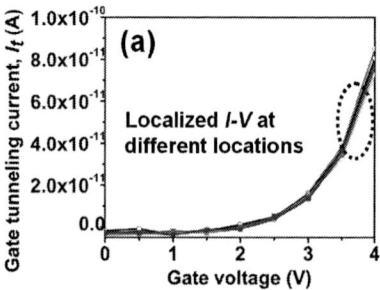

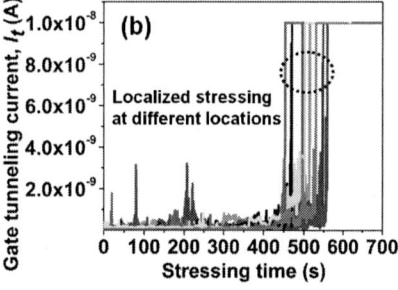

Figure 6. (a) *I-V* characteristics at different locations of LaCe-silicate. (b) Localized evolution of I_t for BD in CVS, +5 V at different locations of LaCe-silicate.

Figs. 6(a) and (b) show localized I_t –*V* plots and I_t evolution with stressing time of LaCe-silicate respectively. More uniform electrical characteristics throughout the dielectric film are evidenced by the narrow span of time to failure. Fig. 7(a) shows the normalized current distribution of the CITS images of CeO_2 and LaCe-silicate dielectrics. Significant reduction in the width of the current distribution in LaCe-silicate compared to that of CeO_2 dielectric is evident, which implies enhanced spatial uniformity of the tunneling current characteristics. *I-V* characteristics of MOS capacitors with LaCe-silicate show much lower I_g as compared to that of the CeO_2 dielectric as in Fig. 7(b).

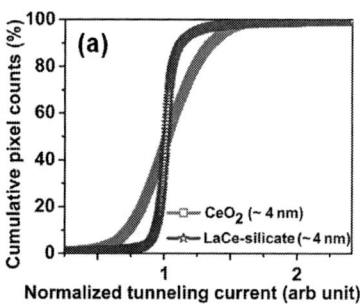

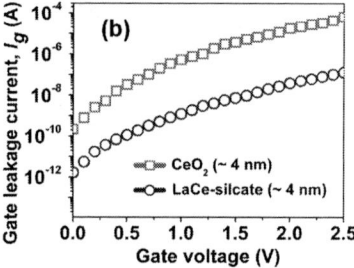

Figure 7. (a) Normalized current distribution plot of CeO_2 and LaCe-silcate dielectrics, as determined by the analysis of the corresponding CITS image (normalized to the current at the 50th percentile). (b) *I-V* curves of 100 x 100 μm^2 MOS structures with CeO_2 and LaCe-silicate as high-κ dielectrics.

B. *HfO_2 HK dielectric analysis and the effect of ZrO_2/HfO_2 dual-layer HK dielectrics*

In polycrystalline HfO_2 dielectrics, variations in the nanoscale characteristics show an impact on the electrical characteristics and reliability [16]. Fig. 8 presents a TEM cross-section of a metal/HfO_2/SiO_x/Si structure showing the polycrystalline HfO_2 dielectric with SiO_x IL of 1 nm. The corresponding STM topographical profile and the CITS map clearly show grain and GB structures as presented in Figs. 9(a) and 9(b), respectively. Correlation between the STM topography and CITS indicates more leakage sites and conductivity along the GBs. Fig. 10 shows localized I_t-*V*, at different grain and GB locations.

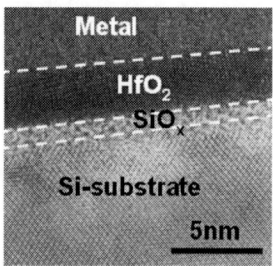

Figure 8. TEM cross-section of a Metal/HfO_2/SiO_x/Si structure showing the polycrystalline HfO_2 dielectric.

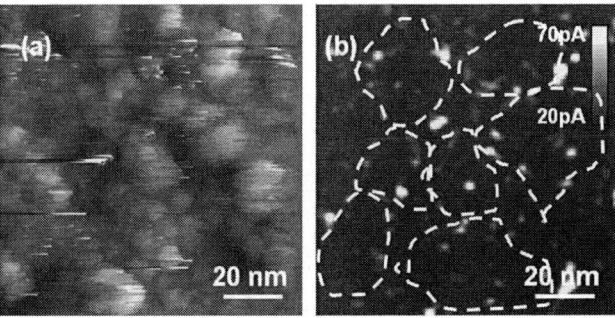

Figure 9. STM (a) topography (bias conditions, +3 V, 30 pA), (b) CITS map (at +3.5 V) of HfO$_2$ (1-layer) dielectric depicting the grain and GB profile.

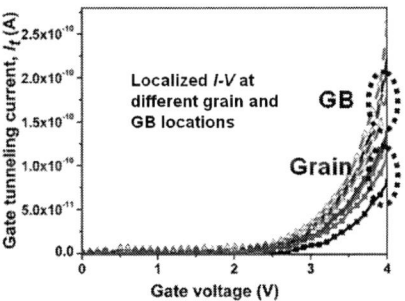

Figure 10. I-V characteristics at grain and GB locations of HfO$_2$ dielectric.

Again, higher leakage current at GBs was observed as compared to locations in grains. Increased concentration of vacancies at GBs act as percolation paths for leakage current [1] and shows the preferential occurrence of failures at the proximity of the GB site. In HfO$_2$–based dielectrics, BD is found to occur primarily at the GB sites [16,17]. In the TEM study of the failed device with HfO$_2$ as a dielectric, BD along the GB region was observed together with dielectric breakdown induced epitaxy (DBIE) microstructural defects [12].

Dual-layer ZrO$_2$/HfO$_2$ dielectric stack shows better electrical characteristics with better layer-to-layer epitaxy, overall lower gate leakage current and higher BD resistance [18]. The crystallization temperature of the individual layers also increases in ZrO$_2$/HfO$_2$ dielectric stack structure [5] and the crystallization of individual layers with different grain size possibly results in mismatch in GB alignment in the dual-layer dielectric stack [19]. A TEM micrograph of a ZrO$_2$/HfO$_2$ dual-layer dielectric stack is shown in Fig. 11. It has been reported that dual-layer started with HfO$_2$, has a strong tendency to be amorphous in thin films and suppresses the crystallization of the subsequent ZrO$_2$ layer during the following deposition cycles [5]. At high temperature annealing, larger grain sizes were observed in the ZrO$_2$ layer with HfO$_2$ as a starting layer [5].

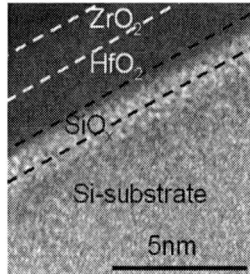

Figure 11. Cross-sectional TEM micrograph of ZrO$_2$/HfO$_2$/IL/Si structure.

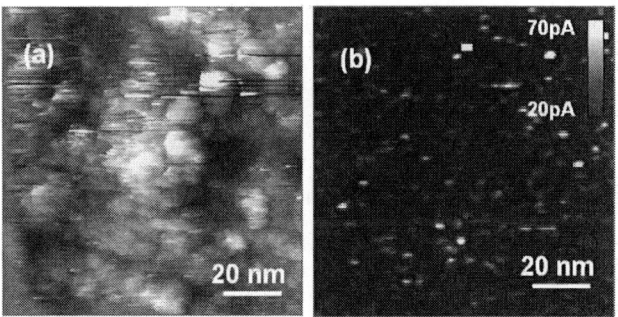

Figure 12. STM (a) topography (bias conditions, +3 V, 30 pA), (b) CITS map (at +3.5 V) of ZrO$_2$/HfO$_2$-dual-layer dielectric stack.

Figs. 12(a) and (b) show the STM topography and CITS maps respectively of a ZrO$_2$/HfO$_2$ dual-layer dielectric stack. The plot of normalized current distribution of the CITS images of HfO$_2$ and ZrO$_2$/HfO$_2$ dual-layer dielectrics is compared (Fig 13(a)). A slight improvement in electrical homogeneity is observed in the ZrO$_2$/HfO$_2$ dual-layer dielectric stack. Misalignment of GBs in a two-step deposited ZrO$_2$/HfO$_2$ dielectric stack could result in reduced gate leakage current and improved performance. Fig. 13(b) shows a schematic of different possible combinations of alignment of grain and GB in the two layers.

The I-V graph of 50 x 50 μm^2 ZrO$_2$/HfO$_2$ based MOS structures with a two-layer material system shows lower I_g than that of one layer HfO$_2$, further supporting the concept of GB misalignment and an increase in the crystallization temperature [5] in the two-layer structure as shown in Fig. 14. This misalignment implies that the number of traps needed to cause BD in a dual-layer dielectric is "more than" that needed for a single layer HK of the same thickness which has more dominant GB paths bridging the gate and substrate. Therefore, the dual-layer HK dielectric stack is more robust to TDDB failures.

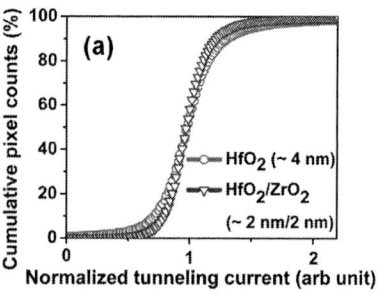

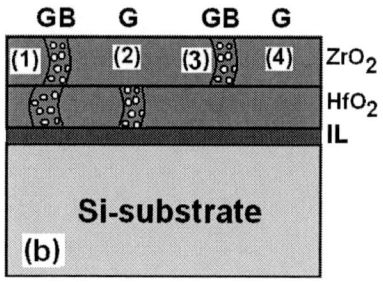

Figure 13. (a) Normalized current distribution plot of HfO₂ and ZrO₂/HfO₂ dual-layer dielectric stack, as determined by the analysis of the corresponding CITS (normalized to the current at the 50th percentile). (b) Schematic of combinations of alignment of grain and GB in two layers.

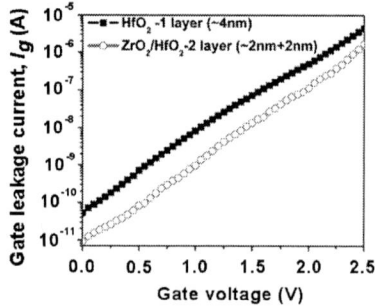

Figure 14. *I-V* curves of 50 x 50 μm² MOS structures with one layer of HfO₂ and ZrO₂/HfO₂ dual-layer dielectric stack.

The crystalline phase of ZrO_2 grown on HfO_2 depends on the thickness of HfO_2 and the identity of the starting layer determines the final grain size and surface roughness of dual-layer gate stacks [5]. In addition, diffusion occurs of hafnium into the zirconia layer and zirconium into the hafnia layer upon post deposition annealing [18]. Hence the quality of the interface and intermixing between each layer depends on the PDA temperature [18]. Kim *et al.* reported that the ZrO_2/HfO_2 dual-layer gate stack shows a clear interface and lesser intermixing between each layer using ALD deposition at 300 °C [5]. Zhang *et al.* have reported a detailed macroscopic analysis of dual-layer gate stacks with different HK materials. Their analysis shows ZrO_2/HfO_2 dual-layer dielectric stack to have the highest breakdown resistance and lowest gate leakage current compared to that of other dual-layer dielectric stacks [20].

IV. CONCLUSIONS

Nanoscale analysis in this study supports the hypothesis of enhanced trap generation and faster degradation of polycrystalline HK dielectrics at GB sites as compared to the bulk grain, hence shorter TDDB lifetime at GB sites. The microstructure of the HK thin film plays an important role in determining the TDDB robustness of MG-HK stacks. The use of multi-layer HK films (La_2O_3 capping layer on CeO_2 (without IL) and ZrO_2/HfO_2 dual-layer dielectric stack that result in GB misalignment and an increase in the crystallization temperature) is proposed as an effective strategy to prolong the lifetime of these polycrystalline dielectric stacks and reduce the likelihood of high GB conduction from gate to substrate.

ACKNOWLEDGMENT

K. Shubhakar is grateful to Nanyang Technological University (NTU) for the Research Student Scholarship (RSS). This work is funded by the SUTD Research Grant SRG ASPE 2010 004. The authors would like to thank Dr. Yee Chong Loke (IMRE) for fabricating metal electrodes of MOS capacitors, Dr. Debbie Seng (IMRE) and Ms. June Ong (IMRE) for TEM sample preparation. The authors also would like to thank Dr. Ramesh Thamankar (IMRE) and Mr. Qin Hailang (NTU) for useful technical discussions.

REFERENCES

[1] K. P. McKenna and A. L. Shluger, " The interaction of oxygen vacancies with grain boundaries in monoclinic HfO₂," *Appl. Phys. Lett.* **95**, 222111 (2009).

[2] K. Kakushima, T. Koyanagi, D. Kitayama, M. Kouda, J. Song, T. Kawanago, M. Mamatrishat, K. Tachi, M. K. Bera, P. Ahmet, H. Nohira, K. Tsutsui, A. Nishiyama, N. Sugii, K. Natori, T. Hattori, K. Yamada, and H. Iwai, "Direct contact of High-κ/Si gate stack for EOT below 0.7 nm using LaCe-silicate layer with V_fb controllability," *Symposium on VLSI Technology, Digest of Techinical Papers*, p.69 (2010).

[3] N. Umezawa, K. Shiraishi, S. Sugino, A. Tachibana, K. Ohmori, K. Kakushima, H. Iwai, T. Chikyow, T. Ohno, Y. Nara, and K. Yamada, "Suppression of oxygen vacancy formation in Hf-based high-κ dielectrics by lanthunum incorporation," *Appl. Phys. Lett.* **91**, 132904 (2007).

[4] W. He, L. Zhang, D. S. H. Chan, and B. J. Cho, "Cubic-structured HfO₂ with optimized doping of lanthunum for higher dielectric constant," *IEEE Elect. Dev. Lett.* **30**, 623 (2009).

[5] H. Kim, P. C. McIntyre, and K. C. Saraswat, "Microstructural evolution of ZrO₂-HfO₂ nanolaminate structures grown by atomic layer deposition," *J. Mater. Res.* **19**, 643 (2004).

[6] P. Patsalas, S. Logothetidis, L. Sygellou, and S. Kennou, "Structure-dependent electronic properties of nanocrystalline cerium oxide films," *Phys. Rev. B.* **68**, 035104 (2003).

[7] S. Zec and S. Boskovic, "Cerium silicates formation from mechanically activated oxide mixtures," *J. Mat. Sci.* **39**, 5283 (2004).

[8] Y. C. Ong, D. S. Ang, K. L. Pey, S. J. O'Shea, K. E. J. Goh, C. Troadec, C. H. Tung, T. Kawanago, K. Kakushima, and H. Iwai, "Bilayer gate dielectric study by scanning tunneling microscopy," *Appl. Phys. Lett.* **91**, 102905 (2007).

[9] R. J. Hamers, R. M. Tromp, and J. E. Demuth, "Surface electronic structure of Si (111)-(7 X 7) resolved in real space," *Phys. Rev. Lett.*, **56**, 1972 (1986).

[10] X. Li, C. H. Tung, and K. L. Pey, "The nature of dielectric breakdown," *Appl. Phys. Lett.* **93**, 072903 (2008).

[11] K. Shubhakar, K. L. Pey, S. S. Kushvaha, S. J. O'Shea, N. Raghavan, M. Bosman, M. Kouda, K. Kakushima, and H. Iwai, "Grain boundary assisted degradation and breakdown study in cerium oxide gate dielectric using scanning tunneling microscopy," *Appl. Phys. Lett.* **98**, 072902 (2011).

[12] R. Ranjan, K. L. Pey, C. H. Tung, L. J. Tang, G. Groeseneken, L. K. Bera, and S. De Gendt, "A comprehensive model for breakdown mechanism in HfO_2 high-κ gate stacks," *Proceedings of the IEEE International Electron Devices Meeting*, p.725 (2004).

[13] J. S. Jur, D. J. Lichtenwalner, and A. I. Kingon, "High temperature stability of lanthunum silicate dielectric on Si (001)," *Appl. Phys. Lett.* **90**, 102908 (2007).

[14] M. Kouda, N. Umezawa, K. Kakushima, P. Ahmet, K. Shiraishi, T. Chikyow, K. Yamada, and H. Iwai*et al.*, "Charged defects reduction in gate insulator with multivalent materials," *Symposium on VLSI Technology*, Digest of Techinical Papers, p.200 (2009).

[15] J. Robertson, "Electronic structure and band offstes of high-dielectric-constant gate oxides," *MRS Bull.* **27**, 217 (2002).

[16] V. Iglesias, M. Porti, M. Nafria, X. Aymerich, P. Dudek, T. Schroeder, and G. Bersuker, " Correlation between the nanoscale electrical and morphological properties of crystallized hafnium oxide-based metal oxide semiconductor structures," *Appl. Phys. Lett.* **97**, 262906 (2010).

[17] G. Bersuker, D. Heh, C. D. Young, L. Morassi, A. Padovani, L. Larcher, K. S. Yew, Y. C. Ong, D. S. Ang, K. L. Pey, and W. Taylor, "Mechanism of high-k dielectric induced breakdown of interfacial SiO_2 layer," *Proceedings of the IEEE International Reliability Physics Symposium,* p.373 (2010).

[18] S. Lhostis, C. Gaumer, C. Bonafos, S. Schamn, F. Pierre, A. Fanton, C. Morin, X. Garros, M. Casse, and C. Leroux, "Crystalline structure of HfZrO thin films and ZrO_2/HfO_2 dual-layers grown by AVD for MOS applications," *ECS transactions*, **13**, 101 (2008).

[19] K. S. Yew, D. S. Ang, K. L. Pey, G. Bersuker, P. S. Lysaght, and D. Heh, "Enhanced electrical uniformity and breakdown of multi-step deposited and annealed HfSiO-Insight by scanning tunneling microscopy," *Proceedings of the International Conference on Solid State Devices and Materials (SSDM)*-2010.

[20] H. Zhang and R. Solanki, "Atomic layer deposition of high dielectric constant nanolaminates," *J. Electrochem. Soc.*, **148**, F63 (2001).

Investigation of Progressive Breakdown and non-Weibull Failure Distribution of High-k and SiO₂ Dielectric by Ramp Voltage Stress

Nilufa Rahim

IBM Microelectronics Division,
Essex Junction, Vermont, United States
phone: (1) –(732)- 692-7959, nrahim@us.ibm.com

Ernest Y. Wu

IBM Microelectronics Division
Essex Junction, Vermont, United States

Durgamadhab Misra

Electrical and Computer Engineering Department
New Jersey Institute of Technology
Newark, New Jersey, United States

Abstract—In this work, the progressive breakdown (PBD) phase and non-Weibull final failure distributions of multi layer high-k and SiO₂ gate dielectric were investigated by voltage ramp stress (VRS) technique. A new hybrid two-stage constant voltage stress/voltage ramp stress methodology was developed to exclusively evaluate the PBD phase. Then the VRS technique was applied to investigate the non-Weibull failure distribution at a specified current (I_{FAIL}) with large sample-size (~1000) experiments. An excellent agreement was achieved in both cases in comparison with the conventional CVS technique, thus demonstrates that VRS is an effective technique to replace the CVS technique for investigation of post-BD and non-Weibull statistics in both SiO₂ and high-k dielectrics.

Keywords-VRS; CVS; TDDB

I. INTRODUCTION

It is known that static random access memory (SRAM) failure can be predicted by the failure-current (I_{FAIL}) based methodology rather than the conventional first breakdown (BD) criterion [1]. Therefore, the final failure distributions (F_{FAIL}) at I_{FAIL} including the progressive breakdown (PBD) phase are non-Weibull [2]. However, conventional constant voltage stress (CVS) measurements are time consuming for data collection of non-Weibull statistics. On the other hand, although voltage ramp stress (VRS) technique is well known for its efficiency, it has only been used in the context of the first BD definition [3,4]. There has been no report of the PBD phase and non-Weibull failure distributions of multi layer high-k and SiO₂ gate dielectric by VRS technique. In this work, a new hybrid dual stage CVS/VRS methodology was developed to exclusively evaluate the PBD phase. Also, it will be shown that the VRS technique can be applied to investigate the non-Weibull failure distribution at a specified current (I_{FAIL}), thus

leading to an efficient experimental methodology for modern TDDB investigation.

II. EXPERIMENTAL SETUP

High-k/metal gate (HK/MG) devices used in this work with hafnium based dielectrics and interfacial layer of SiO₂ were fabricated using conventional CMOS process flow on SOI substrate. The high-k gate stacks had an interfacial layer of ~10Å thickness and high-k layer thickness of <25Å. Both VRS and CVS were performed at 140⁰C. Unit nFET and pFET devices were connected in parallel to construct array structures with larger areas.

The concept of two-stage breakdown naming partial and complete breakdown during Fowler-Nordheim (F-N) stress was reported earlier [5]. The partial breakdown was termed as B-SILC and Ohmic conduction was called as complete breakdown. Based on this concept, Linder *et al.*, has developed two-stage stress both by CVS to study oxide degradation rate [6]. Here we have developed a hybrid dual stage of CVS/VRS techniques to investigate specifically PBD phase [7]. For the new hybrid two-stage stress introduced in this work, a higher constant voltage stress was applied at the first stage with a low current compliance to arrest the first breakdown events. This low current compliance would prevent the dielectrics to go into post breakdown phase. Then at second stage, voltage was ramped until the measured current exceeds a specified failure-current on those samples. Although one would naturally consider the use of VRS technique for arresting BD in the first stage, due to the strong exponential voltage-dependence of tunneling currents, it is practically impossible to discern the occurrence of 1st BD events with a current compliance. This is the key reason for the introduction of hybrid dual stage CVS/VRS technique. Experimentally, it was found that progressive breakdown voltage, V_{PBD} at second stage does not depend on the initial ramp voltage (0 or 1V). Hence, voltage can be ramped from higher than 0 volt. It can also be mentioned that stress condition such as stress voltage in first stage does not influence on V_{PBD} or T_{PBD} derived from second stage.

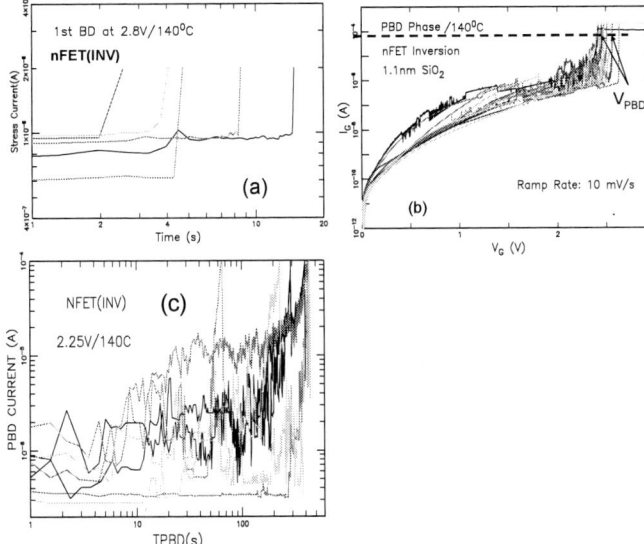

Figure 1. Hybrid two-stage VRS technique. (a) at first stage, BD was detected by CVS with a low current compliance (here 2 μA). For 2nd stage those samples were either subjected to (b) VRS to extract V_{PBD} or (c) CVS for direct T_{PBD} measurements until the gate leakage current exceeds I_{FAIL} of 200 μA).

For small area nFETs with thin oxide, a current compliance of 2μA was applied to arrest BD as shown in Figure 1 (a). High gate voltage of 2.8V was used in 1st stage for shorter 1st BD time. Once BD was detected, these devices were subjected to VRS until gate currents larger than 100μA as shown in Figure 1 (b). It was shown earlier that post-breakdown gate current exhibits exponential dependence on the gate voltage [8]. This model which is based on quantum point contact describes that the experimental post-SBD current can be fitted by exponential law I=A*exp(BV). Here V_{PBD} was extracted based on this exponential voltage dependence of post-BD currents. For comparison purpose, the current transients of CVS was also carried out separately to directly measure T_{PBD} from the same population of samples which were broken at 1st stage shown in Figure 1 (c).

III. RESULTS AND DISCUSSIONS

A. Conversion from Voltage domain to Time domain

The traditional conversion from V_{BD} to T_{BD} (presumably for first BD events) was originally developed based on the cumulative defect generation in time steps during voltage ramp [3,9]. However, with the development of progressive BD in recent years [6,10], time-to-failure at specified failure-current consists of at least two phases: first BD and progressive BD, ie, $T_{FAIL}=T_{BD}(1^{st})+T_{PBD}$. In this work, we extend this V_{BD}-to-T_{BD} conversion procedure to the PBD phase (T_{PBD}) as defined V_{PBD}. Similarly, for the fresh samples without the 1st stage of constant

voltage stress, we refer V_{FAIL} as the voltage at specified I_{FAIL}. For both V_{PBD} and V_{FAIL} conversion, voltage acceleration parameter was extracted from constant voltage stress data.

As it was discussed before, V_{PBD} was extracted based on the exponential relation of progressive breakdown current and voltage shown in Figure 1 (b). Figure 2 shows the conversion of a steeper V_{PBD} distribution into a much shallower T_{PBD} distribution. This conversion requires a relevant voltage acceleration model. Three popular models for BD such as exponential law, power-law and 1/V model have been applied [11-13]. In Figure 3, solid line is for power-law model: $T_{PBD} \sim V_G^{-n}$, dotted line for exponential model: $T_{PBD} \sim exp(-\gamma V)$ and the dashed line for 1/V model: $T_{PBD} \sim exp(C/V)$ have been used for the conversion. The corresponding acceleration factors n, γ and C are 37.6, 16.795 1/V and 84.3V were determined from CVS T_{PBD} data. The conversion between V_{BD} and T_{BD} measurements has been derived based on three models [14]. A good agreement is observed for direct T_{PBD} from CVS and converted T_{PBD} from VRS for all three models. This is because the projected voltage, 2.1 V is close to $V_{PBD,63\%}$ which is 2.51V (not shown here). Differences would be visible between different models if the projection voltage is either sufficiently low or sufficiently high than $V_{PBD,63\%}$. Also, it has been reported earlier that voltage scaling of progressive breakdown time of ultra-thin gate oxide can be modeled by a power-law model as $T_{PBD} \approx T_{PBD0}V_{G,PBD}^{-m}$ [15,16]. Hence, power-law model has been used as the acceleration model throughout this work.

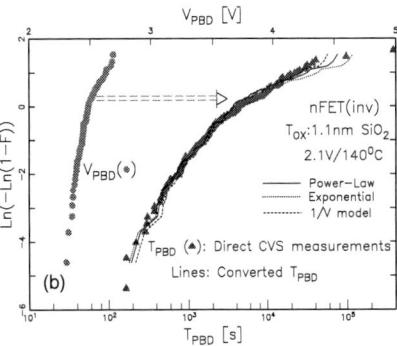

Figure 2. Conversion of V_{PBD} to T_{PBD} distribution. Symbols represent TPBD data from CVS stress. Very steep V_{PBD} distribution is translated to shallow T_{PBD} distribution due to the exponential dependence of T_{PBD} on V_{PBD}. Three popular models as power-law (solid line), exponential (dotted line) and 1/V model (dashed line) have been used as acceleration model for conversion. The T_{PBD} converted from V_{PBD} for a reference voltage of 2.1V agrees quite reasonably irrespective of the choice of the model. The variations among different models would be visible for a larger time window.

It was found that voltage acceleration (n) can be derived from either $V_{PBD,63\%}$ of different ramp rates of VRS or $T_{PBD,63\%}$ of different stress voltages of CVS. For time domain, $T_{PBD,63\%} \sim V_G^{-n}$ and for voltage domain, $R \propto V_{PBD,63\%}^{n+1}$ have been used [3]. From VRS, n was found to be 37.5 and from CVS it was 37.8. Hence both methods yield values which are within statistical uncertainty. Although for VRS, difficulties lie in selecting the range of practical ramp rates. It was reported earlier in the literature that for 1st BD study by VRS, ramp rate higher than 2V/s does not yield correct results [17]. But in this work we have shown that this is not due to fast ramp rate, but

978-1-4244-9113-1/11 $26.00 © 2011 IEEE

due to higher ΔV. As it is known that ramp rate is $\Delta V/\Delta t$ volt/sec. If ΔV is made too large to make ramp rate very fast, then the granularity effect would diverge the converted distribution from actual distribution. The other way to get faster ramp rate is to reduce time step, Δt. But the resolution range of the measuring instrument sets the limit in this case. To expand the ramp rate in the slower region (assuming $\Delta t = 1s$) would be time consuming alleviating the benefit of fast VRS technique. ΔV was always fixed at 1mV and Δt was varied from 1ms to 100ms.

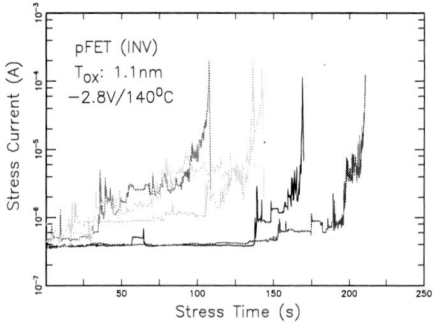

Figure 3. 1st BD detection difficulties in ultra-thin oxide pFETs. The noise in early stress time seen in this figure would impede to fix a low current compliance level for 1st stage of the hybrid stress method.

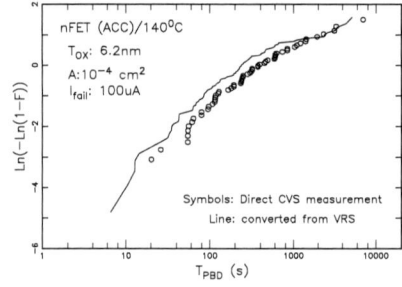

Figure 4. Comparison of residual times (T_{PBD}) for thick (6.2nm) oxide. These devices were initially broken at 7V at 1st stage. For PBD phase or 2nd stage, V_{ref} was also 7V.

B. Progressive Breakdown Time by VRS

To characterize time-to-progressive breakdown (T_{PBD}) with this hybrid method, detection of 1st BD at the first stage is very critical. This issue can be discussed in the context of ultra-thin dielectric of pFET in inversion. When pFET devices are stressed by CVS, background tunneling current and stress-induced leakage current (SILC) due to the generation of defects make gate current very noisy and can easily mask the formation of 1st BD (Figure 3). As the gate currents of ultra-thin pFETs were plotted in Log-Lin scale, some devices show spike in gate current within very short period of stress (<10s). Hence, it becomes challenging to fix a low current compliance which could arrest BD at first stage invariably on a large sample size. If the stress current of pFET devices (Figure 5) is compared to nFETs (Figure 1c) of identical oxide thickness and device dimensions, the difference in gate current noise is clearly visible. Because of this constrain, progressive breakdown study using the hybrid dual stage CVS/VRS technique is limited to the cases where 1st BD events can be clearly arrested such as nFET inversion mode, etc. It is worth

to mention that this is true for PBD time only, not to confuse with time-to-fail (T_{FAIL}) which will be discussed later.

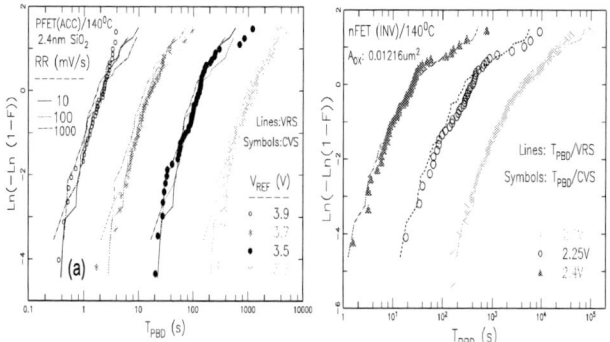

Figure 5. Comparison of residual times (T_{PBD}) for (a) thick oxide, 2.4nm, (b) thin oxide, 1.1nm SiO_2. An excellent agreement between CVS/T_{PBD} and VRS/T_{PBD} can be observed for different reference voltages and 100 μA fail current.

Historically, it is known that thick oxide (>3nm) shows a sudden hard breakdown at TDDB stress voltage. Hence, it is assumed that PBD phase does not exist for thick oxide. This is merely due to the fact that thick oxide has a very short progressive breakdown time compared to 1st breakdown time and detection of that T_{PBD} depends on the time resolution of the test set-up at the stress bias. In this scenario, hybrid VRS technique can separate the two BD phase if the stress bias and current compliance at the first stage are chosen carefully. A very short PBD phase was observed specially at high bias for this 6.2 nm SiO_2 dielectric. Figure 4 shows statistical distribution of progressive breakdown times obtained by the hybrid stress method.

This method was then applied on thinner oxide (2.4nm) pFETs in accumulation. Second stage ramp was performed for three different ramp rates ranging from 1V/s to 10mV/s. V_{PBD} from VRS was converted to T_{PBD} (lines) and compared to the directly measured T_{PBD} by CVS and an excellent quantitative agreement can be observed for all three different ramp rates Figure 5 (a).

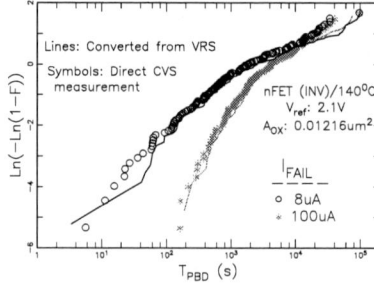

Figure 6. Similar failure current dependence of Time-to-PBD (TPBD) is observed for both VRS and CVS for thin (1.1nm) SiO_2.

The results also include different reference voltages from 3.3 to 3.9 volts. As we can see, T_{PBD} does not depend on the ramp rates. As was discussed earlier, for ultra-thin (1.1nm), only nFETs have been studied. Comparable T_{PBD} were also obtained for ultra-thin oxide based on 100 μA fail current during PBD phase and a non-Weibull distribution is observed (Figure 5 b), similar to the previous finding [18]. It is worth to mention that similar failure current dependence of Time-to-PBD (T_{PBD}) is observed for both VRS and CVS for this thin (11Å) oxide (Figure 6) which indicates that the defects generated by CVS and VRS are essentially equivalent. Also, as I_{FAIL} was increased from 8μA to 100μA, T_{PBD} significantly increases at low percentile, a clear signature of post-BD characteristics since 1st BD does not depend on failure currents.

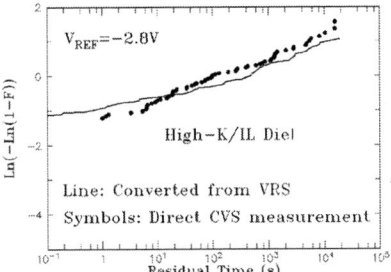

Figure 7. Comparison of residual times (TRES) of high-k/IL dielectric pFET in inversion at 125^{0}C. Constant voltage of -3.1V was used in the first step.

There are reports of the evidence of PBD in high-k gate stack [19, 20]. Figure 7 shows the results of the hybrid two-stage VRS technique for the high-k/IL gate stacks in comparison with CVS technique. The VRS results also yield very shallow distribution of residual time (or PBD time) which is the unique characteristics (β<<1) of post-BD found for high-k/IL gate stacks using CVS method [20].

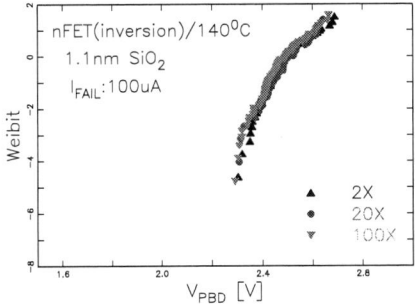

Figure 8. V_{PBD} distributions of ultra-thin oxides extracted during 2nd stage of hybrid stress. Three different area nFETs (2x to 100x, where x is 0.00608 μm^2) were investigated. Area independence of V_{PBD} indicates single spot BD during PBD phase.

C.   Area Independence of Progressive Breakdown Voltage

It has been reported earlier in the literature that ultra-thin nFETs show single spot breakdown and that spot grows during progressive breakdown phase [10]. Therefore, T_{PBD} measured by conventional CVS method is independent of device area as it is a localized degradation phenomenon. In this work, progressive breakdown voltage (V_{PBD}) was measured during second stage ramp for three different areas from 0.01216 to 0.608 μm^2 (2x to 100x). These larger area structures are made by connecting parallel array of unit cells of 0.00608 μm^2 (equivalent to 1x). It was found that V_{PBD} for these nFETs is also area independent shown in Figure 8. This is similar to the results mentioned above about area independent T_{PBD} by CVS. Hence VRS mimics the mechanism of the growth of a single BD spot during progressive breakdown phase of ultra-thin

nFETs by producing area independent V_{PBD}. This again validates that hybrid VRS method can be used to study PBD phase of dielectric breakdown.

D.   Time-to-Fail by VRS

Having established the validity of VRS technique in PBD phase, we then investigate the time-to-final fail (T_{FAIL}) by a single-stage voltage ramp for fresh samples similar to traditional VRS technique. Our goal is to investigate whether the converted T_{FAIL} from V_{FAIL} distribution can yield the non-Weibull distributions in agreement with non-Weibull failure distributions directly obtained by CVS method.

1)   Time-to-Fail of Thick and Thin SiO$_2$

Figure 9 shows time-to-fail (T_{FAIL}) converted from V_{FAIL} by VRS and directly measured by CVS for thick SiO$_2$ (24Å) pFET in accumulation which shows excellent agreement. Here, lines are T_{FAIL} from VRS and symbols represent T_{FAIL} from CVS .The advantage of this VRS method is that one set of V_{FAIL} data can project T_{FAIL} distribution for different stress voltages. This can significantly reduce time and resources. A large sample size study of around ~1000 devices was carried out on thin SiO$_2$ (11Å) for both p- and nFETs in inversion mode. Figure 10 (a) shows that VRS method can effectively reproduce non-Weibull T_{FAIL} distribution extracted by CVS. For nFETs, characteristic life, $T_{FAIL,63\%}$ values agree well even though a disagreement is observed at low percentile which is due to statistical uncertainty in experiment. Figure 11 investigates failure current dependence of time-to-fail for both methodologies. Both VRS (symbols) and CVS (lines) show similar failure current dependence. For low failure current such as 1μA, failure distributions behave more like Weibull distribution because of minimal contribution from progressive breakdown time or for short T_{PBD}, $T_{FAIL} \approx T_{BD}$.

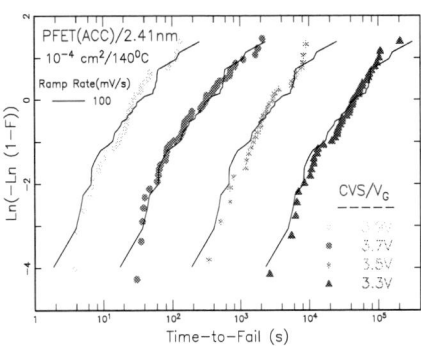

Figure 9. Time-to-fail (T_{FAIL}) extracted from CVS and converted from V_{FAIL} by VRS for thick SiO$_2$ (2.4nm) pFET in accumulation show excellent agreement. Lines are T_{FAIL} from VRS and symbols represent T_{FAIL} from CVS.

As I_{FAIL} was increased to 100μA for example, the low-percentile bending is prominent making this distribution a non-Weibull distribution. This is because as the failure current was increased, for significant T_{PBD}, $T_{FAIL} > T_{BD}$. From this the similarity between breakdown physics of these two breakdown mechanisms is suggested.

978-1-4244-9113-1/11 $26.00 © 2011 IEEE   795

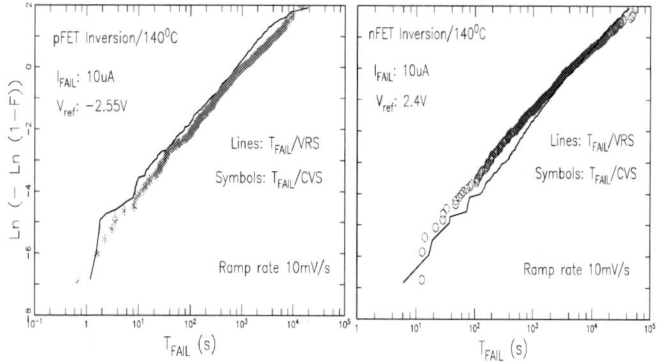

Figure 10. (a) Time-to-fail (T_{FAIL}) extracted from CVS and converted from V_{FAIL} by VRS for ultra-thin SiO_2 (11Å) pFET show excellent agreement on large sample size (~1000 devices each) and (b) For nFET, high percentile data agrees quite well.

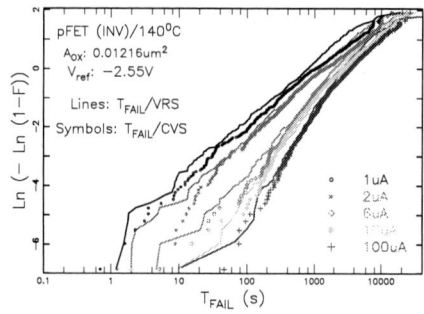

Figure 11. The final failure distributions as a function of fail currents for ultra-thin SiO_2 pFETs. For both VRS (lines) and CVS (symbols), strong failure current dependence of final fail time is observed.

2) Time-to-Fail of High-k/SiO_2 Gate Stack

Poisson area scaling was performed on V_{FAIL} distribution of three different area high-k pFET devices in inversion based on this equation, $Ln(-Ln(1-F_2)) = Ln(-Ln(1-F_1)) + ln(A_2/A_1)$. Here F_1 and F_2 are the failure distribution corresponding to areas A_1 and A_2. This is applicable for weakest-link property and a uniform failure site distribution in the oxide area. So, a non-Weibull distribution can also be scaled using this formula. Both T_{FAIL} and V_{FAIL} distributions follow Poisson area scaling shown in Figure 12 (a), (b). Here a strong bending (or deviation from Weibull distribution which is evident in high percentile) at low percentile distribution is observed. This low-percentile distribution is of paramount importance when studying TDDB reliability of these new high-k gate stacks as it represents product areas relevant to the circuit/chip reliability. It is worth to mention that VRS method can efficiently generate the shallow and steep distributions in high and low percentiles respectively. It is important to point out that the agreement between CVS and VRS is obtained simply using a constant voltage acceleration factor (exponent) to translate the V_{FAIL} data to T_{FAIL} for all the samples. Therefore, we can conclude that voltage acceleration is independent of

distribution percentiles although defect generations in high-k and IL layers can be different [21].

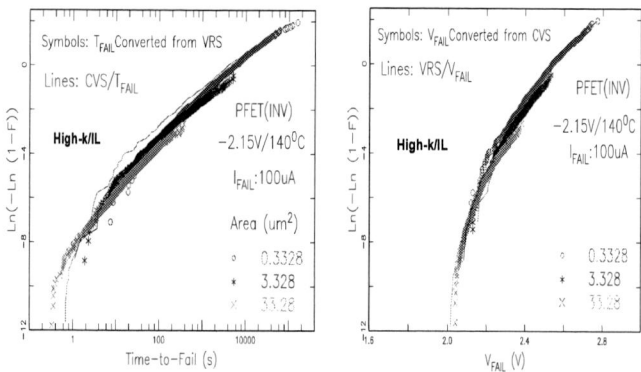

Figure 12. (a) T_{FAIL} distributions for 3 different area high-k/IL pFETs in both time domain. Lines represent T_{FAIL} converted from V_{FAIL} and symbols are direct CVS measurements. (b) Similar comparison between direct and converted V_{FAIL} in voltage domain. Poisson area scaling has been applied in both cases.

$V_{FAIL,63\%}$ and $T_{FAIL,63\%}$ were measured by VRS and CVS at different temperatures from 85^0C to 140^0C on pFETs of $3.328\mu m^2$ area (Figure 13). For CVS, reference stress voltage was fixed at -2.15V. Activation energy was extracted independently from $V_{FAIL,63\%}$ and $T_{FAIL,63\%}$. Power-law exponent (n) of 46 was used for conversion of $V_{FAIL,63\%}$. This acceleration factor was derived from CVS data at 140^0C. Both methods yield E_a~1.15eV. This not only confirms the equivalence of these two methodologies, it also assures that key reliability parameters can be extracted by faster VRS method with sufficient accuracy.

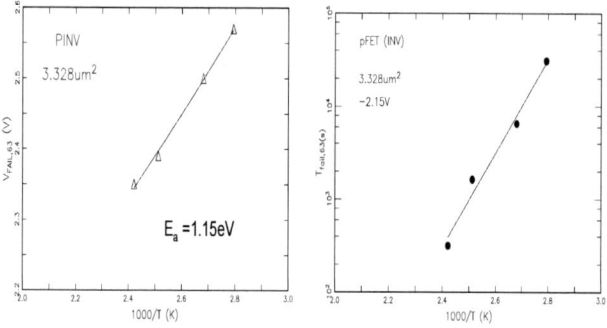

Figure 13. Thermal activation energy, E_a of V_{FAIL} (Lft) and T_{FAIL} (Right) by VRS and CVS measurements. In both cases, E_a was ~1.15eV.

IV. CONCLUSIONS

A new methodology using hybrid two-stage stresses has been developed to study progressive BD phase for high-k and SiO_2. This methodology was then applied to dielectrics of various thicknesses such as 6.2, 2.41, 1.1-nm SiO_2 and high-k dielectric stack as well. It was found that reliability parameters of progressive breakdown time (T_{PBD}) distribution can be efficiently captured by VRS technique for high-k/IL dielectric and other oxides. The voltage ramp stress technique can also reproduce non-Weibull or bending at low percentile

978-1-4244-9113-1/11 $26.00 © 2011 IEEE

distribution of time-to-final fail of high-k/IL gate stacks similar to CVS on large sample size. Finally the activation energies of T_{FAIL} for both methods were consistently similar. So, this study demonstrates that VRS can be used effectively for quantitative reliability studies of progressive BD phase and final BD of high-k and other dielectric materials; thus it can replace the time-consuming CVS measurements as an efficient methodology and reduce the resources and manufacturing cost.

ACKNOWLEDGMENT

The authors would like to thank Dr. John Aitken for the management support and C. Larow, R. Dufrense, T. Merrill, D. Brochu, R. Driscoll for the technical assistance.

REFERENCES

[1] E. Y. Wu, G. Braceras, D. Turner, A. Swift, M. Johnson, J. Sune, S. Tous, B. Li, R. Bolam, G. Massey, and M. Khare, "A viable and comprehensive TDDB assessment methodology for investigation of SRAM V_{min} failure", IEEE Int. Electron Device Meeting, pp.397-400, 2009.

[2] B. P. Linder and J. H. Stathis, "Statistics of Progress Breakdown in Ultra-thin Oxides," Microelectronics Engineer, Vol. 72, pp.24-28, 2004.

[3] A. Kerber, L. Pantisano, A. Veloso, G. Groeseneken, and M. Kerber, "Reliability screening of high-k dielectrics based on voltage ramp stress", Microelectronics Reliability, vol. 47, pp.513-517, 2007.

[4] S. C. Fan, J. C. Lin, and A. S. Oates, "Accurate characterization on intrinsic gate oxide reliability using voltage ramp tests", IEEE Int. Reliability Physics Symposium, pp. 625-626, 2006.

[5] K. Okada, "A new dielectric breakdown mechanism in silicon dioxides", Symposium on VLSI Technology Digest of Technical Papers, pp.143-144, 1997.

[6] B. Linder, J. H. Stathis, D. J. Frank, Salvatore Lombardo, and Alex Vayshenker, "Growth and scaling of oxide conduction after breakdown", IEEE Int. Reliability Physics Symposium, pp.403-408, 2003.

[7] E. Y. Wu, N. Rahim, D. Brochu, T. Merrill, R. Dufresne, and C. Larow, "Dual stage voltage ramp stress test for gate dielectrics", IBM docket no. BUR920100143US1 for patent application.

[8] J. Sune, and E. Miranda, "Post soft breakdown conduction in SiO₂ gate oxides", IEEE Int. Electron Device Meeting, pp. 533-536, 2000.

[9] A. Berman, "Time zero dielectric reliability test by a ramp method", IEEE Int. Reliability Physics Symposium, pp. 204-209, 1981.

[10] F. Monsieur, E. Vincent, D. Roy, S. Bruyere, J. C. Vildeuil, G. Pananakakis, and G. Ghibaudo "A thorough investigation of progressive breakdown in ultra-thin oxides. Physical understanding and application for industrial reliability assessment", IEEE Int. Reliability Physics Symposium, pp. 45-54, 2002.

[11] J. W. McPherson, R. B. Khamankar, and A. Shanware, "Complementary model for intrinsic time-dependent dielectric breakdown in SiO₂ dielectrics", Journal of Applied Physics, vol. 88, pp. 5351-5359, Sept. 2000.

[12] I.-C. Chen, S. Holland, and C. Hut, "A quantitative physical model for time-dependent breakdown in SiO₂", IEEE Transactions on Electron Devices, 1985.

[13] E. Y. Wu, A. Vayshenker, E. Nowak, J. Sune, R.-P. Vollertsen, W. Lai, and D. Harmon "Experimental evidence of T_{BD} power-law for voltage dependence of oxide breakdown in ultra-thin gate oxides", IEEE Transactions on Electron Devices, pp. 2244-2253, 2002.

[14] E. Wu, and J. Sune, "On voltage acceleration models of time to breakdown-part I: experimental and analysis methodologies", IEEE Transactions on Electron Devices, vol. 56, no. 7, pp. 1433-1441, July 2009.

[15] T. Pompl, A. Kerber, M. Rohner and M. Kerber, "Gate voltage and oxide thickness dependence of progressive wear-out of ultra-thin gate oxides", Microelectronics Reliability, vol. 46, pp. 1603-1607, Sept. 2006.

[16] J. Sune, E. Y. Wu, and S. Tous, "Failure-current based oxide reliability assessment methodology", IEEE Int. Reliability Physics Symposium, pp. 230-239, 2008 .

[17] A. Aal, "A comparison between V-ramp TDDB techniques for reliability evaluation", IEEE Integrated Reliability Final Report, pp. 133-136, 2008.

[18] E. Y. Wu, S. Tous and J. Sune, "On the progressive breakdown statistical distribution and its voltage acceleration", IEEE Int. Electron Device Meeting, pp. 493-496, 2007.

[19] G. Bersuker, N. Chowdhury, C. Young, D. Heh, D. Misra, and R. Choi, "Progressive breakdown characteristics of high-k/metal gate stacks", IEEE Int. Reliability Physics Symposium, pp. 49-54, 2007.

[20] E. Y. Wu, J. Sune, B. Linder, R. Achanta, B. Li, and S. Mittl, "Post-breakdown statistics and acceleration characteristics in high-k dielectric stacks", submitted to IEEE Integrated Reliability Final Report, 2011.

[21] T. Nigam, A. Kerber, and P. Peumans, "Accurate model for time-dependent dielectric breakdown of high-k metal gate stacks", IEEE Int. Reliability Physics Symposium, pp. 523-530, 2009.

978-1-4244-9113-1/11 $26.00 © 2011 IEEE

OXIDE DEFECTS GENERATION MODELING AND IMPACT ON BD UNDERSTANDING

[1,2]Y. Mamy Randriamihaja, [1]V. Huard, [1]A. Zaka, [1]S. Haendler, [1]X. Federspiel, [1]M. Rafik, [1]D. Rideau, [1]D. Roy

[1]STMicroelectronics, 850 rue J. Monnet, BP16, 38926 Crolles, France
Phone: (33) –(0)4.38.92.23.21, yoann.mamy-randriamihaja@st.com

[2]A. Bravaix

[2]ISEN – IM2NP, Maison des technologies, 83000 Toulon, France

Abstract—**Microscopic characterization of defects into the depth of the gate oxide is used to compare defects generated during both Hot-Carrier Stress (HCS) and Fowler-Nordheim Stress (FNS). Measured defects are linked to the breakdown (BD) process and there creation rate is modeled.**

Keywords: TDDB, HCS, FNS, oxide defects, interface defects

I. INTRODUCTION

A new approach of MOSFETs' Breakdown (BD) analysis and modeling is proposed. This work is based on microscopic characterization of defects involved in the BD process. To do so, oxide defects N_{ot} (different from interface defects N_{it}) are profiled through the depth of the oxide using Charge Pumping technique (CP) [1]. Low Frequency Noise (LFN) measurement [2], allowing the characterization of N_{ot} is also performed and compared to CP. N_{ot} generated through Fowler Nordheim Stress (FNS) and Hot Carrier Stress (HCS) are compared. A new model of N_{ot} generation, adapted from former HCS models [3] is proposed and compared to measurements. N_{ot} generation field and time dynamic are studied to link those defects to the breakdown phenomenon.

II. EXPERIMENTAL SETUP FOR N_{OT} PROFILING

Various methods can be used to monitor induced defects into the depth of the oxide (N_{ot}) after FNS or HCS. Charge Pumping (CP) allows extracting the N_{ot} profiles through the depth of the oxide [1]. Low Frequency Noise allows monitoring the average N_{ot} created over a particular depth range [2]. TCAD simulation (not shown here) identifies the energy probed zones of those methods, showing common energy probed domain. The same approach was proposed for CP in recent work [4]. This allows comparing N_{ot} creation rate measured with these two methods.

N_{ot} profiling was performed with CP [1] after positive FNS on NMOS device (WxL = 224x4.8μm, gate-oxide thickness T_{ox}=5nm, nominal supply voltage V_{dd}=2.5V), as illustrated in **Fig. 1**. One can already see that the degradation into the oxide depth can be divided into two parts: near the Si-SiO$_2$ interface, a peak is reached (interfacial layer), while a plateau seems to grow deeper into the gate oxide (bulk SiO$_2$).

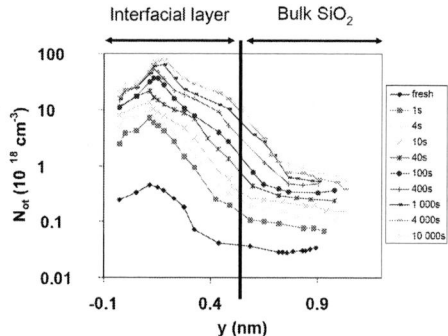

Figure 1. N_{ot} profiles into the depth of the oxide (y direction) extracted after FNS Vg=5.25V, for stress time varied from 0 to 10 000s

N_{ot} profiling and LFN measurements were performed on NMOS devices (WxL = 10x0.28μm, gate-oxide thickness T_{ox}=5nm, nominal supply voltage V_{dd}=2.5V) after HCS (Vg/Vd = 4.3/4.3 V and Vg/Vd = 1.8/4.3 V). The N_{ot} profiles extracted on both FNS and HCS (**Fig. 2**) are normalized and superposed onto the same fresh N_{ot} profile for comparison. This evidences that N_{ot} are generated in both FNS and HCS, as already shown in previous work [5].

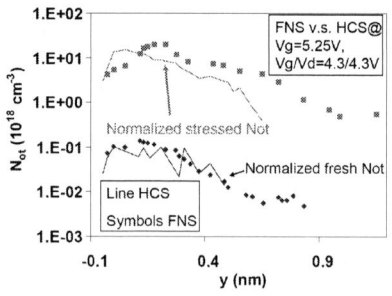

Figure 2. N_{ot} profiles extracted with CP after FNS (@Vg=5.25V) and HCS (@Vg=Vd=4.3V). Fresh N_{ot} profiles are normalized to be superimpose for the eyes. This evidences the generation of N_{ot} for both FNS and HCS.

978-1-4244-9113-1/11 $26.00 © 2011 IEEE

III. LINK BETWEEN N_{OT} GENERATION RATE WITH FNS AND HCS

From our N_{ot} profiles extracted after the various FNS mentioned before, we can reconstruct, at a given position on the plateau, the generation rate of N_{ot} (see **Fig. 3**), using the N_{ot} value in the plateau region. To do so, the stress time at each FNS bias condition was normalized to the bias ratio. We found a time power-law dependence as:

$$N_{ot\_CP\ after\ FNS} \sim A(Vg, Ig).t^{0.5} \Leftrightarrow N_{ot\_LFN\ after\ HCS} \sim A'(Vg,Vd, Id).t^{0.5} \quad (1)$$

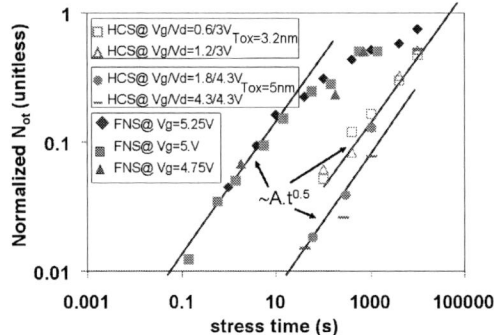

Figure 3. N_{ot} versus stress time, measured with CP for FNS and LFN for HCS, showing the same time power

N_{ot} measurements were also performed with the LFN technique after HCS on two gate-oxide thickness (T_{ox}=5nm and T_{ox}=3.2nm), allowing us to follow its time dependence, as shown in **Fig. 3** and (1).

The time dependence of N_{ot} measured with CP after FNS and LFN after HCS on two gate-oxide thickness, have similar time dynamics. As a consequence, we can assume that defects are generated during FNS and HCS at a similar rate.

IV. OXIDE DEFECTS GENERATION MODELING

In a recent physical framework [3], the N_{it} creation related to carrier-induced bond-breaking excitation has been modeled. A similar formalism was applied to N_{ot} creation under HCS by its relation to the gate-current Ig [10] which is generalized here as following:

$$\Delta N_{ot} \sim K.[R_{ot}.t]^{0,5}, R_{ot}(Vg) = Ig(Vg) . (E_{domi}(Vg) - \varphi_{it})^{pot} (2)$$

where ΔN_{ot} is the amount of created oxide defects (cm^{-3}), R_{ot} is the rate of N_{ot} creation (cm^{-3}.s^{-1}), K is a constant, E_{domi} (eV) is calculated as the "elbow" of the carrier energy distribution function (f(E)), and φ_{it} is the threshold energy of defects creation (1.5eV) [6, 3]. Simulation of f(E), using the deterministic Spherical Harmonic Expansion (SHE) of the Boltzmann equation [8], has permitted to verify the E_{domi} values usually extracted from Ib/Id measurements. p_{ot} calculation is explained below.

V. LINK BETWEEN MEASURED N_{OT} AND BD

N_{ot} profiles were extracted for various FNS's conditions (Vg ranging from 4.25 to 6.25 V with a step of 0.25V, not shown here). For three FNS's conditions (Vg=6.25/6/5.75V),

N_{ot} profiles are measured, and shown before BD in **Fig. 4**. Similar profiles are found, which is in agreement with the hypothesis of a constant amount of defects (N_{BD}) needed to reach the BD [9], and lets us believe that measured N_{ot} may be involved in the BD process.

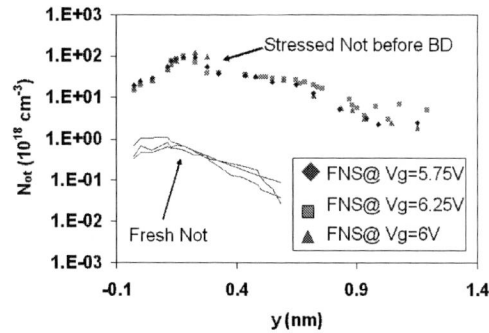

Figure 4. N_{ot} profiles extracted before BD with CP after FNS (@Vg=6.25/6/5.75V). Same profiles found before BD, in agreement with the N_{BD} model [9].

This constant N_{BD} assumption was further checked thanks to statistical measurements : 15 dies were stressed until BD is reached for Vg = 5.75 / 6 / 6.25 / 6.5 V. Partial N_{ot} profiles were performed at the end of both the interfacial layer and the bulk SiO$_2$ which is probed by CP as shown in **Fig. 5**.

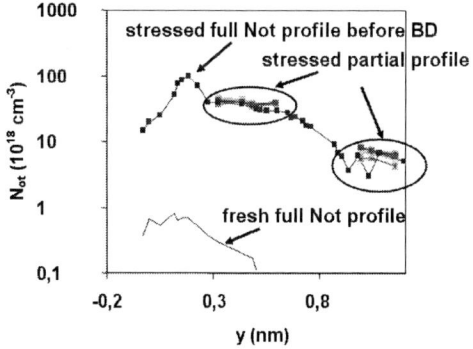

Figure 5. Full N_{ot} profiles extracted before BD with CP after FNS (@Vg=6.25V) superimposed with partial N_{ot} profiles

The statistical N_{ot} measured at the end of the bulk SiO$_2$ probed with CP for the four different stress conditions was found to be similar, as shown in **Fig. 6**. This is in agreement with the constant N_{BD} hypothesis, and gives another indication that measured N_{ot} are involved in the BD process. Furthermore, we decided to follow the Time needed To reach this Degradation (TTD) for a given amount of generated N_{ot} at a given depth. This is illustrated in **Fig. 7**, for three different stressing voltage conditions (Vg=5/5.25/5.75V), where the same amount of degradation (N_{ot}) measured at a certain depth, can be found for each stress conditions after a different stress time (TTD).

978-1-4244-9113-1/11 $26.00 © 2011 IEEE

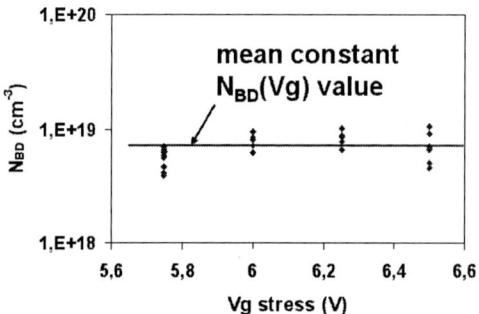

Figure 6. Statistical N_{ot} measured before BD (T_{ox}=5nm) for different stress conditions (Vg=6.5/6.25/6/5.75V). No Vg dependence evidenced

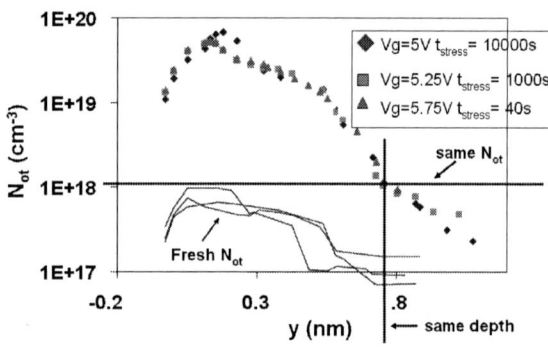

Figure 7. Illustration of TTD: for three stress conditions (Vg=5/5.25/5.75V), the same N_{ot} is reached at a given depth, after different stress time

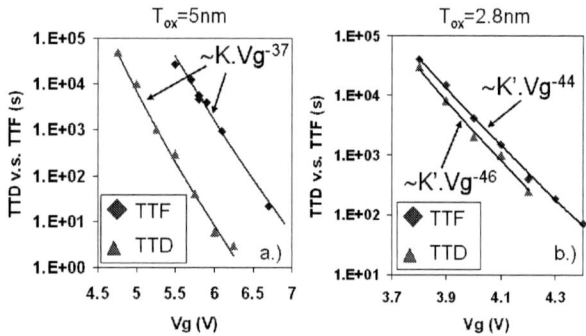

Figure 8. TTD versus Vg, as measured with CP (N_{ot} profiling), compared with the TTF, on a.): T_{ox}=5nm and b.): T_{ox}=2.8nm. TTF versus Vg measured with different MOS areas and normalized after. Similar Vg power dependence for TTD and TTF implying that measured N_{ot} contribute to BD

As seen in **Fig. 8-a,** we found a power-law dependence:

$$TTD_{Not\ with\ CP} \sim K\ [Vg]^{-37} \quad (3)$$

where K is a constant.

The Time To Failure (TTF) was also measured versus Vg. We found a power-law dependence in **Fig. 8-a** with:

$$TTF \sim K'\ [Vg]^{-37} \quad (4)$$

The TTF and TTD measured with CP, on two gate oxide thickness have similar Vg dynamics. This is another indication that measured N_{ot} with CP are the same as those involved in the BD process.

To further verify this assessment, the Vg power dependence of the TTD was measured on two gate oxide (T_{ox}=5nm and T_{ox}=2.8nm), at different depths and for different level of degradation. As illustrated in **Fig. 9**, after a certain depth (~0.6nm), the found Vg power-law dependence is in the range known for NMOS devices. This evidences that only deep defects have the same Vg power-law dependence and may be involved in the BD process.

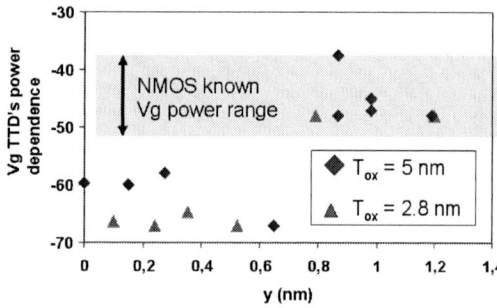

Figure 9. TTD's Vg power dependence measured on two gate oxide thickness, at different depths and degradation levels. The NMOS known Vg power dependence of the TTF is reached after a certain depth (~0.6nm)

Now considering that measured N_{ot} are involved in the BD process, we can rewrite the TTD as follows:

$$TTD \sim N_{ot}/R_{ot}(Vg) \sim N_{ot} / [\ Ig(Vg) \cdot (E_{domi}(Vg) - \varphi_{it})^{pot}\ ] \quad (5)$$

N_{ot} creation measured with LFN after HCS on both gate-oxide thickness was modeled by (5) with p_{ot} ~11, as shown in **Fig. 10**. By extending the E_{domi} range found for those HCS conditions to our range of FNS bias conditions (E_{domi} ranging from 4.5 to 6.5 eV, **Fig. 10**), our extended model gives us:

$$(E_{domi}(Vg) - \varphi_{it})^{pot} \sim E_{domi}^{15} \sim Vg^{15} \quad (6)$$

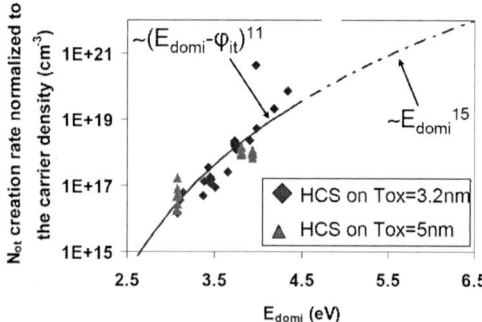

Figure 10. N_{ot} creation rate (normalized to the carrier density) versus E_{domi} measured with LFN after HCS on two oxide thicknesses

Experimental Ig(Vg) show a Vg power-law dependence with an exponent of 22 ($Ig(Vg) \sim Vg^{22}$). Combining (6) and (5) with experimental Ig(Vg) leads us to:

$$TTD \sim N_{ot}.Vg^{-37} \qquad (7)$$

Now if we consider the TTD just before BD, we can rewrite (6) as:

$$TTD_{before\_BD} \sim TTF \sim N_{BD}.Vg^{-37} \qquad (8)$$

This is in agreement with the Vg dependence of TTD found experimentally in (1). This concordance tends to validate our new model of the oxide defect (N_{ot}) generation. This can lead to simple BD modeling based on the calculation of the N_{ot} reached before BD (N_{BD} [9]). This can be further observed in **Fig. 8**: in **Fig. 8-b**, the degradation level is close to the N_{BD}, and so is the TTD close to the TTF, while the opposite is seen in **Fig. 8-a**, allowing, with a same experience time as for classical TTF measurements, to monitor the degradation leading to BD for lower stress conditions.

Concerning the interfacial layer (as shown in **Fig. 1**), what we call the "interfacial layer", is approximately the first two atomic layers. This part of the oxide is a transition zone between bulk Si and SiO_2. Indeed, SiO_2 has not reached its full stoichiometry yet in this zone. This part of the N_{ot} profile can be seen as what is usually called "interface defects" (N_{it}). The fact that it coincides, irrespective of stress condition (see **Figs. 4&7**), only evidences that N_{it} at BD is independent of the stress voltage condition, as was already shown in former work measuring N_{it} at BD for different Vg [11].

Note that N_{ot} profiling using CP was successfully applied to gate oxide thickness of 1.8nm (not shown here). It was also reported to be performed on High K oxide with Effective Oxide Thickness down to 1.3nm [12].

VI. CONCLUSION

This paper uses microscopic defect characterization to gain a better understanding of oxide defects (N_{ot}) generation in correlation to BD mechanism. FNS and HCS are found to generate the same kind of oxide defects, with a similar rate. N_{ot} measured with CP are shown to be deeply involved in the breakdown process. Furthermore, a new model of N_{ot} generation is validated by experiments which have been related to the energy modeling framework, opening a new way of breakdown modeling.

ACKNOWLEDGMENT

The author would like to address special thanks to Gerard Morin and François Dieudonné for their precious help to the measurements.

REFERENCES

[1] D. Maneglia, et al., Jour. of App. Phys., Vol. 79, p. 4187, 1996
[2] M. H. Tsai, et al, Elec. Dev. Let., Vol. 14, p. 256, 1993
[3] C. Guerin, et al, Jour. of App. Phys., Vol. 105, p. 114513, 2009
[4] M. Masuduzzaman, et al, Int. Reli. Phys. Symp., p. 13, 2009
[5] D. J. DiMaria, et al., App. Phys. Lett., Vol. 75, p. 2427, 1999
[6] V. Huard, et al., in proc. IEDM, 2007
[7] S.E. Rauch, et al., Trans. Dev. Mat. Reliab., Vol. 1, p. 113, 2001
[8] S. Jin, et al., SISPAD, 2009
[9] R. Degraeve, et al., Trans. on Elec. Dev., Vol. 45, p. 5757, 1998
[10] A. Bravaix *et al.*,Int. Reliab.. Phys. Symp., p. 531, 2009
[11] D.J. DiMaria, et al., App. Phys. Lett., Vol. 70, p. 2708, 1997
[12] O. Ghobar, et al., Int. Integ. Reli. Work., p. 94, 2007

Comprehensive analysis of charge pumping data for trap identification

D. Veksler, G. Bersuker, A. Koudymov,* C. D. Young, M. Liehr, B. Taylor
SEMATECH Austin, TX 78741 and Albany, NY 12203, USA
phone: 518-649-1141; e-mail: Dmitry.veksler@SEMATECH.org
* Rensselaer Polytechnic Institute, Troy, NY 12180, currently with Sensitron Semiconductor, Deer Park, NY 11729

Abstract— **Analysis methodology for the charge pumping (CP) data, which considers non-elastic electron/hole capturing and releasing processes, is proposed. It is shown that the multi-phonon-assisted rearrangement of the dielectric lattice around the traps, associated with the charge trapping, is important for the interpretation of experimental results and needs to be taken into account. Analysis of the temperature dependent multi-frequency charge pumping data, measured on the MOSFETs with different thickness of the interfacial layer in the high-k dielectric gate stack, allowed to extract the trap energy and spatial profiles and helps to identify the nature of these traps.**

Keywords – charge pumping, trap kinetics, MOSFET characterization, configurational relaxation of traps

I.   INTRODUCTION

Charge pumping (CP) is a widely used technique for characterization of the interfacial trap density in the MOSFETs. With introduction of metal/high-k gate stacks frequently containing a variety of additional capping layers for the work function modulation, it becomes critically important to not only profile the trap distribution through the dielectric thickness but also identify the nature of traps in order to assist the process improvement efforts.

Depth profiling of the bulk dielectric trap concentration is traditionally performed using a frequency-dependent CP (f-CP) method [1-8]. During the CP experiment, the ac signal of a characteristic frequency f and amplitude high enough to achieve both channel accumulation and inversion during the ac cycle is applied to the gate of the MOS structure. The induced DC current between the grounded source-drain electrodes and the body electrode is measured as a function of frequency. This DC current, excluding gate leakage current, is due to the alternate electron and hole trapping on the interface and bulk dielectric traps under the gate metal. The carriers are supplied from the semiconductor body in inversion (Fig. 1a) and from the source-drain electrodes in accumulation (Fig. 1b). Bulk oxide traps located further away from the semiconductor/oxide interface are able to contribute to this recombination current when the signal frequency is reduced. However, the conventional analysis [1-8] of charge pumping data does not take into account the fact that the bulk oxide traps change their atomic configuration when capturing/releasing electrons, which is accompanied by phonon emission [9-12]. Indeed, the characteristic trapping/de-trapping times are controlled, to a

great degree, by these multi-phonon assisted processes (rather than by the carrier tunneling time from/to the substrate), which are uniquely determined by the defect atomic structure. This provides an opportunity for identification of the defect nature from the f-CP data.

In this study, we introduce an analysis methodology for f-CP measurements based on the atomic-level consideration of the physical processes associated with the carrier trapping/detrapping. The bulk defect density (N_T) profiles and structural relaxation properties of the traps are extracted from the frequency and temperature dependencies of the CP currents for the samples with the SiO$_2$ interfacial layers (ILs) in the high-k (HK) gate stacks of different thicknesses (down to a "zero" IL).

II.   MODEL DESCRIPTION

The following key assumptions of the proposed analysis are: (i) The carrier capture/emission involves both the electron (hole) tunneling and multi-phonon assisted structural trap relaxation caused by the Coulomb coupling between the trapped charge and surrounding lattice ions, as illustrated in Fig. 2 (see, for instance, [10,12]).

(i) The bulk traps capture electrons/holes emitted from the conduction/valence bands as well as from the interface states (Fig. 1). Since the characteristic times of the electrons capture by the interfacial states are much faster than that of the f-CP measurements, the former are considered to be always filled, thus providing a constant supply of carriers to the bulk oxide traps.

Total capture rate (per eV of the energy range of the initial states) by a particular bulk oxide trap can be described as follows:

$$\Omega = \Omega_{ELASTIC} \times R,$$
$$R = 1 / \sqrt{4\pi \cdot k_B T \cdot E_{rel}} \exp\left(-\left(E_T - E_{rel}\right)^2 / 4 \cdot k_B T \cdot E_{rel}\right). \quad (1)$$

where $\Omega_{ELASTIC}$ is the rate of the pure elastic capture process, which is determined by the probability of the carrier tunneling to/from the trap (when the interaction between the electronic and phonon subsystems is neglected), R is the probability of the dielectric lattice structural re-arrangement around the trap, which is determined by the electron-phonon interaction (after [11,12]). Here E_T is the trap energy, T is the temperature, k_B is the Boltzmann constant, E_{rel} is the characteristic energy associated with the trap relaxation. We calculate the $\Omega_{ELASTIC}$ rates assuming the trap having 3-D δ-function potential and using the equations derived from the first principles in [13]. These equations describe:

(i) The probability of the carrier transition between the conduction/valence band states and bulk oxide trap (process (1) in Fig. 1a)

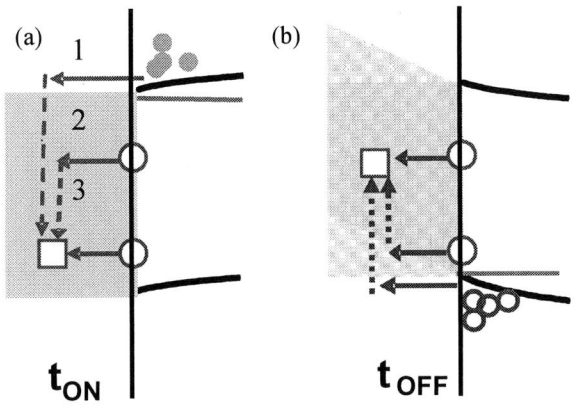

Figure 1. Schematic representation of the processes contributing to the electrons (a) and holes (b) trapping in the bulk oxide defects. The injection from both conduction/valence bands and interfacial states is considered. "1" indicates the electron capture from conduction band accompanied by multiphonon emission, "2" indicate capture from interfacial state with multi phonon emission , and "3" indicate the electron capture throgh the direct tunneling.

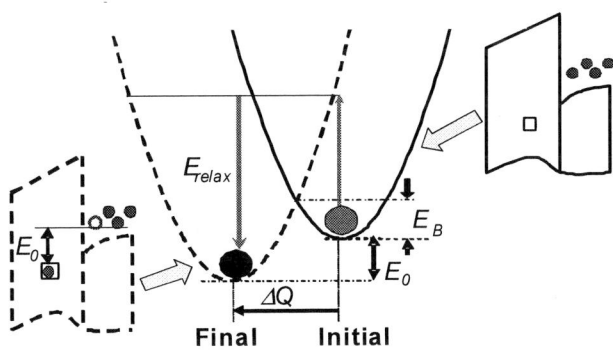

Figure 2. Total energy description of the non-elastic electron capture accompanied by the lattice relaxation (rearrange-ment of the trap atoms). Insets are the corresponding band diagrams using a single-electron description.

The corresponding equation has the dependency on the trap energy and location (depth) in both the pre-factor and exponential term.

(ii) Transition probability between the interfacial states and an oxide bulk trap (processes 2 and 3 in Fig. 1 a). This probability is proportional to $D_{IT}(E)$, the interface state density per area per unit energy. The equations in [13] were derived within the framework of the conventional perturbation theory, and

therefore, they naturally incorporate the dependency of trapping rates on the energy of the initial states. For the case of interfacial states, which are (as opposite to the conduction band states) distributed through the total substrate energy gap, the tunneling barrier height also depends on the initial energy.

The total capture rate of the electrons emitted from both conduction band and interface states is calculated as follows:

$$
\begin{aligned}
c_n(x_T, E_T) = & \int_{E_C}^{\infty} \frac{1}{1 + \exp\left([E - E_F]/k_B T\right)} \cdot \\
& \cdot \Omega_{BAND}(x_T, E) \cdot R(E_T, E) dE \\
& + \int_{-Ev}^{E_C} \Omega_{INT}(x_T, E_{IT}) \cdot R(E_T, E_{IT}) dE_{IT}
\end{aligned}
\tag{2}
$$

where E_F is a Fermi level for the electrons in inversion. Ω_{EAND} and Ω_{INT} are the elastic capture rates from the conduction band and interfacial states, respectively.

The real trap potential can be different from a δ-function. In order to adjust the resulting capture/emission rates we use normalization: We normalize the electron capture rate by the bulk traps having their energy in resonance with the conduction band, $E_T = E_C$, to be $\sigma n v_t$ with σ, n, and v_t to be equal to the previously calibrated parameters [6]. Please note, that the normalization does not affect the functional dependence of the rates vs. distance, energy and temperature. Capture rates for holes (c_p) are calculated in the similar way. Emission rates (e_n and e_p) are calculated using the detailed balance principle [14].

Probability for the trap recharging (which determines the charge pumping currents) is calculated by solving a kinetic equation and using the obtained above electrons/holes capture/emission rates. [6]:

$$
\frac{dF(x_T, E_T, t)}{dt} = (c_n + e_p) - F(c_n + c_p + e_n + e_p)
\tag{3}
$$

Here $F(x_T, E_T, t)$ is the probability for the trap of the energy E_T and located at x_T to be filled at the moment t.

The trap recharging probability during a charge pumping cycle for a particular trap location and energy is $\Delta F = F_{max} - F_{min}$. Charge pumping current can be calculated as:

$$
\begin{aligned}
I_{CP} = & q \cdot f \int_{-\infty}^{\infty} dE_T \int_{0}^{Tox} dx_T \Delta F(E_T, x_T) \cdot D_{BT}(E_T, x_T) + \\
& + q \cdot f \int_{-\infty}^{\infty} dE_T \Delta F(E_T, x_T \to 0) \cdot D_{IT}(E_T)
\end{aligned}
\tag{4}
$$

where D_{BT} is the volume density of bulk traps, D_{IT} is the surface density of the interfacial (ultrafast) traps.

To simplify the calculations, the interfacial traps are considered to be uniformly distributed through the entire energy range of the band gap with the density D_{IT}. On the other hand, the energy spectrum of the bulk traps is expected to be distributed around a certain value, which is specific to an atomic structure of a certain type of an oxide defect (Fig. 3), presumably associated with oxygen vacancies, contributing to the electron/hole capture:

$$D_{BT} = N_{BT}/(\sqrt{\pi}\ \Delta E)\ exp[-(E_{T0}-E_i)^2/\Delta E^2], \qquad (5)$$

Here E_{T0} is the center of the Gaussian distribution of trap energies, and ΔE is the dispersion. E_{T0} can have different values (for example, level "1" and level "2" in Fig. 3), however level "1"-type distribution centered near the midgap, is more effective in the electron/hole recombination process.

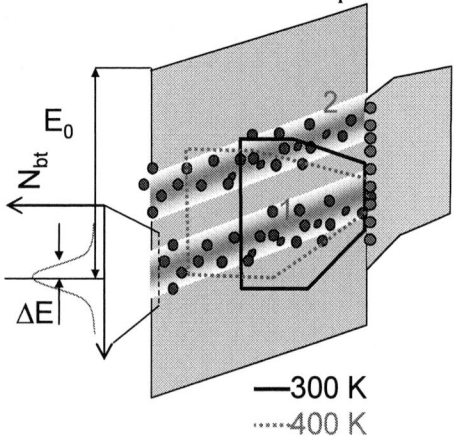

Figure 3. Schematic of the two possible mono-energetic trap distributions,1 and 2, for the traps of a certain given origin (e.g., a certain type of oxygen vacancies). Levels (1) (around mid-gap) and (2) (band edge) would cause different I_{CP} dependencies on frequency and temperature. Contours scematically show the energy and distance ranges of the traps probbed at room (solid curve) and elevated (dasshed curve) temperatures.

III. RESULTS AND DISCUSSION

A. *Role of the dielectric structural relaxation:.*

Figure 4 shows calculated contours of constant trap recharging probability at two different temperatures and two different relaxation energy values, E_{rel} (assuming the interface state density of $N_{it}=10^{10}$ cm^{-2}). One should note that the range of energies and distances for the traps to be recharged during the CP cycle is entirely different with and without considering lattice relaxation. The probing energy range reduces with temperature due to a higher rate of the T-activated carrier escape from the traps closer to the gap edges. Probing depth increases with temperature due to higher n_s and v_t, and, due to increase in the multi-phonon emission probability at higher temperatures (especially for the traps with larger E_{rel} values).

Note that if the lattice relaxation processes are neglected (e.g., only the elastic processes are considered), the electron/hole capture by the bulk traps, which energy distribution is centered within the band gap, can occur only from the interfacial states (which are also distributed through the band gap) – this constitutes a fundamental problem for the elastic trapping description due to a low density of the interfacial traps comparing to density of states in the conduction/valence bands, resulting in a very shallow probing depth, Fig. 4a.

Please note that under a wide range of used CP frequencies and temperatures, the probing depth still lies within a few nanometers from the substrate/oxide interface for any reasonable values of the relaxation energy.

B. *Role of interface traps:*

With the relaxation energies of ≥ 0.5 eV, the contribution of the interfacial traps as carrier suppliers to the bulk traps becomes negligible. However, they do provide a contribution to the charge pumping current, linearly dependent on frequency (Eq. 4).

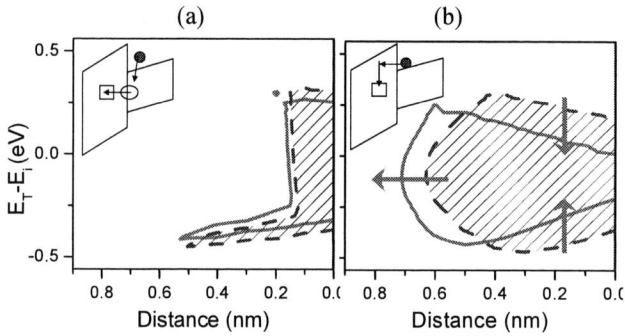

Figure 4. The range of trap energies and distances (counted from Oxide/Semiconductor interface) contributing to an electron-hole recombination. Contours are drawn at the trap recharge probability $\Delta F = 0.65$ assuming different trap relaxation energies (different atomic structure) of traps. (a) Neglecting relaxation: $E_{rel} \sim kT$. (b) With significant relaxation: $E_{rel} = 0.5$ eV. Solid line correspond to $T=400$ K, dashed line – $T=300$K. Arrows show the contour evolution with temperature. Insets show the major trapping process contributing to CP in each case.

C. *Experiment: extracted traps characteristics*

Figure 5 shows the experimental dependence of the charge pumping current vs. frequency at room and elevated temperatures for a 3nm HfO$_2$/1.1 nm SiO$_2$/Si sample. At elevated temperatures the gate leakage current becomes an issue, and the leakage correction is applied to the experimental data.

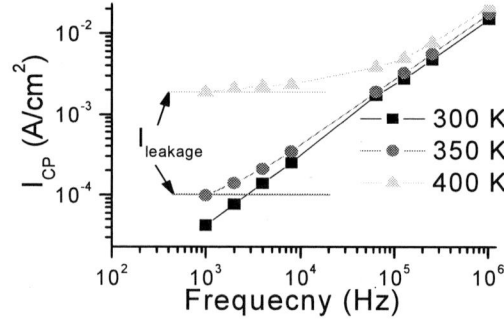

Figure 5. Experimental dependence of the CP current vs. frequency (3nm HfO$_2$/1.1 nm SiO$_2$/Si). Correction for the gate leakage current is taken for the data analysis.

Using the Gaussian distribution for the trap energies (Eq. 5) we performed the N_{BT} extraction from the measurements. The following material and electrical parameters were used in this analysis: the barriers for the electrons and holes, effective

978-1-4244-9113-1/11 $26.00 © 2011 IEEE

masses, band gap, substrate doping, oxide thickness (T_{ox}), oxide dielectric constant ($\varepsilon_{ox.}$), threshold voltage (V_T), and, flat band voltage (V_{FB}). The values for capture cross-sections and thermal velocities for electrons and holes were taken from [6]. The fitting parameters are the trap energy (E_{T0}) and relaxation energy (E_{rel}). Due to the amorphous nature of the interfacial dielectric the broadening of the trap energy level is assumed with $\Delta E \sim 0.5$ eV.

The *T-dependency* of the I_{CP} allows narrowing the range of possible E_{T0} and E_{rel} values. We found that E_{rel}=0.36 eV provides the best match of the cumulative N_{BT} vs. distance dependencies measured at different temperatures (Fig. 6). The same traps in IL were earlier identified as supporting the trap assistant tunneling current in these HK stacks [15].

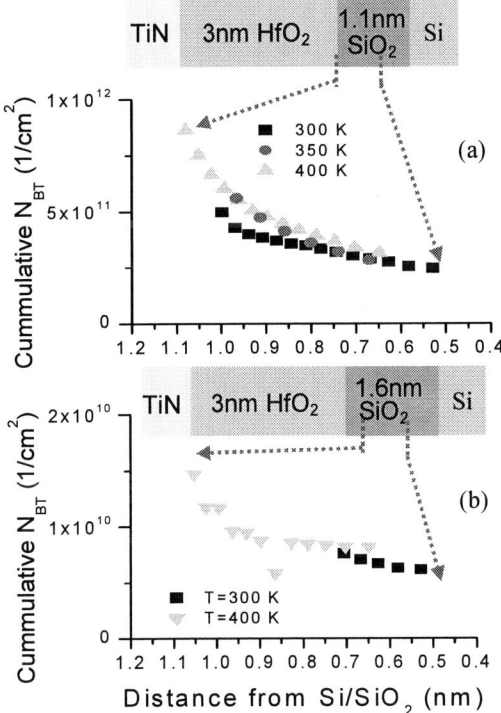

Figure 6. Cumulative bulk trap density profile extracted from the I_{CP} vs. frequency and T dependences in the sample with (a) 1.1nm SiO$_2$ IL (using data in Fig. 5) and (b) 1.6nm SiO$_2$. Distance is counted from the SiO$_2$/Si interface. The trap relaxation energy of E_{rel} =0.36 eV was obtain by matching the trap densities at different temperatures

The volume trap density in the IL of the gate stacks with different thicknesses of the ILs, of 1.1nm and 1.6 nm, are found to be similar at the same distances from the HK/IL interface (Fig. 7), as should be expected when the traps in IL are induced by its interaction with the overlaying HK film [16]. The bulk trap density in IL was found to exponentially decrease with the distance from the HK/IL interface. This fact implies that oxygen extraction from the interfacial layer by the HK layer is energy barrier reduction-driven, rather than diffusion-controlled.

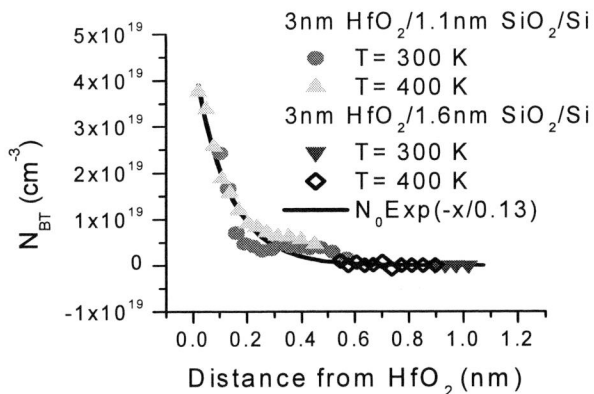

Figure 7. The volume trap density profiles through the SiO$_x$ IL (counting from the HK/IL interface) extracted from the frequency/temperature CP measurements on the gate stacks with 1.1nm and 1.6nm ILs. The N_T decay away from the HK/IL interface can be described by a single exponential dependency = $N_0 e^{-x/\lambda}$, with N_0=4.x10^{19} cm^{-3} and λ= 0.13 nm.

For comparison, different bulk trap density profile was obtained in the devices with the "zero interface" stack of the ~0.3 nm IL (as reported in [17]), Fig. 8. In this case, the CP measurements probe exclusively (besides the Si interface) the HfO$_2$ film, and the trap density was found to be more uniform through the HK thickness, with the values consistent with those of the oxygen vacancies in hafnia [18]. The extracted values of the trap energy of 2.4eV and relaxation energy of 0.65 eV are close to the calculated characteristics of the neutral O-vacancies in monoclinic HfO$_2$ [19].

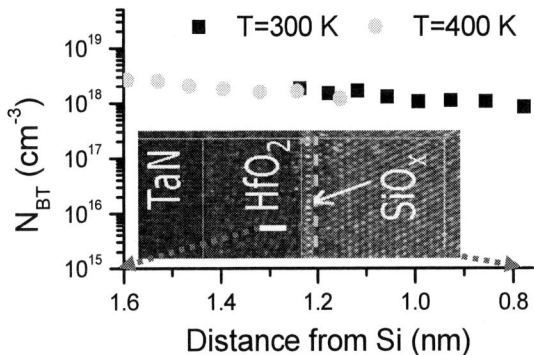

Figure 8. The volume trap density profile in the HK layer in the "zero interface" gate stack. Inset shows a TEM image of the IL-engineered ~0.3nm IL/4nm HfO$_2$ gate stack.

CONCLUSION

We developed a model based on the first principle physical description for the band-to-trap and interface-to-trap tunneling & trapping processes contributing to CP measurements. The analysis shows that trap energy and depth distribution profiles strongly depend on the lattice relaxation determined by the trap atomic structure. In addition to the CP frequency, temperature is found to be an effective factor controlling the probing depth.

REFERENCES

[1] R.E. Paulsen and M.H. White, "Theory and application of charge pumping for the characterization of Si-SiO2 interface and near-interface oxide traps", *IEEE Trans. on Electron. Dev.*, vol. 41, p. 1213, 1994.

[2] Y. Maneglia and D. Bauza, "Extraction of slow oxide trap concentration profiles in metal–oxide–semiconductor transistors using the charge pumping method",*J. Appl. Phys.* 79, pp. 4187-4192 (1996).

[3] R. Degraeve, A. Kerber, Ph. Roussel, E. Cartier, T. Kauerauf, L. Pantisano, and G. Groeseneken, "Effect of bulk trap density on HfO2 reliability and yield," in *IEDM Tech Dig.*, pp.935-938, 2003.

[4] A. Kerber a, E. Cartier , L. Pantisano , R. Degraeve , G. Groeseneken ,H.E.Maes , U. Schwalke, " Charge trapping in SiO2/HfO2 gate dielectrics: Comparison between chargepumping and pulsed ID-VG", *Microelectr. Eng.*, vol. 72, pp. 267-272, 2004.

[5] E. Vogel and D. Heh "Impact of High-K properties on MOSFET electrical characteristics" in *Defects in High-k Gate Dielectric Stacks: Nano-Electronic Semiconductor Devices*, E. Gusev Ed, NATO Science Series II: Math., Phys. and Chem. Springer, 2006, p 85.

[6] D. Heh, C. D. Young, G. A. Brown, P. Y. Hung,, A. Diebold, G. Bersuker, E. M. Vogel, J. B. Bernstein, "Spatial distributions of trapping centers in HfO2 /SiO2 gate stacks", *Appl. Phys. Lett.*, vol. 88, no. 15, p. 152907-3, 2006; D. Heh, C. D. Young, G. A. Brown, P. Y. Hung,, A. Diebold, E. M. Vogel, J. B. Bernstein, G. Bersuker, *IEEE Trans. Electron. Dev.*, vol. 54, no. 7, p1338 (2007).

[7] M. Toledano-Luque, R. Degraeve, M. B. Zahid, L. Pantisano, E. San Andrés, G. Groeseneken, S. De Gendt, "New Developments in Charge Pumping Measurements on Thin Stacked Dielectrics" *IEEE Trans on. Electron. Dev.*, vol. 55, no. 11, pp. 3184-3191 (2008).

[8] M. Masuduzzaman, A. E. Islam, and M. A. Alam, "Exploring the Capability of Multi-Frequency Charge Pumping in Resolving Location and Energy Levels of Traps within Dielectric", *IEEE Trans. on Electron Dev. (T-ED)*, vol. 55, no. 12, pp. 3421-3431, 2008; M. Masuduzzaman, A. E. Islam and M. A. Alam, "Physics and Mechanisms of Dielectric Trap Profiling by Multi-Frequency Charge Pumping (MFCP) Method," *Proc IEEE IRPS*, pp. 13-20, 2009.

[9] Yu. E. Perlin, "Modern Methods in the Theory of Many-phonon Processes," *Sov. Phys- Usp. (Usp Phys Nauk)* vol. 6, p. 542, 1964.

[10] D. Veksler, G. Bersuker, S. Rumyantsev, M. Shur, H. Park, C. Young, et al., "Understanding Noise Measurements in MOSFETs: the Role of Traps Structural Relaxation," *Proc. of IEEE Inter. Rel. Phys. Symp. (IRPS)*, p. 73, 2010.

[11] W. B. Fowler, J. K. Rudra, M. E. Zvanut, F. J. Feigl, "Hysteresis and Franck-Condon Relaxation in Insulator-semiconductor Tunneling," *Phys. Rev. B*, vol. 41, pp. 8313-8317, 1990.

[12] V. N. Abakumov, V. I. Perel, I. N. Yassievich, *Nonradiative recombination in semiconductors,*, North-Holland, 1991.

[13] T. Tewksbury, "Relaxation effects in MOS devices due to tunnel exchange with near-interface oxide traps," Ph.D. dissertation, MIT, May. 1992.

[14] P.T. Landsberg, *Recombination in Semiconductors*, Cambridge and New York: Cambridge University Press, p. 168, 2003.

[15] G. Bersuker, D. Heh, C. Young, H. Park, P. Khanal, L. Larcher, A. Padovani, P. Lenahan, J. Ryan, B.H. Lee, H. Tseng, R. Jammy, "Breakdown in the metal/high-k gate stack: Identifying the "weak link" in the multilayer dielectric," *IEEE IEDM Digest*, pp. 791-794, 2008.

[16] G. Bersuker, C. S. Park, J. Barnett, P. S. Lysaght, P. D. Kirsch, C. D. Young, R. Choi, B. H. Lee, B. Foran, K. van Benthem, S. J. Pennycook, P. M. Lenahan, and J. T. Ryan, "The effect of interfacial layer properties on the performance of Hf-based gate stack devices," *J. Appl. Phys.*, vol. 100, no. 9, p. 094 108, Nov. 2006..

[17] G. Bersuker, D. Heh, C. D. Young, L. Morassi, A. Padovani, L.; Larcher, K.S. Yew, Y. C. Ong, D. S. Ang, K. L. Pey, W. Taylor, "Mechanism of high-k dielectric-induced breakdown of the interfacial SiO2 layer," , *Proc. IEEE IRPS*, , pp. 778–786, 2010

[18] C. Young, Z. Yuegang, D. Heh, R. Choi; B. Lee, G. Bersuker, "Pulsed IV Methodology and Its Application to Electron-Trapping Characterization and Defect Density Profiling*" IEEE Trans. Electron. Dev.* vol. 56, no. 6, p. 1322-1329, 2009.

[19] D. Muñoz Ramo, J. L. Gavartin, and A. L. Shluger, G. Bersuker, "Spectroscopic properties of oxygen vacancies in monoclinic HfO2 calculated with periodic and embedded cluster density functional theory" *Phys. Rev. B*, vol. 75, no. 20 p. 205336, 2007.

978-1-4244-9113-1/11 $26.00 © 2011 IEEE

A Physics-Based Model of the Dielectric Breakdown in HfO$_2$ for Statistical Reliability Prediction

Luca Vandelli, Andrea Padovani, Luca Larcher
DISMI (DIpartimento di Scienze e Metodi dell'Ingegneria)
Università di Modena e Reggio Emilia
42122 - Reggio Emilia, Italy
83688@studenti.unimore.it

Gennadi Bersuker, Jung Yum
SEMATECH, 2706 Montopolis Dr.,
Austin, TX 78741, USA

Paolo Pavan
DII, Università di Modena e Reggio Emilia and IUNET
Via Vignolese, 905
41125 Modena, Italy
paolo.pavan@unimore.it

Abstract— **We present a quantitative physical model describing the current evolution due to the formation of a conductive filament responsible for the HfO$_2$ dielectric breakdown. By linking the microscopic properties of the stress-generated electrical defects to the local power dissipation and to the corresponding temperature increase along the conductive path the model reproduces the rapid current increase observed during the breakdown. The model successfully simulates the experimental time-dependent dielectric breakdown distributions measured in HfO$_2$ MIM capacitors under constant voltage stress, thus providing a statistical reliability prediction capability, which can be extended to other high-k materials, multilayer stacks, resistive memories based on transition metal oxides, etc.**

Keywords-component: TDDB; high-k; breakdown statistics; dielectric breakdown; HfO$_2$, RRAM; forming.

I. INTRODUCTION

Understanding the physical mechanisms governing the temperature (T) and voltage dependencies and the statistics of the degradation and breakdown (BD) of high-k materials is critical for evaluating the reliability of advanced high-k/metal gate CMOS and of emerging resistive memory devices. Despite recent progresses [1]-[4], the key factors governing the degradation and BD of the high-k dielectrics are not unambiguously determined. In this context, a physical model linking the material properties to the degradation mechanisms is highly desirable.

The approach generally adopted for simulating the BD process is based on the percolation model [5]. Although this simple approach has been very helpful explaining the oxide BD statistics, it does not provide insight into the physical mechanisms governing the oxide degradation and BD.

We present here a quantitative physical model, which is capable to describe the prominent features of the oxide degradation process by linking the experimentally observed current evolution under constant voltage stress (CVS) to the generation of defects of the specific atomic characteristics. The different phases of the current increase in the HfO$_2$ dielectric is attributed to a non-uniform localized increase in the density of the defects assisting the electron transport through the dielectric, that eventually results in BD. The model takes into account the microscopic properties of the

HfO$_2$ material, e.g. the physical characteristics of the electrically-active defects at grain boundaries (GBs). The current flow across the dielectric is calculated in the framework of the multi-phonon Trap Assisted Tunneling (TAT) model [6, 7, 8], which allows also to compute the power dissipation at the trap sites and the associated T profile. The field- and T- dependent generation of new defects (i.e. oxygen vacancies) is included through the bond breaking probability description [9, 10]. The proposed model allowed to reproduce the experimental time-dependent dielectric BD (TDDB) distributions obtained in HfO$_2$ capacitors.

II. PHYSICAL MODEL OF THE BD PROCESS

Recent electrical and conductive atomic force microscopy experiments [11] have shown that the electron transport in polycrystalline HfO$_2$ occurs preferentially along the GBs. Therefore, we consider here a GB-driven conduction, in which GBs are modeled as cylindrical filaments of 1nm radius. Since GBs in HfO$_2$ are known to be energetically favorable locations for oxygen vacancy defects [12], we consider the TAT via such defects as the dominant conduction mechanism through the dielectric, in agreement with earlier reports [8,13].

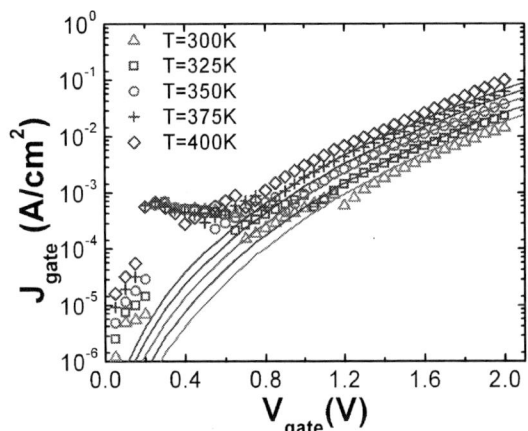

Figure 1. Pre-stress currents at different T measured (symbols) and simulated (solid lines) on a 7nm TiN/HfO$_2$/TiN capacitor using trap energy E$_T$ = 1.7eV-2.7eV and trap relaxation energy E$_{REL}$= 1.2 eV.

978-1-4244-9113-1/11 $26.00 © 2011 IEEE

In order to extract the properties of the GB defects assisting the charge transport, we simulated pre-stress I-V currents measured on TiN/HfO$_2$/TiN capacitors at different T through the statistical multi-phonon TAT model proposed in [6, 7, 8]. In these simulations, the conduction in these large area capacitors, containing many GBs, is simulated by considering a large number of GBs randomly generated across the total device area. As shown in Fig. 1, the experimental IV curves are accurately reproduced by the simulations. The extracted defect parameters (i.e. trap energy, relaxation energy) agree with those derived from ab initio calculations for positively charged O vacancy (V$^+$) defects [14], thus suggesting that V$^+$ defects, which tend to precipitate at the GBs [12], are responsible for the electron TAT transport. Very similar defect parameters were used to reproduce the I-V characteristics of TiN/HfO$_2$/SiO$_2$/Si stacks of various thicknesses in a wide T range [8].

Since GBs constitute a preferential path for the electron conduction, they are also thought to determine the location of the BD filament, in agreement with scanning tunneling microscopy [15] and transmission electron microscopy [16] investigations. The developed electron transport model was applied to simulate the evolution of the current along a single GB under CVS in a HfO$_2$ MIM capacitor. The schematic diagram of the BD simulation flow is shown in Fig. 2. It is comprised of five major steps.

1) Pre-existing defects with the above characteristics (V$^+$) are randomly generated inside the 1-nm radius GB. The initial defect number should be able to reproduce the pre-stress experimental I-V characteristics at different T.

2) The current flow via the GB defects is calculated using the multi-phonon TAT model [6, 7, 8].

3) The energy of the phonons released as a result of the lattice relaxation around each defect contributing to the TAT process is calculated; this energy determines the distribution profile of the power dissipated along the GB. For instance, Fig. 3 shows the 2D profile of the power dissipated during the TAT process via 20 oxygen vacancy defects randomly generated in the GB. Interestingly, the power dissipation occurs primarily in the dielectric region close to the anode, where the average energy released is higher.

4) The calculated power dissipation profile along the GB is then used to calculate the 3D T profile. This is done by solving the Fourier's heat flow equation taking properly into account the heat flow along the GB radial direction (i.e. between the GB and the surrounding HfO$_2$ lattice), as well as between the GB and the metal electrodes, see Fig. 4. The heat transport through the TiN/HfO$_2$ interfaces to the external environment is modeled by two finite thermal resistances, whose values have been estimated according to the model described in [17]. Fig. 5 shows the T profile calculated from the power dissipation case plotted in Fig. 3 considering a 300K external T. As can be seen, T peaks at the centre of the high power dissipation region and decreases significantly approaching the electrodes, since the thermal resistance associated to the heat transfer from the HfO$_2$ interface to the environment is low compared

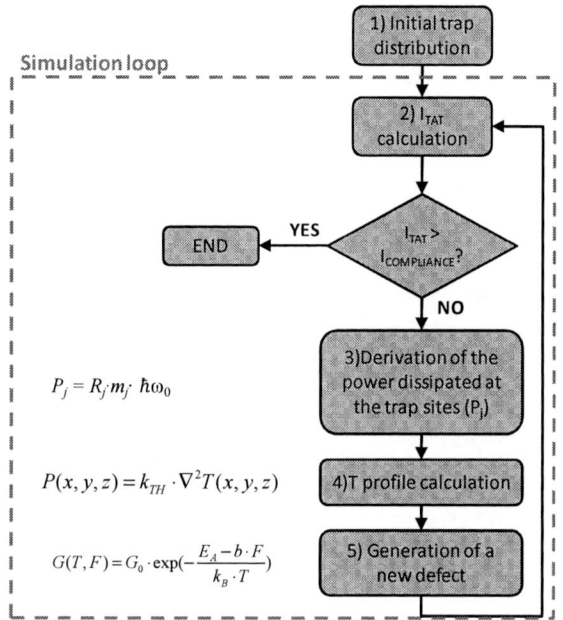

Figure 2. The flow diagram of the model simulating the BD process. P_J is the power dissipation associated to a trap j, R_J the rate of the electron flow via the trap, m_j the number of phonons released to the lattice, $\hbar\omega_0$ the phonon energy. $P(x,y,z)$ is the 3D distribution of the power dissipated at the trap sites, κ_{TH} is the HfO$_2$ thermal conductivity. G is the T- and field- (F) dependent probability of a Hf-O bond breakage (see Sec. II).

to that of the HfO$_2$, whose conductivity is low (~0.005 W/cmK [18]).

5) Once the T profile is calculated, new stress-induced defects are randomly generated employing an effective activation energy description [9, 10]:

$$G(T,F) = G_0 \cdot \exp\left(-\frac{E_A - b \cdot F}{k_B T}\right), \qquad (1)$$

$G(T,F)$ is the T- and field- (F) dependent rate of a Hf-O bond breakage. k_B is the Boltzmann's constant, G_0 is a constant, which has been experimentally derived. E_A is the effective energy required to remove an oxygen atom from its regular position, which depends on the energy of the final location of the removed ion, and it is considered an experimentally determined parameter. The b parameter is determined by the bond polarization factor [9]

$$b = \frac{2+k}{3} \cdot p_0, \qquad (2)$$

where p_0 is the molecular dipole moment and $k \cong 21$ is the HfO$_2$ relative dielectric constant. In this work we considered E_A=4.4eV and p_0=11eA, close to the value theoretically calculated in [9] for cubic HfO$_2$. By using MonteCarlo techniques [19], the model generates the new defect randomly in energy and in position, according to the local generation rates, which depend on the local T and field, see Eq. (1).

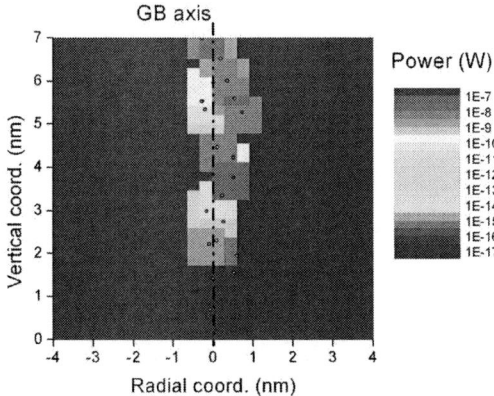

Figure 3. 2D map of the power dissipated by electrons tunneling via a distribution of 20 defects randomly generated in the GB (simulation data refer to V_G=3V; 7nm HfO$_2$ thickness). The map has been derived from the calculated 3D map considering a cut along a plane passing through the axis of the GB. The dots represent the projection of the defects on the plane.

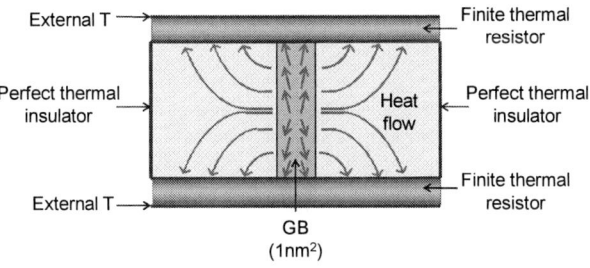

Figure 4. Illustration of the heat flow along a section of the simulated device, with the boundary conditions assumed in the T profile calculation.

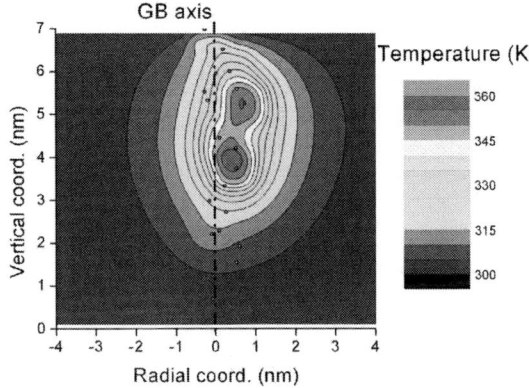

Figure 5. 2D T map calculated for the power dissipation in Fig. 3.

The loop in Fig. 2 is repeated each time a new defect is generated, and the current, the T map and the total simulation time are then updated accordingly. Thus, the T driven positive feedback responsible for the fast current increase observed during the BD experiments is quantitatively reproduced.

The model accounts for the microscopic structural changes related to the O vacancy concentration increase at the GBs, which leads to the formation of a conductive Hf-rich filament. The defect relaxation energy associated with the electron trapping/emission event is reduced when the defects get closer and the energy barrier separating them gets smaller. This relaxation energy reduction is caused by increasingly delocalized nature of the electron wave function, which leads to a smaller lattice re-arrangement to accommodate the trapped charge.

III. SIMULATION RESULTS

The time evolution of the current driven by a single GB simulated under CVS with V_G=2.8V on a 7nm TiN/HfO$_2$/TiN stack is shown in Fig. 6. I_G slowly grows until an abrupt increase corresponding to the BD event is observed. The inset shows the number of defects along the GB conduction path for this specific case at different degradation stages labeled as A, B, C, D. The corresponding evolution of T along the GB is shown in Fig. 7.

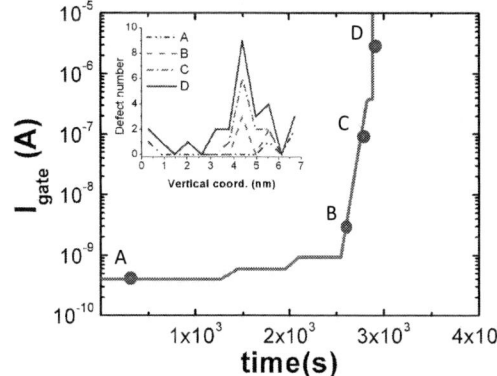

Figure 6. An example of the simulation of the gate current through a single GB in a 7nm TiN/HfO$_2$/TiN stack during CVS (V_G=2.8V). The inset shows the number of generated defects along the GB conduction path responsible for the current at different stress times A, B, C, D.

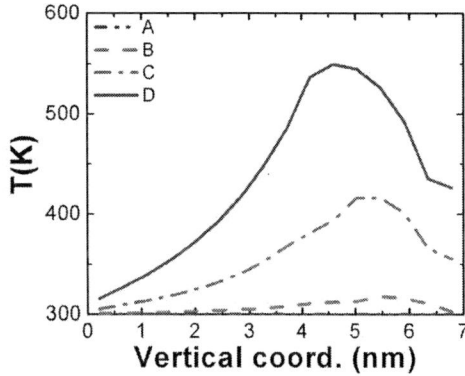

Figure 7. Evolution of the T profile along the GB during the BD simulation. Lables A, B, C and D refer to the different phases of the BD transient shown in Fig. 6.

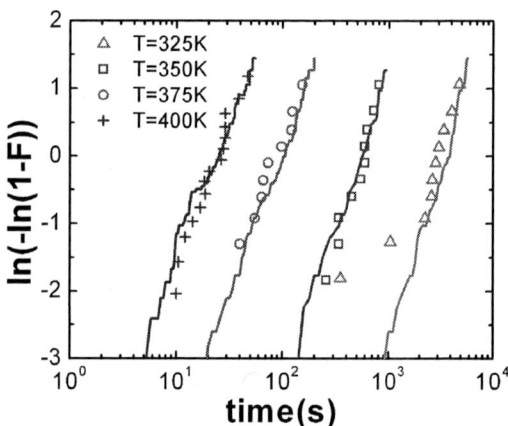

Figure 8. Experimental (symbols) and simulated (lines) TDDB distributions on a 8nm TiN/HfO2/TiN stack at 3.3V at different T.

During the initial period of the stress (point A in Fig. 6) only few pre-existing defects are present in the GB so that the initial TAT current and the related power dissipation are low. As a consequence, the local T along the GB is approximately uniform and equal to the ambient T. Due to the T uniformity the new stress-induced defects are generated in the entire GB volume approximately with the same probability. The process changes when the random generation of additional defects leads to the formation of a particularly favorable percolation path, which strongly enhances the local current (the point B in Fig. 6) and the related power dissipation. In turn, this increases the local T leading to an accelerated defect generation around the high T region. Such process triggers a positive feedback that quickly leads to the BD (points C and D in Fig. 6).

The statistical capabilities of the model, due to the MonteCarlo generation of defects, can be exploited to simulate TDDB distributions. Fig. 8 shows the Weibull plot of TDDB distributions simulated and measured [1] on a 10^{-5} cm^2 8nm TiN/HfO2/TiN stack at different T. These large area devices contain a large number of GBs. Therefore, BD simulations have been performed by calculating the current evolution in a large number of isolated, randomly located GBs, neglecting the interaction between different GBs. The BD is assumed to occur as soon as the current through one of the GBs exceeds a pre-defined compliance value (5µA). As shown in Fig. 8, the simulated TDDB distributions follow the Weibull dependency, indicating that the model correctly describes the "weak link" feature of the BD process. Furthermore, the model reproduces the dependence of the TDDB distribution on the ambient T indicating that the T dependences of both electron transport and defect generation are described correctly.

IV. CONCLUSIONS

We introduce a physical model quantitatively describing the formation of a conductive filament, which leads to the HfO2 dielectric BD. The model connects the current evolution during CVS and the subsequent rapid current increase observed during the hard BD phase to the defect generation process, which is shown to be accelerated by a T increase along the conductive path. The model reproduces the Weibull character of the statistical distribution of the TDDB under different stress conditions demonstrating that it can be used as a powerful tool for investigating the gate oxide reliability and for life-time predictions.

REFERENCES

[1] G. Bersuker et al. "Breakdown in the metal/high-k gate stack: Identifying the weak link in the multilayer dielectric"., IEDM Tech. Dig. 2008, p. 791.

[2] T. Nigam et al. "Accurate model for time-dependent dielectric breakdown of high-k metal gate stacks", Proc. IEEE Int. Rel. Phys. Symp. 2009, p.523.

[3] N. Raghavan et al. "Detection of high-k and interfacial layer breakdown using the tunneling mechanism in a dual layer dielectric stack", Appl. Phys. Lett. 95, 2009, p. 222903.

[4] G. Bersuker et al. "Mechanism of high-k dielectric-induced breakdown of the interfacial SiO2 layer", Proc. IEEE Int. Rel. Phys. Symp. 2010, p. 373.

[5] R. Degreave et al. "New insights in the relation between electron trap generation and the statistical properties of oxide breakdown", IEEE Trans. Electron Devices, Vol. 45(4), pp. 904-911, 1998.

[6] L. Larcher, "Simulation of leakage currents in MOS and Flash memory devices with a new multiphonon trap-assisted-tunneling model", IEEE Trans. Electron Devices, vol. 50 (5), 2003, pp. 1246-1253.

[7] A. Padovani, L. Larcher, S. Verma, P. Pavan, P. Majhi, P. Kapur, K. Parat, G. Bersuker, and K. Saraswat, "Statistical modeling of leakage currents through SiO2/high-k dielectric stacks for non-volatile memory applications," Proc. IEEE Int. Rel. Phys. Symp. 2008, pp. 616-620.

[8] L. Vandelli et al. "Modeling of the temperature dependency (6 - 400K) of the leakage current through SiO2/high-k stacks" ESSDERC Proc. 2010, p. 388.

[9] J. McPherson, J. Y. Kim, A. Shanware, H. Mogul, "Termochemical description of dielectric breakdown in high dielectric constant materials", Appl. Phys. Lett., Vol 82(13), pp. 2121-2123, 2003.

[10] G. Bersuker, Y. Jeon, H. R. Huff, "Degradation of thin oxides during electrical stress", Microelectron. Reliab., Vol. 41(12), 2003, pp. 1923-1931.

[11] G. Bersuker et al. "Grain boundary-driven leakage path formation in HfO2 dielectrics", ESSDERC Proc. 2010, p. 333.

[12] K. P. McKenna, et al. "The interaction of oxygen vacancies with grain boundaries in monoclinic HfO2", Appl. Phys. Lett. 95, p. 222111, 2009.

[13] A. Campera, G. Iannaccone, F. Crupi, "Modeling of tunneling currents in Hf-based gate stacks as a function of temperature and extraction of material parameters", IEEE Trans. Electron Devices, Vol. 54 (1), 2007, pp. 83-89.

[14] D. Muñoz Ramo, J. L. Gavartin, A. L. Shluger, G. Bersuker, "Spectroscopic properties of oxygen vacancies in monoclinic HfO2 calculated with periodic and embedded cluster density functional theory", Phys. Rev. B 75, 2007, p. 205336.

[15] Y. C. Ong et al. "Bilayer gate dielectric study by scanning tunneling microscopy," Appl. Phys. Lett., vol. 91, p. 102905, 2007.

[16] K. L. Pey, R. Ranjan, C. H. Tung, L. J. Tang, V. L. Lo, K. S. Lim, A/L. Selvarajoo, and D. S. Ang, "Breakdowns in high-k gate stacks.of nano-scale CMOS devices" Microelectron. Eng., vol. 80, pp. 353–361, 2005.

[17] S. Lee, S. Song, K. Moran, "Constriction/spreading resistance model for electronic packaging" SME/JSME Therm. Eng. Conference, Vol. 4, pp. 199-206, 1995.

[18] M.A. Panzer et al. "Thermal properties of ultrathin hafnium oxide gate dielectric films" Electron Device Lett., Vol. 30 (1269), p.1269, 2009.

[19] D. T. Gillespie, "A general method for numerically simulating the stochastic time evolution of coupled chemical reactions" J. Comput. Phys, Vol. 22, no. 4, 1976, pp. 403–434.

Advanced 45nm MOSFET Small-signal equivalent circuit aging under DC and RF hot carrier stress

L. Negre[1,2], D. Roy[1], S. Boret[1], P. Scheer[1], D. Gloria[1]

[1]STMicroelectronics

850, rue Jean Monnet, 38926 Crolles, France

phone: (33) – 438-922-630, laurent.negre@st.com

G. Ghibaudo[2]

[2]IMEP-LAHC

3, rue Parvis Louis Neel, BP 257, 38016 Grenoble, France

Abstract—**The continuous CMOS performance improvement enhances the interest of the RF CMOS for millimeter wave application. Hence the extension of DC reliability model in the RF domain is becoming critical. Understanding the MOSFET aging influence on the small signal equivalent circuit is a key concern to integrate the RF reliability simulation at compact model level. In this work, an accurate setup which allows the application of RF stress and the monitoring of DC/RF parameters is detailed. Hot carrier stress is performed on MOSFET and a physical analysis of the small signal equivalent circuit aging is done.**

Keywords-hot-carrier degradation; reliability; aging; load-pull; RF; MOSFET; small-signal; equivalent circuit; extraction.

I. INTRODUCTION

Product aging modeling is becoming a major reliability challenge of modern technologies. It requires accurate reliability models to guarantee relevant circuit lifetime. These models are historically based on the aging of DC MOSFET parameters and used to simulate DC or AC low frequency circuit aging. However, modern technologies are becoming more and more attractive for millimeter wave applications and thus, competitive technology platform requires an extension of the reliability models in the RF domain.

The hot carrier effects were already studied covering DC to millimeter wave domain [1-5], but these analyses did not take into account the impact of hot carrier stress on the small signal equivalent circuit at a set of bias point covering both the extrinsic and intrinsic device elements. The degradation of these elements has to be interpreted and included in the existing age-based model.

In order to develop such a model, accurate and reliable methodology is required, including the essential following steps:

(1) Device stress and key parameters measurements,

(2) Identification of both DC and RF parameters responsible for the device aging and an extraction procedure,

(3) The analysis of the physical origin of the device aging to allow relevant extrapolation at operating condition.

In this work we discuss this three-step methodology with a special attention to the cause of the RF parameters aging. Section II presents the load pull setup adapted for this study in order to perform both DC and RF large signal stresses. In section III, the extraction of the physically-based small signal circuit is developed. Finally the physical origin of the device aging observed on the small signal equivalent circuit is discussed in section IV.

II. EXPERIMENTAL DETAILS

A. Measurement setup

A passive load pull setup [4] is used in order to perform DC or RF stress which is periodically interrupted for DC and RF measurements (Figure 1). Concerning the stress, the parametric analyzer and the synthesizer are used to apply respectively DC and RF stress component on the device under test (DUT). Concerning the measurements, the parametric analyzer and the vector network analyzer (VNA) are used to measure respectively DC parameters and scattering (S) parameters of the DUT. The power meter is used to monitor the power level delivered by the synthesizer and the impedance tuners are used to control the impedances presented to the DUT. External switches allow to automatically calibrate the bench, apply and extract the RF signal, and to measure S-parameters of the DUT. Hence the accuracy of the access characterization is preserved.

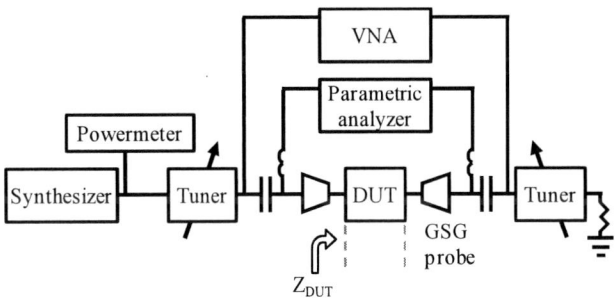

Figure 1. Simplified schematic of the load pull setup used to perform DC/RF stress and monitor DC/RF parameters

B. Stress signal extraction

The complete following methodology is used to extract the actual signal applied to the device and also to accurately monitor DC and scattering (S) parameters:

(1) A calibration of the VNA at both coaxial and probe level is done, so that S-parameters of the input/output access can be extracted by combining measurements of on-wafer standards i.e. short, open and load. Measurements of the de-embedded standards and of the device are also needed in order to analytically extract the impedance (Z_{DUT}) at device level. The impedance is computed using the two port network formula (Eq. 1) from the reflection coefficient. Then, using Z_{DUT}, the time domain voltage and current applied to the device are extracted. The methodology and associated results have been confirmed through Agilent ADS harmonic balance simulation.

$$\Gamma_L = S_{11} + \frac{S_{12}S_{21}\Gamma_O}{1 - S_{22}\Gamma_O} \qquad Z_L = Z_0 \frac{1+\Gamma_L}{1-\Gamma_L} \qquad \text{Eq. 1}$$

where Γ_L and Γ_O are respectively the reflection coefficient at the input and the output of the load, and Z_0 the characteristic impedance.

(2) A full 2-port short-open-load-thru calibration is used and de-embedding is done using open and short structure in order to put the reference plane at the edge of the device [6].

(3) A DC bias is applied to stress the DUT while a RF signal can be applied on the gate or the drain. The stress is interrupted for the DC and RF measurements. Furthermore, it is very important to monitor the DUT input impedance vs. the stress time in order to match the signal in real time.

We have focused our study on hot carrier stress for a nMOSFET with gate length L=0.04μm, total gate width W_{tot}=57.6μm, number of fingers N_{fing}=10. The device has a common source configuration and is embedded in a Ground-Signal-Ground RF test structure.

The stress consisted of a DC gate voltage of V_{GS}=1.8V and a DC drain voltage of V_{DS}=1V, while a RF power of P_{DUT}=9dBm at a frequency of 1GHz is generated on the drain. The equivalent time domain voltage applied on the drain has a peak value of 1.8V and is shown in the inset of the Figure 3. The stress is interrupted five times per decade to measure the threshold voltage (V_T), the maximum of transconductance (G_{mMAX}), the linear (Id_{LIN}) and saturation (Id_{SAT}) current, and S-parameters at a set of bias point.

III. EXTRACTION PROCEDURE

A modeling approach of the equivalent circuit will allow an interpretation of S-parameters drift. The equivalent circuit in Figure 2 is composed of physically-based lumped elements. All capacitances represent the sum of the intrinsic (channel) and extrinsic (fringing and overlap) capacitances. In this schematic, extrinsic series resistances are considered bias independent.

Each lumped element is extracted from the S-parameter measurements using the admittance matrix Y_I and the impedance matrix Z_E representing respectively the part included in the box and the series extrinsic resistances [7].

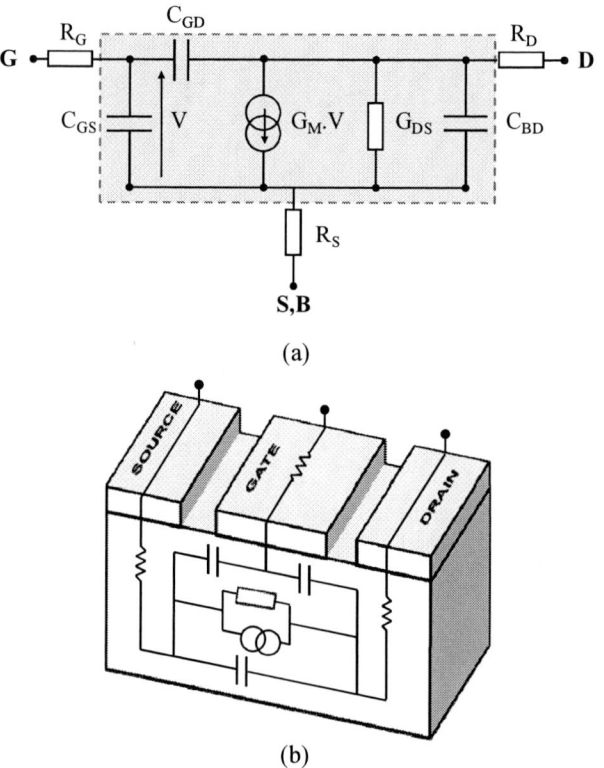

(a)

(b)

Figure 2. (a) Small signal equivalent circuit of a MOSFET in common source configuration including the intrinsic and extrinsic part respectively inside and outside the box. (b)Corresponding physical representation of the small signal equivalent circuit

$$Y_I = \begin{bmatrix} j\omega \cdot (Cgs + Cgd) & -j\omega \cdot Cgd \\ Gm - j\omega \cdot Cdg & Gds + j\omega \cdot (Cbd + Cgd) \end{bmatrix}$$

$$Z_R = \begin{bmatrix} Rg + Rs & Rs \\ Rs & Rd + Rs \end{bmatrix}$$

In cold-FET condition, the intrinsic capacitance and the channel conductance can be neglected so that, the extrinsic parasitic resistances can be evaluated. When all the extrinsic parasitic resistances are known, each element of the equivalent circuits can be calculated from Y_I as follows:

$$Cgd = \frac{\text{Im}(Y_{12}^{-1})^{-1}}{\omega}$$

$$Cgs = \frac{-\text{Im}((Y_{11} + Y_{12})^{-1})^{-1}}{\omega}$$

$$Cbd = \frac{\text{Im}(Y_{22} + Y_{12})}{\omega}$$

$$Gds = \text{Re}(Y_{22})$$

$$Gm = \text{Re}(Y_{21} - Y_{12})$$

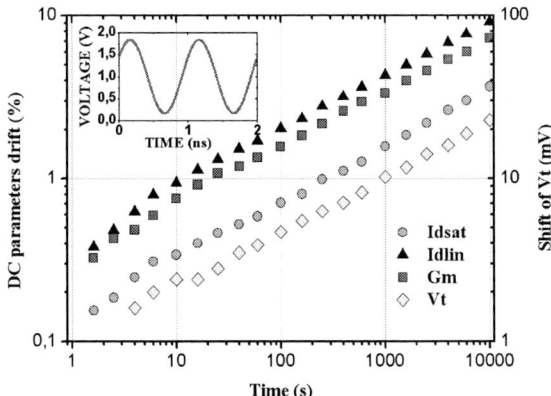

Figure 3. Drift of main dc parameters vs. stress time for rf hot carrier stress (Vgs=1.8V, Vds=1V and a RF signal is applied on the drain). In inset is plotted the rf signal applied on the drain that has been extracted.

IV. RESULTS ANALYSIS AND DISCUSSIONS

A. DC and small-signal parameters degradation

The drift of the main DC parameters are reported in Figure 3. Similar trend of the DC MOSFET parameters drift are reported considering a DC stress or a wide set of accelerated RF stress.

Using the small signal equivalent circuit extraction procedure described in section III, the degradation of each lumped elements has been observed. Figure 4 and Figure 5 clearly reveal that C_{GD}, G_M, G_{DS} are the most impacted elements [8]. Concerning C_{GD}, the degradation is clearly visible at zero drain bias in the inset of Figure 4. At this biasing condition, the extrinsic (overlap and fringing) and intrinsic (channel) contribution can be highlighted.

B. Physical origin of the degradation

We have already demonstrated that most of the degradation of the small signal equivalent circuit's elements can be explained by the threshold voltage, the mobility and the drain series resistance variations [8]. In addition, this work confirms the validity of these corrections for RF stress applied on the gate or the drain (Figure 4 and Figure 5).

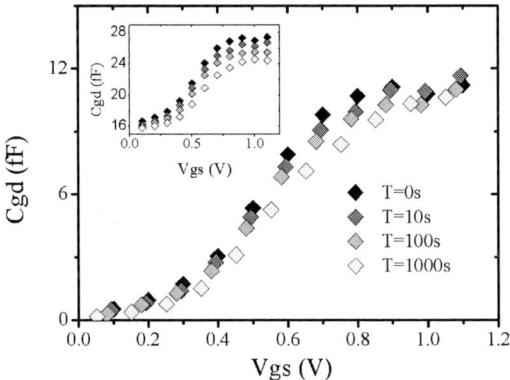

Figure 4. Intrinsic Cgd vs. Vgs for different stress time at Vds=0V including threshold voltage, mobility and extrinsic series resistance correction (uncorrected data are plotted in inset)

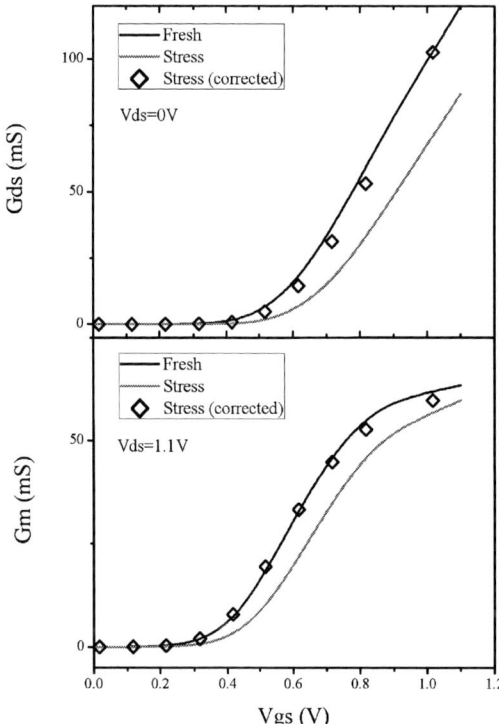

Figure 5. Gds and Gm vs. Vgs for specific Vds biasing. Black lines represent fresh devices and red lines represent stressed devices. Symbols represent stressed devices including threshold voltage, mobility and extrinsic series resistance correction.

The evolution of the extrinsic C_{GD} was revealed. The degradation of C_{GD} for V_{GS}=0V (i.e. overlap and fringing capacitance) and of C_{BD} for V_{GS}=0V (i.e. junction capacitance) shown in Figure 6 allow to emphasize the overlap capacitance drift. This can be explained by the shift of the local flat band voltage ($V_{FB,OV}$) in the n+ poly/n- LDD overlap region. The shift of $V_{FB,OV}$ is confirmed by modeling the V_{GD} drift using the BSIM model [9]. A good agreement is obtained between the proposed model and experimental data (Figure 7).

$$C_{gd}(V_{gd}) = C_{gdo} + C_{gdl}\left[1 + \frac{v}{\sqrt{(V_{gd}+\delta)^2 + 4\delta}}\left(1 - \left(1 - \frac{4v}{ckappad}\right)^{-\frac{1}{2}}\right)\right]$$

$$v = \frac{1}{2}\left(V_{gd} + \delta - \sqrt{(V_{gd}+\delta)^2 + 4\delta}\right) \quad \delta = 0.02V$$

where C_{GDO} is the fringing part of the capacitance, C_{GDL} and ckappad are parameters associated to the bias dependent part.

Thus, as shown in Figure 4 and Figure 5, this physical explanation of the small signal equivalent circuit degradation is confirmed when the fresh and stressed characteristics of the device are plotted by taking into account the threshold voltage, the mobility, the overlap flat band voltage and the drain series resistance drift. In fact, the stressed characteristics well merge with the initial fresh one.

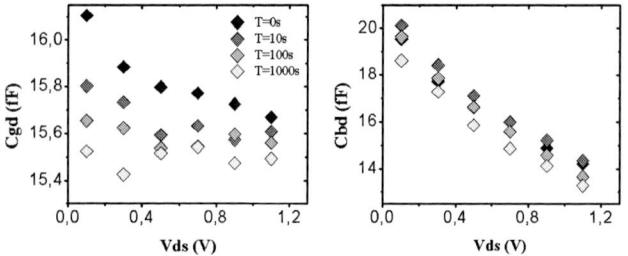

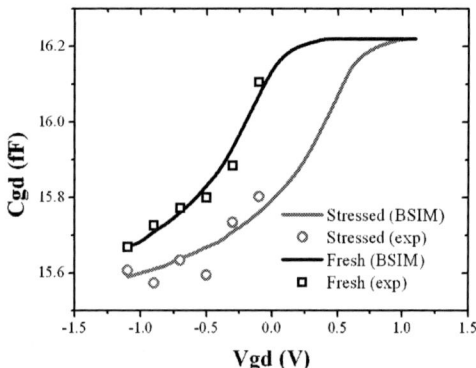

Figure 6. Extrinsic Cbd and Cgd vs. Vds for different stress time at Vgs=0V for different stress time.

Figure 7. Comparison between experimental data and BSIM model of the overlap gate to drain capacitance vs. gate to drain voltage for a fresh and stressed device. A shift of 400mV has been induced after the hot carrier stress and is included in the BSIM model to interpret the aging of Cgd.

Furthermore, extended quantitative lateral profile measurements (Figure 8) have been carried out combining Lateral Profiling with Charge Pumping (LPCP) and Lateral Profiling with Id(Vd) (LPIV) techniques [10]. Concerning LPCP, the drift of pumped drain current is measured while a pulse of increasing amplitude with constant low level is applied over the stress time. For LPIV, the drift of the saturation drain current is measured over the stress time. Combining these two measurements result in an extraction of the interface states density from the channel to the LDD side. Under the stress condition defined previously, these lateral profiling experiments have been performed and infer that the majority of defects generated is located in the gate to drain overlap area.

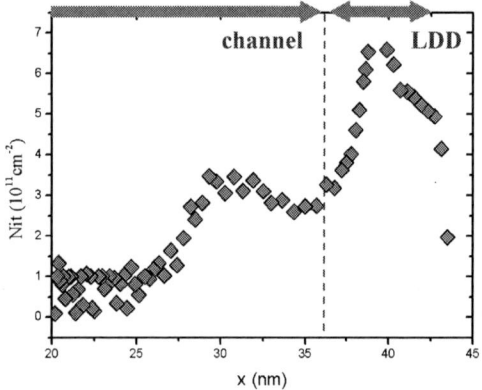

Figure 8. Lateral profile extracted from charge pumping measurements shows an important degradation localized on the LDD (right part of the x-axis)

V. CONCLUSION

The impact of RF hot carrier stress on DC and RF performances has been analyzed. Owing to equivalent circuit modeling, a good description of all RF performances degradation has been achieved using physical parameters. As demonstrated for DC case, RF hot carrier stress induces degradation on the G_{DS}, G_M, C_{GD} that can be linked to the DC parameters degradation. In addition, the degradation of the extrinsic component of the C_{GD} has been explained by the generation of interface states in the gate to drain overlap area. Excellent agreement has been obtained with experimental data, which makes this approach promising in order to integrate RF reliability assessment in industrial compact model.

REFERENCES

[1] J. T. Park, B. J. Lee, D. W. Kim, C. G. Yu, and H. K. Yu, "RF Performance Degradation in nMOS Transistors due to Hot Carrier Effects," *IEEE Trans. Electron Devices*, vol. 47, no. 5, pp. 1068-72, May 2000.

[2] C. Yu and J. S. Yuan, "MOS RF Reliability Subject to Dynamic Voltage Stress - Modeling and Analysis," *IEEE Trans. Dev. Mat. Rel.*, vol. 52, no. 8, pp. 1751-8, Aug. 2005.

[3] G. T. Sasse, F. G. Kuper, and J. Schmitz, "MOSFET Degradation Under RF Stress," *IEEE Trans. Electron Devices*, vol. 55, no. 11, pp. 3167-74, Nov. 2008.

[4] D. Stephens, T. Vanhoucke, and J. J. T. M. Donkers, "RF Reliability of Short Channel NMOS Devices," in *IEEE Radio Frequency Integrated Circuits Symp.*, 2009, pp. 343-6.

[5] C. H. Liu, R. L. Wang, Y. K. Su, C. H. Tu, and Y. Z. Juang, "DC and RF Degradation Induced by High RF Power Stresses in 0.18-µm nMOSFETs," *IEEE Trans. Dev. Mat. Rel.*, vol. 10, no. 3, pp. 317-23, Sep. 2010.

[6] M. C. A. M. Koolen, J. A. M. Geelen, and M. P. J. G. Versleijen, "An improved de-embedding Technique for on-wafer High-Frequency Characterization," in *IEEE Bipolar Circuits and Technology Meeting*, 1991, pp. 188-91.

[7] D. Lovelace, J. Costa, and N. Camilleri, "Extracting Small-Signal Model Parameters of Silicon MOSFET Transistors," in *MTT Digest*, 1994, pp. 865-8.

[8] L. Negre, et al., "Hot carier impact on the small signal equivalent circuit," in *IEEE Int. Integrated Rel. Workshop*, 2010.

[9] *BSIM4.6 Model, User's Manuel.* 2006.

[10] Y. Mamy Randriamihaja, et al., "Multiple Microscopic Defects Characterization Methods to Improve Macroscopic Degradation Modeling of MOSFETs," in *IEEE Int. Integrated Rel. Workshop*, 2010.

Characterization of Hexagonal Rare-Earth Aluminates for Application in Flash Memories

M.B. Zahid, R. Degraeve, M.Toledano-Luque*, J. Van Houdt, *Senior Member*, IEEE

imec, Kapeldreef 75, B-3001 Leuven, Belgium, *Dpto. Física Aplicada III,
Universidad Complutense, Madrid, Spain
Phone: +32-16-287-787 Fax: +32-16-281-844 E-mail: zahidm@imec.be

Abstract - We use Gate-Side Trap Spectroscopy by Charge Injection and Sensing (GS-TSCIS) and Post Program-Post Erase Discharge (PPD-PED) to characterize crystalline Al_2O_3 and hexagonal rare-earth aluminates. The results show that high concentrations of electron traps are present close to the Al_2O_3/metal gate interface, whereas, in the hexagonal aluminates, the electron defect density is reduced, but the lower energetic barrier of these aluminates promotes a parasitic electron injection from the gate during erase.

Index Terms- Rare-Earth Aluminates, Al_2O_3, Spatial profile, Gate- Side Trap Spectroscopy by Charge Injection and Sensing (TSCIS).

I. INTRODUCTION

TANOS charge trapping-based devices are promising candidates to overcome the scaling limitation of Floating Gate based NAND flash memories. In particular, the discrete nature of the storage element makes it more resistant to SILC and coupling interferences issues [1]. The key element of this device derived from the SONOS concept [2] is the use of Al_2O_3, a higher-κ value material for top dielectric. Indeed the higher dielectric constant reduces the field in the top dielectric and prevents electron injection from the gate. However, Al_2O_3 still has a moderate κ-value (~9), and further improvement requires a material with higher dielectric constant. Previous papers have successfully used GdAlO [3], LaAlO [4], or HfAlO [5] high-k top dielectrics. However, in general, the program / erase window is still quite small for multi-level applications. In particular, the erase saturation level is not deep enough, probably because of a too small energy bandgap (E_g) of the top dielectric, which promotes parasitic injection of electrons from the gate during erase. Furthermore, electrically active defects in the top layer can interfere with the program and erase operations, either by charging/discharging or through trap-assisted leakage. Defects also play an important role in the reliability characteristics of TANOS.

In this paper, we investigate the electrical defects in hexagonal aluminates as LaAlO, LuAlO and GdAlO intended as replacement of Al_2O_3 top dielectric in TANOS stacks. We use a recently developed Gate-Side Trap Spectroscopy by Charge Injection and Sensing (GS-TSCIS) shown in Fig. 1 [6] to monitor electron trapping occurring at the aluminates/metal gate interface and Post-Program / Post- Erase Discharge (PPD / PED) shown in Fig. 2 [7] to investigate the involved charging/discharging mechanisms.

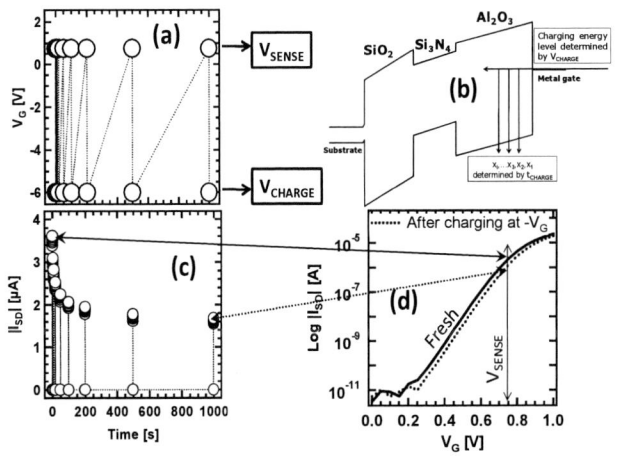

Figure 1. The measurement principle of GS-TSCIS. (a) By applying $V_{CHARGE} < 0$ at the transistor gate, Al_2O_3 traps are charged by direct tunneling from the metal gate as illustrated in the band diagram in (b). In order to measure the trapped charge density, the change of the source-drain current (with $V_D=0.1V$) is measured at V_{SENSE} for a short time of 3 ms, shown in (c). The change of I_{SD} is converted to a V_{TH}-shift using an initially measured I_{SD}-V_G characteristic, illustrated in (d).

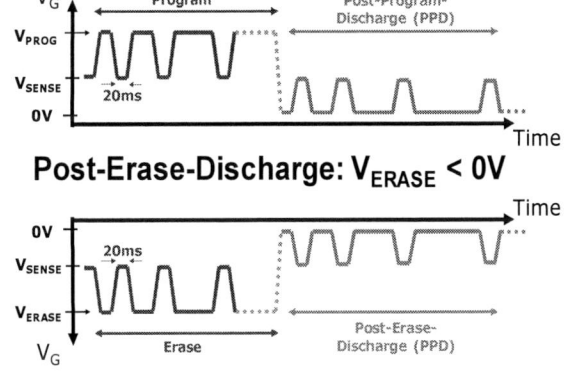

Figure 2. A positive voltage pulse V_{PROG} with exponentially increasing length from $t_{PROG}=20ms$ to 1s is applied to the gate. In between the pulses, the gate voltage V_G switches during a short time of 20ms to a sense voltage V_{SENSE}, chosen such that the gate capacitance C_G presents strong voltage dependence. Afterwards, V_G is fixed to 0V and regularly switched to V_{SENSE} for monitoring the fast transients immediately after program. At V_{SENSE}, the measured C_G is converted to V_{TH}-shift.

978-1-4244-9113-1/11 $26.00 © 2011 IEEE

II. DEVICES

Table 1 shows the samples studied in this work. Large TANOS capacitors 50x50 μm² were fabricated in order to focus on the intrinsic properties of the gate stack materials, without any process integration related issues. For the gate stack, 4nm ISSG SiO_2 tunnel oxide was grown, followed by a 6 nm LPCVD stoichiometric nitride trapping layer. Water based 10 nm ALD Al_2O_3, annealed at 1000 °C or 1100 °C in diluted N_2 or O_2 atmosphere was used as a reference top dielectric [8]. 10 nm PVD-TaN was used for metal gate. In the different splits of this experiment, the Al_2O_3 reference layer was replaced by GdAlO (13 nm for 25% Gd and 15 nm for 50% Gd), LaAlO (15 nm for 25% La and 24 nm for 50% La) and LuAlO (15 nm for 25% Lu and 24 nm for 50% Lu) deposited by ALD. Annealing after top dielectric deposition was performed in N_2 atmosphere at 1000°C, targeting respectively hexagonal phases.

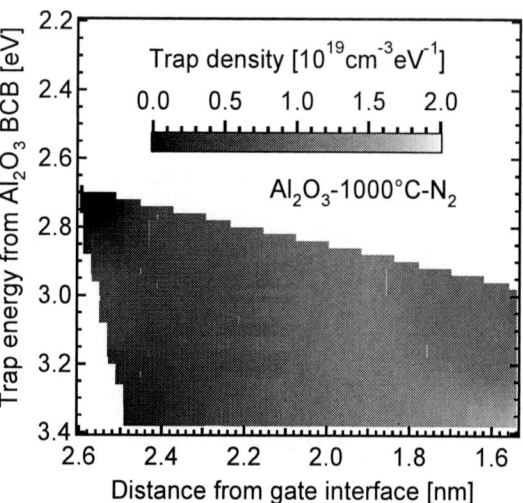

Figure 2. GS-TSCIS trap density plots as a function of the trap energy below Al_2O_3 conduction band (BCB) and spatial position in crystalline Al_2O_3 with TaN metal gate. Scanning further away from the interface, the trap density drops gradually. This high trap density is most probably due to an interface interaction (Al_2O_3/TaN).

Sample	SiO₂	SiN	Al₂O₃ /Aluminates	PDA , 1min	Gate
1			10 nm Al₂O₃	1100 °C – O₂	
2			10 nm Al₂O₃	1100 °C – N₂	
3			10 nm Al₂O₃	1000 °C – N₂	
4			15 nm - 25% La	1000 °C – N₂	
5	4 nm	6 nm	15 nm - 50% Gd	1100 °C – N₂	PVD TaN
6			13 nm - 25% Gd	1000 °C – N₂	
7			13 nm - 25% Gd	1100 °C – N₂	
8			24 nm - 50% Lu	1000 °C – N₂	
9			15 nm - 25% Lu	1000 °C – N₂	
10			15 nm - 25% Lu	1100 °C – N₂	

Table 1 Summary of the samples studied in this work

III. RESULTS AND DISCUSSIONS

Figs.2-5 show the trap density as a function of the distance from the gate interface and the energy below the Al_2O_3 and aluminates bottom of conduction band (BCB).

Note that the energy scale in Fig. 2 is not the same as in Figs. 3-5 due to the difference in barrier for Al_2O_3 and aluminates samples. The maps are extracted using the procedure describe in [6], [9]. Fig. 6 summarizes the electron trap density at 0.2 eV from the gate work function for all the samples characterize in this work. The circle ('o') represents the trap density close to the Al_2O_3 –Aluminates/ metal gate interface and the triangle ('▼') represents the trap density at 2.6 nm from the gate interface.

A. Gate-Side Trap Spectroscopy by Charge Injection and Sensing (GS-TSCIS)

Figs. 2-5 show the trap density map for crystalline Al_2O_3 and hexagonal aluminates. In Fig. 2 and 6, Al_2O_3 shows high trap density close to the Al_2O_3 / metal gate interface. Scanning further away from the interface, the trap density drops gradually. This high trap density is most probably due to an interface interaction (Al_2O_3/TaN).

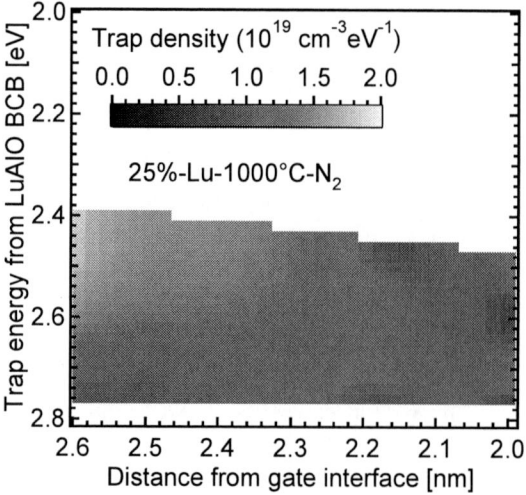

Figure 3. GS-TSCIS trap density plots as a function of the trap energy below LuAlO conduction band (BCB) and spatial position in hexagonal LuAlO with TaN metal gate. Note the different energy scales compared to Fig. 2. High electron defect density is present close to the metal gate interface.

In Figs. 3-5 hexagonal aluminates are shown:

(i) Fig. 3 show the trap density of 25% LuAlO sample, similar as in Al_2O_3, a high electron defect density is present close to the metal gate interface, when scanning further away from the interface, a weak increase in the defect density is observed (Figs. 3 & 6).

(ii) Fig. 4&5 show the trap density of 25% GdAlO and LaAlO sample respectively. These two samples show a very low trap density over the scanned range (Fig. 6). Possible reason can be that the interaction with the metal gate is reduced. Alternatively, the trap energy levels may be outside of the scanned energy range.

978-1-4244-9113-1/11 $26.00 © 2011 IEEE

Summarized, Al_2O_3 show a large amount of electron defects at the gate interface, reducing when scanning further. This indicates an interface interaction is occurring. In the case of the aluminates the trap density is more or less constant over the scanned range, which means that the defects are intrinsically in the bulk.

We use in this work also Post Program and Post Erase Discharge (PPD-PED) as a technique complementary to GS-TSCIS. PPD and PED are based on fast time-resolved V_{FB}-shift measurement and provide more insight in the charging/discharging mechanisms in rare-earth aluminates.

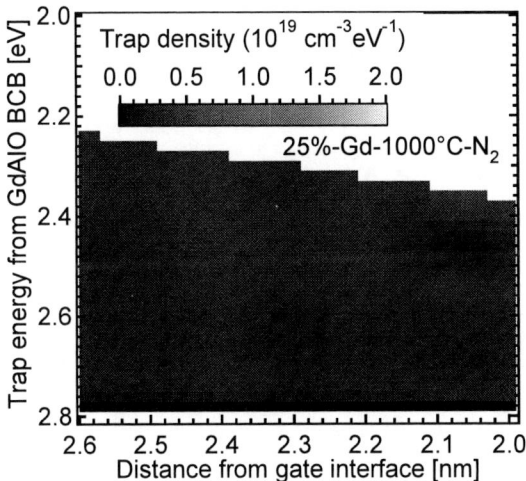

Figure 4. GS-TSCIS trap density plots as a function of the trap energy below GdAlO conduction band (BCB) and spatial position in hexagonal GdAlO with TaN metal gate. Note the different energy scales compared to Fig. 2. Very low trap density over the scanned range is observed.

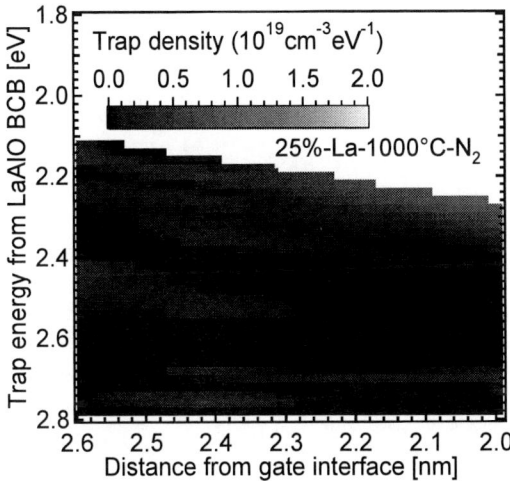

Figure 5. GS-TSCIS trap density plots as a function of the trap energy below LaAlO conduction band (BCB) and spatial position in hexagonal LaAlO with TaN metal gate. Note the different energy scales compared to Fig. 2. Very low trap density over the scanned range is observed.

B. Post Program and Post Erase Discharge (PPD-PED)

In PPD and PED technique, one measures with high time resolution the evolution of the flat band (V_{FB}) voltage immediately after applying a positive (V_{PROG}) or negative (V_{ERASE}) voltage pulse. In fact, PPD/PED is a short integrated version of TSCIS with three additional advantages: (i) it does not need dedicated structures with thin SiO_2 interface layer and can be used on any stacks, (ii) the measurement is significantly less time-consuming, and (iii) it also allows scanning electron traps below the Si-conduction band. Disadvantage is that all extracted information remains qualitative and no deconvolution between energy and space can be carried out. When applied to a memory stack (floating gate or TANOS), PED/PPD reveal in fact the transient behavior of a room temperature retention characteristic. Whereas retention is a long term experiment that shows the charge loss by steady-state leakage mechanisms through the tunnel or top oxide layers, PED/PPD reveal only the discharging phenomena of the dielectrics. Detail description of this technique can be found in [7].

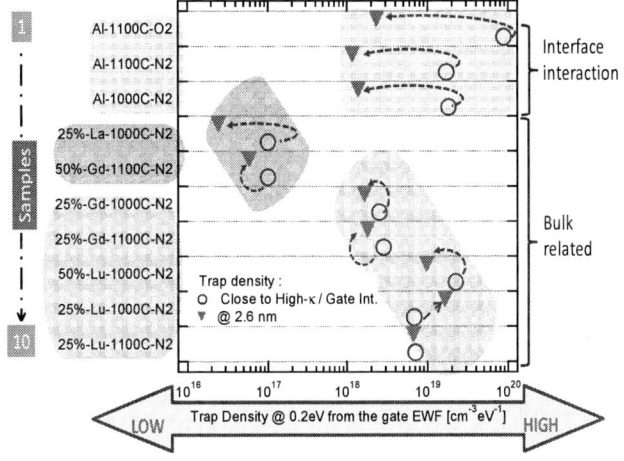

Figure 6. Trap density at 0.2eV from the gate work function (EWF) for all the samples studied in this work. The circle ('○') represents the trap density close to the Al_2O_3 –Aluminates/gate interface and the triangle ('▼') represents the trap density at 2.6 nm from the gate interface. The arrows indicate the evolution of the trap density from interface to bulk.

Note that the voltage apply to perform PPD/PED is not the conventional one used during normal program/erase characteristic in flash memories. For fair comparison, due to the different κ-value and thicknesses, the V_{ERASE} and V_{PROG} are adjusted in order to keep the same electrical field in the Al_2O_3 and aluminates. Fig. 7 shows the Post-Program and Post-Erase Discharge results (PPD-PED) from the different sample studied in this work. During V_{PROG} ('○') and V_{ERASE} ('◊') operations one can see that all Al_2O_3 stacks show the widest P/E window, whereas devices with aluminates dielectric show similar behavior on the program operation, but the erase levels are insufficient or even positive. This is explained by the lower energetic barrier (~2 eV) of aluminates compared to Al_2O_3 (~ 2.6 eV) which promotes parasitic electron injection from the gate during erase [10]. After 1s V_{PROG} and V_{ERASE}, the

978-1-4244-9113-1/11 $26.00 © 2011 IEEE   817

gate voltage is switched to 0 V for 1000 s and PPD or PED are continuously monitored.

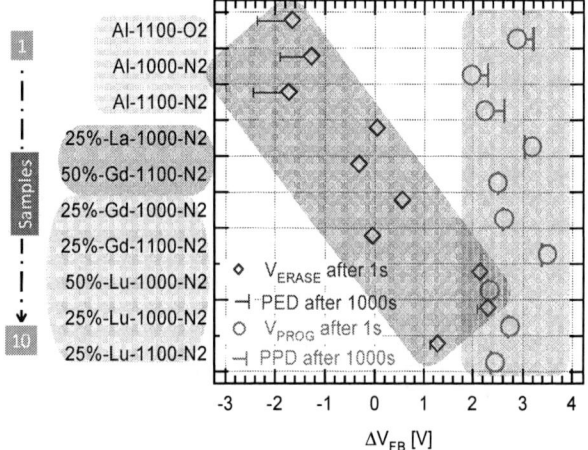

Figure 7. Flat band voltage shift after a 1 s pulse at either V_{PROG} ('○') or V_{ERASE} ('◇'), together with the PPD ('⊢') and PED ('⊣') after 1000 s, respectively. Note that, due to the different k-values and thicknesses, V_{ERASE} and V_{PROG} need to be adjusted to keep the same electrical field in the Al_2O_3 and aluminates.

(i) During PED we can observe that Al_2O_3 stacks show an extra V_{FB}-shift ~ -0.4 V (symbol ('⊢')), which is caused by the rapid discharging of electrons from the near gate Al_2O_3 traps [6], [11]. In the case of the aluminates stacks no extra or very weak V_{FB}-shift is observed. These results confirm the results extracted by GS-TSCIS, where no high amount of electron defects is present at the gate interface in aluminates stacks (Figs. 3-6).

(ii) During PPD, electron trapping to the Al_2O_3 deep traps close to the gate produces an extra positive V_{FB}-shift. This shift is possibly reinforced by redistribution of trapped electrons in the SiN layer, which shift towards the substrate due to the strong band bending. A total V_{FB}-shift of ~0.1 V (symbol ('⊣')) is observed for Al_2O_3 (Fig. 7). Aluminates on the other hand show a very weak electron detrapping during PPD and the V_{FB}-shift remains more or less constant after 1000 s. In summary, the electron traps in Al_2O_3 cause a program and erase instability, which is strongly suppressed in aluminates.

IV. CONCLUSION

In this work we have investigated the defect density close to the metal gate interface using Gate-Side Trap Spectroscopy by Charge Injection and Sensing (GS-TSCIS) and charging/discharging mechanisms using Post-Program-Post and Post-Erase Discharge (PPD-PED) in hexagonal aluminates that are used as top dielectrics in TANOS as a replacement of Al_2O_3. The method of trap energy/depth profiling by using GS-TSCIS allows scanning from ~1.6 nm up to 2.6 nm from the metal gate interface and 0.5eV above the metal work function. For TANOS devices with Al_2O_3, we have shown that high concentration of electron traps exists close to the Al_2O_3/metal gate interface for crystalline sample, whereas in the hexagonal aluminates, the electron defect

density is reduced, but these high-κ materials suffer from low energetic barrier which promotes the parasitic electron injection from the gate during erase. This phenomenon limits the program / erases window and retention from high program level, as required for multi-level application.

ACKNOWLEDGMENTS

This work was funded partly by the EC in project GOSSAMER "Giga-scale oriented soli-state Flash memory for Europe", partly by IMEC's Industrial Affiliation Program on Advanced Flash Memory. M. Toledano-Luque's stay was supported by the Spanish Ministry of Education and Science under contract TEC2007-63318/MIC, and the grant program 'José Castillejo' (ref. JC2009/00052).

REFERENCES

[1] Yoocheol Shin; Jungdal Choi; Changseok Kang; Changhyun Lee; Ki-Tae Park; Jang-Sik Lee; Jongsun Sel; Kim, V.; Byeongin Choi; Jaesung Sim; Dongchan Kim; Hag-ju Cho; Kinam Kim; "A novel NAND-type MONOS memory using 63nm process technology for multi-gigabit flash EEPROMs", IEEE, IEDM, pp 327-330, 2005

[2] Frank R. Libsch and Marvin H. White., "Charge transport and storage of low programming voltage SONOS/MONOS memory devices", Solid-State Electronics, Volume 33, Issue 1, January 1990, Pages 105-126

[3] Jing Pu, Chan, D.S.H., Sun-Jung Kim., Byung Jin Cho., "Aluminum-Doped Gadolinium Oxides as Blocking Layer for Improved Charge Retention in Charge-Trap-Type Nonvolatile Memory Devices", Trans. Elec. Dev., Vol. 56, No 11, pp. 2739 - 2745 ,2009

[4] Wei He, Jing Pu, Chan, D., Byung Jin Cho., "Performance Improvement in Charge-Trap Flash Memory Using Lanthanum-Based High-k Blocking Oxide" IEEE Trans. Elec. Dev., Vol. 56, No 11, pp. 2746 – 2751, 2009

[5] Yan Ny Tan., Chim, W.K., Wee Kiong Choi., Moon Sig Joo., Byung Jin Cho., "Hafnium aluminum oxide as charge storage and blocking-oxide layers in SONOS-type nonvolatile memory for high-speed operation", IEEE Trans. Elec. Dev. Vol. 53, No4 , pp.654-662, 2006

[6] M.B. Zahid, A. Arreghini, R. Degraeve, B. Govoreanu, A. Suhane, J. Van Houdt, "Electron Trap Profiling near Al_2O_3/ Gate Interface in TANOS Stack Using Gate-Side-Trap Spectroscopy by Charge Injection and Sensing" Accepted for publication in EDL

[7] M. Toledano-Luque, R. Degraeve, M. B. Zahid, B. Kaczer, J. Kittl, M. Jurczak, G. Groeseneken, and J. Van Houdt, "Resolving Fast VTH Transients After Program/Erase of Flash Memory Stacks and Their Relation to Electron and Hole Defects", 2009 IEDM Tech. Dig., p. 749

[8] A. Rothschild, L. Breuil, G. Van Den Bosch, O. Richard, T. Conard, A. Franquet, A. Cacciato, I. Debusschere, M. Jurczak, J. Van Houdt, J. A. Kittl, U. Ganguly, L. Date, P. Boelen, and R. Schreutelkamp, "Post Deposition Anneal of Al2O3 Blocking Dielectric for Higher Performance and Reliability of TANOS Flash Memory," in Proc. ESSDERC 2009, Athens, pp. 272-275 (2009)

[9] R. Degraeve, M. Cho, B. Govoreanu, B. Kaczer, M. B. Zahid, J. Van Houdt, M. Jurczak, and G. Groeseneken, "Trap Spectroscopy by Charge Injection and Sensing (TSCIS): A quantitative electrical technique for studying defects in dielectric stacks," in Proc. IEDM Tech. Dig. pp. 775 (2008).

[10] L. Breuil, C. Adelmann, G. Van den bosch, A. Cacciato, M.B. Zahid, M. Toledano-Luque, A. Suhane, A. Arreghini, R. Degraeve, S. Van Elshocht, I. Debusschere, J. Kittl, M. Jurczak, J.Van Houdt , "Optimization of the Crystallization Phase of Rare-Earth Aluminates For Blocking Dielectric Application In TANOS Type Flash Memories", Accepted in ESSDERC 2010

[11] R. Degraeve, M. Zahid, G. Van Den Bosch, P. Blomme, L. Breuil, B. Kaczer, M. Mercuri, A. Rothschild, A. Cacciato, M. Jurczak, G. Groeseneken, J.Van Houdt, "Explanation of anomalous erase behaviour and the associated device instability in TANOS Flash using a new trap characterization technique," Proc. 2009 SSDM, p.428

Charge gain, NBTI recovery and Random Telegraph Noise in localized-trapping NVM devices

Meir Janai, Ilan Bloom and Yael Shur

Spansion Israel Ltd, Netanya 42504, Israel

(+972)-(9)-892-8437, Meir.Janai@Spansion.com

Abstract—**Three different physical reliability processes – charge gain (CG) in EEPROM nonvolatile memory devices, the recovery of negative bias temperature instability (NBTI-R) and random telegraph noise (RTN) are linked together to interpret the CG effect. CG in nitride trapping devices is shown to exhibit discrete detrapping steps similar to NBTI recovery in MOS devices. The height of the RTN-like steps of the CG process can be tuned by modulating the electric field that controls the point of hole injection during the negative erase bias of the device.**

Keywords: NVM, RTN, NBTI, EEPROM, Flash MirrorBit, NROM, Charge Trapping storage

I. INTRODUCTION

Charge Gain (CG) and Charge Loss (CL) are instabilities of the threshold voltage of Electrically-Erasable-Programmable (EEPROM) nonvolatile memory devices under storage conditions. CG is generally observed after erase operation, while CL is observed after program operation. In EEPROM Flash memory devices based on n-channel transistors the device is programmed by injecting electrons into the storage layer under positive gate bias, and it is erased by ejecting the electrons or by injecting holes into the storage layer under negative gate bias. Programming increases the cell's Vt (i.e. reduces the cell's current), and erasure reduces the cell's Vt (i.e. increases the cell's current).

Vt instabilities in EEPROM devices were studied intensively over the last decade [1,2]. Three different mechanisms were reported in the literature for the CG process:

The first CG mechanism, originally reported in NOR floating-gate (FG) devices, affects only a small fraction of the bits, but these bits exhibit relatively large Vt shift after erasure [3]. This mechanism is enhanced by Program/Erase (P/E) pre-cycling. It is also enhanced by applied electric field, but it is nearly invariant to storage temperature. This mechanism was attributed to Flash-SILC (Stress Induced Leakage Current). According to the Flash-SILC model, cycling-induced defects in the tunnel oxide form a trap-assisted tunneling percolation path via the oxide layer, manifested in CG [3, 4].

The second CG mechanism, reported in NAND FG devices, is also enhanced by P/E pre-cycling but is not enhanced by electric field, and, unlike the first mechanism, it is thermally activated with activation energy of 1.1 to 1.2 eV [1]. This mechanism was attributed to thermally-activated detrapping of charges captured in the tunnel oxide layer during P/E cycling.

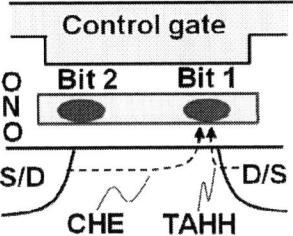

Fig. 1: Schematic illustration of a MirrorBit™ 2-bit nitride trapping Flash memory cell. The cell is programmed by Channel Hot Electorn (CHE) injection, and erased by Tunnel Assisted Hot Hole (TAHH) injection. Bit 2 can be programmed/erased independently by alternating the Source/Drain voltages.

A third CG mechanism was originally characterized in 2-bit nitride storage flash memory devices of the type illustrated in Fig. 1. It was reported to be independent of temperature [2] like the first CG mechanism. However, unlike the first mechanism, the entire Vt distribution is affected uniformly, and, unlike either the first or second mechanism, it is not enhanced by P/E pre-cycling [5].

We reported recently that CG of the third kind has similar phenomenology to the recovery of Negative Bias-Temperature Instability (NBTI) in CMOS devices [6]. NBTI is the instability of the threshold voltage of a CMOS device under prolonged negative bias, and NBTI recovery is the Vt instability after the negative bias is removed [7]. According to our model, the negative bias during erasure results in a significant NBTI effect [7]. The NBTI effect may be obscured by the large Vt shift due to hole injection into the storage layer as the cells get erased. However, it should be realized that the large Vt shift from the programmed state to the erased state during erasure of the non-volatile memory device has two components: its large part is due to hole injection into the storage layer, and a smaller part is due to NBTI. When the negative erase bias is removed the recoverable NBTI component leads to Vt rise which is the observed CG. According to the proposed model, CG of the third kind results from capture of channel electrons by the oxide traps or detrapping of holes from the oxide. The CG rate is controlled by a non-equilibrium Random Telegraph Noise (RTN) process, similar to the detrapping seen during NBTI recovery [8,9].

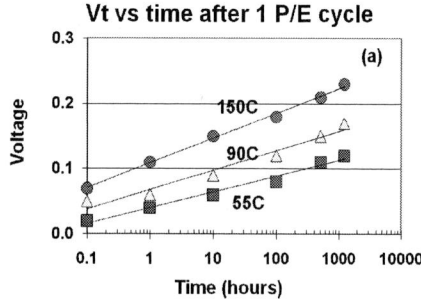

Vt vs time after 1 P/E cycle

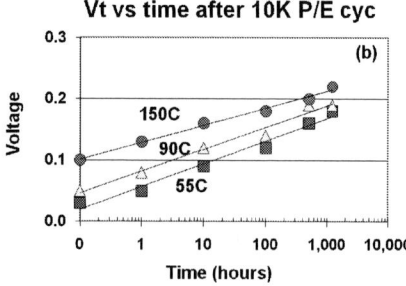

Vt vs time after 10K P/E cyc

Fig. 2: Charge Gain vs. storage time for storage temperatures 55C, 90C and 150C, (a) non-cycled product and (b) 10,000 P/E pre-cycled product. The vertical axis displays the Vt at time t at storage temperature T relative to Vt of same units measured 5 min after last erase at 25 C.

In this paper we provide additional evidence for the relation between CG and RTN, and by that its relation to NBTI recovery. We demonstrate that the point of hole injection and location of charge trapping can be tuned by selection of the drain and gate voltages during erasure, thus modulating the magnitude of the RTN-like step-heights of the CG process.

II. EXPERIMENTAL RESULTS

CG of the third kind may be observed in Flash localized-trapping devices. These devices are immune to Flash SILC [10] and hence the first CG mechanism does not interfere with the measurements. The experiments described below were performed on NROM arrays of products from different manufacturers, and on single-cell test structures embedded in wafers of MirrorBit[TM] NOR Flash products.

Fig. 2 shows typical behavior of the Vt of a localized nitride-trapping product after sector erase, for a non-cycled sector and a 10Kc pre-cycled sector. The plots exhibit the average CG of an ensemble of 0.5Mb of NROM cells manufactured at 0.18 μm technology node, with cell's bottom oxide thickness of 5 nm. The erase conditions during cycling and prior to the CG measurement were Vg=-7V, Vd=7V, Vs float, substrate voltage =0V, and pulse duration = 1 msec. The erase voltage was applied over a dielectric layer of effective oxide thickness of 22 nm. After pattern erase different groups of devices were stored at 55 C, 90 C and 150 C. Vt measurements before storage and at storage intervals were taken at room temperature. The observed CG exhibits a few distinct features:

(a) It has a lin-log time dependency over many time decades. Previous studies showed the lin-log dependency holds from 10 μsec to 2000 hrs, namely, over 11 orders of magnitude [2,5];

(b) The slope of the CG plots is independent of storage temperature [2]. The product in Fig. 2 exhibits CG of about 30 mV per time decade in the temperature range 55 to 150C.

(c) The magnitude of the instability is hardly cycle-count dependent. In Fig. 2, no practical difference can be observed between devices that were programmed and erased only once and devices that were cycled 10,000 times. Previous studies on other products of similar technology have shown that the cycle-count dependence might exhibit a mild increase at low cycle counts, but will then decrease at high cycle counts [6,11,12].

The later feature indicates that CG of the third kind cannot be attributed to cumulative cycling damage of the EEPROM device. It is rather the result of the charging effect caused by an individual erase operation. It further suggests that each program operation in between two erasures re-sets the device by discharging the centers that are responsible for CG. This feature is very different from the behavior of CL in same class of devices, where the magnitude of CL increases monotonically with P/E cycle count [13].

The similarity of the first two features of CG to the behavior of NBTI recovery [14,15] led us to re-examine the behavior of CG in small-scale individual cells. As was recently reported, NBTI recovery in small scale structures exhibits a cascade of RTN-like steps [8,9].

Fig. 3 exhibits a typical CG trace of an NROM single cell produced on 90 nm MirrorBit[TM] wafer. The X axis presents the logarithmic time scale from end of the negative erase bias pulse. The trace shows a monotonic log-time staircase of Vt steps, with occasional RTN signal superimposed on that staircase. The staircase trace is quite similar to the NBTI recovery traces reported previously [8,9]. Fig. 3(a) shows that CG occurs in discrete steps of about 0.10±0.04 μA in height. As seen from Fig. 3(c), the steps obey (stochastically) the equation

$$dN/dt = kN_o/t \qquad (1)$$

where N is the CG step number counted from the first step of trace 3(a) and t is the time. The dimensionless rate constant kN_o equals 3.54. The RTN signal in Fig. 3(a) has amplitude of 0.3μA with mean frequency of 1/65 sec[-1] (about 1 event per min) in the range 0 to 2400 s.

Experiments on different single-cell devices exhibited different RTN amplitudes and frequencies, while some cells did not exhibit any RTN within the experimental time. The consecutive CG steps of a given device showed some spread as can be seen from Fig. 3. The different RTN amplitudes in different cells might be attributed to different locations of the traps that capture/release charges with respect to the locations of dopant atoms in the channel [16]. We point out that in all experiments the CG steps were smaller in magnitude than the RTN amplitudes. The reason for the variance in CG step heights and the reason the CG steps are smaller than the RTN amplitude are explained in section III.

978-1-4244-9113-1/11 $26.00 © 2011 IEEE   820

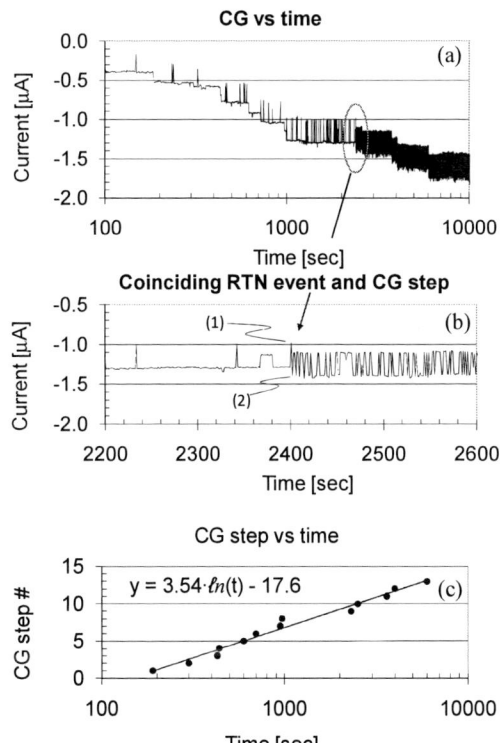

CG vs time

(a)

Coinciding RTN event and CG step

(1)

(2)

(b)

CG step vs time

$y = 3.54 \cdot \ln(t) - 17.6$

(c)

Fig. 3: (a) Typical Charge Gain trace after one program/erase cycle. (b) Blowup of the event at 2400 s; (c) Timing of the CG steps of Fig. 3(a). The field during the negative erase pulse was 6 MV/cm, applied for 1 ms at RT (see Fig. 1 for erase geometry)

An interesting event is seen on the chart at 2400 s from end of the erase pulse (Fig. 3(b)). At this point of time an RTN and a CG step coincided. Concurrently, the frequency of the RTN signal changed: the electron capture time decreased by a factor of 20 and the time to release increased by factor of 4. Our interpretation of this event is given below.

III. ANALYSIS OF EXPERIMENTAL RESULTS

A. Location of hole injection and its effect on CG steps

The smaller amplitudes of the CG steps relative to the RTN amplitudes suggest that the oxide traps responsible for CG are located away from the main current path in the channel, while the traps responsible for RTN are closer to the main current path.

Figs. 4-7 explain the origin of this phenomenon. Fig. 4 shows the sensitivity function of the cell current to trapped charges under similar read conditions to those used for Fig. 3. The figure shows that trapped charges over the channel have a larger effect on the cell's current than charges trapped over the junction edge or over the drain area. The simulation conditions used to derive Fig. 4 are somewhat simplified relative to real cells that were programmed and erased few times. Nevertheless, it can be seen that a ratio of 2 between CG

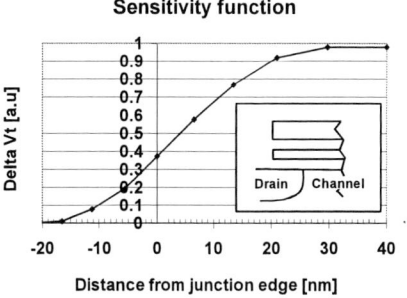

Fig. 4: NROM cell sensitivity function.

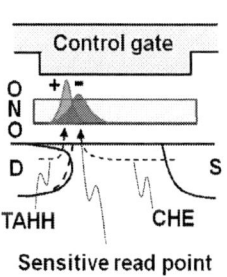

Fig. 5: Schematic illustration of TAHH (hole) injection and electron injection relative to the sensitive read point.

amplitude and RTN amplitude can easily result from the RTN trap being located >10 nm from the metallurgical junction towards the channel while the traps responsible for CG are located at the drain side of the metallurgical junction.

To confirm the last point we erased the same cell with different erase voltages. As shown schematically by Fig. 5, the location of hole injection under increased negative gate bias bends backwards towards the drain. The more negative the gate bias the further backwards will the maximum field be, and the further away will be the point of hole injection. Hence the charge capture responsible for CG will occur further away from the point of maximum cell-current sensitivity, resulting in smaller CG steps.

Fig. 6 shows three CG traces for two different erase conditions. The injections for the three plots were performed intermittently on the same cell with alternating erase conditions, to discriminate aging effects. The figure shows that higher negative gate bias reduces the magnitude of the CG steps, which in turn reduces the total CG magnitude. Fig. 7 summarizes the effect of increased negative gate bias. The figure shows that the amplitude of the CG steps increases when the injection point is moved towards the channel. All experimental points of Fig. 7 were derived with gate/drain voltage combinations that produce same net Vt shift of the erase operation. We conclude that negative erase pulls the maximum field point backwards away from the sensitive read area, thus reducing the magnitude of the CG steps, which are the manifestation of the NBTI recovery steps.

978-1-4244-9113-1/11 $26.00 © 2011 IEEE

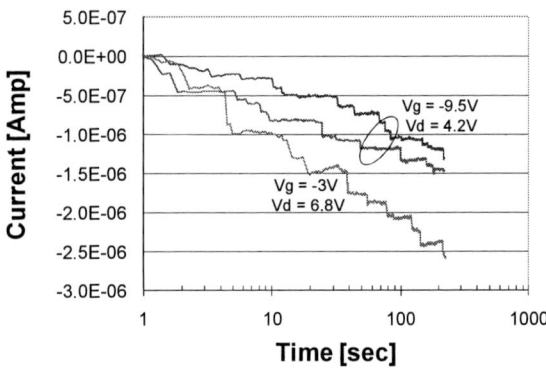

Cell current vs time after erase for different erase bias conditions

Fig. 6: Cell current following two different erase conditions.

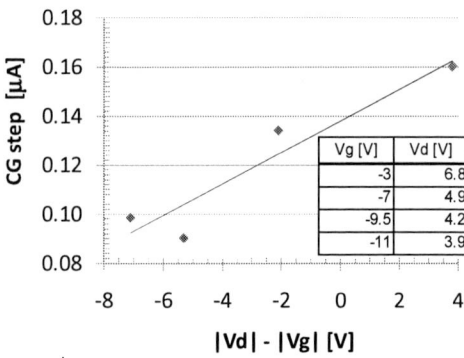

Average CG step height as a function of negativity of erase

Vg [V]	Vd [V]
-3	6.8
-7	4.9
-9.5	4.2
-11	3.9

Fig. 7: Average CG step height as a function of the negativity of the erase bias. The values of Vg and Vd as well as the erase pulse length of each experimental point were selected so the same ΔVt is obtained by the respecive erase operation.

Lateral charge distribution similar to that shown in Fig. 5 above was recently indicated by an RTN-based profiling technique that probed the location of injected charges in similar devices [17].

Our interpretation of the CG effect is further supported by previous study showing that hole injection with negative gate bias {Vg=-10V, Vd=3V} injects the holes' peak 5 nm to the drain side of the metallurgical junction, while positive bias combination {Vg=0V, Vd=8V} injects the holes' peak 4 nm to the channel side [18]. According to the sensitivity function of Fig. 4 this spread in injection locations will account for a factor of 2 in ΔVt per unit charge capture. Indeed Fig. 7 exhibits a change by a factor of 2 of the average height of CG steps for the specified variation of erase negativity.

While we use the results of [18] to support our model, we point out that the final conclusion of [18] regarding the root cause of CG is not supported by the current study. Reference [18] concludes that the lateral gradient of hole concentration in the nitride layer results in lateral hole migration, and the holes in the nitride migrating from over-the-drain to over-the-channel, via the reported sensitivity function, raise the Vt of the cell and cause the observed CG phenomenon. However, had the interpretation of [18] been correct, one would have expected to see a monotonically inclined CG Vt plot. The observed RTN step-like behavior of CG indicates that vertical charge exchange takes place between the silicon and the oxide, contrary to the interpretation of [18].

B. Interaction of CG and RTN events

According to the RTN theory, change in frequency of the RTN indicates a configuration transformation of the oxide bonds at the location of the RTN trap, leading to a change of the trap energy and hence changing its capture and release time constants [19]. The concurrent CG step and RTN event followed by RTN frequency change, as seen in Fig. 3(b), suggests that the combined CG and RTN events, implying a

concurrent 2-electron capture, was associated with an oxide bond transformation in the vicinity of the RTN trap. The same height of the RTN signal before and after the transformation suggests that it is the same trap location before and after the event. While changes of RTN frequency are documented [19], the interaction between a CG step and RTN transformation was not reported before. It suggests that the mechanisms of CG electron capture and RTN electron capture are physically identical, establishing further the link between CG of the third kind and NBTI recovery. Oxide structural transformations were recently proposed in few NBTI studies [9,20,21].

We cannot rule out, though, that the event of Fig. 3(b) is a random coincidence. However, the probability of such random coincidence is low; in our experiments we have not observed any RTN frequency change other than that occurring in coincidence with the CG step. Furthermore, with our sampling rate (1 read per second) and the spacing between CG steps in the vicinity of the frequency change (1500 sec in Fig. 3) we conclude that the probability of the two events to coincide randomly is 1/1500.

Similar relation to Eq. (1) was derived previously for charge gain based on the tunneling front model [2,22]. However, as was previously claimed, the tunneling front model cannot explain NBTI recovery, because NBTI recovery is observed also in thin (~2nm) gate oxides, and this thickness is too small to produce log(*t*) behavior according the tunneling front model over the wide time range observed in NBTI experiments [14]. As pointed out above, log(*t*) dependence of CG is observed over 11 orders of magnitude in devices with tunneling oxide thickness of only 5 nm, which makes the tunneling front model an unlikely explanation for the log(*t*) dependence of CG in trapping-base EEPROM devices. Structural oxide transformations that accompany charge capture / charge detrapping as postulated for NBTI seem a more likely explanation.

978-1-4244-9113-1/11 $26.00 © 2011 IEEE

IV. SUMMARY AND CONCLUSIONS

Our proposed CG model aligns well with a recent NBTI recovery model [8-9]. The relative P/E cycling independency of the CG process in nitride trapping flash memory devices indicates that the oxide traps involved in this mechanism do not accumulate but rather get populated and depleted by each program/erase cycle, i.e. no cumulative damage is accrued in this process.

The similarity of CG phenomenology to NBTI recovery teaches us that the basic recovery physics after high negative bias is the same whether the stress prior to the recovery measurement involved hot hole injection or not. NBTI experiments generally use stress conditions that minimize or prevent hot hole injection. On the contrary, our experiments involved massive hot hole injection through the stressed oxide. The similarity of results indicates that the hot hole injection accompanying the erase conditions does not alter the basic physical process. We further conclude that the erase conditions in trapping-based NVM devices do not create a new damage mechanism other than that used in the interpretation of NBTI experiments, namely, oxide transformations [9,20].

We also learn from the above experiments that 10 years of unbiased storage of nitride trapping NVM devices may cause the Vt of erased bits to rise by up to 300 mV above the level measured 10 ms after erase, irrespective of storage temperature. This level of Vt instability should be mitigated by selecting properly the product operation window.

REFERENCES

[1] N. Mielke, H. Belgal, I. Kalastirsky, P. Kalavade, A. Kurtz, Q. Meng, N. Righos and J. Wu, "Flash EEPROM Threshold Instabilities due to Charge Trapping During Program/Erase Cycling", IEEE Trans. Device and Mat. Rel. 4 (2004) pp 355-344.

[2] T. Wang, W.J. Tsai, S.H. Gu, C.T. Chan, C.C. Yeh, N.K. Zous, T.C. Lu, Sam Pan, and C.Y. Lu, "Reliability Models of Data Retention and Read-Disturb in 2-bit Nitride Storage Flash Memory Cells" IEDM Tech Digest (2003) pp. 169-172

[3] A. Modelli, F. Gilardoni, D. Ielmini and A. S. Spinelli, "A new conduction mechanism for the anomalous cells in thin oxide Flash EEPROMs" 39th International Reliability Physics Symposium (2001) pp. 61-66

[4] T. Wang, N. K. Zous, J. L. Lai and C. Huang, "Hot hole stress induced leakage current (SILC) transient in tunnel oxides," IEEE Elect. Dev. Lett. 19 (1998) pp. 411-413.

[5] A. Shappir, Y. Shacham-Diamand, E. Lusky, I. Bloom, B. Eitan, "Lateral charge transport in the nitride layer of the NROM nonvolatile memory device" Microelectronic Engineering 72 (2004) pp. 426-433.

[6] M. Janai and I. Bloom, "Charge gain, NBTI and Random Telegraph Noise in EEPROM Flash memory devices" IEEE Elect. Device Lett. 31 (2010) pp. 1038-1040.

[7] J. H. Stathis and S. Zafar, "The NBTI in MOS Devices: a Review", Microelec. Reliab. 46 (2006) 270

[8] H. Reisinger, T. Grasser and C. Schlünder, "A study of NBTI by the statistical analysis of the properties of individual defects in pMOSFETS", IEEE International Reliability Workshop (2009) pp. 30-35.

[9] T. Grasser, H. Reisinger, W. Goes, Th. Aichinger, Ph. Hehenberger, P.J. Wagner, M. Nelhiebel, J. Franco and B. Kaczer "Switching oxide traps as the missing link between negative bias temperature instability and random telegraph noise" IEDM (2009) pp. 729-732.

[10] M. Janai and B. Eitan, "The kinetics of degradation of data retention of post-cycled NROM nonvolatile memory products ", in proceedings 43rd Int. Rel. Phys. Symp. (2005) pp. 175-180.

[11] W.J. Tsai, N.K. Zous, C.J. Liu, C.C. Liu, C.H. Chen, T. Wang, S. Pan, C.Y. Lu and S.H. Gu, "Data retention behavior of a SONOS type two-bit storage flash memory cell", IEDM Tech Digest (2001) pp. 719-722.

[12] Y. Roizin, E. Pikhay, and M. Gutman, "Suppression of erased state Vt drift in two-bit per cell SONOS memories," IEEE Electron Device Letters, 26 (2005) pp. 35-37.

[13] M. Janai, B. Eitan, A. Shappir, E. Lusky, I. Bloom and G. Cohen, "Data retention reliability model of NROM nonvolatile memory products", in IEEE Tran. Device and Materials Rel. 4, (2004) pp. 404-415.

[14] H. Reisinger, O. Blank, W. Heinrigs, W. Gustin, and C. Schlünder, "A comparison of very fast to very slow components in degradation and recovery due to NBTI and bulk hole trapping to existing physical models", IEEE Tran. Device and Materials Rel. Vol. 7, No. 1, (2007) pp. 119-129

[15] A. Aichinger, M. Nelhiebel and T. Grasser, "On the temperature dependence of NBTI recovery", Microelec. Reliab. 48 (2008) 1178.

[16] P. Fantini, A. Ghetti, A. Arinoni, G. Ghidini, A. Visconti and A. Marmiroli, "Giant RTS in nanoscale floating gate devices", IEEE Elect. Device Lett. 28, 12 (2007) pp. 1114-1116.

[17] Y.L. Chou, J.P.Chiu, H.C. Ma, T. Wang, Y.P. Chao, K.C. Chen and C.Y. Lu, "Use of random telegraph signal as internal probe to study program/erase charge lateral spread in a SONOS flash memory", IEEE 48th International Reliability Physics Symposium (2010) pp. 960-963

[18] A. Shappir, D. Levy, Y. Shacham-Diamand, E. Lusky, I. Bloom and B. Eitan, "Spatial characterization of localized charge trapping and charge redistribution in the NROM device", Solid State Electr. 48 (2004) pp. 1489-1495.

[19] M. J. Kirton and M. J. Uren, "Noise in solid-state microstructures: A new perspective on individual defects, interface and low-frequency (1/f) noise", Advances in Physics, 38, no. 4, (1988) pp.367-468.

[20] D. Ielmini, M. Manigrasso, F. Gattel and M.C. Valentini, "A new NBTI model based on hole trapping and structural relaxation in MOS dielectrics", IEEE Trans. Electron Devices 56, 9 (2009) pp. 1943-1952.

[21] S. Mahapatra and M. A. Alam, "Defect-generation in p-MOSFETs under negative-bias stress: An experimental perspective," *IEEE Trans. Device Mater. Rel.* 8, 1 (2008) pp. 35–46.

[22] S. Manzini and A. Modelli, "Tunneling discharge of trapped holes in silicon dioxide", in Insulating Films on Semiconductors, Edit. J.F. Verweji and D.R. Wolters, (Elsevier Science Pub. B.V. North-Holland 1983) pp 112-115.

A Highly Reliable Embedded P-Channel SONOS Memory using Dynamic Programming Method

Ying-Je Chen, Cheng-Jye Liu, Chun-Yuan Lo, Yun-Jen Ting, T. H. Hsu, and Wein-Town Sun

Device Engineering Department, Technology Development Division,
eMemory Technology Inc.
8F, No. 5, Tai-Yuan 1st St., Jhubei City, HsinChu County 30265, Taiwan
Phone: (+886) –(3)- 560-1168 ext. 5870, yjchen@ememory.com.tw

Abstract— In this paper we propose a dynamic programming scheme in p-channel SONOS operation in order to achieve high reliability and scalability. By adopting the new method, cell can perform better writing efficiency and suffer less oxide degradation than the conventional CHE programming. Besides, the low programming current under the low program bias also makes simple circuitry design and small charge pumping area.

Keywords; SONOS; P-channel; dynamic programming;

I. INTRODUCTION

More recently, there has been a dramatic increase in the number of applications in embedded flash memory. However, few solutions are available in advanced technology nodes because of process complexity and reliability constraint. A simple and logic based single poly p-channel SONOS technology, NeoFlash® technology, was proposed to fulfill the requirement of embedded non-volatile memory (NVM) applications [1].

With the adoption of channel-hot-electron (CHE) injection programming method, a relatively higher bias on drain side junction to initiate impact ionization is necessary. This often leads to the engineering effort for the optimization of junction breakdown and hot electron efficiency. Besides, without finding the suitable bias condition for higher hot electron injection efficiency, larger channel conduction current by CHE injection programming method results in greater power consumption. Furthermore, the memory cell biased with poor hot electron injection efficiency is sometimes injected by hot hole. Endurance and data retention of this kind of memory cell would become worse.

In this paper, a dynamic programming scheme is presented and demonstrated with 0.13μm P-Channel NeoFlash® memories (with top oxide / Si₃N₄ / bottom oxide thickness = 45/45/35 Å) which are used as a development vehicle as shown in Fig.1 With a programming pulse modification, cell performance with high programming efficiency, low programming current and low bit-line bias operation is achieved. The bottom oxide can suffer less stress by this new approach during writing. As a result, cells perform higher endurance than being written with the conventional approach.

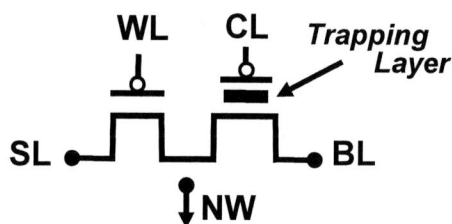

Figure 1. P-channel NeoFlash cell with an ONO charge trapping layer.

II. METHOD

The conventional channel-hot-hole induced hot-electron (CHHIHE) programming methodology was first reported in 1992 [2]. The authors demonstrated that the gate bias of p-channel devices operated near threshold voltage has proper charge injection efficiency as shown in Fig.2. However, as the memory cell is programmed by fixed gate bias, the hole current will also be generated. It is believed that part of reliability degradation is caused by hot-hole injection [3] and become a major concern of charge losses in advanced technology.

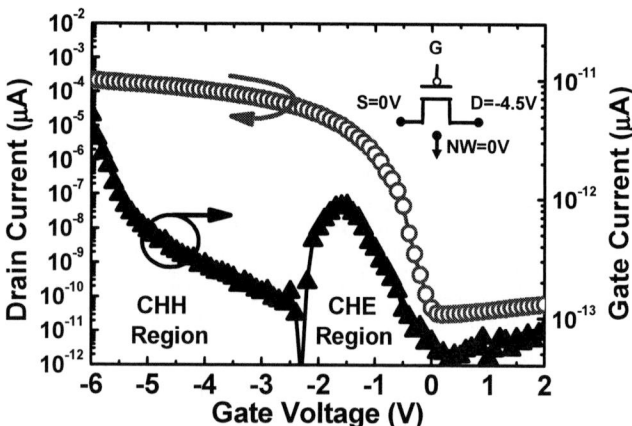

Figure 2. Drain and gate current characteristics of pure PMOS transistors at various gate voltages.

The conventional programming method is applied by a fixed voltage near threshold voltage of erase state on control-line (CL). During a short programming period, cell-Vt becomes more positive and then the gate current moves from electron-favored region toward hole-favored region. That is, the writing efficiency and film reliability are depressed by the large channel hole current. Instead of a fixed bias, a ramped pulse applied on the control-line is to improve the efficiency and reliability.

Figure 3 points out the gate current of conventional method and new approaches. The peak of gate current is gradually moved to the positive value as program bias is proceeding. For conventional method, fixed program bias is applied on CL, so the injected charges include both hot electrons and hot holes as shown in Fig.3 (a). On the other hand, Fig. 3(b) shows the injected charges mainly include hot electrons in new approach because the program bias on CL is dynamically increasing to trace Vt of memory cell during programming.

Figure 4 sketches the waveforms of the dynamic programming which further explain the behavior shown in Fig. 3(b). Different from the conventional operation [1], the CL is applied by a set of step ramping pulses. The typical range of CL voltage is from -0.5V to +2.5V which is dependent on cell-Vt state and limited by the pumping circuitry. The programming procedure would be accomplished within 7us (typical) or 14us (for high endurance operation with program verification) which is relatively shorter than the conventional one (~30us).

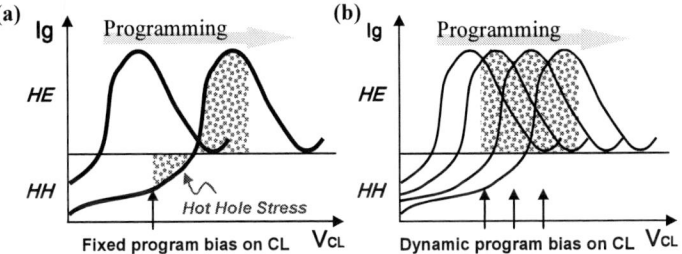

Figure 3. Injection charges diagrams of (a) the conventional programming which induced hot-hole stress (b) the dynamic programming which is always filled by hot-electrons.

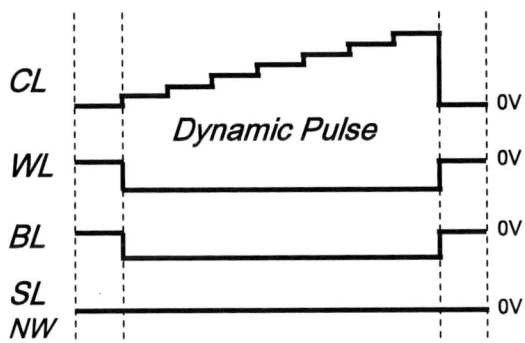

Figure 4. Dynamic programming waveforms.

III. RESULT

The threshold voltages corresponding with programming time indicate that the programming speed of the new approach is much faster than convention as shown in Fig. 5. Besides, the programming current measured by Transient-IV technique [4] also reveals that several times of power consumption is saved

compared with the conventional one shown in Fig. 6. The total program time of new method is equal to 7us (1us for each step). The program was operated at the maximum hot-electrons injection point (as drawn in Fig.2) for each 1us step. The programming current by this new method behaved low while the dynamic pulses ramped.

Due to higher writing efficiency, the bit-line bias can be decreased above 1V to achieve the same programming/erasing Vt window (ΔVt=3V). Programming current of a cell can be decreased from 150uA (BL=-5.5V for the conventional one) to 80uA (BL=-5V for the new one) or even lower value for more optimum program conditions.

Owing to low program current property of the new approach, the word-line (WL) bias can be further decreased for low voltage application. Fig. 7 is the Id-Vd characteristic of word-line transistor under various gate voltages. During program operation, the channel current of the word-line transistor needs to be larger than the program current of memory cell as shown in Fig. 6. Since the program current is reduced to 80uA in new approach, the required WL voltage can be also reduced to -2.5V.

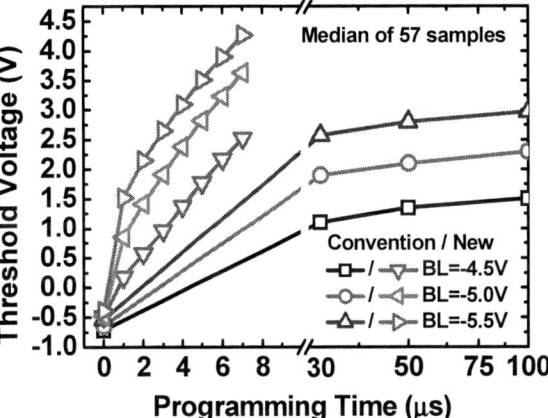

Figure 5. Program transients of the conventional and the dynamic methods at various bit-line voltages.

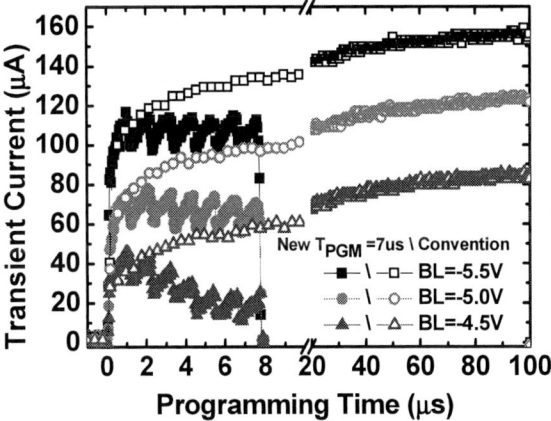

Figure 6. Dynamic PGM current with 7us programming time compared with convention at various bit-line voltages.

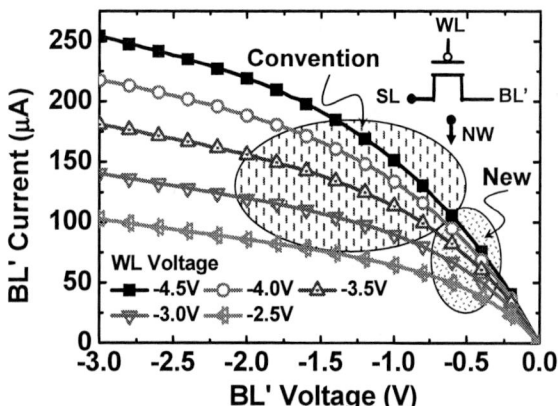

Figure 7. Id-Vd characterization of 1T-WL transistors at various WL voltages.

The cell endurance characteristics of both conventional method and new method are illustrated in Fig. 8. Conventional method means the memory cell is programmed with fixed program pulse width (100us) and higher BL bias (-5.5V). In new method, the memory cell is programmed with shorter pulse width (gradually increased from 7us to 14us through cycling) but lower BL bias (-5.0V). In conventional method, more channel-hot-holes were generated to cause the stress on interface state and make the trans-conductance (Gm) degraded as shown in Fig. 9. Therefore, the Vt level of program state is decreasing during cycling in conventional method because of worse interface state [5]. By using new approach, the endurance characteristic can be kept without obvious degradation owing to less channel hot holes generated to stress the interface state.

Data retention performance under 85°C is shown in Fig. 10. There are 2 major mechanisms responsible for the charge loss of SONOS flash memory cell, one is thermal emission and the other is trap-to-band tunneling through bottom oxide [6]. If the bottom oxide and top oxide are thicker enough or still with good quality after cycling to prevent the stored electrons from tunneling out, thermal emission mechanism will be dominant. The charge loss of 1K cycles by conventional method is worse than the charge loss of fresh cell in Fig.10. This implies some electrons tunneling out through degraded bottom oxide. By using new approach, the charge loss after 10k cycles is relatively smaller than 1k cycles by the conventional one. These findings lead us to believe that there are significant reliability improvements in cells by the new approach.

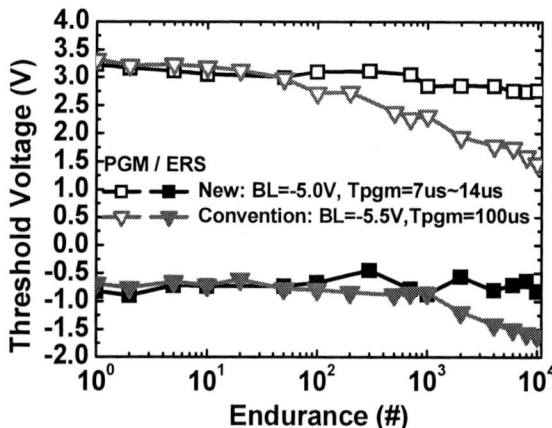

Figure 8. Endurance characterization of the dynamic and the conventional programming methods.

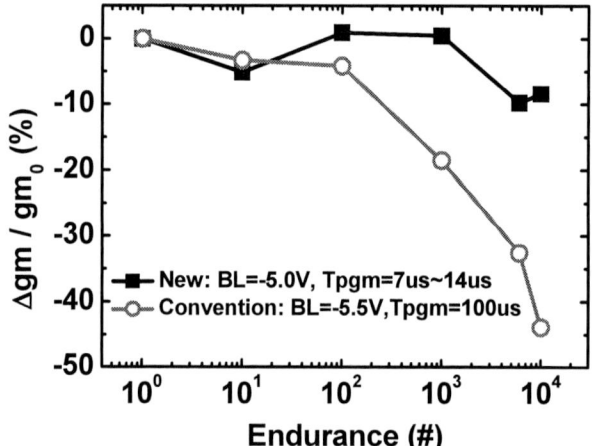

Figure 9. Comparisons of gm degradation while proceeding P/E cycles.

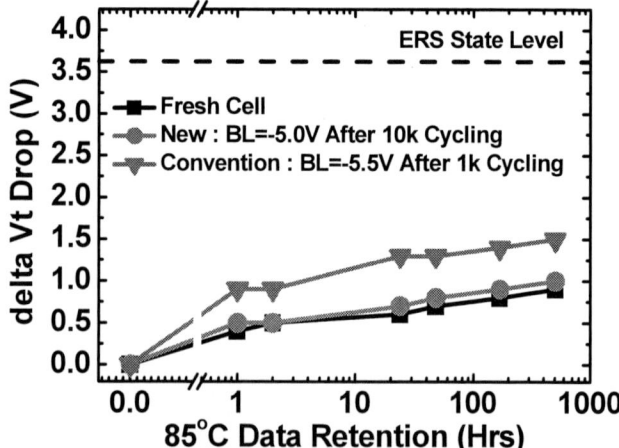

Figure 10. Data retention capability of the dynamic and the conventional methods and used fresh cell which was only stressed by 1-time program as a reference.

978-1-4244-9113-1/11 $26.00 © 2011 IEEE

Figure 11 shows the programming trend on chip-level. With a set of multiple pulses applied on control-line, program distribution after each pulse shows no tail-bits. Figure 12 shows the chip endurance characteristic, and the results are remarkably consistent with the device data above. Even with lower bit-line bias and shorter programming time, the cell current of program state in array by the new method is still higher and there is no obvious current drop compared with the conventional one.

A benchmark table between the new approach and the conventional method was summarized in Table 1. This preliminary study reveals a great reliability improvement by using the new method which provides more applications in growing demands.

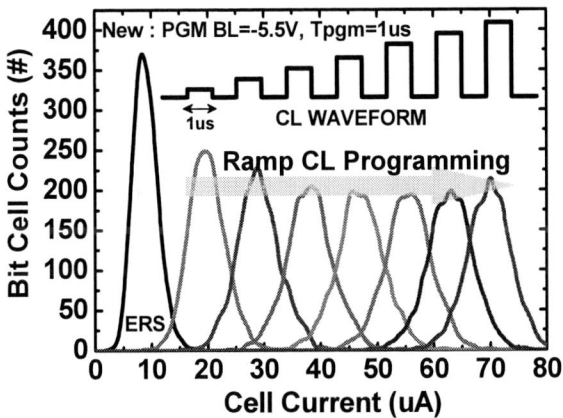

Figure 11. Dynamic programming trend in 2Mb chip array.

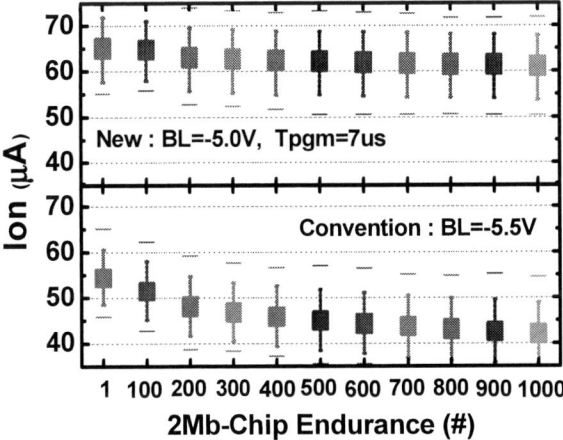

Figure 12. Endurance characterization in 2Mb-Chip without program verification.

IV. CONCLUSION

In this work, a dynamic CHHIHE programming method is proposed. The new scheme successfully leads to larger programming window and suppression of tail-bit issue. Besides, the lower programming current and voltage level greatly prevent the oxide damage. With the current/voltage reduction, the power consumption is greatly reduced as well.

Furthermore, the charge pumping circuitry and chip area could be much saved. Without the effort of changing any existing process or mask layer, the new programming scheme provides a promising approach for embedded non-volatile memory.

Table I. Benchmark Table

Item	Convention	New Approach
Program Power	— (> 150uA/cell)	Low (< 100uA/cell)
Program Speed	— (~ 30us @ TT)	Fast (< 7us @ TT)
Device Scalability	— (BL > 5.5V @ PGM)	Good (BL < 5.5V @ PGM)
Word-Line Effect	— (> 1V)	Small (< 0.5V)
Endurance	— (> 1k Cycling)	Better (> 10k cycling)
Retention	—	Better (Tend to Fresh Cell)
Reliability	—	Better (Less Damage)
Circuit Design	—	Simple & Small

— : Baseline

ACKNOWLEDGMENT

The authors would like to thank Global Foundry for wafer preparing.

REFERENCES

[1] H.M. Lee, L. Lim, S. M. Jung, S.T. Woo, H. M. Chen, C. Y. Lin, R. Shen, C. D. Wang, C. C.-H. Hsu, and S.C. Sun, "NeoFlash – True logic based 0.18um single poly embedded SONOS flash," SSDM, IEEE, 2005, p.196

[2] C.C.-H. Hsu, A. Acovic, L. Dori, B. Wu, T. Lii, D. Quinlan, D. DiMaria, Y. Taur, M. Wordeman, and T. Ning, "A high speed, low power p-channel flash EEPROM using silicon rich oxide as tunneling dielectric," SSDM, IEEE, 1992, p.140

[3] A. Bravaix, D. Goguenheim, N. Revil, and E. Vincent, "Comparison of low leakage and high speed deep submicron PMOSFETs submitted to hole injcetions," IRW Final Report, IEEE, 2002, p.14-20

[4] Ying-Je Chen, Yun-Jen Ting, Cheng-Jye Liu, Wein-Town Sun, and Rick Shen, "Precision programming power control in embedded p-channel SONOS flash using transient-IV method," SSDM, IEEE, 2009

[5] Sung-Rae Kim, Kyung Joon Han, Kin-Sing Lee, Pavan Singaraju, Rophina Li, Patty Liu, Yingbo Jia, Ben Schmid, Yu Wang, Fethi Dhaoui, Frank Hawley, and Huan-Chung Tseng, "Cycling impact on the gm degradation and GIDL current of 65nm 2T-embedded flash memory," NVMTS, IEEE, 2009, p.77-79.

[6] A. Arreghini, N. Akil, F. Driussi, D. Esseni, L. Selmi, and M.J van Duuren, "Long term charge retention dynamics of SONOS cells," Solid-State Electronics, 2008, p.1460-1466.

Analysis of Edge Wordline Disturb in Multimegabit Charge Trapping Flash NAND Arrays

Cristian Zambelli, Andrea Chimenton* and Piero Olivo
Università degli Studi di Ferrara, Dipartimento di Ingegneria
Via Saragat 1, Ferrara, 44122, Italy
Email: cristian.zambelli@unife.it

Abstract—The Edge Wordline Disturb (EWD) represented a reliability issue on traditional Flash NAND memories, evidenced as an unwanted positive threshold voltage shift of all the cells belonging to the first wordline (WL0) connected to the Ground Select Transistor (GSL). In this work, throughout the experimental characterization of Multimegabit arrays it has been investigated the presence and the physical nature of the EWD in Charge Trapping (CT) NAND Flash, emphasizing its dependency on parameters such as the programming voltage, the inhibit voltage and device aging.

I. INTRODUCTION

The constant scaling of traditional Floating Gate (FG) Flash memories is facing several limitations, ascribed both to architectural and physical issues [1]. One of the promising candidates for replacement of high density FG Flash NAND memories is represented by the Charge-Trapping (CT) memories [2]. However, such a technology has to overcome some reliability issues that have been a sore for the traditional FG memories, such as the so-called Edge Wordline Disturb (EWD) [3] occurring during programming.

In fact, in a standard NAND architecture (see Fig. 1), a disturb affects the cells belonging to the first wordline (WL0) connecting the cell strings to the string selector GSL. This disturb is evidenced as a difference between the average threshold voltage $< V_{T\_WL0} >$ of the cells belonging to WL0 and the average threshold voltage $< V_T >$ of all other cells.

The difference between cells belonging to WL0 and the other cells can be ascribed to three effects: i) different potentials at their terminals with respect to the other cells depending on the specific WL selected for programming; ii) a different cell geometry due to the fact that these cells are located between a cell and a transistor (differently from cells belonging to WL1 ÷ WL30), therefore with a different field underneath their channels and a modified programming dynamics; iii) the possible presence of a large GIDL (Gate Induced Drain Leakage) current generated at the drain edge of GSL transistors due to their drain potential raised by channel boosting [4]: such a field can efficiently trigger electron-hole pair generation followed by an acceleration of the electrons toward the channel of WL0 cells. These electrons can be

*Present address: Intel Corporation, USA.

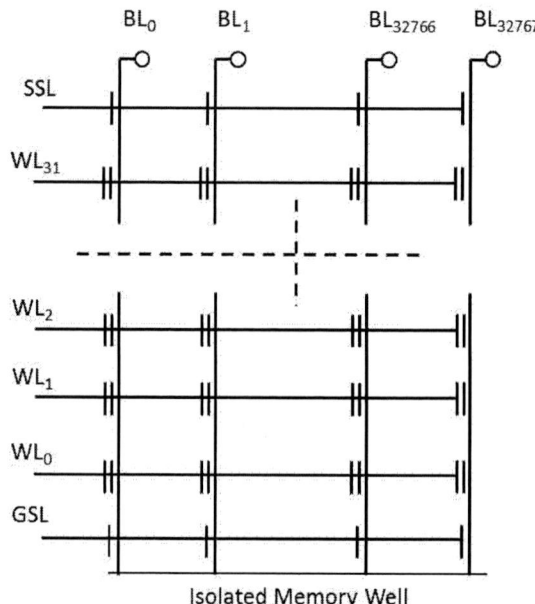

Fig. 1. Standard NAND array architecture. The figure represents a single block of the CT array considered in this work.

injected into the floating gate of these cells, thus provoking an undesired increase of their threshold voltages.

In this work we have experimentally analyzed the EWD in CT NAND Flash arrays during programming. In particular EWD has been measured by comparing $< V_{T\_WL0} >$ with respect to $< V_T >$ as a function of the programming voltages applied to the selected WL, the device aging represented as the number of write/erase cycles, the number of partial programming operations (NOP) and the voltage applied to unselected cells to inhibit their programming.

II. EXPERIMENTAL SETUP

The experimental data have been obtained by testing 4Mbits CT NAND Flash arrays with a CT version of the Active Technologies RIFLE-SE Automated Test Equipment (ATE) suitable for applying arbitrary waveforms, thus fully controlling the voltages applied during write and read operations. The array is

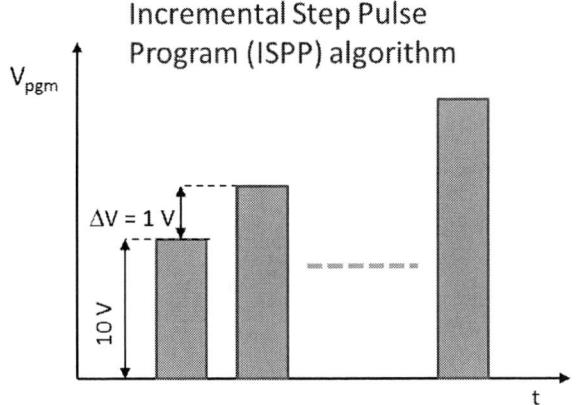

Fig. 2. ISPP Pulse characteristics exploited in this work. The duration per pulse is $4\mu s$.

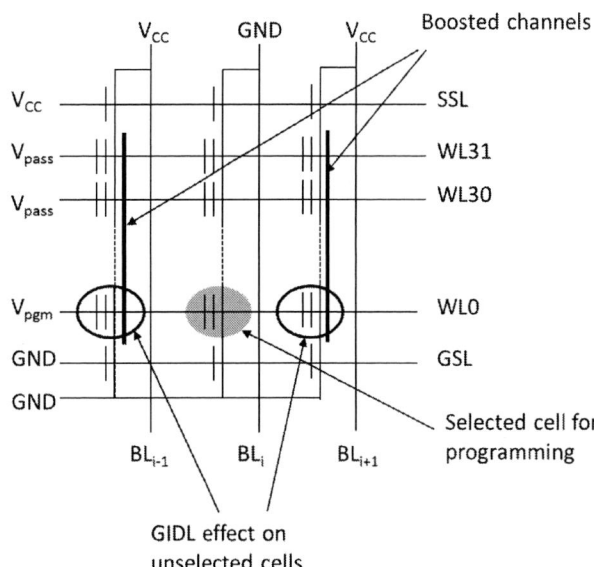

Fig. 3. Bias conditions possibly activating the GIDL effect from GSL transistors belonging to columns BL_{i-1} and BL_{i+1}.

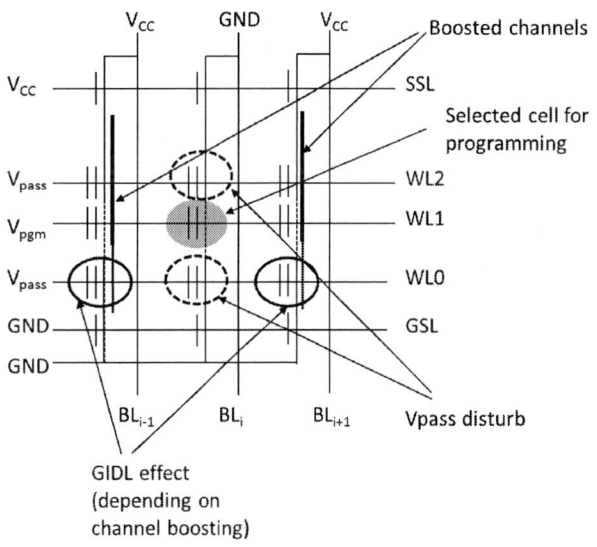

Fig. 4. Bias conditions possibly activating the GIDL effect from GSL transistors belonging to columns BL_{i-1} and BL_{i+1} and the V_{pass} disturb in unselected cells belonging to columns BL_i.

arranged into four blocks, each one sized 1Mbits and organized with 32 wordlines and 32768 bitlines. The memory cells feature a p-Si/SiO$_2$/Si$_3$N$_4$/Al$_2$O$_3$ stack overwhelmed by a high work function TaN/Ti/TaN metal gate. The program operation is performed page-wide (mapped as an entire 32Kbits word-line) by using an Incremental Step Pulse Program (ISPP, see Fig. 2) algorithm with a starting voltage of 10V and 1V steps with $4\mu s$ duration, whereas keeping the unselected wordlines at a constant intermediate V_{pass} voltage. In order to minimize interbitline capacitive coupling noise during a page Program, a dual bitline scheme has been used, in which only the even or the odd bitlines are activated simultaneously in the array. When WL0 is selected for programming, the bias situation is that depicted in Fig. 3 and the unselected cells are affected by the GIDL phenomenon. When other bitlines are selected for programming (see for example the case depicted in Fig. 4 where WL1 is selected for programming), the cells belonging to WL0 are affected either by the GIDL effect or by a V_{pass} disturb. In the latter case, however, the boosted channel voltage may be different for cells above or below the selected one, depending on the actual voltage applied to the selected WL. These effects, controlling the actual program/inhibit dynamics, will be detailed in the next section.

The erase operation is performed block-wide (1 Mbits) with a single voltage pulse featuring 19V amplitude and 100 μs duration. All the cells within the array are in the erased state before every experiment.

III. EDGE WORDLINE DISTURB DEFINITION

As introduced in the previous paragraphs, the Edge Word-line Disturb (EWD) magnitude, indicated as ΔV_{T_WL0}, has been measured as:

$$\Delta V_{T\_WL0} = <V_{T\_WL0}> - <V_T> \qquad (1)$$

where $<V_{T\_WL0}>$ is the average threshold voltage of all the cells belonging to row WL0, and $<V_T>$ is the average threshold voltage of all the other cells of the array.

As stated in the introduction, a first contribution to EWD is attributed to the different programming dynamics affecting cells belonging to WL0 with respect to all other cells and related to their different geometries, therefore to a different field distribution that allows cell programming with lower ISPP voltages.

Cells belonging to WL0 are also characterized by a peculiar field distribution in the space region between their source diffusion and the drain diffusion of the GSL selector making

978-1-4244-9113-1/11 $26.00 © 2011 IEEE

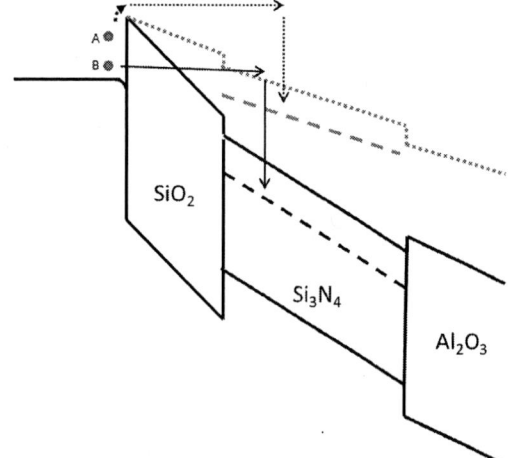

Fig. 5. Band structure of the CT cells (excluding the TaN/Ti/TaN metal gate) considered in this work. For high biases electron B is free to tunnel into SiO_2 and then gets trapped into Si_3N_4 layer. For low biases electron A is still able to get trapped, although the energy levels of the bands are shifted upwards.

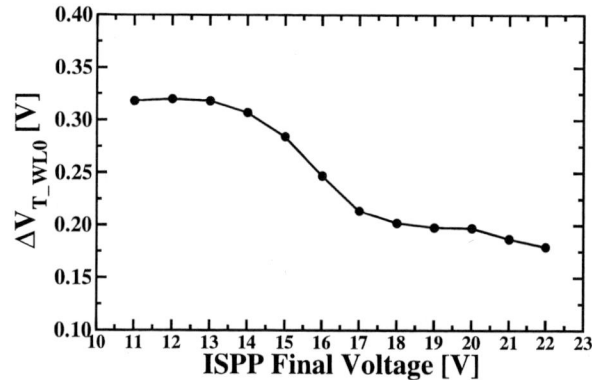

Fig. 6. EWD dependence on the exploited ISPP voltages during a program operation. V_{pass} has been fixed at 8 V for the unselected wordlines.

possible the presence of a GIDL current. The GIDL is a major leakage mechanism occurring in OFF MOS transistors when high voltages are applied to the Drain and it is attributed to tunneling taking place in the deep-depleted or even inverted region underneath the gate oxide. Such a phenomenon has been revealed in FG NAND Flash during the Program operation [5] and it is theoretically present also in CT NAND Flash, as the basis of the Program operation are shared.

The GIDL effect may interest the cells belonging to WL0 in two different bias conditions: i) when WL0 is selected for programming and the channel of the inhibited bitlines is locally self-boosted (see Fig. 3); ii) when WL0 cells act as pass transistors and the bitlines are inhibited through channel self-boosting (see Fig. 4). Under these bias conditions large GIDL current may be generated at the drain edge of GSL transistor because the potential at the drain node of GSL transistor is raised by channel boosting [5]. Then, electron-hole pair generation follows and the generated electrons are accelerated at the GSL-WL0 space region which can be hot enough to be injected into the floating gate in FG NANDs or trapped into the Si_3N_4 layer of the CT NANDs WL0 cells. This causes a positive shift of the average threshold voltage of the cells belonging to WL0, whereas keeping unmodified the average threshold voltage of all the other cells of the array. The electron injection and trapping depend on the relative values of the WL0 voltage and that of the self-boosted channel. In Fig. 5 it is shown the band diagram for the CT NAND devices considered in this work. For high gate biases (i.e. when WL0 is programmed) the electrons tunnel through the SiO_2 and get trapped into the Si_3N_4 layer, whereas for low gate biases (i.e. when WL0 is inhibited) it is still possible for high energy electrons to pass the SiO_2 barrier and subsequently get trapped.

The difference on the impact of the EWD on the mem-

ory reliability between standard FG and CT Flash NANDs principally relies on the fact that FG NANDs depend only on tunneling efficiency, whereas CT NANDs depend also on trapping efficiency.

The last marginal contribution to EWD is caused by the V_{pass} disturb that affects cells that are inhibited for programming (Gate voltage at V_{pass} and bitlines at GND), as in Fig. 4. The impact of V_{pass} disturb, that affects all cells acting as pass transistor because of an unwanted soft programming, is different for cells belonging to WL0 because of their different geometries and, therefore, coupling ratios.

IV. EXPERIMENTAL RESULTS

Different experiments have been carried out to analyze the EWD in CT NAND arrays, and to expose its dependence on bias conditions.

A. Dependency on WL0 programming

In order to discriminate from the presence of other indeterminate sources of measurements perturbation, a bulk Program operation has been executed on a whole memory block, by varying the final voltage of the ISPP algorithm.

Fig. 6 shows the dependency of EWD on the voltage steps exploited in ISPP retrieved by measuring the threshold voltage of all cells within a block. The plot evidences that for the first ISPP steps the disturb magnitude is in the range of 300mV, and it decreases for higher programming voltages. At low voltages indeed, the large EWD is justified by the geometrical and the coupling ratios mismatches that accelerate the program dynamics in WL0 cells. At high voltages, on the contrary, when the programming transient approaches a saturating trend, EWD is justified by the GIDL effect.

B. Dependency on WL0 inhibition

The presence of EWD has been analyzed when WL0 is driven into an inhibit condition. However, rather than analyzing all possible cases, the analysis has been focused on a repeated programming of WL1. In fact, in some applications such as multimedia and embedded systems, the multiple writing of a single wordline is allowed in NAND memories in

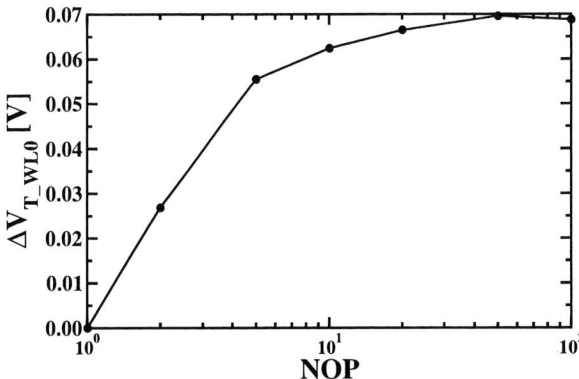

Fig. 7. EWD dependence on the number of performed NOP on WL1. V_{pass} has been fixed at 8 V for the unselected wordlines.

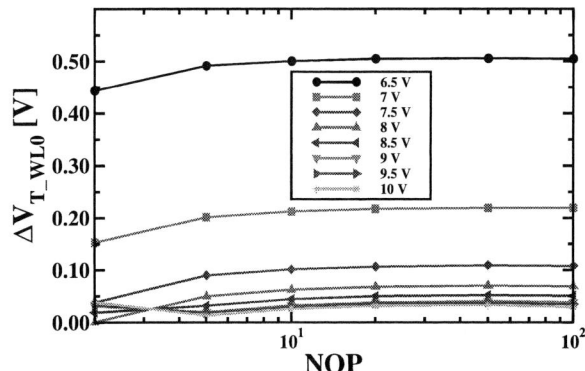

Fig. 8. EWD as a function of the NOP number. V_{pass} has been varied in order to evaluate the dependence of the phenomenon on the pass voltage.

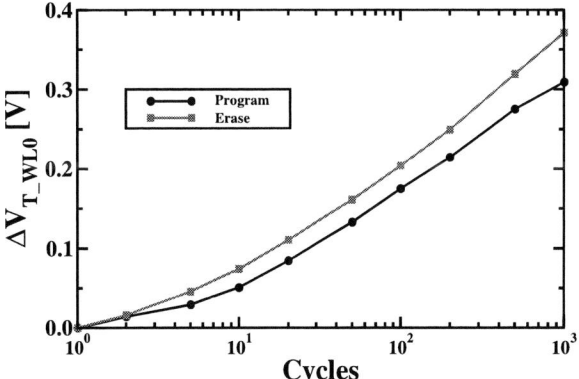

Fig. 9. EWD dependence on the number of write cycles performed on the CT NAND array

order to optimize the usage of the memory addressing space. This policy is called *partial programming* and the number of multiple writing operations sustained by the memory is called NOP [5]. The impact of EWD on CT NAND Flash arrays has been analyzed with respect of NOP number, by applying the bias configuration indicated in Fig. 4.

By programming only cells belonging to WL1, while keeping all other wordlines at V_{pass}, it is possible to evaluate two different EWD causes: GIDL and V_{pass}-induced disturb. Fig. 7 shows EWD magnitude as a function of NOP number. In this case the EWD has been calculated as:

$$\Delta V_{T\_WL0} = <V_{T\_WL0}> - <V_{T\_WL2}> \qquad (2)$$

where $<V_{T\_WL2}>$ is the average threshold voltage of the cells belonging to WL2. The choice of limiting the reference cells on which calculate EWD to those belonging to WL2 is justified by the need of maximizing the differences between the two bitlines adjacent to the selected one. Results in Fig. 7 show that by applying a number of 200 NOP on cells belonging to WL1, whereas keeping V_{pass} voltage at 8V, an EWD of about 70mV is detected. The dependency on NOP number shows a logarithmic trend on the very first operations performed on WL1, whereas it reaches a steady state after a large number of operations.

C. Dependency on V_{pass}

Starting from the theoretical base of the previous experiments, the EWD can also be analyzed as a function of V_{pass}. As in the previous experiment, only the cells belonging to WL1 are programmed (up to 200 NOP) and a V_{pass} bias is applied to the other wordlines. After each set of 200 NOP, the entire array is erased and the experiment is repeated by applying a different V_{pass} voltage. The V_{pass} ranges from 6.5 V to 10 V with 0.5 V step.

As shown in Fig. 8, a maximum EWD of 500 mV has been measured for the lowest V_{pass} bias whereas it decreases for higher V_{pass} voltages. A detailed EWD analysis must take into account the actual band structure within the region including the GSL selector and the 3 cells in series belonging to WL0,

WL1 and WL2, together with the band structure modification induced by coupling effects and charge trapping.

D. Dependency on cycling

The EWD dependency on cycling has been evidenced by stressing the CT array by a sequence of 1000 bulk Program/Erase cycles with intermediate Read operations for retrieving the threshold voltage values of all the cells. The EWD has been calculated by using equation (1), since bulk programming bias configuration has been considered. A measure of EWD is available both for the Programmed state and the Erased state of the cells belonging to WL0, as the performed Read measurements took place after each operation.

As shown in Fig. 9, the EWD initially increases the charge trapped in the Si_3N_4 layer: such charge cannot be efficiently removed by subsequent Erase operation and a cumulative voltage shift in cells belonging to WL0 is evidenced. No saturating trend can be depicted after 1000 cycles.

After cycling the array the experiment for evaluating the EWD dependency on the WL0 inhibition has been repeated, in order to compare the EWD impact on cycled devices with respect to fresh devices. As shown in Fig. 10, the EWD impact is greater on cycled devices. This can be attributed to the

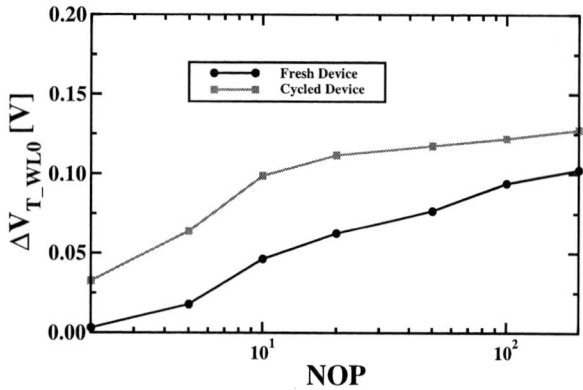

Fig. 10. EWD as a function of the NOP number. A dependence on the device aging is evidenced. V_{pass} has been fixed to 8 V for the unselected wordlines.

higher number of traps generated during cycling, which allows for a higher number of electron to be trapped thus causing a further upward threshold voltage shift.

V. CONCLUSIONS

In this work we have analyzed the presence of the Edge Wordline Disturb in CT NAND Flash arrays, by measuring its effect through a set of dedicated experiments. The results evidenced that such a phenomenon depends on different causes such as the GIDL effect, geometrical and coupling ratio mismatches and actual bias conditions.

The dependency of EWD in CT NAND Flash on the

programming voltage, the WL0 inhibition, the V_{pass} and aging of the devices has been extensively analyzed, showing that such a phenomenon could represent a long term reliability issue especially when the inhibit policies for the Program operation are not opportunely optimized and the device becomes heavily cycled. Future works will be targeted to device simulations of CT structures, in order to qualitatively support the statements about EWD's nature and to quantify its impact on scaled technologies.

ACKNOWLEDGEMENTS

The authors of this paper would like to thank P. Tessariol for helpful discussion.

This work was partially supported by European Commission under the FP7 research contract 214431 GOSSAMER.

REFERENCES

[1] R. Micheloni, L. Crippa, and A. Marelli, *Inside NAND Flash memories.* Springer-Verlag, 2010.
[2] C. Lee, J. Choi, C. Kang, Y. Shin, J. Lee, J. Sel, J. Sim, S. Jeon, B. Choe, D. Bae, K. Park, and K. Kim, "Multi-level nand flash memory with 63 nm-node tanos (si-oxide-sin-al2o3-tan) cell structure," in *VLSI Symp. Tech. Dig.*, 2006, pp. 21–22.
[3] K. Park, S. Lee, J. Sel, J. Choi, and K. Kim, "Scalable wordline shielding scheme using dummy cell beyond 40 nm nand flash memory for eliminating abnormal disturb of edge memory cell," *Jpn. J. Appl. Phys.*, vol. 46, pp. 2188–2192, 2007.
[4] T. Melde, M. Beug, L. Bach, A. Tilke, R. Knoefler, U. Bewersdorff-Sarlette, V. Beyer, M. Czernohorsky, J. Paul, and T. Mikolajick, "Select device disturb phenomenon in tanos nand flash memories," *Electron Device Letters, IEEE*, vol. 30, pp. 568–570, 2009.
[5] J. Lee, C. Lee, M. Lee, H. Kim, K. Park, and W. Lee, "A new programming disturbance phenomenon in nand flash memory by source/drain hot-electrons generated by gidl current," in *Proc. NVSM Workshop*, 2006, pp. 31–33.

Investigation of the Programming Accuracy of a Double-Verify ISPP Algorithm for Nanoscale NAND Flash Memories

Carmine Miccoli, Christian Monzio Compagnoni, Alessandro S. Spinelli* and Andrea L. Lacaita*

Dipartimento di Elettronica e Informazione, Politecnico di Milano–IU.NET,
piazza L. da Vinci 32, 20133 Milano, Italy; e-mail: miccoli@elet.polimi.it
* also with IFN-CNR, Milano, Italy

Abstract—**This paper presents a detailed investigation of the performance of a double-verify algorithm for accurate programming of deca-nanometer NAND Flash memories. In order to minimize the programmed threshold-voltage distribution width in presence of discrete and statistical electron injection, a weakened programming step is applied to cells if their threshold voltage falls between a low- and a high-program-verify level during incremental step pulse programming. Clear improvements are shown with respect to the single-verify case, with minimal burdens on programming time and complexity.**

Index Terms—**Flash memories, program verify, electron injection statistics, semiconductor device modeling.**

I. INTRODUCTION

Accurate programming of NAND Flash memories is usually obtained by the incremental step pulse programming (ISPP) algorithm [1], consisting in the application to cell control-gate (with grounded bulk and channel) of short programming pulses of equal duration τ_s and increasing amplitude. This algorithm allows very tight threshold voltage (V_T) distributions to be obtained when a constant increase V_s is given to the control-gate pulses, leading to an average V_T variation per step ($\Delta V_{T,s}$) rapidly converging to V_s [1]–[4]. In this case, inserting a verify operation after each pulse and stopping the algorithm when cell V_T exceeds the desired program-verify (PV) level, a maximum width of the programmed V_T distribution equal to V_s should theoretically be obtained, regardless of the width of the neutral V_T distribution. However, this result was shown to be compromised by the electron injection statistics (EIS), introducing a statistical spread in $\Delta V_{T,s}$ and allowing the cells to be displaced from the PV level more than V_s [2], [3], [5], [6]. The severe scaling of the NAND technology, and in particular of the control-gate to floating-gate capacitance (C_{pp}) [7], has in fact increased the impact of single electrons stored in the floating gate on cell V_T, reducing, in turn, the number of electrons to be transferred to accomplish the program operation [6], [8], [9]. As a consequence, the statistical process ruling the granular electron injection into the floating gate during programming has become a major source of dispersion for the final cell V_T [2], [3], [5], [6]. The possibility to overcome the single-verify (SV) accuracy limitations by means of more complex double-verify (DV) algorithms [10], [11] has never been clearly assessed so far.

In this paper we investigate an ISPP-based DV algorithm, considering its ability to tighten the V_T distribution in presence of EIS. The algorithm compares cell V_T with two PV levels, namely a high-PV (HPV), used to determine the end of the program operation, and a low-PV (LPV) level: in the case cell V_T falls between LPV and HPV, a positive bit-line bias (V_{BL}) is applied to the selected string to reduce the V_T growth when the next ISPP pulse is applied. In so doing, the programmed V_T distribution can be tighter than V_s, trading-off the programming speed with a better programming accuracy. In order to correctly evaluate the algorithm performance, the programmed V_T distribution width is studied by means of Monte Carlo (MC) simulations for the electron injection process, therefore carefully accounting for the EIS spread. Results show clear improvements in the programmed V_T distribution width, with minimal burdens on the programming time and complexity.

II. DOUBLE-VERIFY ISPP ALGORITHM

A. Algorithm description

The basic features of SV ISPP of NAND arrays are schematically shown in Fig. 1: when a sequence of programming pulses whose amplitude has a constant increase V_s is applied to the selected word-line, cells below the PV level (namely, **A** and **B** in the figure) display an average V_T increase per step $\overline{\Delta V_{T,s}} = V_s$, while cells having $V_T >$ PV (namely, cells **C**) preserve their V_T state thanks to the inhibit bit-line voltage $V_{BL} = V_{CC}$ applied for channel boosting [12]. In so doing, the width of the programmed V_T distribution is mainly set by cells **B**, *i.e.* those that are closer to the PV level before overcoming it, on average, by V_s.

The basic idea for a DV ISPP algorithm [10], [11] is to reduce the V_T shift of the cells that come in close proximity to the PV level. To this aim, the conventional ISPP algorithm is modified as schematically depicted in Fig. 2: cell V_T is compared against two PV levels, namely HPV and LPV=HPV-αV_s, with $0 < \alpha < 1$, and three possible values of the bit-line bias are applied at the next programming pulse: (1) $V_{BL} = 0$ V for cells with $V_T <$ LPV (region **A** of the V_T axis in Fig. 2), (2) $V_{BL} = (1-\beta)V_s$ for cells with LPV $< V_T <$ HPV (region **B**) and (3) $V_{BL} = V_{CC}$ for cells with $V_T >$ HPV (region **C**). Cells in region **A** are considered sufficiently far from the end

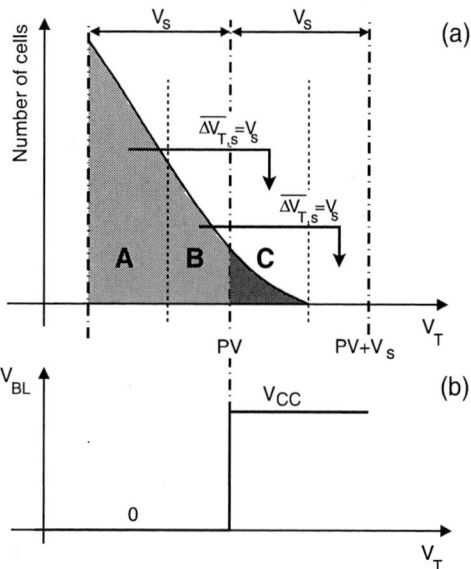

Fig. 1. Schematic for the single-verify algorithm (a) and bit-line bias applied during the next programming pulse as a function of V_T (b): the bit-line is kept to ground for cells below the PV level (region **A** and region **B** of the V_T axis), while V_{BL} is raised to V_{CC} for program inhibit when cells overcome the PV level (region **C**).

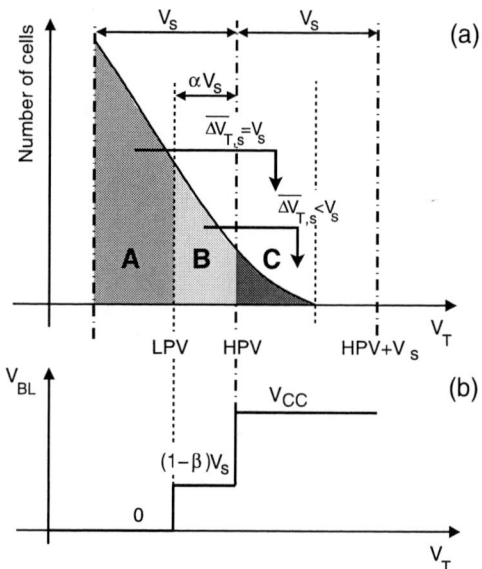

Fig. 2. Schematic for the double-verify algorithm investigated in this work (a) and bit-line bias applied during the next programming pulse as a function of V_T (b): the bit-line is kept to ground for cells that fall in region **A** of the V_T axis, while $V_{BL} = (1 - \beta)V_s$ is applied if cell V_T is intermediate between the LPV and HPV levels (region **B**). When cells overcome the HPV level (region **C**), V_{BL} is raised to V_{CC} for program inhibit.

of programming, obtained when V_T overcomes the HPV level, to withstand an average increase of their V_T equal to V_s. As a consequence, their bit-line is kept to ground when the next word-line pulse is applied as in the standard ISPP algorithm. Cells whose V_T is above the LPV but below the HPV level (region **B**) are instead close to the end of programming: a positive bit-line bias $V_{BL} = (1 - \beta)V_s$ (with $\beta \leq 1$) is applied in this case during all the following programming pulses. The bit-line bias applied to cells **B** aims at reducing the increase of their tunnel-oxide electric field during the next pulses and should, however, be sufficiently low to keep the string-select transistor ON during program. In so doing, these cells will experience an average V_T increase that is lower than V_s, therefore limiting the maximum displacement they can have from the HPV level when they overcome it. Finally, cells having V_T beyond the HPV level (region **C**) have reached the end of their programming transient and no further V_T increase is needed. Therefore, their bit-line starts to be biased at V_{CC} to inhibit the effect of the following word-line pulses.

B. Effect of a bit-line bias on the ISPP transients

Figs. 3-4 display the average behavior out of 10^3 MC simulations for the V_T transient during ISPP, in the case of $\beta = 1$, 0.5 and 0. The MC model, originally presented and validated against experimental data in [2] for SV ISPP, reproduces the discrete electron injection during programming, correctly describing the tunnel-oxide field variations due to (1) electron storage in the FG, (2) word-line bias increase and, in the case of the DV algorithm, (3) bit-line bias increase (more details on the simulation procedure will be given in

the next Section). The same neutral V_T was initially assumed for the cells. Fig. 3 shows the staircase waveform applied to the selected word-line ($V_s = 500$ mV) and the simulated average V_T evolution in the case $\beta = 1$ (circles), clearly displaying $\overline{\Delta V_{T,s}} = V_s$. Note, in fact, that $\beta = 1$ represents the reference programming transient, with no bit-line bias applied, and should be compared to the $\beta = 0.5$ and $\beta = 0$ cases, corresponding to $V_{BL} = 250$ mV and 500 mV. The bit-line bias was always applied since step 18 in the figure: in the following steps, the V_T transients for $\beta = 0.5$ and $\beta = 0$ depart from the reference transient, first showing a slower programming for some steps and then growing parallel to the $\beta = 1$ curve, with only a horizontal shift depending on the selected β value. This is also clearly highlighted in Fig. 4, where the average $\Delta V_{T,s}$ abruptly moves from V_s to a quite lower value after step 18. The reduction of $\overline{\Delta V_{T,s}}$ is due to the application of the bit-line bias, increasing the channel potential and decreasing the tunnel oxide field. This is, however, only a transient effect, with $\overline{\Delta V_{T,s}}$ that increases and converges again to V_s after some programming steps. This is due to the constant increase V_s of the word-line pulses, leading to the recovery of the stationary working point for the tunneling current [3].

C. Algorithm design parameters

The reduction of $\overline{\Delta V_{T,s}}$ immediately after the application of the bit-line bias in Fig. 4 can be exploited to tighten the programmed V_T distribution. In fact, if the cells overcome the HPV level during the transient phase during which $\overline{\Delta V_{T,s}} < V_s$, their maximum displacement from this level at the end

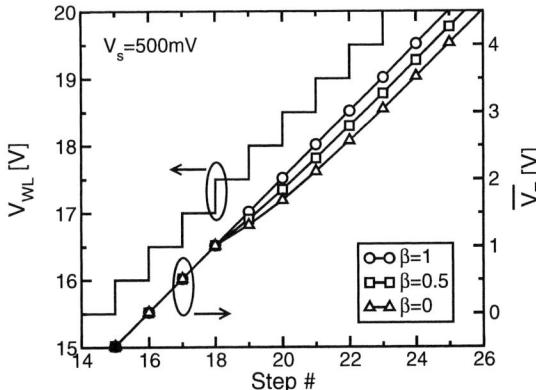

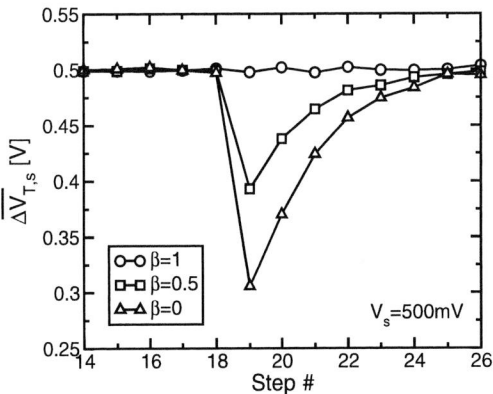

Fig. 3. Word-line bias applied during ISPP of the NAND cells (V_s = 500 mV). The resulting average V_T evolution from 10^3 MC simulations is also shown for the case $\beta = 1$ (conventional ISPP algorithm), $\beta = 0.5$ and $\beta = 0$, starting applying the bit-line bias at step 18.

Fig. 4. Average $\Delta V_{T,s}$ from the transients shown in Fig. 3, as a function of the programming step. At step 18 the bit-line bias is applied, giving rise to a clear reduction of $\overline{\Delta V_{T,s}}$ in the next steps with respect to V_s.

of programming can be reduced. This can be obtained when the HPV is not too far from the LPV level, determining the initial step for the application of V_{BL}. Otherwise, if the HPV level is not overcome during the transient V_T growth following the application of V_{BL}, the same width of the V_T distribution of the conventional ISPP algorithm is obtained, due to the convergence of $\overline{\Delta V_{T,s}}$ to V_s. In addition to that, note that the achievable reduction of $\overline{\Delta V_{T,s}}$ in Fig. 4 depends on β, with lower β giving a more abrupt decrease of $\overline{\Delta V_{T,s}}$ and the possibility for tighter V_T distributions to be obtained. All these considerations suggest that the programmed V_T distribution width obtained by the DV algorithm can be optimized by a careful selection of both α and β. For this optimization, the maximum number of pulses required by cells receiving the bit-line bias to overcome the HPV level is also a very important parameter. In fact, if very low $\overline{\Delta V_{T,s}}$ are obtained but the HPV is too far from the LPV level, a large number of programming pulses may be required by the cells to reach the end of the program operation, compromising the programming speed. All these points should be carefully considered for a proper design of the DV programming scheme.

As a final remark, note that the optimization of the algorithm in terms of α and β requires the width of the programmed V_T distribution to be investigated including all the sources of statistical spread that may compromise the algorithm accuracy. Spread sources may impact either the verify operation or the charge transfer to the floating gate during the programming pulses: precision of V_T sensing, stability of the PV levels and RTN [5], [13]–[18] are among the main accuracy constraints of the former group, while the latter includes EIS and erratic behaviors [2], [3], [19], [20]. In the next Sections we will focus our attention only to EIS, as this is related to device physics and not to the sensing circuitry, and, differently from erratic phenomena, impacts the programmed V_T distribution at high probability levels [21], [22].

III. PROGRAMMING ACCURACY IN PRESENCE OF EIS

A. Simulation methodology

In order to investigate the programming accuracy of the DV ISPP algorithm, we referred to a 32 nm NAND technology, simulating the ISPP operation from the erased cell state in a MC fashion, taking into account EIS limitations. Each MC run consisted in the extraction of initial cell V_T from a gaussian distribution with average value equal to the erased level and standard deviation corresponding to the dispersion of neutral cell V_T of the technology [23]. We then simulated the electron transfer process to the floating gate extracting the time delay between one electron injection to the next from an exponential distribution with average value q/I_t, where I_t is the tunneling current through the tunnel oxide, which is a function of the floating gate potential V_{FG}. In order to carefully reproduce the electron injection process during ISPP, V_{FG} and I_t were updated both after each electron storage in the floating gate and at the beginning of each programming pulse, when the word-line and, in some cases, the bit-line bias is modified [2]. In so doing, simulations account for the non-homogeneous nature of the Poisson process ruling electron injection from the substrate to the floating gate [3], [24], and can be used to extract the EIS contribution to programming dispersion for arbitrary V_s. In order to correctly reproduce the DV algorithm, V_{FG} is calculated taking into account the applied bit-line bias since the beginning of the first programming step at which cell V_T becomes larger than the LPV level and the application of the programming pulses is stopped when V_T overcomes the HPV level.

B. Simulation results

Fig. 5 shows the simulation results for the V_T distribution obtained after ISPP with $V_s = 500$ mV and $\alpha = 0.5$, assuming $\beta = 1$, 0.5 and 0. The former case gives the reference curve corresponding to the SV algorithm, showing the EIS impact making the distribution width larger than V_s, while the other two cases represent possible implementations of the

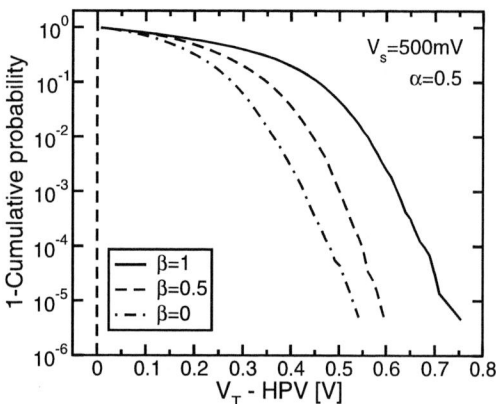

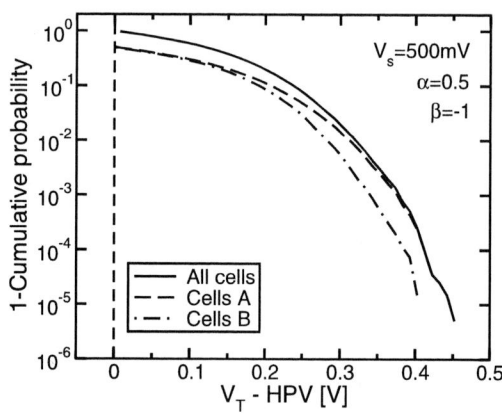

Fig. 5. Simulated cumulative distributions of V_T obtained after ISPP programming with $V_s = 500$ mV, in the case of the conventional SV algorithm ($\beta = 1$) and of two implementations of the DV algorithm ($\beta = 0.5$ and $\beta = 0$).

Fig. 7. Simulated cumulative distribution of V_T programmed by the DV algorithm in the case $V_s = 500$ mV, $\alpha = 0.5$ and $\beta = -1$. Results including all the cells, only cells **A** or only cells **B** of Fig. 2 are shown.

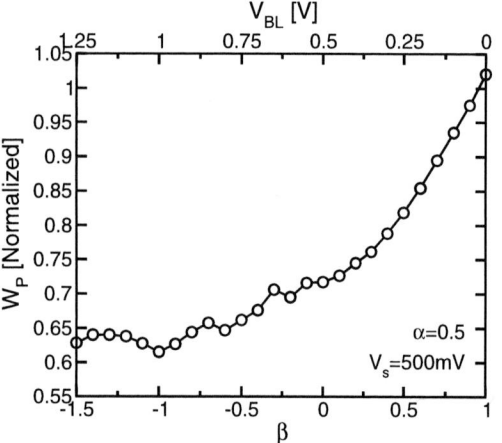

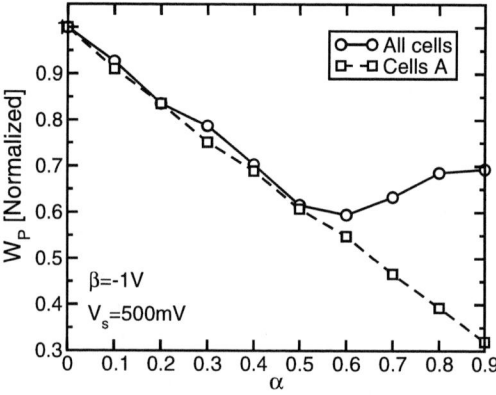

Fig. 6. Simulated V_T distribution width at a probability level of 10^{-4} as a function of β for $V_s = 500$ mV and $\alpha = 0.5$. Results are normalized to the distribution width obtained from the SV algorithm.

Fig. 8. Simulated V_T distribution width at a probability level of 10^{-4} as a function of α for $V_s = 500$ mV and $\beta = -1$. Results are normalized to the distribution width obtained from the SV algorithm.

DV algorithm. Results confirm that the algorithm presented in Section II can actually narrow the programmed V_T distribution with respect to the conventional SV programming scheme, pointing out also that the improvements strictly depend on the value of β. In order to discuss the results more quantitatively, the V_T distribution width (W_P) was extracted at a reference probability level of 10^{-4}. This level was chosen to have reliable results from the 10^5 MC simulations used to obtain the distributions of Fig. 5, nevertheless not being too much higher than the ECC level for NAND Flash. For $V_s = 500$ mV and $\alpha = 0.5$, Fig. 6 shows that a strong reduction of W_P with respect to what obtained from the SV algorithm can be obtained, revealing a 35% decrease when β is reduced from 1 to -1, with a clear saturation of W_P for $\beta < -1$.

The reduction of W_P with β in Fig. 6 reveals that the V_T distribution width after program is determined by cells **B** of Fig. 2, receiving the bit-line bias during the last programming

pulses of the ISPP algorithm. The reduction of β from 1 ($V_{BL} = 0$) to -1 ($V_{BL} = 2V_s$) makes, in fact, cells **B** reduce their $\overline{\Delta V_{T,s}}$ in the steps immediately following the application of V_{BL}, as shown in Fig. 4. This, in turn, allows cells **B** to lower their maximum displacement from the HPV level at the end of programming, compacting the V_T distribution when cells **A** have a lower displacement from the HPV level. In this way, the best achievable accuracy is then obtained when the statistical distribution of cells **B** over the HPV level becomes narrower than the statistical distribution of cells **A**. These cells, in fact, do not receive any bit-line bias during programming and their final V_T distribution does not depend on β. Fig. 7 shows that for $V_s = 500$ mV and $\alpha = 0.5$ this condition is obtained when $\beta = -1$, as for this value the high-V_T tail of the programmed distribution is given by cells **A**, with cells **B** being closer to the HPV. This result means that for $V_s = 500$ mV and $\alpha = 0.5$, $\beta = -1$ represents the optimum conditions for the reduction of the programmed V_T distribution width, as confirmed in Fig. 6 by the saturation of W_P for β values

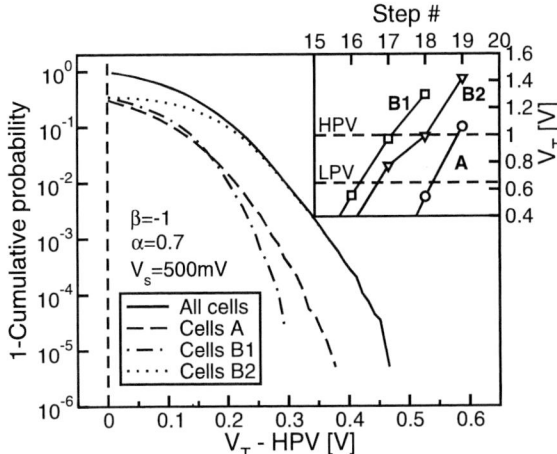

Fig. 9. Simulated cumulative distribution of V_T programmed by the DV algorithm in the case $V_s = 500$ mV, $\alpha = 0.7$ and $\beta = -1$. Cells **B** are shown considering those requiring only 1 (**B1**) or more than 2 (**B2**) steps to overcome HPV since V_{BL} is applied (see inset for an example).

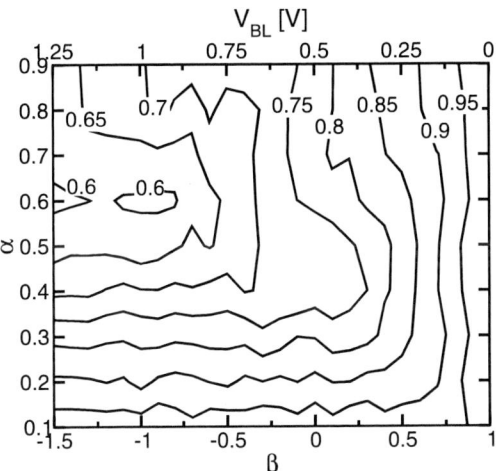

Fig. 10. Contour plot for the simulated W_P at a probability level of 10^{-4} as a function of α and β in the case $V_s = 500$ mV. W_P levels are normalized to the distribution width obtained from the SV algorithm. The bit-line bias $V_{BL} = (1 - \beta)V_s$ applied to cells **B** (see Fig. 2) is quoted on the upper axis.

lower than -1.

For $\beta = -1$, Fig. 8 shows that W_P is limited by and therefore decreases with the programmed V_T distribution of cells **A** for α ranging from 0 to 0.5, while for larger values of α, W_P is limited by cells **B** and grows for $\alpha > 0.6$. This is due to the larger number of cells of group **B** requiring more than two steps to overcome the HPV level after the application of V_{BL} (namely, **B2**) for larger α: as resulting from Fig. 4, these cells display a $\overline{\Delta V_{T,s}}$ at their last programming pulse that is larger than that of cells completing program in one pulse (**B1**) and, therefore, are the main limitation to W_P for large α, as shown in Fig. 9.

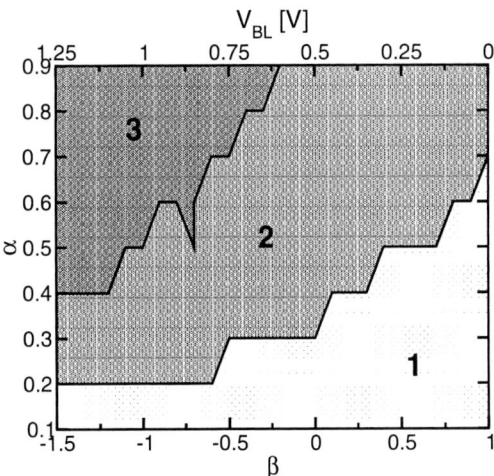

Fig. 11. Contour plot for the maximum number of steps required by cells **B** to overcome the HPV level since the application of the bit-line bias, in the case $V_s = 500$ mV. Results have been extracted from a statistics of 10^5 MC simulations.

C. Algorithm optimization

In order to explore the possibility for a further reduction of W_P when changing α from the 0.5 case addressed in Fig. 6, Fig. 10 shows a contour plot for the simulated W_P as a function of α and β. Results show that the parameters $\alpha = 0.5$ and $\beta = -1$ are near the optimum value for the DV algorithm in the case $V_s = 500$ mV, with only a slightly better W_P obtained when α approaches 0.6. In the optimal conditions, Fig. 10 reveals that a reduction nearly equal to 40% can be obtained from the DV with respect to the SV algorithm. Note that this large improvement in W_P does not require a significant increase in the number of steps needed to complete the program operation. Fig. 11 shows, in fact, the maximum number of steps required by cells **B** to overcome the HPV level since the application of the bit-line bias, for $V_s = 500$ mV. Results have been extracted from a statistics of 10^5 MC simulations, considering the worst case cell for each value of α and β. From Fig. 11, in the case $\alpha = 0.5$ and $\beta = -1$, a maximum number of 3 steps are required for a cell to overcome the HPV level since the application of V_{BL}. This does not represent a critical delay of the programming speed when considering that a larger dispersion of the number of programming pulses is determined by the statistical spread of neutral cell V_T. This is also confirmed by Fig. 12, showing the minimum W_P and the corresponding maximum number of steps required by cells **B** to complete program as a function of V_s, keeping the additional constraint of $V_{BL} < 1$ V in order to ensure that the string-select transistor is ON during program.

IV. CONCLUSIONS

This paper presented a detailed investigation of the accuracy of a DV algorithm for deca-nanometer NAND Flash memories. In order to account for EIS, optimization of the algorithm was studied by means of MC simulations for the electron

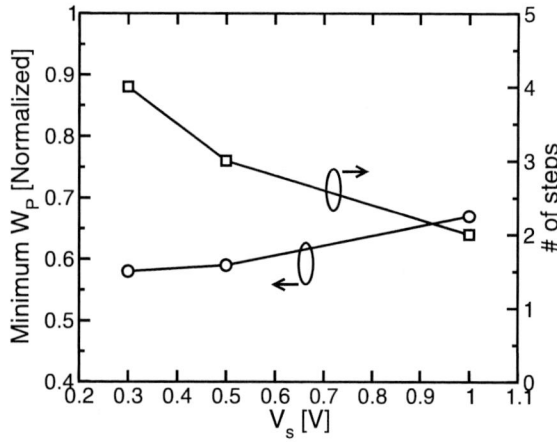

Fig. 12. Minimum W_P (normalized) achievable by the DV algortihm as a function of V_s, for $0 < \alpha < 1$ and limiting β to require a bit-line bias lower than 1 V. The maximum number of steps required by cells **B** to complete the program operation since the application of the bit-line bias is also shown.

injection process, showing that quite large V_T distribution narrowing can be obtained (*e.g.* for $V_s = 500$ mV and the optimal conditions $\alpha = 0.6$ and $\beta = -1$, corresponding to $V_{BL} = 1$ V applied to cells **B**, a narrowing nearly equal to 40% is obtained). These results are of fundamental importance for future NAND Flash technologies, especially for multi-level memory devices.

V. ACKNOWLEDGMENTS

Authors would like to thank P. Cappelletti, E. Camerlenghi, R. Bez, M. Robustelli, A. Visconti and S. Beltrami from Numonyx now Micron for discussions and support. This work has been partially supported by the European Commission under the Project MODERN.

REFERENCES

[1] G. J. Hemink, T. Tanaka, T. Endoh, S. Aritome, and R. Shirota, "Fast and accurate programming method for multi-level NAND EEPROMs," in *1995 Symp. VLSI Tech. Dig.*, pp. 129–130, 1995.

[2] C. Monzio Compagnoni, A. S. Spinelli, R. Gusmeroli, S. Beltrami, A. Ghetti, and A. Visconti, "Ultimate accuracy for the NAND Flash program algorithm due to the electron injection statistics," *IEEE Trans. Electron Devices*, vol. 55, pp. 2695–2702, Oct. 2008.

[3] C. Monzio Compagnoni, R. Gusmeroli, A. S. Spinelli, and A. Visconti, "Analytical model for the electron-injection statistics during programming of nanoscale NAND Flash memories," *IEEE Trans. Electron Devices*, vol. 55, pp. 3192–3199, Nov. 2008.

[4] A. Chimenton, P. Pellati, and P. Olivo, "Constant charge erasing scheme for Flash memories," *IEEE Trans. Electron Devices*, vol. 49, pp. 613–618, Apr. 2002.

[5] C. Friederich, J. Hayek, A. Kux, T. Muller, N. Chan, G. Kobernik, M. Specht, D. Richter, and D. Schmitt-Landsiedel, "Novel model for cell-system interaction MCSI in NAND Flash," in *IEDM Tech. Dig.*, pp. 831–834, 2008.

[6] K. Prall and K. Parat, "25 nm 64 gb MLC NAND technology and scaling challenges," in *IEDM Tech. Dig.*, pp. 102–105, 2010.

[7] C. Monzio Compagnoni, C. Miccoli, A. L. Lacaita, A. Marmiroli, A. S. Spinelli, and A. Visconti, "Impact of control-gate and floating-gate design on the electron-injection spread of decananometer NAND Flash memories," *IEEE Electron Dev. Lett.*, vol. 31, pp. 1196–1198, Nov. 2010.

[8] Y. Shin, "Non-volatile memory technologies for beyond 2010," in *2005 Symp. VLSI Tech. Dig.*, pp. 156–159, 2005.

[9] G. Molas, D. Deleruyelle, B. De Salvo, G. Ghibaudo, M. Gely, S. Jacob, D. Lafond, and S. Deleonibus, "Impact of few electron phenomena on floating-gate memory reliability," in *IEDM Tech. Dig.*, pp. 877–880, 2004.

[10] T. Tanaka and J. Chan, "Non-volatile semiconductor memory device adapted to store a multi-valued data in a single memory cell." U.S. patent 6 643 188 B2, November 4 2003.

[11] V. Moschiano, G. Santin, T. Vali, and M. Rossini, "Non-volatile multilevel memory cell programming." U.S. Patent 7 692 971 B2, April 6 2010.

[12] K.-D. Suh, B.-H. Suh, Y.-H. Lim, J.-K. Kim, Y.-J. Choi, Y.-N. Koh, S.-S. Lee, S.-C. Kwon, B.-S. Choi, J.-S. Yum, J.-H. Choi, J.-R. Kim, and H.-K. Lim, "A 3.3 V 32 Mb NAND flash memory with incremental step pulse programming scheme," *Solid-State Circuits, IEEE Journal of*, vol. 30, no. 11, pp. 1149 –1156, 1995.

[13] C. Monzio Compagnoni, M. Ghidotti, A. L. Lacaita, A. S. Spinelli, and A. Visconti, "Random telegraph noise effect on the programmed threshold-voltage distribution of Flash memories," *IEEE Electron Dev. Lett.*, vol. 30, pp. 984–986, Sep. 2009.

[14] H. Kurata, K. Otsuga, A. Kotabe, S. Kajiyama, T. Osabe, Y. Sasago, S. Narumi, K. Tokami, S. Kamohara, and O. Tsuchiya, "The impact of random telegraph signals on the scaling of multilevel Flash memories," in *2006 Symp. VLSI Circ. Dig.*, pp. 140–141, 2006.

[15] H. Kurata, K. Otsuga, A. Kotabe, S. Kajiyama, T. Osabe, Y. Sasago, S. Narumi, K. Tokami, S. Kamohara, and O. Tsuchiya, "Random telegraph signal in Flash memory: its impact on scaling of multilevel Flash memory beyond the 90-nm node," *IEEE J. Solid-State Circuits*, vol. 42, pp. 1362–1369, 2007.

[16] K. Sonoda, K. Ishikawa, T. Eimori, and O. Tsuchiya, "Discrete dopant effects on statistical variation of random telegraph signal magnitude," *IEEE Trans. Electron Devices*, vol. 54, pp. 1918–1925, Aug. 2007.

[17] C. Monzio Compagnoni, R. Gusmeroli, A. S. Spinelli, A. L. Lacaita, M. Bonanomi, and A. Visconti, "Statistical model for random telegraph noise in Flash memories," *IEEE Trans. Electron Devices*, vol. 55, pp. 388–395, Jan. 2008.

[18] A. Ghetti, C. Monzio Compagnoni, A. S. Spinelli, and A. Visconti, "Comprehensive analysis of random telegraph noise instability and its scaling in deca-nanometer Flash memories," *IEEE Trans. Electron Devices*, vol. 56, pp. 1746–1752, Aug. 2009.

[19] T. C. Ong, A. Fazio, N. Mielke, S. Pan, N. Righos, G. Atwood, and S. Lai, "Erratic erase in ETOX^{TM} Flash memor array," in *1993 Symp. VLSI Tech. Dig.*, pp. 83–84, 1993.

[20] A. Chimenton, P. Pellati, and P. Olivo, "Analysis of erratic bits in Flash memories," *IEEE Trans. Device and Materials Reliab.*, vol. 4, pp. 179–184, Dec. 2001.

[21] S. Gregori, A. Cabrini, O. Khouri, and G. Torelli, "On-chip error correcting techniques for new-generation flash memories," *Proceedings of the IEEE*, vol. 91, pp. 602 – 616, April 2003.

[22] N. Mielke, T. Marquart, N. Wu, J. Kessenich, H. Belgal, E. Schares, F. Trivedi, E. Goodness, and L. Nevill, "Bit error rate in nand flash memories," in *Proc. IRPS*, pp. 9–19, 2008.

[23] A. Spessot, A. Calderoni, P. Fantini, A. S. Spinelli, C. Monzio Compagnoni, F. Farina, A. L. Lacaita, and A. Marmiroli, "Variability effects on the V_T distribution of nanoscale NAND Flash memories," in *Proc. IRPS*, pp. 970–974, 2010.

[24] P. A. W. Lewis and G. S. Shedler, "Simulation methods for Poisson processes in nonstationary systems," in *Proc. of the 10th conference on Winter simulation - Volume 1*, pp. 155–163, 1978.

Precise understanding of data retention mechanisms for MONOS memories: Toward simultaneous improvement of retention and endurance performances by SiN engineering

Shosuke Fujii, [1]Ryota Fujitsuka, [1]Katsuyuki Sekine, and Naoki Yasuda

Advanced LSI Technology Laboratory, Corporate R&D Center,
[1]Advanced Memory Development Center, Semiconductor Company,
Toshiba Corporation
8, Shinsugita-cho, Isogo-ku, Yokohama 235-8522, Japan
Phone: (+81) -(45)-776-5926, shosuke.fujii@toshiba.co.jp

Abstract— **We investigate the charge leakage path during data retention through the evaluation of its temperature dependence. As a result, it is experimentally demonstrated for the first time that the main leakage path of trapped charge changes depending on retention time. Furthermore, the direction of leakage path rather than trap energy profile in the SiN layer determines the temperature dependence of data retention characteristics. In addition, it is found that cycling degradation of data retention is due to increase in the charge loss through the tunnel layer. Based on the accurate understanding of data retention mechanisms, we show the possibility to achieve both of data retention and endurance improvements by SiN engineering.**

Keywords; MONOS, TANOS, Data Retention, Charge leakage path, Cycling degradation

I. INTRODUCTION

Metal-oxide-nitride-oxide-semiconductor (MONOS) type devices are candidates to replace conventional floating-gate non-volatile memory devices because of their low program/erase (P/E) voltage and reduced cell-to-cell interference effects. However, for applying MONOS devices in such application, it is necessary to improve erase performance without sacrificing data retention and endurance properties. Whereas we have already reported that both of erase and endurance improvements can be achieved by employing Si-rich SiN as charge trapping layer[1], Si-rich SiN is known to degrade data retention characteristics severely. It has been widely suggested that the data retention degradation by Si-rich SiN is due to its shallower trap levels[2-4]. On the other hand, recent atomistic simulations reported that the energy depth of trap levels is insensitive to SiN composition[5]. Thus, the physical origin of data retention degradation in Si-rich SiN is still controversial.

In this study, it is experimentally demonstrated for the first time that the main leakage path of trapped charge changes depending on retention time. Furthermore, we find that the leakage path rather than trap energy profile in the SiN layer determines the temperature dependence of data retention. These findings are different from previous models, where the energy profile of trap sites was extracted from the temperature dependence of data retention characteristics.

II. EXPERIMENTAL

MONOS capacitors with n^+diffusion layer on p-type Si substrates were fabricated. The 5nm-thick tunnel oxide was thermally grown, and the 5nm-thick Si-rich SiN layer was deposited by atomic layer deposition (ALD) method using dichlorosilane (DCS) and NH_3. The composition (N/Si ratio) of the SiN was modulated by controlling the DCS/NH_3 gas supply ratio. Refractive indices (R.I.) of the SiN layers are described in Tables I and II. Since R.I. increases with decreasing the N/Si ratio[6], higher R.I. indicates more Si-rich composition. After the SiN deposition, Al_2O_3 block layer (13nm or 15nm) and TaN gate electrode were formed. Areas of the capacitors are all $100 \times 100 \mu m^2$. Equivalent oxide thicknesses (EOTs) of the MONOS devices are estimated by C-V measurements.

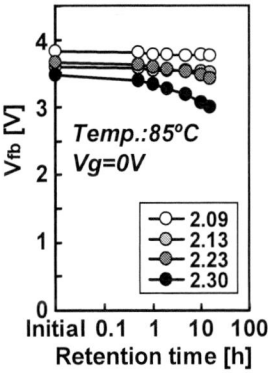

Figure 1 Data retention properties for the MONOS devices. Measurements were performed at 85°C.

TABLE I. MONOS DEVICES USED FOR THE EVALUATION OF DATA RETENTION MECHANISMS.

Sample name	Tunnel ox	Charge SiN	Block Al_2O_3	EOT
2.09		R.I.=2.09 5nm		12.9 nm
2.13	SiO_2 5nm	R.I.=2.13 5nm	13nm	13.1 nm
2.23		R.I.=2.23 5nm		12.9 nm
2.30		R.I.=2.30 5nm		13.2 nm

978-1-4244-9113-1/11 $26.00 © 2011 IEEE

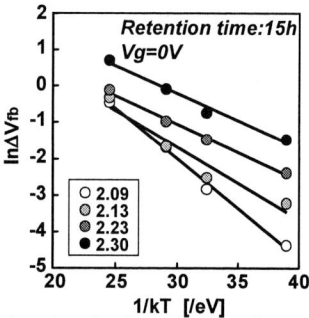

Figure 2 Arrhenius plots for data retention characteristics after 15h.

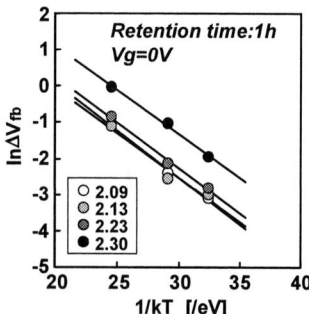

Figure 3 Arrhenius plots for data retention characteristics after 1h.

III. EVALUATION OF CHARGE LEAKAGE PATH DURING DATA RETENTION

Fig.1 shows data retention characteristics of the MONOS devices. Data retention characteristics degrade with more Si-rich composition. Arrhenius plots for the data retention after 15h are shown in Fig.2. Extracted activation energy decreases as the SiN layer becomes more Si-rich composition. Previous reports suggested that this reduction of activation energy is attributed to shallower trap energy in Si-rich SiN layer[2-4]. However, we found that short-term data retention exhibits different trend. Fig.3 shows Arrhenius plots for the data retention after 1h. Different from the result after 15h, activation energy is almost the same irrespective of the SiN composition. Fig.4 compares the activation energies for the data retention after 1h and 15h as a function of the SiN composition. For the MONOS devices of "2.13", "2.23", and "2.30", activation energies of short-term retention are much higher than those of long-term retention. This result for Si-rich SiN cannot be explained only by previous models[3,4] where electrons trapped at shallower energy levels detrap in shorter term. Therefore, another mechanism must be involved in the data retention characteristics.

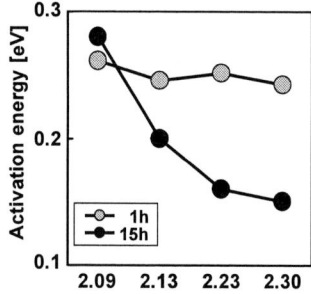

Figure 4 Comparison of activation energies for data retention after 1h and 15h as a function of the SiN composition.

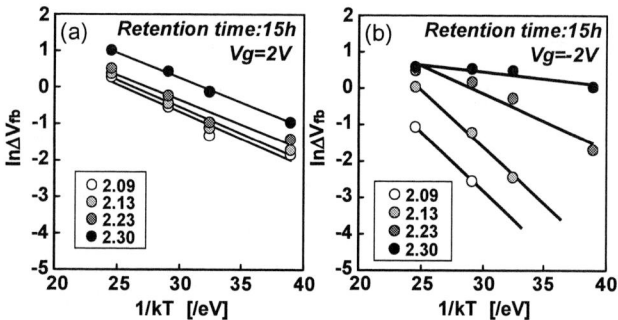

Figure 5 Arrhenius plots for data retention characteristics under (a)Vg=2V and (b)Vg=-2V.

To clarify the origin of the temperature dependence, we performed data retention measurements by applying gate biases so that we can evaluate the charge loss through the tunnel layer and through the block layer separately. Fig.5 illustrates Arrhenius plots for data retention with applying gate biases. Activation energy for the charge loss through the block layer (Fig.5a) is almost constant over all the MONOS devices, whereas that through the tunnel layer (Fig.5b) drastically decreases with more Si-rich composition. These results are interpreted as follows. The activation energy for detrapping through the block layer is determined only by the current conduction in the block layer, since the location of trapped electrons during programming operation is found to be around the SiN/Al2O3 interface (Fig.6), which is in agreement with previous reports[7,8], and the electrons do not need to move across the SiN layer when leaking through the block layer. In contrast, trapped electrons have to move across the SiN layer in order to detrap through the tunnel layer, leading to a strong dependence of activation energy on the SiN composition. Thus, charge leakage path has strong impact on the temperature dependence of data retention.

Here, we re-examine the results of Arrhenius plots for short-term and long-term retention (Fig.2 and Fig.3). In the case of short-term retention, activation energy is insensitive to the SiN composition, which is similar to the result in Fig.5a, indicating that the charge loss proceeds through the block layer. Activation energy of long-term retention, in contrast, decreases with more Si-rich composition, which is in agreement with the result in Fig.5b, revealing that the charge loss mainly occurs through the tunnel layer. Therefore, these results demonstrate that the main leakage path of trapped charge changes depending on retention time, as schematically illustrated in Fig.7.

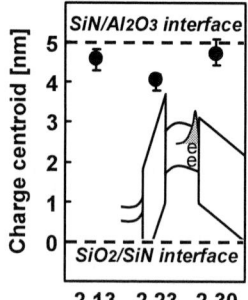

Figure 6 Charge centroid of trapped electrons measured by our developed method[7].

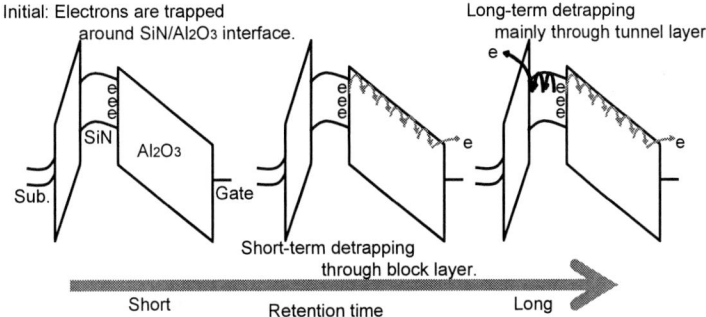

Figure 7 Schematic illustration for explaining the data retention mechanisms. Main leakage path changes depending on the retention time.

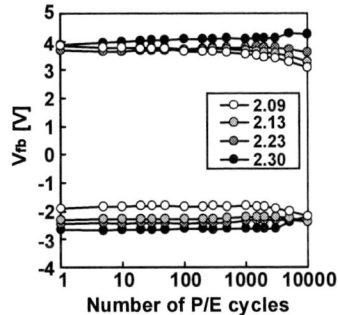

Figure 8 Endurance characteristics for the MONOS devices.

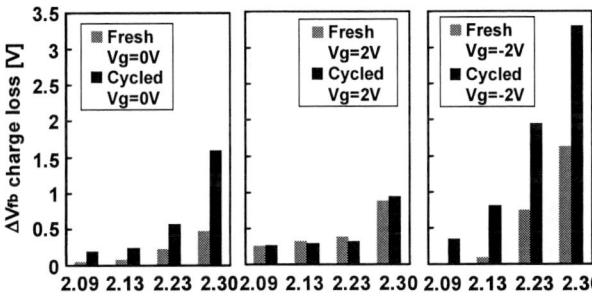

Figure 9 Comparison of data retention characteristics before and after P/E cycling. Gate biases (Vg=(a)0V, (b)2V, and (c)-2V) were applied during retention. Measurements were performed at 85°C.

IV. DATA RETENTION DEGRADATION AFTER P/E CYCLING

Endurance characteristics of the MONOS devices are shown in Fig.8, and we investigate the data retention properties after 1.2k P/E cycles. Fig.9 shows the results of data retention measurements with applying gate biases. Data retention properties for Vg=0V (Fig.9a) and Vg=-2V (Fig.9c) degrade after the P/E cycling, whereas that for Vg=2V (Fig.9b) exhibits no significant degradation. Fig.10a shows Arrhenius plots for the data retention with Vg=2V. The activation energy is almost the same as that before P/E cycling (Fig.5a and Fig.10a), indicating that the degradation of the block layer by P/E cycling is extremely small. On the other hand, activation energy for detrapping through the tunnel layer decreases after P/E cycling (Fig.5b and Fig.10b). From these results, it is concluded that the degradation of data retention after P/E cycling is caused by the increase in detrapping through the tunnel layer.

V. RELIABILITY OPTIMIZATION BY LAMINATED SiN STRUCTURE

We have demonstrated previously that hole injection during erase operation is the main cause for cycling degradation, and that the erase mechanism via electron detrapping can suppress the damage to the tunnel layer[1]. Thus, the electron detrapping erase in Si-rich SiN is effective for suppressing the data retention degradation because of the decrease in damage to the tunnel layer. On the other hand, electrons trapped in Si-rich SiN are relatively mobile even during data retention as suggested by Fig.5b. Hence, Si-rich SiN is not favorable for the preservation of trapped charge.

Considering the fact that the electrons injected during program operation are trapped around SiN/Al₂O₃ interface, MONOS with a laminated SiN structure listed in Table II is expected to suppress charge loss during retention while maintaining the erase mechanism of electron detrapping[9]. The laminated SiN consists of a stack of SiN layers, where more Si-rich (R.I.=2.25) and less Si-rich (R.I.=2.09) SiN layers are close to the tunnel layer and the block layer, respectively. Fig.11 shows J-E_{ox} curves of MONOS devices listed in Table II, extracted by our developed method[1]. For both Si-rich SiN and the laminated SiN, current flowing during erase operation is larger than the theoretical hole current from Si substrate, indicating that the erase operation proceeds by electron detrapping.

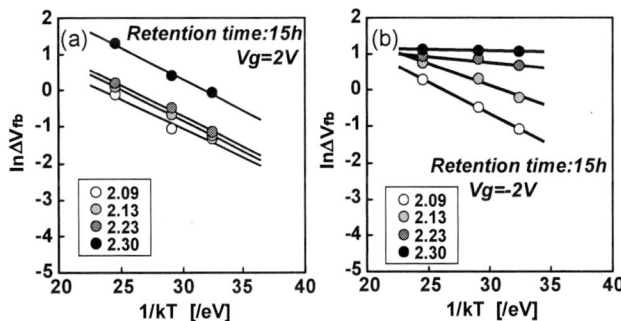

Figure 10 Arrhenius plots for data retention characteristics after P/E cycling. Gate biases (Vg=(a)2V and (b)-2V) were applied during retention.

978-1-4244-9113-1/11 $26.00 © 2011 IEEE   841

TABLE II.   MONOS DEVICES USED FOR DEMONSTRATING
RELIABILITY OPTIMIZATION.

Sample name	Tunnel ox	Charge SiN	Block Al$_2$O$_3$
2.25+2.09	SiO$_2$ 5nm	R.I.=2.25 2.5nm / R.I.=2.09 2.5nm	15nm
2.25		R.I.=2.25 5nm	

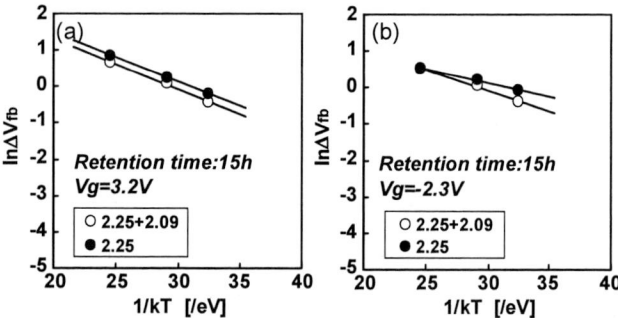

Figure 12   Arrhenius plots for data retention characteristics of MONOS devices listed in Table II. Gate biases (Vg=(a)3.2V and (b)-2.3V) were applied during retention.

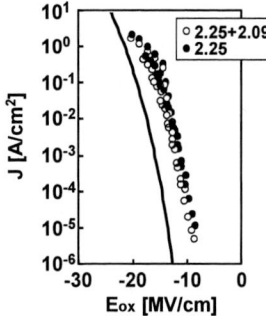

Figure 11 J-E$_{ox}$ characteristics during erase operations extracted by our developed method[1]. Theoretical hole current from Si substrate, which is calculated by WKB approximation, is also plotted as a solid line.

Fig.12 illustrates Arrhenius plots for data retention with applying gate biases. MONOS device with the laminated SiN structure shows similar tendency of temperature dependence to the MONOS with a single SiN layer: Charge loss towards the gate is determined by the conduction in the block layer (Fig.12a) while that through the tunnel layer shows SiN composition and/or structure dependence (Fig.12b). Moreover, charge leakages through both the block and tunnel layers are suppressed relative to the MONOS with the Si-rich single SiN layer. Therefore, MONOS with the laminated SiN can suppress charge loss during retention while maintaining the electron erase. These results show the possibility of simultaneous improvement of data retention and endurance performances.

VI. CONCLUSIONS

In this work, we investigated the charge leakage path during data retention through the evaluation of its temperature dependence. We demonstrated experimentally that the main leakage path changes from through the block layer to through the tunnel layer, depending on the retention time. Furthermore, the direction of leakage path rather than trap energy profile in the SiN layer determines the temperature dependence of data retention characteristics. In addition, it was found that cycling degradation of data retention is due to increase in the charge loss through the tunnel layer. Finally, we showed the possibility to achieve both of data retention and endurance improvements by SiN engineering. These findings will be helpful to design reliable MONOS structures based on the accurate understanding of data retention mechanisms.

REFERENCES

[1]   S. Fujii, R. Fujitsuka, K. Sekine, J. Fujiki, and N. Yasuda, "Transition of erase mechanism depending on SiN composition and its impact on cycling degradation," IRPS, (2010) p.956.

[2]   C. Sandhya, U. Ganguly, N. Chattar, C. Olsen, S. M. Seutter, L. Date, R. Hung, J. M. Vasi, and S. Mahapatra, "Effect of SiN on performance and reliability of charge trap flash (CTF) under Fowler–Nordheim tunneling program/erase operation," IEEE Electron Dev. Lett. 30 (2009) 171.

[3]   T. H. Kim, I. H. Park, J. D. Lee, H. C. Shin, and B.-G. Park, "Electron trap density distribution of Si-rich silicon nitride extracted using the modified negative charge decay model of silicon-oxide-nitrideoxide-silicon structure at elevated temperatures," Appl. Phys. Lett. 89 (2006) 063508.

[4]   A. Suhane, A. Arreghini, R. Degraeve, G. Van den bosch, L. Breuil, M. B. Zahid, M. Jurczak, K. De Meyer, and J. Van Houdt, "Validation of Retention Modeling as a Trap-Profiling Technique for SiN-Based Charge-Trapping Memories," IEEE Electron Dev. Lett. 31 (2010) 77.

[5]   E. Vianello, L. Perniola, P. Blaise, G. Molas, J. P. Colonna, F. Driussi, P. Palestri, D. Esseni, L. Selmi, N. Rochat, C. Licitra, D. Lafond, R. Kies, G. Reimbold, B. De Salvo and F. Boulanger, "New insight on the charge trapping mechanisms of SiN–based memory by atomistic simulations and electrical modeling" IEDM, (2009) p.83.

[6]   T. Makino, "Composition and structure control by source gas ratio in LPCVD SiNx," J. Electrochem. Soc. 130 (1983) p.450.

[7]   S. Fujii, N. Yasuda, J. Fujiki, and K. Muraoka, "A new method to extract the charge centroid in the program operation of MONOS memories," SSDM (2009) p.158.

[8]   C. Sandhya, A. B. Oak, N. Chattar, A. S. Joshi, U. Ganguly, C. Olsen, S. M. Seutter, L. Date, R. Hung, J. Vasi, and S. Mahapatra, "Impact of SiN composition variation on SANOS memory performance and reliability under NAND (FN/FN) operation," IEEE Trans. Electron Dev. 56 (2009) 3123.

[9]   R. Fujitsuka, K. Sekine, A. Sekihara, A. Fukumoto, J. Fujita, F. Aiso, and Y. Ozawa, "Engineering of Si-rich nitride charge-trapping layer for highly reliable MONOS type NAND flash memory with MLC operation," SSDM (2009) p.861.

978-1-4244-9113-1/11 $26.00 © 2011 IEEE

Variability of Resistive Switching Memories and Its Impact on Crossbar Array Performance

An Chen and Ming-Ren Lin
Strategic Technology Group
GLOBALFOUNDRIES
Sunnyvale, CA 94086, USA
1-408-373-9800, an.chen@globalfoundries.com

Abstract— **Metal oxide based resistive switching memories (also known as RRAM for Resistive Random Access Memory) often show large variability, due to the stochastic nature of the switching process. This paper discusses the variability of key RRAM parameters with the focus on the resistance variation. The dependence of resistance variation on operation conditions is analyzed, using Cu_2O-based RRAM as an example. The impact of device variability on the sensing margin of crossbar RRAM arrays is studied by statistical modeling. The variability of the selected device contributes more to the signal degradation in crossbar arrays than the variability of unselected devices.**

Keywords – Resistive switching memory, RRAM, variability, crossbar arrays

I. INTRODUCTION

Metal oxide based resistive switching devices are considered promising candidates for next-generation non-volatile memory and programmable logic applications [1-14]. They are also known as Resistive Random Access Memory (RRAM). These devices can be electrically switched between a high-resistance state (HRS, or off-state with resistance of R_{off}) and a low-resistance state (LRS, or on-state with resistance of R_{on}). They are usually made in a two-terminal metal-insulator-metal (MIM) structure and have demonstrated promising scalability. RRAM can be integrated in CMOS or other novel architectures (e.g., crossbar arrays). Resistive switching behaviors have been observed in numerous metal oxides. Although the switching mechanisms are not yet fully understood, research has shown that the switching is related to defects, e.g., charge traps, mobile ions, oxygen vacancies, *etc.* If not controlled well, these defects may cause considerable variability. Although the on/off ratio from ~ 10 to above 10^3 has been reported in many RRAM papers as proofs of large memory operation window, device variability and its impact are often neglected. This paper will discuss the variability of RRAM devices with the focus on the variation of LRS and HRS resistance, using Cu_2O-based RRAM as an example.

Although it has not been emphasized, RRAM variability is not uncommon in literature. Fig. 1 plots the LRS and HRS resistance variation for some metal oxides reported in literature [1-12]. The X and Y error bars show the variability of LRS and HRS, respectively. Fig. 2 shows the variability of these materials more directly by plotting resistance variation ratio *vs.* on/off ratio for LRS (blue) and HRS (red). The resistance

variation ratio is calculated by dividing the upper value by the lower value of the resistance variation range. HRS variation is generally larger than LRS variation, although exceptions exist. The dashed line marks where the variation ratio is equal to the on/off ratio. In many devices, the HRS variation is even larger than the on/off ratio, i.e., the noise (i.e., resistance variability) exceeds the signal (i.e., on/off ratio). Typically, LRS variation ranges from 2× to 10× and HRS variation from 5× to 100×.

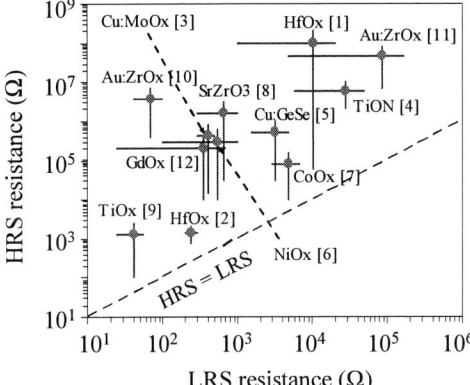

Fig. 1 LRS and HRS variation of metal oxide based RRAM devices reported in literature.

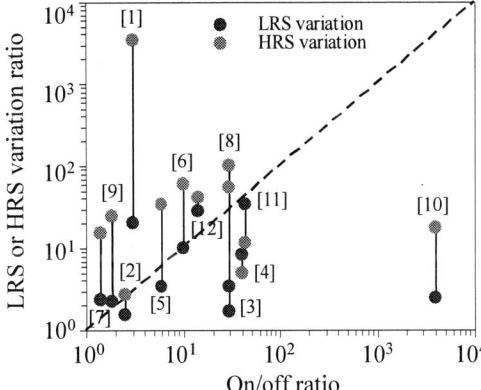

Fig. 2 LRS and HRS variation *vs.* on/off ratio reported in literature.

978-1-4244-9113-1/11 $26.00 © 2011 IEEE

II. EXPERIMENTS

Cu$_2$O-based devices are used to study RRAM variability quantitatively. 64kbit memory arrays are fabricated with standard 0.18 μm CMOS process, and array tests provide the statistics needed to characterize variability [14]. Each device is connected with a transistor to form a 1-transistor-1-resistor (1T1R) structure, where the transistor provides both device selection and current-limit functions. Typical DC switching current-voltage (I-V) curve is shown in Fig. 3, where set and reset are defined as the switching from HRS to LRS and that from LRS to HRS, respectively. Notice that there are two types of variability: device-to-device variability that characterizes the uniformity of memory arrays and cycle-to-cycle variability that characterize device stability.

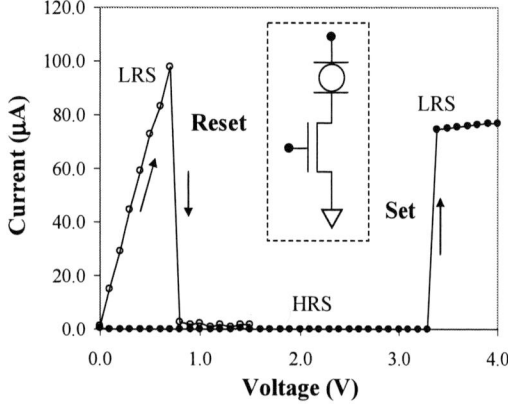

Fig. 3 Typical switching I-V of Cu$_2$O-based resistive switching device.

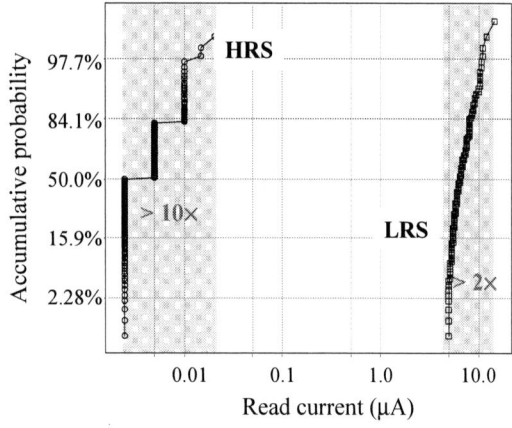

Fig. 4 Distribution of read current in LRS and HRS of Cu$_2$O-based resistive switching devices.

Fig. 4 shows the distribution of LRS and HRS resistance of Cu$_2$O-based devices where device-to-device variability can be measured. LRS shows slightly larger than 2× variation and the HRS variation is > 10×. The stair-like shape of HRS is due to tester resolution. The uniformity of R$_{on}$ distribution can be improved by the operation method [14]. Cycling of Cu$_2$O-based devices is plotted in Fig. 5, where cycle-to-cycle

variation can be determined: variation > 2× for LRS and ~ 50× for HRS. The LRS variability shows an increasing trend over cycles, indicating an "aging" effect where variability may deteriorate during cycling.

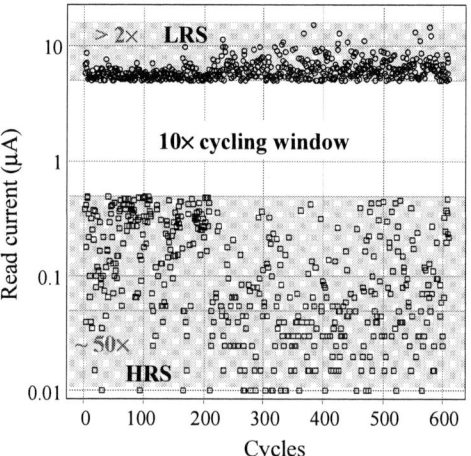

Fig. 5 Cycling of Cu$_2$O-based resistive switching devices.

III. VARIABILITY DEPENDENCE ON OPERATION

It has been shown that LRS of RRAM devices can be modulated by set operation conditions, whereas HRS is less controllable. Therefore, the dependence of LRS variability on operation conditions is studied on Cu$_2$O-based devices. Many RRAM devices show strong LRS dependence on set current limit I$_{limit}$ (controlled by transistor V$_g$ in a 1T1R structure). Lower V$_g$ leads to higher R$_{on}$ and lower reset current, which favors low-power operation. Fig. 6 shows that the LRS variability increases with decreasing set V$_g$, indicating a tradeoff between low-power operation and variability control.

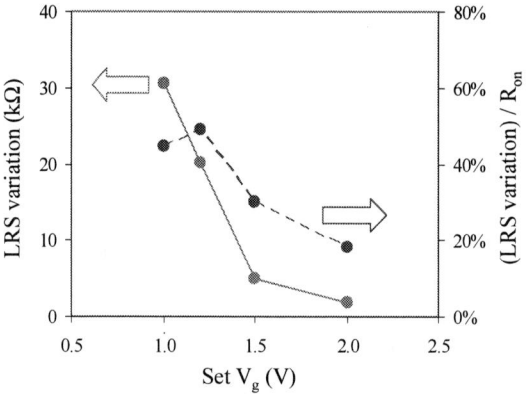

Fig. 6 LRS variation *vs.* set transistor V$_g$ for Cu$_2$O-based resistive switching devices.

The variability of LRS also increases with the decrease of set pulse width, which is undesirable for fast-speed operation (Fig. 7). Although resistive switching is known to be as fast as several ns, it is found that long pulse can continuously reduce both R_{on} and its variation in a certain range, analogous to post-switching "electrical annealing". With shorter switching pulses, resistance states may be unstable and exhibit larger variation. The variation-I_{limit} tradeoff in Fig. 6 and the variation-speed tradeoff in Fig. 7 both show that variability cannot be ignored in the discussion of low-power and fast-speed operation of RRAM.

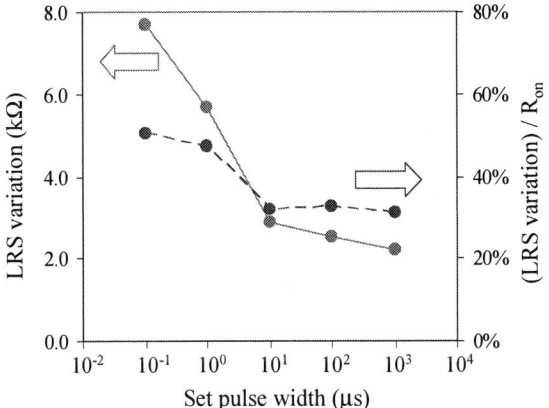

Fig. 7 LRS variation *vs.* set pulse width for Cu$_2$O-based resistive switching devices.

Fig. 8 shows that LRS variability increases at higher temperature, especially after certain critical temperature range. LRS degrades (i.e., R_{on} increases) with increasing temperature as a result of retention loss. Because of non-uniformity, not all devices lose retention at the same rate at high temperatures; therefore, variability also increases.

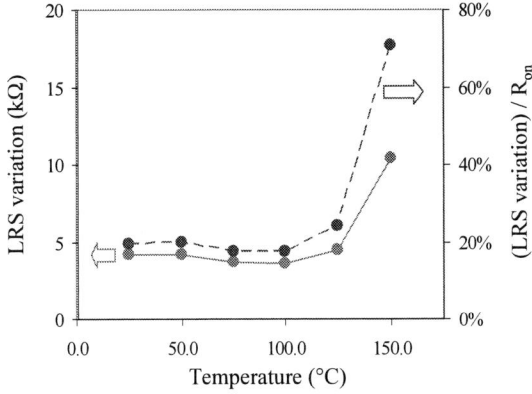

Fig. 8 Temperature induced LRS degradation and variation for Cu$_2$O-based resistive switching devices.

IV. VARIABILITY IMPACT ON CROSSBAR ARRAYS

Most RRAM arrays demonstrated so far are integrated in CMOS with the 1T1R structure, where the transistor may help reduce variability. This is because R_{on} of RRAM has shown strong dependence on the I_{limit} during set switching and I_{limit} can

be "adaptively" adjusted by changing transistor V_g to drive the RRAM resistance into a target range. However, a promising advantage of RRAM devices is their suitability for crossbar array architectures to achieve high device density, where transistors as selection devices may not be applicable. Fig. 9 shows a crossbar array with a suggested voltage configuration for reading. The RRAM device is built at the junctions between the horizontal wordlines (WLs) on the top and the vertical bitlines (BLs) at the bottom. WLs are biased to V_{dd} through a pull-up resistor (R_{pu}) and the sensing signal (V_{out}) is the voltage of the selected WL. Ideally without the parasitic leakage paths formed by the unselected devices, R_{pu} and the selected device form a voltage divider between V_{dd} and ground; therefore, V_{out} is determined by the state of the selected device. However, in reality large amount of parallel leakage paths exist, which may significantly degrade the sensing margin (i.e., the V_{out} difference for the selected device in HRS and LRS).

Analysis has shown that by optimizing the pull-up resistors (R_{pu}) of the selected and unselected lines separately, it may be possible to read crossbar arrays without selection devices. When parasitic access resistance to the WLs/BLs can be ignored and WLs/BLs are accurately biased to V_{dd} or ground, the crossbar circuit can be greatly simplified and the minimum sensing margin can be obtain analytically using the worst scenario analysis. The existence of finite access resistance will cause the WL/BL voltage to vary depending on the random resistance pattern in the array, and the sensing margin cannot be calculated as analytical equations. The sensing signal has to be solved using a set of Kirchhoff equations, Ohm's equations, and current continuity equations. For linear RRAM devices with two resistance states (R_{on} and R_{off}), the equations can be organized in a matrix format and solved in Matlab. This generic approach is applicable to any resistance patterns; therefore, a large amount of resistance patterns can be randomly generated in the simulation and the statistical distribution of the sensing signal can be calculated. To study the impact of device variation, certain amount of variation of R_{on} and R_{off} is randomly applied on each device in the simulation.

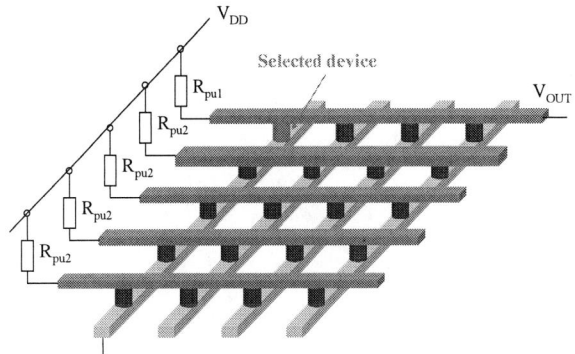

Fig. 9 A schematic of crossbar array and a suggested sensing voltage configuration (bias all top lines to V_{dd}, ground the selected bottom line, and float the other lines).

Fig. 10 compares the calculated V_{out} with and without variation applied only on unselected devices. Although the assumed variability of 2× for LRS and 10× for HRS is not small, V_{out} shows surprisingly little change and there still exists a clear sensing window. This can be explained by the fact that V_{out} is already significantly degraded by the large amount of parasitic leakage paths formed by unselected devices, as shown by the spread of the V_{out} distribution, especially for the case of the selected device in LRS. The assumed variability slightly broadens the distribution and decreases the sensing margin, but is not large enough to change the sensing margin significantly. However, when the variability of unselected devices is increased to 5× for LRS and 100× for HRS, the original sensing margin disappears, as shown in Fig. 11. Variation at this magnitude has been observed in reported RRAM devices (Fig. 1 and 2); therefore, variation-induced signal degradation is a realistic concern for crossbar arrays.

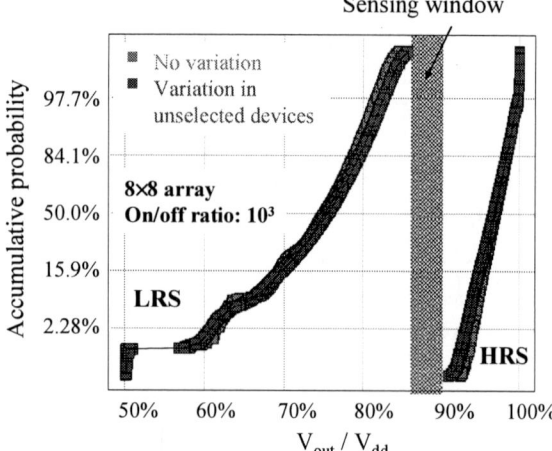

Fig. 10 Distribution of V_{out} (10,000 samples) for the selected device in LRS and HRS, without variation (red) or with variation (blue) on unselected devices; assume variation of 2× for LRS and 10× for HRS on unselected devices and no variation on the selected device.

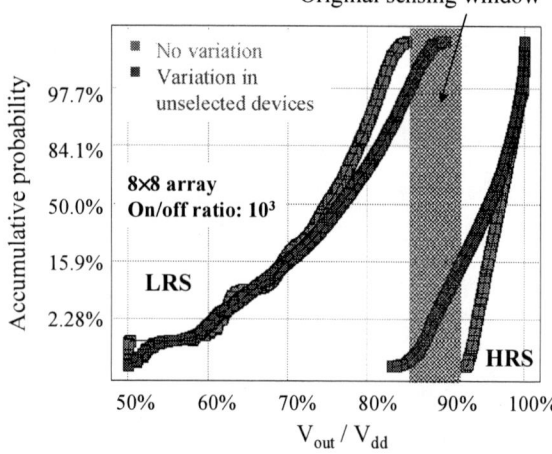

Fig. 11 Distribution of V_{out} (10,000 samples) for the selected device in LRS and HRS, without variation (red) or with variation (blue) on unselected devices; assume variation of 5× for LRS and 100× for HRS on unselected devices and no variation on the selected device.

Notice that in the simulation in Fig. 10 and 11, only variation of unselected devices is considered. When variability is applied directly on the selected device, the sensing signal V_{out} varies significantly due to the variability of the selected device in LRS, while variability of devices in HRS has almost no impact on V_{out} (Fig. 12). Therefore, the variability of the selected device has much more impact on the sensing margin of crossbar arrays than that of unselected devices.

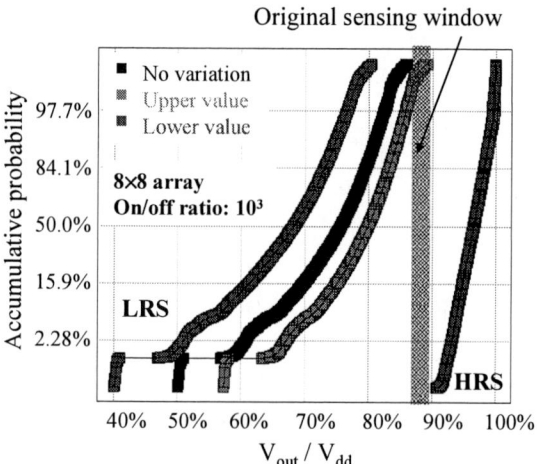

Fig. 12 Distribution of V_{out} for the selected device in LRS and HRS; compare the selected device without variation (black) and with variation to upper value (red) and lower value (blue); assume variation of 2× for LRS and 10× for HRS on both selected and unselected devices.

V. SUMMARY

Variability is an important characteristic for RRAM, typically in the range of 2× - 10× for LRS and 5× - 100× for HRS. Experimental results have shown that tradeoffs exist between variability reduction and low-power fast-speed operations. Large variability causes signal degradation in crossbar arrays, which is affected more by the selected devices than by the unselected devices.

REFERENCES

[1] S.S. Sheu, *et al*, Symposium VLSI Tech., 82 (2009).

[2] K.C. Liu, *et al*, Thin Solid Films **518**, 7460–7463, (2010).

[3] D. Le, *et al*, IEDM Tech. Digest (2006).

[4] Y.H. Tseng, *et al*, IEDM Tech. Digest, 109 (2009).

[5] S.Z. Rahaman, *et al*, VLSI-TSA, 134 (2010).

[6] I.G. Baek, *et al*, IEDM Tech. Digest 587 (2004).

[7] S. Kawabata, *et al*, IMW (2010).

[8] M.H. Lin, *et al*, J. Phys. D **43**, 295404 (2010).

[9] W.Y. Chang, *et al*, APL **95**, 042104 (2009).

[10] Q. Liu, *et al*, ESSDERC, 221 (2009).

[11] W. Guan, *et al*, APL **91**, 062111 (2007).

[12] K.C. Liu, *et al*, Microelectronics Reliability **50**, 670 (2010).

[13] Y.S. Chen, *et al*, IEDM Tech. Digest, 105 (2009).

[14] A. Chen, *et al*, IEDM Tech. Digest, 764, (2005).

[15] A. Chen, *et al*, APL **92**, 013503 (2008).

Statistical Analysis of Retention Behavior and Lifetime Prediction of HfO$_x$-based RRAM

Lijie Zhang[1,2], Ru Huang[1]
Institute of Microelectronics,
Peking University,
Beijing, China
86-10-62752546, zhanglijie@ime.pku.edu.cn.

Yen-Ya Hsu[2], Frederick T. Chen[2], Heng-Yuan Lee[2], Yu-Sheng Chen[2], Wei-Su Chen[2], Pei-Yi Gu[2], Wen-Hsing Liu[2], Shun-Min Wang[2], Chen-Han Tsai[2], Ming-Jinn Tsai[2]
Electronics and Optoelectronics Research Laboratories,
Industrial Technology Research Institute (ITRI),
Taiwan

Pang-Shiu Chen[3],
Department of Chemical and Materials Engineering,
Ming Shin University of Science and Technology,
Taiwan

Abstract—In this paper, statistical measurements on the retention behavior of the stable HfO$_x$-based RRAM under various thermal/voltage/cycling stresses are investigated. Testing results show that, data retention of high resistance state (HRS) of a RRAM is insensitive to temperature and cycling-aging. An empirical equation involving the voltage/thermal/cycling-stress acceleration is given for lifetime prediction. 10 years lifetime can be obtained with a constant read voltage of 0.2 V even at 160 °C. Also the set time of the RRAM extrapolated by the empirical equation coincides with the experiment value. In addition, the shallow Weibull slope of the retention time can be improved when the variations of the initial resistance is well controlled.

Keywords- lifetime; data retention; breakdown; RRAM;

I. INTRODUCTION

Recently, RRAM as an emerging nonvolatile memory has attracted great attention due to its fast speed, excellent endurance and low operating voltages [1, 2]. As for embedded and vehicle applications, reliability issue should be evaluated scientifically. Reported works have shown that the RRAM, unlike PCM, which suffers from data loss at high temperature due to the crystallization, exhibits good retention property at high temperatures during the limited testing time [3, 4]. Since the switching of RRAM is trap/defect related, voltage/temperature acceleration test is an effective way for lifetime prediction which has been usually used for reliability characterization of gate oxide in the past years [5]. Up to now, lifetime predictions based on simulation or retention data at zero bias for one time testing have been reported in several works [6, 7]. However, systematic investigation into the data retention based on a large amount of testing data is still lacking.

In this paper, constant voltage stress (CVS) is applied on HfO$_x$-based RRAM at various temperatures. Data are collected and analyzed with Weibull statistics. The conduction mechanism of HRS which is critical to the breakdown/set

process of the RRAM device is analyzed. The relationship between the HRS conduction and the breakdown process is obtained with an empirical lifetime projection model based on the large amount of measured data.

II. EXPERIMENTAL DETAILS, RESULTS AND DISCUSSION

A. Experimental Details

Fig. 1 (a) and (b) respectively show the structure of the tested device and the testing configuration. Details of the device structure and fabrication process are reported in the previous work [1]. As can be seen from Fig. 1 (a), the thickness of the HfO$_x$ film of the fabricated device is about 8 nm. CVS is applied to the RRAM with a transistor as the current limiter to avoid the current overshooting during the soft breakdown-like process. When the soft breakdown happened, the device was reset to HRS again with sweep mode for another retention testing cycle. 50 retention time data were collected for a single device under a thermal/voltage stress. Since the device can be easily switched for more than 10^6 times, the degradation induced by the 50 successive retention testing cycles can be neglected.

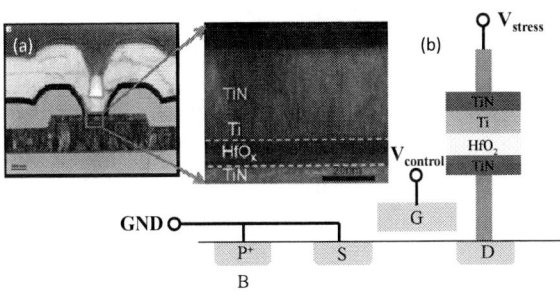

Fig. 1 (a) Cross sectional TEM image of RRAM, (b) Retention testing configuration for 1T1R cell.

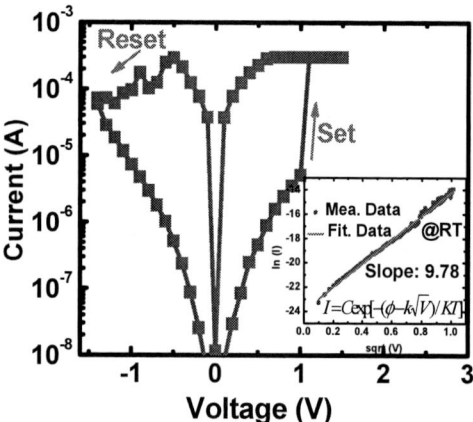

Fig. 2 Typical I-V curve of HfO$_x$-RRAM, the inset shows log$I$ has a linear relationship with $\sqrt{V}$.

B. Results and Discussion

1) Conduction mechanism of HRS

It is well known that RRAM with larger R$_{HRS}$ (resistance of HRS) exhibit higher set voltage and good immunity to voltage disturb. Therefore, it is necessary to investigate into the conduction mechanism of the HRS before HRS-failure analysis. Fig. 2 is the typical I-V curve of 1R device. The device exhibits symmetric current conduction with both positive and negative voltage bias. The inset figure shows that log$I$ has linear relationship with $\sqrt{V}$, satisfying the Schottky emission theory or the Poole-Frenkel theory [8]. It is hard to determine which conduction mechanism dominated just by the I-V symmetry or the slope. But it is certain that log$I$ has a linear relationship with $\sqrt{V}$ for different devices under various conditions including temperatures, cycle-aging and thickness as shown in Fig. 3. Therefore, the current of the HRS can be estimated by:

$$I = C \exp[-(\phi - k\sqrt{V})/KT] \qquad (1)$$

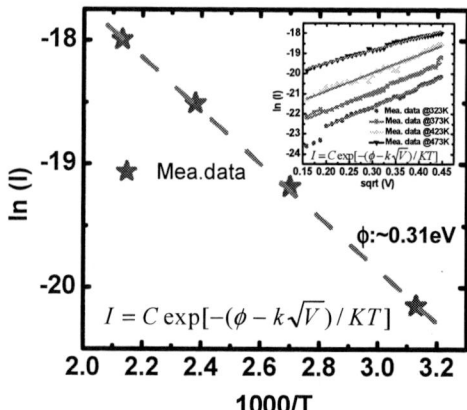

Fig. 4 Temperature dependence of current of the HRS @ 0.2V, the barrier height is about 0.31eV. The inset shows I-V curves of HRS at different temperatures.

where ϕ is barrier height of electron transport.

The HRS conduction at various temperatures is also investigated and the results are shown in the inset of Fig. 4. I-V curves of the device under different temperatures follow (1) well, further verifying the correctness of the equation (1). The conductance of the device under different temperatures follows the Arrhenius law as shown in Fig. 4. From the slope of the ln$I$ vs $1/T$, the barrier height of the electron transport in HRS can be extrapolated. The extrapolated barrier height is about 0.31 eV, which is much lower than the ideal conduction band offsets between the electrode and the oxide while is very close to the experimental extracted trap depth or the calculated value in many HfO$_2$-related works (0.2~ 0.35 eV) [9-12]. Therefore, the HRS conductance may be dominated by trap-assistant tunneling.

2) Voltage/temperature acceleration retention test

CVS ranging from 0.35 V to 0.8 V is applied to devices at various temperatures (25 $^\circ$C, 150 $^\circ$C, 180 $^\circ$C). For each stress

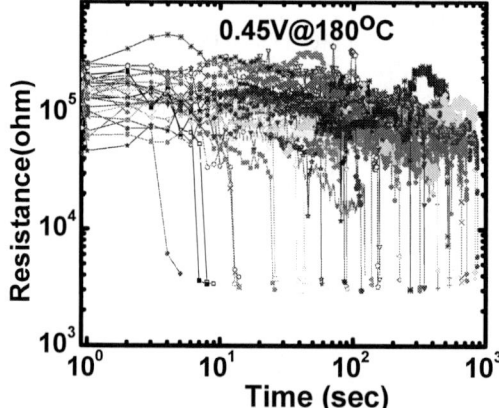

Fig. 5 50 cylces of CVS retention test for a device at 180.$^\circ$C.

Fig. 3 I-V curves of HRS of different devices under variours conditions including temperature, cycle-number, and thickness of HfO$_x$.

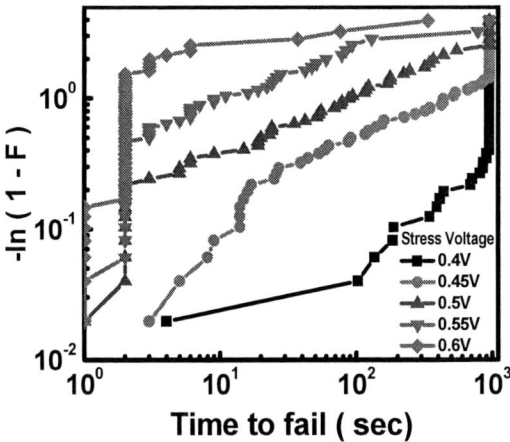

Fig. 6 Weibull distribution of t_{bd} with different voltage stress at 180℃.

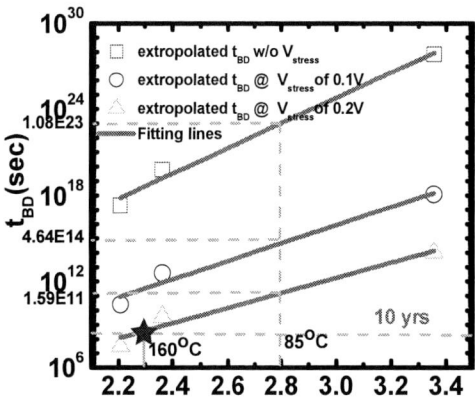

Fig. 8 Temperature dependence of the extrapolated t_{BD} with different applied voltage stress.

condition, retention behavior of one device is measured for 50 times. The retention behaviors of one device with the stress of 0.45V under 180℃ are shown in Fig. 5 as an example. As shown in this figure, breakdown time (t_{bd}) of the device varies randomly for each testing, which is similar to oxide-breakdown [5], exhibiting the statistical properties of the soft-breakdown of HRS. Many groups of data have been measured with different CVS under various temperatures. After the data collection, the data were treated with Weibull statistics, and the result of the device under 180 ℃ with different voltage is shown in Fig. 6. The Weibull shape factor of the t_{bd} is relatively small（0.4~0.5）, which is possibly caused by the variations of HRS before every testing-cycle, and will be discussed about below. The t_{bd} at 63% of the cumulative probability is extracted to assess the voltage/temperature acceleration effect. t_{BD} under different temperatures with various voltage stress can be obtained in the same way.

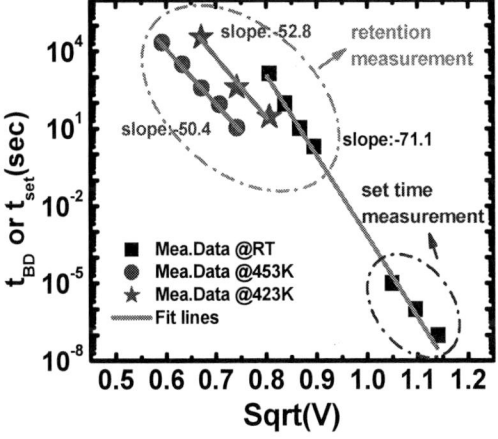

Fig. 7 Exponential relationship of t_{BD} or t_{set} with $\sqrt{V}$ at different temperatures.

Fig. 7 shows that t_{BD} under different temperatures exhibit the negative exponential relationship with $\sqrt{V}$, indicating that breakdown or set process is possibly triggered by the HRS conduction current during the stress. An empirical equation between t_{BD} and $\sqrt{V}$ can be given as follows：

$$t_{BD} = A\exp((E_a - k_{BD}\sqrt{V})/KT) \qquad (2)$$

where E_a is the activation energy, A and k_{BD} are constant.

The slopes which indicate the voltage acceleration factor decrease with temperature as equation (2) indicates. The set time of the RRAM with high set voltages extrapolated by the empirical equation coincide with the experiment values as shown in Fig. 7, indicating the soundness of the equation, which links the gap between the fast switching speed and the long retention time.

Since only the voltage (V) and temperature (T) are variables, based on the experimental value shown in Fig. 7, t_{BD} under different temperatures with various voltage stress can be extracted and the result is shown in Fig. 8. Lifetime projection for 10-years with CVS of 0.2 V can be obtained even up to 160 ℃ as shown in Fig. 8. E_a for the breakdown or set process is extracted from the slope of t_{BD} vs $1/T$ without voltage stress in Fig. 8 and its value is about 1.8 eV, which is about 6 times of ϕ in (1). It is found that k_{BD} is also about 6~7 times of k in equation (1) when comparing Fig. 7 with Fig. 4. And thus t_{BD} has a negative power-law relationship with the current, which can be explained as follows. The breakdown or set process of the RRAM is possibly triggered by the current conducted through the oxide under the voltage stress. Defects or traps are generated in the oxide under the impact of the electron current. The newly generated defects are beneficial for the current conduction, resulting in the increase of current, which will further induce more defects. This positive feedback process lead to the quick breakdown or set process of the RRAM.

The retention behavior of the RRAM after different switching cycles is also investigated. The devices are cycled for

978-1-4244-9113-1/11 $26.00 © 2011 IEEE

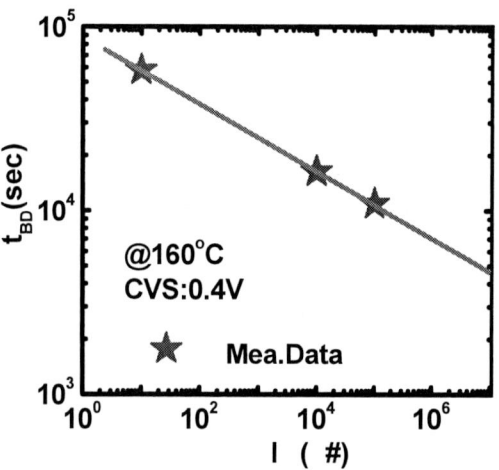

Fig. 9 Linear relationship between t_{BD} and cycle-number in log-log scale. t_{BD} decreases with swithicng cycles.

different times with pulse mode. Then the retention behaviors of three devices are investigated in the same way as that used in voltage/temperature acceleration retention test. Power law relationship between t_{BD} and cycle-number (N) can be obtained, which is similar to flash memory [13]. With switching cycle increasing, more defects will exist in the oxide layer of RRAM, the activation energy decrease due to the increase number of the traps, resulting in the decrease of the retention time.

Considering the cycling-aging effect on the retention time, the lifetime projection equation is finally written as:

$$t_{BD} = A \times N^m \exp((E_a - k_{BD}\sqrt{V})/KT) \quad (3)$$

where m is the cycling-aging acceleration factor.

Based on this equation, the life-time of a device with different activation energy can be projected under different

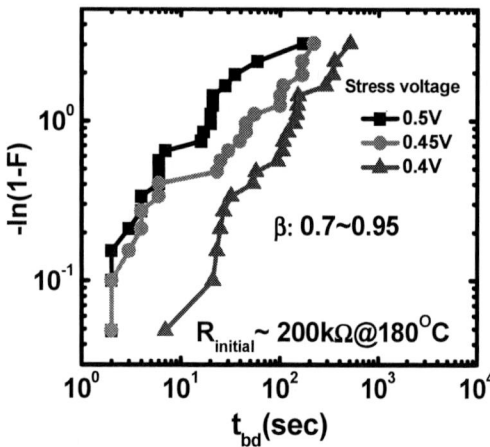

Fig. 10 Improved Weibull slope of t_{bd} with the verification method.

temperatures with different cycles. This equation can also be used to explain the large distribution of t_{bd} during the retention testing. As (1) and (3) indicate, variations of ϕ which relates to E_a will cause the variations of R_{HRS}, resulting in the

variations of current under the voltage stress, and thus leading to the large distributions of t_{bd} as shown in Fig. 6.

If the variation of the HRS can be controlled, the distribution of the t_{bd} can be tightened. As can be seen in Fig. 6, without verification method, the fluctuation range of the HRS before retention test in different testing-cycles is about one order. When RRAM was reset to the similar value before every retention test, the Weibull slopes are improved as shown in Fig. 10.

III. SUMMARY

The data retention of RRAM under different thermal/voltage/cycling stress is statistically analyzed. An empirical equation for lifetime prediction based on the HRS conduction mechanism is given, which explains both good retention behavior and fast speed of the RRAM. Furthermore, the variations of t_{bd} can be suppressed when a verification operation of initial HRS is used.

ACKNOWLEDGMENT

This work is financially supported by Ministry of Economic Affairs, ROC

REFERENCES

[1] H. Y. Lee, P. S. Chen, T. Y. Wu, Y. S. Chen, C. C. Wang, P. J. Tzeng, C. H. Lin, F. Chen, C. H. Lien, and M.-J. Tsai, "Low power and high speed bipolar switching with a thin reactive Ti buffer layer in robust HfO₂ based RRAM," in *IEDM Tech. Dig.*, 2008, pp. 297-230.

[2] Z. Wei, Y. Kanzawa, K. Arita, Y. Katoh, K. Kawai, S. Muraoka, S. Mitani, S. Fujii, K. Katayama, M. Iijima, T. Mikawa, T. Ninomiya, R. Miyanaga, Y. Kawashima, K. Tsuji, A. Himeno, T. Okada, R. Azuma, K. Shimakawa, H. Sugaya, T. Takagi, R. Yasuhara, K. Horiba, H. Kumigashira, and M. Oshima, "Highly reliable TaO$_x$ ReRAM and direct evidence of redox reaction mechanism," in *IEDM Tech. Dig.*, 2008, pp. 293-296.

[3] Y. S. Chen, H. Y. Lee, P. S. Chen, P. Y. Gu, C. W. Chen, W. P. Lin, W. H. Liu, Y. Y. Hsu, S. S. Sheu, P. C. Chiang, W. S. Chen, F. T. Chen, C. H. Lien, and M.-J. Tsai, "Highly scalable hafnium oxide memory with improvements of resistive distribution and read disturb immunity," in *IEDM Tech. Dig.*, 2009, pp. 105-108.

[4] Y. S. Chen, T. Y. Wu, P.-J. Tzeng, P.-S.Chen, H. Y. Lee, C. H. Lin, F. T. Chen, and M.-J. Tsai, "Forming-free HfO₂ bipolar RRAM device with improved endurance and high speed operation," in *VLSI-TSA.*, 2009, pp. 36-37.

[5] E. Y. Wu, A. Vayshenker, E. Nowak, Sune. J, R.-P.Vollertsen, W. Lai, and D.Harmon, "Experimental evidence of T$_{BD}$ power-law for voltage dependence of oxide breakdown in ultrathin gate oxides," *IEEE TED*, pp. 2244-2253, 2002.

[6] P. Zhou, H. J. Wan, Y. L. Song, M. Yin, H. B. Lv, Y. Y. Lin, S. Song, R. Huang, J. G. Wu, M. H. Chi, "A systematic investigation of TiN/Cu$_x$O/Cu RRAM with long retention and excellent thermal stability," in *IMW*, 2009.

[7] C. Cagli, D. Ielmini, F. Nardi, and A. L. Lacaita, "Evidence for threshold switching in the set process of NiO-based RRAM and physical modeling for set, reset, retention and disturb prediction," in *IEDM Tech. Dig.*, 2008, pp. 301-304.

[8] John G. Simmons, "Poole-Frenkel effect and Schottky effect in Metal-Insulator-Metal systems," *Phys. Rev.*, vol. 155, pp. 657-660,1967.

[9] P. Broqvist, and A. Pasquarello, "Oxygen vacancy in monoclinic HfO₂: A consistent interpretation of trap assisted conduction, direct electron

injection, and optical absorption experiments," *Appl. Phys. Lett.*, vol. 89, p. 262904,2006.

[10] G. Ribes, J. Mitard, M. Denais, S. Bruyere, F. Monsieur, C. Parthasarathy, E. Vincent, and G. Ghibaudo, "Review on high-k dielectrics reliability issues," IEEE *Trans. Device Mater. Reliab.* vol. 5, pp.5-19, 2005.

[11] G. Bersuker, B.H. Lee, H.R. Huff, J. Gavartin, and A. Shluger, "Mechanism of charge trapping reduction in scaled high-k gate stacks," *NATO Science Series*, vol. 220, pp. 227-236, 2006.

[12] Ch. Walczyk, Ch. Wenger, R. Sohal, M. Lukosius, A. Fox, J. Dąbrowski, D. Wolansky, B. Tillack, H.-J. Müssig, and T. Schroeder, "Pulse-induced low-power resistive switching in HfO_2 metal-insulator-metal diodes for nonvolatile memory applications," *J. Appl. Phys.*, vol.105, p.114103, 2009.

[13] "Failure mechanisms and models for semiconductor devices," in JEDEC PUBLICATION, p.20, 2009.

Behaviors and Physical Degradation of HfSiON MOSFET Linked to Strained CESL Performance Booster

Kidan Bae[1], Minjung Jin[1], Hajin Lim[2], Lira Hwang[1], Dongseok Shin[2], Junekyun Park[1], Jinchul Heo[2], Jongho Lee[2], Jinho Do[2], Ilchan Bae[1], Chulhee Jeon[1] and Jongwoo Park[1]

Technology Reliability[1], Technology Development[2], System LSI division, Samsung Electronics
San #24 Nongseo-Dong Giheung-Gu, Yongin-City, Gyeonggi-Do, Korea 446-711
82-31-209-1344 (phone), 82-31-209-4312 (fax), jongwoo.s.park@samsung.com (email)

Abstract

The propensity of HCI and BTI degradation of HfSiON MOSFET on strained SiN-CESL performance booster is meticulously investigated. It is found that HCI and BTI lifetime of HfO based n/p MOSFET devices depend on hydrogen, initial Dit and plasma charging inherently related to the stress type of CESL fabricated with PECVD. In case for tensile CESL, n/p MOSFET devices far exceed reliability targets for both HCI and BTI. While compressive CESL on n/p MOSFET drastically depresses HCI and BTI lifetime. (Keywords: CESL, FTIR, BTI, HCI, Plasma, CP, Reliability)

I. INTRODUCTION

Since 90nm technology below, strained silicon technology used to boost up device performance becomes essential particularly for high speed and low power CMOS. Even in high-k based technology nodes, the contact etch stop layer (CESL) is one of the most feasible engineering remedies for improving device performance in comparison to other candidates [1]. However, its influence on device reliability that includes HCI and BTI is still controversial associated with SiN-CESL stress types, entangled with physical properties of SiN film and effect of intrinsic mechanical stresses [2-4].

In this study, a systematical approach is taken to refine the propensity of HCI and BTI degradation on the CESL stress type from physical characterization of CESL to meticulous reliability assessments of n/p MOSFET along with electrical characterization of charging pumping before and after stress.

II. EXPERIMENTAL

HfSiON nMOSFET and pMOSFET devices having a nitride oxide thickness in a range of 34~37A are processed with PECVD SiN-CESL (contact etch stop layer) in a way of varying pressure, RF power and gas flow rate at below 500°C in order to provide different intrinsic mechanical stress types, such as tensile and compressive stress. In sequence, HCI and BTI test are conducted at room and 140°C with voltage stress conditions, respectively. In case for BTI, ΔVt is carefully measured by known as a fast measurement to avoid recovery during measurement. Figure 1 intends to illustrate adopted strain technologies. As shown in Fig. 2, intrinsic mechanical stress induced by CESL rapidly increases as technology node shrinks. Figure 3 shows ~10% of nMOSFET device improvement with tensile CESL (tCESL) in comparison to the neutralized.

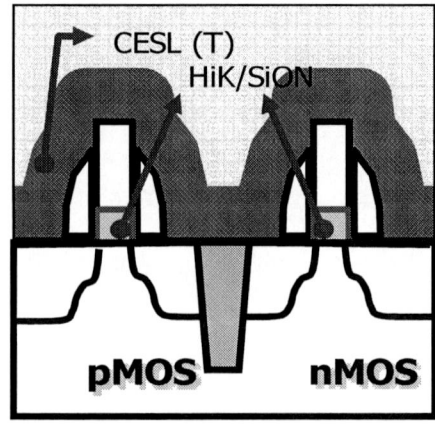

Fig. 1. Schematic diagram of strained silicon technology used for n/pFET.

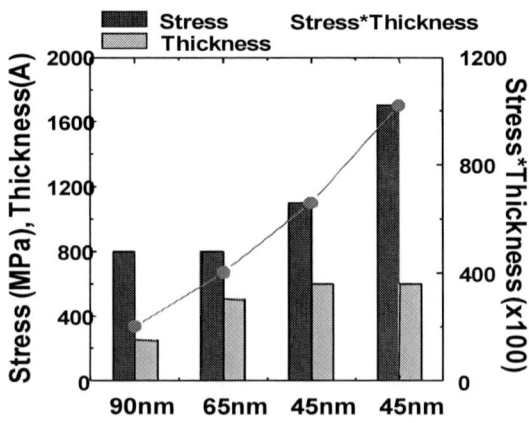

Fig. 2. Effects of CESL on device performance from since 90nm.

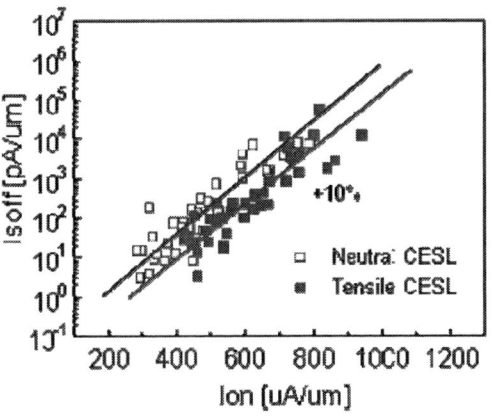

Fig. 3. Increase of carrier mobility with tensile CESL on nMOSFET

III. PHYSICAL CHARACTERIZATION AND PLASMA CHARGING

Figure 4 reveals FTIR spectra on the cCESL and tCESL fabricated by PECVD. As shown, cCESL has more Si-H bonds than tCESL. Since N-H bond is thermodynamically more stable than Si-H bond, MOSFET device would not expect to be degraded in the presence of N-H bond in tCESL film. It is manifested that the total amount of hydrogen that includes Si-H and N-H bond decreases with increasing mechanical stress from compressive to tensile stress (see Fig. 5). Such fragile Si-H bonds can diffuse into the gate oxide and channel region then create trap, resulting in severe Idsat and Vth shift of MOSFETs. It is, in fact, found that n/p MOSFET HCI and NBTI reliability strongly depend on the magnitude of initial Dit and plasma charging (see Fig. 6). No doubt, the larger plasma charging from cCESL fabrication, the short NBTI and HCI lifetime will be. Accordingly, tCESL is more reliable than cCESL, which is in well agreement with literature [5]. However, it is also reported that cCESL increases HCI immunity particularly for nMOS device, although device performance is improved with tCESL [4].

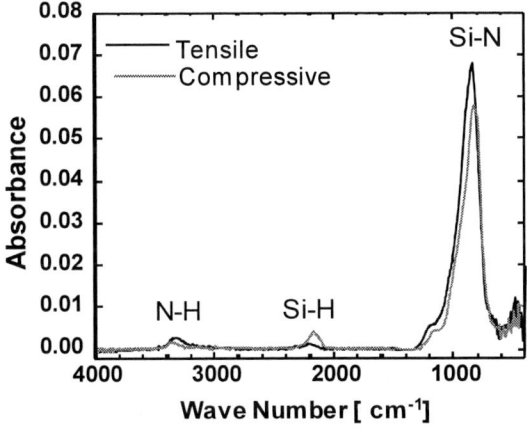

Fig. 4. FTIR spectra on CESL deposition by PECVD (compression vs. tensile).

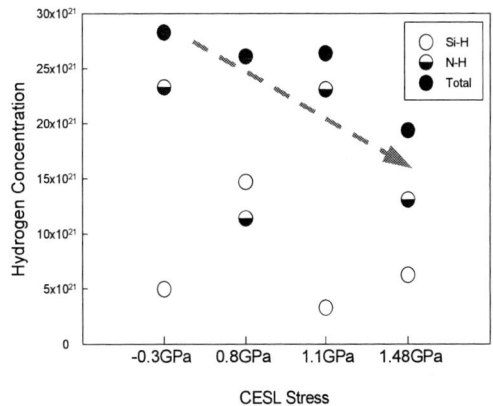

Fig.5 . Hydrogen concentration as a func of SiN-CESL film stress

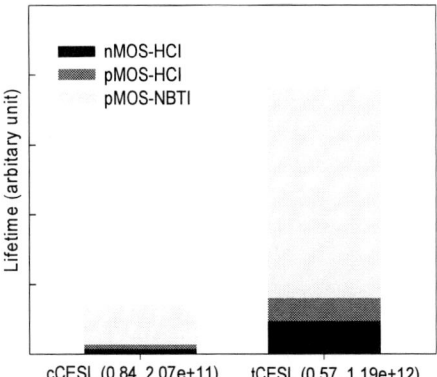

Fig. 6. Effects of plasma charging and initial Dit on HCI and NBTI (plasma, $eV^{-1} \cdot cm^{-2}$)

IV. EFFECT OF CESL ON RELIABILITY

Figure 7 shows less ΔIdsat for n/p MOS devices with tCESL than with cCESL. In turn, tCESL provides a better HCI immunity for both n/pMOS devices. Although cCESL yields initially improved interface quality, n/pMOS HCI degradation becomes worse due mainly to a higher breakage rate of Si-H bonds at the interface from hydrogen rich cCESL (see Fig. 5). Indeed, n/p MOSFET HCI lifetime with tCESL is > 5× and >7× better than with cCESL. Figure 8 shows ΔIdsat of nMOS with cCESL induced by HCI under low and high Vd stress. Surprisingly, ΔIds of nMOSFET stressed at lower Vd meets the target of >AC 10 years. This implies retarded dissipation reaction of Si-H bonds at lower Vd stress. As such, HCI lifetime prediction can be misled when the data is garnered then extrapolated from higher stresses. However, such phenomenon is not observed from pMOS due presumably to polarity effect of plasma charging [6]. Since mechanical stress discrepancy between cCESL (-0.3GPa) and tCESL (+0.8GPa) is shallow, it is posited that the effect of mechanical strain on degradation would be negligible especially for channel length larger than 0.1µm [7]. Figure 9 supports that tCESL can suppress ΔVth shift induced by BTI more efficiently than cCESL, at least > 5×.

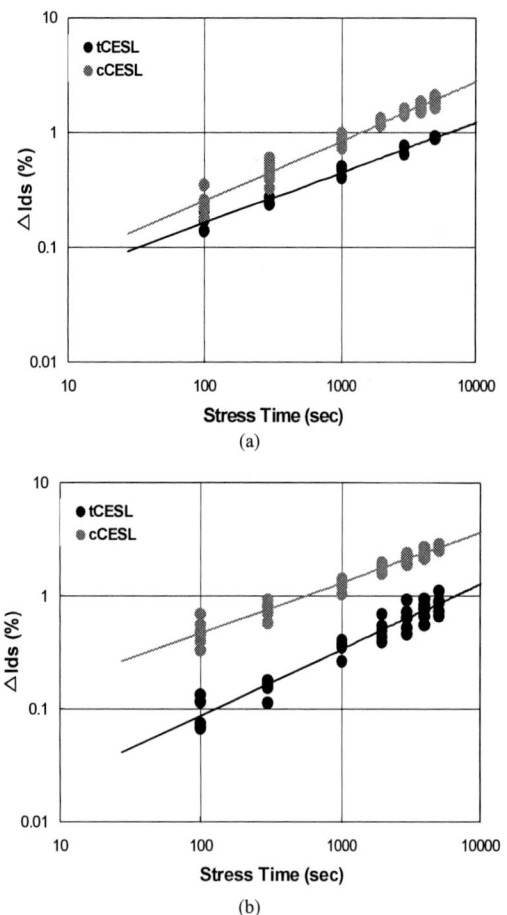

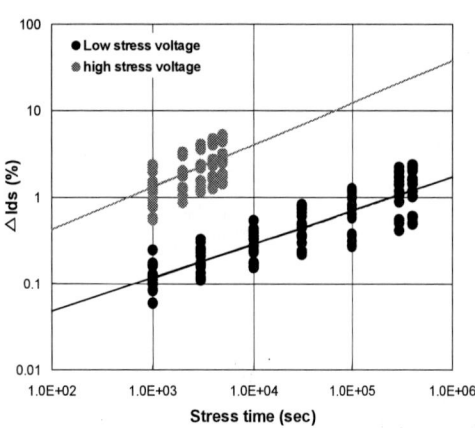

Fig. 7. ΔIds degradation of nMOS (a)/pMOS (b) with tCESL (0.8Gpa), and cCESL(-0.3GPa) resulting from HCI stress test at ambient with Vg=0.5× Vd for nMOS and Vg=Vd for pMOSFET

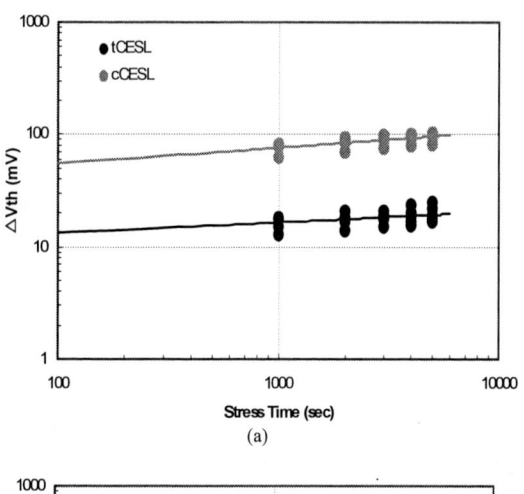

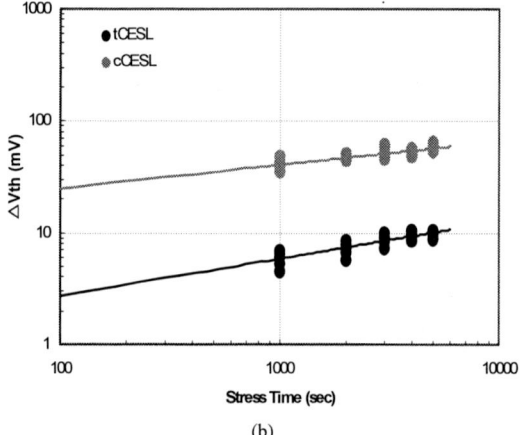

Fig. 9 ΔVth degradation of nMOS (a) and pMOS (b) with tCESL (0.8GPa) and cCESL (-0.3GPa) resulting from BTI stress test at 140℃ with Vg= -2.9V for nMOS and Vg =3.3V for pMOS

Under NBTI stress, the bulk trap can catch holes or H+ released from the broken Si-H bonds then, in turn, results in the increase of interface trap so as to enhance Vth shift. Hence, cCESL is more susceptible than tCESL against HCI and BTI because of larger ΔDit (see Fig. 10). If ΔVth shift results from the creation of interface trap known as Ea of ~0.15eV, both CESL should have a similar Ea. However, Fig. 11 shows different Ea, implying another physical model involved for ΔVth shift. In spite, relatively lower ΔNit from tCESL with less hydrogen causes erroneous data would not be excluded. When ΔVth and ΔNit are measured, the subtraction of interface trap to total Vth shift is assumed as trapped hole density on oxide trap (Not). In Fig. 12, ΔNit increases as temperature augments especially for cCESL, while little dependency appears on tCESL. Apparently, more ΔNit and ΔNot exist on cCESL than tCESL. This is attributed to the higher breakage rate of Si-H bonds and plasma induced hole trapping. As results, cCESL causes more pMOS ΔVth shift than tCESL during BTI.

Fig. 8. Behaviors of ΔIds of nMOSFET with tCESL at lower and higher Vd stress for HCI

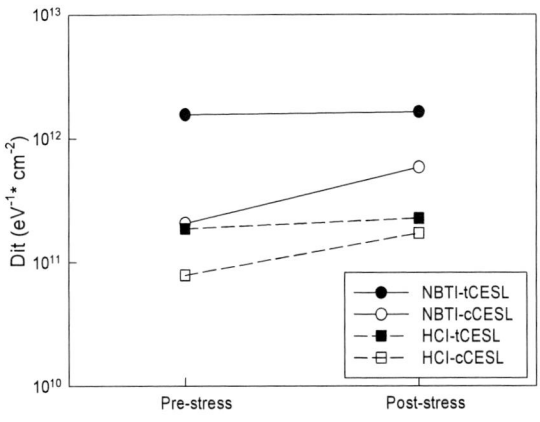

Fig. 10. Changes in ΔDit between tCESL and cCESL resulting from HCl/ NBTI stress test on n/pMOS

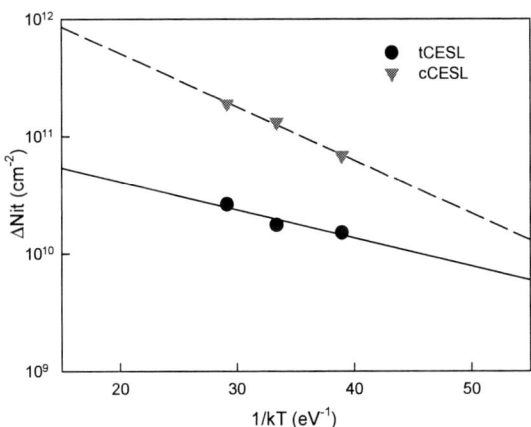

Fig. 11. Ea of tCESL (0.71eV) and cCESL (0.11eV) estimated from 25C, 75C and 125C by using charge pumping with respect to changes in ΔVth

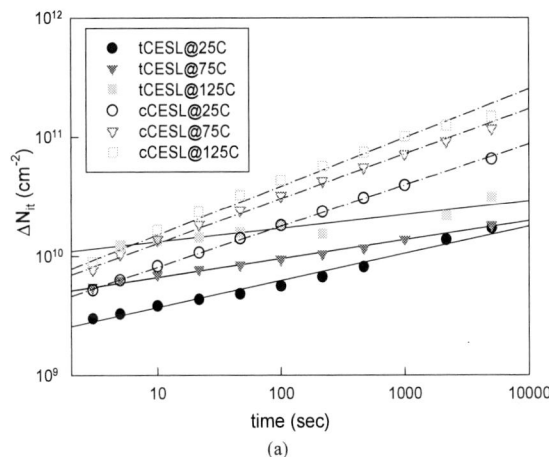

(a)

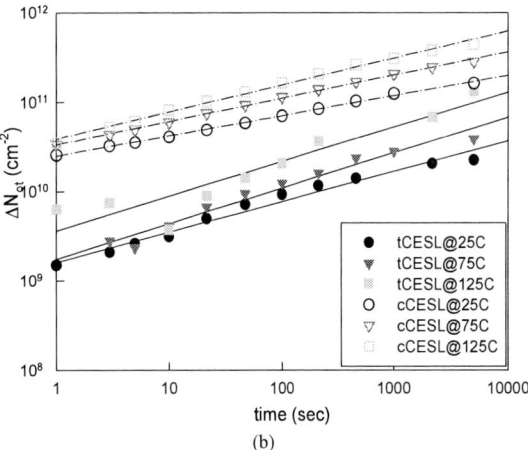

(b)

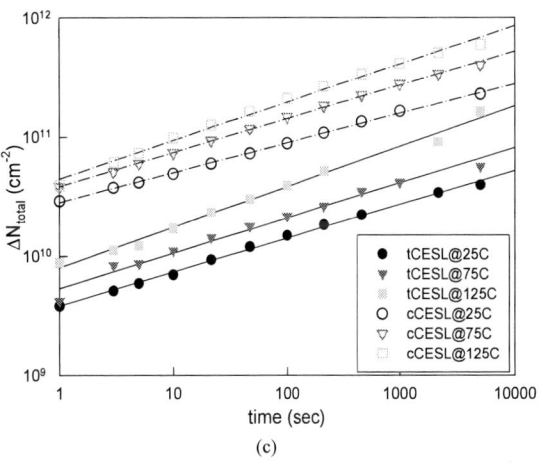

(c)

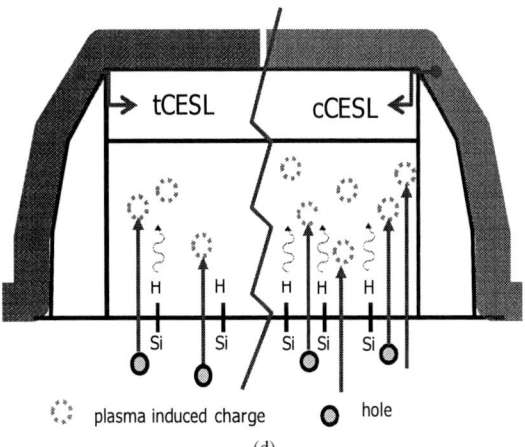

(d)

Fig. 12. Changes in ΔN total [=ΔNit +ΔNoxide trap (ot)] between tCESL(solid) and cCESL (open) from pMOSFET exposed to NBTI stress at 25C, 75C and 125C: ΔNit (a), ΔNoxide trap (b), and ΔN total (c) measured by charging pumping with respect to changes in ΔVth and the schematic diagram of hole trap induced by plasma charging (c).

978-1-4244-9113-1/11 $26.00 © 2011 IEEE

V. CONCLUSION

It is found that HCI and BTI degradation cohesively adhere to hydrogen concentration, initial Dit and plasma charging during PECVD SiN-CESL deposition. Apparently, tCESL is more reliable than cCESL for both n/p MOSFET with respect to physical and electrical aspects. As such major reliability concerns of HCI and BTI degradation for HfSiON devices associated with CESL deposition can be efficiently controlled through the optimization of the front-end-of-line process.

REFERENCES

[1] S. Chung, E. R. Hsieh, D. C. Huang, C. S. Lai, C. H. Tsai, P. W. Liu, Y. H. Lin, C. T. Tsai, G. H. Ma, S. C. Chien, S. W. Sun, "More Strain and Less Stress- The Guideline for Developing High-End Strained CMOS Technologies with Acceptable Reliability," *IEDM Tech. Dig.*, 2008, PP. 1-4

[2] A. Shickova, B. Kaczer, P. Verheyen, G. Eneman, E. San Andres, M. Jurczak, P. Absil, H. Maes, G. Groeseneken, "Negligible effect of process-induced strain on intrinsic NBTI behavior," *IEEE Electron Device Letters*, vol. 28, no.3, 2007, pp. 242-244.

[3] H. Rhee, H. Lee, T. Ueno, D. Shin, S. Lee, Y. Kim, A. Samoilov, P. Hansson, M. Kim, H. Kim, N. Lee, "Negative bias temperature instability of carrier-transport enhanced pMOSFET with performance boosters", *IEDM Tech. Dig.*, 2005, pp. 692-695

[4] I. Han, H. Ji, O. You, W. Choi, J. Lim, K. Hwang, S. Park, H. Lee, D. Kim, H. Lee, "New observation of mobility and reliability dependence on mechanical film stress in strained silicon CMOSFETs", *IEEE Trans. on Electron Devices*, 2008, vol. 55, no. 6, pp. 1352-1358.

[5] T. Irisawa, T. Numata, E. Toyoda, N. Hirashita, T. Tezuka, N. Sugiyama+, and S. Takagi, "Physical Understanding of Strain Effects on Gate Oxide Reliability of MOSFETs," , *VLSI symp. Tech. Dig*, 2007, pp. 36-37.

[6] S. Krishnan, A. Amerasekera, S. Rangan, S. Aur, "Antenna Device Reliability for ULSI Processing", *IEDM Tech. Dig.*, 1998, pp. 601-604

[7] J. Liao, Y. Fang, Y. Hou, C. Hung, P. Hsu, K. Lin, K. Huang, T. Lee, M. Liang, "Strain effect and channel length dependence of bias temperature instability on complementary metal-oxide-semiconductor field effect transistors with high-k/SiO2 gate stacks", Appl. Phys. Lett., 93, 2008, pp. 092101

Experimental Study on Origin of V_{TH} Variability under NBT Stress

Yuichiro Mitani

Corporate Research & Development Center, Toshiba Corporation
8 Shinsugita-cho, Isogo-ku, Yokohama235-8522, Japan
phone: (+81) –(45)- 776-5943, yuuichiro.mitani@toshiba.co.jp

Akira Toriumi

Department of Materials Science, The University of Tokyo
7-3-1, Hongo, Bunkyo-ku, Tokyo 113-8656, Japan

Abstract— The origin of NBTI variability was investigated experimentally using the recovery by hydrogen annealing after NBT stressing. In the case of hydrogen-annealed devices after low voltage NBT stress, ΔV_{TH} and ΔI_{CP} values including these distributions completely coincide with those in the case of the first NBT stress, irrespective the number of both the stress and the hydrogen annealing. This result indicates that the defects generated by applying low voltage NBT stress can be completely recovered by hydrogen incorporation. On the contrary, in the case of the hydrogen-annealed devices after high voltage NBT stress, the variability of ΔV_{TH} under low-voltage NBT stress after hydrogen annealing becomes more marked compared to that for fresh device, while the relationship between the mean values and the variability of ΔI_{CP} coincides with that of fresh device. Based on the experimental results, we conclude that the variability of the quantity of hole-trapping precursors in gate oxide films is the dominant origin for the V_{TH}-shift distribution, rather than the variability of the quality of the gate dielectric interface.

Keywords-NBTI; variability; hydrogen; SiON; interface trap; hole trap;

I. INTRODUCTION

Issues concerning the reliability of gate dielectrics constitute one of the most serious challenges in the scaling of ultra-large scale integrated (ULSI). In particular, negative bias temperature instability (NBTI) in pMOSFETs has become increasingly serious in the context of efforts to realize highly reliable integrated analog and digital CMOS devices [1-6]. Moreover, the statistical distribution of NBTI is also important from the viewpoint of determining the lifetime of the devices. This is because the ULSI consists of over one billion transistors and the device reliability cannot be decided only by average characteristics. NBTI variability and its origin have been already discussed in the previous reports [7-9]. According to the previous reports, the distribution of the number of generated interface traps mainly causes the variability of NBTI. On the other hand, it has been reported that NBTI arises from both hole trapping to defects in gate oxide and hydrogen release from Si/SiO$_2$ interface subsequent to this hole trapping (so-called 2-stage model) [10]. Therefore, it is expected that the hole trapping also relates to the distribution of NBTI. The

purpose of this paper is to further investigate experimentally the variability of NBTI and to discuss the influence of the hole trapping on the variability.

II. DEVICES AND EXPERIMENTAL PROCEDURE

The devices used in this work were pMOSFETs fabricated on Si (100) substrate having oxynitride (SiON) films as the gate dielectrics. The SiON films were nitrided base gate oxides using nitric oxide (NO) gas. The gate oxide thickness was 1.9 nm.

We have already investigated that NBTI can be recovered by hydrogen annealing [11]. The hydrogen annealing was performed to the devices after NBT stress. As a result, the mean values of ΔV_{TH} and ΔI_{CP} were completely recovered. In this paper, we also used this technique.

III. EXPERIMENTAL RESULTS AND DISCUSSION

A. Hydrogen Anealing after Low Voltage NBT Stress

First of all, the effect of hydrogen annealing on the recovery of Low-voltage NBTI is shown using the process flow in figure 1. In this experiment, the identical MOSFETs were measured before and after hydrogen annealing.

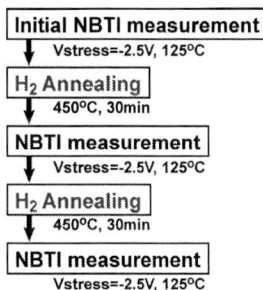

Figure 1   Process Flow for NBTI recovery by hydrogen annealing. NBTI is measured using the same MOSFETs.

978-1-4244-9113-1/11 $26.00 © 2011 IEEE

Fig. 2 shows the stress time evolution of the distributions of charge-pumping current, I_{CP}, and threshold voltage, V_{TH}, in fresh devices, and these recoveries by annealing in nitrogen (N_2) or hydrogen (H_2/N_2) ambient at 450 °C for 30 minutes. The devices were applied NBT stress with V_G=-2.5 V at 125 °C. After that, the annealing was performed, and V_{TH} and I_{CP} were measured again using the same devices. Not only the average values of I_{CP} and V_{TH} but also these distributions are fully recovered by hydrogen annealing, while nitrogen annealing could not recover the devices to the values of pre-stressed conditions.

not only median value but also standard deviation increase with stress time. On the other hand, median value of ΔV_{TH} also increase monotonously with stress time, while no increase of standard deviation of ΔV_{TH} is observed as shown in figure 5.

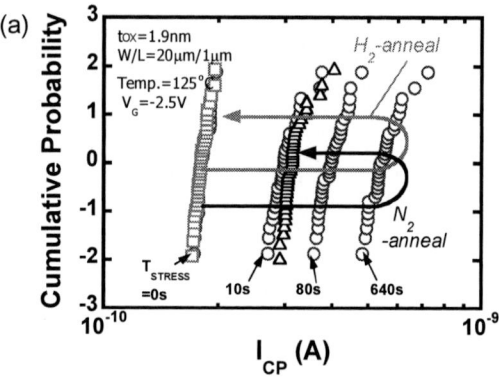

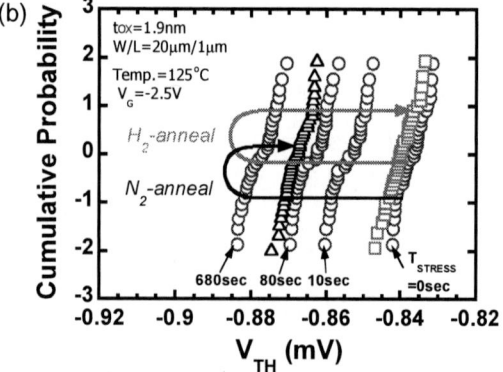

Figure 2 Recovery of NBTI (I_{CP}, and V_{TH}) by hydrogen or nitrogen annealing. When the devices were annealed in nitrogen, V_{TH} and I_{CP} values were recovered by nitrogen atmosphere. However, V_{TH} and I_{CP} values did not return to the former values. On the other hand, V_{TH} and I_{CP} values were completely recovered by annealing in diluted hydrogen atmosphere.

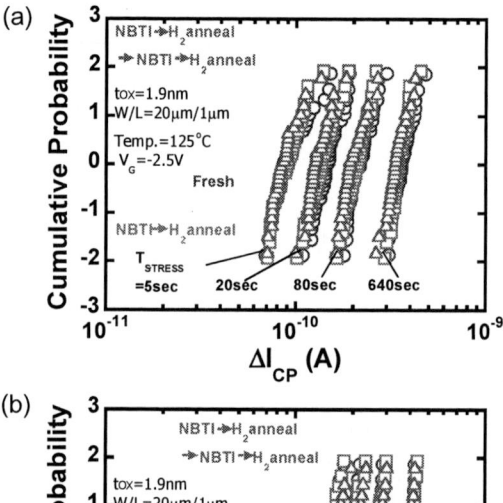

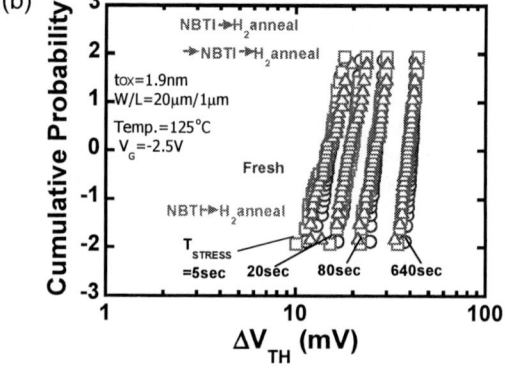

Figure 3 Log-normal plots of ΔI_{CP} and ΔV_{TH} of fresh, 1st and 2nd hydrogen-annealed devices. Both the distribution and mean value after hydrogen annealing are completely identical to those in the case of the first NBT stress.

Furthermore, figure 3 show the log-normal plot of ΔI_{CP} and ΔV_{TH}. The applying of NBT stress was repeated to the devices which were completely recovered by hydrogen annealing. In this procedure, the NBT measurement and hydrogen annealing were repeated alternately as shown in figure 1. It should be noted that the degradation behaviors including the variability of ΔI_{CP} and ΔV_{TH} under 2nd and 3rd NBT stress following hydrogen annealing are completely the same as that under 1st NBT stress. Figs. 4 and 5 show the stress time evolution of median values and standard deviations for ΔI_{CP} and ΔV_{TH} in fresh, 1st and 2nd hydrogen-annealed devices. In the case of ΔI_{CP},

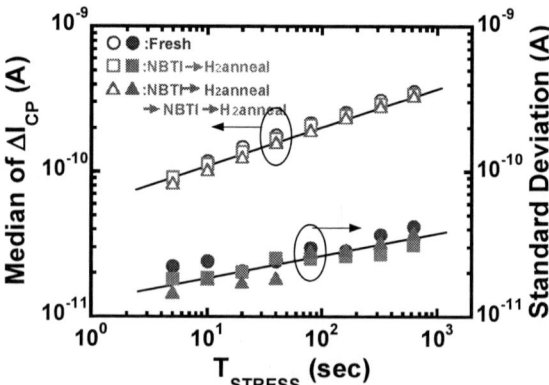

Figure 4 Stress time evolutions of median values and standard deviations for ΔI_{CP} in fresh, 1st and 2nd hydrogen-annealed devices. ΔI_{CP} of hydrogen-annealed devices completely correspond to those in the case of 1st NBT stress irrespective of the number of the NBT stress and the hydrogen annealing.

978-1-4244-9113-1/11 $26.00 © 2011 IEEE   858

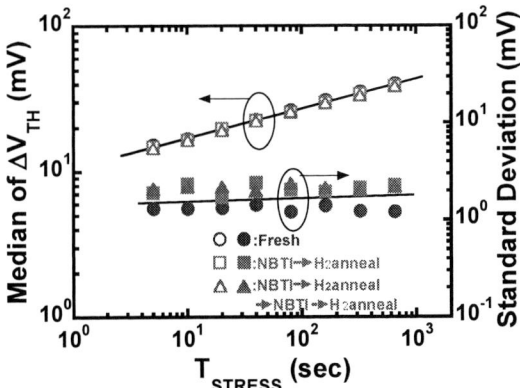

Figure 5 Stress time evolutions of median values and standard deviations for ΔV_{TH} in fresh, 1st and 2nd hydrogen-annealed devices. Both values of ΔV_{TH} for hydrogen-annealed devices completely correspond to those in the case of 1st NBT stress irrespective of the number of the NBT stress and the hydrogen annealing.

These results coincide to the previous report [6]. It should be noted that both median values and standard deviations of ΔI_{CP} and ΔV_{TH} almost correspond among fresh, 1st and 2nd hydrogen-annealed devices. These results imply that the NBTI and its distribution after low-voltage NBT stress can be recovered completely by supplying thermal energy and hydrogen.

B. Valiability of NBTI under High Voltage Stress

Next, hydrogen annealing is performed to the devices which are applied high-voltage NBT stress as shown in figure 6(a). The purpose of this high-voltage NBT stress is that not only interface-states but also bulk defects increase, which might function as hole-trapping precursors. By the high-voltage NBT stress, I_{CP} markedly degrade as shown in figure 6(b). Furthermore, it is found that the high-voltage NBT stress could change the trend of the ΔI_{CP}-ΔV_{TH} correlation compared to the case under low-voltage NBTI as shown in figure 7. And, as can be seen in figure 6(b), the degraded devices are also fully recovered as same as fresh devices by hydrogen annealing.

NBTI behaviors after hydrogen annealing are shown in figure 8. Fig. 8(a) shows the log-normal plots of ΔI_{CP} for fresh devices and hydrogen-annealed devices after high-voltage NBT stress. It is found that the increase of ΔI_{CP} of the hydrogen-annealed devices after high-voltage NBT stress is more marked than that for fresh devices. Furthermore, the degradation of ΔV_{TH} for the hydrogen-annealed devices is also larger than that for the fresh devices as shown in figure 8(b). However, it should be noted that the slop of these ΔV_{TH} distributions in the case of hydrogen-annealed devices seem to be gentler. This result implies that the variability of ΔV_{TH} is degraded by high-voltage NBT stress.

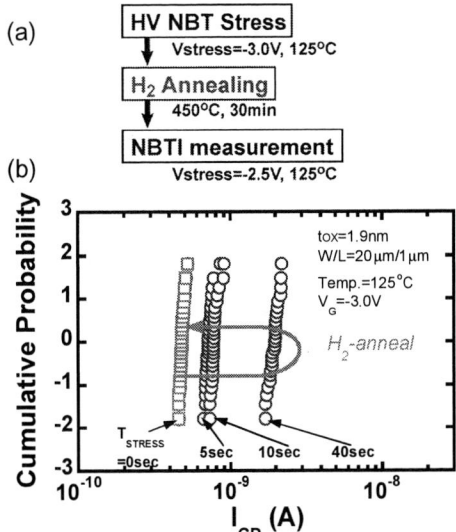

Figure 6 (a) Process Flow for NBTI recovery by hydrogen annealing. NBTI is measured after applied high-voltage NBT stress and H_2 annealing. (b) Time evolution of I_{CP} under high-voltage NBT stress and recovery by hydrogen annealing after the stress.

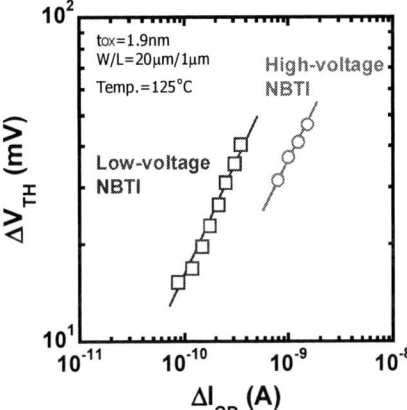

Figure 7 ΔI_{CP}-ΔV_{TH} correlations in low-voltage NBT stress case and high-voltage NBT stress case. ΔI_{CP}-ΔV_{TH} correlation can be changed by high-voltage NBT stress, which is expected to be changed the generated interface-state density and the amount of hole trapping compared to the case of low-voltage NBT stress.

Here, the resulting total ΔI_{CP} and ΔV_{TH} of hydrogen-annealed devices were found to be uncorrelated with the ΔI_{CP} and ΔV_{TH} just after applying high-voltage NBT stress as shown in figure 9. These results indicate that the NBT degradation of the respective devices after applying high-voltage stress has no influence on the variability of NBTI after hydrogen-annealing.

978-1-4244-9113-1/11 $26.00 © 2011 IEEE

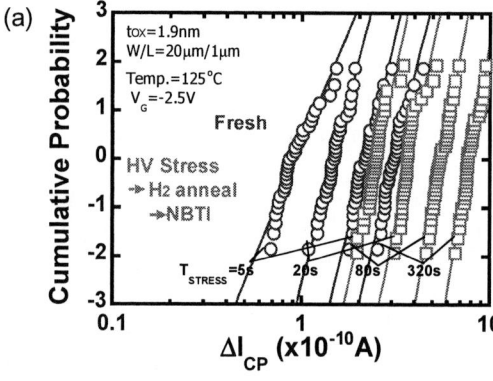

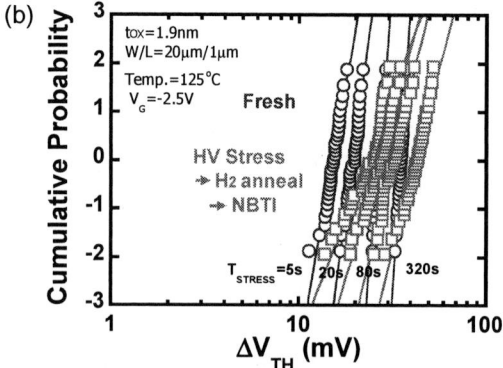

Figure 8 Log-normal plots of ΔI_{CP} and ΔV_{TH} of fresh devices and hydrogen-annealed devices after high-voltage NBT stress. The mean value of ΔI_{CP} is degraded markedly, and the variability of ΔV_{TH} is definitely degraded in the case of hydrogen-annealed devices after high-voltage NBT stress.

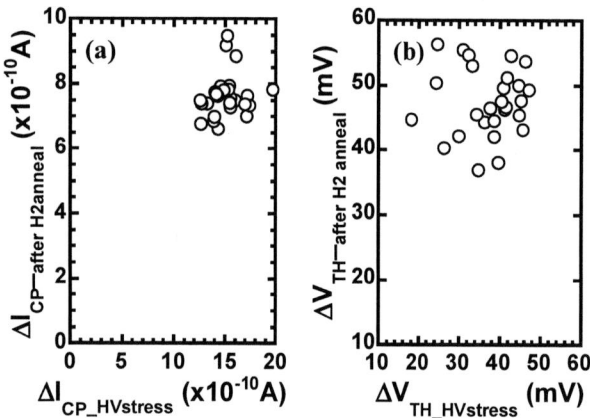

Figure 9 Correlation between high-voltage NBTI and low-voltage NBTI after hydrogen annealing. No relationship between them is observed.

Fig. 10 shows NBT stress time evolution of the median value and standard deviation of ΔI_{CP}. Both median and standard deviation of ΔI_{CP} for the hydrogen-annealed devices after high-voltage NBT stress is more marked than those for fresh devices.

In the same way, the median and standard deviation of ΔV_{TH} for the hydrogen-annealed devices is also larger compared to those for the fresh devices as shown in figure 11. In particular, standard deviation is degraded markedly by high-voltage NBT stressing.

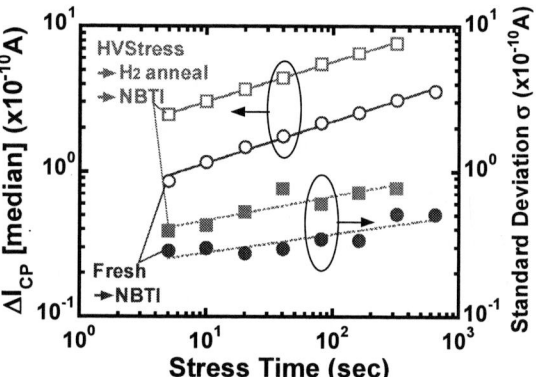

Figure 10 Median and standard deviation (σ) of ΔI_{CP} as a function of NBT stress. Median and σ of the hydrogen-annealed devices after high-voltage NBT stress is more degraded compared to the fresh devices.

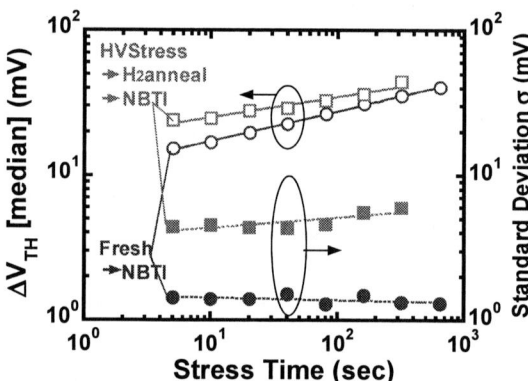

Figure 11 Median and standard deviation (σ) of ΔV_{TH} as a function of NBT stress. These Median and σ of the hydrogen-annealed devices after high-voltage NBT stress is also degraded markedly.

Figs. 12 and 13 show the correlation between the median values and standard deviations of ΔI_{CP} and ΔV_{TH}, respectively. It can be seen in figure 12 that the correlation for the hydrogen-annealed devices after high-voltage NBT stress shows the same tendency with that for fresh devices. From this result, it is inferred that the variability of interface-state generation arises from the quality of the gate oxide interface, irrespective of the stress condition. On the other hand, the correlation between median and the standard deviation of ΔV_{TH} for the hydrogen-annealed devices after high-voltage NBT stress does not coincide with that for fresh devices, and the degradation of the variability becomes more markedly by applying high-voltage NBT stress.

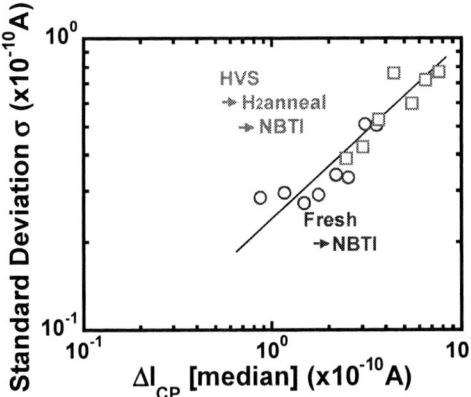

Figure 12 Correlation between median and standard deviation (σ) of ΔI_{CP}. The relationship for the hydrogen-annealed devices after high-voltage NBT stress is plotted on the same line of those for fresh devices.

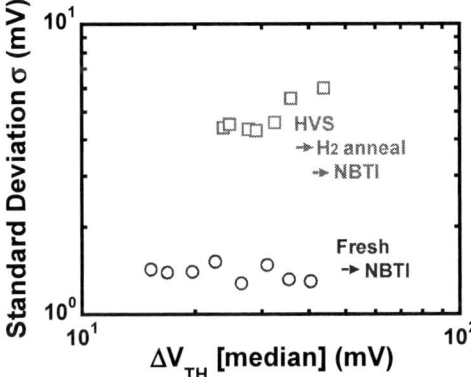

Figure 13 Correlation between median and standard deviation (σ) of ΔV_{TH}. The trend of the median-σ relationship does not coincide between that for the hydrogen-annealed devices after high-voltage NBT stress and that for fresh devices. The degradation of s becomes worse by applied high-voltage NBT stress.

Finally, in order to discuss the influence of the variability of interface-state generation on ΔV_{TH} variability, the median and the standard deviation of ΔV_{TH} is plotted as a function of those for ΔI_{CP} as shown in figure 14. It is found that both the median and deviation for the hydrogen-annealed devices after high-voltage NBT stress does not show the same tendency comparing to those for the fresh devices. From these results, it can be concluded that the variability of ΔV_{TH} arises from the variability of the quantity of hole-trapping precursors existing in gate oxide film rather than the variability of the quality of the interface.

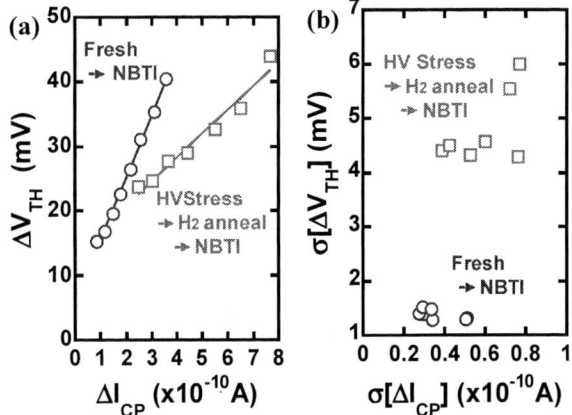

Figure 14 Relationships of median and σ between ΔI_{CP} and ΔV_{TH} of the hydrogen-annealed devices after high-voltage NBT stress. No correlation in median or σ is observed.

IV. CONCLUSIONS

In this paper, the origin of the variability of NBTI has been investigated using hydrogen annealing after NBT stress. As results, not only the median values of ΔI_{CP} and ΔV_{TH} but also these distributions can be fully recovered by hydrogen annealing in the case of low-voltage NBTI. Furthermore, NBTI behavior including its variability after hydrogen annealing corresponds to that for fresh devices. On the other hand, in the case of high-voltage NBTI applying, both the median and standard deviation after hydrogen annealing degrade more markedly comparing to those for the fresh devices, though the high-voltage NBTI also seems to be fully recovered as same as the fresh devices by hydrogen annealing. In addition, the deterioration of ΔI_{CP} distribution is less marked than that of ΔV_{TH}. From these experimental results, it is inferred that the variability of the quantity of hole-trapping precursors in gate oxide films is one of the dominant origins for the ΔV_{TH} distribution, rather than the variability of the quality of the gate oxide interface.

ACKNOWLEDGMENT

The authors would like to thank Drs. K. Kato, Y. Nakasaki, I. Hirano, M. Miyata and K. Matsuzawa for their thoughtful discussions and comments. In addition, I thank Dr. J. Koga for encouragement and support throughout this work.

REFERENCES

[1] N. Kimizuka, T. Yamamoto, T. Mogami, K. Yamaguchi, K.Imai, T.Horiuchi, "The impact of bias temperature instability for directtunneling ultra-thin gate oxide on MOSFET scaling," in Symposium on VLSI Technology Digest of Technical Papers, 1999, p.p. 73-74.

[2] N. Kimizuka, K. Yamaguchi, K. Imai, T. Iizuka, C. T. Liu, R. C. Keller, T. Horiuchi, "NBTI enhancement by nitrogen incorporation into ultrathin gate oxide for O.10-~m gate CMOS generation," in Symposium on VLSI Technology Digest of Technical Papers, 2000, p.p. 92-93.

[3] G. La Rosa, F. Guarin, S. Rauch, A. Acovic, J. Lukaitis, E. Crabbe, "NBTI-channel Hot Carrier Effects in PMOSFETs in Advanced CMOS Technologies," in Proc. IEEE International Reliability Physics Symposium proceedings, 1997, p.p. 282-286.

[4] K. Uwasawa, T. Yamamoto, T. Mogami, "A new degradation mode of scaled p+ polysilicon gate pMOSFETs induced by bias temperature (BT) instability," in IEEE International Electron Devices Meeting Technical Digest, 1995, p.p. 871-874.

[5] T. Yamamoto, K. Uwasawa, T. Mogami, "Bias temperature instability in scaled p+ polysilicon gate p-MOSFETs," IEEE Trans. Electron Devices, 46, 1999, p.p 921-926.

[6] Y. Mitani, M. Nagamine, H. Satake, A. Toriumi, "NBTI mechanism in ultra-thin gate dielectric - Nitrogen-originated mechanism in SiON," in IEEE International Electron Devices Meeting Technical Digest, 2002, p.p. 509-512.N. Kimizuka et al., in Symposium on VLSI Technology, p. 73 (1999).

[7] S. E. Rauch, "Review and Reexamination of Reliability Effects Related to NBTI Statistical Variations," IEEE Trans. Dev. Mat. Rel. 7, 2007, p.p. 524-530.

[8] V. Huard, C. Parthasarathy, C. Guerin, T. Valentin, E. Pion, M. Mammasse, N. Planes and L. Camus, "NBTI Degradation: From Transistor to SRAM Arrays," in IEEE International Reliability Physics Symposium proceedings, 2008, p.p. 289-300.

[9] B. Kaczer, T. Grasser1, Ph. J. Roussel, J. Franco, R. Degraeve, L.-A. Ragnarsson, E. Simoen, G. Groeseneken, H. Reisinger, "Origin of NBTI Variability in Deeply Scaled pFETs," in IEEE International Reliability Physics Symposium proceedings, 2010, p.p. 26-32 (2010).

[10] T. Grasser, B. Kaczer, W. Goes, Th. Aichinger, Ph. Hehenberger, and M. Nelhiebel, "A Two-Stage Model for Negative Bias Temperature Instability," in IEEE International Reliability Physics Symposium proceedings, 2009, p.p. 33-44.

[11] Y. Mitani and H. Satake, "Experimental Investigation of Related / Non-related Mechanisms with Released Hydrogen from SiON/Si Interface in NBT Degradation," in Ext. Abst. Solid State Devices and Materials (SSDM), 2004, p.p. 212-213 (2004).

Multiple Cell Upsets Tolerant Content-Addressable Memory

Syed Mohsin Abbas[1], Sanghyeon Baeg, Sungju Park
Hanyang University, ERICA Campus, Ansan, Korea
[1]smak85pk@ mslab.hanyang.ac.kr

Abstract — **Multiple cell upsets (MCUs) become more and more problematic as the size of technology reaches or goes below 65 nm. The percentage of MCUs is reported significantly larger than that of single cell upsets (SCUs) in 20nm technology. In SRAM and DRAM, MCUs are tackled by incorporating single-error correcting double-error detecting (SEC-DED) code and interleaved data columns. However, in content-addressable memory (CAM), column interleaving is not practically possible. It has been previously proposed that Hamming distance based approaches are good for SCUs but are not effective for MCUs. These schemes require a large number of extra parity bits for mitigating MCUs, and so they are not a practical solution for CAM devices. A novel error correction code (ECC) scheme is proposed in this paper that will cater for ever-increasing MCUs. This work demonstrated that *m* parity bits are sufficient to cater for up to *m*-bit MCUs, with an understanding of the physical grouping of MCUs. The results showed that the proposed scheme requires 85% fewer parity bits compared to traditional Hamming distance based schemes**

Keywords: Error correcting code; multiple cell upsets; soft-error rate; single-error correcting codes; parity bits; MCU confinement

I.   INTRODUCTION

Content-addressable memory (CAM) is very important due to its fast search capability, and is typically used for routing and policing in network routers and associating instructions in cache memories [1]. Due to disturbances from high-energy neutron particles, semiconductor memories are susceptible to soft errors [2]. Soft errors can be classified as single cell upsets (SCUs) or multiple cell upsets (MCUs). Baeg et al. observed MCU trends in 45-nm, 65-nm and 90-nm technologies [3]. Their observations clearly show that as technology is scaled down, MCUs become more and more problematic. Ibe et al. [4] simulated neutron-induced soft errors in SRAMs from a 250 nm to a 20 nm process; the results showed that the soft-error rate in SRAM increases by a factor of six to seven between 130 nm and 22 nm processes. The ratio of MCU to SCU increases by as much as 46% as the process technology node shrinks from 250 nm to 22 nm [4].

The interleaving approach has been used in memory devices to convert physical MCUs into logical SCUs; single-error correcting (SEC) codes can then be used to correct the logical SCUs. However, the interleaving approach is not practically possible for CAM due to the tight

coupling of the hardware structures from both cells and comparison circuit structures [5]. Hamming distance based error correction code (ECC) schemes are usually applied to CAM in order to protect data against SCUs and MCUs [6] [8]. However, these Hamming distance based schemes require greater overheads in terms of parity bits, and typically provide protection for up to one or two bit upsets. Single-error correcting double-error detecting (SEC-DED) codes are the most commonly used for CAM; however, they can only correct 1-bit errors and can only detect double-bit errors. The mitigation of MCUs requires a large number of additional parity bits. Therefore, an alternate scheme, which can mitigate MCUs at a lower cost in terms of parity bits, is required.

Pagiamtzis et al. [6] proposed applying Hamming distance based coding schemes to CAM in order to mitigate soft errors. They proposed the application of (81, 72) code, using nine parity bits along with 72 bits of data to form an 81-bit code word. Their scheme was limited to handling SCUs, meaning that MCUs are still problematic.

Noda et al. [7] proposed using embedded DRAM along with TCAM in order to mitigate SCUs and MCUs. In this approach, data entries are stored in both eDRAM and TCAM redundantly and a Hamming code ECC decoder/encoder is implemented at the interface between eDRAM and TCAM. In order to mitigate MCUs, during the refresh period of eDRAM, data are checked using ECC circuitry and corrected data are overwritten into the TCAM. One drawback of this approach is that it has redundancy and area overheads along with complex timing constraints.

Dutta et al. [8] proposed an ECC methodology for the correction of adjacent double-bit errors. Their method is an extension of SEC-DED codes and uses the same number of bits as SEC-DED codes. Their proposed code can detect adjacent double-bit errors, but is not effective for larger MCUs. In short, to the best of our knowledge, previous works have primarily targeted SCU-related issues with a limited handling of MCUs; no simple and efficient technique exists for the mitigation of MCUs in CAM.

In this research, a novel ECC scheme is proposed which can mitigate MCUs. Our scheme requires 85% fewer parity bits compared to previous Hamming distance based schemes. The rest of the paper is organized as follows. Section II briefly describes CAM architecture and MCU confinement. The proposed ECC scheme is presented in section III. The results and discussion are presented in section IV, and the conclusions are presented in section V.

This research was supported in part by the "GRRC" Project of Gyeonggi Provincial Government, Republic of Korea and the National Research Foundation of Korea (NRF) grant (MEST) (No. 2010-0026822).

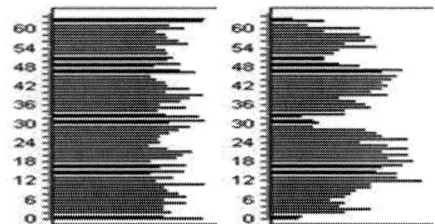

Fig. 1: Distribution of SCU (left) and 8-bit MCU (right) for the SRAM with well tapping every 32 bits in the Y-direction. The errors are folded every 64 bits for both SCU and MCU cases.

II. PIPELINED CAM ARCHITECTURE AND MCU CONFINEMENT

An accelerated test was performed for an SRAM device. The device had well tapping every 32 bits of SRAM cells in the Y-direction. Fig. 1 shows the distribution of SCU and 8-bit MCU. All error events were collected at intervals of 64 bits and overlaid in Fig. 1. The SCU events to the left of Fig. 1 were randomly distributed over the entire space of SRAM. However, the effect of the well tapping is noticeable for the case of MCU to the right of Fig. 1; MCU distribution is mostly distributed between the well-tapped points.

MCU occurring at the boundary of well-tapped points does not necessarily cross over well tapping. Fig. 2 shows one example of a 40-bit MCU, which was a very rare case. We could see that there are MCUs in the Y-direction in columns 33 and 34. The left-hand case is between well-tapped points. The right-hand case shows that the MCU crossed over the well tapping. However, most of the upset was on one side of the well-tapped points.

With the moderate efforts from substrate and design engineering practices, MCU can be confined within a segment such as the bits between well tapping. In pipelined CAM architecture, a match-line is divided into multiple segments [9]. Each segment can evaluate the match result independently from the other segments, as shown in Fig. 3. This work develops a new error-correction scheme with two fundamental observations: MCU is confined in a segment with the aid of substrate and design engineering, and the pipelined CAM architecture can be used to independently process a segment.

III. PROPOSED ERROR-CORRECTION SCHEME

Fig. 4 shows the physical CAM structure with N data segments and one parity-bit segment, which is the underlying architecture in this work. The boundary of two adjacent segments is structured, to prevent an MCU from crossing over, with the aid of the substrate engineering practices. In the proposed coding scheme, data bits are divided into M groups and a parity bit is calculated for each group, which make M parity bits. The data bits are then placed at data segments in the way that no two bits from a group belong to the same segment and each segment contains M data bits. All parity bits are placed in the parity-bit segment.

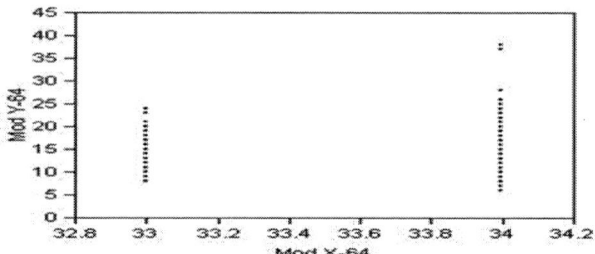

Fig. 2: Distribution of 40-bit MCU (X-axis shows the column of a memory and Y-axis shows the bit positions)

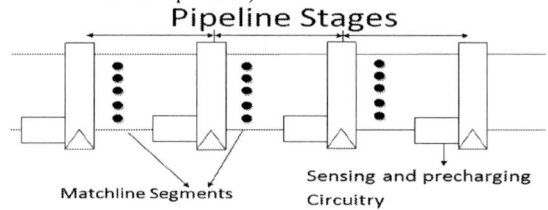

Fig. 3: Pipelined match-line: each segment can evaluate the match result independently from other segments.

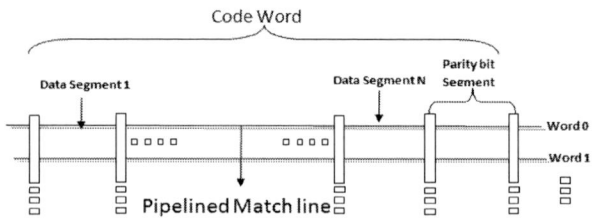

Fig. 4: Pipeline CAM architecture: the pipeline is divided into multiple segments, each segment storing a fixed number of data bits.

A theorem is developed for generating parity bits; the theorem is based on the fact that one parity bit can detect single-bit errors in a data group. n is the number of bits in a data word and m is the size of the MCU. n bits of data are divided into m groups and k is the number of bits in each group.

Theorem: One parity bit is generated to detect a single-bit error in a k-bit data group. The m parity bits from m data groups can detect up to m-bit MCU errors in a segment, where $m < n$.

Proof: Let n bits of data be divided into m groups, each group having k bits such that $k = n/m$. Then a parity bit is independently calculated for each group. Let p_1 be the parity bit for the first group, p_2 for the second group and p_m is the parity bit for the last mth group; as a result, there are m groups of data and m parity bits. If only one bit changes from any group, then from the corresponding parity bit we can detect this 1-bit error.

The bits in m groups are assigned to N physical segments in a word such that only one bit from m groups is exclusively mapped to one segment. For example, the first bits from m groups are mapped to the first segment; this concept is similar to interleaving technique in memory design, and each segment contains m data bits. All parity bits are placed in the parity-bit segment.

Since each segment contains a bit from m groups, an MCU in a segment will be translated as 1-bit error in m

groups; the parity bits can detect the single-bit error by definition. Therefore, m parity bits are sufficient to detect any MCU as long as the MCU is contained within a segment of m bits.

Suppose that 15 bits of a data word are divided into three groups of five bits each, so that $n = 15$, $m = 3$, $k=5$. The first group, D_1 is composed of five data bits, a_1, a_2, a_3, a_4, a_5 as shown in Fig. 5. Similarly the second group, D_2 is composed of b_1, b_2, b_3, b_4, b_5 and the third, D_3 is composed of c_1, c_2, c_3, c_4, c_5. Parity bits are calculated for each group. p_1 is the parity of the first group , p_2 is the parity of the second group and p_3 is the parity of the third group, as shown in Fig. 5. If only one bit erroneously changes from any group, then its corresponding parity bit will detect this 1-bit error.

Now we have CAM with N ($N=5$) data segments and one parity-bit segment. The data bits are then placed at data segments in such a way that no two bits from a group belong to one segment and each segment contains m ($m=3$) data bits as shown in Fig. 6. If an MCU occurs and it is limited to one segment only, with the help of corresponding parity bits, the MCU will always be detected. This is because, in a segment, each bit is from different group S_1, S_2, S_3 .

Due to parity bits, each valid code word differs from other code word by at least 2 segments. During the search operation, the search key is compared with all of the code words stored in the CAM. The search key is also a code word having data bits concatenated with parity bits. All of the groups of the search key are compared with all of the segments of a stored code word. If more than one segment is a mismatch, then the search key is considered to be mismatched with the stored code word; otherwise, it is considered as a match. In other words, if the search key and the stored code word differ from one another by only one segment, this difference can be considered to be originated from soft errors and the search can be declared to be successful. When more than one segment of the search key and the stored code words do not match, it is declared to be a mismatch. For example, the data stored in CAM are depicted in Table 1, comprising four bits of data and two parity bits. This 6-bit code word is divided into two data segments and one parity-bit segments, as shown in Table 1. Suppose that a

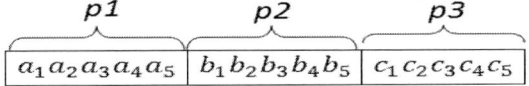

Fig. 5: A parity bit is generated for each group in a data word: p_1, p_2, and p_3 are parity bits for the first, second, and third groups of a data word respectively.

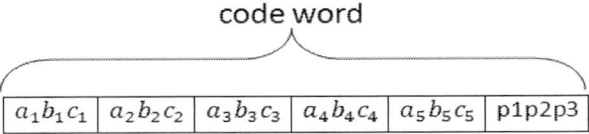

Fig. 6: Data bits are placed on data segments such that each segment contains bits from different groups and parity bits are placed in parity-bit segment.

Table 1: Data bits are stored in CAM with a word length of six. Each word is divided into three segments, with two bits in each segment.

Seg. 1	Seg. 2	Parity
00	00	00
00	01	01
00	10	10
00	11	11
01	00	01
01	01	00
01	10	11
01	11	10
10	00	10
10	01	11
10	10	00
10	11	01
11	00	11
11	01	10
11	10	01
11	11	00

neutron hits and the fourth code word 001111 is changed to 000011; it is considered a two-bit MCU.

When the search key 001111 is applied across CAM, the fourth code word differs from the search key by just one segment; all of the other stored entries differ by at least two segments. As long as only one segment is corrupted, it is considered as a match. In this way, m bit MCU can be tolerated adding only m parity bits.

IV.   RESULTS AND DISCUSSION

The proposed ECC scheme has been compared with previous Hamming distance based schemes [6] [8]. For the creation of Hamming distance between code words, BCH encoding is used, and the required number of parity bits is calculated. The parity bits required for Hamming distance based schemes were then compared with the parity bits required for our proposed ECC scheme. For any integer $i \geq 3$ and $m \leq 2^{i-1}$ there exists a primitive BCH code with the following properties [10].

Block length: $n = 2^i - 1$
Parity check bits: $n - k \leq mi$
Minimum distance: $d \geq 2m + 1$

This code can correct m or fewer errors over a span of $2^i - 1$ bits.

For 512 x 72 CAM, $i = 7$, and so minimal block length becomes 127 bits. Therefore a truncated BCH code set is used for calculating parity bits, as used by Cheng et al. [12] for their W-ATM protocol design. A particularly popular SEC-DED (72, 64) code is a truncated (127, 120) Hamming code [11]. Table 2 shows the calculation of parity bits for different MCU sizes; the last column shows the truncated BCH codes.

The parity bits required for Hamming distance based schemes are then compared with the parity bits required for our proposed ECC scheme, as shown in Fig. 7. The X-axis shows the size of MCU to be mitigated, and the Y-axis shows the parity bits required. Long gray bars show the number of bits required by Hamming distance based schemes and short black bars show parity bits required by our schemes. For example, if a 5-bit MCU is to be tolerated, then according to Hamming distance based schemes a

978-1-4244-9113-1/11 $26.00 © 2011 IEEE

Hamming distance of 11 is needed between code words. In order to create a Hamming distance of 11 between code words, 35 parity bits are needed for the truncated BCH code (72, 36), if the word length is assumed to be 72 bits for CAM. On the other hand, by using the proposed ECC scheme, only five parity bits are needed to mitigate a 5-bit MCU, provided the MCU is confined to a segment of five bits in size.

By applying the proposed ECC scheme, the number of parity bits required is reduced up to a maximum of 85% as shown in the example case. This saving of parity bits comes from the assumption that the MCU is confined in one segment. Based upon this assumption, the number of parity bits required for MCU mitigation is equal to the segment size.

In the application of the proposed ECC scheme, a trade-off exists between the segment size and the number of segments. Segment size is directly related to MCU size; the greater the segment size, the greater the size of MCU which can be tolerated. On the other hand, the number of segments is inversely proportional to segment size. Fig. 8 shows the relationship between segment size and the number of segments. The gray line shows the number of segments and the black line shows segments size; as we can see, with the increase in segment size, the no. of segments required is decreased. For example, if we have 512 x 72 CAM, and each segment is nine bits, then we require eight segments along one word line and we are able to mitigate MCU of up to eight bits.

Table 2: Calculation of parity bits using a truncated BCH coding scheme

MCU size (no. of bits) m	Hamming distance $2m + 1$	Parity bits required mi	Code word (n,k)	Truncated code word (n,k)
1	3	7	(127, 120)	(72, 64)
2	5	14	(127, 113)	(72, 57)
3	7	21	(127, 106)	(72, 50)
4	9	28	(127, 99)	(72, 43)
5	11	35	(127, 92)	(72, 36)
6	13	42	(127, 85)	(72, 29)
7	15	49	(127, 78)	(72, 22)
8	17	56	(127, 71)	(72, 15)

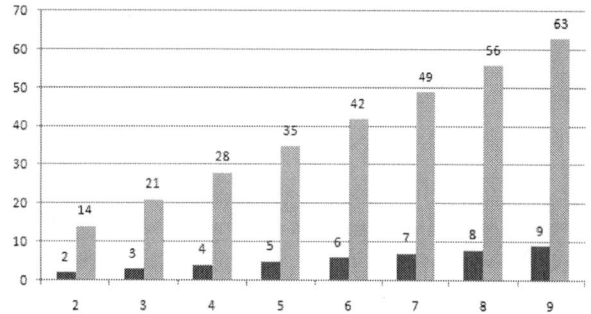

Fig. 7: Parity overhead comparison (X-axis shows the MCU size to be mitigated and Y-axis shows parity bits required.)

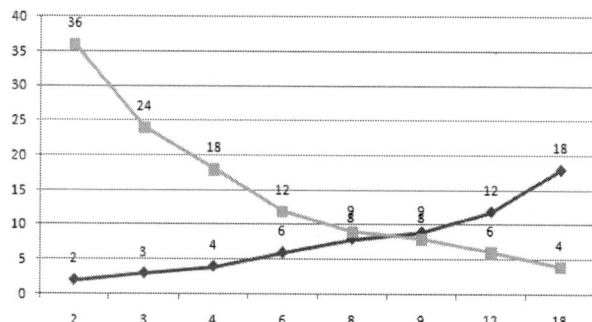

Fig. 8: Comparison of the number of segments required for a particular segment size (X-axis shows the segment size and Y-axis shows the number of segments required)

MCUs corrupting the data in CAM either result in a false hit or a false miss during search operation. The application of the proposed scheme can guarantee elimination of false miss cases; however some cases of false hits might result. One example of a false hit can be the case in which an MCU crosses the boundary between segments. An accelerated test performed for an SRAM device shows that some MCUs can cross the boundaries between segments, as shown in Fig. 2. Although these events are believed to be rare, an MCU crossing the boundary can corrupt the code word in such a way that it can become another valid code word; such code words are called false positives. False positives cannot be detected by any ECC scheme and result in the failure of the ECC scheme. In future we wish to analyze the failure probability of our proposed scheme in case of MCUs crossing the boundary and false hits.

V. CONCLUSION

In this work, a novel ECC scheme has been proposed to cater for the ever-increasing MCU problem in CAM. A 512 x 72 CAM was used to analyze this ECC scheme. The parity bits required by previous Hamming distance based schemes were compared with the parity bits required by the proposed ECC scheme. The application of the proposed ECC scheme resulted in a saving of 85% of the parity bits required. This saving of parity bits came from the assumption that the MCU would be confined to one segment due to best practices applied in substrate and design engineering. This scheme resulted in a trade-off in terms of calculating segment size and the number of segments, and this trade-off was also estimated in this research work.

ACKNOWLEDGMENTS

The authors would like to thank ShiJie Wen and Rick Wong for providing soft-error test data. The efforts of the Higher Education Commission (HEC) in Pakistan in providing funding to students pursuing higher education in Hanyang University, Korea are also acknowledged.

REFERENCES

[1] Kostas Pagiamtzis and Ali Sheikholeslami, "Content-Addressable Memory (CAM) Circuits and Architectures: A Tutorial and Survey", IEEE Trans. on Solid-State Circuits, vol. 41, no. 3, 2003, pp. 712–727.

[2] http://selse5.selse.org/Papers/selse5_submission_7.pdf.

[3] Sanghyeon Baeg, ShiJie Wen and Richard Wong, "SRAM Interleaving Distance Selection with a Soft Error Failure Model", IEEE Trans. on Nucl. Sci., vol. 56, no. 4, Aug. 2009, pp. 2111–2118.

[4] Eishi Ibe, Hitoshi Taniguchi, Yasuo Yahagi, Ken-ichi Shimbo and Tadanobu Toba, "Impact of Scaling on Neutron Induced Soft Error in SRAMs from an 250 nm to a 22 nm Design Rule", IEEE Trans. on Electron Devices, vol. 57, no. 7, July 2010, pp. 1527–1538.

[5] Sanghyeon Baeg, ShiJie Wen and Richard Wong, "Minimizing Soft Errors in TCAM devices: A Probabilistic Approach to Determining Scrubbing Intervals", IEEE Trans. on Circuits and Systems, Reg. papers, vol. 57, no. 4, Apr. 2010, pp. 814–822.

[6] K. Pagiamtzis, N. Azizi and F.N. Najm, "A Soft-Error Tolerant Content-Addressable Memory (CAM) Using an Error-Correcting-Match Scheme," Proc. IEEE Custom Integrated Circuits Conf. (CICC '06), 2006, pp. 301–304.

[7] Hideyuki Noda, Katsumi Dosaka, Fukashi Morishita and Kazutami Arimoto, "A Soft-Error-Immune Maintenance-Free TCAM Architecture with Associated Embedded DRAM", Proc. IEEE Custom Integrated Circuits Conf., 2005, pp. 451–454.

[8] Avijit Dutta and Nur A. Touba, "Multiple Bit Upset Tolerant Memory Using a Selective Cycle Avoidance Based SEC-DED-DAEC Code", Proc. 25th IEEE VLSI Test Symposium, May 2007, pp. 349–354.

[9] Kostas Pagiamtzis and Ali Sheikholeslami, "Pipelined Match-Lines and Hierarchical Search-Lines Content-Addressable Memories", Proc. IEEE Custom Integrated Circuits Conf., 2003, pp. 383–386.

[10] http://cwww.ee.nctu.edu.tw/course/channel_coding/chap5.pdf.

[11] http://en.wikipedia.org/wiki/Hamming_code.

[12] Fang-Chen Cheng and Jack M. Holtzman, "Wireless Intelligent ATM Network and Protocol Design for Future Personal Communication Systems", IEEE Trans. on Selected Areas in Communication, vol. 15, no. 7, 1997, pp. 1289–1307.

An Automated Approach to Isolate Dominant SER Susceptibilities in Microcircuits

James Castillo, David Mavis, Paul Eaton, Mike Sibley, Don Elkins, Rich Floyd
Microelectronics Research Development Corporation
Albuquerque, NM. USA
Phone: (505)-294-1962, email: james.castillo@micro-rdc.com

Abstract —With the ever decreasing feature sizes of modern integrated circuits (IC), new test and analysis approaches are needed to isolate dominant soft error rate (SER) susceptibilities. A new test capability which provides SER raster scanning of microcircuits with a collimated heavy-ion beam, having spatial isolation as small as 10 microns, is presented. The system termed the Milli-Beam^{TM} provides, through post processing, three-dimensional surface plots showing the location of error counts over an entire IC. A new cross section measurement technique that accounts for beam variations and uncertainties independent of laboratory dosimetry is also presented.

Keywords- Heavy-ion testing; Milli-Beam; SER; Soft Error Rate; SEU; Single Event Upset; SEE; Single Event Effect; SET; Single Event Transient

I. INTRODUCTION

As microcircuit feature sizes decrease, single event upset (SEU), single event transient (SET), and multiple bit upset (MBU) effects, due to cosmic-rays, dominate the radiation response of these circuits in space applications. In high-altitude and terrestrial applications, cosmic-ray neutron recoil byproducts can easily produce an unacceptable soft error rate (SER) in modern microcircuits. These cosmic-ray showers are formed when an energetic heavy ion from space undergoes a nuclear reaction in the atmosphere. As these charged particles pass through microcircuits, they produce electron-hole pairs which may be collected by the junctions in the device. If a sufficient amount of charge is deposited on a given node, an error either in the form of a flipped bit or a transient pulse may occur. [1, 2, 3, 4]

Process modifications and engineered substrate attempts have not provided significant levels of single event effect (SEE) mitigation. Circuit-level hardening approaches have, however, proven effective in mitigating a number of important heavy-ion related effects. The size and speed penalties associated with these circuit hardening techniques often cannot be tolerated in commercial product designs. For this reason, experimental SEE characterization is necessary to identify dominant response mechanisms so that critical circuits can be identified and hardened with minimal impact on overall integrated circuit (IC) performance and permit the most effective trade-off between SER and the area/speed overhead.

Broad-beam heavy-ion testing, such as that performed at the Lawrence Berkeley Laboratory (LBL) 88-inch cyclotron [5], is often used to measure the SEE response of various ICs. Characterization of SEE basic mechanisms is usually performed using specially designed test chips while qualification testing is often performed using large ICs containing complex designs. For complex designs, conventional broad-beam test data provides limited data to isolate the exact cause of observed errors and little insight into potential design improvements.

New test methods are needed to address these issues and help isolate dominant circuit susceptibilities of complex modern microcircuits. This paper provides an overview of the Milli-Beam™ test apparatus and associated data acquisition software that provides rapid SER raster scanning to physically isolate dominant circuit susceptibilities. The paper will then illustrate a method for implementing an independent dosimetry technique to correct for variations in beam focus and position. Finally, results from testing high performance microcircuits will be described. While the Milli-Beam is presently used at the LBL 88-inch cyclotron facility, it can in principle be used at any heavy-ion test facility.

II. DESCRIPTION OF THE MILLI-BEAM SYSTEM

The Milli-Beam system diagram is shown as it currently exists for use at the LBL heavy-ion test facility in Fig. 1. Two X-Y stages are mounted on a shelf attached inside of the Berkeley vacuum chamber wall where the four inch beam line entrance port is located. Each stage holds a small square aperture that can be maneuvered in both the horizontal (X) and the vertical (Y) direction with better than 1 μm accuracy. The primary aperture is the critical component of the system as it defines, through collimation, the beam size and position on the device under test (DUT). The secondary aperture is used to prevent edge-scattered ions from the primary aperture reaching the DUT. Each stage can be independently controlled using either LabVIEW or a Perl/Tk based software developed by Micro-RDC for improved automation of the whole system.

The authors wish to express appreciation to the Defense Threat Reduction Agency (DTRA) for its support of the design and development of the Milli-Beam heavy-ion test capability described in this paper.

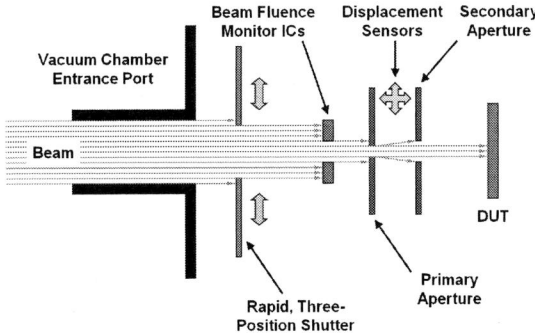

Figure 1. Milli-Beam hardware schematic diagram

The Milli-Beam differs from other beam manipulating devices, such as microbeams, because it is not ion dependent and can be used at any heavy-ion particle accelerator facility as a result of its portability. Very high intensities, nearly 10^9 cm$^{-2}\cdot$s^{-1} flux, have been achieved at the LBL facility which allows for a rapid automated raster scan over a DUT since each raster position only needs to be in place for a few seconds. Abundant selections of aperture configurations are also available ranging from 10 to 250 microns slit widths. Additionally, all the Milli-Beam components have been designed to withstand operation in a vacuum [6].

A. X-Y Stages

Each X-Y stage is constructed from two linear actuator stages which can be positioned with sub-micron resolution and high repeatability [7]. The linear actuators controlling the horizontal (X) position are mounted directly on the mounting table while the vertical (Y) actuators are mounted on the X-actuators. The machined aluminum alloy stages are also designed to be vacuum compatible and use linear guide bearings to provide a smooth motion with maximum load specifications well above the implemented application. The stages also employ a specialized drive nut that offers minimal backlash operation that automatically adjusts for wear and is further accounted for by the controlling software. Separate connectors for motor power and limit/encoder signals are used for ease of setup and operation. Limit switches are also incorporated in each stage to define the travel and protect the high torque motors from driving the carriage into itself. The beam apertures are then mounted on their respective X-Y actuator.

B. Apertures

The beam apertures are formed by back-to-back horizontal and vertical slits constructed from steel feeler-gauge material. Various slit sizes have been constructed (10, 25, 50, 75, 100, and 250 µm) for both horizontal and vertical beam definition. This approach enables square collimation (equal width and height), rectangular collimation (different width and height), or simple slits (either horizontal or vertical).

The aperture assemblies are secured to the X-Y actuators using a slit mounting system designed for any variation of the collimating configuration. The mounting system retains the collimator assemblies in a slotted holder. Retention is achieved using a pressure plate that derives its forces from a system of Neodymium rare earth magnets. Alignment is reproducibly achieved by abutments between the right and bottom collimator assembly edges with the right and bottom holder slot edges. The beam operator can use a simple set of pliers to remove and replace collimator assemblies from a safe distance so that beam activation of the stages is not an issue.

C. Shutter

A three position rapid beam shutter is incorporated as part of the Milli-Beam system. The shutter is controlled by the same software package that controls the raster scanning. During a normal raster scan the shutter is initially blocking the beam over the starting position. The shutter then moves to the middle position to expose the DUT coordinate to be irradiated until a preset fluence level. Once the fluence level has been reached, the shutter receives a command from the software to actuate into the third position to once again block the beam. The X-Y stages then maneuvers the beam collimator into the next coordinate and the shutter sequence is repeated for all locations programmed into the preset raster scan using position three as the initial position. This alternating sequence ensures symmetric exposure times on each raster position and a beam fluence monitor is utilized to correct for any variations in beam flux. If it proves necessary to jump over a sensitive circuit, such as a PLL (phase locked loop) or a control mode register, not relevant to the measurement being taken, the beam shutter can be actuated and held in position to block the beam during the raster scan steps.

D. Beam Monitor Fluence ICs

Micro-RDC has developed specialized beam monitor chips that were fabricated in the IBM 9LP 90 nm bulk CMOS processes. Physically each chip is less than 1 mm × 1 mm in size and contains eight separate set reset flip-flop (RSFF) chains. Each RSFF chain has 1024 stages, with a total measured SEU saturated cross section on the order of 5×10^{-6} cm^2. Three select bits control the total number of chains that are active, providing a variable saturated cross section between 5×10^{-6} cm^2 and 4×10^{-5} cm^2 in 5×10^{-6} cm^2 increments. A 3 mm hole is bored through the center of a 180 PGA ceramic package and four beam monitor chips are attached symmetrically around this hole within 2 mm of the center of the primary aperture. The chips are wire bonded to the package and the package is mounted in a ZIF socket mounted on a PCB and attached to the slit mounting hardware bracket described above, also depicted in Fig. 2. Four of these beam monitor chips placed symmetrically around the primary Milli-Beam aperture therefore has a maximum total saturated cross section on the order of 1.6 $\times10^{-4}$ cm^2.

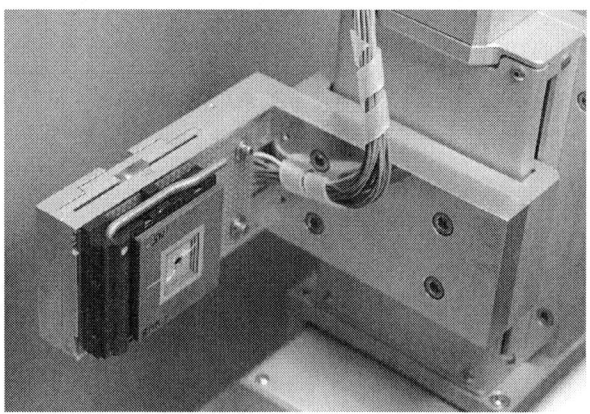

Figure 2. Beam monitor assembly fixed to the X-Y stage along with the slit holder, pressure plate and primary aperture.

Figure 3. Upstream view of primary aperture, shutter, actuator and beam entrance port.

The dead time of the beam monitor is proportional to the sum of the following processing times:

- The mean transit time of an RSFF SEU to the end of its chain,

- The subsequent propagation time through the logic selecting and combining errors from different chains,

- The delay through the IC output pads driving the FPGA monitoring software,

- The time needed by the FPGA to sense the error and generate a reset pulse, and

- The beam monitor on-chip reset delay.

To assess the amount of dead-time in the measurements, so that it could be corrected in our analysis, we performed Monte-Carlo simulations of the circuit and readout circuitry. These simulations provide dead time correction (DTC) factors to account for events that might occur during the processing of a previous event, and subsequently not be counted [8]. The DTC factor will have a value of 1.0 for the case that no events are lost (very low count rates) and values >1.0 for higher count rates. To correct the beam monitor value, one simply multiplies the measured number of beam monitor counts by whatever DTC factor is appropriate for the event rate at which the device is operating.

SPICE simulations as well as electrical bench measurements were used to estimate each of these processing times [9]. Monte-Carlo simulations were then performed at various beam monitor count rates expected for the particle fluxes achievable at Berkeley. At a Berkeley flux of 1×10^7 cm$^{-2} \cdot$s^{-1} and a maximum device cross section of 16×10^{-5} cm^2, about 1600 counts/s should be observed in the IC.

From the simulations, at this flux, the correction factor was found to be no more than 1.00006 and can subsequently be ignored. This however suggests making an experimental measurement of the DTC factor as a function of available IC event rates would be next to impossible due to the small effect expected. However, the FPGA could control the amount of dead time that contributes to the total dead time (item 4 in the above list of contributing processing times). The FPGA test software was therefore modified to include this capability to be able to experimentally verify the Monte Carlo simulation. Measurable values are considered in the range of 10% to 20% loss of counts, corresponding to DTC values of 1.10 to 1.20, respectively. Simulations were performed at various beam monitor error rates. At an error rate of 200 counts per second, which was estimated for a beam flux of 10^6 cm$^{-2} \cdot$s^{-1}, a 1 ms FPGA delay should provide a DTC value on the order of 1.20. The observed agreement between the experimental data and Monte-Carlo simulations proved the DTC estimates were correct for the beam monitor IC.

Routine recalibrations of the beam monitors have been executed to measure any performance changes overtime such as total dose effects. However, none of these effects have been found to contribute a significant alteration to the beam monitors performance. The beam monitors also have a linear response to changes in beam fluence.

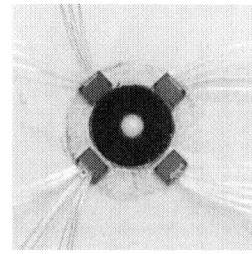

Figure 4. Zoomed view of four wire bonded beam monitors

E. Software

Two separate software packages have been developed for the control and automation of the Milli-Beam system. The initial programming language was based in National Instruments LabVIEW [10]. Due to encountered complexities, a duplicate software package was created using Perl and interfaced with Tk. The software is responsible for computing coordinate transformations, setting the beam position on the DUT, controlling and monitoring the runs, and communicates with the FPGA test board. The software program stops data acquisition at preset fluences as measured by a microprocessor controlling the beam monitor chips, steps the Milli-Beam aperture position, updates the FPGA test board with the new coordinate position, and resumes data acquisition. All raster scan data is contained a single ASCII data file, with each observed error having an X-Y position tag. Data reduction to extract and display positional information is finally performed by Perl post-processing scripts.

III. INITIAL DEVELOPMENT

While developing the Milli-Beam system, several preliminary measurements stepping over simple SRAM devices to calibrate the system were performed [11]. In the course of these measurements (which should display a uniform error cross section) error rate variations well outside statistical uncertainties were observed. These variations were easily attributed to beam flux variations. Similar beam calibration problems in the past can be attributed to the calorimetry technique used at Berkeley. Scintillators monitor the beam flux at the extreme edges (top, bottom, left, and right) and a periodic calibration establishes the intensity ratio between the beam center (as measured by a removable center scintillator) and the edge scintillators. This ratio can be quite sensitive to the broad-beam profile and can change over time with variations in beam tuning and focusing. If the beam is focused tighter the flux has a higher concentration in the center than the edges, resulting in the edge scintillators predicting a lower flux. Likewise, if the beam defocuses, the center has a lower flux concentration than the edges and is predicted to be higher. Beam focus is likely to change whenever ions are switched during a test. Past experiments have shown well over 10% variations in error cross sections made in back-to-back measurements for identical fluences.

For this reason independent beam monitors are incorporated as part of the Milli-Beam apparatus. These beam monitors can control the run time at each raster step using a desired fluence level. Fig. 5 demonstrates the accuracy achievable using the four detector Milli-Beam fluence monitoring system. The data points with error bars represent beam monitor counts, proportional to the fluence, at the primary aperture location. The solid line labeled prediction is computed from the four curves labeled Monitor 1 through 4. These results indicate that for a total accumulated fluence of 10^7 cm^{-2} at each raster scan position, the fluence normalization will be accurate to 1.6%, based on the Poisson counting statistics in the four monitor chips. The calibration procedure for these monitors will be described later in the paper.

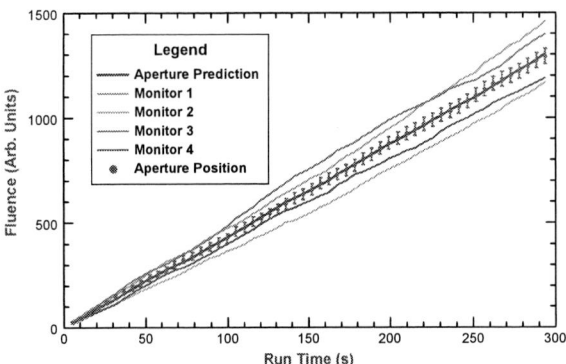

Figure 5. Accuracy of the Milli-Beam fluence monitor.

A. Ion Beam Considerations

The finite emittance of the heavy-ion beam along with edge scattering from defining slits and collimators are the major factors that must be considered in constructing and using the Milli-Beam technique. Beam emittance is defined as the area an ellipse in Rθ phase space containing all particles in the beam at any given point along the beam axis where R is the radius of a particle in the beam, θ is the angle dR/dz of the particle trajectory, and z is the beam axis coordinate [12]. The emittance area is invariant as the beam is transported which means that particle trajectories become more parallel as the beam diameter is made larger. To ensure beam uniformity over the area of an IC, the beam diameter has historically been made as large as possible (~10 cm) as it enters the chamber. This fortunately means that the angular spread of the beam particles in minimized since as θ will decrease inversely as R is increased to maintain a constant emittance ellipse area.

The angular spread of the Berkeley 88-inch cyclotron beam as it enters the vacuum chamber was measured and a Gaussian distribution of angles with a sigma of 0.0025° was found. This angular spread means that for a defining collimator 40 cm before the DUT, all edges of beam will degrade on the order of 18 μm. Thus, a 10 μm diameter beam will blow up to ~46 μm by the time it reaches the IC and will not have well defined edges. This can be seen in Fig. 6 as the rounded edges of the beam profile of a 100 μm square beam.

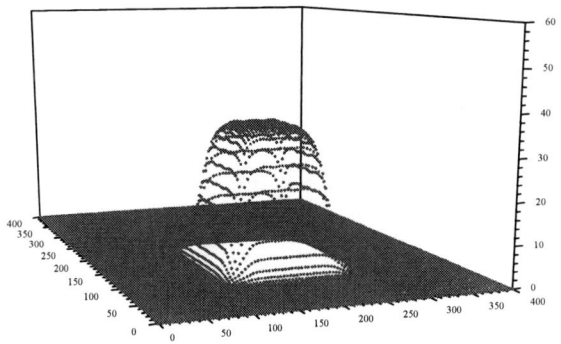

Figure 6. SRAM 40 cm from the defining aperture.

By placing the DUT within 5 cm to the defining apertures, the edge degradation is reduced to less than 2 μm, as seen by the sharper edges of the profile in the plot of Fig 7.

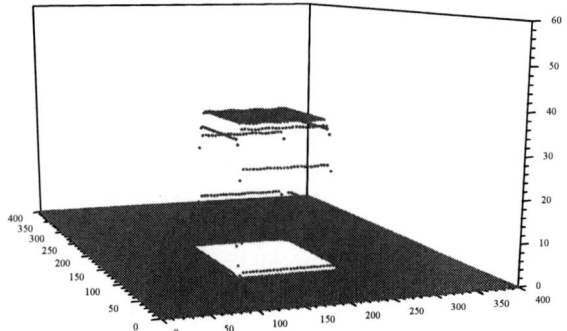

Figure 7.   SRAM 5 cm from the defining aperture.

The surface plots in Figures 6 and 7 were obtained by performing a least-squares fit to the errors produced in an SRAM positioned 40 cm and 5 cm, respectively, from the Milli-Beam primary aperture. The fitting function consisted of a 2-dimensional convolution of a Gaussian product $Z(X)*Z(Y)$ with a box in X-Y-Z space. The X and Y dimensions represent the width and height of the Milli-Beam, respectively, and the Z dimension represents the number of observed errors. The parameters of the fitting function included total number of errors under the convolution integral, the Milli-Beam width and height (before edge washout by the Gaussian), the center location of the beam (the mean X and Y), and the edge washout parameters σ_x and σ_y (the sigma values for the $Z(X)$ and $Z(Y)$ Gaussians, respectively).

While this edge degradation does not present many problems for larger beams (~100 μm to 200 μm diameter), it does mean that the defining apertures must be much nearer the DUT for smaller beams (~10 μm to 20 μm diameter). Beam scattering from the edges of the defining apertures now presents new problems that must be addressed in constructing the Milli-Beam apparatus. For this reason we use the second cleanup aperture to remove these scattered particles so that heavy-ion strikes will not occur outside the targeted region on the DUT.

B.   Raster Scan Coordinate Calibration

In any particular heavy-ion Milli-Beam test, the displacement and rotation of the DUT relative to a calibration SRAM on a test board must be measured. These measurements must determine three parameters to describe the relative displacement (both X and Y) and the relative angle of rotation between the devices. By finding a known coordinate on the SRAM, the parameters allow areas of interest on the DUT to be reliably and quickly targeted during a test.

The calibration SRAM and DUT can be mounted on the same test board or can be mounted in separate packages that share the same socket on the test board. In the first case, measurements are made of the DUT relative to the SRAM. In the second case, each chip location is measured separately to obtain displacement and angle parameters relative to two known positions on the board. The extracted results for each chip are then subtracted to provide the final DUT to SRAM displacement vector and relative rotation angle.

These measurements were made using a stationary high powered (150× objective) microscope with the DUT/SRAM test board mounted on an accurate (to 1 μm) X-Y stage. Supporting Perl scripts extract these parameters at each of four different test board rotations and subsequently combine the results to remove systematic errors associated with any non-orthogonality of the optical axis with the plane of X-Y stage. These Perl scripts give both values and expected uncertainties of the DUT and SRAM relative displacement vector as well as the angle of rotation of the DUT relative to the calibration SRAM.

Two additional matrix parameters then need to be extracted at the heavy-ion facility. The first of these is the angle between the calibration SRAM Y-axis and the Milli-Beam Y-axis. The second is the angular deviation from orthogonal of the Milli-Beam X and Y stages. (It should be noted that physical systems, such as right angle mounting brackets, cannot be machined with sufficient precision and must therefore be measured and accounted for in the various matrix transformation operations.)

This is done by making Milli-Beam ΔY steps with ΔX=0 and extracting the relative angle (θ_y) of the Y stage with respect to the SRAM die. By holding ΔY=0 and making ΔX steps gives a measurement from which one can extract both the X stage angle with respect to the IC and determine the angle of non-orthogonality between the Y actuator and the X actuator ($\theta_\perp$). These values are subsequently used, through the use of appropriate coordinate transformations, to target known positions on the DUT.

As in any data reduction and error analysis effort, establishing estimates of final parameter variances is just as important as determining the final parameter values themselves. The expected uncertainties are computed using the standard error propagation formula,

$$\sigma_y^2 = \sigma_u^2 \left(\frac{\partial f}{\partial u}\right)^2 + \sigma_v^2 \left(\frac{\partial f}{\partial v}\right)^2 + \cdots \qquad (1)$$

where σ_y is the desired parameter uncertainty, σ_u and σ_u etc are the input value uncertainties, and the partial derivatives are obtained numerically using the parameter extraction function directly [13].

Using an SRAM rotation matrix transformation, a Milli-Beam axis skew transformation, a DUT displacement translation, and a final rotation of the DUT relative to the SRAM finally allows one to place the Milli-Beam over the coordinate system origin on the DUT itself. Using similar

978-1-4244-9113-1/11 $26.00 © 2011 IEEE

inverse matrix transformations, additional software then computes the Milli-Beam actuator positions needed to strike the specified X-Y locations on the DUT. This software also provides an estimate of the variance, again based on Equation 1, for each targeted DUT coordinate.

IV. BEAM FLUENCE MONITOR

An absolute calibration of the Milli-Beam fluence monitor system was accomplished by finding the fluence using a silicon particle detector and the known primary aperture area.

A. Independent Dosimetry

A particle detector system was used to calibrate the Milli-Beam fluence monitor. The system consists of a 300 mm^2 partially depleted Silicon particle detector and associated electronics. The detector depletion depth is 500 µm minimum, adequate to measure all beams in the 10 MeV/nucleon cocktail without punching through.

The fluence values used for these data analyses were measured independent of the Berkeley dosimetry system using a ~100µm × ~100µm Milli-Beam aperture whose size was determined by performing a least-squares fit to a beam profile measurement on an SRAM with 3x3 micron bit cell size. The length and width of the aperture were each determined to be 101.6 µm ±0.3 µm, giving an aperture area of 1.032×10^{-4} cm^2 ±0.004×10^{-4} cm^2. Placing our Silicon particle detector directly downstream of this aperture then allowed us to count individual heavy ions passing through the known sized aperture, thus providing an accurate fluence determination.

A total of five separate runs were made at each ion in the 10 MeV/nucleon cocktail and each run accumulated data for a total fluence of 1×10^8 $ions/cm^2$ as measured by the Berkeley data acquisition system. All five runs together therefore accumulated ~50,000 counts in the particle detector which, together with the aperture area uncertainty, gave a measured fluence variance of only 0.6%.

B. Absolute Calibration of the Mill-Beam Fluence Monitor

The calibration procedure was conceptually straightforward. The total events detected by the four chips in our beam monitor hardware were counted as a function of LET for known beam fluences. The total beam monitor event count divided by the fluence gave the beam monitor system cross section which is used to compute raster scan fluence values.

Computational uncertainties in this calibration arose from Poisson counting statistics in both the particle detector and the beam monitor chips as well as the accuracy with which the area of the aperture can be measured. The ~100 µm × ~100 µm aperture size was chosen so that in saturation, the particle detector counts would be roughly

equal to the beam monitor chip counts (about 10,000 for a run fluence of 1×10^8 $ions/cm^2$).

Table 1 gives the resulting cross section as a function of LET in units of cm^2 per RSFF cell. The total cross section of each beam monitor chip is this cross section multiplied by 8,192, the total number of RSFF cells on each of the four chips. The total cross section for the entire beam monitor system is therefore the cross section shown in Table 1 multiplied by 32,768.

LET (MeV-cm²/mg)	σ (cm²)	dσ (cm²)
58.78	4.52×10^{-9}	0.033×10^{-9}
48.15	4.07×10^{-9}	0.033×10^{-9}
30.23	3.09×10^{-9}	0.027×10^{-9}
21.17	2.41×10^{-9}	0.022×10^{-9}
14.59	2.00×10^{-9}	0.019×10^{-9}
9.74	1.65×10^{-9}	0.024×10^{-9}
6.09	1.32×10^{-9}	0.018×10^{-9}
3.49	9.38×10^{-10}	0.118×10^{-10}
2.19	4.98×10^{-10}	0.049×10^{-10}
0.89	6.89×10^{-12}	0.356×10^{-12}

Table 1. RSFF Cell Cross Section Versus LET

Fig. 8 shows this same data as a function of LET. Also shown on the figure are two least squares fits to the data. The first fit is to a standard lognormal function while the second fit is to a conventional Weibull distribution. The lognormal clearly does not fit as well as the Weibull, particularly near the LET threshold.

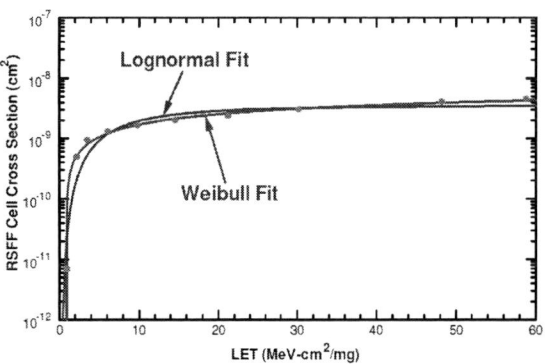

Figure 8. Beam monitor cross section and least-squares fits

V. APPLICATION EXAMPLES

The two following sections demonstrate simple application capabilities achievable by the Milli-Beam. The first shows an alternate method for finding a cross section of a DUT independent of laboratory dosimetry. The second example illustrates a method for finding positional error susceptibilities of a phase lock loop.

A. Cross Section Measurements Using the Milli-Beam Fluence Monitor

A variant of the described Milli-Beam fluence monitoring system has been established in order to more accurately find the cross section of an SRAM. The test procedure is similar to a conventional broad beam heavy-ion SEU measurement since it uses an extremely large Milli-Beam aperture (~15 mm x ~15 mm). Unlike a conventional Berkeley experiment, the Milli-Beam fluence monitoring system is used to accurately establish the total fluence at each LET. This alternate fluence monitoring system still utilized the simultaneous positioning of the beam monitors and SRAM in the heavy-ion beam.

The cross-section measurements are performed at normal incidence (0°) for each heavy ion in cocktail. The experimental cross-section values, along with a variance of each data point, is computed from the measured number of error counts and the total beam fluence as measured by the beam monitor. This process eliminates the dependence on laboratory dosimetry which has been previously shown to neglect beam variations over time. The cross section values, as a function of LET are fit using a least-squares parameter extraction software developed by Micro-RDC. The threshold, shape, and cross section parameters (and their variances) are then extracted using both Lognormal and Weibull distribution fitting functions.

The cross-section fit for the calibrated beam monitor in Fig. 8 using a Weibull cumulative distribution is given by the closed form

$$F(T; \sigma_{sat}, \gamma, \eta, \beta) = \sigma_{sat} \left(1 - e^{-\left(\frac{T - \gamma}{\eta} \right)^{\beta}} \right) \quad (2)$$

where, T is the LET, σ_{sat} is the saturation level, γ is the location parameter, η is the scale parameter, and β is the shaping parameter. The cross section for the SRAM may then be computed independent of the Berkeley fluence by

$$\sigma_{RAM} = \frac{N_{RAM}}{N_{BM}} F(T; \sigma_{sat}, \gamma, \eta, \beta) \quad (3)$$

where N_{RAM} and N_{BM} are the error counts of the SRAM and beam monitor respectively.

The variance of each data point is then computed using the propagation of errors formula derived for the SRAM cross section given by Eq. 4.

$$\delta\sigma_{RAM} = \sigma_{RAM} \sqrt{ \left(\frac{\sqrt{N_{RAM}}}{N_{RAM}} \right)^2 + \left(\frac{\sqrt{N_{BM}}}{N_{BM}} \right)^2 + \left(\frac{\partial F / \partial \eta}{F} \right)^2 + \left(\frac{\partial F / \partial \beta}{F} \right)^2 + \left(\frac{\partial F / \partial \gamma}{F} \right)^2 + \left(\frac{\partial F / \partial \sigma_{sat}}{F} \right)^2 } \quad (4)$$

B. Positional Error Mesurements

A phase lock loop (PLL) is a relatively complex mixed signal device containing a number of circuits susceptible to SEE. Charge pumps, voltage controlled oscillators, and phase detectors are subject to transient effects while dividers and multipliers are subject to SEU. Any observed loss of lock in a broad-beam experiment would be difficult to attribute to any given functional block. The Milli-Beam has therefore been used to isolate and attribute observed malfunctions to specific blocks within a PLL.

Micro-RDC has previously designed a PLL on the IBM 90nm CMOS process. The PLL was tested using the Mill-Beam raster scan with a 100 μm square beam to determine primary elements causing an SEE. The physical layout, as seen in Fig. 9, was used to correlate and map the observed errors of specific circuit elements during the experiment. The lock signal was monitored and the measured recovery time was also measured.

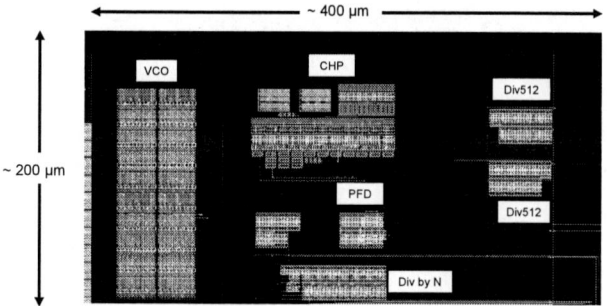

Figure 9. Layout of the Micro-RDC PLL

During a Milli-Beam raster scan of the PLL, the PLL X-Y beam coordinate was recorded along with a Vertex FPGA data log file. The FPGA monitored the PLL under test for loss of lock which was interpreted as an SEE induced malfunction. The FPGA then, for each malfunction, generates an error log file entry containing information describing the error, a time stamp for the event, and the X-Y coordinate on the PLL. These errors were logged throughout the entire raster scan of the PLL and plotted as error contours in Fig. 10 which is a zoomed view of the switching current charge pump (CHP) and the voltage controlled oscillator (VCO).

The gathered data was accumulated during a raster scan pattern of a 7 x 13 grid. The overall run took two hours but upon further improved automation of the system the same procedure can now be repeated under an hour.

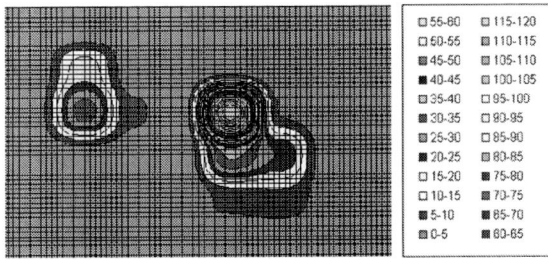

☐ 55-80	☐ 115-120
☐ 50-55	▦ 110-115
▦ 45-50	▨ 105-110
■ 40-45	☐ 100-105
☐ 35-40	☐ 95-100
▦ 30-35	☐ 90-95
▨ 25-30	☐ 85-90
■ 20-25	▨ 80-85
☐ 15-20	■ 75-80
☐ 10-15	▨ 70-75
▨ 5-10	▨ 65-70
▨ 0-5	▨ 60-65

Figure 10. Contour plot of error rates for the Micro-RDC PLL.

Since any given circuit larger than 100 μm in either the X or Y direction could contribute to multiple scan position measurements, a smoothing algorithm was applied so that the centroid of the errors could be better estimated. Fig. 11 shows a full surface plot of the resulting error rate analysis as a function of location.

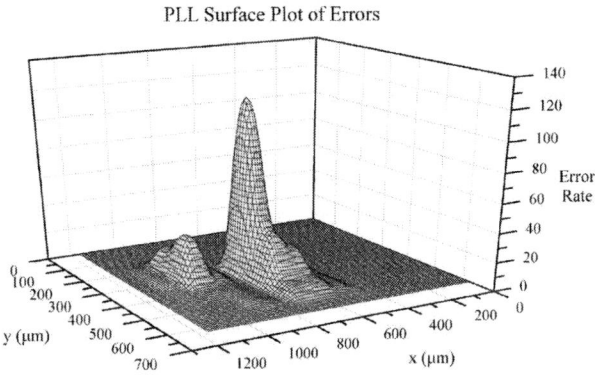

Figure 11. Surface plot of the Micro-RDC PLL errors.

The largest error rate corresponds to the switching current charge pump. Smaller, but still substantial error contributions are observed in the partially hardened VCO and the phase frequency detector (PFD).

VI. SUMMARY

A new test capability termed the Milli-Beam, which compliments existing heavy-ion test methods, has been described. By raster scanning complex designs with any beam presently available at the LBL cyclotron, the spatial location associated with each temporally observed error can be determined. This provides detailed information that can be used to isolate the exact cause of each observed error as well as providing maximum insight into potential design improvements for the mitigation of SER and SEU in microelectronic devices.

REFERENCES

[1] F. W. Sexton, "Measurement of Single-Event Phenomena in Devices and ICs", IEEE Nuclear and Space Radiation Effects Conference Short Course Text, 1992.

[2] L. Massengill, "SEU Modeling and Prediction Techniques", IEEE Nuclear and Space Radiation Effects Conference Short Course Text, 1993.

[3] E. L. Peterson, "Single-Event Analysis and Prediction", IEEE Nuclear and Space Radiation Effects Conference Short Course Text, 1997.

[4] E. L. Peterson, P. Shapiro, J. H. Adams, Jr., and E. A. Burke, "Calculatons of Cosmic Ray Induced Soft Upsets and Scaling in VLSI Devices", IEEE Transactions on Nuclear Science, Vol. 29, No. 6, December 1982, pp. 2055-2063.

[5] M. A. McMahan, "The Berkeley Accelerator Space Effects (BASE) Facility," Proceedings of the Nuclear Space Conference 2005, June 5-9, 2005, San Diego, CA, Paper 1011.

[6] J.A. Pellish, R.A. Reed, D. McMorrow, G. Vizkelethy, V. Ferlet-Cavorois, J. Baggio, et al., "Heavy Ion Microbeam- and Broadbeam-Induced Transients in SiGe HBTs", IEEE Nuclear and Space Radiation Effects Conference 2009.

[7] *Newmark Systems Inc*, Internet: http://www.newmarksystems.com/, 2010

[8] G. F. Knoll, *Radiation Detection and Measurement, 4th* ed. John Wiley & Sons, Inc., 2010

[9] L. W. Nagel and D. O. Pederson, "Simulation Program with Integrated Circuit Emphasis (SPICE)", Electronics Research Laboratory, Technical Report Number ERL-M382, University of California, Berkeley, 1973.

[10] *NI LabVIEW,* Internet: http://www.ni.com, 2010

[11] P. E. Dodd and F. W. Sexton, "Critical Charge Concepts for CMOS SRAMs", IEEE Transactions on Nuclear Science, Vol 42, No. 6, December 1995, pp. 1764-1771

[12] P. H. Rose, R. P. Bastide, N. B. Brooks, J. Airey, and A. B. Wittkower, "Description of a Device Used to Measure Ion Beam Emittance," Review of Scientific Instruments, Volume 35, page 1283, (1964)

[13] P.R. Bevington, D.K. Robinson, *Data Reduction and Error Analysis for the Physical Sciences,* McGraw-Hill, New York, 1969.

Bit Error and Soft Error Hardenable 7T/14T SRAM with 150-nm FD-SOI Process

Shusuke Yoshimoto, Takuro Amashita, Shunsuke Okumura, Kosuke Yamaguchi,
Masahiko Yoshimoto*, and Hiroshi Kawaguchi

Graduate School of System informatics, Kobe University, Japan   *JST, CREST
Phone/Fax: +81-78-803-6629
E-mail: yoshipy@cs28.cs.kobe-u.ac.jp

Abstract — **This paper presents measurement results of bit error rate (BER) and soft error rate (SER) improvement on 150-nm FD-SOI 7T/14T (7-transistor / 14-transistor) SRAM test chips. The reliability of the 7T/14T SRAM can be dynamically changed by a control signal depending on an operating condition and application. The 14T dependable mode allocates one bit in a 14T cell and improves the BER in a read operation and SER in a retention state, simultaneously. We investigate its error rate mitigating mechanisms using Synopsys TCAD simulator. In our measurements, the minimum operating voltage was improved by 100 mV, the alpha-induced SER was suppressed by 80.0%, and the neutron-induced SER was decreased by 34.4% in the 14T dependable mode over the 7T normal mode.**

[Keywords: SRAM, single-event upset (SEU), soft error rate (SER), bit error rate (BER), alpha particle, neutron particle]

I. INTRODUCTION

Static random access memory (SRAM) is compatible with CMOS process technology. It is used as on-chip cache and embedded memory in processors and system-on-a-chip (SoC) applications. These days, SRAM occupies an area of 90% or more of a silicon die [1], meaning that SRAM is the device that is most sensitive to process variation and soft error. An SRAM on a silicon-on-insulator (SOI) substrate has one-half the threshold voltage (Vth) variation and one-third to one-fifth soft error rate (SER) of a bulk CMOS [2–3], but it has still crucial problems to the SER and bit error rate (BER):

- BER: The variation in process parameters hinders low-voltage operation. Disturbance to a static noise margin (SNM) in read operations particularly degrades the BER.

- SER: The critical charge (Qcrit) decreases with process scaling [4]. An ionizing particle passing through a Si (silicon) substrate generates electron-hole pairs, which possibly flip a datum in SRAM. A neutron does not ionize Si directly, but generates electron-hole pairs via secondary ions as nuclear reactions [5]. The SOI SRAM collects a lesser electron charge by the funnel effect, but the collected charge is amplified by the parasitic bipolar effect, which causes a single event upset (SEU) [6]. The SEU is suppressed by an error correction code (ECC) [7] or triple modular redundancy (TMR) [8]. However, the ECC and TMR have large area overheads and access time penalties.

Many types of SRAM cells have been proposed to overcome these problems. Refs [9–10] introduce a 10T or 8T cell which eliminates the disturbance to read/write operations.

Other papers present SER-improved SRAM using a self-feedback mechanism [11–12]. However, these SRAM cells cannot address the bit error and soft error problems, simultaneously.

We have proposed a 7T/14T (7-transistor / 14-transistor) SRAM with a dependable mode for low-voltage operation [13]. Figure 1 portrays the 7T/14T SRAM structure. As well as the conventional 6T cell, the normal mode allocates one bit in a 7T cell while the dependable mode does so in a 14T cell by enabling a CTRL signal (CP0 and CP1 are activated). In the 14T dependable mode, either wordline, WLA or WLB, is asserted, which enlarges an SNM because a β ratio (a ratio of the driver transistor's size to the access transistor's size) is doubled. Thus, the reliability of the proposed cell can be dynamically changed depending on an operating condition. For instance, the 7T/14T SRAM is able to be implemented as on-chip cache memory on a processor with dynamic voltage and frequency scaling (DVFS) [14].

In this paper, with 150-nm FD-SOI 7T/14T SRAM test chips, we present not only a BER improvement but also mitigating alpha- and neutron-induced SER. We will show experimental results that the 14T dependable mode is superior to ECC and TMR in terms of BER. The respective alpha- and neutron-induced SERs in the 14T dependable mode are suppressed by 80.0% and 34.4% over the 7T normal mode because Qcrit in the 14T mode is increased by 70%, at a supply voltage of 0.3 V. Namely, the proposed 7T/14T SRAM is a dynamically hardenable device.

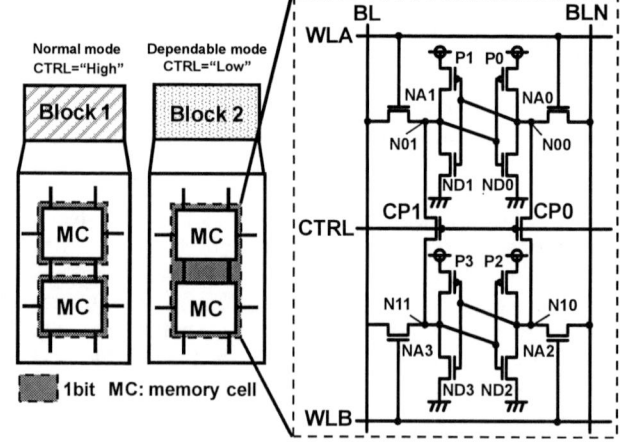

Figure 1.   Structure of 7T/14T SRAM cell [13].

978-1-4244-9113-1/11 $26.00 © 2011 IEEE

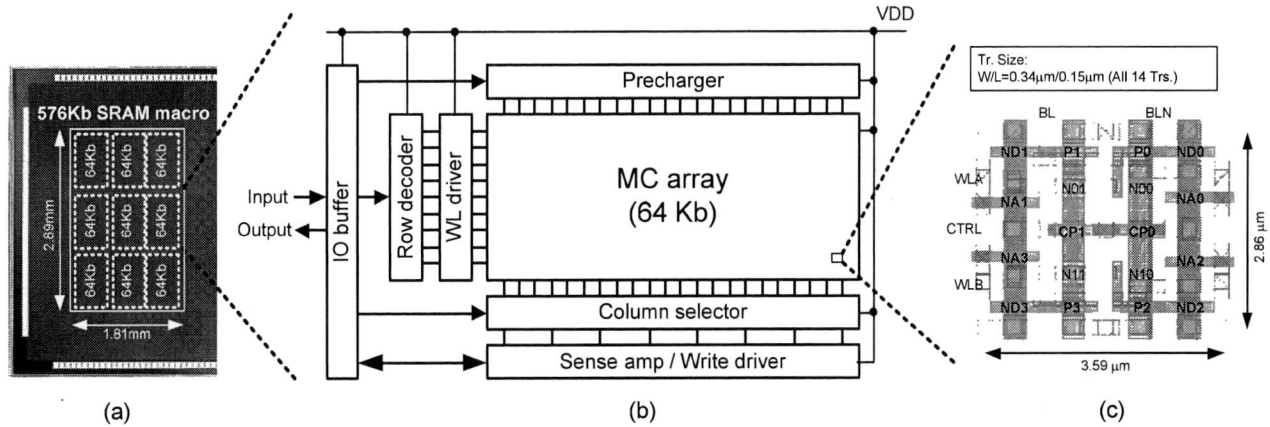

Figure 2. (a) Chip micrograph, (b) block diagram of 64-Kb bank, and (c) 7T/14T SRAM cell layout (based on logic rule).

II. SRAM STRUCTURE

We designed and fabricated a 576-Kb SRAM macro (512 rows x 128 columns x 9 banks) with a nominal supply voltage of 1.5 V in a 150-nm FD-SOI process. Figure 2 illustrates the implemented chip micrograph, a block diagram of a 64-Kb bank, and 7T/14T SRAM cell layout. The SRAM adopts bit-interleaving technique for preventing multiple-cell upset (MCU) [15]. The 7T/14T memory cells are accessed by bitlines (BL/BLN), wordlines (WLA/WLB), and control signal (CTRL). Since the memory cell is inter connected by layer-1 metal, the bitlines vertically tracks with layer-2 metals; then the wordlines and control signal are horizontally connected using layer-3 metals. The layout is based on a logic rule, and its area overheads are 9.5% and 119% in the 7T mode and 14T mode over the minimum 6T memory cell.

Figure 3 shows operating waveforms of the 7T/14T SRAM in the write, read, and standby cycles. When the CTRL signal is disabled (CTRL is "high"), the SRAM acts as the 7T normal mode. In this case, the SNM and Qcrit are quite similar to those in the conventional 6T cell. In the 14T dependable mode, the both wordline, WLA and WLB, are simultaneously activated in the write cycle, and a single wordline, WLA, is enabled in the read cycle; thus, the write margin and SNM are enlarged, compared to the 7T mode. The CTRL signal is kept active (= "low") during a standby state in the 14T dependable mode, which improves a retention margin.

III. SIMULATION RESULTS

In this section, we will present mixed-mode simulation results that the 14T dependable mode is superior to the 7T normal mode at SEU tolerance. Figure 4 portrays the 7T/14T SRAM cell circuit that has a Synopsys TCAD [16] model (ND1) and FD-SOI SPICE models (P0–3, ND0, ND2–3, and CP0–1). The device profile and SPICE models are provided by a foundry.

We simulate that an alpha particle whose LET is 0.1 pC/um perpendicularly strikes the center of a gate. Figure 5 shows waveforms of internal nodes (N00 and N01) in the 7T/14T SRAM cell. When N01 (N00) is high (low), the ND1 transistor that has the N01 node is the most sensitive to the SEU; it is pulled down to the ground if a heavy ion strikes the ND1 transistor. The datum in the 7T cell (7T per bit) is possibly flipped by this disturb current. However, that in the 14T dependable cell can be sustained by compensating current flowing through the connecting pMOS, CP1.

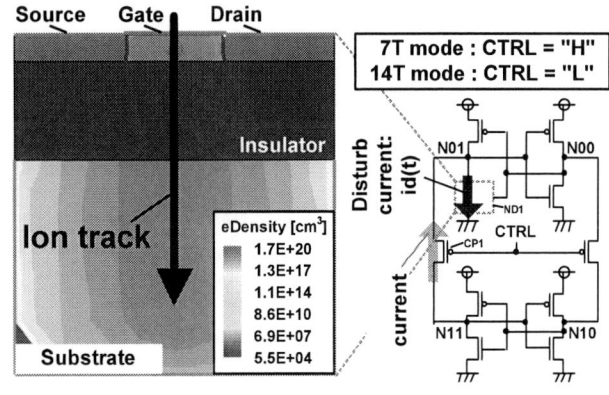

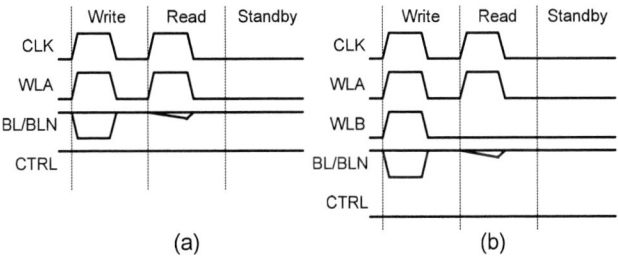

Figure 3. Operating waveformes of (a) 7T normal mode and (b) 14T dependable mode.

Figure 4. (a) Cross section of nMOS TCAD model, and (b) 7T/14T SRAM cell circuit for mixed-mode simulation using tool suite of Synopsys Sentaurus package [16].

978-1-4244-9113-1/11 $26.00 © 2011 IEEE

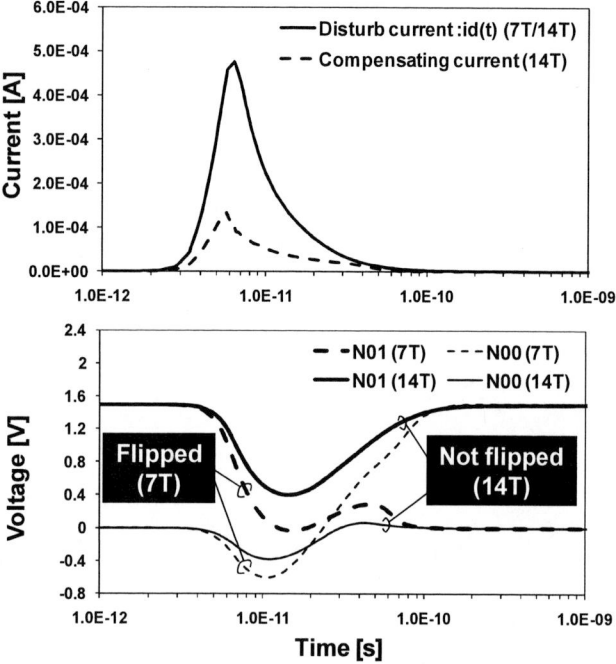

Figure 5. Mixed-mode simultion results on the structure presented in Figure 4. The 14T dependable cell is not flipped due to compensating current flowing through CP1. The heavy ion's LET is 0.1 pC/μm.

We investigated the LET threshold (LET$_{th}$) and Qcrit; they were calculated by the following equations (1–4). LET$_{th}$ is a minimum LET in which a memory cell is flipped. Qcrit is defined as a deposited charge which is an integral of the disturb current:

$$LET_{th} = \min(LET\,|_{SEU\ happens}) \tag{1}$$

$$Qcrit = \min(Deposited\ charge(LET)\,|_{SEU\ happens}) \tag{2}$$

$$Deposited\ charge\,(LET) = \int_{0}^{10e-9} id(t, LET)\,dt \tag{3}$$

$$LET = 10k\ [fC/\mu m]\,(1 \le k \le 15, k \subset N) \tag{4}$$

We simulate the SEU effect over a LET range of 10-150 fC/μm. Figures 6 and 7 portray the simulated LET$_{th}$ and Qcrit under 0.3-1.5 V operating voltages (VDD). The LET$_{th}$ is improved by 10-50% and the Qcrit is increased by 10-70%. Because an alpha and neutron SERs are logarithmically proportional to Qcrit when Qcrit is small [4], the SERs is expected to be rapidly decreased on that condition.

IV. EXPERIMENTAL RESULTS

A. Bit Error Rate (BER) measurement

Figure 8 portrays the measured BERs in the 7T normal and 14T dependable modes. We took three steps: 1) normal write operation at 1.5 V, 2) dummy read operation for disturbance to SNM at low voltage (VDD < 1.5 V), and 3) read operation at 1.5 V. We confirmed that the BERs in the write and retention

modes have much margins than that in the read disturb. So, the read operation is critical. The curves of the 7T with ECC and 7T with TMR can be calculated according (5) and (6), and shown as well in the figure.

$$BER_{ECC} = BER_{7T} \times \left[1 - \left(1 - BER_{7T}\right)^{37}\right] \tag{5}$$

$$BER_{TMR} = 3 \times \left(BER_{7T}\right)^{2} - 2 \times \left(BER_{7T}\right)^{3} \tag{6}$$

The minimum operating voltages are 0.49 V, 0.45 V, 0.44 V, and 0.39 V in the 7T normal, 7T + ECC, 7T + TMR, and 14T dependable modes, respectively. The 14T dependable mode reduces the minimum operating voltage by 0.1 V, compared to the 7T normal mode.

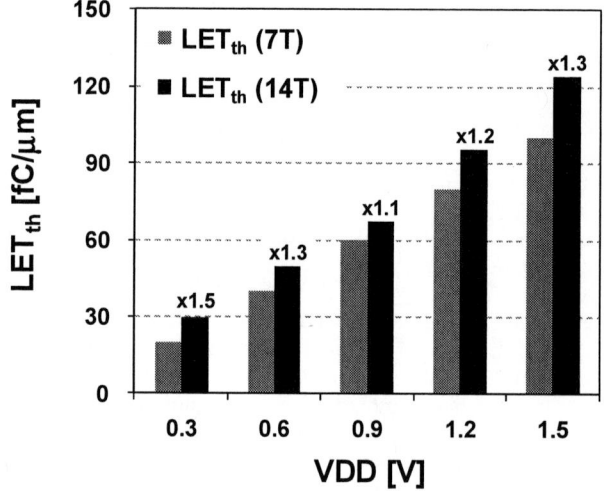

Figure 6. Simulated LET$_{th}$ when VDD is varied. LET$_{th}$ in the 14T dependable mode is improved by 10-50% over the 7T normal mode.

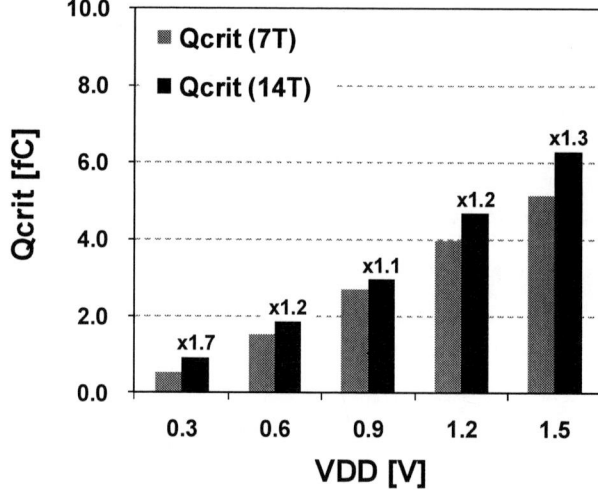

Figure 7. Simulated Qcrit when VDD is varied. Qcrit in the 14T dependable mode is improved by 10-70% over the 7T normal mode.

978-1-4244-9113-1/11 $26.00 © 2011 IEEE

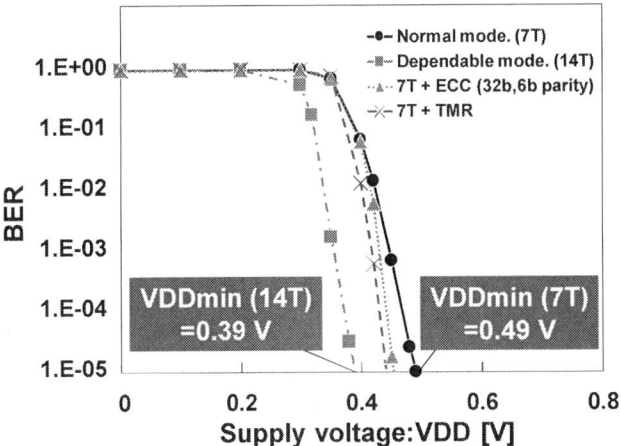

Figure 8. BER curves: The 14T dependable mode has the smallest BER.

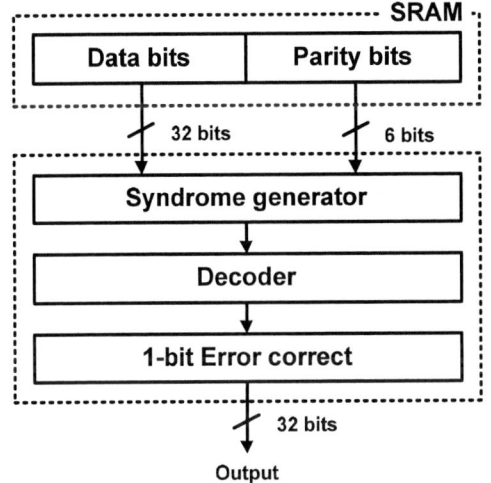

Error correction code

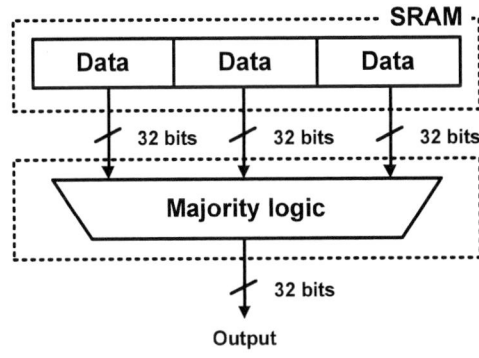

Triple modular redundancy

Figure 9. Error correction code (ECC) and triple modular redundancy (TMR).

Figure 9 illustrates overviews of 1-bit correct ECC and TMR configurations when a word width is 32 bits. The ECC includes a syndrome generator, decoder, and error correction, which needs 158% speed penalty and 18% area overhead [17]. Because the TMR simply decides its outputs by majority voting, the speed penalty is less than that in ECC. However, the area overhead is 200% due to the triple redundancy. Note that the proposed 7T/14T SRAM can also adopt these classic methods. The combination with the proposed 14T dependable mode will realize higher reliability.

B. Soft Error Rate (SER) measurements

Figures 10 and 11 show diagrams of the alpha and neutron accelerated tests. For an alpha particle source, Am-241 foil was used; its flux was 9×10^9 cm^{-2}h^{-1}. The foil was placed above the package [18]. A neutron irradiation experiment for SER verification was conducted at the Research Center for Nuclear Physics (RCNP), Osaka University. A neutron white beam generated by a 400-MeV proton beam irradiates a measurement board for six hours, on which three sample chips were placed (Figure 12). The neutron flux was normalized to 3.6×10^{-3} N/cm^2-s at the ground level in New York City.

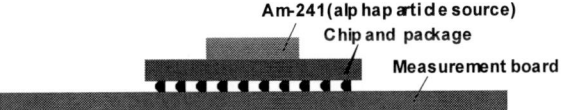

Figure 10. Experiment diagram of alpha accelerated test.

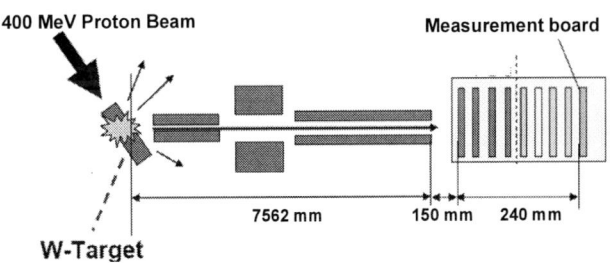

Figure 11. Experiment diagram of neutron accelerated test.

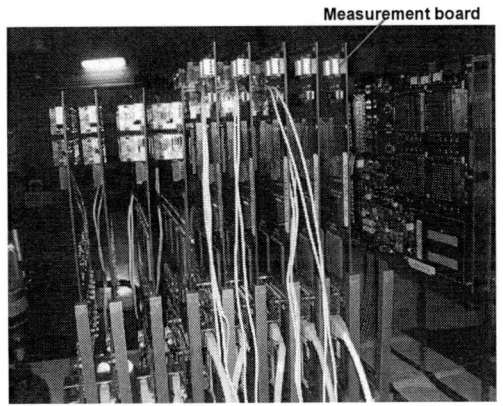

Figure 12. Photograph of neutron accelerated test.

978-1-4244-9113-1/11 $26.00 © 2011 IEEE

Figures 13 and 14 illustrate the alpha and neutron SERs, respectively. In these experiments, any MCU did not occur. The supply voltage was fixed to 1.5 V according to the specification of the measurement board. The alpha-induced SER is improved by 80% because Qcrit is increased in the 14T dependable mode. The average neutron SERs in the three samples are 131 FIT/Mb and 86 FIT/Mb in the 7T normal and 14T dependable modes, respectively (34.4% reduction). The standard deviation is decreased from 9.06 FIT/Mb to 5.87 FIT/Mb. The 14T dependable mode has a double area but achieves a less SER than to the 7T normal mode.

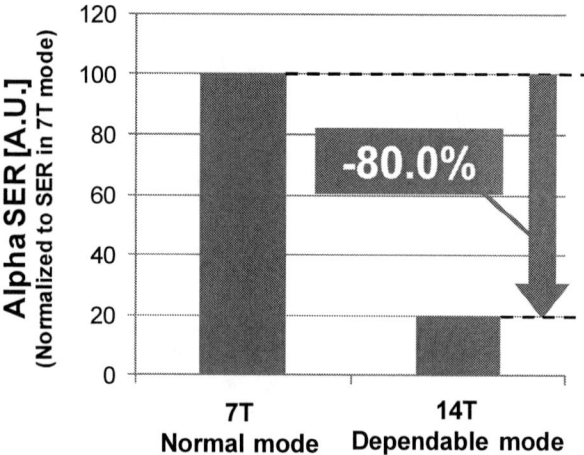

Figure 13. Measured alpha-induced SERs in the 7T/14T SRAM.

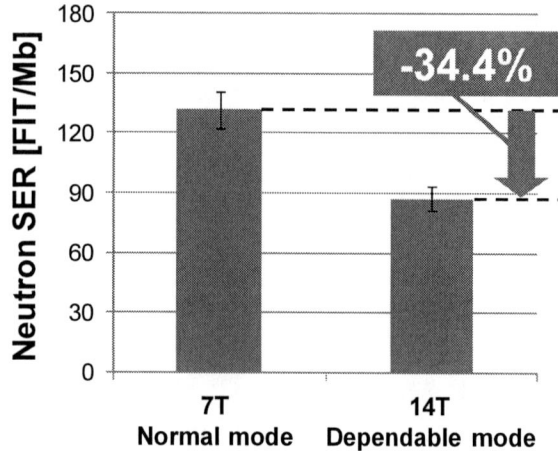

Figure 14. Measured neutron-induced SERs in the 7T/14T SRAM.

V. CONCLUSION

We measured BERs and alpha-particle/ neutron accelerated SERs using a 150-nm 576-Kb 7T/14T FD-SOI SRAM. We confirmed that the 14T dependable mode improves the minimum operating voltage to 0.39 V from 0.49 V in the 7T normal mode. The BER in the 14T dependable mode is superior to those in the TMR and ECC. The respective alpha- and neutron-induced SERs in the 14T dependable mode are 80.0% and 34.4% less than that in the 7T normal mode. We

observed 10-70% increase of the 14T mode's Qcrit in a range of 0.3-1.5 V, by using Synopsys TCAD tool. Because the proposed 7T/14T SRAM can dynamically change its BER and SER, users can take a tradeoff between the reliability and area (cost), which is useful and effective to various applications.

ACKNOWLEDGMENTS

This work was supported by KAKENHI (20360161), VLSI Design and Education Center (VDEC), the University of Tokyo in collaboration with Cadence Design Systems, Mentor Graphics and Synopsys, Inc. The test chip was fabricated by Oki Electric Industry Co. Ltd. We thank Dr. Satoshi Kuboyama with Japan Aerospace Exploration Agency for technical discussion and Dr. Koji Nii with Renesas Electronics Corporation for measurement setup.

REFERENCES

[1] International Technology Roadmap for Semiconductors (ITRS) Report. (http://www.itrs.net/Links/2010ITRS/Home2010.htm)

[2] A.V-Y Thean, Z-H Shi, L. Mathew, T. Stephens, H. Desjardin, C. Parker, T. White, M. Stoker, L. Prabhu, R. Garcia, B-Y. Nguyen, S. Murphy, R. Rai, J. Conner, B.E. White, S. Venkatesan., "Performance and Variability Comparisons between Multi-Gate FETs and Planar SOI Transistors," *IEEE International Electron Devices Meeting* (IEDM), pp. 1-4, 2006.

[3] Ethan H. Cannon, Daniel D. Reinhardt, Michael S. Gordon, and Paul S. Makowenskyj, "SRAM SER in 90, 130 and 180 nm Bulk and SOI Technologies," *IEEE International Reliability Phisics Symposium* (IRPS), pp. 300-304, 2004.

[4] David F. Heidel, Kenneth P. Rodbell, Phil Oldiges, Michael S. Gordon, Henry H. K. Tang, Ethan H. Cannon, and Cristina Plettner, "Single-Event-Upset Critical Charge Measurements and Modeling of 65nm Latches and Memory Cells," *IEEE Trans. Nuc. Sci.*, vol. 53, pp. 3512-3517, 2006.

[5] F. Wrobel, J.M. Palau, M.C. Calvet, O. Bersillon, H. Duarte, "Incidence of multi-particle events on soft error rates caused by n-Si nuclear reactions," *IEEE Trans. Nucl. Sci.*, vol. 47, pp. 2580-2585, 2000.

[6] J. R. Schwank, V. Ferlet-Cavrois, M.R. Shaneyfelt, P.E. Dodd, "Radiation Effects in SOI Technologies," *IEEE Trans. Nuc. Sci.*, vol. 50, pp. 522-538, 2003.

[7] R. Baumann, "The Impact of Technology Scaling on Soft Error Rate Performance and Limits to the Efficacy of Error Correction," *IEEE International Electron Devices Meeting* (IEDM), pp. 329-332, 2002.

[8] C. H. Chen, and A. K. Somani, "Fault-Containment in Cache Memories for TMR Redundant Processor Systems," *IEEE Trans. on Computers*, vol. 48, no. 4, pp. 386-397, April, 1999.

[9] L. Chang, D.M. Fried, J. Hergenrother, J.W. Sleight, R.H. Dennard, R.K. Montoye, L. Sekaric, S.J. McNab, A.W. Topol, C.D. Adams, K.W. Guarini, and W. Haensch, "Stable SRAM Cell Design for the 32 nm Node and Beyond," *IEEE Symp. VLSI Technology, Dig. Tech. Papers*, pp. 128–129, 2005.

[10] I. J. Chang, J. J. Kim, S. P. Park, and K. Roy, "A 32kb 10T Subthreshold SRAM Array with Bit-Interleaving and Differential Read Scheme in 90nm CMOS," *IEEE International Solid-State Circuits Conference* (ISSCC), pp. 398-399, 2008.

[11] T. Calin, M. Nicolaidis, R. Velazco, "Upset Hardened Memory Design for Submicron CMOS Technology," *IEEE Trans. Nuc. Sci.*, vol. 43, No. 6, pp. 2874-2878, 1996.

[12] S.M. Jahinuzzaman, D.J. Rennie and M. Sachdev, "A Soft Error Tolerant 10T SRAM Bit-Cell With Differential Read Capability," *IEEE Trans. Nuc. Sci.*, vol. 56, No. 6, pp. 3768-3773, 2009.

[13] H. Fujiwara, S. Okumura, Y. Iguchi, H. Noguchi, Y. Morita, H. kawaguchi and M. Yoshimoto, "Quality of a Bit (QoB): A New Concept in Dependable SRAM," *IEEE International Symposium on Quality Electronic Design* (ISQED), pp. 98-102, 2008.

[14] Y. Nakata, S. Okumura, H. Kawaguchi and M. Yoshimoto, "0.5-V Operation Variation-Aware Word-Enhancing Cache Architecture Using 7T/14T Hybrid SRAM," *ACM/IEEE International Symposium on Low Power Electronics and Design* (ISLPED), pp. 219-224, 2010.

[15] G. Gasiot, D. Giot and P. Roche, "Multiple Cell Upsets as the Key Contribution to the Total SER of 65 nm CMOS SRAMs and Its Dependence on Well Engineering," *IEEE Trans. Nuc. Sci.*, vol. 54, No. 6, pp. 2468-2473, 2007.

[16] (2010) Synopsys Sentaurus TCAD tools. [Online], Available: http://www.synopsys.com/Tools/TCAD/DeviceSimulation/Pages/default.aspx

[17] T. Suzuki, Y. Yamagami, I. Hatanaka, A. Shibayama, H. Akamatsu and H. Yamauchi, "A Sub-0.5-V Operating Embedded SRAM Featuring a Multi-Bit-Error-Immune Hidden-ECC Scheme," *IEEE Trans. Solid-State Circuit* (JSSC), vol. 41, no. 1, pp. 152-160, 2006.

[18] JEDEC standard JESD89, "Measurement and Reporting of Alpha Particles and Terrestrial Cosmic Ray-Induced Soft Errors in Semiconductor Devices."

Pulsed Laser-Induced Transient Currents in Bulk and Silicon-On-Insulator FinFETs

F. El-Mamouni[1], E. X. Zhang[1], R. D. Schrimpf[1], R. A. Reed[1], K. F. Galloway[1]
[1] Department of Electrical Engineering & Computer Science, Vanderbilt University, Nashville TN, USA
email: farah.el.mamouni@vanderbilt.edu

D. McMorrow[2],
[2] Naval Research Laboratory, Washington DC, USA

E. Simoen[3], C. Claeys[3]
[3] Imec, Kapeldreef 75, B-3001 Leuven, Belgium

S. Cristoloveanu[4]
[4] IMEP - INP Grenoble MINATEC, BP 257, 38016 Grenoble Cedex, France

W. Xiong[5],
[5] Texas Instruments Inc. Dallas, TX, USA

Abstract: **Pulsed laser-induced current transient experiments are used to understand the mechanisms of single-event effects in bulk and fully depleted silicon-on-insulator p-channel FinFETs. The drain current transients are significantly larger in the bulk FinFETs than in the SOI devices. Bulk FinFETs collected 270 times more charge than SOI FinFETs. 98% of the charge collected in the bulk FinFETs is generated in the substrate. The rest of the collected charge (2%) is generated in the fins. Most of the collected charge in the SOI FinFETs is generated in the fins.**

Index Terms—Silicon-on-insulator (SOI), multi-gate transistors, FinFET, fin, charge collection, charge generated, single event effects (SEEs), transient measurements, single photon absorption (SPA), laser energy.

I.   INTRODUCTION

Planar silicon-on-insulator (SOI) technologies exhibit compelling benefits over their bulk-silicon counterparts [1]. The incorporation of the buried oxide layer dramatically reduces the parasitic drain-substrate and source-substrate capacitance, enhancing the performance of SOI-based devices [1]. For 22 nm technologies [2] and beyond, however, non-planar structures may be required to provide adequate gate control [3-5]. FinFETs are among the most promising devices for these future technology nodes, particularly in low power applications [6]. SOI planar devices have excellent resistance to radiation-induced single event effects because the buried oxide (BOX) reduces the collection volume compared to bulk silicon devices [7, 8]. However, the charge-collection characteristics of SOI FinFETs have not been compared to those of bulk FinFETs. With no buried oxide layer, bulk FinFETs offer lower fabrication cost and more ease of integration with established CMOS technologies [9].

In this work, we use the single photon absorption (SPA) pulsed laser technique [10] to investigate radiation-induced transient currents in both bulk-silicon and SOI FinFETs. Charge collection is reported as a function of drain bias and laser energy for both bulk and SOI FinFET technologies.

II.   EXPERIMENTAL DETAILS

A.   *Device details*

The PMOS bulk FinFETs investigated in this work were fabricated at Imec, Belgium. Fig. 1 shows an SEM picture of one of the investigated bulk FinFETs. A 2.5 nm gate oxide (HfSiON) was grown on top of a 1 nm SiO_2 interfacial layer. The FinFET gate oxide was topped by a 100 nm poly-silicon layer. The fin height and width are 65 and 160 nm, respectively.

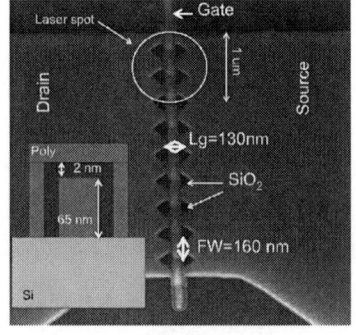

Fig. 1: SEM picture of a 10-fins bulk FinFET. The figure shows the fin width (FW) and the gate length (Lg) parameters. The inset shows the cross section of the bulk FinFET corresponding to one fin.

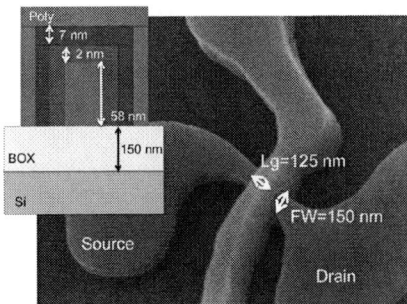

Fig. 2: SEM picture of a single-fin SOI FinFET. The figure shows the fin width (FW) and the gate length (Lg) parameters. The inset shows the cross section of the FD SOI FinFET.

Devices with 10 fins and a gate length of 130 nm were investigated. SiO_2 was used for fin-to-fin isolation.

The PMOS SOI FinFETs were manufactured by Texas Instruments. Fig. 2 contains an SEM picture of one of the fully depleted (FD) SOI FinFETs used in this work. A 2 nm gate oxide (SiO_2) was grown by steam oxidation. 7 nm of TiSiN was deposited and capped by 100 nm of poly-silicon. The fin height and width are 58 and 150 nm, respectively. Single-fin SOI FinFETs with gate length of 125 nm were investigated.

A. Experimental setup

Topside single photon absorption (SPA) pulsed laser experiments were performed at the Naval Research Laboratory in Washington DC [2]. SPA experiments are frequently used as a surrogate test for ion irradiation to simulate single event effects (SEEs) in microelectronic devices. The photon pulse energy defines the amount of charge that is liberated in the semiconductor. This charge produces effects that are similar to those produced by ion tracks, although the details of the charge distribution differ. Fig. 3 illustrates the experimental setup used to measure the transient events. Devices were mounted in high-speed (50 GHz) packages, which were fixed on an (xyz) stage with a step size of 100 nm. The stage helped to optimize the focus and the location of the laser beam on top of the tested FinFETs. Optical pulses with wavelength of 590 nm and pulse width of 1 ps with a repetition rate of 1 MHz were used. The absorption coefficient of silicon for laser beam with 590 nm wavelength is $5.83 \ 10^3$ cm^{-1} [11]. A 100X microscope objective was used to focus the laser beam on the device under test (DUT) producing a laser spot size of 1.1 μm. Devices were irradiated in the off state configuration with a drain bias of -0.6 V, source and drain grounded. The body in SOI FinFETs was floating. FinFETs were biased through a 50 GHz bias tee. A 12 GHz TDS6124C scope with 50-ohm impedance was used to visualize and record the transients. The single shot scope was externally triggered by the laser source itself.

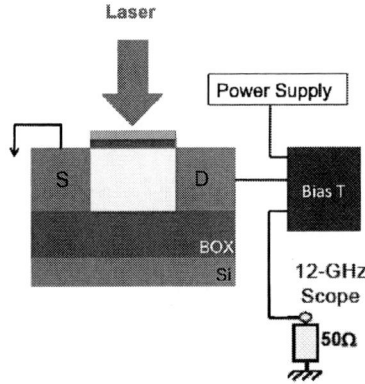

Fig. 3: Schematic diagram of the experimental setup used to record transients in SOI FinFETs. The same setup was used for bulk FinFETs. Transients were visualized and recorded in a 12 GHz sampling scope.

Current transients were captured when the laser beam was focused on the top of the fins for different laser energies and varying drain biases.

III. EXPERIMENTAL RESULTS AND DISCUSSION

Laser energy dependence

The laser spot diameter (1.1 μm) is bigger than the single-fin SOI sample dimensions (W/L = 150/125). Thus, charge is generated in the entire active region of the SOI sample (i.e., fin). However, the bulk FinFET has ten fins, as shown in Fig. 1. The laser spot covers approximately four fins of the bulk device. Fig. 4 illustrates the recorded current transients at the drain terminals for both bulk and SOI FinFETs for a laser energy of 22.4 pJ. Larger and longer transients are exhibited by the bulk FinFETs. A full width at half maximum (FWHM) of 80 ps was obtained for the SOI samples. Taller and faster current transients (67 ps) were recorded for gamma-shaped SOI n-channel FinFETs with thinner fins (25 nm) [12] than those used in this work (58 nm). The higher amplitude current transient for the bulk FinFET is partially due to the additional three fins contributing to charge collection in bulk samples. A FWHM value of 310 ps was recorded for the bulk FinFETs, which is three times larger than the value obtained for the SOI FinFETs. Fig. 4 also shows a tail extending over more than 4 ns. This is caused by carrier diffusion from the substrate to the fins where they are collected. This carrier's diffusion is eliminated in the case of SOI FinFETs where the active region (fins) is isolated from the substrate.

Fig. 5 illustrates the charge collected for both FinFET technologies as a function of the laser energy. At lower laser energy (2 pJ), the charge collected in the bulk FinFETs is one order of magnitude larger than that collected in the SOI FinFETs.

978-1-4244-9113-1/11 $26.00 © 2011 IEEE

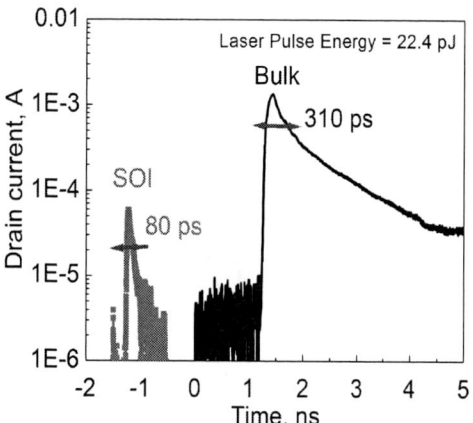

Fig. 4: Semi log drain current transients generated by the pulsed laser in both SOI and bulk FinFETs. The laser energy is 22.4 pJ. A voltage of -0.6 V was applied to both samples during irradiation. The full width at half maximum (FWHM) is illustrated for both devices on the figure.

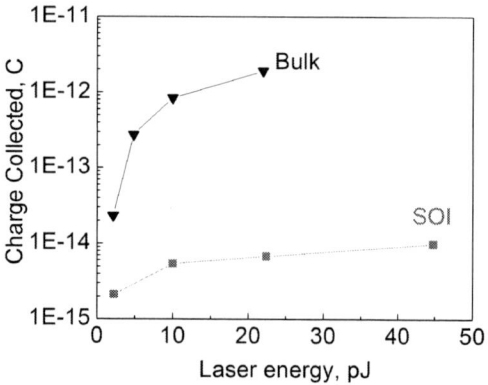

Fig. 5: Semi log curves showing charge collected (C) for bulk (triangles) and SOI (squares) FinFETs as a function of laser energy.

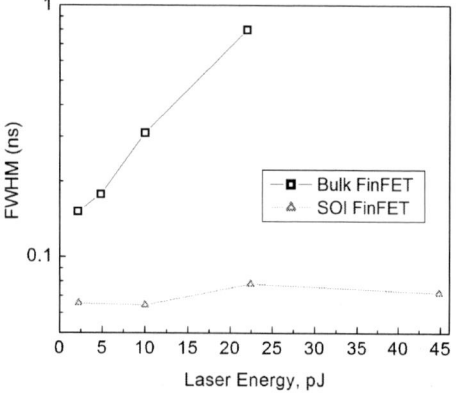

Fig. 6: Semi log plot showing FWHM as a function of laser energy for bulk (empty squares) and SOI FinFETs (empty triangles).

This difference aggravates (more than two orders of magnitude) as the laser energy further increases (starting from 10 pJ). For a laser energy of 22.4 pJ, the bulk FinFETs collect 270 times more charge than SOI samples.

Although the number of irradiated fins is only about four times greater (~4 vs. 1), the contribution of the additional three fins to charge collection in the bulk samples is not likely to be the main mechanism responsible for this significant difference. Indeed, in bulk devices, charge is collected from the substrate, as well as from the body (fin). The total amount of charge generated in the fins in bulk FinFETs was calculated using the Beer's law expression [11]:

$$N(z) = \frac{\alpha}{\hbar\omega}\exp(-\alpha z)\int_{-\infty}^{\infty} I_0(z,t)\,dt \qquad (1)$$

where N(z) is the density of laser generated carriers, is the absorption coefficient of the material (silicon in this case), $\hbar\omega$ is the photon energy, z is the penetration depth and I_0 is the laser irradiance [11]. The dependence of I_0 on z is neglected in this work. For a fin height z = 65 nm, a total charge of 8.67 fC was obtained in one fin. In other words, the charge generated in 4 fins of the bulk FinFET is 34 fC. This value is 1.8% of the total charge collected in the bulk FinFET. In fact, 98.2% of the charge collected in the bulk FinFET was due to collection from the substrate rather than the four fins. On the other hand, similar calculation for the SOI FinFET at a laser energy of 22.4 pJ results in a total charge generated in the fin of 7 fC. This value is very close to the collected charge as illustrated in Fig. 5. The contribution of the fins to the charge collection in SOI FinFETs is limited because of the finite dimensions of the fins.

Fig. 6 illustrates the FWHM as a function of laser energy for bulk and SOI FinFETs. A minor increase in the FWHM with increasing laser energies was observed for SOI FinFETs. A significant increase in the FWHM with increasing laser energy was obtained for bulk FinFETs. Bulk FinFETs exhibit slower transients for higher laser energies. Indeed, carriers generated in the substrate diffuse to the drain terminal (or the fins) where they eventually get collected. This diffusion of the carriers is possible in bulk FinFET because of the direct contact of the substrate with the fins (and the drain terminal) of the device. Carrier's diffusion from the substrate necessitates additional time, which explains the dramatic increase in the FWHM for bulk FinFETs. On the other hand, in SOI FinFETs, charge gets collected only from the fins since the buried oxide isolates the active region from the substrate.

Drain bias dependence

Transients were captured for different drain biases while irradiating the FinFETs. Transients with larger amplitudes were obtained with increasing drain biases for both bulk and SOI FinFETs. In fact, higher drain biases produce higher electric fields in the fins, which increases the amount of carriers drifting toward the drain terminal, resulting in larger current transient amplitudes.

Fig. 7 illustrates the rise time of current transients for varying drain biases in bulk and SOI FinFETs. Fig. 7 show no significant change in the rise time of the transient for SOI FinFETs, whereas, a notable decrease in the rise time of transients in the bulk FinFETs is obtained for increasing drain biases. This may be caused by the increased electric field in the depletion region under the drain for higher drain biases in the bulk FinFETs. TCAD simulations produce a trend that agrees with the experimental data, with shorter rise times obtained for higher drain biases. This result can be a signature of the substrate effects in charge collection in bulk FinFETs.

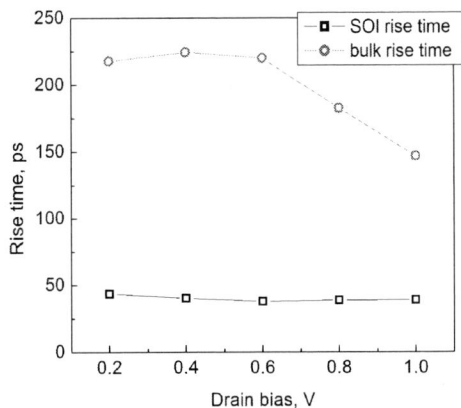

Fig. 7: Rise time in ps for bulk and SOI bulk FinFETs current transients for different drain biases.

IV. CONCLUSIONS

SOI FinFETs are more suitable than their bulk counterparts for rad hard circuit applications because of the BOX layer that isolates the active region from the substrate and consequentely reduces the collection volume. This result may have a considerable impact on the strategy for developing FinFET-based memory and logic devices on SOI or bulk.

Acknowledgments:

This work was supported by the Air Force Office of Scientific Research (AFOSR) under the MURI program.

REFERENCES

[1] J. P. Colinge, *Silicon-On-Insulator Technology: Materials to VLSI*. Boston: Kluwer Academic Publishers, 1997.
[2] http://www.conceivablytech.com/513/science-research/intel-confirms-production-of-22nm-processors-for-late-2011/.
[3] J.-P. Colinge, M. H. Gao, A. Romano-Rodriguez, H. Maes, and C. Claeys, "Silicon on Insulator "Gate All Around" Device," in Proc. IEDM Tech. Dig., 1990, pp. 595-598.
[4] X. Huang, W.-C. Lee, C. Kuo, D. Hisamoto, L. Chang, J. Kedzierski, E. H. Anderson, H. Takeuchi, Y.-K. Choi, K. Asano, V. Subramanian, T.-J. King, J. Bokor, and C. Hu, "Sub-50 nm P-channel FinFET," IEEE Trans. Electron Dev., Vol. 48, No. 5, pp. 880-886, May 2001.
[5] C. Jahan, O. Faynot, M. Casse, R. Ritzenthaler, L. Brevard, L. Tosti, X. Garros, C. Vizioz, F. Allain, A.-M. Papon, H. Dansas, F. Martin, M. Vinet, B. Guillaumot, A. Toffoli, B. Giffard, and S. Deleonibus, " FETs transistors with TiN metal gate and HfO$_2$ down to 10 nm," in Proc. Symp. VLSI Tech., 2005, pp. 112-113.
[6] A. Bansal, S. Mukhopadhyay, and K. Roy, "Device-Optimization Technique for Robust and Low-Power FinFET SRAM Design in NanoScale Era", IEEE Trans. Nucl. Sci., Vol. 54, No. 6, pp. 1409-1419, JUNE 2007.
[7] V. Ferlet-Cavrois, P. Paillet, M. Gaillardin, D. Lambert, J. Baggio, Member, IEEE, J. R. Schwank, G. Vizkelethy, M. R. Shaneyfelt, K. Hirose, E. W. Blackmore, O. Faynot, C. Jahan, and L. Tosti, "Statistical Analysis of the Charge Collected in SOI and Bulk Devices Under Heavy Ion and Proton Irradiation Implications for Digital SETs," IEEE Trans. Nucl. Sci. Vol. 53, No. 6, pp. 3242-3252, DECEMBRE 2006.
[8] V. Ferlet-Cavrois, P. Paillet, D. McMorrow, A. Torres, M. Gaillardin, J. S. Melinger, A. R. Knudson, A. B. Campbell, J. R. Schwank, G. Vizkelethy, M. R. Shaneyfelt, K. Hirose, O. Faynot, C. Jahan, and L. Tosti "Direct Measurement of Transient Pulses Induced by Laser and Heavy Ion Irradiation in Deca-Nanometer Devices," IEEE Trans. Nucl. Sci. Vol. 52, No 6., pp. 2104-2113, DECEMBRE 2005.
[9] K. Okano, T. Izumida, H. Kawasaki, A. Kaneko, A. Yagishita, T. Kanemura, M. Kondo, S. Ito, N. Aoki, K. Miyano, T. Ono, K. Yahashi, K. Iwade, T. Kubota, T. Matsushita, I. Mizushima, S. Inaba, K. Ishimaru, K. Suguro, K. Eguchi, Y. Tsunashima and H. Ishiuchi, "Process Integration Technology and Device Characteristics of CMOS FinFET on Bulk Silicon Substrate with sub-10 nm Fin Width and 20 nm Gate Length", IEDM Technical Digest, IEEE International, 2005.
[10] D. McMorrow, J. S. Melinger, S. Buchenr, T. Scott, R. D Brown, and N. F. Haddad,"Application of a Pulsed Laser for Evaluation and Optimization of SEU-Hard Design," *IEEE,Trans. Nucl. Sci.*,Vol. 47,No. 3, pp. 559-565, 2000.
[11] D. McMorrow, W. T. Lotshaw, J. S. Melinger, S. Buchner, R. L. Pease, "Subbandgap Laser-Induced Single Event Effects: Carrier Generation Via Two-Photon Absorption", IEEE Trans. Nucl. Sci., Vol. 49, No. 6, pp. 3002-3008, 2002.
[12] M. Gaillardin, P. Paillet, V. Ferlet-Cavrois, J. Baggio, D. McMorrow, O. Faynot, C. Jahan, L. Tosti, S. Cristoloveanu, "Transient Radiation Response of Single- and Multiple-Gate FD SOI Transistors", IEEE Trans. Nucl. Sci, Vol 54, No. 6, p. 2355-2362, 2007.

Neutron- and Alpha-Particle Induced Soft-Error Rates for Flip Flops at a 40 nm Technology Node

Srikanth Jagannathan[1], T. D. Loveless[1,2], Z. Diggins[1], B. L. Bhuva[1,2], S-J. Wen[3], R. Wong[3], L. W. Massengill[1,2]

[1]Department of EECS and [2]Institute for Space and Defense Electronics, Vanderbilt University, Nashville, TN USA
[3]Cisco Systems, Inc., San Jose, CA, USA

Abstract—**Flip-flop designs fabricated in a 40 nm bulk technology node with a wide range of soft-error hardness, area, power, and speed have been tested for neutron and alpha single event upsets. Neutron results show that the error rates of flip-flop designs that were considered hardened at older technologies are comparable to that of the conventional D-flip-flop. The soft-error rates (SER) of all the flip-flops consistently increase with reduction in supply voltage and increase in ambient temperature.**

Keywords-component; DICE, D-FF, flip flop, SER, SEU, SET, alpha, neutron, Radiation hardening.

I. INTRODUCTION

Due to scaling of CMOS technology (smaller node capacitances, increased clock frequency, and lower supply voltage), integrated circuits (ICs) are more susceptible to soft errors caused by alpha particles and neutrons [1-4]. When these particles are incident on a silicon substrate, they create electron-hole pairs which may be collected by p-n junctions via drift and diffusion mechanisms. The resulting perturbation in the output node voltage is described as a single-event transient (SET). Typically, a flip-flop experiences a logic bit upset (also called single-event upset or SEU) when the SET created at one of the storage nodes is longer than the feedback loop of the flip-flop. Since SET pulse width is dependent on technology node, circuit design, and operating parameters [1-4], soft-error rates are expected to vary for these designs as a function of temperature and supply voltage. For ICs operating in the field, there will be variations in operating temperature and supply voltages; thus, it is important to experimentally verify the soft-error rates for flip-flop designs for accurate performance predictions at the system level. These results will also provide insight into the effectiveness of conventional hardening approaches and novel approaches for mitigating soft errors at advanced technology nodes.

There have been many different design approaches to mitigate soft errors in flip-flops - such as Triple Mode Redundancy (TMR), the temporal latch [5], or the Dual Interlocked Cell (DICE) [6]. These techniques are based on the assumption that an incident ion affects only one circuit node. However, in nanoscale technologies, this assumption is not valid, and charge-sharing between multiple nodes is prominent [7, 8]. This combined with lower supply voltage, lower critical charge and reduced drive-strength of transistors leads to increased susceptibility of hardened flip-flop designs to soft errors. Flip-flops designed under the assumption that only a single node may be perturbed at a given moment, and

deemed radiation-hardened at older technologies, may display an increase in vulnerability by orders of magnitude.

This paper presents the neutron- and alpha-particle-induced soft-error rate (SER) of 14 different flip-flops designs fabricated in a 40 nm bulk CMOS technology node. The flip-flop designs were chosen with a wide range of radiation hardness, area, power consumption and speed of operation. Experimental results show increased vulnerability of hardened flip-flops across the whole range. The hardened flip-flop designs (including DICE design) show a maximum of 4X improvement over conventional D-flip-flop (DFF) designs in the case of neutrons [9] and 100X improvement in the case of alpha particles. Supply-voltage and temperature variations for alpha-particle exposure show increased vulnerability for all flip-flops.

II. TEST CHIP AND EXPERIMENT DETAILS

A single test chip containing 14 different flip-flop shift registers with an on-chip error detection circuit was designed and fabricated in a 40 nm bulk CMOS process. The shift registers contained 8000 stages and could operate at multi-GHz frequencies. The clock for the shift registers was provided by an on-chip PLL capable of providing up to 3 GHz frequency. Figure 1 shows the block diagram of the shift register–error detection circuit for one flip flop design. The SEUs due to particle hits in any of the flip-flops in the shift register chain were identified using an error detection circuit and buffered out serially using a 6-bit counter.

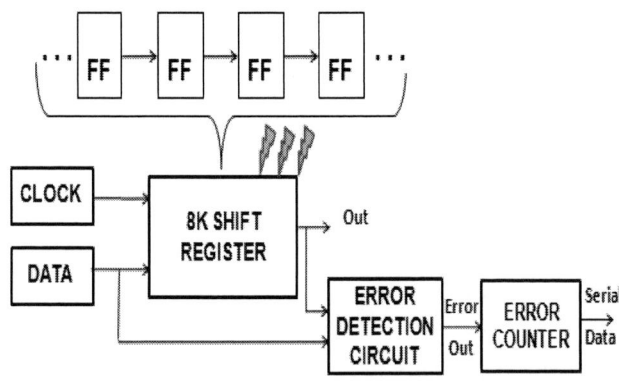

Figure 1. Block diagram of a single 8K shift register with error detection circuit to measure the FIT rate of each flip-flop design.

This work was supported in part by Cisco Systems, Inc. and the Defense Threat Reduction Agency (HDTRA1-09-C-0038).

978-1-4244-9113-1/11 $26.00 © 2011 IEEE

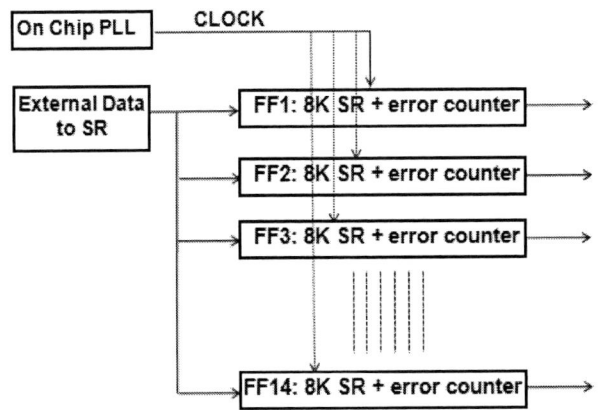

Figure 2. Block diagram of the test chip implemented in TSMC 40nm process.

Figure 2 shows the block diagram representing the entire test chip with all the 14 flip flop designs. A single PLL unit generated the clock for all 14 shift register blocks. Each shift register block had its own error detection and counter circuit. The 14 counters were controlled externally and independently thus providing the option to test all the 14 designs at the same time.

Table 1 contains the area specifications and brief descriptions about the 14 different flip flop designs. DFF1 was the standard D-FF (master-slave design) with inverters and transmission gates as shown in Figure 3. DFF2 through DFF10 were variants of D-FF with design variations (NAND-based designs, NOR-based designs, layout variations, low power versions, and high speed versions). The DICE flip-flop design is the conventional design with two DICE latches, each with four storage nodes implemented with interconnected inverters. The schematic of a DICE latch is shown in Fig. 4. Hard-1 through Hard-3 were flip-flops employing various proprietary hardening techniques using redundancy techniques (such as DICE-like designs) or capacitive enhancements for lengthening the feedback-loop delay.

Specific details in regard to the alpha- and neutron- -particle exposures are as follows:

1) Alpha source: Experiments were performed using an Americium-241 5 MeV alpha source whose activity is 10 Ci. The input to the shift registers was a fixed logic HIGH (or LOW) value with operating frequency at 225 MHz. The supply voltage was varied from 0.81 V to 0.93 V. Nominal supply voltage for this technology is 0.9 V. Temperature measurements were taken at room temperature and at elevated temperatures. The test procedure was compliant with the JEDEC SER test standard labeled JESD89 [10].

2) Neutron source: Experiments were performed at the TRIUMF neutron facility at the University of British Columbia, Vancouver. Figure 5 compares the neutron flux spectrum at TRIUMF with that at the ground level. The energy range of neutrons used for testing was

between 0.1 MeV to 400 MeV. Seven devices were simultaneously tested at 225 MHz to a total neutron fluence of $11.7 \times 10^9/cm^2$. The input to the shift registers was a fixed logic HIGH value.

Table 1: 14 flip flop designs with area specification and description

Flip Flop Design	Area μm^2	Description
DFF1	4.4	Standard – Inverter, TG
DFF2	4.2	High speed– low V_{th} devices
DFF6	6.2	High speed
DFF9	5.5	High speed
DFF4	6.6	Low power – with pass gates
DFF7	5.3	Low power
DFF8	4.3	Low power
DFF5	4.1	NAND gates
DFF3	11.6	NOR gates
DICE	11.8	Hardness by Redundancy
Hard-1	8.6	Hardness by Redundancy
Hard-2	13.8	Hardness by Redundancy
Hard-3	15.3	Hardness by TMR
Hard-4	9.8	Hardness by Capacitance

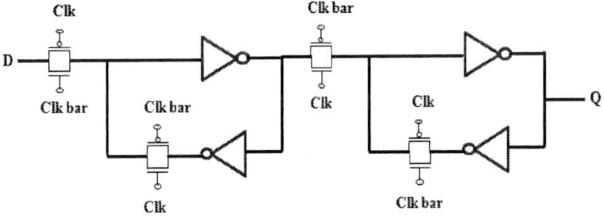

Figure 3. Circuit diagram of standard D flip-flop (DFF1).

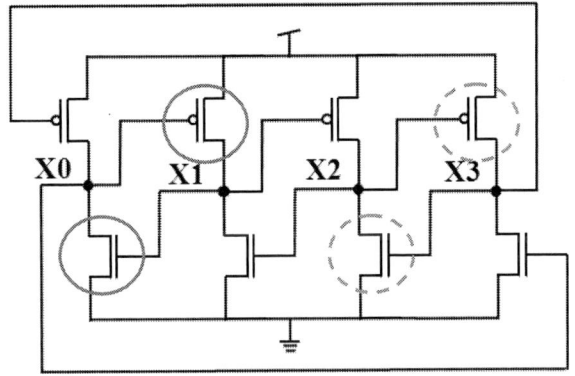

Figure 4. Circuit diagram of standard DICE latch.

| Hard-4 | 1 | 0.95 |

In previous technologies, the DICE flip-flop was considered relatively immune to soft errors showing orders of magnitude difference in FIT rates when compared to D-FF [12]. At the 40 nm technology, the difference in error rates between the DICE-FF and D-FF is only ~1.4X in the case of neutrons [9], while it is 10X in the case of alpha particles. Also, from the neutron experiments, it is observed that the most hardened design (Hard-1) is ~4X less susceptible than D-FF, while it is ~100X in the case of alpha particles.

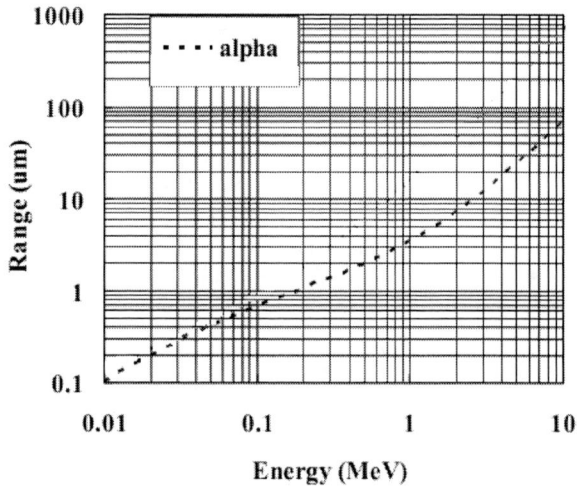

Figure 6. Range of alpha particle in Si [11].

Previous researchers [7, 8, 13], have predicted a dramatic increase in SER of redundancy based hardened devices due to charge sharing. Sensitive nodes in the flip-flop, upon sufficient charge collection (a.k.a. critical charge), will cause the circuit to erroneously flip its state. In order to estimate the critical charge of the nodes due to multi-node charge collection, simulations were performed on designs implemented in TSMC 40 nm technology using Cadence Spectre circuit simulator. In this paper, the struck node is denoted as the primary node and the adjacent node that collects the charge induced by the strike as the secondary node. First, for each of the flip-flop designs the sensitive node pairs were identified. For example, the two pairs of sensitive nodes for DICE are marked in Figure 5. Then, the critical charge (charge required to cause an upset) of the primary node was obtained for a given amount of charge deposited on the secondary node by connecting current sources based on 3D TCAD simulations [14], to both the nodes simultaneously.

Figure 7 shows the critical charge (Q_{crit}) of the primary node of the three flip-flops (DFF1, DICE and Hard-1) as a function of charge deposited on the secondary node. The data points on the curve denote the charge required to be deposited on both the nodes simultaneously to cause an upset.

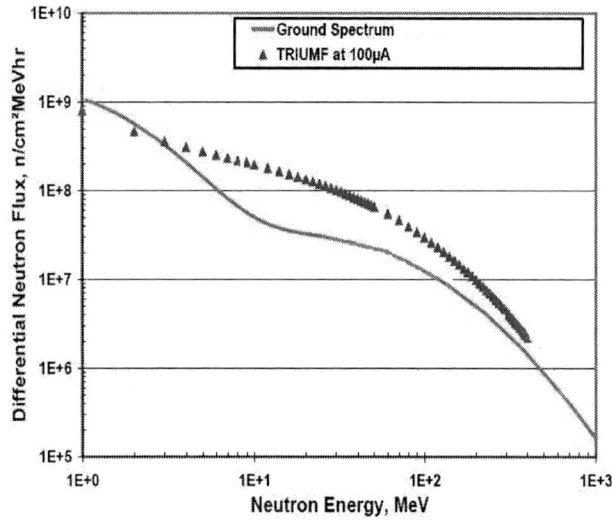

Figure 5. Comparison of neutron flux spectrum at TRIUMF with that at ground level.

III. NEUTRON/ALPHA EXPERIMENT RESULTS AND ANLYSIS

Table 2 shows the relative failure-in-time (FIT) rates of the different flip-flop designs (relative to DFF1) due to neutrons and alpha particles. It is observed that the relative FIT rates due to alpha particles are different from that due to neutrons. This is because of the difference in energy, charge deposition mechanism and the range of alpha particles in Silicon wafers [11]. The range of alpha particles is shown in Fig. 6 and varies with energy of the particle. For example, the range of an alpha particle is 24 μm at energy of 5 MeV. The energy of the alpha particles that reach the active region is less than 5 MeV as they lose energy when travelling through the silicon over-layers. Energy of particles is directly related to the amount of charge deposited and hence to the FIT rate [11].

Table 2: 14 flip flop designs with relative FIT rates due to neutron and alpha particles

Flip Flop design	Relative FIT = $\dfrac{FIT_{FF}}{FIT_{DFF1}}$	
	Neutron	Alpha
DFF1	1	1
DFF2	0.91	0.95
DFF6	0.87	0.98
DFF9	1	1.31
DFF4	0.68	0.23
DFF7	0.79	0.82
DFF8	1.04	0.94
DFF5	1.08	0.87
DFF3	1.04	0.75
DICE	0.68	0.1
Hard-1	0.23	0.01
Hard-2	0.28	0.08
Hard-3	0.3	0.09

978-1-4244-9113-1/11 $26.00 © 2011 IEEE

a) Q_{crit} due single- node strikes:

The point of intersection of the curve and the axis in Figure 7 gives the amount of single-node charge required to cause an upset. It is observed that only ~2 fC of charge is required on the primary node of DFF1 to cause an upset (single node upset). Since the the hardened devices (DICE and Hard1) did not upset for single-node strikes, the curves in figure 12(b) do not intersect the axes.

b) Q_{crit} due multi- node strikes:

In Figure 7, the non-zero values for the charge deposition on the primary node and secondary node denote the multi-node strikes. Any combination of charge deposition that falls in the region above the curve will cause an error whereas any combination that falls in the region below the curve does not cause an upset. Only the most vulnerable node pairs for each of flip-flop design are shown. From the plot, the charge requirement for multi-node upsets was found to be lesser that for single-node upsets, which implies that even if two nodes collected charge less than the Q_{crit} of single node, there is still a chance for an upset to occur.

In the region marked on Figure 7, it is observed that there could be an upset in both DFF1 and DICE for charge deposition of less than 1 fC on the sensitive nodes. The by-products of nuclear reaction due to interaction of high energy neutrons with Silicon could easily deposit this amount of charge in the primary and secondary nodes [10] causing an upset. Hence, multi-node upset is more probable in the case of high energy neutrons than the 5 MeV alpha particles. As a result, we see the increase in susceptibility of DICE to neutrons at 40 nm technology when compared to previous technologies. Also, from the figure it is seen that the amount of charge required by the sensitive nodes to cause an upset in Hard1 design is higher when compared to DICE. This results in increase in hardness of Hard1 when compared to DFF1 and DICE.

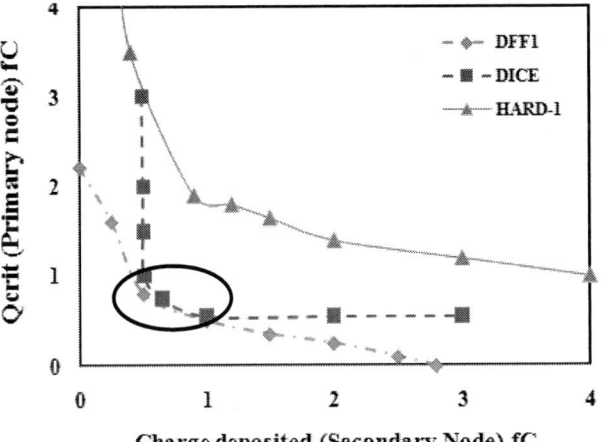

Figure 7. Soft error upset on primary node due to charge collection on secondary node for DFF1, DICE and Hard1.

FIT vs. Supply Voltage

The critical charge of a node decreases as the supply voltage is scaled thereby increasing the probability of an upset [1] due to radiation. The dependence of alpha-particle-induced SER on the supply voltage for all 14 flip-flops is shown in Figure 8. The supply voltage ranges from 0.93 V to 0.81 V. The nominal supply voltage for this technology is 0.9 V. All the designs consistently show an increase in SER due to supply voltage reduction. A maximum of 60% increase in SER is observed for DFF4 for the given voltage range.

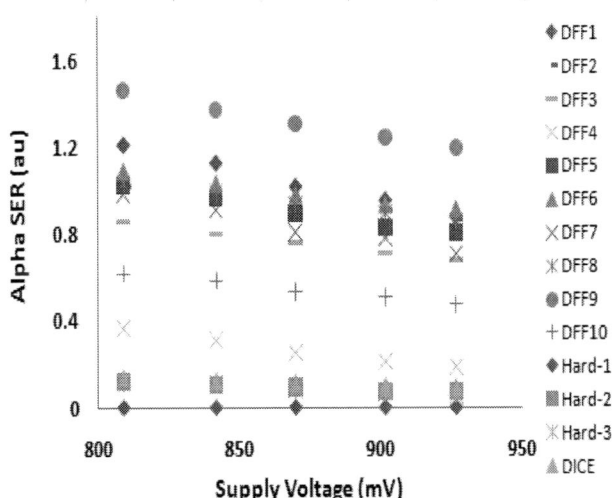

Figure 8. Dependence of alpha SER with supply voltage. SER of all the flip flops inreases with reduction in supply voltage.

FIT vs. Temperature

The radiation-induced charge carriers are mainly collected at circuit nodes through drift, diffusion, and parasitic bipolar processes [15]. The Drift process is a strong function of carrier mobility, which decreases as temperature rises, resulting in decreased charge collection at short times [16]. Diffusion is a strong function of minority carrier lifetime and diffusion coefficient, which move in opposite directions as temperature rises [17]. Parasitic bipolar transistors exhibit increased amplification at elevated temperatures [18]. The overall effect of all these factors has been predicted by simulation to lengthen SET pulses in sub-micron bulk technologies [17].

Figure 9 shows the effects of temperature variations on error rates for alpha-particle exposure. Alpha experiments were conducted at 3 different temperatures (25°C, 50°C and 75 °C). All the flip flop designs consistently show an increase in vulnerability (higher FIT rate) at higher temperature.

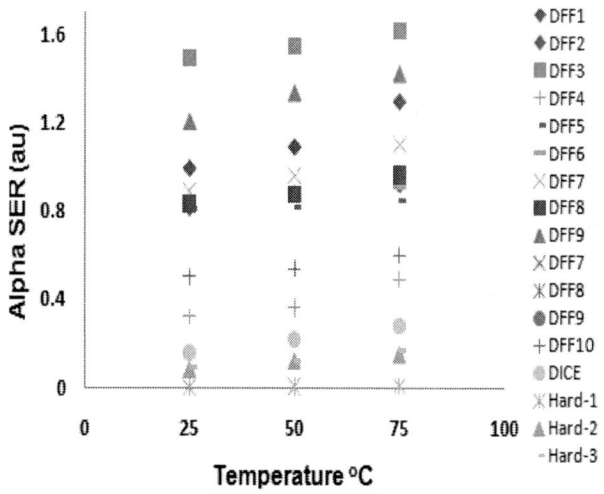

Figure 9. Dependence of alpha SER with temperature. SER of all the flip flops inreases with reduction in temperature.

IV. CONCLUSION

The soft error susceptibility of circuits increases due to technology scaling. At 40 nm technology node, the results show varying degrees of SER between the neutrons and alpha particles. The hardened flip-flops are more susceptible to the neutrons than the low-energy alpha particles. The difference in the rates between the hardest flip-flop (Hard-1) and the standard D-flip-flop is 4X in case of neutrons and 100X in the case of alpha particles. The DICE flip-flop in the case of alpha particles is 10X harder than D-FF while is shows only 1.4X improvement in the case of neutrons. This indicates that hardening techniques in nanoscale technologies need to be carefully selected keeping in mind the radiation source and environment. For all the 14 flip flop designs considered, the soft error rate increases with supply voltage scaling and increase in temperature.

REFERENCES

[1] R. C. Baumann, "The impact of technology scaling on soft error rate performance and limits to the efficacy of error correction," in *IEDM Tech. Dig.*, 2002, pp. 329–332.

[2] A. H. Johnston, "Radiation effects in advanced microelectronics technologies," *IEEE Trans. Nucl. Sci.*, vol. 45, no. 3, pp. 1339–1354, Jun. 1998.

[3] T. P. Ma, and P. V. Dressendorfer," Inonizing Radiation Effects in MOS Devices and Circuits," John Wiley & Sons Inc, 1989.

[4] M J. Gadlage, J. R. Ahlbin, B. Narasimham, B. L. Bhuva, L. W. Massengill, R. A. Reed, R. D. Schrimpf, G. Vizkelethy, "Scaling Trends

in SET Pulse Widths in Sub-100 nm Bulk CMOS Processes," *IEEE Trans. Nucl. Sci.*, vol. 57, no. 6, pp. 3336–3341, 2010.

[5] D. G. Mavis and P. H. Eaton, "Temporally Redundant Latch for Preventing Single Event Disruptions in Sequential Integrated Circuits," *U.S. Patent No. 6 127 864*, Oct. 2000.

[6] T. Calin, M. Nicolaidis, R. Velazco,, "Upset hardened memory design for submicron CMOS technology," *IEEE Trans. Nucl. Sci.*, vol.43, pp. 2874–2878, Dec. 1996.

[7] O. A. Amusan, A. F. Witulski, L. W. Massengill, B. L. Bhuva, P. R. Fleming, M. L. Alles, A. L. Sternberg,, "Charge Collection and Charge Sharing in a 130 nm CMOS Technology," *IEEE Trans. Nucl. Sci.*, vol. 53, no. 6, pp. 3253-3258, 2006.

[8] O. A. Amusan O. A. Amusan, A. L. Steinberg, A. F. Witulski, B. L. Bhuva, J. D. Black, M. P. Baze, L. W. Massengill,, "Single event upsets in a 130nm hardened latch design due to charge sharing," *in Proc. 45th Int. Reliability Physics Symp.*, Arizona, 2007, pp. 306–311.

[9] T. D. Loveless, S. Jagannathan, T. Reece, J. Chetia, B. L. Bhuva, L. W. Massengill, S-J. Wen, R. Wong, D. Rennie, "Neutron- and Proton-Induced SEU Error Rates for D- and DICE-Flip/Flop designs at a 40 nm Technology Node," *presented at the 11th European Conference on Radiation Effects on Components and Systems*, Langenfeld, Austria, September 2010.

[10] JEDEC Standard Measurement and Reporting of Alpha Particles and Terrestrial Cosmic Ray-Induced Soft Errors in Semiconductor Devices, JESD89 Arlington, VA: JEDEC Solid State Technology Association.

[11] R. C. Baumann, "Radiation-induced soft errors in advanced semiconductor technologies," *IEEE Trans. Device Mater. Rel.*, vol. 5, no. 3, pp. 305–316, Sep. 2005.

[12] J. E Knudsen and L. T. Clark, "An Area and Power Efficient Radiation Hardened by Design Flip-Flop," *IEEE Trans. Nucl. Sci.*, vol. 53, no. 6, pp. 3392–3399, Jun. 2006.

[13] N. Seifert, B. Gill, V. Zia, Ming Zhang, V. Ambrose, "On the Scalability of Redundancy based SER Mitigation Schemes," *ICICDT 2007*, pp. 1-9, 2007

[14] S. Dasgupta, "Trends In Single Event Pulse Widths And Pulse Shapes In Deep Submicron CMOS," 2007.

[15] P. E. Dodd and L. W. Massengill, "Basic mechanisms and modeling of single-event upset in digital microelectronics," *IEEE Trans. Nucl. Sci.*, vol. 50, no. 3, pp. 583–602, 2003.

[16] M. Gadlage, "Impact of Temperature on Single-Event Transients in Deep Submicrometer Bulk and Silicon-On-Insulator Digital CMOS Technologies," *PhD. Dissertation*, 2010.

[17] Gang Guo, T. Hirao, J. S. Laird, S. Onoda, T. Wakasa, T. Yamakawa, T. Kamiya,, "Temperature dependence of single event transient current by heavy ion microbeam on p+/n/n+ epilayer junctions," *IEEE Trans. Nucl. Sci.*, vol. 51, no. 5, pp. 2834-2839, Oct. 2004.

[18] Chen Shuming, Liang Bin, Liu Biwei, Liu Zhen, "Temperature dependence of digital SET pulse width in bulk and SOI technologies," *IEEE Trans. Nucl. Sci.*, vol. 55, no. 6, pp. 2914-2920, Dec. 2008.

[19] M. L. Alles, L. W. Massengill, S. E. Kerns, "Effect of temperature-dependent bipolar gain distribution on SEU vulnerability of SOI CMOS SRAMS," *Proc. 1992 IEEE Int. SOI Conf.*, pp. 96-97, Oct. 1992.

Analysis of Multiple Cell Upsets due to Neutrons in SRAMs for a Deep-N-Well process

Nihaar Mahatme, Bharat Bhuva
Department of Electrical Engineering and Computer Science
Vanderbilt University, Nashville TN, USA
nihaar.n.mahatme@vanderbilt.edu

Y-P Fang, Anthony Oates
Taiwan Semiconductor Manufacturing Company
Hsinchu, Taiwan
{ypfanga, aoates}@tsmc.com

Abstract— **This work accounts for the single-bit and multiple-cell upset phenomena due to neutron strikes in highly scaled SRAMs implemented in a Deep-N-well process. 3D TCAD simulations are used to explain test results, upset mechanisms and implications for ECC.**

Keywords— *SRAMs, Soft-Error Rate, single-bit upset (SBU), multiple-cell upset (MCU), multiple-bit upset (MBU)*

I. INTRODUCTION

Soft-error rates for static Random Access Memories (SRAMs) are strongly influenced by the cell layout and subsequently the vulnerable area per bit. As technologies advance, decreased vulnerable area will reduce the soft-error rate per memory bit. However, reduced distances between transistors make multiple transistors vulnerable to a single ion strike resulting in multiple-cell upsets. Thus, multiple-cell upsets in SRAMs due to terrestrial neutrons have increased significantly, in spite of a decrease in Soft Error Rate (SER) per bit. Data in Figure 1 shows the increasing contribution of multi-cell upsets to overall SER for recent technology nodes [1-3]. Since the multiple-cell soft-error rates are a strong function of layout strategies, it is imperative to evaluate different layout strategies for a given technology and cell design. In case of highly scaled commercial SRAM arrays, designers only have control over well contact placement and the error correction code (ECC) to be implemented. This paper investigates both these factors using 3D TCAD simulations for 40 nm bulk CMOS technology node.

The multiple-cell upsets in 40 nm bulk CMOS Deep-N-Well (DNW) SRAMs chips due to ground level neutrons were investigated through testing at Los Alamos Neutron Science Center (LANSCE) neutron test facilities. The primary focus of this work was the number of word-line upsets involved in a single multiple-cell upset (MCU) event as this determines the interleaving and Error Correction Code (ECC) strategy to be adopted. Simulations have been used in this work to identify the mechanism responsible for the multiple-upset phenomena in the vertical (bit-line) direction

and horizontal (word-line) direction, to evaluate the effects of well-contact placement, and to suggest ECC strategies to reduce word-line errors. Predictive simulations for sub 40 nm technology nodes show that such an analysis needs to be carried out for each future technology nodes to prevent significant increases in multiple-bit soft-error rates.

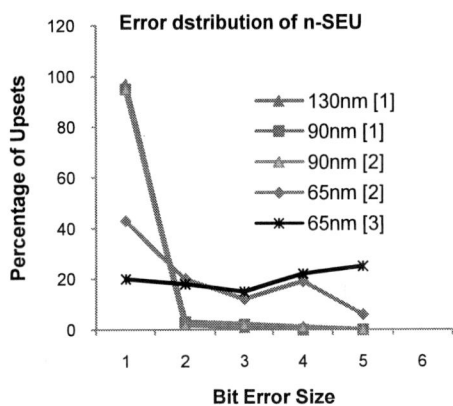

Fig. 1. Single Bit Upset (SBU) and MCU probabilities from previously published results

II. TEST SETUP AND RESULTS

All_0, All_1 and checkerboard (CHB) patterns were written into the 40 nm SRAM chips implemented in Deep-N-Well (DNW) technology and irradiated at the Los Alamos Neutron Science Center (LANSCE) neutron test facilities with varying supply-voltage and temperature. The greatest interest was in the MBU/MCU patterns in the word-line and bit-line directions to understand the underlying failure mechanisms and to analyze the appropriateness of the layout and ECC strategy used. Experimentally, maximum observed MCU in the horizontal (word-line) direction was 4 while those in the vertical (bit-line) direction were 15. For this test, single-bit upsets ranged from 38-59% for different operating voltages and temperatures as seen in Figure 2. Comparing this data with those from other technologies shown in Figure 1, it is clear that the multiple-cell upsets now account for a large percentage of the soft error rate

978-1-4244-9113-1/11 $26.00 © 2011 IEEE

(SER) of memories and must be mitigated to achieve lower SER.

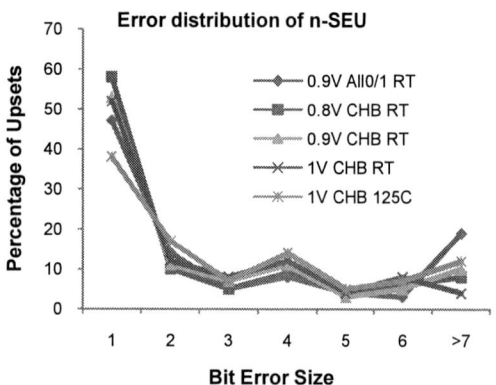

Fig. 2. SBU and MCU probabilities for DNW SRAMs observed after neutron exposure (this work).

Typically observed failure patterns from the error map are shown in Figure. 3. Pattern (a) was observed when all-0 or all-1 pattern was used. Patterns (b) and (c) were observed when checker-board pattern was used. The MBU/MCU double-column pattern represented by Figures 3 (a), 3 (b), and 3 (c) was most common. This was attributed to the presence of smaller p-well regions in DNW process resulting in charge confinement and turn-on of NMOS-based parasitic bipolar transistors [4]. A single-column MCU/MBU pattern was never observed. Since the PMOS transistors are fabricated in larger n-wells this mechanism was not found to be dominant. Thus, NMOS transistor upsets dominate the overall SER of the DNW SRAMs.

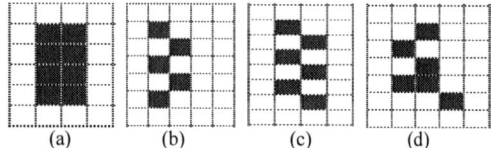

Fig. 3. Typical SRAM upset patterns observed after neutron exposure (this work).

In a large percentage of the MCU events, the number of cells involved was even. For this SRAM design, the cells share a well in the vertical direction with transistors from two cells adjacent to each other as shown in Figure 4. As a result, the de-biasing of the well will affect an even number of cells more often for any ion hit in the vertical direction. For MCU's, this will result in a higher probability for an even number of bit errors than an odd number of bit errors. This is seen in Figure 2 as well.

These observations clearly show a data-dependent effect on SER for multi-cell upsets. 3D TCAD simulations were carried out to understand the underlying mechanisms for MCUs. In the next sections, 3D TCAD simulation methodology and the failure mechanisms responsible for MCU upsets are discussed in detail.

III. SRAM CELL LAYOUT AND PLACEMENT

A representative layout for the 6T-SRAM cell in a DNW process is shown in Figure 4 (a). The SRAM cell itself is flipped and then laid in the horizontal direction. This is shown in Figure 4 (b).

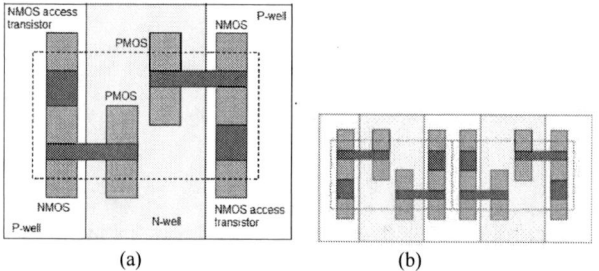

(a)   (b)

Fig. 4 (a) Representational layout placement of transistors for DNW SRAMs [3]
Fig. 4 (b) SRAM cells flipped and placed horizontally

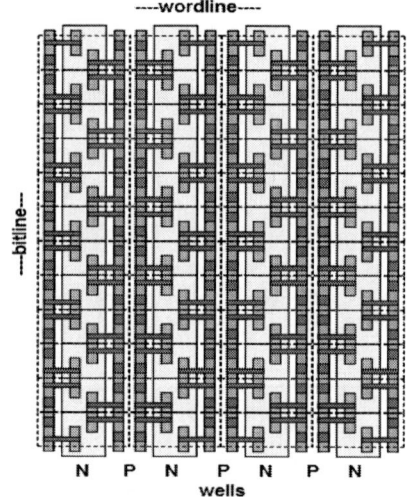

Fig. 5. Section of the layout of an SRAM array with alternating wells.

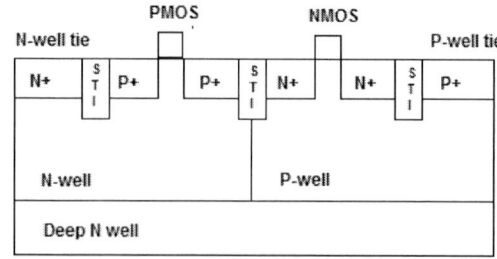

Fig. 6. DNW structure used for simulation [3]

978-1-4244-9113-1/11 $26.00 © 2011 IEEE

For this design, wells run in the vertical direction with well contacts placed 64 cells apart vertically. NMOS transistors of adjacent cells are placed in the same well. Within a word (in horizontal direction), two adjacent cells will have half of the NMOS transistors within one p-well. Thus, all PMOS transistors within a cell are placed in one n-well, while NMOS transistors are placed in two separate p-wells. PMOS (and NMOS) transistors from different rows share a common well in the vertical direction. The wells are oriented in long strips, top to bottom in Figure 5. For clarity, well contacts are omitted in Figure 5.

The cross-sectional view of an SRAM cell is shown in Figure 6. In this paper, multiple-cell upsets (MCUs) will be exclusively used to refer to bit-line errors in the vertical direction along the wells, and single word multiple-bit-upsets (MBUs) will be used to refer to errors in the horizontal (or word line) direction.

IV. TCAD SIMULATIONS FOR MCU VERIFICATION

The chief objective of simulation efforts was to evaluate the MCU phenomena rather than SBUs using TCAD simulations. The parasitic bipolar mechanism has been shown to be the dominant mechanism resulting in MCU events in a well, i.e., in the vertical direction. If the perturbation in the well-potential near the ion hit location is large enough to cause minority carrier injection across the body- (base of the parasitic bipolar) source (emitter of the bipolar transistor) junction, then the parasitic bipolar transistor will turn-on. The carrier movement after the ion strike is in effect the base current for this parasitic bipolar transistor. The well/substrate p-n junctions have the highest probability of charge collection, and they constitute the largest single event currents [5].

This allows for the parasitic bipolar mechanism to dominate as compared to the charge collection due to diffusion and drift of carriers. When this transistor turns on, large voltage perturbations may result at the drain (sensitive) node, causing upsets. On the other hand, MCU events in the word-line direction result due to charge deposition by ionizing particle tracks that cross the wells horizontally. The first part of the simulations focuses on vertical or bit-line MCUs and the second part focuses on horizontal or word-line MCUs. Full-cell TCAD simulations were carried out to mimic the 3D structure of the DNW SRAMs. In addition, the single-event response of DNW SRAMs was compared with conventional dual well structures. All simulations were carried out with appropriate layout and placement dimensions.

Neutron secondary-reaction products with varying LET ranging up to 10.2 MeV-cm^2/mg with path lengths up to 4 μm were used for simulations. These represent the maximum LETs and path lengths from the high end of secondary reaction products of the neutron energy spectrum [6, 7]. The parasitic bipolar transistor turn-on mechanism was observed to occur beyond particle LET values of 2 MeV-cm^2/mg. This value of particle LET corresponds to about 20 fC/μm of charge deposition by the incident particle. Below this particle LET

value, the upsets were caused by charge collection due to conventional drift and diffusion mechanisms. In other words, the charge deposited is low enough for well contacts to effectively remove excess charges, thus preventing bipolar action. For secondary particles with higher LET values, well contacts can't effectively remove all deposited charges, resulting in lowering of well-potential and parasitic-bipolar turn-on.

Most previous studies have used direct hits on drain regions, and the resulting charge collection due to drift and diffusion mechanisms, to study upsets in SRAMs. These efforts are accurate for single-bit upset studies. However, for multiple-bit upsets, the dominant mechanism for charge collection is the turning on of the parasitic bipolar transistor caused by a perturbation in the well potential. To accurately model such an event, profiles of realistic ion paths that do not necessarily cross the drain regions were used for simulations. The trajectory of secondary particles was modeled as passing under the OFF transistors in the vertical direction at varying depths for bit-line upsets and across the wells for word-line upsets.

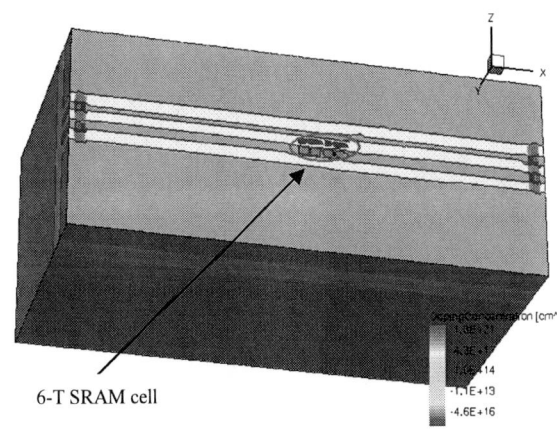

6-T SRAM cell

Fig. 7. 6-T SRAM cell modeled in 40nm triple well process. The cell position was varied to study the effect of resistance modulation between

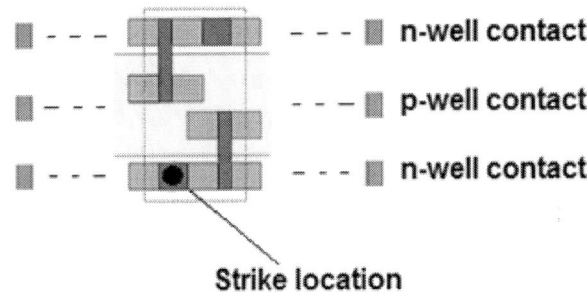

Fig. 8. Shows the relative well contact placement and strike origin under the NMOS channel

The TCAD structure used for the simulations is shown in Figure 7. The SRAM cell seen in the center of the structure shown in Figure 7 is expanded in Figure 8 for a clear view.

The two main factors that influence the upset probability of SRAMs due to parasitic bipolar action are the well depth and the distance between the SRAM cell and the well contact. To evaluate the first, the ion strike location representing the origin point of secondary neutron reaction products was chosen at the center of the well. The distance between well contact and SRAM cell was varied during simulations to evaluate effectiveness of well contact density for SE mitigation. Similarly, well-depth was varied while keeping all other parameters constant to understand the effects of well-potential perturbation (larger wells will have smaller well-potential perturbations for identical conditions). For all simulations, the charge collected due to parasitic bipolar transistor was recorded.

A. Well depth dependence

For simulations to study the well-depth dependence, the location of ion hit was chosen to be at the center of the p-well for all simulations. This ion-strike location also represents the origin of secondary reaction products of neutrons.

Two different charge deposition values were considered; the first about 2 MeV-cm²/mg and 10 MeV-cm²/mg. Two cases were examined: in the first case, the well-depth was the same as the value used for the 40 nm DNW process. In the second case, this well depth was increased by a factor of two to evaluate the charge collection.

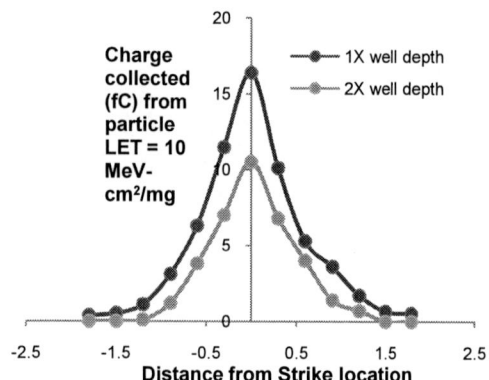

Fig. 9. Charge collected due to parasitic bipolar as a function of distance from center of well strike, for 1X and 2X p-well depth (LET = 10 MeV-cm²/mg)

Simulation results showed that charge collection reduces by over 40% for an identical ion hit when the well-depth is doubled. The main reason for this improvement is reduced source-injected current for the parasitic bipolar transistor. This current, which is a function of the voltage drop across the source-body junction, was reduced by 75%. This is mainly caused by the reduced well resistance between the well contact and the SRAM cell. Figures 9 and 10 show the effects of well depth on collected charge for two different values of particle LET. For the LET values of secondary reaction products chosen, the parasitic bipolar is responsible for 8-10 upsets vertically around the center of the well. This agrees well with

the test results. However this number reduces to 3-6 in the case where the well depth is doubled. The range of cells that upset due to strikes in the p-well is shown in Figure 11.

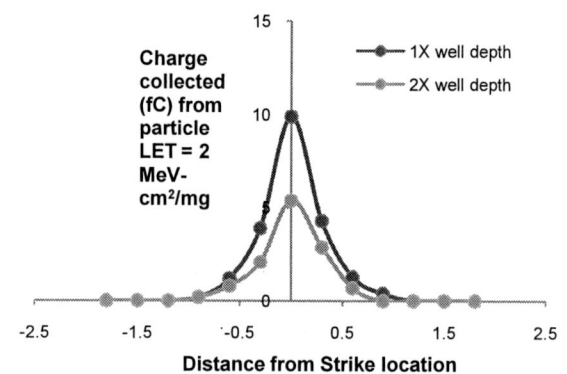

Fig. 10. Charge collected due to parasitic bipolar as a function of distance for 1X and 2X p-well depth (LET = 2 MeV-cm²/mg)

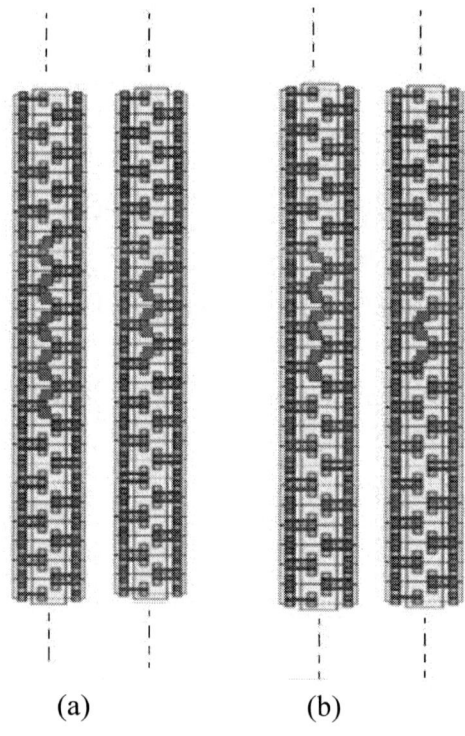

(a)　　　　(b)

Fig. 11 (a) Shows the number of cells that upset about the center of the well for 1X and 2X well depth for LET 10 MeV-cm²/mg
Fig. 11 (b) Shows the number of cells that upset about the center of the well for 1X and 2X well depth for LET 2 MeV-cm²/mg

The red squares represent the cells that upset. Figure 11(a) represents the number of cells that upset in the vertical direction along the well for 1X and 2X well depth respectively for particle LET 10 MeV-cm²/mg. Clearly, when the well depth is doubled, the number of SRAM cells that upset about the same position decrease. Figure 11(b) illustrates the same for a lower particle LET.

978-1-4244-9113-1/11 $26.00 © 2011 IEEE

Increasing the well depth will be a powerful mitigation strategy for MCUs within a single well. However, this will increase the fabrication process complexity. Process and design engineers will need to weigh the penalty and associated performance improvement for such an approach.

B. Well contact dependence

In order to evaluate the effects of the distance of the SRAM cell from well contacts, the cell position was varied about two different locations in the well.

For all simulation runs, the ion strike was always placed in the center of the n-well. The ion-hit location was chosen to be 15 µm or 10 µm away from the well-contact. The source-injected charge collected by NMOS drains of adjacent SRAM cells was recorded. Figure 12 shows the comparison of charge collected due to source injection for these conditions. As expected, collected charge decreases as distance between well-contact and SRAM cells decreases.

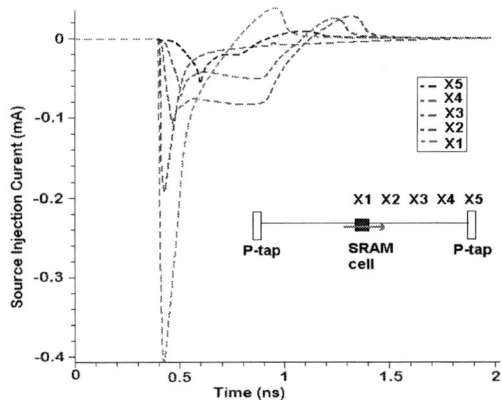

Fig. 13. Source injection currents due to parasitic bipolar turn on at off NMOS drain as a function of SRAM cell location from well-tap

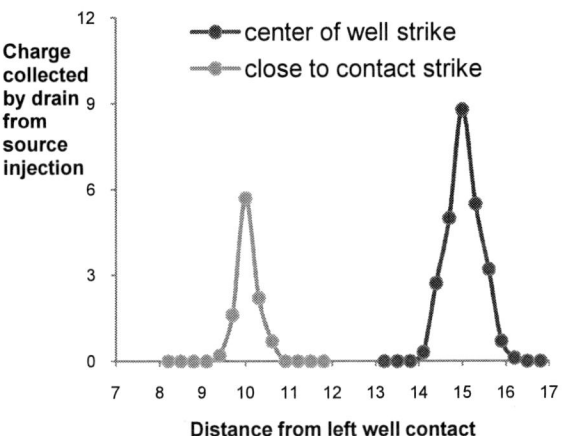

Fig. 12. Charge collected due to parasitic bipolar as a function of distance from well contact

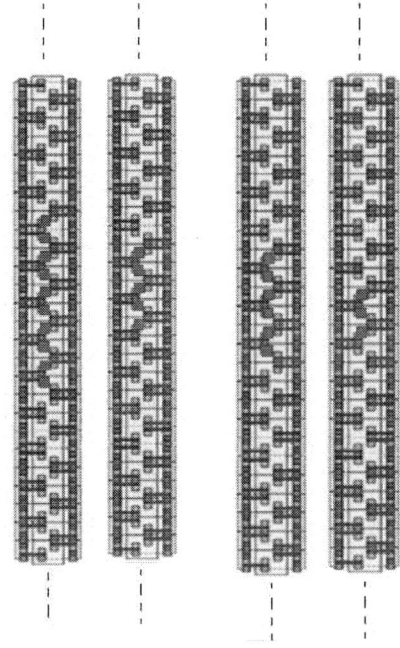

Fig.14 (a) Shows the number of cells that upset about the center of the well for 1) 10 MeV-cm^2/mg and 2) 2 MeV-cm^2/mg charge deposition

Fig.14 (b) Shows the number of cells that upset about a point approximately 10 µm form the well contact for 1) 10 MeV-cm^2/mg and

This is so because the base current for the bipolar is provided by direct charge deposition [5]. As the distance between the cell and the strike location increases the base current of the bipolar decreases. The voltage drop across the base of the parasitic bipolar and well contacts also decreases. This results in weaker source injection in the case of the SRAM cells closer to the contacts as seen in Figure 13. The range of cells that upset due to strikes in the p-well resulting in well perturbation is shown in Figure 14. Figure 14 (a) represents the number of cells that upset in the vertical direction along the well for two different LET values about the center of the well. Cells closer to the well contacts have lower upset probability. This is seen in Figure 14 (b). These results were experimentally observed for the 40 nm test IC. Experimental results in Figure 15 show that the error count for SRAM cells close to the well contacts is about half that of those close to the center of the wells.

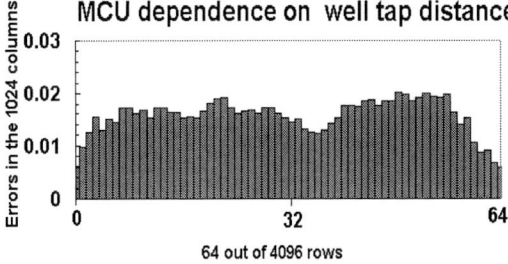

Fig 15. Distribution of MCUs as a function of distance from well taps

C. MCU events in the word-line direction.

For the SRAM sizes and dimensions evaluated for the present technology generation, at most 4 cells were observed to upset in the horizontal direction with realistic values of neutron secondary ion LET and path lengths. All of these SRAM cells belong to a single word in memory. Since, the horizontal SRAM cell dimension is 3X longer than the vertical direction; an ion-hit in the horizontal direction will affect a smaller number of cells. But the affected SRAM cells belong to different words stored in memory. As a result, interleaved memory with a distance of 4 will work fine for SRAM design used in this study. However predictive simulations carried out for the 32 nm technology node, with appropriately scaled dimensions, suggest that more than 4 cells upset horizontally, for identical LET values and simulation techniques adopted for 40 nm SRAMs. Since all bits in a word are stored in the horizontal direction, current interleaving strategy of MUX-4 with ECC may have to be extended to 8 bits, adding to decoding complexity, area and delay overhead for 32 nm technology [8].

V. CONCLUSION

Neutron charge deposition simulations for DNW devices show that the MCU upset probability depends on the location of the ion track and bipolar turn-on mechanism. The NMOS parasitic bipolar transistor turn-on is the dominant mechanism for MCU related upsets.

Efforts must be made to increase the hardness of the NMOS transistors for such a layout. The use of dual well technology with a similar layout may reduce the MCU reduce rates by half because the bipolar transistor turn-on mechanism would affect PMOS transistors more than the NMOS transistors. But unlike NMOS transistors which are shared between two adjacent SRAM cells, only a single column of PMOS transistors would upset. Layout thus strongly influences upset patterns and number of bits that upset. Well contacts can be used effectively to reduce the number of multi-cell upsets. An interleaving strategy of 8 bits with ECC may be required for the next technology node based on predictive simulations.

REFERENCES

[1] Maiz J. et al., "Characterization of multi-bit soft error events in advanced SRAMs", IEDM Tech. Digest, pp 21.4.1-4, 2003.
[2] Georgakos G. et al., "Investigation of Increased Multi-Bit Failure Rate Due to Neutron Induced SEU in Advanced Embedded SRAMs" IEEE Symposium On VLSI Circuits, pp 80-81, 2007.
[3] Gasiot G. et al., "Multiple Cell Upsets as the Key Contribution to the Total SER of 65 nm CMOS SRAMs and Its Dependence on Well Engineering", IEEE TNS, pp 2468-2473, 2007.
[4] Osaka K. et al., "Cosmic-Ray Multi-Error Immunity for SRAM, Based on Analysis of the Parasitic Bipolar Effect", Proc. Of VLSI Symposium on Circuits, 2003.
[5] Black J., PhD Thesis, Vanderbilt University, 2008.
[6] Zhu. W. et al., IEEE TNS, pp 1378-1385, 1999.
[7] Deglahal V., PhD Thesis, Pennsylvania State University, 2003.
[8] Gold B. et al., "Mitigating Multi-bit Soft Errors in L1 Caches Using Last Store Prediction", Workshop on Architectural Support for Gigascale Integration (ASGI-07), June 2007.

Impact of Ion-Induced Transients on High-Speed Dual-Complementary Flip-Flop Designs

Dolores A. Black, *Student Member, IEEE*, Robert A. Reed, *Senior Member, IEEE*, William H. Robinson, *Senior Member, IEEE*, Jeffrey D. Black, *Member, IEEE*, Daniel B. Limbrick, *Student Member, IEEE* and Kevin D. Dick, *Student Member, IEEE*

Department of Electrical Engineering and Computer Science
Vanderbilt University School of Engineering
Nashville, TN 37235
1-615-322-2962, dolores.black@vanderbilt.edu

Abstract— This paper describes the single event performance of a dual-complementary D-type Flip-Flop (DC-DFF) implemented similarly to Dual Interlocked Cell (DICE-DFFs), but without pass-gates. Circuit-level modeling indicates that the DC-DFF is resistant to single event transient (SET) capture of errant signals on the data lines while increasing the operating speed, as compared to the DICE-DFF. However, the simulations also predict that the DC-DFF is susceptible to internal single events during data transitions. This susceptibility is not present in basic DICE designs, but is present in standard DFF designs. Heavy ion testing verified the simulations of the internal single-event clock-dependent mechanism in the DC-DFF design. This dynamic clock-dependent mechanism is described in detail.

Keywords-built-in self-test (BIST); Dual Interlocked Cell (DICE); dynamic upsets; heavy ion testing; single event transient

I. INTRODUCTION

Dual Interlocked Cell (DICE) circuit hardening of sequential logic gates was developed with redundant storage nodes so that an ion strike to a single node would not change the contents of the logic gate [1]. The DICE design has been shown to be susceptible to co-incident ion strikes on two or more circuit nodes [2], multiple node charge collection due to well-collapse mechanisms [3, 4], and clock-rate-dependent charge collection mechanisms [5-7]. This paper describes the Dual-Complementary DFF (DC-DFF) design for higher speed circuits, similar to DICE, and explains new insights into its susceptibility to clock rate effects.

Previous research has described two different mechanisms for clock-rate-dependent single event upsets (SEUs). The first mechanism is a circuit response to an SET propagating to the data input of the DFF [5, 8]. The SET in this case must arrive during the window of vulnerability, i.e., the time when the data input must be stable prior to or after the clock transition. SETs outside of this window will not be captured by the DC-DFF and will not show up as an SEU on the output. The second mechanism is a circuit response to a single event on: (1) the clock input, (2) internal clock buffering, or (3) other controls such as set, reset, preset, and clear [9]. Ion strikes affecting the clock can sample data at the wrong time and thus capture

This work was supported in part by the NASA Electronic Parts and Packaging Program under Grants HDTRA1-08-1-003.

input data before it settles to the correct value. Ion strikes affecting the controls can change the DC-DFF state to a particular state, such as setting the DC-DFF to logic state 1.

In this paper, we propose a new clock-rate-dependent SEU mechanism for the DC-DFF, where an ion strike inside the cell prevents it from writing the correct data into the storage node. This new mechanism is demonstrated in both circuit-level simulation and heavy ion test data collected on a test structure.

The paper is organized as follows. Section II describes a unique DC-DFF cell design, which reduces the likelihood of capturing transients propagating on data signal paths. Then, results of single event simulation of the DC-DFF cell are presented in Section III, followed by a discussion of the circuit used to evaluate the DC-DFF cell and the results of heavy ion testing in Section IV. In Section V, we describe how this new mechanism can be used to explain heavy ion data collected on the DC-DFF. This mechanism would exist in standard DFFs, but not necessarily in traditional DICE-DFFs due to their unique design. Section VI summarizes the paper.

II. DUAL COMPLEMENTARY DFF (DC-DFF)

The basic DICE design is given in [1]. The DC-DFF is a modification of the typical DICE design, implemented to increase the speed of its operations due to faster switching between the complementary logic. It also makes use of its four internal nodes to transmit data.

A. Cell Description

The operation of the DC-DFF can be better understood when several are connected in a shift register format. Typical DFFs are designed with one data input, **d**, and two complementary outputs, **q** and **nq**. The distinguishing characteristic of the DC-DFF shift register configuration is that each complementary pair that constitutes the four storage nodes (**qa**, **qb**, **nqa**, and **nqb**) are output from one DC-DFF to the four input ports (**da**, **db**, **nda**, and **ndb**) of the next DC-DFF (similar to Fig. 1). All connections internal to an individual DC-DFF are also dual complementary. The DC-DFF is designed as a master-slave DFF and is created by connecting two D-latches in series. It is called master-slave because the second latch in the series only changes in response to a

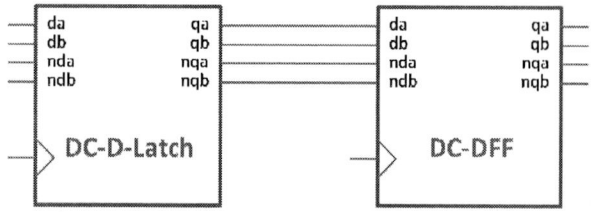

Figure 1. Block diagram of circuits for single event simulation of DC-D-Latch.

change in the first (master) using a non-overlapping clock. Fig. 2 shows the input circuit for either the master or slave of the DC-DFF and is followed by the memory circuit (Fig. 3). Both of the figures combined (32 transistors in all) form a single D-latch and are one-half of the total DC-DFF cell.

B. Cell Write Operation

The nominal operation of the cell can be understood by examining the procedure for overwriting the state of the memory portion of the DC-DFF. Assume that the memory circuit in Fig. 3 (illustrated by simple switches in Fig. 4) is holding **qa** and **qb** high (logic state 1) and **nqa** and **nqb** low (logic state 0). In the hold state, the clock input into the input circuit (Fig. 2) is low, driving the internal connections (**n1**, **n2**, **n3**, and **n4**) high or logic state 1; this configuration is demonstrated with closed switches in Fig. 4. This makes the

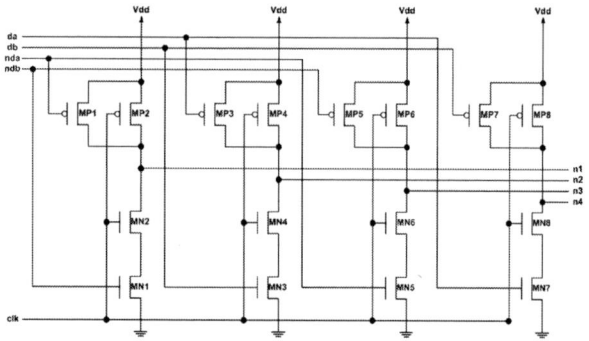

Figure 2. Dual-Complementary DFF input circuit showing internal connections.

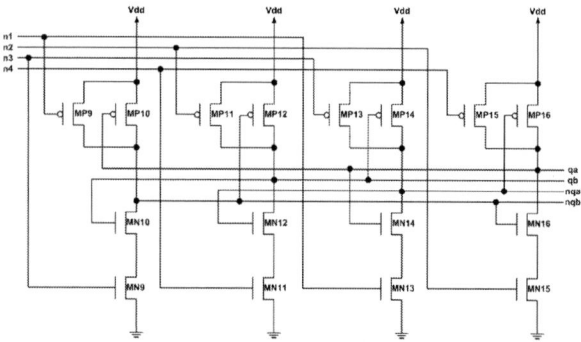

Figure 3. Dual-Complementary DFF memory circuit showing internal connections.

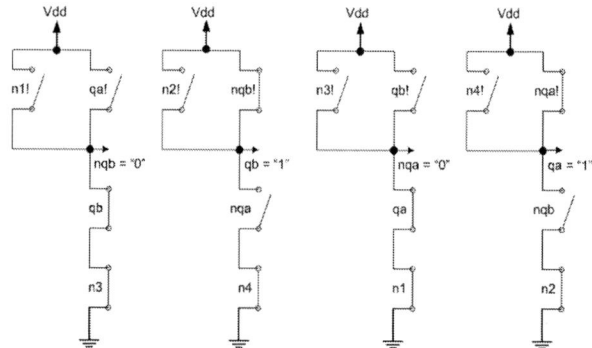

Figure 4. Memory Circuit - start state qa/qb holds logic state 1.

even-numbered transistors in Fig. 3 (MP10/MN10, MP12/MN12, MP14/MN14 and MP16/MN16) or their equivalent switches in Fig. 4 (**qa!/qb**, **nqb!/nqa**, **qb!/qa** and **nqa!/nqb**) in the memory circuit operate in traditional DICE storage operation. The odd numbered transistors in Fig. 3 (MP9/MN9, MP11/MN11, MP13/MN13 and MP15/MN15) are otherwise known as the access transistor pairs of the traditional DICE storage, with their equivalent switches in Fig. 4 (**n1!/n3**, **n2!/n4**, **n3!/n1** and **n4!/n2**). These switches provide the control to determine if the memory circuit should hold its existing state or sample a new state through the input circuit of the DC-DFF.

Table I and Fig. 5 show the input controls, output signals, and the results of the transistor conditions to overwrite the stored data value. The first row (Hold) is the condition just before the clock input goes from logic state 0 to logic state 1. The next five rows show the progression of how the transistors change. The red items in each row are the ones that changed from the previous row. The final row (Stable) is such that the clock could return to low, and the new state will remain in the memory circuit. The table shows that changing the stored value from logic state 1 to logic state 0 begins via closing the switches of **n1!** and **n3!** (i.e., pulling up the internal nodes **nqa** and **nqb**) and conversely opening the switches **n1** and **n3**, thereby changing the output signals. Similarly, changing the memory state from logic state 0 to logic state 1 is accomplished in reverse.

TABLE I.   INPUT CONTROLS, OUTPUT SIGNALS, AND TRANSISTOR CONDITIONS REQUIRED TO CHANGE DC-DFF MEMORY CIRCUIT FROM LOGIC STATE 1 TO LOGIC STATE 0.

Step	Input Control				Output Signal			
	n1	n2	n3	n4	qa	qb	nqa	nqb
Hold	1	1	1	1	1	1	0	0
Change-1	0	1	0	1	1	1	0	0
Change-2	0	1	0	1	1	1	1	1
Change-3	0	1	0	1	1	1	1	1
Change-4	0	1	0	1	0	0	1	1
Stable	0	1	0	1	0	0	1	1

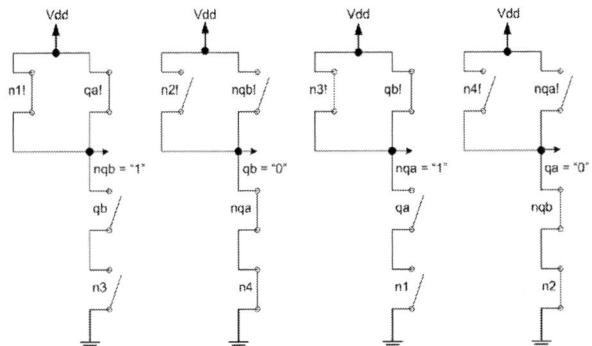

Figure 5. Results from memory circuit normal operation - circuit will settle to logic state qa/qb = 0, (i.e., no error).

III. SINGLE EVENT TRANSIENT CIRCUIT SIMULATION

The input and memory circuits (Fig. 2 and Fig. 3) were simulated in Cadence Spectre with various stimuli to define the input ports (**clk**, **da**, **db**, **nda**, and **ndb**) and a full DC-DFF connected to the output ports, as shown in Fig. 6. This simulation captures the behavior of a shift register. In this model, all transistors in the input and memory circuits could be simulated with a current source between the drain and body, which emulated a heavy ion strike to the transistor. The circuit simulation determined: (a) if an SET generated in either the input or memory circuit would result in a captured SEU in the following DC-DFF, (b) if an SET on the clock input resulted in a captured SEU, or (c) if an SET produced by an ion strike in either the input or memory circuit could prohibit the propagation of the correct data down a shift register.

A double exponential current source was used to represent SETs in all simulations. The rise of the current pulse was defined with a damping factor of 50 ps. The length of the pulse was defined to guarantee errors between clock edges. The amplitude of the current was defined to be 1 mA so that the effect of the current was saturated in the circuit level simulation. The tail of the current pulse had a damping factor of 500 ps and is illustrated in Fig. 7.c (IO_sink).

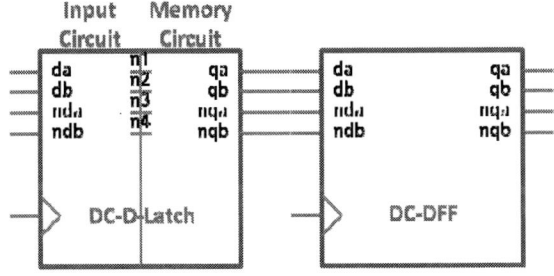

Figure 6. Block diagram of circuits for single event simulation of Dual-Complementary D-Latch. The current sources for single event modeling are placed in the input and memory circuits.

A. Data Line SET

SETs on the data line were simulated with the input and memory circuits in a defined static mode. The SET was generated at defined nodes and simulated to determine if it was latched into the following DC-DFF. Several conditions were evaluated including: (1) an initial logic state 0 in the DC-DFF and a follow-on logic state 0, (2) an initial logic state 0 and a follow-on logic state 1, (3) an initial logic state 1 and a follow-on logic state 0, and (4) an initial logic state 1 and a follow-on logic state 1. The first condition represents a constant logic state 0 data being propagated down the shift register and the last condition represents a constant logic state 1. The middle two conditions represent alternating data propagating down the shift register. Applying these simulation conditions, no errors were observed in the output of the DC-DFF. An example of the alternating data set of simulations results is shown in Fig. 7. Shown in Fig. 7.a is the input clock – **clk**, Fig. 7.b is one of four alternating input data signals – **da**, Fig. 7.d output signal – **nqb**, Fig. 7.e output signal – **qb**, Fig. 7.f output signal – **nqa**, Fig. 7.g output signal – **qa** and Fig. 7.h shifted output data signal – **shift_qa**. This series of waveforms is consistent throughout the paper.

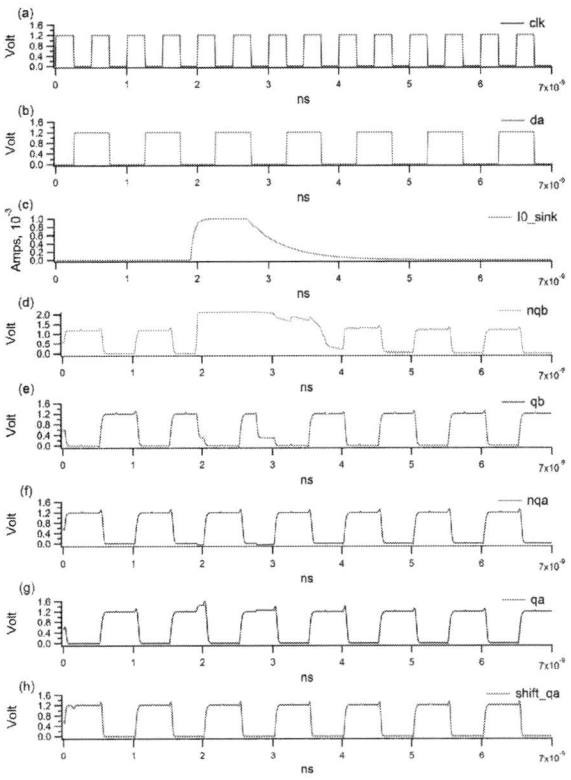

Figure 7. Complementary data - no error (b) matches (h), change of logic state 0 to 1, or logic state 1 to 0; a.input clock pulse b.input state c. ion strike d. through g. internal storage nodes h. output data.

978-1-4244-9113-1/11 $26.00 © 2011 IEEE

899

B. Clock Line SET

SETs on the clock line were analyzed for this circuit (the clock line was not hardened to SETs). Clock-line SET-induced upsets will not occur if the input data signal is held constant (logic state 1 or 0). However, when alternating logic state from 1 to 0 back to 1 (or vice versa), a clock-line SET occurring at the incorrect time may change the stored state early or cause a miss of the change of data logic state. In general, SETs on the clock lines would show up when the data value was alternated, but not when the data value was constant. A comparison of bit errors that occur during switched input versus constant input allows one to determine errors induced on the clock distribution circuit. If more upsets are observed when using alternating data as opposed to constant data, then clock-line SETs can be suspected of contributing to this increased upset rate.

C. Internal DC-DFF Single Event

SETs internal to the DC-DFF that prohibit the memory circuit from loading the proper state were simulated similarly to the SETs on the data lines. The difference was that the circuit in Fig. 6 was operated in a shift register fashion. As previously stated in Section III.A, there were no errors observed in the SET on the data line simulations (Fig. 7); errors observed here result from ion strikes in the DC-DFF, causing the DC-DFF input circuit to be unable to write the correct data into the memory circuit. Once again, we simulated all four conditions, constant input of logic state 0 or 1 and two alternating data logic states 1 and 0. There were no errors in the conditions where the data were constant, meaning that the design of the DC-DFF was hardened against SETs when the propagating data was unchanging. However, there were many sources of error in the propagation of alternating data patterns. Once again, if more upsets are observed when using alternating data as opposed to constant data, then internal DC-DFF single events can be suspected of contributing to this increased upset rate. We propose these internal DC-DFF single events as new clock dependent mechanisms discussed in greater detail below.

1) Input Circuit Single Events

In the alternating data case, every single event on every transistor in the input circuit (Fig. 2) could block the change of state in the memory circuit. To illustrate one of these errors, consider an SET affecting input control **n1**. A single event on the transistors driving **n1** could prohibit the node from pulling down and remain at logic state 1. Table II, Fig. 8, and Fig. 9 show what occurs in the memory circuit under this condition. The items highlighted in blue (or bold/italic font) show the differences from the non-single-event case shown in Table I. The easiest way to understand this mechanism is to examine the input controls to the memory circuit. Only one control signal changes, i.e., **n3**, on the DC-DFF. DICE-like cells are resistant to change from perturbations on a single node by design [1]. The net result is one internal node being in contention (C) and one node floating (F). For purposes of this paper, contention is defined as a short from power, Vdd, to ground caused by both PMOSFETs and NMOSFETs being turned on at the same time. In this example, the DC-DFF circuit will return to its original state. Thus, there is no error if the access transistors are not trying to change the state, while

TABLE II. INPUT CONTROL, OUTPUT SIGNAL, AND TRANSISTOR CONDITIONS REQUIRED TO CHANGE DC-DFF MEMORY CIRCUIT FROM LOGIC STATE 1 TO LOGIC STATE 0. A SINGLE EVENT TO THE INPUT CIRCUIT KEEPS N1 AT LOGIC STATE 1.

Step	Input Control				Output Signal			
	n1	n2	n3	n4	qa	qb	naq	nqb
Hold	1	1	1	1	1	1	0	0
Change-1	*1*	1	0	1	1	1	0	0
Change-2	*1*	1	0	1	1	1	*C*	*F*
Change-3	*1*	1	0	1	1	1	*C*	*F*
Change-4	*1*	1	0	1	*1*	*1*	*C*	*F*
Unstable	*1*	1	0	1	*1*	*1*	*C*	*F*

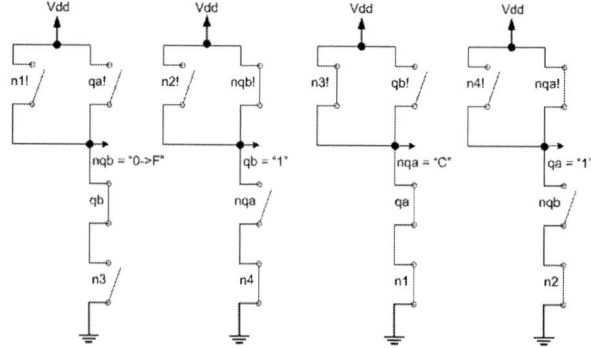

Figure 8. Results from input circuit single event - circuit will settle to logic state qa/qb =1, which is an error.

an error occurs (Fig. 9) if the intent is to change the storage state.

The simulations showed both types of transistors in the input circuit, i.e., PMOSFET or NMOSFET contribute to the prevention of the input data being written to the DC-DFF when struck by a single event. The simulations further showed that this would only occur in one direction, either on the data transition from logic state 0 to logic state 1 or from logic state 1 to logic state 0. This was observed to be dependent on the location of the transistor in the input circuit design. In Fig. 9, the input data transition from logic state 0 to logic state 1 at the time of the single event (at 2.5 ns) was output correctly. However, on the next data input transition from logic state 1 to logic state 0, the output data was blocked and therefore, incorrectly output.

2) Memory Circuit Single Events

The mechanism in the memory circuit was not the same as the input circuit. Once again, the constant input conditions do not cause an error. However, in the condition where the data value was changing, single events on the PMOSFETs in the memory circuit do not cause any errors, and single events on the NMOSFETs in the memory circuit all caused errors. Table III, Fig. 10, and Fig. 11 show the simulation results for a single event to an NMOSFET in the memory circuit, i.e., MN13 (Fig. 3). In the example, the transistor is struck while it was on. When the two input controls, **n1** and **n3**, pull down, the internal

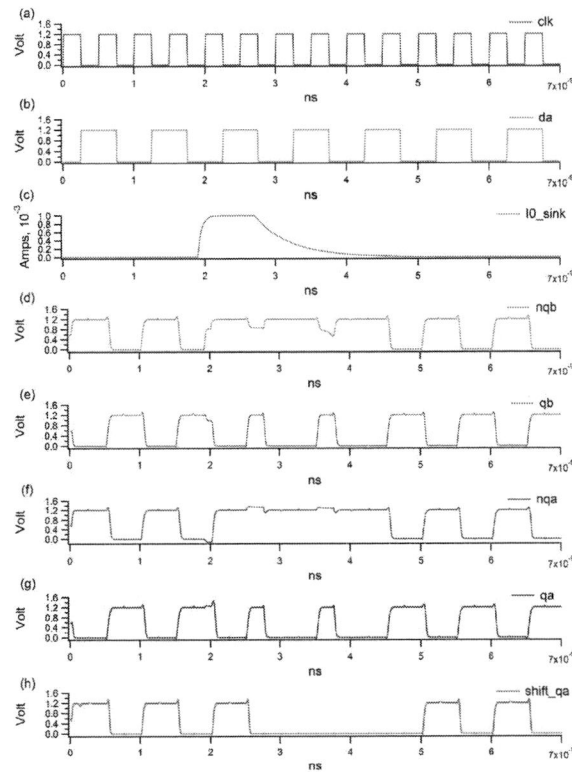

Figure 9. Complementary input data with error located between 3 ns and 5 ns - single event on input circuit.

node **nqa** goes into contention (C). As the transition continues, another internal node goes into contention (C), **qa**, and one floats (F), **qb**. Therefore, only one storage node is correct. This state leads to a very unstable condition for the cell and will require some time to recover. Because various nodes are in contention, the recovery could be very complicated.

The simulations on the memory circuit showed a difference in the PMOSFET and NMOSFET single event response. No errors were observed for single events occurring on PMOSFETs in the memory circuit. However, if a single event occurred on the NMOSFETs, an error resulted in all simulations. Changing the stored value begins by pulling up the

TABLE III.   INPUT CONTROL, OUTPUT SIGNAL AND TRANSISTOR CONDITIONS REQUIRED TO CHANGE DC-DFF MEMORY CIRCUIT FROM LOGIC STATE 1 TO LOGIC STATE 0. A SINGLE EVENT TO THE MEMORY CIRCUIT KEEPS TRANSISTOR MN13 ON, THEREBY KEEPING QA CLOSED.

Step	Input Control				Output Signal			
	n1	n2	n3	n4	qa	qb	nqa	nqb
Hold	1	1	1	1	1	1	0	0
Change-1	0	1	0	1	1	1	0	0
Change-2	0	1	0	1	1	1	C	1
Change-3	0	1	0	1	1	1	C	1
Change-4	0	1	0	1	C	F	C	1
Unstable	0	1	0	1	C	F	C	1

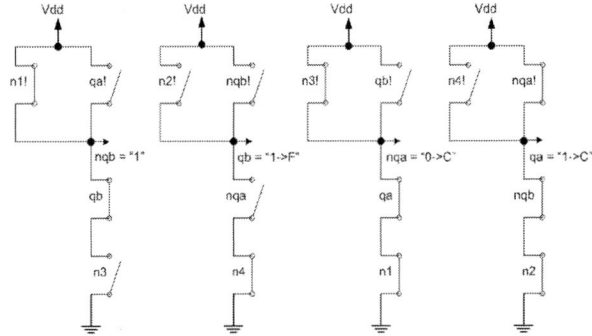

Figure 10. Results from memory circuit single event – circuit will settle to logic state qa/qb = 1, which is an error.

internal nodes. Errors result because ion strikes to NMOSFETs block the pull-up action, while ion strikes to PMOSFETs will not block that action.

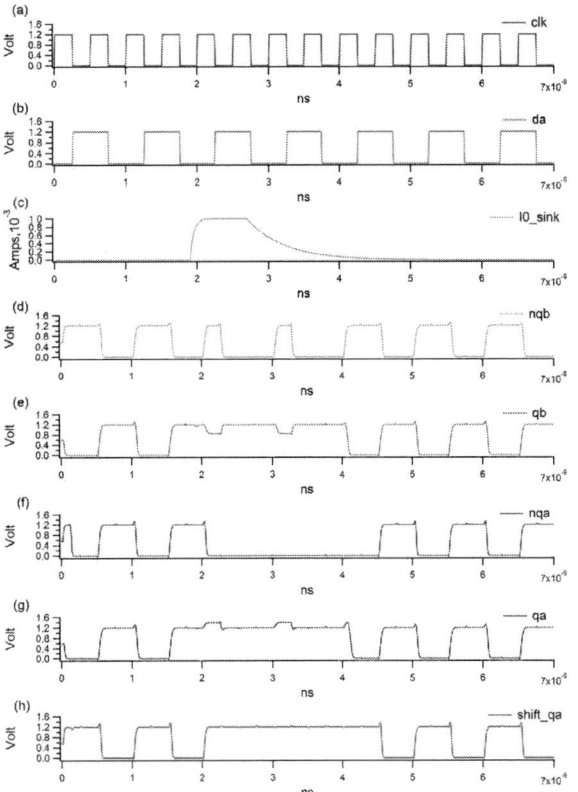

Figure 11. Complementary input data with error between 2.5 ns and 4.5 ns – single event on memory circuit.

978-1-4244-9113-1/11 $26.00 © 2011 IEEE

IV. EVALUATION OF SINGLE EVENT EFFECTS IN DUAL-COMPLEMENTARY DFFS

We evaluated the DC-DFF cells by developing a shift register design in IBM's CMOS9SF process. The circuit includes built-in self-test (BIST) to evaluate the DC-DFF at high frequencies [6, 10]. In this section, we describe two circuit layouts evaluated during single event effects testing, a standard layout and a guard-band layout.

A. V-CREST Test Chip Design

The DC-DFF was designed in a shift register for single event testing. We based the shift register design on the Circuit for Radiation Effects Self Test (CREST) as originally described by Marshall et al. [10]. This original concept implements built-in self-test (BIST) of SiGe flip-flops (FFs) subject to single event effects (SEE) experiments. It enabled SEE experiments at high operating speeds, reduced challenges with test chip input/output pad design and packaging, and removed requirements for high frequency cabling at SEE facilities. It accomplished these features by integrating the clock and the data pattern from a 127-bit pseudo-random number generation onto the test chip. It also moved the error/upset detection circuit onto the test chip. This significantly reduced the switching frequency needed for the test chip input/output.

Vanderbilt University expanded the original CREST design to accommodate bulk complementary metal-oxide-semiconductor (CMOS) FFs; this will be referred to as V-CREST. A block diagram of V-CREST is shown in Fig. 12. Since bulk CMOS FFs SEU sensitive area is smaller than heterojunction bipolar transistors (HBTs) SiGe FFs, the shift register chain needed to include a large number of FFs. Also, more circuits can be operated at once on the V-CREST design. The shift registers were implemented using 584 FFs, which were designed in groups of four, where all four types of shift registers were powered and clocked together.

The V-CREST circuit was designed to enable separation of errors caused by SETs in the clock path versus those in the data path [3, 6]. The clock distribution was designed so that its last buffer drives the four FFs. Single events to the clock distribution would affect all four FFs more or less equally, so upsets to all FFs would be seen as an SET in the clock distribution. The error detection circuit in the V-CREST was designed to ignore such errors. If all four FF shift registers showed an error, then the error flag was not triggered. Finally, the clock distribution design included both the clock and its inverse so that clock signals were not generated within a FF.

Two different DC-DFF layouts were evaluated. The physical difference between the two layouts was one utilized guard-bands and the other did not [3, 11]. There was no physical difference in the placement of the transistors. The size of each layout was also exactly the same.

B. Heavy Ion Test Results

The heavy ion facility for these tests was the 88-inch cyclotron at Lawrence Berkeley National Laboratories. Prior and subsequent to beam exposure, electrical measurements

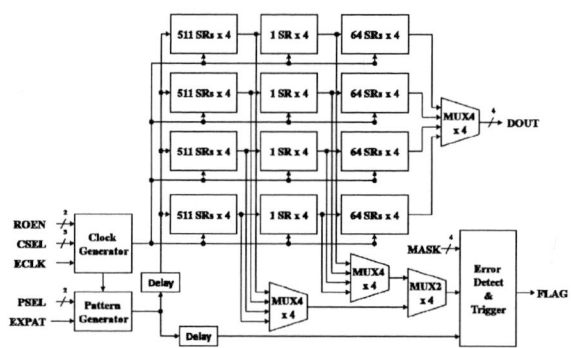

Figure 12. V-CREST block diagram.

were made to verify the samples were functional and within electrical parameter limits. The de-lidded test devices were exposed to a range of beam conditions (energy and LET) while executing test routines that exercise various device functions. During each exposure the device outputs and supply current were monitored for erroneous conditions.

The fluence for each test was selected so that at least 30 errors were observed for each exposure. In test runs where no errors were observed and the accumulation of total dose permitted, the minimum fluence was 10^7 particles/cm^2. The ions used in heavy ion testing of the DC-DFF are summarized in Table IV. Finally, the overall cross-section of the experiment and the cross-section per bit were calculated and plotted in Fig. 13. There are not many data points on the constant input data, though the additional points tested were consistent with the one presented here. These data were from different devices under test and cannot be compared directly. A plot of the data, cross-section versus LET, is shown in Fig. 13.

V. DISCUSSION OF CLOCK DEPENDENT MECHANISMS

We analyzed the susceptibility of the input and memory circuits by simulating ion strikes to all circuit nodes. The simulations showed the DC-DFF was not susceptible to data line SETs but was susceptible to clock line SETs and internal DC-DFF single events. The heavy ion testing of the DC-DFF shows a much higher cross-section when propagating the alternating input data versus constant input data when clocked. Therefore, the higher cross-section can only be explained by clock line SETs and internal DC-DFF single events.

TABLE IV. HEAVY IONS, LETS AND ION ENERGIES USED TO TEST DC-DFFS.

Ion	LET	Energy, MeV
Ne	5.76	90
Ar	14.33	180
Cu	30.04	284
Kr	38.25	387
Xe	68.50	612

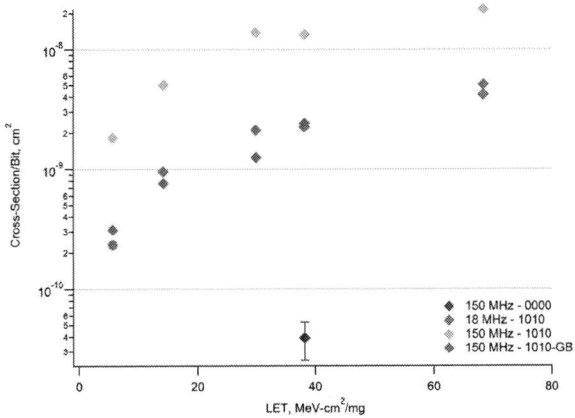

Figure 13. Upset cross-section versus LET for DC-DFF layouts at two different clock frequencies.

The SEU cross-section, Fig. 13, for the guard-band design (150 MHz – 1010-GB) is lower than that for the non-guard-band design (150 MHz – 1010). The guard band was not applied to the transistors in clock distribution circuitry. Therefore, for the standard design, the internal DC-DFF single event upsets are the largest contributor to the cross-section. If the clock line SETs were the larger contributor, then the cross-sections would be approximately equal because the clock distribution network on the chip was exactly the same. Therefore, we can conclude that internal DC-DFF single event upsets are clock-dependent and are a significant contributor to the error cross-section.

To give a feel for the upset cross-section of the internal DC-DFF single events, consider the simulations of the cell. The simulations revealed that all 32 of the NMOSFETs and PMOSFETs in the two input circuits, and all 16 of the NMOSFETs in the two memory circuits, were susceptible to this mechanism; therefore, 48 potential transistors factor into the upset cross-section for this mechanism. Even with the sharing of drain regions in layouts, there will still be significant upset cross-section in DC-DFFs, at least 24 independent drain regions per FF.

VI. SUMMARY

We have used a DC-DFF cell to demonstrate a clock dependent upset mechanism due to ion strikes internal to the DC-DFF. The mechanism prevents the cell from writing new data into the cell. The mechanism showed susceptibility when the data in the DC-DFF cell was alternating from logic state 0 to logic state 1 and vice versa, and no susceptibility when the data in the DC-DFF was constant at logic state 0 or logic state 1. Simulations and heavy ion testing verified the dominance of this mechanism in this design. While the DC-DFF cell is not a practical circuit to use in chip design, it was useful in verifying this mechanism. Simulations further showed traditional DICE-DFF designs did not exhibit this mechanism, but standard DFF designs did exhibit the mechanism.

ACKNOWLEDGMENT

The authors would like to thank Michael Fritze (Defense Advanced Research Project Agency), Lewis Cohn (Defense Threat Reduction Agency), Warren Snapp and Anthony Amort (Boeing Phantom Works) for the support in producing the test chips. The authors would also like to thank Kenneth LaBel, Hak Kim and Mark Friendlich (National Aeronautic Space Administration, Goddard Space Flight Center) for heavy ion test support.

REFERENCES

[1] T. Calin, M. Nicolaidis, and R. Velazco, "Upset hardened memory design for submicron CMOS technology," *IEEE Trans. Nucl. Sci.*, vol. 43, no. 6, pp. 2874-2878, Dec. 1996.

[2] K. M. Warren, B. D. Sierawski, R. A. Reed, R. A. Weller, C. Carmichael, A. Lesea, M. H. Mendenhall, P. E. Dodd, R. D. Schrimpf, L. W. Massengill, T. Hoang, H. Wan, J. L. Long, R. Padovani, and J. Fabula, "Monte-Carlo based on-orbit single event upset rate prediction for a radiation hardened by design latch," *IEEE Trans. Nucl. Sci.*, vol. 54, no. 6, pp. 2419-2425, Dec. 2007.

[3] J. D. Black, A. L. Sternberg, M. L. Alles, A. F. Witulski, B. L. Bhuva, L. W. Massengill, J. M. Benedetto, M. P. Baze, J. L. Wert, and M. G. Hubert, "HBD layout isolation techniques for multiple node charge collection mitigation," *IEEE Trans. Nucl. Sci.*, vol. 52, no. 6, pp. 2536-2541, Dec. 2005.

[4] O. A. Amusan, A. F. Witulski, L. W. Massengill, B. L. Bhuva, P. R. Fleming, M. L. Alles, A. L. Sternberg, J. D. Black, and R. D. Schrimpf, "Charge collection and charge sharing in a 130 nm CMOS technology," *IEEE Trans. Nucl. Sci.*, vol. 53, no. 6, pp. 3253-3258, Dec. 2006.

[5] J. Benedetto, P. Eaton, K. Avery, D. Mavis, M. Gadlage, T. Turflinger, P. E. Dodd, and G. Vizkelethyd, "Heavy ion-induced digital single-event transients in deep submicron processes," *IEEE Trans. Nucl. Sci.*, vol. 51, no. 6, pp. 3480-3485, Dec. 2004.

[6] J. R. Ahlbin, J. D. Black, L. W. Massengill, O. A. Amusan, A. Balasubramanian, M. C. Casey, D. A. Black, M. W. McCurdy, R. A. Reed, and B. L. Bhuva, "C-CREST Technique for Combinational Logic SET Testing," *IEEE Trans. Nucl. Sci.*, vol. 55, no. 6, pp. 3347-3351, Dec. 2008.

[7] K. M. Warren, A. L. Sternberg, J. D. Black, R. A. Weller, R. A. Reed, M. H. Mendenhall, R. D. Schrimpf, and L. W. Massengill, "Heavy ion testing and single event upset rate prediction considerations for a DICE flip-flop," *IEEE Trans. Nucl. Sci.*, vol. 56, no. 6, pp. 3130-3137, Dec. 2009.

[8] P. E. Dodd, M. R. Shaneyfelt, J. A. Felix, J. R. Schwank, "Production and propagation of single-event transients in high-speed digital logic IC's," *IEEE Trans. Nucl. Sci.*, vol. 51, pp. 3278–3384, Dec. 2004.

[9] M. P. Baze, J. Wert, J. W. Clement, M. G. Hubert, A. Witulski, O. A. Amusan, L. Massengill, and D. McMorrow, "Propagating SET Characterization Technique for Digital CMOS Libraries," *IEEE Trans. Nucl. Sci.*, vol. 53, no. 6, pp. 3472-3478, Dec. 2006.

[10] P. Marshall, M. Carts, S. Currie, R. Reed, B. Randall, K. Fritz, K. Kennedy, M. Berg, R. Krithivasan, C. Siedleck, R. Ladbury, C. Marshall, J. Cressler, G. Niu, K. LaBel, and B. Gilbert, "Autonomous bit error rate testing at multi-gbit/s rates implemented in a 5AM SiGe circuit for radiation effects self test (CREST)," *IEEE Trans. Nucl. Sci.*, vol. 52, no. 6, pp. 2446-2454, Dec. 2005.

[11] B. Narasimham, B. L. Bhuva, R. D. Schrimpf, L. W. Massengill, M. J. Gadlage, O. A. Amusan, W. T. Holman, A. F. Witulski, W. H. Robinson, J. D. Black, J. M. Benedetto, and P. H. Eaton, "Effects of guard bands and well contacts in mitigating long SETs in advanced CMOS processes," *IEEE Trans. Nucl. Sci.*, vol. 55, pp. 1708–1713, Jun. 2008.

Stability Improvement of a-ZIO TFT Circuits Using Low Temperature Anneal

Aritra Dey and David R Allee

Flexible Display Center at Arizona State University

Tempe, Az-85281

Tel: +1 480-3990216; email: adey3@asu.edu

Abstract— Long duration of low temperature thermal anneals show performance and stability enhancement for low-temperature fabricated amorphous zinc-indium-oxide (a-ZIO) thin-film-transistors (TFTs). The turn-on voltage (V_{on}) of 50 hour annealed TFTs shifts by 1.5 V for a positive gate bias stress period of 10^4 s when compared to a 2.2 V shift for the un-annealed TFTs. The performance and stability improvements are attributed to a reduction of the interface trap density and removing of defects states in the band-gap of the a-ZIO.

Keywords: ZIO, TFT, V_{on}, anneal, subthreshold swing, V_{th}, D_{it}, LCD, OLED.

I. INTRODUCTION

Amorphous metal oxide TFTs are emerging as strong candidates for replacing hydrogenated amorphous silicon (a-Si:H) as the dominant thin film device [1-4], due to their superior mobility and much higher stability than the former. They find potential application in Liquid Crystal Display Systems (LCDs) and active matrix backplane for organic light emitting diodes (OLEDs). Among the metal oxide TFTs, zinc indium oxide (ZIO) is one of the dominant candidates [1-4]. Although extensive studies of gate bias stress have been reported [4-8], not much work can be found on effects of low temperature anneals of a-ZIO TFTs.

Low temperature fabrication of these TFTs results in a high defect density leading to instability and poor performance like low mobility [1-2]. These are mainly caused due to dangling bonds and amorphous nature of these materials. In this paper, we report on the stability and performance improvement by low-temperature (150◦ C) long-time (25-50 hrs) anneals of a-ZIO TFTs. These long time anneals improve the stability by a reduction of the defect density and the number of interface traps, formed during fabrication [9]. Low temperature anneal has also been used as a technique to understand the different mechanisms that cause the instability in these TFTs.

This paper is divided into the following four sections: Section I gives an introduction of a-ZIO TFTs and brief motivation of writing this paper. Section II describes the detailed TFT structure and low temperature fabrication process at the Flexible Display Center (FDC). Section III gives the experimental set up and discusses the results obtained. Finally we present the conclusions in section IV.

This research was sponsored by the Army Research Laboratory and was accomplished under Cooperative Agreement W911NG-04-2-2005.

II. TFT STRUCTURE AND FABRICATION

The a-ZIO TFTs have a bottom gate inverted staggered structure fabricated at 180˚ C. Molybdenum is used as gate and is sputtered and patterned (thickness = 150 nm). A stack of SiO$_2$ (thickness = 100 nm), ZIO (thickness = 50 nm) and SiO$_2$ (thickness = 100 nm) is deposited next. The ZIO deposition is done between 77 and 91 °C and top layer of SiO$_2$ is deposited at 180 °C. The ZIO is patterned and molybdenum is sputtered to form source/drain contacts, followed by inter-layer dielectric (ILD) deposition. Indium-tin-oxide (ITO) is patterned as connecting metal to the electrophoretic material in an active matrix display. Finally the TFTs are annealed in an atmosphere of nitrogen for 1 hour. Fig. 1 shows the schematic of the device with the different layers.

Arrays of TFTs with W/L equal to 96/9 were fabricated. An average threshold voltage (V_{th}) of -0.4V, linear mobility varying between 4-5 cm^2/Vs, and a subthreshold swing variation of 0.61-0.81 V/dec were obtained with the TFTs. The typical I_{on}/I_{off} ratio was 10^6. The linear mobility was calculated using the following formulae

$$\mu = \frac{g_m (= \partial I_{DS} / \partial V_{GS})}{(W/L) C_{ox} V_{DS}} \quad (1)$$

where g_m is the transconductance, C_{ox} is the oxide capacitance per unit area, V_{GS} is the gate to source bias, and V_{DS} is the drain to source bias. The values of V_{GS} and V_{DS} used to measure μ were 20 V and 1 V respectively. Fig. 2a and b show the typical I_{DS} versus V_{DS} and I_{DS} versus V_{GS} characteristics of these devices. All measurements were made with Keithley 4200-SCS.

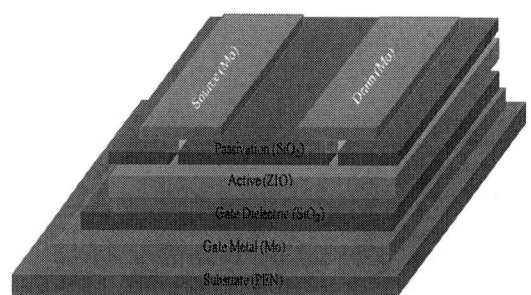

Figure 1. Schematic of the TFT used in this paper

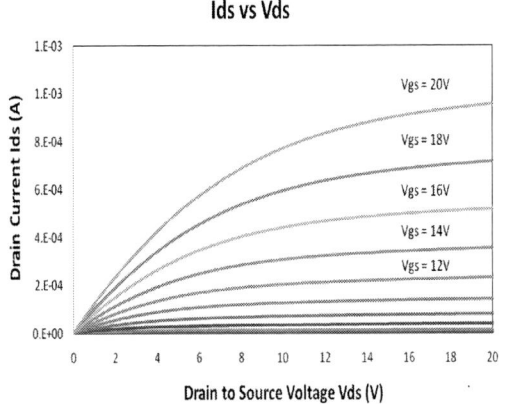

Figure 2a. Output characteristics of one of the TFTs

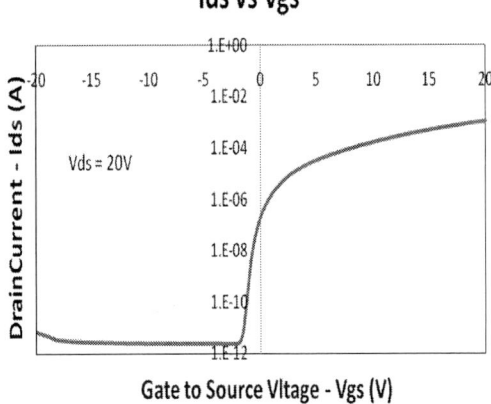

Figure 2b. Transfer characteristics of one of the TFTs

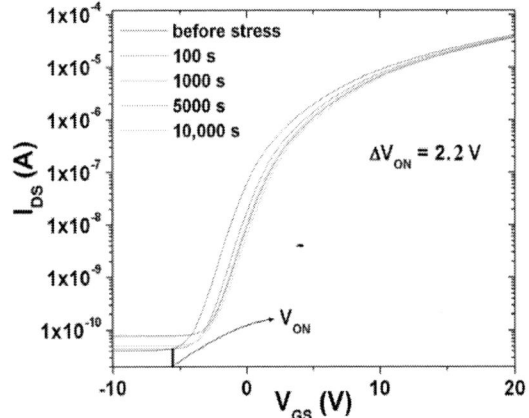

Figure 3a. Drain current (I_{DS}) vs gate to source bias (V_{GS}) curves under gate bias stress (without anneal)

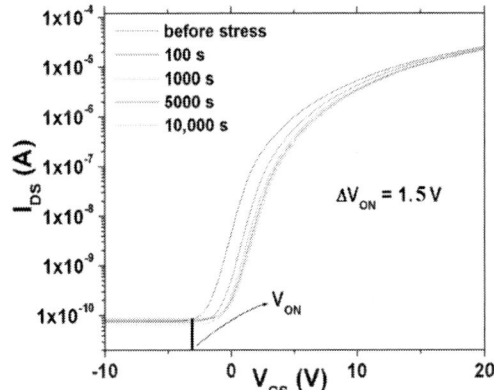

Figure 3b. I_{DS} vs V_{GS} curves under gate bias stress (with 50 hr anneal)

I. EXPERIMENTS AND RESULTS

For the present study the a-ZIO TFTs were annealed at 150 °C for various times before any electrical or thermal stress. Fig. 3 shows the transfer characteristics for the un-annealed (Fig. 3a) and 50 hr annealed TFTs (Fig. 3b) after a V_{GS} stress of 20 V for 10^4 s. The V_{DS} was kept at 0 V during stress and 1 V during measurement. An almost parallel shift of the transfer characteristic in the positive V_{GS} direction is observed. The post stressed V_{on} shifts by 2.2 V (un-annealed TFT), and 1.5 V (50 hr annealed TFT). Such improvements in gallium indium zinc oxide (GIZO) TFTs have also been reported after 65 hrs post thermal anneals at 250 °C [10]. The shift in V_{th} after 65 hrs annealing was only 2.2 V, compared to the shift of 8.5 V for the as fabricated device, when stressed for an hour. Since the characteristics are not exactly MOSFET like, we use V_{on}, which is defined as the V_{GS} at the onset of the initial sharp rise in I_{DS}, instead of V_{th}.

Figure 4 shows the variation of the relative subthreshold swing (S) with bias stress time for both un-annealed and annealed devices (under the same stress condition as above).

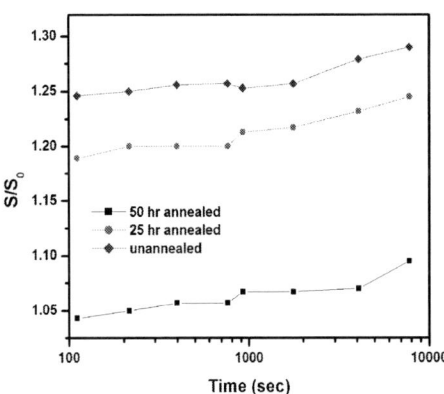

Figure 4. Degradation of subthreshold swing (S) under gate bias stress for different anneal times

Although instability in metal oxide semiconductors is generally attributed to charge trapping [7-8], the degradation in S for longer stress time indicates state creation also playing a role in the degradation. [11]. However, it can also be a cause for field dependent charge/discharge of pre-existing traps that the TFTs almost return to their virgin state when un-stressed for long time [12].

To gain further insight into the mechanism of degradation we subjected the devices (both annealed and un-annealed) to elevated temperatures (370 K). Fig. 5 shows the relative change in linear mobility μ with temperature. The degradation of μ is mainly due to increase in band tail states (D_{it}), via charge trapping in the increased states. Mobility in presence of charge trapping is given by [14]

$$\mu = \mu^0 \left(\frac{t_{free}}{t_{trap} + t_{free}} \right)$$
(2)

where t_{free} and t_{trap} are the average time spent by carriers in free and trap states respectively and μ^0 is the free electron mobility.

Fig.6 shows how the relative S degrades with elevated temperature for both annealed and un-annealed TFTs. The degradation is obviously more for the un-annealed devices. The degradation of S is attributed to increase in D_{it}, rather than charge trapping, as well as calculated from the slope of activation energy (E_a) versus V_{GS}. The E_a was extracted

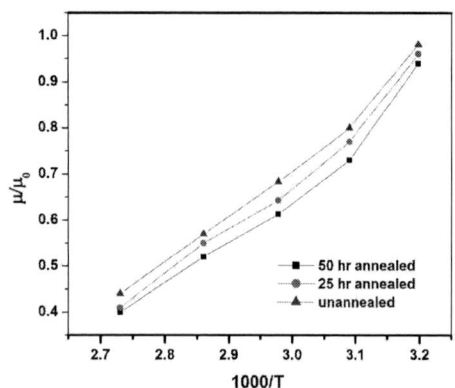

Figure 5. Relative change in mobility with temperature T (K) for annealed and un-annealed devices.

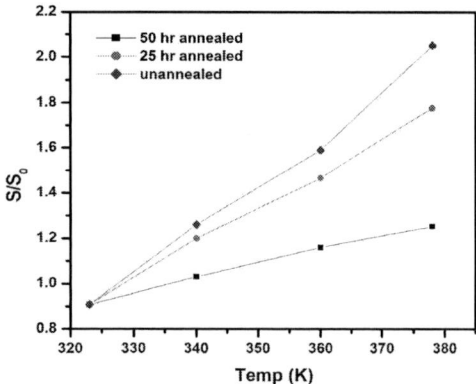

Figure 6. Degradation of subthreshold swing (S) under high temperature for different anneal times

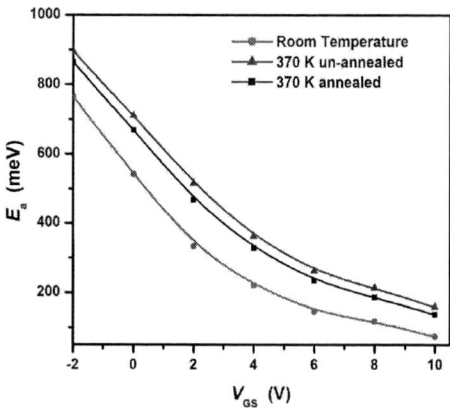

Figure 7. Measurement of activation energy E_a vs V_{GS} for both annealed and un-annealed TFT.

from the Arrhenius plot of log I_{DS} versus $1000/T$ plots for both un-annealed and annealed devices. To measure E_a we obtained I_{DS}-V_{GS} curves from 25°C to 125°C, at an interval of 25°C. It is well known that the slope of E_a versus V_{GS} is inversely proportional to D_{it} [14]. Measurement of the slope as in [15] shown in Fig. 7, showed D_{it} values of 1.55×10^{12} at room temperature. At 370 K the values were $3\text{-}4 \times 10^{12}$ and 2×10^{12} for un-annealed and annealed TFTs respectively. These values were also confirmed using the following formulae [16]

$$N_{it} = \left(\frac{S \log(e)}{kT/q} - 1 \right) \left(\frac{C_{ox}}{q} \right).$$
(3)

where k is Boltzmann's constant, T is the absolute temperature, and q is the electronic charge. We believe that the improvement is due to fraction of the band tail states in the ZIO band-gap because defect states are reduced by annealing the TFTs for long times (even calculated above). Hence, low-temperature, long-time anneals not only improve the electrical performance; but also improve their stability at elevated operation temperatures.

II. CONCLUSIONS

In conclusion, the effect of low-temperature anneals on the stability of a-ZIO TFTs has been studied. The electrical performance and elevated temperature stability of these TFTs has been significantly improved by low-temperature, long-time anneals due to the reduction in trap density at the semiconductor/insulator interface, and/or defects state reduction in band gap of a-ZIO. Finally it also showed that our TFT process is also much more stable than previous ones [10,12].

REFERENCES

[1] B. Bayraktaroglu, K. Leedy, and R. Neidhard, "Microwave ZnO thin-film transistors", *IEEE Electron Device Lett.*, vol. 29, no. 9, pp.1024-1026, Sep. 2008.

[2] Y. Lin Wang, F.R.W. Lim, D.P. Norton, S. J. Pearton, I. I. Kravchenko and J. M. Zavada, "Room temperature deposited indium zinc oxide thin film transistors," *Appl Phys. Lett.*, vol. 90, pp. 232103.1-232103.3, June 2007.

[3] D.P. Heineck, B. R. McFarlane, and J. F. Wager, "Zinc Tin Oxide Thin-Film-Transistor Enhancement/Depletion Inverter," *IEEE Electron Device Lett*, vol. 30, no. 5, pp. 514-516, May 2009.

[4] R. Hoffman, T. Emery, B. Yeh, T. Koch and W. Jackson, "Zinc Indium Oxide thin-film transistors for active-matrix display backplane", in *SID Symp. Int. Tech. Papers*, pp. 288-291, May 2009.

[5] A. Suresh and J. F. Muth, "Bias Stress Instability of Indium Gallium Zinc Oxide Channel Based Transparent Thin Film Transistors," *Appl. Phys. Lett.,* vol. 92, pp: 033502.1-033502.3, Jan, 2008.

[6] K. Kaftanoglu, S. M. Venugopal, M. Marrs, A. Dey, J. R. Wilson, E Bawolek, D. R. Allee and D. Loy, "Flexible Electrophoretic Display using Low temperature ZIO TFT Backplane on PEN substrates", *IEEE Journal of Display Technology,* in press.

[7] R. B. M. Cross and M. M. DeSouza "Investigating the stability of thin film transistors with Zinc Oxide as the channel layer," *Proc. IEEE Int. Reliability Physics Symposium, Phoenix 2007,* pp: 467-471.

[8] R.B.M Cross and M.M. DeSouza, "Investigating the Stability of Zinc Oxide Thin Film Transistors," *Appl. Phys. Lett.,* vol. 88, pp: 263513.1-263513.3, Dec. 2006.

[9] C. Casteleiro, H. L. Gomes, P. Stallinga, L. Bentes, R. Ayouchi, and R. Schwarz, "Study of Trap States in Zinc Oxide (ZnO) Thin Films for Electronic Applications," *J. Non-Cryst. Solids,* vol. 354, 2519- 2522, May 2008.

[10] J. S. Jung *et al*, "Stability Improvements of Gallium Indium Zinc Oxide Thin-Film Transistors by Post Thermal Annealing," *ECS Trans.*, vol. 16, pp. 309-310, Oct. 2008.

[11] K. Hoshino, D. Hai, Q. Chiang, and J.F. Wager, "Constant-Voltage-Bias Stress Testing of a-IGZO Thin-Film Transistors," *IEEE Trans. Electron Devices*, vol. 56, pp. 1165-1170, May 2009.

[12] R. B. M. Cross and M. M. DeSouza, "The effect of gate bias stress and temperature on the performance of ZnO thin film transistors," *IEEE Trans.,Device, Matter Reliability,* vol. 8, pp.277-282, June 2008.

[13] R.A. Street, J. Kakalios, and M. Hack, "Electron Drift Mobility in Doped Amorphous Silicon," *Phys. Rev. B*, vol. 38, pp. 5603, Sep. 1988.

[14] T. Globus, H. C. Slade, M. Shur and M. Hack, "Density of Deep Bandgap States in Amorphous Silicon from the Temperature Dependence of Thin Film Transistor Current," *in Mat. Res. Soc Proc.*, vol. 334, pp: 823-828, Nov. 1994.

[15] A.Dey, A.Indluru, S.Venugopal, D R. Allee and T. Alford, " Effect of electro-mechanical and mechanical stress on amorphous ZIO TFTs," *IEEE Electron Device Lett.*, vol. 31, no. 12, pp.1416-1418, Dec, 2010.

[16] J. Kanichi, and S. Martin, *Thin Film Transistors*, C. Kagan, and P. Andy, Eds. New York: Marcel Dekker, pp. 87- 89, 2003.

Low-Frequency Noise Behavior of La-Doped HfSiON/Metal Gate nMOSFETs

Do-Young Choi, Min Sang Park, Chang Woo Sohn, Hyun Chul Sagong, Eui-Young Jung, Jeong-Soo Lee, and Yoon-Ha Jeong
Department of Electronic and Electrical Engineering
Pohang University of Science and Technology (POSTECH)
Pohang, Korea
+82-54-279-2897, ddoei@postech.ac.kr

Chang Yong Kang
SEMATECH
Austin, USA

Abstract—We investigate the low-frequency noise characteristics of HfSiON/metal gate nMOSFETs with and without La-doping and report new findings on the impact of La-doping on low-frequency noise of the nMOSFETs. The La-doped devices show lower noise intensity than the un-doped devices and it is attributed to the reduced trap density (N_t) and tunneling attenuation length (λ) caused by the La-doping. In the case of submicron devices, however, the La-doped devices show additional mobility-fluctuation noise at low-field condition and the additional noise intensity increases as the gate length decreases. These results indicate that the advantage of low noise of La-doping technique can decrease or even disappear in case of short-channel devices used for analog applications.

Keywords-La-doping, HfSiON/metal gate, low-frequency noise, trap density, tunneling attenuation length, dipole

I. INTRODUCTION

The low-frequency noise is becoming a major concern for advanced gate stack CMOS devices because excessive low-frequency noise and fluctuations could lead to serious limitation of the functionality of the analog and digital circuits [1], but low-frequency noise of nMOSFETs with La-doped Hf-based dielectrics has been investigated by only a few groups [2], [3]. In addition, the impact of La-induced dipole on low-frequency noise characteristics is still not clear while many research groups have reported that the La-induced dipole have considerable effect on device reliability [4], [5]. In this study, we investigate the low-frequency noise of La-doped and un-doped HfSiON/metal gate nMOSFETs and analyze the impact of La-doping on low-frequency noise which is mainly generated by the La-induced dipole.

This research was supported by BK21 program, System IC 2010 project, the National Center for Nanomaterials Technology (NCNT), and WCU (World Class University) program through the National Research Foundation of Korea funded by the Ministry of Education, Science and Technology (R31-2008-000-10100-0).

II. EXPERIMENTS

About 0.7 nm SiO_2 interfacial layer (IL) was grown by rapid thermal oxidation on a HF-cleaned p-type Si wafers and followed by HfSiO film deposited by atomic layer deposition (ALD). A 0.5 nm La_2O_3 cap layer was deposited by physical vapor deposition and a post-deposition annealing was done in N_2 ambient to incorporate La into the dielectric layers. TaN gate electrodes were deposited by ALD. For comparison, co-processed nMOSFETs without La_2O_3 cap layer were also fabricated. The effective oxide thickness (EOT) of both devices was about 1 nm. All noise measurements were performed in linear operation at drain voltage of 50 mV.

III. RESULTS AND DISCUSSION

The La-doped devices showed a lower threshold voltage (V_{th}) than the un-doped devices by about 200 mV, which is due

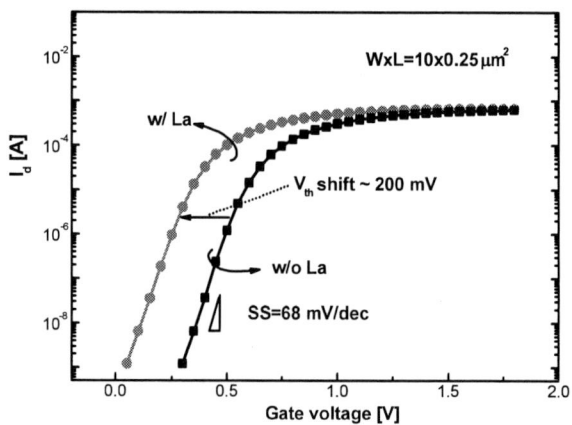

Fig. 1. I_d-V_g characteristics for the nMOSFETs with and without La-doping. The La-doped devices showed a lower V_{th} than the un-doped devices by about 200 mV.

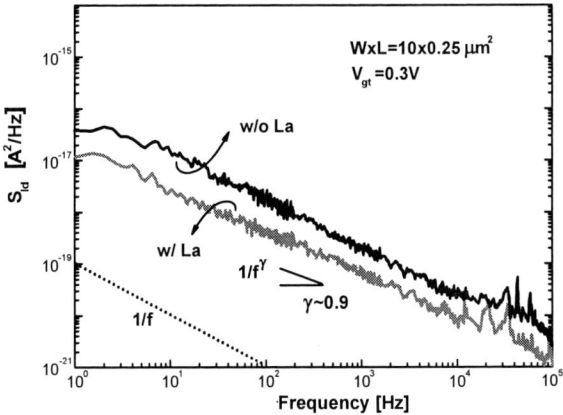

Fig. 2. Drain-current noise spectra for nMOSFETs with and without La-doping. The La-doped devices showed lower noise intensity than the un-doped devices.

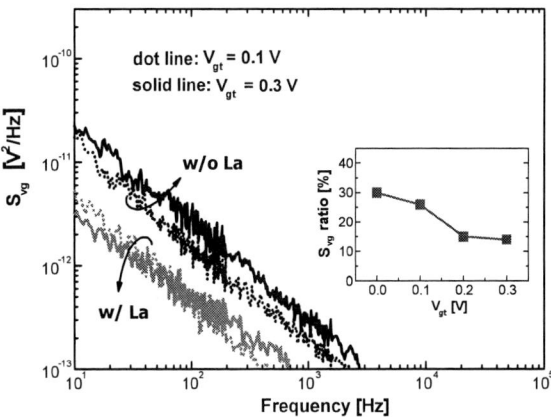

Fig. 4. The S_{vg} of both devices at $V_{gt} = 0.1$ V and $V_{gt} = 0.3$ V. The La-doped devices showed lower S_{vg} than the un-doped devices but the ratio of the S_{vg} of La-doped devices to that of un-doped devices was quite different at each bias point (inset of Fig. 4).

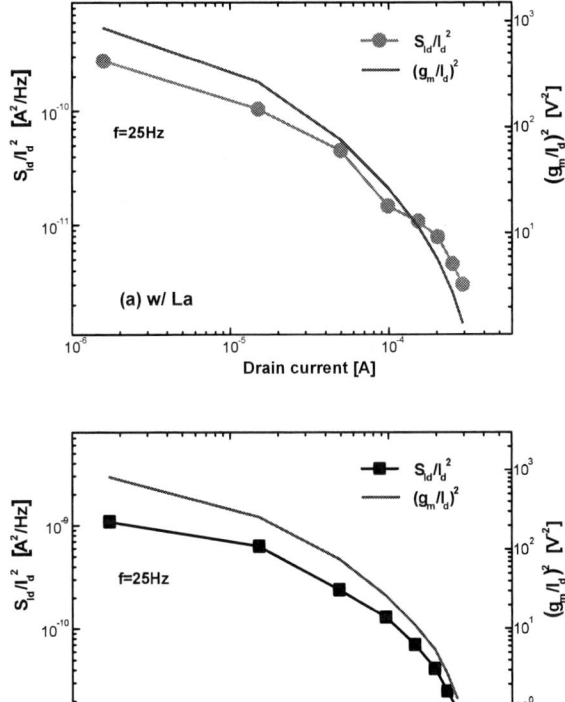

Fig. 3. S_{Id}/I_d^2 at the frequency f = 25 Hz is plotted with $(g_m/I_d)^2$ for (a) La-doped devices and (b) un-doped devices. For both devices, S_{Id}/I_d^2 and $(g_m/I_d)^2$ followed the same trend.

Both devices showed a typical $1/f^\gamma$-like noise with the frequency exponent (γ) close to 1 and the La-doped devices showed lower noise intensity than the un-doped devices.

The 1/f noise is a conductance fluctuation in the channel and there are two mechanisms which generate the noise in MOSFETs. One is number fluctuation of channel carriers and the other is mobility fluctuation of the carriers. In order to figure out the main origin of the noise, the normalized drain-current noise spectral density (S_{Id}/I_d^2) at the frequency f = 25 Hz is plotted with transconductance to drain current ratio squared ($(g_m/I_d)^2$) in Fig. 3. For both devices, S_{Id}/I_d^2 and $(g_m/I_d)^2$ followed the same trend, which indicates that the noise was mainly originated from the number fluctuation of channel carriers [1]. A number-fluctuation noise occurs by trapping and de-trapping of channel carriers at traps in dielectric layers and drain-current noise spectral density is expressed as [1]:

$$S_{Id} = \frac{q^2 kT \lambda N_t}{f^\gamma WLC_{ox}^2} g_m^2 \qquad (1)$$

where kT, λ, N_t, W, L, C_{ox}, and g_m are thermal energy, tunneling attenuation length for channel carriers penetrating into the dielectric layer, occupied trap density, gate width, gate length, gate dielectric capacitance per unit area, and transconductance, respectively.

The input-referred gate-voltage noise (S_{vg}) was calculated from the measured S_{Id} by $S_{vg} = S_{Id}/g_m^2$. Fig. 4 shows the S_{vg} of both devices at $V_{gt} = 0.1$ V and $V_{gt} = 0.3$ V. The La-doped devices showed lower S_{vg} than the un-doped devices probably due to a lower N_t but the ratio of the S_{vg} of La-doped devices to that of un-doped devices was quite different at each bias point (inset of Fig. 4). The S_{vg} ratio increased notably at $V_{gt} = 0.1$ V.

To investigate the bias-dependent noise characteristics, we

to the La-induced dipole formation (Fig. 1) [6]. Both devices showed same subthreshold slopes (SS) of 68 mV/dec. The drain-current noise spectra of La-doped and un-doped devices at the gate voltage overdrive (V_{gt}) of 0.3 V are shown in Fig. 2.

978-1-4244-9113-1/11 $26.00 © 2011 IEEE

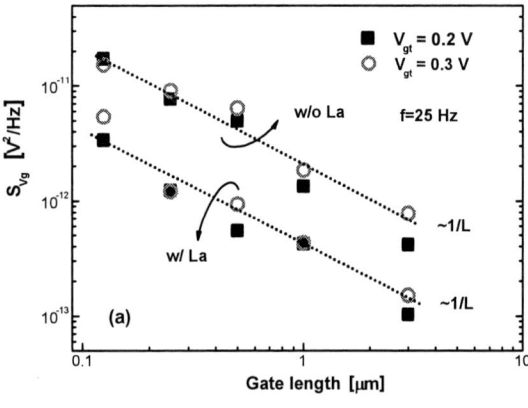

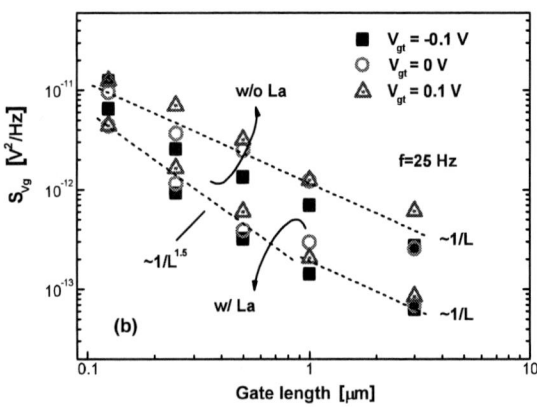

Fig. 5. The gate-length dependence of S_{vg} at various bias points. (a) For $V_{gt} \geq 0.2$ V, both devices showed 1/L-dependent S_{vg}. (b) For $V_{gt} \leq 0.1$ V, both devices showed 1/L-dependent S_{vg} above 1 μm however in the case of submicron devices, the La-doped devices showed $1/L^{1.5}$-dependent S_{vg}.

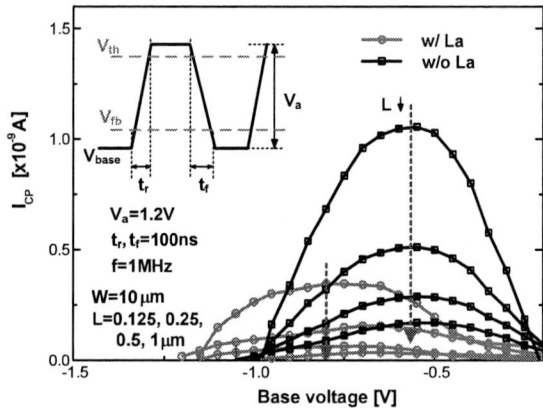

Fig. 6. CP measurement results of both devices with various gate length and same gate width W = 10 μm. The La-doped devices showed lower I_{CP} than the un-doped devices.

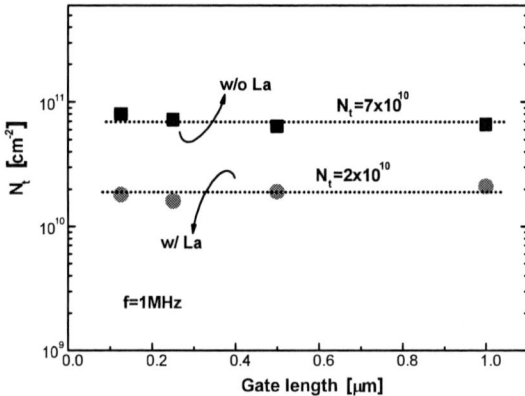

Fig. 7. The extracted N_t from the CP measurement results. The La-doped devices showed lower N_t than the un-doped devices and the N_t was independent of the gate length for both devices.

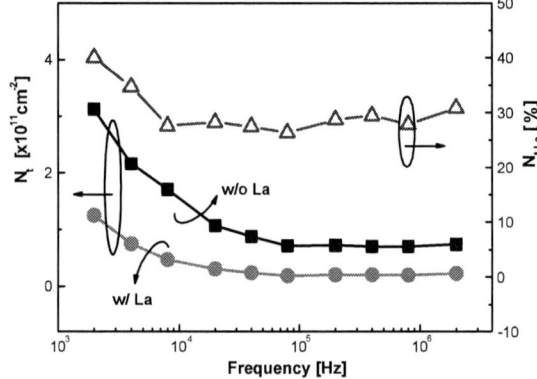

Fig. 8. Frequency dependent CP results in the frequency range of 2 KHz to 2 MHz and the N_t ratio of La-doped devices to that of un-doped devices. The La-doped devices showed at most about 70 % lower trap density than the un-doped devices.

analyzed the gate-length dependence of S_{vg} at various bias points as shown in Fig. 5. For $V_{gt} \geq 0.2$ V (Fig. 5(a)), both devices showed 1/L-dependent S_{vg}. The S_{vg} difference was also 1/L-dependent and the S_{vg} ratio was almost constant for all gate length. These results are consistent with the Eq. (1) of number-fluctuation noise. Fig. 6 shows the charge-pumping (CP) measurement results of both devices [7] with various gate length and same gate width W = 10 μm under the following conditions: pulse amplitude V_a = 1.2 V, rising time t_r = 100 ns, falling time t_f = 100 ns, pulse frequency = 1 MHz, duty cycle = 50 %. The La-doped devices showed lower CP current (I_{CP}) than the un-doped devices and N_t is proportional to the maximum of I_{CP}. The extracted N_t from the CP measurement results is shown in figure 7. The La-doped devices showed lower N_t than the un-doped devices and the N_t was independent of the gate length for both devices. Therefore, the lower noise intensity of La-doped devices was attributed to the reduced N_t caused by the La-doping [8] and 1/L dependence of S_{vg} for both devices was well explained by the Eq. (1) with the gate-length independent N_t. Fig. 8 shows frequency dependent CP results in the frequency range of 2 KHz to 2 MHz and the N_t

ratio of La-doped devices to that of un-doped devices. As the pulse frequency is reduced, electrically active traps farther away from the dielectric/substrate interface can be detected [9]. They showed at most about 70 % lower trap density than the un-doped devices while the La-doped devices showed about 85 % lower S_{vg} than the un-doped devices (Fig. 5(a)), which means the λ of La-doped devices was reduced and also contributed the lower S_{Vg}. The λ is expressed as [10]:

$$\lambda = \sqrt{\hbar^2/(8m^*\varphi_B)} \qquad (2)$$

where ħ, m*, and φ_B are Plank's constant normalized by 2π, effective mass of carrier, and tunneling barrier height seen by channel carries. The effective barrier height of La-doped devices for substrate injection is increased because of the La-induced dipole formation between the interface and high-k layer [4]. The increased barrier height of the La-doped devices resulted in the reduction of λ and hence reduction of S_{vg}. Therefore, the reduced λ caused by the La-induced dipole is responsible for the lower noise intensity of La-doped devices.

In Fig. 5(b), for $V_{gt} \le 0.1$ V, the both devices showed 1/L-dependent S_{vg} above 1 µm however, in the case of submicron devices, the La-doped devices showed $1/L^{1.5}$-dependent S_{vg}. As a result, the S_{vg} ratio of submicron La-doped devices was larger than that for $V_{gt} \ge 0.2$ V and increased rapidly as the gate length decreased. The noise level of La-doped devices was comparable with that of un-doped devices at around L = 100 nm. Because the noise of La-doped devices was mainly generated by the number-fluctuation of channel carriers (Fig. 3(a)), the $1/L^{1.5}$-dependent S_{vg} of submicron La-doped devices could be explained by an additional noise depending on the gate length added to the dominant number-fluctuation noise.

The additional noise could not be explained by the extracted N_t because the additional-noise intensity varied with the gate-length of the devices. In other words, the additional noise was not originated from the number fluctuation of channel carriers through the trapping and de-trapping at the traps in dielectrics. Therefore it was mobility-fluctuation noise which is originated from the carrier scattering. Mobility-fluctuation noise is closely connected to mobility degradation phenomenon through identical scattering mechanism [11]. Tatsumura et al. [12] and C. Y. Kang et al. [4] have reported that in La-doped HfSiON/metal gate nMOSFETs, La-doping caused mobility degradation at low field conditions and it was attributed to remote Coulomb scattering by La-induced dipole at high-k/IL interface. Therefore, the additional mobility-fluctuation noise of submicron La-doped devices for $V_{gt} \le 0.1$ V may be attributed to remote Coulomb scattering by the La-induced dipole.

IV. CONCLUSION

The La-doped devices showed lower noise intensity than the control devices and it is attributed to the reduced N_t and λ of the devices cased by the La-doping. In the case of submicron

devices, however, the La-doped devices showed additional mobility-fluctuation noise which increased as the gate length decreased. The La-doped devices have advantage of lower noise comparing with the un-doped devices but that merit is reduced at low-field conditions in case of short-channel devices due to additional mobility-fluctuation noise and this phenomenon is getting more severe as the gate length decreases. The additional mobility-fluctuation noise of short-channel La-doped devices should be considered to use the devices in analog, mixed-signal, and RF circuits.

ACKNOWLEDGMENT

The authors thank to SEMATECH for device fabrication.

REFERENCES

[1] G. Ghibaudo, T. Boutchacha, "Electrical noise and RTS fluctuations in advanced CMOS devices," Microelectron. Reliab., vol. 42, no. 4/5, pp. 573-582, Apr./May 2002.

[2] H. S. Choi, S. H. Hong, R. H. Baek, K. T. Lee, C. Y. Kang, R. Jammy, B. H. Lee, S. W. Jung, and Y. H. Jeong, "Low-frequency noise after channel soft oxide breakdown in HfLaSiO gate dielectric," IEEE Elec. Dev. Lett., vol. 30, no. 5, pp. 523-525, May 2009.

[3] E. Simoen, A. Akheyar, E. Rohr, A. Mercha, and C. Claeys, "Low-frequency noise analysis of the impact of an LaO cap layer in HfSiON/Ta2C gate stack nMOSFETs," ECS Tans., vol. 25, no. 7, pp. 237-245, Oct. 2009.

[4] C. Y. Kang, C. D. Young, J. Huang, P. Kirsch, D. Heh, P. Sivasubramani, H. K. Park, G, Bersuker, B. H. Lee, H.S. Choi, K.T. Lee, Y-H. Jeong, J. Lichtenwalner, A. I. Kingon, H-H Tseng and R. Jammy, "The impact of La-doping on the reliability of low Vth high-k/metal gate nMOSFETs under various gate stress conditions, " in IEDM Tech. Dig., 2008, pp. 115-118.

[5] J. Huang, P. D. Kirsch, D. Heh, C. Y. Kang, G. Bersuker, M. Hussain, P. Majhi, P. Sivasubramani, D. C. Gilmer, N. Goel, ^M.A. Quevedo-Lopez, C. Young, C.S. Park, C. Park, P. Y. Hung, J. Price, H.R. Harris, B .H. Lee, H.-H. Tseng and R. Jammy, "Device and Reliability Improvement of HfSiON+LaOx/metal gate stacks for 22 nm node application," in IEDM Tech. Dig., 2008, pp. 45-48.

[6] P. D. Kirsch, P. Sivasubramani, J. Huang, C. D. Young, M. A. Quevedo-Lopez, H. C. Wen, H. Alshareef,1 K. Choi, C. S. Park, K. Freeman, M. M. Hussain, G. Bersuker, H. R. Harris, P. Majhi, R. Choi, P. Lysaght, B. H. Lee, H.-H. Tseng, R. Jammy, T. S. Böscke, D. J. Lichtenwalner, J. S. Jur, and A. I. Kingon, "Dipole model explaining high-k/metal gate field effect transistor threshold voltage tuning," Appl. Phys. Lett., vol. 92, no. 9, pp. 092901-1-092901-3, Mar. 2008.

[7] G. Groeseneken, H. E. Maes, N. Beltran, and R. F. De Keersmaecker, "A reliable approach to charge-pumping measurements in MOS transistors," IEEE Trans. Device Mater. Rel., vol. 31, no. 1, pp. 42-53, Jan. 1984.

[8] M. Sato, N. Umezawa, J. Shimokawa, H. Arimura, S. Sugino, A. Tachibana, M. Nakamura, N. Mise, S. Kamiyama, T. Morooka, T. Eimori, K. Shiraishi, Y. Yamabe, H. Watanabe, K.Yamada, T. Aoyama, T. Nabatame, Y. Nara, and Y. Ohji, "Physical model of the PBTI and TDDB of La incorporated HfSiON gate dielectrics with pre-existing and stress-induced defects," in IEDM Tech. Dig., 2008, pp. 119-122.

[9] C. D. Young, D. Heh, A. Neugroschel, R. Choi, B. H. Lee, and G. Bersuker, "Electrical characterization and analysis techniques for the high-k era," Microelectron. Reliab., vol. 47, no. 4/5, pp. 479-488, Apr./May 2007

[10] K. K. Hung, P. K. Ko, C. Hu, and Y. C. Cheng, "A unified model for the flicker noise in metal-oxide-semiconductor field-effect transistors," IEEE Trans. Electron Devices, vol. 37, no. 3, pp. 654-664, Mar. 1990.

[11] M. von Haartman, B. G. Malm, and M. Östling, "Comprehensive study on low-frequency noise and mobility in Si and SiGe pMOSFETs with

high-k gate dielectrics and TiN gate," IEEE Trans. Electron Devices, vol. 53, no. 4, pp. 836-843, Apr. 2006.

[12] K. Tatsumura, T. Ishihara, S. Inumiya, K. Nakajima, A. Kaneko, M. Goto, S. Kawanaka, and A. Kinoshita, "Intrinsic correlation between mobility reduction and Vt shift due to interface dipole modulation in HfSiON/SiO2 stack by La or Al addition," in IEDM Tech. Dig., 2008, pp. 25-28.

NEUTRAL INTERFACE TRAPS FOR NEGATIVE BIAS TEMPERATURE INSTABILITY

Z. Chen, X. Zhou, Y. Z. Hu, and K. S. Machavolu

Nanyang Technological University, School of Electrical & Electronic Engineering, Singapore 639798

E-mail: zh_chen@msn.com

Negative Bias Temperature Instability (NBTI) is a critical reliability issue and becoming more and more seriously in the modern CMOS technology. Many models have been proposed to account for the NBTI phenomena, but most of the models are based on the popular hydrogen reaction-diffusion (R-D) mechanisms. For the first time, in this work, a neutral interface-trap model due to the random variations of bond angles and lengths of Si••Si, Si••O and Si••N bonds is applied to explain the NBTI phenomena. The unified compact model with the neutral interface traps can very well characterize stressed device performance. The neutral interface traps with energy distribution in the silicon energy gap can very well account for the root cause of NBTI fast recovery. The model is verified by TCAD simulation and experimental measurement with excellent agreement.

NBTI is an important reliability concern for modern digital and analog technology. In recent years, several NBTI model have been proposed [1–4]. Although the inconsistency [5], hydrogen-based reaction-diffusion (R-D) model is the most popular interpretation of NBTI phenomenon. It is generally believed that NBTI degradation is due to the generation of interface traps along SiO_2/Si interface channel during the stress. The generation of interface traps is due to the broken Si••H bonds, and subsequently the hydrogen diffuses away from the interface. In this work, a neutral interface-trap model with physics basis is applied to account for the NBTI phenomena. The interface-trap behavior is accurately captured in the I_{DS}-V_{GS} characteristics of the stressed MOSFETs. The bias-dependence of neutral interface traps with energy distribution in the silicon energy gap can well account for the fast recovery of NBTI problem.

NEUTRAL INTERFACE TRAPS A Coulomb potential energy is generated when an impurity atom is introduced into a periodic lattice. The periodic potential of the lattice is perturbed by the introduced impurity atom. Fig. 1 shows the Slater's electronic traps [6] from bulk impurity ions: Fig. 1(a) shows that an n-type impurity atom generates a bound electron state, by shifting a conduction band state downwards into the energy gap, and Fig. 1(b) shows that a p-type impurity atom generates a bound hole state, by shifting a valence band state upwards into the energy gap. Slater's impurity potential model in the bulk is applied to the interfaces of two crystalline lattices [7]. Different from Slater's electronic bulk traps, the interface-trap model is a physical defect model or lattice mismatch model. The physical defects are the random variations of bond angle and length of Si••Si, Si••O and Si••N bonds. These random variations generate electrical-neutral perturbation potential wells with random depth and random diameter or size. The neutral-potential well acts as the neutral interface trap. Since both neutral electron traps and neutral hole traps are due to the perturbations, respectively, from conduction and valence bands, many random perturbations would give a U-shaped energy distribution of interface traps in the silicon energy gap as shown in Fig. 2.

MODEL FORMULATION The neutral electron interface trap is defined as an electrically neutral trapping potential well that can bind only one electron. Thus, the neutral electron traps have charge states 0 (i.e., empty) and −1 (i.e., capturing an electron). Similarly, the neutral hole interface trap is defined as a neutral trapping potential well that can bind only one hole. Thus, the neutral hole traps have charge states 0 (i.e., empty) and +1 (i.e., capturing a hole). Eqns. 1 and 2 respectively shows the neutral electron and hole interface-trap equations. The total neutral interface-trap charge is the sum of the charges at the neutral electron and hole interface traps over the silicon energy gap as shown in Eqn.7. Thus, the total neutral interface-trap charge not only depends on the interface-trap density but also the applied gate voltage because charge trapping concentration is directly related to the surface carrier concentration which is modulated by the gate voltage. Eqns. 8–15 show neutral interface traps are added in the unified compact model [8], in which I_{RE} is recombination current, I_{DD} is drift and diffusion current and I_D is drain-terminal current. The total interface-trap charge is sum of the trapped charges at the neutral interface traps, N_{NT}, interface traps due to the hydrogen-broken bonds, N_{HT}, and oxide traps, N_{OT}, as shown in Eqn. 9. During regular operations in pMOSFETs with n-type substrate and p-type source and drain contacts, the source-body p/n junction is zero biased while the drain-body p/n junction is forward biased. Thus, electron-hole recombination is inherent along the channel region due to the trapping electrons and holes at the interface traps. The recombination currents along the channel region greatly modify the subthreshold characteristics, not only the slope but also great distortion of the drain-terminal current.

RESULTS AND DISCUSSION Fig. 4 shows the gate-voltage dependence of interface-trap charges at various interface-trap densities. As anticipated by Eqn.1, the charge trapping concentration is closely related to the surface carrier concentration which is a function of the surface potential, and hence, the applied gate voltage. Fig. 5 shows the effect of the interface-trap density dependence on drain-terminal current I_D in short-channel MOSFETs. Because interface traps shift the threshold voltage which reduces the useful gate voltage at a given condition, the lineshapes of I_D are distorted increasingly as the neutral electron-trap density is increased. Good agreement is achieved between the model and Medici simulations. In order to further check the neutral interface-trap model, Fig. 6 shows the comparison of neutral interface traps on I_D-V_{GB} in short-channel MOSFETs between model calculation and experimental data, which is measured using ultra-fast switching method (UFS) [9]. The model of interface traps on I_D-V_{GB} shows very good agreement with the UFS experimental data. Fig. 7 shows the threshold voltage variation ΔV_t as a function of stress time. Fig. 8 shows that the ΔV_t is measured using ultra-fast switching method (cycle symbol) and DC method (Square symbol). Device degradation involves not only oxide traps and the interface traps due to the broken Si••H bond, but also neutral interface traps due to the random variation of bond angles and lengths of Si••Si, Si••O and Si••N bonds near the SiO_2/Si interface. The bias-dependent neutral interface traps with energy distribution in the silicon energy gap can account very well for the fast recovery of NBTI phenomena. When the stress voltage is removed, the low energy portion of neutral interface traps, such as interface trap with energy levels near the conduction- and valence-band edges, would quickly lose charges and become neutral, which can account for the fast recovery phenomena of NBTI problem. The oxide traps and the interface traps due to the broken hydrogen bonds would slightly increase as the stress time to account for the permanent part of NBTI problem. For a U-shaped energy distribution of neutral interface traps, if the low-energy traps with $E_{TI} < 0.20$ eV lose charges during NBTI cycles, $\Delta V_t = 0.811$ V for a device with $T_{OX} = 1.7$ nm and $N_{NT} = 1.1 \times 10^{12}$ cm^{-2}. $Q_{NT} = 0.1893 \times 10^{-6}$ C/cm^2 under the NBTI stresses while $Q_{NT} = 0.01992 \times 10^{-6}$ C/cm^2 during the measurement. Thus, the recovery gives 89% variation of neutral interface-trap charge.

CONCLUSION A neutral interface-trap model with energy distribution in the silicon energy gap is applied to characterize device performances and account for the fast recovery of NBTI phenomena. The bias and energy dependence of interface-trap model can very well account for the root cause of fast recovery of NBTI problem. With the neutral interface traps, the unified compact model can very well characterize the interface-trap effect on transistor performance. The modeling approach has been verified by the TCAD simulation and experimental data with excellent agreement.

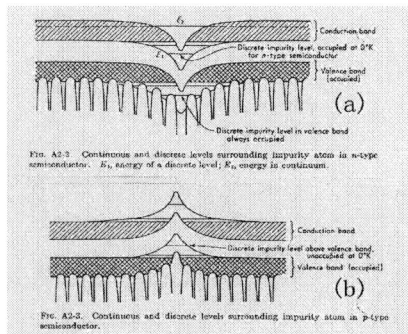

FIG. 1 SLATER'S ELECTRONIC TRAPS FROM BULK IMPURITY IONS: (A) AN N-TYPE IMPURITY ATOM GENERATES A BOUND ELECTRON STATE; (B) A P-TYPE IMPURITY ATOM GENERATES A BOUND HOLE STATE.

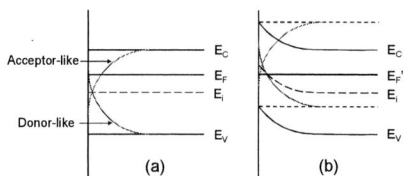

FIG. 2 ENERGY BAND DIAGRAM OF A P-MOSFET WITH A U-SHAPED ENERGY DISTRIBUTION OF INTERFACE TRAPS: (A) NEGATIVE INTERFACE-TRAP CHARGES AT FLATBAND AND (B) POSITIVE INTERFACE-TRAP CHARGES IN INVERSION. NEUTRAL INTERFACE TRAPS ARE DUE TO RANDOM VARIATIONS OF BOND ANGLES AND LENGTHS OF SI••SI, SI••O AND SI••N BONDS AT THE SIO$_2$/SI INTERFACE.

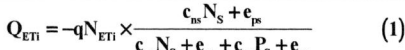

$$Q_{ETi} = -qN_{ETi} \times \frac{c_{ns}N_S + e_{ps}}{c_{ns}N_S + e_{ns} + c_{ps}P_S + e_{ps}} \quad (1)$$

$$Q_{HTi} = +qN_{HTi} \times \frac{c_{ps}P_S + e_{ns}}{c_{ns}N_S + e_{ns} + c_{ps}P_S + e_{ps}} \quad (2)$$

$$N_S = n_i \exp(U_S - U_N) \quad (3)$$

$$P_S = n_i \exp(U_P - U_S) \quad (4)$$

$$e_{ns} = c_{ns}n_i \exp(+U_{TI}) \quad (5)$$

$$e_{ps} = c_{ps}n_i \exp(-U_{TI}) \quad (6)$$

$$Q_{NT} = \int_{E_V}^{E_C} (Q_{ETi} + Q_{HTi}) dE_{TI} \quad (7)$$

Symbols: N_{ETi} and N_{HTi} are the neutral electron-trap and hole-trap densities (cm^{-2}eV) at the ith energy level, respectively. c_{ns}, c_{ps}, e_{ns}, and e_{ps} are the electron-hole capture-emission coefficients. N_s and P_s are the electron and hole surface concentrations. U_{TI} is interface-trap energy potential.

$$V_{GB} = V_S + \Phi_{MS} - Q_{IT}/C_{ox} + \varepsilon_{Si}E_s/C_{ox} \quad (8)$$

$$Q_{IT} = Q_{NT} + Q_{HT} + Q_{OT} \quad (9)$$

$$R_{SS} = \frac{c_{ns}c_{ps}N_SP_S - e_{ns}e_{ps}}{c_{ns}N_S + e_{ns} + c_{ps}P_S + e_{ps}} N_{IT} \quad (10)$$

$$I_{RE} = q\iiint R_{SS}(V_{GB}, y) dy dz dE_{TI}$$
$$= \frac{q(c_{ns}c_{ps})^{1/2}n_i W}{2} \int_y \int_{E_V}^{E_C} \frac{[\exp(U_{PN}(y)) - 1] N_{IT}(E_{TI}) dE_{TI} dy}{\exp(U_{PN}/2)\cosh(U_S^*) + \cosh(U_{TI}^*)} \quad (11)$$

$$U_S^* = U_S + \frac{1}{2}[\ln(c_{ns}/c_{ps}) - (U_P + U_N)] \quad (12)$$

$$U_{TI}^* = (U_T - U_I) + \frac{1}{2}\ln(c_{ns}/c_{ps}) \quad (13)$$

$$I_{DD} = \mu_{eff0}C_{ox}\frac{W}{L}\left(V_{GB} - \Phi_{MS} + \frac{Q_{IT}}{C_{ox}} - \frac{V_{S,eff} + V_{D,eff} + v_{th}}{2}\right)V_{DS,eff} \quad (14)$$

$$I_D = I_{DD} + I_{RE} \quad (15)$$

Symbols: Φ_{MS} is the work-function difference between the gate and semiconductor, R_S is steady-state electron-hole recombination-generation rate, ϕ_s^* and U_{TI}^* are the effective surface potential and interface-trap energy potential, respectively. I_{RE} is generation current, I_{DD} is drift and diffusion current and I_D is drain-terminal current.

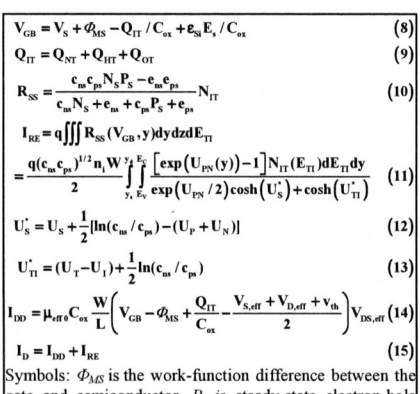

FIG. 4 GATE-VOLTAGE DEPENDENCE OF INTERFACE-TRAP CHARGES AT VARIOUS NEUTRAL INTERFACE-TRAP DENSITIES $N_{ET} = N_{HT} = 1.0 \times 10^{10}$, 1.0×10^{11}, 1.0×10^{12}, 3.0×10^{12}, AND 5.0×10^{12} CM^{-2}.

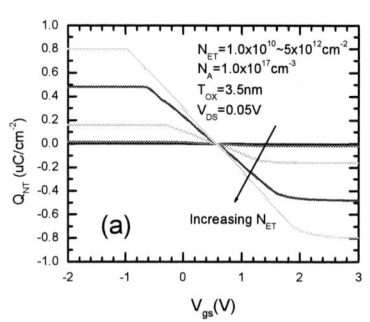

FIG. 5 COMPARISONS OF I_D-V_{GB} CURVES BETWEEN MEDICI SIMULATION AND MODEL CALCULATION WHICH EMPLOYS THE NEUTRAL INTERFACE TRAPS. SYMBOL: MEDICI SIMULATION, LINE: MODEL. NEUTRAL INTERFACE-TRAP DENSITY $N_{ET} = 0$, 1.0×10^{12}, 2.0×10^{12}, AND 5.0×10^{12} CM^{-2}.

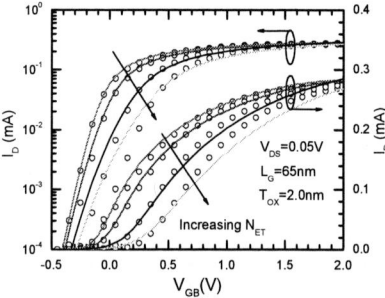

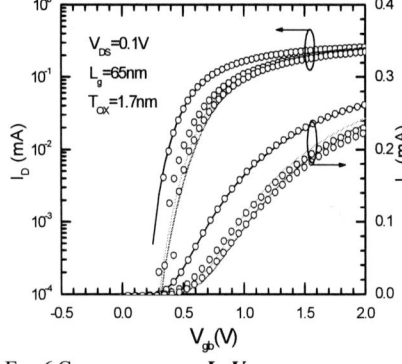

FIG. 6 COMPARISONS OF I_D-V_{GB} CURVES BETWEEN EXPERIMENTAL DATA AND MODEL CALCULATION WHICH EMPLOYS THE NEUTRAL INTERFACE TRAPS. EXPERIMENTAL DATA IS MEASURED USING ULTRA-FAST SWITCHING METHOD. THREE DATA CURVES ARE RESPECTIVELY CORRESPONDING TO THE STRESS TIME $T = 0$, 405 AND 10005S AT STRESS VOLTAGE $V_{GS} = 2.55$S.

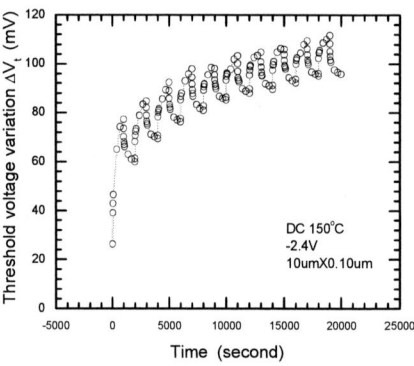

FIG. 7 THRESHOLD VOLTAGE VARIATION MEASURED USING DC METHOD. THE THRESHOLD VOLTAGE VARIATION ESSENTIALLY IS A CONSTANT DURING THE STRESS AND RELAXIATION CYCLES. STRESS CONDITION: $V_{GS} = -2.4$V, $T = 150$°C.

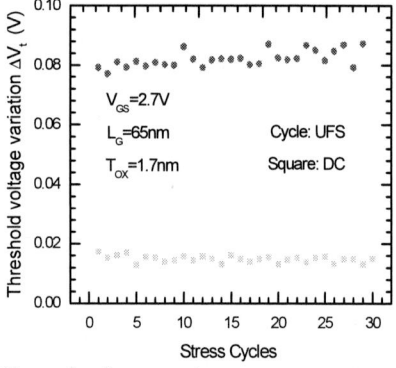

FIG. 8 STRESS-CYCLE DEPENDENCE OF THRESHOLD VOLTAGE VARIATIONS ARE MEASURED USING ULTRA-FAST SWITCHING METHOD (CYCLE SYMBOL) AND DC METHOD (SQUARE SYMBOL). BIAS DEPENDENT OF NEUTRAL INTERFACE TRAPS WITH ENERYG DISTRIBUTION IN SILICON ENERGY GAP CAN VERY WELL ACCOUNTE FOR THE FAST RECOVERY OF NBTI PROBLEM.

REFERENCES

[1] Huard et al., *IEEE Trans. Electron Dev.*, pp. 558–570, 2007; [2] Mahapatra et al., *IEEE Trans. Electron Dev.*, pp. 35–46, 2008; [3] Grasser et al., *IRPS* 2009 pp. 33–44; [4] Ielmini et al., *IEEE Trans. Electron Dev.*, pp. 1943–1952, 2009; [5] Teo *et al.*, *IEDM* 2009, pp. 737–740; [6] Slater, vol.3, pp. 292–307, 1967; [7] Chen *et al.*, *JAP*, pp. 115511–11, 2006; [8] Zhou et al., *IEDST*, 2009, pp. 1–6; [9] Du et al., *IEEE Trans. Electron Dev. Lett.*, pp. 275–277, 2009;

Atomistic approach to variability of bias-temperature instability in circuit simulations

B. Kaczer[1], S. Mahato[1,2], V. Valduga de Almeida Camargo[1,3], M. Toledano-Luque[1,4], Ph. J. Roussel[1], T. Grasser[5],
F. Catthoor[1,6], P. Dobrovolny[1], P. Zuber[1], G. Wirth[3], G. Groeseneken[1,6]

[1]imec, Kapeldreef 75, B-3001 Leuven, Belgium
[2]TU Munich, Germany & NTU, Singapore
[3]UFRGS, Brazil
[4]UC Madrid, Spain
[5]TU Wien, Austria
[6]KU Leuven, Belgium

phone: +32 16-281-557, e-mail: kaczer@imec.be

Abstract—A blueprint for an atomistic approach to introducing time-dependent variability into a circuit simulator in a realistic manner is demonstrated. The approach is based on previously proven physics of stochastic properties of individual gate oxide defects and their impact on FET operation. The proposed framework is capable of following defects with widely distributed time scales (from fast to quasi-permanent), thus seamlessly integrating random telegraph noise (RTN) effects with bias temperature instability (BTI). The use of industry-standard circuit simulation tools allows for studying realistic workloads and the interplay of degradation of multiple FETs.

Bias-temperature instability (BTI), random telegraph noise (RTN), time-dependent variability, single-carrier effects, circuit simulations

I. INTRODUCTION

It is generally accepted that systematic and statistical variability will have to be considered in the design of future ULSI circuits [1]. This process relies on describing the as-fabricated device parameters in terms of their statistical *distributions*. The challenge is to extend the same concept to circuit operation [2-6]. To that end, the *time dependence of the parameter distributions* during circuit operation, after being thoroughly understood, will need to be inserted into circuit simulators. Reliability assessment of future applications can thus be seen as *time-dependent variability* analysis.

We have previously shown that a large portion of BTI degradation and recovery is due to charging and discharging of gate oxide defects [7]. Here we argue that incorporating *the stochastic* nature of these defects and their *individual impact on FET operation* into existing industry-standard circuit simulation tools should *naturally* guarantee realistic time-dependent variability results under actual circuit workloads. Because only a handful of defects will be present in future nm-sized devices [8], tracking the occupancy of individual defects becomes a feasible approach to circuit reliability assessment. *Aging and degradation is thus viewed as "merely" following the occupation of each defect in time, while considering the impact of the individual charged defects on the respective FET's operation.*

II. PROBLEM STATEMENT

Large devices employed in the past technologies were assumed to behave identically during BTI stress. Consequently, one device was typically sufficient to project a well-defined lifetime at each stress condition (Fig. 1a) [9]. With the downscaling of device dimensions, the number of defects per device decreases. As a consequence of that and of the stochastic nature of the defects, the projected lifetime spread drastically increases (Figs. 1bc). The example in Fig. 1 is calculated for constant stress only using a custom toolkit. The challenge, addressed here, is to translate this approach to an arbitrary circuit employing arbitrary workloads using industry-established simulation tools.

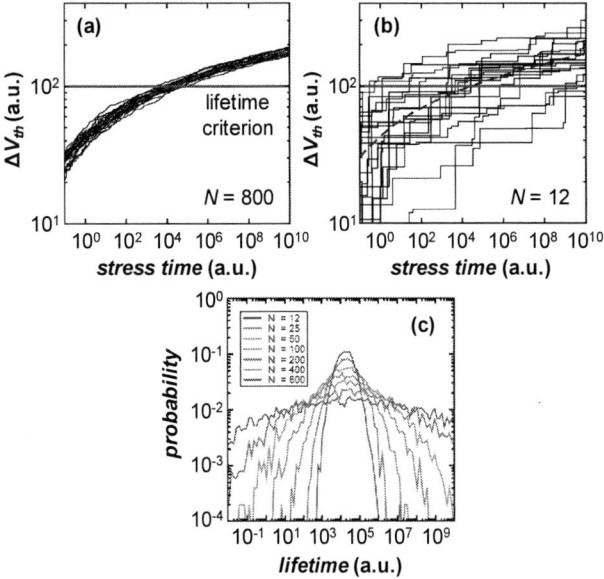

Figure 1.   (a) The random properties of many defects *N* in large devices average out, resulting in a well-defined lifetime while (b) the stochastic nature of a handful of defects in deeply-scaled devices becomes apparent, resulting in large variation in the lifetime, shown in (c). Only capture events are shown for the sake of simplicity.

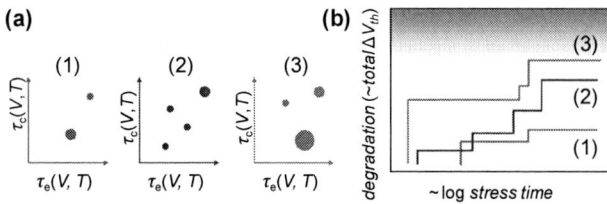

Figure 2. (a) Parameters affecting degradation schematically illustrated for three FET instances (1, 2, and 3). Each device is characterized by the number of defects (circles), their capture and emission times τ_c and τ_e, and their impact on the device *when charged* (demarked by the size of the circle). (b) Schematic showing the progress of degradation during constant stress in the three devices. As in Fig. 1, only capture events are shown.

III. APPROACH

A. Assumptions

The general assumptions are schematically illustrated in Fig. 2a [9]. Each device is characterized by the number of defects, their capture and emission times τ_c and τ_e, and their individual impact on the device when charged. Fig. 2b then schematically illustrates the calculation of Fig. 1b for the 3 devices in Fig. 2a.

These assumptions are more formally given in Fig. 3. Each instance of a FET with gate dimensions $L_G \times W_G$ is initiated with a specific number of defect precursors n taken from a Poisson distribution with mean $N = L_G W_G N_{ot}$. Here, N_{ot} is the gate oxide defect precursor areal density. Each defect can be independently initialized as empty (uncharged) or occupied (charged). In the proof-of-concept version of our simulation framework, only the impact of defects on the FET threshold voltage V_{th} is assumed in the form of a voltage shift ΔV_{th}, taken from an exponential distribution with the mean value η [9]. The scaling of η with device dimensions is presently under discussion [8,10]—we use the most simple scaling $\eta \propto 1/L_G W_G$.

Furthermore, each defect is characterized by its capture and emission times τ_c and τ_e. Due to the inelastic nature of the charging mechanism, τ_c and τ_e are strongly voltage and temperature dependent [7,11,12]. The voltage dependence of each time constant is simplified in the case of digital circuits to two values, at low (L) and high (H) gate biases. Hence each defect is described by 4 time constants: $\tau_{c,H}$, $\tau_{c,L}$, $\tau_{e,H}$, and $\tau_{e,L}$. The latter is assumed to be uniformly distributed on the log scale between 10^{-9} and 10^{9} s. $\tau_{c,H}$ and $\tau_{e,H}$ are taken to be weakly correlated with $\tau_{e,L}$, with $<\tau_{c,H}> \sim 0.01 <\tau_{e,L}>$ [13] and $<\tau_{e,H}> \sim 100 <\tau_{e,L}>$ [11]. $\tau_{c,L}$ is assumed to be large [11]. Values of τ's for intermediate voltages are interpolated from the supplied L and H constants following the behavior given in Fig. 13 of Ref. 11. Impact of temperature is not assumed in this first iteration.

The initial (time-zero) variability is captured by assuming a Gaussian distribution for the initial threshold voltage V_{th0} of each FET. The parameters of this distribution are the mean $<V_{th0}>$ and the variance σ^2.

---

for each FET instance with gate length L_G and width W_G:
 initial values:
  V_{th0}: from Gaussian($<V_{th0}>$, σ)
 oxide defects:
  for each defect $i = 1 .. n$ from Poisson(N), where $N = L_G W_G N_{ot}$:
   initial occupancy state P_{c0} (0 or 1): Eq. 4
   $\Delta V_{th,i}$ when occupied: from Exponential(η), where $\eta \sim 1/L_G W_G$
   $\tau_{c,H}$, $\tau_{e,H}$, $\tau_{c,L}$, $\tau_{e,L}$:
    $\tau_{e,L}$ uniformly distributed on log scale [13]
    capture and emission τ's weakly correlated [13]
    L and H τ's connected via voltage dependence [11]

Figure 3. Prescription for including initial (time-zero) and time-dependent variability in the trap-enhanced FET model.

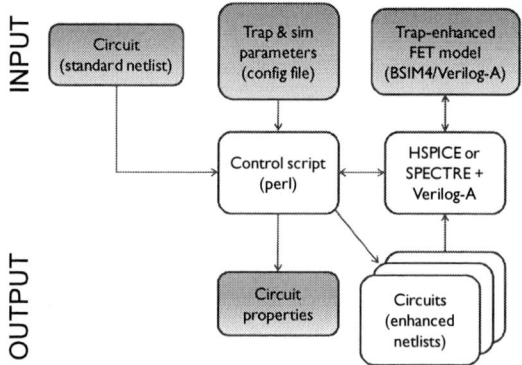

Figure 4. Simulation setup to study time-dependent variability of circuits based on industry-standard tools.

B. Implementation

The simulation framework (Fig. 4) accepts an arbitrary circuit in the form of a standard netlist. Based on the configuration file the control script generates multiple instances of the circuit, *each enhanced with unique random values of defect parameters for each FET* as per Fig. 3 and submits them to the HSPICE or SPECTRE solvers. The crucial component of the framework is the Verilog-A-based BSIM4 FET model *enhanced to simulate the impact of individual defects on the FET's behavior*, specifically, *its ability to follow the occupancy of each defect in every degrading FET*. Its details are discussed in the following section. Finally, the resulting circuit parameters from all instances are statistically analyzed.

C. Time-dependent defect occupancy

To a good approximation, the capture and emission of charge can be described by first-order kinetics. The probability of a defect capturing charge P_c is then described by the differential equation

$$\frac{dP_c}{dt} = \frac{1-P_c}{\tau_c} - \frac{P_c}{\tau_e}. \qquad (1)$$

The two terms on the right-hand side of Eq. 1 correspond respectively to capture and emission probabilities. The general solution of this equation is

$$P_c = \frac{\tau}{\tau_c} + \left(P_{c0} - \frac{\tau}{\tau_c}\right) \exp\left(-\frac{t}{\tau}\right), \qquad (2)$$

where $\tau^{-1} = \tau_c^{-1} + \tau_e^{-1}$ and t stands for the time elapsed since the trap was occupied with the probability P_{c0}.

Our setup determines each defect's occupancy transitions *during the circuit simulation* using

$$P_{p,V} = \frac{\tau_V}{\tau_{p,V}}\left[1 - \exp\left(-\frac{\Delta t}{\tau_V}\right)\right]. \tag{3}$$

Here p stands for process (capture or emission), V the gate voltage (H or L), $\tau_V^{-1} = \tau_{c,V}^{-1} + \tau_{e,V}^{-1}$, and Δt the simulation time step (at voltage V) [11, 14]. Eq. 3 is implemented in Verilog-A (see Fig. 4) and used to update the occupancy (i.e., *instantiate it* with 0 or 1 based on a randomly generated number) of every trap at every simulation step. The process p is selected depending on the state P_{c0} (0 or 1) of the defect at the beginning of the simulation step. *In this way our setup directly and naturally simulates the impact of individual charge trapping events (such as RTN) on FETs, and hence on the circuit operation.*

D. Initialization of defect occupancies

It is, however, computationally unfeasible to simulate the circuit after long operation t_s (e.g. 1 year). To circumvent that, we have derived and experimentally verified [15] an analytical description of trap occupancy after stressing with a square periodic signal with frequency f, duty cycle (or duty "factor") DF, and duration t_s

$$P_c = \frac{B}{A}[1 - \exp(-At_s)]. \tag{4}$$

Here A and B are in general functions of f, DF, t_s, and arbitrary $\tau_{c,H}, \tau_{c,L}, \tau_{e,H}$, and $\tau_{e,L}$, elaborated in Ref. 15. Eq. 4 is valid for both "slow" (i.e., τ's $\gg 1/f$) and "fast" (i.e., τ's $\ll 1/f$) traps. Fig. 5 graphically shows the behavior of Eq. 4 for a particular set of trap parameters, t_s, DF, and f. Instantiated P_c (i.e., 0 or 1 decided with the probability P_c) is input into the circuit simulation as the initial state of each trap.

For the more realistic case of *irregular* workloads, Eq. 4 can be also used to initialize the simulation at time t_s if *workload-equivalent DF* [2,4] is used. Fig. 6a shows that Eq. 4 is still a good approximation for "slow" defects. As will be illustrated below, such defects become dominant at longer operating times. *The framework is thus also capable of integrating the impact of defects with larger time constants into the circuit simulation.* Eq. 4 is obviously not strictly correct for "fast" defects in case of irregular workloads (Figs. 6b and 6c) but may be essentially irrelevant as their occupancy will change rapidly during simulation.

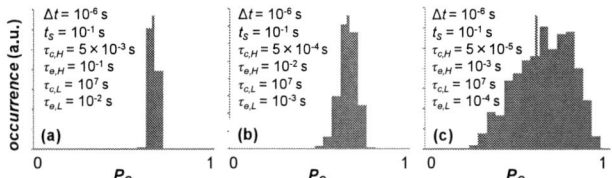

Figure 6. Occupation probabilities P_C (histograms) calculated using Eq. 3 by following 5000 irregular random signals with mean $f = 25$ kHz and mean $DF = 50\%$. (a) Slow defects generally follow the analytical description for a periodic signal (red vertical lines, Eq. 4), while (b) 10× faster and (c) 100× faster traps do not.

IV. DEMONSTRATION

Our approach is demonstrated on the simple case of a CMOS inverter—the basis of many circuit elements, such as the SRAM cell. Publicly available PTM 16 nm Metal Gate / High-k / Strained-Si technology model card is used [16]. Initial V_{th0} variation is assumed in both FETs of the inverter, while, for demonstration purposes, only the inverter pFET is assumed to be degrading. The studied inverter is sandwiched between two inverters with fixed parameters (Fig. 7) to condition the digital signal.

We first confirm that the threshold voltage shift distribution generated by the framework for a single pFET agrees with the analytical prediction (Fig. 8) [9].

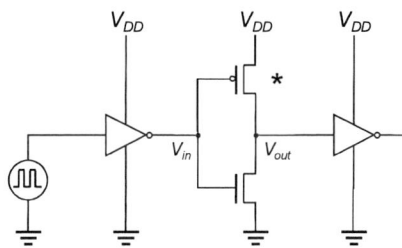

Figure 7. The circuit used in the demonstration consists of a 3-inverter chain. Only the pFET in the middle inverter (denoted with an asterisk) is assumed to be degrading. Parameters: $V_{DD} = 0.8$ V, $f = 1$ GHz, $DF = 50\%$, $L_n = L_p = 22$ nm, $W_n = 45$ nm, $W_p = 90$ nm, $V_{th0} = \pm0.25$ V, $\sigma = 25$ mV, $\eta = 16$ mV, $N_{ot} = 5\times10^{11}$ cm^{-2}.

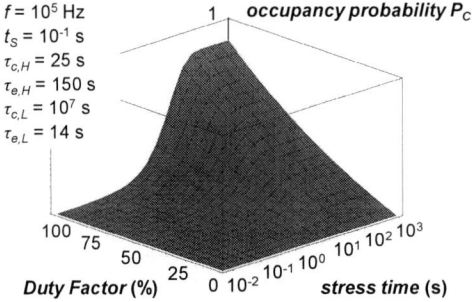

Figure 5. Graphical representation of the analytical description (Eq. 4) of a single trap occupancy probability P_C after periodic AC stress with duty cycle DF lasting t_S for given emission and capture times.

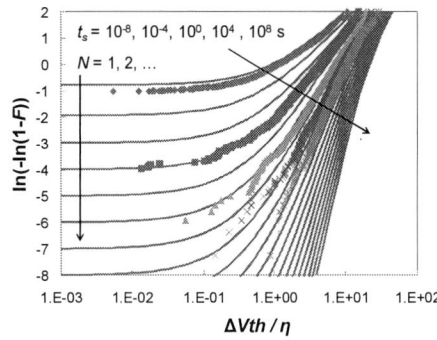

Figure 8. Threshold voltage shift distributions for a single pFET at different stress times t_S generated by the framework (symbols) agree with the analytical description for the average *charged* defect number N (lines) [9].

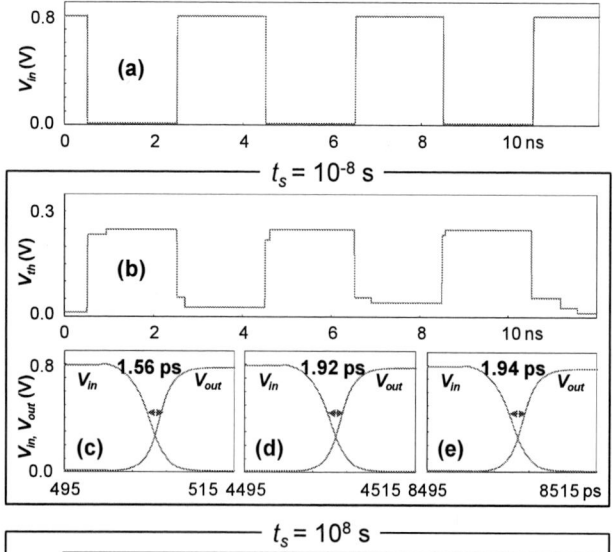

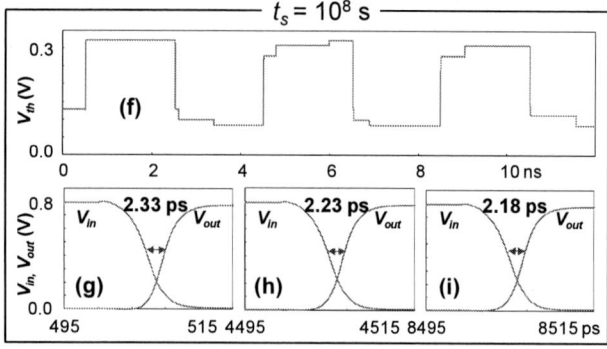

Figure 9. Inverter degradation simulation transient snapshots. (a) Three periods of a $f = 250$ MHz input signal of the studied inverter. Corresponding time-dependent pFET threshold voltage V_{th} (b) after 10^{-8} s and (f) 10^8 s. Two fast defects modulate the V_{th} in both cases. Note that due to short channel effects, V_{th} is also modulated by V_D, i.e., the output signal of the inverter. (f) V_{th} after 10^8 s is futher degraded with respect to (b) due to the capture of charge in slow defects. Switching delay of the studied inverter during the three periods (c-e) after 10^{-8} s and (g-i) 10^8 s.

Fig. 9 then shows snapshots of one instance of the inverter simulation at the beginning of its operating lifetime (Figs. 9b-e) and after 10^8 s (Figs. f-i). Fig. 9b shows the pFET threshold voltage V_{th} behavior at the beginning of the circuit operation. V_{th} is changing as single holes are captured in two fast "RTN" defects when the inverter input is low (i.e., the pFET gate is stressed), and subsequently emitted when the inverter input is high (i.e., pFET $V_G = V_S$). The same two defects are still active 10^8 s into the circuit operation (Fig. 9f); note, however, the pFET V_{th} is further degraded with respect to its initial value (Fig. 9b) due to the charge capture in slow defects. This latter behavior thus naturally emulates the "classical" BTI degradation.

The inverter switching transients are also illustrated in Fig. 9. It is apparent that there is a variation in the inverter switching delay from period to period (see Figs. 9c-e and 9g-i), resulting in the so-called delay jitter. The overall slowdown of the inverter after 10^8 s of operation is also apparent (cf. Figs. 9c-e and 9g-i).

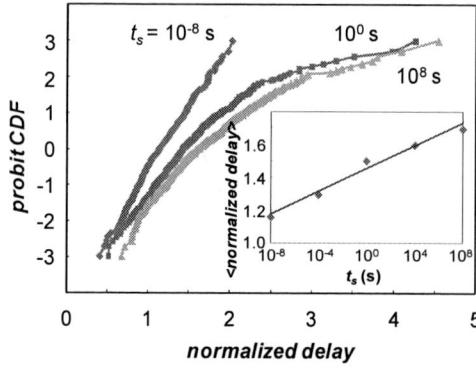

Figure 10. Distributions of normalized inverter delay for three points of the circuit lifetime. Inset: the average values increase as log t_S, in agreement with the assumptions (Section III.A).

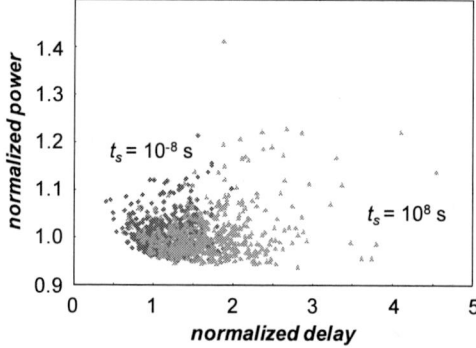

Figure 11. Normalized delay-power plot at the beginning (diamonds) and end (triangles) of circuit life. Shift in the distribution cloud due to increased trap occupation is apparent.

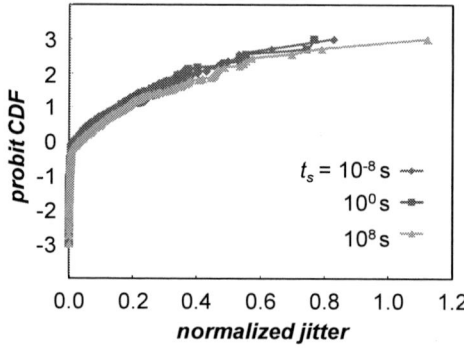

Figure 12. Normalized inverter delay jitter at three points of the circuit lifetime. The delay jitter is defined as the difference between the maximum and the minimum inverter delays simulated over 100 periods. The values are normalized by the nominal delay. Analysis of the impact of fast traps *during* circuit operation (e.g. delay jitter) is possible within the proposed framework.

The ability of the framework to generate statistical distributions of circuit parameters is demonstrated next. 500 instances of the circuit in Fig. 7 are generated at each of several points of the circuit operation t_s. Fig. 10 shows the increasing variability of the inverter delay distribution with ongoing circuit operation [3]. Here, the delay is averaged over 100 periods of operation of each circuit instance. All circuit parameters are normalized to the corresponding value of the

978-1-4244-9113-1/11 $26.00 © 2011 IEEE

inverter without either initial (i.e., V_{th0}) or time-dependent variations (i.e., defects). The same simulation results are shown in the energy-delay diagram in Fig. 11. Again, increased variability of the circuit parameters with ongoing operation is apparent.

The ability of our simulation setup to investigate variations in circuit parameters *during operation* at different points of the circuit lifetime is documented in Fig. 12. Unlike the results Figs. 10 and 11, the jitter does not change as the circuit ages. This is because the jitter is caused by fast traps, which are assumed (see Section III.A) to be present in the pFET from the beginning.

Finally, we note that some penalty in terms of simulation speed is incurred by moving to a Verilog-A-based FET description (Table I). However, the additional penalty of following individual defects is minor.

SPECTRE 7.1.1 + BSIM4.4	SPECTRE + Verilog-A	SPECTRE + trap-enhanced Verilog-A
16.5 s	69 s	95 s
24%	100%	138%

Table I: Comparison of circuit simulation speed. Circuit in Fig. 7 is used, with 15 defects in *each* of its 6 FETs (i.e., in total 90 defects) simulated in the "trap-enhanced" case.

V. CONCLUSIONS

We have demonstrated a blueprint for an atomistic approach to introducing time-dependent variability into a circuit simulator, based on the stochastic properties of individual defects and their impact on the FET. The framework is capable of following defects with widely distributed time scales (from fast, i.e., RTS-like, to slow and quasi-permanent, i.e., BTI-like) in a unified manner. The employment of existing industry-standard circuit simulator tools ensures correct combination of the deterministic workload-dependent component with the stochastic modeling aspect while simultaneously incorporating interactions among different devices. The refinement and expansion, be it by other types of defects or by their impact on FET behavior, can be further explored. We expect the framework should prove useful for investigation of the reliability of ULSI circuits.

VI. ACKNOWLEDGEMENTS

This work was carried out as part of IMEC's Industrial Affiliation Program funded by IMEC's core partners. V.V.A.C. thanks CNPQ Brazil for financial support. M.T.L. thanks the IberCaja Postdoctoral grant program and Spanish MEC (TEC2010-18051) for financial support.

REFERENCES

[1] M. Miranda Corbalan, B. Dierickx, P. Zuber, P. Dobrovolny, F. Kutscherauer, P. Roussel, and P. Poliakov, "Variability aware modeling of SoCs: from device variations to manufactured system yield", *Int. Symp. Quality Electronic Design (ISQED)*, San Jose, CA, March 2009.

[2] S. V. Kumar, C. H. Kim, and S. S. Sapatnekar, "NBTI-Aware Synthesis of Digital Circuits", *Design and Automation Conference (DAC)*, 2007.

[3] Tong Boon Tang, A. F. Murray, Binjie Cheng, and A. Asenov, "Statistical NBTI-effect prediction for ULSI circuits", *Proc. IEEE Int. Symp. Circuits and Systems (ISCAS)*, p. 2494, 2010.

[4] Wenping Wang, Shengqi Yang, S. Bhardwaj, S. Vrudhula, F. Liu, and Yu Cao, "The Impact of NBTI Effect on Combinational Circuit: Modeling, Simulation, and Analysis", *IEEE T. VLSI Systems* **18**, p. 173, 2010.

[5] V. Huard *et al.*, "Managing SRAM reliability from bitcell to library level", *Proc. Int. Reliab. Phys. Symp. (IRPS)*, p. 655, 2010.

[6] H. Kufluoglu, V. Reddy, A. Marshall, J. Krick, T. Ragheb, C. Cirba, A. Krishnan, and C. Chancellor, "An Extensive and Improved Circuit Simulation Methodology For NBTI Recovery", *Proc. Int. Reliab. Phys. Symp. (IRPS)*, p. 670, 2010.

[7] T. Grasser *et al.*, "Recent Advances in Understanding the Bias Temperature Instability", *Int. Electron Dev. Meeting (IEDM) Tech Dig.*, p. 82, 2010.

[8] A. Asenov, R. Balasubramaniam, A. R. Brown, and J. H. Davies, "RTS Amplitudes in Decananometer MOSFETs: 3-D Simulation Study," *IEEE T. Electron Dev.* **50**, p. 839, 2003.

[9] B. Kaczer, T. Grasser, Ph. J. Roussel, J. Franco, R. Degraeve, L.-A. Ragnarsson, E. Simoen, G. Groeseneken, and H. Reisinger, "Origin of NBTI Variability in Deeply Scaled pFETs", *Proc. Int. Reliab. Phys. Symp. (IRPS)*, p. 26, 2010.

[10] A. Ghetti, C. M. Compagnoni, A. S. Spinelli, and A. Visconti, "Comprehensive Analysis of Random Telegraph Noise Instability and Its Scaling in Deca–Nanometer Flash Memories," *IEEE T. Electron Dev.* **56**, p. 1746, 2009.

[11] T. Grasser, H. Reisinger, P.-J. Wagner, F. Schanovsky, W. Goes, and B. Kaczer, "The Time Dependent Defect Spectroscopy (TDDS) for the Characterization of the Bias Temperature Instability", *Proc. Int. Reliab. Phys. Symp. (IRPS)*, p. 16, 2010.

[12] T. Grasser, H. Reisinger, P.-J. Wagner, and B. Kaczer, "Time-dependent defect spectroscopy for characterization of border traps in metal-oxide-semiconductor transistors", *Phys. Rev. B.* **82**, p. 245318, 2010.

[13] H. Reisinger, T. Grasser, W. Gustin, and C. Schlünder, "The statistical analysis of individual defects constituting NBTI and its implications for modeling DC- and AC-stress", *Proc. Int. Reliab. Phys. Symp. (IRPS)*, p. 7, 2010.

[14] J. Martin-Martinez, B. Kaczer, M. Toledano-Luque, R. Rodriguez, M. Nafria, X. Aymerich, and G. Groeseneken, "Probabilistic Defect Occupancy Model for BTI", accepted to *Proc. Int. Reliab. Phys. Symp. (IRPS)* 2011.

[15] M. Toledano-Luque, B. Kaczer, Ph.J. Roussel, J. Franco, T. Grasser, C. Vrancken, N. Horiguchi, and G. Groeseneken, "Response of a single trap to AC Negative Bias Temperature Stress", accepted to *Proc. Int. Reliab. Phys. Symp. (IRPS)* 2011.

[16] http://ptm.asu.edu/ .

Probabilistic defect occupancy model for NBTI

J. Martin-Martinez[1,*], B. Kaczer[2], M. Toledano-Luque[2,3], R. Rodriguez[2], M. Nafria[2], X. Aymerich[2], G. Groeseneken[2,4]

[1]Universitat Autònoma de Barcelona. Dept. Enginyeria Electrònica Edifiçi Q. 08193 Bellaterra (Spain)
[2]imec, Kapeldreef 75, 3001 Leuven (Belgium)
[3]Dpto. Fisica Aplicada III, Universidad Complutense de Madrid, 28040 Madrid (Spain)
[4]Katholieke Universiteit Leuven, 3000 Leuven, Belgium
*corresponding author: javier.martin.martinez@uab.es

Abstract—A new Negative Bias Temperature Instability (NBTI) model based on the probability of charge/discharge of defects in devices is presented. The model correctly describes the main experimental NBTI features, such as the threshold voltage shift (ΔV_T) evolution under DC or AC stresses, its frequency and duty cycle dependence and the statistical ΔV_T distribution in small transistors. Finally, the model is used to explain the dependence of NBTI degradation on the stress/relaxation times, frequency and duty cycle in terms of defect occupancy.

Keywords: NBTI, CMOS technologies, Monte Carlo simulation, modeling, Reliability.

I. INTRODUCTION

NBTI is one of the most relevant failure mechanisms in pFETs with ultrathin gate dielectric. Extensive work has been done in the past in order to characterize and model the phenomenology behind the NBTI [1-4]. In small FETs the NBTI reveals a stochastic behavior, which has been attributed to the charge/discharge of defects during stress/relaxation [5,6]. The analysis of these isolated defects provides very valuable information for the development of future BTI models [6-8]. In this work, a new model based on the description of the defects that cause the threshold voltage shift in the NBTI damage is presented. After the experimental details explained in Section II, while in Section III a model to calculate the probability of occupancy of a single defect as a function of the applied DC or AC stress is proposed. In Section IV the model is extended to account for the effect of several defects that can exist in a device. The simulations are compared with measurements or previous experimental results showing that the model is able to reproduce the most important NBTI features observed under DC or AC stresses. Finally, in Section V, the model is used to describe the dependence of ΔV_T with stress time, frequency and duty cycle in terms of defect occupancy.

II. EXPERIMENTAL

The samples used in this work were ultrascaled pFETs (nominal WxL=90x70nm^2) with SiON as gate dielectric (EOT=1.8nm). The devices (75 in total) were stressed for 2s by applying -2.3V to the gate and -0.1V to the drain with the other terminals grounded. The NBTI relaxation was induced by applying a gate voltage close V_T and a small drain voltage. Measuring the channel current evolution with the relaxation time, ΔV_T can be easily obtained as explained in [9].

This work has been partially supported by the Spanish MICINN (TEC2007-61294/MIC and TEC2010-16126) and the Generalitat de Catalunya (2009SGR-783).

III. DEFECT OCCUPANCY PROBABILITY MODEL

In small pFETs, the V_T recovery after the NBTI stress appears as discrete voltage drop events that take place in a wide range of relaxation times. An example of this effect is shown in Fig. 1a where the ΔV_T measured on several of the stressed pFETs is plotted as a function of the relaxation time. These V_T drops are attributed to the discharge of defects that have been previously charged during the stress.

We start by proposing a model to describe the trapping/detrapping of charge in/from a single defect in the device. The basis of our single defect model is that every instant Δt it has a charge capture probability $P_c = \Delta t / \tau_c$ when empty and a charge emission probability $P_e = \Delta t / \tau_e$ when charged, τ_c and τ_e ($>\Delta t$) being the capture and emission times, the time parameters of the model. These parameters are voltage and temperature dependent and uncorrelated [6]. In this work we will restrict the analysis to rectangular waveforms, so that only two values of τ_e and τ_c are necessary, which will depend on the waveform voltage; namely $\tau_e(V_H)$ and $\tau_c(V_H)$ when the voltage is high (V_H) and $\tau_e(V_L)$ and $\tau_c(V_L)$ when the voltage is low (V_L). The last parameter that characterizes the defect, μ, corresponds to the change in V_T produced by the defect when charging, that can be obtained from the discharge events, as those observed in Fig. 1a.

To determine when a defect is occupied or empty Monte Carlo (MC) simulations have been performed following the flow diagram illustrated in Fig. 2. The occupancy of a defect is

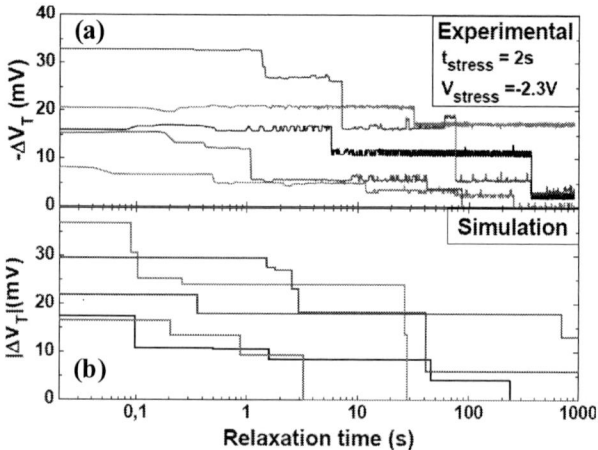

Figure 1. (a) Experimental recovery traces in pFETs (WxL=90x70nm^2). The discharge events are evident. (b) Simulated recovery traces obtained using the model presented here when the same conditions of stress and relaxation than in top are considered.

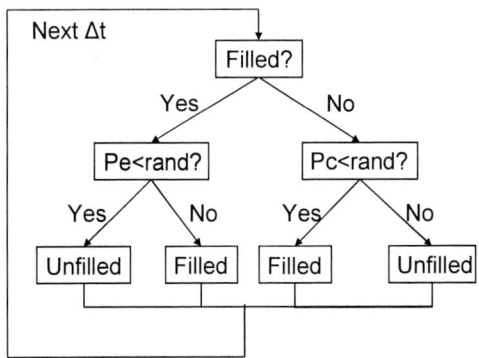

Figure 2. Simulation flow designed to determine the occupancy of a defect. The emission (P_e) and capture (P_c) probabilities are compared with a random number uniformly distributed from 0 to 1 (rand).

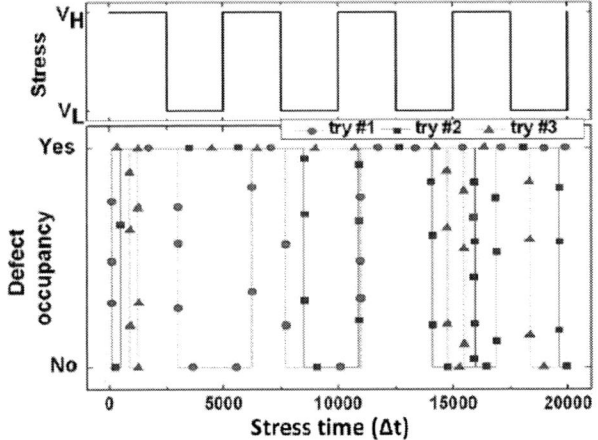

Figure 3. Three typical MC simulated traces of the occupancy for the same defect (parameters: $\tau_c(V_H)=10^3 \cdot \Delta t$, $\tau_e(V_H)=5 \cdot 10^3 \cdot \Delta t$, $\tau_c(V_L)=10^{20} \cdot \Delta t$ and $\tau_e(V_L)=2 \cdot 10^3 \cdot \Delta t$) obtained under the pulsed stress at the top of this figure. The MC traces have been calculated following the simulation flow presented in Fig. 2.

determined at every instant Δt from the comparison of P_e or P_c with a random number uniformly distributed between 0 and 1. As an example of these simulations, Fig. 3 shows a simulated rectangular stress waveform (top) and three traces of the occupancy evolution with time (in units of Δt) for a defect with time parameters $\tau_c(V_H)=10^3 \cdot \Delta t$, $\tau_e(V_H)=5 \cdot 10^3 \cdot \Delta t$, $\tau_c(V_L)=10^{20} \cdot \Delta t$ and $\tau_e(V_L)=2 \cdot 10^3 \cdot \Delta t$ (bottom). The defect occupancy probability (P_{occ}) can be numerically evaluated from the average of many simulations as those shown in fig. 3. Fig.4 shows P_{occ} calculated from the average of 10, 100 and 1000 trials of the defect considered in Fig. 3 under the same pulsed waveform. The thick line in Fig. 4 shows that P_{occ} can be analytically described by Eq.1

if ($V_{stress}=V_H$):

$$P_{occ}(t) = P_{occ}(t_i) + \left(\frac{\tau_e(V_H)}{\tau_e(V_H)+\tau_c(V_H)} - P_{occ}(t_i) \right) \left(1 - \exp\left(\frac{-(t-t_i)}{\tau_H} \right) \right)$$

if ($V_{stress}=V_L$): $\qquad$ (1)

$$P_{occ}(t) = \frac{\tau_e(V_L)}{\tau_e(V_L)+\tau_c(V_L)} + \left(P_{occ}(t_i) - \frac{\tau_e(V_L)}{\tau_e(V_L)+\tau_c(V_L)} \right) \cdot \exp\left(\frac{-(t-t_i)}{\tau_L} \right)$$

where $\tau_H^{-1} = \tau_e^{-1}(V_H) + \tau_c^{-1}(V_H)$, $\tau_L^{-1} = \tau_e^{-1}(V_L) + \tau_c^{-1}(V_L)$ and t_i is the time at which the i-th transition of the pulsed stress occurs. In [10] a simplified version of this equation was validated from experimental data; Using Eq. 1 P_{occ} only has to be calculated at every voltage transition of the pulsed waveform, and not every Δt, saving an important amount of calculation time in the simulations.

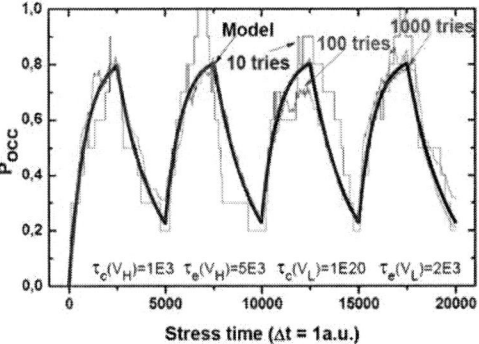

Figure 4. Thin lines: occupation probability of a defect obtained from the average of several occupancy traces as those presented in Fig. 3. Thick line: the occupation probability can be described using Eq.1, which only has to be calculated at the voltage transitions of the stress waveform.

IV. ΔV_T OF A DEVICE

In this section, the model developed for one defect is extended to calculate ΔV_T in a device, where several defects can coexist. Simulations with stress configurations similar to the typical experimental conditions in the NBTI studies are performed to check the validity of the model. ΔV_T is calculated from the simulation of N defects in the device contributing each one of them with an amount μ to the ΔV_T of the device when charged. Then, the total ΔV_T caused by NBTI can be expressed using Eq. 2

$$\Delta V_T(t) = \sum_{j=1}^{N} k_j(t) \cdot \mu_j \, , \qquad (2)$$

j being an index that denotes each defect. k_j will be equal to 1 or 0 if the j-th defect is occupied or empty respectively. In order to calculate ΔV_T the values of N, k_j and μ_j must be determined.

To obtain k_j, it is necessary to calculate the occupation probability of each defect, $P_{occ,j}$ which will be dependent on the $\tau_{c,j}$ and $\tau_{e,j}$ parameters of the particular defect. Previous works in large area devices show that the recovery component of NBTI increases during the stress and decreases during relaxation following a log(t) law in a wide range of times [9]. This result suggests that the characteristic times $\tau_{c,j}(V_H)$ and $\tau_{e,j}(V_L)$ are uniformly distributed on a log scale [11]. Therefore, these two parameters are obtained in our model by MC simulation assuming their corresponding distributions in a time interval larger than the typical experimental window. In our simulations we have taken $\tau_{e,j}$ and $\tau_{c,j}$ values ranging from 10^{-8}s to 10^6s. Moreover, we consider $\tau_{c,j}(V_H)$ and $\tau_{e,j}(V_L)$ uncorrelated [6], and, for the shake of simplicity, null probabilities of capture at V_L and emission at V_H are assumed. That is, in the model is imposed $\tau_{c,j}(V_L)$ and $\tau_{e,j}(V_H) \rightarrow \infty$. This simplification will not imply relevant changes in the

978-1-4244-9113-1/11 $26.00 © 2011 IEEE

conclusions of the work. Finally, $\tau_{c,j}(V_H)$ and $\tau_{e,j}(V_L)$ are introduced into Eq. 1 to calculate $P_{occ,j}$. k_j is obtained from the comparison of $P_{occ,j}$ with a random number.

The μ parameter of each defect, that indicates the V_T shift caused in the device when the defect is charged, can be directly extracted from the ΔV_T drops observed in small devices during the relaxation, as those shown in (Fig. 1a). Fig. 5 shows the μ distribution obtained from V_T relaxation traces of the stressed devices where only μ values larger than 1mV are considered because of the resolution limit of the measurement technique. The μ probability can be described with an exponential distribution [6] with mean value of $\langle\eta\rangle=4.75$mV in this particular case. So that, the μ_j parameter of each defect can be calculated from MC simulation assuming the distribution showed in Fig. 5.

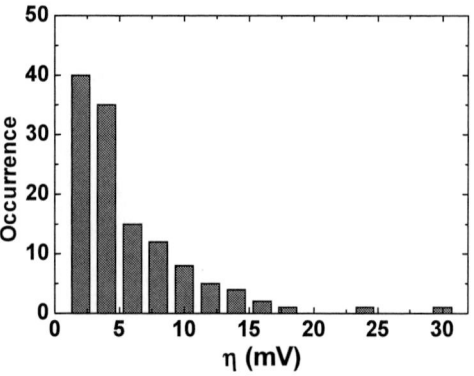

Figure 5. Occurrence distribution of μ obtained from the relaxation traces measured in small pFETs (Fig. 1a). Values of $\mu<1$mV can not be registered because of the resolution limit of the measurement technique.

The number of defects, N, follows a Poisson distribution [2], which is included in our simulations, therefore, the number of defects varies from device to device. Finally N, k_j and μ_j are introduced into Eq. 2 to calculate ΔV_T.

To validate the model, ΔV_T has been simulated for different stress conditions and the results have been compared with well known experimental results. Firstly, we have considered ultrascaled devices, that is, with a small number of defects (in this case, the average number of defects of each transistor was $\langle N\rangle=9.2$). The ΔV_T evolution with stress and relaxation times has been simulated at the same conditions that the measurements in Fig. 1a. Some of the relaxation traces obtained from these simulations are plotted in Fig. 1b, where it is observed that the stochastic nature of the ΔV_T caused by NBTI is well reproduced, both in time and in magnitude.

The next step has been to study the variability of threshold voltage produced by NBTI. It has been done by the simulation of ΔV_T in 1000 devices at different stress time and a fixed relaxation time of 100ms. Fig 6a shows the ΔV_T distributions obtained in the simulations. The shape of the distribution and its evolution with stress time is well reproduced when compared to experimental results obtained at similar stress conditions (Fig. 6b [6]). Fig. 6c shows the ΔV_T distribution simulated at the stress time of 1900s and several relaxation times, which reproduces the evolution of the ΔV_T distribution experimentally obtained again in [6] (fig. 6d).

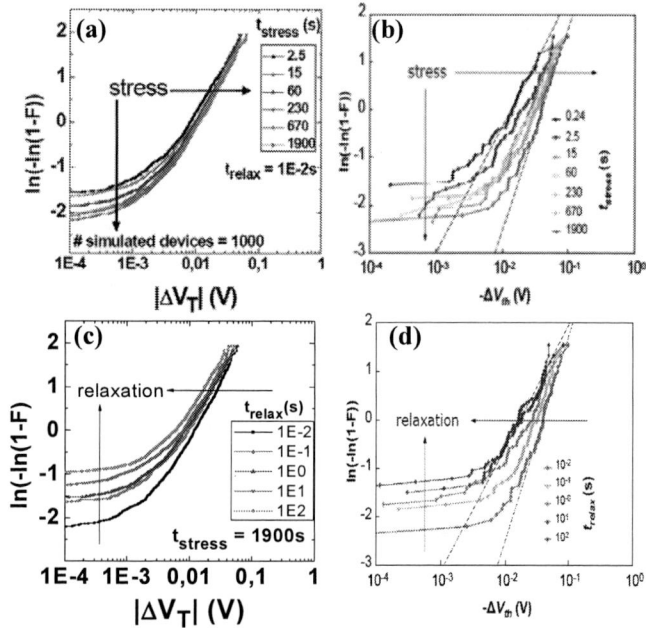

Figure 6. ΔV_T distributions simulated in 1000 devices for different stress (a) and relaxation (c) times. The shape and time dependence match with experimental results observed in devices under similar measurement conditions (b and d) [6].

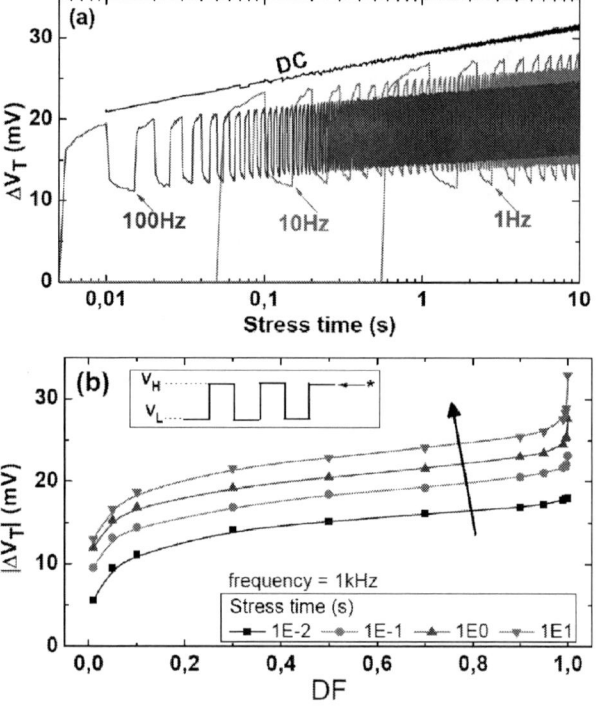

Figure 7. (a) Simulations of ΔV_T under DC and AC stress at different frequencies reproduce the prediction of the R-D model [3] and experimental observations [13]. (b) The distinctive duty factor dependence of NBTI after long enough stress time is correctly reproduced by the model [9]. ΔV_T is calculated when the input is at V_H just before the transition to V_L, as marked in the inset.

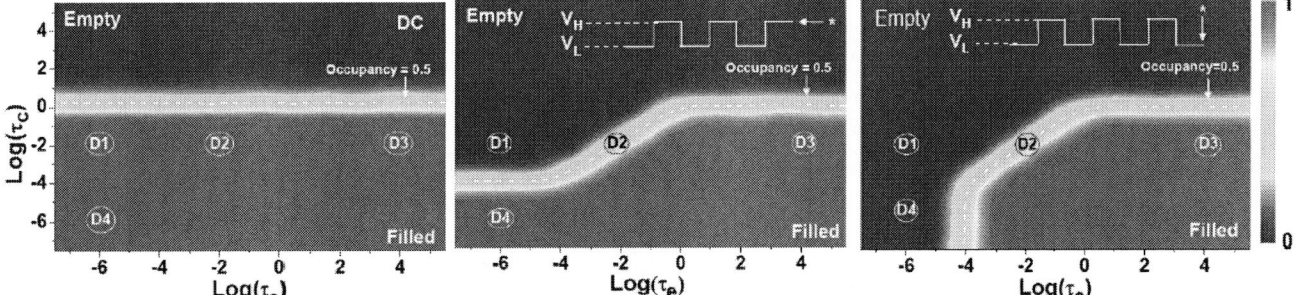

Figure 8. Occupancy as a function of τ_c and τ_e at DC (left) and AC stresses when at high (middle) and low (right) voltages, t_{stress}=1s, DF = 0.5, f=10kHz. D1-D4 correspond to the defects of interest which have been used to describe the effect of traps on ΔV_T for the different stress conditions. The dashed lines indicate occupancy = 0.5. Area below this line is aprox. proportional to ΔV_T.

The NBTI behaviour predicted by the model for AC stress has been also studied. Fig 7a shows the simulated ΔV_T dependence with stress time calculated from the average of 1000 devices stressed with pulsed waveforms for three different frequencies. For comparison, the ΔV_T evolution when the devices are DC stressed is also included. As can be observed, the simulation reproduces the oscillations previously described by the R-D model [3] and experimentally observed at low frequencies [13]. Finally, ΔV_T has been simulated for different duty cycles (in this work duty factor, DF) and different stress times. Fig. 7b shows the results obtained for ΔV_T calculated when the input is at V_H but just before its transition to V_L (marked with an * in the inset). Our proposal improves the R-D model in the sense that it correctly reproduces the NBTI distinctive DF dependence [9,14].

In summary, the model is able to reproduce the main experimental features observed in NBTI studies, and this is done from the calculation of all the defects occupancy in the device. In the following this property will be used to study NBTI from the point of view of the defect trapping predicted by the model.

V. NBTI DESCRIPTION IN TERMS OF TRAPS OCCUPANCY

In this section the defect occupancy as a function of the parameters τ_e and τ_c has been calculated at different stress conditions. The simulations have been performed considering a large number of defects (N>5×10^4) with τ_c and τ_e ranging from 10^{-8}s and 10^6s and calculating the occupancy P_{occ} for each defect at different stress conditions. Fig 8 shows the occupancy probability as a function of τ_c and τ_e obtained in DC (left), AC at V_H (middle) and AC at V_L (right) stresses. The stress conditions were t_{stress} for DC and t_{stress}=1s, period=0.1ms and DF=0.5 for AC. The dashed line in Fig. 8 indicates occupancy=0.5. Since the τ values of the defects are uniformly distributed on a log scale, then the area below this dash line is approximately proportional to the number of charged defects and to ΔV_T. Four defects of interest, labeled as D1-D4, have been chosen to describe the trapping process during the stresses, whose τ_e and τ_c values are given in table 1. At DC stress (Fig. 8 left) there is no relaxation, and the four defects are charged (all of them τ_c<<t_{stress}). For the AC stress, for both V_H and V_L voltages, D1 is empty due to the high probability of emission in the relax phase ($\tau_{e,D1}$<<period). D3 will always be filled because its negligible emission probability ($\tau_{e,D3}$>>t_{stress}). D2 (period << $\tau_{c,D2}$ = $\tau_{e,D3}$ << t_{stress}) has the same probability to

Defect	Time parameters
D1	$\tau_e = 1\mu s$; $\tau_c = 10ms$.
D2	$\tau_e = \tau_c = 10ms$.
D3	$\tau_e = 10ks$; $\tau_c = 10ms$.
D4	$\tau_e = \tau_c = 1\mu s$.

Table I Parameters of the defects chosen to describe the trap occupancy process at different stress conditions.

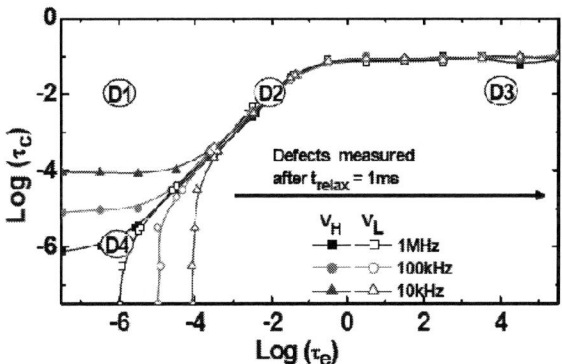

Figure 9. Occupancy = 0.5 lines in the situations of Fig.8 middle and right as a function of frequency. Increasing the frequency decreases the number of defects that trap and detrap every pulse (defects around D4). Defects with τ_e>t_{relax}, and therefore ΔV_T, do not depend on frequency [14].

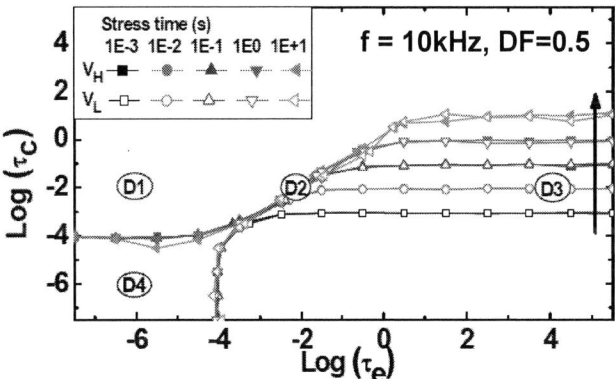

Figure 10. Occupancy = 0.5 lines in the situations of Fig.8 middle and right as a function of t_{stress}. Trapping is produced in defects with τ_c<t_{stress} and τ_e>t_{stress} (region close to D3). Occupancy of D2 remains constant for the longer stresses.

978-1-4244-9113-1/11 $26.00 © 2011 IEEE

trap a charge at V_H than to detrap at V_L, resulting in an occupancy = 0.5. D4 has very short capture and emission times ($\tau_{e,D4} = \tau_{c,D4} \ll$ period), so that it trap (Fig. 8 middle) and detraps (Fig. 8 right) every pulse. Then, defects near D4 explain the oscillations observed in Fig. 7a. The evolution of the 0.5 line occupancy with frequency has been calculated at different frequencies (Fig. 9), when at V_H (filled symbols) and V_L (open symbols). Defects with τ_c and τ_e lower than the period of the stress waveform can trap and detrap with each pulse (region around D4), explaining the decrease of the amplitude of the oscillations with frequency in Fig. 7a. However, these defects only are occupied if ΔV_T is measured at short relaxation times. After $t_{relax}>$period, as for example $t_{relax}=1$ms marked in Fig. 9, only those defects whose τ_c and τ_e are larger than the period contribute to ΔV_T. The occupancy of these defects is independent of frequency, explaining why no change of ΔV_T with frequency is observed in [14]. Fig. 10 shows the defect occupancy obtained for AC stresses (f=10kHz and DF=0.5) for different stress times. The occupancy of those defects in the region near D3 ($\tau_e \gg t_{stress}$) increases with stress time. Defects close to D1 and D4 have no relevant influence on the increase of ΔV_T with stress time under AC conditions because of their very low τ_e; if they were charged at any moment of the stress, they will rapidly discharge during the relaxation. Note that D2 occupancy depends neither on frequency nor on t_{stress}. However D2 occupancy depends on DF, as shown in Fig. 11, where the D2 occupancy with stress time is plotted for different DF. Occupancy increases with stress time until it starts to oscillate around a constant value that depends on DF, but also on the ratio τ_e/τ_c. In Fig. 12 the occupancy calculated from a finite distribution of defects with period$<\tau_e<t_{stress}$ and period$<\tau_c<t_{stress}$ (defects around D2, inside the dashed rectangle of the upper inset) is plotted versus DF, for different t_{stress}. As t_{stress} increases, the obtained curve is more similar to the ΔV_T dependence with DF. Therefore, the results of this simulation suggest that defects with period$<\tau_e<t_{stress}$ and period$<\tau_c<t_{stress}$ are the ones responsible for the peculiar relation between ΔV_T and DF.

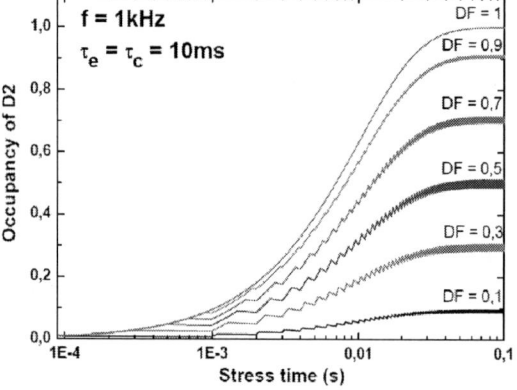

Figure 11. Simulation of the occupancy of D2 as a function of t_{stress} for different values of DF. D2 occupancy increases with t_{stress} until it reaches oscillation around a constant value which depends on DF.

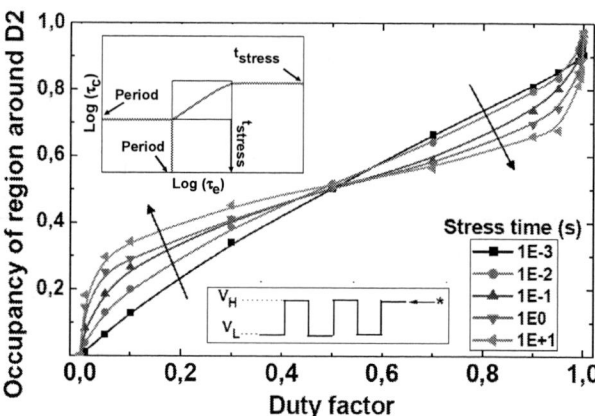

Figure 12. Dependence of the occupancy of defects around D2 (period$<\tau_e<t_{stress}$ and period$<\tau_c<t_{stress}$, area inside the dashed rectangle of the upper inset) with DF. ΔV_T has been calculated when the input is at V_H immediately before changing to V_L (marked with * in the lower inset). For longer stress times the occupancy at these conditions has a dependence on DF similar to ΔV_T.

VI. CONCLUSIONS

This work presents a new model for NBTI based on the experimental evidence of the phenomena in FETs. The model uses two parameters, the charge capture and emission time (τ_c and τ_e) to describe the occupancy probability of a defect under DC or AC stresses and one more (μ) to describe the change in V_T caused by the defect trapping. The model is used to calculate the V_T shift (ΔV_T) in devices under different stress conditions. The number of defects and the distribution of their parameters (τ_e, τ_c and μ) in the device are obtained from our measurements or previous works. The simulations show that the model can describe the stochastic nature of NBTI and the ΔV_T distribution observed in small pFETs. Moreover the model can reproduce the ΔV_T evolution with stress time under DC and AC conditions at different frequencies and the dependence on the duty factor. All these results allow concluding that the model proposed in this work is able to describe the NBTI from the point of view of defect trapping. From the simulations performed, the experimental ΔV_T caused by NBTI at different stress conditions is explained in terms of trap occupancy. Our calculations indicate that frequency in AC stresses only affects to defects with τ_c and τ_e lower than the period, explaining the oscillations observed during the AC stress and the independence of ΔV_T on frequency after short relaxation times. Simulations have been performed to calculate the occupancy of the defects with τ_e and τ_c values larger than the stress waveform period but smaller than t_{stress}. The results have shown that for long enough stress times the occupancy of those defects and ΔV_T have qualitatively the same DF dependence. Then, according to our model, these defects are the cause of the ΔV_T distinctive DF dependence experimentally observed.

REFERENCES

[1] J. H. Stathis and S. Zafar, "The Negative Bias Temperature Instability in CMOS devices: A review". Microelectronics Reliability, vol 46. pp. 270-286, 2006.

[2] S. E. Rauch, "Review and Reexamination of Reliability Effects Related to NBTI Statistical Variations", IEEE T. Dev. Mat. Rel. 7, pp. 524-530, 2007.

[3] M.A. Alam, H. Kufluoglu, D. Varghese and S. Mahapatra, "A comprehensive model for PMOS NBTI degradation: Recent progress". Microelectron. Reliab., **46**, pp. 853-862, 2007.

[4] J. Martin-Martinez, R. Rodriguez, M. Nafria, X. Aymerich, B. Kaczer, G. Groeseneken, "An equivalent circuit model for the recovery component of BTI," Proc. ESSDERC, pp. 55-58, 2008.

[5] B. Kaczer, T. Grasser, Ph. J. Roussel, J. Franco, R. Degraeve, L. A. Ragnarsson, E. Simoen, G. Groeseneken, H. Reisinger. "Origin of NBTI variability in deeply scaled pFETs," *Proc. Int. Reliab. Phys. Symp.*, pp. 26-32, 2010.

[6] T. Grasser, H. Reisinger, P.-J. Wagner, F. Schanovsky, W. Goes, and B. Kaczer, "The Time Dependent Defect Spectroscopy (TDDS) for the Characterization of the Bias Temperature Instability", *Proc. Int. Reliab. Phys. Symp.*, pp. 16-25, 2010.

[7] H. Reisinger, T. Grasser, W. Gustin, and C. Schlünder, "The Statistical Analysis of Individual Defects Constituting NBTI and its Implications for Modeling DC- and AC-Stress," *Proc.Int. Reliab. Phys. Symp.*, pp. 7-15, 2010.

[8] B. Kaczer, S. Mahato, V. Valduga de Almeida Camargo, M. Toledano-Luque, Ph. J. Roussel, T. Grasser, F. Catthoor, P. Dobrovolny, P. Zuber, G. Wirth, and G. Groeseneken, "Atomistic approach to variability of bias-temperature instability in circuit simulations," presented at *Int. Reliab. Phys. Symp.*, 2011.

[9] B. Kaczer, T. Grasser, Ph. J. Roussel, J. Martin-Martinez, R. O'Connor, B. J. O'Sullivan, and G. Groeseneken, "Ubiquitous Relaxation in BTI stressing—New Evaluation and Insights", *Proc. Int. Reliab. Phys. Symp.*, pp. 20-27, 2008.

[10] M. Toledano-Luque, B. Kaczer, Ph. J. Roussel, T. Grasser and G. Groeseneken. "Temperature dependence of the capture and emission times of SION individual traps after positive bias temperature stress," J. Vac. Sci. Technol. B, in press.

[11] B. Kaczer, T. Grasser, J. Martin-Martinez, E. Simoen, M. Aoulaiche, Ph. J.Roussel, and G. Groeseneken, "NBTI from the Perspective of Defect States with Widely Distributed Times," *Proc. Int. Rel. Phys. Symp.*, pp. 55-60, 2009.

[12] M. F. Bukhori, S. Roy, A. Asenov, "Simulation of statistical aspects of reliability in nano CMOS," *Proc. Int. Integ. Rel. Workshop*, pp. 82-85, 2009.

[13] M. A. Alam, "A critical examination of the mechanics of dynamic NBTI for PMOSFETs," *Int. Electron Devices Meeting Tech. Dig.*, pp. 345-348, 2003.

[14] Fernandez, R.; Kaczer, B.; Nackaerts, A.; Demuynck, S.; Rodriguez, R.; Nafria, M.; Groeseneken, G., "AC NBTI studied in the 1 Hz -- 2 GHz range on dedicated on-chip CMOS circuits," *Int. Electron Devices Meeting Tech. Dig.*, pp. 1-4, 2006.

MOSFET'S HOT CARRIER DEGRADATION CHARACTERIZATION AND MODELING AT A MICROSCOPIC SCALE

[1,2]Y. Mamy Randriamihaja, [1]A. Zaka, [1]V. Huard, [1]M. Rafik, [1]D. Rideau, [1]D. Roy
[1]STMicroelectronics, 850 rue J. Monnet, BP16, 38926 Crolles, France
Phone: (33) –(0)4.38.92.23.21, yoann.mamy-randriamihaja@st.com

[2]A. Bravaix
[2]ISEN – IM2NP, Maison des technologies, 83000 Toulon, France

Abstract—**Microscopic characterization of interface defects along the channel length is used to monitor the HC induced defect generation. The modeling of HC degradation is adapted to a microscopic scale and is found to be consistent with obtained lateral profiles.**

Keywords: HCS, interface defects, degradation modeling

I. INTRODUCTION

Recent work on Hot Carrier Stress (HCS) modeling has considerably improved the understanding of HC induced defects [1]. More-over, most of these models are based on macroscopic quantities (Id, Gm ...) which embrace the additive effect of the non-uniform defects generated during the stress. In this paper, we focus on the degradation modeling at a microscopic level with several defects characterization methods and advanced Monte Carlo (MC) simulation tools. Based on this methodology, various modes are explicitly monitored along the channel. In particular, it is shown that similar macroscopic degradation can lead to very different microscopic situations.

II. EXPERIMENTAL SETUP FOR LATERAL PROFILING

Various methods can be used to monitor HC Induced defects along the channel length. In [2] it has been shown that the combination of Gated Diode (GD) [3], Lateral Profiling with Charge Pumping (LPCP) [4], [5] and LP with Id(Vd) (LPIV) techniques can provide a complete picture for defects localization in the channel and in the LDD-region. In particular, new method using Id(Vd) measurement in the saturation regime is proposed in [2]. In this method, stressed forward Id(Vd) curves tend to converge toward the fresh ones when going from early saturation toward high saturation regime (**Fig. 1**). This method allows an accurate delimitation of the stress induced defects in the pinch off region.

Lateral Profiling (LP) measurements were performed on NMOS structures featuring a 0.28μm gate length (L), 10 μm gate width (W), 5nm gate oxide thickness (T_{ox}) and a nominal

supply voltage of 2.5 V (V_{dd}). Varying HCS time (0 to 10 000 s) and bias (Vg/Vd = 3.3/3.3, 4.3/4.3, 1.8/4.3 V) conditions are applied. **Fig. 2** summarizes the methods used to monitor the interface trap generation (N_{it}) and their probed domain, after HCS Vg=Vd=3.3V. Also shown in the Figure are the extracted Lateral Profiles for various stress times obtained with LPCP and LPIV.

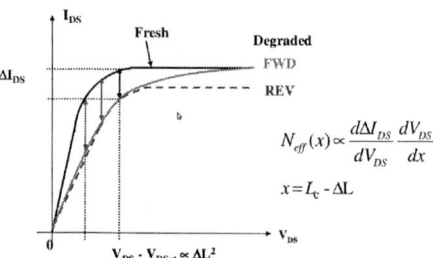

Figure 1. Schematic of the LPIV principle, where $N_{eff}(x)$ is the effective interface trap's density, and x is the lateral position

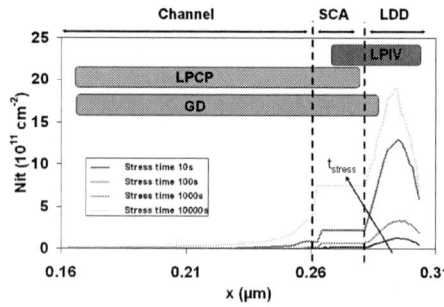

Figure 2. Extended quantitative lateral profiles obtained by combining LPIV and LPCP, after HCS Vg=Vd=3.3V. The probed zones by LPCP, GD and LPIV techniques are also represented above

978-1-4244-9113-1/11 $26.00 © 2011 IEEE

III. MACROSCOPIC DEGRADATION MODELING ADAPTED TO MICROSCOPIC LEVEL

In a recent physical framework [1], the N_{it} creation related to carrier-induced bond-breaking excitation has been divided into three modes: 1) a single carrier has enough energy to break the bond (Single Vibrational Excitation, SVE) or, 2) some carriers gain enough energy through Electron-Electron Scattering (EES), or 3) numerous carriers with reduced energy cooperate (Multi Vibrational Excitation, MVE) to create a defect. As this work focuses HCS in input/output devices, the modeling results mainly in mode 1 and 2 disregarding the MVE mode involved in core devices [1]. Thus, the formalism of [6] expressed in (1) is coupled with the modeling of [7] in (2-3) to determine the degradation of the first two modes (see **Fig. 3**) where:

$$\Delta N_{it}(t) \sim [\, R_{it} . t\,]^n \qquad (1)$$

The amount of ΔN_{it} (cm^{-2}) versus time t (s) follows a power law (n ~ 0.25 experimentally), where R_{it} stands for the rate $(cm^{-2}.s^{-1})$ of N_{it} creation, and can be related to SVE and EES modes by:

$$R_{it-SVE} = f(E_{domi}).S_{it-SVE}(E_{domi}) = f(E_{domi}).(E_{domi} - \varphi_{it})^{pit} \,(2)$$

$$R_{it-EES} = f(E_{domi})^2.S_{it-EES}(E_{domi}) = f(E_{domi})^2.(1,8.E_{domi} - \varphi_{it})^{pit} \,(3)$$

where $f(E_{domi})$ is the carrier distribution function $(cm^{-3}eV^{-1})$ at the energy E_{domi} (defined below), S_{it} is the probability of creating a defect, φ_{it} is the threshold energy of defects creation (1.5eV) [8], K is a constant, and p_{it} is found around 11 [1].

The energy distribution functions f(E) were obtained using the deterministic Spherical Harmonic Expansion (SHE) of the Boltzmann equation [9], [10]. For a given f(E), E_{domi} has been calculated as the "elbow" of the distribution function at each lateral position close the Si-SiO$_2$ interface [7].

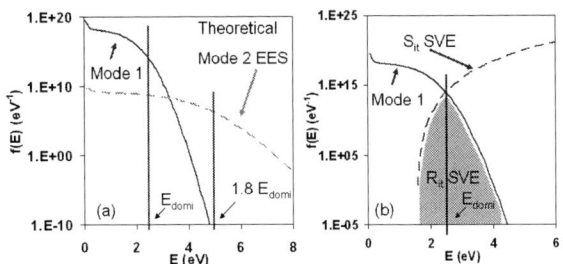

Figure 3. (a): carriers energy distribution function at a given position along the channel. Mode 1 (SVE), as simulated, and the theoretical mode 2 (EES), taken here as the square root of mode 1, shifted by 1,8.E_{domi}. (b): carriers energy distribution function superposed with the probability of N_{it} creation for mode 1 (SVE). The N_{it}'s creation rate is represented as the shaded area

Stress conditions corresponding to Vg = Vd = 4.3 V (mode 2 at a macroscopic level), and Vg = 1.8 V, Vd = 4.3 V (mode 1 at a macroscopic level) have been considered. The obtained f(E) and the E_{domi} extractions are shown in Fig.3 (a). It ought to be noted that the f(E) of mode 2 is presented as an illustration (taken here equal to the square root of mode 1, shifted by 1,8.E_{domi}), as described by [7], since we didn't simulate the EES. **Fig. 3-(b)** shows the subsequent R_{it} determination. This

enables us to determine the ΔN_{it} calculation for each stress condition simulated.

Figs. 4, 5 show LPs obtained for both HCS's conditions at t_{stress}=400s. At this stress time, one can observe that the two conditions reach the same 1/Gm degradation, with different dynamics (not shown). On the same figures, ΔN_{it} creation calculated by using the described methodology has been also reported. Both ΔN_{it} were scaled with the K constant to match the LP extracted through measurements. It is accurately checked that each mode of degradation (SVE, EES) complete each other to lead to the overall macroscopic degradation (**Fig. 4**) which extends from the LDD towards the spacer according to energetic carriers in mode 1 [1].

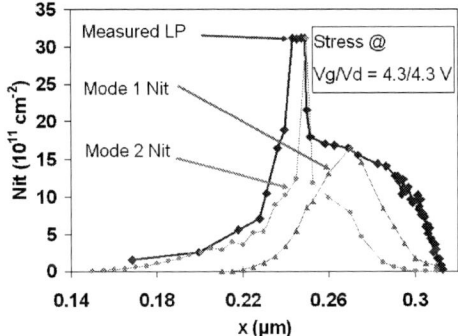

Figure 4. LPs obtained by combining LPIV and LPCP after HCS (Vg=Vd=4.3V, t_{stress}=400s) superposed with simulated N_{it} created with mode 1&2

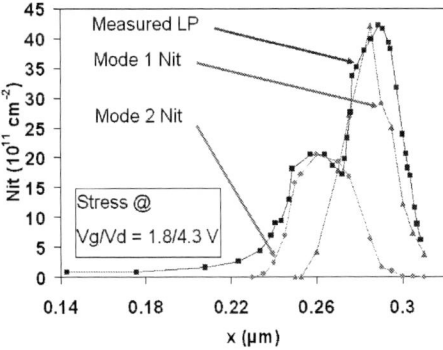

Figure 5. LPs obtained by combining LPIV and LPCP after HCS (Vg=1.8V, Vd=4.3V, t_{stress}=400s) superposed with simulated Nit created with mode 1&2

In contrast, HCS at the maximum substrate current in **Fig.5** extends further towards the channel due to EES (mode 2). Even though macroscopic degradation is similar for the considered conditions, the defects localization is radically different. Therefore degradation modeling based on macroscopic parameters can be misleading. Furthermore, even if macroscopic parameters degradation can be modeled with one mode, we demonstrate that all modes coexist at a microscopic level and contribute to the degradation process.

Fig. 6 shows LP obtained for HCS at Vg = Vd = 3.3 V, after 10 000s of stress to illustrate the reproducibility of the method.

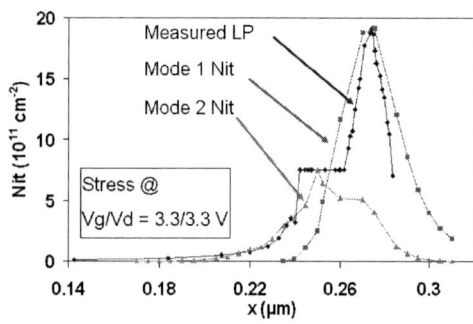

Figure 6. LPs obtained by combining LPIV and LPCP after HCS (Vg=Vd=3.3V, t_{stress}=10 000s) superposed with simulated N_{it} created with mode 1&2

IV. CONCLUSION

This paper shows that the combination of different defects characterization methods is useful to obtain a better understanding of defect generation. This is the necessary basis to model the degradation process from a microscopic point of view. Each one of those methods has its own 1) probed area, 2) probing resolution and 3) physical way of probing. By combining them, an extended lateral profile was obtained. The lateral profile already refine the understanding of the degradation process (LDD and Space Charge Area (SCA) are the mainly degraded zones) compare to macroscopic parameters such as V_{th} and Gm. Those obtained lateral profiles were compared with simulations of degradation rate, adapted at a microscopic scale, and were found to be consistent.

ACKNOWLEDGMENT

The author would like to address special thanks to Gerard Morin and François Dieudonné for their precious help to the measurements.

REFERENCES

[1] C. Guerin, et al., Jour. of Appl. Phys., Vol. 105, p 114513, 2009
[2] Y. Mamy Randriamihaja, et al, IIRW, in press
[3] A. Asenov, et al, Microelec. Eng., pp. 445-448, 1991
[4] R. Lee, et al, Trans. on Elec. Dev., Vol. 43, p 81, 1996
[5] C. Chen, et al., Trans. on Elec. Dev., Vol. 45, p 512, 1998
[6] Y. Hu, et al., Trans. on Elec. Dev., Vol. 32, p 375, 1985
[7] S.E. Rauch, et al., Trans. Dev. Mat. Reliab., Vol. 1, p 113, 2001
[8] V. Huard, et al, in proc. IEDM, 2007
[9] S. Jin, et al., SISPAD, 2009
[10] A. Zaka, et al., Sol. Stat. Elec., Vol. 54, p 1669, 2010

Simultaneous extraction of threshold voltage and mobility degradation from on-the-fly NBTI measurements

Rodolf W. Herfst[1,2], Jurriaan Schmitz[1], and Andries J. Scholten[2]

[1]MESA+ Institute for Nanotechnology, University of Twente, P.O. Box 217, 7500 AE Enschede, The Netherlands

[2]NXP-TSMC Research Centre, High Tech Campus 37, 5656 AE Eindhoven, The Netherlands

Abstract—Conventional on-the-fly characterization of NBTI translates measured changes in drain current to a threshold voltage shift only. In this paper, we show how to extend this method to the simultaneous determination of threshold voltage and zero-field-mobility degradation by. This is achieved by using a Vector Network Analyzer for OTF characterization of g_{ds} and g_m. For the technology under study, we have found that degradation in the zero-field mobility is responsible for at most 10% of the drain current change. Effective mobility, on the other hand, does change as a direct consequence of the threshold-voltage shift.

Index Terms—NBTI, On The Fly, Reliability, RF

I. INTRODUCTION

Negative Bias Temperature Instability (NBTI) is a serious reliability issue in today's CMOS technologies. Its characterization is complicated by the fact that the device (partly) recovers when the gate stress is removed. Degradation and recovery occurs over a wide range of time scales. This makes the traditional "measure-stress-measure" (MSM) method unsuitable for NBTI characterization, and has led to a host of new measurement methods, amongst which the "on-the fly" (OTF) method [1], which is used by a number of different groups [1]–[10].

With OTF NBTI characterization the degradation of I_d is monitored while continually applying a gate stress voltage. More advanced variations also monitor the transconductance g_m by measuring the drain current response to small excursions ΔV_g from the applied stress voltage. Whichever variation of OTF is used, a key step in the interpretation of measurement results is the translation of changes in I_d (and if available g_m) to a change in threshold voltage V_{th}.

Of course, V_{th} is not the only transistor parameter that can change due to degradation of the device, e.g. carrier mobility might also degrade under BTI stress. The question whether this happens, and if so, how large the effects are, has received relatively little attention in the literature [8], [11]–[13]. While some mobility-degradation-related corrections have been proposed for OTF ΔV_{th} extraction [11], [13], [14], actual mobility degradation results have only been published in the context of MSM techniques [12].

In this paper, we will first review the underlying assumptions and validity of some of the existing methods for extracting ΔV_{th} (section II). A new method for simultaneously extracting both V_{th} shifts and changes in mobility from g_{ds} and

g_m data is introduced in section III. In section IV we apply this method to measure the change in zero-field mobility with the help of on-the-fly small-signal two-port RF measurements.

II. EXISTING EXTRACTION METHODS

During an OTF BTI test, degradation of the current (or of g_{ds}) is measured while a certain stress voltage V_{gs} is applied to the gate. To calculate ΔV_{th} from ΔI_d, assumptions must be made regarding the (unstressed) $I_d - V_{gs}$ characteristic around V_{stess} or V_{meas}. One of the widely used methods calculates $\Delta V_{th}(t)$ as [2]–[4], [10]

$$\Delta V_{th}(t) = -\frac{\Delta I_d(t)}{I_d(0)}\left[V_{gs} - V_{th}(0)\right]. \tag{1}$$

The (implicit) assumption here is that there is a linear relation between measured drain current and applied gate voltage, i.e. that at $t = 0$ the IV curve can be modeled as a straight line from $(V_{th}, 0)$ to $(V_{stress}, I_d(0))$. This is of course never true, and especially wrong for currents measured at higher voltages typically used during accelerated life-time tests. Here current is reduced with respect to the linear relation by mobility reduction (dominant in long channel devices) and series resistance (dominant in short channel devices).

A more accurate result can be obtained if g_m at $t = 0$ is also measured. In this case, instead of relying on the initial V_{th} and $I_d(0)$, the IV characteristic is assumed to be locally linear with slope $g_m(0)$ at $V = V_{stress}$. It is derived from a first order Taylor expansion of current with respect to ΔV_{th}:

$$\Delta I_d \simeq \frac{\partial I_d}{\partial V_{th}} \cdot \Delta V_{th} = -\frac{\partial I_d}{\partial V_g} \cdot \Delta V_{th} = -g_m(0) \cdot \Delta V_{th}. \tag{2}$$

$\Delta V_{th}(t)$ is then given by [9]:

$$\Delta V_{th}(t) = -\frac{\Delta I_d(t)}{g_m(0)}. \tag{3}$$

This method is accurate for small values of ΔI_d, but will deviate for large changes in V_{th} due to the curvature of the IV characteristic (i.e. this is only accurate for $|\Delta V_{th}| \ll |2g_m/g_{m2}|$).

In the most advanced version of the OTF method, not only the drain current I_d, but also the transconductance g_m is monitored by measuring the drain current response to small excursions ΔV_g from the applied stress voltage [1], [7], [8],

978-1-4244-9113-1/11 $26.00 © 2011 IEEE

TABLE I
COMPARISON OF VARIOUS ΔV_{TH} EXTRACTION FORMULAS.

Extraction Method	Valid for nonlinear $I_d - V_{gs}$?	Valid for large ΔV_{th}?	Valid for $\Delta \mu_0 \neq 0$?	Assessment
$\Delta V_{th} = -\frac{\Delta I_d}{I_d(0)}\left(V_{gs} - V_{th}\right)$	–	–	–	Never true
$\Delta V_{th} = -\frac{\Delta I_d}{g_m(0)}$	✓	–	–	Use only for small ΔV_{th}
$\Delta V_{th} = -2 \cdot \sum_{i=1}^{n} \frac{I_d(t_i)-I_d(t_{i-1})}{g_m(t_i)+g_m(t_{i-1})}$	✓	✓	–	Good method, but all ΔI_d attributed to ΔV_{th}
This work, Eq. (8)	✓	✓	✓	Most general OTF ΔV_{th} extraction method

[11], [15]. The threshold voltage shift ΔV_{th} is subsequently obtained using [11]

$$\Delta V_{\text{th}} = -2 \cdot \sum_{i=1}^{n} \frac{I_d(t_i) - I_d(t_{i-1})}{g_m(t_i) + g_m(t_{i-1})}. \tag{4}$$

If the time steps used are small enough, the above method (to which we will refer as "conventional method" hereafter) is essentially correct, also when the $I_d - V_{gs}$ relationship is nonlinear and ΔV_{th} is large. However, it still tacitly assumes that the entire drain current degradation is due to a threshold voltage shift, while it is conceivable that not only the threshold voltage, but also the carrier mobility degrades under NBTI stress conditions.

A summary of the different methods, the requirements for their validity, and an assessment of them is given in Table I. Also included is the new extraction method that is introduced in the next section, which can also be used to extract changes in low-field mobility.

III. NEW EXTRACTION METHOD

NBTI degradation is usually measured in the linear region of operation (small V_{ds}), where the current is accurately described by [8], [13]

$$I_d = \frac{\beta \cdot V_{ds} \cdot \left(V_{gt} - \frac{1}{2}V_{ds}\right)}{1 + \theta_1 V_{gt}}, \tag{5}$$

where $V_{gt} = V_{gs} - V_{th}$ is the gate overdrive voltage, $\beta = \mu_0 C_{ox} W_{\text{eff}}/L_{\text{eff}}$ is the gain factor (which is a direct measure of zero-field mobility μ_0), and θ_1 is the vertical-field mobility reduction coefficient. From Eq. (5), transconductance g_m and output conductance g_{ds} are easily derived by differentiating to V_{gs} and V_{ds} respectively:

$$g_{ds} = \frac{\beta \cdot (V_{gt} - V_{ds})}{1 + \theta_1 V_{gt}} \tag{6}$$

$$g_m = \frac{\beta \cdot V_{ds} \cdot \left(1 + \frac{1}{2}\theta_1 V_{ds}\right)}{\left(1 + \theta_1 V_{gt}\right)^2} \tag{7}$$

Eq. (6) is divided by (7) to eliminate β and is then solved for V_{gt}. This V_{gt} is then used to derive β. This all results in:

$$V_{gt} = \frac{-1 + \theta_1 V_{ds} + \sqrt{(1 - \theta_1 V_{ds})^2 + 4\theta_1 V_{ds}\left[1 + \frac{g_{ds}}{g_m}\left(1 + \frac{1}{2}\theta_1 V_{ds}\right)\right]}}{2\theta_1} \tag{8}$$

and

$$\beta = g_{ds}\frac{1 + \theta_1 V_{gt}}{V_{gt} - V_{ds}}. \tag{9}$$

In the above, the parameter θ_1 is assumed to be constant. Although this is an approximation, one should note that this is still a step beyond the conventional method which tacitly assumes both β and θ_1 to be constant. One could think that extending our method to use I_d in addition to g_{ds} and g_m would help to determine θ_1 as well. However, in the linear regime of operation, $I_d \approx V_{ds}g_{ds}$, so that I_d and g_{ds} carry essentially the same information. So, keeping one parameter of Eq. (5) constant is inevitable.

The initial values of V_{th}, β, and θ_1 are found as follows: we determine $V_{\text{th}}(t = 0)$ using a pre-stress $I_d - V_{gs}$ curve, which is measured up to limited V_{gs} to avoid any NBTI degradation prior to the experiment. The parameters $\beta(t = 0)$ and $\theta_1(t = 0)$ are determined from g_m and g_{ds} at $t = 0$ of the actual stress test.

In our new method, we have chosen to use the set of parameters (g_{ds}, g_m), which are conveniently measured using a network analyzer. However, the method can also be applied to the data sets (I_d, g_m) from a conventional OTF experiment. The equations needed for this case can easily be derived, similar to the derivation of Eqs. (8) and (9). This is detailed in the appendix.

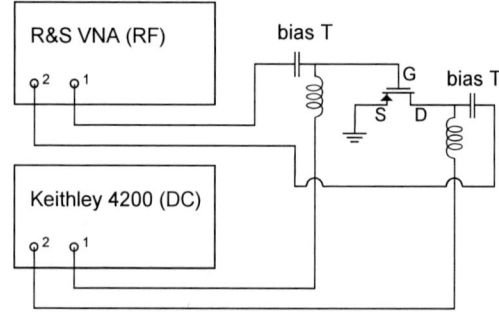

Fig. 1. Schematic representation of the measurement setup. IV capability of a Keithley 4200 is combined with a Rohde&Schwarz VNA to obtain g_{ds} and g_m from RF Y-parameters. The RF frequency is 2 GHz, and the RF power -25 dBm (25 mV RMS).

Before applying our method to real measurement data, we verified our equations to simulations. MOS Model 11 was used

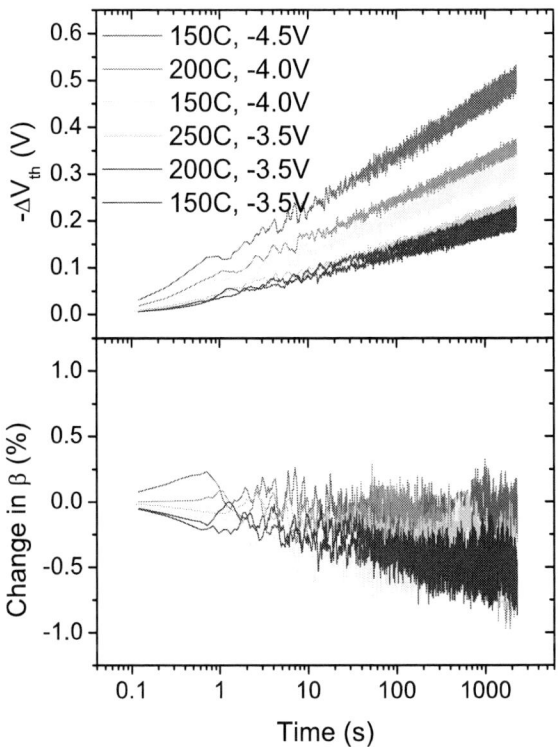

Fig. 2. Threshold voltage shift and change in β as function of time for several NBTI experiments (thin oxide). Used devices: PMOS, gate-length 160 nm, (nitrided) oxide thickness 2.9 nm.

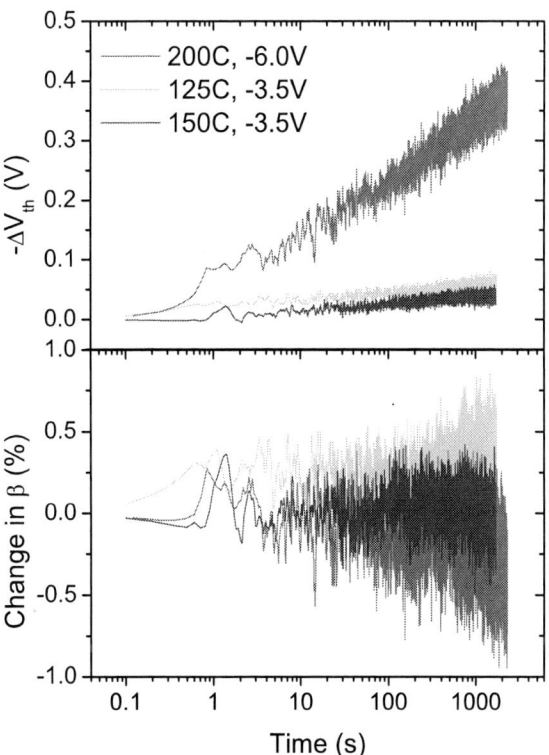

Fig. 3. Threshold voltage shift and change in β as function of time for several NBTI experiments (thick oxide). Used devices: PMOS, gate-length 322 nm, (nitrided) oxide thickness 7.2 nm.

to simulate realistic IV curves with various threshold voltage shifts as well as degradation of low-field mobility. This was used as "measurement data". The four different V_{th} extraction methods were then used to check how well the applied ΔV_{th} is extracted by each of the methods.

When Eq. (1) is used, ΔV_{th} is significantly underestimated. Depending on the voltage at which current change is translated into threshold voltage change, ΔV_{th} can easily be off by a factor of two. As expected, for small ΔV_{th} Eq. (3) correctly extracts the applied ΔV_{th}. It can however deviate quite significantly for higher applied ΔV_{th}, as explained in section II. Provided $\Delta\beta = 0$, Eq. (4) and (8) give identical results, as expected. However, if β degradation is also present, only the new method retrieves the applied ΔV_{th} correctly.

IV. RESULTS

Using the experimental setup of Fig. 1, DC NBTI stress was applied to PMOSFETs with nitrided oxide of two different thicknesses, while its Y-parameters were recorded on-the-fly. The frequency f was set to 2 GHz. For our devices this is low enough for $\mathrm{Re}(Y_{21})$ and $\mathrm{Re}(Y_{22})$ to be frequency-independent, and $g_m = \mathrm{Re}(Y_{21})$ and $g_{ds} = \mathrm{Re}(Y_{22})$ holds. Changes in g_m and g_{ds} were translated into changes in V_{th} and mobility, using Eqs. (8) and (9), respectively. This is shown for thin (2.9 nm) oxide devices in Fig. 2, and thick (7.2 nm) oxide devices in Fig. 3.

The V_{th} degradation that we observe shows all the well-known dependencies on stress voltage, temperature, and oxide thickness. Just as in for example Ref. [4], it follows a logarithmic time dependence. Contrary to the large V_{th} degradation in our experiments, we only observe sub-1% shifts in β. These small shifts of β do not correlate with either stress settings or threshold voltage shifts, and do not point at true β degradation. Instead, they are most probably due to small instabilities in the measurement equipment, as well as to the fact that Eq. (5) cannot perfectly describe the behavior of the device. Analysis of these results shows that –at most– 10% of the drain current degradation can be attributed to β degradation, justifying the use of Eq. (4) for this technology.

It is noteworthy that since the zero-field mobility does not change significantly, a threshold voltage shift will actually *increase* the *effective* mobility at constant bias voltage, since $\mu_{eff} = \mu_0/(1+\theta_1 V_{gt})$. This is also explains why g_m is observed to increase during stress.

V. SUMMARY

We have presented a method to extract both threshold-voltage and mobility changes from on-the-fly NBTI degradation measurements. For the technology under study, only a small part of the drain current degradation is due to zero-field mobility degradation for both thin and thick oxide devices. This confirms that –for the technology at hand– the threshold voltage shift is by far the most significant effect of NBTI stress.

ACKNOWLEDGMENT

The authors acknowledge financial support from the Dutch Ministry of Economic Affairs / Agentschap NL through the knowledge workers program "Resilience for Automotive".

APPENDIX: EXTRACTING ΔV_{TH} FROM I_{D} AND g_m

In our work, we use a two-port Y-parameter measurements for determining g_m and g_{ds}, from which ΔV_{th} and β are extracted. However with conventional OTF I_{d} is measured in stead of g_{ds}, so that V_{gt} must be obtained from I_{d} and g_m. V_{gt} and β can be then derived as follows. Again, we start with the equations for I_{d} and g_m:

$$I_{\text{d}} = \frac{\beta \cdot V_{\text{ds}} \cdot \left(V_{\text{gt}} - \frac{1}{2}V_{\text{ds}}\right)}{1 + \theta_1 V_{\text{gt}}} \qquad (10)$$

$$g_m = \frac{\beta \cdot V_{\text{ds}} \cdot \left(1 + \frac{1}{2}\theta_1 V_{\text{ds}}\right)}{\left(1 + \theta_1 V_{\text{gt}}\right)^2}. \qquad (11)$$

We divide I_{d} by g_m to get

$$\frac{I_{\text{d}}}{g_m} = \frac{\left(V_{\text{gt}} - \frac{1}{2}V_{\text{ds}}\right)\left(1 + \theta_1 V_{\text{gt}}\right)}{1 + \frac{1}{2}\theta_1 V_{\text{ds}}}. \qquad (12)$$

From this we can easily solve V_{gt}:

$$V_{\text{gt}}(t) = \frac{-A + \sqrt{A^2 + 4\theta_1 \left[\frac{1}{2}V_{\text{ds}} + \frac{I_{\text{d}}(t)}{g_m(t)}\left(1 + \frac{1}{2}\theta_1 V_{\text{ds}}\right)\right]}}{2\theta_1}, \qquad (13)$$

where A is equal to

$$A = 1 - \frac{1}{2}\theta_1 V_{\text{ds}}. \qquad (14)$$

Using $V_{\text{gt}}(t)$, we can also calculate $\beta(t)$:

$$\beta(t) = I_{\text{d}}(t) \frac{1 + \theta_1 V_{\text{gt}}(t)}{V_{\text{ds}} \left(V_{\text{gt}}(t) - \frac{1}{2}V_{\text{ds}}\right)}. \qquad (15)$$

REFERENCES

[1] M. Denais, C. Parthasarathy, G. Ribes, Y. Rey-Tauriac, N. Revil, A. Bravaix, V. Huard, and F. Perrier, "On-the-fly characterization of NBTI in ultra-thin gate oxide PMOSFET's," in *IEEE 2004 International Electron Devices Meeting, IEDM Technical Digest.*, 2004, pp. 109–112.

[2] S. Mahapatra, V.D. Maheta, A.E. Islam, and M.A. Alam, "Isolation of NBTI Stress Generated Interface Trap and Hole-Trapping Components in PNO p-MOSFETs," *IEEE Transactions on Electron Devices*, vol. 56, no. 2, pp. 236–242, Feb. 2009.

[3] V.D. Maheta, E.N. Kumar, S. Purawat, C. Olsen, K. Ahmed, and S. Mahapatra, "Development of an Ultrafast On-the-Fly Technique to Study NBTI in Plasma and Thermal Oxynitride p-MOSFETs," *IEEE Transactions on Electron Devices*, vol. 55, no. 10, pp. 2614–2622, Oct. 2008.

[4] S. Mahapatra, K. Ahmed, D. Varghese, A.E. Islam, G. Gupta, L. Madhav, D. Saha, and M.A. Alam, "On the Physical Mechanism of NBTI in Silicon Oxynitride p-MOSFETs: Can Differences in Insulator Processing Conditions Resolve the Interface Trap Generation versus Hole Trapping Controversy?," in *Proceedings of the 45th annual IEEE international Reliability physics symposium, 2007*, 2007, pp. 1–9.

[5] A.E. Islam, V.D. Maheta, H. Das, S. Mahapatra, and M.A. Alam, "Mobility degradation due to interface traps in plasma oxynitride PMOS devices," in *Proceedings of 2008 IEEE International Reliability Physics Symposium*, April 2008, pp. 87–96.

[6] M. Denais, A. Bravaix, V. Huard, C. Parthasarathy, C. Guerin, G. Ribes, F. Perrier, M. Mairy, and D. Roy, "Paradigm Shift for NBTI Characterization in Ultra-Scaled CMOS Technologies," in *Proceedings of 44th Annual IEEE International Reliability Physics Symposium, 2006*, 26-30 2006, pp. 735–736.

[7] Zhigang Ji, L. Lin, Jian Fu Zhang, B. Kaczer, and G. Groeseneken, "NBTI Lifetime Prediction and Kinetics at Operation Bias Based on Ultrafast Pulse Measurement," *IEEE Transactions on Electron Devices*, vol. 57, no. 1, pp. 228–237, Jan. 2010.

[8] V. Huard, M. Denais, and C. Parthasarathy, "NBTI degradation: From physical mechanisms to modelling," *Microelectronics and Reliability*, vol. 46, no. 1, pp. 1–23, 2006.

[9] J.F. Zhang, Z. Ji, M.H. Chang, B. Kaczer, and G. Groeseneken, "Real V_{th} instability of pMOSFETs under practical operation conditions," in *2007 IEEE International Electron Devices Meeting*, 10-12 2007, pp. 817–820.

[10] S. Deora, V.D. Maheta, A.E. Islam, M.A. Alam, and S. Mahapatra, "A Common Framework of NBTI Generation and Recovery in Plasma-Nitrided SiON p-MOSFETs," *IEEE Electron Device Letters*, vol. 30, no. 9, pp. 978–980, Sept. 2009.

[11] Jian F. Zhang and Mo H. Chang, "An Assessment of Mobility Variation during Negative Bias Temperature Stress," *ECS Transactions*, vol. 6, no. 3, pp. 301–311, 2007.

[12] A.E. Islam, E.N. Kumar, H. Das, S. Purawat, V. Maheta, H. Aono, E. Murakami, S. Mahapatra, and M.A. Alam, "Theory and Practice of On-the-fly and Ultra-fast VT Measurements for NBTI Degradation: Challenges and Opportunities," in *Proceedings of 2007 IEEE International Electron Devices Meeting*, Dec. 2007, pp. 805–808.

[13] T. Grasser, P.-J. Wagner, P. Hehenberger, W. Gos, and B. Kaczer, "A rigorous study of measurement techniques for negative bias temperature instability," in *Integrated Reliability Workshop Final Report, 2007. IRW 2007. IEEE International*, Oct. 2007, pp. 6–11.

[14] J. B. Yang, T. P. Chen, S. S. Tan, and L. Chan, "A Simple Negative Bias Temperature Instability Characterization Methodology to Minimize the Immediate Recovery Effect during Measurement," *Japanese Journal of Applied Physics*, vol. 45, no. 8A, pp. 6137–6140, 2006.

[15] Y.Z. Hu, D.S. Ang, and G.A. Du, "An improved methodology for monitoring NBTI induced threshold voltage shift of scaled p-MOSFETS," in *Proceedings of 2008 IEEE International Reliability Physics Symposium*, April 2008, pp. 743–744.

Analysis of recoverable and non-recoverable NBTI and PBTI using AC and DC stresses

Frederic MONSIEUR, Eduard CARTIER[+], James STATHIS[+]

STMicroelectronics, [+] IBM Research Division, Albany Nanotech, 257 Fuller Rd., Albany, NY 12203,
phone: 518-292-7232; fax: 518-292-7385; e-mail: frederic.monsieur@st.com / fmons@us.ibm.com

INTRODUCTION

Much effort continues to be devoted to the modeling of the NBTI and PBTI instabilities in MOSFETs. Most analyses are based on Eq. 1, which describes the threshold voltage shift, Δ, under uniform gate stress [1],

$$\Delta = \frac{\Delta_{max}}{1 + B \cdot \left(\frac{tr}{ts}\right)^q}. \qquad (1)$$

The parameter, Δmax, is the maximum Vt shift possible for a given stress voltage, Vs, and relaxation voltage, Vr. This parameter depends on the trap occupancy and the energy distribution of the bulk oxide traps. The parameter, B, acts as a time acceleration factor and modulates both, the stress time, ts, and the relaxation time, tr. Finally, the parameter, q, is a constant.

Other authors, following [2], fit the recovery behavior with a log-law according to,

$$\Delta(tr) = Ar - Br \cdot \ln(tr). \qquad (2)$$

The parameter, Br is known as the recovery rate, while Ar is the intercept which represents the Vt shift for $t_r = 1$ s.

This paper is dedicated to linking both Equations in order to extract important physical information from the basic experimental parameters Ar and Br. We will show that both NBTI and PBTI behave very similarly. They both show permanent and recoverable traps. The recovery rate is a good estimator of the amount of recoverable traps enabling to decouple smartly both kinds of traps.

EXPERIMENTAL

n- and p-channel metal gate MOSFETs with Hafnium-based high-k dielectrics were fabricated utilizing a conventional CMOS process flow. The DC data were collected during an uninterrupted relaxation phase that followed an uninterrupted stress phase. During the relaxation, $Idlin$ is sensed at the relaxation voltage and is then converted into a Vt shift by projecting $Idlin$ onto the fresh Id-Vg characteristics. For each stress voltage, Vs, the relaxation was measured for ts from 300 ms up to 100 ks.

DATA ANALYSIS

If the BTI data are parameterized in terms of Eq. 1, the local log-slope is

$$Br = -\frac{\partial \Delta}{\partial \ln(tr)} = q \cdot (\Delta_{max} - \Delta) \cdot \frac{\Delta}{\Delta_{max}}. \qquad (3)$$

Eq. 3 relates the recovery rate, Br, as defined in Eq. 2, to the local Vt shift, $\Delta = \Delta loc$ (Δloc is the average local Vt shift around which Br is fit to Eq. 3). The relation is a polynomial of the second order in Δ without a constant term. It is important to notice that B does not appear in Eq. 3 any longer. As a result, this new approach allows the investigation of the Δmax dependences independently of the time acceleration parameter, B. In Fig. 1, the recovery rate, Br, obtained by fitting experimental data to Eq. 2 is plotted as a function of Δloc. The functional dependence can be directly compared to Eq. 3. As predicted by Eq. 3,

for low Vt shift $\Delta loc \ll \Delta max$, a linear relation between Δloc and Br is obtained. The slope is equal to the parameter, q, independent of the stress voltage, Vs.

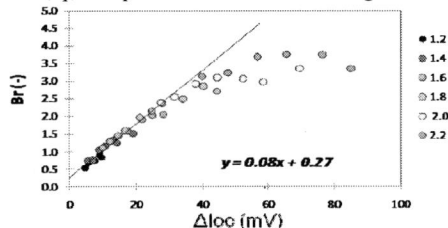

Figure 1. *Br-Δloc correlation plot (NBTI data, Vr=-0V5, 125C)*

Furthermore, when $\Delta loc = \Delta max/2$, Br reaches its maximum at $Br_{max} = q.\Delta max/4$, also supported by the data in Fig. 1, where a deviation from the linear dependence is found at high Vt shift. Finally, there is an indication in Fig. 1 that Br decreases for the larges Δloc values as expected by Eq. 3. The range near $\Delta loc = \Delta max$ where $Br = 0$ cannot be observed experimentally. Careful inspection of the data suggests that, Br_{max} depend on Vs, at high values of Δloc, implying that Δmax is Vs dependent.

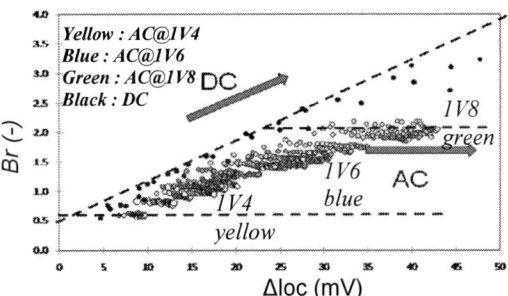

Figure 2. *Br-Δloc plot (NBTI data, Vr=-0V5, 125C). AC for 3 stress voltages is compared to the DC data from Fig. 1. Dashed line and arrows are guides.*

The same samples were also subjected to low frequency AC stresses ($f=10mHz$). In this case, the power-off part of each pulse is fit according to Eq. 2 and thus a local recovery rate is associated with a local Vt shift, pulse after pulse. The resulting Br-Δloc correlation is shown in Fig. 2 for 3 stress voltages. Data for various duty cycles from 1% to 99% is shown for each Vs. The DC data for $\Delta loc \ll \Delta max$ is included in Fig. 2 for comparison. As expected, the first pulse of each AC stress agrees with the DC data. However, if for the following pulses the Vt shift increases continuously, we show that the recovery rate remains constant or decreases slightly, in disagreement with Eq. 3. This behavior can be explained if a permanent Vt shift contributes to Δloc in addition to a recoverable component. The existence of a permanent component for NBTI is hardly a new concept and it was demonstrated several times that interface states contribute to the Vt shift and this contribution does not recover in the time window investigated experimentally [3]. The existence of a permanent component can also be seen when plotting the total Vt shift with respect to the ratio tr/ts as shown in Fig.

978-1-4244-9113-1/11 $26.00 © 2011 IEEE

3a. The recovery traces merge into a universal curve as expected from Eq. 1 only if a permanent component is subtracted from each trace, as illustrated in Fig. 3. The permanent component obtained by this method from DC data with a large range of ts and Vs values, behaves as described in Fig 3b and is independent of Vr. This result is consistent with different previous works [3-4].

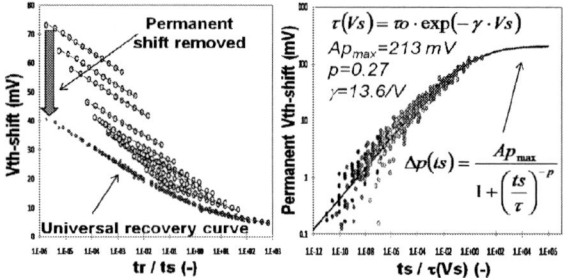

Figure 3. Left: Extraction procedure of the permanent component (NBTI data). Right: The permanent component extracted according to the left-plot procedure. Data are time-scaled in voltage.

When the data in Fig. 1 and Fig. 2, is corrected for the permanent component, the Br-Δloc correlation becomes universal as shown in Fig. 4a. To correct the AC data, the permanent component was calculated from the total elapsed stress time. As can be seen from Fig. 4b, the correction results in higher q value and leads to the correct intercept of the Br-Δloc correlation.

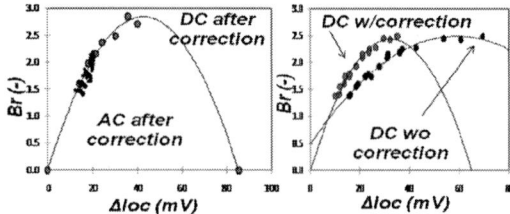

Figure 4. Br-Δloc plot for NBTI data (Vs=-1V8, Vr=-0V4, 125C). The left plot shows a comparison of AC and DC data after permanent shift correction. The right plot shows the same DC data w/ and w/o correction (Vs=-1V6, Vr=-0V4, 125C).

We now switch to PBTI data for which Br and Δloc can be extracted in the same way. Results are shown in Fig. 5 where Br-Δloc correlation is plotted for various Vs values.

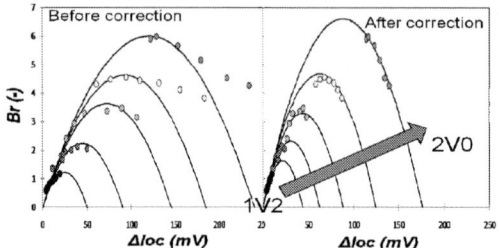

Figure 5. Br- Δloc plot for PBTI data (various Vs, Vr=+0V4, 125C). The left plot shows Δloc as measured and the right plot shows Δloc after permanent component correction.

Again, in Fig 5a, the data deviates from a second order polynomial and it cannot be fit by Eq. 3. However, if a permanent component is subtracted by the same method as used for NBTI, a good fit to Eq. 3 is obtained (Fig. 5b). The predicted decrease of the recovery rate is quite apparent in this PBTI dataset. Fitting the extracted PBTI permanent component as in Fig. 3 yields the parameters $p = 0.4$ and, $\gamma = 11.7$ per Volt. These numbers are quite

different from the values obtained for NBTI, showing that defects may be different from those responsible for the NBTI permanent component. For clarity, AC stress results are shown in Fig. 6 for one particular voltage and several duty cycles. Similar to the NBTI result, when Vt increases during AC stress, the recovery rate decreases in contradiction to Eq. 3. Again however, the AC and DC Vt shifts merge together after correction for the permanent component as demonstrated in Fig. 6.

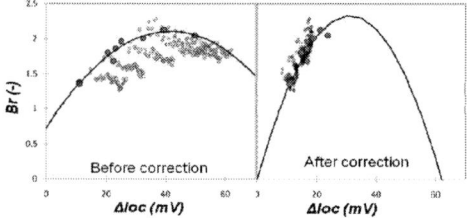

Figure 6. Focus on PBTI data $Vs=1V$, $Vr=+0V4$, 125C. AC and DC comparison without (left) and with (right) permanent shift correction.

DISCUSSION

We have shown that the Vt shift for NBTI and for PBTI is composed of a permanent and a recoverable component. The permanent component was extracted with the assumption of a symmetrical trapping-detrapping process according to Eq 1. This is strongly supported by Fig. 7 where B, the parameter as defined in Equ. 1, does not depend on Vr as simulated in [5].

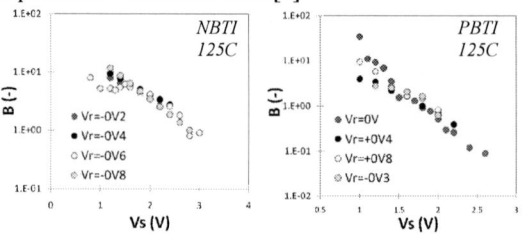

Figure 7. B parameter after Equ. 1 plotted with respect to Vs for various Vr. NBTI shown on the left and PBTI data shown on the right.

For NBTI, the permanent component is attributed to interface states but for PBTI permanent traps are still to identify. For all conditions studied, the recovery rate was shown to be correlated to the magnitude of the recoverable Vt shift for both NBTI and PBTI. The recovery rate is the signature of recoverable traps and is the best indicator of the recoverable Vt. During low frequency AC stress permanent traps become greater and it may dominate the Vt shift at end of life. In any case, the amount of recoverable traps is capped by Δmax. If at nominal voltage, Δmax is below technology specification (which is likely according to Fig. 5) then permanent traps would consequently drive the product end of life.

AKNOWLEDGEMENTS

This work was performed by the Research Alliance Teams at various IBM Research and Development Facilities.

REFERENCES

[1] T. Grasser, IEEE IRPS 2007. [2] Tewksbury, IEEE Journal of solid-State Circuit, Vol.29, No.3 [3] V. Huard, IEEE IRPS 2010 [4] A. McMahon, IEEE IRPS 2001. [5] T.Grasser, IEEE IRPS 2009

On the Evolution of the Recoverable Component of the SiON, HfSiON and HfO$_2$ P-MOSFETs under Dynamic NBTI

Y. Gao, A. A. Boo, Z. Q. Teo, and D. S. Ang[*]

School of Electrical and Electronic Engineering
Nanyang Technological University
Singapore 639798
[*]E-mail : edsang@ntu.edu.sg

Abstract—**The evolution of the recoverable (R) component of negative-bias temperature instability (NBTI) is examined, as a function of the number of stress and relaxation cycles, for the SiON, HfSiON, and HfO$_2$ p-MOSFETs. At typical NBTI oxide fields (~7 MV/cm), a steady and substantial *decrease* of the R component in the case of the HfO$_2$ p-MOSFET is observed, while the R component of the SiON and HfSiON p-MOSFETs are found to remain constant. A decrease in the R component of the SiON and HfSiON p-MOSFETs is observed only at much higher oxide fields (> 10 MV/cm). Evidence shows that the decrease in the R component is due to a greater tendency for the hole traps in the HfO$_2$ to be transformed into a permanent form (P) under a given oxide field. The result therefore implies that, under typical NBTI oxide fields, the R and P components could share a common defect origin in the case of the HfO$_2$ p-MOSFET. On the other hand, the R and P components are likely to have originated from different defect precursors in the case of the SiON and HfSiON p-MOSFETs. The existence of different oxide fields at which the transformation of the R component into a permanent form occurs for different gate dielectrics implies that the nature of the defect precursors responsible for the R component is *intrinsic* to the gate dielectric material.**

Keywords — Bias temperature instability, recovery, ultra-fast measurement, pulsed I-V, hole trapping, interface states

I. INTRODUCTION

It has been commonly shown that NBTI of the oxynitride (SiON) gate p-MOSFET is made up of a recoverable (R) and a relatively permanent (P) component [1], [2]. There has been considerable attention on the R component as its transient nature could pose significant challenges to both device and circuit reliability assessment and modeling. Recently, it was shown that the R component remains constant, regardless of the number of times the p-MOSFET was stressed and relaxed under typical NBTI oxide fields (~7 MV/cm) and temperatures (~100 °C) [3]-[5]. This behavior is inconsistent with the notion that dynamic NBTI is driven by hydrogen diffusion[1] [6], [7].

This is because even during relaxation, some hydrogen still diffuses away from the Si/SiO$_2$ interface. Thus, in the context of the hydrogen diffusion model [6], [7] the R component should decrease gradually as the p-MOSFET is being repeatedly stressed and relaxed. But this is not observed experimentally for the SiO$_2$ and SiON gate dielectrics [3]-[5]. In fact, apart from the R component remaining constant, it is also found to exhibit a cyclic behavior, i.e. the amount of parametric shift which is recovered during a relaxation cycle is nearly equal to the amount generated by the prior stress cycle. This behavior implies that the repetitive charging and discharging of oxide (hole) traps are responsible for the R component [3]-[5], [8], [9]. Studies based upon electron spin resonance measurement confirmed that these "switching" hole traps are related to oxygen vacancy defects in the gate dielectric [10], [11].

In this work, we examine the evolution of the R component under dynamic NBTI for p-MOSFETs employing the HfO$_2$ and HfSiON gate dielectrics, in addition to the SiON gate dielectric. There have been limited reports on this aspect for the high-κ gate stack despite the importance of this technology for future CMOS applications. Under typical NBTI oxide fields (~7 MV/cm), the R component of the HfO$_2$ p-MOSFET is found to decrease gradually with repeated stressing and relaxation. Evidence shows that the decrease is a consequence of the transformation of the hole traps into a more permanent form. This behavior is unlike those of the SiON and HfSiON p-MOSFETs where the R component remains constant. A decrease in the R component of the SiON and HfSiON p-MOSFETs is, however, also observed at much higher oxide fields (> 10 MV/cm). Implications of these new experimental observations are discussed.

II. EXPERIMENTAL DETAILS

Fig. 1 is the schematic illustration of the ultra-fast switching (UFS) measurement set-up used in our study. The method is based on the commercial hardware for the pulsed current-voltage measurement, which incorporates two bias-tees,

---

[1] It is also recognized that detrapping of the holes trapped in the oxide traps during stressing could give rise to a substantial R component, in addition to the presumed defect annealing mechanism driven by hydrogen diffusion. But

some authors have regarded the former as an extrinsic mechanism and argued that it should not be analyzed as part of NBTI (see later discussion).

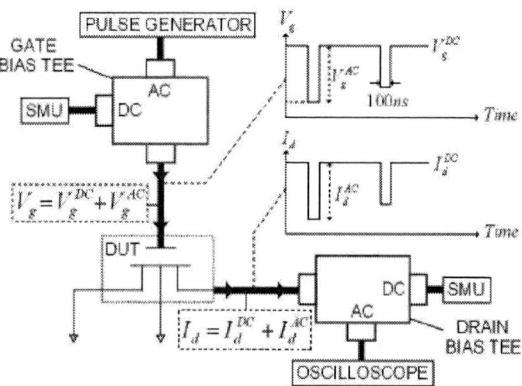

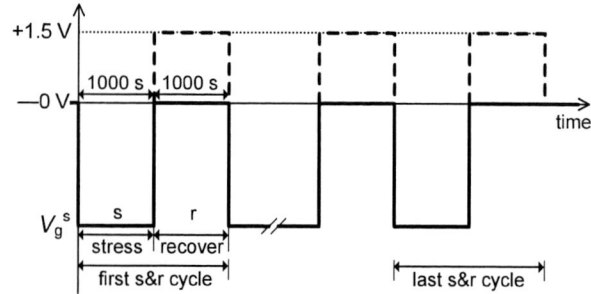

Fig. 1. Schematic diagram of the UFS measurement set-up. The bias-tee at the gate terminal couples the dc input (the stress voltage V_g^s during the stress stage) and the ac input (a narrow positive pulse from the pulse generator to bring the gate voltage to the measurement level). The bias-tee at the drain terminal decouples the ac drain current (due to the gate pulse) from the dc current (drain current corresponding to the gate stress voltage V_g^s).

Fig. 2. Schematic diagram of the gate voltage waveform during the NBTI stress and relaxation (s&r) cycles. The p-MOSFET was stressed at V_g^s and was recovered either at a 0 V or +1.5 V (dashed line) gate voltage.

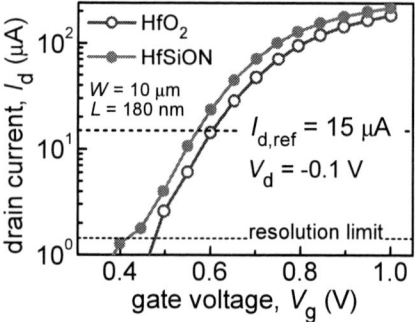

Fig. 3. Examples of linear I_d-V_g curves measured at a selected interval of the stress/relax cycling experiment, using the pulsed V_g method [12], [13]. The good current resolution (~1 μA) of the measurement set-up enabled part of the subthreshold I_d regime to be accurately measured. The threshold voltage was extracted using the constant subthreshold I_d method ($I_{d,ref}$ = 15 μA).

one each at the gate and drain terminal. At the gate terminal, the bias-tee couples the dc and ac voltages applied at the respective inputs. The dc input (V_g^{DC}) was set to either the stress voltage or the recovery voltage (cf. Fig. 2). The ac input was connected to a pulse generator which functioned only during the measurement. At specific intervals, the linear drain current versus gate voltage (I_d-V_g) curve was measured by pulsing the gate voltage from the stress level to different measurement voltage levels. At the drain terminal, the bias-tee decoupled the dc and ac components in the I_d. The dc component corresponded to the gate stress voltage V_g^s and the ac component was the change in I_d arising from the gate-pulse injection. The latter was measured by a high-speed oscilloscope. Each measurement pulse lasted a short duration of 100 ns [12]. Repeated pulsing of the gate to a given voltage was made, at a duty cycle of 0.1%, to reduce measurement noise in the drain current. This was shown to have negligible cumulative recovery effect during the entire I_d-V_g measurement process, which lasted ~2 s [13].

The waveform of the dc gate input, V_g^{DC}, during the stress and relaxation cycles is shown in Fig. 2. The stress and relaxation cycles lasted 1000 s each. The dc drain voltage was held at −0.1 V constantly throughout. During relaxation, the gate voltage was set to 0 V unless stated otherwise. Examples of the linear I_d-V_g curves measured via our UFS method are shown in Fig. 3. The resolution of the UFS method is 1 μA. Threshold voltage shift $|\Delta V_t|$ was extracted using the constant subthreshold drain current ($I_{d,ref}$ = 15 μA) method. The time-zero reference V_t was measured before applying the stress.

We tested p-MOSFETs with three different gate dielectrics: 1) a 1.7 nm SiON formed via conventional decoupled plasma nitridation; 2) a 2-nm HfSiON (formed via metal organic chemical vapor deposition) on a 1-nm chemical oxide; 3) a 4-nm HfO$_2$ (formed via atomic layer deposition) on a 1-nm chemical oxide. The temperature was set at 100 °C and the gate stress voltage was adjusted to achieve similar oxide fields, taking into account the different physical thicknesses of the gate stacks.

III. RESULTS AND DISCUSSION

Fig. 4(a) depicts the evolution of $|\Delta V_t|$ under repeated stress and relaxation cycling, for the HfSiON and HfO$_2$ p-MOSFETs. The amount of $|\Delta V_t|$ recovery per cycle or the R component, determined as the difference between $|\Delta V_t|$ at the end of relaxation ($|\Delta V_t|^{eor}$) and that at the end of the prior stress cycle ($|\Delta V_t|^{eos}$; Fig. 4(b)) are shown in Fig. 5(a), as a function of the number of stress/relaxation cycles, N. While the R component of the HfSiON p-MOSFET remains constant throughout the 30 stress/relaxation cycles, in agreement with the observation for the SiO$_2$ and SiON p-MOSFETs [3]-[5], a progressive decrease, by up to ~35 % of the initial value at N = 1 (Fig. 5(b)), is evident for the case of the HfO$_2$ p-MOSFET. It should be emphasized that the decrease of the R component is not unique to the HfO$_2$ p-MOSFET used in this study; a similar decrease is also observed for HfO$_2$ p-MOSFETs from other sources (not shown). Also shown in Fig. 5(b) is the constant R component of the SiON p-MOSFET [4], [5].

Based on the result of the HfO$_2$ p-MOSFET *alone*, the gradually decreasing R seems to agree with the hydrogen diffusion model, as discussed above. However, one needs to be cautious about making such an inference since this trend is generally not observed for the SiO$_2$ [3], SiON [4], [5] and HfSiON devices, for which the R component remains constant. A possible argument for the discrepancy is that the

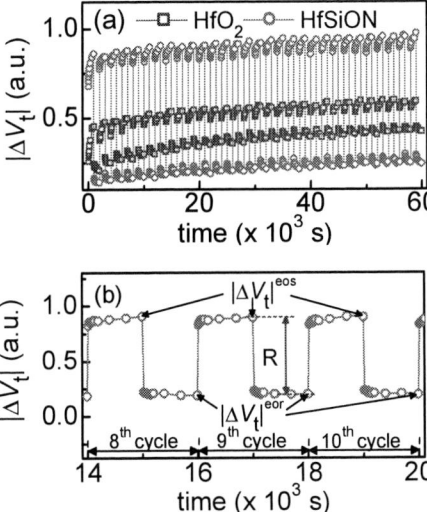

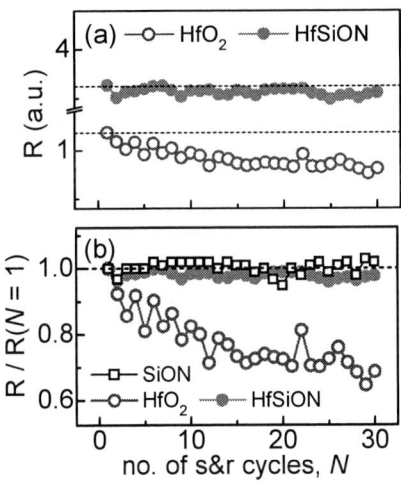

Fig. 4. (a) Evolution of threshold voltage shift $|\Delta V_t|$ of the HfO$_2$ and HfSiON p-MOSFETs under stress/relax cycling. The stress and recovery cycles are each 1 × 10^3 s. The oxide stress field was −8.5 MV/cm. (b) Zoom-in view of specific cycles showing the definition of $|\Delta V_t|^{eos}$ and $|\Delta V_t|^{eor}$ at the end of the stress and recovery cycle, respectively. The recoverable component R per cycle is the difference between $|\Delta V_t|^{eor}$ and $|\Delta V_t|^{eos}$ of the preceding stress cycle.

Fig. 5. (a) Evolution of the recoverable (R) component as a function of stress/relax cycling. (b) The R component is normalized to that of the first relax cycle, showing a substantial decrease of ~35 % in the case of the HfO$_2$ p-MOSFET after 30 stress/relax cycles. Also shown in (b) is the data for the SiON p-MOSFET, exhibiting a constant normalized R component throughout.

hydrogen diffusion effect may be particularly strong in the case of the HfO$_2$ p-MOSFET such that its effect on R is more readily observed compared to the other devices. But this argument is not supported by the further analysis of the data shown below. To probe the reason behind the reduction in the R component of the HfO$_2$ p-MOSFET, we examine the increase in $|\Delta V_t|^{eos}$ or $\delta|\Delta V_t|^{eos}$, and the increase in $|\Delta V_t|^{eor}$ or $\delta|\Delta V_t|^{eor}$, measured with respect to the $|\Delta V_t|^{eos}$ and $|\Delta V_t|^{eor}$ of the $N = 1$ cycle, respectively, as a function of N in Fig. 6. In the context of the hydrogen diffusion model, the increase in the

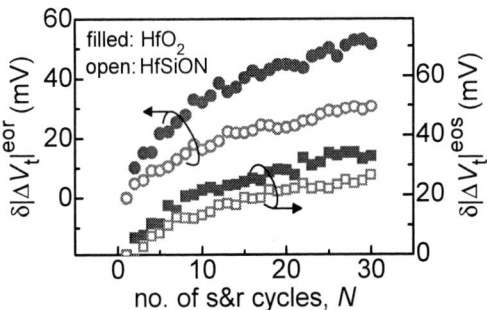

Fig. 6. Increase in $|\Delta V_t|^{eos}$ and $|\Delta V_t|^{eor}$, denoted as $\delta|\Delta V_t|^{eos}$ and $\delta|\Delta V_t|^{eor}$, respectively, measured with respect to the $|\Delta V_t|^{eos}$ and $|\Delta V_t|^{eor}$ of the first stress/relax cycle. The $\delta|\Delta V_t|^{eos}$ of both the HfO$_2$ and HfSiON p-MOSFETs (right axis) are comparable, i.e. the $|\Delta V_t|^{eos}$ (= R + P) in both cases increase at approximately the same rate. The much larger increase in $\delta|\Delta V_t|^{eor}$ of the HfO$_2$ p-MOSFET (left axis) indicates unambiguously that the decrease in the R component (cf. Fig. 4) is a result of its transformation into a more permanent form (P).

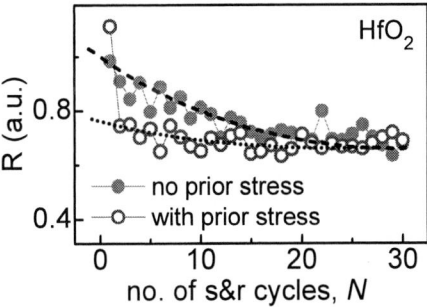

Fig. 7. A *prior* continuous stressing of the HfO$_2$ p-MOSFET at the same gate voltage for 2 × 10^4 s is observed to reduce the rate of decrease of R in a subsequent stress/relax cycling experiment. This confirms that during the prior stressing stage, a substantial part of R is already transformed into P, and hence the decrease in R during the subsequent stress/relax cycling is less pronounced than the case when no prior stressing is carried out.

interface state density is controlled by the speed at which hydrogen species diffuse away from the interface under a given oxide field. Hole trapping in a direct tunneling gate dielectric, on the other hand, is presumed to saturate rapidly and remains a constant. Subtracting $|\Delta V_t|^{eos}$ of the first stress cycle from the rest should thus yield an incremental change $\delta|\Delta V_t|^{eos}$ that is sensitive to the hydrogen diffusion speed if this is indeed the mechanism that drives the long term evolution of $|\Delta V_t|$. Fig. 6 shows that the $\delta|\Delta V_t|^{eos}$ of the HfSiON and HfO$_2$ p-MOSFETs both increase at similar rates. If the hydrogen diffusion model were valid, the speed of hydrogen diffusion must necessarily be comparable for both devices. But this inference is unable to explain the apparent difference in the behavior of the two R components (cf. Fig. 5(b)). The evidence thus does not support the argument that the decrease in R is due to a stronger hydrogen diffusion effect in the HfO$_2$ p-MOSFET.

On the other hand, $\delta|\Delta V_t|^{eor}$ of the HfO$_2$ p-MOSFET rises more significantly than that of the HfSiON p-MOSFET. Given that the $\delta|\Delta V_t|^{eos}$ of both devices increase at comparable rates, this observation implies that the decrease in the R component of the HfO$_2$ p-MOSFET is because of its transformation into a more permanent form. In other words, the switching hole traps in the HfO_2 p-MOSFET have a greater tendency to be rendered permanent, as compared to those in the HfSiON counterpart, under a given stress condition. To further check this inference, a HfO_2 p-MOSFET was stressed continuously for 2×10^4 s before it was subjected to the repetitive stress/relaxation cycling test depicted in Fig. 2. In this case, the decrease in R for the latter is suppressed compared to the case without the prior continuous stress (Fig. 7). This observation corroborates our inference that part of the hole traps in the HfO_2 p-MOSFET is gradually transformed into a more permanent form during stressing.

NBTI studies on the SiON gate p-MOSFET has shown that a fraction of the hole traps responsible for the V_t shift have relatively deep energy levels [14], [15]. It is believed that the trap energy levels are situated above the intrinsic Fermi energy or even the conduction band edge E_C of the n-Si such that these hole traps could not readily discharged under a 0-V gate bias via electronic charge exchange with the substrate. As their emission time could be very long, these deep-level hole traps would manifest themselves as a relatively permanent V_t shift. But their influence on the V_t shift could be revealed by a positive gate voltage, which lowers the trap energy levels below the E_C of the n-Si, thus allowing them to be spontaneously neutralized by electron injection from the substrate. It is believed that the trapped hole sites are subsequently annihilated through bond reformation [8], [9] as they are no longer observed electrically upon returning the gate voltage to 0-V. To investigate whether the R to P transformation of the HfO_2 p-MOSFET seen in Fig. 5 is due to a possible shift in the hole trap energy levels from shallow to deep, the HfO_2 p-MOSFET was subjected to a positive gate voltage (+1.5 V) during the relaxation phase and its effect on the evolution of R is recorded in Fig. 8. Compared to the 0-V case, the R component is clearly increased. This behavior is similar to that reported for the SiON p-MOSFET and could be explained by the annihilation of deeper hole trap states [14], [15]. However, the difference between the R components of the 0-V and +1.5-V recovery is constant, independent of the

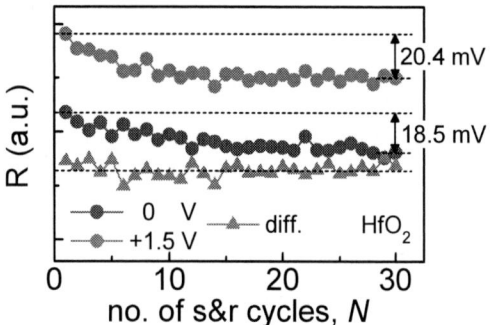

Fig. 8. Comparison of the R component under a 0-V and a +1.5-V gate voltage during the relaxation phase. The gate stress voltage was kept the same. Although the R component is increased in the latter, the amount of decrease is similar — the difference in the R component of the two cases is constant throughout the whole stress/relax cycling experiment. These observations show that the decrease in R for the case of the 0-V recovery is not a consequence of a portion of the positive trapped charge being converted into deeper trap states.

number of stress and relaxation cycles. This observation implies that in both cases, the R component is decreasing at the same rate, i.e. the reduction is not because of an increase in the density of deep-level hole traps. The results hence indicate that the defects have already been transformed into a relatively permanent form during stress and the process is not reversible by a positive gate recovery voltage.

Fig. 9(a) shows that the R component of the HfO_2 p-MOSFET decreases more significantly at a higher oxide stress field. This dependence of the R component on the oxide stress field prompted further study of the SiON and HfSiON p-MOSFETs at higher oxide stress field. The R component of the HfSiON p-MOSFET is found to decrease by ~8 mV after 30 stress and relaxation cycles at an oxide field of 10 MV/cm, while no change could be observed at lower oxide fields (Fig. 9(b)). A similar decrease in the R component is apparent for the SiON p-MOSFET (Fig. 10(a)). In this case, the p-MOSFET was not stressed at the large negative oxide field for 1000 s, as doing so would cause the ultra-thin gate oxide to breakdown before the stress period was completed. To examine the effect of a large negative voltage on the evolution of the R component, each NBTI stress and relaxation cycle (except for the very first) was preceded by a large negative gate voltage stressing lasting a short time (50 s). The impact of the large negative oxide field stressing was checked by measuring the R component of an ensuing NBTI stress and relaxation cycle. The sequence was then repeated several times. Just as in the case of the HfO_2 p-MOSFET, the decrease in the R component (square symbol in Fig. 10(a)) corresponds to a faster rise in $|\Delta V_t|^{eor}$ or the P component in relation to $|\Delta V_t|^{eos} - R(N = 1)$ (Fig. 10(b)). This observation is again indicative of the fact that the R component has transformed into a more permanent form. More importantly, the oxide stress field at which such a decrease becomes apparent is found to be highest for the SiON p-MOSFET. This is followed by the HfSiON p-MOSFETs and lastly the HfO_2 p-MOSFET.

Some authors have presumed that the R component arises from the trapping and detrapping of holes at *preexisting bulk (oxide) traps* [16], [17]. A constant R component and its cyclic behavior observed in our earlier studies on the dynamic NBTI of the SiON p-MOSFET [3]-[5] in fact appear to corroborate this presumption. A caution that the phrase "preexisting oxide traps" could be misleading as it tends to imply an extrinsic mechanism. The decrease in the R component, particularly prominent at high negative oxide field, runs contrary to the presumption. It is widely known that a high oxide field results in the generation of more oxide traps. If the presumption were correct, one should observe a larger R component when the density of the oxide traps is increased. The transformation of R to P implies that the former arises from *transient electronic traps induced by the applied gate stress voltage.* Moreover, the different oxide fields at which the transformation of R to P occurs for different gate dielectrics implies that the defect precursors responsible for the R component are intrinsic to the type of gate oxide materials. These observations indicate that the R component is an intrinsic part of NBTI.

Based on recent *ab-initio* simulation studies on the oxygen vacancy defect V_O ($\equiv$Si–Si$\equiv$) [18], [19], we provide a plausible physical framework for the transformation of R to P. The tendency for R to transform into P is believed to be strongly dependent on the degree by which V_O structurally relaxes (i.e. by how much the distance between the two Si atoms increases) after the capture of a hole. From a simulation study of the growth of the SiO_2 using molecular dynamics, Nicklaw *et al.* [18] showed that a majority (~90 %) of V_O's exhibit little structural relaxation upon the capture of a hole. In fact, the increase in the Si–Si distance is so marginal that the remaining unpaired spin is still equally shared between the two Si atoms [18]. This makes it possible for the Si–Si bond to spontaneously reform once an electron is recaptured, thus annihilating the trapped hole site after the stress is terminated. For V_O's which undergo more significant structural relaxation, bond reformation may not be possible and these trapped hole sites would manifest as a permanent V_t shift. But these V_O's constitute a minor fraction of the total defect density. A higher oxide field would increase the distortion in the oxide network, promoting greater structural relaxation of the trapped hole upon the capture of holes. Thus, the decrease in the R component becomes more apparent only at high oxide fields. The fact that the R component remains constant while the P component increases substantially under nominal NBTI oxide field implies that the two have different origins. This inference is in accordance to earlier temperature dependence study of the SiO_2 and SiON p-MOSFETs, showing different activation energies for the R and P components [3], [20], [21]. The latter is generally ascribed to Si dangling bond defects or the P_b centers [22].

Unlike the SiO_2 or the SiON, the greater ionic character of the HfO_2 tends to promote the structural relaxation of V_O's after the capture of holes [19]. This explains (i) a larger extent of R to P transformation at a given oxide field and (ii) a lower

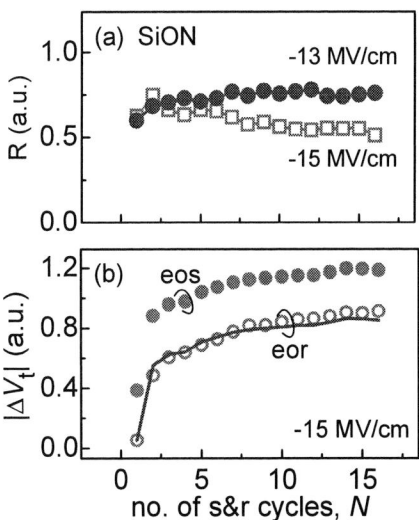

Fig. 10. (a) Evolution of the R component of the SiON p-MOSFET. A high negative oxide-field stress (value indicated) lasting 50 s was inserted before each stress/relax cycle, except for the first. The gate voltage was −1.8 V and +1.5 V during the stress and recovery cycle, respectively. (b) Evolution of $|\Delta V_t|^{eos}$ and $|\Delta V_t|^{eor}$. The solid line is obtained by subtracting from $|\Delta V_t|^{eos}$ the R component of the very first relax cycle, i.e. it depicts the evolution of $|\Delta V_t|^{eor}$ under the assumption of a constant R component.

oxide field for the onset of R to P transformation (Figs. 5 and 9). The transformation means that in the HfO_2 p-MOSFET, both the R and P components share a common defect origin. The faster rise in the P component as a result of this transformation (Fig. 6) also poses a challenge to the long-term NBTI reliability of the device.

IV. SUMMARY

A detailed study on the evolution of the R component of NBTI under repetitive stress and relaxation is performed for the HfSiON and HfO_2 p-MOSFETs. While the R component of the former remains constant, independent of the number of stress and relaxation cycles – just like the SiON counterpart – it was observed to progressively decrease for the HfO_2 p-MOSFET under nominal NBTI oxide fields. The R component of the SiON and HfSiON p-MOSFETs are also observed to decrease, but this happens only at much higher oxide fields (> 10 MV/cm). Evidence clearly shows that the decrease in the R component is a result of its transformation into a more permanent form. An explanation involving structural relaxation of the trapped hole sites was proposed as the underlying mechanism for the transformation. The greater tendency for structural relaxation of the trapped holes sites to occur in the HfO_2 (due to its greater ionicity) is believed to account for the more apparent R to P transformation at lower oxide fields.

ACKNOWLEDGEMENT

This work is supported in part by a Singapore Ministry of Education research grant MOE2009-T2-1-050. Y. Gao would like to thank NTU for the award of a Ph.D. scholarship. A. A. Boo and Z. Q. Teo would like to thank the Singapore

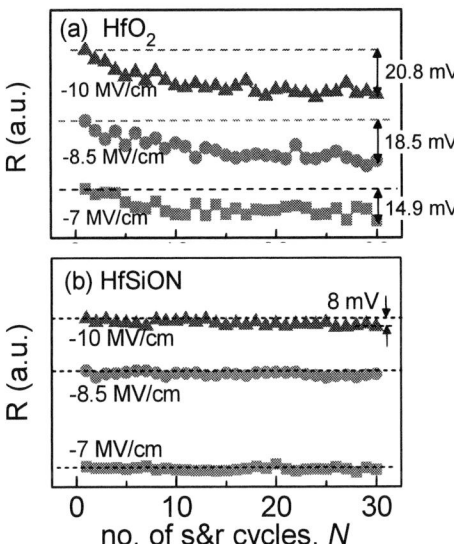

Fig. 9. A greater reduction in the R component is observed at a higher oxide stress field. After accounting for EOT difference, the decrease in the R component of the HfSiON p-MOSFET is ~11 mV for an oxide stress field of 10 MV/cm. The much smaller decrease in the R component in this case implies that the oxide field at which the transformation of R to P occurs is higher for the HfSiON p-MOSFET.

978-1-4244-9113-1/11 $26.00 © 2011 IEEE

Economic Development Board and GLOBALFOUNDRIES Singapore for a joint Ph.D. scholarship grant.

REFERENCES

[1] C. Shen, M.-F. Li, C. E. Foo, T. Yang, D. M. Huang, A. Yap, G. S. Samudra, and Y.-C. Yeo, "Characterization and physical origin of fast V_{th} transient in NBTI of pMOSFETs with SiON dielectric," in *IEDM Tech. Dig.*, 2006, pp. 333-336.

[2] T. Grasser, B. Kaczer, P. Hehenberger, W. Gos, R. O'Connor, H. Reisinger, W. Gustin, and C. Schunder, "Simultaneous extraction of recoverable and permanent components contributing to bias-temperature instability," in *IEDM Tech. Dig.*, 2007, pp. 801-804.

[3] Z. Q. Teo, D. S. Ang, and K. S. See, "Can the reaction-diffusion model explain generation and recovery of interface states contributing to NBTI?" in *IEDM Tech. Dig.*, 2009, pp. 737–740.

[4] Z. Q. Teo, D. S. Ang, and C. M. Ng, ""Non-hydrogen-transport" characteristics of dynamic negative-bias temperature instability," *IEEE Electron Dev. Lett.*, vol. 31, pp. 269-271, Apr. 2010.

[5] Z. Q. Teo, D. S. Ang, and C. M. Ng, "Separation of hole trapping and interface-state generation by ultrafast measurement on dynamic negative-bias temperature instability," *IEEE Electron Dev. Lett.*, vol. 31, pp. 656–658, July 2010.

[6] K. O. Jeppson and C. M. Svensson, "Negative bias stress of MOS devices at high electric fields and degradation of MNOS devices," *J. Appl. Phys.*, vol. 48, pp. 2004-2014, May 1977.

[7] M. A. Alam, "A critical examination of the mechanics of dynamic NBTI for PMOSFETs," in *IEDM Tech. Dig.*, 2003, pp. 345-348.

[8] A. J. Lelis and T. R. Oldham, "Time dependence of switching oxide traps," *IEEE Trans. Nucl. Sci.*, vol. 41, pp. 1835-1843, Dec. 1994.

[9] T. Grasser, B. Kaczer, W. Goes, Th. Aichinger, Ph. Hehenberger, and M. Nelhiebel, "A two-stage model for negative bias temperature instability," in *Proc. IRPS*, 2009, pp. 33-44.

[10] J. P. Campbell, P. M. Lenahan, A. T. Krishnan, and S. Krishnan, "Identification of the atomic-scale defects involved in the negative bias temperature instability in plasma nitrided p-channel metal-oxide-silicon field-effect transistors," *J. Appl. Phys.*, vol. 103, art. no. 044505, Feb. 2008.

[11] J. T. Ryan, P. M. Lenahan, T. Grasser, and H. Enichlmair, "Recovery-free electron spin resonance observations of NBTI degradation," in *Proc. IRPS*, 2010, pp. 43-49.

[12] G. A. Du, D. S. Ang, Z. Q. Teo, and Y. Z. Hu, "Ultrafast measurement on NBTI," *IEEE Electron Dev. Lett.*, vol. 30, pp. 275–277, Mar. 2009.

[13] Y. Z. Hu, D. S. Ang, and Z. Q. Teo, "Threshold voltage and mobility extraction by ultrafast switching measurement on NBTI," *IEEE Trans. Electron Dev.*, vol. 57, pp. 2027–2031, Aug. 2010.

[14] D. S. Ang and S. Wang, "Recovery of the NBTI-stressed ultrathin gate p-MOSFET: The role of deep-level hole traps," *IEEE Electron Dev. Lett.*, vol. 27, pp. 914-916, Nov. 2006.

[15] D. S. Ang, S. Wang, G. A. Du, and Y. Z. Hu, "A consistent deep-level hole trapping model for negative bias temperature instability", *IEEE Trans. Dev. & Mat. Reliab.*, vol. 8, pp. 22–34, Mar. 2008.

[16] J. H. Lee, W. H. Wu, A. E. Islam, M. A. Alam, and A. S. Oates, "Separation method of hole trapping and interface trap generation and their roles in NBTI reaction –diffusion model," in *Proc. IRPS*, 2008, pp. 745-746.

[17] G. Kapila, N. Goyal, V. D. Maheta, C. Olsen, K. Ahmed, and S. Mahapatra, "A comprehensive study of flicker noise in plasma nitrided SiON p-MOSFETs: Process dependence of pre-existing and NBTI stress generated trap profiles," in *IEDM Tech. Dig.*, 2008, pp. 103-106.

[18] C. J. Nicklaw, Z.-Y. Lu, D. M. Fleetwood, R. D. Schrimpf, and S. T. Pantelides, "The structure, porperties and dynamics of oxygen vacancies in amorphous SiO_2," *IEEE Trans. Nucl. Sci.*, vol. 49, pp. 2667-2673, Dec. 2002.

[19] Y. P. Feng, A. T. L. Lim, and M. F. Li, "Negative-U property of oxygen vacancy in cubic HfO_2," *Appl. Phys. Lett.*, vol. 87, art. no. 062105, Aug. 2005.

[20] D. S. Ang, S. Wang, and C. H. Ling, "Evidence of two distinct degradation mechanisms from temperature dependence of negative bias stressing of the ultra-thin gate P-MOSFET," *IEEE Electron Dev. Lett.*, vol. 26, no. 12, pp. 906-908, Dec. 2005.

[21] D. S. Ang and S. Wang, "On the non-Arrhenius behavior of negative-bias temperature instability," *Appl. Phys. Lett.*, vol. 88, no. 9, art. no. 093506, Feb. 2006.

[22] J. P. Campbell, P. M. Lenahan, A. T. Krishnan, and S. Krishnan, "Direct observation of the structure of defect centers involved in the negative bias temperature instability," *Appl. Phys. Lett.*, vol. 87, art. no. 204106, Nov. 2005.

New Observations on the Physical Mechanism of Vth-Variation in Nanoscale CMOS Devices After Long Term Stress

E. R. Hsieh[1], Steve S. Chung[1], C. H. Tsai[2], R. M. Huang[2], C. T. Tsai[2], and C. W. Liang[2]

[1]*Department of Electronics Engineering, National Chiao Tung University, Taiwan* [2]*United Microelectronics Corporation (UMC), Taiwan*

Abstract—**A new effect, called random trap fluctuation(RTF), is proposed to study the impact of hot carrier stress on the device variability. It was found that not only the popular random dopant fluctuation (RDF), but also the traps, caused by the HC stress or FN-stress, induce the V$_{th}$ variation. After the FN stress, it was found that V$_{th}$ variation is worse in pMOSFETs due to stress-induced interface traps. While, under the HC stress, different Vth variations were found for nMOSFETs and pMOSFETs. The V$_{th}$ variation is enhanced in pMOSFETs due to RTF and reduced in nMOSFET as a result of the Trap Blocking Effect (TBE). RTF in pMOSFET might be the dominant factor of CMOS reliability for future generations.**

Keywords-hot carrier effect, random dopant fluctuation, random trap fluctuation

I. INTRODUCTION

Moore's Law has driven CMOS devices scaling for several decades. However, the scaling limits continue to be a great challenge. One of the most significant issues in the scaling is the variability induced by the process and device design [1-4], especially the random dopant fluctuation(RDF)[1] which is the major source of V$_{th}$ variation. To solve RDF, carbon co-implant [5], FDSOI or FinFET with undoped or lighter channel [6-8], and gate dielectric with high-k material [9] have been proposed to be effective. However, so far, only RDF induced V$_{th}$ variation has been reported. Rare has been reported for another similar effect, the **random trap fluctuation** (RTF), caused by the HC stress. As a consequence, we are interested in finding the correlation between hot carrier effect and variability.

The basis of the V$_{th}$ variation can follow the Takeuchi plot in Table 1 [10], i.e.,

$$\sigma V_{th} = B_{VT} \sqrt{T_{inv}(V_{th} + 0.1V)/LW}. \quad (1)$$

The slope, B$_{VT}$, is an indicator of the V$_{th}$ variation. Moreover, to explain the trap-induced V$_{th}$ variation, two sophisticated profiling techniques, L^2-GD[11] and IFCP[12] have also been demonstrated.

II. DEVICE PREPARATION

The technology node and EOT (SiON) of planar CMOS devices are 40nm and 11^{0}A respectively. V$_{th}$ is measured in several areas of the device to gather the statistics. The values of V$_{th}$ were determined by the G$_{m,max}$ method.

III. RESULTS AND DISCUSSION

A. Evolution of Vth Variation After FN stress

Figs.1 and 2 show the cumulative probability of fresh and FN-stressed devices respectively. The stressed one exhibits a broader distribution of V$_{th}$. Fig. 3 depicts the Takeuchi pot for

fresh nMOSFETs and pMOSFETs. The B$_{VT}$ of pMOSFETs is much smaller than that of nMOSFETs. Thus, V$_{th}$ variation for fresh nMOSFETs is more serious caused by the RDF. However, after the FN stress, Figs. 4 and 5, the changes of B$_{VT}$ for nMOSFETs and pMOSFETs are reversed; specifically, pMOSFET demonstrated a much larger increase of B$_{VT}$. Fig. 6 describes the mechanism, in which the green balls are the interface traps, and the blue arrows are the carrier-transporting paths. The individual carrier-transporting path will be scattered by the different numbers of interface-traps depending on how many traps the carrier pass through in terms of the random distribution of interface traps. As a result, the V$_{th}$ variation becomes worse for both devices after the FN-stress. Furthermore, in Figs. 4 and 5, we found the increasing rate of B$_{VT}$ for pMOSFETs is faster than that for nMOSFETs and assume it is related to interface traps. The traps extracted by charge pumping measurement (Figs. 7) have been used to validate the results. Fig. 8 shows that *B$_{VT}$ is proportional to traps*. Moreover, the slope of B$_{VT}$ versus traps for pMOSFETs is larger than that for nMOSFETs, which is consistent with the previous results, Figs. 4 and 5.

B. Evolution of Vth Variation After HC stress

Figs.9 and 10 show the Takeuchi plots after HC stress for nMOSFETs and pMOSFETs respectively. The value of B$_{VT}$ for pMOSFETs increases, but that of nMOSFETs after HC stress *decreases*. The physical mechanism is illustrated in Fig.11. For pMOSFET after HC stress, Fig.11(a), the generation of traps is inefficient and the distribution of traps is sparse. σV_{th} is also enhanced. In comparison, for nMOSFET after the HC stress, Fig. 11(b), traps are highly localized near the drain; thus, σV_{th} reduces as a result of Trap Blocking Effect(TBE). To identify the trap distribution, gated-diode (GD) measurement was performed [11]. Three peaks in the GD-curve indicate damaged regions (Fig.12). In nMOSFETs, traps are indeed highly localized near the drain, Fig. 13, confirming our assumption in Fig. 11(b). *In short: (1) The trap density and distribution dominate stressed V$_{th}$ variation, (2) RTF causes and increasing σV_{th}; TBE causes a decreasing σV_{th}, and (3) with the further scaling and long term operation, RTF might be an important factor affecting the CMOS reliability.*

In summary, a new random trap fluctuation(RTF) effect is proposed to study its impact on the device variability for HC and FN-stressed devices. Trap density and distribution determine the variation of V$_{th}$, in which different physical phenomena were observed between n- and p-channel MOSFETs. The V$_{th}$ variation will be worse after the stress due to the random trap fluctuation effect (**RTF**). However, an interesting phenomenon was observed, i.e., the V$_{th}$ variation was reduced after HC stress in nMOSFETs because of the trap-blocking effect (TBE). RTF in pMOSFET might be the dominant factor of CMOS reliability for future generations.

Acknowledgments This work was sponsored by the National Science Council, under contract number NSC-96-2221-E009-280.

References:

[1] F. Yang et al., *Symposium on VLSI Tech.*, p. 208, 2007.
[2] A. Asenov, *Symposium on VLSI Tech.*, p. 86, 2007.
[3] T. Matsukawa, *Symposium on VLSI Tech.*, p. 118, 2009.
[4] T. Tsunomura et al., *Symp. on VLSI Tech.*, p. 156, 2008.
[5] T. Tsunomura et al., *Symp. on VLSI Tech.*, p. 110, 2009.
[6] A. V-Y Thean, *Tech. Dig. IEDM*, p. 881, 2006.
[7] K. J.Kuhn, *Tech. Dig. IEDM*, p. 471, 2007.
[8] O. Weber et al., *Tech. Dig. IEDM*, p. 245, 2008.
[9] S. Kamiyama, *Tech. Dig. IEDM*, p. 431, 2009.
[10] K. Takeuchi et al., *Tech. Dig. IEDM*, p. 467, 2007.
[11] S. S. Chung et al., in *Tech. Dig. IEDM*, pp. 513-516, 2002.
[12] S. S. Chung et al., *Symp. on VLSI Tech.*, pp. 19-20, 2002.

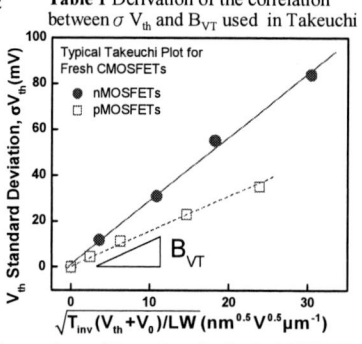

Table 1 Derivation of the correlation between σV_{th} and B_{VT} used in Takeuchi plot.

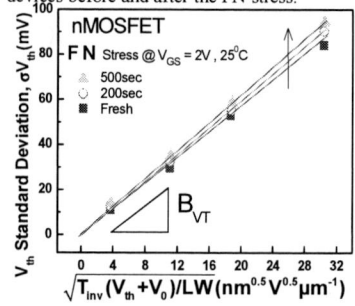

Fig. 1 Cumulative probability of Vth for nMOS devices before and after the FN stress.

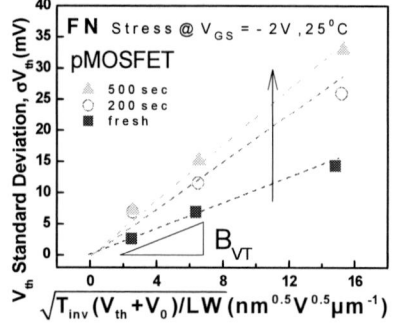

Fig. 2 Cumulative probability of Vth for pMOS devices before and after FN stresses.

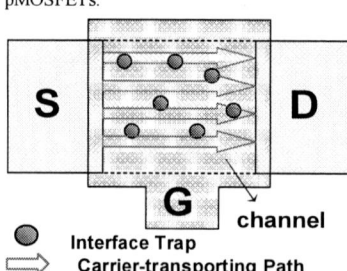

Fig. 3 Comparison of B_{VT} values for fresh nMOSFETs and pMOSFETs.

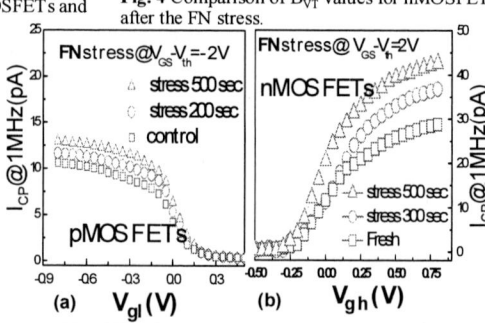

Fig. 4 Comparison of B_{VT} values for nMOSFETs after the FN stress.

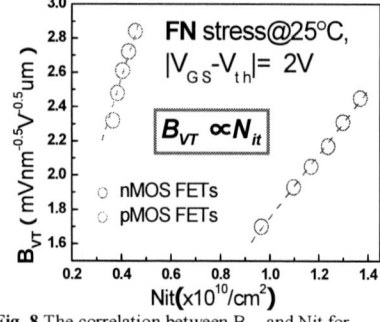

Fig. 5 Comparison of B_{VT} values for control pMOSFETs after the FN stress.

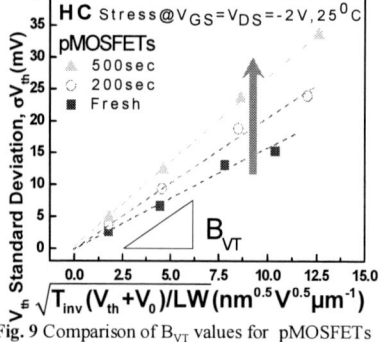

Fig. 6 The schematic of RTF induced Vth variation after FN stress., traps are uniformly distributed in the channel.

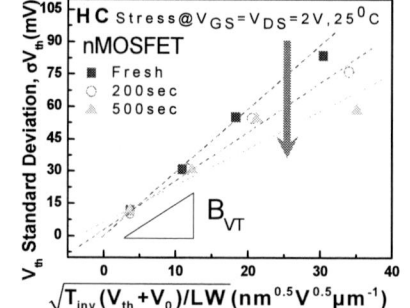

Fig. 7 The charge pumping current after FN stress for pMOS devices (*left*) and nMOS devices (*right*).

Fig. 8 The correlation between B_{VT} and Nit for nMOS and pMOS devices under FN stresses. Note that B_{VT} is proportional to the interface trap, Nit.

Fig. 9 Comparison of B_{VT} values for pMOSFETs after the HC stress.

Fig. 10 Comparison of B_{VT} values for nMOSFETs after the HC stress.

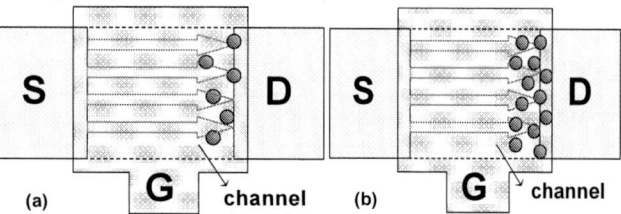

Fig. 11 Schematic of the mechanisms for Vth Variation after different stress schemes: (a) HC stress for pMOSFET, and (b) HC stress for nMOSFETs. Note that the generated traps in pMOS are sparsely distributed, while it is highly localized near the drain in nMOSFET.

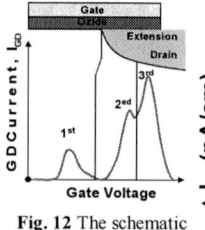

Fig. 12 The schematic of gated diode measurement; the three peaks indicate the HC-stress damaged regions.

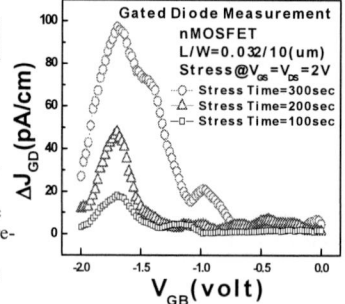

Fig. 13 Density of interface traps dominates the Vth variation, which is highly localized in nMOSFETs.

978-1-4244-9113-1/11 $26.00 © 2011 IEEE

On the Cyclic Threshold Voltage Shift of Dynamic Negative-Bias Temperature Instability

Z. Q. Teo[+], A. A. Boo, D. S. Ang, and K. C. Leong[*]

School of Electrical and Electronic Engineering
Nanyang Technological University
Singapore 639798
E-Mail[+]: TEOZ0019@ntu.edu.sg

[*]GLOBALFOUNDRIES Singapore Pte. Ltd.
Singapore 738406

Abstract—Based on new experimental evidence for the cyclical threshold voltage shift (ΔV_t) under dynamic NBTI and a recent *ab-initio* study on the oxygen vacancy defects (hole traps) in the SiO_2, an improved physical hole-trapping model for dynamic NBTI involving the E_δ' center is proposed. This model stipulates that the hole-trap precursor (i.e. the Si-Si dimer) responsible for the cyclic ΔV_t only undergoes marginal structural relaxation under typical NBTI stress condition, such that the Si-Si bond is completely re-formed when the stress is terminated. This framework is subtly different from an existing one based on the earlier HDL model. The latter assumes that the switching hole traps are oxygen vacancy defects that have undergone significant structural relaxation and that the switching behavior is due to the repetitive transitions between the positively charged state and the charge-compensated state. Experimental results obtained from higher oxide-field stressing in fact *do not* support this proposition.

Keywords — *Bias temperature instability, recovery, ultra-fast measurement, pulsed I-V, hole trapping, interface states*

I. CURRENT STATUS

A *cyclic* behavior for the increase and decrease of the magnitude of the threshold voltage (V_t) was revealed in recent studies on dynamic NBTI (DNBTI) [1], [2]. This cyclic behavior is shown to be inconsistent with the hydrogen transport model [3]-[4], which stipulates that the decrease or recovery of V_t must reduce progressively under dynamic NBTI. Electron spin resonance (ESR) measurement on NBTI [5]-[7] has shown that oxygen vacancy defects (or hole traps) are the origin of this cyclic threshold voltage shift ($|\Delta V_t|$). The present hole-trap model for NBTI [8] is based on the Harry Diamond Laboratories (HDL) framework [9]. It stipulates that a switching hole trap is formed from the structural relaxation that follows the capture of a hole at the oxygen vacancy defect ($1 \rightarrow 2$). The cyclic $|\Delta V_t|$ is explained by the alternate transition of the switching hole trap between a positive state (2) and a charge-compensated state (3) (see Fig. 1).

II. PURPOSE OF THIS WORK

In this paper, we present characteristics of the cyclic $|\Delta V_t|$ which do not completely correspond to the HDL model. The

results imply that a majority of the oxygen vacancy defects behind the hole trapping process must inevitably involve only minimal structural relaxation such that near-complete Si-Si bond reformation always happens during the recovery. Transformation of the defects into a more permanent form, which involves more substantial structural changes, only occurs under very severe stress conditions. An improved model involving the E_δ' center [10] is shown to better explain the behavior of the cyclic $|\Delta V_t|$ under typical NBTI conditions.

II. EXPERIMENTAL DETAILS

Test devices (DUTs) were p[+] polysilicon gate p-MOSFETs employing 2.4 nm SiO_2 gate dielectrics. DNBTI test was carried out at 100 °C. An ultrafast switching method, based on the pulsed current-voltage setup, was used in the measurement of the linear I_d-V_d curve [11], [12]. The narrow timing (100 ns)

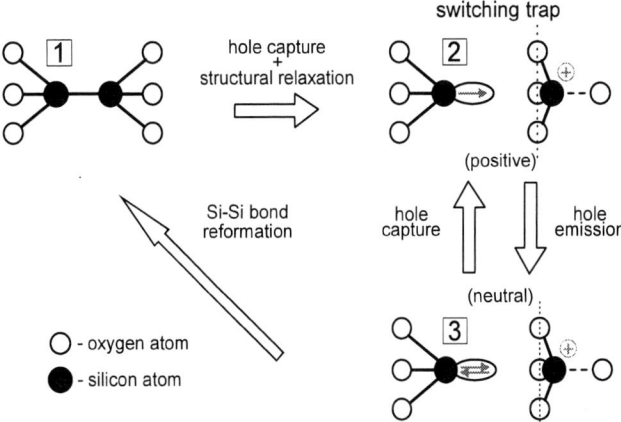

Fig. 1. Schematic illustration of the $E_{\gamma 1}'$ center model for NBTI [8]. An important feature of this model is the significant asymmetrical structural relaxation which occurs upon hole capture ($1 \rightarrow 2$), which results in one of the Si atom moving past the plane containing the oxygen atoms and getting back-bonded to a neighboring oxygen atom. This significant structural relaxation reduces the likelihood of the reformation of Si-Si bond ($3 \rightarrow 1$). Transition from state 3 to 2 (during stressing) and from state 2 to 3 (during relaxation) are believed to explain the cyclic shifts observed in the threshold voltage $|\Delta V_t|$.

978-1-4244-9113-1/11 $26.00 © 2011 IEEE

and short duty-cycle (0.1 %) gate pulses ensured that recovery was minimized. $|\Delta V_t|$ was extracted by a constant subthreshold drain current method.

III. RESULTS AND DISCUSSION

The main idea behind the HDL (or E_γ' center) model [9] is that strained Si-Si bonds broken upon the capture of holes structurally relax into the stable $\equiv$Si$\cdot$ $^+$Si$\equiv$ configuration (Fig. 1), which in turn acts as a switching hole trap causing the V_t fluctuations of DNBTI. Because structural relaxation and the resultant back-bonding of the Si$^+$ to a neighboring O atom has

stabilized the defect cluster, most of the immediate $|\Delta V_t|$ recovery (Fig. 2(a)) is ascribed to the charge compensation effect (Fig. 1; 2 → 3). Reformation of the Si-Si bond is deemed unlikely and would perhaps only account for the recovery at very long term. Alternate transition of the switching hole trap between the positive (2) and charge-compensated (3) states appear to explain the cyclical behavior (Fig. 2 (b); $|\Delta V_t|^r \approx |\Delta V_t|^s$) and highly repetitive evolution (Fig. 2(c)) of $|\Delta V_t|$ under alternate stressing and relaxation cycles.

To further test this model, a DUT was first stressed at $V_g^s =$ −1.15 V (~4 MV/cm) and relaxed at 0 V repeatedly for 4 cycles. The test was repeated with V_g^s increased (negatively) by −0.25 V each time until it reached −2.65 V (~10 MV/cm) – "ramp-up". This was followed by decreasing (positively) V_g^s by the same 0.25 V step back to −1.15 V, and the 4 stress/relaxation cycles applied at each V_g^s step – "ramp-down". Fig. 3(a) shows the evolution of $|\Delta V_t|$ for the test sequence. During the ramp-up phase, additional hole trapping occurs, as evident from the larger cyclic $|\Delta V_t|$ as well as the increase of the relatively permanent $|\Delta V_t|$. In spite of this, the cyclic $|\Delta V_t|$ at lower V_g^s did not register an increase on the ramp-down phase (Fig. 3(b)).

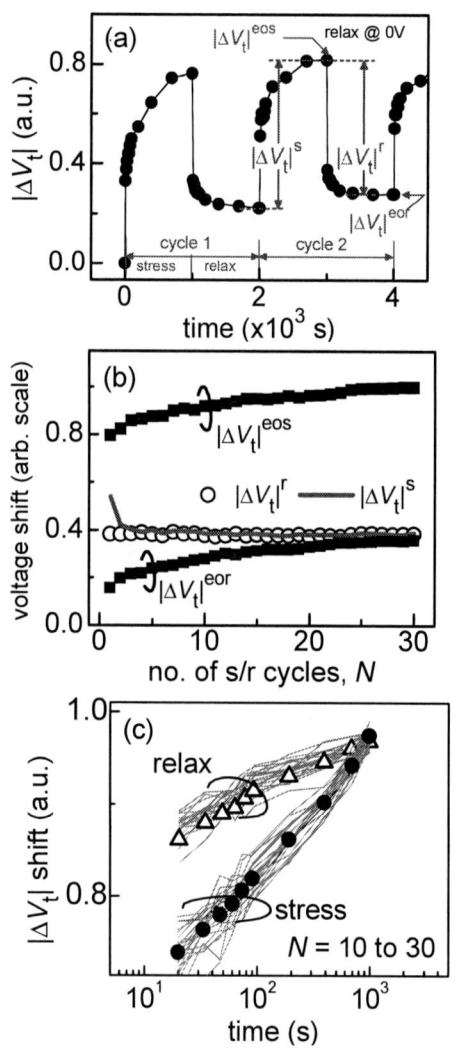

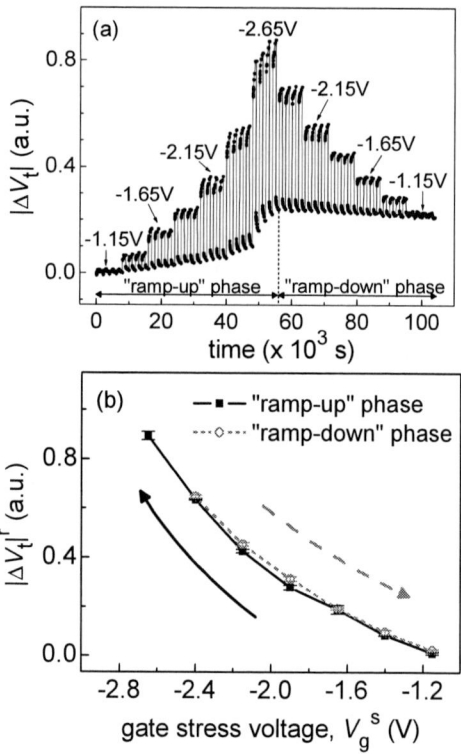

Fig. 2. Typical evolution of threshold voltage shift, $|\Delta V_t|$ during the stress (−2 V) and relaxation cycles of dynamic NBTI. (b) $|\Delta V_t|$ at the end-of-stress (eos), end-of-recovery (eor), and the amount of $|\Delta V_t|$ shift during each stress and relaxation cycle, $|\Delta V_t|^s$ and $|\Delta V_t|^r$, respectively as a function of the number of stress/relaxation (s/r) cycles, N. The constant $|\Delta V_t|^r$ in conjunction with $\Delta V_t|^s \approx |\Delta V_t|^r$, ($N > 10$) clearly depict a cyclic $|\Delta V_t|$ shift behavior. (c) This is confirmed by the highly repetitive $|\Delta V_t|$ shift over many stress and relaxation cycles (thin lines). Symbols denotes the average.

Fig. 3. (a) A fresh DUT was subjected to 4 stress (@ −1.15 V) and relaxation (@ 0 V) cycles. Each s/r cycle lasted 1 ×10^3 s. The test was then repeated on the same DUT with V_g^s increased (negatively) by −0.25 V each time, until $V_g^s = -2.65$ V. The V_g^s was then decreased (positively) each time by the same voltage step and the test repeated. Evolution of $|\Delta V_t|$ for the entire test sequence is shown. (b) Despite having subjected the DUT to a higher stress oxide field, there is no change in the $|\Delta V_t|^r$ at less negative gate stress voltages.

978-1-4244-9113-1/11 $26.00 © 2011 IEEE

There are two possible explanations for the observation. First, new switching hole traps were generated but they had fully recovered and hence were not manifested in the ramp-down phase. We used a relaxation period of 1×10^3 s; it is inevitably arguable whether this is long enough for full recovery to happen, but the non-negligible increase in the relatively permanent part of $|\Delta V_t|$ suggest that recovery is incomplete. In imposing full recovery of the switching hole traps, one must necessarily presume a different defect generation mechanism for the permanent $|\Delta V_t|$ increase. This inference is supported by temperature dependence study [1], which showed that the activation energy for the permanent $|\Delta V_t|$ increase was different from that of the cyclic $|\Delta V_t|$. An alternative explanation is that no new switching hole traps were generated, i.e. the increase of the cyclic $|\Delta V_t|$ in the ramp-up phase was due to the activation of a greater number of *preexisting* charge-compensated switching hole traps (Fig. 1; 3 → 2).

To test the second explanation, we deliberately introduced a very high oxide-field (> 13 MV/cm) stress prior to every DNBTI test. The idea is to generate more switching hole traps through the very high oxide-field stressing and to reveal their effect on the cyclic $|\Delta V_t|$ through a subsequent DNBTI test (Fig. 5(a)). A separate experiment which monitored the gate leakage current on a similar DUT registered a significant increase, confirming the generation of a substantial number of new oxide traps (Fig. 4). Evolution of the cyclic $|\Delta V_t|$ under this alternate DNBTI and high-field stressing sequence is shown in Fig. 5(b). Instead of an anticipated increase in the cyclic $|\Delta V_t|$, a *decrease* is observed despite the now higher oxide trap density. The result does not support the explanation that the cyclic $|\Delta V_t|$ is a consequence of the activation of pre-existing charge-compensated switching hole traps (3 → 2), since a higher density of switching hole traps should correspond to a larger cyclic $|\Delta V_t|$ (which is not observed).

Fig. 5(c) compares the increase of $|\Delta V_t|^{eos}$ and $|\Delta V_t|^{eor}$ as a function of the number of alternate DNBTI and high-field stress to the case without a preceding high-field stress. The

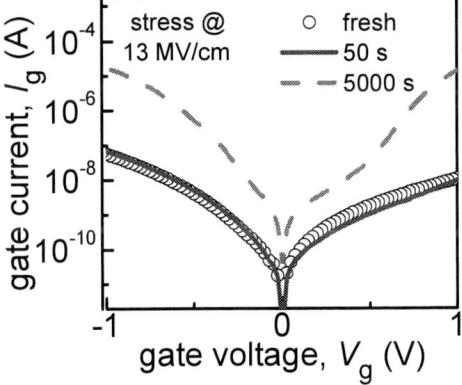

Fig. 4. Significant increase in the gate leakage current following high oxide-field stressing indicates substantial generation of oxide traps (including E' centers).

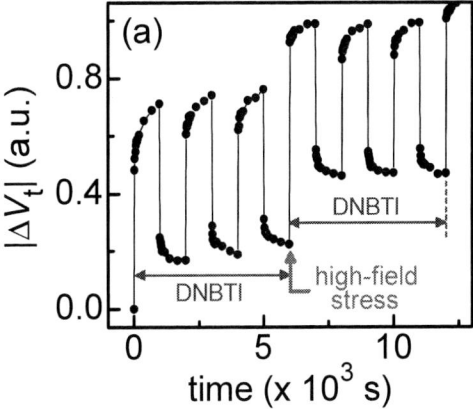

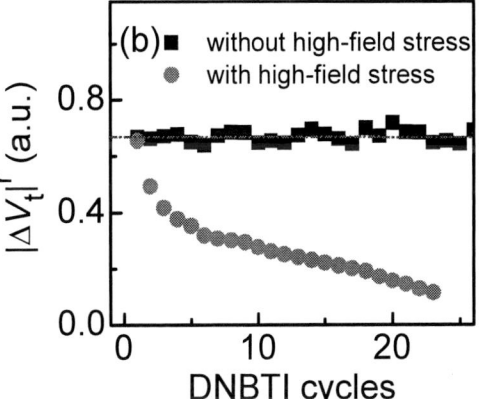

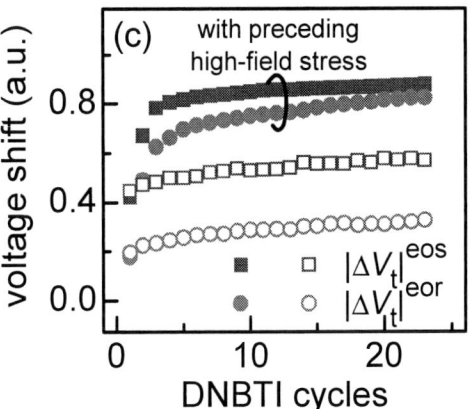

Fig. 5. (a) A DUT was first subjected to 3 NBTI stress (@ −1.8 V) and relaxation (@ 0 V) cycles. This was followed by a high oxide-field stressing (>13 MV/cm) for 50 s. The 3 stress/relaxation cycles were reapplied to check the effect of the high oxide-field stressing on the cyclic $|\Delta V_t|$. Transition between the dynamic NBTI and high oxide-field stressing was done on-the-fly. (b) Instead of an expected increase due to oxide trap generation (cf. Fig. 4), the cyclic $|\Delta V_t|$ registers a decrease. (c) The $|\Delta V_t|^{eos}$ or the total trapped hole density for both the case with and without a preceding high oxide-field stressing increase with comparable rates. The reduction in the cyclic $|\Delta V_t|$ is therefore due to a portion of it being transformed into a permanent component by the high oxide-field stressing.

$|\Delta V_t|^{eor}$ for the case with a preceding high-field stress rises much faster (and approaches $|\Delta V_t|^{eos}$ at the later stage) as compared to the case without. This observation shows that following each high-field stress, the amount of recoverable $|\Delta V_t|$ in the ensuing DNBTI test is reduced, indicating that the preceding high-field stress has transformed the oxygen vacancy defects responsible for the cyclic $|\Delta V_t|$ into a more *permanent* form. It is worth mentioning that recent ESR measurements have made parallel observations [5]-[7]. For moderate stress, a large E' signal was observed *during* stressing but it almost entirely disappeared upon removal of the stress [7]. Under more severe stressing, a weak nevertheless stable E' signal was observed even after a long time after the stress was removed [13]. The latter ESR observation and the reduction in the recoverable $|\Delta V_t|$ (Fig. 5(b)) or the rapid rise in $|\Delta V_t|^{eor}$ relative to $|\Delta V_t|^{eos}$ (Fig. 5(c)) are inevitably possible only if the switching hole traps are rendered more permanent under severe stressing. Our findings support the first explanation, i.e. the cyclic $|\Delta V_t|$ one sees under DNBTI is a result of the generation of new hole traps which, however, fully recover (i.e. broken bonds are reformed) when the gate stress voltage is removed.

The characteristics of the cyclic $|\Delta V_t|$ may be reconciled by an alternate E_δ' center model (Fig. 6). First-principles simulation using molecular dynamics has shown that the E_δ' center accounts for > 90% of all the oxygen vacancy defects in the gate oxide, while the E_γ' center accounts for the rest [10]. Unlike in the case of the E_γ' center which is formed following significant structural relaxation and with the trapped hole localized on one Si atom, a unique feature of the E_δ' center is that the trapped hole is shared equally between the two Si atoms [10]. The latter is inevitably possible only if there is minimal increase in the separation of the two Si atoms upon hole capture, which in turn implies a much higher likelihood of spontaneous Si–Si bond reformation once the center is charge compensated. Given the predominant density of the E_δ' center, this could explain why $|\Delta V_t|$ is cyclic and its magnitude is not affected by a preceding stress at a higher V_g^s (provided the oxide field is kept to a moderate level; cf. Figs. 2(b), 2(c) and 3(b)). At a much higher oxide field, distortion of the oxide network may give room for greater structural relaxation upon hole capture, thus transforming a fraction of the E_δ' centers into the more permanent E_γ' centers (cf. Figs. 5(b) and (c)).

IV. SUMMARY

New experimental observations which show that the E_γ' center model is inadequate for DNBTI are presented. An alternative model involving the E_δ' center and its subsequent transformation into the more permanent E_γ' center under more severe stress condition are shown to be able to reconcile the observations.

ACKNOWLEDGEMENT

This work is supported in part by a Singapore Ministry of Education research grant MOE2009-T2-1-050. Z. Q. Teo and A. A. Boo would like to thank the Singapore Economic

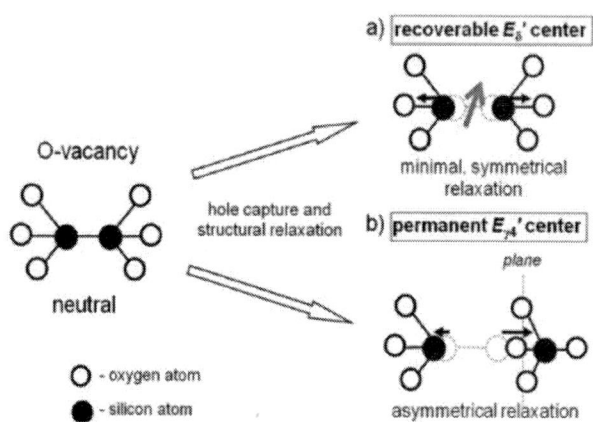

Fig. 6. An alternative oxygen vacancy defect model for dynamic NBTI. First-principles simulation [10] has shown that a large fraction (> 90 %) of the E' centers that form upon the capture of holes are the E_δ' center which involves *minimal* symmetrical structural relaxation. The positive trapped charge (or unpaired spin – see red arrow) is equally shared between the two Si atoms, implying that the likelihood of reformation of the Si-Si bond is very high when an electron is captured. It is thus believed that the E_δ' centers are mainly responsible for the cyclic $|\Delta V_t|$ observed under dynamic NBTI. The E_γ' centers, however, are present in much smaller amount and the significant asymmetrical relaxation renders subsequent Si-Si bond reformation unlikely. Hence, the E_γ' centers are responsible for part of the more permanent component (the other constituent is the P_b center [13]). The percentage of E_γ' centers may, however be increased under very high oxide field where network distortion enhances structural relaxation. This increase is reflected through a decrease in the number of the E_δ' (cf. Fig. 5(b)), for a given density of oxygen vacancy defects.

Development Board and GLOBALFOUNDRIES Singapore for a joint Ph.D. scholarship grant.

REFERENCES

[1] Z. Q. Teo, D. S. Ang and K. S. See, "Can the reaction-diffusion model explain generation and recovery of interface states contributing to NBTI?," in *IEDM Tech. Dig.*, 2009, pp. 689-692.

[2] Z. Q. Teo, D. S. Ang and C. M. Ng, ""Non-hydrogen-transport" characteristics of dynamic negative-bias temperature instability," *IEEE Electron Dev. Lett.*, vol. 31, no. 4, pp. 269-271, Apr. 2010.

[3] K. O. Jeppson and C. M. Svensson, "Negative bias stress of MOS devices at high electric fields and degradation of MNOS devices," *J. Appl. Phys.*, vol. 48, no. 5, pp. 2004-2014, May 1977.

[4] M. A. Alam, "A critical examination of the mechanics of dynamic NBTI for PMOSFETs," in *IEDM Tech. Dig.*, 2003, pp. 345-348.

[5] J. P. Campbell, P. M. Lenahan, C. J. Cochrane, A.T. Krishnan, and S. Krishnan, "Atomic-scale defects involved in the negative-bias temperature instability," *IEEE Trans. Dev. & Mat. Reliab.*, vol. 7, no. 4, pp. 540-557, Dec. 2007.

[6] J. P. Campbell, P. M. Lenahan, A. T. Krishnan, and S. Krishnan, "Identification of the atomic-scale defects involved in the negative bias temperature instability in plasma-nitrided p-channel metal-oxide-silicon field-effect transistors," *J. Appl. Phys*, vol. 103, art. no. 044505, Feb. 2008.

[7] J. T. Ryan, P. M. Lenahan, T. Grasser, and H. Enichlmair, "Observations of negative bias temperature instability defect generation via on the fly electron spin resonance," *Appl. Phys. Lett*, vol. 96, art. no. 223509, Jun. 2010.

[8] T. Grasser, B. Kaczer, W. Goes, Th. Aichinger, Ph. Hehenberger, and M. Nelhiebel, "A two-stage model for negative bias temperature instability," in *Proc. IRPS*, 2009, pp. 33-34.

[9] A. J. Lelis and T. R. Oldham, "Time dependence of switching oxide traps," *IEEE Trans. Nucl. Sci.*, vol. 41, no. 6, pp. 1835-1843, Dec. 1994.

[10] C. J. Nicklaw, Z.-Y. Lu, D. M. Fleetwood, R. D. Schrimpf and S. T. Pantelides, "The structure, properties, and dynamics of oxygen vacancies in amorphous SiO_2," *IEEE Trans.Nucl. Sci.*, vol. 49, no. 6, pp. 2667-2673, Dec. 2002.

[11] G. A. Du, D. S. Ang and Z. Q. Teo, "Ultrafast measurement on NBTI," *IEEE Electron Dev. Lett.*, vol. 30, no. 7, pp. 275-277, Jul. 2009.

[12] Y. Z. Hu, D. S. Ang and Z. Q. Teo, "Threshold voltage and mobility extraction by ultrafast switching measurement on NBTI," *IEEE Trans. Electron Dev.*, vol. 57, no. 8, pp. 2027-2031, Aug. 2010.

[13] J. P. Campbell, P. M. Lenahan, A.T. Krishnan, and S. Krishnan, "Observations of NBTI-induced atomic-scale defects," *IEEE Trans. Dev. & Mat. Reliab.*, vol. 6, no. 2, pp. 117-122, Jun. 2006.

2012 IEEE INTERNATIONAL
RELIABILITY PHYSICS SYMPOSIUM

April 15 - 19, 2012 • Hyatt Regency Orange County • Anaheim, CA, USA
CALL FOR PAPERS

IRPS addresses state-of-the-art developments in the reliability physics of devices, materials, circuits, and products used in commercial, industrial, and harsh or unusual environmental conditions, such those found in space, automobiles, renewable energies, and medical applications. At IRPS, a specific problem is posed, extent of reliability risk is identified, and often a novel solution is presented to minimize exposure to such risk. IRPS is the venue where important reliability challenges to integrated circuit reliability are first described, especially those that emerge as a consequence of dimensional scaling, new materials introduction, new processes or integration strategies, and/or fundamentally new device architectures. Past work includes important contributions on transistor reliability, interconnects, failure analysis, circuit reliability (including soft-error and ESD/LU), and package reliability. Thus, IRPS provides the necessary reliability baseline in microelectronics systems so that vertically integrated reliability - from process to final product - can be achieved for electronic components or applications.

The IRPS technical program includes • technical sessions, • keynote and invited talks on timely and important subjects, • tutorials, • workshops, • evening poster session, • year-in-review seminar, and • equipment demonstrations.

Awards: IRPS bestows awards for Best Paper, Outstanding Paper, Best Poster and Best Student Talk.

YOUR ORIGINAL PAPERS AND POSTERS ARE SOLICITED IN THE FOLLOWING AREAS:

Circuits and Products

Circuits – Reliability in Designs and Circuits for Digital and Analog including Simulation.

Component/System Reliability – Product Reliability, Board-level Reliability, Component Reliability, System-level Reliability, Reliability Metrics for ICs.

ESD and Latch-Up – Novel Structures including SOI and Bipolar; Modeling & Checking.

Extreme Environment – Device Reliability in Deep Space, Ultra-high or -low Temperatures, High Reliability for Mission Critical Applications.

Foundry and Fabless – Reliability for Designers, Reliable Process Solutions for various IC Applications.

Health and Biological Systems – Reliable Electronics for *in-vivo* Health Monitoring, Medical Equipment/Tools.

Industrial/Harsh Conditions – Device and Component Reliability and Failure Mechanisms in Challenging Ambient such as in Automobiles/Smart Transportation, Solar Energy, Wind, Nuclear, Energy Grid and Storage.

Product Reliability, HTOL & Burn-in – Chip-level Reliability; Defect Detection; Technology Model Predictions.

Simulation/Modeling Techniques – Circuit and Device Simulators for Reliability; Reliability Checkers.

Soft Error – Neutron and Alpha SER for Scaled CMOS including Multi-Bit and Logic SER/SEU; Mitigation Techniques, Simulation & Modeling.

Device, Process, Chip and Materials Reliability

Analog – Reliability of Bipolar Devices; Power Devices including Silicon and Compound Semiconductors; Passives.

Advanced Device Technologies – Reliability of Nanoelectronics, FINFETs, CNTs, Graphene Devices, Molecular Devices, and Hybrid Technologies, Novel Transistors.

Assembly and Packaging – Chip-Package Interaction, Multichip Modules, 3D Integration including TSV; Thermomechanical Stress Modeling; Bump/Solder EM & Stress.

BEOL Dielectrics – Low-k Dielectric Breakdown; Inter/Intra-Level Reliability including Impact of Process Variation.

Failure Analysis – Evidence of New Failure Mechanisms; Advances in Failure Analysis Techniques; New Technologies.

Gate Dielectrics – Breakdown Mechanisms & Modeling; Novel Dielectric Materials Reliability, Post-Breakdown Modeling.

Interconnects – Wearout in Cu and Al Metallization; Electromigration, Joule Heating, Stressmigration.

MEMS – Reliability of New Structures, Sensors, Actuators; Reliability Testing, Analysis & Modeling.

Non-Volatile Memory – Unique Reliability Phenomena and Failure Mechanisms in Non-Volatile Memory Devices.

Photoelectronics – Photovoltaics/Solar Cells, μm and mm ICs, Lasers, Optoelectronics; Novel Si and Compound Semiconductors.

Process – Reliability-Driven Process Interactions; New Process Related Reliability Issues.

Transistor – Hot Carrier Phenomena; Random Telegraph Noise; Bias-Temperature Instability; Transistor Scaling Issues.

Abstracts Must Be Received By: OCTOBER 3, 2011.

Abstract/Paper/Poster Submission: All submissions are done electronically at www.irps.org. Please do not email or mail your abstract.

Your original abstract submission (≤ 2 pages) should clearly and concisely state specific results, why they are important, and how they relate to prior work. An IRPS document template is available at www.irps.org. Following abstract review and paper acceptance, full manuscripts of accepted papers will be due before the conference.

Late Paper Submission: Full-length manuscripts with late breaking news *(See www.irps.org)* may be considered for inclusion in the conference/proceedings - space permitting - and must be completed and submitted by **January 9, 2012.**

Technical Program Chair:	General Chair:	Sponsored by:
Prasad Chaparala	**Ennis Ogawa**	
Alta Devices, Inc.	Broadcom	
Tel: +1-408-886-9381	Tel: +1-949-926-5507	
Email: prasadc@altadevices.com	Email: etogawa1@yahoo.com	

Find us online

AUTHOR INDEX

Abbas, Syed Mohsin 863
Abe, Kenichi 381
Achanta, R. 1, 134
Achanta, Ravi 665
Aghassi, Jasmin 342
Agostinelli, M. 533
Agrawal, Manish 342
Ahlbin, J. R. 258
Ahn, Jae-Young 126
Aichinger, Th. 605
Aitken, J. 134, 696
Alaeddine, A. 674
Alam, M. A. 196, 207, 353, 557, 614
Allee, David R. 904
Altimime, L. 99
Amashita, Takuro 876
Anand, Sindhu 220
Ang, D. S. 935, 943
Anghel, L. 347
Angyal, M. 134, 312, 317
Aoulaiche, M. 99
Arie, H. 503
Aritome, Seiichi 641
Arnaud, L. 297, 347, 746
Arora, Rajan 461
Athikulwongse, Krit 65
Augustine, Charles 353
Aymerich, X. 920
Backes, Benjamin 527
Badami, D. 696
Bae, Ilchan 852
Bae, Kidan 17, 852
Baeg, Sanghyeon 91, 863
Baek, Dong-Cheon 734
Bagatin, M. 751
Bai, P. 533
Bai, Shawn 706
Baik, Seung Jae 650
Baker, Michael 220
Balasubramanian, S. 41
Ball, D. R. 258
Banerjee, Kaustav 280
Banerjee, Sanjay K. 539
Bansal, Aditya 47, 696, 700
Barabadi, Banafsheh 271
Bari, Daniele 113, 566
Bashir, Muhammad 65
Baumann, F. 134
Baumann, Robert C. 247
Beltrami, S. 751
Bersuker, Gennadi 802, 807
Berthold, J. M. 56
Bertuccio, M. 751
Beyer, G. P. 142, 321
Bhuva, B. L. 258, 467, 886

Bhuva, Bharat 891
Bieler, Thomas R. 573
Black, Dolores A. 897
Black, Jeffrey D. 897
Blair, Lauren 573
Bloom, Ilan 819
Boit, Christian 514, 774
Bolam, R. 696
Bonilla, G. 134
Boo, A. A. 935, 943
Boret, S. 811
Bosman, M. 182, 786
Brain, Ruth 61
Bravaix, A. 704, 798, 926
Brown, Thomas M. 566
Bultreys, D. 500
Burdeaux, David C. 435
Burger, Wayne R. 435
Byun, Jun Young 656
Cacho, Florian 430, 454
Camargo, V. Valduga De Almeida 915
Campbell, J. P. 202
Cao, Yu 36
Carbonell, L. 142, 321
Carroll, Barry 476
Cartier, E. 41, 372, 387, 696, 933
Carulli, John 36
Cassell, Alan 280
Castillo, James 868
Catthoor, F. 915
Cester, A. 113
Cester, Andrea 566
Cha, Seon-Yong 95
Chanda, Kaushik 317
Chang, Soon W. 335
Chang, Tang-Long 717
Chang, Yao-Wen 396
Chappaz, C. 347
Cheffah, Saad 704
Chen, An 843
Chen, Chia-Yu 190
Chen, F. 134
Chen, Frederick T. 847
Chen, K. C. 645
Chen, K. F. 645
Chen, Kuang-Chao 396
Chen, Lu-An 392, 714
Chen, M. S. 645
Chen, Min 36
Chen, Pang-Shiu 847
Chen, Wei-Su 847
Chen, Willis 160
Chen, Yan-Yu 396
Chen, Ying-Je 824
Chen, Yu-Sheng 847

AUTHOR INDEX

Chen, Z. 913
Cheng, C. H. 645
Cheng, C. M. 160
Cheng, Chun-Min 661
Cherman, Vladimir 592
Cheung, K. P. 202
Chevallier, Remy 704
Chia, Pierre 91
Chiang, Puo-Yu 444
Chikaki, Shinichi 130
Chikyow, Toyohiro 335
Chimenton, Andrea 171, 828
Chin, H. W. 636
Cho, Hyun-Jin 190
Cho, M. 624
Cho, Moonju 207
Choi, Do-Young 449, 908
Choi, Gil-Heyun 307
Choi, Hanmei 126
Choi, Seung-Man 734
Choi, Siyoung 307
Choi, Wonsup 650
Chong, L. H. 645
Chou, Bruce 409
Christiansen, Bradley D. 680
Christiansen, Cathryn 312, 317
Chu, W. T. 636
Chu, Y. S. 636
Chua, C. M. 780
Chuang, Ching-Te 47
Chudzik, Michael 387
Chung, Chilhee 307
Chung, Steve S. 941
Ciappa, M. 28
Cirak, J. 113
Claeys, C. 882
Clark, Brett M. 476
Classe, F. C. 685
Cohen, S. 134
Collaert, N. 99
Compagnoni, Christian Monzio 833
Concannon, A. 405
Coutu, Ronald A. 680
Cressler, John D. 461
Cristoloveanu, S. 882
Croes, K. 142, 321, 592
Crupi, F. 624
Curello, G. 533
Czeppel, L. T. 751
Dadgour, Hamed F 280
Dalal, Vikram L. 546
Daoud, K. 674
Darbandi, Payam 573
Das, K. 696
De Wachter, B. 99

Degraeve, Robin 7, 207, 815
Del Alamo, Jesús A. 422
D'Emic, Christopher P. 630
Deora, S. 614
Deshpande, H. 533
Dey, Aritra 904
Di Carlo, Aldo 566
Dick, Kevin D. 897
Diggins, Z. 886
Diop, Malick 430
Dixit, Anand 486
Do, Jinho 852
Dobrovolny, P. 915
Dongaonkar, Sourabh 557
Donoval, D. 113
Dosseul, Franck 592
Dunn, S. 51
Dutton, Robert W. 190
Eaton, Paul 868
El Farhane, R. 347
Elkins, Don 868
El-Mamouni, F. 882
Eneman, G. 624
Eng, Genghmun 177
Ermisch, Karsten 597
Falk, John 75
Fang, Y.-P. 891
Federspiel, Xavier 454, 798
Ferreira, P. J. 500
Fichtner, W. 28
Fleetwood, Daniel M. 426
Floyd, Rich 868
Foran, Brendan 509
Franco, J. 364, 605, 624
Frank, David J. 630
Frank, T. 347
Freeman, Greg 461
Frei, Michel 557
Fronheiser, J. 202
Fujii, Shosuke 839
Fujitsuka, Ryota 839
Fukutani, Katsuyuki 335
Furuta, Jun 471
Gaddamraja, Sesha 685
Gaddi, Roberto 171
Gadlage, M. J. 258
Galand, R. 297, 746
Galloway, K. F. 882
Ganesh, K. J. 500
Gao, Y. 935
Geer, Robert E. 527
Gerardin, S. 751
Ghibaudo, G. 811
Gloria, D. 811
Gordon, Michael 476

AUTHOR INDEX

Gossner, Harald 342
Grasser, T. 364, 597, 605, 624, 915
Green, Ronald 756
Groeseneken, G. 7, 99, 364, 624, 915, 920
Gu, Pei-Yi 847
Guillorn, Michael A. 630
Gustin, W. 56, 597
Gyun, Byung Gu 656
Habersat, Daniel 756
Haendler, Sebastien 454, 798
Haensch, Wilfried 630
Hafez, W. 533
Hamanaka, Chikara 471
Han, J. H. 202
Han, Jeong-Uk 582
Han, T. T. 645
Harada, Ryo 253
Harm, Gregory J. 740
Hashimoto, Masanori 253
Hattori, N. 503
Hau-Riege, Christine 588
Hauser, M. 696
He, Chieh-Wei 396
He, Jun 61
He, Yi 476
Hekmatshoar, Bahman 562
Heller, Eric R. 680
Heo, Jinchul 852
Herfst, Rodolf W. 929
Hicks, Jeff 61
Higgins, R. 51
Hillenbrand, R. 493
Hirose, Y. 503
Ho, Dah-Chuen 444
Ho, Paul S. 264
Hoffman, M. 56
Hoffmann, T. Y. 624
Hong, Heebum 105
Hong, Sung-Joo 95, 641
Hopstaken, Marinus 562
Horiguchi, N. 364
Horii, Sadayoshi 335
Horikawa, Tsuyoshi 335
Hsieh, E. R. 941
Hsu, C. L. 160
Hsu, C. M. 160
Hsu, Chia-Lin 670
Hsu, Chi-Mao 661
Hsu, H. K. 160
Hsu, T. H. 824
Hsu, Yen-Ya 847
Hu, Y. Z. 913
Huang, C. C. 160, 661
Huang, Clement 731
Huang, Climbing 670

Huang, Hsin-Fu 661
Huang, I. J. 645
Huang, J. S. 645
Huang, R. M. 941
Huang, Ru 847
Huang, Rui 264
Huang, Y. C. 22
Huang, Yu-Hui 444, 690
Huard, Vincent 430, 454, 704, 798, 926
Huber, A. J. 493
Hussain, Muhammad M. 280
Hwang, Kihyun 126
Hwang, Lira 17, 852
Ibe, Eishi 239
Ighilahriz, Salim 430
Im, Jay 264
Ioannou, D. P. 696
Islam, A. E. 614
Islam, Ahmad 207
Ito, Kyosuke 710
Ito, Shuu 335
Iwai, H. 786
Jack, Nathan 409
Jagannathan, Srikanth 886
Jain, A. 614
Jakabovic, J. 113
Jan, C.-H. 533
Janai, Meir 819
Jang, Tae-Su 95
Jenkins, Keith A. 47
Jeon, Chulhee 852
Jeong, Eui-Young 449
Jeong, Jae-Goan 95
Jeong, Tae-Young 734
Jeong, Yoon-Ha 449, 908
Jesudoss, Pio 224
Jha, N. K. 22
Jin, Minjung 17, 852
Joh, Jungwoo 422
Johnston, Allan 165
Joo, Moon Sig 656
Joshi, K. 614
Joshi, Yogendra 271
Juan, Alex 731
Jung, Eui-Young 908
Jung, Eunji 307
Jung, Hye Kyung 307
Jurczak, M. 99
Kaczer, B. 364, 605, 624, 915, 920
Kadi, M. 674
Kakushima, K. 786
Kane, Terence 312
Kang, A. C. 636
Kang, Chang Yong 449, 908
Kang, Chang-Jin 126

AUTHOR INDEX

Kang, J. F. 196
Karthik, Y. 557
Kato, Ichiro 149
Kauerauf, Thomas 7, 182
Kawaguchi, Hiroshi 876
Kayaba, Yasuhisa.......................................130
Keilmann, F. 493
Keller, Robert R. 740
Kendig, Dustin 725
Ker, Ming-Dou 717
Kerber, Andreas..............................41, 190, 372, 387
Kikkawa, Takamaro....................................130
Kim, Andrew T. 734
Kim, Bio 126
Kim, Chris H. 47
Kim, Dae Hyun 65
Kim, Dae-Hyong..111
Kim, Dong Woo 126
Kim, Hyungseok 641
Kim, Hyunjin 17
Kim, Jae-Joon..............................47, 696, 700
Kim, Jung Nam 656
Kim, Jungin 17
Kim, Kyoung-Hwan 582
Kim, Kyung-Do 95
Kim, Sook Joo 656
Kim, Taeh Wan 656
Kim, Tae-Hyoung...47
Kim, Wan Gee 656
Kim, Won..656
Kim, Yong Tae 582
Kim, Yongshik...105
Kim, Yong-Taik 95
Kindereit, Ulrike 514
King, Everett E. 81
Ko, Yongsun 126
Kobayashi, Kazutoshi.........................471, 710
Koh, L. S. 780
Kohmura, Kazuo 130
Komeyli, K. 533
Kopanski, J. J. 202
Kouda, M. 786
Koudymov, A. 802
Koutsoureli, M. 290
Kovac, J. 113
Koyama, T. 503
Koyanagi, Mitsumasa..................................328
Krishnan, A. T. 51
Krishnan, S. 36, 51, 387
Ku, S. H. 645
Kumar, Satish...271
Kumari, Sangita 740
Kurtz, Sarah 551
Kushvaha, S. S. 786
Kwasnick, Robert..75

La Rosa, G. 696
Labie, Riet 592
Lacaita, Andrea L. 833
Lafonteese, D. 405
Lai, Tai-Hsiang 392, 714
Lamontagne, P. 297, 746
Larcher, Luca 807
Lauzeral, Olivier 476
Leduc, P. 347
Lee, C. H. 645
Lee, Chienying 645
Lee, Heng-Yuan 847
Lee, Hyun-Bae 307
Lee, Jack C. 449
Lee, Jeong-Soo 908
Lee, Jian-Hsing 690
Lee, Jong Myeong 307
Lee, Jongho 852
Lee, Kyong Taek 17
Lee, Miji 734
Lee, Seokkiu 641
Lee, Shou-Chung 155
Lee, Tackhwi 539
Lee, Tae-Kyu 573
Lee, Y. H. 22, 444, 636
Lelis, Aivars 756
Leong, K. C. 943
Lepla, Keith 476
Leung, Martin 177
Li, Baozhen 1, 134, 312, 317
Li, Jinglong 765
Li, Zhao 546
Liang, C. W. 941
Liang, James W. 731
Liao, P. J. 22
Liehr, M. 802
Lim, Hajin 852
Lim, Koeng Su 650
Lim, Seung-Hyun 126
Lim, Sung Kyu 65
Lim, Sun-Me 105
Limbrick, Daniel B.......................................897
Lin, C. P. 690
Lin, Chun-Yu 717
Lin, Hung Sung..520, 770
Lin, J. F. 160
Lin, Jack...160
Lin, Jin-Fu 661
Lin, Kun-Hsien 661
Lin, Ming-Ren 190, 843
Lin, Mingte 731
Lin, W. C. 160, 661, 670
Linder, Barry P. 1, 47, 372, 696
Litt, Brian 111
Liu, C. C. 160, 444

AUTHOR INDEX

Liu, Cheng-Jye 824
Liu, Kuo-Chuan 573
Liu, T. 121
Liu, W. H. 182, 847
Liu, Yan-Chun 661
Liu, Yang 190
Liu, Ziyuan 335
Lo, Chun-Yuan 824
Lofrano, M. 321
Lorut, F. 347
Loveless, T. D. 258, 886
Lowrie, Anthony 61
Lu, Chih-Yuan 396, 645
Lu, Kuan-Hsun 264
Lu, Pong-Fei 47
Lu, Tao-Cheng 396
Lu, W. P. 645
Lu, Z. 99
Machavolu, K. S. 913
Maeda, H. 503
Mahapatra, Souvik 557, 614
Mahatme, Nihaar 891
Mahato, S. 915
Maheta, V. D. 614
Maize, Kerry 725
Malandruccolo, V. 28
Maloney, Timothy J. 409
Marckmann, Jennifer 476
Martin-Martinez, J. 920
Mason, Maribeth 177, 509
Massengill, L. W. 258, 886
Massey, G. 696
Masuduzzaman, Muhammad 196, 207
Mathewson, Alan 224
Matsumoto, Takashi 710
Mavis, David 868
McDonough, Colin 527
McGahay, Vincent 312
McLaughlin, Paul 134, 665
McMahon, W. 41
McMorrow, D. 882
McNally, Brendan 476
Mendenhall, Marcus H. 247
Meneghesso, Gaudenzio 113, 566
Mercha, A. 99
Miccoli, Carmine 833
Michalas, L. 290
Miki, Hiroshi 630
Milor, Linda 65
Min, Hong Kook 582
Misra, Durgamadhab 792
Mitani, Yuichiro 857
Mitard, J. 624
Mitsuyama, Yukio 253
Mittl, S. 1, 696

Molzer, Wolfgang 342
Monsieur, Frederic 933
Moquillon, Laurence 454
Moreau, S. 347
Moss, Steven C. 509
Motohiko, Masuda 765
Mukhopadhyay, Saibal 47
Murakami, E. 503
Muthuswamy, Jit 220
Mutihac, Oana-Mihaela 514
Nafria, M. 920
Nakanishi, N. 503
Nakanishi, Toshiro 126
Nam, Hyeowoo 91
Nam, Jongik 17
Narasimham, B. 467
Nayfeh, Hasan M. 461
Negre, L. 811
Nelhiebel, M. 605
Ney, D. 297
Nielen, Heiko 597
Nigam, T. 41
Nishii, Koji 239
Nishizawa, Shinichi 710
Oakley, Jennifer 317
Oates, A. S. 155, 467, 706, 891
O'Connor, Robert 7
Ogawa, Arito 335
Ohmi, Tadahiro 381
Okandan, Murat 220
Okumura, Shunsuke 876
Olivo, Piero 171, 828
Olson, Nicholas 719
Ong, T. C. 636
Onodera, Hidetoshi 471, 710
Onoye, Takao 253
Osborn, Jon V. 81
O'Shea, S. J. 786
Paccagnella, A. 751
Padovani, Andrea 807
Pagano, Carlo 774
Pantelides, Sokrates T. 426
Pantuovaki, M. 142
Papaioannou, G. 290
Papathanasiou, Athanasios E. 75
Park, Dae-Gyu 630
Park, Jongwoo 17, 105, 533, 734, 852
Park, Junekyun 17, 852
Park, Min Sang 449, 908
Park, Se Yeoul 582
Park, So Ra 582
Park, Sung Ki 656
Park, Sungju 863
Park, Sungkye 641
Parks, C. 134

AUTHOR INDEX

Patterson, K. 467
Pavan, Paolo 807
Penna, Stefano 566
Petitprez, E. 297, 746
Pey, K. L. 182, 786
Phang, J. C. H. 780
Phoa, K. 533
Pimparkar, N. 41
Pobegen, G. 605
Poh, John Chung Hang 461
Poling, Brian S. 680
Pourboghrat, Farhang 573
Prabhumirashi, Prad 61
Presser, Nathan 509
Puzyrev, Yevgeniy S. 426
Quemerais, Thomas 454
Rafik, M. 798, 926
Raghavan, N. 182, 786
Ragnarsson, Lars-åke 7
Rahim, Nilufa 792
Rahman, A. 533
Rajasekhara, S. 500
Rakowski, M. 99
Ramaswami, Sesh 524
Ran, Qiushi 190
Randriamihaja, Y. Mamy 798, 926
Ranjan, R. 22, 444, 636
Rao, R. 696
Rao, Rahul M. 47, 700
Rao, V. Ramgopal 342
Rashid, Al 75
Read, David T. 740
Reale, Andrea 566
Reddy, Vijay 36, 51
Reed, Robert A. 247, 882, 897
Reilly, Matthew 75
Reisinger, H. 56, 597, 605
Ren, Zhibin 630
Rentala, Vijay 36
Rhoton, Brent 231
Rideau, D. 798, 926
Robinson, William H. 897
Roche, Philippe 476
Rodriguez, R. 920
Rogers, John A. 111
Roh, Jae Sung 656
Rosa, Giuseppe L. 461
Rosenbaum, Elyse 409, 719
Rothleitner, H. 28
Roussel, Ph. J. 7, 364, 624, 915
Roy, D. 798, 811, 926
Roy, Kaushik 353
Roy, Tania 426
Ryan, J. T. 202
Ryu, Suk-Kyu 264

Sadana, Devendra 562
Sagong, Hyun Chul 449, 908
Sahhaf, Sahar 7
Saito, Tomoya 335
Saito, Wataru 417
Sakamoto, Mitsuhiro 149
Savage, Timothy Scott 213
Scarpulla, John 81
Scheer, P. 811
Schepens, Cor 171
Schluender, Christian 597
Schlünder, C. 56
Schmitz, Anthony 61
Schmitz, Jurriaan 929
Scholten, Andries J. 929
Schrimpf, Ronald D. 247, 426, 882
Schulz, Thomas 342
Sekine, Katsuyuki 839
Seo, Soonok 641
Seth, Sachin 461
Shahrjerdi, Davood 562
Shakouri, Ali 725
Shaviv, Roey 740
Shealy, Jeffrey B. 680
Shih, J. R. 22, 444, 690
Shim, Byung Sup 582
Shimbo, Ken-Ichi 239
Shimizu, Tatsuo 149
Shin, Dongseok 852
Shin, Hyunjung 650
Shinosky, M. 134
Shrivastava, Mayank 342
Shubhakar, K. 786
Shukla, Vrashank 719
Shur, Yael 819
Shusterman, Yuriy 61
Sibley, Mike 868
Sierawski, Brian D. 247
Siew, Y. K. 321
Simoen, E. 882
Simon, A. 134
Sin, Yongkun 509
Singh, Navab 280
Slayman, Charles 476
Smith, Charles 171, 231
Sohn, Chang Woo 449, 908
Sokolsky, M. 113
Song, Grace 765
Spinelli, Alessandro S. 833
Srinivasan, Venkatesh 36
Stam, Frank 224
Stathis, James 372, 387, 696, 933
Stupian, Gary 177
Sturm, J. C. 121
Su, David 690

AUTHOR INDEX

Su, Kuan-Cheng 392, 401, 714, 731
Suehle, J. S. 202
Sugawa, Shigetoshi 381
Sun, Wein-Town 824
Sunagawa, Hiroki 710
Suñé, J. 1
Sung, Min Gyu 656
Sutanto, Jemmy 220
Sutton, Akil K. 461
Tagliaferro, Roberto 566
Tam, Nelson 247, 467
Tang, Tien-Hao 392, 401, 714
Taniguchi, Hitoshi 239
Taniguchi, Yoshio 239
Taylor, B. 802
Tega, Naoki 630
Teo, J. K. J. 780
Teo, Z. Q. 935, 943
Teramoto, Akinobu 381
Thuaire, A. 347
Tian, Li 765
Tilak, V. 202
Tillack, Bernd 514
Ting, Yun-Jen 824
Toba, Tadanobu 239
Tökei, Zs. 142, 321
Toledano-Luque, M. 364, 815, 915, 920
Torii, Kazuyoshi 630
Toriumi, Akira 857
Tsai, C. H. 533, 847, 941
Tsai, C. T. 941
Tsai, Ming-Jinn 847
Tsai, T. C. 160, 670
Tsao, W. C. 160
Tsuchiya, Hideaki 149
Tsukasa, Matsuda 307
Tucker, Mike 476
Twomey, Karen 224
Uno, Satoshi 149
Vaidyanathan, Balaji 706
Van Houdt, J. 815
Vandelli, Luca 807
Vandevelde, Bart 592
Varghese, D. 51
Vasanth, Karthik 231
Vashchenko, Vladislav 405, 725
Veksler, D. 802
Vetury, Rama 680
Via, G. David 680
Vilchis, M. 467
Visconti, A. 751
Viventi, Jonathan 111
Vrancken, C. 364
Wagner, P.-J. 605
Wagner, S. 121

Waltz, P. 297
Wang, C. 202, 636
Wang, Chang-Tzu 401, 714
Wang, Dapeng 557
Wang, Shih-Yu 396
Wang, Shun-Min 847
Wang, Tahui 645
Wang, W. 22, 527, 765
Wang, Y. H. 636
Wang, Yun Yu 312
Wang, Z. R. 786
Webers, Tomas 592
Weigmann, J.-M. 56
Weiland, R. 493
Weller, Robert A. 247
Wen, S. J. 91, 247, 467, 886
Wilde, Markus 335
Wilkinson, Jeffrey D. 476
Wilson, C. J. 321
Winters, Christophe 592
Wirth, G. 364, 915
Wittborn, J. 493
Witters, L. 624
Witulski, A. F. 258
Wohlgemuth, John H. 551
Wong, Richard 91, 247, 467, 476, 886
Wood, Alan 476
Wrachien, Nicola 113, 566
Wright, William M. D. 224
Wu, Chunlei 765
Wu, E. 1, 792
Wu, Guan-Wei 396
Wu, J. Y. 160, 661, 670
Wu, Kenneth 22, 444, 636, 690
Wu, Miao 765
Wu, Mong Sheng 520
Wu, Tommy 476
Wu, X. 182
Xia, Feng 61
Xiong, W. 882
Xu, J. 533
Xu, Xiaochen 231
Xu, Y. 467
Yamaguchi, Kosuke 876
Yang, I. C. 645
Yang, J. Q. 196
Yang, Sangryol 126
Yang, Seung Jin 582
Yao, Shaoning 312
Yasuda, Naoki 839
Yau, You-Wen 588
Yeh, J.-Y. 533
Yeo, Myung-Soo 734
Yeoh, Terence 177
Yew, T. Y. 22

AUTHOR INDEX

Yokogawa, Shinji 149
Yokoyama, Yoshiyuki...774
Yoo, Hyunjun 650
Yoo, Jong Hee...656
Yoo, Min-Soo...95
Yoshimoto, Masahiko 876
Yoshimoto, Shusuke...876
Young, C. D. 802
Yu, H. Y. 786
Yu, Joe 765
Yu, L. C 202
Yu, Nick 588
Yu, Sunil 105
Yum, Jung 807
Yun, Jong-Ho 307
Zahid, M. B...207, 815
Zaka, A. 798, 926
Zaknoon, Bashir 75
Zambelli, Cristian 171, 828
Zamiri, Amir 573
Zhang, E. X..426, 882
Zhang, F..202
Zhang, Lijie 847
Zhang, Ming 105
Zhao, K. 372, 696, 700
Zhao, L..142
Zhou, Bite 573
Zhou, X. 913
Zhu, Xiaowei 231
Zhu, Yu 630
Zous, N. K. 645
Zuber, P. 915

CURRAN ASSOCIATES INC.
proceedings
.com

9781424491131